2nd International Congress on Materials and Structural Stability (CMSS-2017)

MATEC Web of Conferences Volume 149 (2018)

Rabat, Morocco
22 – 25 November 2017

Part 1 of 2

Editors:

A. Diouri
A. Boukhari
L. Ait Brahim
L. Bahi
N. Khachani
M. Saadi
J. Aride
A. Nounah

ISBN: 978-1-5108-5259-4

Printed from e-media with permission by:

Curran Associates, Inc.
57 Morehouse Lane
Red Hook, NY 12571

Printed by Curran Associates, Inc. (2018)

For additional information, please contact EDP Sciences – Web of Conferences
at the address below.

EDP Sciences – Web of Conferences
17, Avenue du Hoggar
Parc d'Activité de Courtabœuf
BP 112
F-91944 Les Ulis Cedex A
France

Phone: +33 (0) 1 69 18 75 75
Fax: +33 (0) 1 69 28 84 91

contact-edps@webofconferences.org

Additional copies of this publication are available from:

Curran Associates, Inc.
57 Morehouse Lane
Red Hook, NY 12571 USA
Phone: 845-758-0400
Fax: 845-758-2633
Email: curran@proceedings.com
Web: www.proceedings.com

TABLE OF CONTENTS

PART 1

PART 2

Concrete casting robustness improvement due to active rheology

Wolfram Schmidt, Hans-Carsten Kühne, Sarah Leinitz
Bundesanstalt für Materialforschung und -prüfung (BAM), Department „Safety of Structures", Berlin, Grmany

Abstract. With ongoing innovation in process technology, the challenges of concrete technology are more and more focused on the rheological optimisation for these processes, since improper mixture stability or poor compaction ability negatively affect the concrete homogeneity and quality. However, along with the increasing complexity of today's concrete mixture compositions, concrete becomes more prone to failure regarding the casting process. Variable properties of the raw materials typically cause changing workability. The reasons can be found among others in scattering water contents, physical or chemical properties of the cement or varying environmental temperatures. Robustness in the delicately adjusted rheology, however, is of utmost importance for modern and future process technology, from sprayed concrete over pumpable concrete towards 3D-printing, with regard to the long-term strength, the function and the durability. Typically, material induced changes cannot be identified easily due to the complex interactions of concrete constituents. Therefore, a precise and prompt counteraction is impossible. However, it is known that the yield stress can be controlled by addition of supplementary superplasticizer or stabilising agent. In combination with computerized process observation tools that can rapidly interpret and react on changes in the rheology, it is therefore thinkable, that only these two admixture types can adjust the rheology steadily and permanently, regardless of the actual root cause for observed macroscopic rheology change. The presentation will firstly give a comprehensive overview of effects at the interface between pore solution, particles and hydrates, which affect the rheology of fresh concrete. Secondly, ways are recommended how the rheology can be actively manipulated before eventually computerized methods are demonstrated that help to actively and rapidly assess and counteract performance scatter during steady casting processes.

1 State of the art

1.1 Relevance of robust rheology and effective process control

Concrete construction today has become more sophisticated, innovative, future oriented, and performance based. Compared to the past, where cement, water, and aggregates were major components, today supplementary additions and chemical admixtures have become important. Admixtures eventually facilitated the development of new concrete types such as self-compacting concrete, high performance concrete, strain hardening cement-based composites and a variety of special mortars and concrete types by uncoupling the workability from the water-cement ratio (w/c).

The price for the technological potentials are more sensitive systems that are prone to failure, particularly in terms of the rheological properties. Future additive automated process technologies will further increase the demands on the rheology of concrete. Table 1 provides an overview of some recent and future technologies and the specific rheological challenges that lie within.

Table 1. Rheological challenges for different concrete manufacturing technologies.

Manufacturing technology	Rheological challenge with relevance for unction, safety and durability
3D-printing of concrete elements	In-situ measurement of the workability and in-situ counter control required
Self-compacting concrete	Effective avoidance of segregation and stagnation
Ready-mixed concrete	Good workability properties despite transport delays
Grouting, esp. Offshore-Grouting	Better consistency to avoid blockages in pipes which lead to technical problems
Pumping concrete	Optimised rheology can save energy costs and increase pumping height
Fibre concrete/SHCC	Differences in the rheology of pastes can cause poor fibre distribution and agglomeration of fibres
Sprayed concrete	Small changes to the water/additive content influences rebound and run out

Today, a variety of binders are used from ordinary Portland cement (OPC) to largely calcium free geopolymers and everything in between. At the same time, modern construction chemicals found their way into concrete daily routine, which, however, often react sensitive towards changes in the environmental conditions as well as scatter in the quality of raw materials. These parameters were relevant already in the past with normal concrete but they could be compensated typically by increased compaction effort, and in addition the expectations on the performance were lower. For many high performance modern concrete types and particularly those that will come up in future in conjunction with additive manufacturing, the rheological performance is crucial for the manufacturing process, the function of the concrete as well as eventually the durability. The higher complexity of binders, admixtures and additions is hence a thread to high performance concrete technology and automated processing. Hence, robust rheology is the key challenge of the future, but maintaining a permanent rheology without large scatter is difficult to achieve in daily practice since:

- Due to the complexity of concrete, which is a multiple scale (nm to cm), multiple phase (various organic and inorganic components), and time-dependent system, the root cause for changing and scattering rheology performance is difficult to find. E.g. changes can be induced by the water absorbed by aggregates, the gypsum in the cement, the fineness of the powder constituents, variations in the particle size distribution, and many more.
- Rheology affecting influences can be induced uniquely or systematically. Upon occurrence of a rheology change in concrete production, it is difficult to assess rapidly whether a change is systematic or has just happen within the normal range of scatter.
- When counteractions against unwanted rheology are identified (e.g. more or reduced superplasticizer) it is still difficult to find the relevant dosage without overshooting the target. This is aggravated by the complexity of the system and the typically unknown root cause for scatter and deviation.

1.2 Process control and decision making methods

Due to the relevance of robust rheology, particularly in automated processing, process observation and control tools are getting more and more relevant in the future. In production control, most of the time people observe process values (e.g. in concrete technology this would be compressive strength, air pore content or density). In order to better assess process data, control charts were developed, where the process data are plotted over the course of time and often compared to indicator values. The objective is a steady production with regard to the observed process value and as low as possible scatter around a target value. Largely known and well established are so called Shewhart charts, which were developed in the 1920s by W.A. Shewhart [1]. The relevant process values are plotted consecutively in comparison to a deviation of the twofold and threefold standard deviation. In case the twofold standard deviation line is touched by the process values it indicates that the process is at risk to run out of control, and corrective actions should possibly be taken, but still the process is within the 95% confidence interval. A crossing of the threefold standard deviation line indicates that the process significantly runs out of control. Counteractions have to be taken instantly to bring the process back into control. However, Shewhart charts react quickly on dramatic and instant changes of a process, but they react relatively slowly on smaller but systematic process deviations. In order to identify a process deviation. Small systematic changes cannot be detected at all by Shewhart charts. A user has to observe the chart for a long period of time to identify the changed production, so that counteractions can only be taken after a long period of time. In order to have a more efficient tool that helps to react quicker on systematic changes in processes, Cusum control charts were developed in the 1950s by E. S. Page [2, 3]. Cusum stands for "cumulative sum".

Cusum charts look at the deviations from a target value. Cusum charts react more rapidly on systematic changes in processes than other charts such as Shewhart charts, which look at the process values. It has been used in various fields successfully, e.g.

- As a learning curve for the evaluation of different medical treatments, anaesthesia, or intubation [4-7]
- Documentation of the complication rates occurred in chirurgical practice and competence assessment [8]
- Evaluation of patient data with regard to laboratory errors and more rapid error identification [9]
- For the improvement of the predictability of chirurgical protocols [10]
- Surveillance and evaluation of fish stocks and balances without and with little historic data information [11, 12]
- For the monitoring of production events [13]

A cusum can be derived according to the equation:

$$C(n) = \Sigma(A_i - A_{target}) \qquad (1)$$

Where: C(n) = cusum function; A_i = last process value; A_{target} = target value

Hence, instead of observing process values, cusum shows trends in the deviations from the target value and the following conclusions can be made based on eq. (1):

- An upward trend indicates that real process values are higher than the targeted value. This is often an indicator for uneconomic processes and should be a trigger for measures that help reducing the production costs (e.g. products are too thick → material can be saved; or strength is higher than required → cement can be saved).
- A horizontal trend indicates that the process runs as expected and no changes in the process or mixture composition should be made.
- A downward trend indicates that the real process values are lower than the targeted value. This is often an indicator for production that is technically not appropriate (products are too thin → more material is needed to fulfil the technical specifications; or strength too low → more cement required).

Table 2. Sample for cusum calculation.

	Target value f_{cta} = 45.0	Production data	Target deviation	Cusum
	[MPa]			
	Seq. Number	f_{ci}	f_{ci}-f_{cta}	$\Sigma(f_{ci}$-$f_{cta})$
Production below target value	1	40.1	-4.9	-4.9
	2	45.0	0.0	-4.9
	3	41.1	-3.9	-8.8
	4	41.9	-3.1	-11.9
	5	42.1	-2.9	-14.8
	6	43.5	-1.5	-16.3
	7	42.2	-2.8	-19.0
Production according to plan	8	43.1	-1.9	-20.9
	9	48.3	3.3	-17.7
	10	41.5	-3.5	-21.2
	11	47.3	2.3	-18.9
	12	45.4	0.4	-18.5
	13	46.2	1.2	-17.3
Production above target value	14	47.5	2.5	-14.8
	15	48.5	3.5	-11.3
	16	45.5	0.5	-10.8
	17	44.6	-0.4	-11.2
	18	44.2	-0.8	-12.0
	19	47.2	2.2	-9.8
	20	48.1	3.1	-6.7

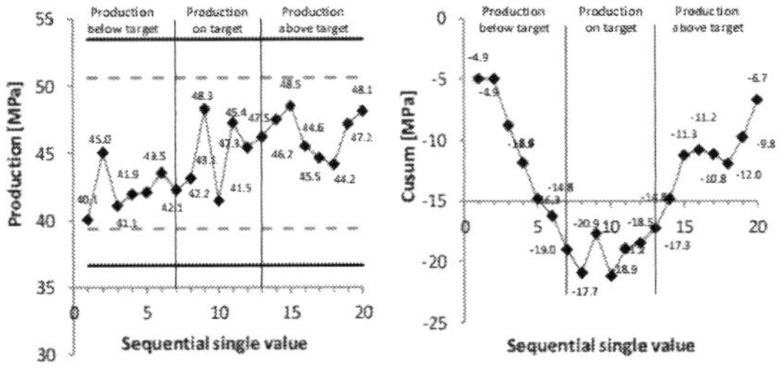

Fig. 1. Control charts for data from Table 2; left: sample shewhard chart with control lines at twofold (dashed) and threefold (full line) σ; right: respective cusum chart.

Error! Reference source not found. show a cusum calculation example with the respective data plot as a Shewhart-chart (left) and as a cusum chart (right). While the Shewhart chart cannot clearly indicate systematic process changes, the cusum plot indicates the values below target (1-7) directly by a downwards trend, and the values above target (14-20) by a steep increasing slope.

Although the cusum delivers a vivid picture of process deviations, it remains difficult to assess quickly, whether corrective actions are required or not. An effective process control should avoid under-control by indicating deviating processes as early as possible to minimise defective production, but it should also avoid over-control, where counteractions are taken although the process deviates within normal scattering ranges. Hence, decision making is always a compromise between over-control and under-control.

A so called V-mask helps to assess whether trending curves show critical trends or not [14-16]. It ideally provides very clear decision rules whether counteractions must be taken or not. V-masks are composed of V-shaped decision lines and a decision interval (DI) that can be superimposed retroactively on the cusum curve. No intervention is necessary as long as the cusum curve is located within the area delineated by the limbs of the V-mask (decision lines). As soon as a decision line is intersected this clearly indicates that immediate corrective actions are required. DI, slope of the decision line (SL), and observation interval form a unit and should be adjusted to the process according to the following considerations:

- The decision interval determines the reliability of the detection.
- The slope of the decision line determines the size of change to be detected.
- The observation interval should be appropriate for the process.

1.3 Cusum process control in concrete technology

In concrete technology, the use of cusum has become popular for the quality control in the UK, South Africa, and Australia [18-23]. In other countries, it is not well established. It has mostly been applied for strength observation, but also as multi-parameter method for e.g. air content and concrete temperature.

Cusum in combination with a V-mask evaluation has also entered the European standards recently since it has been approved to become a method to assess the conformity of the compressive strength in EN 206 as an alternative for the existing mean value criterion. However, cusum was fundamentally developed for process control and not for conformity control, which follows different objectives. The conformity is regulative, has to be observed by third parties, and counteractions in case of non-conformity are mandatory, but not required before the conformity is at risk. The process optimization is not mandatory, but lies within the interest of the producer to produce cost efficient and technologically satisfying. Corrective actions should be taken as early as possible in the interest of the producer. Caspeele and Taerwe proved that depending upon the parameters of the V-mask both conformity control and quality control can be achieved, but a tailoring of the V-mask to the production target value is required [16]. However, these recommendations were not considered in EN 206, so that the recommended standard method of EN 206 is not a feasible tool for process optimization [14, 24].

1.4 Cusum process control for robust rheology

Cusum cannot only be used effectively for conformity control but it can be a viable option for process control. For the concrete production, it has not been used to date. It has largely been applied for mechanical properties of concrete, but strength relevant data typically occur only after a while, so that the strength of the cusum method to react rapidly on systematic changes cannot be efficiently used, even when predictive early strength values are applied.

The strength of cusum is its capacity to identify systematic process changes very rapidly. Therefore, the system is an ideal candidate for the process control and robustness enhancement of the rheology of concrete – much more than for any other concrete property.

The V-mask for cusum can deliver very clear decision lines for the deviation of the slump flow, the V-funnel time, the yield stress, the plastic viscosity, or any other value related to rheology, depending upon the available equipment. However, once the V-mask indicates counteractions, they need to be taken effectively. Ideally, there should be only two options – one possible counteraction per limb of the V-mask.

1.5 Challenges for decision making

The major challenge for the application of a cusum in concrete technology lies in identifying counter actions for changed rheological properties. In concrete, the root causes for process deviations can be vast, and similarly vast can be the possible counteractions. The rheology of concrete can be affected by a vast variety of parameters. With increasing complexity of cementitious binders and considering the complex interaction between cement and rheology modifying admixtures on the micro scale [25-34] the variety of possible rheology changing influences is becoming significantly worse.

Practically this means, segregation can be effectively counteracted among many others by alternatively reducing the water volume in the mix, by increasing the specific surface of the sand, by changing the cement type, by varying the superplasticizer (SP), by adding stabilising agent (STA) or fines, or many other options.

Hence, cusum process control can only effectively work with regard to rapid rheology adjustment, in case very clear and effective counteractions can be defined. Due to the complexity of the overall system, the corrective actions need to be strong rheology modifiers with a good performance predictive capacity. Their effect has to manipulate the rheology of the concrete regardless of the root cause. Due to their strong effect on the flowability, superplasticizers based on polycarboxylate ethers (PCE) and polysaccharide stabilising agents could be potential agents to actively control the rheology and to re-adjust the flowability target value when scatter in the constituents occurs. This study assesses the feasibility of these agents to control the rheology of self-compacting concrete actively by using a cusum control chart in conjunction with V-mask decision making.

2 Experimental setup

The objective of the study is to observe whether a cusum evaluation with V-mask can respond quickly on a variety of influences resulting from modified parameters in the mixture composition. In order to obtain realistic values for mean value changes and standard deviations induced by the changes, experiments were conducted with self-compacting mortar. The reference had slump flow (SF) of 237.7 mm 15 minutes after the mixing.

- The following stiffening manipulations were made, each variation of which was observed with 0.00%, 0.20% and 0.40% of supplementary PCE in addition to the reference dosage of 1.8% PCE. The solid content of the PCE was 30%.
  - -10% water addition
  - +10% cement
  - +10% limestone filler (LSF)
- The following liquefying manipulations were made, each variation of which was observed with 0.00%, 0.03% and 0.06% of ST based on potato starch.:
  - +10% water addition
  - -10% cement
  - -10% limestone filler (LSF)

For all variations arithmetic means (AM) and standard deviations (σ) for the normal processes and manipulated processes were determined as well as for the same processes with different dosages of superplasticizers and stabilising agents. Each test was repeated three times for the determination of the σ and the AM. This allowed assessing and predicting performances of manipulated SCC systems with and without PCE or ST, as well as of normal SCC systems with overdose of PCE or unnecessary dosage of ST. The data was used to derive linear regression functions to predict the required admixture dosages to bring the deviating SF performance back to the set target SF of 237.7 mm, and to predict the performance of the reference mixture with different admixture dosages.

3 Statistical data evaluation

The experimentally derived data were used to generate representative normally distributed random values. The scenario shown in Table 3 was observed for the manipulations described in the experimental setup. The general purpose, standard, truncated V-mask according to BS 5703-2 [17] that was applied every ten values has a DI of 5σ, an SL of 0.5σ, and an observation interval of 10 process values.

Table 3. Observed scenarios with 50 production values.

Prod. values	Situation	Observation target
1-10	Normal production	-
11-20	Defective production	Can defective production be identified quickly?
21-30	Defective production	Can PCE or STA bring a defective production back to target?
31-40	Normal production	Can the return to normal production be identified?
41-50	Normal production	-

3.1 Stiffening effects

The results of the data analysis for the stiffening effects are shown for exemplary data sets in Figs. 2-4. In all cases, the slope of the cusum changes dramatically, after the first manipulation is applied. The reduction of the SF causes a downward trend, since the production values are lower than the target. The limbs of the V-mask are clearly touched, indicating corrective action requirement. After the correction using additional PCE, the cusum trends horizontally again and stays within the limbs of the V-mask until the manipulation is gone but the corrective actions are still taken. This situation can also be clearly identified by a distinct upward slope change of the cusum and by crossing the limbs of the V-mask. After the no longer required corrective actions are stopped, the cusum runs horizontally again.

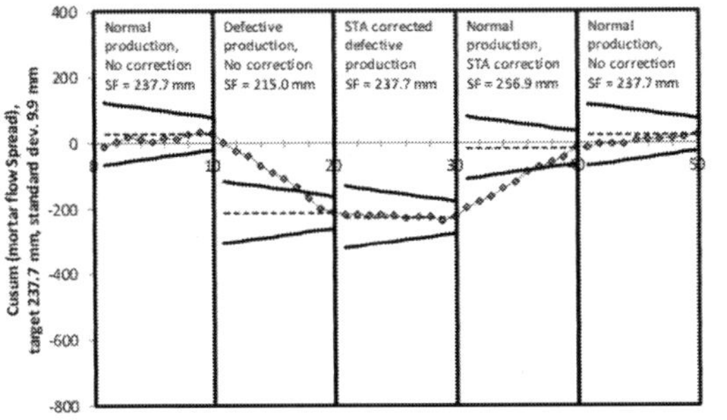

Fig. 2. Sample cusum + V-mask for +10% cement betw. values 11 and 30 and corrective actions (PCE) betw. values 21 and 40.

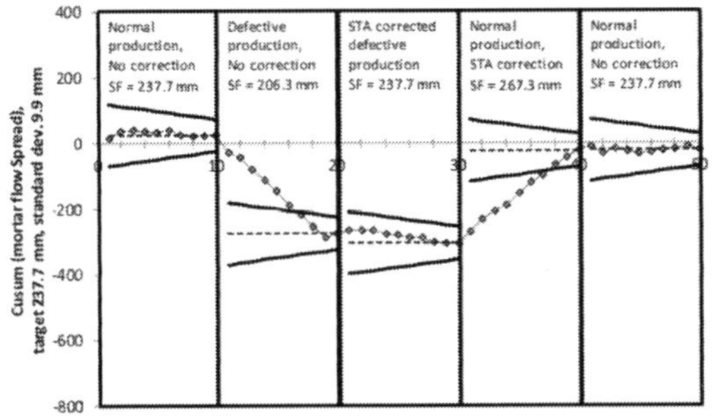

Fig. 3. Sample cusum + V-mask for +10% LSF betw. values 11 and 30 and corrective actions (PCE) betw. values 21 and 40.

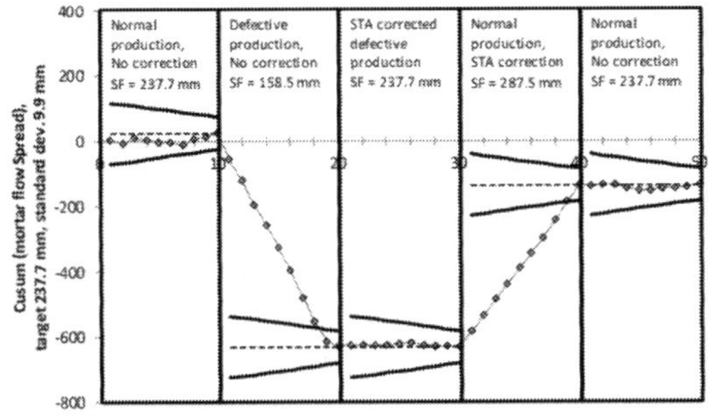

Fig. 4. Sample cusum + V-mask for -10% water betw. values 11 and 30 and corrective actions (PCE) betw. values 21 and 40.

All stiffening manipulations had a very strong effect on the rheological properties. Hence, the slope change of the cusum is quite strong in all cases. The strongest effect can be observed for -10% water addition.

3.2 Liquefying effects

Exemplary data set plots for the liquefying manipulations are shown in Figs. 5-7. The liquefying effects are generally less pronounced than the stiffening effects. However, for less cement and less LSF, the cusum and V-mask clearly shows process changes clearly within the relevant periods. The effect of limestone filler was significantly more pronounced than the effect of cement. For the case of extra water (Fig. 7), the liquefying effect cannot be identified clearly by the standard truncated V-mask, although a prominent change of the slope is clearly visible. Therefore, for small effects as in the case of additional water addition, the standard truncated V-mask according to [17] might not be the best choice. And alternative V-masks should be observed.

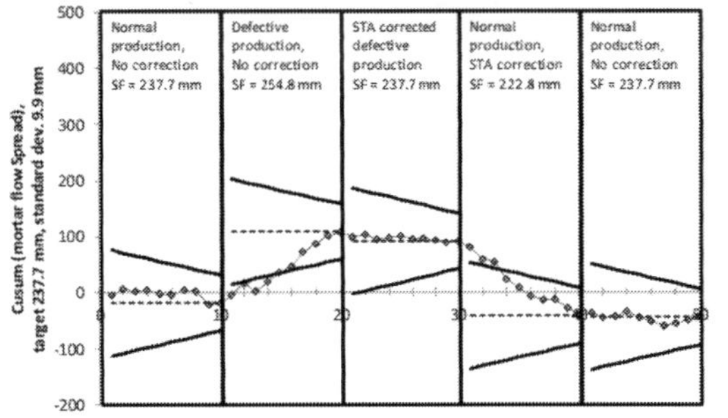

Fig. 5. Sample cusum + V-mask for -10% cement betw. values 11 and 30 and corrective actions (STA) betw. values 21 and 40.

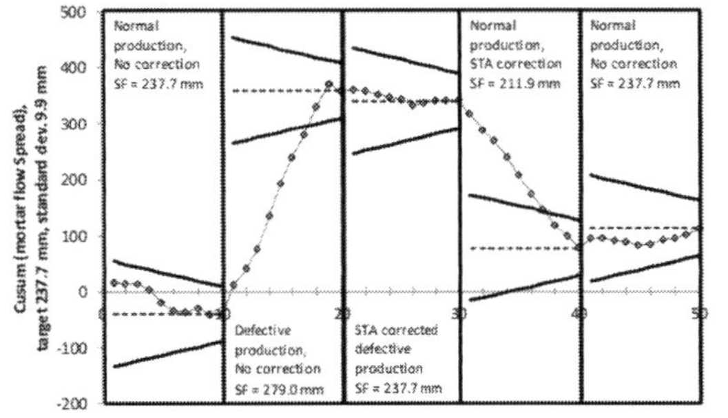

Fig. 6. Sample cusum + V-mask for -10% LSF betw. values 11 and 30 and corrective actions (STA) betw. values 21 and 40.

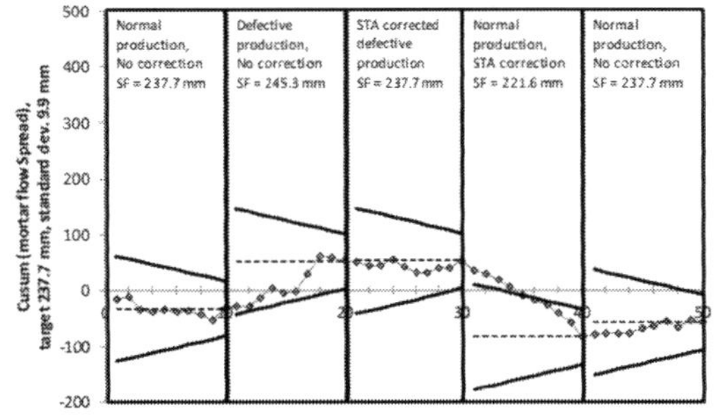

Fig. 7. Sample cusum + V-mask for +10% water betw. values 11 and 30 and corrective actions (STA) betw. values 21 and 40.

3.3 Influence of decision interval and slope

The Figs. 8-9 show cusum curves of the mean values of normally distributed random values with +10% water addition between value 11 and 30. However, different V-mask parameters are observed. Fig. 8 shows that decision intervals of 2.5σ and 1σ can clearly indicate the effect of supplementary water on the rheology clearly and effectively.

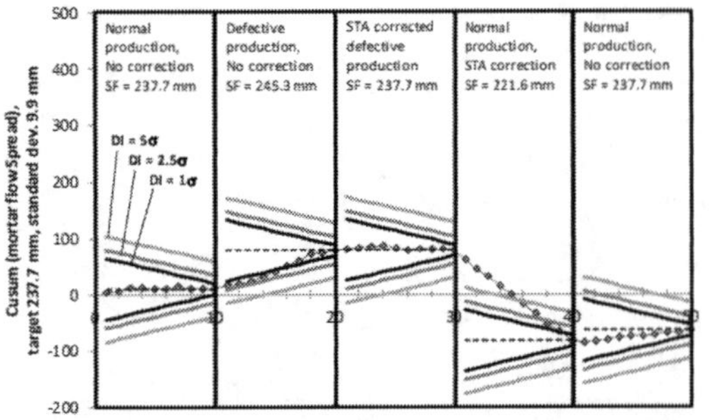

Fig. 8. Cusum of mean values of 10 random normally distributed cusum functions for +10% water betw. values 11 and 30 and corrective actions (STA) betw. values 21 and 40 with different decision intervals (DI) for the V-mask.

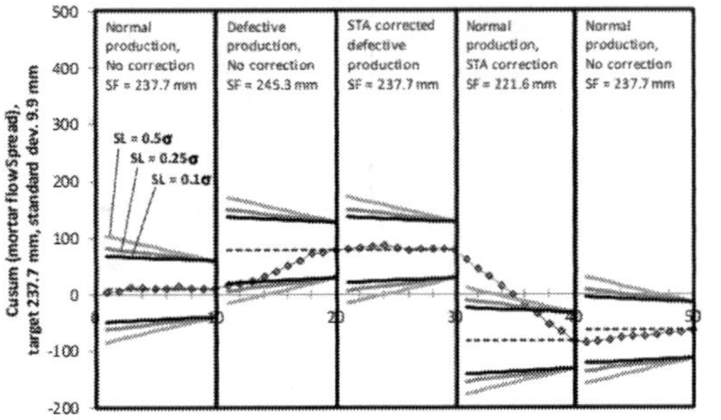

Fig. 9. Cusum of mean values of 10 random normally distributed cusum functions for +10% water betw. values 11 and 30 and corrective actions (STA) betw. values 21 and 40 with different slopes (SL) for the V-mask.

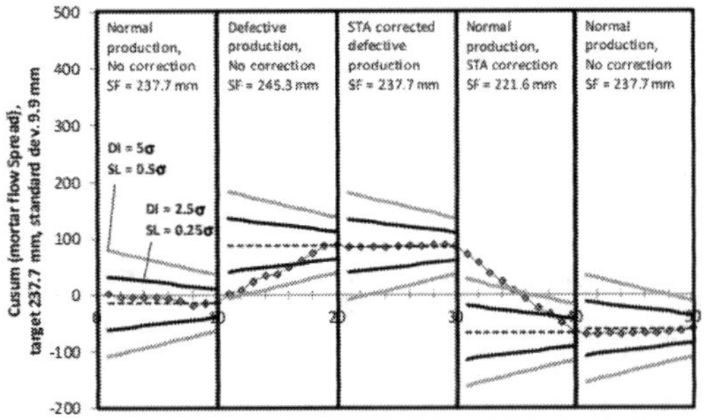

Fig. 10. Comparison of standard V-mask and optimised V-mask for a cusum of mean values of 10 random normally distributed cusum functions for +10% water betw. values 11 and 30 and corrective actions (STA) betw. values 21 and 40

Fig. 9 shows the same situation for different slopes. A slope of 0.1s would have identified the effect of the extra water on rheology also clearly, but compared to the effect of the decision interval, the effect is less pronounced.

The evaluation with an optimised V-mask with DI = 2.5σ and SL = 0.25σ is shown in Fig. 10 compared to the standard V-mask. The optimised V-mask reacts on changes very quick, so that process failures could be indicated already between the first 4-5 defective values. An even more rapid indication would be possible with a narrower V-mask, but it has to be considered that the use of mean values for Figs. 8-10 causes less scatter for individual values. Therefore, a narrower V-mask would increase the risk that a normal production could be falsely interpreted as defective production.

4 Summary and conclusions

Today, robust process control becomes increasingly important for concrete manufacturing. Due to the complexity of concrete mixture compositions, it is difficult to identify root causes for process changes. However, with the use of superplasticizers and stabilising agents it is possible to control the rheology independent of the root cause.

A cusum data processing with V-mask for decision making helps to identify systematic changes in processes rapidly. The results have shown that identification of systematic changes is possible with a standard V-mask, and that defective processes can be corrected effectively with PCE or STA. However, when the process is well understood, tailored V-mask parameters can help to minimise the reaction time. In the present case, a decision interval of 2.5σ and a slope of 0.25σ seemed to be ideal.

With increasing use of computational tools, perspectively it is possible to use the method for in-situ rheology control in complex automated processes such way, that rheology changes are identified in the beginning of the process using cusum and V-mask, so that still during the processing corrective actions can be taken so that at the final casting step the rheology fully achieves the target value.

5 References

[1] W. A. Shewhart, *Economic control of quality of manufactured product*. New York: Van Nostrand, 1931.

[2] E. L. Grandt and R. S. Leavenworth, *Statistical Quality Control*, 6 ed. New York, 1988.

[3] E. S. Page, "A Test for a Change in a Parameter Occurring at an Unknown Point," *Biometrika*, vol. 42, pp. 523-527, 1955.

[4] L. Dessolle, D. J. Biau, V. de Larouzière, C. Ravel, J.-M. Antoine, E. Daraï, and J. Mandelbaum, "Learning curve of vitrification assessed by cumulative summation test for learning curve (LC-CUSUM)," *Fertility and Sterility*, vol. 92, pp. 943-945, 9// 2009.

[5] D. J. Biau, S. M. Williams, M. M. Schlup, R. S. Nizard, and R. Porcher, "Quantitative and individualized assessment of the learning curve using LC-CUSUM," *British Journal of Surgery*, vol. 95, pp. 925-929, 2008.

[6] O. D. Aguirre Ospina, Á. M. Ríos Medina, M. Calderón Marulanda, and L. M. Gómez Buitrago, "Cumulative Sum learning curves (CUSUM) in basic anaesthesia procedures," *Colombian Journal of Anesthesiology*, vol. 42, pp. 142-153, 7// 2014.

[7] A. Young, J. P. Miller, and K. Azarow, "Establishing Learning Curves for Surgical Residents Using Cumulative Summation (CUSUM) Analysis," *Current Surgery*, vol. 62, pp. 330-334, 5// 2005.

[8] A. M. Parikh, A. M. Park, and J. Sumfest, "Cumulative summation (CUSUM) charts in the monitoring of hypospadias outcomes: A tool for

quality improvement initiative," *Journal of Pediatric Urology,* vol. 10, pp. 306-311, 4// 2014.

[9] M. L. Sampson, V. Gounden, H. E. van Deventer, and A. T. Remaley, "CUSUM-Logistic Regression analysis for the rapid detection of errors in clinical laboratory test results," *Clinical Biochemistry,* vol. 49, pp. 201-207, 2// 2016.

[10] R. J. Novick, S. A. Fox, L. W. Stitt, T. L. Forbes, and S. Steiner, "Direct comparison of risk-adjusted and non–risk-adjusted CUSUM analyses of coronary artery bypass surgery outcomes," *The Journal of Thoracic and Cardiovascular Surgery,* vol. 132, pp. 386-391, 8// 2006.

[11] D. G. Pazhayamadom, C. J. Kelly, E. Rogan, and E. A. Codling, "Self-starting CUSUM approach for monitoring data poor fisheries," *Fisheries Research,* vol. 145, pp. 114-127, 8// 2013.

[12] D. G. Pazhayamadom, C. J. Kelly, E. Rogan, and E. A. Codling, "Decision Interval Cumulative Sum Harvest Control Rules (DI-CUSUM-HCR) for managing fisheries with limited historical information," *Fisheries Research,* vol. 171, pp. 154-169, 11// 2015.

[13] L. Qu, Z. Wu, M. B. C. Khoo, and P. Castagliola, "A CUSUM scheme for event monitoring," *International Journal of Production Economics,* vol. 145, pp. 268-280, 9// 2013.

[14] W. Schmidt, H.-C. Kühne, and F. Haamkens, "Shewhart- und CUSUM-Kontrollkarten zur gleichzeitigen Prozesssteuerung und Konformitätsbewertung - Neue Nachweisverfahren der Konformität in der zukünftigen EN 206," *Beton- und Stahlbetonbau,* vol. 109, pp. 417-427, 2014.

[15] W. Schmidt and H.-C. Kühne, "Cusum control charts with V-mask in process control," *CPI International,* 2015.

[16] R. Caspeele and L. Taerwe, "Combined production and conformity control of concrete with acceptance cusum control charts," in *Proceedings of the 7th International Probabilistic Workshop,* Dresden, Germany, 2009.

[17] "BS 5703-2:2003, Guide to data analysis, quality control and improvement using cusum techniques - Part 2: Introduction to decision-making using cusum techniques."

[18] "Monitoring concrete strength by the cusum system," The Concrete Institute, Midrand, South Africa2013.

[19] K. W. Day, "Economical control of concrete quality," *Reprint from Australian Civil Engineering published via* http://www.kenday.id.au/4.1-4.7.html, 1969.

[20] K. W. Day, "Quality Control of Concrete: Philosophy and Implementation," presented at the Seminar on Quality Control of Concrete, Hong Kong, 1980.

[21] K. W. Day, "Quality Control of 55 MPa Concrete for Collins Place Project, Melbourne, Australia," *CONCRETE INTERNATIONAL,* vol. 3, pp. 17-23, 1981.

[22] K. W. Day, *Concrete Mix Design, Quality Control and Specification*, 3 ed.: Taylor & Francis, 2006.

[23] I. Gibb and T. Harrison, "Use of control charts in the production of concrete," ERMCO2010.

[24] W. Schmidt and H.-C. Kühne, "EN 206:2014: Cusum method or mean value criterion? / EN 206:2014: Kusum-Verfahren oder Mittelwertkriterium?," *BFT international,* vol. 80, pp. 24-31, 2014.

[25] M. Palacios, R. Flatt, F. Puertas, and A. Sanchez-Herencia, "Compatibility between Polycarboxylate and Viscosity-Modifying Admixtures in Cement Pastes," in *10th International Conference on Superplasticizers and Other Admixtures in Concrete*, Prague, Czech Republic, 2012, pp. 29-40.

[26] R. Flatt and I. Schober, "Superplasticizers and the rheology of concrete," *Understanding the Rheology of Concrete,* pp. 144-208, 2012.

[27] R. J. Flatt, I. Schober, E. Raphael, E. Lesniewska, and C. Plassard, "Polymer Physics and Superplasticizers," in *Ninth ACI International Conference on Superplasticizers and Other Chemical Admixtures in Concrete*, Seville, Spain, 2009, pp. 113-122.

[28] D. Lowke and C. Gehlen, "Effect of Pore Solution Composition on Zeta Potential and Superplasticizer Adsorption," in *Eleventh International Conference on Superplasticizers and Other Chemical Admixtures in Concrete*, Ottawa, 2015, pp. 253-264.

[29] A. Zingg, F. Winnefeld, L. Holzer, J. Pakusch, S. Becker, and L. Gauckler, "Adsorption of polyelectrolytes and its influence on the rheology, zeta potential, and microstructure of various cement and hydrate phases," *Journal of Colloid and Interface Science,* vol. 323, pp. 301-312, 2008.

[30] J. Plank and B. Sachsenhauser, "Impact of Molecular Structure on Zeta Potential and Adsorbed Conformation of α-Allyl-ω-Methoxypolyethylene Glycol - Maleic Anhydride Superplasticizers," *Journal of Advanced Concrete Technology,* vol. 4, pp. 233-239, 2006.

[31] J. Plank and C. Hirsch, "Impact of zeta potential of early cement hydration phases on superplasticizer adsorption," *Cement and Concrete Research,* vol. 37, pp. 537-542, 2007.

[32] J. Stroh, M. C. Schlegel, W. Schmidt, Y. N. Thi, B. Meng, and F. Emmerling, "Time-resolved in situ investigation of Portland cement hydration influenced by chemical admixtures," *Construction and Building Materials,* vol. 106, pp. 18-26, Mar 1 2016.

[33] W. Schmidt, S. Peters, and H.-C. Kühne, "Effects of Particle Volume Fraction and Size on Polysaccharide Stabilizing Agents," in *SP-302 Eleventh International Conference on Superplasticizers and Other Chemical Admixtures in Concrete*, V. M. Malhotra, Ed., ed, 2015, pp. 39-52.

[34] W. Schmidt, H. J. H. Brouwers, H.-C. Kühne, and B. Meng, "Influences of superplasticizer modification and mixture composition on the performance of self-compacting concrete at varied ambient temperatures," *Cement and Concrete Composites,* vol. 49, pp. 111-126, 2014.

Rheology of Cementitious Materials: Alkali-Activated Materials or Geopolymers

F. Puertas[1], M.M: Alonso[1], S. Gismera[1], M. Lanzón[2] and M.T. Blanco-Varela[1]
[1] Cements andRecycling Materials Dpt. Eduardo Torroja Institute for Construction Sciences (IETcc-CSIC), Madrid, Spain
[2] Arquitectura y Tecnología de la Edificación Dpt. Universidad Politécnica de Cartagena, Cartagena (Murcia), Spain

Abstract. A clear alternative to reach the goal of sustainable development in the Construction Sector is the development of alternative building materials to Ordinary Portland Cement (OPC) in a more energetically as well as environmentally eco-efficient way. Alkaline cements (Alkali-Activated Materials, AAMs) and geopolymers meet these requirements; and they are based on the alkali activation of aluminosilicates (mainly waste and industrial by-products, such as blast furnace slag, fly ash and ceramic waste) in highly alkaline solutions. AAMs cements and concretes are notable for being very durable and mechanically resistant. However, to date their rheological behaviour is not well controlled and there is little understanding of it, with very disparate experimental data. Despite this, their rheological behaviour is not fully understood and little is known on the disparate data obtained in AAM pastes. Moreover, the common additives used in the preparation of OPC concretes and the rheology modifiers/controllers are also unstable in the AAMs systems.
Understanding and controlling the rheology of the AAMs systems will ultimately determine whether they can be implemented in the market, and will open up greater competitive possibilities in a crisis-affected sector. A systematic study of the factors that affect the rheological properties of AAMs (pastes, mortars and concretes) is therefore necessary in order to ultimately develop more resistant and durable materials.

1 Introduction

Rheology is the study of the fluidity and deformation of matter. An understanding of the rheological properties of cement pastes, mortars and concretes is essential to ascertaining their consistency and workability and consequently their ease of casting or placement. In addition, the rheological properties of cement-based materials have a strong influence on their microstructure, mechanical features and durability [1].

A number of methods are in place for determining the fluidity of binders, including slump tests, the compactness factor and the Vebe consistometer. The advantages of these primarily empirical procedures are their low cost, on-site applicability and simplicity. These measurements, may lead, however, to erroneous conclusions, such as assuming that the same slump value indicates similarity between two pastes that may, in fact, be different. Another drawback of these methods is their high sensitivity to the equipments, technicians and test preparations. As a consequence of this, appropriate methods are needed to characterise, with sufficient accuracy, the rheological behaviour of these materials. Cement system fluidity can be determined from physical parameters, such as viscosity or yield stress measured with viscometers or rheometers. Many authors have studied Portland cement paste rheology as a preliminary to determining the rheology of the respective mortars and concretes [2-8].

Rheological testing on Portland cement pastes, normally conducted with viscometers, consists of subjecting the pastes to shear at a given rate to break down the flocs formed mainly due to the hydration of cement grains. The shear rate (rotor speed) is then gradually lowered. As a rule, the down ramp on the shear rate vs shear stress (τ) curve is a straight line that fits the Bingham model (Equation 1): i.e., where the slope is the plastic viscosity (μ) and the yintercept the yield stress (τ_0). Plastic viscosity is related to the number and size of the flocs, while the yield stress is a measure of the strength and number of inter-floc interrelations that are broken down when shear is applied [4].

$$\tau = \tau_0 + \mu\gamma \qquad (1)$$

The literature [1] nonetheless describes other models for cement paste rheological behaviour, such as the Ostwald de Waele and Herschel-Bulkley models (Equations 2 and 3, respectively).

Ostwald de Waele model $\tau = K\gamma^n$ (2)

Herschel-Bulkley model $\tau = \tau_0 + K\gamma^n$ (3)

where K is the consistency coefficient (Pa·s^n) and n is the dimensionless fluidity index.

There are many studies that relate the rheology of OPC concrete with their final performance and behaviour [9]. But very few studies have been conducted on alkaline activated cement paste, mortar or concrete rheology, however.

Alkaline cements (Alkali-Activated Materials, AAMs) and geopolymers cover a wide variety of materials whose differences depend on the origin and composition of the starting aluminosilicates. Alkali-activated slag and fly ash cements are among the most widely known for their availability and high performance. Due to chemical differences in the starting materials (vitreous blast furnace slag and silicoaluminate fly ash) the reaction products and the microstructure of these cements vary considerably. These features determine the properties of the resulting cements and concretes, which have been shown to exhibit suitable strength development and good durability [10-12]. Those properties, along with drying shrinkage, have likewise been proven to depend heavily on the nature of the activating solution [13, 14].

In general terms, the AAMs are characterized by a good mechanical strength and durability against different aggressive agents [12, 15-20]. However, to date little is known on the physical behaviour of AAMs pastes as well as the intrinsic variability observed in rheological tests. What is more, numerous additives and rheology modifiers widely used in OPC pastes are either unstable or show poor performance in AAMs systems.

In the following section, a brief description of important rheological aspects of AAMs systems (pastes, mortars and concretes) will be analysed. Likewise, the best conditions of dosage and preparation of these materials will be described.

2 Rheology of AAMs paste, mortar and concretes

In alkali-activated slag (AAS) cements, the nature of the activator is a key factor in the rheological behaviour of the paste [21]. When the activators are NaOH (N) or NaOH/Na2CO3 (NC) solutions the exhibited consistently (or shear stress) is lower than OPC pastes ones. But when the activator is a waterglass (Wg) solution (sodium silicate hydrate dissolution), the rheology of the pastes is related to the SiO2/Na2O ratio and Na2O concentration. Probably, a primary C-S-H gel is formed in this case due to the interaction between silicate ions and Ca2+ ions from the activator and the slag, respectively. This reaction might explain the fast set of alkali-activated pastes prepared with waterglass. In Figure 1, the variation of shear stress versus time shows the different behaviour observed in OPC and AAS pastes.

Considering the above mentioned facts, longer mixing times could be beneficial (or even necessary) when preparing handling Wg-ASS mixtures.

The rheology of alkali-activated fly ashes (AAFA) pastes depends on temperature; at room temperature the rheology of these pastes is quite similar to those of OPC. An increase of the concentration of alkaline activator leads to an increase of the viscosity of the pastes.

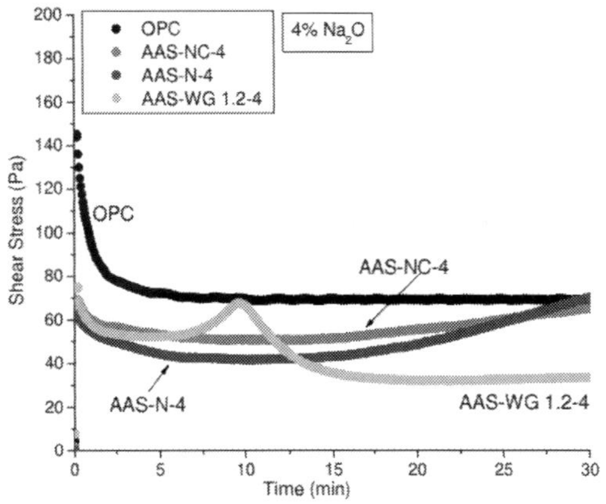

Fig. 1. Shear stress versus time in AAS pastes tested at a constant shear rate of 100 s-1: effect of the nature of the activator at constant concentration (4 % Na2O by slag mass)

According to different studies, the workability of AAS and AAFA could be more sensitive than that observed in OPC mortars. The effect is more evident when the Liquid/Solid ratio or the amount of aggregate is modified in AAS and AAFA mixtures. Moreover, workability is much more dependent upon the nature of the activator in AAS than in AAFA mortars [22].

OPC, AAS and AAFA mortar rheology was shown to be comparable, although certain considerations around AAMs, such as precursor and activator nature, must be borne in mind. The presence of waterglass (Si) in the activator played a significant role in the rheology of AAS and AAFA mortars, which should be taken into consideration when alkaline-activated materials are casted in place.

At a dry consistency, OPC mortars were found to be more workable than AAS and AAFA mortars at the same L/S ratio. In contrast, after reaching plasticity, AAS and AAFA mortars activated with silicon-containing solutions had greater workability than the respective OPC mortars. In other words, the fluidising effect of silicon was observed in plastic mortars, but not in dry consistency mortars. Under the latter conditions, factors such as particle packing and the physical properties of the precursors and activators (viscosity and surface tension, for instance) appeared to prevail.

Previous studies on pastes and mortars demonstrate the positive effect of mixing time on the rheological behaviour of AAS systems. The slump and rheological behaviour of OPC concrete and AAS concrete are different, especially when waterglass was the alkaline activator. As in pastes and mortars the nature of alkaline activator is the main factor in the rheological behaviour of AAS concretes. Longer mixing protocol has a negative effect on rheological parameters (e.g. dynamic and static yield stress) of OPC and NaOH-AAS concretes. That mixing protocol may produce irreversible coagulation among cement particles that lead to a lower workability. In contrast, in Waterglas-AAS concrete, longer mixing times is beneficial to obtain workable mixtures. In this case, longer mixing times would be required to break down flocs within the microstructure, thus improving their rheological behaviour of the pastes and concretes [23].

3 Final remarks

The rheology of AAMs is quite different to OPC ones, and it is strongly influenced by the particular nature of the silicoaluminate which is used as precursor. When the precursor is a ground vitreous blast furnace slag (AAS), the nature of the alkaline activator is the key factor in the workability of the system. If waterglass is used as activator, additional mixing time is required to obtain adequate slump and workability. No changes are needed when the activators are NaOH or NaOH/Na2CO3 solutions. Finally, the rheology of pastes, mortars and concretes prepared with siliceous fly ash as precursor, is comparable to that observed in OPC systems.

Acknowledgements

The authors thank the Spanish Ministry of Economy, Industry and Competitiveness for funding the Project BIA2013-47876-C2-1-P.

References

1. G.H. Tattersall, P.F.G. Banfill The reology of fresh concrete. Ed, Pitman, Advanced Publishing Program (1983).
2. J. Miranda, V. Flores-Alés, J. Barrios, J. Mater. Construcc. **50**, 257, 47-55 (2000).
3. C.F.Ferraris, K.H. Obla, R. Hill, R. Cem Concr. Res. 31, 245-255 (2001).
4. P.F.G. Banfill, Proc. XI ICCC (Durban, South Africa), pp. 50-62 (2003).
5. J. Golaszewski, J. Cement Wapno Beton. **1**, 17-28 (2006)
6. O. Burgos-Montes, M. Palacios, P. Rivilla, F. Puertas, F. Constr Build Mater **31**, 300-309 (2012).
7. M.M: Alonso. M. Palacios, F. Puertas, A.G. de la Torre, M.A.G. Aranda Mater Construcc. **57** 286: 65-81 (2007
8. M.M. Alonso, M. Palacios, F. Puertas Cem. Concr. Comp, **35** 151-162 (2013).
9. K. Kovler, N. Roussel, N. Cem Concr Res **41**, 775-792 (2011).
10. C. Shi, A. Fernández-Jiménez, A. Palomo Cem. Concr. Res., **41** (7), 750-763 (2011).
11. C. Duran Atiş, C. Bilim, Ö. Çelik, O. Karahan. Constr.Build. Mater **23** (1) 548-555 (2009).
12. A. Fernández-Jiménez, J.G. Palomo, F. Puertas, Cem. Concr. Res., **29** (8), 1313-1321 (1999).
13. M. Palacios, F. Puertas Cem. Concr. Res. *37* (5) 691-702 (2007).
14. F. Puertas, M. Palacios, T. Vázquez J. Mater. Sci., 41, 3071–3082 (2006).
15. A. Favier, G. Habert, J.B.D'Éspinose De Lacaillerie, N. Roussel Cem. Concr. Res., **48**, 9-16 (2013).
16. F. Pacheco-Torgal, J.A. Labrincha, C. Leonelli, A. Palomo, P. Chindaprasirt Handbook of Alkali-activated cements, mortars and concretes, Woodhead Publishing (2015).
17. C. Shi, P. Krivenko, D. Roy. Alkali-Activated Cements and Concretes, Taylor and Francis, London and New York. (2006).
18. J.L. Provis, A. Palomo, C. Shi Cem. Concr. Res. **78** 110–125(2015)
19. J.L. Provis, Mater. Struct. **47,** 11–25 (2014)
20. J.L. Provis, S.J. Van Deventer. Alkali-Activated Materials. State of Art Report, RILEM TC 224-AAM. Ed. Springer (2014).
21. F. Puertas, C. Varga, M.M. Alonso Cem Concr Com, **53**, 279-288 (2014).
22. M.M. Alonso, S. Gismera, M. T. Blanco, M. Lanzón, F. Puertas. Construc Build Mat, **145**, 576-587 (2017)
23. F. Puertas, M.M. Alonso, M. Torres-Carrasco, B. González-Fonteboa, I. González-Taboada, G. Rojo, F. Martínez-Abella. 10th ACI/RILEM Int Conf Cement Mat and Alternative Binders for Sustainable Concrete. Montreal (2017)

X-ray diffraction, cements and environment, three worlds in one.

Angeles G. De la Torre[1], Diana Londono-Zuluaga[1,5] Jesus D. Zea-Garcia[1], Marta García-Maté[2], Gema Álvarez-Pinazo[2], Miguel A. G. Aranda[1,3], Isabel Santacruz[1], Ana Cuesta[3], Laura León-Reina[4], Francisco Franco[1], Jorge I. Tobón[5].
[1] Universidad de Málaga, Departamento de Química Inorgánica, Cristalografía y Mineralogía, Málaga, Spain
[2] X-Ray Data Services S.L., Málaga, Spain.
[3] ALBA Synchrotron, Barcelona, Spain
[4] Universidad de Málaga, Servicios Centrales de Apoyo a la Investigación, Málaga, Spain
[5] Universidad Nacional de Colombia, Grupo del Cemento y Materiales de Construcción, CEMATCO, Medellín, Colombia

Abstract. This keynote lecture will be focused on the strategies for reducing CO_2 emissions in the cement production. Concretely, the production of ecocements with optimised formulations that yield reductions in CO_2 emissions of up to 25%, when compared to OPC production. Phase assemblage has to be carefully optimised to be competitive and these new ecocements should develop compressive strengths of at least 50 MPa at 28 days of hydration. Optimised compositions of several ecocements will be discussed, but all of them are ye'elimite or calcium sulphoaluminate containing ones: belite-ye'elimite-ferrite (BYF), belite-alite-ye'elimite (BAY) and ye'elimite rich ones (CSA). The clinkering temperature of BYF and BAY has to be established to obtain the targeted phase assemblages. Moreover, the stabilisation of alpha-forms of belite is needed to develop high mechanical strengths at early ages. The benefits of the use of waste materials (such as fly ash or slag) as additions to ecocements are three-fold: lower CO_2 emissions due to clinker replacement; valorisation of "useless" products that need a lot of landscape and the consequent efficient consumption of raw materials; and to enhance mechanical properties of the corresponding mortars. The design of appropriate CSA, BYF and BAY mortars, with the final aim of knowing and controlling the hydration mechanisms, will be presented. Particularly, the role of i) type and amount of set regulator (gypsum, anhydrite, etc.), ii) water/cement ratio (w/c); iii) superplasticiser; and iv) pozzolanic additions will be discussed. The role of these parameters in the microstructure and hydraulic behaviour has been investigated through traditional techniques as well as advanced synchrotron characterisation. The formers include laboratory/synchrotron X-ray powder diffraction combined with Rietveld methodology (to obtain phase assemblage), electron microscopy techniques for paste microstructure determination, rheological studies (to control the effect of the different additives, w/c ratio and setting time retarders) and mechanical tests (setting times, compressive strengths and dimensional stability). The latters comprise a group of techniques available at synchrotrons such as: i) high temperature x-ray diffraction for clinkering studies and ii) total scattering data to be analysed by pair distribution function, PDF.

1 Introduction

Ye'elimite containing-cements are considered as eco-cements as less CO_2 is released during their production compared to OPC. This is due to the presence of ye'elimite in their composition, which is a low calcite-demanding phase (0.22 t/t phase). Other phases present in OPC such as alite (0.58 t/tphase), belite (0.51 t/tphase), tricalcium aluminate (0.49 t/tphase) or calcium aluminate ferrite (0.22 t/tphase) are high calcite-demanding phases. For this reason, and their high performances, the research on these cements has recently been increased [1]. The mineralogy of these new binders is of great importance to understand and predict final performances of the derived mortars and concretes.

Ye'elimite, also called Klein's salt or tetracalcium trialuminate sulphate $C_4A_3\hat{S}$, crystallizes as an orthorhombic phase [2]. Moreover, the presence of foreign ions in the structure stabilizes a pseudo-cubic form [3]. Ye'elimite-containing cements can be gathered in three main groups [1]: i) Calcium Sulpho-Aluminate (CSA), ii) Belite Calcium Sulpho-Aluminate or sulphobelite (BYF, from Belite-Ye'elemite-Ferrite cements) and iii) Belite-Alite-Ye'elimite (BAY) cements.

In the last ten years, research has been ongoing to solve the technological problem of BYF, which is the low mechanical strengths developed at early ages [4–6] by stabilising high-temperature belite polymorphs (α-forms) by the addition of minor elements. Another strategy to active these ye'elimite containing cements is the production of a cement with both alite and belite (BAY) [7–9].

The hydration reactions involving ye'elimite have to be fully understood to control and tune-up these type of cements. There is some controversy in the literature data about the hydration behaviour of ye'elimite in absence of other soluble sulphate groups [10–12]. In addition, belite reactivity is important, especially for BYF and BAY cements. Independently on the polymorphism, the hydration behaviour of belite may yield a nanocrystalline C-S-H gel, similar to that arising from the hydration of alite. However, in hydrated BYF systems, belite coexists with a hydrated amorphous aluminate rich environment which favours the formation of an AFm-type phase called stratlingite [4,13,14], which may play an important role in mechanical strengths [15].

2 Clinkering of BYF and BAY.

2.1 Clinkering in-situ Synchrotron X-ray powder diffraction (SXRPD) study of BYF and aBYF.

Clinkering reactions of BYF and aBYF (active BYF) were studied by in-situ SXRPD. Several reactions were followed including α'_H-C_2S → β-C_2S polymorphic transformation, the decomposition of ye'elimite and the melting of C_4AF. Overall, these temperatures decrease with the addition of B_2O_3 [16]. Figure 1 shows the phase evolution, including the melted phase, for aBYF and a Rietveld plot of this clinker at 1586K as an example. The most important results were that i) the α'_H→β polymorphic transformation temperature was reduced from ~1650 K in BYF to 1250 K in aBYF; ii) the decomposition temperature of ye'elimite was also lowered from ~1550 K in BYF to ~1500 K in aBYF. Furthermore, the addition of B_2O_3 also decreased the melting temperature of C_4AF from ~1550 K to ~1430 K.

2.2 Clinkering of BYF and BAY: Activation.

The "medium-scale" synthesis (~2 kg) of BYF and BAY clinkers was performed by mixing the same raw materials commonly used in cement plants.

The targeted composition of BYF clinker was ~50 wt% of C_2S, ~30 wt% of $C_4A_3\hat{S}$ and ~20 wt% of C_4AF and that of BAY clinkers was 25-45 wt% of C_2S, 30-45 wt% of C_3S, ~20 wt% of $C_4A_3\hat{S}$ and ~5 wt% of C_4AF. In the case of BAY clinkers, only CaF_2 or CaF_2 and ZnO (BAYZ) were added, to success in the preparation of a clinker with both alite and ye'elimite phases. Moreover, 2.0 wt% and 0.63 wt% of B_2O_3, added as borax, was used to promote the stabilisation of the high temperature polymorphs of belite, in BYF and BAY, respectively, hereafter aBYF and aBAY.

In all cases, the raw mixture (~3 kg) was homogenised for 30 min in a micro-Deval machine. The resulting mixture was pressed into pellets of about 40 g (55 mm of diameter and ~5 mm of height) and 6 of these ones were placed in a large Pt/Rh crucible. A two-step clinkering process was used for all the clinkers: 900°C during 30 min (heating rate of 5 °C/min) and further heated at 1300°C for 15 min, followed by a quenching with air flow. Table 1 gives the mineralogical phase assemblage of these clinkers.

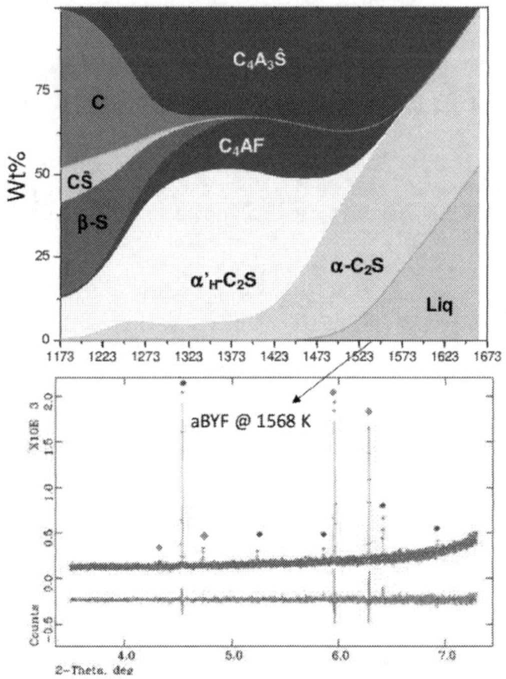

Fig. 1. Normalised phase evolution with temperature aBYF clinkering (top) and Rietveld plot of aBYF at 1568 K (bottom). Orange rhombus stand for α-C_2S and blue circles for ye'elimite.

On the one hand, it is noticeable that the addition of borax to raw mixtures has totally or partially stabilised α'_H-C_2S and pseudo-cubic ye'elimite. On the other hand, the preparation and scale-up of BAY and BAYZ clinkers with the coexistence of alite and ye'elimite phases [17] by adding CaF_2 and ZnO was a success.

Table 1. Mineralogical phase composition of ye'elimite containing clinkers synthetized at medium scale (~2 kg) in the laboratory.

Phase	BYF	aBYF	BAY	BAYZ	aBAYZ
α'_H-C_2S	-	57(2)			18.1(3)
β-C_2S	48.7(6)		59.4(2)	28.4(3)	42.1(2)
γ-C_2S	2.6(5)		1.2(1)		
C_3S			14.3(2)	41.5(2)	15.8(2)
o-$C_4A_3\hat{S}$	15(1)		10.4(1)	9.3(3)	2.9(2)
c-$C_4A_3\hat{S}$	14(1)	31(2)		8.1(3)	4.8(2)
C_4AF	14.9(2)	10.1(6)	6.5(2)	3.8(1)	3.2
Minor phases	5.7	2.1	5.0(1)	9.0(1)	13.1

3 Hydration of ye'elimite containing cements: CSA, BYF and BAY.

3.1 Role of activation of ye'elimite bearing cements.

Mortars prepared with aBYF cement developed higher compressive strengths than BYF at any hydration age and BAY at early hydration time (up to 7 days), and higher strengths than BAY at 28 d, Figure 2, independently of the type and amount of sulphate source [6,18].

Standard BYF clinker contains β-C_2S, as main phase and orthorhombic ye'elimite. It has been demonstrated that β-C_2S in these experimental conditions reacts at a higher pace than alpha forms in active BYF [6]to yield stratlingite. This behaviour may well be justified with the formation of high amounts of ettringite at early hours which implies a concomitant large quantity of amorphous aluminium hydroxide. The availability of amorphous AH_3 promotes the precipitation of stratlingite, C_2ASH_8, from belite reaction. In conclusion, the hydration behaviour of C_2S is likely more dependent on the chemical environment than on its polymorphism. However, it must be underlined that although the phases in BYF react faster than in aBYF, the standard cement develops much smaller compressive strength, Figure 2.

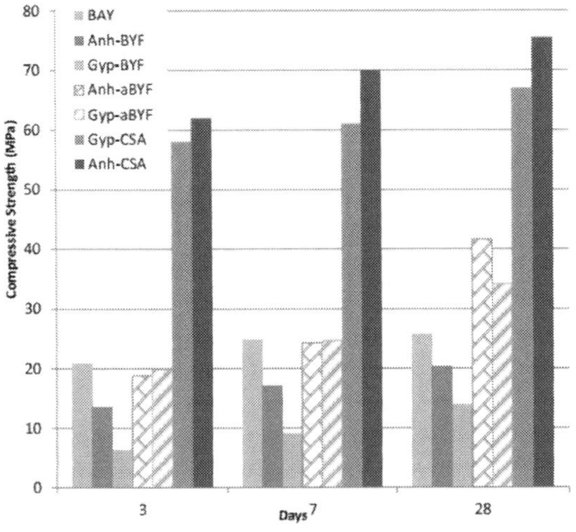

Fig. 2. Compressive strengths of CSA, BYF and BAY mortars.

Furthermore, the activation by synthesising clinkers with alite and ye'elimite, i.e. BAY clinkers, has been a success since BAY mortars show higher compressive strengths than BYF mortars, prepared under similar conditions (anhydrite as setting regulator and w/c) at any hydration time, Figure 2.

None of BYF or BAY cements have developed higher mechanical strengths than CSA cements, Figure 2. However, CSA cements are more expensive due to the need of high amounts of Al_2O_3, mainly as highly pure bauxite. So BYF and BAY are proposed as a suitable cheap environmental solution.

3.2 Role of type of setting regulator: gypsum, bassanite or anhydrite

The effect on hydration behaviour and mechanical strengths of three different setting regulators (gypsum, bassanite and anhydrite) has also been studied. The dissolution rate of the different studied sulphate sources is a key point to control the reactions during hydration of CSA and BYF cements, and hence the mechanical properties of the corresponding pastes and mortars.

The dissolution of anhydrite is the slowest of the three sulphate sources. Cements and mortars prepared with anhydrite, showed higher initial setting times than those prepared with gypsum. Consequently, mechanical strengths after 3 days of hydration, Figure 2, are lower than those of mortars prepared with CSA or BYF with gypsum. However, these longer setting times of anhydrite pastes provides them with high plasticity to accommodate the precipitation of ettringite at later ages, and thus mortars showed similar compressive strength values Figure 2.

CSA or BYF with bassanite pastes showed high water demand, high viscosity values and short setting times with the consequent dramatic effect on the mechanical strength values of the corresponding mortars. This makes necessary the selection of a hydration retarder which also reduces the viscosity of the pastes [18,19], Figure 2.

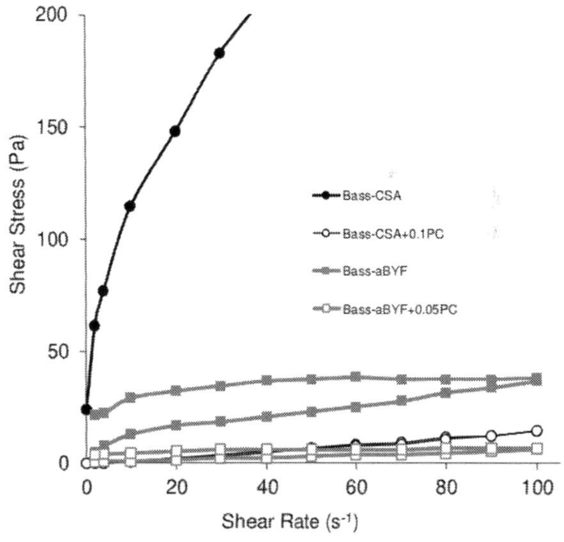

Fig. 3. Flow curves of CSA and aBYF pastes prepared with bassanite as setting regulator without and with (open symbols) a superplasticizer/retarder based on polycarboxylate.

The effect of three additives (polycarboxylate, tartaric acid and phosphonic acid) on CSA pastes and mortars was studied [20]. The amount of each additive was optimised elsewhere [20] and all of them improved the workability of pastes, where the polycarboxylate (PC) was the best additive in reducing viscosity at very early ages. All these parameters could be controlled, making possible the design and preparation of mortars with selected properties for specific engineering applications

[20]. Figure 3 shows the flow curves of aBYF and CSA pastes prepared with bassanite without and with a small amount of PC.

4 Hydration of pure phases related to CSA, BYF and BAY.

4.1 Hydration of ye'elimite.

The crystal structures of orthorhombic ye'elimite [2] and pseudo-cubic ye'elimite [3] were studied. Moreover, the hydration behaviour of both polymorphs with or without sulphate source was studied [21]. These studies unravelled that hydration kinetics of ye'elimite samples not only depends on the water/solid (w/s) ratio and the solubility of the additional sulphate source, but also on the polymorphism of ye'elimite. Firstly, in the absence of additional sulphate sources, orthorhombic ye'elimite reacts slower than pseudo-cubic ye'elimite and the latter produces higher amounts of ettringite than orthorhombic ye'elimite in similar hydrating conditions. Moreover, in this environment, the formation of AFm-type phases from orthorhombic ye'elimite is strongly accelerated and favoured by high w/s ratios. Moreover, in presence of gypsum or anhydrite as additional sulphate sources, orthorhombic ye'elimite reacts faster than pseudo-cubic ye'elimite. In addition, the formation of AFm-type phases at late ages is avoided by the addition of gypsum and anhydrite being ettringite the only crystalline phase formed in this reaction medium.

Furthermore, the particle size [22], water content and the calcium uptake of amorphous aluminium hydroxide gels in ye'elimite eco-cements have been established with Ca contents being minor but not negligible [23].

4.2 Hydration of ye'elimite with ferrite.

The jointly hydration of ye'elimite and ferrite was also studied [24]. The hydration of a sample with ferrite, gypsum and orthorhombic ye'elimite gives a mixture of AFt and AFm phases. However, C_4AF hydration is slightly slowed down and moreover, these hydration products did not contain iron in their structures. The retarder effect of ye'elimite over C_4AF hydration was much stronger with pseudo-cubic ye'elimite and consequently AFt was the main hydration phase.

C_4AF was hydrated with ye'elimite in the absence of gypsum and gave AFm as the unique hydration product, Figure 4. Consequently, it was demonstrated that gypsum (which is a high soluble sulphate) was the main responsible for AFt formation in these systems.

4.3 Hydration of ye'elimite with belite and alite.

The reactivity of ye'elimite in the presence of both dicalcium silicate and alite has also been studied. Both polymorphs, α'_H and β of dicalcium silicate were synthesized [25,26] and mixed jointly with both ye'elimites and gypsum. Moreover, mixtures of alite-ye'elimites and anhydrite were also prepared. The mixtures were hydrated. Table 2 gives the mineralogical composition, including amorphous contents (ACn) and free water (FW) of mixtures after 28 days of hydration of mixtures of C_2S and ye'elimite with gypsum. In these experimental conditions the presence of belite in the mixture has not altered the hydration kinetics of ye'elimite. Pseudo-cubic ye'elimite (with a Dv50 of 5 μm) reacts at a higher pace than orthorhombic (with a Dv50 of 9 μm) yielding higher amounts of AFt, Table 2. This last conclusion is also extensible to the mixtures containing alite.

Moreover, we have demonstrated that the hydration degree of α'_H-C_2S is much larger than that of β-C_2S.

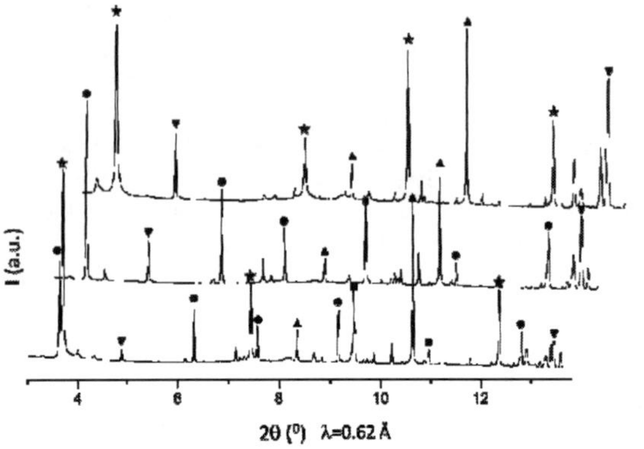

Fig. 4. SXRPD patterns of mixtures of (bottom) orthorhombic ye'elimite/ferrite/gypsum, (medium) pseudocubic ye'elimite/ferrite/gypsum and (top) orthorhombic ye'elimite/ferrite after ~20 hours of hydration. AFt: circle, AFm: star, Quartz: triangle; ye'elimite: square; C_4AF: inverted triangle.

Table 2. Quantitative phase analysis results (wt%) of mixtures of orthorhombic or pseudo-cubic ye'elimite (o-$C_4A_3\hat{S}$ or c-$C_4A_3\hat{S}$), **β-C_2S or α'_H-C_2S** with gypsum after 28 days of hydration obtained from LXRPD.

	β-C_2S-o-$C_4A_3\hat{S}$	**α'_H-C_2S o-$C_4A_3\hat{S}$**	**β-C_2S-c-$C_4A_3\hat{S}$**	**α'_H-C_2S c-$C_4A_3\hat{S}$**
$C_4A_3\hat{S}$	0.9(1)	2.6(1)	-	-
α'_H-C_2S	-	13.9(4)	-	11.8(4)
β-C_2S	20.6(2)	9.3(3)	20.1(2)	9.8(3)
Strat	6.5(3)	6.4(3)	6.7(2)	10.2(3)
AFt	**21.1(2)**	**19.9(2)**	**23.5(2)**	**23.3(2)**
Gib	0.5(1)	0.5(1)	0.5(1)	0.4(1)
ACn	33.0(5)	31.8(6)	32.9(4)	32.7(6)
FW	15.1	14.9	14.3	11.8

Acknowledgements

PhD D. Londono-Zuluaga thanks Colciencias and Enlaza Mundos program PhD grant. Spanish MINECO (BIA2014-57658-C2-2-R, which is co-funded by FEDER, BIA2014-57658-C2-1-R and I3 (IEDI-2016-0079) grants) are acknowledged.

References

[1] M.A.G. Aranda, A.G. De la Torre, 18 –

Sulfoaluminate cement, in: Eco-Efficient Concr., 2013: pp. 488–522. doi:10.1533/9780857098993.4.488.

[2] A. Cuesta, A.G. De La Torre, E.R. Losilla, V.K. Peterson, P. Rejmak, A. Ayuela, C. Frontera, M.A.G. Aranda, Structure, atomistic simulations, and phase transition of stoichiometric yeelimite, Chem. Mater. 25 (2013) 1680–1687. doi:10.1021/cm400129z.

[3] A. Cuesta, Á.G. De La Torre, E.R. Losilla, I. Santacruz, M.A.G. Aranda, Pseudocubic crystal structure and phase transition in doped ye'elimite, Cryst. Growth Des. 14 (2014) 5158–5163. doi:10.1021/cg501290q.

[4] G.S. Li, E. Gartner, High-belite sulfoaluminate clinker: fabrication process and binder preparation, French patent application 04-51586 (publication 2873366), 2006.

[5] A.J.M. Cuberos, A.G. De La Torre, G. Alvarez-Pinazo, M.C. Martín-Sedeño, K. Schollbach, H. Pöllmann, M.A.G. Aranda, Active iron-rich belite sulfoaluminate cements: clinkering and hydration., Environ. Sci. Technol. 44 (2010) 6855–6862.

[6] G. Álvarez-Pinazo, I. Santacruz, L. León-Reina, M.A.G. Aranda, A.G. De la Torre, Hydration Reactions and Mechanical Strength Developments of Iron-Rich Sulfobelite Eco-cements, Ind. Eng. Chem. Res. 52 (2013) 16606–16614. doi:10.1021/ie402484e.

[7] N. Chitvoranund, B. Lothenbach, F. Winnefeld, C.W. Hargis, Synthesis and hydration of alite-calcium sulfoaluminate cement, Adv. Cem. Res. (2016).

[8] Y. Hu, S. Ma, W. Li, X. Shen, The Preparation and Composition Analysis of Alite-Ye'Elimite With Industrial Wastes, Ceram. - Silikaty. 60 (2016) 1–10. doi:10.13168/cs.2016.0027.

[9] S. Ma, R. Snellings, X. Li, X. Shen, K.L. Scrivener, Alite-ye'elimite cement: Synthesis and mineralogical analysis, Cem. Concr. Res. 45 (2013) 15–20. doi:10.1016/j.cemconres.2012.10.020.

[10] F. Winnefeld, S. Barlag, Calorimetric and thermogravimetric study on the influence of calcium sulfate on the hydration of ye ' elimite, J. Therm. Anal. Calorim. 101 (2010) 949–957. doi:10.1007/s10973-009-0582-6.

[11] F. Song, Z. Yu, F. Yang, Y. Lu, Y. Liu, Microstructure of amorphous aluminum hydroxide in belite-calcium sulfoaluminate cement, Cem. Concr. Res. 71 (2015) 1–6. doi:10.1016/j.cemconres.2015.01.013.

[12] A. Telesca, M. Marroccoli, M.L. Pace, M. Tomasulo, G.L. Valenti, P.J.M. Monteiro, A hydration study of various calcium sulfoaluminate cements, Cem. Concr. Compos. 53 (2014) 224–232. doi:10.1016/j.cemconcomp.2014.07.002.

[13] K. Scrivener, R. Snellings, B. Lothenbach, A Practical Guide to Microstructural Analysis of Cementitious Materials, CRC Press, Boca Raton, FL, 2016.

[14] B. Lothenbach, K. Scrivener, R.D. Hooton, Supplementary cementitious materials, Cem. Concr. Res. 41 (2011) 1244–1256. doi:10.1016/j.cemconres.2010.12.001.

[15] I. Mehdipour, A. Kumar, K.H. Khayat, Rheology, hydration, and strength evolution of interground limestone cement containing PCE dispersant and high volume supplementary cementitious materials, Mater. Des. 127 (2017) 54–66. doi:10.1016/j.matdes.2017.04.061.

[16] Á.G. De La Torre, A.J.M. Cuberos, G. Álvarez-Pinazo, A. Cuesta, M.A.G. Aranda, In situ powder diffraction study of belite sulfoaluminate clinkering, J. Synchrotron Radiat. 18 (2011) 506–514.

[17] D. Londono-Zuluaga, J.I. Tobon, M.A.G. Aranda, I. Santacruz, A.G. De La Torre, Clinkering and hydration of Belite-Alite-Ye´elimite cement, Cem. Concr. Compos. 80 (2017) 333–341. doi:10.1016/j.cemconcomp.2017.04.002.

[18] G. Álvarez-pinazo, I. Santacruz, Hydration of belite – ye'elimite – ferrite cements with different calcium sulfate sources, Adv. Cem. Res. 28 (2016) 529–543. doi:10.1680/jadcr.16.00030.

[19] M. García-Maté, A.G. De La Torre, L. León-Reina, E.R. Losilla, M.A.G. Aranda, I. Santacruz, Effect of calcium sulfate source on the hydration of calcium sulfoaluminate eco-cement, Cem. Concr. Compos. 55 (2015) 53–61. doi:10.1016/j.cemconcomp.2014.08.003.

[20] M. García-Maté, D. Londono-Zuluaga, A.G. De la Torre, E.R. Losilla, A. Cabeza, M.A.G. Aranda, I. Santacruz, Tailored setting times with high compressive strengths in bassanite calcium sulfoaluminate eco-cements, Cem. Concr. Compos. 72 (2016) 39–47. doi:10.1016/j.cemconcomp.2016.05.021.

[21] A. Cuesta, G. Álvarez-Pinazo, S.G. Sanfélix, I. Peral, M.A.G. Aranda, A.G. De la Torre, Hydration mechanisms of two polymorphs of synthetic ye'elimite, Cem. Concr. Res. 63 (2014) 127–136. doi:10.1016/j.cemconres.2014.05.010.

[22] A. Cuesta, R.U. Ichikawa, D. Londono-Zuluaga, A.G. De la Torre, I. Santacruz, X. Turrillas, M.A.G. Aranda, Aluminum hydroxide gel characterization within a calcium aluminate cement paste by combined Pair Distribution Function and Rietveld analyses, Cem. Concr. Res. 96 (2017) 1–12. doi:10.1016/j.cemconres.2017.02.025.

[23] A. Cuesta, A.G. De la Torre, I. Santacruz, P. Trtik, J.C. da Silva, A. Diaz, M. Holler, M.A.G. Aranda, Chemistry and Mass Density of Aluminum Hydroxide Gel in Eco-Cements by Ptychographic X-ray Computed Tomography, J. Phys. Chem. C. 121 (2017) 3044–3054. doi:10.1021/acs.jpcc.6b10048.

[24] A. Cuesta, I. Santacruz, S.G. Sanfélix, F. Fauth, M.A.G. Aranda, A.G. De la Torre, Hydration of C4AF in the presence of other phases: A

synchrotron X-ray powder diffraction study, Constr. Build. Mater. 101 (2015) 818–827. doi:10.1016/j.conbuildmat.2015.10.114.

[25] A. Cuesta, E.R. Losilla, M.A.G. Aranda, J. Sanz, Á.G. De La Torre, Reactive belite stabilization mechanisms by boron-bearing dopants, Cem. Concr. Res. 42 (2012) 598–606.

[26] A. Cuesta, M.A.G. Aranda, J. Sanz, Á.G. de la Torre, E.R. Losilla, R. Kastanet, L. León-Reina, M.A.G. Aranda, Mechanism of stabilization of dicalcium silicate solid solution with aluminium, Dalt. Trans. 43 (2014) 2176–2182. doi:10.1039/C3DT52194J.

https://doi.org/10.1051/matecconf/201814901004

Role of Polycarboxylate-ether superplasticizers on cement hydration kinetics and microstructural development

L. Valentini, M. C.Dalconi and G.Artioli
University of Padua, Department of Geosciences, Padua, Italy

Abstract. Polycarboxylate-ether (PCE) superplasticizers are a fundamental constituent of modern cement-based materials due to their impact on the rheology of the fresh mix and mechanical performance of the hardened material. The effect of PCEs on cement hydration kinetics has been known since their introduction in the early 1980s. However, detailed knowledge of the role played by PCE macromolecules on the basic mechanisms of cement hydration (dissolution, diffusion, precipitation) is still lacking. A better understanding of how such mechanisms are influenced by the addition of PCE is no doubt beneficial to the design of novel superplasticizing admixtures. Here, I report on some recent findings about the role of PCE superplasticizers on cement hydration kinetics and microstructural development. The interaction between PCE and C_3S pastes was investigated by an ad-hoc kinetic model based on a combination of generalized forms of the Avrami and BNG (Boundary Nucleation and Growth) models. The model is used to fit the rate of C-S-H precipitation measured by in-situ X-ray powder diffraction combined with mass balance calculations. The results show that a switch from heterogeneous to homogeneous C-S-H nucleation occurs in the presence of PCEs and that the C-S-H growth rate decreases proportionally to the amount of PCE used. The predicted switch to homogeneous nucleation is in agreement with experimental results obtained by XRD-enhanced micro-tomography imaging, showing that, in the presence of PCE, C-S-H preferentially forms in the pore space rather than at the surface of clinker particles.

1 Introduction

The introduction of admixtures for the control of concrete rheology has represented a significant technological advance. Starting from the first generation of superplasticizers, up to the modern polycarboxylate-ether (PCE) molecules, which currently dominate the market, the capability of such admixtures to allow water reduction and better mechanical performance has progressively increased. By providing a detailed description of the mechanisms by which PCE admixtures exert their dispersing action, it has been possible to fine-tune their structure with the aim of providing the desired behaviour as a function of the specific application. For example, the synthesis of PCE macromolecules having different anionic functions can lead to improved affinity for clinker surfaces [1]. However, there are specific situations in which the action of PCE admixtures is hampered. These include the use of PCE with concrete made with cements having large fractions of orthorhombic C_3A [2] and with aggregates having a significant content of clay minerals [3].

From the point of view of hydration kinetics, although the delaying effect of PCE molecules has been well documented, it is still not clear what are the basic mechanisms that control such a delay. Therefore, a detailed investigation of such mechanisms is necessary to achieve a better control of the hydration kinetics in the presence of PCE.

To date, the main invoked mechanisms for the delay of the hydration kinetics are the hindering of dissolution of the clinker phases or nucleation and growth of the hydration products.

Here, I present an investigation of the hydration kinetics of a model C_3S cement, in the presence of different PCE molecules. In-situ XRD data are combined with mass balance calculations and kinetic modelling, with the aim of providing a quantitative description of the modes and rates of C-S-H nucleation and growth.

2 Materials and methods

Cement pastes were prepared by mixing C_3S with deionized water at w/c 0.5, for 1 minute, using a vertical stirrer. PCE admixtures (having 17 repeat units in the side-chains and 3 to 4 repeat units in the backbone for each segment) were added directly to the mix water, at dosages of 0.05% and 0.10% by mass of C_3S.

2.1 XRD measurements

C_3S reaction kinetics, for the blank paste and in the presence of PCE, were monitored by means of in-situ X-ray powder diffraction, using 0.5 mm borosilicate glass capillaries filled with the sample. The measurements were performed in time-resolved mode for 20 hours, with a time resolution of 20 minutes and with a constant temperature of 23 °C. Quantitative phase analyses of the diffraction patterns were performed using the software Topas 4.1 (Bruker). The time-dependent volume of C-S-H was calculated using a mass balance algorithm [4], as implemented in the software RieCalc 3.2 [5].

2.2 Kinetic Model

The kinetics of C-S-H precipitation have been modelled by combining generalized forms of the Johnson-Mehl-Avrami-Kolmogorov (JMAK) and Boundary Nucleation and Growth (BNG) models [6,7]. The following kinetic equation was used to fit the experimental C-S-H rate of precipitation obtained by XRD:

$$X(t) = A\{1 - \exp\left[-X(t)_e^{hom} - X(t)_e^{het}\right]\} \quad (1)$$

Where X is the volume fraction of C-S-H precipitated at time t, A is a scaling constant and the two terms inside the exponential are the extended volume fractions (representing the theoretical volume that would form ifthere was no overlap among particles) formed by homogeneous and heterogeneous nucleation respectively, given by the two following equations:

$$X(t)_e^{hom} = \frac{4}{3}\pi N_v \Gamma^3(t) \quad (2)$$

$$X(t)_e^{het} = O_B \int_0^1 \Gamma(t)\{1 - \exp\left[-\pi N_s \Gamma^2(t)(1-u^2)\right]\}du \quad (3)$$

Where N_v and N_s are the number of nuclei formed by homogeneous and heterogeneous nucleation respectively, Γ is the time-dependent growth rate O_B is the amount of cement surface per unit volume, u is an auxiliary variable expressed as $u = y/\Gamma$, with y being the spatial coordinate perpendicular to the cement surface. An extended version of the above equations can be found elsewhere [8].

3 Results and discussion

Figure 1 displays the fit of the experimental curves for the rate of C-S-H precipitation, by means of the kinetic model outlined in the previous section. The model is capable of fitting the XRD data with excellent accuracy.

The curves displayed in figure 1 show a delay in the beginning of the acceleration stage, proportionally to the amount of PCE dosed, as well established in the literature.

The output from the model suggests a substantial switch from heterogeneous nucleation, to homogeneous nucleation in the presence of PCE. Specifically, the number of nuclei formed by heterogeneous nucleation per unit surface of substrate, as predicted by the model, decreases from 4.55×10^{-11} m^{-2} to less than $3.00 \times 10^{-10} m^{-2}$, with a majority of nuclei forming by homogeneous nucleation, whereas the number of homogeneous nuclei is negligible in the absence of PCE. However, the total number of nuclei formed is nearly halved in the presence of PCE. Such a hindered nucleation rate, in combination with a predicted decrease of the mean growth rate of 30-40% (3.50×10^{-11} m/s for the blank paste to a minimum of 2.10×10^{-11} m/s in the presence of PCE) explains the net decrease in the overall kinetics of C_3S hydration.

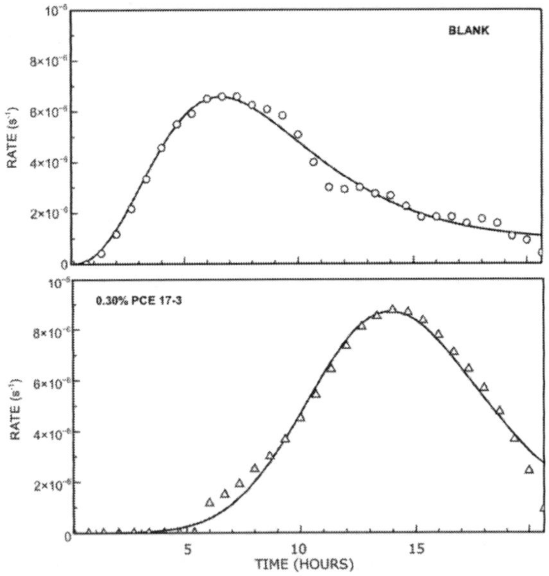

Fig. 1 Fitting of the XRD data (symbols, one in two data points are displayed with the aim of facilitating visualization) by means of the kinetic model (solid curve) for the C_3S paste in the absence (top) and in the presence (bottom) of PCE.

These results suggest that the nucleation and growth of C-S-H exerts a fundamental role in controlling the overall hydration kinetics and that PCE molecules induce a delay by acting on the modes and rates of C-S-H precipitation. The switch to homogeneous nucleation in the presence of PCE had been previously observed based on XRD-tomography measurements [9].

Nonetheless, a better understanding of the action exerted by PCE molecules on hydration kinetics and microstructural development, will require a more refined understanding of the mechanisms of C-S-H precipitation, since classical models of heterogeneous/homogeneous nucleation and growth may be too idealized to provide an accurate picture. Some steps forward have been recently achieved by implementing more complex numerical models for the aggregation of C-S-H nano-particles [10]. A further advance will be obtained by implementing into such models alternative pathways to phase separation, which include the formation of pre-nucleation clusters [11]. Such models are especially relevant in the presence of organic molecules, such PCE [12].

References

1. F. Dalas, A. Nonat, S. Pourchet, M. Mosquet, D. Rinaldi, S. Sabio, Cem. Concr. Res. **67**, 21-30 (2015)
2. G. Ferrari, L. Valentini, V. Russo, M.C. Dalconi, M. Favero, G. Artioli, Constr. Build. Mater.**130**, 83-91 (2017)
3. G. Xing, W. Wang, J. Xu, RSC Adv. **6**,106921–106927 (2016)
4. L. Valentini, M.C. Dalconi, M. Favero, G. Artioli, G. Ferrari, J. Am. Ceram. Soc. **98**, 1259-1264 (2015)
5. L. Valentini, J. Appl. Crystallogr. **46**, 1899-1902 (2013)
6. M. Avrami, J. Chem. Phys. **7**, 1103-1112 (1939)
7. J.J. Thomas, J. Am. Ceram. Soc. **90**, 3282-3288 (2007)
8. L. Valentini, M. Favero, M.C. Dalconi, V. Russo, G. Ferrari, G. Artioli, Cryst. Growth. Des. **16**, 646-654 (2016)
9. G. Artioli, L. Valentini, M. Voltolini, M.C. Dalconi, G. Ferrari, V. Russo, Cryst. GrowthDes. **15**, 20-23 (2015)
10. M. Delhorme, C. Labbez, M. Turesson, E. Lesiewska, C.E. Woodward, B. Jonsson, Langmuir**32**, 2058-2066 (2016)
11. T. Sowoidnich, T. Rachowski, C. Roessler, A. Voelkel, H.M. Ludwig, Cem. Concr. Res. **73**, 42-50 (2015)
12. C. Rodriguez-Navarro, L.G. Benning, Elements**9**, 203-209 (2013)

Ground Glass Pozzolan in Conventional, High, and Ultra-High Performance Concrete

Arezki Tagnit-Hamou[1], Ablam Zidol[1], Nancy Soliman[1] Joris Deschamps[1], and Ahmed Omran[1]

[1] Department of civil Engineering, University of Sherbrooke, Sherbrooke (QC) Canada

Abstract. Ground-glass pozzolan (G) obtained by grinding the mixed-waste glass to same fineness of cement can act as a supplementary-cementitious material (SCM), given that it is an amorphous and a pozzolanic material. The G showed promising performances in different concrete types such as conventional concrete (CC), high-performance concrete (HPC), and ultra-high performance concrete (UHPC). The current paper reports on the characteristics and performance of G in these concrete types. The use of G provides several advantages (technological, economical, and environmental). It reduces the production cost of concrete and decrease the carbon footprint of a traditional concrete structures. The rheology of fresh concrete can be improved due to the replacement of cement by non-absorptive glass particles. Strength and rigidity improvements in the concrete containing G are due to the fact that glass particles act as inclusions having a very high strength and elastic modulus that have a strengthening effect on the overall hardened matrix.
Keywords: Ground-glass pozzolan (G), Durability characteristics, Fresh properties, Mechanical performance, conventional concrete, high-performance concrete, and ultra-high performance concrete.

1 Introduction

The use of supplementary-cementitious materials (SCMs) can help cement-based material construction achieving sustainable development by reducing cement production and therefore reduce fossil-fuel consumption and greenhouse-gas emissions. Alternative locally available SCMs are favorable when traditional SCMs are not locally available or costly if transported. Concrete industry is currently seeking additional sources of SCMs to supplement fly ash, slag cement, and silica fume that are currently used in concrete.

There are several million tons of recycled glass available and are widely dispersed across the globe. The ground-glass pozzolan (G) obtained by grinding the mixed-waste glass to same fineness of cement can act as an SCM, especially it is amorphous and pozzolanic material. The mixed colored waste glass cannot be recycled and in the best scenarios, is stockpiled, causing obvious environmental and economic problems. Therefore, valorization of this waste glass as an SCM could provide dual advantages. Over decades, the G showed promising performance in mortar and concrete mixtures both in laboratory and in large-scale field applications with more focus on long-term performance.

The purpose of this paper is to summarize material specifications for the G for use as a pozzolan as a partial replacement of portland cement in concrete. The overall performance (fresh, mechanical, durability, and life-cycle assessment) of incorporating G in cement-based materials (CC, HPC, and UHPC) in both laboratory and field applications are highlighted in this paper.

2 CC and HPC containing G

2.1 Mix design and fresh properties

This section of the paper focuses on the performance of the G in two types of concrete: CC and HPC. The CC is a concrete with water-to-binder ratio (*w/b*) of 0.55 or 0.42 with respective binder content of 350 and 390 kg/m^3. The HPC is a concrete with a *w/b* of 0.30 with relatively higher binder content of 460 kg/m^3. The slump value of the CC is in the range of ±110 mm and that for the HPC is about ±210 mm. The air contents of the two types of concrete varied between 4% and 8 %. These two types of concretes can contain different rates of G in partial substitution of cement.

2.2 Compressive Strength

Figure 1 shows the compressive strength (f'_c) of the CC (Fig. 1-a) and HPC (Fig. 1-b) concrete types. Of course, the HPC presented higher f'_c than CC due to the effect of its low *w/b*. Whether in CC or HPC, the f'_c of concrete mixtures incorporating G are low at an early age but increase significantly in the long term. Similarly to the typical pozzolanic materials, the G concretes have lower f'_c (about 12%) than the control at early age (less than 91 days), but they develop greater f'_c than the control (about 18%) at long-term (beyond the 91 days). The low early-age f'_c results from the dilution effect and the slow reactivity of the G. The improvement of the long-term f'_c results is due the pozzolanic activity of the G [1,2].

2.3 Chloride-Ion Permeability

The results of the chloride-ion permeability of the CC at 28 and 91 days are shown in Fig. (2-a) while the results at 28 and 56 days for the HPC are presented in Fig. (2-b). The incorporation of the G considerably reduces the permeability in both types of concrete. However, the reduction in the CC is even greater with time than that in the HPC. The reduction is not noticeable enough at 28 days due to the slow reactivity of the G. I the CC, at 91 days, the G reduces the permeability by 75% or even 80% depending on the rate of replacement. The corresponding reduction in the HPC is about one third. The effect of G is particularly interesting in the CC by its significant role in the reduction of permeability. Indeed, the CC containing G with a *w/b* as high as 0.55 showed larger reduction in the permeability relative to that with a low *w/b* of 0.30. This demonstrates a greater contribution of the G in decreasing the permeability than a simple reduction in the water content or *w/b*. This important decrease in permeability in the presence of G results from its pozzolanic reaction, which consumes large portlandite crystals to produce additional C-S-H that fills the pores in the microstructure level [3-6]. It is the refining effect of the pores and the grains, which generates an increased tortuosity of the pores reflecting the decrease of the permeability.

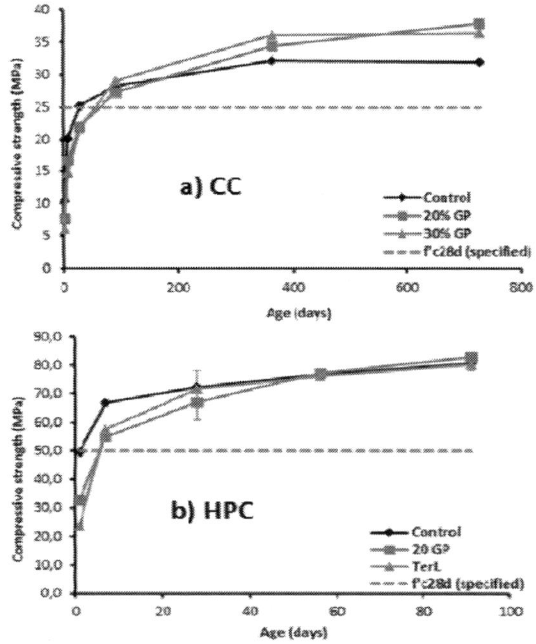

Fig. 1 Development of compressive strength with time for G concrete a) CC with w/b = 0.55, and b) HPC with w/b = 0.30

2.4 Scaling resistance

Figure 3 shows the scaling resistance of the two types of concretes (CC and HPC). For all concrete mixtures containing or not G, the results of the mass losses versus the de-icing scaling cycles are lower than the BNQ 2621-905 standard limits demonstrating good resistance to scaling. However, the concrete mixtures containing G exhibited mass losses slightly greater than those of the control concrete. It is a usual trend observed for the supplementary cementitious materials that seem more susceptible to scaling [7]. The CC usually have mass losses about four times higher than those of the HPC due to the low *w/b* of the latter. Since the low *w/b* would reduce excessive bleeding likely to weaken the skin of the concrete and makes it more vulnerable to scaling.

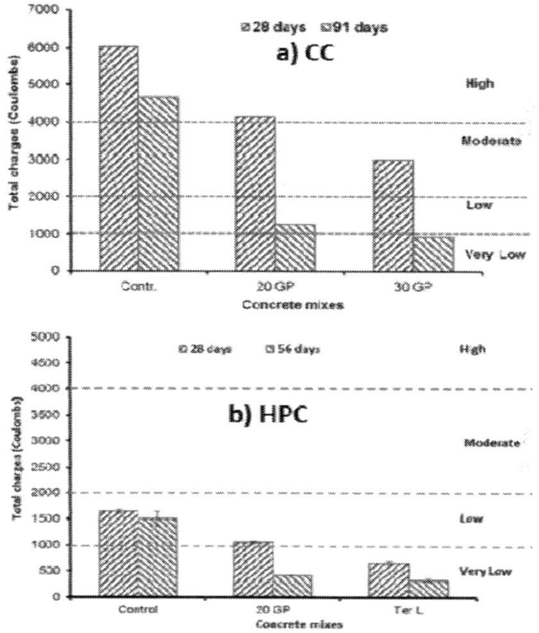

Fig. 2 Chloride-ion permeability of a) CC with w/b = 0.55, and b) HPC with w/b = 0.30

https://doi.org/10.1051/matecconf/201814901005

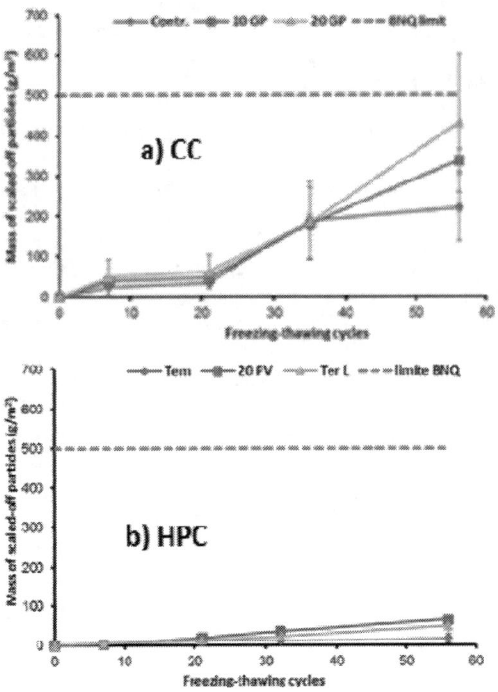

Fig. 3 Scaling resistance of a) CC with *w/b* = 0.42, and b) HPC with *w/b* = 0.30

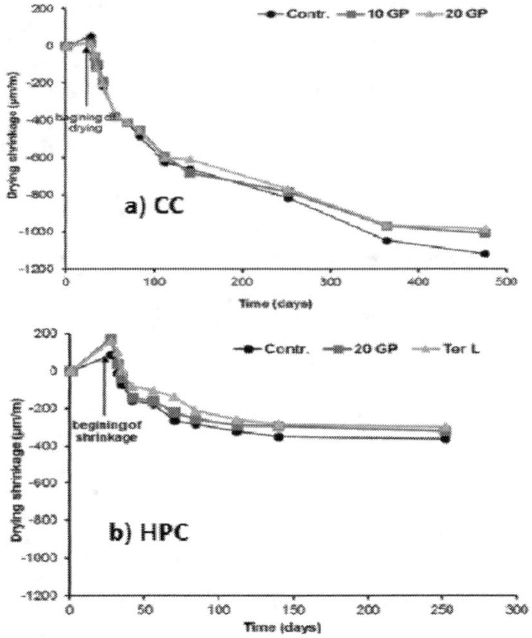

Fig. 4 Drying shrinkage of a) CC with *w/b* = 0.42, and b) HPC with *w/b* = 0.30

2.5 Drying shrinkage

Figure 4 shows the drying shrinkage of the two types of concretes. Whether in CC or HPCs, the mixtures containing G exhibit a drying shrinkage similar to that of the control, suggesting that the G does not enlarge the drying shrinkage of the concrete. The drying shrinkage of the HPC is very low, about one third of that of the CC, due to its low *w/b* that limits the evaporation of free water and correlatively the drying shrinkage. Shayan et al. [8] also reported similar trends. These results suggest that the G concrete may have less or similar deformation and lower risk of cracking due to drying shrinkage than a control.

3 Use of G in UHPGC

3.1 UHPGC classes

Based on the research conducted to develop UHPGC using waste-glass materials [9,10], four different UHPGC classes can be delimited in responding to various construction demands (Table 1). The UHPC mixtures in Class A are characterized by low flowability (using mini-slump flow table) of less than 200 mm but with 2-day f'_c under hot curing conditions (HC) of more than 200 MPa. Concrete mixes in Class A have a *w/b* between 0.150 and 0.190. Highly flowable UHPGC can be obtained, as in Class C, with higher *w/b* between 0.225 and 0.250. The UHPGC mixtures in Class C are characterized by 2-day f'_c under HC between 160 and 175 MPa. The UHPGC mixtures in Class B blend the characteristics of those in Classes A and C. All UHPGC mixtures in Classes A, B, and C are designed with steel fibers. The mixtures in Class D are designed for architectural applications, with the steel fiber replaced by polyvinyl alcohol fiber (PVA). The concretes in this Class D are characterized by higher flowability (260 mm) and moderate strength [91-day f'_c of more than 120 MPa under normal curing conditions (NC)].

Table 1 UHPC with local materials for various construction applications

Characteristics	Class A	Class B	Class C	Class D
Flowability	Semi-flowable	Flowable	Highly flowable	Highly flowable
Average mini-slump flow diameter, mm	200	230	260	260
w/b	0.150–0.190	0.190–0.225	0.225–0.250	0.225–0.250

% of solids in superplasticizer/cement weight	1–3	1–3	1–3	0.225–0.25
Steel fiber (%)	2	2	2	--
PVA fiber	--	--	--	2.5
2-day-HC f'c, MPa	>200	175–200	160–175	--
28-day-NC f'c, MPa	>160	>140	>130	>100
91-day-NC f'c, MPa	>180	>150	>140	>120
Flexural strength, MPa	>25	>20	>15	>10
Modulus of elasticity, GPa	>50	>45	>40	>40
Chloride-ion penetration	Negligible	Negligible	Negligible	Negligible
Relative dynamic modulus of elasticity after 1000 freeze–thaw cycles	100%	100%	100%	100%

3.2 Material properties and mixture proportion

Type high sulfate-resistance (HS) cement formulated with a low C_3A content was selected.. Silica fume (SF) conforming the CSA A3000 Standards "Cementitious materials compendium", and three ground-glass pozzolans ground to different sizes [G, fine G (FGP), and glass sand (GS)] with same Na_2O_{eq} of 13% were used in the UHPGC mix designs. Table 2 provides the properties of these materials. This mixture had more than 400 kg/m^3 of G as cement replacement. A polycarboxylate-based HRWRA with a specific gravity of 1.09 and solid content of 40% was used. The PVA used were 13 mm in length and 0.2 mm in diameter, and had a specific gravity of 1.3 and tensile strength of 400 MPa.

Table 2 Material properties

Property	HS Cement	Silica Fume	G	Fine G	Glass Sand	Quartz Sand
Silica, %	--	99.8	73	73	73	99.8
Specific gravity	3.21	2.20	2.60	2.60	2.60	2.70
d_{max}, µm	<100	<1	<100	<10	<800	<600

Table 3 presents the mixture proportions for seven UHPGC mixtures covering the various concrete Classes described in Table 1. A 2% volume fraction of steel fiber was used for UHPGC-1 to 6 mixtures, while 2.5% PVA fiber was used for UHPGC-7 mixture. As shown in Table 3, the design allowed using G in all mixes in contents varying between 222 and 403 kg/m^3 to replace cement and quartz powder, FGP ranging from 53 to 113 kg/m^3 to replace SF, and GS contents between 25% to 50% replacement of QS.

Table 3 UHPGC mixture compositions for various construction applications (Kg/m^3)

Materials	Class A		Class B		Class C		Class D
	UHPGC-1	UHPGC-2	UHPGC-3	UHPGC-4	UHPGC-5	UHPGC-6	UHPGC-7
Water	196	193	211	186	215	236	224
Cement	812	640	608	790	561	739	544
Silica fume	113	142	158	109	208	205	204
Fine glass powder	113	80	53	109	--	--	--
Glass powder	244	382	380	237	411	222	403
Quartz sand	974	960	684	474	896	443	888
Glass sand	--	--	228	474	--	443	--
Superplasticizer*	13	13	10	17	16	10	16
PVA fiber	--	--	--	--	--	--	32.5
Steel fiber	158	158	158	158	158	158	--

*Solid content

3.3 UHPGC performance

Table 4 provides the workability and the *f'c* after 2 days of HC (*f'c-2d-HC*), 28 days of NC (*f'c-28d-NC*), and 91 days of NC (*f'c-91d-NC*) of the seven UHPGC mixtures. A mini-slump flow of more than 230 mm indicates higher concrete workability. Strength results satisfy the requirements given in Table 1. As shown in the table, the mechanical strength for the samples subjected to 2 days of HC was greater than those obtained under 91-day NC. The concrete mixtures in Class A showed the highest strength results (*f'c-2d-HC* of 210 and 205 MPa for UHPGC-1 and 2, respectively), while those in Class C were the lowest (*f'c-2d-HC* of 169 and 175 MPa for UHPGC-5 and 6, respectively). The concrete in Class B exhibited moderate strength (*f'c-2d-HC* of 200 and 190 MPa for UHPGC-3 and 4, respectively). UHPGC-7 (Class D) tested at 135 MPa for the *f'c-2d-HC* due to the inclusion of the PVA fiber.

Table 4 Performance characteristics of UHPGC mixtures used for various construction applications

	Class A		Class B		Class C		
Materials	UHPGC-1	UHPGC-2	UHPGC-3	UHPGC-4	UHPGC-5	UHPGC-6	UHPGC-7
Mini-slump flow, mm	240	230	260	250	270	290	270
f'c-2d-HC, MPa	210	205	200	190	169	175	135
f'c-28d-NC, MPa	177	170	158	150	130	138	100
f'c-91d-NC, MPa	191	185	175	165	155	162	127

4 Large-scale field applications

4.1 CC and HPC

The G concretes (with replacement ratios between 10% and 30% by cement content) were used to cast various structure elements in Quebec-Canada between 2006 and 2011 including, interior slabs, exterior slabs (sidewalks), and structural wall elements in various environmental conditions – indoors and outdoors). The performance of samples taken from these concretes and subjected to laboratory curing are presented in references [11,12]. The G used had fineness (retained on 45 µm, %) = 0%, maximum-particle size (d_{max}) = 40 µm, mean-particle size (d_{50}) = 12 µm, Blaine specific surface = 382 m^2/kg, and specific gravity = 2.54. In addition to the laboratory-cured measurements, the long-term evaluation of the G-concrete performances were conducted by testing core samples (during the concrete life, they exposed to different environmental and service conditions, consolidation, finishing, curing conditions ..etc.). The core results compared to the concrete performances that were initially sampled at the time of casting and laboratory cured (ages of 28 and 91 days) are presented in Fig. 5. The results showed improved mechanical and durability properties compared to the reference mixtures. The incorporation of G in concrete contribute to enhancing the microstructure of concrete (densify) especially with age allowing a significant reduction in the pore system and generating higher mechanical and durability properties. The use of G generates dense C-S-H gel rim around the G particles and densifies the cementitious matrix. The mechanical strength was improved with time due to the pozzolanic activity of G and the microstructure enhancement.

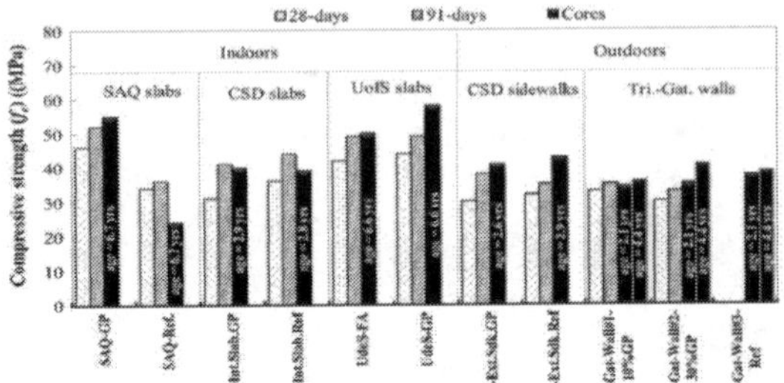

Fig. 5 Compressive strength undertaken on core samples-compared to 28 and 91-days results following casting [14]

4.2 UHPGC

The developed UHPGC using G was used in the construction of two footbridges (Fig. 6) at the University of Sherbrooke showing a potential for the UHPGC to be used in future large-scale projects [13,14]. The G used in the UHPGC mix design for constructing the bridges had maximum particle size of 100 µm and specific gravity of 2.60. The concrete exhibited excellent workability and rheological properties due to the zero absorption of the glass particles and optimized packing density for the entire concrete matrix. The mechanical properties were found to be excellent allowing reductions in the bridge thickness (material saving), and maintenance costs (due to the excellent durability properties of the UHPGC.

Fig. 6 Picture of a pedestrian bridge in the campus of University of Sherbrooke made of UHPC containing G [10,12]

5 Life cycle assessment

In order to assess the environmental footprint of a new technology, or product, life cycle assessment (LCA) is a robust and promoted tool. A study carried out by Quantis for Recy-Quebec [15], demonstrated that the valorization of G into concrete shows significant environmental benefits compared to usual end-of-life scenario of landfilling. This valorization project showed even higher environmental benefits from landfilling than being reused as raw material for glass bottle, glass aggregate or glass wool.

The replacement of cement by G can significantly reduce the carbon footprint of a typical UHPC, as shown in Fig. 7 [10]. The study demonstrated that the use of G can secure a low carbon footprint while maintaining high strength – which is difficult to obtain with other pozzolanic materials (fly ash, slag, and limestone powder).

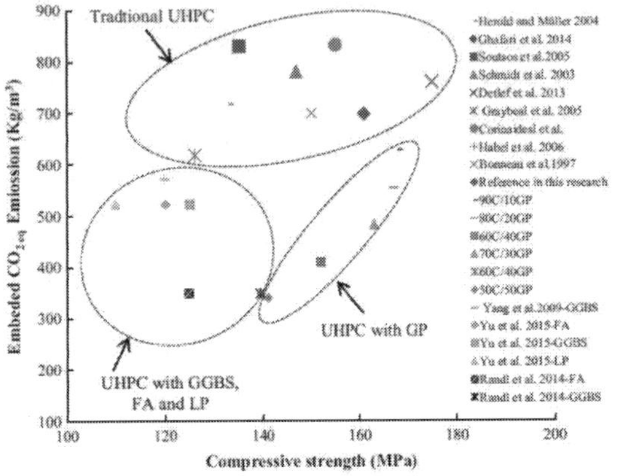

Fig. 7 Relationships between embodied CO_{2eq} for the cement content and 28-day compressive strength under NC [10]

A LCA study by Deschamps et al. [16] compared the environmental life cycle footprint of a conventional UHPC and UHPGC in the case of the pedestrian bridge (Fig 6). For the same concrete volume and service life, results showed that the UHPGC can secure better environmental impacts on all indicators (Fig. 8). It was mainly due to the reduction in cement content with the incorporation of more G. Indeed, cement production appears as a main contributor for most of the environmental indicators of concrete life cycle. It is also important to note that, considering the same volume, UHPGC have a very high environmental impact compared to the conventional concrete. But, considering the volume needed to fulfill the same service life span, UHPGC can reach an environmental impact less important than conventional concrete.

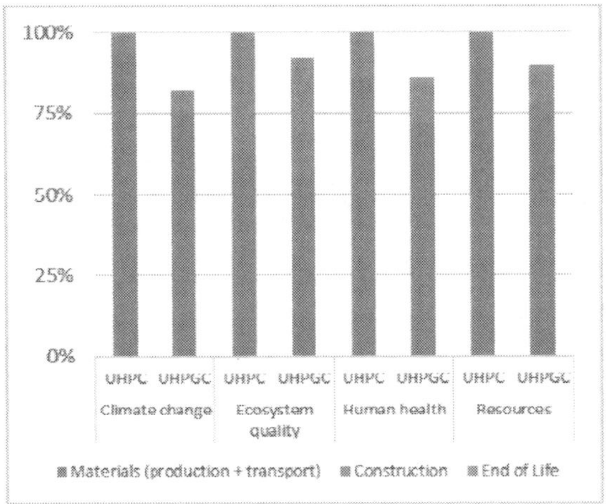

Fig. 8 LCA comparison of UHPC and UHPGC on environmental indicators.

6 Conclusions

Based on the previous finding about the use of ground-glass pozzolans (G) as a partial replacement of cement in concrete, the following conclusions can be drawn:

- The use of G increases the workability due to its non-absorption nature and smoothness surface.
- The use of G can provide better mechanical properties (compressive strength) compared to the reference without G, especially at later ages, provided by the G's pozzolanic activity.
- The use of G can enhance the durability characteristics (resistance to chloride-ion permeability and resistance to de-icing salt scaling) due to the improved characteristics of the pore network, the filling effect of G particles, and the conversion of CH to C–S–H. The G generates dense C-S-H gel rim around the G particles and densifies the cementitious matrix. This effect is more noticeable at later ages. The G reduces the chloride-ion penetrability of concrete to approximately one third, thereby reducing the risk of chloride-induced corrosion of steel reinforcement.
- The use of G can reduce the GHG emission (CO_2 emissions) of concrete associated with cement production and protecting the environment by favoring the valorization of recyclable materials.
- The use of G can save the costs related to the processing and the disposing of the waste glass in landfills as well as reducing the price of concrete.
-
- LEED (Leadership in Energy and Environmental Design) certification says, "using glass in concrete can earn double points through the use of by-products, since glass is considered a post-consumption and locally available material while other materials are considered post-production materials".
- The G is white color material in the natural state and can provide whitish coloring to the concrete products, leading to better reflectance (Albedo effect).

7 References

[1] Zidol, A., Tognonvi, M.T., Tagnit-Hamou, A. (2017) "Effect of glass powder on concrete sustainability, part II", New Journal of Glass and Ceramics, 7, 34-47.

[2] Hussain, S.E. and Rasheeduzzafar (1994) Corrosion Resistance Performance of Fly Ash Blended Cement Concrete. ACI Materials Journal, 91, 264-272. http://esatjournals.net/ijret/2012v01/i03/IJRET20120103042.pdf

[3] Zidol, A. "*Optimisation de la finesse de la poudre de verre dans les systèmes cimentaires binaires,*" Mémoire de maîtrise, Université de Sherbrooke, Qc, Canada, 2009, 168 p.

[4] Aladdine, F. "*Propriétés mécaniques et durabilité des bétons incorporant le verre finement broyé,*" Mémoire de maîtrise, Université de Sherbrooke, Qc, Canada, 2009,124 p.

[5] Zidol, A., Pavoine, A., Tagnit-Hamou, A. (2012) "Effect of glass powder on concrete permeability," International Congress on Durability of Concrete, Trondheim, Norway, June 18 – 21, 2012.

[6] Jain, J. A. et Neithalath, N. (2010) "Chloride transport in fly ash and glass powder modified concrete. Influence of test methods on microstructure," *Cement and Concrete Composites,* 32, 2010 p. 148-156.

[7] Houehanou, E., Gagné, R., and Jolin, M. (2010) "Analysis of the representativeness and relative severity of ASTM C672 and NQ 2621-900 standard procedures in evaluating concrete scaling resistance," Canadian Journal of Civil Engineering, 37(11): 1471–1482. doi:10.1139/l10-091.

[8] Shayan, A. and Xu, A. (2006) Performance of Glass Powder as a Pozzolanic Material a Field Trial on Concrete Slabs. Cement and Concrete Research, 36, 457-468.

[9] Soliman, N., (2017), "Development of ultra-high-performance concrete (UHPC) using waste glass materials — towards innovative eco-friendly concrete", *Doctoral Thesis, Département de génie civil, Faculté de génie, Université de Sherbrooke,* 393 pp.

[10] Soliman, N. A., and Tagnit-Hamou, A., "Development of Ultra-High-Performance Concrete using Glass Powder — Towards Ecofriendly Concrete," *Journal of Building and Construction Materials*, Vol. 125, 2016, pp. 600–612.

[11] Omran, A., and Tagnit-Hamou, A., "Performance of Glass-Powder Concrete in Field Applications," *Journal of Construction and Building Materials*, Vol. 109, 2016, pp. 84–95.

[12] Omran, A., Morin, E.-D., Harbec, D., and Tagnit-Hamou, A., "Long-Term Performance of Glass-Powder Concrete in Large-Scale Field Applications," *Journal of Construction and Building Materials*, Vol. 135, 2017, pp. 43–58.
[13] Soliman, N., Omran, A. F., and Tagnit-Hamou, A., "Laboratory Characterization and Field Application of Novel Ultra-High Performance Glass Concrete," *ACI Materials Journal*, Vol. 113, No. 3, 2015, pp. 307–316.
[14] Tagnit-Hamou, A., Soliman, N., Omran, A., Mousa, M., Gauvreau, N., and Provencher, M., "Novel Ultra-High Performance Glass Concrete," *Concrete International*, Vol. 37, Issue 3, 2015, pp. 41-47.
[15] Quantis, "Analyse environnementale du cycle de vie de projets de commercialisation du verre mixte récupéré via des centres de tri de matières recyclables au Québec", In French, 2015.
[16] Deschamps, J., Tagnit-Hamou, A., Fournier, B., Amor, B., "LCA case study of ultra-high performance concrete with glass powder", International Conference on Cementitious Materials and Alternative Binders for Sustainable Concrete, Montreal, Canada, October 2-4, 2017.

MATEC Web of Conferences **149**, 01006 (2018) https://doi.org/10.1051/matecconf/201814901006
CMSS-2017

Durability and service life prediction for concrete structures – developments and challenges

Mark G Alexander[1]
[1]University or Cape Town, Department of Civil Engineering, Cape Town, South Africa

Abstract. The paper reviews developments in service life prediction for concrete structures. It indicates the difficulties inherent in rational service life design, in view of the multiple factors and variabilities involved in the process. The paper also emphasises the advantages of performance-based approaches to durability prediction, and considers performance testing, which is critical to achieving intended service life. Such approaches allow service life modelling, which the current prescriptive approaches do not. The concept of 'durability indicators' is covered, with a practical example showing how this can be used to improve concrete durability in construction. The paper also stresses the importance of an 'integrated approach' to durability specifications, performance-based predictions, and site quality control.

1 Introduction

Concrete remains the material of choice for the vast majority of construction activities around the world. The reasons are obvious: it is economical, easy to use, can be shaped into endless forms, is excellent structurally, and, despite the claims that it is not a 'green' material, has a very low carbon footprint – see Table 1

Table 1. Environmental indicators of concrete and other materials

	Concrete	Fired clay bricks	Steel
Embodied energy (MJ/kg)	~ 0.95	~ 3.00	~ 35
CO_2 emissions (kg $CO_{2,eq}$/kg)	~ 0.13	~ 0.22	~ 2.80

Concrete is also the 'developmental material' *par excellence*. Global development in infrastructure, housing, and so on is unthinkable in the absence of modern concrete. Further, concrete has today assumed a position of being a sophisticated material when used at the upper end of its technological usage scale, incorporating multiple binders, admixtures, fibres, various aggregates, and the like, to produce a composite material that can be tailored to almost any constructional requirement. At the other end of the technological scale, it remains accessible to the 'common person', helping her or him to improve their lives in meaningful ways.

Nevertheless, there are real challenges in taking concrete forward into future decades as the construction material of choice. One of the most pressing is the question of lack of durability, coupled with 'Design for Service Life', with the associated concepts of 'Performance-based design', particularly for durability.

2 Design for service life

Design for service life implies that we can, rationally, perform the necessary range of design activities that will give reasonable assurance that the structure in question will fulfil its intended function for the required service life. This may be feasible when the service life is 20 or 30 years, or maybe even 50 years, but, practically, how do we design for a 100-y or 150-y service life? This is now being more regularly demanded by infrastructure owners. The following aspects need to be considered:

- Changing exposure environments (for example, global warming and related extreme weather events)
- Rapidly changing materials, especially with the pressure to reduce environmental impacts
- Inadequate knowledge and models
- The pervasive problem of variable construction quality
- Differing perceptions of what 'service life' means, particularly in terms of 'end of service life' criteria
- The physical impossibility of verifying our designs at an age much greater than 50 years!

Take for example the 'Coignet House' in Paris, built in 1853, and thought to be the first reinforced concrete house in the world, shown in Figure 1. It is clearly now in a poor state, with evidence of reinforcing corrosion and other deterioration. The question is: how would this structure have been designed in the 19th Century, to conform to the vastly different requirements of the 21st Century? The house appears to be abandoned, implying that minimal or no maintenance has been carried out over, possibly, many decades. Despite this, the house is still standing – a mute testimony to the longevity of concrete!

Fig. 1. Coignet house in Paris – the world's first reinforced concrete house, 1853

Engineers understand 'design & specification', but usually these are embedded in national codes of practice, standard specifications, and design guidelines or rules. When it comes to service life deign, guidelines are largely non-existent, and the standards are wholly inadequate to provide for meaningful approaches. Since service life design, by definition, involves being able to 'predict' the behaviour or performance of a structure over a long period of time, specifically in terms of its ability to remain serviceable, it is obvious that 'Service Life Design' requires 'Service Life Models' that are adequately *predictive*. Thus, Service Life Modelling implies Service Life Prediction – and vice versa. And this is precisely where the problem lies – our general inability to do adequate service life modelling, because of our inadequate knowledge, but also importantly, because of the range of exposure conditions that might exist even in a single structure. Add to this variable construction quality, and this becomes quite an intractable problem!

Further salient points about 'service life design' are:

- Current approaches are inadequate or very basic
- Current approaches are almost universally 'prescriptive' in nature (see later)
- There is very little true 'durability design' carried out in practice, let alone Service Life Design
- Research has run well ahead of practical implementation

One major problem at present is the prescriptive nature of current approaches – dealt with in more detail later. It is essential that we move to 'performance-based' approaches if we hope to make progress in a reasonable time. First, however, other aspects of service life prediction and modelling are covered.

2.1 Service Life Prediction (SLP) and Service Life Modelling (SLM)

The essential elements in service life prediction and modelling are:

- Understanding of deterioration mechanisms
- Translation of deterioration mechanisms into mathematical models that can describe rates of deterioration and consequent structural damage: SLP is largely an exercise in determining rates of deterioration and damage
- Measurement of key material parameters – over time if needed; this means developing and implementing robust characterisation and quality control tests, which must be routine, easily-carried out, and represent reliable measures of resistance (e.g. to chloride ingress)

2.2 Critique of current service life models

Several general Service Life Models exist, such as DuraCrete, DURACON, LIFE-365, etc., but all still have shortcomings. Current SLMs can broadly be represented as those that are 'universal', such as the *fib* (Model Code) [1] and ISO [2], or those that are more 'local', such as LIFE365, DURACON, etc. The 'simpler' models cover limited exposure conditions, and are mainly related to reinforcement corrosion, being diffusion-based models. As such, they ignore other deterioration mechanisms such as AAR, scaling, sulphate attack, and so on.

The more 'sophisticated' model may cover other (but not all) types of chemical exposures, additional fluid transport mechanisms, and so on, but: they are not necessarily more accurate! Also, their assumptions and algorithms are usually difficult to verify independently.

All current models cannot deal adequately with the effects of construction or design defects, which is a major drawback. Further, they mostly do not fully consider the 'trade-off' between initial construction cost, and later maintenance or rehabilitation costs.

3 Prescriptive vs performance approaches

'Prescriptive' approaches, which are currently the norm, have already been mentioned. This section also briefly introduces the concept of performance approaches. The move to performance-based specifications (PBS) is gathering momentum, for example the NRMCA P2P initiative [3], and the French PERFDUB Programme [4]. Various national jurisdictions also have PBS in various stages of development – e.g. Canada, Spain, Switzerland, and South Africa.

The features of a prescriptive approach are that it sets limiting values, for example mix materials and proportions, it often prescribes construction processes such as curing, and it works on a 'deemed-to-satisfy' approach. The main disadvantages have long been recognised: primarily, that it does not permit, promote, or 'guarantee' a) in-situ verification of specified requirements, b) actual quality of construction, c) prediction for service life requirements, d) economic analysis or maintenance budgeting, and e) innovation.

Further, in prescriptive approaches, there is frequently a major 'disconnect' between the intention of the owner or designer, as set out in the specification, and that actually achieved in construction. Figure 2 shows some results from our lab, plotting oxygen permeability index (OPI) measured on actual structures, against standard cube strength measured on the same concrete as used in the structures. It is clear there is no correlation!

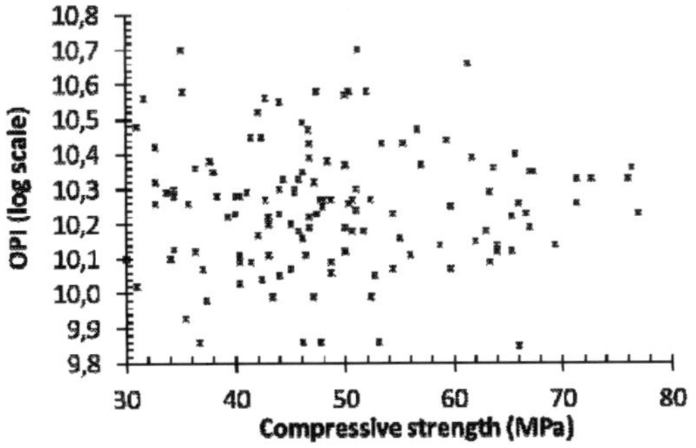

Fig. 2. Oxygen permeability index (OPI (log)) vs. cube strength, actual structures

3.1 Responsibilities in prescriptive vs performance approaches

In the prescriptive approach, since the 'owner' (or their agent such as the architect, engineer, etc.) specifies the mix, materials, properties etc., the owner also assumes full responsibility for the concrete. On the other hand, the performance approach requires the owner, contractor, and materials supplier to cooperate: a) the 'Owner' specifies required *performance* (e.g. concrete properties), b) the supplier is responsible for delivering the appropriate concrete, and c) the contractor is responsible for placing and curing the concrete to achieve the performance properties *in situ*. This implies a proper definition and division of responsibilities between the parties, for achieving concrete quality, see Figure 3.

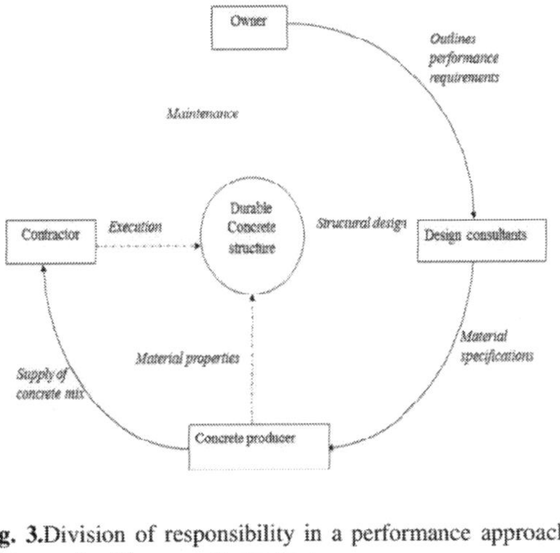

Fig. 3. Division of responsibility in a performance approach, to achieve a durable concrete structure

Note in Figure 3, that it is the *finished concrete structure* that is required to be durable, implying that ways must be found to assess the properties of the *as-built structure*. Many so-called 'performance-based approaches' do not address this critical element, and so should not be called such.

3.2 Shift from prescriptive to performance approaches

A move from 'prescriptive to performance' approaches is absolutely necessary to adopt SLP&SLM methods. However, the shift will not be quick or easy, due to a) concrete industry conservatism, b) a lack of trained personnel to provide technical advice to specifiers and engineers, and c) a serious lack of quick, reliable durability tests, which are essential. Until such tests are developed and verified in practice, we may still have to rely on the "equivalent performance" concept (not covered here).

4 Performance-based design approach

In practice, performance approaches can represent either a partial approach (termed 'hybrid'), or a full performance-ba sed approach. Both encompass key elements such as tests by which to characterise the desired performance, definition and specification of performance limits by which to judge acceptable performance, and, importantly, integration of durability requirements and durability design through service life models to estimate service life of the RC structure [5]. Crucially, in a full performance approach, specified concrete properties should be measurable *in situ* to ensure as-built quality is actually achieved. This leads to questions of performance testing and performance specifications.

4.1 Performance testing and limits

Performance testing requires development of reliable and representative test methods and the imposition of suitable performance limits. Performance tests may be used in different ways: for pre-qualification of mixes prior to construction where only the material 'potential' (in the sense of the ability of the material to potentially be durable) is proved; and for quality control (QC) during construction, where a) tests are carried out on concrete mixes 'as delivered', and b) tests are done on the actual structure, using in-situ tests or samples that are removed from the structure and tested in a laboratory. The object is to assess the actual *as-built* quality of the structure.

4.2 Performance specifications

In any methodology, the performance parameters and criteria for the structure must be explicitly described and quantified, and a conformity scheme must be set up to verify these parameters in practice and ensure the criteria are met. Thus, the performance specification must clearly describe (i) the functional requirements to ensure that the parties involved in their implementation do not interpret them differently; (ii) performance requirements for the structure, so that the concrete producer and constructor work together in the design of the concrete mix that will meet the performance requirements; (iii) means of verifying compliance using tests that are reliable, repeatable, accurate and preferably easily applicable on site; and (iv) a means to enforce compliance with the specifications, e.g. through penalties when specifications are not met.

4.3 Hybrid approaches

A 'hybrid' approach is a mix of prescriptive and performance elements, where greater emphasis should be placed on the performance criteria. The client and/or specifier decide on the desired level of performance in the given exposure conditions and propose relevant 'durability indicators' or other durability parameters which are used to prepare specifications. A hybrid approach may not explicitly address a design service life period; rather, it aims at achieving the relevant durability parameters that demonstrate the suitability of the concrete, including its composition, in relation to the exposure conditions.

4.4 Durability indicators

Durability indicators, or durability indexes, comprise physical, chemical, or electro-chemical parameters that characterise the concrete at an engineering level. They must be a) easily interpreted in an engineering context, such as being linked to a notional service life or rate of deterioration; b) easily measurable and reliably quantifiable to give engineers confidence in use in construction specifications; and c) sensitive to processing and environmental factors such as binder type, influence of SCMs, mix proportions, placing and compaction, and type and degree of curing.

Durability indicators are well-suited to permit characterisation of the concrete in terms of its potential to be durable – largely the constituent materials – and how these are processed to produce a concrete of the desired properties [6]. Durability indicators encompass a wide scope, e.g. direct measures of permeability, conductivity or the like; indirect measures such as calcium hydroxide residualsor porosity indexes; or chemical/electrochemical properties that provide an indication of potential durability. It is critical that the relevant deterioration mechanisms be considered: durability, however described or characterised, relates to a governing deterioration mechanism.

In summary, the following points are important regarding performance-based approaches:

- A means of relating actual performance to quality control tests (e.g. in terms of limiting material parameters)
- Conformity criteria and procedures, which must be calibrated against known performance measures or traditional solutions
- An "Integrated Scheme" in which the relevant material parameters can be used in quality control, service life modelling (prediction), and setting of specification limits
- Finally, the issue of actually assessing*in-situ* durability properties. This is a thorny issue, and likely to be so for many years to come.

5 Practical implementation

5.1 Limit states for durability, and service life design approaches

Practical implementation of a performance-based approach incorporating service life assessment and codification requires the following elements (as already mentioned): (i) quantifiable limit state criteria, with physical meaning; (ii) a defined service life; (iii) deterioration models with parameters that directly or indirectly are linked to the performance criteria; (iv) compliance tests; (v) a strategy for maintenance and repair; and (vi) quality control systems.

Engineers mostly work to codes of practice, making it essential that any usable approach be codified. Structural codes, which include durability provisions, are slow in being updated, so that new knowledge from research and practice takes a long time to find its way into the standards. However, as an example of implementation of performance-based durability design, ISO 13823 [2] outlines a limit-state methodology which is related to different service life design approaches (not covered here); this methodology has been taken up in the 2010 *fib* Model Code [1], with the approaches to service life design being (i) Full probabilistic approach; (ii) Partial safety factor approach, (iii) Deemed-to-satisfy approach, and (iv) Avoidance of deterioration approach.

The *fib* Model Code approach has merit in moving away from the current simplistic approaches. However, for many if not most RC structures, a minimalist approach may be adequate – since many structures are not exposed to severe environments that threaten their longevity, or are not intended to have long service lives. For mild or non-aggressive environments, simple attention to good construction practices, which should include good mix design, compaction and curing, should help to ensure adequate durability.

5.2 Practical example of implementation, using a durability index performance-based approach

In South Africa, we have developed and implemented a 'durability index performance-based approach', for various infrastructure projects. It is based on measurement of so-called durability indexes, of which one is the oxygen permeability index (OPI). For structures located inland, a limit OPI value is provided. In one study [7], OPI test variability for different sub-projects of a large construction project was measured, and compliance with the set limit value of 9.70 (log scale) determined. OPI variability was observed to be high for *in situ* structures, but lower for precast elements - see Figure 4. Clearly, with proper control of construction practices (placing, curing and compaction), the required durability properties of a structure can be obtained. The implementation example serves to stress the importance of as built verification, and the difficulty of controlling quality on site.

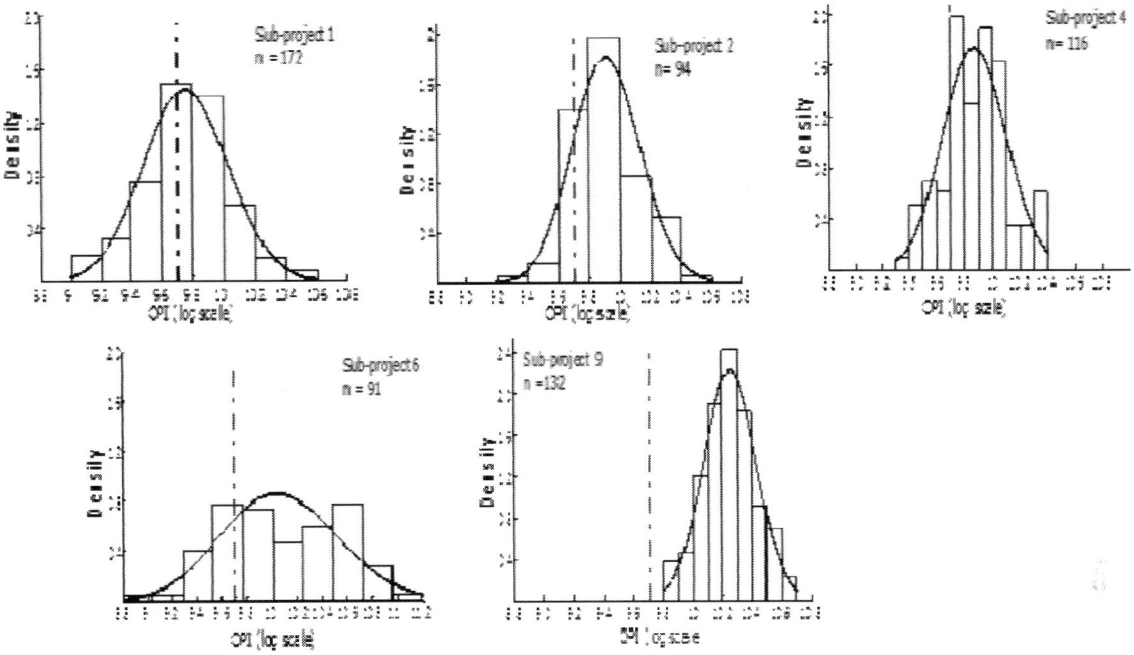

Fig. 4. Illustration of variability in OPI values for different sub-projects. Variability is high for *in situ* structures (sub-projects 1, 2, 4 and 6) and low for precast structures (sub-project 9)

6 Closure

The paper has briefly outlined issues around design for service life in reinforced concrete structures, related to on-going problems in concrete durability. Implementation of 'service life design' cannot occur in the absence of a move to performance-based approaches to design and specification, since the current prescriptive approaches preclude development of more advanced methods. Service life design implies the ability to do service life prediction and modelling; however, at this stage, we are still a long way off comprehensive models that are reliable and accurate. The problem is compounded by the variety of exposure environments for concrete structures, and which are difficult to quantify. However, pragmatic approaches have been developed that can make a significant difference to improving durability performance, and certainly more work is needed at such local and hybrid levels.

References

1. *fib* Model Code. (2010)*Model code for concrete structures*, Ernst & Sohn, 402 pp.
2. ISO 13823-1 *General Principles on the design of structures for durability*. Geneva. International Organization for Standardization, 2008.
3. C. Lobo, L. Lemay, and K. Obla (2005). Performance-based specifications for concrete, The IndianConcrete Journal, **79**(12), 13-17
4. L. Linger &F. Cussigh(2018). PERFDUB: A New French Research Project on Performance-Based Approach for Justifying Concrete Structures Durability. In High Tech Concrete: Where Technology and Engineering Meet.**1**. https://doi.org/10.1007/978-3-319-59471-2
5. M.G. Alexander and M. Santhanam (2013) Achieving durability in reinforced concrete structures: durability indices, durability design and performance-based specifications. *International Conferences on Advances in Building Sciences & Rehabilitation and Restoration of Structures*, Feb 2013, IIT Madras, Chennai, India, 2012, 21pp.
6. M.G. Alexander, Y. Ballim, and K. Stanish (2008).A framework for use of durability indexes in performance-based design and specifications for reinforced concrete structures. Materials & Structures, **41**(5),921-936.
7. G. Nganga, M.G. Alexander and H. Beushausen (2013).Practical implementation of the durability index performance-based design approach. Construction & Building Materials, **45**,251-261.

Vegetable fiber as reinforcing elements for cement based composite in housing applications – a Brazilian experience

Viviane da Costa Correia [1], Sergio Francisco Santos [2] and Holmer Savastano Junior [3]

[1] University of São Paulo, Department of Biosystems Engineering - Research Nucleus on Materials for Biosystems, Pirassununga, SP, Brazil
[2] São Paulo State University, Department of Materials and Technology - School of Engineering, Guaratinguetá, SP, Brazil

Abstract: Vegetable fibers are a hierarchical structure material in the macro, micro and nanometric scales that have been used as reinforcement in cementitious materials. In nanoscale, the nanofibrillated cellulose has the advantage of having good mechanical performance and high specific surface, which contributes to improve the adhesion between fiber and matrix. In hybrid reinforcement, with micro and nanofibers, nanofibrillated cellulose forms bonding with the matrix and acts as stress transfer bridges in the nano-cracking with corresponding strengthening of the cementitious composite. Processing has a strong influence on performance of the fiber cement composite. Two fabrication methods were evaluated: (i) slurry dewatering followed by pressing and (ii) extrusion. The extrusion process strongly depends on the rheological characteristics of the fresh cement material but it can better organize the microstructure of the fiber cement due to the partial orientation of the fibers in the extruder direction. Curing process also plays a key role in the performance of the final product. Accelerated carbonation at early age is a promising technology and a strategy to mitigate the durability problems with the composite materials; it decreases porosity, promotes a higher density in the interface guarantying a good fiber–matrix adhesion and a better mechanical behavior. Alternative MgO-SiO2 clinker free binder is also presented as a suitable alternative to cementitious products reinforced with cellulosic pulps. Finally, mechanical behavior of fiber cement under flexural loading is evaluated by modulus of rupture, fracture toughness, the initial crack growth resistance in cement matrix, and fracture energy that is obtained to evaluate the influence of toughening mechanisms promoted by fibers, such as pullout and bridging, on the mechanical performance of the composites. Degradation during the service life is also crucial for the evaluation of the durability of the resulting materials and components in real applications exposed to different environmental conditions as roofing, partitioning or ceiling elements. It can be concluded that more sustainable and high performance components based on engineered natural raw materials for civil construction can bring valuable contributions for the affordable housing in particular to developing region.

1 Introduction

Fiber-cement is a cement-based material with mineral additions (pozzolans and/or carbonates), without aggregates and reinforced with fibers distributed through the matrix. Normally, in the Brazilian market, fiber-cement is produced with a Portland cement matrix and synthetic fiber combined with cellulose pulp as reinforcement. Hatschek is the most popular processing method followed by curing in saturated air conditions. Asbestos fibers are still used for some products but their application is being progressively reduced in many regions due to occupancy health concerns. Fiber-cement is a versatile material used in different manufacture processing of construction industry products such as corrugated sheets for roofing, water tanks, tubes and flat boards. These products can be applied in the countryside, industries, residences, non-residences and for other applications such as furniture and decoration products [1]-[3].

Fiber-cement has been largely produced since the beginning of the last century with the advent of the Hatschek process. Since 1938, roofing tiles made of fiber-cement have been used in Brazil. Back to the mid-1960s, fiber-cement already had 25% of the roofing area per year in the country. However, it was in the early 1970s that fiber-cement was established in the Brazilian civil construction industry, which continues until now [3].

https://doi.org/10.1051/matecconf/201814901007

In Brazil, fiber-cement products are considered one of the best solutions for housing destined to the population with lower income. It is a lightweight product and have lower cost (taking in account the structure required) compared to other conventional constructive solutions for roofs and flat plates for walls [3]-[4].

Vegetable fibers, as reinforcement of fragile matrices based on inorganic materials, have attracted great interest due their low cost, availability, energy saving and environmental preservation. In the USA market, the main production of fiber-cement is based on autoclaved products reinforced with cellulosic pulp [4]-[6].

2 Vegetable fibers

In Australia, Commonwealth Scientific and Industrial Research Organization (CSIRO) in the early 1970s had active research programs in studying ways to utilize wood fibers as reinforcement in a broad range of composite materials and in modification of the surface and malleability of wood pulp fibers to make them more compatible with various organic and inorganic matrices. The systematic study of the use of vegetable fibers for reinforcement began in England in 1970 [5]. In Brazil, one of the pioneering researches was carried out by the Center for Research and Development (Ceped) in Camaçari, Bahia, which started in 1980 [7]. These materials offer durability and resilience, even in extreme loading conditions, and contribute to infrastructure sustainability via reduced repair events and extended service life. The role of interfaces in relation to design at all scales area corresponds the composites reinforced with vegetable fibers, which have a hierarchical structure and are more sustainable compared to synthetic fibers that are commonly used in the cement industry, such as glass and polypropylene fibers.

Agopyan et al. [3] studied the use of vegetable fibers as reinforcement of fragile matrices and related about 20 fibers potentially useful for the civil construction. The most adequate fibers were selected based on their mechanical properties (tensile strength, Young's modulus and elongation at break), physical characteristics, aspect ratio, cultivation, cost and durability in the natural environment. Softwood pulps, especially those with low lignin content, are already in current use in the Brazilian cement industry. Additionally, hardwood pulps have been studies in countries where their production is abundant [8]-[9].

In addition to cellulosic pulp, vegetable fibers at the nanoscale have also been used as reinforcement for cementitious composites, given that some authors report that nanofibers act as transfer bridges of cracks, which begin at the nanometric scale [10]-[14].

Figure 1 shows an example of bamboo fibers in different scales (macro (a), micro (b) and nano (c)) and their respective micrographs, showing the difference in the morphology of these fibers.

Demonstrating utility of curauá fiber as engineered cementitious composites (ECC) reinforcement would greatly expand the range of applications and would represent a significant step toward the popular use of renewable vegetable fiber in materials mainly for construction applications. The curauá-ECC is well suited to building cladding and façade applications, where lightweight and thermal regulation properties are important. The inherently porous natural fibers provide curauá-ECC with a low density that is theoretically beneficial for both of these aspects and the strain-hardening behavior provides an additional bonus of material durability [15].

Ardanuy et al. [16] used 3.3% (by mass) of nanofibrillated cellulose from sisal as reinforcement of mortar in comparison of mortar reinforced with the same level of cellulosic pulp. The authors reported an increase of the 26.4% in the flexural strength and 41.5% in the modulus of elasticity in the mortar reinforced with nanofibrillated cellulose in comparison to mortar reinforced with the pulp. According to the authors, these results are due to the fact that the nanofibrillated cellulose had higher stiffness and higher intrinsic resistance compared to cellulosic pulp. According to Ardanuy et al. [16], the high specific surface area of the nanofibrillated cellulose promotes a better fiber-matrix bonding, as well as the increase of the hydroxyl groups available in the cellulose to form hydrogen bonds with the cementitious matrix.

Thomson et al. [17]produced hybrid composites, by the slurry dewatering followed by pressing process, reinforced with 8% of cellulosic pulp and the contents of 0.5%, 1.0% and 2.0% of nanocrystalline cellulose to define the best content of nano reinforcement. The mechanical results showed the increase of the modulus of rupture of composites with up to 0.5% of nanofibers and the modulus of elasticity increased with nanofibers content of 2.0%.

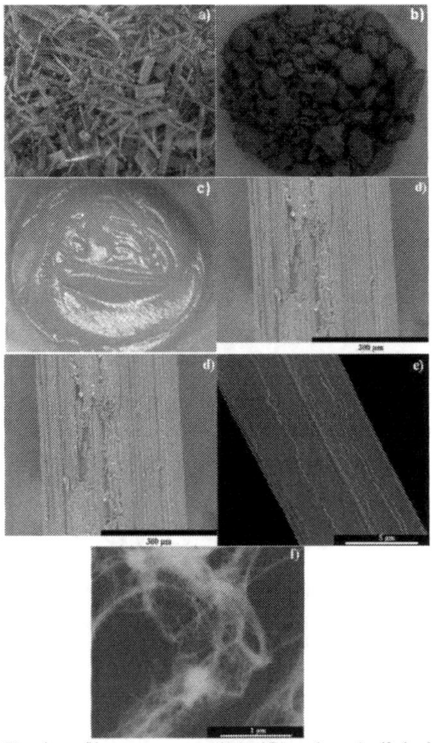

Fig. 1. Bamboo fibers at macro ((a), (d)), micro (cellulosic pulp) ((b), (e)), and nano (nanofibrillated cellulose) scales ((c), (f)) [18].

The results of these studies confirm that the use of nanometric fibers and a hybrid reinforcement is effective for increasing the mechanical properties of cementitious materials.

3 Fiber-cement manufacturing processes

3.1 Hatschek Process

The Hatschek is the most commonly used process for the production of fiber-cement flat and corrugated plates for application in the civil construction industry. The raw materials commonly used in the process are Portland cement, carbonaceous material, silica fume, vegetable fibers in the form of pulp and polymeric fibers (in replacement to the asbestos fiber).

The mainly stages in the Hatschek process are showed in Figure 2. The first step is to prepare the slurry (1), which consists of mixing an adequate proportion of solid materials, with water in a low solid concentration (approximately 20% of total mass). Portland cement (2), reinforcement fibers (3), cellulose fibers, limestone filler (4) and water (5) are the most commonly used materials by the fiber cement technology. The slurry is then transported to the vats (6) with sieve cylinders (7) where wet solid material is deposited. After, the running felt (8) removes the material from the sieve cylinders, thus, forming a green lamina. Vacuum (9) is applied to remove water from the lamina before it is transferred to the formation cylinder (10) where the stacking is performed. Finally, the green sheet (F) is cut (11), shaped (corrugated sheets and accessories, e. g.) (12) and submitted to curing [19]-[21].

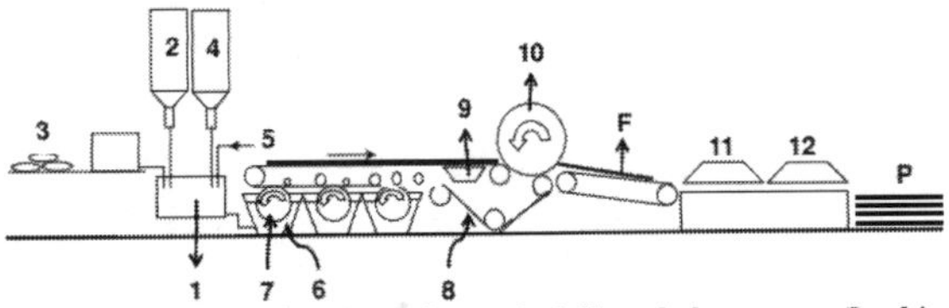

Fig. 2. Schematic drawing of a typical Hatschek process. In this figure F is the green sheet fiber cement and P is a final product (Adapted from Dias et al., [19]; Correia et al., [20]).

During the process the fibers need to meet several variables of the production process, such as: do not sediment and agglomerate when it is mixed with the cement and additions in aqueous suspension; it should be capable of retaining fines during the filtering step and do not damage the felt. The fibers also cannot present a risk to the health of workers and users. In addition, the fibers must be able to achieve components with good mechanical performance and durability.

The products manufactured by the Hatschek process are corrugated roofing tiles, roof ridges, flashings and hips), flat boards, as well as structural roof tiles for large gaps.

3.2 Extrusion Process

The extrusion method displays versatility in the preparation of varied cross-sectional configuration building components, not limited to plates, as in many conventional processes. The shearing and compression forces involved in extrusion process makes the composite more homogeneous and with a better distribution of the fibers in the most favorable direction in order to provide best performance of the fibers, especially in the case of nanofibrillated cellulose. Additionally, extrusion is an efficient process that has the following aspects: the use of a relatively simple equipment; low energy consumption and low water quantity; low investment amount to implement a production line [22]-[23]. Figure 3 illustrates the production of fiber-cement by the extrusion method.

Fig. 3. Single screw extruder used in the extrusion process and the production steps of composite.

The manufacturing of fiber-cement products, in industrial scale, is generally applying the Hatschek process, although this method requires much higher financial investments for a production line. The extrusion process is an alternative process for manufacturing fiber-cement products, for some specific applications such as: tubes, tiles and sidings. The extruder used in the ceramics industry can be adapted for the production of fiber-cement [24].

However, according to Kuder et al. [25], some challenges related to the rheological aspects of the extrusion process require to be studied and improved, such as mixing of raw materials that is important and complex dynamics process related to rheological properties as well as of chemical and physical interactions between the components. For example, the mixing interferes in the agitation of the powders with different bulk densities may result in migration of smaller particles downwards and of larger ones upwards. Besides, the extrusion process requires to study the application of rheological modifiers to adjust the flow behavior (viscosity) and to ensure that the liquid and solid phases of the extrudate do not separated during extrusion, such as those based on high-molecular weight synthetic copolymers, starch ethers, polyacrylamides and clay binders.

4 Accelerated carbonation curing

The main objective of fiber-cement curing thru accelerated carbonation is to minimize the degradation of lignocellulosic fibers in cement matrix and to maintain mechanical performance of composites when they are under aggressive weather conditions [26]-[28].

Tonoli et al. [29] indicated the use of accelerated carbonation after the initial curing of fiber-cement tiles as an effective procedure to minimize the degradation suffered by lignocellulosic fibers. The authors studied the mechanical and physical characteristics of carbonated fiber-cement. The maximum load and toughness of the tiles increased by around 25% and 80% respectively. These results were achieved due to the adoption of accelerated carbonation curing. Water absorption and apparent porosity of composites decreased, while bulk density increased indicating the densification of the material.

Almeida et al. [26] evaluated the effects of accelerated carbonation on the mechanical and physical performance and microstructure of fiber-cement reinforced with only cellulosic pulps at the initial ages, during and after the hydration process. The authors also evaluated the effects of accelerated carbonation in fiber-cement reinforced with cellulosic pulps at the initial ages of hydration, after 200 and 400 soak and dry cycles, and after one year of natural aging. According to the authors, the fiber-cement showed a significant decrease in $Ca(OH)_2$ and a significant increase in $CaCO_3$ and, consequently, low porosity, high density, good fiber-matrix interface and good mechanical performance, according to Figure 4.

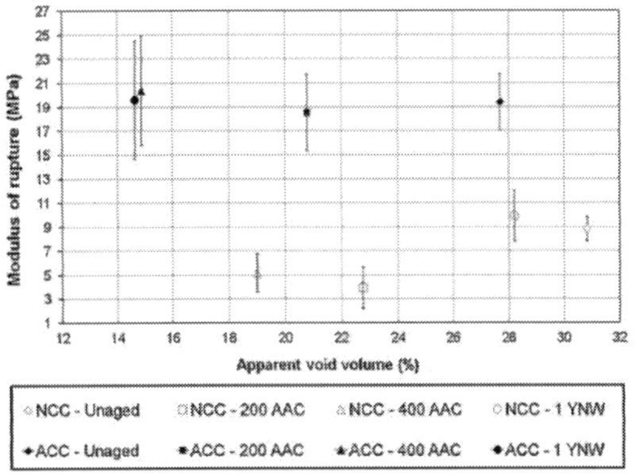

Fig. 4. Average values of mechanical properties vs. apparent void volume (AVV): and modulus of rupture (MOR) of cement based composites: non-carbonated (NCC), carbonated (ACC), after 200 and 400 soak and dry cycles, and after one year of natural aging (YNW).

Santos et al. [28] investigated the effects of supercritical carbonation on extruded fiber-cement reinforced with lignocellulosic fibers (eucalyptus bleached pulp and sisal fibers) after three days of curing. The cure with supercritical carbonation at early ages involves a different chemical kinetics of conventional accelerated carbonation using CO_2 gas, because it uses 100% of liquid CO_2. After 200 immersion and drying cycles, the average fracture energy of the non-carbonated and carbonated composites decreased 56% and 28%, respectively.

Besides the accelerated carbonation curing enhances the durability of cementitious composites reinforced with vegetable fibers, this cure favors the increase of mechanical performance of hybrid composites reinforced with micro and nanofibers. Correia et al. (2015) produced cementitious composites with hybrid reinforcement at micro and nano scales (8% pulp + 1% nanofibrillated cellulose) compared to composites reinforced with only pulp (9% pulp) and observed the effect of accelerated carbonation curing on the physical and mechanical properties of these composites. The authors observed that accelerated carbonation curing contributed to the improvement in the mechanical performance of composites with and without nanofibers, as presented in Figure 5.

The accelerated carbonation curing favored the better performance of the matrix and of the composite in post-cracked condition. This result is attributed to the reactions that occur during carbonation, in which there is the precipitation of calcium carbonate ($CaCO_3$) into the pores of the matrix. As calcium carbonate is denser than calcium hydroxide, there is greater densification of the matrix, reducing pores and improving fiber-matrix bonds and, consequently, improving the mechanical performance of the composite.

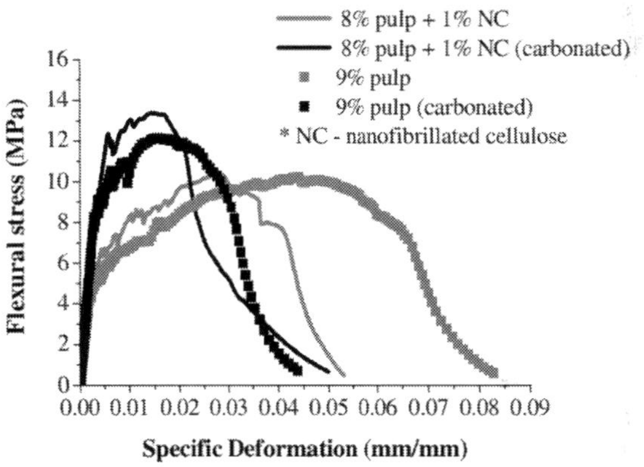

Fig. 5. Typical stress x strain curves at under flexure tests of the composites reinforced with 8% of eucalyptus pulp + 1% NC, and 9% of eucalyptus pulp in carbonated and non-carbonated conditions.

5 Magnesium oxide cement

New types of cement as magnesium oxysulfate (MOS) cements have been successfully used to prevent cellulosic fibers degradation over time. However, the mechanical results of this new kind of cement are far lower than those obtained with Portland cement products. Besides MOS cement presents leaching problems with time. To solve these inconveniencies, a new type of magnesia cement for the production of fiber-cement elements from calcination of MgO powder and reactive SiO_2 ($MgO-SiO_2$ cement) has been developed. Over the last decade, many authors have studied this $MgO-SiO_2$ cement, which hydrates to form magnesium silicate hydrated (M-S-H) gel and magnesium hydroxide (MH), bringing as a result a binding material with low alkalinity and excellent mechanical properties. This cement also has the advantage of offering a chemical composition with irrelevant amounts of calcium, thus avoiding fiber mineralization [30]-[31].

Figure 6 shows the compressive strength results. At 7 days, there is a slight enhancement of the mechanical strength with the increase of the MgO in the samples. This difference is statistically significant when the MgO content varies from 60 to 70%, but at 7 days of curing the samples containing 50 and 60% of MgO and those with 70 and 80% each have statistically similar values. However, at 28 days, all the formulations containing MgO have at least equivalent strength to the control samples. It is also remarkable that the MgO\SiO$_2$ samples present a strength improvement from 7 to 28 days for every formulation, in contrast to the control samples which have no statistically considerable variation [31].

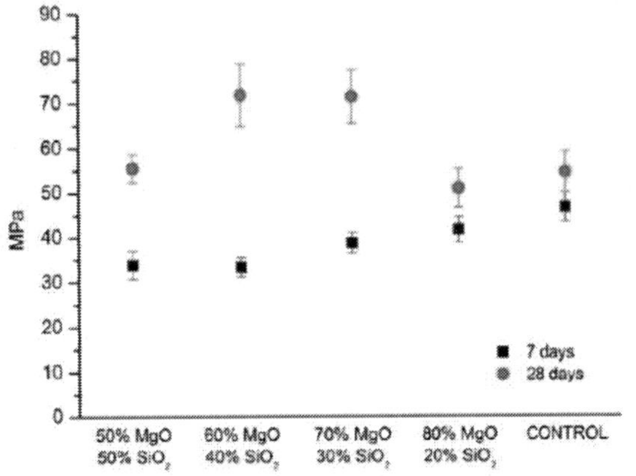

Fig. 6. Axial compressive strength of different MgO\SiO$_2$ pastes and control samples (100% high initial strength Portland cement, CP V-ARI, according to Brazilian Standards NBR 5733) at 7 and 28 days. (Mármol et al., 2016).

Therefore, the use of MgO-SiO$_2$ is presented as a suitable alternative to cementitious products reinforced with cellulosic pulps with potential application in civil construction in view of acceptable performance under binding solicitations and since no appreciable degradation of the fibers takes places after 200 soak and dry accelerated ageing cycles. The 60 wt% MgO - 40 wt% SiO$_2$ system is presented as the optimal formulation, since it is the least alkaline binder with high mechanical strength. Bending tests on pastes reinforced with cellulosic pulps proved the efficiency of this binder, which preserves the reinforcing capacity of the fibers much better than Portland cement pastes after 200 cycles of accelerated ageing [30-31].

Final remarks

The high availability of vegetable fibers in Brazil and in other developing countries in the world justifies the research for the development of new and more sustainable building materials reinforced with vegetable fibers in fragile matrix based on cement. The physical and mechanical characteristics of the vegetable fibers make them a promising material for the partial or total substitution of the synthetic fibers as reinforcement in inorganic matrices.

However, due to low durability of the vegetable fibers in a highly alkaline environment, such as the cementitious matrix, alternative matrices and accelerated carbonation have been developed to enable the production of a more sustainable material with good physical and mechanical performance and with improved durability. That is the case of magnesium based cement products reinforced with cellulosic pulps. Thus, materials reinforced with vegetable fibers are technically possible to be produced and economically viable in regions where the raw material is abundant.

Acknowledgements

The authors would like to acknowledge financial support from the Fundação de Amparo à Pesquisa do Estado de São Paulo, FAPESP, (Grants: 2008/04769-9, 2009/10614-0, 2009/17293-5; 2010/16524-0, 22011/01128-5, 2012/51467-3, 2013/23810-8 and 2015/21079-0) and the Conselho Nacional de Desenvolvimento Científico e Tecnológico, CNPq, (Grants: 472133/2009-8, 305792/2009-1, 142082/2011-2 and 312151/2016-0).

References

[1] M. Ardanuy, J. Claramunt, R.D. Tôledo Filho, Constr. Build. Mater. **79**, 115 (2015)
[2] A. Bentur, S. Mindess, *Fibre Reinforced Cementitious Composites* (Taylor & Francis, 2007)
[3] V. Agopyan, H. Savastano Jr, V.M. John, M.A. Cincotto, Cement Concrete Comp. **27,** 527 (2005)
[4] S.F. Santos, G.H.D. Tonoli, J.E.B. Mejia, J. Fiorelli, H. Savastano Jr, H, Materiales de Construcción **65**, 317 (2015)
[5] R. Coutts, Cement Concrete Comp. **27** 518 (2005)
[6] O. Onuaguluchi, N. Banthia, Cement Concrete Comp. **68**, 96 (2016)
[7] CEPED - Centro de Pesquisas e Desenvolvimento da Bahia. *Cartilha para fabricação de telhão em argamassa armada* (BNH, 1985) (In Portuguese)
[8] G.H.D. Tonoli, U.P. Rodrigues Filho, H. Savastano Jr, J. Bras, M.N. Belgacem, F.A. Rocco Lahr, Compos. Part. A-Appl. S. **40**, 2046 (2009)
[9] G.H.D. Tonoli, E. Fuente, C. Monte, H. Savastano Jr, F.A. Rocco Lahr, A. Blanco, Cement Concrete Res. **39**, 1017 (2009)
[10] O. Galao, F.J. Baeza, E. Zornoza, P. Garcés, Cement Concrete Comp. **46**, 90 (2014)
[11] C.G. Hoyos, E. Cristia, A. Vázquez, Mater. Design. **51**, 810 (2013)
[12] G. Yakovlev, J. Kerienė, A. Gailius, I. Girniene, Mater. Sci. **12**, 147 (2006)
[13] J.M. Makar, J.C. Margeson, J. Luh, *Carbon nanotube/cement composite - early results and potential applications* (International Conference on Construction Materials: Performance, Innovation and Structural Implications, 2005).
[14] G.Y. Li, P.M Wang, X. Zhao, Carbon. **43**, 1239 (2005)

[15] D.G. Soltan, P. Neves, A. Olvera, H. Savastano Jr, V.C. Li, Ind. Crop. Prod. **103** 1 (2017)
[16] M. Ardanuy, J. Claramunt, R. Arévalo, F. Parés, BioResources. **73** 3883 (2012)
[17] S.L. Thomson, D.J. O'Callaghan, J.A. Westland, B. Su, Method of making a fiber cement board with improved properties of the product. Patent US 2010/0162926 AL, (2010)
[18] V.C.,Correia, V. Santos, L.B. Rodier, K. Ghavami, H. Savastano Jr, Green. Mater. **4**, 41 (2016)
[19] C. M.R. Dias, H. Savastano Jr., V. M. John, Constr. Build. Mater. **24**, 140 (2010)
[20] V.C. Correia, S.F. Santos, H. Savastano Jr, World Academy of Science, Engineering and Technology: International Journal of Civil, Architectural, Structural and Construction Engineering **9**, 7 (2015)
[21] V.C. Correia, S.F. Santos, G.H.D Tonoli, H. Savastano Jr, *Characterization of vegetable fibers and their application in cementitious composites* (Woodhead Publishing, 2016)
[22] H. Takashima, K. Miyagay, T. Hashida, V. C. Li, Eng. Fract. Mech. **70**, 853 (2003)
[23] R.S. Teixeira, G.H.D. Tonoli, S.F. Santos, J. Fiorelli, H. Savastano Jr, F.A.R. Lahr, Key. Eng. Mat. **517**, 450 (2012)
[24] G.G. Horst, *Extrusion of FRC products: technology and practical aspects* (Inorganic-Bonded wood and fiber composites materials Conference, 2002)
[25] K.G. Kuder, S.P. Shah, Constr. Build. Mater. **24**, 181 (2010)
[26] E.F.S. Almeida, G.H.D. Tonoli, S.F. Santos, H. Savastano Jr, Cement Concrete Comp. **42**, 49 (2013)
[27] V.D. Pizzol, L.M. Mendes, L. Frezzatti, H. Savastano Jr, G.H.D. Tonoli, Miner. Eng. **59**, 101 (2014)
[28] S. F. Santos, R. Schmidt, A.E.F.S. Almeida, G.H.D. Tonoli, H. Savastano Jr, Cement Concrete Comp. **56**, 84 (2015)
[29] G.H.D. Tonoli, S.F. Santos, A.P. Joaquim, H. Savastano Jr, Constr. Build. Mater. **24**, 193 (2010)
[30] G. Mármol, H. Savastano Jr, Cement Concrete Comp. **80**, 258 (2017)
[31] G. Mármol, H. Savastano Jr, M.M Tashima, J.L. Provis, Mater. Design **105**, 251 (2016)

Connection behaviour and the robustness of steel-framed structures in fire

Ian Burgess
University of Sheffield, Department of Civil & Structural Engineering, Sheffield S1 3JD, United Kingdom

Abstract. The full-scale fire tests at Cardington in the 1990s, and the collapse of at least one of the WTC buildings in 2001, illustrated that connections are potentially the most vulnerable parts of a structure in fire. Fracture of connections causes structural discontinuities and reduces the robustness provided by alternative load paths. An understanding of connection performance is essential to the assessment of structural robustness, and so to structural design against progressive collapse. The forces and deformations to which connectionscan be subjected during a fire differ significantly from those assumed in general design. The internal forces i generally start with moment and shear at ambient temperature, then superposing compression in the initial stages of a fire, which finally changes to catenary tension at high temperatures. If a connection does not have sufficient resistance or ductility to accommodate simultaneous large rotations and normal forces, then connections may fracture, leading to extensive damage or progressive collapse of the structure. Practical assessment of the robustness of steel connections in fire will inevitably rely largely on numerical modelling, but this is unlikely to include general-purpose finite element modelling, because of the complexity of such models. The most promising alternative is the component method, a practical approach which can be included within global three-dimensional frame analysis. The connection is represented by an assembly of individual components with known mechanical properties. Component characterization must include high-deflection elevated-temperature behaviour, and represent it up to fracture.In reality a connection may either be able to regain its stability after the initial fracture of one (or a few) components, or the first failure may trigger a cascade of failures of other components, leading to complete detachment of the supported member. Numerical modelling must be capable of predicting the sequence of failures of components, rather than considering the first loss of stability as signifying building failure. It is necessary to use a dynamic analysis, so that loss of stability and re-stabilization can be tracked, includingthe movements of disengaging members and the load-sharing mechanisms which maintain integrity and stability within the remaining structure, until total collapse occurs.

1 Introduction

Structural steel connections have been extensively investigated over the past three decades to determine their moment-rotation characteristics. However, the importance of tying capacity had been realized even earlier, since the explosion at Ronan Point [1] in 1968 caused progressive collapse of a large part of the building. The UK structural steelwork design code BS5950 [2] now requires connections to have minimum tying capacities. The UK SCI/BCSA design guidance [3] checks the tying capacity as an isolated action, whereas in reality a combination of tying force, shear force and moment usually exists. For individual bolts, resistance to tying force may be affected by co-existence with other forces. For a complete bolted connection combined actions can prevent a uniform distribution of the resultant tying force between the bolts, causing them to fail sequentially, significantly reducing the tying capacity.

In design for fire resistance, the increasing adoption of performance-based design principles means that structures are now treated integrally in structural fire safety design. Connections, as the key components which tie structural members together, are important in maintaining structural integrity and preventing progressive collapse. Evidence from the collapse of the WTC buildings [4, 5] and full-scale fire tests at Cardington [6] have shown that connections are vulnerable to fracture in fire. Only limited research has been done on the performance of connections at elevated temperatures, most of which has concentrated on end-plate connections, and has mainly been confined to moment-rotation behaviour. A further complexity is that interactions between structural members during heating cause continuous changes in the forces and moments taken by the connections.

A recent trend in the design of composite floor systems has been to fire-protect beams on the main column grid, while leaving other beams unprotected. In the early stages of a fire (with structure temperatures typically up to 600°C) unprotected beams expand against restraint from surrounding structure and this creates high compressive forces. Protected beams eventually deflect considerably under the combined effect of high steel temperatures and enhanced loading, shed from the unprotected members, and will impose high tying forces on their connections. In non-composite steel construction, the beams deflect considerably at high temperatures and experience catenary tension, which is transferred to the supporting structure through the connections. Tests by Ding [7] showed that connections were subjected to tying forces varying from 0.65 to 1.6 times their shear force at high temperatures.

Even if the connections survive the heating phase of a fire, when beams contract from their distorted state during cooling this imposes higher tensile forces on connections, and many recorded connection failures observed in full-scale testing have occurred during this phase. These tying forces, either at high temperature or during cooling, together with the local forces on components of the connections imposed by very high rotations, can clearly fracture components, triggering disproportionate collapse of the structure. This was clearly seen in partial connection failures at Cardington. The most dramatic example of the effect of restrained thermal expansion of long-span steel beams in recent years was revealed by the NIST report [5] on the collapse of "7 World Trade" on 11 September 2001 in the immediate aftermath of the collapse of the Twin Towers. Thermal expansion of very long steel secondary beams caused the connections of the supporting primary beams to columns to be pushed-off, initiating progressive collapse of the building.

2 Component modelling of connections

Figure 1[8] schematically illustrates the layout of components within a component-based element.

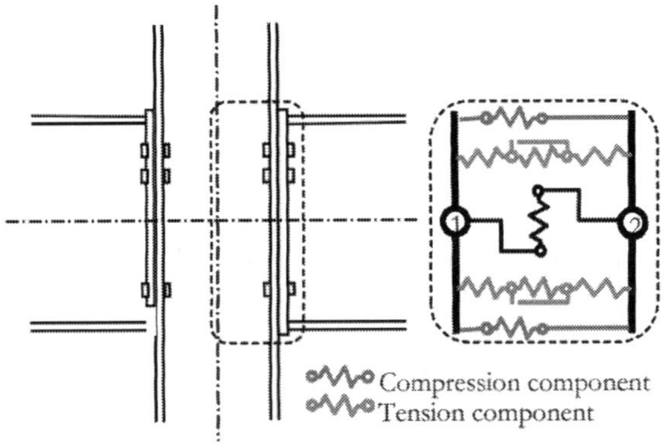

Fig. 1. Component assembly.

The assembled element has 2 external nodes. Although only two component-based "springs" are shown in the figure the model can consist of appropriate numbers of tension (bolt) rows and compression rows. Node 1 is located at the intersection between the beam and column reference axes. Node 2 is the end-node of the beam. Column-face shear components can be included in this assembly, but are commonly assumed to be rigid in the vertical shear direction.

2.1 Tension components

Each tension bolt row includes three tension components, which work in series. The middle component in each series is designed to represent the bolt in tension. The other two tension components are: the column flange in tension and end-plate in tension for flush end-plate connections; for other types of connection there will certainly be at least two outer components, although there may be additional effects to be represented by further component springs in each row.

The three components in each tension bolt row are combined into one effective spring at each temperature step, as is illustrated in Figure 2. The force-displacement curves of the tension bolt rows are used to derive the connection's local force and stiffness.

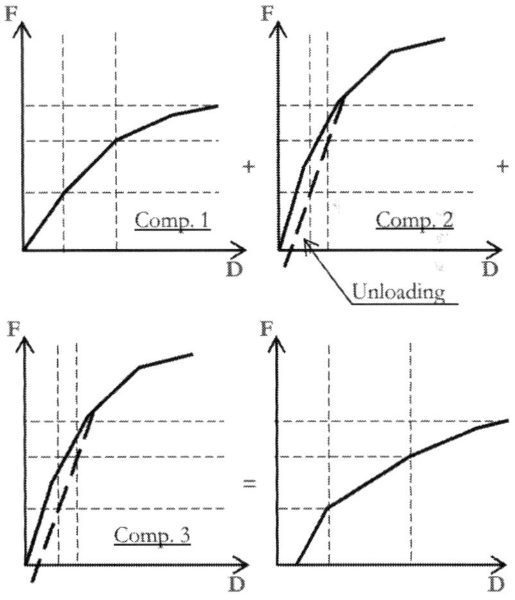

Fig.2. Assembly of the components of a tension bolt row.

After the global analysis reaches a converged stable equilibrium, the forces in the tension bolt rows are established, and the displacements of each tension component are calculated. The related information, such as each component's permanent deformation, is then updated. The maximum resistance of the effective spring is defined by the weakest component in this series. Any force above the weakest component's ultimate resistance is ignored. At each force level, the effective spring's displacement is the total of the components' displacements under this force level.

2.2 Compression components

Compression components clearly do not include bolts, since these are in clearance holes, and become unloaded when the net force at their level is compressive. They may of course include bolts in pure shear in cases where there is a plate connected to the beam web. However, the major effect which needs to be taken into account in the compression zone is hard contact between the beam and column flanges, which may be present at all times (end-plate connections) or may occur when the initial clearance between the beam bottom flange and the column-face closes completely. This hard contact generates a stiffness which is much higher than those of the tension components.

As a result of a project conducted by the Universities of Sheffield and Manchester on the capacity and ductility of steel connections at elevated temperatures [9-12] a large number of high-temperature tests were conducted on four common connection types under combined tension and moment. These were accompanied by detailed finite element analyses. The outcome was that many component types were characterized in terms of their load/deformation behaviour. An example of the component arrangement and characteristics (at ambient temperature) is given in Figure 3.

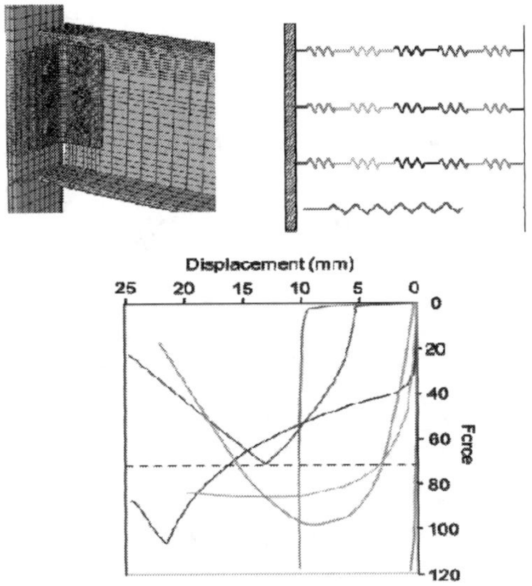

Fig.3. Component characteristics for a web-cleat joint.

3 Analytical implications

Because of the nature of conventional quasi-static analysis, an analysis of a structure in fire which includes component-based connection elements can only trace the behaviour of a connection up to the point where its first component fails. In reality a connection may either be able to regain its capacity after the initial fracture of a component, or the first failure may trigger a cascade of failures of other components, leading to complete detachment of the connected member. This possibility should be considered in performance-based design when a structure is being tested for robustness. If connections are to avoid the possibility of becoming detached from members, this numerical modelling must be capable of predicting the sequence of failures of components, rather than simply the first loss of stability. A numerical procedure in which the whole behaviour, from first instability to total collapse, can be modelled effectively, has been developed in *Vulcan*.

The *Vulcan* model combines alternate static and dynamic analyses [13], in order to use both to best advantage. Static analysis is used to follow the behaviour of the structure at changing temperature until instability happens; beyond this point an explicit dynamic procedure is activated to track the motion of the system until stability is regained. When combined with the parallel development of general component-based connection elements, this procedure can effectively track the behaviour of connections, from the initial fracture of a component, via the failure of successive bolt-rows, to final detachment from the column. This sequence is illustrated schematically in Figure 4(a); static analysis is initiated at the start of a loading process, and this continues until a loss of stability is detected. The analysis then switches to explicit dynamic, so that the dynamic motion of the unstable structure is tracked until re-stabilization occurs; the static analysis is then re-started. This is capable of modelling the partial fractures of partial-depth end-plates, of the kind shown in Figure 4(b),which were seen in cooling of some Cardington tests.

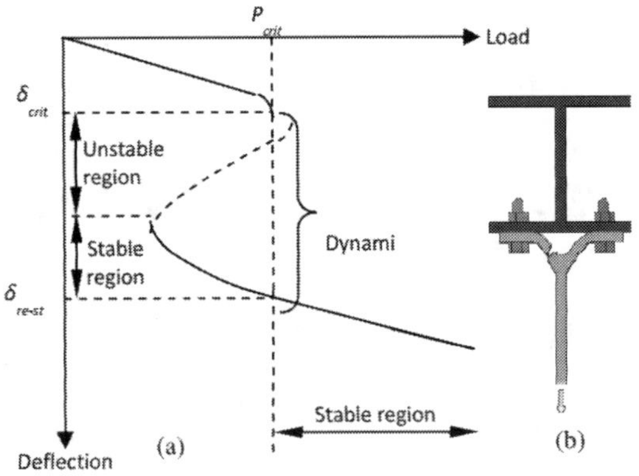

Fig.4. Component characteristics for a web-cleat joint.

In fact the analysis of a simple frame model, depicted as a sequence of deflected shapes in Figure 5, carries on beyond connection fracture row-by-row, within which the actual fractures in each row were of the components representing the bolts. Logically, this occurs sequentially, starting at the top bolt row and progressing downwards rapidly; the final bolt row essentially simply carries the catenary tension without any of the leverage that applies to the upper rows, and so its curve differs from the others. Detachment of the heated beam does not, in fact cause structural collapse;this occurs due to column buckling at a higher temperature.

4 The potential of ductile connections

It seems unlikely that the problem of robustness in fire can be solved using the normal solution of increasing the strength of the vulnerable structural elements, in this case the connections.

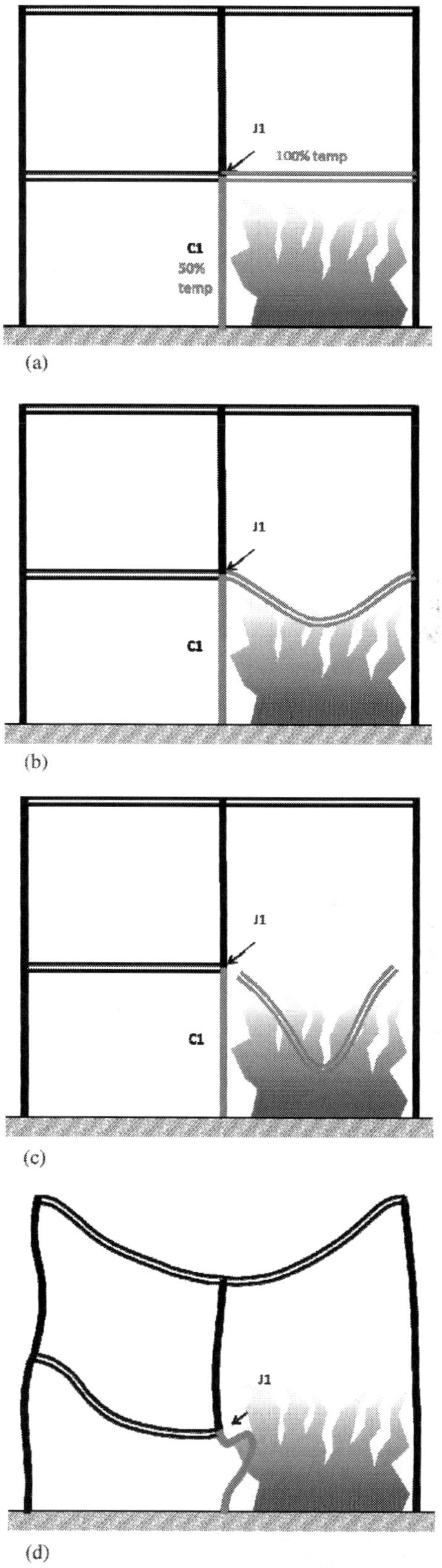

Fig. 5. Plane frame fire scenario: (a) Initial heating stage, (b) as connection components fracture,(c) connections detached, (d) final collapse due to column buckling.

The variations of component forces at each bolt row of the connection J1, and the vertical movement of the central column, with the beam temperature, are plotted in Figures 6(a) and 6(b) respectively.

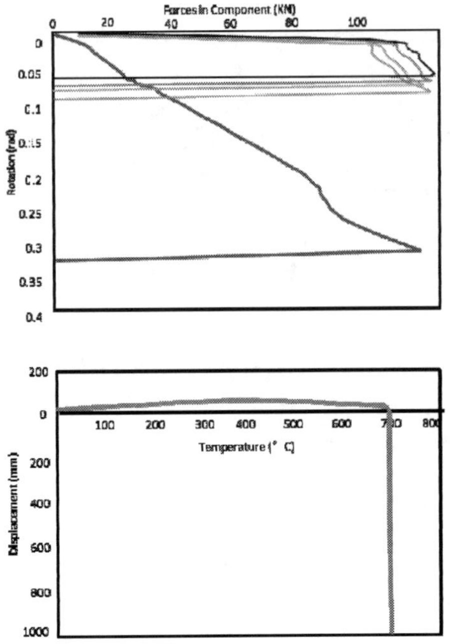

Fig. 6. Plane frame fire scenario: variation of (a) tensile forces in bolt rows of connection, (b) column vertical displacement, with key temperature.

Long-span beams cannot be prevented from generating either large axial expansions or high axial forces in fire, depending on the extent of their restraint from surrounding structure. At relatively low temperatures restrained expansion forces can cause integrity failures in concrete slabs, while at high temperatures the catenary tension forces in beams have magnitudes which are inversely dependent on the amount of axial movement of the beam ends. The latter can even fracture the heated beam in pure tension when its tensile strength is reduced to a few percent of its normal value.

A potential strategy to improve the robustness of connections against fire is to devise connection details which possess the ductility to deform in push-pull without fracturing, while also remaining rigid with respect to vertical shear deformation. In the early stages of heating this will prevent large forces from being transmitted to the surrounding structure. At high temperatures it will allow sufficient movement of the beam ends to reduce the catenary tension in the beam (and hence on the connection) to a level where no fracture occurs at the highest beam temperature. The exact levels of movement required constitute the "ductility demand" of the structure in the fire limit state. The ductility demand is very largely dictated by the characteristics of the beams which are supported by the connections. If the connection is considered as an additional element attached between the beam-end and the column-face then the key demands for a non-composite steel beam are illustrated in Figures 7(a) and 7(b). During the beam expansion phase the ideal situation occurs when the maximum movement at the beam-ends does not cause hard contact with the column faces. In the contraction phase the net movement should be sufficient to reduce the catenary tension below that which causes connection fracture at the highest elevated temperatures. There is a secondary requirement that the additional tension caused by cooling of the beam can be sustained at all stages as the strengths of components are restored.

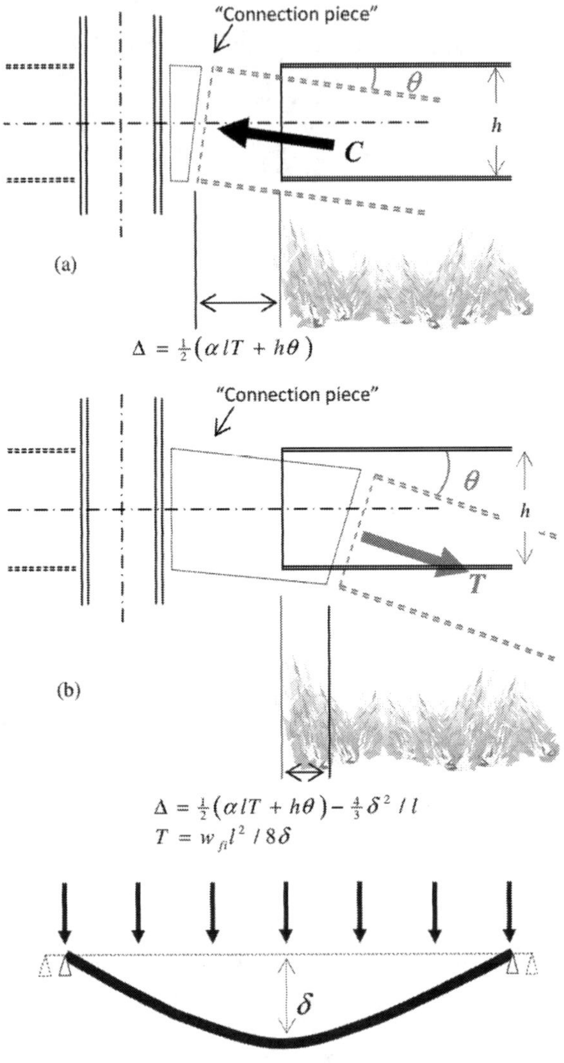

Fig. 7. Movements of beam-end at different stages of heating: (a) Expansion at low temperatures, (b) Shortening due to large deflection, (c) Deflected shape.

5 Conclusions

Connections within a structural sub-frame, if heated together with the beams that they support, will initially be subjected to compressive force due to the restrained thermal expansion of the beams. The magnitude of this force depends on the span of the beam and the axial restraint stiffness provided by the connections themselves and adjacent structure. Some connections, such as fin plates, can fail due to this force. Although this has been suggested as the cause of failure of WTC 7 [5], it has never been observed in the UK, probably because the multi-storey composite-framed structures typical of UK practice provide less axial restraint to beams. End-plate connections cannot fail under compressive forces; the compression will continue to increase until the beam reaches its limit capacity under the combined effect of the bending moment and compression. It then experiences a rapid increase of deflection, which attenuates the compression force to a limiting value as the thermal expansion is accommodated by the deflection. With further temperature increase the progressive reduction of steel strength decreases the compressive force, to the extent that the axial component eventually becomes tensile. This tension increasingly takes over from the bending resistance of the beam in carrying the loads by catenary action. At this stage the upper bound to the tensile force is given by the lower of the reduced strengths, at the appropriate temperatures, of the beam or its connections. This has been demonstrated in small-scale structural frame tests by Ding [7]. Ductile design of connections is important because the catenary force is reduced with increase of the deflection which is allowed by movement at the ends of the beams; some of this movement can be allowed by the connections themselves. The capacities of a connection in terms of moment, tying force and rotation are completely inter-dependent. Both moment and tying capacity are based on the tensile behaviour of each bolt row. The rotation of a connection, in terms of movement of the beam-end relative to the column face, is the most important influence on its strength and ductility. Semi-rigid or rigid connections, which have higher moment resistances, generally have lower rotational capacity than simple connections, which may limit their ability to develop catenary action. It is not necessary to consider these three parameters directly in order to establish the limit state of a connection in fire. In most cases a component-based model can provide a sufficiently accurate and practical solution to the modeling of connections in fire. Previously component-based models have been developed mainly for end-plate connections at ambient temperature, in order to generate rotational stiffnesses and moment capacities for semi-rigid frame design. The Sheffield group has now conducted several successive research projects on steel connection behaviour in fire, culminating in the component characterization reported here. The behaviour of most components of the four connection types tested has been represented in simplified high-temperature non-linear spring models. Because of the need to emphasize the issue of robustness in fire, it is advantageous for these models to have two innovative characteristics:

- A model of the pre-peak part of the load-displacement curve for a component is needed, since uniform distribution of displacement to all the bolt rows is unlikely.
- Formation of a yielding mechanism is not necessarily synonymous with fracture. The behaviour of each component up to large deflection or actual fracture is necessary.

Following these basic principles, components have been characterized for use in modelingof four types of connection, and these have been shown to predict the connection behaviour with satisfactory accuracy. A general-purpose component-based connection element has been assembled which can accommodate the appropriate components within its bolt-rows, and this development has been made in parallel with a static/dynamic solution process for the Vulcan software. It has been shown that the behaviour of a structural frame can be modelled throughout the duration of a fire with this combination, so that transient failures of parts of connections do not cause an end to the analysis, and re-stabilization is shown if it occurs. This kind of analysis will be necessary in future for true performance-based analytical design of framed buildings against fire, so that potential disproportionate collapse can be predicted and the design of the structure, including that of its connections, can be adjusted to reduce this possibility.

References

[1] Bignell, V., Peters, J., and Pym, C., Catastrophic failures, Open University Press, Milton Keynes, 1977.
[2] BS5950, Structural Use of Steelwork in Building-Part 1: Code of practice for design-Rolled and welded sections, British Standards Institution, London, 2001.
[3] SCI & BCSA, Joints in steel connection, Simple connections, The Steel Construction Institute and British Constructional Steelwork Association, UK, 2002.
[4] NIST, Final Report on the Collapse of the World Trade Center Towers, National Institute of Standards and Technology, USA, 2005.
[5] NIST, Final Report on the Collapse of World Trade Center Building 7, U.S. Department of Commerce and National Institute of Standards and Technology, USA, 2008.
[6] Newman, G.,M., Robinson, J.,T., Bailey, C.,G., Fire Safety Design: A New Approach to Multi-Storey Steel-Framed Buildings, The Steel Construction Institute, Ascot, 2004.
[7] Ding, J., Behaviour of Restrained Concrete Filled Tubular Columns and Their Joints in Fire, Ph.D Thesis, University of Manchester, UK, 2007.
[8] Block, F.M., Development of a Component-Based Finite Element for Steel Beam-to-Column Connections at Elevated Temperatures, PhD thesis, University of Sheffield, 2006.
[9] Yu H.X., Burgess I.W., Davison J.B., and Plank R.J., Experimental and Numerical Investigations of the Behaviour of Flush Endplate Connections at Elevated Temperatures, *Journal of Structural Engineering, ASCE*,137 (1), (2011) pp 80-87.
[10] Hu, Y., Davison, J.B., Burgess, I.W. and Plank, R.J., 'Component Modelling of Flexible End-plate

Connections in Fire ', *International Journal of Steel Structures*, 9, (2009) pp 29-38.

[11] Yu, H.X., Burgess, I.W. Davison, J.B., and Plank, R.J., Experimental Investigation of the Behaviour of Fin Plate Connections in Fire, *J. Construct. Steel Research*, 65, (2009) pp 723–736.

[12] Yu H.X., Burgess I.W., Davison J.B., and Plank R.J., Tying Capacity of Web Cleat Connections in Fire Part 1: Test and Finite Element Simulation, *Engineering Structures*, 31 (3), (2009) pp 651-663.

[13] Sun, R.R., Huang, Z. and Burgess, I.W., 'Progressive collapse analysis of steel structures under fire conditions', *Engineering Structures*, 34, (2012), pp 400-413.

Atmospheric Weathering of Historic Monuments and Their Related Conservation Issues

Emine N. Caner-Saltık

Graduate Program in Conservation, Department of Architecture,

Archaeometry Program, Graduate School of Natural and Applied Sciences,

Middle East Technical University (METU), Ankara

E-mail: canersal@yahoo.com.tr

Abstract. Atmospheric environment affects the materials of historic monuments and their structure starting from the time of their construction. Daily and seasonal changes in temperature and humidity, wind, snow and rainfall, soluble salts carried by water, biological agents, pollutant gases and particulate matter are some of the agents in atmospheric environment that introduce weathering by physical, chemical and biological processes in the materials of the monuments such as natural building stones, bricks, mortars and plasters, mud brick etc. The weathering processes need to be well diagnosed by identification of main mechanisms of decay and major responsible agents, degree and depth of deterioration expressed with measurable parameters of physical, physicomechanical properties, and micro structural changes together with their distribution on the monument. Success of conservation treatments strongly depend on those diagnostic studies and compatibility of the treatments with the deteriorated and relatively sound parts of the historic materials. Current approach to materials conservation is to be able to make minimum intervention to historic material by targeting the conservation treatment to the deteriorated area for the purpose of controlling the deterioration factors and achieving compatible and durable conservation of historical material. In this presentation, two examples of diagnostic research and conservation treatments based on and guided by the diagnostic results are summarized concerning historic stone monuments exposed to atmospheric environment since more than two thousand years. The first example is on the marble walls of Temple of Augustus in Ankara exposed to polluted urban atmosphere. The second example is on the limestone statues of Nemrut Mount Monument in Adıyaman-Turkey, exposed to rural atmosphere with harsh climatic conditions. Finally, a brief discussion on current research issues related to historic materials conservation in atmospheric environment is made.

Introduction

Materials of historic monuments and their structure are affected by the atmospheric environment they live in since the time of their construction. Daily and seasonal changes in temperature and humidity, wind, snow and rainfall, soluble salts carried by water, biological agents, pollutant gases and particulate matter are some of those agents in atmospheric environment that introduce some changesby physical, chemical and biological processes.In time, those weathering processes cause considerable changes in microstructure of stone, such as increase in porosity, decrease in mechanical properties, quite a number of changes in chemical composition etc. starting from exterior surfaces towards interiors of the stone. Those changes are also visible by change in color, detachments as scales and flakes, crack formation, material loss as powdering, granular disintegration, outbursts etc. Diagnostic analyses aim at finding out the main process of decay and its responsible decay agents as well as the degree and extent of decay on the material and its distribution on the monument. They serve as a guide for the selection and development of conservation treatments. The approach in diagnostic analyses of monuments is similar to the diagnostic analyses in medical field without which no curative treatment can be done, although human life is very short compared to the life of monuments many of which surviving since centuries. It should not be forgotten that conservation treatment should not endanger the life of the monument and its survival capacity but help its survival in its aesthetic unity.

In this study, diagnostic studies and conservation treatments research about two important monuments in Turkey, namely marbles of Temple of Augustus and limestone statues of Nemrut Mount Monument are summarised and discussed.

Temple of Augustus

The Temple of Augustus is one of the oldest standing monuments in Ankara from Roman period. The Roman Baths and a part of Roman city walls are situated not very far from the Temple.The Temple of Augustusis not only a standing monument from Roman Period but also a written document. It carries on its marble stone walls the only original text of Emperor Augustus in Greek and Latin letters. It is essential to preserve that monument in its present condition with its historic and aesthetic integrity. In order to be able to do that, one had to clarify the point about what it was going wrong with the monument.

That work is a diagnostic study to determine what the main decay agents are, and how much damage they have produced on the monument. Those questions were investigatedby the use several analytical tools with minimum representative sampling (1).

The air pollution damage was drastic in the years starting from 1970 up to mid-eighties, while the population in the city had rapid growth, coal was being used in central heating and traffic started being heavy. Climatic conditions of Ankara atmosphere with high average relative humidity values in winter months and during the nights through the year even in summer times favored adsorption of both nitrogen oxides and sulphur dioxide on allstone surfaces.Those gases are able to react with the materials of the monument at high relative humidity conditions thus causing great damages on historic monuments.The pollutants reaching the materials surfaces as particulate matter carried by air currents or gases are called dry deposition mechanisms while the pollutants in acid rain directly wetting and washing the surfaces of materials are called wet deposition mechanisms.Nitrogen oxides from traffic are known to act as catalysts in the reactions of sulphur dioxide with the calcareous stones during the conversion of calcite to gypsum. Damage is known to be more extensive on the monuments constructed of calcareous building stones such as marble, limestone and travertine.

A project was started for the diagnosis of the current materials and structural problems of Temple of Augustus and development of its conservation treatments (2-6).

The visual forms of the decay on the monument are observed as grey, black, yellow, brown colour change and depositions, detachments as scales, detachments as granular disintegration and powdering, material loss by loss of scales, grains, and powders. Photogrammetric drawings of the Temple were used for mapping those main decay forms observed on the Temple. Forms and degrees of the weathering at the monument do not show an equal distribution. The decay forms range between severe decays and light decays. Those differences were thought to be due to differences of the microclimatic conditions on the surfaces of the stone,different wetting conditions with rain water since there was no roof on the monument for its protection from rainwater,and probably the different types of marbles used for the construction of the monument. IR thermography studies on the surfaces of the monument have proved to be a useful non destructive investigation to reveal the state of deterioration together with the USV measurements

Mineralogical and micro structural properties of the marble samples from the Temple were investigated by optical microscopic analyses of cross sections and thin sections, SEM-EDX analyses of cross sections, XRD and FTIR analyses of powdered samples.The marbles of Temple of Augustus were medium to coarse grained. The sizes were 20-200 microns for medium grains, and up to 1mm for coarse grains. It was found that the marbles of the walls having latin inscription consist of calcite minerals together with dolomite minerals while the marbles of the walls having greek inscription consisted of calcite. SEM-EDX analyses showed that the main mineral calcite was transformed to gypsum mineral ($CaSO_4.2H_2O$). XRD and FTIR analyses confirmed that the main decay product is gypsum together with biominerals such as calcium oxalate.That.has proved the presence of biodeterioration on the surface of Temple of Augustus together with gypsum formation as a result of air pollution. Biological activity and its distribution was further investigatedby qualitative and quantitative analyses with fluorescein diacetate (FDA) (6).

The formation of gypsum was not only present on the surface but also distributed towards the interiors of the marble structure. Transformation of calcite to gypsum included volume expansion.The analyses indicated that the characteristic compact structure of marbles was changed by deterioration. The grains were separated from one another and spaces of various sizes were produced in between. The pore sizes varied from capillary size of few microns to a few hundreds of microns. The pore structure of the deteriorated marble had to be taken into account in the conservation treatments.

Then it comes the solutions about how to control and stop the main decay agents, and the description of treatments to be applied to the decayed materials based on the current conservation methods whether those are applicable or adaptable to that case or a new treatment is necessary to be developed.

Since gypsum is soluble in water (~2.2 g/l), the Temple needed a roof to protect it from wetting by rain penetration.Calcium oxalate minerals are quite insoluble . their solubility product is much less than calcite.Preesence of calcium oxalates, although itwas not in the form of a continuous layer but in a scattered formation, decreased the solubility of gypsum to some extent. The SO_2 pollution in Ankara is now diminished since the use of coal is stopped and natural gas is being used instead.Traffic is tried to be eliminated at least in the vicinity of the monument.

It was necessary to strengthen the weak marble by consolidation and sticking the scales with compatible repair mortars each of which required the laboratory and in situ research on development of suitable methods. Here, marble consolidation treatments will be mentioned (5).

Aim of marble consolidation treatments was to improve the properties of decayed stone to such an extent that a compatible relationship is formed with the decayed and the relatively non-decayed parts after the treatment. It was aimed to form a structure with similar physico-mechanical properties and compatible with the chemical and mineralogical properties of the stone.. In other words, deteriorated, less deteriorated and not deteriorated parts of the marble should be connected to get a compatible and durable unit. For those purposes some consolidation treatments were performed based on the information gained on the types and degree of decay of the marbles in the Temple of Augustus. It was avoided to use synthetic polymers of any sort since they form a new structure in decayed stone that differ a lot in its response to the changes of atmospheric temperature and humidity and separate itself from the rest of the stone in a short period of time.

In this research, consolidation treatments were developed by the preparation of nanodispersive solutions of $Ca(OH)_2$ in ethyl alcohol. Efficient growth of calcium carbonate crystals in the pores and capillaries by incorporating them with the present calcite network of marble were tried to be achieved with nanodispersive solutions of calcium hydroxide. Precipitation of micritic calcite was observed in the fine cracks of marble that were formed as a result of consolidation treatment. After the consolidation treatments the ultrasonic velocity values of the marbles have increased and their porosities have decreased.

The method was found to be successful, however the different nanodispersive solutions needed to be prepared in accordance to the locations with different types and degrees of deterioration. For this reason, samples with different deterioration problems were studied in detail with several analytical techniques before and after treatments.

Long term performance of consolidation treatments continue to be monitored by nondestructive tests of ultrasonic pulse velocity measurements and quantitative infrared thermography imaging in situ and in laboratory.

Views from the Temple of Augustus

Nemrut Mount Monument

Nemrut Mount Monument is one of the most spectacular archaeological sites situated on top of a high Mountain, more than two thousand meters high having very harsh climatic conditions with strong winds, covered with snow during almost half of the year. Monument is a temple with giant statues of Gods and King Antiochos around his Tumulus, built in around two thousand years ago during Kommagene Kingdom. The site is in World Heritage List since 1986.Nemrut Mount belong to the large Midyat Formations of Eocene age, precipitated in marine environment. Limestone statues were selected and carved from the limestone of the mountain itself. They are karstic limestone with visible and invisible karstic veins. Quite a number of decay forms are visually seen on the limestone statues. Abundance of decay forms as fragmental disintegration through some karstic veins of limestone, colored biological depositions and pitting were observed at the site, both on limestone statues and geological limestone deposits. In this study, the causes and mechanism of fragmental disintegration formation, and extent of biological weathering was examined in order to figure out what type of conservation treatments were to be done (7-8). Examinations were done on the small fragments of samples that were lost from the statues and on the samples from the nearby geological formations that showed similar deterioration forms with the limestone statues. Samples from geological formations were cut to 5 cm cubes and were used for artificial weathering by salt crystallization in the laboratory. Progress of stone deterioration was examined in the laboratory by salt crystallization tests. Unlike most other techniques the sodium sulphate crystallization test was a comparative test to obtain data for monitoring and discovering the weak zones of limestone.Both naturally and artificially decayed limestone were examined for their physical, physico-mechanical and microstructural properties. Physical and physico-mechanical properties were examined by bulk density, effective porosity, and ultrasonic velocity measurements (8). Microstructural properties of deteriorated limestones were observed by examination of cross sections with SEM-EDX, by analyses of thin sections with optical microscopy and by examination of powdered samples scraped from exterior surfaces, crack surfaces and interiors of the stone using XRD and FTIR analyses. Distribution of biological activity was determined by qualitative and quantitative analyses with FDA (6). The petrographical analysis of limestones indicated that the limestones had micritic and microsparitic particles and they bonded with sparitic calcite crystals. Some submicroscopic opaque minerals as iron oxides and fossils were also observed in thin sections of limestones. Physico-mechanical studies indicated that weathered limestone samples taken from the monument had bulk density ~2.67g/cm^3 and porosity ~0.8%. Samples collected from geological formations in the vicinity were a bit denser and had slightly lower porosity. Some karstic zones of limestone were found to be preferred sites of decay for the start of surface dissolution and for the formation and widening of cracks. Presence of clay minerals kaolinite, illite, chlorite in those decay zones, iron oxides that move through those zones, as well as biological activity that found those sites favorable and started to live in there were found to be closely related to fragmental disintegration phenomena. Swelling of chlorites and increase in their swelling by weathering were expected. The swelling characteristics of kaolinite and illite were considered to be very small. However, they may take part in deterioration by several other mechanisms. The mobility of clay minerals in the aqueous media caused the transportation of the clay minerals in the fine capillaries of the stone. The role of clay minerals in deterioration by other mechanisms such as ion exchange capacity, acting as reservoirs for nucleation of new crystals etc. might also be important in the growth of cracks.

The most important deterioration type observed in limestones was fragmental disintegration that proceeded through micro cracks in the structure of limestone and enlarged in changing atmospheric conditions. The changes in atmospheric conditions introduced wetting and drying cycles of the clay minerals and iron oxides in the micro cracks. At the same time, calcium carbonate dissolution and recrystallization reactions took place in the cracks of the limestone. Some clay minerals accumulated from outside and settlement of micobiological growth and increase in biological activity accelerated fragmental disintegration process. To prevent the widening of the cracks and fragmental disintegration, it was necessary to decrease the activity of clay minerals and control the wetting and drying in the micro cracks. Nano dispersive $Ca(OH)_2$ solutions were used to decrease the activity (cation exchange capacity) of the clay minerals in the cracks of limestones. After treatment with nano dispersive $Ca(OH)_2$ solutions, a considerable decrease was observed in the cation exchange capacity of the clay minerals. The decrease in clay minerals activity was measured by methylene blue absorption method (8). Decreasing the activity of clay minerals and controlling their swelling by the use of nano dispersive $Ca(OH)_2$ solutions was expected to control the widening of the cracks and loss by fragmental disintegration. The control of wetting and drying in the cracks were done by further treatments of the cracks with nano dispersive $Ca(OH)_2$ solutions that would form a calcite network in the cracks similar to the structure of the original limestones. Carbonation of nano dispersive $Ca(OH)_2$ solutions injected in the cracks were accelerated by providing high relative humidity and higher CO_2 concentrations in the treatment areas by using temporary covers. Filling of cracks with calcite network would prevent rain penetration through them. Therefore, the study was focused on the preparation of variable concentrations of nano dispersive $Ca(OH)_2$solutions with some additives that were able to penetrate through the deep cracks in the limestone (8-9). Pilot treatment areas at the site are monitored to verify the longterm success of the treatments.

A View from the Nemrut Mount Monument (7)

Research topics and trends in conservation of historic building materials exposed to atmospheric environment

Recent research topics in conservation science concerning materials of historic monuments exposed to atmospheric environment are now concentrated on the development of treatments with nano dispersive solutions and repair materials including nano powder additives. Compatibility and durability criteria for the treatments and improvements of the methods in their evaluation are topics of high interest.

Many methods of conservation are stopped being used due to their damaging effects that developed in time and due to their irreversibility and incompatibility.

While studies on the conservation treatments develop, the need for the diagnostic studies that describe in detail, the degree and the process of deterioration in historic material is increased. Better the diagnosis, better the chance for adjustment of the proper dose of the treatment that serves the purpose.

Advances in analytical techniques contribute a lot to the advances in diagnosis and treatments; small representative samples now become sufficient enough to do the identification of decay problems.

Researchers are in need of describing the interaction of new treatments with the historic material in time. Non-destructive techniques are developing to be used in diagnostic studies as well as in monitoring the efficiency of conservation treatments. Quantitative Infrared Thermography (QIRT) and Ultrasonic velocity measurements (USV) are very useful to follow the success of conservation treatments in time and measure their compatibility and durability with the original structure.

European standards on the investigation and conservation of historic materials and structures are being developed (CH-EN).

Conservation studies of historic monuments involve multidisciplinary work. That fact presents some coordination and management problems to be overcome. Another important topic is to fill the gap between conservation science research results and current conservation practice.

Importance of research laboratories specialized on conservation of historic monuments is increased. Research laboratories on historic monuments can only serve their purpose if well-educated conservation scientists are responsible and in charge of conservation issues.Planning of a diagnostic and conservation research, its execution, its monitoring need to be in the responsibility of highly qualified researchers who have Ph.D on conservation science issues. Multidisciplinary nature of research team is the key to the success. Scientists from the fields of chemistry, physics, biology, construction science, materials science, archaeometry, engineers and architects who are knowledgeable on functional systems of buildings, all who are further qualified by Ph.D in conservation science are badly needed. All those are common European assets and are being supported by a number of different collaborations.

References

1. E. N. Caner, H. Göktürk, A. G. Türkmenoğlu, G. Eseller,. Effects of air pollution on the monuments of Ankara, Case Study: Temple of Augustus. Durability of Building Materials, **5**, 463-473. (1988)

2. E. H. Göktürk, M. B. Al-Badawi, S. Aygün, E. N. Caner, "Determination of Nitrate and Nitrite by Ion Selective Electrodes in Relation to Gypsum Formation on Some Travertine Buildings in Polluted Atmosphere of Ankara" Intern. J. Environ.Anal. Chem. **40** 47-57. (1990).

3. E. N. Caner, H.Böke, "Occurence of Calcium Oxalates on Marble Monuments in Anatolia" *Proceedings of International Symposium on; The Oxalate Films: Origin and Significance in the Conservation of Works of Art;*G. Alessandrini Ed., Centro CNR "GINO BOZZA" Milano, 299-307, (1989)

4. Temple of Augustus Conservation Project: Control of Stone Deterioration, Structural Maintanence and Site Presentation.Coordinator: Prof. Dr. Emine N. Caner-Saltık. METU, World Monument Fund. Ministry of Culture: General Directorate of Cultural Heritage and Museums. (2006)

5.Development of new methods of conservation for marble in archaeological sites and monuments: A study to develop combined treatments for marble conservation using nanoparticlesand biomineralization, Turkish-French University Cooperation Project 2004-2005, METU-MCL and LRMH France. Turkish Team Coordinator: Prof. Dr. Emine N. Caner-Saltık, (2005)
6. M. C. Üstünkaya, Biological Decay and Its Control by Biomineralisation in Calcareous Stones, Unpublished Master Thesis, METU. Ankara, (2008)
7. Commagene-Nemrut Conservation and Development Program: Program for Preparing a Management Plan, Materials and Structural Conservation of Nemrut Mountain, Tumulus and Its Monuments. Coordinators: Prof. Dr. Emine N. Caner-Saltık & Assoc. Prof. Dr. Neriman Şahin Güçhan. METU, Ministry of Culture: General Directorate of Cultural Heritage and Museums. (2010)
8. E. Caner, E. N. Caner-Saltık, G. Orial, J. Mertz, "Deterioration Mechanisms of Historic Limestone and Development Of Its Conservation Treatments With NanodispersiveCa(OH)2 Solutions" In: 8th International Symposium on the Conservation of Monuments in the Mediterranean Basin: Monument Damage Hazards & Rehabilitation Technologies, Eds: M. Koui, F. Zezza, P. Koutsoukos, Technical Chamber of Greece, Athens, V.3, 54-72.(2013)
9. P. Lopez-Arce, L. Gomez-Villalba, S. Martinez-Ramirez, M. Alvarez de Buergo, Powder Technol., **205**, 2

Prestressing Concrete with CFRP Composites for Sustainability and Corrosion-Free Applications

A. Belarbi[1], M. Reda[1], P. Poudel[1], H. Tahsiri[1], M. Dawood[1], and B. Gencturk[2]
[1]University of Houston, Civil and Environmental Engineering, Houston, TX, USA
[2]University of Southern California, Civil and Environmental Engineering, Los Angeles, CA, USA

Abstract. Advancement in material science has enabled the engineers to enhance the strength and long-term behavior of concrete structures. The conventional approach is to use steel for prestressed bridge girders. Despite having good ductility and strength, beams prestressed with steel are susceptible to corrosion when subjected to environmental exposure. The corrosion of the prestressing steel reduces load carrying capacity of the prestressed member and result in catastrophic failures. In the last decades, more durable composite materials such as Aramid Fiber Reinforced Polymer (AFRP), Glass Fiber Reinforced Polymer (GFRP) and Carbon Fiber Reinforced Polymer (CFRP) have been implemented in concrete structures as a solution to this problem. Among these materials, CFRP stands out as a primary prestressing reinforcement, which has the potential to replace steel and provide corrosion free prestressed bridge girders. Despite its promise, prestressing CFRP has not frequently been used for bridge construction worldwide. The major contributing factor to the lack of advancement of this promising technology in the United States (U.S.) is the lack of comprehensive design specifications. Apart from a limited number of guides, manuals, and commentaries, there is currently no standard or comprehensive design guideline available to bridge engineers in the U.S. for the design of concrete structures prestressed with CFRP systems. The main goal is to develop design guidelines in AASHTO-LRFD format for concrete bridge girders with prestressing CFRP materials. The guidelines are intended to address the limitation in current AASHTO-LRFD Bridge Design Specifications which is applicable for prestressed bridge girders with steel strands. To accomplish this goal, some of the critical parameters that affect the design and long-term behavior of prestressed concrete bridge girders with prestressing CFRP systems are identified and included in the research work. This paper presents preliminary results of an experimental study that is part of a National Highway Co-operative Highway Research Program (NCHRP) project.

1 Introduction

Prestressed reinforced concrete for highway bridges has been in progress since 1951 [1]. At present, around 27% of all the highway bridges in the United States consist of prestressed concrete. Nearly all the prestressed beams use prestressing steel. The problem with steel prestressing reinforcement is corrosion (Figure 1). Prestressed steel beams have prestressing steel tendons under high stress with the relatively small cross-sectional area. The possibility of brittle fracture due to stress corrosion or hydrogen embrittlement is higher in higher strength steel. This may lead to the sudden failure of a bridge without any warning [3-4]. These have initiated the use of non-metallic tendons such as Glass Fiber Reinforced Polymer (GFRP), Aramid Fiber Reinforced Polymer (AFRP) and Carbon Fiber Reinforced Polymer (CFRP) collectively referred to as Fiber Reinforced Polymers (FRPs). Among these materials, CFRP stands out as a primary prestressing reinforcement. The initial cost of CFRP is higher than that of GFRP and AFRP. Furthermore, all three FRPs are comparable regarding their tensile strength. The one thing that stands out the most for the CFRP is its high tensile stiffness. The durability properties of the CFRP are also excellent in comparison to other FRPs [5].

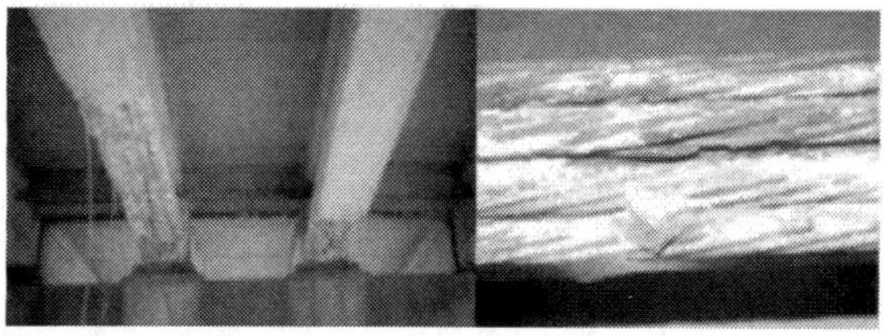

Fig. 1. Corrosion of the prestressing steel [2]

CFRPs are becoming a recognized alternative to traditional construction materials with a wide range of applications in current civil engineering practice. A novel example of these applications is the use of CFRP strands as prestressing reinforcement for prestressed concrete bridge girders, especially in aggressive environments where steel strands are susceptible to corrosion. CFRP materials have specific advantages such as high strength-to-weight and stiffness-to-weight ratios, and resistance to electrochemical corrosion. Despite their promise, however, prestressing CFRP strands have not frequently been used for bridge construction in the United States (US). The major contributing factor to the lack of advancement of this technology in the US is the lack of comprehensive design specifications.

2 Previous Studies

Many researchers have proven that CFRP can be an alternative to steel strands in prestressed concrete bridge applications, especially in aggressive environments where corrosion resistance is essential.

The flexural behavior of CFRP-prestressed beams with bonded tendons has been extensively studied in the past. Test results consistently indicate that while the behavior of CFRP-prestressed and steel-prestressed concrete beams are similar in many regards, there are several important differences which must be considered. These differences are primarily due to the linear-elastic nature of CFRP reinforcement. On the other hand, the flexural behavior of post-tensioned beams with unbonded CFRP tendons has been less studied comparing to the bonded prestressed beams with CFRP tendons. The failure of unbonded post-tensioned beams is generally attributed to concrete crushing. The fact that the cables are unbonded leads to release of stresses at critical sections and average it out along the tendon's length, so the load in the tendon will be less than their bonded counterpart. Therefore, the ultimate strength of the bonded specimens is higher than the unbonded specimens regardless of the tendon type.

Beams that are prestressed with CFRP can exhibit failure due to rupture of the tendons before crushing of the concrete, whereas the flexural design of prestressed steel beams is always defined by crushing of the concrete regardless of the reinforcement ratio.

The initial behavior of the beams prestressed with prestressing CFRP or steel is comparable. All beams exhibit a linear response before cracking. The tendon type had little impact on the uncracked stiffnesses and the cracking load, as the beam geometry, total prestressing force, and eccentricity of the prestressing force are comparable. After cracking, the beams prestressed with CFRP exhibited slightly lower stiffnesses than their steel-prestressed counterparts. This is due to the slightly lower modulus of elasticity of the CFRP tendons as compared to steel tendons. Abdelrahman et al. tested four beams with two different types of prestressing CFRP under monotonic and fatigue loading. These four beams had the same span-to-depth ratio and 1:3.3 scale as the girders of first smart highway bridge built in Calgary, Alberta, Canada. They reported that the beams pre-tensioned with prestressing CFRP exhibit a significant deformation before failure. The beams tested under fatigue loading with a load range of 70% of cracking to cracking load survived 2 million cycles with little effect on beam stiffness. The capacity of the beams after fatigue loading was comparable to the capacity of the beams under monotonic loading. The length of the tested beams was 13.5 m. [6]

Dolan et al. tested one beam pretensioned with harped prestressing CFRP and one beam pretensioned with harped prestressing steel under fatigue loading. The load range of the fatigue was approximately between 11% and 35% of the ultimate monotonic load which induces a stress range of $0.05\sqrt{f'_c}$ and $0.6\sqrt{f'_c}$ on the bottom concrete fiber, respectively. The tendon stress varied from $0.28f_{pu}$ to $0.36f_{pu}$ for prestressing CFRP. They found that the fatigue behavior of the beam pre-tensioned with steel and prestressing CFRP is similar. Both beams were found to experience similar amounts of permanent deflection, the same crack distribution and the same strength due to fatigue load. They also found that the beam pre-tensioned with prestressing CFRP experienced no reduction in the capacity due to the fatigue loading. They also pointed out that special consideration must be given to the CFRP prestressed beams to avoid brittle failure, especially when the prestressing CFRP are harped. [7]

Heo et al. tested seven simply supported post-tensioned beams with internally unbonded CFRP tendons (four rectangular and three T-beams). The test parameters were the prestressing reinforcement ratio, the initial prestressing, the loading type (three-point or four-point loading) and the section shape. Unstressed auxiliary bonded reinforcements (steel for the T-beams and CFRP for rectangular beams) were added to the bottom of the beam. Regardless of the reinforcement ratio, all the beams failed in flexure due to concrete crushing without tensile rupture of the unbonded prestressed tendons. Also, all rectangular beams maintained residual strength after the peak load, which is equal to section capacity of the beams after removing the top concrete cover. In addition, the type of loading affected the failure mode of the T-beams as the beam tested in four-point loading failed in an explosive manner that took out the constant moment region without any residual capacity, however, the beam tested under 3-point loading showed a residual capacity after the peak. Moreover, the prestressing ratio affects the brittleness of the compression failure mode, as the higher the prestressing ratio, the higher the brittle compression failure will happen. Comparing the ductility of beams with auxiliary bonded steel bars with bonded CFRP tendons for rectangular sections, the ductility of the beams with auxiliary bonded steel bars were higher. Moreover, they evaluated the ductility of post-tensioned beams with CFRP tendons based on the absorbed elastic and inelastic energy under the load-deflection curve. The results showed that the calculated ductility overestimates the measured ones and need further refinement. [8]

Braimah et al. tested five unbonded post-tensioned concrete beams with two different prestressing reinforcement types (prestressing CFRP and steel) under fatigue flexural loading. The steel post-tensioned beams had bonded non-prestressed steel reinforcement, while the CFRP post-tensioned beams had bonded non-prestressed GFRP rods. A severe test regime was chosen that induced a stress range of about 100 MPa in the

https://doi.org/10.1051/matecconf/201814901010

bottom prestressing reinforcement. The beams were initially loaded past the cracking load and then unloading to obtain an initial load–deflection response. The fatigue tests were performed under four-point loading with a frequency of 1 Hz. None of the post-tensioned beams survived the 2.0 million cycles. The failure of the post-tensioned steel beams started by fracture of the non-stressed steel then the seven-wire prestressed strands. However, the failure of CFRP post-tensioned beams was initiated by the failure of the prestressed CFRP tendons, and non-prestressed GFRP rods didn't fail during the fatigue cycles. The fracture of the CFRP tendons occurred at the anchor's location. [9]

Kakizawa et al. investigated the behavior of beams fully and partially prestressed with CFRP. The prestress force and bonding properties were the test parameters. The results showed that the failure mode and deformation depend on the reinforcement system. In addition, the absorbed energy is affected by the reinforcing system, but little by the failure mode. The partially prestressed beams tended to have greater energy absorption than the fully prestressed beams. The fully prestressed beams showed no energy absorption after the failure of the FRP cable. However, the fully unbonded beams that failed by concrete crushing showed the same energy absorption after reaching the maximum load. [10]

Grace et al. investigated the behavior of multi-span CFRP prestressed continuous concrete bridges with external longitudinal post-tensioning using draped tendons and bonded transverse post-tensioning. The ductility of such a bridge system increased and mid-span deformation decreased compared to the behavior of a simply supported beam constructed with similar member properties. At the ultimate load stage, the progressive failure of CFRP tendons was observed. [11]

3 Objectives

The objectives of the current study are to; (1) study the behavior of prestressed beams with different prestressing CFRP systems under monotonic and fatigue loadings, (2) compare the behavior of beams post-tensioned and pre-tensioned with prestressing CFRP.

4 Experimental Program

4.1 Beam Design

Four full-scale prestressed beams were constructed at a precast plant. The beams are 711 mm deep AASHTO Type 'I' with a 203 mm cast-in-place composite deck. The span of the beams is 12 m. All of the beams were reinforced in shear with steel stirrups to ensure a flexural dominant failure.

The post-tensioned beams "CPouDF" and "CPoDM" were stressed with five straight 19.3 mm. twisted-wire prestressing CFRP cables. The cables inside the duct of the post-tensioned beam "CPoDM" were grouted with high strength cementitious grout. However, the cable on the other post-tensioned beam "CPouDF" left un-grouted (unbonded). The two pre-tensioned beams, "CPrSM" and "BPrSM", were stressed with eight straight 15.2 mm. twisted-wire CFRP cables and twelve straight 12.7 mm round CFRP bars, respectively. Figure 2 shows the beam's dimensions, shear reinforcement distribution details, deck steel reinforcement details, and prestressing elements' locations. The CFRP cables and bars were stressed to the 60% of the guaranteed breaking load, provided by the manufacturer. The properties of the CFRP, based on conducted laboratory tests, are presented in Table 1.

Table 1. Prestressing CFRP properties

Beam ID	Diameter (in.)	Breaking Load (kN)	Young's Modulus (GPa)	Ultimate Strain (%)
Cable	15.2	327.8	155	2.0
Cable	19.3	467.9	153	1.7
Bar	12.7	226.5	144	1.3

4.2 Fabrication

All the beams were fabricated at the precast plant. The production of the beams was conducted in different batches. The girders were poured and stressed, and then a composite deck was added on the top.

The pre-tensioning is applied through the steel strands and transferred to the prestressing CFRP with couplers, as shown in Figure 3 a-b. On the other hand, the post-tensioning of the prestressing CFRP cables was done in sequence using a single hydraulic jack, as shown in Figure 3c.

For bonded post-tensioned beam, a high strength cementitious non-shrinkable grout was used for grouting the prestressing CFRP. The grouting was performed at the precast plant, immediately after the post-tensioning process.

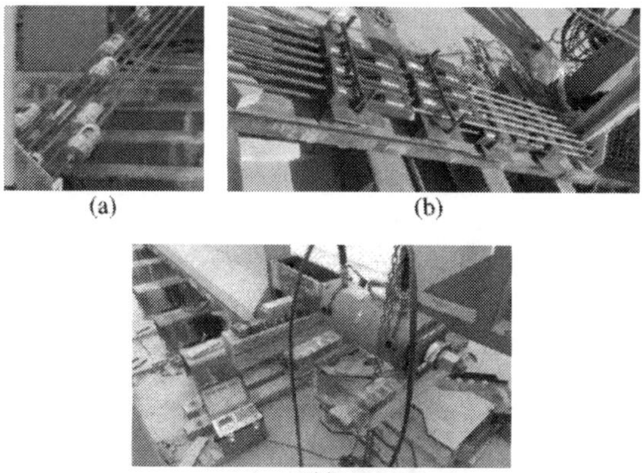

(a) (b) (c)

Fig. 3. Pre-tensioning and post-tensioning of prestressing CFRP cables and bar

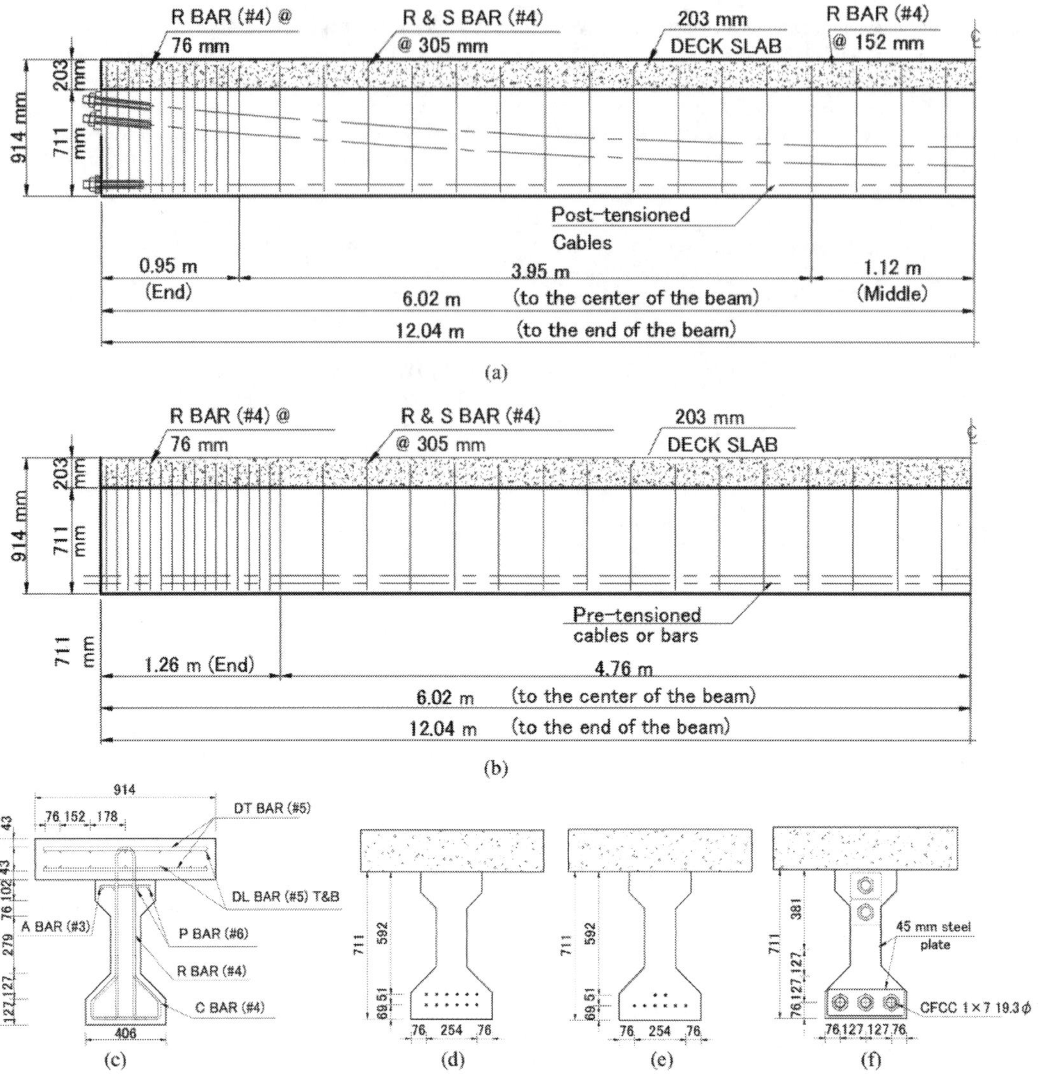

Fig. 2. Schematic drawings of the four prestressed beams: elevation and reinforcement details of a) post-tensioned beam with straight and draped cables b) pre-tensioned beams, c) typical reinforcement distribution, and typical dimensions for prestressed beams d) CPrSM, e) BPrSM, and f) CPouDF and CPoDM.

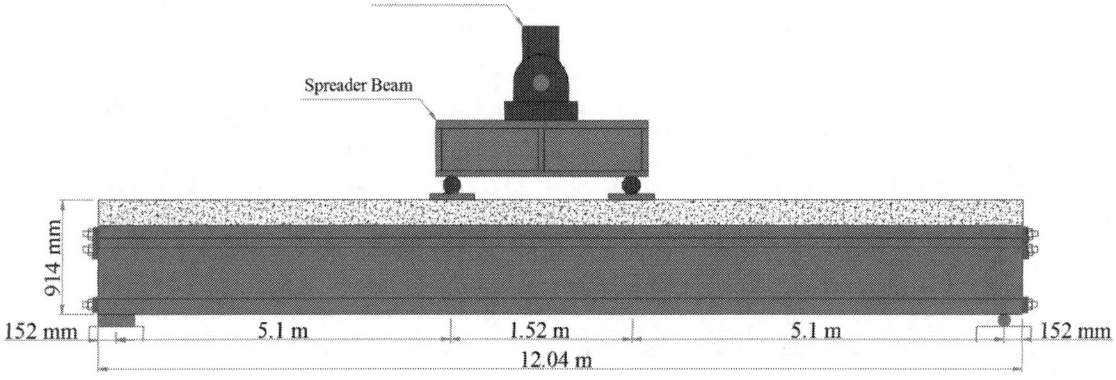

Fig. 4. Test setup for flexural testing

4.3 Test Set-up

The beams were tested under four-point monotonic flexural loading. The applied load was measured with a calibrated load cell attached to the hydraulic actuator with 1045 kN capacity as shown in Figure 4. Twelve string potentiometers along the length of the beam were used to measure the deflections of the beams.

4.4 Testing protocols

The testing of the beams in monotonic flexural loading was performed in displacement control mode. Prior to cracking, the beams were loaded with a rate of 0.84 mm/min. After cracking, the loading rate was increased to 4.24 mm/min. up to 75% of the estimated capacity and 3.05 mm/min. afterwards until failure. A 66.7 kN seating load was applied at the beginning to reduce the initial slack of the load path.

4.5 Results

4.5.1 Load-deflection curves

Figure 5 shows the load-deflection curves of the four CFRP prestressed beams. The displacements presented in the figure are the average of two potentiometers at mid-span, and the applied load is measured from the load cell on the actuator. The presented load does not include the weight of the beam and the composite deck.

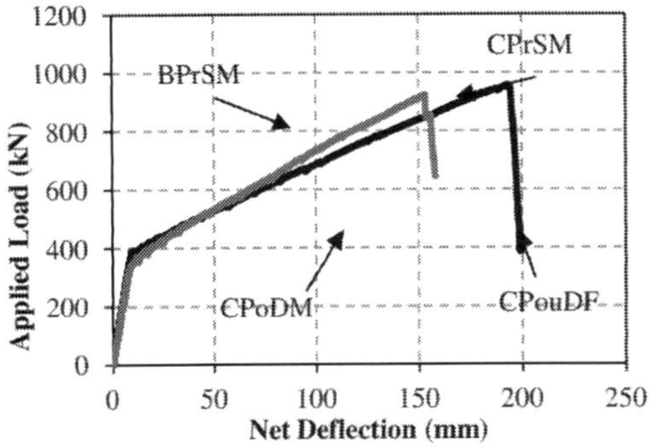

Fig. 5. Experimental load-deflection response of CFRP prestressed beams

Table 2. Summary of the test results

Beam ID	Cracking Load (kN)	Ultimate Load (kN)	Deflection (mm)	Failure Mode
CPrSM	320	936	196	Cable rupture
BPrSM	338	920	152	
CPoDM	362	778	132	
CPouDF	320	636	226	Concrete crushing

The overall behavior of the beams were piece-wise linear with two distinct stiffnesses up to failure. The initial stiffnesses of the beams reduced after the crack openings and remained almost constant up to ultimate load. For the pre-tensioned beams and bonded post-tensioned beam, the failure was due to rupture of the cables, and the load carrying capacity dropped to zero immediately. However, for the unbonded post-tensioned beams, the failure was due to concrete crushing and once the beams reached the ultimate load, there was a drop, but the beams had a reserved load carrying capacity. A summary of the test results is presented in Table 2.

4.5.2 Failure Modes and crack distributions

The monotonic tests of the full-scale beams were done in stages to, 1) draw the crack propagation and 2) measure the crack width. The cracks of the unbonded post-tensioned beams were concentrated in few wide cracks which forked with the loading during the test. This behavior is the same as unbonded post-tensioned with prestressing steel. However, for bonded prestressed beams, the cracks were spread over more than a quarter of the length from the center of the beam on each side. Figure 6 shows the failure modes and crack distribution at the peak load for the prestressed beams with CFRP systems.

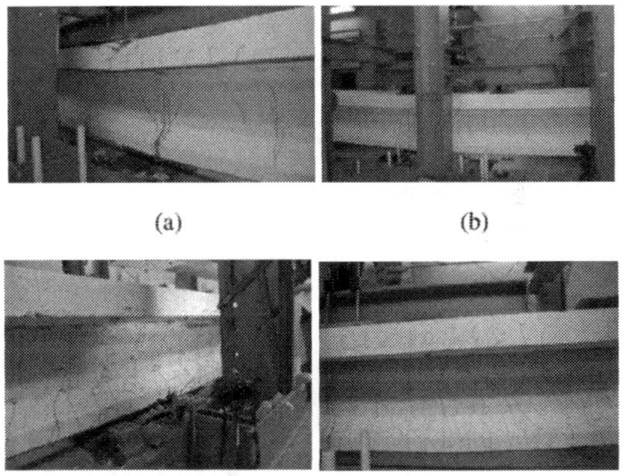

(a) (b)

(c) (d)

Fig. 6. Crack distribution and failure mode at the ultimate load for; a) CPouSM, b) CPoDM, c) CPrSM, and d) BPrSM.

4.5.3 Crack width

The crack widths were manually measured in addition to continuously recording by using non-contact instruments during the monotonic tests. The maximum crack opening against the applied load near the mid-span is shown in Figure 7, for all tested beams. As shown in the figure, the unbonded beams have a higher cracking width than the bonded case (up to seven times the crack width of bonded beams).

https://doi.org/10.1051/matecconf/201814901010

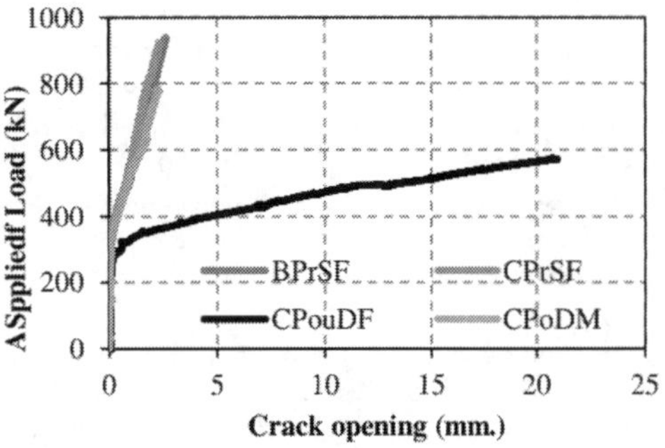

Fig. 7. Maximum crack opening measurements for the beam prestressed

4.5.4 Deflection profile

The deflection profile is obtained from the string potentiometers located along the beam length. In the locations where having potentiometers on both sides, the average displacement between the two potentiometers was taken while calculating the deflections. Figure 8 shows a comparison between the deflections of the tested beams at the ultimate load.

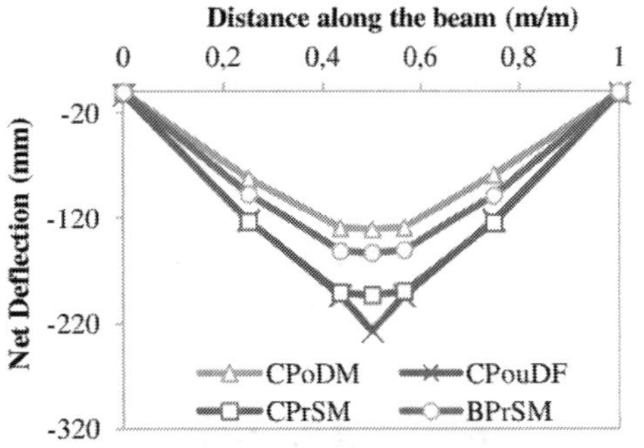

Fig. 8. Comparison of deflection profiles for the full-scale prestressed beam at the ultimate

4.5.5 Cable load

During the experimental tests, the cable force was recorded through a load cell attached to the dead end of some of the unbonded cables while the strain of the bonded cables was measured through a pre-attached strain gauge.

Figure 9 shows the change in the force of the middle bottom cable in the post-tensioned beams for two cases; unbonded in the beam CPouDF and bonded in the beam CPoDM. The results are plotted against the applied load.

At a same applied load level, the bonded case has less increase in cable force compared to the unbonded cables, however, at the same net deflection the increase in cable load is higher for the bonded case.

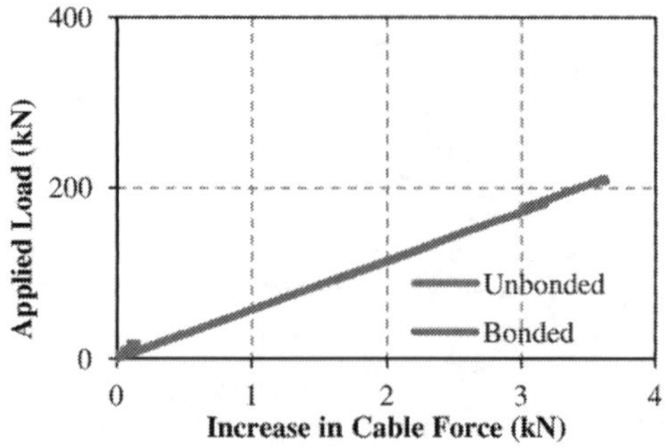

Fig. 9. Increase in cable force for the middle bottom cable for post-tensioned beams against the applied load for the bonded and unbonded case.

5 SUMMARY AND CONCLUSION

The preliminary results from the experimental tests showed that the CFRP prestressed beams demonstrate enough warning prior to failure. In addition, the deformability of the post-tensioned beams with unbonded prestressing CFRP cables are higher than the other CFRP pretensioned beams. However, the ultimate strength was lower than all the other bonded beams.

6 ACKNOWLEDGEMENTS

This research Project was funded by National Co-operative Highway Research Program (NCHRP). The views expressed in this paper are mostly of the authors and do not necessarily reflect the views of the funding agency.

7 REFERENCES

1. J. Warwaruk, M.A.Sozen, C.P. *University of Illinois at Urbana-Champaign*,(1960)
2. H.C. Ozyildirim, S.R. Sharp, *Virginia Concrete Conference*, **March**, (2013)
3. W. Podolny Jr, PCI Journal, **37(5)**, 34-55 (1992)
4. M. Nagi, D. Whiting, *Special Publication*, **151**, (1994)
5. J.J. Myers, T. Viswanath, *Structures Congress 2006: Structural Engineering and Public Safety*, 1-8 (2006)
6. A.A. Abdelrahman, G.Tadros, S.H. Rizkalla, ACI Structural Journal, **92**, 4 (1995)
7. C. W. Dolan, H. R. Hamilton, C. E. Bakis, and A. Nanni, *Report DTFH61-96-C- 00019*, May(2000)
8. S. Heo, S. Shin, C. Lee, Journal of Composites for Construction, **17**, 2 (2012)
9. A. Braimah, M.F. Green, T.I. Campbell, Canadian Journal of Civil Engineering,**33**, (2006)
10. T.Kakizawa, S.Ohno, T. Yonezawa, *Special Publication*, **138**,(1993)
11. N.F. Grace, B.M. Tang, and G.A. Sayed, *International Conference on FRP Composites in Civil Engineering*, (2001)

Ternesite as a component of sulfobelitic cements

M.T. Blanco-Varela and P.M. Carmona-Quiroga
Instituto de Ciencias de la Construcción Eduardo Torroja, Cements and Recycling Materials Department, Madrid, Spain

Abstract. The cement industry is presently devoting substantial effort to the development of less energy intensive, lower emission cements. Calcium sulfoaluminate (CSA) cements, one of the most promising alternatives, are known as 'eco-cements' in light of their lower energy costs and GHG emissions. Whilst their clinker composition may vary, all these cements lie in the $CaO-A_{12}O_3-SiO_2-Fe_2O_3-CaSO_4$ system, with constituent minerals that include C_2A, C_2S, $C_{12}A_7$, CA, $C\hat{S}$, and C_4AF. Ternesite ($C_5S_2\hat{S}$), a phase compatible with ye'elemite ($C_4A_3\hat{S}$) and C_2S, can also be found in that system.Recent studies show that it is activated and hydrated in the presence of AH_3 and gypsum, yielding ettringite and C-S-H gel, the main hydration products of CSA and OPC, respectively, as well as C_2ASH_8.This study analyses ternesite reactivity in different aluminate blends.

1 Introduction

The environmental impact of Portland cement manufacture has prompted the industry to implement a number of strategies geared to lowering energy and natural raw material consumption while reducing greenhouse gas, particularly CO_2, emissions. The use of secondary cementitious materials (SCMs, often industrial by-products), alternative fuels, grinding additives, along with improvements in combustion, heat exchange and grinding technologies have diminished both consumption and emissions with no detriment to binder properties [1].

Another strategy consists in developing calcium sulfoaluminate cements (CSA) based on ye'elemite ($C_4A_3\hat{S}$), a phase initially used to produce expansive cement and subsequently stable high-performance binders. CSA cement manufacture is less energy-intensive (at 1250 °C, the clinkering temperature is 200 °C lower than in OPC) and, depending on the composition, emits 25 % to 40 % less CO_2 than OPC production [2,3]. Whilst not presently used for structural purposes for want of the necessary regulation, these cements are commercialised as components in special mortars and concretes (such as repair or self-levelling concrete, sealing mortar or shotcrete).

The composition of the clinker in these cements may differ, although all lie in the $CaO-A_{12}O_3-SiO_2-Fe_2O_3-CaSO_4$ system and have lower calcium and silicon and higher aluminate and sulfate phase contents than portland cements. The minerals present in the clinker include $C_4A_3\hat{S}$, C_2S, $C_{12}A_7$, CA, $C\hat{S}$ and C_4AF, with a predominance of aluminates [4,5].

Ternesite, $C_5S_2\hat{S}$, a phase formed in the reaction between anhydrite and beliteand compatible with ye'elemite ($C_4A_3\hat{S}$) and C_2S, also lies within this system.

Ternesite was found as a natural mineral at Ettringer Bellerberg, near Mayen, Eifel, Germany, in the nineteen nineties. Its synthetic form had been identified earlier, however, as a component of cement [6] or lime [7] kiln crust, the sulphur for which was sourced from the fuel, as well as a minority component in some CSA cements [8,9,10].

Ternesite was chemically, physically, structurally and petrologically characterised between 1960 and 2000 [6, 7, 11-13]. Regarded as a non-hydraulic material [14-15] until it was recently proven to be activated by aluminium hydroxide [16], it has since become a target for research. This paper describes some of the characteristics of ternesite and its hydraulic activation by other mineral compounds.

2 Physical and structural characteristics

Natural [11] and synthetic ternesite, which are identical, are isostructural with silicocarnotite, $Ca_5(PO_4)_2SiO_4$ [12,13].The unit-cell parameters of this orthorhombic mineral,Pnma space group crystal are a= 6.863(1) Å, b=15.387(2) Å, c=10.181(1) Å, Z=4.Its structure consists in isolated SiO_4 and SO_4 tetrahedral units connected by octa- or heptahedrally coordinated Ca ions. Both S and Si tetrahedrally coordinated with oxygen occupy different tetrahedral positions in the structure, in which double layers of silicate tetrahedral alternate with a layer of sulphate tetrahedral. Natural ternesite is optically biaxial and negative, with refractive index $n_x = 1.630$, $n_y = 1.637$ and $n_z = 1.640$ and an optical angle, $2V_x = 63.5°$.

3 Thermal stability of ternesite

$C_5S_2\hat{S}$ thermal stability has been studied by several authors, some of whom position it in a narrow range of temperatures (1100 °C to 1180 °C [16] or 1200 °C to 1298 °C [15]), above which it would decompose further to the equation:

$$Ca_5(SiO_4)_2\ SO_4 \rightleftharpoons 2Ca_2SiO_4 + CaSO_4. \quad (1)$$

The decomposition temperature depends on whether it is determined in an open (≈1180 °C) or closed (1289 °C) system at a total pressure of 1 atm. [17,18].Similar findings (≈1290 °C at 1 atm.) were reported by Hanein et al. [19], who determined the thermodynamic constants of ternesite to lie between 1000 °C and 1400°C, based on previously published vapour pressure measurements of several sulphates obtained between those temperatures [22]. Those studies showed that the vapour pressure of ternesite, like that of anhydrite, rises with temperature, while ternesite exhibits a pressure lower than anhydrite at 1000 to 1290 °C and higher above that range.

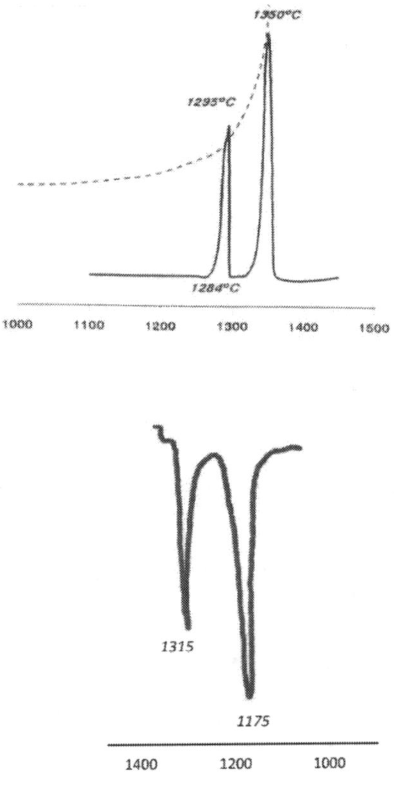

Fig. 1.DTA/TG curves for ternesite, heating (top) and cooling (bottom)

The decomposition reaction of ternesite would also depend, then, on the rate of SO_2 loss and for high temperatures could be described by the equation:

$$Ca_5(SiO_4)_2\ SO_4 \rightleftharpoons 2Ca_2SiO_4 + CaO + SO_2 + \tfrac{1}{2}\,O_2, \quad (2)$$

in which ternesite thermal decomposition would be a function of temperature and the partial pressure of SO_2 and O_2.

For a given temperature and partial pressure of O_2(which must always be higher than in SO_2 for SO_3 to form), there is a maximum SO_2 pressure above which ternesite does not form [19].

On the DTA curve (10 °C/min in CO_2-free air), ternesite thermal decomposition induces an endothermal peak at around 1290 °C. As the temperature rises, a second endothermal peak appears due to the fusion of the C_2S-$CaSO_4$ eutectic, which accounts for 10 %-20 % of the composition of $CaSO_4$ [18].Mass loss intensifies on the TG curve at temperatures of over 1200 °C. During cooling, the DTA curve exhibits two exothermal peaks, respectively denoting the solidification of the C_2S-$CaSO_4$ eutectic (1310 °C) and the α to β polymorphic transformation of $CaSO_4$ (1175 °C).No signal whatsoever is detected that would indicate new ternesite formation along cooling. FTIR and XRD identify $\alpha'C_2S$ and anhydrite (traces of CaO on the XRD pattern), but no ternesite, an indication that at the cooling rate applied (10 °C/min) the decomposition reaction is not reversible.

Table 1. Minerals forming after ternesite thermal treatment

T (°C)	Time	Phases
1315	20 min	α_L' C_2S; $CaSO_4$; CaO
1400	20 min	α_L' C_2S; $CaSO_4$; CaO
1315 and 1000	20 min and 24 h	α_L' C_2S; $Ca_5S_2\hat{S}$; CaO
1400 and 1000	20 min and 24 h	α_L' C_2S; $CaSO_4$; CaO

The XRD findings for ternesite exposed to different thermal treatments (Table 1) suggest that once it decompóses, its recrystallization depends on the cooling rate and the temperature reached. Recrystallization is possible after reheating to 1000 °C when the temperature reached does not exceed the melting point of the eutectic; otherwise, recrystallization does not appear to occur.

According to the literature, ye'elemite, ternesite and anhydrite clinkers at a ratio (by weight) of 1:1:0.5 can be obtained from traditional raw materials or industrial by-products at temperatures of around 1200 °C [14].Two-stage clinkerización has also been successfully applied to produce such cements, clinkering the raw meal at 1250 °C for 1 h, followed by gradual cooling to 1100 °C and quick cooling below that temperature [10,20,21].Single-stage belite and ternesite-rich calcium sulphoaluminate (BY(F)T) have been obtained at a maximum temperature of 1260 °C in a pilot plant where the partial pressure of O_2 and $SO_2$was controlled in the kiln [19].

4 Compatibility between ternesite and CaO-SiO_2-Al_2O_3-$CaSO_4$ system phases

The following is a discussion of some of the features of the equilibrium phases of the three subsystems most closely related to the chemistry of cement, namely:CaO-SiO_2-$A_{12}O_3$; CaO-SiO_2-$CaSO_4$and CaO-$A_{12}O_3$-$CaSO_4$.

The equilibrium phases of the system CaO-SiO_2-$A_{12}O_3$have long been known in the fields of cement chemistry and ceramics [23]. In addition to mullite, the calcium silicates (C_3S, C_2S, C_3S_2, CS) and calcium aluminates (C_3A, $C_{12}A_7$, CA, CA_2, CA_6), the system includes two ternary phases: gehlenite and akermanite.C_2S is compatible with C_3A, $C_{12}A_7$, CA and gehlenite, as well as with C_3S and C_3S_2.

In system CaO-SiO_2-$CaSO_4$, $CaSO_4$ is incompatible at 1100 ºC with C_2S, with which it reacts to form ternesite. Consequently, the equilibrium phases in the CaO-rich zone would be: i) CaO-$CaSO_4$-$Ca_5S_2\hat{S}$; ii) CaO-C_2S-$Ca_5S_2\hat{S}$; iii) C_2S- $Ca_5S_2\hat{S}$-C_3S_2; iv) $CaSO_4$- $Ca_5S_2\hat{S}$-C_3S_2 [18].

A ternary phase, $C_4A_3\hat{S}$, present in system CaO-$A_{12}O_3$-$CaSO_4$, is positioned in between the $CaSO_4$ and the C_3A, $C_{12}A_7$ and CA. The phases in the Ca-rich zone of the system that are compatible at temperatures under 1350 ºC are: i) CaO-$C\hat{S}$-$C_4A_3\hat{S}$; ii) CaO-C_3A- $C_4A_3\hat{S}$; iii) C_3A- $C_{12}A_7$- $C_4A_3\hat{S}$; iv) $C_{12}A_7$-CA- $C_4A_3\hat{S}$ [24].

Studying system CaO-SiO_2-$A_{12}O_3$-$CaSO_4$ at temperatures between 950 ºC and 1150 ºC, Pliego-Cuervo [9] found ternesite to be incompatible with $C_{12}A_7$ and CA, and both C_2S and C_3A to be compatible with ternesite, ye'elemite or both. The compatible phases in the CaO-rich zones of the system were identified as: i) i) CaO-$C\hat{S}$-$C_4A_3\hat{S}$- $C_5S_2\hat{S}$; ii) CaO-C_3A- $C_4A_3\hat{S}$-$C_5S_2\hat{S}$; iii) CaO-C_2S-C_3A-$C_5S_2\hat{S}$; iv) C_3A- C_2S-$C_5S_2\hat{S}$-$C_4A_3\hat{S}$; v) C_2S-$C_{12}A_7$-C_3A- $C_4A_3\hat{S}$; vi) C_2S-$C_{12}A_7$-CA- $C_4A_3\hat{S}$.

Strigac [23] broadened the field of study, identifying two new sets of compatible phases in which ye'elemite and ternesite would co-exist: vii) C_2S-C_3S_2-$C_4A_3\hat{S}$-$C_5S_2\hat{S}$; viii) C_2AS-C_3S_2-$C_4A_3\hat{S}$-$C_5S_2\hat{S}$.

Ye'elemite and ternesite can co-exist, then, not only in calcium-rich zones of the system but also in areas with less CaO and more $CaSO_4$ and SiO_2, although obviously at temperatures below the ternesite decomposition threshold.

5 Ternesite activation

Although a number of authors have concluded that ternesite is a non-hydraulic phase [14-15], as early as 1992Berteka et al. [25] synthesised clinkers consisting primarily in ye'elemite and ternesite, with anhydrite as a residual phase. Produced at 1200ºC, they exhibited very high strength after hydration. Later studies [16] showed that ternesite hydration would be moderate and very slow.

According to Ben Haha et al. [26], ternesite can be activated with aluminate hydroxide to yield ettringite, stratlingite and C-S-H, the proportions of which depend on the proportion among reactives and their degree of reaction. Bullerjahn et al. [27] reported that ternesite contributed more to early age cement properties than C_2S in the CSA cements they synthesised.

Montes et al. [28] showed that ternesite can also be activated by other aluminates, such as C_3A, $C_{12}A_7$, CA and $C_4A_3\hat{S}$, and established the order of activating effectiveness as: $C_{12}A_7$> CA> C_3A>>>>$C_4A_3\hat{S}$.

The calorimetric curves for hydrated aluminate pastes contain a first dissolution peak, followed by an induction period and an intense reaction peak in pastes CA, $C_{12}A_7$ and $C_4A_3\hat{S}$ [29-31]. In 2:1 ternesite: aluminate blends hydrated at 25 ºC for 7 d, the presence of ternesite in the pastes was reported to alter the calorimetric curves of four aluminates, shortening the induction period and bringing the reaction peak forward [28].

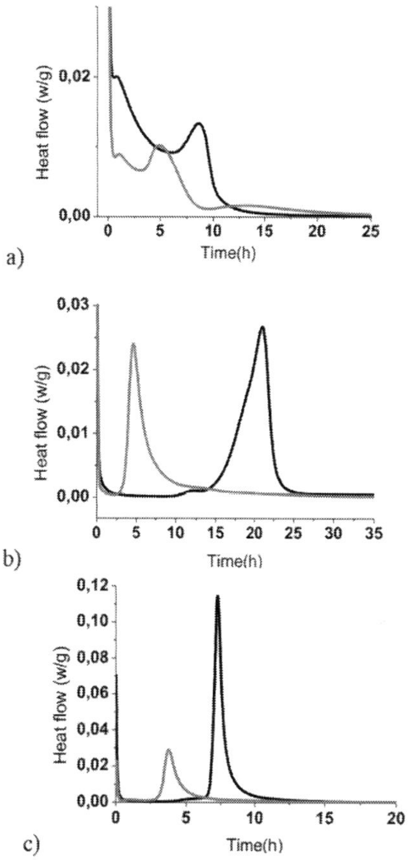

Figure 2.Heat flow curves for: a) $C_{12}A_7$ (black), $C_{12}A_7$+ternesite (grey); b) CA (black), CA+ternesite (grey); c) $C_4A_3\hat{S}$ (black), $C_4A_3\hat{S}$+ternesite (grey)

The presence of ternesite alters the hydration products of the aluminates. Pastes obtained after calorimetric studies of the ternesite/aluminate blends contain $C_4A\hat{S}H_{12}$ as a hydration product.$C_2ASH_8$is also found in pastes containing $C_{12}A_7$ or CA. Stratlingite forms neither in ternesite + C_3A pastes nor in the pastes with $C_4A_3\hat{S}$, in which amorphous AH_3 is present [29].

Acknowledgements. This paper was funded under research project BIA2016-76466-R and by the Regional Government of Madrid and the European Social Fund (Geomaterials Programme2 S2013/MIT-2914).

References

1. J.S. Damtoft, J. Lukasik, D. Herfort, D. Sorrentino, E.M. Gartner, "Sustainable development and climate change initiatives",Cem. Concr. Res. **38** (2) 115–127 (2008)
2. E.M. Gartner," Industrially interesting approaches to "low-CO_2" cements",Cem. Concr. Res. **34**, 1489–1498, (2004).
3. M.C.G. Juenger, F. Winnefeld, J.L. Provis, J.H. Ideker, "Advances in alternative cementitious binders", Cem. Concr. Res. **41**, 1232–1243, (2011).
4. M. Carmen Martín-Sedeño, A.J.M. Cuberos, A. G. De la Torre, G.Álvarez-Pinazo, L. M. Ordónez; MilenGateshkic, M.A.G. Aranda. "Aluminum-rich belitesulfoaluminate cements: Clinkering and early age hydration".Cem.Concr. Res. Volume **40**, (3), March 359–369, (2010).
5. K. Morsli, A.G. de la Torre, M. Zahir, M.A.G. Aranda. "Mineralogical phase analysis of alkali and sulfate bearing belite rich laboratory clinkers".Cem.Concr. Res., **37** 639–646, (2007)
6. N.Sundius, Peterson (1960) "Doppelvereinigung von Sulfat und SilikataussogenanntenSulfatringen in den Zementtfen in Limnham".Radex-Rundschau. 100-103 (1960).
7. M.W. Pryce "Calcium sulphosilicate in lime-kiln wall coating". Min Mag **38**:968-971, (1972)
8. M.Marroccoli, F.Montagnaro,M.L. Pace, A. Telesca, G.L. Valenti, "Synthesis of calcium sulfoaluminate cements from blends of coal combustion ashes with flue gas desulfurization gypsum", *PTSE, Ischia*, Italy, 2010.
9. Y.B. Pliego-Cuervo, F.P. Glasser, "The role of sulphates in cement clinkering: subsolidus phase relations in the system CaO–$A_{12}O_3$–SiO_2–SO_3".Cem.Concr. Res., **9**, 51–56, (1979).
10. F.Bullerjahn, D.Schmitt, &M.BenHaha, "Effect of raw mix design and of clinkering process on the formation and mineralogical composition of (ternesite) belite calcium sulphoaluminate ferrite clinker".Cem.Concr. Res., **59,** 87-95, (2014).
11. E. Irran, E. Tillmanns, and G. Hentschel."Ternesite, $Ca_5(SiO_4)_2SO_4$, a new mineral from the EttringerBellerberg/Eifel, Germany" Mineralogy and Petrology **60:**121-132, (1997).
12. B. Dickens,WE. Brown "The crystal structure of $Cas(PO_4)2SiO_4$ (silico-carrlotite)".Tschermaks Min Petr Mitt 16:1-27, (1971).
13. PD. Brotherton ,JM Epstein , MW.Pryce,AH.White "Crystal structure of 'calcium sulphosilicate', $Ca_5(SiO_4)_2SO_4$".Aust J Chem **27:**657-660, (1974).
14. N. Sherman, J. Beretka, L.Santoro and G.L. Valenti"Long-term behaviour of hydraulic binders based on calcium sulfoaluminate and calcium sulfosilicate"Cem. Concr.Res, Vol. 25. No. 1, pp. 113-126,1995
15. Su, Muzhen, W.Kurdowski, F. Sorrentino, Developpement in non Portland cements". *IX. ICCC*, Vol.l, p.317, New Delhi (1992).
16. Sahu,"Developpements in Low energy cements". Ceramics-Silikáty.**38** 191-, (1994)
17. W. Gutt, M.A. Smith, "A new calcium silicosulphate," Nature **210**, 408–409, (1966).
18. W. Gutt, M.A. Smith, "Studies of sub-system CaO-CaO. SiO_2-$CaSO_4$, Trans Br."Ceram.Soc. **66** 557-567, (1967).
19. T. Hanein, I. Galan, F. P. Glasser, S.Skalamprinos, A. Elhoweris, M.S. Imbabi, M.N. Bannerman, "Stability of ternesite and the production at scale of ternesite-based clinkers,"Cem. Concr. Res. **98,** 91–100 (2017)
20. F. Bullerjahn, D. Schmitt, M. B. Haha, "Method for producing ternesite," *US9073785 B2* (2015).
21. W. Dienemann, D. Schmitt, F. Bullerjahn, "Belite-calciumsulfoaluminate-ternesite (BCT)-a new low carbon clinker technology,"Cem. Int. **11**, 100–109, (2013).
22. Gang-Soon Choi and F.P. Glasser"The sulphur cycle in cement kilns: vapour pressures and solid-phase stability of the sulphate phases"Cem. Concr. Res. **18,** . 367-374, (1988)
23. J. Strigac, S. Sahu, M. Lopasovska, R. Durovcikova, V. Kovar, J. Majling, Phase Compatibility in the system CaO-SiO_2-$A_{12}O_3$-SO_3," Ceramics **42** 90–98, (1998).
24. I.Kaprálik , F.Hanic,"Phase changes in the system CaO-SiO_2-$A_{12}O_3$-Fe_2O_3 MgO-$CaSO_4$-K_2SO_4 in air up to 1300°C referred to sulphoaluminate cement clinker"Br.Ceram.Trans.J. **85**, 131 (1986)
25. J.Beretka, L.Santoro, N.Sherman, G.L.Valenti, "Synthesis and properties of low energy cements base don C_4A_3s". *XI. ICCC*, Vol.3, p.195, New Delhi 1992
26. M. B. Haha, F.Bullerjahn, MaciejZajac"On the reactivity of ternesite" 14th ICCC, Pekín 2015
27. Frank Bullerjahn • MaciejZajac •M.B. Haha "CSA raw mix design: effect on clinker formation and reactivity" Materials and Structures **48**:3895–3911, (2015)
28. M. Montes; M.T.Blanco Varela "Can calcium aluminates activate ternesite hydration?"Cem. Concr.Res. **103,** 204-2015 **(**2018).
29. B. Raab, H. Poellmann, "Heat flow calorimetry and SEM investigations to characterize the hydration at different temperatures of different 12CaO·$A_{12}O_3$ ($C_{12}A_7$) samples synthesized by solid state reaction, polymer precursor process and glycine nitrate process Thermochimica"Acta **513**, 106–111, (2011).
30. M.J. Sánchez-Herrero, A. Fernández-Jiménez, A.Palomo, "$C_4A_3\hat{S}$ hydration in different alkaline media,"Cem.Concr.Res. **46**, 41-49, (2013).
31. S.R. Klaus, J. Neubauer, F. Goetz-Neunhoeffer, "Hydration kinetics of CA_2 and CAInvestigations performed on a synthetic calcium aluminate cement,"Cem.Concr.Res. **43** 62–69, (2013).

Application areas of phosphogypsum in production of mineral binders and composites based on them: a review of research results

Leonid Dvorkin[1], Nataliya Lushnikova[2] and Mohammed Sonebi[3]
[1], National University of Water and Environmental Engineering, Department of Building Products Technology and Materials Science, Rivne, Ukraine
[2] National University of Water and Environmental Engineering, Department of Architecture and Environmental Design, Rivne, Ukraine
[3]School of Natural and Built Environment, Queen's University Belfast, Belfast, Northern Ireland (UK)

Abstract. The increase of the consumption of gypsum products in construction industry with a limited amount of natural gypsum deposits requires alternative sources of gypsum-containing raw materials. In some countries which have fertilizers industry plants, the problem can be solved using industrial wastes, e.g. phosphorgypsum – a byproduct of fertilizers' production. Kept in dumps over decades, phosphorgypsum is subjected to the chemical changes due to washing out impurities with rain and other natural factors. However, there are observed deviations of harmful impurities in dumped PG depending on its age., Phosphorgypsum of any age requires chemical treatment to neutralize remains of phosphorus and sulfuric acids, fluorine compounds. According to our researches one of the most simple and effective method of neutralization the impurities is using lime-containing admixtures. The paper presents results of laboratory tests of phosphorgypsum as a component of clinker and non-clinker binders. There were investigated the impact of phosphorgypsum as admixture for clinker binders to substitute natural gypsum. Neutralized phosphorgypsum can be applied as mineralizing admixture in calcination of Portland cement clinker. Adding 2 to 2.5% of phosphorgypsum as setting time regulator resulted in a similar physical and mechanical properties compared to mix made with natural gypsum. Another important area of phosphorgypsum application is sulphate activatoion of low-clinker blast-furnace slag cement (clinker content is less than 19%). According to results, the incorporation of phosphorgypsum as sulphate activator in cement has the better effect as natural gypsum. Other development has been carried out to modify the phosphorgypsum binder properties. Complex additive consisted of polycarboxylate-based superplasticizer and slaked lime permitted an increase mechanical properties of hardened phosphorgypsum binder due to significant a reduction of water consumption. Such modified binder can be used as partial or complete replacement of gypsum binder for filling cements and finishing plasters. It can substitute gypsum in non-clinker binders like supersulphated cements. There were also developed compositions of supersulphated cements based on low-alumina blast furnace slag and phosphorgypsum. Supersulphated cements were tested in normal-weight and light-weight concrete.

1 Introduction

Synthetic gypsum is applied as an alternative for natural gypsum raw in production of construction materials [1], in the areas of location of fertilizing plants. at insufficient level of wastes recycling and reuse. For instance, the only fertilizing plant located in Western Ukraine accumulated around 10 mln tones of accumulated phosphorgypsum (PG) wastes – a by-product of phosphoric acid production, small part of it is offered as agricultural product. According to data of Florida Institute of Phosphate Research, for each ton of phosphoric acid produced by wet processes in fertilizing plant - approximately 4.5 tons of gypsum generates. Annual world generation of PG is around 100-280 Mt. [2].

At gradual exhausting of the deposits of natural gypsum [3] calcium sulphate containing waste in few decades will become a main raw source for gypsum products.

The main producers of phosphate rock and phosphate fertilizers are in the USA, the counties of former USSR, China, Africa and the Middle East [2]. The leading positions in PG recycling belong to Japan, which does not have natural gypsum deposits [4].

The main disadvantages of PG are rather energy-consuming need in removing harmful admixtures and drying. There are different types of cleaning, some of them are rather resources and energy-consuming [5, 6]. Due to the wet technological processes PG also requires drying.

Another issue is radioactivity of natural raw for fertilizing plants and PG [5, 6]. Activity concentrations of the radionuclides in PG vary worldwide [7].

There are suggested different areas of PG use in the construction industry, including rather challenging areas like road construction [8, 9], cellular gypsum [10]. One of the most common applications for different PG binders [8] is mortars and different types of concrete [11].

2 Aim and scope of the research

The aim of the research was revealing the areas of PG application in production of binders and composites made with PG.

Regarding the high energy consumption α-hemihydrate and anhydrate, there research was oriented on production of PG β-hemihydrate as one of the most low-energy environmentally friendly binders.

There are presented the key results in testing clinker based binders with phosphogypsum as mineralizing additive and setting time regulator for Portland cements of I and II type (65-94% of clinker, which we consider as high-clinker); PG as sulphate activator for low-clinker blast furnace slag cements (5-19% of clinker) and non-clinker PG binders and supersulphated cements (Fig. 1).

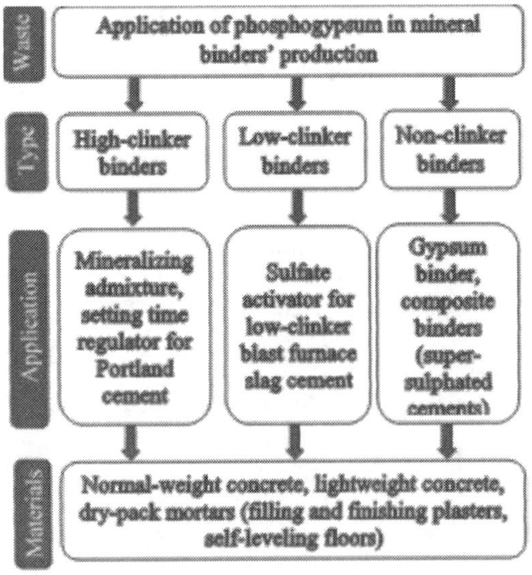

Fig.1. Research areas of phosphorgypsum application in mineral binders' production.

3 Materials and methods applied

Composition of phoosphogypsum (PG), blast-furnace slag (BFS), gypseous stone (GS), ordinary Portland cement (PC, I type) used in research are shown in Tab. 1.

The phosphogypum of one of the largest fertilizing plant in Western Ukraine PJSC «Rivneazot» (OstChem) was used. The Portland cement of local plants and local gypseous stone were employed.

To apply PG in construction materials it must be odorless and do not have toxic impurities (arsenium, cromium), harmful evaporations, minimum radioactivity [2]. As it can be seen from the Tab 1. the chemical composition of PG accumulated in dumps changes with time. Fluorine and phosphor oxides contents reduces due to washing out with rains.

Table 1. Chemical oxide composition of materials (% by mass)

Component, %	**PG fresh**	**PG aged (>15 years)**	**BFS**	**GS**	**PC**
SiO_2	-	-	39.51	8.54	22.38
Al_2O_3	-	0.34	6.47	0.70	5.26
Fe_2O_3	-	0.16	0.14	0.40	4.09
CaO	40.0	38.30	47.19	29.88	66.25
CaO free		–	–	–	0.41
MgO		0.004	3.12	0.41	0.63
SO_3	57.0	59.1	1.76	41.85	0.46
MnO		–	1.14	–	–
P_2O_5 total	**1.2**	**0.69**	–	–	–
P_2O_5 water soluble	**0.6**	**0.04**	–	–	–
R_2O	-	–	–	–	0.29
F^-	**0.4**	**0.14**	–	–	–
Cl^-		0.01	–	–	–
L.O.I.	-	–	0.59	17.87	–
IR	-	–	–	–	0.21

Dumped PG demonstrates multi-disparity of grains with prevalence of size ranges 0.1 to 0.4 mm, 0.05 to 0.1 mm and less 0.5 almost in equal proportions [12].

According to the results of sanitary testing, radioactivity of PG is within the required limits and does not exceed 370 Bq/kg and it does not contain toxic elements [13]. Local standards determine the content of basic components of PG (see Tab. 2). As it can be seen form Tab. 1 and 2, PG meets the required standards as $CaSO_4$ content is 96.13 to 97%.

Table 2. Requirements to synthetic gypsum made of phosphogypsum [14]

Content, % by dehydrated product	**Mineralizing admixture**	**Setting time regulator**	**Raw for gypsum binder**
$CaSO_4$, more than	80	80	90
P_2O_5 (total), less than	1.8	1.4	1.5
P_2O_5 (soluble), less than	0.7	0.18	0,15
F (total), less than	0.9	0.36	0.4
W_{hyh}, less than	14.5	14.5	14.5

Water consumption of binders was tested according setting time Vicat needle test [15], normal consistency of cement [16]. For determination compressive and bending strength [17], prisms of binder sand mortar in 1:3 ratio 40x40x160mm were tested. Quartz sand used for preparing the PC, BFSC and supersulphated cement (SSC) based mortar specimens has the following properties: fineness modulus 1.9 to 2.0, content of washed out admixtures about 1.7 %, real density - 2.69g/m^3, bulk density – 1.38 g/m^3, voidage – 48.7%. The average value for 3 specimens was determined. Basic properties were considered in comparison to those of control specimens without PG.

4. Preparation of phosphogypsum to application in production of construction materials

According to preliminary research there have been selected parameters of PG production stages: drying, neutralizing harmful admixtures, fineness of grinding. Applying statistical modeling method, it was determined, that temperature of thermal treatment (drying) is 150 to 160 ^{0}C, percentage of the neutralizing admixture (slaked lime) 2 to 2.5% (by PG weight); PG binder was ground up to maximum sieve residue #02 up to 5% [18]. PG neutralization with lime milk was made. For complete neutralization and providing sufficient homogeneity, it was kept for 1 to 3 days (subjected to ageing). Before calcination PG was dried to a residual humidity – 1 to 2%, drying temperature was up to 100 ^{0}C. For obtaining the binder, neutralized PG was calcined during 2 h, after that it was grounded in the laboratory ball mill.

Neutralizing with lime milk led to increase of PG moisture content by 4 to 5% and required additional drying of the material. For practical technology, dry methods for PG neutralizing are the most interesting, using powders of slaked lime. At the same time, dry methods are possible when moisture content of PG is 15 to 20%. At thorough mixing of PG with a neutralizing additive, introduced in an amount of 2 to 3% (in terms of CaO), after keeping for 2 to 6 hours the pH value increases from 4 to 5 to 8 to 9 [13].

To give the PG a marketable form and high processability it is necessary either to granulate or briquette it. The known methods of granulation are complex, require the use of bulky drying-roll-up devices (drying drums, bowl or drum granulators), in addition, they do not provide required strength of granules and their safety at transport and technological operations. To our opinion, briquetting of the material is more promising.

Table 3. Properties of phosphogypsum briquette

Parameters	Value
Diameter of briquette, mm	60
Thickness, mm	10-15
Humidity, %	7-12
Compacting pressure, MPa	20-40
Compressive strength, MPa	6-10

Briquetting of neutralized PG was worked out to prevent dusting, improve transportation conditions, dosage and storage. The experiments were carried out on a briquette-roll press. After laboratory tests, industrial approbation was carried out. The optimal pressing parameters, at which the compressive strength of the briquette was from 6 to10 MPa, correspond to the moisture content of the material 7 to 12%, pressure 20 to 40 MPa. To prepare a mass containing PG and lime to it is necessary to dry the dump PG. The characteristics of a briquette of industrial production are given in Table 3.

5 PG as admixture for Portland cement

5.1 Phosphogypsum as mineralizing admixture

The use of PG as mineralizer at Portland cement clinker calcination is known in many countries (Japan, Russia, Lithuania and others). Adding 2 to 3% of PG (1% of SO_3) allows to reduce fuel consumption by 1% and improve grindability of cement [13].

The reason for adding of mineralizing admixture in local plant was caused by lime saturation factor of the PC sludge and presence of coarse-grained silica in charge mixture.

There have been determined the amount of bound CaO depending on the type and dosage of additives (blast furnace slag and PG) (Tab. 4) and the temperature of calcination (Fig. 2). As follows from the data obtained, at calcination in a laboratory furnace during 10 min, the introduction of PG in an amount up to 1% SO_3 allows to accelerate the binding of free lime both without slag and with slag. At a temperature of 1400 ° C, almost complete lime binding was observed (Tab. 4, Fig. 2). PC clinker used has lime saturation factor LSF= 0.93, silica ratio (SR)=2.4; alumina ratio AR=1.3.

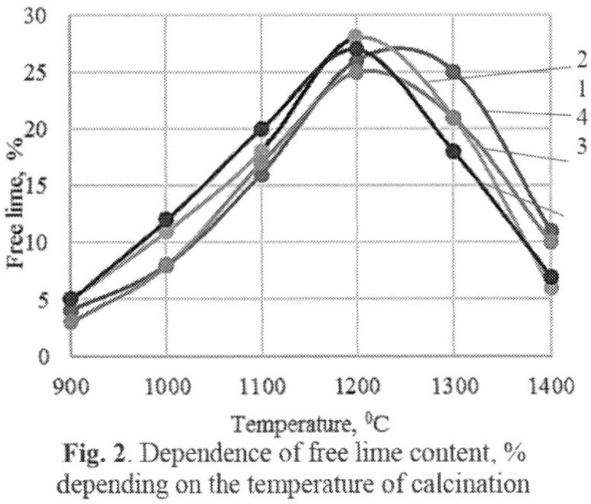

Fig. 2. Dependence of free lime content, % depending on the temperature of calcination

Table 4. Proportion of sludge, %

#	Dry sludge	BFS	PG (SO_3)
1	100	0	0
2	98	0	2 (1)
3	96	4	0
4	94	4	2 (1)

At temperature of calcination 1400°C for sludge without admixture CaO content was 11% (#1), at adding PG- 6% (#2), when slag was added these values were 10% (#3) and 7% (#4).

These data demonstrate mineralizing effect of PG, which can be applied in cement plant rotary kilns. It is rational to add PG into the kiln in the loose state with the slag.

5.2 Phosphogypsum as setting time regulator

Phosphogypsum was tested as a substitute of natural gypsum for setting time regulation of cement. Compositions, water consumption of cement paste and W/C to provide flow spread diameter of tested cement sand mortars 110 to 116 mm are given in Tab.5.

As it can be seen from the results of setting time testing (Fig. 3), there is no direct dependence of setting time change and GS or PG dosage variation.

Table 5. Testing phosphogypsum as setting time regulator

#	PG, $\%SO_3$	GS, % SO_3	#008 sieve residue	Water demand, %	W/C	Flow spread, mm
1	-	3	10	27,0	0.40	115
2	1.5	-	9	27,5	0.40	115
3	2.5	-	10	27,2	0.40	115
4	0.5	2.5	8,6	26,0	0.42	110
5	1.5	1.5	8,7	26,5	0.43	111
6	2	1	9,2	27,0	0.40	110

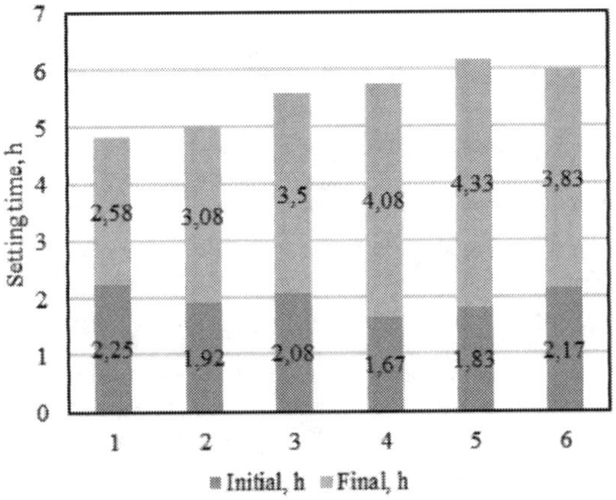

Fig. 3. Setting time of PC with PG and GS as regulators

Fig. 4 and Fig. 5 present kinetics of bending and compressive strength of cement mortar. As it can be seen the compressive strength of cement with PG additive is compatible to that with natural gypsum and meets the requirements of the standards for CEM I 42.5 R. Therefore, PG can replace natural gypseous stone. The optimum dosage of PG is from 2 to 2.5% SO_3.

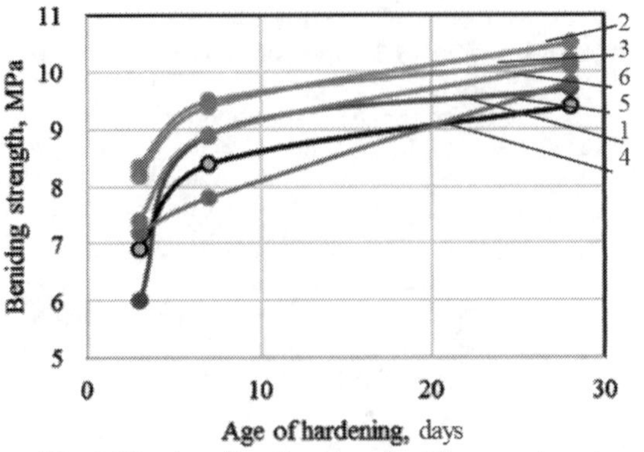

Fig, 4. Kinetics of bending strength of the cement mortar depending on the type and dosage of setting time regulator (# of compositions according to Tab. 5)

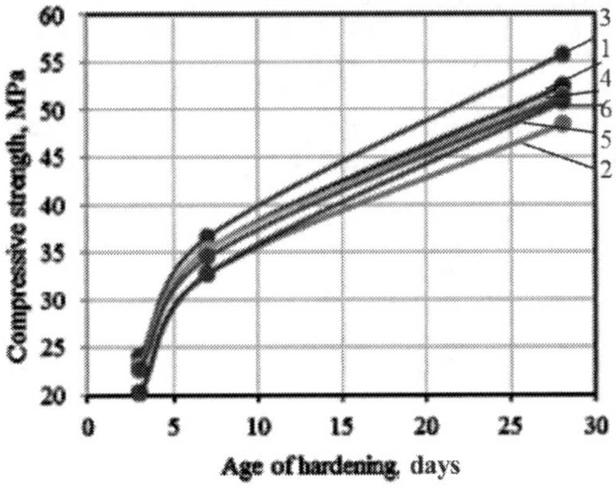

Fig, 5. Kinetics of compressive strength of the cement mortar depending on the type and dosage of setting time regulator (# of compositions according to Tab. 5)

6 PG as sulphate activator to low-clicker binders

Sulphate activation of low-clinker binders can be made by different modifications of calcium sulphate: calcium sulphate dihydrate (gypseous stone and PG), calcium sulphate hemihydrate and anhydrite.

There was applied aged PG (stored more than 1 year in the dump) and gypseous stone for comparison setting time and strength parameters of low-clinker binder – BFSC. The content of sulphate component varied from 3.5 to 5.5% SO_3. Components ratio of binder: BFS – 88%, clinker- 12%. Specific surface area of the binder was 450 m^2/kg. Water consumption of the low-clinker BFSC depending on the type of sulphate component and its dosage is shown in Tab. 6.

To reduce water consumption for all the compositions naphthalene formaldehyde superplasticizer was used (1% by weight of binder). The research results of setting time and strength of cement mortar at the age of 28 days are shown on Figs. 6 to 8.

Table 6. Composition and water consumption of low-clinker BFSC

#	PG, %	GS, %	Water demand, %
1	3.5	-	22.7
2	4.5	-	23.3
3	5.5		24.5
4	-	3.5	23.5
5	-	4.5	25.5
6	-	5.5	26.2

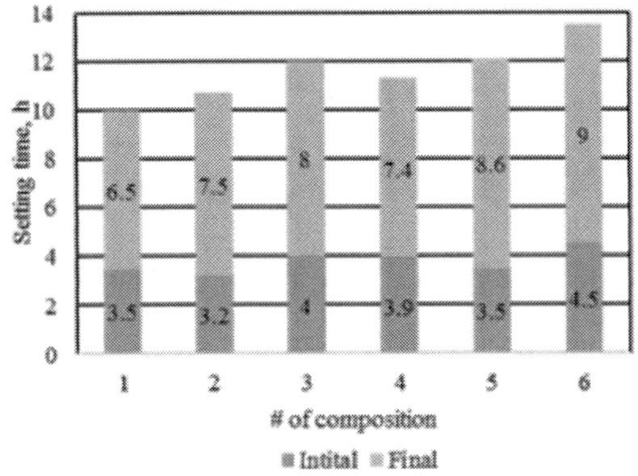

Fig. 6. Setting time of low-clinker BFSC

According to the obtained data, we can conclude that PG is more effective sulphate activator than natural gypsum. Probable reason for this are the structural features of the PG. It is characterized by porous structure, preserved even after grinding, high dispersity of PG and the presence of impurities in its composition. Fluorites, mainly fluorspar CaF_2, dominate, and it is the accelerator of hardening of slag-based binders [19, 20]. PG is also the cheapest sulphate component; its use contributes to solving environmental problems of storage and disposal of potentially hazardous industrial waste. However, it should be noted that the increase of sulphate component dosage of more than 7.5%, both PG and gypseous stone, led to a

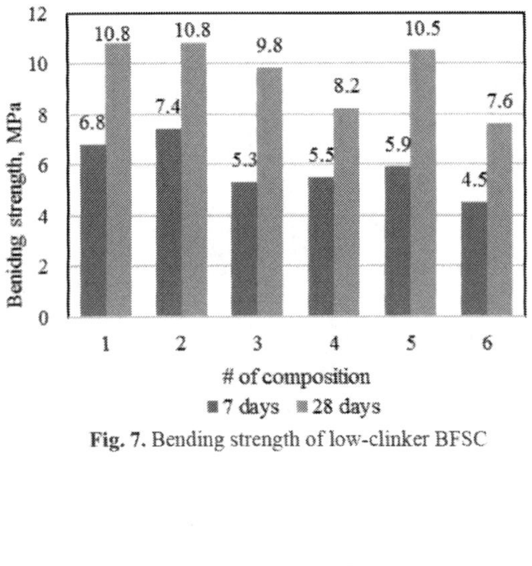

Fig. 7. Bending strength of low-clinker BFSC

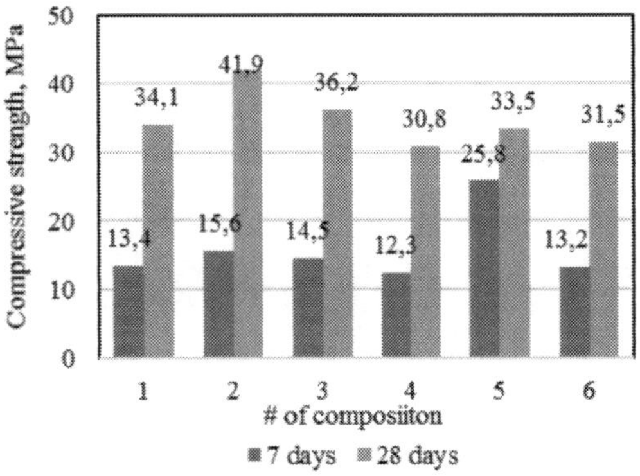

Fig. 8. Compressive strength of low-clinker BFSC

decline in the compressive strength of low-clinker BFSC [21]. The reason is that the dual activation of slag by calcium hydroxide, which is formed by hydration of clinker, and sulphate component, destructive processes are possible due to recrystallization of hydrosulphoaluminates. That led to decreasing hardening rate and compressive strength reduction. Therefore, the optimal proportion of activators should be considered. As it can be seen from Fig. 7 and 8, the optimum content of PG is 4.5% (SO_3). That permits achieving maximum bending and compressive strength. Comparing to the strength of BFSC with GS, PG permits increase bending strength on 15 to 20% and compressive strength on 10 to 25%.

Preliminary tests proved the possibility of application low-clinker BFSC in production of normal-weight concrete, fiber-reinforced concrete, dry pack mixes for masonry mortars [21].

7 Modified gypsum binders

One of the important disadvantages of phosphorgypsum is its high-water consumption, which is limens its application as substitute for calcium hemihydrate binder. Water gypsum ratio of PG at normal consistency by Suttard cylinder is 0.9, comparing to gypsum paste on natural gypsum is 0.6. That causes low strength of hardened gypsum.

Therefore, the ways of reduction water consumption with subsequent strength increasing are vital. The effective way of modifying PG binder is application of superplasticizers (SP). However, most of known high-range water reducers, produced for alkaline based binders like Portland cement, have lower effect in neutral gypsum environment [22]. Previous researches proved the effectiveness of polyacrylates like Dynamon and modified polycarboxylate ethers as Melflux.

To increase the water reducing effect slaked lime added. It provided formation of alkaline environment and both steric and electrostatic effect of molecules of modified polycarboxylate ether was more evident [22, 23]. Series of researches proved that stability of gypsum plaster under the influence of steric effect is better than that caused by electrostatic repulsive force [12, 24]. It means that there is no need in using setting retarders for such binders.

Superplasticizer used and compositions of modified binders are given in Tab. 7. The dosage of SP was selected to provide minimum W/G ratio at forced mixing the paste. Application of "forced" mixing during 2 to 3 min at adding modifier SP+lime led to achieving higher homogeneity, higher water-reducing effect and high strength parameters (Fig. 9 and 10). Small portions of binder for laboratory research have been mixed by hand for 3 to 6 min. Consumption of lime as a part of multifunctional modifier is about 3% (2% of CaO).

Table 7. Compositions of modified PG binder with different SP (flow spread of Suttard cilinder– 180-220 mm)

#	Type of SP	SP content, % by PG weight	Lime content, % by PG weight	W/G
1	Dynamon SP3	0.66	3	0.34
2	Melfulx 1641 F	0.60	3	0.32
3	Melfulx 2651 F	0.60	3	0.31

For complex additive Melflux 1641F+lime the optimal setting time is within 18 to 25 min, for Melflux 2651F+lime – 50 to 70 min, for Dynamon SP3+lime setting time was 15 to 20 min. Modified PG binders based are slow-hardening gypsum binders.

The data in Figs. 9 & 10 demonstrate that mechanical properties of PG binder containing polyacrylate SP are close to the strength of specimens containing polycarboxylate SP. The effectiveness of Melflux 2651 F is higher than Melflux 1641 F. It can be explained by longer side chains of the molecules; therefore, the steric effect is higher.

SEM micrographs (Fig. 11) show that morphology of $CaSO_4 \cdot 2H_2O$ is not related to the type and portion of admixtures comparing to the PG binder specimens. However, at the application of optimal values of SP and slaked lime denser disposition of crystals is observed.

There have been conducted research on properties of plasters, filling cements, self-levering floor mortars with PG binder and PG filler. As fine aggregate quartz sand was used, water retaining admixture of Tylose and citric acid as setting time regulator were applied [25]. Bending strength of the mortars was 1.5 to 2.5 MPa, compressive strength 2.5-6.5 MPa.

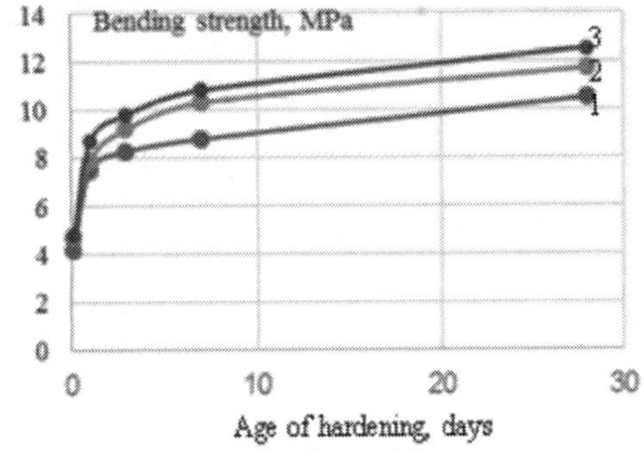

Fig. 9. Kinetics of bending strength of modified PG binder depending on the type of SP

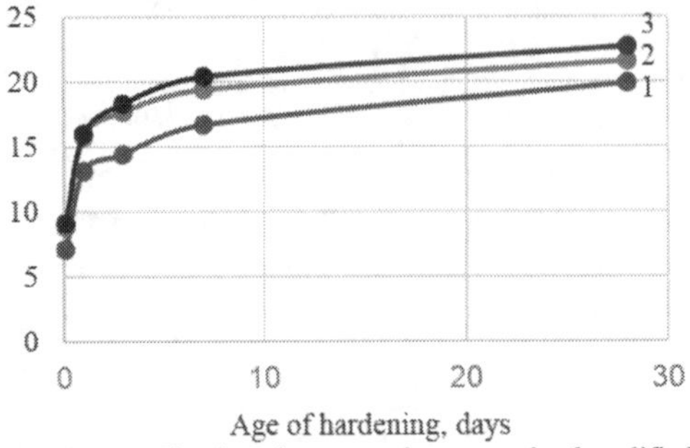

Fig. 10. Kinetics of compressive strength of modified PG binder depending on the type of SP)

Fig 11. Micrographs of non-modified PG binder (W/G=0.9, left) and modified PG binder specimens (0.6% Melflux + 3% lime, W/G=0.32, right)

8 PG as a component of SSC

Supersulphated cement (SSC) is low-energy and low-resource consuming non-clinker cement [26]. It is obtained by combined intergrinding of blast furnace slag, sulphate and alkaline activators in small relative quantity. Proportion of the binder under research selected as optimum was [12] was as follows: 85% of blast furnace slag, 10% of sulphate activator and 5% of alkaline activator (PC). Table 8 shows the components of SSC with different types of sulphate components.

Table 8. Sulphate components of SSC

#	Type of sulphate component	W/C	Specific surface of SSC, m^2/kg
1	Gypseous stone	0.4	610
2	β-hemihydrate (building gypsum0	0.4	615
3	Phosphogypsum	0.4	620

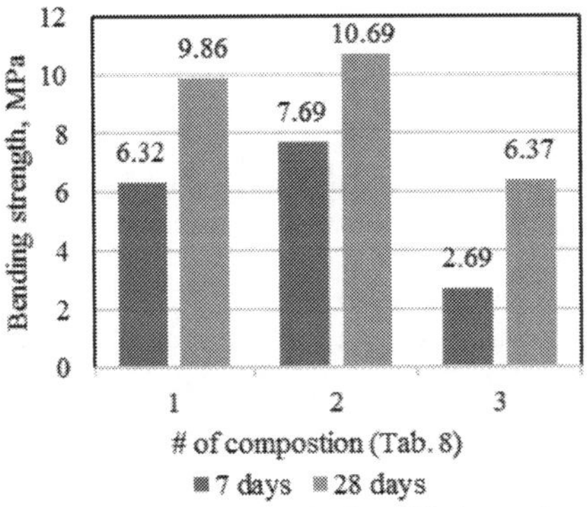

Fig. 12. Bending strength of modified SSC depending on type of sulphate component

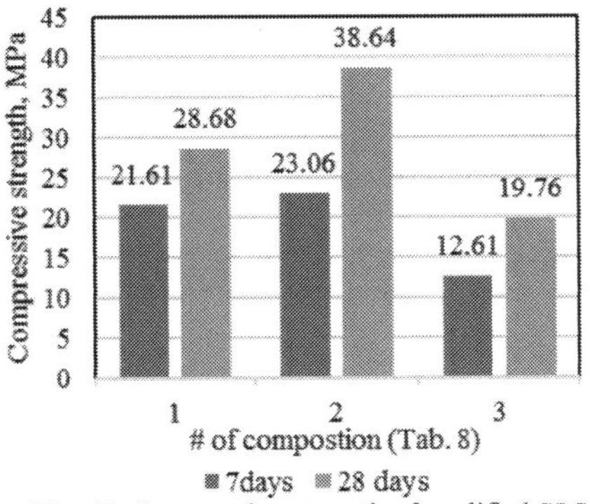

Fig. 13. Compressive strength of modified SSC depending on type of sulphate component

Table 9. SSC based on PG modified with superplasticizers

#	Superplasticizer	SP content, % SSC weight	W/C
1	Melflux–2651F	0.2	0.37
2	Melflux–2651F	0.4	0.33
3	Melflux–2651F	0.6	0.29
4	Dynamon SP3	0.2	0.38
5	Dynamon SP3	0.4	0.34
6	Dynamon SP3	0.6	0.30

As it can be seen in Figs. 12 and 13, PG is the most effective sulphate component of SSC. Strength of SSC with building gypsum sulphate slag binders using plaster was as almost twice lower as with PG. It can be explained by different rate of calcium sulphate dihydrate and ettringite crystallization and formed by the interaction of aqueous plaster with water. Crystallization of $CaSO_4 \cdot 2H_2O$ is rapid and ends no later than in 2 hours after mixing with water, while formation of ettringite lasts few days, thus destroying the crystal structure of hardened gypsum and hardened SSC structure.

For reduction of water consumption of SSC based on PG there was suggested adding SP (Tab. 9).

As it can be seen from Fig. 14 the bending strength is almost in direct proportion to SP dosage.

Superplasticizers make strong effect on water demand of SSC, and permit to reduce W/C from 0.4 to 0.3 at keeping the same consistency of fresh mortar. The compressive strength of the binder (Fig. 15) increased respectively from 1.2 and 1.5 times comparatively to control composition at dosage of SP 0.4%.

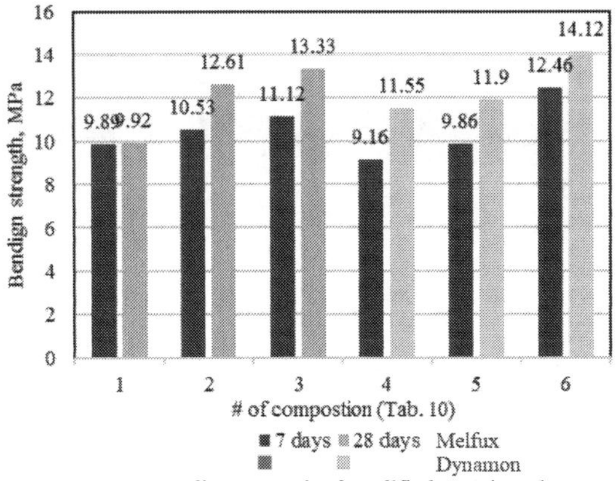

Fig. 14. Bending strength of modified SSC based mortars

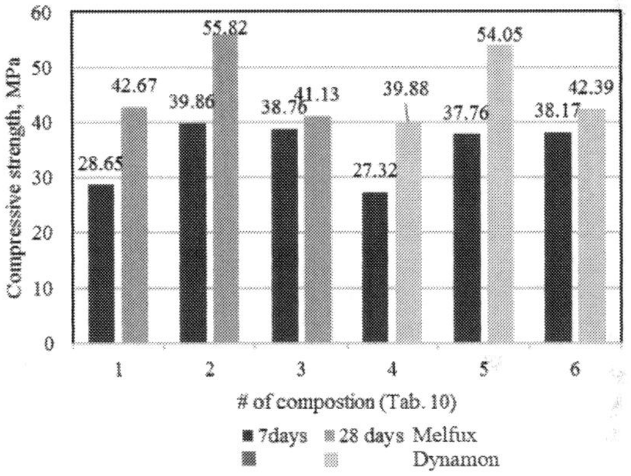

Fig. 15. Compressive strength of modified SSC based mortars

Reducing water-binder ratio to 0.29-0.3 resulted in a reduction of strength, despite the preservation of a sufficient level of mortar workability. We can assume that water content was insufficient for completion the transfer of existing $CaSO_4 \cdot 2H_2O$ in hydrosuphoaluuminates, in particular ettringite (Fig. 15). Therefore, the optimal content SP to maximize compressive strength was about 0.4% by weight of SSC at W/C = 0.3-0.34).

The setting time of SSC was not affected by type of sulphate component as well as by the presence of SP. Water consumption and setting time corresponds to those for ordinary Portland cement (Tab. 10).

Table 10. Normal consistency and setting time of SSC

#	Components, %				SP, %	Setting time, h-min	
	BFS	PG	PC	GS		initial	final
1	85	10	5	-	-	3-10	7-40
2	85	10	-	5	-	3-30	7-50
3	85	10	5	-	0.2	4-20	7-40
4	85	10	5	-	0.4	4-40	7-10

Microstructure of SSC specimens at the age of 7 days is shown on Fig. 16.

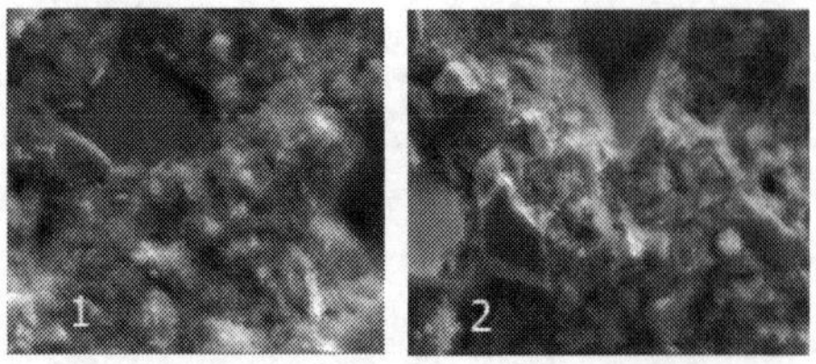

Fig. 16. Micrographs hardened SSC binder (BFS-85%, PG-10%, PC-5%) : 1) W/C=0.4 , without admixtures (left) and W/C=0.33 ; with 0.4% Melflux (right) x1000

Comparison of SEM images specimens of SSC demonstrate the presence of large number of needle-like prismatic crystals of ettringite and lamellar crystals of calcium sulphate dihydrate. The use of superplasticizer affected of new formations. Prismatic structures that can be attributed to the calcium sulphate dihydrate gypsum and hydrosulphoaluminate. Fibrous structures are likely to indicate the presence of significant amounts of low-basic calcium hydrosilicates.

SSC can be applied as binders of normal-weight and light-weight concrete, like sawdust concrete and expanded clay aggregate concrete [27].

Conclusions

Phosphorgypsum as alternative to natural gypsum can be applied in wide range of the construction materials. Special treatment on drying and neutralizing harmful admixtures is required. Using slaked lime (2-3% by PG weight) is one of them most cheap ways for it.

Results of laboratory research confirms the possibility to use PG of Western Ukraine fertilizing plant as mineralizing and setting time regulator for Portland cement. It can also serve as sulphate activator for low-clinker blast furnace slag cement and supersulphated cement.

To reduce water consumption of PG superplasticizer with steric effect are applied. There have been determined the optimal content of SP. For PG binder, the water consumption can be reduced at 30 to 40%, whereas compressive strength increases by 2 to 2.5 times.

The durability aspects of PG-based materials require further research.

References

1. N. Lushnikova and L. Dvorkin. The sustainability of lightweight aggregates manufactured from clay wastes for reducing the carbon footprint of structural and foundation concrete. In book: Sustainability of Construction Materials. Ed. by. J. Khatib. Cambridge: Woodhead Publishing, 643-681(2016).
2. H.Tayibi, M.Choura, F.A. López, F.J. Alguacil, and A.L. Delgado. Environmental impact and management of phosphogypsum (Review). Journal of Environmental Management 90(8):2377-86 May (2009).
3. *Mineral commodity summaries*: U.S. Geological Survey, 196 p. (2015). Available at: http://minerals.usgs.gov/minerals/pubs/mcs/2015/mcs2015.pdf
4. *Phosphogypsum*. Introduction. Proceedings of International Symposium on Phosphorgypsum. Utilization and/or Disposal of Phosphogypsum. Potential Barriers to Utilization. Ed. by David P. Borris and Patricia W. Boody Bartow, Florida. Reprinted, p III-VI (19870.
5. L.Y. Dvorkin, A.V. Myronenko, and T.O. Polishchuk-Gerasimchuk, New technologies of phosphogypsum binder made of dump phosphogypsum./ Resourses saving materials, elements, structures and buildings, 16 (1), 27-34 (2008) (In Ukrainian).
6. A.P. Kouloheris. Chemical nature of hosphorgypsum as produced by various wet processes phosphoric acid processes./ Phosphogypsum. Proceedings of International Symposium on Phosphorgypsum. Utilization and/or Disposal of Phosphogypsum. Potential Barriers to Utilization. Ed. by David P. Borris and Patricia W. Boody Bartow, Florida. Reprinted, p.7-34 (1987)
7. S.K. Sahu, P.Y. Ajmal, R.C. Bhangare, M. Tiwari, G.G. Pandit Natural radioactivity assessment of a phosphate fertilizer plant area. Journal of Radiation Research and Applied Sciences 7, 123-128 (2014).
8. G.S. Bandgar, M.B. Kumthekar, A.B. Landage, A Review of Effective Utilization of Waste Phosphogypsum as a Building Material/ International Journal of Engineering Research. Jan, No.5, Issue Special 1. pp.: 277-280. Doi: 10.17950/ijer/v5is1/065 (2016).
9. E. Prandi, Use and valorization of phosphogypsum in road construction and civil engineering/ *Phosphogypsum*. Proceedings of International Symposium on Phosphorgypsum. Utilization and/or Disposal of Phosphogypsum. Potential Barriers to Utilization. Ed. by David P. Borris and Patricia W. Boody Bartow, Florida. Reprinted p. 269-278 (1987).
10. P. Pichat and R. Shinn Developments performed by A.P.C. – CdF Chimie in the field of phosphogypsum/ *Phosphogypsum*. Proceedings of International Symposium on Phosphorgypsum. Utilization and/or Disposal of Phosphogypsum. Potential Barriers to Utilization. Ed. by David P. Borris and Patricia W. Boody Bartow, Florida. Reprinted p 269-278 (1987).
11. S.S. Bhadauria, Utilisation of phosphogypsum in cement mortar and concrete. Rajiv Gandhi Proudyogiki Vishwavidyalya, Bhopal (M. P.) India Rajesh B. Thakare, JSPM's B. N. College of Engineering, Pusad (M. S.), India, 9 p. 31st Conference on our world in concrete and structures: 16 - 17 August, Singapore (2006).
12. L. Dvorkin, A. Myronenko, O. Dvorkin, N. Lushnikova Effect of Low-Alumina Blast Furnace Slags and Phosphogypsum Dihydrate on Properties of Supersulfated Cements. Journal "Scientific Israel-Technological Advantages", Vol.15, # 2, 16-29(2013).

13. L.Y. Dvorkin, V.L Shestakov, A.A Ishchuk. Complex use of phosphogypsum I ncement production/ Resourses saving materials, elements, structures and buildings. Ed. by Y.M. Babych. — Rivne:, Issue 16. Vol 1, pp. 35-40 (2008) (In Russian).
14. DSTU B V 2.7-3-93. Building materials. Synthetic gypsum made phosphogypsum. Technical conditions. (In Ukrainian).
15. ASTM C191 – 13. Standard Test Methods for Time of Setting of Hydraulic Cement by Vicat Needle.
16. ASTM C187 – 16. Standard Test Method for Amount of Water Required for Normal Consistency of Hydraulic Cement Paste
17. ASTM C348 - 14 Standard Test Method for Flexural Strength of Hydraulic-Cement Mortars
18. L. Dvorkin, N. Lushnikova, M. Sonebi, J. Khatib Properties of modified phosphogypsum binder./ 2nd International Conference on Bio-based Building Materials & 1st Conference on ECOlogical valorisation of GRAnular and FIbrous materials. June 21th - 23th. Clermont-Ferrand, France (2017) [accepted to publication].
19. L. Kacimi, A. Simon-Masseron, A. Ghomari, Z. Derriche, Reduction of clinkerization temperature by using phosphogypsum. J. Haz. Mat. B137, 129–137 (2006). http://dx.doi.org/10.1016/j.jhazmat.2005.12.053.
20. L. Kacimi, A. Simon-Masseron, A. Ghomari, Z. Derriche, Influence of NaF, KF and CaF2 addition on the clinker burning temperature and its properties. C. R. Chimie. 9, 154–163 (2006). http://dx.doi.org/10.1016/j.crci.2005.10.001
21. L.Y. Dvorkin, O.L. Dvorkin, Y.A. Stepasiuk , Ways of activization of hardening low-clinker blast furnace slag/ Drypack mortars (Stroitel'nyie Smesi). Moscow,. Issue 3. – pp. 35-39 (2015). (In Russian)
22. L.Y. Dvorkin, N.V. Lushnikova, Properties of Gypsum Binders Modified with Complex Admixtures. Ibausil, Internationale Baustofftagung, Tagunsbericht Weimar, Institut für Baustoffkunde, Band I: 1-0701 – 1-0707 (2009).
23. M. Marschetzky, H.-B. Fischer, H.-M. Ludwig, Superplasticizers in the calcium sulfate system – plasticizing action and influence on hydration. ZKG International. January 66(2):42-54 (2013).
24. J. Peng, J. Qu, J. Zhang, M. Chen, T. Wan, Adsorption characteristics of water reducing agents on gypsum surface and its effect on the rheology of gypsum plaster. Cement and Concrete Research. 35(3): 527–531 (2005).
25. L.Y. Dvorkin, A.V. Myronenko, T.O. Polishchuk-Gerasimchuk, Drypack gypsum and phosphogypsum mixes with effective modifiers. / Building mateirals and products (Stroitel'nyie materialy i izdelia)., Issue 6, pp. 20-23 (2012). – Available at: http://nbuv.gov.ua/UJRN/smii_2012_6_10 (In Ukrainian).
26. M.C.G Juenger, et al. Advances in Alternative Cementitious Binders. Cement and Concrete Research, 41 (12), 1232-1243 (2011).
27. L. Dvorkin, N. Lushnikova, M. Sonebi, J.M. Khatib, Light weight claydite and sawdust concrete based on supersulphated cement. / 2nd International Conference on Bio-based Building Materials & 1st Conference on ECOlogical valorisation of GRAnular and FIbrous materials. June 21th - 23th. Clermont-Ferrand, France (2017) [accepted to be publication].

Hydration characteristics and structure formation of cement pastes containing metakaolin

Leonid Dvorkin[1], Nataliya Lushnikova[2], Oleksandr Bezusyak[1], Mohammed Sonebi[3] and Jamal Khatib[4,5]
[1]National University of Water and Environmental Engineering, Department of Building Products Technology and Materials Science, Rivne, Ukraine
[2] National University of Water and Environmental Engineering, Department of Architecture and Environmental Design, Rivne, Ukraine
[3]School of Natural and Built Environment, Queen's University Belfast, Belfast, Northern Ireland, UK
[4]Faculty of Engineering, Beirut Arab University, Beirut, Lebanon
[5]Faculty of Science and Engineering, University of Wolverhampton, Wolverhampton, UK

Abstract. Metakaolin (MK) is one of the most effective mineral admixtures for cement-based composites. The deposits of kaolin clays are wide-spread in the world. Metakaolin is comparable to silica fume as an active mineral admixture for cement-based composites. In this paper, the rheological and mechanical properties of cement paste containing metakaolin are investigated. The effect of MK is more evident at "tight" hydration conditions within mixtures with low water-cement ratio, provided by application of superplasticizers. The cement is replaced with 0 to 15% metakaolin, and superplasticizer content ranged from 0 to 1.5% by weight of cementitious materials (i.e. cement and metakaolin). An equation is derived to describe the relationship between the metakaolin and superplasticizer content and consistency of pastes. There is a linear dependence between metakalolin content and water demand. Second-degree polynomial describe the influence of superplasticizer content. The application of SP and MK may produce cement-water suspensions with water-retaining capacity at 50-70% higher than control suspensions. The investigation of initial structure forming of cement pastes with SP-MK composite admixture indicates the extension of coagulation structure forming phase comparing to the pastes without additives. Crystallization stage was characterized by more intensive strengthening of the paste with SP-MK admixture comparing to the paste without admixtures and paste with SP. Results on the porosity parameters for hardened cement paste indicate a decrease in the average diameter of pores and refinement of pore structure in the presence of metakaolin. A finer pore structure associated with an increase in strength. X-ray analysis data reveal a growing number of small-crystalline low-alkaline calcium hydrosilicates and reducing portlandite content, when MK dosage increases. Scanning electron microscopy (SEM) data confirm, that hardened cement paste containing MK has crystalline structure with dominance of partially crystalized hydrosilicates and gel-like formations.

1 Introduction

Metakaolin (MK), a pozzolanic admixture of N type [1] consists mostly of amorphous aluminosilicate, obtained by firing kaolin clays with subsequent grinding. One of the common area of its application in construction is high-performance concrete, self-compacting concrete [2-5]. Self-compacting high-strength concrete is characterized by basal structure with 'floating' aggregate particles. In such a structure, grains of aggregate are considerably separated and almost do not interact with each other. Thus, properties of fresh and hardened concrete depend on the rheological, structural and mechanical properties of the cement-matrix.

Incorporation of MK as partial replacement of cement leads to significant changes in chemical composition and pore size distribution [6]. Studying of poor Greek Kaolin clays proved positive effect on the cement strength in different age of hardening. However, water demand increased significantly. The $Ca(OH)_2$ content of is decreases from 7 and 28 days. MK content of 10% is considered the optimal replacement [7].

Replacement of Portland cement (PC) by MK up to 20% elongates the initial and final setting times of cement pastes due to coating effect of MK particles on the cement grains [8].

The rheology parameters of cement grouts containing MK, viscosity-modifying admixtures and polycarboxylate superplasticizer proved that MK reduces fluidity and increases yield stress and plastic viscosity [9].

The studying of Portlandite consumption in cement paste with metakaolin addition from 0 to 15% at constant water-binder ratio of 0.55 by thermogravimetric analysis proved that maximum removal due to pozzolanic reaction occurs at the ages of 14 days during the 365 days period of research. This corresponded to the maximum value of relative strength. After the age of 14 days, the reaction retarded [10].

MK adding proved to lead to refinement of the pores at the same water-binder. The proportion of pores with radii smaller than 20nm increased as the level of PC with MK increased [11].

Most of the researches on cement pastes do not consider mutual influence of MK and superplasticizer (SP). However, the presence of SP at subsequent water demand reduction permits to create "tight" conditions of hydration. There is limited information about the properties of cement pastes with equal flowability.

2 Aim and scope of the research

The aim of the research was studying fresh and hardened cement paste properties with metakaolin in superplasticizer as if in this combination they usually applied in concrete.

There have been investigated the peculiarities of hydration and early structure formation.

The research was conducted in three main stages:

1 – studying of hydration processes at mixing with water: determination of water consumption (normal consistency) and water retention;

2 – studying of initial structural formation of pastes at setting and early age of hardening;

3 – studying of hydration and chemical composition of hardened cement pastes.

3 Materials and methods applied

The materials were used as follows. Locally produced PC CEMI 42,5, DIN 1146, EN-196 was used. The properties of cement are given in Tab. 1 and 2.

Table 1. Physical-mechanical properties of cement

No.	Property, units	Value
1	Fineness (008 sieve fraction content), %	5
2	Specific surface by Blaine, cm²/g	3,300
3	Normal consistency, %	24%
4	Hardening - begins at - ends at	 1 h 35 min 3 h 45 min
5	Uniformity of volume change	Meets the requirements
6	Ultimate strength at 28 days, MPa: - flexural - compressive	 8.75 54.0
7	Gypsum content (SO_3), %	3.5

Superplasticizer (SP) of naphthalene formaldehyde type, high range water reducers – admixtures of F type [12].

The properties of metakaolin are given in Tab.3 and 4.

Content of SP and MK was calculated as % of binder weight (PC+MK+SP). SP content varied in the range of 0 – 1.5%, MK content – 0 – 15%.

Table 2. Chemical composition of cement, % by weight

No.	Composition	Value
1	CaO	67.15
2	SiO_2	21.70
3	Al_2O_3	5.36
4	Fe_2O_3	4.10
5	FeO	-
6	MgO	0.74
7	Cl^- ion content	-
8	Loss on ignition	0.34
9	Insoluble residue	0.28

Table 3. Chemical composition of metakaolin, % by weight

SiO_2	Al_2O_3	Fe_2O_3	TiO_2	CaO	MgO	MnO	Na_2O	K_2O	L.O.I.
52.5	42.20	0.34	0.70	0.30	0.25	0.01	0.10	0.90	0.50

Table 4. Physical and chemical properties of metakaolin

No.	Property, units	Value
1	Specific surface by Blaine, cm²/g	18,000
2	Density, g/cm³	2.50
3	Bulk density, kg/m³	350
4	Pozzolanic activity, mg/g (by CaO absorption)	25
5	Normal consistency of metakaolin paste, %	46

There was tested normal consistency, water retention, setting time, water absorption according to standard methods. The other methods used are described in relevant parts of the article.

For each composition, the tests were conducted two times.

The adequacy of equations obtained was checked by calculating the adequacy dispersion, the design value of Fisher's criterion (F_d – criterion) [13] and comparing the last with a theoretical one F_t. The regression equation is adequate for the given probability level if $F_d < F_t$.

4. Normal consistency and water retention of cement pastes

Water consumption of the cement pastes is determined by their normal consistency and homogeneity.

Normal consistency is an important indicator of the physical characteristics of the cement paste. It has linear dependence on ultimate water retention and depends on cement composition and dispersion, as well as the type and quantity of additives.

To determine the impact of composite admixture of superplasticizer and metakaolin on normal consistency and water retention of pastes, experiments were performed two times for each composition. Compositions of paste and mean values of normal consistency are presented in Table 5.

Table 5. Cement pastes compositions and normal consistency

#	SP, % of binder weight	MK, % of binder content	Normal consistency (NC), % experimental value	calculated value
1	0		0.240	0.240
2	0.5	0	0.216	0.216
3	1.0		0.197	0.197
4	1.5		0.185	0.184
5	0		0.252	0.251
6	0.5	5	0.227	0.227
7	1.0		0.210	0.208
8	1.5		0.197	0.195
9	0		0.262	0.262
10	0.5	10	0.239	0.238
11	1.0		0.223	0.219
12	1.5		0.206	0.206
13	0		0.272	0.273
14	0.5	15	0.251	0.249
15	1.0		0.228	0.230
16	1.5		0.215	0.217

Experimental data are approximated by the following equation:

$$NC = (A \cdot {}_{1}^{2} - B \cdot X_1) + (NC^0 + C \cdot X_2 / 100) \qquad (1)$$

where A, B – coefficients, determined by the type superplasticizer: for used SP: $A = 0.011$; $B = 0.053$;
X_1 – SP content, % by binder weight;
NC_0 - normal consistency of cement paste without additives, $NC^0 = 0.24$;
X_2 - metakaolin content, % by binder weight;
C - coefficient, $C = NC_{MK} - NC_0$, where $NC_{MK}=0.46$ (see Tab. 4, normal consistency of metakaolin paste): $C=0.46-0.24 = 0.22$;
Eq. (1) is adequate at 95% confidence probability:
$F_d = 2.03 < F_t=2.46$.

Fig. 1 demonstrates the influence of MK and SP content on normal consistency. As it can be seen from eq.(1) and Fig. 1, the dependence of NC on SP content is can be approximated by parabola, which means that effectiveness of SP decreases as its content grows. It corresponds to known experimental data related to silica fume [14]. Effect of metakaolin is linear and does not depend on the SP content. According to Akhverdov [15], the minimum amount of water required for the formation of coagulation structures in cement paste, is as follows:

$$NC^{min} = 0.876\ NC \qquad (2)$$

When water content is equal to NC, water is contained mainly in solvate shells. As follows from Tab.5 and Fig.1 when SP is added water consumption can be less than NC^{min}. At this case effect of dilution caused by the action of SP leads to significant reduction in the thickness of the solvate shells and the formation of additional quantities of free water. NC of cement paste with SP is close to theoretical W/C, required for the hydration, which is 0.21 ... 0.23 [16]. Introduction of MK which has a higher normal consistency than cement (see Tab. 4), leads to regular growth of water demand of cement paste and decreasing of water-reducing effect of SP. However, as shown in Fig. 1, the introduction of high dosages of SP (1 ... 1.5%) with MK leads to appreciable water content reduction compared to cement paste without additives. Replacing 5% of PC by MK at SP=1.5%, decreases normal consistency at 20%, for replacing PC with 15% of MK water reduction is 10%.

Water-binder ratio (W/B), which corresponds to ultimate water retaining capacity of paste is from 1.60·NC to 1.65·NC. However, considering some possible vibrations at transportation it is considered from 1.30·NC to 1.35·NC [17, 18].

There was determined influence of MK and SP on water-retaining capacity of cement suspensions.

There was determined water bleeding coefficient (K_b, %) of cement suspensions with the SP and MK at water-binder ratio (W/B) equal to 1. K_b represents the volume of water separated:

$$K_b = \frac{a-b}{a} \cdot 100 \qquad (3)$$

where a is an initial volume of cement suspension cm^3;
b is the volume of cement of settled suspension, cm^3;
Then the water-retaining capacity suspensions is as follows

$$WR = 1 - K_b \qquad (4)$$

WR, divided by 100, is W/B of settled suspension. Mean values of bleeding coefficients and water-retaining capacity is shown in Tab. 6.

Table 6. Water bleeding coefficient and water retaining capacity of cement suspensions

#	SP, %	MTK, %	Water bleeding coefficient, % / Water retaining capacity, % after, hours: 0	0.5	1	1.5	2	2.5	3	3.5	4
1	0	0	0 / 100	25.4 / 74.6	36.8 / 63.2	41.9 / 58.1	44.2 / 55.8	45.2 / 54.8	45.6 / 54.4	45.8 / 54.2	45.9 / 54.1
2	0	10	0 / 100	13.2 / 86.8	19.0 / 81.0	22.0 / 78.0	23.9 / 76.1	25.5 / 74.5	25.9 / 74.1	25.9 / 74.1	-
3	0.5	10	0 / 100	17.6 / 82.4	25.5 / 74.5	29.0 / 71.0	30.0 / 70.0	31.1 / 68.9	32.0 / 68.0	32.4 / 67.6	32.5 / 67.5
4	1	10	0 / 100	21.0 / 79.0	30.0 / 70.0	35.2 / 64,8	37.8 / 62,2	38.0 / 62,0	38.7 / 61,3	39.4 / 60.6	39.5 / 60.5
5	1.5	10	0 / 100	23.5 / 76.5	34.5 / 65.5	41.6 / 58,4	43.8 / 56,2	44.0 / 56,0	45.5 / 54,5	45.6 / 54,4	45.7 / 54,3
6	1	0	0 / 100	30.0 / 70.0	45.7 / 54.3	54.5 / 45,5	56.0 / 44,0	57.5 / 42,5	58.9 / 41,1	58.9 / 41.1	-
7	1	5	0 / 100	25.0 / 75.0	38.5 / 61.5	43.0 / 57.0	47.0 / 53.0	48.5 / 51.5	48.8 / 51.2	48.9 / 51.1	49.0 / 51.0
8	1	10	0 / 100	19.0 / 81.0	30.0 / 70.0	34.5 / 65.5	38.0 / 62.0	38.3 / 61.7	38.9 / 61.1	39.0 / 61.0	-
9	1	15	0 / 100	14.5 / 85.5	24.0 / 76.0	25.5 / 74.5	27.0 / 73.0	28.5 / 71.5	28.8 / 71.2	28.9 / 71.1	-

The influence of water retention on time of settlement can be described by the hyperbola dependence with horizontal asymptote as following

$$WR=100-(A+BX_1+CX_2)\cdot(1-1/D^t), \quad (5)$$

where X_1 – SP content, % binder weight; X_2 – MTK content, % binder weight; t- time of sedimentation, h; *A, B, C* and *D* are coefficients, which depend on properties of materials applied (MK, SP); in current research: $A=46; B=13; C=-2; D=5$.

Eq. (5) is adequate at 95% confidence probability:
$F_d = 1.35 < F_t=1.36$.

Fig. 2 shows dependencies of water retaining capacity on the SP and MK content. With increasing SP content ordinate asymptote decreases (Fig. 2 a); as MK dosage increases - ordinate asymptote increases too. (Fig. 2 b). The second part of the equation describes the dependence of water-retaining capacity on the time.
As time of settlement of suspension grows, water retaining capacity reduces close to asymptote. The presence of MK reduces bleeding of suspension due to higher sedimental stability of smaller particles of MK comparing to Portland cement.

According to the data Tab. 6, the use of additives SP and MK, allows to obtain at the end of experiment suspensions with WR= (2.5-3)·NC. At the same time, the control suspension without additives at the end of the experiment has WR≈2 NC, that value is close to the theoretical limit of water-retaining capacity of cement (upper limit of thixotropy) [14]. That is caused by higher water demand of MK comparing to PC.

Thee water-retaining capacity of suspensions made with binders of equal water consumption was compared.

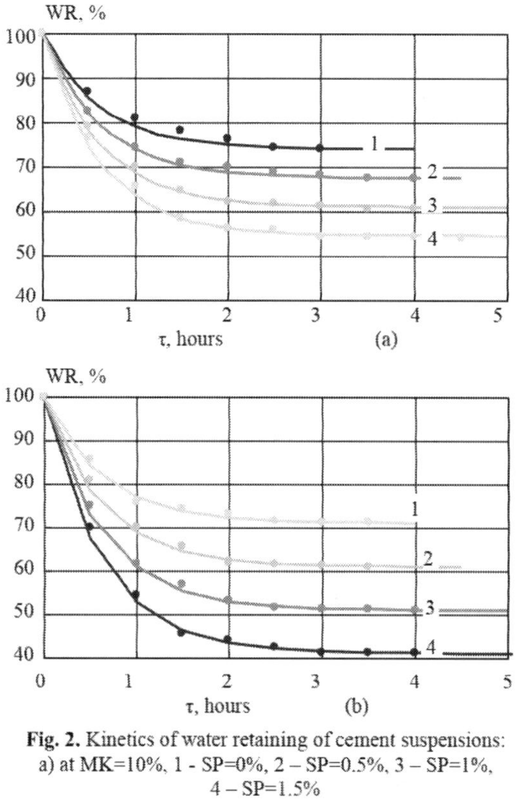

Fig. 2. Kinetics of water retaining of cement suspensions: a) at MK=10%, 1 - SP=0%, 2 – SP=0.5%, 3 – SP=1%, 4 – SP=1.5%
b) at SP=1%, MK=15%, MK=10%, MK=5%

It was assigned NC=0.24 (as for control composition). Varying MK content from 5 to 15%, SP content was calculated according to formula (1). According to formula (4) the water retention was determined. The results are shown in Fig. 3.

As it can be seen increasing of MK content at substantial SP dosage increase leads to significant growth of water retaining capacity. At replacing 5% of PC by MK WR=61.3% (W/B=0.613), at 15% - WR=74.7% (W/B=0.747).

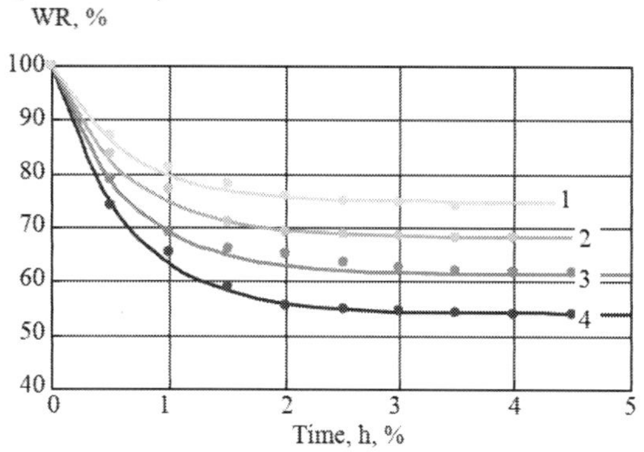

Fig. 3. Kinetics of water retaining of cement suspensions with equal normal consistency 24% calculated by formula (4): 1 - SP=0.72%, MK=15%; 2 – SP=0.45%, MK=10%; 3 – SP=0.21%, MK=5%; 4 – SP=0%, MK=0%

In the case of self-compacting high-strength concrete W/C is normally lower than 0.4. Taking into consideration water consumption of fine and coarse aggregate for their moistening, actual W/C of cement paste in concrete is within the range 0.25-0.35 [19]. Taking into consideration ultimate water retention of control cement paste water-binder ratio selected for further research was $W/B=1.35NC^0= 1.35\cdot 0.24= 0.33$ (5).

5 Peculiarities of initial structure forming of cement pastes

The flowability of the pastes tested in the research was determined by the flow spread diameter of the Vicat apparatus cone (upper Ø70 mm, bottom Ø100 mm, h=40mm). The diameter of paste was assigned for compositions #2-5 as (200±1) cm. In this case we received self-levering paste like that in self-compacting concrete. The compositions are shown in Tab. 7.

Table 7. Cement pastes compositions

#	W/B	Flow spread diameter, mm	MK, % by binder weight	SP, % by binder weight	NC (according to formula (1))
1	0.33	100	0	0	0.240
2		200	0	0.32	0.224
3			5	0.46	0.229
4			10	0.64	0.232
5			15	0.95	0.233

For determination of initial structure formation of cement paste there were used standard setting time test and ultrasonic pulse velocity. Comparing setting time of the compositions is shown in Tab 8.

Table 8. Setting time of cement pastes according to Vicat apparatus and ultrasonic pulse velocity test

#	NC, %	Setting time (Vicat apparatus), h-min				Setting time by ultrasonic pulse velocity test
		W/B=NC		W/B=0.33		
		initial	final	initial	final	initial
1	24.0	1 - 35	3 - 45	4 - 25	6 - 45	4 - 00
2	22.4	3 - 15	4 - 35	6 - 30	7 - 50	6 - 10
3	22.9	1 - 25	4 - 20	4 - 40	8 - 00	4 - 30
4	23.2	1 - 45	4 - 35	4 - 50	8 -25	4 - 25
5	23.3	2 - 15	4 - 55	5 - 00	8 - 40	4 - 50

As cement had relatively low-volume of C_3A, SP do not adsorb intensively by aluminates. Therefore, SP retards setting of cement paste due to lower thickness of the solvate shells comparing to high-alumina cements and releasing of additional mixing water. The water can adsorb both at the grains of unhydrated cement and hydrated new formations. At adsorption, diffusive resistance increases and dissolution of cement grains inhibits [14, 19]. At MK addition and SP there is a "tight" hydration conditions [20] there is close contact between the solid particles due to the increase of its concentration and higher surface area of MK. Because of the increase of physically and chemically bound water, the thickness of water interlayers reduces. This leads to acceleration of initial setting time. With time, the intensity of new formations growth reduces, and the final setting time of pastes #3-5 is close to paste #2. As MK content increases, structure forming processes of paste retards due to adsorption of SP on particles of MK and blocking of cement hydration. Similar phenomena are observed for pastes with W/B = 0.33. However, because of their higher water content, processes initial structure forming retard at 3-5h (Tab 8).

The ultrasonic velocity can determine the end time of inductive stage in structure formation after mixing with water [21], which corresponds to initial setting time. The method of through passage of longitudinal ultrasonic pulse was used during 24 h with measuring each 30 min. Results of the test are given in Tab. 8 and Fig. 4 (for compositions #1, 2 and 4, W/B=0.33).

As shown in the Fig. 4, curves of ultrasonic pulse velocity have sections, which correspond to the stages of structure forming. At the initial stage (the inductive period) C slightly increases (marked as horizontal section). Subsequent formation of low-strength crystallization structure of aluminate components significantly affects C: $dC/d\tau$ is maximum at this stage. At the stage of silicates crystallization, which are the main carriers of strength, the rate of velocity growth reduces.

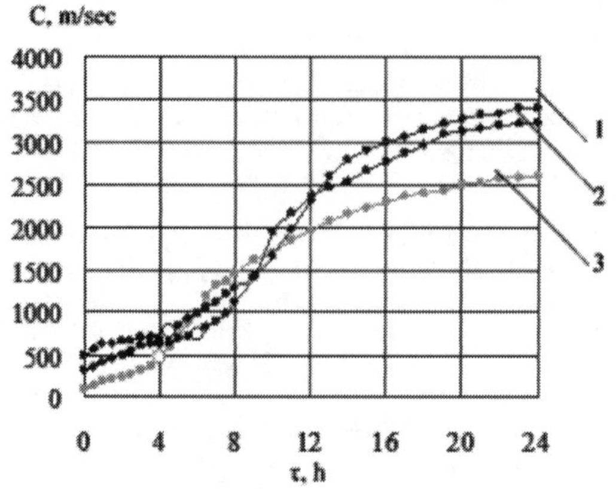

Fig. 4. Ultrasonic pulse velocity at early structure formation of cement pastes:
1 – SP=0.64%, MK=10%; 2 – SP=0.32% 3 - control

The initial stage of coagulation structure forming is characterized by horizontal sections, which ends by inflection point (white point). It corresponds to the initial setting time. Growth of MK content leads to intensifying of the ultrasonic pulse passage.

This research method is more informative for initial structure forming than setting time test. Setting time test permits to estimate the period of coagulation. Whereas ultrasonic pulse velocity test – to estimate the physical and chemical changes that lead to the formation of microstructure of the pastes. However, the values of initial setting time determined by these both methods are correlated with each other. The introduction of the SP and MK retards initial structure formation due to the of adsorptive films of SP, which prevent hydration.

Crystallization structure stage is characterized by intensive strengthening of paste with SP and MK due to the active binding of portlandite with metakaolin and the formation of additional quantities of stable low-basic crystalline hydrate CSH (I). For more detailed testing of structure formation processes X-ray SEM analysis were applied.

6 Peculiarities of structure of hardened cement paste

There were determined influence of MK and SP on porous structure of cement paste and its phase composition. To provide uniformity and compatibility of output parameters all the pastes had equal consistency (flow spread diameter 200 mm) and W/B=0.33.

The peculiarities of the modification of cement paste with MK and SP additions is determined by the parameters of its pore structure. One of the most simple and universal methods of investigation pore structure of cement is water absorption method [22], which allows to determine how integrated (apparent porosity) and differential parameters of pore structure (the average pore size and uniformity of their size).

Numerous experiments of water absorption of multicapillary materials such as cement stone and concrete, can be described by exponential function. In general, water absorption curves of cement pastes are approximated by three-parameter exponential function of the form:

$$W_\tau = W_{max}\left(1 - e^{-\lambda \cdot \tau^\alpha}\right) \quad (6)$$

where W_τ is water absorption after τ hours, %; W_{max} – conventional value of maximum water absorption, %; λ – coefficient, which determines average diameter of capillary pores; α – coefficient, which determines uniformity of capillaries size; τ – time of saturation of specimen, h.

For determination of porosity paraments according to kinetics of water absorption cubic specimens with 7 cm sedge were prepared. After curing at normal hardening chamber for 28 days, the specimens were dried at t=105…110°C. The water absorption was determined during 2 h after saturation and in 24 h of saturation, (which corresponds to conventional total open porosity). Kinetics of water absorption is shown on Fig. 5.

For all the pastes (see Tab. 7) Eq. 6 is adequate at 95% confidence probability: composition #1: $F_d=1.94<F_t=4.19$; #2: $F_d=1,77<F_t=4,19$; #3: $F_d=2,05<F_t=4,53$; #4: $F_d=1,66<F_t=3,84$; #5: $Fd=1,83<F_t=4,1$.

There was determined weight water absorption, %:

$$W_m = \frac{m_{24} - m_0}{m_0} \cdot 100, \quad (7)$$

And volumetric water absorption, % (open porosity):

$$W_0 = \frac{m_{24} - m_0}{m_{24} - m_{24}^{в}} \cdot 100, \quad (8)$$

where m_0 is a mass of the specimen in dry state (weight in air), kg; m_{24} is a mass of specimen after 24h of saturation (weight in air), kg; m^{w}_{24} is a mass of specimen after 24h of saturation, (weight in the water) kg.

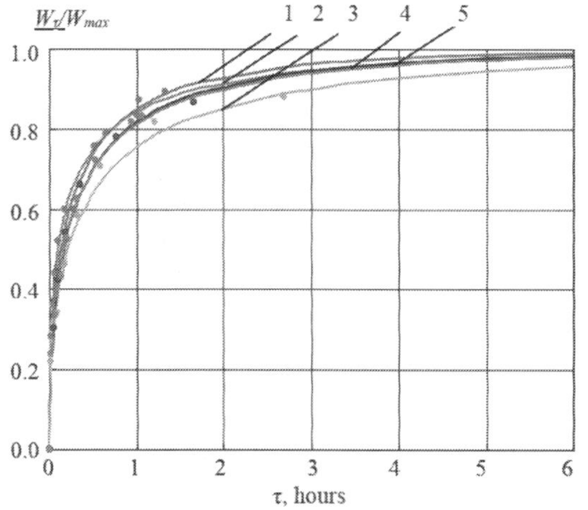

Fig. 5. Kinetics of water absorption of cement paste:
1 - MK=0%, SP=0%; 2 - MK =0%, SP=0.32% ;
3 - MK =5%, SP=0.46%; 4 - MK =10%, SP=0.64%;
5 - MK =15%, SP=0.95%

As it can be sen from the Tab. 9, the growth of MK content in binder leads to reduction of average diamenter of pores (coefficient λ) and increasing their uniformity (coefficient α). Water absorption and open prosity reduces.

Table 9. Parameters of porous structure of cement paste

#	SP, %	MTK, %	α	λ	Water absorption (mass) W_m, %	Water absorption (volume) W_0, %
1	0	0	0.494	1.889	9.80	19.39
2	0,32	0	0.417	1.847	9.61	18.72
3	0,46	5	0.491	1.690	7.44	14.55
4	0,64	10	0.476	1.663	7.52	14.01
5	0,95	15	0.451	1.401	7.06	13.20

It confirms the data about formation of fine-crystalline structure with prevalence of low-base calcium silicates CSH(I) [24]. In accordance to the general laws of structure formation [23], the introduction of ultra-fine materials leads to a change in the balance between the gel (and capillary pores in favor of the first and cement structure is more dispersed.

There were conducted X-ray and SEM study of cement at the age of 28 days, at early stages of structure formation with the addition of ultrafine reactive materials "physical factor" dominates, which " fill the volume between coarse particles of cement and form numerous low-strength coagulation contacts" [24]. "Chemical factor", which is expressed in changing the balance of hydration new formations to growth of number of stronger and more stable compounds, has significant value in later ages of hardening.

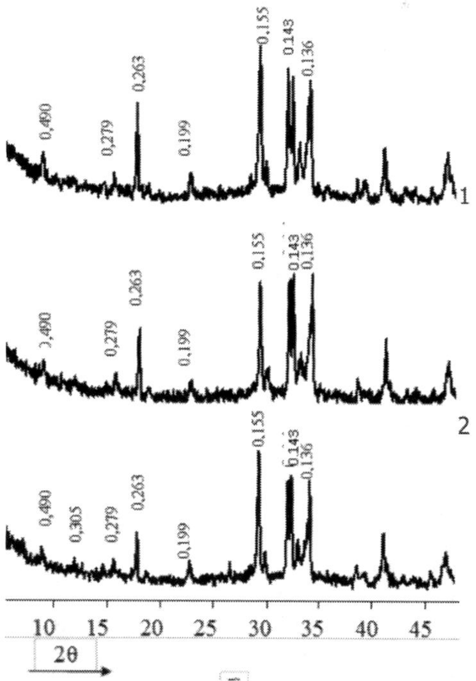

Fig. 6. X-ray of cement stone at the age of 28 days:
1 - MK =0%, SP=0.32%; 2 - MK =5%, SP=0.46%;
3 - MK =10%, SP=0.64%

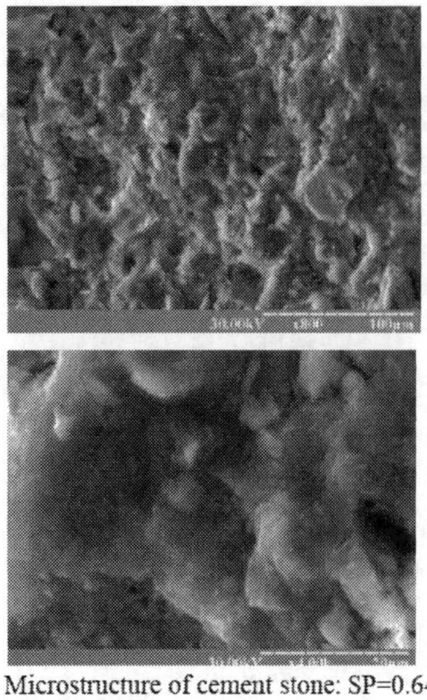

Fig. 7. Microstructure of cement stone: SP=0.64%; MK=10% (x800 upper, x 4000 - bottom)

The results of X-ray analysis at the age of 28 days (Fig. 6) For all pastes indicate the presence of crystalline crystallized tobermorite-like calcium hydrosilicates CSH(I) with variable structure: (1-1.5) $CaO{\cdot}SiO_2{\cdot}H_2O$ (d=0,305; 0,279; nm). As the dosage of MK grows – the content of the minerals increases, as it can be seen from increased intensity of the diffraction peaks. Portlandite $Ca(OH)_2$ content (d=0,490; 0,263; 0,143 nm), decreases with increasing portion of MK. Small amount of ettringite (d = 0,155 nm) is present in both early and late terms of hardening, the content of it is proportional to the amount of metakaolin and tobermorite-like calcium hydrosilicate CSH (II) $2CaO{\cdot}SiO_2{\cdot}H_2O$ (d = 0.199 nm).

The results of SEM analysis of cement paste structure (Fig. 7) show dominance of fine crystalline hydrates and gel-like products. Overall cement structure is dispersed.

Conclusions

1. The following conclusions are based on the results of this investigation. The normal consistency depends on the content SP and MK, which indicates a positive linear relationship between the water demand increase and MK content growth. The influence of SP content is described by polynomial of the second degree that confirms reduction of SP efficiency superplasticizer when the content in the paste over 1 ... 1.5%.
2. It is shown that the use of SP and MK allows to obtain cement suspensions with water-retaining capacity up to (2.5-3.0) NC, while for the control suspension it is (1.65-1.75) NC. It is caused by high water-retaining ability of metakaolin.
3. The structure of cement pastes with admixtures show elongation of coagulation stage structure compared to control paste. Crystallization stage is characterized by more intensive strength growth of pastes with SP and MK comparing to that with SP indicated by ultrasonic pulse velocity.
4. There is a decrease in the total diameter pores and improve in their homogeneity when SP and MK are added.
5. X-ray data give evidence of the growing number of small-crystalline low-base calcium hydrosilicates and reduction of Portlandite in cement, at increasing the MK content. According to SEM analysis cement paste with SP and MK has crystalline structure that has fine crystalline structure with partly crystalized hydrosilicates and gel-like formations.

References

1. ASTM Standard C618-89a. Standard Specification for Coal Fly Ash and Raw or Calcined Natural Pozzolan for Use in Concrete.
2. S. Wild. J.M. Khatib..A. Jones. Relative strength, pozzolanic activity and cement hydration in superplasticised metakaolin concrete. Cement and Concrete Research. 26(10), October 1996, 1537-1544. https://doi.org/10.1016/0008-8846(96)00148-2
3. C.S. Poon, S.C. Kou, L. Lam. Compressive strength, chloride diffusivity and pore structure of high performance metakaolin and silica fume concrete. Construction and Building Materials. Volume 82, 1 May 2015, Pages 133–141. https://doi.org/10.1016/j.conbuildmat.2005.07.001
4. Dvorkin L., Bezusyak A., N. Lushnikova, Ribakov Y. Using mathematical modeling for design of self compacting high strength concrete with metakaolin admixture. Construction and Building Materials. Volume 37, December 2012, Pages 851–864. https://doi.org/10.1016/j.conbuildmat.2012.04.019
5. E.G. Badogiannis, I.P. Sfikas, D.V. Voukia, K.G. Trezos, S.G. Tsivilis. Durability of metakaolin Self-Compacting Concrete. Construction and Building Materials. Volume 82, 1 May 2015, Pages 133–141. https://doi.org/10.1016/j.conbuildmat.2015.02.023
6. N.L. Coleman, C.L.Page Aspects of the pore solution chemistry of hydrated cement pastes containing metakaolin. Cement and Concrete Research 27(1):147-154 · January 1997 DOI: 10.1016/S0008-8846(96)00184-6
7. E. Badogiannis, G. Kakali, G. Dimopoulou, E. Chaniotakis, S. Tsivilis. Metakaolin as a main cement constituent. Exploitation of poor Greek kaolins. Cement & Concrete Composites 27 (2005) 197–203.
8. H. El-Diadamony, A.A. Amer, T.M. Sokkary, S.El-Hoseny. Hydration and characteristics of metakaolin pozzolanic cement pastes. HBRC Journal Available online 27 May 2016 In Press, https://doi.org/10.1016/j.hbrcj.2015.05.005
9. Influence of the type of viscosity-modifying admixtures and metakaolin on the rheology of grouts. M. Sonebi, W. Schmidt, J. Khatib Chemistry and Materials Research, Vol.5 2013. Special Issue for International Congress on Materia ls & Structural Stability, Rabat, Morocco, 27-30 November 2013.
10. S. Wild,.J.M. Khatib. Portlandite consumption in metakaolin cement pastes and mortars Cement and Concrete Research Volume 27, Issue 1, January 1997, Pages 137-146.
11. J.M. Khatib, S. Wild Pore size distribution of metakaolin paste. Cement and Concrete Research. Volume 26, Issue 10, October 1996, Pages 1545-1553.
12. ASTM C494 / C494M – 11. Standard Specification for Chemical Admixtures for Concrete.
13. Ž.R. Lazić. Design of experiments in chemical engineering: A practical guide. Weinheim: WILEY-VCH Verlag GmbH & Co, KGaA; 2004.

14. V.G. Batrakov. Modified concrete. Theory and practice. 2nd edition, Moscow, 1998. – 768 p.
15. I.N. Akhverdov. Concrete Physics Fundamentals, Moscow, 1981. (in Russian)
16. O.P. Mchedlov-Petrossyan. Chemistry of non-organic building materials. – M.:Stroyizdat, 1988. – 304 p. (in Russian)
17. V.S. Demianova, V.I. Kalashnikov, I.E. Ilyina. Comparative estimation of the influence of local and abroad superplasticizers on the properties of cement compositions. // Strointelnye materialy (Building Materials). 2002 (9) 4-6. (in Russian)
18. V.R. Falikman, A.Y. Vainer, N.F. Bashlykov. New generation of superplasticizers. // Beton i Zhelezobeton (Concrete and Reinforced Concrete). 2000 (5).5-7. (in Russian)
19. V.P. Sizov. Strength of concrete based on low water demand binders. // Beton i Zhelezobeton (Concrete and Reinforced Concrete) 1990 (10) 14-15. (in Russian)
20. M.M. Sychev Perspectives of increasing of the strength of cement stone. // Cement, 1987 (9) 17-19. (in Russian)
21. Y. Lin, C.-P. Lai, and T. Yen. Prediction of Ultrasonic Pulse Velocity (UPV) in Concrete. Materials Journal. Volume: 100 Issue: 1: 21-28. (in Russian)
22. Sheikin, A.E. Structure and properties of cement concrete / A.E. Sheikin, Y.V. Chekhov, M.I. Brusser. - M.: Stroyizdat, 1979. - 344 p. (in Russian)
23. S.Kaprielov, A. Sheinfeld Influence of Silica Fume / Fly Ash / Superplasticizer Combinations in Powder-like Complex Modifiers on Cement Paste Porosity and Concrete Properties. //Sixth CANMET/ACI International Conference on Superplasticizers and other Chemical Admixtures in Concrete. Nice, France, October 2000. Proceedings, pp.383-400.
24. Massazza F. Chemisry of pozzolanic admixtures and composite cements. // Proc. 7th International congress of the chemistry of cements. Vol.3. – Mockow , 1976 – p. 209-211.

Challenges of the growing African cement market – environmental issues, regulative framework, and quality infrastructure requirements

Wolfram Schmidt[1], Inès L. Tchetgnia Ngassam[2], Gerhard Breitschaft[3], and Stephen Virchow[4]
[1]Bundesanstalt für Materialforschung und -prüfung, Department "SafetyofStructures", Berlin, Germany
[2]University of Cape Town, Department of Civil Engineering, Cape Town, South Africa
[3]Deutsches Institut für Bautechnik (DIBt), President, Berlin, Germany
[4]Physikalisch-Technische Bundesanstalt (PTB), Department "International Cooperation", Braunschweig, Germany

Abstract. The African cement, concrete and construction business is growing at rapid pace. The cement sales are expected to grow rapidly until 2050. The number of newly built cement plants increases dramatically and in addition more cements are being imported from outside the continent, e.g. from Turkey, Pakistan, Indonesia, and China, driven by overcapacities in the countries of origin. This causes a high number of potentials and challenges at the same time. Newly built cement plants can operate directly at best technological state of the art and thus incorporate more sustainable technologies as well as produce new and more sustainable products such as cements blended with sustainable supplementary cementitious materials such as calcined clays, and industrial or agricultural by products. At the same time the new variety of binding agent as well as the international imports, which are driven by price considerations, make the cement market prone to quality scatter. This puts pressure on the quality control regulations and institutions to ensure safety of construction, healthy application, and environmental safety for the population. The paper presents possible solutions to build up the rapidly increasing African cement production more sustainably than in the rest of the world as well as the related challenges and obstacles that need to be overcome. Based on experiences with a series of pan-African cement testing laboratory proficiency schemes conclusions are made on technical, regulative and political level.

1 Cement and construction technology from a global point of view

Concrete construction has always been a local business of global concern. Due to the indispensable nature of concrete, globally and locally the material has an enormous ecologic, economic and social dimension.

Concrete makes out 50% of everything the world produces [1]. It consists of the elements that form 98% of the outer Earth's crust [2]. Regardless of their possible performance, materials based on the complementary 2% are not feasible for building global infrastructures and habitats, due to the limited volume of required resources.However, it has also pointed out that compared to other construction materials concrete shows the lowest embodied energy and the lowest embodied carbon [3].

The global cement demand is predicted to increase from 4.6 Gt/a [4] in 2015 to about 6-13.5 Gt/a until 2050 [3, 5]. Based on the high global production volume, concrete today is responsible for 5-10% of the global CO_2 exhaust [1]. In 2050 it can be as much as one third [4].

Concrete is affordable and can be produced everywhere in the world from mainly local resources.It is often called a "democratic" material, since despite its global business, it always involves a many local businesses and players.It empowers 545,000 employees worldwide. The annual turnover is 17.5 billion EUR [5].

Eventually, concrete is inevitable for infrastructures that are lifelines for local developments and triggers of global networking. And, concrete is of highest importancefor adequate housing that gives shelter and safety, and thus lets the populace participate in the wealth and growth of economies. It therefore provides the foundation for adequate life and future perspectives, which contributessignificantly to reduction of global social problemse.g. related to health, social unrest, or migration.

Hence, cement and concrete technologies are inevitable in the world with an increasingly important role for developing economies. Due to the enormous global impact, it will become more important to use cement and concrete more sustainably. Recommendations for a more eco-efficient cement and concrete based industry were made available recently in a UNEP report [6]. However, despite the environmental and economic aspects, it is also important to build with cement and concrete in a safe and qualitatively sound way.

2 Challenges in the growing African cement and construction market

2.1 Perspectives and peculiarities of the African cement and concrete market

Many African countries are booming with steady and increasing velocity. Along with the economic upturn come the construction activities. Without any doubts, the numerous challenges that can be identified regarding housing and infrastructure can only be mastered by opting for cement and concrete as building material.

The African continent has numerous regions with different boundary conditions, and not every region provides the best potential for cement production [7]. The lack of typical industrial by-products such as fly ash (FA) and ground granulated blast furnace slag (GGBS) makes saving of ordinary Portland cement a challenge. However, less established local alternatives do exist and will be discussed later in this paper.

Despite the great regional differences in Africa a few peculiarities can be observed nearly everywhere over the continent. The variety of cements available in sub-Saharan Africa is generally limited. While modern cement technology considers cement blending as mean to individualise the cement performance and increase the number of marketable products, the variety of cements on the sub-Saharan African market is rather limited to OPC, and limestone filler or pozzolana blended cement. South Africa with huge deposits of FA and certain availability of GGBS is an exception. In order to serve the market with GGBS or fly ash blended cement types, the raw materials are sometimes shipped in, although from global perspective shipments of huge materials volumes does not make sense regarding the global carbon emissions.

Cement plants are being rapidly built today in Africa, but to date the density of cement plants is still low. The related transportation costs cause high local prices for cement, so that building with concrete is extremely expensive for local builders. This encourages global cement exporters to ship materials from all over the world into Africa. The implications are:

- High pressure on local cement producers
- A vast variety of cement types and producers
- Challenges in quality and safety regulations

While cement prices expressed in local purchasing power are significantly higher, labour costs are typically low in Africa compared to other parts of the world, since in construction business workers are commonly paid upon requirement and on daily basis. The cost positions for the calculation of concrete constructions thus significantly diverge from European calculations.However, not only the costs diverge, but also the entire boundary climatic and infrastructural framework:

- Most African regions have high temperature and often high relative humidity, while most standards were developed in moderate climate.
- Transportation distances for raw materials are long, time intensive, and costly.
- Neither pre-cast nor ready-mixed concrete are well established. Mostly cement, sand, and aggregates are stored and mixed on the construction site.
- Poker compaction and automated mixing is not always and everywhere available. Many process steps during casting are largely conducted manually.

However, this does not generally apply. Many large scale construction projects such as dams, airports, or central infrastructures are operated at a high technological level. The result is that, eventually, cement and concrete technological regulations need to cope with a much wider variety than e.g. in Europe, from very simple to very sophisticated. Due to the vast differences between the African boundary framework and the framework of the European continent it is questionable whether a European regulative framework can be an ideal role model for countries in Africa[8].

2.2 Sustainability potentials in Africa

While in the past concrete performance was linked to the content of OPC and the water-cement ratio, today, high quality concrete can be produced with low clinker cement types and due to the use of chemical admixtures, the water-cement ratio can be minimised, thus allowing to optimise the cement use in concrete. Therefore today'slow per capita cement consumption in most African countries can turn out to become an enormous potential for Africa to spearhead sustainable cement and concrete technologies [9], since the predicted growth of the construction market can be conducted at best level of available knowledge without repeating past mistakes and without using unnecessarily high amounts of cement.

2.3 Experiences from a pan-African cement testing proficiency scheme

Although there is a distinct demand for high quality products in the African society, the infrastructure to meet the demand mostly needs improvement. Particularly in the growing construction market with an increasing number of materials in use, it is most important to develop efficient quality assessment and control schemes, since this market is crucial for Africa's development, and affects the safety of the entire population.

Results from a pan-African cement proficiency testing scheme (PACE-PTS), which is conducted as cooperation between BAM Federal Institute for Materials Research and Testing and the German Metrology Institute (PTB) with support of the University of Cape Town illustrate the challenges related to quality control procedures in the cement and concrete area. In the PACE-PTS, laboratories conduct cement, binder and aggregate testing with identical reference materials provided by BAM according to formerly agreed upon procedures and report back their results. Furthermore, the laboratories were requested to conduct some of the tests in addition with local cement types.

This gives opportunity for public and private laboratories to self-assess their performance and to increase their network and skill in a final evaluation workshop. In the 2016/2017 scheme 55 laboratories from 25 countries, 22 of which from Africa, have participated (Figure 1). Most laboratories were from sub-Saharan Africa with strong participation from S. Africa, Nigeria, Ghana, Angola, and Kenya.

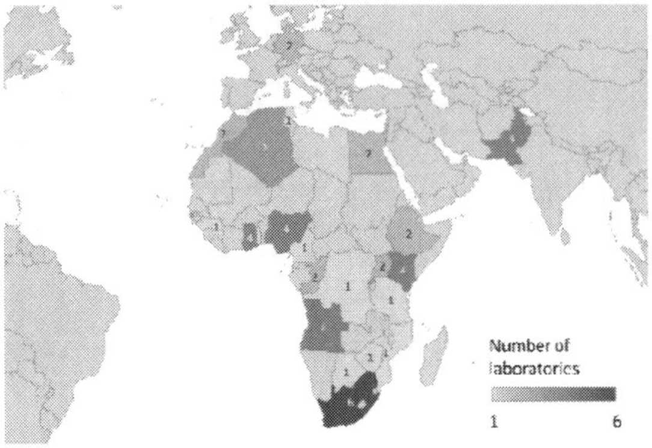

Figure 1: Distribution of registered laboratories for the PACE-PTS in 2016/2017.

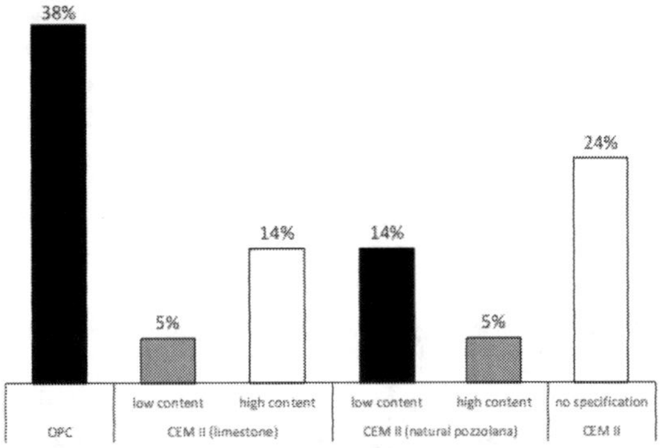

Figure 2: Local cement types used and percentages.

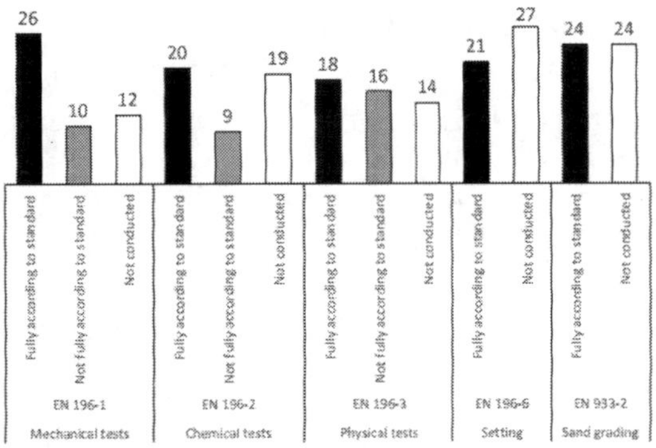

Figure 3: Distribution of registered laboratories for the PACE-PTS in 2016/2017.

Figure 2 shows the usage of different cement types for the African laboratories. It can be found that 72% of the local cement types were blends, the majority of which blended with limestone filler or natural pozzolana. A high percentage of cements were only classified as CEM II without further specification. The diversity of different cement types used, and the fact that limestone filler and natural pozzolana dominated, both of which are prone to performance scatter, indicates that proficiency in testing becomes very important for the safety of construction, but as shown in Figure 3, in many experiments less than half of the laboratories could conduct the testing fully according to the standards. The reasons were manifold, such as lack of equipment or lack of possibility to adjust and maintain the required laboratory climate.

2.4 Challenges for quality control

In order to make use of the existing potentials for more sustainable cement and concrete construction technologies, in the future, mixture compositions have to be more complex than in the past[6, 10]. Besides the classical components for concrete, cement, water and aggregates, further components, such as chemical admixtures and mineral additions are used.

Particularly industrial and agricultural waste products as well as organic compounds can become a threat to the population, and shall be carefully used, while it is also necessary to see the potential of immobilising certain compounds in concrete structures.These changes and the increasing need for the associated quality control lead to an even higher requirement of well-equipped laboratories with well-schooled staff members. Both requirements are not always met as the results of arecent and past pan-African cement testing proficiency schemes (PACE-PTS) have indicated [11, 12].

Other challenges for efficient quality control in the African cement and construction market include the lack of legally binding requirements to guarantee users' security. Besides, functioning market surveillance schemes are rare in most countries in the region which could lead to serious accidents such as collapsing buildings. Customers' lacking awareness of the importance of quality control in the market additionally challenges the development of laboratories in this market, since the demand might be too weak, though past proficiency testing schemes indicated a definitively increasing demand for quality control.

2.5 Challenges for a safe construction framework

To supply African markets with reliable construction products in the long run, it is important to have appropriate legal provisions. Unfortunately, there have been various buildings that collapsed due to inadequate execution of works and improper construction products recently, causing injury and worse.The required safety can only be provided by government initiatives.

As a first step, governmentshave to draw up the requirements for construction works that shall be binding in their countries. These requirements first and foremost include safety issues such as stability, fire protection, health and environmental protection as well as secondly building physical properties such as acoustic and thermal insulation - where needed. Legal requirements can be kept quite general in the law, as long asthey are specifiedfor users and appliers in more detail through technical specifications.

For example, it is important to define acceptable failure probabilities for load-bearing components or necessary fire resistance durations for certain building sections, or tolerable elution to the groundwater. These requirements then lead to specifications for construction products. To make sure that construction products meet thespecifications permanently and reliably, it is necessary to establish a system of certifications, control and inspections. This system does not inevitably have to be run under government responsibility, but it can also be private run. In either case, it needs to be tailored to the economic and political situation of each country. It is not advisable to adopt European or American systems and apply them in Africa without reasonable and careful adaption.

2.6 Standard challenges

Developing standards is vital to guarantee consistent and reproducible quality for construction products and construction works. As mentioned before, climatic and economic differences make it necessary for Africa to develop own standards or adapt those from other continents carefully and reasonably based on expert knowledge.Particularly the adjustment of a uniform temperature seemed to be a major challenge during the PACE-PTS scheme. The reported testing temperatures varied between 10 °C and 30 °C. A reasonable standard has to take this into account, and still provide uniform test specifications, tolerances, evaluations, calculation methods as well as implementation rules etc. to ensure a reliable system, provided it be monitored adequately.

3 Solution strategies

3.1 Way forward

The boundary framework for cement and concrete construction varies greatly from conditions in Europe. Existing solutions that were developed elsewhere cannot master the challenges of Africa in an ideal way and the full potentials in Africa cannot be tapped. To create the best available solutions, construction technologies and materials need to be developed in Africa based on short supply chains, tailored regulative frameworks and policies. While international experts can support processes, the final decisions have to be made by local experts.

3.2 Solutions for alternative materials

Today, it is undisputed that binder has to be reduced in OPC clinker in order to minimise the carbon emissions related to cement production [6]. One of the most promising ways to develop sustainable cement types is the use pozzolanic materials, or calcined clays. Limestone calcined clay cement have pointed out to be very eco-efficient and sustainable [1, 13-16], the resources for which are abundantly available all over Africa. Since a high number of cement plants will be required in the future all over the continent, the limestone calcined clay technology can directly be implemented with a reasonable market acceptance and an adequate regulative framework.

However, while clay materials will assumedly have an enormous global impact, alternative solutions might have a high potential locally. Africa has a huge and growing agricultural sector [17], which creates a large number of by products that are not adequately used to date [18]. For example, due to its high amorphous silica content, rice husk ashespoint out to be most excellent cementitious materials that can outperform OPC by far[18, 22, 24-28]. Further alternatives exists, such as ashes from cassava peels [19], sugar cane bagasse, groundnut shells, locust beans or bamboo leaves[20], which have shown to be suitable for both low-end and high-end performance applications [21-23]. Their use is extremely promising, considering that Africa provides60% of all globally yet unused arable land [17]. Particularly Africa's economic growth is closely linked to the agricultural sector. The development of plant based cementitious binders and concrete additions is therefore future oriented and sustainable with enormous growth potential.

Sustainable construction also requires to use cement in concrete more efficiently. Today, various petrol based chemical admixtures help to produce higher strength concrete with less cement [18, 19]. The use of polycarboxylate ether superplasticizers has become quite common in concrete technology due to their versatility [10, 30]. However, their uncomplicated and cost efficient availability is limited in most countries in Africa [22, 24]. They have to be shipped in from overseas, which creates additional global warming potential.

Locally easy accessible alternatives are available. Lignosulphonates are by-products of the cellulose industry and they can well enhance the concrete properties [9, 22]. Furthermore, Acacia gum from the Karoo and Arabic gum from Sudan can have positive influence on set retardation as well as flow performance[9, 31, 32]. These are just examples of many alternatives, which do not need complex processing to become ready to use. Furthermore, fibres can be used to enhance the ductility. For structural applications, typically steel fibres are used. Alternative fibres can be found again from bio-based sources such as sisal an coconut fibres [23]. They have to become more brittle due to ageing. However, their deterioration can be reduced by immersion in silica fume slurry [33, 34]. Since this may not be the most cost effective treatment, in the future, it might be interesting to observe whether a solution in silica from rice husk ash can have a similar effect.

Therefore, the African continent provides a high number of materials, which seem to be technologically promising and significantly more sustainable than established materials. However, their usage is typically not covered in existing standards. The materials and

 https://doi.org/10.1051/matecconf/201814901014

technologies have potentials to become globally similarly or more sustainable than existing technologies. Their rapid implementation can help to overcome some of the most pressing challenges in Africa, that are housing and infrastructure, while in parallel new local value chains can be created thus fostering the economic growth and the independence of international monetary flows.

Hence, the African markets need to develop rapidly quality surveillance infrastructures that form a reasonable compromise between support of the verve for innovation and provision of adequate safety.

3.3 Enhanced quality infrastructure

In order to improve the whole system of quality control, an enhancement of national and regional systems of quality infrastructure is necessary. Testing is part of the national quality infrastructure, which is understood as a complex network of metrology, standardisation, testing, certification, and accreditation. To guarantee the reliability of their results, testing laboratories must ensure that their devices are properly calibrated and thus the results are traceable to national and international measurement standards [35]. Furthermore, laboratories should use quality management systems and work according to defined procedures. Proficiency testing schemes are one efficient tool to compare installed measures.

The current state of national quality infrastructure systems in Africa is heterogenous. On the one hand, there is a positive development throughout Africa. Since 2014, improvements in 32 national systems could be registered. On the other hand, there is still room for improvement. Around half of African states have national quality infrastructure systems which are not in accordance to international requirements. [36]

While the need for qualified testing laboratories is obvious, the current state of quality infrastructure in most African countries does not necessarily support such laboratories. Consequently, it is difficult for them to calibrate their instruments in a cost and time efficient way. Furthermore, support for the introduction of quality management systems and accreditation is not always easy to get. Besides measures to enhance the national quality infrastructure, another solution might be the division of labour on a regional level. This might include e.g. regional reference laboratories, regional market surveillance schemes as well as regional capacity development.

3.4 Research capacity

The series of 25 tests, which were selected by the participants of the first PACE-PTS series in 2012/2013 [37] and amended by further investigations over the course of the years, were divided in"standard cement tests" and "advanced tests". The first group comprises of the mechanical, chemical and physical testing according to the standards, the latter group comprises of further mechanical tests,sand evaluation, andmore sophisticated rheological testson interactions between cement and superplasticizers following a procedure developed at BAM.

Table 1 presents the numbers and percentages of laboratories that conducted part of the tests or all tests. It can be found that many laboratories do not provide the adequate equipment to conduct all required tests. 20% of the laboratories registered for the PACE-PTS were from cement producers (Figure 4). These laboratories are typically well equipped to perform the required testing. However, about 80% of the participating laboratories were either university laboratories, governmental laboratories or private laboratories. Since these laboratories ensure the public safety, that lack of equipment for standardised tests is alarming. Considering that university and private laboratories are furthermore driving forces for innovation, it is getting obvious that investment in research capacity must include investment in up to date equipment.

A way to finance the relevant investments could be to create central excellence laboratories that cooperate with universities and private laboratories as well as more public private partnerships.

Table 1: Numbers and percentages of tests conducted by the African laboratories for the PACE-PTS.

	Type and number of tests		Laboratories that conducted tests		Laboratories that conducted all tests	
Standard cement tests	Mechanical tests	**6**	37	77%	28	58%
	Chemical tests	4	42	88%	23	48%
	Physical tests	**6**	46	96%	18	38%
Advanced tests	Mechanical tests	4	38	79%	20	42%
	Rheological tests	4	38	79%	25	52%
	Sand test	1	26	54%	26	54%

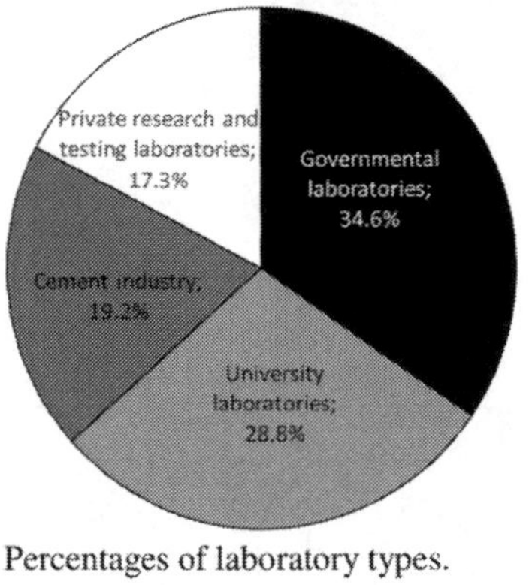

Figure 4: Percentages of laboratory types.

3.5 Improved framework for safe construction

Establishing a functioning system for controlling and inspecting construction products and the execution of construction works is costly and time intensive. However, such a system is inevitable for keeping the general public safe from damage caused by collapsing buildings, fire or hazardous materials and products. Each society determines which risks it is willing to take in this area, and how much money can be spent on control systems.

Europe uses different ways to ensure safety[38], depending on whether the construction products and construction works are standardised or new and innovative; whether they serve an important function within the building or are less relevant. These control and inspection systems can only serve as a rough model for Africa. They have to be adapted to the local circumstances. A network of laboratories is necessary as well as independent organisations (state-run and/or private) which check each other's results to ensure that only high-quality products enter the markets. Independence and mutual control are vital to prevent mistakes, fraud and corruption. This applies just as much to Europe as it does to Africa.

4 Summary and conclusions

The entire African continent provides huge potentials to develop a more sustainable cement and concrete construction market than in the rest of the world. However, the African continent also has its own challenges and peculiarities, which hinder timely tapping of the full potentials.

Africa provides a variety of local materials that can be used in cement and concrete construction that are comparable or better than existing materials used in other parts of the world. The short supply chains ensure local market growth combined with low environmental impact. Besides natural mineral resources, agricultural wastes provide manifold potentials regarding use as supplementary cementitious materials, chemical admixtures, and fibrous constituents.

For the implementation of these future oriented technologies, it is required to re-think the respective quality infrastructure and normative framework:

- The regulative framework and the quality infrastructure should have a strong focus on African demands.
- Concepts have to be innovation driven and target the African markets, yet ensuring public safety.
- In order to cope with the much higher variety of future cements and materials, research capacity and adequate equipment are inevitable.
- Central excellence laboratories can help keeping cost for equipment low, and public private partnerships can support sustainable quality infrastructures, when independence can be guaranteed.

Acknowledgements

The project is financed by the Federal Ministry for Economic Cooperation and Development

5 References

[1] K. Scrivener, "Realistic options to meet the challenge of increasing concrete demand in Africa (Presentation)," presented at the 2nd International Conference on Advances in Cement and Concrete Technology in Africa, Dar es Salaam, Tanzania, 2016.

[2] P. Wray and K. Scrivener, "Straight talk with Karen Scrivener on cements, CO2 and sustainable development," *American Ceramic Society Bulletin,* vol. 91, pp. 47-50, Jun-Jul 2012.

[3] L. Barcelo, J. Kline, G. Walenta, and E. Gartner, "Cement and carbon emissions," *Materials and Structures,* vol. 47, pp. 1055-1065, 2014.

[4] K. Carstensen and O. Rapf, "A blueprint for a climate friendly cement industry," WWF, Nürnberg, Germany2008.

[5] CEMBUREAU, "Cement & concrete: key facts & figures," t. E. c. association, Ed., ed, 2015.

[6] K. Scrivener, V. M. John, and E. M. Gartner, "Eco-effcient cements: Potential, economically viable solutions for a low-CO2, cement-based materials industry," United Nations Environment Programme2016.

[7] W. Schmidt, N. N. M. Hirya, D. Bjegovic, H. C. Uzoegbo, and S. G. Kumaran, "Cement technology in sub-Saharan Africa—Practical and scientific experiences," *American Ceramic Society Bulletin* vol. 91, p. 82, 2012.

[8] W. Schmidt, A. Radlińska, C. Nmai, A. Buregyeya, W. L. Lai, and S. Kou, "Why does Africa need African concrete? An observation of concrete in Europe, America, and Asia – and conclusions for Africa," in *International Conference on Advances in Cement and Concrete Technology in Africa*, Johannesburg, South Africa, 2013, pp. 1139-1148.

[9] W. Schmidt, "Why Africa can spearhead innovative and sustainable cement and concrete technologies globally," in *KEYS Knowledge Exchange for Young Scientists: Valorisation of Industrial By-products for Sustainable Cement and Concrete Construction – Improvement of Solid Waste Management*, N. S. Msinjili and W. Schmidt, Eds., ed Accra, Ghana: BAM, 2016, pp. 7-19.

[10] W. Schmidt, M. Sonebi, H. J. H. Brouwers, H.-C. Kühne, and B. Meng, "Rheology modifying admixtures: The key to innovation in concrete technology - a general overview and implications for Africa," *Chemistry and Materials Research,* vol. 5, pp. 115-120, 2013.

[11] W. Schmidt and G. Gluth, "Knowledge transfer and quality assurance - key factors for the future prospects of building with cement in Africa / Wissentransfer und Qualitätssicherung - Schlüsselfaktoren für die Zukunftsperspektiven der Bauweise mit Zement in Afrika," *ZKG International,* vol. 65, pp. 34-38, 2012.

[12] W. Schmidt, H. C. Uzoegbo, N. Bella, G. Rongai, H. C. Kühne, and T. Diergardt, "Cement Testing in Africa – Conclusions from the First Africa-Wide Proficiency Testing Scheme," *Civil and Environmental Research,* vol. 4, 2013.

[13] S. S. Berriel, A. Favier, E. R. Dominguez, I. R. S. Machado, U. Heierli, K. Scrivener, F. M. Hernandez, and G. Habert, "Assessing the environmental and economic potential of Limestone Calcined Clay Cement in Cuba," *Journal of Cleaner Production*, vol. 124, pp. 361-369, Jun 2016.

[14] F. Avet, R. Snellings, A. Alujas, M. Ben Haha, and K. Scrivener, "Investigation of the reactivity of various grades of kaolinitic calcined clays used as partial clinker substitute," in *Advances in Cement and Concrete Technology in Africa*, Dar es Salaam, Tanzania, 2016, pp. 137-142.

[15] L. M. Vizcaino-Andres, S. Sanchez-Berriel, S. Damas-Carrera, A. Perez-Hernandez, K. L. Scrivener, and J. F. Martirena-Hernandez, "Industrial trial to produce a low clinker, low carbon cement," *Materiales De Construccion*, vol. 65, 2015.

[16] A. Alujas, R. Fernandez, R. Quintana, K. L. Scrivener, and F. Martirena, "Pozzolanic reactivity of low grade kaolinitic clays: Influence of calcination temperature and impact of calcination products on OPC hydration," *Applied Clay Science*, vol. 108, pp. 94-101, May 2015.

[17] A. Perry, *The Rift: A New Africa Breaks Free*: Little, Brown and Company 2015.

[18] N. S. Msinjili, W. Schmidt, A. Rogge, and H. C. Kühne, "Optimising available resources for production of good concrete properties," in *Advances in Cement and Concrete Technology in Africa*, Dar es Salaam, Tanzania, 2016, pp. 323-332.

[19] K. A. Olonade, "Cassava Peel Ash: An Emerging Pozzolanic Material," in *1st Symposium on Knowledge Exchange for Young Scientists (KEYS) - Sub-Saharan African Standards for Cement and Concrete Research and Raw Materials, Quality Control and Maintenance of Cementitious Products*, Dar es Salaam, Tanzania, 2015, pp. 77-81.

[20] I. C. Mayowa, "Making Durable Concrete Through Inhibition of Chloride Ion Penetration by Pozzolanic Reaction," in *1st Symposium on Knowledge Exchange for Young Scientists (KEYS) - Sub-Saharan African Standards for Cement and Concrete Research and Raw Materials, Quality Control and Maintenance of Cementitious Products*, Dar es Salaam, Tanzania, 2015, pp. 145-150.

[21] T. T. Loan and P. Navard, "Treatments of plant biomass for cementitious building materials – A review," *Construction and Building Materials*, pp. 131-176, 2016.

[22] W. Schmidt, N. S. Msinjili, H. C. Uzoegbo, and J. K. Makunza, "Admixture Concepts for the Sub-Saharan African Environment with Indigenous Raw Materials," in *SP-302 Eleventh International Conference on Superplasticizers and Other Chemical Admixtures in Concrete*, V. M. Malhotra, Ed., ed, 2015, pp. 491-505.

[23] W. Schmidt, N. S. Msinjili, S. Pirskawetz, and H. C. Kühne, "Efficiency of high performance concrete types incorporating bio-materials like rice husk ashes, cassava starch, lignosulfonate, and sisal fibres," in *First International Conference on Bio-based Building Materials*, Clermont-Ferrand, France, 2015.

[24] W. Schmidt and N. S. Msinjili, "SCC: a solution to the challenges of concrete construction in Africa," *Concrete trends - Journal of the African cement and concrete industry*, pp. 16-20, 2014.

[25] J. L. Akasaki, A. R. B. Vasconcelos, J. Payá, D. R. O. Assi, M. V. Borrachero, M. M. Tashima, J. L. Melges, and C. F. Fioriti, "Durability Of High-Performance Concrete With Addition Of Rice Husk Ash And Rubber Tire," in *13th International Congress on the Chemistry of Cement*, Madrid, Spain, 2011.

[26] G. C. Cordeiro, R. D. Toledo Filho, and E. M. R. Fairbairn, "Influence of Rice Husk and Sugar Cane Bagasse Ultrafine Ashes on Hydration Characteristics of Cement Based Pastes," in *13th International Congress on the Chemistry of Cement*, Madrid, Spain, 2011.

[27] G. Rodríguez de Sensale, C. Romay, and F. Costa, "Waste Valorization From the Burning of Rice Husk," in *13th International Congress on the Chemistry of Cement*, Madrid, Spain, 2011.

[28] W. Wilson and A. Tagnit-Hamou, "Workability and Hydration of Superplasticized Cementitious Mixtures with Rice Husk Ash," *Aci Materials Journal*, vol. 111, pp. 491-500, Sep-Oct 2014.

[29] W. Schmidt, N. S. Msinjili, and H.-C. Kühne, "Rheological Optimisation for Flowable Mixture Compositions Specified for African Boundary Conditions," in *International Conference on Advances in Cement and Concrete Technology in Africa*, Johannesburg, S. Africa, 2013, pp. 359-366.

[30] W. Schmidt, H. J. H. Brouwers, H. C. Kuehne, and B. Meng, "Influence of Environmental Temperatures on the Performance of Polymeric Stabilising Agent in Fresh Cementitious Materials," *Polymers in Concrete*, vol. 466, pp. 97-104, 2011.

[31] R. Mbugua, R. Salim, and J. Ndambuki, "Effect of Gum Arabic karroo as a Water-Reducing Admixture in Concrete," *Materials*, vol. 9, Feb 2016.

[32] R. Mbugua, R. Salim, and J. Ndambuki, "Utilisation of Gum Acacia Karoo as Set-retarding Water-reducing Admixture in Cement Mortar and Concrete at Optimum Dosage," in *1st Symposium on Knowledge Exchange for Young Scientists (KEYS) - Sub-Saharan African Standards for Cement and Concrete Research and Raw Materials, Quality Control and Maintenance of Cementitious Products*, Dar es Salaam, Tanzania, 2015, pp. 97-101.

[33] R. D. Toledo, K. Ghavami, G. L. England, and K. Scrivener, "Development of vegetable fibre-mortar composites of improved durability,"

Cement & Concrete Composites, vol. 25, pp. 185-196, Feb 2003.

[34] R. D. Toledo, K. Scrivener, G. L. England, and K. Ghavami, "Durability of alkali-sensitive sisal and coconut fibres in cement mortar composites," *Cement & Concrete Composites,* vol. 22, pp. 127-143, Apr 2000.

[35] C. Sanetra and R. M. Marbán, "The answer to the global quality challenge: a national quality infrastructure.," 2007.

[36] PTB, "PAQI, 2017: Pan-African Quality Infrastructure. Stocktaking Document 2017 (in preparation)," 2017.

[37] W. Schmidt, "Start of round-robin test for cement testing in Africa / Start für Ringversuch zur Zementprüfung in Afrika," *BFT international : concrete plant + precast technology,* vol. 08, pp. 54-55, 2012.

[38] G. Breitschaft, "Safe Construction Works Made of Reliable Construction Products," presented at the 2nd International Conference on Advances in Cement and Concrete Technology in Africa, Dar es Salaam, Tanzania, 2016.

Non-Destructive Testing for Building Diagnostics and Monitoring: Experience Achieved with Case Studies

Ayşe Tavukçuoğlu
Middle East Technical University, Faculty of Architecture, Department of Architecture, Ankara, Turkey

Abstract. Building inspection on site, in other words *in-situ* examinations of buildings is a troublesome work that necessitates the use of non-destructive investigation (NDT) techniques. One of the main concerns of non-destructive testing studies is to improve *in-situ* use of NDT techniques for diagnostic and monitoring studies. The quantitative infrared thermography (QIRT) and ultrasonic pulse velocity (UPV) measurements have distinct importance in that regard. The joint use of QIRT and ultrasonic testing allows *in-situ* evaluation and monitoring of historical structures and contemporary ones in relation to moisture, thermal, materials and structural failures while the buildings themselves remain intact. For instances, those methods are useful for detection of visible and invisible cracks, thermal bridges and damp zones in building materials, components and functional systems as well as for soundness assessment of materials and thermal performance assessment of building components. In addition, those methods are promising for moisture content analyses in materials and monitoring the success of conservation treatments or interventions in structures. The in-situ NDT studies for diagnostic purposes should start with the mapping of decay forms and scanning of building surfaces with infrared images. Quantitative analyses are shaped for data acquisition on site and at laboratory from representative sound and problem areas in structures or laboratory samples. Laboratory analyses are needed to support in-situ examinations and to establish the reference data for better interpretation of in situ data. Advances in laboratory tests using IRT and ultrasonic testing are guiding for in-situ materials investigations based on measurable parameters. The knowledge and experience on QIRT and ultrasonic testing are promising for the innovative studies on today's materials technologies, building science and conservation/maintenance practices. Such studies demand a multi-disciplinary approach that leads to bring together knowledge on materials science and building science.

1 Introduction

Building inspection on site, in other words *in-situ* examinations of buildings is a troublesome work that necessitates the use of non-destructive investigation (NDT) techniques. The quantitative infrared thermography (QIRT) and ultrasonic pulse velocity (UPV) measurements have distinct importance in that regard. One of the main concerns of those non-destructive testing methods is to improve their quantitative use on site for diagnostic and monitoring purposes. The knowledge achieved on QIRT and ultrasonic testing is presented here, mostly with a focus on case studies conducted on historical materials and structures. Those case studies are, in fact, research studies shaped to develop in-situ use of QIRT and ultrasonic testing methods for specific topics related to building inspection and monitoring as well as to discover their potentials and limitations in this regard.

2 Infrared thermography and ultrasonic testing in the context of building diagnostics and monitoring

In-situ NDT studies for diagnostic purposes start with field observations composed of mapping of decay forms with visual analyses and IR scanning. Quantitative analyses are shaped for data acquisition on site and at laboratory taken from representative sound and problem areas in structures or laboratory samples. Laboratory analyses are needed to support in-situ examinations and to establish the standard/reference data for better interpretation of the in-situ data.

In order to enhance the accuracy of non-destructive investigations, particularly the in-situ ones, there is necessity of using more than one NDT testing method and supporting the in-situ investigation with laboratory tests. Advances in laboratory tests using IRT and ultrasonic testing are promising to give the hints of using those methods on site for quantitative analyses, in other words, allow making in-situ examinations based on measurable parameters.

Infrared (IR) thermography is commonly used for detection of building defects, such as thermal bridges, air leakages or moist spots, particularly in the context of energy conservation. It measures thermal radiation emitted by the material and depicts the examined area as an image in colours corresponding to a predefined temperature scale. The application of a hot or cold source to a specific area results in the warming up or cooling down of the surface area under examination at varying rates. Final surface temperatures differ depending on the thermal properties, specifically thermal resistance and thermal inertia characteristics, of the surface and sub-surface layers. For instance, entrapped moisture in a porous material increases its thermal conductivity, therefore decreases its thermal resistance and creates a kind of thermal bridge. Such a defect is visible in infrared images as cooler areas. In addition, exposure of wet surfaces to mild wind or sun increases the evaporation rate, therefore accelerates evaporative cooling. It means that heating the surfaces or wind effect may also enhance the visibility of moist areas in infrared images.

The IR scanning is useful and time-saving for the in-situ check-up of the overall structure, especially with a focus on visible and invisible materials defects, moisture and thermal failures, and various materials use. However, during the in-situ examinations the quantitative analyses of the representative areas are obligatory for the correct interpretation of the problem areas detected in single IR images. Thermal monitoring of the problem area by means of sequential thermal imaging is a favourable technique for quantitative analysis of the problem area. This technique allows taking infrared images in sequences from the target area during the periods when the target area is exposed to heating and then cooling conditions as well as producing differential thermal images which show surface temperature differences between the initial and the last IR images for heating or cooling period. Due to the changes in physical and thermal properties of the defect area, the impact of any failure can easily be followed in differential images. In addition, the temperature evolution in time under heating and/or cooling exposure conditions can be examined by the graphs showing changes in surface temperature as a function of square root of time. The slope of the linear regression presents the rate of warming up (R_W) or the rate of cooling down (R_C) for each target area. Those rates are the measurable parameters related with the thermal inertia characteristics of materials. Thermal inertia characteristics of the problem area can also be interpreted relatively by comparing the warming up/cooling rates of problem area in reference to the rates of sound material (reference area).

In short, that investigation technique is sensitive to thermal characteristics of materials. Since thermal characteristics are very related with the physical, physicomechanical and/or mechanical properties of materials, changes in those properties can be monitored precisely by quantitative IR thermography.

Ultrasonic testing of building materials is based on measuring the travel time (transit time) of ultrasonic waves passing along a solid material through a certain distance between the transmitter and receiver. The measurable parameter of ultrasonic testing is the velocity of ultrasound waves propagating through the building material. The ultrasonic pulse velocity (UPV) of a material is related with its physical and mechanical properties as well as the state of deterioration, moisture/water content and presence of discontinuity. Any discontinuity within a material or increase in porosity is expected to increase the travel time and, consequently, decrease the pulse velocity. The UPV measurements can be made by positioning two transducers (one transmitter and one receiver) on opposite faces (direct transmission mode) or on the same surface (indirect transmission mode) of the material sample.

The UPV measurements taken in direct transmission mode (UPV_{DIRECT}) are used for assessing the state of deterioration of a material and non-visible failure/discontinuity at deeper layers, such as invisible cracks. The UPV measurements in indirect transmission mode ($UPV_{INDIRECT}$) provides precise data on estimating the depth of a visible crack while investigating surfaces in layers only accessible from one side. Difficulty in access to mutually-perpendicular faces of a building component limits the applicability of direct UPV measurements on site. In such a case, there is necessity to establish the reference data obtained from control samples and correlate the UPV data taken in direct and indirect transmission modes. That correlation makes it possible to interpret the in-situ UPV data correctly in reference to the control UPV data.

The combined use of infrared thermography and ultrasonic testing enhances the accuracy of the non-destructive in-situ examination, especially the studies on soundness assessment. A good correlation is determined between the state of deterioration of stone and its thermal inertia characteristics. The deteriorated stone samples present lower UPV values and faster warming up and cooling down rates than the sound ones.

Some research fields where QIRT and ultrasonic testing can be useful are summarized as follows:–

- detection of visible and invisible defects/failures in materials and structures, such as deep and surface cracks, detachments [1, 2];
- detection of different materials use hidden behind the surfaces or buried within the section of building components [3];
- assessment of the state-of-deterioration of building materials and their distribution in the structure [2, 4 5];
- failures in functional systems of historic structures, such as water supply and drainage systems [6-9];
- assessment of thermal performance of structures as well as thermal and moisture failures in structures [3, 10, 11];
- in-situ monitoring of the existing conditions before and after treatments, success of conservation treatments in historic structures [2, 12, 13].

The advances in laboratory tests using QIRT are also promising to determine the thermal properties of materials and building walls while give the hints for in-situ QIRT investigations to assess moisture content in materials [3, 11, 14, 15].

3 Some case studies: prominent findings and advancements

Here, some case studies are summarized under respective subheadings with a focus on the prominent results of those studies and guiding remarks related with non-destructive building inspection.

3.1 Soundness assessment of structural timber elements on quantitative basis

A comprehensive study is needed for the in-situ soundness assessment of structural timber elements that involves:

- *preliminary laboratory tests* on control samples (deteriorated and non-deteriorated timber samples) to produce the reference data on UPV and thermal characteristics,
- *field measurements* taken from representative problem areas, and
- *evaluation of in-situ data in reference to the control data*.

Such a study was conducted on two types of structures [16-18]:

- The 19th century traditional timber dwellings in Ankara belonging to Ottoman Period, namely Ayaş House in the town of Ayaş and Istiklal House in Istiklal district: The timber post and beams forming the timber frame wall of traditional dwellings were examined by in-situ UPV measurements taken in direct and indirect transmission modes and by sequential IR imaging during cooling conditions.
- Aslanhane Camii, 13th century mosque, belonging to Seljuks Period: The timber pillars supporting the roof and timber ceiling of the structure were examined by in-situ UPV measurements taken in direct transmission mode and IR scanning.

Reference data established by the laboratory analyses on control timber samples were used for the precise interpretation of in-situ data.

3.1.1 The reference data achieved at laboratory

The old and new pine samples, visually-deteriorated and sound ones, collected from the structures were examined in terms of density, equilibrium moisture content at various RH conditions, UPV taken in direct and indirect transmission modes and warming up and cooling rates.

The UPV values of timber samples decrease at high moisture conditions [17]. This means that physicomechanical properties of timber elements weaken when they get wet (figure 1). The relationship between the UPV values and relative humidity conditions show that microclimatic data is an important input for correct interpretation of the UPV data, particularly for the in-situ ultrasonic investigations.

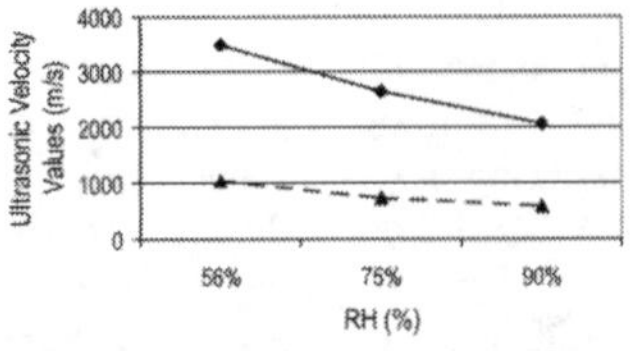

Fig. 1. Changes in ultrasonic velocity values of timber samples due to the changes in moisture content [17].

The UPV measurements exhibit the anisotropic behaviour of timber in relation to fiber direction [16-18]. The UPV_{DIRECT} values parallel to fiber direction are considerably higher than (up to 3 to 3.5 times higher) the ones perpendicular to fiber direction. The $UPV_{INDIRECT}$ values are much lower than the UPV_{DIRECT} values with a ratio of 0.32 in average for the measurements taken in parallel to fiber direction. Based on these relationships, the $UPV_{INDIRECT}$ values measured on site from timber surfaces in parallel to fiber direction can be used to predict the UPV_{DIRECT} values, therefore, the soundness of timber.

In literature, the UPV_{DIRECT} values for sound timber are given in the range of 1000 to 2000 m/s [19, 20]. The ultrasonic data for the sound timber (control) samples measured in perpendicular-to-fiber direction correspond with the data in literature with the UPV_{DIRECT} value of 1040 m/s in average at 56% RH (figure 1) [17,18]. The same control samples have the $UPV_{INDIRECT}$ value of 1439 m/s in average, measured in parallel to fiber direction [16].

When the timber deteriorates, its density decreases and its warming up/cooling rates increases. The sound pine samples (control samples) have the lowest warming up/cooling rates and the highest density, indicating their higher thermal inertia. Depending on severeness of deterioration, the visually-deteriorated pine samples warm up or cool down faster than the sound one (control sample) in the range of 1.2 and 1.7. Measuring the changes in warming up or cooling down rates of timber samples allows assessing their state of deterioration in comparison to sound (reference) samples.

3.1.2 Evaluation of in-situ data with respect to reference data

Some timber beams and posts in timber framed houses had $UPV_{INDIRECT}$ values in the range of 1337 to 1176 m/s taken in parallel to fiber direction [16]. Evaluation of in-situ UPV data with respect to the reference data has shown that those timber elements are still sound. That result was also confirmed with in-situ UPV_{DIRECT} measurements taken in perpendicular to fiber direction wherever cross arrangements for the transducers could be made. Those UPV_{DIRECT} measurements in the range of 1400-1660 m/s are in agreement with values given for the sound timber material in literature [19, 20]. Partial deteriorations on some parts of timber posts and beams could be detected with $UPV_{INDIRECT}$ values below the acceptable range. In short, in-situ $UPV_{INDIRECT}$ and UPV_{DIRECT} measurements have assured the soundness of structural timber elements in some parts and extensive deterioration in other parts of the structures under examination.

The sound timber elements present even surface temperature distribution in differential IR images while the deteriorated ones show heterogeneous temperature distribution (figure 2). In addition, the cooling rates of deteriorated and severely deteriorated timber post/beam were found to be 1.55 and 2.48 times faster than the cooling rate of the sound post/beam, respectively [16]. This meant that thermal inertia of deteriorated timber decreases due to decrease in density and increase in porosity.

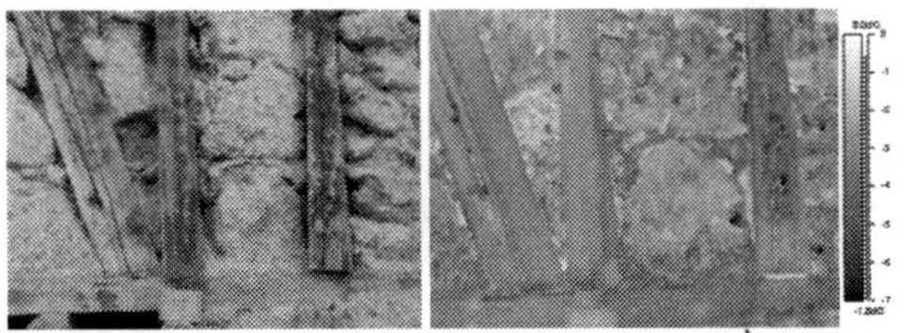

Fig. 2. Partial view from the timber frame wall of İstiklal House (at the left); the differential IR image showing the temperature difference between the initial and the last IR images during the cooling period of 603 seconds [16].

The UPV_{DIRECT} values taken in perpendicular to fiber direction from the timber pillars in Aslanhane Camii were measured in the range of 706 to 1753 m/s [17, 18]. Those data falling into the reference UPV_{DIRECT} values obtained for sound timber show that the timber pillars of the mosque are still sound.

3.2 Crack depth assessment in stone masonry

An in-situ examination on structural cracks with a focus on depth assessment was conducted on a 16th century Ottoman mosque in Ankara, Cenabi Ahmet Paşa Camii [1]. This stone masonry structure with a brick upper structure suffers from serious cracks observed at its walls and dome due to the differential settlement of the clayey ground following the extremely dry seasons of recent years [21, 22].

By the study, a non-destructive investigation method was developed for the estimation of crack depth in a structure by the combined use of QIRT and ultrasonic testing [1]. The depths of cracks in accessible areas were predicted by quantitative analyses of $UPV_{INDIRECT}$ data taken parallel to the stone surfaces (figure 3). The thermal inertia characteristics of those cracks with known depths were defined by quantitative analyses of surface temperature data. The cracks which are not accessible for ultrasonic testing were then able to be monitored by thermographic analyses to identify whether they are deep or superficial cracks.

The superficial and deep cracks have different thermal responses to exposed conditions which made them easily distinguishable by QIRT analyses (figure 4). For instances:-

- The superficial cracks associating with/neighbouring plaster detachments had thermal response similar to detached surfaces, having faster warming up and cooling down rates than the sound stone surfaces.
- The deeper cracks at masonry wall, on the other hand, presented noticeably slower warming up and cooling down rates than the sound stone surfaces.
- The deepest cracks allowing air leakage through the wall section had the coldest initial surface temperature and slightly cooled down during the heating period due to the accelerated evaporative cooling in the crack cavity.

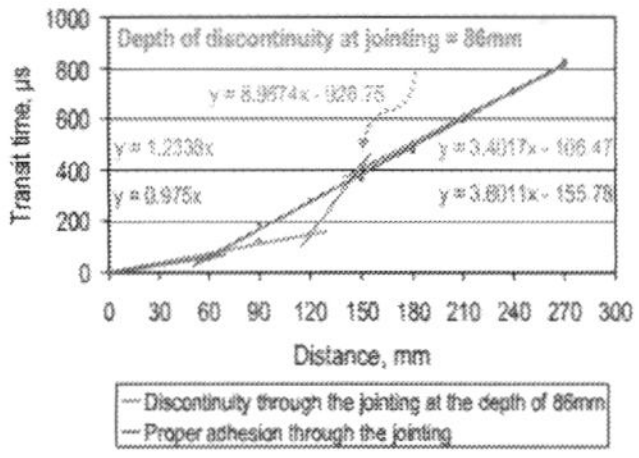

Fig. 3. The slope of regression lines for the tuff stone followed by proper jointing without any discontinuity (lines in green) and the change in the regression slope corresponding to the depth of discontinuity at jointing (lines in orange) [1].

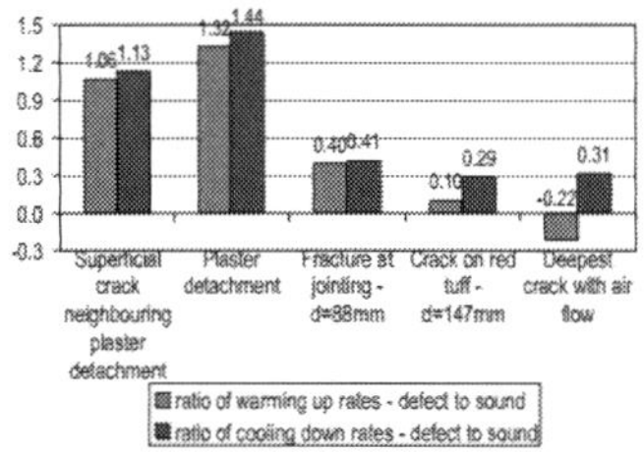

Fig. 4. The ratios of the R_W (or R_C) of crack/defect to the R_W (or R_C) of sound/reference stone surface [1].

3.3 Identification of emergency areas that needed conservation treatments and their monitoring

A comprehensive research involved QIRT and ultrasonic testing supported by the maps of visual decay forms and laboratory tests was conducted on Nemrut Dağ Monument, which is an archaeological site located in eastern Turkey in the province of Kahta- Adıyaman and positioned at the top of Nemrut Dağ at 2150m altitude. The site is in the list of World Cultural Heritage since 1987. The study was focused on the in-situ examination of the limestone and sandstone statues in Nemrut Dağ Monument to assess their state of deterioration and to determine the target areas that needed urgent conservation treatments [2, 4].

The reference data on UPV and thermal inertia characteristics of sound and deteriorated limestone and sandstone samples were produced (figure 5). The relationship between the state of deterioration of stone surfaces and their thermal inertia characteristics was then used for the assessment of problem areas on statues surfaces. The deteriorated stone samples present lower UPV values and faster warming up (R_W) and cooling down (R_C) rates than the sound ones. In other words, the weakening in physicomechanical properties causes the thermal inertia of the deteriorated stone samples to decrease.

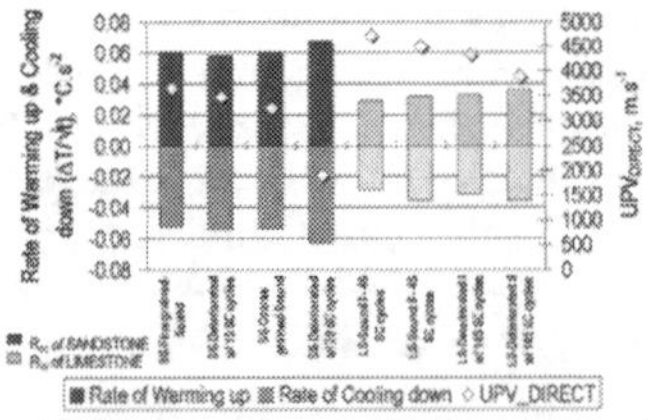

Fig. 5. The UPV_{DIRECT}, R_W and R_C values of sound and artificially-weathered sandstone and limestone samples showing that the deteriorated samples had lower UPV_{DIRECT} and faster R_W values than the sound ones [4].

The decay forms that required urgent conservation treatments are the material loss and detachment areas by fragmental disintegration in limestones and the scales detaching from the sandstones. The fragments and scales detaching from the rock body could easily be detected in differential IR images due to their fast warming up or cooling down behaviour under exposed conditions (figure 6). The soundness of the fragments and scales were assessed on site by $UPV_{INDIRECT}$ measurements taken from stone surfaces. Most stone fragments/scales separating from the rock body were determined to be still sound (figure 7). The deep and superficial crack patterns on limestone surfaces could be distinguished in differential IR images (figure 8). The fragments to be detached from the limestone body were mostly next to the deeper cracks belonging to a crack network. The deep and superficial cracks could not be well-distinguished in individual IR images (figure 8 - at the right middle) while the deeper cracks became visible as the warmer traces in differential IR image when sufficient heating condition was achieved on site (figure 8 - at the right bottom).

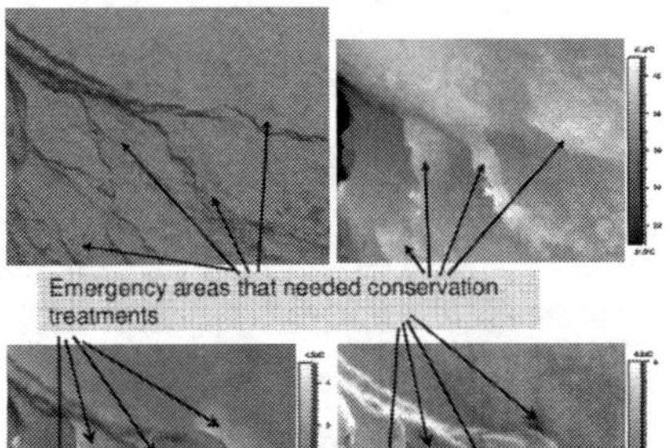

Fig. 6. The differential IR images processed for the warming up and cooling periods showing that the detaching scales on the sandstone stele warm up and cool down faster than the non-detached surfaces.

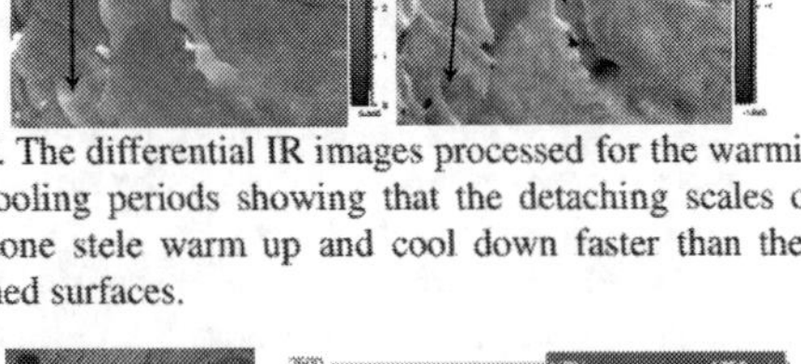

Fig. 7. Partial view showing the locations where $UPV_{INDIRECT}$ measurements were taken (*at the left*) and the chart showing the $UPV_{INDIRECT}$ values for the non-detached sound surfaces and detaching stone fragments.

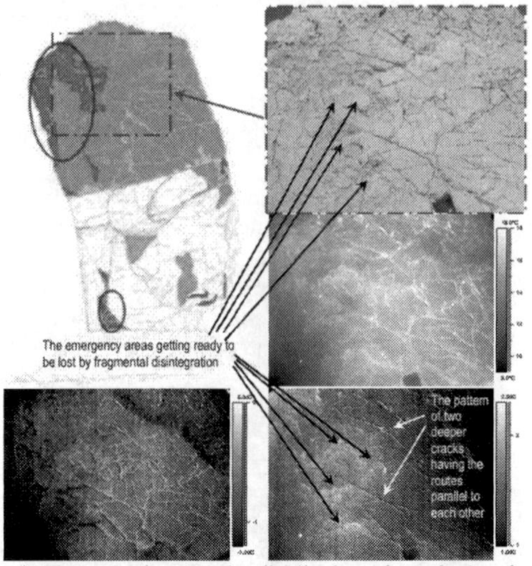

Fig. 8. The map showing the distribution of weathering forms (*at the left top*), the digital photo of the area under QIRT examination (*at the right top*), and the IR image of the selected area at the climatic conditions of 8.6°C and 63% (*at the right middle*), its differential images processed for the cooling period of 191s (at the left bottom) and heating period of 1426s (*at the right bottom*).

The use of QIRT and ultrasonic testing on site complemented each other to point out the emergency areas that needed conservation treatments. Those emergency areas had priority for the conservation treatments, such as for the deep cracks to be treated against water penetration, the prevention of losses due to scales or fragmental disintegration and control of decay/detaching mechanisms in the rock body.

The in-situ data obtained from the emergency areas of limestone and sandstone statues/steles before treatment have vital importance for monitoring the progress in those emergency areas after treatment in terms of $UPV_{INDIRECT}$ values and warming up/cooling down rates..

3.4 Determination of critical moisture content in porous materials – Recommendation for a standard laboratory test

Quantitative IR Thermography (QIRT) can be used for the determination of critical moisture content (Θ_c) level for porous materials [14, 15]. The weathering conditions, such as the cycles of freezing-thawing, wetting-drying and salt crystallization, are more damaging for wet porous materials when the moisture content is above the critical moisture content (Θ_c) [23, 24]. For in-situ diagnostic studies, it is therefore, essential to determine whether moisture content in damp zones of a structure is above or far below the Θ_c.

A procedure based on QIRT was developed for precisely-analysing the transition stage between the saturated and dry phases of a porous material in order to determine its effective Θ_c, as a physical property [14]. The Θ_c level for some building materials, such as repair brick and tuffs, were examined in laboratory by QIRT and the Standard Gravimetric Method (SGM).

During the process of moisture desorption at certain boundary conditions, the weight loss of water-saturated brick and tuff samples together with the thermographic view of their evaporating surface were recorded with 2s intervals. The data obtained were then analysed in terms of evaporation rate (R_E) and a new informative parameter called Evaporative Thermal Index (ETI) (figure 9) [14]. The standard test of evaporation rate is based on the sample weight loss change in time during the drying process, knowing the evaporating area [25]. ETI is related to the surface temperature and defined as the normalized temperature difference between actual (evaporating) and reference (dry) surface [14]. Then, R_E and ETI values were computed as a function of the moisture content in the sample. Both R_E and ETI curves as a function of Θ were produced for their joint analysis in terms of knickpoint at time t_c (time critical) when Θ_c has been reached (figure 9). A good correlation was achieved between the values of R_E and ETI.

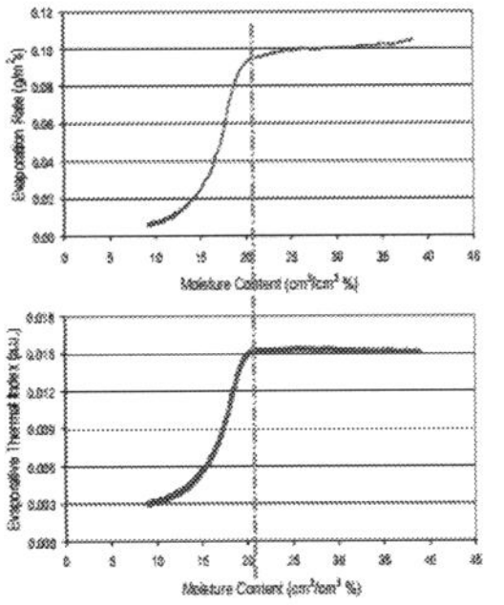

Fig. 9. A good correlation was achieved between the R_E and *ETI* values: The knick-points indicating the θ_c for the materials are the same for both Standard Gravimetric Method and Quantitative IR Thermography.

The study exhibited that the monitoring of a wet surface during the drying process for a certain period of time by means of IR sequences is useful to decide on whether the moisture content in the material is above or below Θ_c. In addition, the results obtained in the laboratory are promising to develop in-situ methods for quantitative determination of moisture content, only by using temperature measurements.

3.5 Assessment of thermal transmittance value of existing building walls

In today's constructions, the emphasis is given to the design of energy efficient buildings and improvement of existing buildings in terms of their energy efficiency. The recent studies of QIRT focus on development of standard laboratory and in-situ tests:

- to determine thermal insulation characteristics of building walls
- to identify thermal failure types,

in terms of measurable thermal parameters [3, 11].

The joint use of calibrated hot box method and IRT at laboratory conditions allowed assessment of thermal transmittance value (U, W/m^2K) of the perforated brick wall samples [11]. The experimental setup and the results achieved also gave the hints of in-situ testing methods in order to measure the real thermal transmittance value of existing building walls by thermal monitoring of wall surfaces at the cold side while inside is heated.

Considering the knowledge above, another QIRT study was carried out with a focus on in-situ thermal performance assessment of some commonly-used wall sections in terms of thermal transmittance value [3]. The infrared images were taken during winter when inside was warm while outside was cold in steady state condition. The actual thermal transmittance value (U_{INSITU}, W/m^2K) of existing walls were determined by using the in-situ surface temperature and microclimatic data together with relevant equations defined in the literature [26, 27]. Therefore, the QIRT methods are promising for the studies on *thermal performance certification of building walls* and *energy efficiency rating of buildings*. The common use of infrared thermography in building diagnostics, therefore, need to be encouraged.

4 Conclusion

The advances in NDT techniques are promising for the in-situ examinations and monitoring of historic and contemporary structures on quantitative basis. The combined use of infrared thermography and ultrasonic testing has potential for detecting the problem areas on building surfaces and monitoring the changes in thermophysical and physicomechanical properties of problem areas before and after treatments/interventions.

Ultrasonic testing, particularly the combined use of UPV measurements taken in direct and indirect transmission modes, is useful for predicting the depth of cracks at stone masonry and state of soundness/deterioration of building materials. This method can also be used for detecting non-visible failure(s) at deeper layers. However, UPV measurements on site is really demanding since data acquisition is restricted to the levels available for human body and requires long working hours at site due to the repetition of UPV measurements for several times to achieve reliable data. Therefore, its use is more reasonable for representative sampling instead of its extensive use in the structure. The QIRT survey, on the other hand, is easier and more practical method to achieve reliable and precise data in a short period of time in case that the method of data acquisition on site has been carefully-thought and well-organized beforehand. It requires long working hours at office for the quantitative analyses of the surface temperature data. Here, the selection of sensitive thermographic equipment with high image quality allows its extensive use in the structure, especially for imaging the unattainable regions.

The knowledge on the combined use of QIRT and ultrasonic testing are promising for innovative studies on today's materials technologies, building science and conservation/maintenance practices. Such innovative studies involving non-destructive investigations demand a multi-disciplinary approach that leads to bring together knowledge on materials science and building science.

References

1. A. Tavukçuoğlu, S. Akevren, E. Grinzato, In-situ examination of structural cracks at historic masonry structures by quantitative infrared thermography and ultrasonic testing. *Journal of Modern Optics, 57*(18), 1779-1789 (2010).
2. A. Tavukçuoğlu, E.N. Caner-Saltık, K.G. Akoğlu, M. Isıkoğlu, E. Caner E. In-Situ Examination of Nemrut Dağ Sandstone and Limestone Statues by NDT Methods. In: Proceedings of International Conference on "Jardins De Pierrres – Conservation of stone in Parks, Gardens and Cemeteries", Paris, France, 22-24 June 2011 (pp. 204-213). France: SFIIC – Institut National du Patrimoine (2011).
3. M. Sayın, A. Tavukçuoğlu. Quantitative assessment of thermal transmittance in building walls by in-situ infrared thermography. In A. Tavil and O.C. Celik (Eds.): *Interdisciplinary Perspectives for Future Building Envelopes - ICBEST 2017 International Conference on Building Envelope Systems and Technologies*, 15-18 May 2017, Istanbul, Turkey (pp. 202-215). İstanbul Technical University (2017).
4. A. Tavukcuoglu, E.N. Caner-Saltik, Quantitative Infrared Thermography and Ultrasonic Testing for InSitu Assessment of Nemrut Dag Stone Statues, *Atti della Fondazione Giorgio Ronchi, Vol: LXIX No :4* (459-472). Fondazione Giorgio Ronchi (2014).
5. A. Tavukçuoğlu, E.N. Caner-Saltık, F. Erol, Ö. Çetin, F. Karahan-Dağ, Non-destructive Monitoring of Salt Weathering in Dolostone. In A. Tavil and O.C. Celik (Eds.): *Interdisciplinary Perspectives for Future Building Envelopes - ICBEST 2017 International Conference on Building Envelope Systems and Technologies*, 15-18 May 2017, Istanbul, Turkey (pp. 242-256). İstanbul Technical University (2017).
6. A. Tavukcuoğlu, A. Düzgüneş, Ş. Demirci, Ş., E.N Caner-Saltık, The assessment of a roof drainage system for an historical building. *Building and Environment, 42,* 2699–2709 (2007).
7. A. Tavukcuoğlu, A. Düzgüneş, E.N. Caner-Saltık, Ş. Demirci, Use of IR thermography for the assessment of surface water drainage problems in an historical building, Ağzıkarahan (Aksaray), Turkey. *NDT & E International, 38*, 402–410 (2005).
8. G. Disli, A. Tavukcuoglu, L. Tosun, E. Grinzato, The assessment of roof drainage system of a historical Turkish bath: Sengul Hammam. In: *11 DBMC International conference on Durability of Building Materials and Components*, 11-14 May 2008, İstanbul, Turkey, Vol 3. (pp. 1287-1297). İstanbul, Turkey: İstanbul Technical University Press (2008).
9. G. Disli, A. Tavukcuoglu, L. Tosun, An investigation on water consumption and waste water discharge systems for historical Turkish baths. In: *8th International HVaC-R Technology Symposium*, 12-14 May, 2008, Istanbul, Turkey (pp. 556-565). Turkish Society of HVAC & Sanitary Engineers (2008).
10. A. Tavukcuoglu, A., P. Cicek, E. Grinzato, Thermal analysis of an historical Turkish bath by quantitaive IR thermography. *QIRT (Quantitative InfraRed Thermography) Journal*, 5(2), 151-173 (2008).
11. R. Tuğla, A. Tavukçuoğlu, M. Arslan, Examination of thermal properties and failures of brick walls by the use of infrared thermograpy and hot box method; In: *RENEWBUILD 2013 –1st International conference & exhibition on "Application of efficient & renewable energy technologies in low cost buildings and construction*, Ankara, Turkey, 16-18 September 2013 (pp. 180-199). Ankara, Turkey: Gazi University (2013).
12. F. Karahan Dağ, Ç.T. Mısır, S. Çömez, M. Erdil, A. Tavukçuoğlu, E. N. Caner-Saltık, B. A. Güney, E. Caner, Assessment of Dolomite Conservation by Treatment with Nano-Dispersive Calcium Hydroxide Solution. In J.J. Hughes, T. Howind (Eds.): Science and Art: A Future for Stone: *Proceedings of the 13th Int. Congress on the Deterioration and Conservation of Stone, Vol. 2*, 6 - 10 September 2016, Paisley, Scotland (pp. 839-846). University of the West of Scotland, Paisley, Scotland (2016).
13. A. Tavukçuoğlu, E. Caner-Saltık, A. Güney, Ö. Çetin, F. Erol, F., F. Karahan Dağ, F., Assessment of Nanostructured Inorganic Consolidants Used in Stone Conservation Practice. In: *XIV DBMC – 14th International Conference on Durability of Building Materials and Components,* 29-31 May 2017, Ghent, Belgium (pp. 85:1-85:12). Ghent University (2017).
14. A. Tavukcuoglu, E. Grinzato, Determination of critical moisture content in porous materials by IR thermography. *Quantitative InfraRed Thermography Journal, 3(2),* 231-245 (2006).
15. A. Tavukcuoglu, E. Grinzato, Determination of critical moisture content level for porous materials by quantitative IR thermography. In, J.W. Lukaszewicz, P. Niemcewicz (*Eds.): Proceedings of 11th International Congress on Deterioration and Conservation of Stone*, September 15-20, 2008, Torun, Poland. Vol.1. (pp. 529-538). Poland. Nicolaus Copernicus University Press (2008).
16. A. Kandemir, Soundness assessment of historic structural timber by the use of non-destructive methods, *Unpublished Ph.D dissertation,* Graduate Program in Restoration, Department of Architecture, Middle East Technical University, Ankara, Turkey (2010).
17. A. Kandemir Yucel, A. Tavukcuoglu, E. N. Saltik, In situ assessment of structural timber elements of a historic building by infrared thermography and ultrasonic velocity. *Infrared Physics & Technology, 49,* 243–248 (2007).
18. A. Kandemir Yucel, A. Tavukcuoglu, E. N. Saltik, Non-destructive analyses of Historic timber elements to assess their state of preservation. In F. W. Broker (Ed.), *Proceedings of the 14th International Symposium on Nondestructive Testing of Wood,* Germany, May 2-4, 2005 (135-144).Germany: Shaker Verlag (2005).
19. F.C. Beall, Overview of the use of ultrasonic technologies in research on wood properties, Wood Science and Technology, 36 197-212 (2002).

20. D.E. Bray, D. McBride, Nondestructive Testing Techniques, Canada, John Wiley & Sons, Inc. (1992).
21. K.O. Cetin, E. Canbay, Cenabi Ahmet Pasa Camii, Temel Zeminleri Geoteknik Değerlendirme Raporu. *Unpublished research report*, submitted to General Pious Foundation. ODTU Doner Sermaye Protokol No: 2008.03.03.1.00.17. March 2008.
22. E. Canbay, Tetkik Raporu - Cenabi Ahmet Paşa Camii Değerlendirme Raporu. *Unpublished research report*, submitted to General Pious Foundation. April 2008.
23. H. Garrecht. Corrosion of Building Materials Caused by Microclimatic and Weathering Attack. In C. Sjöström (Ed.), *Durability of Building Materials and Components 7*, Vol.1. pp.150-159. London: E. & FN Spon (1996)
24. G. Massari, M. Massari *Damp buildings, old and new*. Rome: ICCROM (1993).
25. RILEM, "Tentative Recommendations, Commission – 25 – PEM, Recommended tests to measure the deterioration of stone and to assess the effectiveness of treatment methods", *Materiaux and Construction, 13,* pp.173-253 (1980).
26. R. Albatici, A. M. Tonelli, Infrared thermovision technique for the assessment of thermal transmittance value of opaque building elements on site. Energy and Buildings, 42(11), 2177-2183. (2010).
27. R. Albatici, R., A. M. Tonelli, M. Chiogna. A comprehensive experimental approach for the validation of quantitative infrared thermography in the evaluation of building thermal transmittance. Applied Energy, 141, 218-228 (2015).

Durable protection of the surface of wood used outdoors: material constraints, problems and approaches to solutions

A. MERLIN[1], B. GEORGE [1], L. MALASSENET[2] and L. PODGORSKI[2]

[1] Université de Lorraine – LERMAB : Laboratoire d'Etudes et de Recherche sur la Matériau Bois EA 4370 USC INRA, Vandoeuvre les Nancy, France
[2] InstitutTechnologique FCBA: Forêt Cellulose Bois-construction Ameublement, Bordeaux, France

Abstract. The aesthetic durability of wooden structures is a major challenge for the use of this material in construction. Wood is used for its technical performances but also for its architectural qualities and its aesthetic perception. The premature aging of the wooden structures is detrimental because these disorders, even if they do not affect the strength of the structures, are mostly irremediable. The surface protection of wood is generally ensured by the use of a finish, whose essential role is to protect wood from climatic aggressions (water, solar radiation, oxygen, ...). The secondary wood processing industry consists of a series of manufacturing and processing activities, each containing a portion of the added value of the product. The application of a finish on a wood-based work is usually the last and most visible step in this value chain.In outdoor use, the protection of the wood surface with transparent finishes is not yet sufficiently durable to be able to compete with materials used in industrial carpentry such as PVC or aluminum. Opaque finishes generally provide more durable protection but they mask the appearance of the wood sought by users.With the aim of positioning wood in this construction sector, research on transparent finishes has focused on the efficiency and improvement of the durability of the protection of the surface appearance of structures. Faced with climatic aggressions, the optimum conservation of a structure is not only linked to the performance of the finish but also to the characteristics of the wood material. In particular, in order to fulfill its protective function, the finish film must be able to follow the dimensional variations of the wood it covers without breaking and without detachment. In addition to the criteria for the effectiveness of finishes in the protection of structures, the environmental impact must be considered with increasing attention. Currently, more than 80% of composite or solid wood products are still protected with solvent-based liquid products, which are an important source of VOC emissions. Does the solution come from photo-polymerizable systems that can be formulated with 100% dry matter either in liquid phase in reactive diluents or powdered?

Keywords: Wood, Durability of surface appearance, Surface coatings, Photopolymerizable finishes

1 Introduction

Le bois a toujours été un matériau indispensable pour l'homme ; avec la terre c'est le matériau le plus ancien qui ait été mis en œuvre. Ses utilisations ont été variées et diversifiées au cours du temps que ce soit pour la fabrication d'objets, comme source d'énergie ou encore dans la construction. L'utilisation du bois en construction remonte à des temps ancestraux. Dès le moyen-âge, avec la forte fréquence des incendies dans les villes, le bois sera de plus en plus remplacé par la pierre pour les constructions urbaines. Au XIX° siècle, avec l'apparition du béton et la déforestation, les maisons en bois vont pratiquement disparaître dans les pays du sud de l'Europe. Ce n'est pas le cas du nord de l'Europe, des Etats Unis et du Canada où la couverture forestière est très importante. Ainsi, en Amérique du Nord et en Scandinavie, le bois est employé en structure dans plus de 90% des maisons individuelles. En France, depuis quelques années, on constate une progression de la part du bois dans le bâtiment à la suite de la crise de l'énergie et des matières premières non renouvelables. Cette évolution est liée à la qualité environnementale du bois et à de nouvelles réglementations thermiques qui peuvent être respectées dans des conditions économiques compétitives en associant le bois à des matériaux isolants.

Si de nombreux arguments plaident en faveur de l'utilisation du bois en construction :

- sa ressource abondante et renouvelable,

- son faible coût énergétique,
- sa présentation sous multiples formes couvrant l'ensemble des utilisations potentielles,
- son rapport poids / résistance favorable pour assurer une construction légère,
- sa capacité isolante (12 fois plus isolant que le béton),
- sa résistance au feu,
- son esthétisme,

il présente quelques faiblesses liées à son caractère naturel :

- son anisotropie et son hétérogénéité
- sa sensibilité à l'humidité et son instabilité dimensionnelle
- sa sensibilité aux agents biologiques, physico-chimiques et climatiques

duPour pouvoir concurrencer d'autres matériaux utilisés en construction comme les polymères synthétiques, il faudra soit traiter le bois soit le modifier pour améliorer sa durabilité. En effet, face aux matériaux concurrentiels comme le PVC ou l'aluminium, un handicap dont pâlit encore le bois réside dans la durée de vie trop faible de ses finitions et la nécessité de rénovations fréquentes. Si la garantie décennale, tant du point de vue durabilité du bois que du point de vue de la fonctionnalité de l'ouvrage ne pose pas de problème, l'impossibilité d'assurer en finition, surtout lorsque les produits sont transparents, une garantie d'aspect de surface pénalise considérablement la menuiserie extérieure en bois. Le vieillissement naturel du bois résulte d'une combinaison complexe de divers phénomènes physiques, chimiques et biologiques qui modifient la surface du matériau. En pratique le bois perd sa couleur d'origine et grisaille progressivement. Ce grisaillement s'accompagne d'une érosion très lente, de l'ordre de quelques millimètres par siècle, des couches superficielles du bois. C'est un phénomène purement visuel qui n'entraine aucune dégradation d'ordre mécanique. Cette modification d'aspect de surface est essentiellement due à l'action du rayonnement UV de la lumière solaire, au lessivage de la surface de l'ouvrage et à l'augmentation de la rugosité de surface liée à la dégradation plus importante du bois initial. Parallèlement à cette évolution d'aspect, le bois en ambiance extérieure est soumis à de brusques changements d'humidité dus aux variations climatiques. Si le bois n'est pas protégé, ces variations rapides d'humidité créent des tensions importantes dans les ouvrages et sont à l'origine de variations dimensionnelles, déformations, fentes, ... L'application d'une finition a donc deux objectifs essentiels : protéger le bois du rayonnement solaire et limiter les variations dimensionnelles du bois en réduisant les échanges entre le bois et l'atmosphère. Il ne faut pas confondre finition et préservation. La finition constitue uniquement une protection de la surface du bois. Elle ne confère pas de protection contre les insectes et les champignons à une essence non durable naturellement et ne peut en aucun cas se substituer à l'opération de préservation dont le but est de rendre durable biologiquement un bois qui ne l'est pas naturellement pour l'emploi auquel on le destine. L'industrie de la deuxième transformation du bois est composée d'une chaîne successive d'activités de fabrication et de transformation qui contiennent chacune une partie de la valeur ajoutée du produit. L'application d'une finition est généralement la dernière étape et la plus visible dans cette chaîne de valeur ajoutée. En plus de l'aspect décoratif, de la protection contre les agressions et de la limitation des reprises d'humidité du bois, des fonctions additionnelles peuvent être apportées : facilité de nettoyage, propriétés auto-nettoyantes, antisalissures, propriétés optiques, effets barrière… On parle alors de finition « intelligente ».

2 Les facteurs affectant les performances des systèmes bois-finitions

Les performances des finitions sont tributaires de facteurs liés au bois brut avant l'application de la finition, de facteurs climatiques et de facteurs qui affectent la finition elle-même.

2.1 Les facteurs liés au bois.

Les propriétés de surface sont déterminées par l'anatomie et la composition chimique de l'essence de bois. Les cernes annuels et leurs propriétés affectent la façon avec laquelle une finition va être appliquée et la durée de vie de cette finition. En effet un bois va être classé comme plus ou moins facile à revêtir selon la grosseur des cernes et la transition entre les cernes du bois initial et du bois final. Par exemple, le peuplier, le hêtre, l'érable ou le bouleau sont des feuillus dont les cernes présentent des pores de taille et de répartition relativement uniformes. Ces bois ont une texture fine ce qui les rend facile à revêtir. Par contre, le chêne est un feuillu caractérisé par des pores du bois initial plus gros que les pores du bois final ce qui nécessite une préparation de la surface avant l'application d'une finition. Les extractibles, qui représentent 12% de la masse sèche du bois pour les essences résineuses et jusqu'à 20% pour des essences tropicales sont responsables de l'odeur, la couleur naturelle ou la résistance aux agents de dégradation biologique. Les extractibles changent de couleur lorsqu'ils sont exposés au rayonnement solaire. Cette dégradation par la lumière des extractibles n'est pas empêchée par un revêtement de surface transparent. Les extractibles peuvent aussi affecter la finition elle-même en retardant son durcissement. Par ailleurs, les nœuds des résineux peuvent exsuder de la résine ce qui confère une fragilité à la finition pouvant aller jusqu'à la rupture du film. L'état de surface après usinage a également une influence sur l'accrochage d'une finition et donc sur la durabilité de la protection.

2.2 Les facteurs climatiques

L'impact climatique affecte différemment les composants du bois que sont les carbohydrates (cellulose, hemicelluloses), les lignines et les extractibles. La lumière solaire est constituée des rayonnements ultraviolets, visibles et infrarouges. Les rayonnements UV qui ne représentent qu'une faible proportion du rayonnement solaire sont les plus impliqués dans les réactions de photodégradation du bois qui induisent des altérations de l'aspect de surface des ouvrages. Ce sont des chromophores présents dans la structure moléculaire des lignines et des extractibles qui absorbent la lumière solaire et créent des espèces radicalaires qui sont le point de départ des réactions de photodégradation oxydative. Si les carbohydrates sont des polymères résistants aux UV, ils sont très sensibles à l'humidité. Les cycles d'humidification du bois puis séchage provoquent des variations dimensionnelles du bois, on dit que le bois travaille. L'eau aura également un impact sur la performance de la finition en agissant sur les liants présents dans les formulations qui sont souvent hydrolysables ce qui peut provoquer une perte des propriétés mécaniques du film. Pour un bois revêtu d'une finition, les variations d'humidité dépendent de la perméabilité (à l'eau liquide et vapeur) du film de finition .En rendant le système imperméable à l'eau, la finition empêche le développement de moisissures.

2.3 Les facteurs liés à la finition

La pigmentation a une influence sur la durabilité de la finition aux intempéries. Les pigments apportent la couleur en absorbant ou en réfléchissant les rayonnements UV-visible du rayonnement solaire, ce qui limite la photodégradation des liants. Ainsi pour un même liant, un revêtement pigmenté a une durabilité deux à quatre fois supérieure à celle d'un revêtement transparent. Toutefois, en se dégradant, le liant va permettre aux pigments de se libérer et provoquer le farinage de la finition. Ce farinage se produit avec l'action conjuguée des UV, de l'oxygène et de l'eau. Même si les revêtements pigmentés sont plus durables pour la protection des ouvrages en bois, l'intérêt se porte plus sur les revêtements transparents qui laissent visible la texture du bois mais pour lesquels les performances en extérieur sont à améliorer. Pour les revêtements de surface réticulant par un mécanisme radicalaire, l'oxygène réduit non seulement la vitesse de séchage mais affecte également les propriétés mécaniques des films séchés à l'air. La perméabilité à l'eau liquide et à la vapeur est un paramètre important qui fait l'objet des normes EN 927-5 et ISO 7783. La souplesse de la finition est un paramètre clé pour la durabilité du revêtement appliqué sur le bois. Le manque de souplesse ne va pas permettre au film de s'adapter aux variations dimensionnelles du bois ce qui va entrainer des craquèlements. Le manque d'adhérence qui peut provenir de l'essence de bois ou de la formulation de la finition provoque un écaillage du revêtement de surface.

3 Les finitions à séchage UV pour le bois : une solution pour la limitation des composés organiques volatils (COV) ?

Les recherches sur les finitions appliquées sur les ouvrages en bois destinés à un emploi intérieur ou extérieur ont essentiellement porté sur l'efficacité et sur la durabilité de la protection. S'il est vrai que ces deux paramètres peuvent conditionner la survie de l'utilisation du bois en menuiserie industrielle, la prise en compte des facteurs environnementaux doit être considérée avec une attention croissante. Pour se mettre en conformité ou pour devancer une législation de plus en plus contraignante sur la limitation des émissions de COV et pour satisfaire une demande des utilisateurs de plus en plus soucieux des aspects hygiène, santé et environnement lors de l'élaboration et de la durée de vie des matériaux, les formulations des résines de finition pour le bois doivent être reconsidérées dans leur ensemble. Les finitions de demain devront allier performance et respect des contraintes environnementales sans pour autant alourdir le prix de la menuiserie. La diminution des COV dans un atelier de finition peut être atteinte par des actions portant sur les procédés d'application, sur les procédés de nettoyage des appareillages et surtout sur les formulations en favorisant des systèmes à faible taux d'émission des COV. Le durcissement sous rayonnement UV permet l'utilisation de formulations à 100% de matière sèche par dilution des oligomères dans des diluants réactifs (monomères multifonctionnels photopolymérisables) qui ne s'évaporent pas et qui s'incorporent dans le film en participant à la réticulation. L'ajout de ces diluants permet de diminuer la viscosité des formulations et ainsi de les rendre compatibles avec les procédés d'application. Ce type de formulations séduisantes au niveau environnemental présente des inconvénients quant aux propriétés mécaniques du film durci. Les diluants réactifs conduisent à des films peu flexibles qui ne conviennent pas à la protection de supports instables dimensionnellement comme le bois. Pour de tels matériaux, la mise au point de formulations photosensibles sans diluant réactif serait une solution et trois voies de recherche sont explorées :

- revêtements liquides à 100% de matière sèche à base d'oligomères de faible viscosité,
- revêtements UV aqueux. Pour ces systèmes, il faut évaporer l'eau avant la réticulation,
- revêtements « poudre » : l'oligomère est un solide pulvérulent à température ambiante qui, par chauffage, fond en formant un film qui est réticulé par exposition aux UV.

A l'heure actuelle, les finitions à réticulation sous rayonnement UV sont industriellement appliquées sur des ouvrages en bois destinés à des utilisations intérieures (ameublement, parquets, panneaux décoratifs...) et uniquement en finition transparente ou peu pigmentée. Le développement des vernis photoréticulables dans ce secteur est attribué à certains paramètres technologiques et écologiques :

- grande souplesse des caractéristiques physiques des vernis avant et après l'application,
- faible toxicité due à l'absence de solvant,
- grande stabilité thermique des formulations photoréticulables,

- vitesse de réticulation élevée à température ambiante,
- surface plus faible des installations avec économie d'énergie et d'investissement.

Si l'utilisation des résines photosensibles présente un intérêt au niveau environnemental, elle n'est pas sans poser des problèmes qui ne sont pas résolus, à l'heure actuelle, de façon définitive.

4 La recherche de solutions pour améliorer la durabilité en utilisation extérieure des systèmes bois-finition photoréticulable transparente

On peut distinguer des difficultés liées d'une part à la nature photochimique de l'amorçage et d'autre part à la nature du bois.

4.1 Difficultés liées à l'amorçage

Actuellement le séchage par un rayonnement UV ne peut être réalisé qu'en atelier de fabrication. De ce fait, l'utilisation des finitions photoréticulables pour la protection des menuiseries extérieures est délicate car elle nécessite la pose sur site de l'ouvrage préalablement revêtu en atelier. L'impossibilité d'appliquer ces finitions sur site pose également le problème de la rénovation. Pour lever ce verrou lié à la nécessité d'utiliser le rayonnement UV, des recherches récentes s'orientent vers le développement de formulations réticulant sous la lumière naturelle dans des conditions adaptées pour le séchage d'un revêtement de surface appliqué sur du bois. Pour obtenir une finition régulière, il faut assurer un flux photonique homogène sur l'ensemble de la pièce à vernir. De ce fait, la finition d'éléments assemblés est techniquement difficile. Avec les amorceurs photochimiques courants dont l'efficacité est maximale dans le domaine UV, le durcissement des finitions pigmentées ne peut être obtenu que sur des faibles épaisseurs. La mise au point de systèmes amorceurs complexes fonctionnant dans les fenêtres de transmission des pigments ont permis le séchage d'un film pigmenté de 100 □m sur le bois.

4.2 Difficultés liées à la nature du bois

Les finitions à séchage UV ont fait l'objet de recherches poussées dans le domaine de la protection des surfaces métalliques utilisées en extérieur, en particulier pour le traitement des carrosseries automobiles. Des formulations très performantes quant à leur tenue sont disponibles sur le marché. Appliquées sur le bois, ces mêmes formulations optimisées pour le traitement des surfaces métalliques n'assurent plus une durabilité de la protection suffisante pour concurrencer des menuiseries en PVC ou en aluminium. Cette perte d'efficacité doit être imputée à des caractéristiques propres du matériau bois.

4,2.1 Le bois est un matériau poreux :

La pénétration du rayonnement proche UV dans le bois est de l'ordre de 70 □m. En conséquence, la fraction de la résine qui migre plus en profondeur dans la structure poreuse du bois ne sera pas atteinte par les photons nécessaires à l'amorçage. Pour certaines essences poreuses, cette perte de résine peut devenir importante et ainsi alourdir le prix du traitement de surface. L'augmentation de la viscosité de la formulation pour limiter la migration peut devenir incompatible d'une part avec le procédé d'application et d'autre part avec les propriétés mécaniques du film durci.

4,2.2 Le bois est instable dimensionnellement :

Lors d'un vieillissement climatique de systèmes bois/finition photoréticulable, les dégradations n'ont pas pour origine une perte de cohésion du film de finition. La tenue face aux agressions physico-chimiques de l'environnement d'un revêtement de surface est garantie par l'efficacité des additifs de stabilisation (absorbeurs UV, antioxydants...) entrant dans les formulations des résines pour extérieur. La dégradation se manifeste par une fracture à l'interface bois/finition provoquant un décollement du film. En ambiance extérieure, le bois subit des variations dimensionnelles importantes liées aux variations d'humidité. Pour remplir sa fonction de protection de surface, la finition doit être capable de suivre sans rupture et sans décollement les variations dimensionnelles de l'ouvrage qu'elle recouvre. Différentes stratégies ont été explorées pour améliorer la durabilité de la protection :

- formuler des formulations souples : la flexibilité du film de finition et la conservation de cette flexibilité durant la période d'utilisation de l'ouvrage sont des paramètres influents sur la rupture à l'interface bois/finition. La flexibilité est obtenue en abaissant la température de transition vitreuse T_g du film durci en dessous de la température d'utilisation. Dans le cas des revêtements liquides photosensibles, l'ajout de diluants réactifs pour ajuster la viscosité conduit à des degrés de réticulation trop importants pour pouvoir abaisser suffisamment la T_g. L'application de revêtements UV aqueux pourrait paraître séduisante ; ces résines peuvent être formulées à faible viscosité tout en conduisant à des films flexibles. Toutefois, l'application de ces formulations aqueuses sur le bois se heurte au problème de l'élimination de l'eau nécessaire à la réticulation qui se fait essentiellement par migration dans la structure poreuse du bois. De ce fait, la résine réticule sur une surface gonflée par l'eau. Un passage sous microonde pour éliminer l'eau avant l'excitation UV donne des résultats intéressants. Ce procédé est industrialisé pour la menuiserie intérieure (technologie MOS Gardina). De plus, l'eau d'une part favorise la migration en surface de substances extractibles colorées qui provoque des taches peu esthétiques et d'autre part provoque un relevage des fibres qui modifie l'aspect de surface et le toucher des ouvrages.

- Assurer la stabilité dimensionnelle du bois avant l'application de la finition : des traitements thermiques ou chimiques permettent de limiter de façon durable les variations dimensionnelles du bois soumis aux contraintes physico-chimiques de l'environnement. Ces traitements modifient l'énergie de surface et diminuent l'accrochage des revêtements de surface.

Pour corréler ces deux approches, nous avons proposé d'intercaler entre le bois et le film de finition un composite de surface (sous-couche) qui, par ses caractéristiques de flexibilité amortit les variations dimensionnelles du bois et ainsi réduit les sollicitations du film de finition. Les résultats obtenus sur des formulations simplifiées lors d'un vieillissement accéléré d'échantillons de différentes essences ont montré la faisabilité et l'intérêt de ce procédé qui n'est pas appliqué industriellement jusqu'ici. La durabilité de la protection dépend de la flexibilité du composite de surface et de l'adhérence de la couche de finition sur le composite.

4,2.3 Le bois est un matériau réactif :

Il est connu des applicateurs que le séchage des finitions alkydes ou polyesters qui réticulent par voie radicalaire est retardé sur les essences riches en substances extractibles caractérisées par une forte teneur en composés phénoliques. Une étude fondamentale de l'influence des molécules phénoliques sur les caractéristiques d'une réaction de réticulation amorcée par voie photochimique nous a montré que la vitesse de séchage et les propriétés mécaniques du film durci dépendent du mode de fonctionnement de l'amorceur et de la réactivité des molécules phénoliques du bois sur les espèces transitoires et radicalaires intervenant dans le mécanisme de polymérisation.

4,2.4 Le bois ne présente pas une pérennité de son aspect de surface :

Même revêtu d'une finition transparente, l'aspect d'un bois n'est pas immuable et il est susceptible de varier sous l'action de facteurs liés à l'environnement. Le vieillissement de systèmes bois/finition transparente a mis en évidence le rôle protecteur limité des vernis. Les effets principaux du revêtement transparent sont de filtrer une partie du rayonnement proche UV, de faire barrière à l'eau et à l'oxygène. Ainsi, il a pu être montré qu'un bois recouvert d'une finition transparente subit qualitativement les mêmes altérations de l'aspect de surface que le bois brut exposé sous atmosphère d'azote, à la partie visible du rayonnement solaire. De ce fait, la conservation optimale d'un système bois/finition transparente n'est pas seulement tributaire des performances en terme de stabilité de la finition elle-même mais également de la durabilité de l'aspect de surface du bois qu'elle recouvre. Pour proposer un système de finition assurant une pérennité de l'aspect d'un ouvrage en bois qui puisse concurrencer l'utilisation du PVC ou de l'aluminium en menuiserie, il est nécessaire de stabiliser la couleur du bois par un traitement de surface avant l'application de la finition transparente.

5 Les finitions « poudre » pour la protection du bois

Le développement de produits de finition en phase aqueuse ou à haut extrait sec contribue à diminuer les émissions de COV mais ces produits ne répondent pas toujours aux attentes du marché quant à la qualité du film de finition. De plus, leur utilisation est fortement consommatrice d'eau et génère de nombreux déchets solides et liquides. L'arrivée sur le marché des résines thermodurcissables a fait évoluer la technique des finitions poudre qui sont actuellement largement utilisées dans l'industrie de la métallurgie. Les avantages des finitions poudre sont nombreux : compatibilité avec l'environnement, facilité d'application, rapidité de changement de couleur, faible consommation d'énergie, récupération facile par aspiration de l'excédent de poudre, réduction des risques sanitaires… Cependant, ce procédé reste peu développé pour les substrats comme le bois. En effet, les formulations poudre utilisées pour la protection des surfaces métalliques ne sont pas applicables sur des matériaux thermosensibles ; la fusion nécessaire pour la formation du film et la réticulation exigent des températures et des temps de chauffe incompatibles avec le bois. C'est avec l'apparition récente de produits polymérisant à basse température ou sous rayonnement UV qu'une application sur le bois est rendue possible. Actuellement, les finitions « poudre » sont uniquement utilisées pour la protection intérieure d'ouvrages en bois ou en matériaux dérivés. Pour les menuiseries extérieures, les formulations commercialisées conduisent à des films pas suffisamment flexibles pour suivre les variations dimensionnelles du support. L'optimisation du procédé et des formulations est un enjeu crucial pour le développement et la promotion d'une telle technologie dans l'industrie du bois.

6 Conclusion

La finition est l'étape la plus importante pour ajouter de la valeur aux ouvrages en bois et protéger aussi efficacement que possible leur aspect de surface. La

https://doi.org/10.1051/matecconf/201814901016

durée de vie d'un produit en bois est souvent associée à la durabilité du revêtement de finition. Les progrès les plus marquants dans les finitions transparentes ont été observés en ameublement intérieur ; des formulations performantes alliant durabilité de la protection et respect de l'environnement ont pu être mises au point. La durabilité de la protection des ouvrages en extérieur et en revêtement transparent reste insuffisante pour assurer une concurrence efficace de la menuiserie bois face aux polymères synthétiques. Alors que le matériau bois offre en construction des avantages recherchés par les particuliers, les performances des revêtements de surface restent un des freins au développement de la construction en bois. En plus des propriétés de protection de la surface des ouvrages et pour satisfaire les attentes et les exigences de la clientèle, il devient nécessaire de proposer des finitions dites intelligentes, qui auront la capacité de défendre activement et durablement le bien-être, la santé, le confort... Le développement de ces finitions innovantes représente un saut technologique important dans le domaine de la finition des matériaux en général et du matériau bois en particulier et contribue au développement de produits à plus haute valeur ajoutée. Certaines techniques font déjà l'objet de brevets, mais très peu sont destinées à une application sur du bois. En ce qui concerne l'industrie du bois, le panel des différentes fonctions envisagées est important: parfum, désodorisant, antibactérien, nettoyage facile, absorption des COV, modification de la couleur en fonction de la température, ignifugation, haute résistance thermique (jusqu'à 450°C), préservation biocide, anti microbien, haute conductivité électrique (résistivité inférieure à $10^4\Omega$), forte isolation électrique (résistivité supérieure à $10^{13}\Omega$), résistance chimique particulière (résiste aux hydrocarbures, acides, bases), anti dégazage (pour des supports difficiles à traiter), flexibilité (film post-formable)....

Références :

1. R.A. Williams, M.T. Knaebe, W.C. Feist, *Finishes for Exterior Wood: selection, application and maintenance*, Forest Products Laboratory (1996)
2. R.J. Tichy, *Interior Wood Finishing*, Forest Products Laboratory (1997)
3. P. Dulbecco, D. Luro, *L'essentiel sur le bois*, Document CTBA (1998)
4. M.L. Roux, *Finition des ouvrages en bois dans le bâtiment*, Document CTBA (1994)
5. Association des Ingénieurs de l'Ecole Supérieure du Bois, *Manuel de l'ingénierie Bois*, Eyrolles (2013)
6. L. Podgorski, A. Merlin, *Systèmes bois-finitions extérieures. Vieillissement, comportement.* In : DIROL Daniel, DEGLISE Xavier. Durabilité des bois. Paris : Hermès Science Publication, (2001)
7. Y. Irmouli, *Etude de systèmes de finitions pour bois : application de nouvelles formulations, vieillissement et amélioration des performances.* Thèse Sciences du bois et des fibres, Université de Lorraine (2007)
8. J. Bessière: *Les finitions poudres appliquées sur le bois et matériaux dérivés : optimisation du procédé industriel, étude des relations finition/substrat et fonctionnalisation.* Thèse Sciences du bois et des fibres, Université de Lorraine (2012)
9. M. Sylla, *Rôle des molécules phénoliques dans la photopolymérisation : application à des résines de finition pour le bois.* Thèse Sciences du bois et des fibres, Université de Lorraine (2001)
10. F. Aloui, *Rôle des absorbeurs UV inorganiques sur la photostabilisation des systèmes bois-finition transparente.* Thèse Sciences du bois et des fibres, Université de Lorraine (2006)
11. F. Mauguière-Guyonnet, *Photopolymérisation radicalaire sous lumière visible : application aux revêtements pour bois.* Thèse de l'Université de Haute Alsace, Chimie des matériaux (2004)
12. V. Maurin, *Revêtements poudres UV – mécanismes de polymérisation et étude des relations structure/propriétés.* Thèse de l'Université de Haute Alsace, Chimie des matériaux (2012).

Research on the use of Ferro-Chrome slag in civil engineering applications

Khalifa S. Al-Jabri
Sultan Qaboos University, Department of Civil and Architectural Engineering, Al-Khod, Oman

Abstract. Over recent decades there has been rapid increase in the industrial waste materials and by-products yields due to the progressive growth rate of population, development of industry and technology and the growth of consumerism. With the growing environmental pressures to reduce waste and pollution, Intensive research studies have been conducted to explore all suitable reuse methods. Wastes such as construction waste, blast furnace, steel slag, coal fly ash and bottom ash have been approved in many places as alternative materials in bridges, roads, pavements, foundations and building construction. The use of industrial solid waste as a partial replacement of raw materials in construction activities not only saves landfill space but also reduces the demand for extraction of natural raw materials. Ferrochrome slag is a by-product from the production of chrome. There are environmental and economic advantages in seeing slags as a potentially useful resource rather than as waste products. Slag management at ferrochrome producing companies has been influenced by the limited space available and financial cost implications of the slag dumps. Internationally, e.g. South Africa, India, Norway, Turkey, East Europe, China, Sweden and USA, ferrochrome slag is used commercially in the road and construction Industries. This material is being used for road construction, as aggregates in concrete industry, brick manufacturing, and in pavement construction as engineering fill and has recently been tried in cement. This paper presents an overview of the recent advances of the use of ferrochrome slag in various civil engineering applications such as road construction, and cement and concrete industries.

1 Introduction

The greatest user of natural resources is the rapid developing sector of the construction industry. This raises very important concerns on the depleting of these resources at an alarming rate which causes serious threats to the environmental balance. Due to the increase in the world populations, the rapid growing of the industrial, domestic, commercial and technological activities, the associated generation of the solid wastes are increasing substantially creating an additional burden on the environment. The solid waste materials generating from industries are not only occupying valuable land mass but also polluting the environment and creating real challenges to the safe disposal. The best strategy for solid waste management is to work towards achieving the 5Rs of reduction, recovery, recycling, reuse, and research [1]. Industrial By-products generated from different industries are creating environmental problems associated with disposal and pollution. However, such materials can be used in manufacturing fireproof materials, used in concrete as partial substitutes for cement and/or sand or they can be added to clay and heat-treated to produce building materials such as ceramic tiles, refractory and insulation bricks and as a material in road construction.

The generated industrial by-products in Oman have increased substantially in the last decade due to the rapid development in the industrial projects, and mining industries. Accumulation and lack of proper disposal strategies of such materials are posing environmental, pollution and health problems. Copper slag (CS), Cement By-pass Dust, CBPD (also called Cement Kiln Dust), Spent Catalyst (SC), and Ferrochrome (FeCr) slag are industrial by-products produced from the process of manufacturing copper and Portland cement, in oil refinery's processes, and chrome, respectively. In Oman, large quantities of copper slag (60,000 tons/year), cement by-pass dust (25,000 tons/year), spent catalyst (7300 tons/year) and ferrochrome slag (355,000 tons/year) are produced, most of which is not effectively utilized and disposed on-site without any reuse. The generation of such materials are expected to increase in the future.

The use of the industrial slag products such as FeCr slag in beneficial applications such as materials in building and road construction reduces the depletion of the natural resources and it results in positive environmental effects. The FeCr slag possesses physical properties similar to natural aggregates which makes it a suitable material for civil engineering applications. This paper presents an overview of the recent advances of the use of ferrochrome slag in various civil engineering applications such as road construction, and cement and concrete industries.

2 Ferrochrome (FeCr) Slag

Ferrochrome (FeCr) slag is by product from the production of Ferrochrome, an essential component in stainless steel industry [2]. An amount of 1.1 to 1.6 ton of slag is produced for each ton of FeCr [3]. In Oman, there are two FeCr plants that produce about 355,000 tons of FeCr slag annually [4].

FeCr metal is produced in electric-arc furnaces by a physical –chemical process from the oxide of chromium ore with coke as a reducing agent at temperatures between 1,500°C to 1,700°C. Both the molten FeCr and the slag flow out into ladles. After gravity separation from the metal, the molten slag, slowly cools in the air, forming a stable and dense crystalline product having an excellent mechanical properties similar to basalt [5,6]. The granulated slag (size less than 4.75mm) samples result from water cooling of the slag whereas lumped slag (size between 8 to 20 mm) samples produced when slag is air-cooled [7]. The FeCr slag is classified as ferrous slag under iron-alloy slags [6].

Fairly small amount of this material is used in useful applications whereas the vast majority of the material produced annually is disposed of site without further reuse.

3 Properties Ferrochrome (FeCr) Slag

The chemical composition of FeCr slag includes three major elements: Silicon (Si), Aluminium (Al), and Manganese (Mg). Together with their oxides, these components make up 83% of the slag. The slag also includes oxides such as Cr, Ni, Fe and Ca. Analysis of FeCr slag generated in Oman showed that it contains 5-6 wt% Cr_2O_3, 23-24 wt % Al_2O_3, 22-23 % MgO, 34-35 wt.% SiO_2, 1-4 wt % CaO and 4-5 wt % NiO [4,8]. The free and combined limes contribute to nearly 3% of the chemical composition of FeCr slag and the summation of silicon, aluminium and iron oxides is about 60% which less than 70% which is required in order to classify this material as a pozzolanic material. This indicates that FeCr is not chemically reactive material to be used as cementitious material in lieu of cement since sufficient quantity of lime must be available in order to reach the required rate of hydration and to achieve the required early age strength.

Also, FeCr slag contains high percentage of Magnesium oxide (MgO) of about 34-35% which can have an adverse effect in concrete made with FeCr slag as cement substitute and limit the use of FeCr slag in rigid pavements [9]. The existence of MgO in large quantities can retard the initial hydration of cement and increase its setting time [10] as well as causing large changes of volume of about 10% in concrete and asphalt mixes [11]. However, this adverse effect can be eliminated by slag weathering in atmospheric conditions [11].

Studies [5,6,8,12-15] reported that FeCr slag possesses excellent physical properties compared to other aggregate since it has good adhesion and good abrasion, has a rough and porous surface (figure 1) and it lacks clay and organic ingredients in tis composition. However, the water absorption rate is relatively high due to the porous nature of the slag.

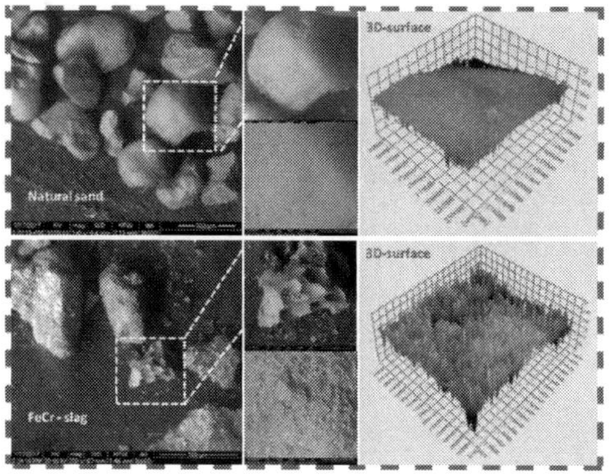

Fig. 1. SEM micrographs and 3D-surface topography images of sand and FeCr-slag [8].

Results from tests conducted by several researchers to determine the physical properties of FeCr slag indicated that the specific gravity of the FeCr slag ranges between 2.84 and 3.15 g/m^3 compared with 2.65 g/m^3 for limestone aggregate. While it has water absorption values between 0.25% and 2.3% in comparison with 0.38% for limestone aggregate. The Los Angeles Abrasion and CBR values of FeCr slag are between 14-18% and 107-140% compared with 21% and 103% for limestone aggregate, respectively.

Table 2. Physical properties of FeCr slag generated in Oman and natural aggregates [4].

Property	FeCr Slag	Aggregates	
		Coarse	Fine
Specific Gravity g/cm^3	2.84	2.83	2.7
Water Absorption (%)	0.63	0.20	0.73
Flakiness Index (%)	9.83	18.50	
Elongation Index (%)	10.50	13.5	
Impact Value (%)	11.00	15.30	
Crushing Value (%)	17.89	20.10	
Abrasion Resistance (%)	18.19	18.60	

Also, figure 2 shows that the XRD pattern of FeCr-slag indicates the presence of dominant mineral phases like metalic phase of magnesiochromite ($MgCr_2O_4$), chromoferide, silicate phases like forsterite (Mg_2SiO_4), and Fayalite (Fe_2SiO_4) and the broadening from 2θ: 15-35 points to the presence of amorphous active silica [8].

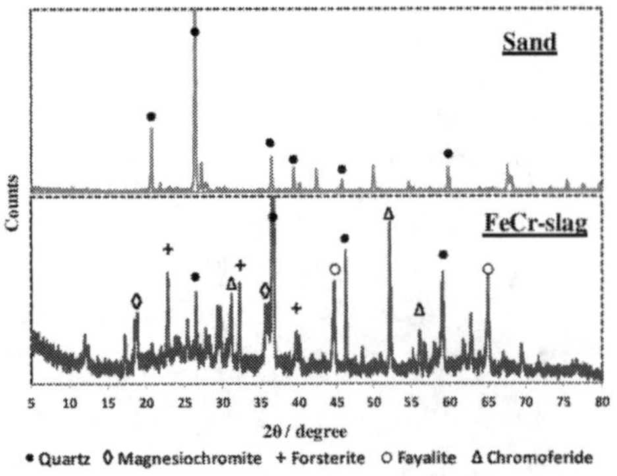

Fig. 2. XRD patterns of aggregates used (natural sand and FeCr-slag) [8].

From the physical properties test results it can be revealed that there are several potential uses of FeCr slag: 1) it can be used as an aggregate in concrete and hot mix asphalt and concrete pavements, 2) as a base or sub-base material, and 3) as an embankment and landfill material.

4 Use of Ferrochrome (FeCr) Slag in concrete

Very limited studies were conducted to study the effect of FeCr slag on the properties of cement mortars and concrete. These studies were focused mainly on the use of FeCr slag as aggregate substitute. Zelić [12] investigated the properties of concrete pavements prepared with ferrochromium slag as concrete aggregate. The results showed that the 28-day compressive strength of the concrete made with original un-fractioned slag and with standard limestone as aggregate reached the values of 57 MPa and 36.7 MPa, respectively. The properties of concrete was investigated by Gencel et al. [13] under combined effects of fly ash as cement replacement and ferrochrome slag as aggregate substitute. Cement was replaced with fly ash at the ration of 10, 20, and 30wt% whereas coarse limestone aggregates were replace with coarse ferrochromium aggregate at the ratio of 25, 50, and 75wt%. The results from the study revealed that ferrochromium aggregates increase the strength of concrete and also the abrasive wear resistance while it has negligible influence on the porosity and water absorption of concrete. A Study [6] which was carried out on the effect of microsilica addition within a low cement castables matrix of calcined bauxite and ferrochrome slag showed that slag containing castables achieved good thermal properties such as thermal shock resistance, permanent linear change, and pyrometric cone equivalent.

A number of researchers [16-19] studied the properties of normal and high strength concrete made with FeCr slag as an aggregate. The results indicated that replacement up to 75% of FeCr slag in lieu of fine aggregates improved the strength of concrete compared with conventional concrete. In a recent study, Acharya and Patro [15] investigated the effect of using ferrochrome ash (FCA) and lime dust in concrete properties. Up to 40% of FCA was used as cement substitution in concrete mixtures in four different substitutions at an interval of 10%. FCA was modified with lime dust to enhance the performance of concrete in which substitution of lime was considered as 7% after studying its effect on blended cement based concrete. The results from Acharya and Patro study indicated that replacement of OPC up to 47% by FCA (40%) and lime dust (7%) has comparable positive impact on the 28 days strength and appreciable impact on long term strength properties.

The author and his colleagues [8] has recently studied the reuse of waste ferrochrome slag (FeCr-Slag) generated from a chrome production plant in Sohar, Oman, as a partial replacement for sand in the production of cement mortars with improved thermal and mechanical properties. Sand was replaced by FeCr-Slag in different percentages ranging from 0 up to 20% (by weight). Compressive strength, flexural strength, drying shrinkage, thermal conductivity and specific heat of the blended mortars were determined in accordance with ASTM standards at 3,7 and 28 days of curing. X-ray diffraction (XRD) was used to study the phase composition. The microstructure characteristics of the hardened samples were investigated by scanning electron microscope (SEM).The experimental results showed considerable improvements in both compressive and flexural strengths, enhancements of about 33 and 39 %, respectively were attained at 20% FeCr-Slag replacement after 28 days of curing. The drying shrinkage generally decreased with increasing replacements of sand by FeCr-Slag. FeCr-slag acts to increase both the thermal conductivity and specific heat effectively. FeCr-slag led to noticeable modifications in the microstructure characteristics of the hardened blends. The results from the study are presented in figures 3-5.

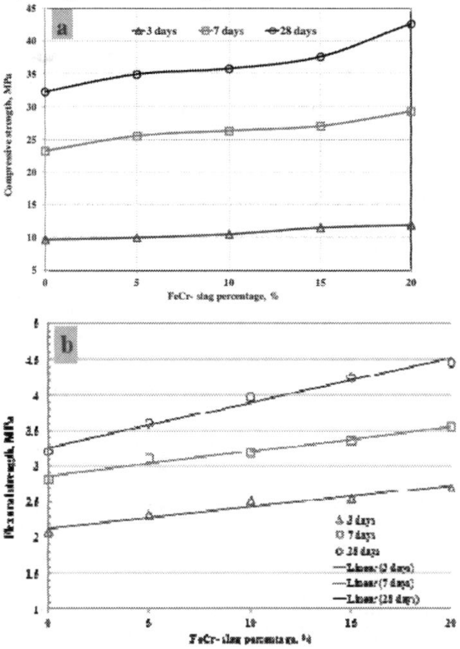

Fig. 3. Variations in the (a) compressive strength and (b) flexural strength of cement mortars incorporating 0, 5, 10, 15 and 20% FeCr-slag hydrated for 3, 7 and 28 days [8].

https://doi.org/10.1051/matecconf/201814901017

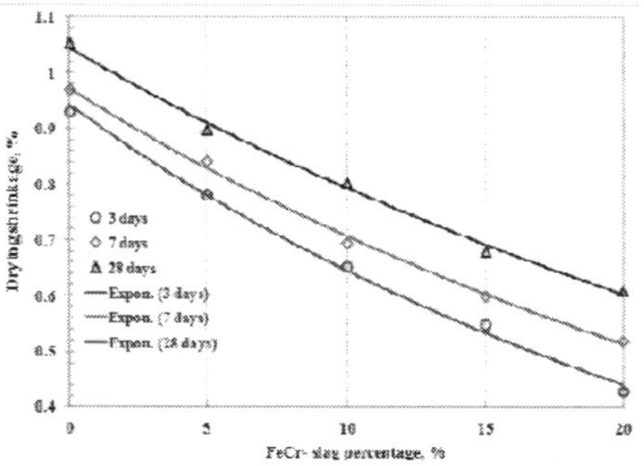

Fig. 4. . Variations in drying shrinkage of cement mortars incorporating 0, 5, 10, 15 and 20% FeCr-slag hydrated for 3, 7 and 28 days [8].

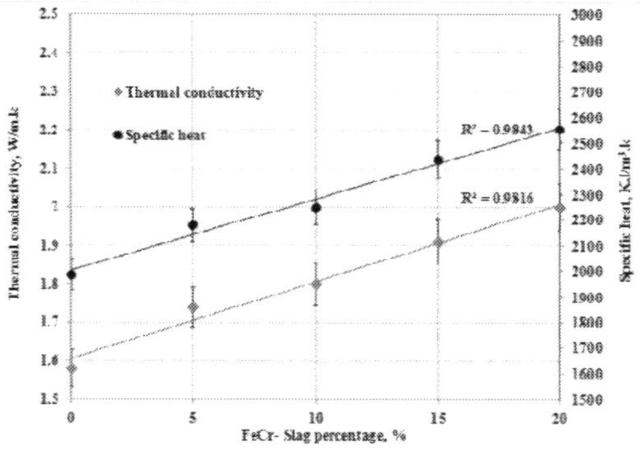

Fig. 5. Variations in thermal conductivity and specific heat of the cement mortars incorporating 0, 5, 10, 15 and 20% FeCr-slag at 28 days of hydration [8].

5 Application of Ferrochrome (FeCr) Slag in Road Construction

Different types of slag find wide acceptance in road construction industry as an aggregate in flexible and rigid pavements, hot asphalt mixes and as a material in base and sub-base layers. Slags possess superior properties such as volume stability, high volume mass, good abrasion resistance to wear and crushability which make it a suitable material for road construction [12]. The first use of slag as an aggregate in asphalt mixtures dates back to 1969 [11], when a trial road section was built in Toronto, on which steel slag was used as an aggregate in base course and road asphalt surfaces. The studied asphalt mixtures have demonstrated very good properties in terms of bearing capacity, resistance to external impacts, and durability.

In the last decade, the potential application of FeCr as an alternative aggregate in pavements and road construction has been investigated [2], [3], [5], [7], [12], [14], [21]. In 2001, Lind et al. [2] investigated the environmental impact of FeCr slag in road construction. The results indicated that FeCr slag is safe to use as a material in road construction. The study conducted by Zelić [12] on the use of FeCr slag as concrete aggregate in concrete pavement concluded that the reinforced slag concrete is suitable for wearing courses of concrete pavements for traffic classes 1 and 2 where carbonate stone material does not meet the standard technical requirements for cement concrete slab pavements according to the Croatian standard. Yilmaz and Karaşahin [5] studied the mechanical properties of FeCr slag in granular layers of flexible pavements. The experimental program consisted of two parts: 1) study of the physical and chemical properties of slag, and 2) study of the mechanical properties of specimens made with FeCr slag and limestone as aggregate. Test results indicated that the physical properties of the slag such as the LA and CBR values, and high frost resistance meet the requirements of the aggregates for granular layers of flexible pavements which qualify FeCr slag as aggregate suitable for pavement layers. In another stud, the authors [14] verified that mixtures of FeCr slag and Portland cement and FeCr and silica fume can be utilized as a road base layer material in specific mixture ratio. The study recommended conducting field tests to verify the slag use in cement-bound layers of pavements. The potential use of FeCr slag as an embankment and granular subbase was evaluated by Das [21].

The use of FeCr slag as an aggregate in road construction is practiced in many countries such as Finland, Sweden, South Africa and India [4]. Finland is one of the leading countries in the world that utilizes FeCr slag commercially in road construction industry [20]. The FeCr slag products have been used in Finland since early seventies for many different purposes [3]. Outokumpu [20], Finland is the major producer of FeCr slag, used the FeCr slag in road construction as shown in figure 6. Due to its superior physical properties over the natural aggregates, roads can be built with thinner bases and sub-bases than when using natural aggregates (figure 7) [4]. FeCr slag has similar applications in other countries such as India, Sweden, and South Africa.

Fig. 6. Use of FeCr slag in road construction in Finland [20].

The outcomes from the previous studies concluded that FeCr slag have the potential to be utilized as a pavement base layer material in similar applications where traditional natural aggregates are used. It can also be used as a base or sub-base material in highway pavements and as an embankment material. The use of FeCr slag as aggregate in pavement layers reduces the requirement for natural aggregates and saves the existing natural resources.

https://doi.org/10.1051/matecconf/201814901017

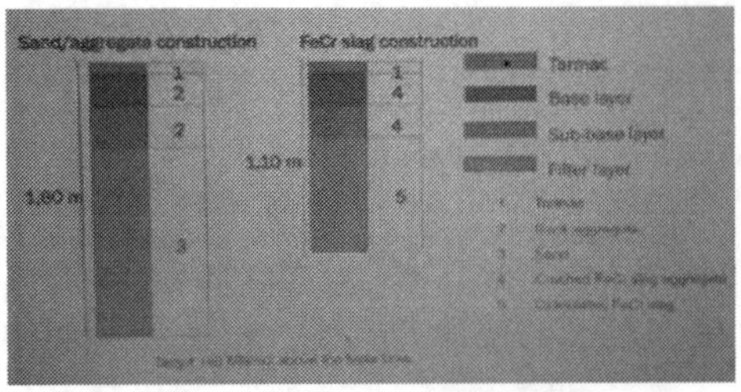

Fig. 7. The advantage of using FeCr slag in road construction [4].

6 FeCr Slag, Is It Environmentally Safe?

Chemical analysis on various FeCr slag samples have detected traces of heavy metals specially chromium oxide, Cr (VI) [2-4], [6-7], [11], [14], [17]. The waste slag contains about 6–12% chromium as chromium oxide and it has the potentiality of releasing hazardous chromium compounds. The Cr (VI) compounds are very toxic and highly leachable. This leaching characteristic may classify the slag as hazardous waste and restricts its use and disposal. To evaluate the potential hazards that may arise from the use of FeCr slag, leaching tests was conducted by various researchers according to the Synthetic Precipitation Leaching Procedure (SPLP). The results confirmed that the FeCr slag is not pollutant and it can be used safely. In all studies, the SPLP test results demonstrated that except for the Cr(VI), the leaching of other heavy elements is negligible in terms of environmental impact. The Cr (VI) concentrations were slightly above target for inert waste meet the US EPA limits of 5mg/l for chromium [22]. This implies that FeCr slag is not classified as a hazardous substance with respect to chromium. An environmental impact study on FeCr slag in road construction was conducted by Lind et al. [2]. The results from an in situ leachate test revealed that there was a low migration of particles from the slag to the underlying soil and that the leaching into the groundwater was also low for all the elements analysed.

7 Conclusions

This paper presented an overview of the recent research on the use of ferrochrome slag in various civil engineering applications such as road construction, and cement and concrete industries. The study concluded that the physical and mechanical properties of FeCr are as good or better than those of natural aggregate which make it suitable material to be used as aggregate replacement in concrete and road construction applications such as in rigid and flexible pavements and base or sub-base materials. Environmental impact assessment studies proved that FeCr slag is non-hazardous material and environmentally friendly to be used as a green construction material in lieu of natural aggregates.

Waste is a source of secondary raw materials but at the same time can negatively impact the natural environment and public health. The production of industrial by-products in Oman is expected to increase rapidly in the coming few years due to the industrial development in the country which will have a detrimental impact on the environment unless waste management policies and legislations are developed. Successful utilization of industrial by-products in the construction industry would require more detailed economic and environmental analyses as well as engineering design specifications.

References

1. N. Su, Z. Chen, H. Fang, Cem. Concr. Comp. **23** (2001)
2. B.B. Lind, A.M. Fallman, L.B. Larsson, Waste Manage. **21** (2001)
3. P. Niemelä, M. Kauppi, INFACON XI (2007)
4. A. Das, M. Lobo, AL TAMMAN INDSIL FERRO CHROME LLC (unpublished presentation) (2017)
5. A. Yilmaz, M. Karaşahin, Mater Struct **43** (2010)
6. P.H. Kumar, A. Srivastava, V. Kumar, M.R. Majhi, V.K. Singh, J. Asian Ceram. Soc. **2** (2014)
7. P. Chittaranjan, PhD Thesis (2015)
8. K. Al-Jabri, H. Shoukry, I.S. Khalil, S. Nasir, H.F. Hassan, J. Mater. Civil Eng. (under review)
9. J. Zhang, Ch. Gong, L. Lu, W. Shoude, P. Hou, Ceramics – Silikáty **59**, 2 (2015)
10. I. Zheng, C. Xuehua, T. Mingshu, J. Cem. Concr. Res. **22** (1992)
11. I. Barišić, S. Dimter, I. Netinger, Technical Gazette **17**, 4 (2010)
12. J. Zelić, Cem. Concr. Res. **35** (2005)
13. O. Gencel, F. Koksal, C. Ozel, W. Brostow, Constr. Build. Mater. **29** (2012)
14. A. Yilmaz, M. Karaşahin, Turkish J Eng Env Sci **37** (2013)
15. P.K. Acharya, S.K. Patro, J. Clean. Prod., **131** (2016)
16. M. Nadeen, A.D. Pofale, Open Journal of Civil Engineering **2** (2012).
17. C.R. Panda, K.K. Mishra, K.C. Panda, B.D. Nayak, B.B. Nayak, Constr Build Mater 49 (2013)
18. S.M. Susheel, S.R. Sathwik, T. Vinayak, S. Darshan, J. Sanjith, A. Ranjith, IJISET **3**, 6 (2016)
19. S.R. Sathwik, J. Sanjith, G.N. Sudhakar, AJER **5**, 9 (2016)
20. V. Kallio, T. Holappa, A. Tikkakoski, Outokumpu Report, http://www.morenia.fi/wp-content/uploads/sites/7/2015/02/OKTO_rakennustuotteet_eng.pdf (accessed 20/06/2017).
21. B.B. Das, MSc Thesis, National Institute of Technology Rourkela (2014)
22. W.A. Gericke, INFACON7 (1995)

Corrosion behaviour of reinforced concrete constructions in acid aggressive environments: case studies, laboratory tests and corrosion mechanisms

Comportement à la corrosion des constructions en béton armé dans des milieux agressifs acides: études de cas, tests de laboratoire et mécanismes de corrosion

Ioan Pepenar[1]
[1]ICECON GROUP, Centre de Recherche CERTINCON, Bucarest, Roumanie

Abstract. The paper presents the results of research on the corrosion behaviour of reinforced concrete elements which compose the bearing structure of some constructions subjected to the corrosive action of acid aggressive agents, after a long-term service in a strongly aggressive environment. To evaluate the damage state of reinforced concrete elements/constructions, both "in situ" investigation of constructions and laboratory tests on concrete samples extracted from some elements damaged by corrosion, using a specific investigation methodology were performed. The results of the research revealed the existence of a severe corrosion-induced damage of the structural elements, which affected the resistance, stability and durability of the constructions. Based on case studies on the service behaviour of reinforced concrete elements, there are pointed out the specific damages generated by the corrosive action of the acid aggressive agents, the causes of their occurrence and presents considerations on the mechanisms of the corrosion processes of concrete and steel reinforcement in acid aggressive environments. Finally, there were proposed intervention measures in order to assure normal service conditions of the reinforced concrete constructions located in such aggressive environments.

Résumé. Le papier présente les résultats des recherches sur le comportement à la corrosion des éléments en béton armé qui constituent la structure de résistance des certaines constructions soumises à l'action corrosive des agents agressifs acides, après un service de longue durée dans le milieu très agressif. Afin d'évaluer l'état de dégradation des éléments/constructions en béton armé, suite à l'action corrosive des agents agressifs acides, on a utilisé une méthodologie spécifique d'investigation et de diagnostic, y comprenant des investigations "in situ" des constructions et des tests de laboratoire sur des échantillons de béton prélevés des certains éléments dégradés par corrosion. Les résultats des recherches ont révélé l'existence des dégradations graves par corrosion des éléments structuraux qui affectent la résistance mécanique, la stabilité et la durabilité des constructions. Sur la base des études de cas sur le comportement en service des éléments en béton armé, on met en évidence les dégradations spécifiques provoquées par l'action corrosive des agents agressifs acides, les causes de leur apparition et on présente des considérations sur les mécanismes des processus de corrosion du béton et des armatures en acier dans les milieux agressifs acides. Finalement, on présente des mesures d'intervention proposées pour assurer les conditions normales de service des constructions en béton armé situées dans tels milieux agressifs.

1 Introduction

La recherche du comportement des constructions en béton soumises à l'action des milieux agressifs, particulièrement des milieux acides et la réhabilitation des constructions dégradées par corrosion est d'actualité et d'une grande importance, tant du point de vue de la protection du milieu construit existant, de même que de la nécessité du perfectionnement continu de la conception, du calcul, de l'exécution et de la durabilité des constructions en béton armé et béton précontraint qui seront placées dans tels milieux agressifs.

Au cours des décennies récentes, ils ont rapporté de nombreux cas de durabilité insatisfaisante des constructions en béton situées dans des milieux agressifs, respectivement de dégradation des éléments structuraux sous l'action des agents agressifs après différents intervalles de service des constructions. L'expérience acquise après un grand nombre d'études et de recherches sur le comportement en service des constructions a montré que les plus fréquentes et les plus dangereux cas de dégradation par corrosion, avec des conséquences importantes sur leur résistance mécanique et stabilité se rencontrent dans les constructions de l'industrie chimique et d'autres industries où se déroulent des processus technologiques de nature chimique [1].

Par rapport à d'autres milieux agressifs, les milieux industriels contenant des agents agressifs à caractère acide sont parmi les milieux les plus agressifs et les plus complexes. Ces milieux présentent une agressivité spécifique et complexe sur les constructions en béton armé et béton précontraint, affectant tant le béton de même que l'armature en acier [2, 3].

Dans le présent papier sont synthétisés les résultats des recherches sur le comportement à la corrosion acide des éléments en béton armé qui constituent la structure de résistance des certaines constructions après une service de longue durée dans des milieux très agressifs, les mécanismes des processus de corrosion du béton et des armatures en acier et les mesures d'intervention proposées pour assurer les conditions normales de service des constructions en béton armé situées dans tels milieux agressifs.

2 Étude de cas

On présente deux études de cas sur la dégradation par corrosion des constructions en béton armé situées dans milieux agressifs acides, l'un sur un bâtiment placé dans un milieu à base d'acide sulfurique et l'autre sur un bâtiment placé dans un milieu à base d'acide chlorhydrique.

Afin d'évaluer l'état de dégradation des éléments/constructions en béton armé, suite à l'action corrosive des agents agressifs acides, on a utilisé une méthodologie spécifique d'investigation et de diagnostic, y comprenant des investigations "in situ" des constructions et des tests de laboratoire sur des échantillons de béton prélevés des certains éléments dégradés par corrosion [4].

2.1 Bâtiment de fabrication dans un milieu à base d'acide sulfurique

Cette étude de cas se réfère à la dégradation par corrosion des éléments en béton armé d'un bâtiment industriel étagé, à l'intérieur duquel se déroule le processus technologique de fabrication du dioxyde de soufre liquide.

La structure de résistance du bâtiment est constituée de cadres transversales (poteaux et poutres) en béton armé monolithe, plancher intermédiaire technologique et éléments de toiture en béton armé. Indépendante de structure du bâtiment, il y a deux plateformes technologiques en béton armé de soutènement des équipements technologiques. Les plateformes sont réalisées sous forme des plaques qui s'appuient sur deux poutres longitudinales et des poteaux en béton armé avec propres fondations. On a effectué toute une série de réparations des dégradations subies lors le service du bâtiment, mais sans l'efficience attendue.

Les éléments en béton armé de l'intérieur du bâtiment de fabrication du dioxyde de soufre liquide ont été soumis à l'action de long durée d'un milieu fortement agressif constitué principalement de gaz et de vapeurs de dioxyde de soufre (SO_2), dont la concentration a été comprise entre 27 et 60 mg/m^3 d'air, en présence de l'humidité élevée de l'air (supérieure à 75% pendant l'activité de production dans de l'intérieur du bâtiment), ainsi que des solutions concentrées d'acide sulfurique (H_2SO_4) et d'acide sulfureux (H_2SO_3). Suite aux dégradations du revêtement de sol en briques antiacides et des canalisations, ces solutions corrosives se sont infiltrées dans le sol, en affectant les fondations du bâtiment et en provoquant son tassement. Dans ces conditions, le milieu de l'intérieur du bâtiment présent une agressivité très fort sur les éléments en béton armé, conformément aux réglementations techniques en vigueur [5].

L'investigation de l'état technique du bâtiment de fabrication du dioxyde de soufre liquide, après plus de 30 ans de service dans le milieu très fortement agressif, a révélé l'existence des dégradations graves par corrosion de la structure de résistance entière, qui ont conduit à la réduction significative ou même à la perte de la capacité portante des certains éléments de construction et qui constituent un risque potentiel d'effondrement du bâtiment et d'accidents.

Les dégradations par corrosion des éléments de construction en béton armé consistent principalement en:

- *aux poteaux du bâtiment* (figure 1):

- la dégradation en profondeur du béton par corrosion acide, notamment à la base des poteaux, en dépassant le niveau des armatures dans plusieurs cas et en laissant découvertes les armatures en contact direct avec le milieu agressif;

- la corrosion avancée des armatures longitudinales et transversales (étriers), avec des réductions significatives de section, jusqu'à leur disparition complète à la base des certains poteaux, par corrosion généralisée ("dissolution").

Fig. 1. Poteaux: dégradation en profondeur du béton par corrosion acide; réduction de la section de béton (plus de 50% ;

corrosion avancée des armatures, leur disparition à la base des poteaux et au champ; armatures "flottantes".

Certains poteaux sont dans un état grave de dégradation, très dangereux, pour lesquels la réduction importante de la section de béton (plus de 50%) est accompagnée de la disparition des armatures longitudinales et transversales (étriers) sur grandes zones/hauteurs, certaines armatures étant "flottantes". Il est évident que ces poteaux ne peuvent plus prendre les sollicitations sous des actions sismiques.

• *aux poutres des cadres et des planchers*:
- la dégradation en profondeur du béton par corrosion acide, en dépassant le niveau des armatures dans certains cas, pour les poutres des cadres et des planchers;
- la corrosion avancée des armatures longitudinales et transversales (étriers), avec des réductions significatives de section, jusqu'à la disparition des armatures, sur certaines zones;
- la dégradation complète des nervures du plancher de toiture dans les zones au-dessus des équipements technologiques;
- dégradations locales avancées de la plaque du plancher de toiture et du plancher technologique, surtout dans les zones à trous technologiques, à cause des infiltrations de solutions agressives; la dégradation en profondeur du béton sous l'action des gaz et vapeurs à caractère acide;

• *aux plateformes technologiques* (figure 2):
- dégradations graves des poteaux de soutènement des plateformes technologiques: la dégradation en profondeur du béton par corrosion acide, notamment à la base des poteaux, en dépassant le niveau des armatures dans certains cas; la corrosion avancée des armatures longitudinales et transversales (étriers), jusqu'à la disparition des armatures, sur certaines zones; il est évident que ces poteaux ne peuvent plus prendre les sollicitations sous des actions sismiques;
- dégradations graves des poutres et de la plaque des plateformes: la corrosion en profondeur du béton due aux écoulements des solutions acides, en dépassant le niveau des armatures dans plusieurs cas, jusqu'à la disparition des armatures, sur certaines zones.

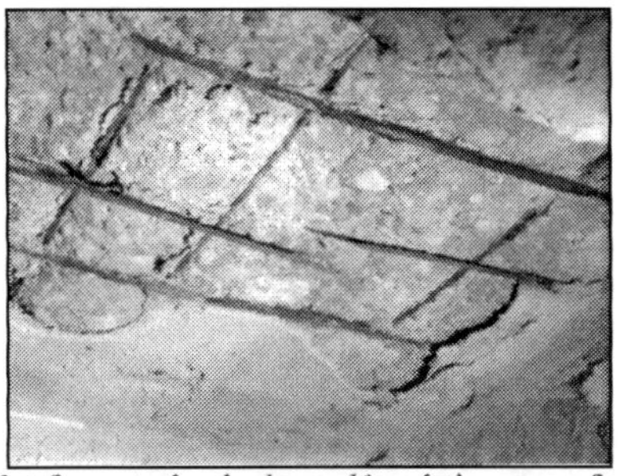

Fig. 2. Plateforme technologique: dégradation en profondeur du béton; corrosion avancée des armatures jusqu'à leur disparition, sur certaines zones.

La sécurité d'utilisation des plateformes technologiques est complètement compromise et il y a le risque de leur effondrement.

Les dégradations graves existantes sont dus au service de longue durée des éléments de construction dans un milieu fortement agressif, suite au fonctionnement défectueux des installations technologiques, y compris celles de capture et d'évacuation des agents agressifs (gaz, vapeurs et solutions à caractère fortement acide).

L'apparition et le développement des dégradations ont été aussi favorisé par l'absence des mesures efficientes de protection anticorrosive des éléments de construction par rapport à la nature et au degré d'agressivité du milieu de l'intérieur du bâtiment ainsi que de l'inefficience des travaux de maintenance du bâtiment et des installations technologiques.

Les processus de dégradation par corrosion des éléments de structuraux en béton armé sont dus à l'action spécifique des agents agressifs à caractère fortement acide, en affectant tant le béton (par corrosion acide) que les armatures en acier (par corrosion généralisée) et ayant comme résultat final la destruction complète des deux matériaux de construction.

Dans le cas du béton, sur l'effet destructive des ions d'hydrogène H^+ (corrosion acide) a été superposée l'effet agressif des ions SO_4^{2-} (corrosion sulfatique), en provoquant des effets importantes de corrosion par l'expansion.

Étant donné l'état de dégradation très avancée du bâtiment, qui affectent la résistance mécanique, la stabilité et la sécurité d'utilisation du bâtiment, le niveau élevé d'infestation du sol avec des solutions fortement acides (H_2SO_4), le degré avancé d'usure physique et morale de l'installation de fabrication du dioxyde de soufre liquide, on a proposé des mesures radicales d'intervention consistant en la démolition du bâtiment existant et la réalisation d'un nouveau bâtiment, conçu et construit conformément aux exigences fondamentales applicables aux ouvrages de construction [6] imposées par le processus technologique de fabrication, qui doit être modernisé également.

2.2 Bâtiment de fabrication dans un milieu à base d'acide chlorhydrique

Cette étude de cas se réfère à la dégradation par corrosion des éléments en béton armé d'un bâtiment industriel, à l'intérieur duquel se déroule le processus technologique de fabrication de l'acide chlorhydrique de synthèse par la combustion du chlore dans l'hydrogène sous vide.

La structure de résistance du bâtiment est constituée de deux cadres en béton armé monolithe, composés de deux poteaux avec section variable et d'une poutre de toiture, deux planchers technologiques et d'éléments de toiture en béton armé.

Les éléments en béton armé de l'intérieur du bâtiment de fabrication ont été soumis à l'action de long durée d'un milieu fortement agressif constitué principalement de gaz et de vapeurs de chlore (Cl_2) et d'acide chlorhydrique (HCl), dont la concentration a variée entre 7,30 et 8,75 mg/m^3 d'air, en présence de l'humidité atmosphérique. En dehors des ces agents agressifs de l'intérieur du bâtiment, les éléments de construction ont été soumis à l'action d'agents agressifs dégagés de l'installation d'acide chlorhydrique située en plein air à proximité immédiate du bâtiment. Dans ces conditions, le milieu de l'intérieur

https://doi.org/10.1051/matecconf/201814901018

du bâtiment présent une agressivité très fort sur les éléments de construction en béton armé, conformément aux réglementations techniques en vigueur [5].

L'investigation de l'état technique du bâtiment de fabrication du l'acide chlorhydrique de synthèse, après plus de 35 ans de service dans le milieu très fortement agressif, a révélé l'existence des dégradations graves par corrosion des éléments structuraux en béton armé, qui affectent la résistance mécanique, la stabilité et la sécurité d'utilisation du bâtiment et qui constituent un risque potentiel d'effondrement du bâtiment et d'accidents. Les dégradations par corrosion des éléments de construction en béton armé consistent principalement en:

• *aux poteaux du bâtiment* (figure 3):

- la dégradation en profondeur du béton par corrosion acide, notamment à la base des poteaux, en dépassant le niveau des armatures dans plusieurs cas et en laissant découvertes les armatures en contact direct avec le milieu agressif;

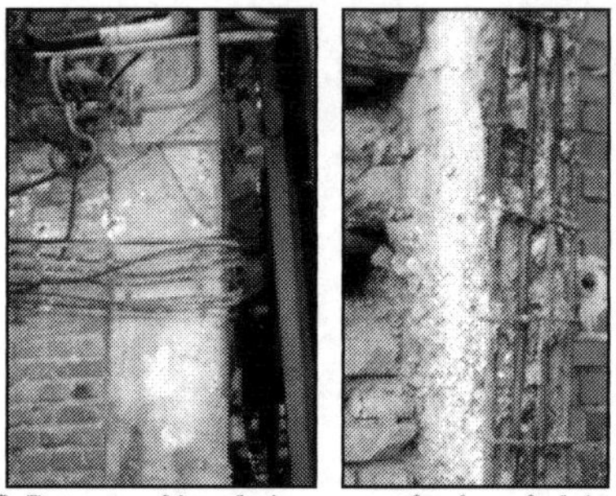

Fig. 3. Poteaux: dégradation en profondeur du béton par corrosion acide; corrosion avancée des armatures, avec réductions significatives de section/leur disparition sur certaines zones; armatures "flottantes".

- la corrosion avancée des armatures longitudinales et transversales (étriers), avec des réductions significatives de section, jusqu'à la disparition complète des armatures, sur certaines zones; aux certains poteaux, les armatures longitudinales sont devenues "flottantes" sur toute l'hauteur du poteau, suite à la disparition des étriers et la destruction par corrosion du béton d'enrobage;

• *aux poutres (plancher de toiture, planchers technologiques)* (figure 4):

- la fissuration du béton au long des armatures et l'éclatement du béton d'enrobage dus à la corrosion des armatures en acier;
- la corrosion avancée des armatures longitudinales et transversales (étriers), avec des réductions significatives de section, jusqu'à la disparition des étriers, sur certaines zones, par corrosion acide généralisée ("dissolution");

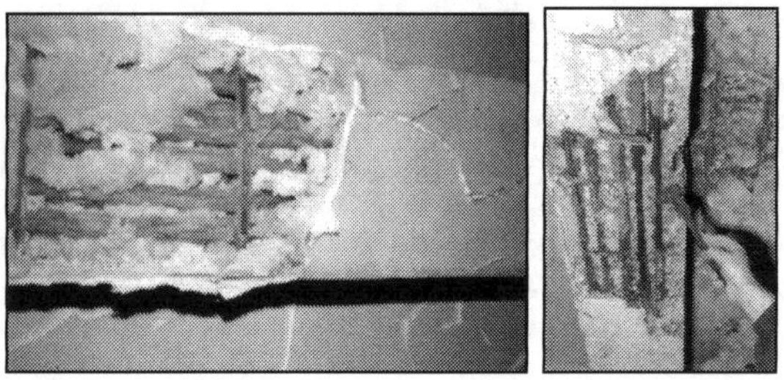

Fig. 4. Poutre de toiture/plateforme technologique: fissuration et éclatement du béton d'enrobage; corrosion avancée des armatures, avec des réductions significatives de section, jusqu'à la disparition des étriers, sur certaines zones.

• *aux planchers technologiques*:

- la fissuration et l'éclatement du béton d'enrobage dus à la corrosion avancée des armatures en acier, avec des réductions significatives de section, jusqu'à la disparition des étriers, sur certaines zones; corrosion acide du béton (figure 5);
- la dégradation complète du béton due à la corrosion acide dans la zone d'appui de la plaque sur le mur en maçonnerie (figure 6); corrosion avancée des armatures, avec des réductions importantes de leur section, jusqu'à la disparition des étriers, sur certaines zones;

Fig. 5. Plancher technologique: fissuration et éclatement du béton d'enrobage; corrosion avancée des armatures, avec des réductions significatives de section, jusqu'à leur disparition.

Fig. 6. Plancher technologique: dégradation complète du béton par corrosion dans la zone d'appui de la plaque sur le mur en maçonnerie; corrosion avancée des armatures, avec des réductions significatives de section.

• *au plancher de toiture:*

- dégradations prononcées par corrosion acide, tant de la plaque que des poutres en béton armé: la corrosion en profondeur du béton; corrosion avancée des armatures en acier qui a conduit à l'éclatement du béton d'enrobage.

Les dégradations graves existantes sont dus au service de longue durée des éléments de construction dans un milieu fortement agressif, constitué principalement de gaz, vapeurs et solutions acides (HCl), suite au fonctionnement défectueux des installations technologiques (y compris la ventilation) et de la préoccupation insuffisante pour la réalisation des travaux de maintenance du bâtiment et des installations technologiques.

Les dégradations ont été favorisées et accentuées par un certain nombre de facteurs, tels que:

- l'utilisation à l'exécution du bâtiment des éléments en béton armé monolithe à défauts initiaux d'exécution (porosité élevée du béton - nids de cailloux, creux, cavernes) (dans le cas des poutres) et des éléments avec l'épaisseur insuffisante de la couche d'enrobage en béton par rapport au degré d'agressivité du milieu (dans le cas

des tous éléments) ainsi que l'utilisation à l'exécution des certains éléments des bétons la classe de résistance et le degré d'imperméabilité inferieurs (dans le cas des poteaux);
- l'existence des trous technologiques dans la plaque des planchers, à travers lesquelles ont eu lieu des fuites de solutions agressives;
- l'absence des mesures efficientes de protection anticorrosive des éléments de construction par rapport à la nature et au degré d'agressivité du milieu de l'intérieur du bâtiment ainsi que l'inefficience des travaux de maintenance du bâtiment et des installations techniques.

Les processus de dégradation par corrosion des éléments structuraux en béton armé sont dus à l'action spécifique des agents agressifs à caractère fortement acide de l'intérieur du bâtiment, en affectant tant le béton (par corrosion acide) que les armatures en acier (par corrosion généralisée) et ayant comme résultat final la destruction complète des deux matériaux de construction.

Étant donné l'état de dégradation par corrosion très avancée de la construction, qui affectent la résistance mécanique, la stabilité et la sécurité d'utilisation du bâtiment, en existant le risque d'effondrement à toute surcharge accidentelle ou même en leur absence, on a proposé des mesures radicales d'intervention consistant en la démolition du bâtiment existant.

3 Tests de laboratoire

Les résultats des tests chimiques de laboratoire sur les échantillons de béton prélevés des éléments de construction dégradés par corrosion acide, ont montré les résultats suivants:

3.1 Bâtiment de fabrication dans un milieu à base d'acide sulfurique

- le pH de la suspension aqueuse des échantillons de béton dégradé a varié entre 1.0...2.5, les valeurs indiquant la désalcalinisation et destruction complète du béton par corrosion acide, ce qui confirme les résultats des tests effectués in situ;
- la concentration des ions sulfate (SO_4^{2-}) solubles dans l'eau dans les échantillons de béton dégradé a variée entre 2.11... 19.2% SO_4^{2-}, les valeurs auxquelles les ions SO_4^{2-} ont un effet fortement corrosif, tant sur le béton que sur les armatures en acier.

3.2 Bâtiment de fabrication dans un milieu à base d'acide chlorhydrique

- le pH de la suspension aqueuse des échantillons de béton dégradé a varié entre 4.0...6.0, les valeurs indiquant la désalcalinisation totale du béton sous l'action de agents agressifs à caractère acide, ce qui confirme les résultats des tests effectués in situ;
- la concentration des ions Cl^- solubles dans l'eau dans les échantillons de béton dégradé a variée entre 0.23... 2.98% Cl^-, les valeurs auxquelles les ions Cl^- ont un effet fortement corrosif, spécialement sur les armatures en acier.

4 Mécanismes de corrosion

L'agressivité des milieux acides sur des éléments de construction en béton armé se manifeste complexe, en affectant tant le béton que l'armature en acier, par des mécanismes de corrosion différents.

4.1 Corrosion du béton

Les processus de corrosion du béton sous l'action des milieux acides se manifeste à un pH < 6,5 et se produit par réactions de double échange entre les constituants du milieu agressif et les composés hydratés du ciment, par la formation de certains produits de réaction très solubles (qui sont lessivés) ou peu solubles/insolubles (qui précipite sous forme de masse des gels, sans propriétés liantes).

Les acides forts tels que *HCl* dissolvent tous les composants de la pâte de ciment durci, par la formation des sels de calcium, aluminium et fer, selon les réactions chimiques suivantes:

$$2HCl + Ca(OH)_2 \rightarrow CaCl_2 + 2H_2O \quad (1)$$

$$4mHCl + m(2CaO.SiO_2.aq) \rightarrow 2mCaCl_2 + mSi(OH)_4 \quad (2)$$

$$mSi(OH)_4 \rightarrow SiO_2.n\,H_2O + 2(2m\text{-}n)\,H_2O \quad (3)$$

$$8HCl + 4CaO.Al_2O_3.13H_2O \rightarrow 4CaCl_2 + 2Al(OH)_3 + 14H_2O \quad (4)$$

Il en résulte que par la réaction directe des solutions acides avec les produits d'hydratation du ciment se forment des sels de calcium ($CaCl_2$ qui est très soluble) et, à la corrosion complète, des gels d'acide silicique, l'hydroxyde d'aluminium et l'hydroxyde de fer. L'hydroxyde d'aluminium et l'hydroxyde de fer réagissent avec des acides forts (par formations des sels) ou peuvent exister en tant que tels dans le cas d'acides faibles.

Un cas particulier est l'acide sulfurique H_2SO_4, dont l'action a un effet cumulatif sur la corrosion du béton: la corrosion acide, due aux ions d'hydrogène H^+ et la corrosion sulfatique, dues aux ions de sulfate (SO_4^{2-}).

Dans le cas de H_2SO_4, se forme le sulfate de calcium qui précipite sous forme de gypse, par la réaction chimique directe de la forme:

$$H_2SO_4 + Ca(OH)_2 \rightarrow CaSO_4.2H_2O \quad (5)$$

Le sulfate de calcium résultant peut cristalliser dans le béton (à des concentrations élevées de H_2SO_4) ou peut réagir avec les aluminates de calcium (C_3A) pour former des sulfo-aluminates de calcium hydratés (ettringite), ayant pour effet l'expansion du béton:

$$3CaSO_4.2H_2O + C_3A + 26H_2O \rightarrow C_3A.3CaSO_4.32\,H_2O \quad (6)$$

La corrosion sulfatique se produit par un mécanisme d'expansion et a lieu par les réactions entre les sulfates du milieu agressif et l'hydroxyde de calcium libre et les aluminates de calcium hydratés de la pâte de ciment durci. Suite à ces réactions chimiques, se forment tant le sulfate de calcium (gypse) que plus particulièrement, le trisulfo-aluminate de calcium hydraté (ettringite), qui cristallise avec 32 molécules d'eau, ce qui produit une augmentation significative du volume (d'environ8 fois) et l'apparition de phénomènes forts d'expansion dans le masse du béton. Finalement, se produit la fissuration et, puis, la destruction rapide du béton. À des concentrations élevées de sulfates, précipite aussi le gypse, qui présente également une augmentation de volume.

De même agit l'acide sulfureux H_2SO_3, l'hydrogène sulfuré H_2S, le dioxyde SO_2 et le trioxyde de soufre SO_3, dans une atmosphère humide, parce que toutes ces substances peuvent être oxydés, pour former H_2SO_4.

4.2 Corrosion de l'armature en acier

Le processus de corrosion de l'armature en acier est de nature électrochimique, étant le résultat du déroulement simultané de deux types de réactions aux électrodes, anodiques et cathodiques. Simplifié, le mécanisme de corrosion des armatures en acier est décrit par les deux réactions [7]:

- la réaction anodique, de dissolution (oxydation) du fer:

$$Fe \rightarrow Fe^{2+} + 2\,e^- \qquad (7)$$

- les réactions cathodiques, de réduction des ions d'hydrogène ou d'oxygène (en milieu acide):

$$2\,H^+ + 2\,e^- \rightarrow H_2 \qquad (8)$$

$$O_2 + 4H^+ + 4e^- \rightarrow 2\,H_2O \qquad (9)$$

Dans la plupart des cas, le processus de corrosion de l'acier dans des milieux acides se déroule avec une dépolarisation d'hydrogène.

La corrosion de l'armature dépend tant des propriétés du béton et de l'acier, que la nature et le caractère du milieu dans lequel est situé l'élément de construction.

Dans des conditions normales, l'armature d'acier est protégée dans le béton sain par le phénomène de passivation grâce au pH élevé (pH = 12 ... 13) de la solution interstitielle contenue dans les pores du béton. À ces valeurs de pH, la formation d'une couche protectrice microscopique sur la surface de l'acier ne permet pas la corrosion (dissolution du fer), même en présence d'humidité et d'oxygène.

La couche de passivation de l'acier dans le béton peut être détruite, respectivement la dépassivation et la corrosion de l'acier deviennent possibles dans diverses conditions:

- la réduction de l'alcalinité du béton par l'action d'agents agressifs à caractère acide (ex. carbonatation par CO_2) jusqu' aux valeurs du pH du béton inférieur à environ 9;
- la présence des anions dépassivants (Cl^-, SO_4^{2-}, NO_3^- etc.) à concentrations critiques au niveau de l'armature;
- la destruction par corrosion de la couche d'enrobage en béton;
- la présence des fissures dans le béton etc.

Dans le cas des gaz et vapeurs acides (Cl_2, HCl, etc.) qui diffusent facilement à travers du béton, la corrosion de l'armature peut se produire avant la carbonatation / désalcalinisation totale de la couche d'enrobage en béton. L'aspect extérieur de la corrosion du béton armé provoquée par Cl_2 et HCl gazeux (en présence de l'humidité atmosphérique), ainsi que par les composés du chlore en général, est la fissuration du béton au long des armatures à cause de la pression des produits de corrosion de l'acier (rouille) sur la couche d'enrobage en béton.

Dans le cas des solutions acides, la corrosion de l'armature d'acier se produit après la désalcalinisation complète de la couche d'enrobage en béton.

En général, dans les milieux agressifs acides se produit une corrosion généralisée ou localisée de l'acier, avec des réductions significatives de section des armatures (parfois même leur "dissolution" complete). Une dernière forme de corrosion est induite par l'action simultanée de la corrosion et d'une sollicitation mécanique. Il s'agit de corrosion fissurante sous tension induite par l'hydrogène (fragilisation par l'hydrogène) qui est caractéristique pour les aciers à haute résistance pour la précontrainte du beton dans certains milieux agressifs spécifiques, tels que les milieux contenant ions SH^-, SCN, H^+ etc.). Cette forme insidieuse de corrosion, caractérisée par la fissuration et la fracture brusque de l'armature en acier prétendue, sans striction et sans perte visible de métal, est très dangereuse parce que peut provoquer l'avarie des éléments/constructions en béton précontraint sans l'apparition au préalable des signes extérieurs avertissant le danger [8].

Dans le cas de l'action des anions corrosifs, tels que des ions sulfate (SO_4^{2-}), la couche protectrice existante sur la surface de l'armature dans le béton est détruite localement et provoque la corrosion localisée de l'acier. Étant donné que la zone anodique est beaucoup plus réduite par rapport à la zone cathodique se produit une diminution significative de la section de l'armature dans cette zone. Par rapport aux ions Cl^-, les ions SO_4^{2-} ont une action corrosive inférieure sur l'armature en acier; le sulfate de fer formé initialement se transforme en rouille en présence de l'oxygène et de l'eau. À des concentrations élevées des ions SO_4^{2-} se produit la corrosion généralisée de l'armature en acier.

La diminution de section de l'armature en acier et le gonflement simultané de la rouille entraînent une diminution plus ou moins notable de l'adhérence entre l'armature en acier et le béton.

5 Conclusions

Les résultats des recherches sur le comportement en service de longue durée des éléments en béton armé qui constituent la structure de résistance des certaines constructions soumises à l'action corrosive des milieux agressifs acides ont révélé l'existence des dégradations graves par corrosion des éléments / constructions, qui ont conduit à la réduction significative ou même à la perte de la capacité portante des certains éléments structuraux et qui constituent des risques potentiels d'effondrement des constructions et d'accidents.

Les dégradations par corrosion ont été causées par les conditions sévères de service des éléments / constructions dans des milieux acides fortement agressifs, étant favorisées et accentuées par l'absence des mesures efficientes de protection anticorrosive des éléments de construction par rapport à la nature et au degré d'agressivité des milieux de l'intérieur des constructions ainsi que de l'inefficience des travaux de maintenance des constructions et des installations technologiques.

Afin d'intervenir avec des mesures efficientes de prévention de la dégradation des éléments de construction soumis à l'action des milieux agressifs acides est nécessaire de bien connaître les mécanismes des processus de corrosion des matériaux de construction (béton, armatures en acier).

Étant donné l'état de dégradation très avancée par corrosion des constructions, qui affectent leur résistance mécanique, stabilité et sécurité d'utilisation, on a proposé des mesures radicales d'intervention consistant en démolition des constructions dégradées existantes et la réalisation de nouvelles constructions, conçues et construites conformément aux exigences fondamentales applicables aux ouvrages de construction imposées par les processus technologiques de fabrication, qui doivent être modernisés également.

References

1. I. Pepenar, *Conferinta Nationala Comportarea in situ a constructiilor*, Romania, 57 (1994)
2. I. Pepenar, Constructii, **3**, 40 (1977)
3. I. Pepenar, Constructii, **11**, 33 (1979)
4. I. Pepenar, *Durability of Building Materials and Components 8*, Ottawa, Canada, 231 (1999)
5. NE 012/1, *Normativ pentru producerea betonului si executarea lucrarilor din beton, beton armat si beton precomprimat, Partea 1: Producerea betonului, Anexa I*, Romania (2007)
6. Règlement (UE) N° 305 du Parlement Européenne et du Conseil, JO L 88 du 4.4.2011 (2011)
7. I. Pepenar, G. Grimaldi, A. Raharinaivo, Études et recherches des LPC, OA 14, 17 (1994)
8. I. Pepenar, *Durability of Construction Materials*, Paris, France, 1388 (1987)

Towards more sustainable construction–application of superabsorbent polymers in cementitious matrices with reduced carbon footprint

Agnieszka J. Klemm[1] and Fernando C.R. Almeida[1]
[1]Glasgow Caledonian University, School of Engineering and Built Environment, Glasgow, United Kingdom

Abstract. Construction industry is constantly searching for sustainable innovations to mitigate negative environmental impacts. Ground granulated blast-furnace slag (GGBS) is a well-known supplementary cementitious material which contributes to reduction of energy and CO_2 emissions from cement industry. However, its use in cementitious systems leads to materials with high cracking susceptibility due to their greater autogenous shrinkage triggered by self-desiccation processes. This problem is even more pronounced when concrete is exposed to severe dry-hot weather conditions, such as in North Africa. In order to mitigate this negative effect of cracking, internal curing agents in the form of Superabsorbent polymers (SAP) can be successfully used. This approach leads to more durable cement based materials and in turn more sustainable constructions.

1 Introduction

Construction industry is constantly searching for sustainable innovations, which aim to reduce environmental impact and propose more durable solutions for buildings. It is widely accepted fact that Supplementary Cementitious Materials (SCM) reduce energy and CO_2 emissions from cement production, improve material's performance as well as reduce wastes from other industries [1–3]. Ground granulated blast-furnace slag (GGBS), a by-product of iron industry, is one of the most popular SCMs. Due to its latent hydraulic property, activated by alkalis/lime present in Portland cement (PC), it can effectively replace up to 95% of cement [4–8].

One of the most common problems in concrete is its premature drying (self-desiccation), which in turn often results in micro-cracking and a subsequent deterioration. This problem is even more relevant to the concrete structures in regions with high temperatures/low humidity such as those in the North Africa [9–11]. Superabsorbent polymers (SAP) may eliminate or substantially reduce this effect by supplying setting and hardening concrete with additional water resources for internal curing [12–16]. Although significant efforts have been made during the last couple of decades there is still a lack of consensus on mechanisms ruling SAP actions in cementitious materials [17–19].

This paper therefore aims to present the current state-of-the-art and to provide an overview of SAP as a new admixture, highlighting its effects in GGBS-PC matrices.

2 Superabsorbent polymers

Superabsorbent polymers (SAP) are hydrogels formed by macromolecular networks of hydrophilic polymer chains, with the ability to absorb aqueous fluids. SAPs are lightly crosslinked polyelectrolytes, generally based on ionic monomers, such as acrylamide and acrylic acid, which are partially neutralized with hydroxides of alkali metals (usually sodium). This lightly crosslinked three-dimensional structure displays a high capacity to swell by water absorption (Fig. 1); SAPs can absorb as much water as about 10-1000 g/g [12,20].

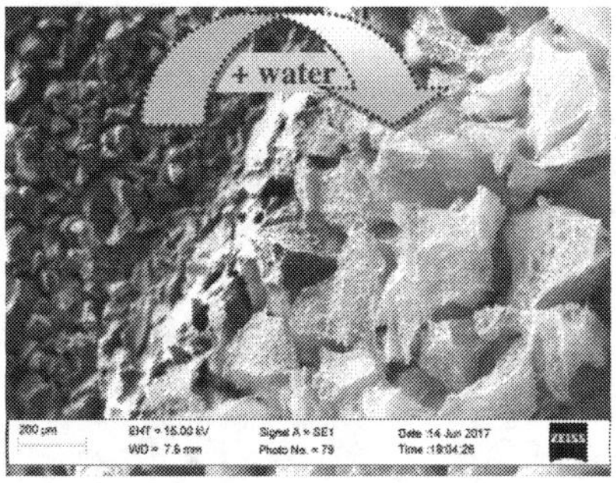

Fig. 1. SAP in contact with water: dry particles (left) and swollen hydrogel (right).

Although SAPs are mainly used in personal disposable hygiene products they also have several other applications, such as in agriculture, pharmaceutics, separation technology, fibers/textiles, water-swelling rubbers, soft actuators/valves, electrical and construction

[20]. In 2014, 3.12 million tons of SAP was produced worldwide, and its consumption is estimated to be expanded by 20% by 2018. Six enterprises are responsible for more than 80% of the global SAP production, concentrated mainly in the USA, Germany, China and Japan [21].

Different SAP applications may lead to different sorption behaviour. The swelling capacity and strength of the polymer network depends not only on the crosslinking density and chemical structure of monomers, but also on the external stimuli surrounding SAP [22]. Highly crosslinked SAPs made of less hydrophilic monomers have lower absorption capacities. On the other hand, polymers with lower anion density may store the absorbed liquid for longer and, eventually, provide very moderate release of solution [20,23].

Such a property is desired especially when SAP is used as internal curing admixture in cement-based materials. SAPs absorb water from fresh mix and release it over time during cement hydration processes, resulting in enhancement of some concrete properties. In this context, both sorption capacity and the kinetics of absorption are highly influenced by pH and ionic concentration of a pore solution. The complex interaction of di- and trivalent ions (e.g. Ca^{2+} and Al^{3+}) from cement with carboxylate groups results in an additional crosslink and hence reduces SAP absorption capacity [12,23]. Thus, raw materials, in particular type of cement, play an essential role in determining SAP efficiency in concrete and mortars. The use of SCM, in turn, may change the behaviour of absorption and desorption capacity. Fig. 2 illustrates the difference of water absorption capacity over time (up to 24h) in Portland cement (PC) and GGBS filtrates, by means of tea-bag method [23,24]. Although SAP achieved the max absorption capacity in the first minutes for both solutions (around 40 g/g), GGBS systems seems to lead to a faster water release.

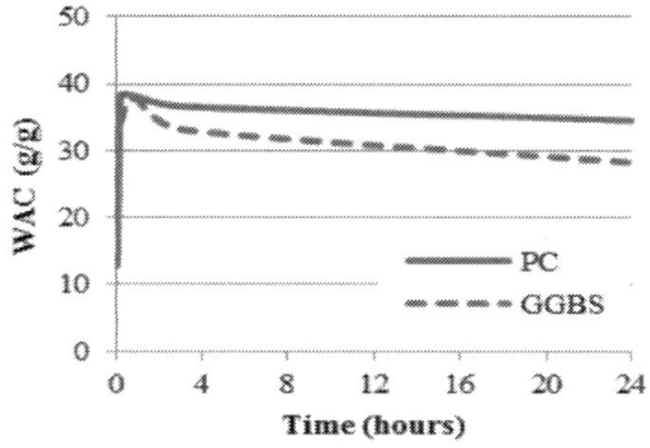

Fig. 2. SAP water absorption capacity in PC and GGBS systems.

In an attempt to show the effect of SAPs on PC-GGBS systems, two types of polymers with different water absorption capacities has been considered and presented in this paper. SAP A and SAP B (with water absorption capacities of 10 g/g and 25-30 g/g in cement paste solution, respectively) have been used in the proportion of 0.25% of binder mass. Four levels of PC (CEM I 52.5N [8]) replacement by GGBS has been considered: 0%, 25%, 50% and 75%. SAP effects on GGBS-PC matrices were assessed by analysing mortars with binder:sand ratio 1:2 and water/binder ratio 0.5. Samples were named by type of SAP (A, B, Ref – no SAP) and GGBS content (0, 25, 50, 75).

2 Workability and setting times

Fresh state properties can be significantly altered by SAP addition, since it absorbs water almost instantly from fresh mix. Thus, sorption characteristics of SAPs are crucial in the determination of early age properties of cementitious matrices. As mentioned previously, many factors can affect SAP absorption capacity and the kinetics of absorption/desorption processes. These include type of polymer (chemical composition and molecular structure), SAP characteristics (shape and size of particles) and type of medium (ions concentration and type of cementitious matrix – e.g. paste, mortar or concrete).

For example, regarding SAP particle sizes, for the same amount of internal curing water, finer particles provide higher values of yield stress and plastic viscosity (drier mixes). This is because of greater total absorption surface of these particles and, consequently, more rapid achievement of equilibrium. On the other hand, larger SAP particles lead to slower water absorption, resulting in less consistent mixes during the first minutes [25].

Moreover, the type of raw materials used in the mix composition can affect rheological properties. The use of finer materials in concrete, such as silica fume, leads to a "competition" between SAP absorption and adsorption of water on finer particles. Thus, it becomes more difficult for SAP to uptake water resulting in reduced workability [25].

Overall, SAP reduces workability of concrete and mortars for the same w/b ratio. This reduction is evidenced by the experimental results of flow table test [26] showed in Fig. 3. Mortars modified by SAPs exhibited lower values than the reference samples (below 140mm), for all GGBS contents. Also, due to the fineness of GGBS, a slight reduction in flow values was noticed for higher PC replacement levels.

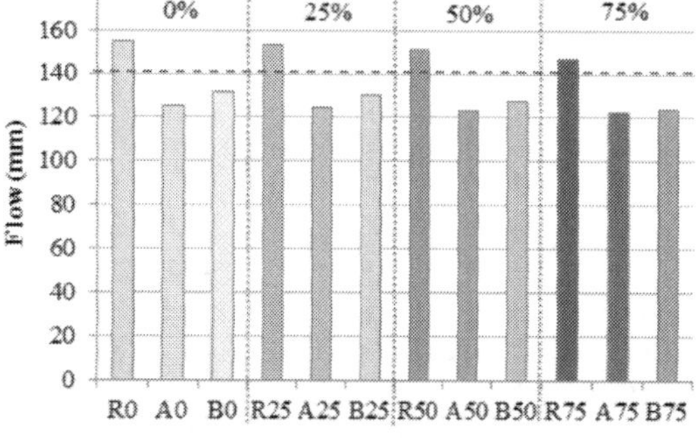

Fig. 3. Flow results of PC-GGBS mortars modified by SAPs.

Nonetheless, this drawback in reduced workability by SAPs can be minimized or even eliminated by using additional water and/or superplasticizer [13,25,27].

However, any of these additional efforts, as any other chemical admixture, should be carefully evaluated beforehand. For instance, increasing w/b ratio may improve workability but also result in higher water absorption by SAP, and consequently, in higher relative changes of rheological parameters [25]. Also, the use of superplasticizer can alter ionic concentration of the medium and hence change SAP sorption behaviour.

Regarding setting time, there is still no consensus whether SAPs increase or decrease this property. Significant scatter in experimental results can be due to variations of raw materials, type and particle size of SAPs, accuracy of test methods [28] and to the fact that SAPs can leach out some monomers into the fresh mix [13,25,29]. Fig. 4 supports this statement. Cement composition and type of SAP can lead to different values of setting times compared to the reference mortars.

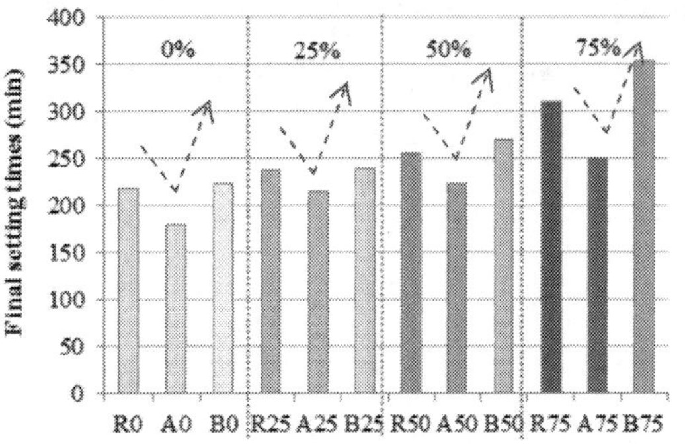

Fig. 4. Setting time results of PC-GGBS mortars modified by SAPs.

While SAP A reduced setting time, SAP B had an opposite effect. This difference is even more considerable for an increased GGBS content in the mix (above 50%).

3 Cracking susceptibility

The main and most well accepted application of SAP in cementitious materials is for the reduction of cracking susceptibility triggered by self-desiccation process. SAP capacity to store and supply water over time can significantly mitigate the effects of volume changes in concrete due to shrinkage, in particular those related to autogenous deformations. Fig. 5 shows that all SAP samples displayed reduced autogenous shrinkage compared to the reference samples. Results cover the period between final setting time and 7 days [30].

Autogenous shrinkage is a result of cement hydration process during hardening of concrete. The use of SAP as an internal curing agent leads to changes in thermodynamics and kinetics of cement reaction due to additional quantity and availability of water. This effect can also contribute to reduction of other types of deformations, such as those provoked by plastic and drying shrinkage [12].

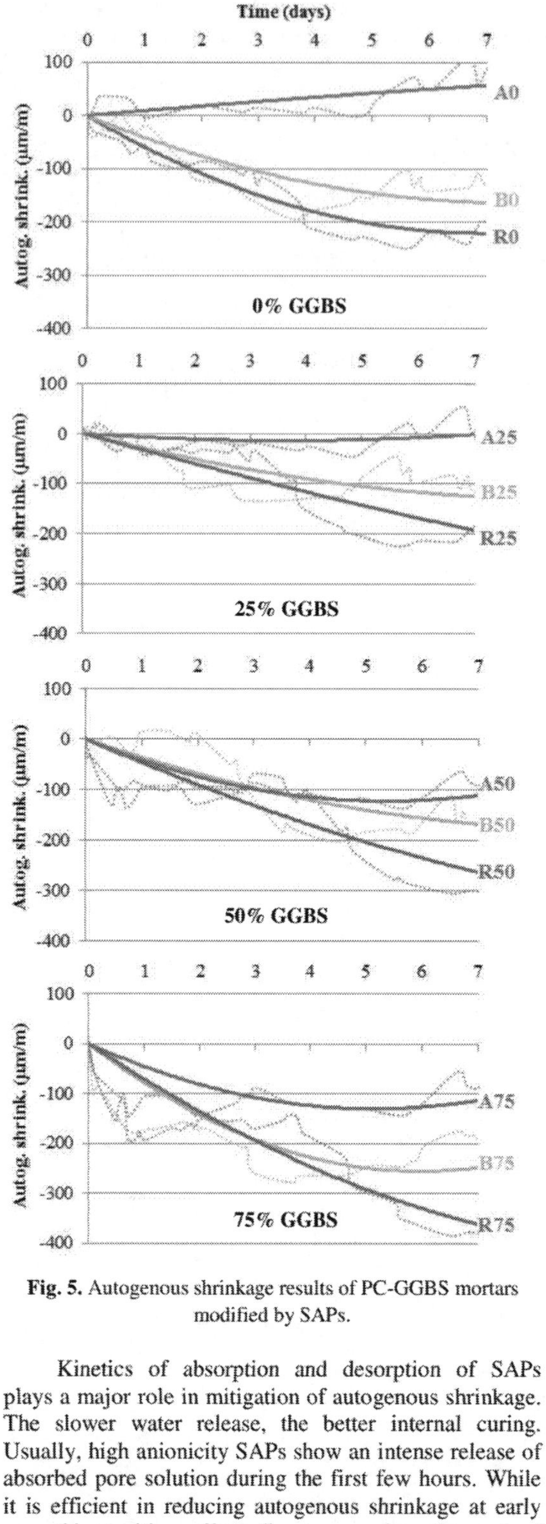

Fig. 5. Autogenous shrinkage results of PC-GGBS mortars modified by SAPs.

Kinetics of absorption and desorption of SAPs plays a major role in mitigation of autogenous shrinkage. The slower water release, the better internal curing. Usually, high anionicity SAPs show an intense release of absorbed pore solution during the first few hours. While it is efficient in reducing autogenous shrinkage at early age, this positive effect disappears when the rest of internal curing water is consumed. Thus, the increase in autogenous shrinkage starts to take place with the same pace as in SAP-free concrete. However, SAPs with low anion concentration, which do not exhibit any, or just a very moderate release of pore solution, can be used to reduce autogenous shrinkage for longer periods of time [13,23].

SAPs can also have slower water intake extending even after setting time (from bleeding water). In this case, an expansion can be observed, which counteracts an expected shrinkage in cementitious matrices (e.g. SAP A for low GGBS levels in Fig. 5). This expansion is also a function of the cement chemistry and the admixture dosage; SAP increases the portlandite oversaturation level in solution, which can result in higher crystallization stresses, and hence lead to a material expansion [18,31].

Concrete composition and type of raw materials also contributes to the SAP behaviour in reducing shrinkage. High performance concrete (lower w/b ratio), for example, is very prone to early age cracking induced by self-desiccation and autogenous deformation [13,15]. High content of GGBS also leads to increased chemical shrinkage due to its filler effect and greater hydration degree [1,4,5]. Thus, SAPs showed themselves as a great shrinkage reducer in GGBS-PC matrices (Fig. 5).

Thus, cracking susceptibility by shrinkage deformation is very dependent to the amount of water available for hydration and curing of GGBS cementitious systems. In this context, creating mechanisms that contribute to controlling of water supply is essential for the material durability, especially in hot and dry climates due to high surface evaporation rate. For example in regions such as in North Africa, day temperature may reach 43°C in summer with amplitude between day and night often reaching about 18°C. The relative humidity (RH) level of air can reach 24% (Table 1) [9].

Table 1. The monthly mean climatic data in North Africa (Ouargla Algeria region) [9].

	Jun	Jul	Aug	Sep	Oct	Nov
T_{max} (°C)	39.8	43.2	42.2	37.6	30.7	23.6
T (°C)	31.5	34.3	33.6	29.8	23.4	16.6
T_{min} (°C)	23.2	25.3	25.0	22.0	16.0	9.6
RH (%)	28.5	24.4	26.4	34.4	47.6	56.5

It has been well documented that in this type of environment, conventional concrete and mortars exhibit high level of shrinkage and cracking [9–11].

Therefore, the higher shrinkage triggered by GGBS, especially in critical wheatear conditions, the greater the need for SAPs. It helps to keep durability and sustainability in acceptable levels for cementitious matrices, particularly in avoiding crack formation.

4 Mechanical properties

The use of SAP with additional water generally can lead to losses in strength. However, it can be partly recovered with time due to ongoing hydration facilitated by the internal curing mechanisms [17,23]. SAP promotes creation of a dense network of CSH in a collapsed state, and hence leading to pore closures and increased compressive strength [16,19,32].

This recovery depends on SAP efficiency in water supply during cement hydration. Premature release of water (before setting) leads to a decrease in compressive strength. This is attributed to some increase in effective w/b (due to additional water for internal curing) and subsequently higher porosity of concrete. In contrast, for mixes with SAP that do not exhibit an early water release, no significant decrease in strength is observed due to its capacity to release water in later stages [13,23].

However, other factors can contribute to strength level, such as, shape and size of SAP particles. Spherical SAPs and smaller particles lead to higher compressive strengths in comparison to mixes with irregular and larger particles. This is because higher stress concentration in combination with larger voids left by collapsed SAP has a strong impact on mechanical properties in the hardened state [13].

In case of GGBS-PC matrices, it is expected that the hydration processes last longer due to slower reaction of slag compared to Portland cement. This delay in GGBS hydration is related to its chemical composition, fineness, glass content, temperature, and to the fact that GGBS has to be activated by alkalis/lime to start to form CSH [2,6]. In this case, the presence of SAP plays an important role in providing water for longer period.

However, the type of cement can also have an effect on the SAP performance (see Fig. 2). This effect on compressive strength can be observed in Fig. 6 [33].

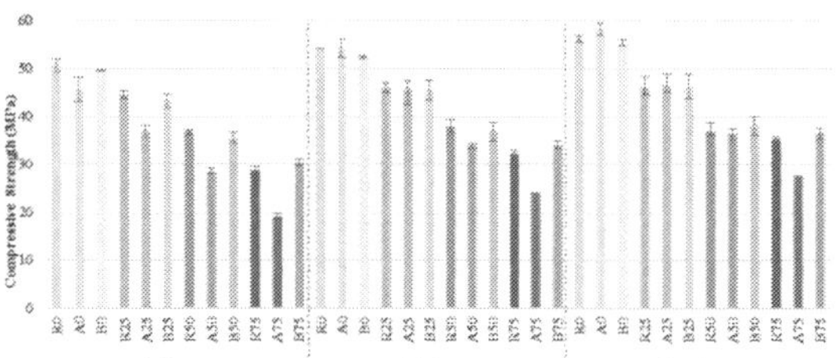

Fig. 6. Compressive strength results of PC-GGBS mortars modified by SAPs.

For low GGBS contents, the effect of SAP on compressive strength is very limited. However, for higher GGBS contents, the effect of SAP type is more pronounced. SAP B, with higher absorption capacity, is able to store and provide more water for hydration for any PC-GGBS system. Thus, it can lead to higher densification of cementitious matrix resulting in a slight increment in compressive strength up to 28 days. On the other hand, SAP A, with lower WAC, has significantly decreased compressive strength for high GGBS contents. This can be related to the instability of the polymer that is not able to retain water for longer (after one week), leaving behind pores that negatively affect compressive strength. This effect on pore formation is discussed in the following paragraph.

5 Microstructure alterations

When SAP absorbs water from fresh mix, it turns to a gel and swells in the mixture. This gel is therefore responsible for storage and release of water for internal curing in cementitious matrix. In other words, this water desorption happens (or should happen) simultaneously with the hardening process leading to a substantial modification of the skeleton formed by hydrated products. Such alterations can be related to the voids created by collapsed SAPs, when gel polymer releases absorbed water and, in consequence, leave behind pores in the hardened microstructure. Although these pores can be filled with later hydration products (especially for GGBS-PC), microstructure alterations are directly dependent on SAP absorption, storage capacity and desorption rate in each cementitious reacting system. Fig. 7 schematically represents the modification of cementitious systems by SAP.

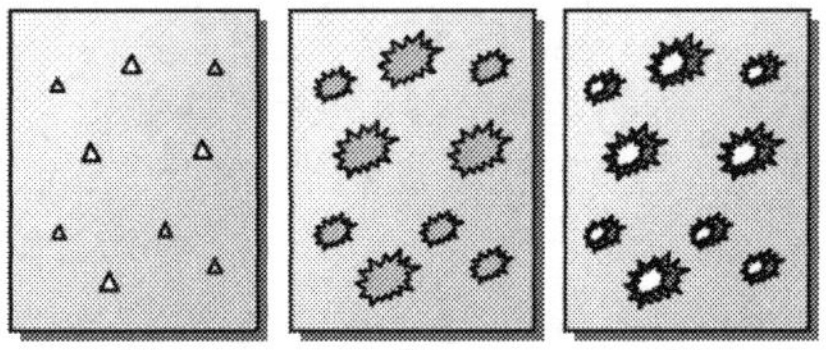

(a) (b) (c)

Fig. 7. SAP modifications in cementitious systems. (a) dry mix with dry SAP particles; (b) fresh matrix with swollen SAP particles filled with water (gel); (c) hardened matrix with collapsed SAPs and voids partially filled with cement hydration products.

When capillary pores (10nm-10μm) are analysed, it is expected that SAPs increase total porosity due to macro-pore formation, especially when additional water is considered [34]. In the case of smaller pores, additional amounts of water do not significantly affect micropore range (< 2 nm) and slightly increase mesopore range (2-50 nm) [35].

Reduction of smaller capillaries is observed in matrices modified by SAPs without extra water. This is due to the filling of existing pores with hydration products and the reduction of initial micro cracks caused by now mitigated autogenous shrinkage. Thus, the microstructure around SAPs is denser due to internal curing and unconnected macropores [32,35,36].

GGBS can also have an impact on the microstructure of cement based matrices. Its ability to slow down cementitious reactions and induce a filler effect leads to a denser CH-laden interfacial zone and formation of finer capillaries, in particular, those in the mesopore range [1,2,6,36].

Fig. 8 shows an example of how different types of polymers can influence porosity of GGBS-PC matrices. SAP A is able to create more pores, since it was not completely filled with hydrated products in the collapsed state. This polymer has presented the highest peaks for all samples.

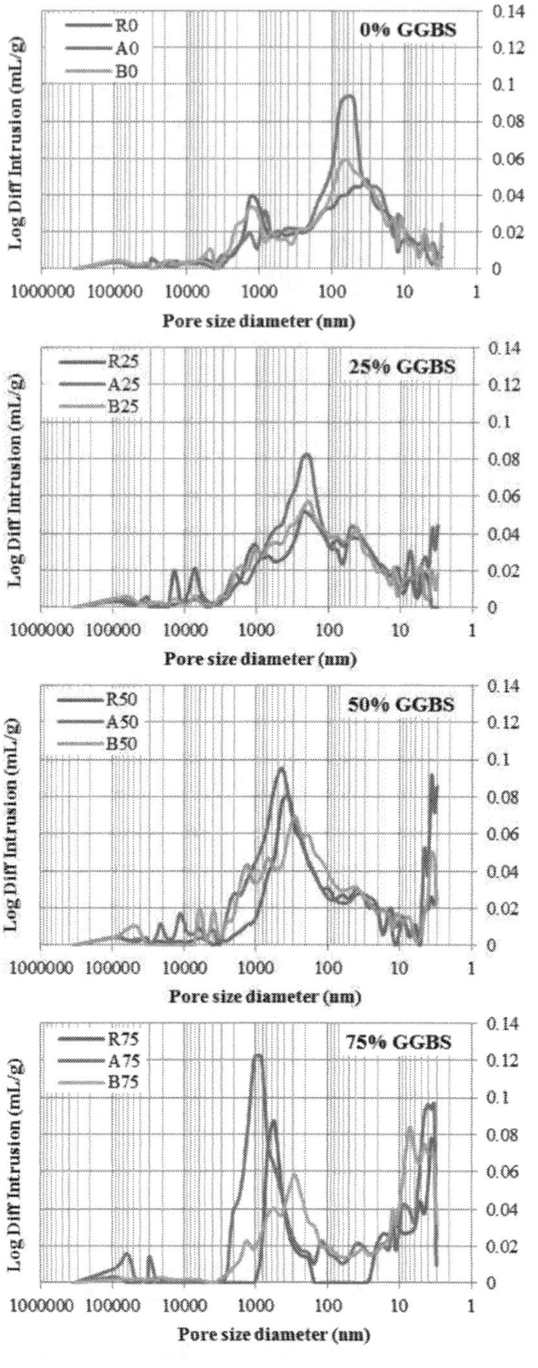

Fig. 8. Pore size distribution of PC-GGBS mortars modified by SAPs at 28 days.

However, for higher GGBS contents (from 50% onwards), both SAPs have created larger capillary pores (above 1μm) when compared to the reference sample; SAP A formed the largest pore sizes (curves shifted to the left) indicating the weakness of this polymer in sustaining slower and later GGBS hydration. In turn, SAP B (with higher water absorption capacity) showed more efficiency in decreasing the amount of larger pores for higher GGBS contents.

6 Final remarks

Superabsorbent polymer (SAP) is a promising new admixture particularly in cementitious materials with reduced carbon footprint (GGBS-PC). Its efficiency in providing water for internal curing results in reduced autogenous shrinkage and, consequently, lower susceptibility for cracking. However, its application in GGBS-PC matrices has to be carefully evaluated, since the effects of different types of SAPs depend on different GGBS concentrations. Overall, SAPs reduce workability and may increase total porosity of concrete in hardened state. Its ability in re-filling the pores (left by collapsed SAPs) and maintaining similar mechanical properties depends on its capacity to store and release water for longer, especially when GGBS is considered (slower reaction rate). SAPs can reduce smaller capillaries with further hydration products, creating a denser and poorly interconnected microstructure. This, in turn, may lead to more durable cement based materials and more sustainable constructions.

Acknowledgments

The authors acknowledge CNPq (National Council for Scientific and Technological Development – Brazil) for the financial support, Hanson Cements for CEM I and GGBS supply, and BASF for SAPs supply.

References

[1] K.L. Scrivener, B. Lothenbach, N. De Belie, E. Gruyaert, J. Skibsted, R. Snellings, A. Vollpracht, Mater. Struct. 48 (2015) 835–862.

[2] M. Thomas, Cem. Concr. Res. 41 (2011) 1224–1231.

[3] R. Siddique, M.I. Khan, Supplementary Cementing Materials, Springer, Berlin, 2011.

[4] K.L. Scrivener, P. Juilland, P.J.M. Monteiro, Cem. Concr. Res. 78 (2015) 38–56.

[5] K.M. Lee, H.K. Lee, S.H. Lee, G.Y. Kim, Cem. Concr. Res. 36 (2006) 1279–1285.

[6] R. Siddique, R. Bennacer, Resour. Conserv. Recycl. 69 (2012) 29–34.

[7] M. Ben Haha, B. Lothenbach, G. Le Saout, F. Winnefeld, Cem. Concr. Res. 42 (2012) 74–83.

[8] BS EN 197-1 (2000).

[9] A. Kriker, G. Debicki, A. Bali, M.M. Khenfer, M. Chabannet, Cem. Concr. Compos. 27 (2005)

[10] K.. Hassan, P.. Robery, L. Al-Alawi, Cem. Concr. Compos. 22 (2000) 453–458.

[11] A.A. Alekrish, S.H. Alsayed, , Cem. Concr. Compos. 16 (1994) 299–307.

[12] V. Mechtcherine, H.-W. Reinhardt, eds., Springer, RILEM, 2012.

[13] V. Mechtcherine, M. Gorges, C. Schroefl, A. Assmann, W. Brameshuber, A.B. Ribeiro, D. Cusson, J. Custódio, E.F. Silva, K. Ichimiya, S. Igarashi, A. Klemm, K. Kovler, A.N. Mendonça Lopes, P. Lura, V.T. Nguyen, H.-W. Reinhardt, R.D.T. Filho, J. Weiss, M. Wyrzykowski, G. Ye, S. Zhutovsky, Mater. Struct. 47 (2013) 541–562.

[14] A.J. Klemm, F.C.R. Almeida, Concr. Plant Int. J. (2016) 38–46.

[15] S. Zhutovsky, K. Kovler, A. Bentur, Cem. Concr. Res. 41 (2011) 981–986.

[16] A.J. Klemm, K.S. Sikora, Constr. Build. Mater. 49 (2013) 134–143.

[17] H. Beushausen, M. Gillmer, M. Alexander, Cem. Concr. Compos. 52 (2014) 73–80.

[18] D. Snoeck, O.M. Jensen, N. De Belie, Cem. Concr. Res. 74 (2015) 59–67.

[19] A.J. Klemm, F.C.R. Almeida, K.S. Sikora, Concr. Plant Int. J. (2016) 50–58.

[20] M.J. Zohuriaan-Mehr, H. Omidian, S. Doroudiani, K. Kabiri, J. Mater. Sci. 45 (2010).

[21] Market Research Reports, (2015). http://www.rnrmarketresearch.com/global-and-china-superabsorbent-polymers-sap-industry-report-2014-2018-market-report.html.

[22] W. Siriwatwechakul, J. Siramanont, W. Vichit-Vadakan, Int. RILEM Conf. Use Superabsorbent Polym. Other New Addit. Concr., RILEM, Lyngby, 2010: pp. 253–263.

[23] C. Schröfl, V. Mechtcherine, M. Gorges, Cem. Concr. Res. 42 (2012) 865–873.

[24] C. Schroefl, V. Mechtcherine, P. Vontobel, J. Hovind, E. Lehmann, Cem. Concr. Res. 75 (2015) 1–13.

[25] V. Mechtcherine, E. Secrieru, C. Schröfl, Cem. Concr. Res. 67 (2015) 52–65.

[26] BS EN 1015-3 (2006).

[27] L. Senff, R.C.E. Modolo, G. Ascensão, D. Hotza, V.M. Ferreira, J.A. Labrincha, Constr. Build. Mater. 95 (2015) 575–584.

[28] BS EN 196-3 (2005)

[29] D. Snoeck, D. Schaubroeck, P. Dubruel, N. De Belie, Constr. Build. Mater. 72 (2014) 148–157.

[30] ASTM-C1698-09 (2009).

[31] G. Sant, B. Lothenbach, P. Juilland, G. Le Saout, J. Weiss, K. Scrivener, Cem. Concr. Res. 41 (2011) 218–229.

[32] F. Wang, J. Yang, S. Hu, X. Li, H. Cheng, Cem. Concr. Res. 81 (2016) 112–121.

[33] BS EN 1015-11 (2006).

[34] V. Mechtcherine, C. Schroefl, M. Wyrzykowski, M. Gorges, P. Lura, D. Cusson, J. Margeson, N. De Belie, D. Snoeck, K. Ichimiya, V. Falikman, S. Friedrich, Mater. Struct. 50 (2017).

[35] D. Snoeck, L.F. Velasco, A. Mignon, S. Van Vlierberghe, P. Dubruel, P. Lodewyckx, N. De Belie, Cem. Concr. Res. 77 (2015) 26–35.

[36] F.C.R. Almeida, A.J. Klemm, in: 14th Int. Conf. Durab. Build. Mater. Components, PRO 113, Ghent, 2017: pp. 205–216.

Mineralogy of C-S-H belite hydrates incorporating Zn-Al-Ti layered double hydroxides

F. Amor[1], A. Diouri[1] and A. Boukhari[1]
[1]Laboratory of Applied Solid State Chemistry, Faculty of Sciences, Mohammed V University, Rabat, Morocco

Abstract. Recently, the belitic cements with low alite content were the subject of several research works which aimed to replace the Ordinary Portland Clinker (OPC) for ecological reasons (reduction of CO_2 emissions), so to understand the reactivity of this cement, the hydration study of the C_2S "dicalcium silicate" phase is primordial research step. As well for a clean environment, the TiO_2 photocatalyst has been extensively applied in the science of building materials because of its ability to degrade the cement surface pollutants. New photocatalyst based layered double hydroxides (LDH) associated with zinc, aluminium and TiO_2 was introduced to increase the compatibility with mortars. The present work is subjected to investigate the effect of the layered double hydroxides on the hydration of C_2S in following the evolution of hydration by X-ray diffraction at 2, 7, 28 and 90 days and analyzing the calcium/silicon ratio of different formed hydrates.

1 Introduction

The ordinary Portland cement is a complex multi-phase system, the hydration includes several simultaneous processes of dissolution/precipitation [1]. As soon as the cement is contacted with the gauging water, the silicate phases C_3S and C_2S react rapidly with H_2O to form less soluble hydrates than these phases. Nevertheless, despite of the almost immediate forming of the first hydrates, the hydration may continue for many years to come [2]. When the cement is hydrated, the reactions during the gauging form the Calcium Silicate Hydrate C-S-H (notation in the cement industry C = CaO, S = SiO_2, and H = H_2O) by releasing Ca^{2+} and OH^- ions [3], these two species combine and form Portlandite. Dicalcium silicate C_2S gives the same products with low quantities and slow kinetics compared to C_3S. In Portland cement the dicalcium silicate has three polymorphic forms α-C_2S, β-C_2S and γ-C_2S (metastable) with a composition which varies between 20% and 25% in the clinker. The slower hydration of C_2S, after 28 days, leads to improved long-term strength. Tests on alite and belite showed that long term strengths are comparable and that only the hydration kinetics is different [4].

C-S-H is the predominant phase resulting from the hydration of dry cement, it is directly responsible for the evolution of mechanical properties. In the literature, several studies show that the tobermorite-like structure is the closest to that of C-S-H [5]. There are three isomorphic structures of tobermorite which are distinguished by their basal space (distance between the two basic plans of two first lamellas neighbors): 9.3 Å, 11 Å [6, 7] and 14 Å [8]. According to the ratio of water/cement and Ca/Si, the C-S-H can adopt one of the isomorphous structures. Tobermorite is generally considered as a valid analogy for the C-S-H with low ratios Ca/Si, while jennite structure is used to describe the structure of C-S-H with high ratios Ca/Si [9,10]. Similarities between C-S-H and tobermorite were also deduced from atomic simulations [11], although this model is the subject of criticism [12]. The chemical composition and the unit cell of the clinotobermorite are closely linked to those of the tobermorite, it is very likely that the clinotobermorite is a stable polymorph at a lower temperature than the tobermorite [13]. Studies of the structure and the kinetic of C-S-H are essential so as to understand the process of the cement setting and the "sticking" mechanism at the microscopic level, their chemical composition is variable. In particular, the ratio (Ca/Si) of calcium-silicon structure is commonly assumed to vary from 0.6 to 2.3, the highest ratio being found in the clean Portland cement, and the lowest in cements containing products such as fly ash or metakaolin [14,15].

The layered double hydroxides LDH have spurred the interest many researchers, they are more likely useful as additions in cements. Cement based materials containing the nanoparticles have shown several benefits under distinct perspectives for sustainable construction practice, as well as, the kinetics of hydration acceleration [16]. Photocatalytic phenomena have become an attractive field of studies in the last decade because of its great potentials for the purification of the environment and the ability to break down aquatic and air pollutants [17, 18].

Furthermore and as part of research on the reduction of pollutants in the atmosphere, the Portland cement is also studied for this aim by adding TiO_2 which allows the degradation of air pollutants by photocatalysis. The photocatalytic principles are based on the free-radical reaction initiated by light onto a concrete surface. Its efficiency depends on the mobility of electron–hole pairs that determines the probability of electrons and holes to achieve contact with active sites on the photocatalyst surface [19]. The most commonly used photocatalyst is TiO_2 due to its strong oxidizing power, its photo-stability and its non-toxicity. This oxide is already widely marketed for its properties and use in surface state [20, 21]. The titanium oxide properties can be improved when using double hydroxides lamellar based Zn-Al-Ti [22].

The introduction of the photocatalytic active inorganic–inorganic nanocomposites to the cement-based mortars improved overall mortar properties (micro-hardness and crystallinity) and showed that the synergetic effect between TiO_2 as traditional photocatalytic and Zn-Al-LDH contributes to the overall photocatalytic performances, improving also the compatibility of the photocatalytic active phase with the mortar matrix [23]. Zinc was selected as constituent LDH metal because of its photocatalytic and antimicrobial activity with the intention to possibly contribute to the overall activity of novel Ti-Zn-Al nanocomposite. ZnO is frequently looked as an alternative to TiO_2, since it can absorb a larger energy fraction of the solar spectrum and more light quanta [24]. In our work, the LDH compounds after synthesis are added under a calcined form with a quantity of $ZnTiO_3$, the compounds, thus introduced under a dried active form, are more likely to react to form again LDH structures that are themselves trapped in hydrates dicalcium silicate. Thermal treatment provides important physicochemical properties to LDH compounds [25]: a "memory effect" of the hydroxide lattice, which allows different anionic species to be incorporated into the LDH interlamellar space, a larger surface area, increasing adsorption of anions and elimination of the interlayer carbonate (CO_3^{2-}), which strongly hinders anion exchange processes in LDHs [26, 27].

In this paper we seek to use LDH compounds and exploit their memory effect by heat treatment and addition to the C_2S phase to study the evolution of hydrates formed in these mixtures. The purpose is to make a mineralogical study of the hydrates of dicalcium silicate in the presence of varying amounts (1% and 3 wt%) of layered double hydroxides Zn-Al-Ti, monitoring of phases used is carried out by XRD after 2, 7, 28 and 90 days of hydration to compare the polymorphism of phase formation and determine the Ca/Si in the formed phases.

2 Materials and methods

2.1 Synthesis of C_2S

The synthesis of the C_2S phase was performed by firing a silica source SiO_2 finely ground with calcite ($CaCO_3$) and of $(NH_4)H_2PO_4$ used as a dopant to stabilize the β-C_2S variety, the mixture is subjected to a heat treatment at different temperatures 500 °C, 800 °C and 1000 °C for 24, 12 and 8h respectively, followed by rapid air cooling. The heat treatments are interspersed by milling with the addition of ethanol whose role is to increase the reactivity of the products.

2.2 Synthesis of Zn-Al-Ti Layered Double Hydroxides

In a beaker an acid solution of $Zn(NO_3)_2 6H_2O$, $Al(NO_3)_3 9H_2O$ precursors and a basic solution of Na_2CO_3 and NaOH were simultaneously added (4cm3/min) so as to adjust a constant pH between 9 and 9.5 at a constant temperature of 45°C for 10 hours. A Zn-Al LDH white precipitate is formed after aging from 12h at 100°C in a stove and calcined for 5 hours at 500 °C. The wet impregnation of TiO_2 on the Zn-Al layered double hydroxide was used for the preparation of Zn-Al-Ti LDH. The wet impregnation process was carried out using TiO_2 suspension diluted (3% by weight) in a base solution 0.67M Na_2CO_3 and loaded on the calcined powder Zn-Al LDH. The excess water was removed in a stove at 100°C. The impregnated sample is dried for a second time 12 hours at 100°C and calcined for 5 hours at 500 °C to finally obtain the powder Zn-Al-Ti LDH.

2.3 Hydration of blended samples

The resulting binders and the hydrated samples were analyzed by the X-ray diffraction performed by a Siemens D5000 diffractometer. This unit uses the mounting BRAGG-BRENTANO ($\theta/2\theta$) and a radiation $\lambda_{K\alpha}Cu$ = 1.5406 Å. The spectra 2θ interval is between 10 and 60 ° (with a step of 0.04°).

Two samples are prepared from the synthesized C_2S phase and the addition of 1% and 3% by weight of Zn-Al-Ti LDH. The adopted nomenclature is given in table 1. The designation for example of sample C2S3LDH7d means that the addition of 3% LDH is performed to C_2S and hydration of sample was followed for 7 days.

Table 1. Nomenclature of prepared samples

Hydration time	C_2S+ 1% Zn-Al-Ti LDH	C_2S+ 3% Zn-Al-Ti LDH
2 days	C2S1LDH2d	C2S3LDH2d
7 days	C2S1LDH7d	C2S3LDH7d
28 days	C2S1LDH28d	C2S3LDH28d
90 days	C2S1LDH90d	C2S3LDH90d

3 Results and discussion

3.1 X-ray diffraction of the anhydrous samples

The C_2S phase is synthesized by reacting of $CaCO_3$ and SiO_2 at 800°C and 1000°C. The α-C_2S form of orthorhombic variety with a large amount of CaO is manifested at 800°C. At 1000°C the intensity of CaO peaks decreases with the appearance of a second form of C_2S (β-C_2S) which has a monoclinic crystal system.

The X-ray diffractograms of the Zn-Al LDH and Zn-Al-Ti LDH before and after calcinations are shown in Figures 1, 2, 3 and 4.

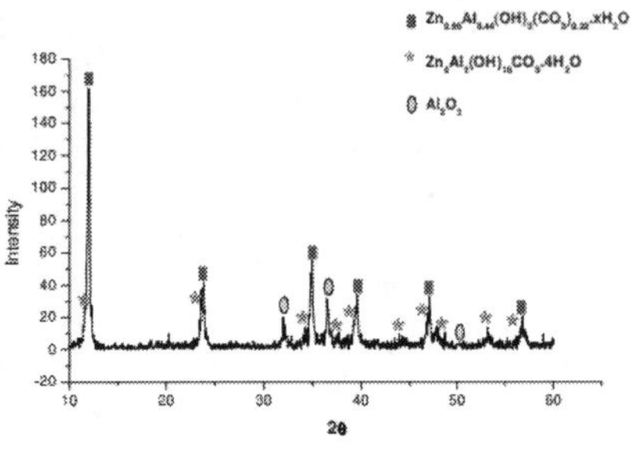

Fig. 1. XRD pattern of Zn-AL LDH phase before calcinations

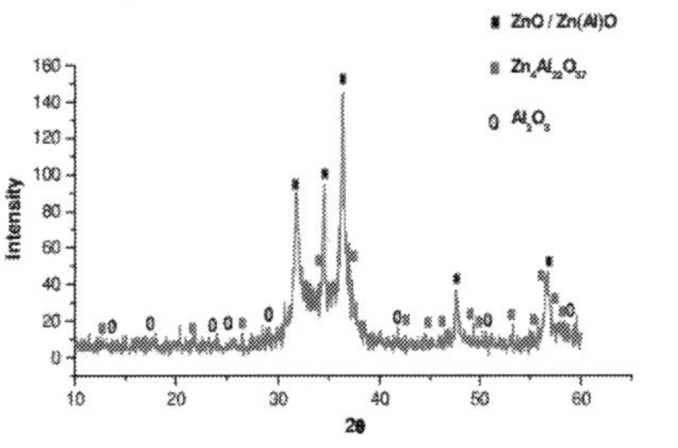

Fig. 2. XRD pattern of the Zn-Al LDH phase calcined at 500 °C

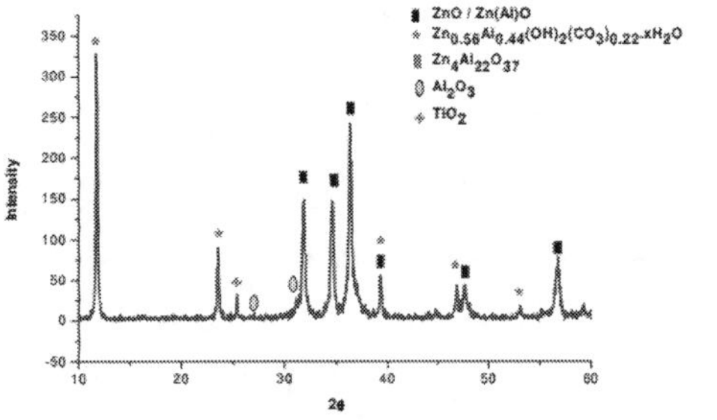

Fig. 3. XRD pattern of the Zn-Al-Ti LDH phase

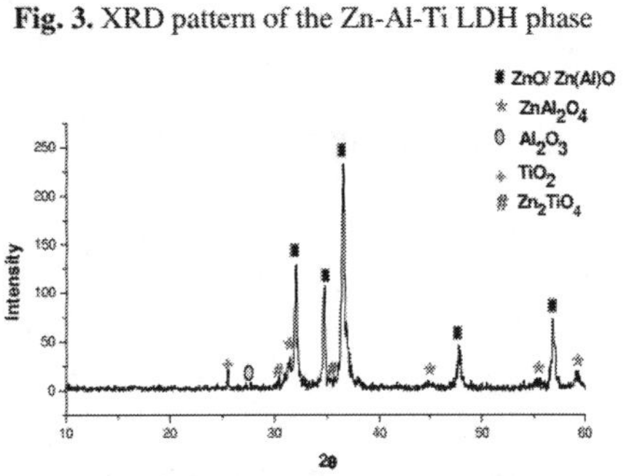

Fig. 4. XRD pattern of the Zn-Al-Ti LDH phase calcined at 500°C

The diffraction peaks demonstrates a LDH double layered hydroxides by forming two LDH phases: A stoichiometric phase $Zn_6Al_2(OH)_{16}CO_3.4H_2O$ and a non-stoichiometric phase $Zn_{0.56}Al_{0.44}(OH)_2(CO_3)_{0.22}.xH_2O$ was detected as the dominant phase. During the first calcination of Zn-Al LDH phase at 500°C we can note the disappearance of the diffraction peaks corresponding to LDH, the structure is destroyed which resulted in the appearance of ZnO, Al_2O_3 and $Zn_4Al_{22}O_{37}$ and the disappearance of carbonates (figure 2). After impregnation of TiO_2 onto Zn-Al LDH phase, TiO_2 was not detected in diffractograms because of its low concentration. However the appearance of zinicite phase ZnO and aluminum oxide Al_2O_3 as well as $Zn_{0.63}Al_{0.37}(OH)_2(CO_3)_{0.185}.xH_2O$ LDH phases were due to the hydration following the impregnation process (figure 3). After the second calcination the dominant phase detected is assigned to the mixed oxides of ZnO and Al_2O_3, since the origin of the ZnO phase is the Zn-Al LDH [28, 29], we also noticed the disappearance of the LDH phase such as what occurred in the first calcination (Figure 2), which resulted in the formation of stable phases of ZnO, $ZnAl_2O_4$ and Zn_2TiO_4. The reflections of the TiO_2 anatase phase were detected, probably due to the titanium oxide that did not react with ZnO (Figure 4). These results are consistent with the work of Milica Hadnadjev-Kostic et al. [30].

The interest of the calcination of $Zn_6Al_2(OH)_{16}CO_3.4H_2O$ is to form Zn-Ti LDH by the reconstitution method, the formed LDH were calcinated at 500°C to remove nitrates and form the ZnO and Zn_2TiO_4 which is the essential final product for the photocatalytic activity [30]. The reconstruction of the LDH structure took place during the impregnation process under the effect of the mixed oxides of ZnAl who have the ability to re-establish the hydroxide lamellar structure when are exposed to water and anions [31], which will be used as an addition to the hydration of different samples of C_2S.

3.2 X-ray diffraction of hydrated samples

Figure 5 shows the diffraction patterns of C_2S samples + 1% of Zn-Al-Ti LDH hydrated for 2, 7, 28 and 90 days. Table 2 summarizes the semi-quantitative analysis of the formed hydrates and their Ca/Si ratios as a function of time. The global Ca/Si observed for total C-S-H phases is calculated as the sum of all proportions of each observed C-S-H phase multiplied by its own factor Ca/Si.

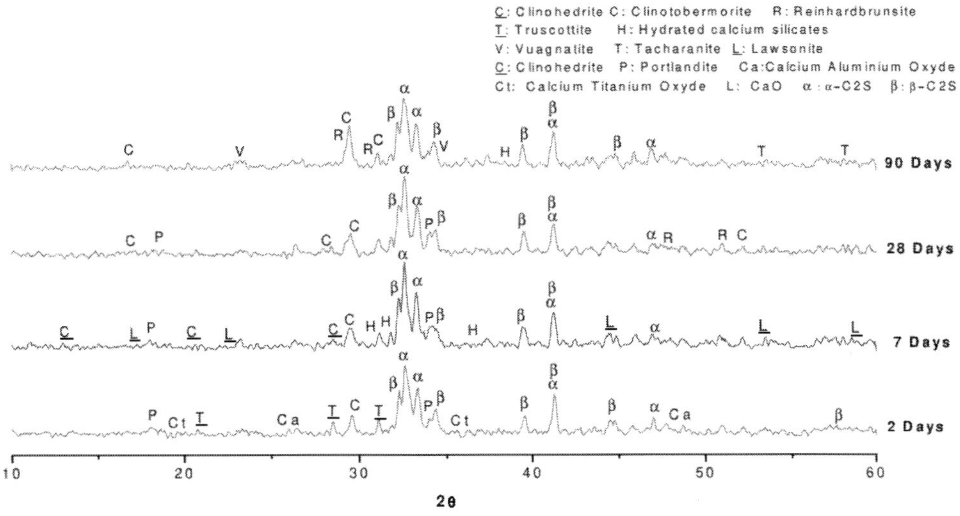

Fig. 5. XRD patterns of C_2S + 1% Zn-Al-Ti LDH hydrated samples

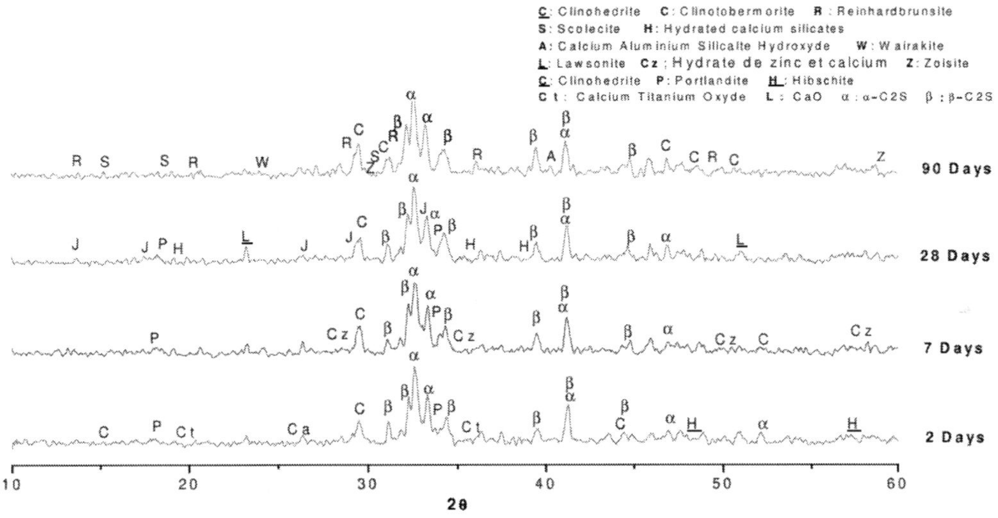

Fig. 6. XRD patterns of C_2S + 3% Zn-Al-Ti LDH hydrated samples

https://doi.org/10.1051/matecconf/201814901020

Table 2. Semi-quantitative analysis of hydrated samples with 1% Zn-Al-Ti LDH

	Phase nomenclature	Formulas	Percentages of hydrates				Ca/Si
			2 Days	7 Days	28 Days	90 Days	
Anhydrous phases	Dicalcique silicate	Ca_2SiO_4	41.27	34.59	23.62	20.13	-
	Calcium Titanium Oxyde	$Ca_4Ti_3O_{10}$	1.03	1.17	-	-	-
	Calcium Aluminium Oxyde	$Ca_3Al_2O_6$	6.05	-	-	-	-
Hydrated phases	Portlandite	$Ca(OH)_2$	8.80	9.24	13.22	-	-
	Lawsonite (C-S-H-Al)	$Ca_4Al_8Si_{36}O_{48}.16H_2O$	-	13.73	-	-	0.11
	Clinotobermorite (C-S-H)	$Ca_5(Si_6O_{17})(H_2O)_5$	15.52	13.63	20.47	17.65	0.83
	Truscottite (C-S-H)	$Ca_{14}(Si_{24}O_{58})(OH)_8(H_2O)_2$	9.8	-	-	-	0.58
	Clinohedrite (C-S-H-Zn)	$CaZn(SiO_4)(H_2O)$	-	12.22	-	-	1
	Hydrate calcium silicates (C-S-H)	$Ca_6Si_3O_{12}.H2O$	-	6.17	-	14.8	2
	Reinhardbraunsite (C-S-H)	$Ca_5(SiO_4)_2(OH)_2$	-	-	12.78	10.38	2.5
	Tacharanite (C-S-H-Al)	$Ca_{12}Al_2Si_{18}O_{51}.18H_2O$	-	-	-	6.7	0.67
	Vuagnatite (C-S-H-Al)	$CaAlSiO_4(OH)$	-	-	-	4.1	1
	Global Ca/Si observed for total CSH phases		0.181	0.373	0.489	0.787	

Table 3. Semi-quantitative analysis of hydrated samples with 3% Zn-Al-Ti LDH

	Phase nomenclature	Formulas	Percentages of hydrates				Ca/Si
			2 Days	7 Days	28 Days	90 Days	
Anhydrous phases	Silicate bicalcique	Ca_2SiO_4	47.32	40.80	38.74	29.84	-
	Calcium Titanium Oxyde	$Ca_4Ti_3O_{10}$	10.03	-	-	-	-
	Calcium Zinc Titanium Oxyde	$Ca_2Zn_4Ti_{15}O_{36}$	7.8	-	-	-	-
Hydrated phases	Portlandite	$Ca(OH)_2$	8.019	10.4	11.06	-	-
	Clinotobermorite (C-S-H)	$Ca_5(Si_6O_{17})(H_2O)_5$	16.77	19.93	19.20	15.29	0.83
	Hibschite (C-S-H-Al)	$Ca_3Al_2(SiO_4)_{1.53}(OH)_{5.88}$	10.05	-	-	-	1.96
	Silicates de calcium hydratés (C-S-H)	$Ca_2SiO_4.H_2O$	-	17.4	-	-	2
	Hydrate de zinc et calcium (C-S-H-Zn)	$CaZn_2(OH)_6.2H_2O$	-	11.28	-	-	-
	Lawsonite (C-S-H-Al)	$Ca_4Al_8Si_{36}O_{48}.16H_20$	-	-	11.84	-	0.11
	Jennite (CSH)	$Ca_9H_2Si_6O_{18}(OH)_8.6H_2O$			10.95	-	1.5
	Hydrate calcium silicates (C-S-H)	$Ca_6Si_3O_{12}.H_2O$	-	-	8.32	-	2
	Zoisite (C-S-H-Al)	$Ca_2Al_3Si_{36}O_{12}(OH)$	-	-	-	14.97	0.05
	Wairakite (C-S-H-Al)	$CaAl_2Si_4O_{12}.2H_20$	-	-	-	11.56	0.25
	Reinhardbraunsite (C-S-H)	$Ca_5(SiO_4)_2(OH)_2$	-	-	-	9.77	2.5
	Scolecite (C-S-H-Al)	$CaAlSi_3O_{10}.3H_20$	-	-	-	9.50	0.33
	Calcium Aluminium Silicalte Hydroxyde (C-S-H-Al)	$Ca_3Al_2SiO_4(OH)_8$	-	-	-	9.04	3
	Global Ca/Si observed for total CSH phases		0.336	0.513	0.622	1.103	

Analysis of results shows that the main products of hydration of the C_2S phase + 1% Zn-Al-Ti LDH are α-C_2S and β-C_2S phases, Portlandite and hydrated calcium silicates. We notice that the C-S-H phases are present in forms of different solid solutions including different proportions of Al and Zn. On the 2nd day of hydration, we have a small amount of C-S-H phases whose presence is of formulas $Ca_5(Si_6O_{17})(H_2O)_5$ et $Ca_{14}(Si_{24}O_{58})(OH)_8(H_2O)_2$. On the 7th day of hydration, there is the appearance of other hydrates of formulas $Ca_6Si_3O_{12}.H_2O$ and $CaZn(SiO_4)(H_2O)$ with decreasing percentage of Clinotobermorite, while the content of Portlandite continues to increase. After 28 days of hydration, we observe fluctuations in the amounts of this phase accompanied by appearance of other compounds such us hydrates incorporating the LDHs. These last phases have been identified in the forms of $CaZn(SiO_4)(H_2O)$ and $Ca_4Al_8Si_{36}O_{48}.16H_2O$. We observe also the formation of the $Ca_5(SiO_4)_2(OH)_2$ phase with a percentage of 12.78% with a slight decrease of C_2S phase and the total consumption of the Portlandite $Ca(OH)_2$, these results are consistent with the work of H. El-Didamony et al. [32]. The hydrated phases formed at a young age, before 7 days, are essentially constituted by cristalline phases with low ratio Ca/Si less than 1 while in the medium term, 28 to 90 days, the identified phases become morerich in calcium and the Ca/Si ratio increases and may exceed 2. The hydration of C_2S shows the presence of crystallized C-S-H compounds with percentages that increase and become more abundant in the 90th day, this is explained by the known character of this phase whose hydration is slow and participates in the creation of long-term resistances [2].

Figure 6 shows the diffraction patterns of C_2S samples + 3% of Zn-Al-Ti LDH hydrated for 2, 7, 28 and 90 days. Table 3 summarizes the semi-quantitative analysis of the formed hydrates and their Ca/Si ratios as a function of time. The observations for the sample with 3% Zn-Al-Ti LDH addition are similar to those attributed to the sample with 1% of the added. From the spectra of Figure 6 and table 3 we can see the formation of the clinotobermorite $Ca_5(Si_6O_{17})(H_2O)_5$ C-S-H type and Portlandite $Ca(OH)_2$ which are the main hydration products calcium silicates. Depending on the hydration time, the content of the clinotobermorite is variable and accompanied by the formation of jennite $Ca_9H_2Si_6O_{18}(OH)_8.6H_2O$ (C-S-H). The Zn-Ti-Al LDH phases also interacted in this hydration by hydrate formation where Zn and Al are incorporated, these hydrates appear since the second day of hydration, they have formulas of $CaZn_2(OH)_6.2H_2O$ and $Ca_4Al_8Si_{36}O_{48}.16H_2O$.

From the tables 2 and 3 we plot versus time the Ca/Si ratios of different hydrates formed (Figure 7), which clearly shows that in dependence of time the global Ca/Si ratio increases with time hydration, and that the Ca/Si ratios with the addition of 1% of Zn-Al-Ti LDH are lower than those with 3% of addition. These results agree well with several studies showing that the variation of the stoichiometry of C-S-H in terms of the ratio Ca/Si varies with the concentration of lime in the solution in which it is in equilibrium, so the higher lime concentration, the ratio Ca/Si of C-S-H is higher [33,34].

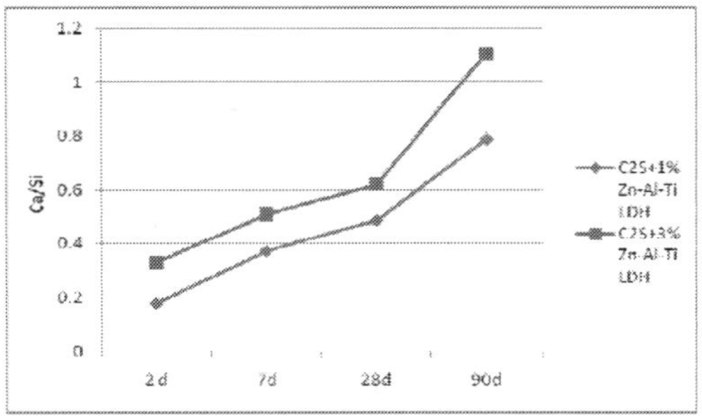

Fig. 7. Evolution of the global Ca/Si ratio of hydrated samples as a function of time

Conclusion

This study examines the mineralogical aspects of hydration of belite phase incorporating Zn-Al-Ti layered double hydroxides in time when the proportions of Zn-Al-Ti LDH varied of 1% and 3%. The results can be summarized as follows:

- As a function of the hydration time we observe a slow increasing of the formed C-S-H and a decrease in the Portlandite phase $Ca(OH)_2$, this may contribute to the increase of the mechanical resistances of the C_2S phase known having a slow hydration character and which participates essentially in the creation of long-term resistance.

- The basic elements of Zn-Ti-Al LDH interacted in the C-S-H phases that are presented in forms of different solid solutions including different proportions of Al and Zn.

- The hydrated phases formed (before 7 days), are essentially constituted by crystalline phases with low ratio Ca/Si less than 1 while in the medium term, 28 to 90 days, the identified phases become more rich in calcium and the Ca/Si ratio increases and may exceed 2 such as in the formation of Reinhardbraunsite $Ca_5(SiO_4)_2(OH)_2$.

References

1. H. F. W. Taylor, *Cement Chemistry, 2 ème edition, Edition Thomas Telford.* Londres (1997)
2. N.B. Singh, Hydrothermal Synthesis of β-Dicalcium Silicate (β-Ca_2SiO_4). Prog. Cryst. Growth. Ch. **52**, 77-83 (2006)
3. S. Goñi, F. Puertas, M. S. Hernández. Quantitative study of hydration of C_3S and C_2S by thermal analysis. Evolution and composition of C–S–H gels formed. J. Therm. Anal. Calorim. **102** 965-973 (2010)
4. H.F.W. Taylor, Proposed structure for calcium silicate hydrate gel. J. Am. Ceram. Soc. 69(6) 464–467 (1986)

5. M. Medala. Etude des interactions entre les phases minérales constituants le ciment portland et des solutions salines concentrées. PhD thesis, Université de Bourgogne, UFR des Sciences et Techniques, (2005)
6. S. A. Hamid, The crystal-structure of the 11 Å naturel tobermorite $CA_{2.25}[Si_3O_{7.55}(OH)_{1.5}].1H_2O$. Zeitchrift fur kristallographie, **154** (3-4) 189-198 (1981)
7. S. Merlino, E. Bonaccorsi and Armbruster. The real structure of tobermorite 11 Å: normal and anormalous forms. Eur J Mineral. **13** 577-590 (2001)
8. E. Bonaccorsi, S. Merlino and A. R. Kumpf. The crystal structure of tobermorite 14 Å (plombierite), a C-S-H phase. J. Am. Ceram. Soc. **88** 505-512 (2005)
9. I.G. Richardson, The calcium silicate hydrates. Cem. Concr. Res. **38** 137–158 (2008)
10. Taylor, H. F., Proposed structure for calcium silicate hydrate gel J. Am. Ceram. Soc. **69** 464–467 (1986)
11. R. J. Pellenq, A. Kushima, R. Shahsavari, K. J. Van Vliet, M. J. Buehler, S. Yip, F. J. Ulm, Proc. Natl Acad. Sci. USA, **106** 16102–16107 (2009)
12. I. G. Richardson, The importance of proper crystal-chemical and geometrical reasoning demonstrated using layered single and double hydroxides. Acta Cryst. B **69** 150–162 (2013)
13. C. HENMI, I. KUSACHI, Clinotobermorite, $Ca_5Si_6(O,OH)_{18}.5H_2O$, a new mineral from Fuka, Okayama Prefecture, Japan. *Mineralogical magazine* 353-358 (1992)
14. A. Girao, I. Richardson, R. Taylor & R.M.D. Brydson, Composition, morphology and nanostructure of C-S-H in 70% white Portland cement-30% fly ash blends hydrated at 55 °C. Cem. Concr. Res. **40** 1350–1359 (2010)
15. I. Richardson, The nature of C-S-H in hardened cements Cem. Concr. Res. **29** 1131–1147 (1999)
16. L. Senff, JA. Labrincha, VF. Ferreira, D. Hotza , WL. Repette, Effect of nano-silica on rheology and fresh properties of cement pastes and mortars. Constr Build Mater. **235** 2487–91 (2009)
17. G. Sivalingham, K. Nagaveni, M.S. Hegde, G. Madras, Photocatalytic degradation of various dyes by combustion synthesized nanoanatase TiO_2, Appl. Catal. B **45** 23–38 (2003)
18. M. Hadnadjev, J. Ranogajec, S. Petrovic, S. Markov, V. Ducman, R. Marinkovic- Neducin, Design of self-cleaning TiO_2 coating on clay roofing tiles, Phil. Mag. **90** (22) 2989–3002 (2010)
19. G. Flores, J. Carrillo, J.A. Luna, R. Martínez, A. Sierra-Fernandez, O. Milosevic, M.E. Rabanal, Synthesis,characterization and photocatalytic properties 5 of nanostructured ZnO particles obtained by low temperature 6 air-assisted-USP, Adv. Powder Technol. **25** 1435–1441 (2014)
20. K.-J. Hwang, J.-W. Lee, W.-G. Shim, H.D. Jang, S.-I. Lee, S.-J. Yoo, Adsorption and photocatalysis of nanocrystalline TiO2 particles prepared by sol–gel method for methylene blue degradation, Adv. Powder Technol. **23** 414–418 (2012)
21. A. Fujishima, X. Zhang, D.A. Tryk, TiO_2 photocatalysis and related surface phenomena, Surf. Sci. Rep. **63** 515–582 (2008)
22. O. Rudic, D. Rajnovic, D. Cjepa, S. Vucetic, J. Ranogajec, Investigation of the durability of porous mineral substrates with newly designed TiO_2-LDH coating. Ceram. Int. **41** 9779–9792 (2015)
23. T. Vulic, M. Hadnadjev-Kostic, O. Rudic, M. Radeka, R. Marinkovic-Neducin, J. Ranogajec, Improvement of cement-based mortars by application of photocatalytic active Ti–Zn–Al nanocomposites. Cement&Concrete Composites. **36** 121-127 (2013)
24. S. Sakthivel, B. Neppolian, MV. Shankar, B. Arabindoo, M. Palanichamy, V. Murugesan, Solar photocatalytic degradation of azo dye: comparison of photocatalytic efficiency of ZnO and TiO_2. Solar Energy Mater Sol Cells. **77** 65–82 (2003)
25. H. Pfeiffer, E. Lima, V. Lara, J.S Valente, Thermokinetic Study of the Rehydration Process of a Calcined MgAl-Layered Double Hydroxide. Langmuir. **26**(6) 4074–4079 (2010)
26. S.B Ghorbel, F. Medina, A. Ghorbel, A.M Segarra, Phosphoric acid intercalated Mg–Al hydrotalcite-like compounds forcatalytic carboxylation reaction of methanol in a continuous system. Appl Catal., A. **493** 142–148 (2015)
27. X. Cheng, X. Huang, X. Wang, B. Zhao, A. Chen, D. Sun, Phosphate adsorption from sewage sludge filtrate using zinc–aluminum layered double hydroxides. J. Hazard. Mater. **169** 958–964 (2009)
28. M. Hadnadjev-Kostic, T. Vulic, D. B. Zoric, R. P. Marinkovic-Neducin, The influence of the UV irradiation intensity on photocatalytic activity of ZnAl – layered double hydroxides and derived mixed oxides, Chem. Ind. Chem. Eng. Quarter. **18** (2) 295–303 (2012)
29. D. Carriazo, M. del Arco, E. Garcia-Lopez, G. Marci, C. Martin, L. Palmisano, V. Rives, Zn, Al hydrotalcites calcined at different temperatures: Preparation, characterization and photocatalytic activity in gas–solid regime, J. Mol. Catal. A: Chem. **342–343** 83–90 (2011)
30. M. Hadnadjev-Kostic , T. Vulic, R. Marinkovic-Neducin. Solar light induced rhodamine B degradation assisted by TiO_2-Zn-Al LDH based photocatalysts. Adv. Powder Technol. **25** 1624–1633 (2014)
31. J. Liua, G. Zhang, Recent advances in synthesis and applications of clay-based photocatalysts: a review, Phys. Chem. Chem. Phys. **16** 8178–8192 (2014)
32. H. El-Didamony, A.M. Shararab, I.M. Helmy and S. Abd El-Aleem, Hydration characteristics of β-C2S in the presence of some accelerators. Cem. Concr. Res. **26** 1179-1 187 (1996)
33. M. W. Grutzek, A. Benesi, and B. Fanning, 29Si magic angle spinning nuclear magnetic resonancestudy of tricalcium silicate hydrate. J. Am. Ceram. Soc. **72** 665-668 (1989)
34. H.F.W. Taylor, Hydrated calcium silicates. Part 1. Compound formation at ordinary temperatures .J. Chem. Soc 3682-3690 (1950)

Valorization of glass and shell powders in the synthesis of Belitic clinker

A.Bouregba*[1], A. Diouri[1], F.Amor[1], H. Ez-zaki[1] and O. Sassi[2]
[1]Mohammed V University, Faculty of science, Rabat, Morocco
[2]Mohammed V University, ENS of Rabat, Morocco

Abstract. In recent years, waste recycling has become a key issue in construction materials science for achieve the goal of sources of raw materials. The aim of this study is to use waste glass as source of SiO_2 oxide and mussel shells consisting essentially of calcium carbonate as starting materials to synthesize belite rich cement. The elaborated clinker is chemically and structurally related to β-Ca_2SiO_4 by heat solid state activation. After grinding, the raw materials are mixed and heated up to 1050°C. The samples are characterized by X-ray diffraction (XRD) and Fourier Transformed Infrared spectroscopy (FT-IR). The mechanical strength of some sample was measured for 28 and 72 days. The results show the different mineralogical stapes of formation of clinker, the state reaction method allows the elaboration of a cementitious material by a simple heat treatment at relatively low temperature and low release of CO_2. The hydration products are also identified by XR diffraction at 2, 7, 28 and 72 days.

1 Introduction

Each year, a large amount of coast shell is being picked up around all the costs sea and oceans of the world, also many factories produce a high amount of waste glass. These wastes become a serious environmental problem. The best solution to overcome the environmental impact of these wastes is to reuse them. Many researchers prove the benefit derived from reusing various waste materials, such as, fly ash [1], paper sludge ash [2], clay aggregate [3], tire crumb [4], siliceous and aluminous materials, as well as some natural pozzolanic materials [5] in the production of cement clinkers, and it has been proposed that these wastes have high potential for being alternative raw materials [6]. The use of by-products or residues generated from other activities seems to be a good approach to solve both economic and environmental problems related to cement production [7].

The use of shells and waste glass are widely studied in literature in many fields. Shells are used to improve mechanical proprieties of mortar [1], monetite [8] and engineering properties of controlled low-strength materials [9]. Shells also used to synthesis bioactive Hydroxyapatite [10]. Waste glass can be utilized in the elaboration of cement and concrete [11], can impact cement hydration [12], and can be used also to improve and enhancing durability of mortar [5] and to reduce and minimize cement content in rendering of mortar [13]. These studies proved that the use of coast shell and waste glass in cement can improve the mechanical properties and durability of elaborated cements by beneficially influencing the material microstructure.

The syntheses of β-Ca2SiO4 in moderately low temperature (< 1400°C) attract a lot of interest of many researchers In recent years. The total or partial replacement of conventional Portland cement by belite cement presents many positive aspects, such as energy and raw material reducing and savings as well as the possible higher durability [7].

Belite is one of the essential components of the Portland cement. It constitutes 15 to 30 wt. % of clinker and Exists in five forms: α, α'_H, α'_L, β and γ [14, 15]. The β form is metastable under 500 °C but is hydraulically active. However, the γ stable form at ambient temperature is inactive towards water reaction. The α, α'_H, α'_L forms are stabilized at room temperature by the incorporation in their structures of foreign ions [15, 16]. The belite phase C2S is generally present in the β form at room temperature [15, 17].

The seashells were composed of Ca as the essential metal with a small amount of Mg, Al and Si. Generally, Ca should present in the seashells as CaCO3 [18], similar to that for other shells resumed in researches [19]. Like marine clam mussel is an aquatic bivalve mollusk. It characterize by two shells connected by a hinge-like muscle [19]. Mussel shells are considered as a calcium-rich resource that can be utilized to produce calcium oxide [20]. This latter can be utilized in several different ways in environmental technologies [21].

Glass is resulting from fusion of several inorganic mineral raw materials, which after undergoing a process of controlled freezing becomes a hard, homogeneous, amorphous and isotropic material [22]. With the exception of Al2O3 and CaO, the rates of the main constituents of different types of glass are similar [23]. For soda lime glass is characterized by an approximately composition: 70% of silica, 13–17% Na2O and 10% CaO [24]. However, glass has a composition that favors the development of pozzolanic reactions [25]. Furthermore, conferring to the chemical requirements in ASTM C 618 glass can be considered as a pozzolanic-cementitious material [26].

In this work, we have used mussel shells and glass which predominantly composed respectively of calcium carbonate and silica as starting materials in elaboration of dicalcium silicate C_2S by solid state reaction at 1050°C without any stabilizers.

2 Materials and method

2.1 Materials

Dicalcium silicate powder (C2S) was elaborated by solid state reaction of the mixture of mussel shells and glass powders as starting material. The mussel shells are gathered from the coast of the Atlantic Ocean of Rabat-Morocco, then dried and finely milled (<40 µm). Glass powder is gotten by crushing the White glass beverage bottles and cups glass (<40 µm). In Table 1, the chemical composition of the starting materials shows that mussel shell contained a huge amount of calcium oxide 55.24 % by weight. It also contained a littler amount of impurities, primarily in the form of silicon, sodium, aluminum and magnesium oxides. The loss on ignition of 43.51 % was ascribed to the loss of carbon dioxide due to the partial decomposition of calcium carbonate. White Glass comprises basically of silicon 72.49 % and lime 11.98 %. Other minor elements as aluminum and sodium oxides are detected.

Table 1: Chemical composition of raw materials (wt.%)

Oxides	Mussel shells	Glass
SiO_2	0,35	72,49
Al_2O_3	0,12	1.50
Fe_2O_3	0.08	0.17
CaO	55.24	11.98
MgO	0.08	0.51
SO_3	0.20	0.32
K_2O	0.02	0.17
TiO_2	0.00	0.03
Na_2O	0.08	2.76
P_2O_5	0.04	0.01
LOI*	43.51	9.97

2.2 Method

After crushing and grinding, shell and glass powders are mixed with a proportion of 68 %wt and 32 %wt respectively, then heated up to 500 °C for 24 h followed by milling with ethanol after quick removal from the furnace. This procedure was repeated by increasing the temperature until reaching a heat treatment of 1050 °C for 4 h in which we have gotten a C_2S phase. Table 2 shows the following heat treatment of the raw materials.

Table 2: Heat treatment of mixed materials

Temperature °C	500	600	700	800	1050
Time	24	24	12	8	4

The heated sample was performed with a Siemens D5000 diffractometer operating with 40kV and 20mA, equipped with a copper anticathode and a secondary monochromatic (λ = 1.5406 Å). The powder is further characterized by Infrared (FT-IR) which is used to study the functional groups having in the synthesized powder. The FT-IR spectra were obtained on the region of 400-4000 cm^{-1}.

After this, the synthesized C_2S was hydrated with water to solid ratio of W/S = 0.5, and then stored at the room temperature (21 ± 1°C). A mineralogical characterization was performed on hydrated C_2S powder with different curing time. On the other hand, cylindrical paste (diameter 15 mm × height 30 mm) based on C_2S were prepared with W/S = 0.5. Afterwards, specimens were stored in a moisture room at 21 ± 1°C until the days of mechanical determinations (28 and 72 days). The compressive strength was obtained on cylindrical samples with the DY22 machine. The average of three values was obtained.

3 RESULTS AND DISCUSSION

3.1 Characterization of synthesized C_2S

Figure.1 shows the X-ray diffraction patterns of the synthesized powders. The obvious sharp peaks and low backgrounds suggest that the C_2S (S1) powder is highly crystalline. The corresponding reflections have been indexed using the Diffraction Data (ICDD)-Card No. 01-070-0388 (JCPDS) that are assigned to the β-C_2S Phase with small peaks of Wollastonite-2M, corresponding to JCPDS card No. 00-027-0088.

Figure 2 shows the FT-IR spectra of the synthesized powder at 1050°C. The main bands corresponding to the silicate groups are due to the SiO_4^{4-} tetrahedral. According to Puertas et al [27], the absorption bands of different polymorphs of dicalcium silicate are assigned to (υ1) (symmetric stretching) located around 800-900 Cm^{-1}, (υ3) (antisymmetric stretching) are typically located around 800-1000 cm^{-1} and those of (υ4) (the triply degenerated out of plane bending) located around 400-500 Cm^{-1}.

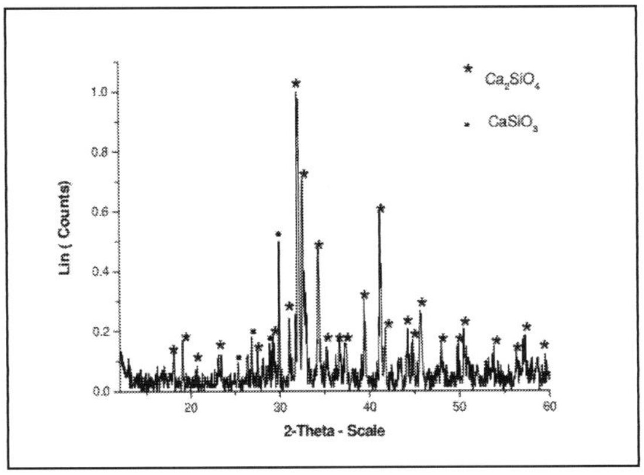

Fig. 1. XRD pattern of C_2S powder synthesized at 1050 °C

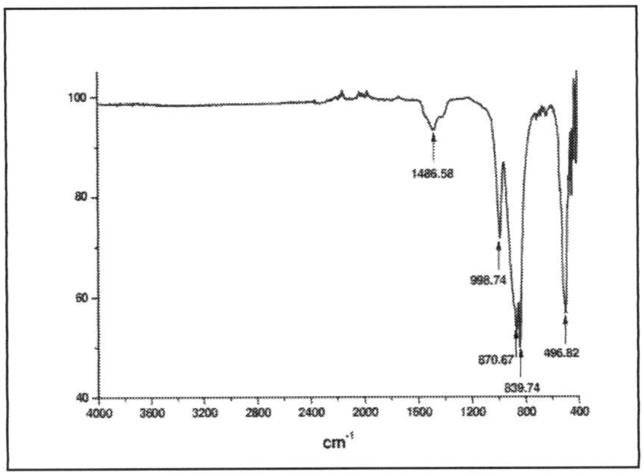

Fig. 2. FTIR spectra of C2S powder synthesized at 1050 °C

3.2 Mineralogical analysis of hydrated C_2S

Figure.3 shows the evolution of hydrated C_2S powder with different curing time. In general, the intensity of C2S peaks decreases with the curing time from two days to 72 days. Besides, we notice that the growth of hydrated products showed a variety of C-S-H phases as Clinotobermorite $Ca_5(Si_6O_{17})(H_2O)_5$, Reinhardbraunsite $Ca_5(SiO_4)_2(OH)_2$ and Suolunite $Ca_2Si_2O_5(OH)_2 \cdot H_2O$ after two days of hydration and then increased with the curing time. The hydration of synthesized C2S during time shows the formation of the crystallized C-S-H compounds which increase and become more abundant after 72 days. On the other hand, the portlandite appears with a low intensity. This can be explained by the known mechanism of this phase whose the hydration of C_2S is slow and contributes to the creation of long-term resistance [28]. It is reported that the quantity of portlandite produced by C_2S hydration is one third the amount from the C3S hydration in a pure clinker [29].

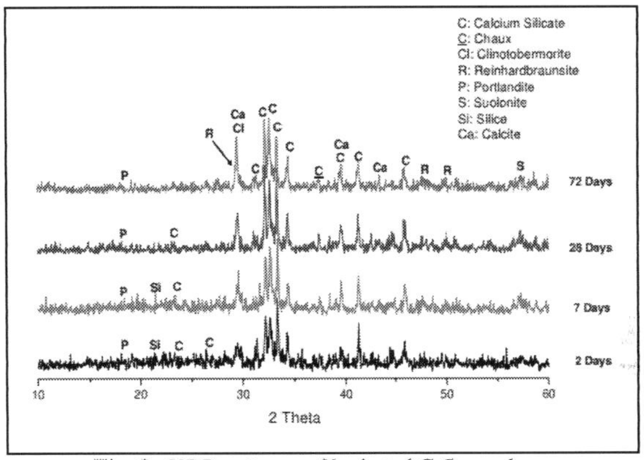

Fig. 3. XRD patterns of hydrated C_2S powder

3.3 Compressive strength

Cylindrical samples are tested with compressive tests after 28 and 72 days of curing. The obtained results are summarized in Table 3.

Table 3: Compressive strength (MPa) at 28 and 72 days

Curing time (days)	28	72
Compressive strength (MPa)	0.625	7.695

The results showed that the compressive strength was improved from 28 to 72 days with a value of 0.625 to 7.695 MPa. It is noted that the strength development for the pure clinker minerals is not similar [30, 31]. The C2S hydrates sluggish and contributes mostly to later age strength. A complete hydration of 30% at 28 days was reported for C2S, and 90% completion at one year [29]. This means that the strength increases gradually with time and can reach high final values after one year.

CONCLUSION

This paper demonstrate the feasibility of using mussel shell as a source of calcium carbonate and waste glass as source of silica as starting materials to synthesize belite rich cement by heat solid state activation. After grinding, the raw materials are mixed and heated up to 1050°C.

The result show that the elaborated clinker is chemically and structurally related to β-Ca_2SiO_4, with a good mechanical strength, which prove that the state reaction method allows the elaboration of a cementitious material by a simple heat treatment at relatively low temperature and low release of CO_2.

References

1. Her-Yung Wang , Wen-Ten Kuo, Chih-Chung Lin, Chen Po-Yo , Construction and Building Materials 41 (2013).
2. V. Ferrándiz-Mas, T. Bond, E. García-Alcocel, C.R. Cheeseman, Construction and Building Materials 61 (2014).
3. Miguel C.S. Nepomuceno, Pedro D. Silva, Construction and Building Materials **63** (2014).
4. FarahNora AznietaAbd. Aziz, Sani Mohammed Bida, Noor AzlineMohd. Nasir, Mohd Saleh Jaafar, Construction and Building Materials **73** (2014).
5. Ana Mafalda Matos, Joana Sousa-Coutinho, Construction and Building Materials **36** (2012).
6. Ying-Liang Chen, Chien-Jung Lin, Ming-Sheng Ko, Yi-Chieh Lai, Juu-En Chang, Cement & Concrete Composites **33** (2011).
7. F.A. Rodrigues, Cement and Concrete Research **33** (2003).
8. Huan Zhou, Timothy J.F. Luchini, NarimanMansouriBoroujeni, Anand K. Agarwal, Vijay K. Goel, Sarit B. Bhaduri, Materials Science and Engineering C **50** (2015).
9. Wen-Ten Kuo, Her-Yung Wang, Chun-Ya Shu, De-Sin Su, Construction and Building Materials **46** (2013).
10. A. Bouregb and A. Diouri, Materials Letters **183** (2016).
11. Yahya Jani, William Hogland, Journal of Environmental Chemical Engineering **2** (2014).
12. Nathan Schwarz, Narayanan Neithalath, Cement and Concrete Research **38** (2008).
13. Renata Oliveira, Jorge de Brito, Rosário Veiga, Journal of Cleaner Production **95** (2015).
14. D.K. Smith, A. Majumdar, F. Ordway, Acta Crystallogr.**18** (1965).
15. W. Eysel, T. Hahn, Z. Kristallogr, Bd. **31** (1970).
16. K. Rajczyk, 10th ICCC, Gothenburg, Sweden, **2** (1997).
17. My.Y. Benarchid, A. Diouri, A. Boukhari, J. Aride, J. Rogez, R. Castanet, Cement and Concrete Research **34** (2004).
18. Siyada Jaiyen, Thikumporn Naree, Chawalit Ngamcharussrivichai, Renewable Energy **74** (2015).
19. Shengyang Hu, Yun Wang, Heyou Han, biomass and bioenergy **35** (2011).
20. Jun Haslinda Shariffuddin, Mark Ian Jones, Darrell Alec Patterson, chemical engineering research and design **91** (2013).
21. Currie, J.A., Harrison, N.R., Wang, L., Jones, M.I., Brooks, M.S, J. Chem. Eng. **2** (2007).
22. Ortiz SA, O Vidro e as Embalagens de Vidro. Campinas, Brasil: CETEA/ITAL; (1996).
23. Miao Liu. Constr Build Mater (2011).
24. Renata Oliveira, Jorge de Brito, RosárioVeiga, Journal of Cleaner Production **95** (2015).
25. Fragata, A., Paiva, H., Velosa, A.L., Veiga, M.R., Ferreira, V.M., Sustainable Construction, Materials and Practices & Challenges of the Industry for the New Millennium, Lisbon, (2007).
26. ASTM C618-12 standard specification for coal fly ash and raw or calcined natural pozzolan for use in concrete.Philadelphia; (2012).
27. F. Puertas and F.Triviño, Cement and Concrete Reserch. **15** (1985).
28. N.B. Singh, Prog. Cryst. Growth. Ch. **52** (2006).
29. P. Klieger, Significance of tests and properties of concrete and concrete making materials. Philadelphia, Pa., (1994).
30. Metwally abd allah Abd elaty, HBRC Journal **10** (2014).
31. P. C. Hewlett, Lea's Chemistry of Cement and Concrete. Oxford: Elsevier Butterworth-Heinmann, (2004).

Study of the reinforcement of structure members by polypropylene fibres waste

Etude de renforcement des éléments de structure par un déchet de fibres de polypropylène

Khadra Bendjillali[1] and Mohamed Chemrouk[2]
[1]University Amar Telidji, Faculty of Civil Engineering and Architecture, Civil Engineering Department, Laghouat, Algeria
[2]University Houari Boumediene, Faculty of Civil Engineering, Algiers, Algeria

Abstract. The valorisation of industrial waste in the field of construction became a very interesting axis of research from scientific, economic and environmental point of view. We have conducted this work to study the effect of the addition of polypropylene fibres waste on the mechanical behaviour of reinforced concrete beams subjected to a simple flexural loading, with and without transversal reinforcement. The used fibres are coming from the waste of the fabrication of domestic brushes and sweeps; they have an average diameter of 0.47 mm and a length between 40 and 60 mm. Two weight dosages of fibres are used, 0.25 and 0.5 %. The experimental results showed that the incorporation of polypropylene fibres waste into the concrete affects negatively its workability, but its flexural and compressive strength are improved. The fibers have presented a significant contribution on the shear behavior and the cracking of beams, particularly in absence of transversal bars. The waste used in this work as fibrous reinforcement has not only increased the ductility of reinforced concrete beams, but it have also provided a perfect cracking distribution on the concrete surface and it has participated in a considerable way in the reduction of cracks number and dimensions, which allows to ensure the material durability and then the structure longevity. The reinforcement of concrete beams with 0.5 % of polypropylene fibers waste with a minimal steel bars can ensure an excellent mechanical behavior in shear, as in flexion.

Résumé. La valorisation des déchets industriels dans le domaine de la construction est devenue actuellement un axe de recherche très intéressant, non seulement du point de vue scientifique, mais encore du point de vue économique et environnemental. C'est alors dans cette optique que nous avons mené ce travail qui consiste à étudier l'effet de l'addition d'un déchet en fibres de polypropylène sur le comportement mécanique des poutres en béton armé soumises à la flexion simple en présence ou non du ferraillage transversal. Le renforcement employé est un déchet fibreux provenant de la fabrication des balais et des brosses domestiques, de diamètre moyen de l'ordre de 0.47 mm et de longueur variant entre 40 et 60 mm. Deux dosages massiques de fibres sont employés 0.25 et 0.5 %. Les résultats expérimentaux ont montré que bien que l'incorporation du déchet en fibres de polypropylène dans le béton a affecté négativement l'ouvrabilité du matériau, mais leur effet sur sa résistance à la flexion et à la compression est très considérable. Les fibres employées ont présenté également un apport très significatif sur le comportement au cisaillement des éléments poutres, particulièrement en absence d'armatures transversales et aussi sur leur état de fissuration. Le déchet utilisé comme renforcement fibreux non seulement a augmenté la ductilité des poutres en béton armé, mais il a aussi offert une parfaite distribution des fissures sur la surface du béton et il a participé d'une façon très appréciable à la réduction du nombre de fissures et de leur dimensions, ce qui permettra d'assurer une durabilité du matériau et par la suite une longévité de la structure. Le renforcement des poutres en béton avec un pourcentage massique de 0.5 % de déchet de fibres de polypropylène avec un minimum d'armatures d'acier peut leur assurer un excellent comportement mécanique au cisaillement, comme en flexion.

1 Introduction

Le mode de rupture et la fissuration des poutres en béton armé dépendent de leurs dimensions, du mode de chargement, des propriétés mécaniques du béton et de l'acier et de la quantité et le type du ferraillage employé. Si la rupture par flexion de ces poutres est bien maitrisée et bien prise en compte par les différents textes normatifs, la rupture par effort tranchant plus complexe demeure mal appréhendée. Des armatures transversales sont disposées pour reprendre une partie des contraintes de cisaillement et coudre les fissures d'effort tranchant. La présence de fibres dans les poutres peut aider le béton notamment dans les zones les plus sollicitées et ainsi soulager le ferraillage conventionnel. La question qui se pose actuellement est la suivante: peut-on substituer ce ferraillage en acier par des fibres? Cette question est d'un intérêt économique énorme car avec moins d'armatures transversales, non seulement le prix de la construction est réduit, mais aussi le temps d'exécution [1]. Certains auteurs [2, 3] ont trouvé que le comportement mécanique des poutres renforcées par des fibres et sans ferraillage conventionnel est meilleur que le comportement des poutres sans fibres et avec des armatures transversales. Selon Halvax et Lublóy [1], les performances des poutres renforcées uniquement de fibres d'acier sont pratiquement identiques à celles des poutres avec uniquement un ferraillage transversal, avec une légère augmentation de la ductilité et de la charge ultime de cisaillement dans le cas des poutres renforcées de fibres d'acier. Selon les mêmes auteurs, la rupture par cisaillement des poutres peut être évitée en incorporant un volume de 1 % de fibres d'acier. Dans une autre étude [4], il était constaté que les fibres d'acier sont capables de contrôler parfaitement la fissuration, sans influer la ductilité et la résistance du matériau. La chose la plus sûre selon toutes les études de la littérature et relatives à ce sujet, est que les fibres sont capables de retarder la fissuration engendrée par l'effort tranchant et de contrôler son ouverture. Cependant, leur effet sur le mode de rupture et sur la résistance à l'arrachement est conditionné par plusieurs paramètres, tels que leurs caractéristiques géométriques et leurs dosages, leurs orientations et leur adhérence avec la matrice cimentaire. Selon certains chercheurs [1, 4, 5], les fibres ne peuvent pas remplacer complètement le ferraillage transversal, mais elles sont capables de remplacer une partie de sa section, notamment dans les zones nodales, où il y'a trop d'aciers qui constituent un obstacle pour le passage du béton, ce qui empêche d'avoir un bon remplissage des coffrages. L'emploi de fibres de polypropylène dans le béton a connu ces dernières années une grande évolution grâce à leurs performances, telles que la légèreté, la non absorptivité d'eau, la non réactivité avec le ciment et la facilité de dispersion dans la matrice. Les fibres de polypropylène améliorent la ductilité du béton, réduisent significativement la largeur des fissures [6] et peuvent retarder la dégradation des structures en réduisant la perméabilité, le retrait et l'expansion du béton [7].

Dans ce travail expérimental, nous allons étudier l'effet du renforcement des poutres en béton armé par un déchet de fibres de polypropylène issu de l'industrie des brosses et des balais ménagères sur leur comportement mécanique et leur état de fissuration. Nous allons tester ce comportement en présence ou non d'armatures transversales, pour voir si nous pouvons remplacer le ferraillage conventionnel par des fibres. Par ailleurs, l'emploi d'un déchet de fibres comme renforcement dans les structures en béton est une tentative de recyclage de déchet afin de réduire le coût de la construction et protéger notre environnement.

2 Programme experimental

2.1 Matériaux

Le béton utilisé dans la présente étude est à base de:

- Ciment portland composé CEM II/A-L 42.5R (C) importé du Portugal, ayant une finesse de Blaine de 3520 cm^2/g et une densité absolue de 3.13.
- Eau de gâchage (E) provenant du réseau de distribution d'eau potable.
- Granulats calcaires de concassage, un sable (S) 0/3 mm et deux graviers (G1) 3/8 mm et (G2) 8/15 mm provenant de la station de concassage Ouazane de Laghouat, de densité absolue au voisinage de 2.50 et de coefficient d'absorption égal respectivement 4.71, 1.50 et 1.80 %. Les courbes granulométriques des granulats utilisés sont présentées sur la figure 1. Le sable possède un équivalent de sable de l'ordre de 66 %.
- Déchet de fibres synthétiques en polypropylène (F) de diamètre moyen de 0.47 mm, de longueur variant entre 40 et 60 mm et de résistance à la traction de l'ordre de 218 MPa (figure 2).
- Superplastifiant (SP) du nom commercial SIKAVISCOCRETE TEMPO 12 fabriqué par la société SIKA EL DJAZAIR.

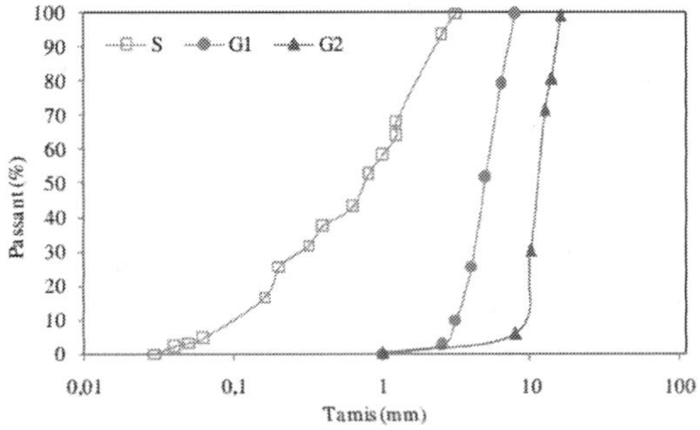

Fig. 1. Distribution granulaire des granulats.

Fig. 2. Déchet de fibres de polypropylène.

2.2 Formulation et préparation des échantillons

Les compositions du béton témoin sans fibres B0 et les deux bétons fibrés B0.25 et B0.5 renfermant respectivement un dosage massique de 0.25 et 0.5 % de fibres sont données dans le tableau 1. Les rapports (E/C) et (SP/C) pour ces trois bétons sont constants et valent respectivement 0.57 et 1.4 %. La formulation du béton témoin a été déterminée par la méthode pratique de Dreux-Gorisse [8]. Les composants secs sont mis en premier dans une bétonnière et après un bon malaxage, nous ajoutons la moitié de la quantité d'eau mélangée avec le superplastifiant, puis la deuxième moitié qui reste. Les fibres sont introduites manuellement en dernier afin d'assurer une bonne dispersion et d'éviter la formation des oursins. Nous avons confectionné des échantillons cubiques 10x10x10 cm pour l'essai de compression simple, la valeur retenue est la moyenne de six résultats et d'autres échantillons prismatiques 7x7x28 cm pour l'essai de flexion à trois points, dont la résistance est calculée par la moyenne de trois valeurs. Des poutres 10x10x50 cm avec le ferraillage schématisé sur la figure 3 sont également confectionnées. Les échantillons de béton sont démoulés 24 heures après leur confection puis ils sont conservés sous la toile de jute mouillée à une température variant entre 20 et 30 °C jusqu'au jour de l'essai à 28 jours d'âge du béton.

Tableau 1. Composition des bétons (kg/m^3).

Béton	C	E	S	G1	G2	F	SP
B0	350	199.5	670	186	899	0	4.9
B0.25	350	199.5	670	186	899	5.8	4.9
B0.5	350	199.5	670	186	899	11.6	4.9

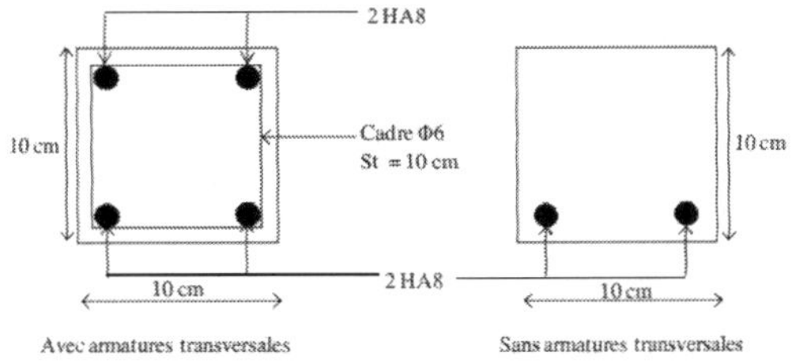

Fig. 3. Ferraillage des poutres.

3 Résultats expérimentaux

3.1 Maniabilité

La maniabilité du béton diminue avec l'augmentation du dosage en fibre de polypropylène. En effet, le béton témoin B0 présente un affaissement mesuré au cône d'Abrams de 9.3 cm, alors que les bétons de fibres B0.25 et B0.5 présentent respectivement un affaissement de 6 et 1 cm (figure 4). Avec la présence de fibres de polypropylène, le béton devient plus ferme et sa mise en œuvre plus délicate. Ainsi afin de maintenir une bonne plasticité du béton et une bonne distribution des fibres, l'augmentation du dosage de fibres nécessite une augmentation du dosage de superplastifiant. Pour cette raison, nous avons limité le dosage massique de fibres à 0.5 % afin d'éviter la formation d'oursins (figure 5) qui constituent un problème très délicat dans la fabrication des bétons de fibres. Les études de la littérature [9-13] montrent que l'addition des fibres dans les matrices cimentaires engendre une réduction de leur plasticité.

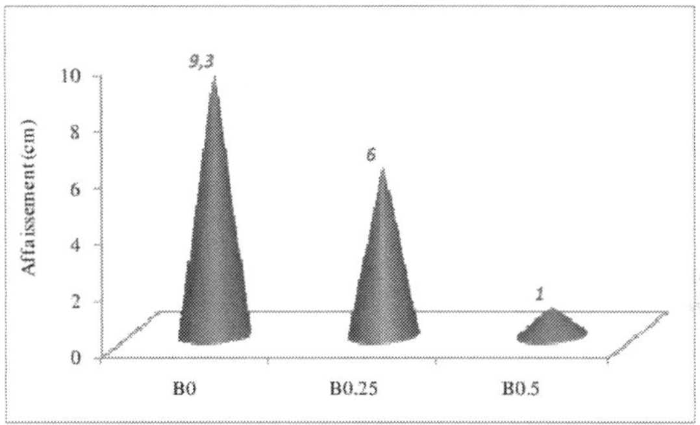

Fig. 4. Affaissement au cône d'Abrams des bétons étudiés.

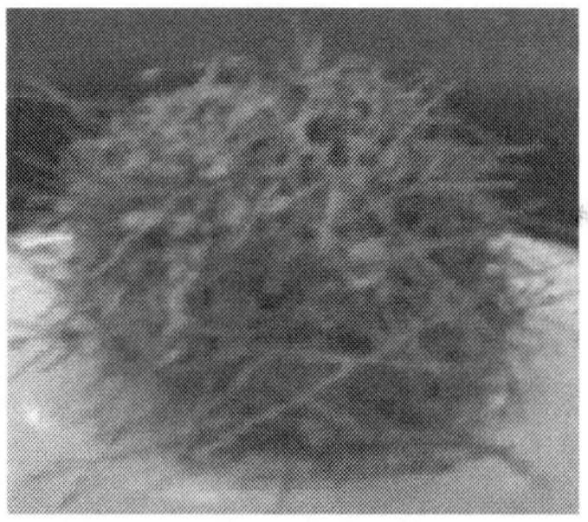

Fig. 5. Oursins formés.

3.2 Comportement à la flexion

Le comportement des bétons en flexion s'améliore avec l'addition de fibres de polypropylène (figure 6). L'amélioration augmente avec le dosage de fibres. Par rapport au béton témoin sans fibres B0, les bétons renforcés par 0.25 et 0.5 % en masse de fibres conduisent à des augmentations de la résistance à la traction par flexion respectivement de l'ordre de 9 et 30 %. Les fibres n'interviennent qu'après la rupture du béton, elles transfèrent les contraintes à travers la fissure, en offrant ainsi au matériau une certaine rigidité et un contrôle parfait de l'ouverture des fissures. La majorité des recherches faites sur les matériaux cimentaires renforcés de fibres [14-17] confirment l'efficacité des fibres sur le comportement en flexion des matériaux. Nous avons constaté que l'effet des fibres sur le comportement en flexion est conditionné par leur distribution efficace dans le béton. Certains chercheurs [17] confirment qu'une mauvaise distribution des fibres conduit à une réduction de cette résistance.

3.3 Comportement à la compression

La résistance à la compression des bétons à 28 jours augmente également avec l'ajout de déchet de fibres de polypropylène (figure 7). Cette augmentation est moins importante que celle observée avec la résistance à la traction par flexion puisqu'elle varie entre 2 et 16 % respectivement pour les bétons B0.25 et B0.5 par rapport au béton témoin sans fibres B0. Ce résultat est concordant avec d'autres travaux [18-21], même si d'autres auteurs ont constaté une réduction de la résistance à la compression après l'ajout de fibres [22-25]. L'effet positif des fibres de polypropylène sur le comportement à la compression peut être expliqué par leur flexibilité et leur souplesse qui leur permettent de se loger facilement entre les granulats sans les perturber et par conséquent d'augmenter la compacité du matériau, et donc améliorer son comportement en compression. Certains auteurs [26] voient que l'efficacité des fibres sur la résistance à la compression dépend de la géométrie et du dosage de fibres et non de leur type.

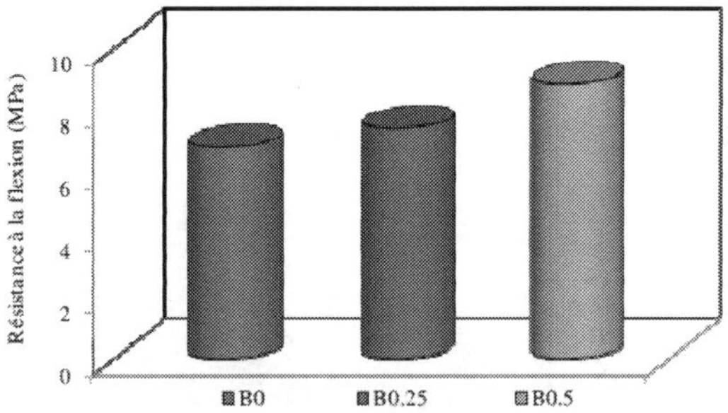

Fig. 6. Résistance à la flexion à 28 jours des bétons.

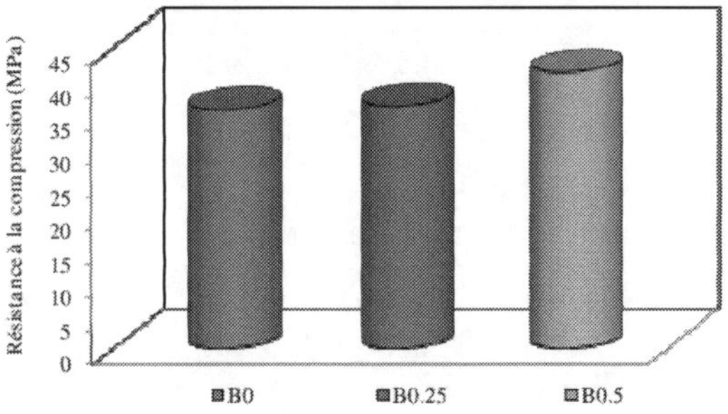

Fig. 7. Résistance à la compression à 28 jours des bétons.

3.4 Comportement mécanique des poutres

Dans les poutres en béton témoin sans fibres B0, la rupture se fait par cisaillement si aucune armature transversale n'est prévue (figure 8-a). Mais avec la présence de déchet de fibres de polypropylène, aucune fissure d'effort tranchant n'est observée sur les poutres, même en absence du ferraillage transversal (figure 8-b). L'apport de déchet de fibres de polypropylène est beaucoup plus apprécié en l'absence d'armatures transversales, du faite que dans ce cas, le renforcement au cisaillement est assuré par les fibres seules qui ont présenté une grande capacité de distribution de contraintes et donc une aptitude de déformation et d'absorption d'énergie plus élevée. La même constatation était faite par d'autres chercheurs [27]. Des fissures de flexion sont localisées au voisinage de la mi-portée des poutres et qui sont plus fines en présence des armatures transversales (figures 8-c, 8-d). La rupture des poutres est alors due au moment fléchissant.

a. Poutres en B0 sans ferraillage transversal.

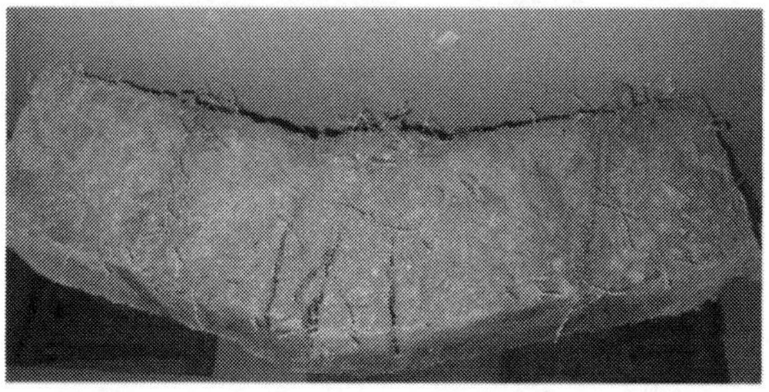

b. Poutres en B0.50 sans ferraillage transversal.

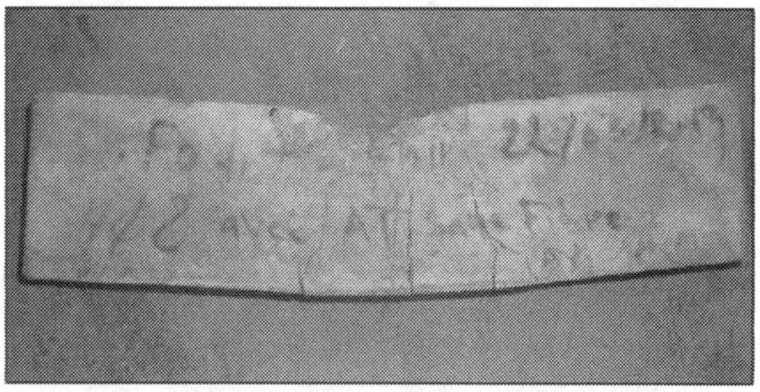

c. Poutres en B0 avec ferraillage transversal.

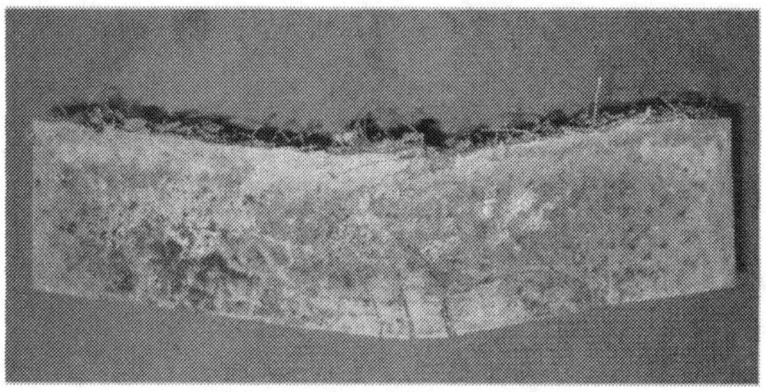

d. Poutres en B0.50 avec ferraillage transversal.

Fig. 8. Mode de rupture des poutres.

Les figures 9 et 10 schématisent la variation des valeurs du taux d'amélioration de la résistance à la flexion des poutres en béton de fibres respectivement à la rupture et à la première fissure. Nous pouvons clairement observer que le déchet de fibres de polypropylène a un effet positif sur le comportement mécanique des poutres même sans ferraillage transversal. Le taux d'amélioration du comportement des poutres à la rupture apporté par un dosage massique de 0.25 % de fibres passe de 8 % en présence d'armatures transversales à 17 % sans armatures transversales; alors que dans les poutres renforcées par 0.5 % de fibres, le taux d'amélioration varie entre 13 et 20 % respectivement dans les poutres avec et sans ferraillage transversal (figure 9). Cependant qu'à la première fissure, l'amélioration varie entre 1 et 12 % dans les poutres renforcées par 0.25 % de fibres et entre 15 et 19 % dans les poutres renforcées par 0.5 % de fibres respectivement avec et sans ferraillage transversal (figure 10). En présence de fibres de polypropylène, les fissures qui sont apparues sur les poutres sont plus fines et mieux dispersées même dans les poutres sans ferraillage transversal. Dans certains travaux [28], il était constaté que les fibres n'affectent pas le processus de fissuration, mais plutôt la longueur et le nombre des fissures, qui deviennent plus courtes et multiples. Nous avons remarqué aussi un effet positif des fibres de polypropylène sur le comportement au cisaillement des poutres, qui est en présence de 0.5 % de fibres et sans armatures transversales meilleur que celui des poutres sans fibres et avec des armatures transversales. Plusieurs travaux [29, 30] ont confirmé l'efficacité des fibres synthétiques à l'amélioration de la résistance des poutres au cisaillement. Furlan et de Hanai [31] ont confirmé selon leur essais expérimentaux que les poutres renforcées avec 2 % de fibres d'acier et sans armatures transversales ont montré une fissuration semblable à la fissuration des poutres avec armatures transversales et sans fibres. Un bon comportement mécanique des poutres peut être obtenu avec la combinaison de fibres et un minimum d'armatures d'acier transversales.

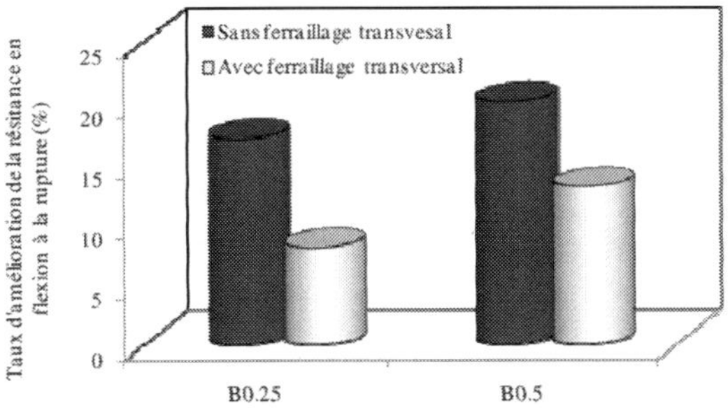

Fig. 9. Amélioration de la résistance en flexion des poutres à la rupture.

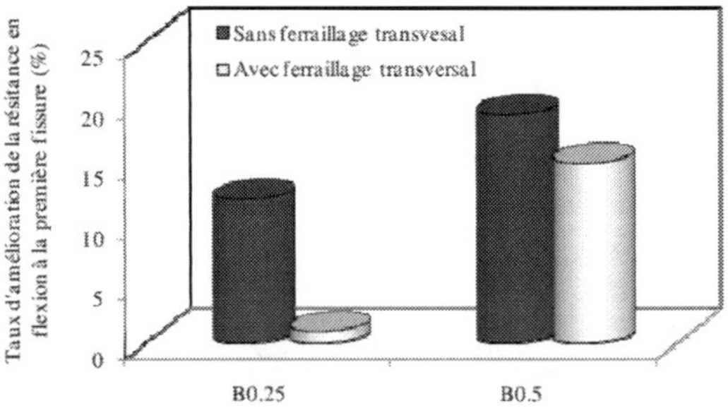

Fig. 10. Amélioration de la résistance en flexion des poutres à la première fissure.

4 Conclusion

À travers les résultats obtenus nous avons pu tirer les conclusions suivantes:
- L'ajout du déchet de fibres de polypropylène au béton diminue sa maniabilité.
- Les résistances à la flexion et à la compression simple du béton subissent un effet positif de l'incorporation de déchet de fibres de polypropylène. Avec un dosage massique de 0.5 % en fibres, l'amélioration des résistances à la traction par flexion et à la compression est respectivement de 30 et de 16 %.
- L'incorporation du déchet de fibres de polypropylène dans le béton a considérablement amélioré le comportement des poutres au cisaillement même sans présence du ferraillage transversal.
- Le comportement en flexion à la rupture et à la première fissure des poutres est amélioré suite à leur renforcement par le déchet des fibres de polypropylène.
- La présence du déchet de fibres de polypropylène dans les poutres a diminué considérablement la fissuration, même sans armatures transversales.

Références

[1] K. Halvax, É. Lublóy, Steel Fiber Reinforced Concrete–Shear Strength, *Acta Technica Jaurinensis*, **5(2)**, pp 149-166, (2012).
[2] A. Meda, F. Minelli, G. A. Plizzari, P. Riva, Shear behaviour of steel fibre reinforced concrete beams, *Materials & Structures Journal*, **38(3)**, pp 343-351, (2005).
[3] T. Nyomboi, H. Matsuda, A. Demizu, K. Makino, Experimental and analytical study on shear capacity in steel fiber and stirrup RC beam, *Journal of Structural Engineering*, **56A**, pp 13-22, (2010).
[4] Y. Fritih, T. Vidal, A. Turatsinze, P. Gérard, Flexural and shear behavior of steel fiber reinforced SCC beams, *KSCE Journal of Civil Engineering, Structural Engineering*, **17(6)**, pp 1383-1393, (2013).
[5] D. H. Lim, B. H. Oh, Experimental and theoretical investigation on the shear of steel fibre reinforced concrete beams, *Engineering Structures Journal*, **21(10)**, pp 937-944, (1999).
[6] D. Suji, S. C. Natesan, R. Murugesan, Experimental study on behaviors of polypropylene fibrous concrete beams, *Journal of Zhejiang University Science A*, **8(7)**, pp 1101-1109, (2007).
[7] S. Kakooei, H. M. Akil, M. Jamshidi, J. Rouhi, The effects of polypropylene fibers on the properties of reinforced concrete structures, *Construction & Building Materials Journal*, **27(1)**, pp 73-77, (2012).
[8] G. Dreux, J. Festa, *Nouveaux guide du béton et de ses constituants*, 8ème ed, Eyrolles, (2002).
[9] C. D. Johnston, *Fiber-reinforced cements & concretes*, Pub Gordon & Breach Science, Amsterdam, (2001).

[10] A. Sivakumar, M. Santhanam, A quantitative study on the plastic shrinkage cracking in high strength hybrid fibre reinforced concrete, *Cement & concrete Composites Journal*, **29(7)**, pp 575-581, (2007).
[11] H. Y. Aruntaş, S. Cemalgil, O. Şimşek, G. Durmuş, M. Erdal, Effects of super plasticizer and curing conditions on properties of concrete with and without fiber, *Materials Letters Journal*, **62(19)**, pp 3441-3443, (2008).
[12] K. Bendjillali, S. Rezzoug, M. S. Goual, M. Chemrouk, A. Zaidi, Mechanical behaviour of polypropylene fibre concrete beams, *Inter Cong CMSS-2013*, Rabat, Morocco, p 65, (2013).
[13] T. A. Söylev, T. Özturan, Durability, physical and mechanical properties of fiber-reinforced concretes at low-volume fraction, *Construction & Building Materials Journal*, **73**, pp 67-75, (2014).
[14] A. García-Santos, J. M. Rincón, M. Romero, R. Talero, Characterization of a polypropylene fibered cement composite using ESEM, FESEM and mechanical testing, *Construction & Building Materials Journal*, **19(5)**, pp 396-403, (2005).
[15] P. Ramadoss, K. Nagamani, Tensile strength and durability characteristics of high-performance fiber reinforced concrete, *The Arabian Journal for Science & Engineering*, **33(2B)**, pp 307-319, (2008).
[16] Z. Sun, Q. Xu, Microscopic, physical and mechanical analysis of polypropylene fiber reinforced concrete, *Materials Science and Engineering A Journal*, **527(1-2)**, pp 198-204, (2009).
[17] V. Alamshahi, A. Taeb., R. Ghaffarzadeh, M. A. Rezaee, Effect of composition and length of PP and polyseter fibres on mechanical properties of cement based composites, *Construction & Building Materials Journal*, **36**, pp 534-537, (2012).
[18] A. Hashem, A. Zakiah, Characterization of natural fibre concrete using destructive and non-destructive testing methods, *The 6th Inter Conf on Concrete Technology for Developing Countries*, Amman, Jordan, pp 551-560, (2002).
[19] Z. J. Grdic, G. A. Toplicic Curcic, N. S. Ristic, I. M. Despotovic, Abrasion resistance of concrete micro-reinforced with polypropylene fibers, *Construction & Building Materials Journal*, **27(1)**, pp 305-312, (2012).
[20] A. Belferrag, A. Kriker, M. E. Khenfer, Improvement of the compressive strength of mortar in the arid climates by valorization of dune sand and pneumatic waste metal fibers, *Construction & Building Materials Journal*, **40**, pp 847-853, (2013).
[21] H. Cifuentes, F. Garcia, O. Maeso, F. Medina, Influence of the properties of polypropylene fibres on the fracture behaviour of low, normal and high-strength FRC; *Construction & Building Materials Journal*, **45**, pp 130-137, (2013).
[22] A. Sadrmomtazi, A. K. Haghi, Properties of cementitious composites containing polypropylene fiber waste, *Composite Interfaces Journal*, **15(7-9)**, pp 867-879, (2008).
[23] M. S. Meddah, M. Bencheikh, Properties of concrete reinforced with different kinds of industrial waste fibre materials, *Construction & Building Materials Journal*, **23(10)**, pp 3196-3205, (2009).
[24] A. A. Ramezanianpour, M. Esmaeili, S. A. Ghahari, M. H. Najafi, Laboratory study on the effect of polypropylene fiber on durability and physical and mechanical characteristic of concrete for application in sleepers, *Construction & Building Materials Journal*, **44**, pp 411-418, (2013).
[25] T. Tahenni, M. Chemrouk, T. Lecompte, Effect of steel fibers on the shear behavior of high strength concrete beams, *Construction & Building Materials Journal*, **105**, pp 14-28, (2016).
[26] L. A. Pereira-de-Oliveira, J. P. Castro-Gomes, M. C. S. Nepomuceno, Effect of acrylic fibres geometry on physical, mechanical and durability properties of cement mortars, *Construction & Building Materials Journal*, **27(1)**, pp 189-196, (2012).
[27] C. E. Chalioris, E. F. Sfiri, Shear Performance of Steel Fibrous Concrete Beams, *Procedia Engineering*, **14**, pp 2064-2068, (2011).
[28] M. Hamrat, B. Boulekbache, M. Chemrouk, S. Amziane, Flexural cracking behavior of normal strength, high strength and high strength fiber concrete beams, using Digital Image Correlation technique, *Construction & Building Materials Journal*, **106**, pp 678-692, (2016).
[29] A. Yazdanbakhsh, S. Altoubat, K. A. Rieder, Analytical study on shear strength of macro synthetic fiber reinforced concrete beams, *Engineering Structures Journal*, **100**, pp 622-632, (2015).
[30] D. R. Sahoo, K. Maran, A. Kumar, Effect of steel and synthetic fibers on shear strength of RC beams without shear stirrups, *Construction & Building Materials Journal*, **83**, pp 150-158, (2015).
[31] Jr. S. Furlan, J. B. de Hanai, Shear behaviour of fiber reinforced concrete beams, *Cement & Concrete Composites Journal*, **19(4)**, pp 359-366, (1997).

The benefic effect of Moroccan oil shale's ash on blended cement (CMII)

Nour El Houda Hamsi, Khadija Nabih* and Redwan Barbach
Mohammed V University, Faculty of Science, Department of chemistry, Rabat, Morocco
*khadijanabih@yahoo.fr

Abstract. Recycling of Tarfaya's oil shales (OSA) was conducted for two essential reasons: First, to follow the change of their mineralogical composition with the heat treatment and second to investigate the possibility of their use in building materials. The mineral compositions and the mineralogical transformations of the oil shale during their thermal treatment were followed respectively by X-ray fluorescence (XRF), by X-ray diffraction analysis (XRD) and Fourier transform infrared (FTIR). The XRF showed that the mineral composition of the ash burned at 850°C is similar to that of Portland clinker and the DRX showed the presence of quartz, mayenite, gehlenite, anydrite, hematite and lime as well as the existence of some characteristics phases of the clinker as C_2S, C_3A and C_4AF. The Infrared spectrum showed the various functional groups that exist in OSA. The substitution of cement CMII by OSA increased the content of free lime and Blaine specific surface. The ash influenced beneficially the mechanical strength of the blended cement. The best results were obtained at all times of curing (2, 7, 28 and 91 Days) for the substitution of CMII by 7 and 10 % of OSA.

1 Introduction

The substitution of a part of cement by thermal power plants wastes as fly ash or oil shale's ash provides multiple benefits. It reduces the consumption of non-renewable natural raw materials [1] and the energy due to the fabrication of clinker, limits greenhouse gas emissions (CO_2), improves the quality of concrete and mortar by giving them a dense and compact microstructure, low permeability, low porosity and very good mechanical properties [2]. New technologies use OSA as additive or substitute to clinker or cement because of its availability as a by-product from the energy industry [3].

In this study, we are considering the use of Moroccan oil shale in cement. These shales are considered one of the largest energy and chemical resources in the world. They are estimated at several billions of oil barrels potentially exploitable. Their combustion for energy production will have a serious impact on the environment since one ton of oil shale produces 450-550 kilograms of ash [4, 5, 6]. Tarfaya's OSA is very promising to be used as additive to cement because its chemical composition is similar to that of Portland clinker [7].

2 Materials and methods

The raw materials used in the preparation of blended cements in this study are clinker, limestone, gypsum and OSA. These materials were crushed separately by a jaw crusher (old generation Fritsh pulverisette 01-703 type) to minimize their size.

The clinker was recovered from the cooler outlet of the plant. Limestone and gypsum were respectively dried in an oven respectively at 110°C for four hours and at 40°C for two hours.

The oil shales used in this study were taken from Tarfaya deposit (south of Morocco). The samples were donated by the National Office of Hydrocarbons and Mines (ONHYM). They were extracted into blocks of several kilograms, crushed and finely ground using a disc Mixer mill (Siebtechnik). The product was burned at 850°C for 30 minutes in a tubular laboratory furnace under atmospheric pressure, cooled down and sealed in plastic bags until its use in this study.

Clinker and raw materials included in the preparation of cements have been provided and prepared according to the method of standardized plant Asment (CIMPOR) [8].

The cement CMII was prepared in the laboratory through the normal processes involved in the plant. 75% clinker, 22 % limestone and 3 % gypsum were ground by a ball mill to fine powder in order to avoid any influence of the fineness on the mechanical properties of mortars. The refusal to 80 nm of cement was fixed between 1.8 and 2.7%.

In our case, the refusal of CMII was of 2.6 %. We varied the percentages of the shale ash in the cement by the substitution method (partial replacement of cement from 0% to 20% by the OSA). The oil shales and their ash were characterized by XRD, XRF. The Blaine surface areas were measured by the Blaine permeability apparatus. This measurement was based to determine the time required for a fixed volume of the air flow through a bed of powder. The density was determined by using Le Chatelier flask apparatus. This method was necessary to know the density as a part of the determination of the specific surface area.

The mechanical tests consisted in determining the resistance to compression and bending of prismatic specimens with dimension 4 x 4 x16 (cm) according to the standards of the Asment plant [8]. These specimens were molded from a plastic mass containing a batch of mortar cement portion (450g), three parts by weight of standard sand (1350g) and a half part of water (225g). The water / cement ratio was 0.5.

The mortar was prepared by mechanical mixing with a mixer (5L capacity, Vmin = 140 tr/min and Vmax = 285 tr/min), filled into a mold 4 x 4 x 16 and stored in a damp atmosphere (T = 20 ± 1°C and humidity ≤ 90) for 24 hours. After removal, the specimens were stored under water (wet cabinet) until resistance tests. The mechanical tests were carried at 2, 7, 28 and 91 days.

The free lime (CaO_{free}) determination was accomplished by using chemical analysis (Ethylene glycol method).

Analyses of the chemical and the mineralogical compositions of the all materials used in this study were investigated using respectively X ray fluorescence with Philips spectrometer PW 16660 device and X-rays diffraction with a Siemens D 5000 diffractometer.

The IR spectrum was recorded with a FTIR Bruker model Vertex 70 spectrometer by using the KBr pellet technique in the infrared region of 4000– 400 cm^{-1}.

Measurements of the expansion are performed using the needles of the normalized Le Chatelier according to the standard specifications [8]. The cement studied was tempered to normal consistency and introduced into small molds which were kept for 24 hours in water (20 ± 1°C). In order to observe the expansion of the cement in a short time, the reactions of hydration were accelerated by thermal treatment of the paste at 100°C for 3 hours.

3 Results and discussion

3.1 Chemical analysis by X-ray fluorescence (XRF)

3.1.1 Chemical composition of raw materials

The XRF analyzes showed the chemical composition of different materials used in the preparation of the blended cement studied. Table 1 shows the composition of the oil shale heated at 850°C. The OSA is composed of major oxides, predominantly calcite and quartz with moderate amounts of Al_2O_3 and Fe_2O_3. This composition is almost similar to that of clinker CIMPOR whose composition is that of an ordinary Portland clinker. We also observed that the percentage of SO_3 is very high for OSA. The same chemical analysis was performed for the gypsum to examine its degree of purity. The result showed that the purity of gypsum ($CaSO_4$, H_2O) used in this study was equal to 98.48 %, while its content of SO_3 does not exceed 45.80%.

3.1.2 Chemical composition of different blended cements

In this study, we varied the percentage of the addition of OSA in CMII by the partial substitution method to study their effects on the physicochemical properties of mortar.

From Table 2, we observed that the contents of the various major oxides (SiO_2, CaO, Al_2O_3 and Fe_2O_3) in the blended cements were quite similar to that of CMII. These cements contained also minor oxides such as MgO (less than 5%), TiO_2, K_2O, P_2O_5, SO_3 (less than 4% [8]) and very low contents on MnO and Na_2O.

These minor oxides play an important role in the clinker burning process and they have an influence on the mechanical and physicochemical properties of cements. The metal ions of these oxides are present in different phases of the clinker and are therefore in the network of silicates, aluminates and alumino-ferrites either replacing the four major oxides or in interstitial sites [9].

The free CaO content was between 1.064 and 2.3 % (must be less than 2% according to the standard Moroccan 10.1.004-2003). This explained the large speed of combination of CaO with other components of the mixture including silica [10]. The increase of free CaO can modify the mechanical properties of the cement. The formation of $Ca(OH)_2$ can cause the swelling of the cement and consequently lower the mechanical resistances [11]. The addition of oil shale in the blended cement decreased the loss on ignition (L.O.I) which can improve the cement strength (the higher the L.O.I, the less strength the cement will develop (Table 2).

3.2 Chemical Analysis X-ray diffraction (XRD)

Figure 1 showed the X-ray diffraction analysis of raw oil shale (a) and the calcined oil shale at 850°C (b). It can be seen from the figure 1 that, in the raw material there was a predominance of calcite ($CaCO_3$), followed by quartz (SiO_2) and dolomite ($CaMg(CO_3)_2$). The latter occured in small quantity regarding the small peak and the small content of MgO confirmed by of X-ray fluorescence (Table 1).

At 850 ° C, we observed a complete decomposition of dolomite, the appearance of new phases as hematite (Fe_2O_3) and anhydrite $CaSO_4$.

The decomposition of the dolomite gave rise to MgO and CaO oxides. The later was also produced by the decomposition of calcite according to the equations [7]:

$$CaMg(CO_3)_2 \rightarrow CaCO_3 + MgO + CO_2 \quad (1)$$

$$CaCO_3 \rightarrow CaO + CO_2 \quad (2)$$

Table 1. Chemical analysis of raw materials given by x-ray fluorescence.

Oxides	SiO_2	Al_2O_3	Fe_2O_3	CaO	MgO	SO_3	K_2O	TiO_2	MnO	P_2O_5	LOI[a]
Limestone	11.14	1.16	1.42	47.44	0.52	0.01	0.15	0.16	0.04	0.29	37.51
Clinker CIMPOR	20.11	4.86	3.40	64.96	3.17	1.60	0.88	0.34	0.09	0.25	0.22
OSA (850°C)	21.64	3.69	1.68	62.17	1.4	7.44	0.72	0.18	0.01	0.43	2.49

a. loss on ignition

Table 2. Chemical analysis of different blended cement by x-ray fluorescence.

Cements	SiO_2	Al_2O_3	Fe_2O_3	CaO	MgO	SO_3	K_2O	TiO_2	MnO	P_2O_5	LOI	% of CaO_{free}
0% OSA	18.76	4.48	2.91	58.05	2.91	2.06	0.70	0.33	0.08	0.22	9.43	1.064
5% OSA	18.03	3.99	2.87	59.67	2.42	2.21	0.66	0.30	0.08	0.27	9.08	1.13
7% OSA	18.65	4.24	2.94	58.90	2.45	2.26	0.68	0.30	0.08	0.27	9.11	1.23
10% OSA	18.39	4.43	2.83	59.03	2.40	2.45	0.66	0.29	0.07	0.27	8.82	1.736
15% OSA	18.27	3.92	2.75	59.43	2.34	2.72	0.68	0.29	0.07	0.28	8.64	2.07
20% OSA	18.76	4.02	2.76	59.37	2.32	2.82	0.70	0.29	0.07	0.28	8.4	2.3

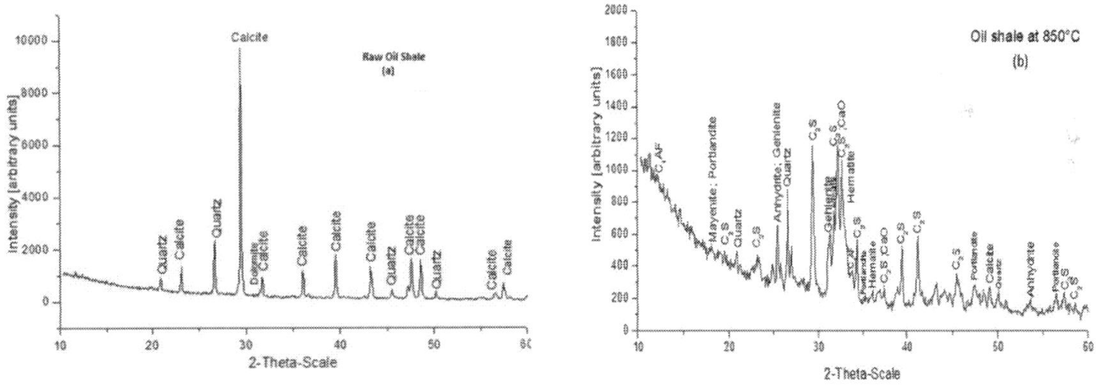

Fig. 1. X-ray Diffraction spectra of oil shale: (a) the raw oil shale and (b) the calcined oil shale ash at 850°C.

The presence of anhydrite $CaSO_4$ was due to the reaction of calcium oxide with sulfur dioxide that was released during the decomposition of the kerogen. The reaction equation is as follows:

$$CaO + \frac{1}{2}O_2 + SO_2 \rightarrow CaSO_4 \quad (3)$$

The formation of $CaSO_4$, allows us to confirm that the sulfur was captured by calcium oxide. This result has been mentioned by other researchers [12-13]. They have reported that oil shales contained a large amount of calcite have a beneficial effect on the environment because CaO acted as a trap of SO_2.

A great reduction of calcite and a considerable quartz peaks reduction were also observed.

These two oxides reacted together, and their reaction was confirmed by the existence of peaks characteristic of C_2S (belite) in the X-rays pattern.

We remark in the same spectre the presence of portlandite $Ca(OH)_2$. The existence of this phase was probably due to exposure of the sample to atmospheric moisture.

The equation for the reaction between the free lime (formed because of the decomposition of calcite) and water is: $CaO + H_2O \rightarrow Ca(OH)_2$ (4)

we could also identify the presence of other phases namely hematite (Fe_2O_3), the gehlenite ($Ca_2Al(AlSi)O_7$) that normally forms around 800°C and there was also the appearance of the mayenite ($Ca_{12}Al_{14}O_{33}$), which was obtained by reacting the aluminum oxide (Al_2O_3) with the calcium carbonate ($CaCO_3$), or with calcium hydroxide ($Ca(OH)_2$) [14].

3.3 Fourier transforms infrared spectra (FTIR)

The chemical groups present in OSA treated at 850°C were characterized by FTIR. As seen in the figure 2, the major bands of silicate group observed at 996, 993, 520 and 425 cm^{-1} were assigned to SiO_2. The very strong peaks at 996, 993 and 520 cm^{-1} were attributed to Si–O stretching vibration. The medium band at 425 cm^{-1} was associated to O–H stretching bands of Si–OH. Those peaks were respectively conducted from C_2S [15,16,17].

The absorption band at 3637 cm^{-1} can be attributed to portlandite ($Ca(OH)_2$), which was well detected in the sample and it was due to the O-H stretching vibration [18, 19]. The spectral bands in the region 3500-3400 cm^{-1} and the sharp peak at 1622 cm^{-1} were assigned to OH vibration of water molecules. The absorption bands at 1518 and 1414 cm^{-1} were attributed to CO_3^{2-} stretching vibrations (the characteristic peak of calcite ($CaCO_3$)). This peak is associated with the one at 847 cm^{-1} [20]. The presence of $CaCO_3$ was attributed to the atmospheric CO_2 absorption [21].

The strong absorption bands observed in 1137, 1124 cm^{-1} and the weak vibration band at 643 cm^{-1} represented the stretching vibration of the sulfate group (SO_4^{2-}) [22,23,24]. These results of FTIR analysis are consistent with the results found in the previous analysis.

3.4 Expansion

The expansion will ensure the stability of cement by determining the swelling of pure cement paste in a humid environment. Cements that contain too much of expansive components such as gypsum, free lime and magnesia can provoke dangerous expansion, cracking, distortion and disintegration with time for the durability of structures [25].

In this study, we can see from the Table 3 that there was an increase of expansion and of free lime by increasing the substitution of CMII by OSA.

The increase of the expansion is either due to the content of free lime or to the SO_3 [26] or to both oxides that increased with the increase of OSA. In this study, the expansion did not exceed 11mm even with 20% of Tarfaya's OSA (the Moroccan standard (NM 10.1.004) value for the expansion is 10mm).

3.5 Fineness of cements with additions of OSA (Blaine)

The measurements of specific surface areas (Table 4) were performed for the ash and for the different cement samples prepared with various percentages of OSA.

From Table 5, we noticed that the values of the specific surface area had increased with the substitution of CMII by OSA that have a low fineness. This Fineness can have an impact on the hydration of the blended cement.

Indeed, the specific surface area plays a decisive role in the evolution of the physical characteristics of cements prepared. The study showed that increase of the fineness increases the wetting of the cement grains, which improved the hydraulic reactivity. It has been shown by several authors that when the Blaine fineness of cement was higher, the hydration speed was faster and the mechanical strengths at early ages were excellent [27]. The plasticity and the cohesion of the fresh cement paste as well as its water retention capacity were influenced by the Blaine fineness [27].

3.6 Bending and compressive strength behaviour

The effects of the substitution of cement CMII by OSA on the bending and compression strength of mortars prepared at different ages 2, 7, 28 and 91 days are shown in the figures 3 and 4.

We noted an increase in the flexural and compressive strengths starting for young ages (2 and 7 days). This improvement of the strength was even greater at 28 and 91 days for various mortars with the addition of 5 to 20% of OSA.

Based on those results, it was concluded that the strengths increased with the increase of the curing time for the various blended cements. This was may be due to the increase of the amount of hydration products especially calcium silicate hydrates [28].

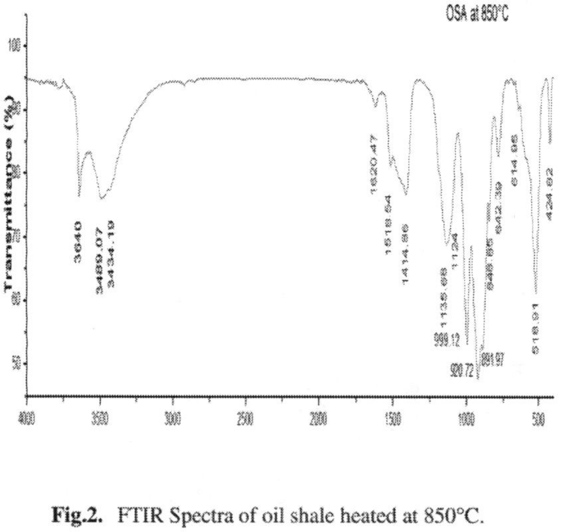

Fig.2. FTIR Spectra of oil shale heated at 850°C.

Table 3. Expansion values for cements as a function of the free lime.

% of OSA	0	5	7	10	15	20
% of CaO free	1.06	1.13	1.23	1.74	2.07	2.3
Expansion (mm)	1	1	3	9	10	11

Table 4. Characteristics of OSA

ρ (g/cm^3)	Density	Surface area (cm2/g)	% CaO free
3.09	3.63	4622	4.256

Table 5. Effects of Substitution CMII by ash on Blaine surface area of blended cements.

% OSA	0	5	7	10	15	20
Surface area (cm^2/g)	3594	4542	4610	4733	5333	5343

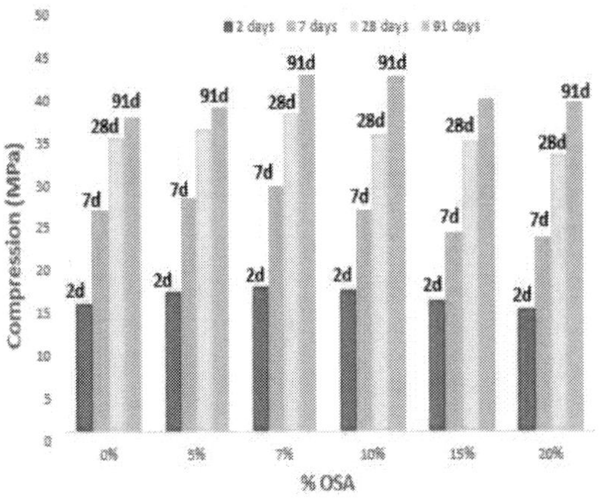

Fig.3. Compression strength of blended cement as a function of ash content.

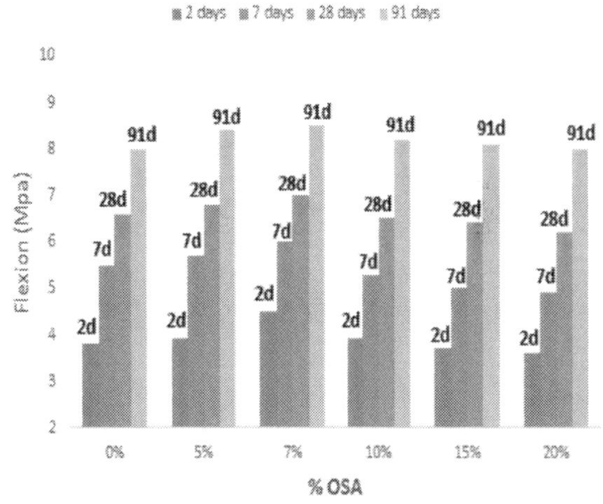

Fig.4. Bending strength of blended cement as a function of ash content.

4 Conclusions

The chemical compositions of the materials used in this study (oil shale burned at 850°C and the raw materials used for the elaboration of CMII) were given by the XRF analysis. From the chemical compositions, we noticed that oil shale's ash has a similar composition to those of Portland clinker, which gave these ashes good hydraulic and pozzolanic properties.

The results obtained by the X-ray diffraction showed that the mineral part of Tarfaya's oil shale was composed essentially of quartz, calcite, and dolomite their ash treated at 850°C contained anhydrite, muscovite, gehlenite, hematite with some of clinker phases such as: belite, alumino-silicate and ferrite with other minor mineralogical phases.

The substitution of CMII by OSA increased free lime content and the Blaine specific surface area and lead to small increase in the expansions without exceeding the one required by the Moroccan standard NM 10.1.004.

This substitution had very good influence on the mechanical properties on the blended cement starting from early ages (2 and 7 days). The mechanical strengths passed through a maximum near 10% of substitution and stayed very good even with 20% of OSA. These strengths were improved considerably at 28 and 91 days for all blended cement.

Acknowledgment

The Authors express their warmest thanks to Asment Temara Plant (CIMPOR) for the technical support and to National Centre for Scientific Research (CNRST), Rabat, Morocco for their analytical instrument facilities.

References

1. L.M. Federico, S.E. Chidiac, Cem. Concr. Compos. **31** 606-610 (2009)
2. W. Ladaoui, Ph.D Thesis., Uni.Tou. III P. Sabatier, (2010) (unpublished)

3. K. Nabih, R. Barbacha, A. De la Torre, O. Sassi, K. Kada, F. Cherkaoui El-Moursli, MATEC web conf. **11 :01046** (2014)
4. M. Uibu, R. Kuusik, Jinyue Yan, Ed., 3rd IGEC, Mälardalens Uni., 1075-1085 (2007)
5. M. Uibu, and A. Trikkel, R. Kuusik, WIT. Trans. Ecol. Environ., **106** 473-483 (2007)
6. A. Shogenova, S. Sliaupa, K. Shogenov, R. Vaher, R. Sliaupiene, Energy Procedia, **1** 2753–2760 (2009)
7. K. Nabih, Ph.D Thesis, UM5/FSR, (1997) (unpublished)
8. NM.10.1.005, The standard technical testing of hydraulic binders (cement), (2008)
9. H.Uchikawa, 8th Int. Cong. Chem. Cem., Rio de Janeiro, **1** 249-280 (1986)
10. W. Mazouzi, L. Kacimi, M. Cyr, and P. Clastres, Cem. Concr. Compo., **53** 170-7 (2014)
11. R. Sersale, 7th Int Cong. Chem. Cem., Paris, IV–1/3-IV-1/21 (1980)
12. R. Bouchta, O. Zemmouri, in Mines, Geo. Energy, **50** 55-69 (1981)
13. A. Baron, Res. Rep. LCPC, **3**-32 (1969)
14. S.N. Ude, Ph.D Thesis, UTK, (2013) (unpublished)
15. K. Nabih, N. Hamsi, Int. CMSS., MATEC web conf., **11** 1-7 (2014)
16. M. Yousuf, A. Mollah, T.R., Y.N. Tsai, D.L. Cocke, Cem. Concr. Res., **23** 773-784 (1993)
17. O.E. Omotoso, R. Mikular, J. Hazard. Mater., **60** 1-28 (1198)
18. Y. Osawa, J.W. Shih, Fuel, **50** 53-57 (1971)
19. S. Chomnanti, P. Deemak, A.F. Gaines, T. Kasomson, K. Keowkamned, V. Lorvldhya, C. Pomtalpong, N. Sucharitakul, V. Vanitsuksombat, W. Wiriyathien, Fuel, **49** 188-196 (1970)
20. B. Čolović, V. Jokanović, N. Jović, Sci. Sinter., **45** 341-350 (2013)
21. M.Y.A. Mollah, L.R. Parga, D.L. Cocke, J. Environ. Sci. Health. A., **27** 1503-1519 (1992)
22. X. Ying-Mei, Q. Ji, H. De-Min, W. Dong-Mei, C. Hui-Ying, G. Jun, Z. Qiu-Min, Oil Shale, **27** 37-46 (2010)
23. T. Nochaiya, Y. Sekine, S. Choopun, A. Chaipanich, J. Alloy. Comp., **630** 1-10 (2015)
24. M.Y.A. Mollah, W. Yu, R. Schennach, D.L., Cocke D.L, Cem. Concr. Res., **30** 267-273 (2000)
25. G.V. Rompaey, Ph.D Thesis, ULB, (2006) (unpublished)
26. A.M. Neville, 5th Ed., Harlow, Essex: Pearson Education (2011)
27. J. Yammine, Ph.D Thesis, ENS Cachan, 262, (2007) (unpublished)
28. M. Lea, 3rd Ed., London: Edward Arnold, 727 (1970)

Durability of the earth mortar: Physico-chemical and mineralogical characterization for the reduction of the capillary rise

Durabilité du mortier en terre: Caractérisation physico-chimique et minéralogique pour la réduction de la remontée capillaire

A.Ammari[1], K. Bouassria[2], N. Zakham[1], M. Cherraj[1], H. Bouabid[1], S. Charif D'ouazzane[3]
[1]LMM, Dept. of physics, Mohammed V University, Rabat, Morocco
[2]LSIMO, Dept. of physics, Ibn Tofail University, Kenitra, Morocco
[3]LMTM, National School of Mineral Rabat, Morocco
[1]abdelmalek09@gmail.com

Abstract. The stabilization of the earth material in the fields related to the earthen construction, in compliance with the standards in force, allows strong results of strength and durability. The chemical and mineralogical elements play an important role, in the presence of an 'optimum' cement dosage, to strengthen the ties between the clays and the grains of the earth.. This approach targets the search for better performances in the use of natural materials resource in an eco-responsible habitat. This study presents the experimental results of the four techniques of mineralogical and chemical analysis on mortar specimens obtained from earth of the city of Fez. The results of the uni-axial compressive tests of the cylindrical specimens for this earth, associated by various percentages 0%, 4%, 7% and 10% by weight of cement, make it possible to analyze the effect of the mineralogical and chemical elements on the mechanical properties, namely Young's modulus, compressive strength and limiting deformation. However, we determine the water absorption coefficient of the mortar for different cement dosages in order to optimize the durability of the mortar against bad weather, rain and / or very wet climates. In the earth mortar of Fez, the strong presence of calcite (CaCO3), quartz SiO2 and dolomite CaMg (CO3)2 amplifies the improvement of the behavior of the material by the addition of cement. In fact, this strong presence of calcite stabilized the clay by cementing quartz and the cement matrix to strengthen the ties between the grains of the earth. In addition, with respect to the capillary rise, the water absorption decreases with the addition of cement. We also note that the evolution of the mechanical properties is of no importance except in the interval [4 to 7%] which represents the zone of effect for cement stabilization and which houses the optimum technico-economic cement dosing.

Résumé. La stabilisation du matériau terre crue dans les domaines liés à la construction en terre, dans le respect des normes en vigueur, permet des résultats de résistance mécanique et de durabilité forts probants. Les éléments chimiques et minéralogiques y jouent un rôle déterminant et permettent, en présence d'un dosage 'optimal' du ciment, de renforcer les liens entre les argiles et les grains de la terre. Ainsi, la connaissance de la nature de la terre crue à utiliser dans la construction, par l'identification scientifique de ses composantes, est incontournable à cet effet. Cette démarche cible la recherche de performances meilleures dans l'utilisation des matériaux naturels et bio ressourcés au niveau d'un habitat éco responsable. Notre recherche présente les résultats expérimentaux des quatre techniques des analyses minéralogiques et chimiques sur des éprouvettes de mortier obtenues à partir d'une terre de la ville de Fès par la diffraction des rayons X, l'Infrarouge, le Microscope électronique à balayage et la fluorescence X. Les résultats des essais de compression uni-axiale des éprouvettes cylindriques de cette terre associée aux différents pourcentages 0%, 4%, 7% et 10% en poids de ciment, permettent d'analyser l'effet des éléments minéralogiques et chimiques sur les caractéristiques mécaniques à savoir le module d'Young, la résistance à la compression et la déformation limite. De plus, nous déterminons le coefficient d'absorption de l'eau du mortier pour différents dosages en ciment afin d'optimiser la durabilité du mortier face aux intempéries, à la pluie et/ou des climats très humides. Dans le mortier de terre de Fès, on constate que la présence forte de la calcite ($CaCO_3$), du quartz SiO_2 et de la dolomie CaMg $(CO_3)2$ amplifie l'amélioration du comportement du matériau par l'ajout de ciment. En effet, cette forte présence de la calcite a stabilisé l'argile par la cimentation de quartz et la matrice de ciment à renforcer les liens entre les grains de la terre. Par ailleurs, en ce qui concerne la remontée capillaire, l'absorption d'eau diminue avec l'ajout du ciment. Nous notons aussi que l'évolution des caractéristiques mécaniques n'a pas d'importance que dans l'intervalle [4 à 7%] qui représente la zone d'effet pour la stabilisation en ciment et qui abrite l'optimum technico-économique du dosage en ciment.

1 INTRODUCTION

Local materials used in construction and civil engineering for the construction of habitats in southern countries represent low environmental and economic costs. The determination of their physicochemical characteristics as well as their control are essential tasks for the good functioning, durability at each stage of the realization of a structure. To reinforce these characteristics, the stabilization of the soil by adjuvants is applied in a systematic manner. To master this practice, we propose a mineralogical and chemical characterization of the earth used based on analysis and mechanical tests [1, 2, 3] and an evaluation of the mechanical characteristics of the earth mortar in two dry and wet cases [4, 5, 6, 15]. We worked with the cement-stabilized earth mortar to determine its mechanical characteristics by simple compression tests on cylindrical specimens in both cases and by the capillary absorption ties to determine the water absorption coefficient [6, 19].
This study also determines the area interval of the technico-economic optimum (OTE) [4, 20].

2 Material and methods

The soil used in this study comes from the region of Fez referenced by **F**. The following table summarizes the granular fraction of this earth:

Table 1. The granular fraction of the earth of Fez

Earth F	Granular fraction (%)			
	Coarse sands	Fine sand	Silt	clays
	41,67	30,55	16,67	11,11

The identification of the earth is completed by the analysis of the clay part to define its states of consistency known as Atterberg limits: liquidity limit (WL), plasticity limit (WP) and plasticity index IP = WL - WP. These limits are given in Table 2. This table also includes the activity coefficient Ca and the specific surface area Sa.

Table 2. Limits of soil consistency F

Earth F	Wp(%)	WL(%)	IP (%)	Ca	Sa (m2/g)
	25,32	35,58	10,26	1,28	48,93

According to table 2, we conclude the earth F is active.

3 Results of mineralogical and chemical analyzes

The chemical and mineralogical of earthen mortar elements are analyzed to determine the proportions of the elements, by four analysis techniques, namely: X-ray diffraction, Infrared, Scanning Electron Microscope and X-ray fluorescence.

3.1 X-ray diffraction (DRX)

The analysis by X-ray diffraction of the earth F shows that it is formed mainly of calcite CaCO3, Of quartz SiO2 And dolomite CaMg(CO3)2 (Figure 1).
The addition of cement to the mortar increases the percentage of calcite and the presence of Silicate Aluminum but in small proportions only is due to the presence of alumina and silica in the soil used and the cement added (Figures 2, 3 et 4).
The presence of calcite and the absence of portlandite show that the binders are fully carbonated. According to Mertens et al. [18] quartz is not a binder and it forms a part of the fine sand fraction of the aggregate.

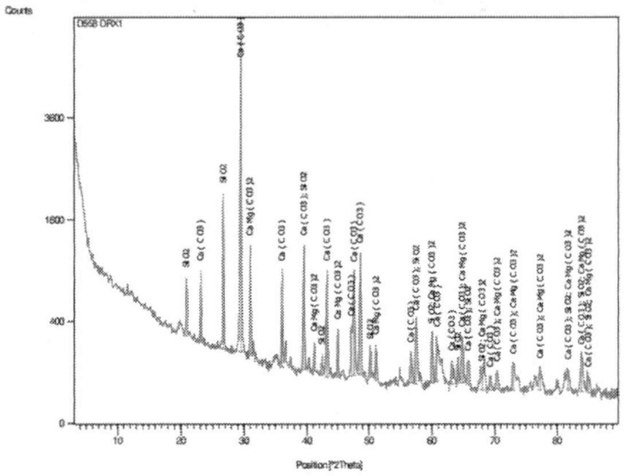

Fig. 1.: X-Ray Diffraction Spectrum of earth

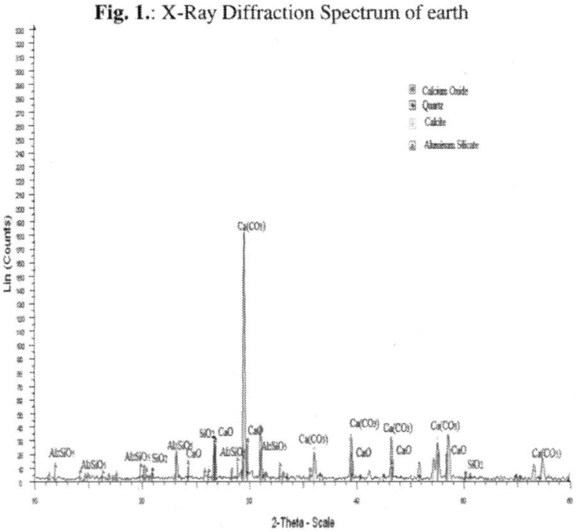

Figure.2. X-ray diffraction spectra at 4% cement

 https://doi.org/10.1051/matecconf/201814901024

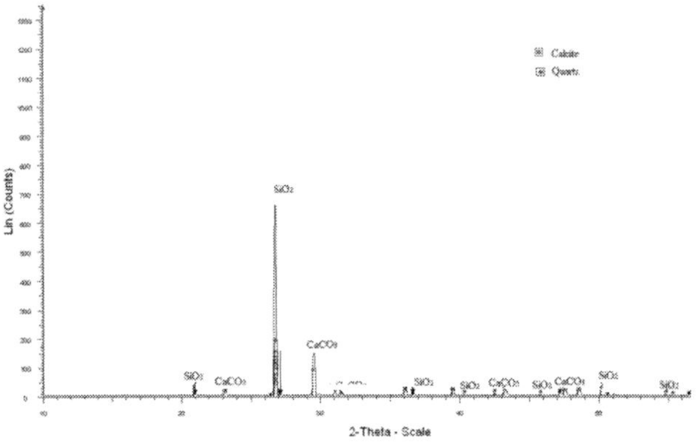

Figure 3. X-ray diffraction spectra at 7% cement

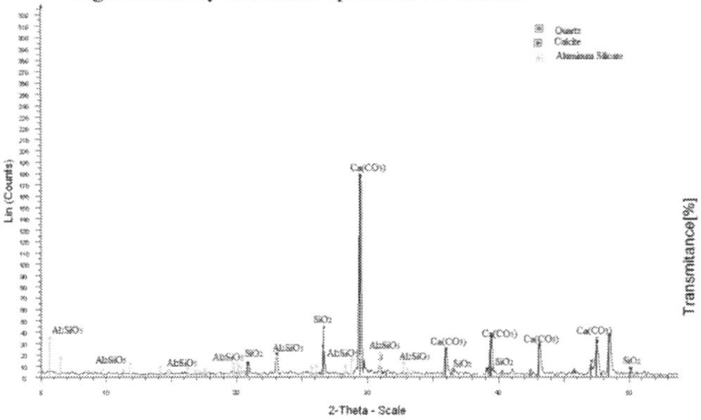

Figure 4. X-ray diffraction spectra at 10% cement

3.2 Infrared Analysis

Infrared analysis is a complementary method to X-ray diffraction analysis to study clays and clay minerals. The infrared spectra of the earth **F** and the specimens of the mortars of different cement dosages are shown in Figures 5, 6, 7 and 8. These spectra are strongly dominated by vibration bands of calcite $CaCO_3$ (1798.1, 1431, 874, 712.6) cm^{-1}, of quartz at (1033.5, 798) cm^{-1} and water at (3416.1 cm^{-1}).

The formation of these carbonates is linked to the biophysical-chemical conditions of the earth. Whereas, the calcite is the product of the carbonation of lime, once mixed with water and sand, it has the property of rapidly setting by agglomerating inert particles, such as quartz [7, 8, 3, 9]. We found bands of humic and fulvic acids that are linked to any root activity at (2516.1, 2876.3, 530.5) cm^{-1}. [7, 8, 3, 9]

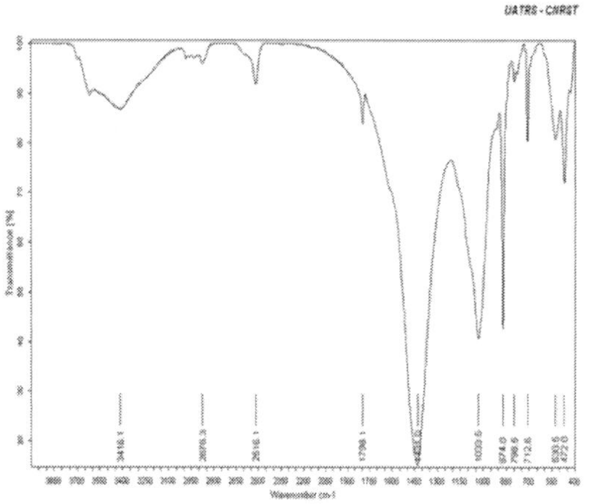

Figure 5. Infrared spectra of the earth of F

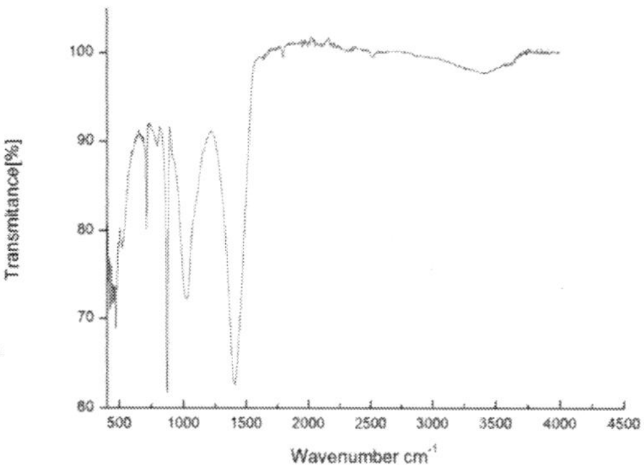

Figure 6. Infrared spectra with 4% cement

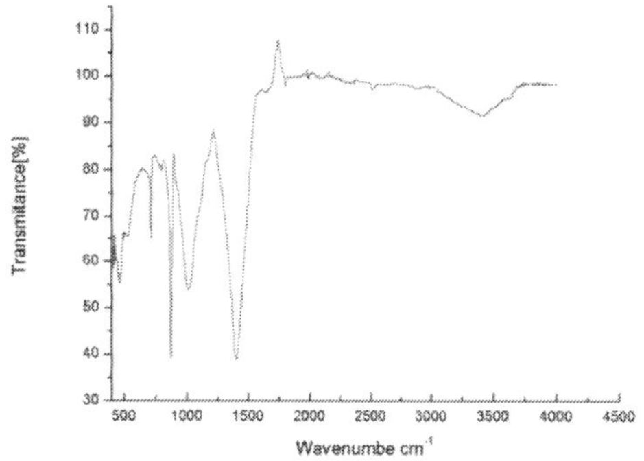

Figure 7. Infrared spectra with 7% cement

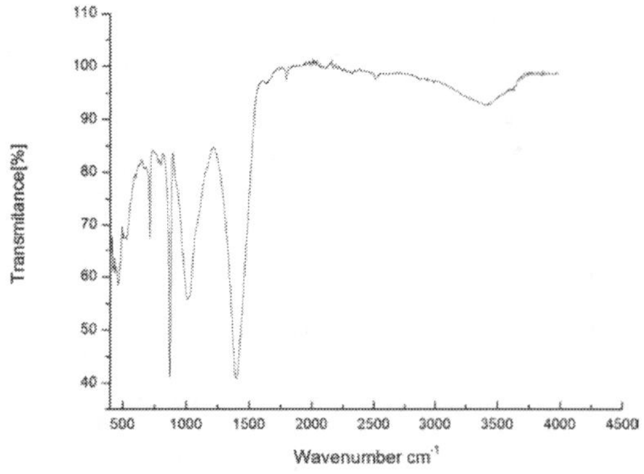

Figure 8. Infrared spectra with 10% cement

3.3 Scanning Electron Microscope (SEM)

SEM is a well-established method that can offer useful information concerning the structure of material. To better understand the evolution of the microstructure of earth samples.
From the images of earth **F** obtained by the SEM (Figure 9), we note that calcite is more dominant than quartz.

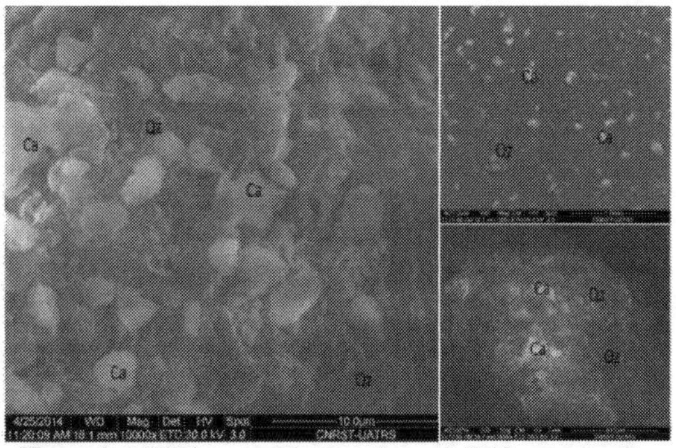

Figure 9. MEB images taken on the earth F (Qz: quartz, Ca: calcite)

3.4 Analysis by fluorescence X

The X-ray fluorescence analysis makes it possible to measure the percentage of chemical elements contained in the earth of sample F and the CPJ cement used in making the earthen mortar. Table 3 summarizes the rate of chemical elements:

Table .3. Chemical compositions of the land of Fez

Composition (%)	Earth F	Cement CPJ 35
SiO2	22,3	12.46
Cao	31,8	57.02
Al2O3	5,96	3.39
Fe2O3	1,72	1.89
MgO	2,13	1.35
SO3	1,29	1.95
K2O	0,536	0.62
TiO2	0,226	0.18
MnO	0,0503	0.03
P2O5	0,753	0.11
Na2O	0,323	-0.03
SrO	0,0125	0.015
PAF (perte au feu à 975°C	32,9	20.90

There is an enrichment of CaO (31.8%) due to the presence of the matrix of carbonate and a high SiO2 content (22,3%), due to the concentration of quartz. The losses on fire are to be linked with the presence of water.

4 Experimental results

4.1 Test mode

The tests are performed on cylindrical specimens in both wet and dry cases, height h = 100 mm and diameter Ø = 60 mm.
First, we immerse the cylindrical specimens so that it is 5 mm below the water level (figure 10). A mortar's mass increases when immersed in water. It is said that the specimen has absorbed water. This increase in mass is due to the penetration of water into the cylinder through the pores. The water absorption represented by the coefficient C_b characterizes the open porosity of a material. The crush test is performed by placing an anti-frettage system composed of neoprene and Teflon between the two plates of the press and the test piece. This system allows better stress distribution thus leading to Homogeneity of applied forces of the cylindrical specimens of earthen mortar. The cement used in this study is CPJ 35. The associated dosages are 0%, 4%, 7% and 10% by weight of cement.
Before the test, all specimens underwent a 28-day cure.
The crushing speed is fixed at 10 mm/min.

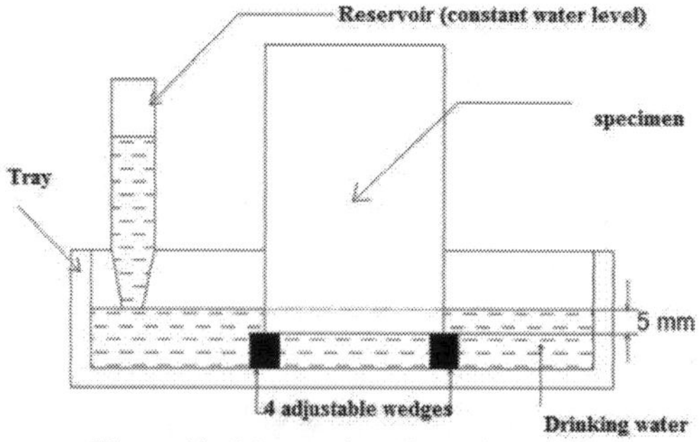

Figure.10. Schema of capillary absorption test

4.2. Durability of mortar

The following figure shows the variation of water absorption in relation to the cement content.

co

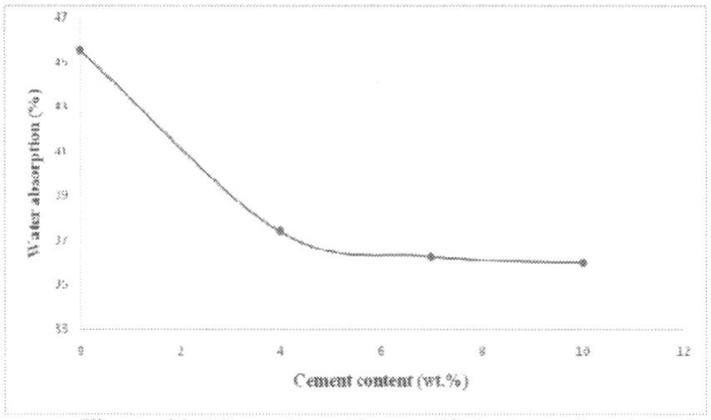

Figure 11. Water absorption coefficient as a function of the cement content

Figure 11 illustrates that the non stabilized specimens absorb a very large amount of water relative to the others which are stabilized. According to the specimens tested, the more cement content increases, the absorption of water by the specimens stabilized decreases. According to this test and to avoid capillary action, if we add a dosage of cement between 4% and 7% with respect to the dry mass of the earth it is sufficient to prevent the absorption of water. The addition beyond [4%- 7%] does not result in any significant gain [4, 14, 20].

4.3 Mechanical Results and Interpretations

The figures.12, 13 and 14 show the gain evolutionof the compressive strength, modulus of elasticity and deformation limit of the test specimens of mortar the earth **F** in both cases (dry and wet).

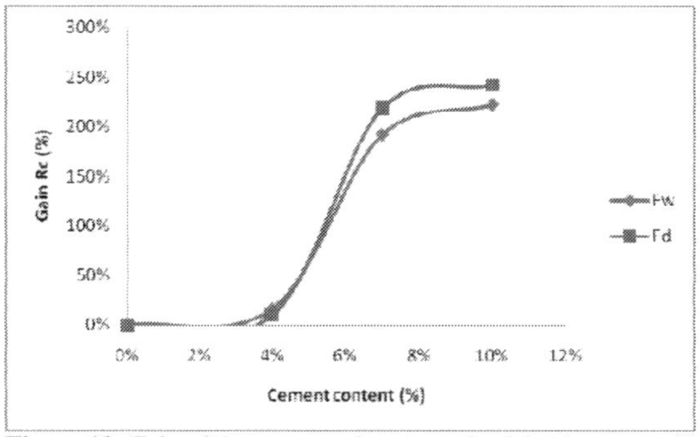

Figure 12. Gain of the compressive strength of the dry mortar F (Fd) and wet (Fw) as a function of the cement dosage

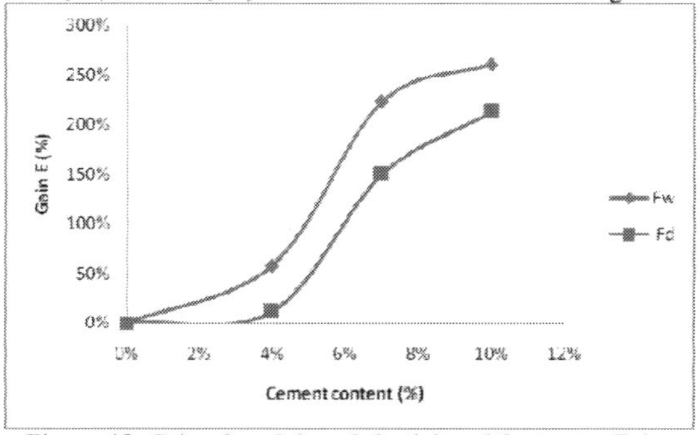

Figure 13. Gain of modulus of elasticity of the mortar F dry (Fd) and wet (Fw) depending on the dosage

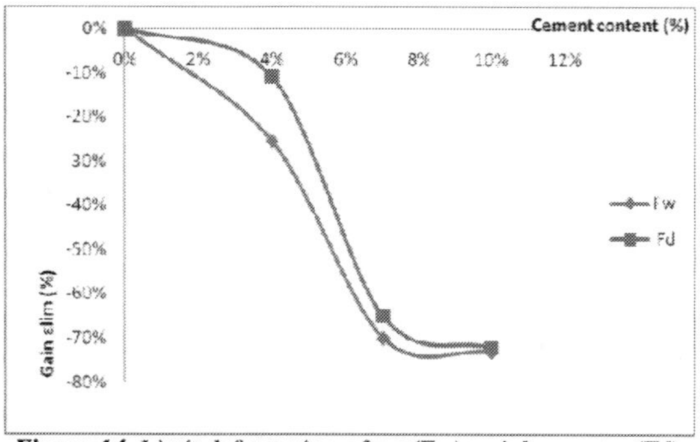

Figure 14. Limit deformation of wet(Fw) and dry mortar (Fd) as a function of the dosage

Figure 12 shows that the compressive strength of the earth **F** increases according to the cement content rate. All recent studies [9, 3, 6, 10, 11, 12, 13, 21, 22] have shown that the mechanical properties increase with the cement dosage without giving the function of this increase. Since this increase is varied according to the particle size and the mineralogical composition of the earth, which impacts the economic investment in cement. There is an increase of the compressive strength gains and the elasticity module according cement content in the dry state of the earth **F**. The percentage, the gain evolution of the compressive strength is irregular: When passing from 0 to 4% of cement dosage it is of the order of 10%, when one passes from 4 to 7% it of the order of 219% and when one passes from 7 to 10% it is of the order of 243%. The same gain evolution is noted for the modulus of elasticity, but this gain in the wet state is high compared to the dry state.

We also note that the limit strain decreases when the cement content increases. In terms of the percentage, the deformation decreases by -11% when passing from 0 to 4% of the dosage, and -65% when passing from 4 to 7%, and decreases by -72% when passing from 7 to 10% of the cement. This comparison shows an increase in the gain of the compressive strength and the modulus of elasticity gain as a function of the wet cement dosage of the same earth **F**. In terms of percentage, the evolution of the resistance in compression is irregular: when one passes from 0 to 4% of cement dosage it is of the order of 16%, when one passes from 4 to 7% it of the order of 192% and when it goes from 7 to 10% it is of the order of 222%. We notice almost the same evolution for the modulus of elasticity. It is also observed that the limiting deformation decreases when the cement dosage increases. In terms of percentage gain, deformation decreases by -25% when passing from 0 to 4% of the assay, and -70% when passing from 4 to 7%, and decreases by -73% 7 to 10% of the cement.

The mineralogical results show that the clay of the earth **F** and their mortar specimens of the different cement dosages are formed mainly of quicklime CaO and quartz SiO_2, which favors the stabilization of the clay by the reaction of quicklime and water. Although the quartz is not a binder, it is constituted by a portion of the fraction of fine sand [16, 17, 18, 7]. The quartz grains of this earth are cemented by the carbonation of the lime.

The addition of low cement content is insufficient to stabilize the sand fraction of this earth, which gives us a small variation in compressive strength from 0% to 4% cement in the both cases dry and wet. It is only when the cement dosage increases to 4% that the value of the compressive strength increases and is equal to 0.100 MPa for the dry state and 0.057 MPa for the wet state. We found that for 7% cement in that earth the silico-aluminous is absorbed (Figure 3), by reacting to calcium hydroxide to form a compound having binding properties. Therefore, all the links are already created, giving a very high compressive strength. On another layer, the hydrated cement matrix reacts in two ways [16]: a strong reaction with the sandy skeleton and a very weak reaction with the clay which is already stabilized by its chemical constituents to improve the compressive strength of mortar specimens.

Conclusion

The mineralogical analysis of the earth of Fez and their mortar shows that it consists mainly of calcite, quartz and dolomite with different proportions.

The chemical analysis carried out on the earth of Fez show an enrichment of CaO (31,8%) linked to the abundant presence of the carbonate matrix, and a content of SiO_2 (22,3%), due to the concentration of quartz.

The matrix in the soil is essentially represented by calcite.

The mechanical behavior of the earthen mortar is strongly influenced by the cement dosage employed, the mineralogy and the chemistry of the earth.

The strength gain and modulus of elasticity improve while increasing the rate of cement dosing. However, the gain of the limiting deformation evolves in the opposite direction. Concerning the capillary rise, water absorption decreases with addition of the cement. We also note that the evolution of the mechanical characteristics does not matter that in the interval [4 at 7%] which represents the area of effect for cement stabilization; consequently, the technico-economic optimum for stabilization will be located in this area [4%, 7%].

References

1. S.L. Pagliolico, S. Ronchetti , E.A. Turcato, G. Bottino, L.M. Gallo, R. DePaolia, (2010)
2. Y. Millogo , M. Hajjaji b, J.C. Morel , Applied Clay Science 52 (2011)
3. Younoussa Millogo, Mohamed Hajjaji , Raguilnaba Ouedraogo , Construction and Building Materials 22(2008)
4. Bouabid H., Thèse de Doctorat d'Etat, Université Mohamed V. Rabat. Juillet (2000).
5. A. Guettala , A. Abibsi , H. Houari , Construction and Building Materials 20 (2006).
6. R. Bahar, M. Benazzoug, S. Kenai b, Cement & Concrete Composites 26 (2004).
7. Abdelilah Dekayir, Marc Amouric, Juan Olives, Claude Parron, Abdelilah Nadiri, Abdelkader Chergui, M. Abdeljalil El Hajraoui,. C. R. Géoscience 336 (2004)
8. F. Gouny, F. Fouchal , P. Maillard, S. Rossignol , Construction and Building Materials 36 (2012).
9. Younoussa Millogo, Jean-Claude Morel ; Materials and Structures 45 (2012)
10. Philip Zak, TahaAshour, AzraKorjenic, Sinan Korjenic, Wei Wu.,Construction and Building Materials. 106 (2016)
11. P. J. Walker., Cemenf & Concrete Composites 17 (1995)
12. Peter Walker, trevvr Stace., Materials and structure/Matériaux et construction .30(1997)
13. C. Jayasinghe, N. Kamaladasa, Construction and Building Materials 21 (2007)
14. H.B. Nagaraj, M.V. Sravan, T.G. Arun, K.S. Jagadish., International Journal of Sustainable Built Environment 3 (2014)
15. Bell. F. G., Eng. Geol. 42(4) (1996)
16. A. Ammari, K. Bouassria, A. Tayyibi, H. Bouabid, M. Cherraj, S. Charif d'Ouazzane, M. Ibnoussina., J. Mater. Environ. Sci. 7 (10) (2016)
17. K. Bouassria, A. Ammari, A. Tayyibi, H. Bouabid, J. Zerouaoui, M. Cherraj, S. Charif d'Ouazzane., J. Mater. Environ. Sci. 6 (12) (2015)
18. Mertens G., Elsen J., Laduron D., Brul R., Revue d'archéométrie. 30 (2006).
19. Bachir Taallah , Abdelhamid Guettala, Construction and Building Materials 104 (2016)
20. M. Cherraj, K. Gueraoui, H. Bouabid, S. Charif d'Ouazzane, R. Bouabid , A. Hammoumi , O Fassi-Fehri., Phys. Chem. News 42 (2008).
21. A. Ammari, K. Bouassria, M. Cherraj, H. Bouabid, S. Charif D'ouazzane, Case Studies in Construction Materials 7 (2017)
22. O. Izemmourena, A. Guettala MATEC Web of Conferences 11,02001 (2014)

 https://doi.org/10.1051/matecconf/201814901025

Effect of the microstructure of mortars with low hydraulicity slag on their behavior in aggressive environments

Effet de la microstructure de mortiers à base de laitier de faible hydraulicité sur leur comportement face à des environnements agressifs

Hadj sadok Ahmed [1]

[1] Ecole Nationale Supérieure de l'Hydraulique (ENSH), GEE, Blida, Algérie
[1]a.hadjsadok@ensh.dz

Abstract. Slag is one of the most used cement additives. Due to its latent hydraulic nature, attached to its hydraulicity, it can contribute to a microstructural modification and an improvement of the durability of the concrete face of aggressive environments. In this study, a low active slag is used in the manufacture of mortars as a substitute for cement, at a maximum rate of 50%. Firstly, a study of the microstructure with mercury porosimetry was used for determination of microstructural parameters (porosity, diameters and volume distribution). The behavior of mortars in aggressive environments (sodium and magnesium sulphate and seawater) was studied later. Despite the low reactivity of studied slag, its presence especially at 50% rate, in the long term, has led to a refinement of the microstructure. This effect, among others, led to better resistivity of the mortars in the sulphate environnements.

Résumé. Le laitier compte parmi les ajouts cimentaires les plus utilisés. De part son caractère hydraulique latent, lié de son hydraulicité, peut contribuer à une modification de la microstructure et une amélioration de la durabilité du béton face à des milieux agressifs. Dans cette étude, un laitier de faible hydraulicité est utilisé dans la fabrication de mortiers comme substituant au ciment, à un taux maximal de 50%. En premier temps, une étude de la microstructure à l'aide de la porosimétrie au mercure a été réalisée permettant la détermination des paramètres microstructurales (porosité, diamètres et distribution volumique). Le comportement des mortiers face à des environnements agressifs (sulfates de sodium et de magnésium et eau de mer) a été étudié en suite. Malgré la faible réactivité du laitier étudié, sa présence particulièrement à un taux élevé de 50%, au long terme, a engendré un raffinement de la microstructure. Cet effet, entre autres, a conduit à une meilleure résistivité des mortiers dans les milieux sulfatiques.

1 Introduction

Le laitier, sous-produit de l'industrie de la fonte, compte parmi les ajouts cimentaires les plus utilisés. Présentant un caractère hydraulique latent, lorsqu'il est présent dans le ciment, son hydratation, liée de son activité hydraulique, conduit à une amélioration de la microstructure (raffinement de la porosité) et à une diminution de quantité de chaux dans la pâte de ciment (réaction pouzzolanique). De ce fait la durabilité du béton contenant du laitier face à des milieux agressifs est généralement améliorée. La dissolution du verre du laitier est assurée par l'attaque hydroxylique (des ions OH- en présence d'eau) provenant de l'hydrolyse de la Portlandite $Ca(OH)_2$, produit d'hydratation du clinker [1]. Néanmoins, ce processus est conditionné par l'hydraulicité du laitier qui est principalement influencée par sa finesse, sa composition chimique et son contenu en verre [2]. Les produits d'hydratation formés dans la matrice cimentaire sont, principalement, des silicates et des aluminates de calcium hydraté (CSH, CAH) additionnelles. La matrice résultante présente une bonne résistance chimique et une structure de pore plus raffiné

[3]. De ce fait les performances mécaniques et de durabilité du béton son améliorées.

En Algérie, le laitier granulé, produit par le complexe sidérurgique d'El Hadjar, est caractérisé par une moyenne réactivité hydraulique [4]. Une étude antérieure [5] a montré son effet positif sur les propriétés de transport, notamment au long terme, dans des conditions de cure humide. Néanmoins, peu d'études ont été menées sur son effet sur la durabilité dans des milieux agressifs, dépendant essentiellement de la résistance chimiques mais aussi des propriétés microstructurales de la matrice.

2 Expérimentation

2.1 Matériaux et mélanges

Le laitier de haut fourneau utilisé est un laitier granulé issu du complexe sidérurgique d'El Hadjar. D'une granulométrie 0/5 mm, sa structure est vitreuse à plus de 95%. Le laitier a été broyé au laboratoire à une finesse d'environ 4200 cm²/g à laide d'un broyeur à boulés. Le ciment utilisé pour la confection des mortiers est un ciment Portland de type CEM I 52,5 N. Les compositions chimiques du ciment et du laitier utilisés sont données dans le tableau 1. Le sable utilisé pour la confection des mortiers est un sable normalisé de granulométrie 0/2 mm. Cette étude est réalisée sur des mortiers composés de trois (03) sables et un (01) ciment avec un rapport E/C de 0,5. Trois (03) mortiers M0, M30 et M50 ont été fabriqués en substituant, respectivement, 0, 30 et 50% du poids de ciment par du laitier.

2.2 Préparation des spécimens et essais réalisés

Les échantillons de mortier sont coulés dans des moules en acier et compactés en deux couches à l'aide d'une table à chocs. Après 24 heures du coulage, les éprouvettes sont démoulées et conservées dans une chambre humide (95% HR) à une température de 20°C ± 2 jusqu'à l'âge de l'essai ou de la mise en solution.

La microstructure des mortiers a été analysée par porosimétrie au mercure après 28, 90 et 360 jours de maturation en cure humide (20°C et 95% d'HR). Un prosimètre Pascal 240 a été utilisé permettant d'atteindre une pression maximale de 200 MPa et ainsi d'accéder aux pores ayant un rayon de 3,7 nm. Les tests sont réalisés sur des échantillons de mortiers d'environ 2 cm3, séchés préalablement à 50°C en présence de gel de silice jusqu'à masse constante ($\Delta M < 0{,}1\%$ en 24 heures). Ainsi la porosité totale et les courbes de distribution volumique des tailles des pores ont été mesurées.

La résistance dans des milieux agressifs est mesurée sur des éprouvettes de mortier (M0, M30 et M50). Après démoulage, les éprouvettes sont conservées en cure humide (20°C et 95% d'HR) pendant 60 jours avant d'être soumises à l'immersion dans deux types de solutions de sulfates (5% de sulfate de sodium ($N\bar{S}$) ou 5% de sulfate de magnésium ($M\bar{S}$)) et une solution d'eau de mer. Une série d'éprouvettes témoins a été conservée en cure humide. Notons que les solutions sont changées chaque mois. L'expansion des éprouvettes prismatique (20x20x160 mm), selon la norme ENV 196-10, a été mesurée à l'aide d'un comparateur de précision de 1µm après chaque mois, durant les 12 mois d'immersions dans les trois solutions. Par ailleurs, l'évolution de la résistance à la compression a été mesurée sur des éprouvettes 40x40x160 mm, après 2, 3, 6 et 10 mois d'immersion dans les sulfates de sodium et de magnésium ainsi que pour les éprouvettes témoins conservées en cure humide.

3 Résultats et discussion

3.1 Microstructure

Le tableau 2 regroupe les valeurs de la porosité volumique totale (P), le rayon moyen de pore "Rm" et le rayon critique "Rc". Le rayon moyen est calculé en se basant sur un modèle de pore cylindrique de rayon Ri et de volume Vi. Si « S » est la surface interne des pores et « V » le volume total, alors Rm est calculé par l'équation 1. Le rayon critique Rcr (ou de percolation) est défini comme le rayon correspondant au pique (ou mode de remplissage) apparent dans la courbe de distribution différentielle des volumes des pores. Ces courbes, obtenues par différentiation logarithmique, sont présentées dans la figure 1.

$$R_m = 2V/S \qquad (1)$$

Au vu des résultats, on peut voir une diminution de la porosité à l'eau avec l'âge, pour tous les mortiers étudiés. Toutefois, on peut constater une porosité plus importante, à tout âge, pour les mortiers au laitier, particulièrement, pour le M50 par rapport au mortier M0. En observant les résultats de plus prêt, on remarque que la diminution de la porosité entre 28 et 360 jours est plus importante pour le M30 et M50 que pour le M0. Cette variation est typique et est la conséquence de l'hydratation latente du laitier dans le ciment Portland. Ceci est largement confirmé dans la littérature [6, 7].

Comme le montre la figure 1, à 28 jours de maturation, la distribution du volume des pores des mortiers au laitier (M30 et M50) est plus grossière que celle du mortier M0. Le rayon critique ainsi que le rayon moyen des pores sont plus élevés pour les mortiers M30 et M50 (Tableau 2). A partir de 90 jours, et particulièrement après 360 jours de maturation, la distribution de la taille des pores des mortiers au laitier a évolué considérablement vers les petits pores. En effet, les rayons moyen et critique des pores sont moins importants pour les mortiers M30 et M50 que pour le mortier M0. Ce raffinement de la porosité, plus apparent pour le mortier avec 50% de laitier est expliqué par l'hydratation du laitier qui se fait par une consommation de chaux (CH), conduisant à la formation des "CSH" additionnels qui contribuent à la densification de la structure poreuse de la matrice [5].

3.2 Résistance dans les milieux agressifs

3.2.1 Expansion

Les résultats d'expansion des éprouvettes de mortier avec et sans laitier après 12 mois d'immersion dans des solutions de sulfates de sodium, de sulfates de magnésium et d'eau de mer sont présentés dans les figure 2, 3 et 4 respectivement.

Au vu de ces résultats, on peut voir une expansion globalement plus importante des mortiers conservées dans la solution de sulfate de sodium, comparée à celle des éprouvettes conservées dans les solutions de sulfate de magnésium et d'eau de mer.

Tableau 1. Composition chimique du ciment et du laitier

	CaO	Al_2O_3	Fe_2O_3	SiO_2	MgO	K_2O	Na_2O	Perte au feu	SO_3
Ciment	61,30	5,60	3,80	18,40	0,90	0,78	0,42	2,2	3,30
Laitier	42,84	9,19	3,44	41,20	2,12	-	-	0,2	-

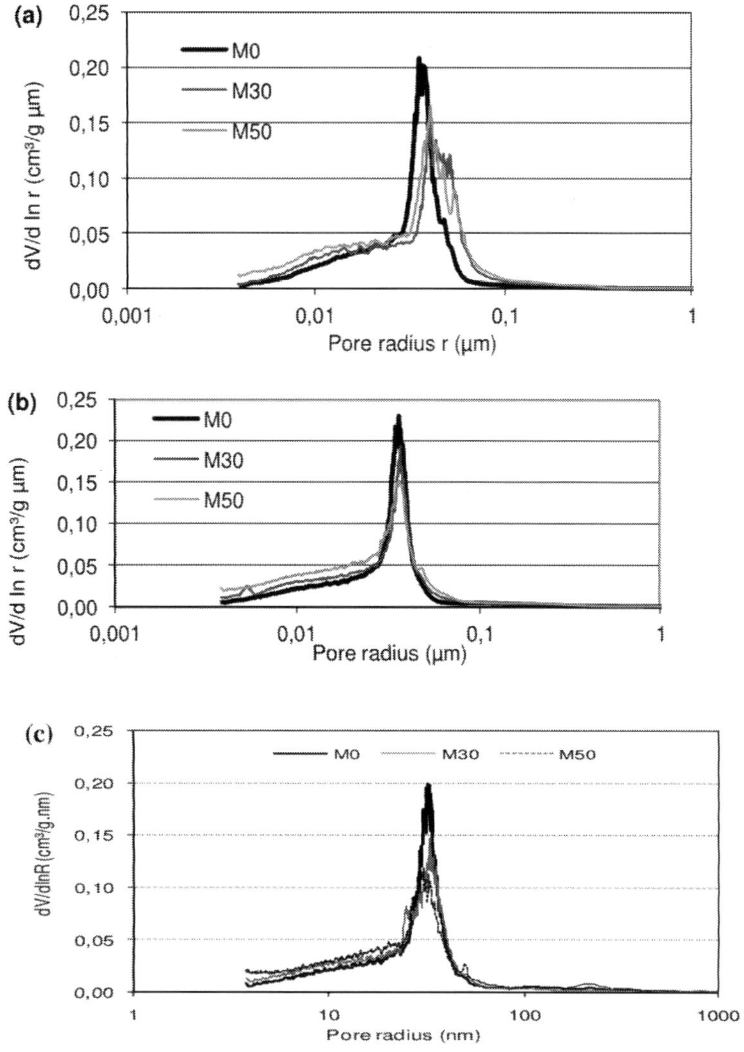

Fig. 1. Distribution des tailles des pores des mortiers au laitier après 28 jours (a), 90 jours (b) et 360 jours (c).

 https://doi.org/10.1051/matecconf/201814901025

Tableau 2. Porosité (P), Rayon critique (R_{cr}) et rayon moyen des pores (R_m) des mortiers étudiés.

	28 jours			90 jours			360 jours		
Mortiers	P(%)	R_{cr} (nm)	R_m (nm)	P%)	R_{cr} (nm)	R_m (nm)	P(%)	R_{cr} (nm)	R_m (nm)
M0	11.20	35,0	35,0	10.90	36,0	32,0	10.40	31,5	30,5
M30	12.50	40,0	38,8	11.50	37,0	30,5	10.50	30,0	28,0
M50	14.15	40,0	35,8	13.45	36,0	27,5	11.60	28,0	25,3

Au regard des figures 2 et 3, on constate que l'incorporation du laitier dans les mortiers, conduit à une diminution significative de l'expansion, particulièrement en solution de sulfates de sodium (5% $N\bar{S}$). Le mortier M0 a atteint une expansion de 0,45% après 12 mois d'immersion dans la solution $N\bar{S}$. A cette échéance, les mortiers M30 et M50 présentent respectivement, une expansion plus faible de 33 et 83% par rapport au mortier M0. Dans la solution de sulfate de magnésium (5% $M\bar{S}$), après 12 mois d'immersion, le M0 et le M30 ont eu une expansion similaire (de l'ordre de 0,12% par contre le mortier M50 n'a subi qu'une faible expansion de 0,02% soit une diminution de 81% par rapport au mortier témoin. Dans la solution d'eau de mer, (figure 4), on observe une faible expansion (0,02-0,03% après 12 mois) et peu variable entre les trois mortiers étudiés avec et sans laitier, en comparaison avec l'expansion observée dans les solutions de sulfates de sodium ou de magnésium. Toutefois, une expansion légèrement supérieure des mortiers au laitier (notamment M50) est constatée.

La variation observée reflète un comportement typique des ciments au laitier face à un environnement sulfatique. En effet, la diminution de l'expansion a été observée par plusieurs chercheurs, surtout pour des taux de laitier supérieur à 50%, particulièrement dans un environnement de sulfates de sodium [8, 9].

La différence d'expansion des mortiers entre les deux solutions (5% $N\bar{S}$) et (5% $M\bar{S}$) indique bien la dissimilitude dans le mécanisme d'attaque. En effet, la détérioration du ciment par les sulfates de magnésium est contrôlée par l'attaque de magnésium - gypse sur le gel de silicate de calcium hydraté (CSH). La formation du gypse et de la brucite (MH) est due aux réactions entre $M\bar{S}$ et CH, CSH [10]. Le gypse formé réagit avec l'aluminate du ciment. La brucite, formée, étant insoluble (sa solubilité est de 0,01 g/l comparée à 1,37 g/l pour le CH) et ayant un bas pH, tend à déstabiliser l'ettringite formée [11, 12]: ceci peut expliquer la relative faible expansion des mortiers dans la solution de $M\bar{S}$. L'expansion élevée mesurée sur les mortiers immergés dans la solution de sulfate de sodium est attribuée à la formation d'ettringite secondaire.

L'expansion plus importante, observée dans les solutions de sulfate comparées à l'eau de mer, peut être attribuée à la différence du taux de diffusion des ions Cl^- et SO_4^{-2}, celui des Cl^- étant le plus élevé. Ceci permet aux ions de Cl^- de pénétrer dans la matrice plus rapidement que les ions de SO_4^{-2}. Les chlorures réagissent ainsi avec les C_3A pour former des chloroaluminates de calcium (Sel de Friedel), diminuant ainsi la quantité de C_3A disponible pour réagir avec les ions sulfates pour former le gypse ou l'ettringite expansive [13]. BICZOK [14] a constaté que l'expansion du béton causée par des ions sulfate en eau de mer est fortement retardée par la présence de chlorures. Il faut toutefois noter que la concentration des sulfates en eau de mer (sulfates de magnésium 5,0 g/l et sulfates de calcium 1,5 g/l) est nettement moins importante que celle des solutions de $N\bar{S}$ et $M\bar{S}$) étudiées (50 g/l).

Dans un ciment au laitier, le taux de C_3A est relativement plus faible que celui d'un ciment sans ajout. Ce taux est d'autant plus faible que le pourcentage du laitier augmente. D'autre part, le laitier de part sa de la réaction pouzzolanique qui consomme de l'hydroxyde de calcium (CH), conduit à une diminution de la formation du gypse et par conséquence, de l'ettringite. Ceci peut expliquer la diminution de l'expansion constatée dans les mortiers au laitier, même si le laitier étudié n'est que moyennement réactif. Cette amélioration peut aussi être expliquée par la microstructure plus raffinée des mortiers au laitier observée précédemment.

https://doi.org/10.1051/matecconf/201814901025

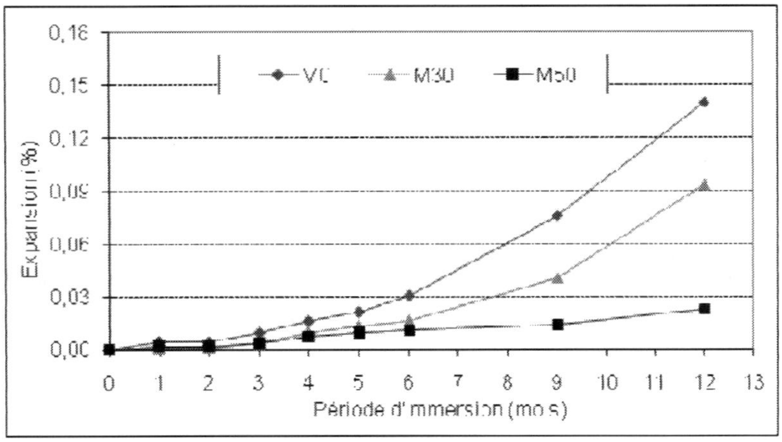

Fig. 2. Expansion des mortiers dans la solution de sulfate de sodium (5% $N\overline{S}$).

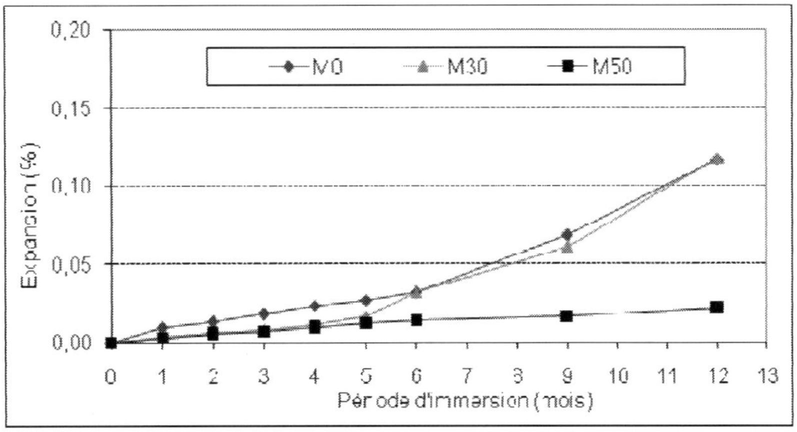

Fig. 3. Expansion des mortiers dans la solution de sulfate de magnésium (5% $M\overline{S}$).

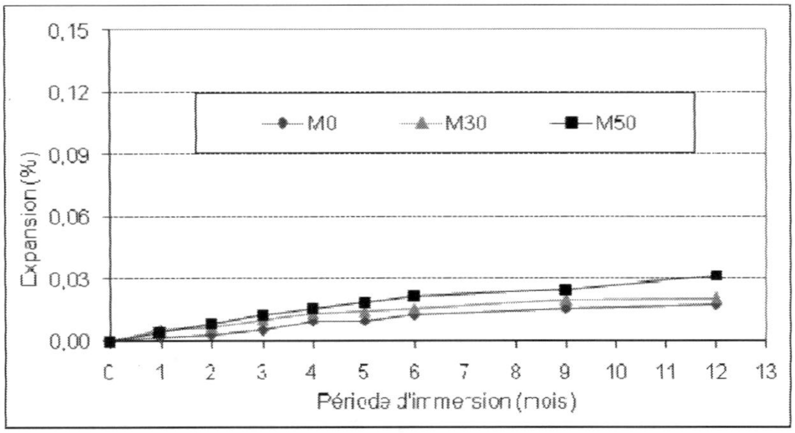

Fig. 4. Expansion des éprouvettes de mortier dans l'eau de mer

 https://doi.org/10.1051/matecconf/201814901025

3.2.2 Résistance mécanique

Le développement de la résistance en compression en fonction du temps d'immersion, dans les sulfates de sodium et de magnésium, des mortiers étudiés est représenté respectivement par les figures 5 et 6. On peut constater une augmentation de la résistance pendant les 2 à 3 premiers mois, suivie d'une diminution, plus importante pour le mortier M0 par rapport aux mortiers M30. Cette variation est observée dans les deux solutions de sulfates, pour tous les mortiers à l'exception du M50 dont la résistance a continué son développement durant toute la période d'immersion dans les sulfates de sodium.

A fin d'estimer la perte de résistance, dans le tableau 3 on présente les rapports des résistances des éprouvettes immergées dans les sulfates par rapport à celles conservées dans l'eau à 20°C. De ces résultats, on constate, dans la solution de sulfate de sodium, une perte de résistance à tout âge pour mortier M0. Cependant, pour les mortiers au laitier, on remarque une légère augmentation jusqu'à 90 jours d'immersion, suivie d'une perte de résistance (de 5 à 35% après 6 mois). Après 10 mois d'immersion, les mortiers M0 s'est pratiquement détérioré. A cette échéance, une perte de résistance relativement faible (18 à 25%) est constatée pour les mortiers M30 et M50. Les mortiers conservés dans les sulfates de magnésium ont eux aussi subi une perte de résistance, notamment après 6 mois d'immersion. Par rapport aux mortiers au laitier (M30 et M50), le mortier sans ajout a accusé une perte de résistance importante de l'ordre de 73% après 10 mois d'immersion.

Par ailleurs, d'une manière générale les mortiers au laitier, ont subi des pertes de résistance plus importante dans les sulfates de magnésium que dans les sulfates de sodium. Plusieurs recherches [15, 16] ont montré une vulnérabilité des ciments au laitier vis-à-vis des sulfates de magnésium. Ceci s'explique par l'attaque préjudiciable des sulfates de magnésium qui se traduit, en plus de par la formation d'ettringite expansive, par la substitution partielle des CSH de la matrice par des MSH (silicates de magnésium hydratés) [17]. Les MSH étant faiblement cohésif, conduisent à une détérioration des couches surfaciques de la pâte durcie de ciment et, par conséquent, à une perte de résistance.

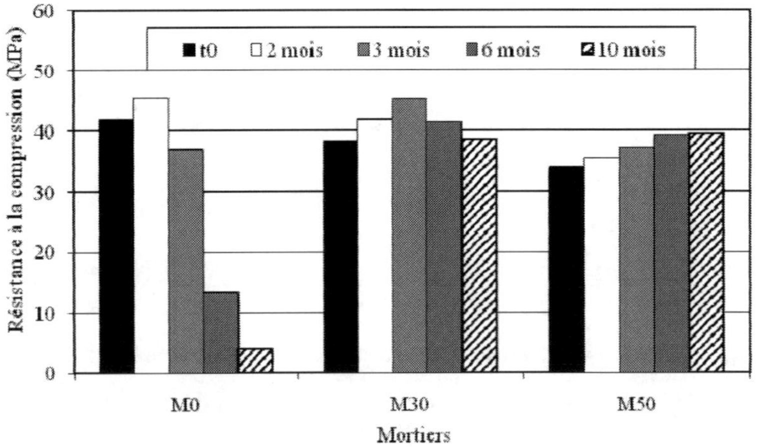

Fig. 5. Développement de la résistance à la compression des mortiers en fonction du temps d'immersion dans les sulfates de sodium (5% $N\overline{S}$).

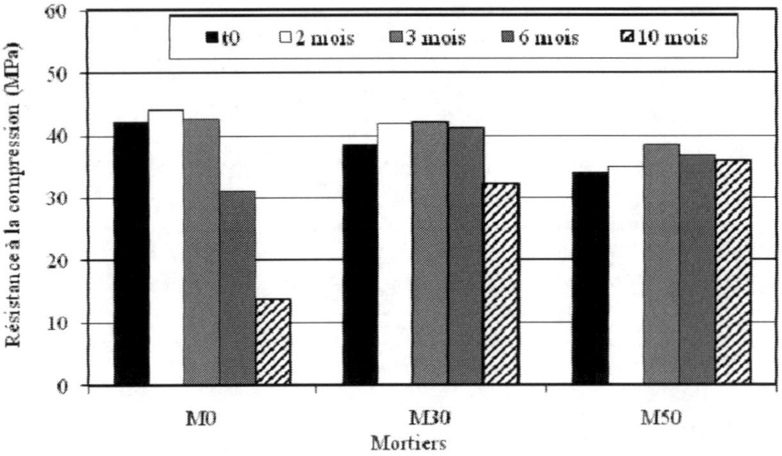

Fig. 6. Développement de la résistance à la compression des mortiers en fonction du temps d'immersion dans les sulfates de magnésium (5% $M\overline{S}$).

Tableau 3 : Rapports des résistances des éprouvettes immergées dans les sulfates par rapport à celles conservées en cure humide à 20°C.

	Sulfates de sodium			Sulfates de magnésium		
Période d'immersion	M0	M30	M50	M0	M30	M50
t_0 *	1,000	1,000	1,000	1,000	1,000	1,000
2	0,974	1,030	1,026	0,942	1,025	1,009
3	0,777	1,051	0,981	0,895	0,976	1,011
6	0,276	0,963	0,840	0,639	0,857	0,848
10	0,080	0,749	0,813	0,268	0,625	0,751

* t_0 correspond à 2 mois de conservation dans l'eau à 20°C

3 Conclusion

De ce travail, les conclusions suivantes peuvent être tirées :

- La porosité des mortiers au laitier est plus importante que celle du mortier sans ajout, par contre, au long terme, les mortiers au laitier développent une distribution de la taille des pores plus raffinée, notamment pour un taux de substitution de 50%.
- Même si le laitier étudié est de faible hydraulicité, les résultats obtenus prouvent que son incorporation aux ciments apporte une modification microstructurale positive conduisant à une amélioration notable de la résistance aux sulfates. Cette amélioration se traduit par une faible expansion et une bonne conservation de la résistance à la compression.
- L'effet bénéfique du laitier sur la résistance des mortiers dans les milieux agressifs est plus apparent dans les sulfates de sodium et de magnésium, que dans l'eau de mer.

Références

1. FP. Glasser, Was. Manag. J. **6,** 159 (1996)
2. S. Pal, A. Mukherjee, S. Pathak, Cem. Concr. Res. **33**, 1481 (2003).
3. M. Han Young, K. Hong Sam, C. Doo Sun, Constr Build Mater. **20**, 725 (2006)
4. A. Bougara, C. Lynsdale, K. Ezziane, Constr Build Mater. **23**, 542 (2009)
5. A. Hadj Sadok, S. Kenai, L. Courard, A. Darimont, Constr Build Mater. **25**, 1018 (2011)
6. A. Cheng, R. Huang, W. Jiann-Kuo, C. Cheng-Hsin, Mater Chem and Phys. **93**, 404 (2005)
7. C. Perlot, J. Verdier, M. Carcass`es, Mater and Struct. **39**, 511 (2006)
8. HT. Cao, L. Bucea, A. Ray, S. Yozghatlian, Cem. Concr. Comp. **19**, 161 (1997)
9. DD. Higgins, Cem. Concr. Comp. **25**, 913 (2003)
10. O.S. Alamoudi, M. Maslehuddin, M. Saadi, ACI. Mater. J. **92** 15 (1995)
11. Rasheeduzzafar, O.S. Al-Amoudi, SN Bduljauwad , M. Maslehuddin, ASCE. J. Mater. Civil Eng, **6**, 201 (1994).
12. FM. Lea, 3rd ed, (ARNOLD, London, 1970)
13. OS. Al-Amoudi, M. Maslehuddin, Y. Abdul-Al, Constr Build Mater. **9**, 25 (1995)
14. I. Biczok, *Concrete Corrosion; concrete Protection* (Akademiai Kiado, Budapest, 1980)
15. RS. Gollop, H. Taylor, Cem. Concr. Res. **26**, 1013 (1996)
16. DD. Higgins, Cem. Concr. Comp. **25** 913 (2003)
17. S. Omar, B. Al-Amoudi, Cem. Concr. Comp. **24**, 305 (2002)

Influence of calcined mud on the mechanical properties and shrinkage of self-compacting concrete

Influence de la vase calcinée sur les propriétés mécaniques et le retrait des bétons autoplaçants

Taieb Fatima[1], Belas Nadia, Belaribi Omar, Belguesmia Khalil and Hadj Sadok Rachid
Laboratoire Construction Transport et Protection de l'Environnement, Département de Génie Civil et d'architecture, Université de Mostaganem, Mostaganem, Algérie
[1]nadakhalifa27@gmail.com

Abstract. The use of SCC has a particular interest in terms of sustainable development. Indeed, their specific formulation leads to a greater volume of dough than for common concretes, thus, a larger quantity of cement. However, for economical, ecological and technical reasons, it is sought to limit their cement content [1]. It is therefore necessary to almost always use mineral additions as a partial replacement for cement because the technology of self-compacting concretes can consume large quantities of fines, in this case calcinated mud issued from dams dredging sediments that can give and/or ameliorate characteristics and performances of this type of concretes. Four SCCs had been formulated from the same composition where the only percentage of calcinated mud of Chorfa (west of Algeria) dam changed (0%, 10%, 20% and 30%). The effect of calcinated mud on characteristics at fresh state of SCC according to AFGC was quantified. Mechanical strengths and shrinkage deformation (total, autogenous, drying) were evaluated. The results show the possibility to make SCCs with different dosages of calcinated mud having strengths that can defy those of the control SCC. The analysis of free deformations indicates the beneficial impact of the mud by contributing to decrease the amplitudes of the shrinkage compared to those of the control SCC.

Résumé. L'utilisation des BAP revêt un intérêt particulier en termes de développement durable. En effet, leur formulation spécifique conduit à un volume de pâte plus important que pour des bétons traditionnels et donc, a priori, à une plus grande quantité de ciment. Or, pour des raisons économiques et des considérations d'ordre écologiques et techniques, on cherche à limiter leur teneur en ciment [1]. Il est donc nécessaire d'utiliser quasi systématiquement des additions minérales en remplacement partiel au ciment car la technologie des bétons autoplaçants est capable de consommer d'énormes quantités de fines en l'occurrence la vase calcinée issue des sédiments de dragage des barrages, qui à son tour pourrait apporter et/ou améliorer des caractéristiques et performances de ce type de bétons. Quatre BAP ont été formulés à partir d'une même composition où seul le pourcentage de la vase calcinée issue des sédiments de dragage du barrage de Chorfa (ouest algérien) change (0%, 10%, 20% et 30%). L'effet de la vase sur les caractéristiques des BAP à l'état frais a été quantifié. Des résistances mécaniques et les déformations de retrait (total, endogène et de dessiccation) ont été évaluées. Les résultats montrent la possibilité de confectionner des BAP avec différents dosages en vase calcinée ayant des résistances qui peuvent défier celles du BAP témoin. L'analyse des déformations libres indique l'impact bénéfique de la vase en contribuant à diminuer les amplitudes du retrait par rapport à celles du BAP témoin.

1 Introduction

L'intérêt des bétons autoplaçants et leur rapide succès auprès des maîtres d'œuvre tient à leur facilité de mise en œuvre qui permet d'améliorer la qualité des travaux en termes de finition notamment pour les ouvrages de formes complexes ou fortement ferraillés [2]. Cependant, le contrôle du retrait de ce type de béton reste un problème majeur non résolu compromettant l'élargissement de son domaine d'utilisation.

L'objectif principal de cette recherche est d'étudier les propriétés des bétons autoplaçants à base de différents pourcentages de substitution en sédiments de dragage traités en l'occurrence, la vase du barrage Chorfa, et ce à l'état durci à savoir les résistances mécaniques en compression et les variations dimensionnelles dues au retrait, et ce en comparant les résultats des essais à ceux des éprouvettes de BAP de référence.

2 Matériaux et Formulations

2.1 Matériaux

2.1.1 Ciment

Un ciment 42.5N ES a été utilisé pour les différentes formulations, sa surface Blaine est de 3200 cm^2/g et sa masse volumique absolue est de 3.17 g/cm^3. La composition minéralogique du clinker est présentée dans le tableau 1.

Tableau 1. Composition minéralogique du clinker selon Bogue.

Les minéraux	C_3S	C_2S	C_3A	C_4AF
%	60.63	17.29	1.82	15.66

2.1.2 Vase

La vase utilisée dans la formulation du béton autoplaçant a été prélevée du barrage de Chorfa II situé au niveau de la wilaya de Mascara dans l'ouest Algérien. Elle a subi un traitement thermique par calcination dans un four lent à température de 750 + 5 C° réglé à 5 C° /minute pendant 5 heures après qu'elle soit étuvée, broyée et tamisée à 80 µm. La surface spécifique et la densité de la vase calcinée sont respectivement 8310 cm^2/g et 2.74 g/cm^3. Ses caractéristiques chimiques sont données dans le tableau 2.

Tableau 2. Composition chimique de la vase calcinée

Oxydes	Teneur (%)
SiO_2	48.39
CaO	17.03
Fe_2O_3	6.82
Al_2O_3	15.63
MgO	2.85
Cl	0.066
K_2O	2.3
SO_3	0.23

2.1.3 Granulats

Les granulats utilisés sont des granulats concassés calcaire provenant de la carrière de Hasnaoui à Sidi Belabbas (Ouest Algérien) de classes 0/3 et 3/8 et 8/15. Leur densité est respectivement 2.65, 2.64 et 2.67. Le sable de mer 0/2 de nature siliceuse provenant de la carrière de Sidi-Lakhdar à Mostaganem (Ouest Algérien) avec une densité de 2.55.

2.1.4. Adjuvant

Un seul type d'adjuvant a été utilisé à savoir le Sika Viscocrète Tempo 12, un super plastifiant haut réducteur d'eau polyvalent de nouvelle génération à base de copolymère acrylique avec une densité de 1.06±0.01 et de 28.0 à 31.0% d'extrait sec.

2.2 Formulations des BAP

Dans l'élaboration des formulations des BAP le ciment a été remplacé partiellement par 10, 20 et 30% de vase calcinée. Au total, quatre BAP ont été confectionnés avec un rapport E/L constant égal à 0.5, leurs compositions sont données dans le tableau 3. La caractérisation à l'état frais des bétons autoplaçants s'est limité aux essais recommandés par l'AFGC (AFGC ,2008) : étalement au cône (SF), écoulement à la boite en L et stabilité au tamis (Sr). Les résultats à l'état frais donnés dans le tableau 3 montrent la possibilité de formuler des BAP avec la vase calcinée qui ont été qualifiés d'autoplaçants selon les spécificités de l'AFGC [3].

Tableau 3. Composition et propriétés à l'état frais des BAP

Designation BAP (kg/m^3)	BAPT	BAP10VC	BAP20VC	BAP30VC
E/L	0.5	0.5	0.5	0.5
Ciment	451	405.9	360.8	315.7
Vase	/	45.1	90.2	135.3
Eau	225.5	225.5	225.5	225.5
Adjuvant	2.3	3.56	4.56	5.70
Sm (0/2)	558	558	558	558
Sc (0/3)	248	248	248	248
G (3/8)	330	330	330	330
G (8/15)	501	501	501	501

SF (cm)	69.75	72.75	68.25	73.5
L-Box (H_2/H_1)	87.25	83.33	81.5	89.8
Sr (%)	6.86	6.64	5.14	1

3 Essais expérimentaux

3.1 Essai de compression simple

Les tests ont été réalisés sur des éprouvettes cubiques 100 mm x 100 mm conservées dans l'eau dès leur démoulage jusqu'aux échéances d'écrasement.

3.2 Retrait libre

Nous savons que l'élément le plus influant sur le retrait c'est la quantité d'eau utilisée. Pour cela et pour mieux cerner le problème, nous avons gardé le rapport E/L = 0,5 fixe et la quantité d'eau 225.5 kg/m^3 fixe.

Les mesures du retrait endogène et du retrait total ont été réalisées 24h après le début du malaxage pour chaque composition, les variations dimensionnelles ont été effectués avec un comparateur sur trois éprouvettes conservées dans une salle maintenue à 20°C et 50% d'humidité relative (retrait total) et sur trois autres éprouvettes protégées par un film adhésif pour empêcher tout échange hydrique (retrait endogène). La différence de ces deux retraits permet d'obtenir le retrait de dessiccation.

4 Résultats et Discussion

4.1 Résistances mécaniques

L'évolution de la résistance en compression en fonction du temps (cure dans l'eau pendant 3, 7, 14, 28, 60 et 90 jours) et du pourcentage de la vase calcinée est donnée par la figure 1. Les résultats montrent que pour tous les BAP étudiés, la résistance en compression augmente avec l'âge en fonction du dosage de vase calcinée.
Les BAP à base de vase affichent de bonnes performances mécaniques par rapport au BAP témoin.

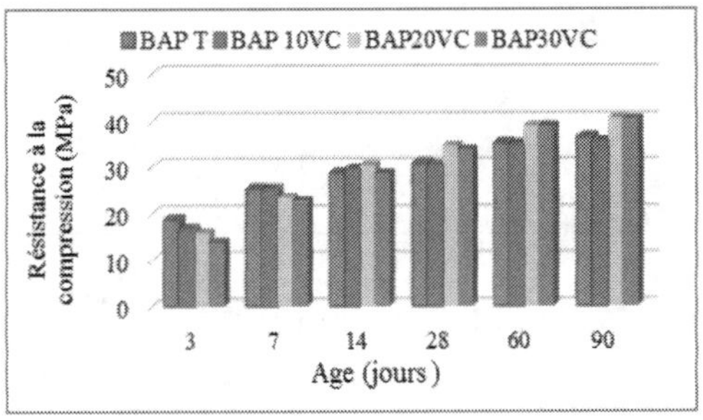

Fig. 1. Evolution de la résistance à la compression des BAP

La figure 1 montre que jusqu'à l'âge de 14 jours les résistances développées pour tous les BAP semblent très proches néanmoins au-delà de cette échéance deux cinétiques se distinguent avec une supériorité pour les BAP à 20% et 30% d'ajout.

A 28, 60 et 90 jours, on remarque déjà une différence de 4 MPa entre les deux BAP (20VC, 30VC) et le BAP témoin, cette différence n'est que la conséquence de l'activité pouzzolanique.
A partir de la figure 2 qui montre les rapports entre les résistances des BAPVC et BAP T, il est à constater que les deux BAP (20VC et 30VC) affichent les meilleures performances par rapport au BAP témoin

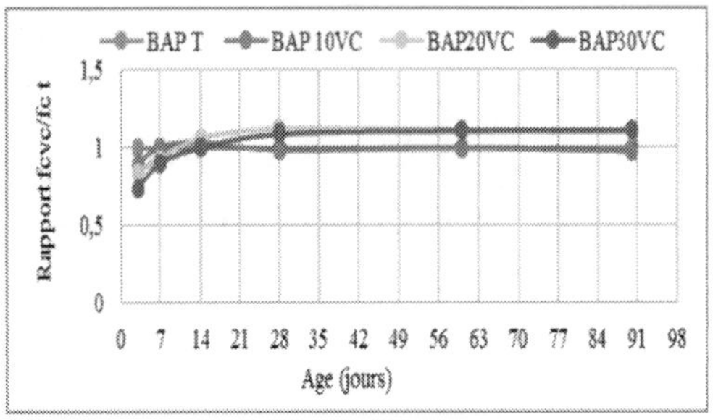

Fig. 2. Variation des résistances en compression des BAPVC/BAPT

En général, la résistance à la compression du BAP à base de vase calcinée est supérieur à celle du BAP témoin pour un même rapport E/L et même quantité d'eau.

4.2 Déformations différées

4.2.1 Retrait endogène

Les évolutions du retrait endogène des BAP mesuré sur des éprouvettes 7x7x7x28 cm^3 sont présentées sur la figure 3.

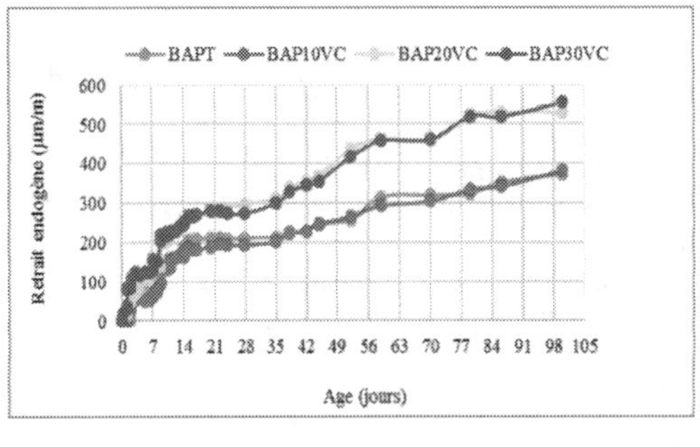

Fig.3. Evolution du retrait endogène en fonction du temps

Il apparaît dans la figure 3, que les cinétiques des déformations du retrait endogène des bétons BAPVC, sont assez semblables, elles présentent des allures similaires et presque identiques aux premiers âges et

commencent à se différencier au fur à mesure que nous avançons dans le temps [4]. Les BAP à base de 20% et 30% de vase présentent les déformations les plus élevées, dès les premiers jours et pour toute la durée des essais. Au contraire les BAP (témoin et à 10% de vase) présentent les plus faibles retraits endogènes. Si l'on se base sur la définition du retrait endogène qui résulte du fait que le volume des hydrates formés est légèrement inférieur au volume des composants initiaux (eau + ciment), ce processus entraîne une diminution de volume après la prise du béton, alors on peut clairement dire que les bétons contenant la vase calcinée nous révèlent la formation d'autres types d'hydrates autres que ceux dus à l'hydratation du ciment, sans négliger l'effet physique de cette même vase suite à sa grande finesse [5]. Les additions minérales et les adjuvants contenus dans la pâte des BAP ont surtout un effet sur la cinétique de la contraction Le Chatelier et peu sur son amplitude finale [6], ce qui est le cas pour nos mélanges à base de 20% et de 30% de vase calcinée.
Il est admis que le retrait endogène est une conséquence des changements d'HR dans les pores [7]. En fonction de la structure des pores et de l'eau disponible ; différents mécanismes sont déclenchés [7].

Le BAP10VC présente le plus faible retrait endogène ; ce qui est remarqué avec les faibles pertes de masse endogène obtenues pour ce mélange. Le retrait endogène présente un même développement en fonction des résistances mécaniques ; il est proportionnel à la résistance en compression du béton [1] ; et ce que nous avons déjà vu dans les résultats de compression.

4.2.2 Retrait de dessiccation

On constate sur la figure 4, que le BAP 10VC présente les valeurs les plus élevées après la première semaine de durcissement. En revanche les BAP à base de 20% et 30% de vase présentent des valeurs semblables et inférieures par rapport à celles du témoin, à 28 jours on constate que le retrait des BAP20VC et 30VC sont inférieurs de 59 à 31 % respectivement par rapport au BAPT, cette phase illustrera le début d'une certaine activité pouzzolanique.
Cela peut être expliqué par le rôle à la fois physique de la vase si l'on considère les résultats des autres chercheurs qui affirment que la présence des ajouts minéraux à des taux convenables fait diminuer le retrait de dessiccation grâce à leur contribution à l'amélioration de la microstructure [5]. On peut dire que l'addition de la vase calcinée réduit le retrait de dessiccation et par conséquent ce matériaux apporte un plus aux problèmes de retrait des BAP [8].

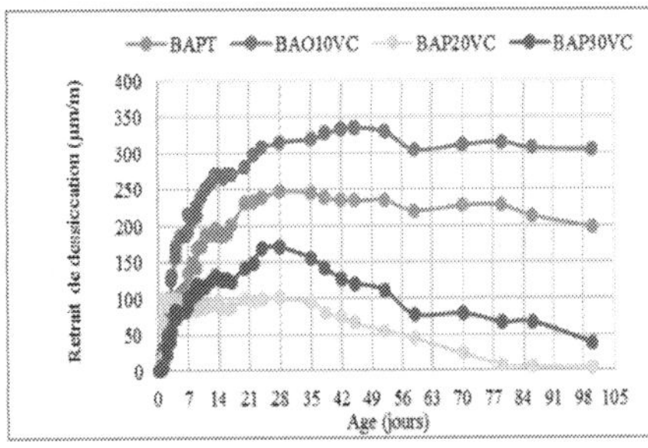

Fig.4. Évolution du retrait de dessiccation en fonction du temps

4.2.3 Retrait Total

Les courbes de la figure 5 montrent l'évolution du retrait total dans le temps des éprouvettes de BAP non confinées.

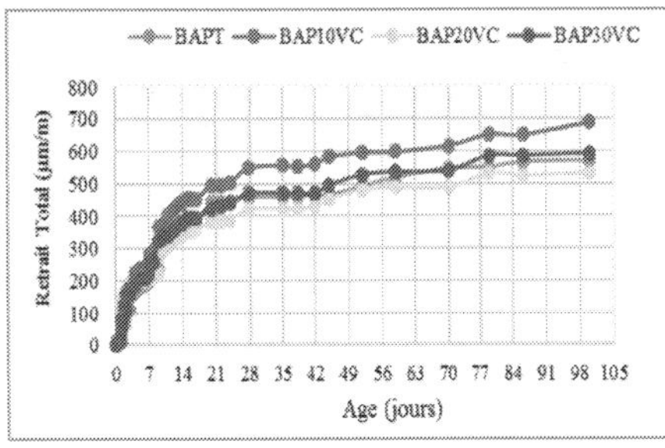

Fig.5. Évolution du retrait total en fonction du temps

Le retrait total évolue très rapidement pour tous les types d'éprouvettes conservées à l'air libre, à cause de la taille de ces dernières qui rend la dessiccation plus favorable [4]. Les évolutions présentent des allures similaires et presque identiques aux premiers âges et commencent à se différencier dans le temps. Le BAP20VC présente le plus faible retrait total. La figure 5 montre une évolution de déformation similaire pour le BAPT et le BAP à base de 30% de vase calcinée, car au très jeune âge nous avons du mal à distinguer entre les graphes représentatifs de chaque béton. Le BAP10VC présente les déformations totales les plus élevées, dès les premiers jours et pour toute la durée des essais.

Si l'on considère le fait que les mesures se font à l'échelle du micromètre alors il est plus facile de dire qu'à long terme la composition des BAP étudiés n'a pas une grande influence sur le retrait total [4].

4.2.4 Perte de masse

Afin de mieux comprendre les interprétations précédentes du retrait nous avons mesuré en parallèle la perte de masse exprimée en %.
Les pertes de masse ont été mesurées sur les éprouvettes destinées aux mesures du retrait total et endogène. Les résultats sont représentés par les figures 6 et 7.

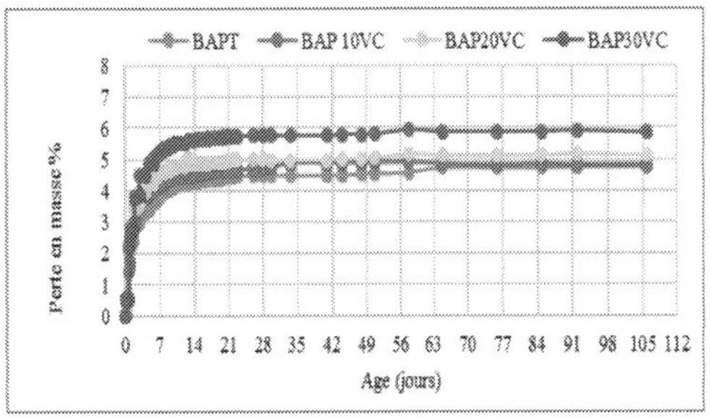

Fig.6. Évolution de la perte de masse totale en fonction du temps

D'après la figure 6, on remarque que le BAPT et les deux BAP de 10% et 20% de vase calcinée exposés au retrait total ont la même cinétique de perte de masse sauf le BAP à base de 30% de vase se détache par rapport aux autres en développant plus de perte. L'analyse des courbes montre que la plus grande partie du retrait a lieu avant 28 jours environ, au-delà de cet âge, il est stabilisé. Ceci s'explique par le fait que la réaction d'hydratation est une réaction exothermique qui participe d'une part au départ d'eau par évaporation d'eau vers l'ambiance extérieure et d'autre part par consommation dans l'action d'hydratation du ciment. Cette échéance, 90 jours, correspond assez bien au ralentissement du séchage, observé sur la perte de masse. Au-delà, le retrait de séchage n'évolue quasiment plus [9].

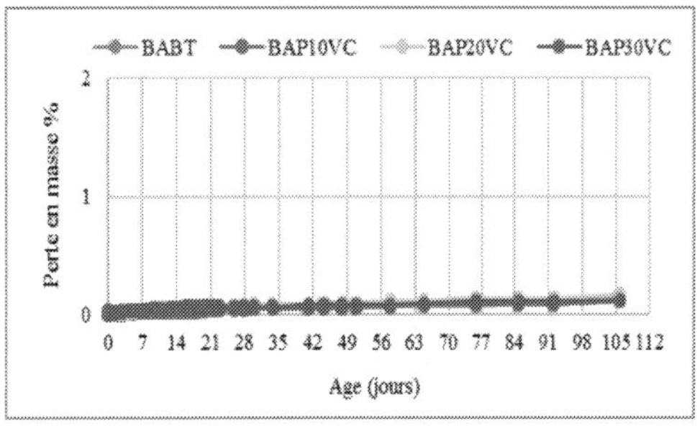

La figure 7 montre le développement en perte de masse de BAP qui est mesuré sur des éprouvettes exposées à un retrait endogène.

Fig.7. Évolution de la perte de masse endogène en fonction du temps

Ces pertes de masse sont causées par la consommation interne d'eau par l'hydratation du ciment et la vase. On observe que tous les BAP évoluent de la même façon en fonction du temps. Les BAP (témoin, 10% et 30% de vase calcinée) croît dans un fuseau très serré jusqu'à l'âge de 60 jours où les valeurs commencent à converger, excepté pour le BAP à 20% qui présente une évolution à part.

Le dosage de 10% semble optimal pour des pertes en masse minimales et par conséquent un retrait réduit et ce que nous avons déjà vu dans la courbe du retrait endogène.

5 Conclusion

Cette étude avait pour objectif de comparer les propriétés mécaniques de bétons autoplaçants à base de différents pourcentages de vase calcinée avec celles du béton témoin utilisé comme référence.

En ce qui concerne les résistances mécaniques, celles-ci augmentent avec l'augmentation du pourcentage de la vase calcinée dans le béton, à long terme.

La présence de la vase à des taux de 20% et 30% fait diminuer le retrait de dessiccation et total, cela est le résultat de leur contribution à l'amélioration de la microstructure.

On peut en conclure qu'une substitution de 30 % du ciment par la vase est le taux le plus intéressant qui s'avère optimal, puisque c'est le taux qui satisfait les deux critères d'amélioration de résistance et réduction du retrait

References

1. Schwartzentruber L et Torrenti J., *Le grand livre des bétons,* Paris, Editions Le moniteur, (2014)

2. Bernardini C., *Les bétons autoplaçants,* Paris., Presses Ponts et Chaussées, Août (2011).

3. AFGC Bétons Autoplaçants Recommandations provisoires, Association française de Génie Civil, (2008)

4. Bouhamou.N, Etude du comportement d'un béton autoplaçant à base de vase de dragage vis-à-vis du retrait, Colloque 01 Ecomatériaux, Matériaux 2014, 24-28 Novembre (2014).

5. Chalarem. N., Caractérisation des BAP à base de sédiments de dragage vis-à-vis du retrait au jeune âge, Mémoire de fin d'étude Master Academique , Université des sciences et de technologie, Mostaganem ,juin ,(2016)

6. Bouhamou.N., « Influence des paramètres de composition sur le comportement du béton autoplaçant vis –à vis du retrait » thèse de Doctorat, (2008)

7. José Oliveira.M, Shrinkage of self-compacting concrete. A comparative analysis CEPAC, University of the Algarve, Faro, Portugal, January (2017), Pages 117-124

8. Mostefa. F, Étude du comportement d'un béton Autoplaçant à base de sédiment de dragage vis-à-vis du retrait, Mémoire de Magister, Université des sciences et de technologie, Mostaganem, 23Février, (2012).

9. Belaribi.O.,Comportement d'un béton autplaçant vis-vis des attaques chimiques, Thèse de Magister, Université des sciences et de la technologie d'Oran Mohamed Boudiaf U.S.T.O Oran,(2009)

Experimental study of a foam concrete based on local Tunisian materials

Etude expérimentale d'un béton mousse à base des matériaux locaux tunisiens

Dorra ELLOUZE[1], Karim MILED[2]
[1]ISET Sfax, Département de Génie Civil, Sfax, Tunisie
[2]Université de Tunis El Manar, Ecole Nationale d'Ingénieurs de Tunis, Laboratoire de Génie Civil, Tunis, Tunisie

Abstract. The building sector in Tunisia is very energy-intensive, the largest share of energy consumption comes from factories of building materials namely brick and cement plants. This work is part of the reduction of the energy bill in the building envelope. Indeed, the foam concrete can be walls in single or double wall with better insulating power. This paper presents an experimental study on the technical problems related to the formulation and manufacture of a new cellular concrete in Tunisia, called "foam" concrete, from Tunisian local materials. Indeed, six varieties of sand of different provenance and grain size will be analyzed, the "good" sand is the one that is best suited for the manufacture of foam concrete. Two clean, fine-grained (0/2mm) rolled grain sands were retained. Then four foam concretes were formulated using each time a single type of sand and varying the density namely 0.8 and 1. These four formulations were tested mechanically and thermally. The results found showed that compressive strengths do not exceed 1.5 MPa at 28 days. Thus, the foam concrete can be used only as a filling concrete in non-load bearing elements such as partition walls. The guarded hot plate method was used to determine the thermal conductivities of the four foamed concretes studied. A low thermal conductivity was found of the order of 0.22 W/m°K which prove the insulating power of foam concrete.

Résumé. Le secteur de bâtiment en Tunisie est très énergivore, la part la plus importante de la consommation énergétique provient des usines des matériaux de construction à savoir les briqueteries et les cimenteries. Ce travail s'inscrit dans le cadre de la réduction de la facture énergétique dans l'enveloppe des bâtiments. En effet le béton mousse peut constituer des murs en simple ou en double cloison avec un meilleur pouvoir isolant. L'objectif de cette étude expérimentale est d'étudier et de résoudre en amont, d'abord, les problèmes techniques liés à la formulation et à la fabrication d'un nouveau béton cellulaire en Tunisie, dit béton "mousse" à partir des matériaux locaux tunisiens. En particulier, six variétés de sable de provenance et de granulométrie différentes seront analysées, le « bon » sable est celui qui convient le mieux pour la fabrication du béton mousse. Deux sables de rivière à grains roulés, propres et fins (0/2mm) ont été retenus. Ensuite, quatre bétons mousse ont été formulés en utilisant à chaque fois un seul type de sable et en variant la densité à savoir 0,8 et 1. Ces quatre formulations ont été testées mécaniquement et thermiquement. Les résistances à la compression trouvées ne dépassent pas 1,5 MPa à 28 jours. Ainsi, le béton mousse peut être utilisé seulement comme un béton de remplissage dans des éléments non porteurs tels que les murs de partition. Enfin, la méthode de la plaque chaude gardée a été utilisée pour déterminer les conductivités thermiques des quatre bétons mousses étudiés. Une faible conductivité thermique a été trouvée de l'ordre de 0,22 W/m°K. Ces résultats prouvent le pouvoir isolant du béton mousse.

1 Introduction

Ce travail s'inscrit dans le cadre de la réduction de la facture énergétique engendrée par les matériaux de construction traditionnels et énergivores tels que le ciment et les briques rouges en terre cuite utilisés dans l'enveloppe des bâtiments, suite aux recommandations de l'Agence Nationale de la Maîtrise de l'Energie (ANME) [1]. Cette recherche propose une étude expérimentale sur la formulation et la fabrication d'un nouveau béton cellulaire en Tunisie, dit béton "mousse" à partir des matériaux locaux tunisiens.

Pour ce faire, on doit identifier les bons matériaux nécessaires dans la confection du béton mousse et notamment le bon sable garantissant une mousse stable dans le mélange. En particulier, six variétés de sable de provenance et de granulométrie différentes seront analysées. Ensuite, on passe à la formulation et la fabrication du béton mousse. Enfin, on réalise les essais mécaniques et thermiques sur les différents échantillons du béton mousse. L'acquisition des matériaux et du matériel nécessaires pour la confection du béton mousse ainsi que les essais mécaniques et thermiques de cette étude expérimentale ont été cofinancés par le Centre Technique Tunisien des Matériaux de Construction, de la Céramique et du Verre (CTMCCV) et l'agence allemande de coopération technique (GIZ).

Le comportement du béton mousse a fait l'objet divers recherches scientifiques à savoir les études de Ramamurthy et al. (2006 à 2012) [2 à 7], de Panesar (2013), de Lukatis et al. (2003) et de Bouvard et al. (2007). [8 à 10]

2 Caractérisation des matériaux

Le béton mousse se compose de ciment, sable, eau et agent mousant. L'eau est celle du réseau d'eau potable de la SONEDE, le ciment est de type CEMI 42.5 provenant de la cimenterie de Jebel Jloud. L'agent moussant désigné par NEOPOR 600 est fourni par la société allemande NEOPOR CLC. C'est un produit organique et naturel, il provient de l'hydrolyse des cornes et des sabots des bovins. Un litre d'agent moussant NEOPOR produit environ 580 litres de mousse. Le "bon" sable du béton mousse doit être propre, ayant des grains roulés et une distribution granulométrique bien déterminée. Selon les recommandations de la société NEOPOR CLC [13], le pourcentage des fines (de diamètre < 80 µm) de ce sable doit être entre 15 et 20%. Au-delà de ces valeurs, les fines consomment une quantité excessive d'eau de gachâge et la réaction d'hydratation du ciment ne sera pas complète, ce qui dècroit la résistance mécanique du béton mousse. Ainsi, pour identifier le bon sable, nous avons analysé six variétés de sable de provenance et de granulométrie differentes:

- Sable S1 : Sable Widhref Chaael de Sfax: sable de rivière, diamètre 0-2mm.
- Sable S2 Sable Widhref Chaael de Sfax: sable de rivière, diamètre 0-5mm.
- Sable S3 Sable de Khlidia : sable de carrière, diametre 0-2mm.
- Sable S4 Sable de Bouargoub : sable de carrière, diamètre 0-2mm.
- Sable S5 Sable de Borj-Hfaiedh: sable de carrière (échantillon sec), diamètre 0-2mm.
- Sable S6 Sable de Borj-Hfaiedh: sable de carrière (échantillon humide), diamètre 0-2mm

Pour ces six sables, nous avons déterminé expérimentalement la teneur en eau naturelle, les masses volumiques apprente et absolue, l'équivalent de sable (ES), la courbe granulométrique par tamisage et le module de finesse (MF). Les résultats sont présentés dans le tableau suivant.

Tableau 1. Caractérisation des six sables étudiés

	Teneur en eau (%)	ES (%)	Densité volumique Apparente	Densité volumique Absolue	MF
S1	1,76	86,95	1,54	2,66	3,42
S2	4,06	78,48	1,48	2,76	2,89
S3	3,55	73,18	1,46	2,63	2,91
S4	3,30	75,24	1,57	2,82	2,36
S5	1,91	91,23	1,57	2,76	2,87
S6	5,23	91,64	1,57	2,76	2,71

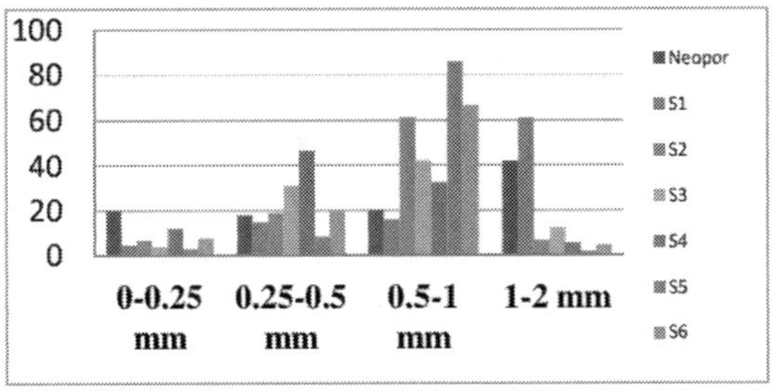

Fig.1. Comparaison des fractions granulométriques des sables étudiés avec celles du sable recommandé par NEOPOR CLC

L'analyse de résultats présentés dans le tableau 1 et la figure 1 décrivant la granulométrie des sables analysés, a montré que les deux sables S1 et S6 respectent le mieux les exigences ou les caractéristiques du "bon" sable décrites précédemment. En effet, S1 et S2 sont deux sables de rivière et le sable S1 s'approche mieux des exigences de propreté et du pourcentage des fines. Comparé aux sables de carrière S3 à S6, le sable S6 est plus propre. Ainsi, nous retenons les deux sables S1 et S6 pour formuler le béton mousse et pour tester l'effet de la forme de grains roulée ou concassée sur le béton mousse.

3 Formulation et confection du béton mousse

Pour formuler le béton mousse, on a utilisé les formules recommandées de la société NEOPOR CLC et qui sont

 https://doi.org/10.1051/matecconf/201814901027

resumées dans le tableau suivant:

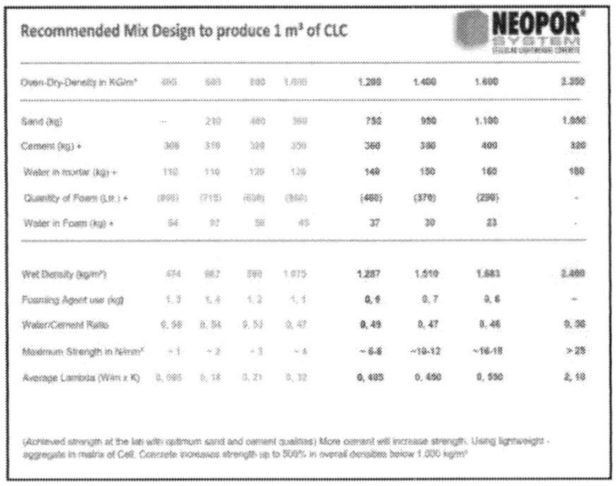

Recommended Mix Design to produce 1 m³ of CLC — NEOPOR SYSTEM

Oven-Dry-Density in KG/m³	[illegible]	[illegible]	[illegible]	[illegible]	1.200	1.400	1.600	[illegible]
Sand (kg)	--	[illegible]	[illegible]	[illegible]	730	950	1.100	[illegible]
Cement (kg) +	[illegible]	[illegible]	[illegible]	[illegible]	360	380	400	320
Water in mortar (kg) +	[illegible]	[illegible]	[illegible]	[illegible]	140	150	160	180
Quantity of Foam (Lit.) +	[illegible]	[illegible]	[illegible]	[illegible]	(460)	(370)	(290)	-
Water in Foam (kg) +	[illegible]	[illegible]	[illegible]	[illegible]	37	30	23	-
Wet Density (kg/m³)	[illegible]	[illegible]	[illegible]	[illegible]	1.287	1.510	1.683	2.400
Foaming Agent use (kg)	[illegible]	[illegible]	[illegible]	[illegible]	0,9	0,7	0,6	-
Water/Cement Ratio	[illegible]	[illegible]	[illegible]	[illegible]	0,49	0,47	0,46	0,38
Maximum Strength in N/mm²	~1	~2	~3	~4	~6-8	~10-12	~16-18	> 25
Average Lambda (W/m x K)	[illegible]	[illegible]	[illegible]	[illegible]	0,405	0,450	0,550	2,10

(Achieved strength at the lab with optimum sand and cement qualities) More cement will increase strength. Using lightweight-aggregate in matrix of Cell. Concrete increases strength up to 500% in overall densities below 1.000 kg/m³

Fig. 2. Formulation du béton mousse NEOPOR [13]

Dans notre étude, deux types de béton mousse de densités 0,8 et 1 et à base de sable S1 ou S6 ont été formulés (voir tableau 2) et ensuite confectionnés.

Tableau 2. Formules des bétons mousse étudiés

Mélange	B2S1	B1S1	B1S6	B2S6
Masse volumique kg/m³	1000	800	800	1000
Ciment (kg/m³) Eau (l/m³) Sable (kg/m³)	350 120 560	320 120 400	320 120 400	350 120 560
Masse volumique mousse g/l	87	88,6	86,9	95,6
Masse volumique kg/m³ béton frais	1059	864	826	1052

La référence [13] décrit la confection de la mousse. Pour ce faire, nous avons utilisé un générateur de mousse en utilisant un émulseur, un compresseur d'air et une pompe à eau. Ensuite, nous incorporonsla mousse dans le malaxeur qui contient le mélange à sec (ciment + sable) déjà malaxé, puis nous ajoutons l'eau de gachâge nécessaire. Nous continuons ensuite le malaxage pendant 2 minutes et nous obtenons enfin le béton mousse. La cure de ce béton se fait à l'air libre et le démoulage est possible après 24 heures.

4 Résultats et discussions

Nous présentons dans cette partie les résultats d'essais de compression et de conductivité thermique.

4.1 Résistance à la compression

Afin de caractériser la résistance à la compression du béton mousse, nous avons commencé par investiger d'abord l'effet de la taille (ou du volume) ainsi que l'effet de la forme de l'éprouvette d'essai sur la résistance à la compression du béton mousse et ce pour déterminer la taille caractéristique de l'éprouvette d'essai à partir de laquelle l'effet d'échelle de volume disparaitera. Ainsi, nous avons réalisé des essais mécaniques de compression simple sur des éprouvettes cubiques en cm³ (10x10x10), cylindriques en cm³ (11x22) et cylindriques en cm³ (16x32). Les résultats trouvés et presentés sur la figure 3 montrent la présence à la fois d'un faible effet de forme et d'un faible effet de volume de l'éprouvette d'essai sur la résistance à la compression du béton mousse. Par conséquent, le cylindre de dimensions en cm³ (16x32) a été retenu pour caractériser la résistance à la compression intrinsèque du béton mousse. Celle-ci diminue quand la densité du béton mousse diminue, autrement dit sa macro porosité (fraction volumique de mousse) augmente, ce qui est prévisible (cf. Figure 4). En outre, les résistances à la compression trouvées sont faibles puisque les résistances les plus élevées, de l'ordre de 1,3 à 7 jours et 1,5 MPa à 28 jours, ont été données par le béton B2S1 de densité 1 et renfermant le sable S1 de Wehdref-Chaael qui est un sable de rivière très fin (0-2mm). Ainsi, ces bétons peuvent être utilisés seulement comme des bétons de remplissage dans des éléments non porteurs bien évidemment.

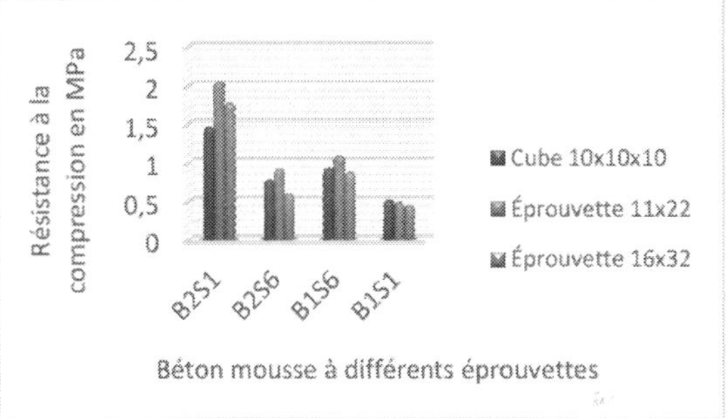

Fig.3. Effet du volume et de la forme de l'éprouvette d'essai sur la résistance à la compression du béton mousse.

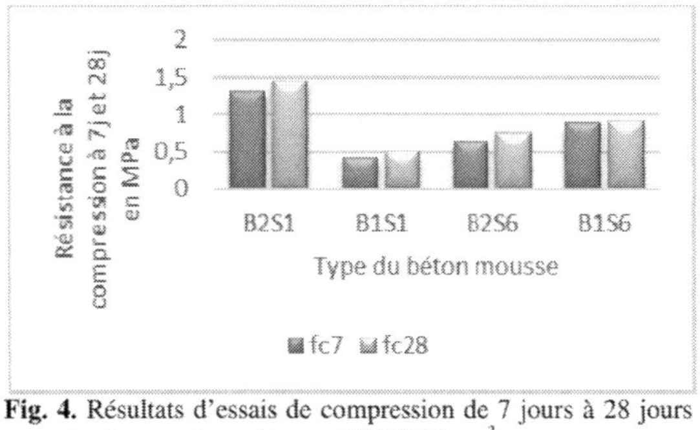

Fig. 4. Résultats d'essais de compression de 7 jours à 28 jours pour des éprouvettes cubiques (10*10*10 cm³).

4.2 Conductivité thermique

Pour évaluer la conductivité thermique du béton mousse, nous avons préparé des éprouvettes prismatiques (des plaques) de dimensions 30x30x4 cm³.La méthode de la plaque chaude gardée a été utilisée pour déterminer les conductivités thermiques des quatre bétons mousses étudiées. Une faible conductivité thermique a été trouvée pour les différents bétons et un effet faible de type du sable utilisé a été observé sur la conductivité thermique du béton mousse. Le béton B1S1 de densité 0,8 et renfermant le sable S1 de Widhref Chaael à Sfax qui est aussi un sable de rivière très fin (0-2mm), possède la plus faible conductivité, de l'ordre de 0,22 W/m°K, et donc le meilleur pouvoir isolant (cf.tableau 3).

Par ailleurs, En comparant ces résultats expérimentaux résumés dans le tableau suivant avec les prédictions théoriques de deux modèles d'homogénéisation classiques dans la littérature et présentés dans les références [11] et [12] , à savoir le modèle de Mori-Tanaka et le modèle différentiel, nos résultats expérimentaux peuvent être prédits par l'expression analytique suivante donnée par le modèle de Mori-Tanaka dans le cas des solides poreux, en fonction de la macro porosité p (fraction volumique de mousse) (cf Figure.5)

$$\frac{\lambda_{bétonmousse}}{\lambda_{mousse}} = \frac{1-p}{1+\frac{p}{2}} \quad (1)$$

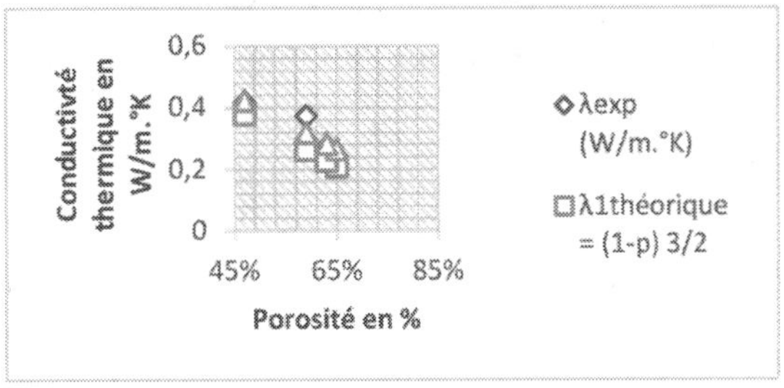

Fig. 5. Comparaison de la variation de la conductivité thermique expérimentale en fonction de la porosité avec les prédictions de deux modèles théoriques.

Le tableau suivant résume les principaux résultats de cette étude expérimentale

Tableau 3. Synthèse des résultats mécaniques et thermiques sur des éprouvettes cubiques

Mélange	B1S1	B2S1	B1S6	B2S6
Masse volumique en kg/m^3	849,22	1222,73	910,64	992,34
Macro porosité en %	65%	47%	63%	59%
f_{c28} (MPa)	0,50	1,45	0,92	0,76
λ_{exp} (W/m.°K)	0,22	0,41	0,27	0,37

5 Conclusion

Dans cette étude expérimentale, nous avons dans un premier temps identifié les « bons » sables qui conviennent le mieux pour la fabrication du béton mousse. Deux sables de rivière à grains roulés, propres et fins (0/2mm) ont été retenus parmi six sables de provenance et de granulométrie différentes. Ensuite, quatre bétons mousse ont été formulés en utilisant à chaque fois un seul type de sable et en variant la densité; deux densités ont été étudiées à savoir 0,8 et 1.

Dans un second temps, ces quatre bétons mousse ont été testés en compression simple. Les résistances à la compression trouvées ne dépassant pas 1,5 MPa à 28 jours. Ainsi, le béton mousse étudié peut être utilisé seulement comme un béton de remplissage dans des éléments non porteurs tels que les murs de partition. Enfin, la méthode de la plaque chaude gardée a été utilisée pour déterminer les conductivités thermiques des quatre bétons mousses étudiées. Une faible conductivité thermique a été trouvée pour les différents bétons et un effet faible de type du sable utilisé a été observé sur la conductivité thermique du béton mousse. Le béton B1S1 de densité 0,8 et renfermant le sable S1 de Widhref Chaael à Sfax qui est un sable de rivière très fin (0-2mm), possède la plus faible conductivité, de l'ordre de 0,22 W/m°K, et donc le meilleur pouvoir isolant.

Références

1. ANME, (Page consultée en Décembre 2016), *La mise en place de la réglementation thermique et énergétique en Tunisie*, [En ligne]. Adresse URL: http://www.anme.nat.tn/index.php?Id=155&L=0
2. .K. Ramamurthy, e.k. kunhanandan nambiar, g. Indu siva ranjani, (2009) *« a classification of studies on properties of foam concrete»*, Cement and Concrete. Composites, Vol.31, pp.388-396.
3. E.K. Kunhanandan Nambiar, K. Ramamurthy, 2006, *« Influence of filler type on the properties of foam concrete*, International Journal of Cement Composites and Lightweight Concrete, Vol 28, pp 475-480
4. E.K. Kunhanandan Nambiar, K. Ramamurthy,« *Sorption caracteristics of foam concret* »,Cement and Concrete Composites, Vol 28, pp 752-760,
5. G. Indu Siva Ranjani, K. Ramamurthy *« Behaviour of foam concrete under sulphate environments »*, Cement and Concrete Composites , Vol. 34 (4), pp.825-834.
6. E.K. Kunhanandan Nambiar, K. Ramamurthy, « Models relating mixture composition to the density and strength of foam concrete using response surface methodology», Cement and Concrete Composites, Vol 28(5), pp 475-480.
7. J. Sathya Narayanan, K. Ramamurthy, (2012), « *Identification of set-accelerator for enhancing the productivity of foam concrete block manufacture* » Construction and Building Materials, Vol.37, pp.144-152
8. D.K. Panesar, (2013), *« Les propriétés du béton cellulaire et l'effet des agents moussants synthétiques et des proteins »*, Construction and building materials, vol. 44, p. 575–84.
9. A.Laukaitis, R.Zurauskus, J Kerien, *The effect of foam polystyrene granules on cement composite properties*, Cement and Concrete Composites, vol 27(1), pp 41-47, 2003
10. D. Bouvard , J.M. Chaix , R. Dendievel , A. Fazekas, J.M. Létang, G. Peix, D. Quenard, *Characterization and simulation of microstructure and properties of EPS lightweight concrete, International Journal of Cement Composites and Lightweight Concrete*, vol 37 , pp 1666–1673
11. Miled K., Limam O., (2016). *Effective thermal conductivity of foam concretes: homogenization schemes vs experimental data and FEM simulations*, Mechanics Research Communications, vol. 76 (2), pages 96-100.
12. Ch. Pichler, R. Lackner, G. Metzler, *Thermo-mechanical optimization of porous building materials based on micromechanical concepts: Application to load-carrying insulation materials*, Composites Part B: Engineering, Vol 43(3), pp 1015-1023, (2012).
13. Documents techniques de NEOPOR, (Page consultée en Janvier 2017) [En ligne]. Adresse URL: http://www.neopor.com

Effects of Basalt Fibres on Mechanical Properties of Concrete

A. M. El-Gelani[1], C.M. High[2], S. H. Rizkalla[3] and E. A. Abdalla[4]
[1,4]University of Tripoli, Civil Engineering Department, Tripoli, Libya
[2]Stweart Engineering, Charlotte N.C., USA
[3]N.C. State University, Civil, Construction and Environmental Engineering Department, Raleigh N.C., USA

Abstract. This paper presents the results of an experimental program carried out to investigate the effects of Basalt Fibre Reinforced Polymers (BFRP) on some fundamental mechanical properties of concrete. Basalt fibres are formed by heating crushed basalt rocks and funnelling the molten basalt through a spinneret to form basalt filaments. This type of fibres have not been widely used till recently. Two commercially available chopped basalt fibres products with different aspect ratios were investigated, which are dry basalt (GeoTech Fibre) and basalt pre-soaked in an epoxy resin (GeoTech Matrix) .The experimental work included compression tests on 96 cylinders made of multiple batches of concrete with varying amounts of basalt fibre additives of the two mentioned types, along with control batches containing no fibres. Furthermore, flexural tests on 24 prisms were carries out to measure the modulus of rupture, in addition to 30 prisms for average residual strength test. Results of the research indicated that use of basalt fibres has insignificant effects on compressive strength of plain concrete, where the increase in strength did not exceed about 5%. On the other hand, results suggest that the use of basalt fibres may increase the compressive strength of concrete containing fly as up top 40%. The rupture strength was increased also by 8% to 28% depending on mix and fibre types and contents. Finally, there was no clear correlation between the average residual strength and ratios of basalt fibres mixed with the different concrete batches.

1 Introduction

Fibre-reinforced concrete (FRC) has been produced using steel, carbon and glass fibres for several years. These existing forms of FRC have proven that the addition of fibres can enhance the mechanical characteristics of concrete. However, basalt fibre-reinforced concrete (BFRC) may also have the potential to improve the mechanical properties of concrete by increasing mainly the concrete's tensile strength, fracture toughness and controlling cracking.

The research program included an investigation of the fundamental properties of BFRC produced using two types of chopped basalt fibre products as additives. This investigation of BFRC was a preliminary study which provided an evaluation of the chopped basalt fibres s an additive to enhance the characteristics of concrete.

Basalt fibres are a material made from extremely fine fibres of basalt. It is similar to carbon fibre and fibreglass, having better physico-mechanical properties than fibreglass, but being significantly cheaper than carbon fibre. Basalt fibres are made from a single material, crushed basalt rock, from a carefully chosen quarry source, and unlike other materials such as fibreglass, essentially no materials are added. The basalt is simply washed and then melted. The manufacture of basalt fibres require the melting of the quarried basalt rock at about 1400°C. The molten basalt rock is then extruded through small nozzles (spinnerets) to produce continuous filaments of basalt fibres. Figure 1 illustrates the stages of producing continuous basalt fibers. There are three main manufacturing techniques, which are centrifugal-blowing, centrifugal-multiroll and die-blowing [1].

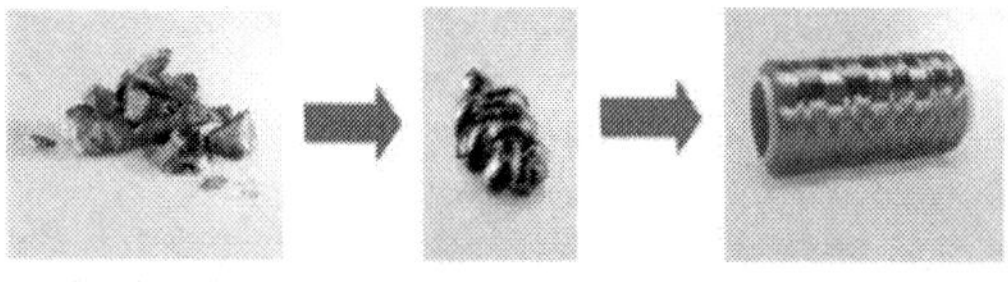

Fig. 1. Manufacturing of basalt fibres.

Basalt fibres are used in several applications, other than as concrete additives, such as military and aerospace applications, in addition to automotive industries and windmill blades. Basalt fibres exhibit considerable tensile strength in the longitudinal direction [2]. Its mechanical properties suggest that adding this type of fibres to concrete may enhance the rupture strength of concrete. However, only a few studies have investigated the mechanical properties of BFRC. This paper aims to determine the potential effectiveness of using basalt fibres as additives to concrete. Borhan studied the compressive and splitting tensile strengths of BFRC with small fibre volume fractions ranging from 0.1% to 0.5% [3]. Test results indicated that increasing the basalt fibre content increased the splitting tensile strength of the concrete without significantly affecting the concrete's compressive strength. That study reports also a decrease in concrete slump as the basalt fibre volume content was increased.

Also, Yang and Lian tested 480 specimens to determine the mechanical properties of BFRC, and to investigate the optimum range of basalt fibre content. This research concluded that adding basalt fibres to concrete can increase the compressive, splitting tensile and flexural strengths of plain concrete. Furthermore, this research suggested that concrete with a basalt fibre volume content between 0.3% and 0.5% resulted in the highest measured strength increases [4].

Another experimental program investigated the mechanical properties of concrete with the addition of basalt fibres that were pre-soaked in epoxy [5]. Three different pre-soaked basalt fibre lengths (10mm, 20mm and 30mm) and three different fibre dosages (3 kg/m^3, 5 kg/m^3 and 7 kg/m^3) were examined in this study. The results showed that as the basalt fibre dosage and fibre length increased, the measured slump decreased. Moreover, according to this study, the presence of the pre-soaked basalt fibres did not significantly affect the compressive strength. The results indicated that the epoxy-presoaked fibres increased the flexural strength of each concrete batch. The splitting tensile strength was increased for each batch with 30 mm long basalt fibres. This study concluded that adding 30 mm basalt fibres at a dosage range of 3 kg/m^3 to 5 kg/m^3 resulted in improved mechanical properties and acceptable workability.

2 Experimental program

The main objective of this paper is to determine the fundamental mechanical properties of BFRC. These include compressive, flexural and average residual strengths. The experimental program included two different concrete mixes, and the basalt fibre type and matrix ratios were varied for each mix. The effect of the basalt fibres was determined by comparing test results to control concrete specimens without fibres.

2.1 Concrete materials and mix design

Test specimens were cast using two different concrete mix designs. Mix design "A" was a standard concrete mix with a target compressive strength of 21 MPa. Mix design "B" contained additional fly ash and admixtures. Table 1 describes the ingredients of each concrete mix design and the weight of each ingredient per cubic meter of concrete.

The SSD specific gravity and percent moisture absorption of the coarse and fine aggregates were determined in accordance with ASTM C127 and C128 respectively [6,7]. The results of the tests are shown in Table 2.

Table 1. Concrete mix designs "A" and "B".

Material	Weight (Kg/m^3)	
	Mix A	Mix B
Cement	307	217
Fly ash	0	145
Water	171	143
Fine aggregate	749	698
Coarse aggregate	988	1016
Air entrainer	0.13 lt	0.17 lt
Superplasticizer	0	0.63 lt
Accelerator	0	0.71 lt

Table 2. Aggregate characterization results.

Material	SSD specific gravity	SSD absorption (%)
Fine aggregate	2.61	0.8
Coarse aggregate	2.62	1.0

Two different basalt fibre products were used as additives to the concrete. The two products are known commercially as GeoTech Fibre and GeoTech Matrix, as shown in Figures 2 and 3, respectively. GeoTech fibre is produced using dry basalt fibres, while GeoTech matrix is produced using basalt fibres pre-soaked in an epoxy resin. The aspect ratio (length of fibre divided by fiber diameter) of the GeoTech fibre is significantly greater than that of the GeoTech matrix.

Fig. 1. GeoTech fibre.

Fig. 2. GeoTech matrix.

For each concrete mix, sub-mixes were generated by varying the basalt content resulting in a total of eight distinct concrete mixes. Table 3 provides an overview of the eight concrete batches and the weight of fibre per cubic meter of concrete for each sub-mix.

Cylinder moulds (100 × 200 mm) and prism moulds (150 × 150 × 510 mm) were used for preparing the concrete specimens for compressive and flexural strength tests respectively. In addition, for average residual strength (ARS) test, prism specimens were cut from the flexure test specimens using an abrasive concrete saw.

Figure 4 shows a finished 100 × 100 × 356 mm ARS specimen.

Table 3. Fibre contents in concrete batches.

Batch designation	Concrete mix	Basalt fiber (Kg/m³)	
		GeoTech Fibre	GeoTech matrix
1A	A (Control)	None	None
2A	A	1.2	0.6
3A	A	1.8	0
4A	A	1.2	1.2
1B	B (Control)	None	None
2B	B	1.2	0.6
3B	B	1.8	0
4B	B	1.2	1.2

Fig. 4. ARS specimen after cutting.

2.2 Test setups and testing procedures

The compressive strength of each sub-mix was measured by testing 100 × 200 mm cylinders at 3, 7 and 28 days in accordance with ASTM C39 [8]. The test specimens were loaded at a constant rate of approximately 0.24 MPa/s. The flexural strength of each sub-mix was determined for 150 × 150 × 510 mm prisms in accordance with ASTM C78 [9]. Flexural specimens were tested using an apparatus suitable for third-point loading, Figure 5. Flexural specimens were turned on their side from the position as cast and loaded at a constant rate of 8 KN/min until failure. Modulus of rupture, R, of each flexural specimen was also calculated.

Fig. 5. BFRC flexural specimen in the loading apparatus.

The average residual strength (ARS) was determined in accordance with ASTM C1399 [10]. ARS specimens, supported by a steel plate, were turned on their sides from the position as cast and initially loaded at a constant displacement rate of 0.635mm/min. The purpose of the steel plate was to control the deflection of the beam once a flexural crack was formed.

The initial loading was terminated once the specimen experienced a mid-span displacement of 0.2 mm relative to the outside supports. Mid-span displacement was taken as the average of two transducer readings positioned on either sides of the test specimen at mid-span. The two transducers were supported by a rigid yoke which rested on top of the specimen directly above the outside supports. Figure 6 shows an ARS specimen positioned in the testing apparatus before the application of the initial loading phase.

Fig. 6. ARS setup for the initial loading phase.

After the initial loading phase, the specimen was unloaded, the steel plate was removed, and the transducers were re-zeroed. The specimen was then reloaded at a displacement rate of 0.635 mm/min until the specimen experienced a mid-span displacement of 1.27 mm. During the initial and reloading phases, applied load and mid-span displacement data were continuously recorded. The ARS was recorded based on the load-deflection data and the dimensions of each specimen.

3 Test results and analysis

3.1 Compressive strength

Figures 7 and 8 show typical failures of tested compressive specimens, while the average measured test results at 3, 7 and 28 days for the mixes A and B are given in Table 4. Only the amount of fibres was intentionally altered among concrete batches of the same mix, however, other parameters influencing the compressive strength of the concrete inevitably varied slightly from one batch to another. Figures 9 and 10 show the average measured 28 days strength of mixes A and B respectively.

An initial assessment of the results indicates the use of GeoTech matrix and GeoTech fibres did not affect the compressive strength of the concrete of mix A. However, test results show some noticeable increase in the compressive strength for concrete sub-mixes of mix B containing GeoTech matrix and Geo-Tech fibres. Table 5 shows the percentage of increase or decrease in the 28 days compressive strength of each sub-mix with respect to the associative control mix.

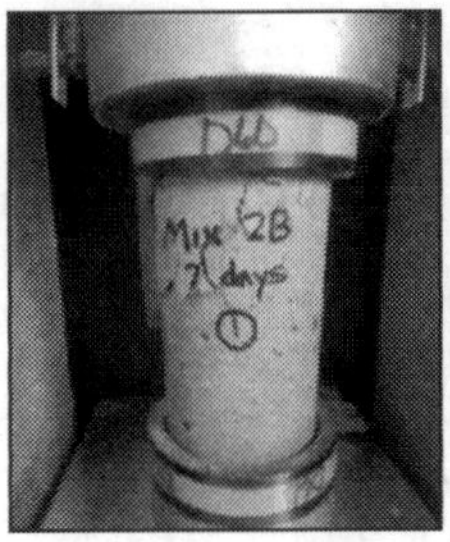

Fig. 7. Typical failure.

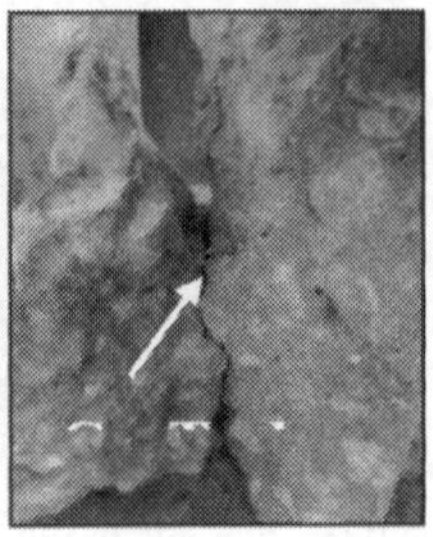

Fig. 8. Exposed GeoTech matrix after failure.

Table 4. Compressive test results.

Batch designation	Compressive strength (MPa)		
	3 days	7 days	28 days
1A	14.6	16.9	23.1
2A	14.6	18.9	23.2
3A	15.3	19.3	24.4
4A	10.6	14.1	18.4
1B	11.2	14.1	21.1
2B	14.7	18.3	25.3
3B	14.3	18.0	24.3
4B	18.5	22.7	29.6

Table 5. Percentage increase or decrease in 28 day compressive strength.

f_c', MPa (%↑↓)	Mix			
	1A	2A	3A	4A
	23.1	23.3 (1.2%↑)	24.4 (5.7%↑)	18.4 (20.3%↓)
	1B	2B	3B	4B
	21.1	25.3 (19.9%↑)	24.3 (15.3%↑)	29.6 (40.5%↑)

The measured compressive strengths at 3 and 7 days were calculated as a fraction of the measured compressive strength at 28 days to determine the effect of the basalt fibre products on the compressive strength at early age, if any.

The analysis indicates that the presence of the basalt additives had little effect on the percentage of compressive strength gained at 7 days for mix A. It can be seen also that the 3 and 7 days fraction of 28 days compressive strength of mix B sub-mixes containing basalt additives are clearly higher than the control batch. Mix 4B gained 10 % more of its 28 days compressive strength at 3 days and 7 days than the control mix. Therefore, the test results suggest that there is some possibility for a correlation between the fibre volume fraction and the development of early strength for mix B. This correlation may be related, as well, to the presence of fly ash in this mix.

Table 6. Early ages compressive strengths as fractions of 28 days strength.

Mix	f_c' 3d/ f_c' 28d	f_c' 7d/ f_c' 28d
1A	0.63	0.73
2A	0.63	0.81
3A	0.63	0.79
4A	0.58	0.77
1B	0.53	0.67
2B	0.58	0.72
3B	0.59	0.74
4B	0.63	0.77

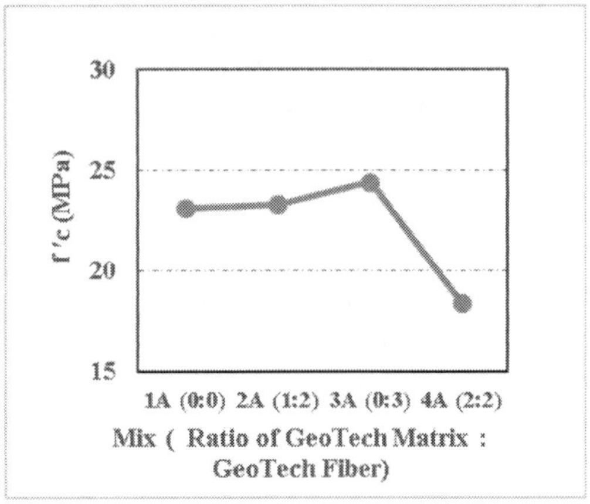

Fig.9. Compressive strength results at 28 days, mix A.

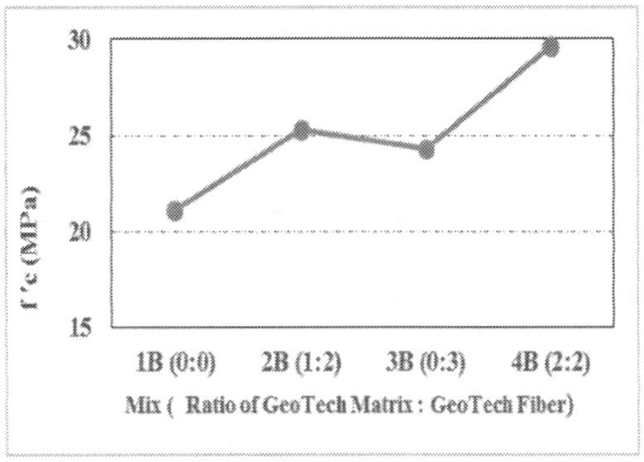

Fig.10. Compressive strength results at 28 days, mix B.

3.2 Flexural strength

Figure 11 shows a typical failure of a flexural specimen, and based on the measurements of three prisms for each batch, Table 7 presents the flexural strength for concrete batches of mixes A and B.

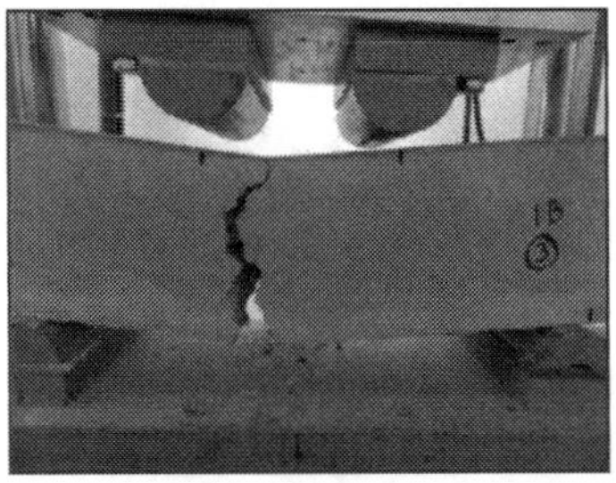

Fig. 11. Typical failure of flexure specimen.

Table 7. Flexural test results.

Batch designation	Flexural strength (MPa)
1A	3.39
2A	3.59
3A	3.95
4A	3.47
1B	2.98
2B	3.91
3B	3.53
4B	3.97

The measured test results of the flexural strength were normalized by the square root of the compressive strength at 28 days for each concrete batch. ACI 318-11 equation (9-10) relates the rupture strength (f_r) and compressive strength ($f'c$) of normal weight concrete as follows [11]:

$$f_r = 0.62\ (f'c)^{1/2} \tag{1}$$

The empirically derived coefficient of 0.62 in the ACI equation providers a lower bound value for comparison with results from this study. However, by comparing the normalized rupture strength values of sub-mixes containing fibres to sub-mixes without fibres, the effects of the basalt additives can be accurately observed. Figures 12 and 13 show the normalized average flexural strength values for sub-mixes A and B, respectively.

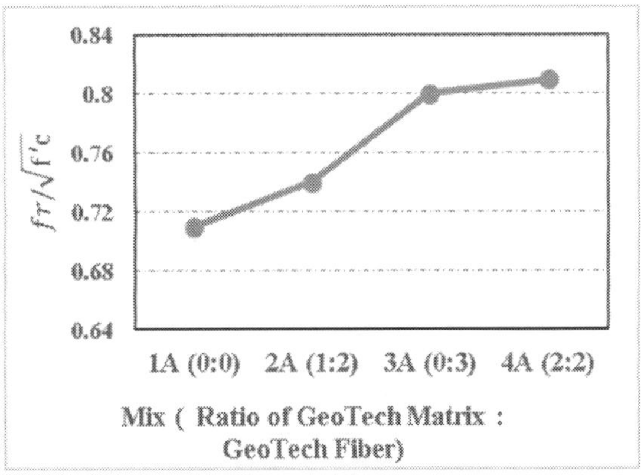

Fig.12. Normalized rupture strength results, mix A.

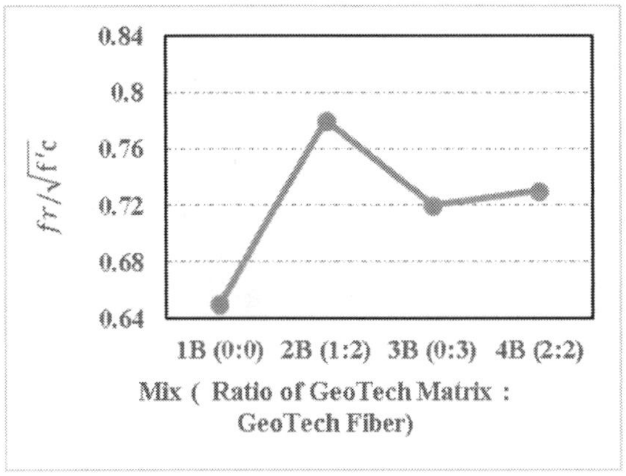

Fig.13. Normalized rupture strength results, mix B.

The normalized rupture strength plots indicate that the addition of basalt fibres increased the rupture strength of mixes A and B. Table 8 shows the percent increase in average rupture strength of each sub-mix with respect to the associative control mix. Test results indicate that the use of both types of basalt fibres provides valuable enhancement to the flexural strength of the concrete which delays the formation of cracks in concrete members.

Table 8. Percentage increase in average rupture strength.

	Mix			
	1A	**2A**	**3A**	**4A**
$\frac{f_r}{\sqrt{f_c'}}$ (% ↑)	0.71	0.74 (4.2 % ↑)	0.80 (12.7 % ↑)	0.81 (14.1 % ↑)
	1B	**2B**	**3B**	**4B**
	0.65	0.78 (20.0 % ↑)	0.72 (10.8 % ↑)	0.73 (12.3 % ↑)

3.3 Average residual strength (ARS)

This test method covers the determination of residual strength of a fibre-reinforced concrete test beam. Figure 14 shows a typical specimen after the application of the initial loading phase and before the application of the reloading phase, while Figure 15 shows the same specimen after the reloading phase. The average residual strength tests results are given in Table 9.

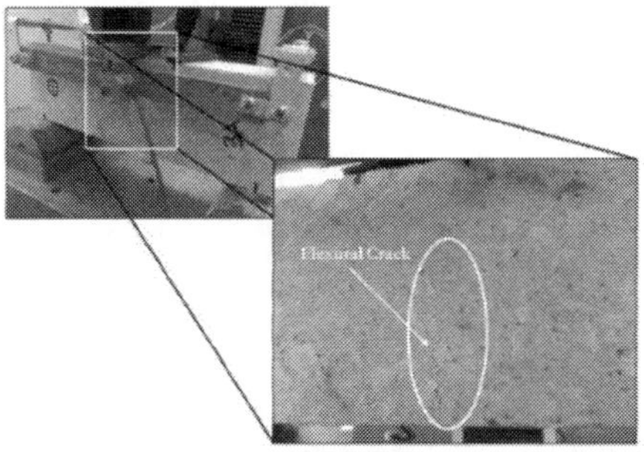

Fig. 14. Typical flexural crack following the ARS initial loading phase.

Fig. 15. Crack propagation and widening after the ARS reloading phase.

Table 9. Average residual strength test results.

Batch designation	Average residual strength (MPa)
1A	-
2A	0.17
3A	0.19
4A	0.19
1B	-
2B	0.25
3B	0.15
4B	0.12

The average residual strength of each specimen was calculated using the following equation given by ASTM C1399-10 [10]:

$$ARS = ((PA + PB + PC + PD)/4) \times k \quad (2)$$

where:
ARS = average residual strength, MPa
$k = L/bd^2$, mm^2
PA + PB + PC + PD = sum of recorded loads at specified deflections (0.50,0.75,1.00 and 1.25 mm, respectively), N
L = span length, mm
b = average width of beam, mm
d = average depth of beam, mm

Measured test results of ARS specimens were used to evaluate performance of basalt fibre-reinforced concrete. By comparing results obtained from specimens containing different fibre types and contents, an optimal mixture of fibre types and proportions can be established. ARS test values were normalized in the same manner as flexural test values, and the results are shown in Figures 16 and 17.

The ARS values of the tested BFRC specimens appear to be low compared to similar studies involving FRP concrete [12]. No control specimens were tested as ARS testing is intended only for comparison of different FRC mixes according to ASTM C 1399-10 [10]. Typical load-deflection curves from an ARS test is shown in Figure 18 in U.S. customary units.

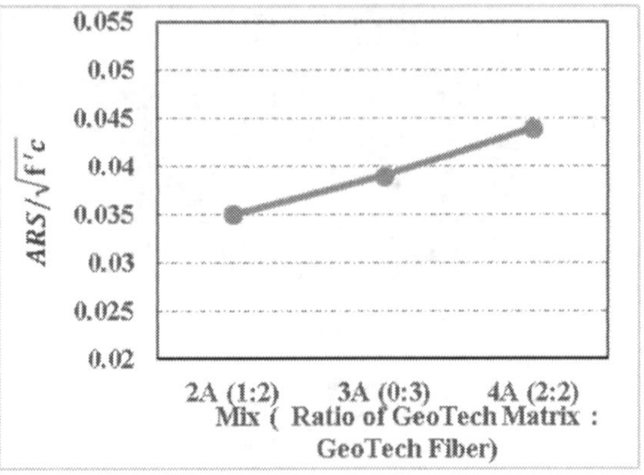

Fig.16. Normalized ARS test results, mix A.

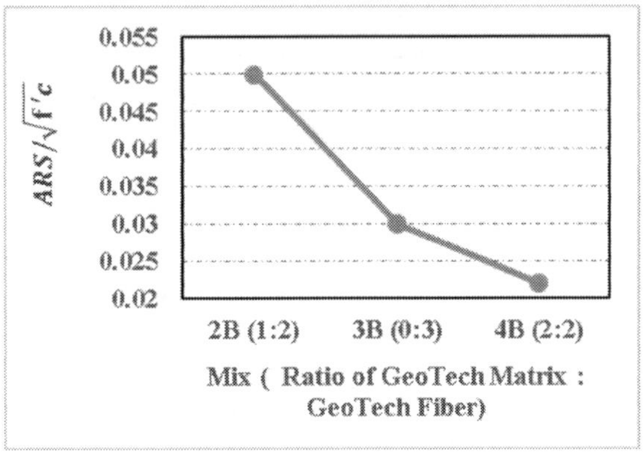

Fig.17. Normalized ARS test results, mix B.

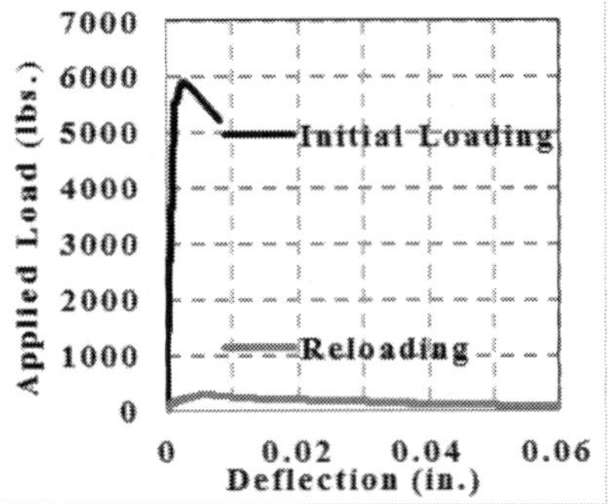

Fig. 18. Typical ARS loading and reloading curves.

As expected, the 'Initial Loading' curve is much steeper than the 'Reloading' curve, because the concrete specimen is initially un-cracked and supported by a 12 mm steel plate.

4 Conclusions

- Results indicated that the use of basalt fibre additives has insignificant effects on the compressive strength of plain concrete at 3, 7 and 28 days, where the increase in strength did not exceed 6 % in the best case. On the other hand, the use of basalt increased the compressive strength of concrete containing fly ash by a percentage ranged from 15 % to 40 % depending on fibre type and volume fraction.

- The research clearly indicated that basalt products increased the rupture strength of the two mixes, 'A' and 'B' with percentages reached 17 % and 33 % respectively.
- The values of average residual strength ranged from 0.12 MPa to 0.25 MPa, lower than similar studies involving FRP concrete.
- The ratios of GeoTech Matrix to GeoTech Fibre had no clear correlation with both the rupture strength and the average residual strength.

5 Recommendations

Testing compressive, flexural and average residual strength values of concrete with increased concentrations of different types of basalt fibres is recommended, in order to determine an optimal mix ratio of the two investigated basalt products. Apparently, fibre volume fractions used in this research, might be too small to have more significant effects on the investigated properties than that obtained in this study.

References

1. 'https://wiki2.org/en/Basalt_fiber', Accessed June 23 (2017)
2. T. Deak and C. Tibor, 'Chemical Composition and Mechanical Properties of Basalt and Glass Fibres : A Comparison', Textile Research Journal, **79**, 7 (2009).
3. T. M. Borhan, International Journal of Civil, Environmental, Structural, Construction and Architectural Engineering, **7**, 4 (2013).
4. Y. X. Yang and J. Lian, 'Basalt Fibre Reinforced Concrete, Journal of Advanced Materials Research, **194-196** (2011)
5. J. Ma, et. al., '*Experimental Research on the Fundamental Mechanical Properties of Presoaked Basalt Fiber Concrete*', Proceedings of the 5th International Conference on FRP Composites in Civil Engineering, Beijing, China, Sep. 27-29 (2010).
6. ASTM International, 'ASTM C127 : Standard Test Method for Relative Density (Specific Gravity) and Absorption of Coarse Aggregate', (2012).
7. ASTM International, 'ASTM C128 : Standard Test Method for Relative Density (Specific Gravity) and Absorption of Fine Aggregate', (2012).
8. ASTM International, 'ASTM C39 : Standard Test Method for Compressive Strength of Cylindrical Concrete Specimens', (2012).
9. ASTM International, 'ASTM C78 : Standard Test Method for Flexural Strength of Concrete (Using Simple Beam with Third-Point Loading)', (2010).
10. ASTM International, 'ASTM C1399 : Standard Test Method for Obtaining Average Residual Strength of Fiber-Reinforced Concrete', (2010).
11. American Concrete Institute, 'ACI 318 : Building Code Requirements for Structural Concrete and Commentary', (2011).
12. C. M. High, 'Use of Basalt Fibers for Reinforced Concrete Members and Concrete Materials', M.Sc. Thesis, North Carolina State University, N.C., USA, (2013).

The application of experimental design models in order to optimize the synthesis of geopolymers

DimitrisKioupis, AggelikiSkaropoulou, Sotiris Tsivilis and GlikeriaKakali
National Technical University of Athens, School of Chemical Engineering, Athens, Greece

Abstract. The aim of this work is to propose the methodology and the mathematical tools for the design and development of geopolymeric materials from industrial wastes and by-products. The applied methodology follows the Taguchi fractional experimental design allowing the investigation of the combining effect of selected parameters in the response of the experimental system by conducting the minimum number of experiments. Two cases of geopolymer synthesis optimization based on the mechanical strength and the apparent density of the products, to which the Taguchi methodology has been successfully applied, are presented in the paper.

1 Introduction

An important part of global industrial production results in production of massive quantities of industrial by-products or other solid wastes. In many cases, these materials contain hazardous compounds and their disposal poses a great social problem with environmental and health repercussions. The best way to handle the above solid waste is through recycling and reusing them into new products.

Recently, a new type of material, geopolymers or inorganic polymers [1], have significantly contributed to both the exploitation and reuse of aluminosilicate wastes and the development of new high added-value construction materials [2,3]. The geopolymer formation involves a chemical reaction between an aluminosilicate material and a silicate solution in an alkaline environment at ambient or slightly elevated temperature and leads to a construction material with rapid compressive strength development [4]. Geopolymers show similar, or even better, properties (mechanical, thermal, insulating, etc.) compared to conventional building materials, while they have minimal environmental impact and contribute to natural resources and energy savings [5]. Most importantly, depending on their design, such materials can exhibit attributes tailor made to the needs of the end-user.

The geopolymerization is a complicated process where the synthesis parameters possess a combined and co-dependent effect and thus their optimization is a difficult procedure. Factors, such as the curing conditions and the composition of the precursor, strongly affect the structure and properties of geopolymers [6-8].

The aim of this work is to suggest the methodology as well as the statistical tools to design and develop geopolymeric materials with a broad range of properties. The applied methodology follows the Taguchi experimental design model, which allows the investigation of the combined effect of selected parameters by conducting the minimum number of experiments [9-11]. In this study, the Taguchi approach was successfully applied in two cases of geopolymeric products development. The first case involves the fabrication of high strength building materials from brick waste while the second one concerns the development of lightweight geopolymer from fly ash with enhanced insulating capacity.

2 Experimental Design: Taguchi Method

The aim of a parameter design experiment is to identify and design the settingsof the process factors that optimize a chosen quality characteristic and are leastsensitive to noise (uncontrollable) factors. When many factors/inputs/variablesmust be taken into consideration, the Taguchi method is a structured approachfor determining the ''best'' combination of inputs to produce a product with optimizedproperties. The Taguchi method utilizes orthogonal arrays from design ofexperiments theory to study a large number of variables with a small number ofexperiments. Using orthogonal arrays significantly reduces the number of experimentalconfigurations to be studied. Furthermore, the conclusions drawn fromsmall scale experiments are valid over the entire experimental region.

The Taguchi experimental design belongs to the fractional factorial designs. In such designs, a simultaneous change of many factors is carried out systematically in order to ensure a reliable and independent study of the effects of the factors on the studied experimental system [12].

The Taguchi method involves the following steps: (i) definition of the problem to be solved, (ii) determination of the target performance of the process under consideration (quality characteristic), (iii) selection of the factors affecting the process and their value range, iv) design and conduction of the experiments, v) analysis of the data obtained from the experiments, vi) interpretation of the results and vii) verification experiments to confirm the results of the previous step.

The major advantage of the Taguchi method and of any fractional factorial design is that a significantly lower number of tests is required to fully describe the system, compared to a full factorial model. This results in a drastic reduction in both time and cost of the entire design. Table 1 shows a test group based on a fractional design.

Table 1.A test group of the Taguchi methodology ($L_9(3^3)$ orthogonal array).

Test	Factor A	Factor B	Factor C
1	1	1	1
2	1	2	2
3	1	3	3
4	2	1	2
5	2	2	3
6	2	3	1
7	3	1	3
8	3	2	1
9	3	3	2

An experimental system examined by the Taguchi method can be illustrated by means of orthogonal arrays (Table 1) [12]. These are basically tables where their columns correspond to the examined factors, while their rows to the factor levels for each experiment.
The orthogonal arrays are denoted by: $L_r(n^k)$, where the letter L is derived from the phrase Latin Squares, the r index is the number of lines, k is the number of the factors and n the number of factors levels. Generally, a large number of orthogonal arrays can be designed in order to meet the needs of the experiment. Table 2 shows some of the most commonly used orthogonal arrays. From Table 2, it is obvious that the selection of the Taguchi method drastically reduces the number of the tests in relation to the corresponding full factorial model.

Table 2. Commonly used orthogonal arrays.

Orthogonal array	No. tests	No. levels	No. factors	No. tests (full factorial model)
$L_4(2^3)$	4	2	3	8
$L_8(2^7)$	8	2	7	128
$L_{12}(2^{11})$	12	2	11	2048
$L_{16}(2^{15})$	16	2	15	32768
$L_{32}(2^{31})$	32	2	31	2147483648
$L_9(3^4)$	9	3	4	81

3 ANOVA analysis

After the implementation of the Taguchi experimental model, the test results are statistically processed using the Analysis of Variance (ANOVA) [13]. The main goal of the ANOVA is to analyze the total dispersion observed in the values of the selected qualitative characteristic and to attribute it to the examined factors in order to derive their significance over the system. The total dispersion also includes the error (balance) which is defined as the random dispersion being present in the results if none of the examined factors affected them. In particular, random dispersion includes the experimental error, the effect of extra-experimental factors and experimental factors considered insignificant. The following relationship summarizes the basis of ANOVA analysis:

$$\begin{pmatrix} Total \\ Dispersion \end{pmatrix} = \Sigma_k \begin{pmatrix} Dispersiondue \\ totheeffect \\ ofkfactor \end{pmatrix} + (Balance) \quad (1)$$

In the ANOVA analysis, the significance of the testedfactors is calculated as the percent contribution of each factor to the selected qualitative characteristic of the process. The contribution of a factor to the system is significant when a change in its value causes a great effect on the response of the system. An appropriate statistical method called significance level is used to characterize the importance of the examined factors [10]. This method classifies the factors as significant or insignificant as to whether the dispersion caused by their variation is random or not.

Once the significant factors of the process have been identified, the optimal levels of the examined factors are calculated. For each factor, the value of the level closest to the desired response is considered as the optimal level. By setting the key factors at their optimum level, the response of the system can be predicted within a confidence interval.

The ANOVA analysis method ends by conducting verification experiments in order to establish whether the experimental design is successful and therefore the resulting conclusions are correct or not. To perform the verification experiments, the combination of the parameters optimizing the qualitative characteristic is selected and then a check if the results fall within the defined confidence interval takes place. If the results verify the predictions, it means that the optimal combination of factor-levels has actually been found and that the experimental system has been properly described. Otherwise, the design has failed to detect the effect of the factors and the system is being re-examined.

4 Examplesof Synthesis Optimization

Two examples of the geopolymer synthesis optimization, where the Taguchi method was successfully applied, are presented in the following paragraphs. Example 1 involves the development of geopolymers from brick waste exhibiting enhanced mechanical properties which can be used as alternative building materials. In the example 2, the objective of the experimental design is to develop low density geopolymers utilizing fly ash from power plants which could find application as construction materials with enhanced insulating capacity.

A potential geopolymer precursor has to contain high amounts of amorphous SiO_2 and Al_2O_3. Table 3 presents the chemical composition of the waste brick and fly ash used in the optimization examples as defined by XRF analysis. Both brick waste and fly ash contain considerable amounts of SiO_2 and Al_2O_3which are necessary for the geopolymerization.

Table 3. Chemical composition of the raw materials (% w/w).

Oxides	Brick waste	Fly ash
SiO_2	51.31	46.96
Al_2O_3	14.62	21.00
Fe_2O_3	8.45	9.83
CaO	6.33	13.74
MgO	8.10	2.20
K_2O	2.28	1.36
Na_2O	0.57	0.16
SO_3	0.44	2.42
TiO_2	0.66	-
$P2O_5$	0.08	-
Cl	0.04	-
L.O.I.	39.7	1.94

In order to define the factors (as well as their value ranges) which took place in the optimization of the synthesis, preliminary experiments conducted with the method of "changing one factor at a time" were performed in both approaches.

4.1 Example 1: Achieving enhanced mechanical strength

In the case of waste brick geopolymers, the performance of the examined system (geopolymer synthesis) was based on the compressive strength (quality characteristic) of the produced specimens. The synthetic factors which were investigated through the Taguchi methodology were the soluble silicon to alkali oxide ($[Si]/R_2O$) molar ratio associated with the amount of dissolved Si in the activation solution, the alkali to aluminum (R/Al) molar ratio expressing the alkalinity of the activation solution and the Na/(Na+K) ratio which is related to the kind of the alkali involved in the geopolymerization reaction. In order to illustrate the experimental design, the orthogonal array $L_9(3^3)$ was chosen. This orthogonal array permits the examination of the three selected factors at three distinct levels (Table 4) and therefore reduces the number of the tests to 9 from 27 of the full factorial model. Compression tests were carried out on a Toni-technik uniaxial testing press, 7 days after the specimens' preparation (load rate 1.5 kN/s, according to the EN196-1 requirements).Three specimens for each experiment were prepared andtested under compression.

Table 5 shows the set of the experiments designed by the Taguchi method along with the corresponding compressive strengths obtained after the tests. The values of the measurements are highly dispersed (2.3-36.4 MPa)indicating that the selected factors and their variation level have a significant effect on the target quality characteristic.

Table 4. Examined parameters and their variation levels (example 1).

Factors	Level 1	Level 2	Level 3
$[Si]/R_2O$	0	1	2
(R/Al)	0.4	0.7	1.0
Na/(Na + K)	0.0	0.5	1.0

Table 5. Synthesis parameters and compressive strength of the produced geopolymers.

Test	$[Si]/R_2O$	R/Al	Na/Na+K	Compr. str. (MPa)
1	0.0	0.40	0.00	2.32
2	0.0	0.70	0.50	14.4
3	0.0	1.00	1.00	12.3
4	1.0	0.40	0.50	14.5
5	1.0	0.70	1.00	24.5
6	1.0	1.00	0.00	36.4
7	2.0	0.40	1.00	6.0
8	2.0	0.70	0.00	8.7
9	2.0	1.00	0.50	33.3

Figure 1 presents the results of the ANOVA statistical analysis by plotting the effect of the selected factors on the compressive strength of the produced samples. At the same time, Figure 1 shows the contribution of each factor in the system response as calculated by the significance level. The parameter with the greatest impact on the system is the R/Al molar ratio (57.75%), followed by the $[Si]/R_2O$ molar ratio with a contribution rate of 35.57%. The kind of the alkali ion (Na/Na+K) has only a marginal effect on the geopolymers properties (6.68%).

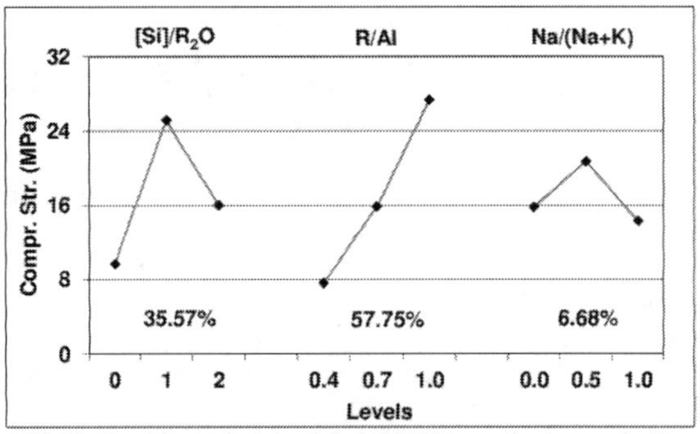

Fig 1. Graphic depiction of the ANOVA results (example 1).

The results analysis of the Taguchi multifactorial model showed that the optimal synthesis conditions for achieving the highest strength are: $[Si]/R_2O$=1.0, R/Al=1.0 and Na/(Na+K)=0.5.

The optimized value of the compressive strength for a 95% confidence interval was predicted to be 39.4 ± 4.8MPa. In order to confirm the model prediction, geopolymer specimens with the optimal conditions were prepared, appropriately cured for 7 days and subjected to compression measurements. The compressive strength (mean of 3 specimens) was found to be 41.0MPa which falls within the predicting range, confirming in this way the validity of the used experimental design model.

4.2 Example 2: Achieving Reduced Density

In the case of fly ash based geopolymers, the qualitative characteristic for the synthesis optimization is the apparent density of the produced specimens. In order to reduce the density of the products, foaming agent in the form of Al powder was introduced into the geopolymeric pastes. The factors selected to be investigated through the Taguchi method are: i) the foaming agent content (Al%w/w on fly ash basis), ii) the alkali to aluminum molar ratio (R/Al, R: Na or K) which associates the alkali quantity in the activation solution with the aluminum content of the raw material and c) the Na/Na+K ratio which is related to the kind of alkali in the activation solution. The experimental design involved the variation of the above factors in four levels and so the $L_{16}(4^3)$ orthogonal array was chosen to describe the experimental system. The apparent density was calculated as the average of three measurements. The variation levels of the selected factors are presented in Table 6.

Table 6. Examined parameters and their variation levels (example 2).

Factors	Level 1	Level 2	Level 3	Level 4
Al % w/w	0.010	0.035	0.065	0.100
R/Al	0.8	1.2	1.6	2.0
Na/(Na+K)	0.00	0.35	0.65	1.00

The apparent density values along with the compressive strength of all the experimental design tests are shown in Table 7. As in the case of brick geopolymers, the system response exhibits a large dispersion of values (0.64-1.56 g/cm^3) indicating the correct selection of the selected factors as well as their value range.

Table 7. Synthesis parameters, apparent density and compressive strength of the produced geopolymers.

Test	Al (%w/w)	R/Al	Na/(Na+K)	density (g/cm^3)
1	0.010	0.800	0.000	1.44
2	0.010	1.200	0.350	1.56
3	0.010	1.600	0.650	1.47
4	0.010	2.000	1.000	1.56
5	0.035	0.800	0.350	1.14
6	0.035	1.200	0.000	1.28
7	0.035	1.600	1.000	1.37
8	0.035	2.000	0.650	1.36
9	0.065	0.800	0.650	0.72
10	0.065	1.200	1.000	1.11
11	0.065	1.600	0.000	1.29
12	0.065	2.000	0.350	1.13
13	0.100	0.800	1.000	0.64
14	0.100	1.200	0.650	0.93
15	0.100	1.600	0.350	1.11
16	0.100	2.000	0.000	1.20

Figure 2 shows the independent effect of the synthesis factors on the apparent density of the produced samples as derived from ANOVA analysis. The significance of each factor is expressed by the percent contribution to the qualitative target characteristic (apparent density). The most significant factor of theexamined system is the Al %w/w content of the composition with a contribution of 66.31%. The R/Al molar ratio is the second most important factor (26.58%), whereas the choice of alkali does not possess a significant effect on the apparent density of the produced geopolymers (7.11%).

In summary, the analysis of the results showed that the delivery of a lightweight product is favored by the following parameter combination: Al 0.1% w/w, R/Al=0.80 and Na/(Na+K)=0.65. In order to confirm the design model, a geopolymer with the above parameters was prepared. The apparent density was measured 0.65 g/cm^3 which falls within the range of predicted values (0.66 ±0.05 g/cm^3) for a 95% significance level.

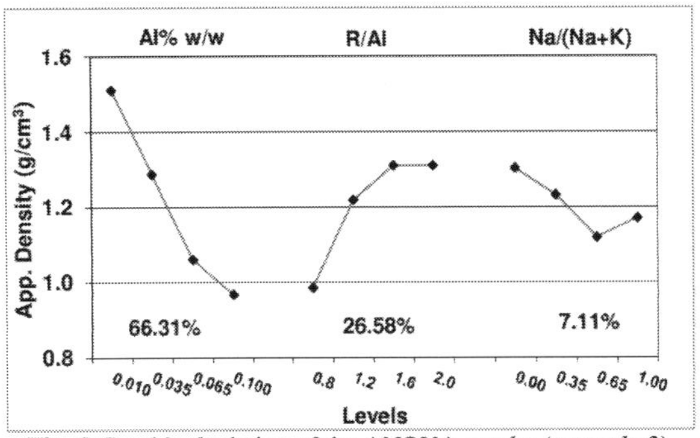

Fig. 2. Graphic depiction of the ANOVA results (example 2).

In order to assess the thermal behavior of the produced geopolymers, the thermal conductivity of the samples was measured. Figure 3 shows the apparent density and thermal conductivity values of the reference geopolymer (55.5MPa and 1.77 g/cm^3) and the optimal composition (2.3 MPa and 0.65 g/cm^3) as derived from the experimental model. It is evident that the addition of metallic Al to the reactant components of the composition results in a drastic reduction in both density and thermal conductivity.

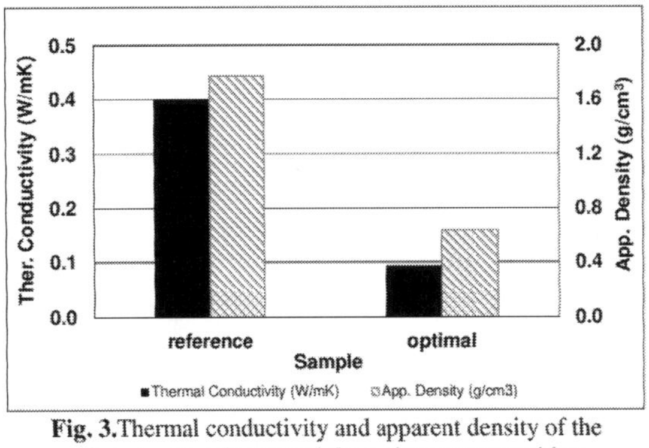

Fig. 3. Thermal conductivity and apparent density of the reference geopolymer and the optimum composition.

Typically, the reference geopolymer exhibits a conductivity value of 0.40 W/mK while the optimal composition has a thermal conductivity value of almost one order of magnitude lower (0.09 W/mK). The above results indicate the successful application of Taguchi fractional factorial design to the development of lightweight geopolymers with thermal insulating properties.

5 Conclusions

In the present study, the application of the Taguchi multifactorial design of experiments is proposed in order to develop geopolymeric materials with desirable properties. In designing the experiment, it is important to define the desired target property (quality characteristic) on which synthesis optimization is based and to select the synthesis factors that have a significant effect on this property as well as their value range. Subsequently, experimental tests are determined by means of appropriate orthogonal arrays. The results of the tests are statistically processed with the help of ANOVA and the levels of the test factors optimizing the target property are determined. Finally, according to the optimum levels of the examined factors, the confirmation test is carried out which verifies or not the results of the analysis. The proposed methodology has been applied to the development of high-strength geopolymers from brick waste and lightweight geopolymersfrom fly ash, with successful results. The produced geopolymers showed compressive strength, density and thermal conductivity values in the range of 2-55 MPa, 0.6-2.0 g/cm^3 and 0.09-0.40 W/mK, respectively.

Acknowledgement

This project has received funding from the European Union's Horizon 2020 research and innovation programme under grant agreement No [723825].

References

1. J. Davidovits, J. Therm. Anal. **37**, 1633 (1991)
2. J.L. Provis,Mater. Struct.**47**, 11 (2014)
3. J.L. Provis, A. Palomo, C. Shi,Cem. Concr. Res. **78**, 110 (2015)
4. J.L. Provis, S.A. Bernal, Annu. Rev. Mater. Res. **44**, 299 (2014)
5. B.Majidi, Mater. Tech. **24**,79 (2009)
6. P.W. Ken, M.Ramli, C.C. Ban,Constr. Build. Mater.**77**, 370 (2015)
7. C. Panagiotopoulou, T. Perraki, S. Tsivilis, N. Skordaki, G. Kakali, Ceram. Eng. Sci. Proc. **29**, 165 (2009)
8. C. Panagiotopoulou, A. Asprogerakas, G. Kakali, S. Tsivilis, Cer. Eng. Sci. Proc. **32**,17-28 (2011).
9. C. Panagiotopoulou, S. Tsivilis, G. Kakali, Constr. Build. Mater. **91**,17-22 (2015).
10. M.W. Weiser, K.B. Fong, Am. Ceram. Soc. Bul, **72**, 87 (1993)
11. G. Taguchi, *System of experimental design (Engineering Methods to optimize quality and minimize costs)*, (UNIPUB/ Kraus Int. Publications, New York, 1987)
12. T.B. Barker,*Quality by experimental design*,(Chapman & Hall/CRC, 2005).
13. R.K. Roy,*Design of experiments using the Taguchi approach*, (John Wiley& sons, New York, 2001)

Thermal treatment of desert sand to produce construction material.

Frank Neumann[1], Manfred Curbach[2]
[1, 2] Technische Universität Dresden, Institute of Concrete Structures, Dresden, Germany

Abstract. The concrete and reinforced concrete industry has refrained from using desert sand. The reasons for this are the rounded form of the particle, its grading and its chemical composition. This article provides an overview of the background, the constructability and technological challenges that the construction industry faces regarding construction sand's shortage. Also, construction aggregates consumption and its influence on fresh and hardened concrete properties are described. The paper concludes by presenting several feasible approaches and provides a feasible technological solution.

1 Demand for construction aggregates

The most common construction material in the world is concrete. About 8 billion m^3 of concrete were produced in 2015 [1]. This means that nearly 5 billion tons of the fine aggregate fraction, denominated as sand was consumed for concrete construction in 2015. In Germany, the consumption is estimated at 253 million tons of sand and gravel [2] per year. In addition to its use as a construction material, sand is often used for artificial land reclamation, especially in China, Singapore and the United Arab Emirates (UAE). The worldwide amount of sand used, along with concrete production included is calculated between 15 and 40 billion tons per year [3, 4].

2 Aggregates for concrete casting

The shortage of construction sand, the economic disadvantage of sand imports and the regional unavailability of sufficient construction sand had led several researchers involved in concrete and mortar production to investigate the feasibility of substituting fine aggregates by desert sands.

Concrete consists of 20 % water, cement and 80 % aggregates. Aggregates with different diameters (Figure 1) are chosen for concrete production to achieve a stable granular structure. The appropriate selection of the material is important especially for fresh concrete as it prevents aggregate separation and sedimentation. To reduce the amount cement and water needed, to minimize pores and to reduce shrinkage and the risk of carbonation of the hardened concrete [5], appropriate grain curve has to be chosen. Because of this, nearly a third up to a half of the used aggregates have a maximum diameter of 2 mm. Figure 1 compares different packing rates and illustrates the effect of stabilization by using different grain sizes for concrete production.

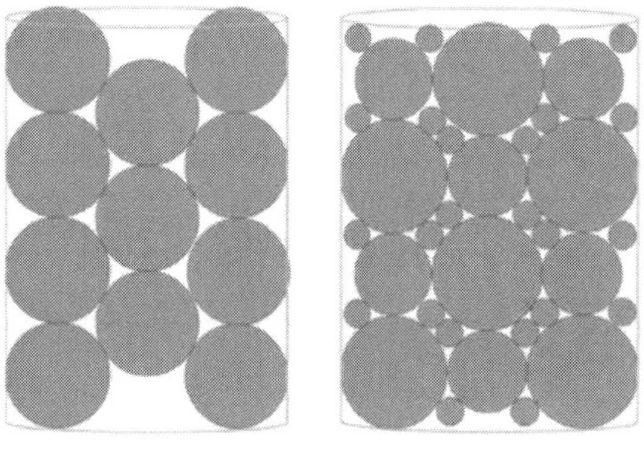

Fig. 1: Structures packed with different tightness to ensure stability within fresh concrete: homogeneous (left) and poorly graded grain (right). (Graphic: F. Neumann)

The specific surface and the shape of aggregates influence the water consumption and workability of fresh concrete. For this reason, the above-mentioned grain curves and aggregates with corresponding surface and shape have to be chosen according to the requirements and properties of the concrete. Desert sand grains are mostly distributed in a monomodal manner. This means grains with one specific diameter are dominating. Thereof only a single fraction of fine aggregates in the ordinary concrete mixture can be substituted to obtain the required grading curve. The shape of desert sand grains, in comparison to aqueous grains, is mostly round. It is caused by aeolian transport effects and varies from equant over elongated to flaky and between very angular and well-rounded. This affects the mechanical interlock between aggregates and thereof the stability of the mixture as well as fresh concretes properties.

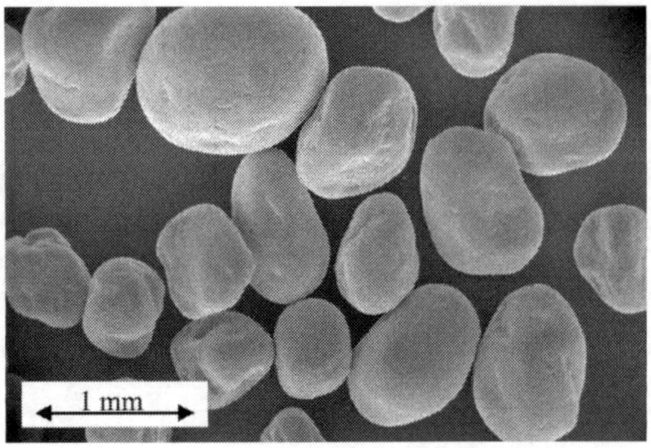

(a) 0.5 mm to 1.0 mm fraction from Libya.

(b) 0.5 mm to 1.0 mm fraction from Algeria.

Fig. 2: ESEM analysis of desert sand grains. (Photos: Simone Hempel, TU Dresden)

Up to a certain extent, the workability is improved with benefits to sedimentation. Figure 2a shows well-rounded desert sand grains from Libya with a diameter between 0.5 mm and 1.0 mm studied via environmental scanning electron microscopy (ESEM). Figure 2b shows grains with the same diameter from Algeria with a slightly edged shape and thereof a higher surface. Concrete properties are also affected by desert sand's chemical composition. The solubility of certain salts contained in aggregates and cement phases depend on the pH value of the pore solution which is affected by adsorption and desorption of dissolved carbonates, chlorides and sulfates [6]. On the one hand, dissolved chloride ions are leaching the portlantide. On the other hand, chloride or sulfate ions that are present dissolute in the mixing water and react with the tricalcium aluminate to give Friedel's salt and the ettringite, respectively [7].

Some researchers investigated concretes made with desert sand aggregates and compared the mechanical properties with those of standard concrete. Although the workability was improved and the variation in air content was negligible while w/c-ratio was constant, a reduction in compressive and tensile strength of 16 % [8, 9, 10] was found. Depending on the substituted amount of fine or very fine aggregates, a reduction of about 21 % was measured. By substituting 100 % of the fine aggregates with dune sand, the compressive and tensile strength of self-compacting concrete (SCC) were reduced to 50 % compared to an SCC made of crushed sand [11].

Another approach to meet the challenge of a construction sand shortage comes along with the idea of reducing the needed amount of concrete to prevent steel reinforcement from corrosion. The layers of concrete surrounding the steel are thicker than statically necessary. The objective of the new material called carbon reinforced concrete, based on substitution of common steel reinforcements with carbon fiber reinforcements, is to reduce the amount of concrete and thereof the amount of cement [12, 13, 14] and aggregates. Along with that, the concrete composition has to be adjusted in order to meet the requirements of carbon reinforced concrete. Typical mesh sizes of textile reinforcements are between 10 mm and 14 mm which prevent ordinary aggregates from penetrating trough the grid. Mostly, the used concrete is based on a grain curve with a maximum grain size of 2 mm.

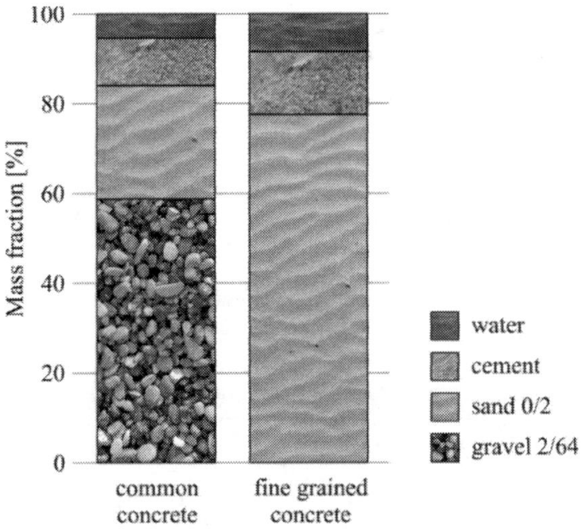

Fig. 3: Comparison of mass fractions between common concrete and fine aggregate concrete. (Graphic: F. Neumann)

In Figure 3, a common concrete mixture for steel reinforced concrete with a maximum grain size of 63 mm is compared to an optimized fine graded concrete mixture for carbon fiber reinforcements with a maximum grain size of 2 mm. The substitution of 20 % of common steel reinforced concrete with carbon fiber reinforced concrete will consequently raise the overall consumption of fine aggregates by about 1 billion tons per year.

However, the substitution of fine aggregates by desert sand comes along with a loss in strength which is acceptable in an economical point of view unless the strength does not fall below a certain value. Nevertheless, most previous studies focus on fine or very fine grains only. Shape as well as chemical composition mostly remain unconsidered although they are affecting fresh and hardened concrete and are varying from deposit to deposit even among single grain sizes.

https://doi.org/10.1051/matecconf/201814901030

3 Approaches for a technical solution

In the past, some technical solutions where developed to modify or reshape desert grains in order to substitute concrete's ordinary aggregates. These technical approaches are mainly based on a melting process followed by a crushing process [15, 16] or on pre-shaping the aggregate with a high voltage electrostatic field [17].

One of the challenges of the 21st century is the construction of buildings on celestial bodies. At this moment the use of local available resources is essential. Several investigations focused on the available aggregates on the moon, called regolith. Such an aggregate compares well with terrestrial ashes [18] with grain sizes between 0.05 mm and 0.1 mm [19]. Lunar base construction materials may be produced by direct sintering of the regolith [20].

The method, patented by Technische Universität Dresden [21], is based on the same process which allows the reshaping of solids by combining single particles (Figure 4). The sintering process, which is known in the field of powder metallurgy, connects two or more particles with each other without melting them.

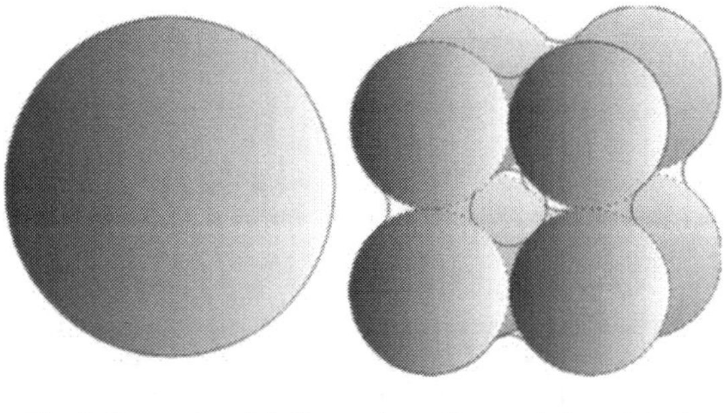

Fig. 4: Concept model. Comparison of a round shaped single grain with a combination of grains. (Graphic: F. Neumann)

The process occurs below the melting point meaning that the crystalline structure of grains is kept and amorphous states are avoided.

In Figure 5, untreated and sintered samples from two different sites in Namibia are compared. Sample I contains flaky, elongated angular grains (a) whereas sample II contains more equant and rounded grains (c) which may have led to the less porous sintered aggregate (d) compared to the porous structure from sintered sample I (b).

(a) Sample I: Sieved grains from Namibia with a diameter between 0.5 mm and 1.0 mm.

(b) Sample I: Sintered product from Namibia.

(c) Sample II: Sieved grains from Namibia with a diameter between 0.5 mm and 1.0 mm.

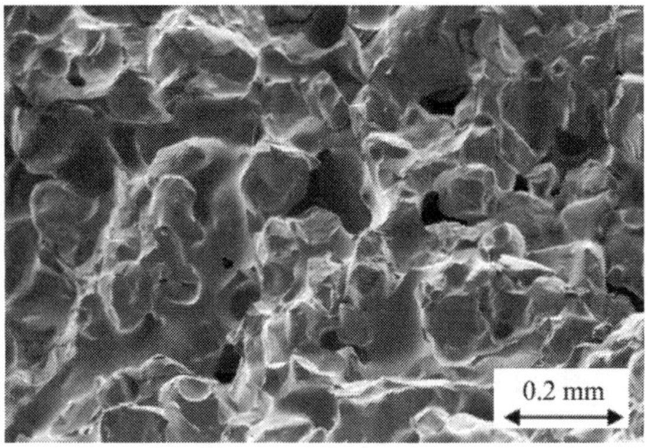

(d) Sample II: Sintered product from Namibia.

Fig. 5: ESEM analysis of different desert sand samples from Namibia. (Photos: Simone Hempel, TU Dresden)

To proof the concept, desert sand samples were gathered and analysed by grading curve, chemical composition, melting point and pH value. The grading curves of unsintered samples from Algeria, Libya and Namibia are given in Figure 6 and are compared with a standard grading curve B 63 according to the German standard DIN 1045-2 [22].

The preliminary tests were conducted with grain size fractions passing the 100 µm sieve. The chemical analysis was conducted to predict the melting point and to ensure that a protective atmosphere while processing is unnecessary (Table 1). Then the melting point was examined via a differential calorimetry scan (DSC) analysis. From all samples, only the ones from Namibia could be analysed due to the maximum operational temperature of 1500 °C of the DCS device. Taking the chemical composition and the melting point into account, the parameters for the sintering process were determined to 1250 °C and one hour under atmospheric pressure without any protective inert gases.

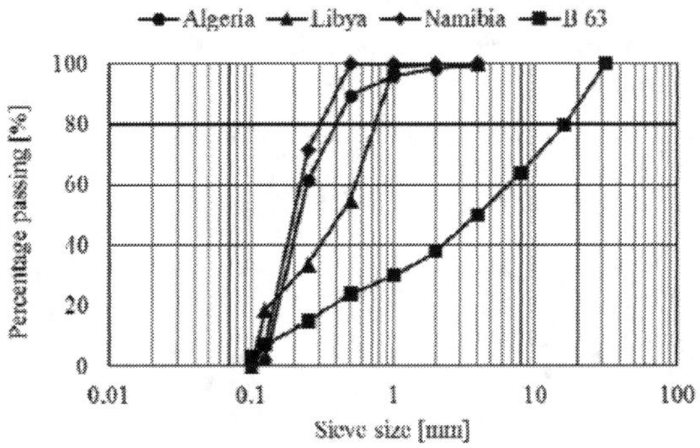

Fig. 6: Desert sand grain size distribution depending on their origin compared to standard grading curve B 63. (Sieve analysis by Simone Hempel, TU Dresden)

After the sintering process, the chemical composition and pH value from produced grains were analysed again to compare them with those from the untreated material. Identified elements and the pH value from raw material grains (a), sintered grains (b) and a reference are compared in Table 1. The reference grain (R) is an unsintered dune sand from the Kharga Oasis in Egypt [8].

Table 1: EDX analysis of sintered and unsintered samples from site I and II in Namibia including pH value comparison. (EDX analysis by Simone Hempel, Institute for Construction Materials and Michal Pejko, Institute of Manufacturing Science and Engineering, both TU Dresden)

No.	Elements [Atomic %]							pH [-]
	O	Al	Si	S	K	Ca	Fe	
Ia	52.6	11.0	24.2	-	3.1	1.3	3.6	5.7
Ib	58.9	7.0	20.4	-	4.0	0.3	2.0	9.0
IIa	58.8	1.5	4.6	13.0	0.5	19.6	1.0	8.1
IIb	53.8	1.8	11.0	na	0.1	28.1	1.4	11.5
R	51.6	5.16	40.3	-	0.1	0.7	0.8	7.5

4 Summary

The Technische Universität Dresden has successfully tested a process, with which different types of desert sand were sintered to produce aggregates. The production of aggregates with varying but distinct grain sizes allows the entire substitution of conventional aggregates with those made from desert sand. Within preliminary tests, the chemical composition of different desert sands was observed via SEM micrograph and EDX analysis before and after the sinter process. It was found, that the pH value of sieved, cleaned and dried samples has changed from acidic to neutral or even alkaline. The process of sintering in combination with the chemical composition of different desert sands affects the porosity. In general, desert sand can be used as a substitute for fine aggregates. The produced aggregates with different porous structures are shown in Figure 7.

Fig. 7: Sintered aggregates produced from desert sand by TU Dresden. (Photo: F. Neumann)

5 Outlook

Based on the presented, first positive results, the Technische Universität Dresden and the German Aerospace Centre (DLR) want to conduct a research project along with participating partners from Argentina, Brazil, Chile and Portugal taking into account the method of physical transformation of sand, patented by TU Dresden. Due to the high energy consumption of the sintering process, the project focuses on a solar sintering technology provided by the DLR. In the project, the partners want to focus firstly on desert sand deposit evaluation and analysis of mono- and multimodal grain sintering. Afterwards, a solar based sintering technology has to be developed. Concrete and mortar development, feasible alternative concrete made materials, developmental and social and techno-economic studies are also conducted.

 https://doi.org/10.1051/matecconf/201814901030

6 References

1. Weimar Institute of Applied Construction Research: http://www.iab-weimar.de/en/termine?view=article&id=206, access on 19.05.2017.

2. O. S. Kaiser, O. Krauss: *Kurzanalyse Nr. 12: Systematische Ansätze zur Steigerung der Ressourceneffizienz im Bauwesen.* VDI Zentrum Ressourceneffizienz GmbH, *(2015).*

3. WeltN24 GmbH: *Die lebensnotwendige Jagt nach mehr Sand.* https://www.welt.de/wirtschaft/webwelt/article127216019/Die-lebensnotwendige-Jagd-nach-mehr-Sand.html, (2014), access on 19.05.2017.

4. P. Peduzzi: *Sand, rarer than one thinks.* United Nations Environment Programme, (2014).

5. J. Stark, B. Wicht: *Dauerhaftigkeit von Beton.* Springer-Verlag Berlin Heidelberg (2013).

6. I.-S. Yoon, T. Saeki, S.-S. Park: *Chloride ions adsorption and desorption in Pore Solution with various pH values.* Second International Conference on Microstructural-related Durability of Cementitious Composites, Amsterdam (2012).

7. U. Dahme: *Chlorid in karbonatisierendem Beton.* PhD thesis. University Duisburg-Essen, (2006).

8. E.-S. S. A. Seif.: *Performance of Cement Mortar Made with Fine Aggregates of Dune Sand, Kharga Oasis, Western Desert, Egypt: An Experimental Study.* Jordan Journal of Civil Engineering, **7** (2013), P. 270–284.

9. F. J. Luo, et al.: *Effect of very fine particles on workability and strength of concrete made with dune sand.* Construction and Building Materials **47** (2013), P. 131-137.

10. A.S. Al-Harthy, et al.: *The properties of concrete made with fine dune sand.* Construction and Building Materials **21** (2007), P. 1803-1808.

11. A. Rmili, et al.: *Incorporation of crushed sands and Tunisian desert sands in the composition of self-compacting concretes Part II: SCC fresh and hardened states characteristics.* International Journal of Concrete Structures and Materials **3** (2009), P. 11–14.

12. Schladitz, F.; Tietze, M.; Lieboldt, M.; Curbach, M.: *Carbon and Concrete – The Future of Construction?* Bridges and Structures Sustainability – Seeking Intelligent Solutions – Proc. of IABSE Conference, 8.–11.5.2016 in Guangzhou (2016), P. 240/241.

13. Brameshuber, W. (Ed.): *Report rep036 : Textile Reinforced Concrete – State-of-the-Art Report of RILEM TC 201-TRC.* Bagneux: RILEM Publ., 2006.

14. T. C. Triantafillou: *Textile Fibre Composites in Civil Engineering.* Woodhead Publishing (2016).

15. D. Behnisch, J. Ikic: *Verfahren zur Herstellung eines, zur Betonherstellung oder als Schüttgut zur Neulandgewinnung geeigneten, Sandes.* Patent DE 10 2014 006 942 B3, (2015).

16. D. Behnisch, J. Ikic: *Method and device for producing artificial crushed sand by means of a thermal treatment using sand in the form of fine sand (fs/fsa) and/or round sand as the starting material.* Patent WO 2015/172765 A1, (2015).

17. E. Rabe: *Formkörper auf der Basis von Quarzsand, ihre Herstellung und Verwendung sowie Vorrichtung zur Durchführung des Verfahrens.* Patent DE 3248537 A1, (1984).

18. G. Heiken, et. al.: *Lunarsourcebook – a user's guide to the moon.* Cambridge University Press (1991).

19. W. D. Carrier: *Particle size distribution of lunar soil.* Journal of Geotechnical and Geoenvironmental Engineering **129** (2003), P. 956-595.

20. S. Willhelm, M. Curbach: *Review of possible mineral materials and production of techniques for building material on the moon.* Structural Concrete **15** (2014), P. 419-428.

21. F. Neumann, M. Curbach: *Verfahren und Vorrichtung zur thermischen Behandlung von Sand.* Patent WO 2017016550 A1, (2015).

22. DIN 1045-2: *Concrete, reinforced and prestressed concrete structures – Part 2: Concrete – Specification, properties, production and conformity – Application rules for DIN EN 206-1* (2008).

The authors like to thank all colleagues, especially Simone Hempel, Michal Pejko, Johannes Trapp (TU Dresden) and Karin Weimann (BAM Berlin) for supporting the idea with trial testing.

Dipl.-Ing. (FH)
Frank Neumann
Research associate
Institute of Concrete Structures
Technische Universität Dresden
01062 Dresden, Germany
frank.neumann1@tu-dresden.de

Prof. Dr.-Ing. Dr.-Ing. E.h.
Manfred Curbach
Professor, institute's director
Institute of Concrete Structures
Technische Universität Dresden
01062 Dresden, Germany

Improvement of the mechanical performance of Fergoug dam sediments treated for reuse in road engineering

LAROUCI Abdelkader[1], SENHADJI Yassine[1], LAOUFI Laïd[1], BENAZZOUK Amar[2]
[1] Université de Mustapha Stambouli de MASCARA, Département de Génie Civil, Algérie
[2] Université de Picardie jules Verne IUT d'AMIENS, Département de Génie Civil, France

Abstract: The phenomenon of siltation is one of the main problems facing many dams throughout the world, and particularly in Algeria. The mud, before and even after extraction, is always harmful. Indeed, mud is a harmful material, first for dams, because it reduces their capacity of water storage, and second, for the environment following the desilting operations that generate large quantities of materials that occupy and sometimes pollute large areas. However, these materials can be exploited differently and used in other fields. The envisaged research work relates to real cases of silted dams, such as the dam of Fergoug (western Algeria) whose siltation rate is very high; it is estimated at 97% of its initial capacity which, according to the National Agency for Dams (Agence Nationale des Barrages - ANB), is equal to 18 million m^3. The large quantities of silt extracted present an environmental problem, and its use as a local raw material for the manufacture of civil engineering materials can contribute to solve this problem. The vulnerable lands of the sub-catchment of Wadi Fergoug (Fergoug River) extend over an area of 122 km^2, from a total surface of 8340 km^2 for the dam catchment area. There is a great diversity of superficial formations with predominantly clay soils from marly formations. The rate of specific erosion has increased to 160 T/ km^2/ year due to irregular annual rainfall resulting from a succession of dry and wet years. Sedimentary materials, which are found in considerable quantities, were collected at the foot of the dam, on its right bank. This study attempts to find a recycling pathway for these sediments. The objective of the present work is to investigate the behavior of silt from the dam of Fergoug (Algeria) for the purpose of using it in road construction (foundation and base layers). The method adopted is to reconstitute, in the laboratory, samples of mixtures containing road aggregates with different proportions of silt. These samples are then subjected to various tests (Normal modified Proctor, Atterberg limits, VBS, Particle size analysis, Sedimentation, DRX etc ...). The results of the tests indicate that some of the prepared mixtures present very interesting properties. These results satisfy the current recommendations of the technical guides; they also corroborate their use in road construction. ***Keywords:*** *Dredged sediments; Valorization; Road construction; California bearing ratio (CBR) index; Mechanical behavior; Environment.*

1 Introduction

Le barrage de Fergoug est situé dans une zone montagneuse caractérisée par un climat agressif avec alternance de période très chaudes et froides, en plus des pluies dévastatrices pour les sols du bassin versant nus de toute couverture végétale. Ainsi les phénomènes d'érosion interne et externe sont très accrus dans la région et par conséquent, le barrage en question est pratiquement rempli de vase.

Pour y remédier à ce problème d'envasement, plusieurs recherches sont lancées dans ce domaine. Certains travaux sont réalisés avant l'envasement dans le but de proposer des solutions réduisant ce phénomène [1,2].D'autres par contre, sont réalisés après envasements dans le but de la réutilisation de la vase dans les différentes activités, notamment dans le domaine du génie civil [3-6].

Cette étude a pour objectif la valorisation des sédiments de barrage dans le domaine routier en vue de leur utilisation dans les corps de chaussées (couches de fondation et couches de base) [7].

2 Intérêts du travail réalisé

Ce travail peut avoir plusieurs intérêts, tant sur le plan économique et environnemental, à savoir :

L'exploitation des quantités importantes de vase évacuées périodiquement par les vannes de fond de barrages dont les effets sont néfastes pour l'environnement.

 https://doi.org/10.1051/matecconf/201814901031

- Par conséquent, la récupération des espaces occupés par ces matériaux pour une éventuelle utilisation dans des autres activités.
- Le renforcement des secteurs du génie civil et des travaux publics par de nouvelles sources de granulats routiers, sachant que les sources habituelles alluvionnaires restent parfois incapables de satisfaire les besoins nationaux.

3 Programme expérimental

3.1 Matériaux utilisés

3.1.1 La vase

La vase est défini comme étant un ensemble d'élément déposes par l'eau, le vent, la glace qui proviennent de l'usure des continents, c'est-à-dire de la destruction de roches ou d'être vivants. La destruction se fait par mécanismes physique produisant la fragmentation des matériaux et des réactions chimiques donnant des solutions de lessivage (altération chimique).

Le terme « vase » englobe une large famille de sédiments représentée par trois phases ; une phase liquide, colloïdale et minérale. Constituée de particules solides dont l'origine peut être naturelle ou anthropique.

3.1.1.1 Caractéristiques physico-chimique

Toute la vase prélevée dans la zone de rejet est convenablement préparée, elle est séchée (60 °C) et réduite en poudre, ensuite tamisée au tamis 1 mm.

Les essais de caractérisation physico chimique sont réalisés au laboratoire LNHC (Laboratoire National de l'Habitat et de Construction) Unité de Sidi Bel Abbes. Elles sont résumées dans le tableau 1.

Tableau 1. Caractéristiques physico-chimique des sédiments du barrage Fergoug

Caractéristiques	Designation	Résultats
Teneur en eau naturelle	(%)	52
Limite de liquidité	W_L (%)	62,54
Limite de plasticité	W_P (%)	29,29
Indice de plasticité	I_P (%)	33,26
Masse volumique sèche	ρd_{OPM} (g/cm^3)	1,779
Teneur en eau de l'optimum Proctor	W_{OPM} (%)	19,30
Valeur du bleu méthylène	VBS (g/100 g du sol)	2,58
Carbonates	$CaCO_3$ (%)	27,94
Matière Organique	(%)	3,15

3.1.1.2 Caractéristiques minéralogique

Cette analyse nous permet d'identifier les minéraux présents dans notre vase, elle a été réalisée au laboratoire de la cimenterie 'Lafarge' d'Oggaz-Algérie.

L'analyse DRX de la vase brute nous montre l'existence de raies correspondants aux minéraux tels que le quartz (SiO_2), les carbonates types calcite($CaCO_3$), et les argiles type Kaolinite ($Al_2O_3.2SiO_2.2H_2O$) et l'illite, ainsi que la Dolomite $CaMg(CO_3)_2$ (Figure 1).

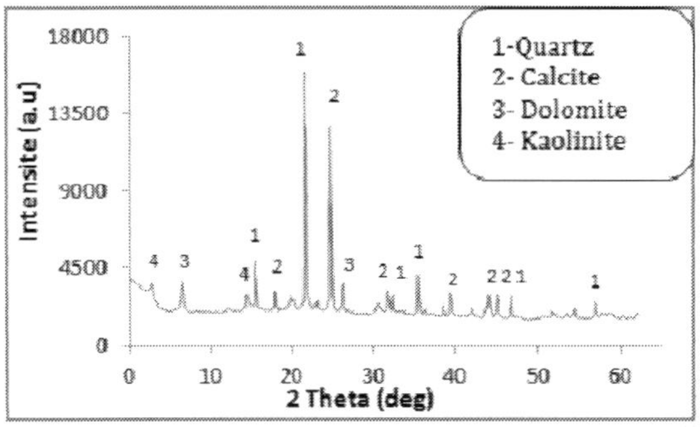

Fig. 1. Analyse minéralogique de la vase par DRX.

3.1.2 Les granulats routiers

Les granulats routiers utilisés sont prélevé dans le tronçon de la pénétrante autoroutière qui relier la ville de Mascara a l'Autoroute Est-Ouest publics ce qui est en cours de réalisation. C'est les granulats utilisés pour la construction routière dans la région Mascara et Oran (Algérie). Cette dernière est caractérisée par la courbe granulométrique présentée sur la Fig. 2. Selon la classification des matériaux utilisables dans la construction des remblais et des couches de forme d'infrastructures routières (NF P 11-300), les granulats routiers utilisés se présentent sous forme d'une grave propre mal graduée, de granulométrie uniforme.

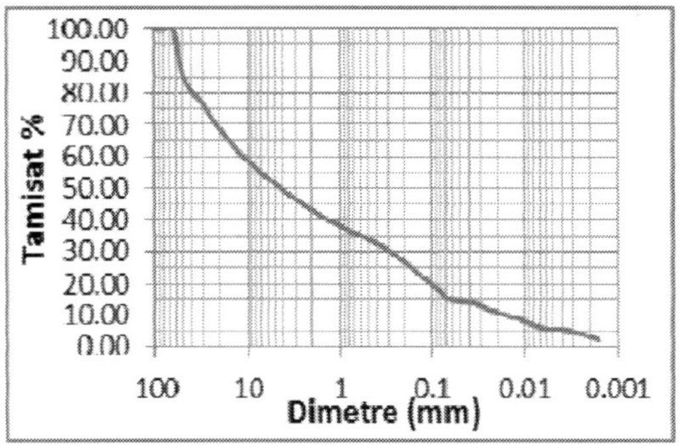

Fig. 2. Courbe granulométrique du matériau utilisé.

4. Echantillons préparés et essais réalisés

Les échantillons soumis aux essais mécaniques sont préparés à base de granulats routiers et de sédiments de barrage, sous forme de mélanges selon les proportions résumé dans le tableau suivants :

Tableau 2. Les mélanges préparés avec le granulat routier et la vase du barrage.

	Granulats routiers (%)	Vase (%)
Echantillon 1	100	0
Echantillon 2	95	5
Echantillon 3	90	10
Echantillon 4	85	15
Echantillon 5	80	20
Echantillon 6	75	25

Afin d'étudier le comportement des mélanges sous les principales sollicitations routières en fonction de leur teneur en sédiments, les échantillons préparés sont soumis aux essais physiques et mécaniques sont déterminés conformément aux normes françaises.

- Essais de limite d'Atterberg (NF P94-051) ;
- Essais de compactage au Proctor Normal et Proctor modifié (NF P94-093) ;
- Essais de portance immédiat (IPI) et essais de portance après immersion (CBR_{imm}) (NF P94-078).

5. Résultats et analyses

5.1 Effet de la vase sur les limites d'Atterberg

Les résultats des essais sur les limites d'Atterberg aux différents pourcentages sont montrés dans le tableau 3 et sur la fig. 3. A travers l'allure de la courbe dans la fig. 3, on peut clairement apprécier la tendance progressive de l'indice de plasticité des différents échantillons, on en déduit que l'augmentation du pourcentage de la vase de barrage à un effet direct sur l'indice de plasticité du matériau composé Granulat routier-Vase.

D'après les résultats trouvés, On montre un changement remarquable de l'indice de plasticité après l'incorporation du vase. Il est évident qu'une addition de 15% et 25% du vase pour une augmentation de 51,84% et 97.07% respectivement. L'augmentation de l'indice de plasticité dans notre échantillon est indiqué une amélioration dans la plasticité du sol. L'indice de plasticité est très fortement lié à la quantité et à la nature des minéraux argileux présents dans le sol et à la surface spécifique des sols.

Tableau 3 : Valeurs des limites d'Atterberg

L'échantillon	W_L	Wp	Ip
1	18.62	13.47	5.15
2	19.41	13.64	5.77
3	20.42	13.83	6,59
4	21.84	14.02	7.82
5	22.81	14.15	8.66
6	24.51	14.46	10,13

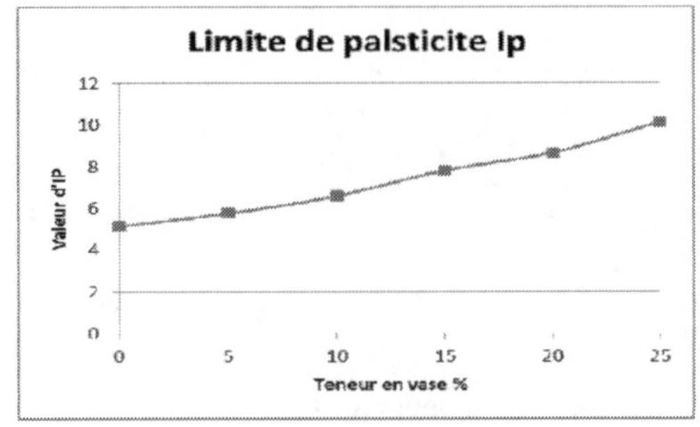

Fig. 3. L'indice de plasticité (Ip) de la vase de barrage de Fergoug

5.2 Essais de compactage au Proctor Normal et Proctor Modifié

Les essais Proctor sont réalisés selon la norme (NF P 94-093). Ils permettent de montrer l'aptitude au compactage des mélanges étudiés en fonction de leurs teneurs en vase. Les courbes de compactage (normal et modifié) correspondant aux différents mélanges sont présentées sur les figs. 4 et 5 (Proctor normal) est les figs. 6 et 7 (Proctor modifié).

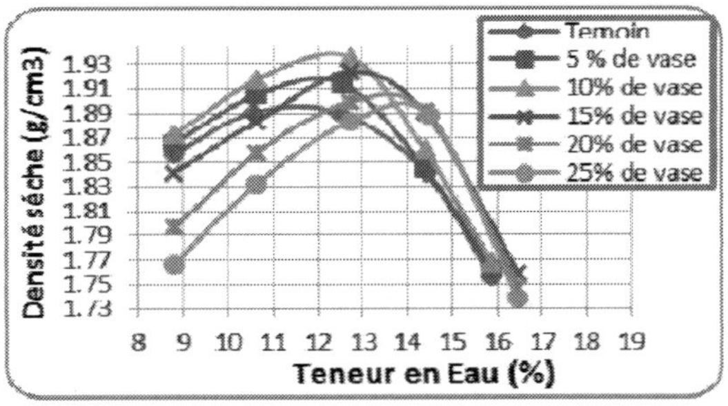

Fig. 4. Courbes de compactage au Proctor normal.

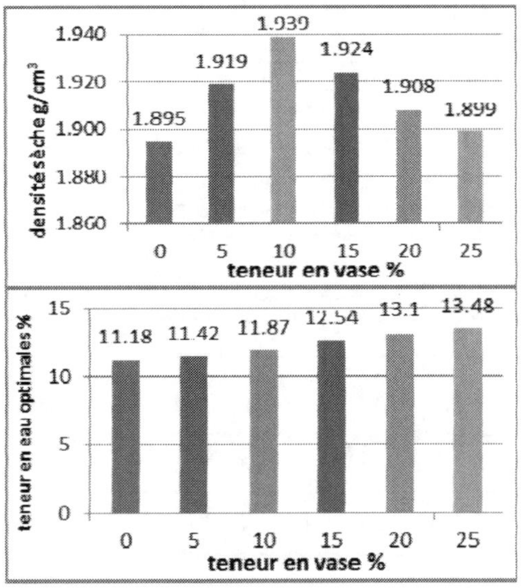

Fig. 5. Caractéristiques de compactage en fonction de la teneur en sédiments (Proctor normal).

https://doi.org/10.1051/matecconf/201814901031

A première vue, nous pouvons constater que les courbes Proctor représentant les différents mélanges se distinguent clairement de celle correspondant aux granulats routiers seuls. Par rapport à cette dernière, nous pouvons remarquer que les courbes représentant les mélanges de 5, 10, 15, 20 et 25% de sédiments sont situées pratiquement du coté humide.

L'évolution de la densité sèche maximale des mélanges étudiés en fonction de leurs teneurs en sédiments est présentée avec plus de précision sur Figs. 4 et 5 (Proctor normal) et Figs. 6 et 7 (Proctor modifié).

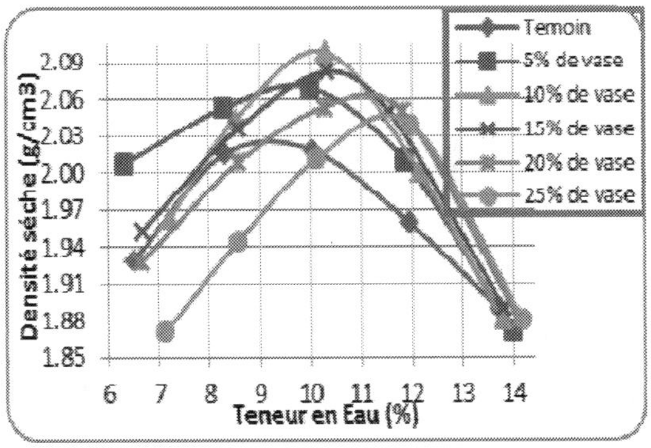

Fig. 6. Courbes de compactage au Proctor modifier.

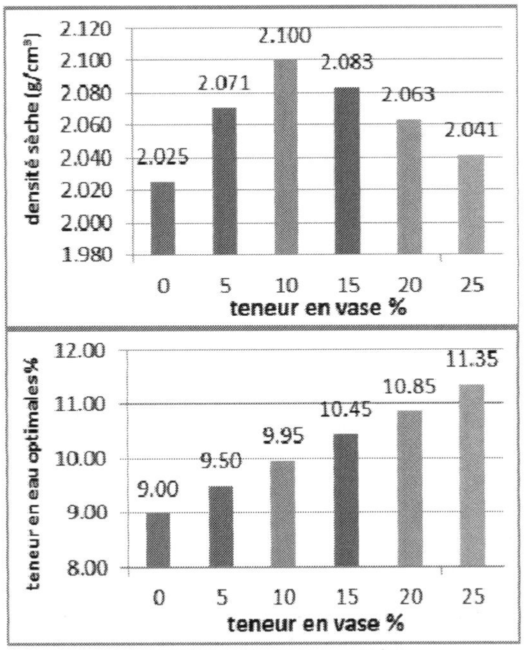

Fig. 7. Caractéristiques de compactage en fonction de la teneur en sédiments (Proctor modifié).

La densité sèche des mélanges s'améliore de plus en plus jusqu'à atteindre une valeur maximale pour une teneur en sédiments de 10%. Au-delà de cette proportion, les mélanges commencent à perdre de leurs densités, toutefois, ces densités restent supérieures aux témoins et pour Proctor modifier reste supérieur a 2, de tels résultats sont assez suffisants pour un matériau routier [8]

Ceci est dû au fait que cette proportion de vase adéquate rentre facilement dans les vides entre les granulats routiers ce qui réduit leur volume (vides) et rend les mélanges plus denses.

L'augmentation de la teneur en eau optimale des mélanges est due à l'accroissement de la surface totale des particules des mélanges par les grains fins de la vase additionnée.

Une proportion supérieure à 10% de sédiments, dépasse le volume des vides existants entres les granulats routiers, la densité sèche elle commence a réduit de plus en plus en fonction de l'ajout de la vase, la raison la diminution de la densité sèche maximum est le faible poids volumique de la vase (1.779 t/m^3) par rapport les granulats routière seul (2.025 t/m^3).

On remarque que l'augmentation de la teneur en sédiments dans les granulats, la courbe de compactage soit à l'OPM ou l'OPN présente de plus en plus une forme accentuée, ce qui explique la sensibilité du mélange à l'eau par rapport à l'état naturel [9].

5.3 Essais de portance immédiats (IPI)

Les courbes représentant les forces en fonction des enfoncements pour les mélanges dont la teneur en sédiments de 5 à 10% se distingue clairement sur la Fig. 8 pour (Proctor normal) fig. 9 (Proctor modifié).

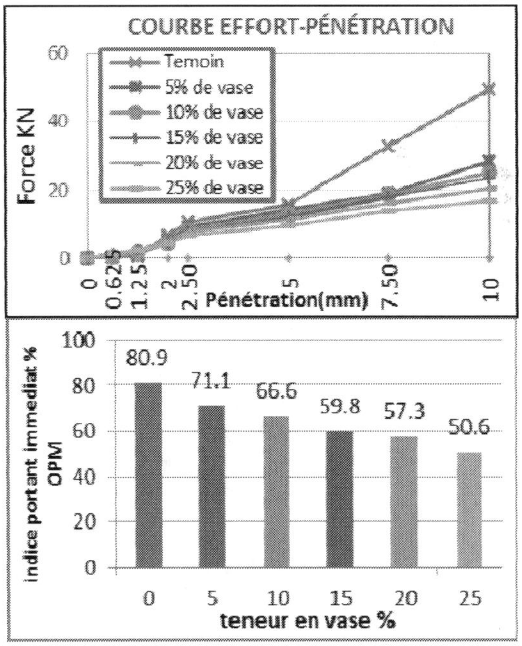

Fig. 8. Effort de pénétration IPI (OPM)

La portance des mélanges empirés dès le premier ajout de sédiments. Néanmoins, les mélanges à 5% de sédiments présentent les portances les plus élevées que les autres mélange (10, 15, 20, et 25%), malgré cela il reste toujours inferieur par rapport des granulats routiers seuls sachant que ces derniers sont constatés les plus denses au cours des essais de compactage.

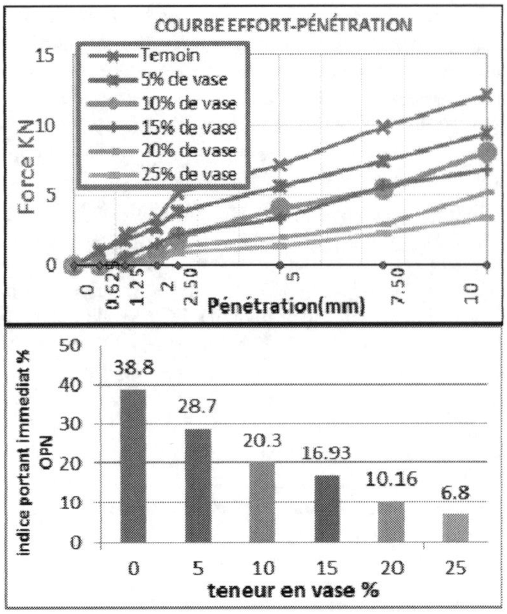

Fig. 9. Effort de pénétration IPI (OPN)

La raison de la diminution de la portance du sol c'est la vase qui entoure les granulats routiers, elle est facile le poinçonnement de plus en plus en fonction de l'ajout des sédiments.

5.4 Essais De Portance CBR Après Immersion

Les essais CBR après immersion réalisés selon la norme NF P 94-078, permettent d'évaluer la portance des mélanges sous la circulation des engins dans les plus mauvaises conditions hygrométriques (Indice CBR après immersion).

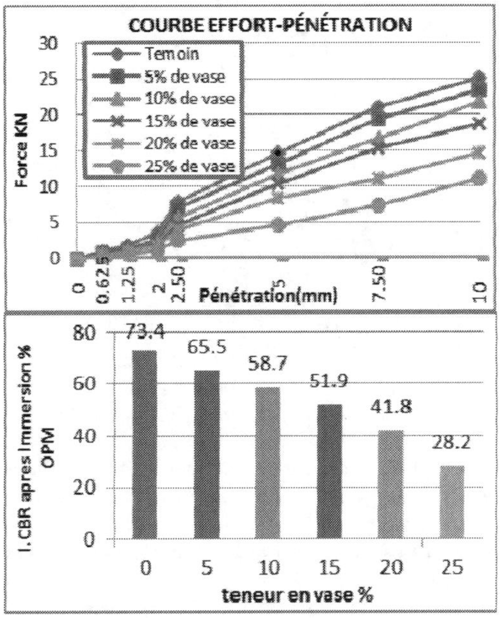

Fig. 10. Effort de pénétration CBR après Immersion (OPM)

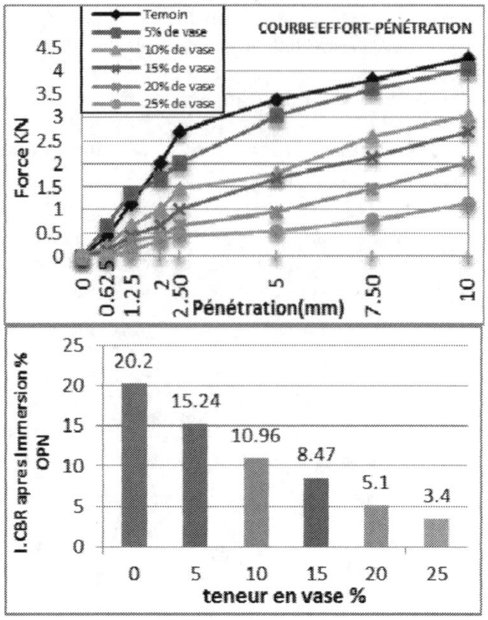

Fig. 11. Effort de pénétration CBR après Immersion (OPN)

La portance des mélanges diminue dès le premier ajout de vase. En effet, les courbes forces enfoncements correspondant aux mélanges (fig. 11 pour Proctor normal et fig. 10 pour Proctor modifié) s'écartent de plus en plus de celle correspondant aux granulats routiers seuls, notamment pour des teneurs en sédiments supérieures à 10% pour le Proctor normal et à 20% pour le Proctor modifier. La vase à l'état humide (immersion) même en faible proportion peut nuire à la portance des mélanges.

En effet, la vase humide présente une très faible résistance au poinçonnement. Les grains entourés de vase évitent facilement les pressions engendrées par le vérin de l'appareillage de poinçonnement en se déplaçant vers les points de faibles pressions, ces déplacements sont de plus en plus favorisés avec l'ajout de vase.

En ce qui concerne le gonflement relatif (G) en présence d'eau ; on peut dire qu'il est négligeable dans tous les cas. En effet le gonflement maximal est enregistrée dans le cas du mélange (granulats routiers + vase), néanmoins, il ne dépasse pas les 0,3% pour le Proctor normal et 0.27% pour le Proctor modifier.

6 Conclusion

Les travaux de recherche présentés dans cet article entrent dans le cadre général de la valorisation sédiments dragué des barrages en Algérie. Notre contribution est essentiellement centrée sur le cas de barrage Fergoug de la wilaya de Mascara ; Cette étude consiste à traiter le comportement des mélanges (granulats routiers + sédiments du barrage de Fergoug) sous sollicitation routière.

L'envasement des retenues de barrages en Algérie a pris une ampleur importante, limitant ainsi leur capacité (diminution importante du volume d'eau stockée). Le dé envasement est un moyen de lutte contre l'envasement néanmoins les conséquences engendrées sont le dépôt sédimentaire et les problèmes liés à son stockage.

L'objectif de valoriser de nouvelles sources de granulats d'une part et de préserver l'environnement d'autre part. Les résultats obtenus ont montré que les mélanges constitués de (granulats routiers +5 à 10% de sédiments) présentent un comportement très satisfaisant sous les différents types de sollicitations appliquées.

Ils peuvent par conséquent, être proposés dans la construction routière, en couches de fondations et en couches de base dans le cas des chaussées de faible à moyen trafic. Ce travail de recherche peut présenter un intérêt capital en géotechnique routière en Algérie, tant sur l'aspect économique et sur l'aspect écologique et environnemental.

Références

1. B. Abdellaoui, A. Merzouk, M. Aberkan, J. Albergel, Revue des sciences de l'eau **15** (2002), 737-748
2. W. Remini, B. Remini, Larhyss Journal **02** (2003) 45-54
3. N. Belas, O. Belaribi, A. Mebrouki, N. Bouhamou, Valorisation des sédiments de dragage dans les bétons, Séminaire International, innovation & valorisation en génie civil & matériaux de construction. INVACO2. Rabat, Maroc, (2011), pp. 23-25
4. A. SEMCHA, Valorisation des sédiments de dragage, Applications dans le BTP, cas du barrage de Fergoug, Thèse, Université de Reims Champagne- Ardenne, (2006).
5. F. Kazi Aoual, M. Ameur, B. Mekerta, A. Semcha, (2014). Caractérisation des sédiments de dragage du barrage de Bouhanifia pour une réutilisation, XIIIèmes Journées Nationales Génie Côtier – Génie Civil, Gestion et valorisation des sédiments marins, Dunkerque, France. pp. 999-1006.
6. Z. Labiod, B. Remini, M. Belaredj. Larhyss Journal, n° 03, Juin 2004, pp 7-12.
7. M. A. Bourabah, Comportement mécanique des sols fins, application à la valorisation des sédiments de barrages en technique routière, Thèse de doctorat en sciences, Université de Tlemcen, Algérie, (2012).
8. P. Autret. (1983). Institut des sciences et des techniques de l'équipement et de l'environnement pour le développement (ISTED), Laboratoire central des ponts et chaussées (LCPC), *pp. 40.*
9. A. Larouci, Y. Senhadji, L. Laoufi & A. Benazzouk, Valorisation de la vase du barrage de Fergoug pour une construction routière. 3rd International Symposium on Materials and Sustainable Development (*ISMSD2017*), 7 & 8 Novembre 2017 – Boumerdes, Algérie.

Influence of recycled polystyrene beads on cement paste properties

Maroua Maaroufi, Akli Younsi, Rafik Belarbi, Armelle Nouviaire
University of La Rochelle, LaSIE UMR CNRS 7356, La Rochelle, France

Abstract. In order to keep up with the requirements of sustainable development, there is a growing interest towards reducing the energy consumption in the construction and rehabilitation of buildings and the promotion of recycling waste in building materials. The use of recycled polystyrene beads in cement-based materials composition constitutes a solution to improve the insulation in buildings. This allows also limiting landfill by reusing the polystyrene waste. The aim of this study is to compare some properties and performances of a cement paste containing polystyrene beads to a reference paste designed with only the same cement. An experimental campaign was conducted and the obtained results showed that adding recycled polystyrene beads to a cement paste improves its hygro-thermal properties. Further studies are however necessary to better understand the real role of the polystyrene beads in the heat and mass transfers.

Introduction

In today's world, the need for quality housing is becoming more pressing, and it has become increasingly common to adopt alternative building materials that use waste or recycled materials, given the overexploitation of natural resources in the construction sector. Moreover, energy consumption in building continues to increase, and another difficulty to overcome is developing suitable materials in order to reduce this consumption. Enhancing the thermal properties of building materials is one way to contribute to improving energy efficiency in buildings.

Lightweight concrete is a good track for alternative building materials in modern construction applications. It greatly improves the technical, economical and environmental aspects of the construction sector. The use of lightweight concrete significantly reduces the load and the weight of the structures. Compared to ordinary concrete, it has many advantages such as: lower density, lower thermal conductivity, better fire resistance, improved durability properties and higher acoustic insulation characteristics. However, this type of concrete is associated with lower mechanical properties and a higher water demand due to the very porous structure of the material. It is generally used to make non-structural insulating items.

Expanded polystyrene (EPS) appears as a good choice to produce lightweight concrete given its good thermal properties, its very low density and its moisture resistance. It is a material commonly used for insulation and packaging of several products for protection purposes due to its flexible structure, impact absorption property and mechanical robustness. It is present in various industrial fields and is most often released into nature after its use. However, being non-biodegradable, it does not disappear by natural means and remains present in nature for many years. For this reason, the recycling of EPS, in its natural form or after some treatments, is beneficial to the economy and for the protection of the environment.

EPS is not poisonous because it does not perform chemical reaction under normal atmospheric conditions. It is not a nutrient for bacteria and fungi. Moreover, EPS is an environmental friendly material as it is recyclable and its components do not give harm to the nature and the ozone layer.

In the recent years, properties of polystyrene concrete were investigated in many studies:

- De Melo [1] found that incorporating EPS aggregates in mortar mixture improves the thermal performances and decreases the mechanical strength.
- Chikhi *et al.* [2] showed that increasing the amount of EPS chips in concrete mixture decreases the density and the thermal conductivity. The authors classify the so-obtained material in the category of capillaro-porous hygroscopic materials.
- Kaya and Kar [3] showed the same positive effect of incorporating EPS in concrete composition. They also found that the water absorption decreases with increasing the amount of EPS in the mixture.

- Li *et al.* [4] and Ferrandiz-Mas *et al.* [5] highlighted that EPS improves the durability of the concrete.

In the present study, two cement-based materials were designed: a lightweight mortar, prepared by mixing EPS beads with cement, and a reference cement paste made with only the same cement. The objective was to assess the influence of EPS on the properties of the so-obtained lightweight mortar and to extend thus the knowledge of polystyrene-based materials by providing informations about the microstructure using the adsorption/desorption isotherms.

1 Materials and samples preparation

Two types of cement-based materials were prepared with a water-to-binder ratio of 0.32: a lightweight mortar containing EPS beads (Fig.1) and a reference cement paste without EPS. Note that EPS contains 98% air and only 2% polystyrene. It has a very low density that varies between 20 and 30kg/m^3 [6]. It is a thermoplastic material that has closed pores. EPS does not absorb liquid water.

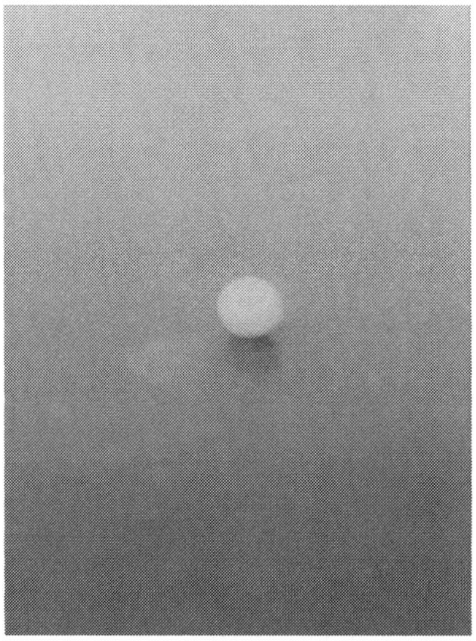

Fig. 1. Polystyrene bead

After casting in different molds, both the lightweight mortar and the reference cement paste were stored in endogenous conditions for 28 days before characterization.

2 Experimental program

2.1 Compressive strength

Compressive strength tests were conducted on 4X4X16cm specimens after 28 days in endogenous conditions. Tests were carried out in a testing machine at a loading rate of 2.4kN/s (Fig. 2).

Fig. 2. Compressive strength testing machine

2.2 Thermal conductivity

The λ-Meter EP500e (Fig. 3) was used to measure the thermal conductivities. The apparatus operates according to the guarded hot plate method described in EN12664 and EN 12667 standards. This is the most commonly used method for measuring low values of thermal conductivity ($\lambda \leq 1 W/m^2.K$), i.e. for insulating materials.

The sample is put between a hot plate and a cold one, and is submitted to a constant temperature gradient. The thermal conductivity is then calculated as follows:

$$\lambda = \frac{Q.d}{A.\Delta T}$$

with Q(W) the electrical power of the measuring heating; d(m) the thickness of the sample; A(m^2) the transfer area and ΔT(K) the temperature gradient.

In our case, 5.3X15X15cm samples were used.

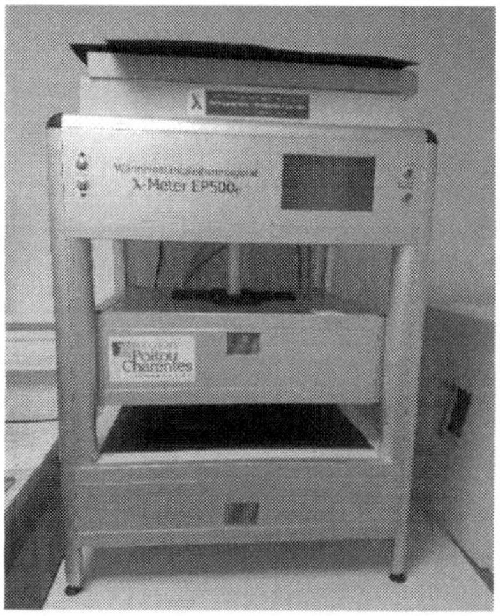

Fig. 3. λ-Meter EP500e

https://doi.org/10.1051/matecconf/201814901032

2.3 Heat capacity

The heat capacity expresses the heat storage capacity of the material. It depends on temperature and was measured using a calorimeter Calvet BT 2.15 (Fig. 4). This technique (3D technique) is more accurate than micro-calorimetry or DSC ones. According to Calvet approach, the sample is completely surrounded by thermocouples. Such a design guarantees the measurement of the total heat exchange between the sample and the surroundings through the calorimetric detector.

Fig. 4. Calorimeter Calvet BT 2.15

2.4 Total porosity

The total porosity was measured according to the French NF P18-459 standard, which is based on the water saturation of the samples under vacuum. However, this method is not suited to lightweight materials, especially those with polystyrene, which is hydrophobic and does not absorb water. For the determination of the polystyrene cement paste porosity, we used an empirical equation established by Kaya and Kar [3].

2.5 Adsorption and desorption isotherms

There are several techniques and devices for measuring water vapor sorption isotherms. The oldest one is the traditional gravimetric method, using saturated salt solutions to regulate the relative humidity. This method requires a long time to provide results. So we used this method with smaller samples and automatic humidity regulator using Vapor Soption Analyser (VSA, Decagon) shown in Fig. 5. Another device was used based on volumetric method (Belsorp Aqua3 in Fig. 6).

Fig. 5. Vapor sorption analyser (VSA)

Fig. 6. BELSORP aqua 3

3 Results and discussion

3.1 Compressive strength

As expected, the compressive strength decreases when EPS beads are added (Fig. 7). Indeed, the 28-day compressive strength of the reference cement paste is equal to 22MPa while it is equal to only 4MPa for the lightweight mortar. This result is in good agreement with the literature since lightweight aggregates, and polystyrene in particular, are known to alter the mechanical properties of cementitious materials.

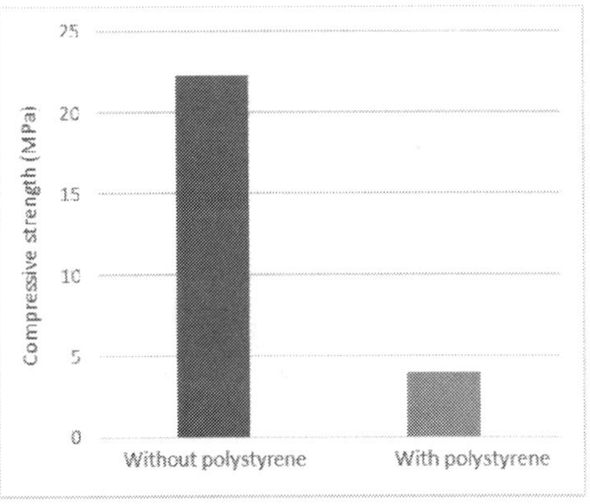

Fig. 7. 28-days compressive strength

3.2 Thermal conductivity

Fig. 8 shows that the thermal conductivity decreases considerably with the addition of EPS beads. At 23°C, the thermal conductivity is equal to 383mW/m°K for the reference cement paste, and it goes down to 142mW/m°K for the lightweight mortar. This can be explained by the insulating behavior of the EPS beads that contain 98% of air and that have a very low thermal conductivity (λ=0.04W/m°K [7]).

Our polystyrene cement paste (lightweight mortar) has a good thermal insulating performance compared to some values from the literature (Table 1).

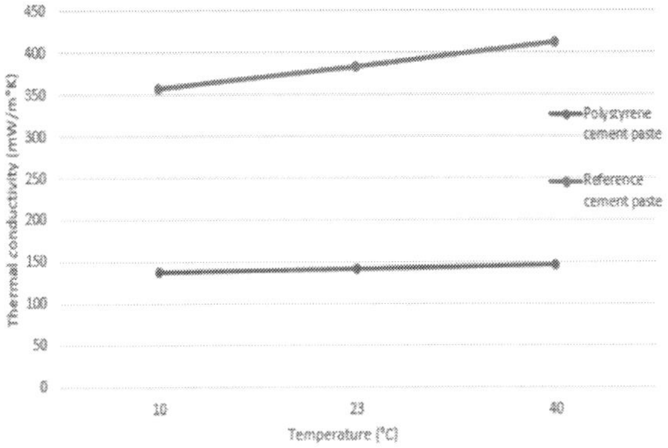

Fig. 8. Thermal conductivity

3.3 Heat capacity

Table 1 shows that the presence of EPS beads improves the specific heat capacity of the cement paste. This means that the lightweight mortar has a better heat storage capacity, and confirms the thermal insulating behavior of this material.

3.4 Total porosity

The obtained water porosities after 28 days in endogenous conditions are given in Table 1. The results show that the porosity increases when EPS beads are added. This is due to the high porosity of the beads which is assessed to be equal to 95%. Note however that the porosity of the polystyrene cement paste was obtained using an empirical equation. Therefore, we should find a more suitable technique to determine the porosity of this kind of lightweight material.

Material	Thermal conductivity (mW/m°K) at 23°C	Heat capacity (J/Kg°K) at 20°C	Total porosity (%)	Literature
Polystyrene cement paste	142	920 - 1030	57	Present work
Reference cement paste	383	900 - 920	33	Present work
Cork based cement paste	270		37.5	[1]
Polystyrene cement paste	230			[1]
Cork based mortar	190 - 320			[8]
Polystyrene foamed concrete	85 - 156			[9]
Polystyrene concrete	930	950	60	[2]
Cork gypsum composite	110 - 124	1267 – 1287		[11]
Hemp concrete	124	1000	76	[12]

Table 1. Properties of our materials and others from literature

3.5 Adsorption and desorption isotherms

The water vapor adsorption/desorption isotherms obtained with the volumetric method (Belsorp) are given in Fig. 9. The two isotherms exhibit the same behavior and tendency. This similitude is due to the hydrophobic property of the EPS beads and to their closed poral structure. The beads do not adsorb water vapor and have no impact on the isotherms.

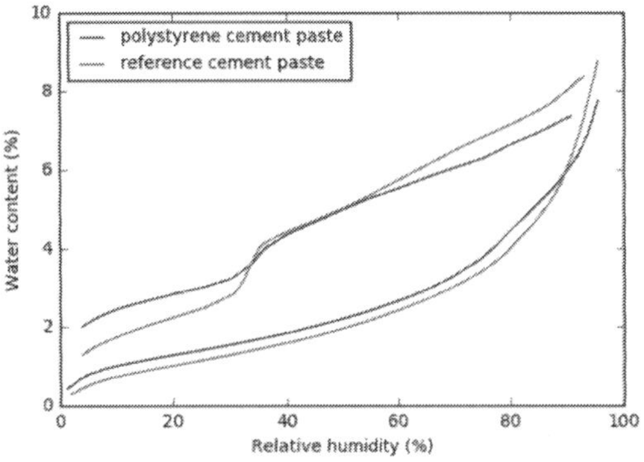

Fig. 9. Adsorption and desorption isotherms (BELSORP)

Sorption isotherms curves of two experimental techniques: the gravimetric method (VSA) and the volumetric method (BELSORP aqua 3) are shown in Figure 10. A good agreement between these results is observed.

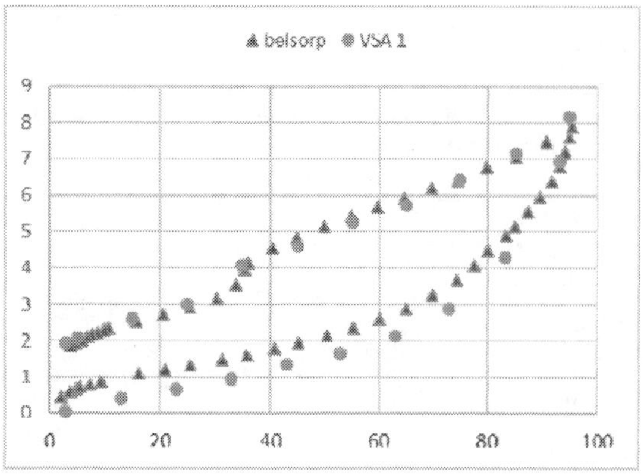

Fig. 10. Comparison of polystyrene cement isotherms with the different techniques

Conclusion

The objective of this work was to assess the influence of the polystyrene beads on the mechanical and hygrothermal properties of cement pastes, along with their microstructure and porosity. The main conclusions are:

- The compressive strength decreases by adding the polystyrene beads, which are lightweight aggregates and composed of 98% of air.
- Thermal conductivity of the polystyrene cement paste is lower than the reference cement paste, due to the insulating behavior of the polystyrene beads.
- The specific heat capacity is higher for the polystyrene cement paste, because the polystyrene has a good heat storage capacity.
- The presence of the polystyrene beads highly increases the total porosity of the cement paste, because of their high porosity.
- The water vapor isotherms do not show a notable impact of the beads on the cement pastes, because of the hydrophobic nature and the closed poral structure of polystyrene.

Acknowledgements

The authors would like to express their sincere thanks to Kamilia Abahri and Rabah Hamzaoui for their kind support and helpful discussions.
This work was funded by the Region Aquitaine, Limousin, Poitou-Charentes through European and National Program CPER-FEDER "Bâtiment Durable 2015-2020".

References

1. Humberto Fernandes Sequeira Borges de Melo, master degree report, technical university of Lisbon, Portugal (2014)
2. A. Chikhi, A. Belhamri, P. Glouannec, A. Magueresse, Journal of building engineering **7**, 183-193 (2016)
3. A. Kaya, F. Kar, Construction and building materials **105**, 572–578 (2016)
4. Y. Li, N. Liu, B. Cheng, Materials and structures **48**, 269–276 (2013)
5. V. Ferrándiz-Mas, E. García-Alcocel, Construction and building materials **46**, 175–182, (2013)
6. P.L.N. Fernando, M.T.R. Jayasinghe, C. Jayasinghe, Construction and building materials **139**, 45–51, (2017)
7. M. Y. Ferroukhi, K. Abahri, R. Belarbi, K. Limam, A. Nouviaire, Heat and mass transfer **52**, 2257–2269 (2015)
8. A. Moreira, J. António, A. Tadeu, Cement and Concrete Composites **49**, 1–8 (2014)
9. A. A. Sayadi, J. V. Tapia, T. R. Neitzert, G. C. Clifton, Construction and building materials **112**, 716–724 (2016)
10. C. Ma, B. Chen, Construction and building materials **123**, 106–114 (2016)
11. A. Cherki, B. Remy, A. Khabbazi, Y. Jannot, D. Baillis, Construction and building materials **54**, 202–209 (2014)
12. D. Lelièvre, PhD thesis, University of Bretagne-Sud (2016)
13. Ecomatériaux et isolation, Construction et travaux publics | Techniques du bâtiment : le second œuvre et les lots techniques (2012)

Thermal characteristics study of light aggregate concrete of the middle Atlas pozzolana

Etude des caractéristiques thermiques de bétons de granulats légers de la pouzzolane du moyen Atlas

A. Bouyahyaoui[1], M. Cherkaoui[1,2], My L. Abidi[1], T. Cherradi[1]
[1]Département de Génie Civil, EMI, Université Mohamed V, Rabat, Maroc
[2]Ecole Nationale Supérieure des Mines de Rabat (ENSMR), Rabat, Maroc

Abstract : We present in this work an experimental study of the thermomechanical characterization of pozzolan (volcanic rock widespread in the region of Azrou in the Middle Atlas in Morocco). The objective of this work is to evaluate the effect of the pozzolan on the thermal conductivity of pozzolanic concrete by using different concrete formulations and this, by substituting various percentages of gravel and usual sands with gravel and sand of the pozzolan. The thermal properties of the samples consist to determine the thermal conductivity and the heat capacity. The results obtained show that the pozzolan is a local material with a high thermal resistance and can be used as a good insulated material in heat transfer and energy efficiency in the building.

Résumé : Dans ce travail, nous présentons une étude expérimentale portant sur la caractérisation thermomécanique de la pouzzolane (roche volcanique très répandue dans la région d'Azrou au Moyen Atlas au Maroc). L'objectif de ce travail est d'évaluer l'influence de la pouzzolane sur la conductivité thermique du béton pouzzolanique en utilisant différentes formulations de béton et ce, en substituant des pourcentages variés de graviers et sables usuels par des graviers et sables à base de la pouzzolane. La caractérisation thermique des échantillons a consisté à déterminer la conductivité thermique et la capacité calorifique. Les résultats obtenus ont montré que la pouzzolane est un matériau local qui présente une résistance thermique importante et peut être considérée comme un bon isolant en efficacité énergétique et thermique dans le secteur du bâtiment.

1 Introduction

Dans le cadre de la transition énergétique, la réglementation thermique de la construction marocaine appliquée depuis 2015, exige que les déperditions dans le bâtiment n'excèdent pas un certain seuil selon le zonage climatique du pays [1-2]. Pour se conformer à ce règlement, deux possibilités peuvent être envisagées, soit d'augmenter les épaisseurs des éléments de l'enveloppe et des séparations, soit de travailler avec des matériaux de faible conductivité thermique. La première solution reste difficile à appliquer pour les maisons de petite surface vu qu'elle entraine une consommation importante d'espace et à fortiori un prix de revient du produit très couteux. De ce fait, la deuxième solution semble être plus acceptable. Dans cette optique, nous proposons d'étudier un matériau de construction élaboré à base de pouzzolane.

La pouzzolane est une roche volcanique du moyen Atlas marocain qui se définit comme un matériau silico-alumineux. Elle se caractérise par sa légèreté et sa grande porosité qui est de l'ordre de 45% [2].

L'objectif de notre travail est de confectionner un béton et un mortier contenant de la pouzzolane et d'évaluer et comparer leurs caractéristiques thermophysiques (la conductivité thermique et la capacité calorifique) à celles du béton et du mortier usuels.

2 Etude expérimentale

Notre étude expérimentale a consisté en l'étude de la conductivité thermique et la capacité calorifique d'échantillons à base de pouzzolane.

2.1 Mesure de la conductivité thermique : méthode du plan chaud

Plusieurs méthodes sont utilisées pour la mesure de la conductivité thermique des matériaux, il s'agit notamment de [3-4], [5-6] :

Plaque Chaude gardée ;
Méthode Fluxmétrique ;
Méthode Flash ;
Fil Chaud ;
Plan Chaud ;
Source Plane Transitoire ;
Source plane modifiée.

Vu la nature de notre matériau et la disponibilité des appareils de mesures, notre calcul a été effectué par la méthode du plan chaud [7-8].

Les échantillons étudiés ont été confectionnés en substituant un pourcentage des graviers usuels par des graviers pouzzolaniques, et une partie du sable de mer par du sable pouzzolanique

Pour la simplification, nous avons utilisé les abréviations suivantes :

- BnGPZ : Béton contenant n% des graviers pouzzolaniques à la place des graviers usuels.
- BnSPZ: Béton contenant n% du sable pouzzolanique à la place du sable de mer.
- MnSPZ: Mortier contenant n% du sable pouzzolanique à la place du sable de mer.

Les mélanges étudiés sont : B25GPZ, B40GPZ, B100GPZ, B40SPZ, M50SPZ et M100SPZ.[2-3]

Les échantillons ont été préparés selon les dimensions normalisées des films chauds, à savoir :

- une forme cylindrique : rayon=10 ±0,4 cm et épaisseur=2±0,4 cm
- une forme parallélépipédique : 10×10×2 cm3

Etant donné que la conductivité thermique est influencée par l'humidité, les échantillons ont été placés dans l'étuve à 70°C jusqu'à ce que leur masse se stabilise.

Technique de mesure

La méthode est basée sur la création d'un flux de chaleur unidirectionnel passant par l'échantillon et le polystyrène à l'aide d'une résistance thermique. Le montage expérimental se compose des éléments suivants :

Un polystyrène, isolant placé au-dessous de la résistance thermique, permet de maximiser le flux passant à travers l'échantillon.

Deux blocs d'aluminium, utilisés pour réduire le temps de la phase transitoire profitant de sa grande conductivité.

Trois thermocouples pour la mesure de la température dans des points particuliers (Fig. 1.).

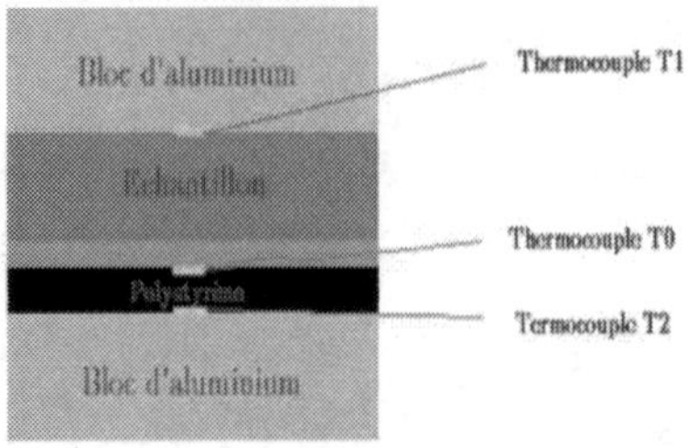

Fig. 1. Méthode du plan Chaud

En effectuant le bilan énergétique de l'échantillon :

Flux induit par effet de Joule

$$\phi_J = RI^2 = R \times \left(\frac{U}{R}\right)^2 = \frac{U^2}{R} \qquad (1)$$

Avec
U : tension aux bornes de la résistance chauffante (V)
R : Résistance chauffante (Ω)

Flux de conduction traversant l'échantillon

$$\phi_E = \frac{\lambda_1 S}{e_1}(T_0 - T_1) \qquad (2)$$

Avec
λ_1 : La conductivité thermique de l'échantillon
S : L'aire de l'échantillon
e_1 : Epaisseur de l'échantillon
T0 : Température de la plaque chauffante
T1 : Température entre l'échantillon el le bloc d'aluminium

Flux de conduction traversant le polystyrène

$$\phi_P = \frac{\lambda_2 S}{e_2}(T_0 - T_2) \qquad (3)$$

Avec
λ_2 : La conductivité thermique du polystyrène
S : L'aire du polystyrène
e_2 : Epaisseur du polystyrène
T0 : Température de la plaque chauffante
T2 : Température entre le polystyrène et le bloc d'aluminium

En utilisant les équations (1), (2) et (3), on trouve que :

$$\phi_J = \phi_E + \phi_P$$

Alors

$$\frac{U^2}{R} = \frac{\lambda_2 S}{e_2}(T_0 - T_2) + \frac{\lambda_1 S}{e_1}(T_0 - T_1) \qquad (4)$$

Finalement

$$\lambda_1 = \frac{e_1}{T_0 - T_1}\left[\frac{U^2}{RS} - \frac{\lambda_2}{e_2}(T_0 - T_2)\right] \qquad (5)$$

2.2 Mesure de la capacité calorifique de la pouzzolane en poudre : Méthode du calorimètre différentiel

La capacité thermique massique à pression constante Cp, est l'une des grandeurs thermodynamiques les plus importantes. Elle permet de déterminer les variations d'enthalpies en fonction de la température pour une phase donnée, le pouvoir de stockage de l'énergie thermique par la chaleur sensible et de déterminer implicitement la diffusivité et l'effusivité thermique d'un matériau.

La pouzzolane a été broyée sous forme de poudre pour la préparation des échantillons de 21 mg.

Technique de mesure

L'appareil utilisé, TA Instrument Q100, constitué de deux enceintes chauffantes et identiques, l'une contient l'échantillon à analyser et l'autre la référence (fig. 2 et fig. 3). Lors de l'utilisation de la DSC (Calorimétrie Différentielle à Balayage), une augmentation linéaire de la température est imposée ; on enregistre alors la différence de puissance électrique W(t) fournie aux deux enceintes (fours) afin de les maintenir en permanence à une température identique (T1(t) = T2(t)) [9-10]

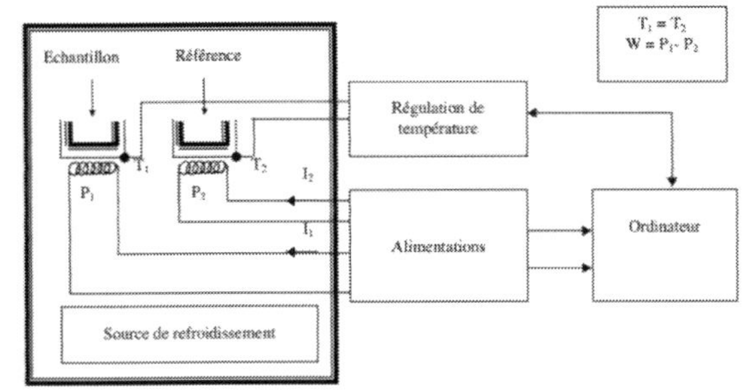

Fig. 2: Principe de mesure du calorimètre différentiel

Fig.3. Calorimètre différentiel

Principe de la détermination du Cp

L'énergie fournie au corps

$$Q = \Delta H = m\, C_p \Delta T$$

$$\frac{dH}{dt} = mC_p \frac{dT}{dt}$$

$\frac{dH}{dt}$: correspond au flux de chaleur que mesure la DSC

$\frac{dT}{dt}$: correspond à la vitesse de montée en température réelle lors de l'essai

$$\text{Donc} \quad C_p = \frac{\frac{dH}{dt}}{m\frac{dT}{dt}}$$

3 Résultats

Les résultats de l'étude de la conductivité thermique des échantillons sont résumés dans les tableaux ci-dessous :

Table 1. Conductivité thermique du béton pouzzolanique

Echantillon	Conductivité thermique (W/K.m)
B25GPZ	0,985
B40GPZ	0,808
B100GPZ	0,528
B40SPZ	1,229

Table 2. Conductivité thermique du mortier pouzzolanique

Mélanges	Conductivité thermique (W/K.m)
M50SPZ	0,506
M100SPZ	0,289

Nous constatons que l'incorporation du gravier pouzzolanique dans le béton permet de réduire la conductivité thermique. Elle passe de 0,985 W/K.m pour un béton contenant 25% de gravier pouzzolanique (B25GPZ) à 0,528 W/K.m pour un béton contenant uniquement du gravier à base de pouzzolane (B100GPZ) (Table 1).

De même pour le mortier, la conductivité thermique est passée de 0,506 W/(°K.m) pour M50SPZ à 0,289 W/(°K.m) pour M100SPZ (Table 2).

En ce qui concerne la capacité calorifique, nous remarquons qu'elle augmente en fonction de la température. Elle passe de 765,7 (J/K.Kg) pour une température de -10°C à 1096 (J/K.Kg) pour une température de 290°C (Fig. 4)

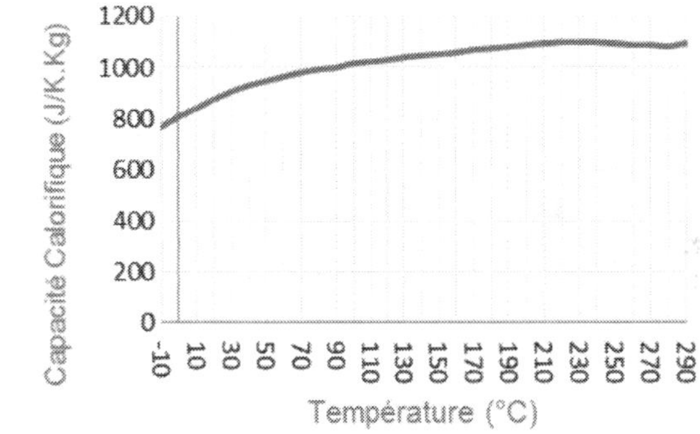

Fig. 4. Capacité calorifique en fonction de la température.

4 Discussion

La pouzzolane se caractérise par une grande porosité, due aux nombreuses bulles d'air qu'elle contient. Son incorporation dans le béton et le mortier permet de réduire la conductivité thermique. Cette réduction est de l'ordre de 76%, elle passe de 2,2 W/(°K.m) pour un béton usuel à 0,528 W/(°K.m) pour un béton à gravier entièrement à base de pouzzolane. Pour le mortier, elle diminue d'environ 78%. Elle passe de 1,3 W/(°K.m) pour un mortier usuel à 0,289 W/(°K.m) pour un mortier à sable entièrement à base de pouzzolane ; ce résultat est conforme aux études bibliographiques [11].

En ce qui concerne la capacité calorifique, elle est de l'ordre de 900 J/°K.Kg pour des températures d'utilisation dans le secteur du bâtiment (-10 °C à 45 °C). Ce qui est en en concordance avec les résultats de la littérature [12].

Le comportement mécanique de ces mélanges a montré une faible résistance à la compression par rapport au béton usuel ce qui limite leur utilisation aux éléments secondaires non structuraux [2].

L'introduction de ces matériaux dans la couche de forme et la couche de protection de l'étanchéité permet d'amortir la variation de la température de la face interne et d'augmenter le déphasage entre les pics des températures.

5 Conclusion

La pouzzolane du moyen Atlas peut jouer un rôle très important dans la politique de la transition énergétique. Son utilisation permet de réduire les déperditions thermiques à travers les éléments de l'enveloppe du bâtiment : les ponts thermiques et la couche d'enduit des façades internes et externes (le mortier).

Enfin, il est recommandé d'étudier le volet économique de ces solutions et de les comparer à l'utilisation d'un isolant thermique existant sur le marché.

Références

1. Réglementation thermique de la construction marocaine RTCM, Novembre 2015
2. A. Bouyahyaoui, M. Cherkaoui, My L. Abidi, T.Cherradi Mechanical and chemical characterisation of pozzolan of middele atlas in morocco accepted, International Journal of GEOMATE, Jan., 2018 Vol.14, Issue 41, pp.126-134
3. B. Ranaivo Rabehaja, Etude des caractéristiques thermique et mécaniques de béton de pouzzolane, Thèse de doctorat, université paris 7, décembre 1986
4. A. Degiovanni, M. Laurent., Nouvelle technique d'identification de la diffusivité thermique pour méthode flash, Journal of Applied Physics, Vol. 21, n°3, pp. 229-237, 1986.
5. Y. Jannot, P. Meukam, Simplified estimation method for the determination of thermal effusivity and thermal conductivity with a low cost hot strip, Measurement Science and Technology, Vol. 15, pp. 1932-1938, 2004.
6. S.E. Gustafsson, Transient plane source techniques for thermal conductivity and thermal diffusivity measurements of solid materials, Rev. Sci. Instrum., Vol. 62, pp.797- 804, 1991.
7. Sébastian Gauthier, Caractérisation thermique de la matière par la méthode 3ω, thèse de doctorat Université Languedoc Roussillon 2012
8. N. OZISIK : *Heat conduction.* John Wiley, Sons inc ., 1993.
9. Jean Grenet, Bernard Legendre, Analyse calorimétrique différentielle à balayage (DSC) techniques de l'ingénieur 10 déc. 2010
10. H.G.E. Hohne, W.F Hemmnger Differential scanning calomerty. Springer 2003
11. S.Azizi, C.Moyne, A.Degiovanni Experimental And Theoretical Approach Of The Thermal Conductivity Of Wet Porous Media International Journal Of Heat And Mass Transfer Volume 31, Issue 11, November 1988, Pages 2305-2317
12. P. Meukam, Y.Jannot, A. Noumowe, T.Ckofane Thermo Physical Characteristics Of Economical Building Materials Construction And Building Materials Volume 18, Issue 6, July 2004, Pages 437-443

Effect of reinforcement fibers on the collapse potential of clayey sands

Souhila Adjabi[1], Mohamed Salah Nouaouria[1] & Cherif Betehi[1]
[1]Laboratory of Civil Engineering and Hydraulics, University 8th May 1945 Guelma, Algeria.

Abstract. The collapse of soils under wetting is a major problem in Geotechnical engineering. The erection of structures on these types of soils, located in arid and semi-arid zones, needs careful treatment of these soils. Soil reinforcement techniques have been rapidly increased during these two decades because of their effectiveness in geotechnical engineering. The aim of this experimental work is to investigate the collapsible soil behaviour in order to improve its characteristics. To achieve this goal, Polyethylene fibers, and Sisal fibers were used as Polyethylene fibers content in mass are varied from 0% (unreinforced samples) to 15%; and Sisal fibers content from 0.5% to 1% . The fiber reinforcement is combined with other processing procedures such as compaction and the addition of CPA cement to decrease the collapse potential.

Keywords: Collapse potential, Polyethylene fiber, Sisal fiber

1 Introduction

Construction of buildings and other civil engineering structures on collapsible soils is highly risky because such soils are susceptible to settle suddenly when are loaded and wetted. It is quite common to find soils that do not fulfill the basic necessities of an engineering project technically, economically or in terms of timing. Economic issues have stimulated interest in the development of alternative materials that fulfill design specifications. The well-established techniques of soil stabilization and soil reinforcement are often used to obtain improved geotechnical materials, either through the addition of cementation agents or through the inclusion of randomly distributed discrete elements such as fibers. Stabilized and reinforced soils are composite materials that result from the combination and optimization of the properties of individual constituent materials.

This paper describes a study of the mechanical behaviour of Kaolinite-sand mixed samples reinforced with randomly distributed polyethylene fibers, and Sisal fibers under oedometer loading conditions. The specific objectives of the present work are to evaluate the effect of fiber insertion on the collapse potential of sand samples mixed with Kaolinite in different proportions such as: 0 %, 5%, 10%, 15%, 20%, 25% and 30% respectively, to determine the optimum content of Kaolinite giving the maximum potential collapse.

In its quest to get healthier drinking water, millions of people are using bottled water. The annual global consumption of bottled water has reached billions of liters. This requires the production of innumerable quantities of bottles which are generally thrown in the nature. This act is harmful to the environment. Only about 20% of plastic bottles are recycled; 80% end up in landfills, or in nature. Plastic bottles begin to decompose in nature only after hundreds of years. In order to reduce the quantity of plastic bottles, we undertake this study based on the use of this material to treat the collapsible soil. This is in order to protect the environment and to address this type of soil problems by finding economic solutions.

The objective of this paper is to determine the effect of randomly distributed short polyethylene-fiber, and Sisal fibers, respectively, on the collapse behaviour of a clayey sand material. A series of compression tests were carried out in the Oedometer apparatus, on soil samples made of sand, fibers and Kaolinite. Various lengths of fibers were respectively used, such as: 5mm, 10mm and 25 mm.

Soil was reinforced with polyethylene fibers contents of 5%, 10% and 15% respectively. For each content three different lengths were used (5 mm, 10mm and 25 mm respectively), and 0.5%, 1% (5mm, 10mm and 25mm long) for Sisal fibers.

2 Literature review

Since the beginning of 1970s, several investigators have studied stress-strain characteristics of reinforced soil using triaxial, direct shear, and plane strain tests. From 1977, extensive experimental work has been performed on geotextile-reinforced sand. Incorporating reinforcement inclusions within soil is also an effective and reliable technique in order to improve the engineering properties of soil. Many investigators have used various types of fibers under different test conditions.

Maher and Ho (1993) evaluated the effect of randomly distributed fiber reinforcement on the response of cemented sand to load. Test results indicated that fiber reinforcement have significantly increased the compressive and shear strength of cemented sand.

Ayadat and Dahili (1998) improved the properties of a collapsible soil, the bituminous treatment of such soil at a depth of less than 4m has been examined, the results indicated that for obtaining non collapsible soil, a minimum water content of 4%, a bitumen content between 10 and 12 % and a high compacting energy (more than 80 blows by layer for a total of two layers in oedometer ring) have been applied.

Consoli et al. (2003a) proposed a field application for soil reinforcement for increasing the bearing capacity of spread foundations when placed on a layer of fiber-reinforced cemented sand built over a weak residual soil stratum.

Yetimoglu and Salbas , (2003), investigated the effect of fiber reinforcement content on the shear strength of soil. The results of the tests indicated that the peak shear strength and the initial stiffness of the sand were not affected significantly. However the fiber reinforcement could reduce soil brittleness providing smaller loss of post-peak strength, and increases the residual shear strength angle of the sand.

Hamidi and Hooresfand (2013) used the conventional triaxial compression tests to determine the effect of cement and polypropylene fiber reinforcement on sand, the addition of fibers increased peak and residual shear strength of cement treated soil and changed its brittle behaviour to a more ductile one.

Botero and Ossa (2015) studied the mechanical behaviour of a silty soil that was reinforced with randomly distributed PET fibers. The results indicated that the reinforced specimens presented increase of shear strength that was associated with the increasing quantities of the PET fiber.

Chegenizadeh and Nikraz (2012) investigated the effect of paper inclusion on the modulus of elasticity of subgrade material; the results indicated that the increase of paper percentage in mass slightly increased the modulus of elasticity.

Al Adili and Azzam (2011) studied the effect of soil reinforced with randomly included papyrus fiber on the strength behaviour. The results of these tests have shown a significant improvement in the failure deviator stress and shear strength parameters of the soil with 10% of papyrus fiber. This addition also reduced the deformation of the soil under loading.

Chauhan and Mittal (2008) investigated the effect of reinforcement with coir fiber and synthetic fiber in the strength of subgrade soil, they reported that the permanent and resilient strains in all materials decreased with confining pressure but increased with the number of load cycles and deviator stress in reinforced and unreinforced conditions.

In comparison with conventional geosynthetics (strips, geotextiles, geogrids, etc.), there are some advantages in using randomly distributed fibers as reinforcement. Firstly, the discrete fibers are simply added and mixed randomly with soil, in the same way as cement, lime, or other additives. Secondly, randomly distributed fibers limit potential planes of weakness that can develop parallel to oriented reinforcement.

3 Materials, equipment and experimental program

3.1 Materials

3.1.1 Sand

The soil used in this investigation was a Kaolinite and sand with the grain size distribution curves shown in Figure 1.

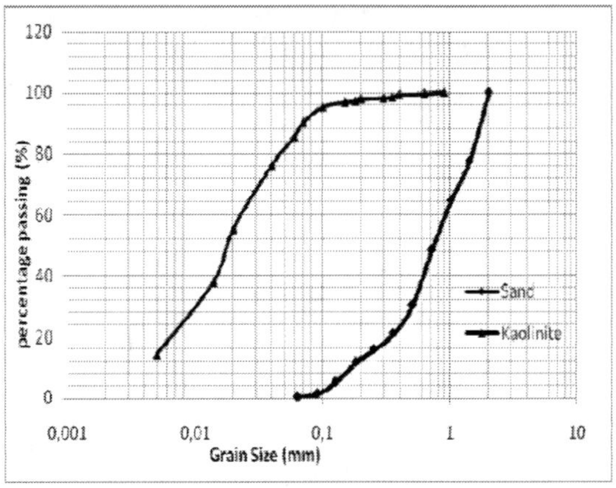

Fig. 1 Particle size distribution curve of sand and Kaolinite

3.1.2 Polyethylene fiber

Polyethylene (abbreviated PET) is the most common plastic material. The annual global production is approximately 80 million tones. Its primary use is in packaging (plastic bag, plastic films, geomembranes, containers including bottles, etc.). Many kinds of polyethylene are known, with most having the chemical formula $(C_2H_4)nH_2$. Thus, PET is usually a mixture of similar organic compounds that differ in terms of the value of n.

3.1.3 Sisal fibers

Sisal, as shown in Figure 2, is a natural fiber fully biodegradable. It is highly renewable resource of energy.

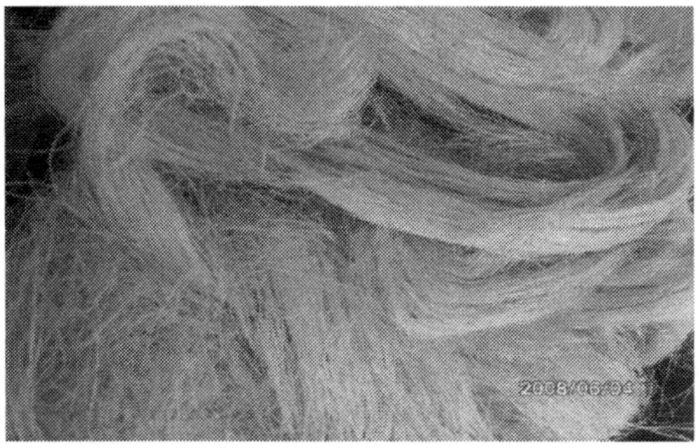

Fig. 2 Sisal fibers

3.2 Equipment

3.2.1 Compacting Device

The compacting device first designed by Ayadat et al. (1998), is consisted of a vertical stem of 200 mm long and a diameter of 12.2 mm. It is welded to a horizontal disc of 5mm thick and a diameter of 50.2 mm. A disc-shaped hammer of 136g, 16,4mm thick and a diameter of 39.2 mm, with a hole of 12.25 mm.

3.2.2 Oedometer

The Oedometer is a laboratory apparatus used in geotechnical engineering. This instrument is generally used to measure the compressibility of a soil sample, preferably intact, under a given applied vertical effective pressure. The operation of loading and unloading is carried out stepwise and is timed in order to determine the compressibility properties of soil.

3.3 Oedometer test:

3.3.1 Sample preparation

In this investigation, the collapse behaviour is assessed with an initial water of 4%.
The first series of tests were conducted on seven reconstituted samples from sand and Kaolinite in various proportions in mass of Kaolinite/Sand, such as: 0/100, 5/95, 10/90, 15/85, 20/80, 25/75 and 30/70 as shown in Table 1.

Table 1 Reconstituted samples

Designation	S1	S2	S3	S4	S5	S6	S7
Kaolinite (%)	0	5	10	15	20	25	30
Normalized Sand (%)	100	95	90	85	80	75	70
Water content (%)	4	4	4	4	4	4	4

The second set of hydrocollapse tests were undertaken on reconstituted samples of sand and Kaolinite. Seven grain size classes of sand are used together with the same content of Kaolinite of 25% and water content of 4%. The reason of taking 25% of Kaolinite is that this content has given the maximum collapse potential in the first set of tests. Based on the tests results to investigate the effect of the Kaolinite content on the collapse potential, it was noted that the Kaolinite content 25% has clearly given maximum collapse potential.
The soil reinforcement was provided by sisal fibers and polyethylene fibers, respectively, in various ratios by mass of fiber/soil, such as: 5%, 10% and 15%. Three different lengths were used: 5mm, 10mm, and 15 mm. The sand is well mixed with Kaolinite and water content of 4 % in order to obtain homogeneous samples. The sample is then placed into the oedometer ring in two layers.
The hydrocollapse tests were carried out based on the procedure of (Jennings and Knight, 1975).
According to Ayadat et al. (1998), the compaction is made with dropping, hammer from a height of H = 150 mm. This hammer slides freely along the vertical rod until it hits the disk that transmits the shock to the sample.

4 Results and discussions

4.1 Effect of Kaolinite content on the collapse potential

The collapse potential is defined by the following expression:

$$CP = (\delta h/H_0)*100 \qquad (1)$$

Where: $\delta h = H_0 - H_1$
H_0 is the sample thickness before saturation, and H_1 is the sample thickness after saturation. The applied vertical stress at the moment of saturation is taken as the nominal pressure, σ_i = 200kPa.
Figure 3 presents the effect of the Kaolinite content on the collapse potential. We note that the collapse potential is clearly affected by the Kaolinite content present in the samples. The maximum collapse potential is achieved with an optimum Kaolinite content of about 25%. These findings confirm with the observations made by Lawton and al. (1992), who stated that the maximum collapse is found with clay content between 10 % and 40 % and are in concordance with the results of the investigation

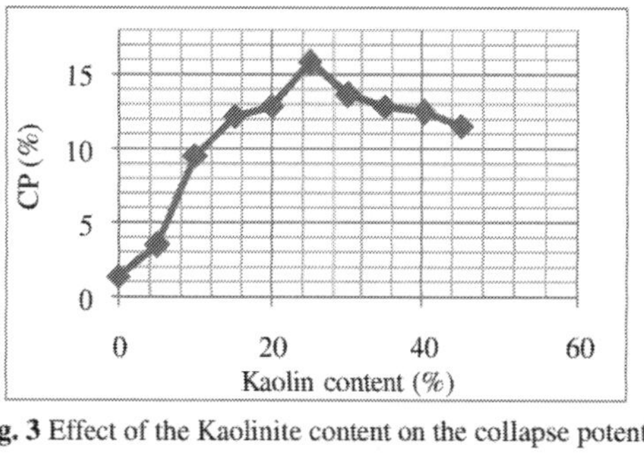

Fig. 3 Effect of the Kaolinite content on the collapse potential

undertaken Nouaouria and al. (2002), who have shown that the maximum collapse potential was obtained with an optimal percentage of about 25% of Kaolinite.

4.2 Effect of sand grain size on CP

A set of tests were carried out in order to show the effect of the sand grain size on the collapse potential (see figure 4). We can clearly see in figure 4 that the potential collapse is approximately ranging between 9.2 % and 13.5 %.The maximum value of CP was found with sand grain size of 500µm. According to Jennings and Knight (1975), this type of soil is classified to cause severe damages.

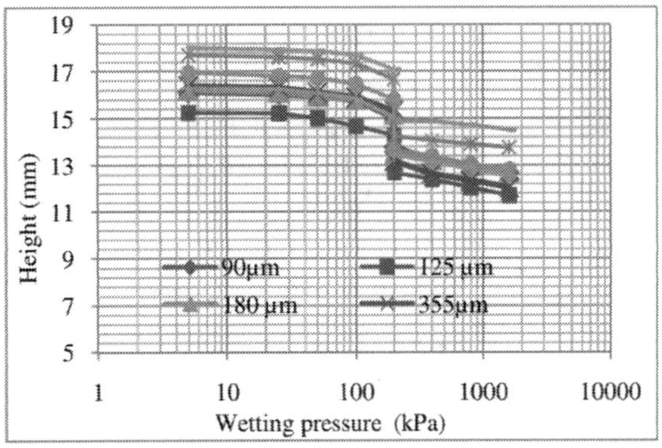

Fig. 4 Hydrocollapse test results

4.3 Effect of the polyethylene fibers reinforcement

The specimens used in this series of tests were prepared by mixing 75% of sand and 25% Kaolinite. The polyethylene fibers were used in three different proportions, namely: 5 %, 10 %, and 15 % respectively, and in different lengths, as follows: 5mm, 10mm, and 25mm respectively.

The results are reported in Figure 5. It is seen in figure 5-(a) that the collapse potential for unreinforced soil is about 17.28%, the addition of 5% of the polyethylene fibers of 5mm long has resulted in a small reduction in collapse potential. Whereas, this collapse potential was reduced to 9.7 with 15% of polyethylene.

Figure 5-(b) shows the treatment of the soil with polyethylene fibers of 10 mm long. In this case, there is a small decrease in the value of collapse potential. Figure 5-(c), with reinforcing fibers of 25 mm long, decreases from 12.76% to 10.20% and then increases with fiber content of 15%. As the collapse is due to the migration of soil particles, during flooding the soil fine particles move through the soil skeleton from one level to another. The increase of polyethylene fiber content beyond 10% prevents the good compaction and results in an open soil structure which will be subjected to high hydrocollapse.

We can note that the minimum value of CP (9.70%) is achieved with 15% of fibers of 5mm long. Using this type of reinforcement the collapse potential has then decreased by approximately 44 %.

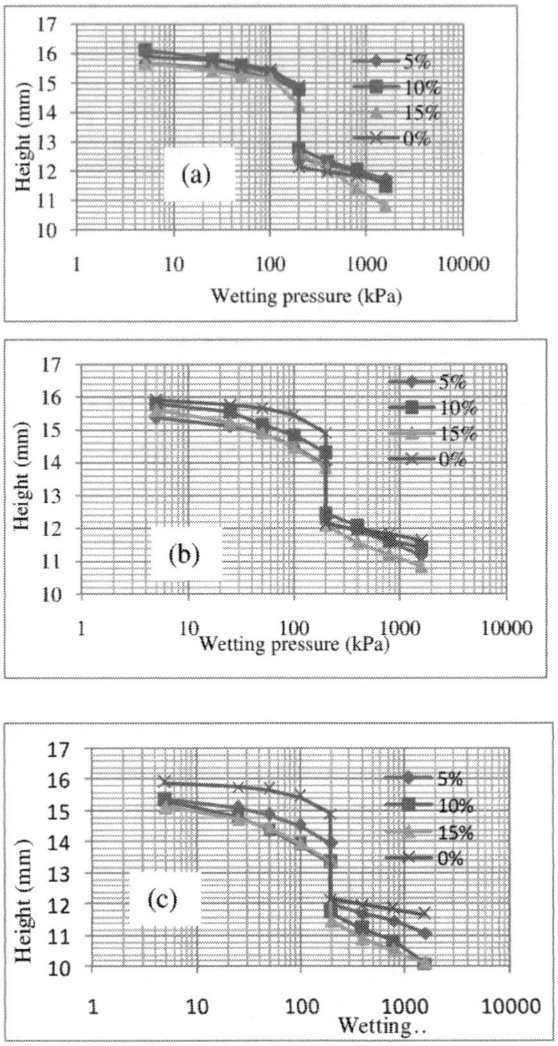

Fig. 5 Effect of Polyethylene fiber on potential collapse
a) length = 5mm, **b**) length = 10mm
c) length = 25mm.

4.5 Effect of the sisal fibers reinforcement

The sisal fibers are used in two different proportions, namely: 0,5 %, and 1% respectively, and different lengths, as follows: 5mm, 10mm, and 25mm respectively.

From the results illustrated in Figure 6-(a), it is noted that the collapse potential increases from 12.48 to 15.42 for a 5mm long of sisal fiber. According to the figure 6-(b), the collapse potential is ranging from 13.06 to 14.64 for a fiber of 10mm long. The only fiber length that decreased the collapse potential was 25 mm long as clearly shown

https://doi.org/10.1051/matecconf/201814901034

figure 6-(c). The soil treatment with sisal fibers gives a reduction of about 31%.

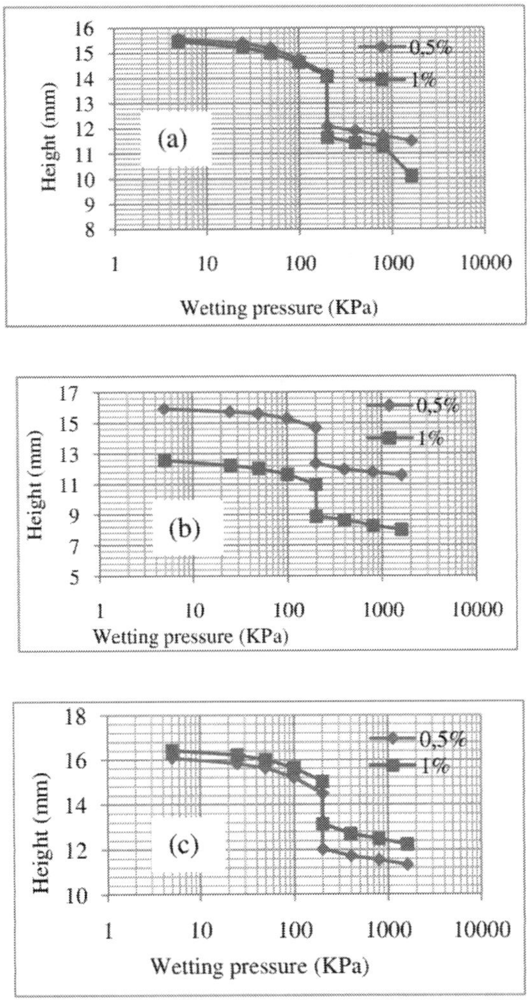

Fig. 6 Collapse potential (a) Sisal fiber length = 5mm
(b) Sisal fiber length = 10mm ,
(c) Sisal fiber length = 25mm .

5 Conclusions

The objective of this study is to improve the behaviour of a collapsible soil with the inclusion of polyethylene fibers and sisal fibers. The following conclusions are drawn:

Reinforcing a collapsible soil with polyethylene fibers reduced considerably the collapse potential.

To preserve the environment from pollution due to used water plastic bottles thrown in the nature, the polyethylene fibers are generally used rather than sisal fibers because this latter does not harm the environment.

References

1. Ayadat T., Belouahri B., Ait Ammar R.., 1998. Migration of fine size particles as collapse mechanism of soils. French geotechnical journal N°83, P 73-81.
2. Botero E. , Ossa .A, Sherwell G., Ovando-Shelley E., 2015. Stress- strain behaviour of a silty soil reinforced with polyethylene terephthalate (PET), Geotextiles and Geomembranes 43, P 363-369.
3. Chauhan M.S., Mittal S., Mohanty B., 2008. Performance evaluation of silty sands subgrade reinforced with fly ash and fibre. Geotextiles and Geomembranes 26. P 429-435.
4. Chegenizadeh A., and Nikraz H., 2012. Investigation on silty sand paper reinforcement. International Journal of Emerging Technology and Advanced Engineering Volume 2
5. Consoli N.C, Prietto P.D.M., Ulbrich L.A,1998. Influence of fiber and cements addition on behaviour of sandy soil, Journal of Geotechnical and geoenvironmental engineering 124 P 1211-1214.
6. Hamidi A., Hooresfand M., 2013. Effect of fiber reinforcement on triaxial shear behaviour of cement treated Sand. Geotextiles and Geomembranes 36, p 1-9.
7. Nouaouria M.S., Harireche O., Rouaiguia A., 2002. Comportement de l'affaissement du loess, Sols et fondations, Anales du batiment et des travaux publics. P 37-41.

Hybrid Latex Particles: Preparation with Miniemulsion Polymerization

F.Bouanani [1,2]
[1]Université d'Oran, Département de Chimie, Oran, Algérie
[2] Ecole Nationale Polytechnique Maurice Audin, Département de Chimie, Oran, Algérie

Abstract. This research developed a novel poly(trifluoropropylmethylsiloxane) (PTMS)/montmorillonite nanoparticles, for technologic applications. PTMS /MMT nanoparticles were prepared by the miniemulsion polymerization method. Montmorillonite clay was encapsulated within a fluorinated cyclosiloxane (1,3,5-tris(trifluoropropylmethyl)cyclotrisiloxane) to give stable water based nanocomposite latex, using miniemulsion polymerization technique. The resulting composite latex was characterized by transmission electron microscopy (TEM), dynamic light scattering (DLS) and RX diffraction. The particles stability was investigated by sedimentation and surface tension measurements. An efficient cationic/nonionic surfactant mixture was used in order to achieve the best compatibility with the monomer. TEM and RX data revealed the partial embedding of montmorillonite in the spherical polymer based nanoparticles. According to DLS measurements, the encapsulated clay particles conserve their size throughout the polymerization process. The melt processing of poly(trifluoropropylmethylsiloxane) matrix to encapsulate the montmorillonite clay was also carried out for comparison purposes.

1 Introduction

Polymer-clay nanocomposites have received growing attention during the last 20 years since they are knownto display much better physical properties than pure polymers or their classical microcomposites [1]. Several kinds of phyllosillicates have been coated [2,3] with a polymer layer or encapsulated in a polymer matrix [4] to enhance their stability and also to improve their particle distribution at a nanoscale. An appropriate process is crucial in order to achieve this nanocomposite. Several strategies have been considered to prepare polymer silicate nanocomposites which include four main processes: exfoliation–adsorption,in situ polymerization, melt intercalation and template synthesis.Under the first one is also gathered the emulsion polymerization synthesis [5]. The polymer layered phyllosilicate nanocomposite
in the second process is developed by swelling the layered phyllosilicate within the liquid monomer so that the polymer formation can occur in between the intercalated sheets [6]. The melt intercalation enables the layered silicate to mix with the polymer matrix in the molten state but compatibility is required to ensure a thermodynamic stability in the final state [7]. The last technique is the in situ hydrothermal crystallisation of the clay in an aqueous gel medium where the polymer often acts as a template for layer formation [8].

In miniemulsion process, the monomer droplets are protected against the monomer diffusion (Ostwald ripening) by an efficient surfactant or a mixture of two surfactants and also by addition of a hydrophobic compound [9,10] allowing the polymerization to occur mainly inside these nanoreactors.

Furthermore, due to their hydrophobic nature platelet clay can be dispersed within the organic phase, i.e. inside the monomer droplets and stabilize the dispersion provided that chemical compatibility and partial clay exfoliation are achieved. In this work, miniemulsion polymerization through an anionic reaction pathway is successfully used to encapsulate a montmorillonite clay within a fluorinated cyclotrisiloxane, the 1,3,5-tris(trifluoropropylmethyl)cyclotrisiloxane (D3F) [11–13]. A cationic/ nonionic surfactant mixture which consists of didodecyldimethylammonium bromide (DDDBr) and FSO-100 whose general formula is $(C2H4O)x(CF2)yC2H5FO$) was employed to stabilize the system. In addition, the melt intercalation strategy was tested from the polymerised D3F, poly(trifluoropropylmethyl)siloxane (PTFPMS) in order to estimate the miniemulsion process efficiency with regards to the classical melt process. Our ultimate goal is to assess silicate dispersion on a nanolength scale.

2 Experimental

2.1 Materials

Montmorillonite clay whose general formula is $Mx(Al_4{-}xMgx)Si_8O_{20}(OH)_4$, was purchased from a local

company (ENOF Maghnia) in Algeria. Prior to employment, an acidified clay form was obtained by shaking the clay in a 0.25M sulfuric acid solution until saturation was achieved at 95 ◦C with continuous stirring during 6 h. The cation-exchanged clay was recovered by filtration and then suspended in de-ionized water.
The monomer D3F was provided by ABCR Company and its nominal purity was larger than 98%. The surfactants DDDBr and FSO-100 were purchased from Acros Organics and Zonyl respectively, and were 99% pure. All compounds were used as received. The solvent is de-ionized and bi-distilled water.

2.2 Characterization

The determination of particle size and particle size distribution was investigated by dynamic light scattering (DLS) using a Malvern Zetasizer-4 with a fixed incident light angle at 90◦.

All samples were diluted when necessary to avoid counter saturation. The latex particle morphology was observed through Philips CM 30 transmission electron microscope (TEM) at a maximum accelerating voltage of 100 kV. The samples were prepared by casting a latex solution drop onto a 200-mesh copper grid covered with carbon film. This drop is absorbed on a filter paper no 41 and deeped immediately in liquid ethane cooled with liquid nitrogen, then transferred inside the electron microscope. The incineration experiments were carried out in programmable Forced-Draft furnace (Fisher Scientific Inc.) at 500 ◦C for 1 h. The ash weight was taken as the clay weight after subtraction of the adsorbed water weight estimated to be 3.5% of the initial sample weight.

Table 1:
Miniemulsion basic components and composition

Component	Total ((w/w)%)
D_3^F	18.0
Montmorillonite	0.75
DDDBr	1
FSO100	0.5
Water	76

3 Results and discussion

3.1 Miniemulsion stability

3.1.1. Surfactant effect

The samples were prepared following the typical procedure described above with the basic components shown in Table 1 except that only the nonionic surfactant FSO 100 was incorporated, the cationic surfactant effect being previously described elsewhere [12]. The variation of the average droplet size ($\bar{R}$) before polymerization deduced from DLS as a function of added FSO 100 amount is represented in Fig. 1. It is noticed that at very low concentration or in absence of FSO 100 the droplet size measured is as high as 800 nm. By increasing the FSO 100 concentration up to 0.15 wt% in the basic recipe, the oily droplet size is reduced down to 350 nm. A further increase of the surfactant content up to 0.3 wt% does not bring any additional size reduction and a leveling off to a plateau-like step is observed.

It is also interesting to mention that at and above 0.15 wt% of FSO100 the system shows remarkable stability over several months.

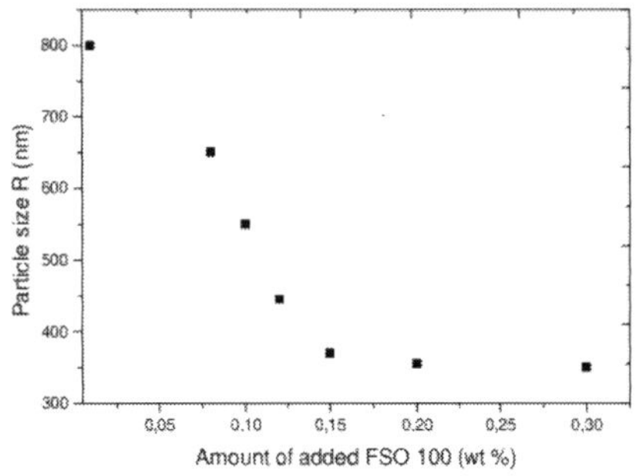

Fig.1. Correlation between particle size and FSO 100 amount added to the aqueous phase before polymerization.

As can be seen from Fig. 2, the particle size change as determined by DLS in the course of the polymerization reaction when combining cationic and nonionic surfactant in optimal proportions as described in Table 1 is minor. This result is consistent with the literature [15] and indicates clearly that a real miniemulsion process has been accomplished. It should be noted that the average $\bar{R}$has about the same value at that obtained above (see Fig. 1). This is in good agreement with the results obtained by Qunhui et al. [16] for polystyrene laponite composites synthesized via miniemulsion polymerization using a nonionic/cationic mixture.

3.1.2. Miniemulsion stability

The miniemulsion surface tension was measured as a function of time just after sonication as shown in Fig. 3. After 50 s, the surface tension remains relatively constant at about 42 mN/cm. Since this value is well above typical values of micellised surfactant aqueous solutions, it can be ascribed to incomplete covering of the latex particle surface by the surfactant molecules [17]. This observation is an additional support for an effective miniemulsion process occurrence.

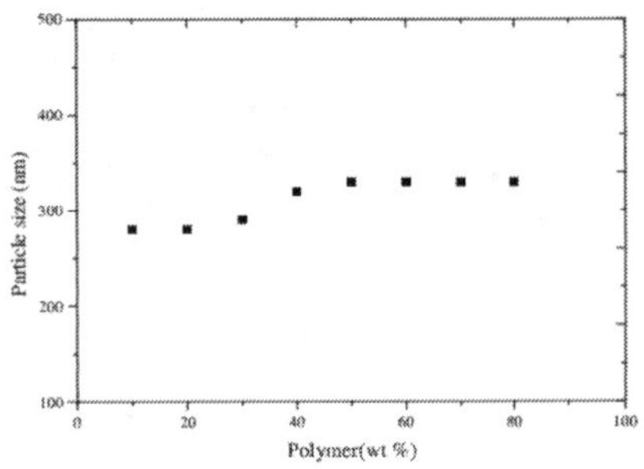

Fig.2. Droplet size evolution during the polymerization reaction.

On the other hand, the constancy for a relatively long time of the surface tension confirms that the nonionic/ cationic surfactant mixture is very efficient in stabilizing the nanocomposite.

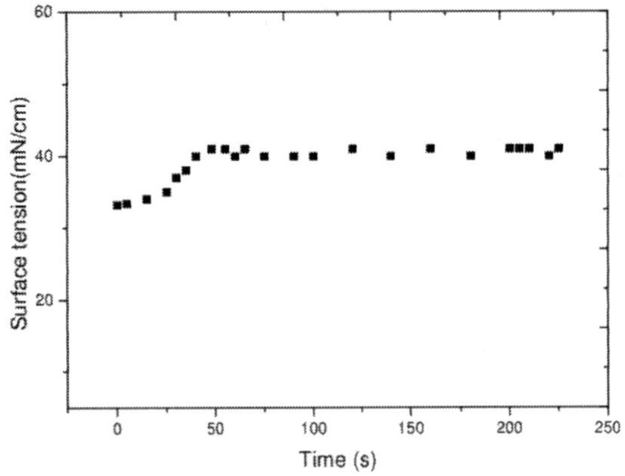

Fig.3. Miniemulsion surface tension time evolution of the D_3F miniemulsion.

3.2 Encapsulation by miniemulsion

3.2.1. Preliminaries

In order to investigate the montmorillonite surface modification efficiency, the clay was stirred with the surfactant in the aqueous phase for 1 h, then the monomer was added under continuous stirring for 24 h. A phase separation was observed after 48 h of free standing. The aqueous and oily phases were separated and clay distribution in each phase was determined by ash weight measurement after mixture incineration. The obtained results show that 65% of the clay are located within the monomer layer. This indicates clearly that the surfactants/clay mixture remained primarily in the oily monomer phase under mechanical shear.

3.2.2. DLS

The monomer and polymer droplet sizes with and without clay added are regrouped in Table 2. The size increase observed upon clay addition is a direct evidence of an effective encapsulation. Indeed, the phyllosilicate hydrophobic nature compatible with the monomer is favourable to the droplet stability. However, the presence of a double population in the montmorillonite polymer sample can be interpreted in terms of a partial encapsulation since some of the particles remain unchanged with regards to their mean size.

Table 2:
Particle size and particle size distribution in the miniemulsions with and without added montmorillonite (from DLS)

Miniemulsion	Without the montmorillonite		With the montmorillonite	
	R (nm)	Standard deviation	*R* (nm)	Standard deviation
Monomer	246	0.04	276	0.17
Polymer	280	0.15	$R_1 = 330$, $R_2 = 276$	0.19

3.2.3. XRD

Spectral information on the clay platelet structure can be obtained by XRD analysis [18]. As depicted in Fig. 4 when the pure montmorillonite XRD pattern (I) exhibits one diffraction peak at diffraction angle $2\theta = 5.6°$ leading to d(001) spacing of 13.5A° , the diffraction profile of the acidified montmorillonite (II) is centered at $2\theta = 4.8°$ with d(001) spacing of 18.01A° . The larger latter value is probably due to the presence of two interlamellar water layers, subsequent to the acid treatment and reflects the change in interlayer cations and their associated hydratation states [19].

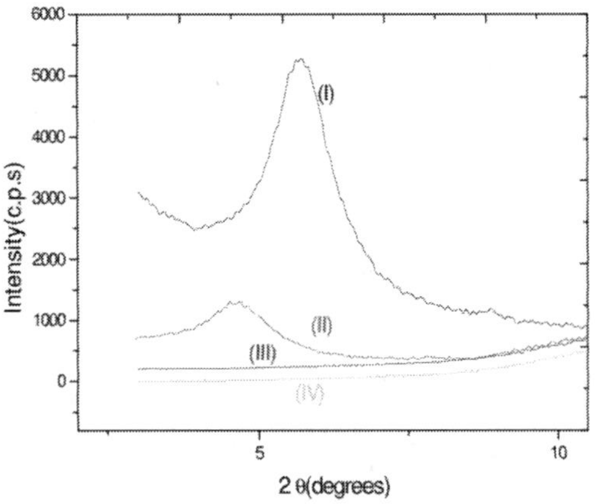

Fig.4. Nanostructure XRD spectra: (I) pure montmorillonite; (II) acidified montmorillonite; (III) nanocomposite before polymerization; (IV) nanocomposite after polymerization.

On the other hand, for both nanocomposites in monomer and latex miniemulsions, no apparent diffraction peak could be detected in the whole available diffraction angle range suggesting a fully exfoliated structure of the montmorillonite clay. This result is also due to the fact that DDDBr pre- modified montmorillonite can easily be exfoliated and encapsulated inside the polysiloxane latex, in agreement with the literature [20,21].

3.2.4. TEM analysis

The structural and composition heterogeneities have been observed by transmission electron microscopy. As seen on the TEM image of the mixed nanocomposite in Fig. 5, most of the particles have a well-defined spherical shape though a small fraction of solid particles co-exist. Hence, the mixed nanocomposite contains both free montmorillonite clay and spherical nanocomposites. In this case, only partial encapsulation is obtained as already deduced from sedimentation data.

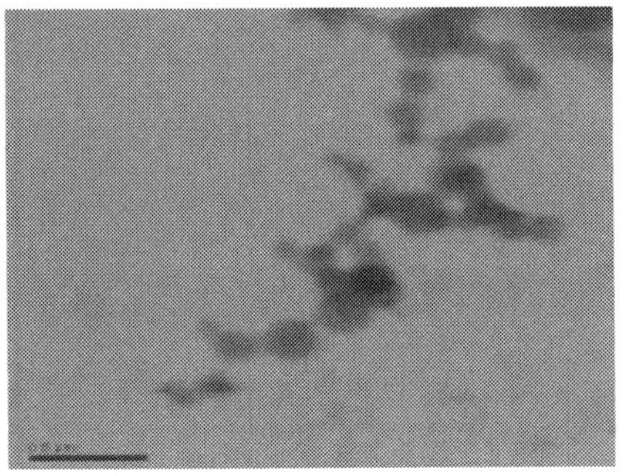

Fig.5. Mixed nanocomposite TEM image

4 Conclusion

It was shown by DLS, TEM and RX diffraction that the miniemulsion process leads to an effective encapsulation of montmorillonite. A novel poly(trifluoropropylmethyl)siloxane / montmorillonite hybrid nanoparticle in the form of a water dispersion was developed with up to 5 wt% clay loading through a miniemulsion polymerization approach. The nanocomposite system obtained in this way was macroscopically homogeneous and stable with clay being embedded within the polymer particles with a fully exfoliated structure. The presence of a cationic/nonionic surfactant mixture together with the sonication process played a cooperative role in stabilizing the miniemulsion.

References

[1] H. Lu, G.-Z. Liang, X. Ma, B. Zhang, X. Chen, Polym. Int. 53 1545 (2004).
[2] E.P. Giannelis, R. Krishnamoorti, E. Manias, Adv. Polym. Sci. 107 138 (1999).
[3] T.J. Pinnavaia, G.W. Beall (Eds.), Polymer-Clay Nanocomposites, vol. 1, John Wiley & Sons, New York, 2000.
[4] M. Zanetti, S. Lomakin, G. Camino, Macromol. Mater. Eng. 279 1–10 (2000).
[5] M. Alexandre, P. Dubois, Mater. Sci. Eng. 28 1–63 (2000).
[6] Y.W. Luo, X.D. Zhou, J. Polym. Sci., Part A: Polym. Chem. 42 2145–2152 (2004).
[7] M. Alexandre, P. Dubois, T. Sun, J.M. Garces, R. J´erome, Polymer 43 2132–2139 (2002).
[8] R.A. Vaia, E.P. Giannelis, Macromolecules 30 7990–7999 (1997).
[9] K.A. Carrado, L.Q. Xu, Chem. Mater. 10 1440 (1998).
[10] (a) J. Delgado, M.S. El-Aasser, J.W. Vanderhoff, J. Polym. Sci., Part A: Polym. Chem. 24 861 (1986).
(b) J.M. Asua, Prog. Polym. Sci. 27 1283–1386; (2002).
(c) S.T. Wang, F.J. Schork, G.W. Poehlein, J.W. Gooch, J. Appl. Polym.Sci. 60; 2069 (1996).
[11] M. Barr`ere, C. Ma^ıtre, F. Ganachaud, P. Hemery, Macromol. Symp. 151 359–367 (2000).
[12] M. Barr`ere, C. Ma^ıtre, M.A. Dourges, P. Hemery, Macromolecules 34 (2001) 7276–7280.
[13] F. Bouanani, D. Bendedouch, C. Maitre, J. Teixeira, P. Hemery, Polym. Bull. 55 (6) (2005) 429–436.
[14] A. Yahiaoui, M. Belbachir, J. Appl. Polym. Sci. 100 ;1681–1687 (2006).
[15] M. Antonietti, K. Landfester, Prog. Polym. Sci. 27 689–757. (2002).
[16] S. Qunhui, Y. Deng, Z. Lin Wang, Macromol. Mater. Eng. 289 288–295 (2004).
[17] K. Landfester, N. Bechthold, F. Tiarks, M. Antonietti, Macromolecules 32 2679 (1999).
[18] T.Yui, H.Yoshida, H. Tachibana, D.A. Tryk, H. Inoue, Langmuir 18 891–899 (2002).
[19] A. Harrane, R. Meghabar,M. Belbachir, Int. J. Mol. Sci. 3 790–798 (2002).
[20] Z. Tong, Y. Deng, Polymer 48) 4337. (2007).
[21] Z. Tong, Y. Deng, Ind. Eng. Chem. Res. 45 2641 (2006).

The use of non-destructive tests to estimate Self-compacting concrete compressive strength

Boukhelkhal Djamila[1], Guendouz Mohamed[1]
[1] LME Laboratory, University of Medea, Medea, Algeria

Abstract. Until now, there are few studies on the effect of mineral admixtures on correlation between compressive strength and ultrasonic pulse velocity for concrete. The aim of this work is to study the effect of mineral admixture available in Algeria such as limestone powder, granulated slag and natural pozzolana on the correlation between compressive strength and corresponding ultrasonic pulse velocity for self-compacting concrete (SCC). Compressive strength and ultrasonic pulse velocity (UPV) were determined for four different SCC (with and without mineral admixture) at the 3, 7, 28 and 90 day curing period. The results of this study showed that it is possible to develop a good correlation relationship between the compressive strength and the corresponding ultrasonic pulse velocity for all SCC studied in this research and all the relationships had exponential form. However, constants were different for each mineral admixture type; where, the best correlation was found in the case of SCC with granulated slag (R^2 = 0.85). Unlike the SCC with pozzolana, which have the lowest correlation coefficient (R^2 = 0.69).

1 Introduction

Self-compacting concrete (SCC), was first developed in Japan in 1986. It is a special kind concrete that can flow through and fill the gaps of reinforcement and corners of molds without any need for vibration and compaction during the placing process [1, 2]. This type of concrete demands a large amount of powder content compared to conventional vibrated concrete to produce a homogeneous and cohesive mix [3]. It was reported that SCC often contains powder in the order of 450–600 kg/m^3 of concrete [4]. One of its disadvantages is its high cost. An alternative for reducing the cost of SCC is the use of fine mineral additions which replace a part of the Portland cement [5-9].

Nondestructive testing of concrete is preferred due to its distinct advantage over conventional compression tests [10-12]. Moreover, ultrasonic pulse velocity (UPV) is one of the popular nondestructive techniques for evaluating the qualitative assessment of homogeneity in concrete. This method is typically based on the measurement of the propagation of velocity, which is closely related to strength of concrete [13]. However, the evaluation by nondestructive methods of the compressive strength of concrete is based generally on empirical relations between this property and nondestructive tests [14, 15].

In the construction industry, self-compacting concrete (SCC) has been increasingly used in ready-mix concrete and in the precast industry due to its technical advantages and to improve several aspects of construction [16, 17]. The specific formulation of these concretes related to their implementation requirements could affect their mechanical behavior in the hardened state, compared to traditional vibrated concrete [18]. Many investigations by various researchers have shown empirical correlation models between the increase of UPV and the increase of the equivalent strength of concrete with age [19-23]. However, these models are only suitable for materials and environment similar to those which they are established.

There are few published studies on the effect of mineral admixtures on relationship correlation models for self-compacting concrete. So, the aim of this work was to study the effect of Algerian mineral admixture such as limestone powder, granulated slag and natural pozzolana on the correlation between compressive strength and corresponding ultrasonic pulse velocity in SCC.

2 Materials and methods

2.1 Materials

Ordinary Portland Cement (CEM I 42.5) with a fineness of 3370 cm^2/g and a specific gravity of 3.15 g/cm^3 was used in this study. Three types of mineral admixtures were used with fineness comparable to that of cement: natural pozzolana (PZ); limestone powder (LP) and granulated slag (GS) their specific gravity are 2590, 2544 and 2857 kg/m^3 respectively. The chemical compositions of Portland cement and mineral admixtures used in this study, are described in table 1.

Table 1. Chemical analysis of cement and mineral admixtures used.

Chemical Constituent (%)	Cement	PZ	LP	GS
SiO_2	20.14	46.40	0.85	40.10
CaO	63.47	9.85	53.08	42.20
Al_2O_3	3.71	17.50	0.93	6.00
Fe_2O_3	4.74	10.50	0.56	2.00
MgO	2.12	3.45	0.23	4.70
SO_3	2.67	0.40	0.60	0.15
Ignition los	1.72	4.31	42.55	0.05

For the fineness modulus of sand used is about 2.50, we combined two sand: roller siliceous sand (S1) and crushed limestone sand (S2). The gravel used is a limestone crushed gravel with two granular classes (3/8 & 8/15). Figure 1 shows the particle size distribution curves of the aggregates used. The specific gravity of the coarse aggregates (G1, G2) and sand (S1, S2) were 2.66, 2.67, 2.59 and 2.63 g/cm^3 respectively; and their absorptions were 0.19, 0.52, 0.04 and 0.38%, respectively. To ensure the self-compacting concrete character, a chemical admixture superplasticizer type was used in this study.

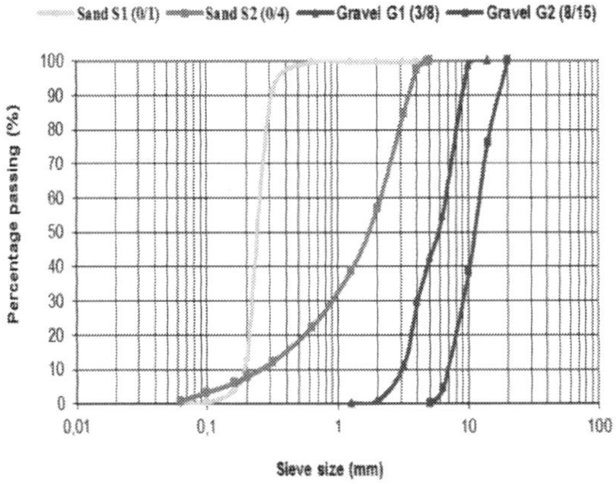

Fig. 1. Grading curves of fine and coarse aggregates used.

2.2 Mix proportions

To determine the quantities of materials: cement, aggregates, water and superplasticizer, we are based on the Japanese method proposed by Okamura [1], with some modifications are made to the sand content in mortar (S/M = 0.5), the water to cement ratio (W/C = 0.42) and the superplasticizer to cement ratio (Sp/C = 1.6) to fit with the requirements of local materials and obtain good characteristics of fresh SCC [24].

Four SCC are prepared in this study with a total powder content of 470 kg/m^3 (Portland cement and mineral admixture), one without any admixture SCC-R and three mixes were prepared by replacing 15% of the cement with one of the three mineral admixtures utilized in this study separately namely: SCC-LP, SCC-GS and SCC-PZ.

2.3 Test procedures

The self compactability of fresh concrete was verified according to tests recommended by the AFGC [25]. Before casting, slump-flow, L-Box and sieve segregation resistance tests were attempted for determining the properties of SCC such as filling ability, passing ability and segregation resistance.

For each mixture, cube specimens (10×10×10) cm^3 were used to determine the ultrasonic pulse velocity and compressive strength at 3, 7, 28, and 90 days in accordance with NF EN 12504-4 [26] and NF EN 12390-3 [27], respectively. All specimens were demolded after 24 h and cured at 20°C and 95% relative humidity. The UPV test method employs the principle of measuring the travel velocity of ultrasonic pulses through a material medium. The pulse velocity equipment consists of an emitter (generating transducer) from which ultrasonic pulses are transmitted, a receiver (receiving transducer) where the pulses are received, and a device for indicating the time of travel from the transmitter to the receiver "Figure. 2.a". The compressive strength is determined by crushing specimen under an applied load until rupture "Figure 2b".

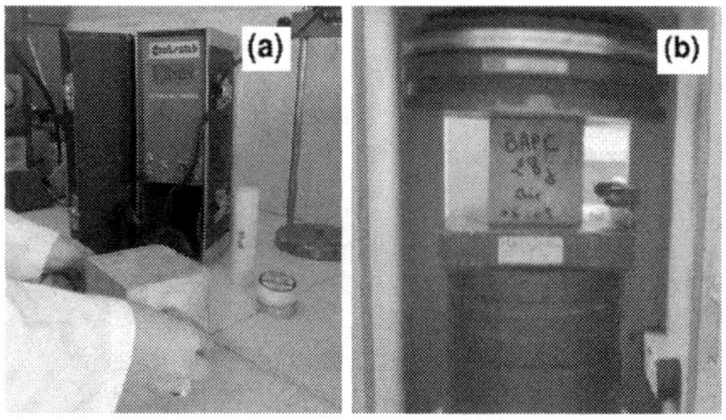

Fig. 2. Tests carried out in the hardened state (a/ ultrasonic pulse velocity test; b/ compressive strength test).

3 Results and discussion

3.1 The effect of mineral admixtures on fresh properties of SCC

The properties of fresh SCCs are presented in table 2 below. The results exposed in this table show that all mixtures, have verified the flow properties of SCC and meeting the requirements of the AFGC recommendations

[25]. We can say also that the incorporation of limestone and blast furnace slag, remarkably improves all the fresh properties of SCC, compared to natural pozzolana which declined slightly their filling and passing capacity, but it increases the segregation resistance. The same results are shown by Uysal and Edamatsu [28, 29].

Table 2. Fresh properties of SCC mixes.

Properties	SCC-R	SCC-LP	SCC-GS	SCC-PZ
Slump flow (mm)	704	784	736	675
H_2/H_1 ratio (L-Box)	0,87	0,97	0,91	0,81
Passing mortar S (%)	4,05	6,25	5,48	2,37

3.2 The effect of mineral admixtures on compressive strength and UPV of SCC

The compressive strength and UPV results of SCCs were determined on the 100 mm cubes for 3, 7, 28, and 90 days. The results obtained for these tests are shown in figures 3 and 4 respectively. From these results, we can see as expected, that UPV progresses with age for all mixes and the increase has slowed after 28 days which is in the same line with the evolution of compressive strength versus time. It is remarked also that all produced SCCs were classified as good as all measured UPV values were greater than 3500 m/s.

Furthermore, lower ultrasonic pulse velocities were observed for SCC-LP compared to those of SCC-R at early and later age, indeed, the same behavior is observed for compressive strength. These results coincide with those found by Mucteba et al. [30] for SCC incorporating 20% of limestone powder. While almost the same UPV and compressive strengths were measured by SCC-R and SCC-PZ for all ages.

As regards the SCC-GS, comparable values to SCC-R were observed with a slight increase in ultrasonic pulse velocity and compressive strength are found after 28 days. In fact, SCC-GS gave the best ultrasonic pulse velocity at later ages compared to the other SCCs, which correlates with the values of compressive strength. The same result was confirmed by Mucteba et al. [30], Türkmen et al. [31] and Guneyisi et al. [32], which found an increase in ultrasonic pulse velocity in SCC with 10 and 20% granulated slag compared to the reference SCC.

3.3 Correlation between compressive strength and UPV for SCC

The different correlation models proposed between the compressive strength by crushing the specimens and the corresponding ultrasonic pulse velocity for different SCC mixes are shown in Figures 5. We can see from this figure, that all SCCs (with and without mineral admixtures) show a good correlation between the ultrasonic velocity and compressive strength (Correlation coefficients between 0,69 and 0,85). Moreover, all the relationships have the exponential form with different constant, which agree with other work for concrete with mineral admixtures [33-36]. We can see also that the best correlation was found for SCC-GS (R^2 = 0.85), while SCC-PZ present the lowest correlation compared to the other SCCs with a correlation coefficient of about 0.69.

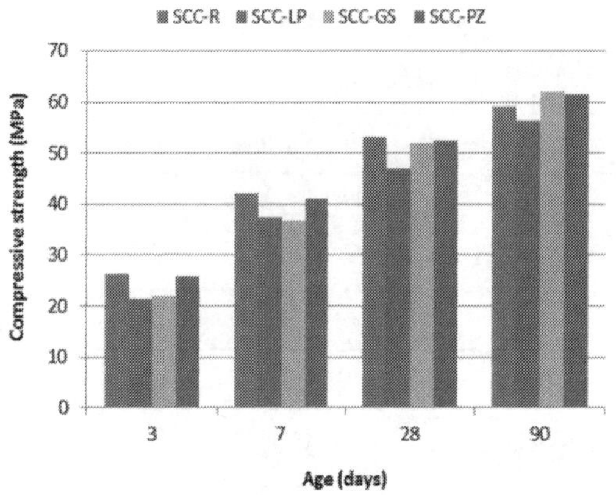

Fig. 3. Compressive strength results for SCC as function of mineral admixture type for different curing periods.

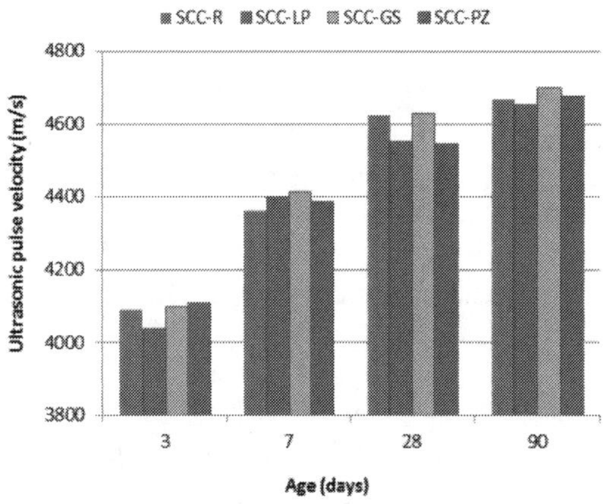

Fig. 4. UPV results for SCC as function of mineral admixture type for different curing periods.

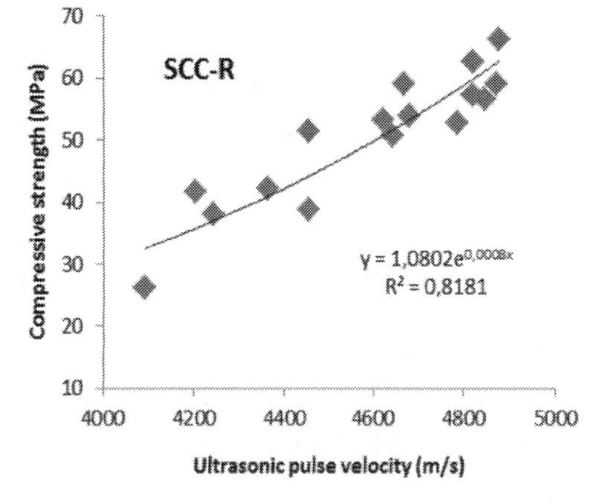

Fig. 5. Relationship between compressive strength and ultrasonic pulse velocity for reference SCC.

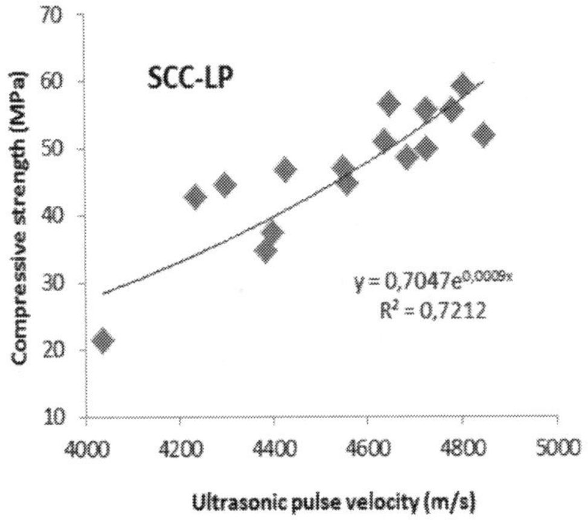

Fig. 6. Relationship between compressive strength and ultrasonic pulse velocity for SCC with limestone powder.

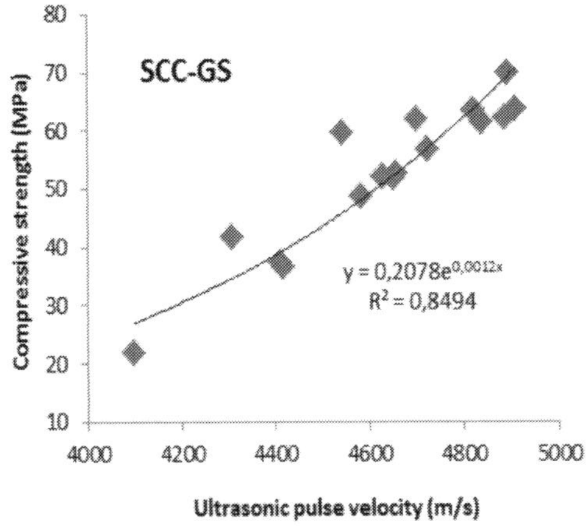

Fig. 7. Relationship between compressive strength and ultrasonic pulse velocity for SCC with granulated slag.

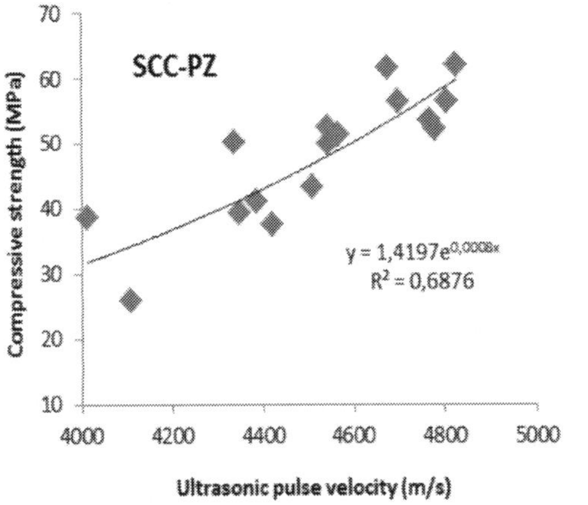

Fig.8. Relationship between compressive strength and ultrasonic pulse velocity for SCC with natural pozzolana.

4 Conclusions

Based on the results obtained through this study, the following conclusions have been drawn:

- It is possible to manufacture self-compacting concrete using Algerian mineral admixtures with acceptable fresh and hardened properties ;
- The use of granulated slag and limestone powder increased the flow properties of SCC, such as filling and passing ability. Unlike the natural pozzolana which has developed the least workable SCC, but it characterized by high resistance to segregation and consequently it leads to produce the more stable SCC ;
- In the context of compressive strength, all the SCCs could be considered to be satisfactory for structural applications. The UPV values increased with increasing curing period for all SCC, which is in the same line with compressive strength ;
- Introduction of LP induced a reduction in strengths and UPV of SCCs for all ages. The strength of SCC with slag and natural pozzolana exceeds that of controlled SCC after 28 days, with almost the same UPV, for all ages ;
- It is possible to develop a good correlation relationship between the compressive strength by crushing and the corresponding ultrasonic pulse velocity for all SCC studied in this research (with and without mineral admixtures) ;
- The correlation between UPV and compressive strength is also exponential for SCC containing mineral admixtures. However, constants were different for each type of mineral admixture in SCCs ;
- The best correlation was found in the case of SCC-GS ($R^2 = 0.85$). Unlike the SCC-PZ, which have the lowest correlation coefficient ($R^2 = 0.69$) ;
- So we can say that the type of mineral addition, can affect the correlation between compressive strength and ultrasonic pulse velocity, and that the establishment of this correlation must be based on appropriate material to the study.

References

1. H. Okamura , M. Ouchi, J Adv Concr Technol, **1**(1):5–15 (2003)
2. D. Bonen D,S.P. Shah S. P, Prog Struct Eng Mater, **7**:14–26 (2005)
3. M.E. Barrak ,M. Mouret , A. Bascoul,J Cem Concr Compos. **31**(1):12–21, (2009)
4. P. Nanthagopalan and M. Santhanam, Const. Build. Mater. **23**(11):3443–9 (2009)
5. Bouzoubaa N and Lachemi M, "*Self-compacting concrete incorporating high volumes of class F fly ash: Preliminary results*", *Cement and Concrete Research*, **31**, 413–420 (2001)

6. M. Guendouz, F. Debieb and E. H. Kadri, 33èmes roncontres de l'AUGC, ISABTP/UPPA, Anglet, 27 au 29 Mai 2015, Bayonne, France, (2015), p. 1-8.
7. P. Dinakar,K G. Babu and M. Santhanam, *Cement & Concrete Composites*, **30**, 880–886 (2008)
8. L. Coppola, T. Cerulli, and D. Salvioni, 11th International Conference on Fracture ICF11, Volume **3**, pp. 2226-2241 (2005)
9. M. Guendouz, F. Debieb, O. Boukendakdji, E. H. Kadri, M. Bentchikou and H. Soualhi, J. Meter. and Env. Scien., **7**(2), (2016), 227-232.
10. A.N. Bin Ibrahim, P. Bin Ismail and M. Forde, International Atomic Energy Agency, Vienna (2002)
11. J. Delacour and J.M. Geoffray, "*Résultats et recommandations du projet national CALIBÉ : La maitrise de la qualité des bétons*", Presses de l'école nationale des Ponts et Chaussées, Paris, France (2004)
12. D. Breysse, M. Larget, Z. Sbartai ,J. Lataste and J.P.Balayssac J. P., 7th International Symposium on Non-destructive Testing in Civil Engineering, Nantes, France, 8p (2009)
13. V.M. Malhotra and N.J. Carino, "*Handbook on Nondestructive Testing of Concrete*", 2nd Edition, .ASTM International, CRC Press (1991)
14. M. Nehdi,H.E Chabib, and A. Naggar, "*Predicting Performance of Self-Compacting Concrete Mixtures Using Artificial Neural Networks*", ACI Mat. J., vol. **98**, no. 5, pp. 394–401 (2001)
15. D. Breysse and O. Abraham, "*Méthodologie d'évaluation non destructive de l'état d'altération des ouvrages en béton*", Presses de l'école nationale des Ponts et Chaussées, Paris, France (2005)
16. M. Yurugi, "*Application of self-compacting concrete in Japan*", Proceedings 23rd OWICS Conference, CI-Premier Ptc Ltd., Singapore, August 25– 26, pp. 29– 42 (1998)
17. O. Petersson , Cem. Concr. Res. Inst. **54** (2), 5 pp (1998)
18. Y. Klug and K. Holschemacher, "*Comparison of the hardened properties of Self-Compacting and Normal Vibrated Concrete*", 3rd International Symposium on Self-Compacting Concrete, Reykjavik,Iceland, pp. 596-605 (2003)
19. J. Blitz and G. Simpson, "*Ultrasonic Methods of Non-Destructive Testing*", London: Chapman and Hall (1996)
20. H.Y. Qasrawi, Cement and Concrete Research, Vol.**30**, pp.739-746 (2000)
21. B. Hobbs and K.M. Tchoketch, Forensic Science International, Vol.**167**, pp.167-172 (2007)
22. A.B. Khodja A. B, "*Corrélation entre essais non destructifs et essais destructifs du béton à faible résistance*", Master theses, University of Hassiba Ben Bouali, Chlef, Algeria (2010)
23. DJ. Boukhelkhal, S. Kenai and F. Debieb, "*Corrélation entre essais non destructifs et essais destructifs de la résistance du béton (Scléromètre & Ultrason)*", International Symposium on Non-destructive Testing in Civil Engineering DIAGNOBETON, Toulouse, France, 6p (2014)
24. O. Boukendakdji,E.H. Kadri ,S. Kenai , Cement & Concrete Composites **34**; 583–590 (2012)
25. AFGC, Association Française de Génie Civil, "*Recommandations pour l'emploi des Bétons autoplaçant*" (2008)
26. NF EN 12504-4 (2005), "*Essais pour béton dans les structures. Partie 4 : Détermination de la vitesse de propagation du son*", AFNOR.
27. NF EN 12390-3 (2003), "*Essai pour béton durci. Partie 3 : Résistance à la compression des éprouvettes*", AFNOR.
28. M. Uysal and K. Yilmaz, Cement and Concrete Composites J., Vol. **33**,(2011),pp.771-6.
29. Y. Edamatsu, T. Sugamata and M. Ouchi , "*A mix-design method for self-compacting concrete based on mortar flow and Funnel tests*". In: Proceedings of 3rd international symposium on self- compacting concrete, Reykjavik, Iceland, (2003) p. 345–55.
30. U. Mucteba, S. Mansur, Construction and Building Materials J., Vol.**25**, (2011),pp. 4112–4120.
31. I. Türkmen,A. Aliöz and A.Cüneyt, , Scientific Research and Essays J., Vol. **5**(15), (2010),pp. 2055-2064.
32. E. Guneyisi and M. Gesoglu, Materials and Structures J., Vol. **41**, (2008), pp. 479–493.
33. K. Tharmaratnam and B.S. Tan , Cement and Concrete Res. **20**: (1990). 335-345.
34. R. Demirboğa and M.R. Türkmen Karakoç, (2004), Cement Concrete Res. **34**: 2329-2336.
35. Z.C. Ulucan, K. Turk and M. Karatas, Russian Journal of Nondestructive Testing, Vol. **44**, N°5, ,(2008) pp. 367–74.
36. C.U. Zulfu, T. Kazim and K. Mehmet, , Russ. J. Nondestructive Test **44**, pp. 367–374, (2008).

Effect of cement fineness and polycarboxylate dosage on the rheological and mechanical behavior of a mortar

Didouche Zahia [1], Ezziane Karim [1].
[1]Hassiba Benbouali University, Civil Engineering Department, Chlef, Algeria.

Abstract. The use of certain organic additives in the production of mortar and concrete influences the workability and the hydration kinetic of mortar. This results in a modification of some properties, namely rheological behavior and mechanical strength. The objective of this work is to evaluate the rheological and mechanical behavior of a mortar by varying the fineness of the cement and using the superplasticizer Polycarboxylate.

1 INTRODUCTION

A l'état frais, la présence des additions minérales modifie la structure du squelette granulaire et les frictions entre les composants solides dans la phase liquide. Certaines additions peuvent réagir chimiquement en milieu cimentaires pour former de nouveaux produits hydratés qui présentent un caractère liant supplémentaire.

La petite taille des particules du filler calcaire comble les vides interstitiels et augmente la densité de la matière. Par conséquent, l`eau attrapée dans les pores granulaires est libérée ce qui augmente la compacité de la pâte, améliore l`ouvrabilité et diminue la demande en eau [1, 2, 3]. Les dosages élevés en filler calcaire augmentent la viscosité de la pâte [1], Le seuil de cisaillement augmente avec la quantité d'addition ultrafine incorporée tandis que la viscosité de la pâte varie avec la nature et la quantité d'addition [4]. D'autre part une augmentation de la finesse offre une ouvrabilité satisfaisante car le clincker adhère mieux au calcaire et la demande en eau diminue par rapport à celle du ciment portland [5]. Aussi la composition chimique d'un ciment est le facteur le plus important qui influence l'hydratation [6]. L`utilisation des fillers augmente le taux d`hydratation [7]. Le calcaire favorise la cristallisation des monocarbonates plus que les monosulfates et stabilise indirectement l`ettringite [8]. Il intervient lors de l`hydratation des C_3A pour former carboaluminate et retarde la transformation d'ettringite en monosulfoaluminate, il constitue des sites de nucléation pour les cristaux de l`hydroxyde de calcium au jeune âge de l`hydratation [9,3], cette addition améliore la réactivité du clincker et exploite son potentiel hydraulique [10], mais le flux de chaleur total après 72h d`hydratation est similaire à celui du ciment portland due à l'effet de dilution [8].

En outre un ciment fin possède une grande surface spécifique avec une large surface de contact avec l'eau ce qui augmente le taux d'hydratation, ainsi les produits hydratés seront de faible épaisseur autour des particules de ciment anhydre et causent une élévation du degré d'hydratation [6]. D'après une étude de recherche, un ciment contenant les particules fines du filler calcaire montre un temps de prise court [9], mais Voglis et al. [4] ont déclaré que l`addition calcaire accélère le début de prise et retarde la fin de prise, par conséquent le temps de prise est prolongé par rapport à celui du ciment portland. Pour d'autres [11], le début et la fin de prise varient en fonction du dosage en calcaire. Une addition de 5% de calcaire augmente la plasticité de la pâte et prolonge le temps de prise, mais pour un dosage en calcaire supérieur à 20 % le début et la fin de prise diminuent.

L'ajout des superplastifiants au mortier améliore ses propriétés dû à la baisse du rapport E/C nécessaire. Les polymères affectent l'interface entre la surface des particules et la solution interstitielle et influencent les propriétés physiques telles que la viscosité et la limite d'élasticité de la pâte [12,13]. Les superplastifiants à base de naphtalène et de mélamine augmentent les propriétés rhéologiques des pâtes de ciment à court et à long termes à cet effet la réduction de la viscosité plastique et le seuil de cisaillement dépendent de la composition du ciment et du temps d'introduction du superplastifiant [14]. Ainsi le type de superplastifiant à une grande influence sur le comportement rhéologique des mortiers de ciment à l'état frais et que les superplastifiants à base de polycarboxylate étaient plus efficaces que les superplastifiants à base de naphtalène [15,13]. Pour Heikal et al. [16] le superplastifiant à base de polycarboxylate adsorbé sur les surfaces des particules du ciment crée une charge négative sur les grains du ciment causant une répulsion électrostatique ce qui augmente les distances entre les particules et améliore la fluidité de la pâte .Si le mortier est considéré comme un ensemble de couches granulaires parallèles soumises à une contrainte tangentielle.

Si une des couches du mortier se déplace par rapport à celle qui lui est sous-jacente en raison du frottement permanent sur les grains de la seconde couche, le mouvement est transmis partiellement à cette dernière en même temps que la vitesse de déplacement de la première couche diminue. Cet effet de retard, provoqué par la friction interne des grains de la couche sous-jacente sur celle de la couche supérieure, est appelé viscosité

μ [Pa.s]. Donc, la viscosité peut être définie comme la résistance au glissement d'un système soumis à une contrainte tangentielle. Elle est caractérisée par un coefficient de viscosité qui joue un rôle essentiel dans la rhéologie des fluides. Pour certains fluides, sa connaissance suffit à caractériser de façon précise son comportement rhéologique [17].

2 Etude expérimentale

2.1 Matériaux utilisés

Le ciment utilisé dans ce travail est un ciment portland composé (CEM II/42.5) contenant 12% de calcaire broyé à des temps variés pour donner quatre types de ciment (Tableau1). Ces ciments sont utilisés pour confectionner des mortiers à base d'un sable normalisé. Un superplastifiant à base de polycarboxylate noté (PC) fourni par l'entreprise GRANITEX et commercialisé sous le nom de MEDAFLOW 30 a été utilisé. Ce superplastifiant a été dilué dans l'eau de gâchage et incorporé dans la composition du mortier avec différents dosages.

Tableau I. Finesse des ciments utilisés.

Ciment	C1	C2	C3	C4
Finesse (cm^2/g)	3026	3298	3668	4000

2.2 Confection des mortiers

Les mélanges destinés à mesurer les paramètres rhéologiques et la résistance mécanique sont confectionnés sous une température constante de 20°C avec différentes teneurs en superplastifiant. En soustrayant la quantité d'eau contenue dans le superplastifiant de la quantité d'eau de gâchage pour chaque dosage utilisé (Tableau 2).

Tableau II . Compositions des échantillons de mortier.

L'essai	E/C	Ciment (g)	Sable (g)	Eau (g)	PC (%)
Rhéologie	0.5	630	1890	315	0, 0.4, 0.8
Résistance mécanique	0.5	450	1350	225	0, 0.4, 0.8

3 Présentation et discussion des résultats

3.1 Rhéologie

La "figure 1" montre la variation du seuil de cisaillement τ_0 des mortiers en fonction du dosage en Polycarboxylate et de la finesse du ciment.

La comparaison du comportement rhéologique des différents mortiers permet de constater en premier lieu que le mortier adjuvanté présente des taux de cisaillement significativement plus faibles que les mortiers non-adjuvantés. Ceci est inhérent avec le rôle dispersant de l'adjuvant et la diminution des forces de frottement entre les grains lors du cisaillement qu'il engendre. D'autre part, les mortiers à base de ciment à finesse élevée ont un seuil de cisaillement plus important que les mortiers confectionnés avec des ciments à faible finesse, ceci peut être expliqué par l'augmentation de la finesse qui comble les vides dans la pâte cimentaire et augmente les forces de frottement entre les grains ce qui conduit à une augmentation du seuil de cisaillement dans la pâte.

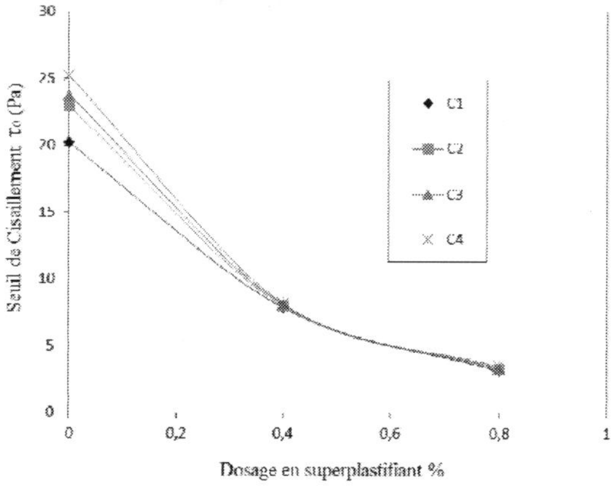

Fig.1. Effet du dosage en Polycarboxylate et de la finesse du ciment sur le seuil de cisaillement des mortiers.

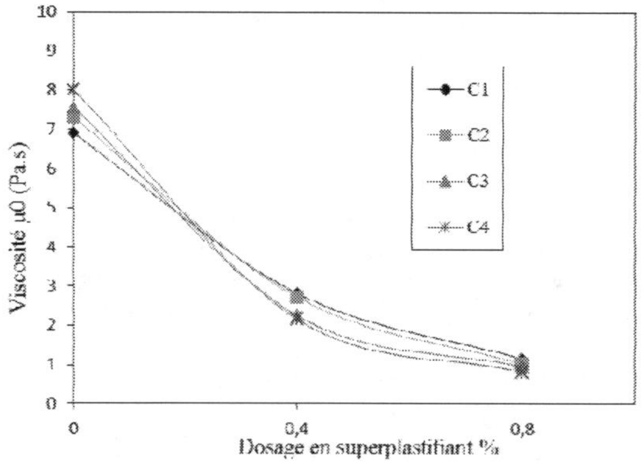

Fig.2. Effet de la finesse et du dosage en plolycarboxylate sur la viscosité des différents mortiers.

Les résultats portés sur la "figure 2" montre une chute considérable de la valeur de viscosité lorsque le dosage en Polycarboxylate augmente, cette chute est autant plus significative pour un dosage supérieur à 0.4% PC. Ce résultat s'aligne aux travaux de Puertas et al. [12] qui ont trouvé qu'un dosage de 0.3% de PC réduit la contrainte tangentielle de la pâte de 70% suivi par une diminution de la viscosité de la pâte.

D'autre part la valeur de la viscosité d'un ciment plus fin est supérieure à celle d'un ciment à faible finesse. Cet effet sur les particules fines a également été étudié par Nawa et *al.* [18] où la viscosité des pâtes de ciment accrue avec la fraction fine (≤ 10 um) de ciment. Cela est dû à l'augmentation du montant de superplastifiant adsorbé lorsque la finesse du ciment augmente.

3.2 Résistance mécanique

De la "figure 3", On observe que la valeur de la résistance à la compression du mortier à base de ciment fin est beaucoup plus élevée que celle d'un mortier à base de ciment à gros grains car les ciments fins s'hydratent rapidement ce qui accélère la formation des hydrates qui offrent à la matrice cimentaire une meilleure résistance mécanique à la compression. Ainsi pour des finesses élevées, le clincker adhère mieux au calcaire [5].

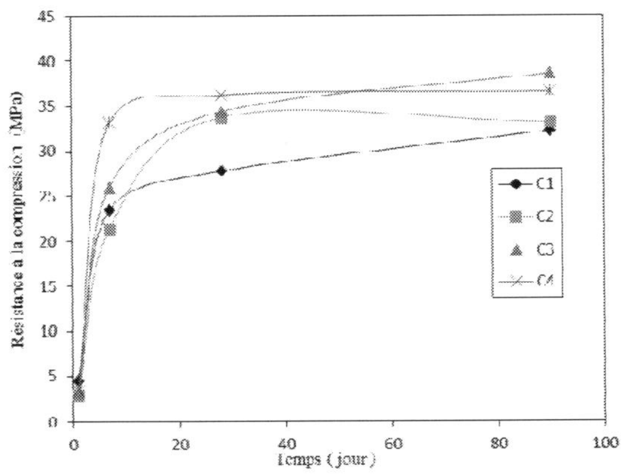

Fig.3. Effet de la finesse sur la résistance mécanique des mortiers non adjuvantés.

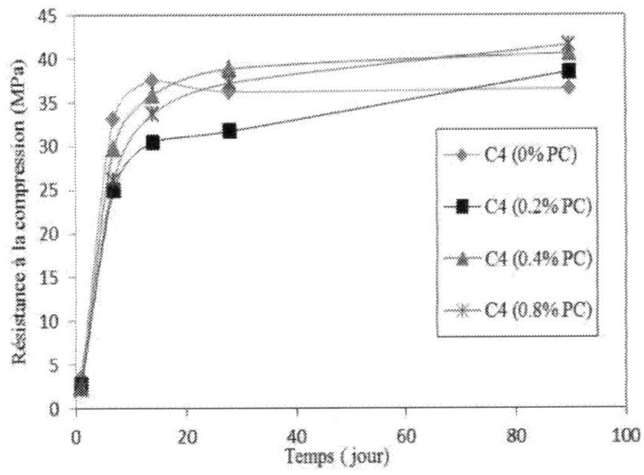

Fig.4. Effet du dosage en polycarboxylate sur la résistance mécanique des mortiers à base de ciment C4.

D'après les résultats illustrés sur la "figure 4", on peut constater qu'au jeune âge la valeur de la résistance du mortier non adjuvanté est presque similaire à celles des mortiers adjuvantés, mais après 28 jours l'effet du superplastifiant améliore les propriétés mécaniques du mortier en enregistrant des résistances mécaniques importantes surtout pour les dosages 0.4% et 0.8% PC.

4 Conclusion

D'après les résultats obtenus nous pouvons constater que l'effet de la finesse du ciment utilisé est très signifiant en présence de l'addition minérale calcaire et du superplastifiant Polycarboxylate Où :

- Le seuil de cisaillement et la viscosité du mortier augmentent lorsque la finesse du ciment passe de 3026 cm^2/g à 4000 cm^2/g et pour des dosages en Polycarboxylate inférieurs à 0.8%.
- L'augmentation de la finesse améliore la résistance mécanique des mortiers testés .
- L'ajout du Polycarboxylate a un effet bénéfique sur le développement des résistances mécaniques surtout pour des dosages inférieurs ou égaux à 0.8 %.

References

1. Yahia A., Tanimura M., Shimoyama Y. , *Réological properties of highly flowable mortar containing limestone filler-effect of powder content and w/c ratio* cement and concrete research, **35**, 532-539,(2005).
2. Bonavetti V., Donza H., Rahhal V., Irassar E., *Influence of initial curing on the properties of concrete containing limestone blended cement* ,Cement and Concrete Research , **30**, 703-708, (2000).
3. Menendez G., Bonavetti V., Irassar EF., *Strength development of ternary blended cement with limestone filler and blast-furnace slag*, Cement and Concrete Composition, **25**, 61–67, (2001).
4. Zhang X., Han J., *The effect of ultra-fine admixture on the rheological property of cement paste,* Cement and Concrete Research, **30**, 827–30, (2000).
5. Tsivilis S., Chaniotakis E., Badogiannis E., Pahoulasa G., Ilias A. , *A study on the parameters affecting the properties of Portland limestone cements*, Cement and Concrete Composites, **21**, 107–116,(1999).
6. Lin F., Meyer C., *Hydration kinetics modeling of Portland cement considering the effects of curing temperature and applied pressure*, Cement and Concrete Research , **39**, 255–265 ,(2009).
7. Soroka I., Stern N., *Calcareous fillers and the compressive strength of Portland cement*, cement and concrete research, **6**, 367-376,(1976).
8. Lothenbach B., Le Saout G., Gallucci E., Scrivener K., *Influence of limestone on the hydration of Portland cements*, Cement and Concrete Research, **38**, 848–860,(2008).
9. Mounanga P. , Irfan M. , Khokhar A. , Elhachem Rana loukili A., *Improvement of the early-age reactivity of fly ash and blast furnace slag*

cementations systems using limestone filler ,Materials and structures , **44**, 437 – 453 ,(2011).
10. Tsivilis S., Chaniotakis E., Kakali G., Batis G., *An analysis of the properties of Portland limestone cements and concrete*, Cement and Concrete Composites, **24**, 371–378,(2002).
11. Heikel M., El-didamony H., Morsy M.S., *limestone filled pozzolanic cement,* Cement and Concrete Research, **30**,1827-1834, (2000).
12. Puertas F., Santos H., Palacios M., Martınez-Ramırez S., *Polycarboxylate superplasticiser admixtures: effect on hydration, microstructure and rheological behaviour in cement pastes*, Advances in Cement Research, **17**, No. 2, 77–89, (2005).
13. Bjornstrom J., Chandra S., *Effect of superplasticizers on the rheological properties of cements* , Materials and Structures / Matériaux et Constructions,**36**, 685-692, (2003).
14. Aiad I., *Influence of time addition of superplasticizers on the rheological properties of fresh cement pastes*, Cement and Concrete Research, **33**, 1229 – 1234, (2003).
15. Golaszewki JG., Szwabowski J., *Influence of superplasticizers on rheological behaviour of fresh cement mortars*, Cement and Concrete Research, **34**, 235-248,(2004).
16. Heikal M., Morsy M.S., Aiad I, *Effect of treatment temperature on the early hydration characteristics of superplasticized silica fume blended cement pastes*, Cement and Concrete Research ,**35**, 680-687, (2005).
17. Courrazé G., Grossiord JL., *Initiation à la rhéologie*, Tec& Doc, (2000).
18. Nawa T., Eguchi H., *Effect of cement characteristics on the fluidity of cement paste containing an organic admixture*, Proceedings of the 9th International Congress on the Chemistry of Cements, **4**, 597-603, (1992).

Experiment-based modelling of the mechanical behaviour of non-hazardous waste incineration bottom ashes treated by hydraulic binder

Approche couplée modélisation-expérimentation pour étudier le comportement mécanique de Mâchefers d'Incinération de Déchets Non Dangereux traités avec un liant hydraulique

Libasse Sow[1,2], Fabrice Bernard[2], Siham Kamali-Bernard[2], Cheikh Mouhamed Fadel Kébé[1]
[1]Ecole Supérieure Polytechnique, Université Cheikh Anta Diop de Dakar, Sénégal
[2]Laboratoire de Génie Civil et Génie Mécanique (LGCGM) INSA Rennes, France

Abstract. Instrumented indentation tests have been carried out on an isolated 25 mm diameter particle of Non-Hazardous Waste Incineration bottom ash. These tests have enabled one to assess the intrinsic mean reduced modulus of elasticity "E_r" of the particles. This result is used as input data for a 3D numerical model of Representative Elementary Volumes (REV) of a road gravel made with this kind of by-products. This numerical model is based on a multi-scale hierarchical modelling strategy. The aggregates treated with cement have been decomposed into two REV at the sub-mesoscopic and mesoscopic scales. The numerical simulations campaign ("virtual laboratory") lead to the following results. At the sub-mesoscopic scale, we determined the input parameters for the Concrete Damaged Plasticity Model (CDPM) used at the mesoscopic scale. At the mesoscopic scale, the mechanical characteristics of the road aggregates usually determined through experiments have been found. The non-hazardous waste incineration bottom ashes treated by hydraulic binder was classified into mechanical classe 3.

Résumé. Des tests d'indentation instrumentée ont été effectués sur particule isolée de Mâchefers d'Incinération de Déchets Non Dangereux (MIDND) de diamètre 25 mm. Ces tests ont permis d'obtenir une estimation du module élastique réduit moyen « E_r » intrinsèque aux particules étudiées. Ce résultat sert de donnée d'entrée à une modélisation numérique tridimensionnelle de Volumes Elémentaires Représentatifs (VER) d'une grave routière à base de ces MIDND. Le modèle numérique est basé sur une stratégie de modélisation multi-échelle hiérarchique. Les granulats traités au ciment ont été décomposés en deux VER : sub-mesoscopique et mesoscopique. A l'échelle sub-mesoscopique, nous avons déterminé les paramètres permettant d'alimenter le modèle d'endommagement plastique (CDPM) utilisé à l'échelle mesoscopique. A l'échelle mesoscopique, les caractéristiques mécaniques de la grave routière traitée au ciment, habituellement déterminées par le biais d'expérimentations, ont été retrouvées. La grave routière à base de MIDND traitée au ciment a été classée en classe mécanique 3.

1 Introduction

La gestion des déchets est un enjeu majeur face à l'augmentation incessante de la quantité produite quotidiennement. Cependant, il est possible de les valoriser sous forme de Mâchefers d'Incinération de Déchets Non Dangereux (MIDND). Les mâchefers répondant aux exigences de la règlementation en vigueur [1] peuvent présenter des caractéristiques permettant leur valorisation en techniques routières. Il n'existe à ce jour aucune connaissance précise sur les caractéristiques élastiques intrinsèques de ces types de matériaux. Par ailleurs, l'approfondissement des connaissances et surtout le développement des matériaux cimentaires à base de produits recyclés ont montré que la structure du matériau était beaucoup trop complexe et aléatoire pour permettre le développement de formules analytiques exactes qui décrivent comment les propriétés et les performances lui sont reliées. Parallèlement à ce constat, le développement incessant des possibilités de calcul numérique et des moyens informatiques a permis le développement de simulations basées sur une description explicite de plus en plus réaliste des matériaux cimentaires. Ainsi, depuis quelques années, le « calcul des microstructures » et les méthodes de changement d'échelles se sont fortement développés dans la communauté mécanicienne au sens large du terme. Dans ce travail, une démarche de modélisation hiérarchique est mise en place à l'échelle mesoscopique afin d'étudier le comportement mécanique d'une grave routière à base de MIDND traitée avec un liant hydraulique. Cette démarche de modélisation nécessite d'être alimentée par des données expérimentales permettant de caractériser les propriétés élastiques intrinsèques des particules de MIDND valorisées. Ces propriétés sont obtenues dans ce travail par le biais d'une campagne d'indentation originale et jamais menée jusqu'à maintenant sur des granulats de mâchefers. L'indentation permet l'obtention du module élastique réduit en utilisant la méthode décrite dans [2].

2 Résultats d'une campagne d'indentation instrumentée sur particule isolée de mâchefer

2.1 Matériaux étudiés

Les mâchefers étudiés proviennent des carrières de la Garenne à Vignoc en Bretagne (France), ils sont sur la plateforme de maturation depuis environ une année et demie. Les granulats de mâchefers étudiés sont présentés à la figure 1.a avec la fraction 12,5/25 mm. Un exemple de particule isolée de mâchefer de diamètre 25 mm, d'apparence « amorphe » découpée en sa ligne médiane et moulée dans de la résine est montrée à la figure 1.b. Les tests d'indentation ont été effectués sur cette particule isolée.

(a)

(b)

Fig.1. Echantillons de mâchefers étudiés

2.2 Détermination des propriétés élastiques intrinsèques des particules de MIDND

Deux séries d'indentation ont été réalisées avec un indenteur sphérique de rayon 140 mm en carbure de tungstène dont les caractéristiques élastiques sont connues. La procédure d'essais est plus précisément décrite dans [3] ; les essais permettent de suivre l'évolution de l'effort appliqué en fonction de l'enfoncement réel de l'indenteur. Avec un indenteur de 140 mm de rayon, un plus grand volume de matériau est testé. Il intègre les différentes phases solides mais aussi la porosité et donne des modules homogénéisés de la particule testée. La déformation du porte-indenteur (correction de la compliance du banc expérimental) est prise en compte. La phase de décharge est considérée suivre la relation puissance de l'équation (1).

$$P = A^*(h - h_f)^m \tag{1}$$

La figure 2 montre un exemple de courbe effort-enfoncement obtenue. A partir des courbes d'indentation obtenues, ont été déterminés les modules élastiques réduits caractérisant les mâchefers étudiés (Tableau 1). Le module réduit d'un matériau s'exprime en fonction de son module d'Young et son coefficient de Poisson ; il est en principe impossible de découpler l'un de l'autre. Toutefois, en considérant la gamme complète de valeurs que peut prendre le coefficient de Poisson, il est possible de déduire que le module d'Young homogénéisé des particules de MIDND est compris entre 15 et 20 GPa.

Tableau 1. Modules élastiques réduits obtenus

N° de test	E_r (GPa)
1	16
2	14
Moyenne	15

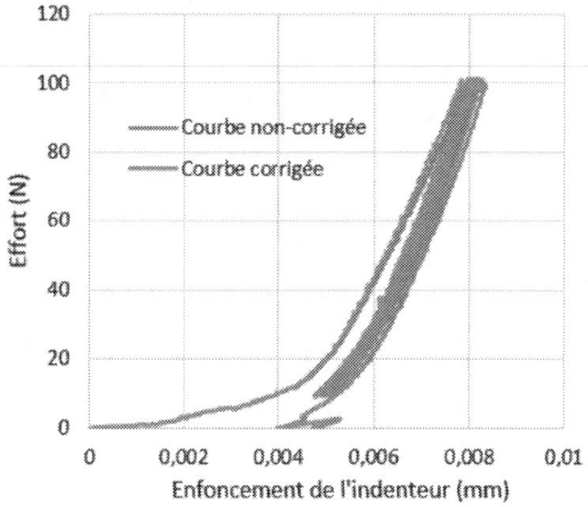

Fig.2. Exemple de courbe d'indentation -Test 1

3 Modélisation numérique hiérarchique multi-échelle tridimensionnelle

3.1 Stratégie spécifique de modélisation

La fraction granulométrique des particules de MIDND (0/25 mm) est trop étendue pour être considérée dans son intégralité dans une modélisation aux éléments finis 3D. Deux échelles d'étude ont donc été considérées : l'échelle sub-mesoscopique contenant la fraction 0/6 mm du squelette granulaire noyée dans une matrice poreuse de liant hydraulique (ou micro-grave-ciment, MGC) et l'échelle mesoscopique contenant la fraction 6/25 mm, noyée dans la MGC. A l'échelle mesoscopique, un modèle d'endommagement plastique (CDPM) est retenu pour caractériser la micro-grave-ciment. Ce modèle utilise une variable scalaire d'endommagement, la surface de charge de Lee and Fenves [4] et le potentiel plastique de Drucker-Prager (version hyperbolique). Il requiert la détermination de paramètres par le biais de simulations menées à l'échelle sub-mesoscopique. Le module d'Young (E), le coefficient de Poisson (ν), la résistance en compression uniaxiale (f_b) sont déterminés à partir des tests uniaxiaux de compression. La résistance en compression biaxiale (f_{b0}) est obtenue avec le test de compression biaxiale. Le degré de déformabilité du matériau (K), le coefficient M_{pic} (rapport entre le déviateur des contraintes au pic et la contrainte moyenne), l'angle de frottement interne (φ) et l'angle de dilatance (ψ) sont obtenus par le biais de tests triaxiaux. Les figures 3, 4, 5 et 6 décrivent respectivement la compression uniaxiale, la compression biaxiale, la traction uniaxiale et le test de triaxial de compression.

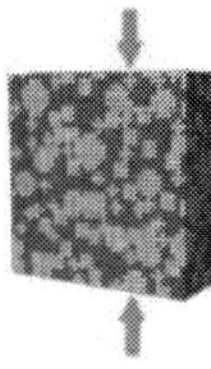

Fig.3. Compression uniaxiale

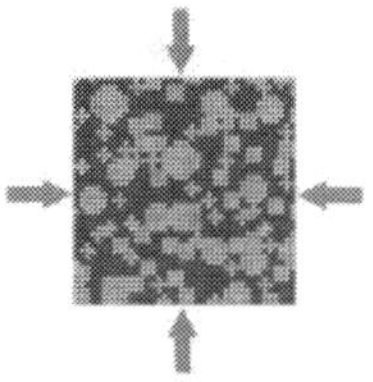

Fig.4. Compression biaxiale

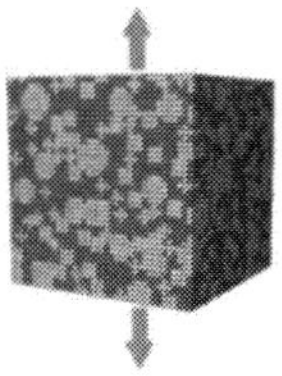

Fig.5. Traction uniaxiale

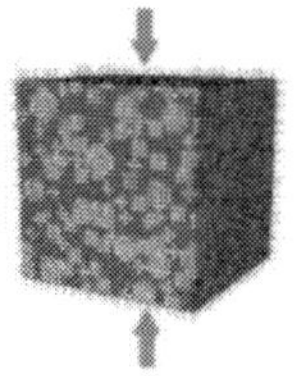

Fig.6. Triaxial de compression

La stratégie de modélisation adoptée est résumée dans la figure 7.

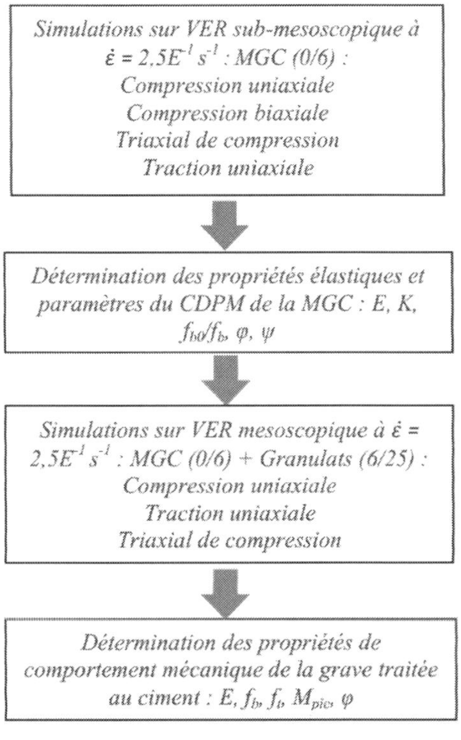

Fig.7. Stratégie de modélisation hiérarchique développée dans ce travail

Avec un test triaxial de compression, il est possible de tracer les courbes déviatoriques et de déformations volumiques ainsi que la réponse du matériau dans le plan (q, p), avec q le déviateur des contraintes et p la contrainte moyenne. A partir de ces courbes, l'angle de dilatance (ψ) et l'angle de frottement interne (φ) sont identifiés par l'intermédiaire des équations (2) et (3).

$$\Psi = arcsin\left(\frac{\frac{d\varepsilon_v}{d\varepsilon_1}}{2+d\varepsilon_v/d\varepsilon_1}\right) \quad (2)$$

$$\varphi = arcsin\left(\frac{3M}{M+6}\right) \quad (3)$$

M : rapport entre le déviateur de contraintes (q) et la contrainte moyenne (p).

Selon [5], le paramètre K caractérisant le degré de déformabilité du matériau peut être déterminé par la relation de l'équation (4). Le paramètre α se détermine quant à lui par la relation de l'équation (5).

$$K = \frac{(1-\alpha)(sin\varphi-3)}{2\alpha(3-sin\varphi)-(sin\varphi+3)} \quad (4)$$

$$\alpha = \frac{1-A}{-2A+1} \quad (5)$$

$A = f_{b0}/f_b$: rapport entre la résistance en compression biaxiale et la résistance en compression uniaxiale.

3.2 Construction des Volumes Elémentaires Représentatifs (VER)

Afin de modéliser la granulométrie étalée 0/25 mm avec la présence d'une pâte de ciment comme liant, il est nécessaire de représenter l'hétérogénéité avec les différentes phases explicites qui la caractérisent : granulats, liant, porosité. Les volumes élémentaires représentatifs aux deux échelles suivent les compositions données aux tableaux 2 et 3.

Tableau 2. Composition du VER sub-mesoscopique

Matrice : 15%	**Squelette granulaire : micro-grave 0/6 à 85%**		
Pâte de ciment 42,5R	40%, Φ=1	31%, Φ=3	14%, Φ=5

Tableau 3. Composition du VER mesoscopique

Matrice : 47,5%	**Squelette granulaire : grave 6/25 à 52,5%**			
MGC	22% de Φ=7,5	14,5% de Φ=12,5	11% de Φ=17,5	5% de Φ=22,5

Φ : diamètre moyen des granulats en mm

La génération de pourcentages de grains aussi grands étant difficile, une méthode du « Take and place » sur un maillage Eléments Finis régulier existant a été utilisée afin de distribuer les grains à l'intérieur des VER. Les particules les plus petites de la MGC sont choisies cubiques pour permettre d'atteindre les pourcentages visés. Les structures obtenues ont ensuite été reprises dans Abaqus pour y effectuer les simulations. A l'échelle sub-mesoscopique, le VER est un cube (40*40*40) mm^3 avec une résolution de 1 mm. A l'échelle mesoscopique, la résolution est de 2,5 mm avec un VER cubique (100*100*100) mm^3. Aux deux échelles, les éléments finis sont cubiques linéaires ; une étude de sensibilité sur la finesse du maillage a été menée.

3.3 Caractéristiques mécaniques des matériaux

3.3.1 La pâte de ciment

Le comportement post-pic de la pâte de ciment, donné par l'évolution bilinéaire de la contrainte post-pic en fonction de l'ouverture de fissure, est déterminé en considérant les résultats de [6]. Un modèle de fissuration diffuse, dont les paramètres sont déterminés par les relations 6 à 10, est alors utilisé.

$$f_1/f_0=\begin{cases}0{,}119\, f_{cap-p} + 0{,}8481 & si\ f_{cap-p} > 0{,}35\\ -22{,}486\, f_{cap-p}^2 + 15{,}99\, f_{cap-p} - 1{,}9525 & si\ f_{cap-p} \leq 0{,}35\end{cases} \quad (6)$$

$$f_0 = 1{,}6673(1 - f_{cap-p})^{1{,}967} \ (MPa) \quad (7)$$

$$w_1 = 0{,}0661 f_{cap-p}^{-0{,}7565} \ (\mu m) \quad (8)$$

$$w_2 = 17{,}102\, f_{cap-p}^{1{,}4863} \ (\mu m) \quad (9)$$

$$E = 35{,}931(1 - f_{cap-p})^{1{,}6973} \ (GPa) \quad (10)$$

Avec : f_0 : résistance en traction ; f_1 : contrainte à un point caractéristique de la forme bilinéaire ; w_1 et w_2 : ouvertures de fissures correspondant à f_1 et lorsque la résistance est nulle ; f_{cap-p} : porosité de la pâte de ciment.

Pour un temps d'hydratation long (α=0,83) correspondant à une porosité dans la matrice de 70 %, les propriétés mécaniques données au tableau 4 ont été obtenues.

Tableau 4. Valeurs caractéristiques de la forme bilinéaire post-pic

E (GPa)	f_0 (MPa)	f_1 (MPa)	f_1/f_0	w_1 (µm)	w_2 (µm)
4,17	0,138	0,138	0,91	0,09	10,45

3.3.2 Les granulats de mâchefers

En raison de l'étendue de la granulométrie, les grains les plus petits (1 et 3 mm) vont percoler d'une face à l'autre du VER. Afin d'empêcher que ce chemin continu ne modifie les résultats, un comportement fragile et donc une possible rupture de cette phase continue ont été imposés. La résistance de cette phase continue a été déterminée par une modélisation simplifiée permettant d'obtenir le comportement de granulats connectés équivalents à des granulats non-connectés. Ce comportement est caractérisé par le déplacement post-pic maximum (u_{ck}), donné dans le tableau 5 pour les différentes sollicitations. Les autres granulats de diamètres 5 mm et plus sont considérés élastiques avec un module d'Young de 20 GPa.

Tableau 5. Caractéristiques des grains de diamètre 1 et 3 mm

Tests	Compr. biaxiale	Compr. Uniaxiale	Triaxial de compr.	Traction uniaxiale
u_{ck} (mm)	3,5E^{-4}	2E^{-4}		
E_{hom} (GPa)	14			

3.4 Résultats obtenus aux deux échelles d'étude

A l'échelle sub-mesoscopique, les paramètres caractérisant le comportement mécanique de la micro-grave-ciment ont été obtenus (tableau 6). A l'échelle de la grave routière, les courbes de comportement obtenues sont données aux figures 8, 9 et 10. Le tableau 7 donne les caractéristiques mécaniques de la grave de MIDND traitée à 3 % de ciment.

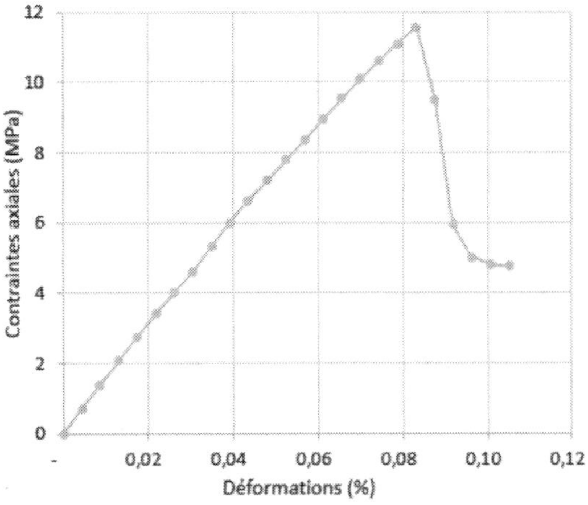

Fig.8. Courbe de comportement en compression uniaxiale

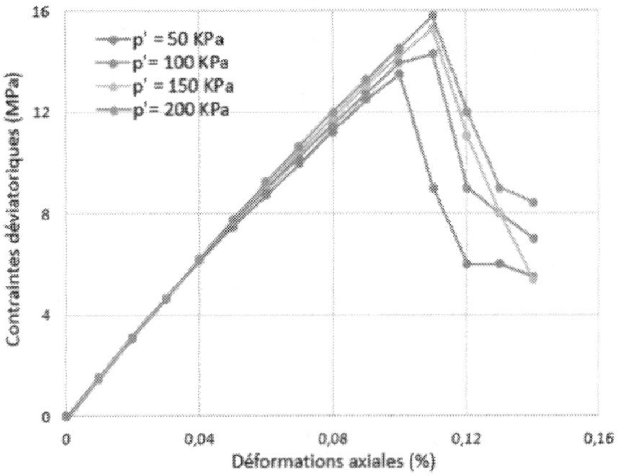

Fig.9. Courbes de comportement en triaxial de compression

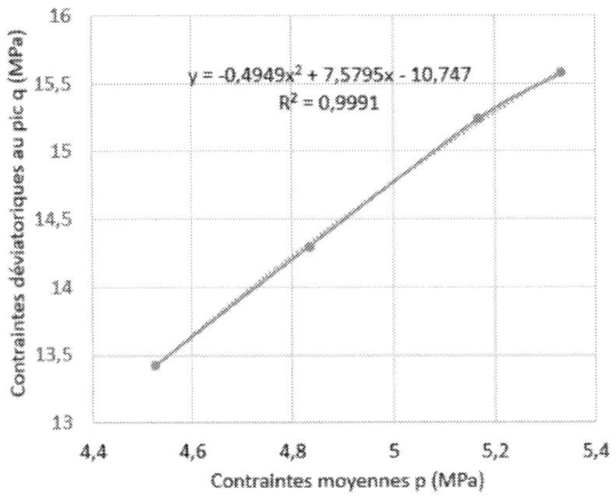

Fig.10. Résultats dans le plan (q, p)

Tableau 6. Paramètres du CDPM caractérisant la MGC

E (GPa)	K	f_{b0}/f_b	ψ (°C)	M_{pic}	φ (°C)
12,63	0,689	1,201	10	10,4	38

Tableau 7. Caractéristiques mécaniques de la grave routière à base de MIDND traitée à 3% de ciment

E (GPa)	M_{pic}	ϕ (°)	$R_c \leftrightarrow f_b$ (MPa)	$R_t \leftrightarrow f_t$ (MPa)
15,8	2,83	74	11,55	0,61

3.5 Classement de la Grave routière à base de MIDND Traitée au Liant Hydraulique

Le classement d'une Grave Traitée au Liant Hydraulique (GTLH) se fait en fonction des valeurs du module d'Young et de la résistance en traction. Le matériau a été classé en catégorie mécanique 3 (cf. figure 11).

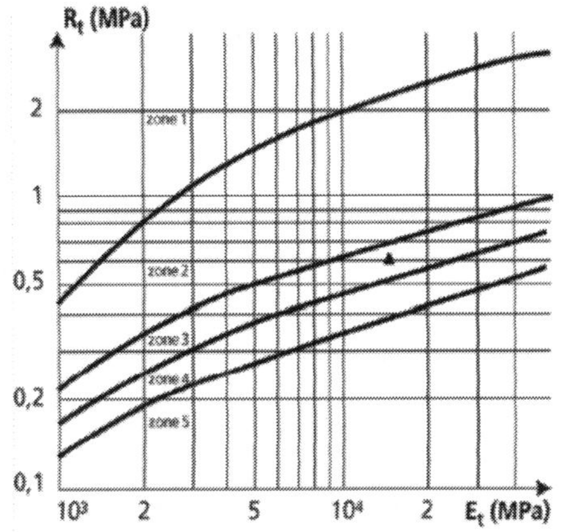

Fig.11. Classement de la Grave routière à base de MIDND Traitée au Liant Hydraulique

4 Discussions et conclusions

Les premiers résultats d'indentation sur particule isolée de mâchefer montrent une faible rigidité homogénéisée. Une étude statistique serait intéressante pour avoir une idée plus précise de l'ordre de grandeur du module élastique de ces types de matériaux du moment que leurs zones de production sont diverses. La modélisation développée a permis de prédire le comportement des mâchefers lorsqu'ils sont traités au ciment. L'angle de frottement interne est comparable à celui déterminé dans les expérimentations de [7] et à celui des graves routières [8]. Les valeurs obtenues s'expliquent par le fait que le matériau granulaire est composé de grains anguleux bien gradués. Le classement mécanique de la grave routière à base de MIDND traitée au ciment en catégorie 3 prouve la faisabilité d'un réemploi de ce type de Sous-Produits Industriels dans ce secteur d'activités.

Remerciements

Les auteurs remercient MM. Gérard Mauvoisin et Olivier Bartier pour leur aide et leur expertise dans la campagne expérimentale de tests d'indentation, ainsi que l'entreprise Eurovia, en particulier M. Loïc Seguin, pour leur approvisionnement en MIDND.

Références

1. Arrêté relatif au recyclage en technique routière des mâchefers d'incinération de déchets non dangereux. Ministère de l'écologie, du développement durable, des transports et du logement, France (18 Novembre 2011).
2. W. C. Oliver, G. M. Pharr, An improved technique for determining hardness and elastic modulus using load and displacement sensing indentation experiments. J. Mater. Res. **7** (1992) 1564-1583.
3. L. Sow, S. Kamali-Bernard, O. Bartier, G. Mauvoisin, F. Bernard, Experimental estimation of the elastic modulus of non-hazardous waste incineration bottom ash aggregates by indentation tests - Microanalysis of particles by scanning electron microscopy. Submitted in Adv. Mater. Res. (Trans Tech Publications).
4. J. Lee, G. L. Fenves, Plastic-damage model for cyclic loading of concrete structures. J Eng. Mech., ASCE, **124** (1998) 892-900.
5. R. Javier, M. Francisco, and M. Joaquín, Concrete Constitutive Model, Calibration and Applications. *Simulia Community Conference* (2013).
6. F. Bernard and S. Kamali-Bernard, Predicting the evolution of mechanical and diffusivity properties of cement pastes and mortars for various hydration degrees – A numerical simulation investigation. Comp. Mat. Sci. **61** (2012) 106-115.
7. F. Becquart, Première approche du comportement mécanique d'un milieu granulaire issu d'un mâchefer d'incinération d'ordures ménagères : valorisation en technique routière. PhD thesis. Université des sciences et technologies de Lille (2007).
8. P. Hornych, A. Kazai, Q. Quibel, Study of the resilient behaviour of unbound granular materials, Proc. *5th Conference on Bearing Capacity of Roads and Airfields*. Trondheim, Norway, **3** (1998) 1277-1287.

Properties of dune sand concrete containing coffee waste

Guendouz Mohamed[1], Boukhelkhal Djamila [1]
[1] LME Laboratory, University of Medea, Algeria

Abstract. In the last years, an increase of coffee beverages consumption has been observed all over the world; and its consumption increases the waste coffee grounds which will become an environmental problems. Recycling of this waste to produce new materials like sand concrete appears as one of the best solutions for reduces the problem of pollution. This work aims to study the possibility of recycling waste coffee grounds (Spent Coffee Grounds (SCG)) as a fine aggregate by replacing the sand in the manufacturing of dune sand concrete. For this; sand concrete mixes were prepared with substitution of sand with the spent coffee grounds waste at different percentage (0%, 5%, 10%, 15% and 20% by volume of the sand) in order to study the influence of this wastes on physical (Workability, bulk density and porosity), mechanical (compressive and flexural strength) and Thermal (Thermal conductivity and thermal diffusivity) properties of dune sand concrete. The results showed that the use of spent coffee grounds waste as partial replacement of natural sand contributes to reduce workability, bulk density and mechanical strength of sand concrete mixes with an increase on its porosity. However, the thermal characteristics are improved and especially for a level of 15% and 20% of substitution. So, it is possible to obtain an insulating material which can be used in the various types of structural components. This study ensures that reusing of waste coffee grounds in dune sand concrete gives a positive approach to reduce the cost of materials and solve some environmental problems.

1 Introduction

The fraction of organic waste in household wastes is large and increases with time. In each country the waste composition is different, since it is affected by socioeconomic characteristics, consumption patterns and waste management programs, but generally the level of organic in waste components is high. Among these wastes is a coffee waste. The valorization of this type of waste can contribute to the reduction of pollution, and to manufacture new building materials [1-3].

Several works has already been done on the use of coffee waste as an aggregate in the preparation of cement mortar [4], concrete [5, 6] and Eco-fired clay bricks [7]. The authors in this works explored the use of light aggregate based on coffee waste, as a composite material reducing the workability, density, the cost, the brittleness, the thermal insulation properties and fire of building materials such as concrete, mortars or bricks. The results found by various authors showed that the properties of concrete containing spent coffee grounds aggregates as lightweight aggregate are similar or even better than those of concrete containing some other commonly used natural lightweight aggregates [5, 6]. The lack of adhesion between the grain of spent coffee grounds and cement paste and the inhibition of cement hydration due to the cellular nature of SCG are the reasons indicated in several studies for the poor mechanical properties of concrete containing SCG [4].

This work focuses on the possibility of recycling waste coffee grounds without any prior treatment (SCG) as a partial replacement of natural aggregate in dune sand, in order to minimize the cost of the final material. The influence of the proportion of waste used (SCG) on properties of the new material has been studied and analyzed.

2 Experimental

2.1 Materials

The sand used in this work is a dune sand (DS) extracted from south of Algeria, with a maximum grain diameter of approximately 2 mm; and the proportion of grains smaller than 0.08 mm is below 5%. The particle size distributions of this sand are shown in Figure 1 and its physical characteristics are presented in Table 1.The X-ray Diffraction analysis of dune sands demonstrates their siliceous nature "Figure 2" and the SEM investigation reveals the rounded shape of its grains "Figure 3".

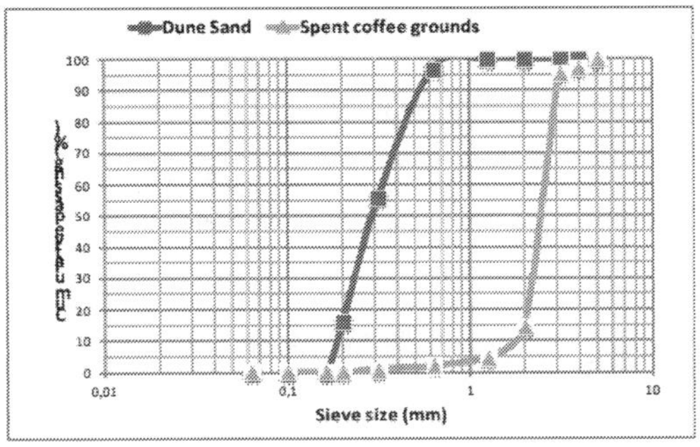

Fig. 1. Particle size distribution of sand and coffee waste used.

Table 1. Physical properties of used sand.

Properties	Sand	Standard
Bulk density (kg/m^3)	1400	NP EN 1097-3
Specific density (kg/m^3)	2580	NP EN 1097-6
Water absorption (%)	2.30	NP EN 1097-6
Sand equivalent (%)	75.50	NP EN 933-8
Fineness modulus	1.16	NP EN 933-1
Compactness (%)	0.54	NF P 18-555
Porosity (%)	0.46	NF P 18-555
Thermal conductivity (W/m K)	1.3-1.4	NF EN ISO 8894-1

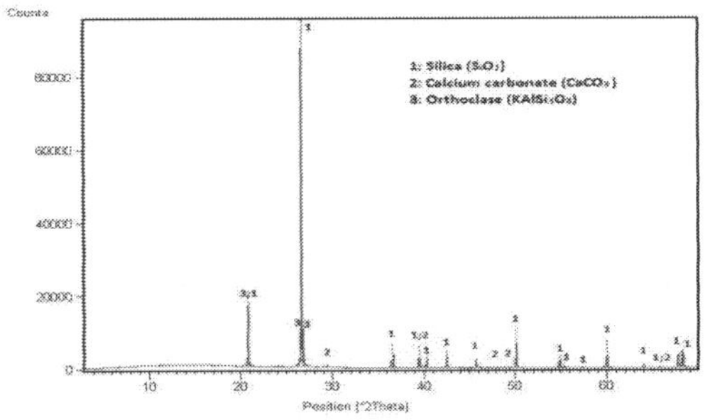

Fig. 2. X-ray diffractogram analysis of dune sand used.

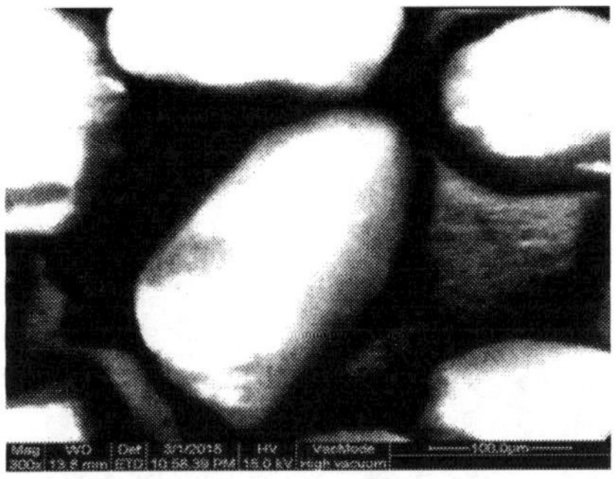

Fig. 3. Scanning electron micrographs of used sand.

Portland cement CEM II/A 42.5 from MASCARA Factory in Algeria with a density of 3100 kg/m^3 was used throughout this study. The chemical analysis of this cement is presented in Table 2.

Table 2. The chemical composition of cement used.

Materials Oxides [%]	Cement
SiO_2	23.83
Al_2O_3	6.05
Fe_2O_3	4.66
CaO	56.35
MgO	2.44
K_2O	0.83
Na_2O	0.58
SO_3	2.37
CaO_{Free}	0.76
P.F	2.23
Total	99.86

The use of fillers in sand concrete composition is essential [8]. Their use is intended to complete the grading curve of sand in its finest part in order to obtain more compact concrete, reduce the cement content and therefore the cost of concrete [9]. In this work the fillers used are the limestone powder. Their specific density is 2.857 kg/m^3, and specific surface area 310 m^2/kg.
The Spent coffee grounds (SCG) used in this work were obtained by recycling a local coffee shop (Arabica type), dried, and powdered after received and oven-dried at a temperature of 105 ° C±2. Figure 4 shows the SCG used in this work. It's a 4 mm maximum size "Figure 1". The chemical analysis of this SCG is presented in Table 3 and their physical and thermal properties are presented in Table 4.

Table 3. Chemical analysis of coffee wastes used.

Elements	Quantity (%)
Carbone (C)	49.70
Hydrogen (H)	s.d
Azotes (A)	2.3

Table 4. Physical and thermal properties of coffee wastes used.

Properties	Spent coffee grounds
Fineness modulus	2.9
Bulk density (kg/m^3)	0.387
Specific density (kg/m^3)	0.8
Compactness (%)	48.38
Porosity (%)	51.62
Water absorption (%)	0.9
Thermal conductivity (W/m K)	0.30

Fig.4. General aspect of used coffee waste.

In this work a polycarboxylate based plasticiser 'MEDAFLUID 104', with a liquid form; a chestnut color; a PH equal to 6, a density of 1.04 ± 0,01 and a content of chlore < 1g/L.
The mixing water used for the different mixes is the distribution drinking water.

2.2 Mix design

In this work, the optimal compositions of the dune sand concrete studied, without SCG, is based on the experimental method proposed from Sablocrete project [9]. The sand is substituted by volume with the SCG at dosages of 0, 5, 10, 15 and 20%. Two types of SC mixtures were made. The mix proportions of each sand concrete are shown in Table 5.

The different manufactured sans concrete mixes will be noted as follows:

- **CSC**: Control sand concrete (Without spent coffee grounds).
- **CSCG**: Sand concrete with spent coffee grounds.

Table 5. Sand Concrete mixture proportion ($1m^3$).

	CSC	CSCG			
		5%	10%	15%	20%
Sand (Kg/m^3)	1320	1254	1188	1122	1056
Cement (Kg/m^3)	350	350	350	350	350
Fillers Kg/m^3)	200	200	200	200	200
Water (l/m^3)	302.5	302.5	302.5	302.5	302.5
SCG (Kg/m^3)	00	19.34	38.64	58	77.36
Pl*(%)	1	1.30	1.35	1.86	2

*plasticizer

All specimens were produced in laboratory environment with 20°C and 50% RH. After 24 h, they were removed from the molds and placed in water at 20°C and 100% relative humidity until the day of testing. This procedure was respected for all compositions and all tests.

2.3 Tests Procedures

The fresh dune sand concrete was characterized by it corresponding slump, determined with means of the Abrams cone test (NF P 18-451), and fresh density (EN 12350-6). For hardened sand concrete mixes, flexural strength was determined by using three points test for each composition on three 4×4×16 cm prismatic samples.

The half-samples resulting from this test were then submitted to compression test on a 4×4 cm test section (EN 196-1). Sample size is significant in comparison with the size of aggregates.

In this study the porosity of dune sand concrete is measured on the base of total immersion test in water; with sample preparation under vacuum.
The thermal properties (conductivity and diffusivity) were measured using the CT-meter device (Fig. 5) on three 40×80×160 mm samples according to NF EN ISO 8894-1.

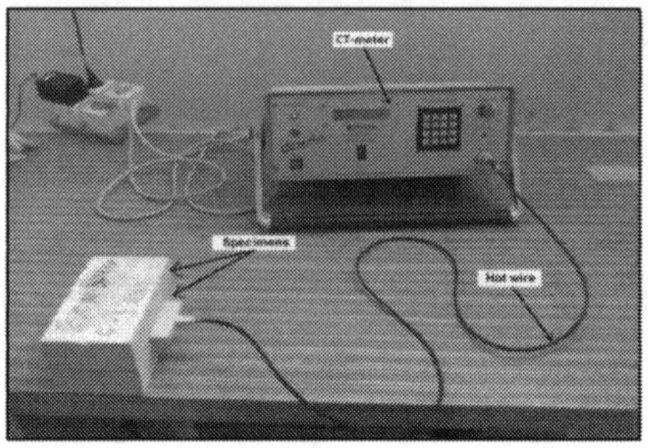

Fig. 5. Measurement of thermal properties.

3 Results and discussion

3.1 Workability

Figure 6 give the results of slump of dune sand concrete mixes as a function of SCG percentages. These results displayed that the use of SCG waste as replacement of sand contributes to decrease the slump about 77% of dune sand concrete. This reduction in value of slump probably due to the absorption of SCG grains compared to natural sand. The same result was obtained by other authors [5], who reported that the use of spent coffee grounds in concrete, involves a significant reduction in the workability.

In order to limit the number of compositions and to be able to compare them in hardened state on a common basis, it has been decided to carry out the research project at a constant workability (slump between 70 mm and 90 mm) and adjusting the workability measure by adding the a quantity of plasticizer for each mixture contains coffee waste.

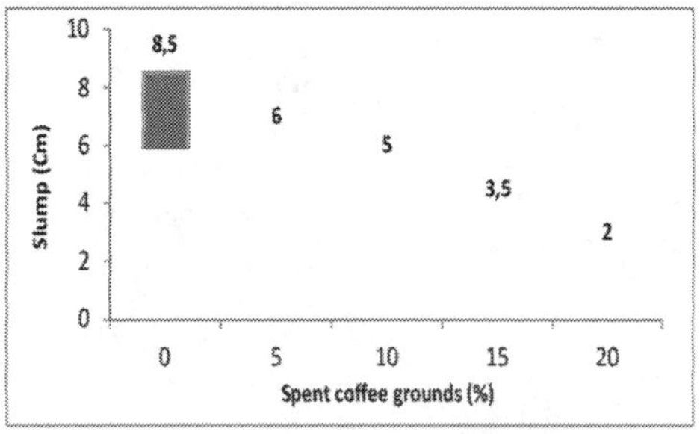

Fig.6. Slump of dune sand concrete as function of spent coffee grounds waste content.

3.2 Bulk density

In figure 7, are presented the results of bulk density for sand concrete mixes as a function of SCG content. The results showed that the use of spent coffee grounds as partial replacement of naturel sand contributes to reduce the bulk density of dune sand concrete for about of 13%. This is probably due to the lower density of the SCG compared with the natural sand. The values of fresh density decrease from 2008 kg/m^3 for mixtures without SCG waste to 1750 kg/m^3 for mixtures containing 20% of SCG for sand concrete. These results agree with those found by other researchers concrete [4, 5,10,11].

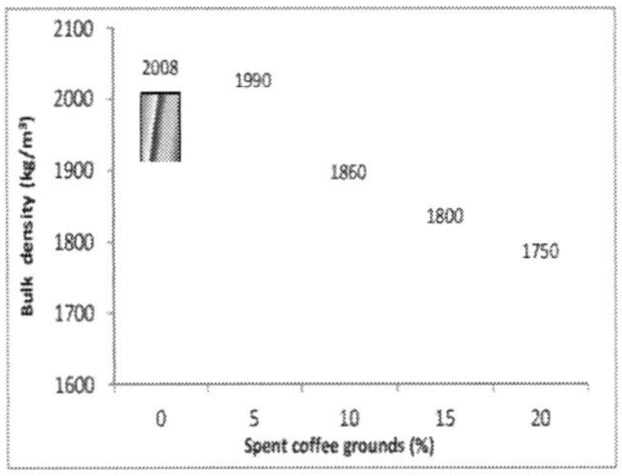

Fig.7. Bulk density of dune sand concrete as function of spent coffee grounds waste content.

3.3 Porosity

Inspecting the results in figure 8, it is possible to observe a difference in terms of porosity between reference sand concrete and all the spent coffee grounds based sand concretes. While an increase on porosity of sand concrete is showed with the spent coffee grounds aggregates content increases. This increase of porosity may be related to a poor adhesion between the spent coffee grounds aggregate surface and binder paste. This result agrees with those observed by authors in [12,13].

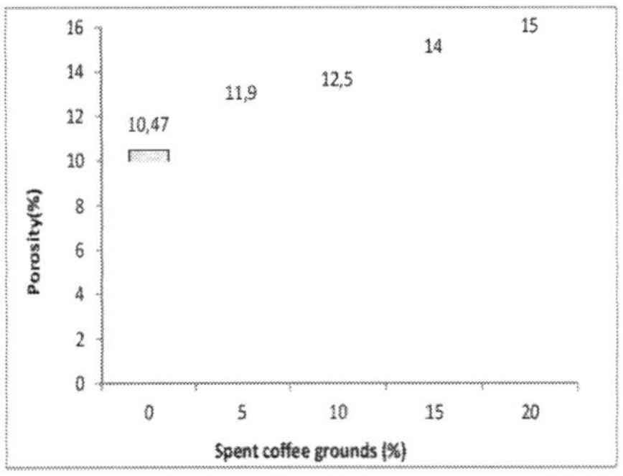

Fig.8. Porosity of dune sand concrete as function of spent coffee grounds waste content.

3.4 Mechanical strength

The compressive and flexural strength of different composites at 28 days of age are presented in Figure 9 and 10 respectively. The results show a decrease in compressive and flexural strength when the spent coffee grounds aggregates content increases (about 44% and 68% respectively with 20 % of SCG content). This trend may be related to the porosity of different dune sand concrete containing SCG, which is likely, contributes to such a strong decay of mechanical properties. The same results are shown by *Lin et al.* [4] in a study on the use of coffee residue in mortar mixture as fine aggregate replacement. For the study of *Velasco et al* [7] on the use of spent coffee grounds in Eco-fired clay bricks, the authors observed a reduction of 65 % in the compression strength which ranges from 38 to 14 N/ mm^2 when 17 % of spent coffee grounds are added.

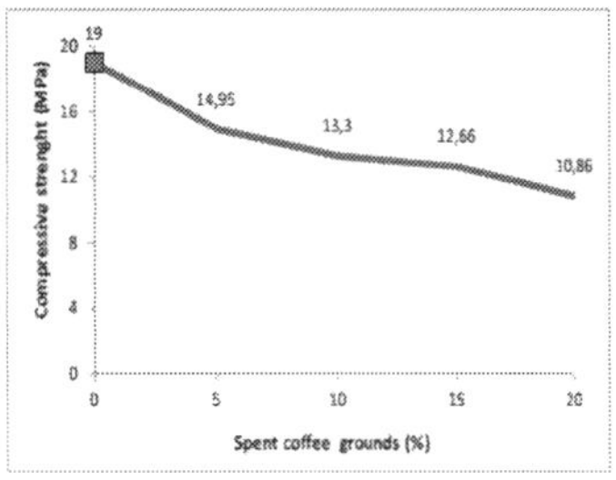

Fig.9. Compressive strength of dune sand concrete as function of spent coffee grounds waste content.

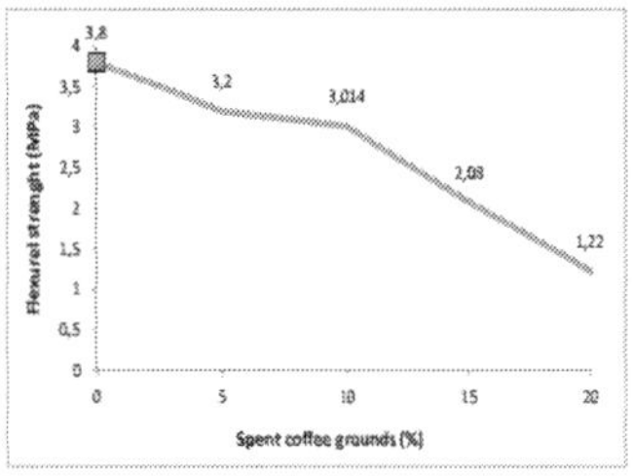

Fig.10. Flexurel strength of dune sand concrete as function of spent coffee grounds waste content.

3.5 Thermal properties

The thermal conductivity values of the specimens measured at 28 days are showed in Figure 11. While the Figure 12 shows the values of thermal diffusivity of different SC. The results show that the thermal properties of dune sand concrete decrease when the percentage of spent coffee grounds increase. Which is certainly due to the lower thermal conductivity coefficient of the SCG aggregate (0.3 W/m K) than natural dune sand (1.4 W/m K) [14]. On the other hand, the decreasing in thermal properties came also from the increase in porosity induced by the SCG. In fact the pores contain air which has a thermal conductivity (0.024 W/mK) much lower than all the other components of the sand concrete [12]. Similar results were observed by several authors in composite systems, *Velasco et al.* [7] in a study on SCG in Eco-fired clay bricks, and it's showed a decreased in value of thermal conductivity from 0.74 W/m K for clay without additive to 0.36 W/m K when 17 % of spent coffee grounds are added.

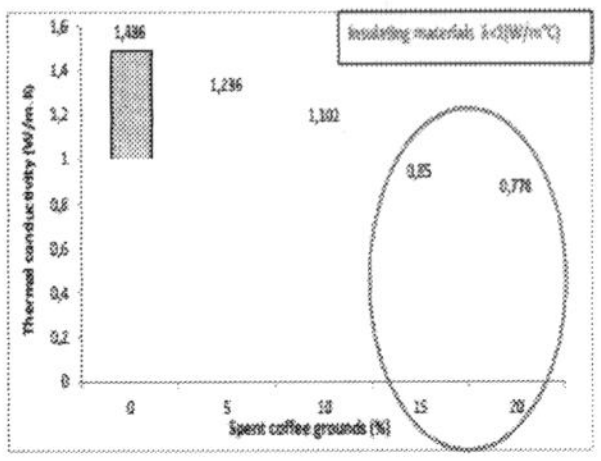

Fig.11. Thermal conductivity of dune sand concrete as function of spent coffee grounds waste content.

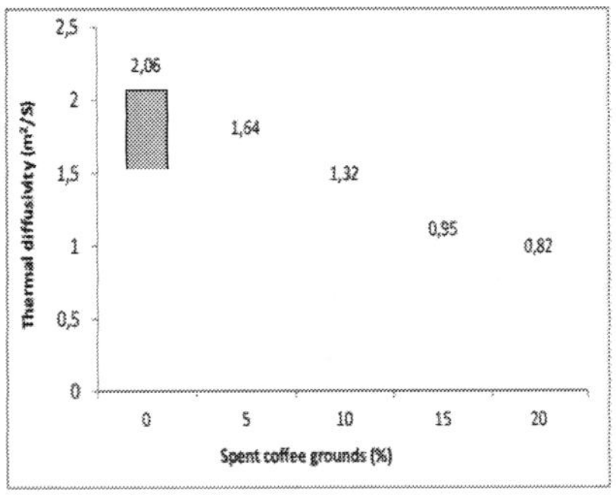

Fig.12. Thermal diffusivity of dune sand concrete as function of spent coffee grounds waste content.

4 Conclusions

This experimental work focuses on the effect of recycling coffee wastes on physical, mechanical and thermal properties of dune sand concrete. The ecological benefit of effectively utilizing this waste material is another prime motivation for this work. From the obtained results of this study, the following conclusions can be drawn:

1-The use of spent coffee grounds (SCG) as partial replacement of sand contributes to decrease the density of dune sand concrete as well as its workability of about 77%.

2-The porosity of dune sand concrete increase with incorporation of coffee wastes.

3-The mechanical properties at age of 28 days of dune sand concrete decrease when the percentage of coffee waste increases. A decrease of 44% and 68% in compressive and flexural strength respectively when 20% of SCG-aggregates are added. But this result was acceptable for lightweight concrete.

4-Spent coffee grounds replacement with natural sand in mixture caused a reduction in the values of thermal conductivity of the sand concrete (48%). The sand concretes containing 15% and 20% of SCG are regarded as insulating materials (λ<1 W/m°C).

Finally, this work insures that the recycling of coffee wastes in dune sand concrete gives a positive approach to conserve the environment and non-renewable natural material resources.

References

1. T. Tokimoto T, N. Kawasaki, T. Nakamura T, J. Akutagawa, S. Tanada. J. Colloid Interf. Sci. **281** ,56-61 (2005)
2. S. Mussatto, I. Carneiro, L. M. Silva , I.C. Roberto, J. Teixeira. *Carbohydrate Polymers*.**83** ,368-374 (2011)
3. R. Cruz , P. Baptista ,S. Cunha ,J.A. Pereira , Casal. *Carotenoids of lettuce*. **17** (2012)
4. Lee- Kuo Lin .,Tsung-Min Kuo ., Yi-Shu Hsu. *J.Mater Cycles Waste Manag*.(2015)
5. LK . Lin.,SI. Yu . J Taiwan Env Eng Assoc. **21,** 1-14 (2010)
6. YS. Hsu. Master Thesis. Institute of Civil and Disaster Prevention. Taipei. (2010)
7. P. Mun˜oz Velasco M., A. Mendı´vil M., P. Morales L., Mun˜oz.,J Mater Stru. **49**,641-650(2016)
8. AFNOR. Concrete – Sand concrete. Projet P 18-500 (1994).
9. Presse de l'Ecole Nationale des Ponts et Chaussées. *'Béton de sable : Caractéristiques et pratique d'utilisation'*. (Projet SABLOCRETE) édition. (1994)
10. M. Guendouz ,F. Debieb ,O. Boukendakdji,E.H. Kadri, M. Bentchikou and H. Soualhi.J Mater and Env Scien, J. Mater. Environ. Sci. 4 (2013) 227-232.7 (2).(2016),p.382-389.
11. M. Guendouz., F. Debieb and E.H. Kadri**.** 33èmes Rencontres de l'AUGC, ISABTP/UPPA, Anglet, 27 au 29 mai 2015 Bayonne, France, (2015)**,**p 1-8.
12 F. Iucolano,B. Liguori, D. Caputo, F. Colangelo, R. Cioffi. J.Mater Desi.**52**,916-922 (2013)
13. S. Akcaozog˘lu, Ku. Akcaozog˘lu , C. Duran Atis. Thermal conductivity WPLA. Cons and Buil Mater, 66,247-258 (2013)
14. K. Hannawi,W. Prince,S. Kamali-Bernard., J.Waste Biomass Valorization **1**,251-9(2010) .

Development of new baked bricks based on clay and sawdust

Omrane Benjeddou[1,2], Chokri Soussi[1,3], Mohamed Amine Khadimallah[4,5], Rayed Alyousef [4], Malek Jedidi[1,2]
[1]Higher Institute of Technological Studies of Sfax, Department of Civil Engineering, Sfax, Tunisia
[2]University of Tunis El Manar, National Engineering School of Tunis, Civil Engineering Laboratory, Tunis, Tunisia
[3]University of Sfax, Faculty of science of Sfax, Georessources, Materials, Environment and Global Change Laboratory, Sfax, Tunisia.
[4]Prince Sattam bin Abdulaziz University, College of Engineering, Civil Engineering Department, Alkharj, Saudi Arabia
[5]University of Carthage, Polytechnic School of Tunisia, Laboratory of systems and applied mechanics, Tunis, Tunisia

Abstract. Current research in the field of building materials is oriented towards the recovery of waste to ensure the lightening of certain building elements and energy saving. The objective of this study is to produce a new bricks by adding sawdust to clay mixtures. The first part of this experimental work consists on detremining the proper manufacturing process and cooking diagram for these new bricks. In the second part, the effect of the maximum cooking temperature, the firing phase time's of the bricks and the sawdust prportion on the physical and mechanical properties of the new bricks have been studied. The studied properties are the absorption ratio, the unit weight and the compressive strength. The results showed that the addition of an increasing amount of sawdust tends to modify the physical and mechanical properties of the cooked samples. The presence of sawdust reduces the number of connections between the clay grains and produces a lighter, more absorbent and less resistant material. Finally, the results showed that the manufacture of bricks based on sawdust is possible.

Résumé. La recherche actuelle dans le domaine des matériaux de construction est orientée vers la valorisation des déchets pour assurer l'allégement de certains éléments de construction et une économie d'énergie. L'objectif de cette étude est de produire des nouvelles briques d'argile plus légères que celles qui sont actuellement sur le marché, en ajoutant des sciures de bois aux mélanges d'argile. Plusieurs paramètres ont été étudiés ainsi que leur effet sur les modifications des propriétés physiques et mécaniques tels que le taux d'absorption, la température maximale de cuisson, la proportion des sciures et la durée de la phase de cuisson de ces briques. Un processus de fabrication, de séchage et de cuisson a été adapté au nouveau produit. Les résultats ont montré que l'addition d'une quantité croissante de sciures tend à modifier sensiblement les propriétés physiques et mécaniques des échantillons cuits. La présence de sciures réduit le nombre de connections entre les grains d'argile et produit un matériau plus léger, plus absorbant et moins résistant. Enfin, les résultats ont montré aussi que la confection des briques à base des sciures de bois est possible.

1 Introduction

La recherche actuelle dans le domaine des matériaux de construction est orientée vers la valorisation des déchets pour assurer d'une part, la pérennité des granulats naturels conventionnels et l'allégement de certains éléments de construction et d'autre part, une économie d'énergie.

L'industrie du bois génère des quantités énormes de déchets (rebus de bois, copeaux et sciures) par an dans le monde. Différentes filières de valorisation de ces déchets existent mais restent très nocives pour l'environnement. Un nouvel enjeu est de trouver des voies et moyens pour une valorisation durable de ces sous-produits. La valorisation surtout des sciures de bois dans la construction représente une solution durable.

L'utilisation des sciures de bois dans l'industrie des briques peut non seulement permettre leur valorisation de façon importante et durable mais aussi permettre la réduction de l'impact environnemental de cette industrie.

Un des objectifs de ce projet est d'élaborer de nouveaux matériaux de construction à base de sciures de bois compétitifs et concurrents des matériaux de construction classiques connus. Il s'agit de développer un nouveau produit : ce sont des briques à base d'argile et de sciures de bois. Ce matériau est obtenu en mélangeant l'argile, la sciure de bois, et l'eau.

Nous avons, à partir d'une série d'essais réalisés au laboratoire Génie Civil de L'Institut Supérieur des Etudes Technologiques de Sfax, sur la pâte d'argile-sciure de bois, déterminé la teneur optimale d'eau et le volume de pâte à utiliser pour bien remplir les moules de la presse de compactage afin d'avoir des briques bien consolidées.

Pour caractériser l'effet de l'utilisation de la sciure de bois sur la qualité des briques, nous avons soumis les briques réalisées, et ce après un cycle de séchage et de

 https://doi.org/10.1051/matecconf/201814901040

cuisson bien déterminé, à des essais tels que l'essai de la résistance à la compression, le comportement hydrique (absorption et tenue dans l'eau).

2 Le protocole expérimental

L'étude en laboratoire se subdivise en plusieurs parties :

- La première partie porte sur la caractérisation des trois échantillons de matières premières prélevées.
- La deuxième partie a pour objectif de se familiariser avec le processus de fabrication et de déterminer la méthode optimale de cuisson des nouvelles briques.
- La troisième partie analyse les modifications subies par les briques d'argile lorsque des sciures de bois sont ajoutées aux mélanges.
- La dernière partie est consacrée à l'étude de l'effet de la variation du ratio de sciures de bois sur les différentes propriétés des briques.

3 Identification des matériaux utilisés

3.1 Identification de l'argile

L'argile est broyée à l'état sec à l'aide d'un concasseur afin de l'homogénéiser.

3.1.1 Teneur en matières organiques de l'argile

L'essai consiste à déterminer la perte de masse d'un échantillon d'argile sèche dans un four à une température de 450 °C. La teneur en matières organiques $C_{MOC} = 1.2 < 3$: L'argile est classée non organique.

3.1.2 Poids volumique des grains solides

La masse volumique des grains solides de l'argile est $\rho s = 2.65$ g/cm^3. Cette valeur de la masse volumique prouve que l'argile est dans la gamme des sols inorganiques.

3.1.3 Limites d'Atterberg

Le tableau 1 présente les limites d'Atterberg, ainsi que l'indice de plasticité, l'indice de liquidité et l'indice de consistance pour l'argile testée.

Tableau 1. Les limites d'Atterberg de l'argile utilisée

Paramètre	Teneur en eau W(%)	Limite de plasticité W_p (%)	Limite de liquidité W_L (%)	Indice de plasticité I_p	Indice de liquidité I_L	Indice de consistance I_c
Valeur	0	25	52	27	1	2

Les résultats montrent que l'échantillon de l'argile a une valeur d'indice de plasticité de 27, ceci permet de les classer parmi les sols très plastiques. Par ailleurs, la limite de plasticité obtenue Wp = 25 donne une valeur approximative sur la quantité d'eau nécessaire lors de façonnage des briques (extrusion),

3.1.4 Analyse granulométrique

Les résultats de l'analyse granulométrique par sédimentation (Norme NF P 94-057) montre que notre argile est constituée à peu près de :
- 40 % de limon fin
- 20 % de limon moyen
- 12% de limon gros
- 18 % sable 0/2

Ceci nous permet de juger à priori que l'ajout du sable comme dégraissant n'est pas nécessaire.

3.1.5 Essai au bleu de méthylène et surface spécifique

Les valeurs du bleu de méthylène et de la surface spécifique de l'argile sont respectivement $V_{BS} = 1.15$ et SSB = 28 m²/g. Ces valeurs prouvent que l'argile est composée majoritairement de limon, mais le calcul de l'activité montre qu'il s'agit d'une argile très active.

3.2 Identification des sciures

3.2.1 Préparation des sciures:

Les sciures de bois sont fabriquées au laboratoire.

Fig. 1. Préparation des sciures de bois au laboratoire

3.2.2 Composition chimique

Quelques soient les essences d'arbres considérées, la composition chimique du bois est approximativement la même : 50 % de carbone, 6 % d'hydrogène et 44 % d'oxygène. La valeur calorifique, par kilogramme, est aussi constante pour toutes les espèces, soit environ 8370 KJ. Quant aux températures de combustion, elles s'établissent entre 300 °C et 500°C, la majeure partie des réactions exothermiques se terminent à des températures voisines de 500°C.

Suite à ces informations très intéressantes, les sciures de bois quelque soit leur origine se comportent de la même manière lors de l'opération de cuisson.

Le caractère fortement hygroscopique des sciures lorsqu'elles sont introduites dans la pate d'argile, fait qu'elles absorbent une partie de l'eau, c'est pourquoi, il faut prendre en compte ce paramètre lors de la préparation de la pate.

3.2.3 Détermination des masses volumiques apparente et absolue :

La masse volumique apparente et la masse volumique absolue des sciures en bois sont respectivement égales à ρa = 0.060 g/cm^3 et ρs = 0.50 g/cm^3.

4 Procédé de préparation des échantillons

Les étapes du procédé de préparation des échantillons sont les suivantes :

4.1 Malaxage à sec du mélange

On commence par ajouter les sciures à l'argile. Puis on mélange manuellement ces deux composantes jusqu'à leur homogénéisation. Enfin, ce mélange est malaxé mécaniquement dans un malaxeur électrique pendant 10 minutes.

Fig. 2. Malaxage à sec des différents composants

4.2 Malaxage de la pate

L'eau est par la suite ajoutée aux autres ingrédients secs. Le tout est mélangé manuellement jusqu'à l'obtention d'une pate homogène. Enfin, le mélange obtenu est malaxé mécaniquement dans un malaxeur électrique pendant 10 minutes jusqu'à l'obtention d'une pate plastique ayant la consistance d'une poudre humide.

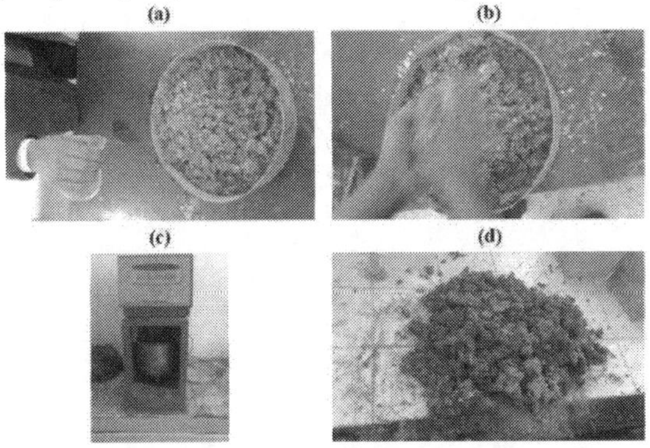

Fig. 3. Malaxage du mélange avec de l'eau

4.3 Mise en forme :

Le mélange, ayant la consistance d'une poudre humide, est versé dans un moule métallique.

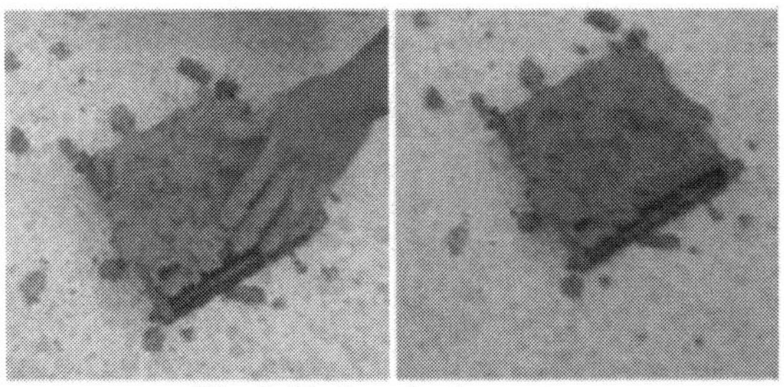

Fig. 4. Moulage des éprouvettes

4.4 Pressage

La pate versée dans un moule métallique est compressée sous une pression de 4MPa à l'aide d'une presse hydraulique de capacité 3000 KN. La pression exercée sur les échantillons lors du moulage favorise le tassement des particules d'argile les unes sur les autres, minimisant ainsi les interstices entre les grains et facilite ainsi la formation de liens céramiques assurant une bonne résistance aux échantillons cuits.

Fig. 5. Pressage des éprouvettes

4.5 Démoulage

Les briques sont alors démoulées en dévissant complètement les parois du moule.

Fig. 6. Démoulage des moules

4.6 Séchage

Les briques ont été séchées dans une enceinte climatique. Le séchage est effectué selon le procédé utilisé par la majorité des briqueteries à une température allant de 40°C jusqu'à 110 °C pendant 4 heures :

Le programme de séchage adéquat est composé des paliers suivants :

- ✓ 0.5 heure à une température de 40°C.
- ✓ 1.5 heure à une température de 60°C.
- ✓ 1 heure à une température de 80°C.
- ✓ 1 heure à une température de 110°C.

5 Détermination du programme de cuisson

5.1 Le programme expérimental

Cette partie de l'étude a pour objectif d'étudier le comportement des briques lors de la cuisson et ceci pour définir le programme de cuisson convenable pour ces briques afin de produire une brique d'argile poreuse de faible densité dont les propriétés répondent aux exigences spécifiées par les normes des briques.

Plusieurs échantillons, composés de l'argile + 4 % de sciures de bois + 20% d'eau, ont été fabriqués et cuits à des températures intermédiaires 300, 550, 850 et 1050°C. Ce programme de cuisson de 23 heures est présenté sur la figure 7.

Pour chacune des températures, la disparition progressive de la coloration noire à l'intérieur des échantillons a été notée, de même que les pertes de poids en fonction du temps de cuisson et de la température.

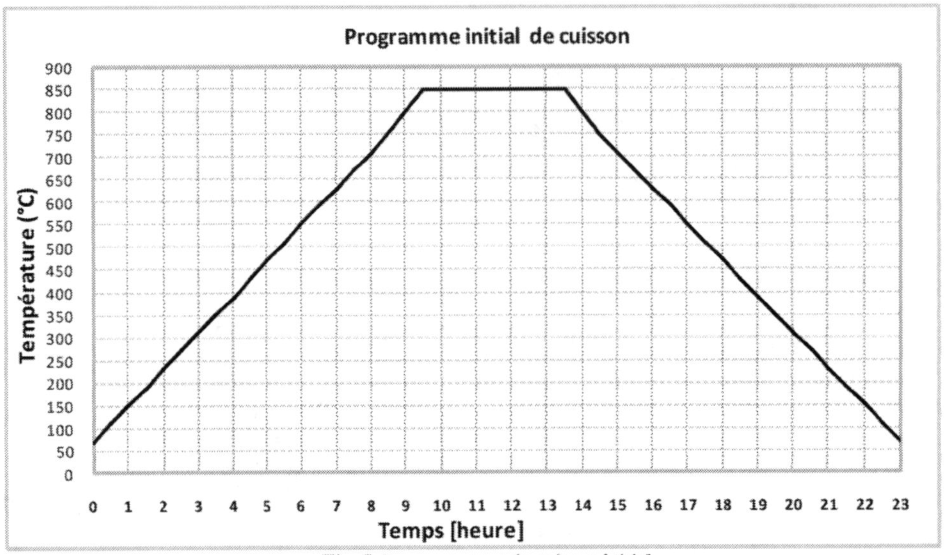

Fig. 7. Le programme de cuisson initial

5.2 Résultats et discussions

5.2.1. Inspections visuelles des échantillons

L'un des changements, immédiatement décelable visuellement, est la modification de la couleur sous l'action de la température. Les échantillons sont passés d'un gris terne après le séchage, à des tonalités de jaune, d'orange et finalement à des teintes de brun-rouge. Les changements de coloration sont habituellement occasionnés par la présence dans l'argile à l'état brut, d'oxydes ferriques (Fe_2O_3) qui changent graduellement de couleur sous l'action de la température.

En ce qui concerne l'incinération des sciures de bois prises isolément, une cuisson d'une heure à 300°C est suffisante pour que 95% du poids initial de sciures soit consumé. Sous l'action de la température les sciures se consument et créent des pores à l'intérieur des briques.

Toutefois, cette combustion est incomplète à des températures inférieures à 850°C. En effet, une teinte noire foncée prédomine, en particulier à la suite de cuisson à des températures de 300 et 400°C. A 500, 600 et 700°C, le noir s'estompe à la surface des échantillons et laisse place graduellement à un jaune-brun. Cette couleur noire est vraisemblablement due à la présence de composés de carbone.

Fig. 8. Début de l'incinération de sciures de bois à 300°C

Les résultats montrent aussi qu'après une cuisson de 4 heures à 300 °C la combustion des sciures de bois est quasi-totale. Un examen visuel indique toutefois que des traces résiduelles subsistent à la suite de cuisson à des températures plus élevées.

 https://doi.org/10.1051/matecconf/201814901040

Fig. 9. Incinération de sciures de bois à 300 °C pendant 4 h

Une inspection visuelle permet de constater que pour un temps de cuisson d'une heure à 550°C réduit progressivement la coloration noire des échantillons jusqu'à ce qu'elle disparaisse complètement après 4 heures.

Fig. 10. Etat des briques après 4 heures de cuisson à 550 °C.

Enfin, le temps et la température de cuisson influencent la combustion des sciures. Plus la température est élevée, plus les sciures se consument rapidement. Les températures ont été de 300, 550 et 850°C avant que la température soit élevée à 1050°C.

Fig. 11. Cuisson des briques à 850 °C

La figure 12 montre l'expansion d'une brique après une heure de cuisson à 1050°C. Cette expansion débute, comme telle, à partir de 850°C et progresse suite à une cuisson d'une heure à 1050°C.

En effet, l'augmentation de la température de 850 à 1050°C environ, affecte la microstructure de la brique. Les pores, généralement ronds, petits et uniformément répartis à la suite d'une cuisson pendant une heure à 1050°C environ, font graduellement place à des pores de plus grosses dimensions et de formes irrégulières. La courbe de la variation de la densité apparente des briques en fonction de la température de cuisson illustre cette progression (Figure 16).

Fig. 12. Expansion d'une brique après 1 h de cuisson à 1050 °C

Des zones noires, clairement délimitées, ont été décelées à l'endos et au centre des échantillons cuits pendant 2.5 heures à 1050°C. Ces zones ont une apparence vitreuse et une surface polie et brillante. Elles sont accompagnées sans exceptions d'une déformation des échantillons telle qu'en témoigne la figure 13. Ce phénomène, dénommé "cœurs noirs" .

Fig. 13. Cœurs noirs au niveau d'une brique

5.2.2 Variation de la résistance à la compression en fonction de la température de cuisson

A partir des courbes présentées sur la figure 15, on peut distinguer deux zones: une zone de croissance de la résistance jusqu'à une température de 850°C; et une zone de décroissance à partir de 850°C.

L'augmentation de la résistance est probablement due au fait que durant la cuisson les oxydes ferreux agissent comme fondants et que l'argile ramollie se solidifie durant la période de refroidissement et lie très fortement les particules d'argile entre elles et entrainant une augmentation marquée de la résistance mécanique des échantillons.

La chute de la résistance à partir de 850°C est vraisemblablement attribuable à la formation de cœurs noirs qui entrainent une augmentation de la porosité, de gonflements et un craquelage en surface. Ces craques rompent la continuité des liens solides établis, et diminuent la résistance en compression des échantillons.

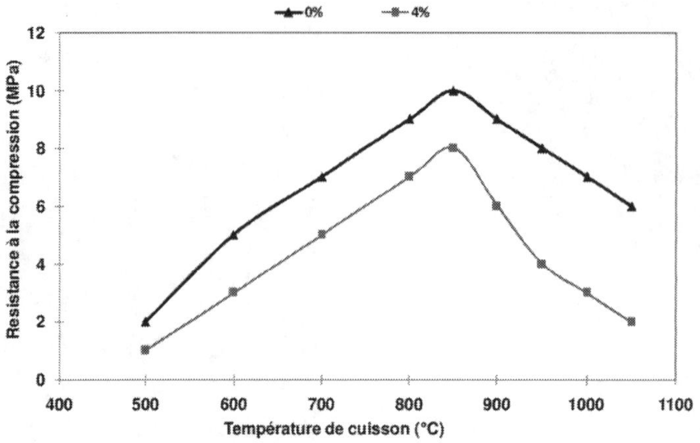

Fig. 15. Variation de la résistance à la compression des briques en fonction de la température de cuisson

5.2.3 Conclusions et recommandations :

Suite aux résultats présentés, on peut conclure qu'en variant la durée de cuisson, les meilleurs résultats ont été obtenus après 3.5 heures de cuisson à 850 °C. (Fig. 16).

Finalement, on peut adopter le programme de cuisson présenté sur la figure 17.

Fig. 16. Etat des briques après 3,5 heures de cuisson à 850 °C : Brique d'argile et brique d'argile et de sciures de bois

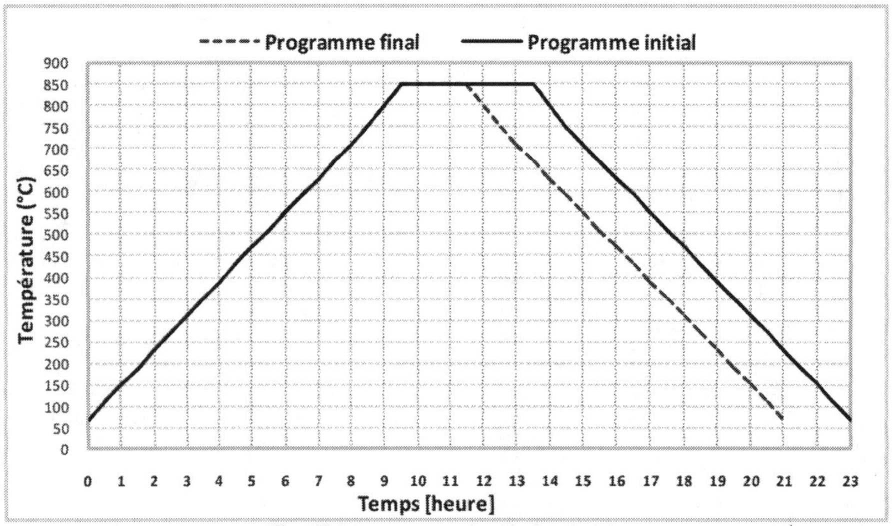

Fig. 17. Le programme de cuisson adopté

6 Variation de la quantité de sciures

6.1 Le programme expérimental

L'objectif de cette partie de l'étude est d'étudier les modifications des propriétés physiques et mécaniques qu'entraine l'addition d'une quantité variable de sciures de bois à l'argile. Pour aboutir à ces objectifs, 6 séries d'échantillons ont été ainsi préparés à partir des compositions présentés dans le tableau 2.

Tableau 2. Compositions des différents échantillons

Série	Proportion de sciures de bois (%)	Proportion d'eau (%)
S1	0	15
S2	2	18
S3	4	22
S4	6	25
S5	8	28
S6	10	30

6.2 Résultats et discussions

6.2.1 Effet de la quantité de sciures sur la densité des briques

La variation de la densité de la brique en fonction de ratio des sciures de bois est présentée sur la figure 18.

D'après cette courbe, on remarque que la densité apparente diminue de façon inversement proportionnelle à la quantité de sciures de bois; passant de 2.09 g/cm^3, pour les échantillons sans sciures, à 1.21 gr/cm^3 pour ceux constitués de 10% de sciures.

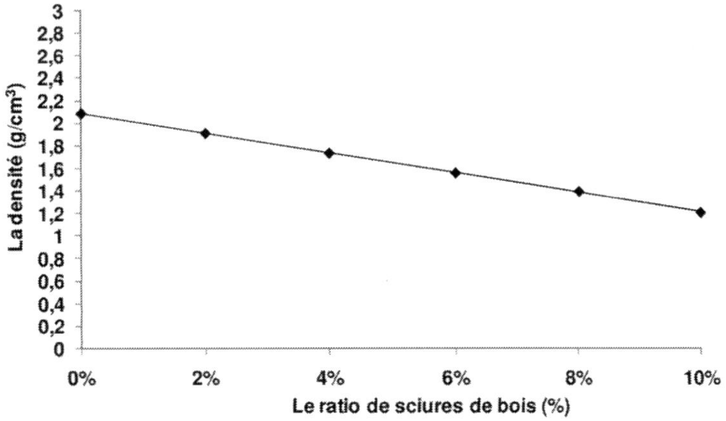

Fig. 18. Variation de la densité des briques en fonction du ratio de sciures de bois

6.2.2 Effet de la quantité de sciures sur le taux désorption d'eau des brique

D'après les résultats présentés sur la figure 19, on remarque que les taux d'absorption de l'eau augmentent de façon linéaire avec le ratio de sciures additionnées. On remarque aussi qu'au delà d'une quantité de sciures de 6%, les pourcentages d'absorption grimpent alors plus rapidement.

L'addition d'une faible quantité de sciures, 2 % ,4% et 6 % crée de petits pores uniformément distribués. Les particules de sciures semblent bien se répartir dans les mélanges et conservent à l'argile cuite sa structure homogène. L'ajout de plus de 6% de sciures provoque la création de gros pores localisés en certains endroits ou encore la formation de pores allongés. Ces grands pores constituent des passages supplémentaires qui facilitent la pénétration de l'eau ce qui explique les augmentations plus importantes des taux d'absorption.

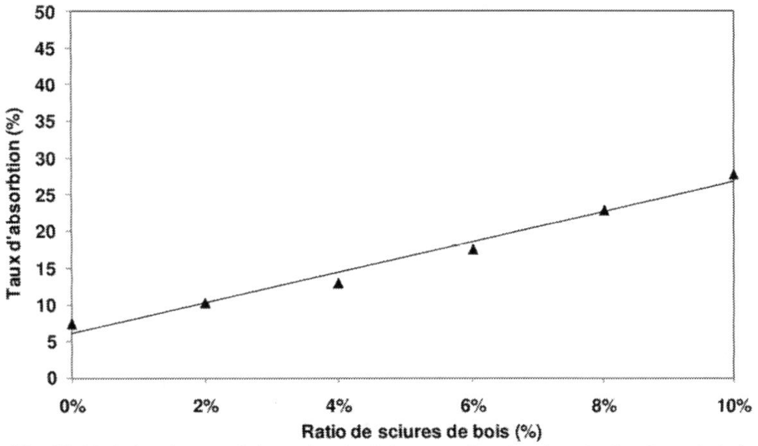

Fig. 19. Variation du taux d'absorption des briques en fonction du ratio de sciures de bois

6.2.3 Effet de la quantité de sciures sur la résistance en compression des briques

La figure 20 présente la valeur de la résistance à la compression en fonction du ratio de sciures de bois additionnés à l'argile. A partir de cette figure, on remarque que les valeurs de la résistance en compression chutent très rapidement en fonction de la quantité de sciures ajoutée aux mélanges.

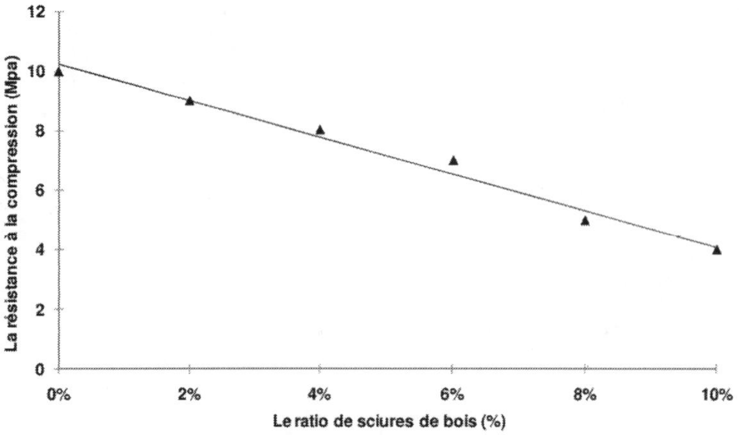

Fig. 20. Variation de la résistance à la compression des briques en fonction du ratio de sciures de bois

La diminution de la résistance lorsque on ajoute de sciures de bois à l'argile peut être expliquée par le fait que plus la quantité de sciures augmente, plus les grains d'argile sont distanciés les uns des autres, et plus la consolidation de ces grains est difficile au moment de la cuisson. En effet, les échantillons cuits offrent alors l'apparence d'une masse de particules non-cimentées qui s'effrite d'ailleurs facilement. Ce phénomène est particulièrement marqué lorsque l'ajout de sciures est plus que 6% par poids.

7 Conclusion

L'objectif principal de cette étude est de produire des briques d'argile plus légères que celles qui sont actuellement sur le marché, en ajoutant des sciures de bois aux mélanges d'argile. Il a été démontré inclusivement que l'ajout de sciures modifie de façon marquée les propriétés physiques et mécaniques des échantillons cuits et que pour optimiser ces propriétés il faut modifier de façon sensible les procédés de fabrication et de cuisson habituels des briques d'argile.

Le processus de fabrication , de séchage et de cuisson était adapté au nouveau produit pour obtenir un abaissement de la densité apparente des produits cuits à base de sciures de bois, tout en minimisant l'accroissement de taux d'absorption et en limitant le plus possible la diminution la résistance en compression.

Une température maximale de cuisson de 850°C a été considérée comme optimale, cette température étant voisine à celle ou la même que pour la température de cuisson des briques à base d'argile.

L'addition d'une quantité croissante de sciures tend donc à modifier sensiblement les propriétés physiques et mécaniques des échantillons cuits. La présence de sciures réduit le nombre de connections entre les grains d'argile et produit un matériau plus léger, plus absorbant et moins résistant.

Enfin, les résultats obtenus dans ce projet montrent que la confection des briques à base des sciures de bois est possible. Les résultats des essais mécaniques confortent cette possibilité d'utilisation.

Références

1. Al-Marahleh, G., 2005. Production of light weight ceramics tiles from local materials. Am. J. Appl. Sci. 2 (2), 778–783.
2. Demir, I., 2005. An investigation on the production of construction brick with processed waste tea. Build. Environ. 41, 1274–1278.
3. Demir, I., 2008. Effect of organic residues addition on the technological properties of clay bricks. Waste Manage. 28 (3), 622–627.
4. Dominguez, E.A., Ullmann, R., 1996. 'Ecological bricks' made with clays and steel dust pollutants. Appl. Clay Sci. 11 (2), 237–249.
5. Dondi, M., Marsigli, M., Fabbri, B., 1997a. Recycling of industrial and urban wastes in brick production: a review. Tile Brick Int. 13 (4), 218–225.
6. Dondi, M., Marsigli, M., Fabbri, B., 1997b. Recycling of industrial and urban wastes in brick production – a review (Part 2). Tile Brick Int. 13 (3), 218–225.
7. Dondi, M., Marsigli, M., Venturi, I., 1999. Microstructure and mechanical properties of clay bricks: comparison between fast firing and traditional firing. Br. Ceram. Trans. 98 (1), 12–18.
8. Pavlola, L., 1996. Use of industrial waste in brick manufacture. Tile Brick Int. 12 (3), 224–225.
9. Cusido, J.A., Cremades, L.V., Gonzalez, M., 2003. Gaseous emissions from ceramics manufactured with urban sewage sludge during firing processes. Waste Manage. 23 (3), 273–280.
10. M. Dondi, M. Marsigli, B. Fabbri, Recycling of industrial and urban wastes in brick production, Tile and Brick International 13 (3) (1997) 218– 225.
11. V. Ducman, T. Kopar, The influence of different waste additions to clay– product mixtures, Materials and Technology 41 (6) (2007) 289–293.

Study of structural short order and surface changes of SiO_2 compounds

A. Oufakir[1,2], L. Khouchaf[2], M.Elaatmani[1], A. Zegzouti[1], G. Louarn[3]and A. Ben Fraj[4]

[1]Laboratoire Sciences des Matériaux Inorganiques et leurs Applications, faculté des Sciences Semlalia, Université Cadi Ayyad Marrakech –Maroc.
[2]IMT Lille Douai, Lille Université, Cité Scientifique, Rue Guglielmo Marconi, BP 20145, 59653 Villeneuve D'Ascq Cedex-France.
[3]Institut des Matériaux Jean Rouxel, UMR 6502, Université de Nantes-CNRS, 2,14 Rue de la Houssiniére 44322, Nantes Cedex, France.
[4]Laboratoire Eco-Matériaux, CeremaldF, 120 Route des Paris – BP 216, Sourdun, 77487 Provins Cedex, France

Abstract. In this work dependence between structure, surface and reactivity of SiO_2 is investigated. Different samples are prepared with and without the presence of calcium cations and are characterized by different techniques. The FTIR frequency shifting of the bridging oxygen stretching vibration Si–O–Si and the binding energy shiftingof the Si2p and O1s photoelectron peaks have been observed and discussed. Results show the effect of presence of calcium on the structural change, surface and reactivity of SiO_2 aggregate. After reaction, a structural and SiO_2 surface changes occur, as the FTIR spectra illustrate depending on the presence of the calcium cations.

1 Introduction

The compounds based on SiO_2, are of interest in several fields: as strengthening additives in the potter's clay materials, insulator in electronic compounds and in glass industry, concrete, Polymers… [1]. Reactivity of SiO_2 depends on the chemical process that occurs between amorphous or poorly crystallized SiO_2 and the matrix, in turn depends on the surface state of SiO_2particles [2-4].SiO_2 surface and short range order is more important in bioactive silica based materials which are characterized by an open structure which enables the accommodation of alkali and alkali-earthcations. The network modifiers such as Ca, K, Na, provoke the disruption of the continuity of starting network dueto the breaking of some of the Si–O–Si bonds leading to the formation of non-bridging oxygen groups (Si–O–NBO).This is the key step of the bioactive process, because the concentration of Si–O–NBO groups controls the dissolution rate of the silica through the formation of silanol groups at the glass surface [5].

Detailed knowledge of the atomic structure of Si-O-Si species and the silanols on surface of SiO_2 compounds and the effect of calcium cations on their atomic structure are essential to understand the nature of fluids–silicates interactions and the dehydration processes in the composite materials such as nanocomposites and concrete [6]. The silanol groups (SiOH) are the main surface reactive sites and their properties (nature, concentration, distribution, accessibility, etc.) will determine the chemical activity of the silica [7]. But there is a strong relationship between reactivity and the structural order of silica compounds [8-9].

The aim of this work is to study the dependence between structure, surface state and reactivity of silica for different heterogeneous SiO_2 compounds with and without the presence of calcium during the reaction.

2 Materials and Experimental

Different type of SiO_2 compounds have been investigated. The materials used in this work are siliceous natural aggregates (called SiO2-1, SiO2-2 and SiO2-3). XRD results show that all compounds are crystalline structure similar to quartz. The results are compared with amorphous a-SiO2 compound from alpfa Aesaras reference.

The experimental protocol was carried out as follows[10]:

The mix of crushed aggregate and calcium oxide $Ca(OH)_2$ is introduced in the autoclavein an oven at 80°C.After 30 min of preheating, 10 ml KOH solution of 0.79 mol/lare added. The mix is then putted in an oven to developreaction under controlled temperature and reaction time. Afterreaction time, we put the autoclave in frozen waterduring 5 min to stop the reaction. The soluble reaction products areremoved by selective acid treatment and filtration. The acidattack is done using 250 ml cold 0.5M HCl solution.The samples are dried by acetone and diethyl ether treatment after the filtration, then kept inside dried atmosphere.

The Transmission Electron Microscope investigations were performed with a FEI Tecnai G2-20 instrument with an accelerating voltage of 200 kV. It is equipped with a filament of lanthanum hexaboride LaB6, a double-tilt holder, an EDS analysis and Gatan digital camera CCD ORIUS.

The XPS measurements were performed using a spectrometer with a monochromatized AlKα X-ray source (hν = 1486.6 eV). The data were collected on the dried and powdered samples which are pressed onto a substrate and is fixed on a sample holder. The samplesare introduced successively in a chamber at a constant pressure (P = 10^{-3} torr) and in an analytical chamber (P = 10^{-9}torr). The ejected electronsphotoelectrons whose binding energies are inferior to the X-ray energy are collected according to their kinetic energies into a hemispherical 150 mm mean radius electron analyzer.

FTIR spectra were acquired with a Brucker VERTEX 70 spectrometer in reflection mode. They were recorded by collecting 100 scans at 4cm^{-1} resolution in the range of 400–5000 cm^{-1}.

3 Results and discussion

Figure 1 shows an image of the sample SiO2-1. Grains of different sizes ranging from ten nanometers to a few micrometers, with angular sides characterizing angles of quartz can be seen. Moreover, other areas of a rather disordered or amorphous appearance are observed. The samples SiO2-2 and SiO2-3 are the same appearance except SiO2-3 which presents more heterogeneous aspect.

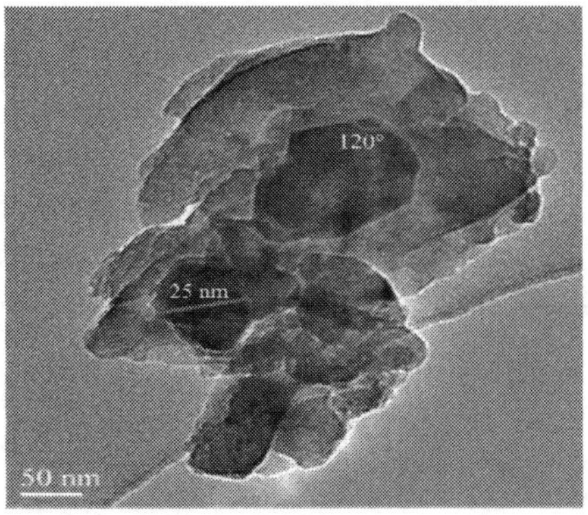

Fig. 1. Transmission Electron Microscope image of SiO2-1 sample.

Analysis of different zones of the samples shows the presence of silicon and oxygen elements as shown in Figure 2. Carbon and copper presence is attributed to the sample holder.

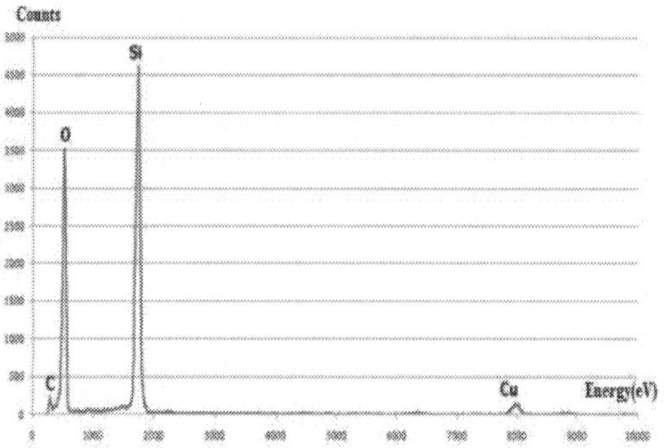

Fig.2. Energy dispersive spectrum of SiO2-1

In order to follow the reactivity of each aggregate, we subjected them to reaction according to the procedure described above. Figure 3 shows the results of the mass recovered after the reaction in the absence of calcium. It is observed that the mass recovered after reaction of the amorphous silica is the lowest, which confirms the expected results since it has the most disordered structure and therefore is the most reactive. In fact, the samples can then be classified in order of increasing reactivity, ranging from SiO2 -2, SiO2 -1, SiO2 - 3 to a-SiO2. These results are in good agreement with those obtained by X-ray diffraction [8].In the presence of calcium, the recovered masses are most important, which shows the reactivity of these compounds decreases in the presence of calcium in agreement with previous study [10].

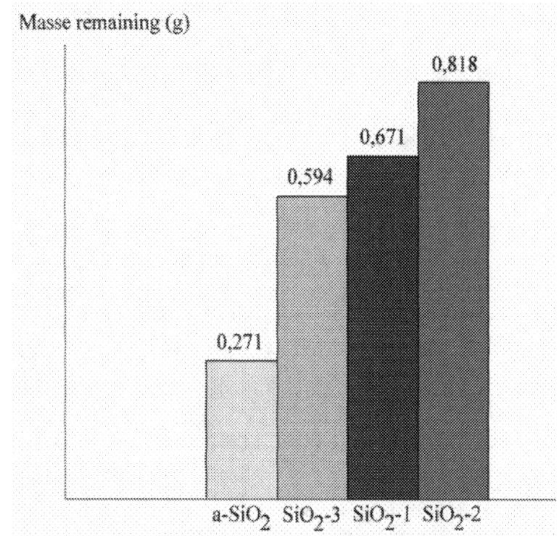

Fig. 3. Evolution of the mass remaining after reaction without calcium.

In order to follow the evolution of the structure as well as the state of the surface, IR spectra of the different samples are recorded. The SiO2-1 as an intermediate case is showed as example.

Figure4 shows the FTIR spectra for SiO2-1 after reaction with and without the presence of calcium compared to SiO2-1 before reaction as reference. These FTIR spectra present the surface species as indicated in the figure.

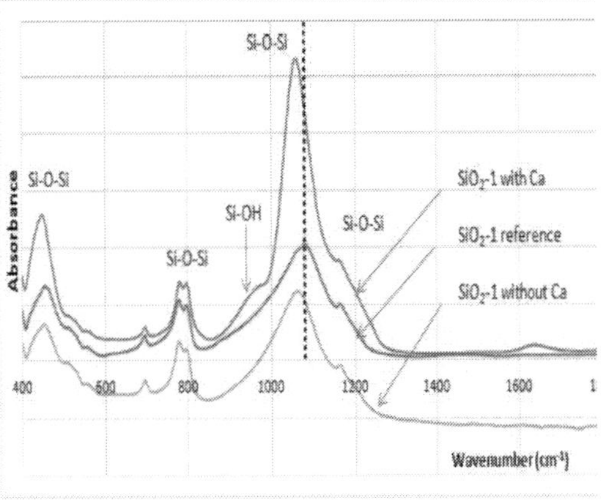

Fig.4. FTIR spectra of SiO2-1 after reaction with the presence and the absence of calcium compared to SiO2-1 reference before reaction.

The infrared spectra of all samples have a broad band between 1000 cm^{-1} and 1300 cm^{-1}. This band consists of a strong peak. The mean peak is associated with the stretching vibration of Si-O-Si [11-12]. In addition, different bending vibrations attributed to Si-O-Si [13-15] are indicated in the figure. The bands located at 555 cm^{-1} and 950 cm^{-1} are associated with Si-O bending vibrations of no bridging Si-OH bonds and stretching vibration respectively Si [13-15].

In addition, we can notice that the mean band of SiO2-1 spectra after reaction with and without the presence of calcium is shifted compared to SiO2-1 before reaction as reference. This pronounced shift may be connected to the structural order change of SiO_2 [13]. Other structural change of Si-O-Si is provided by the band at 800cm^{-1}(Figure5).

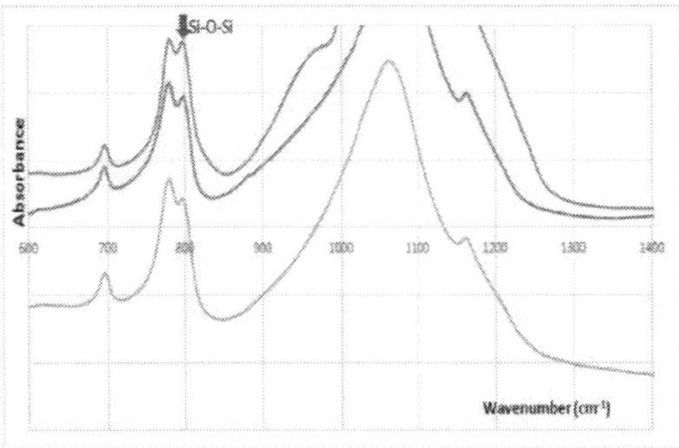

Fig.5. FTIR spectra of SiO2-1 after reaction with the presence and the absence of calcium compared to SiO2-1 reference before reaction.

Furthermore, it can be seen that SiO2-1 after reaction in the presence of calcium has a more intense peakassociated to Si-OH groups and located at 950 cm^{-1} compared to SiO2-1 before reaction as reference and after reaction without calcium. The peak height from silanols groups increases in SiO2-1 with presence of calcium in comparison of SiO2-1 reference and SiO2-1 without the presence of calcium. This result shows that the surface state of SiO2-1 with the presence of calcium is different compared to SiO2-1 reference and SiO2-1 without the presence of calcium.

The XPS spectra were carried out to characterize the surface compositions of samples. Figure **6** presents O1s and Si2pXPS spectra of SiO2-1 after reaction with the presence and the absence of calcium compared to SiO2-1 reference before reaction. Major elements detected are silicon and oxygen for all the samples.

Samples exhibit a clear binding energy difference between the Si2p and O1s lines. In addition, the relative intensities of the Si2p peak and of the O1s peak vary from one sample to another; i.e., the samples do not display the same Si/O atomic ratio. These features evidence that the chemical environment of siliconand oxygen is significantly affected by the presence of calcium more than the SiO2-1 sample with absence of calcium.

It is important to notice that the results represent typically the surface difference between the different samples confirming the changes and the results obtained with FTIR as explained before.

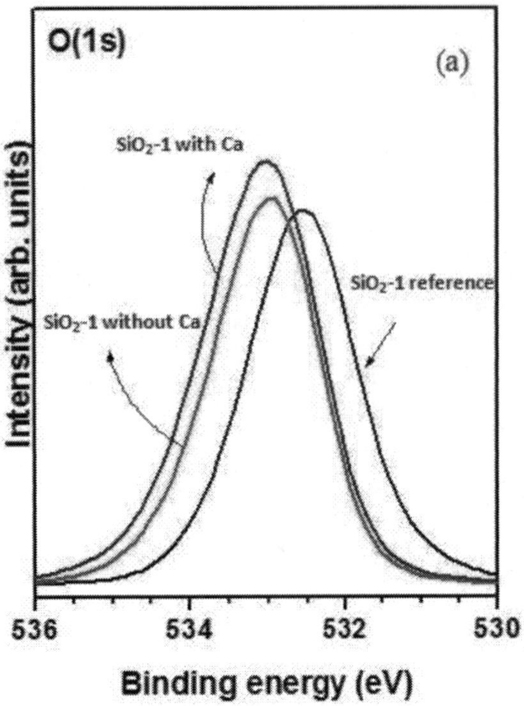

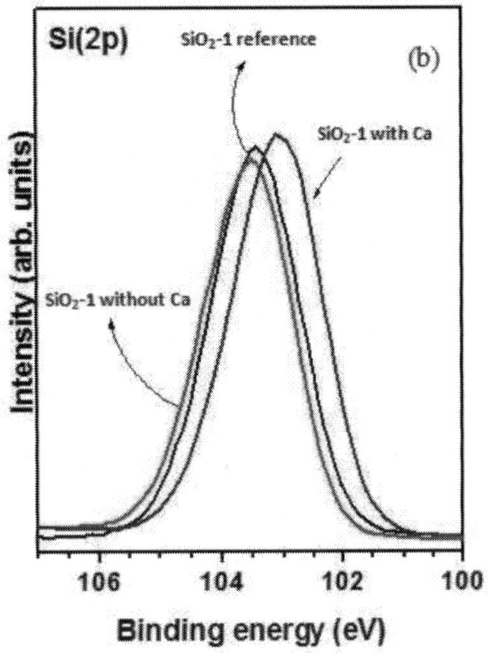

Fig.6. XPS spectra of SiO2-1 after reaction with the presence and the absence of calcium compared to SiO2-1 reference before reaction, (a) O1s, (b) Si2p

The O1s photoelectron spectra for the samples are shown in Fig. 6a. The O1s signal is expressed by a symmetric function and it shifts towards lower energies with the presence of calcium. A similar behavior has been observed for the Si2p photoelectron peak, but shifts towards higher energies as shown in Fig. 6b typically with the presence of calcium.

Crystallized SiO_2 samples are formed by an ordered network with perfect SiO4 tetrahedral connected to the adjacent tetrahedral by bridging oxygen atoms to form a three-dimensional connected network. Consequently, the contribution of Si–OH infrared band is negligible as shown in Figure4. The presence of calciumwithSiO_2 during the reactionchanges the degree of connectivity in the silicon– oxygen disordered network. In fact, the short range structural modifications occur as can be successfully detected by FTIR and XPS spectra.

In fact, the pronounced shift of the position of the peak and the increase of silanols surface species with the presence of calcium show a simultaneous modification of the structure and the surface of the starting aggregate after reaction. These results are interesting in a future study concerning the improvement of the interface properties of some composites materials with the presence of different silica compounds.

4 Conclusion

Dependence between structure, surface and reactivity of SiO_2 is investigated. Different SiO2 samples with and without the presence of calcium cations are prepared and characterized by different techniques. Based on FTIR and XPS spectroscopy, frequency shifting of the bridging oxygen stretching vibration Si–O–Si and the binding energy shifting of the Si2p and O1s photoelectron peaks have been observed and discussed. Results show the effect of presence of calcium on the structural change, surface and reactivity of SiO_2 aggregate. After reaction, a structural order and SiO_2 surface changes occur, as the FTIR spectra illustrate depending on the presence of the calcium cations. TGA/TD and NMR experiments are planned in order to obtain more details about the silanols behavior.

This research was supported by European FEDER and region Nord-Pas-de-Calais.

References

1. T. H. Zhou, W. H. Ruan, J. L. Yang, M. Z. Rong, M. Q. Zhang, Z. Zhang, A nouvel route for improving creep resistance of polymers using nanoparticles, *Composites Science and Technology*, **21**, 539 (2007).
2. F. Boinski, L. Khouchaf, and M-H. Tuilier, Study of the mechanisms involved in reactive

silica, *Materials Chemistry and Physics,* **122,** 311–315, (2010).
3. L. Khouchaf, A. Hamoudi and P. Cordier Evidence of depolymerisation of amorphous silica at medium and short range order: XANES, NMR and CP-SEM contributions, *Journal of H. Materials*, **168**, 1188, (2009).
4. A. Hamoudi, L. Khouchaf, C. Depecker, B. Revel, L. Montagne, P. Cordier, « Microstructural evolution of amorphous silica following Alkali-Silica Reaction » *Journal of non Cryst. Solids*, **354**, 45-46 5074, (2008).
5. T. Kokubo, in: T. Yamamuro, L.L. Hench, J. Wilson (Eds.), *Handbook of Bioactive Ceramics*, CRC, Boca Raton, FL, p. 41,(1990).
6. Qiang Huang, Meiying Liu, Liucheng Maoa, Dazhuang Xu, Guangjian Zeng, Hongye Huang, Ruming Jiang, Fengjie Deng, Xiaoyong Zhang, Yen Wei, Surface functionalized SiO2 nanoparticles with cationic polymers via the combination of mussel inspired chemistry and surface initiated atom transfer radical polymerization: Characterization and enhanced removal of organic dye, *Journal of Colloid and Interface Science,* **499**, 170–179, (2017).
7. Jean-Paul Gallas, Jean-Michel Goupil, Alexandre Vimont, Jean-Claude Lavalley, Barbara Gil, Jean-Pierre Gilson, and Olivier Miserque, Quantification of Water and Silanol Species on Various Silicas by Coupling IR Spectroscopy and in-Situ Thermogravimetry, Langmuir, **25**,5825–5834, (2009)
8. H. El Bahraoui, L. Khouchaf, A. Ben Fraj, "Microscopical and mechanical evaluation of the durability of SiO2 aggregates", *European Physical Journal of Applied Physics,***74,** 2, (2016).
9. N. Tahiri, L. Khouchaf, M. Elaatmani, G. Louarn, A. Zegzouti and M. Daoud, "Study of the thermal treatment of SiO2 aggregate," *IOP Conf. Series: Materials Science and Engineering*, vol. **62**, doi:10.1088/1757-899X/62/1/012002, (2014).
10. J. Verstraete , L. Khouchaf, and M. H.Tuilier , « Contributions of the Environmental Scanning Electron Microscope and X-ray diffraction in investigating the structural evolution of a SiO2 Aggregate attacked by Alkali-Silica Reaction », *Journal of Materials Science*, **39**,**20**, 6221-6226, (2004).
11. G. N.Barbosa and H. P.Oliveira, Synthesis and characterization of V2O5–SiO2 xerogel composites prepared by base catalysed sol–gel method,*J. Non-Cryst Solids,* **352,** 3009-3014, (2006)
12. Z. L.Chen and P. Shen, Thermally activated sintering–coarsening–coalescence-polymerization of amorphous silica nanoparticles, *J. Ceramics International,* **39,** 2365-2373, (2012)
13. M.G. Garnica-Romo, J.M. Yanez-Limon, M. Villicana, J.F. Perez-Robles, R. Zamorano-Ulloa and J. Gonzalez-Hernandez, Structural evolution of sol–gel SiO2 heated glasses containing silver particles,*J. Phys. Chem. Solids,***65,** 1045-1052, (2004).
14. P. Schmidt and F. Fröhlich, Temperature dependent crystallographic transformations in chalcedony, SiO2, assessed in mid infrared spectroscopy,*J. Spectrochimica Acta* A **78,** 1476-1481, (2011).
15. H. El Rassy and A.C. Pierre, NMR and IR spectroscopy of silica aerogels with different hydrophobic characteristics,*J. Non-Cryst Solids,* **351,** 1603-1610, (2005).

Lightweight geopolymer composites as structural elements with improved insulation capacity

Glikeria Kakali, Dimitris Kioupis, Aggeliki Skaropoulou and Sotiris Tsivilis
National Technical University of Athens, School of Chemical Engineering, Athens, Greece

Abstract. This study concerns the development of lightweight fly ash based geopolymers which can be applied as alternatives to the traditional lightweight concrete. Different kinds of expanded polystyrene were used as lightweight agents. The results showed that lightweight geopolymers were successfully prepared, exhibiting compressive strength and density in the range 7.70 – 29.57 MPa and 0.97 – 1.57 g/cm^3, respectively. The product containing 3% w/w of commercial expanded polystyrene possesses low thermal conductivity (0.16 W/mK) combined with sufficient mechanical strengths (11 MPa), excellent stability and fire resistance while its water absorption is comparable to that of conventional construction materials (cement mortars, concrete).

1 Introduction

Lightweight building elements generally exhibit improved thermal insulation and relatively good strength. The main benefits by the application of these elements are associated with the reduction of the total load of the structure making savings in foundations and reinforcement as well as the improved thermal properties. Such elements usually consist of a cementitious binder and a lightweight aggregate (polystyrene beads, expanded perlite etc.). Traditionally, the cementitious material is OPC. Geopolymers or alkali activated materials are produced via the alkali activation of aluminosilicate materials and can be considered as alternatives to OPC [1-3]. The main advantages of geopolymers are the use of wastes and by-products as raw materials as well as the very low embodied energy and CO_2 footprint compared to the conventional cementitious materials [4, 5].

This study concerns the development of lightweight geopolymer composites with enhanced insulation capacity and sufficient mechanical strength, for application as structural elements in energy efficient buildings. For the geopolymers production, a type F fly ash (Megalopolis, Greece) was used as the main aluminosilicate source. In order to reduce the density of the products, different types of polystyrene particles with varying shapes and sizes were tested at a content ranging between 0.5-3.0% w/w (fly ash based). In all cases, compressive and bending strength, apparent density and thermal conductivity were measured. In order to examine the durability of the products, moisture absorption and sorptivity measurements were conducted while the fire resistance was evaluated through flammability tests.

2 Experimental

The fly ash used in this study comes from the power station of Megalopolis in Greece. This material can be classified as Type F according to ASTM C 618, as the sum of the oxides SiO_2 + Al_2O_3 + Fe_2O_3 is higher than 70%. Fly ash was previously ground at a mean particle size (d_{50}) approximately 20 μm. This is the typical fineness of fly ash when used as main constituent in blended cements. The chemical composition of fly ash is presented in Table 1.

Table 1. Chemical composition of fly ash (% w/w).

SiO_2	Al_2O_3	Fe_2O_3	CaO	MgO	K_2O	Na_2O	SO_3
45.52	22.13	8.99	14.75	2.84	1.55	0.72	1.69

The synthesis of the lightweight geopolymers includes three steps: a) the preparation of the activation solution, b) the mixing of the raw materials (fly ash and lightweight agents) with the activation solution and c) the molding and the curing of the specimens. The activation solution contains soluble Si in form of sodium silicate ($3.34SiO_2.Na_2O$, Multiplass Greece) and NaOH (≥98.0%, CAS No. 1310-73-2) in order to provide the alkaline environment that is necessary for the activation of the geopolymerization reactions. The activation solutions were stored for a minimum of 24 h prior to use, to allow equilibrium. In all mixtures, the silicon to aluminum (Si/Al) and the sodium to aluminum (Na/Al) molar ratios as well as the solids to liquids (s/l) mass ratio were kept constant to 2.4, 0.85 and 2.8, respectively [6].

In order to prepare lightweight geopolymers, the incorporation of lightweight additives such as expanded polystyrene (EPS), was investigated. Six different types of expanded EPS were tested. In particular, we used 3 types of commercial EPS and 3 types developed by Fraunhofer ICT institute. The extent of EPS incorporation ranged between 0.5-3.0 % w/w on fly ash basis. Table 2 shows the EPS types along with their characteristics. The lightweight agents were effectively mixed with the fly ash in order to obtain a homogenous blend. Then, this blend and the activation solution were mechanically mixed (standard mortar mixer: Controls 65-L0005) to form an homogenous slurry which was poured to 50x50x50 mm cubic molds and mildly vibrated. The specimens were left at room temperature for 2h and then were transferred to an oven at 70°C for 24h [6]. Compression and bending tests were carried out on a Toni-technik uniaxial testing press, 7 days after the specimens' preparation (load rate 1.5 kN/s, according to the EN196-1 requirements). For each sample three specimens were prepared and tested.

Table 2. EPS types applied in this study.

EPS code	EPS type	Incorporation range (%)
Com1	Commercial - recycled	0.5-3.0
Com2	Commercial - Sphere 1-3mm	0.5-3.0
Com3	Commercial - Sphere 3-5mm	0.5-3.0
FH1	1% polywex / 6% pentane	1.0-2.0
FH2	1% grapheme / 6% pentane	1.0-2.0
FH3	Coperion	1.0-2.0

Geopolymer specimens were submitted to moisture absorption tests at laboratory conditions and in water. In the first case, the examined specimens are exposed to environmental conditions and the mass changes are recorded versus time. The absorption tests in water involve the immersion of the samples into tap water and the record of the mass till variations not greater than 0.2 % are observed.

Sorptivity tests measure the rate of water absorption by capillary suction of the specimens placed in contact with the water [7]. In such tests, the low part of specimens' sides is covered with adhesive tape. Then, the specimens are submerged into water in a 5 mm depth and the mass increase is recorded after 5, 10, 20, 40, 60, 90 και 120 min. In order to export the sorptivity values of the samples, diagrams of water absorption per unit area (g/mm^2) vs square root of time ($t^{0.5}$) were plotted. The slop of the curves indicates the sorptivity magnitude.

The thermal conductivity of the samples was determined by means of the Heat Flow method, on a Netzsch HFM 436 lamda Heat Flow Meter, which meets the EN12667 and ASTM C518 requirements. The samples were left at standard laboratory conditions before testing. The mean temperature was set at 10°, with 10 degrees temperature difference.

For the evaluation of the flame behavior of the lightweight geopolymers, JAR-25 standard (§25.853 (a) and Appendix F, Part I, para (a) (1) (ii), AITM 2.0002 B) was applied. For conducting these tests, three prismatic specimens (30.5x9.5x3cm) for each sample were prepared.

3 Results and discussion

3.1 Mechanical properties

Figure 1 presents the effect of common expanded polystyrene (Com2) on the compressive strength and the density of the produced geopolymers. As it was expected, the incorporation of expanded polystyrene in the geopolymeric matrix favours the production of lightweight materials However, this reduction is accompanied with a more intense decrease of compressive strength values. The addition of 3% w/w of Com2 decreases the density by 40%, with the reduction of strength being much higher (80%). Despite the low content of expanded polystyrene, its volume fraction in the geopolymer is high and the dilution of binding matrix causes a severe decrease of compressive strength. Similar behavior is observed for the whole EPS series. Depending on the kind and percentage of polystyrene, lightweight geopolymers with compressive strength in the range 7.70 – 29.57 MPa and apparent density in the range 0.97 – 1.57 g/cm^3 were developed. The reference geopolymer (0% w/w EPS) exhibits 56 MPa and 1.77 g/cm^3 of compressive strength and density, respectively.

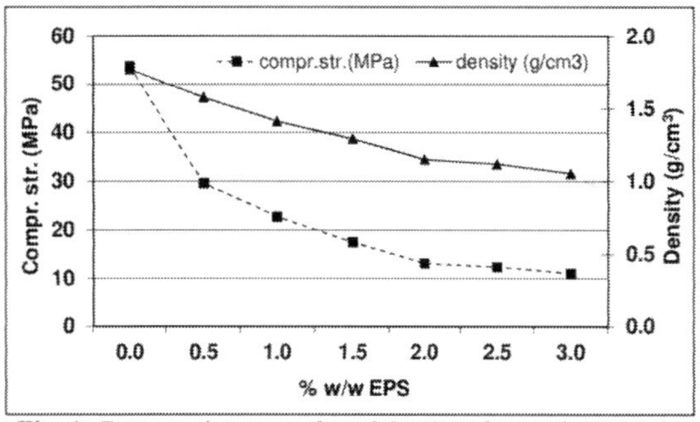

Fig. 1. Compressive strength and density of geopolymers with Com 2 expanded polystyrene.

Figure 2 shows the compressive strength and the density of the geopolymers prepared with the highest incorporation level of the different EPS types. In all cases, the compressive strength of the geopolymers is around 10 MPa showing that the specimens keep a sufficient mechanical strength. The density of the corresponding geopolymers varies from 0.97 to 1.33 g/cm^3. The density of the sample prepared by FH2 EPS is too high for the development of a structural element with enhanced thermal behavior. The lowest density was found in geopolymer with FH1 (1% polywex and 6% pentane). However, the 3% w/w Com2 geopolymer keeps the best combination of properties for the development of a lightweight structural element (C = 11MPa and d = 1.05g/cm^3) and therefore it was promoted for further investigation.

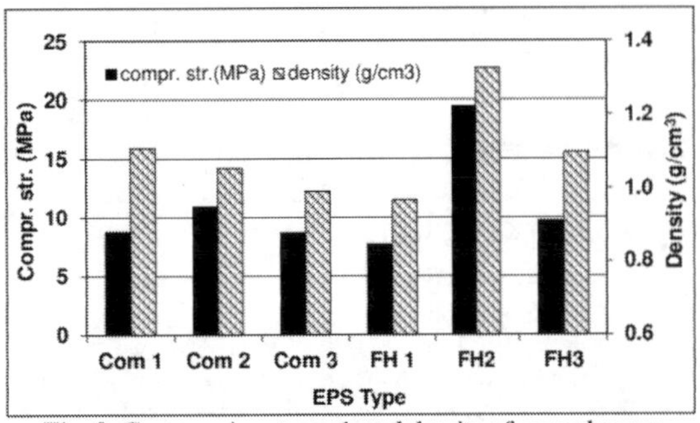

Fig. 2. Compressive strength and density of geopolymers prepared with different types of expanded polystyrene.

The flexural strength of the reference and the 3% w/w Com2 geopolymers was measured 6.03 and 2.15 MPa, respectively.

3.2 Durability tests

Figure 3 presents the mass variations of the reference and the lightweight (3% w/w Com2) geopolymer when exposed to environmental conditions for a period of 370 days. The results showed that both materials exhibit near zero mass variations. In addition, a stabilization of the mass values of the two materials was observed after approximately 150 days. In particular, the reference and the 3% w/w Com2 geopolymer exhibit ±1.05% and ±0.69% mass variations after 370 days of exposure, respectively. The EPS geopolymer presents a greater mass stability in relation to the reference geopolymer. This can be explained in the basis of the tendency of the geopolymer matrix to absorb water. In the case of the EPS product the geopolymer matrix is highly diluted by EPS and so the water absorption is considerably lower.

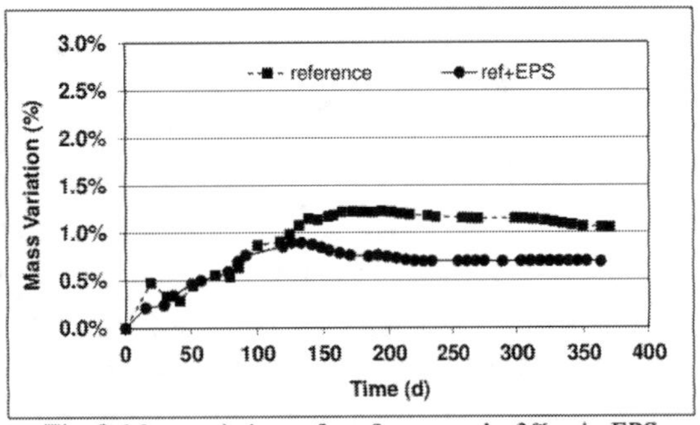

Fig. 3. Mass variations of a reference and a 3% w/w EPS geopolymer when exposed to environmental conditions.

The water immersion tests showed that the lightweight geopolymer exhibits a water absorption of 11.8%. This water absorption value is comparable to conventional building materials. For example, cement mortars present water absorption 5-12%, depending on their compressive strength. Figure 4 presents photos of 3% w/w Com2 specimens which were immersed in water for 270 days. The specimens remained unaffected by water, maintaining their cubic shape with no cracks or surface damages even after 270 days of immersion.

Fig. 4. Photos of 3% w/w Com2 specimens after immersion into water for 270 days.

Figure 5 presents the curves of % water absorption vs. time for the reference and lightweight samples as they were exported from sorptivity tests. As it is seen in figure 5, the lightweight geopolymer (3% w/w Com2) exhibited significant lower water absorption values in comparison to the reference geopolymer. This fact can be explained in terms of EPS nature. It seems that geopolymeric matrix has a higher ability to absorb water than EPS the volume fraction is high (~80%), resulting to significantly reduced water absorption.

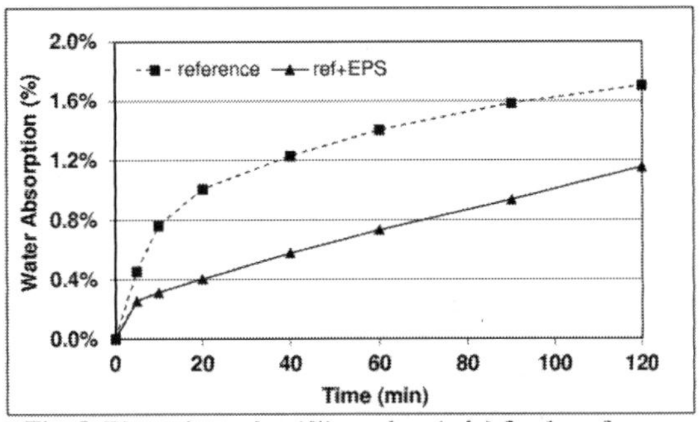

Fig. 5. Water absorption (%) vs. time (min) for the reference and 3% w/w Com2 specimens.

In order to export the sorptivity values of the samples, the curve of water absorption per unit area (g/mm^2) vs square root of time ($t^{0.5}$) were plotted. The slop of the curve indicates the sorptivity magnitude. An indicative sorptivity diagram is showed in Figure 6 for a 3% w/w Com2 specimen. In order to measure the sorptivity of the reference and 3% w/w Com2 geopolymers, three specimens of each synthesis were prepared and tested.

The mean sorptivity value of 3% w/w EPS specimens was calculated at 0.051 mm/min$^{0.5}$ while that of the reference specimens was three times larger (0.150 mm/min$^{0.5}$). These values are comparable or even lower to the typical sorptivity values of concrete (0.09 – 0.17 mm/min$^{0.5}$) [7].

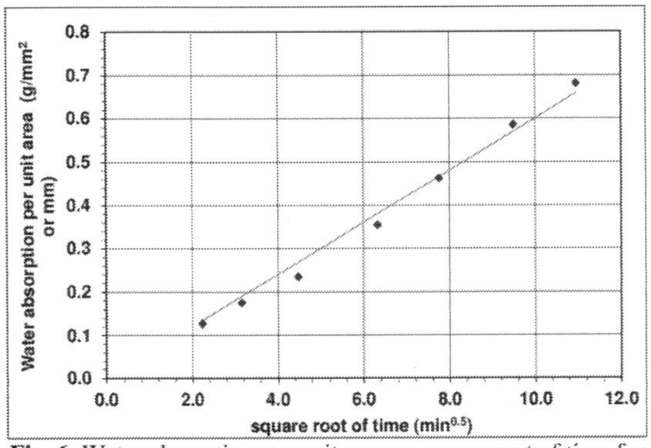

Fig. 6. Water absorption per unit area vs square root of time for the specimen with COM2 (3% w/w)

3.3 Thermal conductivity

Figure 7 shows the density and thermal conductivity of the reference geopolymer and the geopolymer with EPS (3% w/w). It is observed that the values of both thermal conductivity and density, drastically, decrease with the incorporation of the expanded polystyrene. Considering the high volume fraction of EPS in the matrix, the geopolymer volume that can accumulate heat is reduced and so does the conductivity. In particular, the thermal conductivity values of the reference material (0% w/w EPS) and the lightweight geopolymer (3% w/w Com2) were 0.30 and 0.16 W/mK, respectively.

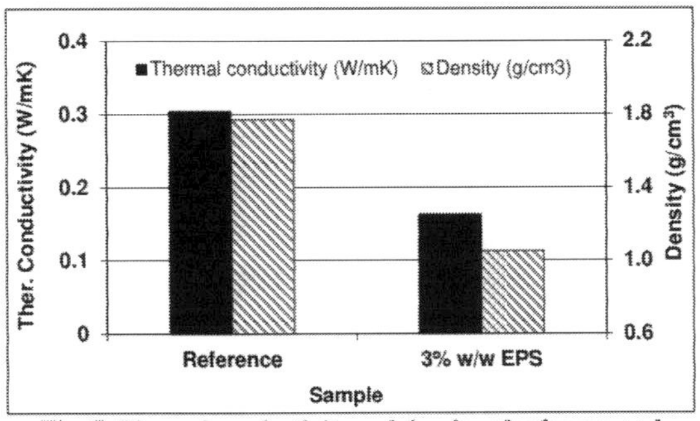

Fig. 7. Thermal conductivity and density of reference and lightweight geopolymers.

3.4 Fire resistance

Figure 8 presents the configuration of the flammability test along with a photo of the burned sample. All geopolymer samples did pass the JAR-25 test. From photo b, it is obvious that even if the flammable EPS was burnt out, the geopolymer matrix prevents any spread of fire. There was no spread of the flame to the rest of the sample.

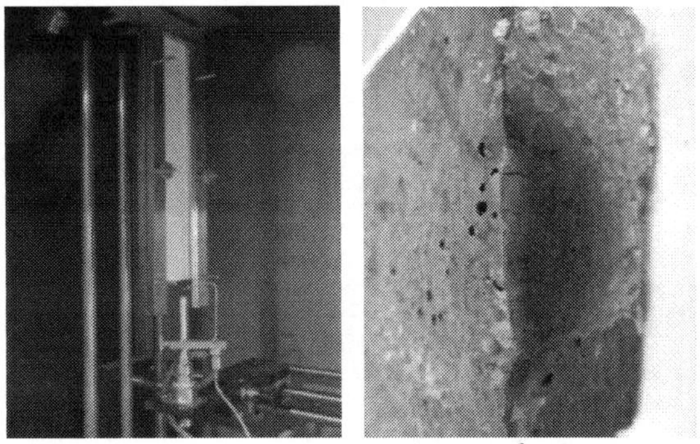

a b

Fig. 8. Configuration of JAR-25 test (a) and burned sample (b)

4 Conclusions

In this study, different types of expanded polystyrene were applied in the production of lightweight geopolymer composites from fly ash. Depending on the kind and percentage of polystyrene, lightweight geopolymers with compressive strength in the range 7.70 – 29.57 MPa and apparent density in the range 0.97 – 1.57 g/cm^3 were developed. The product containing 3% w/w commercial expanded polystyrene can be considered for the development of structural elements with enhanced thermal behaviour since it exhibits low thermal conductivity value (0.16 W/mK) combined with sufficient mechanical strengths (11 MPa). Furthermore, this product possesses excellent stability and fire resistance, while its water absorption is comparable to that of the conventional construction materials (cement mortars, concrete).

Acknowledgments

This research work has been financed by the European Union Seventh Framework Program (FP7-2013-NMPENV-EeB, “Multifunctional facades of reduced thickness for fast and cost-effective retrofitting, MF-Retrofit”, Grant No.: 609345).

References

1. J. Davidovits, J. Ther. Anal. **37**, 1633 (1991)
2. J.L. Provis, J.S.J. van Deventer, *Alkali-activated materials: State-of-the-Art Report* (Springer/RILEM, 2013)
3. P. Duxson, J.L. Provis, J. Amer. Ceram. Soc. **91,** 3846 (2008)
4. J.L. Provis, S.A. Bernal, Ann. Rev. Mater. Res. 299 (2014)
5. P. Duxson, A. Fernández-Jiménez, J.L. Provis, G.C. Lukey, A. Palomo, J.S.J. Van Deventer, J. Mater. Sci. **42,** 2917 (2007)
6. C. Panagiotopoulou, S. Tsivilis, G. Kakali, Constr. Build. Mater. **91,** 17 (2015)
7. A.M. Neville, *Properties of concrete*, (Addison Wesley Longman Limited, England, 1996)

Ecofriendly bricks elaborated from coal waste of Moroccan Jerrada Mining

H. Ez-zaki[1,*], A. Diouri[1], A. Bouregba[1], F. Amor[1], S. Chhaiba[1], O. Sassi[2] and Y. El Rhaffari[3]
[1] Laboratoire de Chimie du Solide Appliquée, Mohammed V University, Rabat, Morocco
[2] Laboratoire de Physico-Chimie des Matériaux, ENS, Mohammed V University, Rabat, Morocco
[3] Equipe thermodynamique-Energétique, Centre d'Energie, Mohammed V University, Rabat, Morocco

Abstract. Solid waste generated during mining is one of the major environmental problems associated with this industrial activity. The best solution to overcome the environmental impact of this waste is to find recycling facilities in mass-produced products that can absorb the large quantities of these available by-products. The present study shows the feasibility of using the coal waste of Moroccan Jerrada mining in the production of ecological brick. The first step consists of consecutive stages of crushing, grinding and heating at 650°C of the coal waste with a small amount of lime in order to promote the reactive products of elaborated binders. The second step of the process consists of mixing treated coal waste with a small amount of marble dust, sand, gravel, and water, then pressed and dried at room temperature to manufacture a laboratory eco-friendly bricks. The mechanical strength and thermal conductivity are investigated.

1 Introduction

In many different parts of the world, coal mine waste rocks generated by coal mining (extraction and processing) affected and damaged the environment due to the huge amount, occupied area, and acid generation [1-7]. However, many researches were focused to increase environmental performance in the mining sector. The reduction, reuse, recycling, and treatment of mineral processing waste are progressively getting more research and development attention for their contribution to enhancing the durability of the mineral industry. Coal wastes are being considered for use in many fields such as filler in hot mix asphalt [8], sulfur recovery [9, 10], fabrication of soils [1] and, as the focus of this work, fabrication of eco-friendly bricks.

Bricks are a structural building and construction material that has been used around the world for a long time. Traditional and usual bricks are fabricated from clay with high-temperature firing or from ordinary Portland cement (OPC) concrete which contains high incorporated energy with a large amount of CO2 generated during production. In many areas of the world, there is already a deficiency of natural source material for the production of the conventional bricks. For environmental protection and sustainable development, extensive research concentrated on the use of different types of industrial waste materials with different methods [11]–[13] has been conducted on the production of bricks from waste material; therefore it became the ideal approach to reduce wastes generation.

Jerada's mine coal production started in 1930 and stopped in 2003. It is located in the Northeast of Morocco. The production left behind large amounts of solid waste consisting mainly of gangue and other coal waste. Currently, twenty slagheaps and dumps of coal waste fill the urban space of Jerada, the largest conical in shape with a base of 450 x 412 m and a height of 70 m and composed of approximately 18 million tones of materials.

This coal waste accumulation poses environmental and economic problems and has become sources of contamination of the soil, water and air [10-14]. The oxidation of sulphuric minerals (pyrite), the danger of draining and leaching is available, creates high concentrations of sulfates in the groundwater and harmful oxides. The presence of high concentration of sulfates in the groundwater is caused by oxidation of pyrite present in these wastes [15].

A suitable way of reducing the accumulation of large amounts of coal waste is the valorisation of these products, which eventually helps to solve urban planning problems, as well as avoiding as possible the environmental impact. In the other hand, coal waste is constituted by a mineral part which is extremely important to reuse them in the construction materials [16].

The aim of this work is to propose a brief procedure to produce ecofriendly bricks from coal gangue based on the mineral processing. The point was to give a useful destination for the huge amounts of coal materials generated from coal processing operations and, simultaneously, to create an appropriate material that enables to use in fabrication of construction materials, also to minimize the environmental impact by absorbing a large amount of coal waste.

2 Materials and methods

2.1 Materials

The coal waste (CW) sample is a mixture of representative soil samples collected from different places on the embankment of Jerada's mine dumps. The CW was grinded under 45 μm, and then characterized by XRF, XRD and DSC/TG. The X-ray fluorescence is a spectrophotometer Philips PW1404 X-Ray tube Sc/Mo LIF200 and the working conditions were 40KV and 70mA. Mineralogical characterization of the samples was carried out with Bruker D8 Advance model X-ray diffractometer. Differential Scanning Calorimetry analysis (DSC) and thermo-gravimetric analysis (TG) were executed using DSC/TG type SDTQ600 with a heating rate of 10°C/min up to 1000°C, and using 100ml/min of dry air free of CO_2 as carrier gas. The table 1, figure 1 and figure 2 show the chemical composition, the x-ray diffraction pattern and DSC/TG curve of CW.

Table 1. Chemical composition (wt. %) of coal waste.

(wt.%)	CaO	Al_2O_3	SiO_2	Fe_2O_3	SO_3	MgO
CW	0,56	21,59	52,18	4,663	0,97	0,99

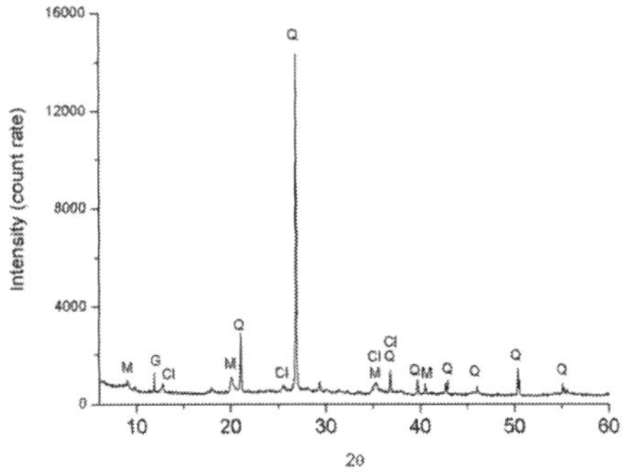

Fig.1. X-ray diffraction patterns for coal waste. Q= quartz; M = Muscovite; G= gypsum; Cl= Clinochlore.

We noticed that coal waste is composed mainly of silicon and aluminium oxides. The remaining compositions are iron, magnesium, sulphur and calcium oxides. The mineralogical analysis showed a large presence of quartz with muscovite, clinochlore and gypsum. In the temperature range 20-650°C (Figure 2), we notice two basic stages: the first endothermic region (20-150°C) is attributed to the evaporation of free and bounded water; the second exothermic region (450-650°C) is attributed to the combustion of the organic matter.

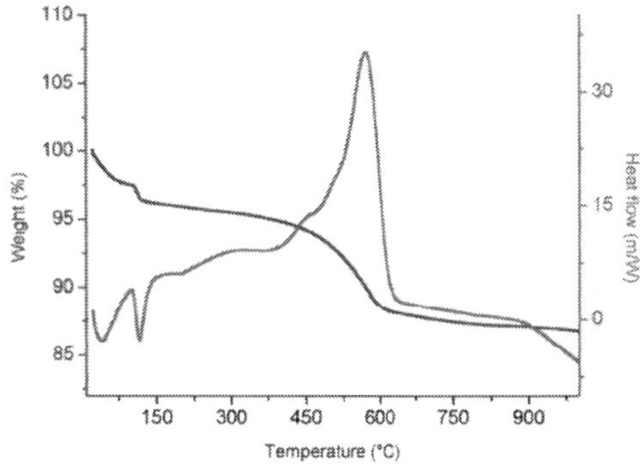

Fig. 2. DSC/TGA curve of coal waste under air atmosphere-heating at 10°C/min.

2.2 Methods

2.2.1 Elaboration of the binder

The sample preparation protocol can be resumed as follow: The first step consists of consecutive stages of crushing, grinding and heating at 650°C of the coal waste (CW) with a little amount of lime in order to advance the reactive products of elaborated binders. The second step of the process consists of mixing treated coal waste (TCW) with a small amount of marble dust, sand, gravel, and water, then pressed and dried at room temperature to manufacture a laboratory ecofriendly bricks. The mechanical strength and thermal conductivity are investigated.

In this study, we focused on heating coal waste (WC) at 650°C for 2 hours to obtain a blend TCW. The choice of this temperature is due to the conversion of clay menials and the remove of the organic matter [17]. Then, we mixed 70 % TCW with 15 % CaO in order to advance the reactive products and 15% marble dust as a filler addition. The table 2 shows the proportion of the binder composition.

Table 2. Proportion of elaborated binder.

(wt. %)	TCW	CaO	Marble dust
Binder	70	15	15

2.1.2 Test specimens

The followed process describes the manufacture of two types of specimens. Cylindrical mortars (7.5 mm Ø × 16 mm) and parallelepiped blocks (70×70×10 mm) are designed based on elaborated binder and then compared to the commercial cement CPj35 samples. Different proportions of elaborated binder, sand (<2 mm) and gravel (<4 mm) are prepared as listed in table 3.

Table 3. Mixes realized to prepare specimens.

(wt. %)	CPj35	Elaborated binder	Sand	Gravel
Mortars:				
MCPj_20	20	-	80	-
M4_20	-	20	80	-
M4_30	-	30	70	-
M4_40	-	40	60	-
Blocks:				
B4_20	20	-	30	50
B4_20	-	20	30	50
B4_30	-	30	25	45

The mixture were homogenized, mixed with adequate amount of water to desired consistency, and then pressed under 5 MPa using a hydraulic press. The samples were dried at ambient temperature until the testing day. A wet spray of specimens was established after one day of drying to promote the hydration reaction of binders, which provides the samples with certain early mechanical strength to avoid the crack in the following tests [18].

After 7, 28 and 56 days of drying, mechanical strength were carried out on cylindrical mortars with DY.22 machine. The thermal properties were characterized after 6 months of drying using CT-meter device. The measurement is based on monitoring the temperature increase for a measurement time of 300s using a ring sensor in which are associated a heating element with a temperature sensor. In order to provide an excellent thermal contact, the ring sensor which has a radius of 15 mm must imperatively be sandwiched between flat faces of two identical parallelepiped blocks.

3 Results and discussion

3.1. Mechanical strength

After each drying time, compressive strength was undertaken on the cylindrical specimens. The values reported in the results represent the average of three strength values. The results for compressive strength as function of curing time are given in figure 3.

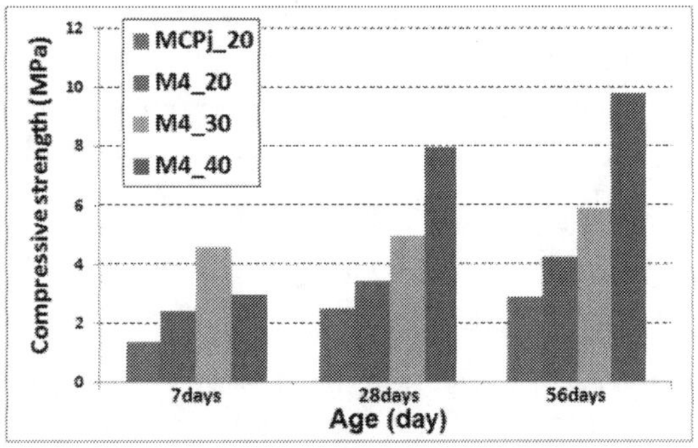

Fig. 3. Compressive strengths of different cylindrical specimens.

The compressive strength of binders was enhanced compared with the control (MCPj_20). It is observed that compressive strength has increased with the increase of binder content at all curing ages. The sample with 40 % by weight of elaborated binder (M4_40) showed the highest compressive strength at almost all curing age. The increase in the strength of the samples with the increase of binder content may be assigned to the pozzolanic activity and the pore structure of the materials.

Studies have shown that the activated coal waste plays an important role in the pozzolanic activity. Activated coal waste showed a high pozzolanicity in which the kaolinite was converted into metakaolinite, and the presence of carbonate as reactive phases [17, 19, 20]. The metakaolinite favors the early pozzolanic activity; while calcium carbonate implies the adhesion between the aggregates and paste [21]. In the other hand, the finer particles of marble dust can involve a filler effect and play a noticeable role in the compressive strength [22]. A study showed that calcium carbonate in marble dust can provide sites for the nucleation and growth of hydration products that leads to further increase in strength [23].

3.2 Thermal properties

Thermal behaviour was investigated on parallelepiped specimens after 6 months of drying at ambient temperature. The thermal conductivity and volumetric heat capacity are presented in figure 4.

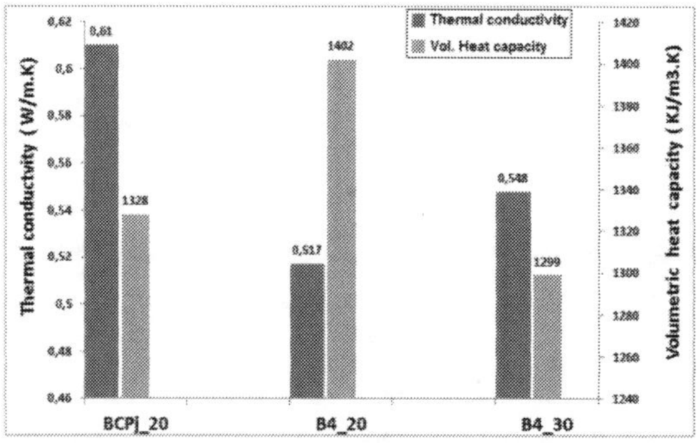

Fig. 4. Data of thermal conductivity and volumetric heat capacity of studied blocks.

As shown in figure 4, the specimen with 20% of Portland cement has the high thermal conductivity which is 0.61 W/m.K, compared to those of specimens with 20% and 30% of elaborated binder. The thermal conductivity increases from 0.517W/m.K to 0.548W/m.K for the shaped blocks with the increase of TCW binder from 20% to 30%. With the increase of elaborated binder addition, the thermal conductivity increases. This may be due to the effect of studied materials.

In addition to the pozzolanic effect of TCW, the incorporation of marble dust can be considered as an effective way to reduce the porosity by the filler effect and improving the mechanical strength. However, the porosity behaves as a barrier to the flow of heat and leads to the decrease of thermal conductivity [24]. The use of high content of TCW and marble dust can cause a lower porosity and an increase in thermal conductivity of blocks. Some theoretical and experimental data showed that thermal conductivity of nanoporous materials decreases with increasing porosity [25]. Rahman et al [24] have reported that the increase of the petroleum coal dust addition conduct to the decrease of mechanical strength and thermal conductivity and the increase of porosity.

It can be seen that volumetric heat capacity of blocks with 20% of binder exhibits the highest value (1402kJ/m3.K) compared to that of control and blocks with 30% of binder (1328kJ/m^3.K and 1299kJ/m^3.K respectively). The high value is considered as a desirable property due to the associated ability to store heat and it combines the specific heat capacity with material's density in which influenced by the content and the types of materials used [26, 27]. Therefore, it can be concluded that the use of a high content of elaborated binder can improve mechanical strength but at the same time deteriorate the thermal insulation of ecofriendly bricks. Thus, the main challenge is to find a balance between the desired thermal and mechanical properties.

4 Conclusions

In summary, coal waste can be used as the main raw material to overcome the environmental impact of this waste. This study showed one of the ways how coal waste as a solid waste can be feasible to produce ecological bricks.

The mixed treated coal waste and marble dust can increase the mechanical strength during curing time with increasing the amount of elaborated binder. The high mechanical strength of samples with high amount of this binder is attributed to the pozzolanic activity of treated coal waste and to the filler effect of marble dust which can improve the porous network and lead to the increase in strength. However, the increase of elaborated binder addition leads to the increase in thermal conductivity by reducing the porosity of materials. It can be explained by the behavior of porosity as a barrier to the flow of heat by storing it, whereof leads to a decrease in thermal conductivity.

Meanwhile, due to the increase in mechanical strength and the drop in thermal insulation by using a high amount of elaborated binder, the main challenge is to find a balance between the desired mechanical and thermal properties.

References

1. B.A. Firpo, J. R.D.A. Filho, I.A.H. Schneider, Minerals Engineering **76**, (2015)
2. N. Gupta,V.V. Gedam, C. Moghe, P. Labhasetwar, Environmental Technology & Innovation (2017)
3. D.M. Franks, D.V. Boger, C.M. Cote, D.R. Mulligan, Resources Policy **36**, (2011)
4. L. Haibin, L. Zhenling, Conservation and Recycling **54**, (2010)
5. H. Si, H. Bi, X. Li, C. Yang, International Journal of Coal Geology **81**, (2010)
6. A.Yu. Stolboushkin, A.I. Ivanov, O.A. Fomina, Procedia Engineering **150**, (2016)
7. D. van Beers, G.D. Corder, A. Bossilkov, R. van Berkel, Minerals Engineering **20**, (2007)
8. A. Modarres, M. Rahmanzadeh, Construction and Building Materials **66**, (2014)
9. C.K. Mbamba, S.T.L. Harrison, J.-P. Franzidis, J.L. Broadhurst, Minerals Engineering **29**, (2012)
10. Y. Taha, M. Benzaazoua, R. Hakkou, M. Mansori, Miner. Eng. **107**, (2017)
11. M. Sutcu, H. Alptekin, E. Erdogmus, Y. Er, and O. Gencel, Constr. Build. Mater. **82**, (2015)
12. J. A. Calabria, W. L. Vasconcelos, and A. R. Boccaccini, Ceram. Int. **35**, 2, (2009)
13. G. Cultrone and E. Sebastián, Constr. Build. Mater. **23**, 2, (2009)
14. Y. Darmane, A. Alaoui, S. Kitane, M. Bennajah, A. Daramy, M. Cherkaoui, Sep. Purif. Technol. **68**, 1, (2009)
15. B. Mounia, B. Mostapha, H. Rachid, B. Hassan, J. Abdelhakim, S. Mohamed, IJEST. **5**, 8, (2013)
16. D. Belkheiri, A. Diouri, M. Taibi, O. Sassi, and J. Aride, J. Mater. Environ. Sci., **6**, 6, (2015)
17. M. Frías, M.I. Sanchez de Rojas, R. García, A. Juan Valdés, C. Medina, Cement & Concrete Composites **34**, (2012)
18. J. Zhou, H. Gao, Z. Shu, Y. Wang, C. Yan, Construction and Building Materials **34**, (2012)
19. Y. Li, Y. Yao, X. Liu, H. Sun, W. Ni, Fuel **109**, (2013)
20. M. Frýas, R. Vigil de la Villa, MI. Sanchez de Rojas, C. Medina, A. Juan Valdes, J. Am. Ceram. Soc. **95**, 1, (2012)
21. I. Vegas, M. Cano, I. Arribas, M. Frías, O. Rodríguez, Construction and Building Materials **99**, (2015)
22. B. Demirel, International Journal of the Physical Sciences **5**, 9, (2010)
23. Taner Kavas, Asim Olgum, Ceramics-Silikáty **52**, 1, (2008)
24. M. H. Rahman, M.T. Islam, I.T. Minhaj, M.A.K. Azad, M.M. Hasan, A.A.M.R. Haque., Procedia Engineering **105**, (2015)
25. I. Sumirat, Y. Ando, S. Shimamura, J. Porous. Mater. **13**, (2006)
26. H. Oktay, R. Yumrutas, A. Akpolat, Construction and Building Materials **96**, (2015)
27. M. Pomianowski, P. Heiselberg, R.L. Jensen, Energy and Buildings **53**, (2012)

Dynamic buckling of laminated composite stringer stiffened CFRP panels under axial compression

Mouhat Ouadia[1], Khamlichi Abdellatif [2], Hasnae Boubel [1], Oumnia Elmrabet [1], Mohamed Rougui [1], El Mehdi Echebba [1] and Ahmed El Bouhmidi[1]
[1] Mohammed V University, Structure LGCE l'EST, Salé -CED de l'EMI, Rabat, MOROCCO
[2] Abdelmalek Essaadi University, Communications Systems and Detection Laboratory, Tetouan, MOROCCO

Abstract. In this work, the dynamic buckling of stiffened panels is evolved numerically through a nonlinear incremental expression through making use of a specific time integration procedure via the finite element software program. the buckling and post-buckling behaviours of hat-stringer-stiffened composite curved panel under axial compression load .Dynamic buckling is extracted from the curve abandoning the very last shortening as a characteristic of time while the shape is subjected with the aid of a square compression pulse movement carried out inside the axial direction. The duration of the heart beat and the amplitude of curvature of decreasing of the cloth inside the band tormented by the warmth, the dynamic buckling motion, are constant. The method approach was proposed to predict the dynamic buckling load of curved panel. Finite element analysis was used to investigate these tests and the FE models were performed by ABAQUS.Approach to determine the reliability of the stiffened panel in dynamic buckling state.

Keywords: Dynamic buckling, stiffened panels, axial compression, finite element method, reliability.

1. Introduction

Stiffened composite curved panel is a model structure widely used in aeronautic industry and spacecraft in the last few years, in the industry efficiency and performance of the structure is the essential objective for the design which requires insistence on the need for strong and light materials. So that the resistance must be high for this in utilised the reinforced polymer fibres for this we find the Stiffened composite curved panel is in several fields of application. it is necessary for the design to study the stability and buckling of the stiff composite panel of the many factors that can affect the buckling characteristics and the post-buckling, Many experimental tests and numerical calculations have been realized to study the behaviour of static and dynamic buckling several researchers who are published in this axis on the static buckling but we are interested in this work by the dynamic buckling.

During the last two decades many articles studied stiffened curved panels under axial compression loading[1-3].Recently, the project that has been the focus of attention in previous years is the project POSICOES and COCO MAT [4] ,Both supported by the European Commission, to study the behavior of composite panels stiffened by all the buckling and post buckling more and more of attention is given to buckling and The aerospace industry has asked to reduce development and operating costs by 20% and 50% in the short and long term. There is a difficulty in preaching the critical load of buckling and the ultimate load by the classical methods. Von Karman et al. [5] a proposed approach "effective width" for the optimization this has been modified by several carters among these loaders Wang Zhenming [6]. There are several experiences of stiffened curved panels under axial compression loading but most are forming I-stiffened panels [7]. And T-stiffened panel [8].

Determination of relative response for the boring load parameters by equating the equations of the motion and trace the selected axial displacement curve as a function of time. The specific value of the charge from this change coincides with the critical value associated with dynamic buckling according to the Budiansky-Roth criterion [9] (Budiansky and Roth, 1962). This criterion seems more practical in the study of dynamic buckling of thin structures.

In the aeronautical and aerospace industries the lifespan of the aircraft is civil or military, generally the aircraft that exceeds 20 years of service or not even, so the buckling performance of stiffened composite panels must also be a big concern for the airplane designer. Scientists now focus on the dynamic buckling that was often ignored composite panels. This article discusses the nonlinear method and dynamic buckling of the hat-stringer-stiffened Composite flat panel subjected to axial compression and made a reliability analysis using the Matlab Monte Carlo method to calculate the probability of failure.

2. Panel description

The stiffened composite panel is composed of three stiffeners. The nominal length and the nominal width of the panel are 952 mm and 757.5 mm. The gap between two stiffeners is 252.5 mm. The two terminals of the panel were recessed ensuring a uniform distributed load calculation. The mechanical properties of the material are defined in Table 1. The layout of the laminate at each zone of the panel is reported in Table 2 and specified graphically in the stiffeners are L-shaped and have the constant thickness $t_w = 2.95\,mm$, height $h_w = 64\,mm$, the thickness of the flange $t_f = 4.3\,mm$ and the height of the flange $b_f = 12\,mm$. Specified graphically in Fig. 1,Nominal ply thickness is 0.185 mm.

In Fig.2. Shows the boundary conditions the configuration of the stiffened panel.

By indicating $u_{\square}$ the displacement coordinates and $\theta_{\square}$ rotations, the boundary conditions examined in the software abaqus are as follows. The lateral ends have the following boundary conditions $u_x = \theta_x = \theta_z = 0$. The border $z = 0$ is supposed to be perfectly wet $u_x = u_y = u_z = \theta_x = \theta_y = \theta_z = 0$, as the uniformly distributed load P_z is applied to the edge $z = a$ with rigid boundary conditions on the side $u_x = \theta_x = \theta_z = 0$.

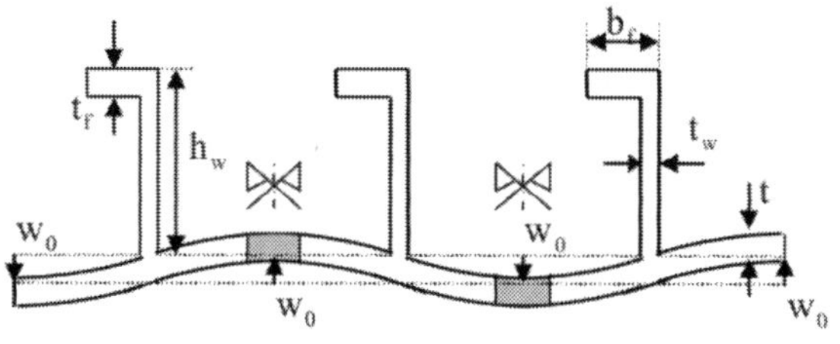

Fig. 1. The initial configuration in of the stiffener panel

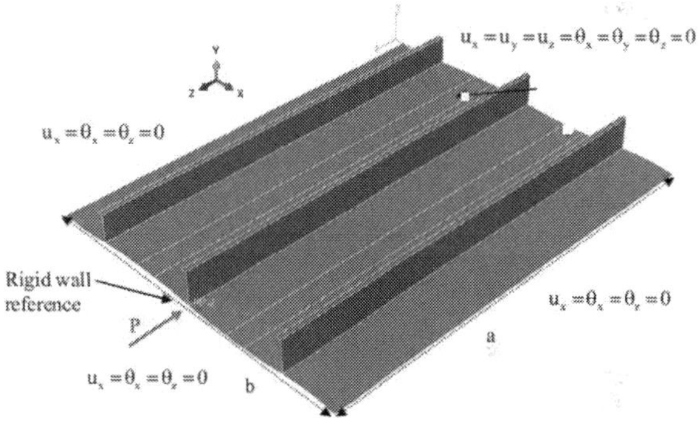

Fig.2. Geometry of panel and boundary condition

Table 1. Mechanical properties

Stiffness		strength	
E_1	141 GPa	X_T	2700MPa
E_2	8.85 GPa	X_C	1600MPa
$G_{12}=G_{13}$	4.57 GPa	Y_T	88MPa
G_{23}	4.2 GPa	Y_C	271MPa
υ_{12}	0.33	S	143MPa

Table 2. Lay-up definition

Skin	$t_w \begin{bmatrix} 45/-45/90/90/0/90/0 \\ /90/0/90/90/-45/45 \end{bmatrix}$
Flange	$t_f\,[45/-45/0/90/90/0/-45/45]$

3. Buckling analysis of stiffened panels

3.1 Composite materials

Composite means in the term composite materials either two or more materials are macerated on a macroscopic scale to form a third useful material. Different materials can be mixing at the macroscopic scale as in the metal alloy, but the resulting materials are, for all practical purposes, macroscopically homogeneous, that is to say that the composite cannot be distinguished with the naked eye and act essentially together.

3.2 Classifications of composite materials

- A laminate consists of a stack of monolayers each a proper orientation with respect to a common frame of reference layers and designated as the laminate reference frame.
- Particulate Composites consist of a matrix reinforced by a dispersed phase in form of particles.
- Fibrous composite materials that consist of fibres in a matrix.
- Combination of some or all of the first three types.

3.3 Stiffness and strength of a lamina

The orientation of the fibres can be randomly in the material; in reality the fibres in the preferred directions can be oriented in expectation of the highest stresses. Such a material is said to be anisotropic, which is different from their physical properties (elasticity moduli, Poisson coefficients, thermal conductivity, etc.) which have different values depending on the spatial orientation of the physical body.

The microscopic level, the properties of these composites are found by the orientation and distribution of the fibres, as well as by the properties of the fibre and matrix substances.

Consider a region of unit dimensions in a material type, which material contains a volume fraction, V_f we chose all orientations in a single direction, with the volume fraction of the matrix is written by this following formula:

$$V_m = 1 - V_f \tag{1}$$

This place can be made in conglomerate all the fibers together, dismissal the matrix to occupy the remaining volume. If a stress σ_l applied along the direction of the fibre, the phases of the fibre and the matrix proceed in similar to support the load. In these parallel bonds the constraints in each period must be the same, so that the strain ε_l in the direction of the fibre can be written in this way as:

$$\varepsilon_l = \varepsilon_f = \varepsilon_m \tag{2}$$

With: the indices L and m denote the lamina and the fibres and the matrix respectively).

For the balance of the total charge in the composite material are added the forces in each phase.

Since the forces in each period are the period constraints times the area (here theoretically equal to the volume fraction), we have:

$$\sigma_l = \sigma_f V_f + \sigma_m V_m = E_f \varepsilon_l V_f + E_m \varepsilon_l V_m \tag{3}$$

The stiffness in the direction of the fibres is found by dividing the stress by the strain:

$$E = \frac{\sigma_l}{\varepsilon_l} = E_f V_f + E_m V_m \tag{4}$$

With: E is the longitudinal Young's modulus.

For the prediction of the mixtures of the global module using this relation which was according to the terms of the modules of hashes which constitutes the volume fractions.

Rule assessments of mixtures for force flow along lines similar to those for rigidity. For example, consider a unidirectional reinforced composite that is stretched to the value at which the fibre begins to break. If the matrix is more ductile than the fibres, then the ultimate tensile strength of the lamina in the equation (3) changed to:

$$\sigma_t^u = \sigma_f^u V_f + \sigma_m^f \left(1 - V_f\right) \qquad (5)$$

When the power u shows an extreme value, and σ_m^f is the stress of the matrix when the fibres break as in fig.3.

It is visible that if the volume fraction of the fibre is not strong, the attitude of the lamina is examined by the matrix.

Mathematically expressed by this following relation:

$$\sigma_l^u = \sigma_m^u \left(1 - V_f\right) \qquad (6)$$

if it is assumed that the lamina is to be used in a practical application, the minimum fraction of fibre volume is deduced this fraction must be added to the matrix, this formula obtained by solving the equations respectively (5) and (6).

Mathematically expressed by this following relation:

$$V_{\min} = \frac{\sigma_m^u - \sigma_m^f}{\sigma_f^u + \sigma_m^u - \sigma_m^f} \qquad (7)$$

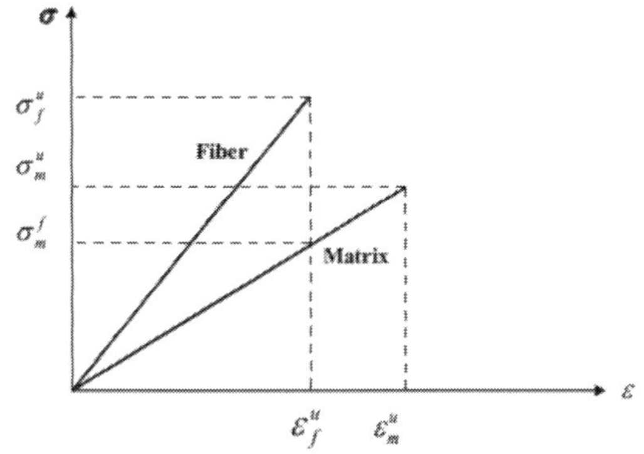

Fig.3.Stress-strain for fiber and matrix

3.4 Nonlinear dynamic buckling analysis

For the dynamic buckling test, the explicit dynamic instruction makes it possible to perform the whole small increment time determination to find a working approximation to the resolution of the problem. This procedure is based on the time integration rule. The increment calculation method is a bit expensive compared to the implicit forward dynamic integration dynamic analysis instruction does not involve any inversion of a linear system. The explicit federation dissimilarity manipulator satisfies the dynamic equilibrium equations at the origin of the time increment t, the increments calculated at that time are used $t + \Delta t/2$ to advance the time-to-time solution $t + \Delta t$ and the on-time move solution.. Generously the principle of preliminary kinematics of an arbitrary time step in acceleration and velocity completions.

Giving the kinematics state at the beginning of a given time step in terms of the acceleration $\ddot{u}_n$ and the velocity $\dot{u}_{n-1/2}$, the equations of motion are integrated using the explicit central-difference integration rule according to

$$\dot{u}_{n+1/2} = \dot{u}_{n-1/2} + \frac{\Delta t_{n+1} + \Delta t_n}{2} \ddot{u}_n$$
$$u_{n+1} = u_n + \Delta t_{n+1} \dot{u}_{n+1/2} \qquad (8)$$

With n refers to the step number.

4. Reliability analysis of composite stringer stiffened panels

To complete a reliability analysis, an analytical model giving an explicit abstract of the problem is developed. This is completed by selecting appropriate application points on the deep division of explored variables according to an experimental design scale (DOE). This table is generated by setting distinguished levels of intermediate factors. Beginning with the objectives received from the simulation generated for these particular points, a model is originated in order to affirm the state function in an explicit form over the whole domain used for the regression. A rectangle polynomial response surface abbreviation is provided.

The variability of the dynamic buckling load is assumed to start from the following three constituent sources which include the magnitude of the geometric imperfection, the Young's modulus, and the square pulse period. All referees geometric and mechanical parameters.

Keeping only these three active variables, the buckling load can be expressed explicitly as

$$P_{cr} = P_{cr}(w_0, E, T) \quad (9)$$

In this anticipation which makes it possible to focus on the fundamental effects governing the problem of dynamic buckling of stiffened panels, the derivation of a predictive model is considered under the adaptation of a rectangle polynomial response surface. P_{cr}.

5. Results and discussion

Table.3. gives the levels of the three parameters selected to perform the reliability analysis. The column T corresponds to the following three values: $T = 0.25T_0,\ 0.5T_0,\ 0.75\,T_0, T_0$ with $T_0 = 12.45$ms

Parameter level	$w_0\,(mm)$	$E\,(GPa)$	$T\,(ms)$
Low	2	64	6.12
medium	3	66.75	9.28
High	4	69.5	12.45

Table 3. Levels of factors considered in the reliability analysis

Table 2. Is used to construct a complete factorial table consisting of 27 combinations involving the levels specified.

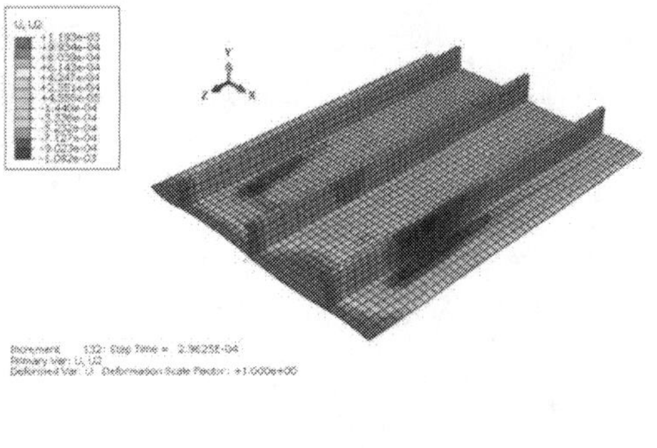

(a)

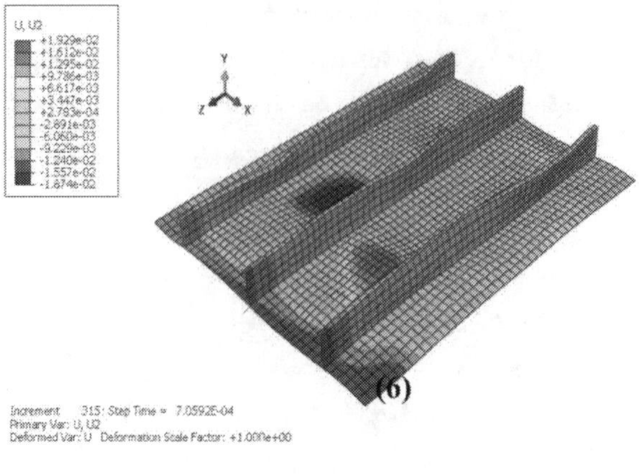

(b)

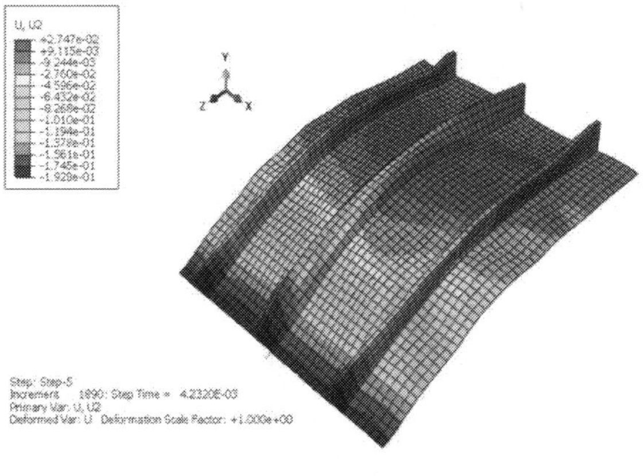

(c)

Fig.4. Three sequences of deformation of the stiffened panel in the form of the iso-values of the transverse displacement.

In order to obtain a response surface that represents the dynamic buckling critical load in an explicit form with coefficients of the same order of magnitude, the factors of Table .3 are dimensioned by asking:

$$\bar{w}_0 = w_0 / w_{0,\max}\ ,\ \ \bar{E} = E / E_{\max}\ \text{et}\ \ \bar{T} = T / T_{\max} \quad (10)$$

The dynamic buckling critical stress and coefficients of this polynomial are obtained by the regstats command of Matlab and the response surface polynomial of the computed system is:

$$\sigma_{cr}(\text{MPa}) = -5.12 + 19.25\tilde{w}_0 - 4.51\tilde{E} - 0.25\tilde{T} - 1.84\tilde{w}_0\tilde{E} + 1.24\tilde{w}_0\tilde{T} - 0.961\tilde{E}\tilde{T} - 8.24\tilde{w}_0^2 + 7.58\tilde{E}^2 + 3.65\tilde{T}^2 \quad (11)$$

The interpolation defined by equation (11) has an excellent coefficient of determination $R^2 = 98.13\%$. It should be noted that even a linear polynomial regression model also gives a very good coefficient of determination that is worth $R^2 = 99.3\%$.

5.1 Analysis of variance

The analysis of variance performed on the results presented in Table .3 using the Matlab anovan command we can notice that the factors and their interactions correctly explain the variability of the dynamic buckling load because the residual error does not exceed. If one works with a linear model then the factors explain by themselves the results with an error limited to 3.4%.

These results are very revealing of the phenomenology of the dynamic buckling which appears in this way as a phenomenon essentially governed by the duration of the impulse loading.

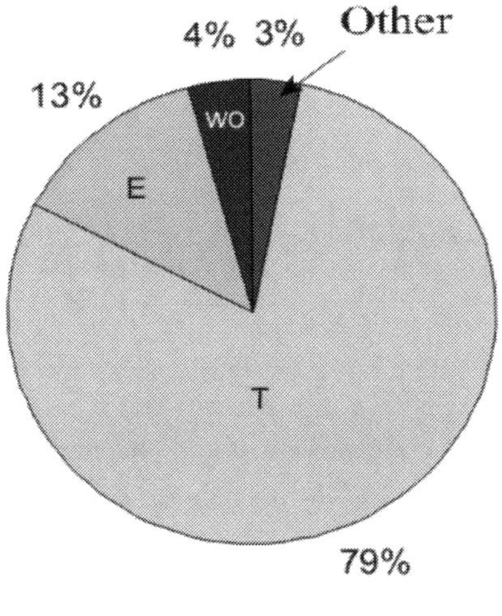

Fig 5. Relative influence of the factors describing the variability of the dynamic buckling critical load; "Other" includes interactions and error

The Fig.5 gives the relative percentage of the influence of each factor on the variability of the dynamic buckling critical load.

Conclusions

This paper discusses dynamic buckling on composite stiffened panels when exposed to axial compression in the manner of a hat and axially reinforced. To obtain a variety of dynamic buckling, panels with different amplitude of curvatures, with different panel thicknesses these models based on ABAQUS software. This allowed the derivation of a surface response representation of dynamic buckling load variations.

References

1. E. Gal ,R. Levy ,H. Abramovich , P.Pavsner . Buckling analysis of composite panels.Compos Struct J.E **73**,179–85 (2006).
2. C.Bisagni, C.Potito. An experimental investigation into the buckling and post-buckling of CFRP shells under combined axial and torsion loading. Compos Struct J.E **60**,391–402 (2003).
3. R.Zimmermann , H.Klein , A.Kling. Buckling and postbuckling of stringer stiffened fibre composite curved panels – tests and computations. Compos Struct ,J E **73** ,150–61 (2006).
4. R.Zimmermann , R.Rolfes . Posicoss-improved postbuckling simulation fordesign of fiber composite stiffened fuselage structures. Compos Structure, J.E **73**,4-171,(2006).
5. T.Von Karman , E.E.Sechler , L.H.Donnel . The strength of thin plates in compression. ASME Appl Mech Trans,J.E **54** (1932).
6. X.M.Wang, W.Cao, C.H.Deng. Experimental and numerical analysis for the post-buckling behavior of stiffened composite panels with impact damage. Compos Struct 2,J.E **133** (2015).
7. F.Caputo , R.Esposito, P.Perugini, D.Santoro. Numerical-experimental investigation on post-buckled stiffened composite panels. Compos Struct ,J.E **55**,347–57 (2002).
8. L.Lanzi , V.Giavotto . Post-buckling optimization of composite stiffened panels: computations and experiments. Compos Struct ,J.E **73**,208–20 (2006).
9. B.Budiansky , J.W.Hutchinson . Dynamic buckling of imperfection-sensitive structures Proc. 11th Int. Congress of Applied Mechanics (Berlin: Springer) ,J.E.**13**,636-651, (1964).

Identification of elasticity modulus by vibratory analysis (Application to a natural composite: Aleppo pine wood)

Identification des modules d'élasticité par analyse vibratoire (Application à un composite naturel: bois de Pin d'Alep)

Abdelhakim DAOUI[1,2], Abdellatif ZERIZER [2]

[1] Département Physique, Faculté des Sciences, Université M'hamed Bougara Boumerdes, Algérie
[2] UR-MPE Unité de Recherche Matériaux Procédé Environnement, Université Bougara, Boumerdès, Algérie

Abstract. In this paper, we present a method for the determination of the elasticity modulus in the vibratory domain of materials. This approach is based on research and interpretation of the spectrum of natural frequencies resulting from natural vibrations based on the theory of elastic beams. The tests consist of classical tests of longitudinal vibrations of natural composite beams (Aleppo pine wood), long enough to observe some natural frequencies. This identification method showed a good correlation between the theoretical and experimental values, notably the evaluation of the modal parameter for the case of the resonant frequencies and the identification of the modulus of elasticity of the materials used.

Résumé. Dans ce papier, nous présentons une méthode pour la détermination du module d'élasticité dans le domaine vibratoire des matériaux. Cette démarche repose sur la recherche et l'interprétation du spectre des fréquences propres issues des vibrations naturelles sur la base de la théorie des poutres élastiques. Les essais consistent en des tests classiques de vibrations longitudinales de poutres composites naturelles (bois de Pin d'Alep), suffisamment longues pour observer quelques fréquences propres. Cette méthode d'identification a montré une bonne corrélation entre les valeurs théoriques et expérimentales, notamment l'évaluation du paramètre modale pour le cas des fréquences de résonance et l'identification du module d'élasticité des matériaux utilisé.

Mots Clés: analyse vibratoire, acquisition de données, module d'élasticité, fréquence de résonance, théorie des poutres élastiques

1 Introduction

Les essais d'évaluation non destructive vibratoire sont basés sur l'analyse spectrale des vibrations naturelles d'un élément de structure (poutre, panneau) soumis à une excitation impulsionnelle. Cette méthode, applicable à bien des matériaux, donne une estimation instantanée des propriétés mécaniques des matériaux testés et permet d'établir une identification mécanique de chaque pièce (Haines and Leban, 1997; Zaveri et Phil, 1984). Les avantages de cette méthodologie, outre le fait qu'elle soit non-destructive, reposent sur sa simplicité, sa répétitivité et traçabilité.

Les données issues de ces essais permettront d'orienter les choix des matériaux de fabrication, d'optimiser les critères relatifs aux propriétés mécaniques, et bien sur d'accroître l'efficacité du classement en se substituant au classement visuel.

Dans les analyses vibratoires, on est toujours amené vers la mesure des modes de vibration (fréquences naturelles de vibration) afin de les comparer à ceux obtenus par des méthodes théoriques. On estime que la confirmation des modes propres de vibration par des essais peut fournir la réassurance pour la validation des résultats des essais effectués (Ewins, 1984).

2 Définition de la méthode d'analyse vibratoire

2.1 Mécanisme d'excitation de structures et d'acquisition de données

La méthode d'analyse vibratoire la plus couramment utilisée dans le domaine expérimental consiste en l'excitation de la structure et l'enregistrement de sa réponse. Les techniques de mesure des sollicitations exercées sur les structures et les réponses de ces dernières pour l'analyse dynamique comprennent trois aspects fondamentaux (DØssing, 1987a; DØssing, 1988b; Dumas et Bennevault, 2001) :

- Le mécanisme permettant d'exciter correctement la structure
- Celui permettant de mesurer les signaux souhaités (forces, accélérations, déplacement…etc.)
- Et enfin celui qui assure l'extraction des informations que l'on désire acquérir.

Les excitateurs peuvent être classés en deux catégories : ceux directement en contact avec la structure étudiée pendant toute la durée de l'essai et ceux qui agissent à distance ou qui ne sont en contact que pendant un certain laps de temps (cas des marteaux d'impulsion pour le cas de test d'excitation transitoire).

2.2 Transformée de Fourier Discrète d'une fonction réelle

Le calcul et la représentation des fonctions nécessitent la connaissance des signaux dans le domaine fréquentiel. En pratique, on a seulement accès aux signaux dans le domaine temporel (réponse d'un accéléromètre par exemple). Le passage du domaine temporel au domaine fréquentiel est réalisé à l'aide de la Transformée de Fourier Rapide (FFT).

L'expression de la FFT d'une fonction réelle $y(t)$ définie pour $0 \le t \le \infty$ a la forme (Lemay, 2001).

$$Y(f) = \int_0^\infty y(t)\exp(-j2\pi ft)\,dt \qquad (1)$$

Pour un signal continu défini par sa transformée de Fourier donnée par l'équation (1), en prenant une fréquence d'échantillonnage f_s et un temps d'observation T on aura alors un échantillon $y(t_n)$ de taille N. Cela revient à observer $y(t)$ à des temps discrets espacés d'un pas $\Delta t = \dfrac{1}{f_s}$; on observe donc un signal aux instants $t_n = n \times \Delta t$ avec $0 \le n \le N-1$.

Ainsi, la transformée de Fourier discrète d'un signal donné s'exprime sous la forme suivante :

$$Y(f_k) = \Delta t \sum_{n=0}^{N-1} y_n \exp(-\frac{j2\pi kn}{N}) \qquad (2)$$

$$\text{avec } k = 0,1,2,\ldots,\frac{N}{2}$$

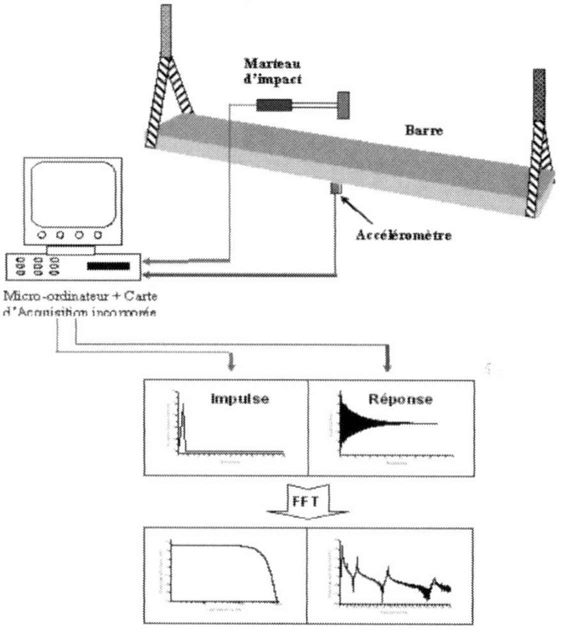

Fig.1 Procédure d'essai de vibration et acquisition de données pour le cas d'une barre libre aux extrémités

2.3 Identification de module d'élasticité

Dans la pratique, le calcul des fréquences de vibrations propres des systèmes ce fait sans tenir compte de la force d'amortissement, cela est dû au fait qui est généralement admis de négligé l'effet de l'amortissement sur les valeurs de la fréquence de vibration vue qu'il est faible par rapport à d'autres facteurs.

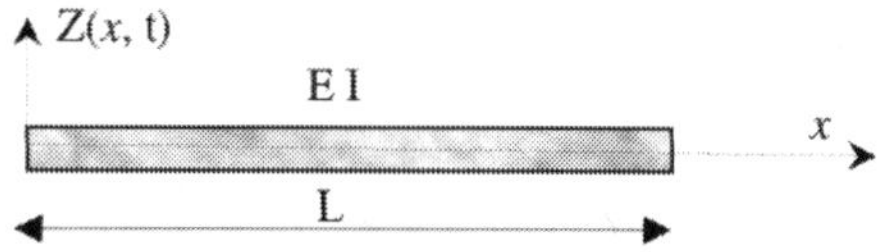

Pour une barre rectangulaire de section constante et de masse uniformément répartie et selon ses conditions géométriques aux extrémités, constitue un système vibrant que les équations du mouvement peuvent être résolues analytiquement par dérivation de l'équation de mouvement obtenue à partir d'une analyse de la

https://doi.org/10.1051/matecconf/201814901045

seconde loi de newton pour les forces et les moments agissant sur la barre (Broch, 1984) :

$$\rho\, S \frac{\partial^2 Z(x,t)}{\partial t^2} + \frac{\partial^2}{\partial x^2}\left(E\, I \frac{\partial^2 Z(x,t)}{\partial x^2} \right) = 0 \qquad (3)$$

E : module de Young
I : moment d'inertie de la barre
ρ : la masse volumique de la barre
S : la section de la barre

A partir de l'expression (3) on peut facilement aboutir a une forme d'écriture permettent de calculer le module de d'élasticité (module de Young) d'un matériau en fonction de la fréquence de résonance du système selon les conditions aux limites :

$$E = 48\,\pi^2 \rho \left(\frac{L^2}{h} \frac{f_n}{K_n^2} \right)^2 \quad N/m^2 \qquad (4)$$

f_n : fréquence de résonance de système.
K_n : constante qui dépend des conditions aux limites de la barre.

3 Procédure expérimentale

Le micro-ordinateur gère, sous le contrôle d'une carte d'acquisition, la totalité du paramétrage du boîtier d'acquisition et du traitement de l'information recueillie à la suite de la percussion appliquée.
L'onde émise est transformée en signal électrique par un capteur de réponse (accéléromètre). Celui ci amplifie, filtre puis échantillonné par l'intermédiaire du boîtier d'acquisition. Puis, après conversion du signal analogique en information numérique, l'enregistrement est transféré directement via le port parallèle dans une mémoire utilisateur du PC. La composition spectrale de l'enregistrement est obtenue par (FFT) transformée de Fourier Rapide (fig.1).

Les matériaux qui ont été soumis à des tests expérimentaux sont des poutres en bois de Pin d'Alep et une poutre en acier.

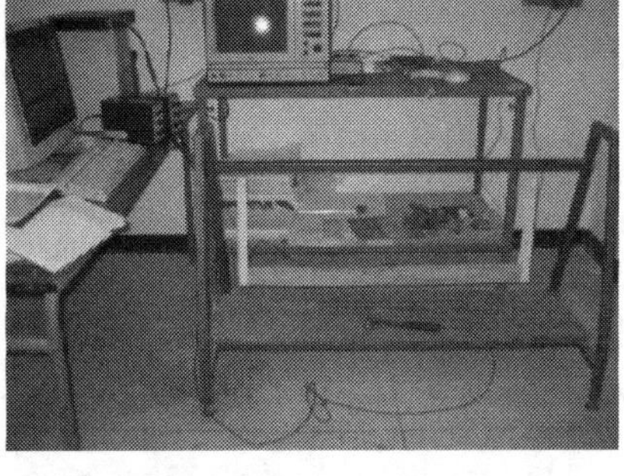

Fig.2. Installation d'équipements pour le cas d'une poutre en bois de Pin d'Alep libre à ses extrémités

Ces poutres en été placé de tel sort quelle constituent un système poutre libre sur ses deux extrémités (fig.1 & 2).

4 Présentation des résultats et discussion

L'acquisition des données nous a permis l'identification des fréquences propre, ainsi que le module d'élasticité longitudinale de chaque mode de vibration des systèmes. En effet, nos systèmes ont été soumis à des excitations impulsionnelles de courte durée. Nous avons enregistré les réponses et les excitations.

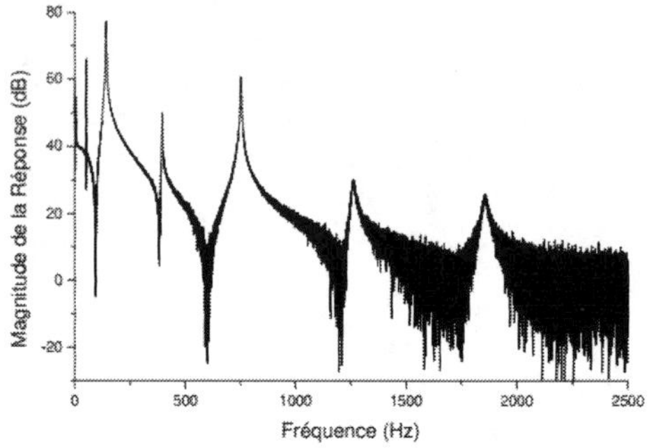

Fig.3. Spectre de réponse de la poutre sur ses deux extrémités

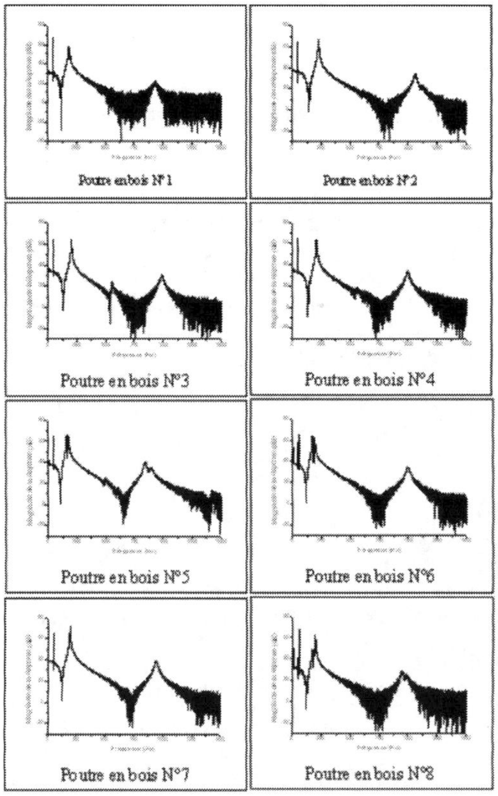

Fig.4. Spectre de réponse de la poutre en bois Pin d'Alep libre sur ses deux extrémités

Ptre	Frq (Hz) E=10000 MPa	Frq (Hz) E=13000 MPa	Frq Exp (Hz)	E ident. (MPa)
1	633.28	631.38	753.37	9400
2	616.67	615.47	607.00	13600
3	609.16	609.16	677.59	10700
4	635.97	635.78	790.91	12300
5	631.95	631.67	788.65	8600
6	627.32	625.89	730.26	9800
7	607.00	607.00	705.99	9800
8	620.18	619.93	718.86	9900

Les graphes obtenus par acquisition, correspondant aux spectres de réponses pour chaque système sont représentés par les figures 3 & 4. Ils correspondent aux transformées de Fourier dans le domaine de fréquence des réponses temporelles. Les fréquences de résonance des modes de vibration sont clairement identifiées et correspondent aux pics présents sur ces spectres. Aussi on peut remarquer que le bruit est prédominant, ce qui empêche l'apparition d'autres modes de vibration.
En effet, pour le cas d'un marteau d'impact, la gamme de fréquence de l'excitation fournie par ce dispositif est déterminée par la rigidité des surfaces de contact entre marteau-poutre et la masse de la tête de marteau.
Après cette gamme, les autres modes de vibration de la structure ne seront pas excités (il n y a pas assez d'énergie pour l'excitation. Ce qui explique la raison pour laquelle nous avons obtenu plus de pics de résonance qui ont été clairement identifiés.

Dans le tableau-1, nous avons présenté les valeurs de fréquence de résonance obtenues à partir des spectres de réponse pour chaque système. Nous avons présenté les valeurs des 4 premiers modes de vibration. L'analyse de ces valeurs, nous a permis d'identifier le module d'élasticité longitudinal de nos éprouvettes.
Par ailleurs, pour estimer la plage des fréquences de résonance du bois de pin d'Alep, nous avons injecté la valeur statique du module d'élasticité de notre matériau. Nous pouvons noter que les valeurs théoriques et celles obtenues expérimentalement vont dans le même ordre.

Pour le cas des poutres en bois de Pin d'Alep, nous présentons (Tab.1) l'ensemble des valeurs expérimentales des fréquences de résonance du premier mode de vibration pour chaque poutre, ces valeurs ont été comparées avec les valeurs théoriques de fréquence de résonance obtenues en admettant une gamme de modules d'élasticité allant de 10000 à 13000 Mpa.

Tableau 1 . Valeurs des fréquences de résonance et de module d'élasticité de flexion de premier mode de vibration pour le cas des poutres en bois de Pin d'Alep.

L'exploitation des valeurs de fréquences de résonance expérimentale a permit le calcul du module d'élasticité pour les poutres de Pin d'Alep. Les valeurs obtenues à l'issue de ce calcul convergent avec celles théoriques (Fig.5).

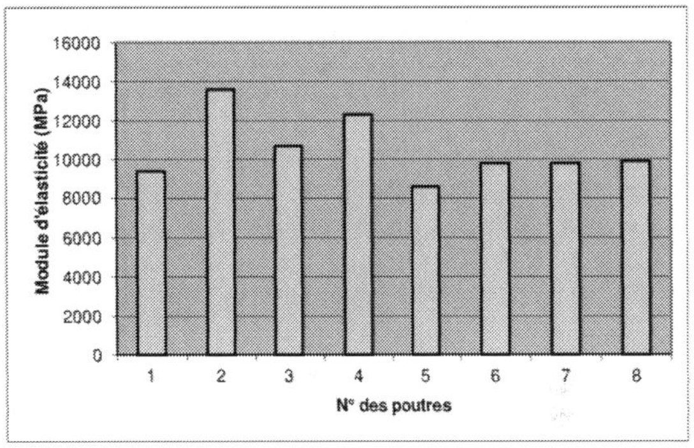

Figure 5 Distribution des modules d'élasticité des poutres en bois de Pin d'Alep

Dans les bois, la répartition de la masse volumique est généralement variable dans les trois sens (radial, tangentiel et longitudinal), et c'est le cas de nos 08 poutres (effet d'hétérogénéité). Cette remarque a été confirmée par la différence des valeurs de module d'élasticité entre ces poutres. Ces différentes valeurs se situent quand même autour de la gamme que peut avoir le bois et notamment le Pin d'Alep.

5 Conclusion

Le but de ce présent travail était de proposer une démarche expérimentale permettant l'identification du module d'élasticité existant dans le matériau. Les résultats expérimentaux obtenus ont montré la concordance des données qui restent en général proches de celles données par les normes aussi bien pour le cas de l'acier que du bois. Au terme de cette intéressante investigation, il y a lieu de noter qu'a travers cette méthode, nous avons pu aboutir a démontrer le caractère variable de module d'élasticité (phénomène reconnu pour le cas des bois). Nous avons utilisé un même type de bois pour remarquer qu'il y avait une différence significative autour de la valeur de module d'élasticité entre ces bois, mais qui reste cependant corrélée avec la gamme que peut avoir ce type de bois.

Par ailleurs, nous avons observé la variabilité du module d'élasticité en fonction du mode de vibration pour le cas de la poutre en acier. Cependant, on n'a pas pu identifier d'autres modes de vibration pour calculer le module d'élasticité pour le cas des poutres en bois. Ceci est du probablement à la nature de la surface de contact entre le marteau d'impact et le bois. Nous pensons que la surface n'est pas aussi rigide pour permettre l'apparition d'autres modes de vibration.

References

[1]-Broch J.T. (1984). *Mechanical vibration and shock measurements*. Brüel&Kjaer. 2nd Edition.

[2]-Dumas J, Bennevault B. (2001). Debuter la mesure vibratoire. *MVI technologies group*. Version Février

[3]- DØssing O. (1987a). *Structural Testing Part 1: Mechanical Mobility Measurements.* Brüel & Kjaer, April.

[4]-DØssing O. (1988b). *Structural Testing Part 2 : Modal Analysis and Simulation.* . Brüel & Kjaer, March.

[5]-Ewins D.J. (1984). *Modal Testing: Theory and Practice.* Research Studies Press, Ltd, England.

[6]-Haines D., Leban J.M. (1997). Evaluation of the MOE of Norway spruce by the resonance flexure method. *Forest Products Journal*, Vol.47 n°10, p 90-94.

[7]-Lemay J. (2001). *Acquisition, traitement de données, cours GMC-64506.* Université LAVAL. Septembre

[8]-Zaveri K, Phil M. (1984). Modal Analysis of Large Structures-Multiple Exciter Systèms. *Brüel &Kjaer,* 1st edition, November

 https://doi.org/10.1051/matecconf/201814901046

Reinforcement of concrete by composite materials

Renforcement du béton par les matériaux composites

Bensaid Boulekbache, Mostefa Hamrat, Nejma Imane Gacem, Kheltoum Halimi
Université Hassiba Benbouali de Chlef, Algérie

Abstract. This work was devoted to the material-scale study of the mechanical behavior of confined concrete by composite materials based on carbon fiber and glass under compression and bending loading. In this context, an experimental work on cylindrical and prismatic concrete specimens has been carried out. The obtained results permitted the identification of the mechanical behavior, strength, deformability and failure modes. The outcome of the present research indicates an increase in the bearing capacity and deformability in compression and in tension. The reinforcement by means of composite materials appears as a pertinent solution and therefore, makes it possible to obtain clear indication on the type of confinement to be considered according to the intended purpose.

Résumé. Ce travail a été consacré à l'étude à l'échelle du matériau du comportement mécanique du béton confiné par des matériaux composites à base de fibres de carbone et de verre sous un chargement de compression et flexion. Dans ce sens, un travail expérimental sur des spécimens en béton de forme cylindrique et prismatique a été réalisé. Les résultats expérimentaux ont permis l'identification comportement mécanique, des résistances mécaniques et la déformabilité et les modes de rupture. Pour l'ensemble des résultats obtenus, on a pu constater un accroissement de la capacité portante et de la déformabilité rn compression et en traction, Le renforcement par les matériaux composite est une solution fiable pour le renforcement et permet de dégager des indications claires sur le confinement à adopter selon l'objectif visé.

1 Introduction

Le béton est un matériau qui se modifie tout au long de sa vie utile, car la structure dont il fait partie subit elle-même des changements. Ces changements proviennent soit de modifications ou de rénovations, soit d'une altération de sa capacité de support par un accroissement des charges. Dans la plupart des cas, ça nécessite un renfort.

Le renforcement d'une structure en béton consiste à améliorer les caractéristiques mécaniques des éléments qui la composent, de manière à ce qu'elle offre une meilleure solidité aussi bien en état de service qu'en état de résistances ultimes. Il s'agit principalement des techniques de renfort avec une incorporation de profiles métalliques, ou bien une modification de la section structurelle et la plus récente qui utilise des matériaux composites en tant que renfort externe.

Si une structure est bien conçue et correctement exécutée, elle n'aura besoin d'aucun renforcement au moment de sa mise en service. Cependant, suite à des erreurs dans le projet ou dans l'exécution, ou suite à diverses circonstances lors de la vie en service (chargements maximums dépassés, changement des conditions d'exploitation, fluage, fatigue, attaques chimiques dues à l'environnement…), des renforts peuvent s'avérer nécessaires. Ce changement de circonstances implique que la nouvelle structure, issue du « projet actualisé », soit plus exigeante que l'antérieure.

Ainsi la notion de réparation vise à restituer les caractéristiques initiales de portance (rigidité–résistance). Le renforcement a pour objectif d'améliorer les performances de la structure vis-à-vis de conditions d'exploitations modifiées ou d'augmenter la durée de vie de l'ouvrage. Le principal avantage du renforcement par

https://doi.org/10.1051/matecconf/201814901046

composites est la manipulation très facile sur chantier sans recours au matériel lourd.

2 Programme expérimental

2.1 Matériaux et matériel utilisés

Les bétons sont confectionnés dans un malaxeur d'une capacité de 50 litres. La formulation du béton testé est donnée dans le tableau 1.

Tableau 1. Composition massique en kg/m^3 du béton.

Ciment	Eau	Sable (0/4)	Gravier 8/15	E/C
350	210	680	1135	0,6

Dans les essais réalisés pour l'étude des bétons, deux types d'éprouvettes ont été utilisés:
- Eprouvettes cylindrique de dimensions 10x20 cm, pour les essais de compression.
- Eprouvettes prismatiques de dimension 7x7x28 cm pour les essais de flexion.

Les écrasements ont été effectués à 28 jours d'âge.

2.2 Matériaux composites

Les matériaux composites utilisés dans cette étude est composé d'une résine époxy et des tissus en fibre de carbone et en fibre de verre (Figures 1 et 2).
-PRFC : Polymère renforcé de fibre de carbone
-PRFG : Polymère renforcé de fibre de verre

Fig. 1. Tissu de fibres de carbone

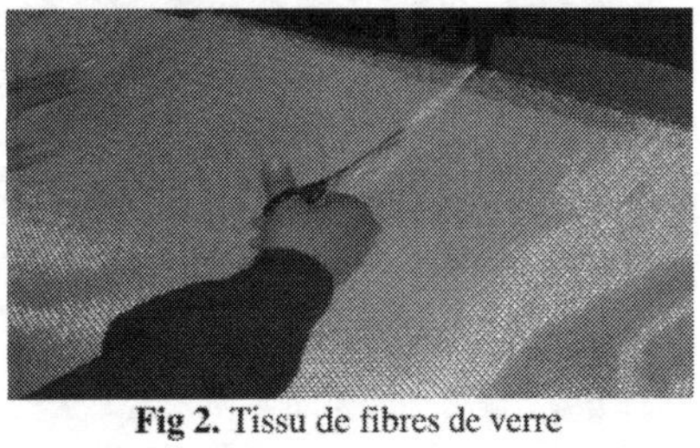

Fig 2. Tissu de fibres de verre

Les propriétés mécaniques fournies par le fabricant sont données dans le Tableau 2.

Tableau 2. Propriétés du PRFC & PRFG et de SIKADUR

	SikaWrap	Verre	Sikadur 330
Résistance à la traction	4300 MPa	2,2 GPa	30 MPa
Allongement à la rupture	1,8%	4,8%	2 à 5 %
Module d'élasticité en traction	234 GPa	72 GPa	4,5 GPa
Epaisseur	0,13 mm	1,1 mm	/
Densité	1,8	2,54	1,3 kg/l

2.3 Préparation des surfaces du béton

Le support doit être propre, sec, exempt de toute trace de graisse, d'huile, de rouille, de produit de cure, d'ancien revêtement ou traitement de surface, de parties non ou peu adhérentes, de laitance.

Avant l'utilisation des fibres (carbone ou verre), on a effectué un ponçage sur l'ensemble des éprouvettes afin d'éliminer la laitance de ciment, les particules friables non ou peu adhérents, et d'ouvrir la porosité de surface. Avant l'application de la résine Sikadur-330, toutes les poussières et autres particules non ou peu adhérentes doivent être complètement éliminées du support de préférence par brossage (papier verre 80) puis par aspiration mécanique (Figure 3).

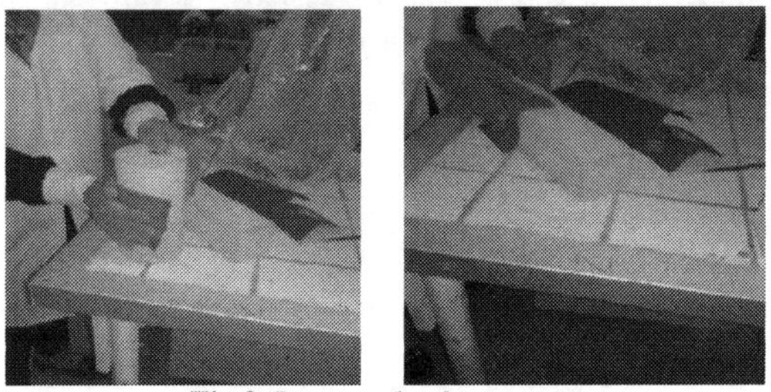

Fig 3. Ponçage des éprouvettes

2.2 Préparation de la résine époxy Sikadur-330

La résine à base d'époxy à deux composantes, résine (A) et durcisseur (B), doit être mélangée peu de temps avant l'application.

La proportion en masse du durcisseur représentait 25% de la masse de la résine (durcisseur/résine = 0,25) conformément aux recommandations du fournisseur. Le malaxage a été réalisé à faible vitesse, afin d'éviter l'inclusion d'air, pendant trois minutes environ jusqu'à disparition complète des traînées de couleur et l'obtention d'un mélange homogène. Le temps d'application de la résine (durée de vie en pot) était d'environ 60 minutes à 10°C ou 30 minutes à 35°C. La durée de vie en pot démarre dès le malaxage des deux composants (résine et durcisseur). Après cette période de temps, la résine est inutilisable. A basse température cette durée de vie sera plus longue qu'à haute température. Plus la quantité de mélange est importante, plus cette durée est courte. Afin d'augmenter cette durée lors de haute température, il est conseillé de répartir le mélange dans plusieurs récipients ou de refroidir chaque composant avant d'effectuer le mélange. Un mûrissement d'environ 5 jours à 23°C est requis pour la cure de la résine, selon les recommandations du fabricant. Ces temps indicatifs peuvent varier suivant les conditions ambiantes particulières à chaque application. Tous les spécimens ont bénéficié d'un temps de cure supérieur ou égale aux recommandations du fabricant (une semaine pour les spécimens).

2.3 Mise en œuvre de PRF

Le renforcement de structures à partir d'un tissu sec se fait directement par la pose de ce dernier sur une couche de résine qui recouvre l'élément à renforcer et qui sert de liant. Après marouflage, une couche de résine de fermeture permet de parfaire l'imprégnation. Les pressions nécessaires appliquées pour cette méthode de renforcement sont faibles par rapport à d'autres méthodes de renforcement. La direction principale des fibres a été orientée dans le sens circonférentiel des cylindres et longitudinal des prismes.

Cette technique permet notamment un suivi parfait de la forme du support ainsi que la maîtrise de l'épaisseur du film de résine. En utilisant des polymères qui polymérisent à température ambiante, il est possible de réaliser une stratification directe, en respectant les étapes suivantes :

- préparation de la surface, application de la couche primaire,
- application d'une première couche de polymère époxyde. La colle est constituée de 2 composants (résine + durcisseur),
- application d'une première couche de tissus secs,
- application d'une seconde couche de polymère époxyde, protection, finition

Ce procédé présente l'avantage d'une mise en œuvre simplifiée, mais le renforcement acquiert toute sa résistance au bout d'une semaine et le contrôle de la qualité de l'application doit être soigné suivant les conditions de mise en œuvre.

3 Résultats et discussions

3.1 Essai de compression simple

Les résultats des essais de laboratoire sont présentés numériquement dans le tableau.

Le Tableau 3 présente les valeurs maximales obtenues des résistances en compression, des déformations axiales, des augmentations en résistance à la compression des cylindres confinés avec PRFG et PRFC et les allongements à la rupture par rapport au béton témoin.

Tableau 3. Résultats des essais sur les cylindres en béton.

Nombre de couches	Résistance à la compression (MPa)	Augmentations des résistances (%)	Allongement A la rupture (‰)
T	20	–	2,1
V1	26	63	11,65
V2	35	120	18,45
V3	45	180	19.6
C1	56	260	24,8
C2	88	450	29,.92
C3	113	607	35

T : éprouvette témoin
V1 : éprouvette renforcée par 1 couche de fibre de verre
V2 : éprouvette renforcée par 2 couches de fibre de verre
V3 : éprouvette renforcée par 3 couches de fibre de verre
C1 : éprouvette renforcée par 1 couche de fibre de carbone
C2 : éprouvette renforcée par 2 couches de fibre de carbone
C3 : éprouvette renforcée par 3 couches de fibre de carbone

Le confinement de 1, 2 et 3 couches de PRFG, en comparaison avec les spécimens non confinés conduit respectivement à une augmentation de la résistance à la compression de 63%, 120 et 180% ainsi que les déformations à la rupture.

Le confinement de 1, 2 et 3 couches de PRFC, en comparaison avec les spécimens non confinés conduit respectivement à une augmentation de la résistance à la compression des spécimens cylindriques de 260%, 450% et 607% ainsi que les déformation à la rupture.

Ces résultats montrent que le confinement améliore considérablement la résistance et la capacité de déformation du béton confiné.

Un cylindre (poteau) soumis à la compression, renforcé par confinement aura l'avantage d'augmenter sa capacité en résistance en même temps que sa déformation. Le confinement circonférentielle des colonnes donne les meilleures performances en matière de résistance à la compression et déformation axiale.

La majorité des éprouvettes en béton confiné avec des matériaux composites PRF ont présenté une allure caractéristique bilinéaire ascendante (Figure 4). Dans la première branche linéaire, l'inclinaison de la courbe contrainte-déformation est semblable à celle du béton ordinaire, ce qui montre que le comportement est peu affecté par l'enveloppe composite. Dans la seconde branche linéaire, le béton est fissuré et la force de confinement du PRF est activée. La contrainte du béton confiné augmente linéairement avec l'augmentation de la contrainte dans le composite et atteint son maximum à la rupture de l'enveloppe PRF.

L'enveloppe en matériaux composites exerce une pression latérale sur le béton pour contrecarrer le gonflement créé par la dégradation de la rigidité du béton. La troisième branche prend naissance lorsque l'enveloppe en matériaux composites est entièrement activée, cette branche est très dépendante de la rigidité de l'enveloppe, elle-même directement proportionnelle au nombre de couches de composites collées.

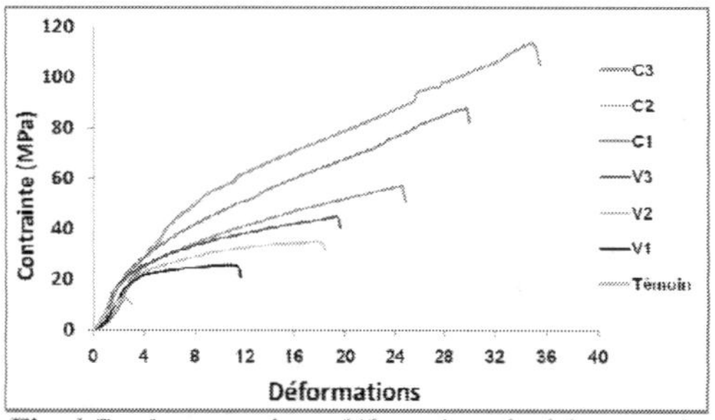

Fig. 4 Courbes contraintes-déformations des bétons testés

3.2 Variation de la résistance à la compression en fonction du nombre de couches

D'après le tableau3, la charge ultime et les déformations axiales augmentent avec l'augmentation du nombre de couches.

D'autre part, les courbes présentées sur la Figure 5 montrent qu'une augmentation du nombre de couches du composite n'affecte pas la pente de la branche initiale de la relation contrainte-déformation, mais elle a un effet significatif sur la pente de la seconde branche.

Les corrélations obtenues à partir de la Figure 4, montrent que la variation de la résistance en compression en fonction du nombre de couches du composite est une fonction linéaire. Les fonctions de lissage ont des coefficients de variation R^2 = 0,998. Ceci montre que la relation entre la résistance et le nombre de couches du composite N est correctement approchable par la fonction :

PRFG : $f_c = f_0(32{,}33N + 20)$

PRFC : $f_c = f_0(9{,}87N + 16)$

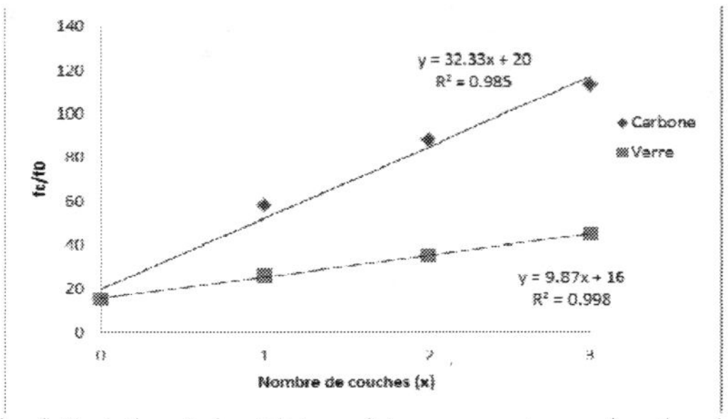

Fig. 5 Variation de la résistance à la compression en fonction du nombre de couches

3.2 Mode de rupture du béton confiné

La rupture des cylindres confinés avec la fibre de verre PRFG a été marquée par une rupture des fibres de verre dans le sens longitudinale du cylindre (Figure 6).

Dans tous les spécimens testés, la rupture a été fragile et d'une manière "brutale" due à l'énergie de déformation emmagasinée dans l'enveloppe composite au cours du chargement. Après la rupture, le béton fissuré suite à l'application de la contrainte de compression reste confiné à l'intérieur de l'enveloppe (Figure 7).

L'examen des spécimens après rupture, montre la présence d'une déchirure du composite le long d'une ligne verticale comme montré sur la Figure 5. Elle apparait pour les spécimens cylindriques dans les zones à faible résistance. Aucune délamination des tissus n'a été constatée et la forme conique du béton après rupture a été retrouvée.

(a) (b) (c)

Fig. 6 Mode de rupture du béton confines par PRFC

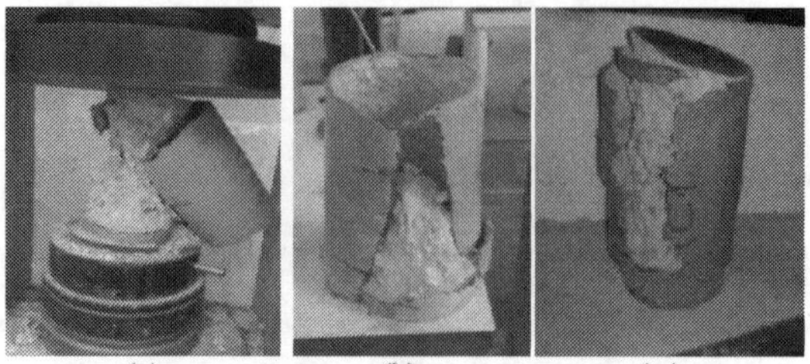

(a) (b) (c)

Fig. 7 Mode de rupture du béton confines par PRFG

3.3 Résistances à la flexion

Les résultats de l'essai de flexion 3 points sur les différents types des corps d'épreuve sont donnés dans le tableau 4.

Tableau 4. Résultats des essais sur les prismes en béton.

N° de couche pour le béton	Résistance à la flexion (MPa)	Augmentations des résistances (%)	Allongement A la rupture (mm)
T	1,7	–	2,14
V1	6,2	265	1,93
V2	6,33	272	2,65
V3	7,01	312	2,97
C1	7,74	355	1,45
C2	8,14	379	1,67
C3	9,27	445	1,85

Lors de l'essai de traction par flexion, on observe une augmentation considérable des forces de rupture pour chaque type de béton confiné (PRFC, PRFG) par rapport à celle du béton témoin. Dans tous les cas, on a réussi à augmenter la résistance et la ductilité.

Une augmentation du taux de confinement de 1, 2 et 3 couches de PRFG, à la flexion avec les spécimens non confinés et confinés avec le PRFC, conduit respectivement à une augmentation de la résistance à la traction par flexion des spécimens prismatiques confinés par PRFG de 265% pour une couche, 272% pour deux couches et 312% pour trois couches (Tableau 4).

Une augmentation du taux de confinement de 1, 2 et 3 couches de PRFC, a la flexion avec les spécimens non confinés et confinés avec le PRFG, tel que montré sur la Figure 7 conduit respectivement à une augmentation de la résistance de traction par flexion des spécimens prismatiques confinés par PRFC de 355%, 379% et 445% tableau 4.

On retient de ces essais que la colle a un effet très limité sur la résistance à la traction du béton et donc à la flexion, tandis que le renforcement par la couche de PRFC en zone tendue augmente la résistance et la déformabilité du béton en flexion.

Le renforcement en PRFC en zone tendue, prolonge la résistance et la déformabilité du béton au delà de la première branche (béton témoin) par une seconde branche linéaire fragile de moindre rigidité. Cette perte de rigidité est due à la fissuration du béton.

Le tissu de verre confère une meilleure déformabilité qui pourra se traduire par une bonne ductilité. Le tissu de verre est 2 fois moins résistant que le carbone mais néanmoins deux fois plus ductile ; il possède un pseudo-palier plastique de ductilité une fois la résistance ultime atteinte.

Cependant, le nombre de couche n'a pas une influence considérable sur l'amélioration de la résistance à la flexion (Figure 8).

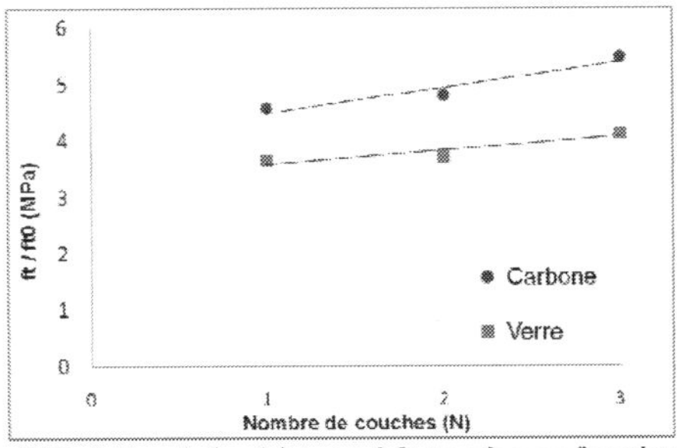

Fig. 8 Variation de la résistance à la traction en fonction du nombre de couches

3.4 Mode de rupture du béton confiné

L'élément renforcé par le tissu de carbone modifie le mode de rupture. On observe une fissure au niveau des appuis puis se propage au milieu, en suite son développement parallèlement à la charge appliquée au fur et à mesure que celle-ci augmente et le béton a été arraché dans les zones de rupture. Lors de l'essai de flexion trois point, on observe des fissures inclinée sur l'élément, au fur et à mesure que l'on augmente le taux de chargement (Figure 9).

La rupture des spécimens renforcés en PRFC est caractérisée par l'apparition de deux fissures quasiment symétriques inclinées d'un angle proche de 45°, la charge est diffusée par le biais de 2 bielles de compression qui soumettent le renforcement PRFC à une traction. Le composite n'a pas atteint sa limite ultime à la rupture. Celle-ci a eu lieu par traction du béton tendu suivi du délaminage de la couche de PRFC.

Concernant le tissu de verre, la rupture est obtenue par une rupture du tissu et la dégradation du béton est plus avancée que celle du béton témoin.

Fig. 9 Mode de rupture de béton de PRFC & PRFG

3 Conclusions

A partir des résultats des essais, on peut tirer les conclusions suivantes :
- Résistance en compression : Le renfoncement par le tissu de verre a donné un gain de 63% pour 1c ouche et 120% pour 2 couches et 180% pour 3 couches. Le tissu de carbone à donné un gain de 260% pour 1 couche et 450% pour 2 couches et 607% pour 3 couches.
- Résistance à la flexion : L'augmentation a atteint 265% pour 1 couche et 272% pour 2 couches et 312% pour 3 couches pour le tissu de verre et 355% pour 1 couche, 379% pour 2 couches et 445% pour 3 couches pour le tissu de carbone.
- L'apport en rigidité du tissu composite aux éléments de structures est largement obtenu. Ceci permettra sans aucun doute, d'entreprendre par la suite des investigations plus poussées pour caractériser le matériau renforcé sous différentes sollicitations.

4 Références

1. A.Avril, "Application des méthodes de mesure de champs a la caractérisation mécanique de poutres en *béton arme réparées par matériaux composites", Thèse de Doctorat, Ecole Nationale Supérieur des Mines de Saint-Étienne et de l'université Jean Monnet, 193 p, 2002.*
2. *S.O Alampalli, J. Connor, A.P. Yannotti. Fiber reinforced polymer composites for the superstructure of a short-span rural bridge", Composites structures, Vol. 58, 2002, pp. 21-27*
3. *Y. H.Chai, , M. J. N. Priestley, et F. Seible "Seismic retrofit of circular bridge columns for enhanced flexural performance", ACI Structural Journal, Vol. 88, No. 5, 1991, pp. 572-584*

Some physical properties of the date palm stem

A. Hakam[1], M. Elkhal[1], M. Ziani[2], A. Famiri[3] and A. Fidah[3]

1. Wood Science Laboratory, Mohammed V University in Rabat, Faculty of Science, PO Box 1014, Rabat, Morocco.
2. National Institute of Archaeology and Patrimony Sciences, PO Box 6828, Rabat, Morocco.
3. Forest Research Center, PO Box 763, Agdal, Rabat, Morocco.

Abstract: The main objective of this study was to test some technological performances of date palm stem (*Phoenix dactylifera* L.) from the region of Marrakech in Morocco with a view to its use in housing construction in southern Moroccan regions. Physical characteristics were determined according to French AFNOR standards. The samples used were respectively 20 mm × 20 mm × 20 mm in the longitudinal, radial and tangential directions. All the results obtained were the average of the measurements on 50 samples. The results obtained showed that the moisture content has an average value of 15 % at the center of stem and 18 % at the periphery and the infradensity has a value of 0.18 g / cm3 in the center and 0.42 g / cm3 in the periphery. For samples from the center, the tangential, radial and longitudinal shrinkage were respectively 2.02 %, 1.77 %, and 0.42 %, while on the periphery of the stem the tangential, radial and longitudinal shrinkage were respectively 3.41 %, 2.98 % and 0.34 %. These results showed that the stem center has better dimensional stability compared to the peripheral region of the stem.

Keywords: Date palm, *Phoenix dactylifera* L., Moisture content, Infradensity, shrinkage.

Quelques propriétés physiques de la tige du palmier dattier

Résumé : L'objectif principal de cette étude était de tester quelques performances technologiques de la tige du palmier dattier (*Phoenix dactylifera* L.) de la région de Marrakech au Maroc en vue de son utilisation dans la construction d'habitat dans les régions du sud marocain. Des caractéristiques physiques ont été déterminées selon les normes françaises AFNOR. Les échantillons utilisés étaient respectivement de 20 mm × 20 mm × 20 mm dans les directions longitudinale, radiale et tangentielle. Tous les résultats obtenus étaient la moyenne des mesures sur 50 échantillons.
Les résultats obtenus ont montré que l'humidité avait une valeur moyenne de 15 % au centre et 18 % à la périphérie de la tige et que l'infradensité avait une valeur de 0,18 g/cm^3 au centre et 0,42 g/cm^3 à la périphérie. Au centre de la tige le retrait tangentiel, radial et longitudinal avaient respectivement des valeurs de 2,02 % , 1,77% et de 0,42% tandis qu'à la périphérie de la tige, les retraits tangentiel, radial et longitudinal avaient respectivement des valeurs de 3,41 %, 2,98 % et 0,34 %. Ces résultats ont montré que le centre de la tige présentait une meilleure stabilité dimensionnelle par rapport à la périphérie.

Mots-clés : Palmier dattier, *Phoenix dactylifera* L., Humidité, Infradensité, Retrait.

1 Introduction

Le palmier dattier (*Phoenix dactylifera* L.) est l'une des plantes fruitières les plus importantes du sud du Maroc. Elle représente une bonne rente pour de nombreux agriculteurs.

Une fois que les palmiers dattier deviennent trop mûrs, leur production de produits traditionnels, tels que les dates, diminue considérablement, ce qui conduit les agriculteurs à rechercher d'autres moyens de valorisation des palmiers dattiers âgés.

Dans le but de répondre à cette demande, plusieurs travaux de recherche ont été consacrés à la valorisation de la tige du palmier dattier [1-4]. Ces travaux ont concernés son utilisation dans le domaine de l'isolation thermique des bâtiments [1,2] et dans le domaine de la construction en général [3], ou dans la fabrication de composites bois-ciment [4].

D'autre part, l'utilisation de la tige du palmier dattier dans la construction nécessite une parfaite connaissance de plusieurs de ses caractéristiques et en particulier ses propriétés physiques [5-6]. Concernant ce point, des études ont été consacrées à la détermination des caractéristiques physique, parmi ces études celle de Fathi [5] qui a montré que l'humidité et la densité varie de manière significative dans la tige du palmier dattier.

Les résultats d'une étude expérimentale sur les propriétés physiques de la tige du palmier dattier de l'oasis de Biskra en Algérie ont été rapporté par Amirou et al., [6]. Selon cette étude, la densité à 12% d'humidité avait des valeurs allant de 0,467 à 0, 616 g/cm^3, avec une valeur moyenne d'environ 0,549 g / cm^3 et des valeurs très élevées des retrais ont été enregistrées.

Des recherches similaires ont été réalisées sur la tige du cocotier (*Cocos nucifera*) [7-9]. Les valeurs des densités moyennes au centre et à la périphérie de la tige du cocotier obtenues par Killmann [7], étaient respectivement de 0,31 et 0,73 g/cm3. Selon cette étude, le retrait tangentiel, radial et longitudinal au centre de la tige avaient respectivement des valeurs de 3,46 % , 3,35 % et 0,18 % tandis qu'à la périphérie de la tige, les retraits tangentiel, radial et longitudinal avaient respectivement des valeurs de 6,33 %, 5,97 % et 0,08 %.

Les études réalisées en 2014 par Fathi et Frühwald [8] sur les propriétés mécaniques du cocotier (*Cocos nucifera*) ont montré que la densité augmente avec la distance transversale du centre de la tige.

Les essais de caractérisation physiques effectués à différentes positions de la hauteur de la tige (c'est-à-dire, en haut, au milieu et en bas) et des positions latérales (c'est-à-dire, centre et à la périphérie) par Rana et al., [9] ont montré qu'Il y avait une différence significative entre les différentes positions de hauteur pour l'humidité contenu, mais il n'y avait pas de différence significative pour la teneur en humidité entre la périphérie et la partie centrale. Selon cette étude, la plus haute teneur en humidité était pour la partie supérieure de la tige et la partie inférieure montrait la plus faible teneur en humidité. Selon cette étude, le retrait moyen dans la direction tangentielle était de 5,25 à 5,81%, de 5,07 à 5,42% dans le sens radial et de 0,88 à 1,20% dans le sens longitudinal. La densité moyenne après séchage à l'ai était de 0,37 à 0,43 g / cm^3.

2 Matériaux et méthodes

2.1 Matériaux

La matière première (rondin de palmier dattier d'environ 1,3 m de long et 24 cm de diamètre) (Figure 1) utilisée dans cette étude a été obtenue de la palmerais de Marrakech (31° 37′ 48″ N - 8° 0′ 0″ W) par le Haut Commissariat aux Eaux et Forêts et à la lutte contre la désertification du Maroc. Par conséquent, nous n'avons aucune garantie sur l'état de la plante avant abatage, l'âge et le temps de séchage à l'air. Son humidité, au moment de la livraison, en mars 2016, mesurée avec un avec un humidimètre KIMO portatif électronique (précision ± 1%) était entre 15 et 18%.

Le sciage a été réalisé avec une scie à ruban pour prélever sept rondelles de 2,5 cm d'épaisseur (Figure 1).

Fig. 1. Rondin et Rondelles du palmier dattier

Le prélèvement des éprouvettes a été réalisé au Centre de Recherche Forestière de Rabat, au Maroc.
Des cubes de 2 cm de chaque coté ont été usinés à partir des rondelles (Figure 2 et 3). Les éprouvettes ont été conditionnées, par la suite dans une chambre de 65 ± 5% d'humidité relative à 20± 2°C pour se stabiliser à 12% d'humidité environ.

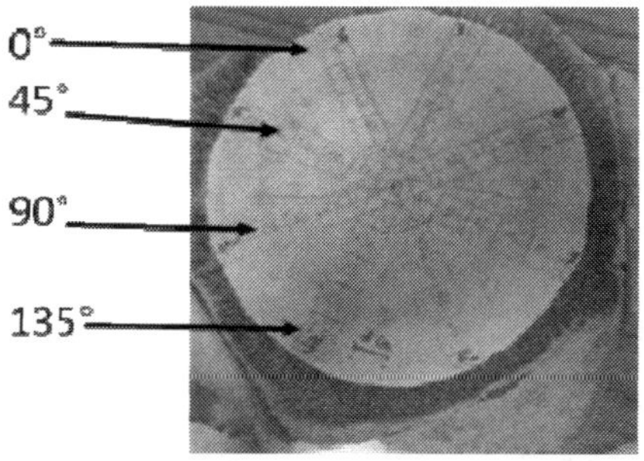

Fig. 2. Préparation des éprouvettes

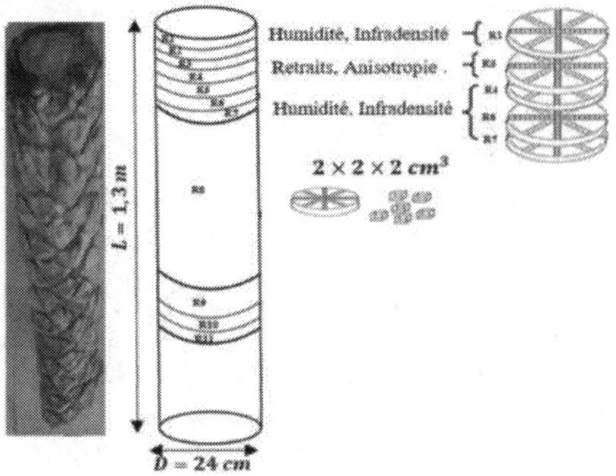

Fig. 3. Prélèvement des éprouvettes

2.2 Méthode

Les caractéristiques physiques ont été déterminées selon les normes françaises AFNOR. Les échantillons utilisés étaient respectivement de 20 mm × 20 mm × 20 mm dans les directions longitudinale, radiale et tangentielle. Tous les résultats obtenus étaient la moyenne des mesures sur 50 échantillons.

2.2.1 Mesure de l'humidité

L'humidité du rondin initial de palmier dattier et des rondelles de 2,5 cm d'épaisseur a été mesurée avec un humidimètre KIMO portatif électronique (précision ± 1%) et l'humidité des éprouvettes qui ont servi pour la détermination des caractéristiques physiques a été mesurée comme décrit dans la norme AFNOR NF B 51-004 [10], à l'aide de la relation 1.

$$H(\%) = \frac{M_h - M_0}{M_0} \times 100 \qquad (1)$$

Avec M_h est la masse de l'éprouvette à l'état humide et M_0 est la masse de l'éprouvette à l'état anhydre. Les masses ont été mesuré avec une balance (précision = 0,1 mg) et les dimensions des éprouvettes avec un pied à coulisse numérique (précision 0,02 mm).

2.2.2 Mesure du retrait

Le retrait est un paramètre qui détermine la stabilité dimensionnelle de l'échantillon. Il a été déterminé selon la norme française NF B51-006 [11], à l'aide de l'équation 2 :

$$R(\%) = \frac{L^{s}_{L,R,T} - L^{0}_{L,R,T}}{L^{0}_{L,R,T}} \times 100 \qquad (2)$$

Avec $L^{s}_{L,R,T}$ est la longueur à la saturation selon la direction considérée (L, R, ou T) et $L^{0}_{L,R,T}$ est la longueur à l'état anhydre. L'anisotropie du retrait est le rapport du retrait tangentiel au retrait radial.

2.2.3 Mesure d'infradensité

L'infradensité est le rapport de la masse anhydre M_0 sur le volume saturé V_S d'un échantillon (Equation 3). Elle est exprimée en g/cm^3 et déterminée selon la norme NF B51-005 [12]. Cette grandeur ne varie pas en fonction de l'humidité comme la masse volumique, car elle est le rapport de deux valeurs constantes pour chaque échantillon.

$$D_b = \frac{M_0}{V_S} \qquad (3)$$

3 Résultats et discussion

3.1 Profil de l'humidité

Quatre rondelles ont été utilisées pour les mesures de l'humidité. Les valeurs Max, Min et Moyenne de 50 éprouvettes de chaque rondelle sont présentées dans le tableau 1.

Tableau 1. Humidité des éprouvettes des quatre rondelles

Rondelle	Max	Min	Moyen	Ecart type
1	19,643	11,789	15,649	1,583
4	19,765	10,959	15,011	1,536
6	19,694	13,144	15,662	1,419
7	22,717	13,804	15,747	1,781

La figure 4 représente le profil d'humidité du centre à la périphérie de la rondelle 7 du palmier dattier pour les quatre orientations (0°, 45°, 90° et 135°).

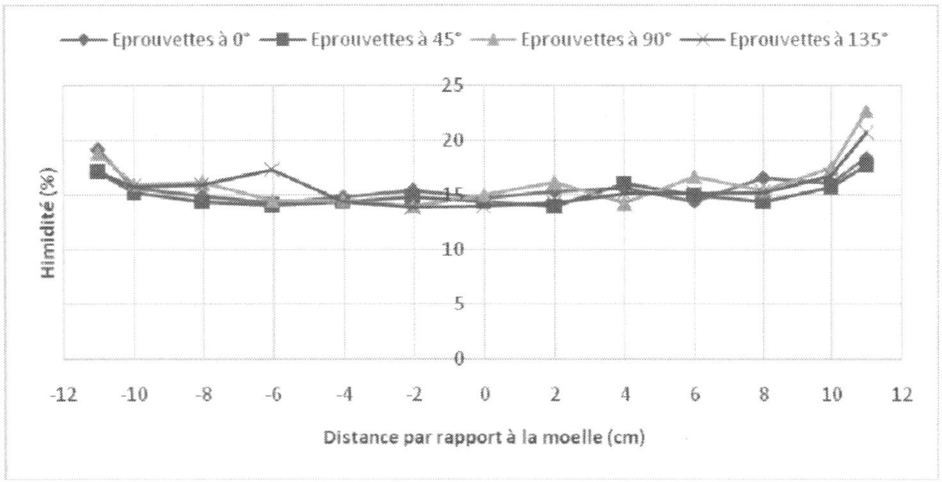

Fig. 4. Profil d'humidité du centre à la périphérie de la rondelle 7 du palmier dattier pour les quatre orientations

D'après la figure 4, l'humidité des éprouvettes proches du centre de la tige est d'environ 15 % et l'humidité des éprouvettes loin du centre ou près de la périphérie de la tige est d'environ 18 %.

3.2 Retrait

Le tableau 2 représente les valeurs du retrait tangentiel (R_T), radial (R_R) et longitudinal (R_L) ainsi que la valeur de l'anisotropie du retrait A = (R_T/R_R) (%), en fonction de la position radiale dans la tige du palmier dattier.

Tableau 2. Retraits du palmier dattier à trois distances du centre de la tige

Position latéral	R_T	R_R	R_L	A
Périphérique	3,41	2,98	0,34	1,14
Milieu	2,14	1,97	0,33	1,08
Centre	2,02	1,77	0,42	1,14
Moyen	2,524	2,240	0,36	1,12
Écart-Type	0,103	0,121	0,034	0,092

D'après le tableau 2, les éprouvettes de la périphérie présentent le plus grand retrait, suivie des éprouvettes du milieu et enfin du centre de la tige du palmier dattier.

3.3 Infradensité

Le tableau 3 représente les valeurs de l'infradensité des éprouvettes de quatre rondelles. D'après ce tableau, la valeur moyenne de toutes les éprouvettes est d'environ 0,308 g/cm^3.

Tableau 3. Infradensité des éprouvettes des quatre rondelles

Rondelle	Max	Min	Moyen	Ecart type
1	0,444	0,119	0,310	0,084
4	0,439	0,139	0,312	0,087
6	0,459	0,143	0,308	0,089
7	0,435	0,127	0,304	0,088

La figure 5 représente le profil de l'infradensité du centre à la périphérie de la rondelle 7 du palmier dattier pou r les quatre orientations (0°, 45°, 90° et 135°).

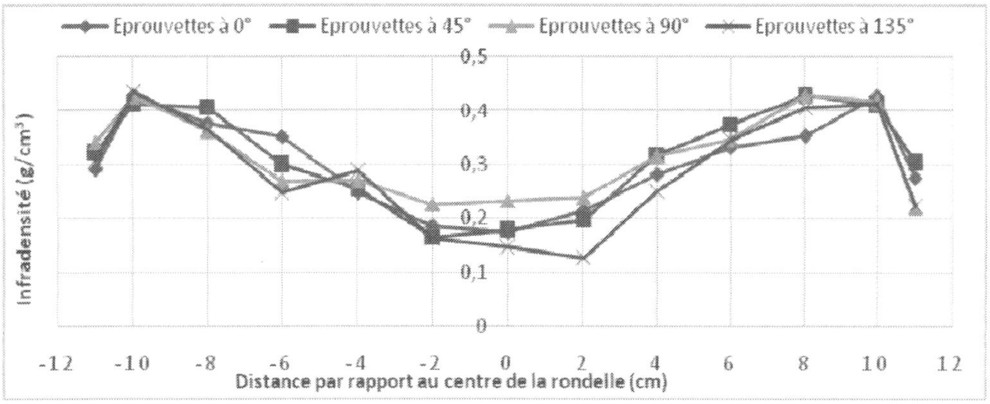

Fig. 5. Profils d'infradensité du centre à la périphérie de la rondelle 7 du palmier dattier pour les quatre orientations

La figure 5 montre que l'infradensité a une valeur d'environ 0,18 g/cm^3 au centre et 0,42 g/cm^3 à la périphérie de la tige du palmier dattier. Le rapport de la valeur de l'infradensité à la périphérie sur sa valeur au centre est d'environ 2,33. Des rapports de 2,35 et de 2 de la densité à la périphérie sur la densité au centre de la tige de cocotier, ont été également trouvés par Killmann [7] et Fathi et Frühwald [8] respectivement.

4. Conclusion

L'objectif de cette étude était de déterminer quelques caractéristiques physiques de la tige de palmier dattier qui cesse de produire, et ce pour mieux orienter son utilisation dans le domaine de la construction des habitats des oasis.

Les variations de l'humidité, du retrait et de l'infradensité selon la position tangentielle et radiale des échantillons dans la tige de palmier dattier ont été étudiées.

Aucune modification des caractéristiques physiques étudiées n'a été observée en fonction de la position tangentielle des échantillons dans la tige du palmier dattier,

Le profil radial de l'humidité à montré que l'humidité à la périphérie était légèrement supérieure à sa valeur au centre de la tige du palmier dattier.

Le profil radial de l'infradensité à montré que le rapport de sa valeur à la périphérie sur sa valeur au centre de la tige du palmier dattier est d'environ 2,33.

Références

1. N. Benmansour, B. Agoudjil and A. Boudenne. *Etude des performances de produits renouvelables et locaux adaptés à l'isolation thermique dans le bâtiment* (XXIXe Rencontres Universitaires de Génie Civil. Tlemcen, 2011)
2. B. Agoudjil, A. Benchabane, A. Boudenne, L. Ibos and M. Fois, Energy and Buildings **43**, 491 (2011)
3. L. Fathi, K. Fruehwald and A. Fruehwald. *The potential use of timber from palm trees for building purposes* (Wood conference on timber engineering, Quebec City, Canada, 2014)
4. F. Touloum, A. Younsi, A. Kaci and A. Benchabane, J. Appl. Eng. Sci. Technol. **2**, 57 (2016)
5. L. Fathi 2014. *Structural and Mechanical Properties of the Wood from Coconut Palm, Oil Palms and Date Palms* (PhD - Thesis, submitted May 2014, Univ. of Hamburg, Germany)
6. S. Amirou, A. Zerizer and A. Pizzi, Materials Testing. Eur J Wood Prod **56**, 236 (2014)
7. W. Killmann, Wood Science and Technology **17**, 167 (1983)
8. L. Fathi and A. Frühwald, Wood Material Science & Engineering **9**, 214 (2014)
9. M.N. Rana, A.K. Das and M. Ashaduzzaman, Bangladesh J. Sci. Ind. Res. **50**, 39 (2015)
10. NF B 51-004, Wood Determination of Moisture (1985), AFNOR, French standard
11. NF B 51-005, Determining the Density of the Wood (1985), AFNOR, French standard
12. NF B 51-006, Determination of Shrinkage (1985), AFNOR, French standard

Parameters estimation of Drucker-Prager plasticity criteria for steel confined circular concrete columns in compression

Walid A. Al-Kutti

Department of Civil and Construction Engineering, College of Engineering, Imam Abdulrahman Bin Faisal University, P.O. Box 1982, Dammam, Saudi Arabia.

Corresponding author: wasalem@iau.edu.sa

Abstract. This paper explores the possibility to use Drucker-Prager model in Steel-Concrete composite section. Numerical simulation was conducted using finite element package to simulate the steel-concrete composite section subjected to uniaxial compressive loading. After calibration with experimental study, parametric study was conducted to evaluate the effect of the friction angle and the cohesion constant c on the stress-strain curve of composite section. Empirical relationship between the friction angle and the confined concrete compressive strength was developed and a range of cohesion constant c between 5-10 MPa was suggested for confined concrete strength range of 25 to 100 MPa, respectively.

1 Introduction

A composite section is an element consisting of two materials with different properties. The idea of using two or more materials to build a composite element takes into account the advantages of both materials. One of the major composite elements is reinforced concrete in which the concrete material combines with steel to form a new composite material known as reinforced concrete, where the new material has the advantage of the high compression strength of concrete and the high tensile strength of steel. Although several experimental and numerical studies have been conducted on composite sections, there is a need to develop a simplified approach in modeling composite sections using Drucker-Prager plasticity criteria.

Several numerical studies have been conducted to simulate the behavior of confined concrete [1, 2]. Jiang et al. [1] used ABAQUS software in their study and adopted the Drucker-Prager plasticity model, and concluded that the plastic dilation angle is a function of the axial plastic strain and the lateral stiffness ratio. Yu et al. [3] developed a modification of the Drucker-Prager plasticity model to predict the strength and ductility behavior of confined FRP concrete. Three features were mentioned in this study: a) the yield criterion including third deviatoric stress, b) a hardening/softening rule, which depends on the confining pressure, and c) the flow rule, which depends on the confining pressure and the rate of confinement increment. The model was implemented in ABAQUS software and a comparison between the numerical and experimental results was conducted.

Another approach was used by Dundar et al. [4] where an experimental study was conducted on the behavior of reinforced concrete columns and concrete-encased composite columns subjected to biaxial bending and axial load. Fifteen reinforced concrete columns were subjected to both biaxial bending and axial load, and the nominal stress- strain relationship was used to model the concrete, steel and the reinforcing bars. The theoretical analysis and the experimental results showed that the compression strength of the concrete and its correspondent strain were the most important parameters on the ultimate strength of the composite columns.

In recent years, several researchers have used the Drucker-Prager model to simulate the behavior of concrete confined by FRP or steel tubes. Dian et al. [5] used the D-P plasticity model to simulate the behavior of a filled steel tube subjected to long term loading. The numerical results matched well with the experimental results. Similar work was reported by Ehab et al. [6] and Fa-xing et al. [7]. The main contribution of this paper will be developing an empirical estimation for the parameters of three Drucker-Prager plasticity criteria in confined concrete subjected to compression. The simplified approach will assist engineers and researchers to estimate the friction angle φ and the cohesion constant c of confined concrete.

2 Drucker-Prager Plasticity Model

The 3-D concrete quarter cylinder was modeled as a

https://doi.org/10.1051/matecconf/201814901048

perfect elasto-plastic Drucker-Prager material model, which includes the effect of hydrostatic pressure on the shearing resistance of the concrete. The Drucker-Prager yield criterion function f can be written as:

$$f(I, \sqrt{J_2}) = \sqrt{J_2} + \alpha I = k \quad (1)$$

where I is the hydrostatic component of the stress tensor, J_2 is the deviatoric stress tensor invariant, α and k are material constants which can be related to the friction angle φ, and cohesion c of the modified Mohr-Coulomb by matching the outer apices of the M–C hexagon with the D–P surface. For 3-D, the relation between the parameters is as follows:

$$\alpha = 2\sin\varphi / \sqrt{3}(3-\sin\varphi)$$
$$k = 6c\cos\varphi / \sqrt{3}(3-\sin\varphi) \quad (2)$$

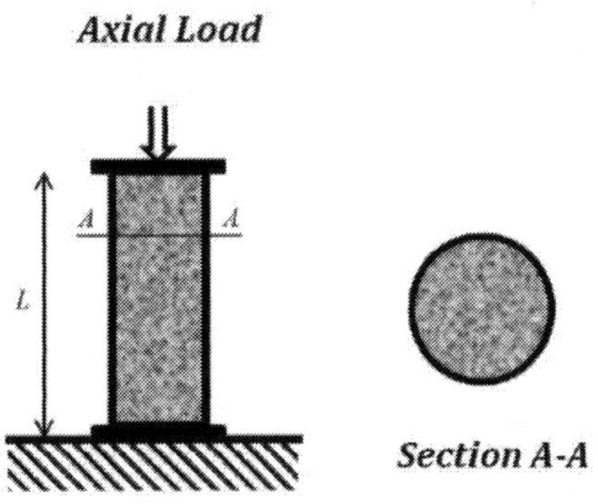

Fig 1. Experimental test setup.

3 Experimental Model for Composite Section

The Drucker-Prager model was calibrated with the experimental results reported by Sakino et al. [8]. Sakino et al. [8] investigated the effect of four parameters in composite sections subjected to uniaxial compressive load; these parameters were: (1) tube shape (circular and rectangular tubes), (2) tube tensile strength, (3) tube diameter-to-thickness ratio, and (4) concrete strength. A total of 114 composite steel short columns were subjected to uniaxial compressive load, and the average longitudinal strains were obtained from four linear variable differential transducers that measured axial deformation. Yielding of steel was the failure mode as reported in this experimental study. Figure 1 shows the experimental test setup used by Sakino et al. [8] including dimensions and boundary conditions of the specimen. The axisymmetric structural mechanics of boundary value problem can be used to simulate the stress-strain behavior of a composite section subjected to compression loading. Experimental results for circular concrete filled steel tubes reported by Sakino et al. [8] are used as a comparison with the numerical simulation using the finite element package software COMSOL and the Drucker-Prager plasticity model. Details of the specimen dimensions and the material properties are shown in Tables 1 and 2.

Table 1. Geometrical properties of the composite section [4]

L (mm)	D (mm)	t (mm)	D/t
450	149	2.96	50.4

Table 2. Mechanical properties of the composite section [4].

σ_y (MPa)	σ_c (MPa)	N_{exp} (kN)	σ_{su} (MPa)	ϵ_{su} (%)
308	40.5	1043	340	1.21

4 Finite Element Model

The axisymmetric structural mechanics of the boundary value problem with only one-eighth of the cylinder was used to simulate the stress-strain behavior of the composite section subjected to compression loading. An elasto-perfect plastic material was used for the steel tube in which the elastic modulus Es was about 200 GPa and the yield strength σy was about 308 MPa. For concrete, a user defined expression function in COMSOL using the Drucker-Prager model, as defined in equations 1 and 2, was used for the concrete. The boundary condition was defined as a symmetrical plane for internal boundaries while it was defined as free for the curve boundary. The cohesion constant c was assumed to be 8.2 MPa, which is about 20% of the compressive strength of concrete σc using the empirical formula for predicting direct shear test of concrete, as reported by Ha Ngoc Tuan et al. [9]; the friction angle φ was 28°. These material parameters were found by calibration of the elasto-plastic solution for the uniaxial compression results experimentally determined by Sakino et al. [8], as mentioned in detail in section 3. The applied load was defined as increment axial displacement in the z direction from the top plane, and constraint conditions were applied in x and y directions. The yielding of steel is the failure mode as reported by Sakino et al. [8], and as predicted from the finite element model.

5 Comparisons between DP Model and the Experimental Results

Figure 2 shows a comparison between the numerical results analysis of the axisymmetric cylinder of a composite section using COMSOL and the Drucker-Prager (DP) model, with the experimental results reported by Sakino et al. [8]. From Figure 2 it can be observed that the DP model matched well the experimental results of the confined can be noted in Figure 3, which shows the cross-sectional compression stress distribution at peak loading. From Figure 3 it can be observed that the maximum compression stress in the steel was found to be 318.5 MPa compared to 340 MPa as shown in Table 1. From Figure 3, the compression stress in the confined concrete is about 55 MPa, which is approximately 1.4 times the compressive strength of unconfined concrete (σc = 40.8 MPa) as shown in Table 1.

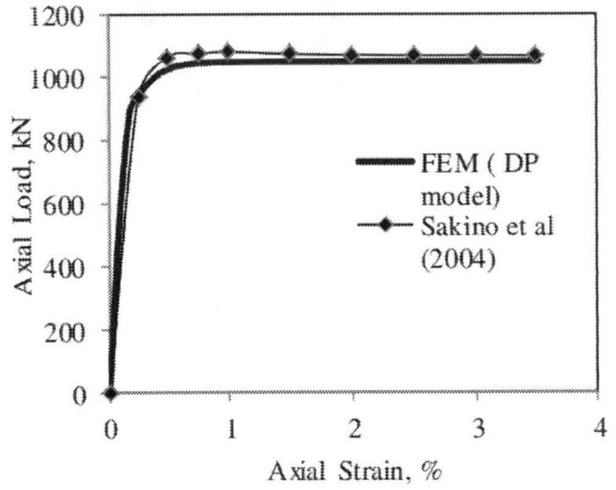

Fig 2. Comparison between numerical and experimental results.

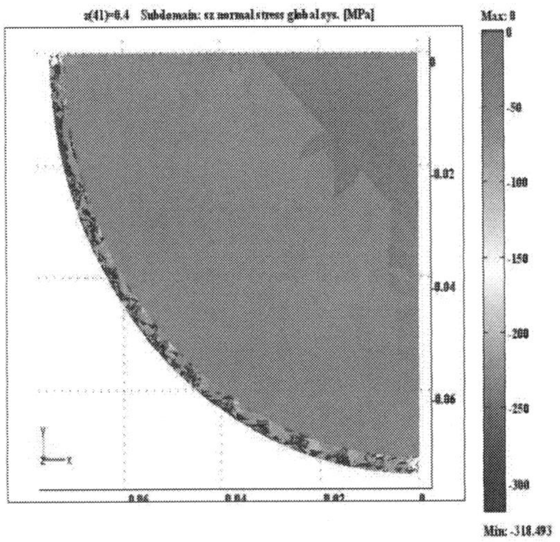

Fig 3. Compression stress distribution at peak loading.

6 Parametric Study of Drucker-Prager Model Parameters

To study the effect of friction angle φ and cohesion c on the confined compressive strength of concrete σcc, a parametric study was conducted. Figures 4 and 5 show the effect of the friction angle φ and the cohesion c on the axial compressive load of the composite section. Figure 4 shows the numerical solution for the axial load-strain of the composite section for constant friction angle φ of 28° and cohesion constant c of range between 1 and 10 MPa. From Figure 4 it can be observed that with the increase in c, the peak load increases. However, for cohesion c of 1 MPa, an elastic-perfect plastic relationship was observed, while for c values between 5 and 10 MPa, a more non-linear relationship can be observed at the end of the elastic zone.

From these results it could be concluded that for normal concrete with compressive strength between 25 and 50 MPa, the cohesion constant c of range between 5 and 10 MPa could be used, which about 20% of the unconfined compressive strength σc.

Figure 5 shows the effect of the friction angle φ on the behavior of confined concrete when using a cohesion constant c of 8.2 MPa. From Figure 5 it can be observed that the Drucker-Prager parameter of friction angle φ affects the non-linearity behavior significantly when compared to the cohesion constant c, as shown in Figure 5. Higher values of φ led to a higher peak load and more tendencies to catch the non-linearity at the end of the elastic zone. Based on the parametric study a relationship was developed between the friction angle φ and the confined concrete compressive strength σcc, as shown in Figure 6. The exponential function was found to fit this relationship as follows:

$$\sigma_{cc} = 14.75exp^{0.05\varphi} \tag{3}$$

For normal concrete with compressive strength between 25 and 50 MPa, the suitable friction angle φ range was found to be between 15° and 28°.

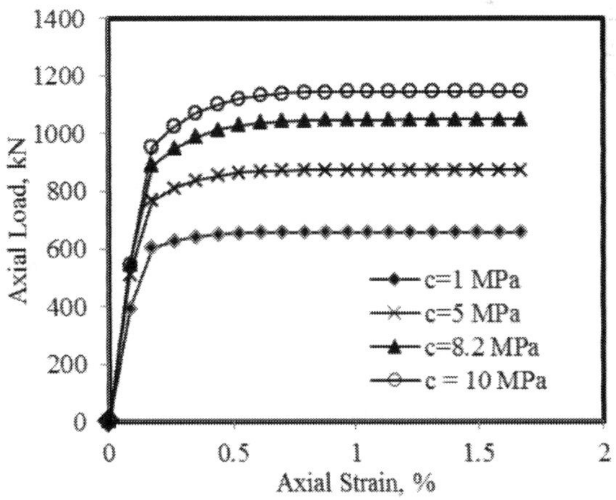

Fig. 4. Numerical simulation of axial load-strain for different values of cohesion constant c and friction angle φ of 28°.

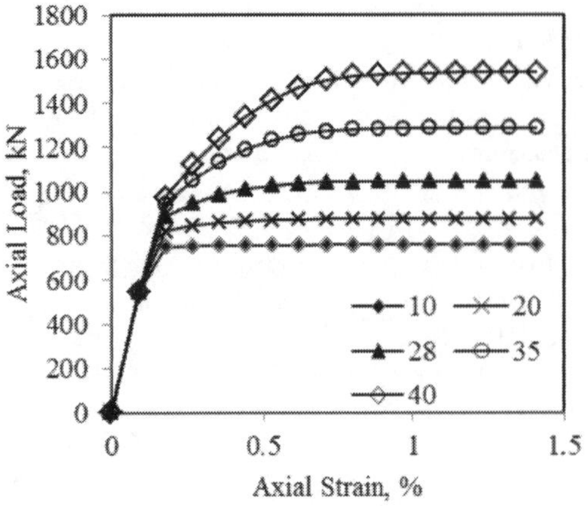

Fig 5. Numerical simulation of axial load-strain for different values of friction angle φ with cohesion constant c of 8.2 MPa.

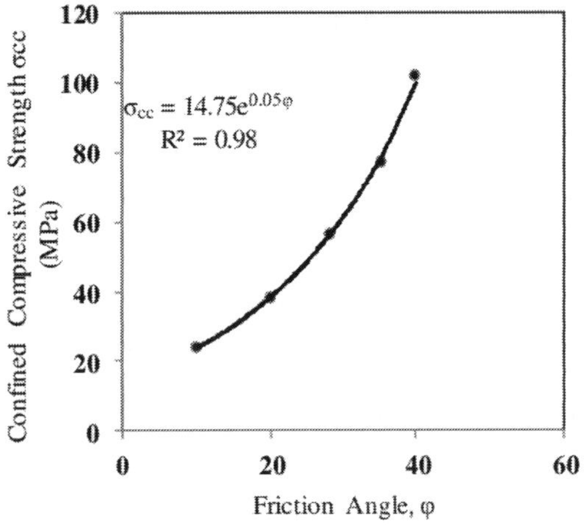

Fig 6. Relationship between the friction angle φ with cohesion constant c of 8.2 MPa and confined concrete compressive strength σ_{cc}.

7 Verification of the D-P plasticity model

To verify the model, eleven concrete-filled-steel composite sections were cast and used for the testing. Eight of these steel tubes were cast with two types of concrete having different compressive strengths (25 and 43 MPa). Three steel tubes were tested without casting concrete to study the mechanical properties of steel tubes. The dimensions of the steel tubes were 300 mm in length, 100 mm in diameter (D) and 6 mm in thickness (t). The steel has yield strength of 360 MPa and ultimate strength of 400 MPa (see Fig.7). A comparison between the numerical results analysis of the axisymmetric cylinder of composite section using COMSOL and Drucker-Prager (DP) model with the experimental results reported is shown in Figure 8. From the data in Figure 8, it could be observed that the DP model matched well the experimental results of confined concrete section. Figure 8 shows that the peak load predicted by the FE analysis is about 10% higher than the experimental results.

a) Curing of composite sections

b) Failure mode

Fig. 7: Confined concrete-filled steel section, a) curing ; and b) failure mode.

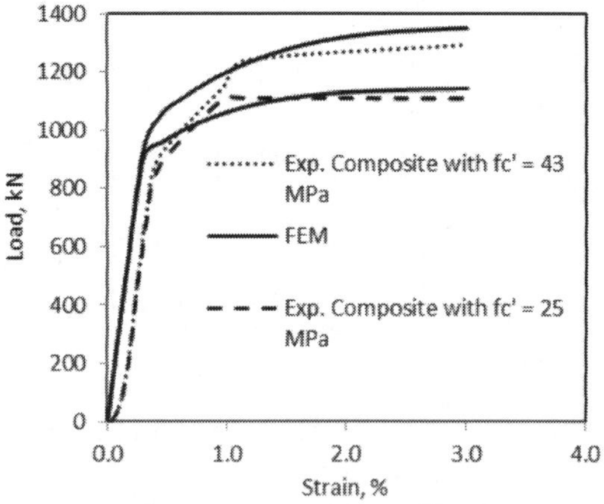

Fig 8. Comparison between numerical and experimental axial load-strain results for fc'=25 and 43 MPa.

8 Conclusions

This paper provides a simple approach to estimate the Drucker-Prager model parameters to simulate the behavior of confined concrete subjected to compressive loading. Numerical simulation was conducted using COMSOL finite element package to simulate the steel-concrete composite section subjected to uniaxial compressive loading. The numerical simulation was calibrated by experimental results available in the literature. A numerical parametric study was conducted to evaluate the effect of the friction angle φ and the cohesion constant c on the stress-strain curve of a steel-concrete composite section subjected to compressive stress. Higher values of φ led to a higher peak load and more tendencies to catch the non-linearity of concrete at the end of the elastic zone. For cohesion constant c of 1 MPa, an elastic-perfect plastic relationship was observed, while for c values between 5 and 10 MPa, a more non-linear relationship was observed at the end of the elastic zone. Based on the numerical analysis, an empirical relationship between the friction angle and the confined concrete compressive strength was developed. For normal concrete with compressive strength between 25 and 50 MPa, the suitable friction angle φ range was found to be between 15° and 28°. and cohesion constant c of range between 5 and 10 MPa could be used, which is about 20% of the unconfined compressive strength σc. Further investigation should be conducted on the effect of the thickness, shape and dimension on the material parameters (c and φ) of the Drucker-Prager (DP) model. Besides, there is a need for further investigation considering high compressive strength.

Acknowledgement

This work was completed under the IAU university funded research project number DSR-2014175. The author acknowledges the support of the IAU, the Department of Civil and Construction Engineering, and Deanship of Scientific Research.

References

[1] J. Jia-Fei, Y. Wu, X. Zhao. Application of Drucker-Prager plasticity model for stress-strain modeling of FRP confined concrete columns. *Procedia Engineering,* **14**, 687–694, (2011)

[2] J. Jia-Fei, Y.F. Wu.. Identification of material parameters for Drucker–Prager plasticity model for FRP confined circular concrete columns. *International Journal of Solids and Structures,* **49,** 445–456, (2012)

[3] T. Yu, J.G. Teng, Y.L. Wong, S.L. Dong. Finite element modeling of confined Concrete-I: Drucker Prager type plasticity model. *Engineering Structures,* **32,** 665-679, (2010)

[4] D. Cengiz, S. Tokgoz, A. K. Tanrikulu, T. Baran. Behavior of reinforced and concrete-encased composite columns subjected to biaxial bending and axial load. *Building and Environment,* **43,** 1109–1120, (2008)

[5] D.J. Zhang, Y.S. Ma, Y. Wang. Compressive behavior of concrete filled steel tubular columns subjected to long-term loading. *Thin-Walled Structures,* **89,** 205–211. (2015)

[6] E. Ehab, Y. Ben. Numerical simulation of concrete encased steel composite columns. *Journal of Construction Steel Research,* **67**, 2, 211–222, (2011)

[7] D. Fa-xing, Y. Zhi-wu, Y. Bai, Y.Z. Gong. Elasto-plastic analysis of circular concrete-filled steel tube stub columns. *Journal of Constructional Steel Research,* **67**, 1567–1577, (2011)

[8] K. Sakino, H. Nakahara, S. Morino, I. Nishiyama. Behavior of centrally loaded concrete-filled steel-tube short columns. *ASCE Journal of Structural Engineering,* **130**, 2, 180-188, (2004)

[9] H.N. Tuan, H. Otsuka, Y. Ishikaw, E. Takeshita. A study on shear strength of concrete under direct shear test. *Proceedings of the Japan Concrete Institute* **28,** 1, 1529-1534, (2006)

Corrosion inhibition of austenitic stainless steel by clay in polluted phosphoric acid with presence of SiC abrasif

S. Skal[1], Y. Kerroum[1], Y. El Aoufir[1], A. Guenbour[1], A. Bellaouchou[1], H. Tabyaoui[1] and H.Idrissi[2]
[1]Laboratoire des Matériaux Nanotechnologie et Environnement, Université Mohamed V Agdal. Faculté des Sciences. Av. Ibn Battouta. BP 1014, Rabat. Maroc
[2]Laboratoire des Matériaux Sciences et Ingénierie MATEIS, INSA Lyon, Bât. B. Pascal, 5° étage - 7, Av. Jean Capelle, 69621 Villeurbanne cedex. France

Abstract. Stainless steels have many properties mechanical and chemical resistances resulting from the formation of the protective layer (passive film) on their surface which prevents the metal to react with corrosive environments such as, phosphoric acid. This acid contains various impurities, including agressive agents and solid particles of gypsum, increase the risk of corrosion damage depending on the type of stainless steel used. In addition, it has been show that abrasion-corrosion causes an acceleration electrochemical process leading to a decrease in the resistance of materials. This work is to find a solution through an ecological inhibitor. That why we have been studied the effect of crude clay on corrosion behavior of Alloy 31 in polluted phosphoric acid with abrasive by electrochemical impedance spectroscopy (EIS) . The clay was characterized by X-ray fluorescence spectroscopy (FX), X-ray diffraction (DRX) and infrared spectroscopy (IR). EIS exhibited that resistance of Alloy 31 increased with increase the concentration of inhibitor.

1 Introduction

Le phénomène de corrosion-abrasion est habituellement causé par un corrodent aqueux ou gazeux qui coule sur la surface métallique ou qui l'empêche. La détérioration mécanique peut être aggravée par la présence d'un corrodent, comme dans le cas d'une usure fretante ou corrosive [1].
La corrosion-abrasion est affectée par de nombreux facteurs, à savoir, la vitesse, la turbulence, l'impact, la présence de solides en suspension, la température et les conditions de cavitation prédominantes [2, 3]. L'accélération de l'attaque est due à la destruction ou à l'élimination du film protecteur par des forces mécaniques, exposant ainsi des surfaces métalliques fraîches anodiques au film voisin non ionisé. Un film dur, dense et adhérent, tel que sur l'acier inoxydable est plus résistant qu'un film fragile et mince, comme celui du plomb. La nature du film de protection dépend en grande partie de la corrosion elle-même [4].
La résistance à la corrosion des aciers inoxydables est déterminée par la formation d'un film protecteur à l'interface métal-solution. Cependant, cette passivité peut être profondément modifiée par l'action électrochimique, mécanique ou physique. Spécifiquement sous l'action de l'effet d'abrasion qui peut conduire à une destruction mécanique partielle ou totale du film passif [5].
Pour la protection contre la corrosion-abrasion, il est possible d'agir sur le matériau lui-même, sur la surface du matériau ou par modification de l'environnement (inhibiteurs de corrosion). Les inhibiteurs de corrosion sont un moyen original pour contrôler la corrosion des métaux [6].
L'inhibiteur utilisé dans ce travail est l'argile. Cette dernière provient de la région de Fès. Les argiles sont écologiques, pas toxiques, et peu coûteuses. L'importance chimique des minéraux argileux est due à leur structure cristalline, à leur surface élevée, à leur capacité d'échange ionique et à de bonnes propriétés de gonflement dans certains cas [7, 8]. Les minéraux argileux sont importants en raison de leur grande échelle d'application dans l'industrie de fabrication de l'acide phosphorique et de la possibilité de modifier leur structure en couches par intercalation [9].
Notre étude porte sur la caractérisation minéralogique afin de l'utiliser comme inhibiteur dans l'acide phosphorique pollué (40% H_3PO_4)

dans le cas d'une corrosion-abrasion des aciers inoxydables.

2 Matériel et méthodes

2.1 Matériaux et solutions testées

L'acier inoxydable utilisé dans cette étude est l'acier allié austénitique (alliage 31). Sa composition chimique est donnée dans le tableau 1. La surface du matériau est de 0,5 cm^2. Avant chaque essai, la surface de l'acier subit un polissage au papier abrasif de granulométrie croissante (de 180 à 1500). Ensuite, les spécimens sont lavés à l'eau distillée et séchés sous flux d'air.
La solution d'acide phosphorique étudiée est de 40% polluée par addition de 4 % de H_2SO_4 et de 0,42 % de KCl, des concentrations typiques pour l'industrie de l'acide phosphorique, en présence de SiC abrasif.
Les concentrations utilisées pour l'argile est de 0.5g/l, 2g/l, 4g/l et 6g/l.
Dans tous les tests, l'angle d'impact est de 90 ° et la concentration des particules abrasives SiC est de 24 g / l.

Tableau 1: Composition chimique de l'acier inoxydable austénitique en alliage 31.

Alliage 31	
Eléments	% massique
Cr	26,75
Fe	31,43
Ni	31,85
Mo	6,6
Mn	1,5
Cu	1,21
N	0,193
Si	0,1
C	0,005
S	0,002
P	0,017

2.2 Mesures électrochimiques

Les expériences électrochimiques ont été effectuées en utilisant un potentiostat Voltalab PGZ 301. La cellule électrochimique consiste en une électrode de travail (alliage 31), une électrode de platine en tant qu'électrode auxiliaire et une électrode au calomel saturé (ECS) comme électrode de référence. Avant les mesures électrochimiques, l'électrode de travail a été initialement réduite potentiostatiquement à -600 mV / ECS pendant 60 s afin d'éliminer les oxydes formés à l'air. Ensuite, l'électrode de travail a été immergée dans une solution d'essai pendant 2 h jusqu'à ce que le potentiel de circuit ouvert stable soit atteint. La spectroscopie d'impédance électrochimique a été effectuée dans la gamme de fréquences de 100 kHz à 100 mHz, avec un signal d'amplitude de 10 mV.

3 Résultats et discussions

3.1 Caractéristiques de l'inhibiteur

3.1.1 Fluorescence des rayons X:

La méthode d'analyse par fluorescence des rayons X est connu par sa rapidité, sa simplicité de mise en œuvre, sa faculté de doser plusieurs éléments sur la même préparation et la grande universalité des éléments dosables [10].
Le tableau 2 résume les différents composants chimiques de l'argile minérale provenant de la région de Fés. D'après le tableau, on remarque qu'il y a trois éléments majoritaires : la silice, le calcium et l'aluminium.

Tableau 2: Composition chimique en éléments majeurs d'argile minérale (Région Fés)

Eléments	% massique
SiO_2	38,86
Al_2O_3	10,76
CaO	16,48
Fe_2O_3	4,42
MgO	3,76
K_2O	1,62
SO_3	0,31
TiO_2	0,55
MnO	0,04
P_2O_5	0,18
Na_2O	0,017
SrO	0,049

Les résultats obtenus montrent une présence de kaolinite en raison du pourcentage très importante de la silice et de l'aluminium. Aussi, le pourcentage massique de calcium est très élevé, d'où la richesse de cette argile en calcite. La présence des oxydes Fe_2O_3, MgO, K_2O et Na_2O montre que l'argile contient des impuretés.

3.1.2 Diffraction des rayons X

Le diagramme d'analyse par la diffraction des rayons X (DRX) est illustré dans la figure 1 :

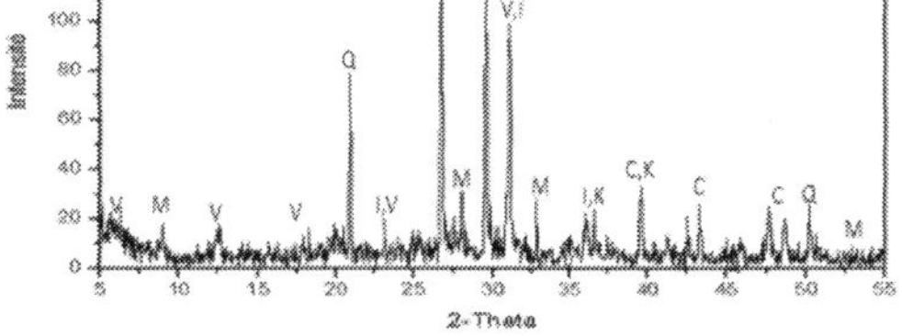

Fig 1. Spectre de la diffraction des rayons X de l'argile brute

L'analyse par DRX révèle que cette argile est composée de quartz (SiO_2), calcite $Ca(CO_3)$, vermiculite $[Mg_{11}Al_5FeSi_{11}O_{42,4}H_2O]$, muscovite $[(K_{0.93}Na_{0.07}Al_{1.66}Fe_{0.18}Mg_{0.16}(Al_{0.82}Si_{3.18}O_{10})(OH)_2]$, illite $[(K,H_3O)Al_2Si_3AlO_{10}(OH)_2]$ et kaolinite $(Al_2Si_2O_5(OH)_4)$ [11].
Dans ce spectre (figure 1), nous remarquons une présence de deux pics intenses, l'un correspond à la présence d'un mélange de Quartz et de Vermiculite et l'autre correspond à la présence de Calcite. Ces résultats montrent que cette argile est une argile hétérogène. Cela confirme les résultats obtenus de l'analyse par fluorescence des rayons X.

3.1.3 Spectroscopie d'infrarouge (IR)

La structure chimique de l'argile étudiée a été identifiée en utilisant la technique IR. Le spectre obtenu est illustré par la figure 2.

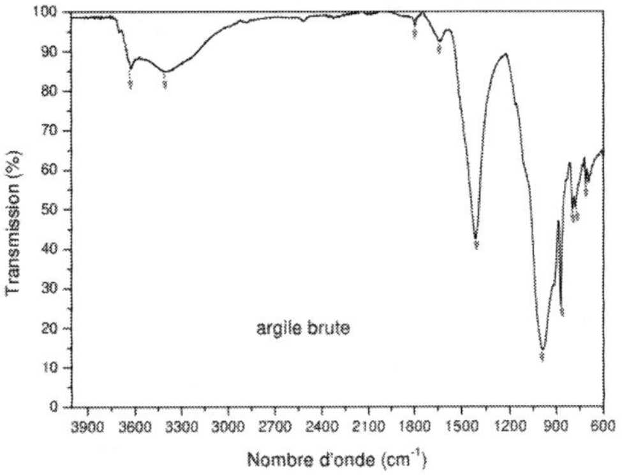

Fig 2. Spectre d'infrarouge de l'argile brute.

On obtient :

- entre 3400 cm^{-1} et 3600 cm^{-1} : vibrations d'élongations du groupement OH ;
- 1800 cm^{-1} : vibrations d'élongations du groupement C=O ;
- 1680 cm^{-1} : vibrations de valences du groupement OH, due à l'eau de constitution de l'argile ;
- 1430 cm^{-1} : vibrations de déformation du groupe CH_3 ;
- 1000 cm^{-1} : vibrations d'allongement de Si-O (kaolinite) ;
- 880 cm^{-1} : présence de liaison C-O carbonate de calcium ($CaCO_3$) ;
- 791 cm^{-1} : liaisons Si-O-Al ;
- 780 cm^{-1} : quartz ;
- 712 cm^{-1} : liaison C-O carbonate de calcium ($CaCO_3$) [12,13].

Ces résultats confirment ceux trouvés dans l'analyse par diffraction de rayon X et cela par la présence de Quartz, Kaolinite et Illite.
Après ce traitement caractéristique, nous avons étudié l'effet inhibiteur pour de cette argile sur le comportement de l'alliage 31 dans le cas de la corrosion-abrasion.

3.2 Efficacité de l'inhibition

3.2.1 Spectroscopie d'impédance électrochimique (EIS)

Pour compléter et comparer les résultats obtenus précédemment, le comportement à la corrosion de l'alliage 31, en solution d'acide phosphorique pollué, avec et sans inhibiteur a été étudié par spectroscopie d'impédance électrochimique (EIS) après 1h d'immersion (Figure 4).

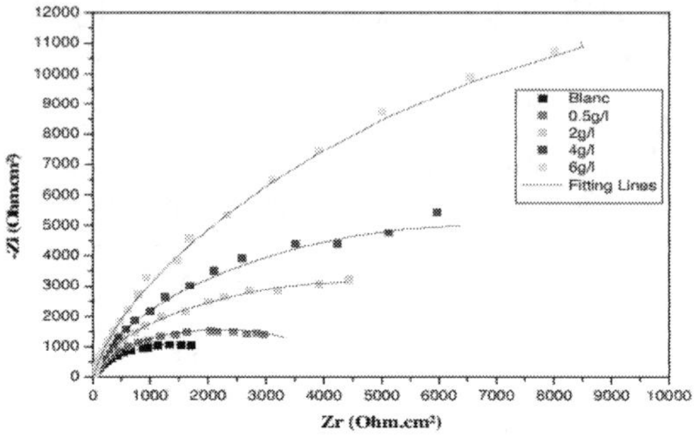

Fig 3. Les courbes de Nyquist de l'alliage 31 dans 40% H_3PO_4 pollué à différentes concentrations d'argile.

Les courbes de Nyquist ont été ajustées à l'aide des schémas électriques équivalents donnés sur la figure 5. Ce circuit se compose de la capacité et la résistance de la double couche (C_1, R_1) placé en parallèle avec la capacité et la résistance du film passif (C_2, R_2). Le symbole CPE (élément de phase constante) est utilisé pour signifier la possibilité d'une capacité non idéale attribuant à la distribution des temps de relaxation associés à des hétérogénéités à la surface de l'électrode.
Son impédance est définie par [14]:

$$Z_{CPE} = \frac{1}{Q(jw)^n} \quad (1)$$

Q est le module, ω et n sont la fréquence angulaire et la phase, respectivement et $j^2 = -1$ le nombre imaginaire. La capacité est considérée comme idéale lorsque n = 1, et non idéale lorsque $0,5 < n < 1$. Le CPE utilisé dans les circuits électriques équivalents de la figure 5 a été convertie en une capacité pure (C) au moyen de l'équation suivante [15] :

$$C = \frac{(QR)^{1/n}}{R} \quad (2)$$

Les principaux paramètres (R_f,R_{dl},C_f et C_{dl}) extraits du diagramme de Nyquist sont regroupés dans le tableau 3. La résistance de polarisation est calculée par addition de la résistance du film et la résistance de la double couche : $R_p = R_f + R_{dl}$. Ainsi l'efficacité de l'inhibition est calculée par l'équation suivante [16]:

$$EI\ (\%) = \frac{R_p - R_p^\circ}{R_p} \times 100 \qquad (3)$$

Où R_p° et R_p représentent respectivement les résistances de polarisation sans et avec inhibiteur.

Table 3 : Les paramètres d'impédance de l'alliage 31 en milieu d'acide phosphorique pollué en absence et en présence de différentes concentrations d'argile:

Milieu	Blanc	Argile			
Concentration en inhibiteur	-----	0.5 g/l	2 g/l	4 g/l	6 g/l
R_2 (Ω.cm²)	522	857	998	1587	3921
C_2 (10^{-4}F/cm²)	3,56	2,96	2,88	2,6	1,67
R_1 (Ω.cm²)	2291	3604	7516	11162	23130
C_1 (10^{-4}F/cm²)	2.54	2,17	2,38	1,65	1,25
R_P (Ω.cm²)	2813	4461	8514	12749	27 051
EI (%)	-----	37	67	78	88

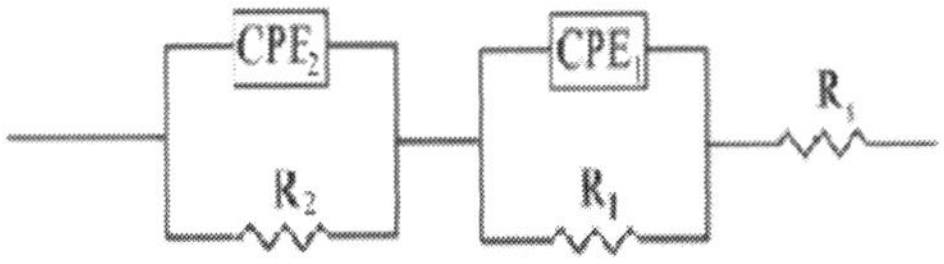

Figure 5. Le modèle de circuit équivalent de l'EIS.

A partir des résultats du tableau 4 on voit bien que l'augmentation de la concentration en inhibiteur engendre un accroissement des valeurs de R_p ainsi qu'une diminution de la capacité de double couche. Cela indique que l'argile inhibe la corrosion-abrasion. La diminution des valeurs C_{dl} en présence de l'argile peuvent être attribuer à la formation du film d'oxyde protecteur sur la surface du métal qui force la zone de passivation en présence de l'abrasif.

$M + H_2O \rightarrow MO + 2H^+ + 2e^-$
$2MO + H_2O \rightarrow M_2O_3 + 2H^+ + 2e^-$

Avec M soit le Fe, Cr ou Ni. [17]
Ceci est confirmé par d'autres auteurs tels que M. Larouj et al., [18] qui a travaillé sur l'effet inhibiteur de AOPT sur le comportement de l'acier au carbone dans une solution de 2M H_3PO_4. Il a conclu que le diamètre de la boucle capacitive augmente avec l'augmentation de la concentration en AOPT, suggérant que le film inhibiteur formé a été renforcé par l'ajout de AOPT. Ainsi M. Boudalia et al., [16] a montré que l'argile s'adsorbe sur la surface du métal et forme un film qui réduit la surface métallique subissant la corrosion.

4 Conclusion

Les techniques de caractérisation de l'argile provenant de la région de Fès ont révélé qu'elle est hétérogène et qu'elle est composée d'un mélange de quartz, muscovite, illite et kaolinite et de calcite.
Les études de polarisation ont montré que l'argile est un inhibiteur de type anodique dans les conditions de corrosion-abrasion qui diminue la dissolution anodique de métal et que son efficacité de corrosion augmente avec la concentration d'inhibiteur. Les résultats d'impédance ont montré que la résistance de la polarisation augmente avec l'accroissement de la concentration en argile. Une meilleure performance d'environ 88% est observée en présence de 6g/L de l'inhibiteur.

References

1. S.S. Rajahram , T.J. Harvey, R.J.K. Wood, Wear **267** 244–254(2009).
2. F.Mohammadi, J.Luo, Corros.Sci **52**, 2994–3001(2010).
3. M. Lindgren, S. Siljander, R. Suihkonen, P. Pohjanne, J. Vuorinen, Wear, 364-365 (2016).
4. A. Guenbour, M.A. Hajji, E.M. Jallouli, A. Ben Bachir, Appl. Surf. Sci. **253**, 2362-2366 (2006).
5. S. Skal, Y. Kerroum, A. Guenbour, A. Bellaouchou, H. Tabyaoui, H. Idrissi, A. Zarrouk, J. García-Antón, J. Mater. Environ. Sci. **8**, 3234-3246 (2017).
6. Y. EL Aoufir, H. Lgaz, H. Bourazmi, Y. Kerroum, Y. Ramli, A. Guenbour, R. Salghi, F. El-Hajjaji, B. Hammouti, H. Oudda, J. Mater. Environ. Sci. **7**, 4330-4347 (2016).
7. L. Jancovic, P. Komadel, J. Catal. **218**, 227 (2003).
8. M. Rouli, A.A. Vassiliadis. J. Colloid Interf. Sci. **291**, 37-44 (2005).
9. D.I. Nistor, N.D. Miron, I. Siminiceanu, Scientific Study & Research **6**, (2006).
10. C. Sadik , I.EL Amrani EL Hassani, A. ALbizane, , *INVACO2*.
11. S. Louati, S. Baklouti, B. Samet, , Hindawi Publishing Corporation Advances in Materials Science and Engineering (2016).
12. A. Qlihaa, S. Dhimni, F. Melrhaka, N. Hajjaji, A. Srhiri,, J. Mater. Environ. Sci. **7**, 1741-1750 (2016).
13. O.Qabaqous, N.Tijani, M.Naciri Bennani, A.El Krouk, J. Mater. Environ. Sci. **5**, 2247 (2014).
14. FB.Growcock, R.J.Jasinski, J.Electrochem. Soc. **136**,(1989).
15. C. Escrivà-Cerdán, E. Blasco-Tamarit, D.M. García-García, J. García-Antón, R. Akid, J. Walton, Electrochimica Acta **111**, 552– 561 (2013).
16. M. Boudalia, A. Guenbour, A. Bellaouchou, A. Zarrouk, Der Pharma Chemica **7**, 301-306 (2015).

17. H. Iken, R. Basseguy, A. Guenbour , A. Ben Bachir, Electrochimica Acta **52**, 2580–2587 (2007).
18. M. Larouj, Y. EL Aoufir, H. Serrar, A. El Assyry, M. Galai, A. Zarrouk, B. Hammouti, A. Guenbour, A. El Midaoui, S. Boukhriss, M. Ebn Touhami, H. Oudda, Der Pharmacia Lettre **6**, 324-334 (2014).

Inhibitive effect of N,N'-Dimethylaminoethanol on carbon steel corrosion in neutral sodium chloride solution, at different temperatures

Mohammed Hassoune[1], Abdelillah Bezzar[1], Latéfa Sail[1] and Fouad Ghomari[1]
[1]Laboratory E.O.L.E, Abou Bekr Belkaid University, Civil Engineering, Tlemcen, Algeria

Abstract. The inhibition of carbon steel corrosion in neutral sodium chloride solution by N,N'-Dimethylaminoethanol (DMEA), at different temperatures, was investigated using weight loss, potentiodynamic polarization and electrochemical impedance spectroscopy (EIS) techniques. The results obtained confirm that DMEA is a good organic corrosion inhibitor for carbon steel in 0.5M of NaCl (concentration encountered in the Mediterranean seawater), over the whole range of temperatures studied. The inhibition efficiency (IE%) increases with increasing DMEA concentration; it reaches highest value for a concentration around 0.125 $mol.L^{-1}$. Potentiodynamic polarization data show that, the compound studied in this research predominantly act as anodic-type inhibitor. The EIS study reveals that the addition of DMEA decreases the corrosion rate of carbon steel in neutral sodium chloride solution, due to the fact that the inhibitor molecules are strongly adsorbed on the active sites following Langmuir isotherm, thus leading to the formation of a stable protective film on the steel surface which is able to keep the metal/solution interface in a passive state. Furthermore, the values of the activation parameters, i.e. ΔHa and Ea obtained in this study indicate that the adsorption process of DMEA is endothermic and could be mainly attributed to chemisorption, respectively.

1 Introduction

Due to low cost and excellent mechanical properties, carbon steel is extensively used in construction and widely applied as the constructional material in large number of industries. However, it gets corroded when exposed to aggressive environments like chloride-rich media, which results in terrible wastage of both resources and money. This is the reason why the corrosion and passivation behavior of carbon steel are quite general and important phenomena in our daily life. Therefore, there is a great economical incentive in developing methods and materials to mitigate the corrosion of carbon steel [1, 2].

The use of corrosion inhibitors is one of the most widely practical methods for protection of metals and alloys against corrosion, this technique has attracted increasing attention in recent years. Different classes of inorganic compounds are used as corrosion inhibitors for carbon steel in various harsh media. However, unfortunately, most of them present health hazards or are very expensive. These features certainly limit the field of their application [3]. Environmentally friendly organic inhibitors containing heteroatoms N, O, and S in the molecules were reported to be effective inhibitors for metals and al alloys in an environment containing aggressive ions [4, 5]. There are many studies deal with the corrosion inhibition effects of these organic compounds in acidic solutions, especially in the treatment of steel and ferrous alloys in hydrochloric or sulphuric acid [6-8]. However, very little literature is reported on the corrosion protection of carbon steel in neutral chloride medium, which leads to extensive general and localized pitting corrosion of carbon steel [9, 10].

The organic inhibitors are typically based on mixtures of alkanolamines and amines. Amino alcohol compounds are frequently used as organic corrosion inhibitors due to their high solubility (easy application) and their low toxicity; they can be more effective when the corrosion degree is relatively low or before the corrosion initiation phase of metal [11]. It is reported that the inhibiting action of this compounds is due to the adsorption and/or formation of physical barrier on the metal/solution interface [12, 13]. This process depends upon the nature and surface charge of the metal, the type of aggressive media, the chemical structure of the inhibitor, and the nature of its interaction with the metal surface [14, 15].

The present work aims to investigate the efficiency of N,N'-Dimethylaminoethanol (DMEA), an amino alcohol compound, used as an organic corrosion inhibitor for carbon steel in 0.5M NaCl solution, using weight loss, potentiodynamic polarization and electrochemical impedance spectroscopy (EIS) methods. The effects of inhibitor concentration and temperature on the efficiency of the DMEA were examined. Moreover, the inhibition mechanism of DMEA in the test solution is studied and discussed.

2 Experimental

2.1 Materials

In order to conduct the gravimetric and electrochemical measurements, specimens of carbon steel were prepared from commercial reinforcing carbon steel (XC38). Its chemical composition is given in Table 1.

Table 1. Chemical composition of tested steel [16].

Chemical Elements	C	Si	Mn	P	S	Cr
Composi- (wt. %)	0.259	0.271	1.25	0.012	0.040	0.05
Chemical Elements	Ni	Al	Cu	V	Su	Fe
Composi- (wt. %)	0.114	0.003	0.29	0.068	0.017	97.65

The N,N'-Dimethylaminoethanol (DMEA) was used as organic corrosion inhibitor in the present study, its chemical structure is shown in Figure 1. The molecular formula of DMEA is $C_4H_{11}NO$, while its molar mass is 89.14 $g.mol^{-1}$.

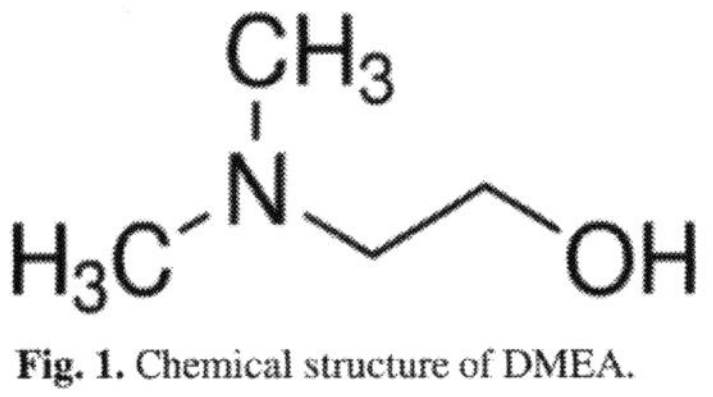

Fig. 1. Chemical structure of DMEA.

2.2 Gravimetric measurements

Prior to the tests, disc-shaped steel specimens (27 ± 2 mm diameter and 2 ± 0.2 mm thickness) were mechanically polished with emery papers (with grade from 120 to 1000), washed with distilled water, degreased with acetone, dried and weighed (m_1) by means of an analytical balance with 0.1mg readability. The clean weighed steel specimens was completely immersed, at an inclined position, in a glass vessel containing 50 ml solution of 0.5M NaCl, without and with addition of different concentrations of DMEA, for 360 hours, at various temperatures (298, 313 and 328 K). Then, the carbon steel specimens were reweighed (m_2) after washing and drying.

Based on the gravimetric measurements, the corrosion rate (CR) in ($mg.cm^{-2}.h^{-1}$) was calculated according to the following equation,

$$CR = \frac{m_1 - m_2}{S.t} \quad (1)$$

where, m_1 and m_2 are the average weights of specimen (mg), before and after immersion, respectively; S is the immersed surface area (cm^2) of the specimen and t is the immersion time (h). The inhibition efficiency (IE%) of the inhibitor was calculated using the relation:

$$IE(\%) = \frac{CR_0 - CR_{inhib}}{CR_0} * 100 \quad (2)$$

where CR_0 and CR_{inhib} represent the corrosion rate, respectively, without and with DMEA.

2.3 Electrochemical studies

The potentiodynamic polarization and electrochemical impedance spectroscopy (EIS), were carried out using a Potentiostat/Galvanostat (Voltalab80 Model PGZ402), driven by (Voltamaster4) software. Conventional three electrode system was used for electrochemical experiments. In this setup, polished carbon steel specimens with 1.05 cm^2 exposed surface area were used as working electrode, platinum electrode as an auxiliary electrode, and saturated calomel electrode (SCE) as a reference electrode. The three electrodes setup was immersed in glass cell, containing 200 ml of the test solution (0.5M NaCl), without and with addition of different concentrations of DMEA.

The potentiodynamic polarization measurements were started from cathodic to anodic direction, from -1000 to -300 mV with respect to the (SCE), with a scan rate of 0.5 $mV.s^{-1}$. The corrosion current (I_{corr}), was obtained by extrapolation of anodic and cathodic regions of the Tafel plots. From (I_{corr}), the inhibition efficiencies (IE%) were calculated using the following relation,

$$IE(\%) = \frac{I_{0corr} - I_{corr}}{I_{0corr}} * 100 \quad (3)$$

were (I_{0corr}) and (I_{corr}) denote the corrosion current densities in the absence and presence of inhibitor, respectively.

EIS measurements were carried out using a signals of amplitude perturbation of 10 mV, over a frequency range from 100 kHz to 50 mHz. The impedance parameters were obtained from the Nyquist plots. The inhibition efficiencies (IE%) were calculated according to the following equation,

$$IE(\%) = \frac{R_{ct} - R_{ct0}}{R_{ct}} * 100 \quad (4)$$

were R_{ct} and R_{ct0} refer to the charge transfer resistances obtained in the presence and absence of inhibitor, respectively.

The potentiodynamic polarization curves and Nyquist plots were recorded after 24 hours of immersion of steel specimens in the test solution, without and with addition of different concentrations of DMEA, at 298K.

3 Results and discussion

3.1 Gravimetric measurements

3.1.1 Effects of inhibitor concentration

The corrosion rate (CR) and inhibition efficiency (EI%), obtained from the gravimetric measurements for carbon steel specimens immersed in the test solution for 360 hours, as a function of inhibitor concentration, at 298K, are shown in Figure 2.

It can be seen from Figure 2 that the DMEA inhibits the corrosion of carbon steel in the test solution, and the inhibition efficiency increases with increasing inhibitor concentration. This can be attributed to the adsorption of DMEA molecules at carbon steel/NaCl solution interface. The maximum efficiency of 88% is obtained at 0.125 mol.L^{-1} of DMEA; and for concentrations higher than 0.125 mol.L^{-1}, no appreciable change in inhibition efficiency was observed. This may be imputed to competitive adsorption between the inhibitor molecules and metal surface, which was already covered with initial layers of molecules from the initial inhibitor concentration (0.125 mol.L^{-1}).

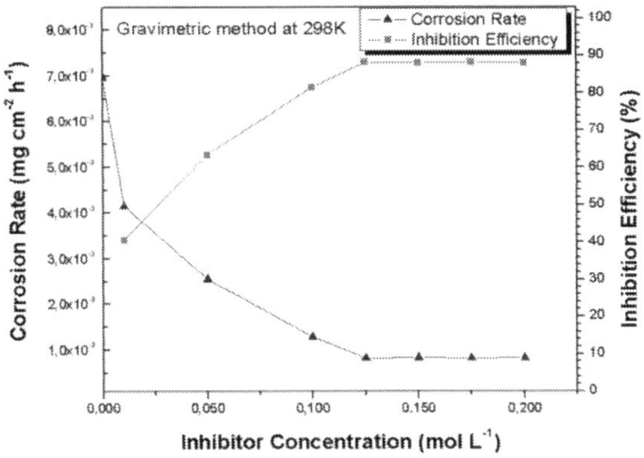

Fig. 2. Corrosion rate and inhibition efficiency of C-steel specimens in the test solution, with and without DMEA.

3.1.2 Effect of temperature

In order to investigate the inhibition mechanism and calculate the activation parameters of the corrosion process, gravimetric measurements were carried out at various temperatures, between 298 and 328 K. The corrosion rate (CR) and corresponding inhibition efficiency (IE%) obtained at studied temperatures are shown in Figures 3 and 4, respectively.

It is clear from Figure 3 that when the temperature increases, the corrosion rate (CR) of carbon steel increased, both in uninhibited and inhibited solutions. The increase in corrosion rate with temperature is more important in the absence or presence of low concentrations of DMEA. Furthermore, Figure 4 shows that the inhibition efficiency increases slightly as the temperature increases from 298 to 328 K. This can be explained by the fact that the increase in temperature leads to higher desorption of water molecules, thus leaving a larger surface area for the adsorption of inhibitor molecules [17].

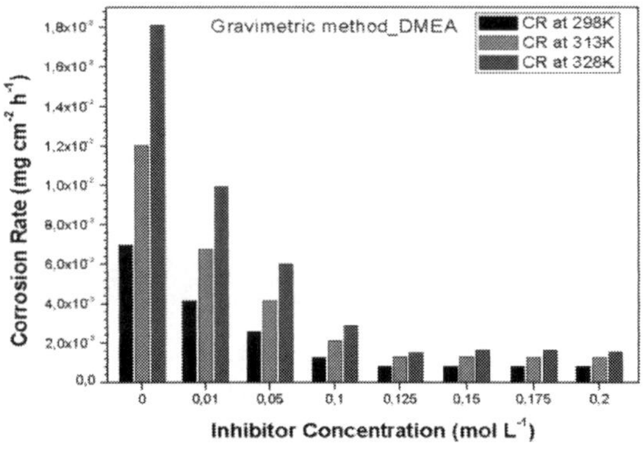

Fig. 3. Effect of temperature on the corrosion rate (CR) of C-steel in the test solution, with and without DMEA.

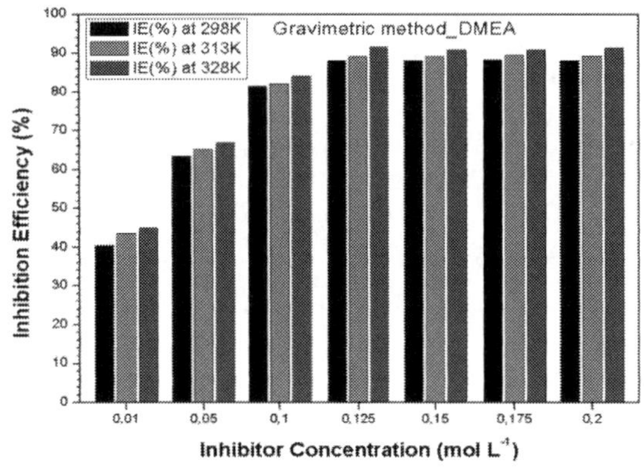

Fig. 4. Effect of temperature on the inhibition efficiency (EI%) of C-steel in the test solution, with and without DMEA.

The degree of surface coverage (θ) was calculated for different inhibitor concentrations by assuming a direct relationship between the surface coverage (θ) and inhibition efficiency (IE%), such as $\theta = IE/100$. The C/θ versus C at the various temperatures were plotted (Figure 5). All plots give straight lines with almost unit slope. This behavior confirmed that the adsorption of DMEA molecules obeys Langmuir adsorption isotherm.

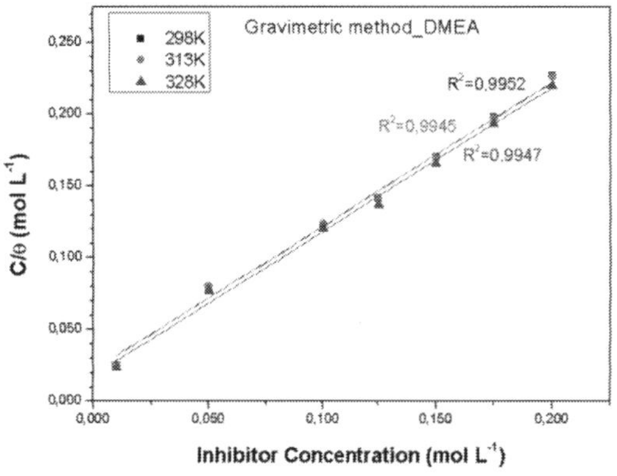

Fig. 5. Langmuir isotherm for C-steel in the test solution, with and without DMEA, at various temperatures.

The activation parameters of the corrosion process in the studied range of temperature were calculated by means of the Arrhenius equation and transition state equation, respectively [18, 19]:

$$Ln(CR) = \frac{-E_a}{RT} + Ln(A) \tag{5}$$

$$CR = \frac{RT}{Nh}\exp\frac{\Delta S_a}{R}\exp\frac{-\Delta H_a}{RT} \tag{6}$$

where E_a is the apparent activation energy, R the universal gas constant, T the temperature, A the Arrhenius pre-exponential factor, N Avogadro's number, h Plank's constant, ΔS_a the entropy of activation and ΔH_a the enthalpy of activation.

The results obtained of activation parameters in the studied range of temperature are shown in Table 2.

An examination of this table reveals that the addition of DMEA decreases the activation energy (E_a); this decline is more significant at concentrations equal or superior to 0.125 mol.L^{-1}. According to Bentiss et al., and Sankarapapavinasam et al. [20, 21], unchanged or lower values of activation energy (E_a) in the inhibited solutions with respect to that of the blank (without inhibitor), suggest charge sharing or transfer from the organic inhibitor molecules to the metal surface, thus forming coordinate covalent bond i.e. chemisorption. Moreover, the positive sign of enthalpy of activation (ΔH_a), as shown in Table 2, reflects the endothermic nature of the metal dissolution process, suggesting that the dissolution of carbon steel is slow in the presence of DMEA [22, 23]. The negative values of entropies of activation (ΔS_a) imply that the activated complex in the rate-determining step represents an association rather than a dissociation step, meaning that there is less disorder when moving from the reactants to the activated complex [24, 25].

Table 2. The values of activation parameters E_a, ΔH_a and ΔS_a for C-steel in the test solution, with and without DMEA.

Concentration (mol L^{-1})	$\mathbf{E_a}$ (KJ mol^{-1})	$\mathbf{\Delta H_a}$ (KJ mol^{-1})	$\mathbf{\Delta S_a}$ (J mol^{-1} K^{-1})
Blank	25.94	23.35	-207.74
0.010	23.67	21.08	-219.70
0.050	23.21	20.61	-225.27
0.100	22.06	19.46	-234.48
0.125	16.46	13.86	-257.04
0.150	18.73	16.14	-249.54
0.175	18.93	16.33	-249.00
0.200	16.97	14.38	-255.35

3.2 Electrochemical measurements

3.2.1 Potentiodynamic polarization

Potentiodynamic polarization curves for carbon steel in the test solution (0.5M NaCl), without and with addition of different concentrations of DMEA at 298K, are shown in Figure 6.

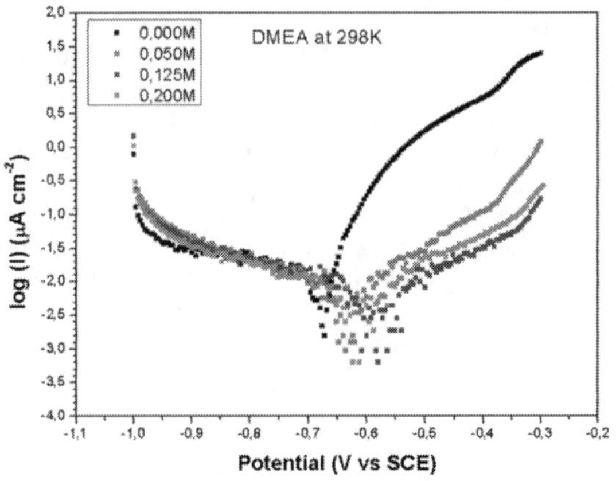

Fig. 6. Potentiodynamic polarization curves for C-steel in the test solution, with and without DMEA, at 298K

The electrochemical corrosion parameters are calculated from this graph; they listed in Table 3.

Table 3. Electrochemical parameters for C-steel in the test solution, with and without DMEA, at 298K.

$\mathbf{C_{inhib}}$ mol/L	$\mathbf{E_{corr}}$ (mV)	$\mathbf{I_{corr}}$ µA/cm^2	**βa** mV/dec	**βc** mV/dec	**IE** (%)
Blank	-693.8	18.52	121.3	581.9	/
0.050	-647.4	6.36	182.1	334.6	65.66
0.125	-591.7	3.59	236.5	456.1	80.61
0.200	-615.0	3.69	238.2	416.8	80.09

It can be seen from this Table 3 that , the value of current density (I_{corr}) for carbon steel in 0.5M NaCl blank solution equals 18,52 µA.cm^{-2}, and decreases with increasing DMEA concentration. This indicates that the investigated organic compounds acts as efficient inhibitor towards carbon steel dissolution in NaCl medium, and the degree of inhibition depends on their concentrations. When 0.125 mol L^{-1} of DMEA was added to the solution, the current density decreased to 3.59 µA cm^{-2} and results in the maximum inhibition efficiency of 80.61%. It can also be noticed from Figure 6 and Table 3 that with the addition of DMEA, both anodic and cathodic currents (βa and βc) change and the corrosion potential (E_{corr}) moves towards a more positive values. According to Ferreira et al. and Cai et al. [26, 27], if the displacement in (E_{corr}) is more than 85 mV in the positive direction with respect to the (E_{corr}) obtained in the solution without inhibitor, then the inhibitor could be considered as an anodic-type. In the present study, the maximum (E_{corr}) shift is more than 100 mV in the positive direction, indicating that DMEA acts mainly as an anodic-type corrosion inhibitor.

3.2.2 Electrochemical impedance spectroscopy(EIS)

EIS measurements were carried out, in order to get more information about the corrosion inhibition phenomenon by DMEA. Nyquist plots obtained for frequencies ranging from 100 kHz to 50 mHz, at corrosion potentials for carbon steel without and with various concentration of DMEA, at 298K are shown in Figure 7.

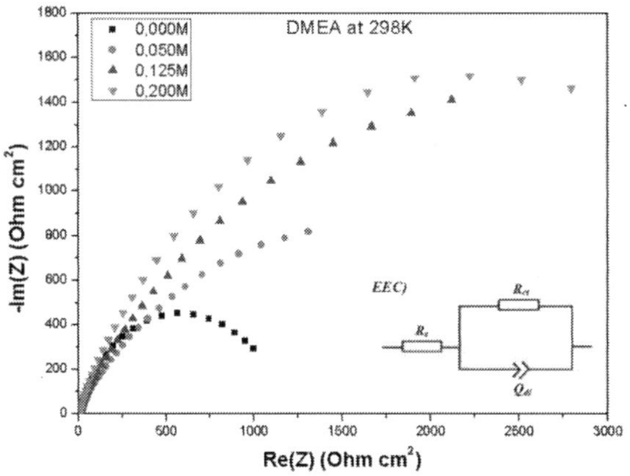

Fig. 7. Nyquist plots obtained for C-steel in the test solution containing different concentrations of DMEA, at 298K.

Nyquist representation of the impedance spectra, can be analyzed through fitting to the equivalent electric circuit (EEC) model, shown in Figure 7. In this model, (R_s) represents the resistance of the electrolyte and (R_{ct}) the charge transfer resistance. The (Q_{dl}) is associated with the double layer capacitance; their values may be obtained by using a constant phase element (CPE), which represents a non-ideal frequency dependent capacitance [28]. There is a variable value (a), associated with the CPE, which generally ranges between 1 and 0.5. It describes the distribution of the dielectric relaxation times in the frequency domain; when (a) is close to 1, the (CPE) represents a capacitor [29]. The corresponding impedance quantitative parameters, derived from Nyquist plots by using the equivalent electric circuit (EEC) shown in Figure 7, and the percentages of the inhibition efficiency (IE%) are given in Table 4.

Table 4. Electrochemical parameters for C-steel in the test solution, with and without DMEA, at 298K.

C_{inhib} (mol/L)	R_s ($\Omega.cm^2$)	Rct ($\Omega.cm^2$)	CPE_{dl}		IE (%)
			$Q_{dl}*10^{-4}$ $F.cm^{-2}.s^{a-1}$	a_2	
Blank	5.744	1069	8.29	0.62	/
0.050	4.576	2455	7.75	0.67	56.46
0.125	5.519	3944	4.87	0.70	72.90
0.200	5.035	4313	3.80	0.75	75.21

It can be seen from this table that, (R_s) value decreases slightly as the concentration of DMEA increased in the test solution. This may be attributed due to more involvement of ions in test solution containing inhibitor, resulting an increase in the solution conductivity [30]. On the other hand, (R_{ct}) value increases while (Q_{dl}) value decreases with increasing inhibitor concentration, and consequently leads to an increase in the inhibition efficiency of DMEA. The increase in (R_{ct}) values may be ascribed to the gradual replacement of water molecules by the adsorbed inhibitor molecules which form an adherent film on the carbon steel surface [31]. Thus, the decrease in (Q_{dl}) values, which may have resulted from a decrease in the local dielectric constant and/or from an increase in the double layer thickness, suggests that the adsorption of the DMEA molecules took place at the metal/solution interface [32, 33]. The maximum inhibition efficiency (EI%) for DMEA measured by (EIS) is 75.21%, obtained at 0.2 mol L^{-1}.

4 CONCLUSION

The inhibition efficiencies obtained in the present study from weight loss, potentiodynamic polarization, and EIS methods are in good agreement. The following conclusions could be drawn:

The DMEA significantly reduces the corrosion rate of carbon steel in 0.5M NaCl solution. The inhibition efficiency increases with increasing DMEA concentration; it reaches highest value for a concentration around 0.125 mol.L^{-1},

The DMEA acts as an efficient organic corrosion inhibitor over the whole range of temperatures studied. The values of the activation parameters, i.e. ΔH_a and E_a obtained indicate that the adsorption process of DMEA is endothermic and could be mainly attributed to chemisorption, respectively.

The adsorption of DMEA molecules obeys Langmuir adsorption isotherm,

The potentiodynamic polarization results reveal that the studied inhibitor is predominantly anodic-type in 0.5M NaCl solution,

The EIS study reveals that the addition of DMEA decreases the corrosion rate of carbon steel in 0.5M NaCl solution, due to the adsorption of inhibitor molecules on the steel surface, thus leading to the formation of a protective film which is able to keep the metal/solution interface in a passive state.

REFERENCES

1. J. L. Yao, B. Ren, Z. F. Huang, P. G. Cao,R. A. Gu, Electrochim. Acta **48**, 1263 (2003).
2. C. A. Melendres, N. CamilloneIII,T. Tipton, Electrochim. Acta **34**, 281 (1989).
3. R. T. Loto, C. A. Loto,T. Fedotova, Int. J. Electrochem. Sci. **7**, 10763 (2012).
4. G. Bereket,A. Pinarbasi, Corros. Eng. Sci. Technol. **39**, 308 (2004).
5. Z. Grubac, R. Babic,M. H. Metikos, J. Appl. Electrochem. **32**, 431 (2002).
6. M. Lebrini, M. Lagrenée, H. Vezin, L. Gengenbre,F. Bentiss, Corros. Sci. **47**, 485 (2005).
7. A. S. Fouda, H. A. Mostafa, F. El-Taib,G. Y. Elewady, Corros. Sci **47**, 1988 (2005).
8. K. S. Jacob,G. Parameswaran, Corros. Sci. **52**, 224 (2010).
9. M. DuduKcu, B. Yazici,M. Erbil, Mater. Chem. Phys. **87**, 138 (2004).

10. E. M. Sherif, J. H. Potgieter, J. D. Comins, L. Cornish, P. A. Olubambi,C. N. Machio, Corros. Sci. **51**, 1364 (2009).
11. T. A. Söylev, C. McNally,M. Richardson, Cem. Concr. Res. **37**, 972 (2007).
12. A. K. Satpati,P. V. Ravindran, Mater. Chem. Phys. **109**, 352 (2008).
13. S. Abd El Rehim, H. Hassan,M. Amin, Mater. Chem. Phys. **78**, 337 (2003).
14. M. Abdallah, Corros. Sci. **46**, 1981 (2004).
15. R. Solmaz, G. Kardas, M. Culha, B. Yazici,M. Erbil, Electrochim. Acta **53**, 5941 (2008).
16. L. Sail, F. Ghomari, A. Khelidj,A. Bezzar, Eur. J. Environ. Civ. En. **18**, 699 (2014).
17. M. Yadav, S. Kumar,L. Gope, J. Adhes. Sci. Technol. **28**, 1072 (2014).
18. H. Zarrok, A. Zarrouk, B. Hammouti, R. Salghi, C. Jama,F. Bentiss, Corros. Sci. **64**, 243 (2012).
19. L. Messaadia, O. ID El mouden, A. Anejjar, M. Messali, i. R. Salgh, O. Benali, O. Cherkaoui,A. Lallam, J. Mater. Environ. Sci. **06**, 598 (2015).
20. F. Bentiss, M. Traisnel, N. Chaibi, B. Mernari, H. Vezin,M. Lagrenée, Corros. Sci. **44**, 2271 (2002).
21. S. Sankarapapavinasam, F. Pushpanaden,M. F. Ahmed, Corros. Sci. **32**, 193 (1991).
22. M. Lagrenée, B. Mernari, M. Bouanis, M. Traisnel,F. Bentiss, Corros. Sci. **44**, 573 (2002).
23. G. N. Mu, X. Li,F. Li, Mater. Chem. Phys. **86**, 59 (2004).
24. S. Sankarapapavinasam,M. F. Ahmed, J. Appl. Electrochem. **22**, 390 (1992).
25. V. Ramesh Saliyan,A. V. Adhikari, Bull. Mater. Sci. **31**, 699 (2008).
26. E. S. Ferreira, C. Giacomelli, F. C. Giacomelli,A. Spinelli, Mater. Chem. Phys. **83**, 129 (2004).
27. J. Cai, C. Chen,J. Liu, Corros. Eng. Sci. Techn. **49**, 66 (2014).
28. H. E. Jamil, A. Shriri, R. Boulif, M. F. Montemor,M. G. S. Ferreira, Cem. Concr. Compos. **27**, 671 (2005).
29. J. J. Shi,W. Sun, Int. J. Miner. Metall. Mater. **19**, 38 (2012).
30. H. S. Ryu, J. K. Singh, H. S. Lee, M. A. Ismail,W. J. park, Constr. Build. Mater. **133**, 387 (2017).
31. S. A. Umoren, J. Adhes. Sci. Technol. **30**, 1858 (2016).
32. E. McCafferty,N. Hackerman, J. Electrochem. Soc. **119**, 146 (1972).
33. S. Jyothi,J. Ravichandran, J. Adhes. Sci. Technol. **28**, 2347 (2014).

Prediction models of mechanical properties for pet-mortar composite in sodium sulphateaggressive mediums

Nabil Kazi Tani[1,4], A.S. Benosman[1,2,3], Y. Senhadji[3], H. Taïbi[2], M. Mouli[3], M. Belbachir[2]
[1]Higher School of Applied Sciences (ESSAT), BP 165, 13000 BelHorizon, Tlemcen, Algeria.
[2]Faculty of Exact and Applied Sciences, Laboratory of Polymer Chemistry LCP, University of Oran 1, Ahmed Benbella, Oran, Algeria.
[3]Department of Civil Engineering, LABMAT, ENPO Maurice Audin, Oran, Algeria.
[4]LCGE Laboratory, Faculty of Mechanical Engineering, USTO-MB, Oran, Algeria.

Abstract. In this research, an investigation was carried out on the effect of sodium sulphate attack on the durability of composites produced with waste polyethylene terephthalate (PET). Experiments were accomplished on limestone sand and cement mortars where the blended Portland cement was partially replaced by various volume fractions of waste PET particles (6%, 12% and 17%). The test solutions used to supply the sulphate ions and cations were 5%sodium sulphate solution. Compressive strengths measured on specimens were used to assess the changes in the mechanical properties of PET-mortars exposed to sulphate attack at different ages, mainly the Young modulus of elasticity. Based on experimental compressive tests on PET-Mortar composite specimens and there densities, the evolution of Young modulus of elasticity has been analyzed in accordance with normative models given by (ACI-318) and (BS-8110) codes of practice. In addition, a comparative study has been carried out for corrosion resistance coefficients K of unmodified mortar to those modified with waste PET particles. It can be noticed that, for the composite immersed in a corrosive Na_2SO_4 solution, the corrosion resistance coefficients decrease with the increase of the immersion period. The corrosion sulphate resistance K based on Young modulus before and after immersion of PET-mortar composites is better than that of the control mortar. Therefore, for safety considerations of PET-mortar composites use, ACI 318 is recommended code for design and investigation works. Also, it can be concluded that adding waste PET by volume fractions (6%, 12% and 17%) to blend Portland cement renders this cement more resistant to the sodium sulphate aggressive medium. Therefore, composites materials based waste PET are often presented as the materials of the future because of their potential for innovation and the advantages they offer. In fact, using waste PET as cement substitutes reduces the energy consumption. These modified mortars address problems related to environmental pollution by CO_2 emissions, and are used to repair various reinforced concrete structures in sodium sulphate aggressive mediums.

1 Introduction

In the last couple years, waste polyethylene terephthalate (PET) is produced within large amount by plastic industry in all over the world [1] andsince PET waste is not biodegradable, it can remain in nature for hundreds of years and causes too many environmental problems.

Various studies [2-6] have been developed so far in order to find ecologic and green ways to dispose of plastics wastes, one of the main solutions proposed by researchers is to incorporate waste PET in building materials technology in order to substitute volumetric amount of cement [7] and/or aggregates [8-10] in concrete and mortars mix-design.

In this research, an investigation was carried out on the effect of sodium sulphate attacks on the durability of composites produced with waste polyethylene terephthalate (PET). Experiments were accomplished on limestone sand and cement mortars where the blended Portland cement was partially replaced by various volume fractions of waste PET particles (6%, 12% and 17 %).

Test solutions used to supply the sulphate ions and cationswere 5% Na_2SO_4 solution. Tap water was used as the reference solution. Compressive strengths measured on specimens were used to assess the changes in the mechanical properties of PET-mortars exposed to sodium sulphate attacks at different ages, mainly the Young modulus of elasticity.After compression testing, X-ray diffraction was conducted on some selected surface fractures to investigate microstructural nature of the sulphate attacks.

https://doi.org/10.1051/matecconf/201814901051

2 Raw materials

2.1 Cement

The cement used was a blended Portland cement type CPJ-CEM II/A42.5 supplied by Zahana factory, located in western Algeria, with 1022 kg/m^3 bulk density; its compressive strength at 28 days was 42.5 MPa. The absolute density of the cement used was 3.15 g/cm^3 and its specific surface area measured with the Blaine method was 3532 cm^2/g. Its initial and final setting times were 170 and 245 min, respectively. Mineralogical and chemical compositions of cement are listed in Table 1.The chemical composition was obtained using an X-ray fluorescence spectrometer.

Table 1. Chemical and mineralogical compositions of cement (wt.%)

Chemical compositions	SiO_2	20.91
	Al_2O_3	5.52
	Fe_2O_3	3.56
	CaO	63.50
	MgO	0.64
	SO_3	2.79
	K_2O	1.23
	Na_2O	0.13
	CaO free	2.35
	LOI	1.19
Mineralogical compositions	C_3S	49.39
	C_2S	22.97
	C_3A	8.61
	C_4AF	10.83

2.2 Sand

The crushed natural limestone sand was obtained from the quarry of Kristel, in Oran, West Algeria. The maximum size of sand grains was 5 mm. The absolute density and absorption coefficient of crushed sand were 2.53 g/cm^3 and 0.5%, respectively. The grading of crushed sand is presented in Table 2, according to standard NF P18-560 [11].

2.3 Waste Polyethylene Terephthalate

Waste PET bottle granules (PET) used as particles were supplied by TRAMAPLAST PET Bottle Plant, in Tlemcen, Algeria. These particles were obtained by collecting the waste PET bottles and washing them; they are then crushed by granules into machines. In addition, they have an irregular shape and a rough texture surface, which enables the adherence of the particle-matrix. The bulk density of the waste PET particles used was 401.4kg/m^3.

After preliminary tests, waste PET particles of size lower than 1 mm were used in this study. The sieve analysis of waste PET particles was carried out according to standard NF P18-560 [11] and is presented in Table 2.

Table 2. Sieve analysis of waste PET-particles and crushed limestone sand.

Sieve size (mm)	Cumulative passing (%)	
	Waste PET	Sand
5	99.92	99.83
2.5	98.16	98.37
1.25	96.82	65.37
0.63	55.78	38.3
0.315	35.48	19.07
0.16	18.28	8.20
0.125	9.56	3.325

3 Composite mixing conditions

The mortar manufactured without waste PET particles was first optimized on the basis of its mechanical criteria and was then used as a reference composite. The composites containing waste PET particles were produced in accordance with the results of the works of Benosman et al. [12]. A massic ratio of 3 between sand (S) and cement (C) was respected. Four different mixtures were prepared (the control mixtures without plastic waste and three PET# mixtures including 6%, 12%and 17% waste PET particles by volume).Mixture name of the different composites were: PET0 (without plastic waste), PET6, PET12 and PET17. The water to binder ratio was kept constant at 0.5. So, after pouring fresh material into the molds(EN 196-1), the samples were stored in a room where hygrometry and temperature were controlled for 24 h (98% relative humidity, and 20 ± 1 °C). After removal from the molds, at 1 day of age, mortar specimens were cured in saturated lime water at 20 ± 1 °C, until the time of testing.

4 Test methods of resistance to sodium sulphate attack

The mortar specimens were cured in water saturated with lime at 20 ± 1°C for 28 days before being exposed to sulphate attack. Three specimens of each mortar and composite mixes (40x40x160 mm^3) were immersed in two types of solutions: distilled water (reference medium) and 5% sodium sulphate solution, Figure 1. According to the standard ASTM C1012-04 [13], the pH of the sulphate solution should be between 6 and 8 and the solution must be renewed each week, which requires huge amounts of sodium sulphate.

For this, Mehta's [14] method and that of Siad et al. [15] were adopted; they all recommend the control of the pH within a range of 6.0–8.0 by adding a suitable amount of sulfuric acid solution (0.1 N H_2SO_4), figure 2.

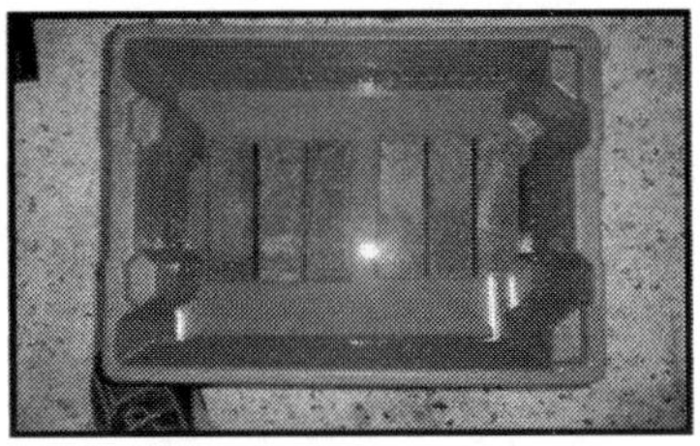

Fig. 1. Immersion of the specimens in the aggressive solution

The correction is performed daily during the first weeks of immersion, and then becomes weekly, for the rest of the test. In addition, the aggressive solutions were totally renewed each 12 weeks.

Fig. 2. Correction of the sulphate solution by adding small quantities of 0.1N H_2SO_4, up to a pH equal to 6-8.

After immersion in sodium sulphate solution (Na_2SO_4) for the required period of time, ASTM C1012-04 [13], the specimens were tested for residual mechanical properties. The Young modulus loss (*YML%*) is calculated as follows:

$$YML(\%) = \frac{E_{cr} - E_{cs}}{E_{cr}} \times 100 \ (1)$$

where$E_{cr}$is the Young modulus of the specimens before immersion (MPa) and $E_{cs}$is the average Young modulus of the specimens after immersion in sodium sulphate solutions for the required period of time (180 days).

5 Results and discussion

5.1 Prediction models of mechanical properties

The static modulus of elasticity E (Young Modulus) represents the one of the most important mechanical characteristics of construction materials (concrete, reinforced concrete, mortars, composites.etc.). This intrinsic property is considering as the basic parameter for the computing strain-stresses in construction structures.

Various countries have been established their design codes based on this empirical relationship between static modulus of elasticity E, and compressive strength of plain concrete at 28 days of curing. The ACI code (ACI-318) [16] defines the relationship between elastic modulus of concrete and compressive strength as:

$$E = w^{1.5}.\,0.043.\sqrt{f_c} (2)$$

The British Code of practice (BS-8110) [17] recommends the following expression for static modulus of elasticity with cube compressive strength of concrete as:

$$E = w^2.\,0.0017.\,{f_c}^{0.33} (3)$$

where, Ec: The static modulus of elasticity (E) at 28 days in MPa

fc : Compressive strength at 28 days, in MPa

w : Air dry density of mortar

Based on experimental compressive tests on PET-mortar composite specimens [18], and there densities, graphs in (figures 3,4) bellow show the evolution of Young modulus evaluated by empirical relationships in accordance to (ACI-318) and (BS-8110) codes [16,17].

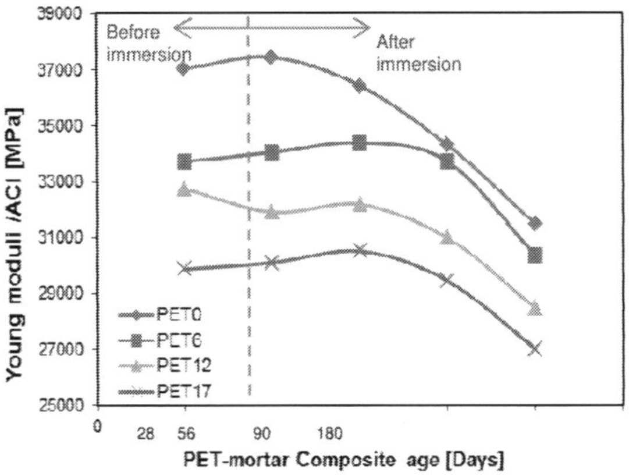

Fig. 3. Evolution of Young modulus with volumetric waste PET rate and composite mortar ages via ACI-318 code before and after 5 % sodium sulphate attacks.

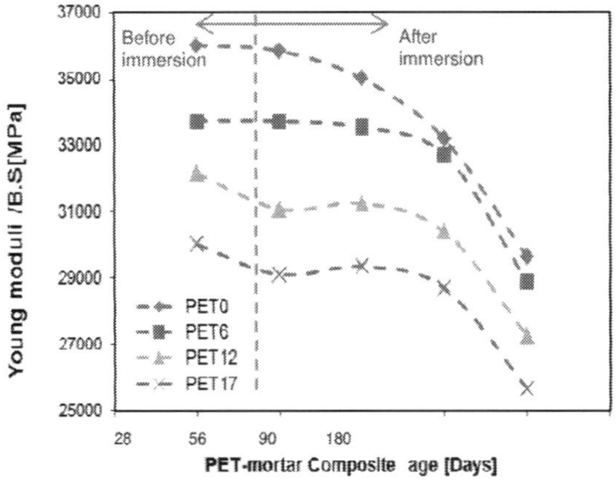

Fig. 4. Evolution of Young modulus with volumetric waste PET rate and composite mortar ages via BS-8110 code before and after 5 % sodium sulphate attacks.

In the other hand, Young modulus loss ratios have been studied for several volumetric waste PET rates and for each model of predicted codes. The results of the Young modulus loss using the specimens immersed in 5% N_2SO_4 solutions (figure5) showed that there are variations in time and group. However, a decrease in

Young Modulus values of all specimens was observed when Young Modulus of elasticity is computed via ACI-318 model. However, the rate of increase varied slightly within the group.

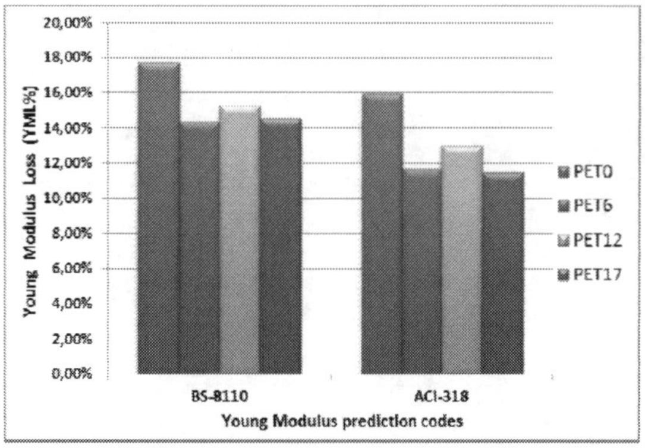

Fig. 5. Young modulus loss (*YML%*) with volumetric waste PET rate and prediction codes (BS-8110 & ACI-318) before and after 5% Na_2SO_4immersion.

It was expected that the mechanical properties loss (*YML%*) values of modified mortars with waste PET plastic particles would be lower than those of unmodified mortar PET0, by 14.35%, 15.28% and 14.53% for PET6, PET12 and PET17, respectively for the case where the prediction model is BS-8110 and respectively by 11.72%, 13.01% and 11.51% when Young Modulus of elasticity is computed via ACI-318 model. It can therefore be concluded that modified mortars by waste PET plastic particles are resistant to the sodium sulphate aggressive mediums conditions often encountered in the field.

Similarly to experimental results obtained by some previous studies [19, 20] in similar fields, it can be stated that, mortar blended within waste polymer particles present a better and promising composite material in building repair against sulphate aggressive mediums (Na_2SO_4, $MgSO_4$).

5.2 Corrosion sodium sulphate resistance

Furthermore, to compare, effectively, the corrosion resistance coefficients K of unmodified mortar (Eq. 4) to those modified with PET particles, as it was used by Benosman et al. [12] and Jiang et al. [21]:

$$K = {}^{Eci}/_{Ecs} \quad (4)$$

Where E_{ci} is the Young modulus of composite mortars immersed in corrosive Na_2SO_4solutions, $E_{cs}$the Young modulus of the normally cured composite mortars. The corrosion resistance coefficients of the specimens with and without waste PETplastic are given in Table 3.

It can be seen from table 3 that, for the composite immersed in a corrosive Na_2SO_4 solution, the corrosion resistance coefficients decrease with the increase of the immersion period. The corrosion sulphate resistance K based on Young modulus before and after immersion of PET-mortar composites is better than that of the control mortar (PET0).

Table 3. Corrosion resistance coefficients of PET-mortar composites in 5% Na_2SO_4

Sodium sulphateattacks	K (BS.8110)		K (ACI.318)	
	90 days	180 days	90 days	180 days
PET0	0.922	0.823	0.927	0.850
PET6	0.970	0.856	1.000	0.900
PET12	0.945	0.847	0.946	0.870
PET17	0.956	0.855	0.986	0.904

For all studied cases, corrosion resistance coefficients K computed via BS.8118 code of practice are lower than the ones computed by ACI 318 model. Therefore, for safety considerations of PET-mortar composites use, ACI 318 is recommended code for design and investigation works. It can be concluded that adding waste PET by volume fractions (6%, 12% and 17%) to blended Portland cement renders this cement more resistant to the sodium sulphate aggressive medium. It is evident that the resistance of cement to sulphate aggression is also related to its content in C_3A[22].

Therefore, composites materials based waste PET are often presented as the materials of the future because of their potential for innovation and the advantages they offer. In fact, using waste PET as cement substitutes reduces the energy consumption. These PET-modified mortars address problems related to environmental pollution by CO_2 emissions, and are used to repair various reinforced concrete structures in sodium sulphate aggressive mediums.Alqahtani et al. [23] reported thatrecovering plastic waste would reduce the CO_2 emissions by 3.8 million tons. So, these results are in agreement with those stated by these authors [23].

Finally, we can suggest the use of thesecurrent composite materials for surface treatment of building and construction façades exposed to marine environment (For example; Hassan II mosque), as well as harbour infrastructures and manufacture of waste water sewer pipes and theirs accessories (junction mortars).

5.3 XRD analysis

Figure 6 exhibits the XRD analysis of PET0 and PET17, after attack by Na_2SO_4solution. The stacking of various spectraconfirms the appearance of:

- Ettringite (E) $Ca_6Al_2(SO_4)_3(OH)_{12}.26H_2O$: 9.73° 2Θ;
- Thaumasite (T), $Ca_6[Si(OH)_6]_2(CO_3)_2(SO4)_2.24H_2O$: 9.73° 2Θ;
- Gypsum (G) $CaSO_4.2H_2O$: 11.59°, 20.72°, 23.40° 2Θ
- Calcite (C), $CaCO_3$: from the crushed natural limestone sand (29.41° 2Θ);
- Portlandite$Ca(OH)_2$ was completely decomposed by Na_2SO_4 solution resulting from the chemical sulphate reactions [6].

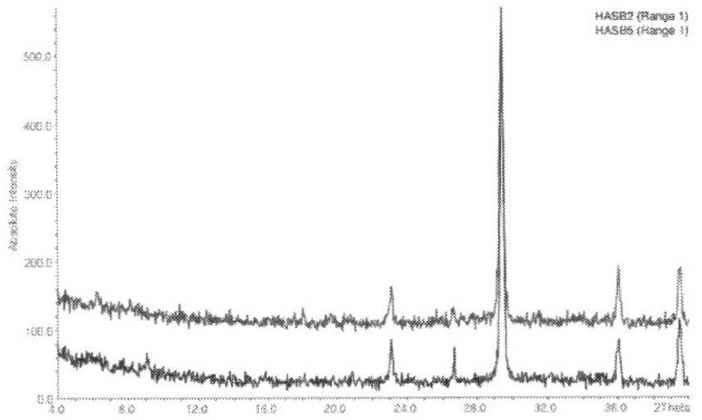

Fig. 6. X-ray diffraction pattern of the specimens under 5% sodium sulphate attacks, after 90 days.
(Blue: PET0; Pink: PET17).

6 Conclusions

With reference to the results presented in this paper, the main conclusions which can be drawn are summarized as follows:

- Composite modified mortars by waste PET plastics present a significant resistance to sodium sulphate attackswhich reduces the consumption energy of modified mortars resulting from PET-cement substitutes. Thecorrosion resistance coefficients decrease with exposure time to sodium sulphate aggressive medium.
- The mechanical properties of PET-mortar composites computed by ACI-318 and BS-8110 codes decrease with the increasing of PET volumetric substitution rate and this, for all studied cases (before and after immersion in 5% Na_2SO_4medium).Prediction model proposed by BS-8110 gives always the lowest values of elastic Young modulus which can be recommended for structural design in terms of safety and mechanical reliability.
- Since PET-mortar composite are resistant to sodium sulphate attack (as concluded above), the mechanical properties loss in terms of static Young modulus of elasticity (*YML%*) of modified mortars are lower than the ones of unmodified mortars (PET0). The optimal minimal values of (*YML%*) are given by ACI-318 code of practice, for the case ofPET-Modified mortars within volumetric rate of 6%, 12% and particularly, 17% (PET17).
- Finally, the better durability properties of PET-mortar composites observed in this study indicate longer life of the repaired structure by using this type of green PET-modified repair materials.

Acknowledgements -The authors acknowledge financial support from the Ministry of Higher Education and Scientific Research of Algeria, under the grants CNEPRU B00L01UN310120130068.

References

1. http://www.planetoscope.com/petrole/989-production-mondiale-de-plastique.html
2. Saikia, N. J. de Brito, Constr. Build. Mater.**34**(2012).
3. R. Sharma, P.P. Bansal, J. Clean. Prod. **112**(2016).
4. I.H. Alfahdawi, S.A. Osman, R. Hamid, A.I. Al-Hadithi, J. Eng. Sci. Technol.**11**,8(2016).
5. L. Gu, T. Ozbakkaloglu, Waste Manage. **51**,19 (2016).
6. A.S. Benosman, H. Taïbi, M. Mouli, *Performances Mécaniques et Durabilité des Composites Mortier-PET,* Recherche et Développement dans la Revalorisation et l'Application des Déchets du PET en Génie Civil. Maison d'édition : Editions universitaires européennes, EUE (2016).
7. Benosman, A.S., Taïbi, H., Mouli, M., Senhadji Y., Belbachir M., Bahlouli I., Houivet D., J. Mater. Environ. Sci. **559** (2015).
8. S. Akcaözoğlu, CD. Atis, K. Akcaözoğlu, Waste Manage. **30**, 2 (2010).
9. N. Saikia, J. de Brito, Constr. Build. Mate. **52**(2014).
10. A. Sadrmomtazi, S.D. Milehsara, O.L. Omran, A.S. Nik, J. Clean. Prod. **112** (2016).
11. NFP 18-560, *Granulats. Analyse Granulométrique par Tamisage*, (Association française de normalisation, AFNOR, 1990).
12. A.S. Benosman, H. Taïbi, Y. Senhadji, M. Mouli, M. Belbachir, M.I. Bahlouli, RAPRA, SMITHERS, Prog. Rubber Plast. Re. J.**33**, 3(2017).
13. C1012-04, *Standard Test Method for Length Change of Hydraulic-Cement Mortars Exposed to a Sulphate Solution*, (Annual Book of ASTM Standards, ASTM, 2004).
14. PK. Mehta, J ACI,**72**, 40(1975).
15. H. Siad, S. Kamali-Bernard, H.A. Mesbah, G. Escadeillas, M. Mouli,H. Khelafi, Constr. Build. Mater. **47**(2013).
16. ACI 318, *Building Code Requirements for Structural Concrete* (ACI 318-05) and Commentary (ACI 318R-05), ACI Committee 318, (American Concrete Institute, Farmington Hills, MI, 2005).
17. BS 8110. *Structural use of concrete. Code of practice for design and construction*. (British Standards Institution, 1997).
18. A.S. Benosman, M. Mouli, H. Taibi, M. Belbachir, Y. Senhadji, I. Bahlouli, D. Houivet, *In the Proceedings of CIMDD'2013*, (UniversitéBoumerdès, Algérie, 2013).
19. A. Aattache, A. Mahi, M. Mouli, A.S. Benosman, *In the Proceedings of CMSS'2013*, (Rabat, Morocco, 2013).
20. A. Aattache, R. Soltani, A. Mahi, Constr. Build. Mater. **146** (2017).
21. L. Jiang, Z. Liu, Y. Ye, Cem. Concr. Res. **34** (2004).
22. S. Miletić, M. Ilić, S. Otović, R. Folić, Y. Ivanov, Constr. Build. Mater.**13**, 3 (1999).
23. Fahad K. Alqahtani, M. Iqbal Khan, G.Ghataora,S.Dirar,J. Mater. Civ. Eng.ASCE, **29**, 4(2017).

Contribution to the damage modelling of reinforced concrete structures

Contribution à la modélisation de l'endommagement des structures en béton armé

Mourad KHEBIZI [1], Hamza GUENFOUD [2], Mohamed GUENNFOUD [2]
[1]Université des frères Mentouri, Constantine, Algérie, E-mail : Mourad_gc@yahoo.fr
[2]Laboratoire de génie civil et d'hydraulique, Université de Guelma, Algérie

Abstract. A two-dimensional multi-layered finite elements modeling of reinforced concrete structures at non-linear behaviour under monotonic and cyclical loading is presented. The non-linearity material is characterized by several phenomena such as: the physical non-linearity of the concrete and steels materials, the behaviour of cracked concrete and the interaction effect between materials represented by the post-cracking field. These parameters are taken into consideration in this paper to examine the response of the reinforced concrete structures at the non-linear behaviour. Two examples of application are presented. The numerical results obtained, are in a very good agreement with available experimental data and other numerical models of the literature.

Résumé. Une Modélisation bidimensionnelle par éléments finis multicouches du comportement endommageant des structures en béton armé sous charge monotone et cyclique est présentée. La non linéarité matérielle est caractérisée par plusieurs phénomène tels que : la non-linéarité physique des matériaux béton et acier, le comportement du béton fissuré, l'effet d'interaction entre les matériaux représentée par le domaine post-fissuration. Ces paramètres sont pris en considération dans cet article pour examiner la réponse des structures en béton armé à comportement non linéaire. Deux exemples d'applications sont présentés. Les résultats numériques obtenus, sont en concordance très favorable avec ceux obtenus par l'expérience et d'autres modèles numériques de la littérature.

1 Introduction

La réponse d'une structure sous un chargement résulte d'une forte interaction entre les effets des matériaux (non-linéarités locales), les effets de géométriques des structures (géométrie, répartition des forces et des raideurs, liaisons) et l'effet d'environnement (interaction sol-structure). Les non-linéarités locales sont notamment liées à la formation, à l'ouverture et la refermeture des fissures d'une part, à la liaison et au comportement des armatures (plastification des aciers) d'autre part. Une bonne description de ces phénomènes est un passage obligé si l'on veut représenter les variations des raideurs de la structure et suivre le comportement jusqu'à la ruine.

Dans cet article nous présentons une méthode de modélisation numérique des structures planes (2D) en béton armé sous chargement statique et cyclique, elle utilise des éléments poutres de type multicouches dont la matrice de raideur est calculée à l'aide d'une discrétisation de la poutre selon la hauteur en couches successives superposées (figure 1). La sommation de ces couches permet le calcul de la raideur d'une manière correcte et la prise en compte des variations du comportement. L'hypothèse cinématique de Bernoulli confère aux différentes couches un comportement uniaxial [1]. Ceci permet de traiter les comportements locaux à travers des lois uniaxiales, pour le béton et l'acier, attribuées à chaque couche. Le calcul des efforts anélastiques s'effectue grâce à une méthode d'itération basée sur la raideur sécante initiale [3]. Un traitement particulier est réservé aux couches comprenant simultanément du béton et d'acier. Le comportement des couches mixtes est homogénéisé par la loi de mélange permettant ainsi de calculer la contrainte de la couche au prorata de chaque matériau [4-6].

$$\sigma_{couche} = (1 - A)\, \sigma_{béton} + A\, \sigma_{acier} \quad (1)$$

Tel que σ_{couche} représente la contrainte axiale de la couche mixte, $\sigma_{béton}$: contrainte du béton dans la couche mixte, σ_{acier} : contrainte d'acier dans la couche mixte et A : représente l'aire relative d'acier au sein de la couche renforcée.

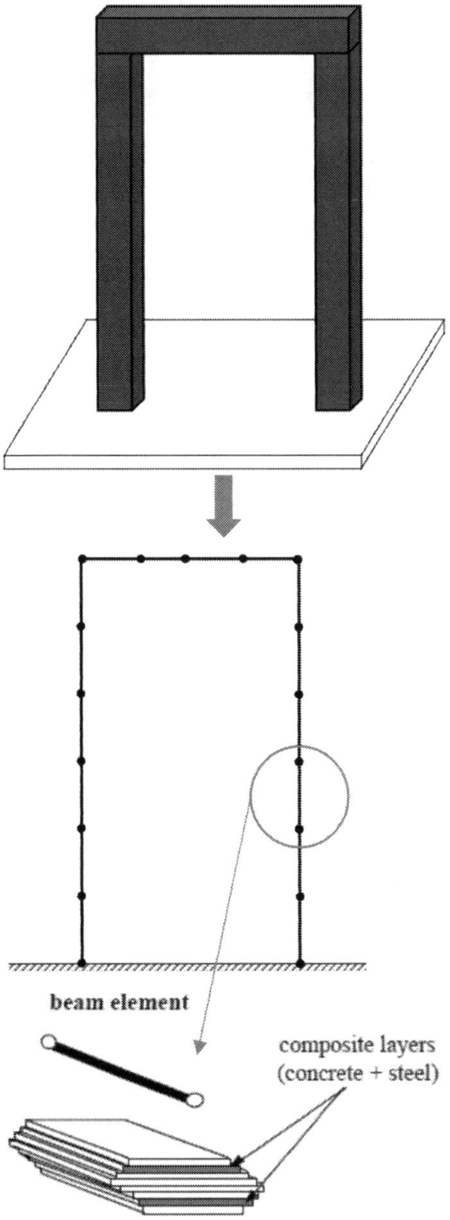

Fig. 1. Principe de discrétisation des structures en béton armé par des éléments poutres multicouches.

2 Modèle d'endommagement pour le béton

Le comportement du béton est décrit dans cet article par le modèle unilatéral de Laborderie [7], il est adapté à la description du comportement endommageable généré par la création de microfissures (abaissement des raideurs) et le fonctionnement lié, au cours de cycles, à leur refermeture (unilatéralité). Deux variables d'endommagement scalaire sont utilisées, l'une en traction D_1, l'autre en compression D_2, agissant sur la rigidité des matériaux et générant des déformations permanentes. La réponse uniaxiale du modèle soumis à un cycle de traction-compression est présentée à la figure 2.

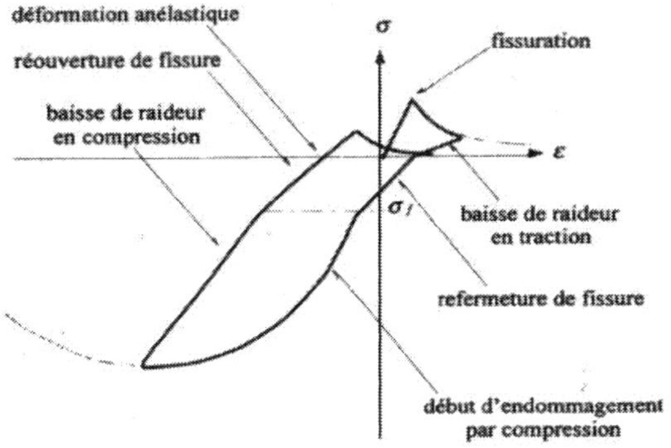

Fig. 2. Réponse uniaxiale du modèle Unilatéral [7]

3 Applications

3.1 Modélisation d'une poutre en béton armé

Dans cette section nous utilisons des éléments finis multicouches pour modéliser la poutre MECA [9]. La géométrie et les conditions de chargement sont présentées dans la figure 3. Vu son élancement, cette structure se comporte comme une poutre dominée par la flexion. L'influence du cisaillement est négligeable et une modélisation de type poutre multicouche de Bernoulli, avec lois de comportement uniaxial pour le béton et l'acier, est suffisante [1]. Le comportement du béton suit le modèle d'endommagement de Laborderie [LAB 91], les caractéristiques mécaniques considérées pour le béton et l'acier sont présentées dans le tableau 1. Dans ce travail nous avons modélisé la structure par 10 éléments poutres à 2 nœuds avec 2 points d'intégration, la section de chaque élément est discrétisée par 10 couches superposées, dont 8 en béton seul et 2 comprenant simultanément du béton et de l'acier (figure 4). Ragueneau [3] a étudié numériquement (3D) la réponse de la même poutre.

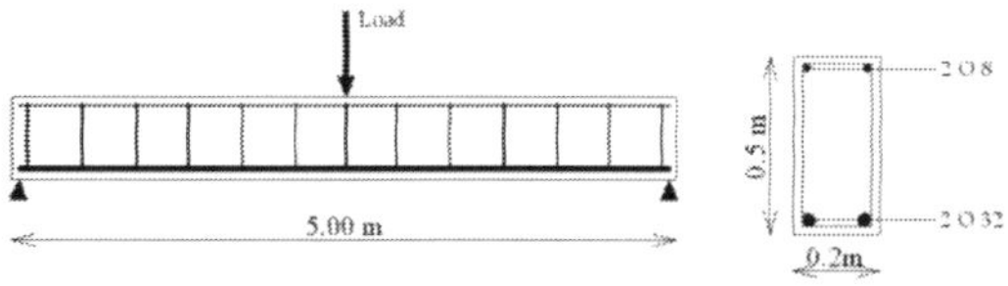

Fig. 3. Poutre en béton armé du MECA benchmark.

 https://doi.org/10.1051/matecconf/201814901052

Table 1. Caractéristiques utilisées pour le modèle de comportement de Laborderie [5].

Paramètre	**valeur**
Module de Young	37272 e^6 Pa
Masse volumique	2400 kg/m3
Seuil d'endommagement en traction	310 Pa
Seuil d'endommagement en compression	7000 Pa
Paramètre d'endommagement en traction	$9e^{-3}$ Pa-1
Paramètre d'endommagement en compression	$5.30e^{-6}$ Pa-1
Paramètre pour la traction	1.20
Paramètre pour la compression	2.00
Activation de la déformation permanente en traction	$1.00e^6$ Pa
Activation de la déformation permanente en compression	$-40e^6$ Pa
Contrainte de refermeture des fissures	$3.50e^6$ Pa

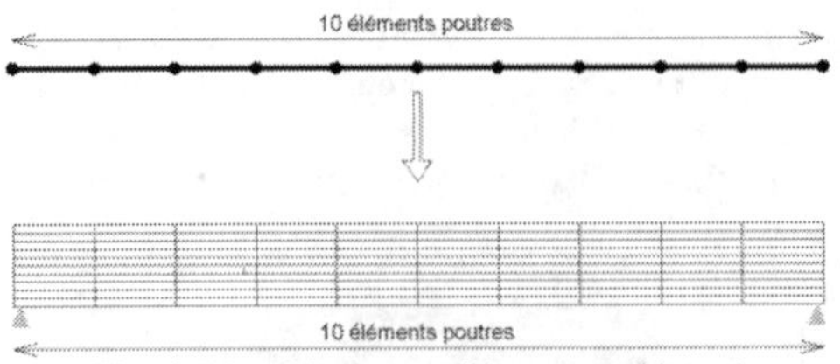

Fig. 4. Discrétisation en éléments multicouches.

La figure 6 représente une comparaison entre les résultats obtenus par notre simulation numérique, ceux de Ragueneau [3] et les résultats expérimentaux extraits obtenus lors du benchmark MECA [9]. Nous voyons que notre modélisation donne dans l'ensemble des résultats comparables avec ceux Ragueneau [3] d'une part et les résultats expérimentaux d'autre part. Cependant, les légères différences entre les résultats peuvent être imputables à un effet de maillage de la section de la poutre.

La figure 5 représente la carte d'endommagement de traction obtenue par la présente modélisation, cet endommagement est localisé au milieu de la fibre inférieure de la poutre, et se propage tout le long de celle-ci.

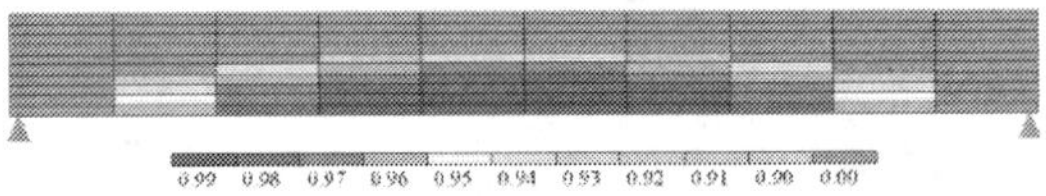

Fig. 5. Carte d'endommagement de traction «D1 ».

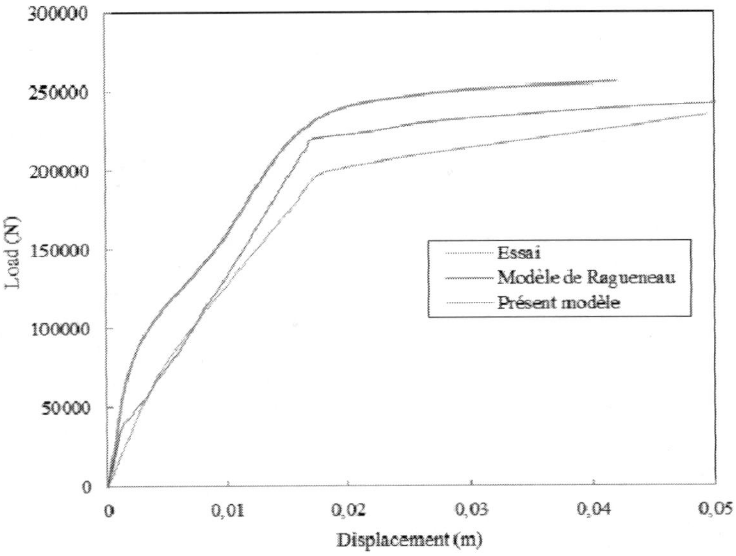

Fig. 6. Courbe charge-flèche de la poutre MECA.

https://doi.org/10.1051/matecconf/201814901052

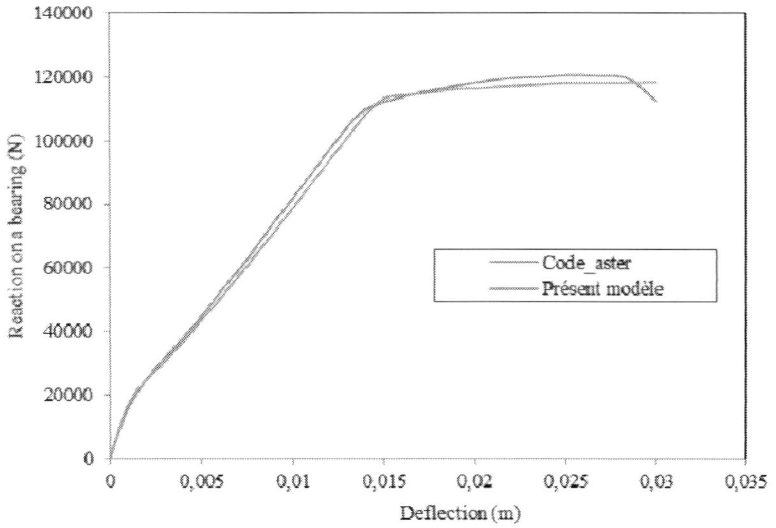

Fig. 7. Courbe réaction d'appui- flèche à mi- travée.

3.2 Modélisation de la réponse cyclique d'une poutre en béton armé

On va modéliser maintenant le comportement cyclique d'une poutre en béton armé dont la géométrie est présentée dans la figure 8. Le chargement est composé d'un cycle d'amplitude 1mm suivi d'un cycle d'amplitude 2mm (figure 9).

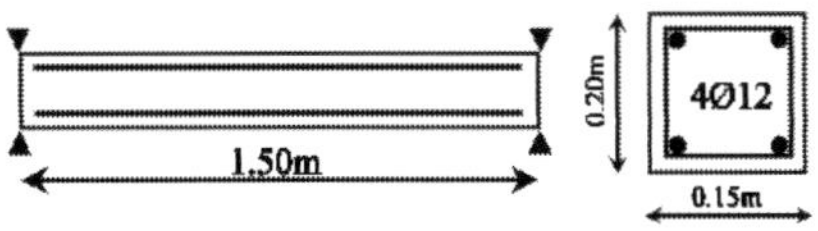

Fig. 8. Géométrie et armatures des poutres.

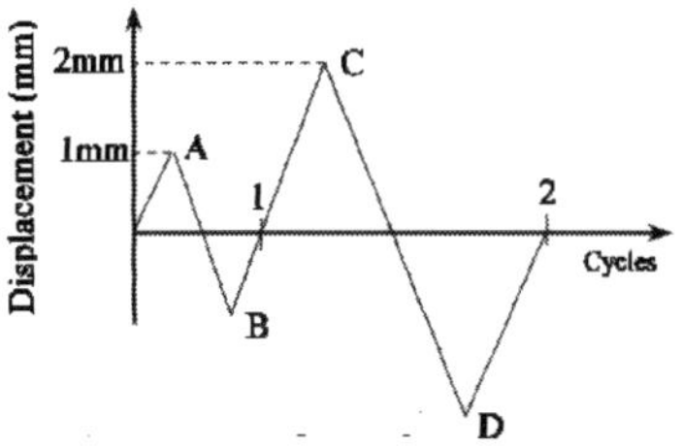

Fig. 9. Histoire de chargement.

La structure est modélisée dans ce travail par 20 éléments poutres à 2 nœuds et 2 points d'intégration, la section de chaque élément est discrétisée par 10 couches superposées, dont 8 en béton seul et 2 comprenant simultanément du béton et de l'acier. Le comportement du béton suit la loi d'endommagement de Laborderie [7], les caractéristiques mécaniques considérées pour le béton et l'acier sont présentées dans [5].

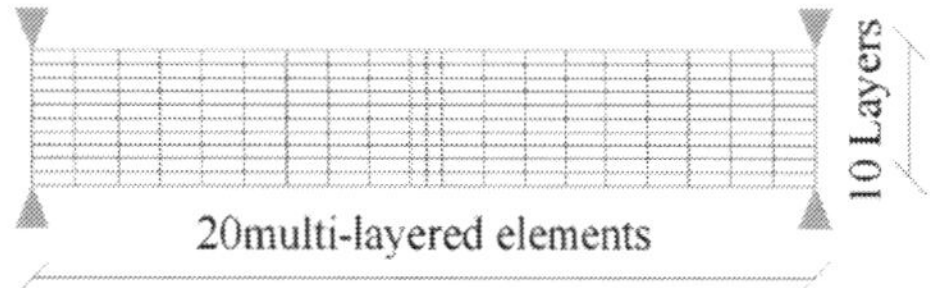

Fig. 10. Discrétisation de la poutre en éléments multicouches.

La figure 11 compare la réponse de la poutre décrite par notre modélisation et celle obtenue expérimentalement par Laborderie [7], on remarque une très bonne cohérence entre les deux réponses. La figure 12 présente également une comparaison de la réponse force-flèche décrite par notre simulation et celle de Matallah [10] obtenue par un modèle de comportement basé sur le couplage de l'endommagement et le l'anélasticité. Les deux modèles numériques donnent des résultats similaires dans le premier cycle du chargement. Cependant, pour le deuxième cycle de chargement, des légères différences sont observées. Les figures 12.a et 12.b présentent la carte d'endommagement de la poutre pour le premier cycle de chargement. A l'état de chargement « A » la partie supérieure de la poutre est endommagée (figure 12.a). L'état de chargement « B » correspond à un chargement opposé, l'état d'endommagement initialement produit est toujours conservé en mémoire alors qu'un nouveau état d'endommagement se crée dans la partie inférieure de la poutre (figure 12.b). Les figures 13.a et 13.b présentent la carte d'endommagement de la poutre au cours du deuxième cycle de chargement.

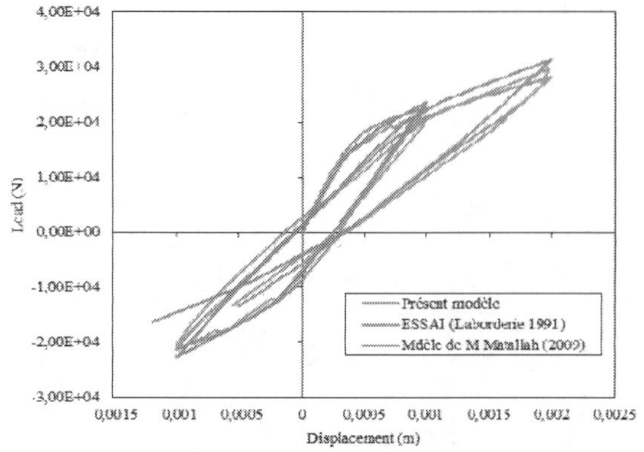

Fig. 11. Comparaison de la réponse force-flèche entre la simulation numérique et l'expérience.

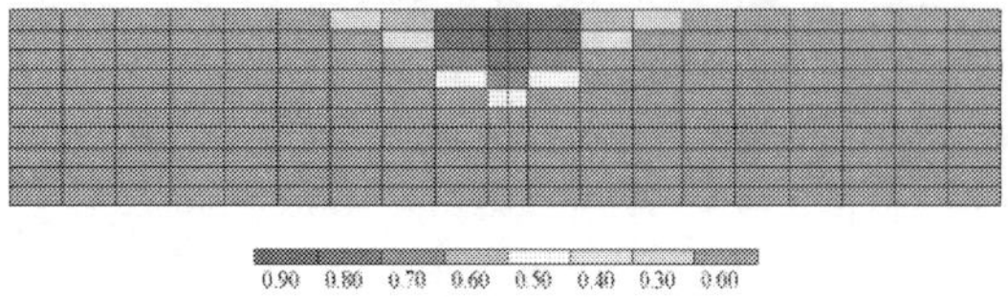

(a) Etat de chargement A

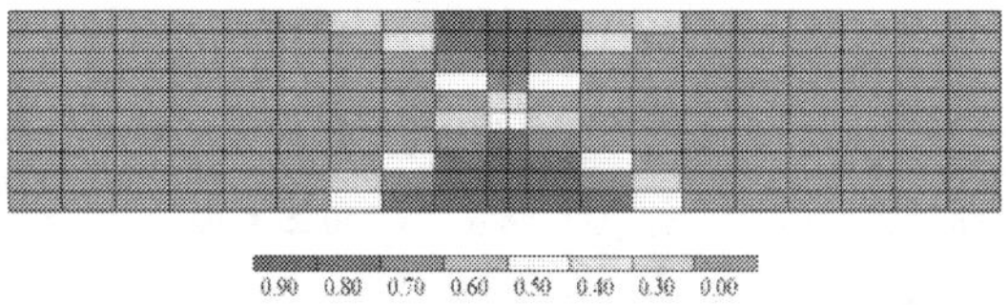

(b) Etat de chargement B

Fig. 12. Carte d'endommagement de traction «D1 » pour le premier cycle de chargement.

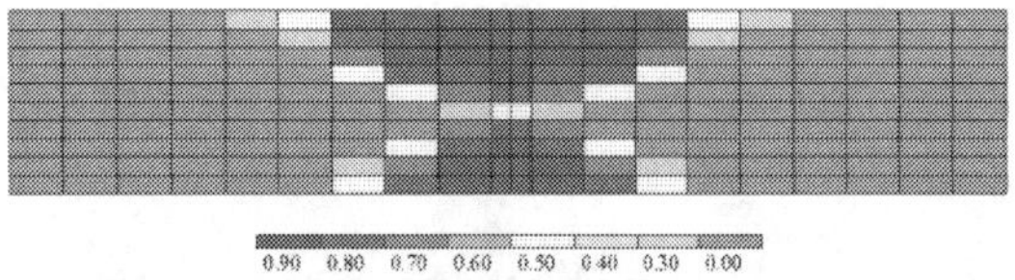

(c) Etat de chargement C

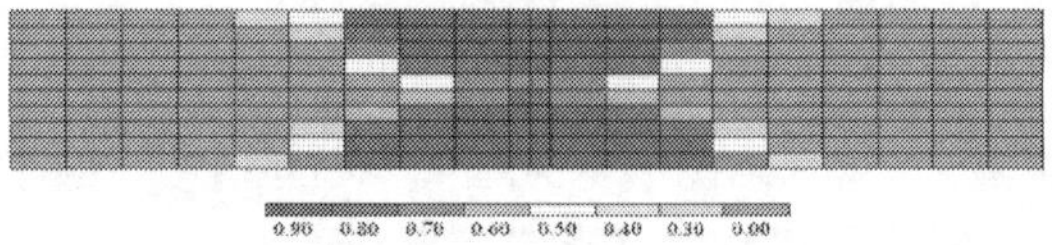

(d) Etat de chargement D

Fig. 13. Carte d'endommagement de traction «D1 » pour le deuxième cycle de chargement.

4 Conclusion

Une modélisation simple du comportement non linéaire des structures en béton armé est présentée. Elle utilise des éléments poutres multicouches obéissant à l'hypothèse de Bernoulli pour conférer aux différentes couches un comportement uniaxial. Elle permet également la description de l'état d'endommagement des structures au cours d'un chargement. Deux exemples d'applications ont été présentés. Le premier est celui d'une poutre soumise à une flexion 03 points (Benchmark MECA), et le deuxième est celui d'une poutre soumise à une flexion cyclique. A la lumière de ces exemples nous avons constaté une très bonne cohérence et concordance entre les résultats de nos simulations et l'expérimentation d'une part et ceux d'autres modèles numériques de la littérature d'autre part. Nous avons constaté aussi, si le matériau est déchargé après avoir subi un état d'endommagement, il restaure sa raideur, les fissures précédemment ouvertes se referment mais la structure interne du matériau reste toujours endommagée.

References

1. P. Kotronis, *Thèse de Doctorat, (*Ecole normale supérieure de Cachan, 2000)
2. P. Kotronis, F. Ragueneau, J. Mazars, Engineering Structures, **27** (2005), P. 1197-1208
3. F. Ragueneau, *Mémoire d'habilitation à diriger des recherches,* (Université Pierre et Marie Curie, Paris 6, 2006)
4. M. Khebizi, H. Guenfoud, M. Guenfoud, Courrier du Savoir, **18**, (2014), P. 111-115
5. M. Khebizi, M. Guenfoud, Computers and Concrete, **15**, 4 (2015), P. 547-562
6. M. Khebizi, H. Guenfoud, M. Guenfoud, Challenge Journal of Structural Mechanics, **3**, 1 (2017) P.52-57
7. C. Laborderie, *Thèse de doctorat,* (Université Paris 6, 1991).
8. C. Laborderie, *Thèse d'habilitation à diriger les recherches,* (Université de Pau et des Pays de l'Adour, 2003)
9. S. Ghavamian, *MECA project benchmark,* (EDP EDF R&D, 2001)
10. M. Matallah, C. Laborderie, Engineering Fracture Mechanics, **76** (2009), P. 1087-1108.

 https://doi.org/10.1051/matecconf/201814901053

Contribution to the requalification of Alkali-Silica reaction damaged concrete: particular respect to the slag effect

Zineb DOUAISSIA[1], Mouloud MERZOUD[2]
[1 et 2] Badji Mokhtar university, Department of civil Engineering, ANNABA , ALGERIA

Abstract. A significant concern when designing durable concretes structures is the risk of alkali silkai reaction (ASR). The cracks formed by ASR increase the permeability of concrete, reduce the strength and lower the economic lifetime of concrete. The use of mineral and chemical admixtures to prevent expansion due to the alkali-silica reaction (ASR) was first reported 40 to 50 years ago. In this paper, a comprehensive experimental research was carried out to prevent the damages and to be able to improve in concrete structures with the reactive aggregates by using the ground granulated blast-furnace slag (GGBFS) in the mixtures. Furthermore, the engineering properties of blast-furnace slag and the test methodology used in the research were presented and the research findings were also discussed.

1 Introduction

In the decade following the discovery of the alkali silica reaction (ASR) by Stanton, recommendations, such as the use of non-reactive aggregate or the use of low alkali cement, were made to avoid or to mitigate the disorders due to this reaction.

At the same time, it was established that blended cements made of OPC mixed with pozzolanic admixtures were efficient in reducing ASR expansion. The use of such admixtures (fly ash, silica fume, ground granulated slag) is now current practice to reduce the risk of ASR.

Many references concern the use of ground admixtures added to mixtures containing a reactive aggregate coming from a different source, particularly Hobbs [1], Peter [2], Moir [3], Swamy[4], Hiroshi Nomachi and al [5] and Yoshihiko Ohama and al [6].

The specification of concrete using ground granulated blast-furnace slag (GGBS) is recognised in several countries [7-9] as an possible measure to reduce the risk of deleterious alkali silica reaction occurring in concrete containing potentially reactive aggregates.

However, the effectiveness of GGBS, particularly at levels of cement replacement below 50%, has been questioned by several workers [10-12].

The aim of this paper is to verify whether ground granulated blast-furnace slag (GGBS) is an efficient and general solution to counteract ASR expansion. Tests were performed on concrete using two reactive aggregates.

2 Materials

2.1 Cement

The binder was an ordinary Portland cement, CEM I 52,5R as specified by European Standard EN 197-1, with a specific weight of 3,15 , a fineness of 350 m^2/kg (Blaine) and a moderately high Na_2O_{eq} content (0,8%). Its chemical composition is presented in Table 1

2.2 Aggregates

Two aggregates were chosen :

– A siliceous limestone (H), mainly composed of calcite with a few percent of dolomite. When observed at microscopic level, diagenetic silica is seen to consist mainly of quartz and traces of feldspars, muscovite and kaolinite. This aggregate is also a reactive rock, with a slow rate of silica dissolution.

– A crushed waste glass (G), resulting from the crushing of window glasses. It was chosen because the reuse of waste glasses has been envisaged by several workers in the last decade [13–17]. It had an amorphous structure as determined by X-ray diffraction.

2.3 Ground Slag

Ground slag (S) is a byproduct of the development of the cast iron in steel complex of El Hadjar (Annaba), it has a glass structure, a vitrification rate of 97% and a low hydraulicity.

it's obtained by crushing and sieving at 80 µm. In order to study the influence of fineness in counteracting expansion, the slag material is ground to different specific surface (Blaine), which are S_1: 100 m^2/kg, S_2: 400 m^2/kg and S_3: 600 m^2/kg, The chemical compositions of this powder appear in Table 1.

Table 1. Chemical composition of cement, aggregates and ground slag (%).

	cement	Aggregates		ground slag		
		H	G	S_1	S_2	S_3
SiO_2	20.1	20.0	72.1	33.0	35.4	36.1
Al_2O_3	5.6	1.3	1.0	11.8	12.7	9.9
Fe_2O_3	2.0	0.6	0.1	1.6	0.30	0.79
CaO	62.5	40.6	6.3	41.3	41.3	41.0
MgO	3.1	1.2	3.7	9.0	06.7	9.5
K_2O	0.9	0.4	0.1	0.51	0.60	0.35
Na_2O	0.2	0.4	12.7	0.32	0.24	0.18
SO_3	3.2	0.3	0.1	0.13	0.14	0.11
L.O.I	1.7	34.7	0.4	0.8	0.6	0.6

3 Experimental methods

3.1. Mortar tests

The mortar test was an accelerated method using an autoclave apparatus (French Standard P18–594 [18]). This test consisted of measuring the length variations of mortar prisms (4 x 4 x 16 cm) doped with alkalis (4% of cement mass) after a treatment lasting 5 h at 127°C and 0,15 MPa. Mortars had a water/cement ratio of 0,5 and an aggregate/cement ratio of 3,0. The first series of tests were performed on mortars containing only cement, water and reactive aggregates (0,16–2,5 mm). For the second series, cement has been substituted with increasing percentage (by volume) of ground slag : 10, 20 and 30% of different fineness from 200 to 600 m^2/kg (Blaine), After this tests, we measured the compressive strength (fc) of each mortar .

3.2 Concrete tests

The concrete samples were made, stored and tested according to French Standard P18-454 [19]. This test consisted of evaluating the performance of a concrete with respect to ASR by measuring its expansion on 7x7 x 28 cm prisms stored at 60°C and 100% relative humidity.

After mixing, the samples were stored in a moist room at 20°C for 24 h. Then, the concrete prisms were placed vertically on grids in watertight containers containing 35 mm of water (the concrete was not in contact with the water).

Each concrete was tested on three prisms and the expansion data represent the average of three measurements. The containers were placed in a reactor generating an atmosphere at 60°C and 100% R.H. Expansion measurements were taken after the containers and their prisms had been cooled for 24 h at 20°C. Immediately after each measurement, the prisms were put back into their containers, which were kept in the reactor at 60°C until the next measurement. The last measurement was taken at 100 weeks. At the same age map cracking was measured in order to quantify the degree of damage due to ASR and compressive strength (fc) was measured on three cubes (10 x10x 10) cm^3.

Details of the concrete mixtures, including the slump (modified Abram's cone of 200 mm) and density, are given in **Table 2**. Each mixture is designated by a capital letter representing the reactive aggregate used, followed by a small letter for mixtures containing ground slag. The water content of the concrete was set between 197 and 212 l/m^3. A significant amount of cement was used (between 442 and 477 kg/m^3) in order to obtain high alkali content. However, considering the value of Na_2Oeq in the cement, it was decided to enhance the alkali content to 35 kg/m^3 by adding NaOH to the mixing water. Ground slag used to replace 20% of the sand (by weight) had finenesses of 200, 400, and 600 m^2/kg.

4 Experimental results and Discussion

4.1. Mortars

Results of the autoclave expansion tests performed on the mortars are given in **Fig.1**, Such as reducing expansion caused by the slag substitution (in% of measured value on the mortar without slag).

They confirm that the two aggregates used can be classified as reactive, since the measured expansions varied between 0,30 and 0,38%. These expansions were much higher than 0,15%, the value under which the aggregate is considered as non-reactive according to French standard P18– 594 [18]. The use of ground slag decreased the expansion of mortars. However, their activity depended on the fineness and quantity used.

Table 2 Composition, density and slump of concrete mixtures

	Composition								Density	Slump
	Cement (kg/m^3)	Water (l)	W/C	Superplasticizer (l)	Ground Slag (kg/m^3)	Sand 0,08-4 (kg/m^3)	Aggregate (kg/m^3)		(kg/m^3)	(cm)
							4 - 8	8 - 16		
H	477	212	0.44	0	0	617	248	929	2445	4
H-S	467	208	0.44	2.49	133.5	454	242	910	2397	7
G	452	201	0.44	0	0	699	200	766	2319	4
G-S	442	197	0.44	4.91	136.8	546	195	749	2272	5

 https://doi.org/10.1051/matecconf/201814901053

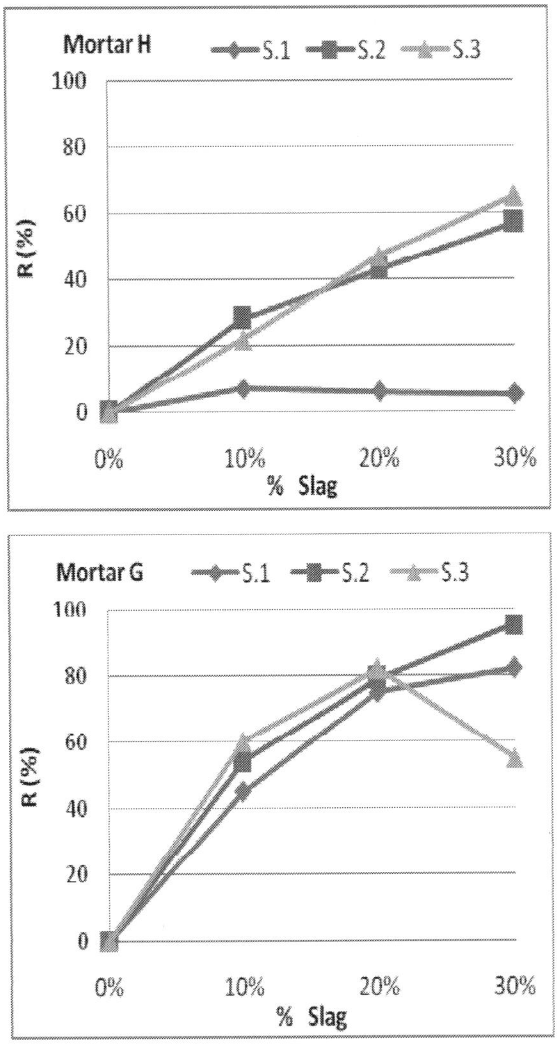

Fig.1. Reduction of expansion (R%) of mortar versus amount and fineness of slag

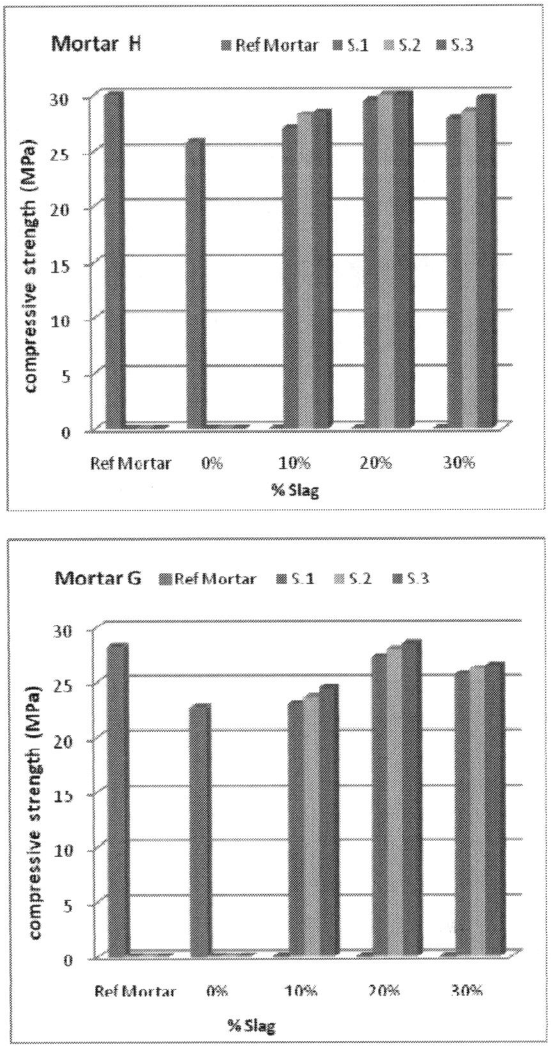

Fig. 2 slag effect on the compressive strength of mortars aged 17 weeks conserved at 60 ° C and 100% RH.

After This test , the compressive strength (fc) of each mortar was measured, the main results from compressive tests are reported in **Fig.2**, it includes the compressive strength for the reference mortars which are undoped mortars (without addition of NaOH) , for the mortars at 0% slag which are doped mortars, for the mortars at different rate of ground slag : 10, 20 and 30% and different finenesses (Blaine), which are S_1: 100 m^2 / kg, S_2: 400 m^2 / kg and S_3: 600 m^2 / kg.

If we are interested in results for the reference mortars which are undoped mortars (without addition of NaOH) and mortars at 0% slag which are doped mortars , we note that :

- The addition of alkali , which causes expansion , occasions a reduction in strength .
- The increasing substitution rate leads to an increase in strength.
- Differences in strength resulting from two antagonistic actions: the alkali-silica reaction that tends to lower the strength, and slag, which probably by a reaction similar to a pozzolanic reaction, increase this strength.
- The slag beneficial effect is measurable by comparing the mortars strength gains with different rate of slag 10%, 20% and 30%, which is 7 MPa, 20 MPa and 16 MPa respectively, mortar strength to 30% of slag is higher than that obtained by the reference mortar did not damaged by alkali-silica reaction.

 https://doi.org/10.1051/matecconf/201814901053

4.2. Concretes

The length variations of the concrete prisms cured for 100 weeks at 60°C and 100% R.H. are shown in **Fig.3**. In a first analysis, it can be seen that the concrete G presents an important expansion during the first week, and that the expansion rate decreases after 30 weeks. It is noteworthy that the ground slag reduced the expansion of all concretes, different behaviors show concrete with ground slag, concrete G-S shows a slow expansion rate.

When the expansion of the prisms had practically stabilized, all faces of the 7 x 7 x 28 cm samples were photographed, Crack patterns are shown on **Fig 4**, The results give information on both the intensity of the damage and the possible isotropic distribution of cracking. We can observed that :

The presence of ground slag dramatically decreased the ASR crack development, as was confirmed by the swelling measurements.

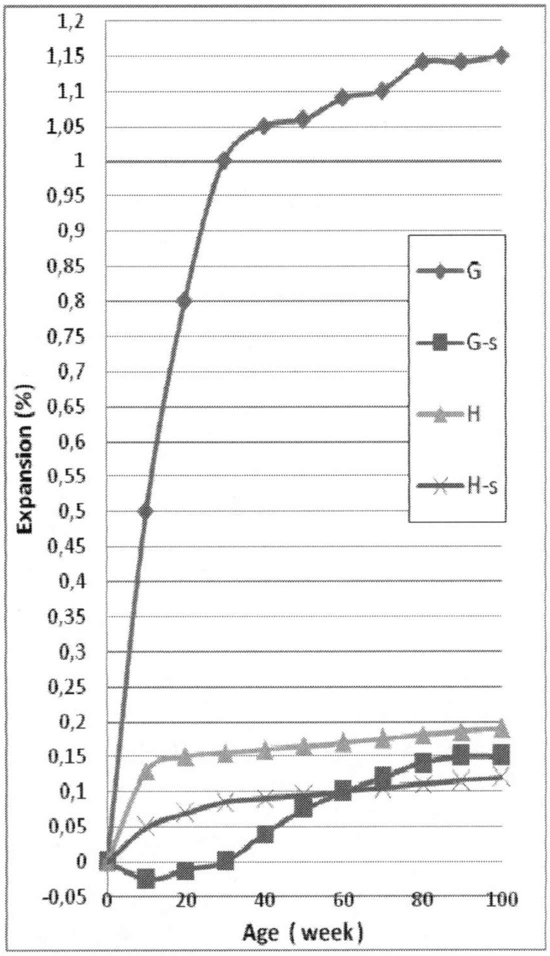

Fig.3. Expansion of concretes versus time, cured at 60°C and 100% RH.

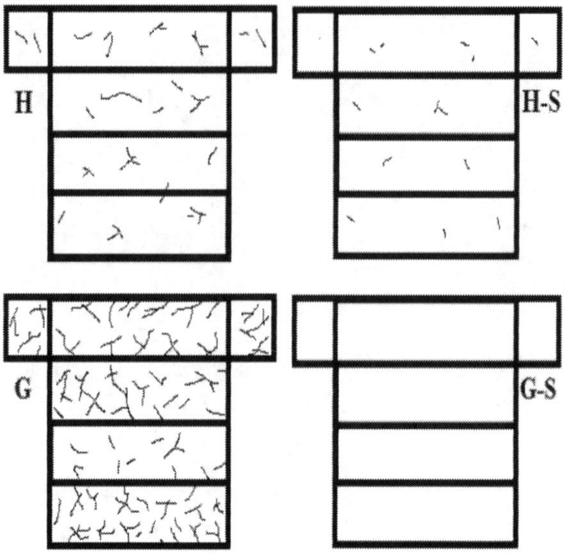

Fig.4 Influence of ground slag on surface cracking of concrete.

At the same age, the compressive strength (fc) of each concrete formulation was measured on three (10 x 10 x 10) cm^3 cubes. These values are reported in **Table 3**. The supplementary data fc = 33 MPa relates to a concrete with same proportioning and curing as the H and G series but with innocuous aggregates. It can be seen that addition of ground slag enhances the compressive strength of all concretes.

Table. 3 Influence of ground slag on compressive strength of concretes (age 34 weeks)

Aggregate	H	G	R[a]
Without ground slag (MPa)	31	44	33
With ground slag (MPa)	37	70	-
Increase in compressive strength (%)	16	37	-

a concrete with innocuous aggregate having the same mixture design and the same curing conditions as other concretes with reactive aggregates

Summarizing, as previously described the selected reactive aggregates present a different kinetics of reaction in accordance with their composition. Then, the manifestations of ASR and the damage produced are quite different. An assessment of the beneficial activity of ground slag against ASR deterioration needs to correlate the different components: expansion as defined by its maximum value and maximum speed, surface cracking and compressive strength. **Table 4** summarizes the main differences of each concrete.

Table 4 Influence of ground slag on the properties of concrete (relative to ASR)

			H	G
Expansion	Without slag	Final expansion (%)	0.20	1.07
		Maximum expansion rate (μm/m/week)	395	1025
	With slag	Reduction of expansion (%)	36	96
		Reduction of expansion rate (%)	63	98
		Acceptance criteria P18-454[a]	no	yes
Cracking	Without slag	Cracking length (cm/cm^2)	0.23	0.10
	With slag	Reduction (%)	91	≈100
Compressive strength	Without slag	Compressive strength (MPa)	31	44
	With slag	Increase in compressive strength (%)	16	37

[a] expansion at 3 months lower than 0.02%

5 Conclusion

The experiments reported in this paper were intended to verify the following hypothesis: the harmful effects of ASR induced by an alkali-reactive aggregate can be counteracted by the beneficial activity resulting from the addition of a finely ground slag.

The work was carried out on concretes incorporating two alkali-reactive aggregates, with or without slag added. The concretes were cured at 60°C and 100% RH and their expansion was measured for up to 88 weeks. In these conditions, which are different from actual conditions in the field, the following conclusions can be drawn :

1. The addition of slag to concrete containing reactive aggregate can reduce or suppress ASR expansion.
2. The efficiency of slag depends on its content and fineness : Generally the reduction rate varies in the same direction as the substitution rate, In most cases the reduction rate varies in the same direction as the finesse; the increase of finesse lowers expansion.
3. Surface cracking is strongly reduced or suppressed when slag is used in concretes.
4. Slag act as pozzolanes do and counteract, more or less, the decrease of compressive strength caused by ASR.

Before any attempt to extrapolate from laboratory to field concrete, our findings have to be verified on other aggregates and more attention must be paid to the influence of the amount of slag on it activity.

Of course it would be interesting to compare the efficiency of this method with that of mitigation due to mineral admixtures such as silica fume, fly ash, pozzolanne and metakaolin.

References

1. Hobbs D.W, Effect of mineral and chemical admixtures on alkali aggregate reaction. 8th ICAAR, Kyoto, Japan, (1989).
2. Peter P. Hudec and Joseph A.Larbi, Chemical treatments and additives to minimize alkali reactivity. 8th ICAAR, Kyoto, Japan, (1989).
3. G.K.Moir and J.S.Lumley, Influence of partial cement replacement by ground granulated slag on the expansion of concrete prisms containing réactive silica. 8th ICAAR, Kyoto, Japan, (1989).
4. R.N.Swamy and M.M. Al-Asali, Effectiveness of mineral admixtures in controlling ASR expansion. 8th ICAAR, Kyoto, Japan, (1989).
5. Hiroshi Nomachi and al, Effects of admixtures on expansion characteristics of concrete containing reactive aggregate. 8th ICAAR, Kyoto, Japan, (1989).
6. Yoshihiko Ohama and al , Inhibiting alkali-aggregate reaction with chemical admixtures. 8th ICAAR, Kyoto, Japan, (1989).
7. Guidelines for alkali-aggregate reactions in concrete. German Building Standars Committee of the Deutsches Institute fur Normunge ev,(1986).
8. Japanese standards Association. ready mix concrete. JSA, Tokyo, (1986).
9. Alkai-silice reaction : minimizing the risk of damage. Guidance notes and model specification clauses. Concrete Society, London, Technical Raport **30**,(1987).
10. Kollak J.J and al. Measurement of OH^- ion concentrations of pore fluids and expansion due to alkali-silica reaction in composite cement mortars. 8 th ICCC, Rio de Janeiro, (1986).
11. Hobbs D.W, Deletarious expansion of concrete due to alkali-silica reaction : Influence of pfa and slag. Mag -concr -res , (1986).
12. Kawamura M. and al , Effet of various pozzolans additives on alkali-silica expansion in mortars made with two types of opaline reactive aggregates. 38 th annual meeting of Japan Cement Association, Japen, Tokyo, 92-95, (1984).
13. Nishikawa T, Takatsu M, Daimon M (1995) Mechanical properties of mortars containing glass powder. 49th annual meeting of Japan Cement Association 49:114–119
14. Polley C, Cramer SM, de la Cruz RV, Potential for using waste glass in Portland cement concrete. J Mat Civ Eng **10**:210–219, (1998)
15. Yixin S, Lefort T, Moras S, Rodrigues D, Waste glass: a possible pozzolanic material for concrete. CANMET/ACI Int Symp on sustainable

development of cement and concrete, Ottawa, 317–326. (1998).

16. Hudec PH, Ghamari C, Ground waste glass as an alkali silica reactivity inhibitor. In: Proceedings of the 11th International Conference on AAR in Concrete, Quebec, Canada, 663–672, (2000).
17. Kojima T, Takagi N, Haruta K, Expanding characteristics of mortar with glass powder produced from waste bottles. In: Proceedings of the 11th International Conference on AAR in concrete, Quebec, Canada, 673–682, (2000).
18. AFNOR NF XP P 18–594, Aggregates –Test methods on reactivity to alkalies, (2004).
19. AFNOR NF P 18–454, Concrete—Reactivity of a concrete formula with regard to the alkali-aggregate reaction - Performance test, (2004).

Study of the Influence of an Air- Entraining Agent on the Rheology of Motars

Nadia Tebbal* [1], Zine El Abidine Rahmouni [1], Lamis Rabiaa Chadi [2]
* [1] Institute of Technical Urban Management, University of M'sila Algeria
[1,2] Geomaterials Development Laboratory, Civil Engineering Department, M'sila University, M'sila, Algeria.

Abstract. The objective of this study is to analyze the effect of the air entrainment on the fresh rheological properties as well as on the compressive mechanical resistances of the mortars. The hardened concrete contains a certain amount of randomly spread air, coming either from a drive during kneading or from the evaporation of the mixing water. The air quantity is in the order of 20 l / m^3, ie 2% of the volume. However, the presence of a large volume of air bubbles causes the mechanical resistances to fall in compression. On the other hand, the use of air entrainment could improve the rheological properties of fresh concrete.
Experimental studies have been carried out to study the effect of air entrainment on compressive strength, density and ingredients of fresh concrete mix. During all the study, water cement ratio (w/c) was maintained constant at 0.5. The results have shown substantial decreasing in water and mortar density followed with decreasing in compressive strength of mortar. The results of this study has given more promising to use it as a guide for mortar mix design to choose the most appropriate concrete mix design economically.

1 Introduction

Air entrained concrete consider one of the important discovery made in concrete technology. It is well-known that AEA can improve the workability of the fresh concrete and improve the frost resistance and permeability of the hardened concrete.

Therefore, it has a practical and theoretical significance to research the influence of air entraining agent on mortar and concrete rheological properties and bubble characteristic parameters.

Y. Jiang, Z., 2006, studied the research of air entraining agent on concrete rheological properties and bubble characteristic parameters effect, it found that air entraining agent decreased concrete compressive strength, slightly added concrete flexural strength, and folding compression ratio increased significantly.

Many researches showed that the addition of air-entraining agent increased the air content up to a saturation level, above which no further increase in air content was observed. With increasing air content, the yield stress increased and the plastic viscosity decreased. (L.J. Struble, 2004).

Frost durability is very important for concrete structures exposed to low temperature because concrete is a porous material that contains water in various forms. To protect the concrete against freezing, the air content and the spacing factor must be taken into account, since it is possible to produce concrete with an air content that meets the specifications but whose spacing factor is not sufficient to protect the concrete from frost (F.Saucier .1991).

The deterioration would not appear in the concrete after a single period of freezing, but it would be the result of progressive damage due to the succession of a greater or less number of freeze-thaw cycles (Carles–Gibergues A. 1992). The use of air entraining aids facilitates the placement of concrete and protects them from frost (Lianxiang Du, 2005). But this use is effective insofar as the bubbles that are added improve the network of air bubbles. This appears to be possible only if the dose of AEA added is quite high (M .Pigeon, 1990), if not there is a risk of only 1% to 2% of air, which is not enough to guarantee an improvement of the network. This increase could just as well be due to the fluidification of the mixture by the addition of water with the AEA.

Generally this fluidization mainly results in large bubbles which are liable to be lost thereafter [Pigeon M, 1987]. Similarly, recent studies in the field such as Nicola Pitrov's (P.Nikola, 2003), have found that the absence of AEA in concrete exposed to de-icing effects causes a rather remarkable decrease in compressive strength, elasticity.

The objective of this paper is to present the results of a research is to formulate a series of air entraining agent dosages incorporated into the mortar mixtures during mixing in order to develop a formulation that provides the mortar with a better durability over time and protecting it in particular at repeated cycles of gels. These formulations of mortars consist in also having a better rheological behavior with respect to the maneuverability without having a significant loss of resistance which exceeds the standards.

2 Experimental program

2.1 Characteristics of the materials used

Cement: This is a Portland cement (CEM II 42.5) has a Blaine specific surface area of 3585 cm^2 / g and a density of 3.2. Chemical and mineralogical composition is given by Bogue's formulas (Table 1).
Sand: The binary sand mixtures aggregates used in this study were SD (dune sand) and CS (crushed sand). The characteristics of binary sand mixtures are given in Table 2. The dune sand used with particles ranging from 0.08 mm to 5 mm in size. This natural sand was taken from Boussâada, Algeria. The granulometric study is performed according to standard NF P18-304 after the treatment process, allows to eliminate a significant portion of clay minerals. In this study the manufactured fine aggregate used is crushed sand generated by the quarry waste.

An improvement of porosity of the binary mixture (crushed and dune sands) and correct the variation of the granulometric composition (particle sizes) using 1/3 of crushed sand blend.
The adjuvant: used is an air entraining agent (MEDA - AIR). It is a solution of pH = 7-8 and a density of 1± 0.01, with ions chlores ≤ 1%. Its normal use scale is fixed by the manufacturer's recommendation of 0.01 to 0.1% of the cement weight. The percentage of the chemical admixture used was 0.08 % by weight of cement for all mixes.
In this study we used the super plasticizer named (MEDAFLOW 30); it was obtained according to NFP 18333 and NFP 18336.

Table 1: Chemical and mineralogical composition of cement

	Compositions (%)	Cement (%)
Chemical compositions	SiO_2	18.48
	CaO	61.04
	Al_2O_3	4.51
	Fe_2O_3	3.21
	MgO	2.16
	SO_3	2.18
	MnO	/////
	Na_2O	0.13
	K_2O	0.56
	Cl^-	
Mineralogical compositions	C_3S	57
	C_2S	18
	C_3A	07
	C_4AF	11
	Gypsum	15

2.2. Experimental procedure

2.2.1 Formulation of Mortar

Fresh mortar mixes were prepared in modified laboratory mixer. Mixtures were prepared from dry aggregates. Cement and sand were dosed by weight, water and chemical admixtures were added by volume.

The mortar formulation tested is performed by formulating's method mortar of the Skramataiv .
The same mixing procedure was used for all mortars:
1. Place the cement and water in the mixer bowl.
2. Mix cement and water 30 seconds min.
3. Add sand 30 second speed min.
4. Mix cement and sand and water 30 seconds fast speeds.
5. Leave to stand for 1 minute 30 seconds.
6. Mix for 1 minute fast speed.

In the case of the use of adjuvants, the latter is diluted in the mixing water. The concrete specimens are preserved in their mould in wet place (20∘C, 95% HR) during 24 hours. Then they undergo immersed in the potable water tap. The physical, mechanical characteristics of the concrete are compared.

In order to investigate the influence of an air-entraining agent on the rheology of mortars, prisms samples with the dimensions of 7x7x14 cm, are produced. After 28 days curing conditions, the samples were used for the determination of the compressive strength.

Freezing thaw test: Six specimens for each type of mortar were tested according to freezing-thawing NF P 18-425 standard. One specimen of each type of mortar was stored at 20 °C and 45 –50% relative humidity for comparison.

It should be noted that the calculation of the composition of the concrete was made by the «Absolute "elaborated by Professor B. Scramtaïv (A.KOMAR 1978). Table 2 present the complete formulation of the 02 mortar mixes for cement dosages of 400 kg / m^3 and for W / C = 0.5, with AEA determinations based on the weight of the cement. The workability was measured by the slump test using the Abrams cone in accordance with the standard NF P 18- 451, which makes it possible to make a connection between the dosage of AEA and the maneuverability.

Table 2: The mortar composition with AEA

Mix (g)	M0	M1	M2
AEA	-	0.36	
Cement (C)	450		
Sand dune (SD)	450		
Crush sand (CS)	900		
Water /Cement (W/C)	0.50		
Super plasticizer	-	-	4.50

3 Results and discussions

3. 1 Behavior of fresh concrete

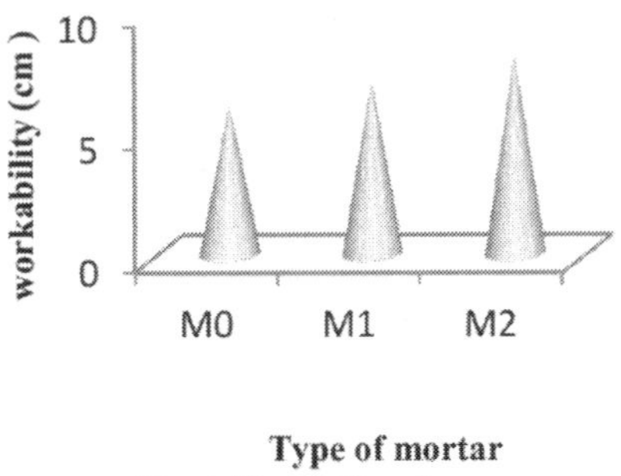

Fig. 1. Workability of mortars

Workability is an essential quality of mortar; it can be defined as the ease of application of the mortar.

Based on the experimental results presented in Fig. 1, the incorporation of adjuvant affects the workability of mortar positively. Indeed, an air- entraining agent or super plasticizer enhances the handling by making a more workable mix of a plastic mortar rheological behavior.

From Fig. 1, it can be observed that all the mortar mixtures with adjuvant gave better workability compared to the control mix at the values ranging from 6 cm slump for the first one and 7 to 8.5 cm slump for the latter. This is in agreement with other results on the subject found previously where there was an improved fluidity of mortars at fresh state of these types with incorporation of blends (Du.Lianxiang ,2005).

3.2 Effect of air entrainment on density of mortar in mortar mix

Figure 2 illustrate the relationship between type mortar and density of concrete.

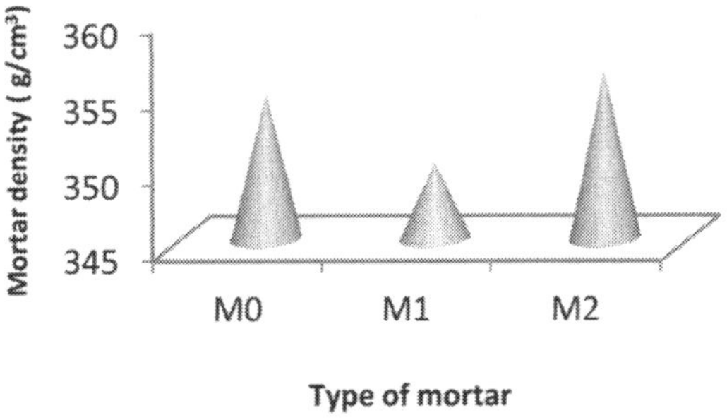

Fig. 2. Effect of air entrainment on density of concrete in concrete mix

Density of non-air entrained mortar is 354.5 g/ cm^3 while air entrained cement mortar from 350 to 356 g/ cm^3 (Fig.2). This confirms the relationship that the bulk density of the mortar is inversely proportional to the amount of air. The (M_1) cement mortar had the lowest bulk density among with air entrained mortars. The highest bulk density has been determined for mortars with Portland cement both non- air entrained and air entrained (M_0, M_2).

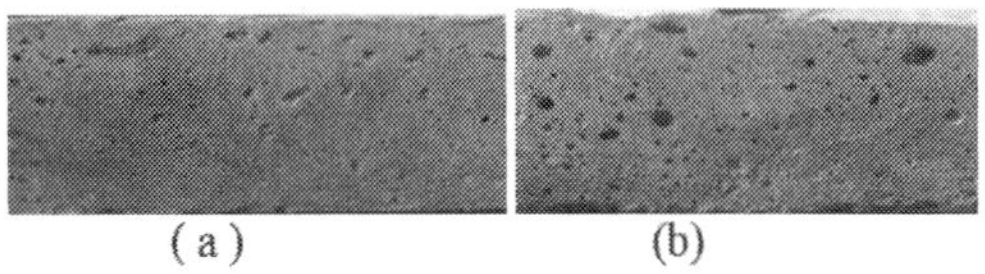

Fig . 3. Mortars with and without AEA

3.3 Compressive strength

Figure 4 shows the evolution of the compressive strength of concretes based on an AEA.

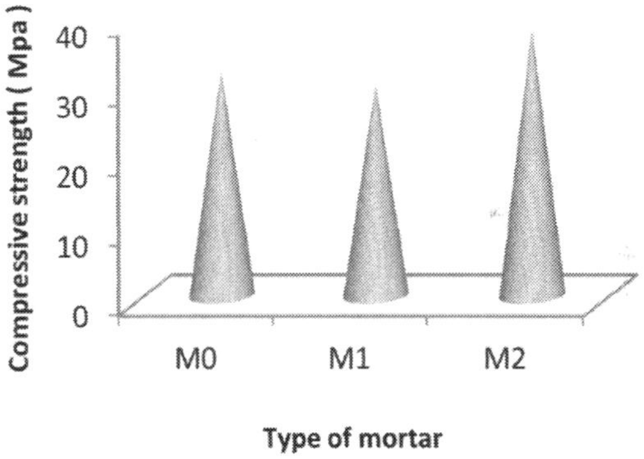

Fig. 4. Influence of the AEA on resistance to compression at the age of 28 days

The results of Fig. 4, it can be concluded that the mortars (M_0 and M_1) has resistances which evolve slightly and which tend towards stable values, whereas for mortar with AEA there is a drop in resistance which depends on the air entrained. The compressive strength of mortar decreases for tests carried out on mortars with E / C ratios of 0.5.

The amount of occluded air measured in the case of these concretes or mortars varies between 1 and 2.8%. It is realized that 1% of air, occupying the same volume as ten liters of water, similarly reduces the compressive strength, it is evident that the air occluded in concrete must always be measured and taken Account (Baron & Ollivier, 1996).

One of the most important effects is certainly the influence of the voids provoked by the air occluded on the compressive strength of the mortar. It is known that the strength of the mortar is a direct function of its compactness and the occluded air voids affect the compressive strength in the same way as voids of any other origin.

3.4 Freezing and thawing

Figs. 5 show the evolution of the compressive strength of mortars based - AEA exposed to freeze-thaw cycles.

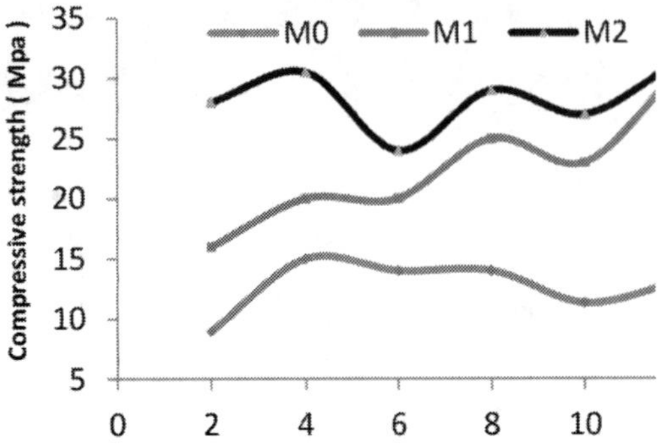

Fig. 5. Compressive strength of the mortar exposed to freeze-thaw cycles

Through the results of Fig. 5, it can be concluded that the mortars without an air-entraining agent have resistances which evolve slightly and which tend towards stable values, whereas for mortars with AEA there is a decrease in resistance which depends on the number of cycles. This reduction in strength becomes quite important when mortars are exposed to severe freeze-thaw cycles, which is consistent with the results of Nicola Pitrov .

3 Conclusions

It has been studied the effect of air entrained on the rheology of mortars. It was noticed that the air bubbles which generated by using either special type of cement or by using an air entrainment agent distributed in the entire mass of the mortar acting as flexible ball bearings. The conclusion of this study can be summarized as following.

➢ The air entrainment can reduce the ingredients of mortar without any effect in the workability of the mortar mix which mean more economical mortar mix can be achieved by using air entrainment agent whoever decreasing of concrete compressive strength by increasing of air entrainment should be considered.

➢ The reduction of density due to the increasing of air entrainment in the mortar mix result in a lower heat of hydration in mass mortar.

References

1. Y. Jiang, Z. Xu, *Engineering performance of polycarboxylate based air entraining agent*, J. of South.Univ (Natural Science Edition), vol. 36, no. 4, pp. 568–571, 2006.
2. L.J. Struble, Leslie J. Struble and Qingye Jiang. *Effects of Air Entrainment on Rheology ACI Mater* J /November-December 2004
3. F.Saucier et al., *Air-void Stability,* Part V, General Analysis and performance Index. ACI Mater J, Vol. 88, No.1, pp 25-36 (Jan.-Feb. 1991).
4. M.A., Carles-Gibergues, A., *La durabilité des bétons face à l'alcali-réaction, La durabilité des bétons* ed. by J. Baron, J-P. Ollivier, Presses Ponts et Chaussées.1992.
5. M.PIGEON, F. SAUCIER et P. PLANTE; « *Air-void stabiliy* ». Part Iv: Retempering, ACI Matér J, pp 252-259, (May/June 1990).
6. M.PIGEON, F.SAUCIER et P. PLANTE, *Production et stabilité du réseau de bulles d'air entraîné dans le béton*». Rapport soumis au Conseil de Recherche en Sciences Naturelles et en Génie et à la Compagnie VERREAULT-FRONTENAC, (Avril 1987).
7. P.Nikola ; *effets combinés de différents facteurs sur l'expansion des bétons causée par la formation différée de l'ettringite. thèse de doctorat* , Sherbrooke (Québec), pp 55-57, Canada (Avril 2003).
8. A.KOMAR 1978 .*Matériaux et éléments de construction* », Version française ; Editions Mir. Moscou, 540, (1978).
9. Du.Lianxiang , *Mechanisms of air entrainment in concrete,* Cem and Conc Research. Vol. 35, Issue 8, August 2005, Pages 1463–1471.
10. J. Baron, et Ollivier, *Les bétons bases et données pour leur formulation*. Edition Eyrolles, Paris.1996.

The effect of the addition of Alumina powder on the confinement properties of a cement mortar

L. Moutei[1]*, Y. Benbrahim[1], T. El ghailassi[2], A. Bouih[2], S. Labied[1], T. Guedira[1], O. Benali[3]
1Laboratoires des matériaux, Electrochimie et environnement, Faculté des sciences, Université Ibn Tofail, BP 133, 14000 Kenitra, Maroc.
2Unité de gestion des déchets radioactifs, Centre National De L'énergie, Des Sciences Et Techniques Nucléaires (CNESTEN) PB 1382, 10001 Kenitra, Maroc.
3Laboratoire des procédés de séparation, Faculté des Sciences, Université Ibn Tofail, BP 133, 14000 Kenitra, Maroc.

Abstract. Organic radioactive waste can occur in solid, liquid or rarely gaseous form. In general, organic components of radioactive waste Generally, the organic components of the radioactive waste have a contaminant effect stronger than most of the inorganic components, because of their low melting point, of their answer to the radiolyse or of their volatilitydu to their low melting point, to their answer to the radiolyse or to their volatility. This property can significantly affect the strategy adopted for their management. Organic radioactive waste results from the use of radioisotopes in the National Center for Energy, Nuclear Science and Technology (CNESTEN) and from medical applications. Organic radioactive waste requires management steps that take into account not only their radioactivity but also their chemical content. The aim of this study is the optimization of the stabilization of radioactive organic effluents in cementitious matrices. Thus, we study the effects of the addition of the alumina powder on the physicochemical and mechanical properties of the cement paste. The mortar was made by mixing Portland cement, sand, water and alumina powder (88 μm). The study showed that the addition of 12% alumina powder improves the compressive strength.

Mots-clés:cement Matrix, alumina powder, compressive strength.

1. INTRODUCTION

The radioactive material is used in scientific studies, medical and industrial applications and in agriculture. However, the use of natural and artificial radioactivities leads to waste, among which many contain important levels of radionucléide. The management approach of the waste radioactive is different from that of the other types of waste [1].The radioactive waste has very different physical natures requiring a special treatmentand a long-term storage. The radioactive organic effluents are mainly producted from the exploitation and the maintenance of research laboratories and from the installations of the nuclear Research center of Maâmoura(CENM) [2].The volume of the organic effluents generated is low Compared with the other radioactive waste; nevertheless, they require anadequate strategy of management for the treatment and the immobilization, which aims at reducing the potential of migration or dispersal of contaminants, including radionucléides. The International Atomic Energy Agency (IAEA) defines the immobilization of waste by the solidification like the conversion of a waste. It facilitates the handling, the transport, the storage and the elimination of the organic radioactive effluents [3-4].However, contaminated waste by organic molecules as oil, cannot be well treated. The addition of organic oil in a cement-based matrix affects negatively its mechanical properties and more exactly its compressive strength[5] .this paper presents a preliminary study to improve the physico-chemical and mechanical properties by realizing acementitious matrix of which the main properties arethe conversion of the liquid radioactive organic waste in a compatible organic solid shape with the cementation to facilitate the long-term conditioning. The mortar as the cementitious material has a low resistance and a limited durability; that's why the improvement of the mechanical properties is necessary to increase its performance as packaging material of the radioactive waste. To surmount these inconveniences, particles are added as fillers incementitious matrix to improve their physico-chemical and mechanics properties. Various studies showed that the use of diverse materials such as SiO2, Fe2O3, Fe3O4, fibers and clays [6-7] improve the compression resistance and in the flexion in the 7th day and in the 28th day the mortars, furthermore, the addition of these particles increases the durability and several other properties.
In this experimental study, the main objective is to study the effect of the addition of the alumina powder on the

physico-chemical and mechanical properties of the mortar.

2. MATÉRIALS AND METHODS

In this work, the influences of Alumina powder (Al_2O_3) on compressive strength of cement matrix have been investigated. The reason for using Al_2O_3 as a partial adds to the cement fraction. A Series of mixtures were prepared by mixing water, fine aggregates (sand) and powder materials (cement and Al_2O_3 particles) in a concrete laboratory.The formulation of cementitious matrix was based on a ratio water/cement (E/C) of 0,4 [8] and an addition of the powder of alumina which varies from 2 to 20 % of the mass of the cement.

2.1. Raw materials

2.1.1. Portland cement

The cement used in this study is a Portland cement CPJ 45 from the AsmentTemara Company which meets the Moroccan standards NM 100.1.004. The composition and the XRF analysis of the cement is indicated respectively in table 1 and table 2.

Composition	%
Clinker	65
Gypse	2,5
Calcaire	32,5
C2S	23,05
C3S	47,85
C3A	5,35
C4AF	9,10

Tab.1. The CPJ 45 Constituent.

SiO2	20,64	K2O	0,67
Al2O3	3,93	TiO2	0,27
Fe2O3	2,99	MnO	0,07
CaO	59,41	P2O5	0,22
MgO	1,40	Na2O	0,02
SO3	2,27	SrO	0,058

Tab.2. The XRF analysis of the CPJ 45 (%)

2.1.2. Sand

The sand used in the cement matrices complies with the NM 10.1.020 (EN 196-1) standard [9]. Characterized by its siliceous morphology, the sand is chemically inert towards the components of the mixture. Based on the previous results [10] which demonstrated, that the use of 0.2 mmsand improves the mechanical resistance of the cementitious matrix. The sand used for the preparation of all formulations has a particle size ranging from 0.08 to1.6 mm (Table 3).

Tamis (mm)	Refus Cumulés %	Tamisât Cumulés %
1,6	0,06	99,94
0,8	0,22	99,78
0,4	4,35	95,65
0,2	95,89	4,11
0,08	99,72	0,28
Fond	100	0

Tab.3. The particle size analysis of sand

2.1.3. Water

The used water in this study is tap water with a pH of 7.38 and an ionic conductivity of 672µs.

2.1.4. Alumina powder

The Al_2O_3particles constitute a fine, uniform, and smooth powder, without aggregates, with a micrometric size. The alumina used in this study was acquired from a Moroccan branch of industry.

matériau	Surface spécifique (m^2/g)	Densité (g/cm^3)	Purity
Al_2O_3	200	0.12	>99.8%

Tab.4. Alumina powder characterisation

2.2. Equipments

The used equipment to measure the compressive strength is a manual type hydraulic press Carver Model (4350.L) S/N (4350-362), it has a capacity that reaches 24000 Pounds.
To mix the components, the apparatus is an automatic mixer, type 3R, this device has a capacity of 5 L, and it ensures an uniform and homogeneous mixture.
The sieving of the sand and the red clay is realized with a column of sieve of 3R type.

2.3. X-Ray diffraction analysis

The Alumina powder was analyzed with a P Analytical diffractometer X PERT POWDER copper anticathode ($\lambda Cuk\alpha$ = 1.5406Å).

 https://doi.org/10.1051/mateconf/201814901055

2.4. X-Ray Fluorescence analysis

XRF is a non-destructive analytical method for materials. It is a quantitative and qualitative analysis of the elements which are present in the samples. An analysis of the Alumina powder, the cement and the cement matrices was carried out by the S1Turbo SD system.

2.5. The experimental protocol

The experimental protocol used in this study to prepare the cementitious matrices is as follows:

- The necessary amounts of cement, sand and Alumina powder are introduced into the bowl and the mixing is carried out dry for 1 min.
- To prevent overflow the required amount of water is added gradually.
- The components are mixed at slow rate for 1 min and at a fast rate for 2 min.
- The cement paste was introduced in cylindrical specimens of 10 × 5,6 cm and vibrated for a few seconds.

2.6. Formulation tests

The main purpose of this study is to manufacture mortars reinforced by alumina powder to study the effects of Alumina powder addition on the compressibility and physicochemical properties of the cement mortars. For that reason, different formulation tests were done (Table 5).

	Ciment %	Sable %	Eau %	Addition d'Alumine en % (*)
Témoin	57,14	20	22,85	0
Test 1	57,14	20	22,85	2
Test 2	57,14	20	22,85	4
Test 3	57,14	20	22,85	6
Test 4	57,14	20	22,85	8
Test 5	57,14	20	22,85	10
Test 6	57,14	20	22,85	12
Test 7	57,14	20	22,85	14
Test 8	57,14	20	22,85	16
Test 9	57,14	20	22,85	18
Test 10	57,14	20	22,85	20

Tab.5. Formulation tests

* The calculation of the percentage of Alumina is adjusted with regard to the mass of the cement.

3. RESULTS AND DISCUSSION

3.1. Particle size distribution of Al_2O_3 powder

The analysis of the Al_2O_3 powder particles is shown in figure 1. According to the results, the majority of the particles constituting the Alumina powder have a size of 88 µm, which satisfy the standards of the use of fillers in the cementitious matrices NF P 18-501 [11].
4.

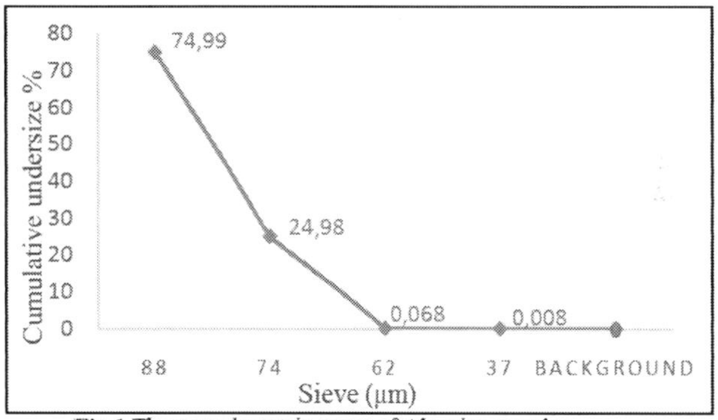

Fig.1. The granulometric curve of Alumina powder

3.2 Quantitative chemical analysis of Al_2O_3

To determine the chemical composition of the Alumina powder, a XRF analysis was performed (Table 6).

Oxydes	Alumine %
Al_2O_3	81,82 ± 1
SiO_2	0,38 ± 0,17
Fe_2O_3	0,070 ± 0,004
CaO	0,18 ± 0,03
K_2O	0,20 ± 0,02
TiO_2	ND
P_2O_5	0,87 ± 0,04

Tab.6.X-Ray Fluorescence of Al_2O_3

3.3 X-ray diffraction of Al_2O_3

The pattern diffraction (Figure 2) shows that corundum (crystalline form of Aluminum oxide Al_2O_3) is the major mineral present in Alumina powder sample.

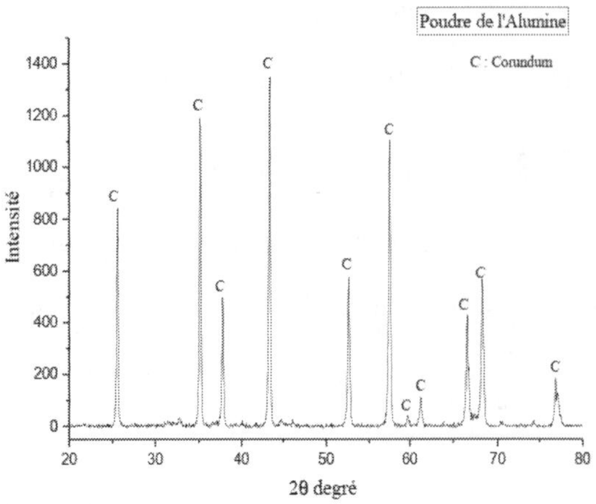

Fig.2 .The diffraction pattern of Alumina powder

3.4 Compressive strength

The Compressive strength of mortars incorporating different content of Alumina powder at different curing periods (7, 14, 21 and 28 days) is illustrated in Figure 3. The results of the compressibility are given in MPa (Mega Pascal).

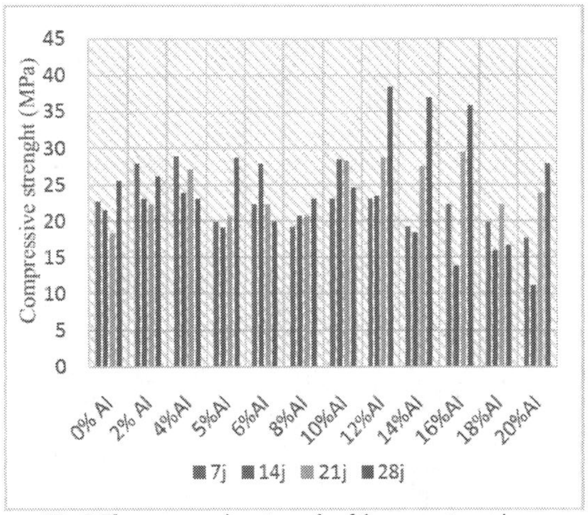

Fig.3. The compressive strength of the cement matrices

According to figure 3, the compressive strength of mortar containing alumina particles was clearly improved. The 28-day compressive strength of matrix was enhanced as much as 51% by adding 12 wt% of cement with alumina powder. The compressive strength increased from 25.48 MPa to 38,43MPa when the content of Al_2O_3 increased from 0wt% to 12wt%. The enhancement of compressive strength of cement matrix containing alumina micro-size particles was probably due to the rapid consuming of calcium hydroxide $Ca(OH)_2$ formed during hydration of Portland cement [12], related to the high reactivity of Al_2O_3 particles [13-14].

$$Ca(OH)_2 + Al_2O_3 \rightarrow CAH$$

Also the Al2O3particles increase the density of the cementitious matrix, and they act as filler which will strengthen the microstructure of the mortar and will reduce the volume of pores in the cement paste [15-16]..
The mechanical properties and frost resistance of matrix containing Al_2O_3 particles in compare with normal witness matrix were improved.
The mechanism allowing to improve the physico-chemical and mechanical properties of cementitious matrix containing the powder of alumina can be interpreted as follows; If the powder of alumina is uniformly dispersed, at the beginning of the hydration, the crystallization will be controlled by limiting the growth of the crystal That $Ca(OH)_2$..
Moreover, the alumina behaves not only as filler, but also as activators which favorites the reaction pozzolanique, and is going to react with the hydroxide of calcium formed from the hydration of the silicate of calcium [17-19]. The rate of pozzolanique reaction is proportional to the available surface quantity for the reaction. consequently, the medium-sized pores in the matrix of cement decrease with the use of particles Al_2O_3 (Picture 1) [20].

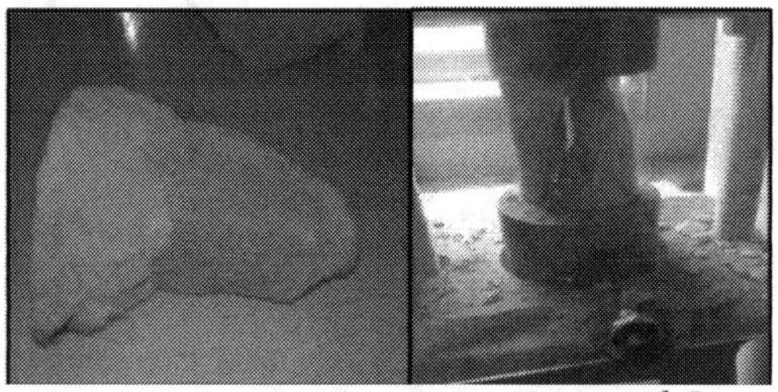

Img.1.Pic.1.cementitious matrix with addedAl_2O_3

This makes the cementitious matrix with an add of a percentage of Alumina powder that ranges from 12% to 20%, as an optimum percentage for addition, more homogeneous and compact, that will give the highest compressive strength compared to the cement matrix without Al_2O_3.

3.5 X-ray fluorescence analysis of cementitious matrix

The x-ray fluorescence analyzes were also performed on the cementitious matrix (Witness, 12% Alumina, 14% Alumina) and the results are indicated in table 7

Oxydes	Mortier témoin	Mortier 12% Al	Mortier 14% Al
Al_2O_3	6,63	16,30	18,45
SiO_2	14,30	12,03	11,25
Fe_2O_3	2,93	2,64	2,58
CaO	57,55	52,70	50,57

K_2O	0,76	0,64	0,58
TiO_2	0,14	0,13	0,14
P_2O_5	2,21	2,04	1,95
MnO	0,07	0,07	0,07

Tab.7.XRF analysis of cementitious matrix

Compared to the witness mortar, the XRF analysis showed an increasing of Al_2O_3 percentages by the addition of 12%Al and 14%Al into the cementitious matrix, next to a decreasing in the CaO content, which will be explained by the pozzolanic reaction that consume more CaO in the presence of Al_2O_3 powder to produce the CSH and CAH phases.

3.6 Analyse par DRX des matrices cimentaires

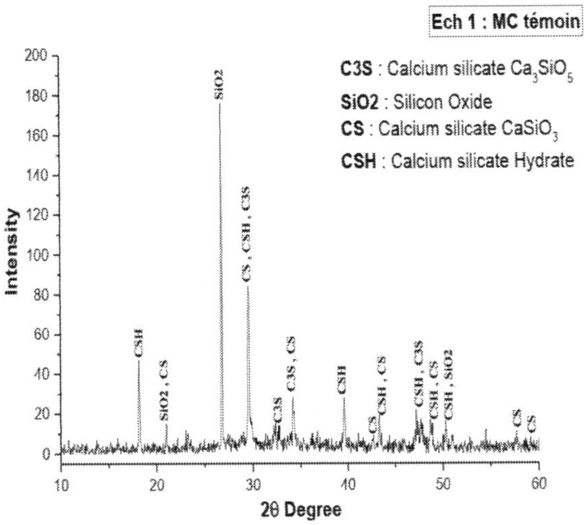

Fig.3. Diffraction pattern of diffraction of matrix without Al.

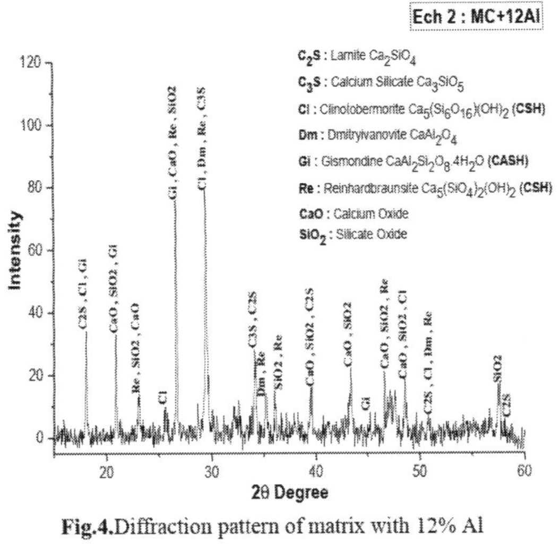

Fig.4. Diffraction pattern of matrix with 12% Al

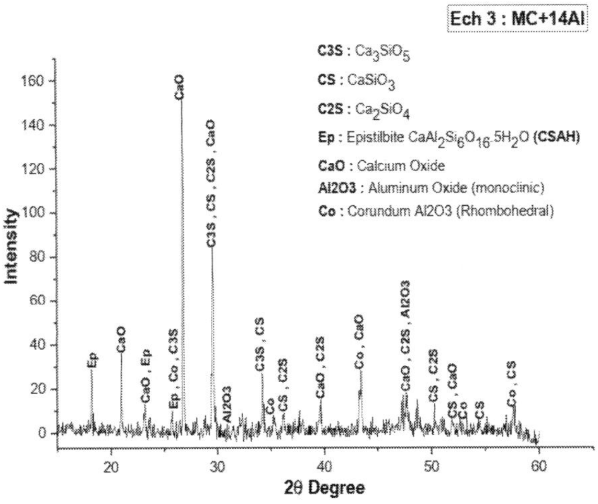

Fig.5. Diffraction patternpattern with 14% Al

The diffractogram of the matrix without Al(fig.3) shows the presence of the phases of silicate of tricalcium, of silicate of calcium hydrated (C-S-H). These crystalline structures constitute the main product of hydration of the Portlandcement and are mainly responsible for the resistance in the cement-based materials.
Matrix containing 12 % and 14 % of alumina are respectively represented on figures 4 and 5, these show that that of products of hydration such as the frosts of aluminate some calcium CAH, result of the reaction pozzolanique between Al2O3 and That (OH) 2, appear. This could strengthen the microstructure of the mortar and reduce the volume of pores in the dough of cement. The introduction of the powder of Alumina in our dough of cement acts on its microstructural properties and consequently strengthens the matrix cimentaire by increasing its compression resistance with regard to the Control mortar[21].

4 CONCLUSION

The result of this study showed that the addition of alumina powder in the cement matrices increases the compressive strength at the age of 28 days compared to the control mortar. An improvement can also be noted by adding 12% to 20% of the alumina to the ordinary mortar matrix. The addition of 12% alumina powder increases the compressive strength of the mortar by 51% compared to the control mortar (without alumina).
This result is explained by the filler effect of the Al_2O_3 powder, the behavior of alumina as an activator which promotes the pozzolanic reaction.

Acknowledgments: The authors would like to express their gratitude to the Radioactive Waste Management Unit (UGDR) of the National Center for Energy, Nuclear

Science and Technology (CNESTEN) of Morocco and the Faculty of Science of Ibn University Tofail, in Kenitra, Morocco, to provide the necessary support for this work, also Pr DIOURI from the Faculty of Sciences of Rabat for his assistance in carrying out the DRX analyzes.

Références

1. Michael I Ojovan, William E. Lee, An Introduction to Nuclear Waste Immobilisation 2, Newnes, 376, (2013).
2. El Ghailassi T., Système de gestion des déchets radioactifs entreposage des déchets radioactifs et combustible Usé Triga Mark II. Rapport de titularisation, 8-9, (2008).
3. IAEA, Predisposal Management of Radioactive WasteSeries No. GSR Part 5, published Tuesday, 19 May, (2009).
4. International atomic energy agency Vienna IAEA-tecdoc-1701, the behaviors of cementitious materials in long term storage and disposal of radioactive waste, 3-20, (2013).
5. D.M. Montgomery.,the Optimization of cement-based stabilization/solidification of organic-containing industrial wastes using organophilic clays,Volume9,Issue 1, 21-34, (February 1991).
6. Behfarnia K., Salemi N., The effects of nano-silica and nano-alumina on frost resistance of normal concrete, Construction and Building Materials 48,580-584, (2013).
7. Sikora P., Horoszczaruk E., Cendrowski K., Mijowska E, The Influence of nano-Fe3O4 on the microstructure and mechanical properties of cementitious composites. NanoscaleResearch Letters11, 182, (2016).
8. El Ghailassi T., AouladBelayachi,A. Bouih, S. Labied, T. Guedira, O. Benali. Mathematical approach for research of new formulation for immobilization of radioactive waste in cementitious matrices, JMES Accepted, (2017).
9. Office national de l'eau potable, Tome 4 Génie Civil 2009-11-16, Morocco, 2-3, (2009).
10. Faiz Z., Fakhi S., Bouih A., Idrissi A., Mouldouira M., J. Mater. Environ. Sci. 3 (6), 1129-1136, (2012).
11. Pineau F., Corcos J., Pitiot A, Spécification technique pour les colis de déchets conditionnés dans une enveloppe durable confinante, à base de liants hydrauliques, destinés au stockage sur un site de surface. ANDRA-ACO-SP-ASQC--95-064/A,1-30, (1995).
12. Herzog A., Mitchell J.K, Reactions accompanying the stabilization of clay with cement. Highw. Res. Board Rec. 36, 146–171, (1963).
13. KiachehrBehfarnia, NiloofarSalemi. , The effects of nano-silica and nano-alumina on frost resistance of normal concrete, Construction and Building Materials 48, 580–584, (2013).
14. Prusinski R., Bhattacharja S, Effectiveness of Portland cement and Lime in Stabilizing Clay Soils.,TRANSPORTATION RESEARCH RECORD 1652, 215-227,(1999).
15. Li H, Xiao H, Yuan J, Ou J., Microstructure of cementmortarwith nano-particles.,
Composites Part B, 35(2), 185–9, (2003).
16. Sneff L, Labrincha JA, Ferreira VM, Hotza D, Repette WL., Effect of nano-silica on rheology and fresh properties of cement pastes and mortars., Constr Build Mater,23(7),2487–91,(2009).
17. A. Nazari, S. Riahi, Abrasion resistance of concrete containing SiO2 and Al2O3 nanoparticles in different curing media, Procedia Eng. 43, 2939–2946, (2011).
18. Li H, Zhang M, Ou J. Abrasion resistance of concrete containing nano-particles for pavement.,260(11–12):1262–6,(2006).
19. Ali N, Shadi R, Shirin R, Fatemeh S, Khademno A. Influence of Al2O3 nanoparticles on the compressive strength and workability of blended concrete. J Am Sci; 6(5), 6–9, (2010).
20. M. R. Arefi, M. R. Javeri , E. Mollaahmadi., To study the effect of adding Al2O3 nanoparticles on the mechanical properties and microstructure of cement mortar, Life Science Journal,8(4), (2011).
21. S0 Mukherjee, S. Barbhuiya, H.Nikraz. Effects of nano-Al2O3 on early-age microstructural properties of cement paste, research gat, (February 2014).

Study of the effect of Kaolin in the mortar of cement matrices by confinement of ion exchange resins

S. Labied[1]*, T. El ghailassi[2], A. Bouih[2], L. Moutei[1], Y. Benbrahim[1], T. Guedira[1], O. Benali[3]

[1]Laboratory of Materials Electrochemistry and Environment, Faculty of Science, University Ibn Tofail, BP 133, 14000 Kenitra, Morocco
[2]Unit of Management of radioactive waste, National Center of Nuclear Energy, Sciences and Nuclear Techniques (CNESTN), Center for Nuclear Studies of Maamoura (CENM).PB 1382, 10001 Kenitra, Morocco
[3]Laboratory of Separation Processes, Faculty of Science, University Ibn Tofail, BP 133, 14000 Kenitra, Morocco

**Corresponding author: E-mail: labiedsoumiya@gmail.com; Phone :(212) 6 01 70 65 99*

Abstract. Radioactive waste arising as a result of nuclear activities should be safely managed from its generation to final disposal in an appropriate conditioned form to reduce the risk of radiation exposure of technical personnel and of the public and to limit contamination of the environment. The immobilization of low and intermediate level radioactive wastes in cementitious matrices is the most commonly used technique to produce inexpensive waste matrix that complies with regulatory requirements in order to protect humans and the environment against nuisance caused by ionizing radiation. Cement based materials are used in radioactive waste management to produce stable waste forms. This matrix constitutes the first build engineering barrier in disposal facilities. In this work, the kaolin is used to enhance the mechanical performance of the matrix of confinement of ion exchange resins by gradually replacing the sand in mortar with kaolin clay. The Kaolin clay sample was a special pure product, sourced from a foreign country. The maximum quantity of resins that can be incorporated into the mortar formulation without the packages losing their strength is 13.915% which results in a better mechanical strength at 6.7686 MPA compression with kaolin.

1 INTRODUCTION

The use of nuclear techniques in various fields such as scientific research, industry and health generates radioactive waste. This waste is composed of different types. Like any developing country, Morocco is responsible for the management of radioactive waste through the National Center for Energy, Science and Nuclear Techniques (CNESTEN). CNESTEN has set up a Radioactive Waste Management Unit (RWMU). This management must be carried out within a rigorous framework in order to guarantee safe solutions for all radioactive waste produced without losing sight of the

permanent requirement for the protection of present and future generations and the risk environment of this waste [1-2].
The low- and intermediate-level radioactive waste conditioning process at the UGDR is immobilization in a cement-based matrix, which is the first barrier of this waste, is the most common technique used to produce radioactive waste packages that comply with regulatory requirements [3-4-5].
As a result of extensive research activities focused on the development of materials, the additives are currently among the most recent developments in cement production, as their uses improve the mechanical properties of cementitious materials (mortar and concrete). Indeed, the incorporation of kaolinite into the mortar obviously produces materials similar to ordinary cement mortar but with best characteristics [6-8].
The aim of this study is to optimize the formulation of a cement matrix by gradually substituting the proportion of sand by kaolinite as a filler. We then follow the impact of this substitution on the compressibility of the test pieces while keeping the E/C ratio constant.

2 MATERIALS AND METHODS

2.1 Materials

2.1.1 Cement

The portland cement used CPJ 35 is a cement whose technical characteristics are in conformity with the Moroccan standard NM 10.1.004.

2.1.2 Sand

The sand used in a laboratory complies with the Moroccan standard NM 10.1.020. We also use this sand for industrial applications for the confinement of radioactive waste.

2.1.3 Water

We used drinking tap water to make mortars

2.1.4 Ion exchange resins

The ion exchange resins [9,10] MDP-15 type, in of clear spherical beads form used in the purification of the reactor water circuits as well as the spent fuel storage pools of the TRIGA MARK II reactor, PUROLITE NRW 37 type, these cations exchange resins are strongly acid, and gel type.
The ion exchange is a process by which the ions contained in a solution are removed to be replaced by an equivalent number of other ions of the same electrical charge. The physical-chemical properties of the resin:
- Hydrocarbon skeleton: Polystyrenic crossed to the gel type DVB
- Functional grouping: R-SO3-
- Physical form: Dark amber balls, translucent
- Ion Shape at delivery: H-
- Moisture content: 51-55% (H + form)
- Maximum swelling: Na + -H +: 5%
- Temperature limit: 120 ° C
- Limit of PH: From 0 to 14
- Apparent density: Approximately 800 g / l
- Actual Density: 1.20 (H + Form)
- Total exchange capacity: Min 1.7 eq / l (H + form)

2.2 Experimental section

The different pastes are prepared in a standardized mixer EN-196-1 [11] following the procedure indicated in standard EN-196-3 [12] relating to the normal consistency of pure pasta.

2.2.1. Study of the influence of the substitution of sand by kaolinite on the mortar

For our experimental approach, four types of mortars were prepared with substitution percentages ranging from 0 to 6% of the kaolinite. The tests were carried out using 10 × 5.5 cm cylindrical specimens, the walls of which were previously washed with oil (Sika type Iron M) which is a demolding agent. the demolding takes place after the following periods: 7 days, 14 days, 21 days and 28 days. The various compositions of the mortars are grouped in Table 1.

Table 1 : Formulation tests

	%/weight (g)	Cement	Sand	Kaolinite	Water
witness	%	57.14	20	0	22.85
	weight (g)	1828.57	640	0	731.43
Test 1	%	57.14	18	2	22.85
	weight (g)	1828.57	576	64	731.43
Test 2	%	57.14	16	4	22.85
	weight (g)	1828.57	512	128	731.43
Test 3	%	57.14	14	6	22.85
	weight (g)	1828.57	448	192	731.43

3RESULTS AND DISCUSSION

3.1 Result of particle size analysis of kaolinite

Figure 1 presents the granulometric analysis of kaolinite which is carried out by series of sieves (88 μm, 74 μm, 62 μm, 37 μm).

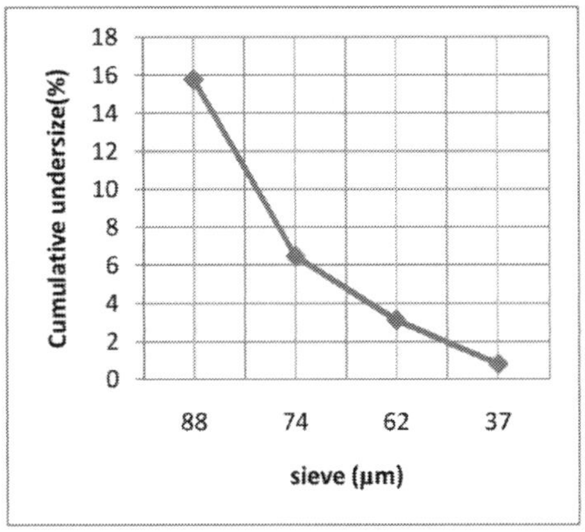

Fig. 1. Grading curve of the kaolinite

3.2 Quantitative chemical analysis of the used materials.

The chemical composition of the samples was determined by The Portable Fluorescence X System with Dispersion of Energy (Bruker S1 Turbo SD). The results obtained are expressed in mass percent.

Table 2. X-Fluorescence Analysis Results

	Chemical components (%)						
	Al_2O_3	SiO_2	P_2O_5	K_2O	CaO	TiO_2	Fe_2O_3
Kaolinite	40.97 ±0.65	55.59 ±0.50	0.78 ±0.04	1.34 ±0.02	0.05 ±0.01	0.22 ±0.01	1.03 ±0.01

3.3 XRD analysis of used materials

Figure 2 shows X Ray diffractogram of kaolinite.

Fig. 2. X-Ray Diffractogram of Kaolinite

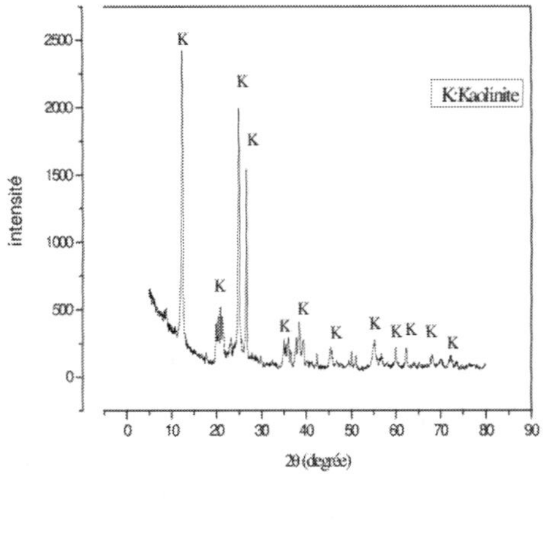

3.4 Formulation tests

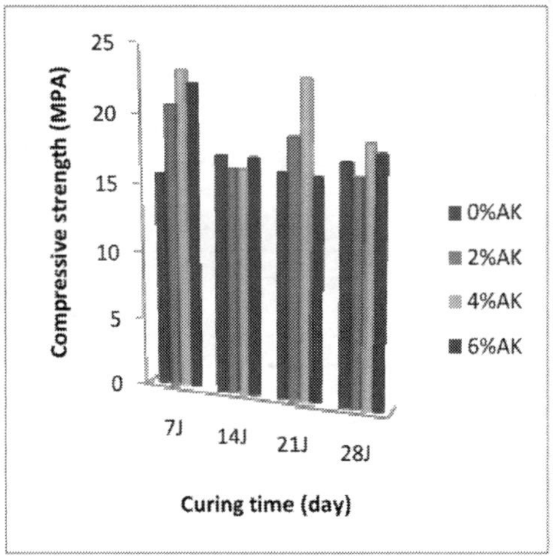

Figure 3. Evolution of the mechanical strength as a function of the curing time of the matrices prepared with different amounts of kaolinite.

The evolution of the resistances as a function of time shows that for seven days as curing time, the resistance corresponding to all the percentages of kaolinite inserted exceeds that of the control mortar.

The evolution of the resistances as a function of time shows that after 28 days the resistance corresponding to 2% kaolinite does not exceed that of the control mortar, while for those of 4% and 6% the resistances increase substantially. Due to the kinetics of the hydration reaction of the constituents of the mixed cement which becomes more and more active.

The evolution of the resistances as a function of time shows that for a cure time of 28 days, the best resistance corresponds to the incorporation of 4% of kaolinite in the mortar.

This improvement can be explained by the double role of kaolinite [13-14-15]:

It permits to fill the tiny voids in the mortar due to its very high fineness.

And there is the pozzolanic effect, the silica It will combine with the portlandite resulting from the hydration of the reactive phases of the clinker to form CSH [16-17].

3.5 Confinement of resins formulation

After finding that optimal formulation corresponds to substitution of 4% of the kaolinite instead the sand, we incorporate a radioactive waste in the matrix in percentages ranging from 3.479% to 13.915% in order to determine the maximum rate of this waste that we can confined in the matrix (Table 3).

Table 3.confinement of resins formulation

	%/weight (g)	Cement	Sand	kaolinite	Water	Wet resin
Test 7	%	57.14	16	4	22.85	0
	weight(g)	1828.57	384	128	731.43	0
Test 8	%	56	15.6	4	20.921	3.479
	weight(g)	1792	499.2	128	669.472	111.328
Test 9	%	54.86	15.2	4	18.982	6.958
	weight(g)	1755.52	486.4	128	607.424	222.656
Test 10	%	53.71	14.8	4	17.044	10.436
	weight (g)	1718.72	473.6	128	545.408	333.952
Test 11	%	52.57	14.4	4	15.153	13.915
	weight (g)	1682.24	460.8	128	484.896	445.28

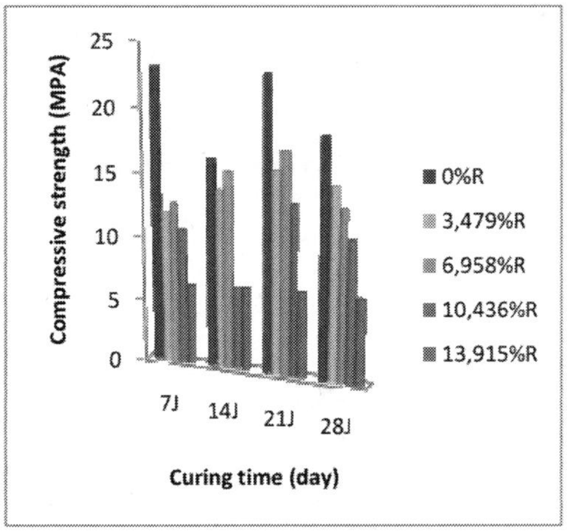

Figure 4 variations of compressive strength for different percentages of resins (R) incorporated in a mortar, according to the age of the test pieces

We observe a decrease in the compressive strength with the increase of the quantity of the ion exchange resins incorporated in the mortar. The maximum quantity of resins which can be incorporated in this formulation without the packages losing their resistances is 13.915 %. Indeed, the compressive strength for this formulation is 6.7686 MPA and seems the most suitable for the conditioning of the resins.

4 Conclusions

The compressive strengths increase with the substitution of sand by kaolinite, thus reflecting the improvement of compactness by three effects which act simultaneously and in a complementary manner: physical, physical-chemical and chemical (pozzolanic) effect.
the main result is that the maximum amount of ion exchange resin that can be incorporated into the cement matrix is 13.95%. This precntage corresponds to a compressive strength of 6.7686 MPa. The package keeps a good resistance.

REFERENCES

1. Collection normes de sureté de l'AIEA, Gestion des déchets radioactifs avant stockage définitifs y compris le déclassement AIEA n° WS-R-2 VIENNE 2004.
2. T. El Ghailassi « Système de gestion des déchets radioactifs entreposage des déchets radioactifs et combustible Usé Triga Mark II » Rapport de titularisation (2008).
3. Céline PERLOT , « Influence de la décalcification de matériaux cimentaires sur les propriétés de transfert : Application au stockage profond de déchets radioactifs» Thèse de doctorat, Université Paul Sabatier de Toulouse U.F.R. P.C.A.(23 Septembre 2005).
4. Junfeng Li, Jianlong Wang, J Hazard Mater, B135 (2006) 443–448.
5. A.Erdal Osmanlioglu Waste Management, Volume 22, Issue 5, August 2002, Pages 481-483.
6.Zheng Guo, Qiming Feng, Weiqing Wang, Yang Huang, Jie Deng, Zhonghui Xu, Procedia Environmental Sciences 31 (2016) 644 – 652.
7.El Ghailassi T., Aoulad Belayachi, A. Bouih, S. Labied, T. Guedira, O. Benali. Mathematical approach for research of new formulation for immobilization of radioactive waste in cementitious matrices, J. Mater. Environ. Sci. (JMES), 8 (11), 3864-3872 (2017)
8. Yingfang Fan, Shiyi Zhang, Qi Wang, Surendra P. Shah, Construction and Building Materials 102 (2016) 486–495.
9. Faiz Z., Fakhi S., Bouih A., Idrissi A., Mouldouira M., J. Mater. Environ. Sci. 3 (6) (2012) 1129-1136.
10. Elhilal B., Lambarki El Allioui T., Bouih A., El Harfi A.,"Characterization of ion exchange resinsused in the purification of nuclearreactor water circuits TRIGA MARK II" article soumis au Journal of Materials and Environmental Science (JMES).
11. International atomicenergyagency, Vienna, Technical Reports Series No. 408: Applications of ions exchange processes for de treatment of radioactive waste and management of spent ion exchanges (2002).
12. «Méthodes d'essais des ciments, partie 3: Détermination du temps de prise et de la stabilité», Août (1995).
13.Geetika Mishra, Anuj Parashar, Shashank Bishnoi, ICSBE2016-100.
14. Abraham Teklay, Chungen Yin, Lasse Rosendahl, Department of Energy Technology, Aalborg University, 9220 Aalborg East, Denmark, Applied Energy (2015)
15. A. Elimbi, H.K. Tchakoute, D. Njopwouo, Construction and Building Materials 25 (2011) 2805–2812.
16. Chahidi ElouazzaniD.,Caractérisation physico-chimique et valorisation en bâtiment et travaux publics des cendres issues de l'incinération des boues de papeterie, thèse de Doctorat , Institut national des sciences appliquées de Lyon, 2005, P 18-21.
17. Z.Faiz ; S. Fakhi ; A. Bouih ; A. Idrissi ; M. Mouldouira J. Mater. Environ. Sci. 3 (6) (2012) 1129-1136.

Rheological study of self-compacting mortars based on ternary cements

Etude rhéologique des mortiers autoplaçants à base des ciments ternaires

Dada Elhadja[1], Belaidi Akram Salah Eddine[1], Soualhi Hamza[2] and Kadri El-Hadj[2]

[1]Laboratoire de recherche en génie civil (LRGC), Université de Laghouat, Algérie
[2]Laboratoire de mécanique et matériaux du génie civil, Université de Cergy-Pontoise, France.

Abstract. Self-compacting concrete (SCC) is able to provide the ability to be easily implemented without vibration and to achieve spectacular structures, by its high fluidity and its rheological stability. By against its formulation requires a large volume of cement, which is necessary to allow its flow. The current environmental considerations lead to reduce the production of clinker however, it is essential to use cementitious additions to replace cement, because of their high availability and their moderate price. Furthermore, their use contributes in a simple and economical way to solve the problems related to the environment. The objective of our work is to study the effects of the incorporation of mineral additions such as: blast furnace slag of El-Hadjar (BFS), and marble powder (MP) on the rheological parameters of self-compacting mortars developed in different combinations in ternary system with a substitution rate ranging from 20% to 60%. According to this study, it been found that the substitution of cement by blast furnace slag and marble powder has negatively affected the rheological behavior of the mixtures. In addition, a considerable decrease in the rheological parameters has been achieved with a substitution rate of 20% of slag and 30% of marble powder. As well as an improvement of workability has been proven to self-compacting mortars and this is due to the increase of ternary cement replacement rate by marble powder from 20% to 30%.

Keywords : blast furnace slag ; marble powder; ternary mortars ; viscosity ; yield stress

Résumé. Le béton autoplaçant (BAP) est capable d'assurer les aptitudes à être aisément mis en œuvre sans vibration et de réaliser des structures spectaculaires, par sa fluidité élevée et sa stabilité rhéologique. Par contre sa formulation sollicite un grand volume de ciment qui est nécessaire pour permettre son écoulement. Les considérations environnementales actuelles incitent à réduire la production du clinker, en revanche il est indispensable d'utiliser des ajouts cimentaires en remplacement au ciment, du fait de leur grande disponibilité et de leur prix modéré. D'autre part, leur utilisation contribue de manière simple et économique à résoudre les problèmes liés à l'environnement. L'objectif de notre travail consiste en l'étude des effets de l'incorporation des ajouts minéraux tels que : le laitier de haut fourneau d'El- Hadjar (LHF), et la poudre de marbre (PM) sur les paramètres rhéologiques des mortiers autoplaçants élaborés selon différentes combinaisons en système ternaire avec un taux de substitution variant entre 20% à 60%. D'après cette étude, on a constaté que la substitution du ciment par le laitier de haut fourneau et la poudre de marbre a affecté négativement le comportement rhéologique des mélanges. Par ailleurs, une diminution considérable des paramètres rhéologiques a été menue avec un taux de substitution de 20% de laitier et 30% de poudre de marbre. Ainsi qu'une amélioration d'ouvrabilité a été prouvée pour les mortiers autoplaçants ternaire et cela est dû à l'augmentation par la poudre de marbre de 20% à 30%.

Mots-Clés : laitier de haut fourneau ; poudre de marbre ; mortier ternaire ; viscosité ; seuil de cisaillement.

1 Introduction

La rhéologie est définie comme étant la science qui traite l'écoulement, les déformations, et plus généralement la viscosité des matériaux sous l'action des contraintes. La rhéologie est capable d'intégrer l'étude de l'ensemble des substances, qu'elles soient liquides ou solides. L'intérêt de la rhéologie grandit chaque jour depuis l'apparition des nouveaux matériaux aux comportements complexes tels que : pâtes, addition, polymères, etc. Parmi ces matériaux les bétons autoplaçants, qui sont des bétons très fluides, qui se mettent en place et se serrent sous le seul effet de la gravité, Cependant, son ouvrabilité dépend fortement de son comportement rhéologique de la phase de mortier dont les propriétés sont dominantes. A cet effet, l'évaluation des propriétés rhéologiques du mortier autoplaçant est une partie intégrale de la formation des BAP, et donc la connaissance des propriétés du mortier est utile [1].

Pour un souci à la fois écologique et économique, il est plus qu'important de formuler des mortiers autoplaçants (MAP) en se basant principalement sur des matériaux locaux et disponibles (naturels ou issus des déchets industriels) tels que : le laitier des hauts fourneaux et la poudre de marbre. Aujourd'hui, le développement des nouveaux matériaux intensifie le besoin de caractériser encore mieux son comportement et son écoulement. [3]. Les tests empiriques conventionnels ne suffisent plus à bien caractériser le comportement lors de l'écoulement et de la mise en place. Le développement de nouveaux appareils basés sur une approche plus fondamentale et rhéologique donne des résultats plus adéquats et reflète bien l'aptitude physique du matériau à se déplacer sous son propre poids. Actuellement le comportement rhéologique des bétons frai et mortier est considéré comme similaire à celui de fluide de Bingham qui est donné par :

$$\tau = \tau_0 + \mu\dot{\gamma} \qquad (1)$$

τ_0 (Pa) est le seuil de cisaillement de Bingham, décrivant la contrainte nécessaire pour commencer l'écoulement, μ (Pas) est la viscosité plastique de Bingham, qui est la résistance de la matière à s'écouler, et τ (Pa) et $\dot{\gamma}$ (1/s) est la contrainte de cisaillement et le taux de cisaillement, respectivement.

Dans le cadre de ce travail de recherche, les changements rhéologiques des mortiers autoplaçants impliqués dans les cas des mélanges composés de ciment plus des ajouts, à savoir le laitier de haut fourneau, et la poudre de marbre en système ternaire a été évaluée par des essais d'ouvrabilité à l'aide de mini-cône et mini V-funnel et par des mesures des paramètres rhéologiques à l'aide d'un rhéomètre

2 Plan expérimental

Le ciment utilisé dans cette étude est de type CEMI 42,5. Les additions minérales utilisées sont le laitier de haut fourneaux (LHF) et la poudre de marbre (PM). Leurs propriétés chimique et physique sont présentées dans le Table 1. Afin de déterminer l'état de surface des additions minérales, une microscopie électronique à balayage (MEB) a été utilisée. La Fig. 1 (a), (b) montre la morphologie du laitier et de la poudre de marbre. Comme on le voit sur ces figures, le laitier présente une forme anguleuse avec une texture de surface rugueuse. Par contre la poudre de marbre présente une uniformité granulaire. Le type de superplastifiant utilisé à base de polycarboxylate de nouvelle génération (SP) avec un poids spécifique de 1,07 g/cm^3 et un extrait sec de 30%. Pour le sable on a utilisé dans ce travail un sable alluvionnaire avec un poids spécifique et de coefficient d'absorption de l'eau 2,7 g/cm^3 et 0,9% respectivement.

Table. 1. Caractéristiques physiques et chimiques du ciment, du laitier et de la poudre de marbre.

Analyse chimique (%)	Ciment (OPC)	Laitier (LHF)	Poudre de marbre
SiO_2	21,6	36,60	0,42
CaO	63,7	37,75	56,01
MgO	1,6	3,77	0,12
Al_2O_3	4,00	6,38	0,13
Fe_2O_3	4,5	6,94	0,06
SO_3	1,9	0,19	0,01
MnO	-	-	-
K_2O	-	-	0,01
T_iO_2	-	-	0,01
Na_2O	0,03	-	0,43
Cl	0,018	-	-
Perte en	0,3	8,37	42,8
Densité	3,15	2,56	2,61
Finesse	3150	4000	3600

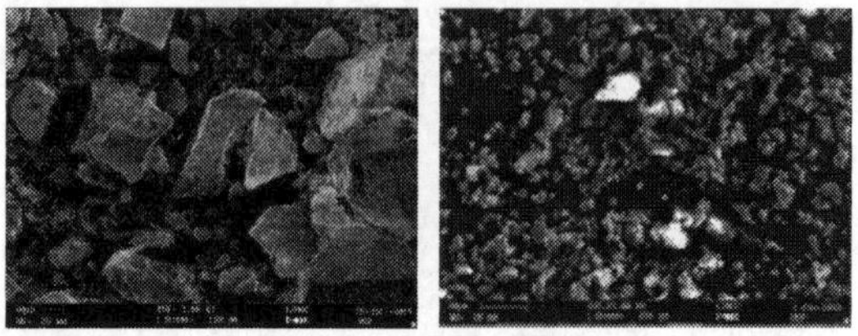

(a) Laitier de haut fourneau (b) Poudre de marbre

Fig. 1. Analyse par microscope électronique à balayage des additions

3 Proportions du mélange

Dans le BAP, la phase du mortier fournit la lubrification entre les particules des gros granulats et la stabilité globale du béton. Les propriétés du mortier sont semblables à celles du BAP lui-même, c'est-à-dire un faible seuil de cisaillement pour assurer l'écoulement sous le seul effet de la gravité et une viscosité plastique suffisante afin d'assurer la non ségrégation du béton durant l'écoulement. Toutefois, l'évaluation des propriétés des mortiers autoplaçant est une partie essentielle de la conception du BAP. Afin d'obtenir les propriétés recherchées à l'état frais du BAP, une proportion élevée de matière ultrafine et l'introduction des ajouts minéraux et chimiques sont nécessaires. Les proportions du mélange sont basées sur la méthode d'Okamura et Ozawa [4], avec des ajustements apportés à la méthode sur la quantité des granulats fins.

Les rapports en volume de sable sur mortier (V_s/V_m), l'eau sur poudre (V_w/V_p) et le superplastifiant sur poudre (SP/P) ont été déterminé par un simple test pour l'évaluation de la contrainte transférabilité du mortier frais [9]. Des essais sur mortier ont été réalisés pour le rapport Eau/poudre (E/P) constant de 0,4 en poids et le rapport sable / mortier (S/M) de 0,5 en volume. Les proportions de mélange de mortier de référence sont présentées dans le Table 2. LHF et PM ont été ajoutés comme une substitution de ciment en masse avec des pourcentages variant de 20% à 60% en système ternaire. Le Table 2 résume les formulations des mortiers ternaires.

La caractérisation rhéologique des mortiers a été effectuée avec un rhéomètre coaxial à palette. Ce rhéomètre a récemment été validé pour les mortiers et les bétons [14]. Les résultats bruts de l'essai sont les couples moment de cisaillement et les vitesses de rotation (M_i, Ωi). Pour exploiter les résultats en forme d'un rhéogramme de comportement rhéologique il faut passer par les équations (2) à (5) permettant de passer du couple

(M_i, Ω_i) au couple contrainte de cisaillement et taux de cisaillement (τ_i, γ_i) et de déterminer les paramètres rhéologiques, la viscosité et le seuil de cisaillement (μ, τ_0) du mortier testé.

$$\dot{\gamma} = 2M\frac{\partial\Omega}{\partial M} \tag{2}$$

$$\frac{\partial\Omega}{\partial M} \cong \frac{\Omega_{j+1} - \Omega_{j-1}}{M_{j+1} - M_{j-1}} \tag{3}$$

$$\tau_j(M) \cong \frac{M_j}{2\pi h R_b^2} \tag{4}$$

$$\tau_i = \frac{1}{2}\left(\tau_{j-1} + \tau_{j+1}\right) \tag{5}$$

Table. 2. Formulation des mortiers à base des ciments ternaires

Mélange		MAP 0	MAP 1	MAP 2	MAP 3	MAP 4	MAP 5	MAP 6	MAP 7	MAP 8	MAP 9	Sable	SP
poudre	Ciments (%)	**100%**	**80%**	**70%**	**60%**	**70%**	**60%**	**50%**	**60%**	**50%**	**40%**		
	Laitier HF (%)	**0%**	**10%**	**10%**	**10%**	**20%**	**20%**	**20%**	**30%**	**30%**	**30%**		
	Poudre de marbre (%)	**0%**	**10%**	**20%**	**30%**	**10%**	**20%**	**30%**	**10%**	**20%**	**30%**		
constituants	Ciments en (g)	1038.41	817.36	709.95	603.96	710.05	603.76	499.01	603.27	498.6	395.7	2045.3	1%
	Laitier en (g)	0	102.17	101.42	100.66	202.87	201.25	199.6	301.64	299.16	296.77		
	Poudre de marbre (g)	0	102.17	202.84	301.98	101.44	201.25	209.4	100.55	199.44	296.77		
	Eau (g)	426.20	420.35	417.27	414.14	416.64	413.66	410.63	413.34	410.3	407.04		

4 Résultats et discussions

Les figures (2-a), (2-b) et (2-c) présentent l'effet combiné de laitier de haut fourneau et de la poudre de marbre sur le comportement rhéologique et la variation des paramètres rhéologiques des mortiers autoplaçants. On remarque que, au-delà de 10 % de substitution par chaque addition, le laitier de haut fourneau en plus de la poudre de marbre affectent légèrement l'ouvrabilité des mélanges composés de ciment ternaire avec une quantité de laitier non élevé.

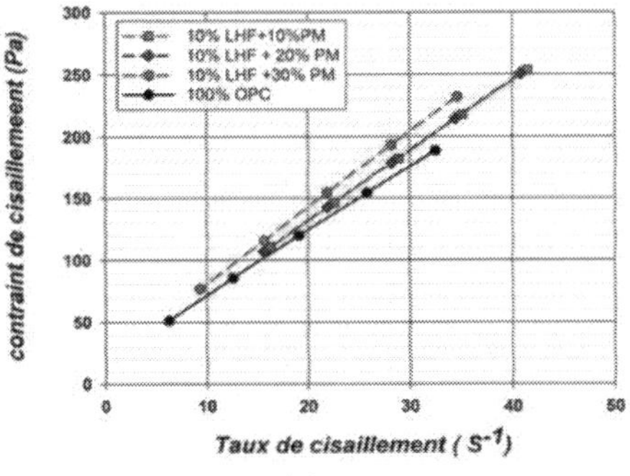

(a)

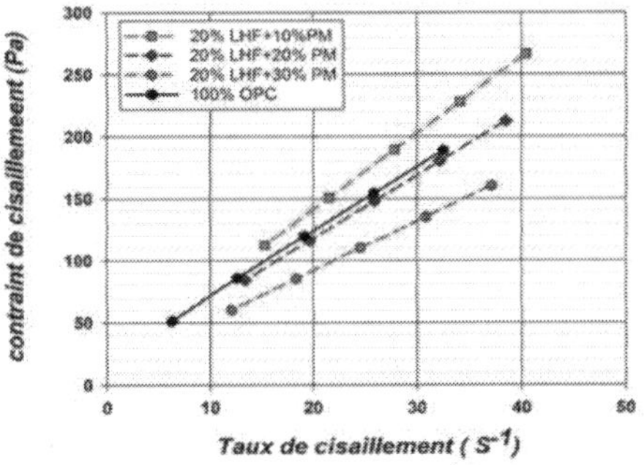

(b)

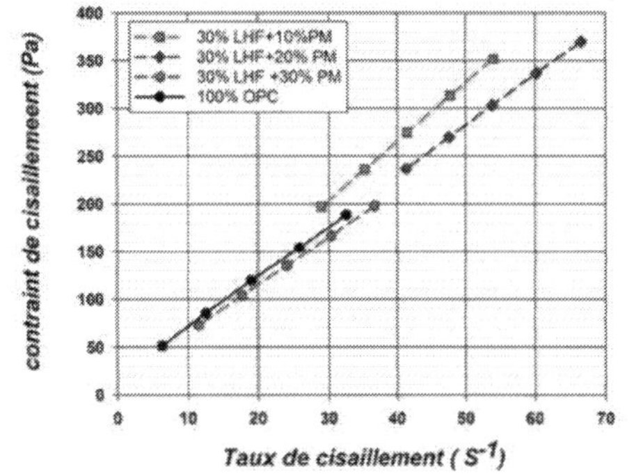

(c)

Fig. 2. Rhéogrammes des mortiers autoplaçants en fonction de dosage LHF et PM

Les figures (2-a et 2-c) indiquent une faible augmentation de viscosité qui est probablement négligeable. Notamment est dû à l'effet de dilution assuré par l'addition de la poudre de marbre de type inerte avec un taux d'incorporation de 20% à 30%. De même, la figure (2-b) présente une diminution considérable de viscosité pour un taux de substitution de 20% LHF et 30% PM. Parallèlement, les figures (3-a, 3-b et 3-c) si dessous présentent la variation de seuil de cisaillement et de la viscosité plastique des MAP en fonction du dosage en LHF+PM. D'où on remarque une diminution générale des paramètres rhéologique τ_0 et μ ou delà de 10 % de substitution de ciment par le LHF et de la PM par rapport au mortier de référence.

 https://doi.org/10.1051/matecconf/201814901057

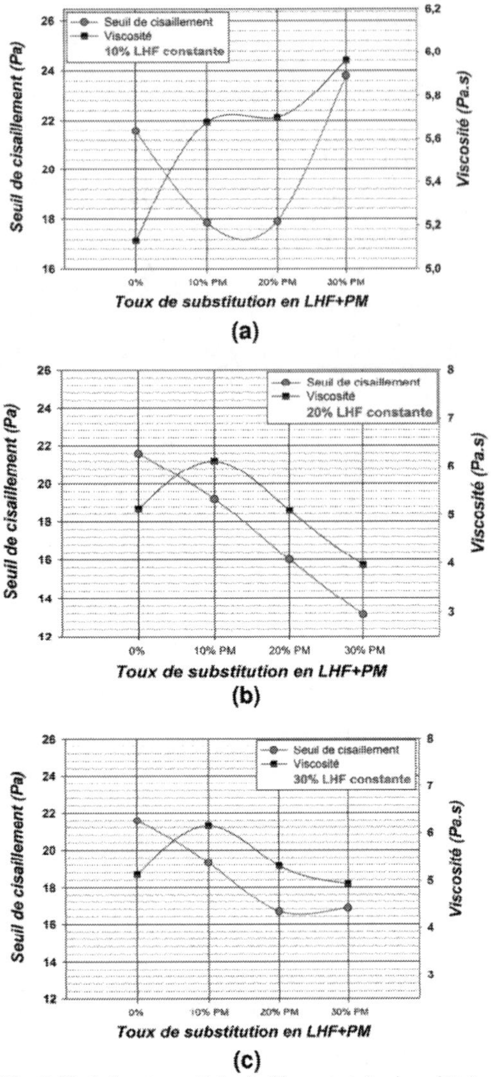

Fig. 3. Variation de seuil de cisaillement et de viscosité des MAP en fonction du dosage en LHF + PM

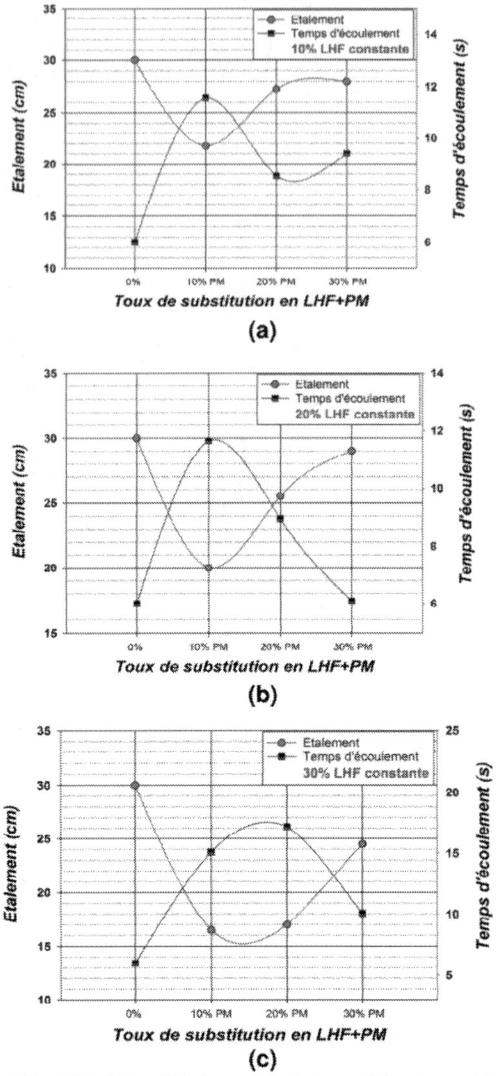

Fig. 4. Variation d'étalement et de temps d'écoulement des MAP en fonction du dosage en LHF + PM

Les figures (4-a), (4-b) et (4-c) présentent les étalements et les temps d'écoulement obtenus avec le mini-cône et le mini v-funnel respectivement. On remarque que les essais l'ouvrabilité des MAP suivent la même tendance avec les résultats obtenu par le rhéomètre. On constate aussi que le remplacement du ciment avec un pourcentage de 20 à 30% de la poudre de marbre et un taux de 10% de laitier, on a une diminution dans l'étalement (figure 4-a) par rapport au mortier de référence, mais cette diminution reste dans l'intervalle d'exigence entre 27 et 33 cm [9].

Par conséquence, le temps d'écoulement par le mini v-funnel, on voit bien qu'on a une augmentation de temps de passage par rapport à celui de référence avec des temps entre 6,9 et 8,3 s. Ces résultats restent toujours dans le domaine d'exigence fixé par Domone [9] entre 2 et 10 s. Cependant, la combinaison ternaire de 20% LHF et 30% PM indiqué dans la figure (4-b) présente un comportement comparable à celui de mortier de référence. Ainsi que la substitution de ciment par 10 à 20% de la poudre de marbre et un taux de 30% de laitier on à une augmentation de temps d'écoulement de 15 à 17 s figure (4-c) Probablement dû à la grande surface spécifique de laitier par rapport au ciment, qui nécessite une demande d'eau supplémentaire.

5 conclusion

Les objectifs fixés dans cette étude étaient de montrer l'influence de la combinaison de laitier de haut fourneau et de la poudre de marbre sur l'ouvrabilité et les paramètres rhéologiques des mortiers autoplaçants à base des ciments ternaires. A la fin de cette étude on peut conclure les points suivants :

- L'utilisation du laitier de haut fourneau combiné avec la poudre de marbre dans les mortiers autoplaçants n'affectent généralement pas le comportement rhéologique et la maniabilité de mélanges à l'exception de quelques compositions ou l'effet de dilution par la poudre de marbre est clairement visible.

- La substitution de ciment par un dosage de 50% d'ajout minéral (20% LH et 30% PM) est idéale pour un comportement rhéologique bénéfique, tout en respectant les exigences en vigueur.

- Ces résultats montrent la possibilité de fabriquer des mortiers autoplaçants à base de ciment ternaire avec un comportement rhéologique et une maniabilité adéquate par un simple ajustement au niveau du dosage en adjuvant.

References

1. A. S. E. Belaidi, « Contribution à l'étude rhéologique et durabilité de béton autoplaçant : apport des ajouts cimentaires ternaires et fillers locaux et industriels » *thèse de doctorat, Université Saâd Dahleb de Blida, Algérie*, (2013).
2. A. S. E. Belaidi, S. Kenai, H. Kadri, H. Soualhi and B. Benchaâ, «Effects of experimental ternary cements on fresh and hardened properties of self-compacting concretes », Journal of adhesion science and Technology, **vol. 30, no. 3**, 247–261, (2016).
3. M. Adjoudj, «Effet des additions minérales et organiques sur le comportement rhéologique du béton », *thèse de doctorat, Université Hassiba Ben Bouali de Chlef, Algerie*, (2015).
4. H. Okamura and K. Ozawa, «*Mix-design for self-compacting concrete* », Concrete library of Jsce, **no. 25**, pp. 107–20, (1995).
5. Park and Noh, «Rheological properties of cementitious materials containing meneral admixture » Cement and concrete research, **vol. 35**, pp. 842–849, (2005).
6. M. Uysol and k Yilmaz, «Effect of mineral admixtures on properties of self-compacting concrete » Cement and concrete composites, **vol. 33**, pp. 771–776, (2011).
7. H. Okamura and M. Ouchi, «Self-compacting concrete », Journal of advanced concrete technology, **vol. 1, no. 1**, pp. 5–15, (2003).
8. A. S. E. Belaidi, L. Azzouz, E. Kadri, and S. Kenai, «Effect of natural pozzolana and marble powder on the properties of self-compacting concrete », Construction and Building Materials, **vol. 31**, pp. 251–257, (2012).
9. P. Domone and J. Jin, «Properties motar for self-compacting concrete » *Procceding international symposium of self-compacting concrete*, pp. 109–20, Sep. (1999).
10. P. Domone, «Self-compacting concret : an analysis of 11years case studies » Cement and concrete composites, **vol. 2, no. 28**, pp. 197–208, (2006).
11. J. Golaszewski, «Influence of superplasticizers on rheological behaviour of fresh cement mortars », Cement and concrete research. **34** 235–248, (2003).
12. C. Lanos, P. Estellé : «Vers une réelle rhéométrie adaptée aux bétons frais », Eur. J. Environ. Civ. Eng. **13,** 457–471, (2009).
13. O. Taleb, F. Ghomari, M. A. Boukli Hacene, H. Kadri & H. Soualhi, « Formulation and rheology of eco-self-compacting concrete (Eco-SCC)», Journal of Adhesion Science and Technology, **Vol. 30, N°3**, 247-261, (2016).
14. F. De Larrard, « Concrete mixture proportioning a scientific approach », Modern Concrete Technology Series. *No 9*, Londres : E & FN SPON ; (1999).
15. H. Soualhi, E.H. Kadri, T. Ngo, A. Bouvet, F. Cussigh, S. Kenai, « A Vane Rheometer for Fresh Mortar Development and Validation », Appl. Rheol, ***24***, 22594, (2014).
16. H. Chong, « Rhéologie des fluides », *thèse de doctorat*, Ecole Nationale des ponts chaussées, France, (2010).
17. P. Estellé, C. Lanos, A. Perrot, « Processing the Couette viscometry data using a Bingham approximation in shear rate calculation », J. Non. Newton. Fluid. Mech. **154,** 31–38, (2008).

Alumina sludge's Influence on the physicochemical characteristics of CPJ_{55} cement

M. Dahhou[1,*], R. Barbach[2] and M. El Moussaouiti[1,*]
[1]Laboratory of Materials, Nanotechnology and Environment, Department of Chemistry, Faculty of Science, Mohammed V University Rabat, Morocco.
[2]Asment Temara (Groupe Votorantim) Ain Atig-Temara, Rabat, Morocco.

*Corresponding authors.
mohammeddahhou@gmail.com (M. Dahhou), m.elmoussaouiti@gmail.com (M. El Moussaouiti).

Abstract. Partial replacement of the Portland cement CPJ_{55} ingredients by various quantities of alumina sludge (AS), produced during drinking water plant sludge, was used in the preparation of mortar mold with dimensions 40×40×160 mm. The characterization of materials is carried out by X-ray fluorescence (XRF), X-ray diffraction (XRD), free lime dosing, and the mechanical tests. Analysis of the chemical composition by XRF shows that the studied alumina sludge is mainly composed of aluminum oxide, silica, which is correlated with the principal mineral phases identified in the XRD analysis results. It is demonstrated that adding 5% of the alumina sludge in Portland cement does not affect the mineralogy of final product. Nevertheless, the compression and flexural strength tests (in 28 days) conducted on mortar sample comprising 5% sludge elucidate that it belongs to cement mortar class of type 32.5 R.

Keywords: Waste management, alumina sludge, cement

1 Introduction

Drinking water treatment plants produce annually thousands of tons of mineral waste called alumina sludge; attributed to the addition of aluminum sulphate during the water coagulation-flocculation step. The latter are mostly discharged to landfills, being the least costly disposal method, which can possibly cause serious environmental problems [1, 2]. The pulverized state of the sludge of about 20% dryness requiring no grinding is a saving in energy costs for the cement industries in particular. In addition, their recycling in this way addresses the environmental protection issues [1-4].

Several studies have examined the replacement of raw precursors in the preparation of building materials by alumina sludge or sewage sludge. It was pointed out that the clinker microstructures and the possible physicochemical properties of ecological cements are strongly influenced by the quantities of sludge [1,5]. According to Zamora et al. [6] it is quite feasible to use alumina sludge from the northern region of Mexico as an additional cementing material and substitute for sand to prepare mortars and concretes for employing in the construction industry. In accordance with Yan et al. [7], the addition of 2.5% by weight of de-inking sludge in the cement mortar does not significantly affect their physical and mechanical characteristics. Although, the replacement of 20% de-inking sludge by weight of cement significantly affects the resistance of the mortars which strongly favors the incorporation potential of the de-inking sludge into cement. Rodriguez, et al. [8] studied the effect of the addition of 5 to 30% of alumina sludge on the physicochemical characterization and the microstructure of the cement mortar 42.5 R. The partial replacement of the cement by the alumina sludge in this case is characterized by the decrease resistance of mortars. Another study by Rodriguez et al. [9] reports that replacement of 20 % constituents of cement type 42.5 N with the sludge enriched with SiO_2, Al_2O_3 and Fe_2O_3 contents produces 32.5 R grade cements. According to El Didamony et al. [10] compressive strength increases with the addition of the alumina sludge up to 5%, and then decreases with its increase to 15% by weight. In this case the addition of the alumina sludge up to 5% can act as a nucleating agent, which gives greater compressive strength at different curing ages. For Piasta and Lukawska, replacement of the cement with 10 or 20% ash from the sewage sludge resulted in an extension of the setting times. The chemical composition suggests that the prolongation in setting times may be due to the appearance of phosphorus in sewage sludge ash. The standard compressive strength of mortars of 10% and 20% ash of sewage sludge is lower than that of reference mortar [11].

This work is a contribution towards exploring the new opportunities to recycle alumina sludge by replacing with it certain proportions of cement ingredients. It will eventually lead to preventing environmental pollution and sustainable development.

2 Materials and methods

Various loadings of alumina sludge were used in our study for the preparation of mortars. In view of comparison, the standard CPJ_{55} cements of type 52.5N were employed. The cements on which we conducted our study were prepared according to the process of production designed in Moroccan cement. A ball mill was used to grind the components (80% clinker, 17% limestone and 3% gypsum), to obtain the required CPJ_{55} cement.

Chemical analyses were carried out by a fluorescence spectrometer X Panalytical PW 4400/24. The samples to be analyzed were prepared as beads using lithium tetraborate as a flux.

Mineralogical compositions were assessed by X-ray diffraction, using a diffractometer X'Pert Pro MPD PANalytical. This diffractometer contains a copper anticathode (Cu-Kα) with wavelength λ = 1.54 Å, supplied with a potential of 45 KV and a current of 40 mA. The recordings were made with a pitch of 0.06 (2θ) and a counting time of 121 seconds per step.

Chemical analysis by conventional dosage with the glycerol-alcohol method was performed to determine the contents of free lime in cement.

Resistance to flexural and compressive strength of the mortar obtained was determined by using prismatic mold with a cross section of 40 mm×40 mm and a length of 160 mm, comprising three horizontal compartments, in order to simultaneously handle three samples. We used normal mortars according to EN197-1 [12] of the following composition:

- 450 g of binder, the additions are always introduced in substitution of cement (C);
- 1350 g of standardized sand (S);
- 225 ml of water (W).

The level of the mixing water was kept constant for all the mixtures W/C = 0.5. The flexural strength and compression were tested after 2, 7 and 28 days [12]. The different mortars studied and their notations are as followings:

-CM: Control mortar: CPJ_{55} cement of type CEM 52.5N;

- CM X: Cement mortar CPJ_{55}+ X% of dried alumina sludge at 105 ° C; X : 5% - 25%.

The mineralogical phases of the cements were calculated by the estimates derived from Bogue's equations according to the expressions (1) - (4) [13]:

$$C_3S\ (\%) = 4.07 \times [C - (0.7 \times \bar{S}) - (1.27 \times (LOI/2) - CaO_f)] - [(7.6 \times S) + (6.72 \times A) + (1.43 \times F)] \quad (1)$$

$$C_2S\ (\%) = 2.87 \times S - 0.75 \times C_3S \quad (2)$$

$$C_3A\ (\%) = 2.65 \times A - 1.69 \times F \quad (3)$$

$$C_4AF\ (\%) = 3.04 \times F \quad (4)$$

Where, C= CaO, S= SiO_2, A= Al_2O_3, F= Fe_2O_3, $\bar{S}$ = SO_3, CaO_f = free lime and LOI: loss on ignition at 975 °C.

3 Results and discussion

3.1 Alumina sludge characterization

Table 1 gives the chemical composition of the alumina sludge obtained by XRF analysis. The main oxides identified were: SiO_2, Al_2O_3 and Fe_2O_3 and CaO. The high alumina contents are attributed to the addition of aluminum sulphate during the water coagulation-flocculation step [1, 2]. The alumina sludge had a very low SO_3 contents and the P_2O_5 concentration was about 0.16% by weight. The loss on ignition of the alumina sludge is 33.14% and the major compounds from the residues of the flocculent and the clay must not undergo considerable variations.

Fig. 1 shows the X-ray diffractogram of the alumina sludge dried at 105 °C. The crystalline mineral phases identified consists mainly of Muscovite ($H_2KAl_3(SiO_4)_3$), Mica (K-Mg-Fe-Al-Si-O-H_2O) and Quartz (SiO_2).

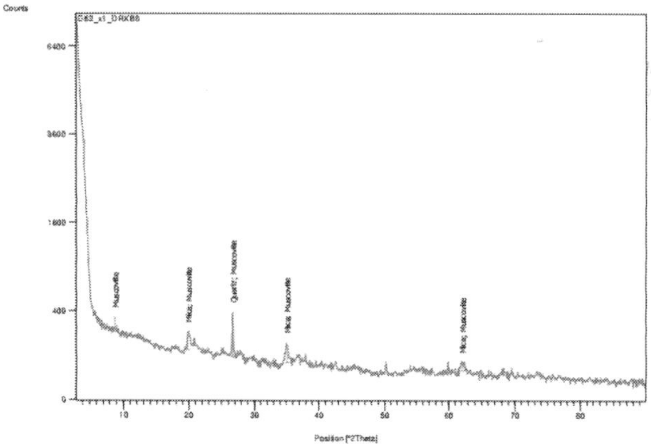

Fig. 1: X-ray diffractogram of alumina sludge dried at 105 °C.

Table 1. Chemical composition of alumina sludge.

	SiO_2	Al_2O_3	Fe_2O_3	CaO	MgO	SO_3	K_2O	TiO_2	MnO	P_2O_5	Na_2O	SrO	L.O.I
Alumina sludge	13.52	47.67	1.13	2.49	0.05	0.18	0.56	0.12	0.90	0.16	-0.04	0.016	33.14

L.O.I: Loss on ignition at 975 °C.

3.2 Influence of additional alumina sludge on physic-chemical characteristics of cement elaborated.

3.2.1 X-ray fluorescence analysis and free lime contents

The chemical and mineralogical composition of various cements obtained with the addition of alumina sludge (5, 10, 15, 20 and 25%) is shown in Table 1. It is found that the replacement of 5% alumina sludge by weight cement 52. N did not induce any qualitative impact on the chemical and mineralogical composition phases formulated by Bogue's equations. Compared with the chemical composition of industrial Portland cement, typical oxides such as SiO_2, Al_2O_3, Fe_2O_3 and CaO are present in concentrations similar to industrial Portland cement [14]. The C_3S and C_2S contents are close to those obtained in a regular Portland cement [15]. On the other hand, the substitution of 10-25% by weight seems to affect the chemical and mineralogical composition of cements. The cements produced by the addition of the alumina sludge were characterized by higher Al_2O_3 contents compared to cement 52.5 N. This is due of course to the high Al_2O_3 proportion which is the main constituent in the alumina sludge. Due to the same reason, higher contents of the C_3A phase were also calculated for these samples.

The Alite (C_3S) contents decreased with the increase in the addition of alumina sludge which justified the increase in C_3A contents. The incorporation of the alumina sludge does not seem to influence the final contents of free lime. The quantity of free lime in our samples remains less than 2% which is in agreement with the quantity of free lime of industrial Portland cement. Moreover, this result is also consistent with the amount of free lime in Portland cement reported in the literature [16].

3.2.2 X-ray diffraction analysis

The cement XRD of 52.5 N references and cements produced by the addition of different percentages of the alumina sludge are given in Fig. 2. As can be seen in Fig. 2 the addition of 5-15% of alumina sludge did not affect the formation of the mineralogical phases and therefore does not entail the formation of new minerals in the cements produced. In these cements, the main mineralogical phases are: C_3S, C_2S, C_3A, C_4AF, SiO_2, $CaCO_3$, $CaSO_4.2H_2O$. The prepared cements were found to have Alite with monoclinic form (Ca_3SiO_5). Belite was found to crystallize in a monoclinic system, similar to the polymorphous form Ca_2SiO_4 (β-Belite) and Celite ($Ca_3Al_2O_6$) also has the monoclinic form. As the addition of the alumina sludge increases the peak intensity of the Alite and calcite decreases and the C_2S lines predominate.

Table 2. Chemical analysis of cements.

	CEM 52.5N	CEM+5 % AS	CEM+10 % AS	CEM+15%AS	CEM+20%AS	CEM+25%AS
Chemical composition						
SiO_2	19.60	18.55	18.98	18.93	18.26	17.69
Al_2O_3	4.33	5.98	8.03	10.02	11.80	13.31
Fe_2O_3	2.99	2.91	2.86	2.81	2.69	2.47
CaO	60.88	58.55	55.46	53.12	49.6	47.38
MgO	1.64	1.56	1.49	1.47	1.32	1.25
SO_3	1.90	1.87	1.54	1.04	0.70	0.38
K_2O	0.66	0.67	0.65	0.64	0.63	0.62
TiO_2	0.32	0.30	0.31	0.29	0.28	0.27
MnO	0.07	0.15	0.24	0.31	0.39	0.48
P_2O_5	0.24	0.23	0.23	0.23	0.23	0.23
Na_2O	-0.02	-0.02	-0.02	-0.02	-0.023	-0.02
SrO	0.064	0.057	0.055	0.054	0.053	0.050
CaO_f	1.96	1.456	0.672	0.504	0.336	0.28
LOI	7.22	9.08	10.08	10.99	13.96	15.78
Estimated mineralogical composition						
C_3S	60.08	47.69	19.15	-1.91	-22.00	-35.58
C_2S	10.86	17.18	39.92	55.68	68.90	77.50
C_3A	6.41	10.93	16.42	21.80	26.74	31.09
C_4AF	9.11	8.87	8.72	8.55	8.17	7.51

LOI: Loss on ignition at 975 °C.

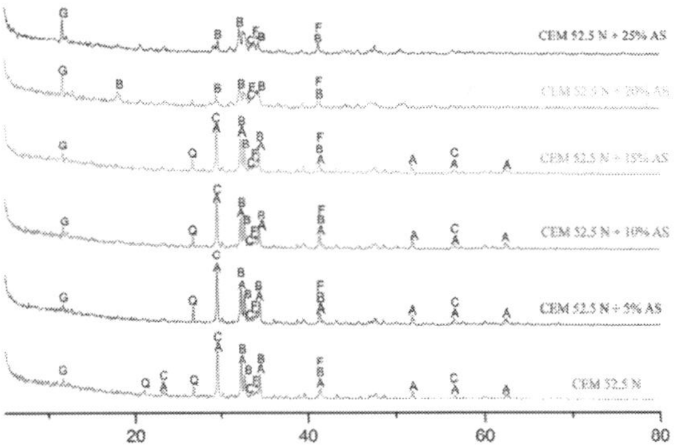

Fig. 2: X-ray diffractogram cements (A: Alite (Ca_3SiO_5), B: Belite (Ca_2SiO_4), C: Calcite ($CaCO_3$), C*: Celite ($Ca_3Al_2O_6$), F: Ferrite ($4CaOAl_2O_3Fe_2O_3$); Q: Quartz (SiO_2) and G: gypsum ($CaSO_4.2H_2O$)).

3.2.3 Physical and mechanical properties of cement mortars produced

The flexural and compressive strengths were measured after being cured for 2, 7 and 28 days, are given in Fig. 3/4, respectively. The mortar based cements produced after 2 days of curing are presented in Fig. 5.

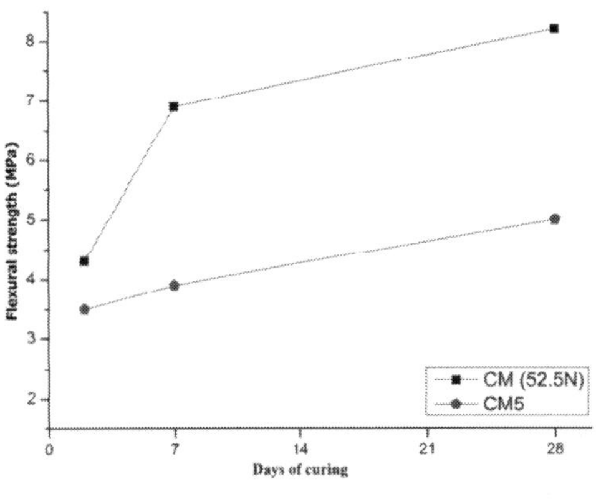

Fig. 3. Flexural strength results of the cement mortar after different curing times.

These compressive strengths were compared with the compressive strength of ordinary Portland cement 52.5 N. Comparison of the results shows that the flexural and compressive strength of the mortar with 5% alumina sludge was lower than that of ordinary Portland cement 52.5 N at all hydration stages. In this case the sludge hydrates slowly and the products obtained during the hydration are reduced when the alumina sludge increases. Mortars based on 10 up to 25% alumina sludge with low resistance are associated with the decrease in the amount of C_3S responsible for the development of resistances during the first four weeks. The results indicate that all mortars with a 5% replacement of alumina sludge can be classified as Portland cements of category 32.5 R according to EN197-1[12].

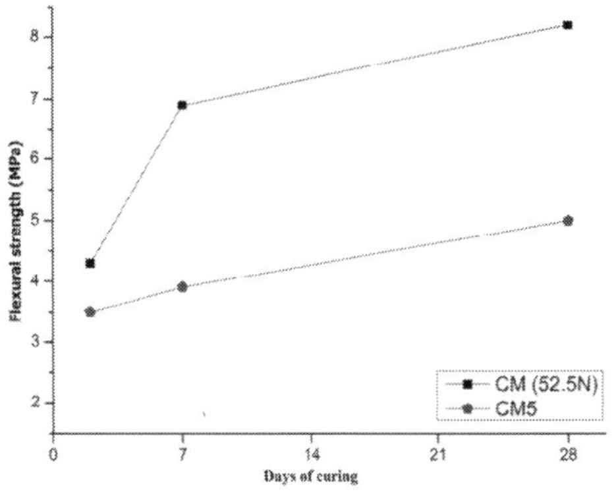

Fig. 4. Compressive strength results of the cement mortar after different curing times.

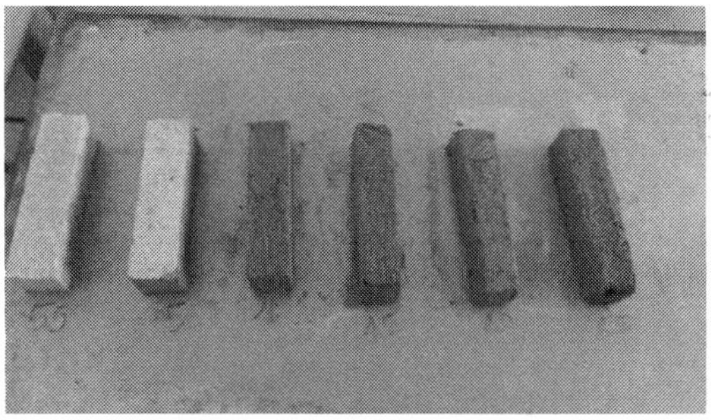

Fig. 5. Images of mortar based cements produced after 2 days of curing. (55: CEM 52.5N (CPJ_{55}); 5: CEM 52.5N + 5% AS; 10: CEM 52.5N + 10% AS; 15: CEM 52.5N + 15% AS; 20: CEM 52.5N + 20% AS; 25: CEM 52.5N + 25% AS.

Conclusion

The direct use of alumina sludge mineral waste as an addition to Portland cement might lead to environmental protection and sustainable development and allows compensating the needs of Al_2O_3 and Fe_2O_3 in cements. It has been found in the present study that in 28 days, all mortars produced with a 5% replacement of alumina sludge can be classified as Portland cements of category 32.5 R. The addition of alumina sludge also contributes to the reduction of the total cost of the cement since the grinding of alumina sludge is much less energy consuming thus contributing in the reduction of the energy requirements of the formation of Portland cement.

References

1. M. Dahhou, M. El Moussaouiti, A. Benlalla, A. El Hamidi, M. Taibi, M. A. Arshad, Waste Biomass Valor. **7**, 5 (2016)
2. M. Dahhou, M. El Moussaouiti, M. El Morhit, S Gamouh, S Moustahsine, J. Taib. Univ. Sci. **11**, 5 (2017)
3. M. Smol, J. Kulczycka, A. Henclik, K. Gorazda, Z. Wzorek, J. Clean. Prod. **95** (2015)
4. W.J. Wang, C.D. Wang, T.C. Lee, C.C. Chang, Constr. Build. Mater. **56** (2014)
5. Y. Lin, S. Zhou, F. Li, Y. Lin, J. Hazard. Mater. **213-214** (2012)
6. R.M.R. Zamora, O.C. Alfaro, N. Cabirol, F.E. Ayala, A.D. Moreno, Am. J. Environ. Sci. **4**, 3 (2008)
7. S. Yan, K. Sagoe-Crentsil, G. Shapiro, J. ENVIRON. MANAGE. **92** (2011)
8. N.H. Rodríguez, S. Martínez-Ramírez, M.T. Blanco-Varela, M. Guillem, J. Puig, E. Larrotcha, J. Flores,. Cement Concrete Res. **40**, 5 (2010)
9. O. Rodríguez, L. Kacimi, A. López-Delgado, M. Frías, A. Guerrero, Constr. Build. Mater. **40** (2013)
10. H. El-Didamony, K.A. Khalil, M. Heikal, HBRC J. **15** (2014)
11. W. Piasta, M. Lukawska, Procedia engineer. **161** (2016)
12. Afnor. EN 197–1, *Cement composition, specifications and conformity criteria for common cements,* (2000)
13. P. Stutzman, A. Heckert, A. Tebbe, S. Leigh, Cement Concrete Res. **61–62** (2014)
14. A. Bordy, A. Younsi, S. Aggoun, B, Fiorio, Constr. Build. Mater. **132** (2017)
15. X. Cai, Z. He, S. Tang, X. Chen, Constr. Build. Mater. **127** (2016)
16. AS-D-G. Vilaplana, V.J. Ferreira, A.M. López-Sabirón, A. Aranda-Usón, C. Lausín-González, C. Berganza-Conde, G. Ferreira, Constr. Build. Mater. **94** (2015)

Influence of cement content on the thermal properties of compressed earth blocks (CEB) in the dry state

N. Zakham[1,2], Y. El Rhaffari[3], A.Ammari[1], M. Cherraj[1,2], H. Bouabid[1], K. Gueraoui[1,2], A. Samaouali[3], A. Mzerd[3]
[1]Mohamed V University, Department of physics, Laboratory of Mechanics and Materials, Rabat, Morocco
[2] Mohamed V University, Department of physics, Team of Modeling and simulating in Mechanics and Energetic, Rabat, Morocco
[3] Mohamed V University, Department of physics, of thermodynamics, Rabat, Morocco

Abstract. The Compressed Earth Block (CEBs) is one of the kinds of building materials which stabilized by cement. Soil is a basic component, a renewable, non-toxic and natural resource. Samples must be stabilized with a limited percentage of cement so that samples do not lose their natural properties including thermal comfort and on other hand offer high mechanical resistance. The objective of this work is to study the effect of cement content on thermal behavior of the building material of CEBs in the dry state, by studying variation of temperature with time, and measuring thermal conductivity and the specific heat, with respect to the various cement ratios added to the samples. This study is mainly an experimental and numerical, to determine how the thermal behavior evolves with the cement content in the samples CEBs. The soil was extracted from the famous city Fez in Morocco, Fez is known for its several historical monuments and buildings. After determining its granulometry and other specific characteristics, the CEBs are made by mixing soil with cement. The samples are put in plastic bags for two weeks, then removed the plastic bags from the samples and let them to dry again for an additional two weeks away from direct sun. The samples CEBs are taken cylindrical form (8 cm diameter with an average height of 12 cm).
The experimental method consists of a hot ring for which a numerical modelization was developed to fit the mathematical equations of heat diffusion and the boundary conditions. For the numerical model Bouabid and Cherraj have developed numerical model which allow, with a good accuracy, to quantify the evolution of the thermal behavior of the earth material in function of cement content.
Indeed, the study provides information on the influence of the cement percentage on the thermal behavior of the samples CEBs, the thermal behavior of samples increase with increase cement content.

Nomenclature

r: Along the radial coordinate.
z: Along the axial coordinate.
T: Temperature [0C].
T: Heating time [S].
ρ: Volume density [kg/ m3].
Cp: Specific heat [J /kg.K].
K: Thermal conductivity [W/m.k].
$\dot{q}$: Heat flux [W.m2].
C(D): The magnitude.
D: The variable of cement content.
C0: Value of the quantity when D =0. The optimal value of 'C' corresponds to the point of inflection of the curve of the evolution.
a_D: The Saturation of cementation limit.
n: Degree of variation (n = 3).
b_D: Constant introduced to approximate the limit (asymptote) by inferior value.

1 INTRODUCTION

In order to ensure obtain thermal comfort in the building, special attention must be in the selection of building materials, and when we identify the properties of building materials used, we can choose the best to get the thermal comfort.

The earth material is considered a construction material which has been used to build thermally comfortable structures in many regions around the world for thousands of years, it is one of the most important and most common in the world, it is suitable as building material and has low thermal properties, and the compressed earth blocks are model for that. In our study, we have chosen compressed earth block stabilized (CEBs) [1, 2] , we carried out experimental and numerical study to understand thermal behavior of the samples compressed earth block stabilized (CEBs) in the dry state when cement is added to it at different percentages (0%, 4%, 7% and 10%). Adding cement to the samples CEBs improves their performance and durability, where we added small amounts of cement (4%,7% and 10%) to the samples, the samples are compressed by machine with pressure 2 MPa.

There are many factors impact to thermal behavior of the samples, such as the humidity in building materials. Our samples have been dried because we need just to study the effect of cement only on the thermal behavior of the samples in dry state.

The instrument for measure thermal conductivity in the experimental study we were used the hot ring method, the hot ring technique is transient method of measurement of the thermal conductivity. The principle of measurement is by a ring at the center of which the temperature rise is measured, the measurement of the temperature rise at the center of the ring allows to determine the thermal conductivity of the samples CEBs [3].

2 The materials used

The materials used for making samples are soil, cement and water. The main component of the samples is soil, soil has been taken from in Fez in Morocco, usually use in traditional building materials in Fez, this the city of the most important cities of Morocco which is famous for its history and its buildings ancient and antique. The granular fraction of the earth Fez is 8% clay, 28% gravels, 12% silt and the sand (main constituent) is composed of 30% large sand and 22% fine sand. The samples (CEBs) are made by mixing earth with cement and water, four ratios of cement stabilizers were used (0,4,7, and 10%), the cement type CPJ35 Portland. The mixture is distributed in molds and pressed by a specialized machine for making compressed earth block with pressure 2MPa, all the samples of CEBs took cylindrical forms, the height of samples (12 cm) and the diameter (8 cm), and left drying the samples in the open air for a month, away from direct sun.

3 Experimental Procedures

After making samples, each sample was divided to two parts, and was dried from the water content (so that the water content becomes in each sample 0%), the drying process has been carried out in an electric oven at temperatures of 60 °C up to the stabilization of the mass of the samples. The drying time was 72 hours, it is very important to dry the samples before the determination of the thermal behavior.

For thermal measurements, there many methods developed for the determination of thermal behavior, in this study used method of hot ring and the measurement of the thermal behavior of the samples is based on heat transfer in regime unsteady. The hot ring method power supply by a hot ring probe (the slab of kapton of the hot-ring apparatus) to measure the thermal conductivity and to estimate the thermal diffusivity of the samples, the principle of the slab of kapton the hot-ring apparatus is placed between two pieces of the sample in the middle, and then we inject current into the ring to generate a heat flux, this heat flux dissipates in the material and arrives after a certain time in the center of this ring, where the measurement is made. There is therefore a time lag between the heating and the measurement, which makes it possible to identify by a mathematical algorithm the thermal diffusivity of the samples (Figures 1 and 2) [9-14].

Fig 1. The slab of kapton of the hot-ring apparatus.

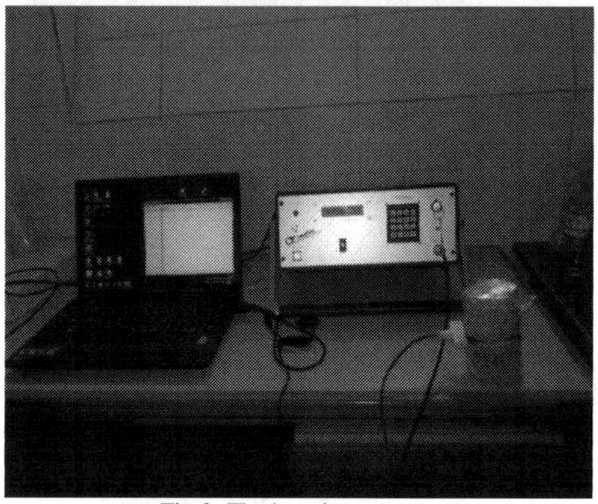

Fig 2. The hot-ring apparatus.

Total heat transfer be determined using the time-dependent heat equation, which for flow in two dimensions can be written as:

$$\rho C_p \frac{\partial T}{\partial t} = k\frac{1}{r}\left(\frac{\partial}{\partial r}\left(r\frac{\partial T}{\partial r}\right)\right) + k\frac{\partial^2 T}{\partial z^2} + \dot{q} \qquad (1)$$

where t is time, ρ is the material density and C_p is the specific heat capacity, the energy required to heat up one kilogram of a given material by one degree, where $\dot{q}$ is the heat flux (heat transfer per unit area), K is the thermal conductivity (heat transfer per unit length for a temperature difference of one degree), T is the temperature, $\dot{q}$ Heat flux [4-6].

The Initial conditions are:

$t \le 0; T(\text{r, z, t}) = T_\infty; \forall r and \forall z$

for all r and z when:

$t \le 0,\ T(r,z,t) = T_{init}$

With the following boundary conditions:

For $r = 0, \forall z \frac{\partial T}{\partial t} = 0$;

$Z = -\frac{H}{2} and + \frac{H}{2}; T = T_\infty$

For $\forall$ Z and r = R, T=T∞;

 https://doi.org/10.1051/matecconf/201814901059

$$r \to \infty \; or \; z \to \infty \; , \; T(r,z,t)=T_{init}$$

4 Numerical modelling

The modeling used for the determination of thermal behavior of the samples CEBs is essential to predict the real thermal behavior of the samples when add a cement. The main characteristic of the proposed model is that it was developed in approximations numerical results with experimental results, Bouabid and Cherraj have developed numerical model which allow, with a good accuracy, to quantify the evolution of the thermal properties of the earth material in function of cement content [7-8]. According to figure 3, one finds that in the areas I and III there is no presence of significant improvement of the magnitude 'C(D)' in function of the stabilization 'D'. Whereas Zone II presents a clear increase of the thermal conductivity and specific heat to the cement content (Figure 3).

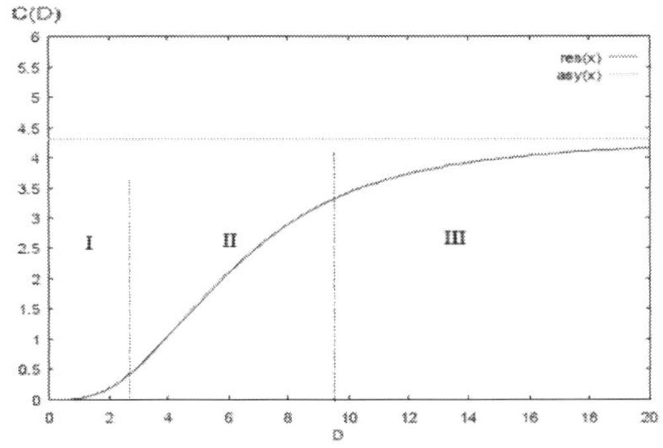

Fig.3. Diagram of the evolution of C as a function of D.

For the single parameter of quantification, the model is reduced based on the following hypothesis:

C(D)= 0 ; D= 0 This is the hypothesis I.
C(D)= C(D)∞ ; D→∞ This is the hypothesis II.

The magnitude 'C(D)', this magnitude supposed to continuous a function of the dosage 'D' in the following relation:

$$C(D)=\frac{a_D * D^n}{b_D + D^n}+C_0 \quad (2)$$

5 Results and Discussions

This work presents the different results obtained in this study and it aims to show the influence of the different percentages cement content on the thermal conductivity, the specific heat and the evolution of the temperature of the samples CEBs in the dry state. The table 1 shows the results of the thermal conductivity and the specific heat of the four samples with different cement content, the thermal conductivity and specific heat increases with high cement content.

Table1. The results of the Thermal conductivity(W/m.K) and Specific heat(J/kg.k) with different percentages of cement.

The samples	Dosage of cement (%)	Thermal conductivity (W/m.K)	Specific heat (J/kg.k)
Sample 1	0	0,662	991,9
Sample 2	4	0,692	1080,9
Sample 3	7	0,746	1205,3
Sample 4	10	0,785	1224,01

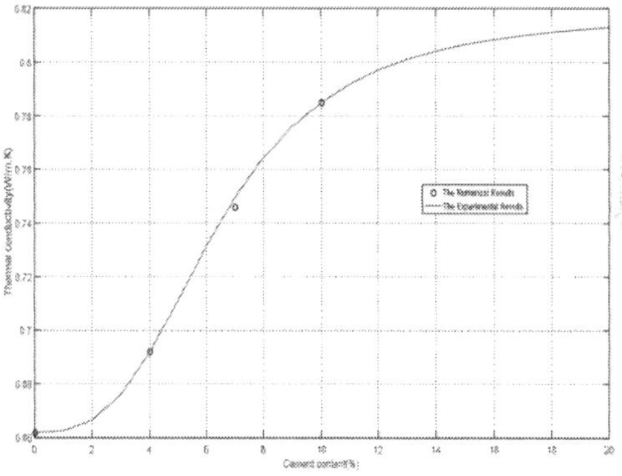

Fig. 4. The thermal conductivity of the samples to increases with an increase in percentage of cement content.

Figure 4 shows experimental and numerical results of the samples CEBs, we can observe that there are three zones as we mentioned in the numerical model. The Induction zone it starts from 0% to 4% of cement content, in this zone can observe the effect of stabilization variable is low, due to the low value of the stabilization variable D (cement content) and therefore does not bring a significant improvement to the magnitude C (thermal conductivity). The effect zone it starts from 4% to 7% of cement content, in this zone can observe the effect of the stabilization is appear clearly, its effect begins to appear and interacts with the samples, and the magnitude C (thermal conductivity) undergoes a clear improvement as a function of the stabilization variable D (cement content). And saturation zone it starts from 7%, in this zone can observe it is beyond a threshold, the magnitude C (thermal conductivity) is less sensitive to the increase of the stabilization variable D (cement content), the contribution of the stabilization is not rewarded by a gain on the magnitude C (thermal conductivity). The experimental results agree with the numerical results, all the experimental results and numerical results illustrate the effect cement content on the thermal conductivity of the samples CEBs in the dry state with C_0 = 0.662, a_D= 0.156084598, b_D= 268.98047.

The results show that the thermal conductivity of the samples varies according to the cement content, in figure 4, the curve obtained shows that the thermal

 https://doi.org/10.1051/matecconf/201814901059

conductivity varies as a function of the cement content, thermal conductivity of the samples CEBs increase with the increasing of cement content. The thermal conductivity as a function of cement content give nonlinear relation.

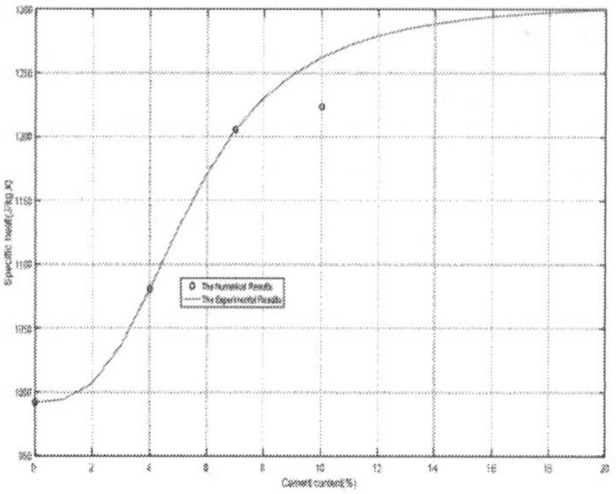

Fig. 5. The specific heat of the samples with different percentage of cement content.

From figure 5 the magnitude C (specific heat) as function of the stabilization variable D (cement content), shows the experimental and numerical results of the samples CEBs of the specific heat with different dosage of cement, we can observe also that there are three zones as we mentioned in the numerical model, as in the curve 4, the experimental results agree with the numerical results, all the experimental results and numerical results illustrate the effect cement content on the specific heat of the samples CEBs in the dry state with C_0 = 991.9, a_D= 314.115, b_D= 161.881.

The term 'specific heat' refers to the heat energy per unit mass (typically 1 kg) required to raise the temperature of a substance by one-degree Kelvin, the higher the specific heat capacity of a substance, the more energy is required to raise its temperature, we observed a characteristic increase of specific heat with increase of cement content.

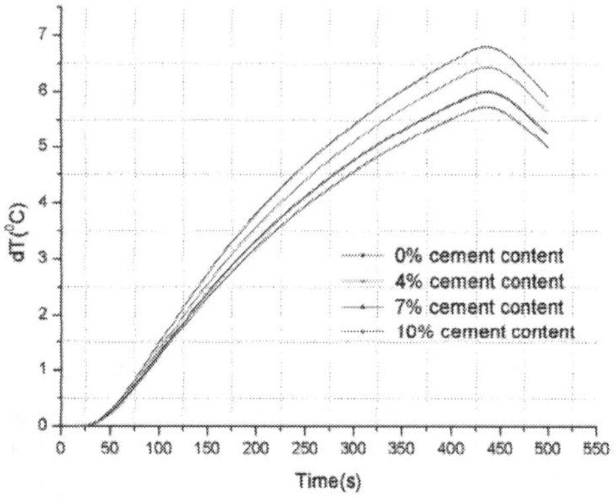

Fig. 6. Evolution of the temperature with time for the samples (dry state) with different percentages of cement (0%, 4%, 7%, and 10 %).

The results illustrated on figure 6 show the experimental results, the cement content (0%, 4%, 7% and 10 %) has effect on the variation of temperature of samples CEBs. We have used the hot-ring apparatus and measured the temperature rise at the center of the ring for each of the CEBs specimens presented. We made temperature rise measurements for the time range 1–500 s, heating time =400s with heating power Q =2 W.

From the figure 6 illustrate thermal mass of the samples is the ability of the samples to absorb and store thermal energy of period time, the samples is heated by the slap kapton of the hot ring method, until time of heating, then it stays warm a few seconds, after that the samples start lose the heating.

6 Conclusion

The compressed earth block stabilized contribute to the improvement of the energy efficiency in building, especially that compressed earth block stabilized owns low thermal conductivity, and can use it for the thermal isolation of buildings.

This work to estimate the thermal behavior of the compressed earth block stabilized by cement and to evaluate the influence of different dosages of cement on the thermal properties of CEBs.

We can observe through the results obtained that thermal behavior of the samples CEBs in the dry state depended strongly on percentage of cement stabilizer, the thermal conductivity and specific heat of CEBs increases with the increase in the percentage of cement.

Thermal conductivity and specific heat of CEBs decrease without add cement also, the addition of cement further increases the thermal conductivity and specific heat of the CEBs.

The results showed that the decrease in cement percentage lowers thermal properties of the samples CEBs, and on the other hand the increase in cement percentage lead to higher thermal properties of the samples CEBs, when cement add to soil was found increase the specific heat and increase the thermal conductivity of the of the samples CEBs.

The mathematical model is used to describe the evolution of the thermal behavior with cement content, the proposed model is that it was developed in approximations numerical results with experimental results, and we have obtained numerical results agrees with experimental results. In fact, this model allows to determine the technico-economic optimum of stabilization which corresponds to the optimum cement rate to be used for a specific earth material. Indeed, this optimum corresponds to the maximum of the thermal behavior with a minimum of economic investment in the stabilizer (cement in this case).

References

[1] Jean-Claude Morel, Abalo Pkla, Peter Walker, Construction and Building Materials 21 (2007)

[2] A. Ammari, K. Bouassria, M. Cherraj, H. Bouabid, S. Charif D'ouazzane, Case Studies in Construction Materials 7 (2017).

[3] Azakine Sindanne S, Ntamack GE, Lemanle Sanga RP, Moubeke CA, Kelmamo Sallaboui ES, Bouabid H, Mansouri K, D'Ouazzane; J. Build. Mater. Struct. **1** (2014).

[4] P. Meukam, Y. Jannot, A. Noumowe, T.C. Kofane, Construction and Building Materials 18 (2004) .

[5] R. Coquard, D. Baillis, D. Quenard , International Journal of Thermal Sciences 47 (2008).

[6] Márcio Buson, Nuno Lopes, Humberto Varum, Rosa Maria Sposto and Paulo Vila Real, fire and materials.Fire Mater. (2012).

[7] M. Cherraj, thèse doctorat, Université Mohammed v-Agdal, Faculté des Sciences, Rabat-Maroc.

[8] H. Bouabid, S. Charif d'ouazzane, O. Fassi-fehri, Rapport Interne, CRR-Bâtiment, LPEE, Casablanca, Déc. (1998).

[9] Hayet Khab, Abla Chaker, Séminaire International sur le Génie Climatique et l'Energétique, SIGCLE'2010.

[10] http://earthblockinternational.com/faq/

[11] P. Meukam, A. Noumowe, T.C. Kofane, thermo physical properties of lateritic soil brick, the Abdus Salam internatinal centre for theorical physics, IC/2002/157.

[12] Joseph Khedari, Pornnapa Watsanasathaporn, Jongjit Hirunlabh, Thailand, Received 19 December (2002); accepted 16 February (2004).

[13] Salim Guettala, Mohamed Bachar, Lakhdar Azzouz, .J Earth Sci Clim Change (2016).

[14] Malahimi Anjorin, Chakirou Akanho Toukourou, Aristide Comlan Houngan, Christophe Awanto, Largum Madougou, Journal of Civil Engineering and Architecture 10 (2016) .

Blast Resistance of Slurry Infiltrated Fibre Concrete with Waste Steel Fibres from Tires

Martina Drdlová[1], Oldřich Sviták[1], Petr Bibora[1], Miloslav Popovič[2], René Čechmánek[1]
[1] Research Institute for Building materials, Applied Research and development, Brno, Czech Republic
[2]SVS FEM s.r.o., Brno, Czech Republic

Abstract. The utilization of waste steel fibres (coming from the recycling process of the old tires) in production of blast resistant cement based panels was assessed. Waste fibres were incorporated in slurry infiltrated fibre concrete (SIFCON), which is a special type of ultra-high-performance fibre reinforced concrete with high fibre content. The technological feasibility (i.e. suitability of the waste fibres for SIFCON technology) was assessed using homogeneity test. Test specimens were prepared with three volume fractions (5; 7.5 and 10 % by vol.) of waste unclassified fibres. SIFCON with industrial steel fibres (10% by vol.) and ultra-high-performance fibre concrete with industrial fibres were also cast and tested for comparison purposes. Quasi-static mechanical properties were determined. Real blast tests were performed on the slab specimens (500x500x40 mm) according to the modified methodology M-T0-VTU0 10/09. Damage of the slab, the change of the ultrasound wave velocity propagation in the slab specimen before and after the blast load in certain measurement points, the weight of fragments and their damage potential were evaluated and compared. Realized tests confirmed the possibility of using the waste fibres for SIFCON technology. The obtained results indicate, that the usage of waste fibres does not significantly reduce the values of SIFCON flexural and compressive strength at quasi-static load - the values were comparable to the specimens with industrially produced fibres. With increasing fibre content, the mechanical parameters are increasing as well. Using of the waste fibres reduces fragmentation of SIFCON at blast load due to the fibre size parameters. Using of low diameter fibres means more fibres in the matrix and thus better homogeneity of the whole composite with less unreinforced areas. Regarding the blast tests, the specimen with waste steel fibres showed the best resistance and outperformed also the specimen with commercial fibres. Using of waste fibres in SIFCON technology can reduce the price of this composite by 70 % by keeping the original SIFCON extraordinary properties, which makes it very competitive material in the concrete area.

1 Introduction

It is estimated that over 4 billion used tires are generated every year [1]. A typical tire consists of approximately 47% rubber, 22% carbon black, 17% steel cords, 5% fabrics, and the remaining percentage consists of some other minor additives [2]. Every passenger car tire thus contains about 1.3 kg of steel fibres, which can be reused. Nowadays, the steel obtained from the recycling process is mostly used as a feedstock in the production of new steel. Using tire waste fibres in concrete technology is limited due to the nature of the obtained fibres – in particular the high diversity of fibre geometry, contamination and bulk nature of the product. Further processing (cutting, selecting and cleaning) of the fibres enables to prepare more suitable material for the concrete reinforcement and brings satisfactory results, but is expensive and demanding. On the other hand, the tire fibres are made from high strength carbon steel with a tensile strength as high as 2,200 – 2,750 MPa [3], so they seem to be promising reinforcing material. Several studies were performed focused on mechanical properties of concrete with classified fibres [1,4,5,6] The pull-out behavior of the recovered fibers may be similar to those of the commercial fibers which may be attributed to the irregular undulations of the fibers as a result of the shredding process that can increase mechanical bonding [6] and [7]. Mechanical properties of concrete containing waste steel fibers may be improved due to the crack-bridging effect of the fibers in concrete similar to the concretes produced with commercial fibers [8], [5] and [9].

Finding the way how to reuse this type of fibers as the concrete reinforcement can save the costs connected with fibreconcrete manufacturing. As the traditional concrete technology is difficult to be adopted when using not classified waste fibres (the fibres tend to stay in bulks and the proper disintegration and homogenous dispersion is difficult to achieve in higher fiber dosage), the slurry infiltration technology was tested in this study. Slurry infiltrated fibre concrete - SIFCON was first developed in 1979 in USA, by incorporating large amounts of steel fibres in cement-based composites [10]. SIFCON is prepared by infiltrating pre-placed fibres with fine grain aggregate mortar. The fibre volume fraction of traditional fibre reinforced concrete is limited, because excessive amount of the fibres affects the workability of the fresh concrete in a negative way. This limits the fibre volume V_f to 1 – 5%, depending on the type of fibre used and the required workability of the mixture. SIFCON specimens can be produced with V_f between 5% and 30% [10], the fibre volume depends on the fibre geometry, length and diameter and vibration of the fibres during their placement process. SIFCON possesses excellent mechanical properties- extraordinary behaviour in flexure and punching shear coupled with very good energy-absorption characteristics [11, 12]. Due to extraordinary ductility of SIFCON, caused by high fibre content, it seems to be very promising material mostly for applications in structures subjected to impact load. The high fibre content is connected with high production costs, so the replacement of the industrial fibres by waste fibres would significantly decrease the price of the whole composite.

The main objective of the work presented herein is to provide more information about the effects of steel fibers recovered from tires on the mechanical parameters of SIFCON, with focus on its behaviour under blast loading. According to the authors knowledge, there is no study dealing with this topic currently available. The modified methodology for blast resistance assessment is also presented.

2 Experimental

2.1 Preparation of specimens and test methods

Waste steel fibers recovered from scrap tires were obtained from recycling plant "RPG recycling". Fibers were partly contaminated by rubber and textiles (see Figure 1), and had variations in their geometrical properties.

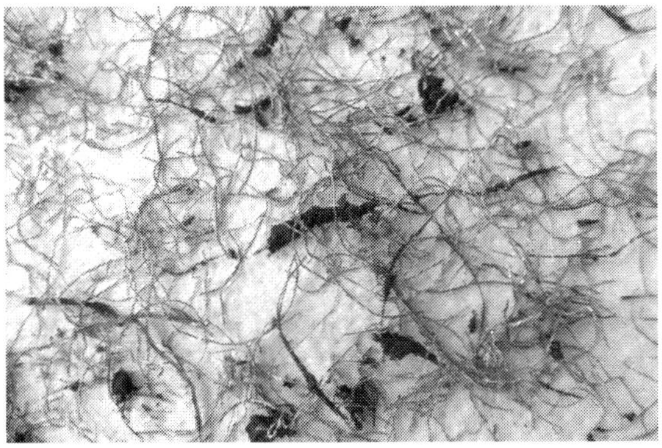

Fig. 1. Waste fibres from recycling process of the tires

The geometrical characterization of the fibers was made by the measurement of their diameters and lengths on samples of randomly selected 100 fibers from three different batches. The distribution of fibre length and diameter is displayed in Figure 2.

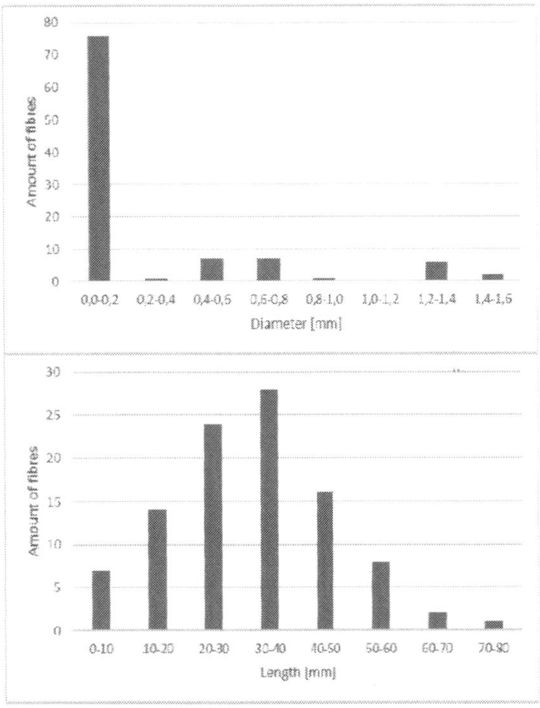

Fig. 2. Distribution of fibre diameter and length

Specimens with three different volume fractions of fibre reinforcement were prepared using waste fibre. The components and mix proportion (in kg.m^{-3}) of the slurry are given in Table 1.

Table 1. Components and mix proportions

Designation/component	SIFCON	UHPFRC
Cement CEM 52.5R Mokra, CR	990	955
Fine aggregate 0-1 mm, Bzenec, CR	890	143
Silica fume Elkem 940U	78	1100
Superplasticizer Glenium ACE 446	12	16
Water	300	230

SIFCON with industrial steel fibres (10% by vol.) and high-performance fibre concrete with 4% by vol. of the fibres were also cast and tested for comparison purposes. The fibres DE 30/06N (KrampeHarex, hooked ends, length 30 mm, diameter 0.6 mm) were used as the reinforcement for the reference specimens, as previous studies revealed their best performance in term of production technology. The manufacturing process of SIFCON involved pre-placing of the steel fibres into the mould with subsequent pouring of the slurry over the pre-placed fibres. The mixing procedure of the reference specimen UHPFRC was as follows: cement, sand and microsilica were dry-mixed, then required quantity of water with plasticizer was added. At the end of mixing, steel fibres were added. The test specimens were demoulded after 24 hours and were cured for 28 days in curing water ponds.

For technological feasibility test, the specimens (100x100x400 mm) with 5-12.5% by vol. (2.5% step) were prepared. After demoulding, the ultrasonic test was adopted to assess the homogeneity of the specimens, using the Proceq Tico device with 54 kHz probe. The velocity of ultrasonic wave was measured at 10 determined spots. The velocity of sound wave in the material can be expressed as: $V = L/t$. [13], where "L" is a travel distance and "t" a propagation time. The measurement of ultrasonic pulse velocity can be used to check the concrete homogeneity and to control the quality of concrete products by evaluation of the velocity variation. The velocity variation coefficient can be calculated according to the following formula: $\Delta = ((v_{max}-v_{min})*100)/v_{min}$, where v_{max}/v_{min} are maximal/minimal values of ultrasonic velocity obtained. For high performance concrete, according to the methodology based on ČSN 73 1371 standard, Δ must not exceed 7.5% for the tested specimen to be evaluated as homogenous.

For the bulk density, compressive and flexural strength investigation the prism specimens of dimensions 400x100x100 mm were cast. The mechanical parameters were obtained using universal strength testing machine TIRAtest 2710, R58/02. The compressive and flexural load was applied in quasi-static conditions at speed of 3 mm/min.

The blast tests were performed using modified methodology M-T0-VTU0 10/09. The methodology was upgraded to cover two tests. First test can be described as follows: The test specimen (500x500x40 mm) is fixed in the frame of the steel stand, which is placed on the solid foundation. The sphere-shaped plastic high explosive Semtex 10, weighting 150 g, is used as testing charge. The weight and distance (100 mm from the test specimen) of the testing charge was adjusted to be strong enough to cause the significant visible damage of the specimens, including fragmentation. Under the stand, the check panel consisting of hardboard-polystyrene-aluminium plate sandwich is placed. Observed and evaluated parameters are the weight of created secondary fragments and their destructive power expressed as the degree of damage of the check panel. Second test covers the loading of the specimen fixed in the stand with the charge 100 g of Semtex 10. The weight for this test was adjusted to cause the low-level damage (small cracks, the slab should remain integral). The change of the ultrasonic

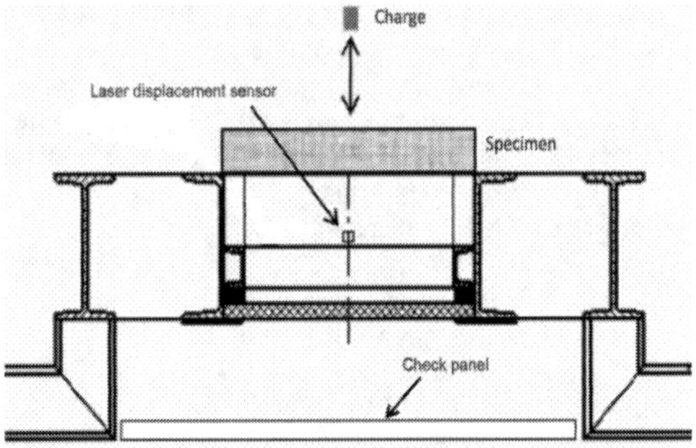

Fig. 3. Schematic view of test rig, arrangement during blast test

wave transit time at four measuring points before and after test was evaluated. The rate of the change of this parameter is directly connected to the damage of material. The scheme of the test rig is depicted in Figure 3. The distance of the charge was same as in the previous test – 100 mm. Additionally, the static and dynamic deflection of the centre of the test slab is measured by the laser sensor optoNCDT ILD2300. The lower the static deflection, the better the performance of the material under the blast load can be expected. The data obtained from both tests corresponds well and give enough information to create good overview about how the particular material performs under blast load; and materials can be compared easily.

2.2 Results and discussion

2.2.1 Technological feasibility

Moulds were filled with 5 – 12.5% by vol. of fibres. 5 and 12.5% of the fibers were evaluated as the threshold values for efficient filling of the moulds – under 5%, the mould was not fully filled, over 12.5%, the structure was so dense, that no more fibres were possible to be placed in the mould even the vibration was adopted. The results of ultrasonic velocity measurements represented by calculated velocity variation coefficient are summarized in Table 2. The specimen with 5.0% of fibres did not achieve satisfactory homogeneity (Δ = 8.4%), due to insufficient fibre amount for creation the uniform skeleton, which corresponds with the findings published in study [10]. All the other specimens showed very good homogeneity-regardless of the measuring point, the ultrasonic velocities of these specimens remain almost constant, with Δ lower than 5%.

Table 2. Calculated velocity variation coefficient

Designation	Velocity variation coefficient [%]
SIF 5.0%	8.4
SIF 7.5%	4.8
SIF 10.0%	4.4
SIF 12.5%	3.1

2.2.2 Quasi-static mechanical tests

Results of the quasi-static mechanical tests are presented in Table 3, the average value of five specimens is presented. As seen in the Table 3, raising of the fibre volume fraction causes increase of the compressive and most noticeable flexural strength - similarly to the SIFCONs with industrial fibres [14].

Table 3. Quasi-static mechanical tests and bulk density

Designation	Bulk density [kg.m^{-3}]	Compressive strength [MPa]	Flexural strength [MPa]
SIF 5.0%	2,488	112.0	18.3
SIF 7.5%	2,510	115.0	20.8
SIF 10.0%	2,550	121.7	23.9
UHPFRC 4%	2,510	146.1	26.2
SIF 10.0% REF	2,585	120.9	25.7

The strength values obtained in the case of specimen SIF 10.0% are very closed to the SIF 10.0% REF with industrial fibres.

2.2.3 Blast tests

The main results of both parts of the blast tests are summarized in Table 4. The test configuration is depicted in Fig. 4, the example of the output of the dynamic deflection measurement shows the graph in Fig. 5. The results of both tests correspond well. Thus, the measurement of the change of the ultrasonic wave transit time can be confirmed as simple, quick and effective method how to assess the blast resistance of the specimens more accurately than just by visual assessment of the damage level of the specimen and check desks. In addition, for this test, less amount of the explosive charge is needed, which reduces the requirements on the testing site and thus costs connected with the testing.

Table 4. Results of the blast tests

Designation	Dynamic /static deflection [mm]	Increase of US wave transit time [%]	Fragment weight [g]
SIF 5.0%	14.17/3.49	112.0	298.3
		238.2	
		380.1	
		311.3	
SIF 7.5%	10.99/3.51	76.4	42.0
		113.7	
		114.4	
		117.3	
SIF 10.0%	11.02/2.87	52.2	23.9
		119.4	
		121.1	
		112.3	
UHPFRC 4%	10.8/2.4	55.92	232.0
		173.58	
		150.1	
		144.1	
SIF10.0% REF	12.5/2.3	120.9	84.0
		181.1	
		100.6	
		98.7	

The blast resistance of the specimens with waste steel fibres rises with increasing amount of incorporated fibres. The specimen with 5% of fibres showed high rate of fragmentation (see Fig. 6 and Table 4). Observed fragments were in form of small aggregates up to 5 mm size. The check panels showed numerous indentations, but without perforation. The increase of the ultrasonic wave transit time was in range 112.0 to 380.0%, which implies the high rate of internal damage. The dynamic deflection was the highest of all tested specimens (14.17 mm). Specimen with 7.5% of waste fibres showed significantly lower fragmentation, dynamic deflection and ultrasonic wave transit time change, compared to the SIF 5% specimen. Ejected fragments´ size was up to 3 mm, indentation but no perforation was observed on the check panel. The best results were obtained in the case of SIF 10.0% specimen. The slab´s integrity didn´t suffer, only some cracks were observed and a few flat splinters were split off from the surface layer (23.9 g), without perforating the check panel. Test slab prepared from UHPFRC with 4% of commercial steel fibres showed different mode of fragmentation – only one fragment (232.0 g) was ejected, creating conic indentation (2 mm depth) without perforation. Internal damage represented

Fig. 4. Test configuration

https://doi.org/10.1051/matecconf/201814901060

by change of the ultrasonic wave transit time was higher compared to slab SIF 7.5% and SIF10.0%. Specimen SIF 10.0% REF with commercial steel fibres showed worse performance compared to the specimen with 10.0% of waste fibres, the weight of the debris was 84.0 g (compared to 23.9 g) and the increase of ultrasonic wave transit time was higher as well.

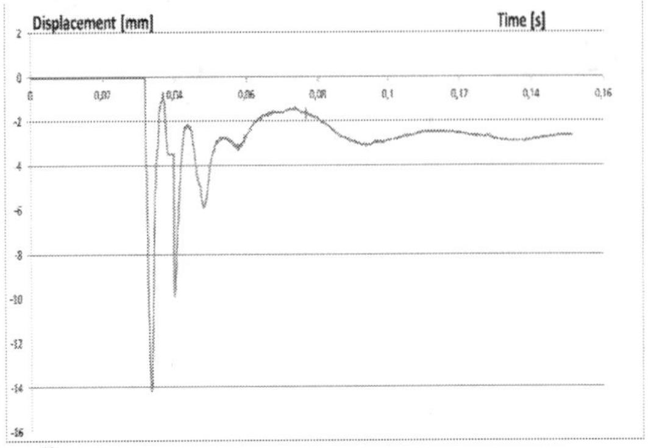

Fig. 5. Example of the output of the dynamic deflection measurement

Comparing the SIFCON specimens, better performance of specimens with waste steel fibres under the blast load was observed, which is probably caused by their hybrid character. Waste fibres from tires are characterized by different shape parameters with high rate of thin shorter fibres (see the distribution in Fig. 2). At quasistatic load, the fracture processes of concrete loaded in tension begin by the formation of numerous fine microcracks; if the tensile load continues to increase, these microcracks connect with each other and form larger cracks, which subsequently leads to failure. Short fibres can bridge microcracks, which increases tensile strength, long fibres bridge macrocracks, so they can provide a stable post-peak response. For the blast load, as it is extremely fast phenomenon, this classical approach cannot be fully adopted, but the hybrid fibre reinforcement is still beneficial, because the same volume content of waste fibres contains more fibres than the commercial ones, so the fibres can be more homogeneously distributed within the concrete, with fewer unreinforced spaces. [15, 16] In this model, thicker and longer fibres help to keep the overall integrity of the material, whereas the shorter and thinner fibres protect the slab from the fragmentation. All the SIFCON specimens (except for SIF 5.0%) outperformed ultra-high-performance concrete, which is caused primarily by the higher fiber amount contained in SIFCON specimens.

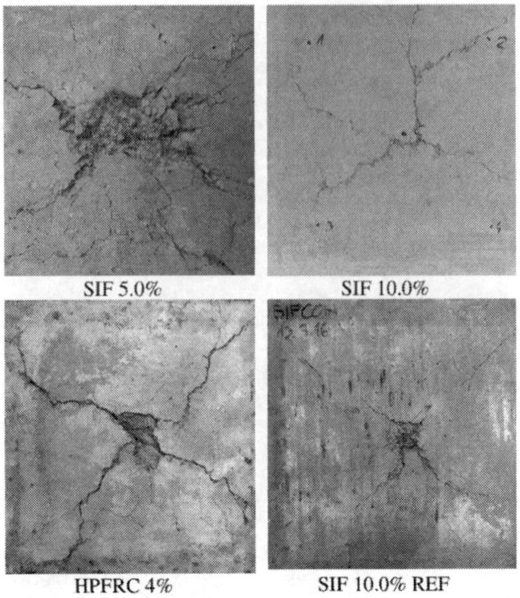

Fig. 6. Specimens after the blast test – 150 g Semtex 10

3 Conclusions

This paper summarizes the results of research on slurry infiltrated fibre concrete reinforced with waste steel fibres from tires. Technological feasibility was assessed and mechanical parameters both at quasi-static and blast load were determined. Obtained results were compared to SIFCON with industrial fibres and HPFRC. The modified methodology for blast resistance assessment is also presented. Following conclusions can be drawn:

- Waste steel fibres have significant variations in their geometrical properties, the length varies between 0.5-80 mm, diameter between 0.03-1.6 mm. Dimension diversity and the bulk nature of the waste fibres were proved not to be obstacle in slurry infiltration technology. Bulk character of the fibres helps to create stable skeleton and fibre mix act like natural hybrid reinforcement. Excellent infiltration level and homogeneity of the specimens can be achieved. Using of unclassified waste steel fibres from tires in SIFCON is feasible from technological point of view.

-The optimal fibre amount for SIFCON technology was determined between 7.5 and 10.0% by vol. The higher the amount of fibres, the better homogeneity is achieved -Mechanical properties (flexural strength, compressive strength and total absorbed energy) increases with raising waste fibre volume fraction.

-Replacement of industrial fibres by waste fibres does not significantly reduce mechanical properties of the SIFCON at quasi-static load.

-Regarding the blast load, SIFCON with waste steel fibres outperformed the SIFCON with industrial fibres, which is attributed to the hybrid nature of the fibres. Thicker and longer fibres help to keep the overall integrity of the material, whereas the shorter and thinner fibres protect the slab from fragmentation. Using of short low diameter fibres means less unreinforced "matrix-only" areas.

-Using of waste fibres in SIFCON technology can reduce the price of this composite by 70% by keeping the SIFCON extraordinary properties, which makes it very competitive material in the area of precast concrete.

Acknowledgement

The authors wish to express their gratitude and sincere appreciation to the authority of The Technology Agency of the Czech Republic, project No. TE02000162 - Centre of advanced materials and technologies for protection and safety enhancement, for financial support.

References

1. O. Sengul, Constr. Build. Mater. **122** (2016)
2. A. Evans and E. Evans, The waste and resources action Programme report (2006)
3. Markets for wire and fiber from waste tires, Contractor´s report, California Department of resources recycling and recovery (2013)
4. G. Centonze, M. Leone and M.A. Aiello, Constr. Build. Mater. **36** (2012)
5. Ch. G. Papakonstantinou and M.J. Tobolski, Cement Concrete Res. **36** (2006)
6. M.A. Aiello, F. Leuzi, G. Centonze and A. Maffezzoli, Waste manag. **29** (2009)
7. A. Caggiano, H. Xargay, P. Folino and E. Martinelli, Cement Concr. Compos. **62** (2015)
8. Y.Wang, H.C. Wu and V.C. Li, Mater. Civ. Eng. **12** (2000)
9. A.G. Graeff, K. Pilakoutas, K. Neocleous, M. Vania and N.N. Peres, Eng. Struct., **45** (2012)
10. D.R. Lankard, Concrete Int., **6**
11. H. Sudarsana Rao and N.V. Ramana, Indian J. Eng. Mater. Sci., **12** (2005)
12. H. Sudarsana Rao, N.V. Ramana and K. Gnaneswar, Indian J. Eng. Mater. Sci. **15** (2008)
13. M. Benachia, O. Jalbaud, X. Roquier, A. H. Alaoui and Y. Burtschell, Alexandria Engineering Journal **54** (2015)
14. Drdlová, R. Řídký and R. Čechmánek, Mater. Sci. Forum **865** (2016)
15. M. Drdlová, J. Buchar, J. Krátký, R. Řídký, Struct. Concrete **16**, 4 (2015)
16. R. Čechmánek, M. Drdlová, M. Boháč, Solid State Phenomena **259** (2017)

 https://doi.org/10.1051/matecconf/201814901061

Influence of initial water content of aggregates on the compressive strength and the shrinkage of High Performance Concretes

Abdelkadir MAKANI [1] and Thierry VIDAL [2]
[1]Université TAHRI Mohamed Béchar, Département de Génie Civil, LFGM (Laboratoire de Fiabilité du Génie Mécanique), BP 417 Béchar (08000), Algérie
[2]Université de Toulouse, UPS, INSA, LMDC (Laboratoire Matériaux et Durabilité des Constructions), 135, Avenue de Rangueil, F-31 077 TOULOUSE Cedex 4, France

Abstract. The time-dependent deformations of concretes, particularly in the case of high performance concrete (HPC) must be taken into account precisely when measurement of civil engineering works. theses deformations is essential for predicting the deflections, the stress distribution in statically indeterminate structures and the loss of pre-stressing force in elements of pre-stressed concrete. However, the experience shows a certain dispersion of values of instantaneous and delayed deformations measured and some significant differences with the values calculated using the building codes. The objective of the present paper focuses on the parameter which is not taken into account in building codes, which could causes the inaccuracies of their predictions: the initial water content of aggregates. The experimental program includes a comparative study of the mechanical behaviours (instantaneous deformations and shrinkage) of HPC with the same basic formulation (water cement ratio, volume of paste), differing principally in the initial water content of aggregates. The experimental results show that HPC made with wet aggregates has a higher resistance than others (dry aggregates). The pre-saturated aggregates would be a water reservoir in which the dough could slow the self-drying due to hydration in the capillary pores at the level of the Interfacial Transition Zone (ITZ). This would increase the amount of hydrates and improve the mechanical behaviour of this interface through better adhesion and less porosity. If the evolution of shrinkage of HPC with wet or dry aggregates can be distinct during the first months, the effect of this formulation parameter is not clear since it differs according to the type of aggregate. Moreover, in the long term, convergent shrinkage shows that this influence becomes insignificant. The comparisons with estimations of Eurocode 2 model were also performed and showed significant differences with the experimental values.

1 Introduction

Les granulats est un des paramètres influant du comportement macroscopique du béton [1-3], plus particulièrement le comportement mécanique instantané et différé à long terme [4]. Cependant, leur influence est mal prise en compte dans les divers règlements en vigueur et pourrait être à l'origine des imprécisions de leurs prédictions. Ceci peut également s'expliquer par la difficulté à estimer et découpler l'impact des nombreuses caractéristiques des granulats sur les propriétés physiques et mécaniques des bétons : nature minéralogique, géométrie/forme, porosité, absorption capillaire, performances mécaniques [5].

L'étude présentée dans cet article a pour objectif d'apporter des informations relatives à l'influence de la teneur en eau initiale des granulats sur les déformations instantanées et différées sans charge des bétons.

Le programme expérimental a été mené sur des Bétons à Hautes Performances (BHP) pour lesquels la connaissance précise du comportement mécanique est primordiale en vue d'un dimensionnement correct des ouvrages. Deux types de granulats (siliceux et calcaire) ont été utilisés dans la production de ces bétons de formulations semblables afin d'étudier l'effet de la teneur en eau initiale des granulats sur la résistance en compression et le retrait.

2 Matériaux et méthodes expérimentales

2.1 Caractérisations des matériaux de base et formulations

Au cours de la caractérisation expérimentale, un sable unique siliceux roulé (SnR), commun à toutes les formulations des BHP étudiés, et deux types de gravillons de nature minéralogique différente ont été testés. Ces granulats seront nommés par la suite selon la terminologie introduite au Tableau 1 qui présente quelques paramètres physiques et mécaniques permettant de caractériser les différents types de granulat.

Tableau 1. Récapitulatif des propriétés physiques et mécaniques des différents granulats utilisés.

	Sable	**Gravillons**	
Forme	Roulé		Concassé
Nature	Siliceux		Calcaire
Classe	0/4	4/10	4/12,5
Nomenclature	SnR	SiR	CaC
Densité	2,8		2,7
Absorption (%)	1,7	1,2	1,3
Porosité totale (%)	8,5	5,3	4,6
Résistance (MPa)	-	-	224
Module (GPa)	-	-	80

Les différents BHP étudiés sont constitués d'un des deux gravillons (SiR et CaC) et d'un même sable de classe 0/4 (SnR) pour l'état sec (S) et humide (H). Le liant est constitué d'un ciment CEM I 52,5R associé à de la fumée de silice (405 + 45 kg/m^3). Pour assurer une maniabilité suffisante, un superplastifiant est ajouté à l'eau. Le rapport E/L (0,35) et de dosage en granulats (gravillon, 1133 kg/m^3 + sable, 944 kg/m^3) sont maintenus constants.

2.2 Méthodes expérimentales

Les échantillons coulés étaient des cylindres 11,8x23,6 cm. Six étaient destinés aux mesures de résistance mécanique en compression pour chacune des échéances, à savoir 3 et 28 jours et six pour le retrait. Après le coulage, les moules sont recouverts par un film plastique afin d'empêcher un départ de l'eau libre du béton par évaporation. Ils sont ensuite conservés dans une salle à 20 °C ± 1 °C et 95 ± 5 % d'humidité relative pendant 24 heures. Le démoulage a été effectué à un jour.

La caractérisation mécanique a été effectuée à 3 et 28 jours à l'aide d'une presse de 3000 kN de capacité maximale, asservie en force avec une vitesse de chargement de 0,5 MPa/s. La résistance en compression a été évaluée suivant les recommandations RILEM 1972.

Pour les mesures de déformations différées sans charge, deux modes de conservation ont été étudiés. Trois éprouvettes sont conservées en mode endogène (sans échange hydrique) dès le démoulage et trois autres en mode dessiccation. Les déformations de retrait sont relevées dès le premier jour après coulage à l'aide d'un rétractomètre (Figure 1).

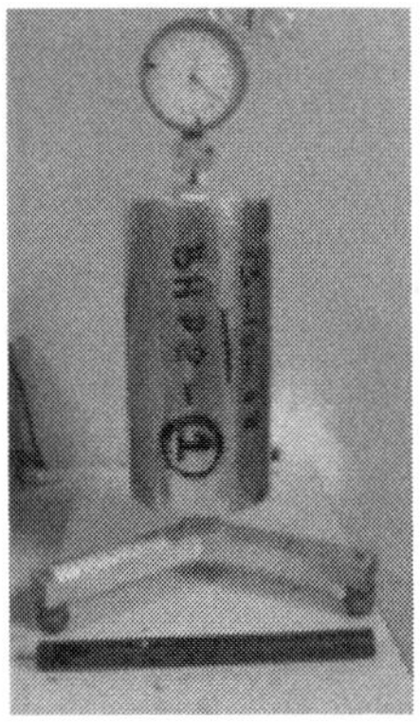

(a)

(b)

Fig. 1. Dispositif de mesure des formations de retrait; (a) en mode endogène et (b) en mode dessiccation.

3 Résultats et discussions

3.1. Comportement mécanique instantané

Les caractéristiques mécaniques instantanées des bétons à 3 jours et à 28 jours sont présentées dans le Tableau 2.

Tableau 2. Récapitulatif des résistances mécaniques en compression.

	Temps (j)	**SiRH**	**SiRS**	**CaCH**	**CaCS**
fcm (MPa)	3	54,5	54,4	66,7	56,2
	28	82,2	73,5	94,7	79,8

Les résistances en compression les plus élevées sont mesurées pour le béton avec granulats calcaires. Ce résultat pourrait être corrélé à la plus forte résistance de ces granulats [6]. L'obtention d'une résistance en compression élevée du béton ne semble donc pas donc uniquement assujettie à celle du granulat utilisé. La qualité de l'interface granulat/pâte de ciment pourrait jouer un rôle [7-9]. Ce constat est relativement proche de celui réalisé par de Larrard et Belloc, à savoir que la résistance en compression des bétons est conditionnée principalement par l'interface entre les deux phases pour les classes de résistance faibles à modérées et par la contrainte ultime en compression du granulat dans le cas où elle se rapproche de celle du béton [6].

La comparaison des résultats des formulations avec granulats préalablement secs ou humides nous permet d'étudier l'influence du paramètre du degré de saturation des granulats sur les performances mécaniques des bétons. Les bétons confectionnés avec des granulats humides (SiRH et CaCH) présentent une résistance plus élevée qu'avec des granulats secs (SiRS et CaCS), cet effet étant plus marqué à 28 qu'à 3 jours. Selon Pigeon et al. [10] et Lopez et al. [11], les granulats saturés représenteraient en quelque sorte un réservoir dans lequel la pâte peut puiser pour ralentir l'auto-dessiccation due à l'hydratation dans les pores capillaires situés au niveau de l'interface. En outre, l'utilisation d'un rapport E/L réduit de nos formulations dans le but d'atteindre des hautes performances mécaniques pourrait entraîner un déficit d'eau nécessaire à l'hydratation localement et notamment au niveau de la zone de transition (ITZ). Dans le cas des granulats humides, l'eau relâchée permettrait d'augmenter la quantité d'hydrates et d'améliorer la résistance de cette interface grâce à une meilleure adhérence. Cette observation conforte la potentielle influence de la qualité de l'ITZ sur la résistance en compression des bétons. Conformément aux cinétiques d'hydratation, cet effet serait donc plus effectif à 28 qu'à 3 jours.

3.2 Comportement différé sans charge en mode endogène

La Figure 2 présente les évolutions des valeurs moyennes des déformations longitudinales de retrait endogène des bétons à base des granulats siliceux et calcaires humides et secs ainsi que les prédictions de l'Eurocode 2 [12].

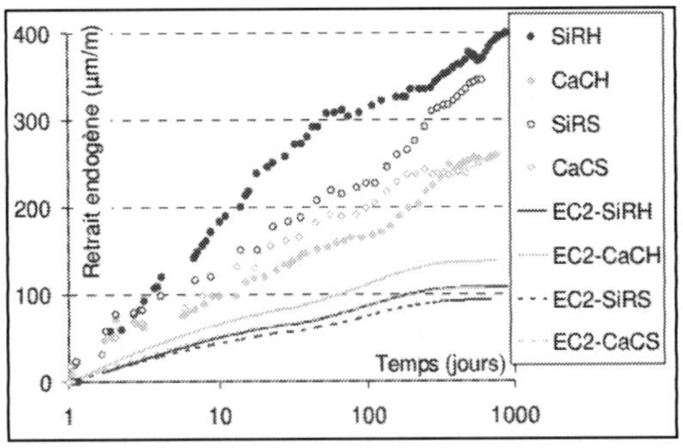

Fig. 2. Effet de la teneur en eau des granulats sur les retraits endogènes des bétons à base des granulats siliceux et calcaires.

On constate des comportements assez différents suivant l'état hydrique initial des granulats. Cependant, l'effet de la teneur en eau n'est pas le même suivant la nature du granulat. Le BHP avec granulat siliceux humide développe le retrait endogène le plus intense au jeune âge comparativement à celui avec granulats secs. Mais l'ordre est inversé pour les BHP avec granulats calcaires. Néanmoins, ces évolutions de déformations assez dispersées durant les premiers mois tendent à se rapprocher aussi bien en termes de cinétiques que d'amplitudes au bout d'un an pour chaque nature minéralogique de granulat.

Il est généralement admis que les granulats, de par leur porosité ouverte, peuvent être considérés comme des réservoirs d'eau pouvant capter ou relâcher de l'eau dans la pâte. Ainsi, les granulats secs absorbent de l'eau pendant le gâchage et une quantité moindre d'eau reste disponible pour l'hydratation à court terme. Ceci implique une structure plus dense de la pâte de ciment contenant des pores plus fins (à un degré d'hydratation donné). Les forces capillaires sont donc plus intenses, ce qui se traduit par un retrait d'auto-dessiccation plus élevé. Une autre explication avancée par certains chercheurs [10-11] et [13] serait que les granulats humides emmagasinent de l'eau et la fournissent plus tard au ciment quand l'eau de gâchage n'est plus disponible pour l'hydratation. Cette eau sera alors utilisée par le ciment pour poursuivre son hydratation. Les deux mécanismes se produisent probablement à des moments différents. Ces transferts d'eau sont probablement complexes et variables car fonctions de plusieurs paramètres interdépendants comme la capacité d'absorption des granulats, la porosité ouverte, la qualité de l'ITZ certainement variable selon la nature minéralogique du granulat (épitaxie pour le granulat calcaire). Ils se produisent au jeune âge principalement durant la phase d'hydratation et des réactions pouzzolaniques de la fumée de silice, lorsque la quantité d'eau non liée est encore importante. Cela engendrerait des différences d'évolutions de retrait endogène. Mais ces phénomènes ne semblent finalement modifier que les cinétiques et non le potentiel de déformation de retrait endogène à plus long terme.

La Figure 2 met en évidence des sous-estimations importantes de l'Eurocode 2. Bien que ce modèle prévoie des écarts dus aux variations des résistances en compression, les différences entre valeurs expérimentales sont plus fortement marquées. On peut en déduire que d'autres paramètres non pris en compte interviennent. Étant donné que les paramètres intrinsèques aux formulations ont été fixés, l'influence du paramètre granulat semble avérée.

3.3 Comportement différé sans charge en mode dessiccation

La Figure 3 présente les évolutions des valeurs moyennes des déformations longitudinales de retrait total des bétons à base de granulats siliceux et calcaires, humides et secs ainsi que les prédictions de l'Eurocode 2[12].

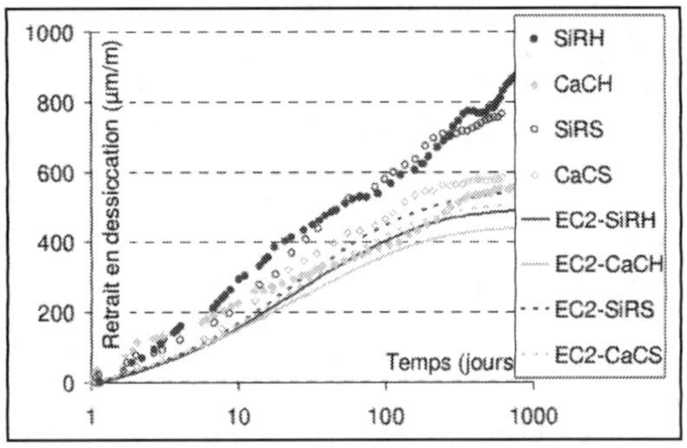

Fig. 3. Effet de la teneur en eau des granulats sur les retraits totaux des bétons à base des granulats siliceux et calcaires.

D'après la Figure 3, les différences entre les évolutions de retrait total sont plus faibles que celles détectées à court terme en mode endogène. A long terme, elles deviennent négligeables quelque soit la nature minéralogique du granulat.

Le fait que les granulats humides fournissent de l'eau à un moment donné au mélange pourrait expliquer les légères variations de comportement durant la phase d'hydratation et des réactions pouzzolaniques. Mais à long terme et à l'échelle de l'éprouvette, cette influence est peu perceptible.

Enfin, l'Eurocode 2 sous-estime le retrait total pour toutes les configurations (Figure 3) et démontre également des imprécisions en n'intégrant que les paramètres de résistance en compression.

4 Conclusions

Cette étude présente le comportement mécanique instantané et différé sans charge des BHP ne se différenciant que par la nature minéralogique et la teneur en eau initiale des granulats. Malgré que les différents bétons aient même formulation de base, les résultats montrent parfois d'importantes différences.

La résistance en compression semble conditionnée par la qualité de l'interface pâte-granulat (ITZ), avec notamment les valeurs plus fortes observées pour le BHP avec granulats calcaires probablement grâce au phénomène d'épitaxie.

Les BHP confectionnés avec des granulats humides présentent une résistance plus élevée que ceux avec des granulats secs. Les granulats saturés représenteraient en quelque sorte un réservoir d'eau dans lequel la pâte pourrait puiser pour ralentir l'auto-dessiccation due à l'hydratation dans les pores capillaires situés au niveau de l'ITZ. Cela permettrait d'augmenter la quantité d'hydrates et d'améliorer la résistance de cette interface grâce à une meilleure adhérence et une moindre porosité.

La comparaison entre les retraits des bétons à base de granulats humides et secs montre que les évolutions de déformations, qui sont assez dispersées durant les premiers mois, tendent à se rapprocher, au bout d'un an, aussi bien en termes de cinétique que d'amplitude pour chaque nature minéralogique de granulat. Le fait que les granulats humides fournissent de l'eau, à un moment donné, au mélange pourrait expliquer les légères variations de comportement durant la phase d'hydratation, y compris les réactions pouzzolaniques. Mais à long terme et à l'échelle de l'éprouvette, cette influence paraît peu perceptible.

De manière générale, les prédictions du modèle Eurocode 2 sous-estiment les déformations de retrait et ne permettent pas de retrouver les écarts entre les amplitudes des différents BHP. Cette étude démontre ainsi les limites de ce modèle dans l'estimation du comportement mécanique à long terme de ces matériaux.

References

1. A. M. Neville, Properties of Concrete. Fourth and Final Edition, *John Wiley & Sons*, Inc, New-York, USA (1996)
2. G. Pons et J. M. Torrenti, Retrait et fluage. La durabilité des bétons, Bases scientifiques pour la formulation de bétons durables dans leur environnement. Sous la direction de Jean-Pierre OLLIVIER et Angélique VICHOT , *Presses de l'ENPC*, Paris, France (2008)
3. A. Makani, T. Vidal, G. Pons et G. Escadeillas, Time-dependent behaviour of high performance concrete: influence of coarse aggregate characteristics, *ICEM14 International Conference on Experimental Mechanics*, Poitiers, France, 4-9 July (2010)
4. A. Makani, T. Vidal et G. Pons, Influence de la nature minéralogique des granulats sur le comportement mécanique différé des bétons, *Colloque International Francophone NoMaD*, Toulouse, France, 19-20 Novembre (2012)
5. A. Makani, Influence de la nature minéralogique des granulats sur le comportement mécanique différé des bétons, *Thèse de doctorat*, INSA, Toulouse, France (2011)
6. F. de Larrard et A. Belloc, The Influence of Aggregate on the Compressive Strength of Normal and High-Strength Concrete, *ACI Materials Journal*, **94**, 5, 417-426 (1997)
7. J. P. Ollivier, J. C. Maso, et B. Bourdette, Interfacial transition zone in concrete, *Advanced Cement Based Materials*, **2**, 1, 30-38 (1995)
8. J. C. Maso, Interfacial transition zone in cement, *E&FN Spon.*, **108**. London, England (1996)
9. I. Gaweska Hager, Comportement à haute température des bétons à haute performance-évolution des principales propriétés mécaniques, *Thèse de doctorat*, ENPC, Paris, France (2004)
10. M. Pigeon, G. Toma, J. Marchand et B. Bissonnette, Étude expérimentale du retrait endogène restreint au jeune âge, *Mater. Struct.*, **36**, 264 ,666-672 (2003)
11. M. Lopez, L. F. Kahn et K. E. Kurtis, Effect of Internally Stored Water on Creep of High-Performance Concrete, *ACI Materials Journal*, **105**, 3, 265-273 (2008)
12. Eurocode 2, EN 1992-2, Design of concrete structures. Concrete bridges, (2006)
13. N. Belaribi, G. Pons et B. Perrin, Delayed behaviour of concrete: Influence of additions and aggregate characteristics in relation to moisture variations, *Cement and Concrete Research*, **27**,9, 1429-1438 (1997)

 https://doi.org/10.1051/matecconf/201814901062

Blended cement hydration assessment by thermogravimetric analysis and isothermal calorimetry

Meriem Meziani[1], Nasser Chelouah[1], Ouali Amiri[2], Nordine Leklou[2]
[1]Laboratoire de Génie de la Construction et d'Architecture, Faculté de Technologie, Université de Bejaia, Algérie
[2]Université de Nantes, GeM, UMR - 6183 CNRS, IUT de Saint Nazaire, France

Abstract. In the present study, the hydration of Portland cement pastes containing 5%, 10%, 15% and 20% tuff, limestone filler and granodiorite was investigated by thermogravimetric analysis coupled with differential scanning calorimetry and microcalorimetry isotherm. The monitoring of the hydration kinetics by thermogravimetric analysis made it possible to quantify the quantity of water combined with the cement (non-evaporable water) and the degree of hydration. By coupling this technique to the differential scanning calorimetry, it was also possible to measure the energy absorbed or released by the material during its decomposition. The results showed that the non-evaporable water content and the degree of hydration of the mixtures containing various mineral admixtures were relatively lower with respect to the reference mixture when as the content of mineral admixture increased. The effect of the evolution of the hydration process on the mechanical properties of mortars was also monitored. The relative variation of the compressive strength to that of the flexural strength was evaluated at 7, 28 and 90 days. Results showed that all the mixtures have a greater contribution in flexion than in compression.

1 Introduction

The constituents of Portland cement undergo a series of chemical and physical processes when mixed with water. The overall process is named as hydration yielding hydrated products. It is known that Portland cement is a multi-component system; therefore its hydration is a rather complex process consisting of a series of individual chemical reactions that take place both in parallel and successively. So, the explanation of its hydration is done, in general, by considering the individual hydrations of the major compounds and their interactions with each other and with other minor phases present in the cement. These considerations affect both the mechanism and the kinetics of hydration. Hydration of Portland cement is affected by a variety of factors such as the compound composition of the cement, the amount of gypsum, fineness, surface area and particle size distribution of the cement, mix composition and cure conditions, presence and amount of minerals and chemical admixtures.

The mechanism and kinetics of the hydration of mineral admixture incorporated cements is of great importance. Some of these materials are considered as inert substances and their effect on hydration is more of a physical nature, some others are of significant chemical reaction with the Portland cement phases. However, mineral admixtures modify the hydration process by both physical and chemical means which are sometimes very difficult to dissociate [1]. Furthermore, the level of the mineral admixtures effect on hydration may fluctuate noticeably, due to the differences in their chemical and mineralogical compositions and physical properties. Study of the hydration process and products in concretes, and even in mortars, is complicated by the presence of the aggregate. But much more detailed information can be obtained from studies on pastes. The hydration of cement is commonly studied using cement pastes which are allowed to hydrate at room temperature.

Although the process of hydration is yet to be fully understood, several modern analytical equipment are being used for the investigation, like transmission electron microscope, scanning electron microscope, quantitative x-ray diffraction, thermogravimetric techniques, conduction calorimetry, and so on.

The progress of the hydration process may be followed by determining the amount of non-evaporable water, determined as the loss on ignition of samples in which free water has been removed first. This method is directly proportional to the degree of hydration and often used as a measure of the fraction of cement that had reacted. The overall kinetics of hydration may be also monitored on the basis of the amount of heat released in the course of hydration.

In this paper, we will restrict our analysis to the examination of the influence of substitution rate and the type of mineral admixture on the performances of the cement pastes and mortars at the hardened state by

https://doi.org/10.1051/matecconf/201814901062

following the evolution of the hydration process (heat and Kinetics of hydration) and its effect on the mechanical properties of mortars.

2 Materials and mix design

The cement used in paste and mortar specimens was a commercially available CEM I 52.5 R according to NF EN 197-1 with a specific gravity of 3.14 and a fineness of 440 m^2/kg [2]. The tuff (T) used is received in bulk and then ground for 20 min until a fineness of 430 m^2/kg, its specific gravity is 2.66. Limestone (LF) contains 97.30% of CaCO3. It is characterized by a fineness of 448 m^2/Kg and a specific gravity of 2.71. The granodiorite (Gr) is received in bulk and then crushed and ground for 12 min until a fineness of 414 m^2/kg, its specific gravity is 2.54. XRD patterns and SEM image of the mineral admixtures were given in **Error! Reference source not found.** and **Error! Reference source not found.**.

Blended cements were obtained by direct replacement of a proportion of Portland cement with tuff (T), granodiorite (Gr), or limestone filler (LF). The samples thus obtained were named according to the type and the rate of substitution of the cement by the mineral admixtures. For example, 5T, 5Gr and 5LF refer to cements containing 5% tuff (T), 5% granodiorite (G) and 5% limestone filler (LF), respectively.

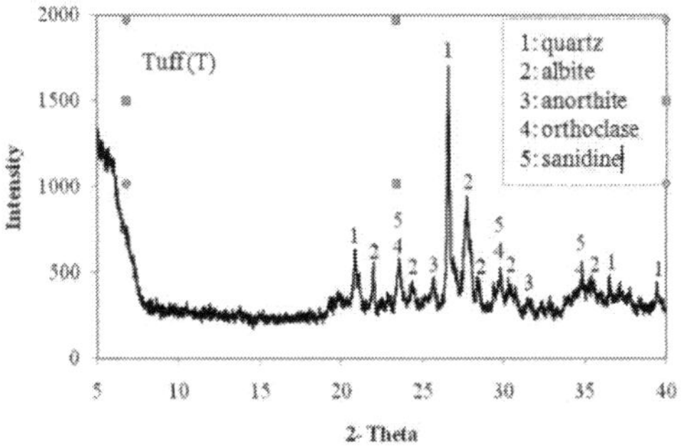

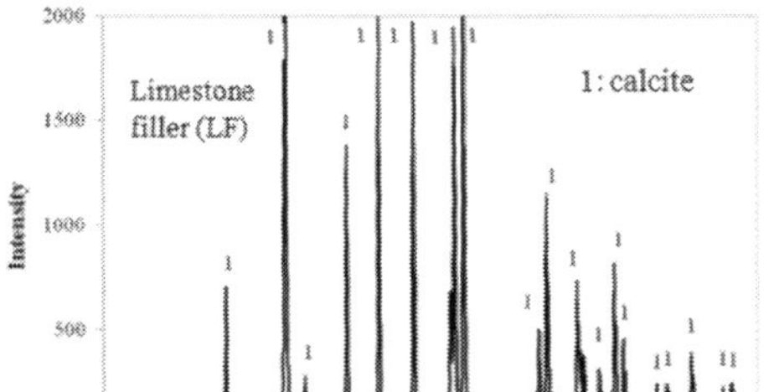

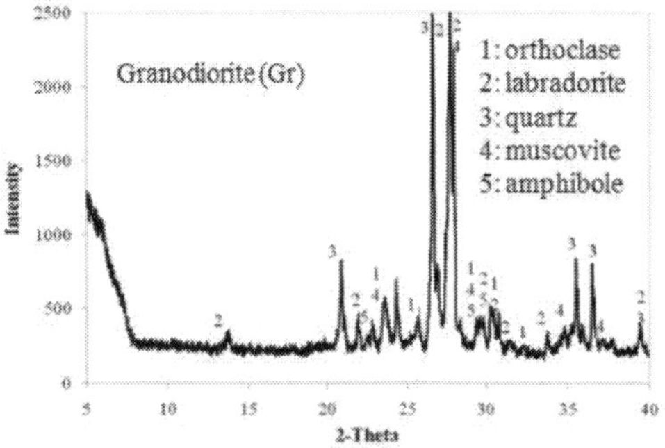

Fig. 1. Caption of the XRD patterns of the mineral admixtures

Fig. 2. SEM image of the mineral admixtures

Cement pastes and normalized mortars were formulated with a partially replacing the cement weight with the corresponding percentage of mineral admixture. Five substitution rates were adopted: 0, 5, 10, 15 and

20%. So, 13 mortars and 13 cement pastes were formulated and mixed according to NF EN 196-1 [3] with a mass ratio of water/binder of 0.35 for pastes and 0.5 for mortars. The content of water is kept constant for all formulations, which leads to changes in the rheological properties of the pastes and mortars. The choice of fixing the quantity of water and varying the rheological properties and not the contrary is based on the fact that the second case affects proportionally the initial porosity of the mixtures. This porosity, consequently, will directly influence the development of mechanical resistances.

3 Techniques of testing

3.1 Heat of hydration

The heat flux due to the hydration reactions of the cement pastes was measured at 20 °C using a multi-channel isothermal microcalorimeter of TAM Air type interfaced by TA instruments software. After mixing, two samples of each paste were immediately taken from the mixture, placed in bottle, weighed and placed in the microcalorimeter with the reference samples. The heat release of the samples characterizing the reaction rate of the binder was measured automatically in order to quantify the influence of the type of the mineral admixture on the hydration and reactivity of cement at early age in the mixtures.

3.2 Thermogravimetric analysis of hydration kinetics

When subjected to high temperatures, the hydration products of the cement decompose releasing water vapor and carbon dioxide. The monitoring and measurement of the different mass losses allowed us to quantify the proportion of water combined with the cement (non-evaporable water) and the degree of hydration. The apparatus used is a Mettler Toledo TGA / DSC 1 type, which measures simultaneously the mass variations of the sample tested and the heat flux between the test sample and a reference crucible. Mass loss calculations and the integration of DTG and heat flux peaks were carried out by the post-processing software associated with the apparatus.

The amount of non-evaporable water and the degree of hydration of the cured cement pastes were determined at 7, 28 and 90 days. At each test age, the hardened pastes were further pulverized to assure that all run through the 80μm sieve and then immediately analyzed between room temperature and 1050 °C at a heating rate of 10 °C/min in a static air atmosphere.

3.3 Amount of non-evaporable water and degree of hydration

Since no drying is carried out, the sample contains both evaporable water and non-evaporable water at the time of the test. The DTG curves show that it is not possible to clearly differentiate the departure of each of the two types of water. It is therefore necessary to define, prior to the analysis of the results, a non-evaporable water evaporation-water boundary temperature. Two temperature values, 105 °C and 145 °C, were defined as the limit between the end of free water departure and the beginning of hydrate decomposition [4]. In view of the literature, we chose a border temperature of 105 °C.

The amount of non-evaporable water was calculated from the following equation [4]:

$$W_n = L_{dh} + L_{dx} + 0{,}41\ (L_{dc} - L_{dca}) \quad (1)$$

W_n (%) : Amount of non-evaporable water

L_{dh} (%) : Weight loss between 105 and 400 °C due to the dehydration of the hydrates

L_{dx} (%) : Weight loss between 400 and 600°C due to the dehydroxylation of calcium hydroxide

L_{dc} (%) : Weight loss between 600 et 800°C due to the decarbonation of calcite

L_{dca} (%) : Weight loss between 600 et 800 °C/weight at 105 °C due to the decarbonatation of anhydrous samples (x% cement + y% mineral admixture).

It should be noted that the weight loss is due to both the decarbonation of calcium hydroxide which has been carbonated after the paste has been made, and the carbonation of the anhydrous material.

The degree of hydration is the percentage ratio between the chemically bound water of a given cement paste and that of a fully hydrated paste. The methodology for calculating the degree of hydration from a thermal analysis differs significantly from one researcher to another.

The degree of hydration α(t) of the cement can be calculated from the following equation [5]

$$\alpha\ (\%) = \frac{Wn}{W\infty} \times 100 \quad (1)$$

W_∞: Water required for a complete hydration of anhydrous cement.

The amount of water required for a complete hydration of anhydrous cement was calculated using the method proposed by S.M. Monteagudo (2014) which is inspired by that of Michaelis-Menten (equation (3)) [4]. The latter is commonly used to calculate the rate of a chemical reaction.

$$V = \frac{Vm \times S}{S + Km} \quad (2)$$

Where V is the reaction rate, Vm is the maximum speed, S is the initial substrate concentration, and Km is the Michaelis-Menten constant.

Applied to the case studied in this study, the equation would be as stated in equation (4).

$$Wn = \frac{W\infty \times t}{t + K} \quad (3)$$

t represents the curing time expressed in hours and K is a constant. To determine the parameters K and W∞ numerically, we have used the linearization of equation (4). It is a change of variables that transforms the initial equation into a linear equation, which can then be adjusted graphically or by linear regression. The linearization method adopted is that of Lineweaver - Burk. This method is considered one of the most accurate in determining the K and W∞ characteristics.

In this approach, the constants K and W∞ were determined by the representation of their inverses (Lineweaver - Burk representation), which is a straight line with the following equation:

$$\frac{1}{Wn} = \left(\frac{K}{W\infty} \times \frac{1}{t}\right) + \frac{1}{W\infty} \quad (4)$$

3.4 Flexural and compressive strength

The mortars were cast in 40 × 40 × 160 mm^3 metal molds. The mixtures were prepared according to the specifications of NF EN 196-1 [3]. After casting, the molds were stored in a humid chamber for 24 hours at 20 ± 1 °C and at a relative humidity of more than 90%. After 24 hours, the samples were demolded and then stored in the humid chamber. At the age of the test, the specimens were first broken in three-point bending flexural test. The half-prisms resulting from the rupture in flexural test were broken in compressive strength test. The data reported for each mortar represent the average values obtained from three flexural strength tests and six compressive strength tests.

3.5 Fourier Transform Infrared Spectroscopy (FTIR)

Fourier Transform Infrared Spectrometry (FTIR) is a method that identifies chemical functions in a product. The wavelengths to which the sample absorbs are characteristic of the chemical groups present in the analyzed material. Infrared analysis measurements were carried out using an IRAFFINITY-1 apparatus between 400 and 4000 cm^{-1}, the scanning number used was 40 and the resolution was 4 cm^{-1}. The traced FTIR spectra were obtained by the KBr pellet technique. The sample of the paste was finely ground and sieved and then mixed with dried KBr, the mixture thus obtained was pressed so as to obtain a pellet as thin as possible so that it is transparent to infrared rays.

4 Results and discussions

4.1 Infrared Spectrometry

Figure 3 shows the IR spectra of hydrated mixtures containing 20% of mineral admixtures compared to the parent mixture. A presence of calcium hydroxide marked by the O–H vibration at 3640 cm^{-1}, but in the case of 20LF this band was loss. The vibration intensities at 3415 cm^{-1} and 1638 cm^{-1} indicates the formation of calcium silicate hydrates (C-S-H). A band at 1114 cm^{-1} attributed to the formation of ettringite; it should be noticed that in the case of 20LF, the IR spectrum showed a total absence of this band. A wide band located at about 970 cm^{-1} related to the presence of C-S-H. A triplet at 1455 cm^{-1}, 875 cm^{-1}, and 713 cm^{-1} characterizing the elongation vibrations of the C-O bond present in calcium carbonate (CaCO3). These three bands were accompanied by harmonics at about 2980 cm^{-1}, 2875 cm^{-1}, 2517 cm^{-1} and 1797 cm^{-1}.

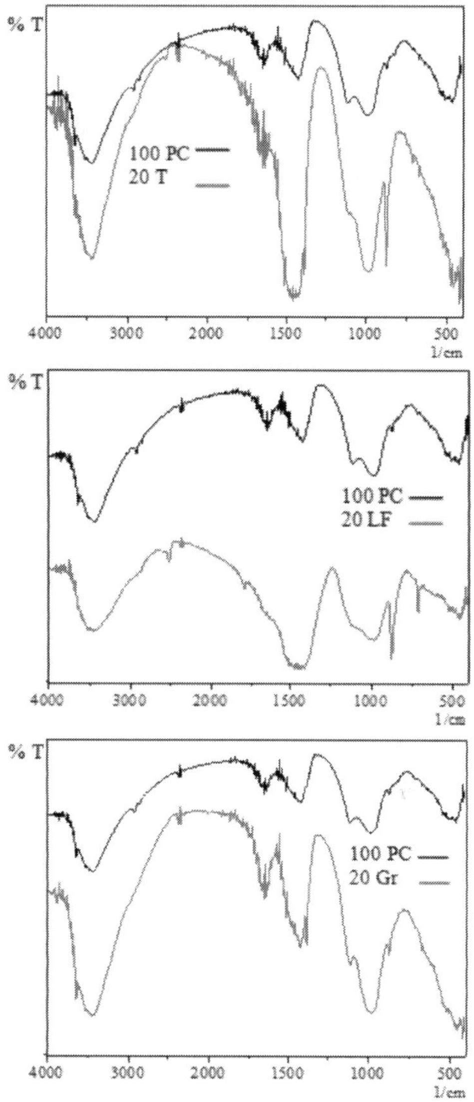

Fig. 3. Infrared spectrometry of the pastes

4.2 Hydration heat by microcalorimetry

Table 1 shows that the heat flux at its maximum decreases with the increase of the mineral admixture amount in the binder. The time taken to reach the maximum peak decreases with the increase of the mineral admixture content, but in the case of the LF the appearance of the peak is delayed.

Table 1 shows also the measured values of the total hydration heat released over a period of 72 hours (exper hydr heat) and the hydration heat calculated for a pure cement (calc hydr heat). The calculated heat of hydration was calculated according to the following expression:

$$Q_{calc\,hydr\,heat} = \frac{Q_{exper\,hydr\,heat}}{100 - \%\,substitution} \times 100 \qquad (5)$$

An increase in the amount of mineral admixture decreases the experimental hydration heat and reduces the total heat released. The experimental hydration heat of 213.39 J/g in the reference paste was reduced to 173.5 J/g, 176.04 J/g and 164.72 J/g for 20% replacement of 20% of cement by LF, Gr and T respectively. As the cement has been partially replaced by the mineral admixtures, this reduction can be attributed to the dilution effect. Indeed, an increase in the rate of mineral admixture is accompanied by a decrease in the cement content, and naturally less cement means less hydration products. This effect is also indicated by M. Tokyay (2016) [1].

If the mineral admixtures used did not have a significant effect on hydration, the heat released in samples containing 20% of mineral admixtures should be equal to half of 213.39 J/g; or 106.7 J/g. This result is fully agree with that of P. Dabic et al (2006)[6]. The limestone filler, which has been considered by several researchers as an inert material, is gradually accepted as contributing to the hydration process by the formation of hydrated calcium monocarboaluminates [7].

If the use of the mineral admixture did not contribute to the release of heat, the hydration heat calculated with pure cement would be 213.39 J/g, although some samples released higher heat. The increased contribution to heat release is particularly marked with the addition of 15-20% limestone filler and 20% granodiorite.

Table 1. Evaluation of the characteristics of the heat flux peak (q_{max}) and the total hydration heat of the different pastes.

	Intensity of hydration peak (mW/g)	Age of hydration peak (h)	Experimental hydration heat at 72h (J/g)	Calculated hydration heat at 72h (J/g)
100PC	4.92	7.76	213.39	213.39
5 T	4.59	7.76	186.60	196.42
10 T	4.62	7.67	185.42	206.02
15 T	4.49	7.53	180.94	212.87
20 T	4.38	7.61	164.72	205.90
5 LF	4.54	8.00	187.76	197.64
10 LF	4.53	8.04	185.25	205.83
15 LF	4.44	8.08	182.65	214.88
20 LF	4.41	8.01	173.50	216.88
5 Gr	4.60	7.78	189.66	199.64
10 Gr	4.59	7.82	186.59	207.32
15 Gr	4.52	7.65	180.97	212.91
20 Gr	4.54	7.38	176.04	220.05

4.3 Non-evaporable water content

Non-evaporable water content is plotted in figures 4–6. The curves have been progressively increased over time but remain below the parent mixture of Portland cement alone at all tests ages. The decrease in the amount of non-evaporable of mixtures containing mineral admixtures was recorded at its maximum for 20% replacement of cement. At 7 days of age, the decrease was 0%, 6.76% and 6.24% for mixtures of Tuff, granodiorite and limestone filler, respectively. However, this decrease increases at 28 and 90 days. It reached 6.39%, 7% and 9.74% at 28 days and 12.27%, 11.7% and 15.46% at 90 days respectively. The results obtained are probably due to the decrease in the quantity of the cement partially substituted by the mineral admixtures, dilution effect [8]. It should be noted that the 5LF mixture exhibited a higher amount of non-evaporable water than that of the parent mixture. Incorporation of 5% limestone filler into the cement matrix resulted in dispersion of the cement particles, thereby reducing the tendency to flocculation and exposing more cement surfaces to hydration. In addition, the dispersion effect leads to a more homogeneous distribution of the water in the cement paste and thus facilitates hydration. However, this effect can be avoided if the mineral admixture is high [9].

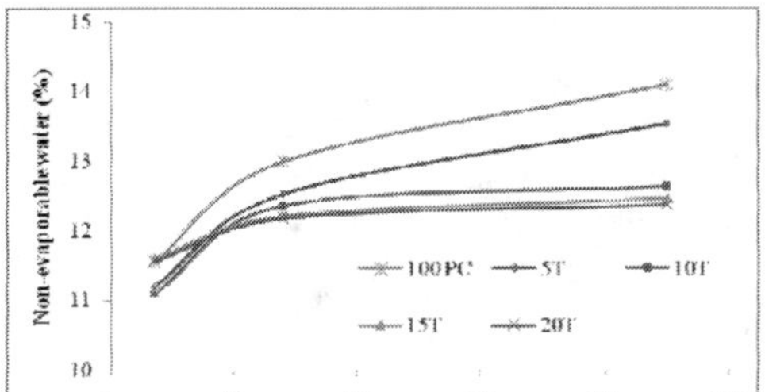

Fig. 4. Non–evaporable water content in the pastes containing tuff

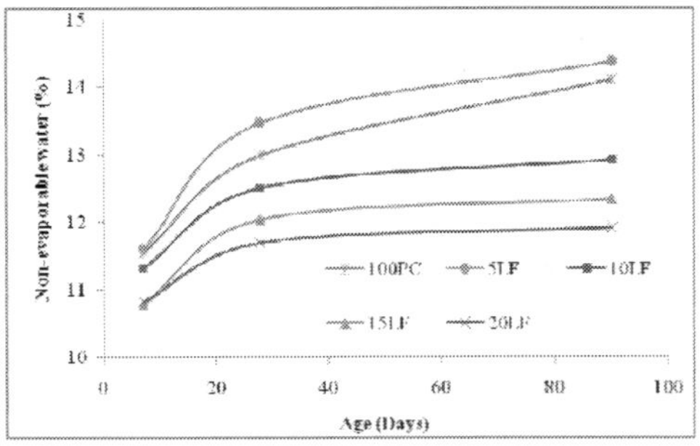

Fig. 5. Non–evaporable water content in the pastes containing limestone filler

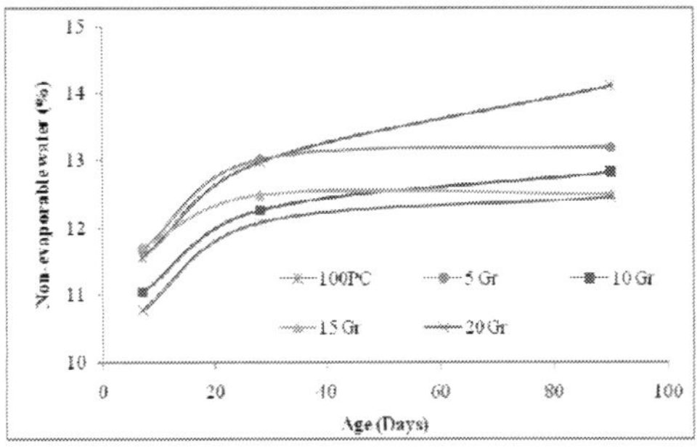

Fig. 6. Non–evaporable water content in the pastes containing granodiorite

4.4 Degree of hydration

The degrees of hydration of the pastes were monitored at 7, 28 and 90 days and presented in table 2. The degrees showed a gradual evolution over time. They were very high and exceed 100% in some mixtures at 90 days. This is a direct result of the number of experimental values considered and from the way of obtaining the interpolated curves. Since the last age for tests was 90 days, mixtures for which hydration has not been completed, hydration values above 100% were obtained because hydration proceeds more slowly and remains in progress. Mixtures containing different percentages of tuff appear to be the most comparable of the parent mixture in terms of degree of hydration.

Table 2. Degree of hydration of the pastes

	Degree of hydration (%)		
	7 days	28 days	90 days
100PC	85.15	95.78	104.02
5T	85.89	96.05	103.80
10T	91.31	94.68	104.00
15T	93.33	98.24	102.40
20T	85.15	96.76	100.05
5Gr	86.29	96.18	97.46
10Gr	81.49	94.63	96.57
15Gr	91.91	96.85	96.44
20Gr	85.42	95.91	98.81
5LF	81.43	94.61	100.94
10LF	87.21	96.46	99.53
15LF	86.10	96.12	98.53
20LF	89.98	97.29	99.15

4.5 Mechanical strength

In order to illustrate the similarities and differences between the flexural strength and the compressive strength, we compared the relative variation (relative to the reference mortar) of the compressive strength to that of the flexural strength. From figure 6 it can be noted that replacement of cement by the mineral admixture resulted in less compressive and flexural strength reduction at the early age (7 days); but at the latter ages of curing, replacement at 5% and 10% produced lower flexural strength improvement. In fact, at 7 and 28 days the tuff mortars line up on the bisector of the diagram. On the other hand, some points diverge from the bisector and present a greater contribution in flexural than in compressive strength namely limestone filler and granodiorite. At 90 days, all the mortars deviate from the bisector and present a greater contribution in flexural than in compressive strength. Similar results were found in the literature [10].

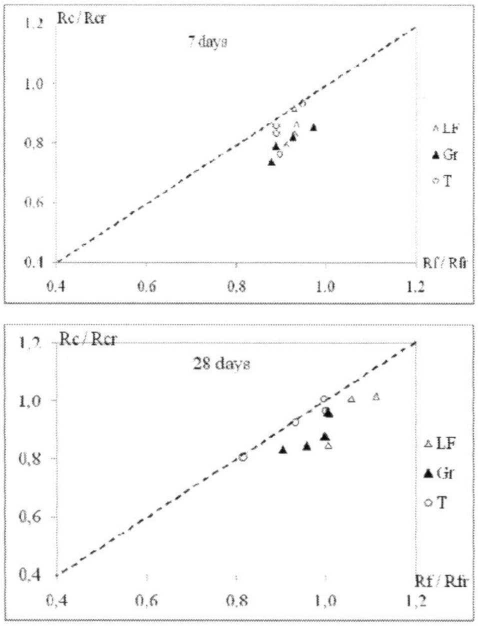

 https://doi.org/10.1051/matecconf/201814901062

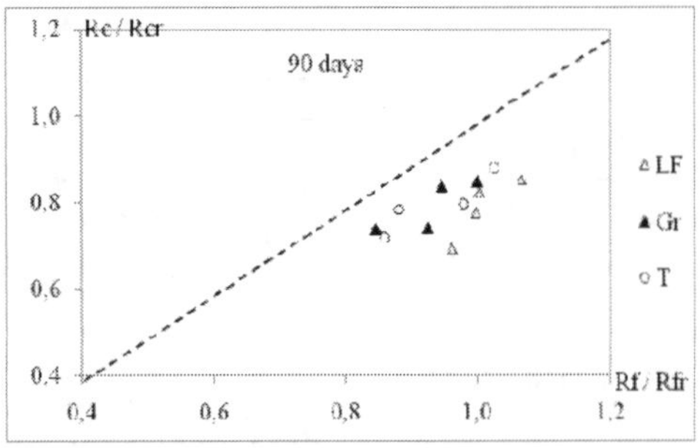

Fig. 6. Relationship between the relative compression and flexural strength of mortars

5 Conclusions

This paper presents the influence of the type and level of mineral admixture in the binder on the evolution of the hydration process and its effect on the mechanical properties. From the results obtained the following conclusions can be drown:

The maximum peak of heat flux decreases when the amount of mineral admixture increases. And, the necessary time to reach the maximum peak decreases with the increase of the mineral admixture content, but in the case of the LF the appearance of the peak is delayed.

An increase in the amount of mineral admixture decreases the experimental hydration heat and reduces the total heat released.

The non-evaporable water content and the degree of hydration of the mixtures containing various mineral admixtures were relatively lower with respect to the reference mixture.

The relative variation of the compressive strength to that of the flexural strength showed that all the mixtures have a greater contribution in flexural than in compressive strength.

References

[1] M. Okyay, *Cement and concrete mineral* admixtures, CRC Press , 2016.

[2] NF EN 197-1 Ciment- Partie 1: Composition, spécifications et critères de conformité des ciments courants, Comité européen de normalisation, Bruxelles, 2001.

[3] NF EN 196-1 Méthodes d'essais des ciments. partie 1: Détermination des résistances mécaniques, (2006).

[4] S.M. Monteagudo, A. Moragues, J.C. Gálvez, M.J. Casati, E. Reyes, The degree of hydration assessment of blended cement pastes by differential thermal and thermogravimetric analysis . Morphological evolution of the solid phases, Thermochim. Acta. **592** (2014) 37–51.

[5] I. Pane, W. Hansen, Investigation of blended cement hydration by isothermal calorimetry and thermal analysis, Cem. Concr. Res. **35** (2005) 1155–1164.

[6] P. Dabic, P. Krolo, A. Bubic, R. Krstulovic, Solidification and stabilization of waste zeolite in cement matrix, in: 16.Internationale Baustofftagung, 2006: pp. 483–490.

[7] R. Lewis, L. Sear, P. Wainwright, R. Ryle, *Cementitious additions*, in: J. Newman, B. Seng Choo (Eds.), Adv. Concr. Technol. Const. Mater., Butterwort, Great Britain,.2003: p. 281.

[8] P. Krolo, R. Krstulovic, D. Pero, A. Bubic, Hydration and leaching of the cement-zeolite composite, Ceram. - Silikaty. **49** (2005) 213–219.

[9] P. Lawrence, M. Cyr, E. Ringot, Mineral admixtures in mortars: Effect of inert materials on short-term hydration, Cem. Concr. Res. **12** (2003) 1939–1947.

[10] C. Poon, L. Lam, S. Kou, Z. Lin, A study on the hydration rate of natural zeolite blended cement pastes, Constr. Build. Mater. **13** (1999) 427–432.

Study of the mechanical characteristics and chemical degradation of concretes based on machining sludge

Souad Kherbache[1], Nedjima Bouzidi[2], Abdelkader Tahakourt[1] and Karim Moussaceb[2]
[1]Laboratory of Construction Engineering and Architecture (LGCA)
[2]Laboratory of Materials Technology and Process Engineering (LTMGP)
University A/Mira, Faculty of Technology, Department of civil engineering, Bejaia, Algeria

Abstract. The main objective of this work is to study the mechanical characteristics and the chemical degradation of mortars and concretes made obtained by substituting the cement by the machining sludge. Several materials were carried out by substituting the cement with different amount of machining sludge (5, 10, 15 and 20%). The obtained materials are compared to those without substitution (ie: 00wt% of machining sludge). Mechanical and physical properties of the two materials (mortars and concretes) as flexural and compressive strengths were studied. Physical properties as shrinkage, the bloating effect, loss in mass, slump concretes and porosity were evaluated. In addition, chemical degradation is studied for the conservation of these concretes and mortars from machining sludge in different aggressive environment. Moreover, the addition of 5wt% of the machining sludge in the mixture of the mortar and the concretes seems to be interesting.

1 Introduction

The common term "machining sludge" includes several types of industrial waste coming from the mechanical industry: rectification sludge, but also sludge from grinding, mass finishing, etc. [1]. Sludge has long been a wasteful and bulky waste for communities and industrialists [2]. Each year, the mechanical industries produce some 50,000 tons of sludge. To date, 80% of these sludges are not reused and end up in landfills. Aware of the enormous potential for recovery of this waste, the Carnot Cetim institute has valorised these muds in bricks [3-4]. For this purpose, the valorisation and management of industrial sludge are topical subjects; the valorization of waste (sludge) as alternative raw materials for the manufacture of ceramics has been studied by J.A. Junkes [5].

Due to the increase in sludge treatment costs, industrial sites must design and implement intelligent methods for the treatment, reduction of sludge from processing and recovery of by-products from production. A sustainable waste management system for sludge effluent treatment was a pressing problem. Recycling is always recommended in terms of environmental sustainability; following a waste recovery approach [6]. There are increasing constraints on regulation. The merger constitutes a technologically and economically value-added [7].

Different studies have been carried out on the processes commonly used for the stabilization and treatment of sludge, as well as the new technologies which have been developed in recent years. It should be noted that very good literature reviews on novelties in the field of treatment, use and disposal of sewage sludge are presented annually in the journal «Water Environment Research» [8-13].

Studies have also shown that neglecting the shrinkage phenomenon leads to an overestimation of the internal diffusion coefficient for convective drying of sludge and sawdust / sludge mixtures [14]. Moreover, the presence of high concentrations of toxic metals in sludge constitutes a significant obstacle [15-18]. Conventional sludge treatment processes, such as aerobic or anaerobic stabilization, are inoperative for the removal of toxic metals [19].

Therefore, the treatment of sludge machining is a recurring problem for mechanical companies. On the other hand, the various industries reject equally large amounts of sludge; the latter occupies an increasingly large storage area, which poses the problem of its evacuation. To remedy this, the sludge was recovered from Ain El Kebira's BCR plant in the city of Sétif, Algeria, and was used for the manufacture of concrete and mortar.

2 Materials and methods

Five (05) specimens of concretes and mortars have been made: a concrete and mortar witnesses without waste addition, and four (04) series of concretes and mortars based on machining sludge in substitution of cement.
The concretes and mortars are prepared in laboratory from the materials listed below:

2.1 Materials

2.1.1 Aggregates

Locally available natural aggregates, a natural sand with the size of 0/3 mm, and two crushed stone aggregates having sizes of 8/15mm and 15/25 mm are used for the manufacture of concretes. The physical properties and chemical composition of aggregates 0/3, 8/15 and 15/25 mm are given in Tables 1 and 2. On the other hand, for the manufacture of mortars, standard sand is used NF EN 196-1, its chemical composition is presented in Table 3.

Table 1. Chemical analysis of sand 0/3.

Element	$CaCO_3$	Nacl	SO_4^{2-}
(%)	**98.00**	**01.17**	**00.00**

Table 2. Physical characteristics of aggregates.

Physical characteristics	Sand 0/4	Aggregate 8/15	Aggregat e 15/25
Density ρ_p (NF P 18-554) (g/cm^3)	**1.44**	**1.40**	**1.28**
Density ρ_s (NF P 18-555) (g/cm^3)	**2.54**	**2.56**	**2.73**
Porosity	**0.43**	**0.45**	**0.53**
Content of impurities (NF P 18-598 , NF P 18-591) (%)	**ES=68**	**0.8**	**0.4**
Module of finesse	**2.92**	/	/
MDE (NF P 18-572) (%)	/	**21**	**22**
LA (NF P 18-573) (%)	/	**22**	**25**

Table 3. Chemical analysis of standard sand.

Element	SiO_2	Al_2O_3	Fe_2O_3
(%)	**97.22**	**nd**	**nd**
Element	CaO	MgO	Na_2O
(%)	**01.34**	**nd**	**nd**

2.1.2 Cement

An ordinary Portland cement (CPA: CEM I 42.5 according to EN 197-1) is used in the concretes and mortars.

2.1.3 Machining sludge

The sludge used in this study (hydroxide sludge), which is an industrial waste from the BCR factory (Ain El-Kebira), which has a green color (figure 1). Its mineralogical composition is given in Table 4.

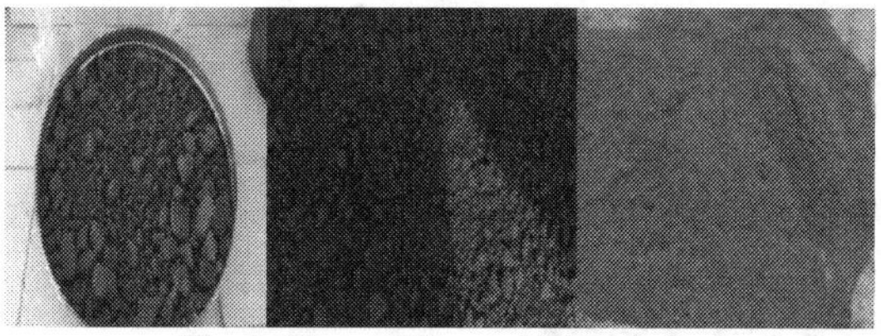

Fig. 1. Photo of the machining sludge.

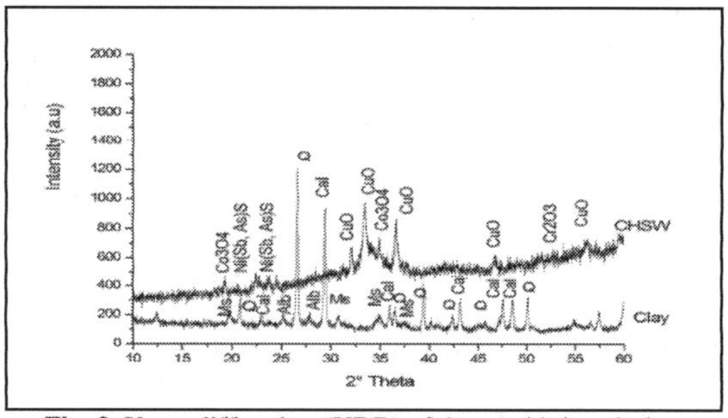

Fig. 2. X-ray diffraction (XRD) of the machining sludge.

Figure 2 presents the XRD analysis of the machining sludge. In the diffractogram to the major peaks of quartz and copper oxide is an indication of the siliceous nature.

Table 4. Mineralogical composition of the machining sludge.

Element	(g/kg)
Ni	1.66053
Cd	0.00867
Cu	12.24374
Zn	0.36480
Cr	20.393

2.1.4 Water

The water used is the drinking water of the city of Bejaia, so it does not require any test. It complies with the requirements of the standard (NF EN 1008).

2.2 Mixture proportions

2.2.1 Mortar

The preparation of formulations of mortar is based on the NF EN 196-1 standard [20]. The formulation is given for five (05) sets of three (03) prismatic test pieces of (40 * 40 * 160) mm^3 and its dosage is shown in Table 5 follows:

Table 5. Compositions of fresh mortars.

Samples	Hydroxide sludge (gr)	Cement (gr)	Standard sand (gr)	Water (gr)
M 00%	**00**	**450**	**1350**	**225**
M 05%	**22.5**	**427.5**	**1350**	**225**
M 10%	**45**	**405**	**1350**	**225**
M 15%	**67.5**	**382.5**	**1350**	**225**
M 20%	**90**	**360**	**1350**	**225**

Figure 3 shows how the specimens are stored for the shrinking measurement.

Fig. 3. Drying of mortars.

2.2.2 Concrete

The formulation method used in our study is that of DREUX-GORISSE [21], figure 4 gives the detail of the method used. Six (06) concrete test pieces of 16 x 32 cm were made; the proportions of the various compositions of the concretes are presented in the Table 6.

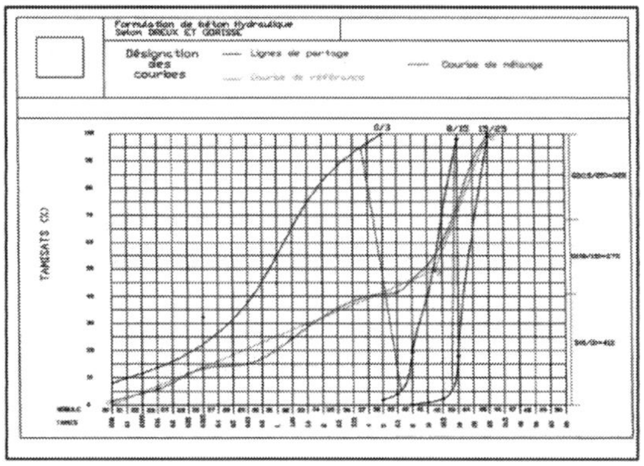

Fig. 4. Curve of mixture of concrete according to Dreux -Gorisse.

Table 6. Compositions of fresh concretes.

Samples	C 00%	C 05%	C 10%	C 15%	C 20%
Hydroxide sludge (kg)	-	**0.448**	**0.897**	**1.346**	**1.794**
Cement (kg)	**8.974**	**8.525**	**8.076**	**7.627**	**7.179**
Sand 0/3 (kg)	**14.907**	**14.907**	**14.907**	**14.907**	**17.907**
Gravel 8/15 (kg)	**14.842**	**14.842**	**14.842**	**14.842**	**14.842**
Gravel 15/25 (kg)	**15.591**	**15.591**	**15.591**	**15.591**	**15.591**
Water (kg)	**4.481**	**4.481**	**4.481**	**4.481**	**4.481**

Fig. 5. Conservation of concretes.

3 Results and discussion

3.1. Compressive strength and flexural strength of mortar

Flexural and compressive strength are presented in figure 6.The value of the flexural strength obtained is the average of three prisms for each series of specimens. The half- prisms of each test-tube obtained break in inflexion will be broken in compression, thus the value of the compressive strength obtained is the average of six half prisms for each series of test-tubes.
From the histogram, it is shown that the compressive strength and traction by inflexion at the age of 28 days decreases progressively with the increasing of the percentage of machining sludge, except for the mortars with 05 and 10 wt% sludge are good resistances. This diminution is due to the presence of the machining sludge, and to the reduction in the quantity of cement in the mixture. Thus, increasing the W/C ratio influences their resistance.

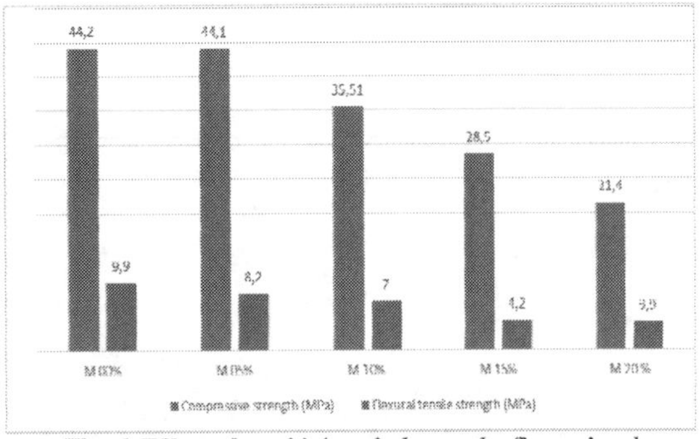

Fig. 6. Effect of machining sludge on the flexural and compressive strength of mortars at 28 days.

3.2 Measure the shrinkage of mortar specimens

It is known that the reaction of hydration is accompanied by a reduction of volume, called shrinkage. Water evaporates of a mortar preserved at the free air causing the shrinkage, which is the consequence of the loss of free water, when this water leaves the material, a contraction (shrinkage) occurs automatically. Results introduced above show a behaviour of shrinkage of test specimens at different ages of measure (1 day, 3 days, 7 days, 14 days, 21 days, 28 days, 60 days, 90 days, 120 days, 150 days and 180 days), and in different percentages of machining sludge (00 wt %, 05 wt %, 10 wt %, 15 wt% and 20 wt %), the volume changes occur after hardening. The variation of shrinkage of mortar specimens at different percentages of sludge has almost the same pace (the representative curves have the same evolution) (Figure 7). However, we note that the shrinkage evolves according to time by evaporation of the water imprisoned in the mortars. The shrinking of the mortar witnesses remains the most important, therefore the use of the machining sludge reduces the shrinkage; this fact is due to the reduction in the quantity of cement which induces the lowering of its hydration what translated by consequent in the reduction of shrinking.
Our results do not coincide with N. Belas work [22], because in their concrete the 20% sludge composite gave a larger shrinkage compared to 0%, 10% and 15%.

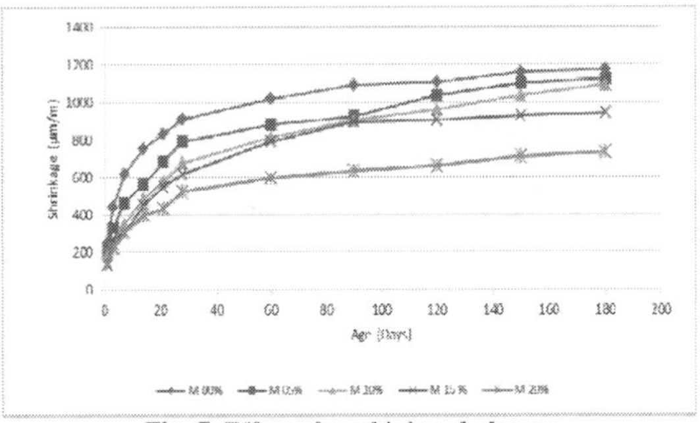

Fig. 7. Effect of machining sludge on the shrinking of the mortars.

3.3 Measure of swelling on specimens of mortar

The figure 8 shows that the swelling increases with time and the percentage of machining sludge in substitution of cement, and it starts to be stabilized as from 120 days. This stabilization and increasing the swelling depending on the sludge can be explained by the saturation of the voids created by the addition of sludge. In fact, by substituting the cement by machining sludge, there will be more voids inside the mortar, who promotes the birth of additional pocket of water that induces more swelling.

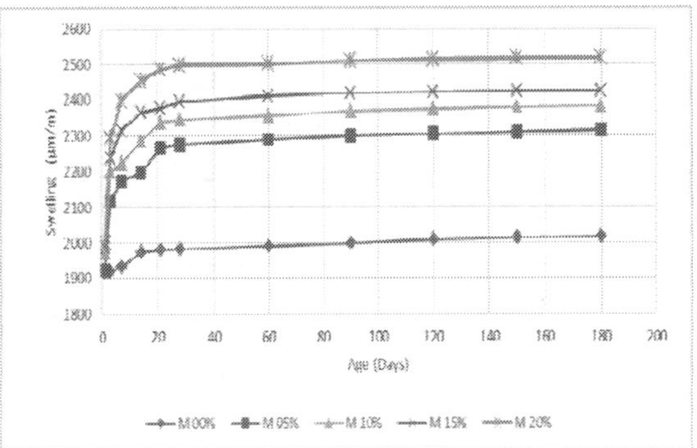

Fig. 8. Effect of machining sludge on the swelling of the mortars.

 https://doi.org/10.1051/matecconf/201814901063

3.4 Evolution of the pH of solutions of conservation of mortars for swelling

Accordingly to the figure 9, all mortars have a basic pH, which varies from 8,4 to 11. This pH increases according to the time and to the increase of the percentage of sludge in the mortars and stabilizes after 150 days. This alkalinity is due to a progressive and continuous salting out elements of chromium (Cr^{3+}) and copper (Cu^{2+}), present in the machining sludge, and which form with the water the alkaline bases.

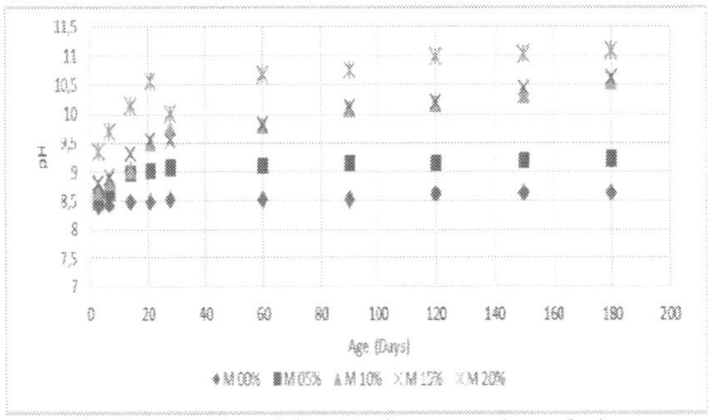

Fig. 9. Evolution of the pH of conservation solutions.

3.5 Compressive strength test of concrete

The compressive strength is an essential characteristic and fundamental parameter of our study. Therefore, measuring at the age of 7 days and at the age of ripening (28 days) was made on different compositions of concrete and the results thus obtained are graphically shown in figure.10. It is observed that the dosage to 00 wt % of machining sludge (concrete witness) gave a better compressive strength for the proportion in 05 wt % of sludge and has a slight decrease in strength relative to the concrete witness.

These results were predictable; because the W / C ratio was increasing, which is respectively: 0.5, 0.52, 0.55, 0.58 and 0.62. Accordingly, the increase of the sludge proportioning in cement substitution, generates a lowering of the compressive strength

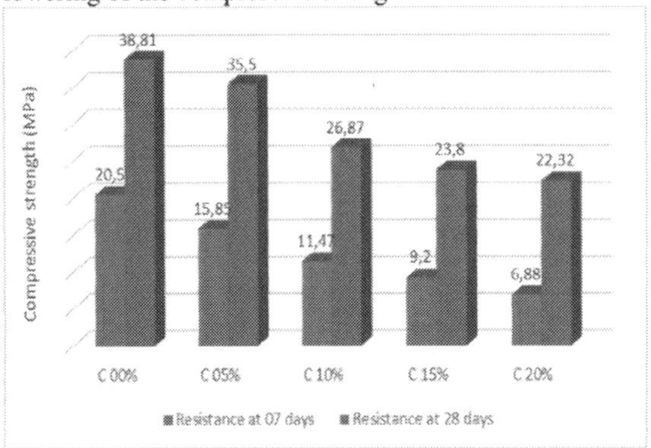

Fig. 10. Effect of machining sludge on the compressive strength of concrete.

3.6 Chemical analyses of preservative solution

After the mechanical tests, monoliths from each mortar sample were recovered and stored in demineralised water for 28 days to evaluate the leaching of existing heavy metals into the sludge.

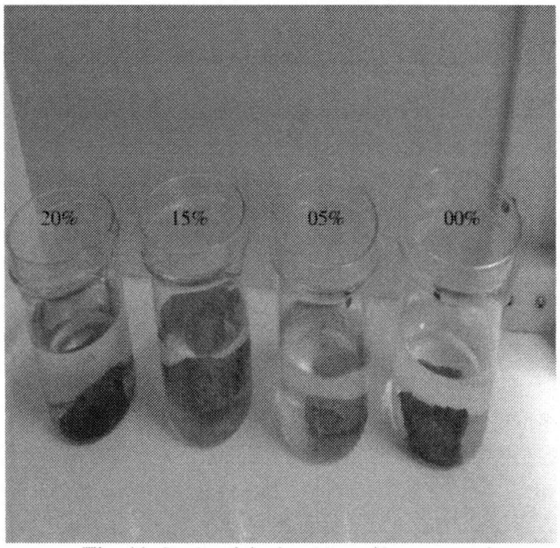

Fig. 11. Study of the leaching of heavy metals of machining sludge on

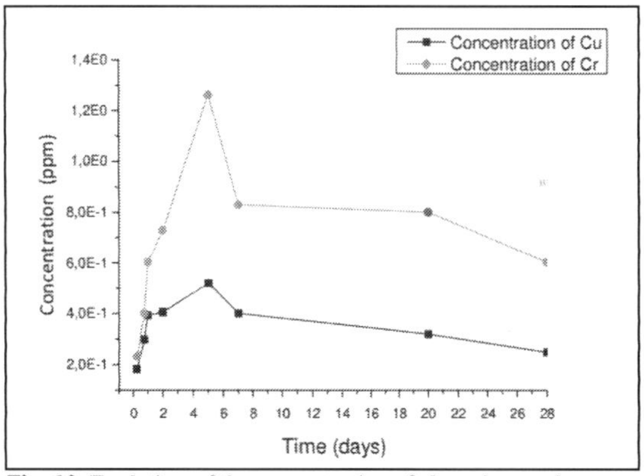

Fig. 12. Evolution of the concentration of chromium and copper as function of time in mortars with 20 wt% of machining sludge.

Mortar leaching with higher sludge content (20%) was studied.

The leaching curves of the Cr and Cu elements contained in the sludge have almost the same variation as in time (Figure 12).

There is an increase in the concentration of these chemical elements in the preservation solution at a young age; up to five (5) days. This behavior can be related to the leaching of these chemical elements. In addition, a reduction of these elements in the solution is observed up to 8 days; this means that these heavy metals have returned to the matrix of the mortar, and there is a trapping of these elements and then stabilization. The curves clearly show that the concentration of

 https://doi.org/10.1051/matecconf/201814901063

chromium metals is higher than that of copper in accordance with the mineralogical composition of the machining sludge.

It can also be said from the results that the leaching rate is low.

4 Conclusion

The main interest of this study is to evaluate the influence of the machining sludge in cement substitution on the mechanical and physico-chemical behavior of mortars and concretes. The results showed that the machining sludge can be used as a substitute for 05 wt % of cement in concrete and mortar, owing to the similar resistance compared to those of mortars and concretes without cement substitution. The use of this machining sludge in cement substitution reduced the shrinkage and the swelling increasing slightly. Therefore, the formulation with 05 wt % of machining sludge in cement substitution is interesting. The results show good compressive and flexural strengths and acceptable shrinkage and swelling. Moreover it is an economic gain concerning cement. The tests of , all mortars have a basic pH , which varies from 8,4 to 11,2; this alkalinity is due to progressive and continuous salting out elements of chromium (Cr^{3+}) and copper (Cu^{2+}) present in the machining sludge, and which form with the water the alkaline bases. But these quantities of salting out remain always weak.

References

1. M. Corboz, Le traitement des boues d'usinage, CT.DEC, Thésame, (2017).
2. J. Bertholon, La gestion durable des boues de station d'épuration, Mémoire académique, université de Cergy Pontoise, (2003).
3. J. Ribeyron, L'expertise technique et scientifique de référence, Technique de l'Ingénieur, Le Cetim valorise les boues d'usinage, (2016).
4. Association des instituts Carnot, la recherche pour les entreprises, (2015).
5. J.A. Junkes, F. Raupp-Pereira, V.P. Della, G.G. Cunto and D. Hotza, Valorization of mineral wastes for ceramic industries, The Minerals, Metals & Materials Society (TMS), (2008).
6. R. Modolo, V.M. Ferreira, M. Rodrigues and I. Coelho, Construction materials as a waste management solution for cellulose sludge, Waste Management (ELSVIER), volume 31, (2011).
7. A. Reynaud, J. Ribeyron, Valorisation des boues d'usinage par fusion Recycling of machining sludge by melting, éditions techniques des industries de la fonderie de Paris, no 192, pp 27-35, 2000.
8. P.T. Bowen, J.M. Entwistle, J.E. Hendrick, J.S. Quilin and U.N. Tyagi, Sludge treatment, utilization, and disposal. Wat. Environment Res. 60(6), 837-843, (1988).
9. P.T. Bowen, J.E. Hendrick, T.A. Woodward, L.S. Mitchell and M. Lahlou Sludge treatment, utilization, and disposal. Wat. Environment Res. 61(6), 821-829, (1989).
10. P.T. Bowen, M.K. Jackson and R.A. Corbitt, Sludge treatment, use, and disposal. Wat. Environment Res. 63(4), 406-414, (1991).
11. P.T. Bowen, M.K. Jackson, R.A. Corbitt and N. Gonce, Sludge treatment, utilization, and disposal. Wat. Environment Res. 64(4), 378-386, (1992).
12. P.T. Bowen, M.K. Jackson, R.A. Corbitt and N. Gonce, Sludge treatment, utilization, and disposal. Wat. Environment Res. 65(4), 360-368, (1993).
13. K.A. Groff, L.A. McLaughlin, Sludge management. Wat. Environment Res. 66(4), 368-374, (1994).
14. J. Li , L. Bennamoun , L. Fraikin , T. Salmon , D. Toye , R. Schreinemachers & A. Léonard, Analysis of the Shrinkage Effect on Mass Transfer During Convective Drying of Sawdust/Sludge Mixtures, Drying Technology, volume 32, (2014).
15. B.J. Alloway et A.P. Jackson, The behavior of heavy metals in sewage sludge-amended soils. Sci. Total Environment 100, 151-176 (1991).
16. A.M. Bruce, et R.D. Davis, Sewage sludge disposal: current and future options. Wat. Sci. Technol. 21, 1113-1128, (1989).
17. L. Korentajer, A review of the agricultural use of sewage sludge: benefits and potentialhazards. Wat. South Afr,17(3), 189-196, (1991).
18. U.S. Environmental Protection Agency (EPA), Standards for the use and disposal of sewage sludge, 40 CFR Parts 257, 403 and 503, Final Rule. U.S. Environmental Protection Agency, Cincinnati, Ohio, U.S.A, (1993).
19. J.F. Blais et J.L Sasseville, Etat de l'art du traitement et de la disposition ou valorisation des boues d'usines d'épuration municipales, rapport de recherche no 429, Institut nationale de la recherche scientifique (INRS-Eau), (1996).
20. R. Dupain, R. Lanchon, et J.C. Saint-Arroman, Granulats, sols, ciments et bétons (Caractérisation des matériaux de génie civil par les essais de laboratoir,éditions CASTEILA,(1995).
21. G. Dreux, J. Fiesta, New Guide of concrete and its constituents, (8th edition); Eyrolles Editions (1998).
22. N. Belas, Valorisation des sédiments de dragage dans les bétons, Séminaire Internationale, Innovation &Valorisation en Génie Civil & Matériaux de Construction, INVACO2, N^0 1P-173, Rabat, Maroc, 23-25 Novembre 2011.

Alkali leaching control of construction and demolition waste based geopolymers

Dimitris Kioupis, AggelikiSkaropoulou, Sotiris Tsivilis and GlikeriaKakali
National Technical University of Athens, School of Chemical Engineering, Athens, Greece

Abstract. This study regards the synthesis of waste brick geopolymers and the evaluation of their alkali leaching potential. Geopolymers with varied alkali to Al molar ratios were prepared and their alkali leaching potential was measured. In order to improve the efflorescence of the waste brick geopolymers, KOH was used instead of NaOH in the activation solution, Ca cations were incorporated through the partial substitution of the brick powder by ground granulated blast furnace slag or extra Al was introduced in the activation solution. The results showed that the leachability of the products is close related to the alkali content of the activation solution while the use of K ions in the starting mixtures reduced the leachability potential. The same effect was observed in the case of the extra Al ions incorporation. The partial substitution of the wastes bricks by slag showeda slight increase of leachability even though denser materials with enhanced strengths were prepared.

1 Introduction

The Construction and Demolition Wastes (CDWs) comprise one of the heaviest and most voluminous solid wastes worldwide generated by the construction industry [1]. A major environmental issue associated with the CDWs is their landfill disposal. A way to release the environmental pressure is by recycling and reusing these materials [2]. Due to the high SiO_2 and Al_2O_3 content of many CDWs (bricks, tiles concrete, ceramics etc), they can be utilized in the production of alkali activated materials or geopolymers [3,4]. Geopolymers are a class of inorganic materials which possess a three-dimensional polymeric-type structure. They are formed by the reaction of an alkaline solution with an aluminosilicate material at ambient or slightly elevated temperature [5].

The alkalis play a dual role in the geopolymerization process. First of all, they generate a high pH solution by releasing hydroxyl ions in order to dissolve the aluminosilicate precursors and to produce small reactive silico and alumino species. Furthermore, the alkali metal cations offer electroneutrality in geopolymers framework by balancing the negative charge in aluminum tetrahedrals. The high amounts of alkalis used in the geopolymers mixes and the weak binding of the alkaline metal cations (Na or K) with geopolymeric matrix lead to products which often suffer from severe efflorescence, when excess alkali oxide remains unreacted in the pore network. The efflorescence influences negatively the appearance of the products and can restrict the variety of geopolymers applications.

Several attempts have been made to reduce the efflorescence of geopolymers either by incorporating an Al-rich [6,7] or Ca-rich source [6,8]or by using potassium hydroxide instead of sodium hydroxide in the activation solution [9]. In addition, hydrothermal curing effectively promote the geopolymerization reactions resulting in less unbound alkali which is able to mitigate efflorescence [6,8].

In this study, geopolymers were synthesized from waste brick and their alkali leachability was evaluated. The effect of the alkali type (Na or K) and alkali content of the activation solution on the alkali leaching was measured. Then, Al-rich and Ca-rich sources in the form of Al_2O_3colloidal dispersion in H_2O and ground granulated blast furnace slag were incorporated in geopolymer synthesis, respectively, in order to control the alkali leaching of brick waste geopolymers. XRD and FTIR were applied to characterize the products. The mechanical strength of the produced specimens was measured through compressive strength tests.

2 Experimental

Waste bricks (WB) collected from various demolished buildings of Attica (Greece), were used as the raw material. In some cases, ground granulated blast furnace slag (GGBFS) was added in the starting materials in order to improve the alkali leaching of the final products.Both materials were pulverized in order to obtain a certain fineness (d_{50} = 20μm) prior to their application. Table 1 shows the chemical composition of the WB and GGBFS in the form of oxides and trace elements, as determined by using X-ray fluorescence spectrometry (XRF).

Table 1. The chemical composition of the raw materials (% w/w).

Oxide	WB	GGBFS
SiO_2	51.31	39.49
Al_2O_3	14.62	10.14
Fe_2O_3	8.45	0.89
CaO	6.33	39.97
MgO	8.10	6.75
K_2O	2.28	0.30
Na_2O	0.57	0.00
SO_3	0.44	0.49
TiO_2	0.66	-
P_2O_5	0.08	-
Cl	0.04	-
L.O.I.	3.97	0.16

The Al and Si leaching of the WB was conducted by mixing 0.5 (± 0.0001) g of solid with 20 ml of alkaline solution under continuous stirring. The concentration of the alkaline solution and the dissolution time were kept at 10 M and 24 h, respectively [10]. After filtering, the liquid part was diluted to 250 ml, the pH was adjusted to pH < 1 by adding concentrated HCl acid and Atomic Absorption Spectroscopy (AAS) was used in order to determine the Al and Si concentrations.

The alkaline solutions were prepared by dissolving NaOH (> 99%, CAS: 1310-73-2) or KOH (86%, CAS: 1013-58-3) anhydrous pellets in distilled water and adding soluble Si in the form of SiO_2(50% in H_2O, colloidal dispersion, CAS: 7631-86-9)and in some cases, soluble Al in the form of Al_2O_3 (20% in H_2O, colloidal dispersion, CAS:1344-28-1).The activation solutions were stored for a minimum of 24h prior to use, to allow equilibrium. Then, the starting materials (WB and / or GGBFS)and the activation solution were mechanically mixed (standard mortar mixer: Controls 65-L0005) to form a homogenous slurry which was transferred to 50x50x50 mm cubic molds and mildly vibrated. The specimens were left at room temperature for 2h and then were cured at 70°C for 48h. Compression tests were carried out on a Toni-technik uniaxial testing press, 7 days after the specimens' preparation (load rate 1.5 kN/s, according to the EN196-1 requirements). For each synthesis, three specimens were prepared and tested under compression.Mixtures were prepared with molar ratios as given in Table 2.

Table 2. Chemical composition of the produced geopolymers.

Mix	WB (%)	GGBFS (%)	Alkali (R)	Si/Al	R/Al	s/l
B1	100	0	Na	3.5	0.1	2.4
B2	100	0	Na	3.5	0.5	3.6
B3	100	0	Na	3.5	1.0	3.6
B4	100	0	Na	3.5	1.5	2.8
B5	100	0	K	3.5	1.0	3.6
B6	100	0	Na	2.9	1.0	2.1
B7	95	5	Na	3.5	1.0	3.6
B8	90	10	Na	3.5	1.0	3.6

The chemical composition of the mixtures was varied by changing the Si/Al molar ratio of the starting mixture, the overall Na/Al molar ratio, the solids to liquids mass ratio (s/l), the GGBFS content (%) and the alkali type (R: Na or K).

In order to determine the alkali leaching ability of the mixes, a modification of the experimental technique developed by Vafaeiet al. [7] which quantifies the free alkali content (alkali which is not structurally bonded) was applied. For this test, each geopolymer samplewas pulverized and passedthrough a 90μm sieve.1.0000 (±0.0001) g of the obtained geopolymer was added to 40 ml of deionized water and stirred for 24 h to leach free-alkali from geopolymerbinder. The solution wasthen filtered to remove solids and titratedwith 0.1 N hydrochloric acid. The pH and electrical conductivity of the filtrate was measured. The free alkali content of the geopolymer was calculated by the volume of the acid used.

The phase distribution of the powders was investigated by X-ray diffraction (XRD), on a Bruker D8 ADVANCE x-ray diffractometer. The data were collected in a 2θ range 2-70°, with 0.01° step size and 1 sec per step and were evaluated using Diffrac.Eva v3.1 software.FTIR measurements were carried out using a Fourier Transform IR Spectrophotometer (Jasco 4200 Type A). The FTIR spectra were obtained by the KBr pellet technique, in the wavenumber range from 400 to 4000 cm^{-1} and resolution 4 cm^{-1}. The pellets were prepared by pressing a mixture of the sample and dried KBr (sample: KBr equals to 1:200, approximately) at 7.5 t/cm^2.

3 Results and discussion

3.1 Extent of waste brick dissolution

Table 3 presents the concentration of Al and Si in the solutions (10M NaOH or KOH) of the WB. The results of the alkaline dissolution experiments showed that WB is susceptible to alkaline attack and, therefore, possess good geopolymerization potential. The degree of dissolution is higher in NaOH than in KOH solution which is in well accordance to the literature [11]. Due to their higher mobility, sodium ions show a higher tendency to attack and dilute aluminosilicon species of the raw material. However, it must be noted, that the leaching behavior of Al and Si cannot predict by itself the quality of the produced geopolymers. Further experiments concerning the preparation and properties of geopolymers from these raw materials have to be conducted.

Table 3. Extent of Al and Si dissolution in relation to alkali metal selection (ppm).

Solution	CDW brick	
	Si (ppm)	Al (ppm)
10 M NaOH	37.1	14.66
10 M KOH	29.3	9.69

3.2 Alkali leachability tests

Table 4 presents the free alkali content (wt. %), the pH and the electrical conductivity of the filtrates as well as the compressive strength measured for the corresponding WB geopolymers.

Table 4. Free alkali content, pH and electrical conductivity of the leachates and the compressive strength of the corresponding samples.

Mix	Free alkali (% wt.)	pH	Electr. Cond. (mS)	Compr. Str. (MPa)
B1	4%	8.8	0.5	0.42
B2	23%	10.0	2.0	7.15
B3	26%	10.4	3.4	16.37
B4	28%	10.6	4.4	1.92
B5	23%	10.3	3.8	20.22
B6	21%	10.4	2.7	3.96
B7	28%	10.4	3.5	16.84
B8	28%	10.8	3.3	23.38

Figure 1 shows, schematically, the free alkali content and the compressive strength of the WB geopolymers varying in their alkalinity level (Na/Al). It is obvious that the gradual introduction of higher amounts of alkalis in the starting mixtures shows a higher extent of alkali leaching, indicating a stronger tendency towards efflorescence. This is related to pore solution chemistry where the higher Na/Al molar ratios generate excess of Na ions. These ions are left unboundfrom the geopolymer network and are trapped to the pores in the form of alkaline solutions.

The sample B1 (Na/Al = 0.1) containa very low value of free alkali content (4%). However, at the same time this sample possesses near zero compressive strength values (0.42 MPa). It seems that the amount of alkali in the activation solution is too little to sufficiently solubilize the Si and Al species of the raw material, resulting in the delivery of a product which is not well geopolymerized.

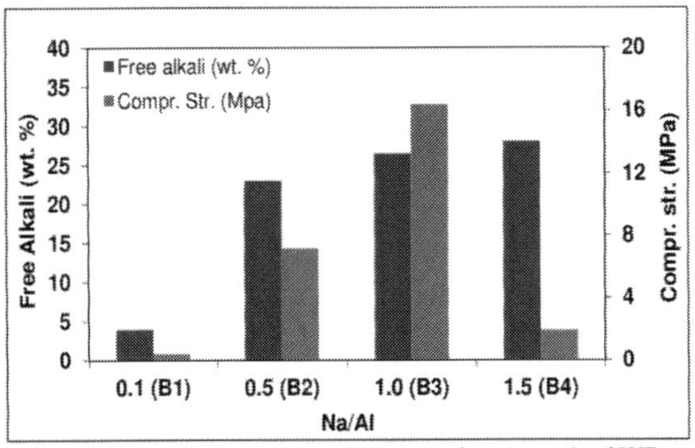

Fig. 1. Free alkali content and compressive strength of WB geopolymers with varying Na/Al molar ratios.

In samples with Na/Al > 0.5, the increase rate of the free alkali content seems to be minimized. This is attributed to the fact that higher Na/Al molar ratios positively affect the extent of the geopolymerization reaction and therefore reduce the matrix permeability[12]. This is also reflected in the mechanical strength of the produced geopolymers which is enhanced.As an example, the sample B3 containing the two-fold amount (Na/Al = 1.0) of alkalis in relation to sample B2 (Na/Al = 0.5) possesses,considerably, higher mechanical strength (16.37 MPa) while the amount of free alkali is not significantly altered. However, this trend is not followed in the case of material B4 (Na/Al = 1.5), which exhibits markedly reduced compressive strengths (1.92 MPa). For the preparation of the particular geopolymer, a large amount of water was required to make the paste workable, resulting in thepH reduction of the gel and therefore the limited dissolution of the WB. Furthermore, a higher porosity is createddue to the removal of the unbound water.

The results of alkali leachability and mechanical strength of the WB geopolymers showed that the highest compressive strength value is not necessarily combined with the lowest free alkali content. This indicates that there is a need to compromise between achieving the optimum mechanical behavior and controlling the unbound alkalis in the pore solutions in order to deliver a durable geopolymer.

Figure 2 presents the effect of the alkali type (Na or K) in the alkali leachability of the WB geopolymers. It is obvious that the incorporation of K ions to the gels favors the formation of high strength products with lower levels of free alkali content. Due to their greater size, K ions are not as mobile as the Na ions and thus create stronger bonding with the aluminosilicate gels, resultingin a lower alkali concentration in the pore solution which causes the efflorescence [13]. Regarding the mechanical behavior of the products, the K based sample exhibits enhanced compressive strength values by about 20% (20.22 MPa) in relation to the Na based product.

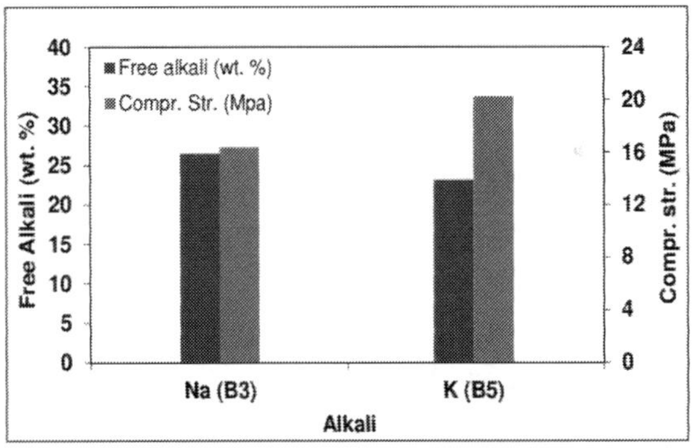

Fig. 2. Free alkali content and compressive strength of WB geopolymers with varying alkali type (Na or K).

As presented in Figure 3, the introduction of an Al rich source, such the one used in this study, led to the improvement of the alkali leachability of the WB geopolymers. The introduction of the additional Al helps the formation of a denser geopolymer molecular network which requires a larger number of Na cations to balance the negative charge of the 4-coordinated Al in the N-A-S-H gels. This results in the promotion of the geopolymerization reaction and the delivery of a structure with a reduced free alkali content [7].

It was expected that the creation of a more stable geopolymeric network will deliver a product with enhanced mechanical strength. However, as it appears in this study, the introduction of aluminum had the opposite effect. This fact can be attributed tothe nature of the starting mixture of the sample B6. As in the case of B4, this mix required a great amount of water to make the paste workable resulting in an increased porosity which adversely affected the strength of the specimens.

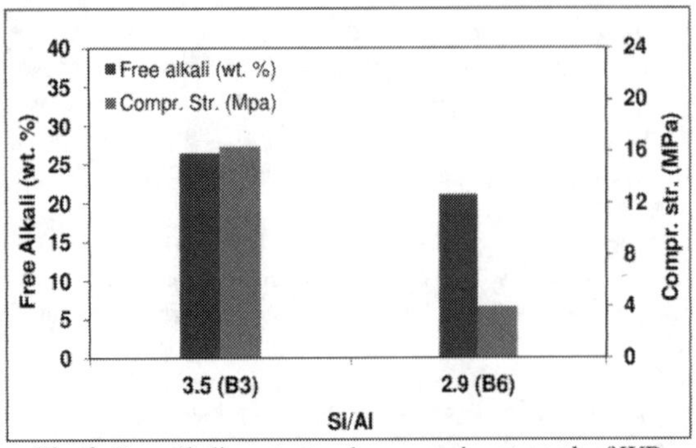

Fig. 3. Free alkali content and compressive strength of WB geopolymers with varying Si/Al molar ratios.

Figure 4 presents the free alkali content and compressive strength of the samples with varying content of GGBFS.It is well understood that the introduction of considerable amounts of calcium in the geopolymerization system forms a homogenous and amorphous sodium calcium aluminosilicate hydrate (N-C-A-S-H) networkin parallel to the sodium aluminosilicate hydrate (N-A-S-H) of the geopolymerization mechanism. The gels containing these extra phases lead to the formation of more compact microstructures which accountfor the high strength achieved in GGBFS containing geopolymers.

As it is observed from Figure 4,the introduction of low levels of GGBFS into the starting mixture does not promote the alkali leachability of the WB geopolymers. On the contrary, the incorporation of Ca ions in the geopolymer network lead to a slight increase of the free alkali content (2%). This is related to the fact that the calcium ions act competitively to sodium ions in their connection to the aluminosilicate network in order to provide charge balance. The increased availability of sodium ions which remain unbound in the gels, contribute to the enhancement of the alkaline pore solution and therefore the free alkali content.

Even though the incorporation of GGBFS favors the formation of more compact structures with higher mechanical strength (e.g. B8 exhibits 23.38 MPa), the alkalinity of the pore solution is higher andthe capacity of pores to hold crystal deposits low.

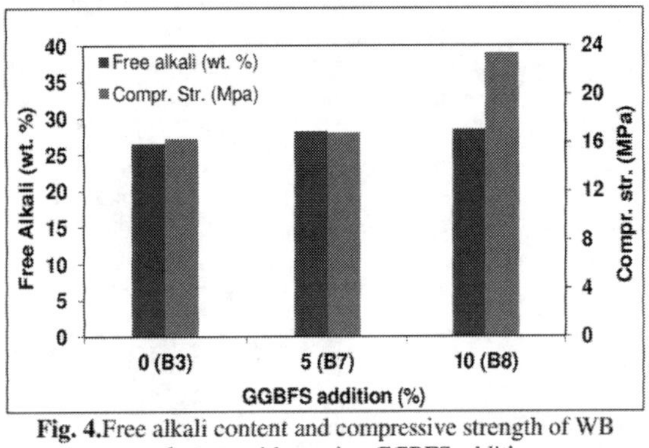

Fig. 4. Free alkali content and compressive strength of WB geopolymers with varying GGBFS addition.

As it seen from Table 4, a proportional relationship between the free alkali content and the values of pH and electrical conductivity is observed in almost all samples.

3.3 Characterizations

The XRD patterns of the WB and some representative geopolymers with varying Na/Al molar ratios are presented in Figure 5. The main mineral phases in the WB are the quartz, albite, diopside, muscovite, microcline, maghemite and traces of carbonates (calcium or potassium). In the case of geopolymers (B2 and B4), the increase of the background, in the range 20-40° 2θ, indicates the formation of the amorphous geopolymer matrix. When the alkali content is high (Na/Al = 1.5), the formation of zeolites, such as chabazite, is favored and this is associated with lower compressive strength (Fig. 5, B4). Besides, it is observed that the increase of R/Al rate favors the enhancement of the alkali carbonates. The unbound alkalis react with the CO_2 of the atmosphere and produces increased amounts of alkali carbonates. All other samples did not exhibit any differences in respect to their crystallographic phase content.

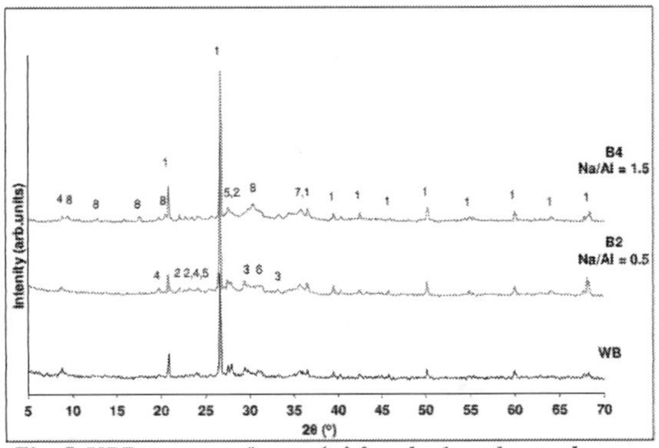

Fig. 5. XRD patterns of waste brick and selected geopolymers. 1: quartz, 2: albite, 3: carbonates, 4: muschovite, 5: microcline, 6: diopside, 7: maghemite and 8: chabazite.

Figure 6 shows the FTIR spectra of WB and selected geopolymer samples with varying Na/Al molar ratios. The region of interest in both WB and geopolymer products lies in the range between 800 and 1300 cm^{-1}. This broad hump exhibits overlapped peaks which are associated with the asymmetric stretching vibrations of Si-O-T (T: tetrahedral Si or Al) bonds. In the case of the WB, the maximum of this hump is at around 1110 cm^{-1} while in the case of the geopolymers this hump becomes narrower and shifts to lower wavenumbers (~1000 cm^{-1}). This alteration is directly linked with the substitution of Si by tetrahedral Al and therefore the formation of an aluminosilicate network [14]. Furthermore, the Si-O-T asymmetric bending vibrations at ~460 cm^{-1}seems to be minimized with the formation of the aluminosilicate network. The band at ~1635 cm^{-1} is related to O-H bending vibrations of molecular water. All samples contain carbonate species pointed out by the presence of the band at around 1450 cm^{-1}, related to anti-symmetric vibrations of CO_3^{2-} ions. Samples with high R/Al ratios (B3 and B4) tend to exhibit more pronounced carbonate ion vibrations. The absorbance at 875 cm^{-1} is related to symmetric stretching vibrations of Al – O and it constitutes sign of geopolymerization. As for 800, 780 and 685 cm^{-1} peaks, these are connected with symmetric stretching vibrations of Si-O-T bonds.

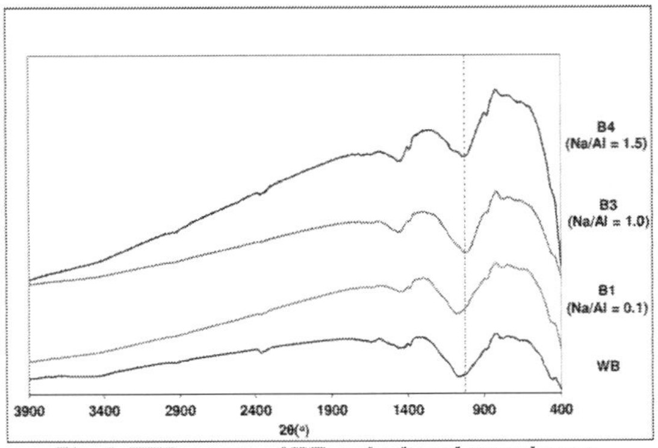

Fig. 6. FTIR spectra of WB and selected geopolymers.

4 Conclusions

This study presents an effort on controlling the alkali leachability of waste brick geopolymers which is responsible for the efflorescence phenomenon. The results showed that the alkalinity of the activation solution plays an important role on the free alkali content. Higher alkalinities tend to increase the leachability but at the same favor the mechanical strength of the products. It seems that the delivery of a durable waste brick geopolymer requires a compromise between the highest achievable mechanical strength and the lowest unbound alkali content. The selection of potassium ions favors the leachability of the products while at the same time promotes their mechanical behavior (20%). The introduction of extra Al ions into the initial gels effectively reduces the free alkalissince more cross-linked geopolymer molecular networks are formed and thus larger amount of Na^+ cations is required to serve as charge balancing. Finally, the substitution of the waste brick by small amounts of ground granulated blast furnace slagenhanced the mechanical strength of the waste brick geopolymers but led to the aggravation of the alkali leachability. Further studies upon the combinational effect of the aforementioned approaches on the alkali leaching ability of the waste brick geopolymers are recommended in order to improve the durability of the products.

Acknowledgments

This project has received funding from the European Union's Horizon 2020 research and innovation programme under grant agreement No [723825].

References

1. European Commission (DG ENV), *Final Report Task 2Service Contract on Management of Construction and Demolition Waste*, 2011 (ENV.G.4/FRA/2008/0112).
2. M. Del Rio Merino, J. Navarro, P. Saez, Open Constr. Build. Technol. J. **5**, 124 (2011)
3. K. Komnitsas, D. Zaharaki, A. Vlachou, G. Bartzas, M. Galetakis, Adv. Powder Technol.**26**, 368 (2015)
4. L. Reig, M.M. Tashima, M.V. Borrachero, J. Monz, C.R. Cheeseman, J. Pay, Constr. Build. Mater. **43**, 98 (2013)
5. J. Davidovits, J. Ther. Anal. 37, 1633 (1991)
6. E.N. Kani, A. Allahverdi, J.L. Provis, Cem.Concr. Compos.**34**, 25 (2012)
7. M. Vafaei, A.Allahverdi,Adv. Powder Technol. **28**,215 (2017)
8. Z. Zhang, J.L. Provis, A. Reid, H. Wang,Cem. Concr. Res.**64**,30 (2014)
9. P. Duxson, J.L. Provis, G.C. Lukey, J.S.J. van Deventer, F. Separovic,Z.H. Gan, Ind. Eng. Chem. Res.**45**,9208 (2006)
10. Ch. Panagiotopoulou, E. Kontori, Th. Perraki, G. Kakali, J. Mater. Sci. **42**, 2967 (2007)
11. X.U. Hu, J.S.J. Van Deventer, Int. J. Miner. Process**59**, 247 (2000)
12. R.R. Lloyd, J.L. Provis, J.S.J. van Deventer, Cem.Concr. Res. **40**,1386 (2010)
13. O. Bortnovsky,J. Dědecek, Z. Tvaruzkova, Z. Sobalík, J. Subrt, J. Am. Ceram. Soc. **1**, 3052 (2008)
14. T. Bakharev,Cem. Concr. Res.**35**, 1224 (2005)

https://doi.org/10.1051/matecconf/201814901065

The use of ultrasonic pulse velocity to estimate the water permeability of concretes

Contribution à l'estimation de la perméabilité à l'eau des bétons par les vitesses des ultrasons

Benouis Abdelhalim[1], Mehamdia Ali[1]
[1] Laboratoire de Génie Civil et d'Hydraulique, Université 8 mai 1945 Guelma (24000), Algérie

Abstract: In this work we investigated the possibility of estimating the water permeability of concrete from the ultrasonic surface wave velocity (Rayleigh's waves). This is a method for the non-destructive permeability diagnosis of the in situ auscultation of a structure. Four ordinary concrete compositions with different W/C ratios and two self compacting concretes SCC were used. This study showed a decrease in of ultrasonic pulse velocity with the increase in the W/C ratio, this is due to the increase in porosity. Curing in air of the concrete specimens produces greater permeability than curing in water. The increase in the permeability with the increase of W/C ratio is more important for curing in water than for the curing in air. SCC1 has a lower permeability than that of SCC2, this difference is respectively 20% and 10 % for curing in air and in water. The study show that permeability estimation with ultrasonic surface waves is more reliable for curing in water mode than tin curing in air. The correlations obtained between the permeability and the indirect ultrasonic velocity are linear, with an inversely proportional relation.

Résumé : Le présent travail consiste à estimer la perméabilité à l'eau par des mesures des vitesses ultrasonores des ondes de surface (ondes de Rayleigh) pour différentes compositions du béton. Ceci dans le but d'un diagnostic non destructif de l'auscultation in-situ d'ouvrage en se substituant aux carottages nécessaires aux essais traditionnels de perméabilité. Quatre compositions de béton ordinaire de différents rapports E/C et deux bétons autoplaçants avec deux ajouts différents ont été utilisés. Cette étude a montré une diminution de la vitesse de surface quand le rapport E/C augmente, ceci est dû à l'augmentation de la porosité. La vitesse des ultrasons pour le BAP1 (fillers calcaires) est 15% supérieure à celle du BAP2 (laitier de hauts fourneaux). La perméabilité des éprouvettes conservées à l'air libre est plus importante que celles conservées dans l'eau. L'augmentation de la perméabilité avec l'accroissement du rapport E/C est plus importante pour les éprouvettes conservées dans l'eau que pour les éprouvettes conservées à l'air libre. Le BAP1 présente une perméabilité inferieure à celle du BAP2, cette différence est respectivement de 20% et 10 % pour les éprouvettes conservées dans l'eau et à l'air libre. Les relations entre les vitesses ultrasoniques et les perméabilités ont montré que l'estimation de la perméabilité à partir des vitesses des ultrasons est plus fiable pour les éprouvettes conservées dans l'eau que celles à l'air libre. Ces relations sont de type linéaires avec une évolution inversement proportionnelle de la perméabilité en fonction des vitesses des ultrasons.

1 Introduction

La détérioration du béton résulte en général de la pénétration d'agents agressifs à travers sa peau, entraînant des dommages irréversibles non seulement à sa surface mais également dans sa structure interne. Donc les bétons employés, aussi bien pour la construction d'ouvrages d'art que pour le stockage des déchets radioactifs, devraient présenter une perméabilité faible. Or, il est admis que la perméabilité d'un matériau est un indicateur fiable de sa durabilité [1].

La perméabilité du béton représente la capacité de ce matériau poreux à être traversé par un fluide sous un gradient de pression. Le flux se produit dans un réseau poreux, comportant des pores initiaux ainsi que dans des vides ou des fissures. Celles-ci, induites par diverses sollicitations, interconnectent généralement les chemins de percolation et diminuent ainsi la résistance du milieu au flux, ce qui provoque une augmentation de la perméabilité. La perméabilité est un paramètre pouvant influencer indirectement la durabilité des structures en béton en gouvernant la vitesse de pénétration d'agents agressifs [2].

Les méthodes ultrasonores sont utilisées depuis les années 1940 pour évaluer le béton. Le principe de ces méthodes est d'estimer d'une façon quantitative les propriétés des matériaux par des mesures de la vitesse et de l'atténuation de l'onde ultrasonore dans une gamme de fréquence donnée. Ces techniques sont largement utilisées pour quantifier les propriétés élastiques des matériaux, ainsi que pour la détection des cavités et des fissures. Plusieurs travaux de recherche ont été menés pour étudier la relation entre les propriétés du béton et les propriétés ultrasonores des ondes acoustiques. Ces études ont mis en évidence la sensibilité des paramètres de propagation à la structure poreuse du béton. Cette méthode permet de mesurer le temps de propagation d'une impulsion ultrasonique traversant le béton [3 - 4].

Le présent travail consiste à estimer la perméabilité à l'eau par des mesures des vitesses ultrasonores des ondes de surface (Onde de Rayleigh) pour différentes compositions du béton. Ceci dans le but d'un diagnostic non destructif de l'auscultation in-situ d'ouvrage en se substituant aux carottages nécessaires aux essais traditionnels de perméabilité. Quatre compositions de béton ordinaire de différents rapports E/C et deux bétons auto plaçant avec deux ajouts différents ont été utilisé dans cette étude.

2 Méthode expérimentale

2.1 Compositions des bétons

Les matériaux utilisés pour la composition des différents bétons sont :

- un ciment CPJ-CEM II/A 42.5,
- un gravier concassé de nature calcaire de dimension 5/15 mm,
- deux sables 0/5 mm (sable de carrière, sable roulé fin),
- un superplastifiant «MEDAPLAST SP40» fabriqué par Granitex,
- deux ajouts (fillers calcaires, laitier de haute fourneaux).

Les compositions des différents bétons et leurs résistances à 28 jours (trois éprouvettes pour chaque composition) sont mentionnées dans le tableau 1.

Tableau 1. Compositions des bétons

Matériaux (kg)	BO 1	BO 2	BO 3	BO 4	BAP 1	BAP 2
Ciment	400	400	400	400	400	400
Eau	210	220	240	260	196	196
Sable carrière	813	813	813	813	/	/
Gravier 5/15	946	946	946	946	865	865
Sable fin roulé	/	/	/	/	765	765
Fillers (20%)	/	/	/	/	80	/
Laitier (20%)	/	/	/	/	/	80
E/C	0.52	0.55	0.60	0.65	0.50	0.50
G/S	1.16	1.16	1.16	1.16	0.88	0.88
Résistance (MPa)	29.39 ± 2.10	25.86 ± 2.09	21.80 ± 0.80	18.98 ± 0.82	25.20 ± 3.85	26.45 ± 0.72

2.2 Mesure de la vitesse ultrasonore de l'onde de surface

La vitesse de l'onde de surface (onde de Rayleigh) est mesurée en utilisant un générateur d'impulsions ultrasonores et deux transducteurs (émetteur, récepteur) de fréquence 54 kHz de diamètre 49.5 mm qui sont mis contact avec la surface des éprouvettes par l'intermédiaire d'un couplant. L'émetteur est fixe, le récepteur se déplace à pas constant (15 cm) (fig.1). Les vitesses ont été déterminées par des mesures du temps d'une impulsion traversant la surface du béton [5].

Les mesures ultrasonores des ondes de surface ont été réalisées sur des poutrelles de dimensions 15x15x100 cm^3 (3 poutrelles pour chaque composition).

La vitesse des ondes de surface est estimée à partir de la pente de la courbe linéaire présentée sur la figure 2 qui donne la variation du temps (acquisition du signal) en fonction de la distance (entre l'émetteur et le récepteur)

$$V = d = \frac{\Delta y}{\Delta t}$$

Où d est la pente de la droite qui donne la variation du temps (acquisition du signal) en fonction de la distance (entre l'émetteur et la récepteur) [3].

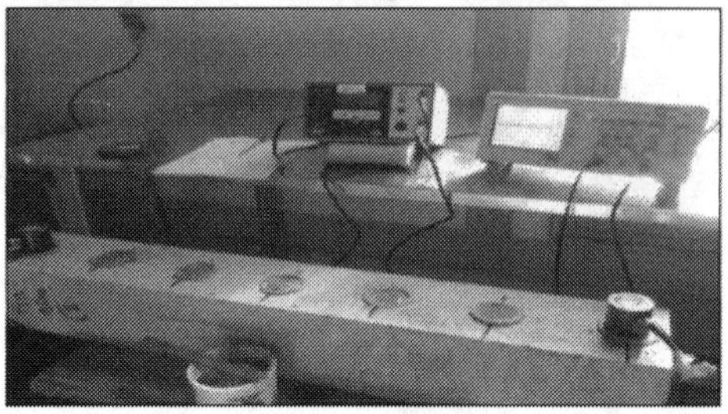

Fig. 1. Dispositif pour la mesure des paramètres de l'onde de surface

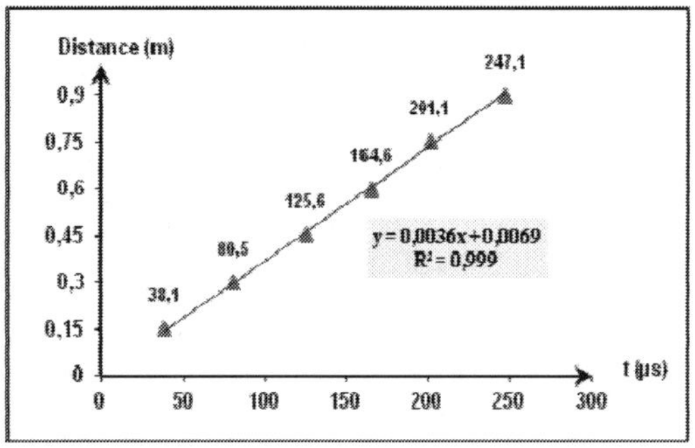

Fig. 2. Estimation de la vi1tesse des ondes de surface

2.3 Mesure de la perméabilité à l'eau

Les essais de perméabilité à l'eau ont été effectués suivant la norme européenne EN 12390-8, cette méthode consiste à déterminer la profondeur de pénétration d'eau sous pression dans un béton durci conservé dans l'eau (EN, 2012) [6].

L'essai a été réalisé sur trois éprouvettes cubiques (15x15x15 cm^3) pour chaque composition de béton à 28 jours avec deux modes de conservation (à l'air libre et dans l'eau). Les perméabilités des éprouvettes conservées dans l'eau et à l'air libre sont étroitement liées, donc ceci nous permet de corréler l'une ou l'autre aux mesures indirectes par les ultrasons. Le montage nécessaire à la réalisation de cet essai est présenté par la figure 3. L'essai consiste à appliquer une pression d'eau de 5 ± 0.5 bar sur la surface des éprouvettes pendant une période de 72 heures, après la rupture des éprouvettes par fendage, on mesure la profondeur de pénétration d'eau (H). L'appareillage utilisé pour mesurer la perméabilité est de type Controlab [7].

Fig. 3. Appareillage pour la mesure de la perméabilité à l'eau (profondeur de pénétration d'eau)

3 Résultats et interprétations

Tableau2. Vitesses des ondes de surface et perméabilités pour les différents bétons.

	Béton	BO 1	BO 2	BO 3	BO 4	BAP 1	BAP 2
	E/C	0.52	0.55	0.60	0.65	0.50	0.50
	V(m/s)	3634 ±100	3534 ± 300	3200 ±200	3167 ± 100	3766 ±300	3200 ±100
H(cm)	conservation à l'air	14.53 ±1.0	14.8 ± 0.6	15 ±0.0	15 ±0.0	11.3 ±2.0	12.5 ±1.2
	conservation dans l'eau	0.17 ±0.1	0.7 ±0.4	1.3 ±0.8	1.4 ±0.3	1.2 ±0.2	1.5 ±0.3

3.1 Influence de la composition du béton sur la vitesse des ondes de surface

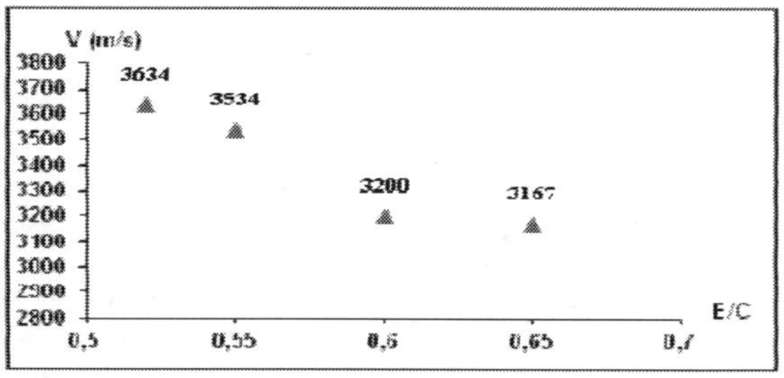

Fig. 4. . Evolution de la vitesse en fonction pour les bétons ordinaires

 https://doi.org/10.1051/matecconf/201814901065

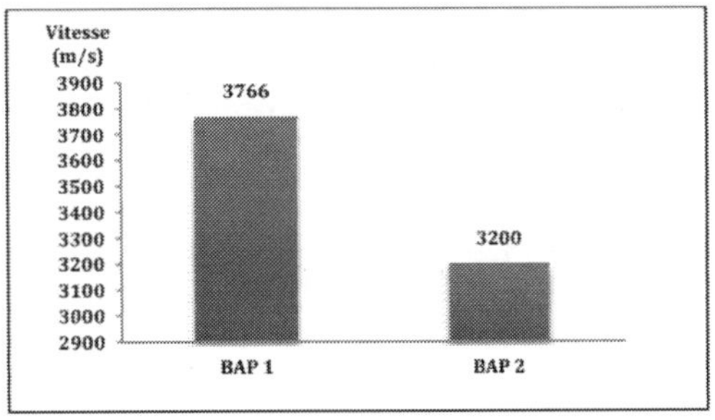

Fig. 5. . Evolution de la vitesse en fonction BAP

La figure 4 présente la variation de la vitesse de propagation de l'onde de surface en fonction du rapport E/C pour les bétons ordinaires. Les résultats montrent une diminution de la vitesse de surface quand le rapport E/C augmente et ceci dû à l'augmentation de la porosité. La diminution de la vitesse est plus importante entre le rapport E/C variant de 0,55 à 0,60 qui se reflète par à une diminution des vitesses des ultrasons de 10% alors pour une variation de E/C de 0,60 à 0,65, la diminution des vitesses n'est que de 1 %. Donc, les vitesses des ultrasons sont beaucoup plus variables dans les bétons plastiques que dans les bétons fluides.

La figure 5 présente la variation de la vitesse de propagation de l'onde de surface pour les deux bétons autoplaçants. Les résultats montrent que la vitesse de l'onde de surface pour le BAP1 (fillers calcaires) est supérieure à celle du BAP2 (laitier de hauts fourneaux). Cette différence (15 %) n'est pas visible en termes de résistance où la différence n'excède pas 5%. Ceci peut s'expliquer par la relation entre les vitesses des ultrasons et la résistance qui n'est de type linéaire.

3.2 Influence de la composition du béton sur la perméabilité

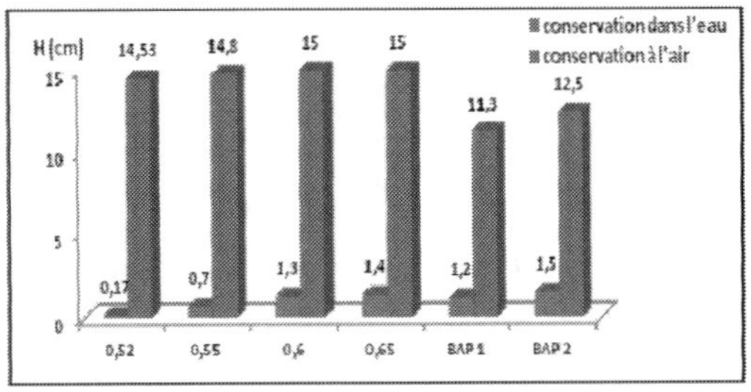

Fig. 6. Evolution de la perméabilité à l'eau (profondeur de pénétration de l'eau) pour les différents des bétons

On remarque sur cette figure que la perméabilité des éprouvettes conservées à l'air libre est plus importante que celles conservées dans l'eau. L'augmentation de la perméabilité avec l'accroissement du rapport E/C est plus importante pour les éprouvettes conservées dans l'eau que pour les éprouvettes conservées à l'air libre. Pour les éprouvettes conservées dans l'eau la profondeur de pénétration de l'eau des différents bétons reste toujours inferieure à la valeur maximale admissible pour les bétons imperméables (5 cm). Le BAP1 présente une perméabilité inferieure à celle du BAP2, cette différence est respectivement de 20% et 10 % pour les éprouvettes conservées dans l'eau et à l'air libre.

Les résultats obtenus concordent bien avec ceux relatés dans la littérature Conz [8] a réalisé des mesures de perméabilité (profondeur de pénétration de l'eau) sur des bétons de différents rapports E/C selon la norme DIN. Il a obtenu des profondeurs d'eau de 12, 15 17 mm respectivement pour des rapports E/C de 0.55, 0.60, 0.65 mm. On remarque que ces résultats évoluent similairement à ceux obtenus par nos essais. il a remarqué que la profondeur de pénétration de l'eau maximal admissible (5cm) est atteinte pour une rapport E/C=0.70.

3.3 Relations entre la vitesse des ultrasons et la perméabilité

On observe sur la figure 7 que la vitesse des ondes de surface présente une augmentation inversement proportionnelle à la perméabilité. L'estimation de la perméabilité à partir des vitesses des ultrasons est plus fiable (R^2= 0.95) pour les éprouvettes conservées dans l'eau que celles à l'air libre (R^2=0.88). Ceci est du à la sensibilité de la transmission des ultrasons à l'état hygrométrique du matériau traversé, les éprouvettes conservées à l'air peuvent présenter des états hygrométriques différentes. Ces relations permettent donc une estimation de la perméabilité des bétons par ces mesures ultrasoniques

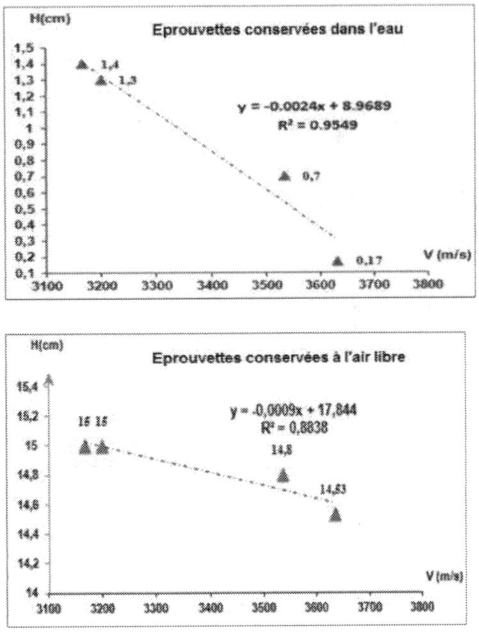

Fig. 7. Evolution de la perméabilité en fonction des vitesses des ultrasons des bétons ordinaires

4 Conclusion

Ce travail a été mené pour étudier la relation entre la perméabilité à l'eau et la vitesse de propagation de l'onde de surface (onde de Rayleigh) dans un matériau à matrice cimentaire. Ceci dans le but de substituer ces essais non destructifs aux essais destructifs de perméabilité réalisés sur des carottes prélevées sur les ouvrages à ausculter. Les mesures des vitesses des ultrasons ont été menées en mode de transmission surfacique pour pouvoir ausculter tous les éléments d'ouvrage (poteaux, poutres, planchers, ...) avec la facilité d'accessibilité. L'étude a concerné quatre bétons ordinaires et deux bétons autoplaçants et a mené aux conclusions suivantes :

- une diminution de la vitesse de surface quand le rapport E/C augmente et ceci dû à l'augmentation de la porosité. La diminution de la vitesse est plus importante entre le rapport E/C variant de 0,55 à 0,60 conduit à une diminution des vitesses des ultrasons de 10% alors pour une variation de E/C de 0,60 à 0,65.

- La vitesse des ultrasons pour le BAP1 (fillers calcaires) est 15% supérieure à celle du BAP2 (laitier de hauts fourneaux).

-La perméabilité des éprouvettes conservées à l'air libre est plus importante que celles conservées dans l'eau. L'augmentation de la perméabilité avec l'accroissement du rapport E/C est plus importante pour les éprouvettes conservées dans l'eau que pour les éprouvettes conservées à l'air libre. Pour les éprouvettes conservées dans l'eau la profondeur de pénétration de l'eau pour un rapport E/C de 0,65 reste toujours inferieure à la valeur maximale admissible (5 cm). Le BAP1 présente une perméabilité inferieure à celle du BAP2, cette différence est respectivement de 20% et 10 % pour les éprouvettes conservées dans l'eau et à l'air libre.

- L'estimation de la perméabilité à partir des vitesses des ultrasons est plus fiable pour les éprouvettes conservées dans l'eau (essais normalisés) que celles à l'air libre. Ces relations sont de type linéaire avec une évolution inversement proportionnelle de la perméabilité en fonction des vitesses des ultrasons.

Références

1. Z. Lafhaj, Caractérisation des propriétés mécanique et acoustique d'un mortier dégradé chimiquement: étude expérimentale, 16ème Congrès Français de Mécanique, Ecole centrale de Lille, France (2005).
2. M. Choinska, Etude expérimentale de l'interaction endommagement-température perméabilité du béton, Thèse de doctorat, Ecole Centrale de Nantes, France, (2006).
3. . Soltani, Etude expérimentale de la propagation de l'onde de surface dans un milieu poreux sec et saturé, COFREND2008, Toulouse, France, (2008).
4. M. Chekroun, Diffusion multiple des ultrasons pour l'evaluation non destructive du beton avec des ondes de surface, COFREND2008, Toulouse, France, (2008).
5. Controls, Instruction manual : Ulrasonic pulse velocity tester, Mod. 58-E0048, Italie, (2002).
6. EN 12390-8, Essai pour béton durci : Partie 8 Profondeur de pénétration d'eau sous pression, AFNOR, France, (2012).
7. Controlab, Manuel d'utilisation Appareil pour mesurer la perméabilité selon la norme EN 12390-8 (AFNOR), Réf.E0651. France (2008).
8. D. Conz, Transport d'humidité dans les éléments de construction en béton imperméable à l'eau (béton WU)", rapport d'étude, Zeugin Bauberatungen AG, Münsingen, Allemagne, (2006)..

Effect of mineral admixtures on the rheological and mechanical properties of mortars

Zineb DOUAISSIA[1], Mouloud MERZOUD[2]
[1 et 2] Badji Mokhtar university, Department of civil Engineering, ANNABA , ALGERIA

Abstract. We propose in this study to determine the role of mineral admixtures on the properties of cementitious materials in order to define a formulation method allowing an optimization of their use and the improvement of the rheological and mechanical properties of the mortars. In order to meet the objectives of this study, and to evaluate the effects that mineral admixtures may have on the properties of mortars in the fresh and hardened state, we have proposed to use a specific method based on progressive volume substitution of cement by the mineral admixtures, in order to preserve the absolute volume of all the solid constituents and the consistency for all the mixtures. The use of the mineral admixtures in the mortars also has for objective to valorize a category of natural materials or coproducts available in Algeria. As a result, the mineral admixtures chosen for this study are Blast furnace Slag, natural pozzolan and silica fume. Then the effect of the mineral admixtures on the rheological and mechanical properties of the mortars was evaluated according to the substitution rate of the cement. We applied to the studied mortars the notion of equivalent binder in order to calculate the coefficient of activity (K) defined by the two theoretical formulas Féret and Bolomey. The specific chemical action of each mixture was evaluated by analysing the variation of the coefficient of activity K as a function of the rate of substitution of the cement.

1 Introduction

Les études récentes qui ont considéré l'influence des aditions minérales sur les propriétés des matériaux de ciment [1-4] ont montré que ces additions par leur finesse et une réactivité plus ou moins significative avec le ciment, peuvent engendrer dans certains cas des modifications significatives dans les propriétés rhéologiques et mécaniques.

La norme européenne EN 206-1 [5] définit les conditions de prise en compte des additions minérales en substitution partielle du ciment dans la formulation des bétons en considérant le concept de liant équivalent et définit un coefficient forfaitaire k de prise en compte des additions minérales qui sont déterminées pour les cendres volantes et les fumées de silice utilisées avec un ciment du type CEM I. Dans la majorité des études publiées et conformément aux textes normatifs, les additions minérales ont été introduites dans les mélanges en substitution massique du ciment. Dans ce cas, le volume absolu du mortier augmente et le rapport volumique (eau/solide) diminue. Mais dans le cas d'une substitution volumique du ciment, le volume absolu du mortier et le rapport volumique (eau/solide) demeure sans changement. Dans ce cas, la compacité de la structure granulaire et la porosité de la matrice cimentaire des mélanges avec additions demeurent comparables à celles du mélange sans additions et seuls les effets des particules d'additions sont pris en compte dans la phase cimentaire.

Pour évaluer l'effet des additions minérales sur les caractéristiques rhéologiques et mécaniques des mortiers, nous proposons d'utiliser cette méthode basée sur la substitution volumique progressive (de 10% jusqu'à 50%) du ciment par du laitier granulé de haut fourneau (La) , la pouzzolane Naturelle de Beni-Saf (Pz) et la fumé de silice (Fs).

Les caractéristiques physico-mécaniques des ciments et mortiers avec la pouzzolane sont étudiées par plusieurs auteurs [6,7]. Demirboga [8] a montré en étudiant l'influence des additions minérales sur la résistance en compression des mortiers, que la densité des mortiers avec additions diminuait avec l'augmentation des additions. Autre part, il avait noté que le laitier granulé de haut fourneau engendrait une diminution de la résistance en compression à 28 jours et une amélioration à 120 jours. Ezziane et al [6] ont montré en étudiant l'efficacité du laitier d'El Hadjer par le concept du liant équivalent, que l'utilisation d'un taux élevé du laitier d'El Hadjar (40%) lui confère un déclenchement d'activité très précoce (2 j), seulement cette activité ne sera jamais achevée à long terme (90 j). dans ce cas, un taux de 15% paraît optimal. Aichouba et al [9] ont montré en étudiant l'effet de la pouzzolane naturelle de Béni-Saf, que la substitution du ciment par la pouzzolane de 10% à 30% engendre une augmentation de la consistance normale, une diminution du temps de prise, une diminution des résistances mécaniques à jeunes âgés et une conservation de ces résistances à long terme. Colak [10] a montré que la substitution du ciment par la pouzzolane engendre une augmentation du besoin en eau des pâtes de ciment à maniabilité constante, une augmentation de la porosité de la pâte avec le taux de substitution du ciment par la pouzzolane et que les résistances en compression des mélanges à 20% de pouzzolanes sont très proches de celles sans additions.

2 Méthodologie

Afin d'évaluer toutes les contributions des additions aux propriétés rhéologiques et mécaniques des mortiers, nous avons été particulièrement intéressés par les propriétés suivantes:

- consistance, temps de prise, masse volumique et maniabilité.
- courbe contrainte-déformation en traction , résistances mécaniques et module d'élasticité.
- les effets des additions minérales sur le coefficient d'activité K.

Pour atteindre cet objectif, nous avons préparé un mortier selon la norme NF EN 196-1 [11], avec une quantité d'eau ajustée pour obtenir une consistance de référence. La consistance de référence a été évaluée en mesurant l'étalement du mortier à l'état frais sur une table à secousses. Pour chaque mortier ayant acquis la consistance de référence, on a préparé des éprouvettes 4x4x16.

Pour cette étude, nous avons indiqué par :

w : le volume d'eau utilisé pour obtenir un mortier avec l'efficacité de référence.

b: le volume absolu du liant dans le mortier (b = c + a).

c : le volume absolu de ciment dans le mortier.

a : le volume absolu du mélange minéral dans le mortier.

2.1 Évaluation des caractéristiques rhéologiques

Le besoin en eau des mortiers avec additions est exprimé par la variation du rapport volumique quantité d'eau par la quantité de liant *(E/L)*. Pour apprécier les temps de début et de fin de prise des pâtes pures des liants préparés, on a effectué les essais avec l'appareil de Vicat conformément à la norme EN196-3 [12]. L'étude de la maniabilité des mortiers est assurée on étudiant deux facteurs qui sont le besoin en eau et l'étalement du mortier avec additions à l'état frais sur la table à secousses selon la norme BS EN 1015-3.

2.2 Évaluation des caractéristiques mécaniques

À 28 jours des essais de traction par flexion et de compression ont été effectué par la machine Zwick/Roe II Z020, dont le pilotage et l'acquisition des résultats se font par ordinateur à partir du logiciel testXpert II, à la fin de chaque essai la collecte de données nous a permis de tracer les courbes contrainte-déformation, de déterminer les résistances à la traction et à la compression ainsi de déduire les modules d'élasticité.

2.3 Evaluation du coefficient d'activité K :

L'évaluation du coefficient d'activité K pour une addition minérale incorporée dans un mélange cimentaire, peut se faire en appliquant aux mortiers étudiés les principes des formules théoriques de Féret et Bolomey.

2.3.1 Application de la formule de Féret :

La formule de Féret exprime la résistance en compression par la relation suivante :

$$R_c = G_F \times f_{mc} \left(\frac{c}{c+e+v}\right)^2 \qquad (1)$$

c, e et *v* désignent respectivement, les volumes du ciment, d'eau et d'air occlus dans le mortier sans additions en (l/m^3).

G_F désigne un coefficient adimensionnel qui dépend de la nature des agrégats.

f_{mc} désigne la résistance en compression du mortier normal au même âge en (MPa).

nous pouvons nous baser sur la notion de liant équivalent définie par la norme européenne EN 206-1, comme suit :

$$L_{eq} = C + kF \qquad (2)$$

C, désigne la quantité de ciment dans le béton (kg/m^3),

F, désigne la quantité d'addition dans le béton (kg/m^3),

k, désigne un coefficient de prise en compte des additions minérales dans le liant.

Le facteur d'activité K_F peut être déduit à partir du modèle prédictif de Féret , comme suit :

$$K_F = \frac{c_r(e+v)\sqrt{R_c/R_{cr}}}{f\left[c_r\left(1-\sqrt{R_c/R_{cr}}\right)+(e_r+v_r)\right]} - \frac{c}{f} \qquad (3)$$

2.3.2 Application de la formule de Bolomey :

En se basant sue le modèle prédictif de Bolomey, la résistance en compression est exprimée par la relation suivante :

$$R_c = G_B . f_c \left(\frac{C}{E+V} - 0{,}5\right) \qquad (4)$$

c, e et *v*, désignent respectivement, la masse du ciment, la masse d'eau et la masse d'eau équivalente au volume d'air occlus dans le mélange en (kg/m^3).

G_B désigne un coefficient adimensionnel qui dépend de la nature des agrégats.

f_c désigne la résistance en compression du mortier normal au même âge en (MPa).

En appliquant aux mortiers étudiés, la notion du liant équivalent, le facteur d'activité K_B peut être déduit à partir du modèle prédictif de Bolomey , comme suit :

$$K_B = \frac{R_c . C_r (E+V)}{R_{cr} . F\,(E_r + V_r)} - \frac{C}{F} + \frac{E+V}{2F}\left(1 - \frac{R_c}{R_{cr}}\right) \ldots\ldots\ldots 5$$

Avec :

$$C_r = \rho_C . C_r$$
$$C = \rho_C . C \qquad (6)$$
$$F = \rho_F . f$$

ρ_C *et* ρ_F désignent respectivement, les masses volumiques absolues du ciment et de l'addition.
c, c_r et *f* désignent respectivement, les volumes de ciment et d'additions dans le mortier.

3 Matériaux

3.1 Ciment

Le ciment utilisé est un ciment Portland du type CPA CEMI 52.5R conformément a la norme NF EN 197-1 [13]. Sa masse volumique et sa surface spécifique sont respectivement 3,15 g/cm^3 et 440 m^2/kg.

3.2 Sable

Le sable utilisé est un sable normalisé, le sable est conditionné en sachets de polyéthylène contenant chacun 1 350 ± 5 g conformément aux exigences de la norme EN 196-1.

3.3 Additions minérales

Pour cette étude, il a été donc utilisé trois additions minérales différentes par leurs natures minéralogiques, morphologiques et chimiques : une addition de pouzzolane (Pz), une addition de laitier granulé vitrifié moulu de haut fourneau (La) et une addition de fumée de silice (Fs), (tableau.1).

Tableau.1: Caractéristiques physiques des additions minérales

Additions		Masse volumique Absolue (Kg/m^3)	Surface spécifique Blaine (cm^2/g)
laitier	La	2800	2900±120
pouzzolane	Pz	2650	3500±200
fumée de silice	Fs	2240	15.10-9

4 Résultats

4.1 Effets des Additions minérales sur les caractéristiques rhéologiques des mortiers

4.1.1 Consistance et temps de prise :

On peut constater sur la figure.1 que l'ajout des additions minérales implique une augmentation de 3 % de la quantité d'eau pour obtenir une consistance normale.

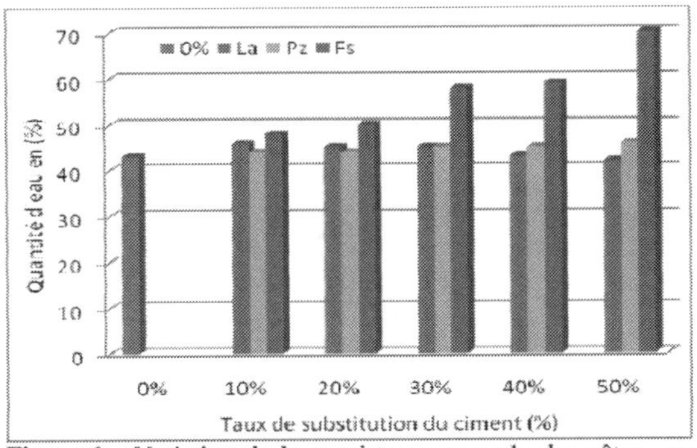

Figure 1 : Variation de la consistance normale des pâtes , en fonction du taux de substitution du ciment par les additions minérales.

Les figures 2 et 3 traduisent les temps de début et fin de prise des pâtes, en fonction du taux de substitution du ciment par les additions minérales. On constate une augmentation des temps de début de prise par rapport à celles de la pâte témoin (0%) a eu pour effet de retarder l'hydratation du ciment (phase de ralentissement), ce retard de la prise est attribué principalement à la baisse de la teneur de ciment en (C_3A) qui est responsable du raidissement précoce.

4.1.2 Masse volumique des mortiers frais :

Les courbes de la variation de la masse volumique des mortiers à l'état frais présentées par la figure 4, sont décroissantes avec le taux de substitution, un optimum est observé pour des taux de 10% et 20% .

En général, les masses volumiques apparentes des mortiers avec additions minérales à l'état frais sont directement liées au besoin en eau et suit une variation opposée.

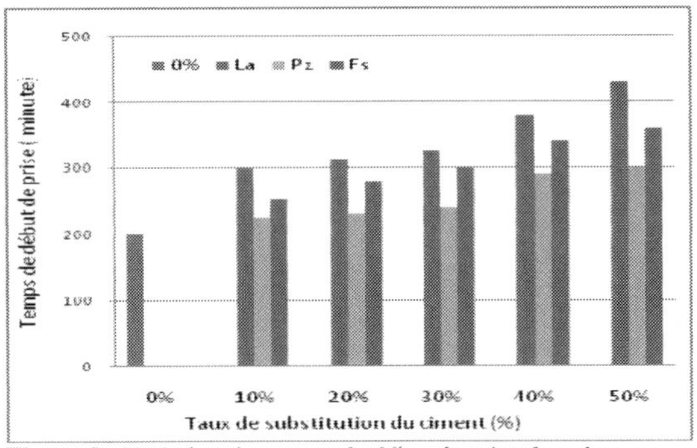

Figure 2 : Variation des temps de début de prise des pâtes, en fonction du taux de substitution du ciment par les additions minérales.

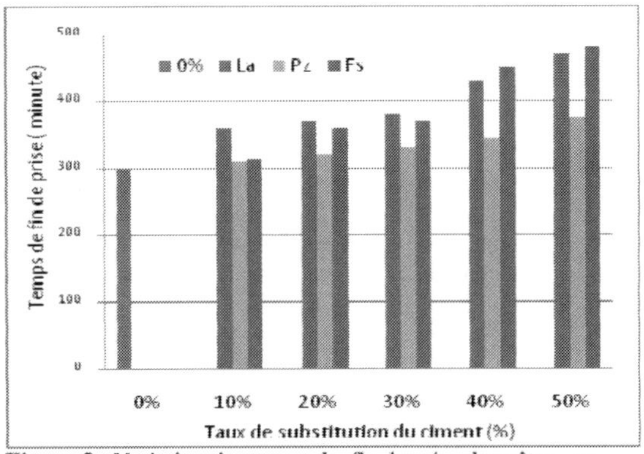

Figure 3 : Variation des temps de fin de prise des pâtes , en fonction du taux de substitution du ciment par les additions minérales.

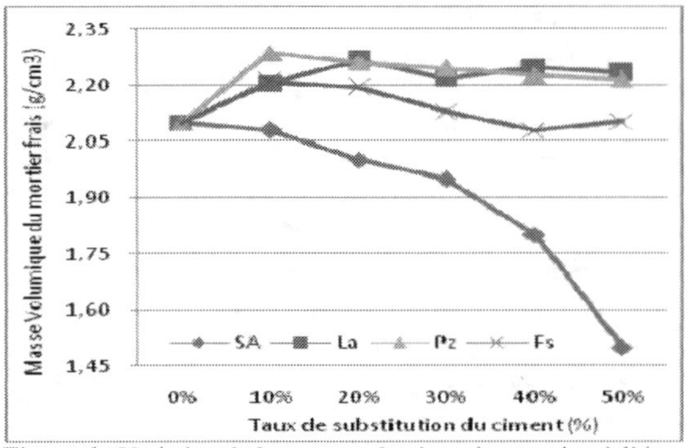

Figure 4 : Variation de la masse volumique des mortiers à l'état frais, à même consistance en fonction du taux de substitution du ciment par les additions minérales.

4.1.3 Maniabilité :

Nous constatons sur la figure 5, que l'étalement du mortier frais avec la pouzzolane (Pz), croît jusqu'à 10 % du taux de substitution, puis décroît avec l'augmentation de la quantité d'addition dans le mélange.
Les mortiers frais avec du laitier (La) et de la fumée de silice (Fs), présentent un étalement qui décroît directement avec l'augmentation de la quantité d'addition dans le mélange. Le mortier sans additions (SA) présente un comportement différent. En effet la substitution du ciment par le sable engendre une amélioration continue de l'ouvrabilité.

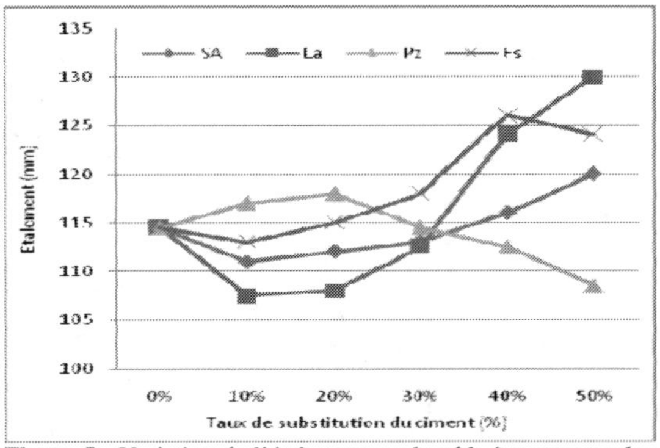

Figure 5 : Variation de l'étalement sur la table à secousses des mortiers, en fonction du taux de substitution du ciment par les additions minérales.

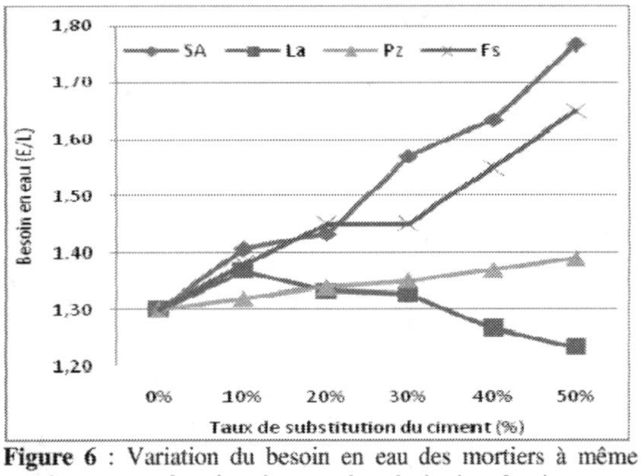

Figure 6 : Variation du besoin en eau des mortiers à même consistance, en fonction du taux de substitution du ciment par les additions minérales

Nous constatons sur la figure 6 que les mortiers avec la pouzzolane présentent un besoin en eau plus faible que celui du mortier de référence pour des taux inférieurs à 20%. À plus forts taux de substitution, le besoin en eau croît avec l'augmentation de la quantité d'addition dans le mélange, mais reste proche de celui du mortier de référence.
Les mortiers avec du laitier présentent un besoin en eau qui croit avec le taux de substitution du ciment. Cette différence de comportement est probablement due à l'irrégularité et à l'aspect poreux de ses particules, qui n'arrivent plus à remplir les pores du squelette granulaire ou absorbent des quantités d'eaux complémentaires qui peut se traduire par un plus fort besoin en eau, à même consistance.
On peut également noter que les mortiers avec fumée de silice présentent les besoins en eau les plus importants en accord avec sa grande finesse, ce besoin en eau est d'autant plus important que le taux de substitution du ciment est grand. Les mortiers sans additions présentent un besoin en eau décroissant.

4.2 Effets des Additions minérales sur les caractéristiques mécaniques des mortiers:

4.2.1 Courbes contraintes - déformations en traction par flexion :

La comparaison des courbes contrainte-déformation en traction par flexion des mortiers est présentée par les figures 7,8 et 9.

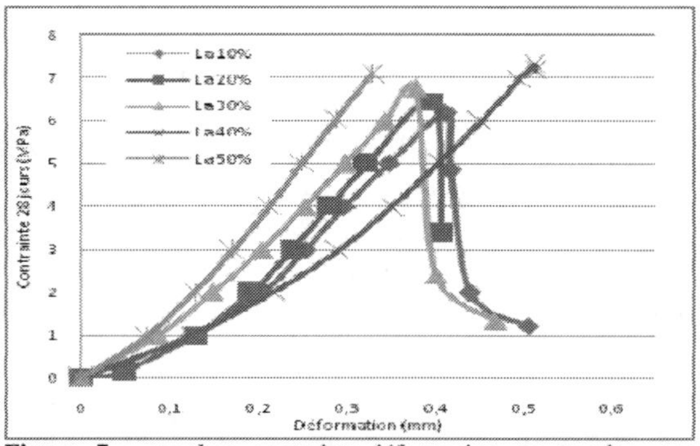

Figure 7 : courbes contrainte-déformation en traction par flexion des mortiers en fonction du taux de substitution du ciment par le laitier.

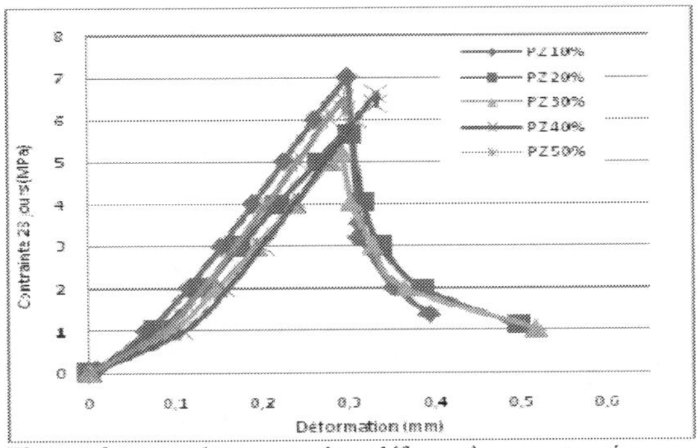

Figure 8 : courbes contrainte-déformation en traction par flexion des mortiers en fonction du taux de substitution du ciment par la pouzzolane

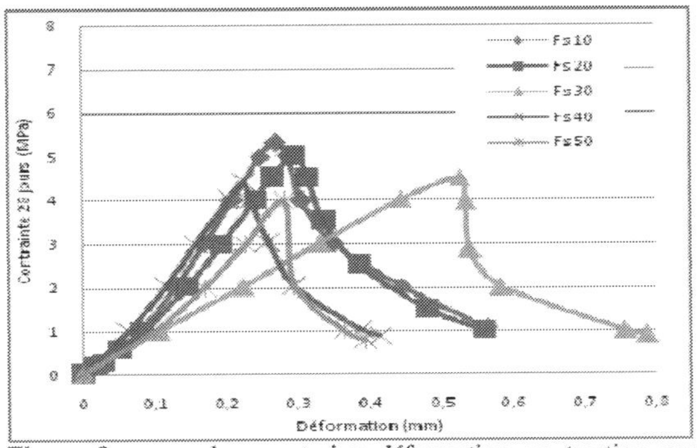

Figure 9 : cou rbes contrainte-déformation en traction par flexion des mortiers en fonction du taux de substitution du ciment par la fumée de silice .

Nous observons que toutes les courbes présentent un comportement linéaire élastique jusqu'à environ 30% de la résistance maximale. Au-delà de la valeur limite d'élasticité, les courbes enregistrent une augmentation graduelle en courbure évaluée autour de 75 à 90% de la résistance ultime. Au-delà du pic, la courbe présente une partie descendante (déformation contrôlée) jusqu'à la rupture. L'addition de fumée de silice améliore considérablement la rigidité des mortiers, tandis que le laitier et la pouzzolane peuvent engendrer une certaine fragilité quoiqu'ils donnent aussi les valeurs de résistances à la traction par flexion les plus ou moins importantes que la fumée de silice.

4.2.2 Résistances mécaniques :

La variation des résistances à la traction et en compression à 28 jours est représentée par les figures 10 et 11.

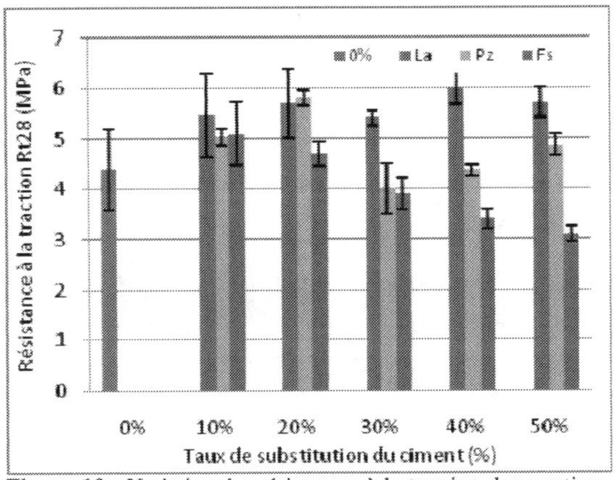

Figure 10 : Variation des résistances à la traction des mortiers, en fonction du taux de substitution du ciment par les additions minérales.

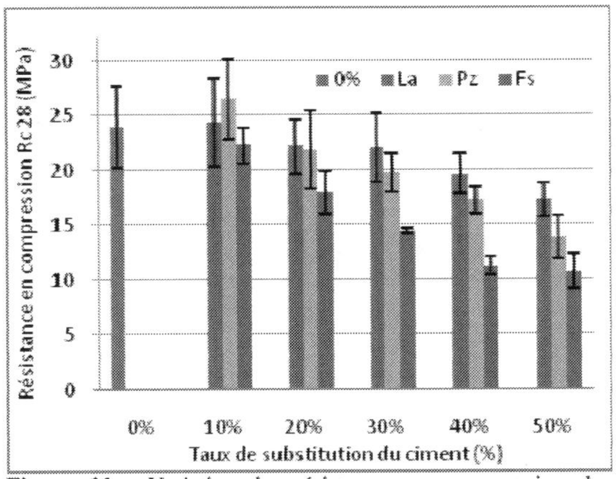

Figure 11 : Variation des résistances en compression des mortiers, en fonction du taux de substitution du ciment par les additions minérales.

Nous remarquons que les résistances à la traction croissent pour les faibles taux de substitution du ciment , 20% pour le laitier et la pouzzolane et 10% pour la fumée de silice, puis commencent à décroître.

Les résistances en compression dépendent de taux de substitution du ciment. Des résistances importantes sont obtenues pour les mortiers avec le taux (10%), avec des valeurs dépassant celui de la référence sans additions minérales. Néanmoins, la tendance générale est au delà de 20% de substitution conduit à une baisse de résistance en compression, principalement en raison de l'effet de dilution, qui correspond bien avec les résultats obtenus par d'autres auteurs [8].

4.2.3 Module d'élasticité E :

Les variations du module l'élasticité à 28 jours sont représentées par la figure 12. Nous remarquons que les modules d'élasticité en compression E_{28} croissent pour les faibles taux de 20% à 30% pour le laitier et 10% pour la pouzzolane et de 10% à 20% pour la fumée de silice, puis commencent à décroître proportionnellement aux taux de substitution.

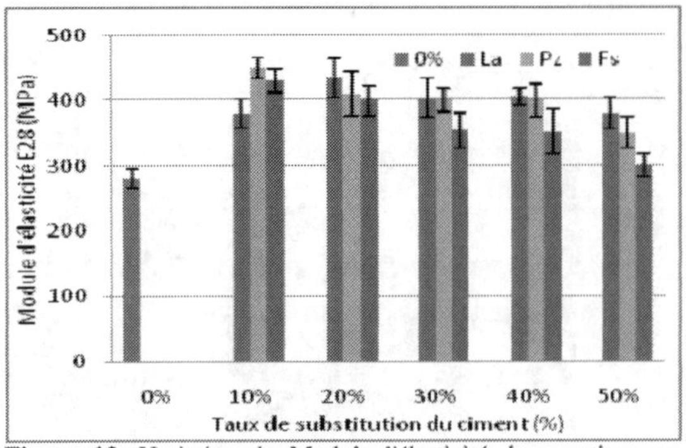

Figure 12: Variation du Module l'élasticité des mortiers, en fonction du taux de substitution du ciment par les additions minérales.

4.3 Effets des Additions minérales sur le coefficient d'activité K:

4.3.1 Application de la formule de Féret :

Les variations des facteurs d'activité chimique des additions calculé par la formule de Féret, est représentée par la figure 13, à faibles taux de substitution, la fumée de silice présente à 10 % facteur d'activité le plus important parmi toutes les additions utilisées. Il atteint une valeur de 1,99. Les facteurs d'activité de la pouzzolane sont les plus faibles en raison de l'atténuation des effets physicochimiques qui sont d'autant plus faibles que la finesse de l'addition est petite. Le laitier présente des facteurs d'activité intermédiaires, avec une meilleure activité à 20 % pour laquelle il atteint une valeur de 1,11 .

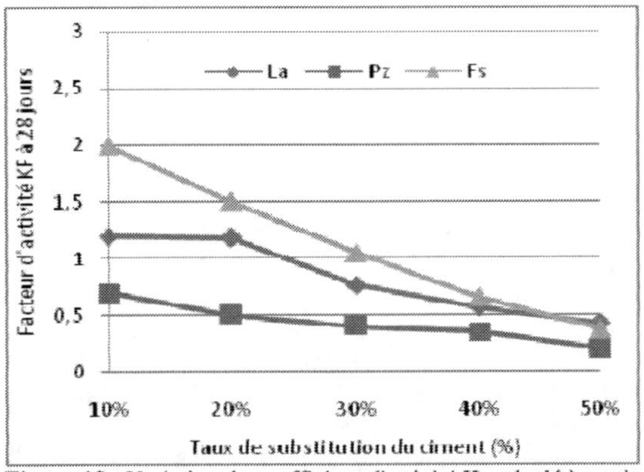

Figure 13 : Variation du coefficient d'activité K, calculé à partir de la formule de Féret, en fonction du taux de substitution du ciment par les additions minérales.

4.3.2 Application de la formule de Bolomey :

Les variations des facteurs d'activité chimique des additions à 28 jours pour la formule de Bolomey, sont représentées par les figures 14.

Dans le cas de la formule de Bolomey, à faibles taux de substitution du ciment, la fumée de silice présente à 10 % du taux de substitution le facteur d'activité le plus important parmi toutes les additions utilisées. Il atteint une valeur de 2,75. Les facteurs d'activité de la pouzzolane sont les plus faibles en raison de l'atténuation des effets physicochimiques. Le laitier présente des facteurs d'activité intermédiaires, avec une meilleure activité à 20 % du taux de substitution du ciment, pour laquelle ils atteint la valeur de 1,30.

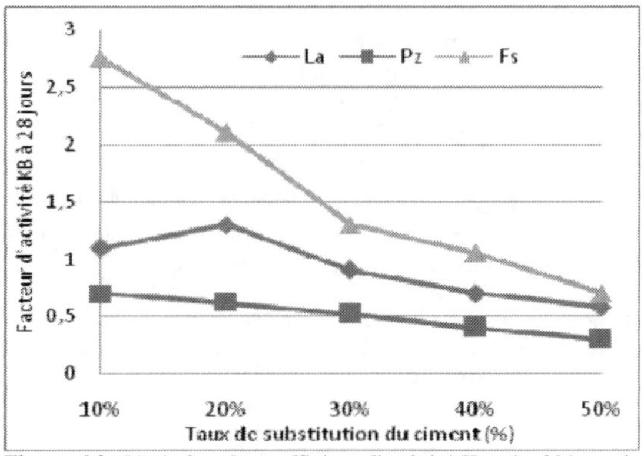

Figure 14 : Variation du coefficient d'activité K, calculé à partir de la formule de Bolomy, en fonction du taux de substitution du ciment par les additions minérales.

Néanmoins, il est important de noter que les facteurs d'activité évalués à partir de la formule de Bolomey sont beaucoup trop surestimés pour le cas de la fumée de silice par rapport à ceux calculés à partir de la formule de Féret.

5 Conclusions

L'approche appliquée consiste en une substitution progressive du ciment par des additions minérales dans des mortiers dont la maniabilité et le volume absolu de toutes les phases constitutives solides sont restés constants.

Les résultats obtenus ont montré que:

- le besoin en eau des mortiers varie sensiblement avec la nature minéralogique et les caractéristiques granulaires du mélange minéral incorporé, et pour les faibles taux de substitution du ciment, elle est directement liée aux phénomènes d'empilage granulaire, donc à la compacité des mélanges frais.
- l'effet des additions minérales se limite principalement à une accélération du processus d'hydratation de ciment à un jeune âge. L'action contraire de la fumée de silice est probablement due aux réactions pouzzolaniques relativement tardives avec le ciment.

Ensuite, les résistances se développent avec des taux de substitution faibles (à proximité de 10%). Ensuite, elles commencent à diminuer, à l'exception du mortier avec de la fumée de silice qui diminue directement sans aucune amélioration.

- L'activité chimique des additions à 28 jours pour les mortiers avec additions, a été évaluée en traçant la variation du facteur d'activité en fonction du taux de substitution du ciment par l'addition, en utilisant la notion du liant équivalent définie par la norme EN 206-1 et en déduisant le facteur d'activité chimique à partir des formules théoriques de Féret et de Bolomey.

Les facteurs d'activités chimiques à 28 jours, des différentes additions sont très variables. Ils dépendent de l'âge du mortier avec additions, de la formule de résistance adoptée, de la nature minéralogique et chimique de la quantité et des caractéristiques granulaires de l'addition incorporée. Ils sont encore plus forts dans le cas de la formule théorique de Bolomey que dans le cas de la formule de Féret.

References

1. Logbi Abdelaziz. "Effet de l'incorporation des ajouts minéraux sur les propriétés physico-mécaniques du béton", Thèse de Magister, l'ENP- Alger (Juin 1999).
2. Chabi. S, Mezghiche. B, et Guettala.H; "Etude de l'influence des additions minérales actives sur le comportement mécaniques des ciments et mortiers" Courrier du Savoir – N°**05**, pp. 03–08, (Juin 2004).
3. A. Kerbouche, M. Mouli, L. Laoufi, Y. Senhadji et S. Benosmane; "Influence des ajouts minéraux sur les résistances mécaniques des mortiers". SBEIDCO -1st International Conference on sustainable Built Environment Infrastructures in Developing Countries-ENSET, Oran-Algeria, (2009).
4. Amouri Chahinez, Houari Hacène; "les performances des mortiers avec differentes additions minerales" SBEIDCO -1st International Conference on Sustainable Built Environment Infrastructure in Developing in Developing Countries ENSET-Oran-Algeria, (2009)
5. AFNOR NF EN 206-1 Béton-Partie 1: spécification performances, production et conformité- (Avril 2004)
6. Ezziane K., Bougara A., Kassoul A., ""Efficacité du laitier d'El Hadjar par le concept du liant équivalent", 2ème colloque Maghrébin en Génie Civil, p. 87-96, Biskara, Algérie, 10-11 (décembre 2002).
7. 7.Boudchicha A., Utilisation des additions minérales et des adjuvants fluidifiants pour l'amélioration des propriétés rhéologiques et mécaniques des bétons, Thèse de doctorat, université de Constantine, (2007).
8. Demirboga R., Influence of mineral admixtures on thermal conductivity and compressive strength of mortar, Energy and building Vol. **35**, p. 189-192, (2003).
9. Aichoube A., Ghrici M., Benaissa A., "Effect de la pouzzolane naturelle de Béni-Saf sur les propriétés d'un ciment à base de calcaire", Colloque Méditerranéen sur les Matériaux CMEDIMAT, Oran, Algérie, 6-7 (décembre 2005).
10. Colak A., "Characteristics of pastes from a Portland cement containing different amounts of natural pouzzolan", Cement and Concrete Research, Vol. **33**, p. 585-593, (2003).
11. AFNOR NF EN 196-1 Méthodes d'essais des ciments- Partie 1 : Détermination des résistances mécaniques, (avril 2006).
12. NF EN 196-3 (P15-473) : Méthodes d'essais des ciments – partie 3 : Détermination du temps de prise et de la stabilité.F. De Lillo, F. Cecconi, G. Lacorata, A. Vulpiani, EPL, **84** (2008)
13. AFNOR EN 197-1 Ciment - Partie 1 : Composition, spécifications et critères de conformité, (février 2001).

Physico-mechanical characterization of composite cementitious mortars modified by resin containing antifoam

Hichem BERKAK[1], Zoubir MAKHLOUFI[1], Madani BEDERINA[1], Abdellah DEMDOUM[2] and Yacine HADJ AISSA[2]
[1] Amar THELIDJI University, Structures rehabilitation and materials laboratory, Laghouat, Algeria
[2] Amar THELIDJI University, Research in civil engineering laboratory, Laghouat, Algeria

Abstract. Nowadays, polymer-modified mortars (PMM) are generally used in the field of building as finishing materials. In general, The properties of the fresh and hardened PMC are affected by several factors that tend to interact each other: the nature of materials used as latex, cement and aggregates; the mix proportions (W/C, polymer/cement ratio, etc.); air content; type and amount of surfactants and anti-foaming; and curing methods. This present experimental work aims to study the influence of the rate of partial substitution of cement by resin containing anti-foaming on the properties of composite mortars in the fresh and hardened state particularly workability, tensile and compressive strengths at different times (3, 7 and 28 days) and module of static elasticity at 28 days. For that purpose, seven mortar combinations were developed from the various substitution percentages (0 to 15% by step of 2.5%). The treatment of the obtained results reveals that this partial replacement of the cement by resin improves, clearly, the workability of mortars and decreases the report W/C at the same time. Besides, the incorporation of resin affects significantly the mechanical strengths, and consequently, the addition by substitution of the polymer-latex improves, slightly, the mechanical behavior in tensile where the rates of increase range from 2.20% to 8.54% for replacement percentages of 2.5% and 15%, respectively. Furthermore, the compressive strengths of cement - resin composite mortars are lower compared to the reference mortar made with the cement Portland (CEM I). The notable decrease in compressive strength of polymer–modified mortars can be justified by the reduction of cementitious part, in hybrid materials, which is an indispensable and responsible part of improving compressive strength.

1 Introduction

La chimie du bâtiment est maintenant basée sur les polymères. L'ingénierie des matériaux polymères est devenue une partie de la construction moderne, car elle permet de fabriquer des matériaux ayant les propriétés désirées et un coût raisonnable [1]. Mais, les mortiers et bétons modifiés par polymères sont utilisés préliminairement comme matériaux de finition ou de réparation [2].

Les différentes combinaisons des dizaines de types des monomères avec des modificateurs produisent beaucoup de variétés de bétons à base de polymères. Parfois, ces bétons ont été utilisés massivement comme matériaux de construction dans certaines applications [3].

La modification des matériaux traditionnels par des polymères devrait augmenter la résistance à la traction ou à la flexion et la ténacité des matériaux, sans induire une diminution importante de la résistance à la compression. En outre, le matériau ciment modifié par polymère présente une bonne liaison et une meilleure adhérence aux agrégats ou ouvrages détériorés déjà existants, il est donc utile pour la réparation des constructions [4, 5] et améliore imperméabilité qui conduit à une plus faible diffusivité des chlorures [6-8], une meilleure résistance au gel [9], une réduction du taux de retrait [10], et finalement améliore la durabilité [11-14].

Les mortiers de ciment modifiés par des polymères sont largement utilisés dans diverses applications comme les infrastructures de transport, les ponts, les murs extérieurs isolants en raison de leur excellente résistance, leur protection de l'environnement et leur meilleure maniabilité [15].

En 1923, le mortier de ciment modifié au latex naturel n'a pas obtenu une place dans l'application pratique à cause de ses compatibilités insatisfaisantes. A l'heure actuelle, une large gamme des polymères synthétiques sont utilisés comme des additifs aux mortiers et bétons cimentaires. Ces additifs peuvent être sous forme liquide ou sous forme de poudre. Lorsqu'ils sont sous forme liquide, on parle de résines comme le Styrène-Butadiène (SBR), le Chlorure de Polyvinyle (PVDC), l'Acrylonitrile-Butadiène (NBR), le Polychloropène (CR), l'Ester Polyacrylique (PAE), l'Acétate de Polyvinyle

(PVAC), le poly (Styrène/Acrylate de n-butyle) (SA) ou encore le poly (Ethylène/Acétate de Vinyle) (EVA). En particulier, les mortiers de ciment modifiés de latex SBR ont présenté une excellente dureté et une réduction d'eau, où ils ont déjà été largement utilisés comme des adhésifs, des matériaux de finition et des matériaux de réparation pour les structures en béton armé [16, 17].

Cet article s'inscrit dans le cadre d'un important programme de recherche au sein du laboratoire SREML portant sur les mortiers calcaires en remplacement partiel du ciment par des additions minérales, en privilégiant l'utilisation simultanée de poly-additions et de polymères. En effet, le but voulu par cette étude est d'élaborer un mortier de sable calcaire additionné de résine afin d'améliorer ses propriétés mécaniques surtout la résistance à la traction. Un autre objectif est de trouver la composition optimale d'un mortier calcaire modifié de résine destiné à la finition, la réparation, la reprise de bétonnage, etc. Cette composition constitue les meilleurs compromis entre les propriétés étudiées, à savoir la maniabilité, les résistances mécaniques à la traction et en compression.

2 Détails expérimentaux

2.1 Matériaux utilisés

2.1.1 Ciment

Le ciment Portland CEM I 42,5 HRS provenant de la cimenterie de LE KARROUBA (Tunisie) a été utilisé. La composition chimique et les caractéristiques du ciment employé sont présentées dans les tableaux 1 et 2 respectivement.

Tableau 1. Composition chimique du ciment

Elément	%
SiO_2	18.83
CaO	61.54
MgO	1.27
Al_2O_3	4.20
Fe_2O_3	5.31
SO_3	1.96
K_2O	0.49
TiO_2	0.20
Na_2O	0.21
P_2O_5	0.29
Perte au feu	5.70

Les constituants minéralogiques de ce ciment, selon la formule de Bogue, sont : le C3S à 66.01%, le C2S à 4.53%, le C3A à 2.16% et le C4AF à 16.14%.

Tableau 2. Caractéristiques du ciment employé

Caractéristique	Norme	Valeur trouvée
Masse volumique absolue (g/cm^3)	NF EN 196-6	3,10
Surface spécifique Blaine (cm^2/g)		3200
Consistance (%)	NF EN 196-3	28
Stabilité (mm)		1

2.1.2 Résine (P)

Au cours de cette expérimentation, la résine utilisée est une résine synthétique nommée TEKWELD provenant de l'entreprise de TECKNACHEM. Cette résine se présente sous la forme d'une solution aqueuse (latex) à haute densité, stable en milieu alcalin, spécialement conçue pour être additionnée à l'eau de gâchage des mortiers, coulis de ciment ou de l'enduit. Les principales propriétés de la résine utilisée sont regroupées dans le tableau 3.

Tableau 3. Principales propriétés de résine TEKWELD

Caractéristiques	Résine TEKWELD
Etat physique	Liquide
Couleur	Blanc
Densité (g/cm^3, à 20°C)	1,040± 0,02
PH (à 20°C)	4± 1
Extrait sec	26 %± 2 %

2.1.3 Sable

Dans le cadre de valorisation des matériaux locaux de la région de l'atlas saharien algérien, le sable calcaire concassé de fraction 0/5mm provenant d'une carrière située au nord de la ville de Laghouat a été utilisé. Les caractéristiques physiques et la courbe granulométrique du sable sont présentées dans le tableau 3 et la figure 1 respectivement. D'après la figure 1, on a remarqué que le sable calcaire est un sable moyennement grossier, à granulométrie continue et étalée et sa courbe granulométrique se situe à l'intérieur du fuseau normalisé.

Tableau 4. Propriétés du sable calcaire

Caractéristique	Norme	Valeur trouvée
Masse volumique absolue (g/cm^3)	NF EN 1097-6	2.59
Masse volumique apparente (g/cm^3)		1.49
Equivalent de sable : SE (%)	NF EN 933-8	66
% des éléments <0.08 mm	NF EN 933-1	7.5
Module de finesse : Mf	NF EN 12620	3.21
Coefficient d'absorption : WA $_{24}$ (%)	NF EN 1097-6	1.90

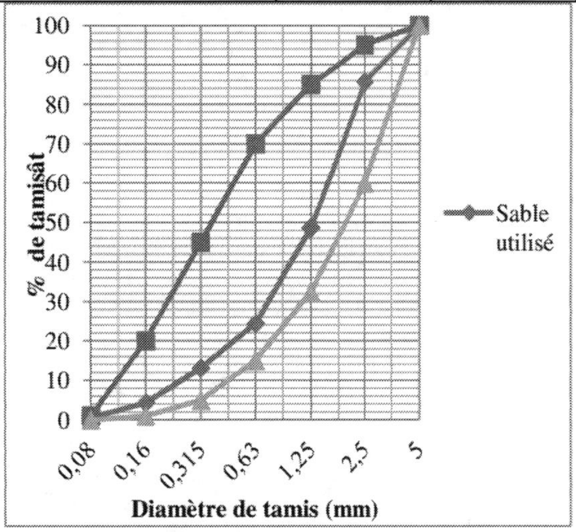

Fig. 1. Courbe granulométrique du sable calcaire

2.1.4 Superplastifiant (Sp)

L'adjuvant utilisé est un superplastifiant fabriqué par la société TECKNACHEM et commercialisé sous le nom de SUPERIOR 9 WG. Il est à base de poly-carboxylates de nouvelle génération qui permettent de réaliser des bétons très fluides avec des rapports E/C très réduits. Il est compatible avec tous les types de ciment. Sa plage normale d'utilisation varie selon la fiche technique entre 0.6% à 2% (en extrait sec) du poids de ciment. Il est sous forme liquide de couleur brune de 33% (±2%) d'extrait sec. Sa densité est de 1.10 (±0.03), et son PH (à 20°C) est de 5.5 (±1) avec une teneur en chlore inférieure à 0.1%.

2.1.5 Eau de gâchage

Il s'agit de l'eau du robinet de l'université de Laghouat exempté d'impuretés.

2.2 Formulation des mortiers composites

Le mode de préparation des échantillons a été effectué selon la norme EN 196-1. Le tableau 3 récapitule la composition des mélanges d'un volume d'un moule de trois éprouvettes de (4x4x16) cm^3. Les essais ont été réalisés sur le mortier de référence et les mortiers contenant différents pourcentages de résine SBR en addition par rapport au dosage en masse du ciment (2.5% à 15 % par pas de 2.5%). Après démoulage, les éprouvettes sont conservées dans l'eau durant 6 jours. Ensuite, elles sont placées à l'air libre au laboratoire à une température de 20±2°C jusqu'à échéances de 3, 7, 28 jours [18, 19]. Les dosages des différents constituants des mortiers élaborés sont donnés dans le tableau 5.

Tableau 5. Dosages des différents constituants des mortiers élaborés

Désignation des mélanges	C (g)	P (%)	Résine (g)	Sable (g)	Sp (%) *	E/C
M réf.	450	0	0	1350	0.8	0.45
MAR-2.5%	450	2.5	11.25		0.8	
MAR-5%	450	5	22.5		0.7	
MAR-7.5%	450	7.5	33.75		0.7	
MAR-10%	450	10	45		0.6	
MAR-12.5%	450	12.5	56.25		0.6	
MAR-15%	450	15	67.5		0.5	

M réf. : Mortier de référence sans addition de résine; MAR-X% : Mortier modifié par X% de résine par rapport de masse du ciment.
*en extrait sec et par rapport du poids de ciment.

3 Résultats et discussion

3.1 Effet de résine sur la maniabilité des mortiers élaborés

La substitution massique simultanée du ciment par la résine engendre une variation de la maniabilité du mélange frais caractérisée par l'écoulement des mortiers à l'état frais. Les essais réalisés, permettant de mesurer le temps d'écoulement au maniabilimètre LCL de mortier type «B».

Les résultats de l'évolution de la maniabilité à l'état frais des mortiers étudiés en fonction de taux de substitution partielle de matrice cimentaire par une matrice polymérique sont représentés sur la figure 2.

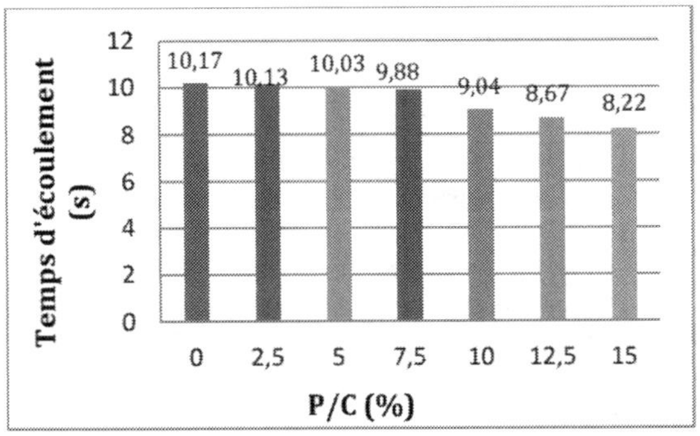

Fig. 2. Variation du temps d'écoulement en fonction du rapport P/C (%)

Les résultats présentés sur toutes les figures montrent une faible variation du temps d'écoulement pour l'ensemble des compositions quelles que soient la teneur de résine. Le constat général qui se dégage est une légère amélioration de la fluidité des mortiers par l'addition de résine malgré que le % du superplastifiant diminue de 0.8% à 0.5% du poids de ciment. Cette constatation est également conforme aux résultats antérieurs [20-22].

Les mortiers contenant un taux élevé en résine montrent une consistance diminuée. Pour une maniabilité fixe, l'incorporation de la résine dans ces mortiers calcaires fait donc décroître la demande en eau.

Nous constatons également que les comportements de tous les mélanges sont proches les uns des autres où le temps d'écoulement de ces mélanges varie de 10.17 à 8.22s.

Enfin, l'ajout du polymère améliore la maniabilité du mélange où le mortier nécessite alors mois d'eau pour une même consistance. De plus, le polymère lui-même réduit la ségrégation [20].

3.2 Effet de résine sur la résistance à la traction

Les figures 3 et 4 montrent l'évolution de la résistance à la traction par flexion des mortiers élaborés, en variant le rapport P/C à 3, 7 jours ensemble et à 28 jours respectivement.

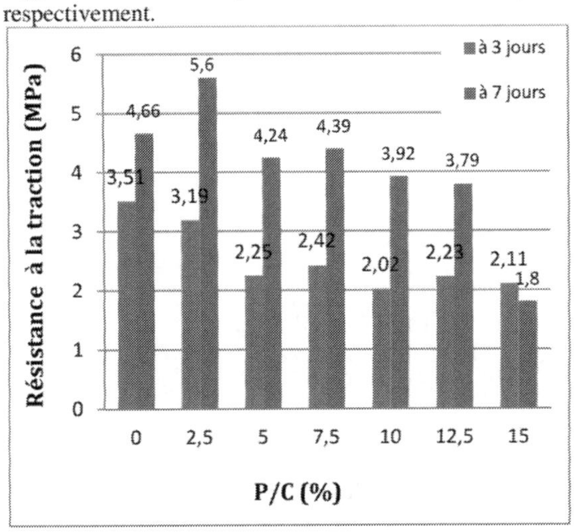

Fig. 3. Effet du rapport P/C (%) sur la résistance à la traction des mortiers calcaires à 3 et 7 jours

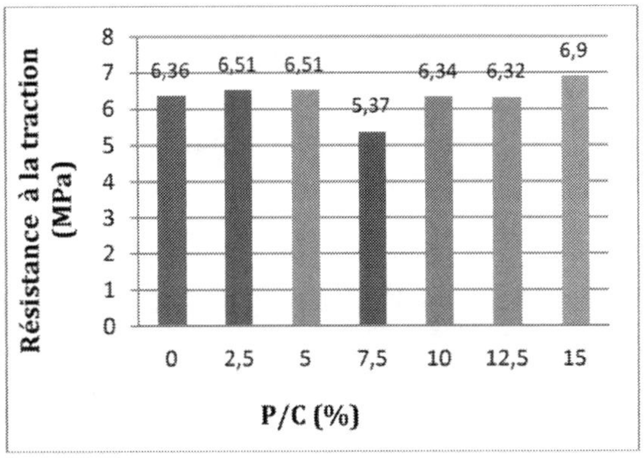

Fig. 4. Effet du rapport P/C (%) sur la résistance à la traction des mortiers calcaires à 28 jours

On constate d'après ces résultats obtenus que, quel que soit le rapport P/C, la résistance à la traction évolue positivement avec le temps. Notons aussi, que la résistance du mortier témoin composé de 100% de ciment Portland ordinaire est constamment la plus élevée que celle des mortiers modifiés de résine à 3 jours. Mais, à 7 jours, le mortier contenant 2.5% de résine semble le plus approprié pour les mortiers modifiés de résine et pour ce rapport, la valeur de résistance à la traction est égale à 5.60 MPa présentant une amélioration de l'ordre de 20% en comparaison avec mortier témoin. Au-delà de 2.5%, la résistance à la traction diminue lorsque le rapport P/C croît à 3 et 7 jours. Ceci est en concordance avec d'autres résultats qui ont montré que la résistance à la traction ou à la flexion augmente avec l'augmentation du taux de polymère latex incorporé aux mortiers jusqu'une quantité seuil qui est égale à 2.5% de résine dans cette étude [27,28].

À 28 jours, ces résistances à la traction continuent évoluer positivement avec le temps. Ainsi que, les incorporations de 2.5% à 5% du latex améliorent légèrement la résistance à la traction en comparant au mortier de référence. En outre, On a également constaté que le mortier additionné de 15% de résine présente le meilleur comportement à la traction et sa résistance est de 6.9 MPa.

En général, ces augmentations peuvent être expliquées par l'amélioration globale de la liaison entre les grains de ciment par le latex, à la formation du film de polymère qui est en partie responsable de l'amélioration mécanique des mortiers modifiés et au pontage des microfissures par le polymère [29-37]. Zhong a observé une augmentation de la résistance en compression et une continuité des valeurs de résistances en flexion, il l'associe donc à la bonne résistance en traction des films de polymère [38].Toutefois, Demiingthon et Justnes ont montré que la présence du film n'est pas nécessaire à l'amélioration des propriétés mécaniques si le latex utilisé est apte à former des complexes avec les minéraux du ciment où la nature et la stabilité de certains hydrates sont modifiées, par exemple l'ettringite formée est plus stable [39].

3.3 Effet de résine sur la résistance à la compression

https://doi.org/10.1051/mateconf/201814901067

Les figures 5 et 6 présentent l'influence du rapport P/C (%) sur la résistance à la compression des mortiers élaborés à 3, 7 jours ensemble et à 28 jours respectivement.

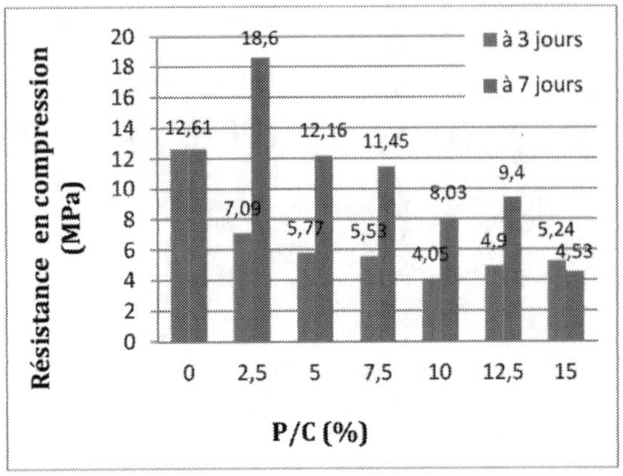

Fig. 5. Effet du rapport P/C (%) sur la résistance à la compression des mortiers calcaires à 3 et 7 jours

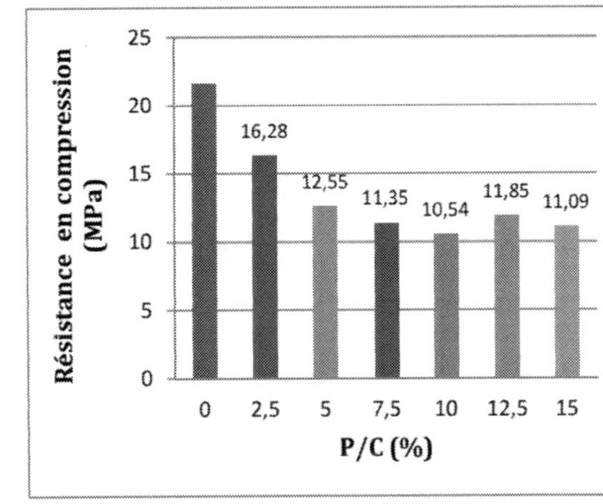

Fig. 6. Effet du rapport P/C (%) sur la résistance à la compression des mortiers calcaires à 28 jours

Il est nettement clair selon les figures 5 et 6 que le mortier contenant 2.5% de résine possède une meilleure résistance à la compression que celle de tous les mortiers élaborés à 7jours. Mais à 28 jours, la résistance en compression du mortier de contrôle est plus élevée que celle des autres mortiers. Ceci confirme les résultats de plusieurs autres chercheurs qui ont travaillé depuis plusieurs années dans ce domaine et qui ont montré que l'addition de polymère a un effet néfaste sur la résistance à la compression des mortiers, où cette résistance reste identique à celle du mortier non adjuvanté [9, 10].

En outre, il a été noté que la résistance à la compression diminue rapidement quand le rapport P/C augmente quel que soit l'âge des éprouvettes des mortiers testés (3, 7 et 28 jours) ; cela est dû à la réduction de la partie cimentaire, dans les matériaux hybrides, ce qui est une partie indispensable et responsable de l'amélioration de la résistance à la compression.

4 Conclusion

L'ensemble des essais effectués dans cette étude a permis de tirer les conclusions suivantes:
*Les résistances à la traction du mortier témoin composé de 15% de polymère latex sont constamment les plus élevées que celles des mortiers élaborés à 28 jours.
*L'addition de polymère a un effet néfaste sur la résistance à la compression des mortiers ou cette résistance reste identique à celle du mortier non adjuvanté.

5 Recommandations et perspectives

Bien que ce travail a permis de donner une idée sur l'influence de résine sur le comportement physico-mécanique des mortiers calcaires contentant un superplastifiant surtout les résistances à la traction et en compression, il devrait être approfondi et enrichi en procédant à:
*Des examens microstructuraux par microscopie électronique à balayage, car ce moyen constitue un outil de choix permettant non seulement de quantifier le degré d'hydratation mais aussi de préciser la structuration des phases ;
*L'identification des phases cristallines, au cours de l'hydratation, par diffractométrie des rayons X ;
*L'étude du degré d'hydratation et son évolution à court et à long terme ;
*La mesure d'autres réponses de mélanges élaborés surtout la nature des matériaux utilisés comme le latex, les types des granulats; le rapport E/C et les méthodes de conservation ;
*La transposition au matériau béton.

References

1. Ch. D. Vacchiano, L. Incarnato, P. Scarfato, D. Acierno, *Conservation of tuff stone with polymeric resins*, Constr. Build. Mater. 22, 855–865(2008).
2. Ohama Y. *Handbook of polymer-modified concrete and mortars*. Park Ridge, NJ. Noyes Publications; (1995).
3. Yao H, Liang N, Sun L, Meng J. *The design of the polymer cement concrete pavement and the analysis of test road.* J Chongqing Jiaotong Univ; 24:83–7; (2005).
4. Islam, MA., Rahman, MM., Ahmed, M., 2011. *Polymer-modified concrete: world experience and potential to Bangladesh,* Point of View, Indian Concrete Journal, pp. 55-63.
5. Barbuta, M., Harja, M., *Experimental study on the characteristics of polymer concrete with epoxy resin*, Bul. Inst. Polit. Iasi, t. LIV (LVIII), f. 1, pp. 53-39 ; (2008).
6. Zhong SY, Chen ZY. *Properties of latex blends and its modified cement mortars.* Cem. Concr. Res.; 32(10):1515–24; (2002).
7. Zhong SY, Shi ML, Chen ZY. *The AC response of polymer coated mortar specimens.* Cem. Concr. Res. 2; 32(6):983 -7; (2002).

8. Yang Z, Shi X, Creighton A, Peterson M. *Effect of styrene–butadiene rubber latex on the chloride permeability and microstructure of Portland cement mortar*. Constr. Build. Mater.; 23(6):2283–90; (2008).
9. Mirza J, Mirza MS, Lapointe R. *Laboratory and field performance of polymer modified cement-based repair mortars in cold climates*. Constr Build Mater; 16(6):365–74; (2002).
10. Wang R, Wang PM. *Function of styrene–acrylic ester copolymer latex in cement mortar*. Mater Struct.;43(4):443–51 ; (2010).
11. Al-Zahrani MM, Maslehuddin M, Al-Dulaijan SU, Ibrahim M. *Mechanical properties and durability characteristics of polymer and cement-based repair materials*. Cem Concr Compos; 25(4–5):527–37; (2003).
12. Ohama Y. *Polymer-based materials for repair and improved durability: Japanese experience*. Constr Build Mater; 10(1):77–82; (1996).
13. Ohama Y. *Principle of latex modification and some typical properties of latex modified mortars and concretes*. ACI Mater J;84(6):511–8; (1987).
14. Sakai E, Sugita J. *Composite mechanism of polymer modified cement*. Cem Concr Res; 25(1):127–35; (1995).
15. L.K. Aggarwal, P.C. Thapliyal, S.R. Karade, *Properties of polymer–modified mortars using epoxy and acrylic emulsions*, Constr. Build. Mater. 21 (2) 379–383; (2007).
16. R. Wang, X.G. Li, P.M. Wang, *Influence of polymer on cement hydration in SBR– modified cement pastes*, Cem. Concr. Res. 36 (9) 1744–1751; (2006).
17. J. Mirza, M.S. Mirza, R. Lapointe, *Laboratory and field performance of polymer– modified cement–based repair mortars in cold climates*, Constr. Build. Mater. 16 (6) 365–374; (2002).
18. Min Wang, Rumin Wanga, , Hao Yao, Shameel Farhan, Shuirong Zheng, Zhujun Wanga, Congcong Du a, Hao Jiang, *Research on the mechanism of polymer latex modified cement*. Construction and Building Materials 111; 710–718 (2016)
19. BERKAK H, MAKHLOUFI Zoubir, BEDERINA Madani, HADJOUDJA Mourad, *Effet de cure sur le comportement physico-mécanique des mortiers calcaires modifiés de résine*. Séminaire International sur les Matériaux Locaux et la Construction Durable (SIMaLoCoD) Adrar, le 04 et 05 décembre (2016).
20. Z. Su. *Microstructure of polymer cement concrete*. Ph D thesis. Delft University Press, the Netherlands, 189 pp; (1995).
21. L.M. Saija. *Water proofing of Portland cement mortars with a specially designed polyacrylic latex*. Cement and Concrete Research, vol. 25 (3), pages 503-509, (1995).
22. G. Barluenga, F. Hernandez-Olivares. *SBR latex modified mortar rheology and mechanical behaviour*. Cement and Concrete Research, vol. 34, pages 527-535, (2004).
23. Pascal S. *Comportement mécanique de composites mortier-polymère*. Thèse de Doctorat, Ecole Centrale de Paris, Paris, 348 p, (2002).
24. Barluenga G., Hernandez-Olivares F. *SBR latex modified mortar rheology and mechanical behavior*. Cement and Concrete Research, vol. 34, p. 527-535, (2004).
25. Y. Ohama. *Polymer based admixtures. Cement and Concrete Composites*, vol. 20, pages 189-212, (1998).
26. F.A. Shaker, A.S. El Dieb, M.M. Reda. *Durability of styrene-butadiene latex modified concrete*. Cement and Concrete Research, vol. 27 (5), pages 711-720.
27. L. Bureau, A. Alliche, P. Pilvin, S. Pascal. *Mechanical characterization of styrene-butadiene modified mortar*. Materials, Science and Engineering A, vol. 308, 2001, pages 233-240, (1997).
28. S. Pascal. *Comportement mécanique de composites mortier-polymère*. Thèse de doctorat de l'Ecole Centrale des Arts et Manufactures, Ecole Centrale Paris, 206 pages, (2002).
29. S. Pascal, A. Alliche, P. Pilvin. *Mechanical behaviour of polymer modified mortars*. Materials Science & Engineering A, vol. 380, pages 1-8, (2004).
30. A. Jenni, L. Holzer, R. Zurbriggen, M. Herwegh. *Influence on microstructure and adhesive strength of cementitious tile adhesive mortars*. Cement and Concrete Research, vol.35, pages 35-50, (2005).
31. W. Jingang, Z. Shuxiang, Y. Haiquin, K. Xiangzheng, W. Xikui, G. Zhongmao. *Study of cement mortars modified by emulsifier-free latexes*. Cement & Concrete Composite, vol. 27, pages 920-925, (2005).
32. S. Zhong, Z. Chen. *Properties of latex blends and its modified cement mortars*. Cement and Concrete Research, vol. 32, pages 1515-1524, (2002).
33. S.P. Dennington, J.W. Thompson. *Proceedings of the 10th International Congress on the Chemistry of Cement*, (1997).

Olive pomace based lightweight concrete, an experimental approach and contribution

Chaabane Lynda Amel [1] and Benferrah Rafik[1]
[1] Djillali Liabes University, Civil Engineering Department, Environment and Civil Engineering Laboratory, Sidi Bel Abbes, Algeria

Abstract. Due to conventional aggregates resources depletion, material recycling has become an economic and ecologic alternative. In this paper, locally available natural residues such as olive pomace were investigated, when partially incorporated in the concrete formulation, since the mechanical characteristics of lightweight aggregate concrete strongly depend on its properties and proportions. Lightweight aggregates are more deformable than the cement matrix because of their high porosity, and their influence on the concrete strength remains complex. The purpose of this paper is to investigate the aggregates properties on lightweight concrete mechanical behaviour through an experimental approach. In addition, the different substitution sequences and the W/C ratio on lightweight concrete behaviour were evaluated, in order to determine the W/C ratio influence on the improvement of the lightweight concrete mechanical properties while knowing that the mixing water quantity gives the cement paste manoeuvrability and mechanical strength effects. The last part of this paper, therefore, was to provide statistical survey for estimating strength and weight reduction through the different natural aggregate substitutions to improve the lightweight concrete properties. The results achieved in a significant olive-pomace lower adhesion with the matrix after the cement setting, making the lightweight concrete mechanical strength weak. However, this work can open several perspectives: Results modeling and correlation with an experimental approach, the evolution and determination of lightweight concrete characteristics when exposed to high temperatures and thermohydric properties.

1 Introduction

In this article, we wish to show that the building sector, being so polluting, can now become an ecological lung in terms of recycling waste to obtain ecological concrete. Moreover, in the construction of a certain number of structures, weight reducing lead to general cost-savings, concrete density may vary between 0.5 to 2 when using lightweight aggregates, but because the concrete has less density, then its compressive strength is lower [1]. Conventional aggregate resources are becoming increasingly depleted over the years, especially in the Saharan areas. These materials are expensive to store and storage sites are also scarce, environmental standards restrict the use of toxic materials. In addition, more processing is required for materials originating from natural environment involving treatment and crushing stations, and therefore, the cost of natural aggregate manufacture is high. That's why part of material recycling remains the best and effective option. [2].

Because the Algerian industry produces huge quantities of olive-pomace every year, lightweight concrete was composed with a variety of olive-pomace aggregates mixtures (731.332t) [3]. The objective of this experimental approach is to demonstrate that the lightweight concrete mechanical behaviour is affected by the aggregate porous properties when inserted in the cement formulation, making it more complex in terms of maneuverability and strength.

2 Experimental materials and procedures

Conventional concrete can be chosen via many methods, in our study case, the method of [4] was applied, which consists of making mixtures with different aggregates. The objective here is to produce an insulating concrete [1]. The S / C ratio remains constant, for all tests we use moulds Ø 16cm x 32cm, filled almost equally in three layers, this is to determine the lightweight aggregates influence on the lightweight concrete mechanical behaviour, 16x32 cm test tubes were made with a variable matrix and a different aggregates percentage.
During material mixing, the subsidence of Abrams cone and density were determined for each batch, density is used to possibly correct the mixtures proportions [5].

- The matrix is made of CPJ-CEM II / A 42.5 cement mortar and non-core sand [6].

2.1 Sand of Sidi Bel Abbès area(Algeria)

The report covers the majority of analysis results led by the Laboratory of the engineering department on the sand quarry of Sidi Bel Abbes area.

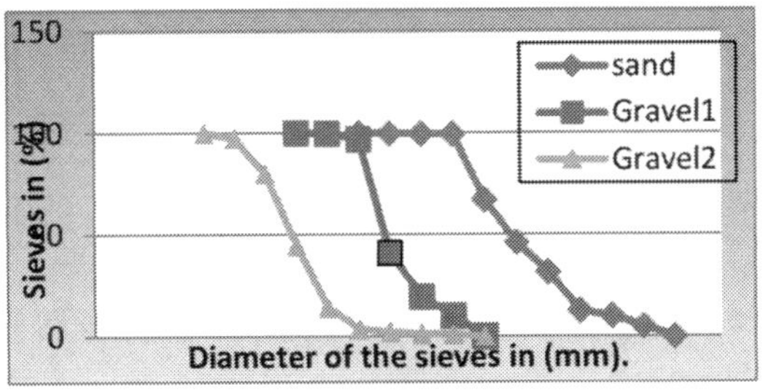

Fig1. Aggregates granulometric analysis.

By curve analysis, we found out that the gravel granulometry and alluvial sand has a continuous shape which will affect the concrete quality, durability and strength.

2.1.1 Physical properties

Alluvial sand from the area of Sidi Bel Abbes (North West Algeria). The equivalent of sand [7] shows that it is clean, and good for a quality concrete. The characteristics are reported in table.1 below.

Table.1. Sand physical characteristics

Characteristics	Unit	Sand
Absolute density	g/cm^3	2.7
Bulk density %	g/cm^3	1.63
Sand Equivalent	%	73.5
Absorption degree	%	1.00

2.2 Cement constituents

The CPJ-CEM II/A 42.5 Portland cement is obtained through the finely ground mixture of clinker and additives, calcium sulphate is added as gypsum to regulate the setting [8].

Composition

CPJ-CEM II/A 42.5 Compound Portland cement consists of:

80 to 94% of Portland clinker.

6 to 20% at maximum of additives (pure limestone).

Minor constituents (0 to 5% Calcium sulphate is added as a setting regulator).

Table 2. Cement physical properties [8]

Cement ECDE-Chlef cement factory (Algeria)	Standard NA442
Initial setting > 60mm Heat stability <7mm Removal 28 days < 650 μ m	Initial setting ≥60mm Heat stability ≤7mm Removal 28 days ≤ 650 μm

2.3.1 Chemical Analysis

Table 3 . Cement chemical composition

Element	SiO_3 (NA442)	Chlorides (NA442)	SiO_3 (ECDE)	Chlorides (ECDE)
%	≤ 3.5	Cl-≤0.1	≤ 2.00	Cl-≤0.08

2.3 Aggregates choice

The choice fell on the olive-pomace based lightweight concrete "fig. 2" from the area of Sig (Wilaya of Mascara - Algeria) , due its abundance and important natural features [9].

Fig.2. Olive-pomace based lightweight concrete.

Knowing that the mechanical properties of lightweight concrete strongly depend on its aggregates proportions present in the concrete formulation, in particular due to their high porosity, lightweight aggregates are more deformable than the cement matrix and have a complex influence on the concrete strength.

As part of the previous research carried out on lightweight concretes, this one aims to understand the behaviour of light aggregates, an experimental part was carried out where the olive pomace was pre-wetted before being introduced into the mixtures, the absorption rate reached 35% during the first days, followed by a long-term stabilization.

Table 4.Olive-pomace physical properties.

Physical properties	Olive-pomace
Density (g/cm^3)	0.9
Mean particle diameter	8mm
Max. particle diameter	12mm

2.4 Adjuvant

The adjuvant is a plasticizer that helps to decrease the water quantity in the cement matrix and to increase the mechanical strength. These additives help reduce the W / C ratio.

3 Results and discussions:

3.1 Lightweight concrete formulation

Through this experimental work, the lightweight concrete is formulated of olive pomace aggregates, using Dreux formulation methods, to determine the substitution rate, rheology and compactness of light concretes.

Table .5: Different compositions of 1m^3 of lightweight concrete.

Concrete	G/S
BLGO1	0.6
BLGO2	0.68
BLGO3	0.75
BLGO4	0.8
BLGO5	1.00
BLGO6	1.2

3.2 Lightweight aggregates effects on mechanical properties.

On the basis of elementary formulations, by varying the W / C ratio as well as olive pomace substitution in the concrete preparation, then correcting the G / S ratio to 1.20, in a way to introduce this ratio in our final composition, in order to achieve objectives.

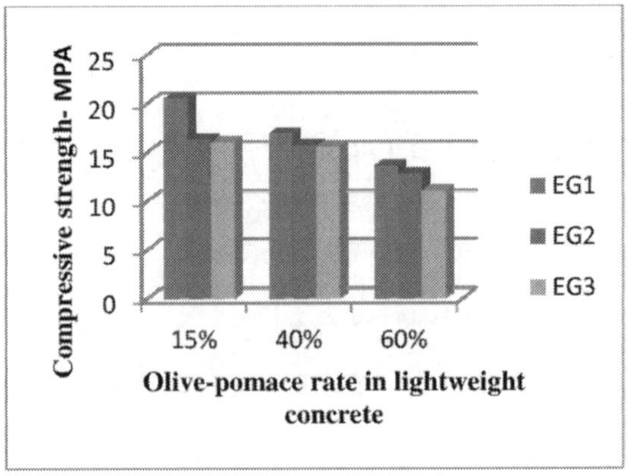

Fig.3. Concrete compressive strength

It can be seen from "fig. 3" that the olive-pomace-based concrete has low compressive strength than conventional (light) concretes, the compressive strength decreased by increasing the olive pomace dosage.
The decreasing resistance throughout 28 days depends on the nature and the substitution rate of the natural aggregates as well as the cement dosage.

The lightweight aggregates concrete strength is determined by the concrete composition, through the W / C ratio "fig. 3", the Aggregate/Sand volume ratio and the Cement content. The concrete strength is higher if the W / C ratio is lower [10]. However, too little water may not be able to completely hydrate the cement and may lead to a concrete strength drop and may limit the variations in the W / C ratio during the concrete manufacture due to high aggregates absorbency, which must be saturated and drained [5].

Fig.4. Test tube crack line subjected to the compression test.

"Fig. 4" illustrates fracture lines corresponding to the simulations carried out by [5]. It demonstrates the the maximum stress of the lightweigh concrete is along Oz axis "Fig. 3" and picture 4", however, the maximum stress of the conventional concrete is along Ox axis, according to [5].

3.3 Void volume variation based on W / C ratio

The notion of compactness, being an intrinsic characteristic of the mixture had to be well defined during the solid suspension elaboration. In practice, they were random mixtures, so the aim was to fill the space with granular materials of different densities. For this purpose, the relationship between minimal porosity and granular extent was investigated.

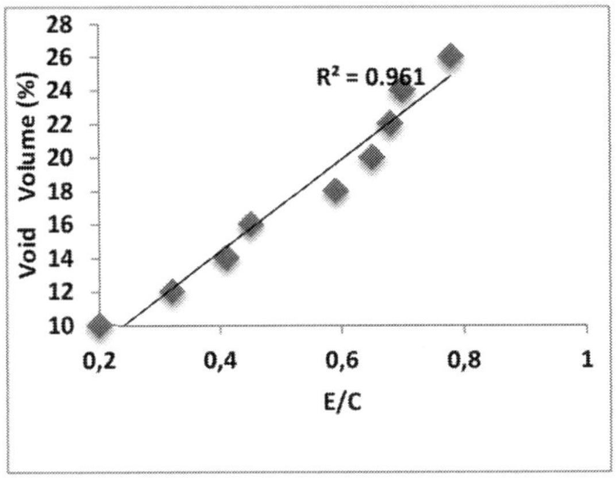

Fig. 5. Void volume variation based on W/C ratio at 60° C.

In order to improve the processing and the compaction, and to reduce the concrete porosity, many tests have been carried out on the water content measurement of lightweight concrete by making hydrostatic weighing tests on olive pomace-based lightweight concrete at 60 ° C, the results shown in "Fig. 5" show that the drying temperature affects the material characteristics, the pre-wetting of the lightweight aggregates proved to be an essential part in the manufacture of lightweight concrete by granting it a long-term fundamental kinetics affecting the hydration of the cement. There is a good correlation between the void volume and the W / C ratio, as demonstrated by [10].

3.4 compressive strength evolution based on W/C ratio.

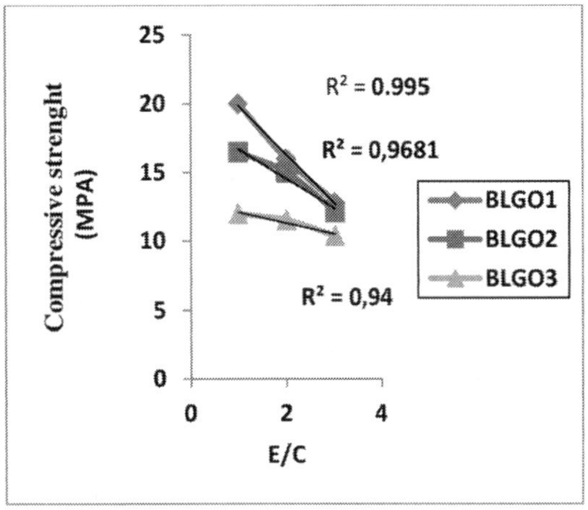

Fig. 6. Compressive strength evolution based on W/C ratio.

The lightweight aggregate concrete resistance is controlled by the following elements:

- the concrete composition, through the W/C ratio "Fig. 6".
- the Aggregate/Sand volume ratio.
- the cement content.

The compressive strength of the concrete is higher than the W/C ratio, that is to say the water quantity is low [11]. However, too little water may not be able to fully hydrate the cement, and may lead to a concrete strength drop, and possibly limit the variations of the W/C ratio during the concrete manufacture high aggregates absorbency, which must be saturated and drained [5].

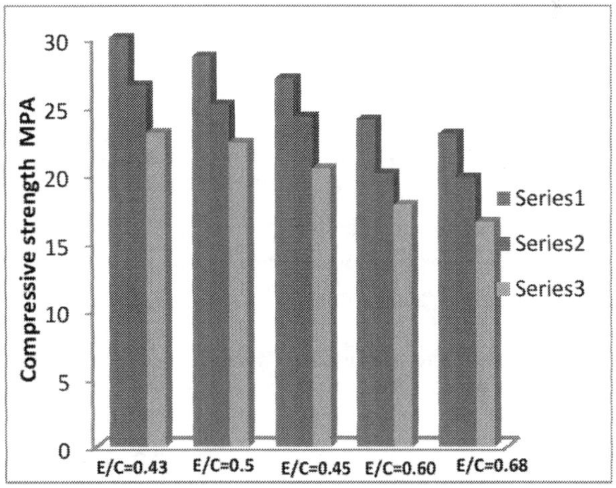

Fig.7. Compressive strength variation based on W/C ratio.

It is seen from "Fig. 7" that the olive-pomace based concrete have lower compressive strength than conventional (light) concrete, and that the strength drop decreases by increasing the cement dosage. The resistence drop throughout 28 days depends on the nature

https://doi.org/10.1051/matecconf/201814901068

and the substitution rate of the natural aggregates as well as the cement dosage.

3.5 Water quantity influence on the concrete compressive strength

The water quantity is a major factor that impacts on the lightweight concrete strength, since lightweight aggregates undergo pre-wetting before they are introduced into the mixture, the amount of water added will be directly related to the effective W/C ratio.

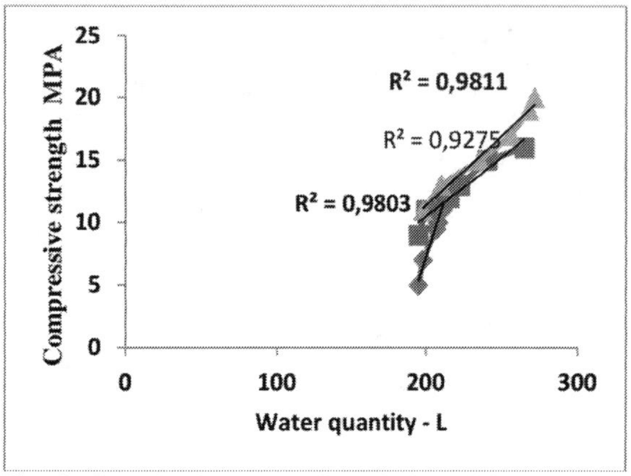

Fig. 8. Compressive strength and the mixing water relation.

"Fig. 8" represents the concrete mixtures with a variation in the W/C ratio, the quantity actually used (mixing water) to increase subsidence in accordance with natural aggregate substitution rate.

"Fig. 8" shows that the evolution according to the quantity of water added to the olive-pomace-based concrete and the conventional lightweight concrete have similar behaviour, this leaves us to conclude that the cement hydration is normal. Finally, the substitution rate of natural aggregate concrete does not exceeding 35%, have an acceptable long-term resistance. As a conclusion, the strength drop rises slightly in the course of time for each lightweight concrete composition.

3.6 Elastic properties of lightweight aggregates

It is hard to describe the grain strength in lightweight aggregates; it has more elastic properties [12].

It can also be seen that their high porosity reduces the wall effect, and the transition zone are different to the normal aggregates. The thin interfacial zone is characterized by mechanical and chemical linkage between the paste and the facial aggregates as defined by different authors [13,14].

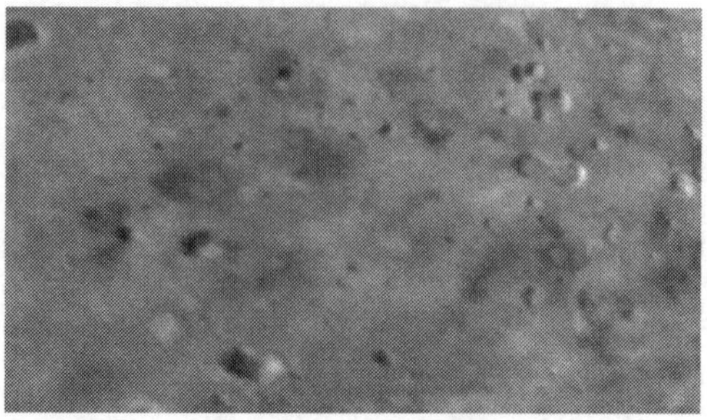

Fig. 9. Picture enlargement of the cement matrix.

3.7. Concrete test tubes subjected to compression

A group of test tubes subjected to the same test should appear more or less identical to the naked eye, however, these test tubes contain structural defects of various sizes, up to the microscopic scale. For this purpose, Weibull distribution is presented in case of uniaxial stress, using it to test material resistance under simple sollicitations. The ability to resist fractures is undoubtedly one of the most important and most studied characteristics of materials. A fracture of a material is the separation into two or more pieces under stress [15].

In many areas such as building, this ability to resist breakage is considered as safety feature. Question is: What is the maximum stress that can be applied so that the probability of fracture is less than 1%.

A statistical survey will be applied on these data in order to reduce the maximum stress on lightweight concrete so that the probability of rupture is less than 1% [15].

It is clearly seen on the obtained curve the fact that Weibull distribution was used:

$$Ln[Ln(1/Ps)]=m[(Ln(\sigma)-Ln(\sigma 0)] \quad (1)$$

Where m: Weibull modulus.

$\sigma 0$: scaling parameter.

σ: The probability of survival.

a- The assumptions

The weibull distribution is based on two simple assumptions, which we present here:

1-The fracture strength of each volume element is modelled by independent variables. The randomness aspect makes allows taking into account the dispersion; Independent sub-volumes imply that there is no interaction between infinitesimal volumes, which means that the fracture is a purely punctual phenomenon [16].

2 - The fracture of the infinitesimal volume containing the first activated fault brings about the ruin of the inspected structure: in other words: as soon as the fracture occurs, the mechanical functions of the whole piece are lost [16].

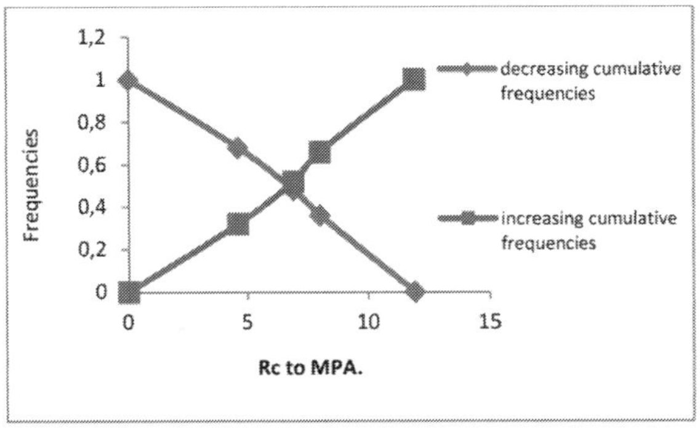

Fig. 10. Cumulative frequencies of lightweight concrete test tubes subjected to compression.

The curve in "Fig. 10" where m=3.1, shows that Weibull modulus is lower, so the dispersion on fracture stress is higher. In a series of test tubes, the median one corresponds to the value chain having a cumulative increasing and decreasing frequency corresponding to a median value class of 0.5, this may result in manipulation errors which reduce the crushing test perfection rate of the lightweight concrete tubes as confirmed by [15].

A concrete element does not break under the effect of a single stress but under the effect of several stresses. The test tubes results can't be taken into account to estimate the concrete strength, only if the stress field is of the same type in both cases [17].

4 Conclusions

The main objective of this study is to technically demonstrate the possibility to use aggregates (olive pomace) in the manufacture of ecological concrete.

The present study shows that olive pomace can reduce the concrete test tubes weight and can be used as a mass substitution additive at an optimum rate of 35%. Throughout this paper, it was concluded that olive pomace had a poor adhesion with the cement matrix after the setting, because of its irregular shape, in addition, the outer side of the olive pomace is rough and let not form granular skeleton that generates the mechanical work of the lightweight concrete.

References

1. El Bakkouri A.Etude thermomécanique du beton allege avec du liege et du beton allege avec des grignons d'olives. *12 èmes Journées Internationales de Thermiques ,France (2005).*

2.E-H.Benaichi .Etude des bétons à base des sables de démolition et ajouts minéraux actifs. *Thèse de Magister, Université mohamed khider Biskra, Algeria (2009).*

3. L. Touati . ,Valorisation des grignons d'olive Etude de cas : Essai de valorisation en Biocarburant .Thèse Magister ,*Université M'hamed Bougara-Boumerdes, Algerie* (2013).

4. G.Dreux ,F.Gorisse. Composition des bétons: méthode Dreux –Gorisse .*Annales de L'ITBTP*, n° **414** (1983).

5. Yang Ke.Caracteristion du comportement mécanique des bétons de granulats legers :Experience et modélisation.*Phd.D.Thesis,université de Cergy-Pontoise ,France* (2008).

6.Y.LOT,Z.CUI H.Effect of porous lightweight aggregate on strength of concrete . *Materials letters* ,**58**,P 916-919 (2004).

7. NF. *Equivalent de sable* P18-597 (1991).

8. http:// www.ecde.dz la norme NA 442-200.

9.H. EZBAKHE. Caractéristiques thermiques de matériaux poreux utilisés comme isolant simples et porteurs . *Phd.D.Thesis,U.C.B.Lyon ,France* (1986).

10. G. COQUILLAT , L. HASNI. Projet national CALIBE –Fabrication du béton, Mesure des teneurs en eau des bétons durcis CEBTP ; *centre expérimental de recherches et d'études et des travaux publics direction de la recherche et des affaires scientifiques et techniques ;DRAST ; Ministère de l'équipement .Saint –Rémy-Les –Chevreuse :CEBTP*(1999).

11. D.MORIN .Sur les bétons légers et leur comportement sous des sollicitations biaxiales . *Phd.D.Thesis ,Université Paul Sabatier de Toulouse ,France* (1976).

12.J. Muller Rochholz .Determination of the elastic properties of lightweight aggregate by ultrasonic pulse velocity measurement ,the international journal of light weigh concrete,**1**,p 87-90(1979).

13.Y.LOT.Y.,Z.CUI H. Effect of porous lightweight aggregate on strength of concrete ; *Materials letters* ,**58**,p 916-919 (2004).

14. Y.LOT.Y.,Z.CUI H. Spectrum analysis of the interfacial zone of lightweight aggregate concrete ,*materials letters*,**58**,p 916-919(2004).

15.J.B LEPRETRE. Compte –rendu de Travaux dirigés Statistique de Weibull , *deuxième année Cours électif CE42 ;Matériaux pour l'ingénieur Année universitaire* (2009-2010).

16. S.GRANGE ,J-L PRENSIER. Le module de Weibull un critère de rupture probabiliste . *Université PARIS-SACLAY ENS de Cachan ,France.* (**Ed** .),(01-07-2006).

17. http://www.e-periodica.ch. Essais statistiques sur mortiers et béton. *Prot, Marcel* 3 (1948).

Characterization and formulation of a new eco-friendly hydraulic binder based on combination of inorganic and organic admixtures

M.H.R. KHUDHAIR[1,2,3,*], M. S. EL YOUBI[3], A.ESSAMRI[1], A. ELHARFI[1]
[1]Laboratory of Agro Resources Polymers and Process Engineering (LAPPE), Team of Macromolecular & Organic Chemistry, Faculty of Sciences, Ibn Tofail University, BP 133, 14000, Kenitra, Morocco.
[2] Laboratory of Cement and Quality Control of Amran Cement Plant of Yemen.
[3]Laboratory of Chemistry of Solid State, Faculty of Science, Ibn Tofail University, Kenitra, Morocco.
*Corresponding author: khudhair.mohammed65@gmail.com

Abstract. This work aims to valorize a mineral and natural resources such as the Limestone Fillers (F-Lime) and the Natural Pozzolan (PN) by incorporating them into the formulation matrix of cement or concrete. In order to minimize the CO_2 emissions into the atmosphere, to reduce the energy and raw materials consumption and as well as, to improve the physical and mechanical properties in fresh cement paste and of mortar or concrete in the hardened state. In this present manuscript, we substituted the clinker by the combination between the F-Lime and NP at 40% by weight of cement with steps of 5% with the admixture of superplasticizers. The influences of the incorporation of these additions on physical and mechanical properties of mortar or concrete in the fresh cement paste and hardened state were evaluated. The obtained results by different formulations elaborated to show that the replacement a part of clinker by the mixing of F-Lime and PN has produced a new hydraulic binder eco-friendly and durable with improved physicochemical, physical and mechanical properties. These results show that we have succeeded to manufacture, characteristic and formulated of new eco-friendly hydraulic binder and sustainable with improved physical, chemical, and mechanical properties while minimizing greenhouse gas emissions on one hand and the reducing the energy raw materials consumption on the other hand.

Keywords: New eco-friendly hydraulic binder, Inorganic admixture, Organic admixture, Limestone Fillers, Natural Pozzolan, chemical admixture, Superplasticizers, Physical properties, Porosity, Capillary absorption, Compressive strength, Durability.

1 Introduction

The Yemen's country contains considerable amounts of mineral admixtures such as the natural pozzolan (PN); of volcanic origin from the Difan-Amran deposit north of Sana'a in Yemen, [1, 2] and the limestone fillers (F.Lime), which is found in several regions of Yemen, namely , Hadramaout, El Maharah, Sana'a, Amran, Etc . [3, 4].

The valorization of these mineral admixtures by incorporation them as an additive in fabrication of cement and various types of concretes has both benefits economic and environmental; to reduce the energy consumption and raw materials used during the manufacture of clinker or cement on one hand and to minimize the greenhouse gas emissions into the atmosphere on the other hand [5–7].

These admixtures by their fineness and their reactivity in presence of cement lead to significant changes in physical and mechanical properties of mortars or concrete in fresh cement paste and hardened state [8–12]. The presence of inorganic or organic admixtures (F.Lime, PN or superplasticizers) in cement or mortar (concrete) matrix modifies the structure of granular skeleton (granular effect) and the friction between the solid components in liquid phase [13–16]. In the same way, during the setting and hardening [17–20], the additions particles interact in cement hydration process while modifying the structure of the hydrated products and for some may react chemically (effects physical-chemical and microstructure) at cementitious environments, in order to form of a new hydrated products which have an additional binding character (improvement of physical properties and mechanical performances).

The interaction between the organic and inorganic admixtures makes it possible to increase the hydration rate of mortars or concretes while modifying the W/C ratio and to improve the mechanical performance [2,21–23]. This improvement finally influences the mechanical of compressive strength. Indeed, when the molecules of admixtures of superplasticizers SP103 are introduced into a suspension of concrete formulated by the combination between the F.Lime and the NP, a large part of them is fixed to the surface of the cement particles (adsorption)[16,24-25].

In this work, different formulations have been elaborated while partially substituting the clinker by the combination between limestone fillers and natural pozzolan at 40% by weight of cement with a step of 5%, in the presence of admixtures of superplasticizers of high water reducer and retarder SP103 of setting. The influence of the incorporation of these admixtures on physical and mechanical properties in fresh cement paste and hardened state were studied. The obtained results from the various formulations elaborated show that the substitution a part of clinker by the mixing of F.Lime the PN in the presence of mixtures of superplasticizers enabled us to production of a new hydraulic binder with improved physicochemical, physical and mechanical properties, while reducing the cost of production, minimizing CO_2 emissions to the atmosphere, reducing the use of mixing water.

2 Materials and methods

2.1. Materials

2.1.1. Cement

The type of cement used in this work is (CMI-42.5), this is a Portland cement as a resulting of simultaneous grinding at (95%) of clinker and (5%) of gypsum according to the standard EN 196-1. The chemical compositions (gypsum and cement) and mineralogical (clinker) are presented in Table 1 and 2.

Table 1: Elementary chemical compositions of clinker, gypsum, and cement in weight of atomic

Chemical name	Chemical formula	Gypsum	Cement
Lime	CaO	33.40	61.29
Silica	SiO_2	0.70	19.99
Alumina	Al_2O_3	0.36	5.57
Ferrite	Fe_2O_3	0.09	2.85
Magnesia	MgO	0.63	1.89
Sulfur trioxide	SO_3	47.20	3.22
Potassium oxide	K_2O	0.03	1.15
Sodium oxide	Na_2O	0.10	0.20
Chlorine	Cl^-	0.01	0.02

Table 2: Mineralogical composition of clinker

Chemical name	Mineral name	Cement nomenclature	Content
Tricalcium silicate	Alite	C_3S	47.70
Dicalcium silicate	Balite	C_2S	25.10
Tricalcium aluminate	Aluminate	C_3A	10.40
Tetracalcium Aluminoferrite	Ferrite	C_4AF	9.10

2.1.2. Inorganic additions

❖ *Limestone Fillers (F.Lime)*

They are mineral materials which spread in several regions in Yemen, such as Hadramaout, Sana'a, Amran and etc.

❖ *Natural pozzolan (PN)*

The natural pozzolan (PN) used is of volcanic origin, extracted from Difan-Amran's deposit of Yemen located in the north of Sana'a.

Physical-chemical characteristics

The results of the analysis by X-ray fluorescence (XRF) of F.Lime and PN after crushing, drying in the oven for 24 hours at a temperature of 50 °C and grinding are represented in the Table 3.

Table 3: Elementary chemical compositions of F.Lime and PN in weight of atomic

Content (%)	F.Lime	PN
CaO	54.96	8.8
SiO_2	0.62	41.43
Al_2O_3	0.12	16.16
Fe_2O_3	0.16	9.41
MgO	0.41	4.79
SO_3	0.08	0.128
K_2O	0.01	0.9
Na_2O	0.00	3.47
Cl^-	0.00	0.04
LOI	43.63	14.87

The physical properties of limestone fillers and natural pozzolan have given in Table 4.

Table 4: Physical properties of F.Lime and PN

Name of addition	Color	Size of particles (μm)	Specific surface Blaine ($cm^2.g^{-1}$)	Density ($g.cm^{-3}$)
F.Lime	White	50	4776	2.13
PN	Black	60	4576	2.81

2.1.3. Chemical admixtures

The admixtures of superplasticizers are polymers in liquid form, especially synthesized for the concrete industry. These superplasticizers are delivered by the company CONMIX Ltd of Sharjah, United Arab Emirates. They are incorporated during the mixing of concrete at doses equal from 0.5% to 4% by weight of cement with a step of 0.5%. In this work, we used a High Range Water Reducing And Accelerating Superplasticiser SP103, Figure 1.

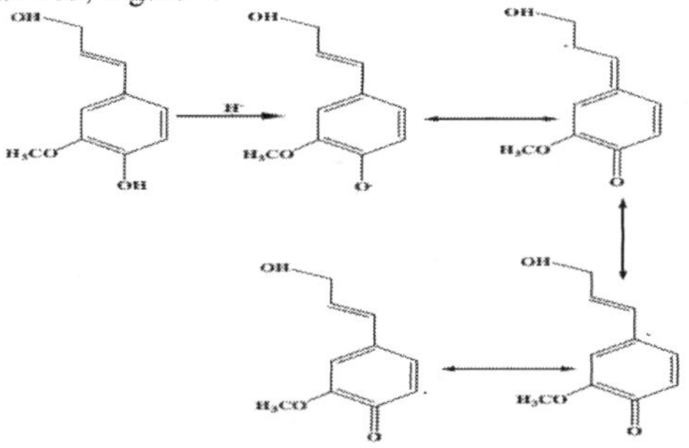

Figure 1: Chemical structure, high purity of lignosulfonate

2.1.4. Mixing water

The water used to mix the mortar or concrete is tap water. The main characteristics of these waters are summarized in Table 5.

Table 5: Main features of the mixing water

Components	Unity	Values
pH	-	7.00
Turbidity	mg/l	450.00
CO_3^{-2}	mg/l	216.00
HCO_3^-	mg/l	0.00
Ca^{+2}	mg/l	56.40
Mg^{+2}	mg/l	52.40
Conductivity	µS/cm	692.00

2.1.5. Sand

To make our mortar and /or concrete, we used standard sand according to the norm EN 196-1, delivered by the new French company of the Littoral. Its particle size analysis is illustrated in Figure 2.

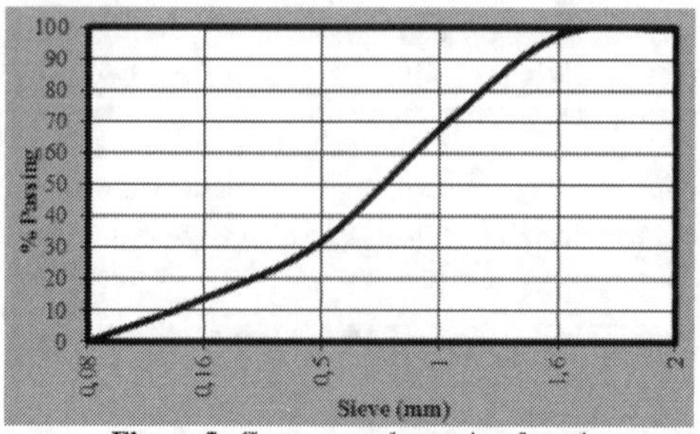

Figure 2: Curve granulometric of sand

According to the granulometric analysis presented in Figure 2, we found that the sand grains used are distributed in a systematic way according to the specifications of the norm EN 196 - 1.

2.2. Methods

To achieve the objective of our study, we have prepared a reference mortar without additions whose compositions are inspired by that of the normal mortar defined by the norm EN 196-1, with a quantity of water adjusted to obtain a reference consistency. And other mortar with mineral additions (PN+ F.Lime) and superplasticizers SP103, always keeping the standardized consistency fixed. For each mortar having acquired this consistency, we have prepared prismatic specimens with dimensions $(40x40x160)\ mm^3$. The compressive strengths were measured at a young age (2 days), in median age (7 days) and long-term (28 days) using a bending test machine to break the specimen into two halves and each party is responsible the subject of compressive using a hydraulic testing machine. The value of the resistance is considered as the average of the crushing stress of three test pieces (6 halves).
The compressive strength was calculated using the equation (1).

$$\text{Compressive strength} = \frac{\text{Load ,in "N"}}{\text{Area, in "}mm^2\text{"}};\text{MPa} \quad \text{Equation 1}$$

The various formulations tests carried out by maintaining the content of gypsum 5% by weight of a total of cement with 40% of the mixture of Fillers Limestone (F. Lime) and Natural Pozzolan (PN) are presented in the Tables 6 and 7.

Table 6: Formulation matrix of cement paste

Content	Cement (%)	F.Lime (%)	PN (%)	W/C Report
1	60	40	0	0.32
2	60	35	5	0.308
3	60	30	10	0.292
4	60	25	15	0.28
5	60	20	20	0.27
6	60	15	25	0.256
7	60	10	30	0.246
8	60	5	35	0.234
9	60	0	40	0.21

Table 7: Formulation matrix of mortar or concrete in the hardened state

Content	Cement (%)	F.Lime (%)	PN (%)	Sand (g)	W/C Report
1	60	40	0	1350	0.54
2	60	35	5	1350	0.52
3	60	30	10	1350	0.51
4	60	25	15	1350	0.5
5	60	20	20	1350	0.49
6	60	15	25	1350	0.47
7	60	10	30	1350	0.46
8	60	5	35	1350	0.43
9	60	0	40	1350	0.42

3 Results and discussion

3.1. Influence on Physicochemical properties

The Table 8 shows the physicochemical properties of fresh cement paste formulated with combination of F.Lime and PN.

Table 8: Physicochemical properties of cement paste

Content (%)	Density "$g.cm^{-3}$"	Fineness "$cm^2.g^{-1}$"	Setting time 'min'		W/C	Expansion
			Initial	Final		
1	3.02	4 050	40	80	0.54	0.5
2	3.02	3 940	70	110	0.52	0.33
3	3.02	3 900	90	130	0.51	0.32
4	3.02	3 830	110	140	0.5	0.325
5	3.02	3 770	120	150	0.49	0.335
6	3.01	3 710	120	160	0.47	0.345
7	3.01	3 680	125	170	0.46	0.376
8	3.01	3 650	125	180	0.43	0.42
9	3	3 630	130	180	0.4	0.45

From Table 7 we have found that the density of cement formulated with combination of F.Lime and PN decrease as a function of the increase in mass fraction of these admixtures. This decrease is generally due to the density of these admixtures, for pozzolan is 2.81 $g.cm^{-3}$, while that of limestone fillers is 2.13 $g.cm^{-3}$, which is usually lower than that of cement. We have noticed that the fineness by the specific surface (Blaine) of cement formulated in combination of F.Lime and PN increases with the increase in mass fraction of these additions. This increase is mainly due to the fineness of F.Lime which is 4776 $cm^2.g^{-1}$ and that of PN is 4576 $cm^2.g^{-1}$. We observed that the initial and final setting times decrease with (40%) of addition of F.Lime. This decrease is due on one hand to the chemical and mineralogical compositions of these fillers which are rich in CaO and poor in Al_2O_3 (physicochemical effect) and on the other hand is due to the fineness of these fillers which fills the voids between cement particles and that of aggregate (granular effect) "(F.Lime gives our material the role of setting accelerator). Subsequently, the setting time increased considerably with the increase in pozzolan content and the decrease in F.Lime proportion. This increase in setting time is generally due to the chemical and mineralogical compositions of NP which is very rich in aluminum oxide (16.16% Al_2O_3) and silicon dioxide (41.43% SiO_2) and low in an oxide of calcium (8.8% CaO), (physicochemical effect). These compositions promote the pozzolanic reaction which occurs between the reactive silica of NP and that of calcium hydroxide $Ca(OH)_2$ released by cement during the hydration of mixture, which slows down the hydration phenomena. (PN gives our material the role of setting retarder). We also found that the W/C ratio of cement paste formulated by combination of F.Lime and PN decreased with increasing in PN proportion and decreased in F.Lime content. Indeed, when the water reacts chemically with the lime released by cement during the hydration of mixture, the water will be absorbed slowly form calcium hydroxide $Ca(OH)_2$ on one hand. On the other hand, the decrease is generally due to the chemical and mineralogical compositions of these additions (PN is poor in CaO, despite, and limestone fillers are rich in CaO). Finally, we distinguished that the expansion of cement paste formulated with combination of F.Lime and PN increases slightly as a function of the mass fraction of these additions. This increase does not exceed the limit (5 mm) given by standard NF EN 196-3 + A1.

3.2. Influence on compressive strength

The Figure 3 illustrates the compressive strength of mortar or concrete formulated by combination of F.Lime and PN as a function of the mass fraction of these admixtures.

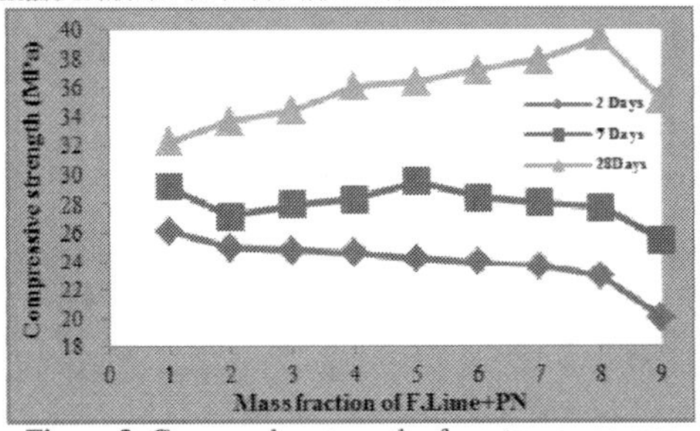

Figure 3: Compressive strength of mortar or concrete

At the Figure 3, we have noted that the compressive strengths at the young age (2 days), the middle ages (7 days) and the long-term (28 days) of all the mortars formulated by the combination of F.Lime and PN increase steadily with age and show no fall. However, the compressive strengths decrease slightly with the addition of 40% of F.Lime. This decrease is generally due to the chemical and mineralogical compositions of F.Lime which is rich in CaO. Then, the compressive strengths are increased with increasing the mass fraction of natural pozzolan and decreasing the proportion of fillers in the formulation matrix. This increase in compressive strength is explained by the fact that the interaction of reactive silica found in the vitreous part of PN and calcium hydroxide Ca $(OH)_2$ released by cement during the hydration mixture, promotes the physicochemical effect of this addition and to contribute of consumption of lime present in cement by subsequently improving the compressive strength.

3.3. Improvement of physical properties and mechanical performance

In order to improve the physical properties and mechanical performance of mortar or concrete formulated by the combination between the F.Lime and the PN, we have incorporated into the different formulation matrix the admixtures of superplasticizers of high water reducer and setting aSP103 with 2.5% by weight of cement.

3.2.1. Influence on porosity

The Figure 4 illustrates the porosity of concrete formulated by the combination between F.Lime and PN as a function of the mass fraction of these additions in the presence of superplasticizer SP103.

https://doi.org/10.1051/matecconf/201814901069

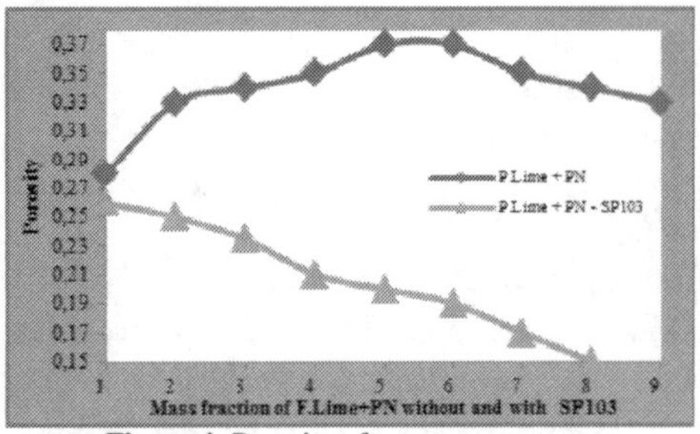

Figure 4: Porosity of mortar or concrete

From the Figure 4, we have found that the porosity of concrete formulated by the combination between of F.Lime, PN, and superplasticizers SP103 increases slightly according to their different percentages. In the second (F.Lime+PN-SP103), we observed a strong decrease in porosity. This decrease is due to the effect of combination of the organic admixtures of superplasticizers SP103 and inorganic admixtures (F.Lime + PN) which disperse the cement grains from one another. On the other hand, it can be explained by the fact that the combination of F.Lime and PN (very fine material) in the presence of superplasticizers fill interstitial voids; between the cement and aggregate particles; (Physicochemical and microstructure effects).

3.2.2. Influence on capillary absorption

The Figure 5 shows the capillary absorption of concrete formulated by the combination of F.Lime and PN as a function of the mass fraction of these additions in the presence of superplasticizers SP103.

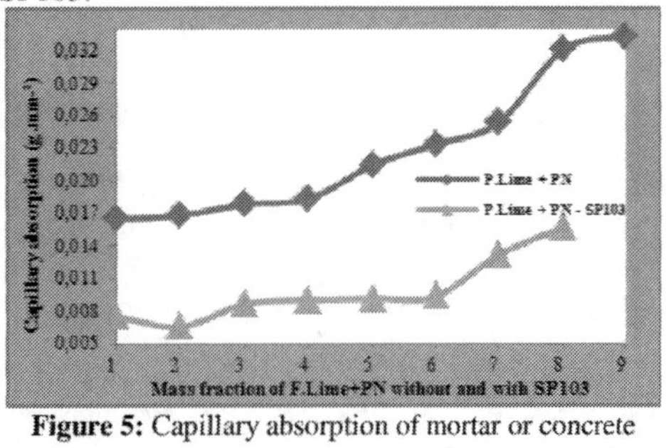

Figure 5: Capillary absorption of mortar or concrete

From the Figure 5, we have observed that the capillary absorption of concrete formulated by the combination between F.Lime and PN (P.Lime + PN) increases with respect to the control mortar. On the other hand, we have found that the incorporation of superplasticizers in the concrete formulation matrix based on combination of F.Lime and the PN (P.Lime + PN-SP103) decreases the capillary absorption. This decrease can be explained on one hand by the fact that the capillary absorption of mortar is influenced by the porous structure and the level of superplasticizers. The latter can contribute to the reduction of the capillarity by the formation of a polymer film and reduce the capillary pressure. On the other hand, it is linked to the initial role of superplasticizers which disperses the cement grains from each other and also due to the fineness of these additions (P.Lime + PN) which fill the voids between the particles by subsequently participating in the reduction of the capillary absorption of the concrete in the hardened state.

3.2.3. Influence on mechanical performance

The Figure 6 shows the evolutions of the compressive strength of mortar or concrete formulated by the combination between the F.Lime and the PN in the presence of superplasticizers SP103.

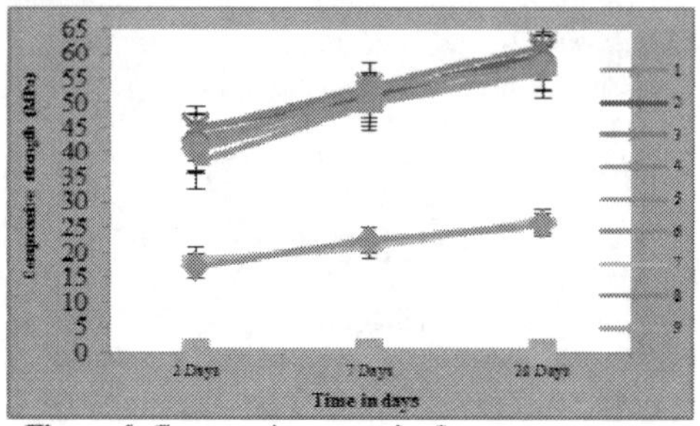

Figure 6: Compressive strength of mortar or concrete with SP103

From the Figure 6, we found that the compressive strengths of all the mortars formulated by the combination between the F.Lime and the PN in combination with superplasticizers SP103 increase steadily and show no drop. These increases can be explained by the fact that the combination between F.Lime and PN (very fine material) with their fine fills fill voids between the cement particles and granules (granular effect) on one hand. And on the other hand, the increases are related to the role of the addition of superplasticizers SP103 which disperse the cement grains from one another (physicochemical effects). In fact, when the organic molecules of SP103 are introduced into a suspension of cementitious material formulated by the combination between F.Lime and PN, a large part of them is fixed by adsorption to the surface of the cement particles (physicochemical and microstructural effect). The latter reduces the attractive forces of interaction between the particles and the atoms of different constituents (possibly chemical effects). The admixture of superplasticizer SP103 by their chemical effects modifies the interparticle forces, that is to say, to minimize the porosity by influencing the increase in the compressive strength.

3.4. Gain in compressive strength at 28 days

We also calculated the gain in compressive strength at 28 days, Figure 6, of mortar or concrete formulated by the combination between the F.Lime and the PN with and without superplasticizers SP103.

https://doi.org/10.1051/matecconf/201814901069

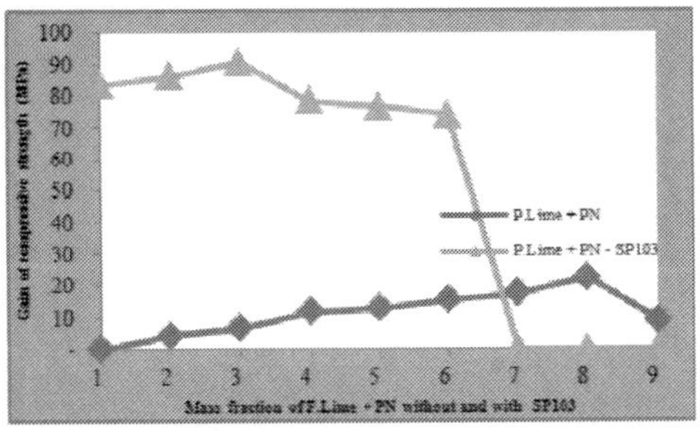

Figure 7: Gain in compressive strength at 28 days

From the Figure 7, we observed that the combination between of F.Lime and PN in the mortar or concrete matrix with superplasticizers SP103 improves the compressive strength.

3 Conclusion

In this work, we studied the influence of partial substitution of clinker by combination of limestone fillers and natural pozzolan (P.Lime + PN) at 40% by weight of cement with a step of 5% in the presence of superplasticizers of high water reducer accelerator setting SP103. The influences of incorporation of these additions on physical and mechanical properties of fresh cement paste and mortar or concrete in hardened state were studied.

The obtained results from the various formulation elaborate shows that granular, physicochemical and microstructural effects have been distinguished. These effects influence the physical-chemical and mechanical properties, namely the reduction of the quantity of water used. Similarly, the density and the expansion have been decreased. In addition, the fineness by surface area and setting time were increased. We found that the porosity and the capillary absorption were decreased (granular and microstructural effects). However, we have observed that the compressive strengths at young ages (2 days), medium ages (7 days) and long-term (28 days) have been improved in the presence of superplasticizers compared to the controls (physical-chemical and mechanical).

Our study contributes to valorize of mineral and natural resources such as natural pozzolan (PN) and limestone fillers (P-Lime) in Amran-region of Yemen, to minimize the greenhouse gas emissions, to reduce the energy and the raw material consumption, and to manufacture of a new ecological hydraulic binder and durable; with physiccochemical and mechanical properties improved.

Acknowledgments- The authors are pleased to acknowledge all engineers and employees of the laboratory of cement and quality control of Amran Cement plant of Yemen for providing the facilities for the research.

References

1. M. H. Khudhair, M. S. Elyoubi, A. Elharfi, J. Mater. Environ. Sci. **8.3**, 902–910 (2017).
2. M. H. Khudhair, M. S. Elyoubi, A. Elharfi, J. Mater. Environ. Sci **.8.6**, 1978–1989 (2017).
3. M. H. Khudhair, M. S. Elyoubi, A. Elharfi, Moroc. J. Chem. **5.1**, 153-163 (2017).
4. M. H. Khudhair, A. Elharfi Int. J. ChemTech Res. **9.12**, 695–704 (2016).
5. M. H. Khudhair, M. S. Elyoubi, and A. Elharfi, J. Mater. Environ. Sci. **8.11**, 3973–3985 (2017).
6. M. H. R. Khudhair, M. S. Elyoubi, A.Elharfi, Mor. J. Chem. **5.3**, 493-504 (2017).
7. M. S. Elyoubi, A.Elharfi, J. Chem. Technol. Metallo. **52.5**, 873-884 (2017).
8. R. Chaid, R. Jauberthie, J. Zeghiche, and F. Kherchi, Eur. J. Environ. Civ. Eng. **15.3**, 427–445 (2011).
9. G. C. Isaia, A. L. G. GASTALDInI, and R. Moraes, Cem. Concr. Compos. **25.1**, 69–76 (2003).
10. R. L. Day and C. Shi, Cem. Concr. Res. **24.8**, 1485–1491 (1994).
11. V. Sata, C. Jaturapitakkul, and K. Kiattikomol, Constr. Build. Mater. **21.7**, 1589–1598 (2007).
12. M. H. Khudhair, M. S. Elyoubi, A. Elharfi ,Res. J. Pharm. Biol. Chem. Sci. **8.3**, 1698–1712 (2017).
13. M. H. R. Khudhair, B. Elhilal, M, S. Elyoubi, and A. Elharfi, J. Mater. Environ. Sci. **8.7**, 2302–2310 (2017).
14. C. Autier, N. Azema, J.-M. Taulemesse, and L. Clerc, Powder Technol. **249**, 282–289 (2013).
15. M. Khudhair, and A. Elharfi. *Éditions Universitaires Européennes*, (2017).
16. M.-H. Zhang, K. Sisomphon, T. S. Ng, and D. J. Sun, Constr. Build. Mater. **24.9**, 1700–1707 (2010).
17. Y. L. Yaphary, Z. Yu, R. H. Lam, and D. Lau, Constr. Build. Mater. **141**, 94–103 (2017).
18. M. H. Khudhair, M. S. Elyoubi, and A. Elharfi, J. Mater. Environ. Sci. **9.1** (2018) 56–65.
19. O. Boukendakdji, E.-H. Kadri, and S. Kenai, Cem. Concr. Compos. **34.4**, 583–590 (2012).
20. J. Baron, J. Olivier, And J. Weiss,. Ed. Eyrolles. **806** (1997).
21. M. Cyr, P. Lawrence, E. Ringot, and A. Carles-Gibergues, Mater. Struct. **33.7**, 466–472 (2000).
22. O. Gheevarhese, C. A. Strydom, J. H. Potgieter, and S. S. Potgieter, Water SA. **28.1**, 45–48 (2002).
23. J. Lanas and J. I. Alvarez-Galindo, Cem. Concr. Res.**33.11**, 1867–1876 (2003).
24. Y. Benachour, C. A. Davy, F. Skoczylas, and H. Houari, Cem. Concr. Res.**38.6**, 727–736 (2008).
25. C. Glotzbach, D. Stephan, and M. Schmidt, Cem. Concr. Compos.**36**, 42–47 (2013)

Performance of self-compacting rubberized concrete

Performance du béton autoplaçant à base de granulats en caoutchouc recyclés.

Bensaci Hamza[1], Menadi Belkacem[1], Kenai Said[1] et Yahiaoui Walid[1]

[1]Laboratoire de Géo-matériaux, Département de Génie Civil, Université de Blida1, Algérie

Abstract: Used tyre rubber wastes present a serious environmental problem of pollution and storage. The recycling of this waste in the industry of construction could be an appropriate solution to produce an eco-concrete and could contribute to the improvement of some of its properties. This paper aims to study the possibility of using tyre rubber waste as fine aggregate replacement in self-compacting concrete (SCC). Fines rubber particles of 0-2 mm of waste tyres were added SCC mixtures as a partial substitution of the total volume of sand at different percentages (5, 10, 15, 20 and 30%). The influence of fines rubber of used tyres on fresh and hardened properties of the SCC was investigated. The fresh properties of SCC were performed by using slump-flow, T50 flow time, L-box, V-funnel and segregation resistance tests. Characteristics of the hardened state were obtained by compressive strength and thermal conductivity. The experimental results showed that the inclusion of fines rubber in SCC decreases the workability, reduced its passing capacity and increases the possibility of blocking. A decrease in compressive strength is observed with the increase in rubber content. On the other hand, the incorporation of the rubber fines aggregates enhances in a remarkably way the thermal conductivity.

Résumé : Les pneus usagés présentent des problèmes de pollution de l'environnement et de stockage. Le recyclage de ces déchets dans l'industrie de construction, s'avère une solution adéquate pour la production d'un éco-béton et peut contribuer à l'amélioration de certaines de ses propriétés. Cet article vise à étudier la possibilité de la valorisation des déchets caoutchouteux des pneus usagés sous forme de granulats dans le béton autoplaçant (BAP). Des particules de caoutchouc de 0-2mm de pneus ont été ajoutées aux mélanges du BAP en substitution partiel du volume total de sable à différents pourcentages (5, 10, 15, 20 et 30%). L'influence du caoutchouc sur les propriétés du BAP, aussi bien à l'état frais qu'à l'état durci, a été analysée. Les propriétés à l'état frais ont été déterminées à l'aide de tests du diamètre d'étalement et du temps d'écoulement, de L-box, du temps d'écoulement à l'entonnoir en V et de la résistance à la ségrégation par tamisage. Les caractéristiques à l'état durci ont été obtenues par l'essai de la résistance à la compression et la conductivité thermique. Les résultats expérimentaux ont montré que la présence du caoutchouc dans le BAP diminue l'ouvrabilité, réduit sa capacité de passage et augmente la possibilité de blocage. Une chute de la résistance à la compression est observée avec l'augmentation de taux de caoutchouc. Par contre, l'incorporation des granulats de caoutchouc améliore d'une façon remarquable la conductivité thermique.

1 Introduction

Les pneus usagés présentent des problèmes de pollution de l'environnement et de stockage. Plus de 275 millions de pneus usagés sont stockés dans les décharges à travers les Etats Unis [1, 2]. En Algérie, le nombre des pneus usagés est estimé à environ 1.5 millions par an [3]. L'élimination de ce type de déchets pose actuellement un sérieux problème pour les pays autour du monde en raison de leur effet direct sur l'environnement et la santé humaine [1, 4]. Le recyclage de ces déchets dans l'industrie de construction s'avère une solution adéquate pour la production d'un éco-béton et peut contribuer à l'amélioration de certaines de ses propriétés. A cet effet ,l'utilisation des particules de caoutchouc des pneus usagés comme des granulats dans le béton a reçu une attention particulière au cours des deux dernières décennies par plusieurs chercheurs en raison de l'accroissement énorme du gisement des pneus usagés.

La majorité des travaux de recherches ont montré que les déchets des pneus usagés (caoutchouc ou fibre) peuvent être utilisés dans la matrice cimentaire comme substitution aux agrégats pour l'amélioration de certaines caractéristiques des bétons.

khaloo et al. [5], ont examiné l'utilisation des granulats en caoutchouc pour fabriquer le béton et une chute de la masse volumique du béton avec l'augmentation du taux de substitution de caoutchouc a été observée. Cependant, Siddique et Naik[6] ont rapporté aussi une augmentation de la teneur en air occlus dans la matrice cimentaire contenant des particules de caoutchouc qui a une réduction de la masse volumique. En outre, Topçu et al. [7], ont étudié des BAP incorporant des agrégats en caoutchouc (G.C) par la substitution des agrégats naturels par des G.C en poids avec différents types d'agent de viscosité. Ils ont constaté queles propriétés rhéologiques des BAP étudiés sont acceptables, quel que soit le taux de substitution du caoutchouc. Güneyisi et al. [8] ont étudié les propriétés mécaniques des bétons de caoutchouc par substitution des granulats naturels par des granulats en caoutchouc (de 2,5 à 50 %) avec et sans l'utilisation de la fumée de silice. Les résultats ont montré une perte d'ouvrabilité ainsi qu'une forte réduction en termes de résistance avec l'augmentation du taux en caoutchouc. D'autre part, les auteurs ont indiqué que pour une teneur de 25% de caoutchouc on peut produire un béton avec une résistance à la compression de 16 à 32 MPa. De plus Eldin et Senouci [9] ont étudié l'influence de l'ajout de caoutchouc comme agrégats par la substitution totale du gravier ou du sable sur la résistance à la compression du béton. Ils ont observé une chute de résistance de l'ordre de 85% et 65% lorsqu'on substitue totalement le gravier et le sable par du caoutchouc, respectivement. Yung et al. [10], ont étudié l'effet du taux et la finesse du granulat en caoutchouc substitué au sable naturel par volume sur les performances du BAP. Ils ont observé une chute de résistance à la compression avec l'augmentation du taux de caoutchouc à l'exception du BAP avec 5% du caoutchouc qui présente une meilleure résistance par rapport au béton de référence. L'incorporation du caoutchouc dans le béton joue un rôle important dans l'amélioration de la capacité de déformation du béton [11]. Afin d'exploiter les gisements du déchet pneumatique et préserver l'environnement dans une vision de développement durable, des études et des recherches intensives et plus détaillé sont nécessaires. Le but de cet article est d'analyser l'effet des granulats de caoutchouc des pneus usagés sur les performances du béton autoplaçant à l'état frais et durci.

2 Matériaux et essais

Pour tous les mélanges de béton autoplaçant réalisés dans la présente étude, nous avons utilisé un ciment Portland CEM II / B 42.5Nde finesse 4640cm^2/kg; deux types de granulats grossiers 3/8 et 8/15 d'origines calcaires concassés de densité 2.53 kg/m^3 et de 1% d'absorption d'eau et un sable fin de dune avec un module de finesse de 2,56, une densité de 2,61 et une absorption d'eau de 1,34%. Les granulats en caoutchouc utilisés sont obtenus à partir du broyage de pneus usagés, leur densité est de 0,77 et un coefficient d'absorption d'eau égale 0,24 et leur taille maximale est de 2 mm. Un plastifiant polycarboxylate (MEDAFLOW 30) de densité spécifique de 1,06 kg/m^{3}est utilisé.

Dans la présente étude, la méthode japonaise est choisie pour la formulation des différents mélanges du béton autoplaçant étudiés. Notre formulation consiste à maintenir un rapport sable/mortier(S/M) de 50%, un super-plastifiant (SP) de 1% en poids de ciment et un rapport E/L=0.42 constant dans tous les mélanges. Six mélanges ont été réalisés avec différents pourcentages de caoutchouc (0%, 5%, 10%, 15%, 20%et 30%) comme substitution du sable en volume. Le détail de la composition des mélanges des différents BAP étudiés sont données dans le tableau 1.

Table 1. Compositions de mélanges des BAPs.

Constituants	(kg/m^3)					
	Réf	**5%**	**10%**	**15%**	**20%**	**30%**
Ciment	442.4					
Gravier	812.9					
Sable	819.7	807	794.5	781.8	769.2	744
Caoutchouc	-	12.6	25.2	37.8	50.4	75.7
SP	4.42					
Eau	200.7					

Pour évaluer les performances rhéologiques des mélanges des BAP, l'essai d'étalement, l'essai à l'entonnoir en V, l'essai L-box et la résistance à la ségrégation par tamisage ont été réalisés sur la base des recommandations d'EFNARC [12].Après la réalisation des essais à l'état frais des BAP, des éprouvettes cubiques de 100 mm de côté et prismatiques (70x70x280 mm^3) ont été confectionnées pour l'essai de résistance à la compression et l'essai de retrait total, respectivement. Les éprouvettes cubiques destinées à l'essai de compression ont été conservées dans l'eau à une température de 20± 2°C pendant 28 jours avant l'essai. Par contre, les éprouvettes prismatiques ont été soumis directement à l'essai du retrait total après démoulage à une humidité de 60% et une température 20 ± 2°C jusqu'à l'âge 28 jours. Pour déterminer les valeurs des coefficients de conductivité thermique des BAPà 28 jours, on a utilisé le dispositif du fil chaud de CT-mètre conforme à la norme NF EN 993-15 sur des éprouvettes cubiques10x10x10 cm.

3 Résultats et discussion

3.1 Propriétés du BAP à l'état frais

Les résultats de l'étalement de tous les mélanges des bétons autoplaçants contenant des granulats caoutchouteux comme substitution partielle du sable sont illustrés sur la figure 1.En comparant, les résultats des différents étalements avec les recommandations du béton autoplaçant[12], on constate que tous les mélanges du BAP présentent un étalement acceptable à l'exception du BAP avec 30% GC.D'après cette figure, on remarque que le diamètre d'étalement diminue avec l'augmentation du taux d'incorporation des granulats en caoutchouc. Cette diminution est de l'ordre de 1,9%, 5,3%, 8,6%, 11,2% et 31.6% pour des taux de substitution de 5%, 10%, 15%, 20% et 30% GC, respectivement en comparaison avec le BAP de référence. Cette réduction peut être attribuée à la

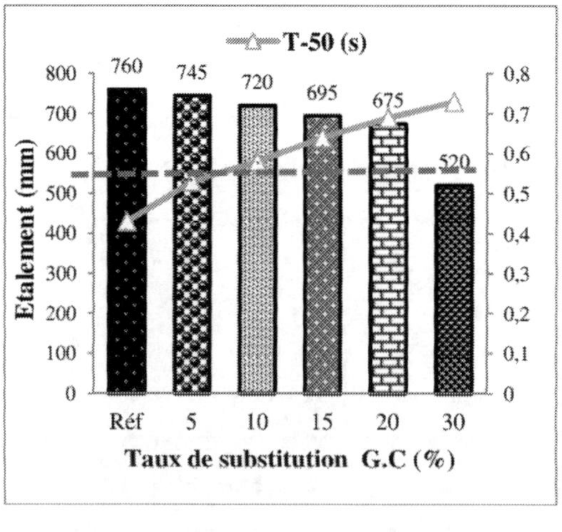

Fig. 1. Effet de l'incorporation du G.C sur l'étalement et le T50.

rugosité de la surface et la forme sphérique des particules de caoutchouc [11].Le temps d'étalement, T50 est inversement proportionnel à l'étalement comme l'indique la figure 1. Les mélanges du BAP avec 20% et 30% de granulats caoutchouteux donnentdes valeurs plus élevées du T50, ce qui indique une viscosité élevée en comparaison avec le BAP de référence. La même tendance est observée pour le temps d'écoulement du V-funnel (Fig. 2). Comme l'indique la figure 2, pour tous les mélanges étudiés le temps d'écoulement est entre 2.2 et 6.4 secondes. Ceci montre que les mélanges du BAP avec et sans granulats en caoutchouc étudiés sont conformes aux recommandations d'EFNAC. Le comportement du béton incorporant des G.C dans une zone ferraillée qui est traduit par l'essai de la boite en L–box est étudié. Cet essai donne une bonne évaluation de la capacité de passage et de remplissage.

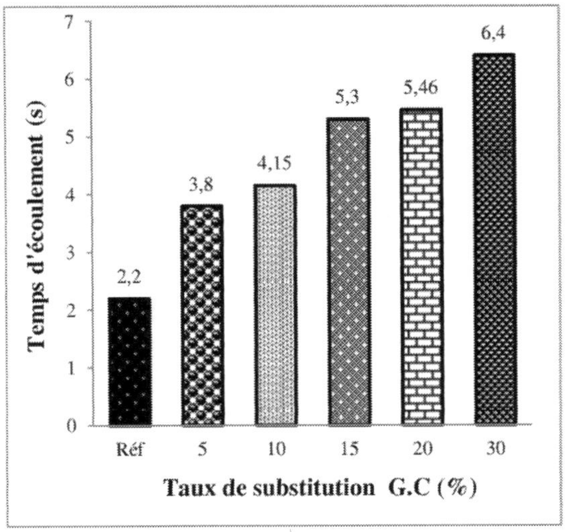

Fig. 2. Effet de l'incorporation du G.C sur le temps d'écoulement en V-funnel.

Les résultats de l'essai L-box exprimés par le rapport H2/H1 de tous les mélanges étudiés dans la présente étude sont illustrés sur la figure 3. Les valeurs mesurées de la capacité de passage (H2/H1) varient de 0.98 à 0.53. D'après la figure 3, on remarque que les mélanges avec 15, 20 et 30% de GC présentent un blocage au niveau des armatures du L-box et ne satisfait pas les exigences recommander pour les BAPs. Les mélanges du BAP avec des taux de substitution du GC supérieurs à 15% présentent une viscosité élevée ce qui empêche le passage à travers les trois barres du dispositif L-box. Par contre, les autres mélanges du BAP présentent des rapports H2/H1 admissibles.

Les résultats de l'effet de la variation du taux d'incorporation des G.C sur la résistance à la ségrégation des BAP sont donnés sur la figure 4. Selon cette figure, on remarque que l'augmentation du taux d'incorporation des G.C s'est traduite par une diminution du taux de ségrégation. Pendant l'essai, aucun signe de ségrégation n'a été détecté pour tous les mélanges des bétons autoplaçants.

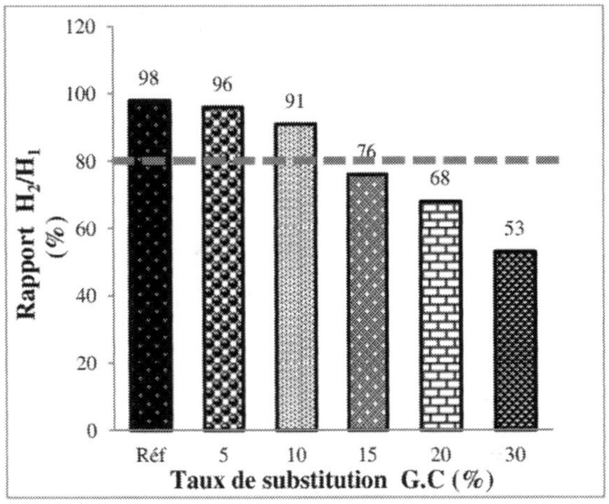

Fig. 3. Effet de l'incorporation du G.C sur le rapport H2/H1.

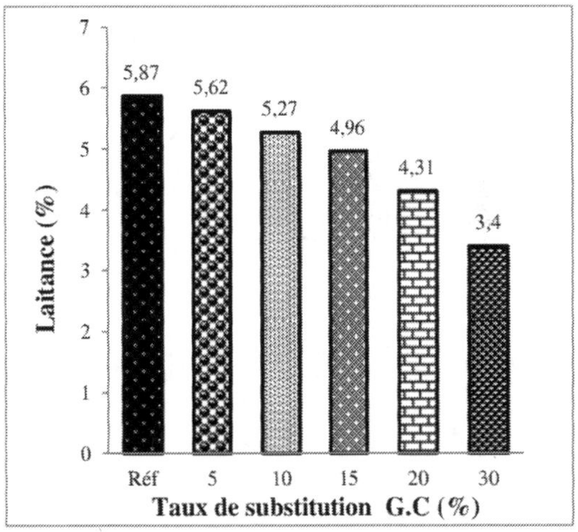

Fig. 4. Résistance à la ségrégation en fonction taux de substitution en G.C.

Ces résultats sont contradictoires à ceux obtenus par Garros [11] et Cuong [13] qui ont trouvé que l'utilisation des gros G.C de classe (0/4mm), entraine un phénomène de ségrégation contrairement à notre cas d'étude ou nous avons utilisé des G.C de classe (0/2 mm).

3.2 Effet de l'incorporation des G.C à l'état durci

3.2.1 Résistance à la compression

Les résultats indiquent clairement (Fig.5) que l'incorporation de G.C est très préjudiciable vis à vis de la résistance à la compression. A titre d'exemple, un taux de substitution en G.C. de 15 % entraîne une chute de résistance remarquable de 45,4 % par rapport au béton de référence. Ceci peut être expliqué par la faible rigidité du caoutchouc par rapport à celle des granulats naturels selon Li et al. [14]. La deuxième raison est le défaut d'adhérence entre le caoutchouc et la matrice cimentaire et la mauvaise qualité de la zone de contact entre ces deux phases. Cet argument a été aussi avancé par Garros [11] qui montre que la zone de transition entre la matrice et le granulat en caoutchouc est peu compacte et présente une porosité importante.

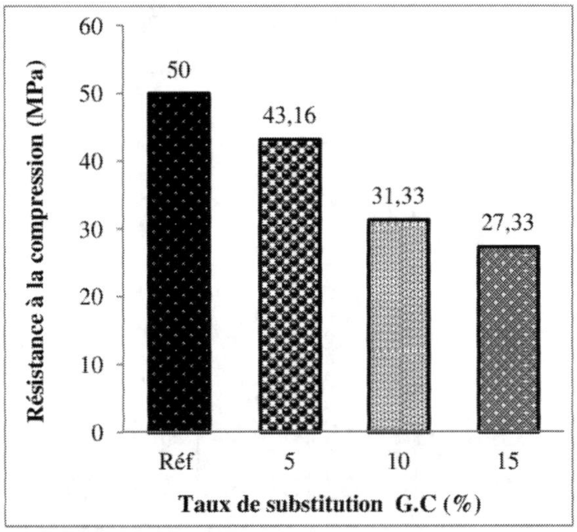

Fig. 5. Effet des G.C. sur la résistance à la compression à 28 jours.

3.2.2 Conductivité thermique

Les résultats des mesures de la conductivité thermique à l'état durci des bétons sont donnés sur la figure 6. Les résultats ci-dessus montrent que la conductivité thermique diminue avec l'augmentation du dosage des granulats caoutchouc. Par exemple, la présence de G.C. entraîne une réduction de la conductivité thermique de 26.50 % lorsqu'on augmente les G.C de 0 à 15%. Cette amélioration peut s'expliquer par la porosité élevée du BAP avec l'adjonction de G.C. D'après Benazzouk [15] l'air entraîné lors du malaxage conduit à une porosité fermée dans la matrice cimentaire, ce qui améliore encore les performances d'isolation du composite cimentaire incorporant des G.C.

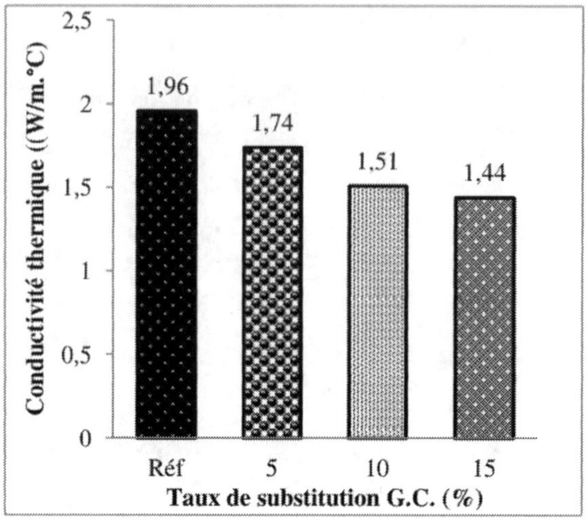

Fig. 6. Influence du dosage en G.C. sur la conductivité thermique.

3 Conclusion

Les résultats obtenus de la présente étude peuvent être résumés comme suit :

-L'incorporation des granulats en caoutchouc diminue les propriétés rhéologiques et conduit à des mélanges des BAP beaucoup plus visqueux à cause de la surface rugueuse de ces granulats.

- L'incorporation des granulats en caoutchouc est très préjudiciable vis-à-vis de la résistance à la compression.

-La conductivité thermique diminue lorsque le taux d'incorporation en G.C. augmente.

-Des bétons autoplaçants conformes aux normes peuvent être fabriqués avec l'utilisation d'un taux de remplacement du caoutchouc (0-2mm) par du sable allant jusqu'à 15%.

References

[1] A.Meddah, M. Beddar, A. Bali,use of shredded rubber tire aggregates for roller compacted concrete Pavement,Journal of Cleaner Production, pp. 187-192, (2014).

[2] M. R. Hall, K. B.Najim, Structural behaviour and

durability of steel-reinforced structural Plain/Self-Compacting Rubberised Concrete (PRC/SCRC), Construction and Building Materials, no. 73 , p. 490–497, (2014).

[3] H. Trouzine, A. Asroun, N. Asroun, F. Belabdelouhab, N. T. Long, Problématique des pneumatiques usagés en Algérie Nature & Technologie, no. 05, (2011).

[4] M. Gesoglu, E. Guneyisi, Permeability properties of self-compacting rubberized concretes,Construction and Building Materials, no. 25, p. 3319–3326, (2011).

[5] A. R. Khaloo, M. Dehestani, P. Rahmatabadi, J.Mechanical properties of concrete containing a high volume of tire–rubber particles,WasteManag. 28, 2472-2482, (2008).

[6] Siddique, R., Naik, T.R,Properties of concrete containing scrap-tire rubber: an overview,Waste Manag. 563-569, 2004.

[7] I. B. Topcu, T. Bilir, Experimental investigation of some fresh and hardened properties of rubberized self-compacting concrete, Materials and Design, vol. 30, no. 123, p. 3056–3065, (2009).

[8] E. Güneyisi, M. Gesoğlu,T. Özturan, Properties of rubberized concretes containing silica fume,Cement and Concrete Research 34, 2309–2317, (2004).

[8] E. Güneyisi, M. Gesoğlu, T. Özturan, Cement and Concrete Research 34, 2309–2317, (2004).

[9] N. Eldin, A.B Senouci, Rubber-Tire Particles asConcrete Aggregate,J. Mater. Civ. Eng, 478-496. (1993).

[10] W. H. Yung,L. C. Yung, L. H. Hua, A study of the durability properties of waste tire rubber appliedto self-compacting concrete,vol. 41, no.122, p. 665–672, (2013).

[11] M .Garros, *Composites cimentaires incorporant des granulats caoutchouc issus du broyage de pneus usagés : optimisation de la composition et caractérisation*, Thèse de doctorat, Université Paul Sabatier de Toulouse, (2007).

[12] EFNARC, *Specifications and Guidelines for Self-Compacting Concrete English ed.*, European Federation for Spec Constr Chem and Concr Syst., (2005).

[13] H. A. Cuong. *Optimisation de la composition et caractérisation d'un béton incorporant des granulats issus du broyage de pneus usagés : Application aux éléments de grande surface* (Thèse de doctorat), INSA de Toulouse,
France,(2010).http://www.theses.fr/2010ISAT0024.

[14] Z. Li, F. Li, Li. J. S. L, Properties of concrete incorporating rubber tyre particles, Magazine of Concrete Research, 50, No4, 297-304, (1998).

[15] Benazzouk. A., *Contribution à la valorisation de déchets de caoutchouc: composites cimentaires à base de caoutchouc compact et cellulaire.* (Thèse de doctorat), université d'Amines, France, (2002).

Experimental Study on Ready-Mix Concrete: Case Study

Étude expérimentale sur un béton prêt à l'emploi: Étude de cas

Dorra ELLOUZE[1], Aida GHAMMOURI[1], Rahma BEN AMAR[1]
[1]ISET SFAX, Département génie civil, Sfax, Tunisie
dorra_ellouze@yahoo.fr

Abstract. Ready-mix concrete (RMC) in Tunisia is becoming more and more in demand in the civil engineering sector thanks to its qualities of handling in the fresh state and resistance in the hardened state, this composite material must respect the quality-price ratio. A RMC with a minimal cost is the object of our work. This research is part of the opening of higher education on professional life, where we optimized the formulation of a RMC. This work has 3 axes. In the first place the resources in building materials were characterized, namely various samples of sand, gravel, cement and water. Subsequently, the adjuvant-cement ratio (A/C) was optimized. Finally, the workability of the concrete as well as its mechanical aptitude at various ages 7, 14 and 28 days were characterized. These examinations have resulted in an appropriate formulation for any type of resource that varies according to the provenance of the quarries (gravel and sand), the effect of the plasticizer-water reducer is found for a very interesting A/C ratio, the mechanical tests for different ages are also conclusive.

Résumé. Le béton prêt à l'emploi en Tunisie devient de plus en plus demandé dans le secteur génie civil grâce à ses qualités de maniabilité à l'état frais et de résistance à l'état durci, Ce matériau composite doit respecter le rapport qualité-prix. Un béton BPE compact avec un cout minimal est l'objet de notre travail. Cette recherche s'inscrit dans le cadre de l'ouverture de l'enseignement supérieur sur la vie professionnelle, où on a optimisé la formulation d'un béton BPE. Ce travail comporte 3 axes. En premier lieu on a caractérisé les ressources en matériaux de construction à savoir de divers échantillons de sable, gravier, ciment et eau, puis on a optimisé le rapport adjuvant-ciment. Enfin on a caractérisé l'ouvrabilité du béton ainsi que son aptitude mécanique à différents âges 7j ,14j, 28j. Ces examens ont abouti à une formulation adéquate pour tout type de ressource qui varie selon la provenance des carrières (gravier et sable), l'effet du plastifiant-réducteur d'eau est trouvé pour un rapport A/C très intéressant, les essais mécaniques pour les différents âges sont aussi concluants.

1 Introduction

Le béton est le matériau le plus utilisé sur le marché grâce à ses performances et sa souplesse d'emploi. Il a mérité sa place pour ses multiples caractéristiques et pour la diversité qu'il permet dans les formes, les teintes et les textures. L'appellation « Béton Prêt à l'Emploi » (BPE) est réservée au béton préparé en usine dans des centrales et transporté jusqu'au lieu d'utilisation. Les centrales de BPE peuvent livrer les bétons les mieux adaptés, dont les caractéristiques font l'objet d'un suivi grâce à des contrôles de laboratoire, qui permettent de s'assurer également de la conformité des constituants. Notre recherche s'inscrit dans ce cadre, elle a pour but de s'ouvrir sur le milieu industriel par le biais d'une convention avec une centrale à béton pour l'élaboration d'une formulation de béton satisfaisant l'exigence des clients de la centrale.

Un bon béton doit être compact et par la suite résistant à l'état durci et ouvrable à l'état frais, notre travail a été basé sur ces deux caractéristiques. Le but étant de trouver un béton qui résiste bien à la compression tout en garantissant une ouvrabilité qui lui permet d'être pompable. La démarche adoptée dans cette étude, est décomposée en quatre phases distinctes. On commencera par une caractérisation des matériaux utilisés afin de connaître leurs propriétés et pour pouvoir les utiliser judicieusement, on passera en deuxième étape à l'optimisation du pourcentage de l'adjuvant utilisé, on présentera dans cette partie le rapport adjuvant-ciment optimal de plusieurs types de ciments pour mettre en évidence l'effet de composition chimique du ciment sur sa réaction avec l'adjuvant. On présentera en troisième lieu une analyse des résultats de la formulation et des essais réalisés. On finira par une conclusion

2 Identification des matériaux

Pour pouvoir utiliser efficacement les matériaux de construction, il est indispensable de savoir leurs caractéristiques.

2.1 Granulats

La caractérisation se fait par la détermination des propriétés intrinsèques, liées à la nature minéralogique de la roche et à la qualité du gisement et des caractéristiques de fabrication, liées aux procédés d'exploitation et de production des granulats. Tous les essais ont été réalisés en se référant à la norme française européenne, on indiquera pour chaque essai le numéro de la norme. Les propriétés physiques et mécaniques des granulats utilisés pour la formulation sont présentées dans le tableau 1.

Essais	Mapp (g/cm^3)	Mabs (g/cm^3)	Es (%)	Micro Deval (%)	LA (%)	Apla (%)	Abs (%)
Norme	NF EN 1097-6 [1]	NF EN 1097-6 [1]	*NF EN 933-8* [2]	NF EN 1097-1[3]	NF EN 1097-2[4]	NF EN 933-3 [5]	NF EN 1097-6 [1]
Sable	1.54	2.5	91	-	-	-	-
Gravier 4/12	1.55	2.56	-	15	21	8	0.5
Gravier 12/20	1.58	2.6	-	15	21	8	0.5

Tableau 1. Propriétés physiques et mécaniques des granulats

L'essai de l'analyse granulométrique est réalisé selon la norme [NF EN 933-1] [6] sur les granulats objet de la formulation, le gravier utilisé est divisé en deux classes granulaires 4/12 et 12/20 ayant la même provenance carrière Fayedh, pour le sable utilisé c'est un sable fin de module de finesse 1.86 et de classe granulaire 0/5 il a comme provenance carrière Wesletia le résultat de l'analyse granulométrique est donné par la figure 1

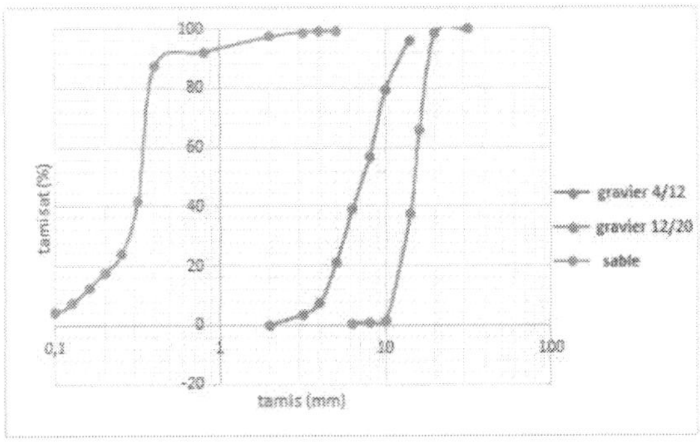

Fig. 1. Courbes d'analyse granulométrique des graviers et du sable

2.1 Ciment

Le choix du type de ciment et son dosage dépendent à la fois des performances recherchées et de la nature des autres composants. On se propose de confectionner un béton courant, en utilisant un ciment dont la provenance est cimenterie Kairouan de type CEM I et de classe de résistance 42,5. Les essais réalisés sur le ciment sont optés pour déterminer sa masse volumique absolue et apparente et pour vérifier sa classe vraie, le tableau 2 fournit les résultats de différents essais réalisés.

Tableau 2. Propriétés physiques et mécaniques du ciment 1

Essais	Masse volumique apparente (g/cm^3)	Masse volumique absolue (g/cm^3)	Classe vraie de la résistance (MPa)
Norme	NF EN 196-1 [7]	NF EN 196-1 [7]	NF EN 196-1 [7]
Résultat	1.04	3.11	42.9

3 Optimisation du dosage en adjuvant

L'ajout d'un adjuvant plastifiant-réducteur d'eau permet d'améliorer l'ouvrabilité du béton tout en réduisant la quantité d'eau. Le but est de déterminer le dosage optimum d'adjuvant pour une quantité de ciment donnée et d'étudier la variation de fluidité en cas de changement de type de ciment puisque le dosage optimal en adjuvant varie selon sa composition chimique du ciment [8].
L'essai Cône de Marsh permet de déterminer le dosage optimum d'adjuvant [9], lorsque la quantité d'adjuvant augmente, la fluidité de la pâte augmente, le but étant de trouver le rapport noté A/C (Adjuvant/Ciment) pour lequel la pâte est saturée en adjuvant et au-delà duquel la fluidité n'augmente plus. Du point de vue économique, on a intérêt à utiliser le moins d'adjuvant possible. L'essai est répété après 60 minutes pour voir l'effet du temps de transport sur la perte de fluidité. On a testé 3 types de ciments de différentes provenances. On a étudié cet effet sur le ciment 1 de provenance cimenterie Kairouan, ciment 2 de provenance cimenterie Jebel Elwest, ciment 3 de provenance cimenterie Carthage. Les résultats sont présentés dans le tableau 3.

	Ciment 1		**Ciment 2**		**Ciment 3**	
%A/C	T0+5 min	T0+60 min	T0+5 min	T0+60 min	T0+5 min	T0+60 min
0	8,18	8,37	8,37	9,05	8,76	8,91
1,6	7,79	8,09	7,99	8,5	7,7	8,2
2,4	7,57	7,89	7,79	8,27	7,16	7,61
3,5	7,4	7,68	7,5	7,97	7,3	7,68
4	7,61	8,02	8,01	8,42	7,79	7,98
4,8	7,58	8	7,89	8,2	7,55	7,8

Tableau 3. Résultats d'Essai cône de MARSH à 5 et 60 min.

 https://doi.org/10.1051/matecconf/201814901071

4 Résultats et interprétation

Cette étude de cas propose une formulation optimale dans les constituants d'un BPE dosé à 350 kg/m^3 en utilisant un adjuvant (plastifiant-réducteur d'eau), il en résulte des performances mécaniques qui dépasse 40 MPa à 28 jours à l'état durci et une meilleure fluidité à l'état frais.

4.1 Formulation du béton BPE

On se propose dans cette partie de déterminer la composition d'un béton prêt à l'emploi satisfaisant l'exigence de la centrale ayant une valeur d'affaissement égale à 16 cm (A= 16 cm) et une valeur de résistance à 28 jours est 30 MPa en utilisant un adjuvant local nommé « SIKA BV40 » qui est un plastifiant-réducteur d'eau. L'étude de la composition d'un béton vise à définir le mélange optimal des différents granulats dont on dispose ainsi que le dosage en ciment et en eau afin de réaliser un béton dont les qualités soient celles recherchées, la méthode utilisée pour la formulation est celle de Dreux Gorisse [10]. Le choix de la méthode est expliqué par sa simplicité à définir une formule de composition adaptée au béton étudié. Nous avons confectionné 12 éprouvettes de dimensions 16x32 en cm pour l'écrasement à différents âges (7 jours, 14 jours et 28 jours). Pour s'assurer de l'obtention de la valeur d'affaissement souhaitée nous avons réalisé l'essai de cône d'Abrams.

4.2 Résultats

Le travail a été élaboré suivant le plan expérimental suivant :
-La formulation d'un béton prêt à l'emploi qui satisfait les exigences souhaitées à l'état frais et durci
-L'optimisation du pourcentage de l'adjuvant utilisé, en étudiant le rapport adjuvant-ciment optimal de plusieurs types de ciments.
Les résultats seront présentés selon cette démarche.

4.2.1 Béton frais

Le béton doit être ouvrable pour faciliter sa mise en place, l'essai de cône d'Abrams permet de renseigner sur la valeur d'affaissement, en tenant compte des résultats souhaités pour le béton et la qualité des granulats utilisés et la nature de ciment pour la formulation, le résultat d'affaissement trouvé A = 16 cm, ce qui vérifie l'exigence de la centrale pour un béton pompable. [11]

4.2.2 Béton durci

La résistance à la compression est vérifiée par l'essai d'écrasement [12], Les résultats de l'essai sur des éprouvettes cylindriques 16x32 en cm sont données par le tableau 4, on note Ep pour désigner une éprouvette.

Tableau 4. Résultats d'essai de compression à différents âges

Age	7j		14j		28j	
	Masse Kg	f_c MPa	Masse kg	f_c MPa	Masse kg	f_c MPa
Ep1	15,99	24,3	15,91	31,5	15,88	46,58
Ep2	15,70	21,6	15,83	32	15,94	44,23
Ep3	15,50	23,6	15,83	32,1	15,89	41,09
		23,73		29,87		43,96

Les valeurs de résistance à la compression obtenues à 7j, 14j, 28 jours montrent l'obtention de bonne valeurs souhaitée et même un dépassement léger, ainsi l'étude expérimentale était concluante, comme le montre la figure 2, une aptitude mécanique à 28j égale 44 MPa, une valeur supérieure à celle proposé dans les hypothèses de formulation (30 MPa), ceci est expliqué par la bonne formulation dans les conditions expérimentales dans un laboratoire, ainsi que l'impact de l'optimisation de l'adjuvant BV40 dans le processus de durcissement.

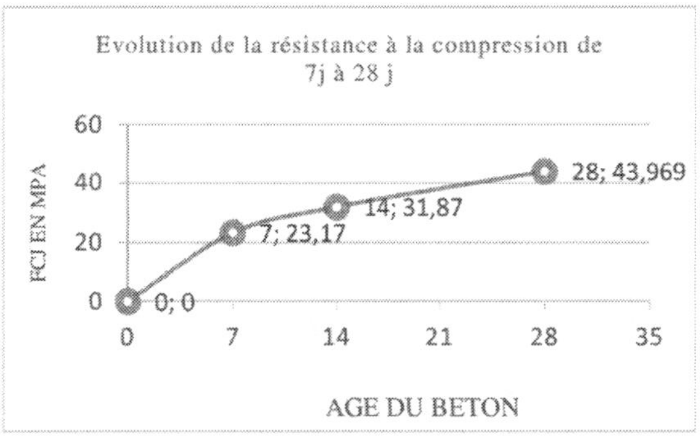

Fig. 2. Evolution de la résistance à la compression de 7 j à 28 j

Avec la formulation décrite précédemment on a obtenu un béton durci ayant un excellent comportement mécanique, en effet la résistance à la compression a atteint presque sa moitié (52%) à 7j, on remarque ainsi une vitesse de durcissement rapide accompagné d'augmentation de la résistance à la compression, 20% de cette résistance est obtenue entre 7j et 14j, et 28% de l'aptitude mécanique est obtenue entre 14 j à 28 jours (cf. Figure 3).

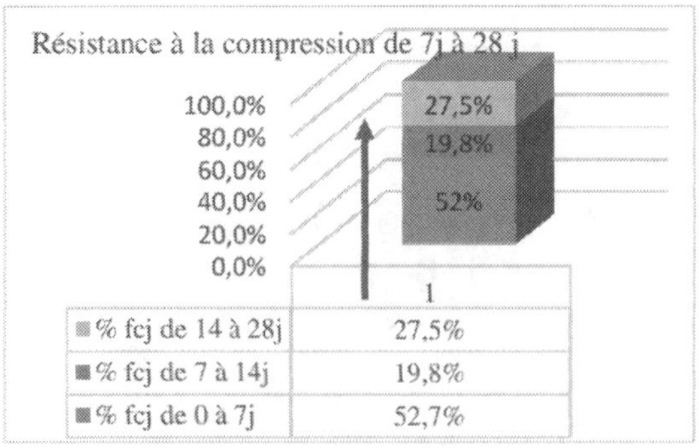

Fig. 3. Pourcentage d'évolution de la résistance à la compression de 7 j à 28 j

https://doi.org/10.1051/matecconf/201814901071

4.2.3 Optimisation de l'adjuvant

Les résultats représentés sur les figures 4, 5 et 6 nous aident à analyser la fluidité d'un coulis de ciment (ciment adjuvant SIkA BV40), la figure 4 présente la courbe d'écoulement d'un coulis de ciment fait avec le ciment 1 de provenance cimenterie Kairouan, les figures 5 et 6 représentent les courbes d'écoulement de deux autres ciments utilisés sur le marché Tunisien, de même type CEM I et classe vraie que la premier ciment (42.5 N), ils différent uniquement par leurs compositions chimiques vue leurs différentes provenances. [7]

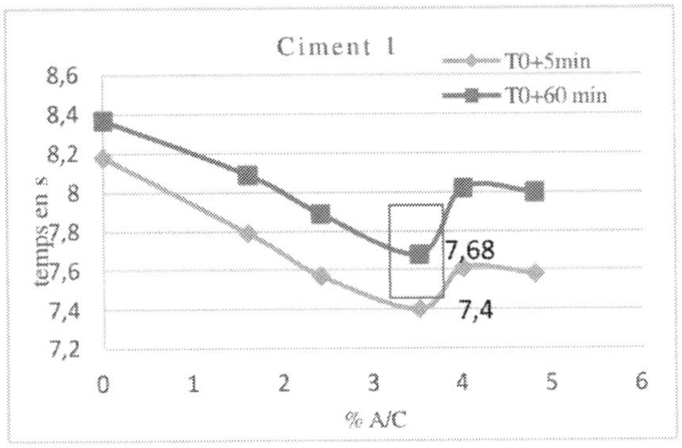

Fig. 4. Résultats essai cône de MARSH pour le ciment 1

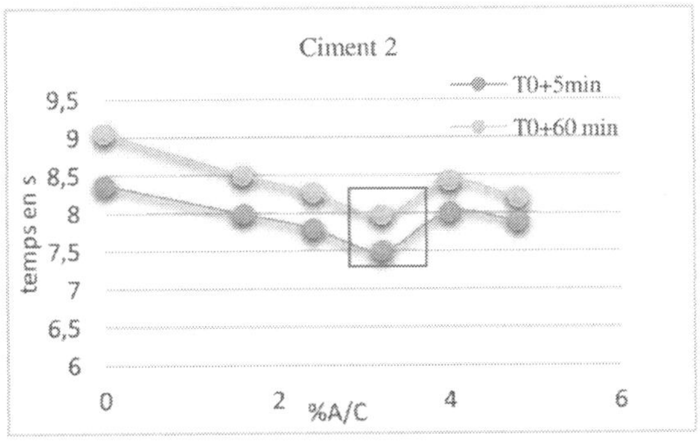

Fig. 5. Résultats essai cône de MARSH pour le ciment 2

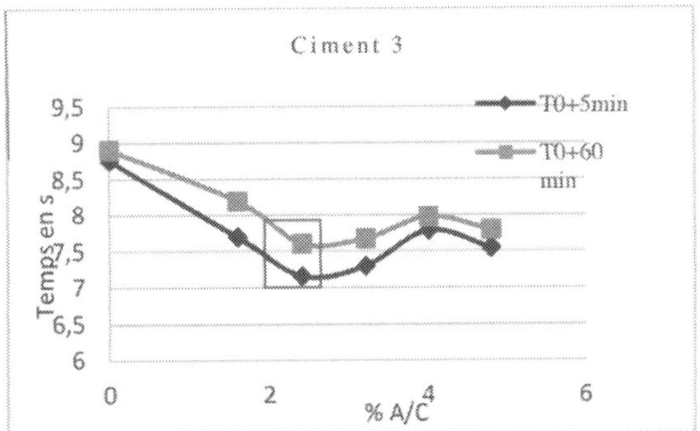

Fig. 6. Résultats essai cône de MARSH pour le ciment 3

Les courbes obtenues dans les figures 4, 5 et 6 nous renseignent sur des critères qui décrivent le comportement rhéologique de la combinaison ciment - adjuvant étudiée.

Le premier critère est que le dosage critique correspondant au point de saturation, c'est une brisure dans la courbe de temps de fluidité/dosage en plastifiant :
Ciment 1 : le dosage critique égale 3,5% (cf. Figure 4)
Ciment 2 : le dosage critique égale 3,2% (cf. Figure 5)
Ciment 3 : le dosage critique égale 2,4% (cf. Figure 6)

Le deuxième critère affirme que le degré de fluidité atteint pour ce dosage critique, dépend en termes de temps d'écoulement du type de ciment utilisé. Autrement dit, pour l'évolution dans le temps de la fluidité de la combinaison ciment-plastifiant et pour les 3 types de ciment, on remarque une position relative des deux courbes à 5 minutes et à 60 minutes.

Un troisième critère parait opportun, c'est que l'adjuvant joue son rôle inverse en cas de dépassement du point critique, il était un accélérateur de prise, il devient retardateur de prise si on dépasse le pourcentage A/C critique.

Un couple ciment-plastifiant est dit compatible lorsqu'à un dosage approprié il atteint une bonne fluidité à 5 minutes et ne subit pas de perte considérable de cette fluidité à 60 minutes [13]. Pour notre cas de formulation, avec le ciment type 1, cette compatibilité est excellente car au voisinage d'un dosage A/C=3,5%, on a une perte de fluidité de l'ordre de 4% (cf. Figure 7). Les ciments 2 et 3 présentent une bonne compatibilité avec l'adjuvant comme c'est illustré à 5 min et à 60min, on trouve une perte qui ne dépasse pas 6% au voisinage du dosage critique A/C.

Pour récapituler, un coulis est dit stable s'il conserve sa fluidité aussi longtemps que possible. La perte de la fluidité d'un coulis est caractérisée par la différence du temps d'écoulement enregistrée à 5 et à 60 minutes (cf. Figure 7). Les résultats illustrés sur la figure 7 montrent que la perte de fluidité diminue avec le dosage en plastifiant et se stabilise avec l'augmentation du dosage en plastifiant. Cela est aussi justifié et trouvé dans divers travaux de recherches fait sur des bétons et des coulis à savoir O. Benjeddou et al.2016, F.taieb et al. 2017 [8, 13, 14,15].

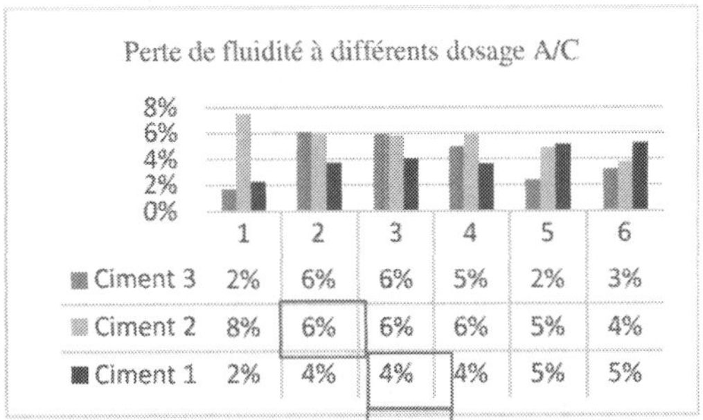

Fig. 7. Perte de fluidité pour différents dosages A/C

5 Conclusions

L'objectif de ce projet est atteint par l'optimisation de la quantité d'adjuvant utilisé dans un BPE d'une centrale à béton sis à Sfax-Tunisie, les résultats trouvés à l'échelle mécaniques sont excellents avec une f_{c28}=44 MPa. Sachant que les méthodes de formulation de béton dans la centrale différent de celle dans un laboratoire, alors il faut tenir compte de l'influence de la teneur en eau des granulats et leurs propretés lors de la formulation. Il faut faire quelques gâchées d'essais pour la confection des éprouvettes permettant d'ajuster au mieux la composition à adopter définitivement.
La méthode de mesure de la fluidité par le cône de Marsh est un moyen efficace pour une première sélection du couple ciment-plastifiant, susceptible d'être utilisé pour la confection d'un béton prêt à l'emploi. La fluidité du coulis est très liée à la composition du ciment et au type de l'addition minérale présente dans sa composition.

Références

1. NF EN 1097-6 (Janvier 2014) *: Essais pour déterminer les caractéristiques mécaniques des granulats et physiques des granulats - Partie 6 : détermination de la masse volumique réelle et du coefficient d'absorption d'eau*
2. NF EN 933-8/IN1 (Juillet 2015) : *Essais pour déterminer les caractéristiques géométriques des granulats - Partie 8 : évaluation des fines - Équivalent de sable*
3. NF EN 1097-1 (Aout 2011) *: Essais pour déterminer les caractéristiques mécaniques et physiques des granulats - Partie 1 : détermination de la résistance à l'usure (micro-Deval)*
4. NF EN 1097-2 (Juin 2010) *: Essais pour déterminer les caractéristiques mécaniques et physiques de granulats - Partie 2 : méthodes pour la détermination de la résistance à la fragmentation*
5. NF EN 933-3 (Mars 2012) : *Essais pour déterminer les caractéristiques géométriques des granulats - Partie 3 : détermination de la forme des granulats - Coefficient d'aplatissement*
6. NF EN 933-1 (Mai 2012) : *Essais pour déterminer les caractéristiques géométriques des granulats - Partie 1 : détermination de la granularité - Analyse granulométrique par tamisage*
7. NF EN 196-1 (Septembre 2016) *: Méthodes d'essais des ciments - Partie 1 : détermination des résistances - Méthodes d'essais des ciments - Partie 1 : Détermination des résistances.*
8. F. Taieb et al, *Effet couplé des fines et de l'adjuvant sur la fluidité du coulis de ciment, Université de Mostaganem.* (2017)
9. NF P18-507 (Novembre 1992) *: Additions pour béton hydraulique - Besoin en eau, contrôle de la régularité - Méthode par mesure de la fluidité par écoulement "au cône de Marsh"*
10. J. Festa, *Nouveau guide de béton et de ses constituants (1998).*
11. *NF EN 12350-2 (décembre* 1999) *: Essai pour béton frais Partie 2 : Essai d'affaissement.*
12. NF EN 12390-*3* (Février 2003*) : Essai pour béton durci Partie 3 : Résistance à la compression des éprouvettes.*
13. O. Benjeddou, M. Jedidi, C. Soussi, M. Benali, *Experimental and theoretical study of the effect of the particle size of limestone fillers on the rheology of self-compacting concrete,* Journal of Building Engineering, Vol 10, Pages 32–41. (2017)
14. A. Hallal, *Effets combinés des ajouts minéraux et des superplastifiants sur la compatibilité des coulis, mortiers et bétons,* Thèse de Doctorat, USTOMB, (2011)
15. Ch. Amouri, *Contribution à l'étude de l'influence des différents ajouts sur les propriétés des matrices cimentaires (caractérisation, Performances, Durabilité),* thèse soutenue à université Mentouri Constantine,158 pages (2009).

 https://doi.org/10.1051/matecconf/201814901072

Feasibility of backfilling mines using cement kiln dust, fly ash, and cement blends

H. Beltagui[1,2], M. Sonebi[1], K. Maguire[2], and S. Taylor[1]
[1]School of Natural and Built Environment, Queen's University Belfast, Belfast, Co. Antrim, BT7 1NN, UK
[2]Quinn Building Products, Derrylin, Co. Fermanagh, BT92 9GP, UK

Abstract. Cement kiln dust (CKD) is an industrial by-product of the cement manufacturing process, the composition of which can vary widely. Recent years of using alternative fuels have resulted in higher chloride and alkali contents within CKDs; as such, this limits the applications in which CKDs can be utilised. Using a CKD containing a high free lime content of 29.5%, it is shown that this CKD is capable of activating pulverized fuel ash (PFA) due to its high alkalinity, which can be utilised in low strength un-reinforced applications. One potential application involves the backfill of mines, reducing the need for continuous maintenance of the mine. This study focuses on the compressive strength achieved by various blends of CKD, PFA, and cement. Samples were hand mixed and compacted in 100 mm x 50 mm diameter cylinders, and unconfined compressive strength measurements taken at 28 and 56 days. The hydration products were assessed through the use of x-ray diffraction and thermogravimetric analysis. Aiming to maximise the use of CKD at a water to binder (w/b) ratio of 0.2, it was found that the maximum CKD content possible to achieve the required strength was 90% CKD blended with 10% cement.

1 Introduction

Cement kiln dust (CKD) is an industrial by-product of the cement manufacturing process, the composition of which can vary widely. Due to the generation of large quantities, it is important both from an environmental and economic view to find suitable applications for this material. Traditionally, a portion of the CKD produced can be returned to the cement production process. However, recent years of using alternative fuels, such as solid recovered fuel (SRF), have resulted in higher chloride and alkali contents within CKDs.

The high alkalinity of these CKDs can be exploited by blending with alkali activated materials, such as pulverised fly ash (PFA) or slags. Typically, Portland cement is also blended to provide high early strength, while the alkali activation can improve the long term strength and durability of the concrete. It has previously been shown that additions of up to 10-15% CKD by weight to high volume fly ash and slag concretes can produce improved strengths to blends with no added CKD [1,2].

The potential for CKDs to activate materials such as PFA and slags typically lies in the form which the calcium oxide takes. While free-lime (CaO) and calcium hydroxide ($Ca(OH)_2$) typically have the capacity for activation, calcium carbonate ($CaCO_3$) is inert and therefore unreactive.

However, the use of such CKDs in reinforced concrete applications is not possible as the chlorides present a high risk of corrosion to the reinforcement. Moreover, the alkalis may result in alkali-silica reactions (ASR) between the pore fluid and the aggregates to take place, causing expansion and cracking of the concrete. Therefore, alternative applications require investigation where the chlorides and alkalis do not present any durability issues.

One potential application for harnessing the alkali activation potential of these blends involves the backfill of mines, reducing the need for continuous maintenance and to extend the life of the mine. While there are several different types of filling techniques used in the mining industry, hydraulic fills using blends of cement and PFA have been used in the past. The blends can be mixed either below or above ground, and pumped into the mine. Typical strength requirements of the material used can be as low as 3 MPa at 56 days, with the chemical composition of the fill material not being of concern, making it an ideal application for the utilisation of such CKDs. Moreover, it can be a cost effective solution for the backfilling of mines.

This study focuses on the compressive strength achieved by various blends of CKD, PFA, and cement. The unconfined compressive strength was measured, and the failure modes recorded. In conjunction, the hydration products were investigated, providing an understanding for the differences in the strengths achieved.

2 Materials and methods

2.1 Materials

The chemical oxide compositions of the CKD, PFA, and cement used, as measured by XRF, are shown in Table 1. The CKD used had a high chloride content of 11.68%, making it unsuitable for many applications, and almost 50% of the CKD was composed of CaO.

Table 1. Oxide composition of materials

Constituent	CKD (%)	PFA (%)	CEM I (%)
CO_2	1.27	5.40	
Water	1.06		
LOI	2.33	5.40	3.00
SiO_2	8.16	51.77	19.83
Al_2O_3	3.68	21.54	4.80
TiO_2	0.12	0.94	
P_2O_5	0.12	0.68	
Fe_2O_3	1.62	6.28	3.02
MgO	1.46	2.06	
CaO	47.98	5.53	63.06
SO_3	7.04	0.84	2.48
Na_2O	3.62	2.55	
K_2O	14.64	2.40	
Cl^-	11.68		

To assess the reactivity of the CKD, the material was analysed by X-ray diffraction (XRD) Rietveld, using the external G-factor method [3], and the resulting mineralogical composition shown in Table 2. It was identified that the main phases present in the CKD were free-lime, sylvite, and belite. Due to the availability of the CaO mainly as free-lime, rather than calcium carbonate, the activation of the PFA becomes possible.

The main crystalline phases of the PFA were identified to be quartz and mullite, both of which are considered unreactive. The reactivity potential of the PFA lies within the amorphous content, which is likely to be composed mainly of aluminosilicate glasses. As quantified by XRD-Rietveld, the amorphous content of the PFA was found to be 73.9%. To ensure consistency between samples, the PFA was dried at 105°C prior to use, while the CKD and cement were used as received.

Table 2. Mineralogical phase composition of CKD

Mineralogical phase	Chemical composition	Phase content (%)
Free lime	CaO	29.5
Sylvite	KCl	21.4
Belite	Ca_2SiO_4	17.5
Dolomite	$CaMg(CO_3)_2$	6.0
Portlandite	$Ca(OH)_2$	3.8
Anhydrite	$CaSO_4$	3.7
Arcanite	K_2SO_4	8.4
Ferrite	$Ca_2(Al,Fe)_2O_5$	5.2
Quartz	SiO_2	3.1
Syngenite	$K_2Ca(SO_4)_2 \cdot H_2O$	1.3

2.2 Methods

The mix designs were selected based on maximising the use of CKD in the mix, with the used mix designs shown in Table 3. In order to achieve sufficient compaction, a dry mix was preferred; as such, a w/b ratio of 0.2 was selected to be used for all mixes. However, two of the mix designs (mixes C and J) were tested at varying w/b ratios, up to a w/b ratio of 0.4, to investigate the effects of moisture content on the compressive strength.

Table 3. Mix designs

Mix Design	PFA	CKD	CEM I
A	90	10	
B	80	20	
C	70	30	
D	60	40	
E	50	50	
F	60	38	2
G	50	47.5	2.5
H	60	36	4
I	50	45	5
J		95	5
K		90	10
L		80	20
M		70	30

The samples were prepared by firstly blending the dry materials, and then hand mixing in a tray with the water added. Following this, the mixes were then passed through a sieve with a 5mm mesh size to ensure that the compaction was not affected by particle size. As the water added was very little, the material remained somewhat dry as shown in Figure 1a.

Fig. 1. (a) dry mix used for sample preparation, and (b) sample wrapped in cling film for storage

The material was then compacted in seven layers into a 100 mm by 50 mm diameter cylindrical mould. After demoulding immediately, each sample was wrapped in cling film to prevent any loss of moisture during curing, as shown in Figure 1b. At the specified test age of 56 days, the unconfined uniaxial compressive strength was measured at a displacement rate of 1mm/min. Each strength measurement represents the average strength measured of 3 replicate samples.

To understand the effects of the mineralogy on the compressive strength, the hydrated phases of four of the mix designs were analysed at 56 days by XRD and thermogravimetric analysis (TGA). These samples were prepared with a higher w/b ratio of 0.35, and were ground to a powder and tested immediately without any ceasing of hydration to prevent any damage to crystalline structures. XRD data were collected using a PANAlytical XPert Powder diffractometer in the Bragg-Brentano geometry with tube operating conditions of 45 kV and 40 mA, and CuKα radiation with a beam wavelength of 1.5418 Å. Thermogravimetric analysis (TGA) was used in combination to identify any amorphous phases within the samples, using a TG 209 F1 Libra, with the sample placed in a flowing nitrogen atmosphere from 20°C to 950°C, and a heating rate of 20C/min.

3 Results and discussion

3.1. Unconfined compressive strength

The unconfined compressive strength of the PFA and CKD blends (mixes A-E) with a w/b ratio of 0.2 at 56 days are shown in Figure 3. It is evident that all samples surpassed the required 3 MPa with very little variation, reaching strengths of between 4.7 and 5.6 MPa, except mix design A which only gained 2.14 MPa. It is likely that in mix A the CKD content was insufficient to activate the PFA.

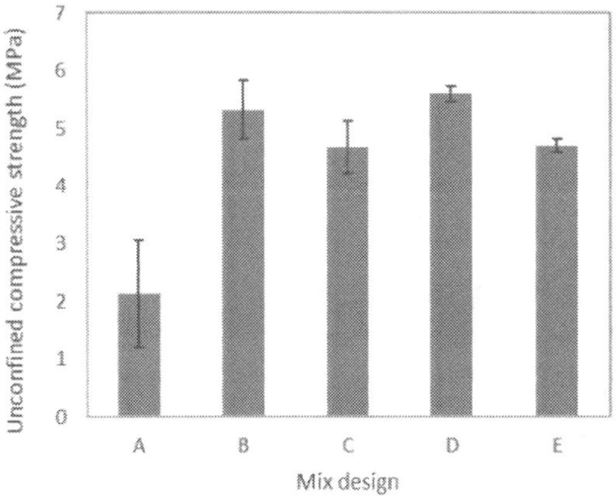

Fig. 3. Compressive strength of PFA and CKD blends with a w/b ratio of 0.2 at 56 days

Figure 4 illustrates the unconfined compressive strength of mixes F – I, which are composed of PFA, CKD, and cement. It is apparent that the addition of small amounts of cement to the mixes did not improve the strength; in fact, a slight drop in the strength was observed. This drop in the strength is likely due to a lack of water availability to fully hydrate the cement, resulting in inert material remaining in the mixes.

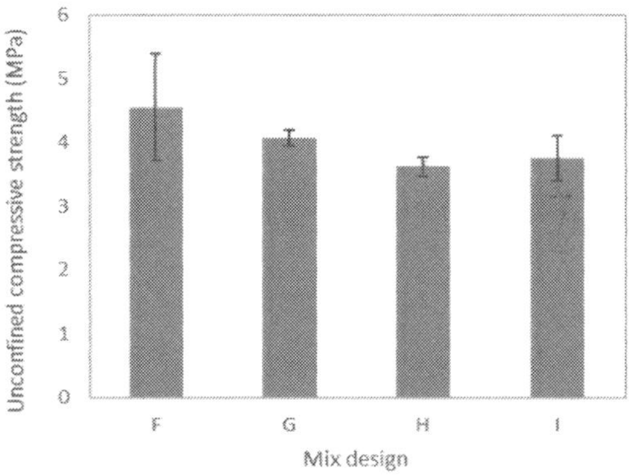

Fig. 4. Compressive strength of PFA, CKD, and cement blends with a w/b ratio of 0.2 at 56 days

In the aim of maximizing the CKD content, CKD and cement only blends were investigated, with the unconfined compressive strengths shown in Figure 5. When blending 95% CKD and only 5% cement, a strength of 2.8 MPa was achieved, which was insufficient to meet the strength requirement of 3 MPa. However, samples with greater than 5% cement surpassed the strength requirement, reaching strengths of up to 5.2 MPa when 30% cement was blended. However, additions of these levels of cement in the blends are unlikely to be economic in the current application.

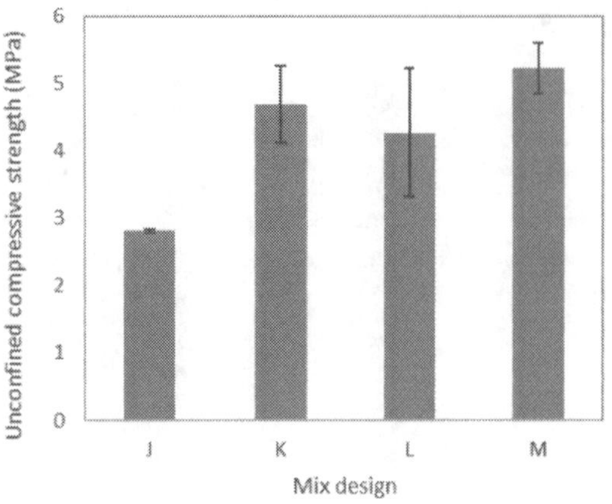

Fig. 5. Compressive strength of CKD and cement blends with a w/b ratio of 0.2 at 56 days

Two of the mix designs, mix C and mix J, were selected to be used for investigating the impact of the w/b ratio on the unconfined compressive strength. Figure 6 illustrates an improvement in the strengths when higher w/b ratios are used. This is more apparent in the mix design containing PFA, where the increased water would result in greater activation of the aluminosilicates in the PFA. On the other hand, increased w/b ratio had less of an impact on the mix design containing cement. As the w/b increases in this mix, it is likely that much of it is uptaken by the free-lime in the CKD, leaving little to fully hydrate the cement in the mix. However, in both cases, it is evident that the water content played a role in the strengths achieved.

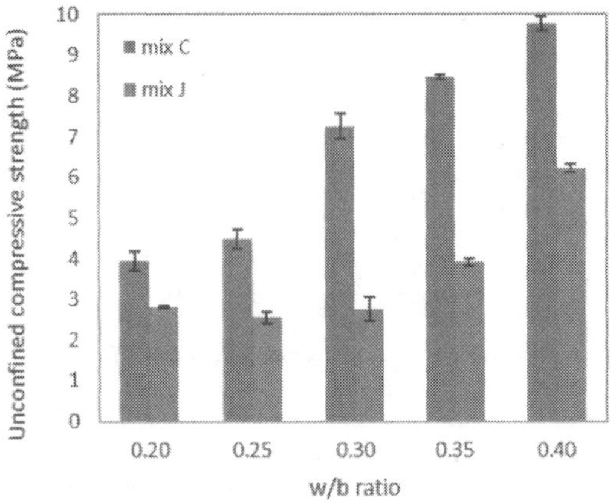

Fig. 6. Compressive strength of mix C and mix J varying w/b ratio at 56 days

3.2 Hydration products

To investigate the activation potential of the CKD, and the further understand the strengths achieved, the hydration products were investigated on four of the mix designs containing CKD and PFA at 56 days, with the XRD patterns shown in Figure 7. The broad "hump" between ~15° and ~35° 2θ, which is associated with the amorphous content is most apparent in mix B, and appears to reduce with increasing CKD content. This is likely a result of two factors: (i) the reduction in the PFA content reduces the final amorphous content of the mix, and (ii) increasing the CKD content provides conditions for higher activation and reactivity of the PFA.

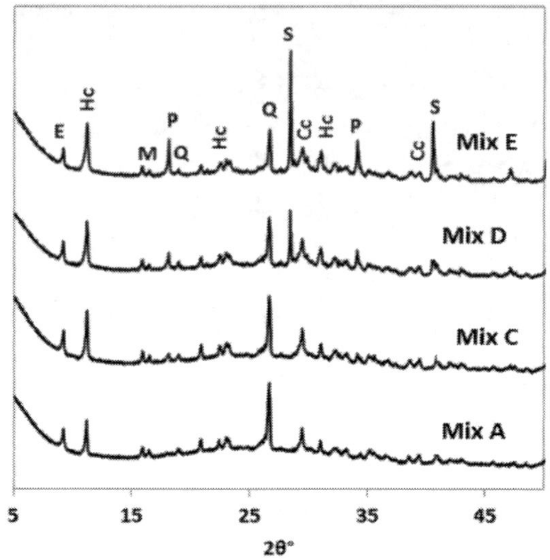

Fig. 7. XRD patterns of hydrated PFA and CKD blends at 56 days, where E = Ettringite, Hc = Hemi-carbonate, M = Mullite, P = Portlandite, Q = Quartz, S = Sylvite, and Cc = Calcium carbonate.

At 56 days, the free-lime contained within the CKD was fully consumed in all cases. Upon initial contact with the mixing water, the free-lime is rapidly converted to calcium hydroxide (Portlandite). Increasing the CKD content results in higher calcium hydroxide formation and higher alkalinity of the pore solution; this provides suitable conditions for the pozzolanic reaction with the aluminosilicate glasses in the PFA to form C-S-H phases. The unconfined compressive strength results were in agreement with this, whereby higher strengths were achieved in the higher CKD content blends with PFA.

However, the XRD patterns show that the amount of residual calcium hydroxide at 56 days increases as the CKD content increased. This suggests that the calcium hydroxide exceeded the amount required to react with the PFA at 56 days. It is possible that the pozzolanic reaction of the PFA will continue to progress slowly with further curing time. On the other hand, the reduction in the strength with the addition of CKD past 40% suggests rather that the optimum CKD for reaction with the PFA has been surpassed; it is likely that the total pozzolanic reaction is close to its full potential.

https://doi.org/10.1051/matecconf/201814901072

Unreactive quartz and mullite remained present in all samples at 56 days. While sylvite was almost completely absent from the samples containing only 20 and 30% CKD, the residual sylvite content at 56 days increased with further CKD addition. At the lower CKD contents, uptake of the small amounts of potassium and chloride ions present into the C-S-H structure may have been possible.

The main crystalline product of CKD-PFA blends has been previously been observed to be ettringite [4], which was also formed in all samples in the present study. Ettringite formation occurs due to reaction of the sulfate bearing phases, anhydrite and arcanite, with the alumina from ferrite or the PFA. Ettringite is known to contribute to strength gain as it has a lower density (1.78 g cm^{-3}) than many other hydrates in cements, and is suggested to have good space filling properties due to the dense packing of its characteristic needle-like crystals [5].

In addition to this, XRD showed the precipitation of hemi-carboaluminate and calcium carbonate in all samples, the carbonate being provided from the PFA.

The TGA-DTG results, shown in Figure 8, confirmed the presence of ettringite and calcium hydroxide, in addition to the formation of amorphous aluminium hydroxide (AH_3) gel. As the CKD content in the binder increased, the aluminium hydroxide gel was observed to increase due to increased dissolution of the aluminosilicates in the PFA.

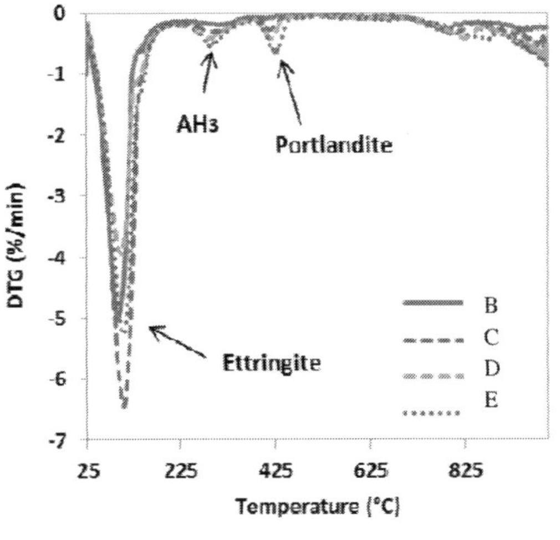

Fig. 8. DTG plots of hydrated PFA and CKD blends at 56 days

The presence of C-S-H could not be directly identified in the DTG plots as the dehydration of this phase typically occurs within the same range of temperatures as ettringite and continues to loss mass across a wide range of temperatures. However, the floating baseline in the DTG plot, and the lack of any crystalline silicate bearing phases in the XRD patterns, suggested that amorphous C-S-H had formed.

4 Conclusions

The present study showed that CKD, containing high free-lime content, has the potential to activate the pozzolanic reaction of PFA and provide sufficient mechanical properties for low strength applications. The main hydration products were identified to be ettringite, calcium carbonate, hemi-carboaluminate, portlandite, and amorphous C-S-H and aluminium hydroxide.

Aiming to maximise the use of CKD at a water to binder (w/b) ratio of 0.2, it was found that the maximum CKD content possible to achieve the required strength was 90% CKD blended with 10% cement. However, it is unlikely to be economic to use cement levels this high, when similar strength can be achieved using only CKD and PFA. On the other hand, when aiming to maximise the strength gained, this was achieved upon increasing the w/b ratio to 0.35, which corresponded to the maximum dry density of the material.

Overall, the results confirm the suitability of using CKD as a backfill material for underground mines. However, some practical considerations should be made regarding its use. Firstly, the high water demand of the mixes due to the free-lime in the CKD, which absorbs water rapidly, produces a large amount of heat which may be difficult to work with in large quantities. In addition to this, the fineness of the material makes it difficult to deal with below the surface in the mines. To deal with this, the mix could be prepared above ground and pumped below, which would require much initial investment. Alternatively, the CKD material can be pre-treated with moisture prior to bringing below ground.

Further work will aim to investigate the microstructure of the hydrated pastes, and to quantify the hydrates formed over time to understand the hydration kinetics of these binders. Moreover, taking into account the practical considerations, early trials on site to investigate the use of the material for the backfill of mines will be carried out.

5 Acknowledgements

The authors wish to acknowledge the Knowledge Transfer Partnership (KTP) 010446 between Quinn Building Products Ltd and Queen's University Belfast.

References

1. D. Bondar, E. Coakley, Construction and Building Mat., **71** (2014).
2. M. Sadique, E. Coakley, Adv. Cem. Res. **28** (2016).
3. D. Jansen, Cem. Conc. Res. **41** (2011).
4. K. Wang, S.P. Shah, A. Mishulovich, Cem. Conc. Res. **34** (2004).
5. F.P. Glasser, L. Zhang, Cem. Conc. Res. **31** (2001).

Mineralogical stabilization of Ternesite in Belite Sulfo-Aluminate Clinker elaborated from limestone, shale and phosphogypsum

K. BEN ADDI[1], A. DIOURI[1], N. KHACHANI[1], A. BOUKHARI[1]
[1] Laboratory of Applied Solid State Chemistry, Faculty of science, University Mohammed V-Rabat, Morocco

*Corresponding authors: khadijabenaddii@gmail.com, abdeljebbar.diouri@gmail.com

Abstract. This paper investigates the mineralogical evolution of sulfoaluminate clinker elaborated from moroccan prime materials limestone, shale and phosphogypsum as a byproduct from phosphoric acid factories. The advantage of the production of this type of clinker is related to the low clinkerisation temperature which is known around 1250°C, and to less consumption quantity of limestone thus enabling less CO_2 emissions during the decarbonation process compared to that of Portland cement. In this study we determine the stability conditions of belite sulfoaluminate clinker containing belite (C_2S) ye'elimite ($C_4A_3\$$) and ternesite ($C_5S_2\$$). The hydration compounds of this clinker are also investigated. The monitoring of the synthesized and hydrated phases is performed by X-Ray Diffraction and Infrared spectroscopy. The results show the formation of ternesite at 800°C and the stabilization of clinker containing y'elminite, belite and ternesite at temperatures between 1100 and 1250°C.

1 Introduction

The literature relating to sulfo-aluminate cement is abundant, but remains much less vast than that dealing with Portland cement. Sulfo-aluminate cements are relatively recent hydraulic binders, these cements are still little known differently than by specialists. They present unquestionable assets and many potential applications. Moreover, the ecological advantages of these cements also arouse the interest of the researchers. Its use thus seems an alternative of choice allowing the reduction of gas emissions for greenhouse effects, and constitutes one of the replacement solutions partial or total Portland cement for some applications.

The production of sulfo-aluminate cements indeed has two advantages compared to that of Portland cement: on the one hand, the temperature of firing is less high (approximately 1250°C) and on the other hand, the manufacturing of this cement requires the consumption of a less quantity of limestone, which reduces the CO_2 emissions during the decarbonation process, compared with that of Portland cement, the manufacturing of sulfo-aluminate cement makes it possible to reduce of almost 35% the carbon dioxide emissions [1]. Energy efficiency is also improved, as the lower burning temperature reduces fuel consumption by roughly 10% [1]. But knowledge on the properties related to its modified chemical and mineralogical composition are less varied and requires advantage research to better understand of the mechanisms, temperatures and phase transformation stages in these cements.

The manufacturing of sulfo-aluminate cement is carried out from a source of aluminuim like bauxite (40%) and of a source of sulphate (20%)[2]. The obtained sulfo-aluminate clinker is mainly composed of ye'elimite ($C_4A_3\$ \approx 65\%$), of bellite ($C_2S \approx 20\%$)(cement notation used throughout the document: A = Al2O3, C = CaO, F = Fe_2O_3, H =H_2O, K = K_2O, M = MgO, S = SiO_2, T = TiO_2, \$ = SO_3).

The formation of solid solutions of $C_4A_3\$$ is hindered by rich concentrations of additives, mainly due to the phosphorus effect that enlarges the stability field of the calcium aluminate phases at the expense of the calcium sulfoaluminate field. My.Y. Benarchid et al. [3] showed that the presence of the Cr_2O_3 and P_2O_5 oxides lowered the decomposition of $CaCO_3$. The enthalpy of decarbonation and the enthalpy of the formation of the solid solutions based on tricalcium aluminate formation decreased when the concentrations of additives increased. In another study [4] the authors showed that the maximum of the initial heat rate of the hydration process of $C_4A_3\$$ decreases with an increase in the concentration of Cr_2O_3 and P_2O_5 oxides. The incorporation of these oxides in the crystal lattice of $C_4A_3\$$ induces a long induction period during the hydration reaction and the principal products of hydration are ettringite, calcium monosulfate and gibsite.

The calcium sulfoaluminate phase, also named as Ye'elimite ($C_4A_3\$$), has been known as hydraulic phase for several decades, but its potential in non-expansive cementitious materials has only recently been reported.

As a result of many researches [5], an optimized formulation based on a clinker containing $C_4A_3\$$, belite (C_2S) and Ternesite ($C_5S_2\$$) was developed [6]. As a consequence, the clinkerization temperature of this type of cement is low (1200°C) compared to Ordinary Portland Cement (1400–1450°C). It is thus currently attracting a great deal of interest worldwide [7, 8]. Its advantages include: i) reduced energy consumption during manufacturing, ii) low CO_2 emission during manufacturing [9], iii) acceptable setting times and high-early strength development, [10], iv) very good durability, particularly in marine constructions [11].

The ternesite phase $C_5S_2\$$ is a calcium sulphosilicate known also as sulphospurrite, its formation is based on the presence of belite and anhydrite [6] according to the formula (1)

$$2C_2S + C\$ \longrightarrow C_5S_2\$ \qquad (1)$$

When ternesite is formed or stabilised at temperatures above 1200 °C, the use of a mineralizer such as phosphates or fluorides is required. Another possibility to form ternesite is a methodology, developed by Frank Bullerjahn and Al. [6], which is based upon a two-step clinkering procedure. In a first step, the clinker is sintered at a temperature of at least 1250 °C. In a second step, a slow cooling or so called "second clinkering" is applied by decreasing the temperature from 1200 to 800 °C.

Accordingly, the production of supposed ternesite belite calcium sulpho-aluminate ferrite clinkers offers facilitate potential for the beneficiation and the advancement of clinkers and bonds in view of calcium sulpho-aluminate. These clinker sorts could be produced utilizing a comparable process to belite calcium sulpho-aluminate clinkers, however inside a temperature scope of 900 to 1300°C.

2 Materials and methods

2.1 Mixture

The raw materials used in this study are: limestone, shale, and phosphogypsum. The chemical compositions determined by XRF are given in Table 1. In order to prepare the basic raw mixture, all of the raw materials are homogenized during 4 minutes in a disc crusher. The final mixtures were homogenized manually by using one ceramics mortar, a rammer and ethanol as a dispersive media to improve homogeneity of sample.

Table 1: Chemical composition of raw materials (Wt %)

	Limestone	Shale	Phosphogypsum
CaO	50,81	5,05	30,45
SiO_2	7,67	35,34	9,50
Al_2O_3	2,41	14,21	2,80
Fe_2O_3	1,15	5,23	0,90
SO_3	0,26	0,15	42,9
MgO	0,57	1,88	0,30
K_2O	0,16	0,30	0,10
P_2O_5	0,29	0,11	0,63
LOI	36,68	37,73	12,42

The basic mixture made by the starting materials was carried out by a first approximation calculating raw believed to have a clinker at an approximate mineralogical composition wherein the produced phases would be within the average values of : 50% of belite (C_2S), 30% of ye'elimite ($C_4A_3\$$) and 20% of ternisite ($C_5S_2\$$). The calculation is based, at a first approximation, on the quantitative distribution of the oxides CaO, SiO_2, Al_2O_3 and SO_3 of starting materials in the C_2S, $C_4A_3\$$ and $C_5S_2\$$ phases planned in the target clinker. Determining the composition of the mixture to achieve, from the chemical composition of raw materials, gives the intended composition shown in Table 2.

The clinkering of the samples were carried out in platinum crucible in a laboratory furnace at different temperatures: 800°C (24h), 1000°C (12h), 1100°C (12h), and 1250°C (6h), with an intermediate crushing in order to increase the homogeneity. The mineralogical evolutions of the various elaborated clinkers and their pastes are carried out using XR-Diffraction and IR spectroscopy. To do this, the hydrated samples were crushed, immersed in ethanol to stop the hydration and then dried in an oven set. In order to investigate the hydraulic behavior of sulfo-aluminate belite clinker, the synthesized clinkers were hydrated for up to 90 days at ambient temperature with the w/c = 0.5.

Table 2: Anticipated composition of raw basic mixture

	Limestone	shale	Phosphogypsum
Wt %	55.04	37.42	7.54

2.2 Characterization

X-ray Fluorescence (XRF) is a widely used technique for both qualitative and quantitative elemental analysis of solids, powders. The pearls were prepared by adding 5g of lithium titraborate (melted) to 1g of crushed and grounded powder, the homogenized mixture was placed into a Platinum crucible and brought up to the complete melting until pearl level; the melt is poured into the mold and cooled up until the pearl detached from the mold.

https://doi.org/10.1051/matecconf/201814901073

X-ray diffraction analysis is performed in order to examine the mineralogical composition of the samples. The diagrams were collected using a Siemens D 500 Powder diffractometer equipped with a copper anticathode (CuK=1.5418 Å) at scanning speed of 0.04°.s-1 from 10 to 60°.

The Infra-red Spectroscopy with Transform of Fourier (or FTIR: Fourier Transform Infra-Red Spectroscopy) is based on the infra-red absorption of radiation by analyzed material. The absorbed frequencies by molecular vibration are ranging between 4000 and 400 cm-1. The used abbreviation to describe studied samples is detailed in Table 3

Table 3: Designation of burned and hydrated samples

Burned Sample designation	**CSA1**	**CSA2**	**CSA3**	**CSA4**
Burning température (°C)	800	1000	1100	1250
Hydrated Sample designation	**CSA5**	**CSA6**	**CSA7**	**CSA8**
Hydration time (Days)	2	7	28	90

3. 3 Results and discussion

3.1 Elaboration and characterization of belite sulfo-aluminate clinker

Different phases are identified by X-ray diffraction for the anhydrous samples CSA1-CSA4. The spectra are presented in figure 1 and their percentages are respectively collected in table 4. These results are obtained by applying the Rietveld method for the determination of semi-quantitative proportions of present phases at each temperature. The observed mineralogy of the synthesized clinker, applying the sintering process described above, displays a composition close to that expected, showing a tendency to form a β-C_2S polymorph of dicalcium silicate as a main formed phase observed at 2θ = 32,50° [12, 13], while $C_4A_3\$$ is present in orthorhombic and cubic forms. Bullerjahn Frank et al. found the same phases using calcite powder, siliceous fly ash, and natural anhydrite as starting materials [6].

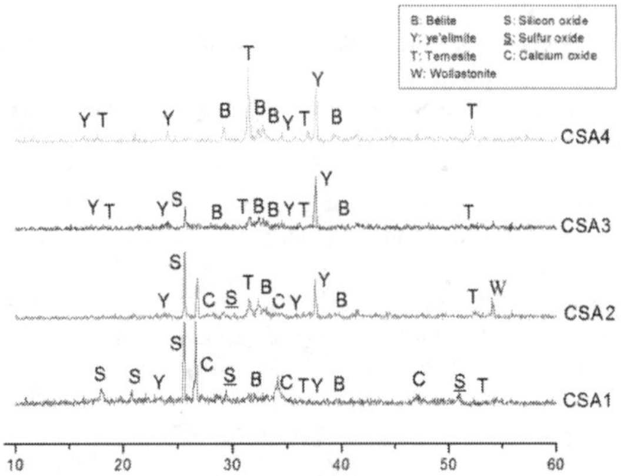

Figure 1: X-ray diffraction spectra of the synthesized clinker CSA1-CSA4

The decomposition of limestone is observed at 800°C, lower temperature compared to those typically known for this reaction (860°C), this result was also observed by Y. WANG et al. [14]. Comparing the amounts of the formed phases during clinkerization we observe that increasing the temperature above 1000°C leads to the reduction and cancellation of the amounts of CaO and SiO_2 and formation of the intermediate phase $CaSiO_3$ between 1000 and 1100°C according to the reaction (2).

$$CaO + SiO_2 \rightarrow CaO.SiO_2\ (CS) \qquad (2)$$

Then, the percentage of C_2S grows beyond 1100°C according to the reaction (3) [15].

$$CaO.SiO_2 + CaO \rightarrow 2CaO.SiO_2\ (C_2S) \qquad (3)$$

Table 4: Semi-quantitative identification of mineral phases in synthesized clinker CSA1-CSA4

	CSA1	**CSA2**	**CSA3**	**CSA4**
C_2S	11,51	53,09	49,01	52,92
$C_4A_3\$$	15,85	22,02	24,96	28,09
$C_5S_2\$$	12,60	17,80	18,45	18,90
CS	_	7,47	0,08	_
CaO	37,76	1,06	_	_
SiO_2	13,70	2,05	_	_
SO_3	8,5	3,01	_	_

The three phases C_2S, $C_4A_3\$$ and $C_5S_2\$$ form in small amounts from the temperature of 800°C, their quantities increase until the order of 53% for C2S, 28% for $C_4A_3\$$ and not exceed 18 to 20% for the ternesite phase, during all the treatment process up to 1250°C. It appears clearly that beyond the temperature of 1100°C, there is a joint stability of the three phases C_2S, $C_4A_3\$$ and $C_5S_2\$$, this result is of agreement with the many previous published results [6, 16-20]. The appropriate burning condition of pure ternesite, is found to be 1200°C for 8 h. The secondary heat treatment between 1100 and 1250°C can significantly promote the formation of ternesite from the reaction of belite and anhydrite in sulfo-aluminate clinkers[18]. The increase of the SO_3 content in the raw meal leads to the replacement of large amounts of C_2S by ternesite, which is stable in the temperature ranging between 900 and 1250°C [6].

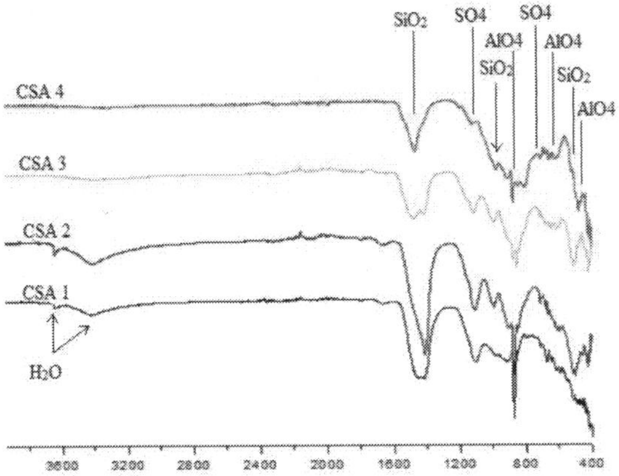

Figure 2: Infrared spectra of CSA1- CSA4 sample

The ternesite phase formed as a second major phase instead of C_2S and CS, in addition to $C_4A_3\$$. Upon hydration, this cement produced very high mechanical strength at early and later ages [20].

The infrared absorption spectra of CSA1-CSA4 samples reported in figure 2 and table 5 showed that the major absorption bands of the clinker phases are in the region 1600 to 400 cm-1. The main vibration bands of the different groups are listed in Table 5. The Si-O2 stretching mode produces an absorption band at 991,879, and 847 cm-1 in C_2S [21], while a broader band with peaks observed at 500 and 1000 cm-1 is assigned to the same stretching mode in $C_5S_2\$$. The SO4 groups are identified at 610, 613, 620 cm-1 in $C_4A_3\$$ [22, 23], while bands with peaks observed at 1100 and 1150 cm-1 are assigned to this vibration in $C_5S_2\$$. The AlO4 groups vibration identified at 420, 600, and 870 cm-1 are assigned to C4A3$ phase.

Table 5: Major vibration bands observed by IR spectroscopy in CSA1- CSA4 samples

	CSA1	CSA2	CSA3	CSA4
Group SiO_2	1000 f	1000 M	1000 M	991 f
Group SiO_2	879 f	879 f	847 f	847 f
Group SiO_2	500 F	500 M	500 M	500 F
Group AlO_4	420 f	420 M	420 M	420 f
Group AlO_4	600 f	600 f	600 f	600 f
Group AlO_4	870 F	870 F	870 M	870 M
Group SO_4	610 f	613 f	615 f	620 f
Group SO_4	1100 M	1100M	1150 f	1150 f
Group H_2O	3620 f	3620 f		
Group H_2O	3400 f	3400 f		

Group CaO	1400 F	1400 F	1450 F	1450 F

3.2 Hydration

The hydrated samples CSA5-CSA8 of the cement pastes has been investigated at 2 days, 7days, 28 days, and 90 days. XRD technique showed the formation of hydration product of ye'elimite, the ettringite phase characterized by the diffraction peaks of maximum intensities observed at 2θ = 9,65 (Figure 3 and Table 6) according to the reaction (4).

$$C_4A_3\$ + 2C\$ + 38H \rightarrow C_3A.3C\$.32H + 2AH_3 \quad (4)$$

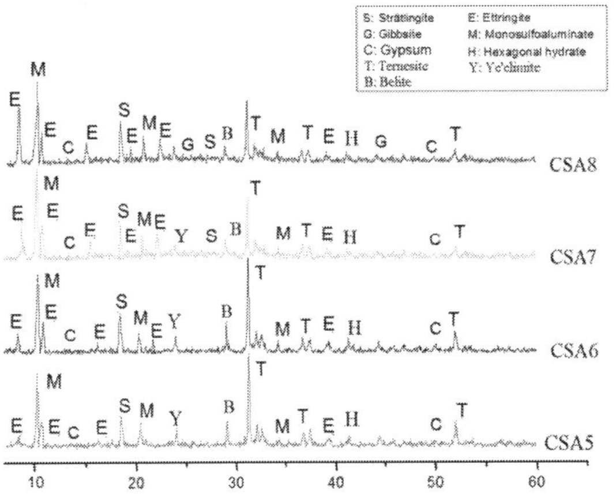

Figure 3: X-ray diffraction spectra of the hydrated samples CSA5-CSA8

The diffraction peaks relating to Gibbsite are identified at 2θ = 44,82 [15], and the peaks of monosulfate are detectable at 2θ = 10,3 , indicating the partial decomposition of ettringite and its transformation into monosulfate. This result is agrees with previous published results [18, 24, 25]. The strätlingite, C_2ASH_3, is formed from the reaction of belite which consumes a portion of AH3 produced by hydration of ye'elimite according to the reaction (5)

$$C_2S + AH_3 + 5H \rightarrow C_2ASH_3 \quad (5)$$

Table 6: Semi-quantitative identification of mineral phases in hydrated samples CSA5-CSA8

	CSA5	CSA6	CSA7	CSA8
Calcuim slicate C_2S	8,68	4,43	1,93	0,53
Ye'elimite $C_4A_3\$$	6,27	3,21	1,32	0,94
Ternesite $C_5S_2\$$	17,22	17,17	17,02	7,85
Ettringite $C_6A\$_3H_{32}$	22,36	28,54	29,93	39,34
Monosulfoaluminate $C_4A\$H_{12}$	8,81	10,41	11,53	20,44
Gibbsite AH_3	16,61	9,26	6,74	1,24
Strätlingite, C_2ASH_3	11,45	19,56	24,43	24,65

Hexagonal hydrates	4,47	4,43	4,34	3,95
Gypsum C$H	4,11	2,98	2,73	1,05

The same results, deduced here, are obtained by Y.Shen and al.[18] which have shown that the pure ternesite can hydrate at a slow rate, and the addition of ye'elimite promotes the hydration of ternesite and the formation of ettringite. The ettringite formation for the clinker containing ternesite mainly depends on the dissolution of gypsum originating from the hydration of ternesite.

Figure 4 and table 7 shows the infrared spectra of the hydrated samples CSA5-CSA8, main infrared vibrations are summarized in Table 7. We identify the vibrations corresponding to the group (SO_4) in the ettringite $C_6A\$_3H_{32}$ at 1100 cm-1 and in the monosulfoaluminate hydrated calcium $C_4A\$H_{12}$ at 420, 630 and 1000 cm-1 like it has been shown by some authors [26, 27]. Absorption localized at 530 cm-1 is due to binding of vibration (AlO_6) in the phase of ettringite, these results are the same obtained by M. Idrissi et al. [15]. on calcium sulfoaluminate phase elaborated by iron inclusion.

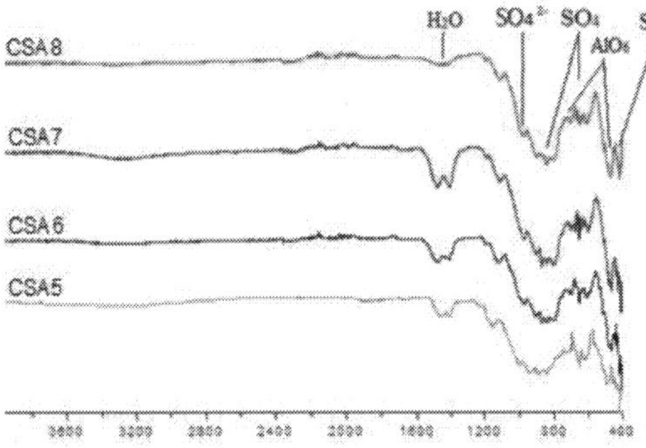

Figure 4: Infrared spectra of hydrated samples CSA5-CSA8

According to Table 6 we found that the percentage of increased ettringite in the sample CSA8 from 22% (at 2 days) to 39% (at 28 days), it may be due to the hydration of ternesite phases beyond 28 days, this result agrees with with previously mentioned results.

Previous studies reported that the ternesite is least reactive in pure ternesite system, the hydration of ternesite does not take place until 90 days, and gypsum has formed as new crystalline phase. The hydration process of ternesite starts by its partial dissolution, and after 28 days of hydration, only traces of ettringite are detectable while gypsum produced from the hydration of ternesite has almost been depleted [18], but in the presence of Yeelimite the ternesite hydration favors also the formation of gypsum and ettringite. [6].

Gallardo et al. [28] and Velazco et al. [29] were shown in the previously published results that the ettringite as the main hydration product although traces of monosuphate phase were also found. The addition of citric acid increased the setting time, modified the morphology of the ettringite needles, changed the microstructural configuration and prevented the decreasing in compressive strength of the samples caused by delayed ettringite formation, also Bazaldúa et al. [30] demonstrated that the main hydration products of supersulphated cements were C-S-H and Ettringite. These products precipitated since the first hours of hydration, forming a porous and weak microstructure; with time and up to one year, the C-S-H and Ettringite consolidated the microstructure and increased the compressive strength. The cements showed hydraulic behavior after curing under water.

Table 7: Major vibration bands observed by IR spectroscopy in CSA5-CSA8 samples

	CSA5	CSA6	CSA7	CSA8
SO_4 corresponding to the ettringite hydrated calcium monosulfoaluminate	410 M	410 M	410 M	410 M
SO_4 corresponding to the ettringite hydrated calcium monosulfoaluminate	630 f	630 f	630 f	630 f
AlO_6 corresponding to the ettringite and strätlingite	520 M	520 M	530 f	530 f
AlO_6 corresponding to hexagonal hydrates	780 M	780 M	770 f	770 f
SO_4^{2-} corresponding to the ettringite	1100 f	1100 f	1100 f	1100 f
H2O	1580 M	1580 M	1580 f	1580 f

Frank Bullerjahn et all. [6] found that the $C_5S_2\$$ phase in the clinker is more reactive than the C_2S, unlike previous studies that considered the ternesite as an inert phase or showing a low reactivity [19, 31-35]. These studies conclude on the low reactivity of ternesite based on the chemical bound water and the low quantities of hydrates formed, which were present in trace amounts only. It has also been reported [36-37] that $C_5S_2\$$ has better hydraulic properties after autoclaving, it transforms to a rapidly hardening and high strength compound [38]. Additionally, the hydration reaction of calcium sulphoaluminate initiates rapidly and forms ettringite and AH_3 [39], contributing to faster setting of CSA cement than Portland cement [40].

All these results indicate that ternesite can be hydrated at a long-term, adding yeelimite not only induces hydration ternesite but also favors the formation of gypsum [6, 18], this gypsum in the pastes leads to ettringite formation and reduced the amount of formed monosulfoaluminate [40].

In this work we have clearly shown the formarion and joint stability at relatively moderate temperatures (1100-1250°C) of the three phases C_2S, $C_4A_3\$$ and $C_5S_2\$$. The hydration of this clinker shows the formation of usually known phases in relatively high proportions for Ettringite $C_6A\$_3H_{32}$ (22%) from the second day and the formation of monosulfoaluminate $C_4A\$H_{12}$ and strätlingite C_2ASH_3 in large amounts at 28 days.

4 Conclusions

The sulfoaluminate cements bring innovating technical solutions, those hydraulic materials find more uses in building and civil engineering by presenting a favorable ecological assessment. This work concerned the sulfoaluminate clinker characterized by the presence of C_2S (belite), $C_4A_3\$$ (Ye'elimite) and of $C_5S_2\$$ (ternesite). The clinker is elaborated by sintering raw materials at a temperature around 1100-1250°C. The blended raw material is made up of a source of alumina (Shale), limestone and a source of sulphate calcium (phosphogypsum derived from phosphoric acid factories). The sulfoaluminate clinker is characterized by a carbonic reduction from approximately 35% due to the use of less limestone and lower clinkerisation temperature. The formation of ternesite is observed at least of 1100°C, that is related to clinkering process. We observe also the stabilization of this phase at 1200°C. The hydration process shows that all hydrate phases are formed in the first two days except the ternesite hydrated phase which takes place at later ages (beyond 28 days).

References

1. E. Gartner. Industrially interesting approaches to "low-CO2" cements, Cement and Concrete Research, 34, 1489–1498 (2004).
2. N. Fukuda. On the Constitution of Sulfo-Aluminous Clinker, Cement Division, 34 [1], 138–139 (1960) .
3. My.Y. Benarchid , A. Diouri, A. Boukhari, J. Aride, R. Castanet, J, Rogezc. Thermal study of chromium–phosphorus-doped tricalcium aluminate, Cement and Concrete Research, 31 [3], 449–454 (2001).
4. My.Y. Benarchid, J. Rogez, A. Diouri, A. Boukhari, J. Aride, Formation and hydraulic behavior of chromium–phosphorus doped calcium sulfoaluminate cement, Thermochimica Acta, 433, 183–186 (2005) .
5. W. Yanmou, S. Muzhen, Microstructural study of the interfacial zone between expansive sulphoaluminate cement pastes and limestone aggregates, Cement and Concrete Research, 26 [5], 805–812(1996).
6. F. Bullerjahn, D. Schmitt, M. Ben Haha, Effect of raw mix design and of clinkering process on the formation and mineralogical composition of (ternesite) belite calcium sulphoaluminate ferrite clinker, Cement and Concrete Research, 59, 87–95(2014).
7. K. Quillin. Performance of belite-sulfoaluminate cements, Cement Concrete Research, 31 [9], 1341–1349(2001).
8. R. J. Mangabhai, Calcium Aluminate Cement, E. & F.N. SPON, London, (1990).
9. J.H. Sharp, C.D. Lawrence, R. Yang, Calcium sulfoaluminate cements-low-energy cements, special cements or what?, Advances in Cement Research. 11 [1], 3–13(1999).
10. Y. Wang, J. Deng, An investigation into cement $CaO-SiO_2-Al_2O_3-Fe_2O_3- SO_3$ system, Proc. of the 8th Inter. Congress on the Chemistry of Cement. Rio. De Janeiro., sub theme 1.3, 11, 300-305(1986).
11. FP. Glasser, L. Zhang, High-performance cement matrices based on calcium sulfoaluminate–belite composition. Cement and Concrete Research, 31 [12], 1881–1886(2001).
12. H.Y. Ghorab, M. Rizk, B. Ibrahim, M.M. Allam, High belite cement from alternative raw materials. Materiales de Construcción, 64 [314], e012 (2014).
13. A. Diouri, A. Boukhari, J. Aride, F. Puertas, T. Vázquez, Elaboration of αL'- C2S form of belite in phosphatic clinker, Study of hydraulic activity. Materiales de Construcción. 48 [249] (1998).
14. L. Zhang, M. Su, Y. Wang, Development of the use of sulfo- and ferroaluminate cements in China, Advances in Cement Research, 11 [1], 35–41 (1999).
15. M. Idrissi, A. Diouri, D. Damidot, J.M. Greneche, M. Alami Talbi, M. Taibi, Characterisation of iron inclusion during the formation of calcium sulfoaluminate phase, Cement and Concrete Research, 40 [8], 1314–1319(2010).
16. F.R.D. Andrade, S.D. Gomes, M. Pecchio, Y. Kihara, F.M.S. Carvalho, J.R. Matos, Effect of sulfur on the polymorphism and reactivity of dicalcium silicate of Portland clinker, Cerâmica, 57, 129–135 (2011).
17. B. Ziemer, B. Altrichter, Effect of SO_3 on formation and hydraulic reactivity of belite, Cement and Concrete Research, 14 [5], 686–692 (1984).
18. Y. Shen, J. Qian, Y. Huang, D. Yang, Synthesis of belite sulfoaluminate-ternesite cements with phosphogypsum, Cement and Concrete Composites, 63, 67-75 (2015).
19. G. Belz, J. Beretka, M. Marrocoli, L. Santoro, N. Sherman, G.L. Valenti, Use of Fly Ash Blast Furnace Slag and Chemical Gypsum for the Synthesis of Calcium Sulfoaluminate-Based Cements, ACI materials journal, 153, 513–530 (1995).
20. N. Sherman, J. Beretka, L. Santoro, G.L, Valenti, Long-term behaviour of hydraulic binders based on calcium sulfoaluminate and calcium sulfosilicate, Cement and Concrete Research, 25, 113–126 (1995).
21. T.L. Hughes, C.M. Methven, T.G.J. Jones, S.E. Pelham, P. Fletcher, C. Hall, Determining Cement Composition by Fourier Transform Infrared Spectroscopy, Advanced Cement Based Materials, 2 [3], 91–104 (1995).
22. P. Zhang, Chen, L. Shi, G. Zhang, W. Huang, J. Wu, Proceedings of the Ninth, The 14th International Congress on the Chemistry of Cement, New Delhi, India. 3, 201–208 (1992).
23. X. Liu, Y. Li, N. Zhang, Influence of MgO on the formation of Ca_3SiO_5 and $3CaO{\bullet}3Al_2O3{\bullet}CaSO_4$ minerals in alite sulphoaluminate cement, Cement and Concrete Research, 32 [7], 1125–1129 (2002).
24. B. Lothenbach, A. Nonat, Calcium silicate hydrates: Solid and liquid phase composition, Cement and Concrete Research, 78, 57-70 (2015).
25. S. Ioannou, L. Reig, K. Paine, K. Quillin, Properties of a ternary calcium sulfoaluminate– calcium sulfate–fly ash cement, Cement and Concrete Research, 56, 75–83 (2014).

https://doi.org/10.1051/matecconf/201814901073

26. J. Bensted, S.P. Varma, Studies of ettringite and its derivatives, Cem. Technol, (1971).
27. R.B. Perkins, C.D. Palmer, Solubility of ettringite Ca6(Al(OH)6) $2(SO_4)_3.26H_2O$ at 5-75°C. Geochim, Cosmochimca Acta, 63,1969–1980 (1999).
28. M. Gallardo, J.M. Almanza, D.A. Cortés, J.C. Escobedo, J.I. Escalante-García, Synthesis and mechanical properties of a calcium sulphoaluminate cement made of industrial wastes, Materiales de Construcción, 64 [315], e023 (2014).
29. G. Velazco, J.M. Almanza, D.A. Cortés, J.C. Escobedo, J.I, Escalante-Garcia Effect of citric acid and the hemihydrate amount on the properties of a calcium sulphoaluminate cement, Materiales de Construcción, 64 [316], e036 (2014).
30. Y. Shen, J. Qian, J. Chai, Y. Fan, Calcium sulphoaluminate cements made with phosphogypsum : Production issues and material properties, 48,67–74 (2014).
31. N. Sherman, J. Beretka, L. Santoro, G.L. Valenti, Long-term behaviour of hydraulic binders based on calcium sulfoaluminate and calcium sulfosilicate, Cement and Concrete Research, 25 [1], 113-126 (1995).
32. T.K. Tadzhiev, T.A. Atakuziev, F.K. Tadzhiev, Hardening of anhydrous calcium sulfoaluminate and sulfosilicate, UDC, 691 [54], 1434–1437 (1973).
33. M.Y. Benarchid, J. Rogez, A. Diouri, A. Boukhari, J. Aride, Formation and hydraulic behaviour of chromium–phosphorus doped calcium sulfoaluminate cement, Thermochimca Acta, 433 (1-2), 183–186 (2005).
34. V. Makhmudova, M. Iskandarova, Y. Ivanova, G. Chernev, N. Ruziev, Synthesis and properties of sulphoferrite calcium clinkers and low temperature cements on their basis, Journal of the University of Chemical Technology and Metallurgy, 46 [2], 151-154 (2011).
35. J. Beretka, M. Marrocoli, N. Sherman, G.L. Valenti, The influence of C_4A_3S content and WS ratio on the performance of calcium sulfoaluminate-based cements, Cement and Concrete Research, 26 [11], 1673–1681 (1996).
36. A.A. Pashchenko, Sulfate-containing Cements in 'New Cements', Kiev, Budivelnik, Russian, (1978).
37. T.K. Tadzhiev, T.A. Atakuziev, F.K. Tadzhiev, Investigation on the process of hardening of water free calcium sulphoaluminate and calcium sulphosilicate, Inorganic Materials, 9, 1634-1637 (1972).
38. N. Sherman, J. Beretka, L. G.L. Santoro Valenti, lqng-term behaviour of hydraulic binders based on calcium sulfoaluminate and calcium sulfosilicate, Cement and Concrete Research.,25 (1), 113-126 (1995).
39. V. Kasselouri, P. Tsakiridis, Ch. Malami, B. Georgali, C. Alexandridou, A study on the hydration products of a non-expansive sulphoaluminate cement, Cement and Concrete Composites, 25 [17], 26–36 (1995).
40. C.W. Hargis, A. Telesca, P.J.M. Monteiro, Calcium sulfoaluminate (Ye'elimite) hydration in the presence of gypsum, calcite, and vaterite, Cement and Concrete Research. 65, 15–20 (2014).

Effect of mechanical activation of fly ash added to Moroccan Portland cement

H. Ez-zaki[1*], A. Diouri[1], M. Maher[2], A. Aidi[2] and T. Guedira[2]
[1]Laboratoire de Chimie du Solide Appliquée (LCSA), Mohammed V University, Rabat, Morocco
[2]Laboratoire de matériaux, électrochimie et environnement, Ibn Tofail University, Kenitra, Morocco
[*]Corresponding Author: ezzaki86@gmail.com

Abstract. Nowadays, the cement industry is the largest emitter of CO_2. In 2015, cement production accounts for roughly 8% of global CO_2 emissions. In order to reduce this impact, cement plants are working on alternative solutions, for instance, producing cement by adding additives like fly ash known for reducing the emissions of CO_2 and minimizing production costs. The thermal power stations in Morocco produce more than 500 000 tons per year. For ecological and sustainable development reasons, it is desirable to recycle these quantities according to beneficial methods to their addition in the cement.
This study aims to investigate the influence of grinding fly ash on the physico-chemical and mechanical properties of fly ash blended CPJ45 cement. The addition of the fly ash particles to the grinder leads respectively to the breakage of the particles and to reduce the agglomeration effect in the balls of cement grinder. Fly ash milling was found to improve particles fineness, and increase the silica and alumina content in the cement. Furthermore, milled fly ash blended cements show higher compressive strength compared to unmilled fly ash blended cements, due to improved fly ash reactivity through their mechanical activation.

1 Introduction

The cement industry is one of the industries that release a significant amount of greenhouse gases. According to the World Business Council for Sustainable Development (WBCSD) [1], emissions from Portland cement manufacturing vary in different regions of the world from 0.73 to 0.99 kg of CO_2 per kg of cement produced [1]. The main sources of carbon dioxide in cement manufacturing are combustion of fossil fuel and limestone calcinations. During the clinkering process, limestone will undergo calcination and CO_2 will be released [2].

With the aim of alleviating their CO_2 emissions, cement manufacturers have implemented new alternatives as adding fly ash to cement. Fly ash is a by-product of the combustion of pulverized coal in thermal power plants. The dust collection system removes fly ash, as a fine particulate residue, from the combustion gases before they are discharged into atmosphere. Most of the fly ash particles are solid spheres; the average particle size is about 10 μm but can vary from less than 1 μm to over than 150 μm [3,4]. Fly ash is entirely composed of silica, alumina, iron and calcium. The magnesium, sulphur, sodium, potassium and carbon are minor constituents [3, 5-7]. The reactive part of fly ash is their amorphous phase which is composed mainly of silicon [8]. When mixed with Portland cement and water, fly ash induces a reaction with the calcium hydroxide released by the hydration of Portland cement to produce calcium-silicate hydrates (CSH) [9-13]. However, there are few limitations of the use of fly ash in cement because of the late pozzolanic reaction that starts only after 7 to 28 days [14-16]. Therefore, several authors have conducted studies on how to activate fly ash trough milling [5, 17, 18]. Kocak et al. [19] reported that the use of fly ash without mechanical activation decreases the compressive strength of blended cements for all fly ash addition rates (5, 10, 15, 20 and 25%) compared to the reference value. However, the authors noticed an improved mechanical strength even below the reference indicating activation of pozzolanic reactions at a high curing time [19]. Mechanical activation of fly ash was conducted by Chen et al. [18] to treat fly ash from a municipal solid waste incinerator. The authors found that the replacement of less than 10 %.wt. of the cement by this activated fly ash leads to a slight improvement of the cement strength. Krishnaraj et al [20] have studied the effect of the replacement of ordinary Portland cement with up to 60 %.wt. of unmilled and milled fly ash in mortar preparation. Results have shown that the replacement of the cement by an amount larger than 15 %.wt. can increase the compressive strength at only 28 days of curing if milled fly ash is used as an addition instead of unmilled fly ash.

In this study, we highlight the effect of mechanical activation of fly ash on the physico-chemical and mechanical properties of fly ash blended Portland cement. An experimental evaluation on the addition of unmilled (UMFA) and milled fly ash (MFA) to the CPJ45 cement was achieved.

2 Materials and method

2.1 Materials

Two materials were used in this study, an industrial CPJ45 cement and fly ash (FA). The milling of FA was achieved using a vibratory disc mill oscillating for 3 minutes. The physical characteristics of the materials were investigated using an Alpine sieve to determine the refusal at 45 and 80 µm sieves, and the ethylene glycol method to estimate the amount of free lime. The chemical composition of the cement, unmilled and milled FA was performed using X-ray fluorescence spectroscopy. The results are reported in Table 1.

Table 1. Physical and chemical properties of Portland cement, unmilled and milled fly ash (wt.%).

(wt%)	CPJ45	Unmilled FA	Milled FA
SiO_2	18,4	48,93	49,38
Al_2O_3	4,65	25,49	26,61
Fe_2O_3	2,85	5,11	5,22
CaO	55,35	7,03	5,76
MgO	1,84	2,37	2,31
SO_3	2,01	0,28	0,13
K_2O	0,78	2,18	2,11
TiO_2	0,28	1,15	1,19
MnO	0,07	0,08	0,08
P_2O_5	0,13	0,70	0,76
L.O.I*	13,51	6,31	6,10
F.L**	1,68	-	-
Rf ***:			
45µm	7,3	31,2	8,6
80µm	0,7	17,3	0,6

*Loss on ignition
**Free lime
***Refusal

The x-ray diffraction analysis of FA was carried out in order to determine its mineralogical composition. The X-ray diffractogram is shown in Figure 1. From the X-ray diffractogram, it can be seen that quartz and mullite make up the majority of crystalline phases in the used FA.

The XRF and XRD results can show that the FA contains a large amount of SiO_2, Al_2O_3, Fe_2O_3, while the content of CaO is poor. This allows us to classify these ashes into class F known as silicoaluminous according to ASTM C 618 specification [21]. The presence of an amorphous halo, located between 21° and 31° in 2θ, indicates the presence of a glassy phase in the used FA [22].

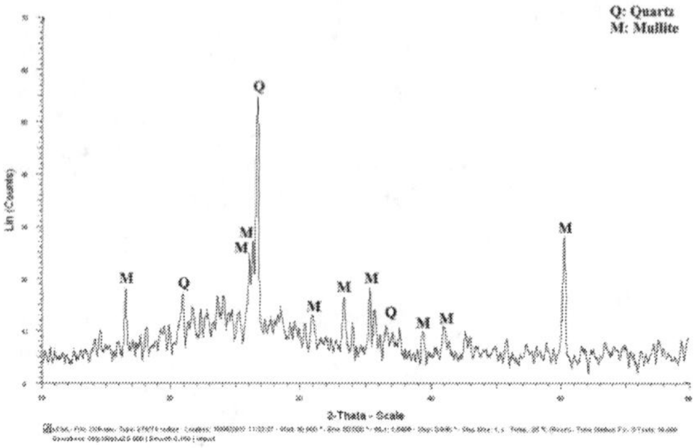

Fig.1. XRD pattern of milled fly ash.

2.2 Method

The addition method consists of introducing 1 kg of industrial cement and fly ash, with different percentages (5, 10 and 20 wt.%), into a plastic bag, then mixed manually for one hour until the mixture became homogeneous. The first series of samples were prepared by mixing the cement with unmilled fly ash. Three blended cements were prepared and noted (C_{UM}). The second series of samples were made by mixing the cement with milled fly ash. The used designations of the three blended cements were (C_M). These six samples were tested and compared to the reference industrial CPJ45 cement, noted C_0. The proportions of prepared composed cements are shown in Table 2.

Table 2. Composition of blended cements.

Samples	CPJ45 (g)	FA (g)
C_0	1000	0
$5C_{UM}$	950	50
$10C_{UM}$	900	100
$20C_{UM}$	800	200
$5C_M$	950	50
$10C_M$	900	100
$20C_M$	800	200

The refusal, the amount of free lime and the chemical composition of the samples were determined using the same methods mentioned above.

Mortars were prepared by mixing one part of cement and three parts of standard sand with water/cement ratio W/C=0.5. The mortar mixtures were then casted in oiled moulds 4x4x16cm and packed with the shock apparatus. Afterwards, the specimens were stored in a moisture room at 20 ± 1°C and relative humidity of ≥90°C. Samples were demoulded after 24h and then cured in water until the days of mechanical determinations (2, 7 and 28 days).

3 Results and discussion

3.1 Physical and chemical characteristics

The physical and chemical characteristics of the tested samples are shown in Table 3. The content of SiO_2, Al_2O_3 and Fe_2O_3 increases proportionally with the increase of fly ash addition. This is due to their high content in the FA used in this study. Contrarily, the low content of CaO in the FA conducts to a decrease of this oxide in the blended cements. Free lime content increases when adding 5% of unmilled or milled FA to the cement, and decreases bellow the reference value for higher adding values of 10 and 20%. However, the loss on ignition decreases bellows the reference value (Table 1) for all additions; unmilled and milled. When unmilled FA was added to the cement, the refusal increased reaching 12.2% and 3.6% for 45 and 80μm sieves respectively. This increase was also noticed when different percentages (3, 5, 7 and 10%) of unmilled FA were added to the cement [23]. On the other hand, when the cement was blended with milled FA, the refusal of the samples was kept the same as the reference.

Table 3. Physical and chemical characteristics of the six tested samples.

(wt%)	$5C_{UM}$	$10C_{UM}$	$20C_{UM}$	$5C_M$	$10C_M$	$20C_M$
SiO_2	19.73	22.16	25.09	20.23	22.4	25.24
Al_2O_3	5.51	7.03	9.34	5.39	7.29	9.35
Fe_2O_3	3.02	3.2	3.34	2.93	3.4	3.36
CaO	53.7	50.07	45.41	52.15	49.14	45.32
MgO	1.86	1.95	1.95	1.85	2.2	1.99
SO_3	1.9	1.8	1.6	1.43	1.13	0.68
K_2O	0.84	0.95	1.05	0.83	0.91	1.06
TiO_2	0.3	0.37	0.46	0.32	0.45	0.47
MnO	0.08	0.08	0.08	0.08	0.08	0.08
P_2O_5	0.16	0.2	0.26	0.17	0.22	0.26
L.O.I	13	12.82	12.29	13.09	12.9	12.09
F.L	1.73	1.56	1.37	1.7	1.54	1.42
Rf.:						
45μm	8.7	10	12.2	7.4	7.6	7.8
80μm	1.4	2.3	3.6	0.7	0.7	0.8

3.2 Mechanical properties

Mortars of the studied samples are tested with flexural and compressive tests after 2, 7 and 28 days of curing. The obtained results are summarized in Figure 2 and 3.

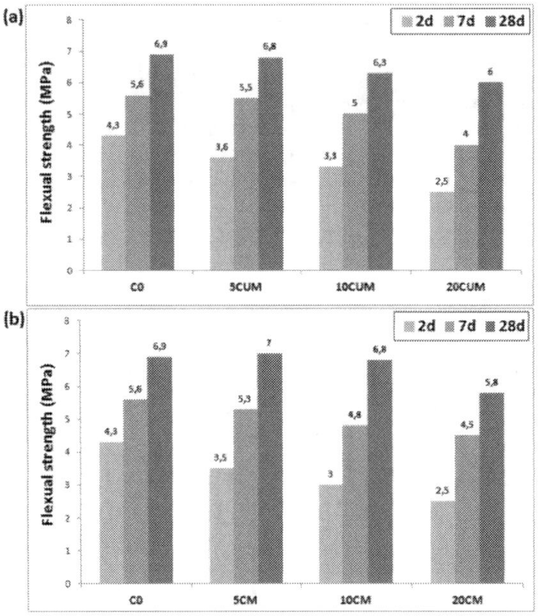

Fig.2. Evolution of flexural strength of (a) UMFA cement mortars and (b) MFA cement mortars with curing age in comparison with control cement.

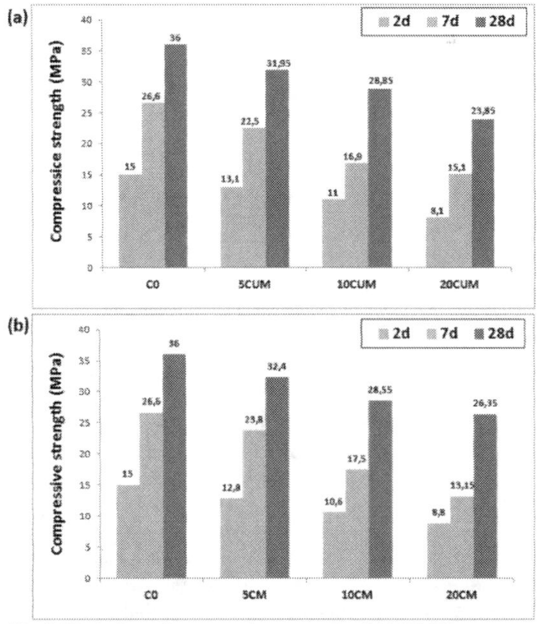

Fig.3. Evolution of compressive strength of (a) UMFA cement mortars and (b) MFA cement mortars with curing age in comparison with control cement.

The Figure 2 shows that flexural strength was affected by the addition of unmilled or milled FA except the mortar with 5% of MFA which displayed a similar value of the reference. Elkhadiri et al. [24] have also reported that the addition of various amounts of fly ash to the cement didn't have a significant effect on its flexural strength. Besides, the compressive strength decreases with the addition of fly ash (Figure 3). These results are in agreement with other studies [23,25], who found that the compressive strength decreases for additions of 5%, 10% and 20% of fly ash, and goes below the cement control values at an addition rate of 7%. This is due to the fact that class F fly ash is a slow reacting pozzolan, which means that it only starts to react noticeably after several weeks at 20°C of curing [26]. According to De Weerdt et al. [27], when fly ash is added to the cement, the amount of the principal reactive component of the cement, which is the clinker, decreases. This is known as the dilution effect. The filler effect of the FA can not compensate for the dilution effect at tested ages (2, 7 and 28 days), hence, the decrease in strength during early ages.

When the compressive strength values of mortars containing FA are compared, it can be noted that mortars made with MFA showed higher strength values than mortars with UMFA. This has also been noticed by J. Payá et al. [28] who carried out a comparative study on the strength development of UMFA/MFA cement mortars, they found that MFA blended cements exhibit higher compressive strength compared to cement mixed with UMFA. They also noticed that greater compressive strength values were obtained for lower replacing percentages (15%) and an increased FA grinding time (40 – 60 min). However, these compressive strength values were bellow the value found for the control mortar containing Portland cement only. In the other hand, the compressive strength of cement mortars with FA milled for prolonged time exceeded the value of the control mortar at all ages, while the UMFA mortars showed a decrease in the compressive strength bellow the reference value [10, 29, 30].

The Figure 3-a shows that when 5% of UMFA is added, the value of the compressive strength at 28 days decreases from 36 MPa, value of the reference, to 31.95 MPa, which means that it only reaches a relative mechanical strength of 88.75% of the control mortar value, and continues to decrease to reach 23.85 MPa which corresponds to 20% of fly ash addition [31]. In the series of mortars made by mixing cement with MFA (Figure 3-b), the compressive strength at 28 days reaches 90% of the reference value with the addition of 5% of MFA, and reaches a value of 73,19% for an addition rate of 20% of MFA.

These mechanical measurements results show that the addition of unmilled or milled FA to industrial cement causes the decrease of the compressive strength at 28 days compared to reference cement. However, the milled FA cement mortars showed higher compressive strength than the unmilled ones. These findings correlate with the results of the study carried out by R. Kumar et al. [17], who found that the mechanical activation of the FA leads to a significant increase in the cement strength compared to cement blended with unmilled FA. According to the same study, the effect of mechanical activation depends on the type of activation device used; attrition mill, vibratory mill, etc.

The content of the amorphous phase of the fly ash, which reacts with $Ca(OH)_2$ to produce C-S-H, increases with a prolonged grinding time. Therefore, the reactivity of milled fly ash is higher than that of the unmilled one [17], which explain the increase in compressive strength.

The milling is performed to improve the reactivity of the FA by reducing the size of the particles and creating new surfaces that serve as precipitation sites for cement hydration products, which is called the "filler effect" [17, 27]. This effect ensures a better distribution of the hydration products in the first days [32].

4 Conclusions

The main conclusions that can be derived from this study are presented as follows:

- The XRF analysis results showed that the mechanical activation of fly ash yielded an increase in the content of sicila, alumina and iron, while the amount of calcium oxide was reduced in both the milled fly ash, and the prepared mortars containing different percentages of milled fly ash.
- The addition of both unmilled and milled fly ash did not have a significant influence on the strength of blended mortars.
- It was found that both unmilled and milled fly ash cement mortars exhibited a decrease in the compressive strength in comparison with the reference at 28 days of curing, due to the fact that we used a class F fly ash, known for being a slow reacting pozzolan, and it only starts to react noticeably after several weeks at the curing temperature we worked with (20°C). This decrease is also a result of the dilution effect of the clinker, which is the essential reactive component of the cement.
- The addition of milled fly ash gave higher compressive strength than the addition of unmilled one as a result of the increase in the fly ash fineness and its reactive components.
- The amount of fly ash that can be incorporated in the cement is limited because of its slow pozzolanic reaction, we found that the optimum addition rate for unmilled and milled fly ash cement mortars was 5%, and they reached a relative compressive strength of 88.75% and 90% respectively.
- The fineness of fly ash acts in the early ages of hydration, it is called "filler effect" while the reactive components act over the long term, thus compensating for the dilution effect.
- It can be concluded that the fly ash fineness effect on compressive strength depends mainly on two parameters; the grinding time and the second parameter is the grinding device used.

References

1. M. Arezoumandi, J. S. Volz, C. A. Ortega, J. J. Myers. Materials and Design **46,** 301-309 (2013).

2. M. Ba-Shammakh, H. Caruso, A. Elkamel, E. Croiset, P. L. Douglas. American Journal of Environmental Sciences **4** (5), 482-490 (2008).
3. A. M. Rashad. International Journal of Sustainable Built Environment **4**, 278–306 (2015).
4. M. Ahmaruzzaman, Prog. Energy Combust. **36,** 327–363 (2010).
5. R. Kumar, S. Kumar, S.P. Mehrotra, Resources Conservation and Recycling **52** (2), 157–179 (2007).
6. A. Upadhyay, M. Kamal, *Characterization and utilization of fly ash.* (National Institute of Technology Rourkela, Orissa, 2007)
7. A. A. Ramezanianpour, *Cement Replacement Materials* (Springer Geochemistry/Mineralogy, Springer-Verlag Berlin Heidelberg, 2014).
8. Lerm info n°26: Fly ashes. http://doc.lerm.fr/lerm-info-n-26-les-cendres-volantes/ (Accessed 05/03/2017)
9. R.C. Joshi, R.P. Lohtia, *Fly Ash in Concrete: Production, Properties and Uses* (CRC Press, 1997).
10. Z.T. Yao, X.S. Ji, P.K. Sarker, J.H. Tang, L.Q. Ge, M.S. Xia, Y. Q. Xi, Earth-Sci. Rev. **141**, 105–121(2015).
11. M. Thomas, *Optimizing the use of fly ash in concrete* (Portland cement association report, 2007).
12. M. Narmluk, T. Nawa, Int. J. Chem. Eng. Appl. **5** (1), 31–35 (2014).
13. S. Goni, M. Frias, I. Vegas, R. Garcia, R.V. de la Villa, Cem. Concr. Compos. **34,** 911-916 (2012).
14. R. Hamzaoui, O. Bouchenafa, S. Guessasma, N. Leklou, A. Bouaziz. Materials and Design **90,** 29–37 (2016).
15. V. M. Malhotra, P. K. Mehta, *High-Performance, High-Volume Fly-Ash Concrete: Materials, Mixture Proportioning, Properties, Construction practice and Case Histories,* (Supplementary Cementing Materials for Sustainable Development Inc, 2002).
16. K.S. Wang, K.L. Lin, B.Y. Tzeng, J. Air Waste Manage. Assoc. **53,** 1340–1346 (2003).
17. M.G. Li, C.J. Sun, S.H. Gau, C.J. Chuang, J. Hazard. Mater. **174,** 586–591 (2010).
18. C.G. Chen, C.J. Sun, S.H. Gau, C.W. Wu, Y.L. Chen, Waste Manag. **33,** 858–865 (2013).
19. Y. Kocak, S. Nas, Constr. Build. Mater. **73,** 25–32 (2014).
20. L. Krishnaraj, P.T. Ravichandiran, R. Annadurai, P.R. Kannan Rajkumar, Int. J. ChemTech. Res. **7** (2), 555–563 (2014–2015).
21. ASTMC618-78: Specification of fly ash and raw or calcined natural pozzolan for use as a mineral admixture in Portland cement concrete, ASTM, Philadelphia, (1978).
22. S. Diamond, Cem. Conc. Res, **13**, 459-464(1983).
23. K. Nabih, A. Quaboul, A. Hamadi, MATEC Web of conferences **2**, 01010 (2012).
24. I. Elkhadiri, A. Diouri, A. Boukhari, J. Aride, F. Puertas, Cem. Conc. Res. **32,** 1597–1603 (2002).
25. K. Nabih, M.K. Kada, M. Hmiri, N. Hamsi. MATEC Web of conferences **11**, 01012 (2014).
26. K. De Weerdt, H. Justnes, K. O. Kjellsen. E. J. Sellevold, Nordic Concrete Research, **42** (2), 51-70 (2010).
27. K. De Weerdt, E. J. Sellevold, K. O. Kjellsen, H. Justnes, Advances in Cement Research. **23** (4), 203-214 (2011).
28. J. Payá, J. Monzó, M.V. Borrachero, E. Peris-Mora, F. Amahjour, Cem. Conc. Res. **30,** 543 - 551(2000).
29. A.A. Zaldívar-Cadena, I. Díaz-Peña, J.R. González-López, F. Vázquez-Acosta, A. Cruz-López, O. Vázquez-Cuchillo, F. Vázquez-Rodríguez, L.M. Serrato-Arias, Adv. Mat. Res. **787**, 286-290, (2013).
30. S. Aydın, Ç. Karatay, B. Baradan, Powder Tech. **197,** 68–72 (2010).
31. M. Georgescu, N. Saca, U.P.B. Sci. Bull., Series B. **71** (3), 11-22 (2009)
32. K. De Weerdt, M. Ben Haha, G. Le Saout, K.O. Kjellsen, H. Justnes, B. Lothenbach.. Mat. and Struc., **45**, 1101-1114(2012).

Influence of Fluorine on Clinker burnability and mechanical properties of CPA Moroccan cement

A. Bouregba*[1], A. Diouri[1], B. Elghattas[2], A. Boukhari[1] and T. Guedira[2]
[1]Mohammed V University, Faculty of science, Rabat, Morocco
[2]Ibn Tofail University, Faculty of science, Kenitra, Morocco

Abstract. It has been found that the addition of certain components, despite their low concentration in raw mix, may accelerate and enhance the reactivity of the cement raw mix. The utilization of mineralizers to facilitate and quicken the process of clinkerization backpedals numerous years, the concept of using such mineralizers to burn normal raw mixes at a much lower temperature with the end goal of decreasing the fuel necessities of the furnace and to enhance the cement proprieties has become to be seriously considered over the most recent couple of years.

The subject of this paper is to investigate the impact of calcium fluoride as mineralizer on addition during the clinkerization process of industrial raw mixtures and its effect on chemical, mineralogical, and mechanical properties of CPA Moroccan cement. Five different raw meals were utilized and were burned with 2% of calcium fluoride. The resulting clinker was analyzed by X-ray diffraction (XRD) and fluorescence spectroscopy to determine the chemical phases of the obtained clinker, and the mechanical properties of resulting CPA cement were determined. The results show that the addition of 2% of CaF_2 to the clinker raw meal induced a decrease in the burning temperature and free lime, while improving the clinker phase formation and the mechanical properties of obtained cement.

1 Introduction

In the recent years many studies have concentrated on the investigation of processing production of clinkers at the lowest temperature than those regularly experienced, while in the meantime producing a better quality product that fulfills Requirements of various standards. The utilization of mineralizers to facilitate the process of clinkerization goes back many years, in the aim of reducing the fuel requirements of the kiln and to improve the cement proprieties [1-5].

Mineralizers are inorganic substances which quicken the reaction process in the solid-liquid interface, liquid phase and in the solid phase [1]. Several works, carried out in different laboratories demonstrate that the addition of a small quantities of these mineralizers have a major impact on the determination of the burning zone clinkerization and the percentage of clinker compound formation. They majorly affect the determination of the burning zone and the clinker compound formation, they can also facilitate and quick the clinkerization process, by decreasing the liquid phase temperature and mineral fusible eutectic in the aim of saving the clinker production energy and improving the cement compound reactivity [2, 3]. They can also improve the hydraulic properties of clinker [3, 6-13]

The aim of this work was to examine the impact of calcium fluoride mineralizer (CaF_2) on the decreasing temperature of burning of Moroccan industrial raw mixtures, and to study it effect on chemical, mineralogical, and mechanical properties of the clinker cement. This mineralizer is added separately to five categories of industrial raw mixtures of Moroccan cement factory.

2 Materials and method

The raw material used in this study is of the industrial type; it comes from deposits operated by the Moroccan cement company. The wines were prepared and characterized. The chemical analysis of five samples (M01-M05) was determined by X-ray fluorescence, their chemical analyses are shown in Table 1.

The mixtures were burned at different temperatures: 1300, 1350, 1400 and 1450 °C. Crystallized phases in the clinker minerals were identified by X-ray diffraction with a Siemens D5000 diffractometer operating with 40kV and 20mA, equipped with a copper anticathode and a secondary monochromator (λ = 1.5406 Å). The acquisition interval in angle 2θ is between 10 and 60 ° in steps of 0.04 ° with an accumulation of a second per step. The cements is obtained by by adding 3% gypsum to the elaborate cements. Mechanical and physical tests on the cements were carried out. The addition effects on the mechanical resistance of these products were determined. The mineralizer effectiveness was compared to the ordinary clinker produced by the Moroccan cement factory

Table 1. Chemical analyse of the raw mixture (wt.%)

	M01	M02	M03	M04	M05
SiO_2	13,49	13,68	13,88	14,47	13,05
Al_2O_3	3,25	3,55	3,25	3,74	3,53
Fe_2O_3	2,33	2,27	2,24	2,17	2,14
CaO	42,45	43,49	44,11	46,28	46,74
MgO	1,34	1,26	1,42	1,28	1,24
SO_3	0,27	0,38	0,49	0,42	0,33
K_2O	0,57	0,67	0,58	0,68	0,67
TiO_2	0,2	0,2	0,22	0,11	0,09
MnO	0,06	0,05	0,06	0,05	0,05
P_2O_5	0,12	0,1	0,12	0,11	0,09
LSF	96	98	100,57	101,97	104
MS	2,51	2,36	2,51	2,39	2,3
A/F	1,53	1,58	1,43	1,61	1,65

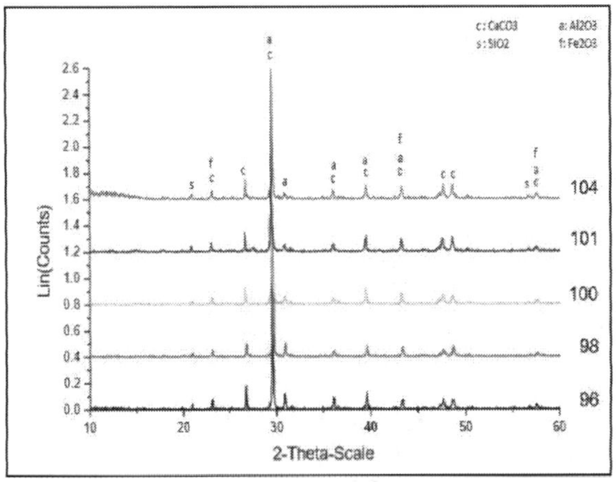

Fig. 1. XRD pattern of five raw mix materials

This investigation studies consists to add 2% of CaF_2 to raw mix and study its impact on the temperature of burning, on the free lime, on the clinker mineral phases formation and on the mechanical strengths

3 RESULTS AND DISCUSSION

3.1 Effect of Calcium fluoride on the free lime

The results of the table 2 shows the evolution of the free lime (CaO_f) content in both case with LSF equal to 98 and 100.57 in addition of CaF_2 (0 and 2%) mineralizer at different temperatures from 1300 °C to 1450 °C. The favorable effect of these substances on the clinkerization process for the two crus studied is clearly compared to the case without mineralization. However, the beneficial effect of adding CaF_2 mineralizers is significativelly remarkable at low temperatures (1300 and 1350 °C) for the both factor lime saturation (LSF). From the table above we observed a decrease of free lime rate when CaF_2 content is increase. this reduction of CaO_f is due to increase in the speed of the solid-state reaction resulting to the easy diffusion effect of fluorine and the acceleration of the necessary reactions for clinker phase formation due to the low-viscosity fluorine melts [6-7].

Table 2. wt.% CaO_f according to the burning temperature, % of mineralizer and LSF

LSF	100.57		98	
% of CaF_2	0	2	0	2
CaO_f (1300°C)	5,84	3,64	5,432	2,912
CaO_f (1350°C)	3,92	2,52	3,864	2,35
CaO_f (1400°C)	3,52	1,792	2,52	1,56
CaO_f (1450°C)	0,95	0,53	0,84	0,46

3.2 Action of CaF_2 on clinker mineral phase's formation

The development of clinker formation was followed mostly on the basis of the free lime (CaOf) contained in the burned samples. It is the most utilized procedure so far because calcium oxide, which formed by dissociation of $CaCO_3$, is gradually consumed by giving clinker phases. The formation of these phases will be reached when the free lime content is much lower, in same case tend to 0 [3].

The clinkers obtained by burning the raw mixtures at 1400 °C without and with 2% CaF_2 and at 1450°C without mineralizer are investigated. The ordinary clinker burned at 1450 °C, has been taken for comparison. The weight percentages of clinker minerals [alite (C_3S), belite (C_2S), $3CaO.Al_2O_3$(C_3A) and $4CaO{\cdot}Al_2O_3{\cdot}Fe_2O_3$ (C_4AF)] were calculated compared to the crystallized mass, by using the following Bogue method:

$C_3S = 4.07\ CaO - 7.602\ SiO_2 - 6.72\ Al_2O_3 - 1.43Fe_2O_3$
$C2S = 2.87\ SiO_2 - 0.754\ C_3S$
$C_3A = 2.65\ Al_2O_3 - 1.69\ Fe_2O_3$
$C_4AF = 3.04\ Fe_2O_3$

Table 3, figures 2-4 summarized the mineral composition of the elaborated clinkers phases after burning in different temperatures. The weight rate of each phase in the studied clinkers is impacted by the temperature of burning and by the presence of Calcium fluoride mineralizer CaF_2. The presence of the latter increases the weight rate of Alite (C_3S), besides it decreases the weight of free lime (CaO_f) and the one of bélite (C_2S), The weight rate of others phases is very close.

The presence of the mineralizer quickens the clinker minerals development. This impact is caused principally by the presence of fluoride anions, characterized by high electronegativity; this latter reacted with the ions (Ca^{2+}, Si^{4+}) on the solid surface. This latter is kept up in a high-energy state, entraining to polarization and rearrangement of atoms, which expand the solid activity either to form new phases, or to melt. However the presences of F^- in the mixture quicken the lime combination by forming the belite phase (C_2S) with the appearance of the amorphous phase, which improves the alite (C_3S) formation [3].

Table 3: Mineralogy of clinkers

Samples	CaO	Ca_3SiO_5 (C_3S)	β-Ca_2SiO_4 (C_2S)	$Ca_4Al_2Fe_2O_{10}$ (C_4AF)	$Ca_3Al_2O_6$ (C_3A)-cubique
CPA	1,96	59,60	16,51	9,64	6,97
K1	3,52	58,10	15,23	10,15	7,37
K2	1,792	78,47	2,21	9,03	7,33
K3	0,95	76,42	3,28	9,36	6,93

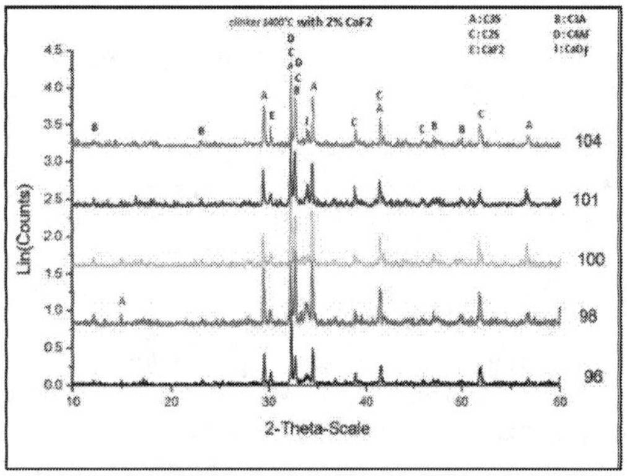

Fig. 2. XRD pattern of Clinker burned at 1400°C with 2% of CaF_2 With different LSF

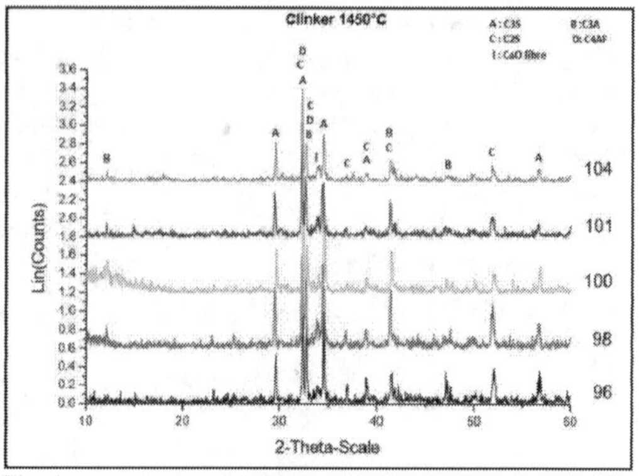

Fig. 3. XRD pattern of Clinker burned at 1450°C without CaF_2 With different LSF

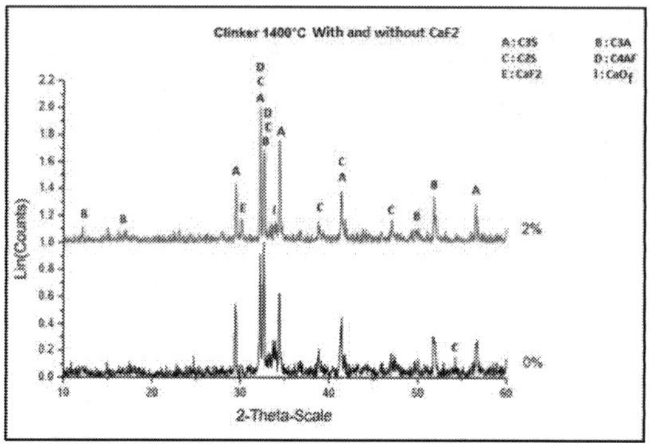

Fig. 4. XRD pattern of Clinker burned at 1400°C with and without CaF_2 (LSF=100.57)

3.3 Mechanical strength

The results of compressive and bending strengths of prepared cements, determined by mechanical tests are given in Fig 5 and 6.

The compressive and bending strengths of clinker cement are affected by CaF_2 addition at short and long times. The mechanical strength values of clinker containing 2% of CaF_2 burning at 1400°C is higher than those developed by the industrial cement, clinker burning at 1450 °C and 1400°C without CaF2. On one hand this is due to fast formation of alite and belite in this clinker and on other hand it is due to percentage weight of alite higher affected by the presence of mineralizer [3].

The presence of CaF_2 in the processes of clinkerization influence the parameter process like temperature of burning, percentage of free lime, clinker phases and also the mechanical strength. L. Kacimi et al., Demonstrated that the presence of alkalines fluorides can affect physical, structural and morphological proprieties. Of a clinker reached at 1300 °C with 1% CaF_2. This latter improve the structural properties of clinker minerals, which prompt to the improvement of the cement hydraulic properties, which increase the C3S rate in the obtained clinker phases wich impact the mechanical strength [3].

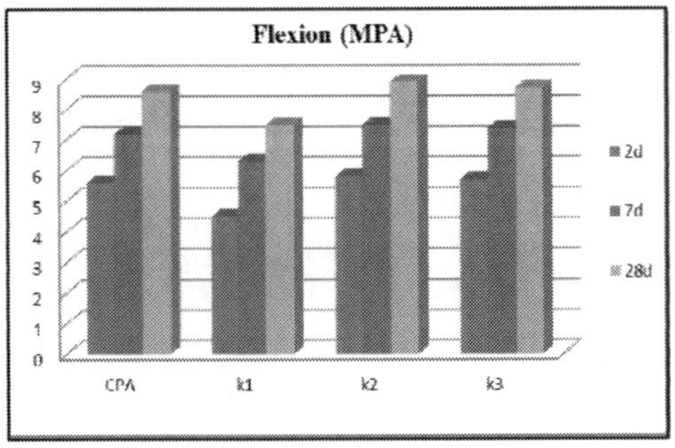

Fig. 5. Flexion strengths of prepared cements

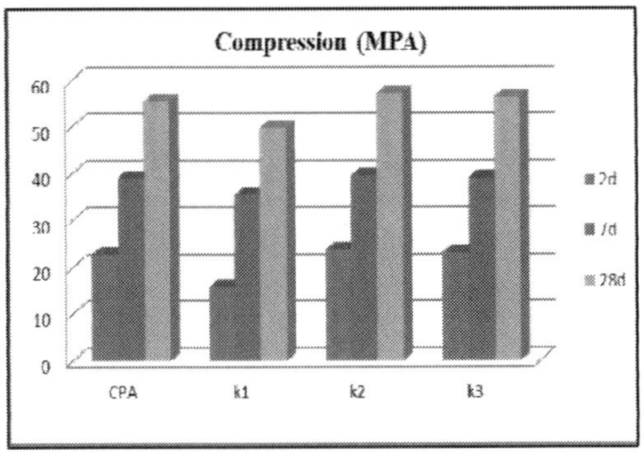

Fig. 6. Compression strengths of prepared cements

Conclusion

In this paper, the chemical and mineralogical analyses are carried out on clinkers from Moroccan cement factory. The investigation of addition a small amount of CaF_2 (2%) during the clinkerization process, demonstrated a satisfactory results without impacting the product quality.

The presence of CaF_2 can decrease the temperature of burning and the percentage of free lime. A clinker obtained at 1400 °C with the addition of 2% of CaF_2 have a high rate of alite which improve the mechanical strength at short and long time in comparison of other clinkers.

References

1. M. Benchanaa, A. Kamali, Ann. Chim. Sci. Mat, **25** (2000)
2. M.T. Blanco-Varela, A. Palomo and T. Vázquez, Cement and Concrete Research, **14** (1984)
3. L. Kacimi A.S. Masseron, A. Ghomari, Z. Derriche, C. R. Chimie, **9** (2006)
4. W.A. Klenm, I. Jawed, and K.J. Holub, Cement and Concrete Research, **9** (1979)
5. I. García-Díaz, J.G. Palomo, F. Puertas, Cement & Concrete Composites **33** (2011)
6. F. Puertas, M.T. Blanco-Varela, T. VBzquez and A. Palomo, Cement and Concrete Research, **26** 9 (1996)
7. I.Akin Altun, Cement and Concrete Research, **29** (1999)
8. C. Gilioli, F. Massazza and M. Pezzuoli, Cement and Concrete Research, **9** (1979)
9. K. Raina and L.K. Janakiraman, Cement and Concrete Research, **28** 8 (1998)
10. G. Kakali, S. Tsivilis, K. Kolovos, K. Choupa, T. Perraki, M. Perraki M. Stamatakis, Ch. Vasilatos, Materials Letters **57** (2003)
11. K.G. Kolovos, Cement & Concrete Composites **28** (2006)
12. I. Jawed and J. Skalny, Cement and Concrete Research, **7** (1977)
13. Salim A. Alimmaryi and Ferenc D. Tamás, Cement and Concrete Research, **10** (1980)

Physical, mechanical and thermal properties of Crushed Sand Concrete containing Rubber Waste

Guendouz Mohamed[1], Boukhelkhal Djamila [1]
[1] LME Laboratory, University of Medea, Algeria

Abstract. Over the past twenty years, the rubber wastes are an important part of municipal solid waste. This work focuses on the recycling of rubber waste, specifically rubber waste of used shoes discharged into the nature and added in the mass of crushed sand concrete with percentage (10%, 20%, 30% and 40%). The physical (workability, fresh density), mechanical (compressive and flexural strength) and thermal (thermal conductivity) of different crushed sand concrete made are analyzed and compared to the respective controls. The use of rubber waste in crushed sand concrete contributes to reduce the bulk density and performance of sand concrete. Nevertheless, the use of rubber aggregate leads to a significant reduction in thermal conductivity, which improves the thermal insulation of crushed sand concrete.

1 Introduction

Rubber waste management is one of the serious environmental problems in the world and it has many challenges for the recycling process. The recycling and disposing of some waste rubber types have many challenges. These are attributed to the difficulty of making separation, type of rubber used, incineration problems, environmental regulations, and increasing the cost of recycling. Use of such waste rubber materials in civil and construction engineering applications has become an attractive alternative to disposal to reduce both the cost of disposal and waste quantities. Many authors [1-9] have reported the properties of concrete with used tyre rubbers and plastics waste. Their results indicate that the size, proportion, and surface texture of rubber particles affect the strength of used tyre rubber contained in concrete [1–9].Eldin and Senouci [3] conducted experiments to examine the strength and toughness properties of concrete made with rubber waste. They noted that the mechanical strength decrease when the rubber waste content increase. Several authors [1,3,4] used tyre chipped rubber containing Portland cement concrete for uses in sound/crash barriers, retaining structures, and pavement structures has been extensively studied. Test results founded that the use of used tyre chipped rubber considerably increases toughness and resistance. Khatib and Bayomy [1] studied the influence of adding two rubbers, crumb and chipped on performances of concrete. They found that the compressive strength of concrete would decrease with increasing rubber content.

This work aims to study the possibility of using waste rubber of shoes in sand concrete without any transformation except grinding, to minimize the cost of the final material. The influence of the proportion of waste used on the physical, mechanical and thermal properties of the new material has been studied and analyzed.

2 Experimental

2.1 Materials

The used sand in this work was extracted from the south of Algeria, near the city of BOUIRA. The particle size distribution is shown in Figure 1, and the physical properties are presented in Table 1. The X-ray Diffraction analysis of used sand demonstrates their essentially limestone nature “Figure 2”.A SEM investigation reveals the crusaded shape of the grains “Figure 3”.

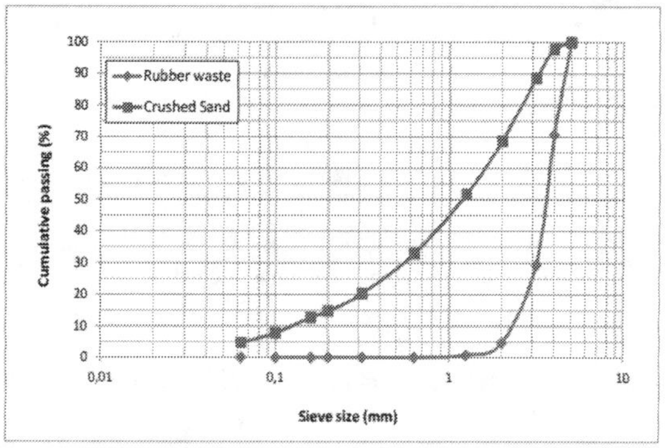

Fig. 1. Particle size distribution of used sand and rubber waste.

Table 1. Physical properties of used sand.

Properties	Sand	Norm
Apparent density (kg/m^3)	1490	NP EN 1097-3
Specific density (kg/m^3)	2500	NP EN 1097-6
Water absorption (%)	1.5	NP EN 1097-6
Sand equivalent (%)	63.50	NP EN 933-8
Fineness modulus	3.14	NP EN 933-1
Compactness (%)	0.60	NF P 18-555
Porosity (%)	0.40	NF P 18-555

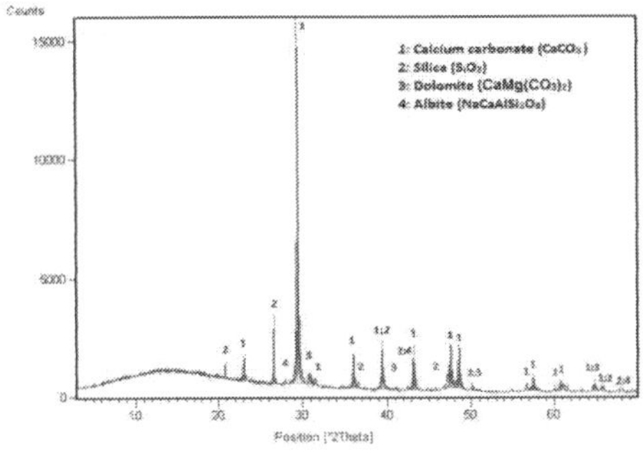

Fig. 2. X-ray diffractogram analysis of used crushed sand.

Fig. 3. Scanning electron micrographs of used crushed sand.

Portland cement CEM II/A 42.5 from CHLEF Factory in Algeria was used throughout this study, with a density of 3100 kg/m^3. Moreover, the chemical analysis revealed the existence of the chemical products shown in Fig. 3 with the proportions presented in Table 2.

Table 2. The chemical composition of used cement (%).

Materials Oxides	Cement
SiO_2	21.89
Al_2O_3	6.15
Fe_2O_3	3.95
CaO	58.76
MgO	1.13
K_2O	0.67
Na_2O	0.57
SO_3	1.62
CaO_{Free}	0.45
P.F	4.67
Total	99.86

The use of fillers in sand concrete composition is essential [10]. Their use is intended to complete the grading curve of sand in its finest part in order to obtain more compact concrete and reduce the cement content and therefore the cost of concrete [11]. In this work the fillers used are the limestone powder. Their specific density is 2,857 kg/m^3, and specific surface area 310 m^2/kg.

The rubber waste used in this work to obtain by the recycling the waste of shoes discharged into the environment after were collected, are washed, compressed, crushed and extruded in the form of grains, and added in the mass of sand concrete with percentage 10%,20%,30% and 40%. Figure 4 shows the general aspect of used rubber waste. The sieve analysis results are given in Figure 1 and the table 3 lists the set of physical, mechanical properties.

Fig.4. General aspect of used rubber waste.

Table 3. Physical and thermal properties of used rubber wastes.

Properties	Rubber waste
Diameter	1.5
Apparent density (kg/m^3)	0.8
Bulk density (kg/m^3)	1.28
Compactness (%)	62.5
Porosity (%)	37.5
Water absorption (%)	0.1
Thermal conductivity (W/m K)	0.15

https://doi.org/10.1051/matecconf/201814901076

The admixture used is an Algerian plasticiser of 'MEDAFLUID 104', liquid form; color chestnut; a PH equal to 6, density 1,04 ± 0,01 and a content of chlore < 1g/L.

The mixing water used for the different mixes is the distribution drinking water.

2.2 Mix design

The optimal compositions of the sand concrete studied, without rubber waste, is based on the experimental method of project of Sablocrete [11]. This gave for cement proportioning of 350 kg/m^3, 1560 kg/m^3 of sand, 130 kg/m^3 of fillers, and a water/cement ratio of 0.75, a percentage of plasticiser of 1% of weight of cement. The rubber waste is added in sand concrete at dosages (0%, 10%, 20%, 30% and 40%). Literary codes identify each mixture in a precise way:

- **CSC**: Control sand concrete (without rubber).
- **SCRW**: Sand concrete with rubber waste.

The specimens produced were cured in air at 20°C and 50% RH. After 24 h, they were removed from the molds and placed in water for another 24 h and then open-air stored until the day of testing. Such curing has been chosen according the normal manufacturing of sand concrete in Algeria. This procedure was respected for all compositions and all tests.

2.3 Tests Procedures

The fresh crushed sand concrete was characterized in what concerns its slump, determined by means of the Abrams cone test (NF P 18-451), and fresh density (EN 12350-6). Flexural strength was determined (using three points) for each composition on three 4×4×16 cm prismatic samples. The half-samples resulting from this test were then submitted to compression on a 4×4 cm test section (EN 196-1). Sample size is significant in comparison with the size of aggregates.

The measurements of thermal conductivity led to the determination of the following parameters: thermal conductivity (k), on three 40×80×160 mm samples using a CT-Meter machine "Figure 5".

Fig. 5. Measurement of thermal properties.

3 Results and discussion

3.1 Workability

Figure 6 give the slump of crushed sand concrete as a function of rubber waste. The slump of sand concrete as a function the content difference of rubber waste is sowed in Figure 3. The use of rubber waste as addition in sand concrete contributes to increase the slump of sand concrete, probably due to the to the presence of more free water in the mixes containing rubber, because the rubber aggregate cannot absorb water during mixing. Several authors [1] reported that the use of rubber waste in the concrete, involves a significant reduction in the workability.In order to limit the number of compositions and to be able to compare them on a common basis, it has been decided to carry out the research project at a constant workability (slump between 70 mm and 90 mm) by the addition of the quantity of admixture for each mixture contains coffee waste.

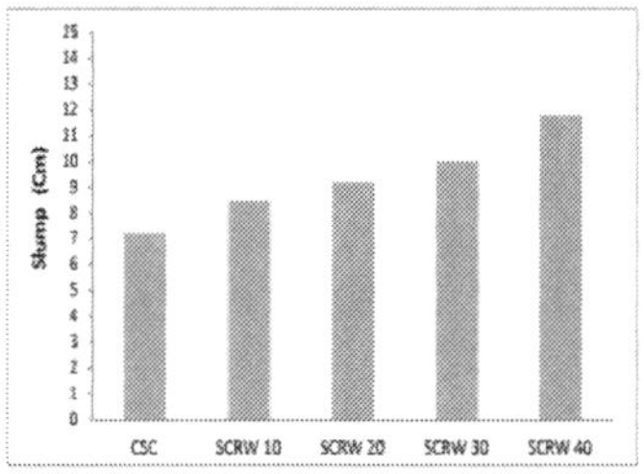

Fig.6. Slump of crushed sand concrete as function of rubber waste content.

3.2 Bulk density

Figure 7 give the bulk density of crushed sand concrete as a function of rubber waste. The results displayed that the incorporation of rubber waste in crushed sand concrete contributes to reduce the bulk density. This reduction probably due to the lower density of the rubber compared with the dune sand and the air content caused by the grain of RW, which in turn reduces the unit weight of the mixtures.Khatib and Bayomy [1] concluded that because of low specific gravity of rubber particles, unit weight of mixtures containing rubber decreases with the increase in the percentage of rubber content.

https://doi.org/10.1051/matecconf/201814901076

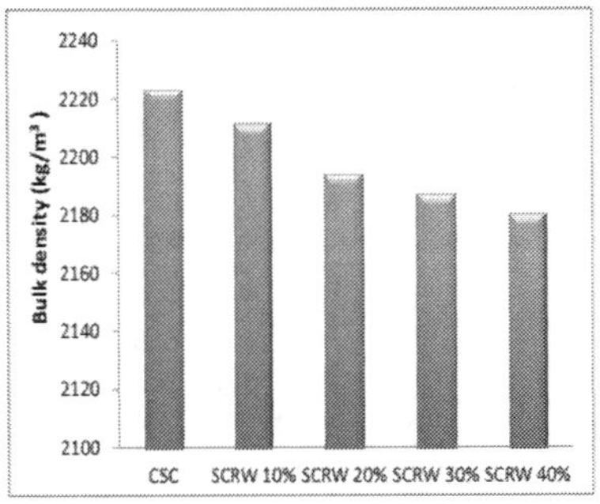

Fig.7 Bulk density of crushed sand concrete as function of spent rubber waste content.

3.3 Mechanical strength

The compressive and flexural strength tests of hardened crushed sand concrete were performed at the age of 28 are showed in figure 8 and 9. The results displayed in that compressive strengths of crushed sand concrete decrease 20 % respectively for addition of 40% natural crushed sand by rubber waste. Even thing in the case of the flexural strengths, which indicates that the addition of 40 % of the rubber waste in a crushed sand concrete represents decrease the performances of new materials. This trend may be related a poor adhesion between the surface of the rubber aggregate and binder paste is likely to have contributed to such a strong decay of mechanical properties. This reduction in performance of SC can be also caused by the great porosity and hardness of grain of rubber waste compared with natural crushed sand. Several authors [3] reported that the addition of rubber waste in the concrete, involves a significant reduction in the mechanical strength of concrete.

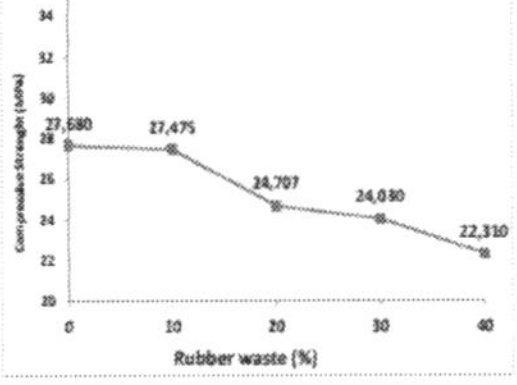

Fig.8. Compressive strength of crushed sand concrete as function of rubber waste content.

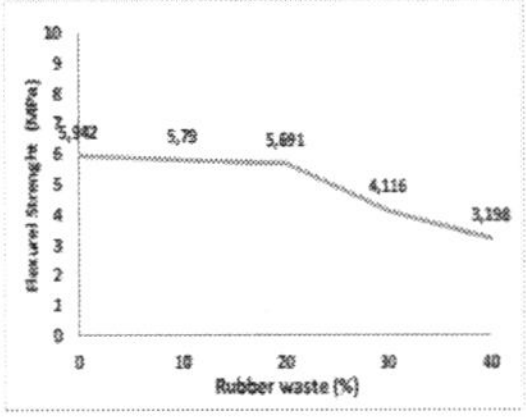

Fig.9 Flexurel strength of crushed sand concrete as function of rubber waste content.

3.4 Thermal properties

The thermal conductivity values of the specimens measured at 28 days are showed in figure 10. The results show that the thermal conductivity of crushed sand concrete decrease when the percentage of addition of rubber waste increase. It is thought that, this situation caused by the lower TC coefficient of the rubber (0.15 W/m K) than natural aggregate (2 W/m K) [12]. So the decreasing thermal conductivity came also from the increase in porosity induced by the rubber waste. In fact the pores contain air which has a TC (0.024 W/mK) much lower than all the other components of the SC. Similar results were observed by several authors in composite systems, for example F. Iucolano et al. [13] in a study on plastic waste in mortar composites, and it's showed a decreased in value of thermal conductivity of mortar containing PET about 50% lower than a traditional mortar.

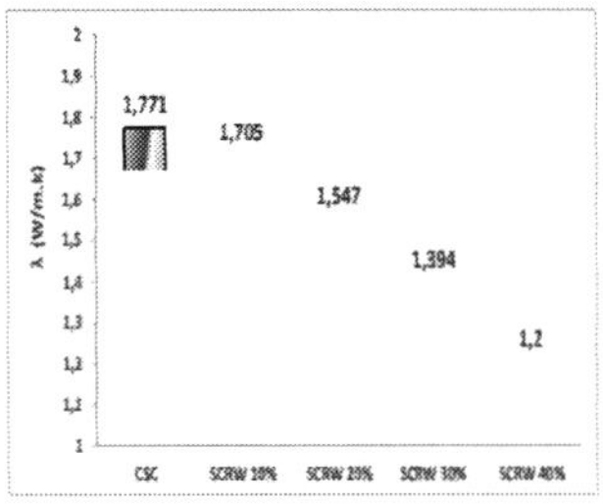

Fig.10. Thermal Conductivity of crushed sand concrete as function of rubber waste content.

4 Conclusions

The results which could be summarized and concluded as:

1- The workability of crushed sand concrete increase when the addition of rubber shoes wastes increase.
2- There is a reduction in the bulk density of crushed sand concrete as the content of rubber waste increases.
3- The compressive and flexural strength values of crushed SC mixtures tend to decrease above the values for the reference SC with increasing the waste rubber ratio. But this result was acceptable for lightweight concrete
4- Rubber waste addition in sand concrete caused a reduction (34%) in the values of thermal conductivity of the sand concrete.

Finally, the use of rubber waste of shoes as aggregates in crushed sand concrete can most of the problems associated with their disposal as well as save natural resources related to aggregates mining.

References

1. ZK . Khatib, FM. Bayomy, ASCE J Mater Civil Eng. **11** (1999),206–13
2. D. Fedroff, S. Ahmad , BZ. Savas , *Washington, DC: Transportation Research Board.* (1996).
3. NN. Eldin., AB. Senouci. ASCE J Mater Civil Eng. (1993),478–96.
4. NA. Ali, AD. Amos., M. Roberts , Scotland (UK): University of Dundee. (2000).
5. Rostami H., Lepore J., Silverstraim T., Zundi I. Scotland (UK): University of Dundee (2000).
6. IB . Topcu , Cem Concr Res.**25** (1995) (2):304–10.
7. NI. Fattuhi , NA. Clark. J Construct Build Mater. (1996).229–36.
8. M. Guendouz ,F. Debieb ,O. Boukendakdji,E.H. Kadri, M. Bentchikou and H. Soualhi.J Mater and Env Scien, J. Mater. Environ. Sci. 4 (2013) 227-232.7 (2).(2016),p.382-389.
9. M. Guendouz., F. Debieb and E.H. Kadri**.** 33èmes Rencontres de l'AUGC, ISABTP/UPPA, Anglet, 27 au 29 mai 2015 Bayonne, France, (2015)**,**p 1-8.
10. AFNOR. Concrete – Sand concrete. Projet P 18-500. (1994).
11. Presse de l'Ecole Nationale des Ponts et Chaussées. *'Béton de sable : Caractéristiques et pratique d'utilisation'*. (Projet SABLOCRETE) édition. (1994)
12. K. Hannawi ,S. Kamali-Bernard, W. Prince , J.Waste Manage.**30** (2010).2312–20.
13. F. Iucolano ,B. Liguori ,D. Caputo ,F. Colangelo , R. Cioffi , Mater and Desi. **52**(2013). 916–922.

Physical and mechanical properties of cement mortar made with brick waste

Guendouz Mohamed [1], Boukhelkhal Djamila [1]
[1] LME Laboratory, University of Medea, Medea, Algeria

Abstract. The development of new building materials is a current problem where researchers are trying to find the right materials for each region and returning cheapest countries. Recycling and recovery of waste are now considered as a promising solution to meet the deficit between production and consumption and protecting the environment. This work focused on the study of the effect of brick waste in the production of cement mortar with substitution rates ranging from 5-30% by weight of cement and to compare its performance with fresh and hardened state with ordinary mortar considered as control mortar. Compressive and tensile strengths up to 28 days of age were compared with those of controlled mortar. Water absorption was also measured at 28 days of age. The test results indicate the beneficial effect of brick waste powder on performance of cement mortar with an optimum of 15% of cement weight substitution.

1 Introduction

The manufacturing of cement, mortar and concrete, as is the case with many industrial materials, consumes a considerable amount of energy, generates large quantities of greenhouse gases (estimated today to 1 tone of CO_2 per 1 tone of cement produced in the world), contributes to the gradual depletion of natural resources and leads to the production of certain residues which must be eliminated.

In order to ensure the growth of these indispensable building materials, the sustainable development approach must be integrated into their production, thus achieving a balance between environmental protection constraints and economic and social considerations [1-5]. Therefore, in view of the increasing need for materials resources and the need to preserve the environment in a sustainable development vision, it has become necessary and relevant to explore all possibilities for reuse and recovery of waste and - industrial products, particularly in the field of civil engineering. However the waste storage disposals are becoming also a serious environmental problem, especially for main cities where disposal sites are lacking. Wastes from construction materials, materials factories, demolition sites, earthquakes and natural disasters are until now rarely used in Algeria: there is hence a need for recycling more and more waste materials.

The need to find new and cheaper binders has led manufacturers to produce various cements containing in addition to clinker, secondary additions with varying proportions. In Algeria, blast furnace slag and pozzolana are used as an additive to cement. Other local materials can also be used and valorized. And it is in this context that our study takes place. It aims to develop new composite cement based on brick waste and having a strength class of 42.5MPa. These by-products of the red products industry, which remain little used or recycled so far, are relatively abundant in Algeria because of the high number of brickyards and the discharge rates (non-compliant or broken bricks) They generate and account for 10 to 15% of their productions [6, 7].

The terracotta products are really very durable and their resistance to the test proves it. Moreover, studies carried out by several researchers have shown the beneficial effect of the use of this type of waste [8-11] in addition to the pozzolanic character of the latter [8, 12-16].

The objective of this work is to study the effect on physical and mechanical properties of mortar made with fine crushed brick waste powder. Cement is substituted by weight by bricks waste powder at rates varying from 0, 5, 10, 15, 20, 25 and 30%. Compressive and tensile strengths are evaluated and compared up to 28 days of age. Capillary water absorption is also measured at 28 days of age

2 Experimental

2.1 Materials

Portland Composite Cement based on limestone (CEM II/A 42.5) with a fineness of 3080 cm^2/g and a specific gravity of 2.93 g/cm^3 was used in this study. Its average compressive strength at 28 days is 46.43 MPa.

https://doi.org/10.1051/matecconf/201814901077

The powder used in the manufacturing of cement mortar derives from the brick waste from the brickworks of Draa Smar in Médéa. Once the waste was received, it was crushed by a laboratory jaw crusher which allowed us to obtain a material whose grain size is 0-5 mm. After they are placed in a disk mill to convert them into fine powder, with fineness comparable to that of the cement used (3500 cm^2/g). Its absolute density has a value of 2.66 g/cm^3.

For the fineness modulus of sand used is about 2.50, we combined two sands: 0/3 roller siliceous sand (S1) from the region of Hassi Bahbah (28%) and 0/4 crushed limestone sand (S2) from the quarry of Mont Guorno (72%). The grain size distribution of roller and crushed sand is presented in "figure 1" and their physical properties are summarized in Table 1. To ensure satisfactory workability of the brick waste-based mortars, a water-reducing plasticizer adjuvant manufactured by the company Granitex-NP and marketed under the name "MEDAFLOW 145" was used in this study which does not exhibit a retarding effect. Its normal range of use is fixed by the data sheet of 0.3 to 2% of cement weight, and its specific gravity is of 1.06 g/cm^3.

Table 1. Physical properties of sand used.

Properties	Roller sand	Crushed sand	Standard
Granular class	0/3	0/5	
Fineness Modulus	1.06	3.45	NF EN 933-1 [17]
Specific gravity (g/cm^3)	2.58	2.59	NF EN 1097-6 [18]
Bulk density (g/cm^3)	1.42	1.48	NF EN 1097-3 [19]
Sand equivalent (%)	61.13	71.31	NF EN 933-8 [20]
Water absorption (%)	0.40	0.30	NF EN 1097-6 [18]

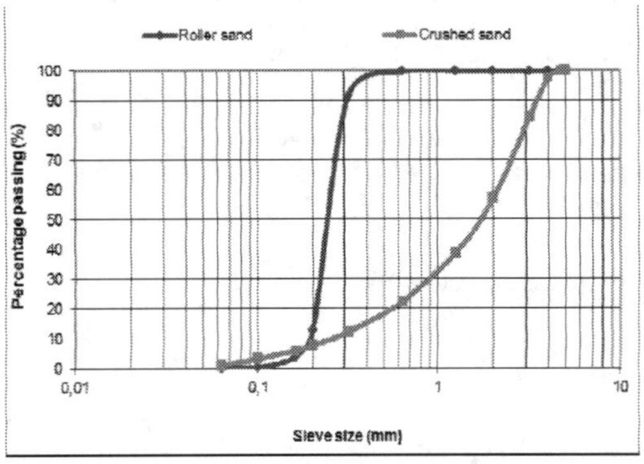

Fig. 1. Particle size distribution of used sand and coffee waste.

2.2 Mix design

For the purpose of this study, a reference mortar is made from 100% cement named MT. And seven (07) other mixtures are prepared by replacing the cement by different % of brick waste (5, 10, 15, 20, 25 and 30%).

In order to determine the quantities of The different materials: sand, cement, water and plasticizer of this mortar, we based on the standard formulation method for normal mortar (S/C = 3 et W/C = 0.5) [21], with some modifications made in the determination of water to cement ratio (W/C = 0.54) in order to adapt the requirements of local materials, taking into account their absorption rate and to get a good plasticity of mortar in fresh state.

The composition adopted for the control mortar and the mortars based on brick waste is that corresponding to the composition which gave satisfactory workability and comparable to that for a normal sand-based mortar (Flow time in LCPC maneabilimeter is close to 10 seconds) [22]. The optimum compositions of the various mixtures are shown in Table 2.

Table 2. Mix proportion of mortar mixes

Constituent (kg/m^3)	C[1]	BW[2]	Sand		W	Ps[3]
			S1	S2		
MT	450	0	978	372	243	0
MBr 5%	427	22.5	978	372	243	0
MBr10%	405	45	978	372	243	0.10
MBr 15%	382	67.5	978	372	243	0.10
MBr 20%	360	90	978	372	243	0.12
MBr 25%	337	112	978	372	243	0.15
MBr 30%	315	135	978	372	243	0.20

[1] Cement
[2] Brick waste
[3] Plasticizer

3 Results and discussion

3.1 Characterization of mortar mixes in fresh state

Preliminary tests on non-adjuvant brick waste mortars have shown that this waste adversely affects the workability of mortar in LCPC maniabilimeter [22], which confirms the results found by several researchers [23-25, 12]. This decrease in workability appears due to the high rate of absorption of brick waste in addition to the irregular shape of its grains [26].

In order to be able to compare the different types of brick waste mortar with the control mortar, the mixtures must have the same workability. This requires the reformulation of the mortars with brick waste. For this, we used a plasticizer admixture to keep the same flow time of the brick mortars with the control mortar.

https://doi.org/10.1051/matecconf/201814901077

The admixture need for the different mortars is shown in "figure 2". It can be seen that the demand of plasticizer in brick waste mortars increases with the increase in the percentage of waste in the mortar.

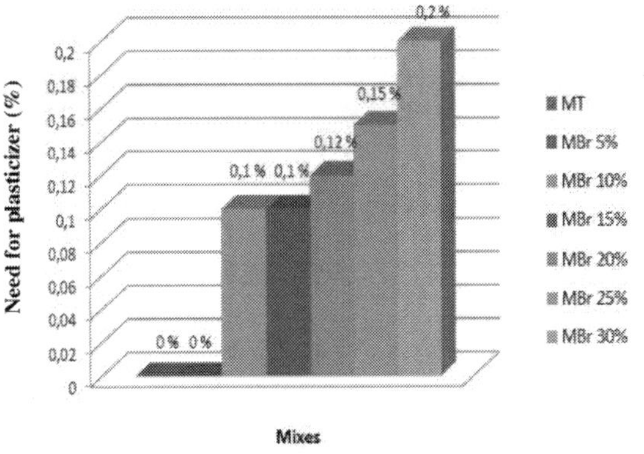

Fig.2. Need of plasticizer admixture for mortar with brick waste.

3.2 Characterization of mortar mixes in hardened state

3.2.1 Mechanical strength

In order to better understand the relative evolution of compressive and tensile strength of the different mortars with age and in function of the rate of substitution of brick waste and to allow comparison with the control mortar, we present this evolution under histogram form given in "figure 3" and "figure 4". respectively.

The first point to be emphasized in these figures is that, at a young age, the resistance of mortars based on brick waste decreases with the increase in the rate of substitution in waste. It is also noted that mechanical performance develops in the early days of hydration. Figure 3 also shows that the resistances of all mortars increase steadily with age and show no fall. It is also observed that the initial resistances of the various mortars (at 3 days) reach 50% compared to that of the 28 days.
At the age of 28 days, the strength of mortars MBr 5%, MBr 10% and MBr 15% exceed that of the reference mortar which is probably due to the pozzolanic activity of this material according other researchers [12-16].

3.2.2. Water absorption

The results of capillary water absorption test for mortars mixes carried out are summarized in "figure 5". This figure shows that the mortar based on the brick waste absorbs less water than the control concrete. This it may be due to the pozzolanic reactions of brick waste which have improved the compactness of mortar structure.

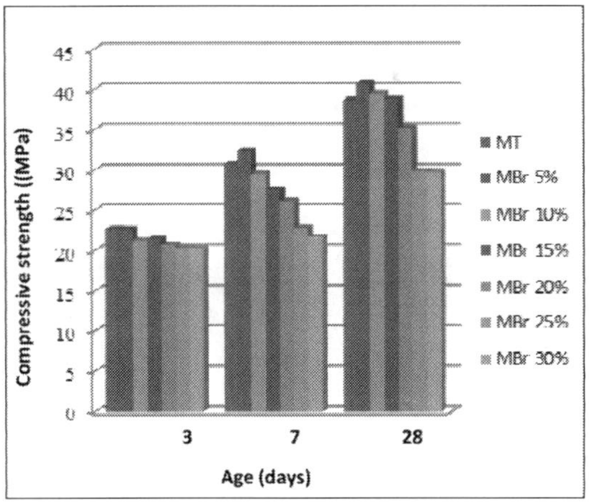

Fig.3. Compressive strength for mortar brick waste.

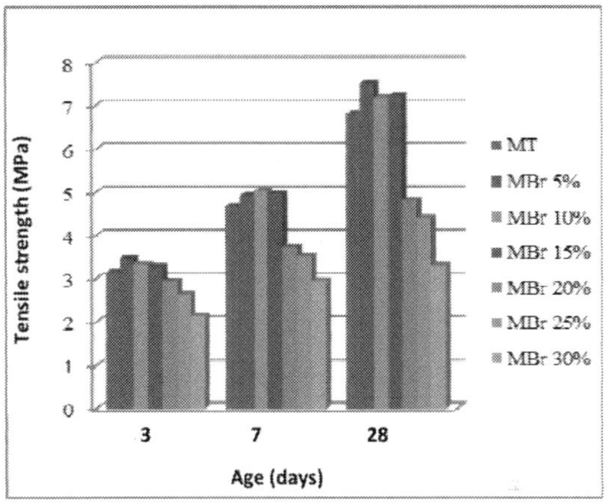

Fig. 4. Flexural strength for mortar brick waste.

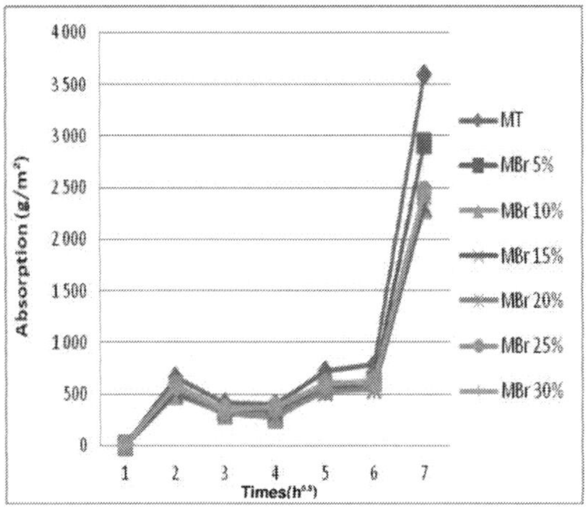

Fig.5. Capillarity water absorption for brick waste based mortars.

4 Conclusions

The main objective of this work was to evaluate brick waste in the manufacture of cement mortars. For this purpose, the study is based on the comparison of the behavior of ordinary mortar with those of mortars based on brick waste powder with constant workability. Based on the results obtained through this study, the following conclusions have been drawn:

- The finely ground brick powder has a negative influence on the workability of the mortar as a function of the cement substitution rate,
- In hardened state the brick waste powder plays both a double pozzolanic role and a densification of the cementitious matrix. Improvements in the mechanical performance of mortars were observed from 28 days of age for substitution rates of 5, 10 and 15% of the cement weight compared to the control mortar,
- The addition of brick waste powder reduces the absorption rate by capillarity;
- The results obtained show that brick waste can be technically used as a cement substitute in the manufacture of ordinary mortars

References

1. L. Courard, *Valorisation des déchets et sous-produits dans le génie civil*, notes de cours, Univ Liège, (1998), 195p.
2. J. S. Damtoft, J. Lukasik, D. Herfort, D. Sorrentino and E. M. Gartner, Sustainable development and climate change initiative, Cem. Concr. Res., **38** (2008), pp. 115–27.
3. M. Guendouz, F. Debieb, O. Boukendakdji, E. H. Kadri, M. Bentchikou and H. Soualhi, J. Meter. and Env. Scien., **7**(2), (2016), 227-232.
4. I. Martínez, M Etxeberria, E Pavón.,N. Díaz, , Constr and Build Mater J., Vol. **49**(2013), pp. 384–392.
5. M. Guendouz, F. Debieb and E. H. Kadri, 33èmes roncontres de l'AUGC, ISABTP/UPPA, Anglet, 27 au 29 Mai 2015, Bayonne, France, (2015), p. 1-8.
6. V. Corinaldesi, M. Giuggiolini and G. Moriconi, Use of rubble from building demolition in mortars, Waste Manage, Vol. **22** (2002) pp. 893–902.
7. Y. Jian ,D. Qiang ,B. Yiwang , Concrete with recycled concrete aggregate and crushed clay bricks, J Constr and Build Mater., Vol. **25**(2011), pp. 1935–1945.
8. Fédération national Belge de la brique, Fabrication de la brique, Brique de parement, Blocs treillis, brique de pavages,…des produits naturels en terre cuite.
9. F. Debieb and S. Kenai, J. Constr. and Build. Mater., **22**(2008), pp. 886-893.
10. S. Kenai and F. Debieb, J. of Materials and Structures, **44**(2011), pp. 815-824.
11. Z. Lan, F. Zhengyue and Z. Shiping , Permeability of recycled aggregate concrete containing fly ash and clay brick waste, Department of Architecture and Civil Engineering, Nanjing Institute of Technology (2014), China.
12. M. S. Dina, Physico-mechanical properties of solid cement bricks containing recycled aggregates, J. Building Materials Research and Quality Control Institute, Housing and Building National Research Center, Egypt (2011).
13. M. Hadjoudja and M. Bederina , *Influence des fillers des déchets de briques sur la durabilité à l'eau du béton de sable de dunes*, Université de Laghouat (2005), Algérie .
14. CIM béton, Fiche technique édité par le centre d'information sur le ciment et ses applications : « Les mortiers », Tome 3 (2008), Paris, France.
15. F. Bektas, K. Wang and H. Ceylan, Effects of crushed clay brick aggregate on mortar durability, Civil Construction and Environmental Engineering, Iowa State University, United States (2008).
16. G. Nigri, Y. Cherait and H. Benouis, Valorisation des déchets de briques dans l'industrie cimentaire, Séminaire International INVACO2, (2011), Rabat, Maroc.
17. NF EN 933-1, Essais pour déterminer les caractéristiques géométriques des granulats. Partie 1: Détermination de la granularité. Analyse granulométrique par tamisage, AFNOR (1997).
18. NF EN 1097-6, Essais pour déterminer les caractéristiques mécaniques et physiques des granulats. Partie 6: Détermination de la masse volumique réelle et du coefficient d'absorption d'eau, AFNOR (2001).
19. NF EN 1097-3, Essais pour déterminer les caractéristiques mécaniques et physiques des granulats. Partie 3: Méthode pour la détermination de la masse volumique en vrac et de la porosité intergranulaire, AFNOR (1998).
20. NF EN 933-8, Essais pour déterminer les caractéristiques géométriques des granulats. Partie 8: Evaluation des fines. Equivalent de sable, AFNOR (1999).
21. NF EN 1008, Eau de gâchage pour bétons: spécification d'échantillonnage, d'essais et d'évaluation de l'aptitude à l'emploi, y compris les eaux des processus de l'industrie du béton, telle que l'eau de gâchage pour béton, AFNOR (2003).
22. NF P 18-452, Béton- mesure du temps d'écoulement des bétons et des mortiers aux maniabilimètre LCPC, AFNOR (1988).
23. NF EN 196-2, Méthodes d'essais des ciment- Partie 2 : Analyse chimique des ciments, AFNOR. 46 p. (2006).
24. B. Étienne and D. Georges-Bernard, *Règlements sur les arts et métiers de Paris*, rédigés au 13e siècle, et connus sous le nom du Livre des métiers d'Étienne Boileau. Crapelet (1837).
25. M. Augustin, Ciment and mortar, Libr. Armand Colin, coll., Génie civil, (1935), p. 2.

26. N. Bourmatte, Granulats recyclés de substitution pour bétons hydrauliques, thèse de magister, Université de Constantine, Algérie (2004).

 https://doi.org/10.1051/matecconf/201814901078

Synthesis and characterization of innovative insulation materials

Aggeliki Skaropoulou, Afroditi Ntziouni, Dimitris Kioupis, Sotiris Tsivilis and Glikeria Kakali
National Technical University of Athens, School of Chemical Engineering, Athens, Greece

Abstract. Insulation elements are distinguished in inorganic fibrous and organic foamed materials. Foamed insulation materials are of great acceptance and use, but their major disadvantage is their flammability. In case of fire, they tend to transmit the flame producing toxic gases. In this paper, the synthesis and characterization of innovative inorganic insulation materials with properties competitive to commercial is presented. Their synthesis involves the mixing of inorganic raw material and water with reinforcing agent or/and foaming agent leading to the formation of a gel. Depending on raw materials nature, the insulation material is produced by freeze drying or ambient drying techniques of the gel. The raw material used are chemically benign and abundantly available materials, or industrial by-products and the final products are non-toxic and, in some cases, non-flammable. Their density and thermal conductivity was measured and found 0.02-0.06 g/cm^3 and 0.03-0.04 W/mK, respectively.

1 Introduction

In Greece, the first thermal regulation was established in 1979, which was replaced in 2010 by the energy efficiency building regulation. Furthermore, the energy consumption for heating and cooling needs in a building with and without insulation is 1:3. The lack of insulation, is both costly, because of increase in energy consumption, and environmental harmful, considering that, in Greece, 82% of buildings are built before 1980 and 45% of total energy consumption is due to the building sector with over than 65% consumed for cooling and heating.

As insulation materials are considered the ones that manage to trap air into their mass, providing energy independence in the spaces and thus ensuring a smooth season transition. They are characterized by the thermal conductivity coefficient. Insulation materials can be either inorganic fibrous elements (glass wool, rockwool, etc.), that are wide used in Europe (cover 60% of European market), or organic foamed ones (expanded polystyrene – EPS, extruded polystyrene – XPS, polyurethane) that cover 80% of the domestic market.

Polystyrene-based insulation materials, are unique for external insulation, due to their low water absorption property. Their major disadvantages consist of the environmental impact and their flammability. During the production process of organic insulation materials, CO_2 emissions are greater compared to the production of inorganic ones, while in case of fire they tend to transmit the flame, producing toxic gases which in most cases are responsible for the human losses [1].

Silica aerogels are a class of materials well known since 1930, with high specific surface area, high porosity, low density and low thermal resistance. Their synthesis involves the formation of a gel, and the removal of the liquid medium through supercritical drying [2, 3]. Over the past 20 years, the research community focuses on the use of various raw materials and the determination of the optimized synthesis conditions. Ambient drying of the formed gel instead of supercritical drying is gaining momentum as it is a safer and less costly procedure. Its synthesis constitutes the mixing of a silica source (TMOS, TEOS, PEDS) with ethanol or acetone in acid or base environment, the formation of gel, its polymerization and, finally, its drying [4-6]. While the received products exhibit excellent properties, the materials that are used as silica source are both expensive and hazardous. Recently, synthesis of insulation materials using as raw materials either clay (clay based aerogels) which are chemically benign and abundant available or industrial by-products (silica based aerogels) gain ground. Synthesis route comprises the mixing of inorganic raw materials with water, reinforcing agent or/and binding agent (polyvinyl alcohol, casein, cellulose fiber, gelatin), and /or foaming agent (sodium dodecyl sulfate) to form a gel. The water is removed either by freeze drying under high vacuum or by ambient drying techniques [7-14]. Depending on the needs and use, products with different properties and structure can occur.

The fabrication of eco-friendly insulation materials, safe to produce and install and with properties competitive to commercial, constitutes a challenge. In this paper, the synthesis and the characterization of innovative, inorganic insulation materials with properties

 https://doi.org/10.1051/matecconf/201814901078

competitive to commercial products is presented. Depending on the raw materials nature and the synthesis procedure, insulation materials with different properties and lamellar or foamed structure can be produced.

2 Lamellar structure insulation materials

From all clays, sodium montmorillonite (Na-MMT), is preferred because of its swelling property, high cation exchange capacity, high aspect ratio and large surface area, rendering it ideal for aerogel formation [14, 15]. This aerogel is synthesized during the freezing and freeze-drying of a gel, which is formed during the mixing of Na-MMT with water. Particularly, during the hydration of Na-MMT, water molecules penetrate between the clay layers, increasing the volume of the clay by up to two times. During freezing the growing ice fronts cause alignment/shearing of the Na-MMT particles along the ice crystal faces. After high vacuum sublimation, air is trapped in the place of ice crystals, forming a lightweight material with lamellar structure [10, 12, 14, 16]. The formed clay aerogel is a material with low density, low thermal conductivity and low mechanical strength, which can be improved by the use of a reinforcement agent [17-21].

Different batches of MMT (Na-MMT, commercial product, Imerys) or organically modified MMT (nanoclay /sigma Aldrich, cloisite /BYK), along with various reinforcement agents, such as polyvinyl alcohol (PVA, 98-99% hydrolyzed, Mw: 31,000-50,000, Acros Organics), casein (Sigma Aldrich) or cellulose fibers (Sigma Aldrich), were investigated. Different freezing temperatures (-40°C, -80°C, gradual freezing, Freezer model: Arctico) were also examined and the produced samples were placed in the freeze dryer of high vacuum (Scientz 50D, ~1Pa). Na-MMT was mixed with deionized water in a slow shear rate mixer and the reinforcing agent was added in the clay suspension. In the case of PVA, a solution was prepared by dissolving PVA in water at 70-80°C until complete dissolution, while the rest reinforcing agents were used in powder form. The mixing of clay suspension with reinforcement agent lasted until fully homogenization of the mixture. The suspension was molded, placed in the freezer and then in the freeze dryer.

Raw materials and samples were characterized through X-Ray Diffraction (Advance B8 Brucker X-ray diffractometer, with CuKα, λ=1,5405Å) and Fourier Transform IR spectrophotometer (Jasco 4200, 400-4000 cm^{-1} wavenumber, KBr technique, 1:200, 8 t/cm^2). Density, thermal conductivity, water vapor absorption and fire resistance was calculated at the final products, while their structure was observed through scanning electron microscope (SEM, JEOL JSM-5600, equipped with OXFORD LINK ISIS 300 EDX) and stereoscope (Zeiss Stemi 2000C, connected with Axio Cam ErcS5). The density of the samples was calculated based on mass and volume measurements. The thermal conductivity was determined by means of the Heat Flow method, on a Netzsch HFM 436 lamda Heat Flow Meter, which meets the EN 12667 and ASTM C518 requirements (mean temperature: 10°C, temperature difference: 10°C). Water vapor sorption was determined according to ASTM C1104/ C1104M, a standard for unfaced insulation materials and fire resistance was ascertained according to DIN 4102, a test which measures the ignitability of building products subjected to direct impingement of flame.

3 Foamed structure insulation materials

Foamed insulation materials of inorganic nature can be produced by the combination of clay or industrial by-products with a binding agent and a foaming agent. From all clays, sodium montmorillonite was preferred, as mentioned in previous section, while from industrial by-products, silica fume, an amorphous, non-crystalline SiO_2, an ultrafine by-product of the silicon and ferrosilicon alloy production was chosen. Raw materials are dispersed in insulation matrix. The matrix is consisted of a binding agent, which ascribes lattice stability, and a foaming agent, which contributes in porosity increase, leading to aerogels with lower density and thermal conductivity. Raw material is dispersed in the matrix, attributing volume stability and enhanced fire resistance due to its inorganic nature. Synthesis route comprises the dissolution of gelatin (binding agent) in water at ~60°C. Subsequently, gelatin solution is mixed with sodium dodecyl sulfate (SDS- foaming agent) until frothed into foam before adding the selected raw material. The foamed mixture is molded and either freezed and the ice is removed via sublimation, or dried at ambient conditions [22]. In every case, a lightweight material with foamed structure is obtained. Properties and characterization of raw materials and final products were defined according to methods mentioned above.

4 Results and discussion

Clay aerogels with Na-MMT and water alone, is very brittle, so PVA, casein or cellulose fibers (CF) were used as reinforcement agents. For the same content, only samples with PVA and casein were received as monoliths, as shown in Figure 1, with the ones with PVA to be more coherent. Cellulose fibers didn't seem to contribute in the specimen's rigidity. Freezing temperature was investigated and found that samples frozen at -40°C showed lower values of density and thermal conductivity.

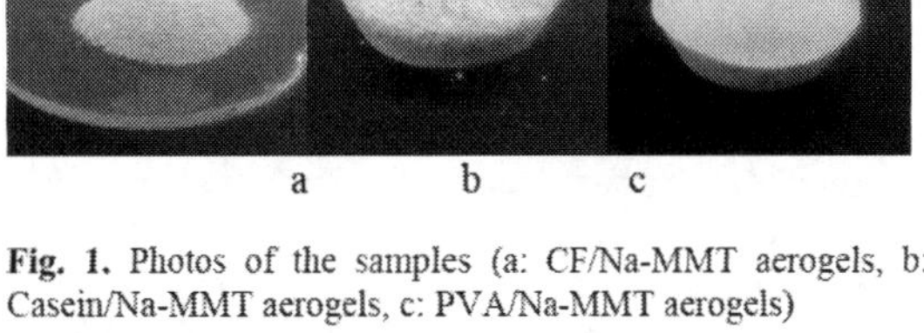

Fig. 1. Photos of the samples (a: CF/Na-MMT aerogels, b: Casein/Na-MMT aerogels, c: PVA/Na-MMT aerogels)

From all results, irrespective of synthesis procedure, it was concluded that increased content of raw materials leads to increased density, lower porosity and higher values of thermal conductivity. Indicatively, the results for the synthesis of lamellar structure insulation materials with PVA as reinforcement agent are presented in Figure 2.

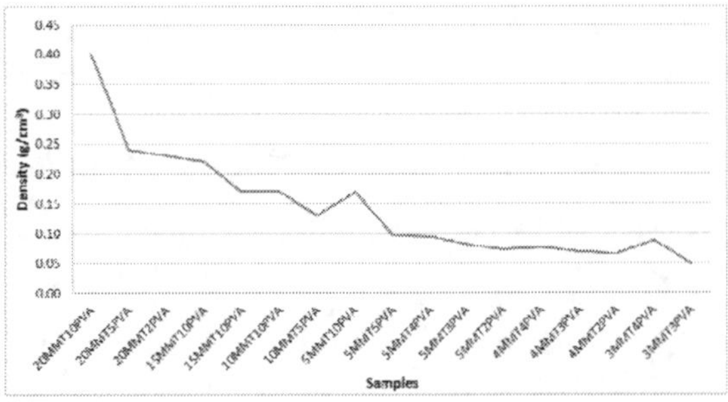

Fig. 2. Density of lamellar structure insulation materials (code name: xNa-MMTyPVA, where x,y: Na-MMT, PVA content)

Synthesis of various samples of lamellar or foamed structure were carried out and the optimized conditions for low porosity and thermal conductivity compared to satisfying rigidity were found. Density and thermal conductivity values of these insulation materials are presented in Table 1, while their stereoscope images are presented in Figure 3.

Table 1. Density and thermal conductivity of synthesized insulation materials

Materials characteristics	Density (g/cm^3)	Thermal conductivity (W/mK)
Na-MMT/Lamellar structure/Freeze drying	0.05	0.035
Na-MMT/Foamed structure/Freeze drying	0.017	0.034
Silica fume/Foamed structure/Freeze drying	0.026	0.033
Silica fume/Foamed structure/ambient drying	0.035	0.035

In the case of freeze drying, foamed structure insulation materials shows an essential decrease in porosity compared to ambient drying, which does not imprint in thermal conductivity values. As it is observed from stereoscope images (Figure 3c, 3d), silica based insulation materials produced via ambient drying are more compact than the ones produced via freeze drying, which is in good accordance with the values of density and thermal conductivity.

Water moisture absorption and fire resistance were examined at materials of all synthesis procedure. Samples were exposed to 100% humidity and it was observed that the samples with lamellar structure had better behavior, as they manage to maintain their shape and their surface integrity. Foamed insulation materials probably because of their great open porosity might absorb greater amount of water vapor causing more intense deformation than the lamellar structure ones.

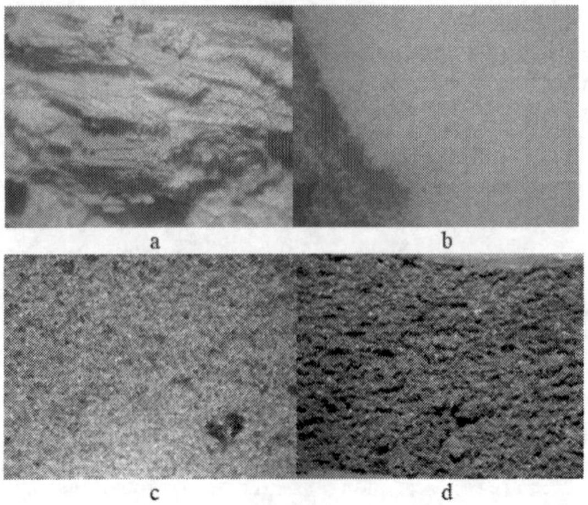

Fig. 3. Stereoscope images, a) Na-MMT/ Lamellar structure/ Freeze drying, b) Na-MMT/ Foamed structure/ Freeze drying, c) Silica fume/ Foamed structure/ Freeze drying, d) Silica fume/ Foamed structure/ Ambient drying

5 Conclusions

Two synthesis procedure of materials with high porosity and low thermal conductivity were investigated. The one involves the mixing of raw material with reinforcing agent, freezing and freeze drying of the suspension. The produced material has lamellar structure. The other procedure include the mixing of raw material with binding or foaming agent, followed either by freeze drying or by ambient drying. In this case, the produced material has a foamed structure, lower density than the lamellar structure products, but similar thermal conductivity values. Optimized synthesis were defined and, depending on procedure, density values were in the range 0.02-0.06 g/cm^3 and thermal conductivity values were in the range 0.03-0.04 W/mK. Clay based aerogels with lamellar structure are inflammable and less susceptible to water vapor sorption compared to foamed structure insulation materials either clay based or silica based.

Acknowledgement

This project was funding from 7th Framework Programme (FP7-2013-NMP-ENV-EeB, "Multifunctional facades of reduced thickness for fast and cost-effective retrofitting, MF-Retrofit".

References

1. A.K. Karamanos, E. Giama, S. Chadiarakou, A.M. Papadapoulos., Heleco '05, 1-11 (2005)
2. J. Fricke, A. Emmerling, J. Am. Ceram. Soc. **75**, 8, 2027-2035 (1992)
3. S.S. Kistler. J. Phys. Chem. **36**, 1, 52-64 (1932)
4. A. Venkateswara Rao, E. Nilsen, M.A. Einarsrud. J. Non-Cryst. Solids, **296**, 3, 165-171 (2001)

5. S.D. Bhagat, , Y,-H Kim, Y.-S Ahn, J.-G. Yeo, App. Surf. Sci. **253**, 6, 3231-3236 (2007)
6. F. Shi, L. Wang, J. Liu, Mater. Lett. **60**, 29–30, 3718-3722 (2006)
7. Y. Wang, M.D. Gawryla, D.A. Schiraldi, J. APPL. POLYM. SCI. **129**, 3, 1637-1641 (2013)
8. B. Lei, K. Shin, I. Jo, Y. Koh, H..Kim, Materials Chemistry and Physics, **145**, 3, 397-402 (2014)
9. S.M. Alhassan, S. Qutubuddin, D.A. Schiraldi, Langmuir, **26**, 14, 12198-12202 (2010)
10. T. Pojanavaraphan, D.A. Schiraldi, R. Magaraphan, Applied Clay Science, **50**, 2, 271-279 (2010)
11. M. Sachithanadam, S.C. Joshi, Procedia Engineering, **75**, 51-55 (2014)
12. H.V. Olphen, 15th Conf on Clays and Clay Materials, **15**, 423-435 (1967)
13. A.P. Perissinotto, C.M. Awano, F.S. de Vincente, D.A. Donatti, A. Mesquita, L.F. da Silva, D.R. Vollet, Microporous and Mesoporous Materials, **223**, 196-202 (2016)
14. S.R. Hostler, A.R. Abramson, M.D. Gawryla, S.A. Bandi, D.A. Schiraldi, Int. J. .Heat .Mass Transfer, **52**, 3–4, 665-669 (2009)
15. Y. Wang, S.M. Alhassan, V.H. Yang, D.A. Schiraldi, Composites Part B, **45**, 1, 625-630 (2013)
16. S.A. Bandi, M. Bell, D.A. Schiraldi, Macromolecules, **38**, 22, 9216-9220 (2005)
17. K. Finlay, M.D. Gawryla, D.A. Schiraldi, Ind. Eng. Chem. Res. **47**, 3, 615-619 (2008)
18. T. Pojanavaraphan, R. Magaraphan, B. Chiou, D.A. Schiraldi, Biomacromolecules, **11**, 10, 2640-2646 (2010)
19. L.S. Somlai, S.A. Bandi, D.A. Schiraldi, L.J. Mathias, AIChE J. **52**, 3, 1162-1168 (2006)
20. A. Skaropoulou, D.Kioupis., S. Tsivilis, G. Kakali. Ashrae *EinB2016*, #27, 1-7 (2016)
21. A. Skaropoulou, S.Tsivilis, G. Kakali. *Ashrae EinB2014*, #110, 1-7 (2014).
22. M. Sachithanadam, S.C. Joshi, J. Mater. Sci., **49**, 1, 163-179 (2014)

Characterization of clay deposits in the Adrar region, with a view to their valorization in the building materials industry

Caractérisation des gisements d'argiles de la région d'Adrar, en vue de leur valorisation dans l'industrie des matériaux de construction

ABBOU Mohammed[1], MOULAY OMAR Hassan[1], SEMCHA Abdélaziz[1], KAZI-AOUAL Fatiha[2]

1 Ahmed Draia University of Adrar, Science and Technology, Adrar, Algeria
2 National Polytechnic School of Oran, Civil engineering, Oran, Algeria

Abstract. In the context of sustainable local development of the Adrar region, one of the largest regions in the Algerian Sahara. The search for local useful substances has been initiated by the Algerian state to cover the need for building materials in the construction industry. However, from a geological point of view, the Adrar zone is located in the extension of the primary chain of the Ougarta which separates two sedimentary basins of Reggane and Timimoun, as well as the basin of Sbâa. In this context, an experimental study is focused on the characterization of clay deposits, with a view to their valorization in the construction materials industry (ceramics) sector, with the aim of contributing to the use of local materials.

Keywords: clay; Ceramic; building materials; Adrar; development.

Résumé. Dans le cadre du développement local durable de la région d'Adrar, l'une des plus vastes régions dans le Sahara algérien, la recherche de substances utiles locales a été entamée par l'état algérien, cela pour répondre au besoin en matériaux de construction dans le BTP. Cependant du point de vue géologique, la zone d'Adrar est située dans le prolongement de la chaine primaire de l'Ougarta qui sépare deux bassins sédimentaires : celui de Réggane et celui de Timimoun, ainsi que la cuvette de Sbâa. Dans ce contexte, une étude expérimentale est focalisée sur la caractérisation des gisements argileux, en vue de leur valorisation dans le secteur de l'industrie des matériaux de construction (céramique), ayant pour objet de contribuer à l'utilisation des matériaux locaux.

Mots-Clés : argile ; céramique ; matériau de construction ; Adrar ; développement.

1 Introduction

Durant ces dernières années le développement de la construction des logements au Sud Algérien, telle que la région d'Adrar, l'une des plus vastes régions dans le Sud algérien a engendré un besoin énorme en matériaux de construction conventionnels. Des efforts ont été orientés vers le développement de nouvelles méthodes de construction en utilisant des matériaux locaux afin d'apporter une réponse adéquate à cette crise.

Dans ce travail, nous présentons la localisation géologique de la région du Touat (Adrar). Par la suite nous ferons une présentation de l'étude expérimentale qui englobera l'identification de l'argile de la région d'Adrar et au préalable une présentation des différentes méthodes utilisées pour identifier notre argile.

Par ailleurs cet article regroupe les caractéristiques céramiques préliminaires de l'argile de la région d'Adrar (Sbaa).

Finalement, nous avons ciblé l'orientation de notre argile dans le secteur de fabrication des produits céramique (Brique de terre cuite).

2 Contexte géologique du site

La région d'étude se situe dans l'Ouest-Sud-Ouest du Sahara algérien au niveau de la plate-forme saharienne (Fig.1). Des ergs (massifs de dunes de sable) et des regs (plateaux caillouteux) constituent les paysages de la zone

étudiée. Seuls les monts d'Ougarta présentent des reliefs significatifs.
Par ailleurs, les affleurements se rencontrent dans le lit des oueds limités par des hamadas plateaux [1].
Dans ce contexte géologique particulier la région d'Adrar dispose d'une grande réserve de dépôts sédimentaires de nature argileuse ainsi que d'un lit de sable de dune d'épaisseur énorme [2].

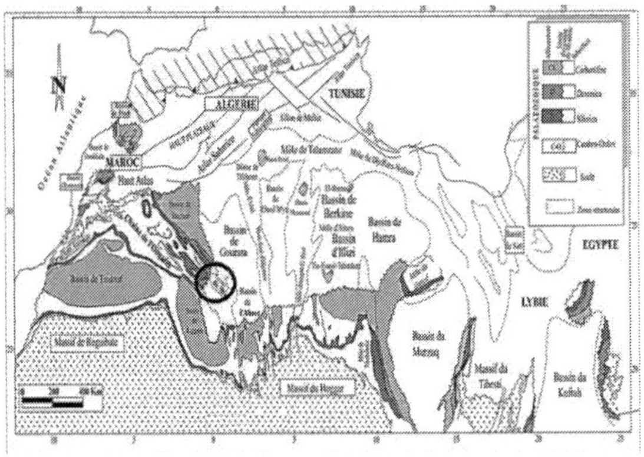

Fig. 1 Carte des principaux éléments morpho-structuraux de la plate-forme saharienne (modifiée d'après Boote et al., 1998)[1].

3 Matériau et méthodes

3.1. Matériau

Les sols sont différents d'une région à une autre, pour cela, une identification de leurs caractéristiques s'impose. L'argile étudiée se localise dans le crétacé inférieur communément appelé continentale intercalaire. Dans la première partie du secteur des sédiments du crétacé inférieur sont recouverts par une épaisse couche de dépôts récents (sable, limon, débris de grés, galets de quartz et localement d'anhydrite).
Le site est situé à quelques kilomètres au nord de la ville d'Adrar (Est de la RN6). Le matériau a une couleur rouge.

3.2 Méthodes

Dans notre étude nous devons identifier le gisement d'argile de la région d'Adrar (Sbâa). La distribution granulométrique a été déterminée par deux méthodes complémentaires, le tamisage par voie humide et une analyse par sédimentation, respectivement selon les normes XP P94-041 [4] et NF P 94-057 [5]. Les propriétés plastiques de la fraction fine, particules de taille inférieure à 400 µm, ont été mesurées comme il est défini dans la NF P94-051 [6]. La masse volumique des particules solides (Gs) a été mesurée en utilisant un pycnomètre NF P 94-054 [7].
Les analyses des compositions chimiques sommaires sont effectuées au laboratoire LTPO à Béchar (Algérie).

Les essais réduits sur l'argile pour déterminer les caractéristiques céramiques ont été réalisés au niveau du laboratoire CETIM à Boumerdès (Algérie).

3.3 Résultats et discussions

3.3.1 Caractéristiques géotechniques et chimiques

Selon les classifications LCPC, cette argile est très plastique (At), et d'activité normale (Ac=Ip/%<2µm). La surface spécifique totale déduite de l'essai au bleu de méthylène par la formule de Tran Ngoc Lan (1977) [3], est de 230,23m^2/g. Le tableau 1 résume les résultats d'identification de l'argile étudiée.

Tableau 1. Propriétés de l'argile utilisée.

Paramètres	Symbole	Valeurs
Poids volumique des grains solides	γs (g/cm^3)	2,60
Limite de liquidité	ω_L (%)	81
Limite de plasticité	ω_p (%)	34
Indice de plasticité	I_P	47
Limite de retrait	ω_R (%)	19,34
Sable	F1 (%)	10
Limon	F2 (%)	54
Argile	F3 (%)	36
Activité des argiles	A_C	1,30
La valeur de bleu de méthylène	VB (%)	11

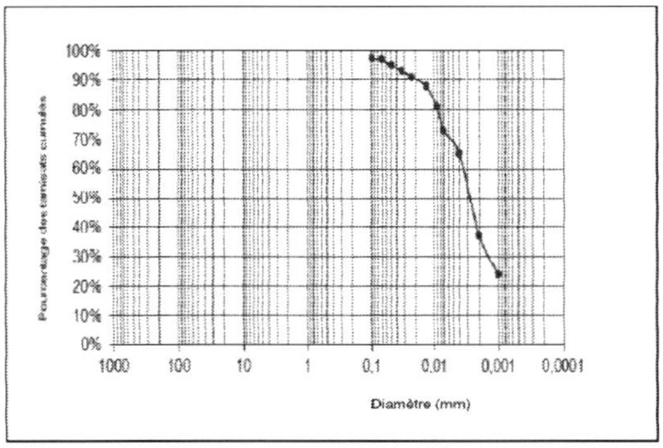

Fig. 2. Courbe granulométrique de l'argile de Sbâa

Suite aux résultats obtenus nous pouvons dire que l'argile est un matériau silico-calcaire faiblement gypseux. Le tableau 3 regroupe les valeurs de composition chimique sommaire de l'argile étudiée.

Tableau 3. Composition chimique sommaire de l'argile.

Paramètres	Valeurs (%)
SO_4^{2-}	0,41
$CaCO_3$	3.6
Cl^-	0,14

Insolubles (SiO_2- Al2O3 - Fe2O3- CaO- MgO)	95,92

3.3.2 Caractéristiques céramiques

- *Propriétés de façonnage et de séchage*

Le tableau 4 ci-après résume les résultats obtenus des essais de façonnage et de séchage.

Tableau 4. Propriétés de façonnage et de séchage de l'argile d'Adrar.

Caractéristiques de séchage				
Régime de séchage	**Humidité de façonnage (%)**	**Retrait sur sec (%)**	**Résistance mécanique en (kg/cm²)**	**coloration**
Lent	23,1	7,8	50,14	Rouge

L'échantillon d'argile d'Adrar (Sbaa) présente un bon aspect physique. Après séchage lent il ressort ce qui suit :

a)Sur le plan céramique :

-Retrait sur sec et humidité de façonnage sont acceptables pour l'argile de Sbâa.

-Résistance mécanique sur sec est admissible.

b) Sur le plan aspect physique :
- Bon aspect physique.
- Coloration : Rouge.

- *Propriétés de cuisson à 900°C*

Les paramètres de cuisson de l'argile sont regroupés dans le tableau.5

Tableau 5. Propriétés de cuisson à 900°C de l'argile étudiée.

Perte au Feu	**Absorption d'eau en (%)**	**Retrait sur cuit (%)**	**Poids volumique (g/dm³)**	**coloration**
10,2	6,8	2,4	1799	Rouge

Le tesson de l'échantillon cuit à 900°C présente les caractéristiques suivantes :

a)Sur le plan céramique
- Le retrait total : Trop élevé
- Absorption : Acceptable

b) Sur le plan aspect physique
-Coloration : Rouge.

4 Conclusion

Dans le contexte géologique, la région d'Adrar est située dans la cuvette limitée par deux grandes bassins de sédimentaire le bassin de Timimoun et le bassin de Reggane. Devant cette formation la région dispose d'une grande réserve de sédiments de nature argileuse continentale intercalaire.

Cette étude a permis :

- d'utiliser des matériaux locaux, et d'identifier les caractéristiques physico-mécanique et chimique de l'argile de la région d'Adrar dans le but de l'utiliser comme matières de base pour la fabrication des produits céramiques tels que brique, tuile et carreaux.

- de présenter les caractéristiques géotechniques, ainsi que la composition chimique sommaire de l'argile d'Adrar.

Nous avons mis en évidence à travers cette étude les caractéristiques céramiques notamment les propriétés de façonnage et de séchage puis les propriétés de cuisson à 900°C qui sont conformes aux recommandations des exigences céramiques.

Enfin, à travers les résultats de cette étude, nous pouvons avancer que :

-L'argile d'Adrar (Sbâa) peut faire l'objet de matière de base pour la fabrication de briques de terre cuite.

-Moyennant un ajout correcteur du retrait total excessif, il reste cependant à étudier la composition minéralogique et chimique détaillées, cela pour bien orienter l'argile d'Adrar (Sbâa) vers l'industrie des produits céramiques.

Références

1. Boote D.R.D., Clark-Lowes D.D. et Traut M.W. *Paleozoïc petroleum systems of North Africa. In : Petroleum Geology of North Africa*, éds. D.S. Macgregor, R.T.J. Moody and D. D. Clark-Lowes, Geol. Soc. Spec. **Pub., 133, 7-68**(1998).
2. Gautier E.-F., Larnaude Marcel. *L'oued Saoura. In: Annales de Géographie,* t. 30, **n°163**, **pp. 50-59,** 1921.
3. Tran Ngoc Lan. "*Nouvel essai d'identification des sols : l'essai au bleu de méthylène*", Bulletin de Liaison des Ponts et Chaussées, 88, **pp. 136-137**, (Mars - Avril 1977).
4. AFNOR. XP P 94-041. Sols : reconnaissance et essais –Identification granulométrique - Méthode de tamisage par voie humide. AFNOR, (1995).
5. AFNOR. NF P 94-057. Sols : reconnaissance et essais – Analyse granulométrique des sols - Méthode par sédimentation. AFNOR, (1992).
6. AFNOR. NF P 94-051 ; Sols : reconnaissance et essais – Détermination des limites d'Atterberg – Limite de liquidité à la coupelle - Limite de plasticité au rouleau. AFNOR, (1993).
7. AFNOR. NF P 94-054 ; Sols : reconnaissance et essais – Détermination de la masse volumique des particules solide des sols - Méthodes du pycnomètre à l'eau. AFNOR, (1991).

The effective complex permittivity stability in filled polymer nanocomposites studied above the glass transition temperature

F. Elhaouzi[1,2], A. Mdarhri[1], M. Zaghrioui[3,4], C. Honstettre[4], I. El Aboudi[1], A. Belfkira[2] and M. El Azhari[1]
[1] FSTG -Cadi Ayyad University, LMCN, Marrakesh, Morocco
[2]FSTG -Cadi Ayyad University, LBCM, Marrakesh, Morocco
[3] François Rabelais University, GREMAN UMR 7347-CNRS,IUT, Blois, France
[4]François Rabelais University, GREMAN UMR 7347-CNRS, UFR Sciences, Tours, France

*Corresponding authors: a.mdarhri@uca.ma and elhaouzifatiha@gmail.com

Abstract. The temperature effecton the dielectric response of nanocomposite at low frequencies range is reported. The investigated samples are formed by a semi-crystalline ethylene-co-butyl acrylate (EBA) polymer filled with three concentrations of the dispersed conducting carbon black (CB) nanoparticles. The temperature dependence of the complex permittivity has been analyzedabove the glass transition temperature of the neat polymer matrix $T_g = -75°C$.For all CB concentrations, the dielectric spectra follow a same trend in frequency range 100-106Hz. More interestingly, the stability of the effective complex permittivity $\varepsilon = \varepsilon' - i\varepsilon''$ with the temperature range of 10-70°C is explored. While the imaginary part of the complex permittivity ε'' exhibits a slight decreasewith temperature, the real partε' shows a significant reduction especially for high loading samples. The observed dielectric response may be related to the breakup of the three-dimensional structurenetwork formed by the aggregation of CB particles causing change at the interfaceEBA-CB.This interface is estimated bythe volume fraction of constrained polymer chain according to loss tangent data of dynamic mechanical analysis.

1 Introduction

The addition of nanoparticles to a polymer matrix has received much attention over several decades[1]. Nanoparticles offer improved mechanical, electrical and thermal properties of polymer systems [2]. Dueto their interesting dielectric properties, nanocomposite materials with polymeric matrix have a great potential for advanced applicationssuch as electromagnetic shielding materials, dielectric resonator antennas, capacitors, electro-active materials to cite a few[3,4,5]. Challenge now is focused on increasing the polymer permittivity without compromising the other characteristics especially ease and quality of polymer processing. For example high permittivity values are needed for reducing the applied electric field to actuate electro-active materials for artificial muscle applications [6]or for increasing the energy density of capacitors [7].
Carbon Black (CB) nanoparticles filled elastomers are among the most commonly nanocomposite widely studied because lightweight, large area, low cost, high flexibility [8], and temperature requirements exist. In addition, the use of nanoscale high dielectric constant fillers make possible to manufacturing polymer thin films at thickness of the order of microns. This study aims at the understanding the temperature dependence of theeffective complex permittivity $\varepsilon = \varepsilon' - i\varepsilon''$ of ethylene-co-butyl acrylatecopolymer (EBA) filled with spherical nanoparticles of CB. Three composite samples containing with fraction volume of CB varying from 6.8 to 19.9% have been examined.Thedirect current electrical conductivityσ_{DC}versus temperature of this series of composites have been already reported by us elsewhere [9].From that study, it is found that σ_{DC}shows a decreasing tendency as the temperature rises indicating a partial breakdown of the CB network in the host matrix. In addition, the CB network restricts the motions of the polymer chains above the glass transition temperature of the polymer and consequently increase the thermal stability as is probed by thermogravimetric analysis tests. Here, attempt has made to investigate the stability of the

electrical properties under dynamic external filed as function of temperature in terms of the effective complex permittivity. Three selected composite samples EBA/CB were studied. The frequency of the applied field wastaken between 100and 10^7 Hz for various temperatures in the range 10–70°C.

2 Experimental

2.1 Materials

The samples of EBA copolymer filled with acetylene carbon black used in this investigation were obtained from Borealis AB (Sweden)[9]. The butylacrylate monomer (EBA) contains butylester side groups, providing a certain polarity and a relatively low crystallinity (about 20% in volume). The glass temperature transition of the neat polymer matrix is $T_g = -75°C$ [10]. A three samples with nominal carbon black volume fraction has been studied.

2.2 Differential thermal analyzes (ATD)

The differential thermal analyzes (ATD) are carried out between temperatures of 20 to 180 °C. 20 mg of sample are placed in a platinum crucible. The acquisition takes place under ambient conditions and the heating rate is programmed at a rate of 10 °C/min.

2.3 Raman measurements

Raman spectra were collected at room temperature using Invia Reflex device (Renishaw). The laser line 514 nm was used. The scattered light was collected in backscattering configuration with a holographic grating (2400 lines/mm) to diffract the light on a CCD detector. The frequency range studied was 900-2000cm^{-1}

2.4 Dielectric measurements

For the dielectric measurements, the samples were prepared as square and the electrical contacts were formed by sputtering gold-palladium of 100 nm thickness on the opposite sites of the sample. The measurements were carried out, in the frequency (F)range from 100to 10^7Hz and temperature from 10 to 70°Cusing an Agilent 4294A precision impedance analyzer. The permittivity may be a complex number quantity, in order to represent both stored energy and losses within the material. Under dynamic external fields, the permittivity $\varepsilon$and conductivity $\sigma$are related via equation:

$$\varepsilon = \varepsilon' - i\varepsilon'' = \varepsilon' - i\left(\varepsilon''_d + \frac{\sigma}{2\pi F\varepsilon_0}\right) \quad (1)$$

The two terms forming the imaginary component represent polarization(ε''_d) and conduction ($\frac{\sigma}{2\pi F\varepsilon_0}$) losses that it is not possible, in practice to be measured separately. $\varepsilon_0 = 8.854 \times 10^{-12}\mathrm{F\,m^{-1}}$is the permittivity of vacuum.

3 Results and Discussion

The differential thermal analysis (DTA) for elastomer matrix filled with different CB volume fractionreported in Figure 1.TheDTA curves show a small peak endothermic between 75°C and 98°C. Melting peak indicates the melting of crystals part on the composite. The peaks have become broader and lower with increasing the filler volume fraction indicating the effect of CB on the thermal stability of the matrix. This result agree with the data obtained from the thermogravimetric analysis (TGA) [9].

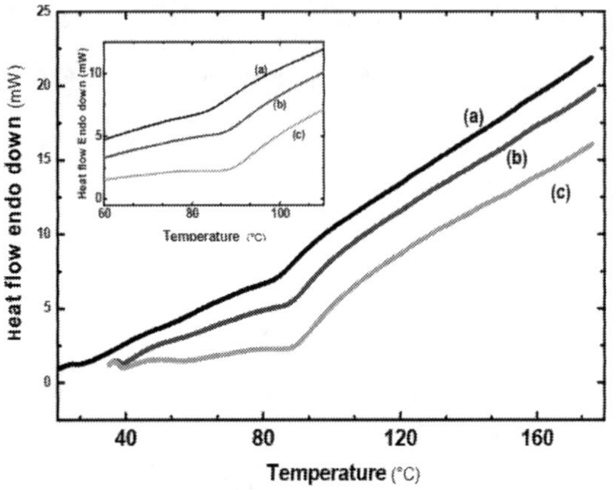

Fig. 1.DTA curves of ethylene-co-butyl acrylate polymer filled with three volume fractions of carbon black: (a) 6.8%, (b) 15.8% and (c) 19.9% versus of temperature.

Raman Spectroscopy was used to assess the affinity between the CB particles and the EBA elastomeric matrix. The Raman spectra of the CB/EBA composite samples are reported in the Figure 2 with twotypical peaks located approximately at 1331 and 1580cm^{-1} that are assigned to the disordered graphite structure (D band) and tangential stretching mode of carbon-carbon bands(G Band) . The Raman spectra exhibit well defined Raman peaks for CB showing that their structural characteristics are not affected when the carbon black are embedded in the polymer matrix [11].

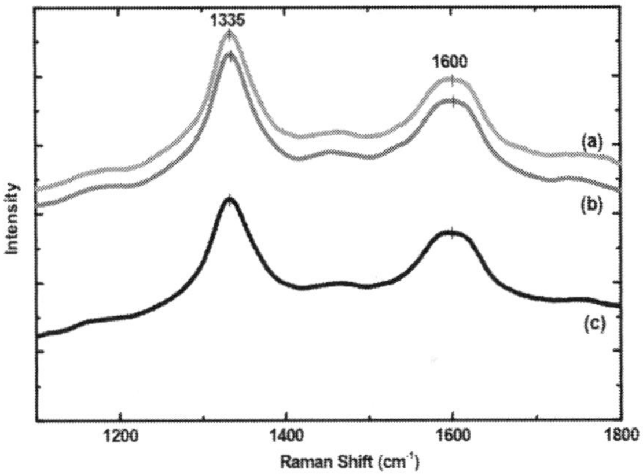

Fig. 2. Raman spectra of ethylene-co-butyl acrylate polymer filled with volume fractions of carbon black: (a) 6.8%, (b) 15.8% and (c) 19.9%

We turn now to the main focus of our study and examine the temperature dependence of the effective complex permittivity. The obtained spectra exhibit high values of both parts of the complex permittivity in the range of frequency considered here indicating a best dispersion of CB into the EBA matrix. In this frequency range, the spectra variation is generally attributed to the interfacial polarization [12] that occurs in heterophase materials. Figure 3 represents thevariation of real ε' and imaginaryε'' parts of the complex permittivity ε versus the temperature for threeCB volume fractions at two frequencies (a) of 10kHz and (b) 1MHz representative of capacitor or actuator applications.Three remarks can be addressed. Firstly, both ε' and ε'' increase with increasing CB volume fractions. This trend is obvious sincethe density of CB/EBA interfaces increases by adding more CB amount. In fact, theattractive interactions between fillers and polymer result in a so-called constrained layer[13]. The volume fraction of constrained polymer chain$V_c$can be estimated according to the dynamic mechanical analysis (DMA)[13].TheV_c values that are 0.342, 0.375 and0.422 are found to increasewith the three CB concentrations 6.8, 15.8 and 19.9%.Secondly, we observea slight decrease of the imaginary part ε'' with temperature for all composites samples as can be seen in Figures 3. This behavior agree with the result reported in the Ref 9 concerning the dc electrical conductivity versus temperature and is attributed to the thermal expansion coefficient (TEC) of the matrix which is larger than that of the CB [14,15].Thermal expansion resulting from temperature increase causes greater separation between the CB particles and hence leading to a decrease of losses.However, the real part ε' behaves differently with temperature. When the temperature exceeds 40°C a significant decrease is clearly observed especially for high loading samples. We suggest that TEC has more effecton the ε' than that on ε'' and consequently reduces the energy stored in the composite material.

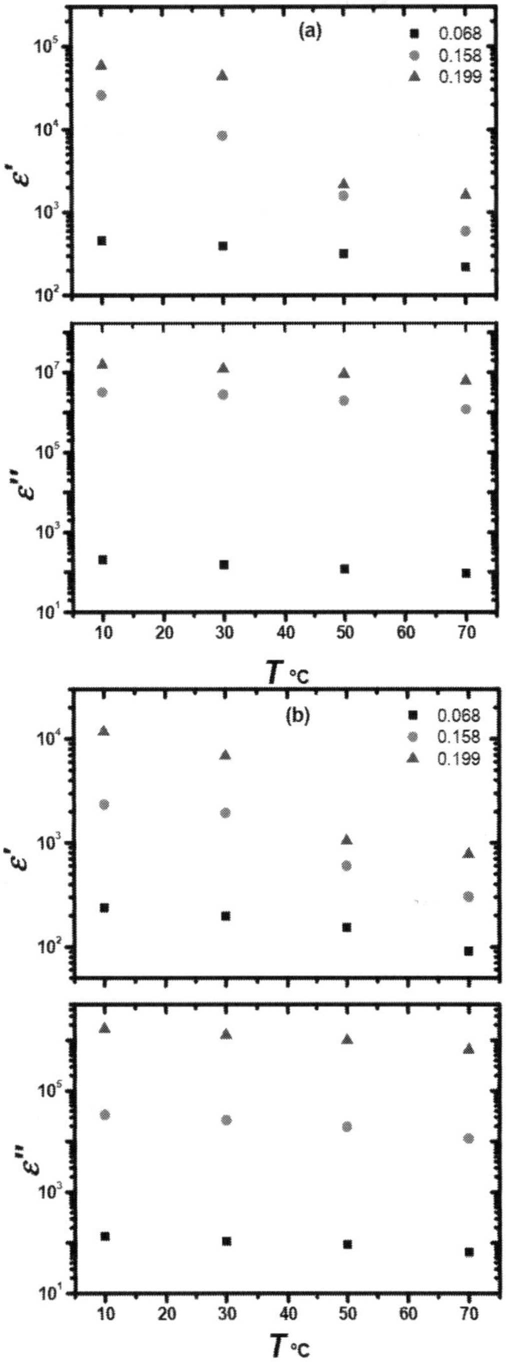

Fig. 3. Temperature dependence dependenceof real and imaginary parts of the effective complex permittivity of EBA/CB for three volume fractions of CB at two frequencies (a) 10 kHz and (b) 1 MHz

4 Conclusion

This paper probesthe dielectric propertiesof nanocomposites formed by incorporating highly conductive carbon black particles in an insulating ethylene butylacrylate matrix. The effective complex permittivity stability in function of temperature is studied in a radio frequency range and above the glass transition temperature (T_g) of the polymer matrix. At the first time, both real ε' and imaginary ε'' parts of the complex permittivity are found to increase with the CB amount. The volume fraction of constrained polymer chain V_c calculated from the DMA tests is used to explain the effect of the CB on dielectric response of the matrix. Secondly, the effect of the temperature on the complex permittivity exhibits a difference behavior between the real and imaginary parts of the permittivity. A small decrease is only observed for the imaginary part ε'' and it is attributed to the difference of the thermal expansion coefficient (TEC) of the matrix and CB particles. However, the real part ε' of the complex permittivity is found to be more sensitive to temperature especially for high loading composite samples. More investigations are needed for a best understanding of the permittivity stability in particularly at wide temperature and high frequency ranges.

Acknowledgments

The authors and especially F. Elhaouzi thank CNRST-Morocco for the Excellence Scholarship.

References

1. H. S. Nalwa, *Handbook of Organic-Inorganic Hybrid Materials and Nanocomposites*, (2003)
2. N. Lombardo,compos scitechnol**67**, 728–736 (2007)
3. T. A. Ezquerra, F. Kremer, and G. Wegner, *in Dielectric Properties of Heterogeneous Materials, Progress in Electromagnetics Research*, edited by A. Priou, ed. (Elsevier, New York, 1992)
4. C. Brosseau, J. Phys. D **39**, 1277 (2006).
5. Mdarhri A, Brosseau C. *Prospects in filled polymers engineering: mesostructure, elasticity network, and macroscopic properties*. Brosseau C, editor. Trivandrum: Research Signpost (2008)
6. V. Bobnar, B. Vodopivec, A. Levstik, M. Kosec, B. Hilezer, and Q .Zhang, Macromolecules, **36**, 4436 (2003)
7. Y. Rao, S. Ogitani, P. kohl, and C. Wong, J. Appl. Polym. Sci. **83**, 1084 (2002)
8. Bibhu Prasad Sahoo, KinsukNaskar, K. AbhinavDubey, Ram N. P. Choudhary, Deba Kumar Tripathy, , J Mater Sci**48**, 702–713 (2013)
9. A. Mdarhri, C. Brosseau, M. Zaghrioui, and I. El Aboudi, J. Appl. Phys. **112**, 034118 (2012)
10. C. Brosseau, A. Mdarhri, and A. Vidal,J. Appl. Phys. **104**, 074105/1-9 (2008)
11. Bokobza, J. Bruneel, M. Couzi, journal of carbon research **1**, 77-94 (2015)
12. A.Patra , D. K.Bisoyi, J Mater Sci**45**, 5742–5748 (2010) 45:5742–5748
13. F. Elhaouzi, A. Nourdine, A. Mdarhri, I. El Aboudi, M. Zaghrioui,Polym Compos. (to be published)
14. A. Sadezky, H. Muckenhuber, H. Grothe, R. Niessner, U. Poschl, Carbon **43**, 1731–1742 (2005)
15. R.N. Rothon, *Particulate-Filled Polymer Composites*, 2nd Edition, (United Kingdom 2003)

Study of the mechanical and chemical properties of potassium manganese phosphate glasses

W. Ahmina[1], M. El Moudane[2], M. Zriouil[1], M. Cherraj[3] and M. Taibi[4]
[1] Université Mohammed V de Rabat, Faculté des Sciences, Laboratoire de Chimie du Solide Appliquée, Rabat, Morocco.
[2] Université Mohammed V de Rabat, Faculté des Sciences, Laboratoire des Matériaux, Nanotechnologies et Environnement, Rabat, Morocco.
[3] Université Mohammed V de Rabat, Faculté des Sciences, LMM, dep of physics, Rabat, Morocco.
[4] Université Mohammed V de Rabat, Ecole Normale, Supérieure Takadoum, Laboratoire de Physico-Chimie des Matériaux associé à l'AUF (LAF 502), Rabat, Morocco.

Abstract. Homogeneous phosphate glasses having the general formula $20K_2O$-xMnO-(80-x)P_2O_5 were synthesized using a melt-quenching method. The amorphous nature of the samples was asserted by X–ray diffraction. The glasses are studied in order to evaluate the influence of MnO addition on the chemical and mechanical properties, such as, dissolution rate, average compressive strength (Rc), Young's modulus (E) and limit strain (ε_{lim}). Some studies show that the addition of MnO significantly reduces the degradation rate (D_R) of soluble phosphate glasses and enhanced chemical stability of both alkali-resistance and acid-resistance. MnO interfere with the chemical structure of the glass, and consequently improve their mechanical properties. Indeed, the mechanical measurements showed that the glass structure was modified by the addition of MnO oxide, becoming more rigid, and as a result, increasing the compressive strength (Rc), Young's modulus (E). These new materials could be an important opportunity for applications in the fields of building, construction, civil engineering, etc…

Keywords: Phosphate glasses, mechanical properties, chemical durability, compressive strength, elastic modulus, strain limit.

1 Introduction

Nowadays, the study of the phosphate glasses represents a special part in the large field of material science because of their applications in the different domains [1-2]. The scientific and technological applications of glasses have encouraged researchers to custom-build the novel glasses for the industry. Glasses are receiving considerable attention due to their unique physical properties like hardness, good strength, transparency, excellent corrosion resistance and high chemical stability [3-5]. These properties of glasses to a large extent are controlled by the composition, structure and nature of the bonds of the glasses formed. Phosphate based glasses have several advantages such as bioactivity; they are degradable and non-toxic and have been used as soluble inhibitors and conversion coating materials [6-8]. However, those glasses present poor mechanical and chemical durability properties when compared to silica based glasses. One approach to increasing these properties is to add various oxides to phosphate glasses, such as PbO, MgO, and TiO_2, and MnO, etc. These oxides lead to the formation of Pb-O-P, Mg-O-P, Ti-O-P and Mn-O-P bonds which replace the easily hydrolysable P-P-P bonds and improve dramatically the chemical durability of the modified phosphate glasses [9-13]. Similarly, the properties of these glasses in terms of mechanical, structural, and electrical properties can be also improved by adding some oxides cited above. The presence of P–O–Mn bands in higher concentration makes the glass more hydration resistant. The addition of metal oxides to the glass leads to a depolymerisation of the network, with the breaking of P–O–P linkages and the creation of non-bridging. Mechanical strength is a crucial factor for structural application of a material. Although all the characters of applied force, temperature, shape, and surface defects significantly influence the strength of a glass product, the mechanical properties of a glass are basically determined by its elastic module values (e.g., bulk, shear, and Young's modulus). To the best of our knowledge, there is no work related to the effect of metallic oxides addition on the mechanical and chemical and properties of a K-containing phosphate glass. The aim of this work was to investigate the effect of MnO addition on the dissolution rates and mechanical properties of potassium phosphates glasses, such as degradation rate (D_R), compressive strength (Rc), Young's modulus (E), and limit strain (ε_{lim}).

2 Materials and methods

The glass samples were prepared from reagent grades of K_2CO_3, $MnCO_3$ and $NH_4H_2PO_4$, mixed according to the molar composition. Were prepared by mixing proportionally weighed raw materials and melted at 1100°C for 1h. Details of the procedure adopted in the preparation of these glasses have been described in our earlier work [13].

X-ray diffraction patterns were recorded at room temperature with a Simens D5000 diffractometer using CuK radiation.

For the chemical durability analysis, samples they were then immersed into a beaker filled with 50 mL of different medium at 25°C for 6, 12 and 24 hours. The dissolution rate (DR) was determined from the weight loss during the aqueous treatment at 25°C. The dissolution rate (DR) was calculated as:

$$DR = \frac{\Delta w}{S.t}$$

Where Δw is the weight loss (g), S is the sample area (cm^2) before the dissolution test and t is the dissolution time (min).

Compression testing of four glasses (V1), (V2), (V3) and (V4) are also performed on cylindrical specimens of dimensions h = 5mm and diameter = 8mm. These specimens are made by a static compaction with an effort of 5MPa associated with the different dosages of manganese oxide 15, 20, 25 and 30 mol % (Figure 1).

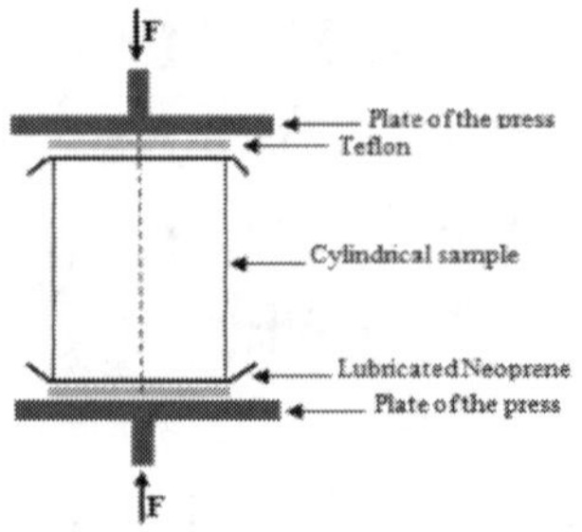

Fig.1. Experimental material of compression test

Glass shows an almost perfect brittle-elastic behavior at temperatures lower than the transformation point. The strain is proportional to the stress according to Hooke's law. If the ends of a rod are exposed to a stress σ, its relative elongation is given by:

$$\sigma = E.\varepsilon$$

Where:

σ: is stress [MPa]

E: modulus of elasticity [MPa]

ε: strain

3 Glass-forming region

The glass forming region is shown in triaxial K_2O, MnO, P_2O_5 diagram by area enclosed by the solid line in fig.2.

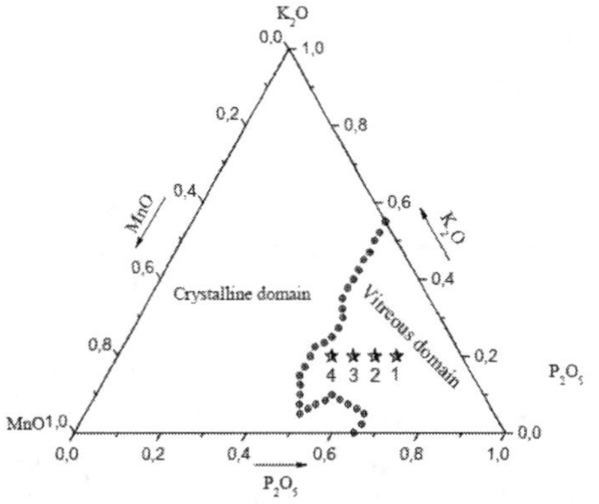

Fig.2. The glass-forming region in the systemK$_2$O- MnO-P$_2$O$_5$ at 1100 °C [12] (see Table 1 for the correspondence between the batch numbers and chemical composition)

Table 1. Composition (mol %), density ρ ($g.cm^{-3}$) of the studied specimens.

Glass no.	K_2O	MnO	P_2O_5	ρ
V1	20	15	65	2.64
V2	20	20	60	2.72
V3	20	25	55	2.81
V4	20	30	50	2.89

3 results and discussion

3.1 X-ray diffraction

As shown in Fig. 3, the X-ray diffraction patterns for all the studied compositions, confirms their amorphous nature, since there is no sharp peak observed in the diffraction pattern.

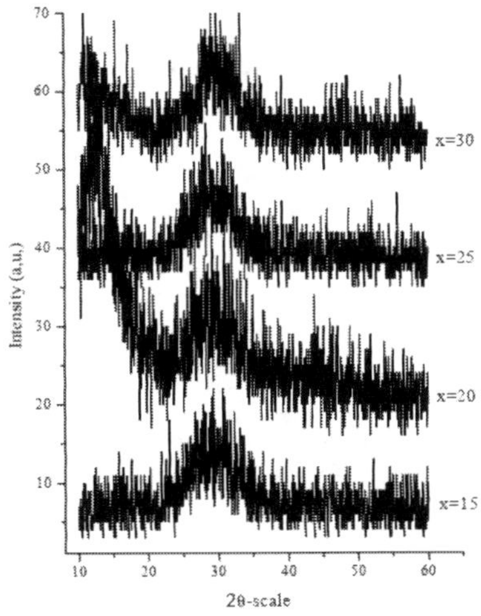

Fig.3. X-ray powder diffraction patterns of the K_2O-MnO-P_2O_5 system.

3.2 Chemical durability

The dissolution rates D_R of the potassium manganese phosphate glasses, measured from their weight loss in distilled water at 25 °C, are given in Table 2 and fig.3.

All the leaching experiments were carried out at pH = 4, 6 and 11. The leaching experiments at ambient temperature were carried out on all glasses from V1 to V4 at all leaching times of 6 h, 12 h and 24 hours, with leaching solutions of HCl (pH = 4), distilled water (pH = 6), and NH_3 (pH = 11).

Table 2 summarizes the values of the dissolution rates DR changes for all of these leaching experiments. In a general manner, all the glasses studied exhibit similar behavior as a function of the pH and whatever the time of alteration. The dissolution rates DR of the leachate decreases in an initially neutral and basic medium whereas it increases in an initially acidic medium.

3.2.1 Variation of the dissolution rates (DR)

Fig. 4 shows the effect of MnO concentration on the dissolution rate at different times of $20K_2O$–$xMnO$–$(80\text{-}x)P_2O_5$ (x=15, 20, 25 and 30) glasses immersed in distilled water at 25◦C. The dissolution

Table 2. Dissolution rate DR ($g.cm^{-3}.min^{-1}$) of $20K_2O$-$xMnO$–$(80\text{-}x)P_2O_5$ glasses in three different middle (pH= 4,6 and 11) at 25◦C for 24 hours.

Initial pH	*Composition (x)*	DR values after an immersion time of 6, 12 and 24 hours		
		6	12	24
4	1 5	$3,89.10^{-6}$	$4,26.10^{-6}$	$3,31.10^{-6}$
	20	$6,91.10^{-6}$	$6,45.10^{-6}$	$5,49.10^{-6}$
	25	$7,76.10^{-6}$	$7,24.10^{-6}$	$6,30.10^{-6}$
	30	$8,51.10^{-6}$	$7,26.10^{-6}$	$6,60.10^{-6}$
6	1 5	$1,69.10^{-5}$	$1,28.10^{-5}$	$1,23.10^{-5}$
	20	$1,54.10^{-5}$	$1,20.10^{-5}$	$9,54.10^{-6}$
	25	$8,91.10^{-6}$	$7,24.10^{-6}$	$6,30.10^{-6}$
	30	$4,17.10^{-6}$	$3,39.10^{-6}$	$2,29.10^{-6}$
11	1 5	$7,08.10^{-6}$	$6,92.10^{-6}$	$6,76.10^{-6}$
	20	$5,50.10^{-6}$	$5,24.10^{-6}$	$4,78.10^{-6}$
	25	$4,78.10^{-6}$	$3,98.10^{-6}$	$3,80.10^{-6}$
	30	$4,46.10^{-6}$	$3,31.10^{-6}$	$2,75.10^{-9}$

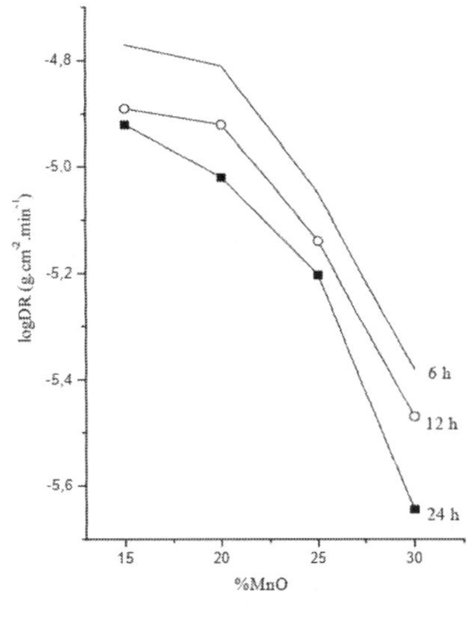

Fig.4. Composition dependence of dissolution rates (DR) for $20K_2O$–$xMnO$–$(80\text{-}x)P_2O_5$ glasses after immersion in distilled water at 25 ◦C for 6 h, 12 h and 24 hours.

rate decreases slightly between 15 and 20 mol% MnO. For MnO additions >20 mol%, the dissolution rate decreases rapidly as MnO concentration increases.

The increasing addition of MnO gives a decrease in solubility. Therefore, in our case we assumed that the higher dissolution rate for less than 20% MnO corresponds to the release of metaphosphate chains into the solution. As it was discussed in our previous works [12,13] from IR and Raman spectra, the addition of less than 20 mol% MnO shows that MnO has a smaller effect on the glass structure with infinite long chains of PO_4 tetrahedra. When increasing MnO concentration, the metaphosphate chains are broken into smaller groups of short chain phosphates such as $P4O_{13}{}^{6-}$, $P_3O_{10}{}^{5-}$ and $P_2O_7{}^{4-}$, which are linked to manganese through P–O–Mn bonds. The formation of P–O–Mn bonds increases the cross-link between the phosphate chains and leads to enhanced chemical durability, in accordance with the increase of Tg [12-13]. These variations can be explained by the effect of the cation substitution on the reticulation of the glasses. The MnO addition causes a shrinking of the phosphate network and creates a more entangled and reticulated metaphosphate chains [14]. In addition, the P-O-K bonds are more sensitive to the hydration process compared to P-O-Mn bonds. These combined effects make the high MnO glasses composition more resistant toward water attack and delay its diffusion into the glasses [15].

3.2.1 Variation of the pH

The table 3 summarizes the pH vales as a function of the time of immersion in three different middles.

Table 3. Composition dependence of pH of $20K_2O$-x MnO–(80-x)P_2O_5 glasses in three different middle (pH= 4,6 and 11) at 25◦C for 24 hours.

Initial pH	Composition (*x mol %*)	pH values after an immersion time of 6, 12 and 24 hours		
		6	12	24
4	15	6.99	7.13	7.17
	20	7.18	7.22	7.32
	25	7.82	7.93	7.98
	30	8.28	8.42	8.78
6	15	8.40	8.75	9.22
	20	8.11	8.53	8.70
	25	7.81	8.02	8.26
	30	7.78	7.96	8.07
11	15	10.55	10.37	10.17
	20	10.58	10.45	10.26
	25	10.61	10.48	10.32
	30	10.79	10.6	10.44

The figure 5 shows the evolution of the pH as a function of the MnO content. It is noticed that the pH of the solution decreases continuously during a times being less than 12 hours, and beyond this value it slows down progressively.

This figure shows that from an initial value of 6 for distilled water, increases continuously with the immersion time to reach a maximum value of 9.22. A similar behavior was observed for the medium of pH = 4.

The increase in the pH of the leaching solutions is related to the inter-diffusion of the hydronium (H^+) ions of the solution and the potassium (K^+) ions of the glass. Thus, the dissolution of potassium manganese metaphosphate glasses can be directly related to the concentration of non-bridging oxygen (NBOs) in these glasses. Indeed, when a water molecule diffuses towards the vitreous network, it becomes immobile in the neighborhoods of non-bridging oxygen (NBOs). The formation of the groups (P-O-H^+) is made simultaneously with the diffusion of the K^+ ion with its hydroxyl (OH^-) co-ion at the glass/solution interface [16].

This process can be schematized as follows:

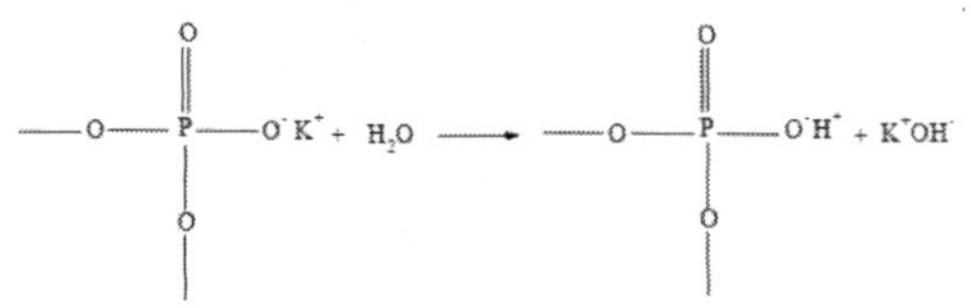

Table 4. Experimental results means the mechanical characteristics of the studied glasses

	E(MPa)	Rc(MPa)	ε_{lim} (%)
V1	417	31.73	28.17
V2	2193	37.45	19.64
V3	2474	42.64	17.73
V4	2759	54.91	15.34

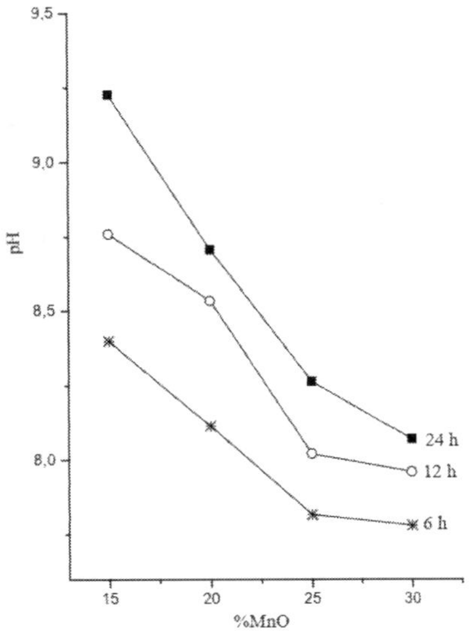

Fig.5. Composition dependence of pH for $20K_2O$–$xMnO$–$(80\text{-}x)P_2O_5$ glasses after immersion in distilled water at 25°C for 6 h, 12 h and 24 hours.

The basification of the pH = 4 and pH = 6 solutions can be attributed to the alkali extraction by ion exchange mechanisms between the protons of the solution and the network modifying elements, this exchange process is responsible for the increase of pH due to the increase of residual OH^- ions in the solution.

For solutions initially basic pH = 11, the evolution of pH over time is less meaningful than for solutions of pH = 4 and 6 for all glasses. Indeed, we notice a drop in pH. It evolves with time and reaches its minimum value at 24 hours certainly due to a consumption OH^- ion.

3.4 Mechanical properties

3.4.1 Effect of the chemical constituents of the glass on the mechanical behavior of the vitreous material

The experimental results of the average compressive strength (Rc), the Young modulus (E) and the limit strain (ε_{lim}) of four glasses, associated with different dosages of manganese oxide, are summarized in Table 4.

Table 4. Experimental results means the mechanical characteristics of the studied glasses

Figs.6 and 7 give the evolution of the compressive strength, the Young's modulus and the limit strain of four glasses (V1), (V2), (V3) and (V4).

As can be seen from Fig. 6, the Young's modulus values increase with MnO content. The influence of the manganese oxide content on the strength characteristics of glasses is illustrated by the data presented in Table 4 and Fig. 6. The glasses contains over 20 mol% manganese oxide, and present a greater rigidity than the glass (V1).The rigidity is generally dependent upon the bond strength and cross-link density. The increase in elastic modulus with an increase in MnO content is attributed to increasing in the cross-link density and rigidity of glass. In addition, MnO enters into the glass network in a position which create a cross-linking between Mn^{2+} ions and the phosphate tetrahedral (Mn–O–P instead of P–O–P). Thus, the enhancement of the structure with addition of MnO beyond 30 mol% results in the increase in modulus, as noticed in Figs. 6-7, and Table 4. That MnO has the ability to enter into the vitreous arrangements.

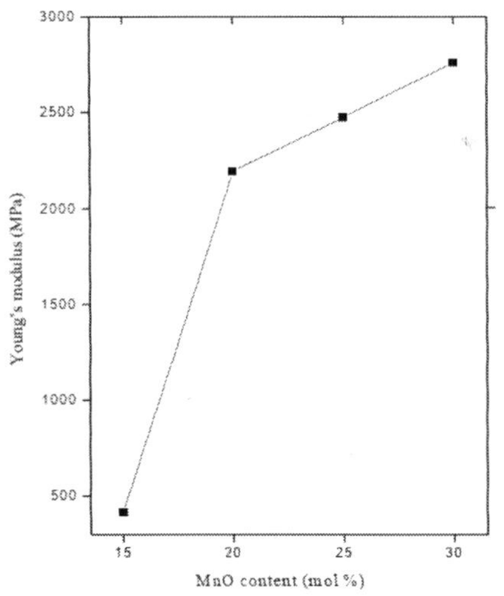

Fig. 6. Variation of Elastic modulus versus MnO content mol%.

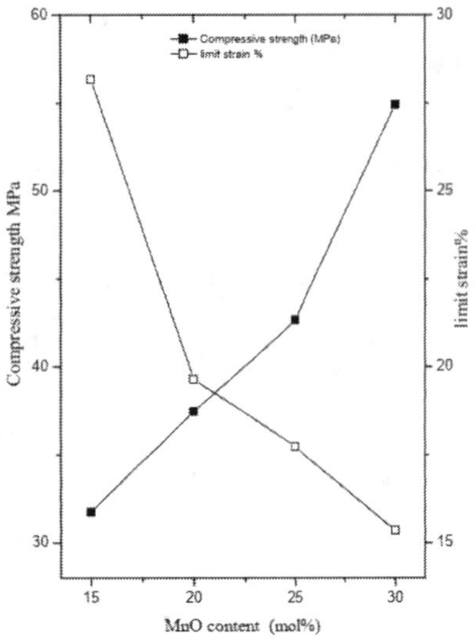

Fig.7. Variation of compressive strength and limit strain versus MnO content in mol%.

and thus, creates a stronger Mn–O–P linkage and increases the O/P ratios in the network are causes depolymerisation in the phosphate network. In consequently, the glass network becomes compact and thereby increases elastic modulus as observed in Figs. 6 and 7. This result is in agreement with our recently studies [12,13]. The addition of manganese oxide to the glass increases the density of glasses which ranged from 2.64 g/cm^3 for (V1) to 2.89 g/cm^3 for (V4) (Table 1). The glass transition temperature also increased with the addition of MnO to the $20K_2O$-xMnO-(80-x)P_2O_5 glasses. Results presented in Tables 1 and 4, and Figures 6 and 7 show that the density and the compressive strength increase with increasing MnO in the glass, stiffens the material and reduces its ability to deform as showed in figure 7.

. This is due to the presence of MnO oxide which provides more density and more strength than K_2O. This indicates that replacement of K^+ ion by Mn^{2+} ion increase the density as well as compressive strength of the of glass.

The potassium manganese phosphate network becomes stronger with increasing MnO content. The improved durability with increased MnO content is attributed to the easy hydrated P-O-P bonds being replaced by more chemically resistant Mn-O-P bonds. The elastic modulus of the glass samples shows an improvement in their mechanical properties with the increase in the MnO content.

This increase in strength is believed to be due to the replacement of monovalent potassium ions by divalent manganese ions which increases the cross-link density in the glass structure and makes the glass stronger. The combination of high strength and good chemical durability of the manganese phosphate glasses, along with the fact that they can be prepared at relatively low temperatures, are valuable advantages for potential technological applications. The observed decrease in limit strain can be attributed to the resulted increase in rigidity and the glass network connectivity. Similar results were found by Hsu et al. [17] and by Hand and Tadjiev [18] when studying, respectively Nb–Sr-phosphate glasses and Mg-silicate glasses.

Conclusions

In this work, we presented the preliminary results obtained for $20K_2O$-xMnO-(80-x)P_2O_5 ($15 \le x \le 30$) glasses investigated by XRD and chemical durability. It was confirmed that the prepared glasses are amorphous. The improved chemical durability with increasing manganese content is attributed to the easily hydrated P-O-P bond being replaced by more chemically resistant Mn-O-P bond. The mechanical properties of potassium manganese phosphate have been investigated. On increasing the amount of MnO in the glass samples, density, Young's moduli and compressive strength were found to increase accordingly. The results revealed that chemical durability is in good agreement with the mechanical properties. Glasses containing more than 20 mol% of MnO had excellent chemical resistance, good mechanical strength and rigidity.

References

1. M.A. El-Ahdal, E.M. Antar, H.H. Mahmoud, F.M. Ezz-Eldin, *J. App. Sci. Res.* **7** (2011) 1434.
2. R. Suwimon, *Sci. Tech.Nuc.Inst.*, **218041** (2014) 5.
3. B. C. Sales, L. A. Boatner, Journal of Non-Crystalline Solids, vol. **79** (1986).
4. Y. B. Peng and D. E. Day, Glass Technology, **32**(1991).
5. N. Aranha, O. L. Alves, L. C. Barbosa, C. L. Cesar, *in Proceedings of the 17th International Congress Glass*, Beijing, China, **7**(1995).
6. I.M. Zin, S.B. Lyon, V.I. Pokhmurskii, Sci. **45** (2003).

7. A.M. Simoes, J. Torres, R. Picciochi, J.C.S. Fernandes, Acta **54** (2009).
8. K. Ogle, R. Bucheit, Conversion coatings, in: A.J. Bard, M. Stratmann (Eds.), *Encyclopedia of the Electrochemistry*, **5**, Wiley-VCH, Weinheim, (2003).
9. R.O. Omrani, S. Krimi, J.J. Videau, I. Khattech, A. El Jazouli, M. Jemal, J. Non-Cryst. Solids **389** (2014) 66.
10. S.T. Reis, M. Karabulut, D.E. Day, J. Non-Cryst. Solids **292** (2001) 150.
11. D.S.L.Weiss, R.D. Torres, S. Buchner, S. Blunk, P. Soares, *J. Non-Cryst. Solids* **386** (2014) 34–38
12. W. Ahmina, M. El Moudane, M. Zriouil, M. Taibi, Phase transitions **89 (11)** (2016) 1051.
13. W. Ahmina, M. El Moudane, M. Zriouil, M. Taibi, J. Mater. Environ. Sci. **7(2)** (2016) 444.
14. D.S. Brauer, Phosphate glasses, in: J.R. Jones, A.G. Clare (Eds.), Bio-Glasses: *An Introduction, John Wiley & Sons*, Chichester, (2012).
15. B.C. Bunker, G.W. Arnold, J.A. Wilder, J. Non Cryst. Solids **64** (1984).
16. M. Abid, M. Et-tabirou, M. Hafid, Journal of Materials Science & Surface Engineering, **5**(4) (2017)
17. S.M. Hsu, J.J. Wu, S.W. Yung, T.S. Chin, T. Zhang, Y.M. Lee, C.M. Chu, J.Y. Ding, J. Non-Cryst. Solids 358 (2012) 14.
18. R. J. Hand, D. R. Tadjiev, J. Non-Cryst. Solids **356** (2010) 2417.

Effect of anionic polyelectrolytes on the flow of activated sodium bentonite drilling mud

Kaci Chalah[1], Abdelbaki Benmounah[1], Khaled Benyounes[2]
[1]Unité de Recherche Matériaux-Procédés et Environnement UR-MPE, Université de Boumerdès, Algérie
[2]Laboratoire Génie physique des hydrocarbures, Université de Boumerdès, Algérie

Abstract. Bentonite is often used in water-based drilling fluids. The xanthan gum is widely used as to increase the viscosity of the bentonite suspension. For the stabilization of the drilled layers, we use filtrate reducers: sodium carboxymethylcellulose low viscosity and cellulose polyanionic low viscosity. The objective of this work is to explain the effect of the polymers on the rheological behavior of the 5% bentonite suspensions. These results will provide practical recommendations for the rational use of different types of additives in water-based drilling muds. Our work is based on rheological trials on a viscometer. The results obtained on the bentonite 5%-xanthane suspension show a rheofluidifying behavior with yield stress conform to the Herschel-bulckly modal. While increasing the concentration of filtrate reducer decreases the yield stress and reduces the viscosity. The effect of CMC LV is more pronounced than PAC L.

1 Introduction

Dans les fluides de forage à base aqueuse, l'eau constitue le milieu de suspension pour les solides. La bentonite confère au mélange un aspect visqueux lors de son mouvement et thixotropique lorsqu'elle est au repos. On ajoute, éventuellement, des polymères pour améliorer la viscosité, la contrainte seuil et le filtrat. Parmi ces ajout, on cite la gomme xanthane et les dérivés cellulosiques : la cellulose polyanionique faible viscosité (PAC L) et la carboxyméthylcellulose à faible viscosité (CMC LV). La principale mission du fluide de forage est de permettre un forage rapide et sûr pendant le forage [1].

La classification des polymères peut être établie en fonction de la nature des charges portées . De ce fait, le CMC, le PAC et le xanthane sont des polyanions. Selon la structure, ils sont classés respectivement en - Polymère linéaire : pour lesquels chaque chaine macromoléculaire est constituée d'un nombre élevé mais fini d'unités monomères (CMC et PAC) [2]. Un tel système correspond à la polymérisation de monomères bivalents - polymère bidimensionnel : ils se présentent sous la forme de feuillets bidimensionnels d'épaisseur comparable à celle des molécules simples (gomme xanthane) [3].

La gomme xanthane est un polyélectrolytique, et comporte un grand nombre de sites ioniques et ayant une continuité des régions d'interactions ioniques. Une fois dissous dans un solvant polaire comme l'eau, le polymère se dissocie, et apparaissent des charges sur son squelette et des contre-ions en solution [4].

Lorsqu'on ajoute un polymère à une suspension d'argile, les interactions mènent à l'adsorption du polymère par les surfaces d'argile. Ce phénomène physique de fixation des chaînes de polymère sur les surfaces de particules d'argile se fait par segments de chaînes appelés "trains". Entre deux trains consécutifs, ces macromolécules forment une "boucle". Le segment de chaîne situé avant le premier train ou après le dernier est appelé "queue" [5].

Une argile en solution aqueuse est reconnue comme étant une suspension fortement anionique [6, 7, 8].

Par conséquent, le mécanisme d'adsorption d'un polymère cationique sur une surface chargée négativement se fait par interactions coulombiennes entre les groupes cationiques du polymère et la surface d'argile [9, 10, 11, 12, 13]

Plusieurs auteurs [14, 15, 16, 17] ont abordé l'étude rhéologique et physico-chimique du mélange eau –argile -polymère anionique en prenant en considération l'influence de la salinité. Feret [15] a étudié l'interaction entre le CMC et l'argile en se basant sur l'évolution de l'isotherme d'adsorption, à différentes salinités et concentrations en argile.

2 Matériels et méthodes

2.1 Matériaux et appareillage

2.1.1 La bentonite

La bentonite utilisée provient de gisement de Hamam Boughrara (Maghnia). C'est une bentonite calcique activée au sodium, de couleur blanche, manufacturée par l'entreprise Bental. Une analyse minéralogique par diffraction des rayons X réalisée sur cette argile a montré une composition cristalline à forte présence de smectite (Na). Nous avons pu identifier également un peu d'illite, des traces de kaolinite, feldspath (K et Na) et du quartz [18].

2.1.2 La gomme xanthane

Le xanthane est un polymère anionique produit à l'échelle industrielle par la fermentation aérobie de la bactérie Xanthomonas campestris. C'est un viscosifiant hydrosoluble, légèrement anionique et à plusieurs branches. Elle est largement utilisée dans le domaine des fluides de forage comme viscosifiant, surtout dans la récupération assistée du pétrole.

Il a un poids moléculaire de 2000 à 3000 Kg/mol. Ces divers groupes fonctionnels (carbonylique, carboxylique, hydroxyle et d'autres) et la longue structure ramifié donnent au xanthane ses propriétés de rhéologie uniques (viscosité apparente et contrainte seuil).

2.1.2 Les réducteurs de filtrats

La carboxyméthylecellulose (CMC LV) est utilisée dans les fluides de forage comme un réducteur de filtrat et il peut être aussi utilisée comme un viscosifiant selon sa masse moléculaire. La CMC LV est qualifié de "technique" ou "purifié" suivant la pureté du produit. Dans cette étude, on a utilisé une CMC LV d'une masse moléculaire de 90kg/mol commercialisé par Sigma-Aldrich, est utilisé.

Le polyanionique de cellulose est disponible en deux catégories (viscosité élevée et faible) qui donnent le même degré de réduction de filtrat mais différent degrés de viscosité. Il est aussi un encapsulant des schistes empêchant le gonflement et la désagrégation. Le PAC L est employé principalement comme réducteur de filtrat pour les boues à base d'eau douce et d'eau de mer.

2.1.2 Viscosimètre

Le viscosimètre utilisé est de type couette, le Viscotester VT550, qui contrôle et caractérise des fluides en mode de rotation déterminant leur propriétés rhéologiques à savoir la viscosité, la contrainte seuil. Il consiste à imposer une vitesse de cisaillement et calculer la contrainte de cisaillement et la viscosité apparente. Un régulateur de température de type (HAAKE DC 30) a pour but de maintenir la température voulue. Dans notre cas, tous les essais ont été réalisés à 20°C ± 1°C. Ce viscosimètre est menu d'un logiciel RheoWin qui nous permet d'avoir les caractéristiques rhéologiques selon le type de modélisation.

2.2 Mode préparatoire

On prépare simultanément une suspension de bentonite-polymère en ajoutant la bentonite à 5% dans l'eau distillée par petites quantités pour éviter le phénomène d'agrégation des particules. Après 4h d'agitation, on ajoute le polymère. Le mélange obtenu est agité pendant 24 heures pour assurer une bonne homogénéisation du mélange et une bonne hydratation du polymère. Un précisaillement de 60 s sera effectué pour garder le fluide bien homogène, suivi d'un repos de 60 s. La vitesse de cisaillement est de 0 à 700 s-1, effectué pendant une durée de cisaillement de 700 s. De la même façon, on prépare les solutions de bentonite et des polymères en maintenant l'agitation pendant 24h.

Le fluide à analyser est placé entre deux cylindres de révolution coaxiaux. Pour l'étude des argiles, on a utilisé la géométrie NV. Cette dernière nécessite un faible volume (9 ml), et elle est destinée pour les fluides à faible viscosité.

Les viscosifiants et les réducteurs de filtrats sont ajoutés à différents concentrations (0.1; 0.3 et 0.5). Dans cette étude, on va mettre en évidence l'action des polyélectrolytes sur le comportement rhéologique du fluide de forage à base d'eau.

3 Résultats et discussions

Nous allons précéder en premier lieu à l'étude du comportement rhéologique des solutions des matières premières : la bentonite, le xanthane, le PAC L et la CMC LV. Les résultats obtenus nous servent de base pour mieux approfondir l'effet des polyélectrolytes sur le comportement rhéologique de la solution de bentonite.

Le comportement rhéologique des suspensions de bentonite de Maghnia est décrit par le modèle de Herschel-Bulkley (éq. 1) notamment pour les fortes vitesses de cisaillement. Il devient plus visible à partir de 3% d'argile [3]. Tandis que les polymères linéaires ont un comportement sans contrainte seuil défini suivant le modèle de Cross (éq. 2).

Le modèle d'Herschel-Bulkley est celui permettant de décrire la plupart des fluides plastiques (fluides à seuil), la courbe d'écoulement de tels fluides finit par devenir rectiligne au-delà d'une certaine contrainte critique appliquée. Il s'agit en fait d'une loi de puissance à laquelle on a ajouté le terme τ_s.

$$\tau = \tau_s + k\,\dot{\gamma}^n \qquad (1)$$

où : K est la consistance, γ ˙ est la contrainte seuil et n l'indice d'écoulement.

Le modèle rhéologique de Cross est l'un des plus populaires en service aujourd'hui. Il peut être trouvé pratiquement dans le logiciel d'un rhéomètre, et il peut être utilisé pour extraire les nombres significatifs de l'ensemble de la viscosité et de la vitesse de cisaillement.

$$(2)= \eta_\infty + \frac{\eta_0 - \eta_\infty}{1+(C\dot{\gamma})^m}$$

où : η_0 est la viscosité au cisaillement zéro, η_∞ est la Viscosité à l'infini de cisaillement, m est connu sous le nom « constante de taux », C est la constante du temps de Cross.

3.1 Solutions de bentonite, de PAC L, de CMC LV et de xanthane

Les résultats d'essais d'écoulements sur les solutions de bentonite, de CMC LV, de PAC L et de xanthane sont illustrés sur la figure 1a et 1b. L'analyse de ces courbes montre que :

- Le rhéogramme de la solution de bentonite à 5% est caractérisé par une contrainte seuil et un comportement rhéofluidifiant. Ce résultat confirme les travaux de Benchabane et Bekkour [19], Luckham et Rossi [20]. En effet le comportement rhéologique des suspensions eau-bentonite est décrit par une loi de puissance de type Herschel-Bulkley.
- Nous présentons sur la figure 1a et 1b les deux rhéogrammes des solutions PAC L et CMC LV à 0.7%. On y observe que, contrairement aux courbes des suspensions de bentonite et de xanthane, les rhéogrammes se présentent en absence de la contrainte seuil. Conformément à la littérature [21], le comportement rhéologique des solutions de PAC L et CMC LV a été décrit d'une manière satisfaisante par le modèle de Cross.
- la solution de xanthane à 0.3% montre un comportement rhéfluidifiant avec contrainte seuil. Au-dessus d'une certaine concentration, les molécules semi-rigides s'enchevêtrent pour former un réseau stabilisé par des liaisons spécifiques, ponds d'hydrogène, et sont à la base de la structuration responsable du seuil d'écoulement et de la résistance à l'écoulement [22,23]. Pour de fortes vitesses de cisaillement, le réseau est déstructuré et les molécules s'alignent dans le sens du cisaillement en favorisant la diminution de la viscosité [23,24]. Le comportement rhéologique est décrit selon le modèle rhéologique Herschel-Bulkley.

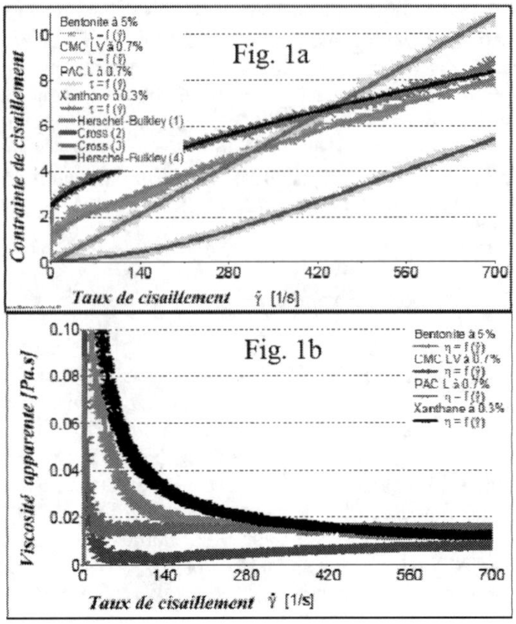

Fig. 1. Comportement rhéologique des Solutions de bentonite de PAC L, de CMC LV et de xanthane : courbes de cisaillement (Fig. 1a) et courbe de viscosité (Fig. 1b)

Les caractéristiques rhéologiques des solutions de bentonite et des polyélectrolytes sont données dans le tableau suivant (Tableau 1). Les courbes de bentonite et de xanthane sont définies par le module d'Herschel-Bulkley tandis que les rhéogrammes de PAC L et de CMC LV suivent le modèle de Cross.

Tableau 1. Caractéristiques rhéologique des solutions de bentonite de PAC L, de CMC LV et de xanthane

Caractéristiques rhéologiques du modèle d'Herschel-Bulkly				
Essais	τ_0	**K**	**n**	**r**
BMg à 5%	1.127	0.0387	0.7920	0.9954
XAN à 0.3%	2.392	0.0723	0.6742	0.9953
Caractéristiques rhéologiques du modèle de Cross				
Essais	η_0	η_∞	**n**	**r**
CMC LV à 0.7%	0.0089	0.0025	-2.2210	0.9989
PAC L à 0.7%	0.0400	-0.0091	-0.0231	0.9995

3.2 Mélanges bentonite (5%)-Polyélectrolytes

3.2.1 Mélanges bentonite (5%)-Xanthane

La gomme xanthane est caractérisée par sa grande masse moléculaire. Son utilisation augmente la cohérence en rendant le mélange plus résistant au cisaillement. La figure 2 met en évidence le comportement rhéologique du mélange bentonite–xanthane.

D'après K. Benyounes et al. [18], les résultats des essais d'écoulements sur les mélanges bentonite (3%)-

https://doi.org/10.1051/matecconf/201814901082

xanthane, à différentes concentrations de xanthane, montrent que la forme de la courbe des mélanges bentonite-xanthane est semblable à celle des solutions de xanthane avec une augmentation de la viscosité apparente et de la contrainte seuil. Ce résultat est en parfait accord avec une étude réalisée sur un tel système à la base d'une bentonite de laboratoire [23]. On remarque également la disparition du plateau de contrainte observé sur la suspension de bentonite [23,18].

La présence du xanthane à différentes concentrations dans la solution bentonite à (5%) a permis d'améliorer les deux paramètres rhéologiques importants à savoir la viscosité et la contrainte seuil. On remarque la disparition de la zone d'agrégation observé sur la suspension de bentonite à 5%, ce qui laisse supposer que le contact entre les particules argileuses n'est plus possible. Dans ce cas, les particules argileuses sont enveloppées par le xanthane et le contact entre les chaînes de polymère va alors former la principale interaction dans le système eau-bentonite-xanthane.

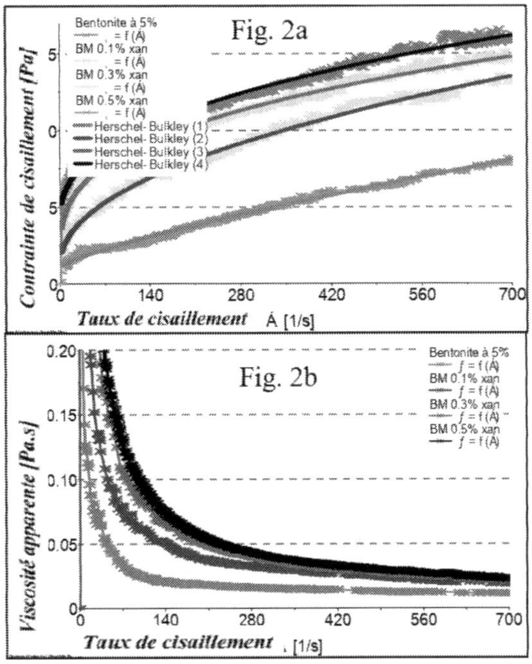

Fig. 2. Comportement rhéologique des Solutions de bentonite-xanthane : courbes de cisaillement (Fig. 2a) et courbe de viscosité (Fig. 2b)

L'ajout de 0.1% de xanthane nous a permet d'améliorer la contrainte seuil de 1,127 à 1.930 Pa et la consistance de 0.038 à 0.530 (tableau 2), ce qui semble intéressant du point de vue technique et économique.

Tableau 2. Caractéristiques rhéologiques du mélange bentonite (5%)-xanthane selon le modèle d'Herschel-Bulkley.

Essais	τ_0	K	n	r
BMg témoin	1.127	0.038	0.792	0.9954
BMg 0.1XAN	1.233	0.530	0.479	0.9952
BMg 0.3XAN	1.644	1.445	0.337	0.9941
BMg 0.5XAN	4.011	0.864	0.403	0.9902

3.2.2 Mélanges bentonite (5%)-PAC L

Le polyanionique cellulose à faible viscosité (PAC L) est utilisé principalement comme réducteur de filtrat. Son effet sur le comportement rhéologique de la solution de bentonite est montré à la figure 3.

On peut classer les interactions en deux grands groupes : une interaction de type électrostatique entre les chaînes anioniques de polymère et les charges positives existant sur les bords des particules d'argile et une interaction entre les charges négatives d'argile et le polymère [5].

A la lumière des résultats obtenus, on note une diminution de la contrainte seuil avec l'augmentation de la teneur en PAC L à cause de la neutralisation des charges des particules par les molécules anioniques.

La viscosité diminue à faible quantité en PAC L (0.3%) à cause de la neutralisation des forces électrostatiques d'attraction des particules argileuses. A une quantité 0.5 à 0.7, la viscosité du système augmente par rapport à la viscosité de la suspension eau-bentonite grâce à l'hydratation des particules colloïdales de PAC L.

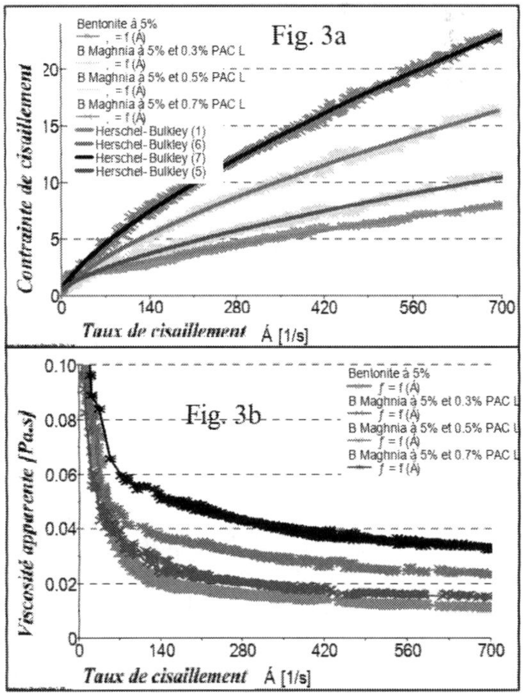

Fig. 3. Comportement rhéologique des Solutions de bentonite-PAC L: courbes de cisaillement (Fig. 3a) et courbe de viscosité (Fig. 3b)

Les caractéristiques rhéologiques du mélange bentonite (5%)-PAC L sont obtenues selon le modèle d'Herschel-Bulkley. Les valeurs de la contrainte seuil, la consistance du fluide et l'indice d'écoulement sont inscrites dans le tableau suivant :

Tableau 3. Caractéristiques rhéologiques du mélange de bentonite (5%)-PAC L selon le modèle Herschel-Bulkley.

Essais	τ_0	K	n	r
BMg témoin	1.127	0.038	0.792	0.9954
BMg 0.3PAC L	0.435	0.1054	0.6949	0.9995
BMg 0.5PAC L	0.091	0.1682	0.6983	0.9977
BMg 0.7PAC L	0.300	0.2017	0.7216	0.9987

3.2.3 Mélanges bentonite (5%)-CMC LV

La carboxyméthylcellulose à faible viscosité est utilisée principalement comme réducteur de filtrat. Les rhéogrammes obtenue sur le mélange bentonite 5%-CMC LV montrent la variation de la contrainte seuil et la viscosité.

Le comportement rhéologique de la suspension de bentonite de Maghnia peut être décrit d'une manière satisfaisante par le modèle de Herschel-Bulkley [18]. Selon A. BENCHABANE et K. BEKKOUR [5], les courbes d'écoulement obtenues avec l'ajout de CMC LV sont décrites par le modèle d'Herschel–Bulkley avec une faible contrainte seuil. Les molécules de CMC LV enrobent chaque particule d'argile, empêchant ainsi tout ré-agglomération après cisaillement. Cela explique la diminution de la viscosité.

Les résultats obtenus montrent une diminution de la contrainte seuil et de la viscosité avec l'ajout de CMC LV dans la solution de bentonite (Tableau 4).
L'ajout de la quantité de CMC LV à 0.3% dans la solution s'accompagne d'une légère augmentation de viscosité, et une forte diminution de la contrainte seuil. Cela s'explique par l'hydratation des particules colloïdales de CMC LV et leur neutralisation des charges des particules argileuses.

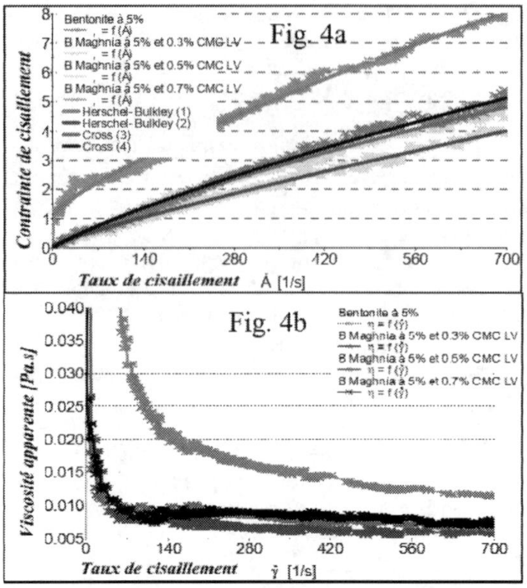

Fig. 4. Impact du CMC LV sur les propriétés rhéologiques de la solution de bentonite à 5%: rhéogrammes de cisaillement (Fig. 4a) et rhéogrammes de viscosité (Fig. 4b)

Les caractéristiques rhéologiques du mélange bentonite (5%)-CMC LV sont obtenues selon le modèle d'Herschel-Bulkley et de Cross. La valeur de la contrainte seuil à 0.3CMC LV est de 0.061 Pa. A partir de 0.5 à 0.7%, les courbent ne présentent pas une contrainte seuil et sont définies selon le modèle de Cross. La viscosité augmente de 0.3 à 0.7% mais reste inférieur à la viscosité de la bentonite sans ajout.

Tableau 4. Caractéristiques rhéologiques de la solution de bentonite (5%)-CMC LV à différentes teneurs en CMC LV

Caractéristiques rhéologiques du modèle d'Herschel-Bulkly				
Essais	τ_0	**K**	**n**	**r**
BMg à 5%	1.127	0.0387	0.7920	0.9954
BMg 0.3CMC LV	0.061	0.0149	0.8491	0.9955
Caractéristiques rhéologiques du modèle de Cross				
Essais	η_0	η_∞	n	r
BMg 0.5CMC LV	0.2527	-0.2342	0.0136	0.9966
BMg 0.7CMC LV	0.1273	-0.1170	0.0252	0.9951

Conclusion

L'étude réalisée sur le mélange bentonite (5%)-xanthane montre que l'ajout du xanthane à la solution de bentonite améliore la viscosité et la contrainte seuil. A 0.1% de xanthane, on note une τ_s = 1.233 Pa avec η100 = 0.063 Pa.s. Tandis que, pour la solution de bentonite, la τ_s = 1.127 Pa avec η100 = 0.027 Pa.s. La 2ème zone de la courbe qui correspond à l'agrégation des particules argileuses a totalement disparue. Le modèle d'Herschel-Bulkley défini bien les courbes obtenues.

Pour étudier l'influence d'un réducteur de filtrat sur les propriétés rhéologiques de la suspension de bentonite, on a utilisé le CMC LV et le PAC L.

- L'augmentation de la concentration en PAC L de 0.3 à 0.7% diminue la contrainte seuil de 1.127 Pa pour la courbe de référence à 0.300 Pa avec 0.7%PAC L. La viscosité se réduit à 0.3%PAC L mais s'améliore avec augmentation à partir de 0.5 à 0.7%.
- L'ajout de CMC LV à la solution de bentonite affecte les propriétés rhéologiques. Les rhéogrammes du mélange de bentonite 5%-CMC LV, avec des concentrations 0.5 à 0.7% en CMC LV, sont définis selon le modèle de Cross. La viscosité augmente de 0.3 à 0.7% mais reste inférieur à la viscosité de la bentonite sans ajout.

References

1. H.C. Darley, G.R.Gray, *Composition and properties of drilling and completion fluids(*Gulf Professional Publishing, Texas,1988)
2. O.H. Houwen, SPE Drilling & Completion **8**, (1993)
3. M. Fontnille, Y. Gnanou, *chimie et physico-chimie des polymères(* Dudon, Paris, 2002)
4. W.Xie, J.Lecourtier , Polym. Degrad. Stab. **38**, (1992)
5. A. Benchabane, K. Bekkour, 39ème Colloque annuel du GFR, pp. 201-205 (2004)
6. Y.Deng, J.B.Dixon, G.N.White, R.H. Loeppert, A.S.R. Juo, Colloids Surf., A **281,** (2006)
7. A. McFarlane, K.Y. Yeap, K. Bremmell, Colloids Surf., A **317,** (2008)
8. P. Mpofu, J. Addai-Mensah, J. Ralston, Int. J. Miner. Process. **71,** (2003)
9. A. Alemdar, N. Öztekin, N. Güngör, ÖI. Ece Colloids Surf., A **252**, (2005)
10. C. Breen, Appl. Clay Sci. **15,** (1999)
11. J. Dau, G. Lagaly, Croatica chemica acta **71**, (1998)
12. S. İşçi, F.S. Güner, ÖI. Ece, N. Güngör, Prog. Org Coat. **54,** (2005)
13. M. Janek, G. Lagaly, Colloid. Polym. Sci. **281,** (2003)
14. A. Alemdar, N. Güngör, F.B. Erim, J. Mater .Sci. Lett. 22,(2003)
15. B. Feret,doctorat de l'Université Pierre et Marie Curie, France,(1992).
16. N. Güngör, A. Alemdar, J. Incl. Phenom. Macro. **33,** (1999)
17. O. M'bodj, N.K. Ariguib, M.T. Ayadi, A. Magnin , J Colloid Interf Sci **273,**(2004)
18. K.Benyounes, A. Benchabane, A. Mellak, Energ Source Part A **32**, (2010)
19. S.Kouloughli, M. B. Lehocine, *Sci.Techn.B* **25**, (2007)
20. P.F. Luckham, S. Rossi, Adv Colloid Interfac **82,** (1999)
21. A. Benchabane, K. Bekkour, Colloid Polym. Sci. **286**, (2008)
22. A. Benchabane, Thèse de doctorat de l'ULP Strasbourg France, (2006).
23. A. Benchabane ,K. Bekkour ,Rheol. Acta **45**, (2006)
24. K.W. Song, Y.S. Kim, G.S. Chang, Fiber Polym **7**, (2006)

Effect of bio-loading on polyvinyl chlorides morphology

Effet du bio-chargement sur la morphologie des polychlorures de vinyle

M.JAMMOUKH[1], K.MANSOURI[1] et B.SALHI[2]
[1]University of Hassan II, Laboratory of Signals, Distributed Systems and Artificial Intelligence, Casablanca, Morocco
[2]University of Mohammed V, Normal Superior School of Technical Education, Rabat, Morocco.

Résumé Permettant d'intégrer la prise en compte des questions environnementales lors de la phase de conception d'un produit, l'éco-conception est une démarche préventive d'amélioration de la qualité écologique de celui-ci tout en vérifiant que le gain environnemental obtenu ne se fait pas au détriment d'autres aspects environnementaux et se focalisant sur la conservation de sa qualité d'usage de ses caractéristiques ou de certaines étapes de son cycle de vie sans faire perdre pour autant la vision globale propre à l'éco-conception. Suite à cette vision, l'utilisation des polymères bio-chargés à base animale, issue d'un mode de production mettant en œuvre des pratiques d'optimisation respectueuses des ressources naturelles de manière durable, s'impose comme changement de manière de penser à remplacer largement l'usage de ressources non renouvelables pour les activités humaines, l'épuisement des gisements de ressources fossiles et à lutter contre l'impact sur le changement climatique généré par cette pratique. Notre contribution tente d'analyser auparavant le comportement mécanique d'une bio-charge et ensuite l'incorporer dans un matériau polymère pour tester l'amélioration des caractéristiques mécaniques et thermiques apportées à ce dernier après addition et optimiser, par la suite, son emploi. C'est une charge, sous forme de poudre ou coupeaux extraits à partir d'une gaine animale. Des essais de traction sont effectués, dans un premier temps, en vue de caractériser mécaniquement le matériau vierge et ensuite le nouveau éco-composite obtenu après l'alliance, à savoir la contrainte maximale à la rupture, la déformation relative et le module de l'élasticité, … etc.

Mots clés: Eco-conception, Polymères bio-chargés, Bio-charge, Eco-composites.

Abstract To integrate the consideration of environmental issues in the design phase of a product eco-design is an ecological quality improvement and preventive approach of it while verifying that the resulting environmental gain is not at the expense of other environmental aspects and focusing on the preservation of its quality of use of its features or certain stages of its life cycle without losing the own global vision of eco-design . Following this vision, the use of bio-laden animal based polymers derived from a mode of production by implementing optimization practices respectful of natural resources sustainably, has become changing way of thinking largely replace the use of non-renewable resources for human activities, the depletion of fossil resources deposits and to fight against the impact of climate change generated by this practice. Our paper attempts to analyze first the mechanical behavior of a bio-load and then incorporated into a polymer material to test improved mechanical and thermal characteristics made to it after adding and optimizing subsequently, its use. This is a load in the form of powder or chips extracted from an animal sheath. Tensile tests are performed in a first step, in order to mechanically characterize the virgin material and eco-composite obtained after the combination, i.e. the maximum tensile strength, the relative deformation and the modulus of the elasticity, etc

1 Introduction

Le choix d'une matière repose, en effet, sur trois principales motivations pour un industriel :
- Optimiser ses coûts et varier son approvisionnement afin d'être toujours compétitif en cas de fluctuation des marchés,
- Obtenir des nouvelles fonctionnalités et propriétés,
- Réduire son empreinte environnementale [1].

Sachant qu'un produit biosourcé ne signifie pas qu'il soit à 100% dérivé de la biomasse, il est fréquent d'avoir des produits mixant plusieurs origines (fossile, minérale, végétale ou encore animale), afin de bénéficier des avantages de chaque constituant [2-7].

Parmi les matériaux basiques choisis pour la caractérisation en perspective, le Polychlorure de vinyle (PVC) souple ou rigide.
Principaux avantages du PVC: Très économique, bonne résistance chimique, intrinsèquement ignifugé.
- PVC souple : Flexible, transparent.
-PVC rigide : Bonne résistance mécaniques, transparence, bonne résistance aux UV [3].

Principaux Inconvénients du PVC : Transformation délicate et peut provoquer une dégradation rapide du matériau, résistance chimique limitée aux solvants, décoloration lors de fortes expositions aux UV, résistance en température limitée [3].

2 Procédure expérimental de l'éco-matériau

Le développement d'un nouveau polymère, sa synthèse à l'échelle industrielle et la mise au point des procédés de sa transformation adéquates sont des étapes longues et difficiles [4]. La voie consiste à mélanger ou à associer des polymères connus pour en combiner les propriétés, à l'instar des métallurgistes qui obtiennent des alliages dotés de propriétés nouvelles en mélangeant différents métaux. Dans une application industrielle, c'est effectivement la compatibilité qui sera un critère déterminant, puisqu'on cherchera à mettre en adéquation les propriétés d'un matériau -éventuellement multiphasé- et un cahier des charges, sans forcément se soucier de la morphologie [5]. Quand deux polymères de natures différentes sont miscibles, on obtient un mélange homogène qui se comporte comme un seul polymère dont les caractéristiques dépendent de celles des polymères « Parents ».

Les véritables mélanges, ceux qui conduisent à des matériaux homogènes, sont des cas d'espèce. Souvent, les deux polymères ne sont pas miscibles (on parlera quand même encore de mélange) et au niveau microscopique, on distingue plusieurs phases [6].

2.1 Elaboration des éprouvettes

2.1.1 Procédé des mélanges

Dans le cas des mélanges, en plus des aspects thermodynamiques et chimiques, il est nécessaire de considérer les aspects mécaniques de la mise en œuvre de mélanges de polymères [7].

Pour des raisons à la fois techniques et économiques, un mélange en solution n'est guère à envisager dans le cadre d'un procédé industriel. Le procédé d'extrusion est bien adapté pour mélanger en continu des matériaux visqueux tels que les polymères [8].

2.1.2 Procédé d'extrusion

En général, les mélanges sont souvent mis en forme à l'état fondu dans des machines plus ou moins standard du type extrudeuses dont le profil de vis devra être choisi avec soin pour "optimiser" la dispersion et la distribution d'un polymère dans l'autre.

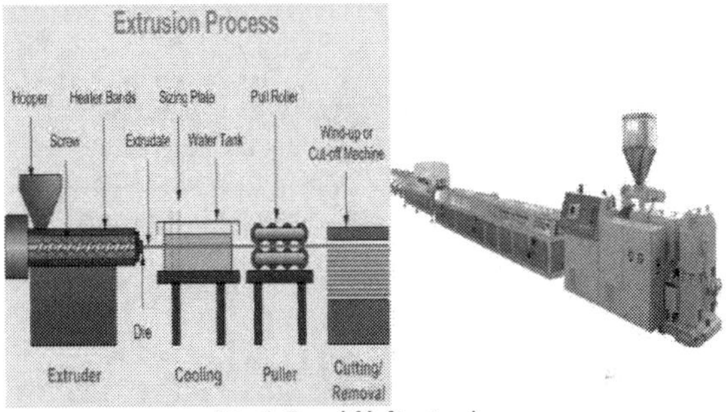

Fig.1. Procédé d'extrusion

Concernant les principes régissant le mécanisme de mélange de deux liquides visqueux non miscibles tels que les polymères, on distingue souvent le mélange dispersif et le mélange distributif dont les caractéristiques sont données ci-après :

Mélange dispersif	Mélange distributif
-Réduit la taille de la phase dispersée. -Fortes contraintes de cisaillement. -Forte viscosité.	-Répartit de façon homogène la phase dispersée dans la matrice

 https://doi.org/10.1051/matecconf/201814901083

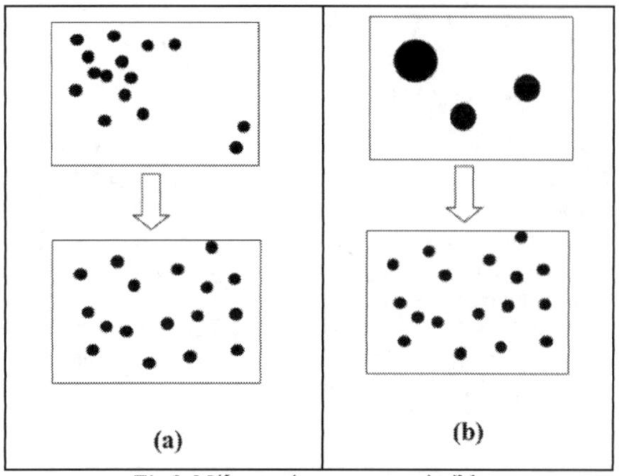

Fig.2. Mélange visqueux non miscible

Les mélanges de polymères compatibles sont rares, toute fois le PVC, par la présence de son hydrogène en α, est miscible avec un grand nombre de polymères [5].
Les mélanges de polymères à base de PVC ont pour but :
• De faciliter la mise en œuvre du PVC lui même, en diminuant à la fois sa température de transition vitreuse (Tg) et sa viscosité.
• D'améliorer sa résistance aux chocs puisque le PVC présente une fragilité importante s'il n'est pas renforcé.
L'utilisation des matériaux conduit à définir les propriétés qualitatives que l'on attend d'eux. Chacune de ces qualités correspond respectivement à une propriété mécanique bien déterminée :

- La ténacité : Résistance à l'extension ou à la compression.
- La résilience : Résistance aux chocs.
- La dureté : Résistance à l'usure.

2.1.3 Formage des éprouvettes de traction

Cette étape permet d'obtenir des éprouvettes pour les essais mécaniques par le procédé de poinçonnage, particulièrement ceux des essais de traction en lots de cinq éprouvettes conformes à la norme en vigueur ISO 527-2, type 1BA et de forme « Haltère ».

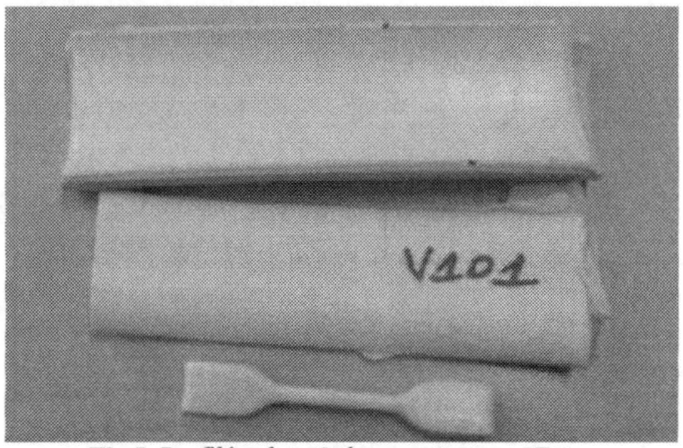

Fig.3. Profilés plats & éprouvette type 1BA
& de forme « Haltère » du PVC rigide vierge

Fig.4. Profilés plats & éprouvette type 1BA
& de forme « Haltère » du PVC rigide bio-chargé

2.2 Mode expérimental

Pour mesurer ces propriétés mécaniques, il est nécessaire d'effectuer des essais. Ces derniers ont pour objet de tester le comportement d'un matériau dans des conditions conventionnelles voisines des situations pratiques [9].

2.2.1 Machine utilisée

Pour l'essai de traction, une machine de traction-compression universelle de type LLOYD Instruments LR50K est mise en œuvre à une vitesse traverse de 1mm/min, une température d'essai de 23°C et une humidité de 50±10%.

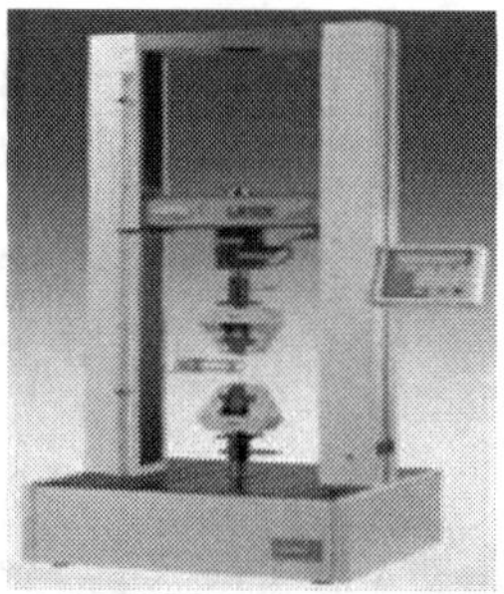

Fig.5. LLOYD Instruments LR50K

Le banc d'essai est équipé de mors auto-serrant et d'une cellule de force ayant une capacité de 5kN. Son pilotage se fait par le logiciel Test expert qui permet à la fois de consigner les paramètres d'essais, d'acquérir et de traiter les données.

 https://doi.org/10.1051/matecconf/201814901083

2.2.2 Résultats des essais pour le PVC dur vierge

Table 1. Relevées enregistrées de l'essai de traction

N° Eprouvette	Contrainte à la rupture (MPa)	% d'allongement à la rupture
1	41,3	50
2	40,0	60
3	39,9	40
4	38,1	50
5	39,1	40
Moyenne	39,7	48
Ecart-type	1,05	7,48

2.2.3 Courbe de traction pour le PVC rigide vierge

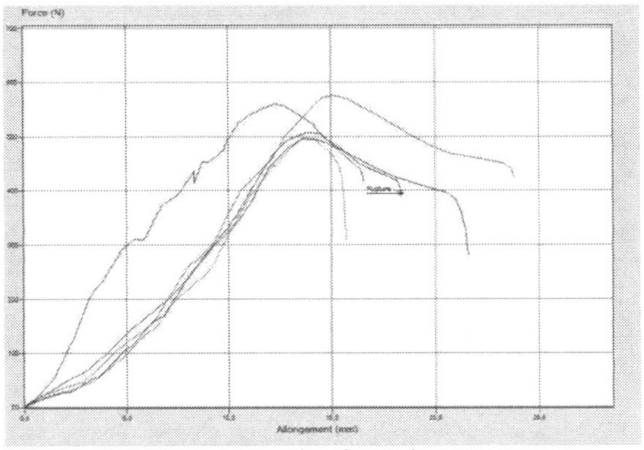

Fig.6. Courbes de traction

2.2.4 Résultats des essais pour le PVC rigide bio-chargé

Table 2. Relevées enregistrées de l'essai de traction

N° Eprouvette	Contrainte à la rupture (MPa)	% d'allongement à la rupture
1	23,7	25
2	26,9	40
3	38	25
4	34,4	25
5	33,6	30
Moyenne	31,3	29
Ecart-type	5,23	5,83

2.2.5 Courbe de traction pour le PVC rigide bio-chargé

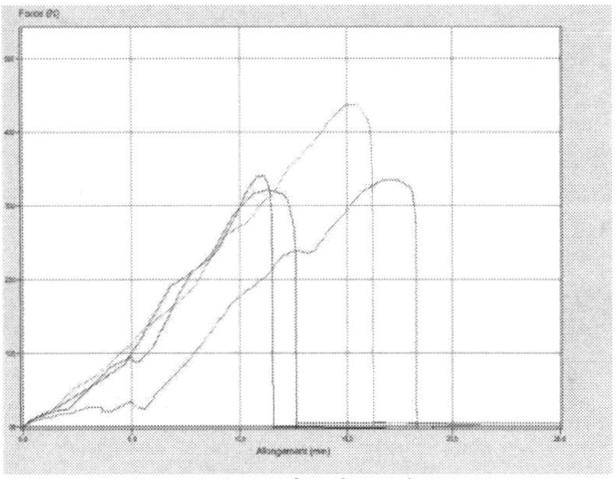

Fig.7. Courbes de traction

2.3 Analyse du comportement élastique

Nous pouvons observer les performances comportementales suivantes du nouveau éco-matériau face aux sollicitations d'extension :

La réactivité du PVC rigide aux sollicitations de traction a généré une variabilité de comportements élastiques intrinsèques, sachant que les comportements enregistrés relatifs aux cinq éprouvettes sont susceptibles de s'être imposés à force de leurs caractéristiques évaluées "Fig. 6 ", "Fig. 7 ".

Au cours de ces essais, les deux types de matériau ont subi plusieurs tests d'extension et on a retenu les valeurs expérimentales les plus significatives à savoir, la résistance maximale à la rupture et le déplacement relatif.

Nous pouvons observer, à travers les résultats des différents essais effectués sur les deux types des matériaux ainsi trouvés face à ces sollicitations que :

Les différentes courbes de traction obtenues présentent un comportement plastique avec seuil très marqué. Outre les paramètres de rupture, on définit une ligne parallèle à la tangente à l'origine des courbes à partir des différentes déformations pour marquer le coefficient directeur appelé module de Young ou module d'élasticité longitudinal caractérisant la dureté du matériau "Fig. 6 ", "Fig. 7 ".

Porté à une température de 185°C, le bio-renforcement à 10% de la charge en cause a contribué à une très sensible amélioration de la rigidité. Sur une même nuance, l'effet de durcissement est d'autant plus significatif que le niveau de dureté initial est faible et l'effet sur les caractéristiques de fragilité est d'autant moins appréciable que l'écart entre la dureté initiale et finale est élevé "Fig. 6 ", "Fig. 7 ".

Pour le lot d'éprouvettes du PVC basique, on remarque un comportement élastique non linéaire suivie d'une déformation plastique irréversible. Ce PVC basique subit une déformation plastique permanente avant de se rompre. C'est un matériau dur et ductile "Fig. 6 ".

Pour le lot d'éprouvettes de l'éco-composite, l'étude a montré un comportement élastique, non linéaire et réversible. Apparition d'une zone semi-rectiligne, pour laquelle les allongements sont relativement proportionnels aux efforts appliqués. C'est la zone élastique qui est réversible, car si l'on supprime l'effort de traction, le retour à l'équilibre s'établit et l'éprouvette conserve un allongement rémanent. La rupture de l'éco-matériau se produit alors qu'il est en régime de déformation purement semi élastique par rapport au cas initial. C'est un matériau rigide et fragile "Fig. 7 ".

Leger abaissement de la résistance à la traction sans aucune influence sur la valeur maximale à la rupture.

Importante chute d'allongement pour cent engendrant une amélioration de module de Young caractérisant une propriété de la matière de nouveau éco-composite accompagnée d'une diminution significative de ténacité "Fig. 7 ".

Variabilité de Comportements élastiques plus intense par rapport au PVC de base due au nouveau composant allié en substance organique qui a rendu le matériau final plus hétérogène. Cela a bien métamorphosé le comportement plus anistropique qu'auparavant "Fig. 6 ", "Fig. 7 ".

Disparition des zones de déformations plastiques du PVC basique après le bio-chargement "Fig. 6 ", "Fig. 7 ". Ce dernier procédé de transformation a permis finalement au matériau final de bénéficier de la propriété de dureté du matériau de base. En fait, la modification de la structure semi-cristalline du PVC bio-chargé est traduite par l'évolution du taux de cristallinité « ϰ ». Cet important paramètre se rencontre souvent dans le cas des matériaux organiques. Il mesure la proportion de matière se trouvant dans l'état cristallin et permet d'apprécier les contributions amorphes et cristallines. Une nouvelle structure qui se manifeste par une grande différence de densité entre phases amorphe et cristalline. Donc, la structure de la phase cristalline est devenue plus compacte [5].

En revanche, une amélioration de l'aspect de la morphologie du nouveau éco-composite à envisager par un ajustement de dosage et de température pour assurer une meilleure compatibilité du mélange. Cette supposition est très probable du fait que le taux de cristallinité massique (ϰm) ou le taux de cristallinité volumique (ϰv), général peu différents, dépend des systèmes cristallins et de l'histoire thermique du polymère parent.

Aussi, il faudra disperser et distribuer les phases, souvent en présence de compatibilisants, ce qui nécessite généralement des études au cas par cas.

Conclusion

Actuellement, les axes de recherches sur Le développement des matériaux plus respectueux de l'environnement concernent bien les matériaux biosourcés, les matériaux biodégradables et les matériaux recyclés.

Dans cette perspective, nous avons pu, dans un premier temps, contribuer à la caractérisation élastique d'un matériau vierge d'origine animal. Ce test de résistance à la traction sur le banc d'essai dans des conditions similaires d'expérience, nous a permis de noter que ce comportement mécanique est jugé favorable à une mise en valeur de la bio-charge en cause pour l'innovation de certains matériaux polymères.

L'objectif recherché était d'analyser expérimentalement l'effet de cette biocharge sur la morphologie du PVC rigide.

A travers Les résultats expérimentalement obtenus, nous constatons que les hypothèses avancés par notre sujet de thèse sont vérifiés.

En termes de développement durable, deux approches principales seront développées dans cet axe de recherche : Le développement du recyclage des organes animaux en fin de vie grâce à l'amélioration d'identification des performances et le développement de matériaux bio-chargés de ressources renouvelables à partir de biomasses organiques.

Obstacles Scientifiques

Le développement de matériaux biosourcés et biodégradables nécessite l'évolution d'essais spécifiques d'évaluation de la biodégradation des matériaux formulés [10]. Ce développement de matériaux à partir de ressources première secondaires nécessite l'obtention de pureté de plus en plus importante afin de viser des applications à fortes valeurs ajoutées [11].

Remerciements

Nos remerciements vont au Laboratoire du Centre Technique de Plasturgie et de Caoutchouc (CTPC), accrédité NM ISO/CEI 17025, pour son accueil, les moyens mis à disposition pour mener à bonne fin cet article. Ce travail a notamment été rendu possible grâce à l'ensemble du personnel technique et administratif de ce laboratoire que nous tenons également à remercier.

References

1. E. Bretagne, J. Bréard, V. Massardier et V. Verney, "*Eco-matériaux: Les matériaux passent au vert*". **367-368,** 100 (2012).
2. CA.Bernardo, AM.Cunha and MJ.Oliveira, the Department of Polymer Engineering, Universidade

do Minho), "The Recycling of Thermoplastics: Prediction of the Properties of Mixtures of Virgin and Reprocessed Polyolefins", 4800 Guimaraes, Portugal, Polymer Engineering and Science, 36 (4) February (1996).

3. D.Dupoux, (Shell Chimie), "*Economie-Ecologie du recyclage des matières plastiques*", Colloque du Groupe Français D'Etudes et d'Applications des Polymères, Douai, **367-368** (1991).
4. A. Kouadri, S. Cretin, "*Des évolutions indissociables* ", (2007).
5. FS.Bates, "Polymer-Polymer phase behaviour", Science, 251 (4996), **898-905** (1991).
6. Y .Agari, M.Shimada. and A.Ueda, Polymer, **2649-2655**, 38 (11) (1997).
7. M. Jammoukh, K. Mansouri et B. Sslhi, « *Etude expérimentale du comportement élastique d'une bio-charge de ressources renouvelables* » RSE-2016. **61-66**, 1 (2016).
8. JC.Angola, Y.Fujita, T.Sakai, T.Inoue, "Compatibilizer-aided toughening in polymer.
9. L. AUGIER, "*Etude de l'élaboration de matériaux composites PVC/bois à partir de déchets de menuiserie : formulation, caractérisation, durabilité et recyclabilité*". **11-12**, 2507 (2007).
10. Allred, *Recycling process to scrap composites and prepregs*, Int. SAMPE Symp.Exhib-1996. **61-66**, 41 (1996).
11. K. Halada, "*Progress of ecomaterials toward a sustainable society, Current Opinion in Solid State and Materials Science*". **209-216-368**, 7 (2003).

Synthesis and crystal structure of two new phosphates with Alluaudite type structure: (M, Mn)$_3$Fe(PO$_4$)$_3$ (M= Ca , Cd)

Ghaleb Alhakmi, Abderrazzak Assani, Mohamed Saadi and Lahcen El Ammari

Laboratoire de Chimie Appliquée des Matériaux, Centre Sciences des Matériaux, Faculty of Sciences, Mohammed V University in Rabat, Avenue Ibn Battouta, BP 1014, Rabat, Morocco.

Abstract. Two new phosphates with alluaudite type structure (M, Mn)$_3$Fe(PO$_4$)$_3$ (M= Ca , Cd), namely $Ca_{1.54}Mn_{1.46}Fe(PO_4)_3$ and $Cd_{0.66}Mn_{2.34}Fe(PO_4)_3$, has been synthesized by a solid state reaction and characterized by single-crystal X-ray diffraction. The two compounds crystallize in the monoclinic system with C2/c space group. Their open framework results from $(Fe1/Mn1)_2O_{10}$ units of edge-sharing $(Fe/Mn)O_6$ octahedra, which alternate with $M(1)O_6$ octahedra (M(1)=Ca1/Mn2 or Cd1/Mn2) that form infinite chains running along [10-1] direction. These chains are linked together through the common corners of PO_4 tetrahedra giving rise to two types of tunnels occupied by bivalent cations Ca^{2+} or Cd^{2+} and Mn^{2+}.

1 Introduction

Depuis la découverte de l'activité électrochimique de $LiFePO_4$ [1-2], la synthèse et la caractérisation structurale des matériaux fonctionnels à base de métaux de transition adoptant des structures en couches ou en tunnel, ont connu un développent considérable, ce qui a ouvert une ligne de recherche prometteuse dans le domaine des sciences des matériaux. Dans ce contexte, notre équipe de recherche a orienté un de ses axes de recherche vers la synthèse et la caractérisation structurale de nouveaux matériaux à base de métaux de transition apparentés plus particulièrement à l'alluaudite [3-4]. Le terme alluaudite se réfère à une grande famille de composés naturels ou synthétiques caractérisées par une formule générale proposée par Moore, A(2)A(1)M(1)M(2)$_2$(PO$_4$)$_3$ [5], où A(1) et A(2) sont deux sites situés dans les tunnels pouvant accueillir des gros cations monovalents et/ou bivalents alors que M(1) et M(2) représentent des sites octaédriques occupés par des cations di et/ou trivalents de petites tailles. Les structures des matériaux de type alluaudite possèdent des charpentes ouvertes présentant des propriétés physiques importantes, en particulier la conductivité ionique [6], l'activité électrochimie [7], le magnétisme [8], l'inhibition de la corrosion [9], la passivation des surfaces métalliques [10-11] et la catalyse [12]. Récemment, un grand nombre de composés phosphatés appartenant à ce type structural ont été élaborés par les chercheurs de notre laboratoire. Nous pouvons citer, à titre d'exemple, $AgMg_3(PO_4)(HPO_4)_2$ [13], $NaMg_3(PO_4)(HPO_4)_2$ [14], $Na_2Co_2Fe(PO_4)_3$ [15] et $Na_{1.67}Zn_{1.67}Fe_{1.33}(PO_4)_3$ [16]. Dans cet article, nous rapportons la synthèse et l'étude structurale de deux nouveaux phosphates, $Ca_{1.54}Mn_{1.46}Fe(PO_4)_3$ et $Cd_{0.66}Mn_{2.34}Fe(PO_4)_3$ de type alluaudite.

2 Synthèse des cristaux

Les deux nouveaux phosphates, $Ca_{1.54}Mn_{1.46}Fe(PO_4)_3$ et $Cd_{0.66}Mn_{2.34}Fe(PO_4)_3$, ont été synthétisés, sous forme de monocristaux, par la réaction de diffusion à l'état solide. Des quantités stœchiométriques de $Ca(NO_3)_2.4H_2O$, CdO, $Mn(NO_4)_2$, $Fe(NO_3)_3.9H_2O$ et H_3PO_4 (85%) correspondant au rapport molaire M (Ca ou Cd) : Mn : Fe : P = 1 : 2 : 1: 3, ont été mélangées dans des béchers contenant 60 mL d'eau distillée. Le mélange réactionnel est homogénéisé par agitation pendant 24 heures. La poudre obtenue après évaporation est broyée dans un mortier en agate et traitée à 300°C pendant une nuit, pour éliminer les gaz volatils. Par la suite, les deux phosphates $Ca_{1,54}Mn_{1,46}Fe(PO_4)_3$ et $Cd_{0,66}Mn_{2,34}Fe(PO_4)_3$ sont portés progressivement à fusion à 1253 K et 1323 K respectivement, dans des creusets en platine et maintenu à ces températures pendant 30 min. Un refroidissement lent avec une vitesse de 5°C/h a conduit à l'obtention de cristaux de couleur marron correspondant à ces nouveaux phosphates.

3 Étude structurale des nouveaux phosphates $Ca_{1,54}Mn_{1,46}Fe(PO_4)_3$ et $Cd_{0,66}Mn_{2,34}Fe(PO_4)_3$

3.1 Collecte des intensités diffractées

A partir des deux préparations, des cristaux marron de bonne qualité et de taille exploitable sont sélectionnés et montés sur des fibres en verre pour les tester au diffractomètre à 4 cercles X8 APEXII de l'UATRS (Unité d'Appui Technique à la Recherche Scientifique) du CNRST (Centre National de la Recherche Scientifique et Technique). La collecte des intensités de diffraction des rayons-X a été effectuée dans les conditions rassemblées dans le tableau (1).

Tableau 1. Données cristallographiques et conditions d'enregistrement des intensités diffractées des phosphates $Ca_{1,54}Mn_{1,46}Fe(PO_4)_3$ et $Cd_{0,66}Mn_{2,34}Fe(PO_4)_3$.

Données cristallographiques		
Formule chimiques	$Ca_{1,54}Mn_{1,46}Fe(PO_4)_3$	$Cd_{0,66}Mn_{2,34}Fe(PO_4)_3$
Système cristallin	Monoclinique	Monoclinique
Groupe d'espace	C2/c	C2/c
Paramètres de la maille	a= 12,3487 (2) Å b= 12,6569 (2) Å c= 6,4650 (1) Å β= 115,673 (1) °	a= 12,1565 (2)Å b= 12,6226 (2),Å c= 6,4311 (1)Å β= 115,200 (1)°
Volume de la maille (Å^3)	910,70 (3)	892,91 (3)
Multiplicité Z	4	4
Densité calculée (Mg m$^{-3)}$	2,886	4,019
Coefficient d'absorption (mm^{-1})	7,24	10,87
Collecte des intensités diffractées		
Diffractomètre	Bruker X8 APEXII	Bruker X8 APEXII
Température (K)	293	293
Longueur d'onde Mo(Kα) (Å)	0,7071	0,7071
Correction d'absorption	multi-scan (SADABS; Bruker, 2005)	multi-scan (SADABS; Bruker, 2005)
Domaine de θ (°)	θ_{max} = 40,10°, θ_{min} = 2.4°	θ_{max} = 38,2°, θ_{min} = 2.5°
Indices de Miller	h = -22→22 k = -23→23 l = -11→10	h = -21→21 k = -21→21 l = -11→10
Nombre de réflexions mesurées	18447	19390
Nbre de réflexions indépendantes	2934	2462
Nbre de réflexions pour I > 2σ(I)	2610	2219

La structure de chacun de ces deux composés a été résolue par la méthode de Patterson à l'aide du programme SHELXS-97 [17]. Cette méthode nous a permis de localiser les atomes lourds Ca, Cd, Mn, Fe et P. Au moyen du programme SHELXL-2014 [18], des séries d'affinements suivies de Fourier-différences nous ont permis de localiser les six atomes d'oxygène indépendants. Divers modèles structuraux ont été testés et les résultats des meilleurs qui respectent le remplissage des sites cristallographiques et la neutralité des charges sont consignés dans le tableau (2). Les coordonnées atomiques et les paramètres de déplacements atomiques isotropes et anisotropes, pour les deux composés, sont rassemblés dans les tableaux 3 et 4.

Tableau 2. Résultats d'affinement des structures des phosphates $Ca_{1,54}Mn_{11,46}Fe(PO_4)_3$ et $Cd_{0,66}Mn_{2,34}Fe(PO_4)_3$.

Composé	$Ca_{1,54}Mn_{1,46}Fe(PO_4)_3$	$Cd_{0,66}Mn_{2,34}Fe(PO_4)_3$
$R[F^2 > 2\sigma(F^2)]$	0,027	0,02

$wR(F^2)$	0,068	0,048
S	1,07	1,07
Nombre des paramètres affinés	99	101

Tableau 3. Coordonnées atomiques et facteurs de déplacements atomiques isotropes équivalents des phosphates $Ca_{1,54}Mn_{1,46}Fe(PO_4)_3$* et $Cd_{0,66}Mn_{2,34}Fe(PO_4)_3$.

Atome	Site	Taux d'occup.	*x*	*y*	*z*	U_{eq}
Fe1 /Mn1*	8f	0,50/0,50	0,2240 (2)	0,1536 (2)	0,6367 (4)	0,0093 (7)
Fe1/Mn1		0,50/0,50	0,2194 (2)	0,15168 (2)	0,6295 (4)	0,0074 (6)
Ca1/Mn2*	4e	0,537 (9)/0,463 (9)	0,0000	0,2594 (4)	0,7500	0,0143 (1)
Cd1/ Mn2		0,305(2)/ 0,695 (2)	0,0000	0,2577 (2)	0,7500	0,0105 (8)
Ca2*	4e	0,192(1)	0,0000	0,0145 (2)	0,2500	0,0108 (4)
Cd2		0,080(1)	0,0000	0,0198 (2)	0,2500	0,0128 (3)
Ca3*	8f	0,404(1)	-0,0277 (9)	0,4910 (7)	0,0456 (2)	0,0160 (2)
Cd3/ Mn3		0,1379(2)/ 0,3221 (2)	-0,02293 (4)	0,5007 (3)	0,0359 (7)	0,0131 (1)
P1*	8f	1	0,2419 (5)	0,1070 (3)	0,1378 (8)	0,0114 (9)
P1		1	0,2415 (4)	0,1085 (3)	0,1358 (6)	0,0084 (8)
P2*	4e	1	0,0000	0,2841 (5)	0,2500	0,0080 (1)
P2		1	0,0000	0,2870 (4)	0,2500	0,0068 (9)
O1*	8f	1	0,1756 (2)	0,1617 (1)	-0,0960 (3)	0,0234 (3)
O1		1	0,1745 (1)	0,1632 (9)	-0,0988 (2)	0,0140 (2)
O2*	8f	1	0,1877 (2)	-0,0018 (1)	0,1287 (3)	0,0302 (4)
O2		1	0,1899 (1)	-0,0016 (1)	0,1287 (2)	0,0213 (3)
O3*	8f	1	0,2279 (1)	0,1769 (1)	0,3203 (2)	0,0145 (2)
O3		1	0,2253 (1)	0,1784 (1)	0,3183 (2)	0,0128 (2)
O4*	8f	1	0,3763 (2)	0,0977 (1)	0,1990 (4)	0,0287 (4)
O4		1	0,3783 (1)	0,1008 (1)	0,1981 (2)	0,0188 (2)
O5*	8f	1	0,0924 (1)	0,3586 (1)	0,2261 (4)	0,0317 (5)
O5		1	0,0965 (1)	0,3624 (1)	0,2360 (2)	0,0190 (3)
O6*	8f	1	0,0515 (1)	0,2137 (1)	0,4646 (2)	0,0151 (2)
O6		1	0,0483 (1)	0,2169 (9)	0,4659 (2)	0,0110 (2)

Tableau 4. Facteurs de déplacements atomiques anisotropes des atomes des phosphates $Ca_{1,54}Mn_{1,46}Fe(PO_4)_3$* et $Cd_{0,66}Mn_{2,34}Fe(PO_4)_3$.

Atome	U11	U22	U33	U12	U13	U23
Fe1/Mn1*	0,0102 (1)	0,0079 (1)	0,0113 (1)	0,00167 (7)	0,0061 (9)	0,002 (7)
Fe1/Mn1	0,0076 (1)	0,0079 (1)	0,0076 (1)	0,0016 (6)	0,0041 (7)	0,0018 (6)
Ca1/Mn2*	0,0128 (2)	0,0191 (2)	0,0148 (2)	0,000	0,0096 (2)	0,000
Cd1/ Mn2	0,0105 (1)	0,0129 (1)	0,0111 (1)	0,000	0,0075 (9)	0,000
Ca2*	0,0051 (8)	0,0168 (1)	0,0086 (9)	0,000	0,0013 (7)	0,000
Cd2	0,0029 (7)	0,0253 (1)	0,0078 (7)	0,000	0,0002 (6)	0,000
Ca3*	0,0157 (4)	0,0088 (3)	0,0162 (4)	-0,0003 (3)	0,0000 (3)	-0,0019 (3)
Cd3/Mn3	0,0126 (2)	0,0056 (2)	0,0131 (2)	0,0010 (1)	-0,0023 (1)	-0,0005 (1)
P1*	0,0201 (2)	0,0068 (2)	0,0082 (2)	-0,0011 (1)	0,0070 (2)	-0,0001 (1)
P1	0,0120 (2)	0,0073 (2)	0,0063 (2)	-0,0002 (1)	0,0042 (1)	0,0004 (1)
P2*	0,0070 (2)	0,0089 (2)	0,0075 (2)	0,000	0,0026 (2)	0,000
P2	0,0063 (2)	0,0075 (2)	0,0055 (2)	0,000	0,0017 (2)	0,000
O1*	0,0464 (1)	0,0110 (6)	0,0092 (6)	0,0025 (6)	0,0087 (6)	0,0025 (4)
O1	0,0230 (6)	0,0099 (5)	0,0069 (4)	0,0010 (4)	0,0042 (4)	0,0019 (4)
O2*	0,0519 (1)	0,0132 (6)	0,0193 (7)	-0,0129 (7)	0,0095 (7)	0,0032 (5)
O2	0,0289 (7)	0,0123 (5)	0,0217 (6)	-0,0052 (5)	0,0099 (6)	0,0046 (4)
O3*	0,0156 (6)	0,0189 (6)	0,0106 (5)	-0,0029 (5)	0,0072 (4)	-0,0046 (4)
O3	0,0118 (5)	0,0192 (5)	0,0083 (5)	0,0002 (4)	0,0051 (4)	-0,0032 (4)
O4*	0,0279 (8)	0,0190 (7)	0,0465 (1)	0,0024 (6)	0,0229 (8)	-0,0083 (7)
O4	0,0154 (5)	0,0159 (5)	0,0282 (7)	0,0026 (4)	0,0122 (5)	-0,0032 (5)
O5*	0,0113 (6)	0,0248 (8)	0,0503 (1)	-0,0021 (5)	0,0053 (7)	0,0239 (8)
O5	0,0105 (5)	0,0149 (5)	0,0273 (7)	-0,0019 (4)	0,0039 (5)	0,0105 (5)

O6*	0,0146 (6)	0,0213 (6)	0,0075 (5)	-0,0029 (5)	0,0031 (4)	0,0032 (4)
O6	0,0105 (4)	0,0156 (5)	0,0068 (4)	-0,0010 (4)	0,0034 (4)	0,0018 (4)

3 .2 Description de la structure des phosphates $Ca_{1,54}Mn_{1,46}Fe(PO_4)_3$ et $Cd_{0,66}Mn_{2,34}Fe(PO_4)_3$

Le tableau (5) rassemble les principales distances et angles interatomiques dans les structures de $Ca_{1,54}Mn_{1,46}Fe(PO_4)_3$-et $Cd_{0,66}Mn_{2,34}Fe(PO_4)_3$.

Tableau 5. Distances (Å) et angles interatomiques (°) dans les structures des phosphates $Ca_{1,54}Mn_{2,46}Fe(PO_4)_3$ et $Cd_{0,66}$ $Mn_{2,34}Fe(PO_4)_3$.

atome	Distances P(1) - O (Å)	$Ca_{1,54}Mn_{2,46}Fe(PO_4)_3$	$Cd_{0,66}$ $Mn_{2,34}Fe(PO_4)_3$
P(1)	P1—O1	1,537 (2)	1,540 (1)
	P1—O2	1,521 (2)	1,517 (1)
	P1—O3	1,543 (1)	1,546 (1)
	P1—O4	1,536 (2)	1,539 (1)
	Distance moyenne	**1,534**	**1,536**
	Angles O - P(1) - O (°)		
	O2—P1—O4	110,39 (1)	109,70 (8)
	O2—P1—O1	109,36 (1)	109,97 (7)
	O4—P1—O1	110,14 (1)	110,24 (8)
	O2—P1—O3	110,84 (1)	111,14 (8)
	O4—P1—O3	108,29 (9)	108,14 (7)
	O1—P1—O3	107,79 (9)	107,61 (7)
	Distances P(2) - O (Å)		
P(2)	P2—O5^{vi}	1,539 (2)	1,543 (1)
	P2—O5	1,539 (2)	1,543 (1)
	P2—O6^{vi}	1,536 (1)	1,536 (1)
	P2—O6	1,536 (1)	1,536 (1)
	Distance moyenne	**1,538**	**1,540**
	Angles O - P(2) - O (°)		
	O6^{vi}—P2—O6	109,09 (1)	109,75 (9)
	O6^{vi}—P2—O5	107,46 (1)	108,04 (7)
	O6—P2—O5	114,28 (8)	113,60 (6)
	O6^{vi}—P2—O5^{vi}	114,28 (8)	113,60 (6)
	O6—P2—O5^{vi}	107,46 (1)	108,04 (7)
	O5—P2—O5^{vi}	104,37 (2)	103,80 (1)
	Distances (Fe1/Mn1) - O (Å)		
Fe1/Mn1	(Fe1/Mn1)—O1^{iii}	2,062 (2)	2,046 (1)
	(Fe1/Mn1)—O2^{i}	1,968 (2)	1,928 (1)
	(Fe1/Mn1)—O3	2,088 (1)	2,061(1)
	(Fe1/Mn1)—O3^{ii}	2,211 (2)	2,230 (1)
	(Fe1/Mn1)—O5^{ii}	2,051 (2)	2,035 (1)
	(Fe1/Mn1)—O6	2,074 (2)	2,060 (1)
	Distance moyenne	**2,076**	**2,060**
	Angles O – (Fe1/Mn1)- O (°)		
	O2^{i}— (Fe1/Mn1)—O5^{ii}	97,73 (8)	94,88 (6)
	O2^{i}—(Fe1/Mn1)—O1^{iii}	85,69 (7)	87,44 (6)
	O5^{ii}— (Fe1/Mn1)—O1^{iii}	108,01 (8)	106,76 (6)
	O2^{i}— (Fe1/Mn1)—O6	100,04 (8)	104,21 (6)
	O5^{ii}— (Fe1/Mn1)—O6	161,00 (6)	160,06 (5)
	O1^{iii}— (Fe1/Mn1)—O6	80,09 (6)	80,17 (5)
	O2^{i}— (Fe1/Mn1)—O3	102,34 (7)	103,97 (6)
	O5^{ii}— (Fe1/Mn1)—O3	86,78 (7)	86,57 (5)

	O1^{iii}—(Fe1/Mn1)—O3	162,26 (6)	161,85 (5)
	O6—(Fe1/Mn1)—O3	82,92 (6)	83,33 (5)
	O2^{i}— (Fe1/Mn1)—O3^{ii}	174,86 (6)	171,66 (5)
	O5^{ii}— (Fe1/Mn1)—O3^{ii}	80,37 (6)	79,23 (5)
	O1^{iii}— (Fe1/Mn1)—O3^{ii}	90,35 (6)	88,63 (4)
	O6— (Fe1/Mn1)—O3^{ii}	82,46 (6)	82,33 (5)
	O3— (Fe1/Mn1)—O3^{ii}	82,37 (5)	81,72 (5)
	Distances (Ca1/Mn2,Cd1/Mn2) - O (Å)		
(Ca1/Mn2, Cd1/Mn2)	(Ca1/Mn2, Cd1/Mn2) —O1^{vi}	2,313 (2)	2,260 (1)
	(Ca1/Mn2, Cd1/Mn2) —O1^{iii}	2,313 (2)	2,261 (1)
	(Ca1/Mn2, Cd1/Mn2) —O4^{v}	2,298 (2)	2,252 (1)
	(Ca1/Mn2, Cd1/Mn2) —O4^{ii}	2,298 (2)	2,252 (1)
	(Ca1/Mn2, Cd1/Mn2) —O6	2,272 (1)	2,204 (1)
	(Ca1/Mn2, Cd1/Mn2) —O6^{iv}	2,272(2)	2,204 (1)
	Distance moyenne	**2,295**	**2,239**
	O6— (Ca1/Mn2, Cd1/Mn2) —O6^{iv}	150,51 (8)	153,02 (6)
	O6— (Ca1/Mn2, Cd1/Mn2) —O4^{v}	119,11 (6)	116,88 (5)
	O6^{iv}— (Ca1/Mn2, Cd1/Mn2) —O4^{v}	85,09 (6)	85,29 (5)
	O6— (Ca1/Mn2, Cd1/Mn2) —O4^{ii}	85,09 (6)	85,29 (5)
	O6^{iv}— (Ca1/Mn2, Cd1/Mn2) —O4^{ii}	119,11 (6)	116,88 (5)
	O4^{v}— (Ca1/Mn2, Cd1/Mn2) —O4^{ii}	76,07 (9)	75,00 (6)
	O6— (Ca1/Mn2, Cd1/Mn2) —O1^{vi}	93,12 (5)	93,02 (4)
	O6^{iv}— (Ca1/Mn2, Cd1/Mn2) —O1^{vi}	70,94 (5)	72,62 (4)
	O4^{v}— (Ca1/Mn2, Cd1/Mn2) —O1^{vi}	85,33 (6)	85,29 (5)
	O4^{ii}— (Ca1/Mn2, Cd1/Mn2) —O1^{vi}	157,45 (7)	156,80 (5)
	O6—(Ca1/Mn2, Cd1/Mn2) —O1^{iii}	70,94 (5)	72,62 (4)
	O6^{iv}— (Ca1/Mn2, Cd1/Mn2) —O1^{iii}	93,12 (5)	93,02 (4)
	O4^{v}— (Ca1/Mn2, Cd1/Mn2) —O1^{iii}	157,45 (7)	156,80 (5)
	O4^{ii}— (Ca1/Mn2, Cd1/Mn2)—O1^{iii}	85,33 (6)	85,29 (5)
	O1^{vi}— (Ca1/Mn2, Cd1/Mn2) —O1^{iii}	115,36 (9)	116,32 (6)
	Distances (Ca2, Cd2) - O (Å)		
(Ca2, Cd2)	(Ca2, Cd2)—O1^{vii}	2,965 (3)	3,004 (2)
	(Ca2, Cd2)—O1^{i}	2,965 (3)	3,004 (2)
	(Ca2, Cd2)—O2	2,758 (2)	2,747 (2)
	(Ca2, Cd2)—O2^{vi}	2,758 (2)	2,747 (2)
	(Ca2, Cd2)—O2^{i}	2,545 (2)	2,554 (2)
	(Ca2, Cd2)—O2^{vii}	2,545 (2)	2,554 (2)
	(Ca2, Cd2)—O6^{vi}	2,815 (3)	2,788 (2)
	(Ca2, Cd2)—O6	2,814 (3)	2,788 (2)
	Distance moyenne	**2,771**	**2,773**
	Distances (Ca3, Cd3/Mn3) - O (Å)		
(Ca3, Cd3/Mn3)	(Ca3, Cd3/Mn3) —O4^{viii}	2,222 (2)	2,248 (1)
	(Ca3, Cd3/Mn3) —O4^{ix}	2,253 (2)	2,279 (1)
	(Ca3, Cd3/Mn3) —O4^{x}	2,377 (2)	2,362 (2)
	(Ca3, Cd3/Mn3) —O5	2,294 (2)	2,285 (1)
	(Ca3, Cd3/Mn3) —O5^{xi}	2,390 (2)	2,348 (2)
	(Ca3, Cd3/Mn3)—O5^{vi}	2,653 (3)	2,668 (2)
	Distance moyenne	**2,365**	**2,365**

Code de symétrie : (i) *x*, -*y*, *z*+1/2; (ii) -*x*+1/2, -*y*+1/2, -*z*+1; (iii) *x*, *y*, *z*+1; (iv) -*x*, *y*, -*z*+3/2; (v) *x*-1/2, -*y*+1/2, *z*+1/2; (vi) -*x*, *y*, -*z*+1/2; (vii) -*x*, -*y*, -*z*; (viii) *x*-1/2, *y*+1/2, *z*; (ix) -*x*+1/2, *y*+1/2, -*z*+1/2; (x) *x*-1/2, -*y*+1/2, *z*-1/2; (xi) -*x*, -*y*+1, -*z*.

Les figures (1) et (2) représentent les unités asymétriques illustrant l'enchaînement des polyèdres ainsi que le déplacement atomique de chaque atome dans les composés $Ca_{1,54}Mn_{1,46}Fe(PO_4)_3$ et $Cd_{0,66}Mn_{2,34}Fe(PO_4)_3$.

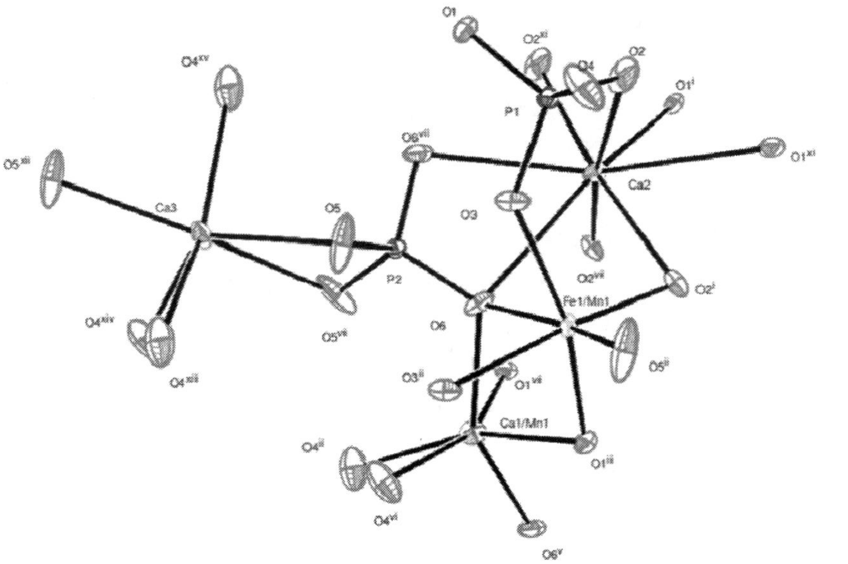

Fig.1 : Unité asymétrique du phosphate $Ca_{1,54}Mn_{1,46}Fe(PO_4)_3$.

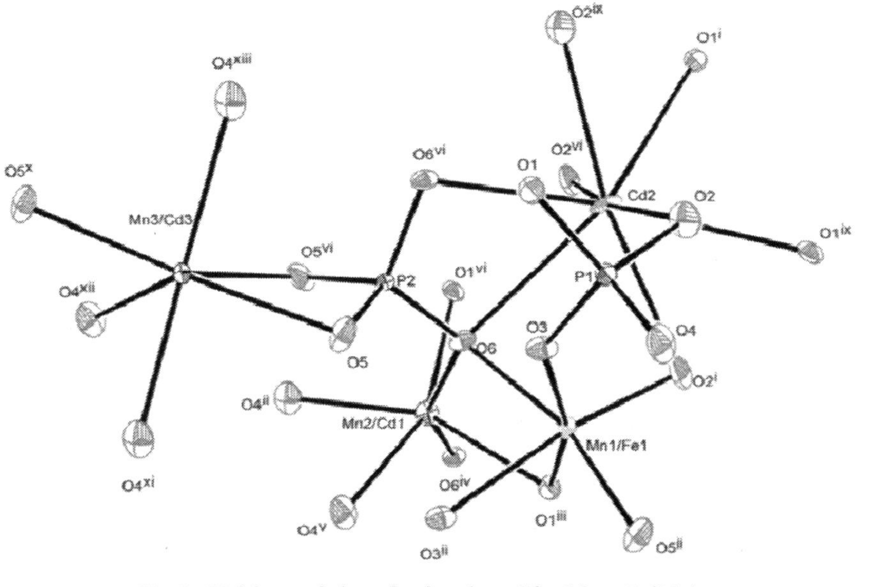

Fig.2 : Unité asymétrique du phosphate $Cd_{0,66}Mn_{2,34}Fe(PO_4)_3$.

La charpente anionique tridimensionnelle est formée de deux type d'octaèdres $M(1)O_6$ occupé par Ca(1)/Mn(2) ou par Cd(1)/Mn(2) selon le composé et $M(2)O_6$ occupé par 0,5Fe/0,5Mn dans les deux phases ainsi que deux tétraèdres $P(1)O_4$ et $P(2)O_4$ dans chaque composé. Les deux types d'octaèdres partagent les arêtes communes formant des chaînes infinies en zigzag parallèle à la direction $[10\bar{1}]$ (figure 3). Les chaînes adjacentes sont connectées entre elles par des tétraèdres PO_4. Ceci conduit à la formation de feuillets parallèles au plan $(\vec{a}, \vec{c})$ (figure 4). La connexion entre deux feuillets est assuré par les sommets partagés des tétraèdres $P(1)O_4$ et des octaèdres (Fe(1)/Mn(1)) pour former ainsi une charpente tridimensionnelle comportant deux types de tunnels parallèles à l'axe *c* (figure 5). Les atomes Ca(2) ou Cd(2) sont localisés dans le premier type de tunnel alors que Ca(3) ou Cd(3)/Mn(3) sont logés dans le deuxième type de tunnel (figure 5).

 https://doi.org/10.1051/matecconf/201814901084

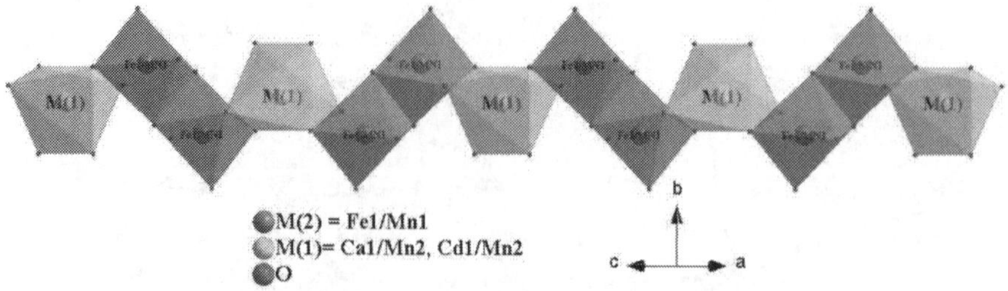

Fig.3 : Enchaînement d'octaèdres $M(1)O_6$ et $M(2)O_6$ suivant la rangé $[10\bar{1}]$.

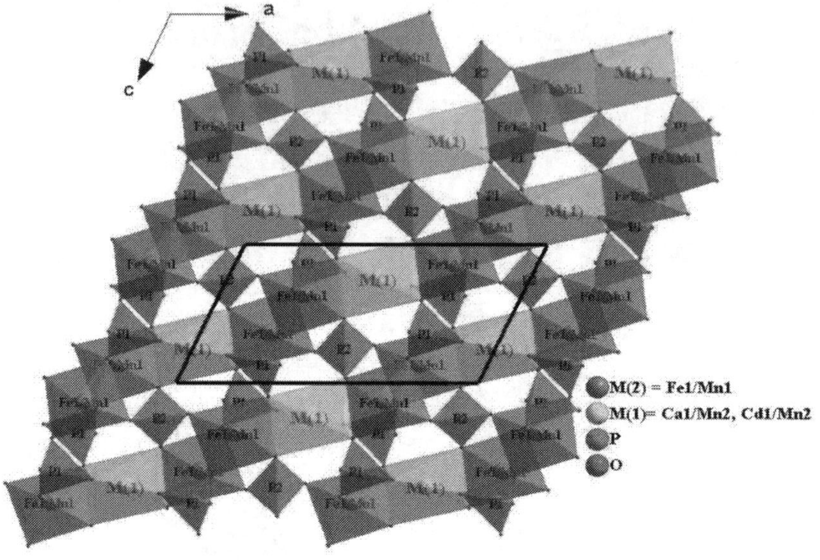

Fig.4 : Représentation d'un feuillet parallèle au plan ($\vec{a}$, $\vec{c}$).

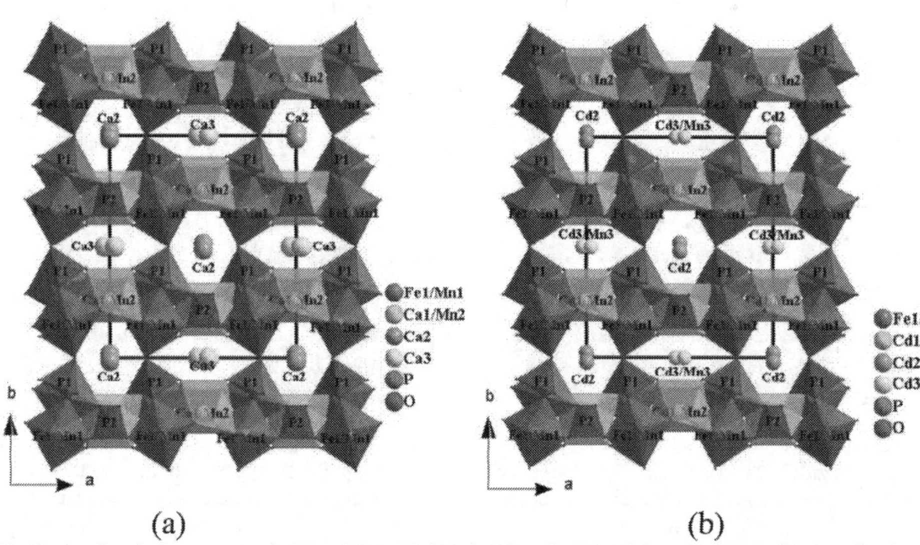

Fig.5 : Projection de la structure de $Ca_{1.54}Mn_{1.46}Fe(PO_4)_3$ (a) et de $Cd_{0.66}Mn_{2.34}$ $Fe(PO_4)_3$ (b) dans le plan (001).

À partir de notre analyse bibliographique, réalisée sur les matériaux de type alluaudite à base de fer, il ressort que le site M(2), correspondant à la position générale 8f, est souvent occupé de manière concomitante par les cations Fe^{3+} et des cations bivalents avec différents taux d'occupation. Plus particulièrement, dans le cas de Mn^{II}, le taux d'occupation du site M(2) est généralement $0{,}5Fe^{III}/0{,}5Mn^{II}$. Les données cristallographiques de certains de ces matériaux sont consignées dans le tableau (6).

Tableau 6. Données cristallographiques des phosphates de type alluaudite avec un site M(2) occupé par 0,5Fe / 0,5Mn.

Composés	Paramètres et volumes des mailles					M(2)	Réf.
	a(Å)	b(Å)	c(Å)	β(°)	V(Å^3)		
$Ag_2Mn_2Fe\ (PO_4)_3$	12.147(3)	12.733(4)	6.499(5)	114.53(3)	907.85(2)	0.5Fe/0.5Mn	8
$AgNaMn_2Fe(PO_4)_3$	12.085(1)	12.684(0)	6.498(4)	114.54(3)	906.25(3)	0.5Fe/0.5Mn	19
$Na_2Mn_2Fe(PO_4)_3$	12.049(5)	12.624(1)	6.507(3)	114.56(3)	900.35(2)	0.5Fe/0.5Mn	19
$Na_2CdMnFe(PO_4)_3$	12.169(5)	12.685(3)	6.524(4)	114.51(2)	916.32(3)	0.5Fe/0.5Mn	20
$K_{0.53}Mn_{2.37}Fe_{1.24}(PO_4)_3$	12.272 (2)	12.606 (2)	6.416 (4)	114.87(2)°	900.5 (6)	0.5Fe/0.5Mn	21
$Na_2CaMnFe(PO_4)_3$	12.283(1)	12.736 (1)	6.494 (5)	114.76 (3)	922.5 (7)	0.5Fe/0.499Mn	22
$Ag_2Mn_2Fe(VO_4)_3$	12.0169(2)	12.990(2)	6.8479(1)	111.614(1)	993.81(4)	0.5Fe/0.5Mn	23
$AgNaMn_2Fe(VO_4)_3$	12.003(4)	12.9766(5)	6.8559(3)	111.693(2)	992.26(8)	0.5Fe/0.5Mn	23
$Na_2Mn_2Fe(VO_4)_3$	11.994(3)	12.952(3)	6.8517(2)	111.904(3)	988.20(5)	0.5Fe/0.5Mn	23
$Ca_{1,54}Mn_{1,46}Fe(PO_4)_3$	12,349 (2)	12,657 (2)	6,465 (1)	115,67 (1)	910,7 (3)	0.5Fe/0.5Mn	Présent travail
$Cd_{0,66}\ Mn_{1,46}Fe(PO_4)_3$	12,157 (2)	12,623 (2)	6,431 (1)	115,20 (1)	892,9(3)	0.5Fe/0.5Mn	Présent travail

4 Conclusion

Nous avons pu synthétiser, par diffusion à l'état solide et fusion directe, deux nouveaux phosphates de type alluaudite de formules chimiques $Ca_{1,54}Mn_{1,46}Fe(PO_4)_3$ et de $Cd_{0,66}Mn_{2,34}Fe(PO_4)_3$. Leurs structures ont été déterminées à partir des données de diffraction de rayons X sur monocristal. Ils cristallisent dans une structure de type alluaudite. La charpente tridimensionnelle de ces phosphates est formée de tétraèdres PO_4 et d'octaèdres $(Fe(1)/Mn(1))O_6$, $(Ca(1)/Mn(2))O_6$ ou $(Cd(1)/Mn(2))O_6$ connectés via des arêtes et des sommets délimitant de larges tunnels occupés par Ca(2) et Ca(3) dans la phase au Ca et par Cd(2) et Cd(3)/Mn(3) dans la phase au Cd.

Références

1. A. K. Padhi, K. S. Nanjundaswamy, and J. B. Goodenough, J. Electro. Soc. **144**, 1188–1194 (1997).
2. A. K. Padhi, K. S. Nanjundaswamy, C. Masquelier, S. Okada, and J. B. Goodenough, J. Electro. Soc. **144**, 1609–1613 (1997).
3. P. B. Moore, Am. Mineral. **56**, 1955–1975 (1971).
4. O. V. Yakubovich, W. Massa, P. G. Gavrilenko, and O. V. Dimitrova, Eur. J. Mineral. **17**, 741–748 (2005).
5. P. B. Moore, J. Ito, Mineral. Mag. **43**, 227–35 (1979).
6. T. E. Warner, W. Milius, J. Maier, J. Solid. State. Chem. **106**, 301–309(1993).
7. J. Kim, H. Kim, K.-Y. Park, Y.-U. Park, S. Lee, H.-S. Kwon, H.-I. Yoo, K. Kang, J. Mater. Chem. A, **2(23)**, 8632–8636 (2014).
8. N. Chouaibi, A. Daidouh, C. Pico, A. Santrich, M. L. Veiga, J. Solid. State. Chem. **159,** 46-50 (2001).
9. W. Meisel, H. J. Guttmann, O. Gütlich, Corros. Sci. **23**, 1373–1379 (1983).
10. S. Attali, B. Vigouroux, M. Lenzi, J. Persia, J. Catal. **63**, 496–500 (1980).

11. M.B. Korzenski, G.L. Schimek, J.W. Kolis, G.J. Long, J. Solid. State. Chem. **147**. 390-398 (1999)
12. M. Kacimi, M. Ziyad, F. Hatert, Mater. Res. Bull. **40**, 682–693(2005)
13. A. Assani, M. Saadi, M. Zriouil, L. El Ammari, Acta Cryst. E**67**, i5 (2011).
14. A. Ould Saleck, A. Assani, M. Saadi, C. Mercier, C. Follet, L. El Ammari, Acta Cryst. E**71**, 813–815 (2015).
15. A. Bouraima, A. Assani, M. Saadi, T. Makani, L. El Ammari, Acta Cryst. E**71**, 558–560 (2015).
16. J. Khmiyas, A. Assani, M. Saadi, T. Makani, L. El Ammari, Acta Cryst. E**71**, 690–692 (2015).
17. G. M. Sheldrick, University of Göttingen. **8** (1997).
18. G. M. Sheldrick, University of Göttingen. (2014).
19. A. Daidouh, C. Durio, C. Pico, M. L.Veiga, N. Chouaibi, A. Ouassini, Solid. state sci. **4(4)**, 541-548 (2002).
20. C. Durio, A. Daidouh, N. Chouaibi, C. Pico, M. L. Veiga, J. Solid State. Chem. **168**, 208–216 (2002)
21. M. Hidouri, M. Ben Amara, $K_{0.53}Mn_{2.37}Fe_{1.24}(PO_4)_3$. Acta. Cry. Sect E: **67(1)**, i1-i1 (2011).
22. S. Jebli, , A. Badri, M. Ben Amara, *Acta Cryst. Sect. E* 72**(12)**, 1806-1808. (2016).
23. H.Ben Yahia, M. Shikano, M. Tabuchi, I. Belharouak, *Inor. chem.*, **55(9)**, 4643-4649 (2016).

Synthesis and structure study of new manganese and indium based phosphate: $(Ca,Ba)_{1.222}Mn_{0.923}In_{1.570}(PO_4)_3$

Elhassan Benhsina*, Abderrazzak Assani, Mohamed Saadi and Lahcen El Ammari
Laboratoire de Chimie Appliquée des Matériaux, Centre des Sciences des Matériaux, Faculty of Sciences, Mohammed V University in Rabat, Avenue Ibn Battouta, BP 1014, Rabat, Morocco.
*Correspondence e-mail: h.benhsina@gmail.com

Abstract. Phosphates are an important class of materials which are well-known due to their structural diversity and their important potential application in many fields of sciences. A member of this family, represented by transition metal phosphates, is in the forefront of important researchers. A new manganese and indium based phosphate is successfully isolated by means of solid state reaction. Its corresponding crystal structure is isotypic to the alluaudite and crystallizes in the monoclinic system with the space group C2/c and cell parameters a=12.4697(5)Å ; b=12.9365(5)Å ; c=6.5625(3)Å and β=115.603(1)°. The framework of this phosphate is built up on the basis of MnO_6, InO_6 and PO_4 polyhedra. The InO_6 octahedra are linked together via common edge, leading to the formation of In_2O_{10} dimers. The In_2O_{10} dimers share edge with MnO_6 octahedra, generating chains along the $[10\bar{1}]$ direction. The PO_4 tetrahedra connect two chains giving arise to a three-dimensional framework with hexagonal tunnels where the barium and calcium are located.

1 Introduction

Les matériaux phosphatés connaissent un développement considérable en raison de leur diversité structurale et de leurs importantes et potentielles applications dans de nombreux domaines de la technologie moderne. Nous pouvons citer, à titre d'exemple, le phosphate $LiFePO_4$ [1] de structure type olivine, dont l'importance est amplement justifiée par son application dans les batteries au lithium commercialisés à l'échelle mondiale. De même, les propriétés électriques de $LiMnPO_4$ [2], de structure alluaudite, avoisinent celles de $LiFePO_4$ et représente l'un des matériaux les plus prometteurs dans le stockage d'énergie en tant que cathode pour les batteries. Par conséquent, l'un des principaux axes de recherche développés dans notre laboratoire est axé sur la synthèse et la caractérisation structurale de nouveaux matériaux phosphatés, apparentés essentiellement à la très convoitée famille alluaudite. Ce choix est justifié par les propriétés physico-chimiques de ces matériaux.

En déployant plusieurs méthodes de synthèse : synthèse par diffusion à l'état solide, synthèse hydrothermale ou synthèse par chimie douce, et en utilisant la diffraction sur monocristal ou sur poudre, nous avons pu isoler et caractériser plusieurs phosphates de type alluaudite. A titre d'exemple, $NaMg_3(HPO_4)_2(PO_4)$ [3], $AgMg_3(HPO_4)_2(PO_4)$ [4], $Na_2Co_2Fe(PO_4)_3$[5] et $Na_{1.67}Zn_{1.67}Fe_{1.33}(PO_4)_3$ [6].

En effet, la structure du premier phosphate de cette famille alluaudite est déterminée par P.B. Moore [7]. Ces phosphates cristallisent dans un système monoclinique avec un groupe d'espace C2/c. Moore a montré que cette famille peut être représentée par la formule générale $A(2)A(1)M(1)M(2)_2(PO_4)_3$ où A représente des sites cationiques pouvant être occupés par des cations de moyen ou grande taille et de faible charge (généralement monovalent ou divalent) alors que les sites cationiques de type M représentent un environnement octaédrique généralement occupé par des éléments de transitions. Toutefois, le site de type M(1) est très distordu alors que celui de type M(2) est presque régulier.

Dans ce travail, nous présentons la synthèse et la caractérisation structurale du nouveau phosphate à base d'indium et de manganèse $(Ca,Ba)_{1,222}Mn_{0,923}In_{1,570}(PO_4)_3$.

2 Expérimental

Le nouveau phosphate $(Ca,Ba)_{1,222}Mn_{0,923}In_{1,570}(PO_4)_3$ a été obtenu par la réaction de diffusion à l'état solide. Les réactifs (nitrates de calcium (Riedel-de haën, 98%), nitrate de baryum (Riedel-de haën, 98.5%), carbonate de manganèse (Riedel-de haën, 99%), iodure d'indium (Ventron, 98.5) et l'acide phosphorique (Riedel-de haën, 85%)) sont dissouts dans l'eau distillée avec quelque goutte d'acide nitrique. Le mélange réactionnel ainsi obtenu est maintenu sous agitation pendant 24 heures avant d'être porté à sec. La poudre obtenue est ensuite broyée dans un mortier en agate et est portée progressivement à la fusion à une température de 1080°C, dans un creuset en platine. Après un refroidissement lent avec une vitesse de 5°C/heure jusqu'à la température ambiante, nous obtenons des cristaux de taille suffisante pour la diffraction des rayons X sur monocristal.

3 Résultat et discussions

3.1 Collection des données et résolution structurale

La résolution structurale a été réalisée en utilisant plusieurs réflexions collectées par le diffractomètre à quatre cercle de type Brüker X8 APEX[8] des Unités d'Appui Techniques à la Recherche Scientifique (UATRS) relevant du Centre National pour la recherche Scientifique et Technique CNRST (Maroc). En effet, seules les intensités répondant au critère statistique $I>2\sigma(I)$ ($\sigma(I)$ étant l'écart type sur l'intensité) ont été utilisées pour localiser les positions des atomes. Ces données ont été corrigées des effets de Lorentz polarisation et des effets d'absorption. Le tableau 1 montre les conditions d'enregistrement et les données cristallographiques de $(Ca,Ba)_{1,222}Mn_{0,923}In_{1,570}(PO_4)_3$. La résolution et l'affinement de la structure ont été réalisés par la méthode directe couplée aux synthèses de Fourier différences successives, à l'aide des programmes SHELXT2014 [9] et SHELXL2014 [10].

Tableau 1: Données cristallographiques et conditions d'enregistrement de $(Ca,Ba)_{1,222}Mn_{0,923}In_{1,570}(PO_4)_3$.

Donnée du cristal	
Formule chimique	$(Ca,Ba)_{1,222}Mn_{0,923}In_{1,570}(PO_4)_3$
Système cristallin	Monoclinique
Groupe d'espace	*C2/c*
Temperature (K)	296
a (Å)	12,4697 (5)
b(Å)	12,9365 (5)
c(Å)	6,5625 (3)
β (°)	115,603 (1)
V (Å^3)	954,68 (7)
Z	4
Type de radiation	Mo $K\alpha$
μ(mm^{-1})	16,79
Collection des données	
Nombre de réflexions mesurées	9990
Nombre réflexions indépendantes	1822
Nombre de réflexions avec [$I>2\sigma(I)$]	1703
R_{int}	0,029
$(\sin \theta/\lambda)_{max}$ (Å^{-1})	0,769
Affinement	
$R[F^2> 2\delta(F^2)]$	0,022
$wR(F^2)$	0,056
S	1,13
Nombre de paramètres	102

Les conditions d'extinction observées pour $(Ca,Ba)_{1,222}Mn_{0,923}In_{1,570}(PO_4)_3$ sont compatibles avec les groupes d'espace C2/c et Cc. La structure de ce phosphate a été résolue dans le groupe centro-symétrique (C2/c). Précisons que la méthode des moindres carrés, appliquée sur le carré des modules des facteurs de structure F^2, a été utilisée pour affiner les différents paramètres structuraux (positions atomiques (x, y, z), paramètres de déplacement isotropes (U_{iso}) ou anisotropes (U_{aniso}) et occupation des sites cristallographiques). Les facteurs d'accord, obtenus à la fin du dernier cycle d'affinement, R(F) = 2,20%, wR(F2) = 5,60% et GooF= S= 1,13. Ces valeurs attestent de la bonne qualité de l'affinement.

Le tableau 2 regroupe les positions atomiques, les déplacements atomiques isotropes équivalents ainsi que les taux d'occupations des sites cristallographiques. Le tableau 3 rassemble les facteurs de déplacements atomiques anisotropes dans $(Ca,Ba)_{1,222}Mn_{0,923}In_{1,570}(PO_4)_3$.

Tableau 2: Positions atomiques, déplacements atomiques isotropes équivalents et taux d'occupations des sites cristallographiques dans $(Ca,Ba)_{1,222}Mn_{0,923}In_{1,570}(PO_4)_3$.

	Wyck.	*x*	*y*	*z*	U_{eq}	Occ.
In1	8f	0,721 (2)	0,3477 (1)	0,6327 (3)	0,009(1)	0,785(1)
Mn1	4e	1,00	0,7375 (1)	0,750	0,010 (1)	0,923 (2)
Ca1	4e	0,500	0,4866 (1)	0,250	0,033 (1)	0,922 (1)
Ba1	8f	0,522 (1)	-0,0006 (1)	0,466 (1)	0,026 (1)	0,151 (1)
P1	8f	0,7397 (1)	0,6104 (1)	0,6341 (1)	0,009 (1)	1
P2	4e	0,5000	0,2114 (1)	0,2500	0,008 (1)	1
O1	8f	0,7263 (2)	0,6795 (1)	0,8136 (2)	0,012 (1)	1
O2	8f	0,6681(2)	0,6628 (1)	0,4025 (3)	0,014 (1)	1
O3	8f	0,5461 (2)	0,2796 (1)	0,4621 (2)	0,014 (1)	1
O4	8f	0,4069(2)	0,1369 (1)	0,2639 (4)	0,023 (1)	1
O5	8f	0,6867 (2)	0,5042 (1)	0,6345 (3)	0,019 (1)	1
O6	8f	0,872 (2)	0,6031 (1)	0,6862 (4)	0,024 (1)	1

Tableau 3 : Facteurs de déplacements atomiques anisotropes dans le phosphate $(Ca,Ba)_{1,222}Mn_{0,923}In_{1,570}(PO_4)_3$.

	*U*11	*U*22	*U*33	*U*12	*U*13	*U*23
In1	0,01137 (9)	0,00810 (9)	0,00950 (9)	-0,00101 (5)	0,00593 (6)	-0,00118 (5)
Mn1	0,0112 (2)	0,0114 (2)	0,0107 (2)	0,000	0,00719 (2)	0,000
Ca1	0,0148 (4)	0,0279 (5)	0,0400 (6)	0,000	-0,0036 (3)	0,000
Ba1	0,0258 (6)	0,0121 (4)	0,0267 (6)	0,0006 (4)	-0,0013 (4)	0,0012 (4)
P1	0,0135 (2)	0,0064 (2)	0,0081 (2)	0,00066 (2)	0,00576 (2)	0,00012 (2)
P2	0,0078 (3)	0,0095 (3)	0,0074 (3)	0,000	0,0031 (2)	0,000
O1	0,0175 (7)	0,0109 (6)	0,0083 (6)	-0,0002 (5)	0,0070 (5)	-0,0016 (5)
O2	0,0252 (8)	0,0104 (6)	0,0080 (6)	0,0017 (5)	0,0081 (6)	0,0019 (5)
O3	0,0120 (6)	0,0211 (8)	0,0066 (6)	0,0015 (5)	0,0027 (5)	-0,0030 (5)
O4	0,0104 (7)	0,0180 (8)	0,0394 (2)	0,0009 (6)	0,0085 (7)	0,0145 (7)
O5	0,0289 (9)	0,0078 (7)	0,0242 (8)	-0.0004 (6)	0,0150 (7)	0,0024 (6)
O6	0,0170 (7)	0,0175 (8)	0,0412 (2)	0,0021 (6)	0,0164 (8)	-0,0042 (7)

Les atomes d'indium et de baryum occupent partiellement les sites 8f avec des taux d'occupation de 0,785(1) et 0,151 (1) respectivement. Quant aux atomes de manganèse et de calcium, ils occupent partiellement les sites 4e avec des taux d'occupation de 0,993(2) et 0,922(1) respectivement. Les autres atomes (phosphore et oxygène) occupent totalement les sites 8f à l'exception de P2 qui est localisé sur le site 4e. Ceci conduit à la formulation suivante : $Ca_{0,922}Ba_{0,300}Mn_{0,923}In_{1,570}(PO_4)_3$. Ce type de non-stœchiométrie est souvent rencontré dans les structures de type alluaudite : $Na_{1,72}Mn_{3,28}(AsO_4)_3$[11], $Ag_{1,49}Mn_{2,90}(AsO_4)_3$[12], $Na_{1,67}Mn_{2,17}(MoO_4)_3$[13], $Na_{2,44}Mn_{1,79}(SO_4)_3$[14].

4 Description de la structure

Les distances et les angles significatifs dans le phosphate $(Ca,Ba)_{1,222}Mn_{0,923}In_{1,570}(PO_4)_3$ sont regroupés dans le tableau 4.

Tableau 4: Distances (Å) et angles interatomiques (°) dans la structure de$(Ca,Ba)_{1,222}Mn_{0,923}In_{1,570}(PO_4)_3$.

Distances P(1) - O (Å)		Distances P(2) - O (Å)	
P1—O5	1,5252 (2)	P2—O3^{xii}	1,5344 (2)
P1—O6	1,5355 (2)	P2—O3	1,5345 (2)
P1—O2	1,5472 (2)	P2—O4	1,5410 (2)
P1—O1	1,5479 (2)	P2—O4^{xii}	1,5411 (2)
Distance moyenne	**1,5389 Å**	**Distance moyenne**	**1,5376 Å**
Angles O - P(1) - O (°)		**Angles O - P(2) - O (°)**	
O5—P1—O6	111,68 (11)	O3^{xii}—P2—O3	154,72 (7)
O5—P1—O2	109,24 (10)	O3^{xii}—P2—O4	90,36 (6)
O6—P1—O2	110,74 (11)	O3—P2—O4	109,79 (13)
O5—P1—O1	109,08 (10)	O3^{xii}—P2—O4^{xii}	114,63 (9)
O6—P1—O1	109,27 (10)	O3—P2—O4^{xii}	107,58 (11)
O2—P1—O1	106,69 (9)	O4—P2—O4^{xii}	102,65 (15)
Distances In - O (Å)		**Distances Mn - O (Å)**	
In1—O5	2,0694 (2)	Mn1—O3^{vi}	2,2660 (2)
In1—O4^{i}	2,1073 (2)	Mn1—O3^{viii}	2,2660 (2)
In1—O2^{ii}	2,1424 (2)	Mn1—O6^{ix}	2,2734 (2)
In1—O1^{iii}	2,1501 (2)	Mn1—O6	2,2735 (2)
In1—O3	2,1605 (2)	Mn1—O2^{x}	2,2905 (2)
In1—O1^{iv}	2,2591 (2)	Mn1—O2^{xi}	2,2766 (2)
Distance moyenne	**2,1481Å**	**Distance moyenne**	**2,3434 Å**
Angles O – In– O (°)		**Angles O – Mn– O (°)**	
O5—In1—O4^{i}	95,63 (7)	O3^{vi}—Mn1—O3^{viii}	152,22 (9)
O5—In1—O2^{ii}	85,41 (7)	O3^{vi}—Mn1—O6^{ix}	116,17 (7)
O4^{i}—In1—O2^{ii}	110,19 (8)	O3^{viii}—Mn1—O6^{ix}	85,77 (7)
O5—In1—O1^{iii}	104,96 (7)	O3^{vi}—Mn1—O6	85,77 (7)
O4^{i}—In1—O1^{iii}	85,63 (7)	O3^{viii}—Mn1—O6	116,17 (7)
O2^{ii}—In1—O1^{iii}	160,45 (6)	O6^{ix}—Mn1—O6	80,23 (9)
O5—In1—O3	103,80 (7)	O3^{vi}—Mn1—O2^{x}	90,83 (6)
O4^{i}—In1—O3	159,40 (6)	O3^{viii}—Mn1—O2^{x}	73,45 (6)

O2^{ii}—In1—O3	78,58 (6)	O6^{ix}—Mn1—O2^{x}	85,01 (6)
O1^{iii}—In1—O3	82,82 (6)	O6—Mn1—O2^{x}	161,49 (7)
O5—In1—O1^{iv}	171,49 (7)	O3^{vi}—Mn1—O2^{xi}	73,45 (6)
O4^{i}—In1—O1^{iv}	80,08 (6)	O3^{viii}—Mn1—O2^{xi}	90,83 (6)
O2^{ii}—In1—O1^{iv}	89,18 (6)	O6^{ix}—Mn1—O2^{xi}	161,49 (7)
O1^{iii}—In1—O1^{iv}	82,15 (6)	O6—Mn1—O2^{xi}	85,01 (6)
O3—In1—O1^{iv}	81,49 (6)	O2^{x}—Mn1—O2^{xi}	111,45 (8)
Distances Ca - O (Å)		**Distances Ba- O (Å)**	
Ca1—O5	2,6005 (2)	Ba1—O4	2,308 (2)
Ca1—O5^{xii}	2,6006 (2)	Ba1—O6^{xv}	2,377 (2)
Ca1—O5^{iii}	2,7446 (2)	Ba1—O6^{v}	2,381 (2)
Ca1—O5^{vii}	2,7446 (2)	Ba1—O4^{xiv}	2,384 (2)
Ca1—O3	2,9577 (2)	Ba1—O6^{iv}	2,480 (2)
Ca1—O3^{xii}	2,958 (2)	Ba—O^{iiv}	2,7204
Ca1—O2	2,9623 (2)	**Distance moyenne**	**2,4417 Å**
Ca1—O2^{xii}	2,9623 (2)		
Distance moyenne	**2,8163 Å**		

Codes de symétrie : (i) *x*+1/2, -*y*+1/2, *z*+1/2; (ii) *x*, -*y*+1, *z*+1/2; (iii) *x*, -*y*+1, *z*-1/2; (iv) -*x*+3/2, *y*-1/2, -*z*+3/2; (v) -*x*+3/2, -*y*+1/2, -*z*+1; (vi) -*x*+3/2, *y*+1/2, -*z*+3/2; (vii) –*x*+1, -*y*+1, -*z*+1; (viii) *x*+1/2, *y*+1/2, *z*; (ix) -*x*+2, *y*, -*z*+3/2; (x) *x*+1/2, -*y*+3/2, *z*+1/2; (xi) -*x*+3/2, -*y*+3/2, -*z*+1; (xii) -*x*+1, *y*, -*z*+1/2; (xiii) -*x*+1, -*y*+1, -*z*; (xiv) -*x*+1, -*y*, -*z*+1; (xv) *x*-1/2, -*y*+1/2, *z*-1/2; (xvi) *x*-1/2, *y*-1/2, *z*; (xvii) *x*, -*y*, *z*-1/2; (xviii) *x*, -*y*, *z*+1/2.

La structure de ce phosphate est formée de tétraèdres PO_4, d'octaèdres InO_6 et MnO_6.

Deux octaèdres d'indium mettent des arêtes en commun pour former un dimer In_2O_{10}. Ces derniers sont liés alternativement par les octaèdres de manganèse MnO_6 pour construire des chaînes parallèle à la direction $[10\bar{1}]$ (Figure 1).

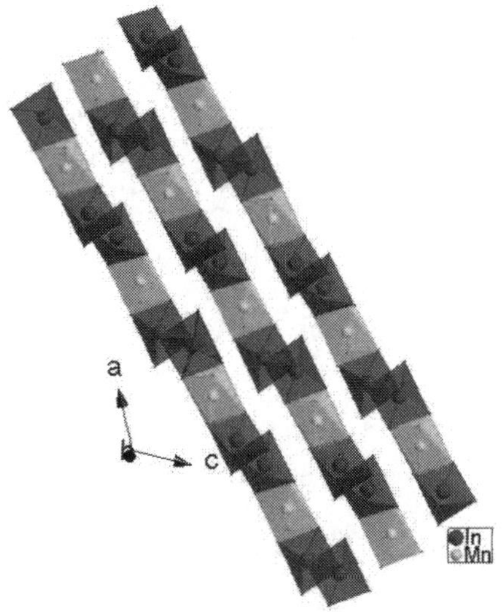

Figure 1: Chaînes formées par les octaèdres MnO_6 et InO_6 parallèle à $[10\bar{1}]$.

Ces chaînes sont connectées via les sommets des tétraèdres $P1O_4$ et $P2O_4$, pour former une charpente tridimensionnelle délimitant des tunnels dans lesquels les atomes de baryum et de calcium se logent (Figure 2).

Figure 2 : Projection de la structure $(Ca,Ba)_{1,222}Mn_{0,923}In_{1,570}(PO_4)_3$ selon l'axe c.

Remerciements :

Les auteurs remercient les Unités d'Appui Techniques à la Recherche Scientifique (UATRS), relevant du Centre National pour la recherche Scientifique et Technique (CNRST) (Maroc) pour la collecte des données de la diffraction de rayons X.

https://doi.org/10.1051/matecconf/201814901085

Références

1. L.-X. Yuan *et al.*, *Energy Environ. Sci.*, vol. 4, no. 2, pp. 269–284, (2011)..
2. J. Kim, H. Kim, K. Park, Y. Park, and S. Lee, vol. 4, pp. 8632–8636, (2014).
3. A. Assani, M. Saadi, M. Zriouil, . & L. El Ammari, Acta Cryst. E67, i5 (2011a).
4. A. Ould Saleck, A. Assani, M. Saadi, C. Mercier, C. Follet, and L. El Ammari, *Acta Crystallogr. Sect. E Crystallogr. Commun.*, vol. 71, no. 7, pp. 813–815, (2015).
5. A. Bouraima, A. Assani, M. Saadi, T. Makani, and L. El Ammari, *Acta Crystallogr. Sect. E Crystallogr. Commun.*, vol. 71, no. Ii, pp. 558–560, (2015).
6. J. Khmiyas, A. Assani, M. Saadi, and L. El Ammari, *Acta Crystallogr. Sect. E Crystallogr. Commun.*, vol. 71, no. Iii, pp. 690–692, (2015).
7. P. B. Moore, Am. Mineral., vol. 56, p. 1955, (1971).
8. Bruker APEX2 and SAINT. Bruker AXS Inc., Madison, Wisconsin, USA. (2005).
9. G. M. Sheldrick, Acta Cryst.A71, 3–8 (2015a).
10. G. M. Sheldrick, Acta Cryst.C71, 3-8 (2015b).
11. B. Ayed, M. Krifa & A. Haddad Acta Cryst. C58, i58, i98-100 (2002).
12. B. Ayed & A. Haddad , Acta Cryst E59, i77-i79 (2003).
13. C. Bouzidi, M. Faouzi Zid, A. Driss & W. Frigui. Acta Cryst. E70, i18–i19 (2014).
14. Debasmita Dwibedi, Rafael B. Araujo, Sudip Chakraborty, Pradeep P. Shanbogh, alini G. Sundaram, Rajeev Ahuja, & Prabeer Barpanda J. Mater. Chem. A, (2015).

Synthesis and structural characterization of new cadmium, zinc and iron based pyrophosphate: $(Cd,Zn)_{2.5}Fe(P_2O_7)_2$

Jamal Khmiyas, Abderrazzak Assani, Mohamed Saadi and Lahcen El Ammari.

Laboratoire de Chimie Appliquée des Matériaux, Centre des Sciences des Matériaux, Faculty of Sciences, Mohammed V University in Rabat, Avenue Ibn Battouta, BP 1014, Rabat, Morocco.

Abstract. The novel compound with the following nominal formulation $(Cd, Zn)_{2.5}Fe(P_2O_7)_2$ has been successfully prepared by the solid state reaction and characterized by single-crystal X-ray diffraction. This phosphate crystallizes in the orthorhombic system with the $C222_1$ space group. Its crystal structure is formed by five types of cationic sites and by two unique $(P_2O_7)^{4-}$ anionic groups with a staggered conformation. In the structure, three sites are fulfilled by Zn^{2+} :(Zn(3), Zn(4)) and $Fe(1)^{3+}$, the fourth site is localized in the Wyckoff position *4a* (site symmetry 2 . .) and is statically occupied by Cd^{2+} : $Zn^{2+}$with ratio 0.586(2) : 0.413(8). The last site, localized in *8c* Wyckoff position (site symmetry 1) is also occupied by a mixture of the two cations Cd^{2+}: $Zn^{2+}$with occupancies of 0.444(6) : 0.555(4).Within this framework, the metallic polyhedra $[MO_n]$ (n = 5 or 6) form corrugated layers stacked along the *b*-axis, and intercalated by P_2O_7 groups.

1 Introduction

La chimie des matériaux phosphatés poreux et à charpente ouverte est en plein essor actuellement [1]. Ceci est dû essentiellement à la grande richesse et la diversité structurales et topologiques affichées par cette classe de matériaux. L'originalité des pyrophosphates tient, en majeure partie, à leurs diverses propriétés physicochimiques remarquables directement liées à leurs structures cristallines. Les pyrophosphates sont recherchées pour diverses applications potentielles à savoir : la catalyse [2], sorption & désorption [3], autant que récepteur et capteur biologique [4], agent thérapeutique en hémodialyse [5], en agroalimentaire [6], conception de composite [7], biocéramique [8], inhibiteur endogène de la bio-minéralisation [9].

Dans ce cadre, on peut citer les pyrophosphates : $Na_{3.12}Fe_{2.44}(P_2O_7)_2$ [10] , $Na_7V_3(P_2O_7)_4$ [11] isotypes de $Na_7M_3(P_2O_7)_4$ (M = Al, Ga, Cr, Fe) [12] et $Na_2MP_2O_7$ (M = Fe, Mn, Co) qui affichent d'excellents taux de capacités et de réversibilités ainsi qu'une grande stabilité thermique [13]. Par ailleurs, la famille des composés $A^IM^{III}P_2O_7$ (A = Li, Na, K, Rb, Cs, Tl, Ag) et (M =Al, Ga, métal de transition), adoptent une structure à cage construite à partir des tétraèdres PO_4 et MO_6 partageant des sommets. Le site crée par cet agencement est occupé par des cations monovalents de coordinence CN = 6 [14].

L'élargissement des dimensions des espaces interstitiels, ainsi que l'occupation totale ou partielle des sites au sein de la structure cristalline est l'un des facteurs favorables à la mobilité des cations monovalents. De même, divers travaux ont été consacrés aux pyrophosphates métalliques et métalliques mixtes : $A^{II}B^{II}P_2O_7$ [15] (A=Ba, Pb ; B= Pb, Co, Mn, Cu), $A^{II}B_2^{III}(P_2O_7)_2$ [16-18] (A=Sr, Pb, Cu ; B= Fe), et $A^IM_3(P_2O_7)_2$ (M = Fe, Co, Ni) [19].Le présent travail relate la synthèse et la caractérisation structurale du pyrophosphate à base de cadmium, de zinc et de fer, $(Cd,Zn)_{2.5}Fe(P_2O_7)_2$.

2 Synthèse et cristallisation

Les cristaux du nouveau pyrophosphate : $(Cd,Zn)_{2.5}Fe(P_2O_7)_2$ ont été synthétisés par réaction de diffusion à l'état solide, à partir des réactifs : CdO (0,1284g), $Zn(NO_3)_2.6H_2O$ (0,5949g), $Fe(NO_3)_3.9H_2O$ (0,4040g) et de H_3PO_4 (85%) (0.20mL), selon les rapports Cd: Zn: Fe: P=1:2:1:3. Dans un bécher contenant de l'eau distillée, le mélange réactionnel est homogénéisé par agitation à l'ambiante pendant 24 heures. Après une évaporation lente, une poudre homogène de couleur jaune est obtenue. Cette poudre est ensuite broyée pendant 30 minutes dans un mortier en agate, puis mis dans un creuset en platine et ~~puis~~ porté lentement à la fusion à 1218K et maintenu à cette température pendant 4 heures.

Après un refroidissement lent avec une vitesse de 5°C/h et un lavage à l'éthanol, dans un bain à ultrasons, nous avons obtenus des cristaux marron relatifs à cette nouvelle phase.

 https://doi.org/10.1051/matecconf/201814901086

3 Etude structurale du nouveau pyrophosphate : $(Cd,Zn)_{2.5}Fe(P_2O_7)_2$

3.1 collecte des intensités diffractées

Un cristal marron de bonne qualité et de taille exploitable, a été sélectionné et monté sur le diffractomètre à 4 cercles X8 APEXII de l'UATRS (Unités d'Appui Technique à la Recherche Scientifique) du CNRST (Centre National pour la Recherche Scientifique et Technique). La collecte des intensités de diffraction des rayons-X, a été effectuée dans les conditions mentionnées dans le tableau(1).

3.2 Résolution et affinement structural

La structure a été résolue par la méthode de Patterson à l'aide du programme SHLEXS-97 et affinée par la méthode des moindres carrées à l'aide du programme SHELXL-97 [20]. La résolution nous a permis de localiser les atomes lourds Cd, Zn, Fe et P. Suite aux séries d'affinements successifs suivies de Fourier-différences, nous avons pu localiser les quatorze atomes d'oxygène indépendants dans l'unité asymétrique. La structure est caractérisée par désordre partiel.

En effet le dernier cycle d'affinement conduit aux résultats suivants : les atomes Cd(1) et Zn(1) sont réparties statiquement sur la position particulière (4a) avec un taux d'occupation de 0,586(6) et 0,414(6) respectivement. De même Cd(2) et Zn(2) occupent la même position générale (8c) avec le taux de 0.445(5) et 0.555(5) respectivement. Le reste des atomes occupent totalement leur site cristallographique.

Cette dernière répartition conduit à la formulation suivante : $Cd_{0.74}Zn_{1.76}Fe(P_2O_7)_2$. L'affinement structural a abouti à de bons facteurs de reliabilité consignés dans le tableau (1). Les coordonnées réduites et les déplacements atomiques isotropes et anisotropes sont rassemblés dans les tableaux (2) et (3), respectivement.

Tableau 1. Données cristallographiques et conditions d'enregistrement des intensités diffractées du nouveau pyrophosphate $Cd_{0.74}Zn_{1.76}Fe(P_2O_7)_2$.

Données cristallographiques	
Formule chimique	$Cd_{0.74}Zn_{1.76}Fe(P_2O_7)_2$
Système cristallin	Orthorhombique
Groupe d'espace	$C222_1$
Paramètres de la maille	a = 8,4761 (1) Å b = 9,7581 (1) Å c = 24,0557 (3) Å
Volume de la maille ($Å^3$)	1989,66 (4)
Nombre de motif par maille Z	8
Densité calculée ($Mg.m^{-3}$)	4.018
Coefficient d'absorption (mm^{-1})	7.934

Collecte des données et résultats d'affinement	
Diffractomètre	Bruker X8 APEXII
Température (K)	296
Longueur d'onde Mo(Kα) (Å)	0,7071
Correction d'absorption	multi-scan (SADABS; Bruker, 2005)
Domaine de θ (°)	θ_{max} = 36.6 θ_{min} = 3.2
Indices de Miller	h = -14→14 k = -16→16 l = -40→35
Réflexions mesurées	16814
Réflexions indépendantes	4898
Réflexions pour I > 2σ(I)	4796
R_{int}	0.021
$R[F^2 > 2\sigma(F^2)]$	0.020
$wR(F^2)$	0.051
S	1.09
Nombre des paramètres affinés	199

3.3 Description de la structure

La structure du nouveau pyrophosphate de cadmium, zinc et de fer $Cd_{0.74}Zn_{1.76}Fe(P_2O_7)_2$ est isotype de : $Fe^{II}_5Fe^{III}_2(P_2O_7)_4$ [21] , $Cd_5MIII_2(P_2O_7)_4$ (M = V, Fe) [22] et $Cd_{4.12}FeII_{0.88}FeIII_2(P_2O_7)_4$ [23]. Les principales distances et angles interatomiques de la structure de $Cd_{0.74}Zn_{1.76}Fe(P_2O_7)_2$ sont reportés dans le tableau (4).

Dans la structure de ce nouveau pyrophosphate, la séquence de polyèdres $(Cd(1)/Zn(1))O_6$ —FeO_6— $(Cd(2)/Zn(2))O_5$—FeO_6—$(Cd(2)/Zn(2))O_5$—$Zn(4)O_6$ permet de construire un anneau à six polyèdres (Figure 1-a). En effet, les cinq premiers polyèdres de cette succession sont liés entre eux par la mise en commun d'une arête, l'octaèdre $Zn(4)O_6$ partage deux de ces oxygènes avec $(Cd(1)/Zn(1))O_6$ et $(Cd(2)/Zn(2))O_5$. De même, chacun des anneaux partage une arête de l'octaèdre $Zn(4)O_6$ avec l'octaèdre $(Cd(1)/Zn(1))O_6$ de l'anneau adjacent, pour engendrer des chaînes d'anneaux infinies évoluant le long de l'axe-a à $z\approx0$ et à $z\approx1/2$ (Figure 1-b).

Les octaèdres $Zn(3)O_6$, qui ne contribuent pas à la formation des anneaux, viennent relier ces deux types de chaînes entre elles pour former des feuillets $[(Cd,Zn)_3Zn_2Fe_2O_{24}]_\infty$ ondulés parallèles au plan (010) (Figure 2a et 2b).Les deux types de groupements P_2O_7 sont isolés (Figure 3) et sont disposés de manière ondulée suivant l'axe c . Les feuillets $[(Cd,Zn)_3Zn_2Fe_2O_{24}]_\infty$ sont inter-connectés par les groupements P_2O_7 pour aboutir à un réseau tridimensionnel (Figure 4).

Signalons que les groupements P_2O_7 adoptent une conformation décalée, également observée dans l'homologue : $Co_5V^{III}_2(P_2O_7)_4$ [24].

Tableau 2. Coordonnées atomiques et facteurs de déplacements atomiques isotropes équivalents (Å^2)

Atome	Site cristallographique	x	y	z	Uiso*/Ueq	Occupation
Cd1	*4a*	0.19839 (4)	0.5	0.5	0.0104 (1)	0,586 (6)
Zn1	*4a*	0.19839 (4)	0.5	0.5	0.0104 (1)	0,414 (6)
Cd2	*8c*	0.31562 (3)	0.64219 (3)	0.86008 (2)	0.01278 (9)	0,445 (5)
Zn2	*8c*	0.31562 (3)	0.64219 (3)	0.86008 (2)	0.01278 (9)	0,555 (5)
Zn3	*4b*	0	0.29054 (5)	0.75	0.00867 (9)	1
Zn4	*4a*	0.08425 (6)	1	0.5	0.0225 (1)	1
Fe1	*8c*	0.07277 (4)	0.65109 (4)	0.62175 (2)	0.00626 (7)	1
P1	*8c*	0.40301 (8)	0.79463 (7)	0.49719 (3)	0.0077 (1)	1
P2	*8c*	0.24181 (8)	0.54279 (7)	0.73124 (3)	0.0075 (1)	1
P3	*8c*	0.37948 (8)	0.44633 (7)	0.62565 (3)	0.0070 (1)	1
P4	*8c*	0.39000 (8)	0.85839 (7)	0.61800 (3)	0.0070 (1)	1
O1	*8c*	0.3317 (3)	0.6361 (2)	0.77110 (9)	0.0133 (4)	1
O2	*8c*	0.3839 (2)	1.0141 (2)	0.62598 (9)	0.0099 (3)	1
O3	*8c*	0.2267 (2)	0.4950 (2)	0.59825 (8)	0.0106 (3)	1
O4	*8c*	0.3991 (3)	0.6393 (2)	0.49590 (9)	0.0127 (3)	1
O5	*8c*	0.3867 (3)	0.2886 (2)	0.62753 (9)	0.0114 (3)	1
O6	*8c*	0.2298 (3)	0.7948 (2)	0.6163 (1)	0.0140 (4)	1
O7	*8c*	0.5060 (3)	0.7911 (2)	0.6573 (1)	0.0149 (4)	1
O8	*8c*	0.1817 (3)	0.4171 (2)	0.7593 (1)	0.0169 (4)	1
O9	*8c*	0.5384 (2)	0.8524 (2)	0.46189 (9)	0.0119 (3)	1
O10	*8c*	0.1177 (3)	0.6188 (2)	0.69813 (9)	0.0129 (4)	1
O11	*8c*	0.5240 (2)	0.5101 (2)	0.5995 (1)	0.0138 (4)	1
O12	*8c*	0.4653 (3)	0.8385 (3)	0.5573 (1)	0.0168 (4)	1
O13	*8c*	0.2462 (3)	0.8551 (3)	0.4830 (1)	0.0201 (5)	1
O14	*8c*	0.3786 (2)	0.4973 (2)	0.68812 (9)	0.0124 (3)	1

Tableau 3. Facteurs de déplacements atomiques anisotropes(Å^2).

	U11	U22	U33	U12	U13	U23
Cd1/Zn1	0.0083 (1)	0.0087 (1)	0.0143 (1)	0.000	0.000	-0.0029 (1)
Cd2/Zn2	0.0072 (1)	0.0115 (1)	0.0197 (1)	-0.00155 (8)	0.00004 (8)	0.00535 (9)
Zn3	0.0086 (1)	0.0076 (1)	0.0099 (1)	0.000	0.0008 (1)	0.000
Zn4	0.0041 (1)	0.0057 (1)	0.0582 (4)	0.000	0.000	-0.0059 (2)
Fe1	0.0069 (1)	0.0055 (1)	0.0065 (1)	-0.0005 (1)	-0.0004(1)	0.0007 (1)
P1	0.0087 (2)	0.0054 (2)	0.0089 (2)	-0.0009 (1)	0.0012 (2)	0.0005 (2)
P2	0.0090 (3)	0.0069 (2)	0.0064 (2)	-0.0010 (2)	-0.0011 (1)	-0.0008 (2)
P3	0.0079 (2)	0.0055 (2)	0.0076 (3)	0.0015 (1)	0.0010 (1)	0.0006 (2)
P4	0.0087 (2)	0.0051 (2)	0.0073 (2)	-0.0011 (2)	-0.0009(1)	-0.0001 (2)
O1	0.0153 (9)	0.0131 (8)	0.0116 (8)	-0.0007 (8)	-0.0041 (7)	-0.0061 (7)
O2	0.0149 (8)	0.0040 (7)	0.0108 (8)	-0.0007 (6)	0.0010 (6)	0.0008 (6)
O3	0.0097 (7)	0.0125 (8)	0.0096 (8)	0.0021 (7)	-0.0016 (6)	0.0015 (7)
O4	0.0202 (9)	0.0062 (7)	0.0118 (8)	-0.0016 (7)	-0.0026 (8)	0.0008 (7)
O5	0.0160 (9)	0.0056 (7)	0.0126 (9)	0.0018 (6)	0.0010 (7)	0.0012 (6)
O6	0.0106 (8)	0.0109 (8)	0.021(1)	-0.0044 (7)	0.0003 (7)	-0.0011 (8)
O7	0.0169 (9)	0.0108 (8)	0.0170 (9)	0.0032 (7)	-0.0062 (8)	0.0023 (7)
O8	0.0225 (1)	0.0127 (9)	0.016 (1)	-0.0042 (8)	0.0020 (8)	0.0045 (7)
O9	0.0109 (7)	0.0151 (8)	0.0097 (8)	-0.0008 (7)	0.0031 (6)	0.0047 (7)
O10	0.0117 (8)	0.0184 (9)	0.0086 (8)	0.0038 (7)	-0.0020 (6)	-0.0002 (7)
O11	0.0110 (8)	0.0134 (8)	0.0170 (9)	0.0004 (7)	0.0058 (7)	0.0040 (8)
O12	0.0148 (9)	0.0223 (1)	0.0132 (9)	-0.0062 (8)	0.0035 (7)	-0.0095 (8)
O13	0.0101 (8)	0.0175 (1)	0.033 (1)	0.0010 (8)	-0.0003 (8)	0.005(1)
O14	0.0102 (7)	0.0168 (8)	0.0102 (8)	0.0017 (7)	-0.0009 (6)	-0.0032 (8)

(a)

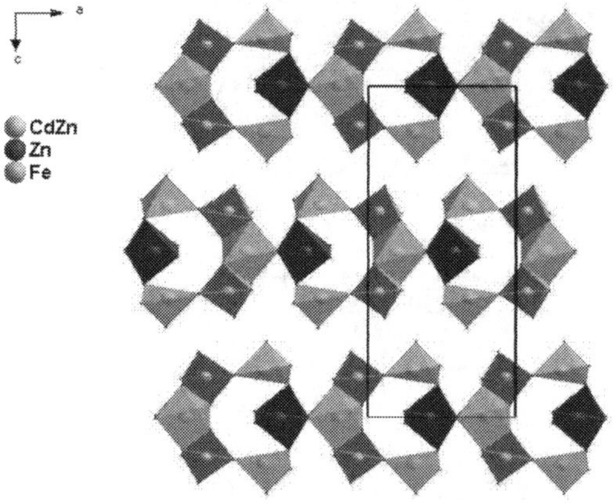

(b)

Fig.1. (a) : Alternance des octaèdres MO_6 (M= Fe^{III} , Zn, (Cd(1), Zn(1)) et la pyramide à base carrée $(Cd(2)/Zn(2))O_5$ formant l'anneau à 6 polyèdres. **(b) :** Succession des chaînes d'anneaux dans le plan (001).

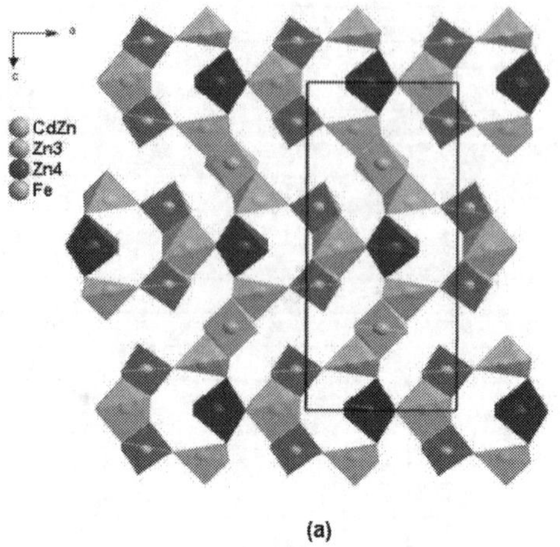

(a)

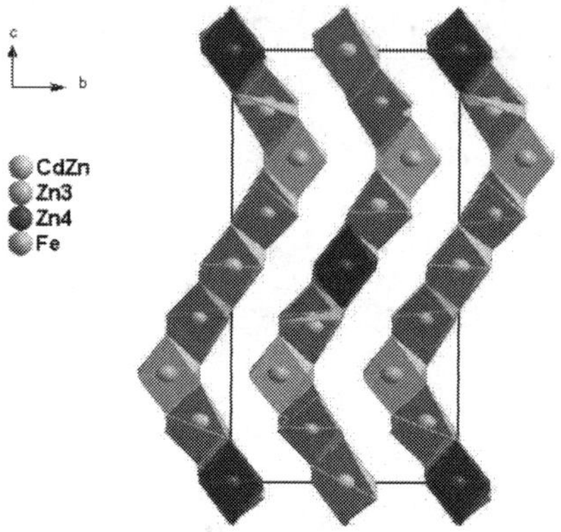

(b)

Fig.2 (a) : Projection d'une couche disposée parallèlement au plan (010).**(b) :** Projection des couches dans le plan (b, c).

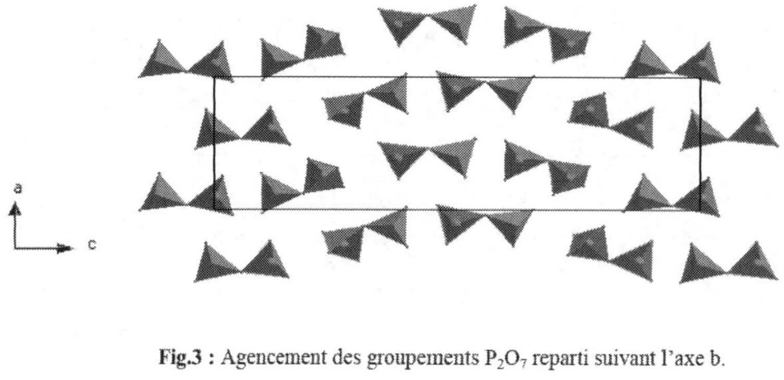

Fig.3 : Agencement des groupements P_2O_7 reparti suivant l'axe b.

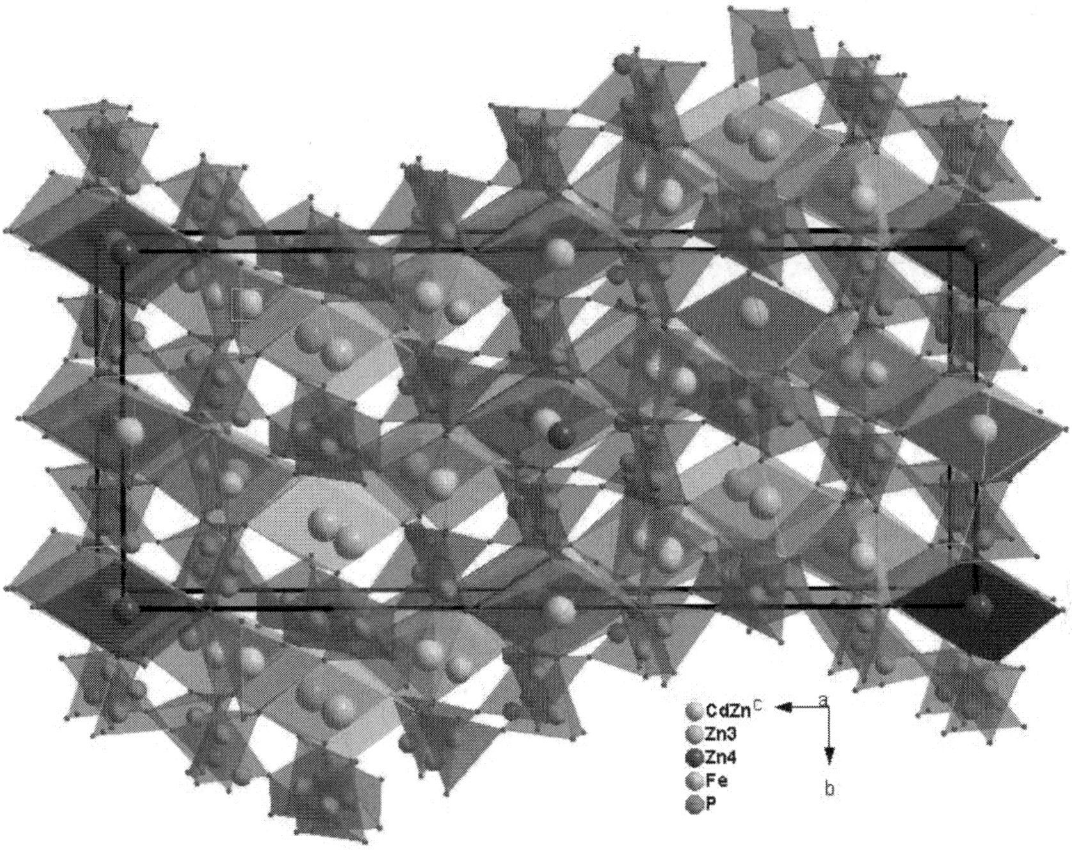

Fig.4. Structure tridimensionnelle du nouveau pyrophosphate $Cd_{0.74}Zn_{1.76}Fe(P_2O_7)_2$.

Tableau 4. Distances (Å) et angles interatomiques (°) pour le pyrophosphate $Cd_{0.74}Zn_{1.76}Fe(P_2O_7)_2$.

Distances Fe(1) - O (Å)	
Fe1—O10	1.903 (2)
Fe1—O6	1.938 (2)
Fe1—O9^{iii}	2.033 (2)
Fe1—O5^{x}	2.075 (2)
Fe1—O3	2.084 (2)
Fe1—O2^{ii}	2.088 (2)
Distance moyenne < Fe(1) – O >	**2.020**
Angles O - Fe(1) - O (°)	
O6—Fe1—O9^{iii}	92.5 (1)
O10—Fe1—O5^{x}	101.20 (9)
O6—Fe1—O5^{x}	93.36 (9)
O9^{iii}—Fe1—O5^{x}	88.17 (9)
O10—Fe1—O3	90.90 (9)
O6—Fe1—O3	94.61 (9)
O9^{iii}—Fe1—O3	79.00 (8)
O10—Fe1—O2^{ii}	90.02 (9)
O9^{iii}—Fe1—O2^{ii}	85.85 (9)
O5^{x}—Fe1—O2^{ii}	80.10 (8)
O3—Fe1—O2^{ii}	91.46 (8)
O10—Fe1—O6	92.7 (1)
O6—Fe1—O2^{ii}	173.29 (9)
Distances Cd(1)/Zn(1) - O (Å)	
Cd1/Zn1—O4	2.179 (2)
Cd1/Zn1—O4^{i}	2.179 (2)
Cd1/Zn1—O9^{ii}	2.181 (2)
Cd1/Zn1—O9^{iii}	2.181 (2)
Cd1/Zn1—O3^{i}	2.376 (2)
Cd1/Zn1—O3	2.376 (2)
Distance moyenne < Cd(1)/Zn(1) – O >	**2.245**
Angles O - Cd(1)/Zn(1) - O (°)	
O4—Cd1/Zn1—O4^{i}	77.4 (1)
O4^{i}—Cd1/Zn1—O9^{ii}	95.32 (8)
O4—Cd1/Zn1—O9^{iii}	95.32 (8)
O9^{ii}—Cd1/Zn1—O9^{iii}	103.1(1)
O4—Cd1/Zn1—O3^{i}	82.14 (8)
O4^{i}—Cd1/Zn1—O3^{i}	88.79 (8)
O9^{ii}—Cd1/Zn1—O3^{i}	70.02 (7)
O9^{iii}—Cd1/Zn1—O3^{i}	117.87 (7)
O4—Cd1/Zn1—O3	88.79 (8)
O4^{i}—Cd1/Zn1—O3	82.14 (8)
O9^{ii}—Cd1/Zn1—O3	117.87 (8)
O9^{iii}—Cd1/Zn1—O3	70.02 (7)
O3^{i}—Cd1/Zn1—O3	168.40 (10)
Distances Cd(2)/Zn(2) - O (Å)	
Cd2/Zn2—O11^{v}	2.110 (2)
Cd2/Zn2—O2^{vi}	2.130 (2)
Cd2/Zn2—O7^{v}	2.138 (2)

Cd2/Zn2—O1	2.146 (2)
Cd2/Zn2—O5^{vii}	2.252 (2)
Distance moyenne < Cd(2)/Zn(2) – O >	**2.1552**
Angles O - Cd(2)/Zn(2)- O (°)	
O11^{v}—Cd2/Zn2—O2^{vi}	94.64 (8)
O11^{v}—Cd2/Zn2—O7^{v}	92.85 (9)
O11^{v}—Cd2/Zn2—O1	113.66 (8)
O2^{vi}—Cd2/Zn2—O1	101.03 (8)
O7^{v}—Cd2/Zn2—O1	77.22 (9)
O2^{vi}—Cd2/Zn2—O5^{vii}	75.32 (7)
O7^{v}—Cd2/Zn2—O5^{vii}	97.65 (8)
O1—Cd2/Zn2—O5^{vii}	101.42 (8)
Distances Zn(3) - O (Å)	
Zn3—O8^{viii}	1.987 (2)
Zn3—O8	1.987 (2)
Zn3—O1^{ii}	2.136 (2)
Zn3—O1^{vi}	2.136 (2)
Zn3—O7^{ii}	2.230 (2)
Zn3—O7^{vi}	2.230 (2)
Distance moyenne < Zn(3) – O >	**2.117**
Angles O - Zn(3) - O (°)	
O8^{viii}—Zn3—O8	103.2 (1)
O8^{viii}—Zn3—O1^{ii}	87.0 (1)
O8—Zn3—O1^{vi}	87.0(1)
O7^{ii}—Zn3—O7^{vi}	179.71 (11)
O1^{ii}—Zn3—O1^{vi}	90.3 (1)
O8^{viii}—Zn3—O7^{ii}	84.44 (9)
O8—Zn3—O7^{ii}	95.38 (9)
O1^{ii}—Zn3—O7^{ii}	104.75 (8)
O1^{vi}—Zn3—O7^{ii}	75.46 (8)
O8^{viii}—Zn3—O7^{vi}	95.38 (9)
O8—Zn3—O7^{vi}	84.44 (9)
O1^{ii}—Zn3—O7^{vi}	75.46 (8)
O1^{vi}—Zn3—O7^{vi}	104.76 (8)
Distances Zn(4) - O (Å)	
Zn4—O13	2.013 (3)
Zn4—O13^{ix}	2.013 (3)
Zn4—O4^{iii}	2.079 (2)
Zn4—O4^{x}	2.079 (2)
Zn4—O11^{iii}	2.451 (2)
Zn4—O11^{x}	2.451 (2)
Distance moyenne < Zn(4) – O >	**2.181**
Angles O - Zn(4) - O (°)	
O13—Zn4—O13^{ix}	94.1 (1)
O13—Zn4—O4^{iii}	93.75 (9)
O13^{ix}—Zn4—O4^{x}	93.75 (9)
O4^{iii}—Zn4—O4^{x}	81.9 (1)
O13—Zn4—O11^{iii}	85.11 (9)
O13^{ix}—Zn4—O11^{iii}	111.7 (1)
O4^{iii}—Zn4—O11^{iii}	82.10 (8)
O4^{x}—Zn4—O11^{iii}	79.78 (8)
O13^{ix}—Zn4—O4^{iii}	164.70 (10)
O13—Zn4—O11^{x}	111.7 (1)
O13^{ix}—Zn4—O11^{x}	85.11 (9)
O4^{iii}—Zn4—O11^{x}	79.78 (8)
O4^{x}—Zn4—O11^{x}	82.10 (8)
Distances P(1) - O (Å)	
P1—O13	1.494 (2)
P1—O4	1.517 (2)
P1—O9	1.535 (2)

P1—O12	1.600 (2)
Distance moyenne < P(1) – O >	**1.5365**
Angles O - P(1) - O (°)	
O13—P1—O4	111.8 (1)
O13—P1—O9	113.2(1)
O4—P1—O9	111.9 (1)
O13—P1—O12	113.4 (1)
O4—P1—O12	107.1 (1)
O9—P1—O12	99.0 (1)
Distances P(2) - O (Å)	
P2—O8	1.490 (2)
P2—O10	1.514 (2)
P2—O1	1.526 (2)
P2—O14	1.618 (2)
Distance moyenne < P(2) – O >	**1.537**
Angles O - P(2) - O (°)	
O8—P2—O10	113.9 (1)
O8—P2—O1	112.2 (1)
O10—P2—O1	112.7 (1)
O8—P2—O14	108.1 (1)
O10—P2—O14	107.2 (1)
O1—P2—O14	102.1 (1)
Distances P(3) - O (Å)	
P3—O11	1.511 (2)
P3—O3	1.528 (2)
P3—O5	1.541 (2)
P3—O14	1.583 (2)
Distance moyenne < P(3) – O >	**1.54075**
Angles O - P(3) - O (°)	
O11—P3—O3	112.3 (1)
O11—P3—O5	113.1 (1)
O3—P3—O5	110.9 (1)
O11—P3—O14	105.6(1)
O3—P3—O14	108.0(1)
O5—P3—O14	106.7(1)
Distances P(4) - O (Å)	
P4—O6	1.494 (2)
P4—O7	1.514 (2)
P4—O2	1.533 (2)
P4—O12	1.604 (2)
Distance moyenne < P(4) – O >	**1.53625**
Angles O - P(4) - O (°)	
O6—P4—O7	115.5 (1)
O6—P4—O2	112.70 (1)
O7—P4—O2	112.0 (1)
O6—P4—O12	106.7 (1)
O7—P4—O12	104.9 (1)
O2—P4—O12	104.3 (1)

Code de symétrie : (i) *x*, -*y*+1, -*z*+1; (ii) *x*-1/2, *y*-1/2, *z*; (iii) *x*-1/2, -*y*+3/2, -*z*+1; (iv) *x*+1/2, *y*-1/2, *z*; (v) -*x*+1, *y*, -*z*+3/2; (vi) -*x*+1/2, *y*-1/2, -*z*+3/2; (vii) -*x*+1/2, *y*+1/2, -*z*+3/2; (viii) -*x*, *y*, -*z*+3/2; (ix) *x*, -*y*+2, -*z*+1; (x) *x*-1/2, *y*+1/2, *z*; (xi) *x*+1/2, *y*+1/2, *z*; (xii) *x*+1/2, -*y*+3/2, -*z*+1

4 Conclusion

Le nouveau pyrophosphate métallique mixte : $Cd_{0.74}Zn_{1.76}Fe(P_2O_7)_2$ a été synthétisé par réaction de diffusion à l'état solide, et a été caractérisé par diffraction de rayons X sur monocristal. Ce nouveau pyrophosphate cristallise dans le système orthorhombique avec le groupe d'espace $C222_1$ et les paramètres de maille : a = 8,4761 (1) Å, b = 9,7581 (1) Å, c = 24,0557 (3) Å, V=1989.66 (4), Z=8. La structure de ce nouveau composé est formée par des polyèdres métalliques déformés : quatre octaèdres MO_6 (M = Zn, Fe, Cd(1)/Zn(1)), et une pyramide à base carrée $(Cd(2)/Zn(2))O_5$, formant des feuillets ondulés parallèles interconnectés par les groupements isolés P_2O_7 le long de (010).

Références

1. C. Bouzidi, W. Frigui, and M. F. Zid, *Acta Cryst. E*,**71**,1,69–72, (2015).
2. S. Loridant, I. C. Marcu, G. Bergeret,J. M. M. Millet, *Phys. Chem. Chem. Phys*,**5**, 20, 4384–4389, (2003).
3. R. J. Xie , A. F. MacKenzie,*Soil Sci. Soc. Am. J*, **54**,1,71–77, (1990).
4. D. H. Lee, S. Y. Kim,J. I. Hong, *Angew. Chemie - Int. Ed*,**43**,36,4777–4780, (2004).
5. A.Gupta,N.B.Amin,A.Besarab,S.E.Vogel , G.W. Divine, J.yee,J.V.Anandan,*Kidney Int*,**55**,5, 1891–1898, (1999).
6. S. Navas-Carretero, A. M. Pérez-Granados, B. Sarriá,M. P. Vaquero,*Nutrition*,**25**, 1,20–24, (2009).
7. Y. Zhang, A. C. Marschilok, E. S. Takeuchi, K. J. Takeuchi,*MRS Adv*,**2**,7,395–400, (2017).
8. A. Bolarinwa, U. Gbureck, P. Purnell, M. Bold,L. M. Grover, *Adv. Appl. Ceram*,**109**, 5, 291–295, (2010).
9. I. R. Orriss, T. R. Arnett, and R. G. G. Russell, *Curr. Opin. Pharmacol*, **28**,7–68, (2016).
10. B. Lin, S. Zhang, and C. Deng, *J. Mater. Chem. A*,**4**,7, 2550–2559, (2016).
11. C. Deng, S. Zhang,B. Zhao,*Energy Storage Mater*,**4**,71–78, (2016).
12. C. Masquelier, F. D'Yvoire, E. Bretey, P. Berthet,C. Peytour-Chansac, *Solid State Ionics*,**67**,3–4,183–189, (1994).
13. G. S. Gopalakrishna, B. H. Doreswamy, M . J. Mahesh, M. Mahendra, M.A. Sridhar, J. S.Prasad, K. G.Ashamanjari , *Bull. Mater. Sci.*,28,1,1–7, (2005).
14. Ģ. Vītiņš, Z. Kaņepe, A. Vītiņš, J. Ronis, A. Dindūne, A. Lūsis, *J. Solid State Electrochem*, **4**,3,146–152, (2000).
15. A. Elmarzouki, A. Boukhari, E. M. Holt, A. Berrada , *J. Alloys Compd*,**227**,2, 25–130, (1995).
16. A. Boutfessi, A. Boukhari, E. M. Holt, *Acta Cryst. Sect.C***51**,3,346–348, (1995).
17. A. Boutfessi, A. Boukhari, E. M. Holt, *Acta Cryst. Sect.***C52**, 1594-1597, (1996).
18. A. Boutfessi, A. Boukhari, E. M. Holt, *Acta Cryst. Sect.***C52**, 1597-1599, (1996).
19. K. H. Lii, P. F. Shih, and T. M. Chen, *Inorg. Chem*,**32**,20,4373–4377, (1993).
20. G. M. Sheldrick, *Acta Cryst. Sect. A***64**, 1,112–122, (2008).
21. B.Malaman, M.Ijjaali , R.Gerardin , G.Venturini, C.Gleitzer,Eur. *J. Solid. State. Inorg . Chem*,**29**, **1**269–1284, (1992).
22. S.Boudin, A. Grandin,Ph. Labbe´ , D. Grebille, N.Nguyen, A. Ducouret, B. Raveau, *J. Solid State Chem*.**121**, 291–300, (1996)
23. A.ELBelghitti ,A.Boukhari,M.E.Holt, *Acta Cryst* , **C50**,1648-1650, (1994).
24. A.Bronova, R.Glaum,C.Litterscheid. *Acta Cryst* , E**69**, i26, (2013).

High efficient photocatalytic activity of Zn-Al-Ti layered double hydroxides nanocomposite

F. Amor[1], A. Diouri[1], I. Ellouzi[2], F. Ouanji[2] and M. Kacimi[2]

[1]Laboratory of Applied Solid State Chemistry, Faculty of Science, Mohammed V University, Rabat, Morocco
[2]Laboratoire of Nanomaterial, Nanotechnology and Environment, Physico-chemical of Material and Catalysis (URAC26), Faculty of Sciences, University of Mohammed V-Rabat, Av. Ibn Battouta, BP 1014

Abstract. This work establishes a simple method for synthesising layered double hydroxides (LDHs) powders with coprecipitation. The characteristics of the samples were investigated y X-ray diffraction (XRD), scanning electron microscopy (SEM) and spectrophotometer UV–Vis (DRS). Non-uniform distribution was shown for LDHs samples by SEM. Photocatalytic efficiencies were tested using methylene blue (MB) dye as a model contaminant under UV irradiation. In particular, Zn–Al-Ti LDH exhibited an excellent performance towards MB degradation compared with commercial TiO_2 nanoparticles. Methylene blue removal percentage was reached at almost 100%, whereas commercial TiO_2 reached a removal rate of only 66% under the same conditions within 20 min. The aim of the current work is to prepare Zn-Al-Ti layered double hydroxides nanocomposite and to evaluate their photocatalytic activity in the removal of methylene blue under UV irradiation.

1 Introduction

Over the past few decades, several researches focusing on photocatalysis have been investigated as an alternative method to eliminate the organic dye pollutant [1–2]. In the last years, the synthesis of layered materials has attracted a considerable attention and was widely used in photocatalysis process due to their stability, their simple preparation method and their low cost [3].

Lamellar double hydroxides (LDH) were synthesized for the first time by Feitknecht [4]. Their development was then amplified by the work of Miyata [5-7] on the $[Zn_3Al(OH)_8][X.\ nH_2O]$ phase. These compounds have been the subject of numerous studies and have been widely described in the literature because of the numerous possibilities of application of these materials in a wide range of fields: catalysis, environment, polymer and health…These materials consist of positively charged sheets containing divalent and trivalent cations. The electro-neutrality of the material is ensured by the presence of intercalary anions, solvated by molecules of water [8]. LDHs associated with zinc, used as an addition because of its photocatalytic activity in order to contribute to the overall activity of the new Ti-Zn-Al nanocomposite [9]. ZnO is often considered as an alternative to TiO_2 because it can absorb a greater energy fraction of the solar spectrum and more quanta of light [10]. Enormousres researches have reported that LDH photocatalysts presented an effective removal performance of organic contaminants [11, 12]. Furthermore, enormous LDH photocatalysts were reported that Zn-Ti LDH exhibited the highest photocatalytic performance in the degradation of organic pollutants [11, 13].

2 Experimental

Zn-Al LDH was synthesized by the simultaneous addition of $Zn(NO_3)_2.6H_2O$ and $Al(NO_3)_3.9H_2O$ ($4cm^3/min$) with a basic solution of 0.67 Na_2CO_3 and 2,25 of NaOH at constant pH 9-9.5 and constant temperature of 45 °C for 10 hours. At the end of the stirring, a white gel is precipitated, aged for 12 hours at 100°C and then calcined for 5 hours at 500 °C. The dried samples were denoted as Zn-Al LDH. To bind the titanium oxide in the previous nanocomposite, a wet impregnation of TiO_2 on Zn-Al LDH was used for the preparation of Zn-Al-Ti LDH; diluted TiO_2 suspension (3wt%) and 0.67 M of Na_2CO_3 basic solution are charged to the Zn-Al LDH calcined powder and removed in a stove at 100°C.The obtained sample was dried for 12 hours at 100 °C, calcined for 5 hours at 500 °C and denoted as Zn-Al-Ti LDH.

The photocatalytic activities of materials were assessed via the photocatalytic degradation of methylene blue (MB) under UV light irradiation using a 125Wlamp.Photocatalytic reaction was carried out in a Pyrex reactor containing10 ppm of MB and 0.5g/L of the photocatalyst. The mixture was vigorously stirred in the dark for 60 minutes in order to reach the adsorption-desorption equilibrium, and then the aqueous samples were filtered. The concentration of the suspension was analyzed using the UV/vis spectrometer.

X-ray diffraction (XRD) was acquired using a Siemens D5000 diffractometer. This unit uses the mounting BRAGG-BRENTANO (θ/2θ). UV/vis spectrophotometer was performed using the UV–vis spectrometer UV2300 Spectrophotometer. The morphology of samples was examined by scanning electron microscopy (SEM) using Jeol JSM-7000F FE-SEM, Japan.

3 Results and discussion

LDHs samples and commercial TiO_2 presented in fig. 1 were analyzed by XRD. ZnO and $Zn_6Al_2O_6$ were the dominant phase detected in Zn-Al LDH and Zn-Al-Ti LDH samples, whereas ZnO phase originates from Zn-Al LDH [14]. After the impregnation, the intensity of TiO_2 peak is weak comparing with others due to their very small amount. The anatase phase in Zn-Al-Ti LDH sample has been detected in DRX diffractogram. Although the formation of Zn_2TiO_4 phase is due to the reaction between ZnO and TiO_2. Formation of very weak peaks intensity were observed in Zn-AL Ti LDH sample calcined at 500 °C, where as Yee-Shin Chang et al., reported that from 700 °C, some peaks correlated to Zn_2TiO_4 were shown in the XRD pattern [15].

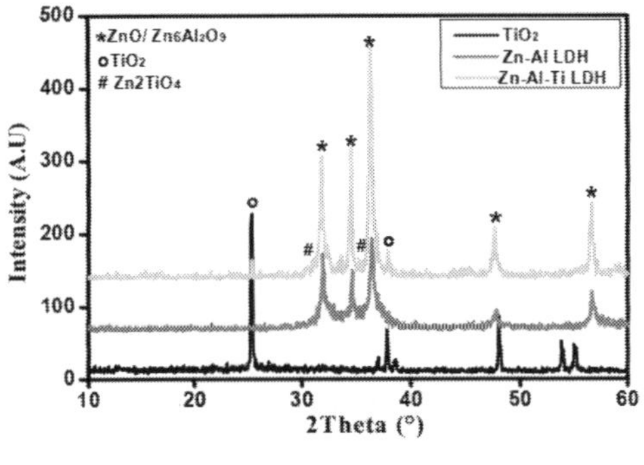

Fig. 1. XRD patterns of Zn-Al-Ti LDH, Zn-AL LDH and TiO_2.

SEM photographs of Zn–Al LDH and Zn-Al-Ti LDH are shown in Figs. 2a and 2b respectively.SEM of LDHs samples showed non-uniform distribution. Furthermore, the images show that all samples were agglomerated and possessed large and lumpy particles.SEM analysis revealed the same morphology for Zn–Al LDH and Zn-Al-Ti LDH.

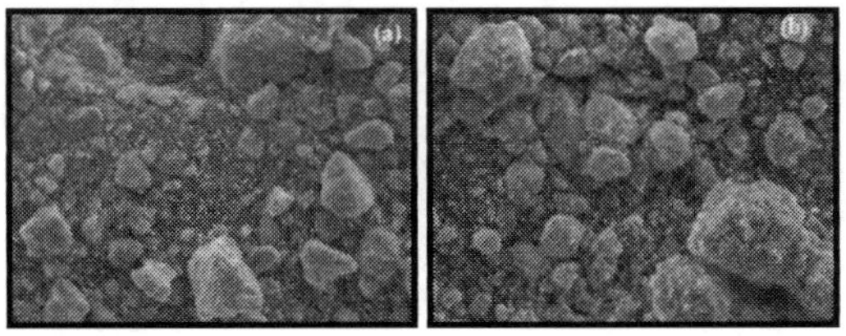

Fig. 2. SEM photographs of (a) Zn-Al LDH and (b) Zn-Al-Ti LDH

The evaluation of optical absorption spectra was recorded by UV–Vis DRS. The optical absorbance versus wavelength plots for as-prepared samples are presented in fig.3. It was observed that the optical absorption of LDH materials was estimated to be 404 nm and 411 nm for Zn-Al LDHs and Zn-AL-Ti LDHs, respectively. Their band gaps were estimated to be 3.06eV and 3.01eV.However, the absorption edge of the commercial TiO_2 is about 3.26 nm [16]. This change of the band gap, indicating that photocatalytic activity enhancement was directly ascribed to the band gap.

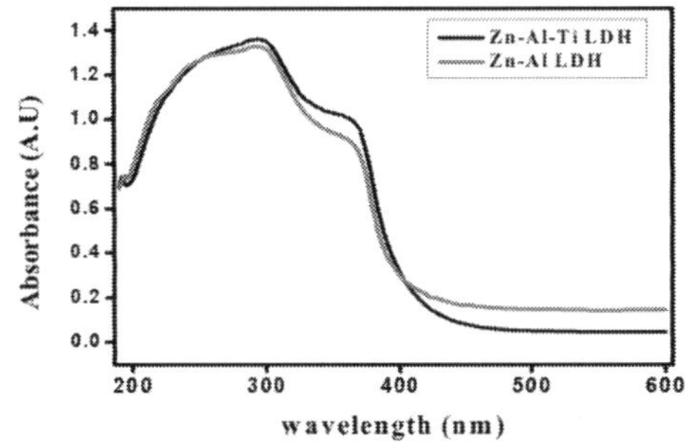

Fig. 3. UV-vis absorption spectra of Zn-Al-LDH and Zn-Ti LDH composites.

The study of methylene blue degradation (MB) was used in order to evaluate the photocatalytic performance of as-prepared samples. The results presented in Fig.4showthat MB concentrations were decreased. Before photocatalytic tests, adsorption performance was conducted.Fig.4 shows the linear plot of MB degradation, which follows a pseudo- first-order kinetic be havior by the equation below [17]:

$$Ln(C/C_0) = -Kt$$

Where C_0 and C are the initial concentration of MB and the concentration at time t, respectively, K is the photocatalytic reaction rate constant (min^{-1}).

Fig. 4 shows the photocatalytic degradation of methylene blue, under UV irradiation. It was remarkable that all the samples exhibited high photocatalytic activity but Zn-Al-Ti LDH present the highest performance. Therefore, it may be concluded that samples prepared using LDH play an important role in the degradation of MB. It has been observed that samples synthesized using LDH exhibit much higher photocatalytic activity than commercial TiO_2.

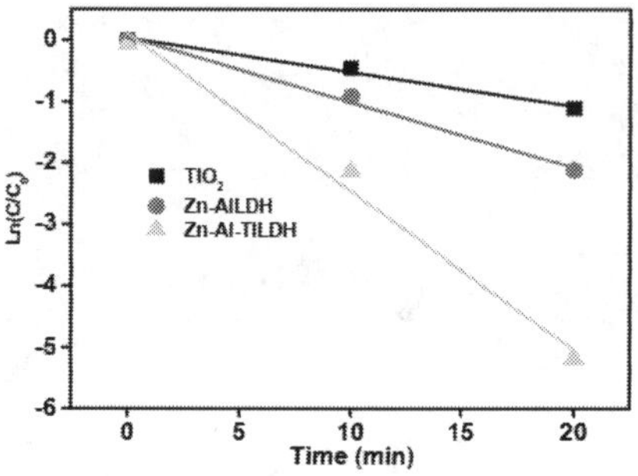

Fig. 4. Photocatalytic performance of as-prepared photocatalysts.

The methylene blue was degraded around 99% in 20 min for Zn-Al Ti LDH. It could be assigned to the fact that LDH materials are rich in OH groups which favor the titania activity. However, at the same photocatalytic time, the degradation rate of Zn-Al LDH and TiO_2 was only 87% and 66%, respectively. The increase in the photocatalytic activity, could be assigned to the rapid electron and hole separation efficiency. Zn-Al-Ti LDHs showed higher degradation which was greater compared with other samples. Many researchers have reported and discussed the photocatalytic performances of Zn type LDH [18, 19]. Zn–Al LDH could have two active species such as superoxide O2•- and hydroxyl radicals OH• under UV irradiation for the removal of MB. Mohapatra et al., and Li et al., [19, 20] have proven thatO2•- plays the key role in the initiation of the photocatalytic process in LDH photocatalytic systems.

Conclusion

Layered double hydroxide was synthesized using simple coprecipitation method. A slight decrease in the band gap of Zn-Al-Ti LDHs was observed by the introduction of TiO_2. All the samples presented a high photocatalytic activity and produced 100% removal of MB. The enhancement of photocatalytic performance could be assigned to the enhanced of the electron–hole pair formation. The photocatalytic process matched pseudo-first-order kinetic model.

References

Here are some examples:

1. Q. Wu R. Krol, J. Am.Chem. Soc, **134** (2012) 9369–9375.
2. R. S. Juang, S. H. Lin, P. Y. Hsueh, J. Hazard. Mater, **182** (2010) 820-826.
3. A. Kaouther, N.SFrini, S.Ezzeddine, Applied Clay Science, 119 (2016) 229–235.
4. W. Feitknecht, G. Fischer, G. Helu, Chim. Acta, **18** (1935) 555- 569
5. S. Miyata and T. Kumura, Chem. Lett, **2** (1973) 843-848
6. S. Miyata, Clays and Clay Minerals, **23**(1975)369-375.
7. S.Miyata, T. Hirose, Clays and Clay Minerals, **26** (1978) 441-447
8. F.Amor, A. Diouri, S. Kamali-Bernard, J. Mater. Environ. Sci, **8** (2016) 2739-2747.
9. T.Vulic, M.Hadnadjev-Kostic, O.Rudic, M.Radeka, R.Marinkovic-Neducin, Cem. Concr.Compos, **36** (2013)121-127.
10. S. Sakthivel, B. Neppolian, MV. Shankar, B. Arabindoo, M. Palanichamy, V. Murugesan, Solar Energy Mater Sol Cells, **77** (2003) 65–82.
11. S.J. Xia, F.X. Liu, Z.M. Ni, J.L. Xue, P.P. Qian, Colloid Interface Sci, 405 (2013)195-200.
12. L. Mohapatra, K.M.Parida, Sep Purif Technol, **91** (2012) 73–80.
13. S.-J. Xia, F.-X. Liu, Z.-M. Ni, W. Shi, J.-L. Xue, P.-P. Qian,Appl. Catal.B—Environ. **144** (2014) 570–579.
14. D. Carriazo, M. del Arco, E. Garcia-Lopez, G. Marci, C. Martin, L. Palmisano, V. Rives, J. Mol. Catal. A: Chem, **343** (2011) 83–90.
15. Yee-Shin Chang, Yen-Hwei Chang, In-Gann Chen, Guo-Ju Chen, Yin-Lai Chai, Te-Hua Fang, Sean Wu.ceramics. **30**(2004) 2183–2189.
16. C. Dette, MA. Pérez-Osorio, CS. Kley, P. Punke, CE. Patrick,P. Jacobson, F. Giustino, SJ. Jung, K. Kern, Nano Lett, **14** (2014) 6533-6538.
17. H. Liu, B.Z. Lin, L.W. He, H. Qu, P. Sun, B.F. Gao, Y.L. Chen, Chem. Eng, J. **216** (2013)396-403.
18. K. Parida, L. Mohapatra , N. Baliarsingh, J. Phys. Chem, C **116** (2012) 22417–22424.
19. L. Mohapatra, K. Parida, M. Satpathy, J. Phys.Chem, C**116** (2012) 13063–13070.
20. H. Li, Q. Deng, J. Liu, W. Hou, N. Du, R. Zhang , X. Tao, Catal. Sci.Technol, **4** (2014) 1028–1037.

Effect of artificial aging at QUV on the color of finish treated woods of *Pinus sylvestris* and *Quercus robur*

Effet du vieillissement artificiel au QUV sur la couleur des bois de Pinus sylvestris et Quercus robur traités par finition

A. Hakam[1], N. El Imame[1, 2], M. El kortbi[1], M. El Matar[2], M. Ziani[1] and A. Merlin[3]

[1]Wood Science Laboratory, Mohammed V University in Rabat, Faculty of Science, PO Box 1014, Morocco
[2]Centre Technique des Industries du Bois et de l'Ameublement, Sidi Maârouf, Casablanca, Maroc
[3]Université de Lorraine, Faculté des Sciences et Technologies, BP 70239 - 54506 Vandoeuvre Les Nancy Cedex, France

Abstract. The durability of exterior finishes on wood is not assured with the systems currently used. Thus, the objective of this study was to evaluate the color change of the finishes applied to wood, the most commercialized in Morocco, in order to find the wood-finishing couple that has the best durability in outdoor use. QUV accelerated aging tests were carried out during 2016 hours, according to the standard (NF 927-6 2006), on Scots pine (*Pinus sylvestris*) and Pedunculate oak (*Quercus robur*) coated with four finishes (two polyurethane finishes F_1 and F_2 and two acrylic finishes F_3 and F_4). Colorimetric measurements in color system CIELab: Gloss (L*), green-red (a*) and blue-yellow (b*) coordinates, color saturation (C*) and tone angle (H*) were performed using an X-Rite Model SP62 spectrophotometer before and at the end of the QUV accelerated aging treatment. The results showed that QUV aging caused a decrease in the L* value and a change in the surface color of all the finishing pairs studied. For wood / finishing couples with *Quercus robur* as a wooden support, the couple *Quercus robur* / F3 has the best resistance to color change, while for couples wood / finish with *Pinus sylvestris* as a wooden support, the couple *Pinus sylvestris* / F1 has the best resistance to color change.

Key words: Wood, finish, gloss, color, accelerated aging, QUV.

Résumé. La durabilité des finitions extérieures sur le bois n'est pas assurée avec les systèmes employés actuellement. Ainsi, l'objectif de cette étude était d'évaluer le changement de couleur des finitions appliquées aux bois, les plus commercialisés au Maroc, dans le but de trouver le couple bois-finition qui présente la meilleure durabilité en utilisation extérieure. Des tests de vieillissement accéléré au QUV ont été réalisés pendant 2016 heures, selon la norme (NF 927-6 2006), sur le pin sylvestre (*Pinus sylvestris*) et le chêne pédonculé (*Quercus robur*) revêtus de quatre finitions (deux finitions polyuréthanes F1 et F2 et deux finitions acryliques F3 et F4). Les mesures colorimétriques dans le système de couleur CIELab: La brillance (L*), les coordonnées vert-rouge (a*) et bleu-jaune (b*), la saturation des couleurs (C*) et l'angle de tonalité (H*) ont été réalisées à l'aide d'un spectrophotomètre X-Rite Model SP62, avant et à la fin du traitement de vieillissement accéléré au QUV. Les résultats ont montré que le vieillissement au QUV a provoqué une diminution de la valeur de L* et une modification de la couleur de surface de tous les couples de finition étudiés. Pour les couples bois/finition avec le *Quercus robur* comme support en bois, le couple *Quercus robur* / F_3 présente la meilleure résistance au changement de couleur, alors que pour les couples bois / finition avec le *Pinus sylvestris* comme support en bois, le couple *Pinus sylvestris* / F_1 présente la meilleure résistance au changement de couleur.

Mots clés : Bois, finition, brillance, couleur, vieillissement accéléré, QUV.

 https://doi.org/10.1051/matecconf/201814901088

1 Introduction

Les principaux facteurs de dégradation des composants en bois à l'extérieure sont l'exposition aux rayons ultraviolets et à l'eau [1-5].

Pour la protection de l'aspect naturel du bois et sa couleur contre les intempéries, des recherches sur les revêtements transparents ou semi-transparents, tels que des vernis, des lasures ou des peintures ont été réalisées [6-9].

Le test de vieillissement au QUV consiste à évaluer la résistance au vieillissement accéléré au QUV du système de revêtement appliqué sur un support en bois. Ce test a l'avantage d'être accéléré par rapport au vieillissement naturel [10-11], qui dure un an et qui présente souvent des problèmes de répétabilité et de reproductibilité.

Le vieillissement artificiel ou accéléré influe sur la couleur du bois traité ou non traité par finition [12-19].

La variation de la couleur de finition observée varie selon la nature du produit de finition, de l'espèce de bois [20-22] et de son traitement, thermique [23-27] ou chimique [28].

D'autres études ont été consacrées à l'effet du vieillissement accéléré sur l'adhérence des finitions [29].

2 Matériaux et méthodes

2.1 Matériaux

Un système Bois / Finition est un échantillon qui est composé d'un support bois et un ou plusieurs produits de finition déposés sur ce bois.

Le choix des systèmes Bois/Finitions a été réalisé en étroite concertation avec les industriels de la filière bois-finition au Maroc.

Le choix du support bois s'est porté sur le Pin sylvestre (*Pinus sylvestris*) et le Chêne pédonculé (*Quercus robur*).

Le tableau 1 donne quelques caractéristiques du support bois utilisé.

Tableau 1. Caractéristiques du support bois utilisé

	(*Pinus* sylvestris)	(*Quercus robur*)
Type de bois	résineux	feuillus
Densité*	~ 0.55	~ 0.74
Structure	Aucun vaisseau	Cernes poreux
Largeur de cerne (mm)	~1,4	~1,4

*Tropix 7 [30]

2.1.1 Confection des éprouvettes

Des éprouvettes, de dimensions 15 x 8 x 1,8 mm, ont été coupées en longueur suivant l'axe de symétrie longitudinal (L), en largeur suivant l'axe de symétrie tangentiel (T) et en profondeur suivant l'axe de symétrie radial (R). Toutes les éprouvettes en bois ont été équilibrées dans une chambre climatique à une humidité relative de 65 ± 5 % et à une température de 20 °C avant le revêtement par la finition. L'humidité d'équilibre des éprouvettes était de 12%.

2.1.2 Produit de finition

Comme pour le bois, les quatre finitions aqueuses commerciales utilisées ont été choisies en concertation avec les industriels de la filière bois-finition au Maroc.

Deux finitions à base de résine polyuréthane (F_1 et F_2) et deux à base de résine acrylique (F_3 et F_4). Le tableau 2 donne quelques indications sur les finitions commerciales utilisées.

Tableau 2. Caractéristiques du support bois utilisé

Finition	Résine de base	Type du produit
F_1	Polyuréthane	Vernis aqueux transparent
F_2	Polyuréthane	Vernis aqueux mate blanc
F_3	Acrylique	Lasure aqueuse Transparente mate
F_4	Acrylique	Lasure Aqueuse opaque blanche mate

2.1.3 Revêtement des surfaces en bois

Les produits de finition choisis ont été appliqués selon les fiches techniques de chaque produit sur une des faces tangentielles de chaque éprouvette. Les autres faces ont été fermées avec un produit scellant, appliqué pendant trois jours, à raison d'une couche par jour.

Avant le début des essais, les éprouvettes ont été laissées sécher pendant une semaine.

2.2 Méthode

2.2.1 Vieillissement accéléré au QUV

L'expérience de vieillissement accéléré au QUV a été réalisée dans un QUV de QLAb avec huit lampes UVA 340 selon la norme EN 927-6 2006. Un cycle d'exposition des éprouvettes dans le QUV implique une première phase de condensation pendant 24 h à T = 45 ± 3°C, suivie d'une phase d'irradiation à l'UV et pulvérisation pendant 144 h à T = 60 ± 3°C, puis une phase d'irradiation de 2,5 h et une dernière phase de pulvérisation pendant 0,5 h. Les échantillons ont été exposés à l'altération accélérée pendant 12 semaines.

2.2.2 Mesures de la couleur

L'analyse colorimétrique des systèmes bois-finition a été effectuée en utilisant un spectrophotomètre X-rite SP 62. Les paramètres mesurés sont : la brillance (L*), les coordonnées chromatiques, vert-rouge (a*) et bleu-jaune (b*) en référence au système de coordonnées de couleur CIELab. La valeur retenue pour chaque paramètre est la valeur moyenne de trois mesures.

A partir des valeurs de L*, a*, et b*, la différence de brillance (ΔL*) et de coordonnées chromatiques (Δa* et Δb*), l'angle de teinte (h), la saturation de couleur (C*) et la différence globale de couleur (ΔE*) ont été calculés en utilisant les formules suivante :

$$\Delta L^* = L^*_f - L^*_i$$
$$\Delta a^* = a^*_f - a^*_i$$
$$\Delta b^* = b^*f - b^*i$$
$$h = \text{arctg}\,(b^* / a^*)$$
$$C^* = (a^{*2} + b^{*2})^{1/2}$$
$$\Delta E^* = (\Delta L^{*2} + \Delta a^{*2} + \Delta b^{*2})^{1/2}$$

où, L^*_f a^*_f b^*_f sont L*, a* et b* des spécimens après essai au QUV et L^*_i, a^*_i et b^*_i sont L*, a* et b* des échantillons avant essai au QUV, respectivement.

3 Résultats et discussion

Le tableau 3 présente les valeurs de L*, a*, b* et E* basées sur les coordonnées de couleur initiales pour chaque couple bois / finition, le tableau 4 présente les valeurs de ΔL*, Δa*, Δb*, ΔE*, C* et H, calculées à partir des valeurs de L*, a*, b* et E* basées sur les coordonnées de couleur initiales pour chaque couple bois / finition et le tableau 5 présente quelques photos des échantillons avant et après vieillissement au QUV.

Comme on peut le voir dans les tableaux 3 et 4, les valeurs de L* de tous les couples de finition diminuent après vieillissement au QUV, mais avec un impact plus important sur les couples (*Pinus sylvestris* / Finition), tandis que les valeurs a* augmentent.

Les valeurs positives de Δa* montrent une tendance de la surface de la finition à devenir plus rougeâtre, sauf pour le couple (*Quercus robur* / F2), où la surface à tendance à devenir plus verdâtre. La coordonnée vert-rouge (a*) des couples (*Pinus sylvestris* / Finition) était plus élevée que celle des couples (*Quercus robur* / Finition) et la coordonnée bleu-jaune (b*) augmente pour les couples avec finition F_1 et le couple (*Quercus robur* / F_4) et diminue pour les couples avec finition F_2 et F_3 et le couple (*Pinus sylvestris* / F_4).

Les couples (*Pinus sylvestris* / finition) présentaient des valeurs plus élevées de saturation des couleurs (C*) que les couples (*Quercus robur* / Finition). Comme on peut le constater dans le tableau 4, les valeurs de la saturation de couleur (C*) des couples avec les finitions F_1 et F_4 augmentent, tandis qu'elles diminuent pour les couples avec les finitions F_2 et F_3, après vieillissement au QUV. La valeur C* résulte des coordonnées a* et b*, et les valeurs les plus élevées de (b*) sont ceux des couples (*Pinus sylvestris* / finition), ce que signifie que la saturation des couleurs (C*) suit les valeurs de (b*) des couples (*Pinus sylvestris* / finition). Le tableau 4 montre également que l'angle de tonalité (H*) diminue pour tous les couples de finition après vieillissement au QUV. D'après les résultats du tableau 4, le classement de la durabilité des systèmes *Quercus robur*/Finition par ordre décroissant est : *Quercus robur* / F_3, *Quercus robur* / F_4, *Quercus robur* / F_1 et *Quercus robur* / F_2 et le classement de la durabilité des systèmes *Pinus sylvestris* / Finition par ordre décroissant est : *Pinus sylvestris* / F_1, *Pinus sylvestris* / F_4, *Pinus sylvestris* / F_2 et *Pinus sylvestris* / F_3.

4 Conclusion

La finition protège le bois contre l'humidité, la température et l'ensoleillement. C'est une étape importante pour valoriser les produits en bois. D'ailleurs, la durée de vie de ces derniers est souvent associée à la durabilité de son revêtement de finition. Dans cette étude, l'effet de l'altération accélérée sur la brillance et la couleur des spécimens de pin sylvestre (*Pinus sylvestris*) et de Chêne pédonculé (*Quercus robur*) revêtus de vernis et de lasure a été étudié en effectuant des tests de la brillance et de la couleur de surface. Le choix des systèmes Bois/Finitions a été réalisé en concertation avec les industriels de la filière bois-finition au Maroc. Deux finitions à base de résine polyuréthane (F_1 et F_2) et deux à base de résine acrylique (F_3 et F_4) ont été choisies. Les résultats ont montré que l'altération provoquait une diminution de la brillance et une modification de la couleur de tous les spécimens étudiés. Cette étude a permis de faire un classement de la durabilité des systèmes bois / finition étudiés.

Le classement de la durabilité des systèmes *Quercus robur*/Finition par ordre décroissant est : *Quercus robur* / F_3, *Quercus robur* / F_4, *Quercus robur* / F_1 et *Quercus robur* / F_2 et le classement de la durabilité des systèmes *Pinus sylvestris* / Finition par ordre décroissant est :

Pinus sylvestris / F_1, *Pinus sylvestris* / F_4, *Pinus sylvestris* / F_2 et *Pinus sylvestris* / F_3.

Tableau 3. Résultats des mesures de la couleur

Finition	Bois	L*		a*		b*		E*	
		Avant	Après	Avant	Après	Avant	Après	Avant	Après
F1	P	64,87	54,43	11,28	14,85	27,56	28,94	71,38	63,41
	C	58,98	51	7,49	10,16	22,78	24,1	63,67	57,32
F2	P	66,12	51,08	11,17	13,8	30,45	22,89	73,65	57,65
	C	55,88	67,77	8,62	5,44	24,34	15,54	61,56	69,74
F3	P	60,85	42,69	13,26	13,47	26,24	20,63	67,58	49,29
	C	53,3	48,51	10,77	11,61	26,76	24,07	60,61	55,38
F4	P	65,45	48,49	11,88	17,32	22,67	21,23	70,28	55,70
	C	63,9	55,09	7,18	13,84	21,64	27,81	67,85	63,24

P : *Pinus sylvestris* et C : *Quercus robur*

Tableau 4. Résultats des mesures de la couleur

Finition	Bois	ΔL*	Δa*	Δb*	ΔE*	C*		H*	
						Avant	Après	Avant	Après
F1	P	-10,44	3,57	1,38	-7,97	29,78	32,53	1,18	1,1
	C	-7,98	2,67	1,32	-6,35	23,98	26,15	1,25	1,17
F2	P	-15,04	2,63	-7,56	-16,00	32,43	26,72	1,22	1,03
	C	11,89	-3,18	-8,8	8,18	25,82	16,46	1,23	1,23
F3	P	-18,16	0,21	-5,61	-18,29	29,4	24,64	1,1	0,99
	C	-4,79	0,84	-2,69	-5,22	28,84	26,73	1,19	1,12
F4	P	-16,96	5,44	-1,44	-14,58	25,59	27,4	1,09	0,89
	C	-8,81	6,66	6,17	-4,60	22,8	31,06	1,25	1,11

P : *Pinus sylvestris* et C : *Quercus robur*

Tableau 5. Photos des échantillons avant et après vieillissement au QUV

Bois / Finition	*Quercus robur* / F_3	*Quercus robur* / F_4	*Quercus robur* / F_1	*Quercus robur* / F_2
Echantillon avant l'essai au QUV	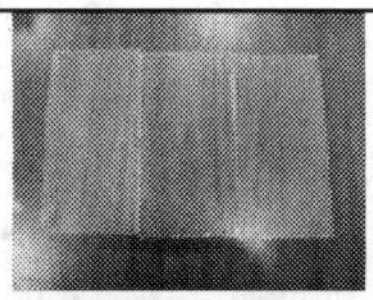	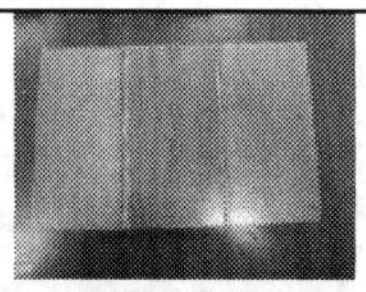		
Echantillon après l'essai au QUV			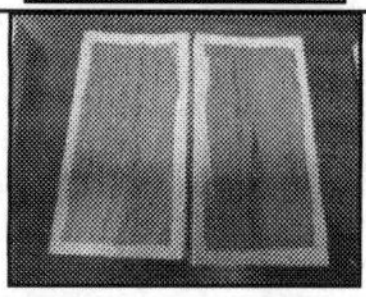	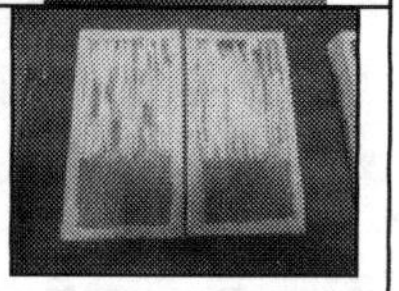

Bois / Finition	*Pinus sylvestris/* F1	*Pinus sylvestris/* F4	*Pinus sylvestris/* F2	*Pinus sylvestris/* F3
Echantillon avant l'essai au QUV	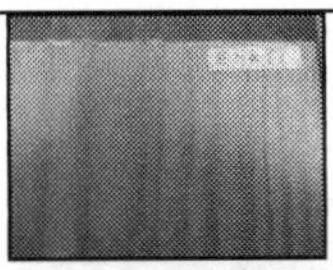	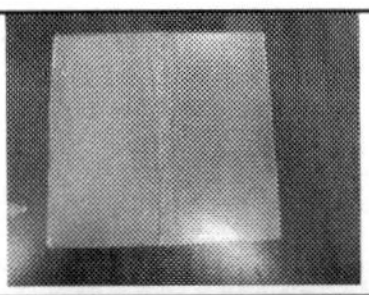	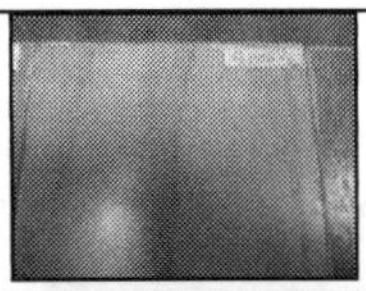	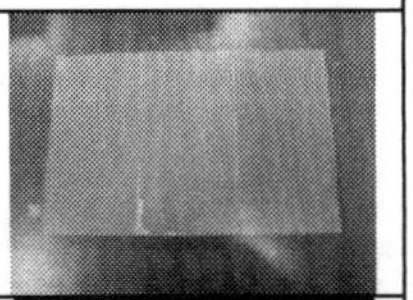
Echantillon après l'essai au QUV		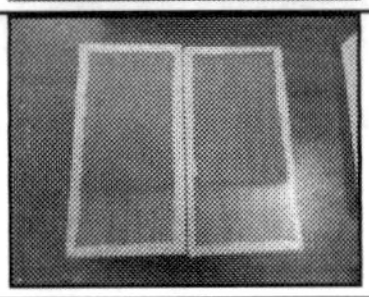		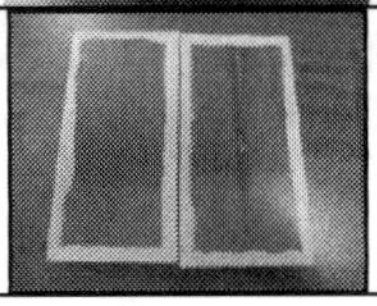

Références

1. W.C. Feist and D.N.S. Hon, *Chemistry of Weathering and Protection* (American Chemical Society, 1984)
2. M.K. Yalinkilic, R. Ilhan, Y. Imamura, M. Takahashi, Z. Demirci, A.C. Yalinkilic and H. Peker, J. Wood Sci. **45**, 502 (1999)
3. J. Zhang and P.D. Kamdem, *Weathering of copper-amine treated wood* (IRG, Kona, Hawaii, 2000)
4. B. George, E. Suttie, A. Merlin and X. Deglise, Polymer Degradation and Stability **88**, 268 (2005)
5. J. Zhang, P. Kamdem, A. Temiz, Applied Surface Science **256**, 842 (2009)
6. V. Jirous-Rajkovic and H. Turkulin, Drvna Industrija **53**, 9 (2002)
7. V. Jirous-Rajkovic, A. Bogner and D. Radovan, Surface Coat International Part B - Coat Trans **87**, 15 (2004)
8. E. Scrinzi, S. Rossi, F. Deflorian and C. Zanella, Prog. Org. Coat. **72**, 81 (2011)
9. G. Grüll, M. Truskaller, L. Podgorski, S. Bollmus and F. Tscherne, Eur. J. Wood Prod. **69**, 443 (2011)
10. T. Turkoglu, E. Baysal, I. Kureli, H. Toker and M. E. Ergun, Wood research **60**, 833 (2015)
11. P.R.Kerber, D.M. Stangerlin, E. Pariz, R.R. de Melo, A.P. de Souza and L. Calegari, Nativa, Sinop, **4**, 303 (2016)
12. A. Temiz , U.C. Yildiz, I. Aydin, M. Eikenes, G. Alfredsen, G. Colakoglu, Applied Surface Science **250**, 35 (2005)
13. S. Zahri, C. Belloncle, F. Charrier, P. Pardon, S. Quideau and B. Charrier, Applied Surface Science **253**, 4985 (2007)
14. N. Çakıcıer, S. Korkut, D.S. Korkut, A. Kurtoğlu and A. Sönmez, International Journal of the Physical Sciences **6**, 1929 (2011)
15. E. Baysal, E.D. Tomak, M. Ozbeyc and E. Altinc, Coloration Technology **130**, 140 (2013)
16. R.F. Teles, A.F. da Costa. *Colorimetric evaluation of three tropical woods after accelerated weathering* (45th IRG, St George, Utah, USA, 2014)
17. T. Krystofiak, B. Lis, M. Muszyńska, S. Proszyk, Drewno, **59**, 127 (2016)
18. E. Baysal, E.D. Tomak, E. Topaloglu, E. Pesman, Maderas. Ciencia y tecnología **18**, 253 (2016)
19. M. Yalcin, H. Pelit, C. Akcay and N. Cakicier, Coloration Technology **133**, 334 (2017)
20. N. Gierlinger, D. Jacques, M. Grabner, R. Wimmer, M. Schwanninger, P. Rozenberg and L.E. Pâques, Trees **18**,102 (2004)
21. A.J.V. Zanuncio, J.P. Motta, T.A. DA Silveira, E.S. Farias and P.F. Trugilho, BioResources **9**, 293 (2014)
22. A.J.V. Zanuncio, A.G. Carvalho, M.T. de Souza, C.M. Jardim, A. de C.O. Carneiro, J.L. Colodette, Maderas Ciencia y tecnología **17**, 857 (2015)
23. N. Ayadi, F. Lejeune, F. Charrier, B. Charrier and A. Merlin, Holz als Roh- und Werkstoff **61**, 221 (2003)
24. P; Bekhta and P. Niemz, Holzforschung **57**, 539 (2003)
25. A.L. Piedade S.M. Pincelli, L.F. de Moura, J.O. Brito, Maderas. Ciencia y tecnología, **14**, 239 (2012)
26. P.H.G. de Cademartori, E. Schneid, D.A. Gatto, D.M. Stangerlin and R. Beltrame. Maderas Ciencia y tecnología **15**, 57 (2013)
27. Z. Demirci, A. Sönmez and M. Budakç, BioResources **8**,1852 (2013)
28. E. Baysal, T. Turkoglu, E. Gavcar, H. Toker, A. Gunduz, M. Kucuktuvek, H. Peker and D. K. Bayraktar, Wood research **62**, 737 (2017)
29. A. Sonmez, M. Budakci and M. Bayram, Scientific Research and Essay **4**, 1432 (2009)
30. J. Gérard, D. Guibal, S. Paradis, M. Vernay, J. Beauchêne, L. Brancheriau and A. Thibaut, (Tropix 7 Version 7.5.1 CIRAD 2011). https://doi.org/10.18167/74726f706978

Study of Local and Distortional Stability of Thin-Walled Structures

Mahi Imene[1], Djafour Naoual[1] Djafour Mustapha[1]
[1]Abou Bekr Belkaid, Civil Engineering, Tlemcen, Algeria

Abstract. Thin-walled structures have an increasingly large and growing field of application in the engineering sector, the goal behind using this type of structure is efficiency in terms of resistance and cost, however the stability of its components (the thin walls) remains the first aspect of the behavior, and a primordial factor in the design process. The hot rolled sections are known by a consequent post-buckling reserve, cold-formed steel sections which are thin-walled elements also benefit, in this case, it seems essential to take into account the favorable effects of this reserve in to the verification procedure of the resistance with respect to the three modes of failures of this type of structure. The design method that takes into account this reserve of resistance is inevitably the effective width method. The direct strength method has been developed to improve the speed and efficiency of the design of thin-walled profiles. The latter mainly uses the buckling loads (for Local, Distortional and Global mode) obtained from a numerical analysis and the resistance curves calibrated experimentally to predict the ultimate load of the profile. Among those, the behavior of a set of C-shaped profiles (highly industrialized) is studied, this type of section is assumed to be very prone to modes of local and distortional instability. The outcome of this investigation revealed very relevant conclusions both scientifically and practically.

1 Introduction

L'utilisation croissante des profilés à parois minces formées à froid réside dans la disponibilité et la facilité des procédés de fabrication qui ont donné la possibilité d'avoir des configurations diverses et plus optimisées afin de répondre aux objectifs structuraux et économiques. L'idée principale derrière l'utilisation de ce type de profilés est l'utilisation de la forme de la section au lieu de l'épaisseur pour résister aux efforts appliqués (1).

L'utilisation d'un matériau mince issu d'un processus de formage à froid induit à des considérations particulières dans la démarche de dimensionnement contrairement à une conception avec les éléments en acier laminé à chaud classique. Parmi les considérations majeures à entreprendre : les modes d'instabilité, l'effet du formage à froid et la résistance post critique (2).

2 l'instabilité locale et distorsionnelle

Étant donné que les composantes individuelles en acier formé à froid sont généralement assez minces par rapport à leur largeur, ces dernières si, elles sont soumises à la compression, ou à la flexion ou au cisaillement, elles peuvent être affectées par les modes d'instabilité à un niveau de contrainte inférieur à la limite d'élasticité, ces modes d'instabilités sont d'ordre local, global et distorsionnel comme le montre la figure1. En l'occurrence le mode local est l'un des principaux facteurs à considérer dans le dimensionnement. Il est bien évident que ce type d'éléments a un comportement et une réserve post critique assez appréciable ; et leur ruine ne se produit pas dès que la charge critique d'instabilité est atteinte, un chargement supplémentaire peut être encore supporté après l'apparition du mode local et avant que la charge ultime de l'élément soit atteinte (3).

Au cours de ces dernières années, le flambement distorsionnel a été considéré comme un état limite important pour la conception des poutres et des colonnes en acier formé à froid, ayant des semelles comprimées renforcées par des raidisseurs. Et par conséquent de nouvelles dispositions de conceptions ont été ajoutées dans les spécifications nord-Américaines actuelles. Le mode distortionnel peut aussi régir le comportement des éléments ayant un centre de gravité et un centre de cisaillement qui ne coïncident pas, ils sont toutefois enclins au flambement flexionnel-torsionnel qui peut être critique vu leur faible rigidité torsionnelle. Des études numériques et des investigations expérimentales ont montré l'existence d'une réserve poste critique pour le mode distorsionnel, mais elle est moins importante que celle du mode local (4).

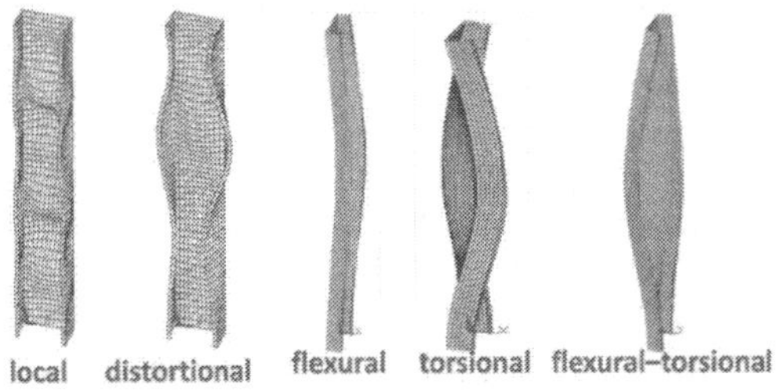

Fig. 1. les modes d'instabilité d'une colonne à parois minces de section en C

2.1 L'instabilité distorsionnelle

C'est un mode qui implique le changement de la forme de la section transversale, cette déformation est manifestée par la rotation et la translation des coins arrondis et des lignes d'intersection entre les plaques formant la section du profilé, l'âme subit une flexion hors de son plan, alors que l'ensemble semelle-raidisseur se déplace en mode rigide. Ce mode se produit à une longueur d'onde intermédiaire entre celle du mode local et global. La figure2 présente la formé déformée du mode distorsionnel pour des sollicitations de compression et de flexion.

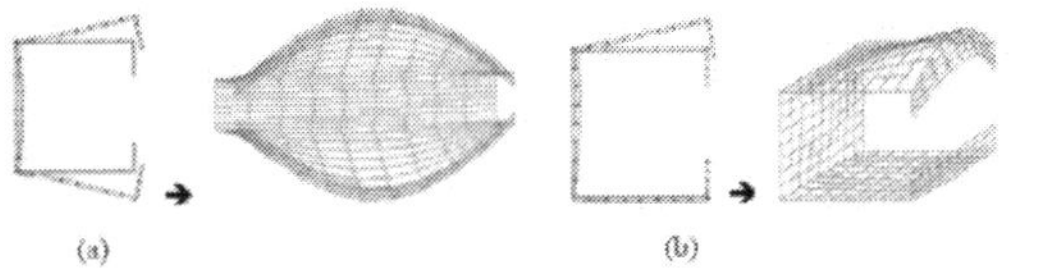

Fig. 2. La déformation d'une colonne en C en mode distorsionnel (a) sous la compression (b) sous la flexion

2.1.1 Le calcul de la contrainte critique du mode distorsionnel

Ce mode est le moins métrisé et le plus compliqué d'ailleurs c'est le dernier qui a été introduit dans les codes de calcul. La mise en évidence de la charge de ruine de ce mode de flambement se doit à Hancock et ses collaborateurs. En l'occurrence la charge critique de cette instabilité est exprimée dans les codes de calculs par les formules (1), (2) et (3) :
Pour le règlement Européen : la contrainte élastique du raidisseur vaut (5) :

$$\sigma_{crs}=2.((k.E.I_s)^\wedge(0.5)) / A_s \qquad (1)$$

K, A_s, Is sont respectivement la rigidité du support, l'aire efficace, et le moment d'inertie de l'aire efficace.
Quant au règlement Américain et australien la contrainte critique est respectivement (6) (7) :

$$F_{cr} = (k_{\phi ef} + k_{\phi ew}) / (k_{\phi gf} + k_{\phi gw}) \qquad (2)$$

$$F_{cr} = E / 2A\,[(\alpha_1 + \alpha_2) - ((\alpha_1 + \alpha_2)^2 + 4\,\alpha_3)^{(1/2)} \qquad (3)$$

Ou $k_{\phi ef}$, $k_{\phi ew}$, $k_{\phi gf}$, $k_{\phi gw}$ sont les rigidités à la rotation élastiques et géométriques de la semelle et l'âme respectivement.

A est l'aire de la section du profilé, E le module d'Young, et α_1, α_2, α_3 sont des paramètres qui tiennent compte des caractéristiques géométriques de la section.

2.2 L'instabilité locale

C'est le mode de flambement le plus fréquent des sections à parois minces, qui se produit en engendrant la déformation des parois planes de la section, sans aucune rotation ou translation des lignes d'intersection, cette déformation se manifeste par la flexion cylindrique de celles-ci (figure3), la longueur de demi-onde de ce mode est le premier minimum sur la courbe signature, définie comme étant plus petite, que la plus grande dimension de la section.

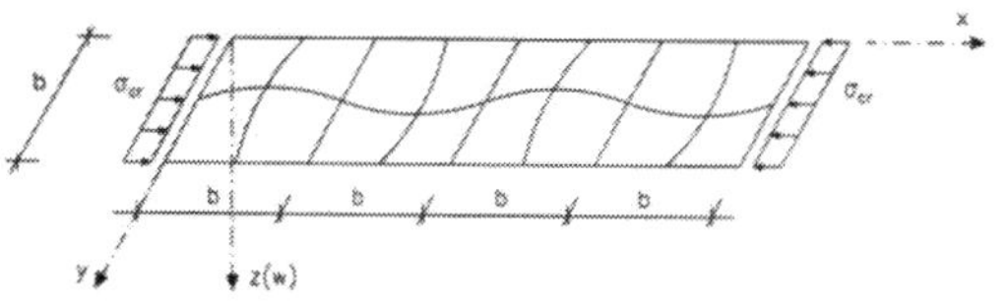

Fig. 3. Le voilement d'une plaque rectangulaire soumise à des contraintes longitudinales de compression

2.2.1 Le calcul de la contrainte critique du mode local

La contrainte critique de ce mode a été investiguée par Timoshenko et Gere 1961, et sa formulation dépend de l'élancement, du coefficient de voilement de la plaque, et de la longueur de demi-onde de ce mode (2).
En adoptant le concept de la méthode énergétique et la mise en œuvre de différentes substituions nécessaires, la contrainte critique du mode local est matérialisée par la formule (4) :

$$F_{crv} = k_V\,\pi^2\,E / 12 * (1 - \nu^2)\,(t / b)^2 \qquad (4)$$

Où Kv est le coefficient de voilement, il dépend des conditions aux limites, du coefficient d'aspect et de la répartition de la contrainte. Plus la plaque est rigidifiée mieux est la valeur du coefficient du voilement et la contrainte du mode local, et par conséquent la performance de l'élément au mode local accroît. A cet effet vient l'intérêt de doter la paroi de raidisseur de bord, qui contribue à sa stabilité tout en empêchant son déplacement.
Du point de vue théorique l'élément se voile dès l'atteinte de sa charge critique, pour un chargement supplémentaire, la ruine ne se produit pas, et cela suite à une redistribution de contrainte, alors une charge additionnelle sera supportée, cette redistribution de contrainte a été idéalisée sur une partie de la parois appelée largeur efficace. C'est le comportement post critique par excellence car le reste de la paroi : les parties inefficaces sont présentes et collaborent également à la résistance. La méthode réglementaire qui prend en considération ce concept dans la procédure de design est la méthode de *la largeur efficace* (MLE) formulée par les équations (5), (6), (7) et (8). Introduite par Von Karman

1932, Winter (1968) a apporté des améliorations et des modifications qui mettent en évidence les imperfections géométriques et les contraintes résiduelles de telle manière que :

$$b_e = b \quad \text{si } \lambda \leq 0.673 \tag{5}$$

$$be = \rho\, b \quad \text{si } \lambda > 0.673 \tag{6}$$

Où ρ est le facteur de réduction de la largeur

$$\rho = (1 - 0.22 / \lambda) / \lambda \text{ et} \quad \rho \leq 1 \tag{7}$$

$$\lambda = 1.052 / (k_v)^{1/2} (b / t)(f / E)^{1/2} \tag{8}$$

f peut être prise égale à f_y, λ est l'élancement réduit, ρ le facteur de réduction b_e b la largeur effective et la largeur de la paroi respectivement.

3 La nouvelle méthode de dimensionnement-la méthode de résistance directe

Quand les sections acquièrent des formes complexes et optimisées, avec des raidisseurs de bords et intermédiaires, le processus de la méthode de la largeur efficace devient extrêmement compliqué et fastidieux en termes de calcul et du temps. En outre, la MLE ne tient pas compte de l'interaction qui existe entre les parois (l'âme et la semelle, semelle raidisseur). Afin de pallier ces problèmes la DSM a été développée par Pekoz et Schafer (1998), son concept consiste à éviter la décomposition de la section par la mise en évidence des courbes de résistances appliquées pour la section complète. Cette procédure s'appuie sur la même hypothèse empirique de MLE : la résistance ultime est en fonction de la charge élastique de flambement et la limite d'élasticité.

Les équations (9), (10), (11), (12), servent à évaluer la capacité portance en compression pour le mode local et distorsionnel

$$\text{Pour } \lambda \leq 0.776 \quad P_{nl} = P_{ne} \tag{9}$$

$$\text{Pour } \lambda > 0.776 P_{nl} = [1 - 0.15(P_{crl}/P_{ne})^{0.4}]\,(P_{crl}/P_{ne})^{0.4} P_{ne} \tag{10}$$

$$\text{Si } \lambda \leq 0.561 \quad P_{nd} = P_y \tag{11}$$

$$\text{Si } \lambda > 0.561\ P_{nd} = [1 - 0.25(P_{crd}/P_y)^{0.6}]\,(P_{crd}/P_y)^{0.6} P_y \tag{12}$$

4 Les méthodes numériques d'analyse du comportement des profils à parois minces

Les méthodes analytiques énoncées dans les standards de calculs sont calibrées par rapport aux résultats numériques et expérimentaux, cependant les ingénieurs et les concepteurs trouvent souvent des difficultés à les mettre en ouvre particulièrement pour des configurations, des chargements, et des conditions aux limites relativement complexes ; cependant les méthodes numériques sont incontestablement l'outil d'analyse structurelle le plus puissant et le plus fiable dans ce sens.

Parmi ces outils numériques on peut citer, la plus populaire la méthode des éléments finis (MEF), la méthode des bandes finies (MBF), la méthode des bandes finies contraintes (MBFc), la théorie des poutres généralisées (GBT) ... etc.

En ingénierie de structure la MEF est la plus robuste, car elle offre une manipulation générale des géométries, du chargement et des conditions aux limites, en revanche elle ne fournit pas l'identification modale (la charge critique d'instabilité et la forme déformée correspondante).

La méthode des bandes finies qui est une variante économique de la MEF (DDL réduits), utilise des harmoniques comme fonctions d'interpolation pour approcher le déplacement dans une direction privilégiée. Popularisée par le professeur Hancock, qui l'a mis en œuvre par courbe de signature qui donne la relation entre le facteur de charge f_{cr} / f_y et la longueur de demi-onde autrement dit, en forçant la structure à se déformer en développant une demi-sinusoïde telle que le montre la figure 4. La méthode des bandes finies est devenue un outil incontournable dans l'analyse de stabilité des profilés à parois minces soumis à des contraintes longitudinales.

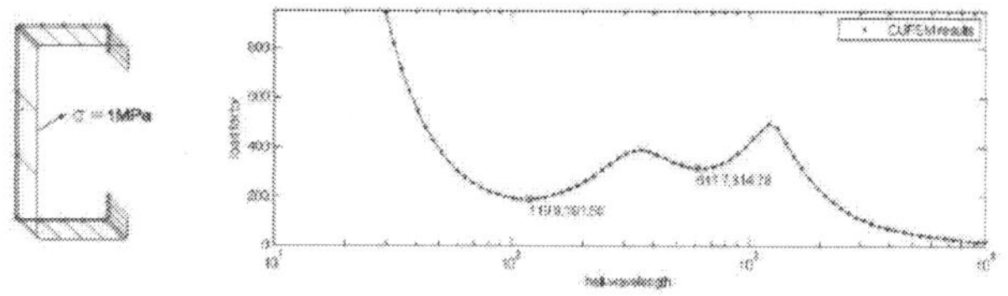

Fig. 4. La courbe signature issue d'analyse par la MBF d'une colonne en C.

5 L'analyse plastique

Finalement, il paraît opportun d'indiquer qu'une méthode de conception adoptant une analyse plastique n'est pas incluse dans les codes de calculs, étant donné que la majorité des sections en acier formées à froid présentent des rapports largeur épaisseur qui dépassent considérablement les limites pour une analyse plastique.

6 Analyse locale et distorsionnelle des profilés en C à parois minces en acier formé à froid

Une étude a été mise en œuvre sur un groupe d'échantillons de profilés de section en c comprimées très utilisés comme pannes et lisses de bardage. Cette investigation consiste à estimer la charge critique élastique locale et distorsionnelle par les formules précédentes et par la suite coordonnées ces données aux résultats numériques de la méthode des bandes finies FSM (finite strip method).

Les propriétés géométriques sont résumées dans le tableau 1 sachant que ''h'' : est la hauteur de l'âme, ''b'' étant la largeur de la semelle, ''c'' : est la profondeur du raidisseur, et ''t'' : est l'épaisseur de la paroi.

Tableau 1. La géométrie des profilés analysés

	h/t	b/t	c/t	Nbr de sections
Min	44	23.33	6.67	20
Max	150	46.67	13.33	

En outre, les caractéristiques mécaniques de l'acier utilisé sont comme suit f_y = 140 MPa, E = 210 GPa, ν = 0.3.

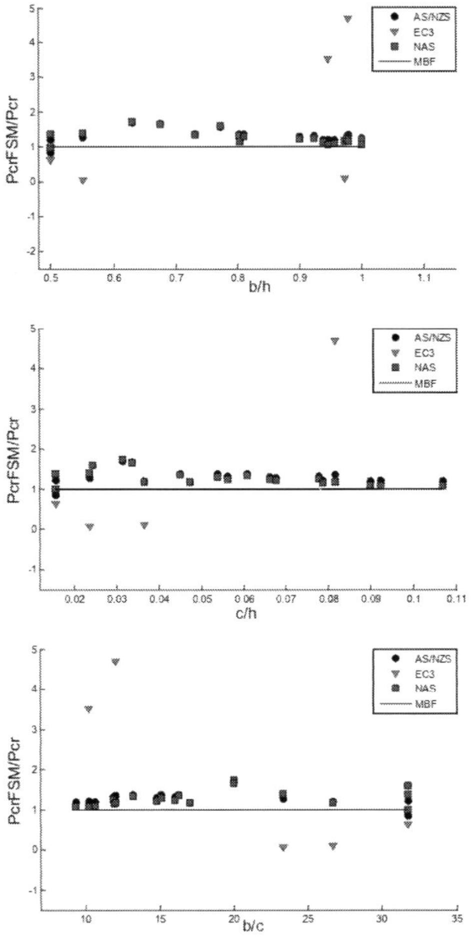

Fig. 5. Comparaison entre les charges critiques distorsionnelles de NAS et ASNZS, EC3 et celles de la MBF pour les rapports h/b, c/h, b/c

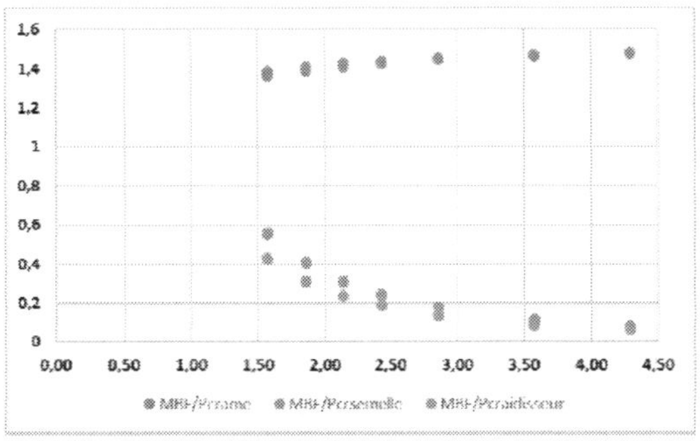

Fig. 6. La performance des trois parois vis-à-vis du mode local évaluée pour le rapport h/b

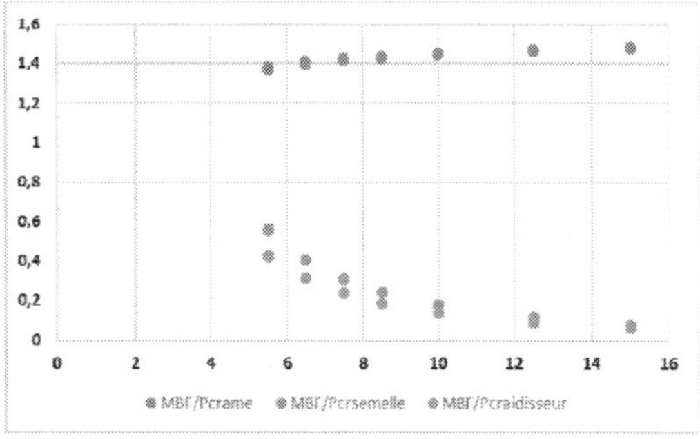

Fig. 7. La performance des trois parois vis-à-vis du mode local évaluée pour le rapport h/c

7 Analyse des résultats

La figure 5 expose le comportement mécanique de la section en fonction de sa géométrie (PcrFSM/PcrAS/NZS, PcrFSM/PcrNAS, PcrFSM/PcrEC3), les modèles américain et australien sont toutes les deux performent vu que le rapport PcrFSM/Pcr NAS ou PcrFSM/Pcr ASNZS sont aux alentours de 1, autrement dit leur dispersion n'est pas importante par rapport à la méthode numérique. Cependant, le modèle européen, exhibe des résultats peu favorables, pour la majorité des valeurs de b/h, c/h et b/c.
L'absence de résultats de certaines sections et la divergence des autres se traduisent par une performance de l'Eurocode qui sont jugées très faibles, et cela est justifié par le manque d'adaptation de ce modèle par rapport à ces géométries, et le comportement de ces sections ne représente pas effectivement la schématisation de l'Eurocode du mode distortionnel.
Quant aux figures 6 et 7 les sections qui sont enclines au voilement de la semelle et du raidisseur sont ceux qui possèdent des rapports h/b et h/c intermédiaires d'où une dispersion minime des charges critiques du voilement de la semelle et le raidisseur par rapport à la MBF, pour des rapports plus importants le voilement de l'âme prend le relais et régit le comportement de la section.

8 Conclusion

La résistance des profilés à parois minces est dans la plupart des cas affectée par le mode local et le mode distosionnel. Le présent article a mis en évidence un état de l'art sur les méthodes analytiques et numériques servant à analyser les modes d'instabilité des profilés à parois minces en acier formé à froid, entre autres la performance de ces méthodes a été estimée pour le mode local et distorsionnel, et les conclusions suivantes ont été révélées :

(i) Plus le raidisseur accroît mieux est la résistance au mode distorsionnel
(ii) Les méthodes américaine et australienne donnent des résultats similaires voire identiques et leurs modèles sont fiables et éprouvent une convergence par rapport à la méthode des bandes finies.

(iii) Plus les rapports h/b et h/c augmentent moins la contrainte critique du mode local de la semelle et le raidisseur est appréciable.
(iv) Le modèle Européen pour le mode distorsionnel, présente des disfonctionnements pour la majorité des sections.
(v) On pourrait avancer que l'EC3 considère le mode distorsionnel comme une continuité du mode local, et en conséquence les résultats divergent par rapport à ceux du deuxième minimum de la courbe signature de la MBF.

Références

1. 4600:2005, AS/NZS. *Australian/New Zealand Standard Cold-Fomred Steel Structures*. s.l. : standards australia standards new zealand, (2005).
2. Institute, American and Iron Steel. *AISI S100-2007, North American Specification for the Design of Cold-Formed Steel Structural Members .(*2007).
3. *Distortional Buckling of Steel Storage Rack Columns*. Hancock, Gregory J. St. Louis, Missouri, U.S.A. : s.n., November 13-14, (1984). Seventh International Specialty Conference on Cold-Formed Steel Structures.
4. *Cold-formed steel structures: Research review (2013–2014)*. Hancock, GJ. 2016, Advances in Structural Engineering, Vol. **19**(3), pp. 393-408.
5. EN1993-1-3, European Standard pr. *EC3. Eurocode 3: design of steel structures*. . s.l. : European Committee for Standardization, (2007).
6. *Effective Width Method for determining distortional buckling strength of cold-formed steel flexural C and Z sections*. Cheng Yu, Weiming Yan. (2011), THIN-WALLED STRUCTURES, Vol. **49**, pp. 233-238.
7. AISI-Specifications for the Design of Cold-Formed, Steel Structural Members. *AISI Manual Cold-Formed Steel Design*. s.l. : American Iron and Steel Institute, (2002).

Effects of the brake shoe friction material on the railway wheel damage

Abdelmajid HAMDAOUI [1], El Houcine JADDI [1]
[1] Mohammed V University, Mohammadia School of Engineers, ERD3M Laboratory, Morocco

Abstract. The wheels of a railway axle are the most critical components of a train. These wheels are subjected to several forms of deterioration, including wear, which significantly influences the safety of traffic as well as the dynamic stability of railway vehicles and the lifetime of wheelsets. The objective of this work is to compare the effect of two friction materials of brake shoes at the damage and the wear rate of the ER8 steel wheels.

1 Introduction

En raison de l'interaction avec le rail et les semelles pendant le freinage, les roues ferroviaires subissent l'usure avec des taux variables dépendant de plusieurs paramètres. Le couple de matériaux roue/semelle est un facteur important dans l'endommagement et l'usure des roues [1]. L'évolution de la forme du profil due à l'usure conduit à des variations des performances dynamiques du train. Par conséquent, les profils originaux sont périodiquement rétablis par reprofilage qui fait diminuer le diamètre et par conséquent la durée de vie de la roue.

La prédiction du taux d'usure des roues devient ainsi importante pour une meilleure planification des interventions de maintenance. Ces opérations fondamentales et périodiquement nécessaires, sont assez couteuses en termes de dépenses et d'indisponibilité du matériel roulant; d'où l'intérêt de réduire leur fréquence et d'optimiser ainsi la durée de vie des roues [2].

2 L'usure des roues d'un essieu ferroviaire

L'usure peut être définie comme étant la perte de matière solide subie par une surface de frottement. L'usure dépend non seulement de la nature du matériau mais aussi de d'autres éléments tels que la géométrie de contact, la topographie de la surface, le chargement, la lubrification et l'environnement. Les mécanismes qui causent de tels dommages sont généralement compliqués et il est difficile de distinguer les uns des autres. Dans [3], environ 60 termes décrivant le comportement et les mécanismes de l'usure sont répertoriés. Certains des mécanismes les plus importants liés aux roues d'un essieu ferroviaire sont mentionnés ci-dessous [4] :

- Usure abrasive: usure causée par des surfaces rugueuses et dures glissant l'une sur l'autre ou une usure causée par des particules dures piégées entre deux surfaces comme les débris d'oxyde dur.
- Usure adhésive: usure causée par le cisaillement des jonctions formées entre deux surfaces en contact, parfois utilisé comme synonyme d'usure par glissement à sec.
- Usure chimique (usure corrosive): usure causée par la formation d'oxyde ou d'autres composants sur les surfaces en raison de la réaction chimique des surfaces avec l'environnement.
- Usure érosive: usure due au mouvement relatif des surfaces de contact en présence d'un fluide contenant des particules solides entre les surfaces.
- Fatigue des contacts roulants (RCF): causée par des variations de contraintes cycliques conduisant à la fatigue des matériaux. Elle en résulte généralement la formation des fissures de surface et de profondeur, des piqûres et l'écaillage.

Dans les travaux de recherche concernant la modélisation de l'usure des roues ferroviaires, trois approches ont été identifiés :

- Mesures sur le matériel.
- Tests de laboratoire.
- Modèles de prédiction théorique.

Différentes techniques ont été donc utilisées pour étudier l'usure des aciers des roues ferroviaires. Des relevés de mesure sur les roues ont été utilisés pour analyser les causes de l'usure. Une grande quantité de données a également été recueillie à partir d'expériences effectuées sur des pistes de test spécialement conçues. Certains essais en laboratoire ont été réalisés à grand échelle et d'autres à échelle réduite [5].

Les travaux de recherche antérieurs effectués pour étudier le comportement d'usure des aciers des roues ferroviaires ont permis d'identifier un certain nombre de régimes d'usure. Les premiers tests ont démontré que deux régimes existaient. Ceux-ci ont été désignés comme légers et sévères. Un troisième régime d'usure qualifié catastrophique a été également identifié [6].

Un essai à deux disques a permis de modéliser trois régimes d'usure pour les roues en acier ER8 : Doux, grave et catastrophique. Ceux-ci ont été classés en fonction du taux d'usure [7].

3 Effet du matériau des semelles de frein sur l'endommagement des roues ferroviaires

L'action de freinage s'effectue par l'appui des semelles de freins sur la table de roulement de la roue comme indiqué sur la Figure 1.

Les semelles de freins sont généralement fabriquées en fonte, en matériaux composites organiques ou en matériaux frittés. Le développement des matériaux pour les semelles au cours des deux dernières décennies a permis l'introduction de nouveaux matériaux de freinage dans l'industrie ferroviaire.

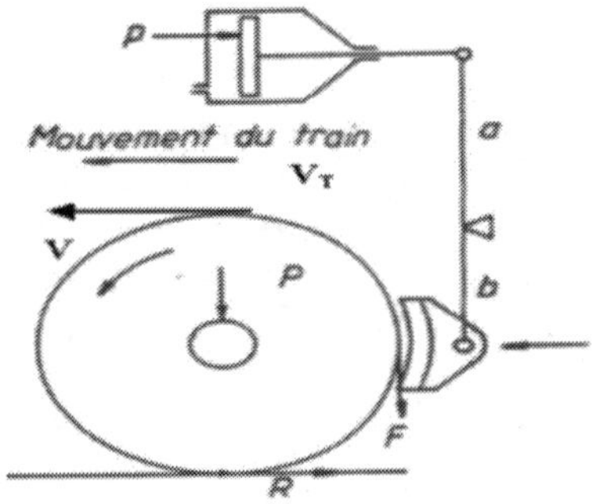

Fig. 1 Freinage par semelles

Des recherches ont été réalisées sur l'influence de divers types de matériaux de semelles sur les performances des freins et les émissions de bruit [8-9]. En conséquence, de nouvelles normes [10] et de nouvelles réglementations pour remplacer les semelles en fonte avec des semelles en matériaux composites sont en cours de développement.

Dans une étude expérimentale récente [11], la comparaison entre les semelles en fonte et les semelles en composite montre que les deux matériaux de semelles testés présentent des différences dans la distribution granulométrique, la morphologie et le contenu élémentaire. Le taux de génération de particules observé est significativement plus important pour la fonte comparée au matériau composite.

Les résultats de d'autres essais confirment une influence significative des semelles en combinaison avec des charges mécaniques sur l'endommagement de la table de roulement [12].

Il a été également prouvé que les semelles en composite produisent des températures plus élevées en par rapport aux semelles en fonte sous les mêmes conditions de freinage. En outre, les semelles en composite imposent moins de dégâts et moins de changement dans les contraintes résiduelles. Tandis que les semelles en fonte de haute teneur en phosphore produisent plus de points chauds sur la table de roulement comparativement aux semelles en fonte ordinaire [13].

Des expériences ont été effectuées par Vernersson pour étudier l'influence de différents types de matériaux en bloc et pour suivre l'élévation de la température (points chauds) dans la bande de roulement [14]. L'étude expérimentale de Vernersson conclut que les semelles en fonte produisent immédiatement des points chauds sur la table de roulement tandis que les semelles en composite nécessitent des vitesses plus élevées et des temps plus importants pour générer ces points chauds. Des essais dynamométriques comparatifs des semelles en fonte et celles en composite montre que les semelles en fonte peuvent augmenter la formation des fissures thermiques dans la table de roulement tandis que les semelles en composite ne changent pas la microstructure de la table de roulement [15].

4 Usure des roues pour deux couples de matériaux Semelle/Roue :

4.1 Matériaux des roues ferroviaires

Les matériaux les plus utilisés dans la production des roues ferroviaires sont les ER7 et ER8, définis dans la norme EN 13262. Ce sont des aciers à structure perlite-ferrite. La jante de la roue est généralement durcie suivant le traitement thermique : 860 ° C / 5h / eau et 520 ° C / 5h / air.

Le Tableau 1 montre les compositions chimiques de ces aciers, avec les teneuses maximales en différents éléments.

Tableau 1. Composition chimique des matériaux ER7 et ER8

Steel grade	C	Si	Mn	P	S	Cr	Cu	Mo	Ni	V	Cr+Mo+Ni
ER7	0.52	0.40	0.80	0.020	0.020	0.30	0.30	0.08	0.30	0.06	0.50
ER8	0.56	0.40	0.80	0.020	0.020	0.30	0.30	0.08	0.30	0.06	0.50

Le Tableau 2 ci-dessous présente les propriétés mécaniques de base qui devraient être atteintes après application du traitement thermique appliqué et avant l'usinage dans l'état final de production.

Tableau 2. Propriétés mécaniques des matériaux ER7 et ER8

Steel grade	Wheel rim			Wheel web	
	ReH	Rm	A5	DRm	A5
	[MPa]	[MPa]	[%]	[MPa]	[%]
ER7	≥ 520	820–940	≥ 14	≥ 110	≥ 16
ER8	≥ 540	860–980	≥ 13	≥ 120	≥ 16

4.2 Comparaison de l'usure des roues en ER8 pour des semelles en fonte et les semelles en composite

Cette comparaison est effectuée par l'analyse de la variation du diamètre des roues en fonction du parcours réalisé par ces roues durant une période prédéfinie. Le taux d'usure est ainsi défini comme la diminution du diamètre de la roue en mm par 10000 Km de parcours. Les résultats de cette analyse montrent que le taux d'usure pour les roues équipées de semelles en composite est de 4mm par 10000Km tandis qu'une valeur inférieur de 2mm par 10000km a été constatée pour les roues équipées de semelles en fonte.

L'écart constaté entre ces taux d'usure est dû d'une part à la différence de leur coefficient de frottement en contact avec l'acier ER8, et d'autre part au nombre de reprofilage subit par chaque type de roue. En effet, la diminution du diamètre est due essentiellement à la diminution du bandage de la roue suite au la reprise du profil conforme d'origine par usinage de la roue. Comparativement aux semelles en fonte, les semelles en composite font évoluer rapidement les cotes caractéristiques du profil, indiqué dans la Figure 2 et Figure 3. , vers des valeurs hors intervalle de tolérance, ce qui impose la reprise du profil original conforme par l'usinage.

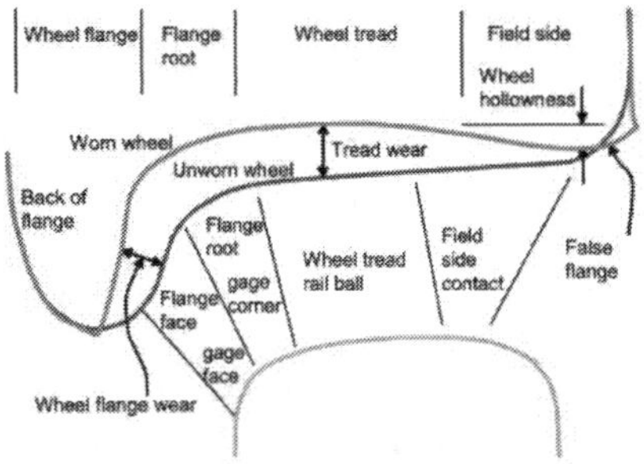

Fig. 2 Evolution du profil par l'usure

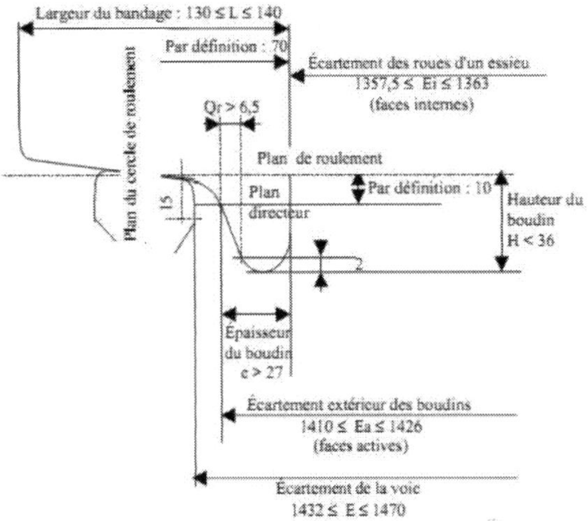

Fig. 3 Cotes caractéristiques du profil de la roue

5 Conclusion

Les effets de chaque matériau de semelles sur la dégradation de l'état des roues ferroviaires en service ont été présentés. Malgré les recommandations du passage de la fonte aux matériaux composites pour les semelles de frein en raison de certains avantages acoustiques et environnementaux, l'usage de ces semelles pour des roues en acier ER8 pénalise leur durée de vie et augmente leurs couts de maintenance à cause du taux d'usure important.

References

1. J. Gerlici, T. Lack, J. Harusinec, Rail vehicles wheels and brake blocks wear laboratory test stand utilization, *Scientific journal „Prace Naukowe (Transport: Analiza i ocena elementów systemów transportowych"* **101** 21-32, (2014).
2. J. Auciello , M. Ignesti , M. Malvezzi , E. Meli & A. Rindi Development and validation of a wear model

for the analysis of the wheel profile evolution in railway vehicles, Vehicle System Dynamics: *International Journal of Vehicle Mechanics and Mobility*, 50:11, 1707-1734, (2012).
3. P.J. Blau (Ed.), Glossary of terms, Friction, Lubrication, and Wear Technology Metals Handbook, vol. **18**, *ASME International, Metals Park*, OH, (1992).
4. [4]Saeed Hossein Nia, Carlos Casanueva, Sebastian Stichel, Prediction of RCF and wear evolution of iron-ore locomotive wheels, *Wear*, Volume **338**, Pages 62-72, (2015).
5. Kumar, S., Rao, D.L.P. Wheel-rail contact wear, work, and lateral force for zero angle of attack - a laboratory study. *Transactions of the ASME, Journal of Dynamic Systems, Measurement, and Control*, **106**, 319-326, (1984).
6. R. Lewis, U. Olofsson, Mapping rail wear regimes and transitions, *Wear*, Volume **257**, Issue 7, Pages 721-729, (2004).
7. Lewis, R. and Dwyer-Joyce, R.S. Wear mechanisms and transitions in railway wheel steels. *Proceedings of the Institution of Mechanical Engineers, Part J: Journal of Engineering Tribology*, **218(6)**, 467-478, (2004)
8. Miyauchi, T., Tsujimura, T., Handa, K., Nakayama, J., and Shimuzu, K. Influence of silicon carbide filters in cast iron composite brake blocks on brake performance and development of a production process. *Wear*, **267**, 833-838, (2009).
9. Bühler, S. Methods and results of field testing of a retrofitted freight train with composite brake blocks. *Journal of Sound and Vibration*, **293**(3-5), 1041-1050, (2006).
10. Hübner, P. The action programme of UIC, CER and UIP "abatement of railway noise emissions on goods trains". *Journal of Sound and Vibration*, **231**(3), 511-517, (2000).
11. Olofsson, U. A study of airborne wear particles generated from the train traffic-block braking simulation in a pin-on-disc machine. In *Proceedings 8th International Conference on Contact Mechanics and Wear of Rail/Wheel Systems*, Florence, Italy, pp 527-533, (2009).
12. Stone, D.H. and Carpenter, G.F., *Wheel thermal damage limits*. Research and Test Department Association of American Railroads: Chicago, IL, USA, pp 57-63, (1994).
13. Stone, D.H. An interpretive review of wheel failure performance with respect to design and heat-treatment. In *Proceedings Joint ASME/IEEE Railroad Conference*, pp 43-53, (1988).
14. Vernersson, T. Thermally induced roughness of tread-braked railway wheels Part 1: brake rig experiments. *Wear*, **236**(1-2), 96-105, (1999).
15. Vukolov, L.A. and Voronchikhin, A.I. New solutions in choice of material and design of brake shoes required by railroad applications. In *Proceedings World Tribology Congress III*, Washington, D.C., USA, pp 197-198, (2005).

Durability of mortars modified by the effect of combining SPA polymers and supplementary cementitious materials

Durabilité des mortiers modifiés par l'effet de la combinaison de Polymère et des matériaux cimentaires supplémentaires

B. Belbachir[1,2], A.S. Benosman[1,2,3*], H. Taïbi[1], M. Mouli[2], Y. Senhadji[2,4], M. Belbachir[1]
[1] Faculté des Sciences Exactes et Appliquées, Laboratoire de Chimie des Polymères, Université d'Oran 1, Ahmed Benbella, BP 1524, El Mnaouer, Oran 31000; Algérie.
[2] Département de Génie Civil, Laboratoire des Matériaux LABMAT, ENPO Maurice Audin, Oran 31000, Algérie.
[3] Ecole Supérieure des Sciences Appliquées, ESSAT, Tlemcen, Algérie.
[4] Faculté des Sciences Techniques, Université Mustapha Stambouli, Mascara, Algérie.

Abstract. Nowadays, the major concern of professionals in the field of building materials is to improve the properties induced by the addition of different additives (polymers) and mineral additions (Supplementary Cement Materials SCMs) and to eventually adapt them to a particular application. This race towards performance has resulted in mortar formulations that are increasingly complex and rich in diversified additions. This is an industry-friendly practice since it generally yields a mortar modified by the combination of a polymer and SCMs, at low cost and low environmental impact, with an improved sustainability in the long term. In order to improve the durability of SCM-modified repair mortars, it seemed interesting to evaluate the influence of adding the styrene polyacrylic (SPA) Latex on the properties of these mortars when exposed to aggressive media such as acids. Composite mortars based on pozzolanic mineral additions, containing different levels of 0.5%, 1% and 2%^{w} latex, were stored in acid solutions, for various periods of immersion. The analysis of the microstructure of these mortars, after exposure to acid attack, was carried out by FTIR spectroscopy. The results obtained allowed to demonstrate the beneficial effect of adding the SPA polymer and the pozzolanic additions to the modified materials and to show their improved resistance to acid attacks, such as HNO_3 and H_3PO_4 solutions at 8%. The best durability properties of the mortars modified by the combination of the polymer and the additional cementitious materials observed in this study indicate a longer service life of the repaired structure when using this type of Latex-modified repair materials.

Résumé. Le souci majeur des professionnels dans le domaine des matériaux de construction est d'améliorer les propriétés induites par l'ajout des différents adjuvants (polymères) et des additions minérales (Matériaux Cimentaires Supplémentaires SCMs) et parfois de les adapter à une application particulière. Cette course à la performance s'est traduite par des formulations de mortiers de plus en plus complexes et riches de ces différents ajouts. Cette pratique est favorable à l'industrie, donnant généralement un mortier modifié par l'effet de la combinaison de l'ajout d'un polymère et des SCMs à faible coût, et à faible impact environnemental, avec l'amélioration de la durabilité à long terme. En vue d'améliorer la durabilité des mortiers modifiés de réparation à base des SCMs, on a évalué l'influence de l'ajout du Latex polyacrylique styrol (SPA) sur les caractéristiques de ces derniers vis-à-vis des milieux agressifs tels que les acides. Des mortiers composites avec des bases minérales pouzzolaniques et contenant différents taux de latex 0.5%, 1% et 2%w sont conservés dans les solutions acides pour diverses périodes d'immersion. L'analyse de la microstructure de ces mortiers après exposition à l'attaque acide est effectuée par spectroscopie IRTF. Les résultats obtenus permettent de mettre en évidence l'effet bénéfique de l'ajout du polymère SPA et des additions pouzzolaniques dans les matériaux modifiés et sur la résistance aux attaques acides telles que HNO_3 et H_3PO_4. Or, les meilleures propriétés de durabilité des mortiers modifiés par l'effet de la combinaison de polymère et des matériaux cimentaires supplémentaires observés dans cette étude, indiquent une plus longue durée de vie de la structure réparé en utilisant ce type de matériaux de réparation modifiés par des Latex.

1 Introduction

Au cours de ces dernières décennies la formulation des mortiers composites a été profondément modifiée par l'introduction de multiples adjuvants [1]. Et depuis bien longtemps, la volonté d'optimisation des propriétés des matériaux cimentaires a poussé les formulations à incorporer divers adjuvants organiques ou minéraux dans leur composition [2]. Les adjuvants polymères sont depuis plusieurs dizaines d'années largement utilisés en génie civil et connaissent un accroissement de leur emploi, du fait de la diversité des propriétés qu'ils peuvent offrir [1,3,4].

Le souci majeur des professionnels dans le domaine des matériaux composites de construction est d'améliorer les propriétés induites par l'ajout des différents adjuvants et des additions minérales (SCMs) et parfois de les adapter à une application particulière. Cette course à la performance s'est traduite par des formulations de mortiers composites de plus en plus complexes et riches de ces différents ajouts (ultrafines, polymères, SCMs [5], etc…). L'utilisation de résidus industriels récupérés et recyclés, tels que les ajouts cimentaires et les gisements de ressources naturelles tels que la pouzzolane et le calcaire, comme produits de remplacement partiel du ciment Portland dans le béton, permet de réduire les émissions des gaz à effet de serre CO_2 et se traduit par la fabrication d'un béton ou mortier non polluant et durable sur le plan environnemental. Cet accroissement du nombre de constituants est, d'un point de vue pratique, très avantageux car il conduit généralement, à l'amélioration très significative des performances mécaniques, physiques et durabilité [6].

Le but de cette étude est de présenter les résultats d'un travail de recherche portant sur la durabilité des mortiers composites à base de l'effet de la combinaison de l'ajout du polymère SPA et des SCMs telles que la pouzzolane naturelle et la fumée de silice, comme substituant au ciment et ceci pour la réalisation des mortiers de réparation conçus au développement durable. Certaines proportions clés sont également étudiées, contrairement à ce qui a été entrepris dans les travaux antérieurs [5,7-12]. Un complément d'information sur la microstructure a été effectué par la spectrométrie IRTF.

2 Conditions expérimentales

Les mortiers composites modifiés par ajout de polymère, ont été réalisés à base d'additions minérales (10% de pouzzolane et 5% de fumée de silice), avec des teneurs en SPA différentes, tel que : P /C = 0, 0.5 et 2 %. Les rapports pondéraux utilisés expérimentalement sont (ciment /sable)= 1/3 et (eau/ciment)= 0.5 constant pour tous les mortiers. On a utilisé un adjuvant superplastifiant (SP) réducteur d'eau type SIKA® Viscocrète® tempo 12 (Tableau 1).

Les échantillons d'essais ont été confectionnés selon la norme NF P15-403 [13] dans des moules prismatiques (4x4x16 cm^3) et vibrés à l'aide d'une table à choc (NF EN 196-1 [14]). Les moules contenant les spécimens sont couverts de film de plastique et conservés dans l'environnement du laboratoire sous une T°C de 20°± 1°C et une humidité relative d'environ 50 ± 5 %.

Tableau 1. Formulations des différents mortiers modifiés.

Code	Ciment (%)	SCMs (%)		SPA (%)	E/C	SP (%)
		Pz	FS			
Témoin	100	-	-	-	0.5	1.3
Pz_{10}	90	10	-	-	0.5	1.3
FS_5	95	-	5	-	0.5	1.3
$Pz_{10}FS_5$	85	10	5	-	0.5	1.3
$Pz_{10}FS_5Pl_{0.5}$	84.5	10	5	0.5	0.5	1.3
$Pz_{10}FS_5Pl_2$	83	10	5	2	0.5	1.3

Témoin: Mortier de Référence; Pz: Pouzzolane naturelle; FS: Fumée de silice; Pl: polymère SPA; SP: Superplastifiant.

Le démoulage est effectué après 24 heures et les échantillons sont conservés dans une eau basique durant 24 heures. Ensuite, les spécimens sont retirés de cette eau et conservés jusqu'aux 60 jours dans une salle de conservation à T°C et hygrométrie (20°±1°C, HR= 50±5 % ; milieu référence). Les échantillons dont les dimensions prismatiques 40x40x50 mm^3, sont pesés pour déterminer Mo, puis ils sont immergés dans deux différentes solutions acides suivantes : solution d'acide nitrique, HNO_3 à 8% et la solution d'acide ortho-phosphorique, H_3PO_4 à 8%. Le choix s'est porté sur ces deux acides car il s'agit des acides majoritaires trouvés dans les zones industrielles.

La résistance des échantillons immergés vis-à-vis des attaques acides, est déterminée selon la norme ASTM C267-01 [15]. Des analyses IRTF ont été effectuées pour les croutes des échantillons immergés et soumis ou non à l'attaque acide.

2.1 Le ciment

Il s'agit d'un ciment Mâtine provenant de l'usine CIBA du groupe LAFARGE sise à OGGAZ (Mascara /Algérie). Ce ciment a une finesse de 4220 cm^2/g, de densité absolue 3.09 et d'une résistance à la compression de 48 MPa à 28 jours. Les compositions chimiques du ciment et minéralogiques de son clinker sont données dans les tableaux 2 et 3.

Tableau 2. Composition chimique élémentaire du ciment, la pouzzolane et la fumée de silice.

Eléments	SiO_2	Al_2O_3	Fe_2O_3	CaO	MgO	SO_3	K_2O	LOI
CM	17.954	4.071	2.750	58.15	1.768	2.428	0.661	10.72
Pz	46.101	16.442	9.140	9.057	5.341	0.059	1.338	7.110
FS	94.609	0.320	1.212	0.434	0.385	0.104	0.527	1.980

Tableau 3. Composition minéralogique du clinker (%).

Eléments	C_3S	C_2S	C_3A	C_4AF
Pourcentage (%)	64	15	8	12.16

2.2 La pouzzolane naturelle "Pz"

La pouzzolane naturelle, qui est une larve rougeâtre extraite du gisement de Bouhamidi région Béni-saf (Nord-ouest de l'Algérie). Elle a une masse volumique absolue mesurée de 2670 Kg/m^3. La pouzzolane est broyée ensuite tamisée à travers un tamis de 80 µm, la poudre résultante donne une surface spécifique de 300 m^2/Kg. Les résultats des analyses chimiques sont présentés dans le tableau 1.

2.3 La fumée de silice "FS"

La fumée de silice, comme poudre amorphe de silice extrêmement fine, est un sous-produit de la fabrication du silicium ou des silico-alliages. L'addition qu'on a utilisée dans nos essais, est fournie par l'entreprise TECKNACHEM (Sidi Bel-Abbes / Algérie). Sa composition chimique est donnée dans le tableau 1.

2.4 Polyacrylique styrol latex "SPA"

Le polymère latex utilisé pour la modification de nos mortiers, se présente sous forme de dispersion aqueuse de couleur blanche laiteuse, nommé TEKWELD fourni par l'entreprise TECKNACHEM, à haute densité et stable en milieu alcalin.

3.2 Le sable

C'est un sable concassé de fraction 0/5 provenant de la carrière de Nedroma (nord-ouest /Algérie). Afin d'avoir une granulométrie conforme au fuseau du sable normalisé, on a procédé à une correction granulaire de ce sable par un sable marin originaire de Tergua (Ain-témouchent /Algérie). Le sable corrigé est 60% sable de carrière, 40% sable de mer.

3. Résultats et discussion

3.1 Attaque par l'acide nitrique (HNO_3)

La Figure 1 exhibe les résultats des pertes de masses pour les spécimens exposés à une solution d'acide HNO_3 à 8%. Ces échantillons n'ont pas la même tendance de détérioration, particulièrement avec le mortier Pz_{10} et le composite $Pz_{10}FS_5Pl_2$. Après 7 jours, la différence devient tout à fait sensible, elle est supérieure à 20%.

Après 77 jours d'immersion, on constate une perte de masse permanente pour l'ensemble des mortiers. En comparant ces composites, on observe que le mortier témoin et le Pz_{10} présentait une perte de masse très importante, elle est de 35% et 42.5%, respectivement. Tandis que les composites modifiés $Pz_{10}FS_5Pl_{0.5}$ et $Pz_{10}FS_5Pl_2$, comparés aux autres mortiers, présentent une grande aptitude à résister à l'attaque acide HNO_3, avec des pertes de masse de 17.6% et 15.2%, respectivement.

A partir de ces résultats, on remarque l'effet bénéfique de la combinaison de l'ajout du polyacrylique styrol (SPA) et des SCMs, ce qui minimise la pénétration des agents agressifs. On a par conséquence résistance et durabilité accrues dans le cas de l'acide nitrique.

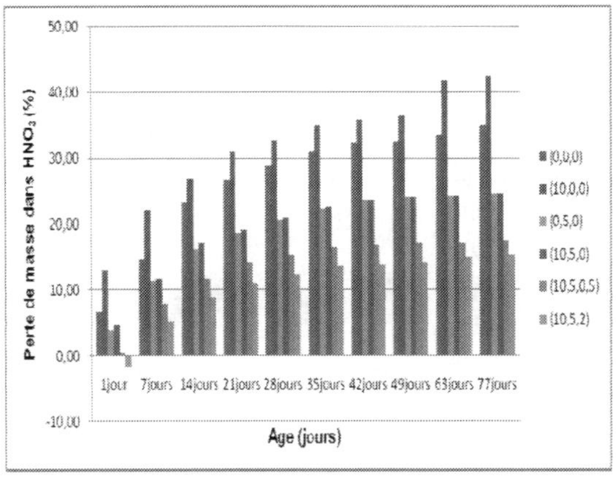

Fig. 1. Effet de la combinaison de l'ajout du polyacrylique styrol et des SCMs sur l'attaque par HNO_3 des mortiers composites.

3.2 Attaque par l'acide phosphorique (H_3PO_4)

Les résultats de la perte de masse des spécimens exposés à la solution d'acide phosphorique à 8%, pour une durée de 1 à 49 jours, sont exhibés dans la Figure 2.

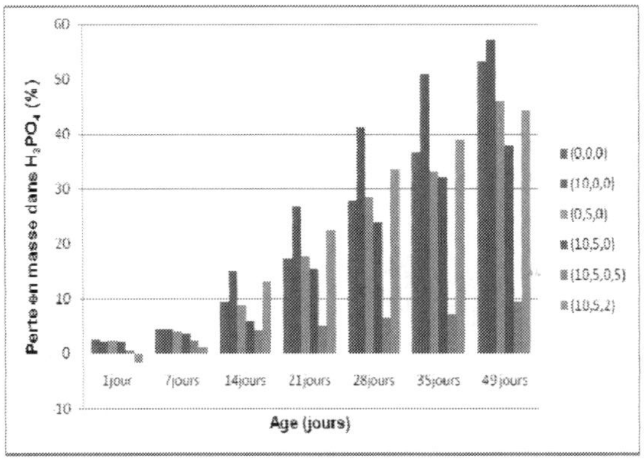

Fig. 2. Effet de la combinaison de l'ajout du SPA et des SCMs sur l'attaque par H_3PO_4 des mortiers composites.

A partir de ces résultats, on observe une perte de masse continue légèrement pour tous les mortiers composites à partir de 7 jours.

Après 14 jours d'immersion, on constate une différence de tendance de détérioration, particulièrement avec le mortier Pz_{10} et le composite $Pz_{10}FS_5Pl_2$ (supérieure à 10%). Tandis que pour le composite $Pz_{10}FS_5Pl_{0.5}$ est inférieure à 4%.

Après 28 jours à 49 jours d'immersion, le mortier témoin, le mortier Pz_{10} et le composite $Pz_{10}FS_5Pl_2$ présentent la grande perte de masse comparée aux autres. Quoique pour le composite $Pz_{10}FS_5Pl_{0.5}$ est inférieure aux autres. On note une perte de masse de 53% pour le mortier témoin, de 57% pour Pz_{10} et de 44% pour le composite avec ajout de 2% de SPA. Par contre, on a seulement une perte de masse qui ne dépasse pas les 9% pour le composite $Pz_{10}FS_5Pl_{0.5}$ à 49 jours d'immersion. Dans cette attaque d'ortho-phosphorique, la diminution de la masse est due à la dissolution des hydrates produits qui mènent à des mortiers poreux. Cependant, seulement l'effet de la combinaison de l'ajout du SPA avec 0.5% et des SCMs augmente la résistance chimique du composite $Pz_{10}FS_5Pl_{0.5}$ vis-à-vis de cette attaque acide.

La Figure 3 montre l'effet de la combinaison de l'ajout du SPA et des SCMs sur les attaques acides des mortiers composites après 28 jours. A cette période d'immersion, on remarque pour l'ensemble des échantillons, que les pertes en masse dues à l'acide phosphorique sont généralement supérieures à celles dues à l'acide nitrique.

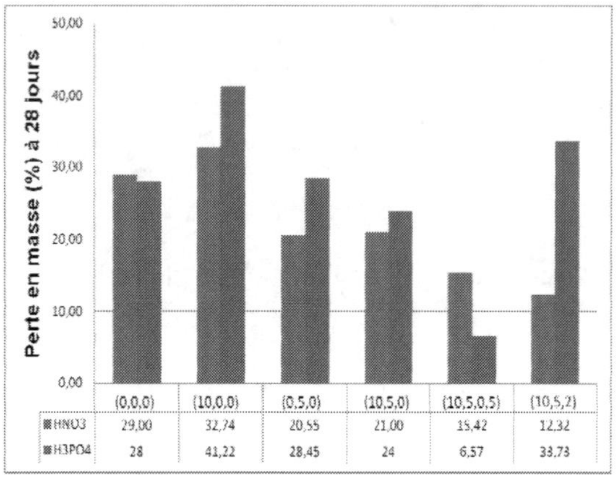

Fig. 3. Effet de la combinaison de l'ajout du SPA et des SCMs sur les attaques acides des mortiers composites après 28 jours.

Avec l'acide phosphorique, il y a formation du phosphate de calcium hydraté $Ca_3(PO_4)_2.6H_2O$, 0.002 g/100ml dans H_2O, qui est un sel peu soluble. Tandis qu'avec l'acide nitrique, la surface du mortier est recouverte du nitrate de calcium $Ca(NO_3)_2.2H_2O$ (di-hydraté) très soluble dans l'eau: 121.2 g/100ml dans H_2O et d'hydroxyde de fer qui est attribué au caractère oxydant de cet acide. Les réactions chimiques correspondantes sont indiquées dans les équations (1 & 2) ci-dessous. Ces résultats sont en accord avec ceux rapportés par Achoura et al. [16], Benosman et al. [17], Omrane et al. [18] et Belbachir et al. [19].

Les résistances mécaniques sont plus ou moins affectées par la concentration mais aussi et surtout par la nature des acides avec dans l'ordre le plus agressif d'après cette étude : l'acide H_3PO_4 puis l'acide nitrique HNO_3 (Figure 3). On propose pour cela la séquence suivante: $H_3PO_4 > HNO_3$.

$$2HNO_3 + Ca(OH)_2 \rightarrow Ca(NO_3)_2.2H_2O \quad (1)$$
$$2H_3PO_4 + 3\,Ca(OH)_2 \rightarrow Ca_3(PO_4)_2.6H_2O \quad (2)$$

Donc on peut conclure que les mortiers FS_5 et $Pz_{10}FS_5$, les composites $Pz_{10}FS_5Pl_{0.5}$ et $Pz_{10}FS_5Pl_2$ ont bien résisté à l'attaque d'acide nitrique. Par contre, on a noté que le composite $Pz_{10}FS_5Pl_{0.5}$ à mieux résister à l'attaque phosphorique.

L'influence positive du pourcentage des SCMs et du latex SPA sur la résistance chimique est traduite d'une part, par la réaction pouzzolanique qui fixe la portlandite libérée lors de l'hydratation du ciment pour former des gels de CSH supplémentaires de 2ème génération et donc l'accroissement de la compacité des hydrates, engendrant une réduction de la porosité et d'autre part, par la formation du film du polymère SPA. Ce qui réduit la pénétration des agents agressifs. Ces résultats sont en accord avec ceux obtenus par Shaikh Ahmed [11], Belbachir et al. [19], Rossignolo et al. [20] et Monteny et al. [21].

3.3 L'examen visuel

L'inspection visuelle montre clairement l'état des différents spécimens avant et après attaques acides : 77 jours d'immersion dans HNO_3 (Figure 4) et 49 jours dans H_3PO_4 à 8% (Figure 5).

Dans l'acide HNO_3, (Figure 4): - les spécimens ne présentent pas une perte remarquable de leur forme, ils gardent toujours leur forme cubique. L'échantillon du témoin a un aspect plus dégradé (plus que la moitié) comparativement aux composites $Pz_{10}FS_5Pl_{0.5}$ et $Pz_{10}FS_5Pl_2$.

- La surface est recouverte d'une mince couche de rouille de couleur jaunâtre différente par rapport à la couleur initiale de la matrice cimentaire et qui semble due à la présence d'hydroxyde de fer Fe_2O_3 mal cristallisé [22].

- La surface des mortiers est très décapée avec un déchaussement important des grains de sable. La solubilité du sel formé, $Ca(NO_3)_2.2H_2O$, après réaction de HNO_3 avec la portlandite $Ca(OH)_2$ libérée lors de l'hydratation du ciment et à sa grande lixiviation à l'extérieur avec décalcification progressive des C-S-H, engendrent ainsi des pertes de masse importantes.

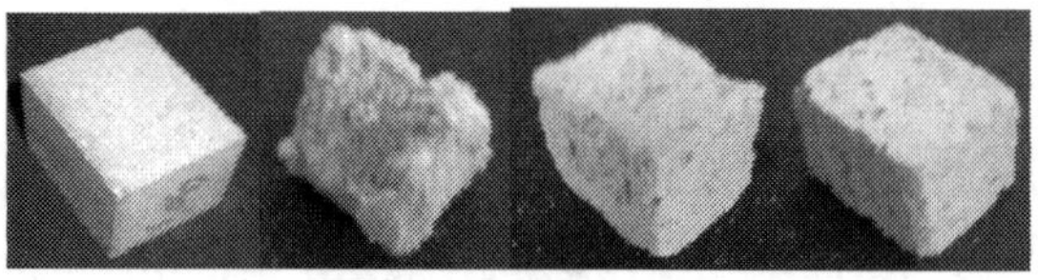

Fig. 4. Détérioration des spécimens après 77 jours d'immersion: (1) témoin, (2) (0;0;0) HNO_3, (3) ($Pz_{10}FS_5Pl_{0.5}$) HNO_3 et (4) ($Pz_{10}FS_5Pl_2$) HNO_3, respectivement (du gauche à droite).

Dans l'acide H_3PO_4, (Figure 5): Réduction importante du volume des échantillons de mortier. En général, les surfaces du mortier sont plus détériorées dans H_3PO_4 que dans HNO_3.

Fig. 5. Etat des spécimens après 49 jours d'immersion dans l'acide H_3PO_4.

Le taux d'agressivité de H_3PO_4 est supérieur à celui du HNO_3. Ces constatations ont permis de confirmer les résultats obtenus par les essais des pertes de masse (Figures 1-3).

3.4 Caractérisation par Spectrométrie IRTF

Le Tableau 4 et les Figures 6-8 présentent les positions et les intensités des bandes IRTF des spécimens {mortier témoin, $Pz_{10}FS_5Pl_2$ et $Pz_{10}FS_5Pl_{0.5}$} exposés aux: milieu référence (HR= 50±5 %), HNO_3 et H_3PO_4. La superposition des différents spectres confirme:

- La formation de trace du nitrate de calcium et du phosphate de calcium hydraté après l'attaque par les acides HNO_3 et H_3PO_4, respectivement, selon les réactions chimiques (Eq.1) & (Eq.2);

- La décomposition totale de la portlandite par la solution agressive, d'où la disparition de sa bande majeure, selon les réactions chimiques (Eq.1) & (Eq.2);

- La diminution de la quantité de la calcite $CaCO_3$ par les différentes attaques acides.

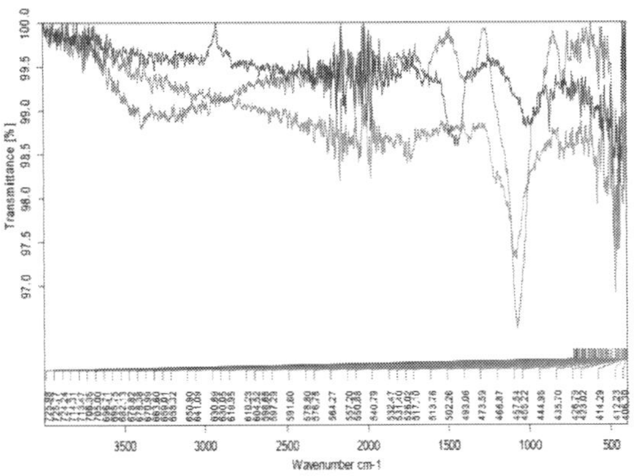

Fig. 6. Spectres IRTF du mortier (0;0;0) dans les différents milieux: Bleu: Référence; Rouge: HNO_3; Rose: H_3PO_4.

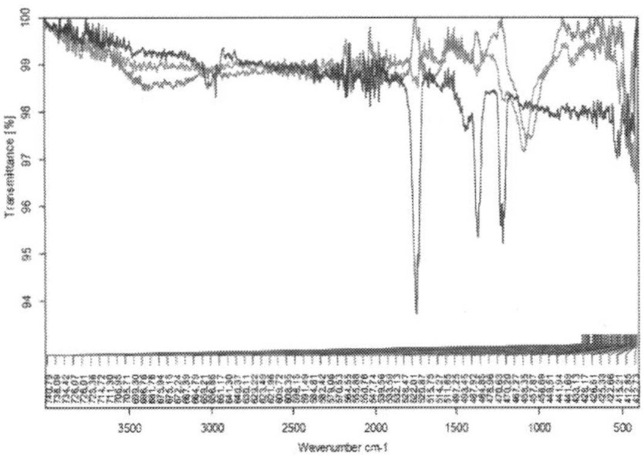

Fig. 7. Superposition des spectres IRTF du composite $Pz_{10}FS_5Pl_2$ dans les différents milieux. Bleu: Référence sans attaque; Rouge: HNO_3; Rose: H_3PO_4.

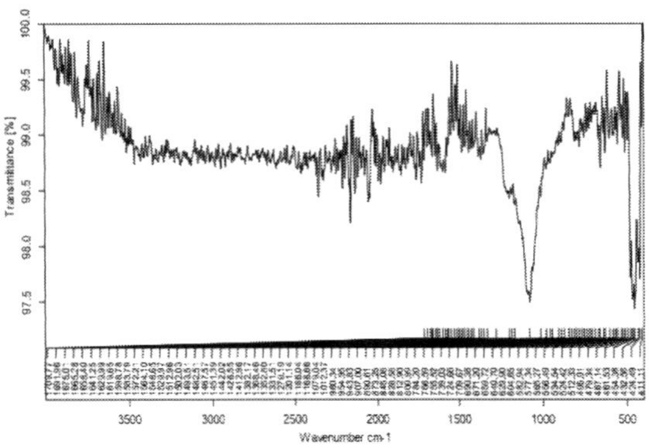

Fig. 8. Spectre IRTF du composite $Pz_{10}FS_5Pl_{0.5}$ dans l'acide H_3PO_4.

Tableau 4. La Table infrarouge à transformée de Fourier des composites avant et après attaque par différents acides.

Positions et Intensités des bandes d'absorption infrarouge (IRTF), 20°±1°C					
(cm^{-1})	**Groupes**	**Composés**	**Référence, HR= 50±5 %**	**HNO_3**	**H_3PO_4**
~ 3000-3600	H_2O, OH, liaison hydrogène	Gypse, CSH	+++	+	+
~ 3638 (Fine)	H_2O, OH, liaison hydrogène	Portlandite, $Ca(OH)_2$	+++	0	0
3435-3452	H_2O, OH	AFm, Aft	++	+	+
1641-1654	H_2O, ν O-H	H_2O	+	+	+
1424-1431	ν C-O	$CaCO_3$ (Calcite)	+	+	+ (F)
872-876	ν C-O	$CaCO_3$ (Calcite)	+	+ (F)	+ (F)
710-714	ν C-O	$CaCO_3$ (Calcite)	+	+	/
~ 1010-1040	ν Si-O	Vibrations de CSH	+ (L)	+ (L)	+ (L)
1030-1080 [23]	ν $Ca(NO_3)_2$	$Ca(NO_3)_2.2H_2O$	0	+	0
1350-1430 [23]	ν $Ca_3(PO_4)_2$	$Ca_3(PO_4)_2.6H_2O$	0	0	++

F: Faible; L: Large; AFt: $C(A,F).3CaSO_4.32H_2O$; AFm: $C_3A.3CaSO_4.12H_2O$.

4. Conclusions

Sur la base de l'étude de la durabilité des mortiers modifiés par l'effet de la combinaison du polymère SPA et des SCMs, on a pu tirer les conclusions suivantes :

- La perte en masse des mortiers non modifiés immergés dans HNO_3 ou H_3PO_4 est supérieure à celles des mortiers composites à base du SPA et des SCMs. La réduction des poids est due à la dissolution des hydrates formés qui mènent à des mortiers poreux.
- On observe que les mortiers FS_5 et $Pz_{10}FS_5$, les composites $Pz_{10}FS_5Pl_{0.5}$ et $Pz_{10}FS_5Pl_2$ ont bien résisté à l'attaque d'acide nitrique. Par contre, on a noté que le composite $Pz_{10}FS_5Pl_{0.5}$ à mieux résister à l'attaque phosphorique.
- L'effet bénéfique de l'ajout du SPA comme addition au mortier modifié sur la durabilité des composites, à savoir $Pz_{10}FS_5Pl_{0.5}$ et $Pz_{10}FS_5Pl_2$. Ceci est dû à l'effet de la combinaison des additions minérales et de l'ajout du SPA. On a la formation du gel CSH de 2ème génération par

réaction pouzzolanique et la formation du film du Latex, respectivement.

- La résistance chimique des mortiers modifiés par l'effet de la combinaison de l'ajout du SPA et des SCMs à faible coût, et à faible impact environnemental, est améliorée vis-à-vis de l'attaque par les différents acides. Les résistances chimiques sont plus ou moins affectées par la concentration et par la nature des acides par ordre d'agressivité. On propose pour cela la séquence suivante : $H_3PO_4 > HNO_3$.
- L'identification des composés formés dans les mortiers composites avant et après attaque acide ont été suivies par spectrométrie IRTF.

Les meilleures propriétés de la durabilité de l'effet de la combinaison du polymère SPA et SCMs (PzN et FS), dans les mortiers modifiés observés dans cette étude, indiquent une durée de vie plus longue de la structure réparée en utilisant ce type de matériaux de réparation verts modifiés par un polymère.

Remerciements - Ce travail a été effectué avec l'apport financier du Ministère de l'Enseignement Supérieur et de la Recherche scientifique de l'Algérie, sous les subventions du projet CNEPRU B00L01UN310120130068. Nos plus vifs remerciements vont aussi à Mr M.T. GOUASMI et Mr O. HASNAOUI Directeur général du Groupe des Sociétés HASNAOUI, La TEKNACHEM Algérie.

Références

1. Y. Ohama, *Hand book of polymer-modified concrete and mortars, properties and process technology*. Noyes Publications, USA, (1995).
2. J. Baron, J.P. Olivier, *Les Bétons Bases et Formulations*. Association de l'industrie des liants hydrauliques, 2ème Edition Eyrolles, (1997).
3. S. Chandra, Y. Ohama, *Polymers in Concrete*, CRC Press, Boca Raton, p 204, (1994).
4. E. Sakai, J. Sugita, Cem. Concr. Res. **25** [1] 127 (1995).
5. Y. Senhadji, G. Escadeillas, H. Khelafi, M. Mouli, A.S. Benosman, Eur. J. Environ. Civ. En. **16**(1), 77 (2012).
6. J. Baron, J.P. Ollivier, *La durabilité des Bétons*. Presses de l'école nationale des ponts et chaussées, Paris, (1992).
7. Y. Senhadji, G. Escadeillas, M. Mouli, H. Khelafi, A.S. Benosman, Powder Technol. **254,** 314 (2014).
8. M. Ghrici, S. Kenai, M. Said-Mansour, E-H. Kadri, J. Asian Archit. Build. Eng. **5**(2), 349 (2006).
9. Y. Benali, F. Ghomari, J. Adhes. Sci. Technol. **31**(3), 219 (2017).
10. B. Belbachir, A.S. Benosman, H. Taïbi, M. Mouli, Y. Senhadji, M. Belbachir. J. Mater. Environ. Sci. JMES, **7**(2), 402 (2016).
11. Shaikh F.U. Ahmed, J. Mater. Civ. Eng. ASCE, **23**(n°9), 1311 (2011).
12. N. Latroch, A.S. Benosman, N. Bouhamou, B. Belbachir, Y. Senhadji, H. Taïbi, M. Mouli, Int. J. Eng. Res. in Africa JERA, **27**, 27 (2016).
13. NF P 15-403. - *Technique des essais. - Sable normal et mortier normal* (homologuée le 31 juillet 1963) (1963).
14. EN 196-1, *Methods of testing cement - Part 1: determination of strength*. European Committee for Standardization, CEN, (2005).
15. ASTM C267-01, *Standard Test Methods for chemical resistance of mortars, grouts, and monolithic surfacing and polymer concretes*, Annual Book of ASTM Standards, (2012).
16. D. Achoura, Ch. Lanos, R. Jauberthie, B. Redjel, J. de Phys. IV France, EDP Sciences, **118**, 159 (2004).
17. A.S. Benosman, M. Mouli, H. Taibi, M. Belbachir, Y. Senhadji, I. Bahlouli, D. Houivet, Environ. Eng. Manage. J., (EEMJ), **16**(7), 1489 (2017).
18. M. Omrane, M. Mouli, A.S. Benosman, Y. Senhadji, Adv. Mater. Res. J., **1064**, 3 (2015).
19. B. Belbachir, A.S. Benosman, H. Taïbi, Key Eng. Mater. KEM, **678**, 123 (2016).
20. J.A. Rossignolo, M.V.C. Agnesini, Cem. Concr. Compos. **26**, 375 (2004).
21. J. Monteny, N. De Belie, E. Vincke, W. Verstraete, L. Taerwe, Cem. Concr. Res. **31,** 1359 (2001).
22. V. Pavlik, S. Uncik, Cem. Concr. Res. **28**(6), 936, (1998).
23. Foil A. Miller, Charles H. Wilkins, *Infrared Spectra and Characteristic Frequencies of Inorganic Ions; Their Use in Qualitative Analysis*. Analytical Chemistry, **24**, No. 8 (1952).

Electrochemical behaviours of Eu(III)/E(II) and Ce(IV)/Ce(III) in H_3PO_4-H_2O media : solvation and complexation reactions

B. Belqat[1, 2] and S. Belcadi[2]

[1] Laboratoire de Chimie Appliquée des Matériaux, Université Mohammed V, Faculté des Sciences, Rabat, Morocco
[2] Laboratoire d'électrochimie et Chimie Analytique, Université Mohammed V, Faculté des Sciences, Rabat, Morocco

Abstract. Many kinds of rare earth elements (REE) such as europium and cerium have been make them essential elements in many high-tech components. The electrochemical studies can be presented as an interesting indication for europium and cerium extraction from phosphoric solutions, including solvation and complexation reactions. The normal redox potentials of Eu^{3+}/Eu^{2+} and Ce^{4+}/Ce^{3+} systems have been determined in H_3PO_4-H_2O media with various phosphoric acid concentration. The solvation of these elements in phosphoric media is characterized by their transfer activity coefficients "f" calculated from the corresponding normal redox potentials. The corresponding solvation increases with increasing the H_3PO_4 concentration. For each REE, the electrochemical properties depend on its number of charges and on its basic properties. Results suggest that solvation and complexation of REE phosphates are important in controlling REE concentration.

1 Introduction

Rare earth elements (REE) are particularly important to emerging technologies in green energy, defense and electronic industries [1-3]. Generally, they are present in sedimentary phosphate deposits, consequently in phosphoric acid. Various phosphoric acid solutions are produced from phosphate ores before fertilizer-grade phosphoric acid using an industrial processing and contain valuable quantities of rare earths elements; hence, it is considered as secondary source for rare-earth elements [4-6]. Only a small proportion of REE deposits are of sufficient size and concentration that they can be exploited economically using existing technology [7]. Rare earth elements exist in phosphorite deposits, which should be extracted using various methods [8-11]. Systematic solvent extraction of rare earths from phosphoric acid by specific reagents aiming to form strong soluble complexes was reported in the literature [12-14]. However, their chemical properties depend not only on their atomic structure, but also on their size and charge.

For the development of a process to extract europium and cerium from phosphoric acid with high current efficiency, it is necessary to understand their electrochemical behavior in produced various phosphoric solutions, which are less developed [15-17]. The acidity function $R_0(H)$ allows to anticipate the thermodynamic constant evolution in phosphoric solutions, involving evening out a concentration gradient and evenly distributing the solute within the water and phosphoric solvents. The improvement of the successive stages in the industrial process to extract europium and cerium from phosphoric solution is possible if the thermodynamic properties such as water activity and solvation properties are known [18-19]. For resolving the solvation properties of Eu^{3+}/Eu^{2+} and Ce^{4+}/Ce^{3+} systems in phosphoric media, the electrochemical study of various concentrations of H_3PO_4 solutions containing the europium and cerium was carried out to determine the reactivity of these species and their solvation in phosphoric acid. This study intends to highlight the most important characteristics of europium and cerium controlling its behavior in H_3PO_4-H_2O media and so closely connected to its electrochemical properties.

2 Theoretical aspects

The solvation transfer coefficients characterize the changes of REE ion reactivity in phosphoric acid-water and consequently help to extract them. Variations of solute solvation are characterized by means of the solvation transfer activity coefficients "f" [20-22].

When f > 1 (or log f > 0) signifies that the REE ion is less solvated in H_3PO_4-H_2O than in pure water; but when f < 1 (or log f < 0) the REE specie is more solvated in the phosphoric solution. The coefficients "f" are calculated from standard potential values of the corresponding redox systems according to the equations (1) and (2):

$$M^{n+}/M: \quad \log f\left(M^{n+}\right) = \frac{n\,(E_o^A - E_o^W)}{0.058} \qquad (1)$$

$M^{n+}/M^{(n-m)+}$: $$\log f\left(M^{n+}/M^{(n-m)+}\right)=\frac{m\,(E_0^A-E_0^W)}{0.058} \quad (2)$$

Where: E_0^w represents the normal potential of the considered redox system in water, E_0^A is its value in a phosphoric medium characterized by the acidity function $R_0(H)$ and the water activity a_{H2O} [22-23], M the metal, Mn^{n+} the metal cation and n the number of electrons exchanged in the redox reaction. Both are referred to the Ferrocene (Fc^+/Fc) as reference system.

3 Experimental

Different H_3PO_4-H_2O solutions are prepared using phosphoric acid (85%) (Riedel De Haen, pure for Analysis). Europium and Cerium ions are introduced in each phosphoric solution by the dissolution of their pure Eu_2O_3 and Ce_2O_3 oxides. Ce^{4+}/Ce^{3+} and Eu^{3+}/Eu^{2+} ions are studied by cyclic voltammetric (potentiostat galvanostat Tacussel type PGS201T) and polarographic method (polaroprocesseur type Pol 150 Polarographic Analyser). The dough of carbon electrode allows covering a large potential area until 2.6 V from -1,0V to 1,6V for large H_3PO_4 concentrations.

To avoid chloride diffusion between the reference electrode and the solution cell, the saturated calomel electrode was connected to the work cell by a capillary, which contained 0.1 M H_3PO_4. During experiments, bubbling of nitrogen was maintained through the solution to avoid the presence of dissolved oxygen. The manufacturing of dough of carbon electrode is made as follows: a known amount of the studied compound was mixed with the graphite powder and the ligand conductor. A mixture of 0.4 - 1.0 mg as electroactif compound, 50 mg of carbon and 40 mg of electrolyte is compressed into glass tube, in which the bottom of vitreous carbon insures the electrical contact. The whole is covered with a fine silica layer or a sintered glass to avoid the dispersion of the dough.

The normal potentials of Eu^{3+}/Eu^{2+}, Ce^{4+}/Ce^{3+} and Fc^+/Fc systems were measured at 25°C by reference to the SCE electrode in a separate compartment containing 0.1M H_3PO_4, where the redox potentials were reported versus Fc^+/Fc electrode at 25°C.

The normal potential values of Eu^{3+}/Eu^{2+} and Ce^{4+}/Ce^{3+} systems were determined directly by using a cyclic voltammetry and polarographic methods respectively as a function of phosphoric concentration (from 0.1 to 14.8 M). The values of acidity function $R_0(H)$ are calculated on the basis of measurements of the hydrogen electrode potential versus the Fc^+/Fc system.

4 Results

Oxidation and reduction processes play a main role in Eu and Ce rare earth elements chemistry requiring many chemical reactions. Their redox potentials can be used to predict the ionic distribution into solution, which may interact with the transfer of electrons. Normal and half-wave potentials of the both redox couples are considered equal, while the diffusion coefficients of cations are similar in the solution and amalgam accompanied with rapid reactions with the transfer number equals to 1. Figure 1 shows a typical cyclic voltammograms for the reversible oxidation and reduction of Ce(IV)/Ce(III) system using gold and carbon electrodes.

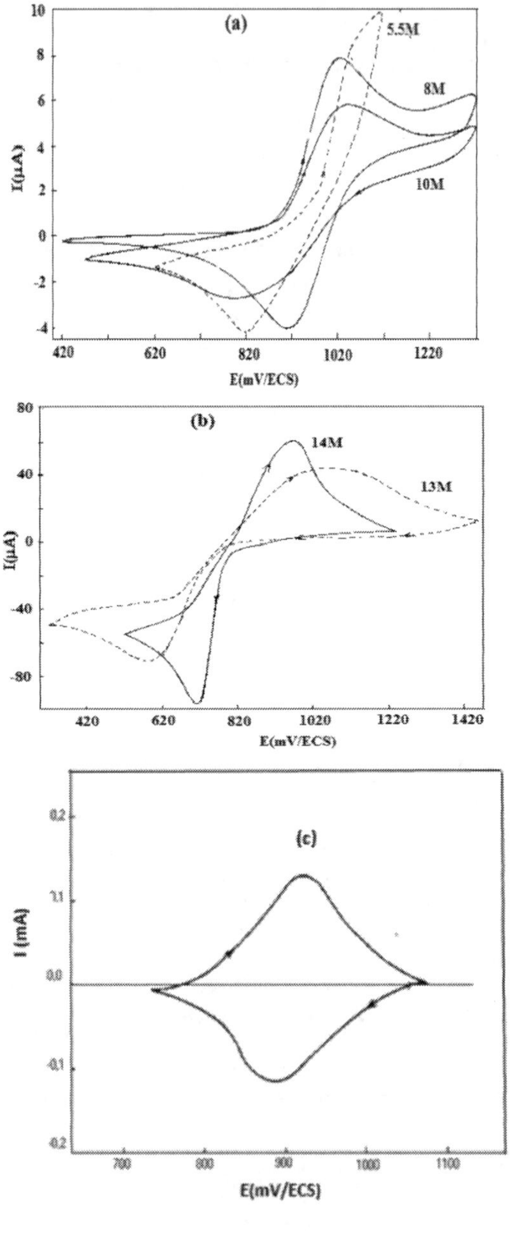

https://doi.org/10.1051/matecconf/201814901092

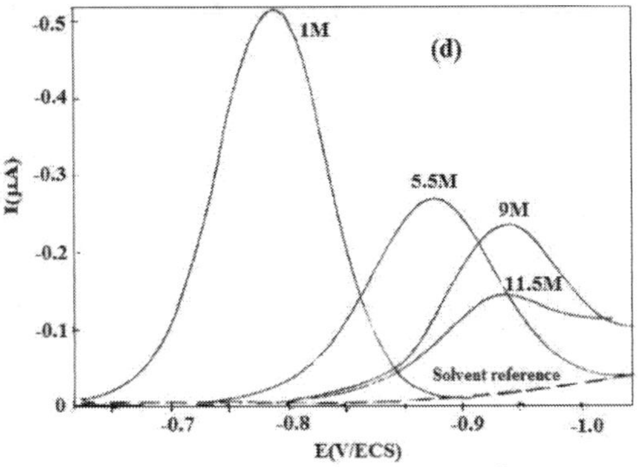

Fig.1. A typical cyclic voltammograms of current verses potential for Ce(IV)/Ce(III) system in (a) medium H_3PO_4 concentration (gold electrode, $[Ce^{3+}]=5.10^{-3}M$, V=20mV/s, initial potential ≈790 mV), (b) high H_3PO_4 concentration (gold electrode, $[Ce^{3+}]=5.10^{-3}M$, V=20mV/s, initial potential ≈790 mV), (c) 8M H_3PO_4 with carbon paste electrode, V=0,5 mV/s and the initial potential of 1050 mV and (d) Pulse polarograms for Eu(III)/Eu(II) system in H_3PO_4 media ($C_{Eu(III)} = 10^{-3}$ M and V = 10 mV/s).

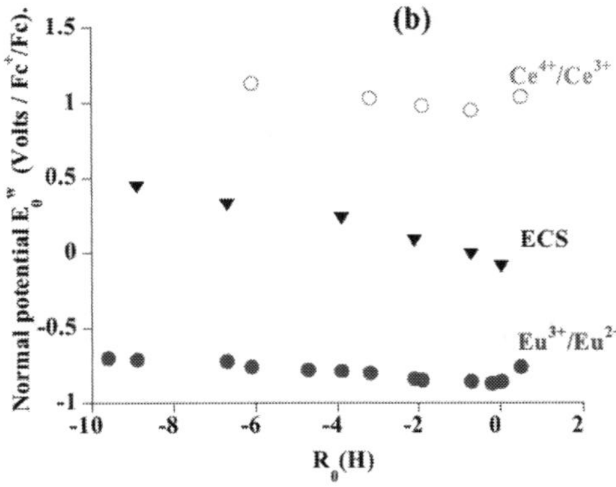

Fig. 2. Normal potential of different redox systems in phosphoric media

The change in electrochemical behavior of Eu(III)/Eu(II) and Ce(IV)/Ce(III) in the function of H_3PO_4 concentration or with $R_0(H)$ are plotted in figure 2. The potential of each reaction depends upon the relative phosphoric concentration and of oxidized/reduced forms. A small change of the normal potential of Eu^{3+}/Eu^{2+} and Ce^{4+}/Ce^{3+} redox systems is observed with the phosphoric acid concentration and a significant variation between the both redox systems, showing a variation in acidity and solvation of REE species. Two factors intervening antagonistic manner: the water activity and "REE-phosphate" complexation, which the last approach becomes significant only in the most phosphoric concentration.

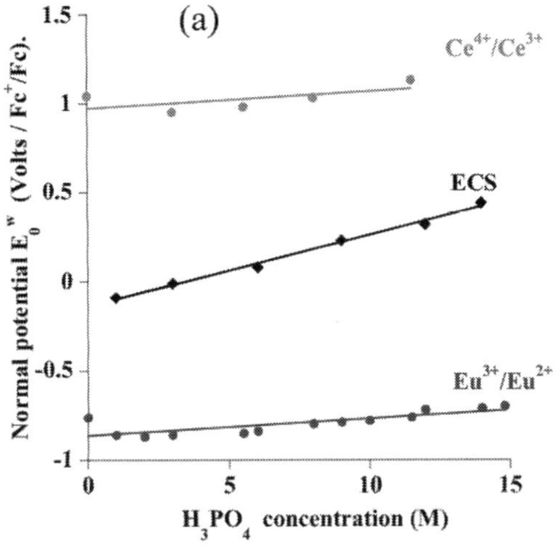

LOUIS and al. [24] have studied the relationship between complexation and solvation of some metal elements in various mineral acids (hydrochloric, perchloric, sulfuric) using acidity function $R_0(H)$ and solvation-transfer activity coefficients. In addition, the electrochemical behaviour of Fe(II) and Fe(III) ions on a gold electrode in H_3PO_4-H_2O solutions has already studied in our previous work [25]. In current study, the transfer activity coefficients "f" were calculated from the normal potential values of redox systems (Figure 2). The thermodynamic data lead to the calculation of the coefficients of solvation transfer activity for Eu^{3+}/Eu^{2+} and Ce^{4+}/Ce^{3+} systems (Figure 3).

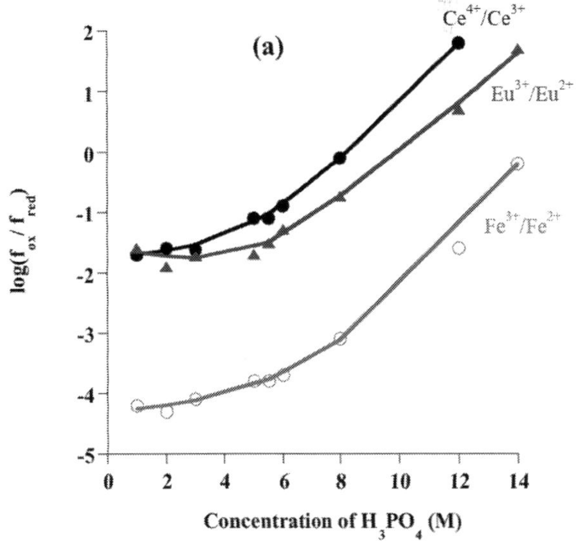

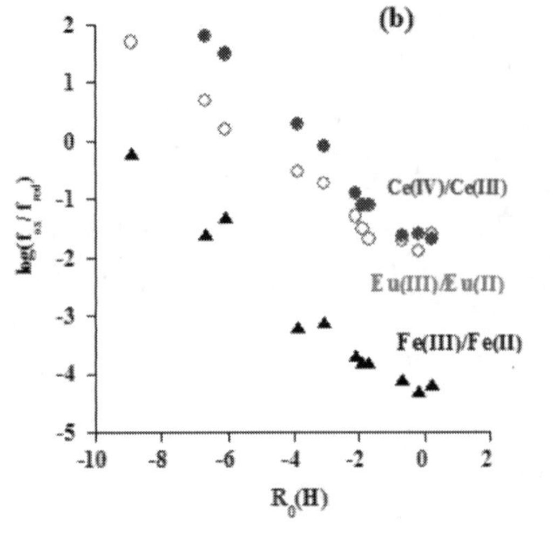

Fig.3. Solvation characteristics $\log(f_{ox}/f_{red})$ of species in H_2O-H_3PO_4 media in a function of (a) H_3PO_4 concentrations and (b) $R_0(H)$.

From figure 2, electrochemical characteristics of Eu^{3+}/Eu^{2+} and Ce^{4+}/Ce^{3+} show a comparable curves shape to that of Fe^{3+}/Fe^{2+} but having a dissimilar solvation character (log f). The behaviour of these cations in H_3PO_4-H_2O media are studied with increasing of the H_3PO_4 concentration (log f). Thus, they are less solvated and more reactive in the high concentration of phosphoric acid. Therefore, $\log(f_{Eu}{}^{3+}/f_{Eu}{}^{2+})$ and $\log(f_{Ce}{}^{4+}/f_{Ce}{}^{3+})$ values are negative, except for those in high concentrated H_3PO_4 acid [>11M], noting that the decomplexation of Eu^{3+} ions by phosphate groups is more important than that of Eu^{2+} ions. The increase of H_3PO_4 concentration favours a decomplexation of cations by a decrease of their solvation and consequently accelerates their reactivity. In 1 M H_3PO_4, the $\log(f_{Eu}{}^{3+}/f_{Eu}{}^{2+})$ =-1.64, as $\log(f_{Ce}{}^{4+}/f_{Ce}{}^{3+})$ = -1.63 expresses that the Eu^{2+} and Ce^{3+} cations are more solvated by the water while Eu^{3+} ions and Ce^{4+} are complexed by phosphate species ($H_2PO_4^-$, HPO_4^{2-}, PO_4^{3-}). This association becomes less at least relevant when the phosphoric concentration increases. When the H_3PO_4 concentration increases until 11 M ($R_0(H)$ = -5.5), the REE-phosphate interactions diminish rapidly and proved by the increase of the normal potential of their redox systems. In addition, the power oxidizing of the Eu^{3+} ions is comparable in water using 11 M H_3PO_4, e.g $\log(f_{red}/f_{ox}) > 0$, indicating that Eu^{3+} and Ce^{4+} cations are less solvated than Eu^{2+} and Ce^{3+} ions respectively, and so their reactivity increases. Similar results found for power oxidizing of Ce^{4+} ions in 8 M H_3PO_4 medium ($\log(f_{Ce}{}^{4+}/f_{Ce}{}^{3+})$ = -0,02). In diluted phosphoric acid, the Ce^{4+} ions {pK ($3Ce^{4+}/4PO_4^{3-}$) = 90.1} are more complexed than Ce^{3+} ions {pK (Ce^{3+}/PO_4^{3-}) = 2}. The analogous phenomenon has been observed for Fe^{3+}/Fe^{2+} and V^{3+}/V^{2+} systems in phosphoric acid media, but Eu^{3+} ions are more complexed {pK(Eu^{3+}) = 8.03} than Fe^{3+} {pK(Fe^{3+}) = 2.2} [26]. This behavior indicates the complexation of Eu^{3+} by means of OH^- groups, which are increasingly formed from the dissociation of water. Bilal and Langer [27] have shown that the formation of hydrolysed species of the lanthanides occurs in solutions having pH = 3.6 at T > 333 °K, even if these solutions contain complexing ligands. They thus propose, the following unique set of hydrolysis constants for the mentioned trivalent ions [28]:

$$M^{3+} + n\,H_2O \rightarrow M(OH)_n{}^{3-n} + n\,H^+ . \qquad (3)$$

The inorganic speciation of M^{3+} cation (M: Eu and Ce) in H_2O-H_3PO_4 media were evaluated primarily to assess the relative importance chelating phosphate {$MH_2PO_4^{2+}$, $MHPO_4^+$, $M(HPO_4)_2^-$ and MPO_4^0}. In comparing the chemical characteristics of Eu^{3+} and Ce^{4+} cations, the desolvation of Ce^{4+} ions {$pK_sCe(OH)_4$ = 50.5} is greater than the Eu^{3+} ions {$pK_sEu(OH)_3$) = 24.5}. However, the existence of the formation of insoluble phosphate containing REE indicates that REE cations are more complexed with phosphates than hydroxide anions. By comparison of the complexation constants of Ce(III) in different acidic media (H_2O-HF, H_2O-HCl, H_2O-H_2SO_4), Ce(III) forms of more stable complexes in H_3PO_4 {pK(Ce^{3+}/PO_4^{3-}} = 23) than in other acids such as in H_2SO_4 {pK(Ce^{3+}/SO_4^{2-}} = 3.6), in HF {pK(Ce^{3+}/F^-) = 4} and in HCl {pK(Ce^{3+}/Cl^-) = 0.5} [29]. Therefore, Ce^{4+} cations form a more stable complexes in H_3PO_4 {pK($3Ce^{4+}/4PO_4^{3-}$) = 90.1} than in H_2SO_4 {pK(Ce^{4+}/SO_4^{2-}) = 4.6} . It has been concluded that the power oxidizing depends therefore of the nature of chelating anions.

Consequently, REE-phosphate complexes tend to dominate the speciation of the REEs in H_3PO_4-H_2O media. Our results indicate that Eu and Ce extraction from the produced phosphoric acid is important in diluted H_3PO_4 concentration. More specifically, elevated phosphate concentration tends to enhance the REE complexation through the formation of stable REE-phosphate complexes and consequently to remove the REE elements from water by the formation of REE-phosphate precipitate.

5 Conclusion

The electrochemical properties of Eu^{3+}/Eu^{2+} and Ce^{4+}/Ce^{3+} are studied in various phosphoric acid concentrations, showing the hydrolysis and the complexation behaviours of studied rare earth elements. These systems are characterized by an equal $R_0(H)$ acidity level and different redox and complexing properties. The relationship between the $R_0(H)$ acidity function, solvation properties and the water activity in phosphoric acid concentrated solutions has demonstrated. These confirm that it is possible to prepare a specific phosphoric acid concentration for extracting europium and cerium metals using a systematic solvent extraction and different classes of cation exchangers.

Acknowledgements

We express our thanks at Professor J. Bessière at Laboratory of Chemistry and Analytic electrochemistry,

Ability of Sciences, University of Nancy, France, for valuable scientific discussions

References

1. F. Xie, T. A. Zhang, D. Dreisinger, F. Doyle, Minerals Engineering, **56**, 18 (2014)
2. X. Huang, Z. Long, H. Li, W Ying, G.Zhang, X. Xue, Journal of Rare Earth, **23**, 4 (2005)
3. A. Jordens, Y. P. Cheng, K.E Waters, Minerals Engineering, **41**, 17 (2013)
4. D. N. Trifonov, *Terres rares* (Ed. Scientifique, Bucarest, 1963)
5. A. Leveque, J. Helgorski, *Techniques de l'ingénieur, Terres rares* J 6020-2353-1.
6. B. Marmodée, J.S. de Klerk, F. Ariese, C. Gooijer, M.U. Kumke, Anal. Chim. Acta, **652** (2009)
7. P. Moreau, S. Colette-Maatouk, P. Vitorge, P. Gareil, P. E. Reiller, Inorganica Chimica Acta, **1**, 7 (2015)
8. J. S. Wu, C. Liao, C Yan, Journal of Rare Earths, **5**, 8 (2004)
9. M. J .Hudson, Hydrometallurgy, **9**, 20 (1982)
10. J. Lu, Z. Wei, D. Li, G. Ma, Z. Jiang, Hydrometallurgy, **50**, 11 (1998)
11. J. S. Preston, A. C. Preez, Journal of Chemical Technology and Biotechnology, **65**, 9 (1996)
12. J. S. Preston, P. M. Cole Preez, A. C. M.H Fox, A. M. Fleming, Part II. Hydrometallurgy, **41**, 24 (1996)
13. D.F. Peppard, G. W. Wason, Kleber, E.V. (Ed.), *Rare Earth Research*. The Macmillan Company, New York, **14** (1961)
14. A. Barkleit, M. Acker, G. Bernhard, Inorganica Chimica Acta, **394**, 6 (2013)
15. C. Louis, J. Bessière. J. Chem. Eng. Data, **31**, 3 (1986)
16. M. Wang, Chem. Abstr, **116** (1989)
17. A. Rollar, Procedia Engineering, **138**, 8 (2016)
18. N. Tahani, Thesis, University of Nançy I, France, (1991)
19. P. A. H. Wyatt, Faraday Discuss. Chem. Soc, **24**, 9 (1957)
20. Z. El Abbassi, S. Belcadi, J. J. Rameau, Electrochimica Acta , **31**, 6 (1986)
21. C. Louis, J. Bessière. Can. J. Chem. **64** (1986)
22. D. Babale. Thesis, University of Yaounde, Cameroun, (1987)
23. F. Coussemant, M. Hellin, B.Torch, *Gordon and Breach* : Paris, London, New York, (1969)
24. C. Louis, A. Bebbaa, ND J. Bessière, Can. J. Chem., **66** (1988)
25. B. Belqat, A. Laghzizil, K. El Kacemi, S. Belcadi. Journal of Fluorine Chemistry, **105**, 5 (2000)
26. B. A. Bilal, Zeitschrift für Naturforschung A, **46**, 9 (1991)
27. B. A. Bilal, P. Langer, Lanthanide and Actinide Research, **3** (1989)
28. V. Neck, J.I. Kim, Radiochim. Acta, **89**, 17 (2001)
29. B. A. Bilal, E. Muller, Zeitschrift für Naturforschung A, **47**, 11 (1992)

Study of the resistance, porosity, and migration of chloride ions of BAP-based crushing sand

Etude de la résistance, la porosité, et migration des ions chlorures des BAP à base de sable de concassage

Benyamina Smain[1,2,3], Siham Kamali-Bernard[2], Kenai Said[3], Menadi Belkacem[3]
[1] Département de Science de la matière, Université de Djillali Bounaama, Kemis Miliana, Algérie
[2] Laboratoire de Génie civil et Génie Mécanique, Institut des sciences Appliquées, Rennes, 35708, France
[3] Laboratoire de Matériaux et de génie civil, Département de génie civil, Université de Blida 1, Algérie

Abstract. Self-compacting concretes (SCC), are hyper-fluid concretes, placed without vibration and are considered as one of the most important innovations of the last decade in construction. SCCs offer many advantages, due to their exceptional characteristics of flow and filling of formwork. Their compositions require a large quantity of fines in order to limit bleeding and segregation. Hence, the use of crushed sand (SC), rich in limestone fines (CF) in the manufacture of self-placing concretes (SCC), can be considered as an alternative source of fillers. These sands reduce the cost of SCC by reducing the high demand for fillers on the one hand and on the other hand, obtaining SCC with good physical and mechanical properties. The main purpose of this paper is to examine the effect of different percentages (0, 5, 10, 15, and 20%) of (CF) in crushed sand on SCC performance. The evolution of the compressive strength, the porosity accessible to water and the migration coefficient of the chloride ions were evaluated. The Okamura method was used for the formulation of all SCC mixtures. Sand/mortar (S/M), water/cement (W / C) ratios and superplasticizer content were kept constant. The results show that (CF) reduce the compressive strength but contribute to the reduction of porosity and migration of chloride ions.

Key word: SCC, crushed sand, limestone, compressive strengths, porosity, migration of chloride ions.

Résumé. Les bétons autoplaçants (BAP), sont des bétons hyper fluides, mis en œuvre sans vibration, constituant l'une des plus importantes innovations de la dernière décennie en matière de construction. Ces bétons offrent de multiples avantages, liés à leurs caractéristiques exceptionnelles d'écoulement et de remplissage des coffrages. Leurs compositions nécessitent une quantité importante de fines afin de limiter le ressuage et la ségrégation. L'utilisation des sables de concassage (SC), riche en fines calcaires (FC) dans la fabrication des bétons autoplaçants (BAP), peut être considérer comme une source alternative de fillers. Ces sables diminuent le coût des BAP par réduction de la demande élevée en fillers d'une part, et d'autre part l'obtention des BAP avec des bonnes propriétés physiques et mécaniques. Le but principal de cette étude, est d'examiner l'effet de différents pourcentages (0, 5, 10, 15, et 20%) de (FC) dans le sable de concassage sur les performances des BAP. L'évolution de la résistance à la compression, la porosité accessible à l'eau et le coefficient de migration des ions de chlorures ont été évalués. La méthode d'Okamura a été utilisée pour la formulation de toutes les mélanges des BAP. Les rapports sable / mortier (S/M), eau / ciment (E/C) et la teneur en superplastifiant ont été maintenus constants. Les résultats montrent, que les fines accélèrent les résistances à la compression aux premiers âges, et contribuent à la diminution de la porosité et la migration des ions de chlorures.

Mots clés : BAP, sable de concassage ; fines calcaires, résistance, porosité, migration des chlorures.

1 Introduction

Les bétons autoplaçants (BAP), sont des bétons extrêmement fluides stables et homogènes, et qu'ils se mettent en place sous l'effet de leur poids propre sans aucun apport d'énergie. Ils peuvent être coulés dans des espaces trop ferraillés et de formes très complexes. Leurs compositions nécessitent certaines spécificités dont un volume de pâte élevé ce qui augmente leur coût. Donc l'utilisation des fines issues de concassage des sables présente un enjeu important pour des raisons à la fois économiques et environnementales.

Les travaux de recherches menés sur des mortiers ou bétons montrent l'effet bénéfique des fines sur leurs comportements [1-4]. Cependant ces études montrent que les interactions multiples et imprévisibles qui peuvent exister entre les fillers et les autres constituants peuvent entraîner des comportements inattendus.

Cette étude contribue à l'examen des performances des BAP, formulés à partir des sables de concassage contenant différentes teneurs en fines. On se basant sur la méthode d'Okamura [5], cinq mélanges de BAP ont été confectionnés dont les rapports E/C=0.4, S/M=0.5, Sp=0.8, soient fixe. Le sable (SC) a été tamisé (80µm) puis substitué par (0, 5, 10, 15 et 20%) de fines calcaire en masse. L'effet de la teneur en fines dans le sable (SC) sur le développement de la résistance à la compression, la porosité accessible à l'eau, la migration des ions de chlorures pour chaque mélange de BAP est évalué.

2 Matériaux et essais

2.1 Matériaux utilisés

Deux sables ont été utilisés, un sable grossier concassé calcaire (SC), et un sable fin siliceux (SS), d'une densité de 2.63, 2.54 et un coefficient d'absorption d'eau de 0.38 et 0.2%, respectivement. Les fines calcaires (FC) sont issues après passage du sable de concassage (SC) dans un tamis de 80µm. Les courbes granulométriques des deux sables utilisés sont présentées sur la figure 1

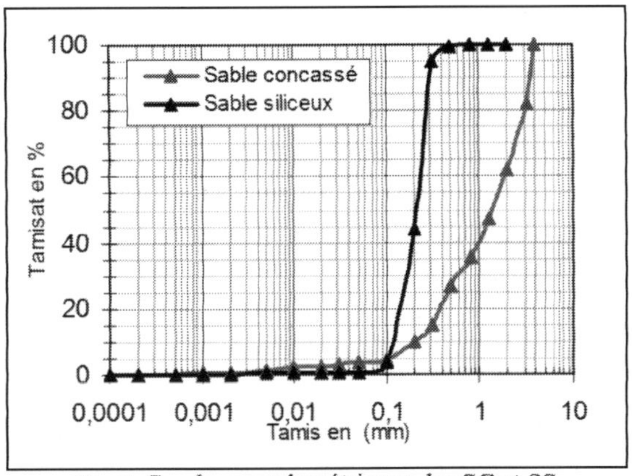

Fig. 1. Courbe granulométriques des SC et SS

L'analyse chimique du sable concassé et les fines calcaires est donné dans le tableau 1. Deux types de graviers ont été employés (G3/8 et G8/15), d'une densité apparente de 2.62 et 2.63 et d'un coefficient d'absorption de 0,20 et 0,30%, respectivement.
Pour la formulation des différents mélanges des BAP, un seul ciment CEM I 42,5, d'une finesse de 3000 cm^2/g et une densité de 3,15 est utilisé.

Tableau 1. Propriétés chimiques des SC et leur FC

(%) des Oxydes	Sable concassé (SC)	Fines calcaires (FC)
SiO2	4.51	1.78
Al2O3	1.09	0.79
Fe2O3	0.54	0.34
CaO	50.68	54.3
MgO	1.01	0.20
K2O	-----	-----
Na2O	-----	-----
H2O	-----	2.8
SO3	traces	-----
Perte au feu	41.04	42.5

Un superplastifiant (MED Flow 145) à base de polycarboxylate avec une densité de 1.05 ± 0.02 g/cm^3 est également utilisé pour toute la formulation des BAP.

2.2 Essais réalisés

La formulation des BAP est réalisée selon la méthode d'Okamura [5]. En total cinq mélanges ont été préparés. Les quantités et dosage pour chaque mélange sont donnés dans le tableau 2.

Tableau 2. Dosage et poids des constituants des BAP

E/C = 0.4, S/M = 0.5 G/S = 0,8, Sp =0,8					
Poids (kg)	BAP0	BAP5	BAP10	BAP15	BAP20
Gravier 8/15	295	295	295	295	295
Gravier 3/8	443	443	443	443	443
Sable (SC)	600	570	540	510	480
Sable (SS)	323	323	323	323	323
FC	0	30	60	90	120
Ciment	490	460	430	400	370
E	196	196	196	196	196
SP	6,86	6,86	6,86	6,86	6,86

Les essais de résistance en compression ont été réalisés sur des éprouvettes cubiques 100×100×100 mm^3. Les éprouvettes testées ont été conservées dans l'eau dès leur démoulage jusqu'à l'échéance concerné (Fig.2).

https://doi.org/10.1051/matecconf/201814901093

Fig.2. Conservation des éprouvettes

Les mesures de porosité accessible à l'eau ont été effectués conformément à la norme NF EN 18-459 [6] qui préconise d'abord une saturation sous vide des échantillons suivi par pesée hydrostatique puis un séchage à 105°C. Chaque essai a été réalisé sur deux prélèvements de 50 mm d'épaisseur découpé au centre d'éprouvettes cylindriques 100x200 mm.

Des essais accélérés de migration des ions chlorures par champ électrique ont été réalisés. La tension appliquée aux bornes des échantillons est de 30 V. Les cellules d'essai utilisées sont illustrées dans la figure 3. Le coefficient de diffusion effectif De (m^2/s) est déterminé selon la norme XP P 18-462 [7].

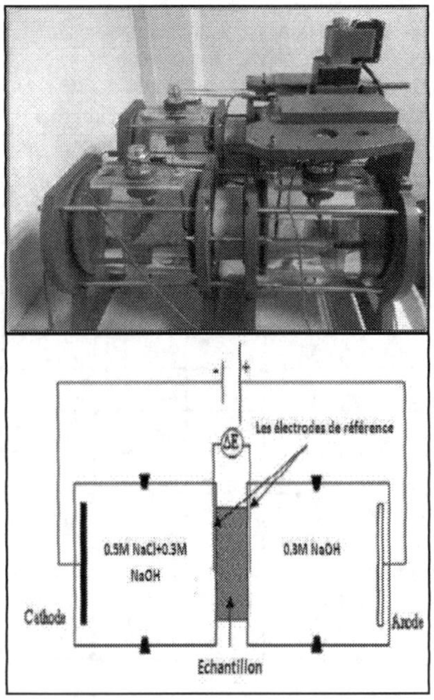

Fig. 3. Essai de migration des ions de chlorures

3 Résultats

3.1. Résistance à la compression

L'évolution de la résistance en compression à l'âge de 28 jours en fonction de la teneur des fines calcaires est donnée en figure 4. Les résultats montrent que pour tous les BAP étudiés, la résistance à la compression diminue avec l'augmentation du taux de fines dans le sable de concassage. Une résistance à la compression de 71,8 MPa est obtenue pour un BAP de 0% de fines calcaires, puis une chute linéaire jusqu'au 65 MPa pour un BAP de 20% de fines. Cette réduction peut être attribuée à l'effet de dilution des fines calcaires [1, 8].

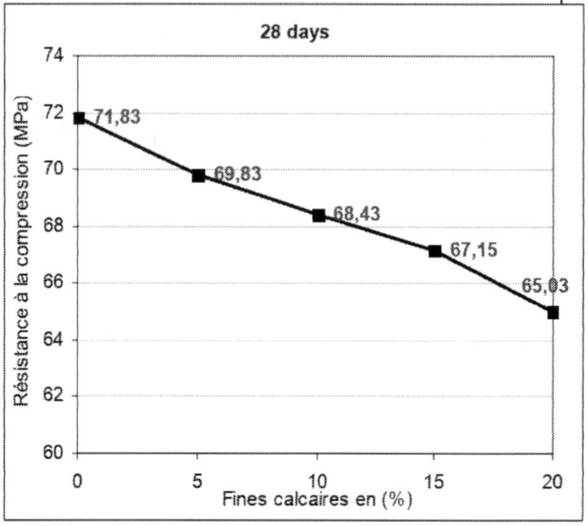

Fig. 4. Evolution de la résistance à la compression en fonction du taux de fines

3.2 La porosité accessible à l'eau

D'après la figure 5, les valeurs maximales des absorptions d'eau sont constatées pour le béton autoplaçant à base de sable de concassage de 0% de fines calcaires, avec une porosité de 15.22 %.

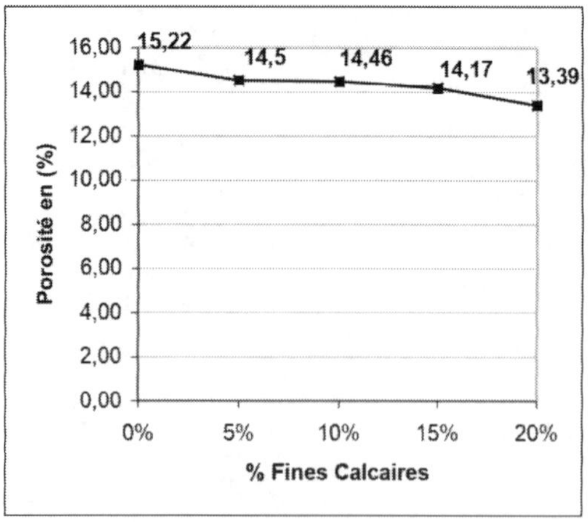

Fig. 5. Essai de migration des ions de chlorures

Cette porosité diminue avec l'augmentation des taux de fines. A titre d'exemple, avec un BAP de 20% de fines, on remarque une réduction de la porosité de 12%. Ce comportement est expliqué par l'influence des fines sur la compacité des BAP et par conséquent sur la porosité.

3.3 Migration des ions chlorures

L'évolution de la concentration en chlorures dans les différents mélanges des BAP en fonction du pourcentage des fines dans le sable concassé est illustrée dans la figure 6.

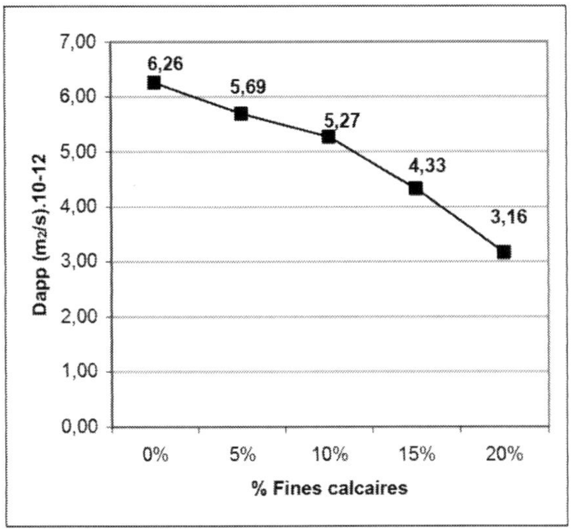

Fig. 6. Essai de migration des ions de chlorures

Les résultats montrent l'effet positif des fines calcaires sur la diminution de migration des ions chlorures dans les BAP. Le coefficient de diffusion obtenu pour le BAP de 0% de fines est égal à égale à $6.26x10^{-12}$ m²/s alors qu'il est égal à $3.16x10^{-12}$ m²/s pour le BAP de 20% de fines. Cette diminution est presque linéaire pour tous les mélanges de BAP. Selon cette étude il est remarqué qu'une augmentation du taux de fines de 0 à 20% en masse dans le sable concassé provoque une diminution du coefficient de migration des chlorures de $3.12x10^{-12}$ m²/s.

4 Conclusion

Selon cette étude qui est portée sur l'influence des sables de concassage notamment les fines calcaires sur les performance des BAP en terme de résistances mécaniques et indicateurs de durabilité à savoir porosité accessible à l'eau et migration des ions de chlorures on peut conclure ce qui suit :

- La substitution d'une partie de sable de concassage par leurs filler calcaires diminue la résistance à la compression. Une réduction de 6.8 MPa est notée pour un BAP avec un remplacement de 20% de fines calcaires en comparaison avec le BAP de référence. L'étude de la porosité accessible à l'eau des différents mélanges des BAP à base des fines issues du sable de concassage montre l'effet positif des fillers calcaires sur la diminution de la porosité. Un taux de 20% de fines dans le sable peut diminuer la porosité de 1.8%.
- Un sable de concassage contenant 20% de fines permet de réduire le coefficient de migration des ions de chlorures dans les BAP.

D'une manière générale, cette étude portée sur les bétons autoplaçants à base de sable de concassage a mis en évidence que ces derniers possèdent des caractéristiques acceptables en termes de résistance et de durabilité.

References

1. B. Menadi, S. Kenai, J. Khatib , A. Ait-Mokhtar, Strength and durability of concrete incorporating crushed limestone sand, Construction and Building Materials 23 (2009):625–633.
2. B._Felekoglu. Utilisation of high volumes of limestone quarry wastes in concrete industry (self-compacting concrete case). Resour. Conserv. Recycling 2007:1–22.
3. E. Meziane, K Ezziane, S Kenai, A Kadri. Mechanical, hydration, and durability modifications provided to mortar made with crushed sand and blended cements, Journal of Adhesion Science and Technology, 29, Issue 18(2015):1987-2005.
4. Y. Benachour, C.A, Davy., F. Skoczylas,_ H. Houari, Effect of a high calcite filler addition upon microstructural, mechanical, shrinkage and transport properties of a mortar. Cement and Concrete Research 38; (2008): 727–736.
5. H. Okamuara, K. Maekawa, K. Ozawa, High performance concrete. 1st ed. Tokyo: Gihoudou Pub.; 1993.
6. NF P18-459. Essai pour béton durci - Essai de porosité et de masse volumique, Mars 2010.
7. XP P18-462.Essai sur béton durci - Essai accéléré de migration des ions chlorure en régime non-stationnaire - Détermination du coefficient de diffusion apparent des ions chlorure. Juin 2012.
8. T. Celik , M. Khaled ,_Effects of crushed stone dust on some properties of concrete, Cement and Concrete Research, Vol. 26, No. 7, pp. 1121-1130, 1996.

Interfacial stresses analysis of FGM porous structures bonded with FRP plates
Analyse des contraintes d'interface des structures poreuse en matériau FGM renforcées par des plaques FRP

A. CHEDAD[1], N. ELMEICHE[12]
[1]Université Mustapha stambouli, département de génie civil, Mascara, Algérie.
[2]Laboratoire de génie civil et environnement, LGCE, Sidi Bel Abbes, Algérie.

Abstract. This paper presents a method for determining the interfacial stresses in an adhesive joint between a strengthening plate and a functionally graded beam. The beam is assumed to be isotropic with a constant poisson's ratio. The present method is based on the deformation compatibility principle taking into account the neutral axis position of the FG beam. A power law for the distribution of the mechanical characteristic with a uniform and non-uniform porosity distribution is presented. A parametric study is presented to show the effect of some parameters on the distribution of stresses in the adhesive joint between the plat and the beam.
Keywords. Interfacial stresses, FGM beam, porosity.

Résumé. Ce travail présente une méthode pour déterminer les contraintes d'interface au niveau de la couche d'adhésive entre une plaque en FRP et une poutre FGM, la poutre est considérée isotropique avec un coefficient de poisson constant. La présente méthode pour modéliser les contraintes d'interface est basée sur le principe de compatibilité des déformations en tenant compte de la position de l'axe neutre de la poutre FGM, la loi qui décrit la distribution des caractéristiques mécaniques de la poutre en considérant une distribution uniforme et non uniforme de la porosité est présentée. Une étude paramétrique est présentée afin d'examiner l'effet des différents paramètres régissant la distribution des contraintes d'interface au niveau de l'interface poutre-composite.
Mots clés. Contraintes d'interface, poutre FGM, porosité.

1 Introduction

La présence de dégradations dans un élément est liée à la perte des caractéristiques mécanique, et pour assurer l'intégrité de la structure les éléments endommagés doivent être remplacé ou réparé, cependant dans plusieurs cas le remplacement de ces éléments est impossible ou couteux, donc le renforcement extérieur des éléments par des plaques semble être la solution la plus efficace pour prolonger la durée de vie de la structure dans ces cas.

Le comportement de l'interface entre une poutre et la plaque FRP peut être influencé par plusieurs facteurs comme les propriétés et la géométrie de la poutre, du FRP et de la couche adhésive. Le rôle de L'interface est le transfert des contraintes de la poutre à la plaque FRP, par conséquent, une compréhension complète sur l'état de contrainte et le mécanisme de transfert de contrainte à l'interface sont nécessaires pour la conception et la mise en œuvre des structures renforcées.

Les contraintes d'interfaces des poutres renforcées ont été étudiées par des méthodes expérimentales et théoriques. Les essais expérimentaux ont été appliqués pour examiner les contraintes d'interface [1-2]. Cependant, l'essai expérimental des champs de contrainte d'interface semble être difficile en raison de la distribution compliquée des contraintes locales.

Plusieurs solutions analytiques ont été développées pendant ces dernières années pour la détermination et la

 https://doi.org/10.1051/matecconf/201814901094

modélisation des contraintes d'interface dans les poutres renforcées par des plaques métalliques ou en matériaux composites. Vilnay[3], Taljsten[4], Malek [5], Smith et teng[6], Tounsi [7] ont considéré la compatibilité de déformation pour déterminer les contraintes d'interface, les contraintes d'interface sont en fonction des déplacements longitudinaux et verticaux de la partie inférieure de la poutre et la partie supérieure de la plaque. Lau et al [8] ont introduit un modèle théorique pour estimer les contraintes d'interface pour différentes orientations des fibres, smith et teng[6] ont présenté une solution applicable pour différents type de chargement et qui semble être précise. Shen et al [9] ont proposé une approche énergétique complémentaire pour la détermination de ces contraintes. Tounsi [7] a développé une nouvelle solution théorique pour la détermination des contraintes d'interface dans une poutre en béton renforcée par une plaque de composite en prenant en compte les déformations des adhérents dues aux cisaillements.

Le présent travail a pour but de développer une procédure pour prédire les contraintes d'interface d'une poutre FGM renforcée par FRP avec une précision raisonnable en se basant sur le modèle de Meradjah [10]. Finalement une étude paramétrique sera présentée pour montrer l'influence de certains paramètres.

2 Hypothèse de la présente méthode

Dans la présente analyse, et en vue de la détermination des contraintes d'interface au niveau de la couche d'adhésif, le modèle théorique développé est basé sur les hypothèses suivantes :

- La poutre en FGM "figure 1", la couche d'adhésif et la plaque FRP ont un comportement élastique linéaire.
- Les contraintes au niveau de la couche d'adhésif sont constantes à travers l'épaisseur de cette couche.
- La poutre FGM est orthotrope dans tous les points
- Les déformations de cisaillement dans la poutre et dans la plaque sont négligées
- La poutre est parfaitement collée avec la plaque.
- L'adhésif a pour rôle le transfert des contraintes de la poutre à la plaque.

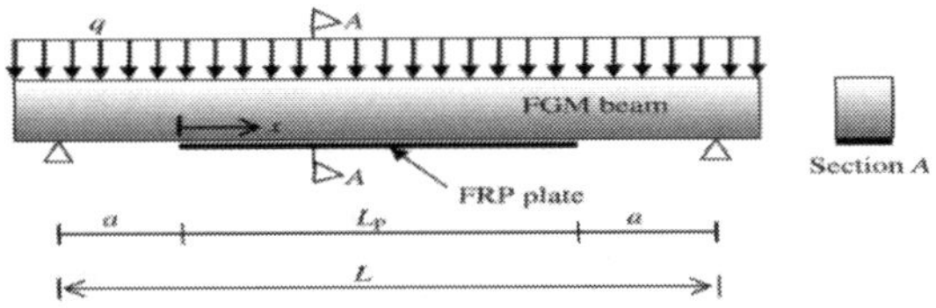

Fig. 1. Poutre FGM renforcée par une plaque FRP.

3 Formulation mathématique

Nous allons suivre la théorie des poutres Euler–Bernoulli qui suppose que les sections planes restent planes après déformation, donc les déplacements sont exprimés comme :

$$u(x,z) = u_0(x) - z\frac{dw}{dx} \text{ et } w(x,z) = w_0(x) \quad (01)$$

Le tenseur de déformation est exprimé comme :

$$\varepsilon_x = u^0_{,x}(x) - zw^0_{,xx}(x) = \varepsilon_0 + zk \quad (02)$$

La relation entre la déformation (équation (02)) et la contrainte est exprimée par :

$$\sigma_x = E(z)\varepsilon_x \quad (03)$$

Le module de Young est supposé variable dans la direction de l'épaisseur et il est exprimé comme :

$$E(z) = E_L + (E_U - E_L)\left(\frac{z}{h} + \frac{1}{2}\right)^P - \frac{(E_U - E_L)\alpha}{2} \quad (04a)$$

$$E(z) = E_L + (E_U - E_L)\left(\frac{z}{h} + \frac{1}{2}\right)^P - (E_U - E_L)\alpha/2\left(1 - 2zh \right. \quad (04b)$$

Ou : L représente inférieur (de l'anglicisme Lower)
U représente supérieur (de l'anglicisme Upper)
P : L'indice de degré d'homogénéité
α : L'indice de porosité
La force axiale et le moment fléchissant résultant N et M, sont définis par la théorie des poutres Euler-Bernoulli:

$$(N, M) = \int_{-h/2}^{h/2} \sigma_x(1, z)dz \quad (05)$$

En substituant l'équation (02) et (03) dans l'équation (05), une relation entre la force, le moment et la déformation est dérivée comme suit:

$$\begin{Bmatrix} N \\ M \end{Bmatrix} = \begin{bmatrix} A & B \\ B & D \end{bmatrix} \begin{Bmatrix} \varepsilon_0 \\ k \end{Bmatrix} \quad (06)$$

Les coefficients A,B et D sont donnés par :

$$(A, B, D) = \int_{-h/2}^{h/2} E(z)(1, z, z^2)dz \quad (07)$$

La relation inverse de l'équation (06) est :

$$\begin{Bmatrix} \varepsilon_0 \\ k \end{Bmatrix} = \begin{bmatrix} A & B \\ B & D \end{bmatrix}^{-1} \begin{Bmatrix} N \\ M \end{Bmatrix} = \begin{bmatrix} A^* & B^* \\ B^* & D^* \end{bmatrix} \begin{Bmatrix} N \\ M \end{Bmatrix} \quad (08)$$

Quand l'effort normal est nul (N=0), la déformation peut s'écrire comme suit :

$$\varepsilon_0 = B^*M, k = D^*M \quad (09)$$

En substituant dans les équations (02) et (03) la contrainte axiale dans la poutre FGM prend la forme :

$$\sigma_x = E(z)M[B^* + zD^*] \quad (10)$$

De l'équation (10) on tire la position de l'axe neutre :

$$z_{NA} = -\frac{B^*}{D^*} \quad (11)$$

3.1 Les contraintes de cisaillement dans la couche d'adhésive

Les contraintes de cisaillement au niveau de la couche d'adhésive est donnés par :

$$\tau_a(x) = \frac{G_a}{t_a}[u_p(x) - u_b(x)] \quad (12)$$

OùG_a,t_a,u_b, et u_p sont respectivement le module de cisaillement, l'épaisseur de l'adhésive, le déplacement longitudinal a la surface inférieure de la poutre FGM, le déplacement longitudinal a la surface supérieure de la plaque .

En dérivant l'équation (12) on obtient :

$$\frac{d\tau_a(x)}{dx} = \frac{G_a}{t_a}[\varepsilon_p(x) - \varepsilon_b(x)] \quad (13)$$

Où

$$\frac{du_b(x)}{dx} = \varepsilon_b(x) \text{et} \frac{du_p(x)}{dx} = \varepsilon_p(x) \quad (14)$$

En utilisant les équations (02) et (08), La déformation à la surface inférieure de la poutre FGM est :

$$\varepsilon_b(x) = -\frac{1}{b}\left[A^* + \left(\frac{h}{2} + z_{NA}\right)B^*\right]N_b(x) + \frac{1}{b}[B^* + \left(\frac{h}{2} + z_{NA}\right)D^*]M_b(x) \quad (15)$$

La déformation à la surface supérieure de la plaque est exprimée comme suit :

$$\varepsilon_p(x) = -\frac{t_p}{2E_pI_p}M_p(x) + \frac{N_p(x)}{E_pA_p} \quad (16)$$

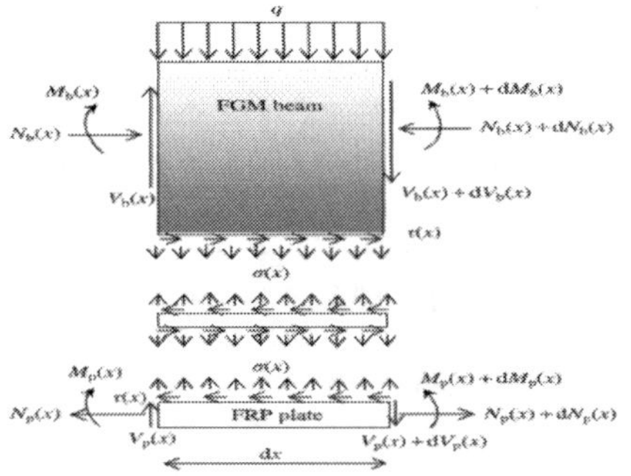

Fig. 2. Segment élémentaire d'une poutre FGM renforcée par une plaque FRP.

En appliquant les conditions d'équilibre on obtient :

$$\frac{dN_b(x)}{dx} = \frac{dN_p(x)}{dx} = b\tau(x) \quad (17)$$

Où

$$N(x) = N_b(x) = N_p(x) = b\int_0^x \tau(x)dx \quad (18)$$

La relation de moment des deux adhérents est exprimée comme :

$$M_b(x) = \frac{b}{D^*E_pI_p}M_p(x) - \frac{B^*}{D^*}N_b(x) \quad (19)$$

L'équilibre des moments dans un segment différentiel "figure 2" donne :

$$M_T(x) = M_b(x) + M_p(x) + N(x)\left[\frac{t_p}{2} + t_a + \left(\frac{h}{2} + z_{NA}\right)\right] \quad (20)$$

Le moment de flexion dans la poutre et dans la plaque de renforcement sont exprimés comme des fonctions du moment total appliqué et de la contrainte de cisaillement d'interface :

$$M_b(x) = \frac{b}{b+D^*E_pI_p}\left[M_T(x) - N(x)\left(\frac{t_p}{2} + t_a + \left(\frac{h}{2} + z_{NA}\right) + \frac{E_pI_p}{b}B^*\right)\right] \quad (21)$$

$$M_p(x) = \frac{D^*E_pI_p}{b+D^*E_pI_p}\left[M_T(x) - N(x)\left(\frac{t_p}{2} + t_a + \left(\frac{h}{2} + z_{NA}\right) + z_{NA}\right)\right] \quad (22)$$

Le premier dérivé des moments fléchissant dans chaque adhérent est :

$$\frac{dM_b(x)}{dx} = \frac{b}{b+D^*E_pI_p}\left[V_T(x) - b\tau(x)\left(\frac{t_p}{2} + t_a + \left(\frac{h}{2} + z_{NA}\right) + \frac{E_pI_p}{b}B^*\right)\right] \quad (23)$$

$$\frac{dM_p(x)}{dx} = \frac{D^*E_pI_p}{b+D^*E_pI_p}\left[V_T(x) - b\tau(x)\left(\frac{t_p}{2} + t_a + \left(\frac{h}{2} + z_{NA}\right) + z_{NA}\right)\right] \quad (24)$$

En substituant l'équation (15) et (16) dans l'équation (13), après la dérivation on obtient :

$$\frac{d^2\tau(x)}{dx^2} = \frac{G_a}{t_a}\left[\frac{1}{E_pA_p}\frac{dN_p(x)}{dx} - \frac{t_p}{2E_pI_p}\frac{dM_p(x)}{dx} + \frac{1}{b}\left(A^* + \left[\frac{h}{2} + z_{NA}\right]B^*\right)\frac{dN_b(x)}{dx} - \frac{1}{b}\left(B^* + \left[\frac{h}{2} + z_{NA}\right]D^*\right)\frac{dM_b(x)}{dx}\right] (25)$$

La substitution de l'effort tranchant (23) et (24) et l'effort normal (18) dans l'équation (25) nous permet d'obtenir l'équation différentielle des contraintes d'interface de cisaillement :

$$\frac{d^2\tau(x)}{dx^2} + \frac{\frac{G_a}{t_a}}{D^*E_pI_p+b}\left[\frac{t_p}{2}D^* + \left(B^* + \left[\frac{h}{2} + z_{NA}\right]D^*\right)\right]V_T(x) + \frac{G_a}{t_a}\left[\frac{b}{E_pA_p} + \frac{t_pD^*b}{2(D^*E_pI_p+b)}\left(\frac{t_p}{2} + t_a + \left(\frac{h}{2} + z_{NA}\right) - \frac{B^*}{D^*}\right) + A^* + \left[\frac{h}{2} + z_{NA}\right]B^* + \frac{b}{b+D^*E_pI_p}\left(\frac{t_p}{2} + t_a + \left[\frac{h}{2} + z_{NA}\right] + \frac{E_pI_p}{b}B^*\right) \times \left(B^* + \left[\frac{h}{2} + z_{NA}\right]D^*\right)\right]\tau(x) = 0 (26)$$

La solution générale de l'équation différentielle (26) est donnée par :

$$\tau(x) = B_1\cosh(\lambda x) + B_2\sinh(\lambda x) + m_1V_T(x) \quad (27)$$

Où

$$\lambda^2 = \frac{G_a}{t_a}\left[\frac{b}{E_pA_p} + \frac{t_pD^*b}{2(D^*E_pI_p+b)}\left(\frac{t_p}{2} + t_a + \left[\frac{h}{2} + z_{NA}\right] - \frac{B^*}{D^*}\right) + A^* + \left[\frac{h}{2} + z_{NA}\right]B^* + \frac{b}{b+D^*E_pI_p}\left(\frac{t_p}{2} + t_a + \left[\frac{h}{2} + z_{NA}\right] + \frac{E_pI_p}{b}B^*\right) \times \left(B^* + \left[\frac{h}{2} + z_{NA}\right]D^*\right)\right] \quad (28)$$

Et

$$m_1 = \frac{G_a}{t_a}\frac{1}{\lambda^2}\frac{1}{b+D^*E_pI_p}\left[\frac{t_p}{2}D^* + \left(B^* + \left[\frac{h}{2} + z_{NA}\right]D^*\right)\right] \quad (29)$$

Pour une charge uniformément reparti, l'effort tranchant est donnée par :

$$V_T(x) = q\left(\frac{L}{2} - x - a\right) \quad (30)$$

B_1 et B_2 sont des constantes déterminées à partir des conditions aux limites.

$$M_b(0) = M_T(0) = \frac{qa}{2}(L - a) \quad (31)$$

En substituant les équations (15) et (16) dans l'équation (13), et en appliquant les conditions aux limites :

$$\frac{d\tau(0)}{dx} = -m_2 M_T(0) \quad (32)$$

Où

$$m_2 = \frac{1}{b}\left[B^* + \left(\frac{h}{2} + z_{NA}\right) D^*\right]\frac{G_a}{t_a} \quad (33)$$

En substituant l'équation (27) dans l'équation (32),B_2 est exprimé par :

$$B_2 = -\frac{m_2}{\lambda}\frac{qa}{2}(L-a) + \frac{m_1}{\lambda}q \quad (34)$$

B_1est donné par :

$$B_1 = \frac{m_2}{\lambda}\frac{qa}{2}(L-a)tanh\left(\frac{\lambda L_P}{2}\right) - \frac{m_1}{\lambda}qtanh\left(\frac{\lambda L_P}{2}\right) \quad (35)$$

Dans le cas ou $\frac{\lambda L_P}{2} > 10, tanh\left(\frac{\lambda L_P}{2}\right) \approx 1$, l'expression de B_1 est simplifiée comme suit :

$$B_1 = \frac{m_2}{\lambda}\frac{qa}{2}(L-a) - \frac{m_1}{\lambda}q = -B_2 \quad (36)$$

En substituant B_1 et B_2 dans l'équation (27), on obtient la fonction qui exprime la distribution des contraintes de cisaillement :

$$\tau(x) = \left[\frac{m_2 a}{2}(L-a) - m_1\right]\frac{qe^{-\lambda x}}{\lambda} + m_1 q\left(\frac{L}{2} - a - x\right) \quad (37)$$

3.2 Les contraintes normales dans la couche d'adhésive

Lorsque la poutre est chargée, une séparation verticale se produit entre la poutre et la plaque. Cela crée des contraintes normales d'interface dans la couche d'adhésif. Cette contrainte normale est donnée par l'expression suivante :

$$\sigma(x) = \frac{E_a}{t_a}\left[w_p(x) - w_b(x)\right] \quad (38)$$

Ou $w_b(x)$ et $w_p(x)$ sont respectivement les déplacements verticaux de la poutre et de la plaque, l'équilibre de la poutre et de la plaque en négligeant les termes de deuxième ordre on obtient :
La poutre :

$$\frac{d^2 w_b(x)}{dx^2} = -\frac{1}{b}B^* N_b(x) - \frac{1}{b}D^* M_b(x) \quad (39)$$

$$\frac{dM_b(x)}{dx} = V_b(x) - \tau(x)b\left[\frac{h}{2} + z_{NA}\right] \quad (40)$$

$$\frac{dV_b(x)}{dx} = -\sigma(x)b - q \quad (41)$$

La plaque :

$$\frac{d^2 w_p(x)}{dx^2} = -\frac{1}{E_p I_p}M_p(x) \quad (42)$$

$$\frac{dM_p(x)}{dx} = V_p(x) - \tau(x)b\frac{t_p}{2} \quad (43)$$

$$\frac{dV_p(x)}{dx} = \sigma(x)b \quad (44)$$

En se basant sur les équations d'équilibre, les équations différentielles des déformations (déflexions) de l'adhérent 1 et 2 en termes des contraintes d'interface normales et de cisaillement sont données comme suit :
La poutre :

$$\frac{d^4 w_b(x)}{dx^2} = -B^*\frac{d\tau(x)}{dx} - \frac{1}{b}D^*\left[-\sigma(x)b - q - b\left[\frac{h}{2} + z_{NA}\right]\frac{d\tau(x)}{dx}\right] \quad (45)$$

La plaque :

$$\frac{d^4 w_p(x)}{dx^2} = -\frac{1}{E_p I_p}\left[b\sigma(x) + \frac{t_p}{2}b\frac{d\tau(x)}{dx}\right] \quad (46)$$

En substituant les équations (45) et (46) dans la quatrième dérivée des contraintes normales d'interface obtenues à partir de l'équation (38), nous aurons l'équation différentielle des contraintes normales d'interface :

$$\frac{d^4\sigma(x)}{dx^4} + \frac{E_a}{t_a}\left(\frac{b}{E_p I_p} + D^*\right)\sigma(x) - \frac{E_a}{t_a}\left(\frac{b}{E_p I_p}\frac{t_p}{2} + B^* - D^*\left[\frac{h}{2} + z_{NA}\right]\frac{d\tau(x)}{dx} + \right)\frac{E_a}{t_a}\frac{1}{b}D^* q = 0 \quad (47)$$

La solution générale de cette équation différentielle de quatrième ordre est :

$$\sigma(x)e^{-\delta x}\left[C_1\cos(\delta x) + C_2\sin(\delta x)\right] + e^{\delta x}\left[C_3\cos(\delta x) + C_4\sin(\delta x)\right] - n_1\frac{d\tau(x)}{dx} - n_2 q \quad (48)$$

Pour des grandes valeurs de «x», il est supposé que les contraintes normales deviennent nulles, et par conséquent d'où la solution générale prendra la forme :

$$\sigma(x) = e^{-\delta x}\left[C_1\cos(\delta x) + C_2\sin(\delta x)\right] - n_1\frac{d\tau(x)}{dx} - n_2 q \quad (49)$$

Où

$$\delta = \sqrt[4]{\frac{E_a}{4t_a}\left(\frac{b}{E_p I_p} + D^*\right)} \quad (50)$$

$$n_1 = \frac{-b\frac{t_p}{2} - B^* E_p I_p + D^* E_p I_p\left[\frac{h}{2} + z_{NA}\right]}{D^* E_p I_p + b} \quad (51)$$

$$n_2 = \frac{D^* E_p I_p}{b\left(D^* E_p I_p + b\right)} \quad (52)$$

Les constantes C_1 et C_2 sont déterminées en utilisant les conditions aux limites appropriées:
En substituant les équations (39) et (42) dans le deuxième dérivé de l'équation (38) on obtient :

$$\frac{d^2\sigma(0)}{dx^2} = \frac{E_a}{t_a}\left(\frac{1}{b}B^* N_b(0) + \frac{1}{b}D^* M_b(0) - \frac{1}{E_p I_p}M_P(0)\right) \quad (53)$$

Dès que $M_P(0) = 0$, $N_b(0) = N_p(0) = 0$ et $M_b(0) = M_T(0)$ au bord de la plaque, la relation (53) s'écrit :

$$\frac{d^2\sigma(0)}{dx^2} = \frac{E_a}{t_a}\frac{1}{b}D^* M_b(0) \quad (54)$$

En substituant les équations (40) et (43) dans le troisième dérivé de l'équation (38) on obtient :

$$\frac{d^3\sigma(0)}{dx^3} = \frac{E_a}{t_a}\left(\frac{1}{b}D^* V_b(0) - \frac{1}{E_p I_p}V_p(0)\right) - \frac{E_a}{t_a}\left(D^*\left[\frac{h}{2} + z_{NA}\right] - \frac{b}{E_p I_p}\frac{t_p}{2} - B^*\right)\tau(0) \quad (55)$$

Dès que $V_P(0) = 0, V_b(0) = V_T(0)$ on obtient :

$$\frac{d^3\sigma(0)}{dx^3} = \frac{E_a}{t_a}\frac{1}{b}D^*V_T(0) - n_3\tau(0) \qquad (56)$$

Où

$$n_3 = \frac{E_a}{t_a}\left(D^*\left[\frac{h}{2} + z_{NA}\right] - \frac{b}{E_pI_p}\frac{t_p}{2} - B^*\right) \qquad (57)$$

Le deuxième et le troisième dérivé de l'équation (49) s'écrivent :

$$\frac{d^2\sigma(0)}{dx^2} = -2\delta^2C_2 - n_1\frac{d^3\tau(0)}{dx^3} - n_2\frac{d^2q}{dx^2} \qquad (58)$$

Et

$$\frac{d^3\sigma(0)}{dx^3} = 2\delta^3C_1 + 2\delta^3C_2 - n_1\frac{d^4\tau(0)}{dx^4} - n_2\frac{d^3q}{dx^3} \qquad (59)$$

Donc C_1et C_2 s'écrivent :

$$C_1 = \frac{E_a}{2\delta^3 t_a}\frac{1}{b}[V_T(0) - \delta M_T(0)]D^* - \frac{n_3}{2\delta^3}\tau(0) + \frac{n_1}{2\delta^3}\left[\frac{d^4\tau(0)}{dx^4} + \delta\frac{d^3\tau(0)}{dx^3}\right] \qquad (60)$$

$$C_2 = -\frac{E_a}{t_a}\frac{1}{2\delta^2}\frac{1}{b}D^*M_T(0) - \frac{n_1}{2\delta^2}\frac{d^3\tau(0)}{dx^3} \qquad (61)$$

Les expressions des constantes C_1 et C_2 ci-dessus sont en terme de moment de flexion $M_T(0)$ et d'effort tranchant $V_T(0)$ au bord de la plaque de renforcement.

4 Résultats et discussions

Une poutre en FGM renforcée par une plaque en FRP est analysée. La poutre ayant une travée de 500 mm, est simplement appuyée et soumise à une charge uniformément répartie d'une valeur de 2 KN/m. La distance de l'appui au bord de la plaque de renforcement est de 50 mm.
Les caractéristiques géométriques et les propriétés des matériaux utilisés sont représentées au tableau 1 :

Table 1. Caractéristiques géométriques et propriétés des matériaux utilisés.

Matériaux	Module d'élasticité (MPa)	υ_{12}	Largeur (mm)	Hauteur (mm)
Plaque FRP	E=100000	0.3	$b_f = 20$	$t_f = 2$
Poutre en FGM	E_{sup}= 200000 E_{inf}= 70000	0.3	$b_b = 20$	$t_b = 30$
Couche d'adhésive	E= 2000	0.35	$b_a = 20$	$t_a = 2$

Les résultats obtenus par le modèle développée par Meradjah [10] ainsi que ceux de la présente méthode pour les contraintes de cisaillement et contraintes normales d'interface, sont illustrés à la figure 3.

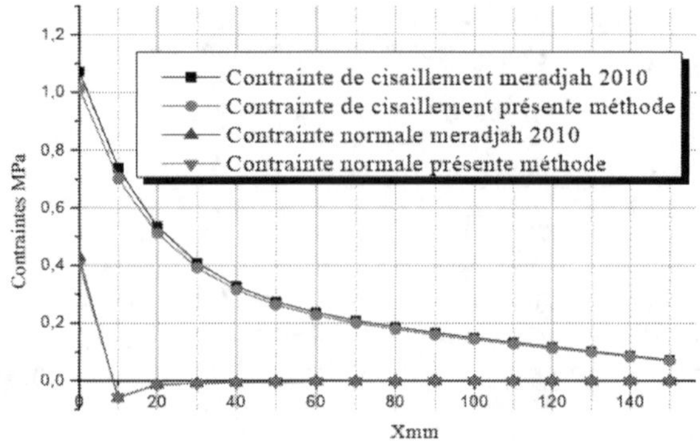

Fig. 3. Comparaison des contraintes d'interface pour une poutre FGM renforcée par une plaque FRP (charge uniformément répartie).

Les résultats obtenus et présentés laissent apparaître clairement, qu'une bonne concordance entre le présent modèle et celui de Meradjah [10] est observée.

4.1 Etude paramétrique

Au cours de ce paragraphe, nous présentons les résultats de la présente méthode afin d'étudier l'effet de divers paramètres sur la distribution des contraintes d'interface d'une poutre en FGM renforcée par une plaque en FRP.
L'effet de la variation de degré d'homogénéité du matériau P-FGM est illustré dans la figures 4, les grandes valeurs de l'indice P correspond à un faible pourcentage de la quantité de céramique , autrement dit, l'augmentation de l'indice de loi de puissance P réduit la rigidité en raison de la grande quantité de métal par rapport à la quantité de céramique et par conséquent les contraintes d'interface augmentent.

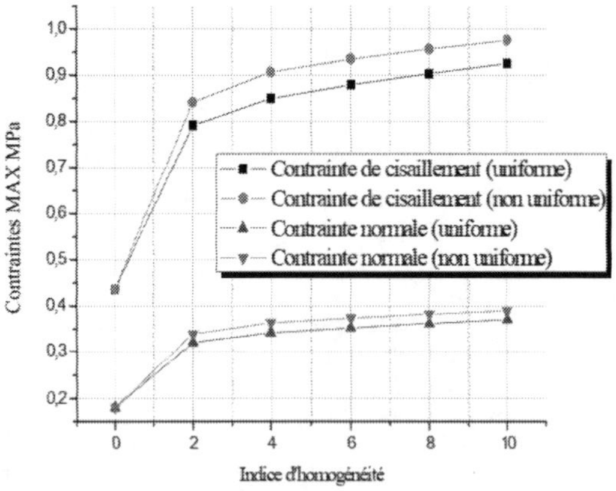

Fig. 4.L'effet de l'indice P (degré d'homogénéité) sur les contraintes normales et de cisaillements d'une poutre FGM renforcée par une plaque FRP.

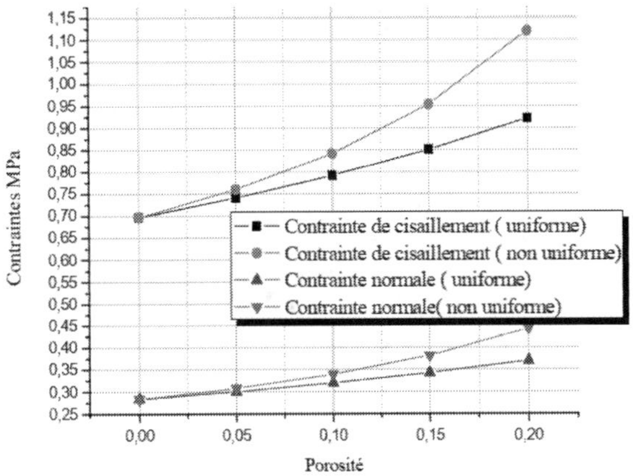

Fig. 5. Influence de la porosité dans la poutre sur les contraintes max pour une poutre en FGM sous charge uniforme renforcée par une plaque en FRP.

L'effet de pourcentage de vide dans la poutre sur la variation des contraintes d'interface est représenté sur la figure 5, la présence du vide dans la poutre affaiblit sa rigidité ce qui mène à une augmentation des contraintes d'interfaces.

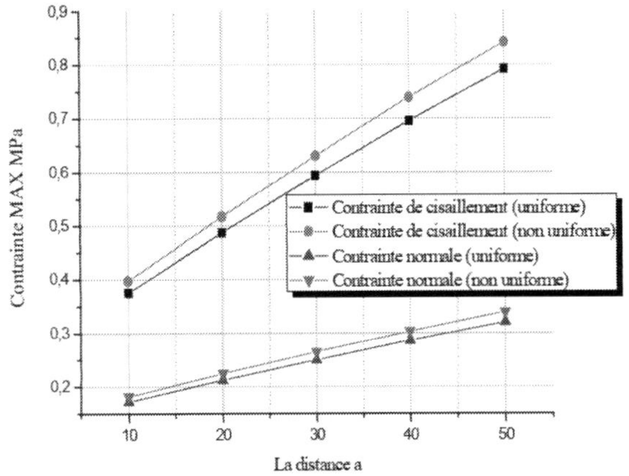

Fig. 6. Influence de la longueur « a » sur les contraintes normales et de cisaillement d'interface pour une poutre en FGM renforcée par une plaque FRP.

L'influence de la longueur non renforcée « a » (longueur entre le bord de la plaque et le support) sur les contraintes d'interface maximales est présentée sur la figure 6. Plus la longueur « a » augmente plus les contraintes d'interface maximales s'accroissent d'une façon significative, donc il est recommandé de prolonger la plaque de renforcement le plus possible aux appuis.

5 Conclusion

Le présent travail étudie la distribution des contraintes normales et de cisaillement au niveau de la colle d'une poutre en FGM renforcée par une plaque en composite FRP, l'élément important de la présente analyse est la prise en considération l'effet de la porosité dans la poutre FGM suivant deux type de distribution uniforme et non uniforme dans la direction de la hauteur, une étude paramétrique a été faite pour étudier l'influence de certains paramètres sur les contrainte d'interface.

References

1. Jones R, Swamy RN, Charif A., Plate separation and anchorage of reinforced concrete beams strengthened by epoxy – bonded steel plates. The Structural Engineer, **66**(5/1): 85 – 94, (1988).
2. Etman E. E., Beeby A. W. Experimental program and analytical study of bond stress distributions on a composite plate bonded to a reinforced RC beam. Cement and Concrete Composites; **22**(4): 281 – 291,(2000).
3. Vilnary O., "The analysis of reinforced concrete beams strengthened by epoxy bonded steel plates". Int. Journal Cement Compo and Light Weight Concrete. **10** (2), 73-78, (1988).
4. Taljsten B., "Strengthening of beams by plate bonding". Journal of Materials in Civil Engineering, ASCE, **9** (4), 206-212, (1997).
5. Malek M, Saadatmanesh H and Ehsani M.R., "Prediction of faillure load of RC beams stregthened with FRP plate due to stress concentartion at the plate end". ACI Structural Journal, **95** (2), 142-152, (1998).
6. Smith S.Tand Teng J.G., "Interfacial stresses in plated RC beams". Engineering Structures, **23** (7), 857-871, (2001).
7. Tounsi A., "Improved theoretical solution for interfacial stresses in concrete beams strengthened with FRP plate". International Journal of Solids and Sructures, **43** (14-15), 4154-4174, (2006).
8. K. T. Lau, P. K. Dutta, L. M. Zhou and D. Hui, Composites, Part B**32** (6), 491 (2001).
9. Shen H.S, Teng J.G and Yang J., "Interfacial stresses in beams and slabs bonded with a thin plate". Journal of Engineering Mechanics, ASCE, **127** (4), 399 – 406, (2001).
10. Meradjah M, Benyoucef S, Tounsi A ,El Abbas A B and Merdaci S,Interfacial Stresses in Plated Beams with Exponentially-Varying Properties, Journal of Adhesion Science and Technology **24** 1063–1081, (2010).

https://doi.org/10.1051/matecconf/201814902001

Earth as Building Material – an overview of RILEM activities and recent Innovations in Geotechnics

Johan Vyncke[1], Laura Kupers[2] and Nicolas Denies[3]
[1] Belgian Building Research Institute, Research and Innovation Department, Limelette, Belgium & President of RILEM
[2] Belgian Building Research Institute, Laboratory of Concrete Technology, Limelette, Belgium
[3] Belgian Building Research Institute, Geotechnical Division, Limelette, Belgium

Abstract. This paper presents an overview of the different earth building techniques, the latest innovations and the normative aspects. The oldest man made earth constructions known to exist date back to 10 000 BC. Since then, earth has remained a popular building material throughout history. With time, different techniques evolved, starting from sundried adobe blocks to cob constructions, rammed earth walls and compressed earth bricks. Today these techniques are still being optimized and alternative binders, specifically adapted admixtures and surface treatments are being developed. Even though nearly one third of the world's population lives in an earth construction, few specific building standards and testing methods exist. Many of the tests used today are based on tests for concrete and thus do not take into account the complex nature of earth constructions, such as their sensitivity to water. RILEM, the union of Laboratories and Experts in Construction Materials, Systems and Structures, set up a new Technical Committee in 2016: TC TCE (Testing and Characterisation of Earth-based building materials and elements). This committee, consisting of an international group of experts on the topic, aim to define testing procedures for earth as a building construction material. To end with, this paper also gives a short introduction to "Deep soil mixing", an "earth" building technique dedicated to geotechnical engineering.

1 Introduction

The oldest manmade earth constructions known were found in Mesopotamia, and date from 10 000 BC, they consist of stacked earth bricks. Remains of rammed earth foundations as old as 5 000 BC were found in Assyria. For these archaeological constructions, mixtures of locally excavated sand, gravel and clay were used [1, 2, 3].

Fig. 1. Fortified village of Aït Ben-Haddou (Photo: uneco.org).

There are many historical monuments made of earth constructions all around the world. Just a few examples are the temple of Ramses II at Gourna, Egypt, the citadel of Bam in Iran, the core of the Sun Pyramid in Teotihuacan, Mexico, the Potala Palace in Tibet, the fortified village of Aït Ben-Haddou in Ouarzazate, Morocco (see Figure 1) etc. [1, 2].

Today earth still remains an important building material. Nearly 30 % of the world's population lives in buildings made of earth [1]. Figure 2 illustrates the worldwide application of earth building.

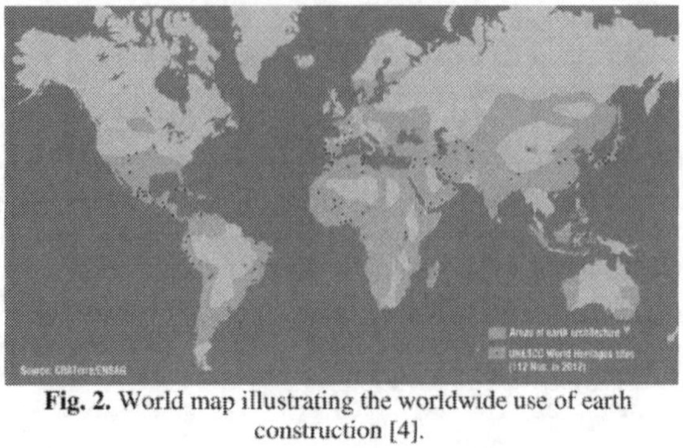

Fig. 2. World map illustrating the worldwide use of earth construction [4].

In developing countries even more than half of the population lives in earth construction. But also in the industrialized countries, building with earth is regaining some interest as it is regarded as an environmentally friendly option due to its low carbon footprint, low thermal conductivity and good hygroscopic features [1, 5, 6].

For modern earth constructions often local soil is mixed with imported materials such as quarry dust, sand and gravel from quarries located at further distance to achieve a suitable mix design [7]. Also additives, such as slaked lime, cement or (synthetic) fibres, are added to the soil mixture in order to achieve higher compressive strength and better durability.

It has been proven that it would be impossible to fulfil the housing demand by the "classical" building materials such as bricks, concrete and steel [2] if earth would not be used. Earth is available in abundance, it is everywhere locally available and cheap [8]. However, some of the drawbacks are the loss of strength due to saturation, its sensitivity to erosion by rain impact and, more generally, its durability [9]. Nevertheless, this can largely be resolved by taking appropriate architectural measures such as large roof-overhangs, high footings and good drainage [10].

Not all soils are suitable for application as building material. Important parameters are the grain size distribution, the clay content, the mineralogical and chemical composition, the organic material content, etc. Depending on the building technique there are different requirements. The raw material must in many cases be prepared before being applied. Preparation usually consists of pulverisation, sieving, and/or drying [2, 8, 11, 12].

2 The different earth building techniques

2.1 Rammed earth

Rammed earth is a building technique in which a wet mixture of mud, sand and gravel is poured in a formwork and is compacted layer by layer [1, 2]. Because of the layer approach, rammed earth constructions are characterized by a horizontally striped pattern. By using different coloured soils this effect can deliberately be emphasized as illustrated in Figure 3.

Originally, the ramming was manually performed but in the 20th century, electric and pneumatic rams were developed [2]. Today the technique has evolved in such a way that the rammed earth constructions can specifically be designed to achieve high compressive strength and fulfil aesthetic requirements. For the execution, sophisticated materials and machinery, such as laminate plywood, aluminium form supports, conveyor belts, pneumatic air compressors, mixing machines and tractors, can be used [7].

Fig. 3. Multi coloured rammed earth wall at Brisbane, Australia (Photo: rammedearthconstructions.com).

2.2 Cob constructions

The great Mosque Djenné, an UNESCO World Heritage Site, is probably the most famous cob construction. Cob is a building technique in which plastic (clayey) soil mixtures are formed in balls and freshly stacked upon each other, as illustrated in Figure 4. The soil mixture typically consists of soil and water and often straw and sand are added. This technique is used a lot in Africa, India and Saudi Arabia. In England and France, this technique was also used a lot in the past and it is now slowly regaining some interest [4].

Fig. 4. Construction of a cob wall by stacking mud balls (Photo: tanzaniagoesmud.wordpress.com).

2.3 Earth blocks

Earth blocks can be divided in three groups, "adobe blocks", "cut blocks" and "compressed earth blocks", depending on the technique used to form the block.

2.3.1 Adobe blocks

Adobe blocks are sun dried bricks. A muddy soil mixture, often containing cut straw, is shaped into blocks by filling a formwork. After shaping the blocks are left to dry in the sun and harden [2, 12].

2.3.2 Cut blocks

In regions where the soil itself is sufficient cohesive, blocks can be cut out directly out of the soil. This technique is referred to as "cut blocks" [12] and is illustrated in Figure 5. Laterite, a soil formed by intense weathering in tropical climates and characterized by a high iron and aluminium concentration giving it its characteristic red colour, is well suited for this form of application. This technique is for example used in India [13] (see Figure 5). Once cut in blocks and exposed to air, the relatively soft soil hardens (indurates) due to dehydration of the iron and aluminium oxides and arrangements of components.

Fig. 5. A laterite block mine in Ratnagiri, India. The blocks are cut out of the soil. (Photo: news.uk.msn.com).

2.3.3 Compressed earth blocks

In the 18th century, the technique to produce "compressed earth blocks" (CEB's) by means of a manual press became popular in Europe. This technique made it possible to work with lower water contents in the soil mixtures than for the adobe blocks and the blocks can be stacked immediately after production. In order to improve the strength of the bricks, cement or lime was added to the soil mixture [2]. Nowadays large industrial presses for producing compressed earth blocks exist, but none the less the manual press is still very popular. Figure 6 illustrates two technicians operating a manual press in the laboratory facilities of the Belgian Building Research Institute (BBRI) where tests on Laterite blocks have been performed for a potential development project in Africa.

The shape of the compressed earth block is determined by the shape of the mold. Most presses can be equipped with a large variety of mold shapes going from different sized rectangular blocks to blocks with rounded angels, blocks with holes (interlocking), sinus-shaped self-blocking blocks to blocks with recess for cables, etc.

3 Mix design and production of CEB's

Just as for concrete constructions, the mix design for earth blocks is important with regard to the final properties (workability, strength, shrinkage…). However, earth constructions have the additional disadvantage that the characteristics of the used earth often vary significantly. For a minimal quality control, the used earth should therefore be submitted to standard testing (Atterberg limits, methylene blue content, organic material content, etc.) during the production process so that the mix design can be adapted if necessary.

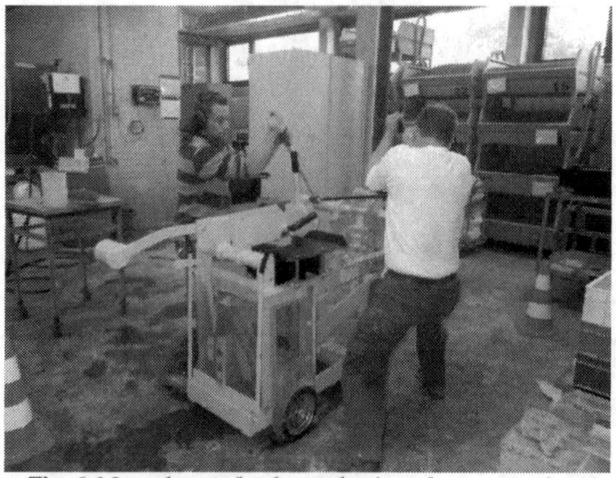

Fig. 6. Manual press for the production of compressed earth blocks (Source: BBRI).

For CEB's, most often a mixture of earth, cement and water is used. The cement/earth and water/cement ratios found in the literature differ greatly - ranging from respectively 0.02 to 0.50, and from 0.5 to 3.9 [14, 15, 16, 17, 18, 19]. These differences can amongst others be attributed to the variation of the used earth and the intended application and desired strength of the blocks.

The cement in the mixture stabilizes the earth by creating a network of hydrated calcium silicate (CSH). A higher cement content results in a denser CSH network thus increasing the compressive strength of the earth block. Sufficient water is needed in the mixture to hydrate the cement and to allow an optimum compression, but a too high water/cement ratio decreases the compressive strength and leads to an increased risk of shrinkage cracks. Sometimes (river) sand is added to improve the particle size distribution or to reduce the percentage of (swelling) clays of the soils [11, 13, 18, 19].

In addition to the mix design, the production process is also important. Experiments show that the degree of compaction and the after-treatment conditions have an important effect on the final compressive strength of the laterite block [18, 20]. After production, the blocks should be able to cure in humid conditions to allow maximum strength development. Tests at the BBRI showed that the difference between a "good cure" (30°C, >90% R.H.) and a "bad cure" (30°C, <60% R.H.) can amount to a loss of strength of more than 35%.

4 Research and innovation regarding earth constructions

In order to improve the strength and durability of earth constructions, different universities and research institutes all around the world are engaged with scientific research in order to develop specific admixtures and surface treatments. Also the use of alternatives, such as eco-friendly binders, are investigated. In the present paragraph, different innovations are listed.

https://doi.org/10.1051/matecconf/201814902001

4.1. Alternative binders

The most common stabilizers for earth constructions are lime and cement. Stabilizing CEBs with lime is possible if the earth used for the mix has a high clay content. If the earth consists dominantly of quartz (sandy soil), a stabilization with cement is recommended [2, 8, 11, 12].

Recently, more and more researches concerning alternative binders are performed. A proposed alternative is, for example, the use of alkaline activation. Alkaline activated materials (AAM's), also called "geopolymers", are considered as more environmentally friendly than cement bonded materials due to substantial reduction in CO_2 emissions [21, 22]. If the earth contains an important clay fraction, the clay can act as a binder. Ternary diagrams, indicating if a resource is a potentially interesting material for alkaline activation, can be found in the scientific literature (see, for example, Figures 7 and 8).

The use of laterite as a potential AAM was tested by Maskell et al [23] and Marques et al [24]. In both cases, the activator NaOH was used. The studies indicated that for an optimal result, a hot curing of the CEBs (at least 55°C) is required. Alkali activated laterite also requires a longer time to obtain its maximum strength than the cement-bonded reference blocks [24].

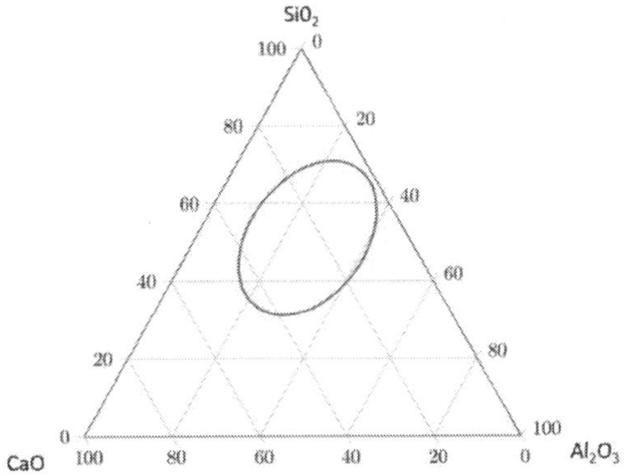

Fig. 7. Ternary Si-Ca-Al diagram showing the ideal chemical composition of potential AAM's [based on 25]

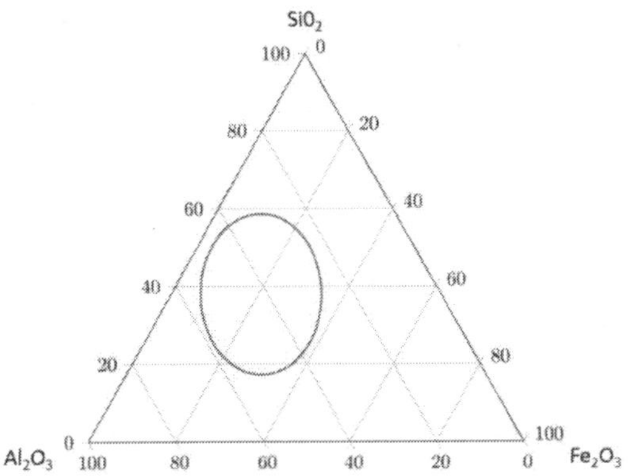

Fig. 8. Ternary Si-Al-Fe diagram showing the ideal chemical composition of potential AAM's [based on 26].

4.2 Admixtures

Specially adapted admixtures can improve the quality of earth constructions by making them less vulnerable to water penetration. For example, adding a low dosage hydrophobic silicone-based admixture could minimise the water absorption without affecting the vapour permeability [9].

4.3 Surface treatments

A surface treatment can significantly improve the surface quality of an earth construction. Many different surface treatments exist including water-repellent treatments, which make the constructions less vulnerable to water penetration, as well as dust suppressant sealing, renders and coatings etc.

The development of nano-acrylic emulsions, similar to the technology used for paint primers, will allow better dust binding without the negative impact on the vapour permeability as it is now the case with the traditional acrylic or latex emulsions. Also still in research phase are fluorinated silicones or fluorinated acrylic polymers sealers for areas that require repeated cleaning such as the kitchen, bathroom or public spaces [9].

5 Standards and testing procedures

Generally, building with earth has not been considered as an "engineered construction technique" in the past and thus building standards were considered unnecessary. However, due to the (re)increasing interest in earth constructions, building standards specifically dedicated to earth constructions have been drawn up the last years. In 2012, thirty-three different standards existed from 18 different countries (Australia, Brazil, Colombia, France, Germany, India, Kenya, Kyrgyzstan, New Zealand, Nigeria, Peru, Spain, Sri Lanka, Switzerland, Tunisia, Turkey, USA and Zimbabwe) which were published by regional, national or local standard bodies.

Nevertheless, there is still no common and internationally accepted terminology for earth building constructions. Moreover, the technical information in the different standards varies widely and no uniform and internationally accepted laboratory tests exist for the moment.

Most of the test procedures described in the existing standards for earth constructions were originally developed for the testing soils or concrete. Thus, they are not always suitable for earthen building materials [27].

6. RILEM activities

RILEM is the Union of Laboratories and Experts in Construction Materials, Systems and Structures. It especially endeavours to advance scientific knowledge related to the use of concrete, masonry, wood, asphalt, and other construction materials and to encourage the worldwide transfer and application of this knowledge. The essential work of RILEM is done in Technical Comities (TC's) which are set up in a Bottom-Up approach, i.e. it are the corporate members and individual members who bring forward the proposals for action. Those are then evaluated on their merits by a technical advisory committee who also provides in monitoring of the progress of the TC's.

Work on earth as building material was initiated in RILEM back in 1996 in the TC 164-EBM where the focus was developing a test method to determine the mechanical strength for compressed earth samples by means of compression and tensile traction tests [28, 29]. Also tests to characterise the durability of the cement-stabilized compressed earth blocks which are subjected to wetting-drying (artificial aging) were performed [30].

Due to the lack of uniform and internationally accepted tests for earth building materials, the TC TCE "Testing and characterisation of earth-based building materials and elements" was initiated in 2016. Their goal is to define testing procedures for stabilized and unstabilized earth as a building construction material in adobe blocks, cob and rammed earth. The required characteristics of the used earth mixture as well as the evaluation of the performances (sustainability, strength, seismic resistance, thermal, moisture buffering...) of the earth building material are treated. A minimal number of laboratory tests allowing an accurate assessment of the mechanical, thermal and hygroscopic performances of the material will be defined. In order to evaluate the repeatability and the robustness of the tests, round robin tests are performed. The experimental developments should serve as a basis for the establishment of international or European standards on earth material testing procedures. The technical committee consists of more than 30 active members from all over the world and all specialized in earth constructions [31].

7. Deep soil mixing

Another, innovative and relatively new application of building with earth is the deep soil mixing. In contrast with the techniques discussed in the previous paragraphs (rammed earth, cob, earth blocks), this technique does not require excavation of the earth. In the deep soil mixing process, the earth is mechanically mixed in place and in depth by an auger or a cutter machine (see Figure 9) while a binder, often based on cement, is injected. Depending on the application, 100 to 600 kg of cement per m^3 of mix is added. After hardening of the earth-cement mixture, further referred as the "soil mix material", different ground improvement patterns can be obtained into the ground: cylindrical columns, rectangular panels, continuous barriers or global mass stabilization.

As reported in Denies and Huybrechts (2015) [32], since several decades, the deep mixing method is known as a ground improvement technique. The deep soil mixing is in this case used for the global stabilization of soft cohesive soils such as silt, clay and peat in order to obtain sufficient bearing capacity (e.g. for the foundation of a structure) or to increase the general stability of the soil (e.g. for the reinforcement of existing dikes or land levees). In this context, 100 to 300 kg of cement per cubic meter of mix are generally used in the design mix.

If originally, the deep mixing method was developed for ground improvement applications, it is currently more and more dedicated to various structural and environmental applications such as alternative to classical foundation solutions, earth-water retaining walls, cut-off walls, slope stabilization, in-situ remediation, liquefaction mitigation, land reclamations...

In Belgium, the soil mixing is mainly used for the construction of earth-water retaining walls within the framework of excavation works. In this particular application, cylindrical soil mix columns or rectangular panels are formed in the ground as illustrated in Figure 10. In order to build soil mix walls for the realization of an excavation, the cylindrical soil-cement columns or the rectangular soil mix panels are placed next to each other, in a secant way such as illustrated in Fig. 10. By overlapping the different soil mix elements, a continuous soil mix wall is built. The soil mix wall is then horizontally stabilized by shoring or anchoring. During execution, steel beams are installed into the fresh soil mix material to resist the shear forces and bending moments due to the earth-water pressure applying on the retaining wall. The role of the soil mix material is to transmit the stresses due to this earth-water pressure to the steel beams by way of an arching effect developing in the soil mix material due to the difference of stiffness between the steel and the soil mix material.

Fig. 9. Typical deep mixing equipment in action on Belgian construction sites: Cutter Soil Mix (CSM) system of Bauer (left) and triple shaft configuration TSM system of Smet F&C (right).

Fig. 10. Soil mix retaining walls within the framework of excavation works in Belgium: soil mix walls built with CSM-panels (left) and with the C-mix shaft system of CVRnv (right).

The success of this new application is based on the recent advances in the deep mixing equipment coupled with the development of binders allowing to reach typical unconfined compressive strength of 1 to 12 MPa for the soil mix material. In Belgium, 200 to 450 kg (sometimes more) of cement per cubic meter of mix are typically used for the construction of soil mix retaining walls.

The soil-mix walls can be used as temporary or permanent geostructural elements and can fulfil different functions (earth-water retaining, bearing and cut-off).

This last decade, the Belgian Building Research Institute (BBRI) performed numerous tests on in-situ soil mix material for the purpose of characterizing its following properties: density, porosity, permeability, unconfined compressive strength (UCS), tensile strength, modulus of elasticity, ultrasonic pulse velocity and adherence between soil mix material and steel reinforcement [33]. Moreover, real-scale soil mix panels and columns were tested in order to obtain the real-scale UCS, the bending stiffness and the bending resistance of soil mix elements. Soil mix panels and columns previously executed on site were excavated after a few days of hardening and transported to the laboratory facilities of the BBRI to be tested [34]. On the basis of these results, the BBRI developed a design methodology, in line with the structural Eurocodes, for soil mix retaining walls. This methodology is described in detail in the BBRI/SBRCURnet soil mix handbook [35]. This handbook also contains information on the hydro-mechanical characteristics of the soil mix material.

Notwithstanding, more research is still required concerning questions such as:

- the influence of the deep mixing equipment on the final product,
- the minimal mixing energy to obtain a homogeneous soil-mix material,
- the durability of the soil mix material,
- the use of reinforcement cage in place of steel beams…

8 EVALUATION

Earth is regaining interest as a building material, not only in developing countries, where due to its abundance and low cost it is a logic and interesting building material for constructing superstructures, but also in modern societies where for reasons of sustainability and limiting transport it has an important advantage opposed to the conventional building materials. In this sense, a relatively new application of building with earth is the deep soil mixing process used within the framework of geotechnical projects. An advantage of this method in comparison with typical concrete solution is its independence from concrete supply on the field. Contrarily to concrete retaining walls, the execution of soil mix walls does not suffer from delayed supply (e.g. due to traffic jams) of the fresh concrete.

With regard to this regaining interest for earth constructions, there is clearly a need for the further development of specific testing techniques to allow further engineering development of earth applications as well as the development of quality standards. RILEM could surely play an important and pioneering role in this context.

References

1. C.S. Costa, F. Rocha & A.L. Velosa. Sustainability in earthen heritage conservation. *Sustainable use of Traditional Geomaterials in Construction Practice,* **416** p 91-100 (2016)
2. G. Minke. Building with earth, Design and Technology of a Sustainable Architecture. *Birkhäuser* (2006)
3. P. Jaquin. History of earth building techniques. *Modern Earth Buildings* p 307-322 (2012)
4. Auroville Earth Institute, Unesco chair earthen architecture. Building with Earth – Technique overview. www.earth-auroville.com (accessed 2017 feb 17)
5. E.C. Hoff. Appraisal of the Sustainability of Compressed Stabilized Earthen Masonry. *Dissertation Architectural Engineering, University of Nebraska – Lincoln* (2016)
6. M.R. Hall, S. Cassey. Hygrothermal behavior and occupant comfort in modern earth buildings. *Modern Earth Buildings* p 17-40 (2012)
7. D. Easton, T. Easton. Modern rammed earth construction techniques. *Modern Earth Buildings* p 364-384 (2012)
8. E.A. Adam, A.R.A. Agib. Compressed Stabilised Earth Block Manufacture in Sudan. *UNESCO* 2001
9. R. Kebao, D. Kagi. Integral admixtures and surface treatments for modern earth buildings. *Modern Earth Buildings* p 256-281 (2012)
10. H. Guillaud, T. Joffroy, P. Odul, CRATerre-EAG. Compressed earth blocks - Volume II. Manual of design and construction (1995)
11. V. Rigassi, CRATerre-EAG. Compressed earth blocks - Volume I. Manual of production (1995)
12. S. Boubekeur, H. Houben. Blocs de terre comprimée – normes. *Guides pratiques Serie Technologies* **N°11** (1996)
13. A.K. Kasthurba, M. Santhanam, M.S. Mathews. Investigation of laterite stones for building purpose from Malabar region, Kerala state, SW India – Part 1: Field studies and profile characterisation. *Construction and Building Materials* **21** p 73-82 (2007)
14. G. Otoko. The use of laterites for production of soil-cement blocks. *International Journal of Engineering and Technology Research* **Vol. 2 n°4** p 1-11 (2014)
15. F. Falade. The significance of source of laterite on strength of cement-stabilized lateritic blocks. *Housing Science* **Vol 15** p 1241-131 (1991).
16. F. Lasisi, E. Ajayi, J. Osunade. Technical notes : Strength characteristics of laterite-cement mortars. *The International Journal of Cement Composites and Lightweigth Concrete* **Vol 6, n° 3** p 201-203 (1984)

17. K. Solomon-Ayeh. Studies of strengths of stabilized laterite blocks and rendering mortars. *Building Research and Information* **Vol 22 n° 3** p 159-166(1994)
18. I. Agbede, M. Joel. Use of Cement-Sand Admixtures in Laterite Brick Production for Low Cost Housing. *Leonardo electronic journal of practices and technologies* **12** p 163-174 (2008)
19. G. Otoko, P. Pedro. Cement stabilization of laterite and chikoko soils using waste rubber fibre. *International journal of engineering sciences & research technology* **3** p 130-136 (2014)
20. J. Ayangade, O. Alake, A. Wahab. The effects of different curing methods on the compressive strength of terracrete. *Civil engineering dimension* **Vol. 11** p 41-45 (2009)
21. M. Estrada. A case study of cob earth based building technique in Matagalpa, Nicaragua – LCA perspective and rate of adoption. *Master Thesis Ecotechnology and Sustainable Development Mid Sweden University* (2013).
22. E. Hoff. Appraisal of the Sustainability of Compressed Stabilized Earthen Masonry. *Dissertation and Student Research Architectural Engineering University of Nebraska-Lincoln* (2016)
23. D. Maskell, A. Heath, P. Walker. Geopolymer stabilisation of unfired earth masonry units. *Key Engineering Materials*, **600** p 175-185 (2014)
24. S. Marques, M. Carvalho, S. Torres, N. Perazzo, K. Gomes, K. Ghavami. Mechanical properties and durability of geopolymer stabilizated earth blocks (2009)
25. A. Buchwald, K. Dombrowski, M. Weil. The influence of calcium content on the performance of geoplymeric binder especially the resistance against acids. *Proceedings of the World Congress Geopolymer 2005* p 35-39 (2005)
26. S. Kumar, R. Kumar. Synergising red (mud) and grey (ash) for greener geopolymers. *Bauxite residue valorisation and best practices, Conference paper* (2015)
27. H. Schroeder. Modern earth builiding codes, standards and normative development. *Modern Earth Buildings* p 72-109 (2012)
28. A. Hakimi, N. Yamani, H. Ouissi. Résultats d'essais de résistance mécanique sur échantillon de terre comprimée. *Materials and Structures* **Vol 29** p 600-608 (1996)
29. M. Olivier, A. Mesbah, Z. El Gharbi, J. Morel. Mode opératoire pour la realisation d'essais de résistance sur blocs de terre comprimée. *Materials and Structures* **Vol 30** p 515-517 (1997)
30. A. Hakami, H. Ouissi, M. El Kortbi, N. Yamani. Un test d'humification-séchage pour les blocs de terre comprimée et stabilisée au ciment. *Materials and Structures* **Vol 31** p 20-26 (1998)
31. RILEM, International union of laboratories and experts in construction materials, systems and structures – Committees. www.rilem.org (accessed 2017 feb 20)
32 N. Denies and N. Huybrechts. Deep Mixing Method Equipment and field of applications. *Book chapter 11 in Ground Improvement case histories, Elsevier* (2015)
33. N. Denies, N. Huybrechts, F. De Cock, B. Lameire, A. Vervoort, G. Van Lysebetten, J. Maertens. Soil Mix Walls as retaining structures – mechanical characterization. *International symposium of ISSMGE – TC211* **Vol. 3** p 99-115 (2012).
34. N. Denies, N. Huybrechts, F. De Cock, B. Lemaire, J. Maertens, A. Vervoort. Large-scale bending tests on soil mix elements. *International Foundations Congress and Equipment Expo 2015, San Antonio (Texas, USA)* p 2394-2409 (2015)
35. BBRI/SBRCURnet. Handboek soilmix-wanden. SBRCURnet/BBRI publication (2016)

Safe Construction Works Made of Reliable Construction Products based on verified structural calculations

G. Breitschaft[1]
[1]President of DIBt (German Institute of Construction Engineering), Berlin, Germany

Abstract: Ensuring the safety of buildings and other construction works is a major public concern. This article gives an overview of the mechanisms and structures put in place in Europe, and more specifically Germany, to ensure safety in construction. The article identifies major risk factors and shows how the European and German governmental and private systems respond to these risks. The questions addressed include: How do regulators ensure the use of safe construction products and designs, and how is the execution of construction works monitored?

1 Introduction

Safety in construction is a key issue for all societies in the world. We want to be sure that our life and health are safe when we enter a building, pass over a bridge or use a street.

However, it is impossible to make construction works 100% safe. This is why each society has to define the level of risk it is prepared to accept. The individual levels of risk vary from one country to another and also strongly depend on economic factors.

Construction works consist of numerous products assembled using manpower and (often) machines. In order to ensure the safety of construction works, the construction products used need to guarantee a certain performance level over time. Also, the construction techniques used need to meet certain minimum standards as well as structural calculations.Establishing and running good quality systems is always expensive, however, rebuilding a collapsed building is much more costly – not to mention the value of people's lives and health.

2 Ensuring the use of reliable construction products

In Germany and in most other European countries, construction products usually need to comply with a standard or another technical specification ("standardised products") before being incorporated into construction works.

"Non-standardised products" generally may not be used. The manufacturer of the construction product has to affix a marking to the product itself or the accompanying documents, indicating which technical specification has been used. In Germany, products complying with the national technical specifications bear the Ü mark (from "Überein-stimmung", the German for "conformity") while products complying with European technical specifications bear the CE marking. At a national level, the applicable rules are laid down in the Building Codes of the German federal states, and at a European level, they are set out in the Construction Products Regulation [1]. The national and European legal frameworks must not conflict. In case of doubt, European law prevails.

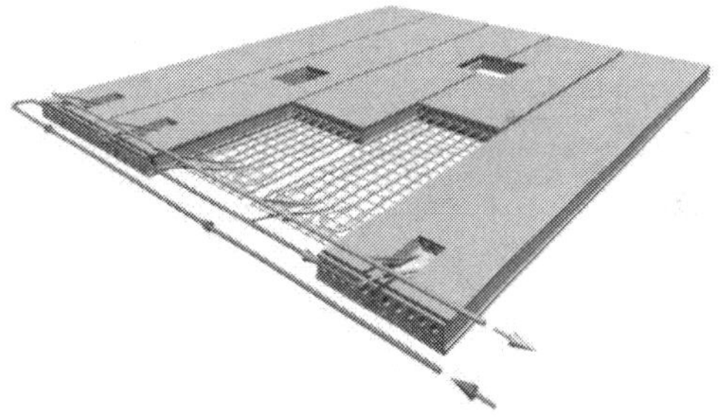

Fig. 1. Technical specifications are used to ensure the reliability of construction products, e.g. in prestressed hollow core slabs (source: Forschungsgesellschaft VMM Spannbetonplatten GbR, Kerpen)

2.1 Technical specifications

Technical specifications are mostly standards, which are developed by European standards organisations. If there is no standard, which is often the case for innovative products, approvals or technical assessments are used. These can only be issued by technical assessment bodies specially recognised for this purpose. For individual projects, it is often possible to apply for a special authorisation with the local authorities. Technical specifications define certain product characteristics and the (test or calculation) methods used in order to assess the performance levels achieved for these characteristics. The technical specifications also contain requirements relating to the production process to ensure that the performance level remains constant over time.

2.1.1 Standards

Most European countries each have one main standards organisation. Standards organisations are usually private sector institutions funded by the industry. Selling the finalised standards is an important source of their financing. However, many of these organisations are still linked to the public authorities via different agreements. New standards projects are usually put forward by industrial stakeholders, but they can also be initiated by public institutions. For example, the European Commission has mandated (and partly financed) the development of all harmonised European standards for construction products. Standards organisations are expected to give all relevant stakeholders the opportunity to participate equally and fairly in the preparation of a standard. The European Committee for Standardization CEN brings together the national standards organisations in Europe.

In addition to the big standards organisations, there are smaller private or public organisations involved in developing technical specifications for speciality areas – usually calling these "guidelines" or "rules" instead of standards.

2.1.2 Approvals and technical assessments

In Germany, if there is no standard (yet), there are two options. The manufacturer can either apply for a national technical approval, or he can apply for a European Technical Assessment (ETA). ETAs are issued on the basis of European Assessment Documents (EADs) which constitute the relevant technical specifications. In other words, within the European legal framework, the EAD sets out the applicable test methods, while the ETA contains the assessment of an individual construction product. With German national technical approvals, both aspects – the technical specification (test methods to be used) and the assessment of the test results – are combined in one document.

The Technical Assessment Bodies entitled to issue ETAs have to be designated by the Member States of the European Union. Germany has designated one single body: DIBt. In total, there are 47 Technical Assessment Bodies across Europe. Some of these bodies are public institutions, like DIBt; others are private sector organisations. An approval or ETA is only issued at the request of the manufacturer, who also has to bear the cost for drafting the document.

2.2 Ensuring the constancy of performance of construction products

Initial testing in accordance with a technical specification is like a snapshot: it certifies that a construction product showed a certain performance at a certain point in time. However, there is also a need for mechanisms to ensure that product performance remains constant over time. Depending on the impact of the construction product on the safety of the building and the risks associated with product failure, more or less stringent assessment systems apply. To clarify, these systems are not designed to assess the durability of construction products once incorporated into the works, but to monitor the manufacturing process. The aim is to make sure that products continue to display the same performance characteristics – let's say two years after the original assessment. In the simplest case, it is left to the manufacturer to ensure the constancy of performance. In this event, the manufacturer simply declares that his products permanently comply with the requirements set out in the technical specification. He also has to implement a production control system called factory production control.

Fig. 2. Example of a highly safety-relevant product (balcony connector; source: Schöck Bauteile GmbH,Baden-Baden)

For highly safety-relevant construction products, the most stringent system applies, meaning that an independent body has to take samples from the production at regular intervals and verify compliance with the technical specification. Furthermore, an independent body has to monitor the factory production control implemented by the manufacturer. In between these extremes, the other systems combine the features outlined above in different ways. It is the task of the public authorities to determine which assessment system applies to which construction product. For products governed by European law, the European Commission determines the applicable system. For products governed by German law, this decision is taken by the national building authorities. The independent testing laboratories and certification bodies involved in the monitoring process have to comply with stringent requirements themselves, and prove their competence and independence. Furthermore, they need to be recognised by a public body. In Germany, DIBt is mandated to recognise these bodies. The manufacturers have to cover the costs for the constancy of performance assessment. The independent testing laboratories and certification bodies have to bear the costs for recognition.

2.3 Market surveillance

By affixing the CE marking to a product or the accompanying documents, the manufacturer guarantees that his product meets the requirements set out in the technical specification and that the constancy of performance assessment has duly been carried out.

The marking allows the contractor, the owner or the site manager to easily determine which construction products may be used on site, and according to which specifications and provisions these products have been manufactured and monitored. Products without such marking may not be stored on site or incorporated into construction works. The marking also gives information about the product's performance characteristics.

To avoid abuse or incorrect use of the marking, a market surveillance system has been put into place as an additional control mechanism. The purpose of market surveillance is to check whether the construction products placed on the market really have the product performance characteristics shown in the marking. Market surveillance is a public responsibility carried out by the authorities. In Germany, DIBt acts as the joint market surveillance authority of the German federal states.

The products and the accompanying documents may be checked at the manufacturing plant, in stores or in transit. The first step is to check if the documents and/or the marking show any evident formal shortcomings. In cases of suspected irregularities, product samples are taken and tested (e.g. for strength, reaction to fire behaviour or environmental effects).

In Europe, a differentiation is made between "active" and "reactive" market surveillance. Active market surveillance means that the market surveillance authorities carry out random checks in accordance with a previously established market surveillance programme. This, however, often only leads to the detection of clerical or formal errors.
Reactive market surveillance means that the market surveillance authorities follow up reports on suspicious products received from competitors or complaints received after an incident. Reactive market surveillance has proved an effective way of detecting fraud. The fraudulent party is liable to sanctions and fines.

The costs incurred by market surveillance measures are covered by the state, as long as the economic operator acted in good faith. However, if a manufacturer, importer or distributor is found to have acted fraudulently, he will have to pay the entire costs and additionally a fine, where applicable.

3 Ensuring reliable design and execution

Apart from ensuring the use of reliable construction products, regulators also have to ensure that the planning, design and construction of buildings and other works meet certain quality standards. During the planning stage, special attention needs to be paid to how the different products interact. As there is no harmonised European legislation in this field, different control mechanisms and measures have been implemented across Europe to ensure the safety of construction works. These systems vary considerably from one country to another. Below, I will therefore focus on the German regulatory system.

Fig. 3. Example of tests to be carried out on construction works (source: Adolf Würth GmbH & Co. KG, Künzelsau)

3.1 Ensuring a reliable structural design

There is a comprehensive range of design standards to cover all types of construction works. The main set of structural design standards are the Eurocodes. This set of European standards covers all common types of construction: concrete, steel, composite steel and concrete, aluminium, timber, masonry and geotechnical design as well as general principles of structural design (including elements of safety theory), actions on structures and seismic design. The Eurocodes contain what are called "nationally determined parameters (NDPs)" to account for different local conditions. These nationally determined parameters are specified in National Annexes to the Eurocodes and are also published as standards. The use of the Eurocodes is mandatory in Germany.
In addition to the Eurocodes, there are several other relevant standards and guidelines in Germany relating, amongst other things, to wind turbines, chimneys, belfries and glass structures. The German building authorities regularly publish lists containing those standards and technical rules which designers and planners should use to ensure safe building. If designers or planners wish to deviate from these technical rules, they have to convince the building authorities that the rules they wish to apply are at

least equivalent to the ones in the official lists (which is not an easy thing to do). If an economic operator wishes to establish a non-standardised design, e.g. a structural design using innovative building materials, he needs to apply to DIBt for approval. Other European Countries use a "voluntary" system. Here, it is not the authorities which require an approval but the insurance companies. In order to get good insurance conditions, the buildings have to comply with certain national specifications.

Structural designs may only be elaborated by experienced structural engineers who need to prove their technical expertise in order to have their names put on special lists kept by the construction industry chambers. Professional liability insurance is mandatory. Additionally, structural calculations and drawings of complex or critical constructions have to be verified by a "certified structural engineer". Certified structural engineers have to demonstrate extended knowledge and experience and pass a public examination. Certified structural engineers are commissioned either by the owner or by the building authorities. The verification services have to be paid for by the owner.

3.2 Ensuring reliable execution of construction works

There are also many standards, guidelines and codes of good practice relating to the execution of construction works. These rules are elaborated by the European Committee for Standardization CEN, the different national standards organisations, such as DIN in Germany, or building trade organisations and associations. They may be made binding at a national level. In Germany, the relevant technical rules are published in official lists established by the building authorities and serve as guidelines to the tradesmen. In addition to this, there are binding legal provisions, especially relating to the vocational training of building tradesmen and to occupational safety and health.

Execution rules specify how construction products have to be assembled to form building components or entire buildings. They also define minimum qualification requirements for tradesmen and certification procedures to check compliance with the relevant standards or the technical expertise of the tradesmen.

Several stakeholders are involved in monitoring execution. Under private law, the owner or the general contractor ensures that contractual obligations are fulfilled. In addition to this, public building controls are carried out by the building authorities. These, however, are limited to infrequent spot checks. For complex or critical projects, the certified structural engineer who verified the structural design has to be involved in the monitoring process, verifying compliance with the plans and the correct execution of the construction works. The private and public stakeholders involved in the monitoring process also ensure that only construction products bearing a conformity marking are used.

Fig. 4. Safe construction products, safe designs and safe execution guarantee safe buildings.

4 Conclusion

As we have said, it is not possible to guarantee absolute safety in construction. However, the measures described above have proven very effective in ensuring a high level of construction safety in Europe. Which lessons can we draw from the European experience? First of all, it can be said that it is insufficient to rely entirely on the civil liability of the manufacturer or the contractor to ensure the safety of construction works. Safety is a public good and public control measures are necessary. Secondly, even though enforcement measures are needed to ensure compliance with the legal provisions, the main focus of the public authorities should be to prevent risks,e.g. through technical assessments before marketing or verification of structural designs before execution. Thirdly, control measures should apply at different stages of the construction process and include the following: initial testing and assessment of the construction product, continuous surveillance of the manufacturing process to ensure constancy of product performance, verification of structural designs and monitoring of the execution of construction works. Finally, it must not be forgotten that construction is an activity. Therefore, proper training, qualification and information of all stakeholders involved are paramount.

References

1. Regulation (EU) No 305/2011, as amended

Major projects in urban sites: Challenge of the Builders

Grands projets en site urbain: Défi et chalenge des Bâtisseurs

Ammar DHOUIB

VINCI, Paris France
61, avenue Jules Quentin 92730 Nanterre cedex. Courriel : ammar.dhouib@vinci-construction.fr

Abstract. Faced with the complexity of urban construction projects and difficulties in the field, engineers must, on the one hand, dimension with rigor and common sense the foundations with all the requirements of today of quality, conformity and respect of deadlines and budget and take into account, on the other hand, safety and environmental requirements and societal and sustainable development criteria, the purpose of this communication is to present concrete projects of foundations and excavation deep in geologically heterogeneous and highly urbanized sites, with monitoring and displacement measurements in order to compare predictions with reality and to promote the "observational method".

Résumé. Face à la complexité des projets de construction en site urbain et aux difficultés de terrain, les ingénieurs doivent, d'une part, dimensionner avec rigueur et bon sens les fondations avec toutes les exigences d'aujourd'hui de qualité, de conformité et de respect de délais et de budget et tenir compte, d'autre part, des impératifs de sécurité et d'environnement et des critères sociétaux et de développement durable, L'objectif de cette communication est de présenter des projets concrets de fondations et d'excavation profondes dans des sites géologiquement hétérogènes et fortement urbanisés, avec suivi et mesures des déplacements afin de confronter les prédictions à la réalité et de promouvoir la « méthode observationnelle ».

1 Préambule

Les projets sont de plus en plus complexes entre tours de plusieurs dizaines, voire plusieurs centaines de mètres de hauteur parfois implantées au milieu d'ouvrages existants sensibles, excavations profondes, ponts en site aquatique et tunnels dans des sols meubles et en milieux urbains sensibles.

A ce défi affronté par les bâtisseurs s'ajoutent les problèmes de terrain où les bons terrains sont, de nos jours, rares. Car, entre friches industrielles et terrains marécageux, le sol est souvent aquifère, mou et/ou compressible et ses déformations sous le poids des ouvrages sont importantes et deviennent, dans ce contexte, souvent incompatibles avec les déplacements admissibles de la structure.

Face à La réglementation, qui a connu un développement considérable et devient de plus en plus exigeante par de nouvelles normes françaises et européennes : missions géotechniques normalisées et « Eurocodes », se limiter à dimensionner, avec rigueur et bon sens, les fondations ne suffit plus en raison, d'une part, des exigences d'aujourd'hui de qualité, de conformité et de respect de délais et de budget et, d'autre part, des impératifs de sécurité et d'environnement et des critères sociétaux et de développement durable.

L'objectif de cette communication est de présenter des projets concrets de fondations d'une grande tour dont le noyau central apporte plus de 1.100.000 kN à un radier général assis sur substratum en forte pente, de fondations superficielles sur semelles, de pieux sous charges verticales centrées, et enfin d'une grande excavation de 70 m de haut en site géologiquement hétérogène et fortement urbanisé.

Ces cas réels présentés ici ont fait l'objet de suivi et de mesures des déplacements, ce qui a permis de confronter les prédictions à la réalité du terrain et de « promouvoir » la « Méthode observationnelle », en guise de démontrer que les aléas et risques sont omniprésents en géotechnique, mais il est possible de les maitriser par des études bien faites, des exécutions soignées et des supervisions adaptées au sens de la norme NF P94-500.

2 Cas de méga-radier d'une grande tour

Dans le cadre d'un projet atypique d'une grande tour d'une quarantaine d'étages sous forme d'amende à façades complexes et lourdes et avec des charges fortement centrées sur un noyau central, le sol est complexe avec un substratum marneux parfois grésifié très compact plongeant de 2,00 à -14,50 NGF vers la mer et partiellement rempli de vases alluvionnaires issues des chenaux qui sillonnent les reliefs montagneux du site et se déversent dans la mer.

Sur le plan mécanique, l'analyse géotechnique initiale d'avant-projet (Mission G12), basée sur les essais pressiométriques, offrait un module pressiométrique moyen (E_M) de 180 MPa dans le substratum marneux. Des études complémentaires en phase d'exécution (Mission G3) avec des essais « haute pression » et des calibrages des sondes pour les corrections des modules pressiométriques selon les règles de la norme NF P94-110 (Dhouib et al., 2002, Dhouib, 2016) ont abouti à un module pressiométrique corrigé moyen dans le substratum marneux de 415 MPa ; ces mesures pressiométriques ont été accompagnées d'essais de compression simple avec mesures des modules de déformation E dans plusieurs échantillons marneux. Les essais complémentaires ont permis de valider en phase EXE une solution de fondations « mixte » :

- Radier général pour le noyau central de la tour, radier fondé équitablement sur le substratum et sur du gros béton remplaçant la vase alluvionnaire, le tout à l'abri d'une paroi de confinement à l'intérieur de la paroi périphérique principale,
- Appuis simples en fondations superficielles très ponctuels aux alentours, et,
- « Barrettes-raidisseurs » contre la paroi périphérique à la fois porteuse et faisant l'office d'enceinte étanche.

La figure 1.a illustre la coupe lithologique issue d'une série de sondages carottés, pressiométriques et destructifs enregistrés ayant permis de déceler avec précision (avec une maille fine de 25 m^2) le toit du substratum marneux. Sur la figure 1.b est présentée l'emprise (en forme d'amende) de la tour avec toit des marnes plongeant vers la mer et position du noyau fondé sur le radier général.

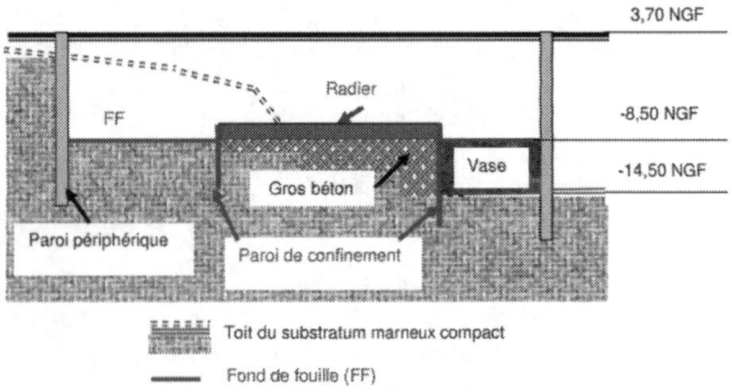

a. Coupe lithologique et fondation de la tour

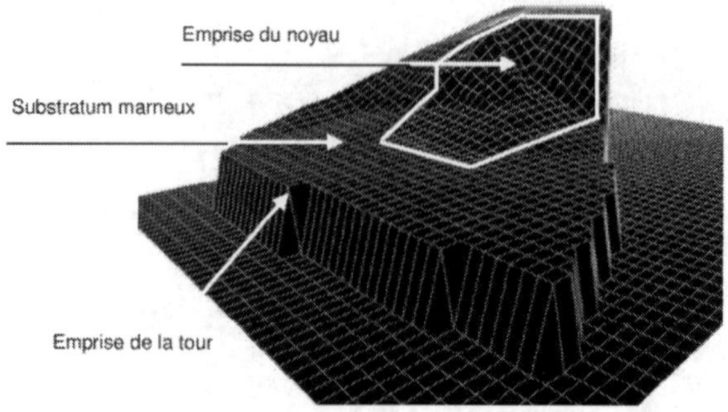

b. Emprise de la tour et position du noyau de la tour

Figure 1 : Contexte géotechnique du site et fondation du noyau de la tour

La figure 2.a met en évidence la synthèse des données pressiométriques E_M corrigés à partir du calibrage des sondes (Dhouib et al., 2002, Dhouib 2016) et des pressions limites nettes p_l* obtenus par des sondes « rocher » où les pressions ont été poussées jusqu'à 10 MPa.

Les tassements de la tour ont été suivis sur une période de 19 mois sur plusieurs cibles, tassements que résume la figure 2.b pour quatre cibles disposées sur la périphérie du noyau en comparaison avec les tassements prédits par des calculs analytiques et numériques en éléments finis en deux dimensions.

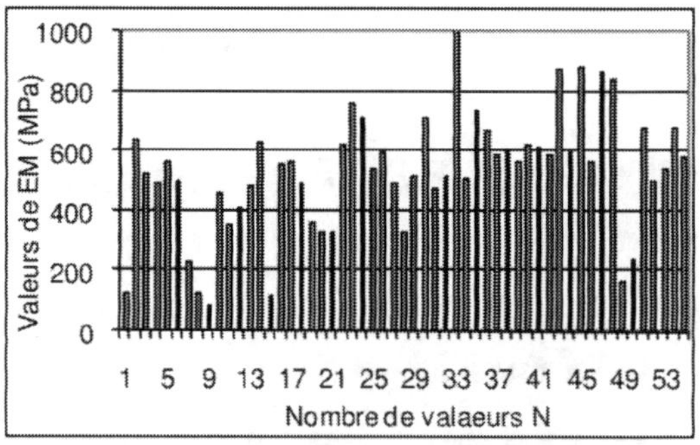

Moyenne harmonique E_M=415 MPa

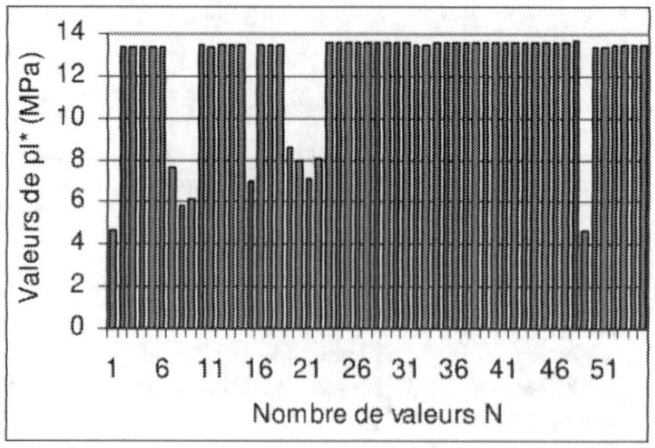

a. Synthèse des données pressiométriques

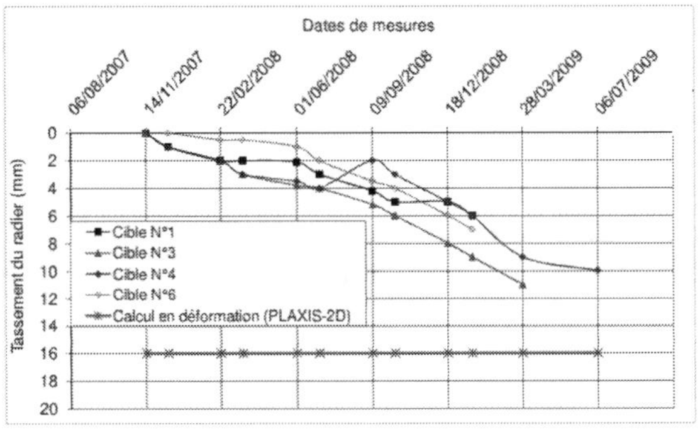

b. Tassements mesurés et prédits du radier général

Figure 2 : Données mécaniques et tassement du radier général de la tour

Commentaires

- *Compte tenu du pendage du substratum, la tour a fait l'objet d'une vérification de la stabilité vis-à-vis du glissement généralisé. Les coefficients de sécurité au glissement vont, selon les combinaisons des charges, de 2,10 à 7,10.*
- *La stabilité du fond de fouille est vérifiée vis-à-vis du « renard-solide » avec un coefficient de sécurité de 16,1.*

3 Cas de semelles filantes

Ce projet de bâtiment de faible hauteur de type RDC + 3 étages sur fondations superficielles filantes a fait l'objet de plusieurs études dans le cadre des missions géotechniques normalisées. La mission, G3 diligentée par l'entreprise en phase EXE, a été clôturée par un rapport de mission G3 – Phase étude.

Pour une emprise au sol du bâtiment de 1800 m^2, 3 sondages initiaux (SP1, SP2 et SP3) dans la cadre des misions G1 et G2 ont été complétés en phase G3 par 2 sondages SP101 et SP102, afin de mieux couvrir l'emprise et respecter le ratio habituel de 1 sondage pour 400 m^2 dans des sols sans anomalies particulières d'origine anthropique (Carrières, marnières, catiches) ou naturelle (dissolutions de gypse, poches karstique). Lorsque le sous-sol du projet se situe dans une zone à anomalies, les sondages doivent être plus serrés jusqu'à un sondage tous les 200 m^2 (selon la notice de l'Inspection Générale des Carrières : IGC, 2003 en région parisienne), voire moins.

A partir des 5 sondages pressiométriques, la synthèse des contraintes aux Etats Limites Ultimes ELU et aux Etats Limites de Service ELS est, par application du DTU 13.12 (1992) pour le dimensionnement et la justification des semelles superficielles ancrées dans des formations de pente (éboulis) sableuses, est portée sur la figure 3.

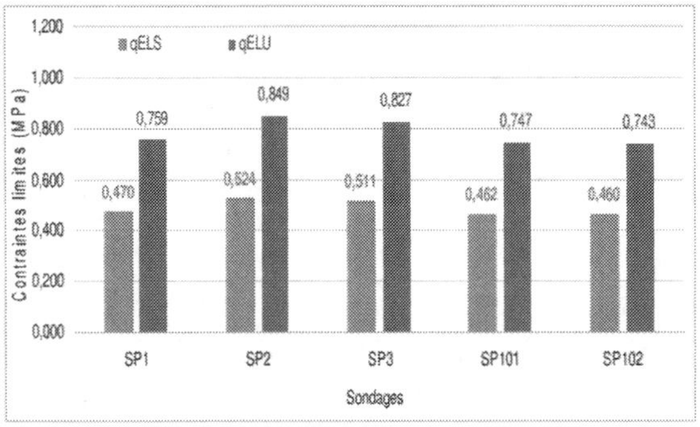

Figure 3 : Contraintes aux Etats Limites (q_{ELU} et q_{ELS})

La contrainte aux Etats Limites de Service ELS varie, d'après les histogrammes de la figure 3, de 0,460 à 0,524 MPa, ce qui permet de retenir raisonnablement une valeur de 0,450 MPa, alors que le rapport initial de mission G1 puis G2 la fixait arbitrairement à 0,30 MPa.

Sous la contrainte aux Etats Limites de Service de 0,450 MPa, les tassements théoriques absolus sont estimés entre 5 mm pour les semelles périphériques moins chargées (SF1 et SF2 de largeur 0,6 à 0,7 m) et 10 mm pour les semelles intérieures plus chargées (SF10 et SF11 de largeur 1,0 à 1,2 m), comme l'illustrent les graphes de la figure 4.a, qui comparent les tassements théoriques absolus aux tassements mesurés pour les semelles d'angle, dont les tassements ont été régulièrement suivis par de simples visées de géomètre pendant 24 mois.

Les tassements différentiels, selon une coupe type et représentative, ne dépassant pas les 0,5 mm entre files espacées de 6,0 m, vérifient largement la règle usuelle de 1/500, ce qu'illustre la figure 4.b.

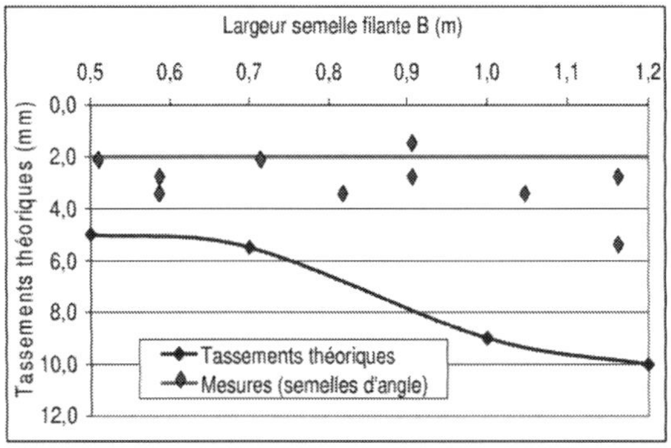

a. Tassements absolus : Mesures et prédiction

https://doi.org/10.1051/matecconf/201814902003

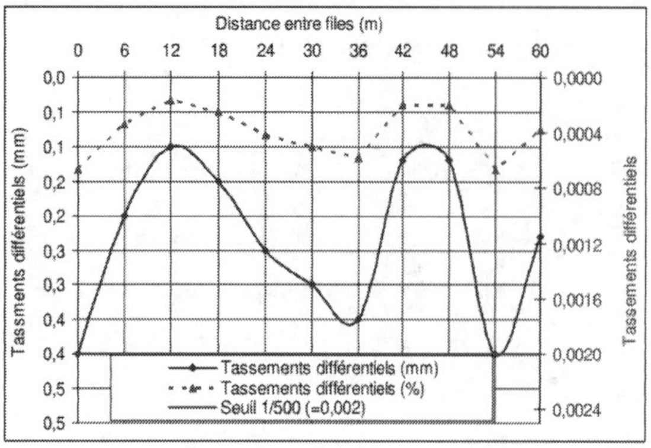

b. Tassements différentiels

Figure 4 : Tassements absolus et différentiels de semelles filantes

Commentaires :

- *Les tassements théoriques absolus calculés ici sont faibles et vérifient parfaitement la règle T0 de Ménard adaptée en très petites déformations (Baguelin et al., 2009). Ils sont compatibles avec la structure, et la contrainte aux Etats Limites de Service ELS de 0,450 MPa est donc admissible.*
- *Lorsque les essais pressiométriques sont bien faits (forage bien calibré) et les charges d'un ouvrage bien réparties entre files, respectant les distances usuelles et les fouilles bien soignées et curées, les tassements, totaux et différentiels, sont généralement faibles et toujours compatibles avec la structure ; c'est ce qui fait la « force » de la méthode pressiométrique de Louis Ménard et son avantage par rapport aux autres méthodes de dimensionnement des fondations.*

4 Comportement de pieux sous charge axiale centrée

Les pieux sous charge verticale ont fait l'objet, par le réseau des laboratoires des Ponts et Chaussées, de plusieurs instrumentations et prédictions des tassements depuis des décennies. Sur la figure 5 sont portés les tassements mesurés en tête de pieux ancrées dans les alluvions de la Seine de la région parisienne et soumis à une charge verticale centrée en comparaison avec les calculs en déformation (Eléments finis) et calculs basés sur le principe des fonctions de transfert (à l'aide du logiciel « PIVER ».

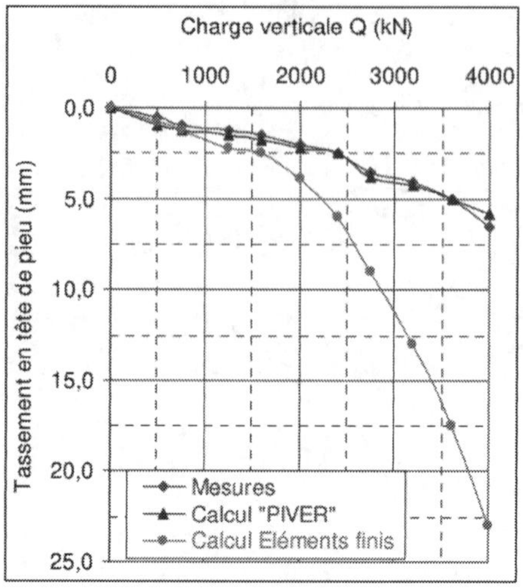

a. Pieu diamètre 650 mm

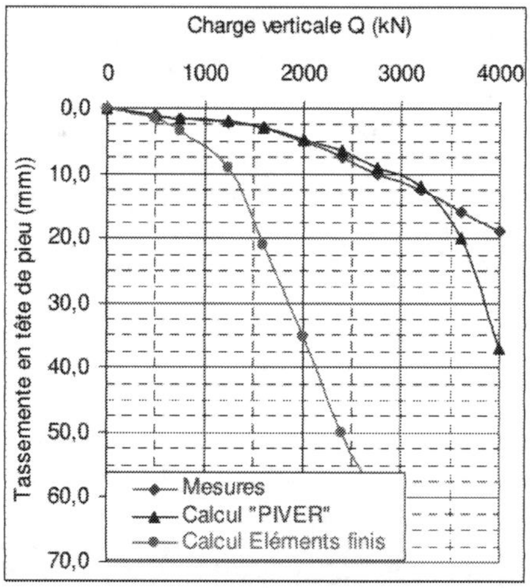

b. Pieux diamètre 1500 mm

Figure 5 : Tassements en tête de pieux : Mesures et prédictions

Notons que, dans les deux cas de pieux de la figure 5, les calculs en déformation par la méthode des éléments finis donnent des tassements beaucoup plus importants que les enfoncements mesurés en têtes de pieux, et ce notamment pour les charges importantes. Ceci est inhérent aux calculs en déformation qui font appel à des modules de déformation, soit mal mesurés, soit mal appréciés. Un autre inconvénient des modèles en éléments finis réside dans la mobilisation de déformations dans les couches profondes qui normalement ne tassent pas. Les calculs utilisant les fonctions de transfert offrent, dans les deux cas de pieux, un parfait ajustement, à l'exception des charges extrêmes pour le fût de 1500 mm de diamètre.

5 Retour d'expérience et leçons à tirer

Les différents cas réels d'ouvrages présentés dans cette présentation montrent que les tassements mesurés sont toujours inférieurs aux prédictions issues des calculs, qu'ils soient analytiques ou numériques. La tendance des jeunes ingénieurs est de courir, de nos jours, aux calculs numériques car il est plus facile d'allumer son ordinateur et de lancer une « boite noire » que constituent notamment les codes en éléments finis que de prendre son crayon, du papier et la calculatrice pour avoir au moins l'ordre de grandeur des phénomènes.

Pour rendre plus crédible leur action en géotechnique, les ingénieurs doivent d'abord analyser avec critique et attention les données issues des mesures, qu'elles soient sur site ou en laboratoire, en faire une synthèse et les vérifier par rapport à des données existantes pour retenir un jeu ou plusieurs jeux de paramètres avant de se lancer dans les calculs. Lorsque les prédictions sont faites, il est important et primordial d'élaborer un vrai programme de mesures et de suivi préalable sur chantier et le mettre en exécution dès le démarrage du chantier. Ce qui n'est pas toujours facile même dans des grandes entreprises de renom car les maîtres de l'ouvrage ne veulent pas aller dans ce sens pour raisons d'économie.

Afin d'enrichir ce retour d'expérience, nous avons apporté sur la figure 6 les tassements mesurés et calculés sous plusieurs ouvrages réels, allant des semelles superficielles, en passant par les radiers généraux, jusqu'aux remblais routiers et autoroutiers.

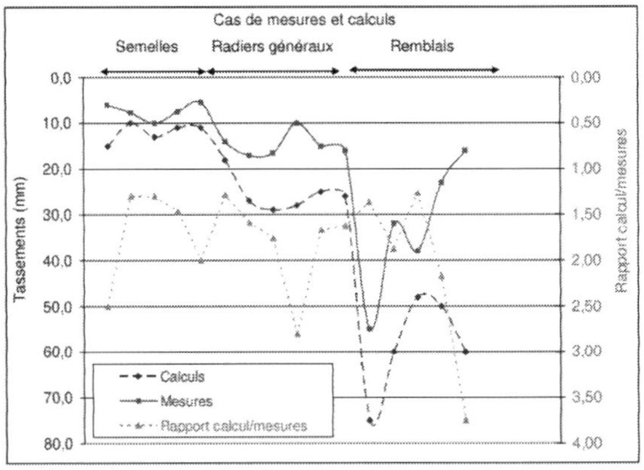

Figure 6 : Exemples de tassements mesurés et calculés sur ouvrages réels

A l'examen des graphes de la figure 6, il y a lieu de noter :

- Les tassements théoriques sont toujours supérieurs aux tassements mesurés. Le rapport des tassements calculés aux tassements mesurés varie de 1,14 (un cas de semelles superficielles) à 3,75 (un exemple de remblai sur sol moyennement compact).
- Sont rares les cas où les tassements mesurés dépassent les tassements prédits par les calculs. Mais il y a, parfois, des exemples atypiques : un remblai de faible hauteur repose sur une puissance de 3 m de limons où divers essais ont été effectués le mois de juillet en plein été. Les calculs des tassements ont été conduits sur la base de caractéristiques mécaniques « estivales » dans des limons secs qui sont trois fois plus élevées que les valeurs réelles « hivernales » dans les limons humides avec des battements de la nappe en relation directe avec un ruisseau tout proche. Les tassements mesurés étaient trois plus importants que les tassements calculés. Un autre cas de remblai de 6 m de haut sur une décharge de 12 m « améliorée » par des inclusions rigides a conduit à des tassements presque 5 fois plus élevés que ceux calculés à partir de mauvais rentrants en raison du caractère très hétérogène des matériaux de la décharge et de points singuliers en termes de modules pressiométriques.
- Dans les divers projets instrumentés et suivis par nos soins ou apportés à notre connaissance par la profession et constatés dans les expertises judiciaires et extrajudiciaires, le rapport des tassements calculés aux mesurés sous fondations (semelles superficielles, radiers et remblais) et entre déplacements calculés et mesurés (écrans et murs de soutènement) de 1,5 à 3 est une valeur courante en géotechnique.
- En ce qui concerne les fondations profondes sur pieux, barrettes et parois porteuses, les enfoncements des fûts dans le sol sont faibles, voire nuls car la pointe des pieux est généralement toujours ancrée (ancrage effectif) dans un sol résistant et compact. Dans ce cas, les tassements des pieux se réduisent aux raccourcissements élastiques sur la hauteur de la couche médiocre. Cependant, il y a eu bel et bien quelques cas, mais rares, de sinistres liés aux tassements excessifs de pieux métalliques ouverts ancrés par battage dans la craie qui se « liquéfie » (« Phénomène de thixotropie ») en cours de battage et de pieux particuliers forés à la tarière, non suffisamment ancrés dans le substratum marneux compact et ayant conduit à un sinistre très couteux en matériel (travaux lourds de reprise par micropieux autour des pieux défaillants) et immatériel (bâtiment non livré à temps et personnel déplacé sur un autre site).
- Les tassements des fondations d'un ouvrage dépendent à la fois de la compacité du sol support et de la contrainte qu'apporte au sol la fondation de l'ouvrage. En fait, les deux paramètres sont liés, car une tour de grande hauteur nécessite un support résistant pour être fondée alors qu'un pavillon a besoin d'un « petit bar (0,1 MPa) » pour asseoir ses fondations dans un limon (la « terre du paysan »).

Sous les contraintes appliquées par l'ouvrage, en partant d'un état initial de faible niveau de déformation caractérisé par un module de déformation du sol maximal E_0 (soit un module de cisaillement maximal G_0), le sol se déforme et sa rigidité diminue lorsque la déformation (ε) sous l'ouvrage croît. Le comportement des ouvrages de construction (fondations, murs et écrans de soutènement, tunnels souterrains, remblais sur sol compressible) et leur stabilité et pérennité dans le temps dépendent du niveau

https://doi.org/10.1051/matecconf/201814902003

de déformation atteint pendant l'application de la sollicitation

La figure 7 présente le domaine conventionnel du niveau de déformation que peut accepter les ouvrages de construction et les variations des modules de déformation (E_s) et de cisaillement (G_s) du sol sous ces ouvrages en fonction de la déformation.

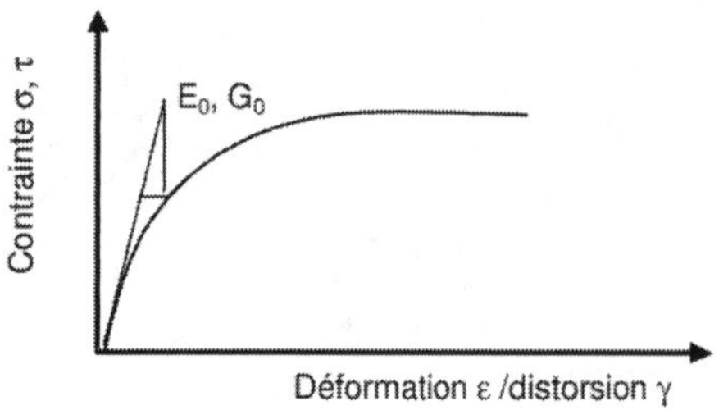

a. Courbe de contrainte-déformation du sol sous les ouvrages

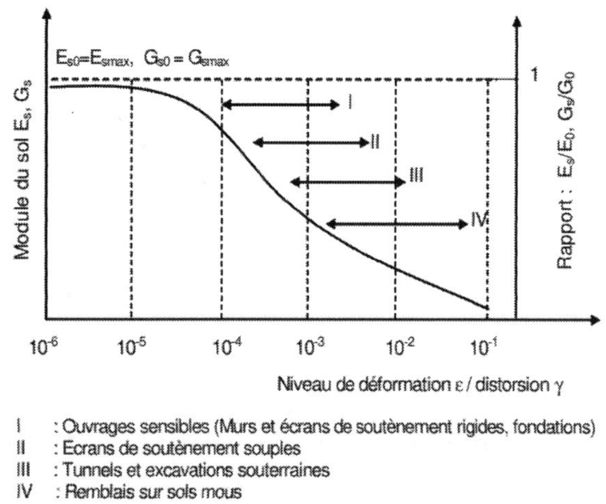

b. Domaine de déformation du sol en relation avec les ouvrages de construction

Figure 7 : Domaine conventionnel du niveau de déformation sous les ouvrages de construction

6 Excavations profondes en site urbain

En France, les calculs, le dimensionnement et les justifications des excavations profondes soutenues par des écrans de soutènement font appel à trois méthodes différentes :

- Calcul à la rupture : détermination du diagramme des pressions et mise en équations d'équilibre,
- Méthode aux coefficients de réaction : principe du ressort « unidimensionnel »,
- Calcul numérique en éléments finis : assemblage de plusieurs ressorts 2D ou 3D.

La méthode basée sur le principe du coefficient de réaction est utilisée, spécifiquement en France, de manière récurrente et elle a fait ses preuves puisque 90% des ouvrages de soutènements sont calculés, dimensionnés et justifiés par cette méthode depuis 60 ans. En effet, cette méthode permet de modéliser le sol en poussée et en butée contre l'écran de soutènement (Figure 8) par des ressorts, comme le montre la figure 9.

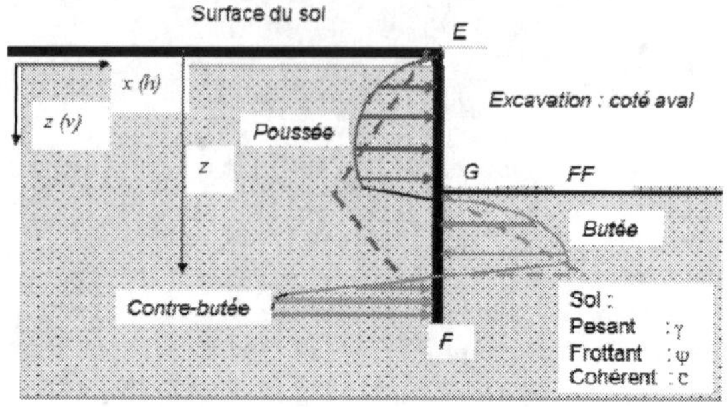

Figure 8 : Mécanisme de poussée-butée contre l'écran de soutènement

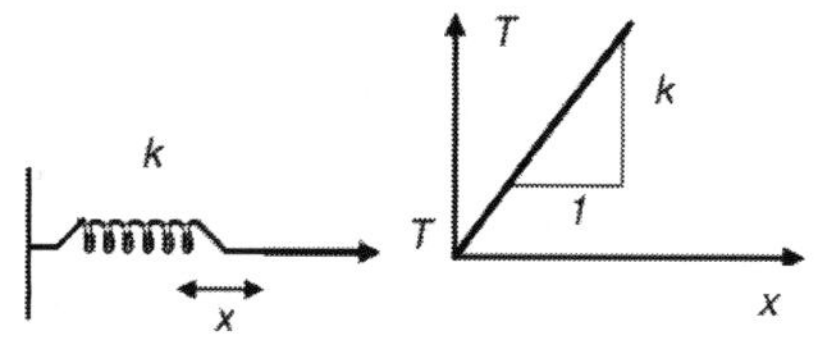

$T = -k.x$ (Winckler, 1868)

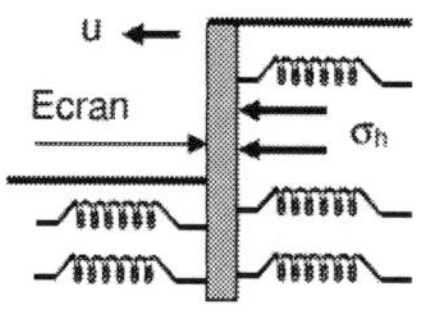

$p = k_s.u$

Figure 9 : Principe de la méthode du coefficient de réaction (Dhouib, 1995)

L'avantage de la méthode aux coefficients de réaction réside dans sa simplicité de mise en œuvre dans les calculs automatiques. Elle a l'inconvénient de supposer le sol « élastique » (Figure 10.a), alors qu'il en est guère (Figure 10.b). En effet, tous les codes de calculs numériques utilisant cette méthode en France sont programmés avec la courbe de réaction bilinéaire de la figure 10.a.

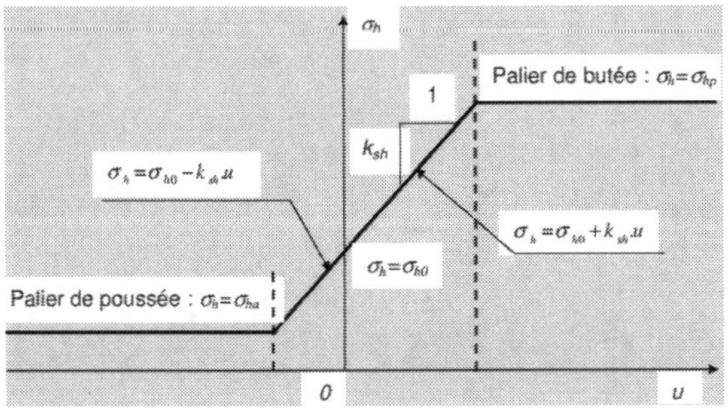

a. Courbe de réaction bilinéaire (Ménard et al., 1964)

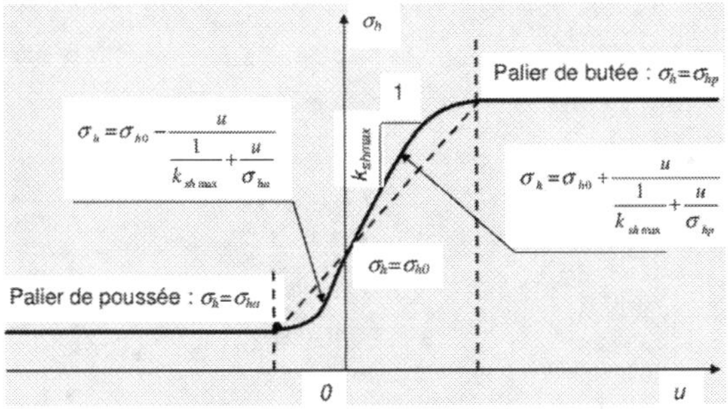

b. Courbe de réaction hyperbolique (Dhouib, 1995)

Figure 10 : Courbes de réaction pour le calcul des soutènements en France.

La figure 11 illustre le cas d'une paroi moulée ancrées par des tirants actifs.

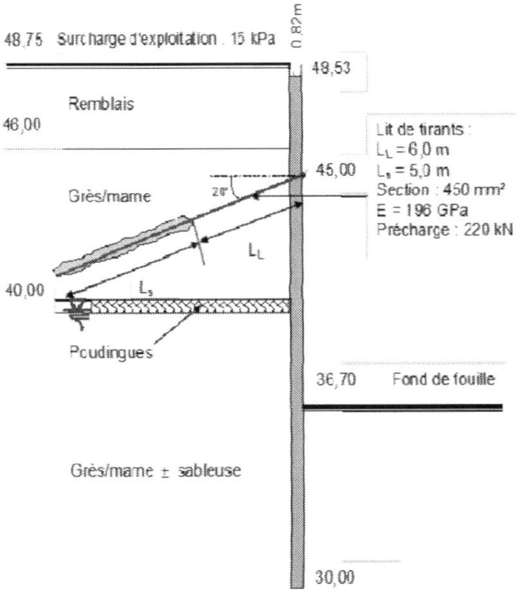

Figure 11 : Cas de paroi moulée ancrée par tirants actifs

La photo de la figure 12.a illustre le panneau de paroi instrumenté et les graphes de la figure 12.b et présentent et comparent les déplacements horizontaux mesurés aux déplacements prédits de la paroi. Les déplacements calculés en éléments finis (code « PLAXIS ») et par la méthode aux coefficients de récation (logiciel « RIDO ») convergent, à ces niveaux de déplacements faibles, avec les mesures (Dhouib et al., 2007).

a. Elévation de la paroi instrumentée

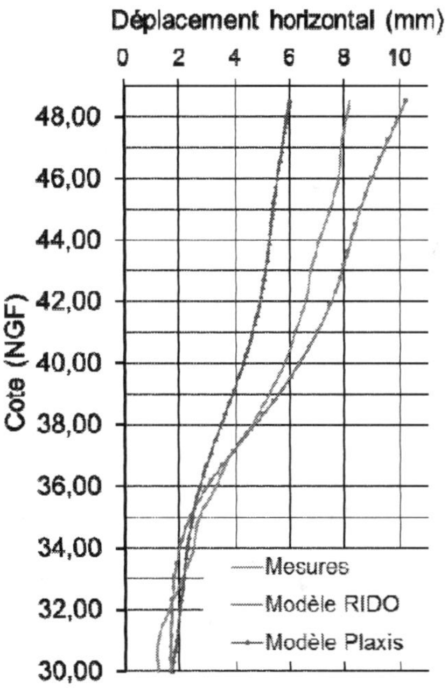

b. Comparaison des déplacements horizontaux mesurés et prédits de la poaroi

Figure 12 : Cas de paroi moulée ancrée par tirants actifs

Une autre paroi exceptionnelle d'un mégaprojet d'une grande tour de 54 étages, fouille de 70 m de hauteur dans un site géologiquement complexe et difficilement accessible a été instrumentée afin de suivre, d'une part,

 https://doi.org/10.1051/matecconf/201814902003

les déplacements de la paroi, et d'autres part, les mouvements de l'environnement immédiat.

La figure 13.a présente le site concerné par le projet et la coupe de la figure 13.b illustre la coupe type des sols et de la fouille de 70 m de hauteur, soutenue par trois parois différentes : micro-berlinoise en tête, puis paroi berlinoise au milieu suivie de paroi moulée pré-fondée en pied, dans laquelle vient s'incorporer la grande tour.

a. Site du projet de la tour

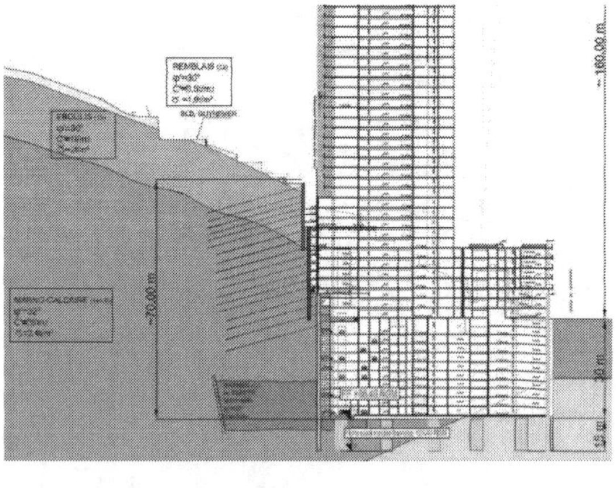

b. Coupe type de la fouille à parois multiples et de la tour de 54 étages

Figure 13 : Cas de fouilles de 70 de hauteur

La sensibilité du projet était si grande que toutes les parois ont fait l'objet de suivi des déplacements et les existants de tassements, voire de soulèvements.

La figure 14 présente les phases d'excavation et les résultats des auscultations de la paroi moulée.

a. Paroi micro-berlinoise en cours d'excavation

b. Fin d'excavations de la paroi berlinoise

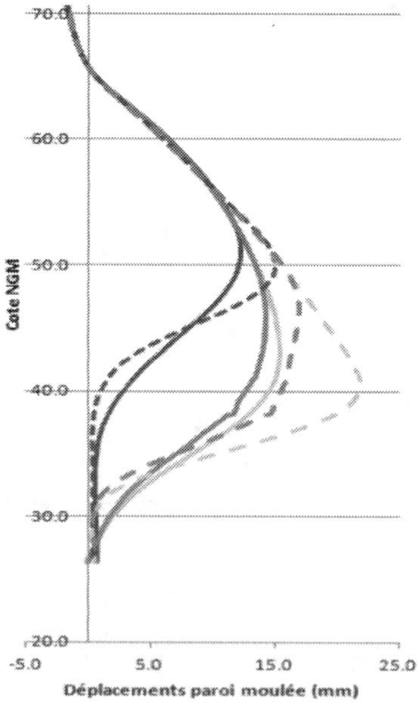

c. Déplacement en fin d'excavation (paroi moulée)

Figure 14 : Résultats des auscultations des parois

7 Conclusions

Les différents cas réels d'ouvrages présentés dans cette communication montrent que les déplacements des ouvrages mesurés sont toujours inférieurs aux prédictions issues des calculs, qu'ils soient analytiques ou numériques. La tendance des jeunes ingénieurs est de courir, de nos jours, aux calculs numériques car il est plus facile d'allumer son ordinateur et de lancer une « boite noire » que constituent notamment les codes en éléments finis que de prendre son crayon, du papier et la calculatrice pour avoir au moins l'ordre de grandeur des phénomènes.

Pour rendre plus crédible leur action en géotechnique, les ingénieurs doivent d'abord analyser avec critique et attention les données issues des mesures, qu'elles soient sur site ou en laboratoire, en faire une synthèse et les vérifier par rapport à des données existantes pour retenir un jeu ou plusieurs jeux de paramètres avant de se lancer dans les calculs.

Lorsque les prédictions sont faites, il est important et primordial d'élaborer un vrai programme d'auscultations et de suivi préalable sur chantier et le mettre en exécution dès le démarrage du chantier. Ce qui n'est pas toujours facile car les maîtres de l'ouvrage ne veulent pas aller dans ce sens pour raisons de « fausses » économies.

Seule cette voie permettrait de confronter les prédictions à la réalité du terrain et de « promouvoir » la « Méthode observationnelle » en guise, en plus, de s'assurer de « *la qualité, la solidité, la stabilité et la pérennité* » des ouvrages de construction.

Références bibliographiques

1. Dhouib, A, "Méthode des coefficients de réaction appliquée en France pour les calculs de soutènements : influence de la variation des paramètres", Revue Française de Géotechnique n° 72, 3ème trimestre 1995, pp. 11-21 (1995).
2. Dhouib, A, Lelièvre S., Idrac, ph., Betoux, O., "Station « La Blancarde » à Marseille : Comportement d'une paroi moulée à ancrages actifs", Revue Travaux n°844, (Septembre 2007).
3. Dhouib, A., "Géotechnique appliquée aux projets de construction", Presses des Ponts et Chaussées, Paris, 1041 pages (2016).
4. Ménard L., Bourdon G., Houy A., "Etude expérimentale de l'encastrement d'un rideau en fonction des caractéristiques pressiométriques du sol de fondation", Sols-Soils n° 9, pp. 11-27 (1964).
5. Ménard L., Bourdon G., "Calcul des rideaux de soutènement : Méthode nouvelle prenant en compte les conditions réelles d'encastrement ", Sols-Soils n° 12, pp. 18-32 (1965).
6. Schmitt P., "Méthode des coefficients de réaction appliquée en France pour les calculs de soutènements : influence de la variation des paramètres", Revue Française de Géotechnique n° 72, 3ème trimestre 1995, pp. 11-21 (1995).
7. Winkler E., "Die Lehre von der Elasticitäs und Festigkeit", Prag. Verlag H. Dominicus (1868).

https://doi.org/10.1051/matecconf/201814902004

Recent Developments in Nanomaterials for Geotechnical and Geoenvironmental Engineering

Mohd Raihan Taha
Institute for Environment and Development (LESTARI) &
Dept of Civil & Structural Engineering,
Universiti Kebangsaan Malaysia, Bandar Baru Bangi, Malaysia

Abstract. Nanotechnology has provided enormous benefits to the industrial and developed world. Most fields of knowledge have been tapping, using, and applying the advantages of nanotechnology to improve systems and living standards. This cannot be said for civil engineering and construction particularly geotechnical engineering. Actually, soil mechanics have been involved with nano scale dimensions and materials for a very long time, i.e. before the term nanotechnology and nanomaterials were coined. From the basic clay structure and the water molecules which give rise to soil plasticity/cohesiveness, geotechnical engineers were exposed to these phenomena since the early days of the subject. Currently, developments in nanotechnology and nanomaterials in geotechnical engineering has been limited to its use as an improvement and clean-up materials. Recent results have shown that the use of a small amount of nanomaterials (<1% of dry weight of the original soil) have resulted in significant change and improvement in soil properties. This is important since efficient use of materials is one of the facets of sustainability related to prolonging the life of our depleting natural resources. Experiences with nano-soil, nano-clay, nano-silica, nano-alumina, nano-magnesium, carbon nanotube and nanofibers, and nano-lime will all be discussed with respect to its use in soil improvement and soil remediation.

1 Introduction

Developments in nanotechnology particularly over the past 15 years has been fully tapped for its advantage by all fields of knowledge. The current state of disciplines such as electronics, manufacturing, medicine, energy, etc. would not have progressed much without nanotechnology. On the other hand, this cannot be much said for geotechnical engineering. However, geotechnical engineers are amongst the earliest "nanotechnologist" in which in the past they have dealt with materials at nano sizes and phenomenon that operates at nano scale [1].

One of the earliest related geotechnical publications was written by Bolt [2] who used the diffuse double layer to study the swelling pressures of clay. Since then these swelling pressure equations (Fig. 1) has been also used by a number of other researchers, notably by Sridharan & Jayadeva [3], Komine & Ogata [4] & [5], Tripathy et al. [6], Tripathy & Schanz [7], and Gurtug [8] to study clay behavior.

This purpose of this paper to provide some latest developments on "nano' related research works in geotechnical and geoenvironmental engineering. Results

$$p=2\eta\kappa T(\cosh u-1) \qquad (i)$$

$$e=G\gamma_w Sd \qquad (ii)$$

$$\int_z^u(2\cosh y-2\cosh u)^{-(1/2)}dy=-\int_0^d d\xi=-Kd \qquad (iii)$$

$$\left(\frac{dy}{d\xi}\right)_{x=0}=\frac{B}{S}\sqrt{\frac{2\pi}{\varepsilon\eta kT}} \qquad (iv)$$

p = repulsive force, η = concentration of cation, κ = Boltzman constant, T = temperature, e = void ratio, G = specific gravity, d = half distance of parallel platelets, y = potential, u = non dimensional mid plane potential, z = non dimensional surface, potential, $K = 1/t_{dl}$, t_{dl} = thickness of double layer, x = the distance from the clay platelets.

Fig.1. Calculations of repulsive pressure from diffuse double layer theory [9].

on the use of nanomaterials such as carbon nanotubes, nanosoil, nanoclay, nanoalumina, etc. for soil improvement will be covered. This serve to show that nanoparticles are the latest generation of materials that can be used for soil treatment. As previously mentioned, geotechnical engineers as one of the major user of materials, should exploit this technology/material to our advantage and provide efficient use of materials for clients/society.

2 Applications of Nanoparticles in Geotechnical and Geoenvironmental Engineering

The use of nanoparticles for geotechnical engineering is still an emerging technology for possibly two fundamental reasons. The first issue is related to cost of the materials. It is generally felt that being nanomaterials, it must be expensive. Generally, the cost of nano materials, for example, carbon nanotubes are much less than it was a decade ago and still reducing due to increase in processing technology. Furthermore, in using nanomaterials, the recommended amount used is quite small, i.e. 0.1-0.5 % of total weight of soil. The use is actually targeted at improving performance of geomaterials not replacing them. The second reason deals with toxicity. It is important to deal with this important subject because it is to be used in the field and directly exposed to the environment. On this concern, much studies have been conducted to evaluate toxicity on animals. While the results indicated that it has some negative impacts, many experts agree that the effects on human is still largely questionable. More studies are required, and safe with responsible use of materials can provide control and minimize risk to human and environment. Nevertheless, nanoparticles are currently already a constituent of much used items [10], for example, skin lotion, and this provide some evidence for supporting the use of nanoparticles against toxicity claims.

3 Nano-Soil

The nano-soil [11] was obtained by grinding a local soil (metasedimentary residual soil within the campus of Universiti Kebangsaan Malaysia, UKM). After the disturbed soil was dug and taken to the laboratory, it was first cleared of all roots and stones, dried in an oven, and sieved. Soil samples passing through 425 mm sieve was then pulverized in a planetary mono mill. In this mill, 20 small (5 mm diameter) balls made of sintered corundum move is a fast planetary/cyclonic fashion to pulverize the soil. The time of pulverization was about 10-13 hours in which during this period, distilled water was added to provide overheating. Fig. 2 show SEM images of the soil sample before and after milling. It can be clearly seen that there is much more soil in the nano range (i.e. 50-60 nm) in the soils after milling. The specific surface of the soil

Fig. 2. SEM images of pulverized UKM soil before (top) and after (bottom) milling process.

obtained from EGME (ethylene glycol monoethyl ether) method improved from 2.4 m^2/g to 3.9 m^2/g. Milling of kaolinite and montmorillonite samples show similar observations.

In order to evaluate its potential for improving soil properties, a soil-cement mix was prepared and the nano-soil was added to the mix. The amount of nano-soil added was 1 and 2 & by weight of the UKM soil. Thus, the overall mixture is 94% UKM original soil, 6 % Portland cement, and nano-soil. The results of compression test on 50 mm^3 samples are shown in Fig. 3 [12]. The results in Fig. 9 show significant improvement in the strength of the nano-soil mixtures over the control test albeit the amount of nano soil added was rather quite low. The strength of the soil-cement mixture almost doubled when 1 % UKM nano-soil was added compared to samples without nano-soil. In addition, the mixture which is usually termed as cement-modified soil in geotechnical engineering will have lower tendency of volume change and PI in addition to increase in load bearing capacity. This study has demonstrated that even a small addition of nanoparticles will show marked enhancement in soil behavior. It is also possible to engineer nanoparticles to improve behavior of soils to suit design and practice requirements.

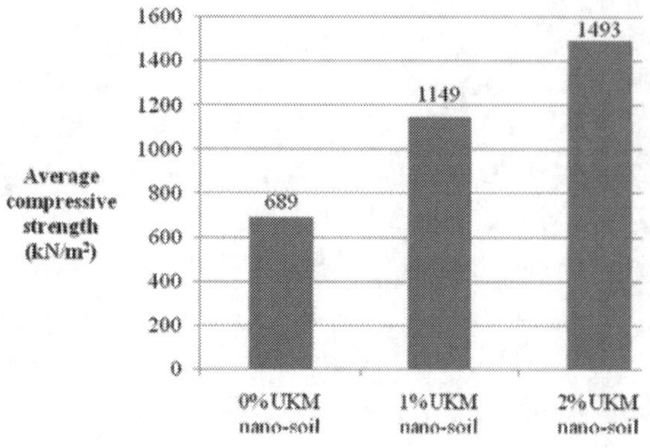

Fig. 3. Compressive strength of soil-cement and nano-soil mix.

4 Carbon Nanotube (CNT)

Carbon nanotube is considered as the first generation nanomaterial in that it is one of the earliest nanoparticles that was produced. Carbon nanotube is so popular and interesting because it has many "super" properties, especially for those working in electronics and superconducting industries. It has high unparalleled strength, high electrical and thermal conductivities, making it ideal for wide range applications [13] & [14]. Thus, it is sometime termed as the "king of nanomaterial" and also "a solution waiting for problem". Interestingly (at least for geotechnical engineers). It resembles a clay mineral, halloysite, which is in the same group as kaolinite, resembles the tubular CNT shape (Fig. 4).

(a) (b)

Fig. 4. (a) Electron photomicrograph of halloysite [5]; and (b) SEM of MWNT (multiwalled nanotube, a type of CNT) [15].

Earlier study has shown that even with small amount of CNT (in the form of multiwalled nanotube from Arkema, France), the changes in basic characteristics are measurable [X]. Increase in CNT led to increase in LL, PL and plasticity index (PI=LL-PL). In these tests, the addition of the fine CNT particles certainly increases the overall specific surface of the mixture. As a result, the need for water, is greater and thus this is indirectly shown by an increase in LL. The increase in PI is due to the fact that the increase in PL is not as dramatic as the increase in LL. It was also found that the values of G_s, in general, also decreases upon addition of CNT due to the low density of CNT.

These results have important significance in geotechnical engineering. In terms of mechanical properties, increase in LL (and increase in surface area) means the soil will have higher capacity for water thus decreasing its strength and increasing its consolidation/compressibility (settlement). However, increase in PI will generally mean a decrease in its hydraulic conductivity. For geotechnical structures (such as foundations and retaining walls) in which strength and compressibility are the main criteria, the use of CNT may not be advantageous. In the case of landfill liners and caps, in which low hydraulic conductivity is very important, its use may thus be considered.

The use of unenhanced CNT was also attempted to adsorb contaminants from the subsurface for application as barrier or filter materials. CNT was used as a remediation material to adsorb DDT from solution. A maximum of almost 90 % adsorption can be achieved (Fig. 5). The relatively high adsorption is due to action of the CNT in which it acts as a sponge due to the strong adsorption forces inside in CNT. Also, because of its small size, the CNT capture and retain small molecules in their interior. Their sorption energy and sorption energy are much higher than that of activated carbon.

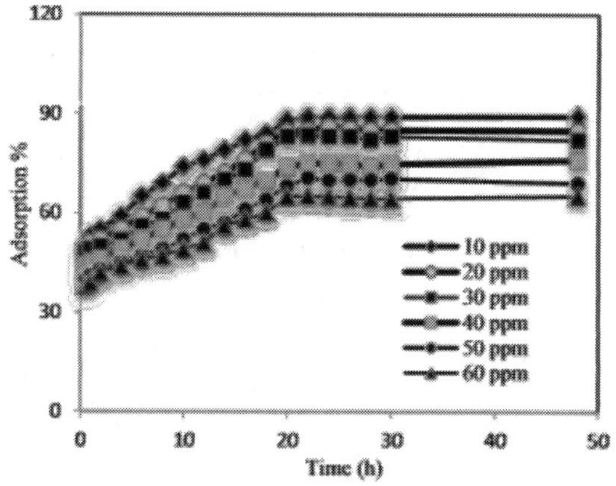

Fig. 5. Adsorption of DDT in various solution by CNT [16]

It must be emphasized that the hydrophobicity and clogging of CNT presents the greatest challenge in mixing soil with CNT. The two test results presented above could be improved if the CNT could be well mixed with the soil. Attempt to mix the soil by sonication have yielded an excellent dispersion of CNT and carbon nanofibers (CNF) within the particles, contributing to bridging effects, which transfer loads across the nano- and micro-cracks.

5 Nano-Clay

Volume change of a compacted residual soil (UKM metasedimentary residual soil) was studied using a natural montmorillonite. The nano-clay material was supplied by Southern Clay Products Inc, Gonzales, Texas, USA under the product name Cloisite®Na^+. Fig. 6 shows the effect of nanoclay on the expansive, shrinkage and total strains. The shrinkage strain slightly decreased at 0.05%, 0.1%, and 0.15% then increase while there was no increased in expansive strain at the these nano-clay content. However, when as the amount of nanoclay is increased, the expansive and shrinkage strain increases leading also to the total volume change increase. The total shrinkage strain after saturation was measured relative to the final volume of sample when become fully dry. The total shrinkage strain is the total volume change and approximately equal to the shrinkage volume change and expansive volume change. The increase in nanoclay more than 0.3% gives a clear increase in total volume change [17].

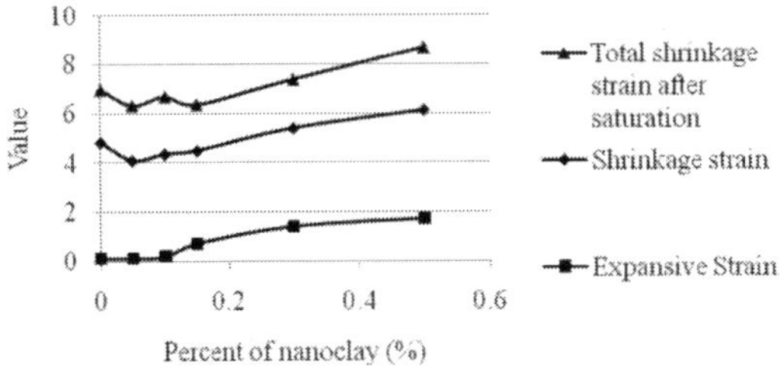

Fig. 6. Influence of nano-clay on the expansive, shrinkage and total strains

Nano-clay gives a noticeable decrease in hydraulic conductivity due to increase in the smaller size of particles. The nanoclay particles probably diffused between the large sedimentary soil particles, closing or minimizing the pores which lead to a decrease in the flow of water in the soil mixtures. This is a very significant results because one of the common materials tested to reduce the shrinkage strain is fiber. Mixing the soil with fiber can reduce the shrinkage strain up to 90% but the fiber caused an increase in the hydraulic conductivity up to 50 times [18]. Thus using nano-clay will improve volume change behavior and at the same time lowering the hydraulic conductivity of soil.

5 Nano-Alumina

Nano-alumina was used to study crack intensity factor (CIF) which is important to assess crack formations around waste containment systems. The nano-alumina was purchased from Inframat Advanced Materials, Manchester, CT, USA. It has a chemical formula Al_2O_3, particle density of 3.6 g/cm^3, surface area of > 150 m^2/g, and a transitional gamma crystal structure [19]. For high plasticity soils (given as S3 and S4 with plasticity index of 18 and 69, respectively), the CIF is minimum at 0.1 % nano-alumina content (Fig. 7). Furthermore, the hydraulic conductivity measurements showed that adding nano-alumina will cause a slight decrease in its value which is always a favorable characteristic for landfill liners and caps. Thus, using nano-alumina for desiccation crack control does not cause detrimental effects on the hydraulic behavior of the soil.

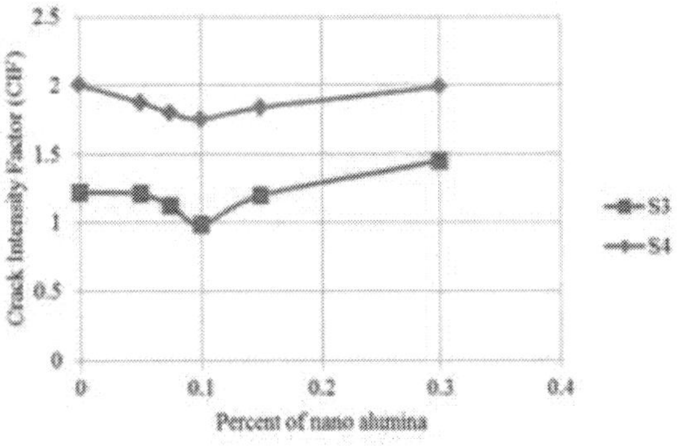

Fig. 7. Effect of nano-alumina on crack intensity factor.

6 Nano-Lime

Lime is a well-established soil stabilization material. The use of nano-lime is possibly aimed at overcoming some of the limitations of traditional lime-based materials, such as the difficulty in achieving a complete carbonation, while maintaining their advantages [20]. Nano-lime powder was imported from Strem Chemicals, Inc., Mulliken Way, Newburyport, Massachusetts, United States. Results in Fig.8 indicated that only about 0.3-0.5% nano-lime is required to achieve UCS strength much higher that without nano-lime. In addition, nano-lime induced noticeable increase in UCS of soil-nano-lime mixtures; that is, the strength of the mixtures with nano-lime was higher than the strength of mixtures with lime. These may be due to higher ability of nano-lime to flocculate and agglomerate the soil particles compared with the lime. Agglomeration and flocculation occurred with the availability of Ca^{+2} cations. Ions exchange mechanism possibly took place between monovalent ions (Ca^{+2}) with divalent (Na^+, H^+) ions leading to reduction of repulsion forces and bridging the clayey particles of negative surface ions. This may again be related to the reactivity of nano-lime and its ability to release more Ca^{+2} ions than lime. This is supported with the purity of nano-lime which is higher than the lime [21].

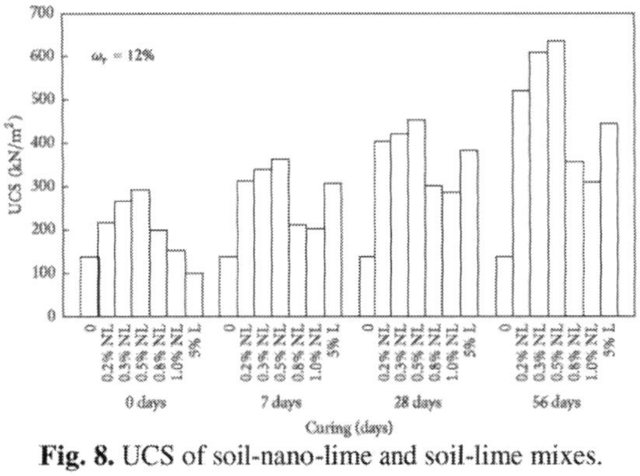

Fig. 8. UCS of soil-nano-lime and soil-lime mixes.

7 Other Nanomaterials

Nano-copper and nano-magnesium are other materials that have been tested to improve the properties of soil [22] [23]. The results are almost similar to most of the studies above in which there is an optimum amount of nanomaterials after which adding more nanomaterials will cause adverse effect to soil properties. These optimum amount normally occurs at less than 1% of the weight of the soil used.

8 Conclusion

In this presentation, laboratory results from work done at Universiti Kebangsaan Malaysia (UKM) on the application of nanomaterials for soil improvements is presented. From the results obtained, it is shown that nanomaterials are potential materials to be innovatively used for enhancement of soil properties. For example, fibers are normally used to improve shrinkage in soils. However, while greatly reducing the shrinkage strain, fibers also significantly increases the hydraulic conductivity of soils. This is where nanomaterials such as nano-clay and nano-alumina becomes handy. The use of these materials will reduce soil shrinkage and at the same time reducing the hydraulic conductivity suitable for waste containment systems. In addition, nanomaterials such as nano-soil and nano-lime significantly increase the compressive strength soil-cement and original soil-lime mix. It can be expected that with time and developments in nanotechnology, more nanomaterials will be produced providing geotechnical and geoenvironmental engineers more options to improve the properties of soils.

References

1. M.R. Taha, E. Ismail, Z. Chik. *2nd Intl Symp on Nanotechnology in Construction*, pp. 373-381 (2005).
2. G.H. Bolt, Geotechnique **6**(2), pp. 86-93 (1956).
3. A. Sridharan, M.S. Jayadeva, Geotechnique **32**(2), pp. 133-144 (1982).
4. H. Komine, N. Ogata. Canadian Geotech J. **33**(1), pp. 11–22 (1996).
5. H. Komine, N. Ogata. Canadian Geotech J. **40**(2), pp. 460–475 (2003).
6. S. Tripathy, A. Sridharan, T. Schanz, Canadian Geotech. J. **41**(3), pp. 437-450 (2004).
7. S. Tripathy, T. Schanz, Canadian Geotech. J. **44**(3), pp. 355-362 (2007).
8. Y. Gurtug, Applied Clay Science **51**, pp. 295-299 (2011)
9. A. Sridharan, *Chemo-Mechanical Coupling in Clays: From Nano-Scale to Engineering Applications*, pp.1-28 (2002).
10. M.R. Taha, *Nanomaterials and Geotechnical Engineering: Big Problems Small Solution* (in Malay). Inaugural Lecture (2012).
11. M.R. Taha, S.Y. Lim, Z. Chik. Jurnal Kejuruteraan, UKM, **21**, pp.1-10 (2009).
12. M.R. Taha, *Nanotechnology in Construction, Proc NICOM3*, pp. 377-382 (2009).
13. G.A. Ozin, A.C. Arsenault, *Nanochemistry: A Chemical Approach to Nanomaterials* (2005).
14. W. Marx, A. Barth, *Carbon Nanotube*, pp. 1-18 (2010).
15. K. McGuire, A.M. Rao, *Carbon Nanotubes – Science and Applications*, pp. 117-136 (2005).
16. M.R. Taha, S. Mobasser, PLoSONE **10**(12), pp. 1-16 (2015).
17. O.M.E. Taha, M.R. Taha, *Proc 5th Jordanian International Civil Engineering Conference*, pp. 417-423 (2012).
18. S.M. Rifai, *PhD Dissertation*, Wayne State University, USA (2000).
19. M.R. Taha, O.M.E. Taha, *NICOM4-Nanotechnlogy in Construction-4 Intl. Symposium*, pp. 1-9 (2012).
20. V. Daniele, G. Taglieri, R. Quaresima, Journal Cultural Heritage, **9**(3), pp. 294–301 (2008).
21. P. Govindasamy, M.R. Taha, J. Alsharef, K. Ramalingam, Applied and Environmental Soil Science, Article **8307493**, 9 pages (2017).
22. M.R. Taha, O.M.E. Taha, J Nanopart Res, **14** :1190 (2012).
23. Z.H. Majeed, M.R. Taha, I.T. Jawad, Research Journal of Applied Sciences, Engineering and Technology, **8**(4), pp 503-509 (2014).

AUTHOR INDEX

AUTHOR INDEX

AUTHOR INDEX

AUTHOR INDEX

AUTHOR INDEX

2nd International Congress on Materials and Structural Stability (CMSS-2017)

MATEC Web of Conferences Volume 149 (2018)

Rabat, Morocco
22 – 25 November 2017

Part 2 of 2

Editors:

A. Diouri
A. Boukhari
L. Ait Brahim
L. Bahi
N. Khachani
M. Saadi
J. Aride
A. Nounah

ISBN: 978-1-5108-5259-4

Printed from e-media with permission by:

Curran Associates, Inc.
57 Morehouse Lane
Red Hook, NY 12571

proceedings.com

Some format issues inherent in the e-media version may also appear in this print version.

Printed by Curran Associates, Inc. (2018)

For additional information, please contact EDP Sciences – Web of Conferences at the address below.

EDP Sciences – Web of Conferences
17, Avenue du Hoggar
Parc d'Activité de Courtabœuf
BP 112
F-91944 Les Ulis Cedex A
France

Phone: +33 (0) 1 69 18 75 75
Fax: +33 (0) 1 69 28 84 91

contact-edps@webofconferences.org

Additional copies of this publication are available from:

Curran Associates, Inc.
57 Morehouse Lane
Red Hook, NY 12571 USA
Phone: 845-758-0400
Fax: 845-758-2633
Email: curran@proceedings.com
Web: www.proceedings.com

TABLE OF CONTENTS

PART 1

PART 2

Multiscale modelling for better hygrothermal prediction of porous building materials

Rafik BELARBI[1], Fares BENNAI[1, 2], MohammedYacine FERROUKHI [3], ChadyEL HACHEM[2], Kamilia ABAHRI [2]
[1]Université de La Rochelle - CNRS, LaSIE UMR 7356, Avenue Michel Crépeau 17042 La Rochelle Cedex 1, France
[2]LMT, ENS-Cachan, CNRS, Université Paris Saclay, 61 avenue du Président Wilson, F-94230 Cachan, France
[3]ENISE, Université de Lyon, LTDS CNRS UMR 5513, 58 Avenue Jean Parot, 42023 Saint Etienne, France

Abstract. The aim of this work is to understand the influence of the microstructuralgeometric parameters of porous building materials on the mechanisms of coupled heat, air and moisture transfers, in order to predict behavior of the building to control and improve it in its durability. For this a multi-scale approach is implemented. It consists of mastering the dominant physical phenomena and their interactions on the microscopic scale. Followed by a dual-scale modelling, microscopic -macroscopic, of coupled heat, air and moisture transfers that takes into account the intrinsic properties and microstructural topology of the material using X-ray tomography combined with the correlation of 3D images were undertaken. In fact, the hygromorphicbehavior under hydric solicitations was considered. In this context, a model of coupled heat, air and moisture transfer in porous building materials was developed using the periodic homogenization technique. These informations were subsequently implemented in a dynamic computation simulation that model the hygrothermalbehaviourof material at the scale of the envelopes and indoor air quality of building. Results reveals that is essential to consider the local behaviors of materials, but also to be able to measure and quantify the evolution of its properties on a macroscopic scale from the youngest age of the material. In addition, comparisons between experimental and numerical temperature and relative humidity profilesin multilayers wall and in building envelopes were undertaken. Good agreements were observed.

1 Introduction

Nowadays, the study of coupled heat and mass transfers in porous building materials becomes an essential part of the methodological arsenal of the modern thermal studies. It became essential step to corroborate the energetic and environmental challenges. In fact, coupled heat air and mass transfers at the scale of porous materials and that ofthe building-environment interface are complex in many respects. It is multi-scale (pore, material, envelope and whole building) and multi-physical (mass transfer, heat with strong interaction). One of the major challenges is to identify the most influential parameters defining the overall behavior of the building (new and old) and the interactions that exist between subsystems in order to propose a design adapted to the built spaces. The knowledge of such a response is important both in terms of predicting the quality of habitable environments and assessing the durability of structures [1].

The preliminary choice of materials used for construction today presents an alternative solution to meet its problems. This certainly reduces the energy and environmental impact, but reveals new challenges, as this could alter summer comfort and indoor air quality.

In addition, the use of materials with low environmental impact, such as bio-based mortars, already known for their thermal insulation properties, acoustic properties and low environmental impact, is increasingly taking place in the construction industry last year's [2].

Moreover, the use of these materials is hampered, in particular, by the unavailability of databases relating to their intrinsic properties and by ignorance of their behaviors over time. This use requires taking into account several criteria that make them vulnerable; the problem of swelling-shrinkage at the cement-fiber interface and fungal growth are the most important criteria for assessing the degradation of these materials.

Indeed, few works in the literature, in the field of civil engineering, take into consideration the effect of local material behavior when modelling the envelope scale ...

Indeed, the macroscopic behavior of these materials is highly dependent on the mechanisms involved on a microscopic scale. In order to predict the behavior of structures on a macroscopic scale, it is essential to know how to capture local behaviors, but also to be able to measure and quantify the evolution of these properties on a macroscopic scale from the youngest age of the material. At the scale of the pore, the physical transfer processes and the topology of the liquid interfaces (bound or free) vapor, air and their interactions with the solid matrix (contact lines, detachment of vapor bubbles from solid surfaces, capillary bridges, ...) make the study of these processes extremely complex. In particular, it is necessary to link the phenomena of mechanical deformations (intra-material swelling and capillary pressure / disjunction) to the equilibrium states of moisture in materials [3].

That why, the present study made recourse to non-destructive method to made refine morphological characterization of the main intrinsic properties of the studied materials.This first step will provide information on the three-dimensional microstructure and also allow to locate any heterogeneity, singularity, porosity, specific surface area and to deduce the representative structure and its deviations via 3D reconstruction techniques.

Then, the implementation of homogenised approachesusing 3D numerical modelling on the real structures of the materials provides more information about the equivalent macroscopic behaviour of the material. These information at the material scale will be then implemented when modelling the envelopes responses to the different hygrothermal solicitations representatives the building environment.

2 Microscopic characterization

2.1 3D reconstruction

The recourse tonon-destructive methods presents some assets for a better microstructural characterization at the local scale of the material.

X-ray tomography is one of these non-destructive imaging techniques that consists in reconstructing the 3D volume of an object from a series of 2D x-rays. This technique was developed and used primarily in the medical field, then in the industry of the director of non-destructive testing and making dimensional measurements. The resolution of tomographic images is function of many parameters, like the distance between the source and the specimen. it is directly related to the dimensions of the specimen.

Fig. 1.3D reconstruction of hemp concrete at a resolution of 31.8μm

The studiedmaterial is hemp concrete, which consists of hemp shiv, water and binder (mixture of: natural lime from Saint-Astier (Hydraulic: NHL and aerial: CL), cement CEM I 52.5, and different adjuvants to improve the rheology and permeability of hemp concretes).

Fig. 2.2D section of a volume of hemp concrete scanned at 31.8μm.

Considering that the material contains hemp shives which are weakly absorbing material, a special treatment was required to perform imaging scans. After performing and analyzing many tests on this material in the tomograph, a choice of using voltage of 100 kV was selected for all the tests treated in this study.

Fig. 2 shows a macro porosity up to a few millimeters in size and random distribution of the hemps shiv. This macro porosity is related to the arrangements of plant particles between them.

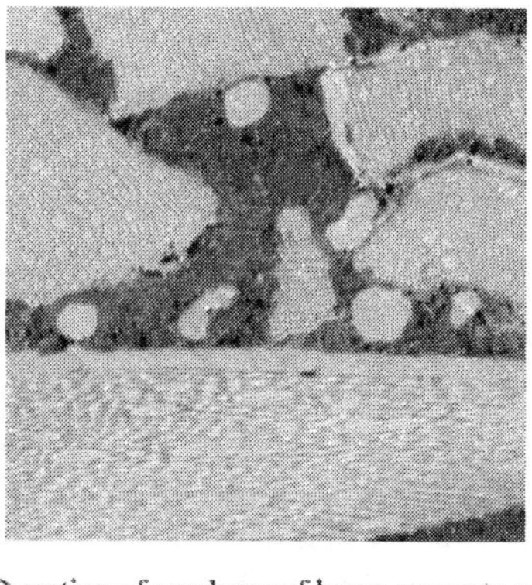

Fig. 3. 2D section of a volume of hemp concrete scanned at 2.3 μm.

Fig. 3 shows an image of the hemp concrete scanned at 2.3μm, where a part of the porosity of hemp shives is observed. the comparison of this figure with the previous figure (Fig. 3) shows that, this very heterogeneous material exhibits a wide distribution of the pore size ranging from the micropores (at the binder and hemp shives) to the macro pores (interparticle porosity). This technique of tomography thus makes it possible to gain access to the heart of the material and to inform on the interface problems in the materialand its complex microstructure. In fact, the hemp shiv is covered by the binder leaving an interparticle porosity between the vegetable particles and the binder.it should be emphasized that the interparticle porosity contributes strongly to the improvement of the hygrothermal properties of the material by natural phase change.

2.2 local properties

It was improved that the local properties are highly dependent on the material morphology especially the fibers orientation. The example cited here concerns the porosity, where the evaluation of the porosity was carried out on a sample of hemp concrete scanned with X-ray tomography at a resolution of 7.5 µm.

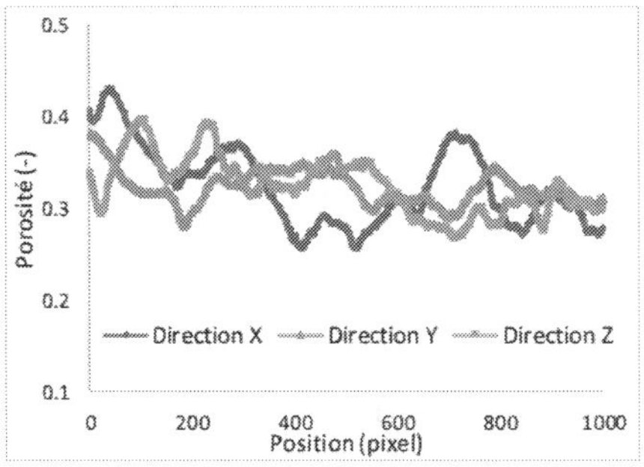

Fig. 4. Distribution of the air volume of the hemp concrete sample in the 3 directions (x, y, z).

A significant heterogeneity of the calculated porosity is observed in Fig. 4, the latter exhibits fluctuations and varies as a function of the position. Indeed, the macro porosity between the binder and hemp shives depends strongly on the orientation and the position of the vegetable particles.

For this, these variabilities were seriously considered when modeling the hygrothamal transfers of such heterogenous material, for a better prediction of the behavior of the materials used in the building envelope.

2.3 hygromorphicbehavior

This material consists of the hemp shives and the binder, which exhibit different behaviors with respect to variations in environmental conditions, capable of causing considerable dimensional changes in contact with the relative humidity of the air. On the microscopic scale, the local deformations are due to the absorption of water by the cell walls of the hemp shiv.

The test consists of scanning the same sample of hemp concrete in the dry state (0% HR) and then in the wet state (85% HR).For this purpose, an original device adapted to the tomograph, in order to preserve a constant relative humidity of the samples during the scans.

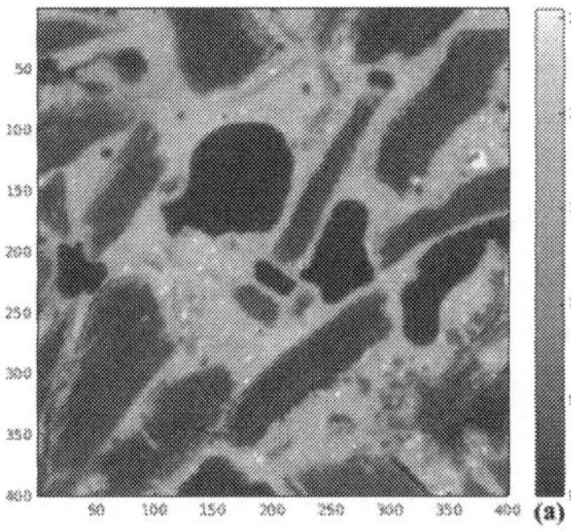

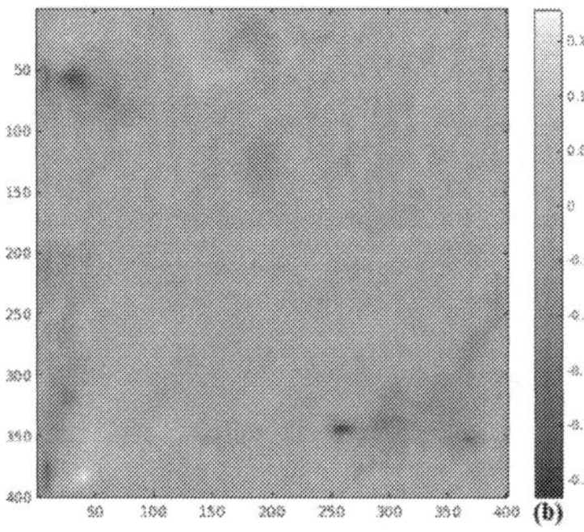

Fig. 5. (**a**)Section in the reference volume, (**b**)Deformations fieldε_{zz}(%)

The volume images correlation allowed to quantify the deformation fields of the studied volumes using Correli-C8R [4]. The comparison between the reference and deformed images illustrated in Fig. 5 shows that the hemp shives undergo swelling due to the phenomenon of water adsorption. This is explained by the yellow dilated zones. This swelling then contributes to the reduction of the porous space which is reflected by the blue narrowed areas.

3 Periodic homogenization

3.1 developed model

A refine periodic homogenization was developed in a previous work in order to predict the hygrothermal behavior of such material. The equations of the mass balance and of thermal equilibrium have been homogenized [5, 6]. Macroscopic descriptions of the different phenomena have been obtained. The homogenized mathematical model for coupled heat, air and moisture transfer in porous building materials can be written as follows:

$$\frac{\partial \theta_l}{\partial t} = div_x(D_\theta \nabla_x \theta_l + D_T \nabla_x T - D_P \nabla_x P) \qquad (1)$$

$$C^{hom}\frac{\partial T^0}{\partial t} = div_x(a_t \nabla_x T + \delta_t \nabla_x \theta_l + \alpha_t \nabla_x P) - h_{lv} m_g \qquad (2)$$

$$h_a \frac{\partial P}{\partial t} = div_x\left(K_{P,P} \nabla_x P + K_{\theta,P} \nabla_x \theta_l + K_{T,P} \nabla_x T\right) - m_g \qquad (3)$$

This model allows us to enfranchise the assumptions of the constancy of the gas pressure and the non-isothermal conditions; as well as take into account the complexity of the internal morphology of the materials. In added, the main physical parameters of this model are non-linear and depend on the water content, temperature, and microstructure of the porous medium.

3.2 case study

Consider a porous material whose microstructure consists of a periodic repetition of a three-dimensional unitary cell (Fig. 6). This cell has been chosen for the purpose of numerically determining the homogenized

https://doi.org/10.1051/matecconf/201814902005

diffusion and thermal conductivity tensors constituting the main input parameters of the developed model. The dimensions of this cell are chosen so as to obtain a low porosity corresponding to the porosity of the cementitious materials, such as concrete

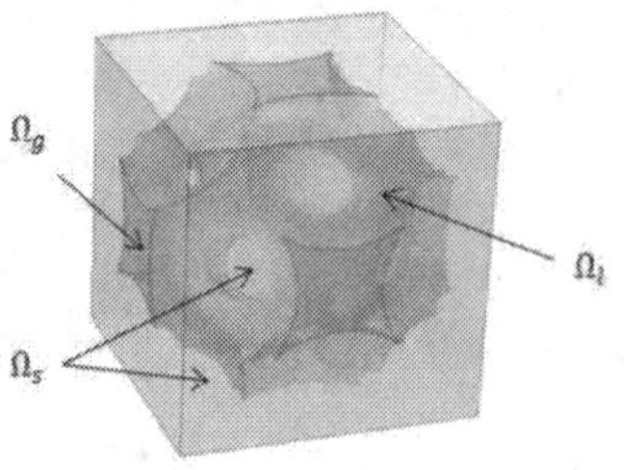

Fig. 6. Elementary cell of periodic medium

- Thermal conductivity tensor homogenized

$$\lambda^{hom} = \lambda_s \begin{bmatrix} 0.78217 & 0 & 0 \\ 0 & 0.78217 & 0 \\ 0 & 0 & 0.78217 \end{bmatrix} + \lambda_l \begin{bmatrix} 0.02188 & 0 & 0 \\ 0 & 0.02188 & 0 \\ 0 & 0 & 0.02188 \end{bmatrix}$$

- Diffusion tensor homogenized

$$D_v^{hom} = D_v \begin{bmatrix} 0.04127 & 0 & 0 \\ 0 & 0.04127 & 0 \\ 0 & 0 & 0.04127 \end{bmatrix}$$

The results showed that these parameters depend strongly on the geometry of the cell. In addition, the experimental validation of the macroscopic diffusion coefficient was carried out in relation to the literature, and the thermal conductivity of the solid phase is thus deduced. This latter is considerably greater than the overall conductivity of the material.

Table1. Conductivité thermique de la matrice solide etcomparaison du coefficient de diffusion

$\lambda_s(W/m.K)$	$D_\theta(m^2/s)$	$D_\theta^{exp}(m^2/s)$	D_θ/D_θ^{exp}
1.9135	$2.507\ 10^{-10}$	$1.01\ 10^{-10}$	2.47

4 Big scale modelling

4.1 Heat, air and moisture transfer in the envelope

To accurately predict the hygrothermal behavior at the building envelope, we opted for a model based on Luikov theory [7, 8]. The developed model considers as driving potentials the temperature for heat transfer, the total pressure for air transfer and the water vapor pressure for the hydric transfer. This allows the avoidance of discontinuity problems at the wall layer interfaces, which is not the case with the water content.

The HAM model is built on heat and mass conservation laws. The first two equationsrepresent bothmoisture (liquid and water vapor) and gaseous balances, and the last equations expresses the heat balance by taking into account the heat conduction and heat advection due to moisture and total pressure gradients. It is noted that the capillary liquid transfer expressed by Darcy's law is taken into account in the HAM model. Indeed, the gradient of capillary suction is reduced to the water vapor pressure gradient and temperature gradient using Kelvin's law. Experimental validation of the proposed HAM model was undertaken in Ferroukhi et al. [9].

$$C_{m_i}(\omega)\rho_{s_i}\frac{\partial P_v}{\partial t} = div\left(k_{m_i}(\omega)\nabla P_v + k_{T_i}\nabla T + k_{p_i}\nabla P\right) \quad (4)$$

$$C_{a_i}\frac{\partial P}{\partial t} = div\left(k_{p_i}\nabla P\right) \quad (5)$$

$$C_{p_i}\rho_{s_i}\frac{\partial T}{\partial t} = div(\lambda_i(\omega)\nabla T + \alpha_i\nabla P_v + \gamma_i\nabla P) + L_v\rho_{s_i}\sigma_i C_{m_i}(\omega)\frac{\partial P_v}{\partial t} \quad (6)$$

4.2 HAM-BES co-simulation platform

The co-simulation approach consists on coupling a dynamic building simulation tool (TRNSYS) with a coupled heat, air and moisture transfer model implemented in COMSOL. The coupling was carried out through MATLAB which represents an integrator tool ensuring the data exchange between TRNSYS and COMSOL. Indeed, TRNSYS is used here because it is widely used by scientific community and by engineers. In addition, its modular architecture is an advantage because it allows extending the modeling to a thermo-hydro-aeraulic modeling. This is achievable by integrating ventilation types in simulations as Comis or Contam. For the hygrothermal transfer simulation in the building's walls, COMSOL was chosen in order to have an enough fine granularity simulation to describe the hygrothermal transfer in 2D and 3D especially in specific envelope components involving multidirectional flows such as thermal bridges.

The COMSOL/MATLAB connection is provided through the LiveLink interface [10]. Thus, the HAM model is defined in MATLAB and differential equations system is solved by the COMSOL solver. Concerning the TRNSYS/MATLAB link, it is established using the type 155 [11].

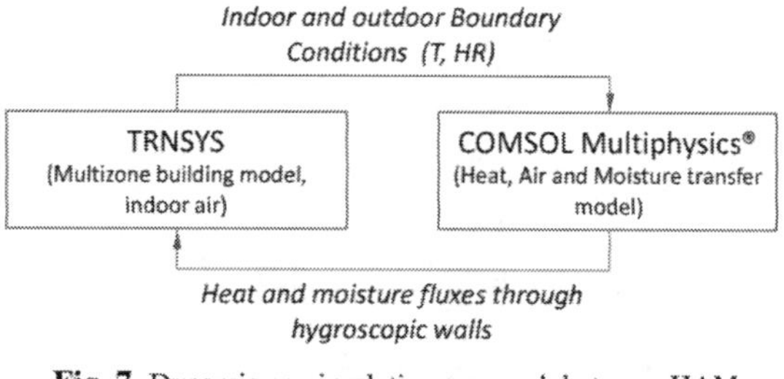

Fig. 7. Dynamic co-simulation approach between HAM (COMSOL) and BES (TRNSYS) models

In this approach, the control of the time step was provided by the BES model (TRNSYS). Consequently, the data exchange between the two coupled codes is made in the post-convergence to the current BES model time step during which the HAM model is solved in COMSOL with a finer time step. The diagram in Fig. 7 summarizes the temporal synchronization approach and data exchange strategy between the BES model in TRNSYS and the HAM model in COMSOL.

4.3 Experimental investigation

As expected previously, an experimental device adapted to study the hygrothermal response of buildings envelopes has been developed in this work. This experimental device is used to validate the proposed HAM-BES co-simulation approach.

The experimental validation was conducted by controlling relative humidity and temperature profiles for a steady state situation.

In this work, a monolayer chipboard wall with a thickness of 6.6 cm was used. The tested thickness envelope was taken enough small to optimize the test time, given the hygrothermal transfer kinetics and especially the moisture transfer which is relatively slow.

The test principle consists on submitting the envelope to two hygrothermal controlled environments. Each compartment was made of a plywood envelope (1.5 cm) with an internal polystyrene insulation (3 cm). The inner surfaces of compartments walls were covered with sheet metal to avoid any moisture exchange with the indoor environment. The temperature is maintained for each ambiance with a thermostatic bath who feeds a copper heat exchange. For the relative humidity, a humidity generator was used. These apparatuses are equipped with deported sensors that allow a downstream hygrothermal control (ie in the two ambiences). Fig. 8 illustrate the developed experimental device.

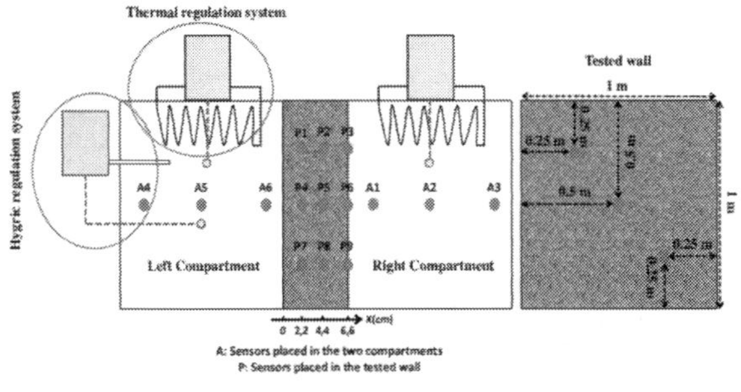

Fig. 8. Experimental device and sensors implementation

4.4 Numerical simulation and validation

In order to validate de developed HAM-BES platform, the numerical resolution obtained by this tool was compared to the experimental data provided from the experimentation previously presented.

It is noted that to perform the numerical simulation task, evidently, the initial and boundary conditions are identical to those of the experiment. For a better understanding of the presented experimental results, the placements of measurement points over the experimental device were illustrated in Fig.8.

The hygrothermal conditions imposed in the two controlled ambiences are represented in Fig.9.The numerical and experimental comparison of temperature and relative humidity distributions in the wall and the relative humidity in the compartment II are shown in Fig.10 and Fig. 11.Focusing on the temperature distribution through the wall, the results show good agreement between the experimental data and numerical simulation.

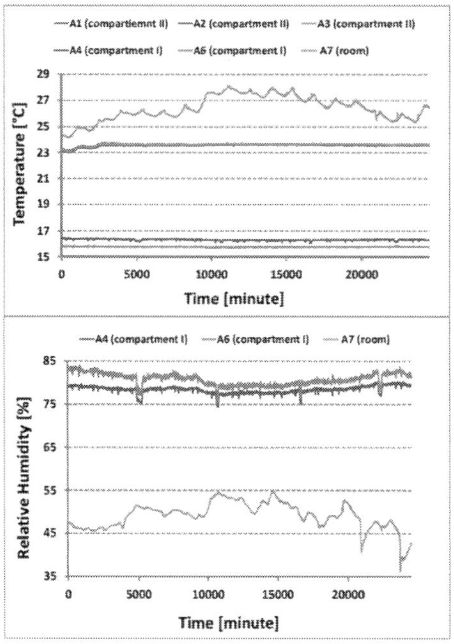

Fig.9. Imposed hygrothermal conditions in the two compartments

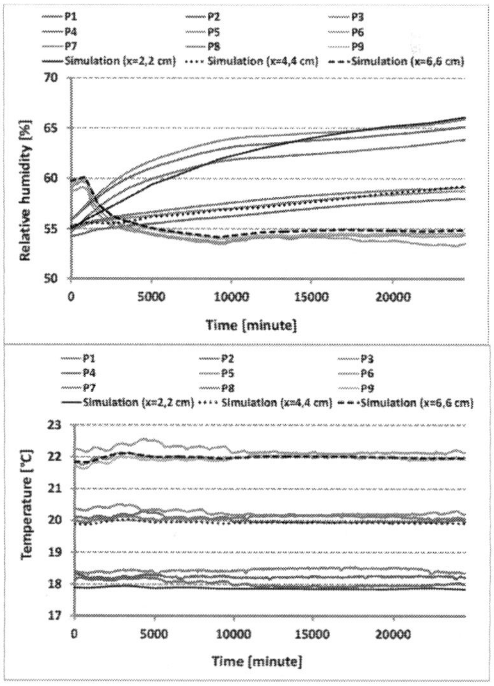

Fig.10. Comparison of numerical simulation with experimental data (temperature and relative humidity at different depths of the wall)

The mean temperature deviation is about 0.2°C at a depth of 2.2cm. At 4.4cm and 6.6cm, the mean deviations are 0.35°C and 0.30°C respectively. Concerning the comparison of the relative humidity distributions, a good agreement between predicted and measured data was observed. The mean variation reaches 2%, 1.5% and 1.3% at 2.2cm, 4.4cm and 6.6 cm of depth respectively. The same observation was noted by comparing the numerical and experimental data of the relative humidity variation in compartment II. The mean deviation over theexperimentation reaches 1.6% with a maximum of 2.7% (see Fig. 11).

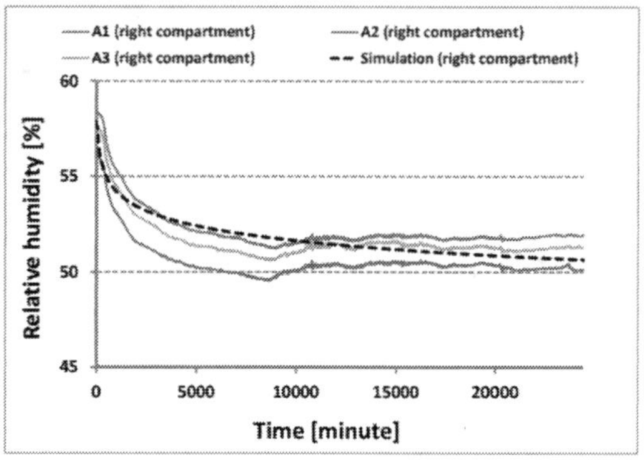

Fig.11. Comparison of numerical simulation with experimental data (Relative humidity variation in compartment II)

5 Conclusions

This paper is a contribution to the multi-scale study of the hygrothermal behavior of buildings to better understand the interactions between the different scales of building envelope materials. The microstructure of the materials was studied using X-ray tomography. This microscopic characterization highlights the complexity of the microstructure of porous building materials and their strong heterogeneities. The 3D reconstructions of the real material scanned with X-ray tomography at different resolutions show that hemp concrete has several scales of porosity, ranging from micro-porosity within the binder and hemp shiv to the inter-particle macro-porosity. In addition,thehygromorphic behavior under hydric solicitations was allowed to follow the deformations on a small scale to better understand the mechanisms of degradation of hemp concrete.

Then, the periodic homogenization method allowed to consider the heterogeneity and the complexity of the porous media and to describe their overall behavior on a large scale. It has been shown that macroscopic behavior of material is strongly dependent on the mechanisms acting at the levels of these constituents.

Finally, a coupled heat, air and moisture transfer model in multilayer walls was developed and examined. Further, an experimental device was designed to study the hygrothermal behavior of several configurations of multilayer porous building materials. Heat and moisture transfer evolutions in the samples were monitored over time, as well as good control of boundary conditions. Comparison of temperature and water vapor pressure profiles resulting from the numerical simulation with those obtained experimentally were undertaken. The results showed good agreement between predicted and measured data. These good results illustrate the advantages of the proposed hygrothermal transfer model, with accessible input data, evaluable experimentally with standardized tests.

Acknowledgments. This work was funded by the Region Aquitaine, Limousin, Poitou-Charentes Region through European and National Program CPER-FEDER "Bâtiment Durable 2015-2020".

References

1. K. Abahri, R. Belarbi, E.H. Chady, Caractérisation macro-hydrique des matériaux biosourcès, *Rencontres Universitaires de Génie Civil.* (2015).
2. S. Amziane, F. Collet, eds., *Bio-aggregates Based Building Materials*, Springer Netherlands, Dordrecht, (2017).
3. C. El Hachem, P. Ye, K. Abahri, R. Bennacer, Fiber's hygromorphic effect on thermal conductivity of wooden fibrous insulation characterized by X-ray tomography, Construction and Building Materials. 758–765. **150** (2017).
4. T. Taillandier-Thomas, S. Roux, T.F. Morgeneyer, F. Hild, Localized strain field measurement on laminography data with mechanical regularization, Nuclear Instruments and Methods in Physics Research Section B: Beam Interactions with Materials and Atoms. 70–79. **324** (2014).
5. F. Bennai, K. Abahri, R. Belarbi, A. Tahakourt, Periodic homogenization for heat, air, and moisture transfer of porous building materials, Numerical Heat Transfer,Part B: Fundamentals. 420–440. **70** (2016).
6. F. Bennai, K. Abahri, R. Belarbi, A. Tahakourt, Convective and conductive thermal homogenization for nonsturated porous building materials: Application on the thermal conductivity tensor, Thermal Science. (2016).
7. B. Remki, K. Abahri, M. Tahlaiti, R. Belarbi, Hygrothermal transfer in wood drying under the atmospheric pressure gradient, International Journal of Thermal Sciences. **57** (2012) 135–141.
8. Luikov, A. V, *Heat and mass transfer in capillary porous bodies*, PergamonLond. (1966).
9. M.Y. Ferroukhi, K. Abahri, R. Belarbi, K. Limam, A. Nouviaire, Experimental validation of coupled heat, air and moisture transfer modeling in multilayer building components, Heat and Mass Transfer. 2257–2269. **52** (2016).
10. Comsol Multiphysics User's Guide, ComsolMultiphysics User's Guide. (2012).
11. Solar Energy Laboratory, TRANSSOLAR Energietechnik GmbH, CSTB – Centre Scientifique et Technique du Bâtiment, et TESS – Thermal Energy Systems Specialists, TRNSYS 17 a TRaNsientSYstem Simulation program, vol. A3. (2010).

Prevention of concrete structures from collapsing

R. Cechmanek[1], B. Nespor[1], M. Drdlova[1], P. Steffan[2] and L. Machan[2]
[1]Research Institute for Building Materials, Department of Applied Research and Development, Brno, Czech Republic
[2]Brno University of Technology, Faculty of Electrical Engineering and Communication, Brno, Czech Republic

Abstract. At the end of the 20th century requirements on using electrical properties of building materials emerged for application in heating of trafficable surfaces, grounding of electrostatic charges in floors, shielding of electro-magnetic fields and diagnosis of concrete structure state in the course of time. For this reason, it was necessary to design special fibre-cement elements able to transfer any mechanical impulse to an electrically-measured signal detected as a change in electrical resistance with computer outputs. Regarding previous research studies, it was concluded that special fibre-cement composites are able to conduct electric current under specific conditions. This property is ensured by using of various kinds of carbon materials. Though carbon fibres are less conductive than metal fibres, composites with carbon fibres were evaluated as better current conductors than the composites with metal fibres. By means of various kinds of carbon particles and fibres it is possible to design cement composites with an ability to monitor changes in electrical conductivity of concretes. The designed composites are assembled with conductive wires and connected with a special electronic equipment for monitoring of changes in alternate voltage passing through the tensometer within mechanical loading of a concrete element in which the composite is integrated. The tensometers are placed preferably into parts of the concrete elements subjected to compression, such as simple reinforced columns or upper parts of longitudinal beams. Several tests of repeated loading and simultaneous monitoring of vertical as well as horizontal prefabricated concrete elements were carried out and evaluated.

1 Introduction

Nowadays, characteristics of composite materials based on cement include also other applications, among others for construction of self-monitoring buildings, bridges, etc. Besides monitoring the usual properties (mechanical stability over the period of time, environmental resistance, design limits or economic profitability) it becomes to be more important and significant for construction of buildings.

One of the most common structural materials used in engineering construction is cement and its mixtures (concrete and mortar). Cement is slightly conducting material, but its electrical conductance, Electro-magnetic interference (EMI) shielding effectiveness and wave absorbing properties are very poor. In order to increase the ability of cement materials to transfer electrons, additional conductive components have to be added.

The strain-sensing properties are achieved by a proper volume amount of conductive filler. In this system, the matrix is made of cementitious material with small amount of silica fume, fly ash, and fine aggregates [1].

Different conductive fillers were tested considering the best strain-sensitivity/material price ratio. The best strain-sensitivity is achieved near the percolation threshold of filler particles [2].

Strain properties of the composite can be evaluated by impedance changing. The impedance changing sensitivity regarding deformation can be widely affected by a proper choice of concrete admixtures [3]. The real part of impedance is strongly affected by deformation and can be used for measurements.

Conventional DC techniques for resistance measurements cannot be used. Electrode system of a sensing element could be damaged by electrolytic corrosion in this case. Square-wave AC technique with an excitation frequency of 1 kHz and an excitation voltage of 1 V were experimentally set [3].

As carbon particles there were used expanded or micronized graphite, fibres were on the base of polyacrylonitrile (PAN) or pitch. The optimal variants were used for preparation of one-dimensional specimens most suitable for measurement of simplified impact of stress and strain.

2 Materials and methods

2.1 Carbon materials for concrete conductivity improvement

In the course of searching for suitable components there were selected mainly various kinds of micronized or expanded graphite with particle size in micro- or nanometers, which is characterized by excellent electrical properties as well as good compatibility with a cement matrix, except for higher demand of batch water due to its bulk specific surface.

Carbon black with similar properties as micronized graphite was used as well. Carbon black has elementary particles generally in a range from 10 to 100 nanometers, but during the production process individual spherical particles agglomerate in chains or clusters. Micronized graphite has carbon content above 80%; carbon black has high carbon content (99%). Expanded graphites have lower carbon content (60 – 96%).

Considering an application based on a relation between electrical behavior and mechanical strain and reverse deformation of concrete elements, two types of fibre reinforcement (carbon and metal) were chosen.

2.2 Composite mixtures with carbon particles and fibres

All mixtures were prepared in a mixer with a stationary drum and forced movement of paddles. Standard fine-grained matrix consisting of cement, sand and fine filler with 3% of dry mixture weight reinforcement by alkali-resistant glass fibres with length 12 mm was chosen for further modifications.

The addition of carbon particles induces good electrical properties in cement-based composites. Though carbon fibres are less conductive than metal fibres, composites with carbon fibres were evaluated as better current conductors than the composites with metal fibres. It is supposed that this is due to extremely fine size of carbon fibres which provides more effective inter-fibre continuity. Therefore, further research was carried out with carbon particles and carbon fibres.

During the tests suitability of fine-grained particle and fibre combination was confirmed. The proportion of carbon particles was expressed as a percentage of a dry mixture weight (cement + sand + fine filler) as a substitution of a part of sand in range of 4 – 10% per weight. Carbon fibres with diameter 18 μm and length 10 mm substituted for glass fibres up to 2% per weight. Basic components stated in Table 1 are as follows: carbon fibre CF, micronized graphite MG, expanded graphite EG and nickel-coated expanded graphite EG/Ni.

Table 1. Basic mixture compositions.

Mixture components	**CF**	**MG**	**EG**	**EG/Ni**
Cement binder	54%	54%	54%	54%
Silica filler	40%	34%	36%	35%
Fine filler	3%	3%	3%	3%
Glass fibre	1%	1%	1%	1%
Carbon fibre	2%	2%	2%	2%
Micronized graphite	-	6%	-	3%
Expanded graphite	-	-	4%	-
Nickel-coated graphite	-	-	-	2%

2.3 Influence on workability

Fibre-cement mixtures were prepared in order to achieve optimal workability and minimal impedance, i.e. maximal electrical conductivity. To compare an influence of the type of carbon matter on impedance of the cement-fibre composite, each carbon powder always replaced the same proportion of dry components.

Carbon origin as well as particle size both affect impedance properties of the final composites, and as well the particle size affects workability of fresh mixtures. It was found that the finer carbon powder, the lower was the impedance and the worse mixture workability. Carbon particles should be properly dispersed among other mixture components without bleeding during transport and molding. With carbon fibres the perfect defibering and ideal anchoring in binder should be reached [4].

2.4 Mixture optimization

Within the optimization process fine fillers were withdrawn from the composition. Despite the positive effect on dense structure and better mechanical characteristics of composites, pozzolanic admixtures increase electric resistance of concrete mixtures [5].

All specimens were cured in laboratory conditions close to real manufacture conditions with temperature 25°C and humidity 55%. The physical-mechanical properties of composites were tested on standard specimens for thin-walled fibre-cement elements with dimensions 250 mm x 50 mm x 10 mm at 28 days.

Higher water/cement ratio causes significantly higher absorption and lower bulk density, associated with low flexural strength and impact strength of the final composites.

2.5 Measurement of electrical properties

Ions in pore solutions cause conducting of electric current in concretes. Conduction of electrical current in cements and concretes is essentially electrolytic. In order to avoid problems of polarization, alternate currents are often used for determining resistivity of electrolytes and therefore also of cements and concretes [6].

Percolation threshold expresses minimal concentration of the certain component that creates the first conductive way through the whole volume of measured composite [4]. In the end percolation threshold of carbon fibres was determined to 0.75% of the dry mixture weight.

3 Application of proposed elements

3.1 Electro-magnetic shielding

Standard glass fibre reinforced concrete inhibits electro-magnetic field up to approximately -5 dB but our modified fibre-cement composite is able to achieve a level up to -35 dB. In comparison to a massive steel reinforced concrete the same shielding effect is achieved with using of incomparably less structural thickness.

In order to approach real conditions as much as possible the measurement was carried out in a special electro-magnetic chamber. Shielding efficiency was proven by the non-availability of any communication network inside the chamber [7]. A modified fibre-cement mixture was used for manufacturing of a concrete channel to demonstrate the application of electro-magnetic shielding. It contains three chambers to separate electric cables with different voltage so that they do not interfere with each other.

3.2 Concrete heating

A fibre-cement slab with dimensions 1,800 x 1,400 mm, thickness 20 mm and circle openings with diameter 110 mm for better incorporation into a concrete panel was cast with the designed composition. An initial setting of openings in square lattice caused heating to occur only in stripes and not across the whole surface, therefore an optimization had to be done. After thermo-camera monitoring an orthorhombic setting of openings was evaluated as the most suitable for uniform heating ability.

A trafficable steel reinforced concrete panel was made with the ability to thaw snow cover or to defrost ice for reduction of slipping hazard on pavements or access ramps. The proposed element combines advantages of the solid bearing steel reinforced concrete and those of a thin fibre-cement slab with the required heating capacity.

4 Monitoring of concrete structures

It was concluded that cement composites with carbon fibres are suitable for monitoring of transition actions caused by change of strain, stress, temperature or humidity. Electrical resistance monitoring is a good way for characterization of cracks in concrete structure [8]. Besides ceiling panels, columns or foundations these elements could be implemented into the overloaded parts of bridge constructions.

Within experimental tests on flexural strength selected specimens with effective carbon addition provided measurable changes in the impedance. For the purpose of monitoring in conditions close to a real application a model of steel reinforced concrete structure was made. The model was used for evaluation of electrical properties of the embedded element as well as an influence of a concrete mass on these properties.

4.1 Description of the measuring device

The device function is based on 16-bit microcontroller and it is equipped with the Secure Digital (SD) card, which can be inserted into a side slot, for saving the measured data. SD card communicates with the microcontroller via Serial Peripheral Interface (SPI). A special feature of used microcontroller is built in Universal Serial Bus (USB) interface connection for transferring the measured data into PC. Microcontroller contains complete physical layer of USB communication device. Specialized software is able to receive this data and display it in a graph.

Under cover, there is a very well readable display panel. Four control buttons are placed from the front. The opposite side is equipped with eight connectors for measuring probes and four connectors for digital sensors. On the left side, there is the main switch and a connector for charging. The right side includes a water and dust proof USB connector with a rubber plug [9].

4.2 Selection of the proper composition for conductive composites

The aim was to design a cement-based composite with the ability to change its electrical properties in dependence on mechanical stress and strain by means of modified cement matrix. For this purpose, there was a search for suitable inorganic additives. 25 carbon materials of different origin were tested, including micro-milled graphites, expanded graphites or high conductive carbon black. For the selected set of carbon particles an optimization process was carried out to reach the percolation threshold, i.e. the proper amount of an additive, for which no significant changes of electrical properties do not take place in the final composite.

Concerning results gathered within static as well as dynamic measurements, the most suitable additives – carbon black and carbon fibres and their dosing were selected. The final composition consists of 50-60% of high-valuable cement, 30-45% of fine aggregate, up to 10% of carbon particles and up to 2% of carbon fibres.

4.3 Testing of modified elements

Planar elements were evaluated as inconvenient for monitoring of electrical response in individual directions of building elements for the reason of combined impact of stress and strain. Therefore, the elements were designed in the shape of prisms with one dominant

 https://doi.org/10.1051/matecconf/201814902006

dimension comparing to others to avoid multi-parametrical loading effect. These specimens were made in two dimensions 40 mm x 40 mm x 160 mm or 20 mm x 20 mm x 100 mm. In one set, there was 1 element with 2 copper contacts ca. 2-5 mm from the end as shown in figure 1. The other 2 elements were intended for testing of physical-mechanical characteristics, i.e. compressive, flexural and tensile strength, bulk density and water absorptivity.

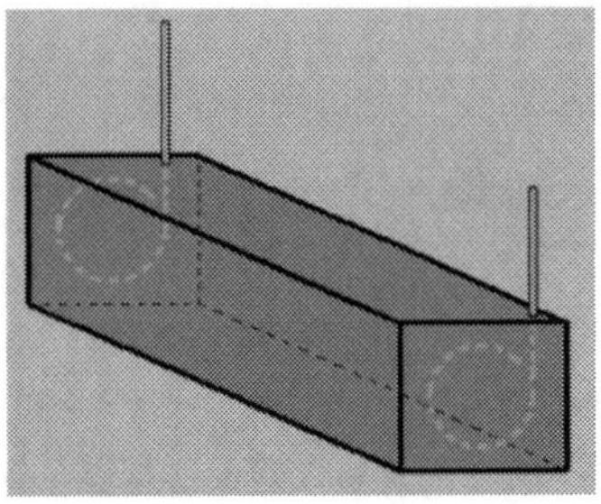

Fig. 1. Design of a modified element.

Cement-based tensometers are prepared from fresh mixture containing all the particular components listed above in accordance with required technological process. The mixture is filled into molds, copper contacts are inserted into it and an isolated part sticks out of it. After 28 days of curing in water the isolation is taken out and the tensometers are provided with long electrical outlets for connection to the measuring device. The tensometers were placed into molds for reinforced structural elements by means of electrically non-conducting straps according to the assembly drawing.

Impedance measurement of individual elements was conducted under following conditions: alternate voltage 1 V and frequency 1 kHz were set to avoid corrosion of used copper electrodes. Maximal loading 1 kN was set within the measurements to avoid unexpected damage of elements.

4.4 Design of a concrete column

Relation of deformation to impedance of composite elements has been already verified [6]. The next logical step was evaluation of this relation in the practical utilization. For evaluation of utilization of integrated composite elements in concrete or reinforced concrete column a model in standard cross section was designed. The final dimensions were 300 mm x 300 mm x 500 mm. There were 2 variants – the first column without steel reinforcement and the second one with steel reinforcement for comparison.

In both variants 3 composite elements were put into the model column. The first one was placed vertically in the column centroid, the second one was placed horizontally above the first one and the third one was placed again vertically at the side of the column next to steel reinforcement. The similar placement was used for the variant without reinforcement. The aim was to compare effect of the steel reinforcement and a different position of individual composite elements.

A mold from water-proof plywood was made according to common praxis of the column production in the horizontal position. The composite elements were fixed in the mold by means of plastic strips to distance elements for the steel reinforcement (see figure 2). The concrete models were made from concrete C30/37 XC4 S4 D22 compacted by inner vibrators.

Fig. 2. Mold with the composite elements and steel reinforcement.

4.5 Measurement of impedance and deformation during loading

Model columns were supplemented by plugs for deformation measurement. Digital deviation indicators were fixed to the plugs [10]. Digital impedance detectors with generated alternate voltage of 1 V were fixed to cable outlets of the composite elements.

Loading of the model columns was simulated by means of the hydraulic press machine. Applied force reached values of 0 to 1,500 kN, which corresponds with pressure 16.7 MPa. Impedance of the composite elements and strain in the model columns was monitored in the course of loading in time of reaching 200 kN (2.2 MPa) by hand operation and 5x 1,500 kN (16.7 MPa) in automatic mode and again de-loading to 200 kN.

In the achieved graphs, it is possible to see clear impedance change of the composite elements corresponding with changes of unit strain in the model columns as it is depicted in figure 3.

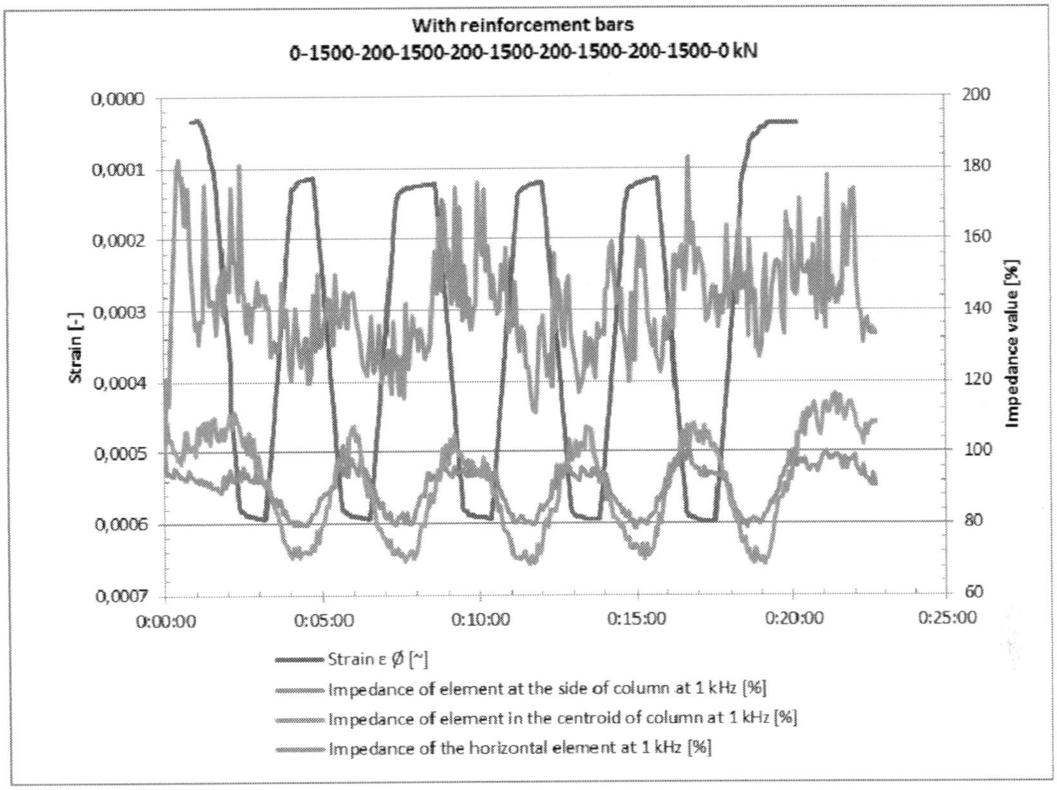

Fig. 3. Graph of strain and impedance in the course of loading for the reinforced column.

The subsequent measurement of relation of mechanical loading of a horizontal element and electrical properties of the composite tensometers was carried out in the specialized testing room without negative effects of the environment (effect of temperature and humidity). The element with dimensions 150 x 150 x 3,400 mm was subjected to 4-point bending. The course of cyclical increase and decrease of the force F = 0-50 kN, the induced deflection and the values of impedance during measurement were depicted again in graphs.

The impedance of tensometers was significantly changed in the case of overloading, when an accumulation of microcracks occurred on the tested element. This state we can describe as the emergency state of the structural elements, when it is necessary to send signal for evacuation of the building.

5 Conclusion

It was concluded that special fibre-cement composites are able to conduct electric current under specific conditions. This property is ensured by using of various kinds of carbon materials in a form of dispersive particles or fibres. Electric conductivity is monitored by means of impedance measurement of the designed elements. These composites could be used for heating of trafficable surfaces or shielding of electro-magnetic fields.

It is also possible to monitor internal processes in building structures and prevent them from collapsing. For this application it is necessary to design one-dimensional elements to simplify the combined loading in the structures. Relation of impedance of the composite elements to unit strain in the model columns was successfully verified thereof. Positions of the composite elements in the models did not significantly affect monitored changes in impedance. It connected only to the absolute change of impedance, but not to the sensitivity to change detection. The ability to transform unit strain to change in impedance seemed to be a material parameter independent on shape and orientation of the composite elements.

Acknowledgement

The research was conducted under the project FR-TI3/485 Engineering structure state monitoring by means of electrically conductive elements with modified cement matrix supported by the Ministry of Industry and Trade of the Czech Republic.

References

1. N. Xie et al, Percolation backbone structure analysis in electrically conductive carbon fiber reinforced cement composites, Composites Part B: Engineering **43** (2012), 3270-3275, ISSN 1359-8368

2. H. Li et al, Effect of compressive strain on electrical resistivity of carbon black-filled cement-based composites, Cement and Concrete Composites **28**, Issue 9 (2006), 824-828, ISSN 0958-9465
3. R. Cechmanek et al, Carbon-Based Composites Enable Monitoring of Internal States in Concrete Structures, International Journal of Civil, Structural, Construction and Architectural Engineering **8**, No.10, Part I (2014), 70-75, eISSN 1307-6892
4. F. Vossoughi, Electrical resistivity of carbon fiber reinforced concrete, *CE241: Concrete Technology* (University of California, Berkeley, 2004)
5. P. Hosseini, M. Moradian, Performance of pozzolanic admixtures on mechanical and durability properties of concrete, *Proceedings of Microstructural-related Durability of Cementitious Composites*, 656-664, ISBN 978-2-35158-129-2, Amsterdam, April 2012 (RILEM, Amsterdam, 2012)
6. P. Nikkanen, *On the Electrical Properties of Concrete and Their Application* (Valtion Teknillinen Tiedotus, Rakennus, 1962)
7. P. Šteffan et al, A New Measuring Method Suitable for Measuring Shielding Efficiency of Composite Materials with Carbon Fibers, *Proceedings of 2010 Fifth International Conference on Systems*, 186-189, ISBN 978-0-7695-3980-5, Menuires, April 2010 (IEEE Computer Society Digital Library, 2010), Available from http://www.computer.org/portal/web/csdl/doi/10.1109/ICONS.2010.39
8. J. Pacheco et al, Relationship between cracking and electrical resistance in reinforced and unreinforced concrete, *Proceedings of Microstructural-related Durability of Cementitious Composites*, 458-466, ISBN 978-2-35158-129-2, Amsterdam, April 2012, (RILEM, Amsterdam, 2012)
9. L. Macháň, P. Šteffan, A New Laboratory Equipment for Characterization of Smart Concrete Materials, *Proceedings of ICONS 2013 The Eighth International Conference on Systems*, 123-126, ISBN 978-1-61208-246-2, Seville, January/February 2013 (IARIA XP S Press, Seville, 2013)
10. J. Malikova, Test protocol No. 060-038, Technical and test institute for construction Prague, Brno (2013)

Prediction of concrete strength by non-destructive testing in old structures: Effect of core number on the reliability of prediction

Boussahoua Youcef[a], Kenai Said[a] and Ali-Benyahia Khoudja[b]

[a]University of Blida1, Civil Engineering department, Blida, Algeria
[b]University of Khemis-Meliana, Civil Engineering department, Khemis-Meliana, Algeria

Abstract. When assessing the quality of concrete on site, it is necessary to base the interpretation on reliable and representative test results. Generally, core tests are the most reliable and effective method for assessing the quality of concrete. Unfortunately, this type of testing is expensive and time consuming and only a limited number of cores can be carried out in practice. Non-destructive tests (NDT) can be used to overcome these drawbacks. The most popular and widely used NDT methods for assessing concrete strength are rebound hammer (RH) and the ultrasonic pulse velocity (UPV). The use of these methods provide unreliable predictions unless their results are correlated to destructive tests. A sufficient number of cores is needed to accurately predict the compressive strength of concrete. Recent researches have been carried out for identifying and optimizing the number of cores able to stabilize the calibration model on data from recent structures and from synthetic data. However, more case studies are needed to draw conclusions. In addition, the effect of the degradation of reinforced concrete elements on the number of cores needed for obtaining a reliable prediction needs to be investigated. In this paper, RH and UPV in conjunction with core tests are used to evaluate the concrete compressive strength in existing structures built in the 1970s and degraded mainly by steel corrosion. More than 234 elements were tested by RH and 86 elements by UPV. Also, 36 cores were drilled and tested under compression. A regression analysis is adopted to establish the correlations between NDT and strength measurements. The accuracy of the predictive assessment was evaluated using two indicators: the root mean square error (RMSE) and the coefficient of determination (r^2). The results of this case study showed that seven to nine cores is the minimal number of cores that guarantees the improvement of concrete strength assessment by combined or single NDT methods.

Key words: non-destructive testing, concrete, cores, prediction, correlation, accuracy.

1 Introduction

Concrete is the most widely used materials worldwide in the construction industry due to its strength, possibility to realize all desired shapes and low cost. However, reinforced concrete structures may suffer significant deterioration that may threaten the stability of construction. There are many reasons for deterioration and degradation of concrete. Deterioration of concrete could be due to mechanical degradation, chemical attack, physical degradations or due to reinforcing steel corrosion. The latter is considered to be the most frequent anomaly encountered in reinforced concrete structures [1]. The compressive strength has a crucial role in the evaluation of concrete in existing buildings. Generally, core tests are the most reliable and effective method for assessing the quality of concrete in existing structures. Unfortunately, this type of testing is expensive and time consuming and also a limited number of cores can be carried out in practice.

In the field of civil engineering, non-destructive testing (NDT) are a valuable method for engineers and structure owners to establish precise diagnosis and to periodically monitor structures to ensure their safety without damaging them [2-4]. However, the use of NDT without cores test does not provide any reliable predictions of concrete strength as it is necessary to carry out a calibration operation to relate the concrete compressive strength of concrete and NDT measurements. In fact, it is difficult to establish this relationship [5], especially when dealing with old concrete structures [6] as the test results are dispersed and not easy correlated. For this reason, many researchers proposed various empirical models to relate RH and UPV measurements (separately or in combination) with the compressive strength [7-10]. The combined methods are known by SonReb methods. Some of these researchers performed their studies in laboratories and others on real structures. It is commonly agreed that none of these models can be used on all concretes and that it's necessary to establish models for each concrete with its own parameters. Core number is considered among the main keys for the accuracy and the reliability of assessment. The challenge on one hand is reducing the amount of core number without structurally affecting the concrete element or the structure and keeping the measurement at moderate costs and on the other hand to acquire a sufficiently complete and reliable understanding of the concrete strength in structures.

For this aim, recent research has been directed toward studying the effect of core number on the reliability of prediction. Authors in [11, 12] carried out an investigation using a synthetic simulation. More recently, Ali-Benyahia et al. [13] analysed these parameters on a recent real case study where 205 cores were tested and more than 155 structural elements (beams, columns and shear walls) were tested by RH and UPV. The main purpose of this study is to investigate the needed number of cores for a reliable strength assessment for old degraded structures.

2 Case study presentation

The case-study subject to the present investigation is a reinforced concrete industrial project composed of 17 different buildings built in the 70s located in the north east of Algeria [14], classified according to the Algerian seismic regulations RPA99 / V2003 as a zone of average seismicity (zone IIa) [15]. The most important anomaly encountered is steel bars corrosion under the effect of carbonation as shown in Figure 1. This degradation is visible as spalling and delamination of concrete and cracks parallel to the steel bars.

Figure 1: Typical degradation by steel corrosion

3 Experimental program

In order to clearly identify the quality of concrete in the structures, in-depth diagnosis was established. A detailed visual inspection was carried out to locate the anomalies and identify the elements to be tested. Then an NDT trial investigation was conducted. More than 230 rebound tests by Schmidt Hammer type N34 were carried out on three locations of each element with at least nine points per location. The UPV was measured by ultrasound in 86 elements by the direct method on at least three locations for each element after the preparation of the surfaces to be tested. Thirty-six (36) cores of 65 mm diameter were drilled and crushed in compression.

The cores were drilled in accordance with ASTM C42 [16]. The preparation of cores extremities and the tolerance of planarity and perpendicularity of the ends of the specimens are also prepared as described in NA 5071 [17].

In order to correct the influence of the ratio L/D (length to diameter) of the core on its compressive strength, and to convert the crushing core result to cylinder strength, the crushing results were multiplied by a correction coefficient (Table 1) according to ASTM C42 [16] if the ratio L/D ≤ 1.75 as shown in the Table 1:

Table 1: Coefficient of correction of core strength [16]

Ratio L/D	Correction Factor
1.75	0.98
1.50	0.96
1.25	0.93
1.00	0.87

4 Statistical analysis

The methodology adopted in this investigation is based on the comparison between two static parameters identifying the reliability of concrete strength prediction: the root mean square error (RMSE) Eq (1) and the coefficient of determination (r²) Eq (2) in order to analyse the effect of cores number on the prediction quality and also to identify the minimum number of cores sufficient to stabilise assessment models.

$$\mathrm{RMSE} = \sqrt{\sum_{i=1}^{n} \left(f_{ci} - \widehat{f_{ci}}\right)^2 / n} \quad (1)$$

$$r^2 = \frac{\sum_{i=1}^{n} \left(\widehat{f_{ci}} - \overline{f_c}\right)^2}{\sum_{i=1}^{n} \left(f_{ci} - \overline{f_c}\right)^2} \quad (2)$$

f_{ci}= measured value; $\widehat{f_{ci}}$ = estimated value;
$\overline{f_c}$ = arithmetic mean value.

https://doi.org/10.1051/matecconf/201814902007

The flowchart presented in Figure 2 summarises the methodology adopted in this study. First, the algorithm is started by the data entry consisting of 36 triplets compromising from 36 crushing cores results and 36 results of the non-destructive tests (RH, UPV), the cores are drilled from the same location where NDT measurement were performed. Afterwards, a random selection was adopted by varying the number of core from 2 to 36 cores, and for each sample, 100 iterations was applied to minimize the error. Next, for each number of cores (NC) a calibration model was established, the rest of the results (N-NC) were used to test the quality of the estimation of the conversion model.

The regression analysis was adopted in our study, it is a statistical process for estimating the relationships among prediction variables and to establish a specific mathematic model. For modelling the relationship between the compressive strength and NDT measurements separately, one variable for the simple regression analysis was used. However, to establish the relationship between the compressive strength and RH and UPV measurements at the same time i.e two variables (combined method), a multiple regression analysis was performed.

Finally, for each approach, RMSE and r^2 both were calculated as these parameters indicate the accuracy of the method. The variability was identified by the calculation of the standard deviation (Std Dev) and the results were verified with in-situ and compared to other methods used for strength assessment. This process is repeated (I) times with a step of one core.

5 Results and discussion

The accuracy of the estimation at the stages of calibration and prediction is presented as a function of number of cores on Figs. 3 to 5, respectively for RH and UPV. The quality of accuracy is tested by two statistical indicators RMSE and r^2. The results presented here are the mean values and their standards deviations of the indicators for the 100 simulations. For the low cores number, the calibration error is low whereas the prediction error is high (i.e the prediction capacity is low). When the cores number increases, the prediction capacity increases. Beyond nine (9) cores, the calibration and prediction errors (represented by RMSE and r^2) stabilize about 6.5 MPa for RMSE and 0.22 for r^2 for RH tests and 5.2 MPa (RMSE) and 0.50 (r2) for UPV.

Increasing furthermore the number of cores does not reduce the error in particular that of statistical origin.

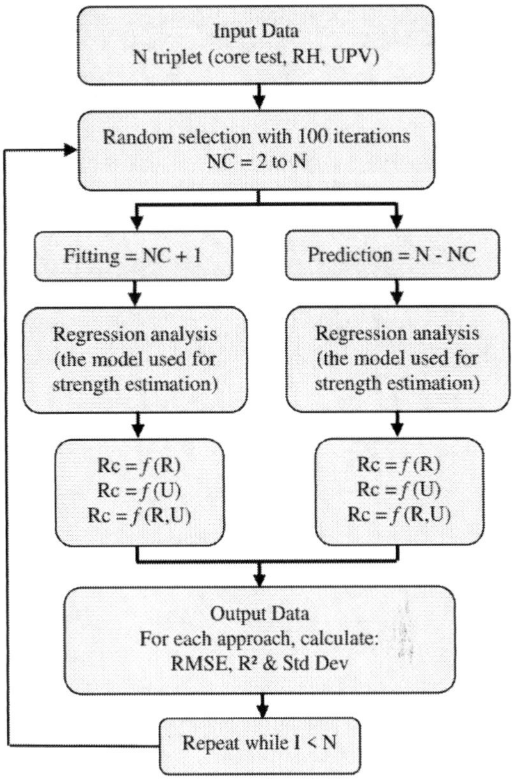

Figure 2: Flowchart represent the statistical analysis performed for strength assessment

These errors result from the variability of material, uncertainty in measurements, model errors, test location and the influence of uncontrolled parameters (humidity, carbonation, cracking, ..).

Alwash et al. [12] have tested the quality of accuracy of various models according to the number of cores using the statistical indicators RMSE for rebound and UPV. This study used synthetic data with three levels of measurement quality (low, medium and high). They concluded that beyond seven to eight cores the added value in terms of accuracy is low. In the case of RH, the error is stabilized at 4.5MPa , 3.5 MPa or 2.5 MPa for low, medium and high NDT measurement, respectively. For UPV, the error is stabilized respectively at 4.7 MPa, 3.8 MPa and 2.8 MPa.

Similarly, Ali-Benyahia et al. [14] analyzed the effect of cores number on the quality of the accuracy of models based on data from a reel case study. The results showed that starting from nine cores, the error is stabilized at about 3 MPa for RH and about 3.5 MPa for UPV.

According to these studies on synthetic data and reel cases studies (old and new constructions), it could be concluded that the number of cores affects considerably the error from statistical origin but for higher number of cores the effect is limited. In addition, the increase of cores number does not affect the non-statistical error (material variability, uncertainty in measurements and variability in uncontrolled parameters).

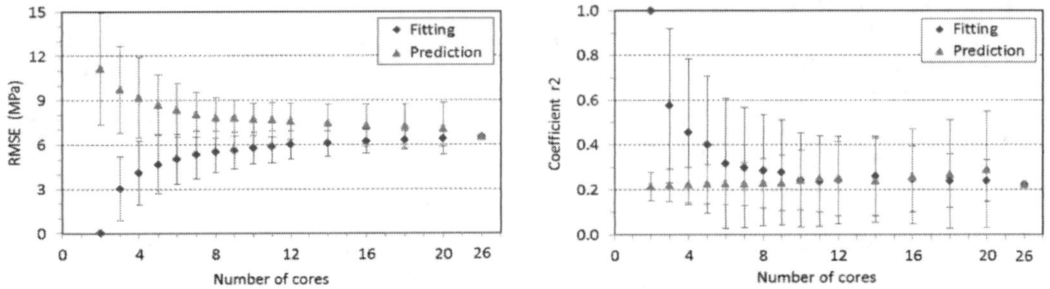

Figure 3: Effect of core number on the reliability of prediction using RH method

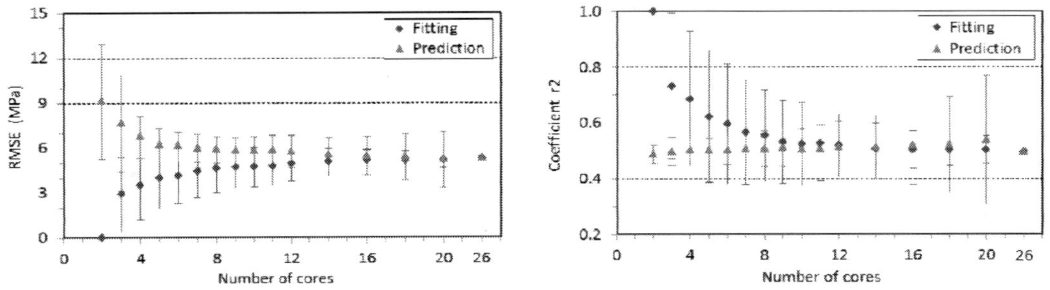

Figure 4: Effect of core number on the reliability of prediction using UPV method

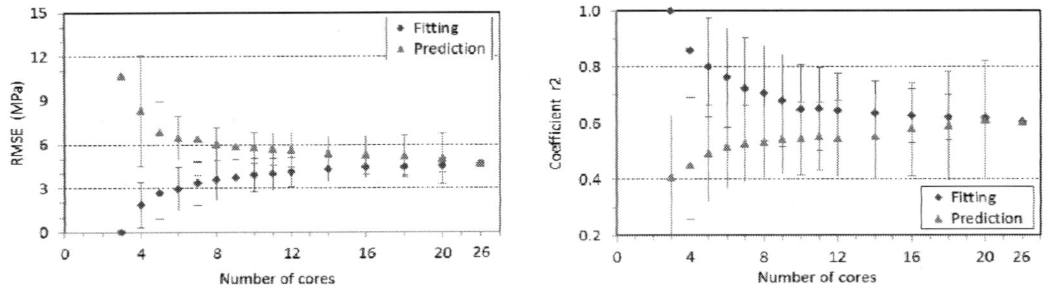

Figure 5: Effect of core number on the reliability of prediction using SonReb method.

6 Conclusion

A companion of destructive tests and NDT tests was carried out on exiting old reinforced concrete structures. A statistical analysis was performed to analyse the effect of concrete aging on NDT measurements and the reliability of strength assessment using two statistical indicators. From the results obtained for this case study, it can be concluded that:

- Seven to nine cores is the minimal number of cores that guarantees the improvement of concrete strength assessment by combined or single NDT methods.
- There is no effect of aging and degradation of concrete on the stability and the reliability of the fit of regression models.

References

1. Y. Boussahoua and S. Kenai, "Diagnostic des Ouvrages en Béton Armé Dégradé par la Corrosion des Armatures: Cas d'Etude," in *3ème Congrès International sur la Technologie et la Durabilité du Béton*, Alger, Algérie, (2016).
2. D. Breysse, "Nondestructive evaluation of concrete strength: an historical review and a new perspective by combining NDT methods," *Construction and*

Building Materials, vol. **33**, pp. 139-63, (2012).

3. D. Breysse and O. Abraham, *Méthodologie d'évaluation non-destructive de l'état d'altération des ouvrages en béton*, France: Association Française de Génie Civil, (2005).

4. D. Breysse, Non-Destructive Assessment of Concrete Structures: Reliability and Limits of Single and Combined Techniques, *RILEM*, (2012).

5. S. Popovics, J. L. Rose and J. S. Popovics, "The behaviour of ultrasonic pulses in concrete," *Cement and Concrete Research*, vol. **20**, pp. 259-270, (1990).

6. J. H. Bungey, S. G. Millard and M. G. Grantham, Testing of Concrete in Structures, London & New York: *Taylor & Francis*, (1992).

7. B. Hobbs and M. Tchoketch Kebir, "Nondestructive Testing Techniques for the Forensic Engineering Investigation of Reinforced Concrete Buildings," *Forensic Science International*, vol. **167**, pp. 167-172, (2007).

8. H. Y. Qasrawi, "Concrete strength by combined nondestructive methods simply and reliably predicted," *Cement and Concrete Research*, vol. **30**, pp. 739-746, (2000).

9. S. R. M. Khan, J. Noorzaei, M. R. A. Kadir, A. Waleed and M. S. Jaafar, "Effectiveness of sclerometer test technique on strength assessment of high performance concrete," *International Journal of Engineering and Technology*, vol. **1**, no. 2, pp. 163-168, (2004).

10. G. Kheder, "A two stage procedures for assessment of in situ concrete strength," *Materials and Structures*, vol. **32**, pp. 410-417, (1999).

11. M. Alwash, D. Breysse and Z. M. Sbartaï, "Non-destructive strength evaluation of concrete: Analysis of some key factors using synthetic simulations," *Construction and Building Materials*, vol. **99**, pp. 235-245, (2015).

12. M. Alwash, Z. M. Sbartaï and D. Breysse, "Non-destructive assessment of both mean strength and variability of concrete: A new bi-objective approach," *Construction and Building Materials*, vol. **113**, pp. 880-889, (2016).

13. K. Ali-Benyahia, Z.-M. Sbartaï, D. Breysse, S. Kenai and M. Ghrici, "Analysis of the single and combined non-destructive test approaches for on-site concrete strength assessment: General statements based on a real case-study," *CSCM Case Studies in Construction Materials*, vol. **6**, pp. 109-119, (2017).

14. S. Kenai and Y. Boussahoua, "Application des technique non destructifs pour l'inspection et diagnostic du béton armé dégradé" in *Diagno-Béton*, Merrakech, (2016).

15. DTR BC, "2 48. (2003). Règlement Parasismique Algérien -RPA 99 version 2003, CGS, Alger.

16. ASTM C 42/C 42M-04, Standard method of obtaining and testing drilled cores and sawed beams of concrete, Philadelphia, PA: American Society for Testing and Materials, (2001).

17. NA 5071, Tests for concrete in structures: Cores sampling, examination and testing compression, Algiers: Algerian Institute of Standardization, (2005).

Feedback of the behaviour of a silo founded on a compressible soil improved by floating stone columns

Ramdane Bahar [1], Omar Sadaoui[2] and Fatma Zohra Yagoub[3]
[1]USTHB, Laboratoire Eau, Environnement, Géomécanique et Ouvrages, Faculté de Génie Civil, Bab-Ezzouar, Algérie
[2]Université Abderrahmane Mira de Bejaia, Département Génie Civil, Bejaia, Algérie
[3]Direction de projets, Groupe Cevital Spa, Bejaia, Algérie

Abstract. The coastal city of Bejaia, located 250 kilometers east of the capital Algiers, Algeria, is characterized by soft soils. The residual grounds encountered on the first 40 meters usually have a low bearing capacity, high compressibility, insufficient strength, and subject to the risk of liquefaction. These unfavorable soil conditions require deep foundations or soil improvement. Since late 1990s, stone columns technique is used to improve the weak soils of the harbor area of the city. A shallow raft foundation on soft soil improved by stone columns was designed for a heavy storage steel silo and two towers. The improvement of 18m depth have not reached the substratum located at 39m depth. The stresses transmitted to the service limit state are variable 73 to 376 kPa. A rigorous and ongoing monitoring of the evolution of loads in the silo and settlements of the soil was carried out during 1400 days that is from the construction of foundations in 2008 to 2012. After the loading of the silo in 2010, settlement occurred affecting the stability of the towers due to excessive differential settlements. Consequently, the towers were inclined and damaged the transporter. This paper presents and discusses the experience feedback of the behavior of these structures. Numerical calculations by finite elements have been carried and the results are compared with the measurements.

1 Introduction

La ville côtière de Bejaia est située à 250 kilomètres à l'est de la capitale Alger, en Algérie. L'expansion urbaine et le développement industriel de la région à partir des années 2000 ont engendré l'occupation des terrains proches de la zone portuaire de la ville. Ces terrains sont constitués de formations d'âge quaternaire, et les reconnaissances géotechniques ont permis de les caractériser dans la catégorie de sols marécageux inondables de faible portance, très compressibles et sujets au risque de liquéfaction [1-2].

En raison des conditions du sol défavorables, la majorité des ouvrages fondés sur des fondations superficielles ont subi des dommages structuraux inhérents aux tassements excessifs nécessitant un renforcement supplémentaire par des travaux de reprise en sous-œuvre très coûteux pour exploiter à nouveau ces structures en toute sécurité [2-3]. Devant ce constat, l'expansion des activités industrielles à partir des années 2000 avait impulsé les concepteurs de projets et les maîtres d'ouvrages à adopter soit des fondations profondes sur pieux fichés dans le substratum, soit de procéder au renforcement des sols pour l'adoption de semelles superficielles.

Parmi les procédés de renforcement de sols, la technique de colonnes ballastées offre une alternative intéressante du fait qu'elle utilise le ballast comme matériau de substitution disponible localement. Ce procédé économique et de réalisation rapide contribue à l'augmentation de la portance des sols médiocres, réduit les tassements et les risques de liquéfaction des sables saturés [4-9]. Cette technique de renforcement a été appliquée pour la première fois en Algérie en 1999 dans le traitement des sols mous sous les silos de stockage de céréales au port de Bejaia.

Après la réalisation de certains ouvrages lourds, une instrumentation a été mise en place en vue de suivre l'évolution des tassements dans le temps en fonction des charges appliquées. Les mesures enregistrées ont dépassé la prévision initiale au stade de projet. Ce qui s'est traduit par interactions entre ouvrages, des pathologies de fondations et des dommages structuraux importants ayant engendré des arrêts d'exploitation silos [10].

L'article traite du retour d'expérience en termes de tassements et de pathologies des fondations d'un silo de stockage de 80 000 tonnes de capacité et de deux (02) tours d'ensilage fondés sur un sol compressible renforcé par des colonnes ballastées flottantes.

2 Site expérimental et description des ouvrages

Le site d'étude est situé à l'embouchure d'oued Soummam et à proximité du port de Bejaia, il s'étend sur une superficie d'environ 8 hectares. La majeure partie des terrains ont été gagnés sur la mer, ils sont sujets à l'influence des courants marins et les fluctuations de la nappe phréatique.

Les ouvrages de stockage ont été conçus pour augmenter les capacités d'emmagasinement en sucre blanc. Ils rentrent dans le cadre du développement des activités du groupe Cevital SPa en matière de traitement et de conditionnement de sucre d'une capacité de 3000 t/j. Les installations constituent l'une des plus importantes en Afrique en termes de hauteur et de capacité de stockage. La superstructure du silo 80 000 tonnes de capacité est conçue en charpente métallique, de 54.3 m de diamètre intérieur, de 63 m de hauteur et de poids à vide de l'ordre de 3000 tonnes (Fig. 1). La coque et le toit sont constitués d'une paroi cylindrique sous forme d'assemblage en double viroles en acier. D'autre part, le radier rigide en béton armé de 58 m de diamètre est fondé sur un gros béton de 0.75 m d'épaisseur, il est constitué de nervures orthogonales (0.7m/3m) et d'une plaque de 0.5 m d'épaisseur. La contrainte due au poids propre est de 73 kPa. La contrainte maximale résultant de la descente des charges à l'état limite de service (ELS) transmise sous le radier est de 376 kPa. Concernant les tours d'ensilage 1 et 2, leurs contraintes en service (ELS) sont respectivement de 242 et 200 kPa. Le sol est renforcé par des colonnes ballastées de 18 m de longueur, réparties uniformément suivant un maillage de 1.6 x 1.6 (m^2) et n'ont pas atteint le substratum situé à 39 m de profondeur. A l'achèvement des ouvrages en 2009, une instrumentation (tassomètres, cibles optiques, capteurs de déformation) a été mise en œuvre avant la mise en chargement du silo survenu en avril 2010.

3 Contextes géologique et géotechnique

Les formations géologiques rencontrées jusqu'à 47m de profondeur sont d'âge quaternaire, elles surmontent la couche marno- calcaire dure du crétacé supérieur qu'on retrouve entre 45 et 50m de profondeur [11]. La formation quaternaire est constituée principalement de dépôts d'alluvions marécageuses de nature argileuse et sableuse lâches et imprégnées d'une matrice vaseuse (Fig. 2). Les investigations géotechniques dénotent une faible résistance du sol au cisaillement et une forte compressibilité, cela est confirmé par certains paramètres mécaniques mesurés sur échantillons intacts. Tenant compte du faible taux de consolidation et de la présence de la nappe et des courants marins, le sol a un comportement évolutif.

L'analyse granulométrique montre que 60 à 95% des passants sont inférieurs à 80µm, ce qui dénote une constitution fine de la majorité des couches de sols. En outre, les essais de limites d'Atterberg ont permis de classifier ces sols dans la catégorie de plastique à très plastique. Le sol est très compressible et présente une tendance à la liquéfaction sur quelques lentilles sablonneuses. Les essais de cisaillement drainés (CD) ont révélé une faible résistance au cisaillement des sols quaternaires ($10 \leq c' \leq 42$kPa et $5 \leq \varphi' \leq 25°$), les valeurs moyennes varient de 18 à 25 kPa pour la cohésion c' et de 10 à 15° pour le frottement interne φ'. Les paramètres mesurés dans la marne entre 39 et 47m de profondeur varient dans les intervalles ($12 \leq c' \leq 52$kPa et $17 \leq \varphi' \leq 33°$), ce qui dénote une bonne résistance. Les essais oedométriques entre 3.5 et 39 m de profondeur ont donné un coefficient de compression c_c variable de 0.09 à 0.27 et un coefficient de gonflement c_g entre 0.01 et 0.08. La formation quaternaire est sous à normalement consolidée et le substratum est surconsolidé à partir de 50m de profondeur. Le niveau moyen de la nappe phréatique est à -2m/NGAL.

4 Suivi des tassements

Les relevés de l'évolution des chargements du silo et des tassements durant 1400 jours ont été effectués d'une manière continue depuis 2008 à 2012. Les tassements ont été mesurés sur 17 points notés (T1,...... à T17) par des tassomètres et des visées topographiques sur réflecteurs fixes (Fig. 3). L'incrément de contrainte de 73 à 260 kPa sous l'effet d'une charge d'exploitation de l'ordre de 50 000 tonnes (60% de la charge maximale) a engendré à partir 2012 des tassements absolus variables de 44 à 53 cm sous le silo et de 27 à 30 cm respectivement sous les tours 2 et 1 (Fig. 4). Les tassements différentiels y afférent varient de 5 à 8 cm sous le silo et de 8 à 18 cm respectivement sous les tours 1 et 2 (Fig. 5). Les premiers indices pathologiques se manifestèrent à partir de février 2010. Ils sont inhérents aux tassements différentiels ayant impacté la stabilité des tours et des locaux techniques. Il s'agit de l'endommagement des transporteurs reliant la tour 1 au silo, de fissures béantes des murs des locaux techniques et l'inclinaison des tours 1 et 2. L'inclinaison des tours a été décelée par des mesures topographiques, un déplacement horizontal maximal au sommet variant de 42 à 51 cm a été mesuré le mois de novembre 2010.

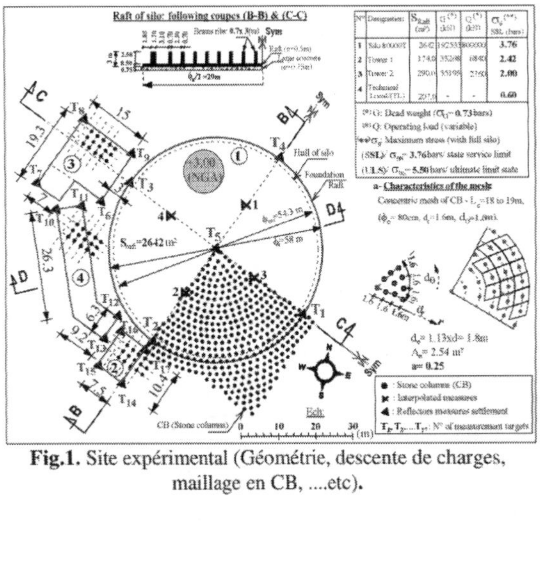

Fig.1. Site expérimental (Géométrie, descente de charges, maillage en CB, ...etc).

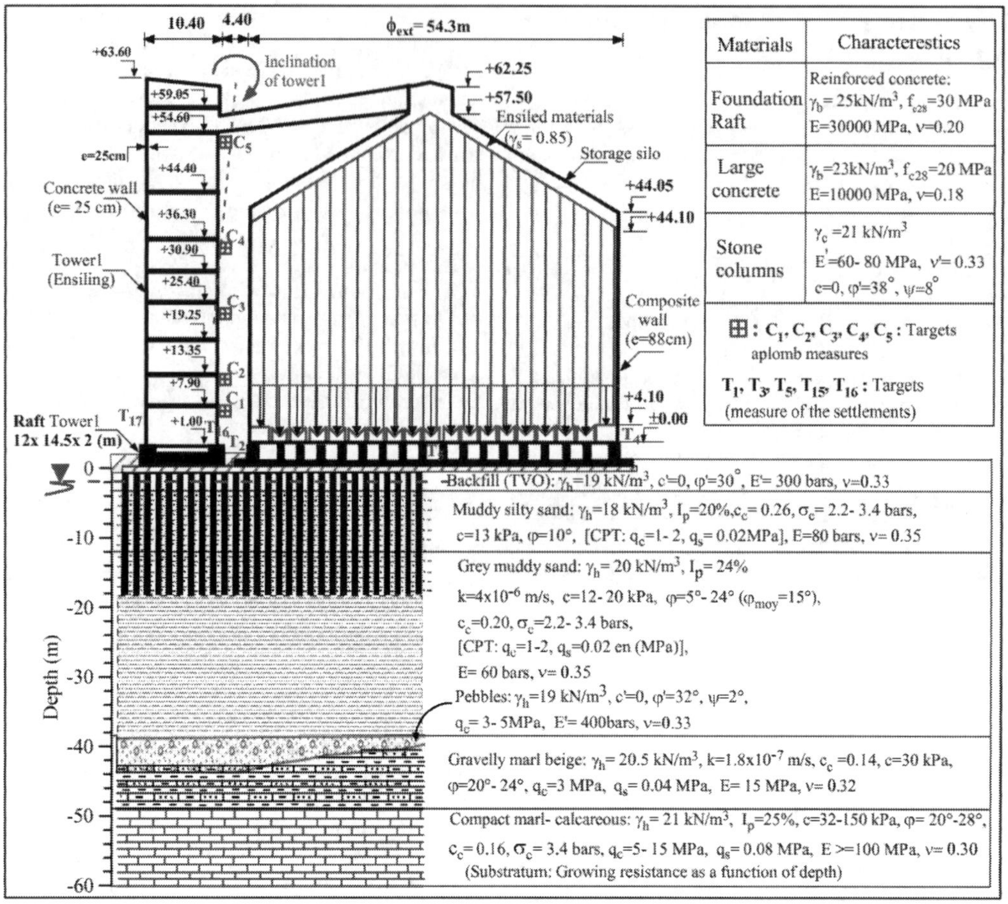

Fig. 2. Profil géotechnique au droit du site expérimental.

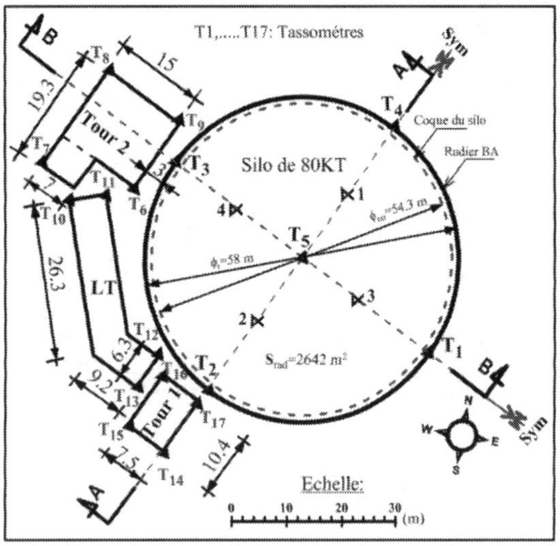

Fig. 3. Vue en plan et dispositifs de mesure des tassements.

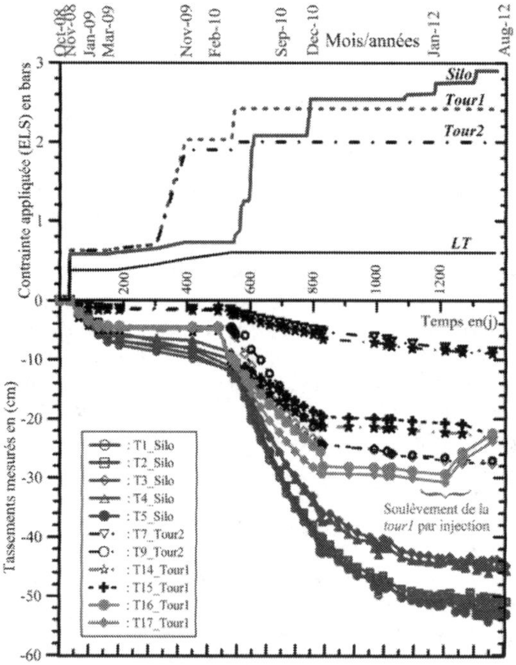

Fig. 4. Evolution des tassements en fonction du temps.

 https://doi.org/10.1051/matecconf/201814902008

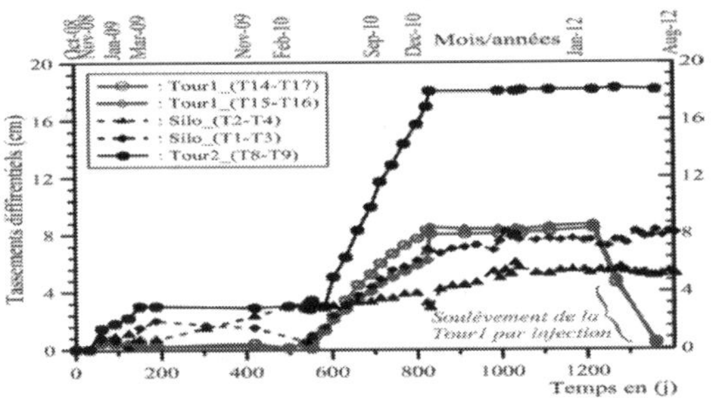

Fig. 5. Evolution des tassements différentiels.

5 Calcul numérique

Les calculs numériques ont été réalisés à l'aide du code d'éléments finis Cesar- LCPC [12] en considérant une loi de comportement élastoplastique avec le critère de Mohr-Coulomb. Le modèle tridimensionnelle (3D) prend en compte 3/4 du radier du silo en intégrant les radiers des tours et les locaux techniques (LT). Il est constitué de 12930 éléments iso-paramétriques hexaédriques H8 (Fig. 6). Les chargements on été appliqués suivant le phasage de la construction des ouvrages et de l'ensilage-déchargement du silo de stockage (Fig. 4). La rigidité (EI) des radiers du silo et des tours est de l'ordre de 34 GPa.m^2/m et celle du radier des locaux techniques est de 2.5 GPa.m^2/m. Les paramètres géomécaniques des couches de sol sont indiqués dans le tableau 1 et ceux de la couche renforcée (0-18m) ont été évalués par la méthode d'homogénéisation de Priebe [13]. Les calculs ont été menés par phasage suivant les contraintes réelles apportées par les ouvrages comme indiqué dans la Figure 4.

Tableau 1. Caractéristiques physiques et mécaniques prises en compte dans le modèle numérique.

Type de sols et des matériaux	EP (m)	γ (kN/m^3)	c' (KPa)	φ' (°)	E_s (MPa)	ν'
Remblai argilo-caillouteux	3.5	19	1	30	20	0.33
Alluvions compressibles	35.5	19	18	13	6 -8	0.35
Marne graveleuse	8.0	20.5	30	23	15	0.32
Marne grise compacte (Crétacé)	≥ 28	21	50- 100	25-30	10^2-10^3	0.30
Colonnes ballastées	19	21	1	38	60-80	0.33
Radier en BA RN30	-	25	-	-	3x 10^4	0.20

Les résultats des calculs numériques par éléments finis de la Figure 7 donnent un tassement maximal de 70 cm sous le centre du radier, soit un écart de 30% par rapport aux mesures. Cet écart est imputable à la variabilité des paramètres géotechniques et aux choix des modules de déformation des sols. Néanmoins, on remarque que l'évolution des cuvettes de tassements mesurés est concordante par rapport aux calculs numériques. Le coefficient de réduction des tassements (β) est un paramètre régissant l'efficacité du traitement, il s'agit du rapport des tassements de consolidation sans et avec renforcement du sol. D'après les calculs à partir de la méthode de Priebe [13] et des mesures, ce paramètre varie de 1.25 à 1.30 dans notre cas.

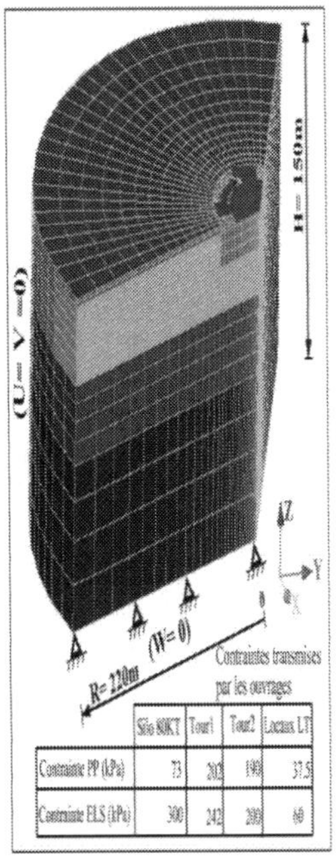

Fig.6. Modèle éléments finis tridimensionnelle à l'aide du logiciel Cesar- LCPC

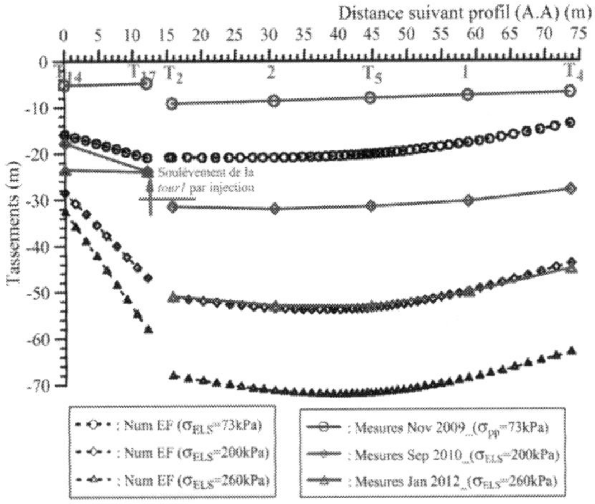

Fig. 7. Evolution de la cuvette des tassements des radiers.

6 Conclusion

La présente étude met en évidence l'inefficacité d'un renforcement par colonnes ballastées flottantes dans des sols mous soumis à l'interaction d'ouvrages à chargement variable. La prévision des tassements au stade de projet était correcte et proche des mesures enregistrées. L'erreur commise dans l'appréciation des effets d'un tassement absolus dépassant les 50cm est très coûteuse. En dépit de la rigidité élevée des radiers conçus pour pallier aux risques déflexions, ils ont accusé des tassements différentiels excessifs qui sont à l'origine de l'instabilité des tours fondées sur des radiers rigides de faibles surfaces (174 à 290 m^2). A noter également le caractère rapide du développement des tassements à cause l'accélération du drainage dû à la forte perméabilité des colonnes ballastées. Tenant compte des risques d'évolution des tassements, la capacité de stockage du silo prévue initialement à 80000 tonnes, n'a pas dépassé 62% depuis 2012 à ce jour.

Remerciements

Nous remercions le personnel technique de la direction projets Cevital SPa Bejaia et les techniciens d'exploitation de la raffinerie 3000T pour leurs contributions dans l'opération d'instrumentation de mesure des tassements et de relevé des chargements du silo. Nous remercions vivement également les cadres techniques du Bureau d'engineering des travaux publics BICS pour la mise à disposition de la station de surveillance topographique et leur assistance technique.

Références

1. M. Khiatine, L. Saci, R. Bahar, O. Sadaoui, Evaluation of dynamics properties and liquefaction potentiel of silty sand in harbour of Bejaia, Algeria. *Proc. 2nd European Conf. on Earthquake Geotechnical Engineering and Seismology,* Istambul- Turkey, pp. 650-659 (2013).
2. R. Bahar, O. Sadaoui, D. Amzal, Diffirenital settlements of cylindrical steel storage tanks- Case of the marine terminal of Bejaia. *7th Int. Conf. on case Histories in Geotechnical Engineering,* Chicago USA, Paper 12, http://scholarsmine.mst.edu/7icchge/session02/12 (2013).
3. R. Bahar, M. Khiatine, O. Belhassani, O. Sadaoui, Settlements observations of industrial structures founded on soft soils: Case of the harbour zone of Bejaia city. ,. *Proc. 3rd Int. Conf. on new Developments in Soils Mechanics and Geotechnical Engineering, Nocosia, North Cyprus,* pp. 149-156 (2012).
4. R.E. Brown, Vibroflotation compaction of cohesionless soils. *ASCE Journal of the Geotechnical Engineering Division 103 (12),*. pp. 1435-1451 (1977).
5. H. B Seed, R.J. Booker, Stabilization of potentially liquefiable sand deposits using gravel drains. *ASCE Journal of Geotechnical Engineering Division 103 (7),*. pp. 757-768 (1977).
6. H.J. Priebe, The prevention of liquefaction by vibro-replacement. *Proc., Earthquake Resistance Construction and Design, Berlin, Germany (1989).*
7. K. Adalier, A. El Gamal . Mitigation of liquefaction and associated ground deformations by stone columns. *Engineering Geology, 72 (1),*. pp. 678-690 (2004).
8. A. Dhouib, F. Blondeau 2005. Colonnes ballastées: Techniques de mise en oeuvre, domaines d'application, comportement, justification, contrôle, axes de recherche et de developpement. *Presses de l'Ecole Nationale des ponts et Chaussées (ENPC), Paris* (2005).
9. S. Lambert. Évaluation de la réduction du risque de liquéfaction par des colonnes ballastées. *Proceedings of the 18th International Conference on Soil Mechanics and Geotechnical Engineering, Paris 2013.* pp. 1-4 (2013).
10. O. Sadaoui, R. Bahar, Field measurements and back calculations of settlements of structures founded on improved soft soils by stone columns, *European Journal of Environmental and Civil Engineering,* http://dx.doi.org/10.1080/19648189.2016.1271358 (2017)
11. R. Bahar, O. Sadaoui, E. Vincens 2010. Settlements of silos founded on compressible soisl reinforced by stone columns. *Proceedings of International Geotechnical Conference- Geotechnical Challenges in Megacities, Moscou (Russie), 07- 10juin 2010, Vol.3.* pp. 965-970 (2010).
12. Itech 2004. Manuel de référence de Cesar et Solveur CLEO 2D & 3D.
13. H.J. Priebe, The design of vibro- remplacement. Reprinted from: Ground Engineering . *Keller Grundbau GmbH, Technical paper 12- 61E. 16pages.* 1995 (1995).

Two-Stage Concrete as a Sustainable Production

H. S. Abdelgader[1], A S. El-Baden[1], H. A. Abdurrahman[2], and A. S.M. Abdul Awal[3]
[1]University of Tripoli, Department of Civil Engineering, Tripoli, Libya
[2]Florida Institute of Technology, Department of Civil Engineering, Melbourne, USA
[3]University of Tun Hussein Onn Malaysia, Faculty of Civil and Environmental Engineering, Johor, Malaysia

Abstract. Two-stage concrete (TSC) is a sustainable concrete which is produced by forcing a flowable cement grout mixture through the voids of a skeletal mass made of compacted preplaced aggregates. From the technical and economic aspects, TSC is particularly useful for construction and repair of concrete structures especially foundations, underwater construction, nuclear reactors, concrete dams, heritage structures and in constructions with closely spaced reinforcement. TSC differs from ordinary concrete in that it contains a higher proportion of stone aggregate and the aggregate stays in point-to-point contact as placed. Thus, the mechanical characteristics of the TSC in failure conditions are distinctly different from ordinary concrete. This paper presents the results of experimental investigations conducted to evaluate the compressive, tensile strength and modulus of elasticity of TSC using grouts with different water-to-cement ratios and admixture contents. It was found that the modulus of elasticity and splitting tensile strength of TSC are equivalent or higher than that of conventional concrete at the same compressive strength. In this method of construction, the splitting tensile strength can be conservatively estimated using the American concrete institute (ACI) equation for conventional concrete.

1 Introduction

The preplaced aggregate concrete (PAC) or two-stage concrete (TSC) gets its name from the method used for placement. Formwork is constructed and the coarse aggregate is densely placed inside the formwork. Grout is then injected through the forms to provide the matrix as shown in Figure 1. Grouting is made at the bottom of the preplaced aggregate [1]. The grout typically consists of sand, cement, water, and an admixture. For each mixture there are optimum amounts of fillers and admixtures which produce the best pumpability or consistency. Proper proportioning of the structural grout mix components is necessary to produce the required strength and durability requirements of the finished preplaced-aggregate concrete [2]. It may be used advantageously on repair jobs of heritage structures, and large concrete repair jobs:

- Where placement by conventional methods is difficult (e.g., massive reinforcing steel and embedded items are present or access to the area is difficult, such as the underpinning of the toe of a dam or apron).
- When low-volume change of the repair concrete is required to avoid cracking caused by excessive tensile stresses in the overlay concrete because of drying shrinkage and restraint provided by the existing concrete.
- Where underwater placement is necessary because dewatering is difficult, expensive, or impractical.

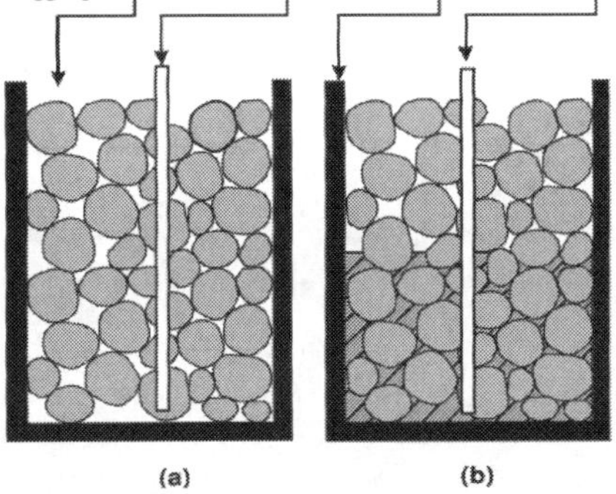

Fig. 1. Fundamental Techniques of TSC

In the last decades, TSC has been involved in many typical applications across the globe. For example, in Poland large-scale two-stage concreting was tested in two applications, namely while laying foundations for an 18 story building in Gdansk (Poland) (about 350 m3) and the repair of damage to a water dam in Czchow near Cracow (Poland) (about 400m3) underwater concreting, on the River Dunajec. In both cases good technological and economical results were obtained [3].
The two-stage concrete has been proved to be essential as radiation shielding in nuclear power plants where heavy weight aggregates are normally used as coarse aggregate. The controversy over the use of nuclear energy fundamentally comes down to three main considerations: its safety of use, ability of nuclear energy generation to reduce greenhouse gas emissions and finally its long term sustainability. In this regard, two-stage method of concrete production appears to be the best choice that promises better environmental performance than conventional concrete in the material production stage by saving a large amount of cement [4, 5].

This paper presents the output results of experimental investigations conducted to evaluate the compressive, tensile strength and modulus of elasticity of TSC using grouts with different water-to-cement ratios and admixtures content. It was found that the modulus of elasticity and splitting tensile strength of TSC is equivalent or higher than that of conventional concrete at the same compressive strength.

2 Experimental program

2.1 Materials and grout proportioning

The coarse aggregate used in the experiment program was sub-angular basalt. A large sample was acquired, washed, and sieved to create the gradation with a maximum size of 50 mm. The coarse aggregate had a bulk loose density 1430 kg/m3, dry specific gravity of 2.687, absorption of 0.34%, and void ratio of 47%. Because of its low absorption and high apparent specific gravity, this basalt is considered a high quality coarse aggregate. The fine aggregate used had an absorption and specific gravity of 1.05 % and 2.63 respectively. The particle size distribution of the sand was divided into three fractions: 2.0/1.0 mm (13%), 1.0/0.25mm (67%), and 0.25/0.0 mm (20%). Its properties were checked in accordance to standards [6]. The cement used throughout the research was ordinary (Type I) Portland cement, having a compressive strength of 41 MPa after 28-days of water curing and Blaine fineness of approximately 3500 cm2/g. The cement properties conformed to standards [7].Water was used directly from the tap and was slightly colder (21OC) than room temperature. The superplasticizer used in the grout was a naphthalene-formaldehyde derivative, trade name 'SikaMent-163' and was mixed at the rate of 2% by weight of cement. The selection of water-cement-sand ratios is more critical in TSC because the amounts of sand and water control the pump-ability and propagation of grout [8]. The void content of the preplaced aggregate was 47% and the bulk density 1430 kg/m3. Five different proportions of water to cement ratio (w/c) (0.42, 0.45, 0.50 and 0.55) were investigated with cement to sand ratios (c/s) of 1/1.5, 1/1, and 1/0.8. Grout preparation was accomplished by combining ingredients with an electric mixer for about three minutes to achieve the desired grout uniformity and consistency. After trials, a w/c ratio of 0.45 was found to be the minimum ratio suitable for grouting. The displacement of aggregate was less noticeable as the w/c ratio was increased; at a w/c ratio of 0.55 there was no displacement.

2.2 Specimen preparation and testing set-up

Three specimens from each one of the three grout mixtures were tested in unconfined compression, split tension and modulus of elasticity at 28 days. A separate set of 27 concrete cubes of size 300x300x300 mm were tested for compressive strength only using grout having w/c ratio of 0.45,0.5 and 0.55 and sand cement ratio of 1/1.5, 1/1 and 1/0.8. Mortar cubes were prepared and tested in unconfined compression as prescribed in standards [9]. Three cubes of each mortar were tested at 28 days. Unconfined compression tests on TSC cylinders were tested in accordance with standards [10]. Splitting tensile tests were also conducted on three specimens of each concrete at 28 days according to the procedures outlined in [11]. Splitting tensile tests were performed using the same machine as the mortar cubes.

3 Strength investigations

3.1 Compressive Strength

Three cylinders from each concrete mix were sub-jected to an unconfined compression test. During the loading procedure, the vertical deformations were measured on three sides of the specimen with respect to axial force increment. On the basis of these results, stress and appropriate strain values were calculated. Some of the test results are presented in Figure 2. Compressive strength results from the cubical shape samples are presented in Table 1.

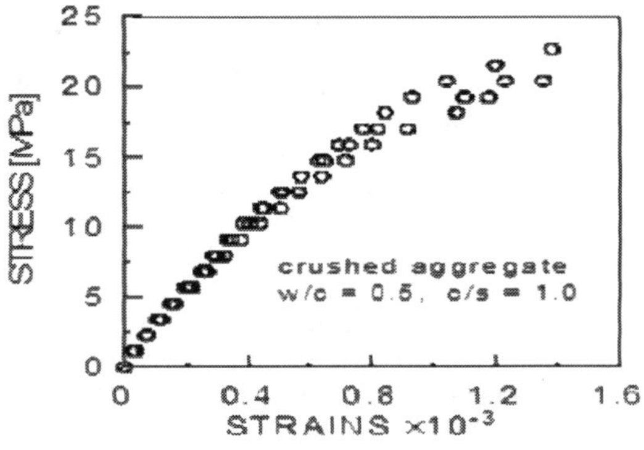

Fig. 2. Stress-strain sample of results for TSC

Table 1. Compressive strength of TSC

W/C	S/C	f_c` (MPa)
0.45	1/1.5	30.82
0.45		27.82
0.45		23.72
0.5	1/1	30.65
0.5		27.10
0.5		23.55
0.55	1/0.8	30.78
0.55		27.23
0.55		23.68

Figure 3 shows the mean and individual cylinder strengths of three specimens per w/c ratio at ages of 28 days. It can be seen that a mean compressive strength of 31.9 is attainable at 28 days with a w/c ratio of 0.45. Figure 3 demonstrates a strength reduction as the w/c ratio increases. Although there is some variation in strength measured per w/c ratio, the strength reduction is approximately linear. This observation is consistent with unconfined compressive strength measurements of TSC cube specimens (300 x 300 x 300 mm), in previous work for the author [12] where the compressive strength (fc`) of TSC at 28 days determined by Equation (1) as shown earlier in Table 1.

$$f_c^- = 62.08 - 71.00(w/c) + 0.52(c/s) \quad (1)$$

Equation (1) is illustrated in Figure 3 for a cement/sand (C/S) ratio of 1.0, and it can be seen that it under-predicts strength of the cylindrical specimens tested in this program. The predicted mean strength from Equation (1) is 87 to 93% of the measured mean strength. Decreasing the multiplicative factor on the w/c ratio in Equation 1 to account for cylindrical specimens yields Equation (2) as the following:

$$f_c^- = 62.08 - 68.00(w/c) + 0.52(c/s) \quad (2)$$

Only c/s = 1.0 for specimens investigated herein. Further research is required to determine the suitability of Equation (2) for a range of c/s ratios. Another significant finding from the compressive strength data was the somewhat limited rate of strength development. This can be explained, in part, because of the fact that no fly ash or other pozzolans were incorporated in the cement grout. The observations show that, although the mechanism of stress transfer is believed to be different from conventional concrete, the mortar strength is a controlling factor in the strength of TSC.

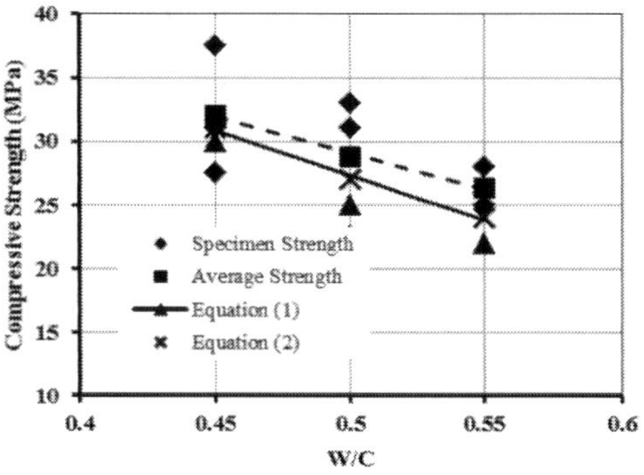

Fig. 3. Compressive strength versus w/c ratio

Table 2 summarizes the average compressive strength of mortar and concrete at 28 days. Firstly, it can be seen that the measured compressive strength at 28 days is sometimes high for both mortar and concrete. Secondly, the ratio of average concrete-to-mortar strengths is reasonably consistent and calculates from 0.50 to 0.51 for all mixtures at 28 days. This observation suggests that the compressive strength of two-stage concrete can be conservatively estimated as one-half of its mortar strength. If this ratio can be substantiated with more mixtures and other sources of coarse aggregate, this simple rule-of-thumb can be adopted in the design of two-stage concrete.

Table 2. Average compressive & tensile strength

W/C	Compressive Strength (fc`) (MPa)	Tensile Strength (f_t) (MPa)
0.45	31.9	3.21
0.45	30	3.11
0.45	25.1	2.73

Table 2 shows the average splitting tensile strengths of three specimens per w/c ratio at age of 28 days. The results indicate strength reduction as the w/c ratio increases. However, there appears to be little difference in strength between specimens produced with a w/c ratio of 0.45 and those produced with a w/c ratio of 0.50. This was also observed with the compressive strengths measured for the same grout mixtures. The actual values of tensile strength at w/c ratios of 0.45 and 0.50 measured from 3.1 to 3.3 MPa, which indicate satisfactory results, especially when one considers the minimum cost of concreting and that no vibration tools are used. Furthermore, excellent results can be expected even when using a high w/c ratio of 0.55, where the mean tensile strength is nearly 2.73 MPa. Test observations show that failure in splitting tension was restricted principally to the line of split and occurs through the mortar and coarse aggregate. Visual assessments of the failed specimens suggest that the percentage of failed aggregate increased in concrete with higher mortar strength.

3.2 Compressive-tensile strength relationships

Knowing the compressive strength (fc`) of concrete the tensile strength (ft) of conventional concrete can be predicted by using ACI code formula as shown in Equation (3). In this investigation Equation (4) was derived from TSC data to predict the splitting tensile strengths (ft`).

$$f_t' = 0.56\sqrt{f_c'} \quad (3)$$

$$f_t' = (0.55-0.56)\sqrt{f_c'} \quad (4)$$

Results from Equation (4) indicated that splitting tensile strength could be well estimated using the ACI equation presented in Equation (3). Comparison between results from this investigation and another investigation conducted by the authors [13] showed a clear similarity as presented in Figure 4. The data from both studies show that the tensile strength of TSC is at least as high as that of conventional concrete, and in fact it can be higher depending on the selection and properties of the coarse aggregate. No causes were apparent for the relatively higher tensile strength in two-stage concrete. However, the greater mechanical interlocking among particles in two-stage concrete could be responsible for the higher tensile strength since factors like aggregate gradation are different from conventional concrete. These observations warrant a much deeper investigation into the influence of coarse aggregate properties on two-stage concrete behaviour in tension. The results obtained in the study corroborate the research findings by Awal [14].

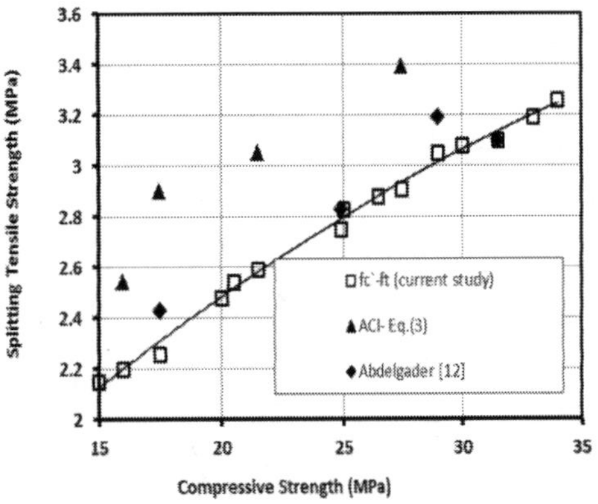

Fig. 4. Compressive strength versus tensile strength of TSC

3.3 Stress-strain relationships

The obtained results are statistically analyzed to estimate equations for the compressive strain-stress relations of various two-stage concrete types. The statistical analysis is performed by use of the MINITAB program [15]. The Minitab Regress Package fits an equation to the experimental data by the least square method. The calculation can be divided into the following steps: assumption of the mathematical relation, estimation of the equation coefficients, testing of the assumed equation and choice of the best model. The statistical analysis of the TSC test data is performed in two successive phases. First, equations for stress-strain relations for type of coarse aggregate and different grout proportions are estimated. In this step the data consist of 27 stress-strain sets of experimental results obtained from cylinder compressive tests [16, 17]. After various trials the following simple relation for all concrete types is:

$$\hat{\sigma} = a\varepsilon_1^3 + b\varepsilon_1^2 + c\varepsilon \quad (5)$$

In the above formulae σ sands for the estimated concrete stresses and ε denotes the concrete strain. The constants a, b, and c in Equation(5) for round aggregate and each concrete type are statistically obtained and presented in Table 3 for c/s=1/1.5 and various w/c ratios. Equation (5) gives reliable results in the strain range from 0 to 0.001. Figure 5 presents typical results of stress and strain relation for round type of aggregate [16].

Table 3. Constants a, b, c of Equation (5)

Aggregate Type	w/c	a	b	c	Correlation Coefficient
	0.45	18.2	-32.5	34.0	0.9941
Round	0.5	1.0	-9.4	28.5	0.9978
	0.55	5.81	-22.0	32.1	0.9861

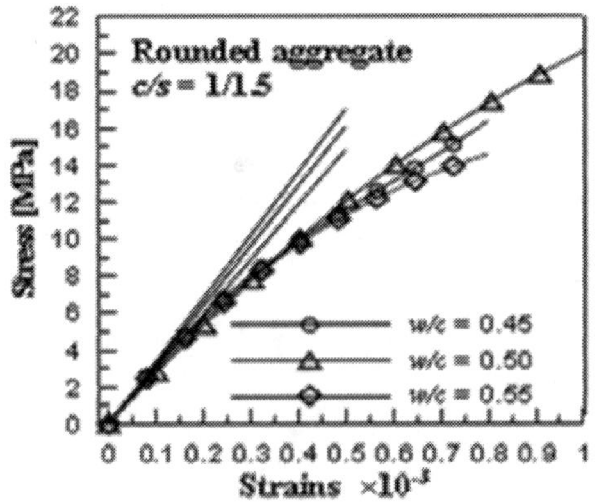

Fig. 5. Graphical presentation of stress-strain results of TSC

In the second step the statistical analysis is used to obtain the TSC stress-strain relations as function of water-cement (ω=w/c) and cement-sand (ς=c/s) ratios. After a complex analysis the functions: $a=a(\omega,\varsigma)$, $b=b(\omega,\varsigma)$ and $c=c(\omega,\varsigma)$ in Equation (5) are assumed for round aggregate as follows:

$$a(\omega,\varsigma) = 14.9 - 17.1\omega + 1.89\varsigma \quad (6\text{-a})$$

$$b(\omega,\varsigma) = -40.8 + 30.1\omega + 5.80\varsigma \quad (6\text{-b})$$

$$c(\omega,\varsigma) = 40.7 - 11.1\omega - 4.46\varsigma \quad (6\text{-c})$$

Some estimation results for concrete stresses according to Equation (5) by use of functions presented in equation (6) is shown in Table 4.

Table 4. Constants calculated for Equation (5)

Aggr. Type	w/c	$a(\omega,\varsigma)$	$b(\omega,\varsigma)$	$c(\omega,\varsigma)$	Correl. Co-eff.
Crushed	0.45	8.47	−21.83	34.49	0.808
	0.5	7.61	−21.78	34.69	0.814
	0.55	6.76	−21.73	34.89	0.735
Round	0.45	10.8	-23.39	32.73	0.827
	0.5	8.24	-21.88	32.18	0.797
	0.55	5.68	-20.38	32.62	0.652

The test observations and the statistical analysis have proved that the linear part of the stress-strain curve in TSC may reach 85% of the ultimate cylinder strength (Figure 5). This may result from a very good contact between the aggregate particles. In consequence the applied load is distributed mainly by the stones. During the laboratory tests it was observed that the skeleton of the two-stage concrete carried the load until the whole specimen collapsed.

It was the reason why the failure of the two-stage concrete was not sudden and explosive but rather gradual. The failure specimen showed extensive lateral expansion in the form of bulging. Thus the specific stress-strain characteristics of the TSC can be explained by a dominant influence of the rigid stone skeleton.

It should be pointed out that the stress-strain relations of traditional concrete result from transfer of stress through the mortar-aggregate uniform matrix. In the TSC there are no such uniform relations because the mechanical properties of TSC are mainly determined by the characteristics of the stone aggregate. Therefore, the subsequent fracture of the two-stage concrete takes place through the stone aggregate particles [17-19]. In the TSC the compressive stresses in the aggregate and in the grout cannot be equal. The compressive stress is influenced by the shear stresses occurring not only in the vertical section but also in the horizontal contact surfaces. The produced shear stress caused the tearing off of the aggregate grains from the grout.

It was also observed that some micro cracking or interfacial cracks were randomly distributed even before the specimens were subjected to load testing. It seems that these cracks were caused by bleeding, settlements of grout, or by shrinkage stresses induced by the drying process [14[. The cracks formed within the grout may be considered to be isolated or discontinuous.

3.4 Modulus of elasticity

The experimental data analysis and the statistically obtained stress-strain relations allow for formulating the relationship between the modulus of elasticity and the compressive strength of the TSC grout. The compressive strength (Table 1) is calculated according to the algorithm for designing the two-stage concrete proposed by the author [16]:

$$\bar{f}_c = \beta_0 + \beta_1 \cdot \bar{f}_g^{\beta_2} \qquad (7)$$

Where: stands for compressive strength of grout (MPa), β_0, β_1 and β_2 are constants obtained from regression analysis and are given in other research by the author [13].The modulus of elasticity of the TSC is mainly affected by the physical properties of the coarse aggregate. The influence of the content of grout in the concrete is rather meaningless. It has been observed that the same factors that affect the compressive strength also alter the elastic modulus of the TSC. The elastic modulus is obtained from the analysis of the stress-strain curves for each type of stone aggregate and the mix proportions. The initial tangent modulus of elasticity of the TSC is found with reference to the tangent line drawn to the stress-strain curve at the starting point as presented in Fig. 5.

The modulus of elasticity of TSC $E_{tsc}(\varepsilon_1)$, determined as a function of strains $\varepsilon(\varepsilon_1 = \varepsilon \cdot 10^3$), can be calculated in GPa from Equation (8) as follows:

$$E_{tsc}(\varepsilon_1) = \frac{d\bar{\sigma}}{d\varepsilon_1} = 3 \cdot a(\omega,\varsigma) \cdot \varepsilon_1^2 - 2 \cdot b(\omega,\varsigma) \cdot \varepsilon_1 + c(\omega,\varsigma) \qquad (8)$$

Assuming that the concrete modulus is estimated as:

$$E_{tsc} = c(\omega,\varsigma) \qquad (9)$$

Equation (9) indicates that the function $c(\omega,\varsigma)$ calculated according to Equation 6, determines the elastic modulus of the TSC. It should be mentioned that the values of the elastic modulus obtained for 30% of the cylinder ultimate strength are almost the same. In this way the initial linear feature of the strain-stress relation of the TSC has been proved. It is possible to elaborate an alternative version of the elastic modulus formulae. Making use of the regression analysis of the experimental data expressions describing the relationship between the elastic modulus and the compressive strength of the TSC the following assumption for round aggregate type can be made:

$$E_{TSC} = 28.7 + 0.08 f_c' \qquad (10)$$

Where: fc` is the cube compressive strengths. The limit value of the cube compressive strengths in equation (10) is:

$$23 \text{ MPa} \le \bar{f}_c \le 32 \text{ (MPa)} \qquad (11)$$

Equation (10) is graphically presented in Figure 6. It should be noticed that the obtained values of the modulus of elasticity are rather scattered. This can be explained by the random allocation of coarse aggregate in the forms. Nevertheless, Equation (10) allows for the approximate determination of the modulus of elasticity for the examined aggregate. It should be pointed out that the obtained results deviate at most ± 10% from the calculated mean values. Similar observations have been made by Awal [20].

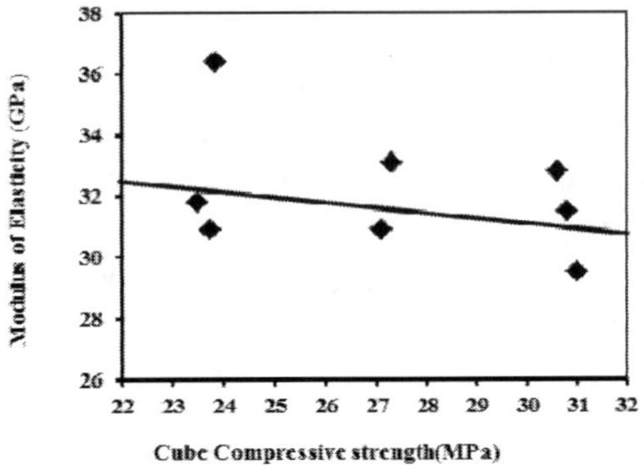

Fig. 6. Modulus of elasticity versus compressive strength

4 CONCLUSIONS

This paper presents sustainable production of two-stage concrete and highlights some physical and mechanical properties. Out of the results obtained and the observations made, the following are the conclusions drawn from the study:

- A grout mixture with water-to-cement (w/c) ratios of 0.45 to 0.55 and a cement-to-sand (c/s) ratios of 1.0/1.5, 1.0/1.0 and 1.0/0.8 optimizes compressive and tensile strength of TSC. Mortar mixed with a w/c ratio below 0.42 is too viscous and does not fully penetrate the voids between coarse aggregate particles, thus creating a honeycombing effect in the hardened concrete. Even when partially bound, however, TSC provides strengths equivalent to that of fully bound concrete with mortar at w/c ratio of 0.45.
- Compressive strength of TSC cylinders can be conservatively estimated as 50% of the mortar cube strength.
- The splitting tensile strength of TSC was found to be similar to that predicted by the ACI equation for splitting tensile strength of conventional concrete. In some cases, the measured tensile strength of TSC is higher than that predicted by the ACI equation.
- The stress-strain relationships for different grout mixes (water/cement ratios and cement/sand ratios) do not show a big difference. The initial stress-strain curves can be estimated by linear relations. This may result from the stresses distributed mainly by the particles of stone aggregate (skeleton of stones). The specific way of stress transmission may also contribute to the initiation and propagation of cracks.
- The modulus of elasticity as a function of compressive strength of the TSC is elaborated. The modulus values for specific type of aggregate are described by linear constant functions which allow engineers to design the TSC according to the algorithm.

References

1. Abdelgader, H., Khatib, J., El-baden, A., Self-Compacting Grout to Produce: Two-Stage Concrete, *Proceeding of The 2nd International Sustainable Buildings Symposium*, 367-372, Gazi University, Ankara, Turkey, (2015).
2. ACI Committee (304.1 R-92), *Guide for the use of pre-placed aggregate concrete for structural and mass concrete applications*: 21-24. Farmington Hills- Michigan: USA, (1997).
3. Abdelgader, H.S., Effect of quantity of sand on the compressive strength of two-stage concrete. Magazine of Concrete Research **48**(3): 353-360, (1996).
4. Liu, C., Ahn, C.R., An, X. and Lee, S. Life-cycle assessment of concrete dam construction: comparison of environmental impact of rock-filled and conventional concrete, Journal of Construction Engineering Management, **139**(12), (2013).
5. Samarin, A. Use of concrete as a biological shield from ionising radiation, Energy and Environmental Engineering, **1**(2): 90-97 (2013).
6. ASTM. C33, *Specification for concrete aggregates. American Society for Testing and Materials.* Philadelphia: Pennsylvania, (1997).
7. ASTM-C150, *Specification for Portland cement. American Society for Testing and Materials.* Philadelphia: Pennsylvania, (1994).
8. Abdelgader, H.S., Elgalhud A. A., Effect of grout pro-portion on strength of two-stage concrete. Structural Concrete Journal **3**(3): 163-170, (2008).
9. ASTM C942. *Standard test method for compressive strength of grouts for preplaced-aggregate concrete in the laboratory.* American Society for Testing and Materials. Philadelphia: Pennsylvania, (2010).
10. ASTM C873. *Standard test method for compressive strength of concrete cylinders cast in place in cylinder molds.* American Society for Testing and Materials. Philadelphia: Pennsylvania, (2010).
11. ASTM-C496/C496M. *Standard test method for splitting tensile strength of cylindrical concrete specimen.* American Society for Testing and Materials. Philadelphia: Pennsylvania, (2004).
12. Abdelgader, H.S., Górski, J. Influence of grout proportions on modulus of elasticity of two-stage concrete. Magazine of Concrete Research **4**(5): 251-255, (2002).
13. Abdelgader, H.S., Górski, J. Stress-strain relations and modulus of elasticity of two-stage concrete. ASCE Materials Journal **4**(2): 251-255, (2003).
14. Awal, A.S.M.A. Manufacture and Properties of Prepacked Aggregate Concrete, M.Eng.Sc. Thesis, University of Melbourne, Australia (1985).

15. Schaefer, R.L. & Anderson, R.B. *The student's edition of MINITAB statistical software adapted for education*: 5-23. New York: Addison Wesley Company, (1989).
16. Abdelgader, H.S., How to design concrete produced by a two-stage concreting method. Cement and Concrete Re-search **3** (29): 331-337, (1999).
17. Iwasaki, N., *Predictions of grouting process in pre-packed concrete by green's function. Proceeding of Japanese Society of Civil Engineering*: 45-50. Tokyo: Japan, (1985).
18. Awal, A.S.M.A. Failure mechanism of prepacked concrete. ASCE Structural Engineering Journal **3**(5): 727-732, (1988).
19. Awal, A.S.M.A., Creep recovery of prepacked aggregate concrete, ASCE Journal of Materials in Civil Engineering, **4**(3): 320-325, (1992).
20. Awal, A.S.M.A., Modulus of elasticity of prepacked aggregate concrete, Journal of Structural Engineering, Madras, India, 15(2) : 65-68, (1988).

Search of the Optimal Position of Sails

Recherche de la Position Optimale des Voiles

MADI Rafik[1], GUENFOUD Mohamed[1]
[1]Université 8 Mai 1945, Génie Civil et Hydraulique, B. P. 401, Guelma, Algerie, madirafik2000@yahoo.fr

Abstract. During its history, Algeria has suffered several destructive earthquakes. According to the observations made during these earthquakes, the Algerian seismic regulation: RPA 82, RPA 83 and RPA 99 revised en 2003, seems insufficient and the buildings have suffered considerable damage. Which explains the presence of failures and insufficiencies in the Algerian seismic code.To resolve this problem, several researchers have advocated the introduction of bracing sails to improve the behavior of structures against the earthquake. In the Algerian seismic regulation, no article is related to this parameter. The objective sought in this study is to study the influence of the position of bracing sails on the behavior of structures across: the ductility demand μ_D, inter-floor displacement D_p and stiffness K_p at the performance point, the elastic stiffness K_e and the overall degradation indicator of the structure I_d. the results obtained showed that the best arrangement of the bracing sails is that which are placed at the end of the structure with perfect symmetry.
Keywords : sails bracing, seismic design, Push-over analysis, modelization, optimization.

Résumé. Au cours de son histoire, l'Algérie a subi plusieurs séismes destructeurs. Selon les observations faites durant ces tremblements de terre, la règlementation parasismique algérienne : RPA 82, RPA 83 et RPA 99 révisé en 2003, parait insuffisante et les constructions ont subi des dommages considérables. Ce qui explique la présence de défaillances et insuffisances dans le code parasismique algérien. Pour résoudre ce problème, plusieurs chercheurs ont préconisé l'introduction des voiles de contreventement pour améliorer le comportement des structures vis-à-vis du séisme. Dans le règlement parasismique algérien, aucun article n'est relatif à ce paramètre. L'objectif recherché dans cette étude est d'étudier l'influence de la position des voiles de contreventements sur le comportement des structures à travers: la demande en ductilité μ_D, le déplacement inter-étage D_p et la raideur K_p au point de performance, la raideur élastique K_e et l'indicateur de dégradation globale de la structure I_d. les résultats obtenus ont montré que la meilleure disposition des voiles de contreventements est celle qui sont placées à l'extrémité de la structure avec une symétrie parfaite.
Mots Clés : voiles de contreventement, conception parasismique, Analyse Push-over, modélisation, optimisation.

1 Introduction

Dans le context sismique, les meilleurs projets en termes de sécurité sont obtenus en respectant les principes de conception. La sécurité est en général mieux assurée dans une structure bien conçue est calculée de manière approchée que dans une structure mal conçue pour laquelle des calculs compliqués sont effectués. Les règlements parasismiques encouragent le concepteur à adopter des dispositions de conceptions en général favorables à un bon comportement de la structure. L'objectif à atteindre des projets de construction parasismiques, est de définir une structure capable de subir, sans s'effondrer, les déformations engendrées par l'action sismique. Cet objectif peut être atteint avec succès, par des projets de différents types d'ossatures et de divers degrés de capacité de dissipation d'énergie. Le respect des principes de construction laisse encore le concepteur du projet devant plusieurs choix à effectuer au préalable, car plusieurs solutions sont possibles qui respectent toutes les principes de conception parasismique. Suite au séisme qui a frappé la région du centre et en grande partie la wilaya de Boumerdes, des conclusions ont été tirées suite aux expertises qui ont été établies, que les constructions en voile en béton armé ont prouvées de leurs efficacités structurelles aux actions sismiques, dans les zones sinistrées [1, 2].

Les structures à voiles en béton armé présentent, particulièrement en zone sismique, plusieurs avantages: leur résistance limite les déformations latérales, leur rigidité permet de protéger les éléments non structuraux et quelques poteaux existants, leur présence permet de s'affranchir du difficile problème posé par la réalisation du ferraillage des nœuds des portiques, elles permettent de ne pas être pénalisées dans le choix du coefficient de comportement en cas de panneaux de remplissage.

Par rapport à d'autres systèmes constructifs tels que les portiques auto stable en poteaux- poutres, des dégâts et effondrements ont été observés pour de multiples raisons: mauvaise conception, sous dimensionnement des poteaux, manque d'armatures transversales dans les zones critiques des éléments structuraux, etc. Ce sont ces raisons qui ont conduit principalement, à pénaliser ce type de système structurel et principalement la limitation de la hauteur des structures. Les critères de vérification prisent en considération dans cette recherche sont les suivants: le déplacement inter-étages maximal D_{max}, la demande en ductilité μ_D, la raideur de la structure au point de performance K_p et l'indicateur de dégradation globale de la structure I_d et le mécanisme de ruine et l'état de dégradation des éléments.

2 Evaluation de la vulnérabilité

L'évaluation de la vulnérabilité d'une construction nécessite l'estimation des dommages potentiels aux différents types de structures selon les intensités sismiques. Les approches pour l'évaluation de la vulnérabilité sont multiples et peuvent être classées soit selon leur degré de difficulté [3]. L'analyse *pushover* est l'une des méthodes d'estimation de la vulnérabilité. C'est une analyse statique non linéaire dans laquelle la structure subit des charges latérales suivant un certain modèle prédéfini, en augmentant l'intensité des charges jusqu'à ce que les modes de ruine commencent à apparaitre dans la structure. Elle est basée sur l'hypothèse que la réponse d'une structure a plusieurs degrés de liberté, peut être assimilée à la réponse d'un système à un seul degré de liberté équivalent. Dans ce cas la réponse est contrôlée par un seul mode de vibration et qui reste constante durant la durée du séisme (figure 1).

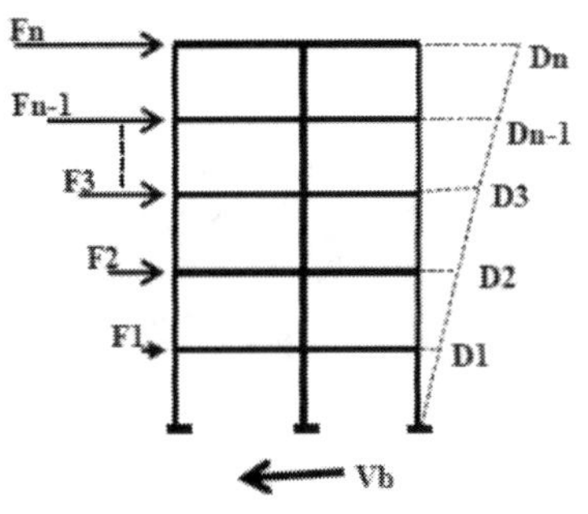

(**a**) Forces et déplacements dans la structure

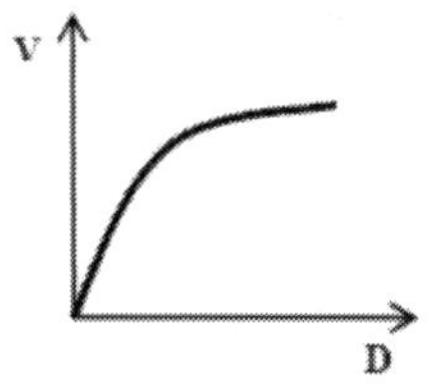

(**b**) Spectre de capacité, format (V_b-D)

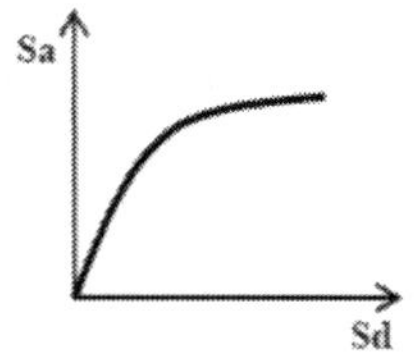

(**c**) Spectre de capacité, format (S_a-S_d)

Fig. 1. Conversion du spectre de capacité.

La force latérale appliquée $\boldsymbol{F}$ et le déplacement $\boldsymbol{D}$ d'un système à un seul de gré de liberté sont donnés par :

$$F=\frac{m_i\,\phi_i}{\sum_{i=1}^{N} m_i\,\phi_i} V_b \tag{1}$$

$$D=\frac{m_i\,\phi_i}{\sum_{i=1}^{N} m_i\,\phi_i} D_i \tag{2}$$

Où :

m_i : la masse concentrée au niveau i.
ϕ_i : Le vecteur déplacement du mode fondamental.
V_b : l'effort tranchant à la base.
D_i : Le déplacement de la masse m_i.

La transformation de l'effort tranchant à la base $\boldsymbol{V_b}$ en accélération spectrale $\boldsymbol{S_a}$, et le déplacement au sommet $\boldsymbol{D}$ en déplacement spectral $\boldsymbol{S_d}$, sont donnés par:

$$S_a=\frac{F}{M_1} \tag{3}$$

$$S_d=\frac{D}{\Gamma_1\,\phi_{t,1}} \tag{4}$$

Où :

M_1 : la masse effective de la construction liée à l'amplitude du premier mode de vibration, et au masses m_i des différents niveaux.
D : le déplacement au sommet lié à l'amplitude du premier mode de vibration.
Γ_1 : le facteur de participation modale correspondant au premier mode de vibration.
$\phi_{t,1}$: amplitude du premier mode de vibration au sommet.

M_1 et Γ_1 sont données par:

$$M_1 = \frac{\left(\sum_{i=1}^{N} m_i \phi_{i,1}\right)^2}{\sum_{i=1}^{N} m_i \phi_{i,1}^2} \tag{5}$$

$$\Gamma_1 = \frac{\sum_{i=1}^{N} m_i \phi_{i,1}}{\sum_{i=1}^{N} m_i \phi_{i,1}^2} \tag{6}$$

Avec ces transformations, nous obtenons la courbe de capacité (figure 2). Le point de croisement entre l'exigence (conversion du spectre ***Sa-T*** au format ***Sa-Sd***) et la résistance (courbe de capacité) fournit le point de performance (fonctionnement) qui décrit l'état d'endommagement de l'ouvrage. Au point de fonctionnement, la demande en énergie à dissiper est gale à la capacité à dissiper de l'énergie et correspond à un état de dommage unique.

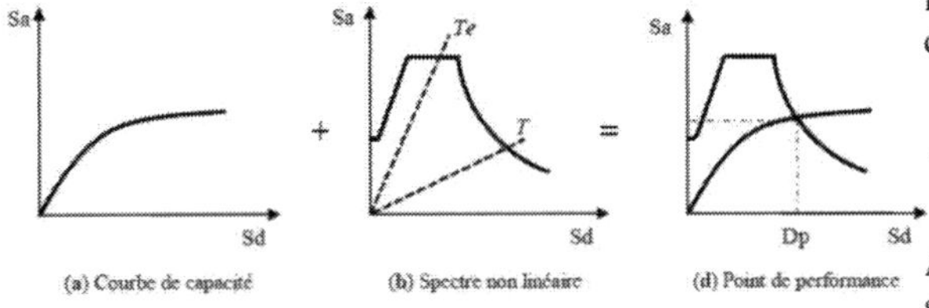

Fig. 2. Point de performance.

La relation force-déplacement utilisée dans SAP 2000 pour définir les zones plastiques est mentionnée sur la figure 3.

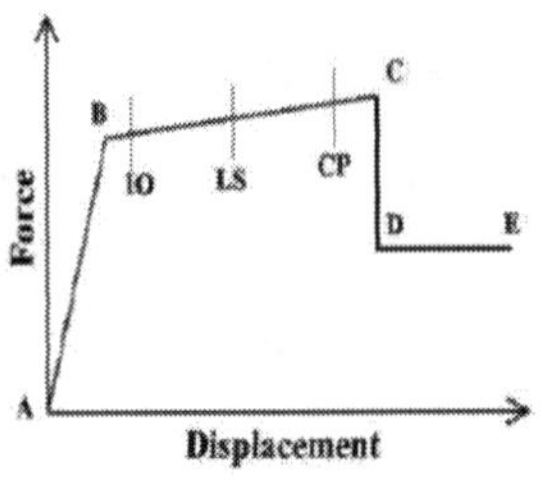

Fig. 3. Relation force-déplacement utilisée dans SAP 2000 pour définir les zones plastiques.

- Le point ***A*** correspond aux conditions initiales (non chargement).
- Le point ***B*** représente la limite élastique de l'élément.
- L'ordonnée du point ***C*** correspond à la résistance nominale et l'abscisse correspondante représente la déformation à partir de laquelle une dégradation marquante de la résistance commence à se produire.
- La baisse manifestant de point ***C*** au point ***D*** représente la rupture initiale de l'élément.
- La résistance aux charges latérales au-delà du point ***C*** est souvent incertaine.
- La résistance résiduelle du point ***D*** au point ***E*** permet à l'élément de supporter les charges gravitaires.
- Au-delà du point ***E*** qui correspond à la capacité maximale de déformation, les charges gravitaires ne peuvent plus être reprises.

Selon l'ATC 40, le déplacement cible D_{cible} en fonction de la hauteur de l'ouvrage H est défini par:

$$D_{cible} = \frac{H}{25} \tag{7}$$

La demande en ductilité μ_D en fonction du déplacement élastique de la structure D_y et le déplacement au point de performance D_p est donnée par:

$$\mu_D = \frac{D_p}{D_y} \tag{8}$$

A mesure que le facteur μ_D est grand, le degré de pénétration de la structure dans le domaine plastique est important.

La raideur initiale (élastique) de la structure K_e en fonction de l'effort tranchant élastique V_y et le déplacement élastique D_y est donnée par:

$$K_e = \frac{V_y}{D_y} \tag{9}$$

A mesure que les éléments de la structure se plastifient, la structure subit des dommages qui se traduisent par une diminution de K_e.

La raideur au point de performance K_p en fonction de l'effort tranchant V_p et le déplacement D_p est donnée par:

$$K_p = \frac{V_p}{D_p} \tag{10}$$

L'indicateur de dégradation global I_d est donné par :

$$I_d = 1 - \frac{K_p}{K_e} \tag{11}$$

3 Modélisation des éléments

Les poutres et les poteaux sont modélisés par des éléments ayant des propriétés élastiques linéaires. Le comportement non linaire des éléments est traduit par l'introduction des rotules plastiques au niveau des sections susceptibles de se plastifier.

Le voile est modélisé par un élément poteau équivalent situé sur l'axe du voile relié à des poutres infiniment rigide. Des rotules plastiques de flexion sont introduites aux extrémités, et une rotule plastique de type (V2 ou V3) au milieu du voile traduit le comportement en cisaillement [4, 5].

4 Etude d'un bâtiment R+5

Il s'agit d'une structure en béton armé R+5. Planchers en corps creux de 20 cm d'épaisseur. Les dimensions sont (30x35) cm pour les poutres, (30x30) pour les chaînages, (30x40) pour les poteaux et 15 cm d'épaisseur pour les voiles de contreventement. La hauteur d'étage est de 3,15 m. L'ouvrage est implanté dans une zone sismique de classe II (RPA 99, version 2003), avec un sol de type meuble. Les caractéristiques des matériaux sont les suivantes (Règle *BAEL 91*):

Béton : $f_{c28}=25\ MPa,\ E_c=32164\ MPa.$
Aciers : *FeE400* pour armature longitudinale,
$E_s=2.10^5\ MPa, f_y=400\ MPa.$
EeE235 pour armature transversale,
$E_s=2.10^5 MPa, f_y=235\ MPa.$

Les charges permanentes ***G*** et les surcharges ***Q*** sont les suivantes:
$G_{terrasse}$= 0,57 tf/m², $Q_{terrasse}$ = 0,10 tf/m², $G_{étage}$ = 0,50 tf/m², $Q_{étage}$ = 0,10 tf/m²

Les différentes variantes de disposition des voiles de contreventement sont mentionnées sur les figures de 4 à 9 pour la variante 1, de 10 à 15 pour la variante 2 et de 16 à 21 pour la variante 3.

Variante 1

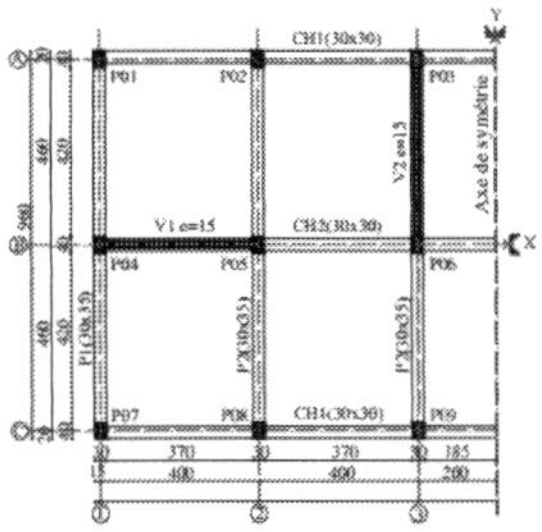
Fig. 4. Variante 1.1

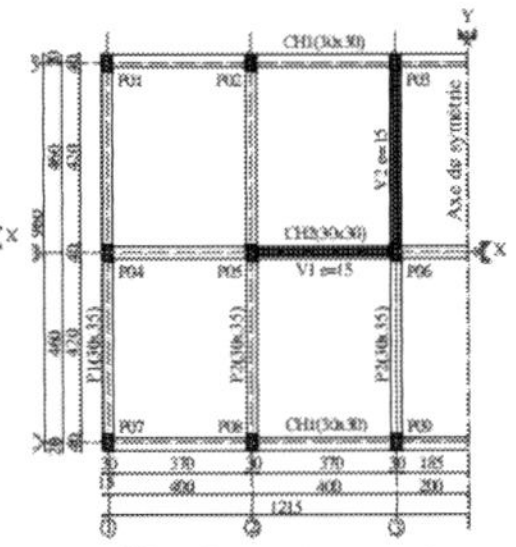
Fig. 5. Variante 1.2

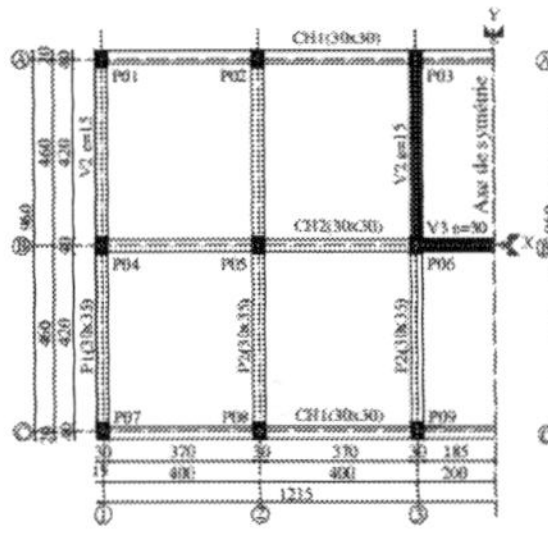
Fig. 6. Variante 1.3

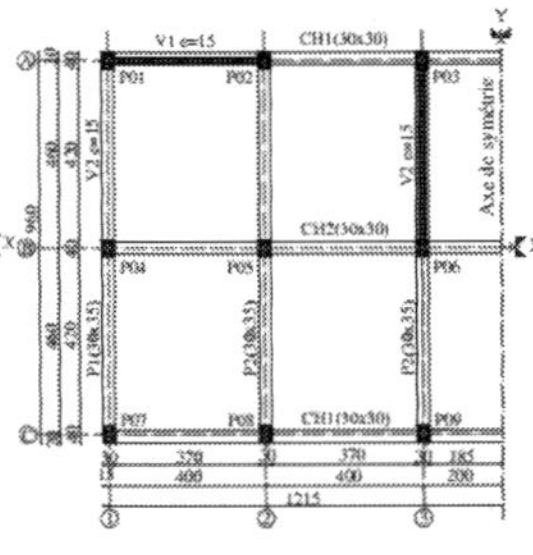
Fig. 7. Variante 1.4

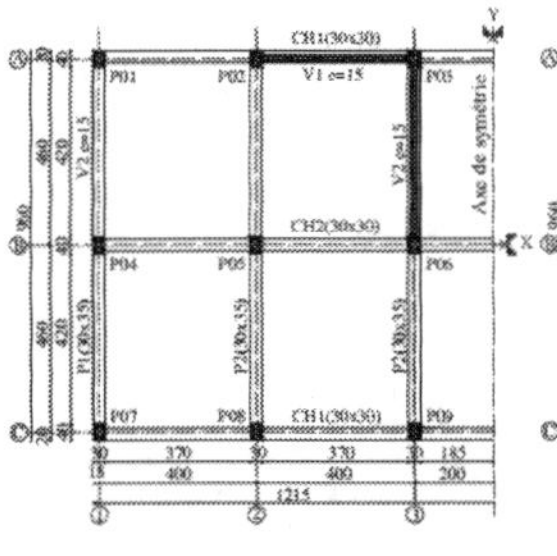
Fig. 8. Variante 1.5

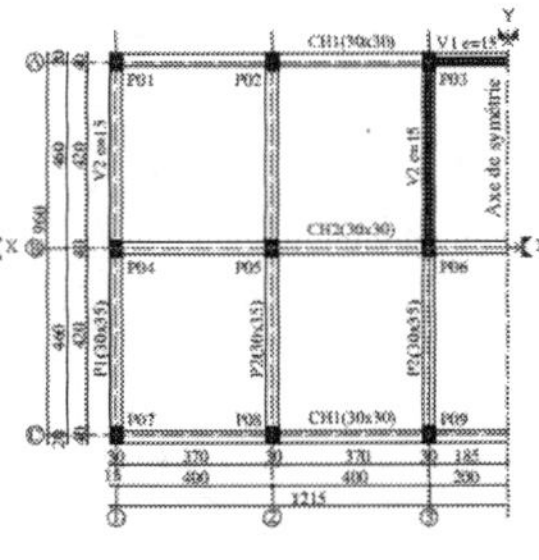
Fig. 9. Variante 1.6

Variante 2

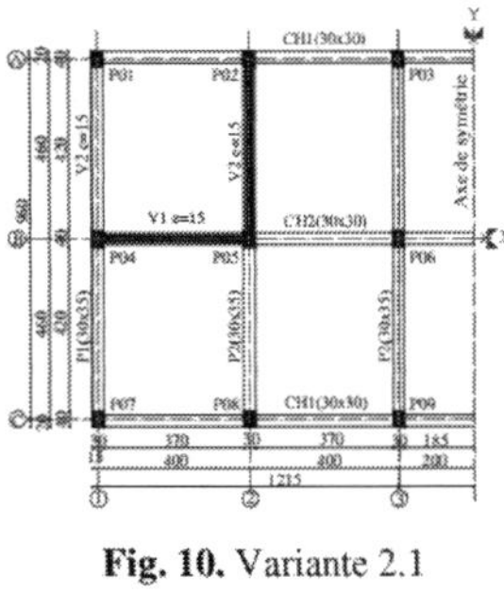
Fig. 10. Variante 2.1

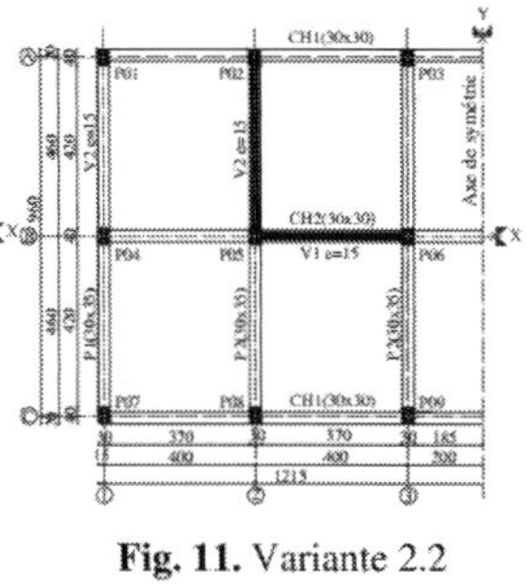
Fig. 11. Variante 2.2

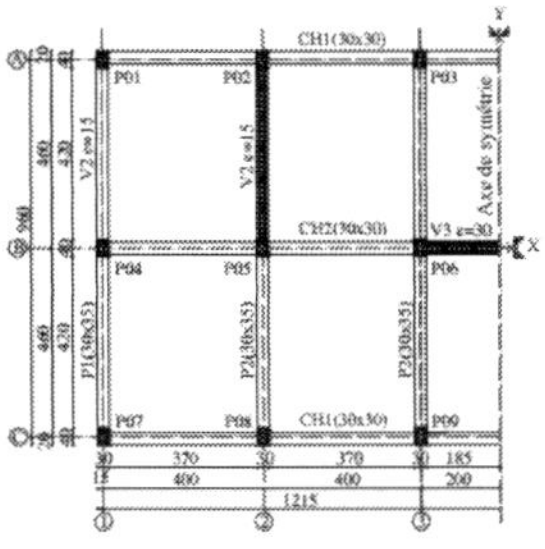
Fig. 12. Variante 2.3

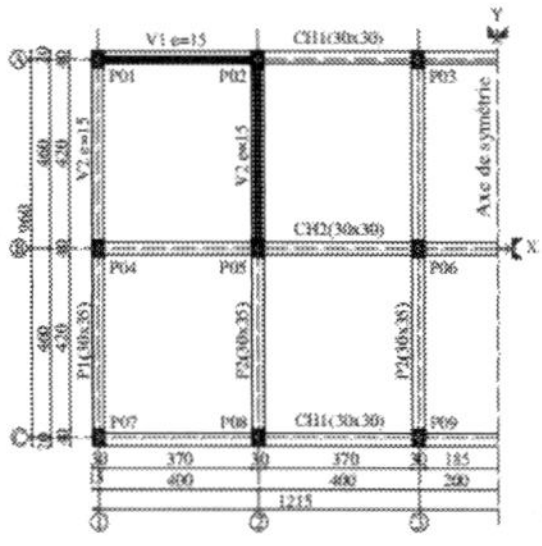
Fig. 13. Variante 2.4

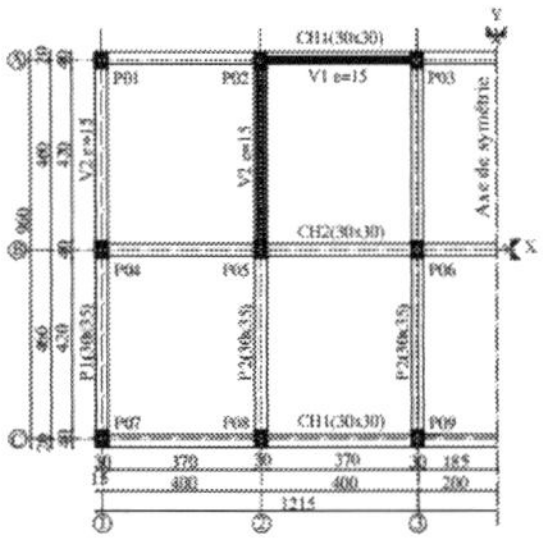
Fig.14. Variante 2.5

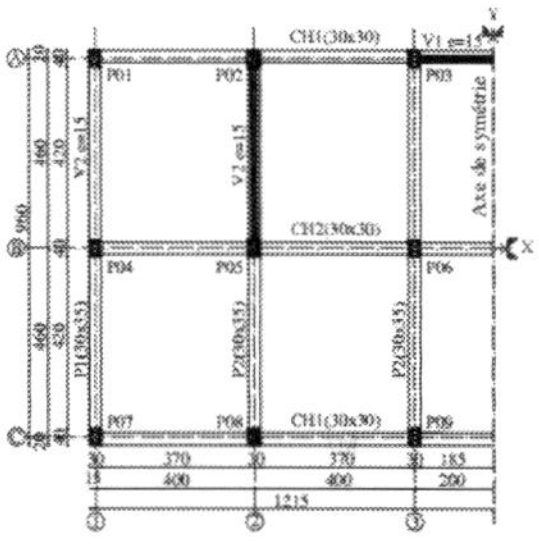
Fig. 15. Variante 2.6

Variante 3

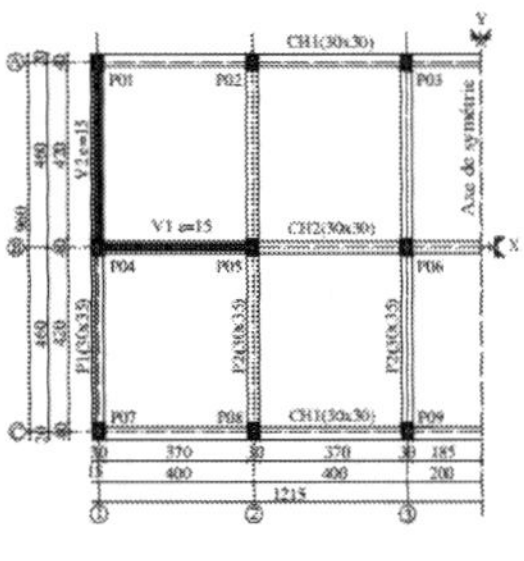
Fig. 16. Variante 3.1

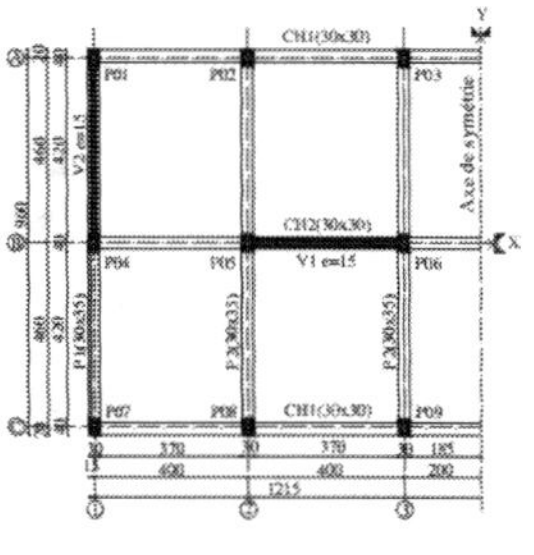
Fig. 17. Variante 3.2

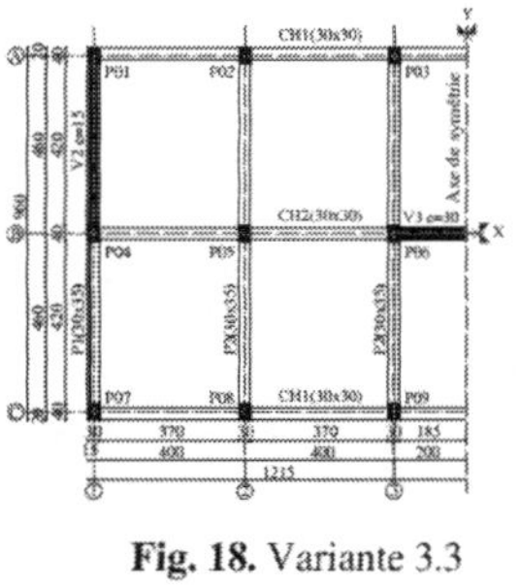
Fig. 18. Variante 3.3

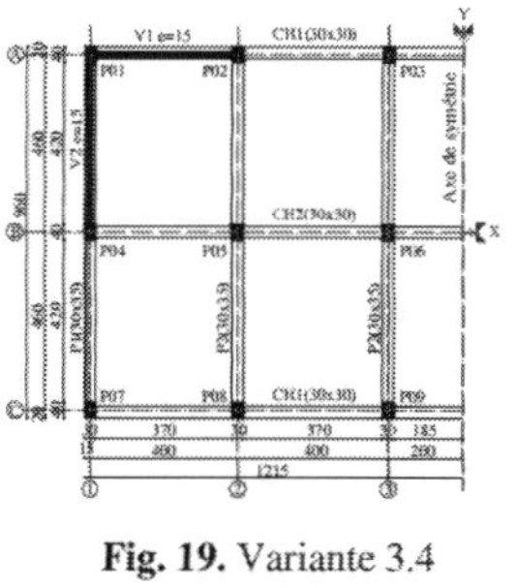
Fig. 19. Variante 3.4

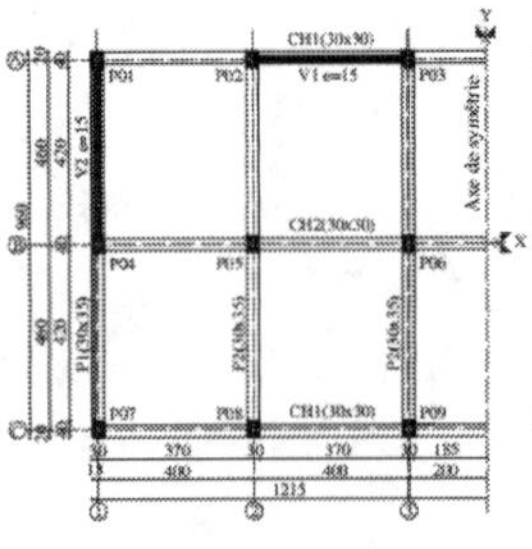

Fig. 20. Variante 3.5

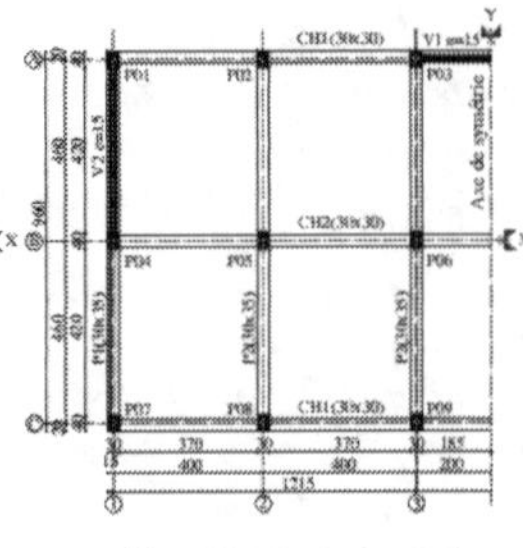

Fig. 21. Variante 3.6

5 Analyse des résultats

Les résultats de l'analyse dynamique et *pushover* sont mentionnés sur les tableaux 1et 2 respectivement.

Pour le comportement dynamique de la structure:

- La participation des masses modales est supérieure ou égale à 90% de la masse totale de la structure,
- Les variantes qui présentent des déplacements purs au premier et deuxième mode de vibrations et une rotation pure au troisième mode sont : V2.6 et V3.1 à V3.6.

Les variantes qui présentent la plus grande participation des masses modales accompagnées de déplacements purs au premier et deuxième mode et une rotation pure au troisième mode sont V3.1 et V3.4, par conséquent la meilleure disposition des voiles de contreventements est celle qui est placées à l'extrémité de la structure avec une symétrie parfaite.

Tableau 1. Résultats de l'analyse dynamique

Variante	Nom Variante	Période fond. (s)	Participation des masses		Délacement au sommet (cm)		Comportement dynamique (T : translation, Rot : rotation)		
			X (%)	Y (%)	X (cm)	Y (cm)	Mode 1	Mode 2	Mode 3
Structure sans voiles	Structure en Portique	1,39	99	99	3,04	3,40	T : x	Rot : z	T : y
Variante 1	V1.1	1,02	90	90	1,13	1,20	Rot : z	T : x et y	T : x et y
	V1.2	0,94	90	91	0,94	1,76	Rot : z	T : x et y	T : x et y
	V1.3	0,93	91	91	1,03	1,03	Rot : z	T : x et y	T : x et y
	V1.4	0,74	90	90	1,15	1,18	Rot : z	T : x et y	T : x et y
	V1.5	0,70	91	91	1,01	1,18	Rot : z	T : x et y	T : x et y
	V1.6	0,56	91	91	0,91	0,93	T : x et y	Rot : z	T : x et y
Variante 2	V2.1	0,58	91	91	0,93	0,95	T : x et y	Rot : z	T : x et y
	V2.2	0,57	91	91	0,87	0,95	T : x et y	Rot : z	T : x et y
	V2.3	0,59	89	90	1,28	1,25	Rot : z	T : y	T : x
	V2.4	0,58	91	91	1,06	1,21	T : x et y	Rot : z	T : x et y
	V2.5	0,57	91	91	0,94	1,02	T : x et y	T : x et y	Rot : z
	V2.6	0,55	90	90	1,13	1,24	T : x	T : y	Rot : z
Variante 3	V3.1	0,58	91	91	0,93	0,99	T : x	T : y	Rot : z
	V3.2	0,55	90	90	1,07	1,24	T : x	T : y	Rot : z
	V3.3	0,59	89	90	1,28	1,24	T : x	T : y	Rot : z
	V3.4	0,57	91	91	0,94	0,98	T : x	T : y	Rot : z
	V3.5	0,55	90	90	1,13	1,26	T : x	T : y	Rot : z
	V3.6	0,55	90	90	1,13	1,24	T : x	T : y	Rot : z

Tableau 2. Résultats de l'analyse pushover

Var.	Sens y-y									
	Courbe de capacité				Analyse Pushover Point de performance		Paramètres			
	V_{max} (tf)	D_{max} (m)	Vy (tf)	Dy (m)	Vp (tf)	Dp (m)	μ_D	Ke (tf/m)	Kp (tf/m)	Id (%)
	Variante 3									
V3.1	262,68	0,057	77,797	0,008	124,593	0,021	2,62	9724,62	5933,00	0,39
V3.2	261,74	0,062	74,915	0,008	123,762	0,02	2,22	9364,37	6188,10	0,34
V3.3	322,17	0,067	76,068	0,008	130,722	0,02	2,50	9508,50	6536,10	0,31
V3.4	374,64	0,09	78,46	0,007	129, 390	0,02	2,85	11208,6	6469,50	0,42
V3.5	384,72	0,102	76,678	0,008	127,548	0,02	2,50	9584,75	6377,40	0,33
V3.6	372,27	0,093	74,712	0,008	127,833	0,02	2,50	9265	6391,65	0,31

6 Conclusion

La participation des masses modales est supérieure ou égale à 90% de la masse totale de la structure pour l'ensemble des variantes. Les variantes qui présentent des déplacements purs au premier et deuxième mode de vibrations et une rotation pure au troisième mode sont: V2.6 et V3.1 à V3.6.

Les variantes qui présentent la plus grande participation des masses modales accompagnées de déplacements purs au premier et deuxième mode et une rotation pure au troisième mode sont V3.1 et V3.4.

Les variantes qui présentent un indice de ductilité élevé et une meilleure pénétration dans le domaine plastique sont V3.1 et V3.4. Par conséquent la meilleure disposition des voiles de contreventements est celle qui sont placées à l'extrémité de la structure avec une symétrie parfaite.

Références

1. Victor Davidovici, *Rapport de mission séisme de Boumerdes*, Ministère de l'habitat et de l'urbanisme, Algérie, (2003)
2. Milan Zacek, *Conception parasismique niveau d'avant projet*, Cahier N°1, Collection conception parasismique, (2004)
3. Arie-José Nollet, *Evaluation de la vulnérabilité des batiments existants*, Département de genie de la construction, (2004)
4. K. Galal, H. El-Sokkary, *Advanced in modelling of RC shear walls*, The 14 th World conference on earthquake engineering, Octoer 12-17, Bejin, China, (2008)
5. Matej Fischinger, Tajana Isacovic, *Benchmark analysis of structural wall*, The 12th world conference on earthquake engineering, Auckland, Nouvelle Zelande, (2000)

Developing a combined Light Detecting And Ranging (LiDAR) and Building Information Modeling (BIM) approach for documentation and deformation assessment of Historical Buildings

Reda Yaagoubi [1] and Yehia Miky [2,3]
[1]College of Geomatics and Surveying Engineering, IAV Hassan II, Rabat, Morocco.
[2]Department of Geomatics, Faculty of Environmental Design, KAU University, Jeddah, Saudi Arabia.
[3]Department of Civil Engineering, Faculty of Engineering, Aswan University, Egypt.

Abstract. Cultural heritage plays a fundamental role in preserving the collective memory of a nation. However, it is noted that many historical buildings suffer from serious deformation that may lead to deterioration or loss. In this paper, we propose an approach for documentation and deformation assessment of historical buildings based on the combination of Terrestrial Light Detecting And Ranging (LiDAR) technology and Building Information Models (BIM). In order to digitally archive the current state of a historical building, classical surveying techniques (Traversing, Levelling and GPS) are integrated with Terrestrial Laser scanner (TLS). A Leica Scan Station C10 is used to accomplish the 3D point cloud acquisition. In addition, Leica GNSS Viva GS15 receivers, a Leica Total Station TCR 1201+ and a Leica Runner 24 are used for classical surveying. The result is a 3D point cloud with high resolution, which is referenced according to the local geodetic reference system Ain el Abd UTM 37N. This point cloud is then used to create a 3D BIM that represents the ideal condition of the building. This BIM also contains some important architectural components of the historical building. To detect and assess the deformation of building's parts that require an urgent intervention, a comparison between the 3D point cloud and the 3D BIM is performed. To achieve this goal, the main parts of the building in the BIM model (such as ceilings and walls) are compared with the corresponding segments of the 3D point cloud according to the normal vectors of each part. A case study that corresponds to a historical building in Jeddah Historical City named 'Robat Banajah' is presented to illustrate the proposed approach. This building was built to serve pilgrims that want to perform the fifth pillar of Islam. Then, it was endowed (waqf) as a charity housing for widows and disabled. The results of assessing deformations of the case study show that some rooms are in a degraded condition requiring urgent restoration (distortions reach up to 22 cm), while other building parts are in a non-critical condition.

1 Introduction

Jeddah city is located on the Red Sea coastline, in the west side of Al Hijaz region. It is considered as theeconomic and the touristic capital of the Kingdom of Saudi Arabia. Also, its locationclose to the two holy cities of Islam, Makkah and Medina, gives this place a fundamental role in the history of Islamic Civilization.

Its old city contains several historical buildings that have their own characters. These buildings take a major place in the collective memory of the Saudi society. In addition, they are considered as attractive landmarks for visitors that come from all over the world. Unfortunately, some monuments inside the old city of Jeddah are under deteriorating conditions due to various human and natural factors (climate, disasters…). Hence, it is necessary to detect and repair any potential deformation at an early stage to preserve the loss of valuable historical buildings. To do so, monitoring and analysis of structural deformations must be performed to check the current condition of such buildings [1].

This study aims to develop an approach for documenting historical buildings and measuring their deformation by combing LiDAR (Light Detecting and Ranging) technology and BIM (Building Information Modeling). To achieve this goal, a comparison between a 3D BIM model (that represents the ideal condition of the building) and a 3D LiDAR point cloud (that represents the current condition of the building) is conducted. In the following sections, an overview of methods used for documentation and deformation assessment ispresented. Then, the proposed methodology based on the combination of BIM and terrestrial LiDAR is described. Finally, a case study that corresponds to a historical building in Jeddah Historical City named 'Robat Banajah' is presented to illustrate the proposed approach.

2 General overview of methods for heritage documentation and deformation assessment

2.1 Heritage documentation methods

The documentation of a historical building may be defined as the process of acquiring, processing, presenting and recording the required data for the determination of the current position and form according to a specific coordinate system [2]. The selection of an appropriate method depends mainly on several factors such as: 1) the required level of details, 2) the accuracy of the acquisition method, 3) the completeness of the model to be generated and 4) the maintainability of the current conditions without any alteration due to the acquisition process.

As mentioned previously, data acquisition is the first step in the documentation process. In the following, three methods thatmay be used for acquiring data in the documentation process will be presented: 1) Total Station, 2) Close-Range Photogrammetry and 3) Terrestrial LiDAR. These three methods do not require any direct contact with the object or the building during the acquisition process.

- Total Station:

A total station is a traditional surveyinginstrument that allows measuring horizontal and vertical angles of a point of interest, in addition to the slope distance to that point. Furthermore, it can also make surveying computation with angle and distance observations. Therefore, the coordinates of the measured point are calculated in real time, according to a specific coordinate system. These functionalities have made total station one of the predominant instruments used in surveying nowadays.

- Close-Range Photogrammetry (CRP):

Close-Range Photogrammetry (CRP) is the art, science, and technology used for obtaining precise mathematical measurements and three-dimensional (3D) data model from two or more images captured with a camera at close range[3]. It allows recording the current position, form and other architectural aspects of the historical building by photographs. If more than two photos are captured, a bundle adjustment solution is possible, including all available measurements on photos at the same time [4]. The 3D model generated from a photogrammetric process may be converted to a set of 3D data points with a very high density [5].

- Terrestrial LiDAR:

Terrestrial LiDAR, also called Terrestrial Laser Scanning, provides remotely sensed 3D data that capture the surrounding environment. Terrestrial LiDAR is based on the use of a continuous laser beam to measure the distance between the target and the scanner. While this scanner rotates around its vertical axis, its mirror also moves the emitted laser beam up and down to cover most of the surrounding area. In general, two main techniques are used to compute the distance to the target: pulse-based and phase-based measurements [6]. Pulse-based LiDAR convert the travelling time that takes some of the emitted energy pulses to return-back to the scanner. Then, this time is converted to distance between the scanner and the target taking into consideration that the laser pulse has a speed of light. Regarding phase-based LiDAR, the scanner measures the phase shift between the emitted and returned signal. This difference in phase is proportional to the traveling time, so it can also be converted to distance between the target and the scanner.

In addition to the distance measurement, additional data including the horizontal angle of the rotating laser and vertical angle of the oscillating mirror are recorded. The scanner automatically combines all these data to calculate an accurate 3D X, Y, and Z coordinates position for each point. The set of acquired points is named 'Point Cloud'. While the scanning with laser is happening, photos of the real world (with true colours RGB (Red Green Blue)) can be taken. These photos are used to assign to each point of the cloud the corresponding RGB value.

2.2 Deformation assessment methods

Deformation assessment methods consist of monitoring the integrity of structures to early detect potential damage and react appropriately in a timely fashion[7]. To carry out deformation analysis, there are three main approaches: 1) Point-to-Point, 2) Point-to-Surface and 3) Surface-to-Surface.

- Point-to-Point analysis:

If the deformation analysis is performed on two triplets of coordinates that represent the same position, but in two different dates, a simple comparison between the two triplets of coordinates gives the corresponding deformation. However, if we have two point clouds representing a scene in two different periods, it is uncertain that the exact same point appears in the two clouds for many reasons (different density, different scanning points...). To overcome this problem, a range image is generated from each point cloud based on the same point of perspective [8]. The value of each pixel of the range image corresponds to the distance to the scanning station. Then, pixel values from the two generated range images are subtracted to assess the deformation.

- Point-to-Surface analysis:

In the Point-to-Surface analysis, a point cloud is compared to a reference surface. This surface may be an existing 3D model of the object of interest, or it may be generated from another point cloud by using specific techniques such as tessellation or surface fitting. The deformation corresponds to the shortest distance that is calculated according the normal of the reference surface [9].

- Surface-to-Surface analysis:

Surface-to-Surface analysis allows comparing between two surfaces, that may come from existing 3D models or that are generated from two different point clouds. This type of analysis requires defining a mathematical parameterization for planar patches of each surface[10]. Then, the plan parameters and their covariance matrices are used to perform Surface-to-Surface analysis for deformation assessment [11].

2.3 Discussion

The use of total station may be considered as a conventional method since it allows using a set of discrete points across the structure (very limited number of points) in the process of deformation assessment. When using a total station in deformation monitoring, the 3D coordinates of specific points on the structure may be measured and compared to their coordinates on a model that represents the ideal condition of the building. In addition, it is possible to make different measurements of the same point at a regular interval of time, so one can deduce information about possible deformation on the structure over the time.

On the contrary, Close Range Photogrammetry (CRP) and Terrestrial LiDAR offer the possibility to acquire a very dense 3D point cloud that may be used in the process of deformation assessment. This 3D point cloud may be compared to an existing model representing the ideal condition of a structure. Also, if we have a set of data corresponding to two different dates, a deformation analysis may be carried out through a comparison between the two point clouds. CRP is very efficient to record the texture of elements and to automate extraction of features in the structure. However, Terrestrial LiDAR is much more appropriate for complex structures where the use of CRP may require complex processing [12]. Terrestrial LiDAR can capture the full surface of an object, even cracks and tiny deformations and map them up to millimeters of accuracy. In addition, the texture is acquired through the photos that are simultaneously taken during the scanning process.

3 The combination of BIM and Terrestrial LiDAR for documentation and deformation assessment

The use of a Terrestrial LiDAR allows a digital archiving of the as-built condition of a building. Indeed, Terrestrial LiDAR provides a very dense and accurate point cloud that captures the current condition of this building. Then, based on the acquired point cloud, a 3D model could be generated. This 3D model may represent either the as-built condition or the ideal condition of such a building.

In this section, we will introduce the concept of Building Information Modelling (BIM) and how it can be used to assess the deformation of historical buildings.

3.1 The importance of BIM in cultural heritage

Building Information Models (BIM) differ from conventional 3D models because BIM are based on knowledge-rich parametric building elements[13]. These parametric objects, which represent the components of the building, are assembled together to automate the generation of building information and to facilitate the creation of rich building models[13]. The parametric objects are defined according to a systemic approach where each object is characterized by its descriptive data and its relationship with other components of the building [14]. The concept of BIM is widely used in the field of cultural heritage preservation [15 - 18]. In [19] a plug-in prototype library, named Historical BIM (HBIM), for parametric objects used to model historical monuments has been developed. HBIM offers interactive parametric objects that represent architectural elements, and it was used to model historical monuments from TLS point cloud [14]. The BIM provides a powerful approach to automate the documentation of cultural heritage thanks to its parametric modelling process. It allows generating highly sophisticated 3D models and offer advanced visualization tools.

3.2 The integration of LiDAR and BIM for deformation assessment

The proposed approach for deformation assessment consist of integrating Terrestrial LiDAR with Building Information modelling. First, we proceed to the scanning of the exterior facades, then the interior components of the building. It is necessary that all scanning stations must be referenced according the same reference system, so we can register all stations in the same point cloud. After cleaning the resulting point cloud, the next step is to create a 3D BIM from this cloud. For this purpose, it is possible to use some commercial software such as Autodesk' Revit and Graphisoft's ArchiCAD [20]. This BIM must represent the ideal condition of the building because it will be compared to the point cloud corresponding to the as-built condition. Once the BIM is completed, the point cloud will be segmented according to the components of the BIM against which deformations will be assessed. Indeed, each segment of the point cloud is compared to the corresponding BIM component through a Point-to-Surface deformation analysis (see section 2.2). The result of such a comparison will provide a useful information about the current condition of the assessed part of the building, and hence if any restoration is needed.

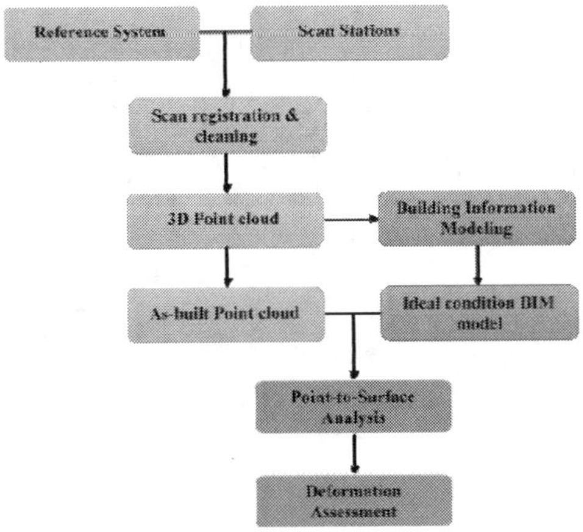

Fig. 1. A typical workflow of the proposed approach.

Figure 1 shows a typical workflow of the proposed approach for documentation and deformation assessment of historical buildings.

4 Case Study: 'Robat Banajah'

'Robat Banajah' is one of the most important historical buildings in Jeddah old city in the Kingdom of Saudi Arabia. This building was built to serve pilgrims that want to perform the fifth pillar of Islam. Then, it was endowed as a charity housing for widows and disabled. This 'Robat' contains 12 rooms on the ground level with a total area of around 430 m^2. The architectural characteristics of the building are very simple with some plaster decorations on the main entrance (Figure 2).

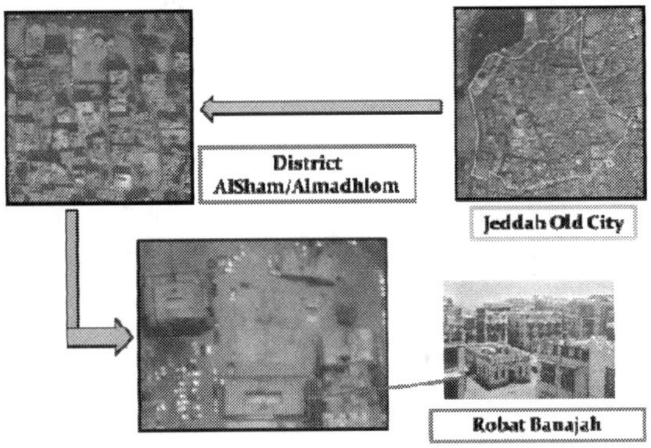

Fig. 2. Robat Banajah.

The acquisition of the current condition of the 'Robat' was performed by the 3D Laser Scanner Leica ScanStation C10. This instrument has a field of view of 360° in the horizontal direction and 270° in the vertical direction. It has a position precision of 6 mm (till 50 m range) and a maximum range of 300 m.

Before proceeding to the scanning, it is necessary to define GCPs (Ground Control Points) according to the local reference system (Ain el Abd UTM 37 North). Leica GNSS Viva GS15 receivers (with a post processing horizontal precision 3 mm + 0.5 ppm) are used in addition to a Leica Total Station TCR 1201+ (0.3 mgon, 1 mm + 1.5 ppm with reflector). The orthometric height values of those GCPs are obtained through a geometric levelling by using a Leica Runner 24 level (2.0 mm per km double-run levelling). The GCPs are used to have a georeferenced 3D point cloud that contains data about the exterior and the interior of the building.

The following figure shows the various surveying instruments that are used in the acquisition of the current condition of the historical building.

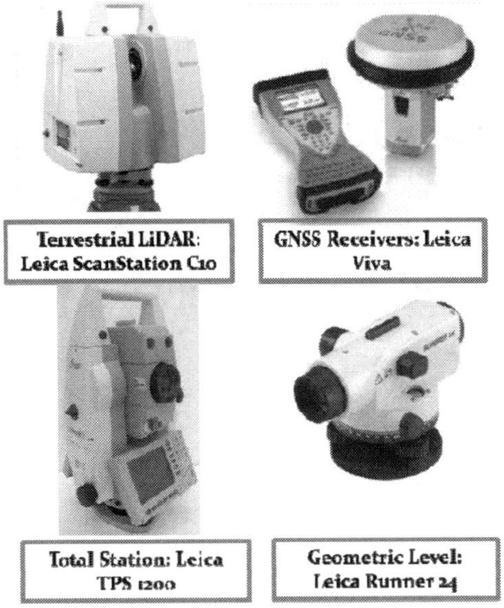

Fig. 3. Surveying instruments used in the acquisition of the current condition of the historical building.

The workflow of the scanning process consists of setting up the laser scanner on a traverse point with known coordinates and to have a known backward azimuth (backward traverse point with known coordinates). Then, we start acquiring forward traverse point with unknown coordinates, in addition to black and white targets that are used in the point cloud registration process. After that, a fully 3D point cloud acquisition is performed from the occupied point by the laser scanner (Figure 4).

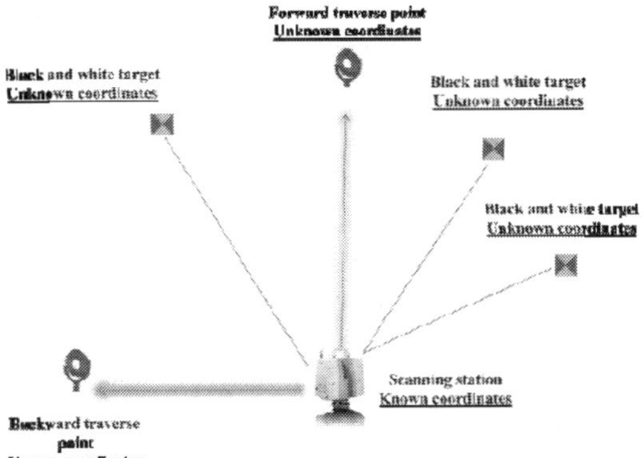

Fig. 4. The Laser scanning workflow.

All scans that come from different scanning stations will be grouped together thanks to common targets. This process is called 'Registration'. The result is a unified point cloud, where each point has 3D coordinates according to the local reference system (Ain el Abd UTM 37 North). The following figure shows the 3D registered point cloud of 'Robat Banajah'.

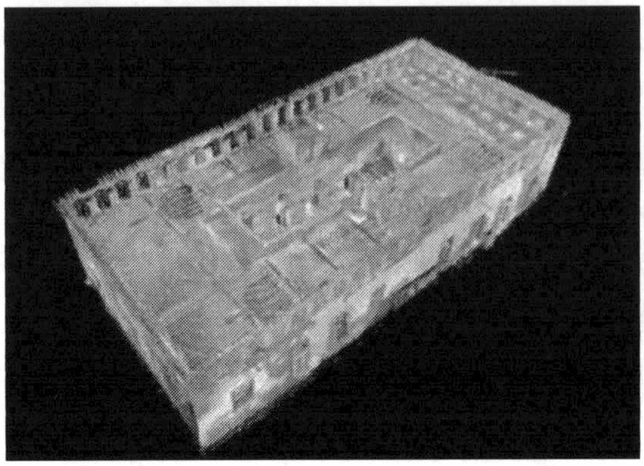

Fig. 5. Registered 3D point cloud of 'Robat Banajah'.

The obtained point cloud is then used to produce a 3D BIM model that represents the ideal condition of the building (Figure 6). This BIM must include the main parts of the building that should be assessed such as walls, ceilings, roofs, etc.

Fig. 6. The 3D BIM model of 'Robat Banajah'.

Before proceeding to deformation analysis, we need to segment the point cloud into parts that will be compared to the corresponding parts of the BIM. In this case study, we are interested in walls and ceilings. So, the first step is to isolate the room (or the set of rooms) to be assessed. Then, the ideal ceiling that comes from the BIM model is used in a Point-to-Surface deformation analysis. This analysis gives a map of differences between the used point cloud and the ideal plan representing the ceiling. The following figure shows the result of deformation analysis on one room of 'Robat Banajah' which has the most deformation values.

Fig. 7. Deformation analysis of a room ceiling.

The deformation analysis of this room shows that the maximum ceiling deflection is about 22 cm. Also, more than 50% of the deflection measurements were more than 10 cm which may require restoration and maintenance works for that ceiling.

It is also possible to make deformation analysis on walls as shown in the following figure.

Fig. 8. Deformation analysis of ceilings and walls.

Conclusion

The aim of this study is to propose an approach for heritage documentation and deformation assessment based on the combination of Terrestrial LiDAR and Building Information Modeling. Terrestrial LiDAR is a very interesting method to provide digital archiving of cultural heritage as a 3D point cloud. This point cloud is characterized by its very high density. In addition, each point of the cloud has very accurate 3D coordinates according to a specific reference system. Also, the point cloud capture information about the texture of the building thanks to realistic photos that are taken during the scanning process.

One of the most important use of Terrestrial LiDAR point cloud is the creation of 3D Building Information models. These 3D models can be used in different applications such as archaeology, history and culture. In this paper, we have highlighted the importance of BIM in the field of deformation assessment. The case study of 'Robat Banajah' shows how a Point-to-Surface analysis can be performed to assess the current condition of an historical building. Results show that some parts of the building have a distortion of about 22 cm, which means that further investigation must be carried out by specialists to decide if some restoration and maintenance interventions are required.

References

1. Roca, P., Cervera, M., &Gariup, G. (2010). Structural analysis of masonry historical constructions. Classical and advanced approaches. *Archives of Computational Methods in Engineering*, *17*(3), 299-325.
2. Callegari, F. (2003). Sustainable development prospects for Italian coastal cultural heritage: a Ligurian case study. *Journal of Cultural Heritage*, *4*(1), 49-56.
3. Zulkepli Majid, Halim Setan, and Albert K. Chong. (2009). *Accuracy assessments of point cloud 3D registration method for high accuracy craniofacial mapping*. Geoinformation Science Journal, 9 (2). pp. 36-44.
4. Grussenmeyer, P., Hanke, K., Streilein, A. Architectural photogrammetry. *in Digital Photogrammetry*, edited by M. KASSER and Y. EGELS, Taylor & Francis (2002), pp. 300-339.
5. Remondino, F., & El□Hakim, S. (2006). Image□based 3D modelling: a review. *The Photogrammetric Record*, *21*(115), 269-291.
6. Wehr, A. (2008). LIDAR: Airborne and terrestrial sensors. In: Z. Li, J. Chen and E. Baltsavias (red.), *Advances in Photogrammetry, Remote Sensing and Spatial Information Sciences: 2008 ISPRS Congress book.*
7. Armesto, J., Arias, P., Roca, J., & Lorenzo, H. (2008). Monitoring and assessing structural damage in historic buildings. *The Photogrammetric Record*, *23*(121), 36-50.
8. Douillard, B., Underwood, J., Kuntz, N., Vlaskine, V., Quadros, A., Morton, P., &Frenkel, A. (2011, May). On the segmentation of 3D LIDAR point clouds. In *Robotics and Automation (ICRA), 2011 IEEE International Conference on* (pp. 2798-2805). IEEE.
9. Barnhart, T. B., & Crosby, B. T. (2013). Comparing two methods of surface change detection on an evolving thermokarst using high-temporal-frequency terrestrial laser scanning, Selawik River, Alaska. *Remote Sensing*, *5*(6), 2813-2837.
10. Lindenbergh, R., & Pfeifer, N. (2005, October). A statistical deformation analysis of two epochs of terrestrial laser data of a lock. In *Proceedings of the 7th Conference on Optical* (pp. 61-70).
11. Mukupa, W., Roberts, G. W., Hancock, C. M., & Al-Manasir, K. (2017). A review of the use of terrestrial laser scanning application for change detection and deformation monitoring of structures. *Survey Review*, *49*(353), 99-116.
12. Soni, A., Robson, S., &Gleeson, B. (2015). Structural monitoring for the rail industry using conventional survey, laser scanning and photogrammetry. *Applied Geomatics*, *7*(2), 123-138.
13. Lee, G., Sacks, R., & Eastman, C. M. (2006). Specifying parametric building object behavior (BOB) for a building information modeling system. *Automation in construction*, *15*(6), 758-776.
14. Murphy, M., McGovern, E., & Pavia, S. (2013). Historic Building Information Modelling–Adding intelligence to laser and image based surveys of European classical architecture. *ISPRS journal of photogrammetry and remote sensing*,*76*, 89-102.
15. Li, Y., Dong, K., & Li, G. F. (2014, November). The Application of BIM in the Restoration of Historical Buildings. In *Applied Mechanics and Materials* (Vol. 638, pp. 1627-1635).
16. Baik, A., Alitany, A., Boehm, J., & Robson, S. (2014). Jeddah Historical Building Information Modelling. *ISPRS Annals of Photogrammetry, Remote Sensing and Spatial Information Sciences*, *1*, 41-47.
17. Di Mascio, D., Pauwels, P., & De Meyer, R. (2013). Improving the knowledge and management of the historical built environment with BIM and ontologies: the case study of the Book Tower. In *13th International Conference on Construction Applications of Virtual Reality* (pp. 427-436).
18. Bogdanova, G. T., &Atanasova, V. (2013). Digitisation and Presentation of Historical Materials in a Virtual Exhibition 'The Image of India in Bulgaria: from the late 19th to the late 20th Cen. *Digital Presentation and Preservation of Cultural and Scientific Heritage*, (III), 212-218.
19. Dore, C., & Murphy, M. (2012, September). Integration of Historic Building Information Modeling (HBIM) and 3D GIS for recording and managing cultural heritage sites. In *Virtual Systems and Multimedia (VSMM), 2012 18th International Conference on* (pp. 369-376). IEEE.
20. Brilakis, I., Lourakis, M., Sacks, R., Savarese, S., Christodoulou, S., Teizer, J., &Makhmalbaf, A. (2010). Toward automated generation of parametric BIMs based on hybrid video and laser scanning data. *Advanced Engineering Informatics*, *24*(4), 456-465.

Development and study of mechanical behaviour reinforcing composites by waste BTP

Elaboration et étude du comportement mécanique de nouveaux composites renforcés par des déchets BTP

M.El kanzaoui[1], A.Hajjaji[2], A.Guenbour[1], R.Boussen[1]
[1]Université Mohamed V Rabat, Faculté des Sciences, Rabat, Maroc
[2]Université Chouaib Doukkali , École Nationale des Sciences Appliquées, El Jadida, Maroc

Abstract. Composite materials are used in many industrial applications for their excellent mechanical and electric properties and their low density compared to metal structures. Most countries are extremely rich waste materials such as white ceramic breakages which represents a potential to be developed. Ceramic breakages have exceptional properties and could be effectively exploited in the manufacture of composite materials for a wide variety of applications. The composite materials reinforced by construction waste materials, such as ceramic breaks which offer significant benefits and gains in strength and stiffness properties (Young's modulus E : a material whose modulus Young is very high is said rigid).This article covers the benefits of breakages as ceramic filler used for reinforcement in composites, as well as improve the mechanical response of these structural elements (test compression).
Key words: composite materials, ceramics, construction materials, mechanical properties, compression test, Young's modulus

Résumé. Les matériaux composites sont utilisés dans de nombreuses applications industrielles pour leurs excellentes propriétés mécaniques et pour leur masse volumique faible par rapport à celles des structures métalliques. La majorité des pays sont extrêmement riches en déchets issue des différents procédés de fabrication industrielle telles que les casses de céramique qui représentent un potentiel à valoriser. Les casses de céramique ont des propriétés exceptionnelles et pourraient être exploitées efficacement dans la fabrication des matériaux composites pour une grande diversité d'applications. Les matériaux composites à renfort par des déchets matériaux BTP, comme les casses céramique qui offrent des avantages notables et des gains en termes de propriétés de résistance mécanique et de rigidité (Module d'Young E : un matériau dont le module d'Young est très élevé est dit rigide).Le présent article couvre les avantages des casses céramique comme charge utilisée pour le renforcement dans les composites, ainsi que d'améliorer la réponse mécanique de ces éléments structurels (Essai de compression).
Mots clés : matériaux composites, céramique, matériaux BTP, propriété mécanique, essai de compression, module d'Young

1. Introduction

Au cours de ces dernières années, les industries des matériaux composites ont bénéficié d'une croissance rapide et régulière, soutenue en particulier par la diversité des applications.

Les matériaux composites sont des matériaux solides hétérogènes et anisotropes, constitués de l'assemblage d'au moins deux matériaux non miscibles et de natures différentes, se complétant et permettant d'aboutir à un matériau dont l'association confère à l'ensemble des propriétés qu'aucun des composants pris séparément ne possède [1]. Ils sont constitués d'une ou plusieurs phases discontinues reparties dans une phase continue. La phase continue est appelée matrice et la phase discontinue, qui présente usuellement des propriétés mécaniques (rigidités et résistances) supérieures à celle de la matrice, est nommée renfort.

Les matériaux composites offrent d'excellents rapports masse/rigidité/résistance en comparaison avec les matériaux métalliques. Ils répondent parfaitement aux exigences mécaniques, thermiques, acoustiques, transparence, résistance aux feux et fatigue [2-3].

Les matériaux composites sont développés et choisis pour leurs caractéristiques mécaniques, thermiques, ou chimiques. Les propriétés obtenues sont généralement meilleures que celles de la matrice seule. Bien évidemment, il est impossible de réunir l'ensemble de ces propriétés dans un même composite mais, par rapport aux caractéristiques recherchées, on peut trouver des solutions optimales.

Les composites ne cessent d'évoluer vers des produits qui sont, soit les moins coûteux possibles, soit les plus performants, ou bien les deux à la fois. De plus, dans un souci de protection de l'environnement et de santé publique, les composites tendent à intégrer un caractère écologique.

De nos jours, l'utilisation des déchets d'industrie matériaux dans les matériaux composites devient de plus en plus fréquente, et ce, du fait que ces ressources sont moins coûteuses et présentent une valeur ajoutée. De plus, les tendances industrielles se penchent vers des produits propres « inertes » et écologiques. D'où l'intérêt d'intégrer des produits à potentiel polluant pratiquement nul, non réactif, non évolutif, non ou peu soluble.

La valorisation des résidus de l'industrie minérale sont devenus extrêmement importants dans l'élaboration des matériaux innovants, écologiques, plus performants, et moins coûteux. L'essentiel de ces déchets est produit sur les chantiers du BTP et bien précisément les résidus de l'industrie céramique « la casse Rouge ».

La casse de céramique rouge représente un déchet solide minéral, non susceptible d'évolution physico-chimique ou biologique. Il ne se décompose pas, ne brûle pas, n'est pas biodégradable.

Les céramiques qui sont en général durs et fragiles à la rupture tout en se caractérisant parfois par des résistances mécaniques très élevée, c'est pour cela qu'ils ont été utilisés en tant que matériaux de constructions depuis l'antiquité.

L'objectif principal de notre travail est la mise en œuvre des composites à matrice renforcée par des casses céramiques « résidu minéral », la caractérisation physico-chimique, la caractérisation morphologique et en fin le comportement mécanique des matériaux ainsi obtenus.

2. Matériels et méthodes

2.1. Matière première

Résine polyester insaturé (UP) thermodurcissable.
Déchet industriel traité :
La casse céramique rouge (région de Casablanca)

2 .1.1 Préparation de la casse rouge

L'échantillon de la casse rouge a subit les opérations unitaires suivantes
-concassage des fragments
-broyage des morceaux dans un broyeur
-tamisage de la poudre

2.2. Elaboration du nouveau matériau composite CR

Le matériau composite a été élaboré avec différentes teneurs en casse rouge brute 15 % ,30% et 50%.

Finement broyée, la casse était ensuite mélangée avec une matrice thermodurcissable (résine).

Le mélange obtenu est versé dans un moule cylindrique puis mis à durcir dans une étuve à 60°C.

Après extraction du matériau on procède à la découpe de plusieurs échantillons avec des épaisseurs variables.

2. 3. Caractérisation des matériaux

Une définition très générale d'une céramique pourrait être la suivante: une céramique est un matériau inorganique polycristallin, présentant une structure complexe de grains et de joints de grains [4].

Le terme générique de céramique recouvre des domaines aussi divers et variés que celui des céramiques traditionnelles (tuiles, briques, carreaux, ...) qui nous intéresse particulièrement, ou celui des céramiques dites techniques à applications thermomécaniques ou électroniques [5].

Les matériaux céramiques présentent une caractéristique essentielle par rapport à d'autres matériaux : les atomes ou constituants de leur réseau cristallin sont en général très solidement liés entre eux par des liaisons fortes (ioniques et covalentes) [6].

Les céramiques jouent actuellement un rôle de plus en plus important mettant à profit leurs propriétés électriques, électromécaniques, isolantes et optiques [7,8].

Notre présent travail s'intéresse particulièrement à des rejets céramiques ou bien les casses céramiques précisément les casses rouges CR.

 https://doi.org/10.1051/matecconf/201814902012

Les céramiques sont souvent des assemblages complexes de cristaux et de phase vitreuse. Seules certaines céramiques techniques, préparées à partir de produits de haute pureté, comportent une phase unique cristallisée. La nature et les proportions des constituants dépendent en premier lieu de la composition chimique du matériau de départ, mais sont aussi influencées par les traitements de cuisson et de refroidissement appliqués. Une céramique se caractérise à la fois par ses constituants et par sa microstructure, c'est-à-dire l'ensemble des paramètres décrivant l'organisation des différentes phases (taille et morphologie des cristaux et des phases inter granulaires, inclusions, porosités ouvertes ou fermées, etc.). [9]

La caractérisation des casses céramiques a été réalisée par spectrométrie IR, Microscopie et par diffraction des rayons X.

2.4. Microscopie optique à haute résolution

L'observation au Microscope nous a permis de visualiser les surfaces, et de mettre en évidence les détails d'une grande finesse. L'observation a été réalisée avec microscope: digital microscope Leica DVM5000 HD.

2.5. Spectroscopie Infrarouge (ATR)

Les spectres infrarouge à transformée de Fourier (IRTF) ont été effectués à l'aide d'un appareil de type VERTEX 70 dans la gamme des longueurs d'onde de 400 cm-1 à 4000 cm-1; la résolution spectrale est de 4,0 cm-1.

2.6. Diffractométrie des rayons X (DRX)

Les phases cristallines ont été identifiées par diffraction des rayons X sur des poudres avec un diffractomètre Siemens d5000 X-Ray à monochromateur arrière en graphite, fonctionnant sous une tension de 40 kV et une intensité de 50 mA avec la radiation CuKα comme source de rayonnement et piloté par un ordinateur muni du logiciel Diffracplus D version 2.2.

2.7. Module d'Young Essai de compression

L'essai mécanique est utilisé pour étudier l'effet positif ou négatif de la charge « casse céramique » sur les propriétés du matériau composite CR.

Les matériaux composites élaborés, avec différentes teneurs en casse rouge 15%,30% et 50%, ont été découpés en des échantillons d'épaisseurs 6, 10 et 15 mm comme décrit dans le tableau 1.

Tableau 1. Gamme d'échantillons matériau composite CR pour essai mécanique compression.

Echantillon / Dimensions	CR 15%	CR30%	CR 50%
Diamètre (D) Epaisseur (E)	D :20mm E :6mm	D:20mm E :6mm	D:20mm E :6mm
	D :20mm E:10mm	D :20mm E :10mm	D :20mm E :10mm
	D :20mm E:15mm	D:20mm E:15mm	D:20mm E:15mm

Pour évaluer le module d'Young, trois échantillons étaient testés et la valeur moyenne déclarée. L'expérience a été réalisée sous charge quasi-statique de compression pour tous les échantillons en utilisant une machine d'essai mécanique normalisée, pilotée par un ordinateur muni du logiciel LM Acq2V version 1.05.

Les propriétés de compression, tel que, le module d' Young obtenu à partir des courbes de compression avec régression linéaire en exploitant le logiciel Statgraphics.

Ces essais nous ont permis de déterminer le module d'Young E donnée par la formule suivante :

$$E = Fe / So \times Lo / \Delta L \quad (1)$$

E=la pente de la droite*facteur de forme
Fe : force appliquée
So : section initiale
Lo : épaisseur initial
ΔL : allongement

3. Résultats et Discussions

3.1. Analyse de la casse rouge

3.1.1. Microscopie optique à haute résolution

La microscopie permet d'observer la texture de la casse rouge et de caractériser les assemblages minéralogiques. L'image obtenue de l'échantillon avec l'agrandissement 200μ, est représentée sur la figure 1.

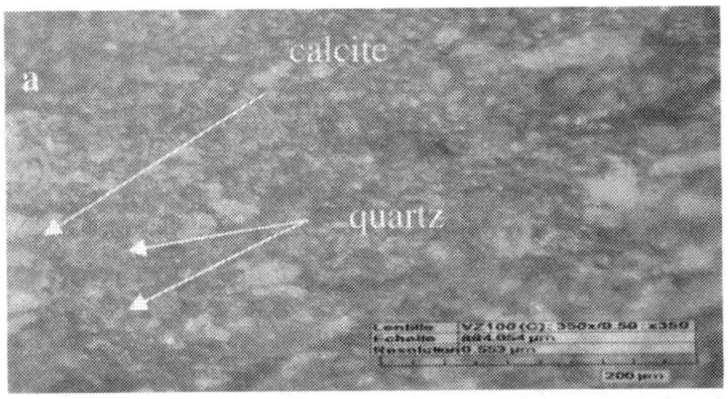

Fig. 1 Observations au microscope optique à haute résolution de la casse rouge.

Les particules argileuses se présentent sous forme d'amas d'agrégats fins et de plaquettes en forme des bâtonnets aux contours irréguliers (fig.1).

Il s'agit d'une morphologie rencontrée aussi bien pour des Kaolinites mal cristallisées que pour des Illites comme l'a observé Konan [10]. L'image de la figure 1 est en accord avec ce que nous avons obtenu en DRX, confirmant ainsi la présence de carbonates et de Quartz dans l'échantillon. Les carbonates (Calcite) se présentent sous forme des agrégats bien visibles et le Quartz se présente sous forme des grains de petite taille [11,12].

3.1.2. Spectroscopie Infrarouge (ATR)

Cette étude a permis de caractériser la casse rouge.

Tableau 2. Principales bandes IR caractéristiques de la casse rouge.

Les bandes (cm-1)	**Interprétations**
1480	(1400 et 1500) cm^{-1} CH3.
1000	(900-1100)cm^{-1} vibrations d'allongement de Si-O (Kaolinite)
795	Liaisons Si-O-Al
400	vibrations de déformation des OH

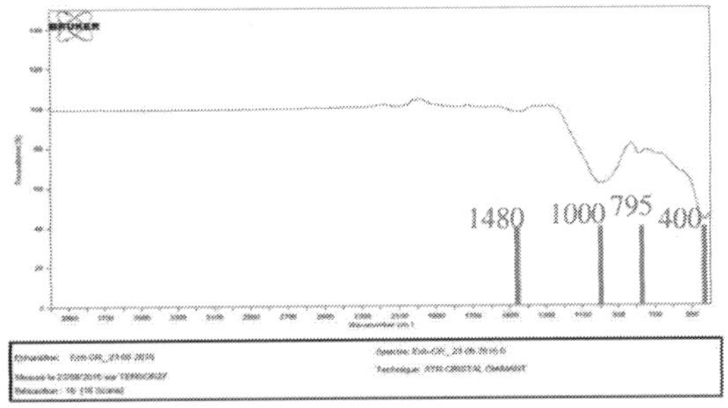

Fig. 2. Spectre infrarouge casse rouge CR.

Les spectres obtenus sont illustrés par la figure 2. On constate: La bande située entre 1400 et 1500 cm-1 est attribuée aux vibrations de déformation des groupes CH3 situés à 1480 cm-1. La liaison Si-O est caractérisée par : - La bande intense située entre 900-1100 cm-1 et centrée vers 1000 cm-1 correspond aux vibrations de valence de la liaison Si-O[13] .

La bande entre 750 et 800 cm-1 , provenant de la liaison Si-O-Al, fait également place à une bande autour de 795 cm-1. La bande d'absorption située à 1000 cm-1 est en accord avec la DRX indiquant la présence de la Kaolinite dans notre matériau CR. La bande observée à 795 cm-1 est attribuable aux vibrations d'élongation des liaisons Si-O-Al et aux hydroxyles perpendiculairement à la surface (OH translationnel) [14]. La bande à 400 cm- 1 est caractéristique des vibrations de déformation des hydroxyles dans les minéraux argileux trioctaédriques en général. Néanmoins, la bande d'absorption à 795 cm-1 peuvent correspondre au Quartz [15].

. Ces résultats sont en accord avec ceux de DRX. Ils confirment la présence du Quartz, de la Kaolinite et de l'Illite dans la casse rouge étudié. Le tableau 1, regroupe les bandes de vibrations et de déformations issues de la figure1.

3.1.3. Diffractométrie des rayons X (DRX)

Les diffractogrammes ont été obtenus à partir des échantillons désorientés (placés directement sous forme de poudre dans un porte-échantillon classique).

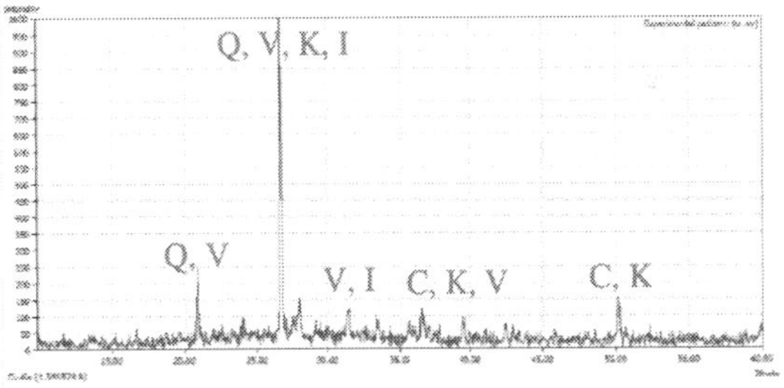

Fig. 3. Diffractogramme X casse rouge CR, avec K = Kaolinite, I = Illite, Q = Quartz, C= Calcite, V= Vermiculite.

Le diagramme des RX de la casse rouge est illustré par la figure 3. L'analyse spectrale indique qu'elle est composée de Quartz(SiO2), CalciteCa(CO3), Kaolinite (Al2Si2O5(OH)4), Illite [(K,H3O)Al2Si3AlO10(OH)2]et vermiculite [(Mg,Al)3(Si,Al)4O10(OH)2,4H2O].

Il révèle principalement la présence de trois pics intenses, le premier correspond à la Quartz et Vermiculite, le deuxième à un mélange de Quartz, Illite, Kaolinite et Vermiculite, le troisième correspond au Calcite et Kaolinite ce qui implique que la casse rouge est hétérogène. Ces résultats sont en parfait accord avec les observations au microscope. La fraction argileuse de notre matériau est constituée de Quartz et de Calcite comme impureté majeure dans notre échantillon.

https://doi.org/10.1051/matecconf/201814902012

3.2. Caractérisation physique du nouveau matériau composite CR

Pour tous nos échantillons, le module d' Young est obtenu à partir des courbes de compression avec régression linéaire en exploitant le logiciel Statgraphics. Nous citons ici l'exemple des échantillons CR à 15% et à différentes épaisseurs : 6mm ,10mm et15mm et les résultats du module d'Young sont regroupés dans le tableau 3.

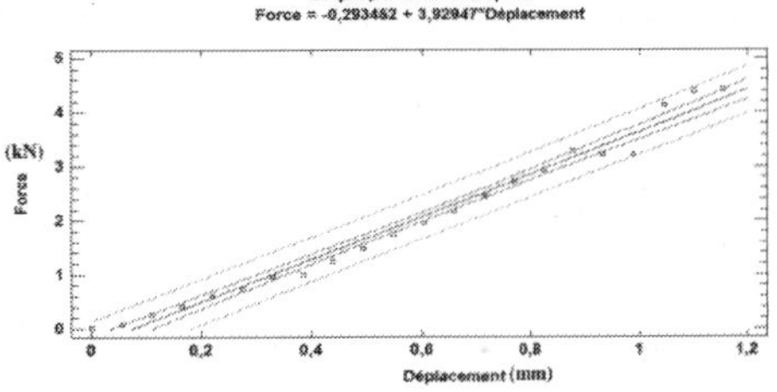

Fig. 4. Graphe du modèle ajusté. Matériau composite CR 15% (D: 20mm /E: 6mm)

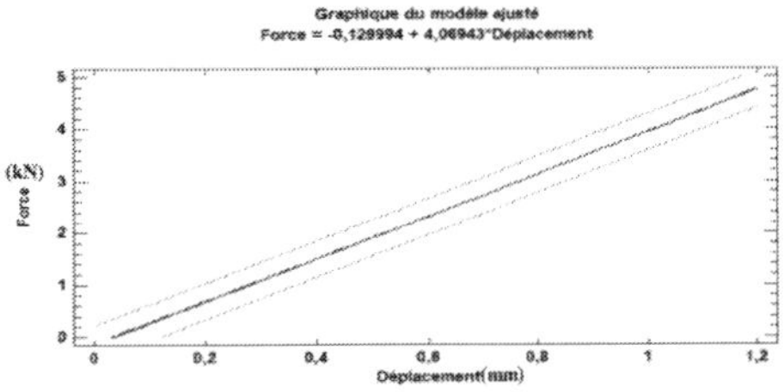

Fig.5. Graphe du modèle ajusté. Matériau composite CR 15% (D: 20mm /E: 6mm)

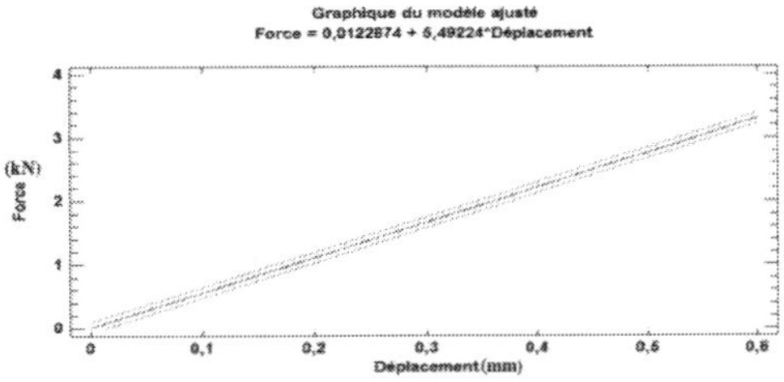

Fig.6. Graphe du modèle ajusté. Matériau composite CR 15% (D: 20mm /E: 6mm)

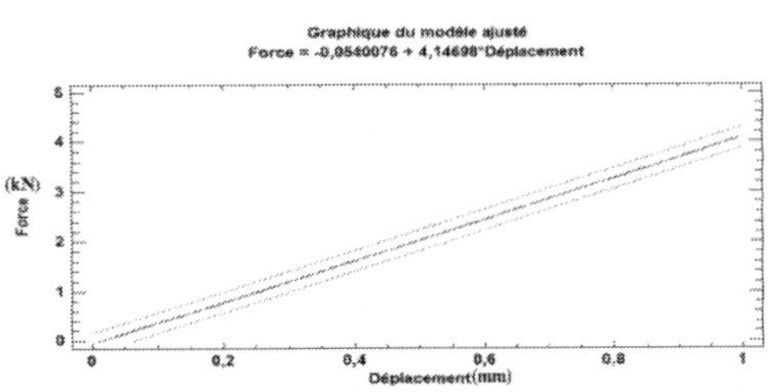

Fig.7. Graphe du modèle ajusté. Matériau composite CR 15% (D: 20mm /E: 10mm)

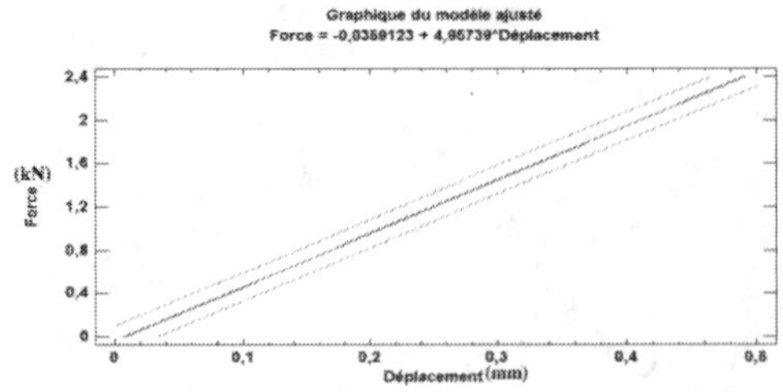

Fig.8. Graphe du modèle ajusté. Matériau composite CR 15% (D : 20mm /E : 10mm)

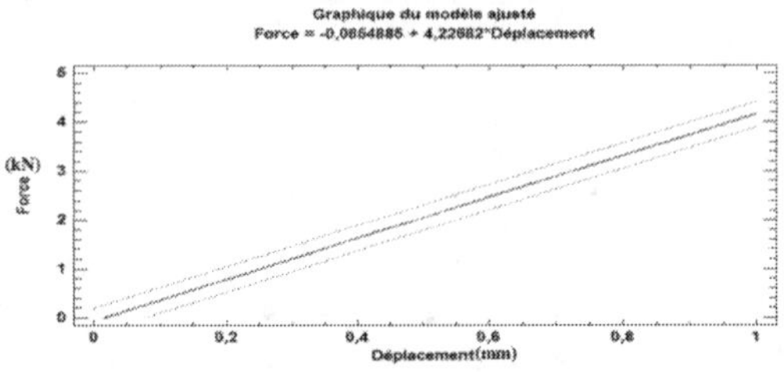

Fig.9. Graphe du modèle ajusté. Matériau composite CR 15% (D: 20mm /E: 15mm)

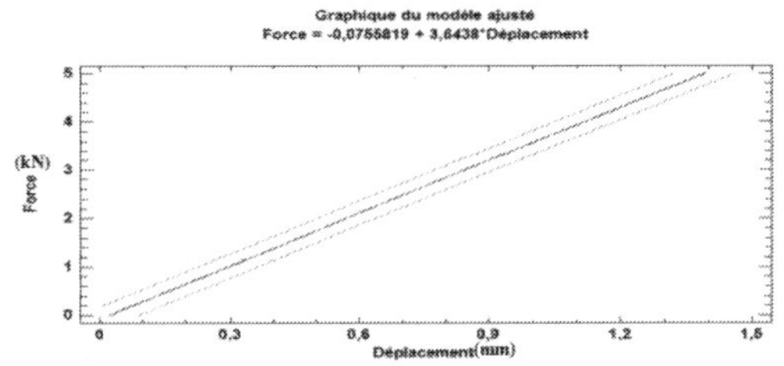

Fig.10. Graphe du modèle ajusté. Matériau composite CR 15% (D: 20mm /E: 10mm)

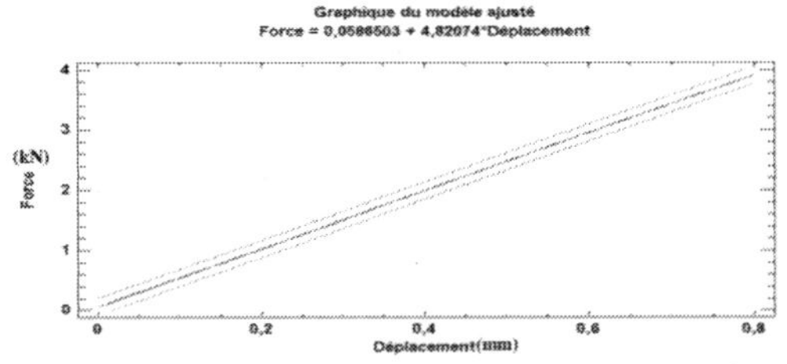

Fig.11. Graphe du modèle ajusté. Matériau composite

CR 15% (D : 20mm /E : 15mm).

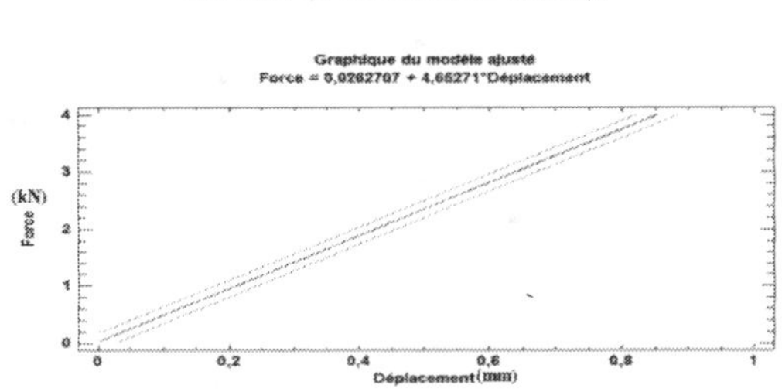

Fig.12. Graphe du modèle ajusté. Matériau composite CR 15% (D : 20mm /E : 15mm)

Tableau 3. Evaluation du module d'Young pour le matériau composite CR15%

Matériau composite / Module d'Young (E) en Mpa	CR 15% Epaisseur 6mm	CR 15% Epaisseur 10mm	CR 15% Epaisseur 15mm
1er Essai	74,90	131,84	173,88
2éme Essai	77,57	157,64	230,25
3éme Essai	104,90	134,39	222,13

Le même processus de caractérisation mécanique a été suivi pour les autres échantillons CR 30% et CR 50%

4 Synthèse

L'évolution du module d'Young a été évalué pour les différents composites CR 15%, CR30% et CR 50%.

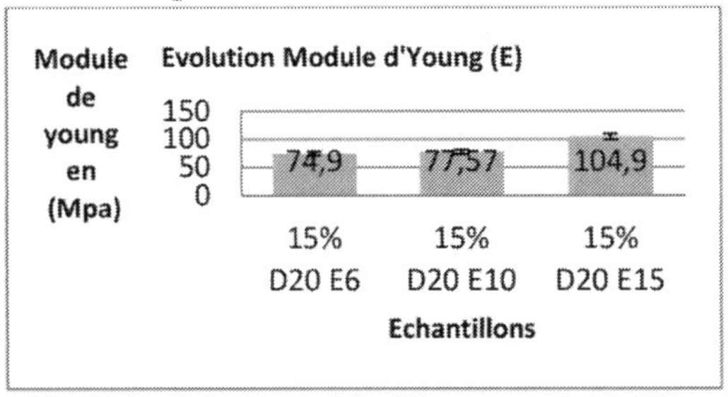

Fig.13. Evolution du module d'Young pour le CR 15%.

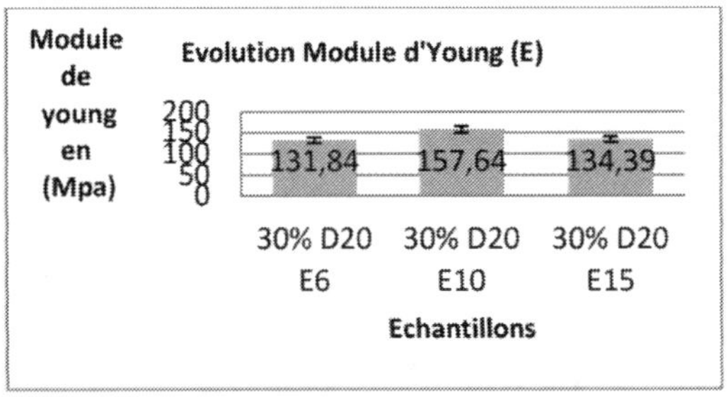

Fig.14. Evolution du module d'Young pour le CR 30%.

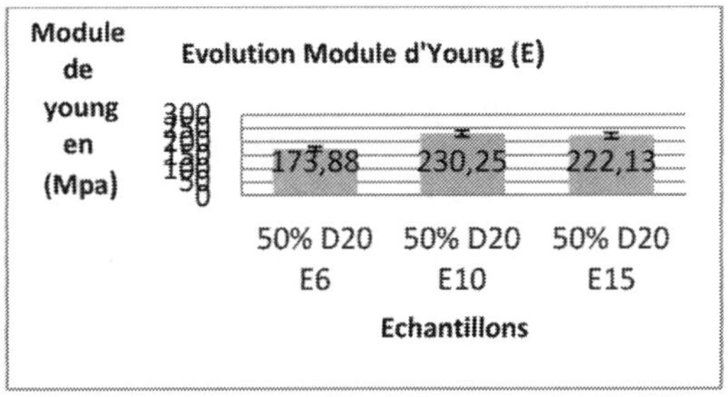

Fig.15. Evolution du module d'Young pour le CR 50%.

D'après ce graphe, on peut conclure que le renforcement par la casse céramique a un effet direct et positif sur les propriétés mécaniques du nouveau matériau composite. Ceci se traduit par un module d'Young élevé.

Conclusion

Les techniques expérimentales utilisées nous ont permis de mettre en évidence les propriétés de ce nouveau matériau composite CR.

Ce nouveau matériau composite CR représente une solution durable et environnementale par la valorisation des déchets BTP.

Les résultats de la présente étude ont mis en évidence la performance et la rigidité du nouveau matériau composite CR traduite par l'évolution du module d' Young. Ce dernier variait progressivement avec la teneur de la charge casse rouge et de l'épaisseur de l'échantillon CR.

References

1. [1] D.Berthelot, Masson 1996
2. [2] J.-J.Barrau, S.Laroze, Collection Résistance des matériaux et structures, Edition Masson, Paris, 1987.
3. [3] B.Harris, Edition B. Harris, University of Bath, UK, 1983.
4. [4] J. M. Haussounne, Technique de l'ingénieur, E1820 , 2 (1996).
5. [5] S. Liebus, Thèse doctorat, Université de Limoges (2003).
6. [6] E. K. Akdogan, A. Safari, J. Appl. Phys, Vol 101, No 6, p 4114 , 4121, (2007).
7. [7] N. M. Hagh, B. Jadidian and A. Safari, IEEE-UFFC Transaction, vol 55, No 1 , p 214 , 224, (2008).
8. [8] V. A. Isupov, Soviet. Phys. Solid. State, Vol 12, No 5, p 1084 , 1088, (1970).
9. [9] Daniel Michel, Bulletin de L'Union des Physiciens N°790 (1997) .
10. [10] K. L.Konan, J.Soro, J.Y.Y.Andji, S.Oyetola, G.Kra, *J. Soc. Ouest-Afr. Chim.* 30,29. (2010)
11. [11] O.Touret, C. H.Pons, D.Tessier, Y.Tardy, *Clay Minerals* 25, 223. (1990)
12. [12] H.Zangue Adija, *Univ. Lorraine. Eco. Doc. Scien. Et Ing des Ress, Proc, Prod et Env.* 132 (2012).
13. [13] A.Aarfane, A.Salhi, M.El Krati, S. Tahiri, M.Monkade, E.K.Lhadi, M.Bensitel, *J. Mater. Environ. Sci.* 5 (6) 1928(2014).
14. [14] A.Arib, A.Sarhiri, R.Moussa, T.Remmal, *C. R.* Gomina, *Chimie* 10,504 (2007).
15. [15] E.Kohler, *Univ. Evry Val d'Essonne. Eco. Doc. Phys. Chim,* (2005).

Analysis of buckling phenomenon of rectangular laminated plates with a hole under different boundary conditions

Ahmed El Bouhmidi[1], Mohamed Rougui[1]
[1]Mohammed V University in Rabat, Department GUE, LGCE High School of Technology- Salé CEDoc-EMI, Morocco

Abstract. In this research, buckling behavior of perforated rectangular plate of symmetric and anti-symmetric laminated composite is investigated based on Finite element analysis. The presence of hole may cause redistribution of stresses in plates with reduction of stability. The aim of the current paper is to find critical buckling load by using the (FSDT) first-order shear deformation theory in concomitance with the variational energy method. The load depends on many factors like as diameter of circular hole, different boundary conditions, lay-up sequences, length/thickness ratio and angle of ply orientation. The result is shown in graphical forms for various boundary conditions.

1 Introduction

Laminated composite plates are widely used in many engineering structures including automotive bridges wind-turbine blades and aviation, due to their lower weights compared to metal structures. Frequently, such plates contain holes of various shapes, for example for access or for weight savings and these perforation or openings often require some type of reinforcing structure to control local structural deformations and stresses near the cutout. In addition, these structures may experience compression loads during operation, and thus their buckling response characteristics must be understood and accurately predicted in order to determine effective designs and safe operating conditions for these structures. Thus in the design context buckling analysis plays a crucial role [4].
Hence, the analytical buckling solutions are inaccessible, and various numerical methods need to be developed for analyzing such plates. In the literature, few published studies investigated the buckling of laminated composite plates with a cutout [4, 7].

This paper deals with buckling of symmetrically and anti-symmetrically laminated composite plates with circular hole under uniaxial static loadings “figure 1”. The FSDT plate deformation theory and the variational energy method are employed in mathematical formulation, and the finite element technique is used for finding critical loads. The effects on critical load by hole size, plate thickness ratio, material modulus ratio, ply-lamination geometry, loading types, and boundary conditions are investigated.

2 Theoretical background

Consider a symmetrically or anti-symmetrically laminated rectangular plate of length *a*, width *b*, central circular hole of diameter *d* and thickness *t* which consists of six orthotropic laminas [12].

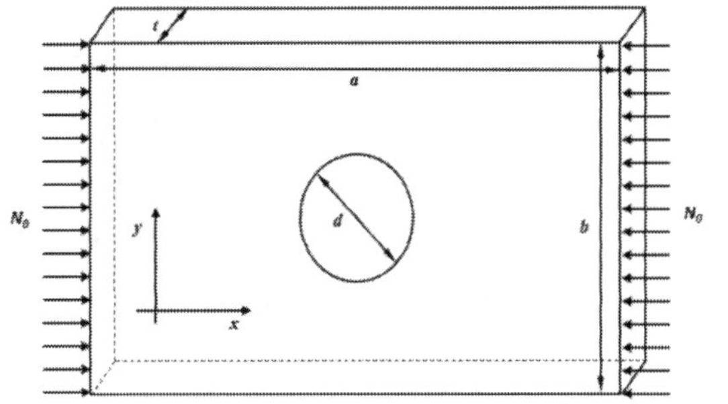

Fig. 1. Geometry of rectangular plate.

In this study, first-order shear deformation theory (FSDT) is employed to analyse the problem and the following displacement field is assumed [8, 10]

$$\begin{cases} u(x,y,z) = u_0(x,y) + z\phi_x(x,y) \\ v(x,y,z) = v_0(x,y) + z\phi_y(x,y) \\ w(x,y,z) = w_0(x,y) \end{cases} \quad (1)$$

where (u_0, v_0, w_0) denote the displacements of a point on the plane $z=0$, and

$$\frac{\partial u}{\partial z} = \phi_x, \frac{\partial v}{\partial z} = \phi_y \quad (2)$$

which indicate that ϕ_x and ϕ_y are the rotations of a transverse normal about the *y*- and *x*- axes, respectively. The strains associated with the displacement field (1) are given by:

$$\varepsilon_B = \begin{Bmatrix} \varepsilon_{xx} \\ \varepsilon_{yy} \\ \gamma_{xy} \end{Bmatrix} = \begin{Bmatrix} \frac{\partial u_0}{\partial x} \\ \frac{\partial v_0}{\partial y} \\ \frac{\partial u_0}{\partial y} + \frac{\partial v_0}{\partial x} \end{Bmatrix} + z \begin{Bmatrix} \frac{\partial \phi_x}{\partial x} \\ \frac{\partial \phi_y}{\partial y} \\ \frac{\partial \phi_x}{\partial y} + \frac{\partial \phi_y}{\partial x} \end{Bmatrix} \quad (3)$$

where ε_B is the bending part

$$\varepsilon_S = \begin{Bmatrix} \gamma_{yz} \\ \gamma_{xz} \end{Bmatrix} = \begin{Bmatrix} \phi_y + \frac{\partial w_0}{\partial y} \\ \phi_x + \frac{\partial w_0}{\partial x} \end{Bmatrix} \quad (4)$$

and ε_S is the transverse shear part.

The resultant force is obtained by integrating the stresses through the thickness of laminas [1, 9]:

$$\begin{Bmatrix} N_{xx} \\ N_{yy} \\ N_{xy} \end{Bmatrix} = \int_{-\frac{t}{2}}^{\frac{t}{2}} \begin{Bmatrix} \sigma_{xx} \\ \sigma_{yy} \\ \tau_{xy} \end{Bmatrix} dz = \sum_{i=1}^{n} \int_{t_{i-1}}^{t_i} \begin{Bmatrix} \sigma_{xx} \\ \sigma_{yy} \\ \tau_{xy} \end{Bmatrix}_i dz$$

$$= \begin{pmatrix} A_{11} & A_{12} & A_{16} \\ A_{12} & A_{22} & A_{26} \\ A_{16} & A_{26} & A_{66} \end{pmatrix} \begin{Bmatrix} \frac{\partial u_0}{\partial x} \\ \frac{\partial v_0}{\partial y} \\ \frac{\partial u_0}{\partial y} + \frac{\partial v_0}{\partial x} \end{Bmatrix} + \begin{pmatrix} B_{11} & B_{12} & B_{16} \\ B_{12} & B_{22} & B_{26} \\ B_{16} & B_{26} & B_{66} \end{pmatrix} \begin{Bmatrix} \frac{\partial \phi_x}{\partial x} \\ \frac{\partial \phi_y}{\partial y} \\ \frac{\partial \phi_x}{\partial y} + \frac{\partial \phi_y}{\partial x} \end{Bmatrix} \quad (5)$$

and we obtained the resultant moment with the same way:

$$\begin{Bmatrix} M_{xx} \\ M_{yy} \\ M_{xy} \end{Bmatrix} = \int_{-\frac{t}{2}}^{\frac{t}{2}} \begin{Bmatrix} \sigma_{xx} \\ \sigma_{yy} \\ \tau_{xy} \end{Bmatrix} z dz = \sum_{i=1}^{n} \int_{t_{i-1}}^{t_i} \begin{Bmatrix} \sigma_{xx} \\ \sigma_{yy} \\ \tau_{xy} \end{Bmatrix}_i z dz$$

$$= \begin{pmatrix} B_{11} & B_{12} & B_{16} \\ B_{12} & B_{22} & B_{26} \\ B_{16} & B_{26} & B_{66} \end{pmatrix} \begin{Bmatrix} \frac{\partial u_0}{\partial x} \\ \frac{\partial v_0}{\partial y} \\ \frac{\partial u_0}{\partial y} + \frac{\partial v_0}{\partial x} \end{Bmatrix} + \begin{pmatrix} D_{11} & D_{12} & D_{16} \\ D_{12} & D_{22} & D_{26} \\ D_{16} & D_{26} & D_{66} \end{pmatrix} \begin{Bmatrix} \frac{\partial \phi_x}{\partial x} \\ \frac{\partial \phi_y}{\partial y} \\ \frac{\partial \phi_x}{\partial y} + \frac{\partial \phi_y}{\partial x} \end{Bmatrix} \quad (6)$$

Equations relating the shear force to the shear strains can be computed as:

$$\begin{Bmatrix} R_{yz} \\ R_{xz} \end{Bmatrix} = k_S \int_{-\frac{t}{2}}^{\frac{t}{2}} \begin{Bmatrix} \tau_{yz} \\ \tau_{xz} \end{Bmatrix} = k_S \sum_{i=1}^{n} \int_{t_{i-1}}^{t_i} \begin{Bmatrix} \tau_{yz} \\ \tau_{xz} \end{Bmatrix}_i dz$$

$$= k_S \begin{pmatrix} A_{44} & A_{45} \\ A_{45} & A_{55} \end{pmatrix} \begin{Bmatrix} \phi_y + \frac{\partial w_0}{\partial y} \\ \phi_x + \frac{\partial w_0}{\partial x} \end{Bmatrix} \quad (7)$$

where $k_S (= 5/6)$ is the shear correction factor [5-6].

Equilibrium equations can be written in terms of the displacement field as follows [10]:

$$D_{11}\frac{\partial^2 \phi_x}{\partial x^2} + (D_{12}+D_{66})\frac{\partial^2 \phi_y}{\partial x \partial y} + D_{66}\frac{\partial^2 \phi_x}{\partial y^2} - A_{55}k_S(\phi_x + \frac{\partial w}{\partial x}) = 0$$

$$D_{22}\frac{\partial^2 \phi_y}{\partial y^2} + (D_{12}+D_{66})\frac{\partial^2 \phi_x}{\partial x \partial y} + D_{66}\frac{\partial^2 \phi_y}{\partial x^2} - A_{44}k_S(\phi_y + \frac{\partial w}{\partial y}) = 0$$

$$A_{55}k_S(\frac{\partial \phi_x}{\partial x} + \frac{\partial^2 w}{\partial x^2}) + A_{44}k_S(\frac{\partial \phi_y}{\partial y} + \frac{\partial^2 w}{\partial y^2}) - N_0\frac{\partial^2 w}{\partial x^2} = 0$$

(8)

Equations (8) are second-order differential equations with (w, ϕ_x, ϕ_y) as the three unknowns.

3 Finite element analysis

The total potential energy Π of a laminated plate under uniaxial loading is equal to strain energy U can be written as [11-12]:

$$\Pi = U_b + U_s + V \quad (9)$$

where U_b is bending term, U_s is the strain energy of shear and *V* denotes the potential energy of in-plane forces. The equilibrium requires that the potential energy Π must be stationary. The equilibrium equations of the cross-ply laminated plate can be derived from the variational principle by using [11]

$$\delta\Pi = 0 \quad (10)$$

The finite element method is applied in this study, by using equations (8), (9) and (10) we obtain the following linear algebraic equations of the form:

$$([K] + \lambda[K_G])\{U\} = 0 \quad (11)$$

where $[K]$ is the stiffness matrix, $[K_G]$ is the geometric stiffness matrix, λ is the eigenvalue and $\{U\}$ is the nodal displacements. The critical buckling load $N_{x,cr}$ is that which corresponds to the least eigenvalue λ determined by:

$$Det([K] - \lambda[K_G]) = 0 \quad (12)$$

The non-dimensional buckling coefficient is given by $n_{x.cr}$ is defined as [3, 7]:

$$n_{x,cr} = \frac{N_{x,cr} b^2}{\pi^2 \sqrt{D_{11} D_{22}}} \quad (13)$$

where $N_{x,cr}$ is critical buckling load and D_{11} and D_{22} are bending stiffness [1].

4 Materials and methods

The critical buckling load for different plates depends on many parameters like as geometry of plate orientation of fiber layers and boundary conditions, have been computed by using ANSYS (Mechanical APDL 17.1); a finite element analysis program [12-13]. It should be noted that the use of this approach will frequently give an imprudent result compared to the real behavior of the structure (experimental).

In this study, the effects of circular cut-out on the buckling load of laminated composite plats have been investigated numerically. The mechanical properties of the E-Glass/Epoxy composite material are shown in Table 1. Perforated laminated plates were analysed for different boundary conditions: simple- simple- simple-simple (SSSS), simple- simple- simple-free (SSSF), simple-clamped- simple-free (SCSF), two different thickness $t=0.23$in; 0.355in, and four width b was varied from 6in, 5in, 4in and 3in and we kept diameter of hole constant $d=0.312in$. Meanwhile, the laminated plates were also analyzed without a hole to compare the influences having a hole and without a hole conditions on buckling loads [2]. Furthermore, the laminated plates were assumed to stack six laminas onto together namely cross-ply $[(90°/+\theta/-\theta)]_{sym}$ for symmetric and $[(90°/+\theta/-\theta)]_{a\text{-}sym}$ for anti-symmetric ply with θ=15°,30°,45° and 60°.

Table 1. Mechanical properties of the E-Glass/Epoxy.

E_1 (Psi)	5.49E+006
$E_2 = E_3$ (Psi)	9.53E+005
$G_{12}=G_{13}$ (Psi)	3.47E+005
$\nu_{12}=\nu_{13}$	0.3001
$\nu_{23}=\nu_{32}$	0.3001

The SHELL 181 element "figure 2" is used for modeling and analyzing thin to moderately-thick perforated plates, it is a four-nodes element with six degrees of freedom at each node: translations in the x, y, and z directions, and rotations about the x, y, and z-axes [13].

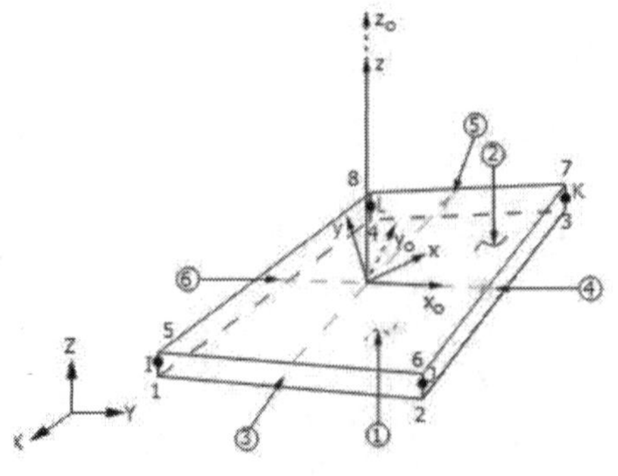

Fig. 2. Shell 181 element geometry [13].

5 Numerical results and discussion

Results obtained are shown in Tables 2, 3 and 4.

It is easily seemed from Tables that while thickness increase the critical buckling load $N_{x,cr}$ increase.

"Figure 3" and "figure4" also refers that critical buckling of the boundary condition (SSSS) plates are approximately 2 times greater than the buckling load of (SSSF) and (SCSF) plates. The effects of a/b ratio on buckling load are illustrated in "figure 5", according to it the buckling loads increase with increasing of a/b ratios. Furthermore, the highest values of buckling loads are calculated for $[(90°/60°/-60°)]_{sym}$ (SSSS) plates, while the lower values are computed angle-ply for $[(90°/+\theta/-\theta)]_{sym}$ (SSSF) plates. The effects of ply orientation symmetric and anti-symmetric on the buckling load can be seen in Tables 2, 3, 4. The results show that the change in buckling load are neglected. The effect of circular cut-out is taken into account. The analysis indicate that the variation of the buckling loads is very sensitive to the presence of hole. It can be seen that the buckling load generally decreases with presence of cut-out "figure 5", we can observe that the predicted buckling loads of plates without hole is 3.5% higher than with hole in (SSSS) plates and 22.5% in (SSSF) plates.

Table 2. Critical buckling load of perforated rectangular laminated plates under the boundary conditions (SSSS)

Plate dimensions (in)				$N_{x,cr}$ (kpsi)							
				Symmetric $[(90°/+\theta/-\theta)]_{sym}$				Anti-Symmetric $[(90°/+\theta/-\theta)]_{a\text{-}sym}$			
a	b	t	d	15°	30°	45°	60°	15°	30°	45°	60°
6	6	0.23	0.312	2.2556	2.4052	2.518	2.5361	2.2531	2.3886	2.528	2.5156
	5			3.4902	3.663	3.8808	3.9489	3.4445	3.6124	3.8311	3.8747
	4			5.5458	5.5539	5.671	5.7	5.5818	5.6488	5.7685	5.7159
	3			9.4055	9.1028	9.4231	9.576	8.9998	9.1516	9.476	9.5089
	6	0.355		7.9823	8.4431	8.6962	8.8701	7.9824	8.4431	8.9297	8.8701
	5			12.307	12.895	13.625	13.764	12.099	12.655	13.418	13.524
	4			18.647	18.675	19.078	19.172	17.295	18.782	19.267	19.182
	3			29.665	29.943	30.971	31.344	29.256	29.599	30.811	31.03

Table 3. Critical buckling load of perforated rectangular laminated plates under the boundary conditions (SSSF)

Plate dimensions (in)				$N_{x,cr}$ (kpsi)							
				Symmetric $[(90°/+\theta/-\theta)]_{sym}$				Anti-Symmetric $[(90°/+\theta/-\theta)]_{a\text{-}sym}$			
a	b	t	d	15°	30°	45°	60°	15°	30°	45°	60°
6	6	0.23	0.312	1.0011	1.0114	1.0087	0.99357	1.0011	1.0184	1.0182	0.99804
	5			1.0744	1.0977	1.0991	1.0741	1.0749	1.1103	1.1164	1.0822
	4			1.1766	1.2454	1.2782	1.2435	1.1851	1.2695	1.3003	1.2466
	3			1.4236	1.5172	1.6469	1.5661	1.4435	1.6114	1.6811	1.5776
	6	0.355		3.5699	3.61	3.5946	3.5269	3.5656	3.6307	3.6288	3.5483
	5			3.8231	3.9132	3.9136	3.8078	3.8211	3.9525	3.9721	3.8411
	4			4.1833	4.432	4.5424	4.4041	4.2098	4.5121	4.6173	4.4183
	3			5.0496	5.5714	5.8299	5.5265	5.1169	5.71	5.9475	5.5706

Table 4. Critical buckling load of perforated rectangular laminated plates under the boundary conditions (SCSF)

Plate dimensions (in)				$N_{x,cr}$ (kpsi)							
				Symmetric $[(90°/+\theta/-\theta)]_{sym}$				Anti-Symmetric $[(90°/+\theta/-\theta)]_{a\text{-}sym}$			
a	b	t	d	15°	30°	45°	60°	15°	30°	45°	60°
6	6	0.23	0.312	1.2129	1.2205	1.1927	1.1515	1.2104	1.2237	1.2001	1.1566
	5			1.4313	1.4363	1.3817	1.3037	1.4291	1.4463	1.3975	1.3109
	4			1.8848	1.9178	1.8306	1.6693	1.8836	1.9302	1.8416	1.6664
	3			3.2651	3.3557	3.1656	2.7395	3.2662	3.3677	3.1688	2.7342
	6	0.355		4.3048	4.3374	4.2351	4.0777	4.2899	4.3429	4.2612	4.102
	5			5.0578	5.0874	4.8951	4.6072	5.0428	5.1132	4.9466	4.6382
	4			6.635	6.7591	6.4548	5.8859	6.6209	6.7905	6.4898	5.881
	3			11.351	11.653	11.02	9.5718	11.344	11.689	11.033	9.56

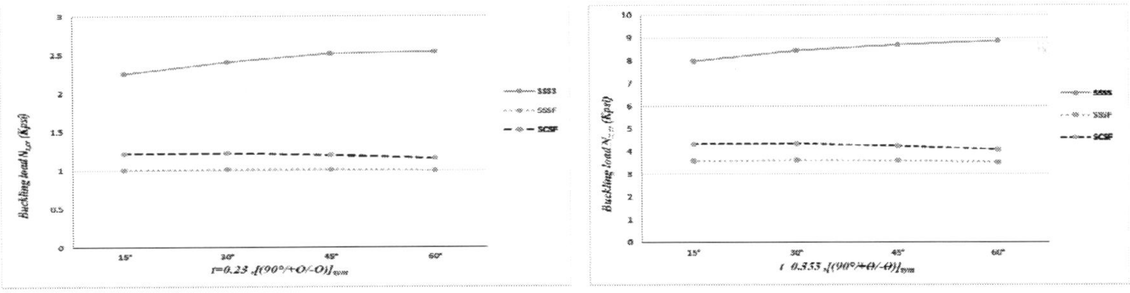

Fig. 3. Variation of the critical buckling load $N_{x,cr}$ compared to symmetric orientation of fibers of a plate 6 ×6 in^2 for t=0.23 in and t=0.355 in

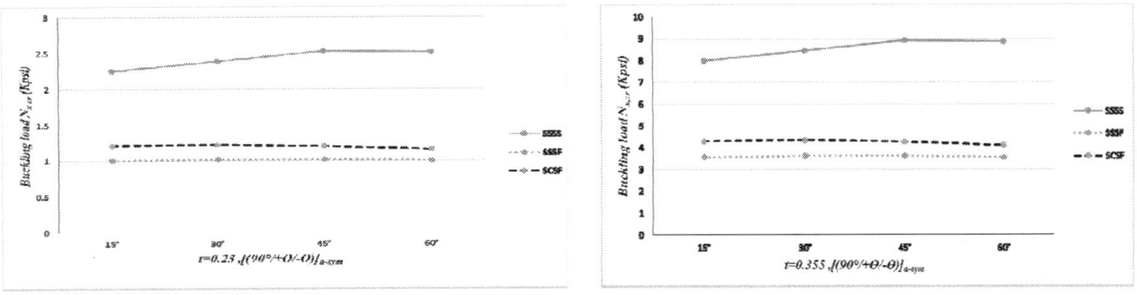

Fig. 4. Variation of the critical buckling load $N_{x,cr}$ compared to anti-symmetric orientation of fibers of a plate 6 ×6 in^2 for t=0.23 in and t=0.355 in

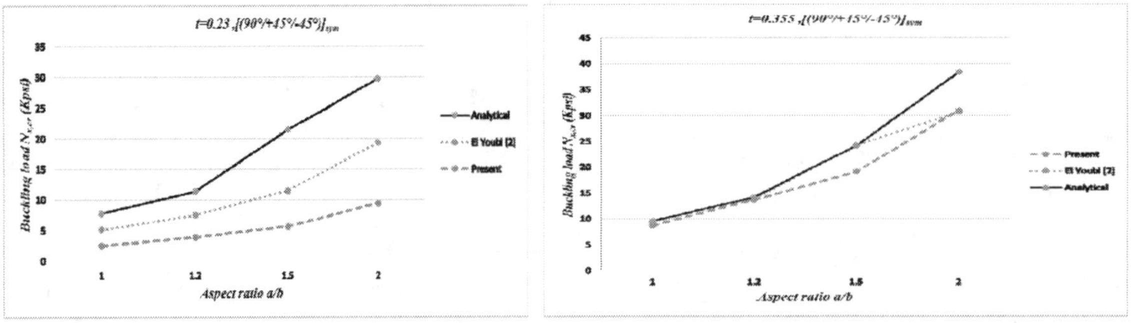

Fig. 5. Variation of the critical buckling load $N_{x,cr}$ compared to aspect ratio $\frac{a}{b}$ for $[(90°/45°/-45°)]_{sym}$ [2]

6 Conclusion

In this study, the buckling load of a E-Glass/Epoxy composite laminated rectangular plate with central circular cut-out is investigated. Additionally, the effect of a/b ratio on buckling loads are calculated. Solutions were done with FEM using ANSYS finite element program. From the present study, the following conclusions can be made:
1. The reduction of the buckling load due to hole is found to be significant, this means big holes cause the weakest plates under the pressure.
2. The increasing of buckling loads by increasing *a/b* ratios.
3. Boundary conditions of composite have the strongest effect on the buckling load, among the other parameters.
4. Buckling is loss of stability due to geometric effects and boundary conditions than the material failure.
Comparison of results of the finite element with test have to be done.

References

1. J.M. Berthelot, *Composite Materials: Mechanical Behavior and Structural Analysis* (Springer-Verlag, New York, 1999)
2. M. El Youbi, M. Rougui, Int. Rev. Mech. Eng. **9**, 3 p.307-313 (2015)
3. M.P. Nemeth, J. Ther. Str. **23**, 9 p.873-916 (2000)
4. B. O. Baba, J. of Reinf. Plast. and Comp. **26**,16 p. 1637-1655 (2007)
5. S. Vlachoutsis, Int. J. for Num. Meth. in Eng. **33**, p.1537-1552 (1992)
6. S.N. Chatterjee, S.V. Kulkarrni, A. I. A. A. J. **17**, 5, p.498-499 (1997)
7. M.F. Altan, M.E. Kartal **30**, p.2243-2249 (2009)
8. W. Kim, J.N. Reddy, L. Amr. J. Sol. Str. **7**, p.201-226 (2010)
9. B.D. Agarwal, L.J. Broutman, K. Chandrashekhara *Analysis and Performance of Fiber Composites* (Wiley, USA, 2006)
10. J.N. Reddy, *Mechanics of Laminated Composite Plates and Shells : Theory and Analysis* (CRC Press, Boca Raton, 2004)
11. J.N. Reddy, *An Introduction to Nonlinear Finite Element Analysis* (Oxford University Press, New York, 2004)
12. E. Madenci, I. Guven, *The finite element method and applications in Engineering using ANSYS* (Springer, New York, 2015)
13. ANSYS Mechanical APDL Theory reference, Release 17.1

Structural evaluation updating based on quality control and proof loads

Mohamed Saad Abbadi, Nouzha Lamdouar
Laboratoire Génie Civil du CEDOC « Sciences et Techniques pour l'Ingénieur », Ecole Mohammedia d'Ingénieurs, Université Mohammed V, Rabat, Maroc

Abstract. In a context of limited available resources, the structural evaluation of bridges is important to avoid the high costs of replacement and repair. As a rational evaluation criterion, the reliability of structures based on probabilistic theories provides an accurate estimate of the current strength and the remaining life time. It also represents an effective approach for analysing structural reliability under uncertainties in parameter estimates during the service life of designed and existing structures, especially during the construction.Where new information concerning the actual bridge conditions is available, which is unknown in the design phase. The certificates of material properties, measurements of actual bridge geometry, collection load data, and results of proof load testing are the major resources of this information. The extra information unknown in the design phase could be used to reduce some uncertainties related to the bridge member resistance and load parameters entering the evaluation process.The aim of this study is twofold: (a) taking into account the uncertainties ofstrength parameters, by presenting some approaches for the efficient inclusion of quality control and proof loads in the structural reliability assessment process.The results of these approaches reveals the actual safety margins in the structure and permits to avoid either unnecessary reinforcement operations or unjustified reduced load limits.

1 INTRODUCTION

The social and economic losses because of the high costs of replacement and repair operations raises the awareness to improve the resilience and sustainability of infrastructures, especially in a context of limited available resources.

Design codes presents safety margins that can be excessive regarding existing bridge assessment, which some characteristics can be defined with more accuracy. The assessment of the structures is relevant as it allow to avoid unnecessarily conservative measures adopted in the design, and reduce some uncertainties related to the bridge member resistance and load parameters.

In particular, some uncertainties can be reduced in the phase post-construction and it's convenient to update the real state of the structure.

Special attention is given to the reliability updating using the two major extra information available after achievement, which are the results of the certificates of materials properties and proof loads.

Proof loading may provide additional knowledge about resistance. For example, a structure which survives a certain load may be considered as more reliable than one which has never been exposed to such a load.

Quality control of material properties during the construction process can be used as a first update of the partial factors used in the design. This should be carried out within the framework of a probabilistic reliability theory, in accordance with the margin of safety intended in the Eurocode.

As a rational evaluation criterion, the reliability of structures based on probabilistic theories provide an accurate estimate of the current strength and the remaining life time. It also represent an effective approach for analysing structural reliability under uncertainties in parameter estimates during the service life of both designed and existing structures.

The present paper intends to provide a semi-probabilistic format in verifications of existing reinforced concrete Bridges, in accordance with the target reliability intended in the Eurocode. Thus, this approach is in order to achieve either of the following goals:

- For a given reliability level, design new partial factors.

- For a given partial factors, estimate the real reliability level.

And permits to efficiently and correctly manipulate information obtained from measurements and inspections.

2 METHODOLOGY:

2.1 Overview on the Assessment methods of reliability:

Depending on the approach to reliability, there are three levels of performance evaluation:

Level 1 consider deterministic approach, the safety margin is introduced through central safety factor (ratio of design and resistance factors). It's simple of use but Inadequate for considering variability of input parameters and generally leads to excessive design. The general formula is expressed in terms of permissible stress as:

$$\sigma \leq \sigma_{adm} = \frac{\sigma_{rupt}}{K} \qquad (1)$$

Where σ is the actual stress on the structure, σ_{adm} is the admissible limit of stress, depending on the properties of materials and the level of safety preconized, $K$is the central safety factor and σ_{rupt} is the ultimate tensile strength.

Level 2, defined as semi-probabilistic approach, consist on the verification of a criterion involving design values of strength R and Load S, noted $R_d$andS_d, Applied in conjunction with the concept of limit states (ultimate, serviceability, fatigue).

$$R_d = \frac{1}{\gamma_{Rd}} R\left(\dots,\frac{f_i}{\gamma_{fi}},\dots\right) \geq S_d = \gamma_{S_d} S(\dots,\gamma_{F_j}F_j,\dots) (2)$$

f_i, F_jCorrespond to characteristic values associated to strength and load variables. γ_{fi} , γ_{F_j} are the related partial safety factors for these variables while $\gamma_{S_d}, \gamma_{Rd}$represent the partial factors covering the modeling uncertainties.

Level 3 require a full reliability analysis to quantify the probability of failure of the structure under various loading scenarios. The acceptance criterion is defined in terms of the closeness of the actual reliability index to the optimum reliability level. Then the basic equation is expressed in terms of probabilities:

$$P_f = P(R < S) \leq P_{f0} \qquad (3)$$

Where P_f is the probability of failure, R is resistance, S is action, P_{f0} is the target probability.

The use of the probabilistic techniques of level 3 for structural assessment is basically not common in civil engineering. But it can be very valuable in the assessment of existing structures, because such approaches can handle uncertainties in a more efficient manner than a semi-probabilistic format as introduced in the Eurocode.

2.2 Partial factor calibration:

The partial factor γ_R of material resistance R is defined, in accordance to Eurocode, as the fraction of its characteristic value R_K and the design valueR_d, as bellow:

$$\gamma_R = \frac{R_K}{R_d} \qquad (4)$$

The characteristic value R_K of a resistance variable R is defined as its 5% fractile. Thus, in the case of a resistance variable R having a normal distribution, then the characteristic value is given as:

$$R_K = \mu_R - 1{,}645 * \sigma_R = \mu_R(1 - 1{,}645 * V_R) \qquad (5)$$

Where V_R is the coefficient of variation of the resistance
$\mu_R$is the mean value of the performance function
$\sigma_R$is itsthe standard deviation

The design value R_d of R can be estimated using the expression:

$$R_d = \mu_R - \alpha_R * \beta * \sigma_R = \mu_R(1 - \alpha_R * \beta * V_R) \qquad (6)$$

Whereα_R is the sensitivity factor of R, taken in the Eurocode equal to 0,8.

The partial factor γ_R for a normal distribution of R can be assessed as:

$$\gamma_R = \frac{(1-1{,}645 * V_R)}{(1-\alpha_R * \beta * V_R)} \qquad (7)$$

Assuming a lognormal distribution of R its characteristic value R_k can be determined using approximate equation:

$$R_K = \mu_R * \exp\,(-1{,}645 * V_R) \qquad (8)$$

Similarly the design value R_d is approximated as:

$$R_d = \mu_R * exp(-\alpha_R * \beta * V_R) \qquad (9)$$

Taking into account equations (14) and (15) it follows that the partial factor γ_R for lognormal distribution can be assessed as:

$$\gamma_R = \exp(-1{,}645 * V_R)\,/\,exp(-\alpha_R * \beta * V_R) \qquad (10)$$

2.3 Truncated method for Updating by proof load:

The truncated method consist on updating the probability density function (PDF) of resistance, after experiencing a load proof with a convincing results in term of deflection and safety. Then, it shows that structural resistance R $>R_{test}$, where R_{test} is the resistance corresponding to the test load applied, then with the truncated method, the $f_R(x)$ is updated and transferred into $f_{R,up}(x)$.

The PDF is described before experiencing the proof load as follow:

$$\int_{-\infty}^{+\infty} f_R(x)dx = F_R(R_{test}) + \int_{R_{test}}^{+\infty} f_R(x)dx \qquad (11)$$

After testing successfully the structure under a proof load, the truncated distribution function is expressed as follow:

$$\int_{R_{test}}^{+\infty} f_{R,up}(x)dx = 1 \qquad (12)$$

From the two equations above, the updated PDF is:

$$f_{R,up}(x) = \begin{cases} \frac{f_R(x)}{1-F_R(R_{test})}, & x \geq R_{test} \\ 0, & x < R_{test} \end{cases} \qquad (13)$$

The updated mean value and variance are respectively:

$$\mu_{R,up} = \int_{-\infty}^{+\infty} x f_{R,up}(x)dx = \frac{\mu_R - \int_{-\infty}^{R_{test}} x f_R(x)dx}{1-F_R(R_{test})} \qquad (14)$$

$$V_{R,up} = \frac{V_R - (\mu_{R,up}-\mu_R)^2 - \int_{-\infty}^{R_{test}} (x-\mu_{R,up})^2 f_R(x)dx}{1-F_R(R_{test})} \qquad (15)$$

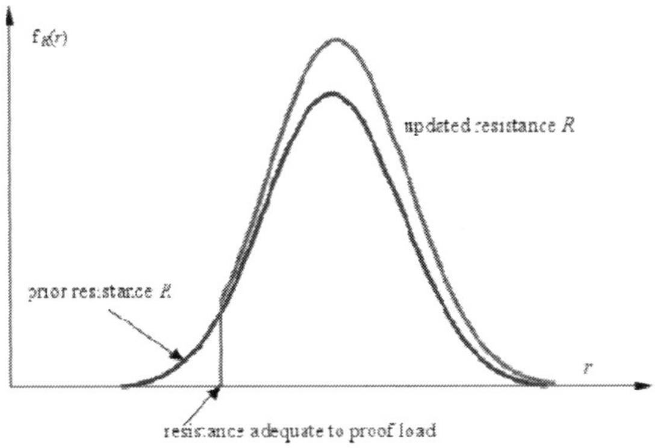

Fig 1.Probability density function of resistance prior to proof loading and after surviving the proof loading.

Loading the structure can be considered as a test, the load may be a service load or an exceptional load. In both cases the proof load level R_{test} can be considered as the lower bound on f_R, and the updating by truncated method can be applied as well as in a case of proof load.

2.4 Updating proof level

This section consider calibration of intensities of proof loads for updated resistance, based on a target reliability index beta

The reliability index of proof load is expressed as:

$$\beta = \frac{\mu_R - Q_{test}}{\sigma_R} \qquad (16)$$

Where:

As the proof load can be fixe with accuracy, its value can be taken as deterministic.

After updating resistance, and preserving the same reliability level, we can consider the expression below:

$$\beta = \frac{\mu_{R,up} - Q_{test,up}}{\sigma_{R,up}} \qquad (17)$$

From the eq and the updated proof level is calculated as follow:

$$\frac{Q_{test,up}}{Q_{test}} = \frac{\mu_{R,up} - \beta * \sigma_{R,up}}{\mu_R - \beta * \sigma_R} = \frac{\mu_{R,up}}{\mu_R}\left(\frac{1-\beta * V_{R,up}}{1-\beta * V_R}\right) \qquad (18)$$

3 RESULTS AND DISCUSSIONS:

3.1 Quality control on material properties:

A sample of n=120 concrete compressive strength measurement, obtained during the construction, is to be used to assess the characteristic value of the concrete strength f_{ck}.

Tab 1. Results of compressive strength measurements

n	f_c	n	f_c	n	f_c	n	f_c	n	f_c
1	35.1	25	27.6	49	38.6	73	35	97	32.6
2	35.6	26	30.1	50	39	74	34.2	98	32.7
3	36.2	27	30.4	51	40.4	75	37.1	99	32.9
4	36.6	28	30.8	52	40.7	76	38.2	100	33.1
5	37.2	29	31.3	53	50.6	77	38.8	101	33.6
6	35.4	30	32.5	54	29.9	78	42.7	102	34.1
7	34.1	31	31.8	55	30.6	79	46	103	33.9
8	37	32	34.4	56	30.3	80	28.6	104	34.7
9	36.8	33	33.3	57	31	81	30.4	105	34.8
10	35.3	34	37.8	58	30.8	82	30.2	106	35.6

11	33.2	35	37.1	59	32	83	30.8	107	34.9
12	33.4	36	40.7	60	32.8	84	30.8	108	36.4
13	32.8	37	38.4	61	32.4	85	30.9	109	37.4
14	32.6	38	26.9	62	33.9	86	31.2	110	37.5
15	33.8	39	30	63	33.2	87	31.3	111	37.4
16	33.7	40	30.5	64	34.6	88	31.3	112	38.3
17	34.1	41	31.5	65	34.9	89	31.7	113	37.9
18	34.3	42	31.9	66	38.3	90	31.5	114	38.6
19	35.4	43	32.3	67	36.4	91	31.9	115	39.8
20	34.9	44	33.2	68	28.2	92	31.8	116	40.3
21	36.1	45	34.4	69	31.9	93	31.9	117	42.4
22	36.3	46	36.3	70	31.1	94	32.2	118	43.1
23	35.8	47	37.4	71	32.6	95	32.1	119	42.6
24	25.9	48	38	72	33.7	96	32.4	120	49.4

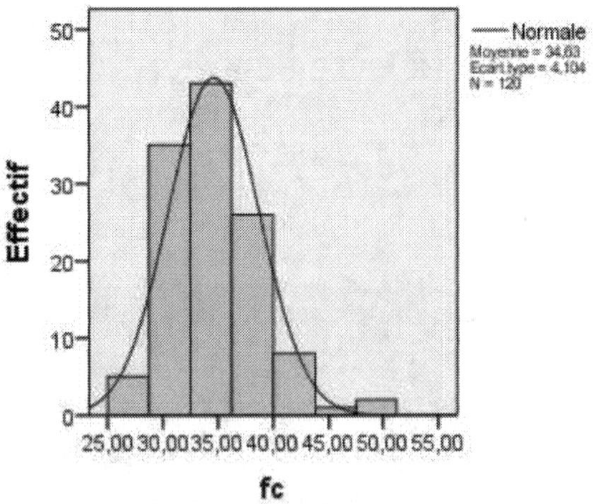

Fig 2. Histogram of compressive strength of concrete showing normal distribution fit

A sample of n=40 yield tensile measurement, obtained during the construction, is to be used to assess the characteristic value of elasticity limit

Tab 2. Results of tensile yield measurements

n	f_y	n	f_y	n	f_y	n	f_y	n	f_y
1	502.1	9	506.2	17	507.3	25	509.4	33	509.5
2	511.6	10	511.7	18	506.8	26	502.9	34	508.1
3	513.2	11	525.3	19	527.4	27	524.5	35	521.6
4	518.7	12	519.8	20	517.9	28	519.1	36	522.2
5	523.3	13	524.4	21	527.5	29	534.6	37	533.7
6	531.8	14	530.9	22	529.1	30	528.2	38	529.3
7	529.4	15	527.5	23	534.6	31	520.7	39	511.8
8	504.9	16	521.1	24	534.2	32	537.3	40	542.8

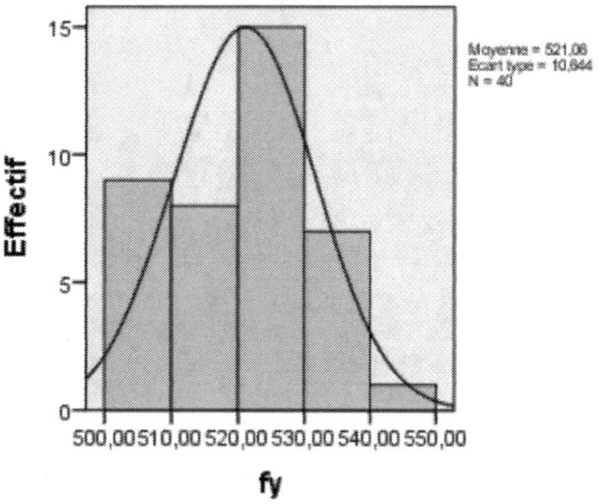

Fig 3. Histogram of yield tensile showing normal distribution fit

The following table resume the statistical results of concrete and yield characteristics:

Tab 3. Statistical results of concrete and yield measurements

 https://doi.org/10.1051/matecconf/201814902014

	Concrete	yield
Mean (MPa)	34.63	521.06
Standard deviation (MPa)	4.10	10.64
Coefficient of variation	0.12	0.02
Characteristic value (MPa)	27.88	503.92
Design value (MPa)	22.15	489.38
Updated Partial factor	1.26	1.03
Initial partial factor	1.50	1.15

3.2 Updating reliabilityof a slab:

Considering a mid-cross section of simply supported slab having the span of 12m, exposed to a bending Moment M_{ELU}, a simple drawing of the cross-section is indicated below, including stress distribution diagrams in the compressive concrete zone (rectangular and triangular) Further, the concrete C25/30 having the characteristic strength $f_{ck} = 25MPa$ and reinforcement bars S 500 having the characteristic strength $f_{yk} = 500\ MPa$

The following table resume the main characteristics of the slab:

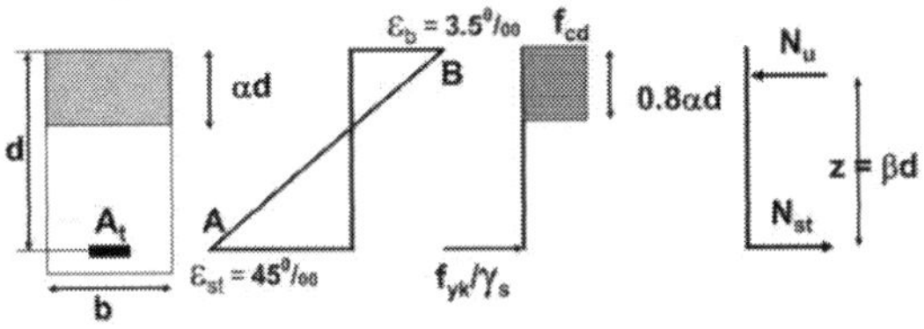

Fig 4. Stress distribution in the reinforced concrete slab

When the rectangular stress diagram is assumed the following equilibrium conditions can be written:

$$N_u = 0,8 f_{cd} \alpha d b \qquad (19)$$

$$M_u = N_u \beta d = N_u (1 - 0,4\alpha) d \qquad (20)$$

$$M_u = 0,8 b d^2 f_{cd} (1 - 0,4\alpha) \qquad (21)$$

Where d denotes the effective depth, αd the depth of the neutral axis, β the distance between $N_u$etN_{st}, b the width of the slab, A_s the area of the reinforcement, f_{cd} the concrete strength and f_{yd} the reinforcement strength (yield point).

$$\mu = \frac{M_{ELU}}{bd^2 f_{cd}} = 0,8\alpha(1 - 0,4\alpha) \qquad (22)$$

$$\alpha = 1,25(1 - (1 - 2\mu)^{\frac{1}{2}}) \qquad (23)$$

$$A_s = 0,8 \alpha b d \frac{f_{cd}}{f_{yd}} \qquad (24)$$

$$M_R = (1 - 0,4\alpha) d f_{yd} A_s \qquad (25)$$

$$M_R = \left(1 - 0,5\left(1 - \left(1 - 2 * \frac{M_{ELU}}{bd^2}\frac{\gamma_c}{f_{ck}}\right)^{0.5}\right)\right) * d * \frac{f_{yk}}{\gamma_s} * A_s \qquad (26)$$

The criterion $\sigma_c < 0,6 f_{ck}$ is not determinant, as well as the crack opening and the fatigue.

After linearizing $M_R$about the design point values, we obtain:

$$M_R \approx -1,736 + 7,13 * 10^{-3} * f_c + 3,12 * 10^{-3} * f_y + 249 * 10^{-4} * A_s$$

(27)

Tab 4. Characteristics of the slab

	Design	Post construction
f_{ck}[MPa]	25	27,88
f_{yk}[MPa]	500	503,92
$A_s[cm^2]$	62,63	64,80
μ_{M_R}[MN.m]	1,562	1,649
σ_{M_R}[MN.m]	$144,02 * 10^{-3}$	$39,47 * 10^{-3}$
β	1,99	9,475

Tab 5. Reliability index before and after updating

Variable	Symbol	Mean	Standard deviation	Distribution function
Span	L[m]	12	0	Deterministic
Load	M_{ELU} [MN.m]	1,275	0	Deterministic
width	b[m]	1	0	Deterministic
depth	d[m]	0,55	0	Deterministic
Concrete strenght	f_c [MPa]	34,63	4,10	Normal
Yield strenght	f_y [MPa]	521,06	10,64	Normal
Area of reinforcement	A_s $[cm^2]$	64,80	0	Deterministic

3.3 Truncated method for Updating by proof load:

A simulation, based on the equations (14) and (15) show the updated mean, by proof load, of the resistance. A comparison of two cases is considered: update based on actualized mean and standard deviation after quality control consideration, and update based on the design characteristics.

 https://doi.org/10.1051/matecconf/201814902014

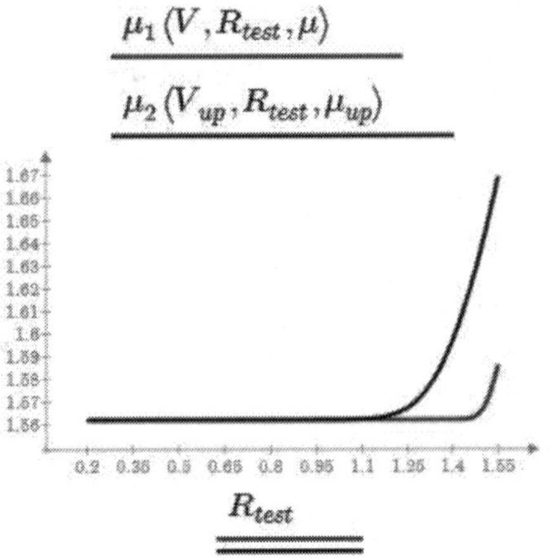

Fig 5. Updated mean v. proof load level in both cases with and without quality control consideration

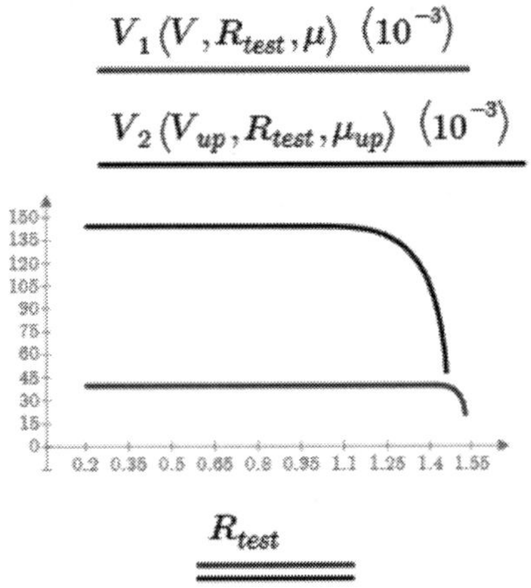

Fig 6. Updated standard deviation v. proof load level in both cases with and without quality control consideration

Structural reliability can be significantly improved by proof loading especially in the cases when V_{up} are smaller than V.

3.4 Updating proof level

Maintaining the same reliability level, the proof load can be updating, in accordance with the equation (18) we obtain:

$$\frac{Q_{test,up}}{Q_{test}} = 1{,}23 \qquad (28)$$

It means that the proof load level can be raised about 23%, and staying with the same safety margin.

Conclusions and perspectives:

Structural evaluation updating have been investigated, taking into account the two extra-information acquired in the construction phase: results of quality control of material properties and proof loads.

The update with the results of quality control gives a better description of the distribution of the variables related to the material properties, permitting a calibration of the partial factors and an assessment of the reliability index.

The update with proof load, consisting on the fact that a successful proof load test demonstrates that the resistance of a bridge is greater than the proof load, reduces uncertainty in the bridge resistance and so increases the bridge reliability, the truncated method constitute a great approach to embody this update, and combined with the update by control quality, gives better results.

As a main results, partial factors of concrete and yield where optimized about 16% and 10%. The reliability index was greatly improved after update by both quality control and proof load.

Another main outcome is the adjustment of proof load level, in accordance with the first update by quality control, as the reliability level remain the same.

In the light of these results,many promising areas of improvement can be conducted, it starts with updating partial factors of actions, based on more accurate analysis of traffic and dead load of the structure, the relationship between the components of a structural system and how a reliability approaches can be established in a context of correlated variables. Also, the reliability of exceptional load passing can be improved, revealing the actual safety margin and permits to avoid either unnecessary reinforcement operations or unjustified reduced load limits.

And finally, prior exceptional loads may be treated as an uncertain proof load, since the fact that a bridge had survived the exceptional trafficking means that the bridge resistance is greater than the load effect.

References

1. T. Shan Lin, A. S.Nowak. Proof Loading and Structural Reliability (1984).
2. W.B Hall, M.Tsai. Load Testing, Structural Reliability and Test Evaluation (1989)
3. M.H.Faber, D.V.Val, M.G.Stewart. Proof load testing for bridge assessment and upgrading (2000).
4. C.Cremona, B.Poulin, A.Colas. Calibration of partial safety factors for the assessment of existing bridges (2013).
5. R.Rackwitz, K.Schrupp. Quality Control, Proof Testing and Structural Reliability (1984).
6. M.Sykora, M.Holicky, J.Markova. Verification of existing reinforced concrete bridges using the semi-probabilistic approach (2013).
7. L.Yuefei, L.Dagang, F.Xueping. Reliability updating and prediction of bridge structures based on proof loads and monitored data (2014)

Comparative study of direct and inverse problems of cracked beams

CHETTAH Mahieddine [1] and LASSOUED Rachid [1]
[1] University of brothers Mentouri-Constantine 1, Department of Civil Engineering,
Laboratory of Materials and Durability of Constructions, Constantine, Algeria

Abstract. In recent decades, the analysis and evaluation of the cracked structures were hot spots in several engineering fields and has been the subject of great interest with important and comprehensive surveys covering various methodologies and applications, in order to obtain reliable and effective methods to maintain the safety and performance of structures on a proactive basis. The presence of a crack, not only causes a local variation in the structural parameters (e.g., the stiffness of a beam) at its location, but it also has a global effect which affects the overall dynamic behavior of the structure (such as the natural frequencies). For this reason, the dynamic characterization of the cracked structures can be used to detect damage from non-destructive testing. The objective of this paper is to compare the accuracy and ability of two methods to correctly predict the results for both direct problem to find natural frequencies and inverse problem to find crack's locations and depths of a cracked simply supported beam. Several cases of crack depths and crack locations are investigated. The crack is supposed to remain open. The Euler–Bernoulli beam theory is employed to model the cracked beam and the crack is represented as a rotational spring with a sectional flexibility. In the first method, the transfer matrix method is used; the cracked beam is modeled as two uniform sub-segments connected by a rotational spring located at the cracked section. In the second method which is based on the Rayleigh's method, the mode shape of the cracked beam is constructed by adding a cubic polynomial function to that of the undamaged beam. By applying the compatibility conditions at crack's location and the corresponding boundary conditions, the general forms of characteristic equations for this cracked system are obtained. The two methods are then utilized to determine the locations and depths by using any two natural frequencies of a cracked simply supported beam obtained from measurements as inputs. The two approaches are compared with a number of numerical examples for simply supported beams including one crack. The theoretical results show that the accuracy of the Rayleigh's method to predict natural frequencies decreases for higher modes when crack depth increases. It is also found that for the inverse problem, the transfer matrix method show a good agreement with those obtained from previous works done in this field.

1 Introduction

In the areas of civil, mechanical and aerospace engineering, the structure must fulfill some requirements such as durability, resistance, working safely and the capacity to satisfy the needs of users during its operational lifetime, which requires a continuous monitoring to detect possible damages (e.g., cracks) that allows an assessment of its state from the current structural conditions also provide a certain forecast of its future performance and information for maintenance, in order to verify its integrity and compliance with applicable regulations. The dynamic behavior of cracked beams has been studied by various analytical, numerical and experimental methods. Dimarogonas [1] and Gasch [2] presented comprehensive reviews of the dynamics of cracked structures. When a structural component is subjected to cracks, it lead to changes in structural parameters (e.g., the stiffness of a structural member such as beam elements), which, in turn, change dynamic behavior (such as natural frequencies and mode shapes).

Damage detection of beam elements has two different aspects: the first is to study the effects of cracks on the structure parameters (mass, damping and stiffness), from a detailed numerical model based on the geometric and material properties of the structure as a direct problem. Different approaches were used to model cracks in beams, such as equivalent reduced cross- section, local bending moment and local flexibility. Ostachowicz and Krawczuk [3] gave comprehensive surveys of crack modeling approaches. Friswell and Penny [4] compare the different approaches to crack modeling, and demonstrate that for structural health monitoring using low frequency vibration, simple models of crack flexibility based on beam elements are adequate. A common approach is representing the crack by a rotational spring, but with different formulas to modeling the effect of a crack on the local flexibility [5-7].

The second can be regarded as how to detect the location and depth of the cracks, from experimentally measured data, by analyzing the changes in the vibration characteristics of the structure (natural frequencies, mode shapes, etc.), as an inverse problem. In the field of inverse problems, both analytical and numerical classes of crack detection problem were usually based on changes of natural frequencies, measuring dynamic flexibility or comparison of mode shapes [8-10]. Among the aforementioned crack identification methods, greater attention has been devoted to solve the crack problem based on changes of fundamental frequencies. Liang et al [11] developed a frequency contour plot method based on measurements of the first three natural frequencies to detect a crack in a beam. Nandwana and Maiti [12] have been extended the method to stepped beams. Lee [13] used a combination of the finite element method and the Newton–Raphson procedure to identify multiple cracks. Attar [14] used a transfer matrix method to solve a similar problem with general form of boundary conditions.

The objective of this paper is to compare the accuracy and ability of two methods to correctly predict the results for both direct and inverse problem of a simply supported beam with one crack. Several cases of crack depths and crack locations are studied. The crack is supposed to remain open. The Euler–Bernoulli beam theory is employed to model the cracked beam and the crack is represented as a rotational spring. In the first method, the transfer matrix method (TMM) is used; the cracked beam is modeled as two uniform sub-segments connected by a rotational spring located at the cracked section. The second one, based on the Rayleigh's method (RM), the mode shapes of the cracked beam is constructed by adding a cubic polynomial function to that of the undamaged beam. By applying the compatibility conditions at crack's location and the corresponding boundary conditions, the general forms of characteristic equations for this cracked beam system are obtained. The two methods are then utilized to determine the locations and depths by using any two natural frequencies of a cracked simply supported beam obtained from measurements as input. The two approaches are compared with a number of numerical examples for simply supported beams including one crack. The results show a good agreement with those obtained from previous works done in this fields.

2 Theoretical model

A simply supported cracked beam as shown in figure 1 has length L and longitudinal uniform rectangular cross section with height of h, width of b and one open traversed crack of depth a located at position $X_1 = L_1$ from the left support of the beam. The beam is modeled as an Euler–Bernoulli in two segments joined by a rotational springs. The transverse deflection function of the beam at axial coordinate X and time T, is denoted by $W_i(X,T)$ for two intervals of $i = 1$ for $0 < X < X_1$ and $i = 2$ for $X_1 < X < L$.

The differential equation of motion for each segment is:

$$EI\,\frac{\partial^4 W_i(X,T)}{\partial X^4} + \rho A\,\frac{\partial^2 W_i(X,T)}{\partial T^2} = 0 \;\; ; \;\; i = 1,2 \qquad (1)$$

E; ρ; $A = bh$ and $I = bh^3/12$ are Young's modulus, density of the material, cross sectional area of the rectangular beam and moment of inertia of the uniform and homogeneous beam, respectively.

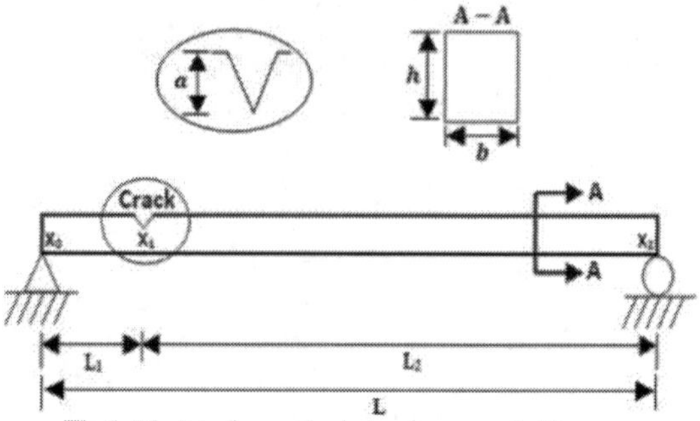

Fig.1. Model of a cracked simply supported beam.

From the above equations, the following quantities are introduced for non-dimensional analysis:

$$w_i = \frac{W_i}{L} \;\; ; \;\; t = T \;\; ; \;\; x_i = \frac{X_i}{L} \;\; ; \;\; l_i = \frac{L_i}{L} \qquad (2)$$

Eq. (1) can be changed to:

$$\frac{EI}{L^4}\,\frac{\partial^4 w_i(x,t)}{\partial x^4} + \rho A\,\frac{\partial^2 w_i(x,t)}{\partial t^2} = 0 \;\; ; \;\; i = 1,2 \qquad (3)$$

The boundary conditions of this simply supported beam as non-dimensional form are:

$$x = 0 \;\begin{cases}\varphi_1(0) = 0\\ M_1(0) = 0\end{cases} ; \quad x = 1 \;\begin{cases}\varphi_2(1) = 0\\ M_2(1) = 0\end{cases} \qquad (4)$$

The continuous conditions at the crack position as non-dimensional form are:

$$x = l_1 \;\begin{cases}\varphi_1(l_1) = \varphi_2(l_1)\\ \theta_1(l_1) = \theta_2(l_1) - \eta\, M_2(l_1)\\ M_1(l_1) = M_2(l_1)\\ T_1(l_1) = T_2(l_1)\end{cases} \qquad (5)$$

φ, θ, M and T are respectively the displacement, slope, bending moment and shear force expressions of the cracked beam.

Where $\eta = \frac{EIc}{L}$, is the non-dimensional cracked beam sectional flexibility, which is related to the ratio of crack depth to the beam height, $\xi = a/h$. It can be expressed as [7]:

$$\eta = \frac{h}{L}\, 6\pi\,(1 - \nu^2)\, f(\xi)$$

$$f(\xi) = 0.6272\,\xi^2 - 1.04533\,\xi^3 + 4.5948\,\xi^4 - 9.9736\,\xi^5 + 20.2948\,\xi^6 - 33.0351\,\xi^7 + 47.1063\,\xi^8 - 40.7556\,\xi^9 + 19.6\,\xi^{10} \qquad (6)$$

Where ν is the Poisson's ratio.

3 Direct problem

3.1 Transfer matrix method

In order to determine natural frequencies and mode shapes, the free vibration solution can be found using the method of separation of variables as:

$$w_i(x,t) = \varphi_i(x)\, e^{j\omega t} \qquad i = 1,2 \tag{7}$$

Using Eq. (7) and applied to Eq. (3) yields to:

$$\varphi_i^{(4)}(x) - \beta^4\, \varphi_i(x) = 0 \qquad i = 1,2 \tag{8}$$

Where:

$$\beta^4 = \frac{L^4 \rho A\, \omega^2}{EI} \tag{9}$$

ω is the natural frequency of the entire uniform beam. The general solution of the ordinary differential equation (9), for each segment, can be expressed as:

$$\varphi_i(x) = A_i \sin[\beta(x - x_{i-1})] + B_i \cos[\beta(x - x_{i-1})] + C_i \sinh[\beta(x - x_{i-1})] + D_i \cosh[\beta(x - x_{i-1})] \tag{10}$$

In which, A_i, B_i, C_i, and D_i are unknown constants associated with the ith segment ($i = 1,2$).
By substituting the boundary and continuous equations (4) and (5) into Eq. (10) for ($i = 1,2$) respectively and calculating the determinant of the matrix, the characteristic equation in this case assumes the form:

$$\eta\, \beta_n[\sinh(\beta_n l_1)\sinh(\beta_n l_2)\sin[\beta_n(l_1 + l_2)] - \sin(\beta_n l_1)\sin(\beta_n l_2)\sinh[\beta_n(l_1 + l_2)]] + 2\sinh[\beta_n(l_1 + l_2)]\, \sin[\beta_n(l_1 + l_2)] = 0 \tag{11}$$

The above characteristic equation is solved with the help of MATLAB software using vpasolve function. Ones the parameters β_n are obtained, the natural frequencies of the cracked beam can be obtained using the following equation:

$$\omega_{cn} = \left(\frac{\beta_n}{L}\right)^2 \sqrt{\frac{EI}{\rho A}} \tag{12}$$

Where, n is the mode range.
The constants A_i, B_i, C_i, and D_i can also be founded directly by the substitution method and normalizing it for the constant A_1.

3.2 The Rayleigh's method

In this section, by extending the Rayleigh's method as used by Fernandez-Saez et al. [15], the fundamental frequencies and mode shapes of a simply supported beam with one crack can be obtained. The transverse deflection of the cracked beam is constructed by adding a cubic polynomial function to that of the un-cracked beam, to represent crack effects on displacement of the section of the cracked beam located to the left and the right of the crack respectively.

The general expression for the mode shapes of cracked beam is:

$$\varphi_i(X) = \sin\left(\frac{n\pi X}{L}\right) + A_i X^3 + B_i X^2 + C_i + D_i \qquad 0 \le X \le L_i \tag{13a}$$

$$\varphi_{i+1}(X) = \sin\left(\frac{n\pi X}{L}\right) + A_{i+1} X^3 + B_{i+1} X^2 + C_{i+1} + D_{i+1} \qquad L_i \le X \le L \tag{13b}$$

By applying the boundary conditions of this simply supported beam and the continuous conditions at the point of crack, the general expression for the mode shapes of cracked beam which is presented in non-dimensional form, can be obtained as:

$$\varphi_i(x) = \sin(n\pi x) + \eta\, [(n\pi)^2 \sin(n\pi l_i)][1 - l_i]\, x \qquad 0 \le x \le l_i \tag{14a}$$

$$\varphi_{i+1}(x) = \sin(n\pi x) + \eta\, [(n\pi)^2 \sin(n\pi l_i)]\, l_i\, [1 - x] \qquad l_i \le x \le 1 \tag{14b}$$

The Rayleigh's method is used, so by equating the maximum values of potential energy U_{max} and kinetic energy T_{max} of the cracked simply supported beam, the natural frequencies can be obtained.
The maximum potential energy can be calculated as:

$$U_{max} = \frac{EI}{2}\left[\int_0^{L_i}\left(\frac{M_i}{EI}\right)^2 dX + \int_{L_i}^{L}\left(\frac{M_{i+1}}{EI}\right)^2 dX\right] + \left[\frac{1}{2}(\theta_{i+1} - \theta_i)M\right]_{X=L_i} = \frac{EI\,(n\pi)^4}{4\,L^3}[1 + \eta\, \gamma_n] \tag{15}$$

The maximum kinetic energy can be calculated as:

$$T_{max} = \frac{\rho A \omega_{cn}^2}{2}\left[\int_0^{L_i}(\varphi_i)^2 dX + \int_{L_i}^{L}(\varphi_{i+1})^2 dX\right] = \frac{\rho A \omega_{cn}^2}{2}\,\frac{L}{2}\left[1 + \eta\, \gamma_n\left[2 + \eta\left(\frac{(n\pi)^4}{3}\right) l_i^2(1 - l_i)^2\right]\right] \tag{16}$$

By equating equations (15) and (16), the nth-natural angular frequency of cracked simply supported beam is:

$$\omega_{cn} = \omega_n\left[\frac{1 + \eta\, \gamma_n}{1 + \eta\, \gamma_n\left[2 + \eta\left(\frac{(n\pi)^4}{3}\right) l_i^2(1 - l_i)^2\right]}\right]^{1/2} \tag{17}$$

Where:

$$\begin{cases} \omega_n = \left(\dfrac{n\pi}{L}\right)^2 \sqrt{\dfrac{EI}{\rho A}} \\ \gamma_n = 2\, sin^2(n\pi l_i) \end{cases} \tag{18}$$

4 Inverse problem

The inverse problem consists of determining the damage parameters (the location and the extent of a crack) when a certain number of measured natural frequencies are known.
The procedures for crack identification for the two methods are given below:

4.1 Transfer matrix method

To obtain the unknown parameters η and l_1, the following steps are used, which are similar to the method used in [16].

- The Eq. (11) is rewritten as:

$$\eta\, \beta_n[\sinh(\beta_n l_1)\sinh(\beta_n l_2)\sin[\beta_n(l_1+l_2)] - \sin(\beta_n l_1)\sin(\beta_n l_2)\sinh[\beta_n(l_1+l_2)]] = -2\,\sinh[\beta_n(l_1+l_2)]\,\sin[\beta_n(l_1+l_2)] \tag{19}$$

- Calculating β_1 and β_2 from Eq. (9) by utilizing the two measured frequencies ω_{c1} and ω_{c2} ;
- Substituting β_1 and β_2 separately into Eq. (19) and dividing these two equations, the non-dimensional crack section flexibility η can then be cancelled out and the following identification equation can be obtained:

$$\frac{\beta_1[\sinh(\beta_1 l_1)\sinh(\beta_1 l_2)\sin(\beta_1) - \sin(\beta_1 l_1)\sin(\beta_1 l_2)\sinh(\beta_1)]}{\beta_2[\sinh(\beta_2 l_1)\sinh(\beta_2 l_2)\sin(\beta_2) - \sin(\beta_2 l_1)\sin(\beta_2 l_2)\sinh(\beta_2)]} = \frac{\sinh(\beta_1)\,\sin(\beta_1)}{\sinh(\beta_2)\,\sin(\beta_2)} \tag{20}$$

- The position of the crack l_1, can be determinate with the help of MATLAB software using solve function;
- After obtaining l_1, the non-dimensional crack sectional flexibility η can be determined from Eq. (11), and the crack depth a can then be obtained from Eq. (6).

4.2 The Rayleigh's method

In this section, the inverse problem can be solved by using Eq. (14) in Ref. [17], as:

$$\frac{\Delta\omega_n}{\omega_n} = \frac{\eta\,\gamma_n}{2} \tag{21}$$

Where:

$$\Delta\omega_n = \omega_n - \omega_{cn} \tag{22}$$

- Calculating ω_1 and ω_2 from Eq. (18) and by utilizing the two measured frequencies ω_{c1} and ω_{c2} , $\Delta\omega_1$ and $\Delta\omega_2$ can be obtained;
- Substituting $\frac{\Delta\omega_1}{\omega_1}$ and $\frac{\Delta\omega_2}{\omega_2}$ separately into Eq. (21) and dividing these two equations, the non-dimensional crack section flexibility η can then be cancelled out and the following identification equation can be obtained:

$$\frac{\Delta\omega_1}{\omega_1}\frac{\omega_2}{\Delta\omega_2} = \frac{\gamma_1}{\gamma_2} \tag{23}$$

- The position of the crack l_1, can be determinate with the help of MATLAB software using solve function;
- After obtaining l_1, the non-dimensional crack sectional flexibility η can be determined from Eq. (11) and the crack depth a can then be obtained from Eq. (6).

5 Results and discussion

In order to validate the study presented in this paper and to show the effectiveness and also the limitations of the two methods, results obtained by applying these two methods are compared with the available data for single open cracked simply supported beam.

The material parameters for an undamaged simply supported beam are: $E = 2.1 \times 10^{11}\, Pa$, $\nu = 0.3$, $\rho = 7860\, kg/m^3$, and its dimensions are: $L = 0.5\, m$, $h = 0.02\, m$, $b = 0.012\, m$.

Table 1 shows comparison of the first three natural frequencies of undamaged beam obtained from the two methods with what computed by [18]. The results show that both methods yield the same exact frequencies and agree well with those obtained from [18].

Table 1. Comparison of frequencies of undamaged beam, obtained by two methods and results given by Ref. [18].

	ω_1	ω_2	ω_3
[18]	1178.160	4712.577	10603.252
1st method	1178.141	4712.566	10603.274
2nd method	1178.141	4712.566	10603.274

5.1 Direct problem

The lowest three natural frequencies of transverse vibration are calculated by the two methods, and compared with what reported by [19] for the damaged beam with one crack. The results for various crack locations and depths, given in Table 2, show a reasonable agreement between the two methods results and those achieved by [19] for all cases of the cracked beam, but the difference between natural frequencies increases for higher order modes.

Table 2. Comparison of first three natural frequencies of cracked beams, obtained by two methods and results given by Ref. [19] for various crack location and depth.

Case number	l_1	ξ	Methods	Natural frequencies, ω_{cn} (rad/s)		
				ω_{c1}	ω_{c2}	ω_{c3}
1	0.1	0.1	[19]	1177.660	4705.604	10573.663
			TMM	1177.709	4706.310	10576.662
			RM	1177.709	4706.234	10574.665
2	0.2	0.3	[19]	1162.672	4557.815	10281.642
			TMM	1164.032	4570.851	10306.810
			RM	1164.036	4518.740	9675.567
3	0.3	0.2	[19]	1165.559	4644.925	10587.306
			TMM	1166.720	4651.012	10588.734
			RM	1166.721	4631.081	10562.019
4	0.3	0.3	[19]	1149.363	4563.174	10568.267
			TMM	1151.851	4575.368	10571.089
			RM	1151.858	4477.435	10428.520
5	0.4	0.2	[19]	1160.876	4686.538	10544.768
			TMM	1162.459	4688.896	10550.062
			RM	1162.460	4677.927	10429.644
6	0.4	0.4	[19]	1105.78	4608.053	10369.596
			TMM	1110.837	4614.974	10384.937
			RM	1110.847	4416.368	8607.153
7	0.5	0.2	[19]	1159.108	4712.566	10436.419
			TMM	1160.849	4712.566	10451.318
			RM	1160.849	4712.566	10105.454
8	0.5	0.4	[19]	1099.580	4712.566	9964.053
			TMM	1104.527	4712.566	10003.928
			RM	1104.528	4712.566	6723.939

https://doi.org/10.1051/matecconf/201814902015

Figure 2 represents plots of the corresponding mode shapes, using the two methods.

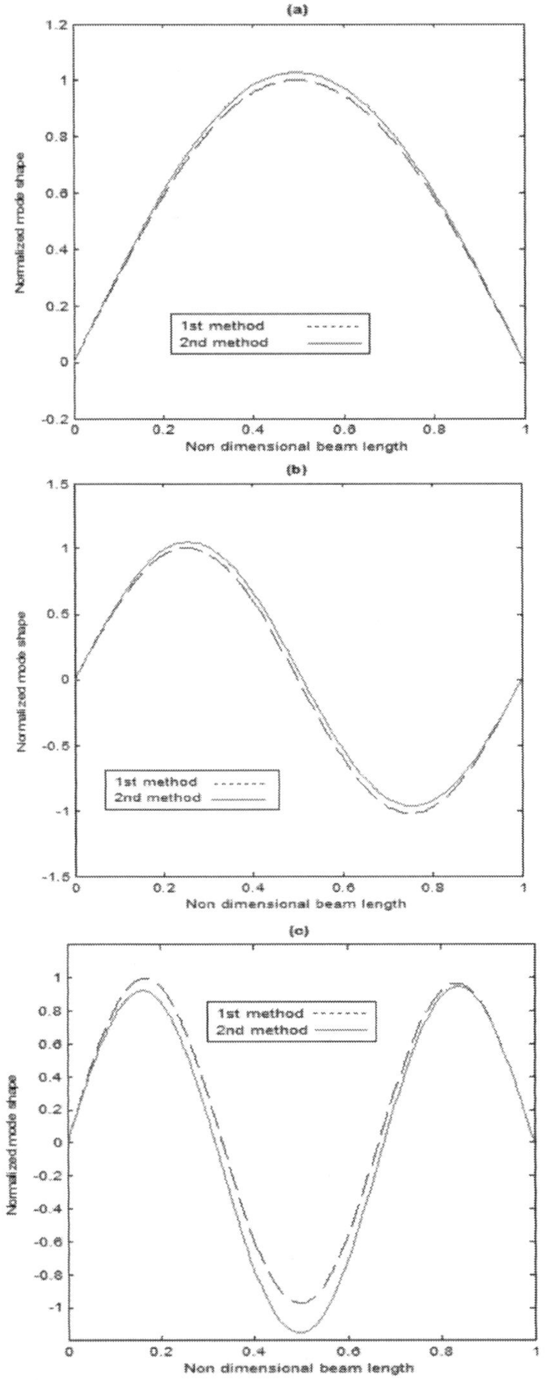

Fig.2. Comparison of the two methods for first three mode shapes of beam having one crack. l_1=0.4 ; ξ=0.2. (a) 1st mode shape, (b) 2nd mode shape and (c) 3rd mode shape.

Table 3 shows comparison of the first three relative errors of natural frequencies (R.E.N.F) obtained by the two methods for various crack location and depth, where:

$$\mathrm{R.E.N.F}\,(\%) = \left|\frac{\omega_{cn1} - \omega_{cn2}}{\omega_{cn1}}\right| \times 100 \qquad (24)$$

ω_{cn1} and ω_{cn2} are the nth calculated natural frequencies by the first and the second methods respectively.
As can be seen, the two methods can approximate first natural frequencies of cracked beam with high accuracy for various crack locations and depths. Also the two methods can approximate second natural frequencies of cracked beam with maximum error less than 5% in the 6th case when the crack location/ depth are respectively (0.4, 0.4), but the difference between natural frequencies increases for the third mode of vibration.

Table 3. Comparison of relative error between first three natural frequencies, obtained by the two methods for various crack location and depth.

Case number	1st R.E.N.F (%)	2nd R.E.N.F (%)	3rd R.E.N.F (%)
1	≈ 0	0.001	0.018
2	≈ 0	1.140	6.124
3	≈ 0	0.428	0.252
4	≈ 0	2.140	1.348
5	≈ 0	0.233	1.141
6	≈ 0	4.303	17.118
7	≈ 0	≈ 0	3.309
8	≈ 0	≈ 0	32.787

5.2 Inverse problem

The inverse problem is solved by using the exact first three frequencies used in [19] as inputs from various cases.

Table 4.a. Comparison of predicted and actual crack parameters for various crack location and depth by using $\omega_{c1} - \omega_{c2}$.

Case	Methods	$\omega_{c1} - \omega_{c2}$			
		l_1^*	Error (%)	ξ^*	Error (%)
1	Method1	0.1002	0.2473	0.1054	5.4543
	Method2	0.1005	0.5028	0.1500	50.0562
2	Method1	0.2000	0.0041	0.3128	4.2935
	Method2	0.2097	4.8554	0.4007	33.5849
3	Method1	0.2999	0.0004	0.2099	4.9612
	Method2	0.3031	1.0639	0.2871	43.5758
4	Method1	0.2999	0.0005	0.3128	4.2947
	Method2	0.3070	2.3661	0.4056	35.2128
5	Method1	0.4000	0.0002	0.2099	4.9622
	Method2	0.4006	0.1726	0.2880	44.0040
6	Method1	0.3999	0.0001	0.4126	3.1733
	Method2	0.4028	0.7158	0.5087	27.1855
7	Method1	0.5003	0.0652	0.2099	4.9610
	Method2	0.5	0	0.2878	43.9407
8	Method1	0.4998	0.0322	0.4115	2.8772
	Method2	0.5	0	0.5073	26.8415

Tables 4a-4c show the comparison of predicted (l_1^*, ξ^*) and actual (l_1, ξ) crack parameters using any two of these three frequencies.
In the majority of cases, the two methods cannot provide a unique value for the non-dimensional crack position l_1^*, which require usually an initial guess closest to the solution. This is due to the symmetric of crack locations

(i.e., 0.2 and 0.8) when the boundary conditions are also symmetric (i.e., simply supported).

Table 4.b. Comparison of predicted and actual crack parameters for various crack location and depth by using $\omega_{c1} - \omega_{c3}$.

Case	Methods	$\omega_{c1} - \omega_{c3}$			
		l_1^*	Error (%)	ξ^*	Error (%)
1	Method1	0.1000	0.0881	0.1056	5.6217
	Method2	0.1005	0.5714	0.1499	49.9553
2	Method1	0.2000	0.0012	0.3128	4.2958
	Method2	0.2081	4.0728	0.4027	34.2600
3	Method1	0.2999	≈ 0	0.2099	4.9609
	Method2	0.3005	0.1801	0.2887	44.3610
4	Method1	0.2999	0.0001	0.3128	4.2944
	Method2	0.3012	0.4000	0.4097	36.5950
5	Method1	0.3999	0.0002	0.2099	4.9624
	Method2	0.3994	0.1495	0.2883	44.1763
6	Method1	0.4000	≈ 0	0.4126	3.1732
	Method2	0.3974	0.6369	0.5105	27.6268
7	Method1	0.2467	50.6406	0.2941	47.0685
	Method2	0.4817	3.6403	0.2883	44.1548
8	Method1	0.2362	52.7556	0.5375	34.3815
	Method2	0.4646	7.0734	0.5093	27.3352

It is observed that the crack identification results for normalized crack location are relatively acceptable. The errors in the 7th and 8th cases are generally the highest for the 1st method.
It is obvious that the crack identification results for normalized crack depth are not satisfactory; this is due to the deferred behavior of the flexibility coefficient.

Table 4.c. Comparison of predicted and actual crack parameters for various crack location and depth by using $\omega_{c2} - \omega_{c3}$.

Case	Methods	$\omega_{c2} - \omega_{c3}$			
		l_1^*	Error (%)	ξ^*	Error (%)
1	Method1	0.1002	0.2473	0.1054	5.4543
	Method2	0.1005	0.5028	0.1500	50.0562
2	Method1	0.2000	0.0041	0.3128	4.2935
	Method2	0.2097	4.8554	0.4007	33.5849
3	Method1	0.2999	0.0004	0.2099	4.9612
	Method2	0.3031	1.0639	0.2871	43.5758
4	Method1	0.2999	0.0005	0.3128	4.2947
	Method2	0.3070	2.3661	0.4056	35.2128
5	Method1	0.4000	0.0002	0.2099	4.9622
	Method2	0.4006	0.1726	0.2880	44.0040
6	Method1	0.3999	0.0001	0.4126	3.1733
	Method2	0.4028	0.7158	0.5087	27.1855
7	Method1	0.5003	0.0652	0.2099	4.9610
	Method2	0.5	0	0.2878	43.9407
8	Method1	0.4998	0.0322	0.4115	2.8772
	Method2	0.5	0	0.5073	26.8415

6 Conclusion

Two methods used to solve the direct and inverse problems of a simply supported beam with one crack are compared in this paper. The Euler–Bernoulli beam theory is employed to model the cracked beam and the crack is represented as a rotational spring. This study allows making the following conclusions:

- The frequencies decreased when the crack is deeper;
- The crack depth has no effects on the second natural frequencies at the mid-span;
- The natural frequencies predicted by the two methods agree well with those obtained from previous work but the error increases for higher modes of vibration;
- The crack parameters predicted in the numerical examples show that both methods are theoretically capable of estimating the location of cracks and with less performance the crack depths.

References

1. A.D. Dimarogonas, Engineering Fracture Mechanics, **55:5** 831-857 (1996)
2. R. Gasch, Journal of Sound and Vibration, **160(2)** 313-332 (1993)
3. W. Ostachowicz, M. Krawczuk, Key Engineering Materials, **204-205** 185-200 (2001)
4. M.I. Friswell, J.E.T. Penny Jet, Structural Health Monitoring, **1:2** 139-148 (2002)
5. W.M. Ostachowicz, M. Krawczuk, Journal of Sound and Vibration, **150:2** 191-201 (1991)
6. P.F. Rizos, N. Aspragathos, A.D. Dimarogonas, Journal of Sound and Vibration, **138:3** 381-388 (1990)
7. T.G. Chondros, A.D. Dimarogonas, J. Yao, Journal of Sound and Vibration, **215:1** 17-34 (1998)
8. O.S. Salawu, Engineering Structures, **19** 718-723 (1997)
9. A.K. Pandey, M. Biswas, Journal of Sound and Vibration, **169(1)** 3-17 (1994)
10. A.K. Pandey, M. Biswas, M.M. Samman, Journal of Sound and Vibration, **145(2)** 321-332 (1991)
11. R.Y. Liang, F.K. Choy, J. Hu, Journal of the Franklin Institute, **328** 505-518 (1991)
12. B.P. NANDWANA, S.K. MAITI, Journal of Sound and Vibration, **203:3** 435-446 (1997)
13. J. Lee, Journal of Sound and Vibration, **320** 482-490 (2009)
14. M. Attar, International Journal of Mechanical Sciences, **57** 19-33 (2012)
15. J. Ferna´Ndez-Sa´Ez, L. Rubio, C. Navarro, Journal of Sound and Vibration, **225** 345-352 (1999)
16. H.P. Lin, Engineering Structures, **26** 427-436 (2004)
17. J. Hu, R.Y. Liang, Journal of the Franklin Institute, **330(5)** 841-853 (1993)
18. J. Xiang, Y. Zhong, X. Chen, Z. He, International Journal of Solids and Structures, **45** 4782–4795 (2008)
19. J.W. Xiang, X.F. Chen, B. Li, YM. He, Z.J. He, Journal of Sound and Vibration, **296** 1046-1052 (2006)

Numerical Derivation of Iso-Damaged Curve for a Reinforced Concrete Beam Subjected to Blast Loading

Yehya Temsah[1], Ali Jahami[1], Jamal Khatib[1,2], M Sonebi[3]
[1]Faculty of Engineering, Beirut Arab University, Beirut, Lebanon
[2]Faculty of Science and Engineering, University of Wolverhampton, Wolverhampton, UK
[3]School of Natural and Built Environment, Queens University of Belfast, Belfast, UK

Email: j.khatib@bau.edu.lb

Abstract. Many engineering facilities are severely damaged by blast loading. Therefore, many manufacturers of sensitive, breakable, and deformed structures (such as facades of glass buildings) carry out studies and set standards for these installations to withstand shock waves caused by explosions. Structural engineers also use these standards in their designs for various structural elements by following the ISO Damage Carve, which links pressure and Impulse. As all the points below this curve means that the structure is safe and will not exceed the degree of damage based on the various assumptions made. This research aims to derive the Iso-Damage curve of a reinforced concrete beam exposed to blast wave. An advanced volumetric finite element program (ABAQUS) will be used to perform the derivation.

1 Introduction

The damage caused by an explosion has been studied after World War 2 by different researchers who derived many formulas. One of the most important formulas used in explosion analysis are the Gurney [1] Equations. They help in calculating the velocity of fragments released during an explosion, and this velocity can help the experts to predict the damage level of the explosion and the fragment penetration of structures.

Cylindrical Charge Equation:

$$\frac{V}{\sqrt{2E}} - \left(\left(\frac{M}{C}\right) + 0.5\right)^{-0.5} \quad (1)$$

Spherical Charge Equation

$$\frac{V}{\sqrt{2E}} = \left(\left(\frac{M}{C}\right) + 0.6\right)^{-0.5} \quad (2)$$

Where:
V = Initial Velocity of Fragment (m/s)
√2E = Gurney Constant for a given blasting materials (m/s)
M_f = Fragment Mass (kg)
M_c = Blasting materials Charge Mass (kg)

Table 1 shows us some of Gurney Constant values (√2E):

Blasting Materials	*Density (Kg/m³)*	*Detonation velocity (m/s)*	*Gurney constant (m/s)*
Comp-B	1.61	7,620	2,774
Comp-C4	1.71	8,200	2,530
HMX	1.91	9,100	2,972
Octal 75/25	1.81	8.640	2,896
PETN	1.78	8,260	2,926
RDX	1.81	8,700	2,926
RDX / TNT 60/40 (Cyclotol)	1.68	7,800	2,402
Tetryl	1.71	7,570	2,499
TNT	1.61	6,900	2,438
Tritonal	1.70	5,480	2,316

Table 1.typical values of Gurney constant (*Gurney 1943)[1]*

Also it's an important to know the range of damage caused by an explosion so we can consider the appropriate steps for safety. The following equations are used to estimate the safety distance required to avoid the

fragments effect. These equations are very efficient in the preliminary design stage since they are easy to use.
When using fragmenting munitions and the demolition area is accessible by public:

$$D = 634(W)^{1/6} \quad (3)$$

When using fragmenting munitions and the demolition area is not accessible by public:

$$D = 444(W)^{1/6} \quad (4)$$

For blasting materials without fragments:

$$D = 130(W)^{1/3} \quad (5)$$

Where:
D = Distance from explosion (m)
W = Total weight of blasting materials (kg)

The vertical danger zone limit is very important for planes flying above blast affected zones. It gives us an indication about the minimum required height to fly in order to avoid the explosion impact.

For single ammunition item only:

$$D = 314(W)^{1/3} \quad (6)$$

For multi-item fragmenting munitions:

$$D = 470(W)^{1/5} \quad (7)$$

Experts are collecting data about explosions and its effect on buildings since World War 2. They used to link the damage level to the standoff distance and the weight of explosion. Many numerical models were derived in the 20th century. One of the most important formulas to estimate the range of damage is illustrated in Equation 8:

$$R_x = (K_x \, . \, M_{exp}^{1/3}) \, / \, (1 + (3175/M_{exp})^2)^{1/6} \quad (8)$$

Where:
R_x = Range for Damage (m)
K_x = Damage Level Constant
M_{exp} = Mass of Blasting materials (kg)

Values for K_x can be obtained from *Jarrett and Gilbert* [2], *Lees and Scilly* [3]. The revised values consider the degree of energy imparted to the primary fragments from the casing which will reduce the air blast energy available.

K_x for Damage Category	Jarrett	Gilbert, Lees and Scilly
A	3.8	4.8
B	5.6	7.1
C_B	9.6	12.4
C_A	28.0	21.3
D	56.0	42.6

Table 2. K_x factors (*Jarrett 1968*)[2]

Category	Definition
A	Houses completely demolished.
B	Houses so badly damaged they are beyond repair and require demolition.
C_B	Houses rendered uninhabitable but can be repaired with extensive work.
C_A	Houses rendered uninhabitable but can be repaired reasonably quickly.
D	Houses requiring repairs to remedy serious inconvenience but remain habitable.

Table 3. Damage category (*Jarrett 1968*)[2]

2 Aim and scope of the research

The aim of this research is to derive the Iso-Damage curve for a reinforced concrete beam exposed to an impact load from an explosion using finite element modeling. The data will be collected from an experimental work done by a Chinese researcher as will be illustrated in the coming sections.

3 Iso-Damage concepts

Assume that we have a structure which is subjected to a quasi-static load pulse. In the elastic range the work done on the system is converted into strain energy. If this work causes a maximum displacement of X_{max}, then:

$$\text{Total work done} = FX \quad (9)$$

And the strain energy "U" is:

$$U = K(X_{max})^2/2 \quad (10)$$

Since the total work is converted into strain energy, then we can conclude the quasi-static asymptote as:

$$\text{Quasi-static asymptote} = \frac{\text{Strain Energy}}{\text{Max displacement}} \quad (11)$$

If the load is of short duration, then from the momentum equation it produces an instantaneous velocity change:

$$v = \frac{I\ (\text{Impulse})}{M\ (\text{Mass})} \quad (12)$$

As a result, the structure gains kinetic energy given by:

$$KE = \frac{Mv^2}{2} = \frac{I^2}{2M} \quad (13)$$

Then the Impulsive asymptote is:

$$Ic = \sqrt{2.KE.M} \quad (14)$$

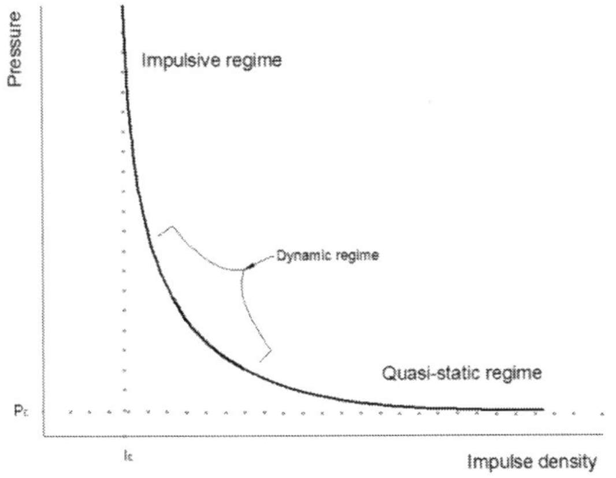

Figure 1. Typical Iso-Damage Curve

4 Data Collection

According to the experiment that was done by Zhang et al (2013) [4] **at** at the National University of Defense Technology in China, there are 3 sizes of reinforced concrete beams with dimensional ratio of 3:4:5. The dimensions are (850mmx75mmx75mm), (1100mmx100mmx100mm) and (1350mmx125mmx125mm). The bars used for tensile, compressive and hoop reinforcement had a diameter of 6mm. The spacing between hoops was 6mm. The number and dimensions of beams are listed in Table 4. The design compressive strength of concrete is 40 MPa whereas the yield strength and ultimate strength of reinforcement steel are 395 MPa and 501 MPa respectively. The blast loading was positioned at the top of the beam and is compressed to a cakey cylinder suspended over the middle of the reinforced concrete (RC) beam, and is ignited by an electronic detonator. The beam is supported using a steel frame as shown in Figure 2. The mid span displacement of the beam is measured using steel needles can move only along the vertical direction.

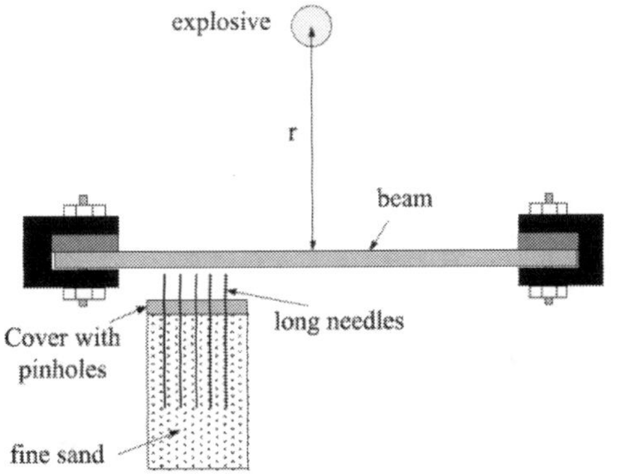

Figure 2. Experimental setup

Beam	*Dimension (mm)*	*TNT (Kg)*	*Standoff distance (m)*
B2-1	100X100X1100	0.36	0.4
B2-2		0.51	0.4

Table 4. Experimental program (Zhang et al 2013)

5 Numerical Modeling

The volumetric finite element program "ABAQUS" was used to perform the nonlinear analysis of the reinforced concrete beams. The model is composed of two main parts: the concrete body that was modeled using a solid element, and steel reinforcement that was modeled as a rebar element. A Dynamic explicit analysis step was chosen for the case with a reasonable time step close to the experimental interaction time.

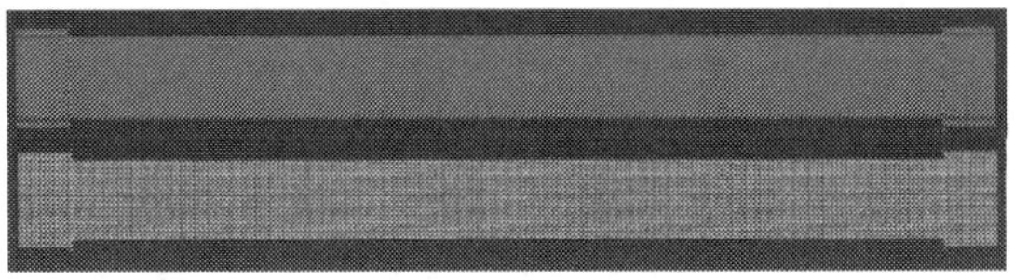

Figure 3. Finite Element Modeling of beam

The Concrete Damage Plasticity Model [5] will be used to define the behavior of concrete. This model works with static and dynamic load conditions. It was derived by Lubliner [6] in 1989 and modified laster by Lee and Fenves [7] in 1998. The model assumes that the concrete behave in a non-linear manner and different input parameters had to be assumed including: inelastic strain, cracking strain, stiffness degradation and recovery, and other parameters. There are many available methods that can model the reinforcing steel. For this simulation, the elasto-plastic behavior of reinforcing steel was considered, and a perfect bond between concrete and steel was assumed.

6 Model Verification

The model was calibrated until an acceptable error was reached. Two verifications were done for this process: comparing mid-span displacements and damage zone length. Table 5 and Figure 4 show the Mid-Span displacements for Beams B2-1 and B2-2. As shown the difference between experimental and numerical analysis is less than 10% (i.e. small). Therefore the verification is acceptable.

Beam Sample	Mid-Span Displacement (mm)		Absolute error %
	Experimental	Numerical	
B2-1	9	8.8	2.22
B2-2	35	32.1	8.29

Table 5. Mid-Span Displacement for B2 samples

https://doi.org/10.1051/matecconf/201814902016

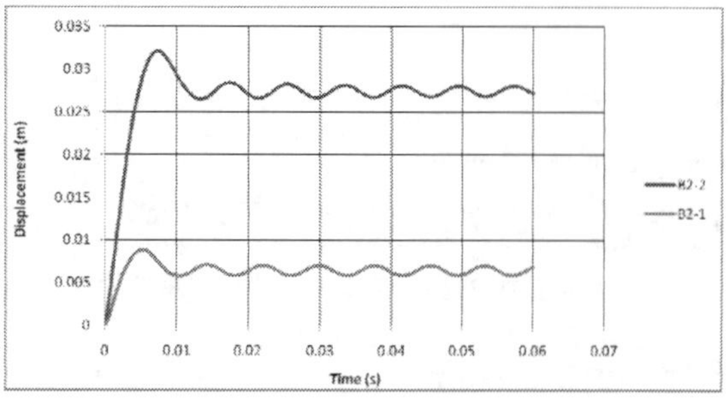

Figure 4. Time-Displacement Curves

7 Iso-Damage Curve

The damage analysis results for beam B2-2 are presented in this section. First the energy curves are shown in Figure 5, which represent the distribution of total work done with respect to denotation time.

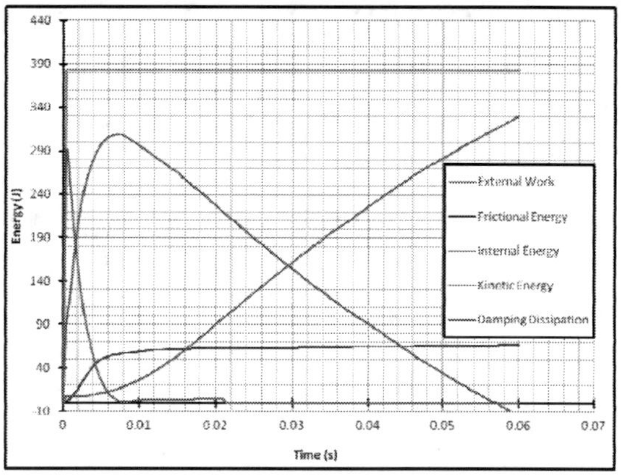

Figure 5. Distribution of Total work for beam B2-2

The Iso-Damage curve for B2-2 is shown below. A four combination of pressure and impulse were tested to check the validity of this curve as will be illustrated in Figure 6 and table 6.

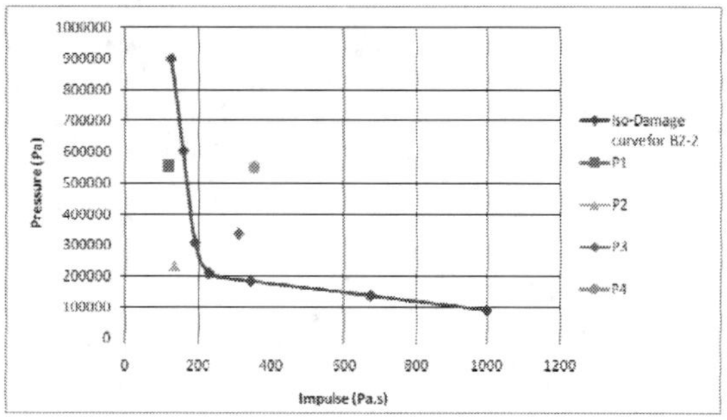

Figure 6. Iso-Damage curve for B2-2

Point	Stand-off distance (m)	TNT mass (Kg)	Pressure (Pa)	Impulse Pa.s
P1	0.3	1	556000	119
P2	1	2	232000	134
P3	10	4	336000	311
P4	8	3	550000	345

Table 6. (Pressure-Impulse) combinations

As shown, Both P1 and P2 will cause less damage compared to the real state of B2-2 (32.1 mm Mid-Span Displacement). On the other hand both P3 and P4 will cause more damage to B2-2. Figure 7 confirms this fact by plotting the Mid-span displacement for these four combinations compared to the real beam state B2-2.

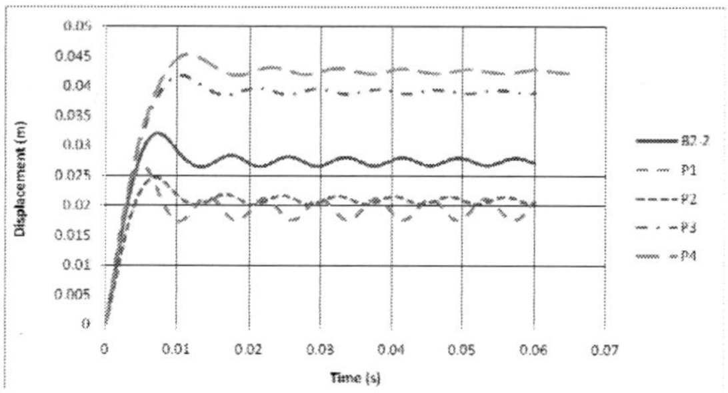

Figure 7. Mid-Span displacement compared to B2-2

8 Conclusions

The outcomes of this investigations lead to the following conclusions:

1- It is possible to form the Iso-Damage curve for any structural elements using Finite Element Program such as ABAQUS.

2- The extend of damage to objects or structures is a function of the both the quantity and distance of blast materials.

9 Further Research

The following points may be considered for further studies:

1- Derivation of Iso-Damage curve for different types of structural elements (columns, walls ...etc).

2- Derivation of Iso-Damage curves using single degree of freedom analysis.

3- The effect of reinforcement ratio and distribution on the damage capacity and Iso-damage curve behavior for structural elements.

4- The possible solutions to increase the range of safety for elements such as new materials.

Reference

1. *Gurney, R. W.* The Initial Velocities of Fragments from Bombs, Shells, and Grenades, BRL-405. Ballistic Research Laboratory, Aberdeen, Maryland. USA. 1943.

2. *Jarrett D E.* Derivation of the British Blasting materials Safety Distances. Annals New York Academy of Sciences, 152, Article 1. 1968.

3. *Scilly N F* and *High W G.* The blast effect of explosions. Loss prevention and safety promotion 5. 1986

4. Duo Zhang , ShuJian Yao, Fangyun Lu, XuGuang Chen, Guhui Lin, Wei Wang & Yuliang Lin. (July 5, 2013). *Experimental study on scaling of RC beams under close-in blast loading*. Retrieved December 7, 2015 from: http: // www.sciencedirect.com /science/ article/pii/ S1350630713002203.

5. Hibbitt, Karlsson, Sorensen, 2011, ABAQUS User's Manual, *Pawtucket, 6th Edition.*

6. Lublinear, J., Oliver, J., Oller, S., Onate, E., 1989. A plastic-damage model for concrete, *Solids and Structures, Vol. 25, No. 3, pp. 299-326.*

7. Lee, J., Fenves, G., 1998. Plastic-damage model for cyclic loading of concrete structure, *Engineering Mechanics, Vol. 124, No. 8, pp. 892-900.*

Behavior of reinforced concrete beams pre-cracked and repaired by composite materials under static loading

Mostefa Hamrat[1], Bensaid Boulekbache[1], Halima Bouziane[1] and Hayet Benkara[1]
[1]Département de Génie Civil, Université Hassiba Benbouali, Chlef, Algérie

Abstract. This work constitutes a contribution to the analysis of the behavior of beams repaired by composite materials. To analyze the overall behavior and failure modes of the beams, an experimental study of nine reinforced concrete beams, pre-cracked and then repaired by composite materials was conducted. Six beams were pre-cracked and repaired in the tensioned part (bending repair) and in the other two beams on the tensioned and lateral parts with strips in the shape of U (shear repair). A comparative study was made between the ultimate moments measured experimentally and those calculated by the theoretical models. Compared to the control beam, the resistance gain for the beams repaired in bending is 50% to 90%, while that of beams repaired in shear is from 120% to177 %. The beams repaired in shear exhibit a ductile rupture in bending. However, the beams repaired in bending were failed by the lift-off of composite or by failure of concrete cover layer (except for beams repaired by fiber glass). BAEL99, EC2-04 and ACI318-08 models give the best prediction of the ultimate moments with a mean value of 1.16 for the ratio of M_{Exp}/M_{theor}. and a mean standard deviation of 0.33.

1 Introduction

La durée de vie des ouvrages en béton armé est de l'ordre de 50-100 ans, mais elle se trouve écourtée du fait de nombreuses pathologies. Un grand nombre d'ouvrages présentent un état de dégradation avancée causé par une exposition prolongée à un environnement agressif, des défauts de résistance à l'effort tranchant ou en flexion due au chargement excessif, ainsi que celle liée à la fatigue de la structure sous chargement cyclique sont à l'origine de la diminution de la durée de vie de ces ouvrages. Face à ces problèmes, la réhabilitation et le renforcement des ouvrages, deviennent un enjeu stratégique.

La réparation des ouvrages en béton armé représente un moyen avantageux d'améliorer leur capacité portante et de prolonger leur durée de vie. En effet, il est souvent beaucoup moins onéreux de réparer certains éléments structuraux que d'effectuer une reconstruction complète de l'ouvrage. Plusieurs techniques de réparation sont disponibles sur le marché, dont celle qui consiste dans la réparation externe avec des matériaux composites. Face à cette situation, il est nécessaire de posséder un certain nombre de connaissances sur le comportement charge-flèche, ductilité et mode de rupture des éléments structuraux en béton armé réparées par des matériaux composites.

Par ailleurs, l'apport des matériaux composites en tant que matériau de réparation soulève de nouvelles questions. Afin d'obtenir une utilisation adéquate de ces nouveaux matériaux dans les tâches de réhabilitation et de renforcement, il est nécessaire de bien comprendre leur interaction avec le béton armé. Toutefois, des modèles ont été formulées par des groupes de travail tel qu'on les trouve dans l'approche de la Fédération Internationale du Béton (FIB) [1] et de l'Association Française de Génie Civil (AFGC) [2], à partir des résultats de la littérature. Ces modèles ont besoin d'être expérimenté et calibré.

2 Programme expérimental

Lors de cette étude, la valeur de la résistance souhaitée du béton a été prise égale à 25 MPa, valeur courante et proche du chantier.

Neuf spécimens de poutres de section 0,10 x 0,15m^2 et de 1.10 m de portée, ont été fabriqués et testés (figure 1). Ces poutres sont armées par deux barres HA8 dans la partie tendue et par deux barres HA6 dans la partie comprimée, ce qui donne une section sous-armée.

Les huit spécimens d'épreuve endommagés jusqu'à un niveau de chargement correspondant à une ouverture de fissure de 0,3 mm [3] (figure 2), vont être réparées, en utilisant le collage des lamelles composite (carbodur) et des tissus (fibre de carbone et fibre de verre).

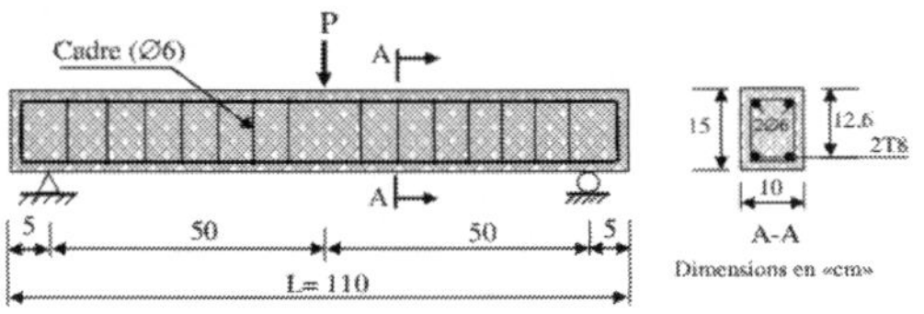

Fig. 1. Dimensions et armatures des poutres.

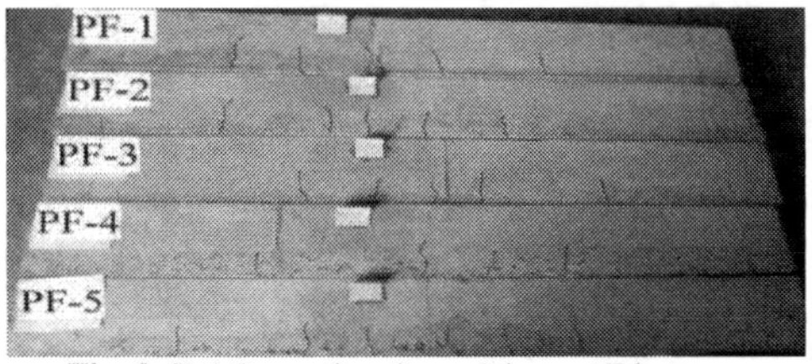

Fig. 2. Pré-fissuration des spécimens d'épreuve.

Deux modes de réparation sont utilisées dans notre étude. Dans le premier mode (A): le composite est collé sur la face tendue avec une largeur égale à celle de la poutre et une distance de 30 mm entre l'extrémité du tissu (lamelle) et l'appui de la poutre. Pour le deuxième mode (B), le composite est collé sur les faces tendue et latérales en forme de U (tableau 1).

Tableau 1. Configuration des Poutres Testées

Désignation poutre	Mode de réparation
Poutre témoin PT	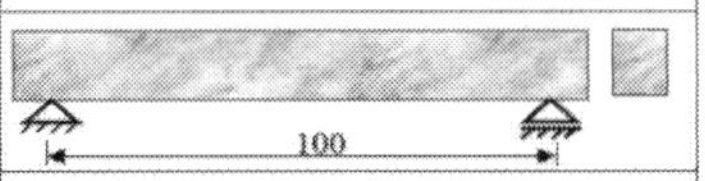
PR-1C, PR-2C PR-3C PR-1V, PR-2V PR-1L (Mode A)	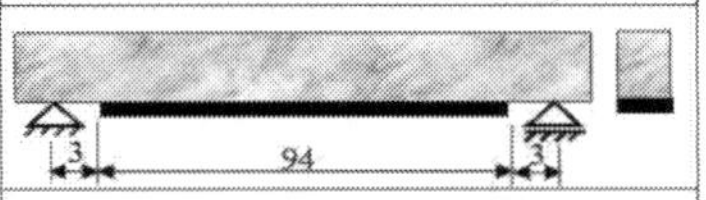
PR-1CU PR-1LU (Mode B)	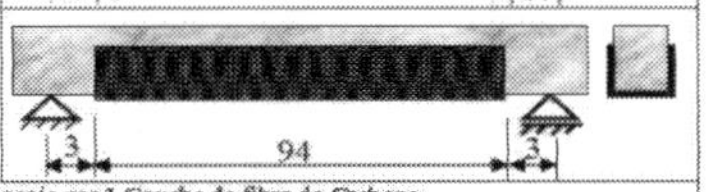
PR-1C : Poutre Réparée par 1 Couche de fibre de Carbone PR-1L : Poutre Réparée par 1 Couche Lamelle carbodur PR-1CU: Poutre Réparée par 1 Couche de fibre de Carbone en forme de U	

3 Résultats et discussions

3.1 Réponse charge-flèche des poutres réparées

La résistance ultime des poutres réparées est principalement dictée par le mode de réparation.

3.1.1 Mode de réparation A

Dans cette série, on a utilisé les fibres (les fibres de carbone unidirectionnelles, les fibres de verre bidirectionnelles et des lamelles carbodur) en 1, 2 et 3 couches. Ces couches sont collées dans la face tendue de la poutre (réparation en flexion).

La figure 3 représente les courbes charge-flèche de la poutre témoin et des poutres réparées selon le mode de réparation A.

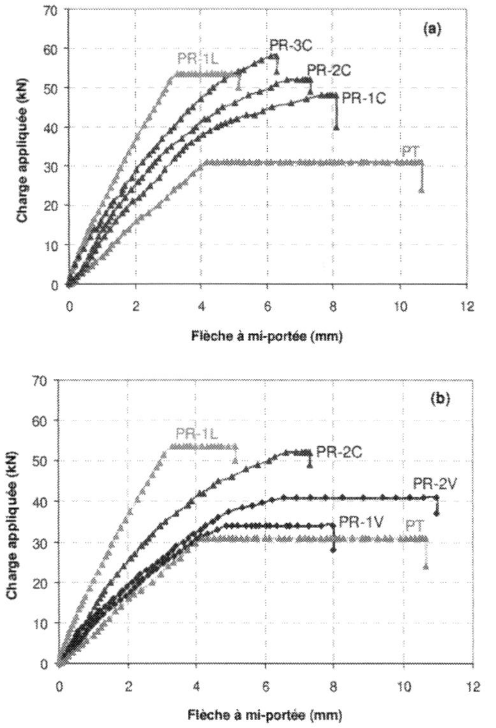

Fig. 3. Courbes charge-flèche des poutres réparées (mode A)

A l'issue de la figure 3, on observe ce qui suit:

-De grandes différences de comportement apparaissent entre la poutre témoin (PT) et les poutres réparées: PR-1C, PR-2C et PR-3C. (figure 3(a)). Les poutres réparées possèdent un pseudo-palier plastique de ductilité une fois la résistance ultime atteinte.

-Les poutres PR présentent un comportement plus rigide avec une réduction significative de la flèche (figure 3(a)). Dans ce sens, les capacités portantes des poutres réparées sont: 48 kN pour PR-1C, 52 kN pour PR-2C et 58 kN pour PR-3C, ce qui représente des gains par rapport à la poutre témoin respectivement de: 55 %, 70 % et 90 % (tableau 2). Les flèches correspondant à ces charges sont de 8,0 mm, 7,3 mm et 6,3 mm alors qu'elle était de 10,6 mm pour la poutre témoin.

-La charge de rupture de la poutre PR-2V est de 78 % la charge de rupture de la poutre PR-2C. Cependant, la poutre PR-2V présentait une flèche de 11 mm à la rupture où la poutre PR-2C s'était avec une flèche de 7,4 mm (figure 3(b)), ceci est principalement lié aux propriétés de la fibre de verre (4,8 % élongation à la rupture contre 1,8 % pour la fibre de carbone) [4].

-L'ajout des lamelles à base de fibres de carbone augmente la charge de rupture et réduit la flèche (figure 3(b)). En effet, la poutre PR-1L a enregistré un gain de résistance de 87 % par rapport à la poutre témoin. De plus, lamelle carbodur augmente d'une manière substantielle la rigidité de la poutre. La poutre PR-1L présentait une charge de rupture presque égale à la charge de rupture de la poutre PR-2C (1 Plaque carbodur ≈ 2 couches de fibres de carbone).

-Les courbes des poutres PR montrent bien le point de plastification qui se situe aux environs de 50 kN. Pour des flèches semblables qui varient de 4 à 6 mm, ceci représente des gains de 40 % environ par rapport à la poutre témoin. C'est donc dire que l'utilisation des matériaux composites en flexion permet de retarder la plastification des armatures longitudinales tendues.

Tableau 2. Résultats expérimentaux à la rupture

Poutre	P_u KN	Δ_u mm	Augm↑ Charge %	Dimin↓ flèche %	Mode de Rupture
PT	31	10,6	-	-	Flexion
PR-1C	48	8,1	55	24	Décollement du composite
PR-2C	52	7,4	70	30	Rupture enrobage du béton
PR-3C	58	6,3	90	41	Rupture enrobage du béton
PR-1V	34	8,0	10	38	Flexion
PR-2V	40,5	11,0	31	4*	Flexion
PR-1L	54	5,2	87	46	Rupture enrobage du béton
PR-1CU	68	9,5	120	12	bris de fibre de carbone
PR-1LU	86	12,1	177	14*	Décollement + Cassure de la plaque

* Augmentation

3.1.2 Mode de réparation B

Cette série comprend deux poutres réparées par tissu de carbone/lamelles en forme de U. Cet essai va nous permettre de comparer le mode de réparation A, avec le mode de réparation B (configuration en U).

La figure 4 représente les courbes charge-flèche des poutres selon le mode de réparation B et la poutre témoin.

D'après ces courbes, on observe ce qui suit:

- La réparation d'une poutre augmente d'une manière substantielle la charge de rupture, en améliorant la rigidité et en réduisant la flèche. On note que la poutre PR-1CU a permis d'atteindre une augmentation de charge significative de près de 120 % par rapport à la poutre témoin (tableau 2). De plus, cette poutre réparée présentait une flèche de 9,5 mm seulement au niveau de charge où la poutre témoin s'était avec une flèche de 10,6 mm (figure 4).

- Pour le mode de réparation B, les courbes charge-flèche ont présenté une très grande rigidité par rapport aux poutres réparées selon mode de réparation A. Les charges de ruptures sont: 68 kN pour PR-1CU et 86 kN pour PR-1LU, ce qui représente des gains par rapport aux poutres réparées uniquement en flexion (PR-1C et PR-1L) de 42 % et 59 %, respectivement (Tableau 2).

En outre, il faut noter que ces deux poutres réparées en cisaillement par des bandes continues en U sur toute leur longueur aient permis d'atteindre une rupture ductile en flexion (figure 8).

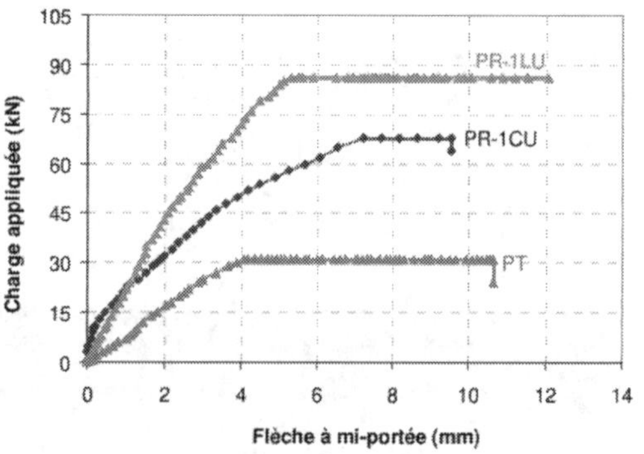

Fig. 4. Courbes charge-flèche des poutres réparées (mode B)

3.2 Faciès et modes de rupture des poutres

La combinaison des matériaux composites comme réparation de poutres en béton armé produit de nouveaux modes de rupture. D'après Meier [5], ces ruptures peuvent se situer à différents niveaux: dans la poutre en béton armé, dans le composite ou dans l'interface béton-composite.

Selon les modes de réparation, les principaux modes de rupture susceptibles d'être observés dans les poutres réparées sont les suivants:

3.2.1 Mode de réparation A

Poutre: PR-1C.

La rupture de la poutre est due au décollement partiel du composite. Ce décollement de composite qui débute aux extrémités de celui-ci, puis se propage vers le centre de la poutre. Le décollement s'est produit entre le béton et l'adhésif du composite (figure 5). Ce mode de rupture provient de la concentration de contraintes à l'extrémité du tissu.

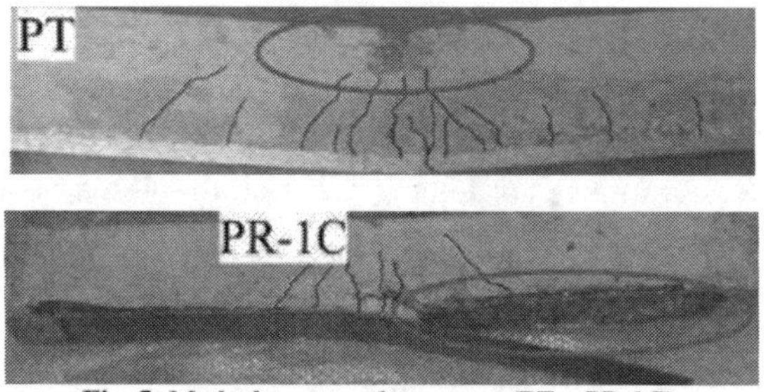

Fig. 5. Mode de rupture des poutres (PT + PR-1C).

Poutres: PR-2C + PR-3C + PR-1L

Il y a eu apparition de fissures de flexion sur la zone centrale, dès le début du chargement. Des fissures verticales apparaissent à l'extrémité où le moment fléchissant est maximal, et sont inclinées vers le vers le point d'application de la charge (figure 6).

La rupture de ces poutres a été fragile et soudaine et s'est caractérisée par décollement du composite avec une séparation complète de l'enrobage du béton sous une fissure diagonale d'effort tranchant (figure 6). Certes, le décollement du composite s'est produit entre le béton et la colle. En effet, la résistance à la contrainte de cisaillement de la colle est plus élevée que celle du béton.

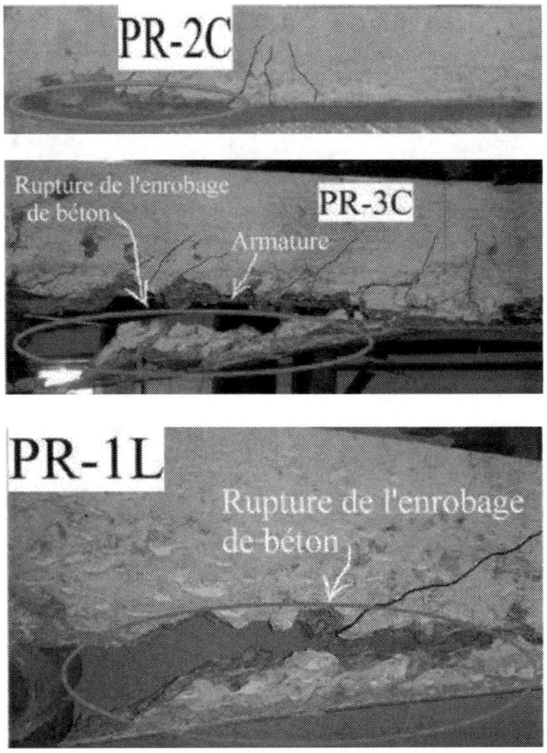

Fig. 6. Mode de rupture des poutres (PR-2C+ PR-3C+ PR-1L).

Poutres: PR-1V + PR-2V

Des fissures de flexion se forment au niveau de la zone tendue et dans la partie centrale de la poutre telle qu'en figure 7. A des niveaux de chargement relativement élevés, des petites fissures diagonales se développent de part et d'autre de la zone centrale de la poutre. La poutre s'est rompue en flexion où la zone comprimée s'est écrasée avec rupture du composite (figure 7).

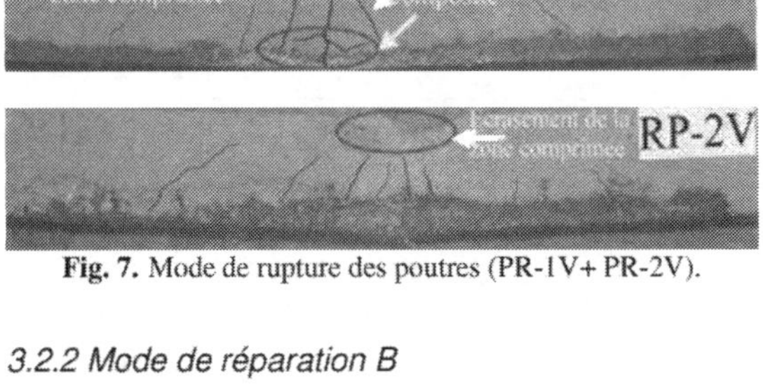

Fig. 7. Mode de rupture des poutres (PR-1V+ PR-2V).

3.2.2 Mode de réparation B

Poutre: PR-1CU

Un bris de fibre de carbone a été observé dans la face tendue et les deux faces latérales. Il y a eu apparition de fissures diagonales sur la face latérale de la poutre (figure 8). La réparation de la poutre par collage du tissu continu en forme de U ne permet pas de "coudre" les fissures. Dans le Mode de réparation B, le composite est sollicité jusqu'à la rupture et que dans le Mode de réparation A, on a assisté au décollement du composite ou rupture de l'enrobage de béton.

La séquence de rupture était moins brutale. En fait, cette poutre présente plus de fissures (durant la phase de pré-fissuration). Ces fissures libèrent l'énergie stockée graduellement de sorte que, à l'état ultime, la quantité finale d'énergie libérée ne soit pas aussi grande, et par conséquent la rupture est moins explosive et suivi d'un bruit assez distinct et particulier dû aux craquements du composite.

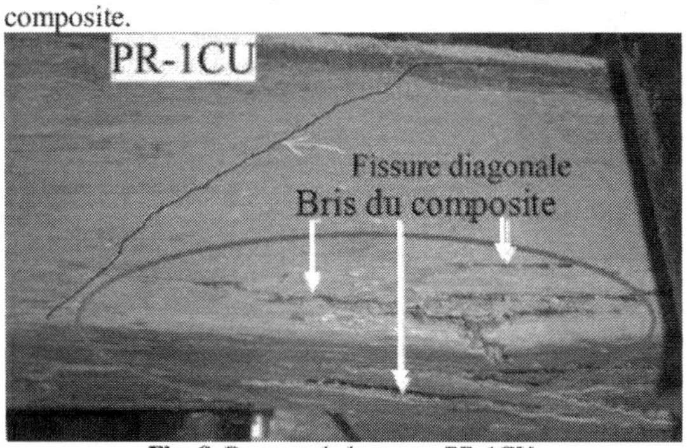

Fig. 8. Rupture de la poutre PR-1CU.

Poutre: PR-1LU

La poutre a été rompue par décollement et cassure de la lamelle collée dans la face latérale sous une sévère fissure diagonale d'effort tranchant à l'extrémité de la plaque. Cette rupture est causée par un déplacement relatif vertical de la face fissurée (figure 9). L'étrier en U fait de lamelles carbodur a recélé toutes fissures de flexion.

Fig. 9. Mode de rupture de la poutre PR-1LU.

3.3 Ductilité des poutres testées

Si la rigidité, la charge et la flèche conditionnent le comportement structurel avant le pic de chargement [6],[7], c'est uniquement la ductilité qui a une influence importante sur le comportement post-pic [8]. Certes, la ductilité d'un élément représente sa capacité à se déformer après la phase élastique. Par souci de simplification, les règlements préfèrent utiliser la notion de déplacement, pour définir l'indice de ductilité comme étant le rapport du déplacement ultime (Δ_u) sur le déplacement limite du comportement élastique (Δ_y). L'indice de ductilité est défini comme suit :

$$\mu_\Delta = \frac{\Delta_u}{\Delta_y} \qquad (1)$$

Les valeurs de l'indice de ductilité (μ_Δ) sont représentées sur la figure 10.

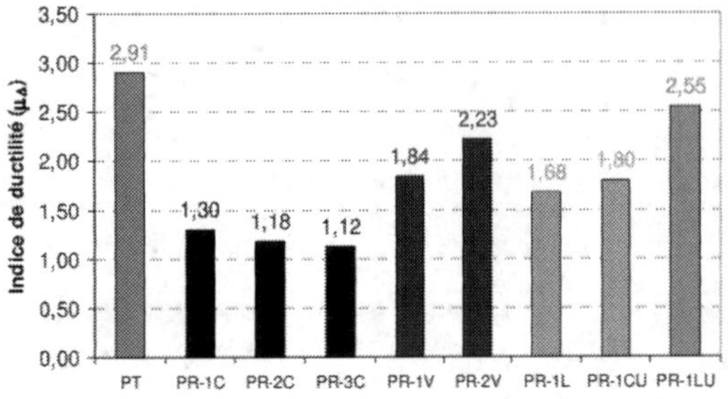

Fig.10. Indices de ductilité des poutres testées.

A partir de la figure 10, on constate que les valeurs de l'indice de ductilité des poutres réparées en cisaillement sont supérieures aux valeurs des poutres réparées en flexion, à l'exception pour les poutres réparées par la fibre de verre où l'indice de ductilité est de 1,84 pour PR-1V et 2,23 pour PR-2V. Ceci prouve que la réparation par plaque carbodur en forme U rend la poutre ductile. Par contre, la réparation en flexion mène à une rupture prématurée, et donc à une rupture fragile (indice de ductilité faible)

3.4 Validation des méthodes de dimensionnement

Le dimensionnement d'une poutre réparée par des matériaux composites à l'aide des méthodes analytiques comprend le calcul du moment fléchissant ultime. Le calcul du moment ultime repose sur les hypothèses conventionnelles du béton armé en tenant compte du comportement élastique linéaire du composite. Ces méthodes sont utilisées sans coefficient de sécurité, afin de comparer les résultats obtenus aux résultats expérimentaux.

En utilisant le modèle de calcul à l'ELU de la méthode FIB [1], nous estimons la résistance ultime au moment fléchissant d'une poutre réparée par des matériaux composites.

Etant donnée la section transversale d'une poutre réparée (figure 11).

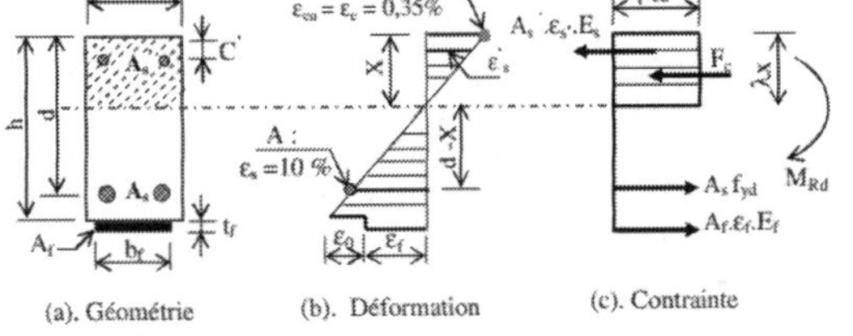

Fig.11. Section transversale d'une poutre réparée.

Le moment fléchissant résistant (M_R) est calculée au centre de poutre par:

$$M_R = A_s.f_{yd}\left(d-\frac{\lambda.x}{2}\right)+A_f.E_f.\varepsilon_f\left(h-\frac{\lambda.x}{2}\right) \qquad (2)$$

$$\varepsilon_f = \frac{h-x}{x}\varepsilon_{cu} \text{ (déformation du composite)} \qquad (3)$$

Les symboles définis pour expliquer les méthodes de base des 3 règlements sont utilisés (figure 12).

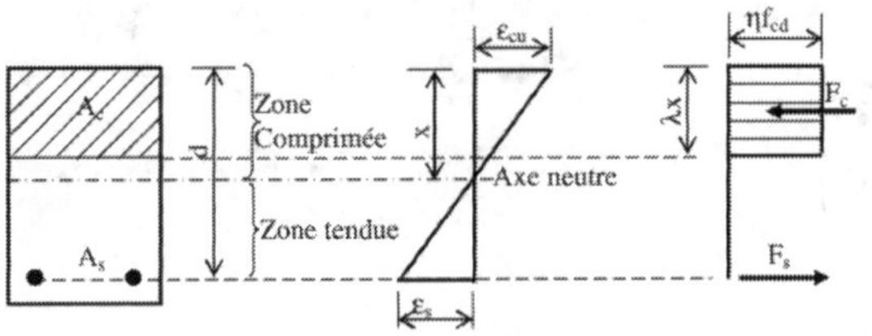

EC2: $0,7 \le \lambda = [0,8-(f_{ck}-50)/400] \le 0,8$

$0,8 \le \eta = [1,0-(f_{ck}-50)/200] \le 1,0$

$f_{cd} = f_{ck}/\gamma_c$

BAEL: $0,7 \le \lambda = [1,0-0,7/(4,5-0,025 f_{ck})] \le 0,8$

$\eta = 1,0$

$f_{cd} = 0,85 f_{ck}/(\theta\gamma_c)$

ACI: $0,65 \le \lambda = (1,05 - f_{ck}/140) \le 0,85$

$\eta = 0,85$

$f_{cd} = f_{ck}$

Fig.12. Diagramme de déformation et diagramme simplifié des forces équivalentes (valeurs λ, η et f_{cd} pour l'EC2, le BAEL et l'ACI318).

Le tableau 3 donne les valeurs du Moment fléchissant résistant prédites par les 3 règlements (BAEL [9], EC2 [3] et ACI318 [10]).

Tableau 3. Moment fléchissant résistant.

Poutre	M_{exp}	Moment calculé			M_{exp}/M_{cal}		
		BAEL	EC2	ACI	BAEL	EC2	ACI
PR-1C	12,1	7,5	7,5	6,7	1,61	1,60	1,81
PR-2C	13,1	11,2	11,3	9,7	1,17	1,16	1,35
PR-3C	14,6	15,6	18,9	15,8	0,94	0,77	0,92
PR-1V	8,6	6,4	6,5	5,8	1,35	1,33	1,47
PR-3V	10,2	9,3	9,5	8,3	1,10	1,08	1,23
PR-1L	13,6	22,2	22,4	18,5	0,61	0,61	0,73
Moyenne					**1,13**	**1,09**	**1,25**
Ecart type					**0,31**	**0,33**	**0,35**
Coef de variation (%)					**27**	**30**	**28**

$M_{exp.}$ = Moment fléchissant expérimental

A l'issue du tableau 3, il est clair que la variation du moment fléchissant entre les valeurs expérimentales et les valeurs prédites par les trois règlements est très faible, avec un coefficient de variation de 28%. Cependant, les trois règlements surestiment le moment fléchissant pour la poutre PR-1L, ceci est principalement lié aux propriétés du composite (module d'élasticité du composite (lamelle carbodur + résine) (E_f) et son épaisseur). En outre, le mode de rupture de cette poutre réparée en flexion est provoqué par flexion et par l'effort tranchant.

4 Conclusions

Les conclusions tirées de cette étude sont les suivantes:

- Les poutres réparées en flexion ont obtenu des gains de résistance, allant de 50 % à 90 %, par rapport à celle de la poutre témoin, et une diminution de la flèche variant de 25 à 45%.

- Les poutres réparées: RP-1V et RP-2V ont obtenu des gains de résistance par rapport à la poutre témoin respectivement de 10% et 31 %. Cependant, ces poutres possèdent un pseudo-palier plastique de ductilité une fois la résistance ultime atteinte. En effet, l'indice de ductilité était de 1,84 pour la poutre RP-1V et 2,23 pour la poutre RP-2V.

- L'ajout d'une plaque carbodur sur la face tendue de la poutre (PR-1L) contribue à augmenter la rigidité de la poutre et à diminuer la flèche. Cependant, on note un gain de charge à la rupture de 87 % par rapport à la poutre de référence.

- L'utilisation d'étriers de fibres de carbone et lamelle carbodur en forme de U permet de réparer et rigidifier la poutre. Un gain de résistance significatif de 120 % a été enregistré pour la poutre PR-1CU, alors qu'un gain était de 177 % pour la poutre PR-1LU. Par ailleurs, l'indice de ductilité était de 2,55 pour la poutre PR-1LU et 1,80 pour la poutre PR-1CU.

- Les poutres réparées en flexion ont été rompues par décollement du composite ou rupture de l'enrobage de béton, à l'exception pour les poutres réparées par la fibre de verre, ces deux poutres (PR-1V et PR-2V) aient permis d'atteindre une rupture ductile en flexion.

- Les trois règlements (BAEL99, EC2-04 et ACI318-08) sont acceptables vis-à-vis du moment résistant ultime des poutres réparées par des matériaux composites, avec un coefficient de variation de 28%.

Références

1. FIB, Externally bonded FRP reinforcement for RC structures. Technical report, Bulletin 14, ISBN 2-88394-055-X, (2001).
2. AFGC, Réparation et renforcement des structures en béton au moyen des matériaux composites, Recommandations provisoires, (2003).
3. Eurocode 2, Calcul des structures en béton -Partie 1-1: Règles générales et règles pour les bâtiments, ENV 1992-1-1, (2004).
4. Guide des Notices Techniques, web:dza.sika.com, Sika El Djazair.
5. MEIER, Carbon Fiber-Reinforced Polymers: Modem Materials in Bridge Engineering, Structural Engineering International, n°. 1, (1992), pp.7- 12.
6. M. Hamrat, Comportement structural du béton à hautes performances (flexion et effort tranchant), Thèse de doctorat, Université USTHB-Algérie/Université européenne de Bretagne-France, (2010).
7. M. Hamrat, B. Boulekbache, M. Chemrouk. S. Amziane, Flexural cracking behavior of normal strength, high strength and high strength fiber concrete beams, using Digital Image Correlation technique, Construction and Building Materials 106, (2016), pp. 678-692.
8. M. Hamrat, B. Boulekbache, M. Chemrouk. S. Amziane, Flexural, Behavior of reinforced concrete beams: Comparative analysis between high-performance concrete and ordinary concrete, MATEC Web of Conferences 11, (2014), 02003.
9. BAEL, règle technique de conception et de calcul des ouvrages et constructions en béton armé suivant la méthode des états limites, (1999).
10. ACI Committee 318, Building Code Requirements for Structural Concrete (ACI 318-08) and Commentary (318R-05), American Concrete Institute, Farmington Hills, MI, (2008), 430 p.

Study of clay soils swelling by the new method based on Laser Interferometry and the classical Odometer test

Etude du gonflement des sols argileux par la nouvelle méthode basée sur l'Interférométrie Laser et l'essai classique par l'Odomètre

Hassan ASSELMAN[1]andAouatefDAOUIDI[1]

[1]AbdelmalekEssaadiUniversity, Science Faculty, Optics and Photonics Laboratory, Tetouan, Morocco.

Abstract. Geotechnical engineering participates in the act of construction, which means that it must meet a double concern for securityand economy. Therefore, an essential part of the engineer's responsibility rests on the recognition of soils in order to determine their nature and properties (taking into account the flow of water). In the present work, we measure the swelling and the permeability by detection of the swelling, by our new optical method based on the interferometry-laser, invented by Hassan Asselman, within the team of Optics and photonics of sciences faculty, Tetouan- Morocco. This new prototype allows us to directly measure the following parameters: the permeability k (m / s), the Young module Eand the swelling index Cs. For the latter parameter, the evolution of the strain as a function of the stresses σ (Pa) is measured for a given degree of saturation (Until saturation). Moreover, we will use the classical odometer test, which reproduces the conditions of deformation of the soils. Using the results of the latter by the graphic methods of Taylor and Gasagrande, it is possible to determine the value of the coefficient of consolidation of the soil Cv. According to the Darcy theoretical modelfor a saturated medium, Cv depends on the permeability, the compressibility coefficient mv (or the inverse of the model of Young odometric) and the voluminal weight of the water γw. These tests will be carried out at the GEORET Geotechnical Laboratory in Tetouan. To perform this work, we chose a sample of claydistrubed, already characterized by X-ray diffraction (whose clay fraction is illite). It is extracted from the so-called "Teffalin" quarry of the Tetouan region, used in the manufacture of pottery. Finally we give a comparison between our new patented method and the classic Odometric test

Résumé. L'ingénierie géotechnique participe à l'acte de construction, ce qui signifie qu'elle doit répondre à un double souci de sécurité et d'économie. Par conséquent, une part essentielle de la responsabilité de l'ingénieur repose sur la reconnaissance des sols afin de déterminer leurs natures et leurs propriétés (en prenant en compte l'écoulement de l'eau). Dans le présent travail, nous mesurons le gonflement et la perméabilité par détection du gonflement, par notre nouvelle méthode optique basée sur l'interférométrie-laser, inventée par Hassan Asselman, au sein de l'équipe d'Optique et photonique à la faculté des sciences, Tétouan-Maroc. Ce nouveau prototype nous permet de mesurer directement les paramètres suivants : la perméabilité k (m/s), le module d'Young E et l'indice de gonflement Cs. Concernant ce dernier paramètre, on mesure l'évolution de la déformation en fonction des contraintes σ (Pa) pour un degré de saturation donné (Jusqu'à atteindre la saturation). Par ailleurs, nous comptons utiliser l'essai classique œdométrique, qui reproduit les conditions de déformation des sols. L'exploitation des résultats de ce dernier par les méthodes graphiques de Taylor et Gasagrande, permettent de déterminer la valeur du coefficient de consolidation du sol Cv. D'après le modèle théorique de Darcy pour un milieu saturé, Cv dépend de la perméabilité, du coefficient de compressibilité mv (ou l'inverse du modèle de Young œdométrique) et du poids volumique de l'eau γw. Ces essais seront effectués au Laboratoire de Géotechnique GEORET à Tétouan. Pour réaliser ce travail, nous avons choisi un échantillon d'argile remanié, déjà caractérisé par diffraction des RX (dont la fraction argileuse est l'illite). Il est extrait de la carrière dite «Téffaline», de la région de Tétouan, utilisé dans la fabrication de la poterie.Enfin nous donnons une comparaison entre notre nouvelle méthode brevetée et l'essai classique Odométrique

1 Introduction

En Génie Civil, les phénomènes de retrait-gonflement de certains sols argileux et des formations géologiques argileuses induisent des tassements différentiels qui se manifestent par des désordres affectant principalement les structures construites en surface (bâtiment, fondation superficielle, ouvrage de soutènement, remblais,...) et les ouvrages enterrés (tunnels, pieux, canalisation, fondation profondes,....) [1].
La détermination des comportements de gonflement et de retrait des matériaux peu perméables, dans notre cas les sols argileux, est une étape incontournable dans la conception de nombreux ouvrages de génie civil.
Pour avoir une vision claire sur ces phénomènes plusieurs paramètres peuvent être déterminés in situ ou en laboratoire. Dans cette deuxième situation, il existe une variété de tests et de méthodes visant à évaluer les propriétés hydrauliques [1]. Dans le présent travail les essais ont été réalisés moyennant la nouvelle méthode inventée dans notre laboratoire [2, 3], par Hassan Asselman[4, 5], qui permet de mesurer : le potentiel du gonflement (ou gonflement libre), le gonflement sous une contrainte et la perméabilité des sols argileux. Ce nouveau prototype possède des qualités, des atouts et des champs d'application très larges qui le rendent très attractif, facile à mettre en œuvre et à commercialisation dans un futur proche. Le RROB (Rapport de Recherche avec Opinion sur le Brevet) élaboré par l'OMPIC indique que le brevet d'invention N°1.586.243 du Français Bernard CORVAISIER déposé le 1er Octobre 1968 sous le titre : « Appareillage pour étudier le gonflement d'un échantillon de sol » (Odomètre), déposant : Etat Français, représenté par le ministre de l'équipement et du logement et laboratoire central des ponts et chaussées, résident en France (Paris) et qui est utilisé par les bureaux d'études du monde entier pour l'analyse des sols Argileux, s'avère efficace, mais très fastidieux et lent. Depuis cette date (1968) il n'y a pas eu d'évolution ! D'où l'énorme importance de notre prototype [3, 4], par son gain de temps incroyable (de quelques heures au lieu de quelques semaines) et sa précision. La conclusion du rapport de préfaisabilité N°LLRA077-12 Moubtakir [6] réalisé par CERIMME (Centre d'étude et de Recherche des Industries Métallurgiques, Mécaniques, Electriques et Electroniques) et financé par l'OMPIC souligne que ce nouveau prototype est techniquement facile à mettre en œuvre et sa rentabilité va dépendre de son prix de revient et de la démarche marketing qui va accompagner sa commercialisation. [7]
Les résultats présentés, s'appuient sur des essais de gonflement à charge variable dans le cas d'un milieu non saturé. Ils ont été réalisés à l'aide d'un système optique et mécanique utilisant le bras de levier.
Rappelons que le phénomène de gonflement est une augmentation de volume du matériau, quelle qu'en soit la cause. Un déchargement mécanique, une imbibition ou une sorption, une exfoliation ou des variations chimiques qui induisent une déformation positive, peuvent être la cause d'un gonflement.
Les analyses minéralogiques montrent que certaines argiles, pour lesquelles les liaisons inter feuillets sont très faibles, ont la propriété de fixer les molécules d'eau entre deux feuillets voisins. L'eau pénètre à l'intérieur des particules et s'organise en couches mono moléculaires, il s'agit alors d'un gonflement intra particulaire ou inter foliaire [1].
Elle est à noter que le phénomène de retrait consiste en une diminution du volume du sol due à l'évaporation de l'eau présente dans ses pores. L'assèchement progressif du sol provoque des tensions capillaires et la formation de ménisques autour des particules du sol. Les tensions capillaires sont à l'origine de contraintes de cisaillement. Plus l'eau s'évapore, plus les tensions capillaires augmentent et plus le rayon des pores diminuent (équation Jurin-Laplace). Parallèlement, la résistance à la traction du sol augmente. Quand cette résistance atteint les contraintes engendrées par les tensions capillaires, le retrait s'arrête, le volume du sol ne diminue plus.
La teneur en eau pour laquelle le retrait cesse correspond à la limite de retrait [1].

2 Matériel et méthode

2.1 Dispositif pour mesurer le gonflement de sols argileux, inventée par Hassan Asselman

Le dispositif expérimental que nous avons utilisé est une adaptation de l'interféromètre de Michelson caractérisé par divers application [8].
Enfin l'apparition de la nouvelle et originale application de cet interféromètre, permettant d'étudier le phénomène de gonflement-retrait ainsi que la perméabilité [3, 4, 7] d'un échantillon de sol argileux. Cette méthode a été présentée pour la première fois dans [9].
Le schéma du dispositif expérimental est représenté sur la figure 1. Il comprend [3] :
Un interféromètre de Michelson
Un laser He-Ne de longueur d'onde λ=632,8nm et de puissance 1mW
Une lentille de courte distance focale (2cm) pour observer les franges d'interférences
Une éprouvette (gouttière) en verre placée verticalement au-dessus de l'échantillon pour contrôler la teneur en eau (distillée).
Un système de détection, constitué d'une caméra reliée à l'ordinateur, un logiciel permettant de suivre au cours du temps, le défilement des franges d'interférences et leur comptage N. Cela nous permet de déterminer la déformation (gonflement).
Pour réaliser ce travail, nous avons choisi un échantillon d'argile remanié, déjà caractérisé par diffraction des RX (dont la fraction argileuse est l'illite) [4]. Il est extrait de la carrière dite «Téffaline», de la région de Tétouan, utilisé dans la fabrication de la poterie. L'échantillon a la forme d'un parallélépipède, séché, de longueur L=73.00 mm, de largeur l=18.00 mm, de hauteur H=6.00 mm et de masse m=20.10g. Il est façonné de manière à retenir les gouttes d'eau à sa surface pour contrôler la teneur en eau (Figure 2)[3]. La fraction de l'échantillon humidifiée de longueur L_H = 53 mm, de largeur l_H = 3 mm et de hauteur H_H = 3 mm.

Pour éviter les vibrations mécaniques, nous avons choisi d'utiliser le bras de levier de l'interféromètre en contact direct avec l'échantillon et à travers lequel on exerce les différentes contraintes σ(Kpa).
L'échantillon est placé sur un support horizontal, perpendiculairement au bras de levier et encastré à l'autre extrémité sur un support vertical fixe. La longueur du bras de levier est de 80 mm.Le miroir mobile est situé à 8 mm de son axe de rotation. Le point de contact de l'échantillon avec ce bras est situé à 48 mm. Nous avons donc une réduction de ce bras de R=1/6. Par conséquent la variation de longueur effective de l'échantillon suite à sa déformation s'écrit :

$$\Delta L = \frac{N\lambda}{2\,R} \qquad (1)$$

N est le nombre de franges d'interférences défilées de même nature.
Pour exercer des contraintes sur l'échantillon, on utilise des masses marquées de quelques centaines de grammes. Elles sont accrochées à in fil horizontal, inélastique et entourant une poulie. Le fil est attaché au bras de levier à 80 mm de son l'axe de rotation du bras de levier.

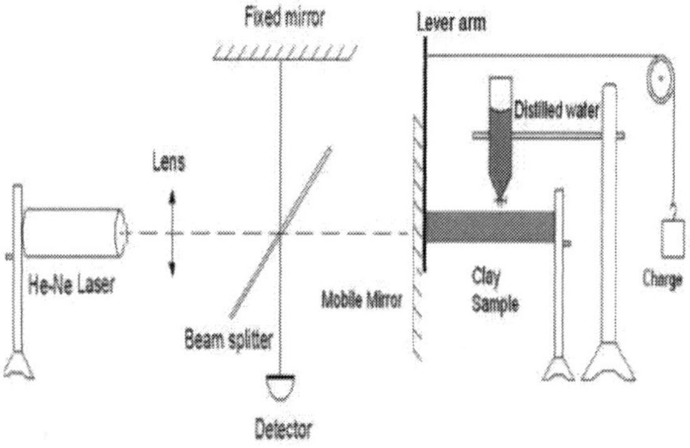

Figure 1: Schéma du dispositifexperimental

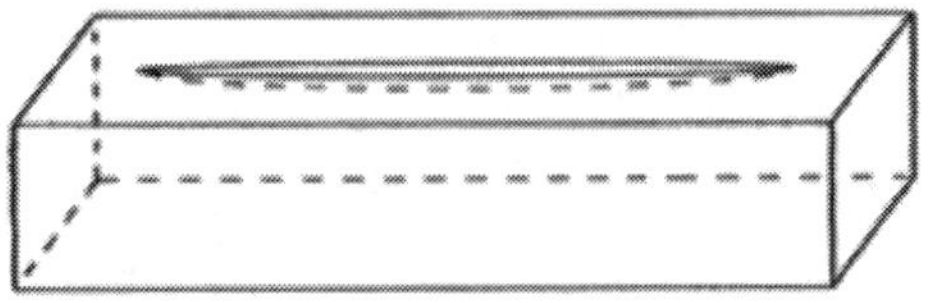

Figure 2: Schéma de l'échantillon argileux remanié

2.2 Procédure de l'expérimentation

La teneur en eau apparente de la fraction humidifiée de l'échantillon est estimée à 62% pour un volume d'eau Ve=1ml. A la fin du processus du gonflement et après avoir effectué la pesée, environs 50% de cette eau s'est évaporé. La teneur en eau effective est de l'ordre de 32% [3].
Lorsqu'on injecte le volume Ve= 1ml, à la surface d'un échantillon de sol argileux, celui-ci se gonfle et le miroir se déplace. On observe le défilement des franges d'interférences et on procède au comptage de leur nombre N durant le temps t que dure le gonflement (environ quelques dizaine de minutes). On peut déterminer le gonflement sous une contrainte constante σ (kPa)..
La contrainte totale (ressort du bras de levier et les masse marquées) appliquée sur la surface de l'échantillon est donnée par :

$$\sigma = \frac{mg}{S}\frac{5}{3} + \sigma_i \qquad (2)$$

σ i = (4.63± 0.02) kPa, est la contrainte exercée par le ressort du bras de levier. Elle maintient en permanence le contact entre le bras et l'échantillon, ce qui permet la détection du gonflement. D'où l'importance de l'utilisation de ce bras.
Nous avons choisi de travailler avec les contraintes (12,35± 0.04) kPa, (20.06± 0.07) kPa et une teneur en eau fixe. Pour chaque valeur deσ nous étudions la cinétique, c'est-à-dire l'évolution du gonflement sous pression au cours du logarithme du temps en seconde [11].
La surface S de l'échantillon subissant la contrainte horizontale est de 108 mm2. Le choix d'une petite surface est dans le but d'atteindre les pressions exercées dans le cas d'un essai odométrique (environ100 KPa) pour une éventuelle comparaison avec cette méthode classique.

Le gonflement sous pression (%) est donnée par :

$$\frac{\Delta L}{L_H} = \frac{N\lambda}{2L_H R}(\%) \qquad (3)$$

L'indice de gonflement est donné par la relation suivante [11]:

$$C_S = \frac{\Delta e}{\Delta Log(\sigma)} \quad \% \quad (4)$$

Δe : est la variation de l'indice des vides de l'échantillon

L'indice des vides e s'écrit :

$$e = e_i + \frac{\Delta L}{L_S} \qquad (5)$$

e_i = 0,66 (indice des vide initial), mesuré par le pycnomètre et l'imprégnation à sec.
$\Delta L = L_0 - L$ (déformation latérale ou gonflement)
L_0 est la longueur initiale de l'échantillon séché
L_s= 0,032m (la longueur du solide)
On utilise le poids volumique du solide γ_s= 26,5 KN/m^3

3 Résultats et discussions

3.1 Gonflement-retrait sous la contrainte σ (kPa)

3.1.1 Teneur en eau fixée (Ve=1ml)

La figure 3, représente nos résultats expérimentaux, concernant l'évolution du gonflement et retrait au cours du temps, pour les contraintes 12,35 kPa et 20,35 kPa avec Ve= 1 ml, T=20°C, Humidité h=90%

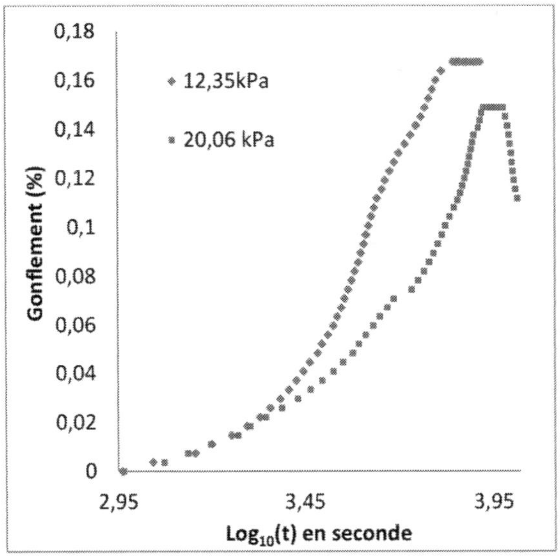

Figure 3: Cinétique du gonflement (%) en fonction $\log_{10}(t)$ (s), pour σ =12 ,35 et σ =20,06 kPa

De nombreux travaux expérimentaux font apparaître ce type de cinétique. Il se traduit par un gonflement primaire et un gonflement secondaire [10].

L'allure de la courbe obtenue relative à σ=20,06 kPa est en bon accord avec celle rencontrée dans la littérature [10, 12]. Elle atteint la stabilisation au bout de 140 min environ. La partie constante de la courbe où le défilement des franges s'arrête, dure environ 18 min. Au-delà des 158 min le phénomène de retrait commence. On observe alors le défilement des franges d'interférences dans le sens contraire.

Sachant que le processus de retrait est très lent. Il dépend des conditions du milieu ambiant (température et humidité), nous avons été amenés à chauffer notre laboratoire à une température voisine de 32°C pour accélérer le processus du retrait et le mettre en évidence. Le retrait complet jusqu'à stabilisation peut être observé à une température ambiante mais cela nécessiterait de laisser le laser allumé pendant quelques heures. Cela permettrait de voir si le processus est réversible.

En ce qui concerne l'effet de la contrainte (KPa) sur le sol argileux humidifie en phase de gonflement. En analysant les deux courbes, nous relevons les observations suivantes :

- Nous constatons une superposition des deux courbes durant la première phase (environ 62min). Le gonflement observé est donc indépendant de la contrainte σ exercée.
- Durant la deuxième phase un décalage temporel Δt (sec) est observé pour chaque valeur de N. Ceci est dû à l'influence de la contrainte σ. En effet la contrainte appliquée provoquerait le ralentissement du processus de pénétration des molécules d'eau entre les feuillets d'argile. La valeur maximale de cet écart Δt est atteinte pour N=32 (Figure 4).

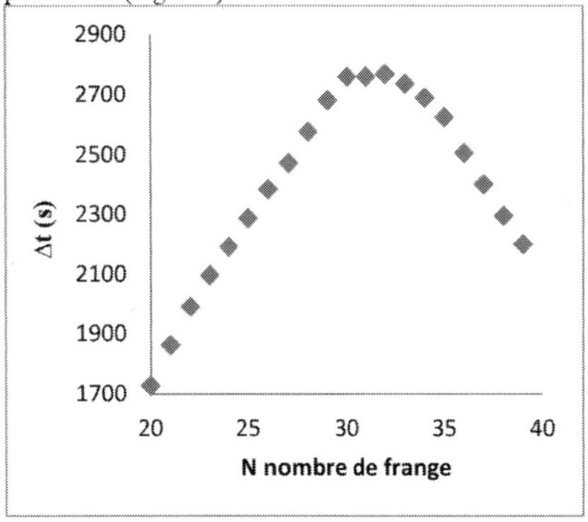

Figure 4 : L'écart temporel Δt (sec) entre les deux courbes en fonction de N

- Lorsque la contrainte augmente, le nombre total des franges défilées diminue sensiblement (tableau 1).

Ce résultat est en bon accord avec la littérature [11].

Le tableau 1 résume nos résultats obtenus.

Tableau 1. Gonflement et sa duration en fonction des contraintes σappliquées

σ(kPa)	Gonflement (%)	Indice des vides : e	Durée du gonflement (minutes)
4,63(0Gr)	-	0,6632681	
12,35 (50Gr)	0,167±0,004	0,66266963	115.
20.06 (100Gr)	0,149±0,004	0,662373	140

La figure 5, représente la variation de l'indice des vides en fonction du Logarithme décimal des contraintes σ.

Nous obtenons la valeur de l'indice de Gonflement C_s= 0,0014 (au voisinage de la saturation), en un temps record d'environs 4 heures (au lieu de quelques semaines par la méthode classique Odométrique). Ce résultat est significatif puisque l'Illite est une argile non gonflante (la distance inter- foliaire d=7Å). Le coefficient de corrélation de la droite R^2= 1.

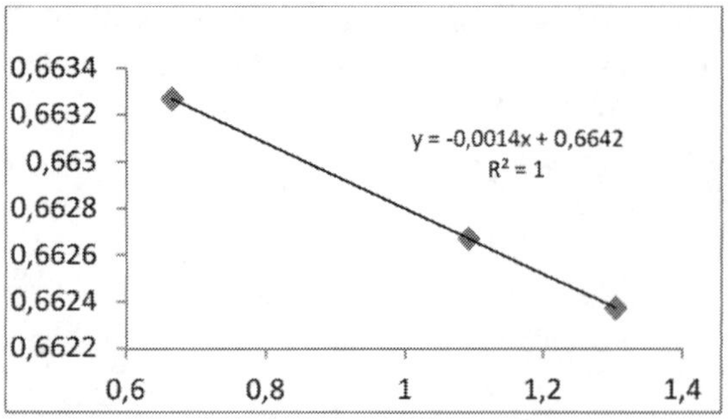

Figure 5 : Indice des vides en fonction de Log10 des Contraintes (kPa)

3.1.2 Teneur en eau fixée (Ve= 0,5ml)

La figure 6, représente nos résultats expérimentaux, concernant l'évolution du gonflement au cours du logarithme décimal du temps en secondes, pour les contraintes 53,29, 58,2 et 66,61 kPa avec Ve= 0,5 ml, T=20°C, Humidité h=90%

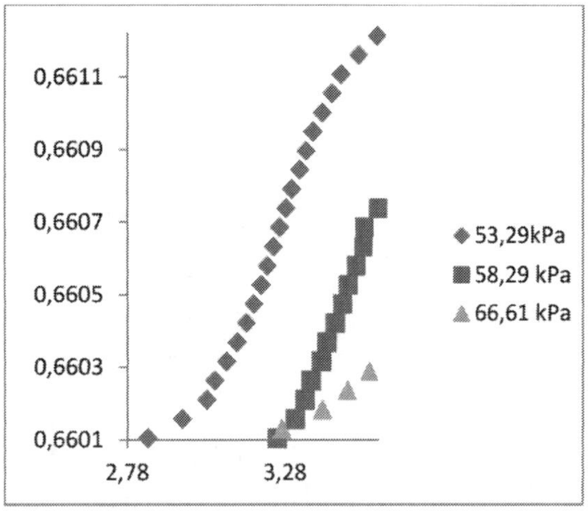

Figure 7 : Indice des vides en fonction du Logarithme décimal du temps (Ve=0,5ml)

La figure 7, représente la variation de l'indice des vides en fonction du Logarithme décimal des contraintes σ. Nous obtenons la valeur de l'indice de Gonflement C_s= 0,015. Le coefficient de la droite est de 0,97. On constate que lorsque la teneur en eau diminue l'indice de gonflement C_s augmente de façon significative ;

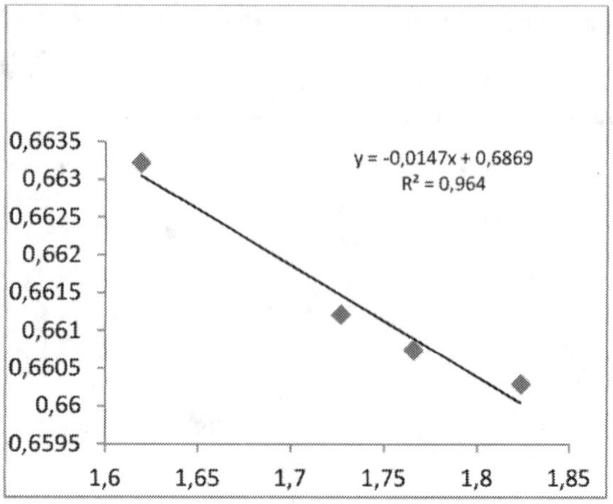

Figure 6 : Indice des vides en fonction de Log10 des Contraintes (kPa)

3.2 Comparaison entre notre nouvelle méthode et la méthode classique

Pour pouvoir faire cette étude comparative importante, il nous a été nécessaire de maitriser la méthode classique sur tous les aspects : procédé de mesure et exploitation des résultats (graphique, signification des paramètres ....)

3.2.1 Exemple d'un essai classique odométrtrique (milieu saturé.

Le bureau d'études géotechniques (GEORET), nous a fourni les résultats d'un essai œdométrique d'une montmorillonite, prélevée à l'ENSA de Tétouan. On obtient les résultats suivants:

C_c= 0,13 (Selon la norme on a un sol moyennant compressible)

C_S = 0,04 (Présence d'une tendance au gonflement)

La valeur de la perméabilité K = 2,5 10^{-11} m/s (milieu saturé) est très faible.

D'après le modèle théorique de Darcy, la Perméabilité d'un milieu saturé s'écrit:

$$k(m/s) = C_v m_v \gamma_w \qquad (6)$$

m_v : Coefficient de compressibilité (ou l'inverse du modèle de Young œdométrique)

C_v : Coefficient de Consolidation

γ_w :Poids volumique de l'eau.

Le résultat de l'essai classique a été obtenu au bout de trente jours!

3.2.2 Perméabilité d'une montmorillonite par la nouvelle méthode (milieu non saturé)

Récemment nous avons mesuré la perméabilité d'une montmorillonite [7]. On a obtenu le résultat suivant: K =3,33 10^{-4} m/s.

Ce résultat a été obtenu au bout d'une heure. Nous avons travaillé avec un échantillon remanié, séché à l'étuve à 105 °C pendant 24 heures.

Dans un milieu non saturé, la Perméabilité est donnée par l'équation Brooks and Corey [13]:

$$K = K_s \left(\frac{\theta - \theta_0}{\theta_s - \theta_0}\right)^{\alpha} \qquad (7)$$

Nous mis en évidence par notre méthode expérimentale, que la perméabilité augmente avec l'augmentation de la teneur en eau. Ceci est en bon accord avec l'équation (7). Lorsqu'on s'approche de la saturation, il se produit une chute brutale de la valeur de K d'un facteur d'environs 10^{-7}

3.2.3 Tableau comparatif entre les deux méthodes expérimentales

La comparaison entre les deux méthodes est indiquée sur le tableau 2

Tableeau 2. Comparaison entre les deux méthodes.

Méthode de mesure	Odomètre	Interférométrie laser
Milieu	Saturé	Non saturé ou presque saturé
Lieu de mesure	Laboratoire	Laboratoireou In-situ
Indice de compression Cc	Oui	Non
Indice de gonflement Cs	Oui	Oui
Perméabilité K(m/s)	Milieu saturé	Non saturé ou presque saturé
Durée de mesure	Quelquessemaines	Maximum quatreheures
Précision	-	3%

On voit clairement que notre méthode possède des avantages surtout le record en temps d'obtention des mesures ainsi que la précision ;

Conclusion

Ce travail nous a permis de mesurer le gonflement-retrait, d'un sol argileux de façon rapide et simple, grâce à l'utilisation de notre nouvelle méthode basé sur l'interférométrie laser. Nous avons réussi à coupler les deux systèmes optique et mécanique, d'où l'originalité de notre travail. On a pu mettre en évidence artificiellement le phénomène de retrait. De plus nous avons observé l'effet de la contrainte σ (kPa) sur un échantillon d'argile humidifié (teneur en eau fixée), s'opposant au phénomène du gonflement de l'échantillon. En effet lorsque la pression sur l'échantillon augmente le nombre de franges défilées diminue et sa durée augmente sensiblement. Ce résultat nous a permis la détermination de l'indice de gonflement C_s dans le cas de l'Illite en un temps record. Ce dernier paramètre dépend de la teneur en eau. D'où l'originalité de notre travail.

Enfin nous donnons le tableau comparatif entre les deux méthodes expérimentales.
Comme perspectives nous souhaitons normaliser notre méthode brevetée en mesurant l'indice de gonflement pour quelques échantillons déjà étudiés par la méthode classique et faire la comparaison. Pour mieux comprendre le phénomène étudié, gonflement-retrait des sols argileux et vu la complexité de ce milieu poreux (phases: liquide, solide et gazeux), nous comptons entamer une étude théorique, adaptée à notre original méthode expérimentale, en se basant sur la méthode des éléments finis (MEF), qui sera validée par nos mesures.

References

1. H. Nowamooz, Sciences de l'ingénieur [physics], *Institut National Polytechnique de Lorraine - INPL, Français*, (2007).
2. H. Asselman, Device for measuring the infiltration velocity the water of deep clay soil, Patent N° **35556**, in OPMIC (*Moroccan Office of Industrial and Commercial Property*), Casablanca-Morocco, (2013).
3. H. Asselman, New optical method to measure the swelling of clay soils by laser interferometry, Patent N° **35229**, in OPMIC (*Moroccan Office of Industrial and Commercial Property*), Casablanca-Morocco, (2012).
4. H. Asselman, A. Daouidi, K. Draoui , et A. Asselman, *MATEC Web of Conferences*, Vol. **11**, (2014).
5. H. Asselman , A. Bakkali, M. Ajdour, and M. Essaaaidi, *MATEC Web of Conferences* **2**, (2012).
6. CERIMME (*Center of Study and Research of Metallurgical Industries, Mechanical, Electrical and Electronic*), Pre-feasibility study, Report N°LLRA**077-12** Moubtakir, Casablanca-Morocco, (2012).
7. A. Daouidi, H. Asselman, *J. Civil Eng. Architect. Res*, Vol. **3** No. 1, (2016).
8. H. Asselman, M. El Addad, *Congers COMPLES'2K7*. ENSA Agadir-Morocco (2007).
9. H. Asselman, *The Second day clays and Sustainable Development*, Faculty of Science, Tetouan-Morocco, (2010).
10. A. Medjnoun, Doctoral Thesis, *Mouloud Mammeri-TiziOuzouUniversity*, (2014).
11. J.-P.Magnan, C. Mieussens, B. Soyez, J. Vautrain. Essais oedométriques. *Ifstar*, (1985).
12. Vizier, J.F., Cah. O.R.S.T.O.M., sér. Pédol., vol. **IX**, n°2, (1971).
13. A. MUSY, Simulation des écoulements en milieu poreux non saturé, *Journal of hydrology, Elsevier scientificpublishingcompany29*, pp 1-21, (1976)

Analysis of some key aspects of soil/foundation interaction- Finite elements modelling

Hamidi Ahmed[1], Bouafia Ali[1]
[1]University Saâd Dahleb at Blida, Faculty of Technology, Dept. Civil. Engineering, Blida, Algeria

Abstract. This paper is aimed to contribute to the analysis of three important aspects of soil-foundation interaction, which are not clearly investigated, by means of a detailed parametric study based on a finite elements modelling.
The first aspect focuses on the effect of the bedrock proximity on the load-settlement behaviour of a continuous or circular shallow foundation. It was found there exists a threshold distance between the foundation base and the top of the bedrock layer beyond which the foundation behaves as in an infinitely deep medium.
The second one deals with the behaviour of shallow foundation on a bi-layered soil where the effect of the underlying layer on the bearing capacity as well as on the settlements depends on the distance between the foundation base and the top of the underlying layer, and beyond a threshold value this effect vanishes.
The third aspect studied was the interference between two strip footings installed on saturated clay. It was shown within a threshold distance between these foundations, an important modification of the foundation behaviour may occur. Finally, the numerical results were fitted and interpreted which allowed suggesting simple practical formulae for shallow foundations design.

1 Introduction

These last four decades were marked by an unprecedented scientific interest in the finite element modelling applied to geotechnical engineering. Earliest work was presented by Duncan and Chang (1972) on the use of a hyperbolic stress-strain formulation in a finite element method for computing the settlements of a model strip footing in sand and a circular footing in a saturated undrained clay. Ever since, availability of several non linear constitutive law-based softwares has made possible to explore soil/structure phenomena besides difficult to access by traditional analytical elastic methods [1].

Although the traditional aspect of shallow foundation design based on the analysis of the bearing capacity as well as on the settlement behaviour, it is recognized some aspects related to the soil/foundation interaction are not yet satisfactorily studied, among which the effect of the bedrock proximity on the load-settlement behaviour, the response of shallow foundations on a bi-layered soil, and the interaction between two closely spaced footings.

Based on a non linear 2D finite element modelling, this paper aims at contributing to understand these three aspects of the soil/foundation interaction as well as at proposing simple formulae for practical purposes.

Due to the limited space allowed to this paper, the main results will be briefly presented.

2 Presentation of the method of analysis

The software CRISP v.5 was used as a computational tool to carry out a detailed parametric study of the load-settlement behaviour of rigid circular foundation as well as rigid strip footing. CRISP is a specialized 2D programme of elastic plastic finite elements analysis offering many computational options. Soil material is assumed in this study following a perfectly elastic-plastic constitutive law obeying the Drucker-Prager's failure criterion and the soil properties are assumed homogeneous in space.

The soil studied is a purely cohesive saturated clayey soil exhibiting an undrained behaviour and submerged by a ground water table at surface. Three consistencies were studied: soft, medium consistent and hard clay.

The foundation is modelled as a rigid reinforced concrete slab with no interface elements between the foundation and the soil.

The FEM model is a plane strain mesh for a strip footing and an axisymmetric mesh for a circular footing. Mesh size is 20Bx20B, B being the width of the strip or the diameter of the circle [1]. Mesh refinement was considered within a zone of 4Bx4B beneath the foundation base whereas the remaining area of the mesh was automatically generated by the software.

The small strain analysis option was considered. Moreover, for nearly incompressible materials like a saturated clay it is possible to avoid an "over-stiff response" of the finite elements by using the technique of integration called the $\bar{\mathbf{B}}$ method [2].

To model the rigid load-settlement behaviour of the foundation, a series of prescribed displacements up to B/10 was assigned at the nodes of the foundation base.

The bearing capacity q_l was conventionally defined as the vertical pressure corresponding to infinite settlement and determined on the basis of the Asaoka's graphical method, whereas the foundation stiffness K_{v0} which is the initial slope of the load-settlement curve was estimated by the Newton's divided differences technique.

3 Foundation on a finite layered soil

3.1. Formulation of the problem

The load-settlement curve parameters K_{v0} and q_l may be formulated according to the following general equation:

$$f(K_{v0}, q_l, D, B, H, E_s, C_u)=0 \quad (1)$$

D, B and H are respectively the foundation embedded length, the width or diameter of foundation and the soil layer thickness beneath the foundation base. E_s, C_u are respectively the elastic modulus and the undrained shear strength. By using the Vashy-Bukingham's theorem of dimensional analysis, this physical equation is transformed into an equivalent dimensionless equation described by (N-k) dimensionless parameters noted π, k being the number of fundamental units in equation (1). Since N=7 and k=3, the equivalent equation may be written as:

$$g(\pi_1, \pi_2, \pi_3, \pi_4) = 0 \quad (2)$$

$\pi_1 = \dfrac{K_{v0}B}{E_s}$ is called factor of settlement influence,

$\pi_2 = \dfrac{q_l}{C_u}$ is the N_c factor,

$\pi_3 = \dfrac{D}{B}$ is the foundation slenderness ratio,

$\pi_4 = \dfrac{H}{B}$ is the normalized layer thickness.

The terms π_1 and π_2 should be independent since the first one corresponds to the small displacements whereas the second one corresponds to the large displacements. Consequently, the equation (2) may by uncoupled to the two following equations [3]:

$$\frac{K_{v0}B}{E_s} = h(\frac{D}{B}, \frac{H}{B}) = I_s \quad (3)$$

$$\frac{q_l}{C_u} = j(\frac{D}{B}, \frac{H}{B}) = N_c \quad (4)$$

3.2. Interpretation of results

3.2.1. Analysis of the N_c factor

In all the cases studied it was found this factor does not depend on the consistency of the clay and slightly increases with D/B. However, sensitive dependency of N_c on H/B, as typically illustrated in figure 1, was noticed in both strip and circular foundations. It was shown the existence of a threshold value noted H_{lim} beyond which the foundation behaves as in an infinitely thick layer. The normalized thickness H_{lim}/B was conventionally determined, as shown in figure 1, as the intercept of the straight lines Δ_1 and Δ_2, which led to the margin of values of 0.7-1.0 depending on D/B for a strip footing with an average value of 0.9, and a margin of 0.67-0.80 with an average value of 0.74 for a circular footing.

By comparison with the Matar-Salençon's method which gives H_{lim}/B equal to 0.79 for a strip footing, this latter fall within the margin found in this study [4].

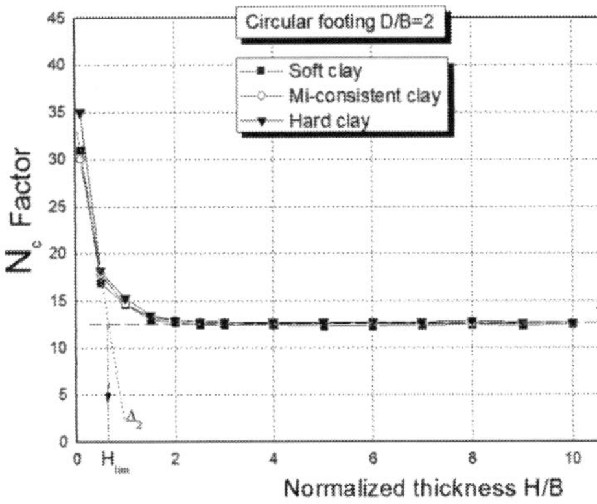

Fig.1. Typical chart giving N_c versus H/B

The function **j** described in equation 4 may be formulated as follows:

$$N_c = \begin{cases} a\left(\dfrac{H}{B}\right)^b & if\ H < H_{lim} \\ N_c^{\infty} & else \end{cases} \quad (5)$$

Coefficients a, b and N_c^{∞} are illustrated in figure 2.

3.2.2. Analysis of the factor of settlement

The factor I_s exhibits similar sensitivity to H/B as N_c, showing a high increase when H/B decreases, which leads to define a threshold normalized thickness (H_{lim}/B) for the settlement beyond which the foundation settles as in an infinitely thick layer, following the same graphical procedure used for N_c. It was found a margin of values of 1.48-1.70 depending on D/B for a strip footing with an average value of 1.6, and a margin of 1.30-1.45 with an average value of 1.40 for a circular footing.

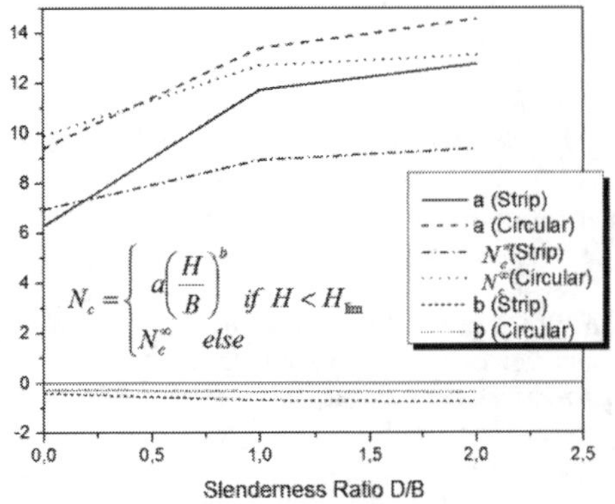

Fig. 2. Chart of coefficients a, b and N_c^∞

Analysis by regression of the values of I_s led to a simple formulation of the function **h** described in equation 3 as follows:

$$I_s = \begin{cases} a\left(\frac{H}{B}\right)^b & \text{if } H < H_{\lim} \\ I_s^\infty & else \end{cases} \quad (6)$$

Figure 3 illustrates a practical chart to compute the coefficients a, b and I_s^∞ in order to use equation (6) and evaluate the stiffness K_{v0}.

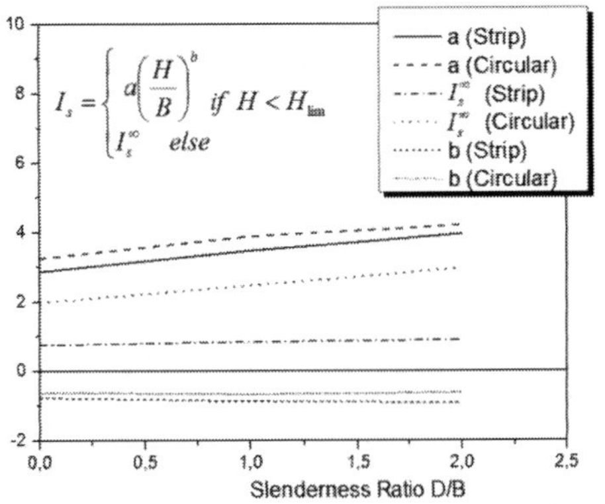

Fig. 3. Chart of coefficients a, b and I_s^∞

4 Foundation on a bi-layered soil

4.1. Formulation of the problem

The general configuration of such a problem is illustrated by the scheme of figure 4. Similarly to the previous part, dimensional analysis led to the following equations:

$$\frac{K_{v0}B}{E_{s1}} = k\left(\frac{D}{B}, \frac{E_{s1}}{E_{s2}}, \frac{H}{B}\right) = I_s \quad (7)$$

$$\frac{q_l}{C_1} = m\left(\frac{D}{B}, \frac{C_1}{C_2}, \frac{H}{B}\right) = N_c \quad (8)$$

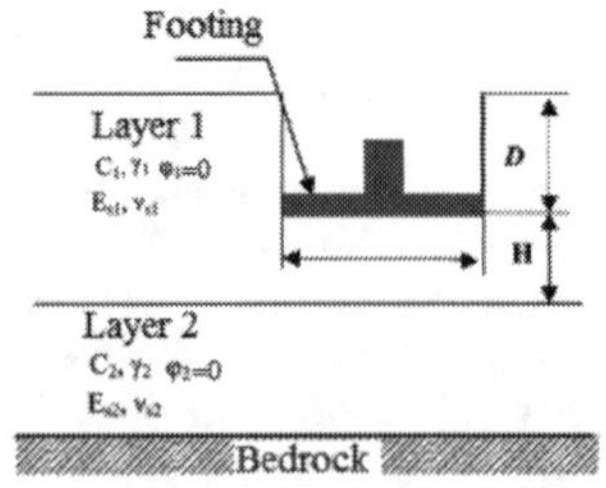

Fig. 4. General scheme of bi-layered soil

4.2. Interpretation of results

4.2.1. Analysis of the N_{c1} factor

It is to be noticed all the cases studied exhibit the same variation of N_c with the normalized distance H/B regardless the shape of foundation and the slenderness ratio. Typical curves illustrated in figure 5 show there exists a threshold normalized distance $H_{\lim}$/B beyond which the effect of the lower layer on the bearing capacity vanishes and the foundation behaves therefore as in a infinitely thick mono-layered soil. This threshold value is about 4 for strip footing and 2.5 for circular footings regardless the ratios C_1/C_2 and D/B.

Figure 5 highlights an important effect of the ratio C_1/C_2 where it can be seen in case of $C_1>C_2$, decrease of H/B results in a decrease in the bearing capacity due to the proximity of the lower layer having a lower undrained shear strength. However, in case of $C_1<C_2$, the bearing capacity increases when H/B decreases.

In case of H< $H_{\lim}$, the effect of proximity of the lower layer should be accounted for. In this regard, the function **m** as described in equation 8 may be written as follows:

$$q_l = f_c N_{c1} C_1 \quad (9)$$

f_c is called the influence factor of the lower layer depending on H/B, D/B and C_1/C_2. It was estimated by least squares regression technique leading to the following formulation:

$$f_c = \begin{cases} a\left(\frac{H}{B}\right)^b & \text{if } H < H_{\lim} \\ 1 & else \end{cases} \quad (10)$$

Values of a and b were found slightly varying with D/B and then tabulated as average values in table 1.

N_{c1} corresponds to the N_c factor of the upper layer considered as infinitely thick and is therefore equal to N_c^∞ illustrated in figure 2.

Table 1. Values of coefficients a and b

Shape	$C_1 < C_2$		$C_1 > C_2$	
Strip	a	1.25	a	0.64
	b	-0.18	b	0.31
Circular	a	1.11	a	0.87
	b	-0.13	b	0.37

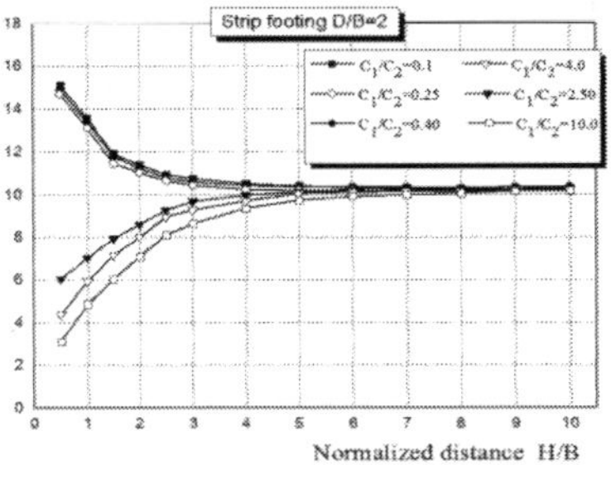

Fig.5. Typical chart giving N_c versus H/B

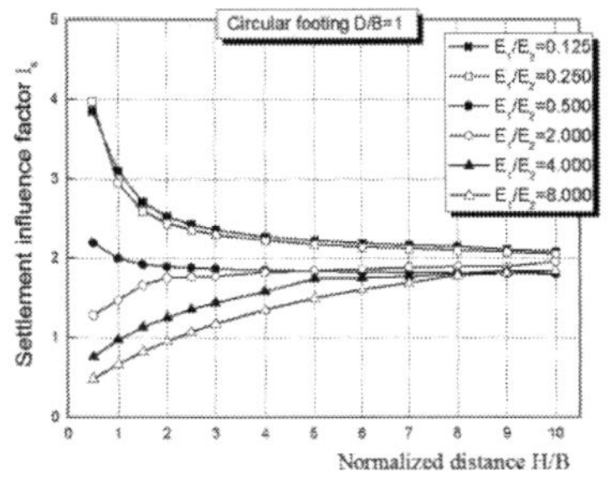

Fig.6. Typical chart giving I_s versus H/B

4.2.2. Analysis of the factor I_s

Similarly to the effect of C_1/C_2 on the bearing capacity, the factor I_s varies sensitively as function of the stiffness ratio E_{s1}/E_{s2}. As shown in figure 6 illustrating a typical variation, in case of $E_{s1}>E_{s2}$ any decrease of H/B results in a decrease of I_s due to the closeness of a looser lower layer. However, in case of $E_{s1}<E_{s2}$, the factor I_s increases when H/B decreases.

It was found the existence of a threshold normalized distance H_{lim}/B regardless the ratios D/B, E_{s1}/E_{s2} and the shape of foundation, beyond which the foundation settles as in an infinitely thick mono-layered soil. This threshold value is about 6.

The function **k** described in equation (7) may be rearranged as follows:

$$K_{v0} = k\frac{E_{s1}}{B} \tag{11}$$

In case of $H<H_{lim}$, the results were fitted by simple power functions and the function **k** may then be formulated as follows:

$$I_s = \begin{cases} a\left(\frac{H}{B}\right)^b & if\ H < H_{lim} \\ I_s^{\infty} & else \end{cases} \tag{12}$$

Values of a and b were found slightly varying with D/B and then tabulated as average values in table 2.

Table 2. Values of coefficients a and b

Shape	$E_{s1}< E_{s2}$		$E_{s1}>E_{s2}$	
Strip	a	1.83	a	0.22
	b	-0.35	b	0.26
Circular	a	2.68	a	1.14
	b	-0.21	b	0.29

5 Interaction of a two closely spaced Identical strip footings

5.1. Formulation of the problem

As illustrated in figure 7, the general configuration of such a problem leads by dimensional analysis to the following equations:

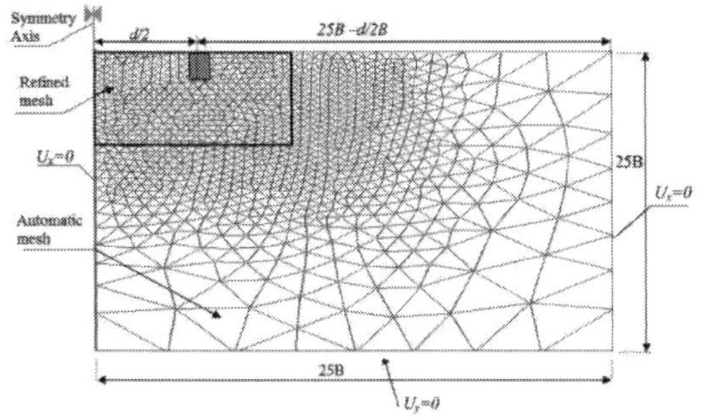

Fig. 7. Typical FEM mesh of two adjacent strip footings

$$\frac{K_{v0}B}{E_s} = n(\frac{D}{B},\frac{d}{B}) = I_s \tag{13}$$

$$\frac{q_l}{C_u} = p(\frac{D}{B},\frac{d}{B}) = N_c \tag{14}$$

d is the horizontal distance between the vertical axes of two identical strip footings embedded at the same depth D. The footings/soil system is identical to that studied in part 3 except for its dimensions which were fixed by gradual increase until the settlement at foundation base nodes start to stabilize. Due to the vertical symmetry only half a plane was modelled.

5.2. Interpretation of results

5.2.1. Analysis of the N_c factor

It was found the consistency of the clay has almost no effect on the N_c factor. Average values of N_c were plotted versus the normalized distance d/B in figure 8 which clearly shows a threshold distance of 3B beyond which both the strip footings behave separately. In order to formulate the function **p** described by equation 14, such curves were fitted by the following hyperbolic function using the least squares regression technique:

$$N_c = \frac{q_l}{C_u} = \frac{\frac{d}{B}}{a+b\frac{d}{B}} \tag{15}$$

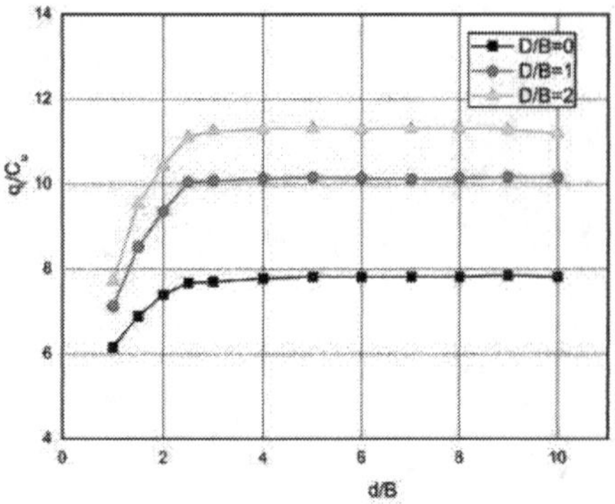

Fig. 8. Chart giving the N_c for different embedments

Table 3 summarized the values of a and b for different values of the slenderness ratio D/B.

Table 3. Values of coefficients a and b

D/B	a	b
0	45.8	0.12
1	43.3	0.09
2	50.5	0.08

5.2.2. Analysis of the factor I_s

In contrast with the N_c factor, the figure 9 shows the factor I_s linearly increases with the normalized distance without exhibiting a threshold value even at distance d of 10B.

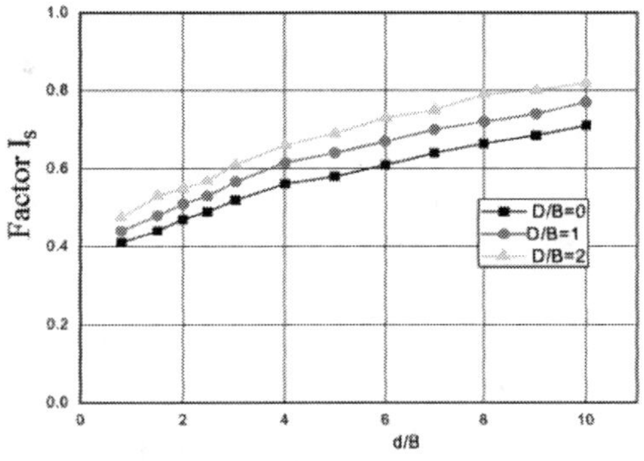

Fig.9. Typical chart giving I_s versus d/B

6 Conclusions

Detailed parametric study based on a non linear 2D finite elements modelling was carried out in order to highlight three important aspects of soil/rigid foundation interaction as well as to suggest simple formulae for practical purposes.

The first aspect is the effect of the proximity of a bedrock on the strip or circular footing behaviour. It was demonstrated the existence of a threshold depth of the bedrock beyond which the foundation behaves as in a infinitely deep layer.

The second aspect is the behaviour of foundations on a bi-layered purely cohesive soil. The results showed also the existence of a threshold normalized distance H_{lim}/B beyond which the effect of the lower layer on the bearing capacity and on settlement vanishes and the foundation behaves therefore as in a an infinitely thick mono-layered soil.

Finally, study of the behaviour of closely spaced identical strip footings on a mono-layered purely cohesive soil illustrated the consistency of the clay has almost no effect on the N_c factor. Moreover, a value of 3B for the distance between the foundations vertical axes is a threshold beyond which both the strip footings behave separately in terms of bearing capacity.

References

[1] H. Zerarka. *Contribution à la modélisation numérique du comportement des ouvrages géotechniques dans les sols de Gibson* (in French), M.S.CE dissertation, University of Blida, 2007.

[2] A. Hamidi. *Etude numérique de certains aspects d'interaction sol-fondation* (in French), M.S.CE dissertation, University of Blida, 2007.

[3] Bouafia, A., Mennaceur,Y. *Finite element analysis of the effect of embedment depth on the immediate settlement of rigid footings on finite layered soil,* Proceedings of the Intl. Symp. on Shallow Foundation *FONDSUP'03*, ENST Paris, 5-7 November, 2003, Presses de l'Ecole Nationale des Ponts & Chaussées, Paris, 9 p.

[4] M. Matar, J. Salençon. *Capacité portante des semelles filantes* (in French), Revue Française de géotechnique No. 9, pp : 51-76, éditions Presses de l'ENPC, Paris, 1982.

MATEC Web of Conferences **149**, 02020 (2018) https://doi.org/10.1051/matecconf/201814902020
CMSS-2017

Influence of granular material characteristics in the behaviour of "Bouregreg Valley" soft ground improved with stone columns

Noura NEHAB[1], Khadija BABA[1], Latifa OUADIF[1], Lahcen BAHI[1]
[1]Mohammadia Engineering School, 3GIE Laboratory, Mohammed V University in Rabat, Morocco

Abstract. The use of finite element analysis has become widespread in geotechnical practice as means of optimizing engineering tasks; it can be easily applied to the treated areas by stone columns, which are a method of improving the soil having low geotechnical properties and likely to deform significantly under load action, by incorporating granular material (commonly called ballast) compacted by remounting passes, so they act mainly as inclusions with a higher stiffness, shear strength than the natural soil. Moreover the stone columns are highly permeable and act as vertical drains facilitating consolidation of the soft soil improving the performance of the foundation. However the characteristics of this granular material influence the behavior of soft soils treated by the stone columns technique, especially: the friction angle, the cohesion, the modular ratio and the constitutive model. The choice of the constitutive model depends on many factors but, in general, it is related to the type of analysis that we intend to perform. Numerical modeling must consider the diversity of the materials nature, the complex geometry of structures- land and the behavior of materials generally nonlinear (permanent deformation). It is a simple and effective alternative to approach the real behavior of soils reinforced by stone columns and the influence of materials characteristics, it allows settlement analysis, lateral deformation, vertical and horizontal stresses in order to understand the behavior of columns and soil. It also has the advantage of integrating the settlements of the underlying layers. This paper aims to study the mechanisms of functioning and interactions of stone columns with the surrounding ground, and vis-à-vis the various parameters characterizing the granular material "ballast" and the surrounding soil, which influence the behavior of the improved soil, The paper presents, in the first part, soil conditions and the parameters associated with columns, is then presented 3D finite element analyses, so the parametric study was carried out varying several properties especially granular material properties.

1 Introduction

La technique des Colonnes Ballastées consiste à incorporer dans le terrain des matériaux granulaires compactés de manière à constituer des inclusions souples, présentant des caractéristiques mécaniques élevées et un fort caractère drainant. Ils sont disposés suivant un maillage régulier et sont coiffées d'un matelas de répartition des contraintes. Elles permettent ainsi d'homogénéiser et de diminuer l'amplitude globale des tassements sous l'ouvrage. Le procédé est adapté pour traiter les sols présentant des caractéristiques mécaniques faibles à très faibles.

Le projet objet d'étude consiste à renforcer ce sol par la méthode des colonnes ballastées.

Wehr (1999) et Muir Wood et al (2000) ont étudié la performance des groupes de colonnes ballastées à l'aide de l'analyse par éléments finis.

Ce travail présente, la modélisation numérique tridimensionnelle d'une colonne isolée et d'un groupe de colonnes en prenant en compte l'influence de leur mise en place.

Les modélisations en 3D sont faites en utilisant le logiciel « Plaxis 3D Foundation ». Outre l'étude de l'expansion latérale lors de la mise en place, on a analysé l'influence dans la modélisation numérique des facteurs tels que l'espacement, le nombre des colonnes ainsi que les propriétés des colonnes (la loi du comportement et l'angle de frottement).

2 Données du sol et caractéristiques des colonnes

2.1 Description de l'aire d'étude

La vallée du Bouregreg sépare les villes de Rabat et de Salé. Elle est située sur la côte Atlantique, au Nord de Casablanca. En coordonnées géographiques, la vallée se trouve à environ 34°01 en Latitude Nord et 6°49 en Longitude Ouest.

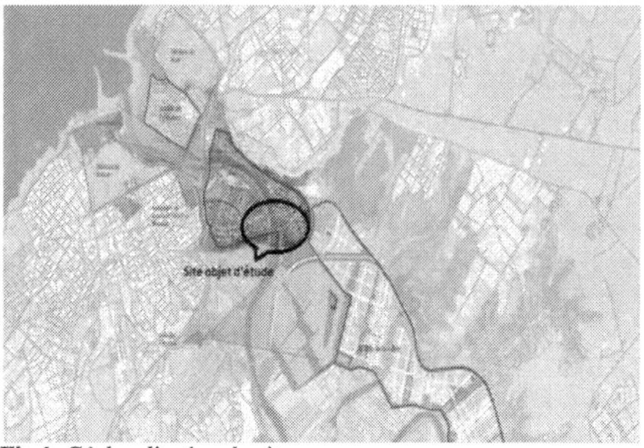

Fig.1. Géolocalisation du site.

La vallée du Bouregreg est relativement encaissée ; c'est le seul accident d'importance du relief du site de l'agglomération. Elle se développe entre les plateaux de Youssoufia Akreuch à l'Ouest et de Kariat Hssaine à l'Est, où le fleuve côtier décrit d'amples méandres. Elle se resserre près de l'embouchure et au niveau de la confluence avec l'oued Akreuch. Le barrage « Sidi Mohamed Ben Abdellah » est construit à la sortie des gorges.

La mauvaise qualité géotechnique des terrains en place contribue à majorer le risque sismique. Les versants de la vallée du Bouregreg présentent par ailleurs des signes d'instabilité par glissement de terrain ou chutes de blocs, ce qui constitue également un facteur aggravant du risque sismique.

2.2 Les conditions géotechniques du sol et les caractéristiques des colonnes

Une campagne de reconnaissance a été mise en œuvre au niveau du site du nouveau pont Moulay Youssef sur « oued Bouregreg ».

Des sondages carottés jusqu'à 30 m de profondeur, avec prélèvement d'échantillons intacts et remanié sont permis d'effectuer des essais de laboratoire (identification et caractérisation des sols) et d'établir une coupe détaillée des terrains.

Fig.2. prélèvement d'échantillons pour effectuer les essais de laboratoire

Des sondages pressiométriques ont été réalisés avec des essais SPT afin de déterminer les caractéristiques pressiométriques des sols. En fin, l'hétérogénéité du site a été mise en évidence par des sondages de pénétration statique (CPT).

Ces essais montrent que jusqu'à 10m on a affaire à des formations sableuses et vaseuses peu consistantes et qu'au-delà de cette profondeur on a du sable consistant.

Compte tenu de la granulométrie du sol et selon le règlement parasismique Marocain, celui-ci est susceptible de liquéfaction.

Vu le risque de liquéfaction ainsi que la présence de la vase jusqu' à 10,5m de profondeur, il faut absolument traiter le sol de la rive droite par les colonnes ballastées.

Nous avons effectué une série de modélisations tridimensionnelles des colonnes ballastées avec une colonne isolée au centre et puis de groupes de colonnes, en utilisant le logiciel PLaxis 3DF.

Le sol considéré est constitué de plusieurs couches compressibles.

Le tableau suivant « Tableau 1 » présente les caractéristiques des couches du sol en place, avec

$$E_{oed} = \frac{E_M}{\alpha} \quad (1)$$

EM : module pressiométrique, et α : un coefficient rhéologique qui dépend du type du sol et de la pression limite Pl.

Table 1. Caractéristiques du sol en place.

Nature du sol	Profondeur (m)	E_M (kPa)	P_l (kPa)	α	E_{oed} (kPa)
silts vaseux	2	1700	200	1/2	3400
vase	6,5	2080	180	1/2	3830
Vase & sable	10,5	15200	1476	2/3	27164
sable fin	17	18800	1902	1/3	49122

 https://doi.org/10.1051/matecconf/201814902020

sable vaseux	18	25000	2600	1/3	75757
sable fin	18,5	24000	2500	1/3	72727
sable vaseux	19,5	26000	2340	1/3	52525
sable fin compact	>20	33840	3006	1/3	85442

3 Modélisation numérique en 2D d'une colonne isolée

Elle permet d'assimiler le comportement de la colonne ballastée, on s'intéresse dans cette partie à l'influence des caractéristiques du matériau granulaire (Ballast) sur l'amélioration du tassement.

Une étude paramétrique a été réalisée afin de déterminer l'influence de certains paramètres du matériau granulaire sur les résultats du tassement.

Nous avons fait varier certains paramètres :
-la loi de comportement du matériau granulaire (Le modèle élastique linéaire et Le modèle élasto-plastique « Mohr-Coulomb »)
-Le module de Young du ballast (de 40 à 100 MPa)
-L'angle de frottement du ballast (de 35° à 45°)

3.1. La loi de comportement de la colonne :

La figure « fig.3 » présente le tassement mesuré déterminé à partir de deux modèles réalisés avec Plaxis 2D. L'un étant avec une colonne à loi de comportement élastique linéaire (Eballast = 70 MPa) « mod lin elast » et une colonne à loi de comportement élasto-plastique « Mohr-Coulomb » (Eballast = 70MPa,φ= 40°) « modMohr C ». Les sols étant à l'état non drainé.

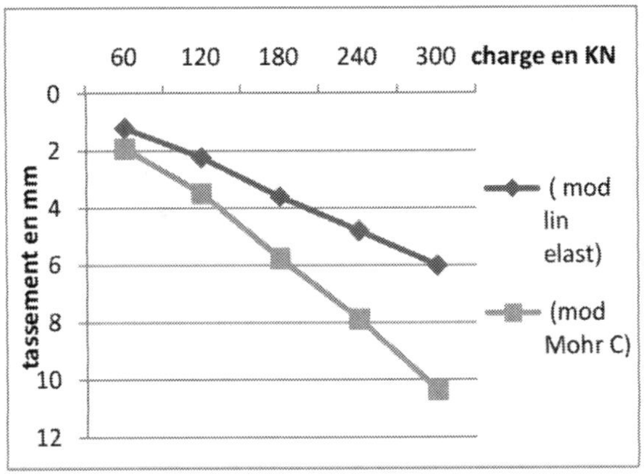

Fig.3. Résultats des tassements obtenus en 2D (colonne à loi de comportement élastique linéaire ou élasto-plastique).

3.2. Le module de Young du ballast (de 40 à MPa) :

Les deux figures ci-après (« fig.4 »et « fig.5 ») présentent les tassements déterminés à partir des deux modèles (comportement élastique linéaire et comportement élasto-plastique) pour lesquels le module de Young du ballast varie entre 40, 70 et 100 MPa,

3.1.1 Comportement élastique linéaire :

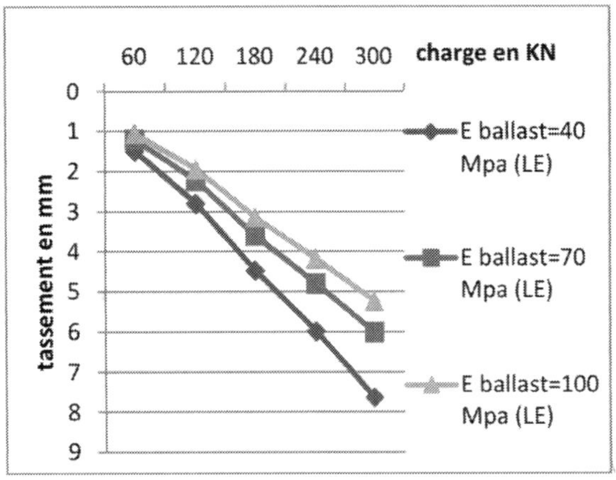

Fig. 4.Résultat des tassements obtenus en 2D (colonne à loi de comportement élastique linéaire avec E ballast=40,70 et 100 MPa).

On constate que :
-Le modèle « E=100MPa » sous-estime les tassements.
-L'amélioration du sol avec l'augmentation du module de Young du ballast.

3.1.2 comportementélasto-plastique :

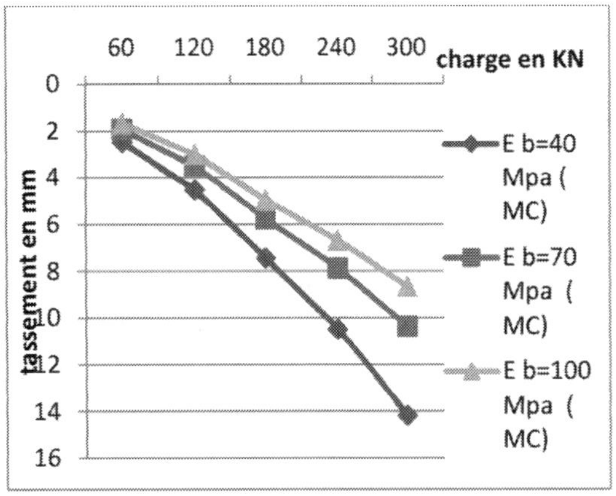

Fig. 5. Résultat des tassements obtenus en 2D (colonne à loi de comportement élasto plastique « MC » avec E ballast=40,70 et 100 MPa).

3.3. L'angle de frottement du ballast (de 35° à 45°):

La figure « fig.6 » présente le tassement de la semelle sur la colonne ballastée, déterminé à partir du modèle réalisé avec une colonne à loi de comportement élasto-plastique « Mohr-Coulomb » (Eballast = 70 MPa) pour lequel l'angle de frottement du ballast varie de 35 à 45°.

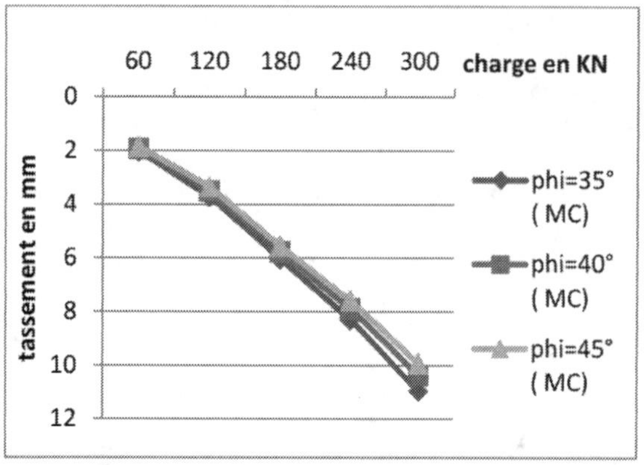

Fig. 6.Résultat des tassements obtenus en 2D (colonne à loi de comportement élasto plastique « MC », E ballast et φ=35°,40° et 45°).

On constate que l'augmentation de l'angle de frottement interne du ballast permet une réduction de tassement.

4 Modélisation numérique en 3D d'une colonne isolée

Cette section analyse l'amélioration du tassement, pour le cas d'une colonne isolée, cette amélioration va être déterminée par le facteur de réduction d tassement, β.
Les paramètres à varier sont :

- L'angle du frottement interne du matériau granulaire.
- La loi du comportement du matériau granulaire

4.1. L'angle du frottement du ballast

Pour examiner l'influence de L'angle du frottement interne du matériau granulaire, et le rapport des surfaces pour une colonne isolée, on prend comme loi du comportement de la colonne « la loi Mohr coulomb ».

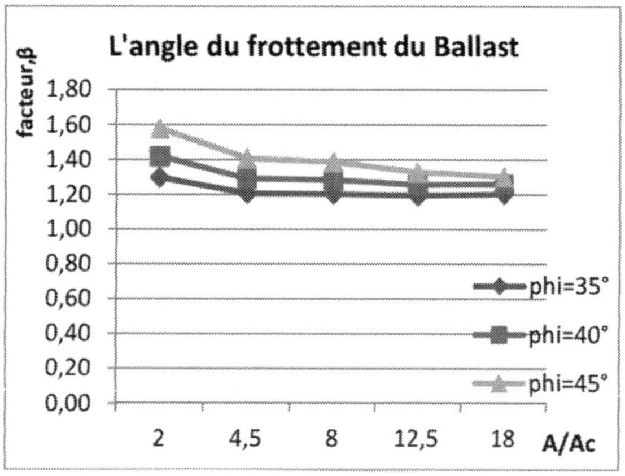

Fig. 7. L' influence de l' angle de frottement interne du ballast et du rapport des surfaces A/Ac.

Les graphes (Figure 7) illustre pour les différentes valeurs de A/Ac, le facteur de réduction du tassement, β pour les différentes valeurs de Φ (35°/40°/45°):

- on observe que l'augmentation de l'angle de frottement interne du ballast permet une réduction de tassement.
- pour A/Ac=2 (petites valeurs du rapport des surfaces); β varie de 1.30 (Φ =35°) à 1.58 (Φ =45°), alors que pour A/Ac=18 (grandes valeurs du rapport des surfaces); β varie de 1.20 (Φ =35°) à 1.30 (Φ =45°).
- Le tassement augmente lorsque A/Ac augmente pour toute les valeurs de l'angle de frottement interne du ballast.
- Les graphes s'approchent pour les grandes valeurs de A/Ac, donc l'augmentation de l'angle de frottement interne du ballast a un effet négligeable pour les grandes valeurs du rapport des surfaces.

4.2. La loi du comportement du ballast

Pour examiner l'influence de La loi du comportement du matériau granulaire, et le rapport des surfaces pour une colonne isolée, on prend comme loi du comportement du sol « la loi Mohr coulomb » et ϕ=45° pour la colonne (MC and HS).

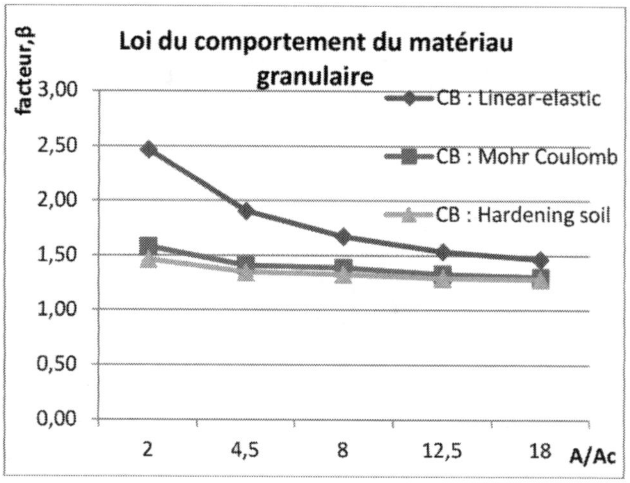

Fig. 8. L'amélioration du tassement pour les différentes lois du comportement du matériau granulaire de la colonne.

Les graphes (Figure 8) illustre pour les différentes valeurs de A/Ac, le facteur de réduction du tassement, β pour les différentes lois de comportement de la colonne (Mohr coulomb, Hardening soil et linéaire élastique):

- Le facteur de réduction du tassement diminue avec l'augmentation de A/Ac.
- Lorsqu'on varie A/Ac de 2 à 18; β varie de 2.46 à 1.46 (pour le modèle linéaire élastique), de 1.46 à 1.28 (pour le modèle Hardening soil) et de 1.58 à 1.30 (pour le modèle Mohr coulomb).
- Les deux lois de comportement de la colonne (Mohr coulomb et Hardening soil) atteignent approximativement le même graphe.
- Les solutions élastiques surestiment le facteur de réduction du tassement.

5 Modélisation numérique en 3D d'un groupe de 3 colonnes

5.1. L'influence de la position des colonnes;

Cette étude paramétrique a pour objectif de montrer l'influence de l'espacement entre colonnes, ou plus précisément, la position relative des colonnes. On considère un groupe de trois colonnes (maille triangulaire), Le chargement est appliqué sur un carré de cote B=8m et l'espacement entre les colonnes est variée de s = 1 à 4 m.

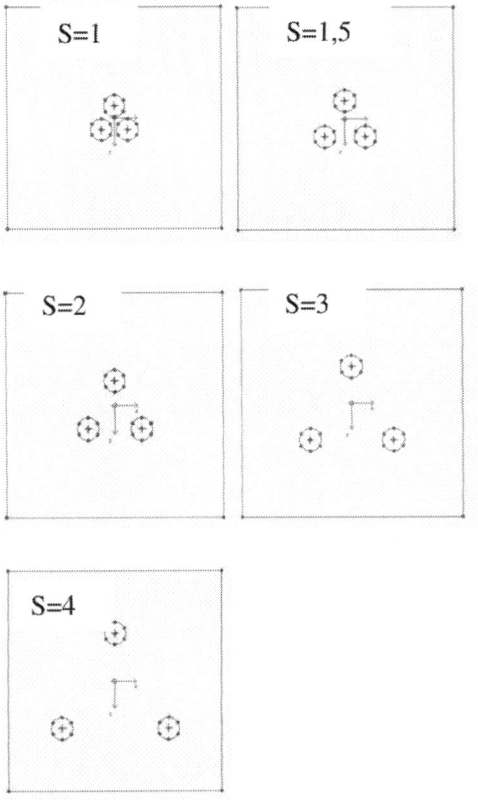

Fig. 9.La variation de l'espacement entre les colonnes pour une maille triangulaire.

5.1.1. Variation de la loi du comportement des colonnes :

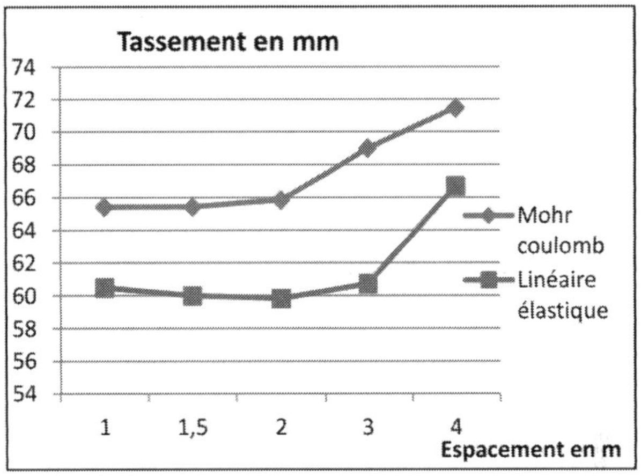

Fig. 10.L'effet de la variation de l'espacement entre les colonnes sur le tassement pour les deux lois de comportement de colonnes (MC et LE).

Le tassement augmente lorsque les colonnes sont plus espacées pour la loi de comportement Mohr coulomb. Alors que pour la loi de comportement linéaire élastique le tassement diminue légèrement entre S=1 et S=2 (mais reste très proche d'une même valeur), puis commence à augmenter.

On constate qu'à partir d'une valeur d'entraxe le tassement augmente d'une manière excessive, donc à partir de cette valeur le renforcement n'est plus bénéfique.

Ces courbes confirment la règle pratique qui recommande un entraxe de colonnes inférieur à 4 Dc ici 4x0,8=3,2 m) pour obtenir un renforcement significatif.

5.1.2.Variation de l'angle de frottement des colonnes :

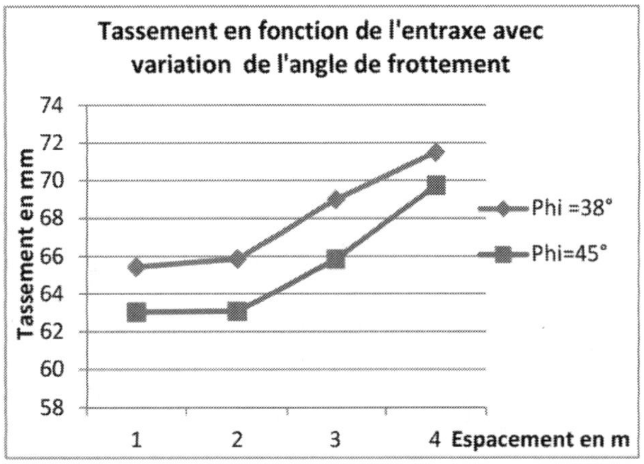

https://doi.org/10.1051/matecconf/201814902020

Fig. 11.L'effet de la variation de l'espacement entre les colonnes sur le tassement pour les deux angles de frottement(phi=38°et phi=45°).

On constate que l'augmentation de l'angle de frottement interne du ballast permet une réduction de tassement.
De même Le tassement augmente lorsque les colonnes sont plus espacées pour les deux valeurs d'angle de frottement interne du ballast.

5.2. L'influence du nombre des colonnes;

Dans cette étude on a utilisé un maillage carré en changent le nombre de colonnes de 1 à 16 colonnes, ainsi que la surface de chargement pour les deux lois de comportement des colonnes Mohr coulomb et linéaire élastique, on a gardé un espacement de 1,5 m entre les colonnes, et un angle de frottement de 38°.

N=4

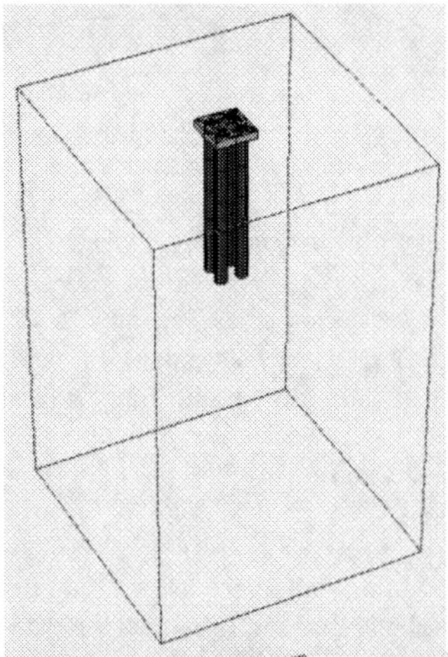

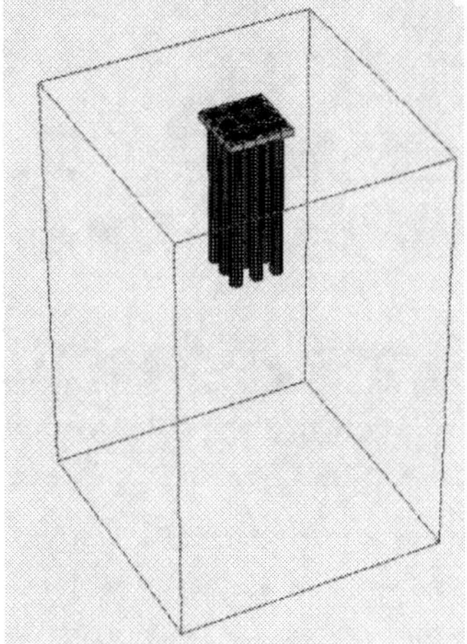

N=16

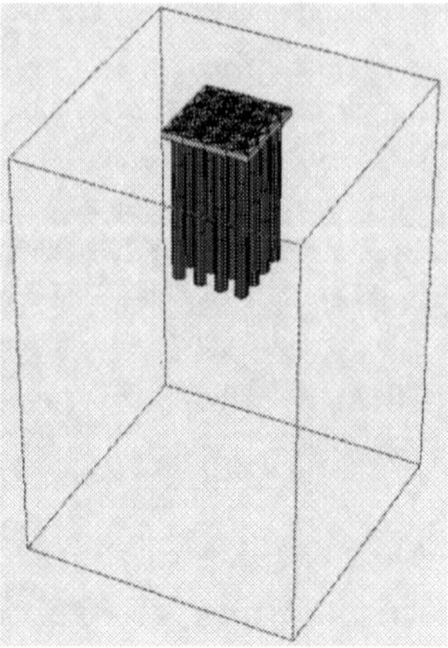

N=9

Fig. 12.La variation du nombre des colonnes pour une maille rectangulaire

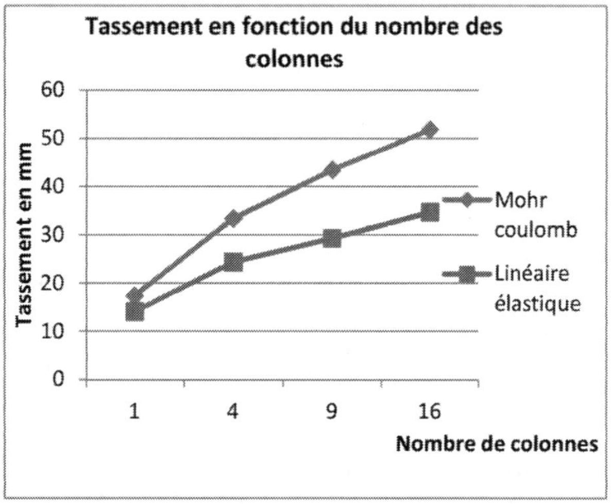

Fig. 13.L'effet de la variation du nombre des colonnes sur le tassement pour les deux lois de comportement de colonnes (MC et LE).

-Les deux courbes se rapprochent dans le cas d'un petit nombre de colonnes et se divergent dans le cas contraire.
-Le nombre de colonnes n'a pas une grande influence sur les tassements, différemment de la littérature qui confirme qu'un nombre croissant de colonnes est considéré comme ayant un effet bénéfique parce que les colonnes sont mieux confinées latéralement [Priebe].

6 Conclusion :

Nous avons effectué une série de modélisations tridimensionnelles des colonnes ballastées avec une colonne isolée au centre, en utilisant le logiciel PLaxis 3DF. Le sol considéré est constitué de plusieurs couches compressibles.

L'augmentation des propriétés du matériau granulaire (l'angle de frottement, le raport modulaire) permet d'améliorer le sol traité en augmentant le facteur de réduction du tassement, l'influence de ces paramètres est liée à l'arrangement des colonnes (la position des colonnes), et à la loi de comportement adopté pour la colonne.
Pour un groupe de trois colonnes, disposées aux sommets d'un triangle équilatéral. La modélisation en 3D permet de montrer que l'espacement entre les colonnes augmente le tassement pour les deux lois de comportement, et à partir d'une valeur d'entraxe le tassement augmente d'une manière excessive et le renforcement n'est alors plus bénéfique. Alors que le nombre de colonnes n'influence pas beaucoup sur les résultats.

References

[1] Wehr J. (2004). Stone columns- single columns and group behavior. In proceedings of the 5th Intconfon Ground Improvement techniques. Malaysia, 329-340.

[2] Wehr, W. (1999). Schottersaulen – das verhalten von einzelnensaulen und saulengruppen.Geotechnik, vol.22,n°1, 40-47.

[3]Killeen M. (2012). Numerical modeling of small groups of stone columns. PhD Thesis. National university of Ireland, Galway.

[4]Brinkgreve, R.B.J., Engin, E. and Swolfs, WM. (2012). Plaxis 3D 2012 Manual.Plaxisbv, the Netherlands.

[5]Baumann V. and Bauer G. E. A. (1974). The performance of foundations on various soils stabilized by the Vibrocompaction process, Rev.canadienne de Géotechnique, Vol.11, n°4, nov.1974, 509-530.

[6] Black, J.V., Sivakumar, V. and McKinley, J.D. (2007). Performance of clay samples reinforced with vertical granular columns. Canadian Geotechnical Journal 44, pp.89-95.

[7]Hughes, J.M.O. and Withers, N.J. (1974). Reinforcing of soft cohesive soils with stone columns.Ground Engineering 7 (3), 42-49.

[8]McKelvey, D., Sivakumar, V., Bell A. and Graham J. (2004). Modeling vibrated stone columns in soft clay.

[9]Proceedings of the Institute of Civil Engineers Geotechnical Engineering 157(3), 137-149.

[10] Muir Wood, D., Hu, W. and Nash, D.F.T. (2000). Group effects in stone column foundations model tests.Géotechnique 50 (6), 689-698.

[11]Najjar, S.S., Sadek, S. and Maakaroun, T. (2010). Effect of sand columns on the undrained load response of soft clays.Journal of Geotechnical and Geoenvironmental Engineering 136(9), 1263-1277.

[12]Nehab, N., Baba , K., Ouadif L., Cherradi C. and Bahi L. (2014). Soft soil strengthening by stone columns: case of the embankment under the bridge "Moulay Youssef" (Rabat/Salé). MATEC Web of Conferences.MatecconfDOI: 10.1051//20141102012.

[13]Shahu, J. and Reddy, Y. (2011). Clayey Soil Reinforced with Stone Column Group: Model Tests and Analyses.

[14]Slocombe, B.C. (2001). Deep compaction of problematic soils, Problematic Soils, Thomas Telford (London),
163-181.

[15]Priebe H., (1976). Abschätzung des SetzungsverhalteseinesdurchStopfverdichtungverbesserte nBaugrundes, Die Bautechnik, Vol. 53 (1976), n°5, 160-162.

[16]Ghionna, V., Jamiolkowski, M. (1981). Colonne di ghiaia. X Ciclo di conferenze dedicate aiproblemi di meccanicadeiterreni e ingegneriadellefondazionimetodi di miglioramento deiterreni, Politecnico di Torino Ingegneria, attidell'istituto di scienza delle costruzioni, n° 507, nov. 1981.

[17]Khabbazian, M., Kaliakin, V. N. and Meehan, C. L. (2012). Numerical Simulation of Column Supported Embankments with Geosynthetic Encased Columns: Influence of Soft Soil Constitutive Model. Conference: GeoCongress 2012. DOI: 10.1061/9780784412121.001

[18]Guetif, T., Bouassida, M. and Debats, J. M. (2007). Improved soft clay characteristics due to stone column installation. Computers and Geotechniques, 34, 104–111.

Study of the degradation of the geometry of the railway between the pk 80 and 105 at the plateau of Settat (Morocco)

Etude de la dégradation de la géométrie de la voie ferrée entre les PK80 et 105 au niveau du plateau de Settat (Maroc)

Mohamed BEN OUAKKASS, Latifa OUADIF, Ahmed AKHSSAS et Lahcen BAHI

3GIE Laboratory, Mohammadia Engineering School, Mohammed V University in Rabat-Morocco

Abstract. The plateau of Settat is formed by Paleozoic age formations surmounted by a subhorizontal cover made up mainly of Cretaceous sedimentary rocks formed of limestones, sandstones and clays with alternating marly limestones and marls. The study area is characterized by a dry to semi-dry climate, an endorheic hydrographic network that flows into the plain, and is affected by a hydraulic fracturing network of WNW-ESE and NE-SW directions. These factors cause repetitive and cyclical disturbances and disorders recorded on the railway section between Settat and Machraa Ben Abbou and are due to the shrinkage-swelling of the clay formations. The statistical and critical analysis of the defects identified by the device of measuring the geometric parameters of the railway, coupled with an exploitation of the Optical and Radar satellite images, made it possible to refine the geological, hydrological and structural study of the plain studied. These approaches have shown that this hazard has been favored by the existence of recent clayey and swelling formations under the influence by the climatic, hydrological and geological contrasts marked by intense hydraulic fracturing of the zone.

Keywords: Swelling clays, railway, remote sensing, radar and optical imagery, Settat, Morocco

Résumé. Le plateau de Settat est constitué par des formations d'âge paléozoïques surmonté par une couverture subhorizontale constituée essentiellement de roches sédimentaires du Crétacé formés de calcaires, grès et d'argiles avec une alternance de calcaires marneux et de marnes. La zone d'étude est caractérisée par un climat aride à semi-aride, par un réseau hydrographique endoréique qui se déverse dans la plaine, et elle est affectée par un réseau de fracturation de directions WNW-ESE et NE-SW, l'ensemble de ces facteurs provoque des dérangements et des désordres répétitifs et cycliques enregistrés sur le tronçon ferroviaire entre Settat et Machraa Ben Abbou de la voie ferrée qui sont dues au retrait-gonflement des formations argileuses. L'analyse statistique et critique des défauts relevés par le dispositif de mesure des paramètres géométriques de la voie, couplée d'une exploitation des images satellitaires Optique et Radar, ont permis d'affiner l'étude géologique, hydrologique et structurale de la plaine d'étude. Ces approches ont fait ressortir que cet aléa a été favorisé par l'existence des formations récentes argileuses et gonflantes sous l'influence des contrastes climatiques, hydrologiques et géologiques marquées par une fracturation intense de la zone.

Mots clé : Argiles gonflant, voie ferrée, télédétection, imagerie radar et optique, Settat, Maroc

1. Introduction

La qualité géométrique de la voie est l'évaluation des variations des paramètres spécifiés par rapport à la moyenne ou aux caractéristiques géométriques de construction, dans le sens vertical ou latéral, qui peuvent mettre en cause la sécurité de circulation des trains [1].

La dégradation de l'état géométrique de la voie dépend, principalement, de la nature et de l'importance du trafic, mais aussi d'autres facteurs locaux à savoir la nature des formations géologiques, les paramètres géotechniques et les conditions climatiques.

Bien que l'armement de la voie, la géométrie de pose et les sollicitations du trafic soient identiques sur une même section, le comportement des zones élémentaires est très différent.

Le tronçon objet de cette présente étude (Figure 1) fait partie administrativement à la province de Settat et situé entre la gare ferroviaire de Settat du point kilométrique (PK) 72 et la gare de Mechraa Ben Abbou du PK 117 de la ligne Casablanca à Marrakech sur une longueur d'environ de 25 km.

La zone traversée actuellement par 18 trains voyageurs dont 16 trains reliant Marrakech à Casablanca et 2 trains Marrakech à Tanger, le trafic marchandise a permet de transporter 234 467 tonnes en 2015. Avec les projets, en cours de réalisation, de doublement complet de la voie entre Casablanca et Marrakech et les zones logistiques de Sidi Bou Othman et Sidi Ghanem, l'offre ferroviaire sera augmentée aussi bien voyageurs que marchandises. Suite à des désordres géométriques répétitifs et cycliques enregistrés dans la zone d'étude, il était nécessaire d'en analyser les causes.

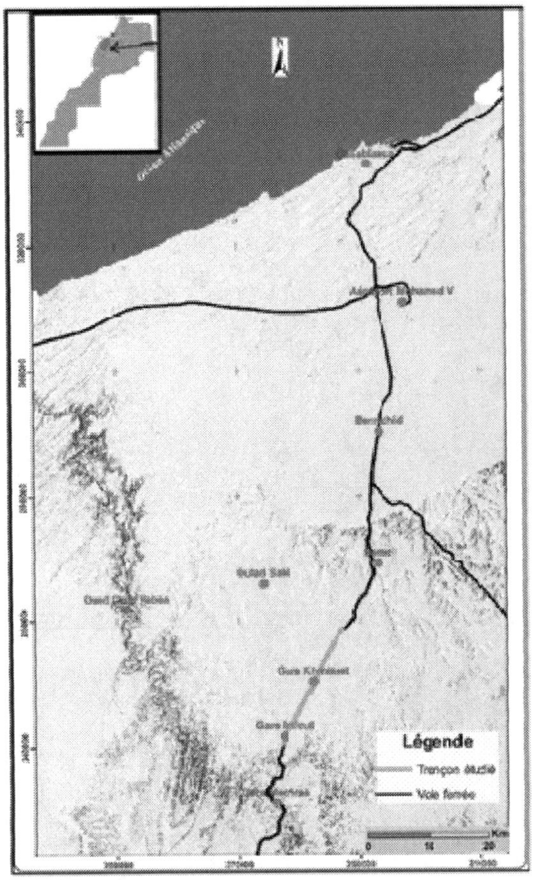

Fig. 1. Situation géographique de la zone d'étude

2. Géométrie de la voie ferrée

La géométrie de la voie se compose du tracé auquel sont superposés les défauts. Le référentiel de la qualité géométrique de la voie a pour objectif de définir les seuils des paramètres géométriques pour permettre une interprétation juste des amplitudes des défauts et des mesures à prendre et pour mieux adapter la nature des corrections à la situation et au degré de gravité des anomalies.

Le tracé de la voie est défini par quatre paramètres : la courbure horizontale, la courbure verticale, le dévers et l'écartement nominal. Quant aux défauts, ils en existent de quatre types suivants : le dressage, l'écartement par rapport à l'écartement nominal, le profil en long et le gauche.

2.1. Défauts géométriques de la voie ferrée

La qualité géométrique de la voie est décrite en utilisant un système orthogonal de coordonnées centré sur la voie. Les indicateurs de qualité de la géométriques sont comme suit [2] :

- Dressage : est le déplacement latéral de la voie dans le plan de roulement, l'écart par rapport à la position horizontale moyenne (flèche) est mesuré sur la base de 10 ml ;
- Ecartement : c'est la distance entre les faces actives des deux files de rail, il est considéré comme étant la plus petite distance entre les rails ;
- Profil en long : correspond à la variation des altitudes consécutives de la table de roulement de chaque file de rail exprimée comme un écart par rapport à la position verticale moyenne (ligne de référence) ;
- Gauche : c'est la différence algébrique entre deux valeurs de dévers mesurées à une distance donnée, exprimée habituellement comme un gradient entre les deux points de mesure. Le gauche est exprimé en mm/m.

2.2. Seuils paramètres géométrique de la voie

Les seuils des paramètres géométriques de la voie sont définis par classe de voie et en fonction de la vitesse limite. Pour chaque paramètre quatre niveaux principaux sont définis [3] :

- Limite d'Alerte (LA) : correspondant à la valeur qui demande l'analyse de la géométrie de la voie et la prise en compte lors des opérations de maintenance programmées ;
- Limite d'intervention (LI) : correspondant à la valeur qui, lorsqu'elle est dépassée, demande une action de maintenance corrective telle que la limite d'action immédiate ne soit pas atteinte avant le prochain contrôle.
- Limite d'action immédiate (LAI) : Les défauts atteignant ou dépassant le seuil « LAI » nécessitent une limitation de vitesse des trains jusqu'à correction du défaut.
- Tolérance de réception (TR) : correspondant à une valeur qui ne doit pas être dépassée après des travaux de pose de voie nouvelle, de renouvellement de voie ou d'entretien de la voie.

Lors de la campagne de mesure, le système informatique de l'engin permet de suivre et de visualiser le déroulement de la mesure et de l'enregistrement de la qualité géométrique de la voie en temps réel sous forme des graphes (Figure 2), qui représentent toutes les dimensions de la géométrie de la voie.

2.3. Mesure de paramètres géométriques de la voie

Les paramètres géométriques de la voie sont mesurés par l'autorail qui permet de calculer des indicateurs représentatifs de la qualité de la voie [4]. C'est est un véhicule automoteur équipé d'un système informatique pour l'enregistrement des mesures et l'analyse des résultats. Le dispositif de mesure se compose d'une partie mécanique et d'une partie électrique.

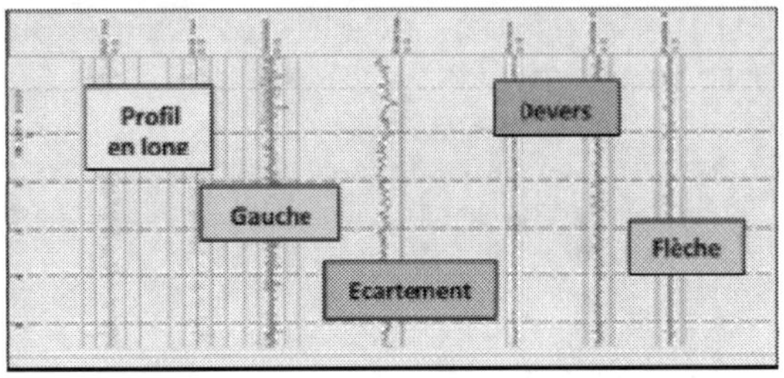

Fig. 2. Graphes de l'enregistrement des indicateurs de la qualité géométrique de la voie ferrée

La périodicité de mesure d'enregistrement est fixée comme suit [3] :

- Mesure et enregistrement en janvier ou février, après la période de froid, pour l'évaluation du bilan de l'année écoulée et élaboration des programmes d'entretien de la voie pour l'année en cours.
- Enregistrement en septembre, après la période de forte chaleur, pour l'appréciation de la qualité des interventions et ajustement des programmes d'entretien de la géométrie.
- Un troisième enregistrement effectué en mai pour l'appréciation de la qualité des interventions et ajustement des programmes d'entretien de la géométrie de la voie.

3. Etude de la dégradation de la géométrie de la voie dans la zone entre Settat et Mechraa Ben Abbou

A trafic donné, il en résulte que la vitesse de dégradation de la qualité géométrique d'une voie peut considérablement varier en fonction de facteurs locaux. L'objectif de l'entretien de la géométrie de la voie est de satisfaire aux conditions de sécurité et de confort des circulations et d'éviter l'aggravation des défauts dont la correction deviendrait par la suite plus difficile et onéreuse. La détection des déformations apparaît comme étant un paramètre essentiel pour l'aménagement et la gestion des risques, notamment en contexte ferroviaire où les enjeux sont importants.

3.1. Analyse statistique des défauts géométriques de la voie

Le rapport ponctuel édité en temps réel par l'engin de mesure de la qualité des paramètres géométriques permet de localiser pour chaque paramètre, les défauts dont l'amplitude atteigne ou dépasse le seuil limite d'action.

La zone d'étude a connu un renouvellement complet de la voie en 2010, la figure 3 représente les résultats de traitement des enregistrements trimestrielles des indicateurs de la géométrie de la voie, notamment le gauche, le flèche et le profil en long, entre 2010 et le deuxième trimestre de l'année 2017.

En tenant compte des opérations de correction des défauts effectuées entre les deux campagnes de mesure consécutives, on constate une variation du nombre moyen des défauts (Gauche, Flèche et profil en long) enregistrés d'un trimestre à l'autre. Nous remarquons aussi que le nombre moyen de défauts mesurés au troisième trimestre, correspondant à la période des fortes chaleurs, est plus élevé par rapport aux autres périodes.

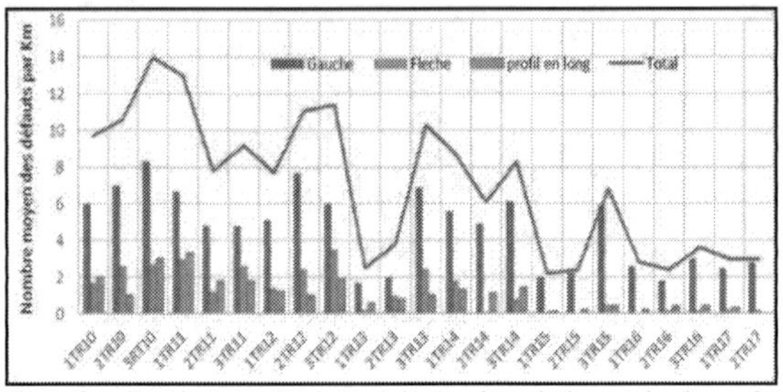

Fig. 3. Evolution de la qualité géométrique de la voie entre les PK 80 et 105 période : 2010-2017

Dans l'objectif d'étudier les variations de défauts géométrique et leur distribution dans l'espace le long de la voie, on a réparti ces anomalies selon les points kilométriques (PK) où elles ont été mesurées. Les figures 4, 5 et 6 représentent la situation géographique des défauts enregistrés par leurs PK et par trimestre entre les années 2012 et 2016. D'après ces représentations on peut distinguer trois zones selon le nombre des défauts enregistrés :

- Zone stable entre les points kilométriques PK 88 et PK 92 : correspond à la zone où on a enregistré moins de défauts, le nombre cumulé ne dépasse pas cinq défauts au premier trimestre et vingt au troisième trimestre ;
- Zones moyennement stables entre PK 80 et PK 87 et entre PK 96 et PK 99 : dont le nombre cumulé de défaut mesuré est de l'ordre de 20 au premier trimestre et 40 au troisième trimestre ;
- Zone perturbante : dans cette zone, on trouve des nombres élevés de défauts qui dépassent 100 défauts au PK 96 et 140 défauts au PK 100.

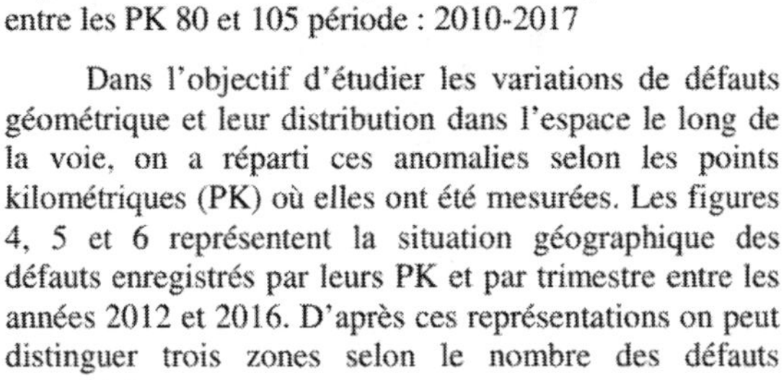

Fig. 4. Localisation des défauts enregistrés au 1er trimestre entre PK 80 et PK 105

D'après les résultats obtenus, les bilans de qualité et l'analyse de la distribution des anomalies dans la zone d'étude, on peut conclure que la dégradation de l'état de la qualité géométrique de la voie ferrée ne dépend pas seulement de conditions internes de la voie comme l'importance du trafic, l'état du ballast et la nature de l'armement de la voie. A cet effet, et pour déterminer l'origine de ces anomalies répétitives et les conditions de leur distribution spatiale, on a recouru aux paramètres extérieurs à la voie à savoir les conditions géologiques, structurales et climatiques.

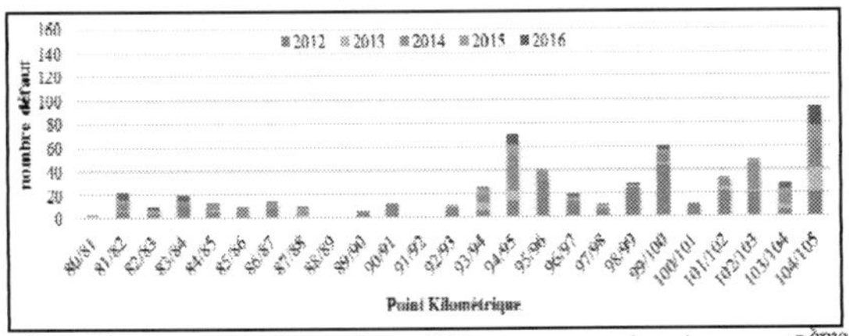

Fig. 5. Localisation des défauts enregistrés au 2ème trimestre entre PK 80 et PK 105

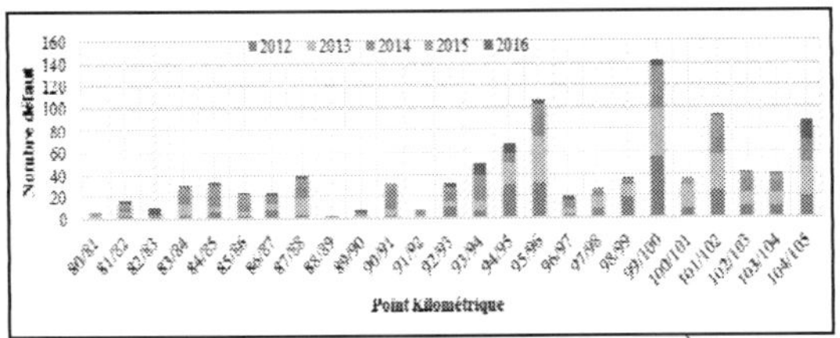

Fig. 6. Localisation des défauts enregistrés au 3ème trimestre entre PK 80 et PK 105

3.2. Climatologie et hydrologie de la région d'étude

Le climat qui règne dans la région de Settat est de type aride à semi-aride [5] à hiver froid [6], la température moyenne annuelle de la région est de 18°C. Le mois de janvier est le plus froid alors que le mois d'août est le plus chaud (Figure 7). On retrouve un gradient croissant des températures du Nord au Sud qui marque un degré de continentalité croissant.

La pluviométrie moyenne annuelle (Figure 7) à l'échelle de la province de Settat est de 372 mm. La période pluvieuse s'étale d'octobre à avril, les mois de décembre, janvier et février sont les plus arrosés. La période estivale est pratiquement sèche avec des précipitations faibles à nulles, les variations interannuelles sont très marquées. Le niveau d'évapotranspiration annuelle est de 1350 mm, il reste très élevé relativement à des précipitations de la région.

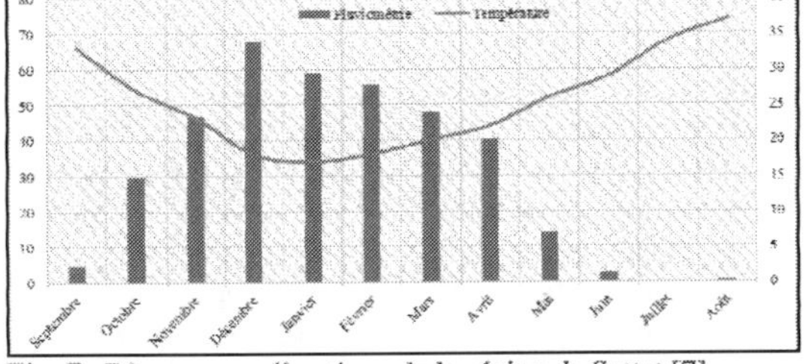

Fig. 7. Diagramme climatique de la région de Settat [7].

Le plateau de Settat est caractérisé par un réseau hydrographique endoréique [8] sous forme d'une série de petits cours d'eau de faible débit et à écoulement irrégulier. L'imagerie satellite optique et radar permet de mettre en évidence ces petits cours d'eau (Figure 8) de caractère saisonnier et de faible calibre qui se perdent en alimentant la nappe plio-quatenaire, ils ne possèdent aucun exutoire superficiel pour les eaux de pluie [9].

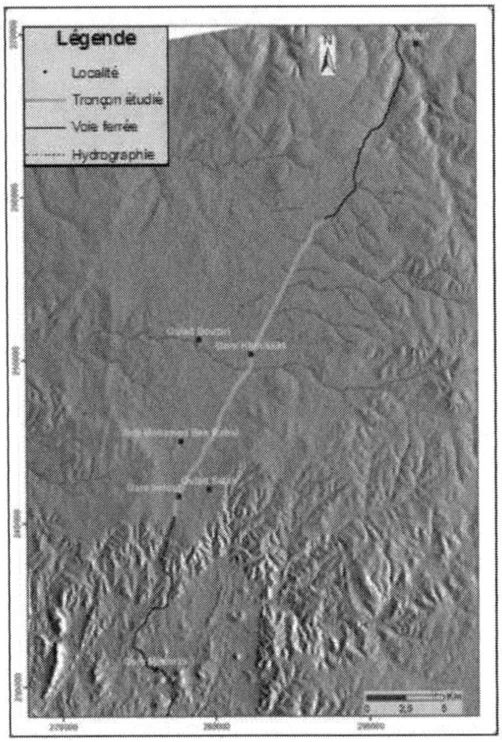

Fig. 8. Réseau hydrographique de la région d'étude

3.3. Contexte géologique

La région d'étude fait partie du plateau de Settat qui constitue la limite Nord-Ouest du plateau des phosphates de la Mesta centrale [10]. Celle-ci est constitué par des terrains du socle hercynien, plissés et granitisés arasés qui affleurent au N et au SW du plateau [11], ils sont recouverts par une couverture subtabulaire [12-13] peu déformés par les phases alpines [14], les formations superficielles quaternaires de la région résultent de l'action combinée des conditions climatiques et des mouvements tectoniques régis par le retrait du Moghrebiein [8] qui a donné naissance à une sédimentation continentale, au moment du soulèvement du plateau.

La série lithologique dans le plateau de Settat (Figure 9) débute par les formations d'âge paléozoïques, essentiellement représentées par des schistes, grès, calcaires récifaux, calcaires métamorphiques, conglomérats, grès quartzitiques et quartzites, plissées et faillées [5]. Le Primaire s'ennoie sous le haut plateau des phosphates [15], il affleure essentiellement au Nord-Ouest du plateau de Settat, sur les bordures Sud-Est de la plaine de Berrechid, au Nord-Ouest du plateau de Ben Ahmed et au Sud aux environs Mechraa Ben Abbou. La couverture subhorizontale constituée essentiellement de roches sédimentaires du Crétacé (Cénomano-Turonien) formés de calcaires, grès et d'argiles avec une alternance de calcaires marneux et de marnes. Les couches les plus récentes sont datées de l'Eocène, l'ensemble de ces formations est couvert par une couche argilo-limoneuse du Quaternaire.

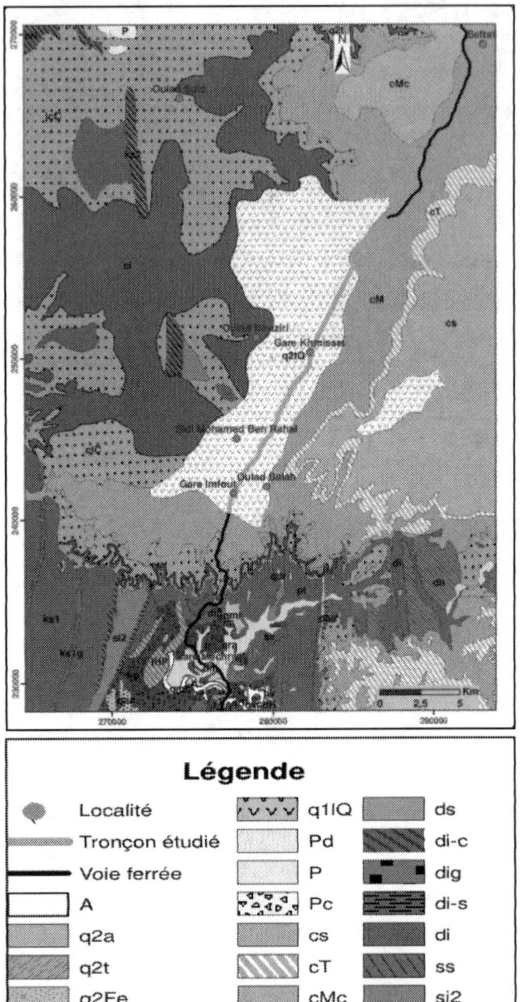

Fig. 9. Extrait des cartes géologiques au 1/200.000 de la meseta entre Settat et Mazagon (Dokkala et chaouia occidental) [16], du plateau du phosphate et la zone synclinale de Tadla [17] et de la meseta entre Mechra BenAbbou et Safi (Abda, Dokkala et Massif des Rehamna) [18] modifiées.

A : alluvions modernes ; q2a : Terrasses limoneuses ; q2t : Tirs ; q2Fe : Limons à concrétions ferrugineuses ; q2l : limons rouges ; q2t ; Q : Croûte ; qra : Terrasses limoneuses ; qma/ q1a : Alluvions anciennes ; qmr : Eluvions ; qar : Quaternaire ancien ; e1 : Eocène inférieur série phosphatée ; q1c : Cônes d'éboulis et de déjections ; q1lQ : Limons inferieurs encroûtés ; Pd : Dunes consolidées attribuées au Villafranchien ; P : Pliocène marin : grès coquilliers, mollasses, lumachelles ; Pc : Conglomérats fluvio-marins pliocènes ; cs : Sénonien : série marneuse ; cT : Turonien : calcaires ; cMc : Cénomanien calcaire ; cM : Cénomanien : série marno-calcaire ; ciC : Néocomien : argiles rouge ; ci : Néocomien marin calcaire ; jcC : Couches rouges de base (sommet de Jurassique) ; dh : Strunien : grès ; dhs : Strunien Schistes ; pt : Permo-Trias : argiles rouges, grès, conglomérats ; hp : Stéphano-Autunien : conglomérats ; hvc : Viséen supérieur : Calcaires ; dm : Dévonien moyen : Calcaires ; ds : Dévonien supérieur : schistes ; di-c : Dévonien inferieur : Calcaires ; dig : Dévonien inferieur : Grès ; di-s : Dévonien inferieur : schistes ; di : Dévonien inferieur ; ss : Gothlandien : schistes à Graptolites ; si2 : Llandeilo : schistes, grès, quartzites ; si1 : Arenig : Schistes ; ks2 : Quartzites d'Imfout ; ks1 : Acadien : Schistes ; xd : Phyllandes d'origine dévonienne ; xsi : Phyllandes d'origine ordovienne ; xk : Phyllandes d'origine cambrienne ; BPT : Basaltes doléritiques du Trias ; dH : Dolérites hercyniennes ; tHP : Trachytes et andésites.

3.4. Fracturation de la zone

Le plateau de Settat a subi les effets de la compression alpine, dont les conséquences sont l'existence quasi permanente d'une légère flexion dans le secteur [5]. Les failles les plus importantes de la région sont de direction NNE-SSW à NE-SW, on trouve aussi une famille de failles d'orientation NW-SE caractérisées par une grande continuité sans décalage apparent, ces failles jouent un très grand rôle dans l'orientation de l'écoulement des oueds du plateau de Settat.

L'imagerie radar et optique fournit des informations fort utiles telle que l'identification des roches et des structures géologiques de surface [19-20], elle permet aussi de mesurer les déplacements de la surface de la terre potentiellement influencée par les changements dans l'humidité du sol [21]. La figure 10 représente le réseau de fracturation du secteur d'étude à l'aide de l'imagerie radar et optique. L'analyse de la rosace des directions de la fracturation montre trois principales directions : ENE-WSW, NE-SW et NNE-SSW qui coïncident avec les familles des failles rencontrées dans l'ensemble de plateau de Settat.

3.5. Analyse des sondages géotechnique

Les neuf sondages carottés de 15 m de profondeur réalisés le long de la voie ferrée permettent de cerner la géologie locale en dégageant l'aspect structural de la zone d'étude ainsi que la nature des formations rencontrées (Figure 11).

En se basant sur les données de ces sondages, le sol en place est formé du marno-calcaire surmonté d'argile et limon tufacés et l'ensemble est coiffé de terre végétale ou remblai hétérogène. Dans le sondage SC36, on a rencontré le substratum sous forme de quartzite à partir de 13,80m de profondeur par rapport au terrain naturel.

Les sols susceptibles de se gonfler sont pour l'essentiel des argiles [22] suite à la succession de périodes humides et sèches en raison des variations saisonnières [23-24], les phénomènes de retrait-gonflement des sols argileux et des formations géologiques argileuses induisent des tassements différentiels [25] qui se manifestent par l'instabilité des ouvrages de génie civil construits [26], l'apparition de fissures et autres désordres dans ces constructions [27], affaiblissement différentiel au-dessous des fondations peu profondes [28]. Les photos la figure 12.**A** illustrent les fissures au niveau des constructions situées à 100 de la voie ferrée dans le secteur d'étude et la figure 12.**B** représente les fissures et glissement de la chaussée de la route nationale n° 9 reliant Settat à Marrakech à 1 Km environ de la voie ferrée.

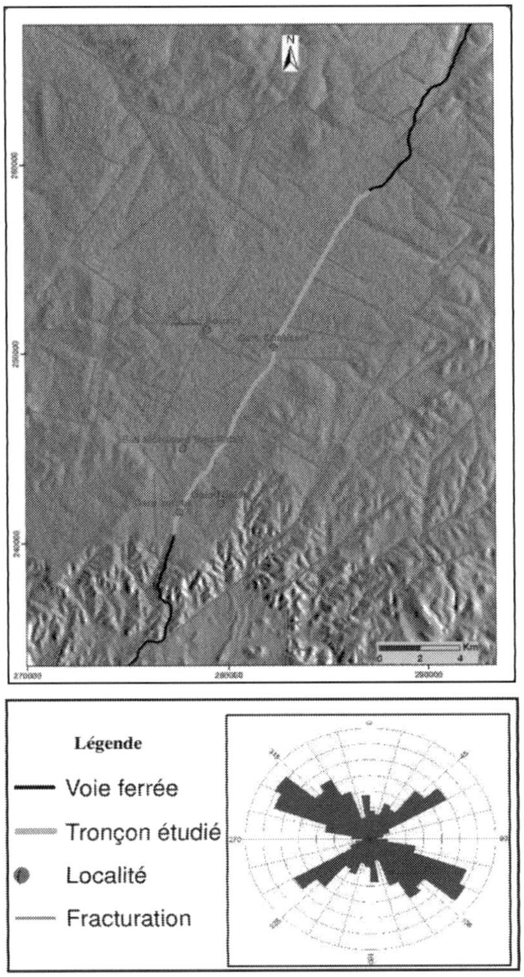

Fig. 10. Réseau et rosace de fracturation

Les problèmes dus aux mouvements des terrains suite à des cycles de retrait-gonflement, reconnues comme un risque naturel coûteux, sont présents partout dans le monde [29], l'abondance de ce sol expansif à l'échelle mondiale a généré trop d'efforts pour mieux comprendre leur comportement [30-31] et d'éviter toutes catastrophes qui peuvent survenir [32]. Ils sont associés à des situations déficientes en eau et dépend d'un ensemble de propriétés mécaniques et physico-chimiques des matériaux [33, 34, 35]. C'est l'une des causes géotechniques les plus courantes [36], et qui a pour conséquence une modification des propriétés physiques du sol entraînant des mouvements de terrain de subsidence et de retrait-gonflement [37].

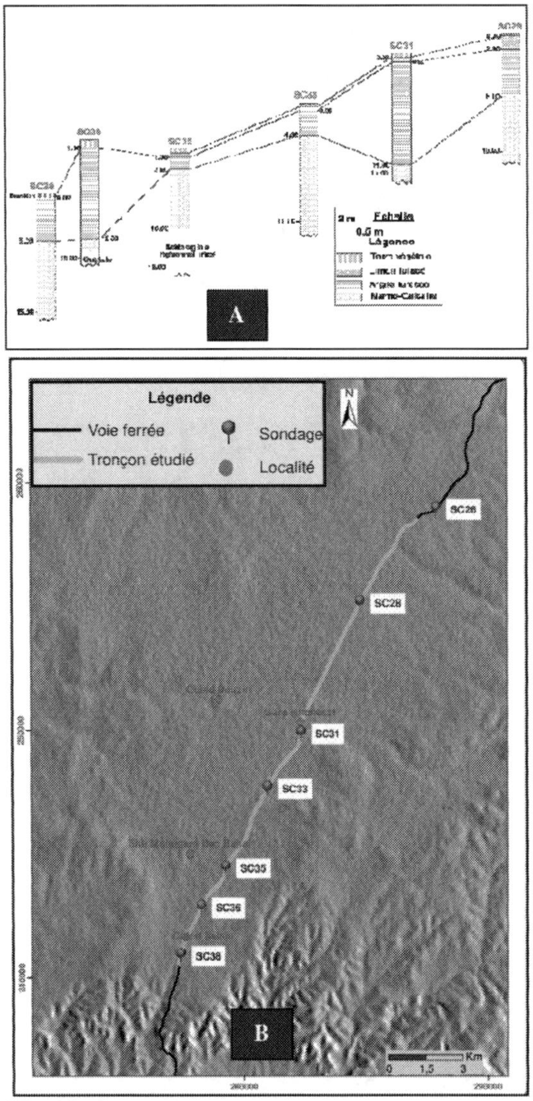

Fig. 11. A : Corrélation et variation latérale des formations à partir des sondages. **B** : Situation géographique des sondages carottés réalisés.

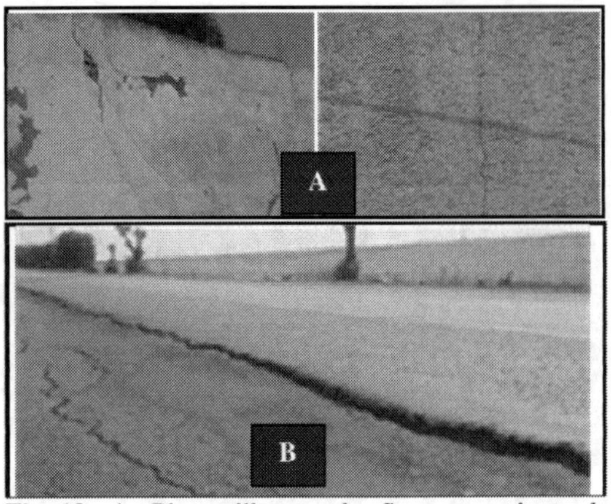

Fig. 12. A : Photos illustrant les fissures au niveau des constructions dans la zone d'étude. **B** : fissures et glissement de la chaussée sur la route nationale n°9

Les formations argileuses et marno-calcaire de la région d'étude sont des matériaux très plastiques [8] et gonflants, qui peuvent causer des problèmes tels que les risques de tassement différentiel vertical et l'ouverture de fissures suite aux variations des teneurs en eau, ce qui favorise la dégradation de la qualité géométrique de la voie. L'aléa retrait-gonflement devra donc être pris en compte dans l'entretien de la voie et dans la planification de projets futurs dans la région.

4. Conclusion

La présente étude a permis d'établir une corrélation entre la dégradation de la qualité géométrique de la voie ferrée et les caractéristiques géotechnique et géologique des sols sous la plateforme ferroviaire. En effet, les formations géologiques du secteur sont constituées d'argiles et marno-calcaire, ce sont des matériaux très plastiques qui changent du volume selon les variations de la teneur en eau : retrait en période de sécheresse, puis gonflement au retour des pluies. Ce phénomène de retrait-gonflement, qui est un risque naturel coûteux, provoque des tassements différentiels verticaux et l'ouverture de fissures qui se manifestent par des désordres affectant principalement les constructions et les infrastructures.

Par ailleurs, le climat aride à semi-aride de zone d'étude, le réseau hydrographie endoréique et les champs de fractures ont favorisé les processus de retrait-gonflement des sols. On a constaté une augmentation de l'amplitude des indicateurs de la qualité géométrique de la voie dans la période d'été et dans zone à forte fracturation.

L'étude géologique, structurale et hydrologique complétée par les données de l'imagerie optique et radar ont permis de mettre en évidence de l'aléa gonflement-rétrécissement qu'est à l'origine de la dégradation de la qualité géométrique de la voie ferrée, par les tassements différentiels verticaux et les pertes de ballast dans les fissures sous la plateforme. Ce phénomène présente un risque naturel qu'il faut prendre en considération dans les opérations de la maintenance et dans les projets futurs dans la région.

Références bibliographiques

1. L. Ricci, Modélisations discrètes et continues de la voie ferrée ballastées. Thèse de Doctorat, Institut Navier-Laboratoire Analyse des Matériaux et Identification (2006)
2. N. Guerin. Approche expérimentale et numérique du comportement du ballast des voies ferrées. Thèse de Doctorat en Structures et Matériaux, Ecole Nationale des Ponts et Chaussées (1996)
3. Office National des Chemins de Fer (ONCF). Instruction relative à la qualité géométrique de la voie. N° d'identification : IN. DMI. R3 009/MVO (2010)
4. N. Rhayma, Contribution à l´évolution des méthodologies de caractérisation et d'amélioration des voies ferrées. Thèse de Doctorat en Génie Civil, Université Blaise Pascal – Clermont Ferrand II (2010).
5. E. El Gasmi, B. El Mansouri et M. Tammael. Les écoulements superficiels dans le plateau de Settat-Ban Ahmed et la plaine de Berrechid : Hydrographie Endoréique. ISSN 2351-8014 Vol. **9** No. 1, 40-53 (2014)
6. E. Hassoune, A. Bouzidi, Y. Koulali et D. Hadarbach, Effets des rejets liquides domestiques et industriels sur la qualité des eaux souterraines au nord de la ville de Settat (Maroc), *Bulletin de l'Institut Scientifique, Rabat, section Sciences de la Vie*, n°**28**, 61-71 (2006)
7. M. El Oumri, A. Osrhiri, R. Moussadek, Z. Moatamid, W. Goebel, A. Tikdirine et A. Ambri. Vocation agricole des terres de la zone de Settat - Rapport et cartes-. Institut National de la Recherche Agronomique (2007)
8. K. El Bouqdaoui. Etude géologique et géotechnique de la Ville de Settat : cartographie géotechnique et substance utile de la Province de Settat, Thèse de Doctorat Géologie, Université Mohammed V, Rabat (1995)
9. G. Choubert et A. Faure-Muret. Plaines et bassins du Maroc atlantique. Ressources en Eau du Maroc Tome 2, *Notes et Mémoires du Service Géologique du Maroc* N° **231**, 1975 (1962)
10. J. P. Ruhard, Chaouia et Plaine de Berrechid. Ressources en Eau du Maroc Tome 2. Plaines et bassins du Maroc atlantique. *Notes et Mémoires du Service Géologique du Maroc*, **231**, 197 p (1975)
11. C. Archambault, M. Combe. J.P. Ruchard, Le plateau des phosphates : ressources en eau. *Notes et Mémoires du Service. Géologique du Maroc*, **231**, 239–258 (1975)
12. A. Piqué, Contribution à la géologie structurale des Rehamna (Meseta marocaine méridionale). Le matériel paléozoïque et son évolution hercynienne dans l'ouest du massif, Thèse de Doctorat en Géologie, l'Université Louis Pasteur de Strasbourg (1972)
13. E. El Gasmi, A. Mridekh, B. El Mansouri, M. Tammal et M. El Bouhaddioui. Apport des données géophysiques et géologiques à la mise en évidence

de nouveaux éléments structuraux associés à la flexure de Settat (Maroc central), *Bulletin de l'Institut Scientifique, Rabat, Sciences de la Terre*, n° **36**, 109-121 (2014)
14. L. Mahdioui, H. Charchari, H. Allali, J. Naja et A. Bakkali. Characterization and development of agricultural land the watershed the Oued Bou Moussa, high Chaouia, Settat, Morocco, IRACST - Engineering Science and Technology: An International Journal (ESTIJ), Vol.**5**, No.2 (2015)
15. A. Michard. Eléments de Géologie Marocaine. *Notes et Mémoires du Service Géologique*, **252**, 426 p (1976)
16. M. Gigout. Carte géologique de la Meseta entre Settat et Mazagon (Dokkala et chaouia occidental) au 1/200000, *Notes et Mémoires du Service géologique du Maroc* (1950)
17. G. Choubert. Carte géologique du plateau des phosphates et de la zone synclinale de Tadla, feuilles Ben Ahmed- El Brouj, au 1/200 000, *Notes et Mémoires du Service géologique du Maroc* n°**137** (1959)
18. M. Gigout. Carte géologique de la meseta entre Mechra BenAbbou et Safi (Abda, Dokkala et Massif des Rehamna) au 1/200 000, Notes et *Mémoires du Service géologique du Maroc* (1950)
19. T. Phan Trong, T. T. Mai, V. L. Ngo, Q. V. Hoang, V. H. Nguyen, V. T. Dinh, D. Closson et A. Ozer. Apport de la télédétection à l'étude des failles actives et des risques sismiques au nord du Viêt-Nam, Revue Télédétection, Vol. **7**, n° 1-2-3-4, p. 465-481 (2007)
20. N. Baghdadi, G. Grandjean, D. Lahondère, P. Paillou et Y. Lasn, Apport de l'imagerie satellitaire radar pour l'exploration géologique en zones arides, *C. R. Geoscience* **337**, 719–728 (2005)
21. J. P. Magnan, H. Ejjaaouani, V. Shakhirev, S. Bensallam, Etude du gonflement et du retrait d'une argile. *Bulletin des Laboratoires des Ponts et Chaussées*, pp 155-170 (2013)
22. S. Zwieback, S. Hensley et I. Hajnsek, Assessment of soil moisture effects on L-band radar interferometry, Remote Sensing of Environment **164**, 77–89 (2015)
23. M. Mrad, Modélisation du comportement hydromécanique des sols gonflants non saturés, Thèse de Doctorat en Génie Civil - Hydrosystèmes – Géotechnique, École Nationale Supérieure de Géologie de Nancy (2005)
24. S. Bensallam, L. Bahi, M. Alaoui, H. Ejjaaouani, V. Shakhirev. The effect of surcharge pressure on the cyclic vertical deformations for expansive clayey soils in Morocco. Geosciences Journal Vol. **18**, No. 1, p. 81 – 87, March 2014
25. H. Nowamooz, Retrait/gonflement des sols argileux compactes et naturels, Thèse de Doctorat en Génie Civil - Hydrosystèmes – Géotechnique, Institut National Polytechnique de Lorraine (2007)
26. A. Muzahim, J. C. Tourayb, F. Bergaya, Une argile modèle pour l'étude du gonflement des sols argileux : la laponite-Na de synthèse. *C.R. Académie des sciences Paris, Sciences de la terre et des planètes*, 329-242 (1999)
27. L. Makki, M. Duc, N. Droniuc, P. Reiffsteck, A. Maloula, J. P. Magnan, Essai de retrait pour une meilleure classification de la sensibilité des sols à la sécheresse, SEC2008, Magnan, Cojean, Cui et Mestat (ed.), Editions du LCPC, Paris (2008)
28. S. Bensallam, L. Bahi, H. Ejjaaouani, V. Shakhirev, Shrinkage Curve: Experimental Study and Modelling. IJE TRANSACTIONS A: Basics Vol. **25**, No. 3, 203-210 (2012)
29. H. Nowamooz, Effective stress concept on multi-scale swelling soils, Applied Clay Science 101, 205–214 (2014)
30. E. Jahangir, Phénomènes d'interaction sol-structure vis-à-vis de l'aléa retrait-gonflement pour l'évaluation de la vulnérabilité des ouvrages. Thèse de Doctorat en Génie Civil - Hydrosystèmes – Géotechnique, Institut National Polytechnique de Lorraine (2011)
31. 27. S. Bensallam, L. Bahi, H. Ejjaaouani, V. Shakhirev, A New Shrinkage Curve Model, Applied to Moroccan Clayey Soil. International Journal of Geosciences, **3**, 507-514 (2012)
32. T. El Bahlouli et L. Bahi, Study On The Swelling Behavior Of The Green Clay Of Gzenaya (Morocco) Compared With Test Conditions. International Journal of Engineering Research and Applications (IJERA), Vol. 2, Issue 6, pp.1413-1423 (2012)
33. S. Fityus, O. Buzzi, The place of expansive clays in the framework of unsaturated soil mechanics, Applied Clay Science 43, 150–155 (2009)
34. S. Bensallamn, L. Bahi, H. Ejjaaouani, V. Shakhirev, K. Rkha Chaham, Clay soil settlement: In-situ experimentation and analytical approach, Soils and Foundations, 54(2):109–115 (2014)
35. H. Ejjaaouani, Interactions des fondations et des sols gonflants : pathologie, calculs et études expérimentales, Thèse Doctorat, Spécialité : Géotechnique, Ecole Nationale des Ponts et Chaussées (2008)
36. S. G. Fityus, D. W. Smith, M. A. Allman, Expansive Soil Test Site Near Newcastle, Journal of Geotechnical and Geoenvironmental Engineering, Vol. **130**, No. 7 (2004)
37. H. Kaveh, Apport de l'interférométrie radar (DinSAR et PSI) pour l'étude des effets de la sècheresse géotechniques : applications à l'Est de la région Île-de-France. Thèse de Doctorat en Sciences de l'Information Géographique, Université de Marne-La-Vallée – Paris Est (2010)

Effect of cement on the stability and the wearrance of soils in earthworks

BENDIMERAD Karam Feth-Allah[1], ZADJAOUI Abdeldjalil[1].
[1]University Abou Bekr Belkaid, Department of civil engineering, Laboratory RisAm, Tlemcen, Algérie.

Abstract. The technique of soil treatment is known for a long time, currently, during the big motorway programs this technique is experiencing important to reach zero imprints and zero deposit in a context of sustainable development or the socio-economic and environmental challenges incite during the work of earthworks. Our work puts highlights the treatment of fine soils in westerns region of Algeria (wilaya of Tlemcen, a project to construct a payment station for the entry of the wilaya (sample n°1), electrical post 400/200 KVA in the locality Ain Fetah (sample n°2)), belonging to the class A_{3h} (very plastic clay) according to the Technical Guide for Earthworks (GTR), whose use in the raw state would involve great risks of stability because of their plasticity (Ip > 20%) and their low capacity (CBR < 25) according to the Specification of the Algerian Loads, Technical Clauses (CSDCA-CT) by the incorporation of the stabilizing agents, for example the variant of locally manufactured CEM II / A42.5N composite cement, in order to define their behaviours by examining the variation of their characteristics determined according to the Algerian standards and at a ambient temperature according to the additive dosage and to contribute to the deduction of the interest of the method demonstrated previously. The results obtained clearly show a tendency the rearrange classification of the soils studied according to the unified system en proportion whit to the additive dosage (0 to 14%) in view of the appreciable change in the plasticity behaviour, which results in a decrease in the plasticity index, going up to 37.7% and the 11% of the maximum dry density accompanied by a satisfactory increase in the suitability of the materials to be supported a load with different moisture content and changes in water regime exceeds 100%. Given the unavailability of other exploitable deposits near the site and the experimental results obtained, the variant of the treatment seems feasible.

KEYWORD: Fine soil, Cement, Soil treatment, Geotechnical, Durability.

1 Introduction

During the last two decades, Algeria has experienced a considerable development in its infrastructure, including transport (road, highway and railroad), this progress has highlighted a technical-economic and environmental in order to interaction to limit the impact of this progress use exclusively have by the exclusive use of soils located in the land reserve of the earthworks project.

This objective requires the use of soils with insufficient properties, given the geological characteristics of the region of western Algeria that is part of an arid to semi-arid zone delimited by the Tell Atlas at the north and the Saharan Atlas at the south is extended of east at west until to the countries bordering the Maghreb, characterized by a cold and wet winter and a hot and dry summer, the majority of these formations are fine-matrix (clay, silt, marl , etc ...) making subject of several construction problems for a road application and pose a compaction problem in the dry side (difficult or impossible) and the wet side (sticky and expansive on the other hand). [4, 11]

So, in this sense, the improvement of the soils end by the incorporation in situ of the stabilizers (additives), for example, cement, remains one of the potential solutions to achieve this goal.

The first works in this field date back more than fifty years, which developed simultaneously with the knowledges of the mechanical performances of improved soils, their short-term and long-term behavior and interaction between the additive and the fine particles of soils which induces a series of complex physicochemical phenomena with very variable kinetics modifying the characteristics of the soil which results in a modification of water content, density of compaction and their lift. [1-3]

The present work is intended as an like a experimental contribution in this field at the macroscopic scale, it's about application of this technique to the case of thes soils of western Algeria with a goal, zero borrowing, zero deposit and which spread over two parts, the identification and analysis of the influence of the treatment on the physical and mechanical properties of the two soils studied.

2 Experimental materials and methods

2.1 Materials and additives

The two soils studied in this work were collected from the wilaya of Tlemcen (M1 at the locality of El Koudia, on the other hand M2 of locality Ain Fetah), extract between 1.50 m and 4.50 m depth and then transported to the mechanical soil laboratory of the University of Tlemcen.

The different physical and mechanical characteristics of the samples studied were determined in accordance with the Algerian standards (NA) and classified according to the GTR, they are reported in table 1.

Table 1. Geotechnical characteristics of the samples studied.

	Standards	M_1	M_2
D_{Max} [mm]	*Particle size analysis : NF P 94-056 _ NA 5232.*	≤ 50	≤ 50
≤ 2 mm [%]		100	85
≤ 80 µm [%]		75	70
Ip [%]	*Atterberg limits : NF P 94-052 _ NA 16213.*	34	28
V.B.S.	*Methylene Blue test : NF P 94-068 _ NA 5288.*	6,15	4,50
W_{opt} [%]	*PROCTOR test : NF P 94-093 _ NA 5262.*	5	7
ρ_d [t/m^3]		1,675	1,725
$CBR_{Immediate}$ [%]	*CBR test : NF P 94-078 _ NA 5252.*	12	15
CBR_{Soaked} [%]		7	9
Classification GTR :		A_{3h}	A_{3h}

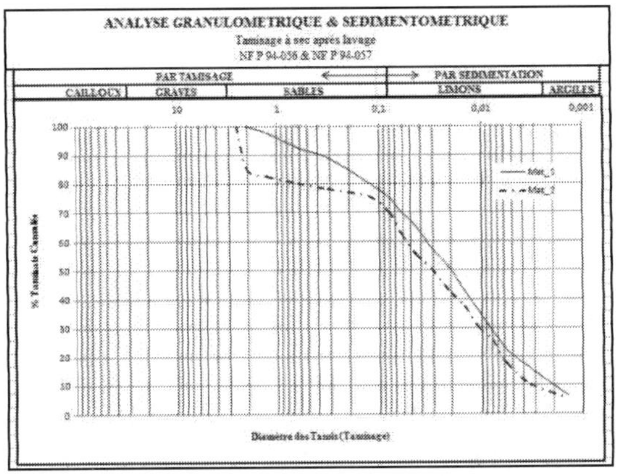

Fig.1. Granulometric curves of the two soils.

The cement used as additive in this study is one of the local products manufactured by the Beni Saf cements company in the western region of Algeria, delivered in a 50 kg bag certified according to the Algerian NA 442 standard.

The technical characteristics provided by the manufacturing plant are summarized in the tables below, tables 2 and 3.

Table 2. Data sheet of cement used.

Type of cement	**CEM II/A-P**
Resistance class	**42.5 N**
Clinker Protland (% by weight)	**80 -94**
Secondary constituents _ P (% by mass)	**6 - 20**

Table 3. Physico-chemical characteristics of cement.

Sulphate content [%] (SO_3)	**≤ 3.5 %**
Magnesium oxide content [%] (MgO)	**≤ 5.0 %**
Start of taking [min]	**T ≥ 60.0**
Compressive strength at 2 days [MPa]	**Li ≤ 12.5**
Compressive strength at 28 days [MPa]	**Li ≤ 42.5**

2.2 Methods

In order to appreciate the variations of the geotechnical characteristics of the samples studied during their interaction with the additive (cement) and in addition to the identification tests, an experimental program was developed in place such as : classical density tests (Proctor compaction), plasticity (Atterberg limits) and immediate and impregnated California Bearing Ratio (CBR), carried out on untreated samples (controls) and on treated samples with various cement content according to the schedule summarized in table 4. [5-10]

Table 4. Different combinations.

-	% Mat.	% Ciment	-	% Mat.	% Ciment
1	100	0	7	90	10
2	98	2	8	89	11
3	96	4	9	88	12
4	94	6	10	87	13
5	92	8	11	86	14
6	91	9			

2.2.1. Sample preparation

After the extraction of the samples, put them in an oven for 24 hours at a temperature of 105 ° C in order to eliminate all traces of moisture, it was then proceeded to their grinding and then to the dry sieving to have a very fine grind.

The cement in turn, went through dry sieving to avoid lumps.

Of the soil and additive quantities were weighed and dry blended according to the combinations mentioned above and stored in dry places to avoid tjeir humidification. [4, 12]

3 Results and discussions

3.1 Results

The following figures from 2 to 6 show the effect of the cement on the clay content, the compaction content, the dry density and the strength of the two materials studied.

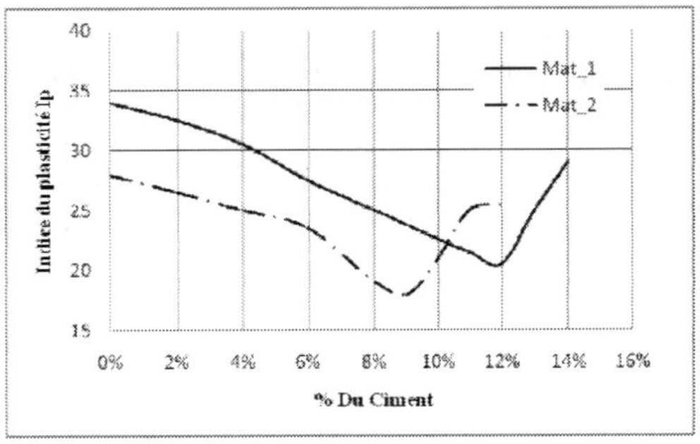

Fig.2. Effect of cement on plasticity.

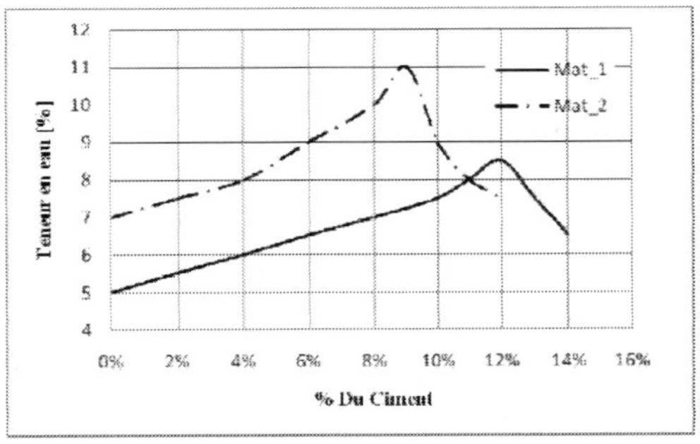

Fig.3. Effect of cement on compaction water content.

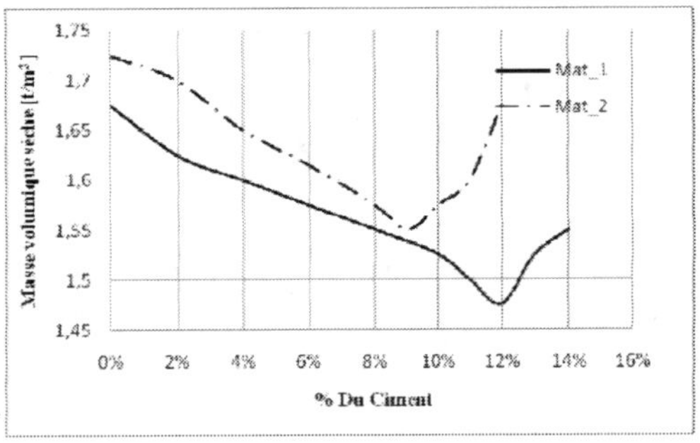

Fig.4. Effect of cement on dry density.

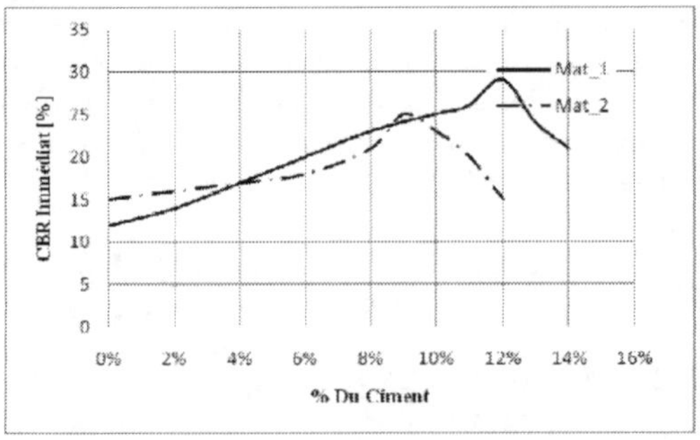

Fig.5. Effect of cement on resistance "CBR Immediate".

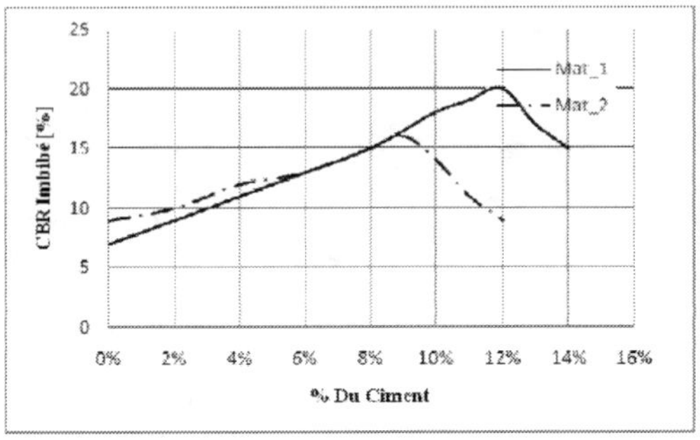

Fig.6. Effect of cement on resistance "CBR Soaked".

3.2 Discussions

The figures above show the results of the tests carried out on the two materials, from where a reshuffling of the geotechnical characteristics is clearly observed in the short term in proportionality with the cement dosage until the opposite effect on the characteristics of compaction (increase in the water content and a lowering in dry density of about 11%), of plasticity (a decrease going up to 37.7% of the plasticity index by aport to its natural state) and the wearrance (an indisputably improvement in CBR indices exceeding thes 100% at various water regimes).

These variations are consistent with the results of several researchers working on different types of soils that can be explained by a series of physico-chemical phenomena with very variable kinetics from the moment of the addition of the additive in the presence of water :

1. Hydration of cement defined as an exothermic reaction that leads to a reduction of the soil water content, concomitantly the variation of plasticity.

$$CaO + H_2O \rightarrow Ca(OH)_2 + Chaleur_{\ (15.5\ kcal/mol\ CaO)} \quad (1)$$

$$Ca(OH)_2 \rightarrow Ca^{2+} + 2OH^- \quad (2)$$

2. Agglomerations of the elementary elements of the soil (flocculation of the clay particles) due to the cation exchange capacity of the clays to fixed the calcium ions Ca^{2+} released during the additive

addition and leads to a modification of the characteristics of the compaction.

3. Pozzolanic reactions (with a more or less slow kinetics) due to the saturation of the minerals in calcium to train of the new hydrates, such as hydrated calcium silicate (CSH), which initiates a second phase in the development of mechanical performances (the bearing of the soil).

These processes are conditioned by the proportion, morphology, specific surface and the cationic exchange capacity of the clay minerals in the soils under study. [1-12]

4 Conclusions

The results obtained in this study give a practical insight into the chemical treatment of soils saw a road application and constitute a important asset when earthworks are being carried out, or the use of cement as a stabilizing material seems to be one of the many methods available allow a durable, stable and economical construction of embankments and layers of form presenting satisfactory characteristics only by using the materials located in the land reserve of the project thanks to the facility of adaptation to the constraints of exploitations and implementation despite the complexity of mechanisms reactions during in the additive soil interaction, what might seem surprising for research that wants to be applied by simple field and laboratory tests, needs precise and rigorous control.

Bibliographical references

1. N. Cabane, *Sols traités à la chaux et aux liants hydrauliques* (Université de Jean Monnet, 2005).
2. V. Ferber, *Sensibilité des sols fins compactés à l'humidification* (Université de Nantes, 2005).
3. L. Saussye, *Traitement des sols aux liants hydrauliques* (Université de Caen Basse Normandie, 2012).
4. W. Khebizi, *Etude de la durabilité des sols argileux améliorés avec des ajouts minéraux* (Université de Chlef, 2011).
5. A. Al-taim, M. Disfani, R. Evans, Procedia engineering J. E. **143**, 615-622 (2016).
6. D. Ribeiro, R. Néri, R. Cardoso, Procedia engineering J. E. **143**, 59-66 (2016).
7. M.A. Ozdemir, Procedia engineering J. E. **143**, 498-505 (2016).
8. G. F. Bell, Engineering geology J. E. **42**, 223-237 (1996).
9. O. Cuisinier, A. Jean-Claude, D. Dimitri, Engineering geology J. E. **123**, 187-193 (2011).
10. L. Saussaye, M. Boutouil, F. Baraud, L. Leleyter, Engineering geology J. E. **189**, 89-103 (2015).
11. A. Mhamedi, M. Khemissa, HBRC J. E. **11**, 82-90 (2015).
12. G. Geoffrey, H. Schad, Otto-Grah Journal J. E. **11**, (2000).

 https://doi.org/10.1051/mateconf/201814902023

Modelling of fine soils railway embankments behaviour under real cyclic loads

Soufiane GUESSOUS, Nouzha LAMDOUAR
Mohammedia School of Engineering (E.M.I) – Mohammed V University in Rabat – Morocco

Abstract. Before the development of regulations, standards and guides for infrastructure building, embankments were generally constructed, by reusing and compacting the excavation materials. Therefore, a large part of existing railway embankments, over the world, is composed of fine compacted soils such as clay or marl. However, the current international trend of the railway undertakings and transport contractors is within the intensification of traffic, and the increase of train's load and speed. Therefore, infrastructure managers are always called upon to evaluate the bearing capacity of existing infrastructures and define the necessary reinforcement solutions. In this perspective, a better awareness and knowledge of the behaviour of fine soils railway embankments is required to define optimal and efficient reinforcement solutions. In that matter, the article will feature the results of finite elements modelling of instrumented railway embankments that were constructed in Morocco in 1920, with marl compacted soil. The model is based on real applied loads and hydraulic conditions monitored for the last three years, which allow to simulate the hydro-mechanical behaviour of the embankment for this period. The article will be also comparing real monitored deformations recorded by the inclinometers equipping the embankments, with the output of the model using different models and adjustments.

1. INTRODUCTION

In the last years, the tendency of the railway undertaking and managers has become within increasing the traffic in order to satisfy the demands of their clients. Therefore, the railway industry has known a very important evolution in the last century within the load and speed of trains, and the weight of superstructure elements (figure 1):

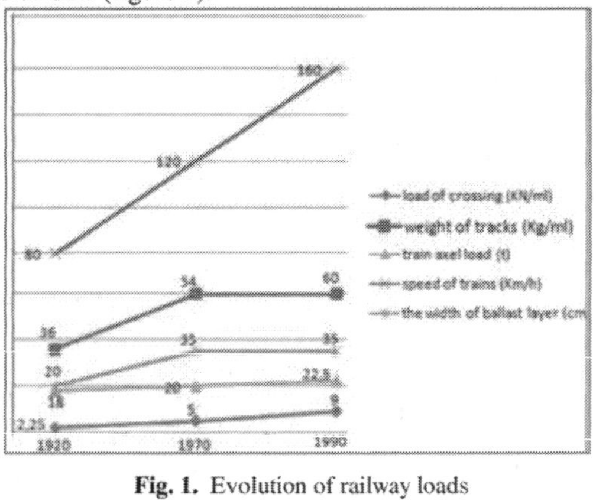

Fig. 1. Evolution of railway loads

However, a large number of existing embankments were constructed in the last 19th and the first of 20th century with clay and marl soils, and submitted millions of hydro mechanical loading cycles, which make the evaluation of their bearing capacity very complicated and difficult.

In that matter, the paper presents the results of a finite elements simulations and modelling of two railway embankments that were constructed in 1920 in morocco using Plaxis 2D version 8.2. These models are based on the geotechnical investigations and the loads monitoring that were conducted in the last three years. The results of the modelling are compared to the deformation recorded for the same period by inclinometers, in order to determine the most convenient materiel model to use in similar modelling and the adjustments to make on the soil characteristics for a better simulation of this type of infrastructures.

2. MODEL'S INPUT :

2.1 Geometry model:

The geometry models were based on topographic cross sections in the axes of the inclinometers implemented at the embankments at Km 244+520 of the line Tanger/Fès and Km 47+960 of the line Fès/Oujda (figures 2. & 3.)

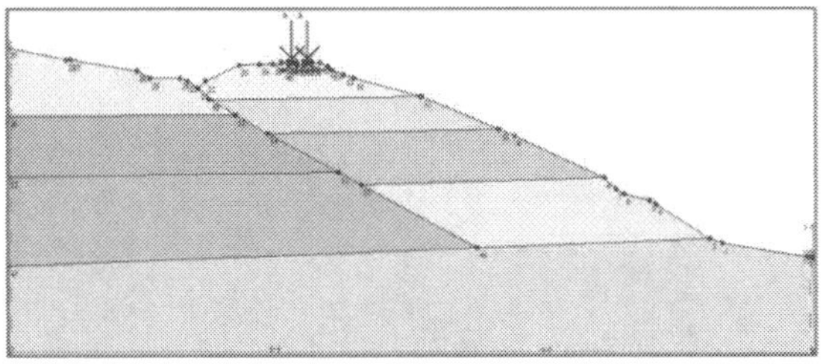

Fig. 2. Geometry model of embankment 1 (47+960)

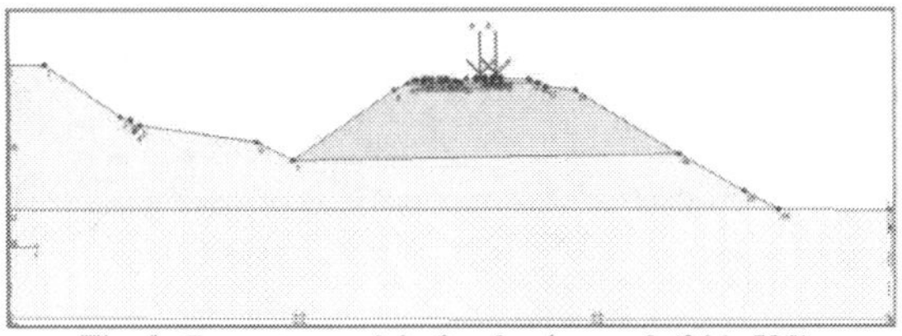

Fig. 3. Geometry model of embankment 2 (244+520)

2.2 Material setting:

2.1.2 Sleepers and ballast characteristics:

The ballast and the sleepers were modelled by a linear elastic model with the following characteristics according to Sauvage G. (1993) [1], and Profillidis (1983) based on the results of the plate tests conducted at derby (1980) [2]

Table 1. Characteristics of ballast and sleeprs as given by Profillidis (1983) [2]

Material	E (MPa)	υ	ρ
Ballast	150	0,2	1,7
Sleepers (split concrete)	30000	0,25	2,4

2.1.2 Soil models and associated parameters:

The material model used to characterise the marl soil is the elastoplastic HSM model.

This model is assumed to be isotropic, depending on the plastic shear and volumetric strains. Unlike the Mohr–Coulomb Model, the stress–strain relationship, due to the primary loading, is a hyperbolic curve in the HSM. The hyperbolic function, as given by Kondner (1963) [3], and later explained by Schanz and al. (1999) [4] and Brinkgreve (2002) [5] is for drained triaxial test:

$$\varepsilon_1=\frac{q_a}{2E50}\frac{q}{q_a-q'} \qquad (1)$$

ε_1: the axial strain,

q : the deviatoric stress.

A total of 10 input parameters are required in the Hardening Soil Model, although some of them have default values at PLAXIS database. The main parameters are γsat, γunsat, Cref, φref, Eur, Eoed E50 and permeability (Kx, Ky). These parameters were based on the pressiometer and oedometric tests results. Eur was considered equal to Pressiometric modulus at first and was adjusted owing to the models results. Eoed and E50 were considered approximately as 1/3 of Eur. The following tables summarize the parameters considered for each model:

Table 2. Material HSM parameters for the model of embankment 47+960

Material	γsat	γunsat	Cref	φref,	Eur =Ep	Eoed	E50
Clay CF	20	16	8	20	4500	1500	1500
Clay CF2	20	16	10	24	8000	2700	2700
CLAY CF 3	20	18	15	28	18000	6000	6000
Clay remblai	18	16	5	22	13000	4300	4300
Clay subs	18	16	15	26	50000	17000	17000
Clay TN	17	15	5	20	3600	1200	1200
Clay TN2	20	18	5	20	8000	2600	2600
Clay TN3	20	18	15	24	13000	4300	4300

Table 3. Material HSM parameters for the model of embankment 244+520

Material	γsat	γunsat	Cref	φref,	Eur=Ep	Eoed	E50
Remblai	20	18	15	24	10000	3000	3000
Clay 2	20	17	8	22	6000	2000	2000
Clay 3	21	18	15	28	35000	11000	11000

3. METHOD OF ANALYSIS :

3.1 Loading :

Real applied railway loads depend on various factors such as the axel static load, the infinitesimal ondulation of rails, track geometry defaults, ovalisation of wheels, wheel/ rail interface...

Many researches were conducted to determine the relationship between Axle load and track deflection, such as the research of Jaco Vorster & Hannes Gräbe

(2010) [6] who used a particle image velocimetry instruments to determine the relationship between Axle load and track deflection, and formulated a logarithmic shape relationship for axel load under 28 t :

$$\mathbf{d} = 0{,}50 \ln(F) - 0{,}29 \qquad (2)$$

F: Axel Load, d: Track deflection

However, the best way to evaluate the axel real applied load is to use instruments and implement loads captors.

Therefore, an in order to reach the precision aimed in the models, the railway real applied loads were recorded for the last two years, by a load monitoring stations that were installed under the tracks near the embankments. This system is based on the measuring of track deflection during the passage of the train's axels.

The monitoring system is composed of twelve fibre optical sensors per track that communicate the measurements to a treatment station.

Fig. 4 Optical bending sensors installed under the track

The recorded loads were introduced in the Plaxis models in a shape of different phases of unloading and reloading using the staged construction box.

3.2 Water conditions:

In order to define the hydraulic factor and integrate its impact on the soil's deformation, water conditions were defined and adapted according to the rainfall and precipitations recorded in the last two years.

To do so, we used the relationship between rainfall and groundwater level described below.

Under the assumption that the rainfall and groundwater level are proportional and the water stored in the soil is gradually accumulated (figure 5), the expression after N times of rainfall is given by Conte and Troncone (2016) [7] as :

$$h(x,t) = \sum_{j=1}^{N} \frac{h_{rj}}{n(1-S_r)} \exp[-k_t i \cos\theta(t - t_{0j})] \qquad (3)$$

Where :
- θ : the angle between infinite slope and horizontal line
- h(x,t) : the change in groundwater level in the vertical direction is due to rainfall infiltration
- Kt = K/b : with K is the permeability of the layer and b is its width
- hrj is the volume of water in the slope (unit area) due to the rainfall infiltration in jth time
- Sr : saturation degree of the soil

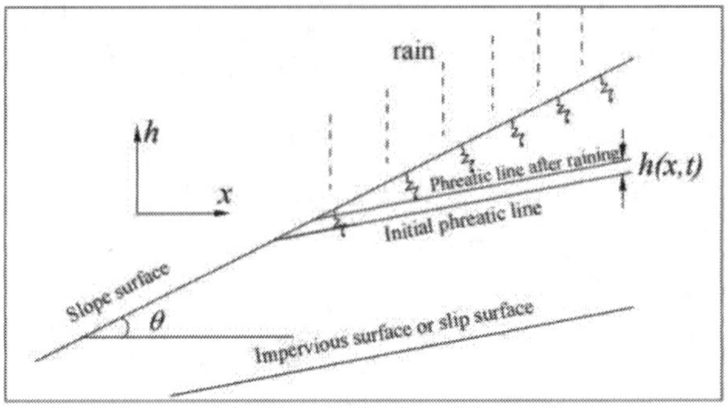

Fig.5 Slope model with an indication of the steady-state groundwater level change owing to rain infiltration h(x,t), and the slip surface (modified from Conte Troncone 2016) [7]

Based on these formulations and results, the phreatic level was calculated and changed in the models owing to the rainfall monthly average.

3.3 Output analysis:

After operating the calculations, the horizontal displacement curve was extracted at a 1,5m depth point from the crest of the slopes in the axis of the implemented inclinometers.

This point was chosen as a reference, because the topper section gives the most significant response to the unloading and reloading cycles and the water conditions fluctuations. Also, the depth of 1,5 m was chosen in order to avoid the impact of soil's erosion

The horizontal displacement curve was analysed from two perspectives:
- ➢ Comparing the results of the models with the real monitored displacements (inclinometers).
- ➢ Evaluating the impact of cyclic variable loads and water conditions fluctuation of the embankment's deformation.

4. RESULTS AND DISCUSSIONS :

4.1. Comparison of the models results with the inclinometers monitored displacements

In order to monitor the real soil displacements, inclinometers were installed at the crests of the embankments with a depth of 30 meters.

The measurements of these instruments for the last two years were extracted and allowed to establish the curve of mensal displacements evolution:

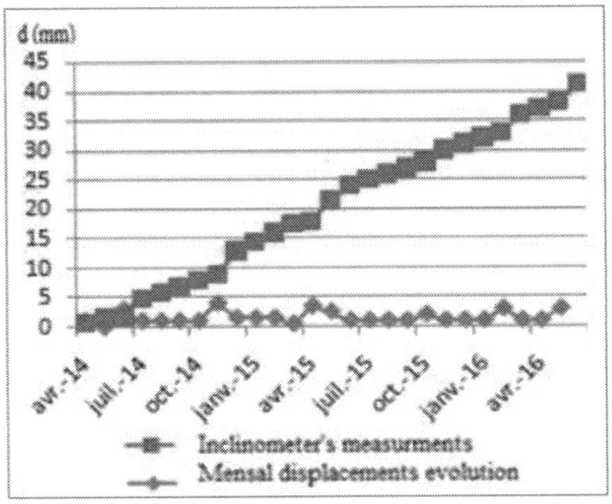

Fig. 6 Inclinometer measurements and mensal displacements evolution for embankment 1

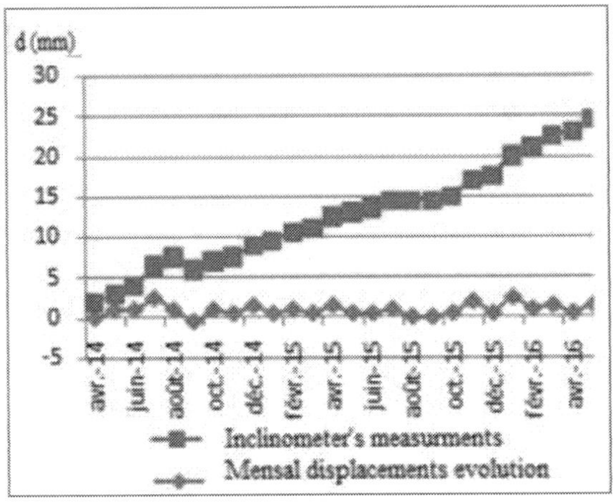

Fig. 7 Inclinometer measurements and mensal displacements evolution for embankment 2

At first,, the unloading reloading stiffness Eur was considered equal to pressiometric modulus in order to have a reference model. Afterwards, the stiffness was modified several times in order to reach and define the optimal Eur/Ep ratio that allows having coherent and similar results with the real measured horizontal displacements. The intervalle considered for the variation of Eur is : (Ep, 2,5 Ep). This interval is based on experiments conducted by Gomes Correia, Armando and Michel Apageo, in 2004 [8] (figure 8).

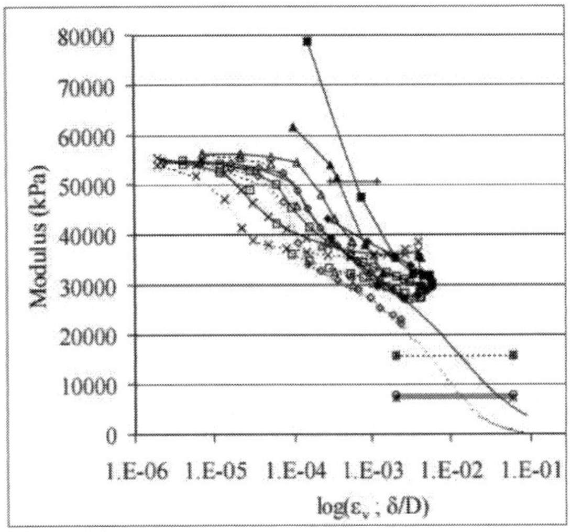

Fig.8 Moduli as a function of strain level for various numerical simulations and test analysis (Gomes Correia, Armando and Michel Apageo, 2004)

The different calculated models showed a convergence between the displacement curve and the measurements of inclinometers for a Eur= 1,5 Ep (fig.8).

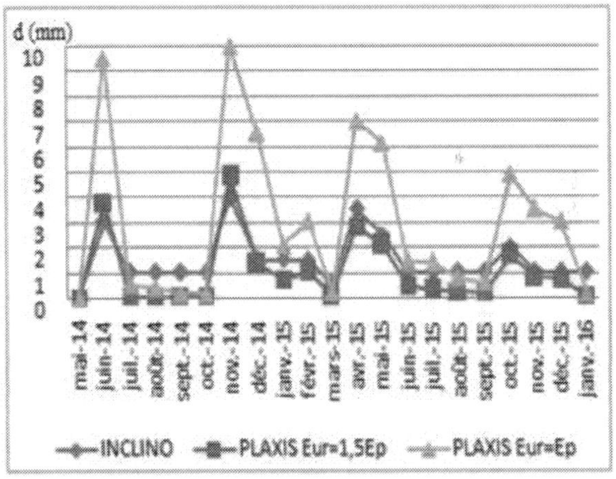

Fig. 9 Inclinometer measurements compared to horizontal displacement calculated with Eur= Ep and Eur=1,5Ep for embankment 1

The same ratio was used in the model of the second embankment and showed a very finite convergence between the measurements and the displacement curve (Fig 9) :

https://doi.org/10.1051/matecconf/201814902023

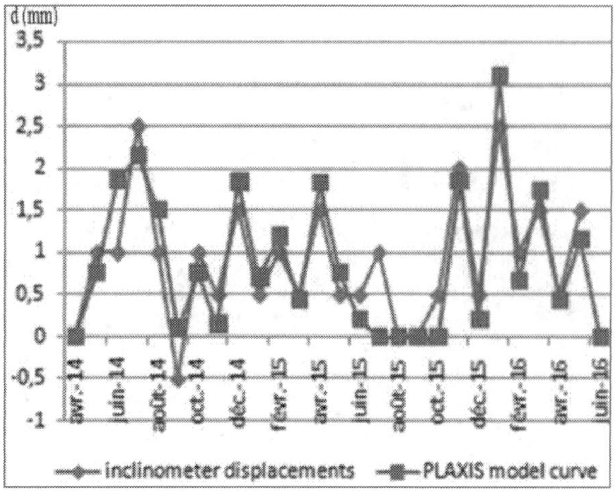

Fig. 10 Inclinometer measurements compared to horizontal displacement calculated with Eur = 1,5Ep for embankment 2

4.2. Impact of variable cyclic loads and water conditions:

The analysis of displacement curves extracted from the models showed that the deformation increases significantly with the groundwater level and the rainfall volume increase. While its much less sensitive to the load increase in the real conditions of railway load fluctuation proportions. Figures 11 and 12 show that after every period of rainfall we have an average displacement of 2-3 mm/month within the embankments:

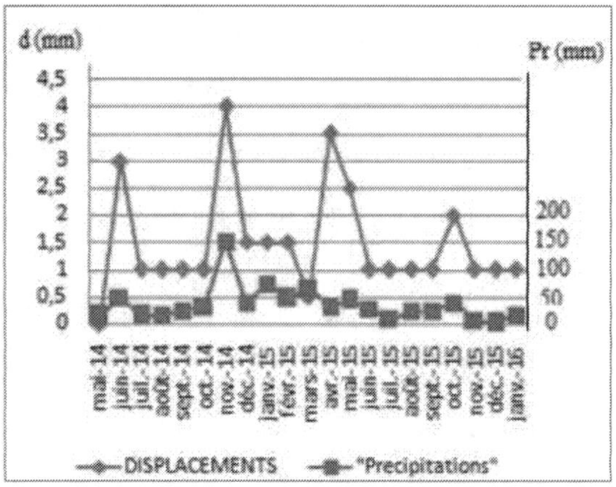

Fig. 11 Mensal displacements and precipitations (embankment 1)

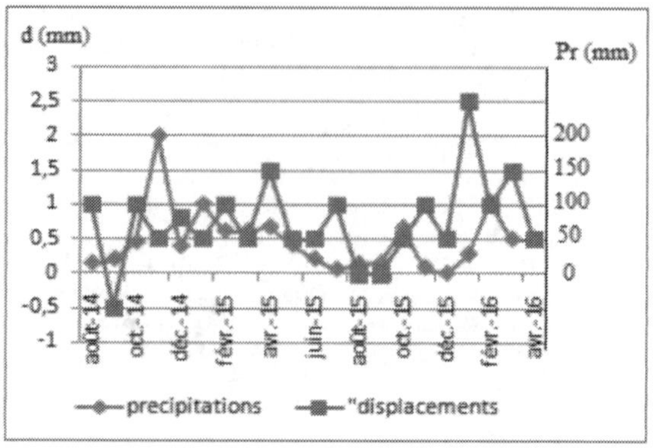

Fig. 12 Mensal displacements and precipitations (embankment 2)

On the other hand, the figure 13 shows that the increase of load impacted the displacements only in two cases where the raise of the load was 7 to 10%.

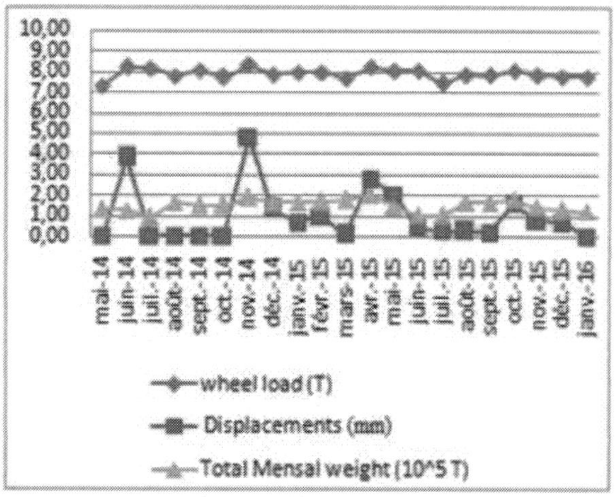

Fig. 13 Evolution of displacements owing the load and total mensal weight.

CONCLUSIONS

The following conclusions can be drawn from the finite model study and the instrumentation approach described in the article:

- The HSM is a very convenient model for simulating the behaviour of fine soil's railway embankments with a history of cyclic loading and hydraulic fluctuations. In which case, The unloading reloading stiffness can be considered as 3/2 of the Menard pressiometer Modulus for a precise approached results.

- The irreversible deformation of soft soils embankments depends mainly of the groundwater level and the rainfall volume, and can evaluate by relatively high values (2 to 3 mm per month). While the influence of the load punctual increase is much less significant in the proportions of train load fluctuation (less than 1mm).

REFERENCES:

1. Sauvage, G., 1993. *Railway track vertical static behaviour*, INRETS/LTN.
2. Profillidis, V., 1983. *La voie ferrée et sa fondation - Modélisation mathématique*. Thèse de doctorat, Ecole Nationale des Ponts et Chaussées.
3. Kondner, R.L., 1963. *Hyperbolic stress–strain response cohesive soils*. Journal of Soil Mechanics and Foundations Division, ASCE 89, 115–143.
4. Schanz, T., Vermeer, P.A., Bonnier, P.G., 1999. *The hardening soil model: formulation and verification, Beyond 2000 in Computational Geotechnics*. Balkema, Rotterdam. (1999)
5. Brinkgreve,R.B.J.,.*PLAXIS Finite Element Code for Soil and Rock Analysis*—version 8.Balkema,Rotterdam. (2002)
6. Hannes Gräbe, Jaco Vorster, *Axle load and track deflection on a heavy haul line* (2010)
7. Enrico Conte, Antonello Troncone, Antonion DONATO, *A simple approach for evaluating slope movements induced by groundwater variations(2016)*
8. Gomes Correia, António Universidade do Minho, Portugal. Antão, Armando Universidade Nova de Lisboa, Portugal Gambin, Michel Apageo, France *Using a non linear constitutive law to compare Menard PMT and PLT E-moduli* (2004)

Estimation of the strength of Nubian sandstone formation from point load test index and other simple parameters

Abdul Karim M. Zein[1] and Mutasim A. Sandal[2]

[1]Building and Road Research Institute, University of Khartoum, Khartoum, Sudan
[2] General Administration for Buildings, Ministry of Physical Planning, Khartoum State Khartoum, Sudan

Abstract: This study investigates the correlation of the uniaxial compressive strength (UCS) and the point load test (PLT) index $I_{s(50)}$, bulk density, water absorption and the RQD properties of the Sudanese Nubian sandstone formation. The UCS being is the rock property needed in engineering practice but its determination is tedious, time consuming and expensive. Alternatively, the UCS may be indirectly evaluated through establishing relationships with rock parameters which are easier, cheaper and more convenient to determine in the laboratory or in the field. An extensive laboratory testing was executed to determine the above rock properties for many NSF samples taken from Khartoum State and other areas. Statistical analysis was performed on the test data and a reliable linear regression equation has been developed with a UCS to PLT $I_{s(50)}$ conversion factor of 10.18 and may be applied to estimate the strength of the Sudanese sandstone formation. The developed correlation is in good agreement with few of the many methods published for similar rock types which indicates that large errors may result in from applying an inappropriate UCS prediction method. Thus, it is important to establish separate correlations or validate published correlations to check their suitability for a specific rock types and local geologies. Useful correlation relationships of reasonable accuracy were also established for rough estimation of the UCS from the bulk density, water absorption and the RQD properties of the Nubian sandstone formation.
Keywords: Nubian sandstone formation, UCS, Point load index $I_{s(50)}$, Bulk Density, absorption, RQD.

1. Introduction

Evaluation of the shear strength characteristics of rocks is an important aspect in most geotechnical investigations related to engineering problems. Rock strength is normally evaluated in terms of the uniaxial compressive strength (UCS) which can be directly and accurately determined from laboratory testing. However retrieval of high quality samples is a prerequisite for testing which is often difficult to obtain for weak and highly weathered rocks and the test is time-consuming and expensive. Alternatively, the UCS may be indirectly estimated from the results of point load test (PLT) or other rock properties. The PLT test can be carried out on irregular shaped specimens and the equipment can be easily transported to the field.

Many empirical methods have been proposed by researchers from different countries to estimate the UCS from the PLT strength Index $I_{s(50)}$. However, the reliability of these relationships depends on the type, location and degree of weathering of rocks.

To the authors knowledge there is no reported research on how the UCS and the PLT strength index $I_{s(50)}$ are related to each other for the Sudanese rock types. The objective of this study is to establish the correlation between the two strength parameters for the Nubian sandstone formation which constitute a main geological feature on which many important structures are founded. The study also investigates the variations of the UCS with the bulk density, water absorption and rock quality designation (RQD) for the same rock formation.

2. Rock strength parameters

The UCS is often used in engineering practice for different applications such as the design of foundations and the stability of rock slopes. The test is typically determined as specified by the ASTM standard D7012-04 [1] from axial loading of unconfined cylindrical rock specimens of flat, smooth parallel ends with a length to width ratio of 2 to 2.5.

The PLT is considered as an attractive indirect substitute to the UCS as it can provide similar data faster and at a lower cost. The ASTM gives a clear justification for the use of the PLT [2]. The test is especially useful when core specimens cannot be extracted from fractured or weathered rock masses. The PLT can be easily performed in the field or the laboratory on samples in the form of cores, blocks or irregular specimens with different diameters which require less or no preparation.

Other rock strength related properties include the "Rock Quality Designation, RQD", bulk density and water absorption. The RQD provides a basis for making preliminary decisions with respect to required depths of excavation for structures foundations. The above properties also can serve to identify potential problems related to bearing capacity, settlement, erosion, or sliding in rock foundations.

2.1. Correlation of UCS and PLT $I_{s(50)}$ index

Many authors have developed empirical relationships between the UCS and the PLT $I_{s(50)}$ with a conversion factor "*k*" expressed by the following linear equation:

$$UCS = k\, I_{s(50)} \qquad (1)$$

It has been shown that the factor *k* is not constant but is rock type dependent; thus, it is important to establish separate correlations for specific rock types and local geologies.

The earliest and most widely used correlation equations were proposed by Broch and Franklin [3] and Bieniawki [4] with a k value of 24 based on data pertaining to different rock types. Since then, many equations have been formulated either for different types of rocks or for local formations. Examples of some of the available UCS-$I_{s(50)}$ correlation relationships are listed in Table 1 with the details given in the corresponding reference. As may be noted from Table 1, most of the UCS-$I_{s(50)}$ equations are linear but some equations are linear with a non-zero intercept or nonlinear in the forms of power or exponential equations. The conversion factor of linear equations varies between 6 and 25 which clearly indicates that using an inappropriate correlation equation may result in erroneous evaluation of the UCS by several times the $I_{s(50)}$ value.

3. Nubian sandstone formation

In Sudan, the term "Nubian Formations (NF)" refers to the cretaceous continental sediments composed of sandstone, mudstone siltstone, conglomerates and pebbles. They are either exposed or overlain by recent quaternary formations and extend between latitude 22° north and 10° south. It constitutes about 45% of the total area of Sudan distributed in the northern, central western and eastern parts as depicted in Fig. 1.

Table 1 Examples of published UCS-$I_{s(50)}$ equations

Correlation equation	Reference name and No.	
$UCS = 23.9\, I_{s(50)}$	Bieniawki (1975)	[4]
$UCS = 18.7\, I_{s(50)} - 13.2$	Singh (1981)	[5]
$UCS = 14\, I_{s(50)}$	Foster (1985)	[6]
$UCS = 18\, I_{s(50)}$	Das (1985)	[7]
$UCS = 24.8\, I_{s(50)}$	Hawkins & Oliver (1986)	[8]
$UCS = 17.4\, I_{s(50)}$	Vallejo et al (1989)	[9]
$UCS = 19\, I_{s(50)} + 12.7$	Ulusay et al (1994)	[10]
$UCS = 24\, I_{s(50)}$	Smith (1997)	[11]
$UCS = 20.6\, I\, s(50)$	Rusnak and Mack (1999)	[12]
$UCS = 3.86(I_{s(50)})^2 + 3.86\, I_{s(50)}$	Quane and Russel (2003)	[13]
$UCS = 7.31\, (I_{s(50)})^{1.71}$	Tsiamaos & Sabatakkis (2004)	[14]
$UCS = 23\, I_{s(50)}$	Koukis et al (2007)	[15]
$UCS = 6.5\, (I_{s(50)})^{1.8}$	Koukis et al (2007)	[15]
$UCS = 11.076\, I_{s(50)}$	Akram and Baker (2007)	[16]
$UCS = 6.21\, I_{s(50)}$	Elkateb (2009)	[17]
$UCS = 21.9\, I_{s(50)}$	Singh et al (2012)	[18]
$UCS = 17.6\, I_{s(50)} - 56.4$	Singh et al (2012)	[18]
$UCS = 5.68(I_{s(50)})^{0.5}$	Salah et al (2014)	[19]

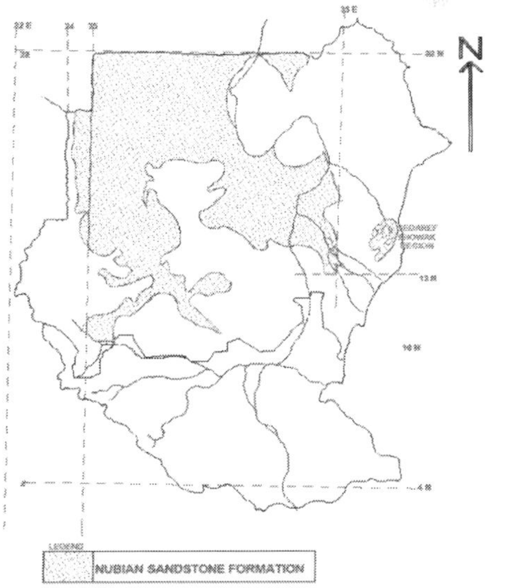

Fig. 1 Distribution of Nubian formation in Sudan

Geologically, the NF uncomfortably rests in Khartoum State on the basement complex with thickness varying between 120 and 420m and is overlain by 7 to 25m thick Tertiary or Quaternary deposits [20]. A number of hypotheses have been advocated about the types of depositional conditions by which the NF has been formed e.g. aeolian, deltaic, estuarine and marine types. Petro-graphic investigations results showed that the NF are essentially comprised of fine to medium sized quartz and rock fragments with an association of opaque minerals including chlorite, crinite and muscovite. Adherence between matrix and grains are weak and therefore the formation can easily be disintegrated. The results of a previous study showed that the NF is jointed, cracked and suffer facies changes and strength characteristics of weak rock and is characterized by variations in color. The sandstone is principally composed of angular to sub-angular quartz of variable sizes floating in a matrix of fine material with kaolin as the basic cementing material [21].

The need for undertaking this study stems from the fact that several important heavy structures such as multi-span bridges across major rivers and high rise buildings, dams, large water front structures pump stations are founded on the NF in Sudan.

4. Research methodology

This study is based on the results of laboratory tests performed on NF sandstone samples taken from different sites of large bridge projects and high rise building towers in Khartoum State and some other parts of Sudan. The bridge sites covered included Al Dabasin and New White Nile bridges on the White Nile, Al Manshiya, Khartoum-Tuti Island, Tuti Island-Khartoum Bahri, Soba, and Almek Nimer bridges on the Blue Nile and Alhalfaya bridge on the main River Nile.

Representative NF sandstone core samples were taken from boreholes drilled to depths reaching 50m below ground elevations. After the sorting and visual examination of rock samples obtained from the boreholes, the RQD was determined then they were cut, trimmed and lapped to standard sizes. Preparation of core specimens and the testing was conducted in accordance to standard methods adopted by the ASTM [1] and the ISRM [22].

The UCS test was performed on sandstone cores with length to diameter ratio L/D more than one. The cores were stressed to fracture, at a very slow rate using a 1500kN compression machine (Fig. 2). The PLT was performed by loading cores against two pointed platens until failure occurred using the apparatus shown in Fig. 3. Specimens having diameters of 80mm, 45mm and 37mm were tested and corrections were applied to calculate the index $I_{s(50)}$ using the equation:

$$I_{S(50)} = \left({}^{D_e}/_{50}\right)^{0.45} \, P_u/D_e^2 \qquad (2)$$

P_u is the failure load and D_e is the effective diameter of the core specimen tested.

Fig. 2 UCS test apparatus used

Fig. 3 Point load test apparatus used

The rock core specimens were trimmed to a regular shape of known volumes, weighed and used in the determination of natural bulk density. Water Absorption of the rock expressed in terms of the degree to which water is absorbed by the core samples was determined according to ASTM D6473-15 method [23] and expressed as a percentage of specimen dry mass.

Table 2 summarizes simple statistical data on the numbers of NF sandstone samples tested and the results of UCS, PLT, RQD, bulk density and water absorption results determined from performing the different test series.

Table 2 Statistical data of the various test results

Test Type	Data Size	Statistical Data		
		Range	Mean	Std. Dev.
UCS (MPa)	369	0.17 -35.88	6.65	6.15
$I_{s(50)}$ (MPa)	141	0.025-3.66	0.64	0.54
γ_b (kN/m^3)	239	0.83 - 3.54	1.89	0.25
A_b (%)	203	2.6 -19.7	12.4	3.3
RQD (%)	204	0 - 100	46.63	25.2

5. Discussion of test results

5.1 General

The results determined from testing of many sandstone samples were analyzed to establish the correlation between UCS and the PLT $I_{s(50)}$ using the statistical regression method. The relationships between the UCS, the bulk density, water absorption and the RQD of the sandstone were also studied using data from different test series.

For the core samples tested the UCS values varied from 0.57 to 18.0MPa and thus the formation may be classified as a rock of low strength. The above UCS values range is consistent with those exhibited in previous investigations carried out Greater Khartoum area. UCS values varying from 0.263 to 17.0MPa were reported at the sites of several engineering structures founded in the same rock formation [24]. The values of the strength index $I_{s(50)}$ determined from point load tests performed on 141 sandstone samples varied from 0.025 to 1.5MPa and thus the formation may be classified as of extremely low to high shear strength.

Relatively high and odd UCS and $I_{s(50)}$ values of about 36 and 3.5MPa respectively were revealed from testing two sandstone samples indicating that they represent a formation of moderate to very high strength. It is believed that the low UCS and $I_{s(50)}$ values determined for 99.5% of the samples tested are pertinent to the locally predominant sandstone type with kaolin as the basic cementing material. On the other hand, the two specimens with much higher UCS and $I_{s(50)}$ values seem to represent sandstone pockets cemented by ferrous oxide or mixed with gravel.

Statistical analysis was performed to establish mathematical models describing the best correlation relationship of the UCS with the $I_{s(50)}$ index, bulk density, water absorption and the RQD of sandstone samples as discussed below.

5.2 Correlation of the UCS and $I_{s(50)}$

To perform regression analysis the test data is normally plotted on a graph which allows visualization of the relationship trends prior to running data analysis. Different curve fitting relationships can be used to analyze the relationship between the dependent and independent variables. The curve fitting produces a measure of variability known as the coefficient of determination (R^2) on one variable that can be accounted for by variability on the other. Once the regression equations and the associated R^2 values are determined the most appropriate curve fit which gives the highest R^2 value is decided.

To establish the relationship between the UCS and PLT $I_{s(50)}$ index of the NF sandstone the data pertaining to 141 sample pairs were plotted as shown in Fig. 4.

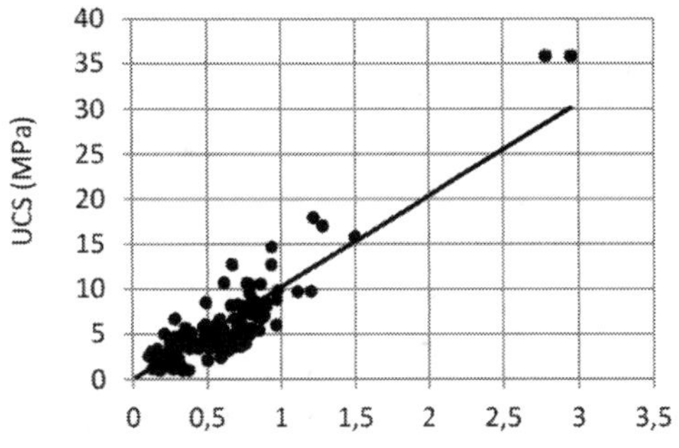

Fig. 4 UCS versus PLT $I_{s(50)}$ relationship

Regression analysis was performed assuming different mathematical models but the following linear equation was found to give the best-fit correlation of the UCS with the $I_{s(50)}$ index:

$$UCS = 10.18\, I_{s(50)} \quad (3)$$

A determination coefficient $R^2 = 0.81$was yielded from data analysis indicating that the correlation of the two rock parameters is highly reliable.

The UCS to $I_{s(50)}$ conversion factor of 10.18 in equation 3 falls well within the range reported in other countries but it is lower than most of values given in the formulae listed in Table 1 for different sandstone rocks. The 10.18 factor value derived for NF sandstones varies from 45% to 180% respectively when compared to the highest and lowest values proposed in the published linear equations of Table 1. The variability in the UCS-$I_{s(50)}$ relationship may be attributed to the intrinsic variability in the physical and mechanical properties of rocks due to their heterogeneous nature [25] as well as their nature and degree of weathering. In addition, part of the variability can be attributed to the inaccuracies in the determination of the true UCS and the $I_{s(50)}$ and the real test type differences. This indicates that a random application of published correlation equations may result in large errors leading to unrealistic prediction of UCS. Such large errors in estimating rock strength from the PLT or other test results are not acceptable and emphasize the necessity of establishing local correlation methods.

The UCS-$I_{s(50)}$ correlation equation developed in this study is in close agreement with that proposed by Akram and Baker [15] for two sandstone types from Pakistan. A good agreement was also revealed between the correlation obtained in this study and the linear and power formulae by Quane and Russel [13] and Salah et al [19] for weak sandstone rocks with UCS and $I_{s(50)}$ values lower than 20 and 1.5MPa respectively as graphically illustrated in Fig. 5.

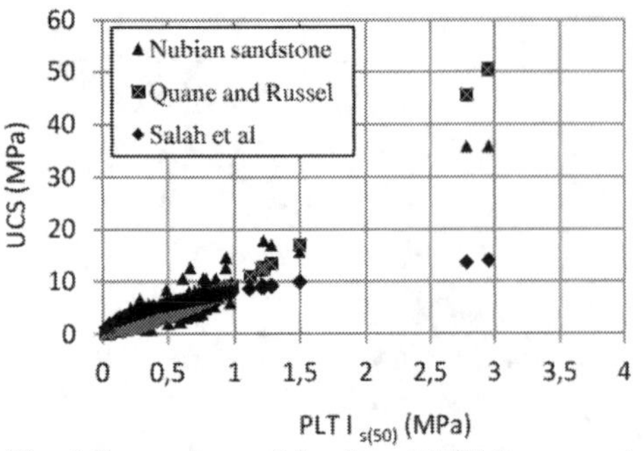

Fig, 5 Comparison of developed UCS-$I_{s(50)}$ equation with previous correlation methods

Equation 3 may be used estimate the UCS of the Sudanese Nubian sandstone formations from known values of the PLT $I_{s(50)}$ index. Though a high R^2 value was yielded from data analysis, the developed relationship accuracy can be further improved as more results are made available from future research works.

5.3 Variations of UCS with bulk density, water absorption and RQD parameters

The UCS and bulk density (γ_b) of the NF sandstone samples were plotted against each other in Fig. 6 which shows that the UCS increases with γ_b in a linear fashion. A sound relationship with an $R^2 = 0.702$ was developed from regression analysis of data pertaining to 153 sample pairs as:

$$UCS\ (MN) = 26.1\ \gamma_b\ (kN/m^3) - 41.68 \quad (4)$$

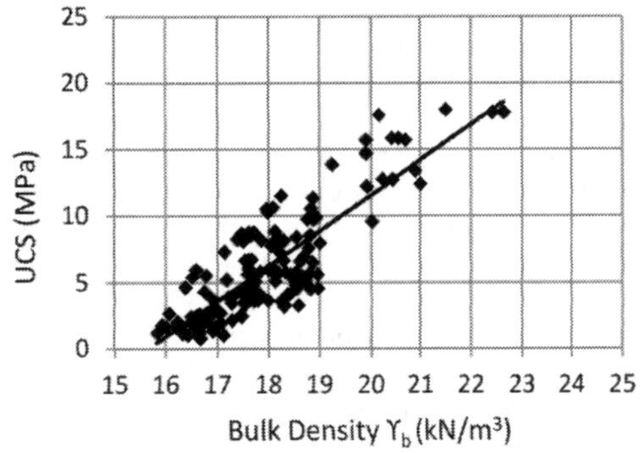

Fig. 6: Plot of UCS against bulk density

Similarly, the rock strength and the degree of water absorption (A_b) by the sandstone material were plotted for as shown in Fig 7 for 150 data points. The figure indicates an inverse linear relationship trend with UCS increasing as A_b is decreasing.

The best-fit line equation given below was established to estimate UCS from A_b with a reasonable degree of accuracy ($R^2 = 0.726$):

$$UCS\ (MPa) = 22.76 - 1.266A_b\ (\%) \quad (5)$$

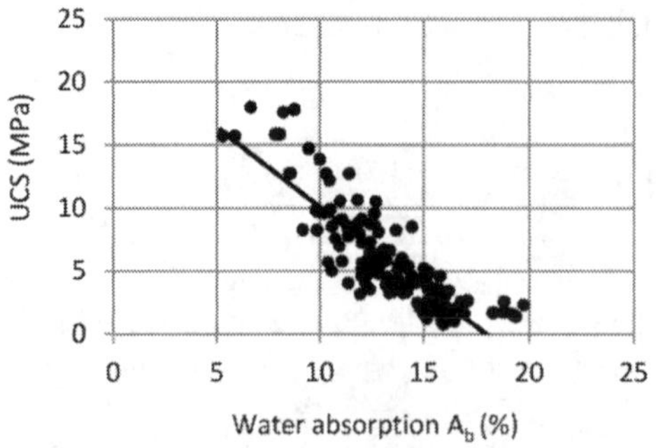

Fig. 7: Plot of UCS against water absorption

A less reliable but perhaps more practically useful correlation was established between UCS and RQD of the NF sandstone using the data points plotted in Fig. 8 which indicates that the two variables increase with each other. The best fit correlation equation given below was established from regression analysis ($R^2 = 0.511$):

$$UCS\ (MPa) = 0.092\ RQD\ (\%) \quad (6)$$

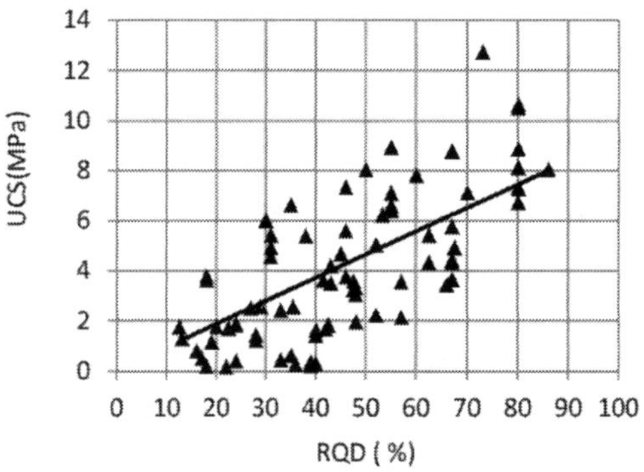

Fig. 8: Plot of UCS against RQD

6. Conclusions

Direct laboratory testing is the most reliable method for the determining the UCS of rocks; however it is often difficult to obtain the quality of specimens required for testing in weak rocks such as NF sandstones. Moreover, the UCS test is tedious, time-consuming and expensive.

The point load test (PLT) represents an alternative for indirect evaluation of the UCS through a correlation with the strength index $I_{s(50)}$. The PLT requires less or no sample preparation and can be easily performed in the laboratory and field using less sophisticated equipment.

Many UCS-$I_{s(50)}$ correlation equations have been proposed worldwide but they are material type and site dependent. Application of such relationships for a different material can result in large errors in rock strength prediction.

A comprehensive study was undertaken to determine the UCS, PLT, bulk density, water absorption and the RQD properties of many Sudanese NF sandstone samples collected from sites wherein they are most predominant.

Analysis the test results was performed using regression method and a reliable linear equation with a UCS to $I_{s(50)}$ factor of 10.18 was developed for the Sudanese Nubian sandstone formation. Similarly, reasonably accurate relationships have been established to estimate the UCS of the NF sandstone from bulk density, water absorption and the RQD parameters. The relationships developed may be used by local engineers to roughly estimate the UCS of the NF sandstone but with some care until they are confirmed or more refined by future research works.

References

1. ASTM International. D 7012-04 Standard test method for compressive strength and elastic modulus of intact rock core specimens under varying states of stress and temperatures, West Conshohocken, USA (2009).
2. ASTM International D 5731-05 Standard test method for determination of the point load index of Rock, West Conshohocken. U.S.A (2009)..
3. Broch, E., Franklin, J.A.. The point load test. Int. J. Rock Mechanics, Mining and Science **9**, 669-697(1972)
4. Bieniawski ZT. The point-load test in geotechnical practice. Engineering Geology; **9**(1):1e11 (1975).
5. Singh, D. P.: Determination of some engineering properties of weak rocks, *Proc. Int. Symposium on Weak Rock*, Tokyo, 21-24. 32) (1981)
6. Forster, I. R. The influence of core sample geometry on the axial point-load test, Int. J. of Rock Mechanics and Mining Sciences and Geomechanical Abstract, **20** (6), 291-295(1983).
7. Das, B. M.: Evaluation of the point load strength for soft rock classification, *Proc., 4th Int. conf. on ground control in mining, Morgantown, WV, 220-226, (1985).*
8. Hawkins, A. B., Olver, J. A. G.,: Point load tests: correlation factor and contractual use. In: Hawkins, A.B. (Ed.), *Site Investigation Practice: Assessing BS 5930,* Geological Society, London, 269-271, (1986)
9. Vallejo, L. E., Welsh, R. A., Robinson, M. K.: Correlation between unconfined compressive and point load strength for Appalachian rocks, *Proc. 30th US Symposium on Rock Mechanics*, Morgantown, 461-468, (1989)
10. Ulusay, R., Tureli, K., and Ider, M. H. Prediction of Engineering Properties of a selected Litharenite Sandstone from its Petrographic Characteristics using Correlation and Multivariate Statistical Techniques, Engineering Geology, **38**(2), 135-157. 36), (1994)
11. Smith, H. J. The point load test for weak rock in dredging applications, In. J. of Rock Mechanics and Mining Sciences, **34**, 295, 3-4, (1997).
12. Rusnak, J. A., Mark, CUsing the point load test to determine the uniaxial compressive strength of coal measure rock, *Proc. 19th Int. conf. on ground control in mining*, **30**, 362-371, (1999).
13. Quane S.L., Russel, J.K. Rock strength as a metric of welding intensity in pyroclastic deposits Eur. J. Mineral., **15**, 855–864. 8. T. N (2003).
14. Tsiambaos G, Sabatakakis N. Considerations on strength of intact sedimentary rocks. J. Engineering Geology; **72**(3e4): 261e73 (2004).
15. Koukis G. t, Sabatakakis N. t, and Papanakli S.l Laboratory testing properties of sandstones Bulletin of the Geological Society of Greece, Vol. **XXXX**, Proc. 11th Int. Congress, Athens, (2007).
16. Akram M. and Bakar M.Z.A. Correlation between uniaxial compressive strength and point load index for salt-range rocks. Pakistani Journal of Engineering and Applied Sciences; **1**: 1e8 (2007).
17. Elkateb T. The point load test in the Persian Gulf carbonate rocks. Proc. 17th Int. Conf. on Soil Mechanics and Geotechnical Engineering, **1**;.372e5. (2009).
18. Singh T.N., Kainthola A and Venkatesh A. Correlation between point load index and uniaxial compressive strength for different rock types. Rock Mechanics and Rock Engineering; **45**(2) 259 e64. (2012).
19. Salah H, Omar M, Shanableh A. Estimating unconfined compressive strength of sedimentary rocks in United Arab Emirates from point load strength index. Journal of Applied Mathematics and Physics; **2**: 296 e303 (2014).
20. Khairalla, M. K., Study of the Nubian Sandstone Formation of the Nile Valley between the 14 N and 17 42 N with reference to the ground water geology. *M. Sc. thesis*, U of K, 283pp, (1966)
21. Yousif, H. O. "Geotechnical Properties of the Nubian Mudstone and Weak Sandstone" *M. Sc. Thesis*, Building and Road Research Institute, University of Khartoum, Sudan, (1985)
22. International Society for Rock Mechanics (ISRM) (1985) *Suggested Method for Determining Point Load Strength*, ISRM Committee on Laboratory and Field Tests, Int. Jour. of Rock Mech, Min. Sci. & Geomech., 22 (2): 53-60.
23. ASTM D6473-15 "*Standard Test Method for Specific Gravity and Absorption of Rock For Erosion Control*"
24. Elsharief A. M., Zein A. K. M. Elarabi H. and Abulgasim R. "Design practice of bored piles in Nubian formation" Sudan Engineering Society Journal, March, Volume 60; No.1, (2014).
25. Idris MA, Saiang D, Nordlund E. Numerical analyses of the effects of rock Mass property variability on open stope stability. In: *The 45th US rock mechanics/geomechanics symposium*, San Francisco, USA., AMRA, pp 1530e40 (2011).

Neural networks and principle component analysis approaches to predict pile capacity in sand

A Benali [1], A Nechnech[2], B Boukhatem[3], M N Hussein[3], M Karry[3]
[1]University of Bounamaa Djilali, Department of Civil Engineering, Khemis Miliana, Algeria
[2] University of Science and Technology, Department of Civil Engineering, Houari Boumediene, Algeria
[3]University of Sherbrooke, Department of Civil Engineering, Sherbrooke, Canada

Abstract. Determination of pile bearing capacity from the in-situ tests has developed considerably due to the significant development of their technology. The project presented in this paper is a combination of two approaches, artificial neural networks and main component analyses that allow the development of a neural network model that provides a more accurate prediction of axial load bearing capacity based on the SPT test data. The retropropagation multi-layer perceptron with Bayesian regularization (RB) was used in this model. This was established by the incorporation of about 260 data, obtained from the published literature, of experimental programs for large displacement driven piles. The PCA method is proposed for compression and suppression of the correlation between these data. This will improve the performance of generalization of the model.

1 Introduction

Pilling has been used for many years as a common foundation solution for different types of civil structures. A large number of design approaches, therefore, have been proposed to predict the ultimate capacities of piles. These approaches range from simple empirical formulations to more sophisticated finite element (FE) analyses, with new methods introduced every few years [1].

Artificial neural networks (ANNs) have been use recently to predict the ultimate capacity of driven piles based on in situ tests [2-4]. However, most of ANNs models available in the literature generally used a limited number of data sets with a lack of more accurate measurements of soils properties from the SPT results.

In this study, two techniques, the ANN and PCA are compiled to develop a new model that would be able to predict the ultimate capacity of large-displacement driven piles embedded in sandy soils based on SPT data.

The performance parameters of the developed model are compared successfully with those obtained from field tests as well as from reliable approaches.

2 SPT N values and failure zone extension

Since the pile capacity depends on the soil compressibility and the SPT is one of most commonly used tests in practices for indicating the in situ compressibility of soils ; the SPT blows count 300 mm (N_{SPT}) are used for the purpose of this study. In most cases, the value of N presents a relatively wide range of variations due to the heterogeneity of soil layers. In order to obtain proper unit shaft and base resistances (the ultimate or total load is the addition of these two components), it is very important to consider the variations of soil resistance properties by presenting an average value for N. Since unit shaft and base resistances are related to the average value of N, this value should be a pertinent representative. As a result of a simple numerical calculate of the studied cases, using the geometrical average method to obtain the logical representative of N values seems to be more accurate and relevant [5]. For example, 26.8 and 15 are the arithmetic and geometric average, respectively of 2, 3, 15, 14, 15, 17, 15, 17, 70, and 100. This example demonstrates that the value of the geometrical average is closer to the predominant values of these numerals, since the arithmetic value is highly affected by the values 70 and 100. It should be noticed that the SPT values used for the geometric average should be at a constant spacing.

In order to obtain the unit shaft resistance of piles from the SPT results, we take the geometrical average of N along the embedded length N_{Shaft}. To account more accurately for the variability of soil properties at the pile base or the unit base resistance from the SPT results, the failure zone should be specified around the pile base. This zone is taken in accordance with the proposal adopted by the UWA method (University of Western Australia) [6] (Table 1, Fig.1)

Table 1 gives the failure zone dimensions, which are noted Zone A and zone B or their corresponding blows number N_A, N_B. For more representation of the tip failure zone, we suggest to take the N value at the pile base N_{Tip}.

Table 1. Proposal used for influence zones for end bearing[6]

Influence zone	UWA method
Zone A	8B
Zone B	(0.7 - 4) B

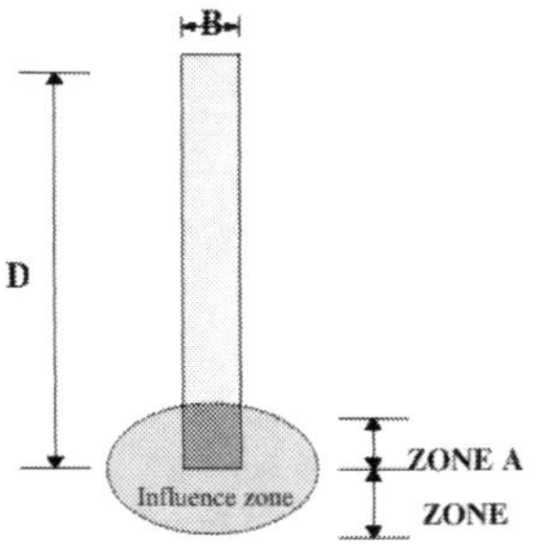

Fig. 1. Influence zone for averaging blows number near the pile base.

The normalized N value to the standard energy ratio of 60% is corrected for a number of effects, is often expressed by the following form [7]:

$$N_{60} = N\ C_E C_B C_S C_R \quad (1)$$

Where C_E, C_B, C_S, C_R are the correction factors for hammer energy efficiency, borehole diameter, sampling method and, length of drill rod, respectively. No available details of the the hammer energy efficiency, borehole diameter, sampling method and, length of drill rod for several studied cases beyond, C_E, C_B, C_S, C_R are taken as unity for all data (Kramer, 1996).

The blows number is then corrected for overburden pressure effect as generally expressed by the following form:

$$N_{corr}=C_N \times N_{SPT} \quad (2)$$

Where, C_N is the correction factor for effective over burden pressure.

There are several expressions for the C_N in literature. However, the expression initially proposed by [8] for sound and later suggested in ASTM D 6066- 96 is used for this study, which is as follows:

$$C_N=(Pa/\sigma_v')^{0.5} \quad (3)$$

Where, σ_v': vertical effective pressure at the SPT test point and, Pa: reference pressure equal to 100 KPa. This correction is not used for clays.

3 Neural networks

A neural network is a system composed of a set of neurons interconnected with each other. A certain disposition of the connection of these neurons produced a neural network model suitable for certain tasks. The Back Propagation Multilayer Perceptron (BPMLP) is the most popular neural network model often used, consisting of three adjacent layers, input, hidden and output [9]. To obtain some desired outputs, weights, which represent connection strength between neurons and biases, are adjusted using a number of training inputs and the corresponding target values. The network error, that is the difference between calculated and expected target patterns, then back propagated from the output layer to the input layer to update the network weights and biases. It arises during the learning process and it can be expressed in terms of mean square error (*MSE*) using:

$$MSE = \frac{1}{p} \cdot \left(\sum_j (t_j - o_j) \right)^2 \quad (4)$$

where: t_j is the target value of jth pattern, O_j is the output value of jth pattern, and P is the number of patterns.

4 Principal component analysis (PCA)

The technique of PCA is a descriptive technique to study the dependencies between variables, for a description or a compact representation of these variables. It was also successfully applied as a technique for reducing the dimensionality of ANN inputs in a variety of engineering applications [10-13]. Mathematically, PCA is an orthogonal projection technique that projects multidimensional observations represented in a subspace of dimension m (m is the number of observed variables) in a subspace of lower dimension ($L < m$) by maximizing the variance of the projections. The estimation of PCA parameters can be summarized in the calculus of eigenvalues and eigenvectors of the matrix Σ. From the spectral decomposition of this matrix it can be written as follows:

$$\sum = p \wedge p^T = \sum_{i=1}^{m} \lambda_i . p_i . p_i^T , \quad (5)$$

where: p_i is the ith eigenvector of Σ and λ_i is the corresponding eigenvalue.

The determination of the number L which represents the number of eigenvectors corresponding to the dominant eigenvalues is very important. Many rules are proposed in the literature to determine the number of L components to retain [14]. In this study, we used the cumulative percentages of the total variance method. The percentage of variance is explained by the first L components and is given by:

$$PCV(L) = 100 . \left(\frac{\sum_{j=1}^{l} \lambda_j}{\sum_{j=1}^{m} \lambda_j} \right) \% . \quad (6)$$

5 Development of the artificial neural network model

5.1 Database construction

In this study, one ANN model that deals with the large displacement piles is developed. The data used to calibrate and validate the ANN model are obtained from the literature and includes 260 static load tests on large displacement driven piles reported by different authors [15]. The conducted tests were performed at different non-cohesive sites.

5.2 Model inputs and outputs

- Nine factors affecting the pile capacity are presented in the ANN as potential model input variables. These include the embedment length (*D*), two corrected blows numbers (N_{Shaft}, $\bar{N}_b$), pile or shaft material (concrete, steel, and aluminium) (SM), and pile classification (large-displacement) (PC). To accurately account the wide variety of pile shapes, four diameters are considered which are; B_{ext} (external or shaft diameter), B_{base} (pile base diameter), B_{int} (internal diameter) and B_{head} (pile head diameter).

- The ultimate pile capacity (Q_t) is defined in this study as the load corresponding to the plunging failure for the well-defined failure cases. For the cases where the failure load is not clearly defined, it is required to determine the failure load from the results of pile load tests through a unique criterion. According to [16], a small-diameter pile is considered to be failed if it experiences a settlement equal to 10% of its nominal dimension. On the other hand, the ultimate capacity of a large-diameter pile is accessed, following the recommendations of [17-18], when the pile settlement (S) equals to:

$$S = (Q.L) / \left(E_p.A_p\right) + B / 30 \,, \quad (7)$$

Where: Q is the test load; L is the pile length; A_p is the cross sectional area of pile and E_p is the modulus of elasticity of the pile material.

5.3 Methodology of implementation of PCA and ANN

This section describes the steps taken to implement the PCA and the ANN approaches. Two types of PCA data processing were implemented in two phases. The first phase is called Pre-PCA, which is responsible for pre-processing the training data matrix and eliminates correlations between them. The second is called Post-PCA, which is used to transform testing and validation data matrix according to its principal components. The implementation and simulation were performed using the MATLAB 7.5 [19] (The Math Works 2007) functions of the neural networks toolbox.

5.3.1. Pre-PCA phase

The input data (Matrix C) were first normalized, so that they had zero and unity variances. Then, the PCA parameters (eigenvalues and eigenvectors) were estimated to calculate the principal components (PC) using the normalized data (Matrix N), the mean and variance values. The uncorrelated components of matrix Ntrans were classified according to their variances. They were then passed to the ANN together with their corresponding target output values for a network training process based on a selected PC variance value.

The nine (9) parameters of the row input matrix can be replaced by seven (7) first principal components based on a chosen PCV value. They were then introduced to the ANN inputs with their desired output. Many ANN were trained using different PCV values to determine the optimal percentage of this value of the total variance in the database. The best model is with PCV equal to 2% and seven principal components.

5.3.2. Post-PCA phase

During each training process of an ANN, validation and generalization performance on testing and validation data sets were evaluated. Each vector of validation or test data was post-processed with the post-PCA before it can be used an ANN to estimate or predict the output. The trained network used these reduced and uncorrelated data with its optimal weights obtained from training process to predict total pile capacity (ultimate load).

5.3.3. Training, testing and network selection

The developed model was trained and tested with its data set for training, testing, and validation using the Bayesian regularization algorithm [20]. We have obtained for the coefficient of correlation respectively: 0.96; 0.94; 0.95, for training, testing and validation.

6 Comparison of ANN model with available SPT-based methods

To examine the accuracy of the large displacement pile model against available methods, the model is compared with four SPT-based methods currently used in practice. Performance of ANN model against available SPT-based methods (Table 2). We can conclude that the developed model gives the better values comparing with the others methods.

Table 2. Performance of ANN model against available SPT-based methods

	Best fit calculations		Arithmetic calculations		Accuracy ±25%
Methods	Q_{fit}/Q_m	R^2	$\mu(Q_p/Q_m)$	$\sigma(Q_p/Q_m)$	Lognormal
Neural network model	0.97	0.96	1.05	0.22	80
[21]	0.92	0.84	0.90	0.30	58
[22]	0.88	0.78	1.14	0.35	54
[23]	0.51	0.26	0.78	0.27	39
[24]	0.91	0.82	0.87	0.27	46

Note: Q_{fit} – pile capacity of the best fit of predicted versus measured pile capacity; Q_p – predicted pile capacity; Q_m – measured pile capacity; μ – mean; σ – standard deviation.

7 Conclusions

In this paper, a comprehensive set of in-situ pile load test results collected from the literature has been utilized to develop an ANN model for capacity predictions of large displacement piles. In order to improve the predictive ability of the developed ANN model, the principal component analyses (PCA) approach was applied. The performance of the ANN model was examined against the most practically used SPT-based pile capacity prediction methods. The results indicate that the developed model was capable of accurately predicting the ultimate capacity of both large displacement piles with high performance parameters (R^2 = 0.96, Mean = 1.05, SD = 0.22).

References

1. M. Iskander, Behavior of pipe piles in sand: plugging and pore-water pressure generation during installation and loading. Springer Series of Geomecha and GeoEngin. Springer Berlin Heidelberg Edition. 250 (2011).
2. M. A. Shahin, Load settlement modeling of axially loaded drilled shafts using CPT based recurrent neural network, Soils and Foundations. **54,** 515–522 (2014) .
3. A. Kordjazi, F. Pooya Nejad, M. B. Jaksa, Prediction of ultimate axial load carrying of piles using vector machine based on CPT data. Comput and Geotech. **55**, 91–102(2014).
4. H. B. Mohammad, S. A. Azizkandi, A. Kashkooli,. Prediction of pile settlement based on cone penetration test results: an ANN approach. KSCE. Civ Eng.J. **19**, 98–106(2015)
5. A. Eslami, B. H. Fellenius, Pile capacity by direct CPT and CPTu methods applied to 102 cases histories. Can Geotech.J. **34**, 886–904 (1997).
6. F. Yu, J. Yang, Base capacity of open ended steel pipe piles in sand. J Geotech Geoenviron. Eng. **138** 1116-1128(2012).
7. A. W. Skempton, Standard penetration test procedures and the effects in sands of overburden pressure, relative density, particle size, aging and overconsolidation. Geotech, **36**,425-447(1986).
8. S. S. Liao, R. V. Whitman, Overburden correction factors for SPT in sand. J Geotech Eng ASCE. **112**. 373-377(1986).
9. G. Dreyfus, J. M. Martinez, M. Samuelides, M. B. Gordon, F. Badran,; S. Thiria,; L. Hérault, Réseaux de neurones – Méthodologie et application. Edition Eyrolles. (1994)
10. M. F. Harkat, Détection et localisation de défauts par analyse en composantes principales. PhD dissertation. Institut National Polytechnique de Lorraine, Centre de Recherche en Automatique de Nancy (2003)
11. K. Kuniar, Z. Waszczyszyn, Neural networks and principal component analysis for identification of building natural periods, Comp in Civil Eng. J. **20**, 431–436(2006)
12. S. W. Shin, C. B. Yun, H. Futura, J. S. Popovics, Non-destructive evaluation of crack depth in concrete using PCA-compressed wave transmission function and neural networks, *Experimental Mechanics* **48**, 225–231(2008).
13. B. Boukhatem, S. Kenai, A. Tagnit Hamou, D. Ziou, M. Ghrici, A system for predicting concrete properties using neural networks (NN) with Principal Component analysis (PCA) technique. *Computer and Concrete* **10**,557–574(2012).
14. S. Valle, L. Weihua, S. J. Qin, Selection of the number of principal components: the variance of the reconstruction error criterion with a comparison to other methods, Indust and Eng Chemistry Research **38**, 4389–4401(1999)
15. A. Benali. Application des réseaux de neurones artificiels pour prédire la capacité portante axiale d'un pieu. *Thèse de doctorat*, (2016)
16. Eurocode 7, *Geotechnical design – general rules*. Thomas Telford, London. (2004)
17. AASHTO (2000) *LRFD bridge design specification*. 2nd *ed.* Washington, D.C., USA.
18. FHWA.1996. *Design and construction of driven pile foundations*. US Department of Transportation, Federal Highway Administration, Florida, Report No. FHWA-HI-97-013.
19. MATLAB 7.5
20. MacKay, D. J. C. 1992. Bayesian interpolation, *Neural Computation* 4(3): 415–447.PHRI Standard (1980)
21. G. G. Meyerhof, Bearing capacity and settlement of pile foundations. Geotech Eng.J. **102**, 1–19(1976).
22. PHRI, Technical standards for port and harbor facilities in Japan. *Bureau of Ports and Harbors, Ministry of Transport*, Japan (1980)
23. Y. Shioi, J. Fukui, Application of N-value to design of foundation in Japan. in *Proceeding of the* 2nd *ESOPT*, , Amsterdam, Netherlands. **1**, 159–164 (1982)
24. Y. Robert, A few comments on pile design. Canad Geotech. J. **34**, 560–567(1997)

Seismic stability analysis of rock slopes by yield design theory using the generalized Hoek-Brown criterion

Mounir Belghali [1,2] and Zied Saada[1]
[1]Université de Tunis El Manar, Ecole Nationale d'Ingénieurs de Tunis, Labo. de Génie Civil (LGC-ENIT), Tunis, Tunisie.
[2]Direction Générale des Études Technologiques, Institut Supérieur des Études Technologiques de Radès, Radès, Tunisie.

Abstract. The stability of rock slope is studied using the kinematic approach of yield design theory, under the condition of plane strain and by considering the last version of the Hoek-Brown failure criterion. This criterion, which is suitable to intact rock or rock mass highly fractured regarded as isotropic and homogeneous, is widely accepted by the rock mechanics community and has been applied in numerous projects around the world. The failure mechanism used to implement the kinematic approach is a log-spiral rotational mechanism. The stability analysis is carried out under the effects of gravity forces and a surcharge applied along the upper plateau of the slope. To take account of the effects of forces developed in the rock mass during the passage of a seismic wave, the conventional pseudo-static method is adopted. This method is often used in slope stability study for its simplicity and efficiency to simulate the seismic forces. The results found are compared with published numerical solutions obtained from other approaches. The comparison showed that the results are almost equal. The maximum error found is less than 1%, indicating that this approach is effective for analyzing the stability of rock slopes. The relevance of the approach demonstrated, investigations are undertaken to study the influence of some parameters on the stability of the slope. These parameters relate to the mechanical strength of the rock, slope geometry and loading.

1 Introduction

Analysis of the rock slope stability was often carried out using the limit equilibrium method and considering the Mohr-Coulomb failure criterion. The success of this approach is mainly due to its simplicity. However, because of the arbitrary simplifying assumptions on which it is based, the solution provided by limit equilibrium method is considered not rigorous in a strict mechanical sense. To avoid this, many researchers have used other methods, particularly the limit analysis method [1-5]. In addition, several works have shown the weakness of the Mohr-Coulomb criterion to adequately describe the resistance of rocks [6-8]. A variety of nonlinear criteria were then proposed in the literature [9-15]. The most practical criterion that has attracted the attention of rock mechanics community is the Hoek-Bown criterion [16]. As soon as it was published in 1980, this empirical criterion has been intensely used and has undergone several improvements in order to correctly evaluate the resistance of different types of rocks. This criterion is applicable for intact rocks and for strongly fractured rock masses, which are assumed to be isotropic [17].

The non-linearity of a failure criterion greatly complicates the study of slope stability, in particular when the limit analysis method is used. To solve this problem, Zhang and Chen [15] proposed a numerical method, which they called the inverse method. Other researchers have proposed methods based on a technique called the tangential technique [18]. Collins et al [18] generalized the work of Chen [1] on the stability of slopes which material's resistance is governed by a linear failure criterion to the case of nonlinear materials. Instead of using Hoek-Brown's original criterion to evaluate the stability factor, they used a tangent Mohr-Coulomb criterion, characterized by a friction angle φ_t and cohesion c_t, and proposed the following expression of the stability factor: $N_n = N_L . c_t/(s^{0.5} . \sigma_c)$. In this expression $N_L = \gamma . H_c/c_t$ is the linear stability factor previously calculated by Chen [1]. At the same period, Drescher and Christopoulos [19] proposed a similar technique for the general nonlinear criterion proposed by Zhang and Chen [15]. Yang et al [20,21] extended the work of Collins et al [18] to the case of materials obeying the Hoek-Brown criterion and proposed the "generalized tangential technique". In their work, the stability factor is calculated from the critical height obtained using the upper bound limit analysis $N_n = \gamma . H_c/(s^{0.5} . \sigma_c)$. Unlike Collins et al. [18], Yang et al. [20,21] do not use the linear stability factor previously calculated by Chen [1].

Earthquakes are one of the sources of slope instability. They can generate destabilizing and cyclical inertia forces or a reduction in shear strength. If these two effects are combined, they induce an overall decrease in the stability of the slope affected by the earthquake. Therefore, it is very important to carry out an assessment of the seismic slope stability. Several methods are available to perform this type of study; the simplest is the pseudo-static method. This approach is widely accepted as a means to assess the seismic slope stability [22-25]. To simulate the potential inertia forces due to ground accelerations in an earthquake, horizontal and vertical static seismic forces are introduced. These forces are assumed to be proportional to the applied forces times horizontal and vertical average seismic coefficients K_h and K_v [26].

This work deals with the study of the static and seismic stability of homogeneous isotropic rock slope obeying the most recent version of the Hoek-Brown failure criterion [27] and using the kinematic approach of yield design theory. This method constitutes a generalization of the limit analysis [28-30]. It is distinguished by taking into account a failure criterion representing the limitation of the stresses and disregards all other information relating to the model of plastic behavior of the material, in particular its flow rule. To carry out the seismic analysis, the shear strength is assumed to be unaffected by the earthquake and the generated inertial forces are simulated using the pseudo-static approach. Only the coefficient K_h is taken into account, the coefficient K_v is neglected. This hypothesis has been adopted by several researchers [22-25]. What distinguishes this work is the direct use, in calculations, of the Hoek-Brown failure criterion. In fact, this will properly reflect the non-linearity of the criterion.

2 Problem description

2.1 Geometry and loading

The slope studied has a height H and two inclinations β and α "figure 1". β is the main inclination belonging to the interval]0,π/2] and α is the inclination of the upper plateau, strictly lower than β. The slope is subjected to gravitational forces $\gamma \underline{e}_1$, a uniformly distributed surcharge $q\underline{e}_1$ applied along the upper plateau and seismic inertial forces. The geometry of the slope and the loading are assumed to be invariable in the direction $\underline{e}_3$. Therefore, the problem is considered to be a plane strain problem.

2.2 Generalized Hoek-Brown Criterion

The material constituting the slope is considered to be homogeneous and isotropic obeying the generalized Howk-Brown nonlinear failure criterion. By adopting the convention of the positive stress in tension, this criterion is written in the following form:

$$f(\underline{\underline{\sigma}}) = \sigma_1 - \sigma_3 - \sigma_c(-m.\sigma_1/\sigma_c + s)^n \leq 0 \qquad (1)$$

where σ_1 and σ_3 are the principal stresses respectively major and minor, σ_c is the uniaxial compressive strength of the intact rock. The parameters of the criterion: m, s and n depend on the geological strength index (GSI) as follows:

$$m = m_i \exp\left(\frac{GSI-100}{28-14D}\right) \qquad (2)$$

$$s = \exp\left(\frac{GSI-100}{9-3D}\right) \qquad (3)$$

$$n = \frac{1}{2} + \frac{1}{6}\left[\exp\left(-\frac{GSI}{15}\right) - \exp\left(-\frac{-20}{3}\right)\right] \qquad (4)$$

Where m_i is the value of parameter m for intact rock. For some typical rocks, Hoek [31] provides approximate values of this parameter. D is a disturbance coefficient that varies from 0.0 for the undisturbed in situ rock masses to 1.0 for very disturbed rock masses.

3 Implementation of the kinematic approach of yield design theory

3.1. Kinematic approach

According to the kinematic approach of the yield design, the slope is stable under the external loading if for any kinematically admissible virtual velocity field $\underline{U}$ (termed failure mechanism in the sequel), defined in the plane $(\underline{e}_1, \underline{e}_2)$, the work of the external loading $P_e(\underline{U})$ and the maximum resisting work $P_{mr}(\underline{U})$ in this velocity field satisfy the following inequality [29,30]:

$$\forall\ \underline{U}\ \text{K.A.} \quad P_e(\underline{U}) \leq P_{mr}(\underline{U}) \qquad (5)$$

The expression of the work of external loading in a kinematically admissible virtual velocity field is:

$$P_e(\underline{U}) = \int_V \gamma.\underline{e}_1.\underline{U}.dV + \int_{\partial V_q} q.\underline{e}_1.\underline{U}.dS + K_h\left(\int_V \gamma.\underline{e}_2.\underline{U}.dV + \int_{\partial V_q} q.\underline{e}_2.\underline{U}.dS\right) \qquad (6)$$

In this expression, V is the volume of the rock mass and ∂V_q is the contour of V on which q is applied.

The maximum resisting work is given by:

$$P_{mr}(\underline{U}) = \int_V \pi(\underline{\underline{d}})\,dV + \int_\Sigma \pi(\underline{\nu}, [\![\underline{U}]\!])\,d\Sigma \qquad (7)$$

Where $\underline{\underline{d}}$ is the strain rate field associated with $\underline{U}$;

$[\![\underline{U}]\!]$ is the jump of $\underline{U}$ when crossing a possible velocity discontinuity surface Σ following its normal $\underline{\nu}$.

The Π-functions are defined by

$$\pi(\underline{\underline{d}}) = \sup_{\underline{\underline{\sigma}}}\left\{\underline{\underline{\sigma}}:\underline{\underline{d}};\ f(\underline{\underline{\sigma}}) \leq 0\right\} \qquad (8)$$

$$\pi(\underline{\nu}, [\![\underline{U}]\!]) = \sup_{\underline{\underline{\sigma}}}\left\{(\underline{\underline{\sigma}}.\underline{\nu}).[\![\underline{U}]\!];\ f(\underline{\underline{\sigma}}) \leq 0\right\} \qquad (9)$$

The Π-functions for a Hoek-Brown failure criterion are given in [32].

The safety factor is defined as:

$$\Gamma = P_{mr}(\underline{U})/P_e(\underline{U}) \qquad (10)$$

A necessary condition of stability is that:

$$\Gamma \geq 1 \qquad (11)$$

It should be noted that this condition of stability requires that the work of the external loading is strictly positive.

The safety factor introduced in the present approach is completely equivalent to the safety factor traditionally defined as the ratio between maximum and mobilized shear stresses. Both can be used to characterize slope stability [29,33].

3.2 Failure mechanism

The failure mechanism selected is a rotational mechanism "figure 1" which has been widely used in the analysis of homogeneous slopes obeying a Mohr-Coulomb failure criterion. It defines two volumes, the first one is rotating about the center Ω with an angular velocity ω and the other one is kept motionless. The curve I_1I_2 separating the two volumes is a log-spiral arc of angle φ and focus Ω defined by the equation $r(\theta) = r_0 e^{(\theta-\theta_1)\tan\varphi}$ in a polar coordinate system (Ω, r, θ). $r_0=\Omega I_1$ is the radius of the log-spiral curve for $\theta=\theta_1$.

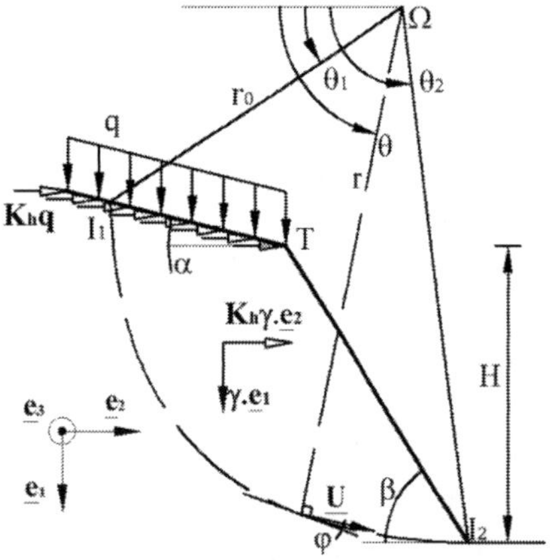

Fig. 1. Loading and slope failure mechanism

Consequently, the discontinuity of velocity at any point of this line is inclined at an angle φ with respect to the tangent to the same point. As shown in "figure 1", the logarithmic spiral curve starts from the upper plateau and passes through the slope toe. Therefore, this spiral arc is completely defined by the three independent parameters: θ_1, θ_2 and φ.

3.3 Safety factor

The slope is subjected to self weight, surcharge and seismic forces. Taking into account the failure mechanism described in the previous section the work of the external forces is:

$$P_e(\underline{U}) = r_0^3.\gamma.\omega.[(f_0 - f_1 - f_2)+K_h.(f_0^s - f_1^s - f_2^s)] + r_0^2.q.\omega[f_3+K_h.f_3^s] \qquad (12)$$

In this expression f_0, f_1, f_2, f_3, f_0^s, f_1^s, f_2^s and f_3^s are dimensionless functions of the parameters θ_1, θ_2 and φ defined by the expressions below:

$$f_0=\frac{e^{3(\theta_2-\theta_1)\tan\varphi}(3\cos\theta_2\tan\varphi+\sin\theta_2)-(3\cos\theta_1\tan\varphi+\sin\theta_1)}{3(9\tan^2\varphi+1)} \qquad (13)$$

$$f_1=D\frac{\sin(\theta_1+\alpha)}{6}\left[\cos\theta_1+X_2^\Omega\right] \qquad (14)$$

$$f_2=H\frac{\sin(\theta_2+\beta)}{6.\sin\beta}.e^{(\theta_2-\theta_1)\tan\varphi}\left[\cos\theta_2\, e^{(\theta_2-\theta_1)\tan\varphi}+X_2^\Omega\right] \qquad (15)$$

$$f_3=\frac{1}{2}D.\left[\cos\theta_1+X_2^\Omega\right] \qquad (16)$$

$$f_0^s=\frac{e^{3(\theta_2-\theta_1)\tan\varphi}(3\sin\theta_2\tan\varphi-\cos\theta_2)-(3\sin\theta_1\tan\varphi-\cos\theta_1)}{3(9\tan^2\varphi+1)} \qquad (17)$$

$$f_1^s=D\frac{\sin(\theta_1+\alpha)}{6}\left[\sin\theta_1-X_1^\Omega\right] \qquad (18)$$

$$f_2^s=H\frac{\sin(\theta_2+\beta)}{6.\sin\beta}.e^{(\theta_2-\theta_1)\tan\varphi}\left[\sin\theta_2\, e^{(\theta_2-\theta_1)\tan\varphi}-X_1^\Omega\right] \qquad (19)$$

$$f_3^s=\frac{1}{2}D.\left[\sin\theta_1-X_1^\Omega\right] \qquad (20)$$

In these expressions:

$$D=TI_1/r_0=\frac{\sin(\theta_1+\beta)-e^{(\theta_2-\theta_1)\tan\varphi}.\sin(\theta_2+\beta)}{\sin(\beta-\alpha)} \qquad (21)$$

$$H=H/r_0=e^{(\theta_2-\theta_1)\tan\varphi}.\sin\theta_2-\sin\theta_1-D\sin\alpha \qquad (22)$$

$$X_1^\Omega=X_1^\Omega/r_0=-\sin\theta_1-D.\sin\alpha \qquad (23)$$

$$X_2^\Omega=X_2^\Omega/r_0=\cos\theta_1-D.\cos\alpha \qquad (24)$$

X_1^Ω and X_2^Ω are the coordinates of the center of rotation Ω.

According to the mechanism adopted, the virtual deformation rate $\underline{d}$ is zero. In addition, the velocity discontinuity occurs along the arc I_1I_2 of the spiral, then (7) is written as follows:

$$P_{mr}=\int_{I_1I_2}\pi(\underline{v},[\![\underline{U}]\!])d\Sigma \qquad (25)$$

The maximum resisting work is then:

$$P_{mr}=r_0^2.\sigma_c\omega.f_4 \qquad (26)$$

The expression of the dimensionless function f_4 is:

$$f_4=\frac{1}{2}\left[\frac{s}{m}+m^{\frac{n}{1-n}}\left(n^{\frac{n}{1-n}}-n^{\frac{1}{1-n}}\right)\left(\frac{1-\sin\varphi}{2\sin\varphi}\right)^{\frac{1}{1-n}}\right].\left(e^{2(\theta_2-\theta_1)\tan\varphi}-1\right) \qquad (27)$$

Taking account of (10), (12) and (26) the safety factor is:

$$\Gamma=\frac{f_4.H}{\frac{\gamma H}{\sigma_c}.[(f_0-f_1-f_2)+K_h.(f_0^s-f_1^s-f_2^s)]+\frac{q}{\sigma_c}.H.[f_3+K_h f_3^s]} \qquad (28)$$

Γ is therefore a function of the parameters: β, α, $\gamma H/\sigma c$, $q/\sigma c$, mi, GSI and D. These parameters are related to slope inclination, gravity level, surcharge and rock strength parameters.

The critical value of the safety factor is obtained by minimization of the objective function Γ with respect to the set of variables θ_1, θ_2 and φ taking into account the following constraints:

$$\begin{cases} 0 < \varphi < \pi/2 \\ 0 < \theta_1 < \theta_2 < \pi\text{-}\beta \\ \sin(\beta+\theta_1) - e^{(\theta_2-\theta_1)\tan\varphi}.\sin(\beta+\theta_2) > 0 \\ r_0 > 0 \end{cases} \quad (29)$$

4 Numerical results

4.1 Comparisons

The validation of the approach presented in this work is carried out using the results published by Yang et al [21]. These authors used the same version of the Hoek-Brown criterion. It should be recalled that in [21] the stability of the slope is characterized by a stability factor calculated from the critical height. Whereas, here, a safety factor that plays this role. Equivalence between these two factors is therefore necessary to make comparisons. It is obvious that if the slope height is critical then Γ=1. In this situation, in the absence of surcharge and taking into account the expression of the stability factor, equation (28) gives:

$$N_n = \frac{f_4.H}{s^{0.5}.[(f_0\text{-}f_1\text{-}f_2)+K_h.(f_0^s\text{-}f_1^s\text{-}f_2^s)]} \quad (30)$$

The minimization of (30) with respect to θ_1, θ_2 and φ and taking account of the conditions (29) gives N_n.

Table 1. Comparison of stability factors

β (deg)	N_n Present Work	N_n [21]	Relative diff. (%)
50	11,56	11,55	-0,09
60	6,53	6,52	-0,15
70	3,79	3,76	-0,80
80	2,28	2,27	-0,44
90	1,41	1,41	0,00

(α=0, q=0, GSI=50, mi=7, D=0 and K_h=0.1)

Finally, to validate this approach, the calculation of a safety factor is transformed into a calculation of a stability factor by stipulating that the height of the slope is critical. Table 1 illustrates a comparison between the numerical results obtained from the minimization of (30) and those published in [21] in the case where m_i = 7. The small relative errors obtained show the effectiveness of the approach described in this paper.

4.2 Effects of parameters α and q/σ_c

In order to study the effect of the inclination α, the safety factor was evaluated in the absence of seismic forces and surcharge considering the following parameters: γH/σc=0.02, GSI=40, D=0 and mi=7. α varies from 0 to 20°. The results obtained are illustrated in Table 2. Examination of these results clearly shows that the effect of the inclination α on the safety factor is insignificant.

The effect of the surcharge was also studied in the following configuration: $\gamma H/\sigma_c$=0.02, GSI=40, m_i=7, D=0, K_h=0 and α=0. Table 3 gives the values of the safety factor when β varies from 50 to 90° and the loading parameter q/σ_c varies from 0 to 2.10^{-3}. From Table 3, it is found that the surcharge has a small effect on the safety factor.

Table 2. Variation of Γ as a function of α

β (deg)	α (deg) 0	5	10	15	20
50	36,14	35,98	35,96	35,55	35,16
60	19,37	19,31	19,28	19,18	19,01
70	10,27	10,23	10,19	10,11	10,07
80	5,60	5,58	5,57	5,51	5,52
90	3,20	3,19	3,18	3,16	3,15

Table 3. Variation of Γas a function of q/σ_c

β (deg)	$q/\sigma_c x10^3$ 0	0,5	1	1,5	2
50	36,14	34,70	33,36	32,06	30,81
60	19,37	18,64	17,90	17,22	16,54
70	10,27	9,87	9,49	9,12	8,76
80	5,60	5,38	5,16	4,96	4,77
90	3,20	3,06	2,93	2,81	2,70

4.3 Effect of gravity level

To illustrate the impact of the gravity level $\gamma H/\sigma_c$ on the safety factor, a study is carried in the absence of surcharge and in the case of undisturbed in situ rock D=0 having the following strength parameters: GSI=50 and m_i=10. The inclination β is varying from 50 to 90°. The obtained results are shown in "figure 2". As expected, Γ is a decreasing function of $\gamma H/\sigma_c$. The decrease in Γ is greater when the slope inclination decreases.

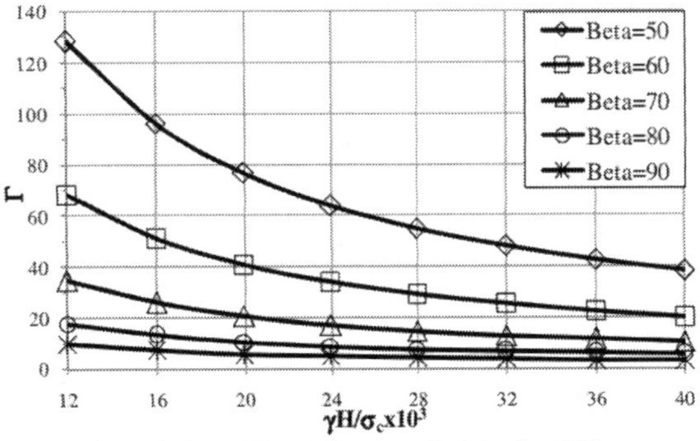

Fig. 2. Variation of Γ as a function of $\gamma H/\sigma_c$ for different inclination of the slope (α=0 and K_h=0)

4.4 Effects of rock strength parameters

This section presents numerical simulations carried out in order to give, in a particular configuration, some preliminary ideas on the individual impact of slope inclination and strength properties on the safety factor. These simulations are performed for three rock types in the following situation: $\gamma H/\sigma_c$=0.02 α=0, q=0 and D=0. The obtained results are presented in Table 4.

https://doi.org/10.1051/matecconf/201814902026

Table 4. Safety factor for m_i = 7, 10 and 15

		GSI=20			GSI=40			GSI=60			GSI=80		
K_h	β	mi=7	mi=10	mi=15	mi=7	mi=10	mi=15	mi=7	mi=10	mi=15	mi=7	mi=10	mi=15
0	50	11,99	18,05	29,01	36,14	51,53	77,99	81,80	113,08	167,01	178,96	242,51	348,90
	60	5,82	8,41	13,19	19,37	26,58	39,20	46,83	61,68	87,64	110,93	140,00	190,85
	70	2,78	3,61	5,18	10,27	12,84	17,42	27,13	33,04	42,64	70,37	81,92	102,24
	80	1,41	1,58	1,88	5,60	6,18	7,21	16,55	18,13	20,35	46,80	50,40	55,96
	90	0,77	0,77	0,77	3,20	3,20	3,20	10,35	10,35	10,35	31,78	31,80	31,80
0,1	50	7,33	10,86	17,25	23,31	32,66	48,93	54,37	73,64	107,00	123,95	161,87	227,54
	60	3,58	4,93	7,45	12,58	16,60	23,62	32,02	40,37	55,06	79,62	96,22	124,95
	70	1,79	2,15	2,79	6,87	8,10	10,11	19,50	22,23	26,81	53,31	59,22	68,65
	80	0,97	1,02	1,10	3,95	4,13	4,47	12,34	12,80	13,47	36,70	37,88	39,39
	90	0,56	0,53	0,49	2,36	2,24	2,08	7,98	7,64	7,18	25,49	24,61	23,41
0,2	50	4,48	6,44	10,03	15,10	20,54	30,17	36,78	48,21	67,87	87,53	109,96	149,16
	60	2,25	2,88	4,06	8,33	10,34	13,85	22,51	26,86	34,32	58,98	68,07	83,09
	70	1,20	1,34	1,54	4,76	5,27	6,01	14,27	15,40	17,34	40,83	43,54	47,50
	80	0,69	0,68	0,67	2,88	2,86	2,80	9,36	9,32	9,19	29,05	28,74	28,63
	90	0,41	0,37	0,32	1,77	1,60	1,41	6,21	5,71	5,10	20,57	19,21	17,47

($\gamma H/\sigma_c$=0.02, α=0, q=0 and D=0)

5 Conclusion

The seismic and static stability of homogeneous and isotropic rock slopes have been studied using the kinematic approach of the yield design theory combined with the pseudo-static method. The strength of the rock is supposed to obey the general Hoek-Brown failure criterion. The use of this approach retains the advantages of the limit analysis and avoids the use of the called tangential methods used to circumvent the difficulty encountered in the calculation of the external work rate and the rate of energy dissipation in the case of a non-linear criterion.

The relevance of the approach has been proved by comparing the results with those published in the literature. Numerical simulations have been carried out to study the influence of some parameters on slope stability.

References

1. Chen W-F., Limit Analysis and Soil Plasticity, 638 (1975).
2. Michalowski RL., Geotechnique, **45**, 11 (1995).
3. Michalowski RL., J Geotech Geoen. Eng , **128**, 5 (2002).
4. Sloan SW., Int J Numer Anal Methods Geomech, 13, 19 (1989).
5. Zhang X. Géotechnique, **49**, 8 (1999).
6. Goodman RE. Introduction to Rock Mechanics, 2nd ed. (1989).
7. Hoek E, Bray JD. Rock Slope Engineering, Third Edition. (1981).
8. Jaeger JC. Géotechnique, **21**, 37 (1971).
9. Barton N., Eng Geol, **7**, 46 (1973).
10. Bieniawski ZT., J South Afr Inst Min Metall, **74**, 9 (1974).
11. Hobbs DW., Int J Rock Mech Min Sci Geomech Abstr, **3**, 32 (1966).
12. Hoek E., J South Afr Inst Min Metall, **64**, 18 (1964).
13. Hoek E., Rock Mech Eng Pract, 26 (1968).
14. Ladanyi B, Archambault G. *American Rock Mechanics Association*, (1969).
15. Zhang XJ, Chen WF., Int J Numer Anal Methods Geomech, **11**, 18 (1987).
16. Hoek E., Brown E.T. *Underground Excavations in Rock*. (1980).
17. Hoek E., *Géotechnique*, **33**, 37 (1983).
18. Collins IF, Gunn CIM, Pender MJ, Yan W., Int J Numer Anal Methods Geomech, **12**, 18 (1988).
19. Drescher A, Christopoulos C., Int J Numer Anal Methods Geomech, **12**, 5 (1988).
20. Yang X-L, Li L, Yin J-H., Int J Numer Anal Methods Geomech, **28**, 10 (2004).
21. Yang X-L, Li L, Yin J-H. Géotechnique, **54**, 7 (2004).
22. R. Richards, D. G. Elms, M. Budhu., J Geotech Eng, **119**, 13 (1993).
23. Saada Z, Maghous S, Garnier D., Int J Numer Anal Methods Geomech, **35**, 25 (2011).
24. Saada Z, Maghous S, Garnier D., Int J Numer Anal Methods Geomech, **37**, 19 (2013).
25. Soubra A-H., J Geotech Geoenvironmental Eng, **125**, 10 (1999).
26. Seed HB. Géotechnique, **29**, 49 (1979).
27. Hoek, E., Carranza-Torres, C., Corkum, B., *NARMS-TAC*, 7 (2002).
28. De Buhan P. *Plasticité et calcul à la rupture*, (2007).
29. Salençon J. *Calcul à la rupture et analyse limite*, (1983).
30. Salençon J. *Yield Design: A Survey of the Theory*, 45 (1993).
31. Hoek E. , Int J Rock Mech Min Sci Geomech,**27**,3 (1990).
32. Saada Z, Maghous S, Garnier D., Comput Geotech, **35**, 11 (2008).
33. Kovarik J-B., Rev Fr Génie Civil, **4**, 45 (2000).

Study of primary creep behavior of clayey soils in landslide area : suction effect

Etude du comportement de fluage primaire des sols argileux dans une zone de glissement de terrain : Effet de la succion

DAHHAOUI Hachimi[1], BELAYACHI Naima[2]et ZADJAOUI Abdeldjalil[1]
[1]Laboratoire RISAM, Université AboubakrBelkaid.. BP 230 Rue Abi Ayad Abdelkrim. 13000. Tlemcen (Algérie)
[2]Université d'Orléans, INSA-CVL, PRISME EA 4229, 8 Rue Léonard de Vinci, 45072 Orléans cedex 2, France

Abstract. Creep behavior of clayey soils plays an extremely important role in the landslide process. The soils that make up these sliding zones are often in unsaturated state. This point indicates the need to take into account the suction effect as hydric parameter on the long-term deformation of clayey soils. In this paper, a primary creep model named Modified Time Hardening (MTH) for unsaturated soils with different matric suction has been built. Based on the literature tests results[1][2], parameters C1 and C2 of the model have relations with suction and deviator stress level respectively. The primary creep strainwill be able to demonstrate unsaturated effect of the soils. comparison between the calculated results and the literature tests results shows a good coherence. The work underway at the university of Orleans will show later the relevance of model used in the present work.
Keywords: Creep behavior, Clay,Suction, Numerical modeling.

Résumé. Lecomportement en fluage des sols argileux joue un rôle essentiel dans le processus de glissement de terrain. Les sols constituants ces zones de glissements sont souvent à l'état non saturé, celaimplique la nécessité et l'importance de prendre en comptel'influence de lasuccion comme paramètre hydrique sur les déformations à long terme des sols argileux.Dans le présent article, un modèle de fluage primaire nommé Modified Time Hardening(MTH)a été considérépour étudier le comportement en fluage des sols non saturés. En se basant sur les résultats d'essais de la littérature[1][2], les deux paramètres C1 et C2 du modèle ont été modifiés en tenant en compte des relations en fonction de la succion et le niveau de déviateurrespectivement. La déformation de fluage primaire ainsi mesurée permettra de mettre en évidence l'effet de la non saturation des sols. La comparaison entre les résultats calculés et les résultats expérimentaux de la littérature montre une bonne cohérence. Les travaux en cours à l'université d'Orléans montrerons par la suite la pertinence de modèle utilisé dans le présent travail.
Mots Clés : Comportement en fluage, Argile, Succion, Modélisation.

1 Introduction

Ladépendance au temps des sols argileux est habituellement associéeà unfluage, c'est-à-dire un comportement sous des contraintes effectives constante, comme l'ont résumé Augustin et al. (2004)[3] et Lingaard et al. (2004)[4].les modèles de fluage peuvent être classés en troiscatégories Selon Purwodihardjo (2004) [5] : (empiriques, rhéologiques monodimensionnels et loi de comportement complète-tridimensionnelle). La plupart des recherches scientifiques utilisent ces modèles dans des conditions de saturation complète, alors que l'état non saturé ne peut être négligé.Le présent article,se veut comme une contribution dans ce sens, c'est-à-dire présenteune étude du comportement au fluage d'un sol non saturé en considérant un modèle de fluage primaire. Dans un premier temps une revue de la littérature sur les modèles de fluage des sols non saturésa été présentée en détail. Dans un second temps le modèle proposé dans ce travail est détaillé. Il présente l'avantage d'être simple et utile dans la pratique de l'ingénieur. Ensuite, les résultats expérimentaux sur un solprovenant d'un grand glissement de terrain dans la région du réservoir des Trois Gorges en Chine ont été exploités pour valider notre modèle. Finalement, la mise en évidence de l'effet de la non saturation sur le fluage primaire a été discutés par les auteurs avec des pistes sur la nécessité d'amélioration des paramètres régissant le phénomène de fluage.

2 Revue des travaux antérieurs

Cette section, présente les modèles de fluage des sols argileux non saturés,développés dans un programme de suivi et d'auscultation d'un grand glissement de terrain dans la région du réservoir des Trois Gorges en Chine. Il s'agit d'un programme expérimental d'essais triaxiaux à succion et déviateur imposés, les auteursLai et al. (2010) [1] (2014) [6] et Zou et al. (2013) [2]ont réussi de prendre en considération la matrice de succion comme variable supplémentaire, favorisant l'amélioration des modèles de fluage classiques.

2.1 Modèlesaméliorés par Lai et al. 2010 et 2014

Les études effectuées par Lai et al. 2010[1] et 2014[6], porte sur l'amélioration du modèle de Singh-Mitchell (1968)[7] et le modèle de Mesri et al. (1981)[8]respectivement.

2.1.1 Amélioration du modèle de Singh-Mitchell

Le modèle de Singh et Mitchell (1968)[7]est une équation phénoménologique à trois paramètres, utilisé pour décrire le taux de déformation axiale des sols argileux soumis à un chargement constant. L'expression du modèle est composée d'une fonction exponentielle décrivant le chargement, et une fonction de puissance tenant en compte de l'effet du temps :

$$\dot{\varepsilon} = A.\exp(\alpha.D^r).\left(\frac{t_1}{t}\right)^m \quad (1)$$

Avec, $(\dot{\varepsilon})$taux de déformation axial,(t)le temps écoulé depuis l'application du chargement,(t_1)le temps de référence désignant un stade précoce dans l'essai de fluage,(D^r)niveau de déviateur désignant le rapport de la contrainte déviatorique imposée (q) sur la résistance au cisaillement induisant la rupture(q_f),(A) $(m)et(\alpha)$sont les paramètres de modèle.

L'intégration de l'équation 1, lorsque m≠1 produit une relation de type déformation-temps :

$$\varepsilon = B.\exp(\beta.D^r).\left(\frac{t_1}{t}\right)^\lambda \quad (2)$$

D'où,(ε)déformation axial,$B = \frac{A.t_1}{(1-m)}$, $\beta=\alpha$,et$\lambda=1-m$.
Dans l'expression 2, la matrice de succion a été intégrée dans une nouvelle variablede niveau de déviateur (D^R) définie comme suite :

$$D^R = D^r.\left(\frac{\sigma_3'}{u_a}\right) \quad (3)$$

Avec, (σ_3') pression de confinement et (u_a) pression de l'eau.

La relation entre la succion et le niveau de déviateur exprimé dans l'équation (3) a été justifié à travers l'analyse des résultats d'essais de fluage triaxiaux a succion contrôlée, dans lesquelles Lai et al. 2010[1]soulignent une relation d'inversement proportionnelle entre la déformation de fluage et la matrice de succion.

L'équation (4) ci-dessous présente l'amélioration du modèle de fluage établi par Lai et al. 2010 [1]

$$\varepsilon = B.\exp(\beta_{DR}.D^R).\left(\frac{t}{t_1}\right)^\lambda \quad (4)$$

Avec,(β_{DR})paramètre correspondant au paramètre(β)dans le cadre la nouvelle variabledéfinie comme(D^R).

Malgré les prévisions raisonnables pour la simulation de fluage des sols non saturés, les différences observées pour les niveaux de déviateursélevées suscitent les auteurs vers d'autres modèles et d'autres investigations.

2.1.1 Amélioration du modèle de Mesri et al

Le modèle constitutif de Mesri et al. (1981)[8] est une équation empirique, composée d'une fonction hyperbolique décrivant la relation contrainte-déformation et une fonction de puissance pour la relation déformation-temps :

$$\varepsilon = \left[\frac{(\sigma1-\sigma3)_f}{E_d}\right]_1 \frac{D_1}{1-(R_f)_1 D_1}\left(\frac{t}{t_1}\right)^n \quad (5)$$

Avec,(ε) déformation axial,(t) le temps écoulé depuis l'application du chargement, *(t_1=60 min)temps* de référence,$(\sigma_1-\sigma_3)_f$ la contrainte déviatorique maximal,(D)niveau de déviateur désignant le rapport de la contrainte déviatorique (q)sur la résistance au cisaillement induisant la rupture (q_f),$(E_d)(R_f)$ et (n)paramètres de fluage,$[(\sigma_1-\sigma_3)_f]_1et(E_d)_1$sont les valeurs de$(\sigma_1-\sigma_3)_f$ et(E_d) a l'instant $t=t_1$respectivement, $(D)_1$et$(R_f)_1$sont le niveau de déviateur(D) et le paramètre(R_f)au même instant.

Selon Mesri et al. (1981)[8], les trois paramètres$(E_d)(Rf)$ et (n) dans l'équation (5)peuvent être exprimés en terme de$\frac{(\sigma_1-\sigma_3)_f}{E_d}$et ε_f, d'où Lai et al. (2014)[6] a déduit que la matrice de succion peut avoirun effet sur$\frac{(\sigma_1-\sigma_3)_f}{E_d}$ et ε_f,cela lui avait permis d'établir une relation entre la matrice de succion et le paramètre R_f:

$$R_f = a.\left(\frac{s}{p_a}\right) + b \quad (6)$$

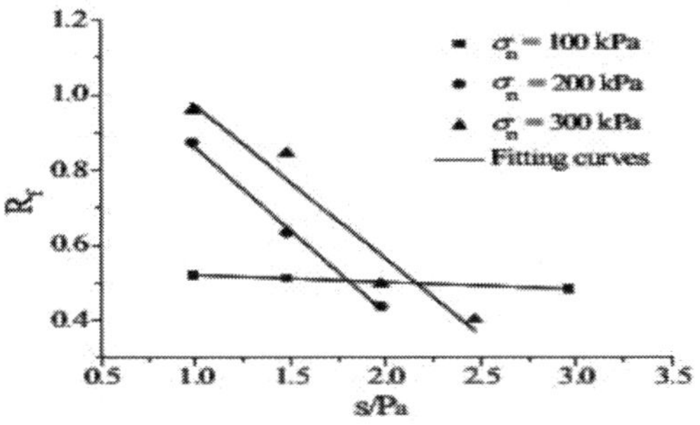

Fig. 1. Relation entre le paramètre$R_f$et le rapports/Pa[6].

Avec,(Pa) pression atmosphérique(101,33 kPa) ;(a)et(b)valeurs constantes obtenue à partir des courbes linéaires illustrés sur la "figure 1".

L'équation (7) présente le nouveau modèle empirique de Lai et al. (2014)[6], pour les argiles non saturées.Ce dernier prend en considération l'effet de la succion sur le comportement en fluage de l'argile étudiée, il s'adapte de manière satisfaisante avec les résultats expérimentaux.

$$\varepsilon = \left[\frac{(\sigma_1 - \sigma_3)f}{E_d}\right]\frac{D_1}{1-[a\,(s/Pa)+b]D_1}\left(\frac{t}{t_1}\right)^n \quad (7)$$

Il convient de mentionner qu'en raison des résultats de tests limités, l'influence de la pression de confinement sur le comportement de fluage de l'argile étudiée n'été pas pris en compte, ce qui est l'une des principales limites de ce modèle.

2.2 Modèle amélioré par Zou et al. 2013

Le modèle de fluage de Burgers,est un modèle rhéologique composé d'un corps de Maxwell relié en série avec un corps de Kelvin "figure 2".

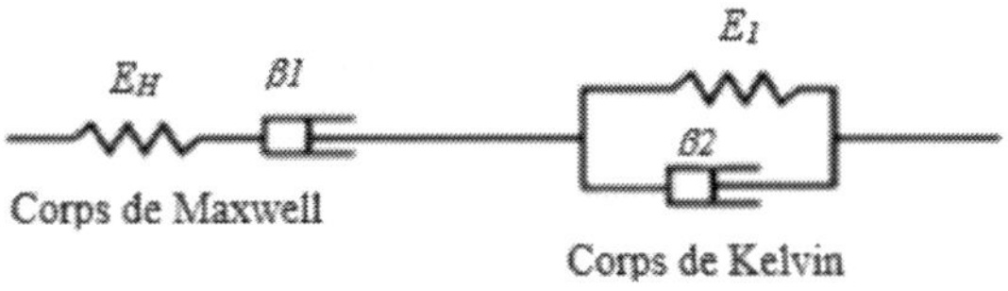

Fig. 2. Modèle de fluage de Burgers[2]

Le travail de Zou et al.2013 [2] porte sur l'amélioration du modèle étendu de Burgers. Ce dernier est développé en insérant un autre corps de kelvin dans le modèle de

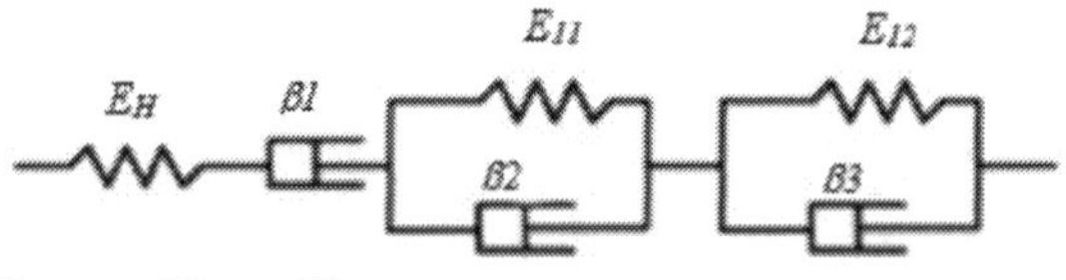

Burgers "figure 3".

Fig. 3. Modèle de fluage « Burgers étendu » (M-2K)[2]

L'équation constitutive de ce modèlepeut être décrite selon Sun. 1999[9]comme:

$$\varepsilon(t) = \sigma.\left\{\frac{1}{EH} + \frac{1}{\beta 1} + \frac{1}{E11}\left[1 - \exp\left(-\frac{E11.t}{\beta 2}\right)\right] + \frac{1}{E12}\left[1 - \exp\left(-\frac{E12.t}{\beta 3}\right)\right]\right\} \quad (8)$$

Avec,$(E_H)(E_{11})$ et (E_{12})modulesélastiques du corps de Maxwell et les corps de Kelvin respectivement,$(\beta 1)(\beta 2)$et$(\beta 3)$présentent la viscosité du corps de Maxwell et les corps de Kelvin respectivement,(t) temps,(σ)contrainte.

A partir des hypothèses de Yan et al.2010[10] l'équation (8) été transformaità un état tridimensionnel pour prendre en compte la matrice de succion dans le modèle :

$$\varepsilon_{ij} = \frac{S'_{ij}}{3K} + \frac{S''_{ij}}{3GH} + \frac{S''_{ij}}{3\eta 1}t + \frac{S''_{ij}}{3G11}.\left[1 - \exp\left(\frac{-G11.t}{\eta 2}\right)\right] + \frac{S''_{ij}}{3G12}\left[1 - \exp\left(\frac{-G12.t}{\eta 3}\right)\right] \quad (9)$$

Avec,(s'_{ij})tenseur de contrainte sphérique,(s''_{ij}) tenseur de contrainte déviatorique,$(G_H)(G_{11})$ et (G_{12}) modules de cisaillement du corps de Maxwell et les corps de Kelvin respectivement,$(\eta_1)(\eta_2)$ et (η_3)les coefficients visqueux du corps de Maxwell et les corps de Kelvin respectivement,(K)module d'élasticité isotrope déterminé par l'équation (10) :

$$K = \frac{2\ G_H\,(1+v)}{3(1-2v)} \quad (10)$$

Avec, (v) coefficient de poisson.

Des relations linéaires ont été trouvé entre les paramètres de l'équation (9) et la succion. Le nouveau modèle de fluage obtenu est comme suit :

$$\varepsilon_{ij} = \frac{S'_{ij}}{3K'} + \frac{S''_{ij}}{3G'H} + \frac{S''_{ij}}{3\eta' 1}t + \frac{S''_{ij}}{3G'11}\left[1 - \exp\left(\frac{-G'11.t}{\eta' 2}\right)\right] + \frac{S''_{ij}}{3G'12}\left[1 - \exp\left(\frac{-G'12.t}{\eta' 3}\right)\right] \quad (11)$$

L'équation (11) présente le modèle de fluage amélioré par Zou et al.2013[2], les nouveaux paramètres$(G'_H)(G'_{11})(G'_{12})(\eta'_1)(\eta'_2)$ et (η'_3)prennent en compte la matrice de succion dans le modèle. La comparaison entre la simulation et l'expérimentation a montré l'efficacité et la capacité prédictive de ce dernier. En outre, même si le modèle prend en considération une large gamme de succion comprise entre 50 et 300 kPa, il reste limité à un seul niveau de déviateur maintenu à 0,55 contrairement au modèles précédents Lai et al. 2010[1]et 2014[6]. Cela, implique la nécessité d'autres essais de fluage a succion contrôlée et à différents niveaux de contraintesdéviatorique.

3 Comportement en fluagedes sols non saturés : Proposition d'un modèle

Dans cette section, le modèle de fluage primaire « Modified Time Hardening » (MTH)[11] est présenté. Ensuite une relation est établie entre l'un des paramètres du modèle et la succion, puis un second paramètreest induit avec le niveau de déviateur appliqué. Sur cette base, le modèle est reconsidéré pour calculer la déformation en fluage en prenant en compte la succion et le niveau du déviateur.

3.1 Modèle de fluage primaire « Modified Time Hardening »[11]

Comme décrit dans la section 2, les modèles développés font déjà preuve d'une bonne efficacité, néanmoins leurs présences dans un code de calculs aux éléments finis n'est pas toujours évidente. Cela implique une procédure d'implémentation qui requiert des connaissances sur le langage de programmation du logiciel utilisé.
Dans ce travail, l'équation de fluage primaire du logiciel ANSYS,intitulé« Modified Time Hardening» (MTH)a été sélectionnée. Le choix de ce modèle est justifié par sa capacité et sa performance pour la représentation de fluage primaire considéré comme réponse macroscopique lors d'un essai de fluage triaxial[12].
L'expression générale du modèle est comme suite :

$$\varepsilon_{cr}(t) = \left(\frac{C1}{C3+1}\right).q^{C2}.t^{C3+1} \quad (12)$$

Avec,(ε_{cr})ladéformation de fluage, (q) lacontrainte déviatorique,(t)est letemps du chargement et Ci (i=1-3)paramètres de fluage.

3.2 Prise en compte de la succion

Pour l'élaboration du modèle de fluage de l'argile non saturé, une méthode alternative consiste d'abord à analyser la prédiction du modèle sur les résultats d'essais de fluage triaxiaux menée par Zou el al[2]. Les tests ont été réalisés sous une pression de confinementσ'_3 maintenu à 100 kPa, et un niveau de déviateurD^r maintenu à 0,55.Un jeu de paramètres pour chacun des essais est répertorié dans le Tableau 1.L'identification des paramètres est réalisée par minimisation de l'écart entre le résultat du calcul numérique et la courbe expérimentale.

Tableau 1. Paramètres du modèle en fonction de la succion

s (kPa)	q (kPa)	D^r	C1	C2	C3
50	103	0,55	$9,5.10^{-5}$	1,64	0,075
100	124	0,55	$6,8.10^{-5}$	1,64	0,075
150	141	0,55	$5,3.10^{-5}$	1,64	0,075
200	188	0,55	$3,4.10^{-5}$	1,64	0,075
250	191	0,55	$2.9.10^{-5}$	1,64	0,075

Sur la base des résultats du Tableau 1, l'augmentation de la succion conduit à la diminution du paramètre C1 d'où le choix d'une relation logarithmique entre C1-s/Pa "figure 4".

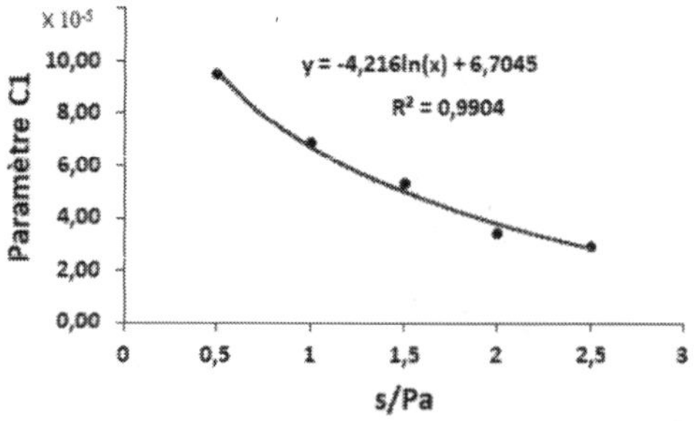

Fig. 4. Relation entre le paramètre C1 et le rapports/Pa.

Cependant, la prise en compte du niveau de déviateur (D^r) dans le modèle ne peut être négligé. Dans le Tableau 2, les paramètres du modèle ont été déterminés à partir des résultats d'essais menée par Lai et al. (2010)[1] sous une pression de confinement σ'3 maintenu à 100 kPa, et des niveaux du déviateur(D^r)différents. L'analyse des résultats, montre la sensibilité du paramètre C2 vis-à-vis leniveau du déviateur.Une relation de puissance a été établie entre C2- D^r "figure 5". En revanche,le paramètre C3 reste constant, ildépend du type de sol et de sa composition.Il vaut 0,075 pourleséchantillons d'argile étudiées.

Tableau 2. Paramètres du modèle pour différents niveaux du déviateur

D^r	q (KPa)	S (kPa)	C1	C2	C3
0,112	39,2	250	$2,9.10^{-5}$	1,04	0,075
0,161	54,9	200	$3,4.10^{-5}$	1,37	0,075
0,170	62,9	300	$2,1.10^{-5}$	1,28	0,075
0,258	90,2	250	$2,9.10^{-5}$	1,30	0,075
0,323	109,8	200	$3,4.10^{-5}$	1,56	0,075
0,340	125,8	300	$2,1.10^{-5}$	1,53	0,075
0,403	141,2	250	$2,9.10^{-5}$	1,45	0,075
0,438	149	200	$3,4.10^{-5}$	1,63	0,075
0,510	188,7	300	$2,1.10^{-5}$	1,77	0,075
0,546	191,2	250	$2,9.10^{-5}$	1,58	0,075
0,553	188,2	200	$3,4.10^{-5}$	1,67	0,075
0,669	227,4	200	$3,4.10^{-5}$	1,76	0,075
0,680	251,6	300	$2,1.10^{-5}$	1,90	0,075
0,784	266,6	200	$3,4.10^{-5}$	1,80	0,075

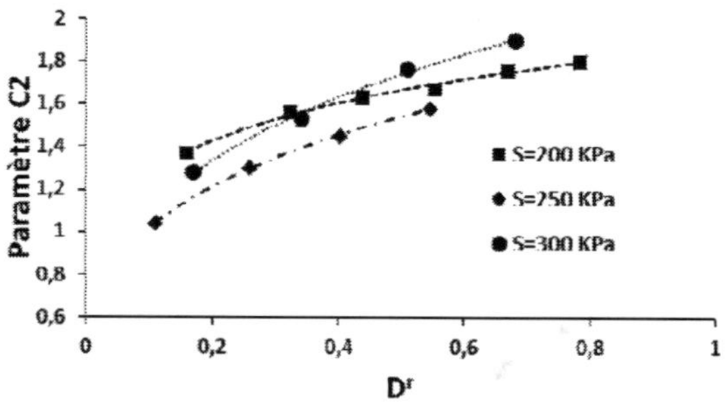

Fig. 5. Relation entre le paramètre C2 et le niveau de déviateur D^r

Compte tenu de la variation du paramètre C1 en fonction de la succion "figure 4", et la variation du paramètre C2 en fonction du niveau de déviateur appliqué "figure 5", le modèle est reconsidéré comme suit :

$$\varepsilon_{cr}(t) = \left(\frac{C'_1}{1,075}\right).q^{C'_2}.t^{1,075} \quad (13)$$

Le calcul des nouveaux paramètres dans l'équation (13) est illustré dans le Tableau 3.

Tableau 3. Conditions et paramètresdu modèle amélioré

Conditions hydriques	Paramètres
Succion [50-250] kPa	$c'_1 = -4.10^{-5} ln(S/Pa)+7.10^{-5}$
Succion [200] kPa	$c'_2= 1,87.D^{r0,17}$
Succion [250] kPa	$c'_2= 1,84.D^{r0,26}$
Succion [300] kPa	$c'_2= 2,12.D^{r0,29}$

3.4Validation

Le but de cette section se veut comme une évaluation du modèle avec la prise en compte des paramètres avec les relations présentées ci-dessus. A partir des conditions illustrées dans le Tableau 3,la matrice de succion et le niveau de déviateur ont été pris en compte pour calculer la déformation de fluage. Les résultats d'essais de fluage triaxiaux publié dans le travail de Lai et al. 2010 [1]ont été simulés, la comparaison entre les résultats montre la capacité prédictive du modèle pour le comportement au fluage des sols non saturés "figure 6".

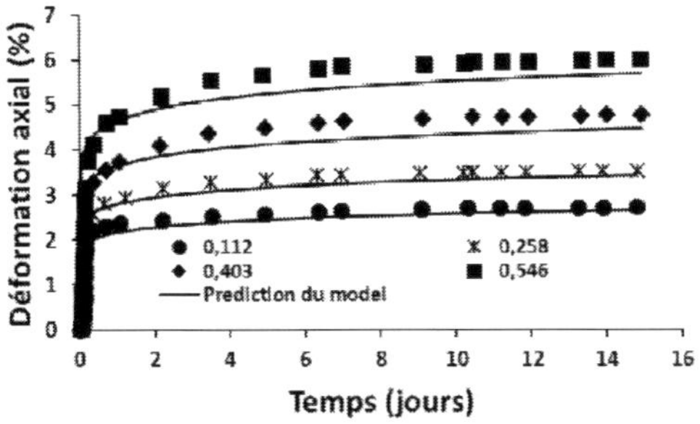

Fig. 6. Simulations d'essais de fluage triaxiaux à succion s= 250 kPa et différents niveaux de contrainte déviatorique

Le modèle est ensuite positionné vis-à-vis les modèles qui ont été présentés dans la section 2.La comparaison illustrée dans la "figure 7"montre une capacité prédictive presque identique entre les différents modèles.

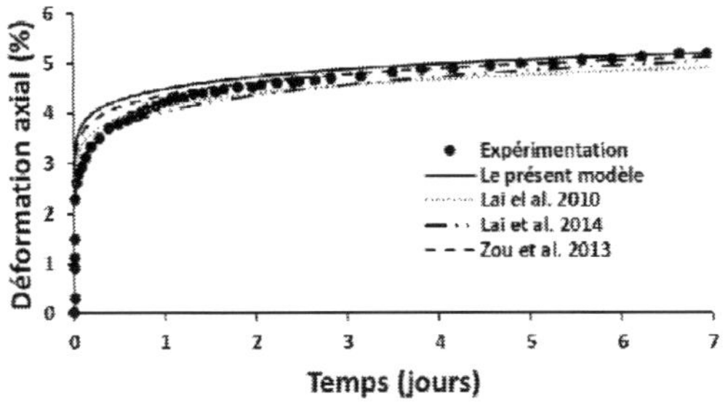

Fig. 7. Simulation d'essai de fluage triaxial à succion s=200 kPa et niveau de contrainte déviatorique D=0.55

Il convient de mentionner qu'en raison du manque des résultats pour d'autres valeurs de succion, le modèle est limité à prédire le fluage pour une gamme de succion comprise entre 200 et 300 kPa. Il convient également de souligner que le modèle est limité à des déformations axiales et ne peut prendre en compte la déformation latérale ou de cisaillement.

4 Conclusion

Un modèle de fluage primaire pour les sols non saturés simulant des tests de fluage triaxiaux a été considéré avec des relations interprétant la sensibilité des paramètres à la succion et au niveau de contrainte. Basé sur une relation logarithmique trouvée entre le paramètre C1 et la succion, et une relation de puissance entre le paramètre C2 et le niveau de déviateur, le modèle prend en compte l'effet dela non saturation et une large gamme de niveau de déviateur, avec des coefficients de corrélation supérieurs à 0,98. Les résultats simulés sont en accord avec les résultats expérimentaux.

L'avantage d'amélioration de ce modèledisponible dans un code d'éléments finis ANSYS, permettra une application étendu pour la simulation du phénomène sous la fluctuation du niveau d'eau cyclique, pour une période de plus long terme. Ceci, permet une base importante pour la prédiction et l'évaluation des glissements de terrain, dansles futures investigations numériques.

5 Références bibliographiques

1. X. Lai, S. Wang, H. Qin, and X. Liu, "Unsaturated creep tests and empirical models for sliding zone soils of Qianjiangping landslide in the Three Gorges," J. Rock Mech. Geotech. Eng., vol. 2, no. 2, pp. 149–154, (2010).
2. L. Zou, S. Wang, and X. Lai, "Creep model for unsaturated soils in sliding zone of Qianjiangping landslide," J. Rock Mech. Geotech. Eng., vol. 5, no. 2, pp. 162–167, Apr.(2013).
3. A. Augustesen, M. Liingaard, P. V Lade, and M. Asce, "Evaluation of Time-Dependent Behavior of Soils," Int. J. Geomech., vol. 4, no. 3, pp. 137–156, (2004).
4. M. Liingaard, A. Augustesen, P. V Lade, and M. Asce, "Characterization of Models for Time-Dependent Behavior of Soils," Int. J. Geomech., vol. 4, no. 3, pp. 157–177, (2004).
5. A. Purwodihardjo, "*Modélisation des déformations différées lors du creusement des tunnels*," Ecole centrale de Lyon, (2004).
6. X. Lai, S. M. Wang, W. M. Ye, and Y. J. Cui, "Experimental investigation on the creep behavior of an unsaturated clay," Can. Geotech. J., vol. 51, no. 6, pp. 621–628, (2014).
7. A. Singh and J. K. Mitchell, "General stress-strain-time function for soils," J. Soil Mech. Found Eng. Div. ASCE, vol. 94, no. 1, pp. 21–46, (1968).
8. G. Mesri, E. Febres-Cordero, D. Shields, and A. Castro, "Shear stress-strain-time behaviour of clays," Geotechnique, vol. 31, no. 4, pp. 537–552, (1981).
9. J. Sun, "Rheological of geotechnical materials and its application to engineering," China Architecture and Building Industry Press, (1999).

10. Y. A. N. Yan, W. A. N. G. Si-jing, and W. A. N. G. En-zhi, "Creep equation of variable parameters based on Nishihara model," Rock Soil Mech., vol. 31, no. 10, pp. 3025–3035, (2010).
11. ANSYS, "*ANSYS Mechanical APDL Theory Reference*," ANSYS Inc, vol. Release15, no. November, pp. 1–909, (2013).
12. R. Plassart, "Modélisation hydromécanique du comportement des ouvrages souterrains avec un modèle élastoviscoplastique," Dr. Diss. Vandoeuvre-les-Nancy, INPL, (2011).

Validation and Verification of Semi-Empirical Methods for Evaluating Liquefaction Using Finite Element Method

Soukaina TOUIJRATE[1], Khadija BABA[1], Mohamed AHATRI[1] Lahcen BAHI[2]
[1]GCE Laboratory, High School of Technology-Salé, Mohammed V University in Rabat-Morocco
[2]3GIE Laboratory, Mohammadia Engineering School, Mohammed V University in Rabat-Morocco

Abstract. Liquefaction is a hazardous and temporary phenomenon by which a soil saturated with water loses some or all of its resistance. The undrained conditions and a cyclic load increase the pores water pressure inside the soil and therefore a reduction of the effective stress.

Nowadays many semi-empirical methods are used to introduce a proposition to evaluate the liquefaction's potential using the in-situ test results. The objective of this paper is to study their ability to correctly predict the liquefaction potential by modelling our case using finite element methods.

The study is based on the data of Cone Penetration Tests experimental results of the Casablanca-Tangier High-Speed Line exactly between PK 116 + 450 and PK 116 + 950 and near of Moulay-Bousselham city. It belongs to the Drader-Soueir basin region which is located in the North-West of Morocco.

This region had a specific soil's formation, the first 50 meters are characterised by the existence of sand layers alternating with layers of clay. These formations are very loose and saturated which suggests the possibility of soil liquefaction.

We present and discuss the results of applying the Olsen method [1], the Juang method [2] and the Robertson method [3], in the evaluation of liquefaction susceptibility.

Apart from the previous empirical analysis to evaluate the liquefaction potential, numerical modelling is performed in this study.

1 Introduction

The soils, during liquefaction, behave as a liquid and no longer support the shearing forces, which causes large deformations which subsequently affect the structures built on this soil.

For assessing the liquefaction potential, many semi-empirical methods are developed. This potential reflects the soil ability to resist the cyclic shear efforts and is mainly dependent on the relative density, the particle size, the soil texture, the saturation degree and the earthquake magnitude.

the case studied is based on the CPT tests data for a section of the Casablanca-Tangier High-Speed Line exactly between PK 116 + 450 and PK 116 + 950, and that is part of the basin Drader-Soueir which is located in the North-West of Morocco.

The aim of this paper is, on the one hand, to evaluate and estimate the liquefaction potential using three semi-empirical methods, namely the Olsen method [1], the Juang method [2] and the Robertson method [3] and, on the other hand, to check and validate these methods using the finite element method.

The numerical modelling of the liquefaction behaviour in this case is carried out in Quake/W software (from GeoStudio) [4], which simulates the soil's behaviour under a cyclic loading by the generalized the finite element method. This modelling will help us to determine the semi-empirical method that predicts the best liquefaction susceptibility.

2 Semi-empirical methods used in the evaluation

The semi-empirical approaches evaluating the liquefaction potential are founded on the earthquakes observation, the liquefaction-induced effects, and the in-situ test results. The tests, which help to estimate the liquefaction potential, are mainly the Cone Penetration Test (CPT) and the Standard Penetration Test (SPT). These approaches can be divided into three categories:

- Approach by cyclic stress;
- Approach by cyclic deformation;
- Energy approach.

In this paper, the evaluation methods [1,2,3] proposed are based on the cyclic stress approach using the CPT test data.

2.1 Evaluating cyclic stress ratio

The CSR estimate for the three methods studied is based on the simplified method of Seed and Idriss (1971) [5] who proposed the following relationship:

$$CSR = \frac{\tau_{av}}{\sigma'_{v0}} = 0{,}65\left(\frac{a_{max}}{g}\right)\left(\frac{\sigma_{v0}}{\sigma'_{v0}}\right) r_d \qquad (1)$$

Where τ_{av} is the average cyclic shear stress; a_{max} is the maximum horizontal acceleration at the ground surface; g = 9,81 m²/s is the acceleration of gravity; σ_{v0} is the initial vertical total stress; σ'_{v0} is the initial vertical effective stress; r_d is the stress reduction factor.

The stress reduction coefficient r_d is expressed in terms of the depth by the following equations (Liao and Whitman 1986) [6]:

$$r_d = 1 - 0{,}00765 \qquad z \leq 9{,}15\ m$$
$$r_d = 1 - 0{,}00765z \qquad 9{,}15\ m < z \leq 23m \qquad (2)$$

The Juang method [2] used the equation (1) to calculate the CSR with the introduction of MSF coefficient that adjusts the CSR value to an earthquake magnitude of 7.5. Equation (1) then becomes:

$$CSR = \frac{\tau_{av}}{\sigma'_{v0}} = 0{,}65\left(\frac{a_{max}}{g}\right)\left(\frac{\sigma_{v0}}{\sigma'_{v0}}\right)\frac{r_d}{MSF} \qquad (3)$$

Where MSF is expressed by the equation (4) according to I.M Idriss 1998 [7]:

$$MSF = \frac{10^{2,24}}{M_w^{2,56}} = \left(\frac{M_w}{7{,}5}\right)^{-2,56} \qquad (4)$$

M_w is the earthquake magnitude.

2.2. Evaluating cyclic resistance ratio

The CRR evaluation in this study is based on the CPT test data, which is among the most common in-situ tests in Morocco.

2.2.1. Olsen method

Olsen has simplified the CRR estimation by the equation (5), based on the study of several parameters influencing soil liquefaction (Olsen 1997) [1]:

$$CRR = 0{,}00128\left(\frac{q_c}{\sigma'_{v0}}\right)^{0,7} - 0{,}025 + 0{,}17R_f - 0{,}028R_f^2 + 0{,}0016R_f^3 \qquad (5)$$

With q_c is the resistance of the tip from the CPT test; R_f (friction ration) is a percentage expressed as:

$$R_f = \frac{f_s}{q_c} \times 100 \qquad (6)$$

f_s is sleeve friction resistance measured by CPT.

2.2.2 Juang method

For the Juang method, the CRR is expressed by (Juang and al., 2003) [2]:

$$CRR = C_\sigma exp\left[-2{,}9 + 1{,}264\left(q_{c1N,cs}/100\right)^{1,25}\right] \qquad (1)$$

Where:

$$C_\sigma = -0{,}016\left(\sigma'_{v0}/100\right)^3 + 0{,}178\left(\sigma'_{v0}/100\right)^2 - 0{,}063\left(\sigma'_{v0}/100\right) + 0{,}903 \qquad (2)$$

$$q_{c1N,cs} = K_I q_{c1N} \qquad (3)$$

$$K_I = 2{,}249(I_c)^4 - 16{,}943(I_c)^3 + 44{,}551(I_c)^2 - 51.497(I_c) + 22.802 \qquad (4)$$

$$I_c = [(3{,}47 - log_{10}q_{c1N})^2 + (log_{10}F + 1{,}22)^2]^{0,5} \qquad (5)$$

F is normalized friction ratio defined as:

$$F = f_s/(q_c - \sigma_v) \times 100 \qquad (6)$$

2.2.3. Robertson method

The Robertson's CRR estimation (Robertson and Wride 1998) [3] appears more difficult than the other two methods. The evaluation steps are summarized in the figure 1.

This method is limited by a standardized the resistance of the tip less than 160 and uses another criterion (liquidity limit and moisture content) to evaluate the liquefaction if Ic is greater than or equal to 2.6.

2.3. Evaluating factor of safety and liquefaction probability

The factor of safety is expressed by the same equation for all methods based on the cyclic stress approach [5]:

$$F_s = CCR/CSR \qquad (7)$$

From the factor of safety coefficient, we can determine the liquefaction probability [8]:

$$PL = \frac{1}{1 + \left(F_s/A\right)^B} \qquad (8)$$

Where:

- A = 1 and B = 2,78 according to Olsen method [1];
- A = 0,96 and B = 4,50 according to Juang method [2];
- A = 1 and B = 3,30 according to Robertson and Wride method [3].

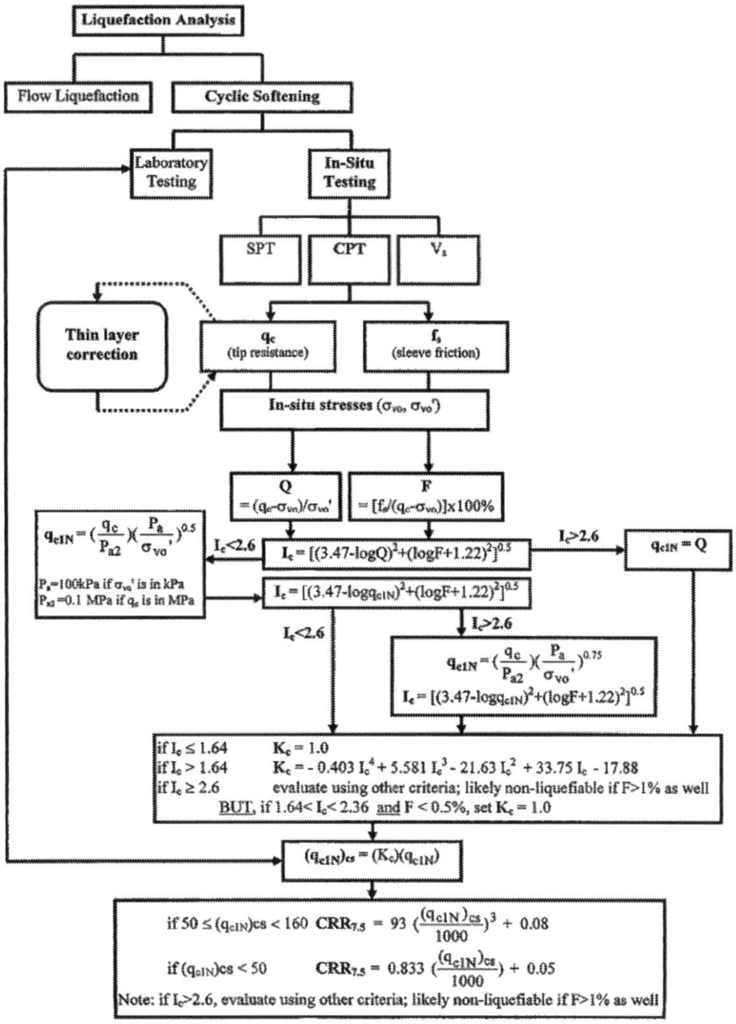

Fig. 1. Method for evaluating Cyclic Resistance Ratio (CRR) (Robertson and Wride 1998) [3]

3 Case study

Our case study positioned at the southern section of the High-Speed Line linking Tangier and Casablanca, it is between the PK 116 + 450 and the PK 116 + 950. Geographically, this section is located in the North-West of Morocco, precisely, 65 km South-East of Moulay Bousselham city.

3.1. Geological framework

Geologically, the section studied is part of the Drader-Soueir basin, which belongs to a unit that extends from the Gharb plain's South to the Oued Loukkos in the North and is limited by the clay and marly pre-Rif outcrops in the East. The unit is composed of the plio-quaternary soils that are quite varied but always more or less detrital (intercalation of sand, sandstone, sandstone limestone, sandy clay) resting on a very clear substratum formed by the Miocene blue marls [9].

The soils in the area are profoundly influenced by the textural nature of their bedrock made of quaternary sands. The granulometric analysis shows that the majority of the soils in this area contain at least 75% of the total sands, the most often 90%. The clay contents are located in the 4 to 6% except in depth where they sometimes reach 18 to 20%. The deep horizon frequently includes calcareous kidneys and needles.

Fig. 2. Satellite view of the section studied

3.2. The region's seismicity

In Oued Drader, the maximum intensity is measured at near VII, but due to the proximity of the area to the critical seismic intensity regions, including a historical seismic site, and for safety measures, considers an intensity of IX.

Many conversion laws between the intensity and the magnitude were developed by Cherkaoui 1991 [10] and Karnick 1968 [11], thus allowing a correlation between the two parameters. For an intensity equal to IX, the Drader-Soueir basin magnitude retained is M = 6.5.

Given that the Oued Drader region is part of the Seismic Zone No. 3 according to the RPS 2000 revised in 2011, it is therefore considered a maximum acceleration of 0.14 g [12].

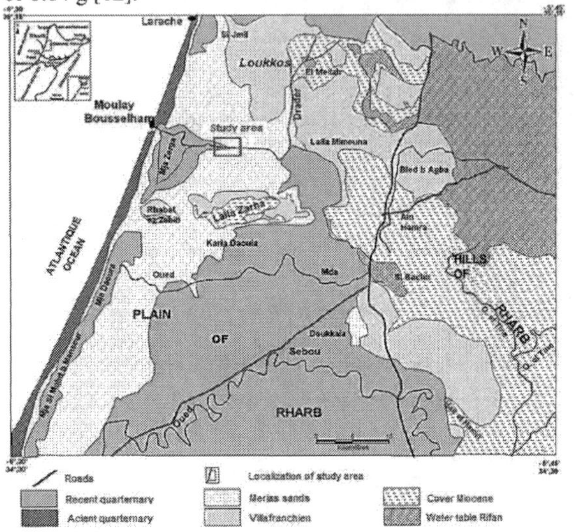

Fig. 3. The Pre-Rif Simplified Geological Map [9]

4 Results and discussions

4.1. Evaluating liquefaction potential

In this study, we evaluate the liquefaction susceptibility using the data from 6 surveys of the test CPT along the section between the PK116 + 450 and the PK 116 + 950 of the High-Speed Line.

This evaluation is made by, on the one hand, the estimation of the safety factor and the liquefaction probability by the 3 semi-empirical methods which are the Olsen method [1], the Juang method [2] and the Robertson method [3] and based on the cyclic stress approach, the results of each one are presented in the figures 4, 5 and 6. On the other hand, by modelling the two surveys executed in the PK 116 + 580 (figure 7) and the PK 116 +720 (figure 8), which aims to analyse and study the soil response when it is subjected to a cyclic loading, and to examine the possibility of a certain generation of an excessive overpressures, which could result liquefaction.

According to the analysis of the different graphs in the figures 4, 5 and 6, it is found the common layers which are the most affected in the depths between 6 and 14 m, with a high liquefaction probability at the silty and the sandy soils such as the silty sand to sandy silt where the factor of security much less than 1 for the three methods [1, 2, 3] in the PK 116 + 950 survey and the silts sensitive also in the PK 116 + 650 survey.

The PK 116 + 780 survey presents a low probability due to the important alternation of clay layers which are less susceptible to liquefaction.

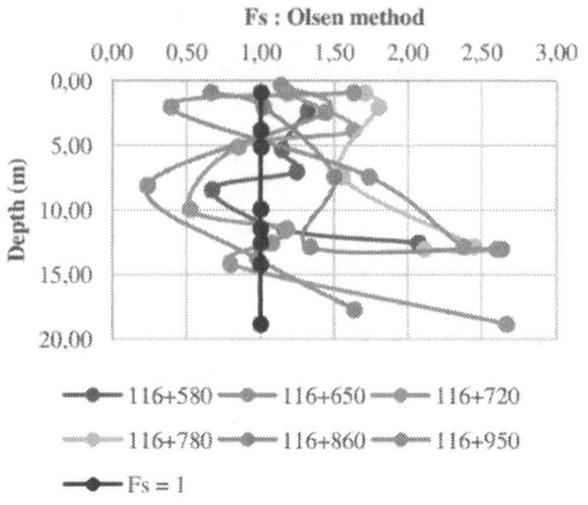

Fig. 2. The Fs Values by the Olsen method

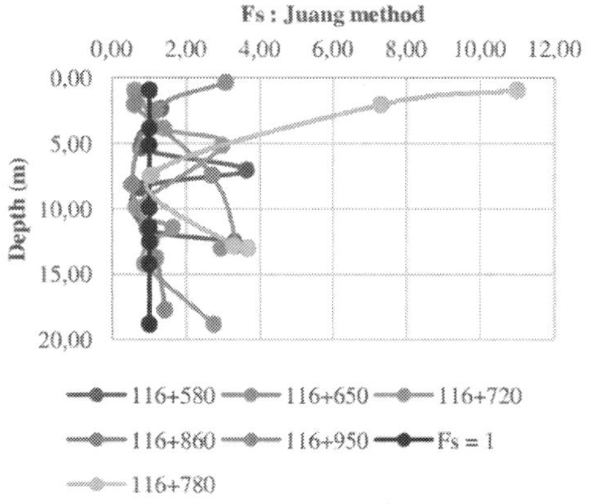

Fig. 3. The Fs Values by the Juang method

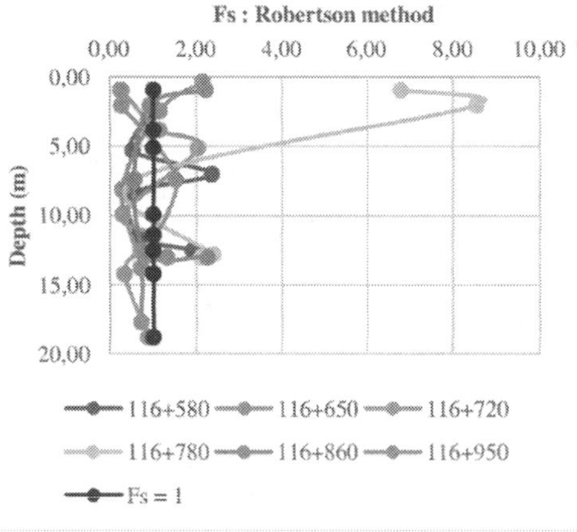

Fig. 4. The Fs values by the Robertson method

https://doi.org/10.1051/matecconf/201814902028

Returning to the numerical modelling results of the two surveys the PK116 + 580 and the PK116 + 720 presented in the figures 9 and 10, we find that the Pore Water Pressure significantly increases, for the first survey, in the sand layer and also in the silty sand to sand and the silty sand to sandy silt layers, and for the second survey, in the three successive layers' silty sand to sandy silt- silty sand to sandy silt - silty sand to sand. This increase, according to the Terzaghi postulate [13], compensates the total stress so it causes the loss of the shear strength and therefore increases the liquefaction risk.

These layers are generally located between 6 and 14 m, which confirms the results preliminary analysis of the semi-empirical methods except that for the fine clayey or sensitive silt layer the pore water pressure is average.

We can conclude that the region of the Drader-Souier basin can be considered as very susceptible to liquefaction, due, on the one hand, to the formations nature in the area, which are characterized by the existence of sand and silt layers alternate with more clayey layers, which are saturated with water because of the basin hydrogeological nature, on the other hand, with the seismic risk present in this region.

4.2. Validation and Verification of Semi-Empirical Methods

As we have already said, the Layers susceptible to liquefaction, from the results of the numerical modelling are the sand layer, the silty sand to sand and the silty sand to sandy silt layers in the PK116 + 580 and the three successive layers' silty sand to sandy silt- silty sand to sandy silt - silty sand to sand in the PK116 + 720, the rest of the layers has an average susceptibility except for the layers more or less clayey that have a low risk.

Comparing these results with the Table 1 results, we find that:

- These results are close to that of the Olsen method [1] even, it does not have an aspect physical well-defined and knowing that it is easy to exploit.
- The Robertson method [3] is also close to modelling results except that it is more restrictive for fine soils.
- The Juang method [2] has an aspect physical robust but in our case study these results are far from that of the modelling except for the layers where the soils are fine and loose soil.

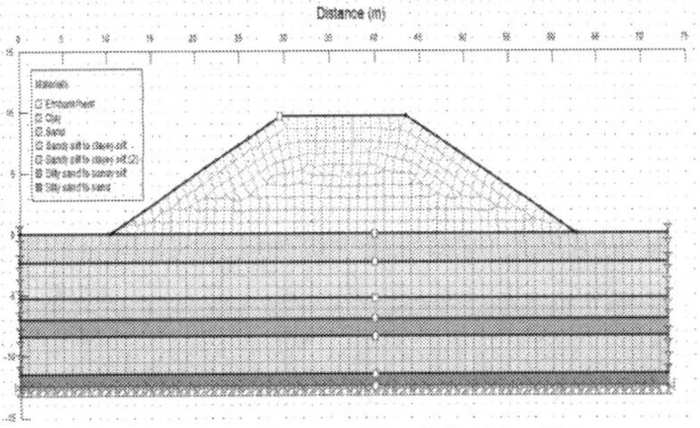

Fig.5. The component layers of the PK 116 + 580 survey

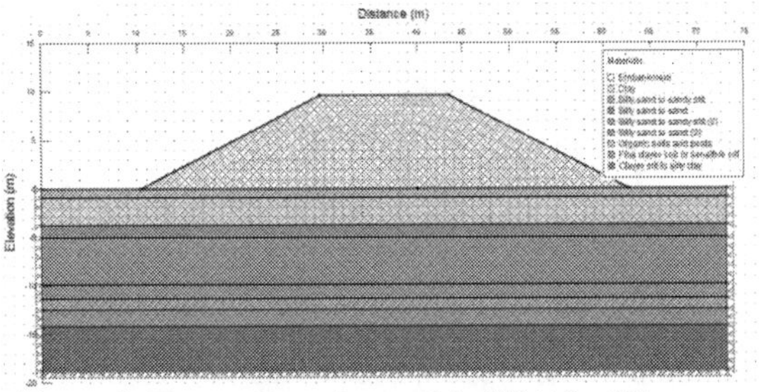

Figure 6 : The component layers of the PK 116 + 720 survey

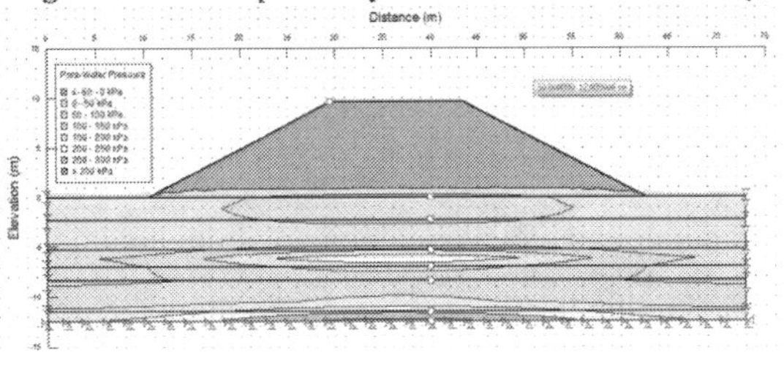

Fig. 7. The PWP distribution in the PK 116 + 580 survey

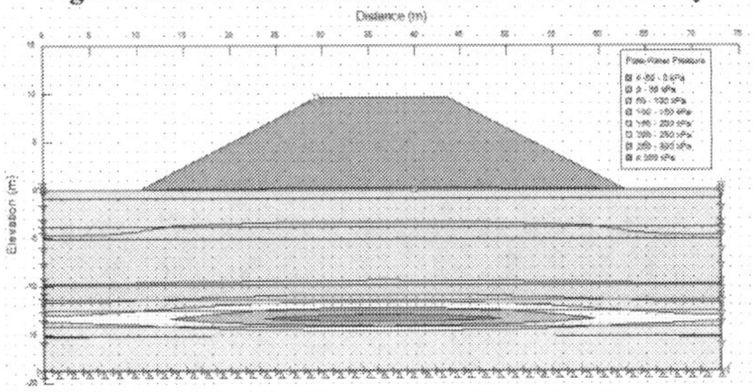

Fig. 8. : The PWP distribution in the PK 116 + 720 survey

Table 1. The liquefaction probability results of the three methods in the PK116 + 580 and PK116 + 720 surveys

	PK	Classification	Olsen	Juang	Robertson
Zone 1	116+580	Sandy silt to clayey silt	0.32	0.20	0.46
		Clay	0.41	0.73	0.90
		Sand	0.35	0.00	0.06
		Silty sand to sandy silt	0.75	0.75	0.87
		Sandy silt to clayey silt	0.40	0.51	0.80
		Silty sand to sand	0.12	0.00	0.11
Zone 2	116+720	Organic soils and peats	0.39	0.89	0.99
		Clay	0.20	0.16	0.40
		Silty sand to sand	0.61	0.01	0.09
		Fine clayey soil or sensitive silt	0.86	0.86	0.98
		Silty sand to sand	0.39	0.08	0.50
		Silty sand to sandy silt	0.45	0.41	0.77
		Silty sand to sandy silt	0.65	0.60	0.97
		Clayey silt to silty clay	0.06	0.01	0.60

Conclusion

Based on the data from six surveys of the CPT test between the PK 116 + 450 and the PK 116 + 950 from the High-Speed Line, the liquefaction potential calculations were carried out using three methods based on the cyclic stress approach which are the Olsen method [1], the Juang method [2] and the Robertson method [3].

The results of these calculations and the numerical modelling with the finite element method allowed us to conclude that the Drader region presents a high risk of liquefaction due the geological, hydrogeological and seismic context of the region.

We also compared the different methods results used in this study and those of the numerical modelling concluded that the Olsen method [2] seems to be the closest and the simplest to operate among those proposed.

We also conclude that the Robertson [3] method is restrictive for the fine soils even if they are not loose and the Juang method [2] results are furthest from the modelling results even if it has an aspect physical robust.

References

1. Olsen RS (1997). *Cyclic liquefaction based on the cone penetrometer test.* In: Youd TL, Idriss IM, editors. Proceedings of the NCEER Workshop on Evaluation of Liquefaction Resistance of Soils. Technical Report NCEER-97-0022, Buffalo, NY: State University of New York at Buffalo; 1997. p. **225–276**.
2. Juang, H., Yuan, H., Lee, D., & Lin, P. (2003). *Simplified cone penetration test-based method for evaluating liquefaction resistance of soils*. Journal of Geotechnical and Geoenvironmental Engineering, ASCE, 129(1), **66—80**.
3. Robertson, P.K and C.E Wride (1998): *Evaluating Cyclic Liquefaction Potential Using the Cone Penetration Test*, Canadian Geotechnical Journal, 35(3), pp **442-459**.
4. *An Introductory QUAKE/W Example*, GEO-SLOPE International Ltd, Calgary, Alberta, Canada, www.geo-slope.com.
5. Seed HB, Idriss IM. *Simplified procedure for evaluating soil Liquefaction potential*. J Soil Mech Found Div, ASCE 1971 ; 97(SM9) :**1249–73**.
6. Liao, S.S.C. and R.V. Whitman (1986): *A catalog of liquefaction and Non-liquefaction occurrences During Earthquakes*, Department of civil Engineering, Massachusetts institute of technology, Cambridge, MA, **117** pp.
7. I.M. Idriss, R.W. Boulanger, *Semi-empirical procedures for evaluating liquefaction potential during earthquakes*, Soil Dynamics and Earthquake Engineering 26 (2006) **115–130**.
8. Juang, C.H., Chen, C.H. and Mayne, P.W. (2008), "*CPTU simplified stress-based model for evaluating soil liquefaction potential*," Soils and Foundations, Vol. 48, No. 6, pp. **755-770**.
9. Safae MAYCHOU, *Étude morphostructurale et cartographie SIG du Rharb Septentrional et du Prérif (Maroc). Analyse sismotectonique et modélisation de la déformation de la région de Moulay Bousselham*, 2009, l'Université Chouaïb Doukkali et L'Université Bordeaux 1.
10. Taj-Eddine CHERKAOUI et Lahcen ASEBRIY, *Le risque sismique dans le Nord du Maroc,* Trav. Inst. Sci. Rabat, sér. Géol. & Géogr. phys., n° 21, 2003, p.**225-232.**
11. KARNIK V. (1969) - *Seismicity of the European area. Part I. (Prague and Dordrecht-Holland);* **364**p.
12. *Le règlement de construction parasismique 2000-version 2011*, Ministre de l'habitat et la politique de la ville.
13. Terzaghi et Peck, *Mécanique des sols appliquée aux travaux publics et aux bâtiments*, Dunod, Paris, 1961
14. Alain PECKER, livre de *Dynamique des sols*, Presse de l'école nationale des Ponts et Chaussées 1984.
15. Anna BOROWIEC and Krzysztof MACIEJEWSKI, *Assessment of Susceptibility to Liquefaction of Saturated Road Embankment Subjected to Dynamic* Loads Studia Geotechnica et Mechanica, Vol. XXXVI, No. 1, 2014 DOI: 10.2478/sgem-**2014-0002**.
16. Francesco Castelli, Antonio Cavallaro, Salvatore Grasso, Michele Maugeri, *Soil liquefaction and risk analysis from in situ tests for the city of Trapani (Italy)*, Fourth International Conference on Recent Advances in Geotechnical Earthquake Engineering and Soil Dynamics 2001.
17. Francesco Castelli, Valentina Lentini, *SPT-Based Evaluation of Soil Liquefaction Risk, 2010* - Fifth International Conference on Recent Advances in Geotechnical Earthquake Engineering and Soil Dynamics.
18. G. Zhang; P. K. Robertson, and R. W. I. Brachman, *Estimating Liquefaction-Induced Lateral Displacements Using the Standard Penetration Test or Cone Penetration Test*,2004, Project Engineer, EBA Engineering Consultants Ltd., **14940-123** Ave., Edmonton AB, Canada T5V 1B4, Journal of geotechnical and geoenvironmental engineering © ASCE.
19. Hudson, M., Idriss, I.M., and Beikae, M. 1994. (1994) "QUAD4M - *A computer program to evaluate the seismic response of soil structures using finite element procedures and incorporating a compliant base*." Davis, CA: Center for Geotechnical Modeling,Department of Civil and Environmental Engineering, University of California, Davis, CA.
20. Idriss and Sun, 1992 Idriss, I. and Sun, J. (1992). *User Manual for SHAKE 91. Technical report, Center for Geotechnical Engineering Modeling, Departament of Civil Engineering*, University of California, Davis. U.S.A.
21. K. Onder Cetin, Raymond B. Seed, Armen Der Kiureghian, Kohji Tokimatsu, Leslie F. Harder Jr., Robert E. Kayen and Robert E. S. Moss, *Standard Penetration Test-Based Probabilistic and Deterministic Assessment of Seismic Soil Liquefaction Potential*, Journal of Geotechnical and Geoenvironmental Engineering © ASCE / December 200, 130 :**1314-1340**.

22. Kramer, S.L., 1996 *Geotechnical Earthquake Engineering Prentice* – Hall Upper Saddle River, New Jersey, pp. **274-275**.
23. Matasovic, N. (1992) "*Seismic response of composite horizontally-layered soil deposits.*" Ph.D. Thesis, University of California, Los Angeles.
24. Muhsin Elie RAHHAL, *Comprendre les méthodes d'évaluation du potentiel de liquéfaction des sols*, 2008, Proceedings of the 4th Canadian Conference on Geohazards: From Causes to Management. Presse de l'Université Laval, Québec, **594** p.
25. Newmark, N. M. [1959]. *A method of computation for structural dynamics*," J. Engineering Mechanics Division **85, 67{94**.
26. Noufail TOUIL, thèse sous le thème « *Analyse expérimentale et numérique du potentiel de liquéfaction des sols sableux de Tanger* » 2012, Université ABDELMALEK SAADI.
27. S. C. Tunal, S. Altun, *Modern Approaches in Soil Liquefaction Analysis*, 6th International Conference on Earthquake Geotechnical Engineering 1-4 November 2015 Christchurch, New Zealand.
28. Youd et al., 2001 Youd, T., Idriss, I. M., Andrus, R. D., Arango, I., Castro, G., &Chistian,J. T. (2001). *Liquefaction resistance of soils: Summary report from the 1996 NCEER and 1998 NCEER/NSF workshop on evaluation of liquefaction resistance od soils*. Journal of Geotechnical and Geoenvironmental Engineering, ASCE, 127(10), **817-833**.

Soil treatment in flood areas: case of the road RP4209, in the province of Kenitra (Morocco)

N. SEBBAR[1], A. LAHMILI[1], L. OUADIF[1], A. AKHSSAS[1], L. BAHI[1]
[1] Mohammed V University, Mohammadia School of Engineering, L3GIE, Rabat, MOROCCO.

Abstract. The Moroccan road network has a linear of more than 57,000 km divided into national, regional, provincial and unclassified roads. Population growth, the transport of goods and the mobility of people require the construction of new links between regions and the maintenance of existing ones. Indeed, several factors, such as rainfall and floods, can cause enormous damage to existing infrastructure. Among the most at-risk areas is the GHARB region, especially the province of Kénitra, represented in this study by Provincial Road RP4209, which includes flood-prone areas with abundant watercourses, associated with poor type of soil. To mitigate these phenomena and to try to reduce their impacts, in particular on the components of a road, one of the appropriate solutions remains the treatment of soils. This process is increasingly practiced in the construction industry, improves the geotechnical and mechanical characteristics of soils without having to go through the contribution of materials from other regions, all for the purpose of a valuation of materials in place by treatment with lime, different hydraulic binders, or industrial products, which will provide an economic and environmental gain

Résumé. Le réseau routier marocain compte un linéaire de plus de 57.000 kms réparties en routes nationales, régionales, provinciales et routes non classées. La croissance démographique, le transport de marchandises ainsi que la mobilité des personnes imposent la construction de nouvelles liaisons entre régions et l'entretien de celles déjà existantes. En effet, plusieurs facteurs, tels que les précipitations et les inondations, peuvent engendrer d'énormes dégâts sur l'infrastructure existante. Parmi les régions les plus à risques, on trouve la région du GHARB en particulier la province de Kénitra, représentée dans cette étude par la route provinciale RP4209, qui regroupe des zones inondables avec l'abondance des cours d'eau, associée à la mauvaise nature du sol. Pour pallier à ces phénomènes et tenter de diminuer leurs impacts, en particulier sur les composantes d'une route, l'une des solutions appropriées demeure le traitement des sols. Ce procédé qu'est de plus en plus pratiqué dans le milieu du BTP, améliore les caractéristiques géotechniques et mécaniques des sols sans avoir à passer par l'apport des matériaux d'autres régions, tout cela dans le but d'une valorisation des matériaux en place par un traitement à la chaux, aux différents liants hydrauliques, ou aux produits industriels, ce qui procurera un gain économique ainsi qu'environnemental [1].

1 Introduction

De nos jours, le transport est devenu une activité primordiale pour le développement de tout pays. Il va de soi, que ce nombre en augmentation quotidienne a, en plus de ses avantages incontestables sur le train de vie actuel des individus, aussi des impacts néfastes que ce soit sur notre environnement ou sur nos infrastructures.
La qualité des chaussées dépend de plusieurs facteurs, entre autre la portance des sols supports. Une faible portance de la plateforme provoquera des déformations qui vont remonter jusqu'à la chaussée et causeront des orniérages et fissurations, parmi d'autres. Cette mauvaise portance est due essentiellement aux caractéristiques de la nature du sol.

L'approche habituelle dans ces cas, c'est d'enlever le mauvais sol (argileux, dans notre cas) et de le substituer par un sol plus résistant et moins fins. Toutefois, cette solution demeure plus onéreuse.

Pour ces raisons économiques et aussi environnementales, le choix de la solution du traitement et surtout avec de la chaux vive, demeure la plus adéquate pour ce type de sols.

La chaux vive permet de réaliser des modifications sur l'état hydrique du sol par la diminution de la teneur en eau du mélange sol-liant, provoquée par une hydratation de la chaux vive et par vaporisation d'une quantité d'eau causée par cette hydratation. Ceci permettra la réduction de la sensibilité du sol à l'eau et aussi du cycle gonflement-retrait de ce sol [2].

Dans cet article, on se penchera en premier lieu, sur une caractérisation de la zone à étudier, par l'identification du sol via une étude géotechnique. Dans cette étude on insistera sur les essais caractérisant l'argilosité du sol (limite d'Atterberg, valeur au bleu de méthylène), tout en passant par une analyse granulométrique dans le but de déterminer la répartition dimensionnelle des grains dans notre sol. Ensuite on se focalisera sur les études de formulation du couple sol-liant. Cette étape consistera en l'obtention d'une meilleure portance.

2 Cadre géographique et géologique

Une caractérisation de la zone à étudier est importante dans le cadre de notre projet, pour nous permettre de situer les zones d'existence des matériaux à traiter, ce qui mènera à une classification géologique et géotechnique des différents matériaux de la zone d'étude, facilitant ainsi la détermination des différents sols favorables pour les divers traitements .

2.1 Cadre géographique

La plaine du Gharb (Figure 1) se situe dans la partie Nord-Ouest du royaume. Elle est limitée à l'Ouest par l'océan Atlantique, au nord par la région de Tanger-Tétouan, à l'Est par les deux régions de Taza-Al Hoceima- Taounate et de Fès-Boulemane et au Sud par les régions de Meknès-Tafilalet et de Rabat-Salé-Zemmour-Zaër.

Fig. 1 : Vue générale de la plaine du Gharb (Ormvag) [3]

La région du Gharb bénéficie d'un climat méditerranéen, caractérisée par des hivers doux et humides et des étés chauds et secs où la moyenne des températures allant de 23°C (au voisinage de la côte) à 35°C (à l'intérieur du bassin). Le réseau hydrographique est représenté par l'une des principales rivières du royaume : la rivière Sebou et ses affluents : Ouargha, Beht et R'dom.

Ce réseau hydrographique complexe avec ses précipitations saisonnières irrégulières ainsi que ses sols argileux, présente un risque d'inondations fréquentes dans cette région. Ces précipitations se situent autour de 500 mm/an à Kénitra dans les 20 dernières années.

Depuis 1933, il y a eu plus que 40 crues d'importance variable causant de graves dégâts aux cultures mais aussi (et surtout) aux infrastructures.

Fig. 2 : Zone d'étude (Google Maps)

Le site du projet se situe dans la région du Gharb, à 40Kms du NE de la ville de Kénitra (Figure 2). Il est reconnu pour ses zones inondables ainsi que ses sols gonflants (Figure 3.a/3.b), conférant ainsi à ce dernier une mauvaise portance.

Fig. 3-a : Inondations région Kénitra (journal Le Matin /2015)

Fig. 3-b : Inondations région Kénitra

https://doi.org/10.1051/matecconf/201814902029

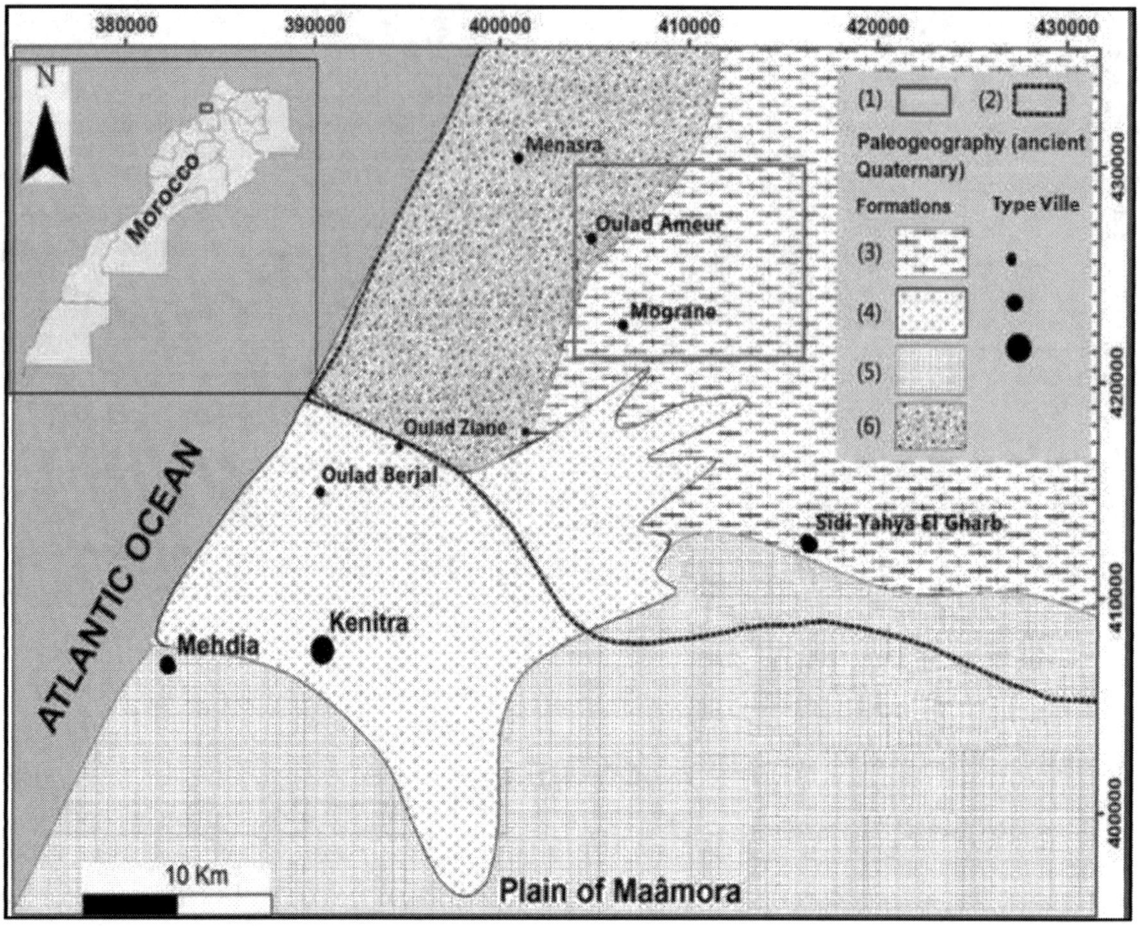

Fig. 4 : Carte de la zone d'étude dans le bassin du Gharb. 1. Limite de la zone d'étude. 2. Limite du bassin du Gharb. 3. Argile et marne de la plaine deltaïque. 4. Biocalcarénite côtière. 5. Sables et colluvions. 6. Sables de dunes et grès.

2.2 Cadre géologique

La plaine du Rharb apparaît comme une cuvette qui fut subsidence depuis le Tortonien moyen et pendant tout le quaternaire. Au Sud, le glacis paléozoïque incliné de la Meseta marocaine est recouvert par des terrains tertiaires et quaternaires constitués de marnes bleues miocènes, de sables pliocènes, puis une formation rouge continental de la Maarnora, appartenant au Villafranchien. Un peu plus à l'Est, au delà de la vallée de l'oued Beht (Dar Bel Hamri), se situe le bassin des Rides prérifaines, plus complexe, constitué de terrains triasiques et jurassiques. Ce sont surtout des calcaires et des marno-calcaires, surmontés par les "grès du Zerhoun" attribués au Bathonien. Les mollasses de base du Miocène enveloppent les plis des rides calcaires [4].

La colonne stratigraphique du Gharb est composée d'un substratum d'âge paléozoïque qui intègre des sédiments et roches métamorphiques : schiste, quartzite et calcaire [17], vaincus par les dépôts argileux rouges de Trias et par les dolomites et calcaires dolomitiques du Jurassique et du Crétacé. Le Miocène qui constitue le plancher imperméable de la nappe phréatique profonde de Gharb [17], est formé par des marnes bleues. Au Pliocène, la sédimentation a une double origine: les premières transgressions sur la côte actuelle ont donné naissance à la sédimentation côtière représentée par le grès et le calcaire. Cette formation de la côte, qui peut atteindre jusqu'à 200m d'épaisseur constitue l'aquifère réservoir de majeur. [5] et les transgressions du quaternaire qui ont été représentées à nouveau par une sédimentation détritique carbonatée en intercalations avec les argiles. Tout au long de cette succession sédimentaire, des argiles turbiditique se sont interstratifiés dans les marnes bleutées et gypsifères. La couverture de surface de Gharb est caractérisée par une extension des sols argileux et sableux favorables à une large gamme de cultures de céréales.

3 Méthodologie

Pour un meilleur résultat au traitement, une caractérisation des matériaux à traiter ainsi que des matériaux de traitement, permettra une optimisation que ce soit technique ou économique. La méthode adoptée est celle mentionnée dans le Guide des Traitement des Sols [6]. Nous procédons par une identification de l'échantillon du sol à traiter, ensuite une étude sur l'aptitude du traitement à effectuer sera réalisée, pour finir par une caractérisation du couple Sol-Liant et une comparaison entre les résultats obtenues avant et après traitement.

3.1. Echantillonnage

Après une reconnaissance géologique et géographique de la zone étudiée, on a procédé à une étude géotechnique du sol par le biais des prélèvements placés dans des sacs hermétiques afin d'éviter de les contaminer par des éléments étrangers. Ensuite ces prélèvements ont été transportés au laboratoire du Centre National d'Etudes et recherches Routière (CNER-RABAT) pour la réalisation des essais d'identification.
Le sol utilisé dans cette étude a été prélevé d'une zone des déblais au niveau de la route provinciale RP4209 entre le PK1+000 et le PK3+000.

3.2 Les essais d'identification

Les essais d'identification [7] ont portés sur: l'analyse granulométrique par tamisage (NM 13.1.008) [8] afin de définir la dimension du plus gros élément présent dans le sol (Dmax) ainsi que la teneur en fines. La limite d'Atterberg (NM 13.1.007) [9] et la valeur au bleu de méthylène du sol VBS (NF P 94-068) [10] pour la connaissance du niveau d'argilosité dans le sol, cette valeur avait un rôle important dans le choix du liant adopté pour le traitement.

Dans le même contexte, on a procédé à la caractérisation de l'état du sol en passant par l'état hydrique déterminée via le rapport Wn/Wopn et par l'indice portant immédiat IPI (NF P94-078) [11]. Ces caractéristiques ont été obtenues après l'exécution de l'essai Proctor selon la norme (NF P94-093) [12]. Suite aux résultats obtenus via ce dernier essai, on a pu réaliser un essai CBR (à 4 jours après immersion dans l'eau), afin de nous statuer sur la portance du sol sans traitement. Ce résultat, sera comparé avec celui obtenu par le même essai, mais cette fois avec l'ajout du liant (chaux vive).

Ensuite, un essai de stabilité des grumeaux, nous a permis dans une durée de temps rapide, d'avoir une idée sur l'aptitude du sol au traitement [13].

L'étape suivant l'essai d'aptitude au traitement, était une étude de formulation : dans ce projet, on a effectué un traitement avec un ajout de 3% de la chaux. Après échantillonnage et étuvage à 50°C, on a procédé à la réalisation de l'essai selon les modalités en vigueur la norme (NF P94-093).

Chaux

La chaux utilisée est une chaux vive CL90-Q, apportée d'une usine de production de Tlat Loulad. Les propriétés physico-chimiques de cette chaux sont récapitulées dans le tableau I.

Tableau I. Propriétés physico-chimiques de la chaux

Nom	Caractéristiques de la chaux
Apparence	Poudre de couleur blanche
SiO2	0 ,6%
Al2O3	0,2%
Fe2O3	0,2%
MgO	0,3
CaO total	≥ 90%
Réactivité T60	≤ 4 minutes

4 Résultats et discussion

4.1. Sans traitement :

Les propriétés physiques et mécaniques de ce sol étudié sont récapitulées dans le tableau II.

Tableau II. Propriétés physiques et mécaniques du sol

Les essais effectués sur le sol	Résultats
Particules ≥ 50mm	0
Particules ≤ 80mm	≥ 90%
Limite de liquidité	61
Limite de plasticité	30
Indice de plasticité	31
Valeur au bleu de méthylène (VBS)	6
Teneur en eau naturelle (Wn)	23,53%
Teneur en eau à l'optimum (Wopn)	22,81%
Densité sèche optimale	1,55
Indice CBR (4jrs)	2 à 3

Compte tenue des résultats obtenus lors de notre étude, on constate que le sol étudié dans cet article est de classe A3 selon la classification GMTR [14]. Une classe dont l'argilosité est moyennement à très argileuse (IP= 31).

Cette conclusion reflète une mauvaise réaction du sol lors du compactage. Les résultats obtenus après la réalisation des essais Proctor, montrent une densité sèche maximale de 1,55 pour une teneur en eau optimale (Wopn) Wopn = 22,81% (Figure 5) et un indice portant immédiat (IPI) entre 8 et 14. Après traitement cette valeur diminue nettement pour atteindre une valeur de 1.36 de densité sèche maximale pour une teneur en eau optimale (Wopn) Wopn = 20,25% (figure 7).

https://doi.org/10.1051/matecconf/201814902029

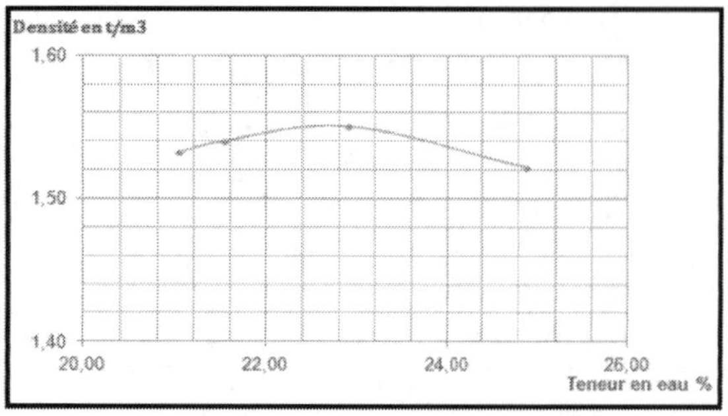

Fig. 5 : Courbe PROCTOR normal (sans traitement)

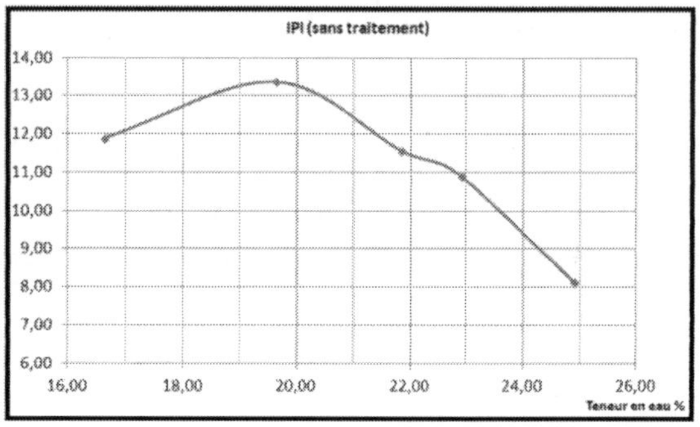

Fig. 6 : Courbe IPI normal (sans traitement)

L'indice CBR_i (à 4jrs) est ≤ 4, ce qui (selon le catalogue des structures types de chaussées neuves CSTCN [15]) confère au sol une portance St0 (portance faible).

4.2. Avec traitement (de 3% de chaux) :

Après un premier essai de traitement à 3% de la chaux, la densité sèche optimale décroit à 1,36 pour une teneur en eau optimale (Wopn) de 20,25). L'indice immédiat portant (IPI) s'est amélioré entre un intervalle de 12 à 21.

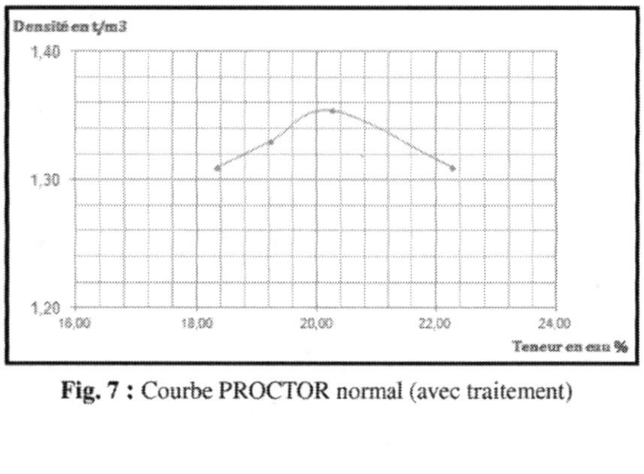

Fig. 7 : Courbe PROCTOR normal (avec traitement)

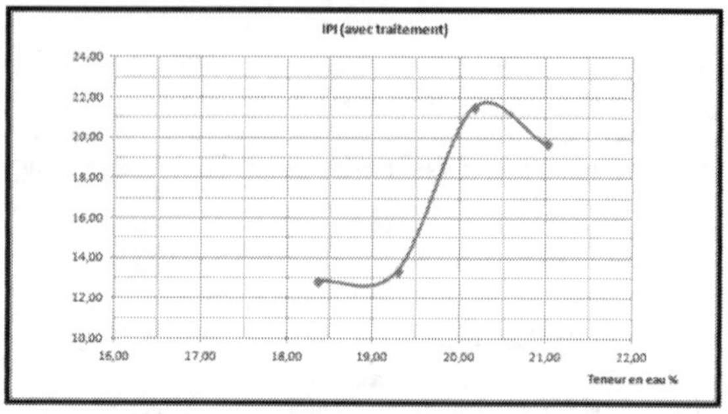

Fig. 8 : Courbe IPI normal (avec traitement)

Une comparaison entre IPI sans traitement (Figure 6) et IPI avec traitement (Figure 8), a permis de montrer une nette amélioration de la portance du sol, ceci est dû principalement à l'effet de la floculation obtenu par l'ajout de la chaux [16].

D'autres essais ont été effectués sur le sol traité pour définir l'Indice de Plasticité. Il en est ressort comme résultat pour IP (avec traitement à la chaux), une valeur de 14%. On remarque que la chaux a un impact important sur la plasticité du sol, impliquant une diminution de la sensibilité de la marne argileuse, soit une réduction la sensibilité à l'eau et du degré du gonflement [16].

L'indice CBR_i (à 4jrs) (avec traitement à la chaux) a augmenté jusqu'à atteindre une valeur de 14, ce qui (selon le catalogue des structures types de chaussées neuves CSTCN) améliorant ainsi la portance du sol (St2).

5 Conclusion

L'étude des marnes argileuses des zones inondables de la région de Kénitra ont de mauvaises propriétés géotechniques. L'étude a permis l'identification des matériaux de la zone du projet, de classer ces marnes et de déterminer leur comportement mécanique.

Ces marnes sont de classe A3 selon la classification GMTR avec une argilosité moyenne à très forte (IP= 31). Ces marnes s'avèrent incapables de supporter toute charge importante.

Le traitement de ces sols par l'addition de la chaux au laboratoire a nettement amélioré leur qualité. En effet, une diminution nette de la densité sèche du matériau qui passe de 1.55 à 1,36, soit une diminution de 12.25% ; une augmentation de la valeur de l'IPI qui évolue de 13 à 22, soit une augmentation de 69.26% et enfin une augmentation importante de la valeur du CBR_i à 4j de 3 à 14, soit un gain considérable qui s'élève à 366.67%.

Ainsi, les résultats prometteux obtenus ont démontré l'aptitude du traitement avec la chaux à améliorer la portance de ces marnes.

Références

1. IDRRIM, Institut Des Routes, des Rues et des Infrastructures pour la Mobilité, *Enseignements de*

TerDOUEST- Propositions de compléments au Guide Traitement des Sols- (Décembre 2015).
2. Magazine « Routes » : *Le traitement des sols à la chaux et/ou aux liants hydrauliques pour l'exécution des remblais et des couches de forme. Rubrique : Documentation technique.* p.7-14, **98**. (Septembre 2004).
3. Ministère de l'Agriculture et de la Réforme Agraire. Royaume du Maroc *Le Périmètre du Gharb; ORMVAG (Office Régionale de Mise en Valeur Agricole du Gharb).* (1997).
4. Mohamed EL Bouhaddioui, Abdelaziz Mridekh, Malika Kili, Bouabid El Mansouri, El Houssine EL Gasmi, Bouchaib Magrane. *Electrical and well log study of the Plio-Quaternary deposits of the southern part of the Rharg Basin, northern Morocco.* Journal of African Earth Sciences. (2016).
5. Combe, M., Bassin du Rharb Maâmora. *In ressources en eau du Maroc. Tome2. Notes et Mem, Serv, Geol, Maroc,* **231**, pp. 93 e128. (1975)
6. Guide technique SETRA-LCPC. *Traitement des sols à la chaux et/ou aux liants hydrauliques- Application à la réalisation des remblais et des couches de forme (GTS).* (janvier 2000).
7. AFNOR *NF P 94-100 : Sols : Reconnaissance et essais. Matériaux traités à la chaux et/ou aux liants hydrauliques- Essais d'évaluation de l'aptitude d'un sol au traitement.* (1999).
8. IMANOR *NM 13.1.008 : Analyse granulométrique des sols par tamisage.* (1998).
9. IMANOR *NM 13.1.007: Essai d'identification : Détermination des limites d'ATTERBERG- Limite de plasticité au rouleau- Limite de liquidité à la coupelle.* (1998).
10. AFNOR *NF P94-068: Sols : Reconnaissance et essais. Mesure de la capacité d'absorption de bleu de méthylène d'un sol ou d'un matériau rocheux par l'essai à la tâche.* (1998).
11. AFNOR *NF P94-078: Sols : Reconnaissance et essais. Indice CBR après immersion- Indice CBR immédiat- Indice Portant Immédiat.* (1997).
12. AFNOR *NF P94-093: Sols : Reconnaissance et essais. Détermination des références de compactage d'un matériau.* (1999).
13. Centre de Recherches Routières : *Mode opératoire des essais en laboratoire pour l'évaluation « de l'aptitude d'un sol à la stabilisation immédiate à la chaux ». Méthode de Mesure* **CRR-MF 33/72.**
14. Direction des routes et de la circulation routière & Laboratoire Public d'Essais et d'Etudes (MAROC) : *Guide marocain pour les terrassements routiers « GMTR »* (2002).
15. Direction des Routes et de la Circulation Routière & Laboratoire Public d'Essais et d'Etudes (MAROC): *Catalogue des structures types de chaussées neuves* (1995).
16. CIMBETON, Centre d'information sur le ciment et ses applications : *Traitements des sols aux liants hydrauliques.* **Collection technique.** (Novembre 2009).
17. MICHARD A. Eléments de la géologie marocaine. Notes et Mémoires, Service Géologique du Maroc, n°252, 420 pp, 1976

Correlation between Rock mass rating, Q-system and Rock mass index based on field data

Amine SOUFI[1], Lahcen BAHI[1], Latifa OUADIF[1], Jamal Eddine KISSAI[1]
[1]Laboratory of Applied Geophysics, Geotechnics, Engineering Geology and the Environment, Mohammadia School of Engineers, Mohammed V University, Rabat, Morocco

Abstract. Throughout the last decades, many authors have published empirical correlations between rock mass classification systems that have arisen from a series of measurements and observations with the special conditions of the work site; this means that the validity of these expressions is strongly dependent on the knowledge of the original data from which they were deduced. Consequently, none of these expressions are universal nowadays.

In recent years, the principal corps B3 "CPB3" of IMITER mine area has witnessed massive mining activities involving a large amount of underground excavation, the rock mass characteristics are undoubtedly the essential requirements for empirical design and numerical modeling. Therefore, the research carried out aims to provide a new specific inter-relation between the most widespread Quality Indexes, Bieniawski's RMR Quality Index, Barton's Q Quality Index and Palmström's RMi Mass index utilizing the data gathered from the jointed volcano-sedimentary rock mass of the "CPB3".

The "CPB3" mining level is located in Imiter silver mine, eastern Anti-Atlas, Morocco, at a depth of 500m. A total of 128 rock blocks were examined for the rock mass quality using the three characterization systems, the outcrop mapping was carried out on freshly parallel exposed faces in the horizontal south to north direction. After processing and plotting the in-situ measured data, several equations of the three indexes has been investigated using regression modeling to analyze the obtained results and find the most suitable equation with the highest correlation coefficients. These relationships were then compared with those reported in the literature.

The proposed regression models reveal strong correlations between RMR, Q and RMi indexes with high values of accuracy coefficients so that they can be used to estimate the "CPB3" underground rock mass quality for the range of RMR between 30% and 80%. The developed mathematical formulations of the geomechanical indexes will certainly offer an effective tool to geologist and geotechnical professionals in the decision-making process, preliminary design phase, stability problems and suggestions of the required supporting system and techniques without the expense of more resources or time.

Key Words: Rock Mass Classification, RMR, RMi, Q-System, Correlation, Regression, Geomechanics

Table 1. Nomenclature used in this study

CPB3	Principal Corps B3	σ_c	Uniaxial compressive strength of intact rock
RMR	Rock mass rating	JP	Jointing parameter
RMi	Rock mass index	jC	Joint condition factor
Q	Rock mass quality	Vb	Block volume, measured in m³
RQD	Rock quality designation	jL	Joint size
RA	Rating adjustment for joints orientation	jR	Joint roughness
Jn	Joint set number	jA	Joint alteration
Jr	Joint roughness number	MAE	Mean absolute error
Ja	Joint alteration number	RMSE	Root mean squared error
Jw	Joint water reduction factor	MAPE	Mean absolute percent error
SRF	Stress reduction factor	Eq No	Equation number

1 Introduction

The last five decades have witnessed the advent of several rock mass classification systems that became a very common practice in underground engineering to make an estimation of the geomechanical characteristics and provides input for stability analysis of the excavated zone to establish the appropriate support system.

Past experience with field examinations and monitoring has provided fairly good correlations with quantitative classifications and these may be used to predict engineering behavior of rock masses with reasonable accuracy. This is the reason that quantitative classifications have become very popular all over the world.

Nowadays, the most widely used classifications are Bieniawski's RMR Quality Index, Barton and al. Q-system, and Palmström's Rock mass index.

2 Rock mass classification systems

2.1 The Rock Mass Rating 1989

Rock mass rating (RMR) was developed by Bieniawski [1], it is an index of rock mass competency based on the rating of six parameters:

A_1	Intact rock strength (IRS)
A_2	Rock quality designation (RQD)
A_3	Joint spacing (JS)
A_4	Joint surface condition (JC)
A_5	Groundwater condition (GW)
RA	Rating adjustment for discontinuity orientation

The first five parameters represent the basic parameters in the classification system. Each of these parameters is given a value. All the values are algebraically summed for the first five given parameters and then adjusted by the sixth parameter depending on the joint and excavation orientation as shown in the following equations:

$$RMR_{basic} = \sum_{i=1}^{5} A_i \quad (1)$$

$$RMR_{89} = RMR_{basic} + RA \quad (2)$$

The RMR value ranges from 0 to 100.

2.2 The Q-system 1993

Barton and al. (1974) of the Norwegian Geotechnical Institute (NGI) proposed a Tunnelling Quality Index (Q) as a classification system for estimating rock support in tunnels [2,3]. It is a quantitative classification system based on a numerical assessment of the rock mass quality. Later, Barton and al. have published several papers on the Q system aiming at extending its applications.

The numerical value of the index Q is defined by six parameters and the following equation:

$$Q = (\frac{RQD}{Jn}) \times (\frac{Jr}{Ja}) \times (\frac{Jw}{SRF}) \quad (3)$$

The symbols in the expressions above represent:

RQD	Rock quality designation
Jn	Joint set number
Jr	Joint roughness number
Ja	Joint alteration number
Jw	Joint water reduction factor
SRF	Stress reduction factor

2.3 The RMi system 1995

The rock mass index is a volumetric parameter indicating the approximate uniaxial compressive strength of a rock mass, it was first presented by Palmström (1995) [3]. It makes use of the uniaxial compressive strength of intact rock (σ) and the reducing effect of the joints penetrating the rock (JP) given as:

$$RMi_{jointed\ rock} = \sigma_c \times JP = \sigma_c \times 0.2\sqrt{jC} \times Vb^D \quad (4)$$

$$jC = jL \times \frac{jR}{jA} \quad (5)$$

$$D = 0.37\, jC^{-0.2} \quad (6)$$

The symbols in the expressions above represent:

σ_c	Uniaxial compressive strength of intact rock
JP	Jointing parameter
jC	Joint condition factor
Vb	Block volume, measured in m^3
jL	Joint size
jR	Joint roughness

3 Review of previous researchers

Throughout the last decades, many authors have published empirical correlations. The most widely used examples are given in Table.2

Table 2. Equations correlating geomechanical classifications from previous works

Approach	Relation	Eq
RMR-Q		
Bieniawski (1984) [5]	$RMR = 9.ln(Q) + 44$	(7)
Rutledge and Perston (1978) [6]	$RMR = 5,9.ln(Q) + 43$	(8)
Cameron Clarke and Budavari (1981) [7]	$RMR = 5.ln(Q) + 60,8$	(9)
Abad and al. (1984) [8]	RMR = 10,5. ln(Q) + 41,8	(10)
RMR-RMi		
Kumar and al. (2004) [9]	$RMR = 5,4.ln(RMi) + 54,4$	(11)
Hashemi and al. (2009) [10]	$RMR = 7,5.\ RMi + 36,8$	(12)
RMi-Q		
Kumar and al. (2004) [9]	$RMi = 0,5.\ Q^{0,93}$	(13)

Hashemi and al. (2009) [10]	$RMi = 1{,}082.\,Q^{0{,}4945}$	(14)

Each of the previous expressions has arisen from a series of specific data that is related to a certain rock mass type.

4 Methodology and case study

The study area is located in Imiter silver mine, eastern Anti-Atlas, Morocco, at a depth of 500m.

- The common rock types in the "CPB3" are volcano-sedimentary rock
- There are prominent foliations over all the rock mass in Est-West direction.
- There are multiple discontinuity sets, many of which intersect each other.

A total of 128 rock blocks were examined for the rock mass quality using the characterization systems from the previous discussion, the outcrop mapping was carried out on freshly exposed faces in the horizontal south to north direction.

After processing and checking the accuracy of the proposed expressions, several types of mathematical equations such as linear, exponential, logarithmic, and power were derived with regression analysis to find the most suitable expression with the highest correlation and lowest error coefficients.

5 Statistical processing

To analyze the results obtained, the R coefficient (Pearson's coefficient for correlation) was used. This coefficient provides information about the degree of relationship between two variables (RMR, Q, and RMi in this case). The mathematical formula for computing R is:

$$R = \frac{n\sum x_i y_i - (\sum x_i \sum y_i)}{\sqrt{n(\sum x_i^2) - (\sum x_i)^2}\sqrt{n(\sum y_i^2) - (\sum y_i)^2}} \quad (15)$$

where, x_i is the input parameter, y_i is the output parameter and n is the number of data.

In order to check the representativeness of the proposed relationships, the Coefficient of Determination (R^2) is used to analyze the results, for example, if a certain correlation has a Coefficient of Determination of X%, it means that the X% of the RMR Index is in direct relation to the Q-Index. So, it represents the proportion of the shared or explained variability.

MAE measures the average magnitude of the errors in the set of rock quality predictions, without considering their direction where all individual differences have equal weight. and is defined by the following expression:

$$MAE = \frac{1}{n}\sum_{i=1}^{n} |A_t - F_t| \quad (16)$$

where, A_t is the actual value and F_t is the forecast value of the output variable.

RMSE is a quadratic scoring rule that also measures the average magnitude of the error and tells how concentrated the data is around the line of best fit. The equation is given by:

$$RMSE = \sqrt{\frac{1}{n}\sum_{i=1}^{n} (A_t - F_t)^2} \quad (17)$$

MAPE is the measure of accuracy in statistics. It usually expresses accuracy as a percentage, and is defined by the following formula:

$$MAPE = \frac{1}{n}\sum_{i=1}^{n} \left|\frac{A_t - F_t}{A_t}\right|$$

(18)

6 Correlation results between the rock mass classification systems

6.1 RMR-Q correlation

The possibility of interrelating RMR and Q Indexes is studied using different types of mathematical expressions that are presented in Table.3

Table 3. Comparison of various regression trendlines between RMR and RMi

Trend-line	Expression	R^2	Eq. No
Linear	$RMR = 1{,}4906Q + 45{,}73$	0.29	(19)
Multi Linear	$RMR = 0{,}24RQD - 0{,}73Jn + 7{,}66Js - 0{,}2Ja + 17{,}17Jw + 19{,}83$	0.79	(20)
Exponential	$RMR = 44{,}953.\,e^{0{,}0304.Q}$	0.27	(21)
Logarithmic	$RMR = 5{,}614.\,ln(Q) + 49{,}395$	0.70	(22)

After testing different types of formulae for fitting, it could be concluded that the multilinear (Eq.20) and logarithmic (Eq.22) models fit well to the available data. By correlating the two sets of Q values with the relevant RMR values, we obtained the figure.1

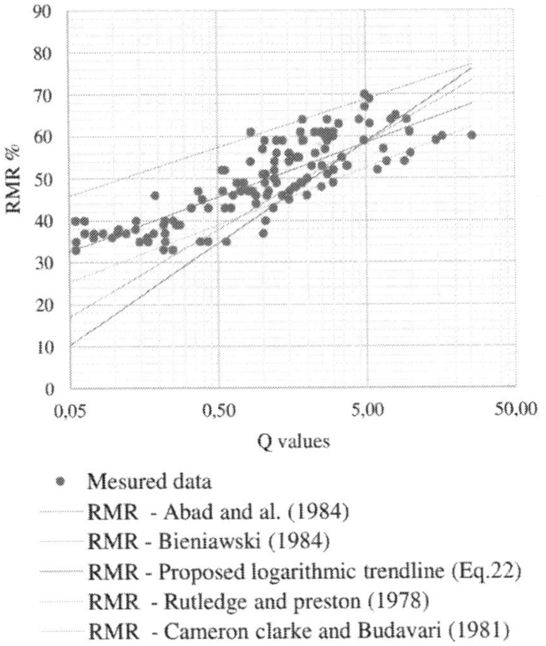

Figure 1. plot of correlations between RMR and Q-system in this case and available literature

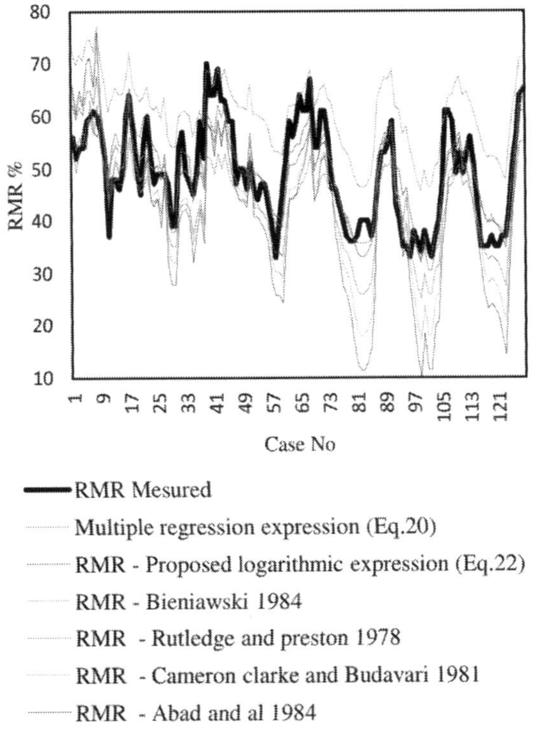

Figure 2. Evaluation of various Correlations between RMR and Q in this case and available literature (128 cases)

Figure.2 summarizes the obtained estimated RMR values from Q compared to other approaches shown in a graphical way

Table 4. Evaluation of correlation and accuracy Coefficients for existing and proposed RMR-Q relations

Approach	R	MAE	RMSE	MAPE (%)	Eq No.
Bieniawski (1984)	0,72	7,5	9,07	22,6	(7)
Rutledge and Perston (1978)	0,72	6,91	8,3	17,14	(8)
Cameron clarke and Budavari (1981)	0,72	11,48	12,62	19,42	(9)
Abad and al. (1984)	0,72	9,67	11,75	37,88	(10)
Proposed regression model	0,72	4,34	5,26	8,81	(22)
Proposed multiple regression model	0,89	3,55	4,37	7,34	(20)

The results of accuracy analysis Table.4 show a satisfactory R and MAPE values for (Eq. 20) and (Eq. 22), the proposed RMR-Q relationships are given by:

$$RMR = 0{,}24RQD - 0{,}73Jn + 7{,}66Js - 0{,}2Ja + 17{,}17Jw + 19{,}83 \quad (20)$$

$$RMR = 5{,}614.\ln(Q) + 49{,}39 \quad (22)$$

6.2 RMR-RMi correlation

The possibility of interrelating RMR and RMi Indexes is studied using different types of mathematical expressions that are presented in table.5

Table 5. Comparison of various regression trendlines between RMR and RMi

Trend-line	Expression	R²	Eq. No
Linear	$RMR = 8{,}2667.RMi + 41{,}77$	0,60	(23)
Multi Linear	$RMR = -0{,}52F - 6{,}32Js - 0{,}04Ja + 0{,}19jL + 5{,}96Vb + 3{,}13jC - 84{,}86D + 107{,}62$	0,87	(24)
Exponential	$RMR = 41{,}519.e^{0{,}1671.RMi}$	0,58	(25)
Logarithmic	$RMR = 7{,}71.\ln(RMi) + 54{,}441$	0,83	(26)

Based on the analysis of the coefficient of determination, the most approximate formulas are the multi-linear (Eq.24) and logarithmic (Eq.26) models.
The obtained results from the correlation between RMi and RMR values are shown in an analytical way by figure.3

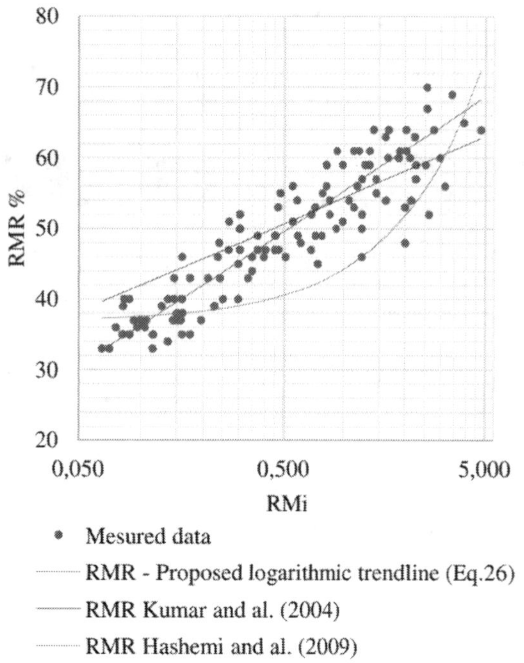

Figure 3. plot of correlations between RMR and RMi in this case and available literature

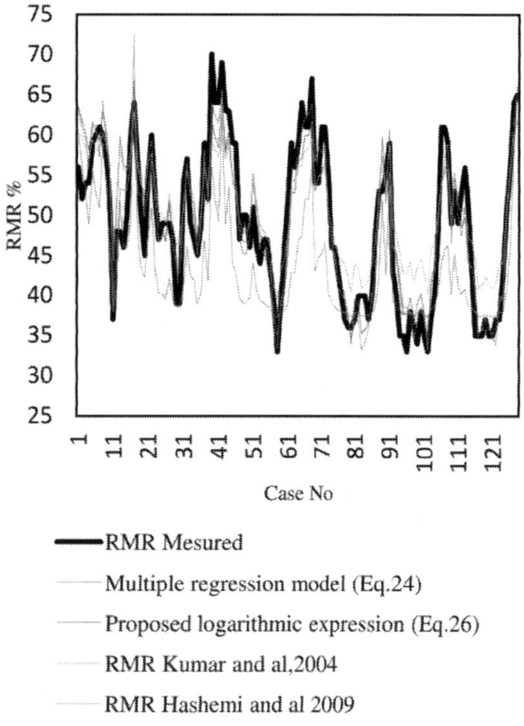

Figure 4. Evaluation of various Correlations between RMR and RMi in this case and available literature (128 cases)

Figure.4 summarizes the obtained estimated RMR values from RMi compared to other approaches shown in a graphical way

Table 6. Evaluation of correlation and accuracy Coefficients for existing and proposed RMR-RMi relations

Approach	R	MAE	RMSE	MAPE (%)	Eq No.
Kumar and al. (2004)	0,89	4,2	4,95	8,56	(11)
Hashemi and al. (2009)	0,81	6,77	8,2	15,47	(12)
Proposed regression model	0,89	3,09	3,99	6,17	(26)
Proposed multiple regression model	0,93	2,88	3,49	5,93	(24)

It can be seen clearly in Table.6 that the estimated value of RMR from RMi using logarithmic and multi-Linear expressions shows one of the biggest values for R and lowest MAPE, the resultants equations for the regression are:

$$RMR = -0{,}52F - 6{,}32Js - 0{,}04Ja + 0{,}19jL + 5{,}96Vb + 3{,}13jC - 84{,}86D + 107{,}62 \quad (24)$$

$$RMR = 7{,}71.\ln(RMi) + 54{,}44 \quad (26)$$

6.3 RMi-Q correlation

The possibility of interrelating RMi and Q Indexes is studied using different types of mathematical expressions that are presented in table.7

Table 7. Comparison of various regression trendlines between RMi and Q

Trend-line	Expression	R^2	Eq. No
Linear	$RMi = 0{,}1962Q + 0{,}444$	0.55	(27)
Multi Linear	$RMi = 0{,}021RQD - 0{,}054Jn + 0{,}445Js - 0{,}129Ja + 0{,}065Jw + 0{,}674$	0.71	(28)
Exponential	$RMi = 0{,}3121.e^{0{,}2092.Q}$	0.41	(29)
Logarithmic	$RMi = 0{,}5073.\ln(Q) + 0{,}9122$	0,63	(30)
Power	$RMi = 0{,}5195.Q^{0{,}7202}$	0,82	(31)

Likewise, a similar analysis was performed for RMi and Q correlation, it could be concluded that the power model (Eq.26) model fits well to the available data.

RMi-Q Correlated data from the "CPB3", along with the other correlations available in the literature, are presented in Figure. 5

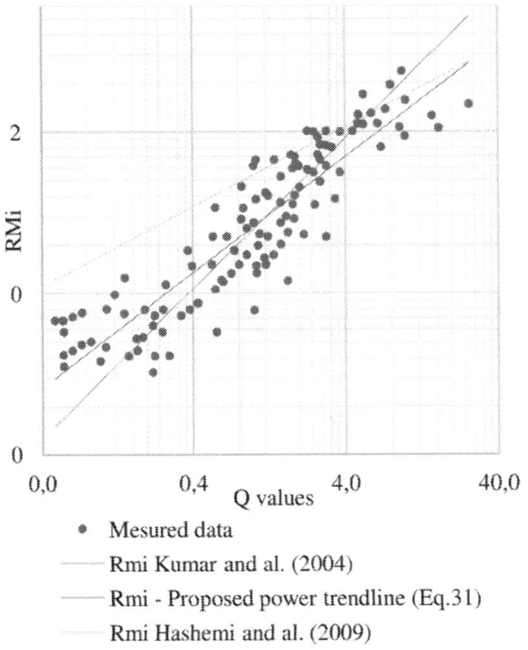

Figure 5. plot of various correlations between RMi and Q-system in this case and available literature

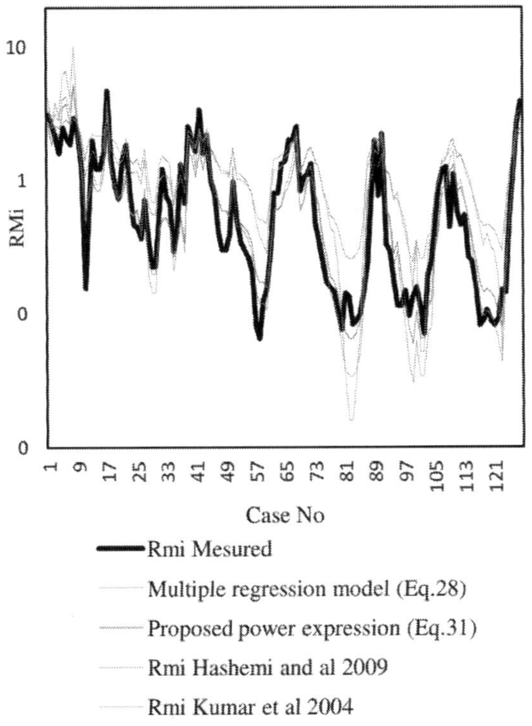

Figure 6. Evaluation of various Correlations between RMi and Q in this case and available literature (128 cases)

Figure.6 summarizes the obtained estimated RMi values from Q compared to other approaches shown in a graphical way

Table 8. Evaluation of correlation and accuracy Coefficients for existing and proposed RMi-Q relations

Approach	R	MAE	RMSE	MAPE (%)	Eq No.
Kumar and al. (2004)	0,93	0,52	0,66	47,93	(13)
Hashemi and al. (2009)	0,72	0,39	0,94	53,21	(14)
Proposed regression model	0,98	0,33	0,53	43,01	(31)
Proposed multiple regression model	0,84	0,52	0,63	60,40	(28)

From a statistical point of view, the power law approximation gives a quite acceptable level of certainty Table.8, we have got the following equation:

$$RMi = 0{,}5195.\, Q^{0{,}7202} \quad (31)$$

Table 9. The recommended relations based on the "CPB3" data

Expression	R	Eq No.
RMR – Q		
$RMR = 0{,}24RQD - 0{,}73Jn + 7{,}66Js - 0{,}2Ja + 17{,}17Jw + 19{,}83$	0.89	(20)
$RMR = 5{,}614.\, ln(Q) + 49{,}395$	0.72	(22)
RMR – RMi		
$RMR = -0{,}52F - 6{,}32Js - 0{,}04Ja + 0{,}19jL + 5{,}96Vb + 3{,}13jC - 84{,}86D + 107{,}62$	0.93	(24)
$RMR = 7{,}71.\, ln(RMi) + 54{,}441$	0.89	(26)
RMi – Q		
$RMi = 0{,}5195.\, Q^{0{,}7202}$	0.98	(31)

7 Conclusion

Most of the available relationships between indexes of rock mass quality have been derived using geological data of Europe, America and Oceania region. In recent years, the "CPB3" of IMITER site has witnessed massive mining activities involving a large amount of underground excavation. Therefore, there is a need to develop specific correlations between the most prevalent rock mass classification indexes for the special conditions of the working site.

The systematic plotting of the graphs using the available relationships reveals a strong potential correlation between RMR, Q and RMi indexes. The proposed regression equations have high-reliability coefficient Table.9 so that they can be used to estimate the "CPB3" underground rock mass quality for the range of RMR between 30% and 80%.

After a preliminary comparison of the different solutions, we recommend using these correlations with extreme prudence at initial stages of a project, being extremely cautious about the origin, compatibility and rating range of the initial data.

References

1. Z.T. Bieniawski, *Classification of Rock Masses for Engineering: The RMR System and Future Trends, Comprehensive rock engineering*, **3**,522-542, (1993)
2. N. Barton, R. Lien, and J. Lunde, *Engineering classification of rock masses for the design of tunnel support*, (1974)
3. N. Barton, International Journal of Rock Mechanics & Mining Sciences, *Some new Q-value correlations to assist site characteristics and tunnel design*, 185–216, (2002)
4. A. Palmström, Journal of Rock Mechanics and Tunnelling Technology, *The Rock Mass Index (RMi) applied in rock mechanics and rock engineering*, **11**, No.2, 1-40, (1996)
5. Z.T. Bieniawski, *Rock Mechanics Design in Mining and Tunnelling*,272,(1984)
6. J.C. Rutledge, R.L. Perston, *Experience with engineering classifications of rock*, In Proc. Int. Tunneling Sym, Tokyo, (1978)
7. I.S Cameron-Clarke, S. Budavari, Engineering Geology, *Correlation of rock mass classification parameters obtained from bore core and in situ observations*, 19–53, (1981)
8. J. Abad, B. Caleda, E.Chacon, V.Gutierrez, E. Hidlgo, *Application of geomechanical classification to predict the convergence of coal mine galleries and to design their supports*, In 5th Int. Congress on Rock Mech., Melbourne, 15–19.(1984)
9. N. Kumar, NK. Samadhiya, R. Anbalagan, Int J Rock Mech, *Application of rock mass classification system for tunneling in Himalaya*, Paper 3B 14, SINOROCK2004 Symposium, (2004)
10. M. Hashemi, Sh. Moghddas, R. Ajalloeian, Rock Mech Rock Eng, *Application of Rock Mass Characterization for Determining the Mechanical Properties of Rock Mass: a Comparative Study*, 305-320, (2010)

Towards building a neural network model for predicting pile static load test curves

A. K. ALZO'UBI, and Farid IBRAHIM
College of Engineering, Abu Dhabi University, Al Ain, United Arab Emirates

Abstract. In the United Arab Emirates, Continuous Flight Auger piles are the most widely used type of deep foundation. To test the pile behaviour, the Static Load Test is routinely conducted in the field by increasing the dead load while monitoring the displacement. Although the test is reliable, it is expensive to conduct. This test is usually conducted in the UAE to verify the pile capacity and displacement as the load increase and decreases in two cycles. In this paper we will utilize the Artificial Neural Network approach to build a model that can predict a complete Static Load Pile test. We will show that by integrating the pile configuration, soil properties, and ground water table in one artificial neural network model, the Static Load Test can be predicted with confidence. We believe that based on this approach, the model is able to predict the entire pile load test from start to end. The suggested approach is an excellent tool to reduce the cost associated with such expensive tests or to predict pile's performance ahead of the actual test.

1 Introduction

Deep foundations such as piles, are the part of a structure used to carry and transfer loads of the superstructure to the bearing ground located at some depth below ground surface. Pile's capacity is one of the most problematic designs in Civil Engineering due to the variability in geomaterials properties and the large number of parameters involved in controlling, or should be involved, in pile's design. In practice, three methods of design are being used: full-scale static load test, analytic methods, and dynamic methods.
There are many factors that contribute to choosing and adapting the most relevant pile design. Some of these factors are size of the project, the significance of the project, and/or the available fund allocated for the investigation, test, and design. Most analytical methods of determining the pile capacity are developed empirically by comparing the results of static load tests with measured soil properties [1]. So, performing the static load tests is essential part of designing these deep foundations or checking their performance especially in the UAE where such test is mandatory.

As a result of the economic boom that has been happening in the United Arab Emirates, construction industry flourished and huge amount of new projects has to be built as fast as possible, keeping in mind the safety of structures such as high rise buildings, malls, and residential developments. With this in mind, engineers relied heavily on the Continuous Flight Auger (CFA) piles (see Figure 1) to speed up the construction process, and over design these piles to avoid failure. The CFA piles are easy and fast to construct in soft ground such as Silty Sand soils which dominates the upper soil layers in the major cities in the UAE. The CFA piles in this region are used to transmit high concentrated loads deep into the ground, mainly by friction. The test load is usually applied in increments up to 1.5 times the design load for working piles load, or more in rare cases, as required by structural designer or geotechnical engineer.

Fig. 1. Piles underneath the future Etisalat centre in Abu Dhabi

In this paper we introduce an Artificial Neural Network framework to build a model that can predict a complete Static Load Pile test. The authors will depend on three factors to reduce the variation of the parameters in the artificial neural Network; the consistency of the geology, the dominant pile diameter, and the number of piles passing the tests. All the tests that are to be used in this research are conducted in three major cities (Abu Dhabi, Dubai, and Al Ain) which are located in the United Arab Emirates. This consistency will reduce the variations in the geomaterial parameters, Figure 2 shows an example of the soil profile encountered typically in the field. Moreover, in this research we choose piles with 500 mm diameter since 80% of about 100 projects that we explore have a diameter of 500 mm and lengths between 6 to 16 m. Also, out of the 100 tests, that we explored, passed the test and only one pile has failed to meet the expectation. This single pile exceeds the limit of the acceptable downward movement of 1 % (of its diameter) by moving 1.1 % at three times not 1.5 times the design load. As a result, we will use successful piles only in this research. Based on the above discussion, it is safe to assume that predicting the pile capacity by means of artificial neural networks can be achieved through training the model on the available data. This approach would allow judging new proposed pile design ahead of the field tests. High prediction accuracy and precision can also be achieved through continuous data feeding in the model by using future tests. This eventually will cut the cost of conducting multiple static load test for the same project. In fact, Coduto [1] pointed out that static load tests are much more expensive than analytic methods, and thus the latter is very attractive. However, the calculated load capacities are not as precise, so designs based solely on analytic methods must be correspondingly more conservative and the final design is more expensive.

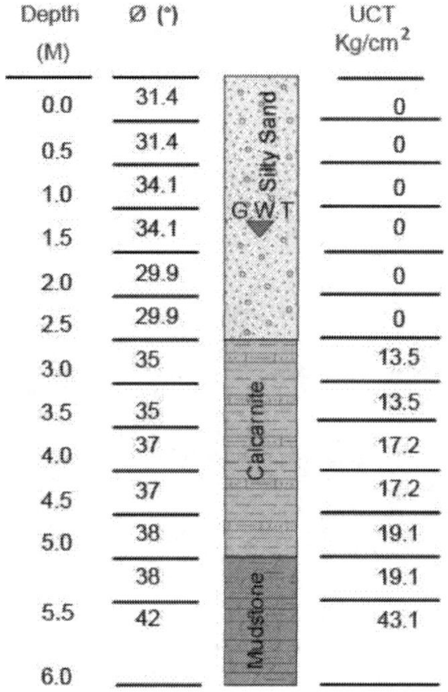

Fig. 2 Static Load test in the future Etisalat centre

2 Artificial Neural Networks

The artificial neural network is the most common training model used in many applications. In this model a set of training records with known output are processed by an algorithm to learn the neural network. It consists of three layers; input, hidden and output layers. Each layer is defined as a set of connected units and weights are associated with each connection. Input attributes are fed simultaneously to form the input layer. The weighted input is propagated forward and passed to the second layer which is the hidden layer; one or more hidden layers may be defined. The number of hidden layers is purely empirical and depends on the experiment result.

The output is calculated for each hidden layer until it reaches the output layer, and in this case it is the machine prediction value. This value is compared to the known output value and the error produced is propagated backward by adjusting the weights and biases at each unit to reflect the error in the network prediction. The machine will continue iterating forward and backward learning the values assigned to the weights and biases until it reaches to a corrected predicted output value that is within an acceptable error difference. Once the model is built, it can be used to process newly input records whose output is unknown. The neural network will use the learned values of the weights and biases to predict the unknown output [2].

3 Related Works

The use of artificial neural networks in predicting the pile capacity started in the 1990's, when Goh [3] used a form of a model to predict an estimate of the friction capacity of driven piles in clay soils. The results were promising when compared to the actual data and some empirical methods. Lee and Lee [4] tried to predict the driven pile capacity by using an artificial neural network model. The error between the predicted and the actual pile test was around 20%. However, they did not attempt to predict the entire static load test. Abu-Kiefa [5] developed three artificial neural network models to forecast driven piles capacity. He compared the results with four empirical formulas, and found that the model that was devoted to forecast the total pile capacity was more accurate than others with 0.95 coefficients of determination. Furthermore, Goh [6] presented a new neural network model to predict pile capacity in sandy soils. The model's pile capacity was satisfactory when compared to other empirical formulas. Shahin [7] discussed different applications of artificial neural network in Geotechnical Engineering; and mentioned different applications including predicting the pile capacity. Benali and Ammar [8] also used artificial neural network to expect the pile capacity; they used a database of 80 cases collected from the literature and corresponding to different sites distributed all over the world. They reported that the networks are feasible for these kinds of problems but they did not attempt to predict the entire static load test. More recently, Maizir and Kassim [9] used artificial neural networks for predicting the axial capacity of a driven pile; the data collected for this study consisted of 300 cases from several projects in Indonesia and Malaysia.

He found that a good prediction was achieved if "both stress wave data and properties of both driven pile and driving system are considered in the input data". In another experiment [10], the pile setup was predicted using an artificial neural network model. The predicted values were compared with those produced by some empirical formulas. It showed that the model produced satisfactory results. A recent paper [11] proposed a framework for a system model that combines the neural network technique to predict the pile capacity with the geographic information system (GIS) data. The proposal suggested building a new layer in the GIS containing the pile load tests.

As we can see in the above mentioned studies, no serious attempt was made to predict the entire static load test; in this paper we demonstrate that static load test can be reasonably predicted by an artificial neural network with enough training data.

4 General Geology

The United Arab Emirates (UAE) lies on the floor of Arabian Gulf which mainly composed of extensive carbonate sediments. Fookes and Knill [12] divided the mountain and piedmont plain of the Arabian Gulf into four sediment depositions. The coast line of UAE, where this study is performed, is located in zone number four or the base plain. The upper layer deposits of this zone consists of sand dunes, loess, and evaporate together with marine sand and silts. Wind and evaporation, due to high temperature, are the principal transporting agents in this region.

Although wind-blown material tends to predominate, as great quantities of silt and sand are moved during periods of high wind, water also plays an important role in this movement and later deposition of high quantities of Silty Sand material. The depth of the Silty Sand layer ranges from 3 to 10 meters and it has a major role in establishing the pile capacity due to friction. Flash floods, last one was witnessed in 2015, are relatively rare, and any water stream actually reaching zone IV is usually short-lived after the rainstorm. However, the water table is very high and it can be as close as 0.5 meter from the ground surface. This water table can dominate the desert processes by limiting the wind erosion to soils above the water table; wind erosion more or less stops when sand and silts are damp. Moreover, capillary pressure forces water movement from the water table to ground surface as the water table is high and evaporation occurs in high rates.

In these conditions a thick salt crust can build and it might affect the pile's capacity. These deposits are common in the urban areas of UAE desert coastal regions and particularly extensive around this coast line. Beneath the first layers of sand and silt, other soil and rock materials exists. The Sand and Silt deposits cover interbedded sandstones (see Figure 2), conglomerates, calcisiltite, limestone, and siltstones; clay deposits may also be encountered. In this research the piles length in all of the projects that will be included ranges from 6 to 15 meter.

5 Classification of the Materials and NN

The field and laboratory test results are summarized for the various ground materials. The sieve analysis and the Atterberg Limits test were performed to classify the soil according to the Unified Soil Classification System and later considered as a parameter in the Neural Network input data to improve the model accuracy. In this research we will classify the soil layers into different layers and then coded to numbers. Table 1 presents example of such coding that shows the soil layers along with the different codes that we used in the artificial neural Network.

Table 1: Example of the coding of soil description that will be adapted in the neural network model.

Soil Description	Soil Profile
Silty SAND (SM): Dry to moist, light brown, fine non-plastic, trace fine gravel	1
Poorly graded SAND with silt (SP-SM): Wet, light brown, fine, and non-plastic.	2
MUDSTONE: Very weak to weak, moderately to highly fractured, moderately weathered.	3
MUDSTONE: Weak moderately fractured, moderately weathered.	4
Crystalline Gypsum: Moderately weak moderately fractured, moderately weathered.	5
Loose to medium dense SAND with trace silt	6
Very loose light brown clayey fine SAND.	7
Dense, light brown fine silty SAND	8

6 Static Load Test

According to the field and laboratory investigations, subsurface conditions, engineering analysis and practical experience, in all of the projects in this research the recommended type of foundation was CFA piles. Pile load tests are considered the most satisfactory method to assess the carrying capacity of a pile, see Figure 3 which shows the details of a static load test conducted in the field. During the foundation construction period, it is recommended that such tests be performed according to British Standard Code of Practice 8004 on specially constructed piles that are installed before the start of the general construction works. The common practice in the UAE is to test the piles after construction for quality assurance and not for design; i.e. to check performance. As a result of this practice CFA piles, in most cases, are over designed to avoid failure in the testing process. However, if testing is carried out for design purposes, significant savings may result from a more economical pile design based upon the specific test data.

Since the theoretical design methods provide an approximate working load, the static load test should demonstrate the pile's performance and its load settlement characteristics. Figure 4, shows the static load test performed in the future Etisalat centre (see figure 1). In the static pile load tests the load-displacement curves were obtained through two cycles of loading-unloading.

As mentioned earlier, the pile load test is usually performed to verify the pile's ultimate capacity. This value is based upon the allowable amount of settlements determined by the requirements of the engineer or a known standard specification. In all of the tests considered in this paper, the BS8004 [13] was used as the testing procedure. The test load is applied in increments up to 1.5 times the design load for working piles load. The load was applied by mean of hydraulic jack as shown in Figure 2, four reaction piles were used to provide support for the loading frame. In addition, four settlement gauges were used to monitor the displacement.

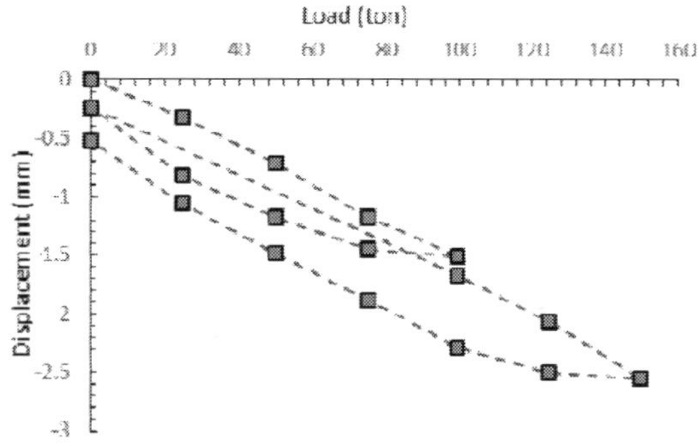

Fig. 3 Field data of a pile static load test

The maintained load method was used in all of the projects. The load was increased in stages and maintained for at least 20 minutes or until the rate of settlement decreased to less than 0.10 mm/hr, whichever is greater. As a matter of fact, in all of the projects in this research, the load was maintained until settlement has seized. In the first loading cycle, increments of 25% of the design load -up to the working load- are imposed to the tested pile and maintained until settlement seized. Settlement readings were taken, for each load increment, every 5 minutes until the settlement stopped. In the first unloading cycle, the load was reduced in inducement of 25% of the testing value until the zero load stage is reached, while settlement was recorded as described above. Reloading was then applied to test pile from zero to the test load (150 % of the design load); first increment from zero to the design load and kept for six hours at which the settlement seized. And then increased in 25 % of the design load until the test load was reached. At last, the unloading occurred in 25 % of the testing load until zero load.

The above mentioned testing procedure is adapted in the majority of tests in the UAE. We rely on this procedure to verify the CFA piles performance ahead of construction. In case the one of the tests failed at a particular site, the CFA piles must be extracted from the ground. This costly solution forced engineers to overdesign the piles to avoid failure in the test. The framework proposed in this paper aims at reducing this cost through predicting the static load test behaviour based on the proposed pile configuration, ground water table, and the soil profile mechanical properties. We believe that those parameters control the CFA piles capacity and performance under loading or unloading.

7 The Network Model

The principle underlying this research is that it is possible to predict the displacement value using an artificial neural network that is trained over data consisting of a combination of the soil profile and the pile configuration, as shown in figure 5. Coduto [1] showed that the soil profile and the pile configuration are essential in determining the pile capacity and behaviour under loading and unloading cycles. Hence, if we assume Si to be the set of soil profile values at site i and whose elements consists of values in the set {v1, v2,…,vj} and Pi be the set of pile configuration at site i, and whose parameters consists of {p1, p2}, then it is possible to predict the displacement D such that F(Si, Pi) = D, where F is a mapping function. This mapping function will be built using a trained artificial neural network.

Fig. 4 Static Load test in the future Etisalat centre

Pile test data are collected from the field and will be implemented in the network. The majority of the points are to be used for training and a certain percentage of them will randomly be selected for validation. The setting of the network is empirical, however keeping the network simple is essential in order to avoid overtraining and consequently overfitting [14, 15]. The model should be able to generalize the mapping between input and output and should be capable of predicting the output for cases not included in the training set [16]. Such network needs substantial amount of data to be trained and verified to guarantee its performance and its ability to predict the displacement for new pile configuration and new soil profile.

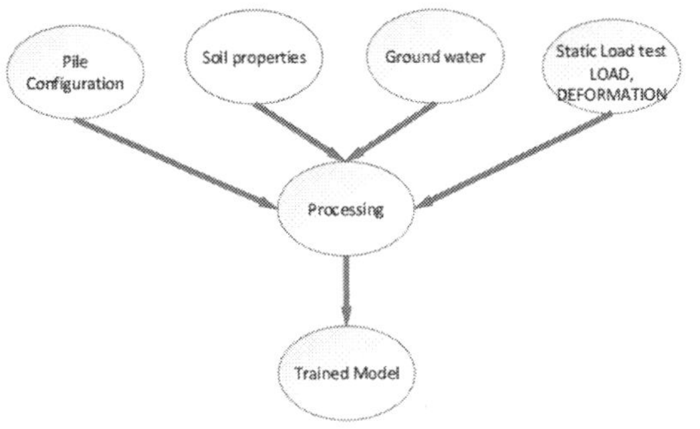

Fig. 5 Training cycle in the artificial neural network

8 The Test Plan

Pile behaviour is a complex process due to the many parameters controlling its performance. Predicting the pile capacity, let alone the entire load-displacement curve, has been a difficult task for geotechnical engineers and that is why the static load test is routinely conducted in the UAE. This paper introduces an artificial neural network model to predict not only the pile capacity, but the two cycles of loading-unloading static load test curves for CFA piles under compression. Although NN has been used, as discussed in section 3 in this paper, to predict the pile capacity; yet no attempt has been made to predict the entire loading-unloading curves of the static load test.

The test plan of our proposed model is that we will randomly pick a number of independent complete pile tests that are not part of the training data. We will input the geotechnical data, ground water level, pile configuration, loading cycle, and load in the model leaving the displacement field to be predicted. After training the model, three piles have 500 mm in diameter, and lengths 6 m, 7 m, and 8.5 m will be used to test the model. These three piles will not be included in the training stage to guarantee the independency of the tests. Figure 6 shows the testing plan that will be adapted in this research. Upon the completion of the testing plan, we will use different statistical methods to compare the predicted data versus the actual field data. Coefficient of determination for example will explain relationship between the two sets of data.

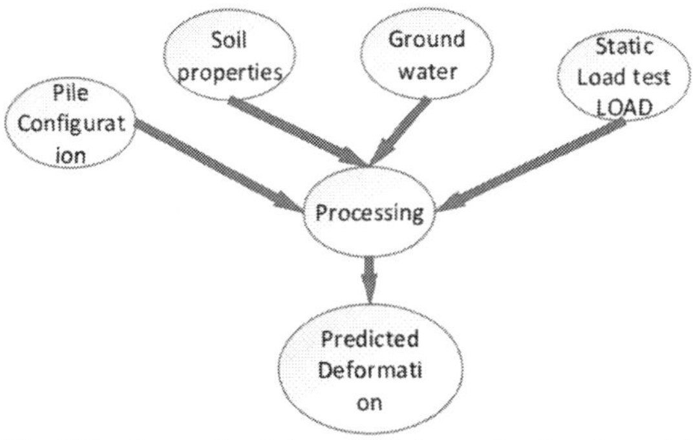

Fig. 6 Testing cycle of the artificial neural network model

Conclusion

In this paper, we introduced a new approach to predict the entire static load tests curve using artificial neural network. This approach is achievable due to three main factors; consistent geology, number of successful tests, and fixed pile diameter. The machine is fed with geotechnical data and the ground water table elevation, pile's diameter, pile's length, and load-displacement curves to train the model. Two loading and two unloading curves are to be obtained through machine learning. We believe that this approach would enable us to predict the static load test ahead of the actual field test for quality control purposes. Statistical methods such as regression models and coefficient of determination will be used to verify the model results.

Acknowledgement

The authors would like to extend their gratitude to the office of scholarship and sponsorship programs at Abu Dhabi University for their financial support of this project.

References

1. Coduto, Foundation design: Principles and practices. New Jersey: Prentice Hall, (2001)
2. Han, J., Kamber, M., Data Mining, Concepts and Techniques, 3rd ed., Morgan Kaufman Publishers, MA, USA, (2012)
3. Goh, A. T. C., Empirical design in geotechnics using Neural Networks, Geotechnique, 45, 709-714, (1995)
4. Lee, I. M., Lee, J. H., Prediction of pile bearing capacity using artificial neural networks, Computers and Geotechnics, 18 (3), 189-200, (1996)
5. Abu Kiefa, M, A., General regression neural networks for driven piles in cohesionless soils, Journal of Geotechnical and Geoenvironmental Engineering - ASCE, 124 (12), 1177-1185, (1998)
6. Goh, A. T. C., Pile driving records reanalyzed using neural networks, Journal of Geotechnical and Geoenvironmental Engineering - ASCE, 122 (6) , 492-495, (1996)
7. Shahin, M. A., Jaska, M. B., and Maier, H. R., Artificial Neural Network applications in geotechnical engineering, Australian Geomechanics, March, 49-62, (2001)
8. Benali, A., Nechnech, Ammar, Prediction of the pile capacity in purely coherent soils using the approach of the artificial neural networks, In: International Seminar, Innovation & Valorisation in Civil Engineering & Construction Materials, N° : 5O-239, University of sciences and technology, Algiers, Algeria, (2011)
9. Maizir, H., Kassim, K., 2013. Neural Network Application in Prediction of Axial Bearing

Capacity of Driven Piles, In: Proceedings of the International Multi Conference of Engineers and Computer Scientists, Hong Kong, (2013)

10. Tarawneh, B., Pipe pile setup: Database and prediction model using artificial neural network, Soils and Foundations, 53 (4), 607-615, (2013)
11. Alzo'ubi, A. K., Ati,M., Ibrahim, F., Smart Framework for Predicting Drilled Shaft Capacity Based on Data Mining Techniques and GIS Data, In: D. Manzanal and A.O. Sfriso (Eds.), From Fundamentals to Applications in Geotechnics, The Pan American Conference on Soil Mechanics and Geotechnical Engineering, 15th PCSMGE / 8th SCRM / 6th IS-BA 2015, 15-18 November, pp. 1909-1915 (2015)
12. Fookes, P.G., Knill, J.L. The application of engineering geology in the regional development of northern and central Iran, Engineering Geology, 3, 81-120, (1969)
13. British standard, BS 8004, Code of practice for foundations, BSI, (2015)
14. Haykin, S, Neural Networks, A Comprehensive Foundation, Macmillan College Publishing Co., New York, USA, (1999)
15. Liu et al., (2008), Optimized approximation algorithm in neural networks without overfitting, IEEE Trans. Neural Networks, 19 (6), pp. 983–995, (2008)
16. Geman, E. et al, Neural networks and the bias/variance dilemma, Neural Computing., 4 , pp. 1–58, (1992)

Time rate of swelling of compacted highly plastic clay soil from Sudan

[1]A M Elsharief, [2]Mai Sufian
[1]University of Khartoum, Khartoum, Sudan
[2]Sudan University of Science and Technology, Khartoum, Sudan

Abstract: This paper investigates the development of swelling with time for a highly plastic and potentially expansive clay from Sudan. Soil samples were prepared in the laboratory at different moisture content values. The prepared samples were placed in the oedometer ring at three density levels and then placed in an oedometer cell which allows one dimensional swelling. Swelling was observed at different time intervals to 48 hours. The data was analyzed to determine the development of swelling with time. The data analysis clearly demonstrated three stages of swelling, initial, preliminary and secondary for all tested samples except the quasi-saturated ones. The swell percent and primary swelling were very sensitive to the initial moisture content and dry density of the tested samples. Most of the swelling took place during the first 24 hours for all the tested specimens. The hyperbolic model was assessed for prediction of the percent swell. The 12 hours data was found to be very successful in predicting the percentage swell.

1 Introduction

Swelling takes place when water percolatesinto the pores of a potentially expansive soil. The potential for a compacted soil to swell is governed by its intrinsic and extrinsic properties. The important intrinsic properties are the soil composition and the mineralogical characteristics of the clay fraction whereas the extrinsic properties are the placement water content, densityand loading [1]. Moisture migration depends on the existence of water source and a potential gradient to drive water into the soil. The percolation potential constitutes three components: gravitational potential, overburden potential and matric suction potential [2]. The first two are negligible compared to the matric suction potential which is responsible for driving water into clay soils [2]. Matric suction of a certain clay is governed by its water content and pore size characteristics which are partly controlled by its density.

Swelling is a time dependent process. It takes place over relatively long time in three distinct stages: initial, primary and secondary [3].The intrinsic and placement factors affect the magnitude and duration of these swelling stages [4]. This paper looks into the different stages of swellingof a highly plastic compacted swelling soil from Sudan.

2 Time rate of swelling for compacted clay soils

The factors controlling the time rate of swelling of compacted clays are: soil composition, density, water content, matric suction and fabric. Dry of optimum compacted clays have flocculated structurewhereas wet of optimum clays have dispersed structure [5].The time rate of swelling varies with the swelling stages, i.e., initial, primary and secondary stages.

The swelling behavior of compacted sand-bentonite mixtures follows a standard 'S' shape[3]. Figure 1 shows typical swell curve for a compacted fine-grained soil plotted on a semi-logarithmic scale. Swelling occurs in three stages: initial or inter-void swelling which comprises 10% of total swelling; primary swelling comprising 80% of total swelling and secondary swelling beyond 90% total swell[3]. Initial swelling is due to hydration of the surface of clay particlessuch thatswelling takes place within the voids of the non-swelling fraction; primary swelling is due to double layer effectsand develops at faster rate when the voids can no longer accommodate further clay particle swelling whereas secondary swelling is explained by the flow processes associated with bimodal pore size distribution for the swelling soil [6]. Secondary swelling is slow compared to the initial and primary swelling stages.

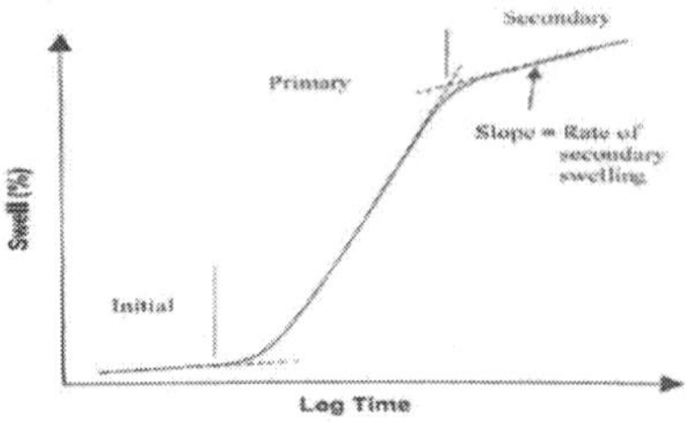

Fig. 1. Schematic of swell versus log of time

When assessing the rate of swelling, the swelling data is often plotted as percentage swell versus time. Time is either plotted on normal scale or on Log_{10} scale. Dakshanamurthy[7] proposed a hyperbolic equation to predict swelling of expansive soils as given below

$$\varepsilon = \frac{t}{(a+bt)} \qquad (1)$$

Where:

ε = percentage swell

t=time

a and b are constants

a and b can be determined by linearizing the non-linear curve through plotting t/ε versus t.

$$t/\varepsilon = a+bt \qquad (2)$$

Here b is the slope of the swelling path and is defined as the coefficient of the rate of swelling [8]. The maximum swell percent can be predicted from equation (1) as follows:

$$\varepsilon max = lim_{t\to\infty}\left(\frac{1}{\frac{a}{t}+b}\right) = 1/b \quad (3)$$

Theoretically the maximum swell percent is the inverse of the coefficient of the rate of swelling. Several researchers have predicted the maximum swell using the hyperbolic model [8].

When ε is plotted versus $\log_{10}(t)$ linear relationships are realized for the different stages of swelling. The time rate of swelling for each stage is given by

$$Cs = ?\varepsilon / ?log_{10}t \quad (4)$$

Where ε and t cover the swelling stage under consideration

3 Soil material and test methods

The soil used in this study is a highly plastic silty clay soil obtained from Alfao in eastern Sudan. The basic soil's physical and engineering properties are as follows: liquid limit is 70, plasticity index is 40, clay fraction 48% ,maximum dry density is 1.43 g/cm^3 and o.m.cis 29.5. The soil is classified, according to the Unified Soil Classification System as highly plastic silty clay (CH).

The objective of the testing program was to study the swelling stages of a compacted clay sample prepared at different water contents and dry densities. The dry densities applied in this research were 1.25 g/cm^3, 1.35 g/cm^3 and 1.45 g/cm^3 (+-0.1 g/cm^3). For each of the three density levels, the soil specimens were prepared at five different water contents;two specimens dry of optimum, one at optimum and two wet of optimum water content. The soil samples were airdried, crushed and pulverized and water was added to attain the target water content.

The tested specimen was manually compacted carefully inside the oedometer ring at the required water content to the target dry density and then mounted into the oedometer cell which allows one-dimensional swelling. To eliminate the effects of loading on the sample and allow free sample swell, very light plastic cover was placed on the sample and distilled water was then added. Swelling was measured at specified intervals for 48 hours.

4 Data analysis

The swell percent is plotted versus time for dry density 1.25 g/cm^3 (Figure 2) and swell percent versus $log_{10}t$,Figures 3, Figure 4 and Figure 5 for dry densities 1.25 g/cm^3, 1.35 g/cm^3 and 1.45 g/cm^3, respectively. The swelling stages are more evident when we look at the swell versus time relationship. Figure2 clearly shows three trends in the flow of the curve; initial fast increase in percentage swell with time and then gradual increase shown by the formation of a curve and finally very small increase of swell with time explained by the asymptotic line.The analysis of data confirms the evolution of the three stages of swelling: initial, primary and secondary. It took less than five minutes for the initial stage of swelling to take place whereas the primary stage continued to 6 to 12 hours (360 to 720 minutes) and the secondary swelling resumes after 12 hours.

Thetime rate of initial swell(Cs in equation 4) is faster for the drier and denser samples and decreases with increase in the molding water content. The first minute witnessed large volume change especially which amounts to more than 25% for the drier and denser sample (Figure 5). The time rate of initial swell decreases with decrease in density and with increase in water content. However, it is evident from the data thatmolding water content has more impact on the amount and rate of swelling when compared to dry density.This is more evidentwhen comparing dry of optimum tests results with the wet of optimum ones. The bottom curves (wet of optimum samples), Figures 3, 4 and 5, are almost identical for the initial stage of swelling and separate from each other during the following stages.The rate of primary swelling,for all densities and water contents, ranges from about 0.02 to about 0.045 whereas the rate of secondary swelling ranges between 0.02 to 0.002, both of them decrease with increasein water content.

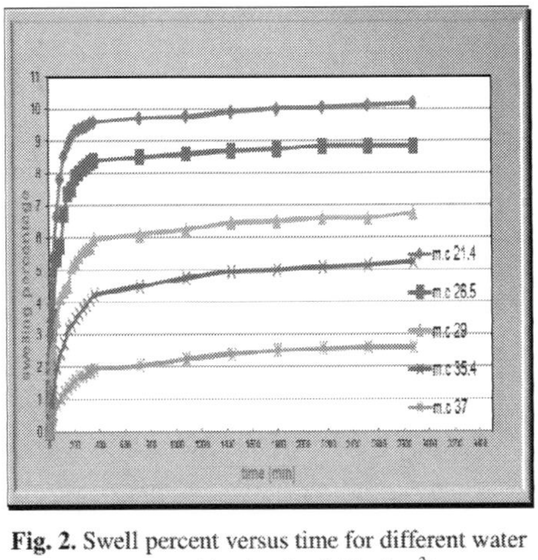

Fig. 2. Swell percent versus time for different water contents (dry density 1.25 g/cm^3)

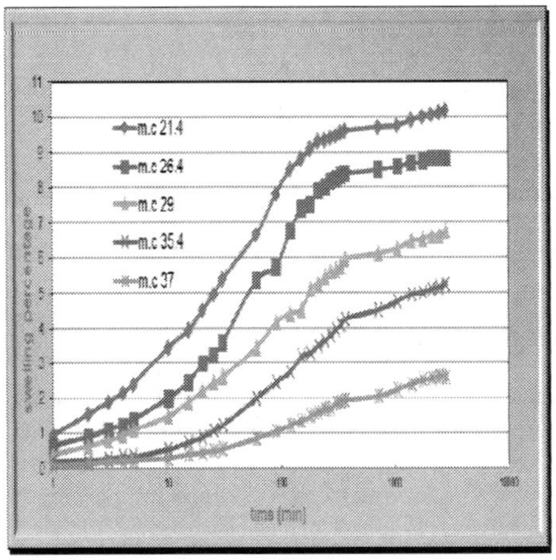

Fig. 3. Swell percent versus log time for different water contents (dry density 1.25 g/cm^3)

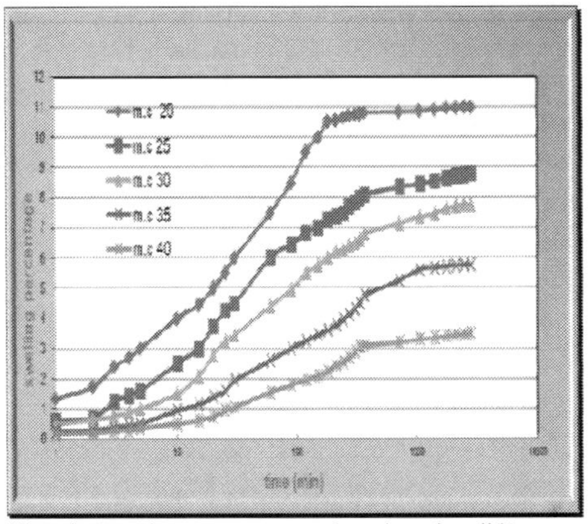

Fig. 4. Swell percent versus log time for different water contents (dry density 1.35 g/cm^3)

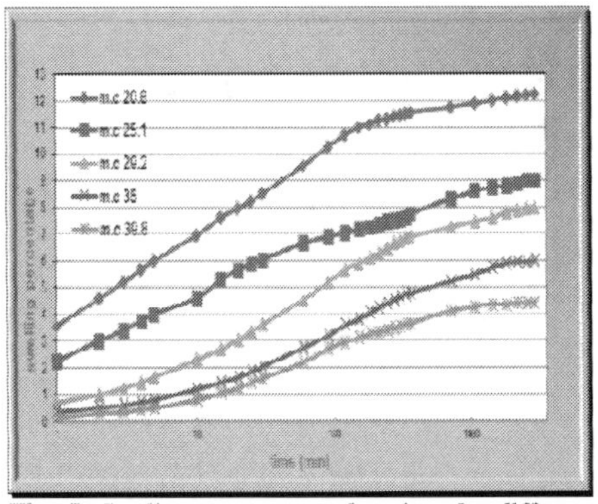

Fig. 5. Swell percent versus log time for different water contents (dry density 1.45 g/cm^3)

The hyperbolic model was used to predict the percentage swell for the test data. From equation (2), $t/\varepsilon{*}t_o$ is plotted,for the whole data, against t/t_o for dry density 12.5g/cm^{3}and all water content values (Figures 6); here t_o is taken as 100. The plots show straight line relationship and very good correlation especially for the primary and secondary swelling stages. The coefficient of correlation is greater than 0.99 for all the tested specimens, i.e. for all water content values. The slope of the straightlineincreases with increase in water content values. The values of a and b in equation (2)were determined. The theoretical maximum percent swell (1/b) was determined for all the tests data and compared with the measured percent swell after 48 hours test time (Figure 7). The slope of the straight line joining the points is about unity, therefore the 48hour percent swell is good approximation of the theoretical maximum percentage swell.

The percent swell was plotted against timefor different time periods, i.e. 180 and 720 for dry density equals 12.5 gcm^3 (Figure 8 and Figure 9). The correlation coefficient was determined for each time period. Excellent correlation (R^2>=0.99) is assured for 720 minutes testingtime. Therefore, the hyperbolic model can be utilized for predicting the maximum percent swell if test data is available for twelve hours. The plots also show that the initial swell data do not fit well to the hyperbolic model, therefore the model best applies to the primary and secondary swell data.

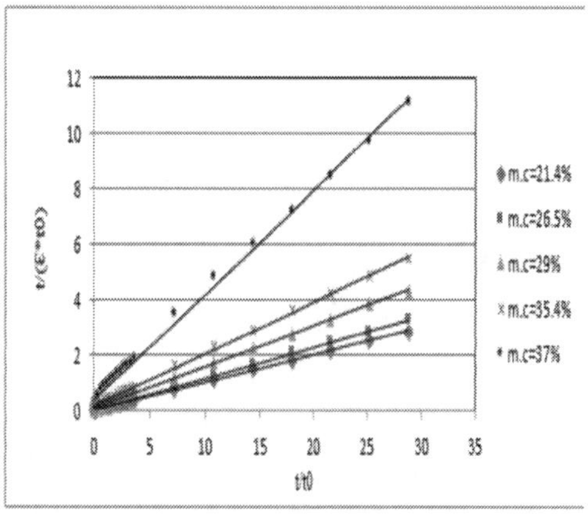

Fig. 6. t/ε versus time for the test data (dry density 1.25g/cm^3)

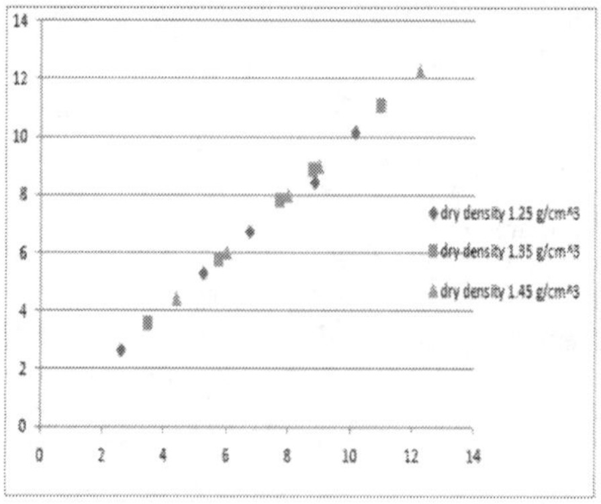

Fig. 7. Computed versus measured percent swell values for all the test data

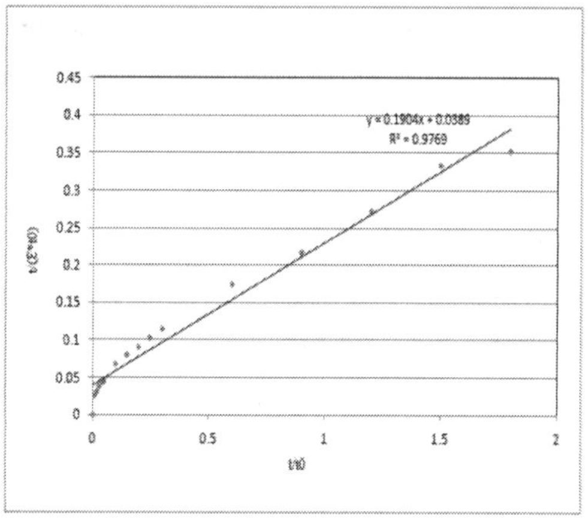

Fig.8. t/ε versus time for the up to 180 minutes of testing (dry density 1.25 g/cm^3)

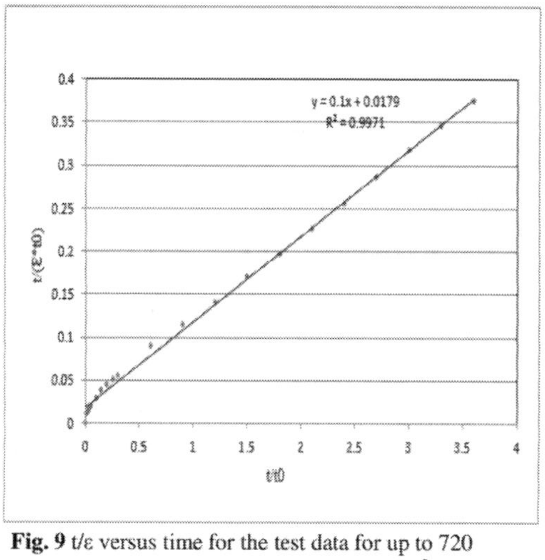

Fig. 9 t/ε versus time for the test data for up to 720 minutes of testing (dry density 1.25 g/cm^3)

5 Conclusions

This paper presented the analysis ofdata of extensive free swell tests performed on a highly plastic silty clay from Alfao in eastern Sudan. The clay is known for its very high potential for swelling. The objective was to study the development of swelling with time, the different swelling stages, amounts and times and determine the swell percent for samples compactedat various densities and water contents. The test specimens were prepared in the oedometer ring at three density levels and dry of optimum, about optimum and wet of optimum water contents.

The swell percent versus time and swell versus log time plots clearly demonstrate three stages of swelling namelyinitial, preliminary and secondary swell. The initial swell took place in less than 5 minutes for all density and moisture content conditions except for the very highwater contents. The secondary swell continues for about 720 minutes whereas the secondary swell dominates after that.

The rate of primary swell decreases as water content increases and is in the range 0.02 to 0.045. The rate of secondary swell is lower than the rate of primary swell and ranges from 0.002 to 0.02, both of them decrease with increase in water content

The hyperbolic model could be used to predict the maximum swell. The 48 hours swell data accurately predicted the maximum swell. The data analysis has shown that the 720 minutes could be used to predict the maximum swell with excellent accuracy for dry of optimum specimens and very good accuracy for wet of optimum specimens.

References

[1] J. D. Nelson and D. J. Miller,*Expansive Soils, Problems and Practice in Foundation and Pavement Engineering*, John Wiley & Sons Inc., New York, (1992)

[2] Sokolov and J. M. Amir, Moisture Distribution in Covered Clays, *3rd International Conference on Expansive Soils*, Haifa pp. 129-136 (1973)

[3] P. V. Sivapullaiah, A. Sridharan, and V. K. Stalin, Swelling Behavior of Soil-Bentonite Mixtures, Canadian Geotechnical Journal, **Vol.33/5**, 808-814 (1996).

[4] A. Sridharan and Y. Gurtug, Swelling Behavior of Compacted Fine-Grained Soils, Engineering Geology **72** 9-18, (2004)

[5] J. K. Mitchell, *Fundamentals of Soil Behavior*, Wiley, London, p.422 (1976)

[6] J. T. Das, *Evaluation of the Rate of Secondary Swelling in Expansive Clay Soils Using Centrifuge Technology*, M.Sc. thesis, University of Texas, Austin,(2014)

[7] V. Dakshanamurthy, A new Method to Predict Swelling Using Hyperbolic Equation, Geotechnical Engineering,**Vol. 9**, pp. 79-87 (1978)

[8] R. Hashim and A. S. Muntohar, Swelling Rate of Expansive Clay Soils, *Recent Advances in Characterization and Treatment*, edited by A. A. Al-Rawas and F. A. Goosen, Taylor and Francis Group pp. 139-148 (2006)

Determination of the yield function and its derivatives of the Barcelona Expansive model

Jamila El Brahmi[1] et Nouzha Lamdouar[1]
[1]Ecole Mohammadia d'Ingénieurs, Rabat, Maroc

Abstract. The Barcelona Expansive Model "BE × M" presented by Alonso and al. in 1999 is one of the best models to correctly describe the hydromechanical behavior of an unsaturated expansive soil in the elastoplastic domain. The implementation of the BE × M in a computer code is, however, complicated by the existence of three yield surfaces (SI, SD and LC) that limit the elastic domain. In order to circumvent this difficulty, the yield surface "F" in the space (p*, q, s) is assimilated to a succession of ellipses along the suction axis "s" inclined by 45° with respect to the two axes (p*, s). Then, the equation of the yield surface corresponds to a given stress state. The partial derivatives of the yield function (with respect to the stress vector « σ », the soil suction « s » and the hardening parameter « p_0^* ») are also determined since they make it possible to calculate the initial stresses $\dot{\sigma}_{ij}^0$ and S_{ij} in the boundary integral equations of the expansive soil displacement.

Résumé. Le modèle de Barcelone «BE×M » présenté par Alonso et al. en 1999 est l'un des meilleurs modèles qui permettent de décrire correctement le comportement hydromécanique d'un sol gonflant non saturé dans le domaine élastoplastique . L'implémentation de BE×M dans un code de calcul se trouve, néanmoins, compliquée par l'existence de trois surfaces de charge (SI, SD and LC) qui limitent le domaine élastique . Afin de contourner cette difficulté, la surface de charge « F » dans l'espace (p*, q, s) est assimilée à une succession d'ellipses suivant l'axe des succions « s » inclinées de 45° par rapport aux deux axes (p*, s). L'équation de la surface de charge correspond alors à un état de contrainte donné. Les dérivées partielles de la fonction de charge (par rapport au vecteur contrainte « σ », à la succion « s » et au paramètre d'écrouissage « p_0^* ») sont également déterminés puisqu'ils permettent de calculer les contraintes initiales $\dot{\sigma}_{ij}^0$ et S_{ij} des équations d'intégrales de frontière du déplacement du sol gonflant.

1 Introduction

Dans le cadre de la modélisation élatoplastiques du comportement hydromécanique d'un sol gonflant non saturé avec le modèle BE×M, nous sommes ramenés à résoudre les équations d'intégrale de frontière du déplacement suivants [1]

$$\dot{U}_j(\xi) = \int_S \left[\dot{t}_i(x) G_{ij}(x,\xi) - F_{ij}(x,\xi) \dot{U}_i(x) \right] dS$$
$$+ \int_V \mathrm{T}\left(1 + 2\ln r\right) y_j \nabla^2 \left(\dot{s}\right) dV +$$
$$\int_S \mathrm{T}\left(\frac{2}{r^2} y_j y_m n_m + (1+2\ln r) n_j \right) \dot{s} dS \qquad (1)$$
$$- \int_S \mathrm{T}\left(1 + 2\ln r\right) y_j \frac{\partial \dot{s}}{\partial n} dS +$$
$$\int_V \frac{\partial G_{ij}(x,\xi_0)}{\partial x_k} \left(\dot{\sigma}_{ik}^0 - S_{ik} ds \right) dV$$

avec:
ξ et x : les coordonnées du point test et du point source respectivement.

$\dot{t}_i$ et $\dot{U}_i$: Incréments de traction et de déplacement sur la frontière du domaine S.

G_{ij} et F_{ij} : Fonctions de Green (solutions singulières).

$\nabla^2\left(\dot{s}\right)$, $\dot{s}$ et $\frac{\partial \dot{s}}{\partial n}$: Laplacien de la succion, les incréments de succion et de gradient de succion.

r: distance entre le point test et le point source donnée par: $r^2 = y_i y_i = \left(x_1 - \xi_1\right)^2 + \left(x_2 - \xi_2\right)^2$ où $y_i = x_i - \xi_i$

T : un scalaire donné par : $\Gamma = \left(\lambda + \frac{2}{3}\mu\right)\frac{\kappa_s}{1+e}\cdot\frac{1}{s+p_{atm}}$

où: λ , μ : les constantes de Lamé

κ_s : coefficient de compressibilité pour des incréments de succion dans le domaine élastique

e: Indice des vides

p_{atm} : pression atmosphérique

S et V : surface de frontière et domaine d'intégrales

$\dot{\sigma}_{ik}^0 - S_{ik}ds$: Contraintes initiales (contraintes plastiques) données par :

$$\dot{\sigma}_{ij}^0 = \frac{\left(\frac{\partial F}{\partial \sigma_{ij}}\right)^T \left(\lambda\delta_{ij}\dot{\varepsilon}_{kk} + 2\mu\dot{\varepsilon}_{ij}\right)\left(\lambda\delta_{ij}\frac{\partial F}{\partial \sigma_{kk}} + 2\mu\frac{\partial F}{\partial \sigma_{ij}}\right)}{H + \left(\frac{\partial F}{\partial \sigma_{ij}}\right)^T \left(\lambda\delta_{ij}\frac{\partial F}{\partial \sigma_{kk}} + 2\mu\frac{\partial F}{\partial \sigma_{ij}}\right)}$$

$$S_{ij} = \left(\frac{\left(\left(\frac{\partial F}{\partial \sigma_{ij}}\right)^T \Gamma\delta_{ij} + \frac{\partial F}{\partial s}\right)\left(\lambda\delta_{ij}\frac{\partial F}{\partial \sigma_{kk}} + 2\mu\frac{\partial F}{\partial \sigma_{ij}}\right)}{H + \left(\frac{\partial F}{\partial \sigma_{ij}}\right)^T \left(\lambda\delta_{ij}\frac{\partial F}{\partial \sigma_{kk}} + 2\mu\frac{\partial F}{\partial \sigma_{ij}}\right)}\right)$$

A condition d'avoir toutes les conditions aux limites requises, toutes les intégrales de la formule de déplacement peuvent êtres déterminées numériquement avec la méthode des éléments de frontière, sauf pour le dernier intégral volumique $\int_V \frac{\partial G_{ij}(x,\xi_0)}{\partial x_k}\left(\dot{\sigma}_{ik}^0 - S_{ik}ds\right)dV$ qui nécessite un traitement particulier. Cette intégrale nécessite d'abord la détermination des contraintes plastiques du problème $\dot{\sigma}_{ik}^0 - S_{ik}ds$ (appelées contraintes initiales) avant de calculer l'intégrale elle-même (qui n'est pas sans complications d'ailleurs).

Le calcul de $\dot{\sigma}_{ik}^0 - S_{ik}ds$ repose sur la détermination de la fonction de charge « F » et de ses dérivées, d'où le présent article.

2 Surfaces de charge du modèle BE×M:

Le modèle de Barcelone « **BE×M Model** » [2, 3] utilise, constitutivement, la contrainte nette $p^* = \frac{\sigma_1 + \sigma_2 + \sigma_3}{3} - u_a$ et la succion matricielle $s = u_a - u_w$ comme variables indépendantes pour décrire le comportement volumique des sols gonflants non saturés. Dans le cas anisotrope (triaxial), une troisième variable (contrainte déviatorique) définie par $q = \sigma_1 - \sigma_3$ s'ajoute aux variables précédentes. Dans le cas général, le comportement du sol est décrit dans le plan (p^*, q, s).

2.1 Surface de charge « LC » dans le plan (p^*, s)

Lors d'un essai de compression à succion contrôlée, on remarque qu'à une succion donnée s, la courbe de compressibilité est formée, dans le plan $(v - \ln p^*)$, de deux droites caractérisées par les pentes κ et $\lambda(s)$ [figure 1]. κ étant le coefficient de compressibilité dans le domaine élastique et λ(s) la pente de la partie normalement consolidée appelée aussi coefficient de compressibilité plastique [4]. Le coefficient κ étant supposé constant, λ(s) diminue quand la succion augmente (le sol tend à se rigidifier et sa compressibilité diminue). Ce dernier paramètre est modélisé par Alonso et al. [4] par l'expression suivante :

$$\lambda(s) = \lambda(0)\left[\left(1-r\right)e^{-\beta . s} + r\right]$$

Avec :

$\lambda(0)$: le coefficient de compressibilité plastique à l'état saturé (coefficient λ du modèle CAM CLAY MODIFIE)

r : Constante liée à la rigidité du sol quand la succion tend vers l'infini ($r = \frac{\lambda(s \to \infty)}{\lambda(0)}$, r <1)

β : Un paramètre qui contrôle la vitesse d'augmentation de la rigidité avec la succion.

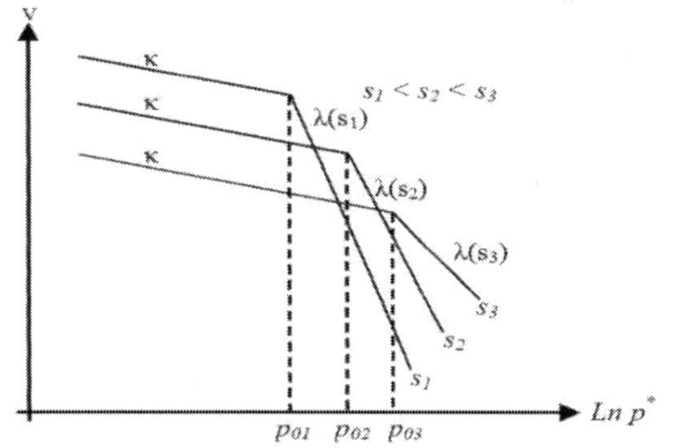

Fig.1. Réponse volumique lors d'un essai de compression à succion contrôlée [4]

La pression de pré-consolidation p_{0s} [figure 1] représente la limite d'élasticité du sol pour une succion donnée. Elle limite la zone où le comportement du sol est élastique (zone sur-consolidée de pente κ) de la zone des déformations irréversibles (zone normalement consolidée de pente $\lambda(s)$).

Cette pression de pré-consolidation augmente avec la succion. La dite augmentation peut être représentée dans le plan (s− p^*) par la courbe LC (*Loading Collapse*) [figure 2]. La courbe LC est, alors, une surface de charge puisqu'elle représente la limite entre la zone élastique et la zone de déformations plastiques. Elle est formulée comme suit :

$$LC: \quad \frac{p_0}{p_c} = \left[\frac{p_0^*}{p_c}\right]^{\frac{\lambda(0)-\kappa}{\lambda(s)-\kappa}} \tag{2}$$

Avec:
p_0: la pression de pré-consolidation apparente sous la succion s ;
p_c : une pression de référence [5]
p_0*: la pression de pré-consolidation à l'état saturé ;

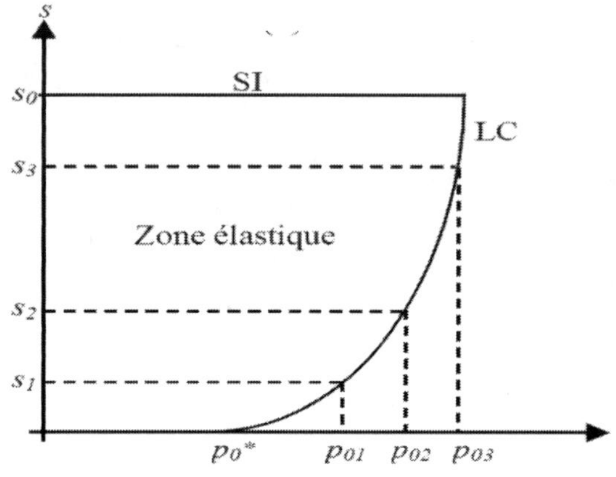

Fig.2. Définition de la surface de charge LC dans le plan (p*-s) [4]

2.2 Surfaces de charge « SI » et « SD » dans le plan (p^*, s)

Un gonflement de la microstructure par humidification (diminution de la succion) peut engendrer une augmentation irréversible de l'indice des vides macrostructural e_M, et inversement, un rétrécissement de la microstructure par dessiccation (augmentation de la succion) peut induire une diminution irréversible de e_M. La limite entre la zone élastique et celle des déformations irréversibles de la macrostructure est définie par deux ligne parallèles à NL « neutral line ». les deux lignes, représentant des surfaces de charges relatives à la succion, sont notées: SI (*suction increase*) et SD (*suction decrease*) [figure 3].

Il est à préciser que la ligne neutre NL est orientée à 45° par rapport aux axes et sépare la zone du gonflement microstructural de celle du retrait microstructural.

Les surfaces de charge SD et SI sont exprimés comme suit :

$$SD: \; p^* + s - s_h = 0 \tag{3}$$

$$SI: \; p^* + s - s_0 = 0 \tag{4}$$

avec:

S_h : est la succion à partir de laquelle les déformations plastiques apparaissent lors de l'humidification à contrainte moyenne nette nulle.
S_0 : est la succion à partir de laquelle les déformations plastiques apparaissent lors de la dessiccation à contrainte moyenne nette nulle.

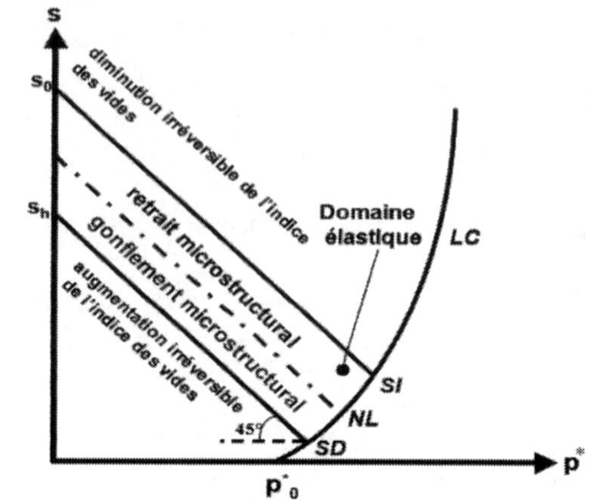

Fig.3. Surfaces de charge « SI » et « SD » du modèle BE×M [3]

2.2 Prise en compte des contraintes déviatoriques

Dans le cas d'un déviateur q, la surface de charge globale pourrait être décrite, dans le plan (p*, q, s), par une succession d'ellipses, suivant l'axe des succions, inclinés de 45° par rapport aux deux axes p* et s [figure 4]. Les projections de ces surfaces de charge dans le plan (p^*, s) et (p^*, q) sont représentées dans [figures 5]. La méthode de détermination de l'équation de cette surface de charge fait l'objet du paragraphe suivant.

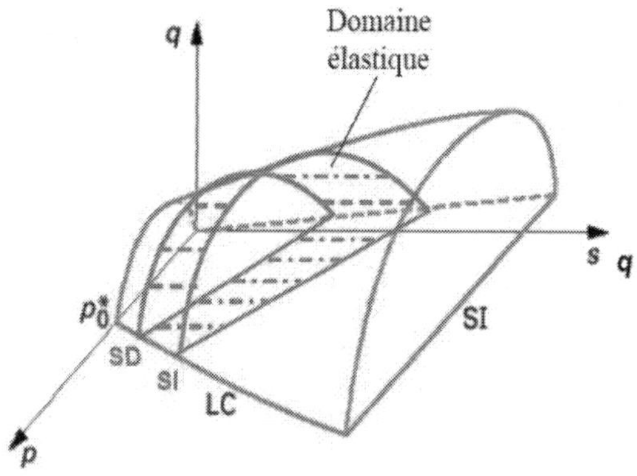

Fig.4. représentation tridimentionnelle des surfaces de charge dans l'espace (p*, q, s) [6]

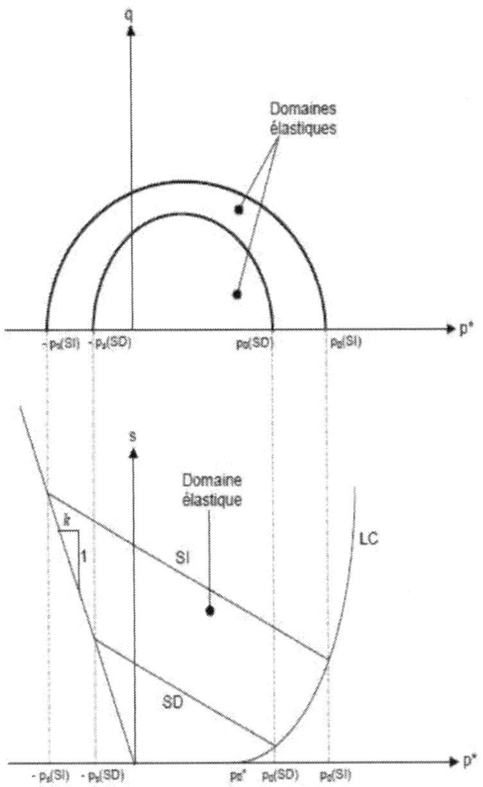

Fig.5. projection des surfaces de charge dans les plans (p^*, q) et (p^*, s) [6]

3 Détermination de la surface de charge globale « F »

La fonction de charge « F » est l'équation de la surface de charge globale en tenant compte de l'ensemble des surfaces de charge mises en jeu. C'est une fonction scalaire de la contrainte $\underline{\underline{\sigma}}$ et de la succion s dont la détermination est difficile mais primordiale pour résoudre le problème élastoplastique du sol gonflant. Vu la complexité du problème, il est plus pratique de calculer l'équation d'une ellipse pour chaque état de contrainte donnée [6].

Soit le sol à un état de contrainte donnée (p_A^*, s_A, q_A) où **A** est le point qui correspond à cet état de contrainte dans le plan (p^*, s, q) et soit **A'** la projection de **A** dans le plan (p^*, s) [figure 6]. La droite passant par le point **A'** et parallèle à SI et SD coupe la droite de cohésion au point **E** et la courbe LC au point **F**. La fonction de charge recherchée « F » est donc une ellipse de grand axe $I = \|EF\|$ et de petit axe $J = \|OM\|$ dans l'espace $(p^* - s - q)$ ($O(p_O^*, s_O)$ étant le milieu de [EF]) et la droite (EF) a pour équation: $p^* + s - b = 0$ où $b = p^*(A') + s(A')$

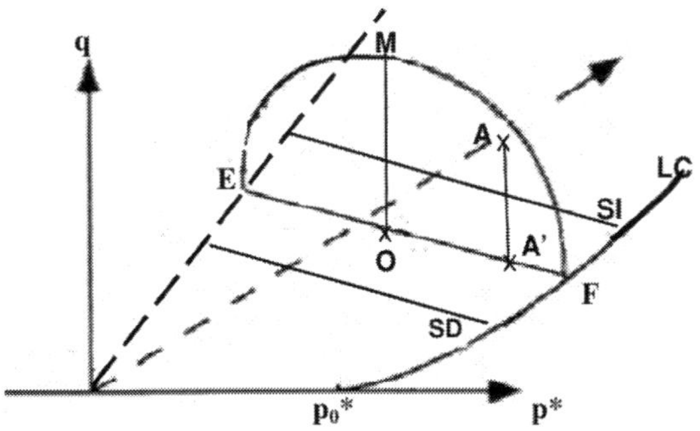

Fig.6. Projection de l'état de contrainte dans le plan (p^*,s) et l'ellipse représentant la fonction de charge

La détermination de la longueur I nécessite le calcul du point d'intersection de la droite (EF) avec la surface de charge LC (point F) et du point d'intersection de (EF) avec la droite de cohésion (point E). Pour ce faire, nous sommes remmenés à résoudre les systèmes d'équations à deux inconnus suivantes :

$$\text{Point E : } \begin{cases} p^* + s - b = 0 \\ p^* = -ks \end{cases} \quad (5)$$

$$\text{Point F : } \begin{cases} p^* + s - b = 0 \\ p^* = p_c \left[\dfrac{p_0^*}{p_c} \right]^{\frac{\lambda(0)-\kappa}{\lambda(s)-\kappa}} \end{cases} \quad (6)$$

L'équation (5) relative au point $E\left(p_E^*, s_E\right)$ est facile à résoudre et les coordonnées du point E peuvent être données comme suit :

$$E : \begin{cases} p_E^* = \dfrac{bk}{k-1} \\ s_E = \dfrac{b}{1-k} \end{cases} \quad (7)$$

La détermination des coordonnées du point $F\left(p_F^*, s_F\right)$ revient à résoudre l'équation $f(s_F) = 0$ suivante :

$$p_c \left[\frac{p_0^*}{p_c} \right]^{\frac{\lambda(0)-\kappa}{\lambda(0)\left[(1-r)e^{-\beta . s_F} + r\right] - \kappa}} + s_F - b = 0$$

ou $b = p^*(A') + s(A')$

et p_0^*, p_c, $\lambda(0)$, κ, r et β sont des paramètres données du modèle BE×M.

si l'on pose :

$$C_1 = p_c$$

$$C_2 = \frac{p_0^*}{p_c}$$

$$C_3 = \lambda(0) - \kappa$$

$$C_4 = \lambda(0)(1-r)$$

$$C_5 = \lambda(0)r - \kappa$$

l'équation $f(s_F) = 0$ devient :

$$C_1 C_2^{\frac{C_3}{C_4 e^{-\beta . s_F} + C_5}} + s_F - b = 0$$

Ce système est insolvable analytiquement. nous sommes donc ramenés à le résoudre numériquement en utilisant la méthode de Newton pour des valeurs données des paramètres p_0^*, p_c, $\lambda(0)$, κ, r et β et pour un état de contrainte donné.

La résolution numérique par la méthode de Newton consiste à donner une solution de départ ($s_{A'}$ par exemple) et faire des itérations ;

$$s_F^{k+1} = s_F^k - \frac{f(s_F^k)}{\frac{\partial f}{\partial s}(s_F^k)}$$

où :

s_F^{k+1} et s_F^k : les solution de l'équation aux rangs $k+1$ et k respectivement

$\frac{\partial f}{\partial s}(s_F^k)$:la dérivée de la fonction f par rapport à la succion donnée par :

$$\frac{\partial f}{\partial s}(s_F^k) = \frac{C_1 C_2^{\frac{C_3}{C_4 e^{-\beta s_F^k} + C_5}} C_3 C_4 \beta e^{-\beta s_F^k} \log C_2}{\left(C_4 e^{-\beta s_F^k} + C_5\right)^2} + 1$$

Il convient de tester la convergence après chaque itération en vérifiant que l'écart entre les valeurs de succion retrouvées aux rang k et k+1 est inférieur à la tolérance fixée par l'utilisateur. Finalement :

$$F : \begin{cases} p_F^* = b - s_F^k \\ s_F = s_F^k \end{cases} \quad (8)$$

Cette méthode de résolution converge vers la solution exacte quelle que soit la solution de départ donnée.

Une fois les coordonnées de E et F (équations (7) et (8)) sont calculées, la longueur du grand axe I est donnée par la formule suivante :

$$I = \|EF\| = \sqrt{\left(p_F^* - p_E^*\right)^2 - \left(s_F - s_E\right)^2}$$

La longueur du petit axe $J = \|OM\|$ ne pourrait pas être déterminée directement vu qu'elle représente la distance du point O (milieu de [EF]) à l'ellipse dont le but final de cet article est de chercher son l'équation! Mais en observant que cette distance est égale à celle du point O à l'ellipse de grand axe [E'F'] [figure 7] dont l'équation est connue (Modèle de Barcelone « *Barcelona Basic Model BBM* » décrit par Alonso et al. [7]:

$$q^2 - M^2\left(p^* + p_s\right)\left(p_0(s) - p^*\right) = 0$$

où :

M: est la pente de la ligne d'état critique CSL dans le plan (p^*, q) supposée indépendante du chemin de chargement et de la succion [8].

$P_0(s)$: est la pression de préconsolidation à succion constante donnée par l'équation (2) ;

P_s : donnée par $p_s = k.s$, avec k un paramètre désignant l'augmentation de la cohésion avec la succion [9]. $p_s = k.s$ est appelée droite de cohésion.

on aura :

$$J = 2q = 2M\sqrt{\left(p_O^* + p_s(s_O)\right)\left(p_0(s_O) - p_O^*\right)}$$

ou p_O^* et s_O sont les coordonnées du point O :

$$O : \begin{cases} p_O^* = \frac{p_F^* + p_E^*}{2} \\ s_O = \frac{s_F^* + s_E^*}{2} \end{cases}$$

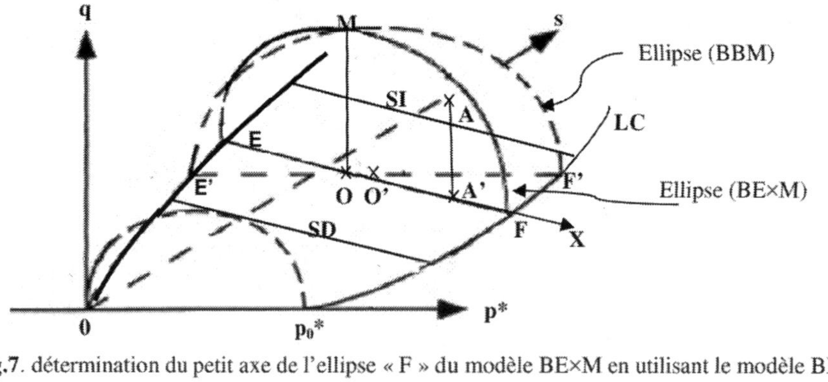

Fig.7. détermination du petit axe de l'ellipse « F » du modèle BE×M en utilisant le modèle BBM

$$p_s(s_O) = k.s_O$$

$$p_0(s_O) = p_c\left[\frac{p_0^*}{p_c}\right]^{\frac{\lambda(0)-\kappa}{\lambda(0)\left[(1-r)e^{-\beta s_O}+r\right]-\kappa}}$$

l'équation cartésienne de l'ellipse dans le plan local (X,q) est alors:

$$\left[\frac{X}{(I/2)}\right]^2+\left[\frac{q}{(J/2)}\right]^2=1$$

qui devient, après transformation des coordonnées, dans le plan (p^*,q) :

$$F=-1+\frac{2\left(p_O^*\right)^2}{\left(\frac{I}{2}\right)^2}-\frac{4p_O^*}{\left(\frac{I}{2}\right)^2}p^*+\frac{2}{\left(\frac{I}{2}\right)^2}p^{*2}+\frac{q^2}{\left(\frac{J}{2}\right)^2}$$

en posant:

$$C_{F1}=-1+\frac{2\left(p_o^*\right)^2}{\left(\frac{I}{2}\right)^2}$$

$$C_{F2}=-\frac{4p_o^*}{\left(\frac{I}{2}\right)^2}$$

$$C_{F3}=\frac{2}{\left(\frac{I}{2}\right)^2}$$

$$C_{F4}=\frac{1}{\left(\frac{J}{2}\right)^2}$$

on peut écrire: $F=C_{F1}+C_{F2}p^*+C_{F3}p^{*2}+C_{F4}q^2$ (9)

4 Détermination des dérivées de la surface de charge « F »

La dérivée de F par rapport au vecteur des contraintes pourrait être calculée comme suit :

$$\frac{\partial F}{\partial\sigma}=\frac{\partial F}{\partial p^*}\frac{\partial p^*}{\partial\sigma}+\frac{\partial F}{\partial q}.\frac{\partial q}{\partial\sigma}$$

avec: $p^*=\frac{\sigma_1+\sigma_2+\sigma_3}{3}-u_a$ et $q=\sqrt{3J_2}$

où J_2 étant le deuxième invariant du tenseur déviateur des contraintes. On peut écrire alors :

$$\frac{\partial p^*}{\partial\sigma}=\begin{pmatrix}\frac{1}{3} & \frac{1}{3} & \frac{1}{3} & 0 & 0 & 0\end{pmatrix}^T$$

$$\frac{\partial q}{\partial\sigma}=3\begin{pmatrix}\frac{1}{2q}\left(\sigma_1-\frac{\sigma_1+\sigma_2+\sigma_3}{3}\right)\\ \frac{1}{2q}\left(\sigma_2-\frac{\sigma_1+\sigma_2+\sigma_3}{3}\right)\\ \frac{1}{2q}\left(\sigma_3-\frac{\sigma_1+\sigma_2+\sigma_3}{3}\right)\\ \frac{1}{q}\sigma_4\\ \frac{1}{q}\sigma_5\\ \frac{1}{q}\sigma_6\end{pmatrix}$$

et $\frac{\partial F}{\partial q}=\frac{2}{\left(\frac{J}{2}\right)^2}q$

pour obtenir finalement :

$$\frac{\partial F}{\partial\sigma}=3\begin{pmatrix}\frac{4}{J^2}\left(\sigma_1-\frac{\sigma_1+\sigma_2+\sigma_3}{3}\right)+\frac{16}{3I^2}\left(p^*-p_O^*\right)\\ \frac{4}{J^2}\left(\sigma_2-\frac{\sigma_1+\sigma_2+\sigma_3}{3}\right)+\frac{16}{3I^2}\left(p^*-p_O^*\right)\\ \frac{4}{J^2}\left(\sigma_3-\frac{\sigma_1+\sigma_2+\sigma_3}{3}\right)+\frac{16}{3I^2}\left(p^*-p_O^*\right)\\ \frac{8}{J^2}\sigma_4\\ \frac{8}{J^2}\sigma_5\\ \frac{8}{J^2}\sigma_6\end{pmatrix}$$

Quant à la dérivée de F par rapport à la succion $\frac{\partial F}{\partial s}$, elle doit être calculée numériquement (méthode de différence finie centrée) vu que la fonction de charge F ne dépend pas explicitement de la succion. Ainsi, et pour un état de contrainte donnée :

$$\frac{\partial F}{\partial s}=\frac{F(s+\Delta s)-F(s-\Delta s)}{2\Delta s}$$

Le modèle BE×M repose sur trois paramètres d'écrouissage ; p_0^*, s_0 et s_h, mais seul le paramètre p_0^* qui intervient dans l'équation de la fonction de charge F. Ainsi, la dérivée $\frac{\partial F}{\partial\chi}$ n'est autre que $\frac{\partial F}{\partial p_0^*}$. Cette dernière peut être calculée numériquement aussi comme ainsi :

$$\frac{\partial F}{\partial p_0^*} = \frac{F\left(p_0^* + \Delta p_0^*\right) - F\left(p_0^* - \Delta p_0^*\right)}{2\Delta p_0^*}$$

5 Conclusion

La détermination de la fonction de charge « F » du modèle BE×M et de ses dérivées est une étape primordiale pour résoudre le problème élastoplastique d'un sol gonflant non saturé. En effet, Ceci permettra de déterminer les contraintes initiales, traiter l'intégrale volumique correspondante et finalement permettre l'implémentation du modèle BE×M dans un code de calcul basé sur la méthode des éléments de frontière.

6 Références

1. J.El Brahmi, M. Zoukaghe, *Numerical Analysis of Nonlinear Coupled Problems, Sustainable Civil Infrastructures*, 262-274 (2017)
2. A.Gens and E.E. Alonso. A framework for the behaviour of unsaturated expansive clays. Canadian Geotechnical Journal, **29**, 1013-1032 (1992)
3. E.E. Alonso, J.Vaunat and A.Gens, Modelling the mechanical behaviour of expansive clays. Engineering Geology, **54**, 173-183 (1999).
4. E.E. Alonso, A.Gens and D.W. Hight. General report – Special problem soils. *Proceedings of the 9th European Conference on Soil Mechanics and Foundation Engineering*, Dublin, **3**, 1087-1146 (1987).
5. S.J.Wheeler, D.Gallipolo and M.Karstunen, Comments on use of the Barcelona Basic Model for unsaturated soils. International Journal for Numerical and Analytical Methods in Geomechanics, **26**,1561-1571 (2002)
6. M.Mrad, Modélisation du comportement hydromécanique des sols gonflants non saturés. Thèse de doctorat, Institut national polytechnique de Loraine, Ecole Nationale Supérieure de Géologie de Nancy (2000)
7. E.E, Alonso, A.Gens, A.Josa, A constitutive model for partially saturated soils. Géotechnique, **40**, n°3, 405-430 (1990)
8. S.J,Wheeler V.and Sivakumar. An elasto-plastic critical framework for unsaturated soil. Géotechnique, **45**, n°1, 35-53 (1995)
9. L.Miao, S.Lius and Y.Lai. Research of soil-water characteristics and shear strength features of Nanyang expansive soil, Engineering Geology, **65**, 261-267 (2002)

Influence of relative density on the cyclic shear strength of sands

Influence de la densité relative sur la résistance au cisaillement cyclique des sables

ARAB A.[1], Marwan Sadek[2,3] and SHAHROUR I.[2]
[1]Laboratory of Materials sciences and Environment
Civil Engineering Department, University Hassiba Ben Bouali, 02000 Chlef, Algeria,
[2]Polytech-Lille, France,
[3]Lebanese University, Lebanon.

Abstrtact. This paper presents a laboratory study of the influence of relative density on the liquefaction potential of a soil. The study is based on undrained triaxial tests that were performed on samples with relative density Id = 0.15, 0.5 and 0.65. The article is composed of three parts. First, we present the materials and characteristics of the studied sands. the second part deals with the procedure and the device used. The third part studies the influence of the relative density on the liquefaction potential of the three sands (Hostun Rf, Chlef and Rass). This study also makes it possible to explore the influence of granulometry on the liquefaction potential. The results of the tests show that concordant results have been obtained which clearly show that the increase of the relative density leads to a significant improvement in the resistance to liquefaction of the sands. This effect is very significant when the initial relative density Id = 0.50 to Id = 0.65.

Résumé: Cet article présente une étude en laboratoire de l'influence de la densité relative sur le potentiel de liquéfaction d'un sol. L'étude est basée sur des essais triaxiaux non drainés qui ont été effectués sur des échantillons ayant une densité relative Id = 0,15, 0,5 et 0,65. L'article est composé de trois parties. Dans un premier temps nous présentons les matériaux et les caractéristiques des sables étudiés. la seconde partie traite le mode opératoire et le dispositif utilisé. La troisième partie étudie l'influence de la densité relative sur le potentiel de liquéfaction des trois sables (Hostun Rf, Chlef et Rass). Cette étude permet également d'explorer l'influence de la granulométrie sur le potentiel de liquéfaction. Les résultats des essais montrent que des résultats concordants ont été obtenus et qui montrent d'une manière très claire que l'augmentation de la densité relative conduit à une importante amélioration de la résistance à la liquéfaction des sables. Cet effet est très significatif lorsque la densité relative initiale Id =0,50 à Id = 0,65.

1 Introduction

La résistance à la liquéfaction d'un site peut-être améliorée si celui-ci est sujet à un haut risque de liquéfaction, en agissant sur l'un de ces paramètres influents sur la résistance. Actuellement les méthodes les plus utilisées pour la stabilisation des sites agissent sur la densité relative et les conditions de drainage du sol. Récemment de nouvelles méthodes de stabilisation des sites liquéfiables ont été envisagées. Ces dernières consistent à améliorer la résistance à la liquéfaction par inclusion de nappes de géotextile, par augmentation de la perméabilité du sol.

La densité relative affecte d'une manière très sensible le potentiel de liquéfaction des sols. La figure 1 donne les résultats obtenus par Tatsuoka et al. (1986b) sur le sable de Toyoura. On constate que la résistance à la liquéfaction augmente de façon linéaire avec la densité relative jusqu'à une valeur de densité relative Dr = 0,70. Après cette valeur, on remarque une augmentation importante de la résistance avec l'augmentation de la densité relative.

Polito et Martin (2001), ont réalisé une série d'essais triaxiaux sur des échantillons de sable (Monterrey et Yatesville) mélangé avec des fines non plastiques. Ils constatent une tendance linéaire entre l'augmentation de la densité relative et la résistance à la liquéfaction pour les échantillons de sol ayant une teneur en fines en dessous du seuil limite (seuil limite des fines étant le passage d'une matrice de sol à prédominance sable à matrice contrôlée par les fines), cependant la résistance à la liquéfaction au-dessus du seuil limite est aussi contrôlée par la densité relative de l'échantillon, cependant la résistance des sols est nettement inférieure par rapport aux

 https://doi.org/10.1051/matecconf/201814902034

même sols qui se trouvent en dessous du seuil avec la même densité relative.

Dans cet article, on présente une étude au laboratoire sur l'influence de la densité relative sur la résistance à la liquéfaction. Ces essais permettent de mieux comprendre l'influence de la densité relative sur le comportement cyclique des sables. L'article est composé de trois parties ; dans la première on présente les matériaux utilisés ; la seconde partie le mode opératoire et le dispositif utilisé ; la dernière donne une analyse des résultats expérimentaux des essais réalisés et discute de l'influence de l'influence de la densité relative sur le comportement cyclique des sables.

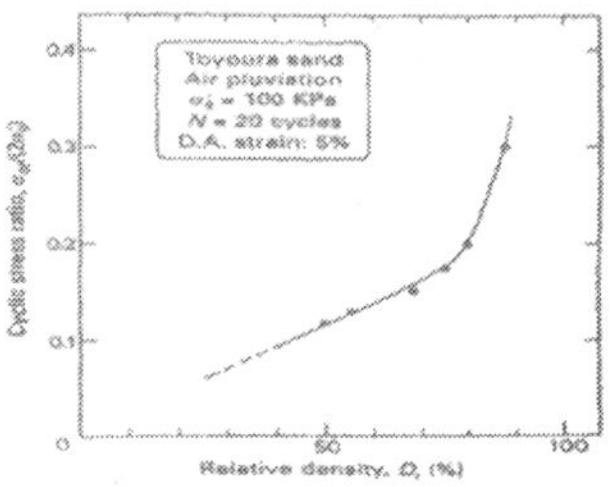

Fig. 1. Effet de la densité relative sur la résistance à la liquéfaction(Tatsuoka et al.,1986b)

2. Matériaux et dispositif expérimental.

2.1. Matériaux

Trois matériaux ont été utilisés dans cette étude à savoir le sable de Chlef et Rass (Algérie) et le sable d'Hostun Rf (France). Le sable de Chlef comme son nom l'indique provient du lit de l'Oued-Chlef qui traverse la ville de Chlef. Le sable d'Oued Rass provient des abords du lit d'Oued Rass (confluent Oued Chlef et Oued Rass). Le sable d'Hostun Rf provient des usines SIKA implantées à 26730 Hostun (Drôme) sur le flanc ouest du Vercors, la carrière d'où provient le sable est implantée dans une série épaisse de sables d'origines Eocène remplissant des poches karstiques. Le sable de Chlef est un sable moyen ayant une forme arrondie (sable alluvionnaire), avec un diamètre moyen D_{50} = 0,61mm. Le sable d'Oued Rass est aussi un sable moyen de forme arrondie (sable alluvionnaire) avec un diamètre moyen D_{50}= 0.39mm. Le sable d'Hostun est considéré comme un matériau de référence dans de nombreux laboratoires en France. De nombreux essais ont été réalisés sur ce sable (Colliat 1986, Fargeix 1986, Flavigny et al. 1990, Biarez et Ziani 1991, Lancelot et al. 1996, Doanh et al. 1997, Al mahmoud 1997, Hoque et Tatsuoka 1998). Il s'agit un sable moyen ayant une forme angulaire avec un diamètre moyen D_{50}=0,471. La figure 2 montre les courbes granulométriques des matériaux utilisés. Ces courbes ont été obtenues par tamisage sur une série complète de tamis. Le tableau 1 donne les caractéristiques de ces sables.

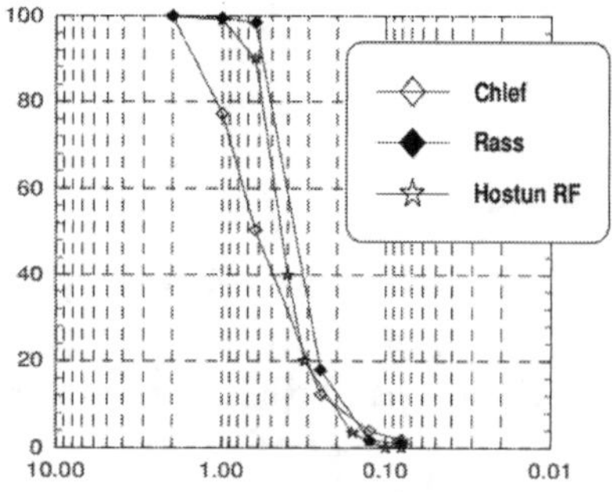

Fig. 2. Courbegranulométrique

Use A4 paper size (210 x 297 mm) and adjust the margins to those shown in Table 1. The final printed area will be 172 x 252 mm.

Tableau 1. Caractéristiques des sables utilisés

	Sable		
Propriétés	Chlef	Rass	Hostun
Masse spécifique $\square_s$ (g/cm$^{3)}$	2,68	2,664	2,654
e_{max}	0,854	0,770	0,983
e_{min}	0,535	0,490	0,622
Cu ($D_{60}/D_{10)}$	3,38	2,42	2,27
D_{10}	0,225	0,227	0,19
D_{50}	0,61	0,39	0,471
Forme des particules	Arrondi e	Arrondie	Angulai re

2.2.Dispositif expérimental

Le dispositif expérimental utilisé est présenté schématiquement dans la figure 3. Il comprend :

- Une cellule triaxiale autonome type Bishop et Wesley (Bishop et Weslay, 1975),
- Trois contrôleurs de pression /volume type GDS (200cc),
- Une pompe à vide relié à un réservoir afin de désaérer l'eau déminéralisée,
- Un micro-ordinateur équipé d'un logiciel permettant le pilotage de l'essai et l'acquisition des mesures.

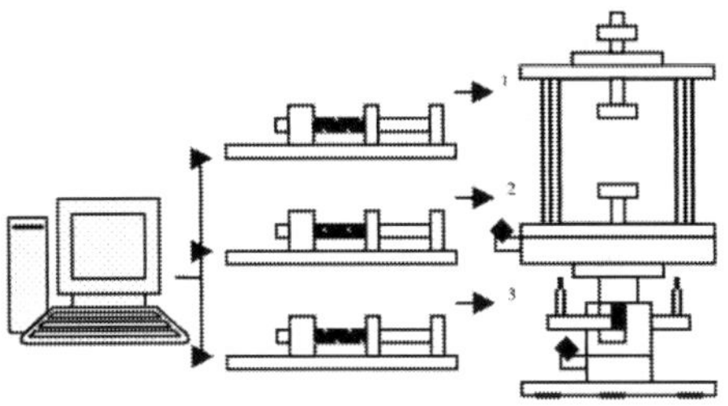

(1) application de la pression dans la cellule
(2) contrôle de la pression interstitielle ou variation de volume de l'échantillon
(3) contrôle du déviateur

Fig. 3. Dispositif expérimental utilisé

Elle est constituée de deux chambres. La chambre supérieure contient l'échantillon et les deux embases servant à appliquer la charge axiale sur celui-ci. La pression maximale que peut supporter cette chambre est de 2 MPa. La partie inférieure consiste en une presse hydraulique. Elle est constituée d'une chambre métallique dans laquelle un piston peut coulisser. L'extrémité du piston est reliée à l'embase inférieure qui applique la charge axiale à l'échantillon. Comme la section du piston est connue, la force et le déplacement axial appliqué par le piston sont déduits directement de la pression et de la variation de volume dans la chambre. Ces dernières sont contrôlées par un GDS qui permet d'appliquer le chargement axial en contrainte ou en déformation contrôlée. L'échantillon est balayé tout d'abord par du gaz carbonique CO_2 pendant 20 minutes ensuite on fait passer l'eau désaérée et déminéralisée à travers l'échantillon pour le saturer. Les échantillons sont consolidés isotropiquement pour atteindre la valeur de contrainte de confinement effective précédent le chargement. La contre pression utilisée pour nos essais est de 400 kPa, la contrainte de confinement est égale à 500 kPa. Le degré de saturation des échantillons estévalué en mesurant le coefficient de Skempton B après consolidation ($B=\Delta\mu/\Delta\sigma$).

3. Résultats des essais

3.1. Essais cycliques sur le Sable d'Hostun

Les résultats de l'ensemble des essais réalisés sont résumés sur les figures 4.a et 4.b. La première illustre l'influence de la densité relative sur le potentiel de liquéfaction. Elle montre d'une manière très claire que l'augmentation de la densité relative conduit à une translation importante des courbes de potentiel de liquéfaction vers la droite. C'est-à-dire, que l'augmentation d la densité relative améliore la résistance à la liquéfaction de ce sable. Ce résultat est cohérent avec l'influence de la densité relative sur l'augmentation de la dilatance du sable.
La figure 5.1b montre l'influence de la densité relative sur la résistance à la liquéfaction qui a définie par l'amplitude du chargement induisant la liquéfaction après 15 cycles. Cette figure montre bien que la résistance à la liquéfaction augmente avec l'augmentation de la densité relative et la diminution de l'amplitude du chargement.

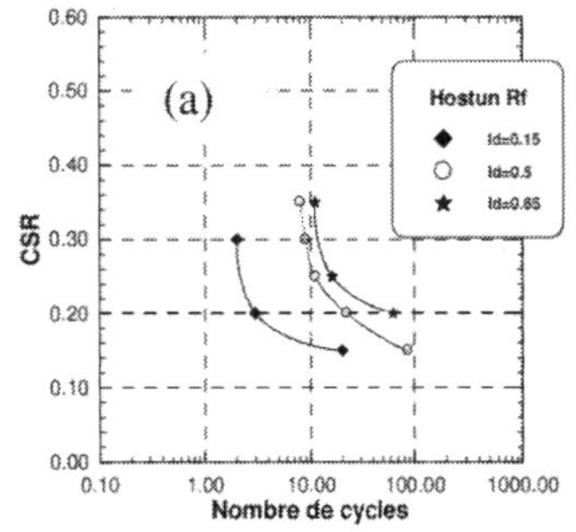

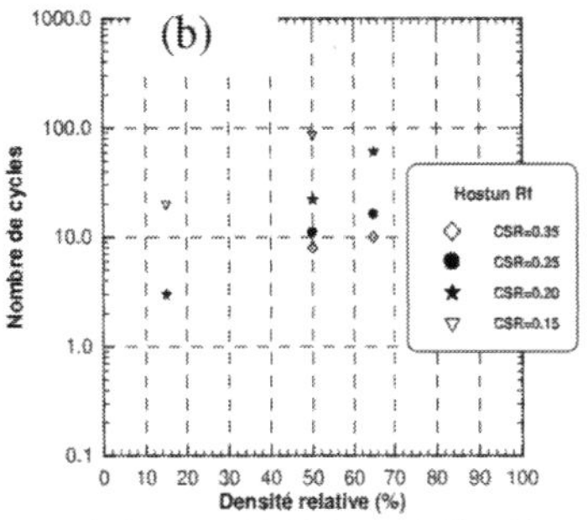

Fig.4. Influence de la densité initiale sur la liquéfactiondu sable d'Hostun Rf
(a) Sur le potentiel de liquéfaction,
(b) sur la résistance à la liquéfaction,

3.2. Essais cycliques sur le sable de Rass

Les résultats de l'ensemble de ces essais réalisés sur les échantillons du sable d'Oued Rass sont illustrés sur la figure 5. La figure 5.a illustre l'influence de la densité relative sur le potentiel de liquéfaction. Elle montre d'une manière très claire que l'augmentation de la densité relative conduit à une importante translation importante de la courbe de potentiel de liquéfaction vers la droite. C'est-à-dire, que l'augmentation de la densité relative améliore la résistance à la liquéfaction de ce sable.

La figure 5.b montre l'influence de la densité relative sur la résistance à la liquéfaction définie par l'amplitude du chargement induisant la liquéfaction après 15 cycles. Cette figure montre bien que la résistance à la liquéfaction augmente avec l'augmentation de la densité relative et la diminution de l'amplitude du chargement. La différence entre la résistance du sable à une densité relative Id = 50 et celle pour Id = 0,65 est très importante en comparaison avec la différence observée sur le sable d'Hostun.

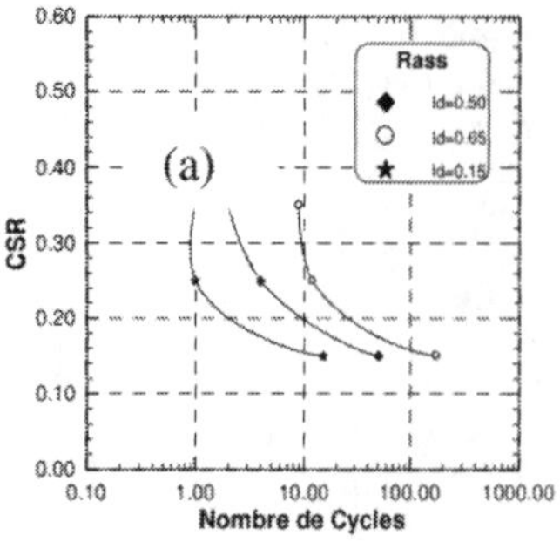

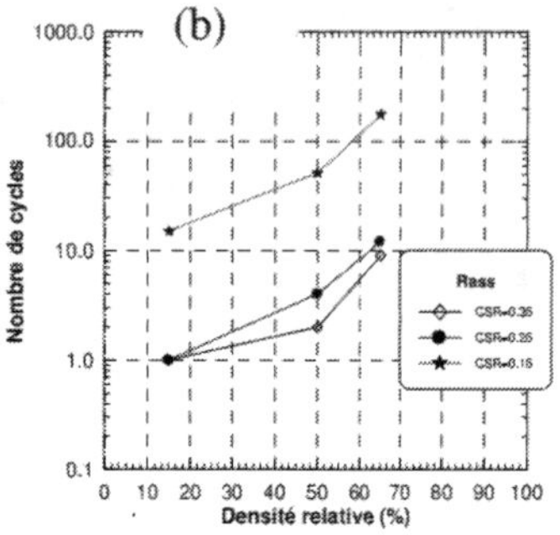

Fig.5. Influence de la densité initiale sur la liquéfactiondu sable de Rass.

(a) Sur le potentiel de liquéfaction,

(b) Sur la résistance à la liquéfaction,

3.3. Essais cycliques sur le sable de Chlef

La figure 6 illustre les résultats des essais cycliques sur le sable de Chlef qui a subi une liquéfaction lors du séisme de 1980. La figure 6.a illustre l'influence de la densité relative sur le potentiel de liquéfaction. Elle montre clairement que l'augmentation de la densité relative conduit à une augmentation de la résistance à la liquéfaction de ce sable. La figure 6.b montre l'influence de la densité relative sur la résistance à la liquéfaction définie par l'amplitude du chargement induisant la liquéfaction après 15 cycles. Cette figure montre bien que la résistance à la liquéfaction augmente avec l'augmentation de la densité relative et la diminution de l'amplitude du chargement. La différence entre la résistance du sable à une densité relative Id = 50 et celle pour Id = 0,65 est très importante ; elle ressemble à celle observée sur le sable de Rass (Figure 5.b).

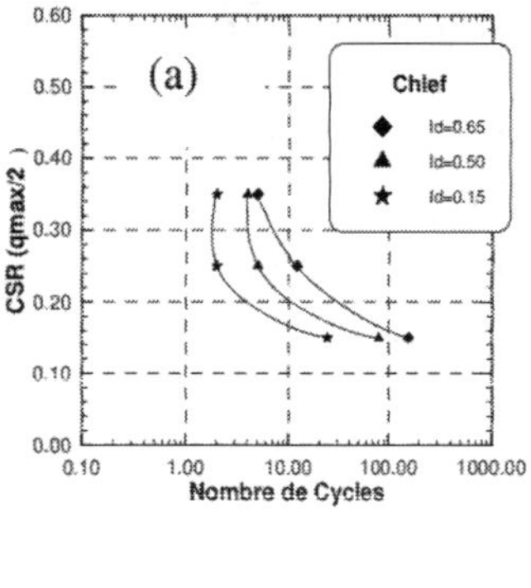

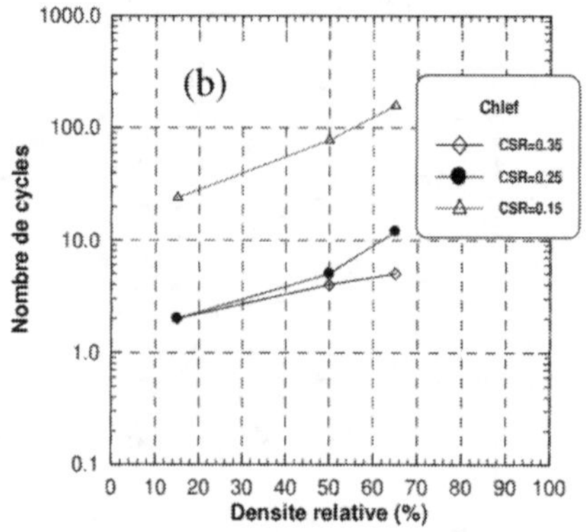

Fig.6. Influence de la densité initiale sur la liquéfaction du sable de Chlef.

(a) Sur le potentiel de liquéfaction,

(b) Sur la résistance à la liquéfaction,

La courbe 7 illustre l'évolution de la résistance à la liquéfaction en fonction de la densité relative (Dr). On déduit à partir de cette figure que le sable d'HostunRf dont les grains ont une forme angulaire présente une résistance à la liquéfaction plus grande que ceux des sables de Rass et Chlef qui sont des sables alluvionnaires et de forme arrondie. La résistance à la liquéfaction augmente linéairement avec l'augmentation de la densité relative.

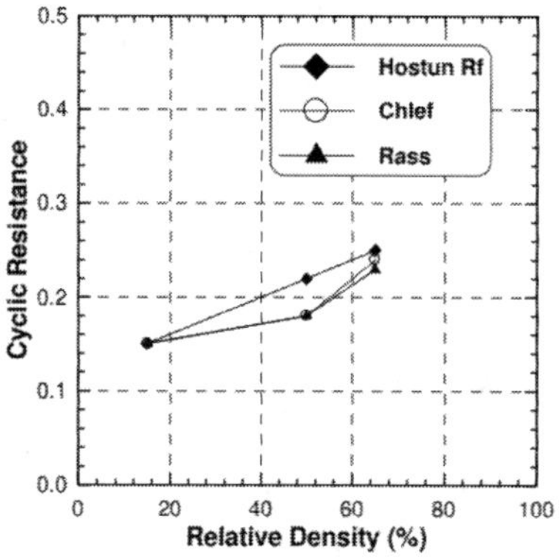

Fig.7. Evolution de la résistance en fonction de la densité relative.

4. Conclusions

Ce travail a été réalisé dans le but de mettre en évidence l'influence de la densité relative sur le potentiel de liquéfaction de trois sables à savoir le sable d'Hostun Rf (France) et les sable de Rass et Chlef (Algérie). Les essais ont été réalisés sur des échantillons lâches, moyennement denses et denses avec des degrés de saturations caractérisés par le coefficient de Skempton B dépassant les 95%.

Les résultats des essais cycliques sur le sable d'Hostun RF dont les grains ont une forme angulaire montre que l'augmentation de la densité relative mène à une augmentation de la dilatance du sable et par conséquent la résistance à la liquéfaction augmente.

Les résultats des essais cycliques sur le sable de Rass dont les grains ont une forme arrondie montrent bien que la résistance à la liquéfaction augmente avec l'augmentation de la densité relative et la diminution de l'amplitude du chargement.

Les résultats des essais cycliques sur la sable de Chlef aussi les grains ont une forme arrondie et peu grossier par rapport au sable de Rass ont montré que la densité relative influe d'une manière significative sur la résistance à la liquéfaction, elle augmente avec l'augmentation de la densité. La différence entre la résistance du sable à une densité relative Id = 50 et celle pour Id = 0,65 est très importante ; elle ressemble à celle observée sur le sable de Rass.

La résistance à la liquéfaction du d'Hostun Rf est supérieure par rapport aux autres sables (Chlef et Rass) et qui peut être expliquer par le faite que le sable d'Hsotun Rf est de caractère dilatant que les autres sables de Chlef et Rass qui sont des sables à caractère contractant.

References

1. Al Mahmoud M. « Etude en laboratoire du comportement des sables sous faibles contraintes», Thèse de doctorat en génie civil à l'USTL, Lille (1997)
2. Biarez J. &Ziani F « Introduction aux lois de comportement des sables très peu denses», Revue Française de Géotechnique, Vol. 54, pp. 65-73(1991)
3. Bishop A. W. & Wesley L.D. « A hydraulic triaxial apparatus for controlled stress path testing», Geotechnique, Vil.25, N° 4, pp. 657-670 (1975)
4. Colliat J.L.: « Comportement des matériaux granulaires sous forte contraintes, influence de la nature minéralogique du matériau étudié», Thèse de doctorat de spécialité, IMG, Grenoble (1986)
5. Doanh, T., Ibraim, E., et Matiotti, R.(1997): " Undrained instability of very loose Hostun sand in triaxial compression and extension.Part 1 : experimental observations.Mechanics of Cohesive-Frictional Materials, 2 :47–70 (1997)
6. Flavigny E., Desrues J.&Palayer B. « Le sable d'Hostun Rf», Revue Française de Géotechnique, N°53, pp.67-70 (1990)
7. Gargeix D. « Conception et réalisation d'une presse triaxiale dynamique-application à la mesure des propriétés des sols sous sollicitations sismiques», Thèse de doctorat, IRIGM, Grenoble(1986)
8. Lancelot L., Shahrour I. & Al Mahmoud M. « Comportement du sable d'Hostun sous faibles contraintes», Revue Française de Geotechnique, N°74, Janvier 1996, pp.63-74 (1996)
9. Polito C.P. & Martin J.R. « Effects of nonplastic fines on the liquefaction resistance of sands», Journal of Geotechnical and Geoenvironmental Engineering, Vol., 127, N°5, pp. 408-415(2001)
10. Tatsuoka F., Miura S., Yoshimi Y., Yasuda S., &Makihara Y. « Cyclic undrained triaxial strength of sand by a cooperative test program, Soils and foundations», 26, 117-128(1986b)
11. Hoque E. &Tatsuoka F. « Anisotropy in the elastic deformation of granular materials », Soils and Foudations, Vol.38, N° 1, pp. 163-179(1998)

Performance-based seismic assessment of vulnerability of dam using time history analysis

Oumnia ELMRABET[1*], Hasnae BOUBEL[1], El Mehdi Echebba[1], Mohamed ROUGUI[1] and Ouadia MOUHAT[1]
[1]University Mohammed V, LGCE High School of Technology- Salé, Rabat, Morocco

Abstract. The current performance-based seismic assessment procedure can be computationally intensive as it requires many time history analyses (THA) each requiring time intensive post-processing of results. Time history analysis is a part of structural analysis and is the calculation of the response of a structure to any earthquake. It is one of the main processes of structural design in regions where earthquakes are prevalent. The objective of this study is to evaluate the seismic performance of embankment dam located on the Oued RHISS in the Province of AL HOCEIMA using the THA method. To monitor structural behavior, the seismic vulnerability of structure is evaluated under real earthquake records with considering the soil-structure-fluide interaction. In this study, a simple assistant program is developed for implementing earthquake analyses of structure with ANSYS, ground acceleration–time history data are used for seismic analysis and dynamic numerical simulations were conducted to study and identify the total response of the soil-structure system.

Keywords. Time History Analysis, embankment dam, finite elements method, soil-structure interaction, nonlinear behavior, earthquake, ANSYS.

1 Introduction

The prediction of the dynamic behavior of dams under seismic loads is a very important issue in the evaluation of their safety in seismic zones. This essentially requires a rational knowledge of the seismic action acting on all the points of connection of the structure with the foundation soil.

The flexibility of soil modifies the dynamic behavior of the supported structure, which in turn generates interaction forces that disrupt the movement of the ground. This phenomenon, designated by soil-structure interaction [1], is a multidisciplinary field of research that remains active given the complexity of the phenomena concerned. The ongoing researches is aimed at solving a multitude of problems encountered both in practice and in numerical modeling.

Moreover, the problem of the application of seismic loading in soil-structure interaction constitutes another difficulty, considering that the incident seismic waves are physically defined in infinite medium and considered as incident plane waves coming from the focus of the earthquake Located in depth and therefore of vertical incidence. Two methods are used in this context; The methods by substructuring [2], applicable in the case of supposed linear behavior of the foundation soil and the global methods to treat the case where the foundation soil has a non-linear behavior [3]. In the latter case, part of the soil is modeled by finite elements, limited to its base by a horizontal surface where the seismic movement is applied. It is in this context that this article is organized.

First, a finite element 3D modeling [4] of the soil-structure system is presented. This modeling takes into account the effects of the non-linear behavior of the structure and the soil, as well as the effects of the SSI. Next, seismic records are loaded from PEER (Strong Motion Database) and seismic analysis files are produced in ANSYS Parametric Design Language (APDL) [5]. The damping value and the direction of the seismic load are introduced into the system. The modal analysis is performed, and the frequency value of the first vibration mode is recorded. The coefficients α and β for the Rayleigh damping are calculated using this value. Then, the time history analysis [6] is executed by applying each acceleration value to the model step by step.

2 Methods of calculation used

2.1 Global method: soil structure interaction

In the seismic calculation of structures, the excitation is defined by one or more acceleration records often measured at the surface of the ground in the absence of any construction. These records give the accelerograms in free fields. In the presence of a structure, the movements near the foundations may differ from those in the free field. Indeed, the forces at the contact between the structure and its foundation disturb the movement of the ground, and the moment of embedment at the base causes the rotation of the latter. This phenomenon is referred to as "Soil-Structure Interaction" or "SSI".

We are led to design models in which the ground is represented either by springs [2] or by finite elements or by other types of modeling allowing to reproduce its behavior.

The global method treats the sol-structure interaction problem as a whole: the numerical model is based on the spatial discretization of structure and soil. This method makes it possible to include in principle any nonlinear behavior of the soil or the structure and any heterogeneity present in the soil.

The matrix system resulting from the discretization of the complete problem is written:

$$M\ddot{u} + C\dot{u} + Fu = Q(t) \quad (1)$$

In equation (1), u, $\dot{u}$ and $\ddot{u}$ are respectively the unknown relative displacements, velocities and accelerations vectors, with respect to the model's base. M is the mass matrix of the soil-structure system, C the damping matrix and the nonlinear term F(u) reduces to Ku – where K is the stiffness matrix of the soil-structure system – in the linear case. Q(t) is a source term, and is equal to the inertial force –M1üg(t) induced by the ground seismic acceleration üg.

2.2 Time history analysis method

The direct integration of the dynamic motion equation of the Dam agitated from the foundation below applied to a finite element model avoids-although at the expense of increased computational effort the disadvantages of model analysis. Hereby the time history of a vibrating system can be analyzed for finite time steps (step-by-step) more suitably and non-linear soil behavior can be modeled.

Below same numerical methods of direct integration shall be discussed. Explicit and implicit integration methods. In the first case for the solutions of the differential equation, the system is considered at time t for the second type of analysis at time $t+\Delta t$ [7].

2.2.1 Method of central Differences

In this explicit method, the unknown displacements $u(t+\Delta t)$ are calculated from the state of equilibrium at time t. Velocity and acceleration are found from:

$$\dot{u}(t) = 0.5\frac{u(t+\Delta t) - u(t-\Delta t)}{\Delta t} \quad (2)$$

$$\ddot{u}(t) = \frac{u(t-\Delta t) - 2u(t) + u(t+\Delta t)}{(\Delta t)^2} \quad (3)$$

2.2.2 Houbolt's Method

Equilibrium at time $(t+\Delta t)$ is considered and the displacements $u(t+\Delta t)$ are found in dependence of $u(t), u(t-\Delta t)$ and $u(t-2\Delta t)$. Velocity and acceleration are given by:

$$\dot{u}(t+\Delta t) = \begin{bmatrix} 11u(t+\Delta t) - 18u(t) + 9u(t-\Delta t) - \\ 2u(t-2\Delta t) \end{bmatrix} / 6\Delta t \quad (4)$$

$$\ddot{u}(t+\Delta t) = \begin{bmatrix} 2u(t+\Delta t) - 5u(t) + 4u(t-\Delta t) - \\ u(t-2\Delta t) \end{bmatrix} / (\Delta t)^2 \quad (5)$$

2.2.3 Method after Wilson

This method of integration developed by Wilson is based on the assumption that acceleration changes in early during the time interval between t and $t + \theta\,\Delta t$.

$$\ddot{u}(t+\Delta t) = \ddot{u}(t) + \frac{\Delta \tau}{\theta . \Delta t}(\ddot{u}(t+\theta.\Delta t) - \ddot{u}(t));\ 0 \le \Delta\tau \le \theta.\Delta t \quad (6)$$

For reasons of computational stability θ must be greater than 1.37.

2.2.4 Newmark's Method

$$u(t+\Delta t) = u(t) + \dot{u}(t)\Delta t + \begin{pmatrix} (0.5-\beta)\ddot{u}(t) + \\ \beta\ddot{u}(t+\Delta t) \end{pmatrix} / (\Delta t)^2 \quad (7)$$

$$\dot{u}(t+\Delta t) = \dot{u}(t) + \begin{pmatrix} (1-\gamma)\ddot{u}(t) + \\ \gamma\ddot{u}(t+\Delta t) \end{pmatrix} / \Delta t \quad (8)$$

Stability can be guaranteed if:

$$\gamma \ge 0.5 \text{ and } \beta \ge 0.25.(0.5+\gamma)^2 \quad (9)$$

Dynamic full-time analysis - universally recognized as the one reproducing the best the behavior of a structure under seismic loads - can be successfully applied to large meshes due to the evolution of software performance of finite element calculations, but also of the machines on which they operate. However, it remains difficult to implement since it depends strongly on a coherent seismic input (series of accelerograms) [8], a correct choice of boundary conditions and constitutive laws adapted in the presence of unsaturated soil. The advantage over the replacement force method lies in the explicit consideration of the non-linearity of the materials and consequently the mobilization of the hidden reserves in the behavior coefficient q present in the calculation of the replacement forces.

3 Finite element computational model

The dam studied is located on OUED RHISS in the province of Hoceima, Morocco. With 84,5m in height and 430 m in width.

All civil engineering structures exhibit a certain degree of interaction or coupling with the surrounding environments. The soil-structure coupling is often treated using the finite element method which allows to treat the case of non-linear behavior.

https://doi.org/10.1051/matecconf/201814902035

In this paper, the dynamic analysis of the dam is performed using the finite element computational model for the three-dimensional dam. We note that the dam-reservoir interaction is considered. The mesh of the dam section is shown in Figure 1. The material properties used in this study are as follow in the table 1.

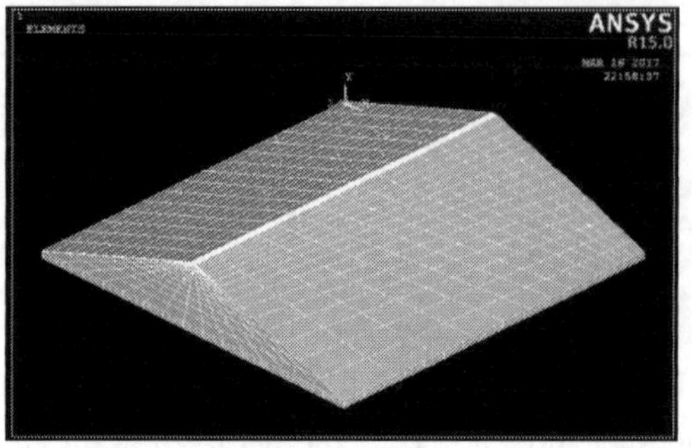

Fig. 1. Mesh of dam.

Table 1. Material properties.

Parameters	Elastic modulus (MPa)	Poisson's ratio	Density (t/m^3)
Foundation	1000	0,25	2,5
Upstream mask	75	0,3	2,3

4 Numerical results analysis

4.1 Modal analysis

4.1.1 Influence of soil-structure interaction on the variation of dam's natural periods and displacements

The analysis of free vibrations of the dam alone and of the dam-reservoir system is the subject of the first step of this study.

A modal analysis is made to follow the period's variation and displacements of the structure with and without interaction of foundation in order to understand its influence on the dam's response. Table 2 lists the values of the periods obtained for the first five modes of vibration.

Table 2. Modal properties of dam without foundation.

Mode	Frequency	Period	U_X	U_Y	U_Z
1	0,39	2,57	8,70E-09	1,19E-05	2,56E-05
2	0,42	2,38	-2,39E-09	-1,58E-05	-3,19E-05
3	0,44	2,27	1,26E-05	8,44E-08	1,91E-07
4	0,45	2,23	-2,11E-05	-4,50E-08	-5,73E-08
5	0,48	2,09	2,53E-05	-1,43E-07	-3,97E-07

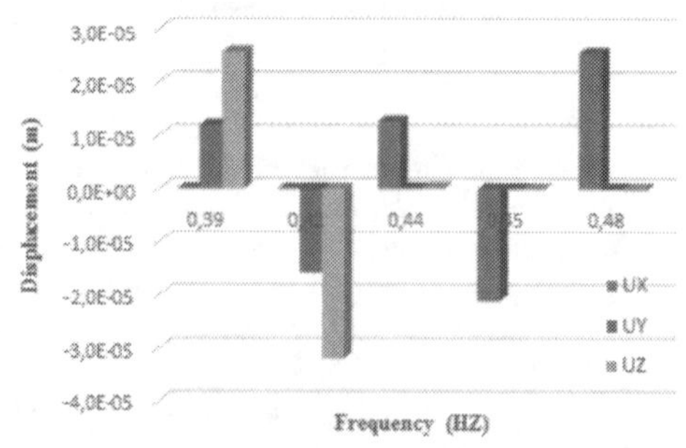

Fig. 2. Variation of displacements.

Table 3. Modal properties of dam with foundation.

Mode	Frequency	Period	U_X	U_Y	U_Z
1	8,07E-02	12,40	1,87E-16	-6,68E-16	-2,49E-06
2	0,10626	9,41	-2,31E-06	9,29E-06	9,10E-07
3	0,13081	7,64	1,07E-06	1,08E-05	2,49E-06
4	0,24669	4,05	-2,59E-06	-2,94E-06	9,69E-07
5	0,2485	4,02	3,28E-07	1,17E-06	7,20E-06

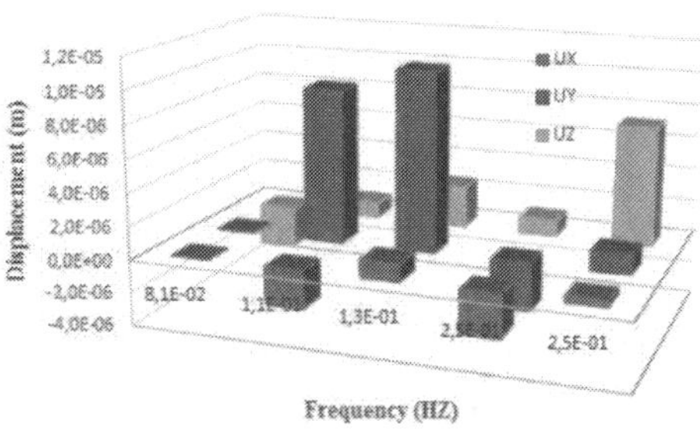

Fig. 3. Variation of displacements.

The interdependence of the mechanical behavior of the soil under the foundation and structure is called " soil-structure interaction." This phenomenon is greatly affected by the stiffness of the structure that depends on the construction materials, geometry and construction method.

The results of modal analysis show that the inclusion of soil-structure interaction is important in the analysis of the dynamic behavior of the dam because it has a remarkable influence in terms of increased natural periods of the structure and decreased displacements. This interaction is more or less important according to the nature of the soil, the characteristics of the structure and mode of foundation. It is in this context that the following section is based.

4.1.2 Influence of foundation's depth

We study the effect of foundation's depth on free vibration characteristics of the dam, we vary the depth as follows h2 >h1. The results obtained are shown in figure 4.

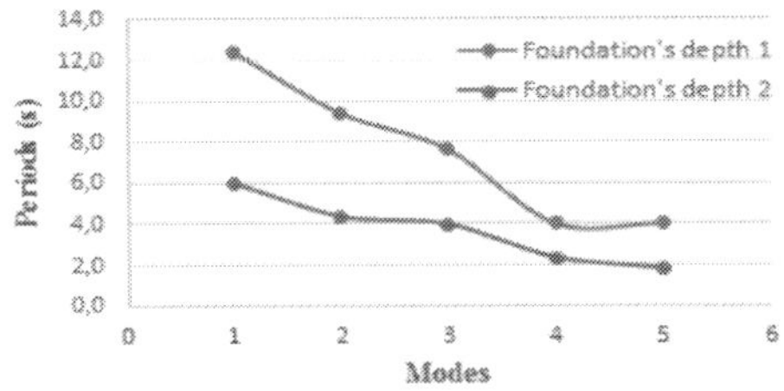

Fig. 4. Comparison of periods.

We have decreasing natural periods of the dam with the increasing depth of the foundation; the influence of the interaction appears significantly.

4.2 Dynamic time history analysis

Time history analysis is a part of structural analysis and is the calculation of the response of a structure to any earthquake. It is one of the main processes of structural design in regions where earthquakes are prevalent.

4.2.1 Seismic record

Al Hoceima region (northeast) experienced an earthquake on February 24, 2004, with a magnitude of 6.3 degrees on the Richter scale [9], causing human and material damage.

Morocco, located on a fracture zone, has already suffered deadly earthquakes. In this unstable region, earthquakes are caused by the rise of the African plate towards the European plate. The Atlas chains thus continue their overlap movements and from time to time the enormous accumulated tensions are released in the form of an earthquake

By exciting the dam by this earthquake (Figure 5), we observed its instantaneous responses. Node (N1933) at the dam crest is selected to represent the time history graphs of vertical movement at the dam crest.

The value of the damping ratio used in this study is 5%. The effects caused by the earthquake such as acceleration, displacements and velocity are summarized in graphs that are discussed in the next paragraph.

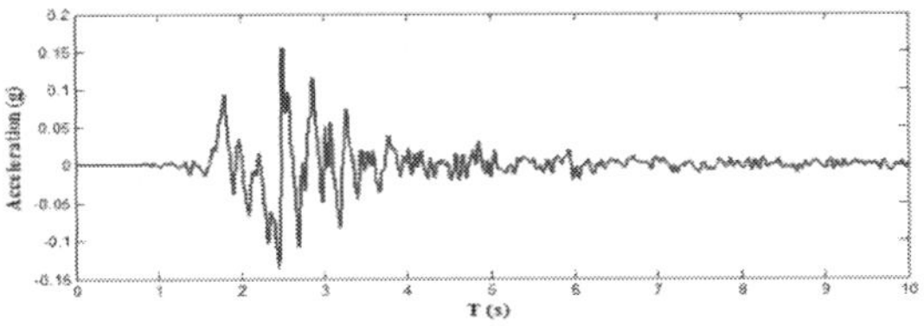

a) Acceleration.

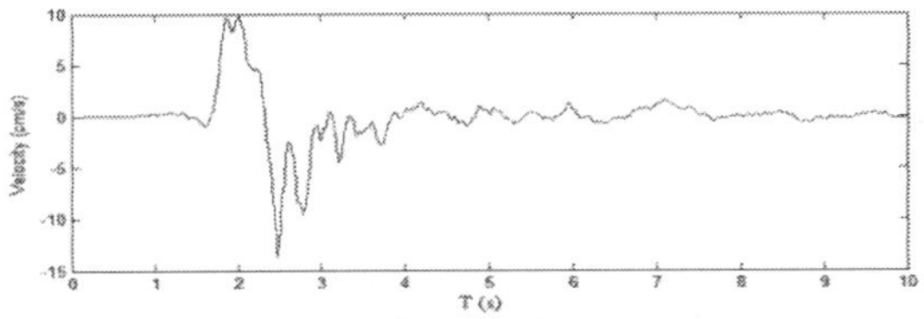

b) Velocity.

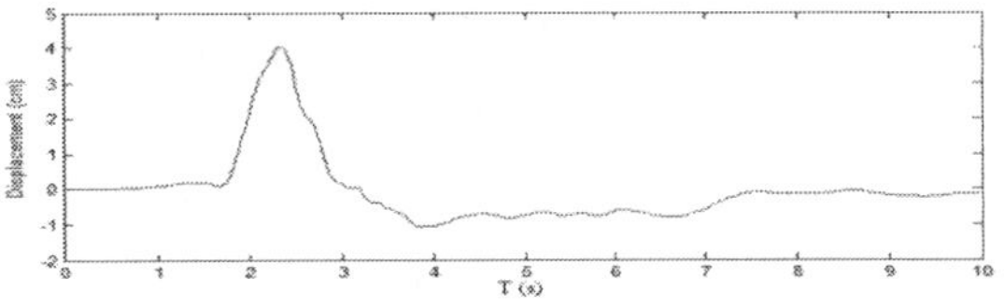

c) Displacement.

Fig. 5. AL HOCEIMA earthquake's record: Acceleration, velocity, displacement.

4.2.2 Analysis results

Figure 6 shows the variation of the acceleration and the velocity as a function of time for the node 1933 situated at the crest of the dam, and an acceleration amplification (amax = 1.4 g) is observed.

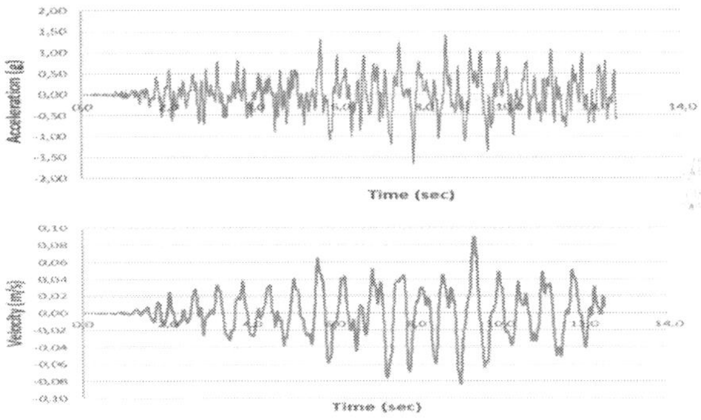

Fig. 6. Acceleration and velocity at the dam crest.

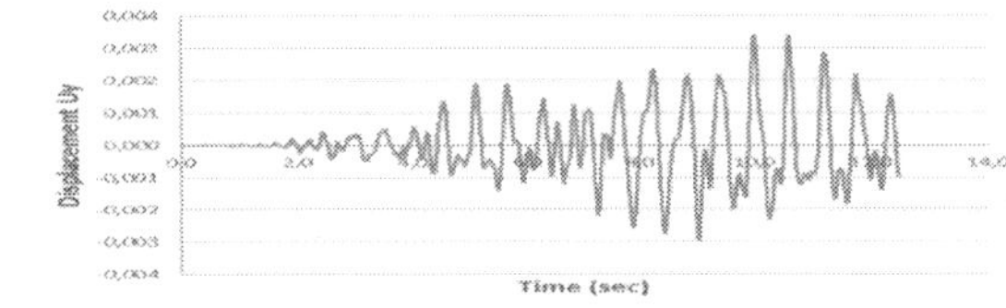

Fig. 7. Time history graphs of the vertical displacements.

Figure 7 show the time history graphs of the vertical displacements of the nodal point (1933) located at the crest of the dam with the considering of the interactions. The maximum value of displacement is 0,34cm.

The soil-structure interaction problem is solved for the embankment dam with a simplified, linear time history analysis, considering three models:

- Model 1: dam only, no foundation, no damping. Eq.1 reduces to :

$$M_{structure}\ddot{u} + K_{structure}u = - M_{structure}1\ddot{u}_g(t) \qquad (2)$$

- Model 2: dam only, no foundation, with damping. Eq. 1 reduces to :

$$M_{structure}\ddot{u} + C\dot{u} + K_{structure}u = - M_{structure}1\ddot{u}_g(t) \qquad (3)$$

- Model 3: dam with foundation (with $\gamma_{rock} = \gamma_{embankment}$ = 23 kN/m3), with damping. Eq. 1 can be rewritten:

$$M_{ot}\ddot{u} + C\dot{u} + K_{tot}u = - M_{tot}1\ddot{u}_g(t) \qquad (4)$$

 https://doi.org/10.1051/matecconf/201814902035

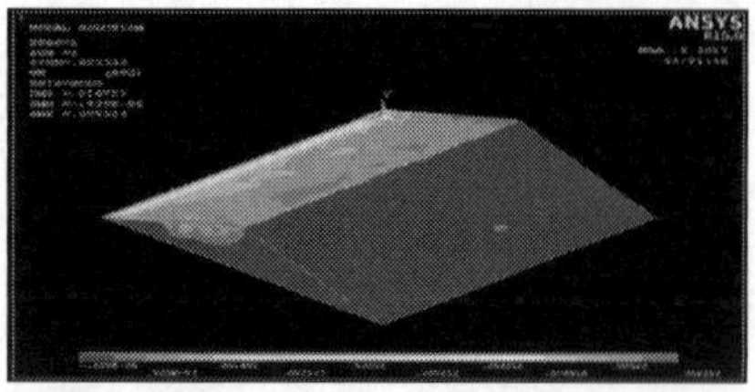

a) Model 1.

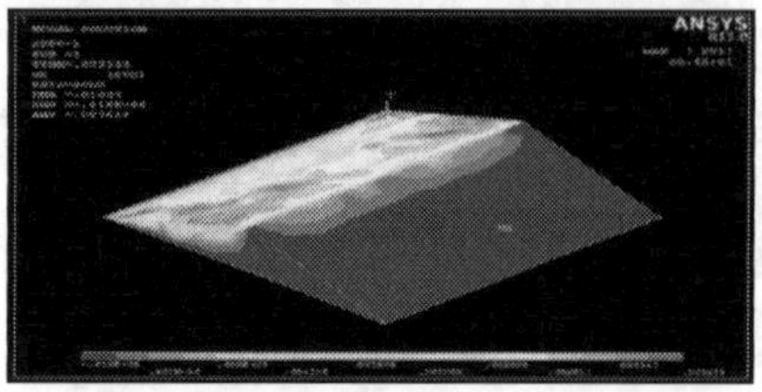

b) Model 2.

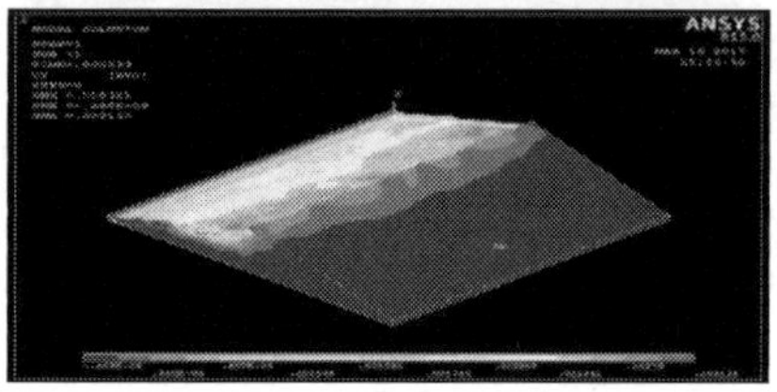

c) Model 3.

Fig. 8. The three considering models for introducing rigorously both the structure and the underlying rock foundation.

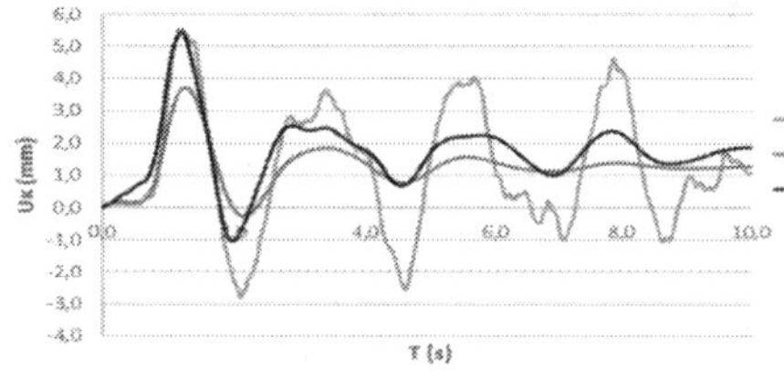

Fig. 9. Displacement time history of the node at the crest of dam.

The radial displacement time history Ux at dam's crest is illustrated in Figure 9. Model 2 (dam without foundation) gives a smaller maximal displacement with respect to Model 1, because of the introduction of the 5% damping.

A significant amplification of the signal can be noticed for Model 3 with respect to Model 2, which is due to the reflection of the waves on the classical boundary conditions. In order to limit this effect in the soil-structure interaction problem, one would have to extend the domain, or use transmitting boundaries, allowing the signal to leave the domain.

5 Conclusion

When a dam and its foundation are subjected to an earthquake, it generates an acceleration field which depends on the shape of the dam and the various constituent materials. Depending on the stiffness of materials and their internal damping capacity, depending on the height and shape of dam in relation to the earthquake spectrum, the response of dam will be more or less strong. This response is also influenced by the characteristics of the foundation and the restraint; Therefore, good consideration of soil-structure and fluid-structure interactions is essential. As in general terms, the characteristics of materials depend on the deformations to which they are subjected (non-linearities) and in particular, on shear deformations or distortions, it is generally necessary to study the response of the dam and its foundation to earthquake.

This paper studies the parametric finite element analysis of an embankment dam with ANSYS software and ANseismic programm. The aim of this work is to contribute to the study of the effects of soil-structure-fluid interactions on the seismic response of civil engineering structures, and more specifically to embankment dams. Another aspect of the work involves a comparative study of numerical models; It has been shown in this paper that when the foundation is rigid enough, simplifying hypotheses allow the use of classical displacement boundary conditions to solve an approximate problem (Model 2 is applicable). Otherwise, in the presence of soft soils, the full soil-structure interaction problem should be solved with the help of transmitting boundaries.

References

1. Seghir, *Contribution à la modélisation numérique de la réponse sismique des ouvrages avec interaction sol-structure et interaction fluide-structure- Application à l'étude des barrages poids en béton*, Université de Béjaia, Faculté de Technologie.
2. X. Zhang, *Modélisation physique et numérique des interactions sol-structure sous sollicitations dynamiques transverses,* Université de Grenoble (2011).
3. C. Liu, L. Zhang, B. Bai, J. Chen, J. Wang, *Nonlinear analysis of stress and strain for a clay core rock-fill dam with FEM.* Proc. Eng. **31** : 497–501, (2012).
4. M. Boumaïza, *Contribution à l'étude et à l'analyse du comportement des barrages en terre et de leur stabilité*, 76-77, (2010).
5. E. H. DILL, *The Finite Element Method for Mechanics of Solids with ANSYS Applications,* (2007).
6. M.A. Hariri-Ardebili, Y. Zarringhalam, H.E. Estekanchi, M. Yahyai, *Nonlinear seismic assessment of steel moment frames using time–history, incremental dynamic, and endurance time analysis methods* Sci. Iran. A (2013) **20** (3), 431–444.
7. K. J. Bathe and E. L. Wilson, *Numerical methods in finite element analysis,* Prentice-Hall, Englewood Cliffs, N. J., (1976), **528**.
8. E. Javelaud, Revue Française de Géotechnique, *Near-field displacement and rotations estimated from strong-motion accelerometers,* **146** (2016).
9. *La réglementation parasismique au Maroc – RPS 2011*, Royaume du Maroc, Ministère de l'Habitat, de l'Urbanisme et de l'Aménagement de l'Espace.

Assessment of Seismic Vulnerability of Reinforced Concrete Frame buildings

CHERIFI Fatiha[1], FARSI Mohammed[2], KACI Salah[1]
[1] University Mouloud MAMMERI, civil engineering department, Tizi-Ouzou, Algeria
[2] National Center of Applied Research in Earthquake Engineering (CGS), Algiers, Algeria

Abstract. The seismic activity remains strong in the north of Algeria since no less than 30 earthquakes per month are recorded. The large number of structures built before the introduction of the seismic standards represents a high seismic risk. Analysis of damage suffered during the last earthquakes highlighted the vulnerability of the existing structures. In this study the seismic behavior of the existing buildings in Tizi-Ouzou city, located in the north of Algeria, is investigated. To make this assessment, a database was created following a building inventory based on a set of technical folders and field visits. The listed buildings have been classified into different typologies. Only reinforced concrete frame buildings are considered in this paper. The approach adopted to estimate structures damage is based on four main steps: 1) construction of capacity curves using static nonlinear method "push-over", 2) estimate of seismic hazard, 3) determination of performance points, and finally 4) deduction of damage levels.

1 Introduction

Lors des tremblements de terre, les constructions subissent des oscillations horizontales, verticales et de torsion. Ces oscillations sont provoquées par des mouvements du sol d'assise qui imposent aux ouvrages des déformations. Contrairement aux oscillations verticales, les oscillations horizontales sont relativement mal supportées par les constructions, plus particulièrement lorsque celles-ci entrent en résonance avec le sol. Les oscillations de torsion, dues à la mauvaise conception des constructions, constituent un des phénomènes les plus destructeurs.

Lors des derniers séismes, il a été constaté que la quasi-totalité des effondrements, dans la ville de Tizi-Ouzou, concerne les structures en portiques auto-stables en béton armé (BA) ; ce qui a causé d'importants dégâts matériels et des pertes en vies humaines. La vulnérabilité de ces structures est en effet élevée à cause de la mauvaise qualité des matériaux et de la mauvaise réalisation.

Cet article traite de la démarche adoptée pour la construction des courbes de capacité de ces structures par utilisation de la méthode push-over, la détermination des points de performance, et la déduction des niveaux de dommages induits par des scenarii préalablement définis.

2 Typologie des bâtiments

Cette typologie s'appuie sur des critères structuraux relatifs au comportement au séisme du bâtiment. La classification des constructions est effectuée selon quatre paramètres prépondérants : l'année de construction (niveau de conception parasismique), le système constructif, le matériau, ainsi que le nombre de niveaux. L'analyse de l'échantillon obtenu (données recueillies et enquête) a fourni le classement des typologies les plus représentatives (environ 94%) du bâti de cette ville :

1. les maisons individuelles, de faibles hauteurs, constituées d'ossatures en béton armé, sans aucune conception parasismique (OBA0L : 46.54%) ;
2. les bâtiments de hauteur moyenne, constitués d'ossatures en béton armé, construits sans conception parasismique (OBA0M : 18.98%) ;
3. les structures en voiles en béton armé, de hauteur moyenne, à faible niveau de conception parasismique (MBA1M : 10.48%) ;
4. les structures en maçonnerie non renforcée, ne dépassant pas 02 niveaux (MNRL : 09.58%) ;
5. les structures mixtes contreventées par des portiques
6. et des voiles en béton armé, de hauteur moyenne, sans conception parasismique (SM0M : 05.24%) ;
7. les structures en voiles en béton armé, de hauteur moyenne, sans conception parasismique (MBA0M : 02.71%).

Notons que les premières règles parasismiques algériennes sont apparues en 1981 (RPA81). Ce

https://doi.org/10.1051/matecconf/201814902036

règlement a connu des révisions successives durant les années 1983 (RPA81 version 83), 1988 (RPA88), 1999 (RPA99) et 2003 (RPA99 version 2003) suite au séisme de Boumerdes du 21 mai 2003. Ces dates, qui définissent les principales périodes d'évolution des règles parasismiques algériennes, correspondent à différents niveaux de conception parasismique.

Les principales caractéristiques du parc immobilier de cette ville sont illustrées sur les Figures (de 1 à 3). Pour mieux visualiser la répartition des différentes typologies sur toute la ville, les résultats sont présentés sur la carte de la Figure 4.

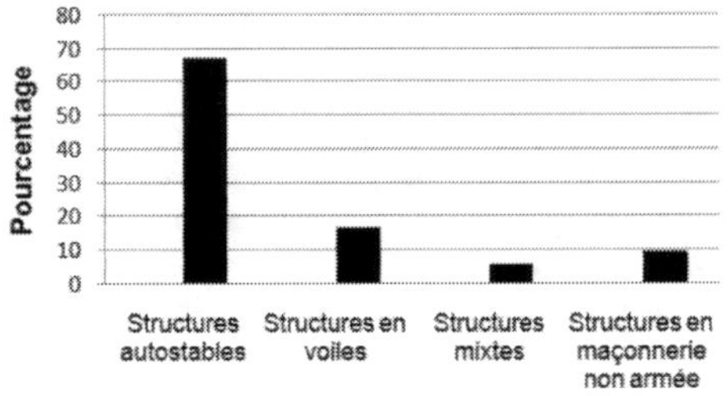

Fig.1. Classification selon le système constructif.

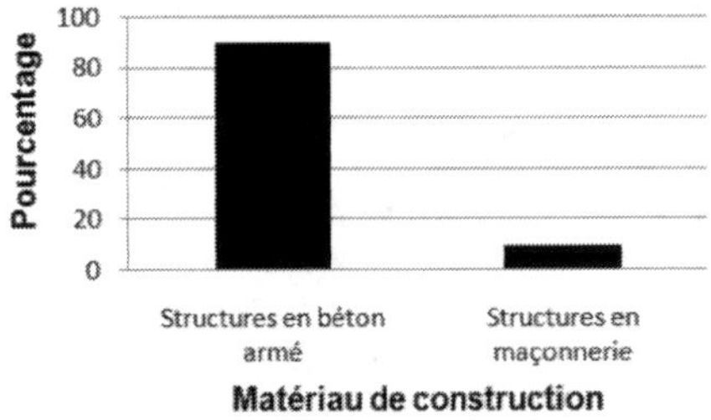

Fig. 2. Classification selon le matériau de construction.

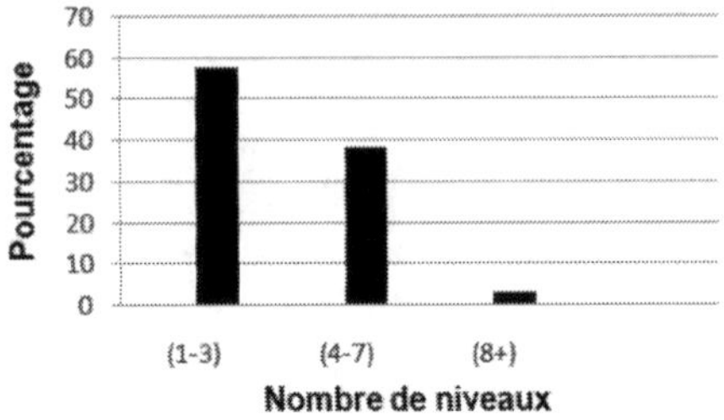

Fig. 3. Classification selon la hauteur du bâtiment.

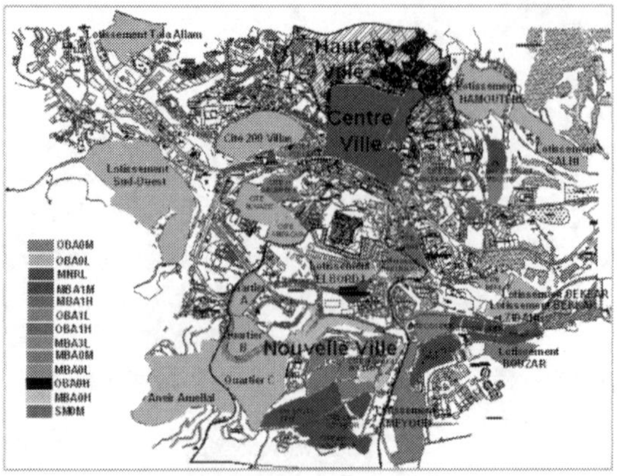

Fig. 4. Répartition des typologies les plus prédominantes de la ville de Tizi-Ouzou. Chaque couleur indique la typologie la plus prépondérante de la zone considérée. Seules les zones inventoriées sont illustrées sur la figure.

3 Hypothèses de calcul

Lors de l'analyse, la structure est considérée comme parfaitement encastrée à sa base, constituée de systèmes résistants reliés par des diaphragmes horizontaux rigides. Des rotules plastiques, introduites aux endroits où la plastification s'effectue de manière localisée, traduisent le comportement non linéaire des éléments. La courbe de capacité (effort à la base - déplacement au sommet) des structures est obtenue à partir d'un calcul statique non linéaire Push-over [1, 2].

L'état d'endommagement des rotules plastiques est fonction des niveaux de déformation [3]. Au niveau *IO*, les dommages structurels sont très limités, et la quasi-totalité de la rigidité initiale est conservée. Au niveau *LS*, la section subit des dommages importants, pouvant entraîner une perte significative de sa rigidité. Le niveau *CP* correspond à des dommages substantiels, au-delà desquelles la section est susceptible de se rompre. Notons que le point *B* correspond à l'apparition de la première rotule plastique.

Les données nécessaires à cette analyse sont résumées dans le Tableau 1.

Tableau 1. Principales données utilisées dans les calculs.

Types de données	Documents
Dimensions des éléments	Plans
Sections d'acier	Plans
Caractéristiques mécaniques du béton	Rapports d'écrasement
Caractéristiques des aciers	Notes de calcul
Charges et surcharges	Documents techniques réglementaires
Caractéristiques du sol	Rapports géotechniques

4 Probabilités d'endommagement

L'approche utilisée pour l'estimation des probabilités d'endommagement repose sur la méthode de capacité spectrale, qui se décompose en quatre étapes principales : la définition de la courbe de capacité, la définition du scénario sismique [4], l'obtention du point de performance (Fig. 5a), et la détermination des courbes de fragilité (Fig. 5b) [5].
Une fois les paramètres de la fonction de fragilité obtenus par la méthode HAZUS, l'on détermine et construit les courbes de fragilité par implémentation numérique de la méthode dans le logiciel MATLAB. Connaissant le déplacement maximal subi par un type de bâtiments, nous déduisons la probabilité d'atteindre ou de dépasser quatre niveaux de dommages (faible *S*, modéré *M*, important *E* et très important *C*) (Fig. 5c).

L'application de la méthode HAZUS nécessite que soit établie une correspondance entre les types de bâtiments considérés par cette méthode et les typologies définies pour la ville étudiée.

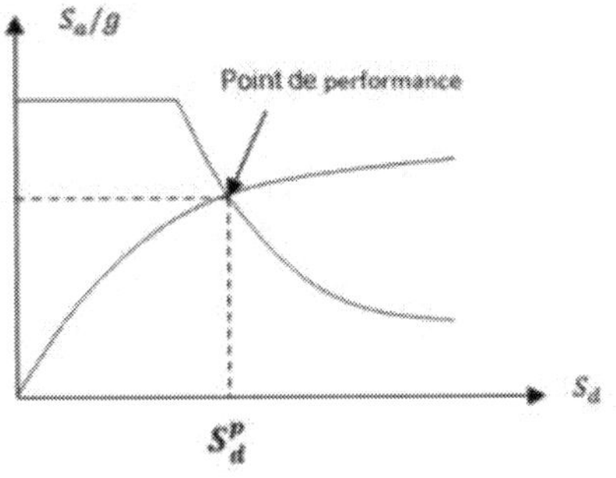

Fig. 5a. Définition du point de performance.

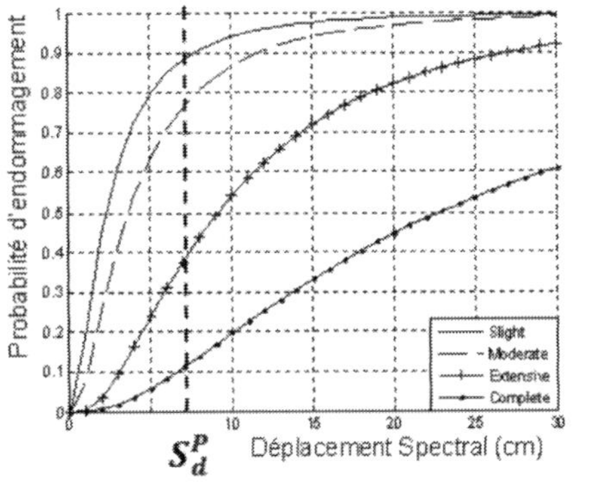

Fig. 5b. Courbes de fragilité selon la méthode HAZUS.

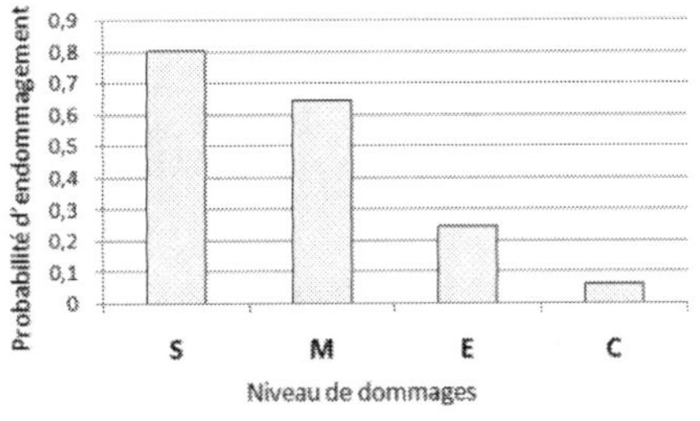

Fig. 5c. Estimation des probabilités d'endommagement.

5 Résultats

5.1 Points de performance

La méthode Push-over consiste à superposer la courbe de comportement de la structure (courbe de capacité) et la courbe de sollicitation du séisme (spectre de réponse) dans un même graphe de type (accélération spectrale S_a - déplacement spectral S_d). Dans ce plan, S_d est le maximum en déplacement de la réponse du modèle équivalent à un degré de liberté qui reproduit le mode désiré, généralement le mode fondamental de vibration de la structure, et S_a est le maximum en accélération de la réponse de ce même modèle.

L'intersection des deux courbes permet d'estimer le point de performance qui définit le déplacement maximal que peut avoir la structure, et donc de prédire les dommages induits. Les points de performance fournis par l'analyse Push-over, sont présentés dans les Tableaux (2 à 9). Pour chaque typologie, nous avons montré l'influence de la qualité du sol sur les points de performance et les niveaux de dommages y correspondant [6].

Tableau 2. Points de performance et niveaux de dommages – Typologie OBA3L.

Type de sol	Point de performance		Type de rotules plastiques
	Déplacement D_p *[m]*	Force V_p *[t]*	
Rocheux	0.048	225.57	**IO**
Ferme	0.070	265.26	**IO**
Meuble	0.087	279.84	**LS**
Très meuble	–	–	–
Nombre de niveaux : **(1-3)**			
Niveau de conception parasismique : **bon**			

Table 3. Points de performance et niveaux de dommages – Typologie OBA2L.

Type de sol	Point de performance		Type de rotules plastiques
	Déplacement D_p *[m]*	Force V_p *[t]*	
Rocheux	0.019	74.68	**IO**
Ferme	0.030	89.76	**LS**
Meuble	0.038	97.44	**LS**
Très meuble	–	–	–
Nombre de niveaux : **(1-3)**			
Niveau de conception parasismique : **moyen**			

Table 4. Points de performance et niveaux de dommages – Typologie OBA1L.

Type de sol	Point de performance		Type de rotules plastiques
	Déplacement D_p *[m]*	Force V_p *[t]*	
Rocheux	0.037	93.41	**IO**
Ferme	0.059	105.45	**LS**
Meuble	–	–	**–**
Très meuble	–	–	**–**
Nombre de niveaux : **(1-3)**			

Niveau de conception parasismique : **faible**

Tableau 5. Points de performance et niveaux de dommages – Typologie OBA0L.

Type de sol	Point de performance Déplacement D_p [m]	Point de performance Force V_p [t]	Type de rotules plastiques
Rocheux	0.064	49.3	**IO**
Ferme	–	–	–
Meuble	–	–	–
Très meuble	–	–	–
Nombre de niveaux : **(1-3)** Niveau de conception parasismique : **aucun**			

Table 6. Points de performance et niveaux de dommages – Typologie OBA3M.

Type de sol	Point de performance Déplacement D_p [m]	Point de performance Force V_p [t]	Type de rotules plastiques
Rocheux	0.051	150.34	**B**
Ferme	0.078	192.53	**IO**
Meuble	0.099	207.13	**LS**
Très meuble	–	–	–
Nombre de niveaux : **(4-7)** Niveau de conception parasismique : **bon**			

Table 7. Points de performance et niveaux de dommages – Typologie OBA2M.

Type de sol	Point de performance Déplacement D_p [m]	Point de performance Force V_p [t]	Type de rotules plastiques
Rocheux	0.089	65.64	**B**
Ferme	0.141	77.33	**IO**
Meuble	0.188	83.75	**LS**
Très meuble	–	–	–
Nombre de niveaux : **(4-7)** Niveau de conception parasismique : **moyen**			

Table 8. Points de performance et niveaux de dommages – Typologie OBA1M.

Type de sol	Point de performance Déplacement D_p [m]	Point de performance Force V_p [t]	Type de rotules plastiques
Rocheux	0.037	82.83	**B**
Ferme	0.058	109.42	**B**
Meuble	0.071	121.48	**IO**
Très meuble	0.116	134.31	**LS**
Nombre de niveaux : **(4-7)** Niveau de conception parasismique : **faible**			

Table 9. Points de performance et niveaux de dommages – Typologie OBA0M.

Type de sol	Point de performance Déplacement D_p [m]	Point de performance Force V_p [t]	Type de rotules plastiques
Rocheux	0.074	59.61	**IO**
Ferme	–	–	–
Meuble	–	–	–
Très meuble	–	–	–
Nombre de niveaux : **(4-7)** Niveau de conception parasismique : **aucun**			

5.2 Dommages

La probabilité d'atteindre ou de dépasser le niveau *C*, où les dommages sont très importants pouvant aller jusqu'à la ruine, atteint 6.0% pour les structures de faible hauteur et ne dépasse pas 10.5% pour les structures de moyenne et de grande hauteur. Ces valeurs sont considérablement réduites pour les bâtiments ayant un bon niveau de conception parasismique.

La probabilité de dépasser le niveau *E*, où les dommages sont importants, est inférieur à 25% pour les structures de faible hauteur et peut atteindre 40% pour le reste des typologies.

La probabilité de dépasser le niveau modéré *M* de dommages atteint environ 88%.

La probabilité de dépasser le niveau faible *S* de dommages varie de 51% (typologies à bon niveau de conception parasismique) à plus de 97% (typologies à niveau moyen de conception parasismique).

5 Conclusion

Les principaux résultats obtenus suite à cette analyse montrent que les bâtiments construits après 2003 présentent une faible vulnérabilité. Les structures de faible niveau de conception parasismique perdent environ 50% de leurs rigidités initiales et subissent le maximum de dommages, pouvant entrainer l'instabilité et la ruine de la structure. Les courbes de capacité obtenues pour les structures construites avant 1981, sans aucune conception parasismique, sont nettement inférieures à la demande sismique et les points de performance sont inexistants. Ainsi, le déplacement engendré par le séisme est supérieur à la capacité de la structure. Il s'ensuit alors l'effondrement partiel ou total de ces structures.

Dans la plupart des structures construites avant 2003, les rotules plastiques apparaissent simultanément dans les poutres et dans les poteaux. L'état d'endommagement des rotules plastiques est plus avancé dans les poteaux et l'effondrement de la structure survient par rupture des poteaux des premiers niveaux. Notons que les sections de ces poteaux sont relativement faibles ; ce qui peut induire une augmentation des efforts et des contraintes et par conséquent des déformations, conduisant à la ruine de ces éléments.

La ville de Tizi-Ouzou dans sa globalité présente une vulnérabilité sismique qui peut être qualifiée de faible pour les structures construites après 2003 et moyenne à importante pour les structures construites avant 2003. Des dégâts structuraux importants peuvent être subis notamment par les structures en ossatures en béton armé sans aucun niveau de conception parasismique.

Références

1. ATC-40, « *Seismic Evaluation and Retrofit of Concrete Buildings* », Applied Technology Council, Redwood City, California, (1996).
2. KRAWINKLER H., « *Procedure and construction of pushover analysis of seismic performance evaluation engineering structure* », vol: 20, edition: Elsever science, Department of civil engineering standford university, U.S.A, (1998).
3. Branson D. E., 《 *Deformation of concrete structures*》, New York (NY): McGraw-Hill, (1977).
4. D.T.R.-B-C.-2.48, « *Règles parasismiques algériennes RPA99 version 2003* », C.G.S., Alger, (2003).
5. HAZUS, « *Earthquake loss estimation methodology* », Technical manuals, National Institute of Buildings Science, Federal Emergency Management Agency (FEMA), Washington, (1999).
6. CHERIFI F., FARSI M.-N., KACI S. « *Évaluation de la vulnérabilité sismique des bâtiments existants en Algérie* », Thèse de Doctorat en génie-civil, Tizi-Ouzou, Algérie, (2016).

Soil-fluid-structure interaction applied to the Oued Taht dam (taking into account the membrane effect).

Krenich Nasreddine[1], Tahar Berrrabah Amina[1], Houmadi Youcef[1], Belharizi Mohamed[2], Mehdeli Mohamed[1], Mohammedi Mohamed el Amine[1]
[1]Centre universitaire Belhadj Bouchaib, Laboratoire des structures Intelligentes, Département de Génie Civil, Ain Témouchent, Algérie.
[2] 37 Impasse Armand, 92160 Antony, France

Abstract. The objective of this work is to analyze the dynamic behavior (modal behavior) of the "Oued Taht" arch dam located at MASCARA, taking into consideration the effect of soil-fluid-structure interaction. The finite element code "Ansys" was chosen for the dynamic modeling of the dam that is the subject of this study. Three hypotheses were used for soil-structure interaction modeling; model with embedded base which corresponds to the case where the phenomenon of interactions soil-structure is neglected, model with ground of foundation without mass which consists in taking into account the kinematic component of interaction soil structure and neglecting the inertial component and the model with foundation soil with mass where the two components of soil-structure interaction are taken into account. For the fluid, the model of added masses (equivalent to the westergaard approach) using the SURF element available in the Ansys code library was used. A comparison between the different models of the "Oued Taht" dam was made; it has been found that the taking into account of the soil-fluid-structure interaction phenomenon modifies the period of the system and that the modeling of the dam with and without fluid gives a very important difference of the periods. The results obtained were compared with those of the "Brezina" dam, which is a gravity dam located in BAYADH. The work has shown that the periods of the "Oued Taht" dam with soil-fluid-structure interaction modeling are very out of phase with the periods without fluid modeling (taking into account only the soil-structure interaction phenomenon). which is not the case for the Brezina dam where the periods for the two models are getting closer. The periods between the two models mentioned before are close to the dam of Brézina because the latter is a dam which participates much more by its own weight than by its vault (thickness of the vault varies between 36.3 m at the base and 5m in crest) which is the opposite for the dam "Oued Taht" which participates by its vault (constant thickness of 7 m) thus the membrane effect is present, which is translated by the shift of the periods between the empty case and the filled case.

1 Introduction

La réponse d'une structure durant un séisme dépend des caractéristiques du signale sismique, du sol de fondation et de la structure elle-même. Le mouvement de la fondation est généralement différent du mouvement en champ libre (la difference du mouvement dépend de la différence de rigidité entre le sol et la fondation), cela est dû au couplage sol-structure durant le séisme, ce phénomène est connu dans la littérature sous le nom « d'interaction sol-structure ». Dans les structures où le domaine fluide existe, le phénomène devient plus compliqué; il s'agit «d'interaction sol-fluide-structure », les barrages en béton font partie de ce phénomène.

L'abondance de la littérature concernant l'interaction sol-structure traduit à la fois la complexité du phénomène et l'intérêt qu'y ont porté nombre de chercheurs [1], [2], [3].

On considère l'interaction sol-structure dans le but d'inclure dans les modes de déformations, les mouvements à la base de la structure, ainsi que pour obtenir une meilleure approximation de la période de vibration du système couplé qui sera plus long que la période du système avec base rigide. Dans la littérature, ils existent trois modèles pour la modélisation du phénomène d'interaction sol-structure: le modèle de la structure à base encastré, le modèle de la structure avec fondation modélisée sans tenir en compte sa masse, et le modèle de la structure avec fondation tenant en compte sa masse.

Par rapport à la modélisation d'interaction fluide-structure, on peut citer trois approches: approche de Westergaard, l'approche de Lagrange et celle d'Euler.

L'objectif du present article est d'effectuer une étude paramétrique pour voir l'effet du phénomène d'interaction sol-fluide-structure sur le comportement du barrage voûte « Oued TAHT » don't l'existence de l'effet membrane. Pour la modélisation des différents cas du barrage en questions, le code en éléments finis ANSYS a été choisi.

2 Présentation du cas d'étude

Barrage Oued TAHT est un barrage voûte en béton, en cours d'exécution, il se situé dans la wilaya de Mascara en Algérie. Les caractéristiques mécaniques du corps du barrage et du sol de fondation ont été données dans le tableau 1.

Tableau 1. Propriétés des matériaux de barrage Oued TAHT

Matériau	Module de young N/m2	Coefficient de poisson	Densité Kg/m2
Corps de barrage	28.5e+09	0.25	2500
Sol de fondation	6.22e+09	0.25	2100

3 Modélisation ansys d'interaction sol-fluide-structure du barrage oued taht

L'effet d'interaction sol-fluide-structure sur le comportement du barrage Oued Taht a été modélisé et étudié utilisant le code ANSYS. Les résultats ont été discutés ci-dessous.

3.1 Modélisation de l'interaction sol-structure

La première application consiste à modéliser l'interaction sol structure, considérant le barrage vide, à cet effet trois hypothèses ont été adoptées :

Barrage à base fixe (encastrement).

Barrage avec sol de fondation, tenant compte de la masse de cette fondation et par conséquent de la composante inertielle de l'interaction.

Barrage avec sol de fondation, négligeant la masse de cette fondation et par conséquent de la composante inertielle de l'interaction.

3.1.1 Modélisation en éléments finis du barrage

L'élément type solid186 [4] disponible dans la bibliothèque du code ANSYS a été utilisé pour modéliser le corps du barrage ainsi que le sol de fondation.

La figure 1 représente le modèle du barrage à base encastrée (c à d sans tenir en compte le phénomène d'interaction sol-structure).

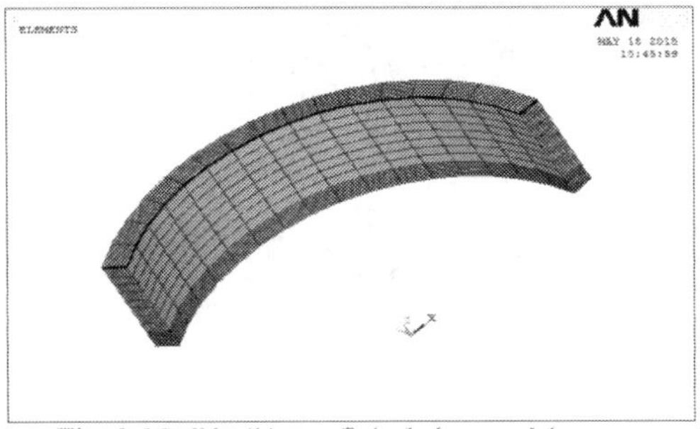

Fig. 1. Modèle élément finis du barrage à base encastré

Fig. 2. Modèle élément finis du barrage avec sol de fondation.

La figure 2 représente le modèle du barrage avec sol de fondation; les dimensions du sol suivant x, y et z ont été choisies après une étude de sensibilité du modèle aux dimensions du sol. Les extensions du sol de fondation sont comme suit :

Suivant x : 50 m de part et d'autre des extrémités du corps de barrage

Suivant y : 176 m de côté aval et amont de l'ouvrage

Suivant z: une extension de -100 m a été choisie,

Il faut signaler que les extrémités du système ainsi que sa base sont encastrées.

3.1.2 Effet d'interaction sol structure sur le comportement modal du barrage Oued Taht

La figure 3, 4 et 5 schématise respectivement le mode fondamentale du barrage seul (à base encastré), barrage avec sol sans masse (non pesant) et barrage avec sol avec masse (pesant).

Fig.3. Schéma du mode fondamentale du barrage seul (mode 2)

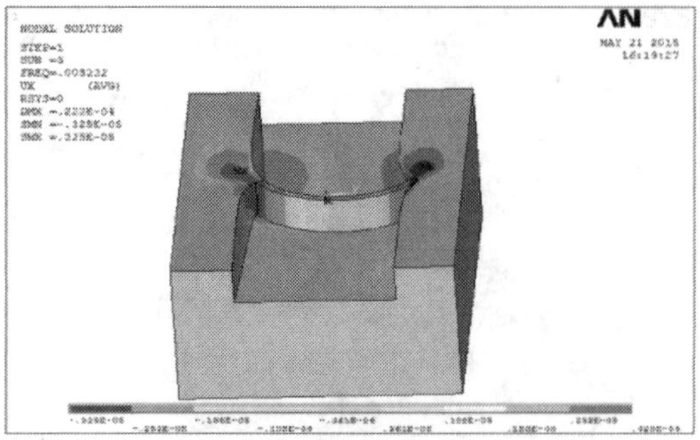

Fig.4. Schéma du mode fondamentale du barrage avec sol sans masse (mode 3)

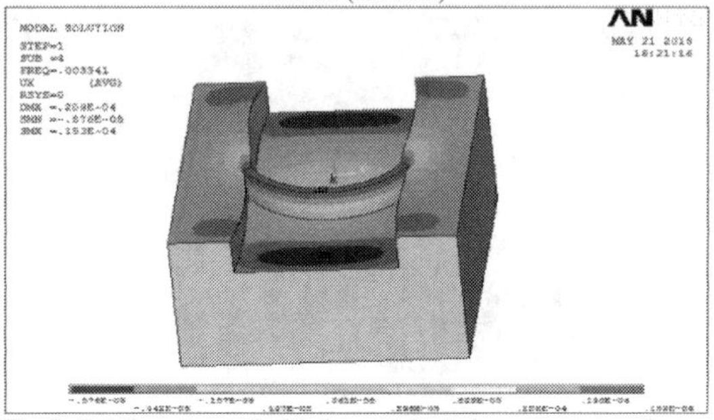

Fig.5. Schéma du mode fondamentale du barrage avec sol avec masse (mode 4)

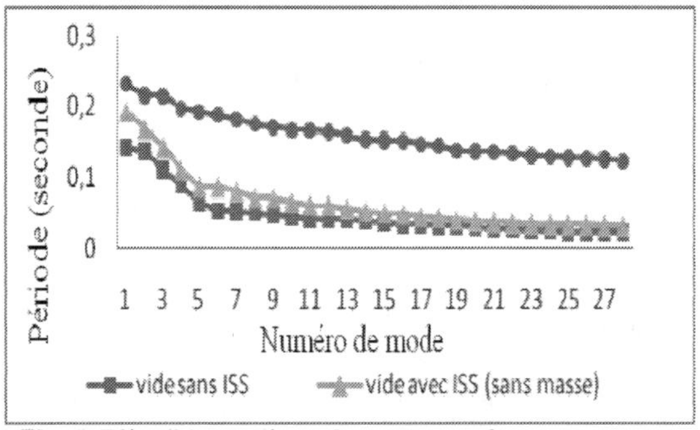

Fig. 6. Effet d' interaction sol-structure sur le comportement modal de barrage Oued Taht

La figure 6 représente une comparaison entre les premières périodes pour chaque modèle, il est à noter que le nombre de modes est choisi de telle façon que la somme des masses entrainées par ces modes atteints les 90% de la masse totale. Il est clair sur la figure 6 que la période du barrage augmente lorsque le sol de fondation est pris en compte, beaucoup plus lorsque le sol de fondation est modélisé avec sa masse. Les même résultats ont été trouvé pour le barrage poids voûte de el Bayadh [5].

Ajouter le sol de fondation à la structure, rend le système flexible et augmente sa masse ce qui engendre une diminution de la fréquence de système et par conséquent une augmentation de sa période.

Suivant les trois figures: 3, 4 et 5, on remarque que la modélisation du sol de fondation modifie non seulement les valeurs des périodes mais aussi le positionnement du mode fondamentale (sachant que le mode fondamentale est le mode qui entraine le maximum de masse). Les mêmes résultats sont trouvés pour le barrage d'el Bayadh [5], [6].

3.2 Modélisation de l'interaction fluide structure

La deuxième application consiste à modéliser l'interaction fluide structure, à cet effet le barrage est considéré plein. Pour chaque modèle d'interaction sol-structure, deux analyses ont été mené avec et sans fluide. L'approche des masses ajoutées a été utilisée pour modéliser le fluide, En utilisant le code ANSYS, l'élément SURF154 disponible dans la bibliothèque ANSYS permet de représenter cette approche. Il s'agit de répartir la masse du fluide sur la surface de contact entre le réservoir et le barrage et/ou sol de fondation.

L'épaisseur de l'élément surf 154 est donnée par l'équation 1:

Epaisseur d'élément SURF154 = $(B/A) \times \rho$ (1)

A=surface de contacte fluide - barrage et/ou fondation (m2)

B=volume d'eau de réservoir (m3)

ρ =masse volumique d'eau (Kg/ m3)

3.2.1 Effet d'interaction fluide structure sur le comportement modal du barrage oued Taht

Les figures 7, 8 et 9 schématisent respectivement le mode fondamentale du barrage seul (à base encastrée), barrage avec sol sans masse et barrage avec sol avec masse tenant en compte la présence du fluide.

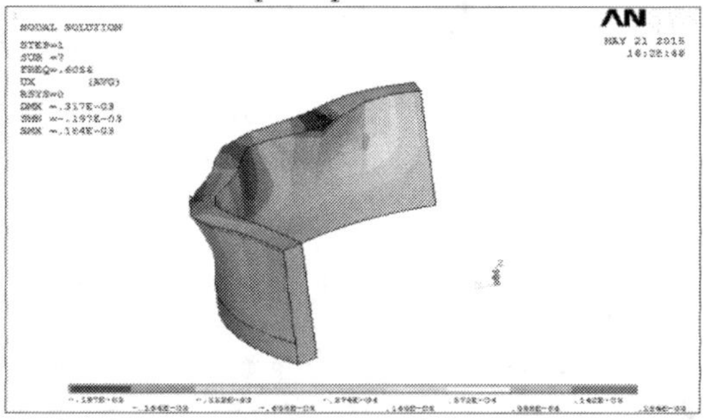

Fig.7. Schéma du mode fondamentale du barrage seul rempli (mode 7)

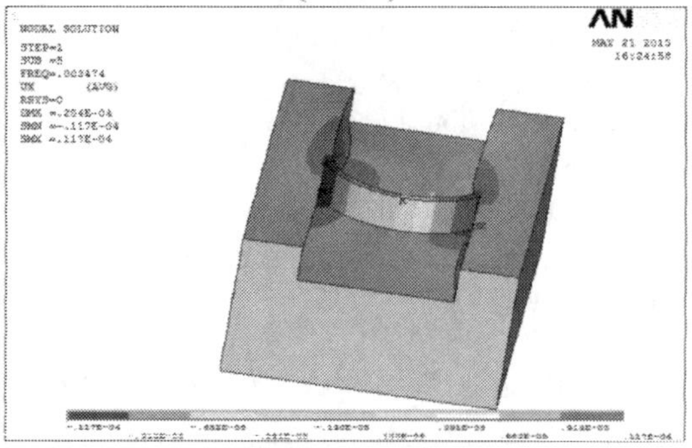

Fig. 8. Schéma du mode fondamentale du barrage avec sol sans masse rempli (mode 3)

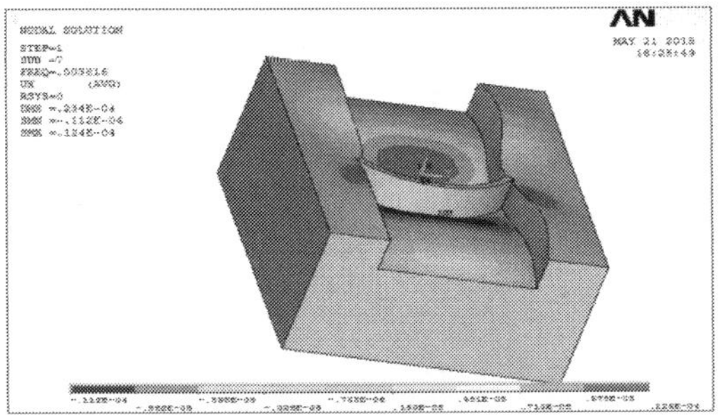

Fig. 9. Schéma du mode fondamentale du barrage avec sol avec masse rempli (mode 5)

3.2.1.1 Barrage seul rempli

La figure 10 présente l'effet de la modélisation d'interaction fluide structure sur le comportement modal du barrage seul c'est à dire sans modélisation d'interaction sol-structure.

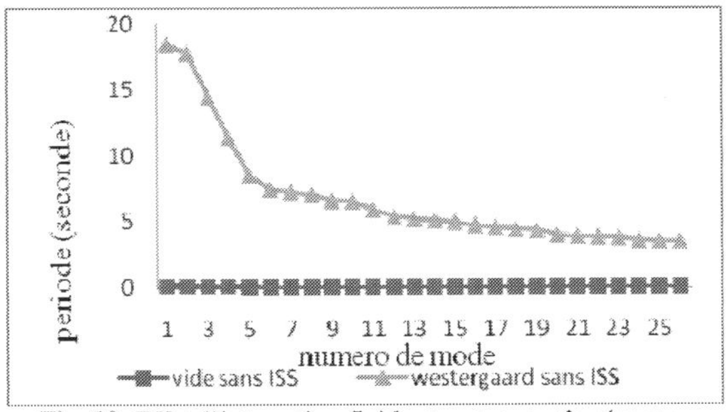

Fig. 10. Effet d'interaction fluide-structure sur la réponse modal du barrage seul (sans interaction sol-structure)

Il est clair sur la figure 10 que la présence du fluide augmente la période du système.

3.2.1.2 Le barrage avec sol sans masse (rempli)

Les mêmes résultats que ceux du barrage à base fixe ont été trouvés pour le cas du barrage modélisé avec sol de fondation sans masse.

3.2.1.3 Le barrage avec sol avec masse (rempli)

Les mêmes résultats que ceux du barrage à base fixe et du barrage modélisé avec sol de fondation sans masse ont été trouvés pour le cas du barrage modélisé avec sol de fondation sans masse.
Il a été constaté que la prise en compte du fluide augmente la période de la structure, c'est les mêmes résultats trouvés pour le barrage à base encastré et le barrage avec sol sans masse. Les mêmes résultats sont trouvés pour le barrage d'el Bayadh [7], la seul différence c'est que la présence du fluide pour ce barrage à un effet faible sur son comportement modale, cela est dû à la géométrie du barrage, c'est un barrage poids voûte, alors que pour le cas de barrage oued Taht, est un barrage voûte, ce qui implique l'effet membrane est présent dans son comportement. L'effet membrane se traduit par la différence de période entre le cas vide et le cas rempli.

3.3 Comportement modal du barrage tenant en considération l'interaction sol-fluide- structure

La figure 11 représente l'effet d'interaction sol-structure et de la présence du fluide sur le comportement modale du barrage Oued Taht, elle montre que la modélisation d'ISS ainsi que celle du fluide ont le même effet sur le comportement modal du barrage, c'est d'augmenter la période du système.

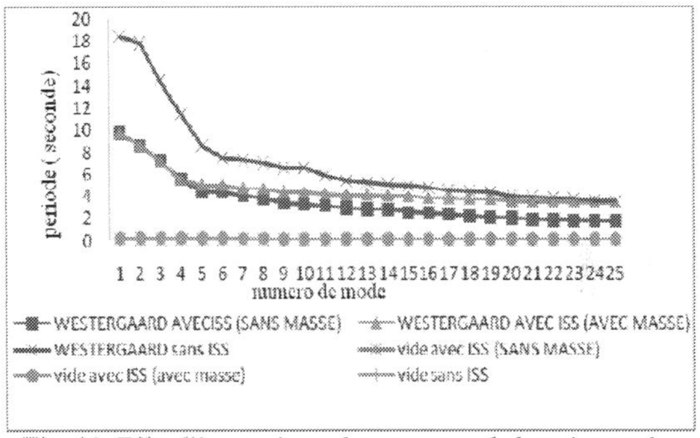

Fig. 11. Effet d'interaction sol-structure et de la présence du fluide sur le comportement modale du barrage Oued Taht

4 Conclusion

Dans cet article, une étude paramétrique a été menée pour le barrage voûte « Oued Taht » utilisant le code élément finis « ANSYS » pour voir l'effet du phénomène d'interaction sol-fluide-structure sur le comportement modale du système. L'analyse modale a permet de conclure que :
La modélisation d'interaction sol structure, modifie la période du barrage, sachant que la période est une caractéristique dynamique à prendre en considération dans les différents types de calcul dynamique (spectrale, transitoire…)
La modélisation d'Interaction fluide structure, modifie aussi la période du barrage.
La modélisation d'interaction sol structure et fluide structure donne une augmentation des périodes du système.
La grande différence de période entre le cas vide et le cas rempli confirme l'effet membrane du barrage Oued Taht, ce qui est n'est pas le cas pour le barrage de Brezina.

Bibliographie

1. LYSMER J. Analytical procedures soil dynamics-state of the ASCE-conference on soil dynamics and earthquake engineering (1978).
2. IDRISS J.M and KENNEDY *R.P.analysis for soil-structure interaction effects for nuclear power*

plants-report (by the AD HOC group, soil structure interaction of the committee on nuclear structures and materials of the structural division of ASCE, 1980).

3. ALAIN PEKER. *Dynamique des sols* (presses de l'école nationale des ponts et chaussées).
4. MANUEL. ANSYS Version **11.0** (2007).
5. A. TAHAR BERRABAH *al.* Three-Dimensional Modal Analysis of Brezina Concrete Arch Dam, *Algeria, Earth Science Research; Volume. 1, No.* **2**. *ISSN 1927-0542 E-ISSN 1927-0550.* (Published by Canadian Center of Science and Education, 2012).
6. A. TAHAR BERRABAH al. Dynamic Soil Structure Interaction Study. *(Jordan Journal of Civil Engineering, Volume 6, No.* **2**, 2012).
7. A. TAHAR BERRABAH al. Fluid–structure interaction of Brezina arch dam: (3D modal analysis, Engineering Structures **84**, 2015)19–28.

Influence of Prior Distributions and Fragility assessment methods in the estimation of the Magnitude of a Historical Seismic Event

Eduardo Charters MORAIS[1], László Gergely VIGH[1] and János KRÄHLING[2]
[1]BME, Department of Structural Engineering, Budapest, Hungary
[2]BME, Department of History of Architecture and Monument Preservation, Budapest, Hungary

Abstract. The production of fragility functions describing the probable behaviour and damage on historical buildings is a key step in a method for the estimation of the magnitude of historical seismic events that uses a Bayes'. The fragilities are estimated by integrating the structural capacity with the seismic demand using either static methods, as the Capacity Spectrum Method (CSM), or dynamic methods, as Incremental Dynamic (IDA) and Multiple Stripes Analysis (MSA). Uncertainties in both resistance, demand, and distance and magnitude models propagate to the posterior magnitude distribution. The present paper studies the effect of uncertainties related both to the production of fragility functions and prior distributions, in the estimation of the magnitude of the 1763 Komárom earthquake (in historical Hungary). In the XVIII century most of the structures in the region were built of earth, adobe, clay or stone masonry, which is complex to model. While micro or detailed macro-modelling strategies are computationally costly, simplified macro-approaches are often more efficient, but require a pre-identification of the failure mode(s) and the determination of the backbone curve. For this study, a simplified macro-model of a Hungarian peasant house archetype is calibrated for CSM and IDA. The physical and geometrical uncertainties are incorporated in the fragilities using Monte-Carlo simulation. Prior magnitude and distance distributions are studied. The final magnitude estimates are presented and discussed.

1 Introduction

The most common methods for the estimation of the magnitude of historical seismic events [1, 2] belong to the domain of seismology, using interdisciplinary knowledge of both geophysicists and historians [3]. Another method applies the background of Probabilistic Seismic Hazard Analysis (PSHA) and uses historical seismic records of damage to translate the nonlinear structural damage into fragility curves [4], obtained either by dynamic methods as Incremental Dynamic and Multiples Stripes Analysis (IDA & MSA) [5] or by static methods as the Capacity Spectrum Method (CSM) [6]. These methods integrate the structural capacity with the seismic demand – usually correlated – and their respective uncertainties, to produce the damage points for fragility function fitting. The magnitude estimation method [4] combines the historical data of the seismic damage with the fragility functions to produce the likelihood of a seismic event (*E*). Then, it applies the total probability and Bayes' theorems using an attenuation model, or ground motion prediction equation (GMPE), and prior distance (r or R_{ep}) and magnitude (m or M_w) distributions – details in subsection 2.3. As a result, both capacity and demand related uncertainties, the GMPE(s) and prior m and r distributions contribute to the reliability of the final magnitude estimates.

The present paper focus both on the impact fragility functions, for historical buildings affected by the 1763 Komárom (nowadays Komárno, Slovakia) earthquake – one of the most destructive in Hungarian history –, and on the weight of assumptions in the final magnitude estimates. Therefore, the modelling tools, their efficiency, as well as the main uncertainty sources in the magnitude estimation method are described and studied. A general overview of the methodologies in study is provided in section 2. In section 3 the empirical capacity curves are studied for a set of historical adobe unreinforced masonry wall panels from a one-storey peasant house archetype [7, 8]. A simplified macro-model that uses a multilinear spring [9] is calibrated for IDA, incorporating physical uncertainties associated with the mechanical parameters defined in literature [10-13], using Monte-Carlo simulation (MCS). CSM is also here implemented with the help of MCS on graphic based approach. In section 4 the weight of the fragilities, GMPE and prior distributions parameters is balanced. The methodologies and results are discussed, and conclusions presented.

2 Methodologies

The methodology presented here is interdisciplinary, requiring the interpretation of historical sources of

https://doi.org/10.1051/matecconf/201814902038

damage and structural data related to the 1763 Komárom earthquake, as well as features of the dynamic modelling and analysis of the historical structures.

2.1. The 1763 Komárom earthquake

The 1763 Komárom earthquake, in the 28th of June, was one of the most destructive in historical Hungary. Although, the estimation of its magnitude is questioned by distinct estimates, in between 5.70 and 6.50 [1, 2, 13]. Its destructive effect led to the production of a vast quantity of documents and depictions, as that in figure 1. In the aftermath, out of 1169 houses in total, 279 completely collapsed, 353 partially collapsed, 213 needed expensive repair and 219 cheap repair [2]. Historical sources identify the most damaged buildings as two storey rigid buildings, made of burnt clay bricks, rather than simple and more flexible peasant houses, made of earthen or adobe masonry. The resulting depictions also helped to identify common failure modes as shear failure [11] enabling the simplification of the modelling work.

Furthermore, the interpretation of the damage sources resulted on the need for the systematized generation of historical building archetypes representing the regional features of XVIII c. West of Hungary [7]. This happened mainly due to the lack damage descriptions of existing representative buildings. Therefore, Archetypes were generated for different building categories, the most representative being tax-payer's houses, as the peasant house [8] represented in figure 2:

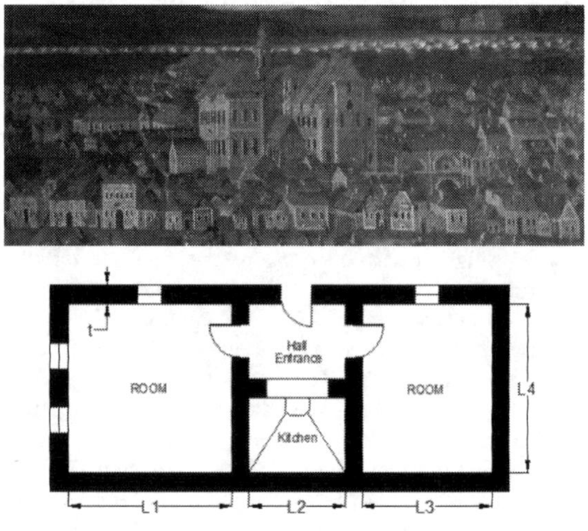

Fig. 1. Anonymous depiction of the damage in the city of Komárom after the 1763 earthquake (up) & plan of a XVIII century Hungarian peasant house archetype, from [5].

2.2. Fragility functions

The fragility functions are produced using dynamic or static structural analysis methods such as IDA and CSM and describe the conditional probability of occurrence of a certain damage measure *DM* implying one or more damage states ds_i ($DM \rightarrow ds_i$) given a set of parameters θ – an intensity measure *im* (may be PGA, PGV, $S_a(T)$), *m* and *r*. They are usually assumed as lognormally fitted CDF functions and described in terms of intensity *im*, with medians μ_{dsi} and standard deviations β_{dsi}:

$$P(ds_i|\theta) = P(ds_i|im,m,r) = \Phi(\ \ln(x/\mu_{dsi})/\beta_{dsi}\ ) \qquad (1)$$

2.3. The Magnitude estimation method

The magnitude estimation method is described in detail in [4] and its flowchart is presented in figure 2:

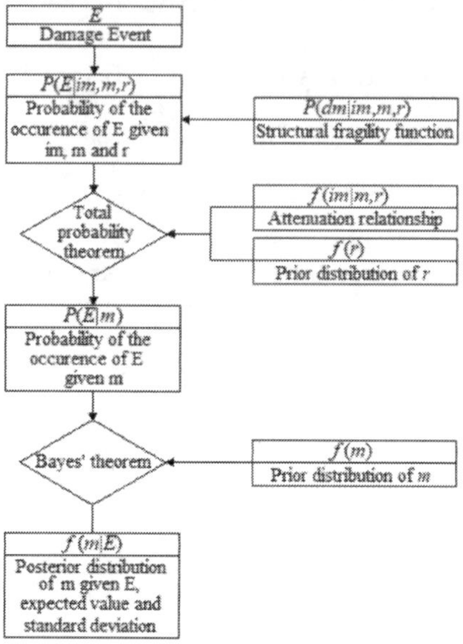

Fig. 2. Flowchart of the magnitude estimation method, from [4].

The original form of the problem consists of stating an occurred seismic event *E*, characterized by an intensity measure *im, or* and *m* and *r*, which caused damage in structures, represented by a damage measure *DM*. So, '*E* caused *DM*' implies here a certain probability of occurrence $P(E|im,m,r)$ described by the combination of fragility functions in the form $P(DM|im,m,r)$ with the number of structures n_{dm} under the damage measures *DM*. Afterwards, the total probability theorem enables to integrate the conditional probability $P(E|im,m,r)$ with the attenuation relationship $f(im|m,r)$ and prior distribution $f(r)$, in order to calculate the conditional probability of *E* given *m*, $P(E|m)$. Then, the posterior distribution $f(m|E)$ is achieved using the Bayes' theorem together with a prior distribution $f(m)$. These operations may be synthesized by expressions (1) and (2), respectively:

$$P(E|m) = \iint P(E|im,m,r) \times f(im|m,r) \times f(r)\ dim\ dr \qquad (2)$$

$$f(m|E) = P(E|m) \times f(m)\ /\ \int P(E|m) \times f(m)\ dm \qquad (3)$$

Despite the use of dynamic methods, applying CSM directly with the help of a GMPE leads to the *im* as a function of *m* and *r*, in the form $im(m,r)$, resulting that the conditional probability of *E* may assume the form $P(E|m,r)$. Thus expression (2) could be simplified:

$$P(E|m) = \int P(E|m,r) \times f(r)\ dr \qquad (4)$$

The direct use of the *m* and *r* inputs via GMPE describes a probability surface of the type $P(E|m,r)$. It is also an economic alternative to directly select the *m* values associated with damage states, if the *r* values are sampled from the prior *f*(*r*), simplifying the numerical integration, resulting on the sole use of expression (3).

3 Static and dynamic based analysis and fragilities

The adopted modelling and uncertainty incorporation strategies and analysis approaches CSM and IDA are here described. Deeper considerations and details about the methods can be found elsewhere [5, 6, 11, 14].

3.1. Complex *vs* simplified modelling

The modelling of historical buildings is a complex issue. The increase of computation capacity has allowed the production of increasingly sophisticated modelling tools, since the 1990's. Different finite and discrete element (FEM and DEM) based approaches, enabled multi-level modelling strategies: detailed micro-, simplified micro- and detailed macro-modelling [15]. Although the computation cost decreases from micro to macro strategies, it is still costly to produce detailed macro-models capable of incorporating at the same time, and using dynamic time-history analysis – IDA or MSA, uncertainties in seismic demand and structural capacity. An alternative is to use a simplified macro-modelling strategy with an SDOF system adapted to simulate the in-plane behaviour of a wall panel by calibration of a nonlinear spring [9] (figure 3, left). In turn, the *F-δ* diagram may be achieved by predicting the F-δ backbone [16] or by using more detailed modelling-levels [15]. This strategy enables to study uncertainties in both structural capacity – as function of compressive strength f_m, Young modulus E_m, internal friction μ and other geometrical parameters θ_G, $R(f_m,E_m,\mu,\theta_G)$ – and the seismic demand – as function of *m*, *r* and other parameters related to fault and soil types θ_{SF}, $D(m,r,\theta_{SF})$.

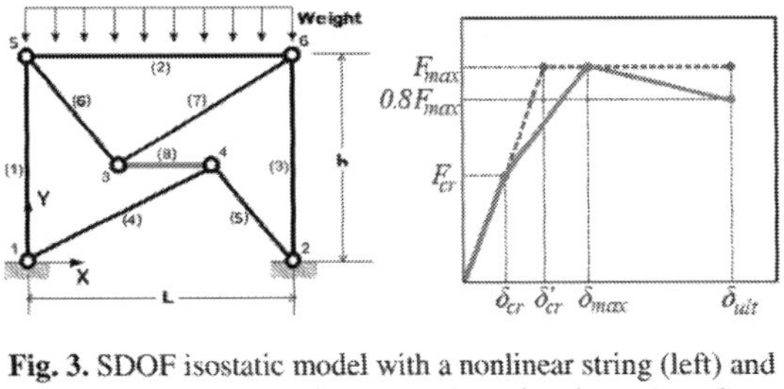

Fig. 3. SDOF isostatic model with a nonlinear string (left) and respective bi- and tri-linear models (right), from [11, 16].

3.2. CSM based fragilities

The CSM involves both the determination of the pushover based capacity curve and demand spectrum, translated in terms of spectral displacements (S_d) and spectral accelerations (S_a). The intersection gives the performance point (figure 4). Details are elsewhere [6, 11]. For the analysis, the median values E_m=0.209 MPa, μ=0.45 and T_{ss}=1.70 MPa [16] are adopted with lognormal distribution, with 25%, 19% and 25% of covariance, respectively. Damage is modelled based on [17]. The final fragilities, presented in table 1. In previous developments, the description of the fragilities has been affected by the *m* and *r* value ranges of interest, resulting that ds_1 resulted to be well described in detriment of ds_2, ds_3 and ds_4 [17]. The previous code has been improved in order to achieve more complete descriptions.

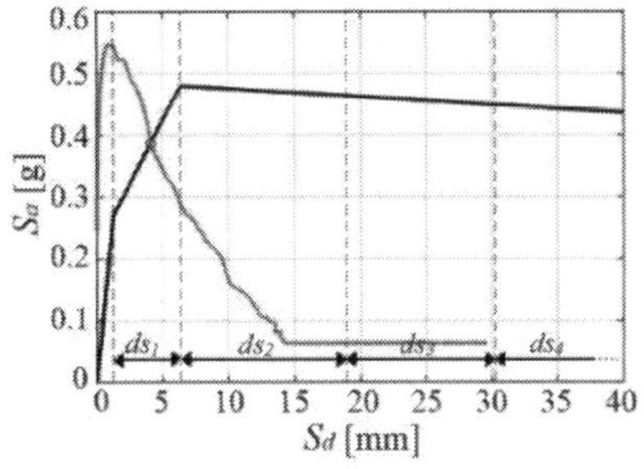

Fig. 4. CSM intersection of structural capacity with the seismic demand in the S_a-S_d axis, based on [11].

Table 1. Fragility moments mean and standard deviation PGA's, in [g], for the walls 1-to-6, from CSM.

Wall	1	2	3	4	5	6
μ_{ds1}	0.21	0.21	0.44	0.23	0.24	0.33
β_{ds1}	0.22	0.37	0.16	0.27	0.28	0.22
μ_{ds2}	0.37	0.38	0.83	0.42	0.43	0.57
β_{ds2}	0.18	0.09	0.21	0.14	0.14	0.16
μ_{ds3}	0.51	0.48	1.26	0.55	0.57	0.81
β_{ds3}	0.08	0.12	0.04	0.08	0.08	0.08
μ_{ds4}	0.69	0.72	1.61	0.84	0.87	1.19
β_{ds4}	0.15	0.15	0.13	0.19	0.19	0.17

3.3. IDA based fragilities

The main idea of IDA is the scaling of the selected *im* values, in the present case peak ground acceleration (PGA), until the collapse of the structure (figure 5). The analysis was carried out with the same tri-linear model and formulation, but using 30 seismic records selected for the region of Komárom [5].

The final IDA based fragilities, presented in table 2, incorporate not only the physical uncertainties, as CSM application in sub-section 3.2, but also the uncertainty related to the frequency and duration associated to the seismic records. Which is not featured by CSM.

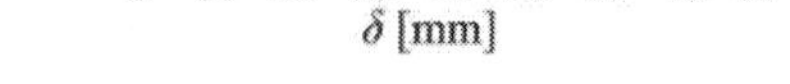

Fig. 5. IDA results for wall 4, based on [11].

Table 2. Fragility moments mean and standard deviation PGA's, in [g], for the wall panels 1-to-6, from IDA.

Wall	1	2	3	4	5	6
μ_{dm1}	0.23	0.19	0.18	0.21	0.20	0.21
β_{dm1}	0.34	0.26	0.24	0.23	0.22	0.22
μ_{dm2}	0.37	0.57	0.30	0.43	0.42	0.41
β_{dm2}	0.26	0.30	0.22	0.19	0.19	0.18
μ_{dm3}	0.68	0.69	1.00	0.70	0.70	0.83
β_{dm3}	0.20	0.32	0.39	0.29	0.30	0.36
μ_{dm4}	0.76	0.91	1.29	0.87	0.88	1.02
β_{dm4}	0.25	0.35	0.50	0.36	0.36	0.41

3.4. FEMA HAZUS based fragilities

FEMA Hazus technical manual [10] makes use of CSM and prescribe *F-δ* diagrams for low and mid-rise buildings (URML & URM, respectively). Therefore, the fragility medians are determined by the *R-D* intersection. The standard deviation is calculated using expression (5).

$$\beta_{ds} = (\mathrm{CONV}[\beta_C,\beta_D]^2 + \beta_{T,ds}^{\ 2})^{1/2} \quad (5)$$

Where β_{ds} is the lognormal standard deviation that describes the total variability for structural damage state, ds_i, β_C=0.30 and is associated with the capacity, β_D=0.50 and is associated with the demand, and $\beta_{T,ds}$=0.40 is associated with the damage state threshold. Assuming the previous values from [10] the total variability β_{ds}=0.43 was achieved. The median values are described with the help of a scaled demand $D(m,r,\theta_{SF})$ given by the attenuation from [18], averaged for the intervals of interest $m\in[5,8]$ and $r\in[0,20]$, with strike-slip fault type and soft soil. The results are presented in table 3.

The elements that transit from CSM and IDA to the magnitude estimation are the fragilities. Despite the small differences evidenced here between CSM and IDA, it is a question how much uncertainty may be reduced with the use of more advanced simulation methods. Other studies suggest that model uncertainties may raise to 0.50 on force side and 0.90 on deformation side [11]. FEMA P695 [12] provides ratings and uncertainty values for different sources: design-requirements, test data and modelling related uncertainties. These are sorted from β_{Mod}=0.10 to 0.50, from high to low 'completeness and robustness', and from high to poor 'confidence in basis of design requirements', but these many sources may not guarantee improvements of the final magnitude estimates.

Table 3. Fragility medians in terms of PGA [g] using FEMA HAZUS suggested capacities for damage states ds_i.

-	ds_1	ds_2	ds_3	ds_4
Low-rise	0.21	0.35	0.80	1.12
Mid-rise	0.19	0.26	0.56	1.10

4 Magnitude estimates

The reliability of the posterior magnitude distribution and estimates depends not only on the fragilities but also on the quality and completeness of the historical data and of the prior distributions and attenuation relationship(s). In this section the impact of the fragilities, the priors and the historical data on structures is evaluated.

4.1. Sensitivity study

The calibration example, developed in [4], consisted on the application of the method to the 1994 Northridge (WES-US) instrumental earthquake using damage data on buildings associated to the FEMA HAZUS fragilities [10]. The code was utilized to study the input parameters and the sensitivity of the method, with input variables and uncertainty sources in: (1) the fragility moments' μ_{dsi} and β_{dsi} (2) the GMPE's μ_{IM} and β_{IM}, (3) the historical records based *ds_i data*, (4) the prior distance *data*, *f(r) type* and *bounds*, and (5) the prior magnitude *data*, *f(m) type* and *bounds*. For immediate purpose of studying the first two topics, the prior *f(m)* is maintained as uniform, the prior *f(r)* is tested in the interval 0 to 20 km, the ds_i array [105 44 42 35 27], as in [19], and the fragilities in table 3, corresponding to a low-rise building are used. The results show a plateau in the magnitude estimate μ_M and respective standard deviation β_M, associated to β_{dsi} and β_{IM} values bigger than 0.3 and 0.5, respectively (figure 6). Both μ_M and β_M are less sensitive to β_{dsi} than to β_{IM}, as the sensitivity factors of 0.11 and 0.99 were approximately estimated in regards to μ_M, and 0.33 and 0.99 in regards to β_M. This suggests that the ground motion uncertainties, rather than the fragility related ones, have a considerable effect on the uncertainty in the magnitude estimates.

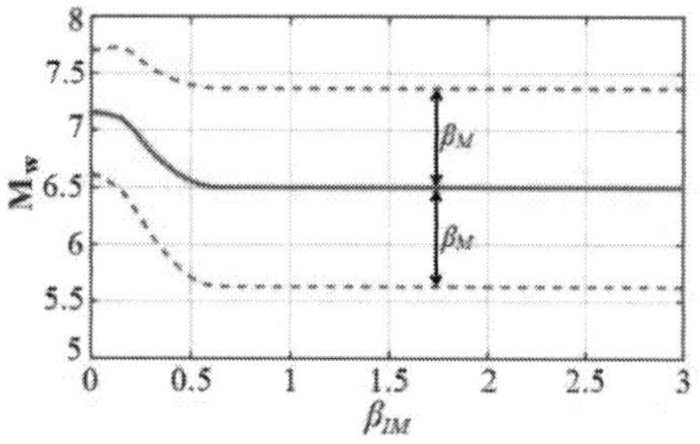

Fig. 6. Variation of $\mu_M \pm\beta_M$ with varying β_{IM}.

4.2. Prior magnitude and distance distributions

In order to achieve the prior distribution of magnitude *f(m)*, historical earthquakes occurred in the regions of Mór, Győr and Komárno (Komárom) between 1615 and 1886, were selected [13]. Their magnitudes were computed as normally distributed with means ranging between 3.8 and 5.7 and standard deviations between 0.33 and 0.50, resulting in a lognormal prior *f(m)*.

Afterwards, the magnitude data was truncated for the range of interest – M_w from 5 to 8 – creating a truncated lognormal prior *f(m)*. In figure 7, six possible *m* priors are shown. They are adapted for the region affected by the 1763 event. The latter were fitted for historical seismic events in the ranges 3.8≤M_w≤5.7 for f_m^{II} and 5.0≤M_w≤5.7 for f_m^{III}. These are lognormally distributed and defined by the moments 4.07±0.17 and 5.41±0.06, respectively. The prior f_m^{I} has a uniform distribution in 5.0≤M_w≤8.0 and the f_m^{IV} to f_m^{VI} priors are described in section 4.3. The suggestions for prior distribution of distance $f(r)$ from [4] are studied together with the sensitivity of the magnitude estimates to the selection of two sets of bounds prior data suggests, in the ranges 0 to 20 km, and 10 and 12 km [2], from Komárom. The priors f_r^I to f_r^{III} are represented in figure 8 in the range 0 to 20 km.

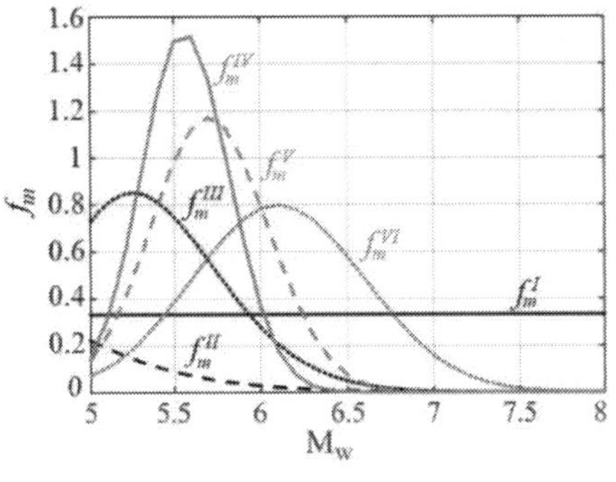

Fig. 7. Prior distributions of magnitude in the range of interest (M_w between 5 and 8).

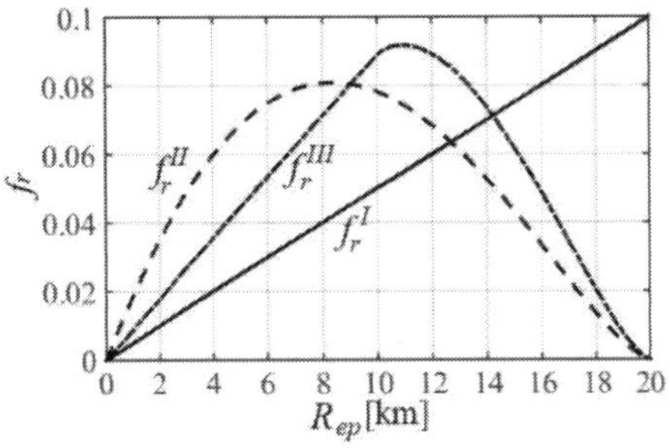

Fig. 8. Prior distributions of distance in the range of interest (R_{ep} between 0 and 20 km), based on [4].

4.3. Comparison of magnitude estimates

The fragilities in 3.2 and 3.3 from CSM and IDA were averaged in order to re-establish new fragility moments, resulting in the values in table 4. Afterwards, in order to compare the magnitude estimates for the different sets of fragilities, damage data and prior distributions, a strategy was set, firstly to compare with previous estimations [19] (table 5, Archetype), secondly, to test the damage descriptions stating the resilience of smaller adobe houses rather than bigger clay buildings (tables 5, respectively), and finally to draw a comparison between the Archetype CSM and IDA and the FEMA HAZUS based fragilities (in table 6). In the last group the R_{ep} range is restricted and three more normally distributed magnitude priors, representing previous magnitude estimations for the 1763 event, are studied: f_m^{IV}=5.56± 0.26, f_m^{V}=5.70±0.34 and f_m^{VI}=6.10±0.50 [13] (figure 7).

Table 4. Averaged fragility moments μ_{LN} and β_{LN}, in terms of PGA [g], for the peasant house walls 1-to-6.

-	ds_1	ds_2	ds_3	ds_4
μ_{LN}	0.23	0.44	0.65	0.95
β_{LN}	0.35	0.31	0.41	0.38

Table 5. Expected $\mu_M \pm \beta_M$ for distance $0 \leq R_{ep} \leq 20$, magnitude priors f_m^I to f_m^{III}, and ds_i= [105 44 42 35 27] for the Archetype (A0), ds_i= [105 219 213 0 0] for Low-rise and ds_i= [105 0 0 353 279] for Mid-rise unreinforced masonry (URML & URM).

-	-	f_r^I	f_r^{II}	f_r^{III}
A0	f_m^I	6.81 ±0.73	6.64 ±0.81	6.69 ±0.78
	f_m^{II}	5.72 ±0.51	5.57 ±0.45	5.63 ±0.46
	f_m^{III}	5.60 ±0.30	5.54 ±0.28	5.57 ±0.28
URML	f_m^I	6.72 ±0.76	6.59 ±0.83	6.62 ±0.81
	f_m^{II}	5.66 ±0.48	5.53 ±0.43	5.57 ±0.44
	f_m^{III}	5.59 ±0.29	5.52 ±0.28	5.54 ±0.28
URM	f_m^I	7.15 ±0.55	7.09 ±0.58	7.15 ±0.55
	f_m^{II}	6.33 ±0.46	6.27 ±0.44	6.33 ±0.46
	f_m^{III}	5.94 ±0.21	5.93 ±0.21	5.94 ±0.28

Table 6. Expected $\mu_M \pm \beta_M$ for distance $10 \leq R_{ep} \leq 12$ and magnitude priors, and ds_i= [105 219 213 353 279].

-	f_m^I	f_m^{II}	f_m^{III}	f_m^{IV}	f_m^{V}	f_m^{VI}
μ_M	7.11	6.55	6.30	6.24	6.34	6.57
β_M	±0.52	±0.34	±0.14	±0.11	±0.15	±0.29

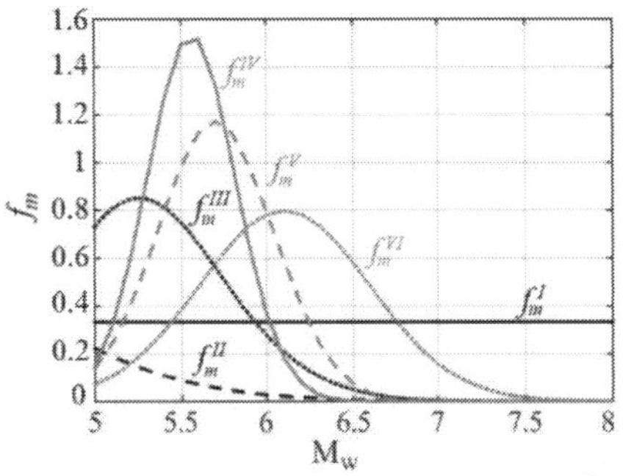

Fig. 9. Posterior magnitude distributions for table 8, considering the priors f_m^I to f_m^{VI}.

4.4. Discussion of the results

The sensitivity study in subsection 4.1 suggest the relevance of β_{IM} rather than β_{dsi} for both μ_M and β_M values, but no effect for β_{IM} bigger than approximately 0.60. According to subsection 4.3, β_M may be reduced by effect of the magnitude priors, as shown by the use of f_m^{III} to f_m^{V} (tables 5 to 8). For the studied R_{ep} ranges, the use of different f_r distributions has a small effect on results, less than 2% on both μ_M and β_M. Otherwise, the application of the priors f_m^{II} to f_m^{VI} highly affects the μ_M values, if compared to the uniform prior f_m^I. A possible reading is that the damage based estimation points to high μ_M values, as can be seen in tables 5 and 6 first rows, while prior information on the seismicity of the region contributes to lower this value. This contribution can be seen as an influence of the median values and inversely proportional to the standard deviation (figure 9).

Additionally, the selected GMPE [18] with m=8.0 and r=0 km produces a PGA=0.42±0.28 g. Consequently, when the likelihood $P(E|im,m,r)$ of the event is estimated, the probability associated to damage states ds_3 and ds_4 for higher PGA's becomes also high, resulting in a poor description of the posterior distribution (figure 9, $f_{m|E}^I$).

Another open point here is the small difference in the fragilities utilized in the analysis, which resulted in small differences between the estimates associated to the two buildings in comparison [8, 10] (table 6). The effect of the variation of fragilities' median values was tested by incrementing them with additional 0.2 g, which resulted in an increase of 10% in μ_M.

5 Conclusions

The present paper focusses on how fragility and GMPE related uncertainties, and assumptions based on prior information, impact the final magnitude estimates, extending earlier works on the simplified assessment of the behaviour of historical unreinforced masonry walls and magnitudes for the 1763 Komárom earthquake.

Despite the poor description of the *im* with CSM, the experimental capacity curves and the FEMA HAZUS manual led to similar fragility moments, although the fragility medians rather than their standard deviations are relevant for the final magnitude estimates. Otherwise, the standard deviation associated to the GMPE has a high impact on the expected value of magnitude and respective standard deviation. The fact that the chosen GMPE leads to limited descriptions for higher *im* values – and thus for higher magnitudes – has been confirmed by the CSM fragilities and the posterior distribution $f_{m|E}{}^{I}$ in figure 9. The expected magnitude and standard deviation are affected mainly by the prior distribution of magnitude rather than by the prior distribution of distance, despite the small studied distance ranges (tables 5 and 6). In addition, the sole consideration of the damage analysis resulted in higher order magnitude estimates (6.6~7.2), while the use of magnitude priors resulted in lower order magnitude estimates of (5.5~6.6). The full use of prior knowledge (table 6, figure 9) led to magnitude expected values in between 6.2 and 6.6.

Improvements may be achieved with further work with other numerical modelling tools and other building archetypes that may be used to produce valid fragilities. Other prior distributions of magnitude and distance may also be tested and discussed in more detail. As to the use of different GMPE(s) associated in a logic tree may improve the *im* descriptions.

Acknowledgements

Proper acknowledgements to CAPES for allowing the present research (Process No. 9178-13-9). This paper was also supported by the János Bolyai Research Scholarship of the Hungarian Academy of Sciences.

References

1. G. Szeidovitz, Earthquakes in the region of Komárom, Mór and Várpalota, *G. T.*, **32**, 20 (1986).
2. P. Varga, G. Szeidovitz, R. Gutdeutsch, Isoseismical map and tectonical position of the Komárom earthquake of 1763, *A. G. G. Hung.*, **36**, 12 (2001).
3. U. Eisinger, R. Gutdeutsch, C. Hammerl, Historical earthquake research, *Abh. Geol. B.-A.*, **1**, 19 (1992).
4. H. Ryu, J.K. Kim, J.W. Baker, A probabilistic method for the magnitude estimation of a historical damaging earthquake using structural fragility functions. *Bull. Seism. Soc. of Am.*, **99**, 18 (2009).
5. E.C. Morais, L.G. Vigh, J. Krähling, Fragility estimation and comparison using IDA and simplified macro-modelling of in-plane shear in old masonry walls, *Spring. Proc. in Math. & Stat.*, **181**, 14 (2016).
6. S.A. Freeman, Development and use of capacity spectrum method, *Proceedings of 6th US National Conference on Earthquake Engineering*, 8 (1998).
7. E.C. Morais, L.G. Vigh, J. Krähling, A methodology for the development of Historical building archetypes for Seismic performance assessment, *P. Periodica*, **13**, 12 (2018).
8. E.C. Morais, L.G. Vigh, J. Krähling, Historical earthquakes and historical building archetypes for dynamic structural analysis using historical surveys, *3rd PROHITECH*, 12 (2017).
9. F. Mckenna, G. Fenves, M. Scott, B. Jeremic, Open system for earthquake engineering simulation (OpenSees)S, (2006).
10. FEMA HAZUS-MH MR-1 Technical Manual, *Fed. Emergency Management Agency*, 699 (2003).
11. E.C. Morais, L.G. Vigh, J. Krähling, Seismic damage analysis of a Hungarian historical peasant house archetype, *9th ENOC*, 6 (2017).
12. FEMA P695 Quantification of building seismic performance factors, *Fed. Em. Man. Ag.*, 216 (2009).
13. AHEAD, *The Eur. Arch. of Hist. Earthquake Data*, DOI: http://doi.org/10.6092/INGV.IT-AHEAD.
14. D. Vamvatsikos, C.A. Cornell, Incremental dynamic analysis, *Earthq. Eng Struct. Dyn.*, **31**, 24 (2002).
15. P.B. Lourenço, Computational strategies for masonry structures, PhD dissert., *Delft University*, 220 (1996).
16. S.H. Rafsanjani, A. Bakhshi, M.A. Ghannad, M. Yekrangnia, F. Soumi, Predictive tri-linear benchmark curve for in-plane behaviour of adobe walls, *Int. Journ. of Arch. Herit.*, **9**, 29 (2015).
17. S. Lagomarsino, S. Cattari, Perpetuate guidelines for seismic performance-based assessment of cultural heritage masonry structures, *Bulletin of Earthquake Engineering*, **13**, 35 (2015).
18. S. Akkar, J.J. Bommer, Empirical equations for the prediction of PGA, PGV, and spectral accelerations in Europe, the Mediterranean Region, and the Middle East, *Seismological Research Letters*, **81**, 12 (2010).
19. E.C. Morais, L.G. Vigh, J. Krähling, Preliminary estimation of the probable magnitude of Komárom 1763 earthquake using fragility functions, *16th WCEE*, 12 (2017).

Qualitative and quantitative evaluation of the ground movement risk on an urban cartographic scale
Example of QUEMADO (Eastern Rif, Morocco)

Evaluation qualitative et quantitative du risque 'mouvement de terrain' à une échelle cartographique urbaine
Cas du versant QUEMADO (Rif oriental du Maroc)

RFIFI Mohamed [1,a], AIT BRAHIM Lahsen [1]
[1] Université Mohammed V, Faculté des sciences, Unité de recherche GEORISK, LGRN, 4 Avenue Ibn Battouta -Agdal, BP 1014, Rabat, Maroc.

Abstract. The morphostructural examination of QUEMADO slope, contributes to distinguish the vulnerability indications that refer to the ground movements. They are supported by the combination of geological, geo morphological, hydro geological parameters. In addition, the seismic activity that is known in this area increases the vulnerability of the coastal areas. Thus, many cracks and locatable indices in the studied area constitute a witnesses of movement risk that threat directly the projects, the installations and the urban constructions of Al Hoceima city, in proximity of the studied cliff.
The present study is consecrated to achieve a comparative study of the evaluated risk "ground movement" by two methodological approaches. The qualitative approach which is most usually employed, and the quantitative approach that requires a scientific effort of adaptation and reproducibility. The risk evaluation is based on using four data components; mainly the topographical slope, the lithology, the geomorphology and the ground occupation. The results were interpreted by a discrete analysis of the unstable areas through the land reconnaissance
Keywords: Morocco, Western Rif, QUEMADO, Ground movements, Vulnerability, Evaluation, Methodological approach.

1 Introduction

Les mouvements de terrain au Rif ont fait l'objet de plusieurs travaux, dont ceux de [1]. D'autres travaux consacrés à des régions précises du Rif, ont invoqué la combinaison la combinaison de plusieurs facteurs provoquant l'instabilité de terrain [2, 3, 4, 5]. Cependant des études sur les mouvements dans les pentes signalent que le facteur sismique peut être à l'origine des surpressions interstitielles [6].

L'objet de la présente étude est de mettre en évidence la conjonction entre divers facteurs déclencheurs d'instabilité à l'échelle de la zone de QUEMADO, l'évaluation de sa vulnérabilité au risque de mouvement de terrain par l'emploi de deux approches d'étude qualitative et quantitative ayant fait l'objet d'études antérieures [7, 8, 9], cette analyse fera l'objet également d'une reconnaissance terrain.

2 Terrain d'étude

2.1 Cadre géologique

La région d'Al Hoceima appartient à la zone interne des unités structurales du Rif résultantes de l'orogénèse de la chaine alpine. Le Rif interne se compose d'un ensemble d'unités allochtones alpines, empilées les unes sur les autres et séparées par des contacts anormaux [10]. Parmi ces unités on trouve la dorsale calcaire externe qui supporte, sous forme de klippe tectonique, les terrains de la dorsale calcaire interne et ceux de la nappe paléozoïque Ghomarides. Ces unités reposent sur les terrains de la dorsale prédorsalienne (semelle tertiaire). L'ensemble de ces formations chevauchent vers le sud les flyschs de Tisirène [11, 12, 13] (figure 1).

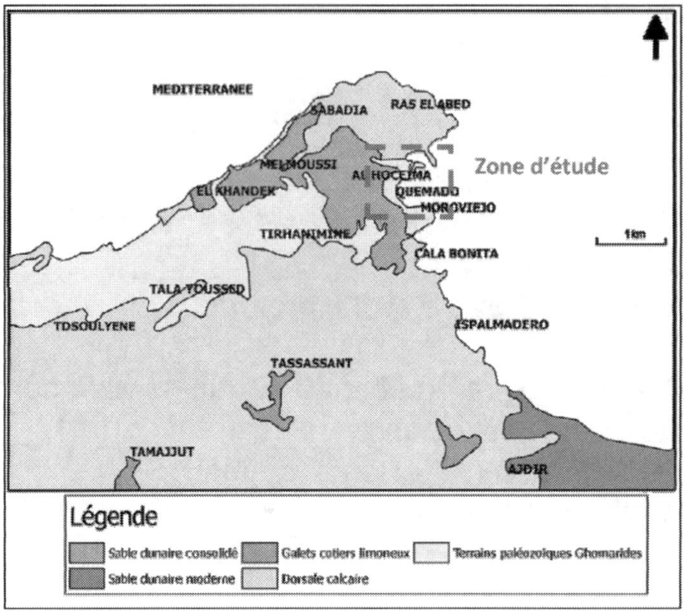

Fig.1. Cadre géologique régional d'Al Hoceima

Les affleurements carbonatés rigides (les calcaires à silex et dolomie) de la dorsale calcaire externe formant des reliefs remarquables (BOUSSEKKOUR, RAS EL ABED et JBEL ASSAGUASAGUANE) suivis des terrains paléozoïques Ghomarides dénommés localement la Klippe d'Al Hoceima. Cette Klippe s'étend d'Al Hoceima à l'oued BOUSSEKKOUR à l'Ouest avec une stratigraphie en schistes satinés et noirs (12) surmontée par des calcaires dévoniens constituant les points géomorphologiques les plus hauts (SIDI MANSOUR, TALA YOUSSEF et la zone étudiée QUEMADO).

2.2 Cadre sismique

La région d'Al Hoceima peut être présentée comme la plus sismogène au Maroc. Par conséquent, elle est parmi les zones actives les plus étudiées le long de la frontière de plaques diffuse Afrique-Eurasie dans la méditerranée occidentale. Les cartes de sismicité de la région montrent un important alignement sismique de direction NE-SW.

Plusieurs tremblements de terre ont pu être considérés dans l'histoire de la région grâce à l'ampleur des désastres qu'ils provoquèrent. On compte trois événements majeurs (dont les deux premiers destructeurs), en 1994, 2004 et 2016 ayant des magnitudes modérées entre 5.9 et 6.3 et des mécanismes au foyer similaires.

2.3 Cadre géomorphologique

L'examen du site étudié montre que le versant possède un relief à pentes irrégulière (figure 2), accentuée par l'existence de trois thalwegs larges et par l'action anthropique des aménagements existants et en cours, ce site est délimité au sommet par les constructions de la ville d'Al Hoceima tandis que le pied du versant abrite des bâtiments relevant d'un complexe touristique.

En dénivelée, le secteur possède des pentes comprises entre la plage méditerranéenne et 65 m d'altitude, les pentes très irrégulières dont les plus fortes dépassent 40°.

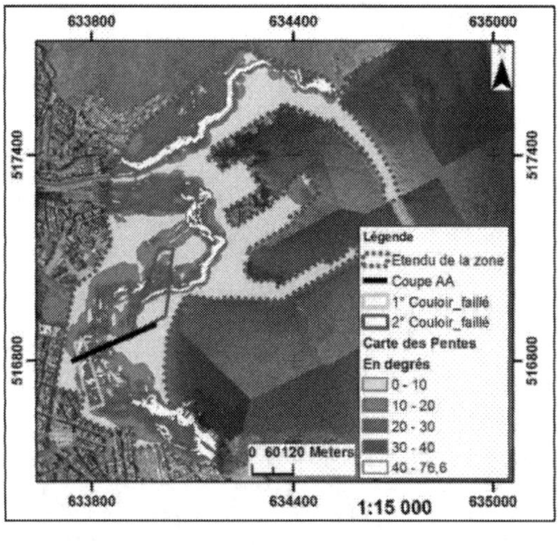

Fig. 2. Cadre morphologique du secteur d'étude

Le Versant QUEMADO présente un cas de figure de la vulnérabilité du Rif au risque « mouvements de terrain ».

Le risque étudié est favorisé par la conjonction de plusieurs paramètres qui relèvent principalement des caractéristiques structurales et topographiques du versant.

L'examen de la zone constituée du versant, reflète l'existence des pélites schisteuses, des calcaires et des dépôts de couverture meubles (figure 3).

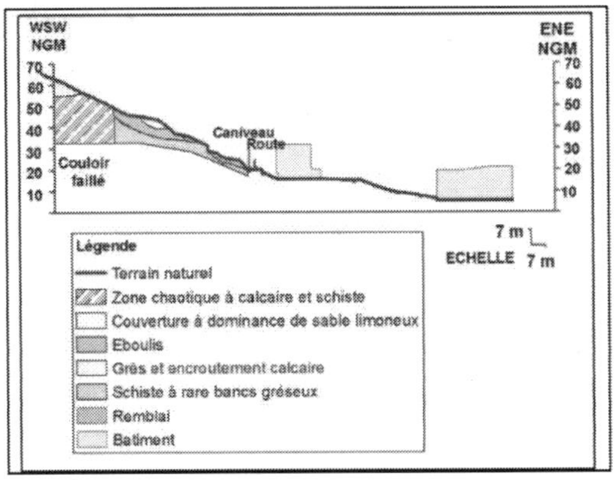

Fig. 3. Coupe géologique (AA)

3 Problématique

Le Versant QUEMADO est sujette au risque « mouvements de terrain », favorisé par la conjonction de plusieurs paramètres qui relèvent principalement des caractéristiques structurales et topographiques du versant.

4 Objectifs de l'étude

Afin de contribuer à une meilleure connaissance du risque dans l'espace d'étude, notre recherche suit une démarche transversale appuyée sur des approches permettant l'évaluation qualitative et quantitative de l'aléa. L'objectif est de caractériser les facteurs de déclenchement du risque et son évaluation à une échelle urbaine.

5 Hypothèses de déclenchement des mouvements de terrain

Les particularités lithologiques et géotechniques de la zone étudiée permettent de définir les hypothèses de déclenchement suivantes :

- Versant à pente raide et sol hétérogène ;
- Présence d'écoulements d'eau ;
- Charge des constructions en amont du versant ;
- Couloirs faillés et sismicité régionale.

6 Démarche d'étude et extraction des paramètres

6.1 Evaluation qualitative du mouvement de terrain

L'évaluation des risques à l'échelle urbaine est un atout [14, 15]. Les trois grandes étapes de l'approche qualitative employée sont présentées ci-dessous.

- Cartographie de la susceptibilité au mouvement de terrain par la prise en compte des paramètres physiques ayant engendré l'aléa. L'intensité de chaque paramètre est exprimée en échelles de valeurs et pondérés par des poids [7] adaptées (Table 1) ;
- Analyse et cartographie des enjeux suivant deux catégories d'enjeux à partir des données d'occupation du sol disponibles (Table 2) ;
- Croisement des cartes de susceptibilité et d'enjeux en multipliant les valeurs de classes (Table 3).

Table 1. Paramètres de prédisposition au risque

Paramètres	**Intitulé du paramètre**	**Pondération**
Topographie	Pente de terrain (P)	4
	Orientation de la pente (O)	2
Hydrographie	Distance du réseau hydrographique (H)	3
Désordres	Distance des désordres naturels (D)	3
Géologie structurale	Susceptibilité au déplacement (G)	4
Structure sismique	Distance des contacts anormaux et failles (F)	4
Confortement	Distance des confortements réalisés (C)	3

Table 2. Valeurs qualitatives d'endommagement attribuées

Carte d'enjeux	**Intitulé du paramètre**	**Valeurs attribuées**
Fonctionnelles et économiques (**FE**)	Aménagements structurants	60
	Voirie	50
	Surfaces artificielles	40
	Terrain nu	20
	Espace vert	12
	Foret	6
	Habitat	100
	Equipements	90
	Tourisme	80
	Dépendances	40
Humaines (**H**)	Forte	100
	Moyenne	90
	Faible	80

Table 3. Croisement des cartes de susceptibilité et d'enjeux

	S1 : Très faible	S2 : Faible	S3 : Moyenne	S4 : Forte	S5 : Très forte
E1 : Très faibles	Très faible	Très faible	Très faible	Très faible	Très faible
E2 : Faibles	Très faible	Très faible	Très faible	Faible	Faible
E3 : Moyens	Très faible	Très faible	Faible	Faible	Moyen
E4 : Forts	Très faible	Faible	Faible	Moyen	Fort
E5 : Très forts	Très faible	Faible	Moyen	Fort	Très fort

6.2 Evaluation quantitative du mouvement de terrain

L'approche d'évaluation quantitative du risque est fondée sur un calcul et une hiérarchisation de la valeur (monétaire ou non monétaire) [16] des éléments exposés, ainsi qu'une évaluation de la susceptibilité par la méthode de régression logistique (ou régression binomiale) [17].

Les cartes résultantes sont classifiées en échelle d'intensité et croisées pour extraire la carte de risque globale. La méthodologie d'évaluation du risque repose sur 2 étapes.

- Etape 1 : Basée sur le calcul de la probabilité d'occurrence des mouvements pour chaque pixel de la zone d'étude :

  $$P = \frac{1}{1+e^{-z}} \quad (1)$$

  Où z représente une combinaison linéaire des paramètres prédicteurs du risque $X_1, X_2, \ldots, X_n$:

  $$z = b + a_1X_1 + a_2X_2 + \cdots + a_nX_n \quad (2)$$

  Avec b le terme libre du modèle et les coefficients a_i de la régression.

  Les paramètres prédicteurs retenus dans le cadre de cette étude ayant un niveau élevé de représentativité sont : la pente (X_1), l'orientation du terrain (X_2), la distance du réseau hydrographique (X_3), la distance des désordres naturels constatés (X_4), la susceptibilité de déplacement des formations géologiques (X_5), la distance des contacts anormaux et couloirs faillés (X_6) et la distance des confortements du sol réalisés (X_7) [18].
- Etape 2 : Basée sur l'identification et l'attribution des valeurs d'endommagement proportionnelles à la valeur monétaire des enjeux exposés (Table 4).

Table 4. Valeurs attribuées des enjeux

	Enjeu structurel (ES)	Enjeu fonctionnel (EF)	Enjeu corporel (EC)
	Valeur monétaire en MAD/m^2 (Valeur hiérarchisé)	Valeurs attribuées	Valeurs attribuées
Aménagements structurants ;	600 (8)	60	50
Voirie ;	400 (5,3)	70	60
Surfaces artificielles.	200 (2,7)	50	45
- Espace vert ;	150 (2)	20	30
- Foret ;	200 (2,7)	40	25
- Eau ;	100 (1,3)	30	15
- Sol nu.	700 (9,3)	20	30
- Plancher habitable ;	5500 (73,3)	100	100
- Plancher d'équipement ;	4000 (53,3)	90	70
- Plancher touristique.	7500 (100)	80	80

7 Résultats

7.1 Approche qualitative

La carte de risque qualitatif (figure 4) est retrouvée par croisement des cartes classifiées de susceptibilité et enjeux qualitatives (Tableau 3).

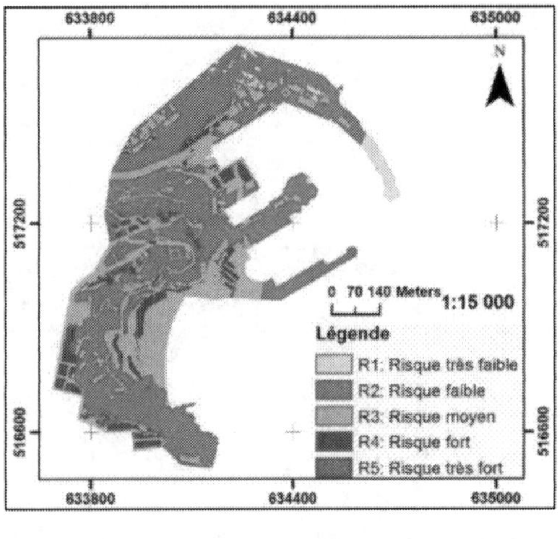

Fig. 4. Carte du risque qualitatif des mouvements de terrain

7.2 Approche quantitative

L'application du de la régression binomiale a permis d'extraire la carte de susceptibilité quantitative aux mouvements de terrain croisée à la carte des enjeux de la zone. La carte de risque quantitatif retrouvée est présentée par la figure 5.

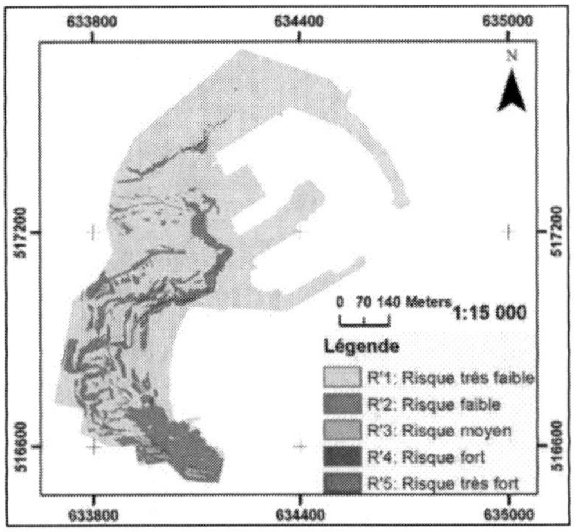

Fig. 5. Carte du risque quantitatif des mouvements de terrain

8 Discussion

La zone d'étude est sujette au risque de mouvement de terrain vue son contexte géologique et morphologique. A cet effet, l'étude a nécessité l'introduction des paramètres prédicteurs du risque.

L'affectation des poids à chaque paramètre de prédisposition est une étape décisive dans la perception de l'aléa et exige impérativement une connaissance des particularités de la zone et le jugement de l'expert.

La comparaison des pondérations intégrées dans le modèle qualitatif aux coefficients calculés par la régression logistique remet en lumière le rôle essentiel que jouent les formations géologiques et la pente topographique dans la répartition des mouvements de terrain à région d'Al Hoceima [21] (Table 5).

Table 5. Coefficients de pondération des deux approches

Paramètres de prédispositions	Coefficients de pondération retrouvés par régression	Coefficients de pondération introduits qualitativement
Pente	0,088	4,0
Orientation de la pente	0,000	2,0
Distance du réseau hydrographique	0,005	3,0
Distance des désordres naturels	0,011	3,0
Susceptibilité géologique	0,255	4,0
Distance des contacts anormaux et failles	0,004	4,0
Distance des confortements	-0,002	3,0

Les résultats acquis par application des démarches présentées au paragraphe 6 ont permis d'établir deux cartographies ayant des proportions du risque différentes.

A cette analyse, une mission d'investigation terrain a été exécutée pour évaluer la pertinence des résultats. L'analyse nous ramène à subdiviser celle-ci en 2 catégories (figure 6) :

- Catégorie 1: L'emprise des talwegs situés en amont des projets touristiques que connait des signes d'instabilité plus ou moins prononcés (désordres, couloirs faillés, caractéristiques mécaniques faibles du sol) avec des pentes raides et moyennes ;
- Catégorie 2 : La partie reliant les talwegs à la limite du versant qui a une apparence relativement stable avec des pentes faibles et un aménagement contribuant à la stabilisation du terrain.

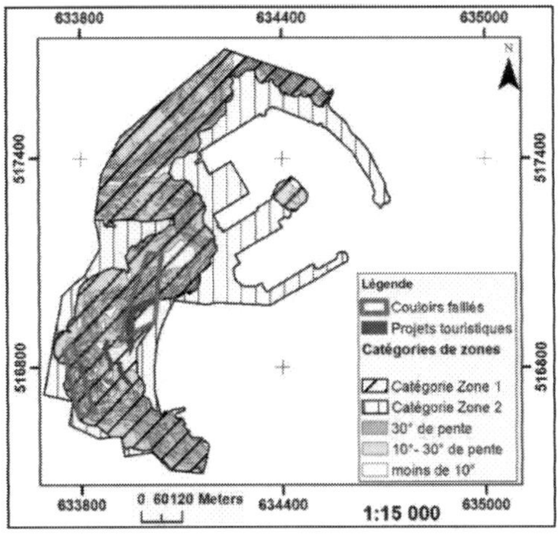

Fig. 6. Constatations et catégories spatiales de risque

La confrontation de cette cartographie aux cartes de susceptibilité précédentes montre que l'approche quantitative est plus proche en résultats, offre un zonage plus fin et des poids objectifs attribués à chaque paramètre de prédisposition.

9 Conclusion

Comme il a été décrit dans des travaux antérieurs (4), la répartition des mouvements de terrain se concentre à proximité des falaises ou le long des versants accidentés des principaux cours d'eau. L'examen du secteur d'étude montre que cette donnée se manifeste clairement et coïncide également avec une proximité des principales failles de la région.

La pertinence des modèles qualitatifs et quantitatifs d'évaluation du risque dépend du type, de la densité et de la qualité des données existantes.

L'approche quantitative est plus objective (calcul des poids de façon probabiliste et qualification des enjeux à la base des valeurs réel des enjeux exposés) et fournie des résultats qui se confirment par les investigations et les relevés terrain. Cette cartographie produite par des données de base au 2000ème en exploitant les fonctionnalités du système d'informations géographiques (SIG) porte un complément reproductible à d'autres zones à risques.

Références

1. Millies-Lacroix, *Les glissements de terrain, présentation d'une carte prévisionnelle des mouvements de terrain dans le Rif.* Mine et Géologie, **27**, pp. 45-54 (1968)
2. B. El Fellah, *Eboulement rocheux dans le Paléozoique du Rif: Présentation d'un cas sur la route Oued Lao-Jebha.* Lisboa, Portugal : 7ème Congré Intern. de l'AIGI, pp. 3927- 3931 (1994)
3. B. El Fellah, O. Azzouz, L. Assebriy, *Exemple de glissement de terrain littoral sur la cote méditerranéenne des Bokoya entre Torrès et Badis, Rif, Maroc.* ORSTOM, Réseau érosion, **16** (1996)
4. K. Margua, *Essai de cartographie des risques naturels: Application à l'aménagement de la région d'Al Hoceima.* Thèse universitaire, Univ. Franche-Comté, 196 p (1994)
5. A. Talhaoui, M. Aberkan, A. Iben Brahim, A. El Mouraouah, N. Jabour, *Contribution à l'étude des risques naturels. Cas de la région d'Al Hoceima* . Oujda : Colloque des Bassins Sédimentaires Marocains (1999)
6. Y. Moriwacki, I.M. Idrissi, T.L. Moses, R.S. Ladd, *A re-evaluation of the "L" Strret Slide in Anchorage during the 1964 Alaska earthquake* . Rio de Janeiro : Comtes rendues XII ICSMFE, **3**, pp. 1583-1586 (1985)
7. M. Levasseur, *Contribution des systèmes d'information géographique à l'analyse quantitative de l'aléa "glissement de terrain" - Exemple d'application au secteur de Ville de la Baie, Québec* . Québec : Université de Québec (2003)
8. F. Léone, *Concept de vulnérabilité appliqué à l'évaluation des risques générés par les phénomènes*

de mouvements de terrain. Grenoble : Rapport de thèse, Université Joseph Fourier (1996)

9. J.P. Malet, H. Thiery, O. Maquaire, A. Puissant, *Analyse spatiale, évaluation et cartographique du risque "glissement de terrain".* Revue Internationale de Géomatique, **10**, pp. 1-10 (2006)
10. M.M. Blumenthal, *Esbozo geologico del Rif en la region de Bokoya* . Bol. Inst. Geol. Min. Espana, Madrid., **44**, pp. 199-352 (1937)
11. J. Andrieux, *La structure du Rif central, étude des relations entre la tectonique de compression et les nappes de glissements dans un tronçon de la chaine alpine* . Notes et Mém. Serv. Géol. Maroc, **235**, 155 p (1971)
12. T. Mouriet, *Etude géologique et structurale du massif des Bokoya* . Trav. Lab. Géol. de l'Afrique . Univ. Paris Sud, (1982).
13. O. Azzouz, *La lithostratigraphie et tectonique des terrains paléozoiques Ghomarides du massif des Bokoya (Rif interne, Maroc).* Thèse 3° cycle .Rabat, 208 p (1992)
14. M. Michael-Leiba, F. Baynes, G. Scott, K. Granger, *Regional landslide risk to the Cairns community.* Natural Hazards, **30**, Issue 2, pp. 233–249 (2003)
15. G. Grünthal, A.H. Thieken, J. Schwarz, K. Radtke, A. Smolka, B. Merz, *Comparative risk assessment for the city of Cologne, Germany – storms, floods, earthquakes.* Natural Hazards, **38**, pp. 21-44 (2006)
16. T. Glade, *Vulnerability assessment in landslide risk analysis*, **134**, pp. 121-138 (2003)
17. Y. Thiery, J.P. Malet, S. Sterlacchini, A. Puissant, O. Mequaire, *Landslide susceptibility assessement by bivariate mehods at large scales : Application to a complex mountainous environment.* Geomorphology, **92**, pp. 38-59 (2007)
18. M.C. Margarint, A. Grozavu, C.V. Patriche, A-M.I. Tomasciuc , R. Urdea, I. Ungurianu, *Evaluation des risques de glissements de terrain par la méthode de régression logistique : Application à deux zones basses de Roumanie* . Dynamiques environnementales. LGPA-EDITIONS, **28** (2011)
19. M. Mejia Navarro, E.E. Wohl, S.D. Oaks, *Geological Hazards, vulnerability, and risk assessment using GIS : Model for Glenwood Springs.* Geomorphology, **10**, pp. 331-354 (1994)
20. S. Aronoff,*Geographic Information Systems:A Management Perspective.*WDLPublications, 294 p (1989)
21. A. Talhaoui, M. Aberkan, Iben Brahim, A. El Mouraouah, *Risques géologiques et activité sismique dans la région d'Al Hoceima (Maroc) : Approche de la quantification des facteurs responsables du déclenchement des instabilités de terrain.* Pangea, **43**, pp. 3-18 (2005)

Relevant lessons learned from the earthquakes El Asnam, 10/10/1980 and Zemmouri, 05/21/2003 on the paraseismic prevention in Algeria

Yamina Ait-Meziane[1], Rania Souici[2] and Farah Lazzali[3]
[1] CGS, National Earthquake Engineering Research Center, Algiers, Algeria
[2] USTHB, University of Science & Technology Houari Boumediene, Algiers, Algeria
[3] UMBB, M'Hamed Bougara Boumerdes University, Boumerdes, Algeria

Abstract. In the field of seismic risk prevention in Algeria, the earthquake of El Asnam, 10/10/1980 is the triggering event and that of Zemmouri 21/05/2003 the accelerating element of the policy already in place. These two major events have been rich in information, as they are among the most studied internationally. The approaches and the development of methods for seismic risk reduction are based largely on feedback. The fundamental goal of seismic risk reduction is to avoid loss of life, which is mainly due to the partial or total collapse of structures. Exploiting lessons learned from past earthquakes is reflected in code requirements for new projects and approaches to vulnerability estimation for existing structures. In this paper, we give a description of the characteristics of the two events. The analysis of pathologies observed and the impact on the Algerian seismic code is discussed.

1 Introduction

Historically, northern Algeria experienced several seismic events [1]. The two major earthquakes that have occurred in Algeria, El Asnam (1980) Ms=7.3 and Zemmouri (2003) Mw=6.8. accounted 3.500 deaths and 60.000 housing units destroyed for El Asnam region, 2.278 deaths and approximately 15.000 buildings destroyed for Algiers-Boumerdes region that is dense urban zone. In urban areas, the casualties due to earthquakes are usually important and often, the deaths and injuries are caused by partial or total collapse of buildings [2]. The results of field investigations and statistical analyses show that a majority of the damaged buildings were masonry buildings and reinforced concrete frame buildings, in part those built without any consideration to the requirements of the seismic code. In the case of the Zemmouri earthquake two wilayates were severely shaken, Algiers whose housing stock is essentially old, and Boumerdes a relatively recent area. In terms of earthquake prevention, the area of El Asnam was classified as a zone B by the AS55, a regulation not adapted to the regional seismic context; The area of Alger-Boumerdes was classified as a zone with moderate seismicity (zone II) by the earthquake-resistant wich standardizes the diagnostic protocol.

2 Expertise protocol

Damage to buildings is closely related to several factors such as: building parameters, geological and geotechnical conditions, and to the severity of the seismic event. The results of the damage observed on the buildings during the two events were carried out using the same post seismic expertise by using the damage evaluation sheet related to buildings in the surveyed area (building address, building use, age, number of stories, soils rupture (fault, liquefaction, landslide and settlement), foundation type, resisting system type and observed damage on structural and non structural elements) is collected. This first phase of buildings investigation is generally carried out very quickly; its objective is to classify globally all buildings into one of the three categories shown in Table 1. The investigator must conclude by given immediate decision about measures to take for the building. Each category (color) corresponds to the damage state of the building and its functionality after the earthquake. Green category is attributed to buildings with very little damage, which can be reoccupied immediately after the earthquake. Orange category is attributed to buildings which need further study before it can be either occupied or condemned (to be repaired or demolished). Red category is attributed to buildings which should be demolished.

Table 1 Categories of building damage and damage degrees according to the EMS-98 scale

Category	Damage state	Damage	Description
Green	Very little damage. Can be reoccupied immediately	Degree 1	Negligible to slight damage (no structural and slight non-structural damage)
		Degree 2	Moderate damage (slight structural and moderate non-structural damage)

Orange	Needs further study before it can be either	Degree 3	Substantial to heavy damage (moderate structural and heavy non-structural damage)
	Occupied or condemned	Degree 4	Very heavy damage (heavy structural and very heavy non-structural damage)
Red	Condemned and should be demolished	Degree 5	Destruction (very heavy structural damage)

3 Zemmouri, may 21st, 2003 earthquake

The latest earthquake that affected Algeria and caused significant damage is the Zemmouri earthquake, on May 21st, 2003. At 19:44:19 local time, the Zemmouri earthquake struck the Algiers-Boumerdes region in northern Algeria, with a magnitude Mw=6.8. Zemmouri is situated approximately 60 km east of the capital city Algiers. The location of the epicenter was at 36.90°N and 3.71°E determined by the U.S. Geological Survey (USGS). The Algerian Research Centre of Astronomy, Astrophysics, and Geophysics (CRAAG) has determined that the epicenter was at 36.91°N and 3.58°E, and at a depth of 8 to 10 km on the coast line near the village of Zemmouri. The maximum acceleration of 0.58g was recorded at Keddara at about 20 km west from the epicenter [3].

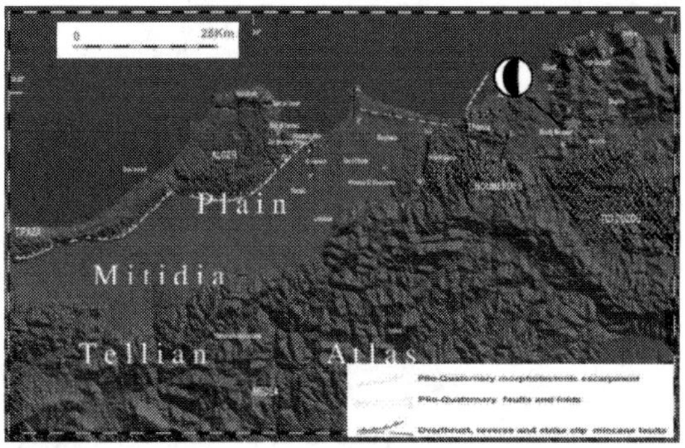

Fig.1. Characteristics of Zemmouri earthquake

The macroseismic study of Zemmouri earthquake [4] allowed establishing the isoseismic map reporting the spatial distribution of the earthquake effects in terms of maximum intensity (Fig. 2).

The earthquake of 2003 has affected the entire Boumerdes Wilaya and a part of capital Algiers. On June 1, the Government of Algeria reported that the earthquake killed 2.278, wounded 11.450, and left approximately 200.000 people homeless. In addition, the earthquake interrupted medical services, water supply lines, electricity, and telecommunications in the region. Many Dairas and municipalities in Boumerdes and Algiers, are declared disaster zones by the central government decree of May 24, 2003.

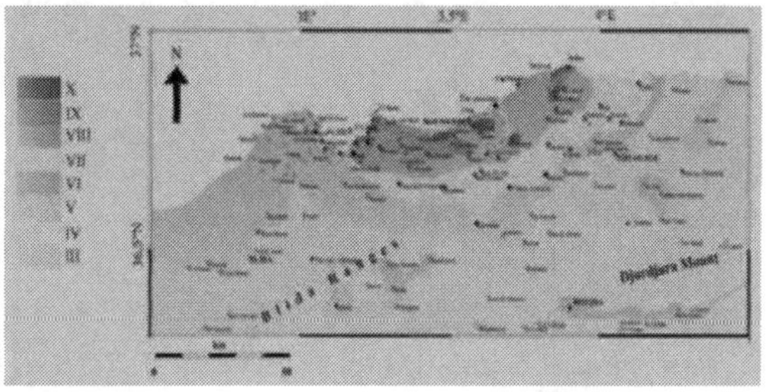

Fig.2. Isoseismal map of Zemmouri earthquake [4]

3.1. Characteristics of buildings and observed damage

In the survey mission following the 2003 earthquake, 728 engineers from the following public institutions and 626 other engineers and architects from private sectors were mobilized in the post-earthquake survey have been deployed throughout the 53 cities as declared disaster zones, in order to classify buildings into one of the three categories of damage.

The predominant category of buildings in the affected regions was the RC frame buildings [5], very few steel and wooden buildings mostly used for industrial purpose.

Based on the field investigations, the number of damaged buildings and the corresponding classification as of the 23 June 2003 are shown in Table 3, respectively [5]. A total of 181.658 buildings were investigated in the affected zones, with 66.383 in Boumerdes region and 105.287 in Algiers. In Boumerdes only, 7.465 buildings were destroyed [6].

Table 2 Buildings investigated and the corresponding classification

Classification	Green	Orange	Red
Algiers	40147 (44, 56%)	42452 (7, 12%)	7497 (8, 32%)
Boumerdes	39054 (59%)	19864 (30%)	7464 (11,2%)

Fig..3. shows the distribution of destroyed and condemned buildings (Red category) in Algiers – Boumerdes region following the Zemmouri earthquake. The High rate of destroyed buildings was recorded in the epicenter area and in the eastern part of Algiers city, dans le view quartier de Belouizdad.

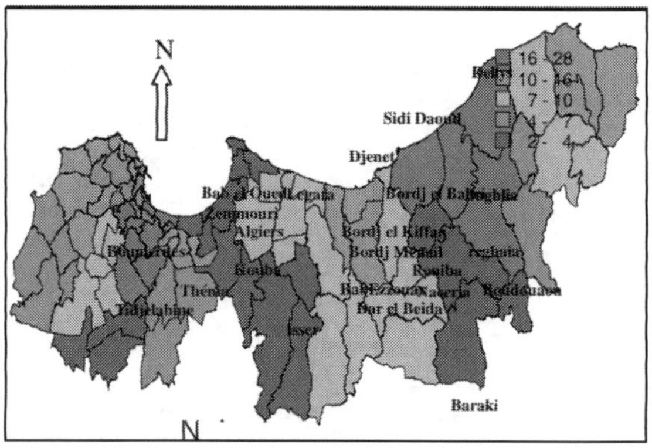

Fig.3. Distribution of destroyed and condemned buildings (Red category) in Algiers – Boumerdes region following the Zemmouri earthquake (rate in %).

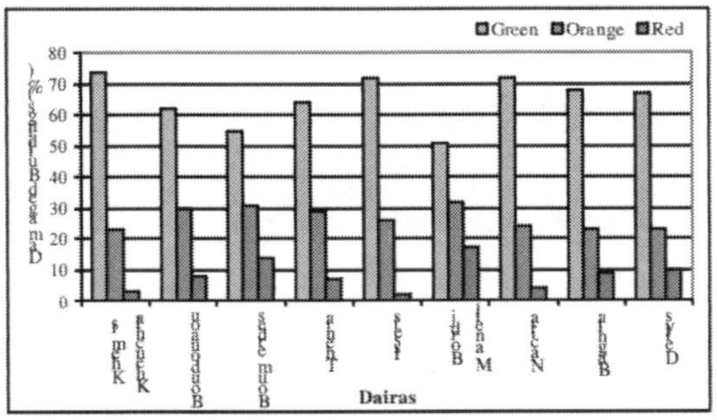

Fig.4. Damage distribution in the affected Daïras of Boumerdes

Observed damage in Algiers-Boumerdes region can be summarized essentially as follow (Fig.5.): collapse of buildings; including total collapse , first soft story collapse, pancake collapse, kneecaps at the top of columns after bursting of masonry walls, collapse of unreinforced masonry buildings shear and compressive failure in columns, rotations at the ends of the columns with cracking and loss of concrete, diagonal cracks, out-of-plane failure with partial or total collapse of infill walls.

Fig.5. Photos showing damage in the affected areas in Boumerdes-Algiers region

4 El Asnam, October 10th, 1980 earthquake

El Asnam (known now by Chlef) earthquake was the first largest seismic event experienced in Algeria since the beginning of the 19 century. On October 10th, 1980 at 13:25:23 local time, El Asnam earthquake struck the Chelf region, at about 200 km west from the city of Algiers, with a magnitude Ms=7.3 (USGS) (Fig. 6). The epicenter was located near the village of Beni-Rached which is15 km north-east of El-Asnam at 36,143°N and 1,413°E at a depth of 10 km. El Asnam earthquake was generated by a thrust fault striking N60°E on 40 km-long rupture zone dipping 45° to NW [7]. The horizontal acceleration of the main shock was estimated in the epicenter area varying in the range of 0.25g and 0.70g [8].

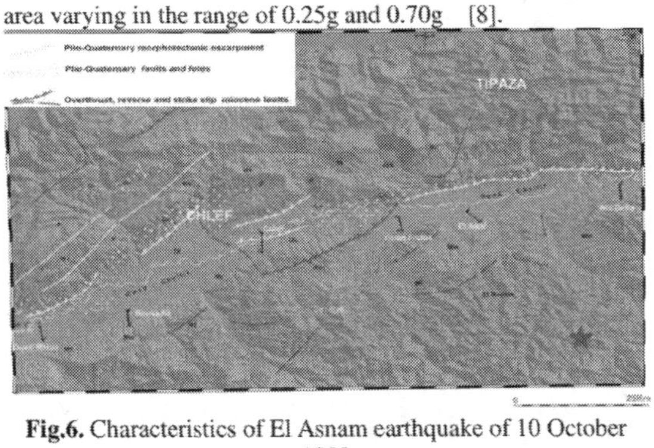

Fig.6. Characteristics of El Asnam earthquake of 10 October 1980

The macroseismic study of El Asnam earthquake [8-9], adapted from [10] seismic hazard study of Algeria, reported the spatial distribution of the earthquake effects in terms of maximum intensity (Fig.7.), based on the damage level recorded in the affected zones. Intensities IX, X and XI (MM) were assigned to the epicenter area and the surrounding, containing sites where many constructions were destroyed or heavily damaged and where there was loss of life. The attenuation of intensity was quite rapid. Algiers, about 200 km east of the epicenter experienced between V and VI. Oran, about the same distance west of the epicenter also experienced similar intensity levels.

The following cities were declared disaster zones after the earthquake are 06 Dairas El Asnam, Ain Defla, Tenes, Boukadir, El Attaf, Oued Rhiou corresponding to 28 municipalities.

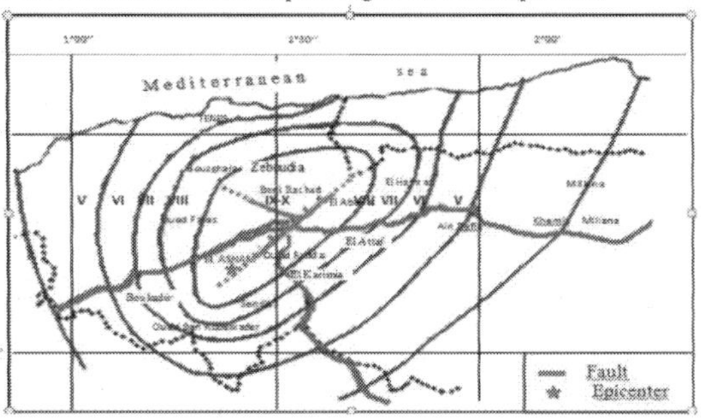

Fig.7. Isoseismal map of El Asnam earthquake (CTC 1981a, adapted from Mortgat and Shah,1978)

The buildings in category 1 are commonly unreinforced masonry buildings. These buildings performed no differently than has been observed in other past earthquakes. Substantial damage and partial collapse were common. A large number of stone and adobe houses that were either on the fault or within 50 meters of the fault collapsed in the village of Beni Rached.

Unreinforced masonry houses and mosques surrounding El Asnam city, has shown great amount of damage. Many were severely damaged and some collapsed. Number of multistory buildings which had been reinforced after the 1954 earthquake with an externally applied concrete frame, performed very well with only non structural damage [8-9]. Buildings in category 2, built during the French period were dominantly concrete frame buildings with masonry infill walls. The performance of these buildings was poor with many total collapses such as the hospital, the schools and the hotels. These buildings appeared to have no rational system of lateral resistance. Collapse was most likely hastened by poor configuration and structural discontinuities [8-9]. In 1980, throughout the environs of El Asnam, large housing projects had been completed or were under construction. Building types varied from 1 story single family units to 5 story apartment buildings.

The distribution of damage to buildings in El Asnam and throughout the disaster area indicates that the most damaged buildings were dwelling and school buildings (Table 3 [8])

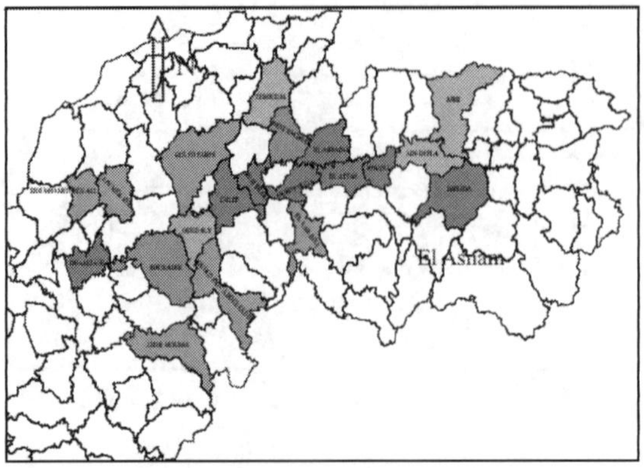

Fig.8. Distribution of destroyed and condemned buildings (Red category) in El Asnam region following El Asnam earthquake (rate in %).

Table 3 Buildings investigated and corresponding classification

Classification	Green	Orange	Red
El Asnam town	1720 (33,41%)	2162 (42%)	1200 (23,31%)
Cities surrounding El Asnam town	592 (43,47%)	579 (42,52%)	191 (14,02%)

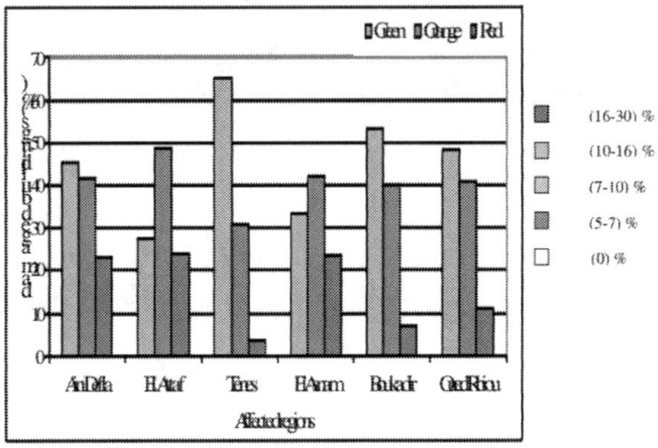

Fig.9. Damage distribution (Green, Orange and Red) in El Asnam town and the surrounding cities

Fig.10. Damages observed in El Asnam earthquake

5. Algerian Seismic codes

The Algerian seismic code was revised in 1999 RPA99 [11], where major modifications were made such as the introduction of soil classification S_1, S_2, S_3 and S_4. The last version of the code was published in 2003 after the Zemmouri earthquake under the title: RPA99 version 2003 [12]. In all three versions before 2003, there were three seismic zones: Zone 0 (no seismicity) in the desert, Zone I (low seismicity), Zone II (moderate seismicity), and Zone III (high seismicity). Boumerdes and Algiers had always been classified as Zone II, with 0.15 g as the PGA for the design of apartment buildings.

In this last version of the code RPA99 Version 2003, the seismic zoning map was revised, including:

-Enlargement of zone III (enclosing before 2003 only El Asnam region). The new values of the seismic zoning factor A with 0.25 g as the PGA for the design of apartment buildings.

-Establishment of two subzones of zone II; IIa and IIb, with two different levels of seismic severity.

-Prohibition of the RC frame system construction in zone III, and strengthening of RC frame structures under construction by addition of concrete walls.

-Strict and severe requirements in construction process.

-Application of the seismic design code became obligatory for private houses.

The Algerian seismic code RPA99 version 2003 defines a static lateral force procedure (simplified method) for determining the seismic actions on standard buildings. These actions do not apply to tall buildings, nuclear power stations, bridges, dams, retaining walls, tunnels and other important structures. The design base shear V is given by the following formula:

$$V = \frac{A\ D\ Q}{R} W \tag{1}$$

Where; A is the seismic zoning factor, D is the dynamic amplification factor, function of site category (S_1, S_2, S_3 and S_4), Q is the quality factor (penalty factor) and R is the behavior factor (17 constructive structural systems are defined in the code with different values of R). The weight W of the building is calculated by assuming that the masses are concentrated at the floor levels. The value of A is given by the code, varying from 0.1 to 0.35, depending of the seismic zone and the building use. The value of D is computed using the fundamental period and the damping. Q is depending of the resisting system organization and construction quality, varying from 1 to 1.35. R is depending of the resisting system type, varying from 2 to 5 for RC buildings.

6 Damage Analysis and discussions

Globally, the analysis in term of damage level, shows that the collapsed and condemned buildings in Boumerdes and Algiers, following the Zemmouri earthquake, and in El Asnam, following El Asnam earthquake, were respectively; 11,2%, 8,15% and 38% (Fig.11). The most affected buildings typologies were the dwelling buildings (Fig.12.), due to their high number compared to schools, hospitals, administrative or commercial buildings. The numbers of dwelling buildings were respectively; 98%, 95% in Algiers and Boumerdes, and 69.75% in El Asnam. In addition, the analysis shows that the number of buildings that suffered significant damage with a possibility of repair (Orange category) was important mainly in Algiers. The dwelling buildings in the affected zones of Algiers, classified in Orange category reached 50%. This can be explained by the fact that buildings in Algiers are old and vulnerable [13-14]

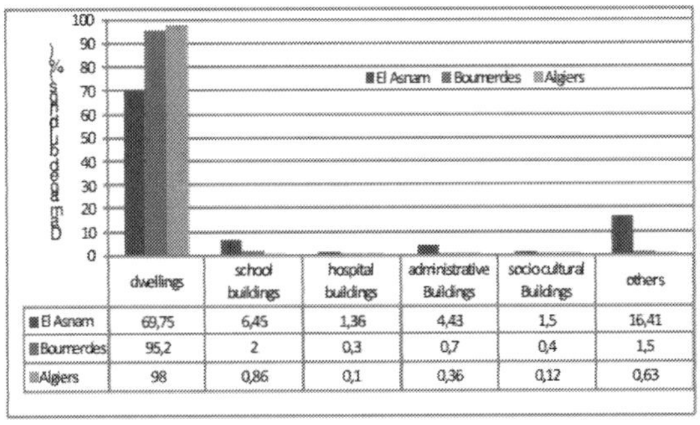

	dwellings	school buildings	hospital buildings	administrative Buildings	sociocultural Buildings	others
El Asnam	69,75	6,45	1,36	4,43	1,5	16,41
Boumerdes	95,2	2	0,3	0,7	0,4	1,5
Algiers	98	0,86	0,1	0,36	0,12	0,63

Fig.11. Damaged buildings by typologies in the affected zones

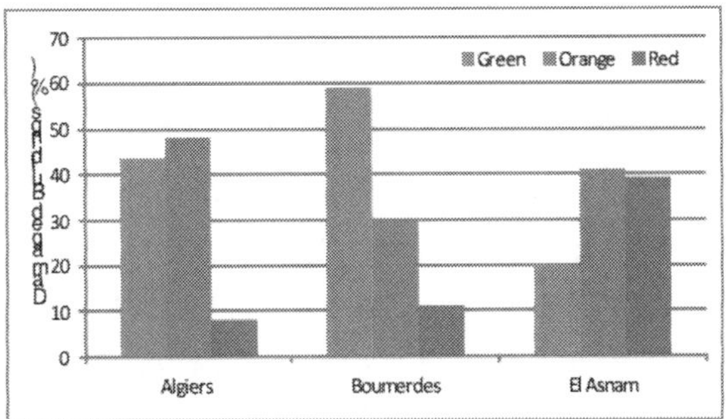

Fig.12. Damaged buildings by category (Green, Orange and Red) in Algiers and Boumerdes following Zemmouri earthquake and in El Asnam following Asnam earthquake

7 Conclusions

In this study, the observed damage distribution during these two strong earthquakes has been analyzed, expressing how buildings suffered severe damage and identifying the deficiencies of the structural systems. The study has also presented the major contributing factors to the observed damage distribution in Algiers-Boumerdes region after the 2003 earthquake and in El Asnam region after the 1980 earthquake, i.e., the structural characteristics and the sites conditions.

Globally, the analysis in term of damage level has shown that the highest rate of collapsed buildings following the two earthquakes was recorded in El Asnam, following El Asnam earthquake with 38%. The most affected buildings typologies following the two events were the dwelling buildings. Among these building 50% located in Algiers were classified in Orange category, corresponding to damage degrees 3 and 4 in the EMS 98 scale, due to the fact that the buildings in Algiers are old and vulnerable. The high rate of damage recorded in the affected zones confirmed the poor behavior of the RC frame system largely used in the region, due essentially to the inadequate sizing and reinforcing of columns and beams and to the bad materials and process quality.

The seismic hazard in Algeria was not sufficiently known and seismic code regulation did not exist before 1980. In all version of the Algerian seismic code published after 1981 and before the 2003 earthquake, the seismic hazard was included in terms of seismic zones and design acceleration. The resistance capacities mandated by the different versions of the seismic code did not differ significantly. The accelerations recorded in Zemmouri 2003 earthquake have confirmed the inadequate level of seismic hazard in the seismic code. Consequently, in the revised version of the seismic code RPA 99 version 2003, seismic zones have been modified and the PGA for the design buildings has been increased.

Decision-makers need adequate information on the expected intensity in the disaster that the cities will face in order to reduce seismic vulnerability.

References

1. (CRAAG) (1994), "*Les séismes en Algérie de 1365 à 1992*". Publication du Centre de Recherche en Astronomie, Astrophysique et Géophysique, Département: Etudes et Surveillance Sismique, ESS, C.R.A.A.G, Algiers.
2. Coburn, A. and Spence, R. (2002), "*Earthquake Protection*", 2nd edition. John Wiley &Sons.Ltd., Chichester, England.
3. Earthquake Engineering Research Institute (EERI) (2003), "*The Boumerdes, Algeria earthquake of May, 21, 2003*", Report, pp. 57. Learning from Earthquakes Program, Oakland, CA.
4. Harbi, A, Maouche, S., Ousadou, F., Rouchiche, Y., Yelles-Chaouche, A., Merahi, M., Heddar, A., Nouar, O., Kherroubi, A., Beldjoudi, H., Ayadi, A., and Benouar, D. (2007), "*Macroseismic Study of the Zemmouri Earthquake of 21 May 2003 (Mw 6.8, Algeria)*", Earthquake Spectra, **Volume 23**, No. 2, 315–332.
5. Belazougui, M. (2008a), "*Boumerdes Algeria earthquake of May 21, 2003: Damage analysis and behavior of beam-column reinforced concrete structures*", 14th World Conference on Earthquake Engineering, 12–17, Beijing, China.

6. Japanese Reconnaissance Team (2004), "*Boumerdes earthquake, May 21, 2003*", 144p.
7. Meghraoui, M., Maouche, S., Chemaa, B., Cakir, Z., Aoudia, A., Harbi, A., Alasset, P.J., Bouhadad, Y., Benhamouda, F. (2004) "*Coastal uplift and thrust faulting associated with the (M_w=6.8) Zemmouri (Algeria) earthquake of 21 May, 2003*", Geophys., Res. Let. **V.31,** 119605.
8. Technical Control of Construction (CTC) (1981b), "*Règles parasismiques Algériennes, RPA81*", Algiers.
9. Earthquake Engineering Research Institute (EERI) (1980), "*El Asnam earthquake, Algeria, 10th October 1980*", Report, 57. Learning from Earthquakes Program, Oakland, CA.
10. Mortgat, C.P. and Shah, H.C. (1978), "*Seismic Hazard Analysis of Algeria*"; Technical Report. The John A. Blume Earthquake Engineering Center, Department of Civil Engineering, Stanford University, Stanford, CA.
11. National Research Centre of Earthquake Engineering (CGS) (1999), "*Règles parasismiques Algériennes, RPA99*", Algiers.
12. National Research Centre of Earthquake Engineering (CGS) (2003), "*Règles parasismiques Algériennes, RPA99 version 2003*", Algiers.
13. Ait-Méziane, Y., Farsi, M. N. and Benouar, D. (2004), "*Seismic vulnerability estimation of representative building in the Algiers district of Bab El Oued (Algeria)*", European Earthquake Engineering, 1, pp. 27-36.
14. Lazzali, F. and Farsi, M.N., (2012), "*Seismic vulnerability assessment of buildings in Algiers area*", International Conference on Environmental and Civil Engineering, WASET, **Issue 61**, pp. 796-800, Dubai.

MATEC Web of Conferences **149**, 02041 (2018) https://doi.org/10.1051/matecconf/201814902041
CMSS-2017

Characteristics of Spectral Responses for a Ground Motion from Mediterranean Earthquake – ZEGHANGHANE Station (6.3Mw) in Morocco, and its Influence on the Structures

Mohamed AHATRI[1], Khadija BABA[1], Soukaina TOUIJRATE[1] and Lahcen BAHI[2]
[1]GCE Laboratory, High School of Technology-Salé, Mohammed V University in Rabat-Morocco.
[2]3GIE Laboratory, Mohammadia Engineering School, Mohammed V University in Rabat-Morocco.

Abstract. The earthquakes are abrupt and sudden, and, in few seconds, they can lead to huge casualties. Nowadays, several approaches can be used to better know the seismic movement than we can expect when and where will have an earthquake. These approaches kind, always begins with a full study of the seismic action: the seismic birth, the wave formation and propagation, accelerometers and associated spectrum representation, etc.

The northern region in Morocco has an important seismic activity, indeed, this region has been struck during its history by many destructive earthquakes. In this paper, we present and discuss the Mediterranean earthquake, January 25, 2016, saved in ZGH station located in Zeghanghane near of Nador city, Morocco.

In this case, we determine the spectral response of the ground motion for ZGH station, and study his influence on the structures as well as make a comparison with the requirements of the Moroccan seismic construction regulations (RPS) 2000 revised in 2011.

1 Introduction

The seismic action can be described as a transitory disturbance due to the aleatory propagation of seismic waves in the soil. The motion in the given site is generally governed by three main factors: the source, the path travelled by the seismic waves and the local geotechnical conditions.

Among the seismic motion characteristics, we have the released energy which propagates in the elastic waves form. These waves are formed near of the focus and propagate towards the ground surface. There are two elastic wave types which are the compression and the shear waves, each one has its own velocity.

The aim of this study, on the one hand, is to construct and examine the spectral response of the ground motion for ZGH station [1], on the other hand, is to study his influence on structure stability which already verified by the regulations.

For that reason, in the first part, a MATLAB code is developed to read the time versus acceleration data of Mediterranean earthquake, which obtained from the Institute National Geophysics (ING) Rabat, then to solve the ordinary differential equation and give various results like accelograms, displacement response spectrum, velocity response spectrum and acceleration response spectrum.

In the second part, we studied the structure stability, which composed of a ground floor, 4 floors and an under floor, with respect to the requirements of the Moroccan seismic construction regulations (RPS) 2000 revised in 2011 [2] and with respect to the real acceleration response spectrum saved in ZGH station [1] in order to perform a comparison between them.

2 Structure Analysis under a seismic action

2.1 Structure design

During an earthquake, the structure behaviour is mainly related to the vibrations generated by the latter. In this sense, the structure seismic analysis allows to determine the seismic forces acting in both the horizontal directions and the maximum deformations that this structure can withstand.

In order to design the building structure for resisting to earthquakes, two approaches exist [3]:

- The static approach equivalent, applicable only for the regular buildings with low and medium height. This approach is based on the static solicitations calculation from the forces system in the earthquake direction, and whose effects are equivalent to those of the seismic action.

- The dynamic approach is based on the maximum response of the structure by means of the response spectra adapted to the construction site and a direct calculation according to the time using an accelerograms adapted to the construction site. This approach is accepted when the building height and the regularity conditions required by the static approach are not satisfied.

2.2 Moroccan seismic construction regulations (RPS) 2000 revised in 2011

Following the Seismic construction regulations (RPS) 2000 revised in 2011, the earthquake evaluation on a structure is carried out using the following parameters:

- The maximum soil acceleration Amax obtained from the seismic zoning (article 5.2.2 of the RPS2000) is: Amax (Z0) = 0.04 g; Amax (Z1) = 0.07 g; Amax (Z2) = 0.10g; Amax (Z3) = 0.14g; Amax (Z4) = 0.18g, with an occurrence probability of 10% in 50 years [2];
- The maximum soil velocity Vmax obtained from the seismic zoning (article 5.2.2 of the RPS2000) is: Vmax(Z0) = 0,05m/s; Vmax(Z1) = 0,07m/s; Vmax(Z2) = 0,10m/s; Vmax(Z3) = 0,13m/s; Vmax(Z4) = 0,17m/s, with an occurrence probability of 10% in 50 years [2];
- A response spectrum in terms of acceleration for the horizontal motion relative to a type of site normalized to the unit acceleration;
- A response spectrum of vertical motion (2/3 of the horizontal spectrum).

The elastic response spectrum or computational spectrum is defined in the RPS2000 revised in 2011 for a relative damping ξ = 5% (reinforced concrete structures and metal frames with heavy external walls and partitions) by: A (T) = (Amax / g) D (T) [2] where g is the acceleration of gravity and D (T) the dynamic amplification factor (art.5.2.3.3 of the RPS2011 [2]). It represents the amplification of the accelerations in the structure compared to that of the ground.

Other parameters are involved in the elastic response spectrum to take account of the site effect (S, art.5.2.3.2 of the RPS2011 [2]), the importance coefficient (I, article 3.1.2 of the RPS2011 [2]) and the damping (μ = (5 / ξ) 0.4, article 5.2.3.3 (d) of the RPS2011 [2]). The inelastic response spectrum is obtained by applying an acceleration reduction factor called the behaviour factor or ductility coefficient (K, Art.3.3.4 of the RPS2011 [2]) as A (T) = (Amax / g) D (T) / K [2].

3 Response spectra by the Duhamel integration method

The response spectra of in acceleration, velocity and / or displacement allows to involve the notion of frequency movement content. Its purpose is to characterize an earthquake based on the response of a simple structure. By definition, an acceleration response spectrum is the curve giving the maximum acceleration of simple oscillators at a variable degree of freedom and natural frequency. Such an oscillator is shown in the figure 1, it is characterized by a mass m, a stiffness k and a damper C.

The motion equation of forced oscillations after the earthquake application in the accelerogram ag(t) form will be the following [4]:

$$ma(t) + Cv(t) + kD(t) = -ma_g(t) \qquad (2)$$

The differential equation resolution makes it possible to know the maximum acceleration undergone by the mass M.

Taking into account the damping definition ζ, the equation can be written in the form [5]:

$$a(t) + 2\omega v\zeta(t) + \omega^2 d(t) = -a_g(t) \qquad (3)$$

Where:

$$T = 2\pi\sqrt{\frac{m}{k}} \qquad (4)$$

$$f = \frac{1}{T} \qquad (5)$$

$$\omega = \frac{2\pi}{T} = 2\pi f \qquad (6)$$

k: The system stiffness

ω : The non-damped system pulsation (in radian / second);

T : Period (in seconds); the motion being periodic, the cycle duration is called the motion period;

f : Frequency (in hertz);

The solution is given by Duhamel's integral. The relative displacement is obtained [5]:

$$d(t) = -\frac{1}{\omega_D}\int_0^t a_g(\tau)e^{-\zeta\omega(1-\tau)} \sin\omega_D(t-\tau)d\tau \qquad (7)$$

$\omega_D = \omega\sqrt{1-\zeta^2}$: Pseudo-pulsation of the free oscillations damped;

τ : Integration variable.

It notes that the displacement value depends only on the pulsation ω, on the damping coefficient ζ and on the ground acceleration a_g.

The structures having always a very low damping, we can consider that $\omega_D = \omega$, which corresponds to a very little damped oscillator; in neglecting the secondary terms, the seismic motion components become:

The relative displacement [5]:

$$d(t) = -\frac{1}{\omega}\int_0^t a_g(\tau)e^{-\zeta\omega(1-\tau)} \sin\omega(t-\tau)d\tau \qquad (8)$$

The relative velocity [5]:

$$v(t) = -\int_0^t a_g(\tau)e^{-\zeta\omega(1-\tau)} \cos\omega(t-\tau)d\tau \qquad (9)$$

The pseudo-acceleration [5]:

$$a(t) = +\omega\int_0^t a_g(\tau)e^{-\zeta\omega(1-\tau)} \sin\omega(t-\tau)d\tau \qquad (10)$$

The formulas below with the earthquake registration (accelerogram) allow to calculate systematically for all simple oscillators possible (i.e. for the whole periods range and possible damping) the maximum response values in the displacements terms [d(t)] max and draw the corresponding graphics, called responses spectra travel. Similarly, we can draw the maximum responses in the velocity terms [v(t)] max and the acceleration [a(t)]max.

https://doi.org/10.1051/matecconf/201814902041

The displacement response spectra [5]:

$$S_d = [d(t)]_{max} = {}^{S_v}/_{\omega} = D_{max} \quad (11)$$

The velocity response spectra [5]:

$$S_v = [v(t)]_{max} = \omega S_d \quad (12)$$

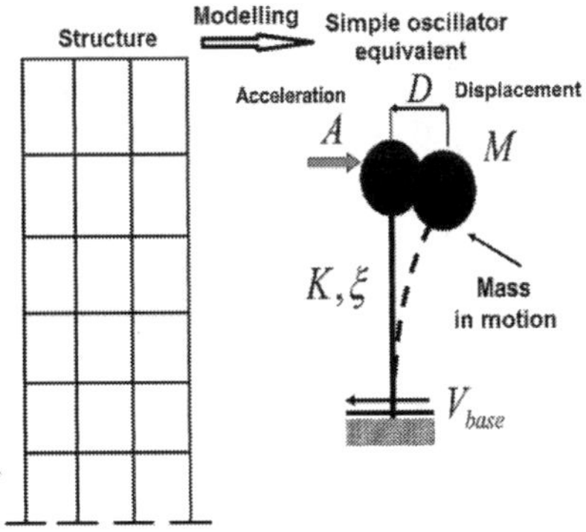

Fig. 1. Simple oscillator [3]

The acceleration response spectra [5]:

$$S_a = [a(t)]_{max} = -\omega^2 d(t) = \omega S_v = \omega^2 S_d \quad (13)$$

3 Case study

A strong earthquake occurred at 04:22 am on 25th January 2016 in the Mediterranean between Morocco and Spain [6], which affected the Nador and El Hoceima coasts [1]. It had a moment magnitude of 6.3 in Richter scale. The epicentre was located at 35.586, -3.690, and the depth was located at 26km [1].

In its initial assessment of the quake's impact, USGS said there was "a low likelihood of casualties and damage".

In this paper, we studied a structure component of a ground floor, 4 floors and an under floor by seismic analysis according to the RPS 2000 revised 2011, as well as by spectral analysis using the Mediterranean earthquake response spectra (ZGH station). We performed both analyses using the Robot Analysis software which works by the finite Element method.

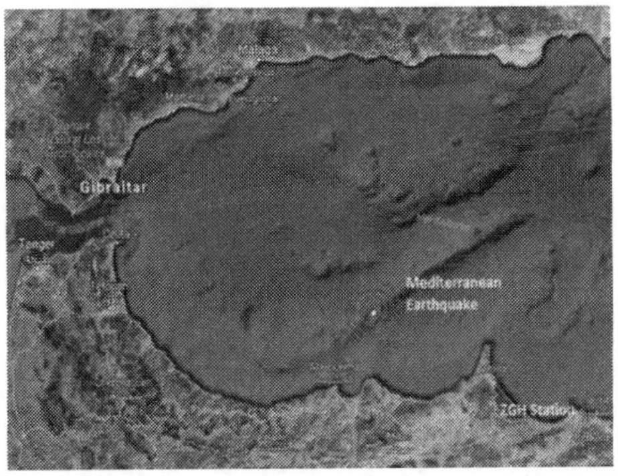

Fig. 2. The locations of the Mediterranean earthquake and the ZGH station

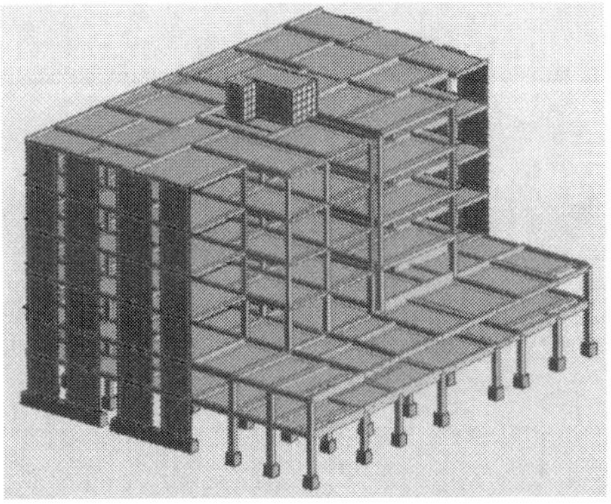

Fig. 3. The structure studied

4 Results et discussion

4.1 Acceleration

Based on the seismic inputs of Mediterranean earthquake (ZGH Station), the acceleration response is obtained.

The figure 1 shows the acceleration history time of the Mediterranean earthquake (ZGH Station).

For East, the maximum acceleration is 0.621g at 17.785s and the minimum acceleration is -0.542g at 18.85s.

For North, the maximum acceleration is 0.458g at 18.245s and the minimum acceleration is -0.3152g at 17.075s.

4.2 The response spectra of the ZGH Station

Using a MATLAB program which developed to solve the ordinary differential equation with the time history records of the earthquake Mediterranean. The response spectra are obtained.

The spectrum taken directly from the accelerogram shows a series of irregular peaks. However, the spectral displacements increase with period, thus, the spectral velocities and accelerations, as expected, show a variation in their tendency with a decrease of the period.

The figures show the response spectra of 5% damping.

From an engineering point of view, the infrastructure designers are more concerned with the structures response in the frequency range between 1 and 10 Hz [7]. Thus, the amplification factors of the ground motion elevated to this frequency range play an important role in seismic disasters.

Finally, depending on the derived equations (11, 12, 13) [5], the differential motion response spectra can be obtained for any time interval.

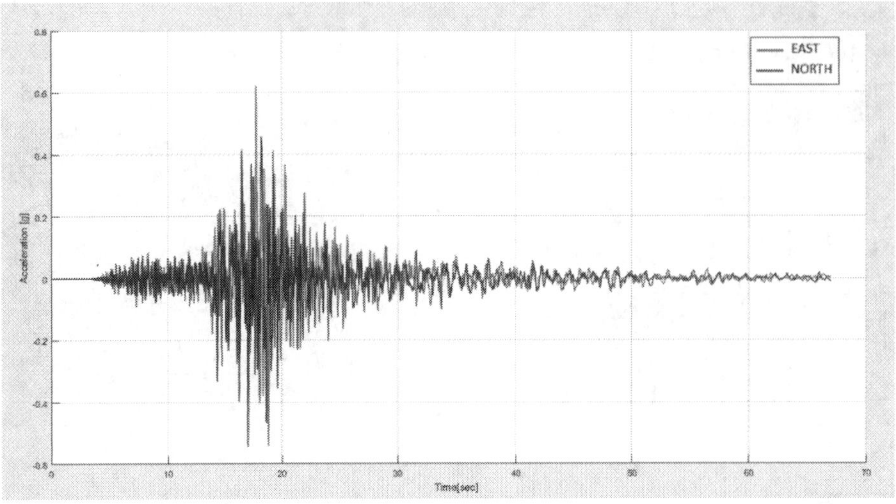

Fig. 4. Acceleration history time

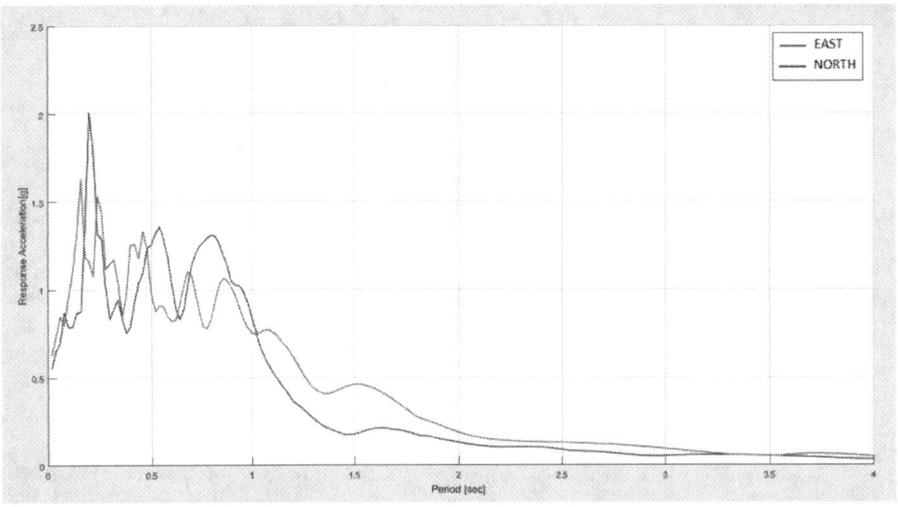

Fig. 5. Acceleration response spectrum (Sa)

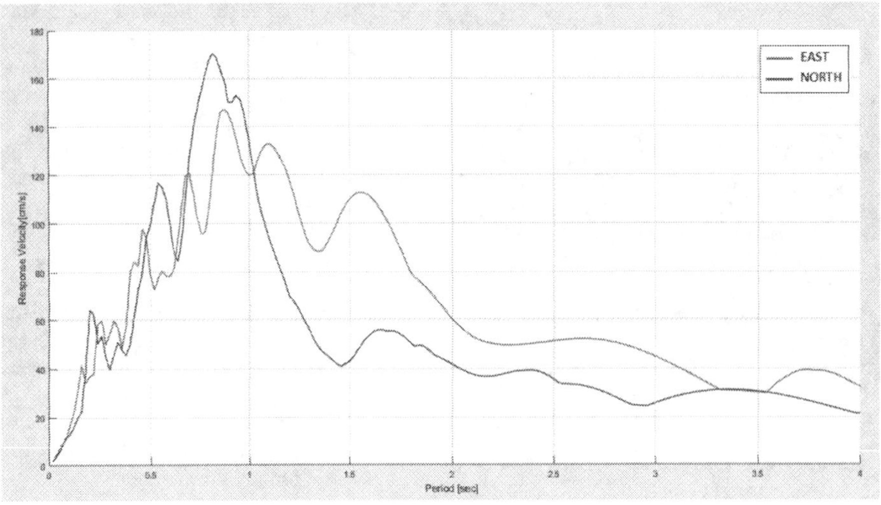

Fig. 6. Velocity response spectrum (Sv)

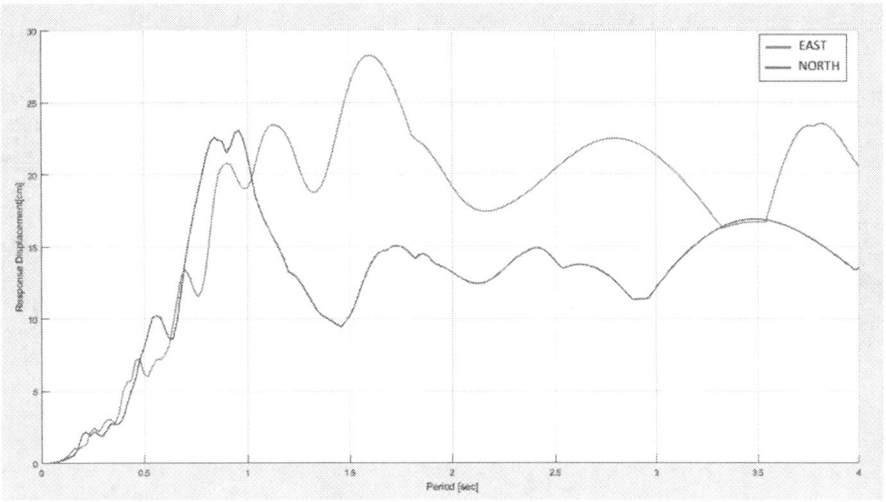

Fig. 7. Displacement response spectrum (Sd)

4.3 The response spectra influence on structures

Using a Robot Structural Analysis software, the displacements, the reactions, the deflections and the forces are obtained.

Using the structure study based on the spectral method by the Duhamel integration, and the study of the same structure according to the RPS2011 by the Robot Structural Analysis software most used in Morocco for the structures design which uses the finite elements method, we had been able to see the differences between the deformations and the solicitations of the elements.

As shown in the tables below, a difference between the total displacements between the study by the seismic regulation 2011 and the study by the spectral analysis on the ZGH Station of 5 cm according to X is 0.9 cm and according to Y we also note a slight difference between inter-floor displacements.

For the whole of deformations and solicitations, the maximum values of the spectral analysis are always lower than the maximum values of the seismic analysis by the RPS 2011, with the exception of the reactions and efforts which in the worst case is the ultimate limit state that depends only on the permanent and farms loads, so it remains the same values for the two types of analysis.

Table 1. Structure deflections values according to the two analysis types

Analysis types	Values	Ux[cm]	Uy[cm]	Uz[cm]
Seismic RPS 2011	Max	0.0	0.9	8.1
	Min	-0.0	-1.2	-5.1
Spectral ZGH Station	Max	0.0	0.4	8.1
	Min	-0.0	-0.6	-5.1

Table 2. Structure forces values according to the two analysis types

Analysis types	Values	Fx[T]	Fy[T]	Fz[T]	Mx[Tm]	Mx[Tm]	Mx[Tm]
Seismic RPS 2011	Max	336.10	17.29	147.82	107.54	104.08	16.87
	Min	-16.42	-0.97	-142.78	-104.46	-187.03	-1.20
Spectral ZGH Station	Max	336.10	9.41	147.82	107.54	104.08	15.05
	Min	-16.42	-9.19	-142.78	-104.46	-187.03	-14.45

Table 3. Structure reactions values according to the two analysis types

Analysis types	Values	Fx[T]	Fy[T]	Fz[T]	Mx[Tm]	Mx[Tm]	Mx[Tm]
Seismic RPS 2011	Max	28.55	13.67	336.10	4.87	5.78	0.38
	Min	-2.33	-4.47	-14.96	-1.13	-1.02	-0.03
Spectral ZGH Station	Max	13.82	7.35	336.10	2.49	3.05	0.20
	Min	-6.88	-6.90	-48.04	-1.61	-1.65	-0.11

Table 4 : Displacements global values according to the two analysis types

Analysis types	Values	Ux[cm]	Uy[cm]	Uz[cm]	Rx[Rad]	Ry[Rad]	Rz[Rad]
Seismic RPS 2011	Max	10.2	4.5	3.1	0.019	0.081	0.014
	Min	-0.7	-0.1	-87.7	-0.171	-0.069	-0.001
Spectral ZGH Station	Max	5.2	3.6	3.1	0.019	0.081	0.008
	Min	-2.8	-3.5	-87.7	-0.171	-0.069	-0.008

Table 5. Displacements inter-floor values according to the two analysis types

Analysis types	Nodes	Ux[cm]	Uy[cm]
Seismic RPS 2011	121	7.0	1.2
	141	7.6	1.5
Spectral ZGH Station	121	2.5	0.4
	141	2.4	0.5

Conclusion

The Robot structural analysis software, based on the finite element method, has been developed and improved for structural calculation, in which 5% damping has been considered for soil and structures.

In our case study, we found a difference between the two analyses, the first being based on the Seismic construction regulations and the second on the response spectrum, this difference in the global and inter-floor displacements, the deflections and the reactions shows that the results of the analysis by introducing a response spectrum of a real case satisfied the requirements of the seismic regulation.

On the other hand, the study shows the importance of studying and analysing the structures by response spectra recorded by accelerograms in addition to the Seismic construction regulations in order to have a general idea on the structures behaviour in an earthquake loading. The important strategic structures such as bridges and long-term administrative buildings, such as hospitals, are the structures most expected to undergo these movements. This is the case of structures near of the earthquake epicentre.

Taking into account the results, it is obvious that other studies of the influence of real response spectra on soils and structures are necessary, in order to improve Moroccan seismic construction regulation.

References

1. ING: Institute National Geophysics, Rabat-Morocco.
2. *RPS2000 révisé 2011*. (Applicable aux bâtiments). Ministère de l'aménagement du territoire, de l'urbanisme, de l'habitat et de l'environnement, Secrétariat d'Etat à l'Habitat, **78 p.**
3. Youssef BELMOUDEN. *Analyse spectrale non itérative des oscillateurs simples sous l'action des tremblements de terre. Bulletin de l'Institut Scientifique,* Rabat, section Sciences de la Terre, 2004, n°26, **1-10.**
4. Alain PECKER, livre de *Dynamique des sols*, Presse de l'école nationale des Ponts et Chaussées 1984.
5. Steven L. Kramer. *Geotechnical Earthquake Engineering*. Prentice Hall. Upper Saddle River, New Jersey **07458.**
6. USGS : US Geological Survey.
7. A.K. Abd El-Aal, *Ground motion prediction from nearest seismogenic zones in and around greater cairo area,* Egypt, Natural Hazards and Earth System Sciences 10 (2010) **1495–1511**.
8. Mohamed I.S. Elmasry. *Response spectra for differential motion of structures supports during earthquakes in Egypt. Housing and Building National Research* Center Journal (2012) 8, **217–227".**
9. B. Madani, F. Behnamfar, H. Tajmir Riahi. *Dynamic response of structures subjected to pounding and structure–soil–structure interaction.* Soil Dynamics and Earthquake Engineering 78 (2015) **46–600.**
10. ALLIN CORNELL. *ENGINEERING SEISMIC RISK ANALYSIS.* Bulletin of the Seismological Society of America. Vol. 58, No. 5, pp. **1583-1606**. October,1968.
11. Chopra A.K. 1995 *Dynamics of structures, theory and applicationto earthquake engineering*. Prentice Hall, Englewood Cliffs,New Jersey**.**
12. Juan C. Reyes, Luis Ardila-Bothia, J. Paul Smith-Pardo, J. Nicolás Villamizar-Gonzalez, Oscar A. Ardila-Giraldo. *Evaluation of the effect of containers on the seismic response of pile-supported storage structures.* Engineering Structures 122 (2016) **267–278.**
13. Xiaolan Pan, Zhi Zheng, Zhenyu Wang. *Estimation of floor response spectra using modified modal pushover analysis.* Soil Dynamics and Earthquake Engineering 92 (2017) **472–487.**
14. A. Douiri, T. Mourabit, A. Cheddadi, M. Chourak, S. Louhibi. *Spectral coincidences of background noise between soil and existing buildings in Imzouren city, Morocco*. Journal of Materials and Environmental Science (2015). **p366-376.**
15. A.F. Drennov, V.I. Dzhurik, S.P. Serebrennikov, E.V. Bryzhak, N.N. Drennova. *Acceleration response spectra for the earthquakes of the southwestern flank of the Baikal Rift Zone.* Russian Geology and Geophysics 54 (2013) **223–230.**
16. A.F. Drennov, V.I. Dzhurik, S.P. Serebrennikov, E.V. Bryzhak, N.N. Drennova. *The basic parameters of acceleration spectra for* $M \geq 5$ *earthquakes in the Baikal Rift Zone.* Russian Geology and Geophysics 56 (2015) **773–781.**
17. Bilal Ahmad and Imtiyaz Akbar Najar. *Comparative Seismic Analysis of EL Centro and Japan Earthquakes using Response Spectra Method.*

International Journal of Current Engineering and Technology. **Vol.6**, No.5 (Oct 2016).
18. Kiureghian, A., Neuenhofer, A. 1992 *Response spectrum method for multisupport seismic excitations.* Earthquake Engineering and Structural Dynamics 21(8) **713-740**.
19. Lukas Moschen, Christoph Adam, Dimitrios Vamvatsikos. *A response spectrum method for peak floor acceleration demands in earthquake excited structures.* Probabilistic Engineering Mechanics 46 (2016) **94–106.**
20. M. Mouzzoun, O. Moustachi, A. Taleb. *Assessment of the behaviour factor for seismic design of reinforced concrete buildings.* Journal of Materials and Environmental Science (2013). **p23-32**.
21. Masumi Yamada; et al. (2010). *Spatially Dense Velocity Structure Exploration in the Source Region of the IwateMiyagi Nairiku Earthquake.* Seismological Research Letters v. 81; no. 4;.Seismological Society of America. pp. **597–604**.
22. Newmark N.M. & Hall W. J. 1982 *Earthquake spectra and design.* Earthquake Engineering Institute.
23. Shinozuka, M. 1970 *Maximum structural response to seismic excitations.* Journal of Engineering Mechanics Division (ASCE) 96 **729-738.**
24. Thomas Wagener, Katsuichiro Goda, Mustafa Erdik, James Daniell,Friedemann Wenzel. *A spatial correlation model of peak ground acceleration and response spectra based on data of the Istanbul Earthquake Rapid Response and Early Warning System.* Soil Dynamics and Earthquake Engineering 85 (2016) **166–178.**
25. Trifunac, M., Gicev, V. 2006 *Response spectra for differential motion of columns paper II : Out-of-plane response.* Soil Dynamics and Earthquake Engineering 26(12) **1149-1160.**

https://doi.org/10.1051/matecconf/201814902042

Weights of evidence method for landslide susceptibility mapping in Tangier, Morocco

Mahfoud BOUSTA[1], Lahsen AIT BRAHIM[1]
[1]Université Mohammed V, Faculté Sciences, Département Sciences de la Terre, Unité de recherche GEORISK: Risques Géologiques, Télédétection et développement durable. Avenue Ibn Battouta Rabat – Agdal, Boîte Postale 1014, Rabat, Maroc

Abstract. Tangier region is known by a high density of mass movements which cause several human and economic losses. The goal of this paper is to assess the landslide susceptibility of Tangier using the Weight of Evidence method (WofE). The method is founded on the principle that an event (landslide) is more likely to occur based on the relationship between the presence or absence of a predictive variable (predisposing factors) and the occurrence of this event. The inventory, description and analysis of mass movements were prepared. Then the main factors governing their occurrence (lithology, fault, slope, elevation, exposure, drainage and land use) were mapped before applying WofE. Finally, the ROC curves were established and the areas under curves (AUC) were calculated to evaluate the degree of fit of the model and to choose the best landslide susceptibility zonation. The prediction accuracy was found to be 70%. Obtained susceptibility map shows that 60% of inventoried landslides are in the high to very high susceptibility zones, which is very satisfactory for the validation of the adopted model and the obtained results. These zones are mainly located in the N-E and E part of the Tangier region in the soft and fragile facies of the marls and clays of the Tangier unit, where landuse is characterized by dominance of arable and agricultural land (lack of forest cover). From a purely spatial point of view, the localization of these two classes of susceptibility is completely corresponding to the ground truth data, that is to say that all the environmental and anthropogenic conditions are in place for making this area prone to landslide hazards. The obtained map is a decision-making tool for presenting, comparing and discussing development and urban scenarios in Tangier. These results fall within the context of sustainable development and will help to mitigate the socio-economic impacts usually observed when landslides are triggered.
Key words: Weights of evidence, landslides, susceptibility map, spatial analysis, predisposing factors, probabilistic modeling, GIS, urban planning, Rif, Northern Morocco.

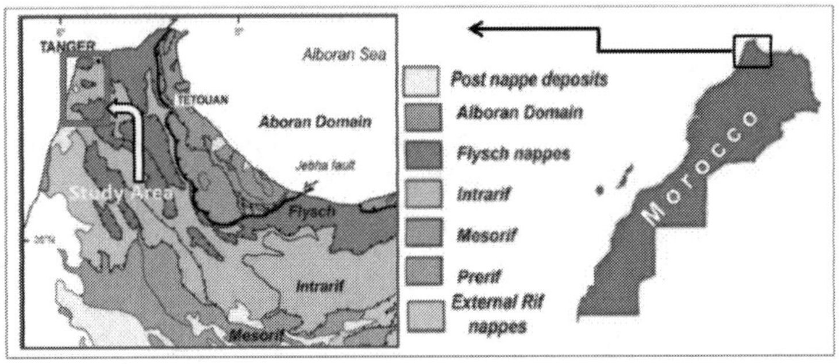

Fig. 1. Study area.

1 Introduction

Landslides belong to natural hazards facing Morocco, especially in the Rif regions [1-7]. It brought about significant environmental and socio-economical losses consequences in recent decades: Road cuts (34%); Subsidence of roads (18%); Destruction of habitat (18%); Collapse of railways (12%); destruction of a large number of hydraulic infrastructures .Every year about 50% (or more) of the total budget allocated to the DPTP (Provincial Commands of Public Works) of the Rif is destined for reinforcement, retaining and repair works of roads because of the occurrence of a large number of mass movements that most often damage the same areas already studied and treated, and it is due to the fact that landslides are not taken in their overall geodynamic context of the watershed, the use of non-adapted methods based on very subjective opinions, the inappropriate choice of scale and especially the pressure imposed by the urgency of reopening the road to traffic to supply the main cities and towns of the Rif. In view of the region's economic development with large land-use planning projects; It is of great interest to develop a landslide susceptibility map of a predictive nature to define the most exposed areas to mass movements. Thus, the main objective of this study is to identify the predictive variables for the assessment and development of this map using the weight of evidence theory [8]. Like other methods [9-15], it is based on the principle that "the past and the present landslide locations are the key to the future" [16]

2 Study area

The Tangier region, located at the NW end of the Moroccan Atlantic coast, is bounded on the north by the Strait of Gibraltar, on the south by Wadi Rharifa, on the west by the Atlantic Ocean and on the east by the plain of Fahs. From a demographic point of view, Tangier is the third largest city in Morocco with 974,000 inhabitants, with an average density of 451 inhabitants / km². The climate ranges between the humid Mediterranean and subhumid in warm winter. The average summer temperature is 28,6 °. The winter is characterized by its softness. The average temperature of the coldest month is 8,5° and the average annual rainfall is 800 mm. Violent winds reach 29% with a speed of between 25 and 50 km/h [17]. In winter, winds are humid from the west, linked to the Azores High which is centered on the Canary Islands, but when the anticyclone is displaced on the Iberian side, the often violent winds are NE direction. In summer it is a warm, dry east wind (Chergui) which blows in gusts (speed over 50kms / h) 2 to 3 days in a row, several times a month.Tangier is located in the external domain of the Rif's chain, where the Rif is the western end of the Maghrebian Alpine chain outcropping along the Mediterranean coast of North Africa from Calabria to the Gibraltar Arc [18-19]. The elevation of the region is characterized by massifs of low mountains bordered by hills with a powerful sandstone mass of the Numidian nappe which generally occupies a high structural position. It forms well marked ridges, armed by the numerous sandstone benches slicing on hills of a Cretaceous substratum. The extreme north of the littoral zone is characterized by a rocky coast. The plain of Fahs is open towards the south with sandy beaches which are sometimes associated with dunes. Such a morphology is the result of the combination of highly contrasted lithology, climatic system, active tectonic regime, and the erosive dynamics giving rise to several mass movements.

3 Landslides and predisposing factors

Using remote sensing imagery, existing maps and field data, we have inventoried landslides, prepared and mapped predisposing factors. Then, we have converted all maps into 30m pixel size's raster format with "Nord Maroc: Lambert Conformal Conic" geographic projection using Arcgis10.Thus, we have mapped 118 landslides that were all checked on the ground. It is difficult to differentiate the typology of mass movements due to the complexity of the land nature and morphology [1]. According to the most dominant morphological traits [20], we recognized four types: Rockfalls, Slides, Mud flows and complex landslides .The typology varies as much as possible the above categories movements combines.According to literature, lithology, faults, elevation, gradient of slope, slope exposure, Drainage network and land use are the most commonly used factors to assess landslide susceptibility in the region. Thus, we used all of them to lead this study. The lithological map (Figure2.b) contains five classes sandstone (35%), marls and argillites (21%), marls sandstone (18%), silts (18%) and sands (8%). The occurrence and spatial distribution of inventoried landslides show that a large number of complex slides and mudflows develop on outcrops of argillite marls and sandstones. The faults shapefile was made by digitizing the 1/50,000th geological maps combined with image processing and field studies. This shapefile was later converted to fault density map by using Kernel method and classified in 5 classes. From ASTER digital elevation model (GDEM), we have derived elevation, slope, and exposure factors. In the elevation map we recognize different geomorphological zones: Low altitudes between 0 and 100m of Fahs, Charf Laquab, Marhar, El Hachef and Asilah plains. Medium altitudes between 100 and 300m corresponding to Dar Zhirou and Haouta Ben Mediar slope, and the high altitudes between 300 and 700m representing the ridge lines of Randa (326m), Haouch Ben Kreaa(643m) and Dar Chaoui (=477m). The slope map (Figure2.e) was classified into five classes. We observed that the most favourable slopes to instability are those whose inclination is moderate. Based on the knowledge of our study area, we have classified the map into five classes: N and flat; NE-E-SE, SW-W; NW. Landuse mapping was carried out by interpreting satellite images and aerial photographs' supervised multi-spectral classification coupled with vegetation cover data o and especially of the forest species made available to us by the HCEFLCD (High Commission for Water and Forests and the Fight against Desertification). This work has been supplemented and updated by collected and/or verified data by our observations at a number of points accessible on the ground (2015-2016). The final document (Figure2.g) has undergone several mergers and adjustments of homogeneous objects in order to make its reading and use more simplified. It then contains the following classes: Buildings and bare land; water and swamps; forests; arable and agricultural land; natural vegetation. The density map of the hydrographical network (Figure2.g) takes into account both the cumulative length of the drainage network and the presence of the hydrographical nodes. The map realized contains 5 classes of density namely: very low; low; moderate; high and very high.

4 Modelling Methodology

Weight of Evidence (WofE) is a log-linear version of the Bayes general theorem, and was originally applied in medicine to diagnose certain diseases. It's a bivariate analysis consists of a relation between the predictive variables of a phenomenon (PV) and the occurrence of this phenomenon (dependent variable, DV: in this case the landslides). The theory is based on three hypotheses [14]: (a) potential landslides will occur under the same conditions as in the past; (b) the predisposing factors are priori known and introduced into the analysis; (c) all landslides are inventoried on the study area. It uses Bayesian statistics to calculate the degree of spatial association between a set of variables DV and predictive factors PV, assigns weights to them, and then uses a mathematical summation technique to summarize the weighted factor maps in one posterior probability map.The a priori probability is the probability that a unit of land (calculation grid) has the dependent variable (VD), calculated from its density on the study area. The posterior probability is estimated after the calculation of the priori probabilities according to the density of DV for each predictive variable PV, in our case each predisposing factor. This association between the DV and the PV will thus make it possible to calculate a positive weight W + and a negative weight W- for each predictive variable, corresponding respectively to a greater or a lower probability of occurrence of DV.

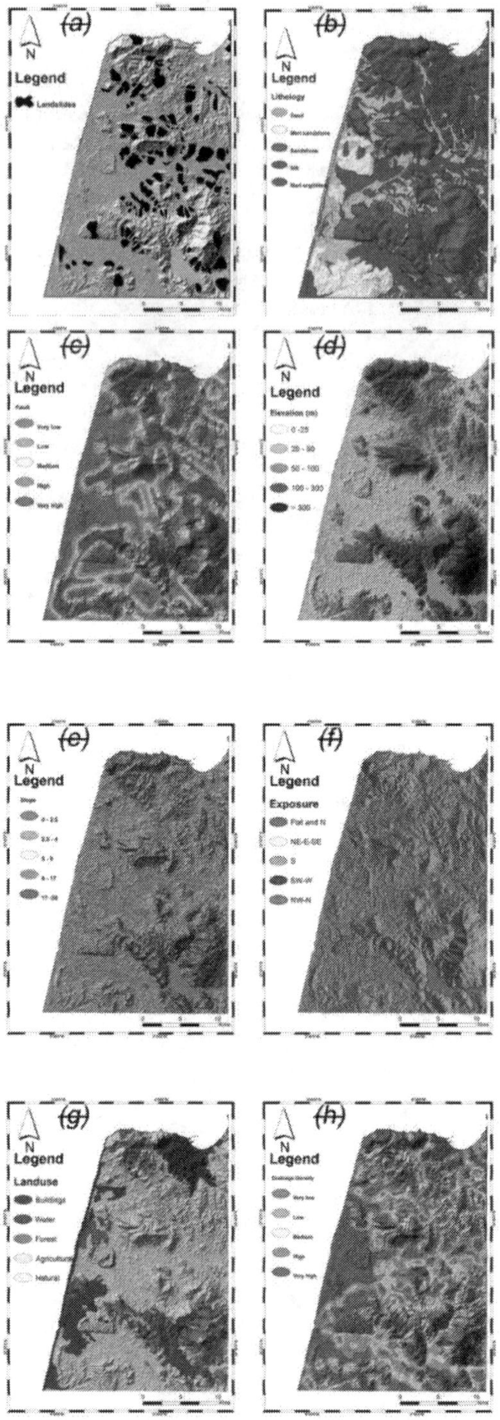

Fig. 2. Landslides and factors maps

$$W^{+} = \ln\frac{P\{PV/DV\}}{P\{PV/\overline{DV}\}} \quad (1)$$

$$W^{-} = \ln\frac{P\{\overline{PV}/DV\}}{P\{\overline{PV}/\overline{DV}\}} \quad (2)$$

Landslide susceptibility and treatments were performed using the ArcGIS' ArcSDM® extension [21-23] Data collection and preparation is done using a multisource approach (ground truth, thematic maps, spatial techniques, etc.). However, the different stages of this approach can be summarized in these points: Preparation of data, WofE application, model validation and landslide susceptibility map generation.

5 Results and discussion

5.1 Predictive factors' weights

Analyzing W +, W- and final weight (W) values assigned to each of the classes of the predictive variables, has allowed us to evaluate and interpret the respective roles of each predisposing factors in the spatial occurrence of landslides. We noted that land use, elevation and fracturing play a major role and that the density of the drainage and the exposure are relatively less influential whereas the slope and the lithology occupy an intermediate place in the triggering of the landslides. Indeed, the value of the maximum weight contrast between the five classes of the drainage density variable is only 0.4, while it is -2.1 for land use and -1.3 for elevation. Slope and lithology variables have a maximum contrast value of 0.92 and -0.95 respectively. Among all the predictive variables, land use is the one that has the greatest impact on the model, notably due to the low susceptibility of forests in landslides' triggering, followed by elevation because of the low susceptibility of high altitudes. Conversely, lithology is characterized by the high susceptibility of marl-argilites and fracturing characterized by the high susceptibility of zones with average fault concentration.

5.2 Conditional Independence assessment

WofE modeling assumes conditional independence (CI) between the layers of evidence (predictive factors). If all themes of evidence are conditionally independent, then the (predicted) calculated number of landslides will be equal to the observed [23]. In practice, when working with geological data, conditional independence (CI) is always violated [22-23] The question is that the expert must assess the degree of the violation (Bonham-Carter, 1994). The ArcSDM® tool provide 3 tests for conditional independence between layers:

1. Conditional independence ratio (CI ratio): The CI ratio is equal to n / T, where values below 1.00 may indicate conditional dependence among two or more sets of data. Bonham-Carter [22] suggests that values <0.85 may indicate a problem..

2. The overall CI test was based on an informal "10 - 15% rule" proposed by Bonham-Carter [22] which suggests that if the predicted number of deposits (T) is more than 10 - 15% greater than the observed number of deposits (n), then there may be an issue with the extant of violation of conditional independence. T is calculated as the sum of the posterior probabilities weighted according to unit cell area [22], while n is simply the number of training points.

3. Agterberg-Cheng (AC) test" [23] tests whether "T-n" is "significantly greater than zero". This is a one-tailed test of the null hypothesis that T-n=0. The test statistic is (T-n)/σT.

Factor	Class	Area (km2)	W+	W-	C	W
Litho	Silt	120.07	0	0	0	-0.26
	Sandstone	234.35	0.105	-0.06	0.165	-0.26
	Marl-sandstone	57.67	-0.9	0.057	-0.96	-0.26
	Marl-argilites	142.45	0.669	-0.26	0.932	0.669
	Sand	121.22	0.119	-0.03	0.147	-0.26
Fault	Very low	155.91	-0.8	0.162	-0.96	-0.8
	Low	95.77	0.271	-0.05	0.323	0.108
	Midium	158.87	0.743	-0.36	1.105	0.743
	High	148.45	-0.01	0.003	-0.01	0.108
	Very high	114.84	-1.61	0.164	-1.78	-1.61
Slope	0 - 2.5	202.7	-0.11	0.043	-0.15	-0.12
	2.5 - 4	60.13	0.809	-0.12	0.925	0.809
	4 - 8	149.65	-0.11	0.031	-0.14	-0.12
	8 - 17	217.62	-0.05	0.022	-0.07	-0.12
	17 - 50	43.75	-0.62	0.033	-0.65	-0.12
Expo	P-N	62.93	-0.27	0.025	-0.29	-0
	NE-E-SE	235.93	0.288	-0.19	0.477	-0
	S	84.11	-0.15	0.019	-0.17	-0
	SW W	172.92	-0.18	0.055	-0.23	-0
	NW	117.96	-0.2	0.038	-0.24	-0
Landuse	Buildings & bareland	53.91	-0.41	0.03	-0.44	-0.29
	Water & swamps	39.07	0	0	0	-0.29
	Forsts	83.78	-1.99	0.125	-2.12	-1.99
	arable & agricultural land	382.59	0.302	-0.57	0.87	0.302
	Natural vegetaion	115.5	0.066	-0.01	0.08	-0.29
Elevation	0 - 25	285.55	0.077	-0.06	0.136	0.137
	25 - 50	133.62	0.353	-0.11	0.458	0.137
	50 - 100	127.43	0.157	-0.04	0.196	0.137
	100 - 300	111.8	-1.17	0.137	-1.31	-1.17
	>300	15.45	0	0	0	0.137
Drainage	Very low	234.04	-0.5	0.198	0.337	-0.5
	Low	223.7	-0.08	0.037	0.304	0.099
	Medium	162.74	0.336	-0.13	0.303	0.099
	High	48.01	0.811	-0.09	0.416	0.811
	Very high	5.35	0	0	0	0.099

Table 1. Weights and contrast(c) for each factor

 https://doi.org/10.1051/matecconf/201814902042

Probability values greater than 95% or 99% indicate that the hypothesis of CI should be rejected, but any value greater than 50% indicates that some conditional dependence occurs [23].

The different tests showed that the model respects the hypothesis of conditional independence, since there is no major difference (T-n = 3.3) between the expected landslides (T) and those observed (n). We note that the value of the CI ratio = 0.94 is greater than the threshold value of 0.85. The 50.1% value of the Agterberg-Cheng test is almost perfect. Thus, the hypothesis of conditional independence is attested to be valid for the model, which is also confirmed by the third Overall test CI = 99.8%.

5.3 Validation

To check the performance of the model, we analyzed the posterior probability map, using the SRC and RPC efficiency curves which are the most important means to evaluate and validate the probability map. We used 50% of all landslides inventoried to create the SRC (success-rate curve) that test how well the posterior probability fits the used. We also used the rest of landslides (50%) that were not introduced into the model to generate the PRC (Prediction-rate curve) which is a blind test that evaluates how well our model predicts unknown landslides. It is therefore the true test of validation of our model.

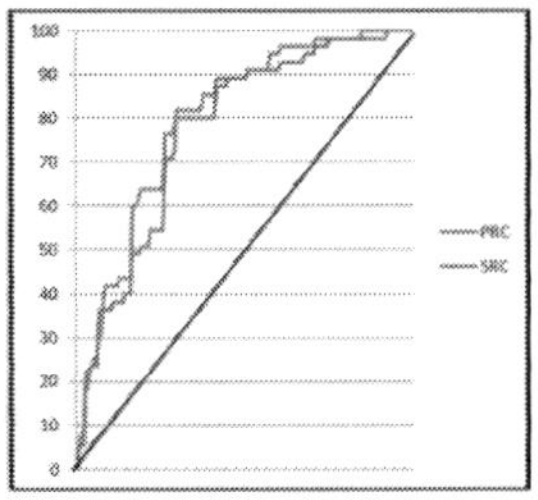

Fig. 3. Success and Prediction rate curves.

Figure3 shows the prediction and success rate curves of the landslide susceptibility map. The shape of the two curves is very similar, which confirms that 50% of the landslides points are sufficient to obtain good results. The two curves show that the model has a success efficiency of 70.0% for known slides distribution and an unknown ones' prediction efficiency of 70.7%.

After validation of the probability map, it was classified into five classes using the reclassification of arcgis' Naural breaks (Jenks).The statistical result of this superposition showed that 60% of inventoried landslides are in the zone of high to very high susceptibility, which is very satisfactory for the validation of the resulting map. High susceptibility zones are located mainly in the NE and E part of the Tangier region, first between the southern slope of Jbel Ronda (spartel) and the north slope of Jbel Zhirou and then north of the slopes of the hills of Haouta Ben Mediar and Haouch Ben Kreaa. They are superimposed on the soft and fragile facies of the marls and clays of the Tangier unit, whose land use is dominated by arable and agricultural land and by bare ground (absence of forest cover), where slope is between 7 ° and 17 ° and the fracturing and the drainage densities are high to very high (with zones of tectonic nodes).From a purely spatial point of view, the localization of its two susceptibility classes is in a complete agreement with the data of the ground truth. That is to say that all the environmental and anthropogenic conditions are met and favourable for this zone to be predisposed to the triggering of the landslides. The zones of low to very low susceptibility present about 20% of instabilities. It covers the areas of the coastal plains between Asilah to the south and the cave of Hercules to the north. The flatlands at the level of the valleys of wadis of Tahadart and El Hachef which can be explained by the very soft slopes generally less than 7°. These slopes, although unfavourable to the genesis of gravity movements, are characterized by severe gullying (badlands) and runoff in marly and clayey formations, especially during periods of occasional floods. Areas of low susceptibility are the reliefs where the most resistant lands of the Tangier region (Numidian sandstones) are partly stabilized by the development of a high density forest.

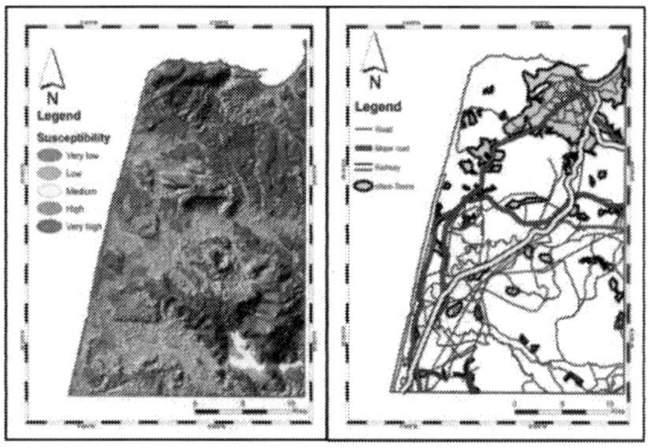

Fig. 4. Landslide susceptibility map(left) and buildings and transportation map(right)

6 Conclusions

In this paper, Weight of Evidence method was applied for landslide susceptibility mapping in Tangier. After classifying the posterior probability, the efficiency of how well the model distributes used landslides has been calculated. It was found to be equal to 71%. Then, the obtained susceptibility map has been furthermore validated by a cross-tabulation with the non-used landslide to evaluate the prediction accuracy of the model (70%). Moreover, the resulting landslide susceptibility map shows that the high to very high susceptibility zones have the highest percentage (60%) of all inventoried landslides in comparison with the other zones. Until 2000 the construction and buildings were focused in low susceptibility areas, but since 2004 the urban extension has began to occupy high susceptibility zones , which tends to increase vulnerability and therefore the risk of mass instability. Thus, the obtained landslide susceptibility map represents a clear view of the present instability and its possible future occurrences. It may helps widely in managing urbanism and construction in Tangier.

References

1. Milles-Lacroix A. Les glissements de terrains. Présentation d'une carte prévisionnelle des mouvements de masse dans le Rif (Maroc septentrional). Mines et Géologie, n° 27. p. 45-55 (1968).
2. Ait Brahim L, Chotin P, Hinaj S, Abdelouafi A, El Adraoui A, Nakhcha C, Dhont D, Charroud M, Sossey Alaoui F, Amrhar M, Bouaza A, Tabyaoui H, Chaouni A. *Paleostress evolution in the Moroccan African margin from Triassic to Present. Tectonophysics*, **Volume 357**, Issues 1-4, Pages 187-205 (2002).
3. El Khattabi J. and Carlier E. Tectonic and hydrodynamic control of landslides in the northern area of the Central Rif, Morocco. Engineering Geology p. 255–264 (2004).
4. El Kharim Y. *Etude des mouvements de versants dans ta région de Tétouan (Rif occidental): lnvenentaire, analyse et cartographie.* Thèse de Doctorat Es-Sciences, Université Abdelmalek Essaadi, Tétouan. Maroc, 250 p(2002).
5. Mansour M. Et Ait Brahim L. *Utilisation de la télédétection pour l'analyse de la fracturation du domaine interne rifain (Maroc): relation avec la répartition des sources.* Télédétection, **Vol. 5**, n°1-2-3, p 95-103(2005).
6. Ait Brahim L., Sossey Alaoui F., Siteri H., Tahri M., Baghdada B., Prise en compte des glissements de terrain dans la quantification des pertes en sols dans le bassin versant Nakhla. Bulletin du Réseau d'Erosion, n° 21, p.50-64 (2002).
7. M.Mastere, B. Van Vliet Lanoë & L. AïtBrahim L, "Land use mapping and its relation to mass wasting and gullying in NorthWestern Rif (Morocco)." Geomorpholog, relief, process, environment, 3, 335–352, (2013).
8. Bonham-Carter G.F, Agterberg F.P, Wright D.F. *Weights of evidence modeling: a new approach to mapping mineral potential. In: Agterberg, F.P., Bonham-Carter, G.F. (Eds.), Statistical Applications in the Earth Science*: Geological Survey of Canada Paper, **vol. 89-9**, pp. 171–183 (1989).
9. Soeters R et Van Westen C.J. Slope instability recognition, analysis, and zonation. In: Turner, K.A., Schuster, R.L. (Eds.), Landslides: Investigation and Mitigation. Transport Research Board Special Report 247. , pp. 129– 177(1996).
10. Guzzetti F, Carrara A, Cardinali M, Reichenbach P. Landslide hazardevaluation:areview of current techniques and their application in a multi-scalestudy, central Italy. Geomorphology 31, 181–216(1999).
11. Dai F.C, Lee C.F, Ngai Y.YLandslide risk assessment and management overview. Engineering Geology 64, 65–87.(2002).
12. Agnesi V, Conoscenti C, Di Maggio C, Iudicello C, Rotigliano E Landslide hazard analysis in the Giardo River Basin (Middle-Western Sicily). Proceedings of the Workshop on Geomorphological sensitivity and system response. Università di Camerino - Università di Modena, Camerino, Italy, pp. 3–11(2003).
13. Agnesi V, Di Maggio C, Fiorito S, Rotigliano E Analytical approach for landslide hazard assessment in El Salvador (C.A). Proceedings of the symposium on Geomorphology: from expert opinion to modelling. CERG Editions, Strasbourg, France (2002).
14. Thiery Y. *Landslide susceptibility in the Barcelonnette basin (French South Alps): morphodynamic cartography, spatial analysis and probabilistic modelling.* PhD Thesis, Caen/basse-Normandieuniversity, Frensh. 445p (2007).
15. Thiery Y, Malet J-P, Sterlacchini S, Puissant A, Maquaire O. Landslide susceptibility assessment by bivariate methods at large scales: Application to a complex mountainous environment. Geomorphology 92 p38–59 (2007).
16. Carrara A, Cardinali M, Guzzetti F and Reichenbach P. GIS-based techniques for mapping landslide hazard. In "Geographical Information System in assessing natural hazard" A. Carrara & F. Guzzetti (èds), KlinverAcadPubl., Dodrecht, pp. 135-176 (1995).
17. El Gharbaoui A., La terre et l'Homme dans la péninsule tingitane. Etude sur l'Homme et le milieu dans le Rif occidental. Travaux de l'Institut Scientifique, série Géologie et Géographie Physique, Rabat, n° 15, 439 p (1981).
18. Didon J., Durand Delga M. Et Kornprobst J. Homologies géologiques entre les deux rives du détroit de Gibraltar (Cordillères bétiques occidentales, Espagne). Bull. Soc.Géol.Fr., (7) XV, 2, p.77-105 (1973).
19. Durand-Delga M. et Mattauer G. sur l'origine ultra-rifaine de certaines nappes du Rif septentrional. Comptes Rendus de la .Société Géologique Française, n° 2, p. 22-23(1960).
20. Varnes D.J. Slope Movement Types and Processes. In Special Report 176: Landslides: Analysis and Control. In: Schuster, R.L., Krizek, R.J., (Eds), Transport Research Board, National Research, p11–33 (1978).
21. Sawatzky, D. L., Raines, G. L., Bonham-Carter, G. F., Spatial Data Modeller (SDM) for ArcMAP 9.3 geoprocessing tools for spatial data modelling using weights of evidence, logistic regression, fuzzy logic and neural networks (2009).
22. Bonham-Carter G.F. Geographic Information Systems for Geoscientists: Modeling with GIS. Pergamon Press, Canada (1994).
23. Agterberg, F.P., and Cheng, Q., *Conditional independence test for weights-of- evidence modeling*: Natural Resources Research, **v. 11**, no. 4, p. 249-255 (2002),.

Evaluating seismic reliability of Reinforced Concrete Bridge in view of their rehabilitation

Hasnae BOUBEL*, Oumnia ELMRABET, Mohamed ROUGUI , Ouadia MOUHAT, Echebba EL Mehdi.

Structure LGCE l'EST, Salé -CED de l'EMI, Mohammed V University in Rabat-MOROCCO

Abstract. Considering in this work, a simplified methodology was proposed in order to evaluate seismic vulnerability of Reinforced Concrete Bridge. Reliability assessment of stress limits state and the applied loading which are assumed to be random variables. It is assumed that only their means and standard deviations are known while no information is available about their densities of probabilities. First Order Reliability Method is applied to a response surface representation of the stress limit state obtained through quadratic polynomial regression of finite element results. Then a parametric study is performed regarding the influence of the distributions of probabilities chosen to model the problem uncertainties for Reinforced Concrete Bridge. It is shown that the probability of failure depends largely on the chosen densities of probabilities, mainly in the useful domain of small failure probabilities.

KEYWORDS: Reliability, ANOVA, Nonlinear, Analysis of variance, finite element method, ANSYS

1 Introduction

Many structures, including bridges, built before 1970 are not adequately detailed for seismic actions. This means bridge structures have historically been vulnerable to seismic loading, with a large number of examples of damage occurring to both superstructure and substructure elements and, in some cases, complete and catastrophic collapse. The turning point event in changing structural design philosophy was the 1971 San Fernando earthquake [1]. That event caused severe damage to major lifelines, including the transportation lifeline. Bridges, an important component of the transportation lifeline, suffered major damage. This prompted almost all the state departments of transportation in earthquake prone areas to upgrade their design specifications and structural details to resist earthquake action, as did the Federal Highway Administration. However, even with this, there are still many existing older bridges that were designed and built prior to 1971. It is critical that these bridges be upgraded to current seismic design standards in regions with strong earthquake potential. While many methods and approaches are available to accomplish this, methods to systematically and rationally compare these methods and select the optimal method are lacking, particularly in terms of target performance levels [2-3]. The aim of this research is to explore one such method based on fragility analyses.

2 Materials and methods

2.1. Modeling the Structure

The bridge studied is located in province Al Hoceima, This structure consists of five simple spans of 35m each. In turn, each bay consists of three equidistant beams transversely. The platform of the structure has a width of 10 m with a floor of 7.80 m and two sidewalks of 1.50 m and 0.70 m. The section of 2.30 m total height deck is formed by three beams spaced 3.40 m.

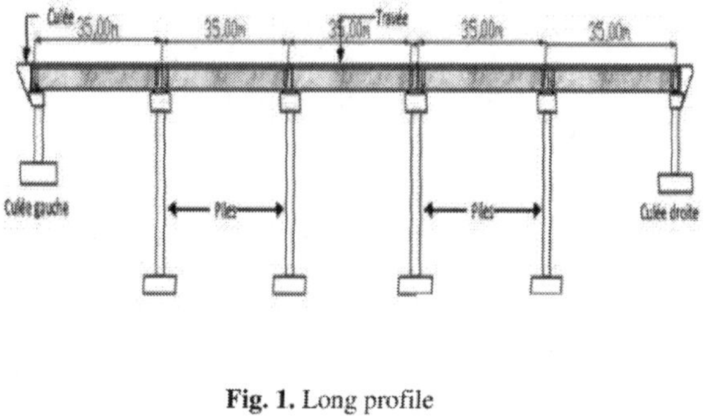

Fig. 1. Long profile

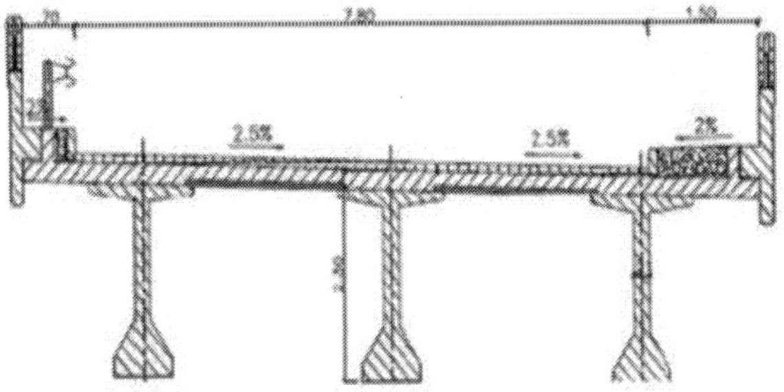

Fig. 2. Cross section of apron

2.2 Loading Idealization

The purpose of this study is to analyze the effect of horizontal and vertical earthquake ground motion on bridges. The focus is on the variations of earthquake loading rather than the gravity load. Loading considers the gravity load (self-weight, superimposed load and traffic load) and earthquake loading.

The dead load of the deck is distributed to the beams based on their respective tributary widths. Superimposed dead loads (wearing course, future wearing surface, railings, barriers), with the exception of footpath loads, are to be distributed equally to all beam lines. Superimposed load is achieved by multiplying the density of material by the area of the elements. The calculation of the superimposed load is as follow:

System A (L):

$AL=\max(0{,}23+36/L+12;0{,}4-0{,}2L/1000)en\ t/m^2$

Case of a single loaded channel, we have: a1 = 1, a2 = 0.96, V = 3.65 m, hence:

$$A\ (lc) = 3.56\ t/ml$$

Case of two charged channels, we have: a1 = 1, a2 = 0.96, V = 7.3 m, hence:

$$A\ (lc) = 7.12t/ml$$

System Bc:

The Bc convoy consists of one or at most two standard trucks per lane. In the transverse direction the number of queues is less than or equal to the number of channels. The characteristics of the Bc convoy are shown in the figure below.

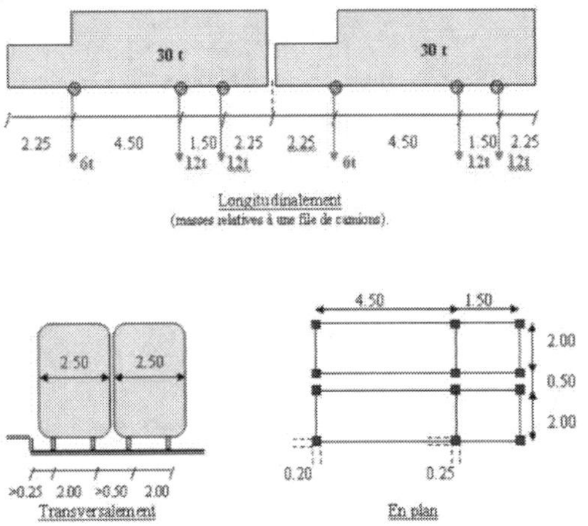

Fig. 3. The charging system Bc

System Bt :

A tandem consists of two axles with pneumatic single wheels. The characteristics of the system Bt are shown in figure 4.

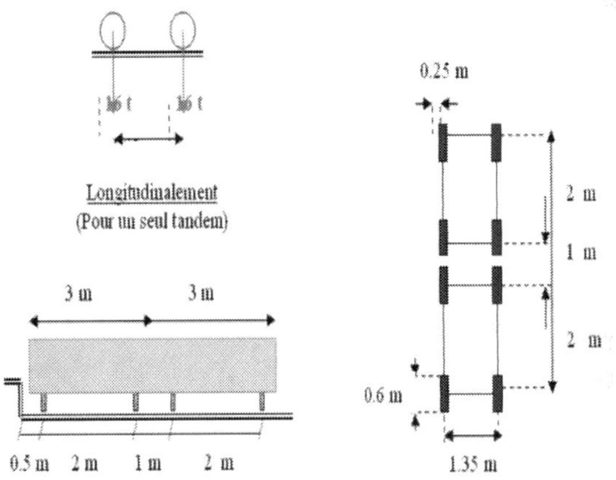

Fig. 4. System Bt

3 Earthquakes - Dynamic Analysis of Bridges

3.1 Fundamentals of Dynamic Analysis

Every structure acts statically and dynamically when subject to displacements or loads. In dynamic responses, the structures will have additional inertial forces which, according to the second law of Newton, are equal to their mass multiplied by acceleration. Thus, if the loads or displacements are enforced very slowly, the inertia forces can be ignored since the time can be assumed as zero and a static load analysis can be justified. Hence, dynamic analysis is a simple extension of static analysis.

Wilson (2002) stated that the force equilibrium of a multi-degree-of-freedom lumped mass system as a

 https://doi.org/10.1051/matecconf/201814902043

function of time can be expressed by the following relationship:

$$F(t)I + F(t)D + F(t)s = F(t) \quad (1)$$

in which the force vectors at time *t* are:
$F(t)I$: is a vector of inertia forces acting on the node masses
$F(t)D$: is a vector of viscous damping, or energy dissipation, forces
$F(t)S$: is a vector of internal forces carried by the structure
$F(t)$:is a vector of externally applied loads
Equation (1) is based on physical laws and is valid for both linear and non-linear systems if equilibrium is formulated with respect to the deformed geometry of the structure.
For many structural systems, the approximation of linear structural behaviour is made to convert the physical equilibrium statement, to the following set of second-order, linear, differential equations:

$$M\ddot{u}(t)a + C\dot{u}(t)a + Ku(t)a = F(t) \quad (2)$$

in which *M* is the mass matrix (lumped or consistent), *C* is a viscous damping matrix (which is normally selected to approximate energy dissipation in the real structure) and *K* is the static stiffness matrix for the system of structural elements. The time-dependent vectors $u(t)a$, $\dot{u}(t)a$ and $\ddot{u}(t)a$ are the absolute node displacements, velocities and accelerations, respectively.
For seismic loading, the external loading $F(t)$ is equal to zero. The basic seismic motions are the three components of free-field ground displacements $u(t)$ that are known at some point below the foundation level of the structure. Therefore, equations can be written in terms of the displacements $u(t)$, $\dot{u}(t)$ velocities and accelerations $\ddot{u}(t)$ that are relative to the three components of free-field ground displacements (Wilson, 2002).

4 Results and discussion

4.1 Analyses de variance (ANOVA)

For each class of bridges, the response of the elements considered as the most vulnerable was analyzed according to the variation of the various parameters [9]. The elements whose response was followed are presented in Table 1.

4.2 Parametric study for the perfect reinforced concrete bridge

A full factorial defined in table 2 and consisting of 3^3 =27 combination was formed. a finite element model was built under ANSYS software. a finite element model convergence, for the case of the reinforced concrete bridge, the calculations were carried out in the Ansys software for the entire combination 27.

Table 1. Range of variation of the considered factors reinforced concrete bridge

Parameter	Height (m)	density	Load(t)
lower threshold	6	2500	1,1E+02
intermediate value	7	3000	1,2E+02
higher threshold	8	3500	1,3E+02

Table 1 is a full factorial design of experiment table consisting of $3^4 = 81$ combinations was formed. Calculations were performed for each of these combinations under Ansys software. The most adverse case was found to be given by the parameters shown in table 2. *Equation II*

Table 2. The obtained stress as function of the combination number.

Combination number	Height (m)	Load (t)	density	stress (Pa)
1	6	1,2E+02	2500	427582
2	6	1,2E+02	3000	427582
3	6	1,2E+02	3500	427581
4	6	1,25E+02	2500	631229
5	6	1,25E+02	3000	631229
6	6	1,25E+02	3500	631229
7	6	1,3E+02	2500	783241
8	6	1,3E+02	3000	783241
9	6	1,3E+02	3500	783241
10	7	1,2E+02	2500	327582
11	7	1,2E+02	3000	337765
12	7	1,2E+02	3500	347544
13	7	1,25E+02	2500	531279
14	7	1,25E+02	3000	521276
15	7	1,25E+02	3500	506547
16	7	1,3E+02	2500	638761
17	7	1,3E+02	3000	603298
18	7	1,3E+02	3500	380765
19	8	1,2E+02	2500	367572
20	8	1,2E+02	3000	357598
21	8	1,2E+02	3500	427598
22	8	1,25E+02	2500	431229
23	8	1,25E+02	3000	421879
24	8	1,25E+02	3500	402971
25	8	1,3E+02	2500	483298
26	8	1,3E+02	3000	473098
27	8	1,3E+02	3500	469876

Table 2 gives the obtained results as function of the Parameters. For the case of the bridge, calculations were performed under Ansys software for all of the 27 combinations. Figure 5 gives the obtained the stress as function of the combination number.

The most adverse case was found to be given by the parameters of combination 3 for which H=7m, L=1.2 E+02tand d=2500, the lowest stress is σ = 327582MPa Then corresponding to the following configuration:

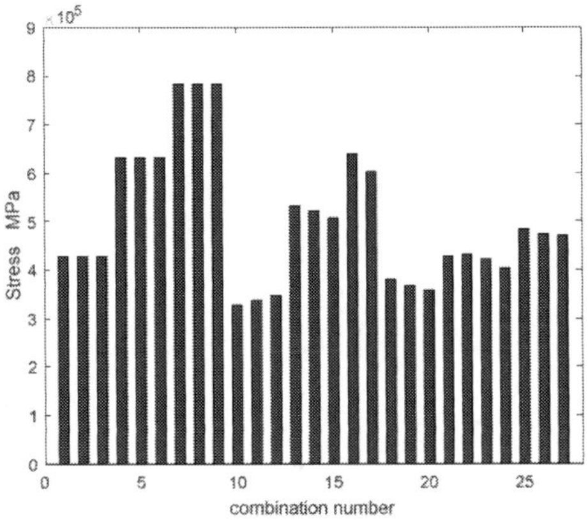

Fig. 5. Perfect shape of the bridge according to the combination (Following x)

These factors were varied for the last case according to Table 2This resulted for each defect configuration case and with Following x. In 27 combinations. Then a DOE full factorial table was considered for the bridge case. The obtained results are shown in Fig. 6, where the non dimensionalized stress defined as the stress is given as function of the combination number as resulting from the following order (H, L and d).

The RSM model as obtained by quadratic polynomial regression using results of the perfect shell in terms of function of parameters x_1=H, x2 = L, and x_3 = d writes:

$$\sigma_{cr}(x_1,x_2,x_3) = 1.9 - 4.2.10^{-10} x_1+2.56.10^{-1} x_2 - 1.3.10^{-10} x_3 -2.2.10^{-20} x_1x_2 + 2.19.10^{-18} x_1x_3 -2.3.10^{-20}x_2x_3 +3.47.10^{-22}x_1^2 +1.8.10^{-25}x_2^2+3.4.10^{-10}x_3^2 \quad (3)$$

The quadratic polynomial regression de fined by Eq. (3) has an R^2=97% value, showing that it is good Enough to represent a RSM for the stress as Function of the considered factors within their ranges as defined in Table 1. The analysis of variance of the obtained results has shown that x_1=H has the greatest influence with Fisher statistics (F= 29228,3), it is followed by x_2=L with (F= 540,8) and then comes the interaction between H and d, x_1*x_3=H*d with (F= 29228, 9).

The associated R-square values are R^2=99.9%, R^2=99.6% and R^2=98%.These indicate that the interpolations are quite good for all the base shears and that the quadratic surface response models can be used to predict theses efforts within the intervals of interpolation.

Let us assume that the design load against buckling is taken to be σ_{lim}. Thus, the performance function for the perfect bridge writes as:

$$G_1(x_1,x_2,x_3)= \sigma_{lim} - \sigma_{cr}(x_1,x_2,x_3) \quad (4)$$

A reliability analysis was performed assuming that x_1,x_2 and x_3 are random variables. Table 3 gives the means and standard deviations for these variables. Two types of probability distribution functions were considered: normal and lognormal. The design load was chosen with a standard deviation of 10%.

Table 3. Means and standard deviation for Reinforced Concrete Bridge design parameters

Random variable	Mean	Standard deviation
x_1	7	0.7
x_2	1250	125
x_3	3000	300

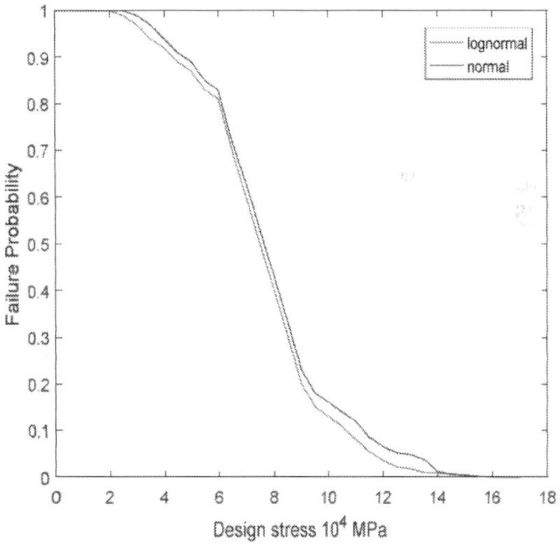

Fig.6. Reinforced Concrete Bridge, variations of failure probability as function of the design Bridge stress

Figure 6 gives the probability of failure P_f as function of the chosen stress design in case of the perfect Reinforced Concrete Bridge.

From Figure 6, we see that the normal and lognormal distributions of the probabilities have a strong influence on the results obtained. The design stress must be lower than 15.10^4 MPa, to avoid cracking of the bridge with great confidence.

This indicates that the identification of the probability distribution functions is very important in order to carry out an accurate analysis of the reliability, particularly in the useful domain of the small probabilities of failure, for example in the art of work on all the building and the bridge.

5 Conclusions

The vulnerability assessment of bridges is useful for seismic retrofitting decisions, disaster response planning, estimation of direct monetary loss, and evaluation of loss of functionality of highway transportation systems. This paper illustrates the results the analysis of the reliability of the bridges against the siege, as this phenomenon could be affected by catastrophic consequences on the structure. Models of these design performance states were obtained by modeling the response surface and quadratic polynomial regression. Then, the reliability analysis of the form was conducted. The results obtained have shown that the distribution of probabilities has a significant influence on the reliability of the x-axis counterpart and in the useful domain of the small probabilities of failure. The dependencies on the chosen distribution of probabilities are not the same for all the situations analyzed.

References

[1] Kircher CA, Wihtman RV,Holmes WT.Hazus earthquake loss estimation methods Nat Hazards Rev 2006;7(2):45-59.
[2] Hazuz-MH.Multi-hazard loss estimation methodology:earthquake model. Washington (DC):Departement of Homeland security,FEMA;2003.
[3] Mackie K, Stojadinovic B.Probabilistic seismic demand model for California highway bridge.J Bridge Eng 2001,6(6):468-81.4
[4] Mackies KR,Stojadimovic B.fragility basis for California highway over pass bridge seismic decision making PEER report 2005/12.Berkeley (CA):Pacific Earthquake Engineering Research center ,University of California:2005.
[5] Powell, G. H., and Allahabadi, R (1988). "Seismic damage predictions by deterministic methods: concepts and procedures."Earthquake Eng. Struct. Dyn, 16. pp. 719–734.
[6] Giberson M (1969). "Two nonlinear beams with definitions of ductility", Journal of Structural Division, ASCE, vol. 95, n° 2. pp. 137-157.
[7] Kondner RL. Hyperbolic stress-stain response: cohesive soils. J Soil Mech Found Div, ASCE 1963;89(1):115–43.
[8] Duncan JM, Chang CY. Nonlinear analysis of stress and strain in soils. ASCE J Soil Mech Found Div 1970;96(M 5):1629–53.
[9] Cardonne, D., Perrone, G. et M., D. (2007). Seismic risk assessment of highway

Effect of spatial variability of ground motion on non-linear dynamic behavior of cable stayed bridges

Mouloud Ouanani[1, 2], Boualem Tiliouine[2] and Malek Hammoutene[2]

[1] Djelfa University, Faculté des Sciences et de la Technologie, Route de Moudjbara, 17000 Djelfa, Algeria
[2] Ecole Nationale Polytechnique, Civil Engineering Department, Laboratory of Earthquake Engineering and Structural Dynamics, 10 Avenue des Frères Oudek, Hassen Badi, BP. 182, 16200, El Harrach, Algiers, Algeria.

Abstract. This present paper summarizes the main results of incoherence of Spatial Variability of Ground Motion (SVGM) component on the non-linear dynamic behavior of a Mila cable stayed bridge. The Hindy and Novack coherence model is developed for the present study in order to examine the SVGM on bridge responses. Nonlinear bridge responses are investigated in terms of transverse displacements and bending moments along the superstructure and substructure of the study bridge, as well as temporal variations of rotational ductility demands at the bridge piers ends under the incoherence SVGM component. The results are systematically compared with those obtained assuming uniform ground motion. As a general trend, it may be concluded that incoherence component of SVGM should be considered for the earthquake response assessments of cable-stayed bridges.

1 Introduction

Cable stayed bridges represent long structures of economic and strategic importance. Analysis of the seismic responses of cable stayed bridges cannot be based on the assumption that all base supports of these structures are subjected to uniform earthquake excitation. In general, the ground motions at different points of an area affected by the earthquake are variable in term of intensity and content of frequencies. This variation noted by Spatial Variability of Ground Motions (SVGM), may induce important quasi-static forces in the extended structures when compared to the case of uniform ground motions. The SVGM is characterized by the coherency function which is mainly represented by four components i) incoherence, ii) wave-passage, iii) site-response and iv attenuation effect. In general, the latter is neglected for long extended structures (e.g. bridges, pipelines).

Many researchers have investigated the response of long span structures under non uniform seismic excitations. Most of them have used relatively simple models of simple and continuous beams (e.g. [2, 3]). More realistic bridge configurations were also studied by various researchers (see e.g. references [4,5]). It was found that the structural response increased substantially when using differential support ground motion, especially for more rigid bridges and bridges with supports on different local soil conditions [6,7]. In reference [7], It was concluded that the effect of spatially varying site conditions on seismic response of cable stayed bridge structures is in general more pronounced than that of uniform earthquake motions. It was found that uniform excitations are generally unacceptable for long-span bridges. In reference [8] the seismic responses of a skewed highway bridge under both spatially variable and identical ground motions using both linear and nonlinear time history analyses have been examined. It was concluded that identical ground motions may predict much lower seismic demands on the bridge than spatially variable ones

In this paper, the main results of a numerical investigations of the Hindy and Novack coherence model component of SVGM on non-linear dynamic behavior of a cable stayed bridge are presented. Nonlinear bridge responses are investigated in terms of transverse displacements and bending moments along the superstructure and substructure of the study bridge, as well as temporal variations of rotational ductility demands at the bridge piers ends under the incoherence SVGM component. The results are systematically compared with those obtained assuming uniform ground motion. Finally, based on the numerical results obtained in the present study, conclusions of engineering significance are given.

https://doi.org/10.1051/matecconf/201814902044

1 Physical and Nonlinear analytical models of bridge structure

The Mila cable stayed bridge of strategic importance is to be built in order to connect the downtowns of Jijel and Mila (North Eastern Algeria) and has an overall length of 502m and consists of three continuous spans in prestressed concrete with a mid-span length of 280m and two end spans of 111m length each as indicated in Fig. 1(a). The superstructure consists of a longitudinally R.C. deck, 13.30m wide with constant height, (see Fig. 1(b)). The bridge is supported by two H shaped concrete towers, double plane semi-harp-type cables and two abutments. The two intermediate towers of heights equal to 111.85m and 142.85m have identical hollow circular cross sections (see Fig. 1(c)). Normal Rubber (N.R.) bearings are located only at right abutment.

In order to include the SVGM effects on the responses of an asymmetric Mila cable stayed bridge, a 3-D FEM model is developed using a small discrete 3-D frame elements for modeling the pylons and the bridge deck and truss elements for the cables modeling. Each adjacent element is connected by a node and at each node six degrees of freedom are considered. The optimal mesh of finite elements of numerical model of the study bridge being obtained based on optimization criteria of the mesh in finite elements [9]. The uncracked section properties are used in the nonlinear model for the bridge piers and their effective torsional moments are determined as 0.2 J_g [10], where J_g is the torsional moment of inertia of the pier gross section.

The physical and mechanical properties of the components of the study bridge are listed in Table 1.

Table 1. Physical and Mechanical properties of the Mila bridge components

Materials	Deck, piers and stiffening girder	Stays
Weight density, (ρ), KN/m^3	25	80
Poisson's ratio, (ν)	0.20	0.3
Modulus of elasticity (E) KN/m^2	3.9×10^7	1.95×10^8
Tensile stress, (σ), KN/m^2		1.24×10^6
Equivalent modulus of elasticity , (E_{eq}) KN/m^2		1.94×10^8

Following Priestley et al. [10], the bi-linear spring element moment-rotation properties were based on the yield moment-rotation (M_y, θ_y) and ultimate moment-rotation (M_u, θ_u) of the pier at the plastic hinge location are obtained from the moment-rotation analysis of the column section based on the moment–curvature relationship (see Fig. 2). The FE software package [11] was used to determine the moment-curvature of the sections of bridge piers. Analysis results are presented in Figs. 2.

The nonlinear analytical model of Mila cable-stayed bridge under SVGM is approximated by the 3-D FEM model (see Fig. 1(a)).

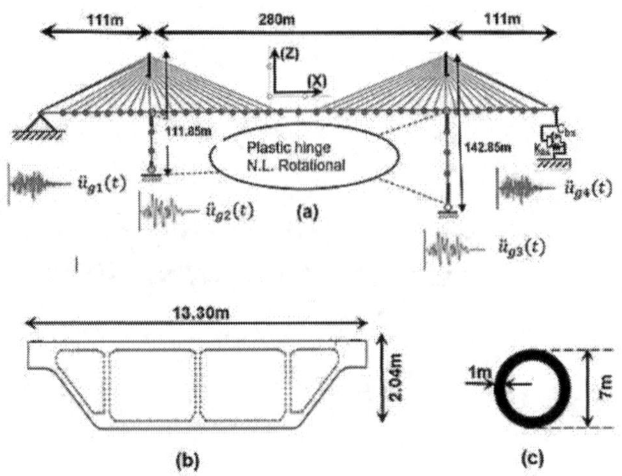

Fig. 1. Description of bridge: (a) Nonlinear analytical model of Mila cable-stayed bridge, (b) Cross-section of segments, (c) Cross-section of pier.

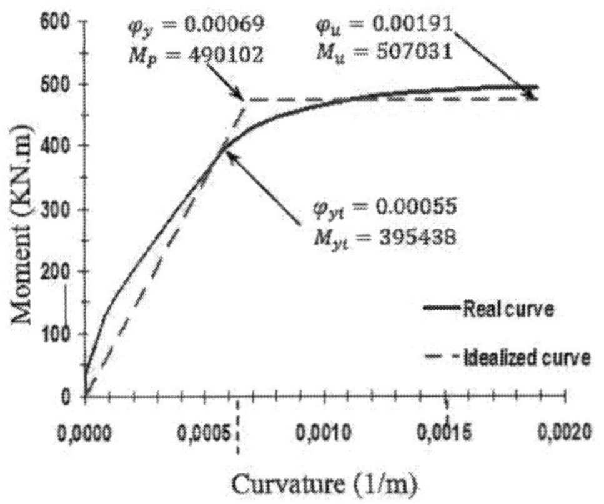

Fig. 2. Pier section analysis results, (a) Moment-curvature curve, (b) P–M interaction diagram

2 Spatially varying ground motion simulation

The power spectral density of a ground motion is a distinctive characteristic of the ground motion. The Clough et Penzien [12] power spectral density of ground acceleration is given by:

$$S_{\ddot{u}_g}(\omega) = S_0 \times H_g(\omega) \times H_f(\omega) \tag{1}$$

Where,

$$H_g(\omega) = \frac{1 + 4\xi_g^2\left(\frac{\omega}{\omega_g}\right)^2}{\left[1 - \left(\frac{\omega}{\omega_g}\right)^2\right]^2 + 4\xi_g^2\left(\frac{\omega}{\omega_g}\right)^2} \tag{2}$$

https://doi.org/10.1051/matecconf/201814902044

$$H_f(\omega) = \frac{\left(\frac{\omega}{\omega_f}\right)^4}{\left[1-\left(\frac{\omega}{\omega_f}\right)^2\right]^2 + 4\zeta_f^2\left(\frac{\omega}{\omega_f}\right)^2} \quad (3)$$

where S_0 is a scale factor, ω_g and ξ_g are the first filter parameters and ω_f and ξ_f are the second filter parameters.

The Spatial Variability of Earthquake Ground Motion (SVGM) is characterized, among others, by the coherency function, which is mainly represented by three main components: incoherence, wave-passage and site-response. Der Kiureghian [13] proposed a general model of spatial seismic coherency function in the following form:

$$\gamma_{kl}(i\omega,\xi_{kl}) = \gamma_{kl}(\omega,\xi_{kl})^{\mathrm{i}} \times \gamma_{kl}(i\omega,\xi_{kl})^{\mathrm{w}} \times \gamma_{kl}(i\omega,\xi_{kl})^{\mathrm{s}} \quad (4)$$

where $\gamma_{kl}(\omega,\xi_{kl})^{\mathrm{i}}$, $\gamma_{kl}(i\omega,\xi_{kl})^{\mathrm{w}}$ and $\gamma_{kl}(i\omega,\xi_{kl})^{\mathrm{s}}$ characterize the incoherence, wave-passage and site-response effects, respectively.

The model expressed by Hindy and Novack [14] is used in the present study for incoherency component.

$$\gamma_{kl}(\omega,\xi_{kl})^{\mathrm{i}} = e^{-(\alpha\omega\xi_{kl})^{\beta}} \quad (5)$$

In which α and β are dimensionless model parameters and ξ_{kl} denote the separation distance between two stations k and l. The model parameters are calibrated to the coherency function computed from simulated ground motion corresponding to the zone IIa earthquake of North Eastern Algeria. Fig. 3 presents the exponential decay of the Hindy and Novack model with frequency along the deck of Mila cable stayed bridge at separation distances of 110, 391 and 502 m for regression parameters, $\alpha = 5.87 \times 10^{-5}$ and $\beta = 1.52$ suggested by [15].

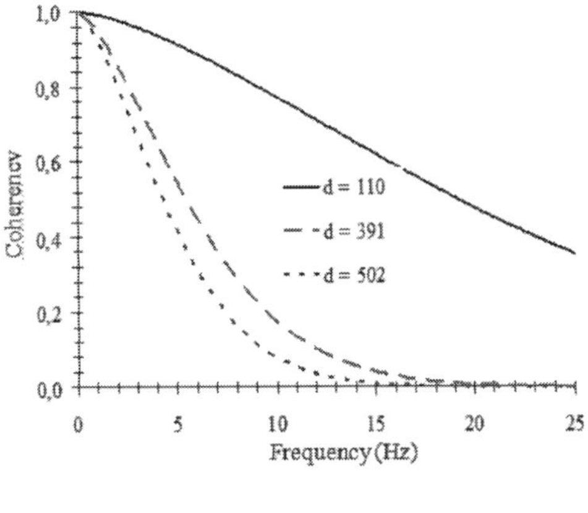

Fig. 3. Variation of coherency models with frequency at separation distances of 110, 391, and 502 m.

3 Numerical results and discussion

In order to further assess the Hindy and Novack coherence model of SVGM component on nonlinear seismic responses of a study, the referential simulated earthquake accelerogram is applied to the left of abutment considered herein as the epicenter of the earthquake and compatible with prescribed scaled to peak acceleration 0.275g supposed equivalent to soil type S_2 (firm site condition) as described in [16].

Nonlinear bridge responses are investigated in terms of transverse displacements and bending moments along the superstructure and substructure of the study bridge, as well as temporal variations of rotational ductility demands at the bridge piers ends under the incoherence SVGM component are evaluated for the study bridge using Newmark direct time integration method with parameters $\alpha = 0.5$ and $\beta = 0.25$ (i.e. using the unconditionally stable average acceleration method). In this study, the two seismic excitation scenarios are simulated:

1) Uniform ground motion characterized by no time delay ($v_{app} = \infty$), no loss of coherence ($\alpha = 0$) and two seat type abutments and the piers are founded on the soil supposed equivalent to soil type S_2 (firm site) in accordance with RPOA [2008] respectively.

2) Only the incoherence effect is considered, the time histories applied are correlated ($\alpha = 5.87 \times 10^{-5}$ and $\beta = 1.52$). Two seat type abutments and the piers are founded on firm site of type S_2 in accordance with RPOA [2008]ely.

3.1 Influence of incoherence effect on absolute maximum seismic responses along the deck

Fig.4 illustrates the variation of absolute maximum vertical displacements and bending moments along the deck of the Mila bridge under uniform ground motion and the Hindy and Novack coherence model of SVGM component.

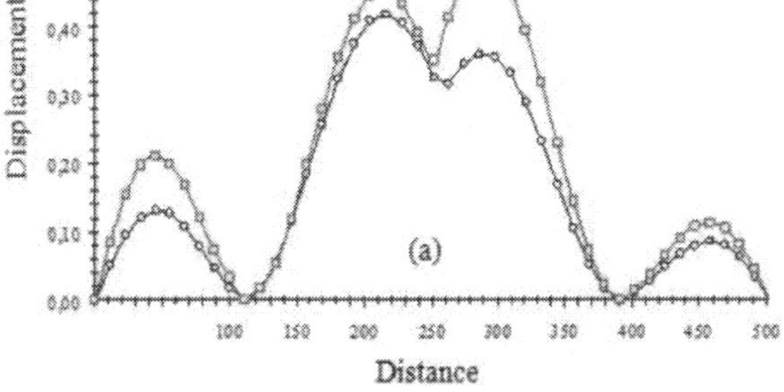

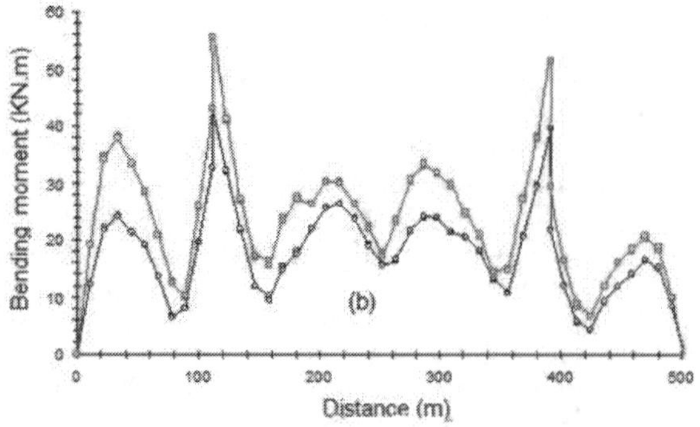

Fig. 4. Absolute maximum seismic responses along the deck of Mila cable-stayed bridge, (a) Vertical displacement, (b) bending moments

From Fig. 4(a) it is clearly seen that, the absolute maximum vertical displacement along the deck of Mila cable stayed bridge induced by incoherence effect are more pronounced than those calculated by the identical seismic excitations. For illustration, the uniform ground motion develops the maximum deck displacement at mid-central span equal to 0.42, which corresponds to significant decreases of more than 14% respectively when the incoherence effects is considered. It is also observed from Fig. 4(b) that the bending moments along the deck are maximums at section on the short pier, large bending moments are obtained along the deck when the incoherence effect is considered. The latter induced a maximum bending moment equal to 55643KN.m, while it is equal to 43087KN when the uniform ground motion is considered.

It is clear for the studied case, the identical seismic excitation model generally produces the lower responses that the uniform ground motion..

3.2 Influence of incoherence effect on absolute maximum seismic responses on pier heights

Figs 5.1 and 5.2 depict the variation of absolute maximum horizontal displacements and bending moments on height of short and tall piers of the Mila cable stayed bridge under uniform ground motion and the Hindy and Novack coherence model of SVGM component.

From Figs. 5.1(a) and 5.1(b) it is clearly seen that, the absolute maximum longitudinal displacements on short and tall towers height of Mila cable stayed bridge are relatively close to each other which corresponds to rigid movement of deck. It is also observed that the incoherence effect induces large values of displacements on towers height in comparison with uniform seismic excitations. For illustration, the maximum deck longitudinal displacement at top of tall tower equal to 0.47m for uniform excitation which is amplified of more than 38%, when the incoherence effect is considered.

It is also observed from the Figs. 5.2(a) and 5.2(b) that the maximum bending moments are maximums at the short pier for uniform ground motions and incoherence effect, very large bending moments values are obtained on tower when the incoherence effect is considered, inducing a maximum bending moment equal to 467447KN.m, while it is equal to 381004KN.m when the uniform ground motion is considered (i.e a substantial reduction of over 18%).

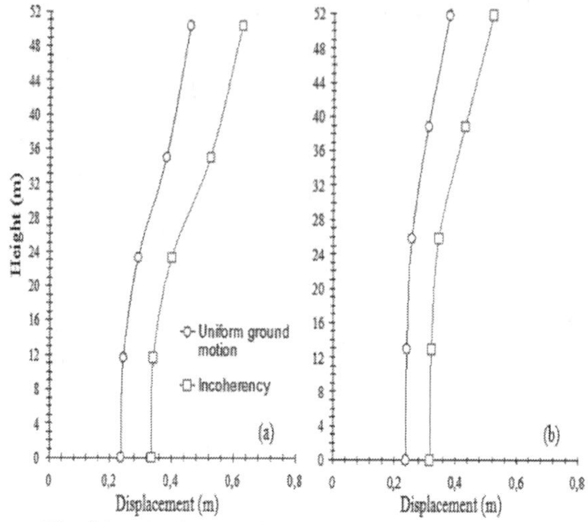

Fig. 5.1. Absolute maximum horizontal displacements, (a) short pier, (b) tall pier

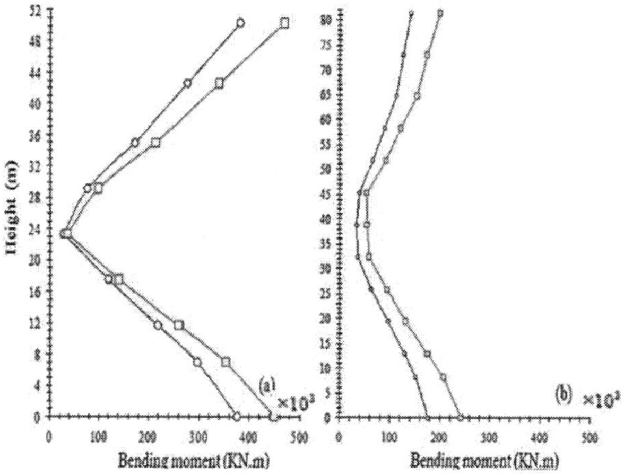

Fig. 5.2. Absolute maximum bending moments, (a) short pier, (b) tall pier.

3.3 Influence of SVGM on rotational ductility demands at pier ends of bridge

Figs. 6.1 and 6.2, below shows the temporal variations in longitudinal direction of rotational ductility demands at short and tall pier ends under uniform ground motions and incoherence effect.

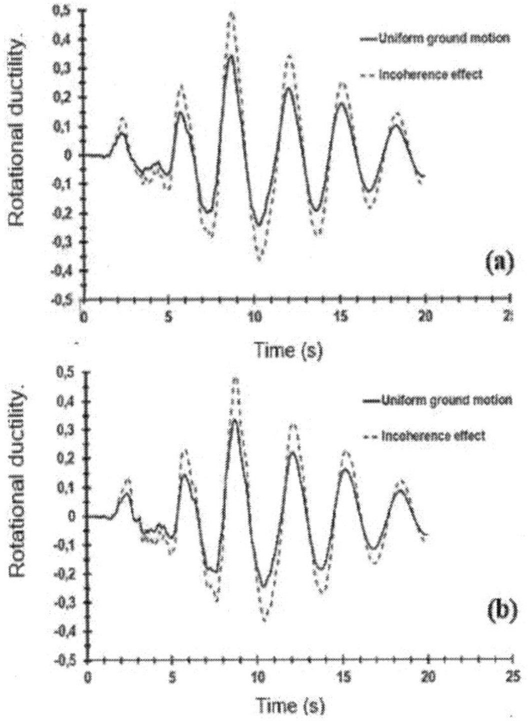

Fig. 6.1. Rotational ductility demands at short pier under uniform ground motions and incoherence effect, (a) Bottom of short pier, (b) Top of short pier.

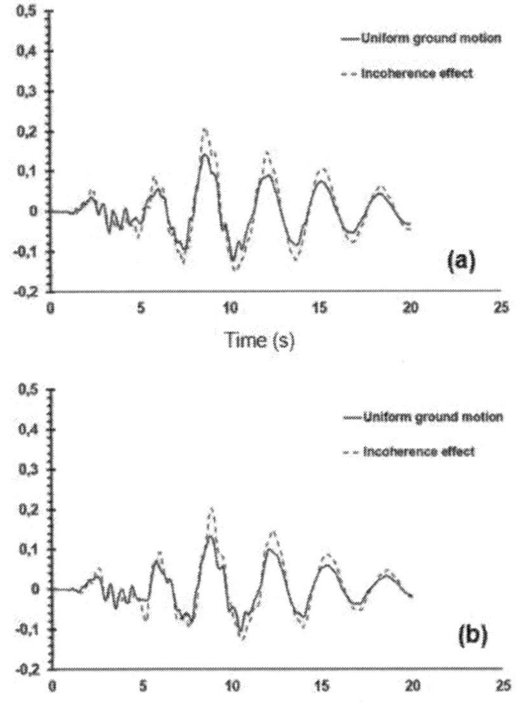

Figure 6.2. Rotational ductility demands at tall pier under uniform ground motions and incoherence effect, (a) Bottom of tall pier, (b) Top of tall

It is noted from Figs. 6.1 and 6.2 that rotational ductility demands at short pier are more important than the tall pier. It is also observed, that the incoherence effect develops rotational ductility demands more important than uniform ground motions. On the other hand, it is clearly seen from Figs. 6.1 and 6.2 that the rotational ductility demands at bottom and top of short and tall piers are well below of elastic ductility limits state and that therefore no damage is likely to occur at the ends of the piers.

As a general trend, it may be concluded that nonlinear analytical models used in the seismic responses of bridge explicitly consider the incoherence effect of SVGM component.

4 CONCLUSION

In this this study, 3-D FEM model of the Mila cable stayed bridge is performed and comparative assessments of the non-linear dynamic behavior of study bridge under incoherence effect and uniform ground motions are presented. Nonlinear bridge response quantities are discussed in terms of displacements and bending moments along the deck and the piers heights of the study bridge as well as temporal variations of rotational ductility demands at the bridge piers ends under incoherence effect and uniform ground motions are evaluated in order to investigate the effect of spatial variability of earthquake ground motion on non-linear dynamic behavior of a Mila cable stayed bridge

From the results obtained in this investigation, the following general conclusions can be drawn:

- Numerical results show clearly the importance of incoherency effect on the seismic response of cable stayed bridge structures.
- The effect of incoherency on seismic response of cable stayed bridge is in general more pronounced than those of excitation ground motion.
- The rotational ductility demands at the bridge piers are more pronounced in the short piers than in the tall piers regardless of the earthquake ground motion model used uniform ground motions and incoherency effect.
- Particularly attention should also be given to higher seismic ductility demands induced in the short piers under incoherency effect that may exceed the rotational ductility limits of piers resulting eventually in global bridge collapse.

References

1 A. Der Kiureghian , A. Neuenhofer, Response spectrum method for multi-support seismic

excitations, Earthquake Engineering and Structural Dynamics, **21**(1992)

2 A. Zerva, AHS. Ang, YK Wen, Study of seismic ground motion for lifeline response analysis, Civil Engineering Studies, Structural Research Series, University of Illinois.

3 A. Zerva, Response of multi-span beams to spatially incoherent seismic ground motions, Earthquake Eng. Struct. Dyn., **19**(1990)

4 RS. Harichandran, W. Wang, Response of indeterminate two span beam to spatially varying seismic excitation, Earthquake Eng. Struct. Dyn. **19**(1990)

5 V. Simeonov, G. Mylonakis, A. Reinhorn, I. Buckle, Implications of spatial variation of ground motion on the seismic response of bridges: Case study', *Proc.FHWA/NCEER Workshop on the National Representation of Seismic Motion*, Tech. Rept, 97-0010; NY, (1997)

6 A. Der Kiureghian, P. Keshishian, Effects of incoherence, wave passage and spatially varying site conditions on bridge response, *Proc. FHWA/NCEER., Report NCEER-97-0010, National Center for Earthquake Engineering Research,* New York, (1997)

7 M. Ouanani, B. Tiliouine, M. Hamoutene, The non-linear dynamic behavior of cable stayed bridges under spatial variability of earthquake ground motion, *2nd International Conference on Material and Structural Mechanics – MSM* (2016)

8 L. Lou, A. Zerva), Effects of spatially variable ground motions on the seismic response of a multi-span bridge, Soil Dyn. Earthqu. Eng. **25**(2005)

9 B. Tiliouine, M. Ouanani, Réponse Sismique 3-D d'un Pont-Caisson à Inertie Variable, *Proceedings of the 8th French National Conference on Earthquake Engineering: Dynamic and Vibratory Aspects in Civil Engineering, AFPS* Paris, France (2011)

10 M. J. N. Priestley, G. Calvi, M.J. Kowalsky, Displacement-based seismic design of structures, *Pavia: IUSS Press*, M.

11 CSI, Bridge, Computer and Structures. Inc. SAP2000 (Nonlinear version 15), *Nonlinear User's Manual Reference*, Berkeley, CA, USA (2012)

12 R. Clough, J. Penzien, Dynamics of structures, *Computers and Structures, Inc.*, Berkeley, USA (1995).

13 A. Der Kiureghian, A coherency model for spatially varying ground motions, Earthquake Engineering and Structural Dynamics, **25**(1996)

14 A. Hindy , M. Novak, Pipeline response to random ground motion, J Eng Mech ASCE, **106**(1980)

15 K. AfifChaouch, B. Tiliouine, M. Hammoutene, R. Sigbjornsson R. Rupakhety, Estimating ground motion incoherence through finite source simulation: a case study of the 1980 El-Asnam, Earthquake, Bull Earthquake Eng, **14**(2016)

16 RPOA (2008). *Partie I: Ponts*, Règlement Parasismiques Applicables au Domaine des Ouvrages d'Arts, Algeria.

 https://doi.org/10.1051/matecconf/201814902045

Extraction of the predisposition parameters for mapping the susceptibility of slides lands, in the region of EL Quola, province of Larache (Nothern Rif)

Ikram El hamdouni[1], Lahcen Aitbrahim[1]

[1] Université Mohammed V, Faculté des Sciences de Rabat, Département des Sciences de la Terre, Laboratoire GEORISK : Risques Géologiques, Télédétection et Environnement. Avenue Ibn Battouta Rabat – Agdal, Boîte Postale 1014, Rabat, Maroc.
ik.elhamdouni@gmail.com; laitbrahim@gmail.com

Abstract. Slide lands (Lands mass movements) are among the natural risks that Morocco faces in the Rifaines regions in particular. They generally constitute small-scale, punctual phenomena. But their diversity and frequency are nevertheless responsible for significant and costly damages and injuries. To deal with this problem, the study of field instabilities must take into account that: i- the ground movements must be taken in a context of the overall dynamics of the watershed, ii- the choice of more objective methods, which is not based on the very subjective opinion of the expert; iii- the choice of an adapted approach to the scale of work. In this article, we used the evidence theory "WofE" to evaluate the sensibility of landslide in the El Quola area. This method is based on the analysis of the relationships between landslides that occurred in the past (the so-called modelable variable Vm) and the spatial distribution of some predictive parameters of instability (Vp). The result obtained is a landslide susceptibility map that is the result of multi-disciplinary and multi-temporal data. This method has been described as one of the most performing by several authors; however, it requires a conditional independence between the different predictive factors. The quality of the performances was evaluated with success curves.

Introduction :

Les glissements de terrain peuvent représenter un enjeu majeur lorsqu'ils interfèrent avec les activités humaines et les biens exposés à ce type de danger. Plusieurs Approches d'analyse spatiale intégrées dans un SIG ont été utilisées pour une meilleure évaluation de la susceptibilité des versants aux glissements de terrain. Les approches probabilistes sont souvent considérées comme les plus performantes par les scientifiques [1]. Ces approches permettent d'obtenir rapidement des cartes issues de la relation statistique entre un certain nombre d'événements spatialisés (Instabilités de Terrain) et un jeu de facteurs de prédisposition. La théorie de l'évidence est l'une des approches les plus adéquates pour une analyse à méso-échelles. Cependant, l'inconvénient de cette approche est le problème de redondance de l'information entre les différentes variables prédictives. Pour répondre à ce problème, il fallait tester et analyser l'indépendance conditionnelle entre les paramètres prédictifs.

L'objectif principal de cet article est de réaliser une carte de susceptibilité au glissement de terrain, la plus proche de la réalité, dans une région très accidentée. Du coup, sept facteurs influençant l'occurrence des instabilités de terrain ont été apportés en tant que couches données de trame dans ArcGIS, et qui ont été utilisés lors de l'analyse de susceptibilité.

1 Cadre géomorphologique et structural :

Située dans le Rif septentrional, entre les latitudes 35°00'_35°45'N et les longitudes 5°30_5°15W, la zone d'étude s'étale sur une surface de 560Km². La commune d'El Quola appartient administrativement à la province de Larache. L'hypsométrie de la région d'El Quola est caractérisée par des reliefs très accidentés et des pentes très abruptes, avec des vallées fortement encaissées et des dénivellations remarquablement fortes [2 ; 3]. Une telle morphologie est le résultat de la conjugaison de plusieurs facteurs: la lithologie complexe, le système climatique, le régime tectonique actif et la dynamique érosive, donnant naissance à plusieurs mouvements de terrain. (Fig1)

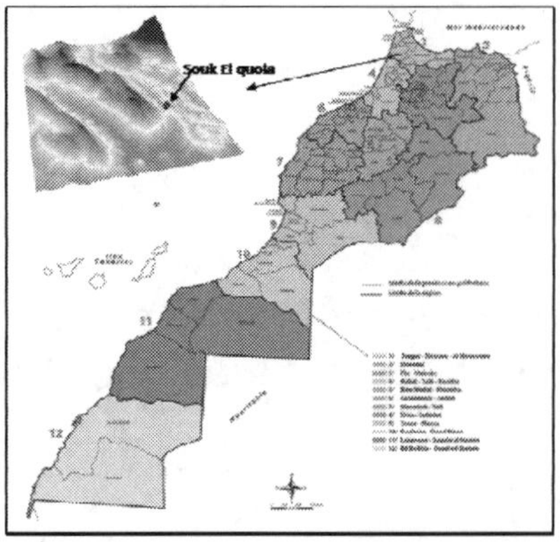

Fig. 1. Situation de la zone d'étude.

La zone d'étude appartient à la bordure SW de la chaine du Rif. Les terrains géologiques drainés sont représentés par des lithologies essentiellement marno-calcaire et marno-quartzitiques appartenant à l'unité de Tanger d'âges Eocène. La structuration de la région s'intègre dans l'évolution tectonique alpine essentiellement méso-cénozoïque [4 ; 5]. Sur ce complexe géologique repose la nappe numidienne caractérisée notamment par une tectonique cassante mio-plio-quaternaires (fig2).

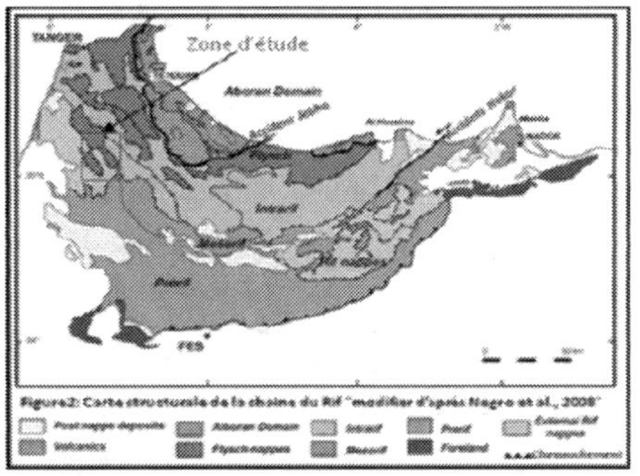

Fig. 2. Cadre géologique de la zone d'étude.

L'ensemble des unités géologique sont séparées par de grands accidents chevauchants orientés NW-SE et à vergence NE, dont l'accident Ejebha qui touche la partie SE de la région [2]. La partie Nord-Est de la région est caractérisée par des reliefs sous forme de crêtes et collines souvent gréseuses, qui appartiennent essentiellement à la nappe numidienne à savoir Jbel Soukna (1603m), Jbel Smia (853m), et Jbel El Kobba (801m).

L'abondance des précipitations associées à la lithologie et l'imperméabilité de la majorité des terrains ainsi qu'aux reliefs accidentés de la région favorisent l'apparition de plusieurs types d'instabilités.

2 Méthodologie et Outils :

La théorie de l'évidence « Weight of Evidence » est une approche quantitative, basée sur le théorème général de Bayes ; elle est fondée sur des règles de calcul statistiques et sur le concept d'unités homogènes [6, 7, 8]. Le principe consiste à définir des relations mathématiques entre les facteurs de prédisposition (variable prédictive, Vp) et l'occurrence spatiale de glissements de terrain (variable dépendante, Vd) afin d'évaluer quantitativement la probabilité de rupture pour des régions non affectées par des glissements de terrain. De ce fait, la méthode suppose que :

- Les glissements de terrain potentiels se déclencheront selon les mêmes conditions que dans le passé ;
- L'ensemble des facteurs de prédisposition sont connus à priori et introduits dans l'analyse ;
- L'ensemble des glissements de terrain qui ont survenu dans le passé sont inventoriés sur l'aire d'étude.

La démarche suivie lors de cette modélisation est résumée sur l'organigramme suivant : (fig.3)

La théorie de l'évidence permet de calculer la probabilité à priori et la probabilité à posteriori. Pour cela, tous les facteurs prédictifs seront pondérés, en calculant un poids positif (W+) (1) et un poids négatif (W-) (1) qui relient les variables prédictives à la variable à modéliser (les mouvements de terrain), ceci selon la formule mathématique suivante [6 ; 7].

$$(1)\ W^{+} = \ln\frac{P\{VP/VM\}}{P\{VP/\overline{VM}\}} \qquad (2)\ W^{-} = \ln\frac{P\{\overline{VP}/VM\}}{P\{\overline{VP}/\overline{VM}\}}$$

VM : Les mouvements de terrain (la variable à modéliser).
VP : Les variables prédictives.
P : la probabilité.

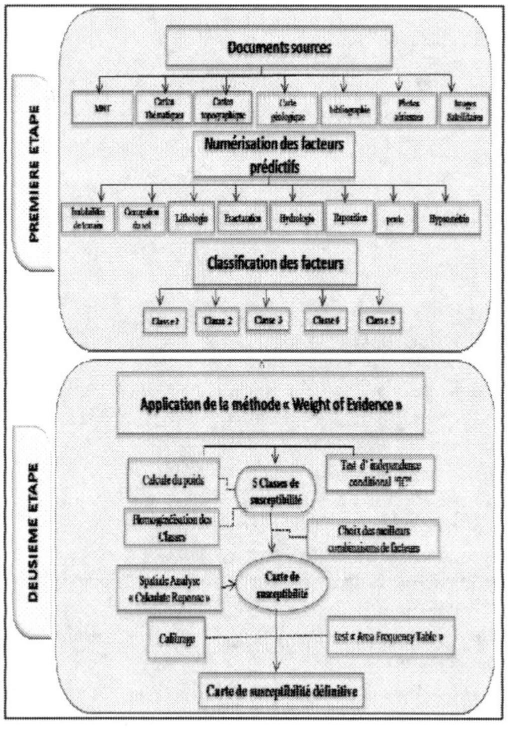

Fig. 3. La démarche de la modélisation.

Grâce aux performances des systèmes d'information géographique (SIG), nous avons pu réaliser une carte de susceptibilité des versants aux glissements de terrain dans la région d'El Quola, la plus objectif possible ; c'est ainsi que nous avons procédé sur deux étapes, **i.** La première étape consiste à la collecte et la mise en forme des données, puis la définition de la meilleure combinaison de variables prédictives, **ii.** La deuxième étape consiste à l'application des modèles de calcul sur ArcSDM, et la validation statistique et experte.

3 Résultats et discutions

4.1 Collecte et mise en forme des données :

Les données indispensables pour effectuer la carte de susceptibilité sont regroupées en cinq catégories [6, 7] (Tableau.1): un inventaire géomorphologique des glissements de terrain, la topographie (gradient de pente, l'hypsométrie, exposition), la lithologie, la densité de fracturation, l'occupation du sol et l'hydrographie.

Tableau.1. Variables prédictives retenues pour élaborer les cartes de susceptibilité.

	Catégorie de variables		Sources
Variable à modéliser	Géomorpho -logie	-Glissement de terrain	Photo-aériennes, observations sur le terrain, cartes topographiques, cartes géomorphologiques
Variables prédictives	Topogra-phie	-Gradient de pente -Exposition -Hypsométrie	MNT issu d'une digitalisation de courbes de niveaux de cartes topographiques au 1/25 000è
	Géologie	-Lithologie -Fracturation	Carte géologique, observations, images L'ansât TM, rapports techniques
	Occupation du sol	-Occupation du sol	Images satellites, littératures antérieures, observations
	Hydrologie	-Réseau hydrographique	MNT, cartes topographiques

4.2 Inventaire des mouvements de versant :

L'inventaire des mouvements de versant a été conduit via l'utilisation des données spatiales satellitaires et aéroportées, des cartes géomorphologiques et géologiques et en se basant sur des données existantes [2 ; 3] ; le tout vérifié sur le terrain (fig4). Les instabilités cartographiées ont été classées en fonction de l'épaisseur et de la nature lithologique des matériaux mis en mouvement et aussi de l'ampleur du mouvement. Quatre catégories, suivant les traits morphologiques les plus dominants, sont distinguées [2 ;11]. Le tableau.2 présente le pourcentage des quatre types de mouvements de terrain cartographiés au niveau de la région d'El Quola.

Tableau.2. Les types d'instabilités en pourcentage.

Types	*Nombre*	*% en Nombre*	*Superficie en Km²*	*% en Superficie*
C.Boueuse	32	34.37%	24.1	36.12%
Ecroulement	5	5.22%	0.52	0.77%
Glissement	17	17.71%	3.24	4.86%
Glissement complexe	41	42.71%	38.87	58.25%

4.3 Caractérisation et détermination des facteurs de prédictions :

Les facteurs de prédisposition sont dérivés à partir : des bases de données existantes, de la digitalisation des cartes, des informations obtenues du traitement d'images satellitaires ou par photo-interprétation, d'interprétation de cartes topographiques et géologiques et d'observations de terrain. Après la numérisation de toutes les données nous avons reclassifié tous les paramètres en cinq grandes classes. La reclassification effectuée a été adoptée grâce aux outils de ArcGIS ; Ceci après plusieurs tests de corrélations entre la distribution spatiale des phénomènes à modéliser (instabilité de terrain) et les différentes classes choisies.

4.3.1 Les facteurs liés à la topographie :

Les paramètres topographiques sont dérivés d'un modèle numérique de terrain issu de la digitalisation des courbes de niveau à partir de la carte topographique au 1/25 000e. Trois paramètres géométriques sont utilisés dans notre modèle : le gradient de pente, l'exposition et l'hypsométrie (fig.5).

4.3.2 Les facteurs géologiques :

Les variables prédictives relatives à la géologie sont la lithologie et la fracturation.

Les principales unités lithologiques de la zone d'étude ont été reconnues sur l'image satellitaire grâce à des traitements spectraux, en se référant à la carte géologique à l'échelle de 1/500 000ème de la chaine Rifaine [12]. Ces interprétations ont été complétées par des observations sur le terrain ; c'est ainsi qu'une carte de facies a été élaborée. Nous avons procédé ensuite à une requalification postérieure des formations géologiques selon leurs affinités lithologiques et mécaniques à l'aide du module « reclassification » (fig.6), [7 ; 11].

La fracturation est un facteur prédictif, étant donné que l'activité tectonique héritée et récente peut intervenir sur un mouvement de terrain dans des zones déjà fragilisées [10]. Tout de même, le réseau de fractures favorise l'infiltration de l'eau, entraînant ainsi une augmentation de la pression interstitielle et une réduction de la résistance au cisaillement des sols. De ce fait, une carte de fracturation du secteur d'étude a été réalisée en se basant sur la numérisation de toutes les failles bibliographiques [5 ; 10] ; l'extraction de la fracturation récente qui a affecté la région basée sur l'analyse morpho-structurale de la carte hypsométrique couplée à celles des anomalies du réseau hydrographique ; l'obtention des principaux accidents d'origine tectonique à partir de l'imagerie satellitaire Landsat7 ETM+, grâce aux rehaussements spatiaux sur ERDAS imagine 11.

L'analyse morpho-structural a permis de déceler trois grandes familles qui, selon leur ordre d'importance, sont de directions : NW-SE à N-S et ENE-WSW à E-W. Ce paramètre a été intégré dans le modèle en tenant compte de la distance par rapport aux failles [1 ; 13]. De ce fait, nous avons réalisé la carte de densité des failles, en utilisant la fonction « Kernel Density » sur ArcGis10. Cette carte a été élaborée selon 5 classes de densité par ré-échantillonnage (fig.7).

4.3.3 Le facteur occupation du sol :

L'occupation des sols joue un rôle primordial dans l'étude de l'aléa glissement de terrain, à travers le caractère stabilisant de la couverture végétale. Lors de son absence ou sa rareté sur les versants, l'effet érosif par ruissellement peut être accéléré. Pour réaliser la carte d'occupation des sols, nous avons eu recours à la conception d'une base de données multi-source. Ces informations ont subi une actualisation par interprétation des données satellitaires et par des données collectées sur le terrain. Les différentes classes sont reconnues par classification multi-spectrale des données Landsat7 ETM+. Afin de simplifier la lecture, plusieurs fusionnements et ajustements ont été effectués. De ce fait, cinq classes sont retenues : les essences forestières, les matorrals, les reboisements, les terrains agricoles et les terres arables. (fig.8)

4.3.4 Le facteur hydrologie :

Certains mécanismes de glissements de terrain correspondent à l'action érosive (sapement de berges) des cours d'eau [9]. Le ruissellement superficiel génère des zones d'intense ravinement (badlands) dans le cas de formations géologiques tendres. Il en résulte le déclenchement d'un nouveau mouvement de terrain ou la réactivation d'un ancien mouvement stabilisé. Le réseau hydrographique sera incorporé dans notre étude via sa densité ; suivie d'une analyse statistique de la répartition des glissements de terrain proches des principaux cours d'eau, ce qui a permis de reconstruire des zones tampons hydrographiques, et après une reclassification, cinq classes ont été retenues. (fig.9)

4.4 Calcul du poids des facteurs :

La fonction Arc-SDM permet de calculer, pour chaque classe de variables, les poids positifs (W^+) et négatifs(W^-), ainsi que leurs contrastes ($C = W^+ - W^-$), qui seront intégrées dans notre modèle. Ce calcul assure l'identification des variables ou des classes de variables qui sont positivement corrélées ($W^+>0$ et $W^-<0$) ou négativement corrélées ($W^+ <0$ et $W^->0$) et les variables non corrélées ($W^+= W^- = 0$) aux glissements de terrain. Ces poids positifs et négatifs sont calculés selon les ratios logarithmiques, fondés sur les théorèmes binaires et résultent du rapport des probabilités conditionnelles suivantes : [6, 7]: (Tableau.3)

Nous avons analysé les poids W+, W- et le contraste attribués à chacune des classes des variables prédictives; ce qui a permis d'évaluer et d'interpréter le rôle respectif de chaque facteurs de prédisposition dans l'occurrence spatiale des glissements de terrain. Nous avons remarqué que l'occupation du sol et l'hypsométrie et le réseau hydrique jouent un rôle majeur dans la susceptibilité aux instabilités de terrain. Alors que la lithologie et l'exposition des versants sont relativement moins influentes. Les variables densité des failles et la pente occupent une place intermédiaire avec une valeur absolue du contraste de 1.05 et 1.08 respectivement. En effet, la valeur absolue du contraste de poids maximum entre les cinq classes de la variable lithologie est de seulement 0.9 ; alors qu'il est de 1.96 pour le réseau hydrographique, de 1.22 pour l'occupation du sol et 1.82 pour l'hypsométrie. Nous avons remarqué aussi que pour la lithologie, les marnes et les argiles sont caractérisées par une haute susceptibilité.

Tableau.3 : Résultats du calcul des poids des facteurs

Facteur	Classes	W+	W-	Contraste
Expo-sition	Nord	-0.02	0.0093	-0.0375
	Est	-0.59	0.0378	-0.2537
	Sud	0.1491	-0.039	0.1879
	Sud-Ouest	0.2632	-0.081	0.344
	Nord-Ouest	-0.67	0.0636	-0.4002
Facies	Grés	0	0	0
	marnes gréseux	0.762	-0.177	0.9394
	limon à galets	0.284	-0.011	0.2954
	Marnes	0.1852	-0.016	0.2015
	Argiles	0.278	-0.311	0.5885
Fractura-tion	0_0,14	-0,76	0.0801	-1.0568
	0,14_0,28	0.4398	-0.213	0.6527
	0,28_0,425	0.3869	-0.194	0.5807
	0,425_0,56	-0.39	0.0881	-0.465
	0,567_0,71	0	0	0
Hypso-métrie (m)	0_185	-1.76	0.1166	-1.8816
	185_415	-0.74	0.0828	-0.8303
	415_655	0.492	-0.494	0.9866
	655_1062	-0.70	0.0767	-0.7866
	1062_1620	0.040	-0.011	0.052
Occupation du sol	Forêt	-0.55	0.3602	-0.9167
	Reboisements	0.7427	-0.077	0.8203
	Matorrals	1.0321	-0.189	1.2213
	T-Agricoles	-0.41	0.0661	-0.4793
	Terres arables	0.2245	-0.056	0.2804
Pente	0_5%	-0.78	0.0884	-0.87
	5%_15%	-1.04	0.0427	-1.0855
	15%_25%	0.3379	-0.099	0.4371
	25%_35%	-0.99	0.1811	-1.1735
	>35%	0.4303	-0.362	0.793
Réseau Hydro-graphyque	0_1,42	-1.72	0.2408	-1.9636
	1,42_2,85	-0.23	0.0776	-0.3099
	2,85_4,27	0.1171	-0.049	0.1659
	4,27_5,70	0.8775	-0.304	1.1811
	5,70_7,21	0.1842	-0.008	0.1919

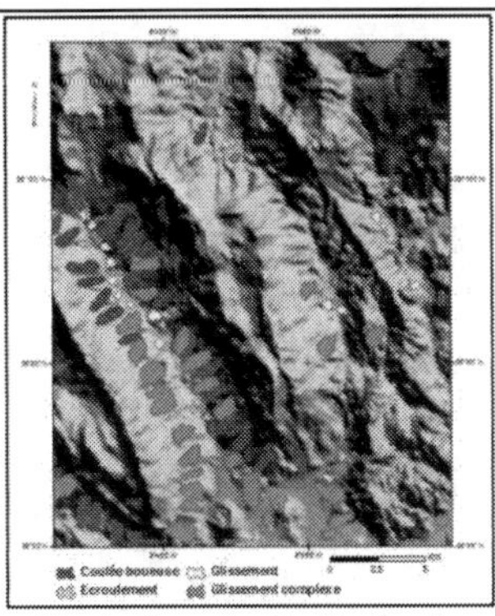

Fig.4 : Carte de distribution des instabilités de terrain

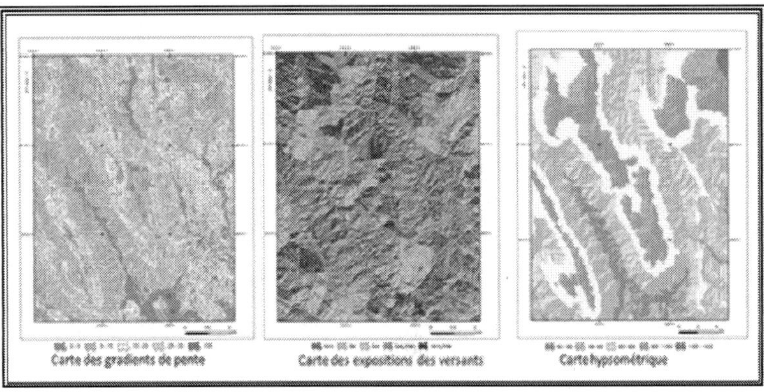

Fig.5 : Facteurs liés à la topographie

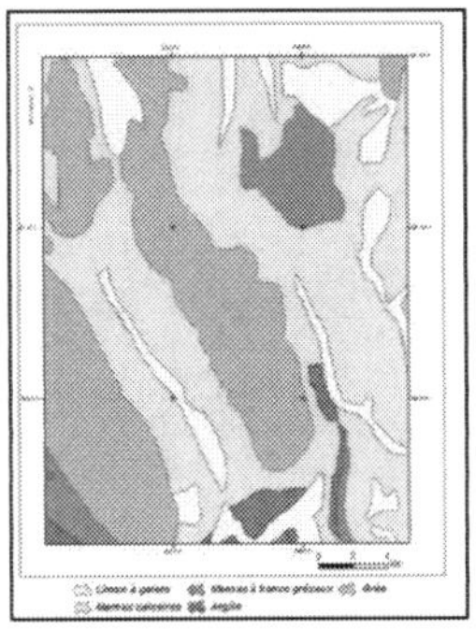

Fig.6 : Carte de facies

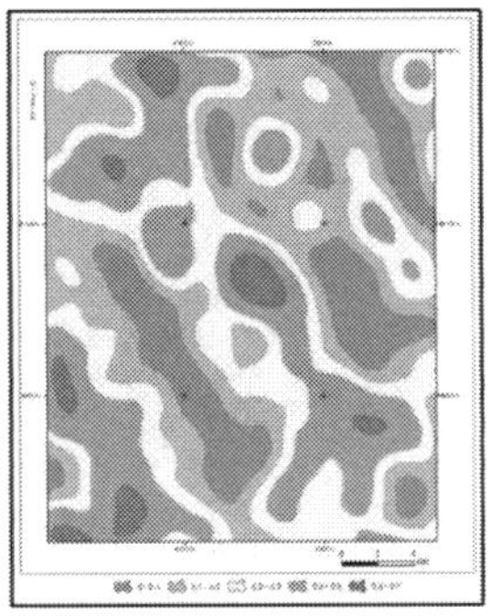

Fig.7 : Carte de densité de fracturation

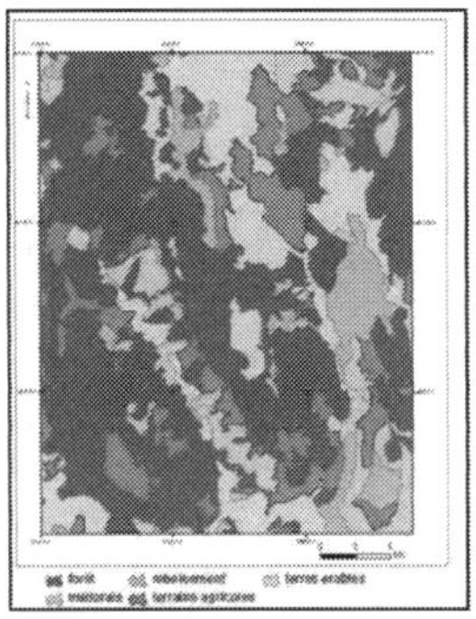

Fig.8 : L'occupation du sol

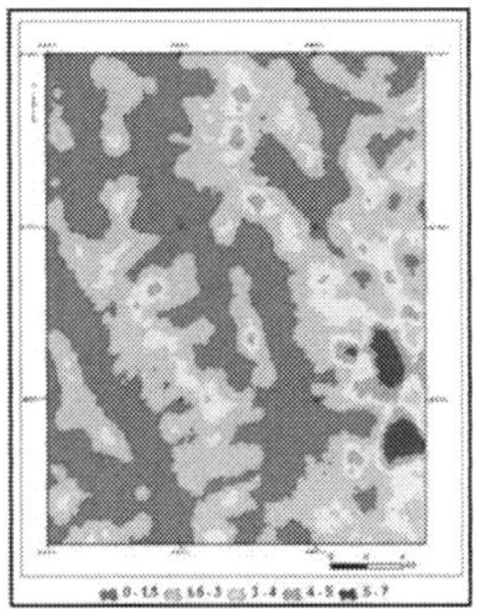

Fig.9 : Carte de densité du réseau hydrographique

4.5 La carte de susceptibilité ;

Les poids relatifs à chaque facteur ont été combinés ensemble afin de calculer la probabilité à posteriori, ce qui permet d'établir la carte de susceptibilité aux mouvements de terrain. Cette carte permet de distinguer cinq classes de susceptibilité (fig.10). La carte de susceptibilité des instabilités de terrain, obtenue par analyse bi variée et en appliquant la théorie de l'évidence, a été reclassée en cinq entités selon le degré de susceptibilité. L'analyse experte montre que plus de 70% des instabilités recensées coïncide avec les classes moyenne et forte à très forte susceptibilité.

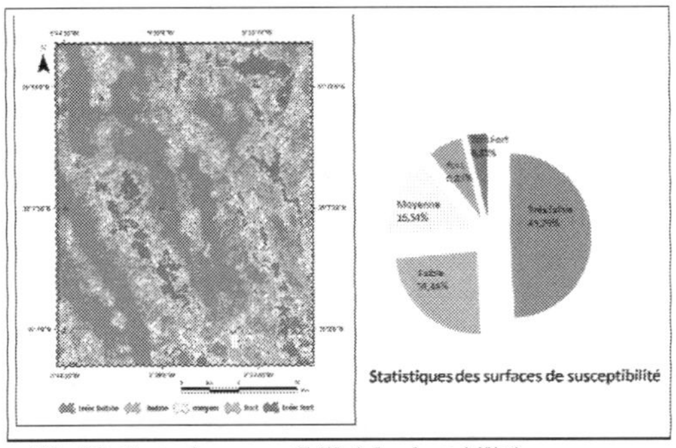

Fig.10. Carte de susceptibilité des instabilités

4 Validation des résultats

4.1 L'indépendance conditionnelle

L'évaluation de l'indépendance conditionnelle a été effectuée en utilisant la fonction « Area Frequency Table » sur l'outil « Spatial Data Modeller Tools ».

Pour le calcul de probabilité et la validation statistique, les 50% de la population totale des glissements de terrain ont été introduits dans le modèle à partir d'un tirage aléatoire. Puis, chaque variable prédictive est ajoutée une à une pour observer son influence sur les simulations [9]. Plusieurs tests indiquent que les meilleures simulations sont obtenues en combinant de manière hiérarchique le gradient de pente, le facies, l'occupation du sol, la densité du réseau hydrographique, la densité de fracturation, l'hypsométrie et l'exposition des versants. Pour la validation experte, les résultats des modèles en termes de surfaces susceptibles aux glissements de terrain modélisés sont comparés à l'inventaire géomorphologique. Les tests de IC utilisés permettent de calculer trois valeurs, celles-ci permettent d'éviter toute surestimation ou sous-estimation de la probabilité [6 ;14]. (Tableau.4)

Tableau.4. Les tests d'indépendance conditionnelle

	Valeurs calculées	**L'intervalle acceptable**
Overall CI simulations (CI)	93.6%	50%<CI≤100%
Conditional Independence Ratio(CIR)	1.01	1≤CIR<2 Valeur acceptable >0.85
Agterberg and Cheng Conditional Independence Test(CIT)	46.8%	CIT<50%

La comparaison des résultats montre que les trois valeurs calculées correspondent parfaitement aux valeurs acceptables théoriquement ; ce qui exclut l'existence d'une dépendance entre les différents facteurs suggérés prédictifs.

4.2 La validation statistique

Pour la validation statistique, nous avons utilisé la fonction « Area Frequency Table » sur l'outil « Spatial Data Modeller Tools» d'ArcGis10, tout en introduisant d'abord les 50% de la population totale, des glissements de terrain qui ont été utilisés dans le modèle. Du coup, nous avons élaboré la courbe du taux de succès (SRC) : Il s'agit d'une représentation graphique définissant la relation existante entre le pourcentage cumulé des surfaces susceptibles et la probabilité à postériori [7]. Le modèle a montré une efficacité d'environ 70%. Dans un deuxième temps, nous avons introduit les 50% de la population totale, des glissements de terrain restants, non utilisés dans le modèle, ce qui a permis de créer la courbe de prévision *(PRC)*; ce test de prévision a montré une efficacité de 68%. (fig11)

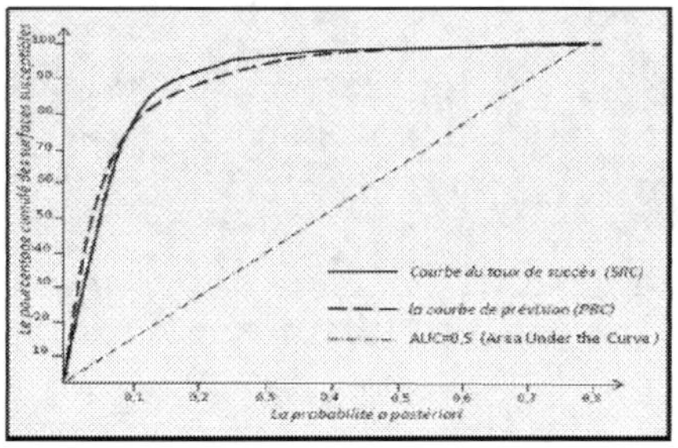

Fig.11. Les courbes de la validation statistique

Conclusion :

L'application de la théorie WoFe a été franchie en utilisant l'outil SDM sur ArcGIS. La plus grande partie de ce projet était consacrée à l'identification des données appropriées, l'amélioration de la qualité des données et à la performance de la prédiction du modèle ; ce qui a permis de préparer une base de données pertinente et adéquate. Ceci était dans le but de diminuer toute subjectivité, lors de la réalisation de carte de susceptibilité. Les résultats indiquent qu'une combinaison de sept variables prédictives (gradient de pente, hypsométrie, facies, occupation du sol, densité de fracturation, densité de réseau hydrographique et exposition) a permis de prédire, pour la meilleure simulation, la localisation spatiale (en termes de susceptibilité) d'environ 70% des aires observées de glissements de terrain, pour l'analyse par la théorie de l'évidence. Le test de prévision montre une efficacité d'environ 68%. Les résultats obtenus sont plus ou moins proches, ce qui nous permet d'affirmer que l'analyse bi variée par théorie de l'évidence semble être une approche toujours adaptable à moyenne échelle considérée (1/50000e).

Références bibliographiques:

1. R. Fell, S. Lacasse, E. Leroi, *A framework for landslide risk assessment and management*. In: Hungr, Landslide Risk Management, 3–26 (2005).
2. A. El Gharbaoui, *La terre et l'Homme dans la péninsule tingitane. Etude sur l'Homme et le milieu dans le Rif occidental*, Travaux de l'Institut Scientifique, série Géologie et Géographie Physique, Rabat, **15**, 439 (1981).
3. G. Maurer, *Les montagnes du Rif Central, étude géomorphologique, travaux de l'Institut Scientifique Chérifien*. Série Géologie ct Géographie Physique, 14 (1968).
4. M. Durand Delga, P. Rossi, P. Olivier, D. Puglisi, *Situation structurale et nature ophiolitique de roches basiques jurassiques associées aux flyschs mahgrébins du Rif (Maroc) et de Sicile (Italie)*, **331**, 29-38 (2000).
5. L. Ait Brahim, *Tectonique cassante et état de contraintes récent du Maroc Nord. Résultat de la cinématique des plaques Afrique-Europe et du bloc d'Alboran*, 233 (1991).
6. G.F. Bonham-Carter, F.P. Agterberg, D.F Wright, Statistical Applications in the Earth Science: Geological Survey of Canada Paper, *Weights of evidence modeling: a new approach to mapping mineral potential*, **89-9**, 171–183 (1989).
7. C.J. Van Westen, *GIS in landslide hazard zonation: a review, with examples from the Andes of Colombia*, Mountain Environments and Geographic Information Systems (1994).
8. P. Aleotti, R. Chowdhury, *Landslide hazard assessment: summary review and new perspectives*, Bull Engineering Geology and Environment **58**, 21–44 (1999).
9. Y. Thiery, J-P. Malet, S. Sterlacchini, A. Puissant, O. Maquaire, *Landslide susceptibility assessment by bivariate methods at large scales: Application to a complex mountainous environment*, Geomorphology (2007).
10. L. Ait Brahim, M. Mansour, F. Sossey Alaoui, I. EL Hamdouni, *Rôle de la fracturation dans la désagrégation mécanique des calcaires et la mise en mouvement de la coulée de pierres d'Amtrasse (Rif, Maroc)*, Edition ISBN 9954-8407-0-2 "L'implication de la géotechnique dans le développement des infrastructures en Afrique ", 69-79 (2003).
11. D.J. Varnes, *Landslide Hazard Zonation, a review of principles and practice*, IAEG Commission on Landslides, UNESCO, Paris (1984).
12. G. Suter, *Carte géologique/structurale de la chaîne rifaine, échelle 1:500,000*.- Serv. Carte géol. Maroc 245a/b (1980).
13. S. Lee, B. Pradhan, *Landslide hazard mapping at Selangor, Malaysia using frequency ratio and logistic regression models*, Landslides **4**, 33–41(2007).
14. F. P. Agterberg, Q. Cheng, *Conditional independence test for weights of evidence modelling*, Natural Resources Research, **11**, 249–255 (2002).

Influence of an absorbers GEM_{PEHD} thermal properties on the propagation of heat in a solar sensor

BABA AHMED Nassim[1], ALIANE Khaled[1], BENMOUSSAT Abderrahim[2]
1Department of Physique, Faculty of Science, Abou Bekr Belkaïd University of Tlemcen, Tlemcen, Algeria
2Materials sciences and Environment Group, Faculty of sciences and Technology, Department of matter sciences, Tamanrasset University center, BP 10034 SERSOUF, Tamanrasset - ALGERIA

Abstract. This paper presents an experimental study to measure some physic-chemical properties of a high-density polyethylene geomembrane (GEM_{PEHD}) in a temperature range from 300 K to 400 K. The results will yield a dependency between temperature and its properties in order to study an energy and process the thermal aging GEM_{PEHD} on the solar collector. In the energy study, we solve the nonlinear equation of unsteady heat with as main non linearity influence of different properties GEM_{PEHD} on the solar collector and one notes that the GEM_{PEHD} presented a maximum exchange Thermal characterized by high absorption and low reflection. Thus, increasing the power absorbed by the GEM_{PEHD} increases its temperature and the useful power received by the air and then the air temperature. Overall losses increase with the increase of the temperature difference between the GEM_{PEHD} and windows and this increases the efficiency of the solar collector and minimize the entropy of the system. Then the study of thermal aging is dedicated to the interpretation of various analyses on GEM_{PEHD} before and after its use in the solar collector. The thermal aging GEM_{PEHD} in the air is a physical loss and chemical consumption immediately followed by a brutal oxidation of the polymer. The losses would be governed primarily by the chemical consumption of antioxidants. After various tests, we conclude that the value of using the GEM_{PEHD} as absorber in the solar collector is doubly beneficial. First, given its very attractive cost compared to other as efficient absorbers (such as metals) gives good thermal performance of the solar collector even reaching 72%. While the life of the GEM_{PEHD} is rather long (several years).

1. Introduction:

Interest in solar air heating has resulted in recent years by the appearance on the market of a number of systems, which include solar collectors. The motivation of this work focuses on the following findings: An environmental statement to limit the intensity of the greenhouse effect and rising temperatures on the surface of the planet by which action plans for emission limitation some greenhouse gases (GHG) have been implemented by many countries. An energy finding where energy needs are growing throughout the four corners of the planet. The industrial sector **[1]** remains the main consumer of energy and the residential and commercial sectors consume mainly because of space heating and coins. The production and consumption of energy contribute to the change in thermal equilibrium at the surface of the earth by producing greenhouse gases that are assigned the current warming of the planet. At that thermal pollution also complemented by a physical or chemical waste pollution **[2]**.

One of problems in solar capture is the choice of materials and their selectivity to optimize absorption and reduce heat transfer through the insulation or transparent materials to reduce reflections and get the maximum radiation converts. We are interested in this study to the polymer material, the high-density polyethylene geomembrane (GEM_{PEHD}) used as a heat absorber in the solar collector energy study that will form during sun exposure. Then, we study the influence of aging on the sustainability and efficiency of the solar collector.

2. Materials and methods:

We use various experimental techniques for physicochemical characterization as well as aging of the samples GEM_{PEHD} to assess the structural and textural properties of our material. In addition, we will present numerical method for modelling the nonlinear equation of unsteady heat. The GEM_{PEHD}, used in this study, is a polymer that provide to us as granules stabilized with antioxidants and process the nature of which has not been

 https://doi.org/10.1051/matecconf/201814902046

specified us. This polymer corresponds to the basic matrix of a commercial GEM. The PE100: it is a PEHD was added 0.4% by weight of a phenolic antioxidant, Irganox B225. Our material has endured aging accelerates high temperature in a ventilated oven in air at atmospheric pressure in order to follow the structural evolution during the period of exposure in the solar sensor.

- Emissivity: The method allows the measurement of the hemispherical directional reflectivity and there after calculate the emissivity of the surface using a portable emissometer **[3, 4, and 5]**. The current equipment design for easy use in the field and better to control of the operation of the infrared source; it allows achieving emissivity measurements in different wavelength bands that the optical detector can be replaced very easily. The method chosen for the characterization of the density is that of the pycnometer. The measurements made with a pycnometer to a balance and small amounts (1 to 4 g) for GEM_{PEHD}. The pycnometer used has a marked capillary to measure a precise volume.
- Thermal Conductivity and Thermal diffusivity: We propose a measuring method enabling the simultaneous characterization of the thermal conductivity (k) and thermal diffusivity (a) a sample of the GEM_{PEHD} at room temperature **[6]**. We modulate the temperature of the sample and measure the temperature variations on each of its faces. The measurements performed under vacuum (pressure < 10^{-4} mbar) to reduce convection losses on the sides.
- ATG thermal analysis of samples conducted used is of type SDT Q600. The analysis performed in air with a temperature ramp of 5 ° C / min in the range of 25 to 1200 ° C for a few samples at 1000 ° C for others.
- UV-visible diffuse reflection analysis and differential scanning calorimetry performed on a Perkin Elmer Lambda 800 flown from a microcomputer.
- Infrared analysis in Fourier transform (FTIR) was performed using an Avatar 320 FT-IR spectrometer THERMO-NICOLET. Spectra were recorded either transmittance (T) or absorbance (A) between 4000 cm^{-1} and 400 cm^{-1}. The X-ray diffraction (DRX) performed with a PANalytical X'Pert diffractometer type Pro goniometer with a theta / theta.
- Solar sensor: The test bench was performed at the University of Tlemcen lying north-west Algeria at latitude 35.28 degrees, a longitude of -1.17° and an altitude of 750 m. It is characterized by lots of sunshine, with an average daily irradiation 6000Wh / m², a sunshine duration of 3000 hours a year and a clearness index K_t average of 0.75 **[7]**. The solar sensor is shown in **Figure 1**, the total area of 2.6 m², and the opening area equal to 2.4 m². It is mounted on a rigid adjustable steel structure with a 45° inclination.

The solar sensor is a parallelepiped shape, constructed with galvanized sheet metal (thickness e = 1.5 mm, length L = 1 m, width = 0.2 m and the thickness of the air gap H = 1 cm). It is internally reinforced by angle irons, covered outside by wooden cork as a thermal insulator between the two there is a glass wool layer of 5 cm thickness on the side surfaces and on the rear surface. Inside the metal sheet is covered by 5 mm thickness GEM_{PEHD} as a heat absorber. Two panes of 4 mm thick form the transparent cover each with a space between two panes of 1.5 cm as recommended in the literature **[8]** because this minimizes the edge effect while maintaining low loss by convection. In this study, we are interested in the front part of the solar collector, the heat from the solar rays that pass through the glass and is absorbed by the inner face of the wall placed on the absorber. The lower face is subjected to a constant heat flux simulating the energy received by the solar radiation on a face of negligible thickness.

Figure 1: Solar collector

3. Results and discussion:

We recap some intrinsic properties of the GEM_{PEHD} in **Table 2**. **Table 2:** Intrinsic properties of the GEM_{PEHD}

Properties	Unit	Value
Volumes' density	$g.cm^{-3}$	0,974 ± 0,01
Thermal conductivity	$W/m \cdot K$	0,28 ± 0,01
Thermal diffusivity	m^2/s	$(0,314 \pm 0,02) \cdot 10^{-6}$
Density*	g/cm	0,949 ± 0,001
Temperature of fusion**	°C	135,7 ± 0,1
Degree of crystallinity**	%	78 ± 2
Weight average molar mass***	g / mol	500,4 ± 0,4
Glass transition temperature	°C	- 125
Coefficient of expansion	C^{-1}	$19 \cdot 10^{-5}$

* Measured by a weighing device on a Mettler TOLEDO in air and in acetone.

** Measured by differential scanning calorimetry with a temperature rise rate of 10 ° C.min-1 in an inert atmosphere.

*** Measured by steric exclusion chromatography at 145 ° C in 1,2,4 trichlorobenzene with a chromatographic system equipped with a triple detection (Viscometer, refractive index and Laser).

The value of the specific heat capacity of the GEM_{PEHD} given by calculating the value of C_p from the values of measured density, conductivity and thermal diffusivity using the following relationship

$$C_P = \frac{k}{a \cdot \rho} \quad (5)$$

 https://doi.org/10.1051/matecconf/201814902046

The value of the following specific heat capacity is $(\mathbf{2300 \pm 0{,}012})\ J/Kg \cdot K$.

Spectral emissivity remains measured at room temperature, approximately 20 °C. The results illustrated in **Figure 2**. This figure shows that the emissivity varies little within this range of temperatures and wavelengths.

Figure 3 shows that the emissivity of the GEM_{PEHD} reaches a constant value for a sample of thickness greater than 0.2 mm. We find the value 0.93 determined above. Beyond this thickness, GEM_{PEHD} may considered opaque with respect to the IR radiation.

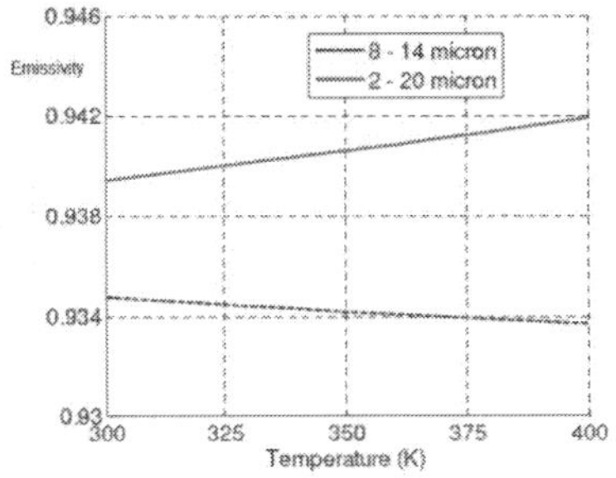

Figure 2: Emissivity of GEM_{PEHD}

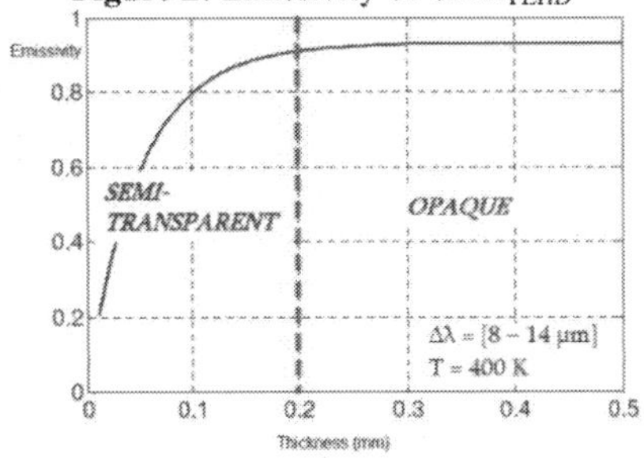

Figure 3: Total Emissivity of GEM_{PEHD}

3.1. *Energy study on the solar collector*: The **figure 4** gives us the intensity of the solar flux for the four days that are respectively 17 January, 15 May, 16 August and 15 October 2015. It noted that the intensity of the solar flux is greater in the days of May 15 and August 16 in which the days are longer.

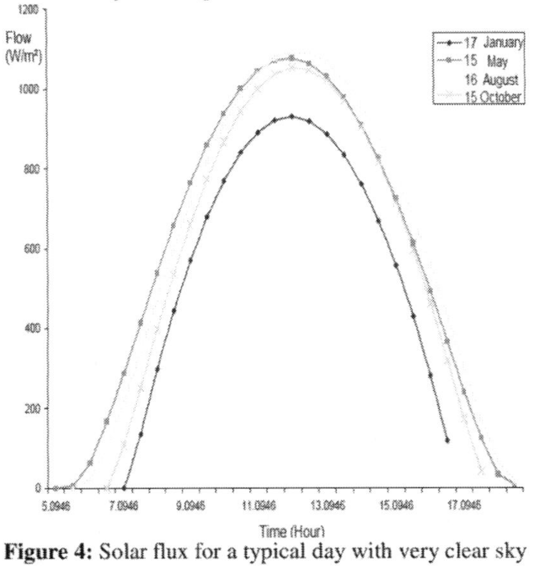

Figure 4: Solar flux for a typical day with very clear sky

To see the effect of climatic parameters on changes in air temperature T_F, weather data was used (direct radiation, wind speed and ambient temperature) measured by the meteorological station in Tlemcen site for day of 16 August 2015. We introduced the values measured in the simulating program of the T_F area, to see the differences compared to the values obtained by semi-empirical models. These data shown in **Figures 5 and 6**. It noted that the numerically values obtained are close to those of the experimental data, the effect of wind speed and room temperature is almost negligible.

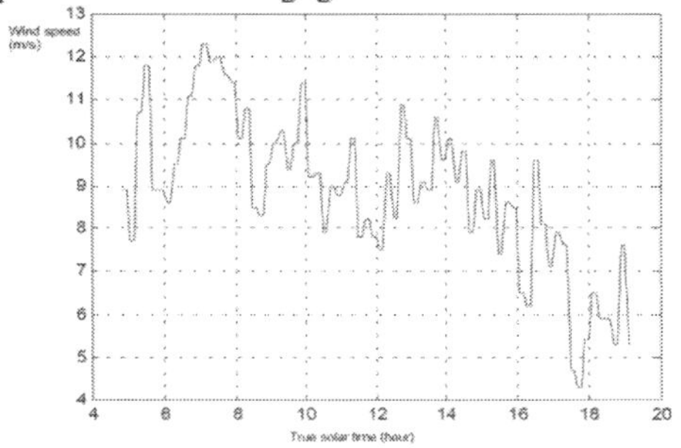

Figure 5: Experimental measurements of wind speed for 16 August 2015

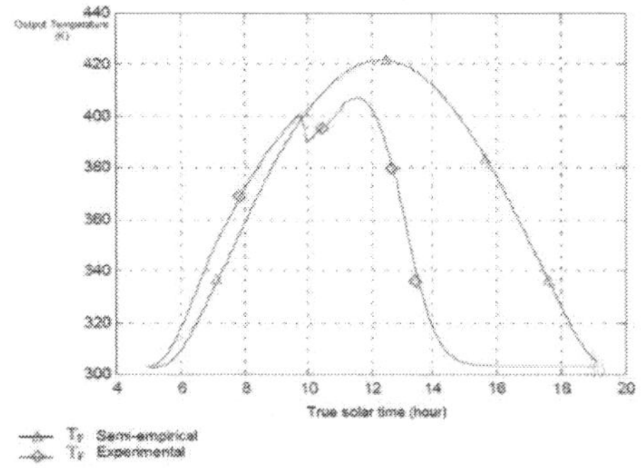

Figure 6: Simulation of the outlet air temperature for 16 August 2015

In the range of airflow values studied from 20 to 200 kg/h, **figure 7** shows the experimental values from previous work **[11]** and are very close to this configuration, in order to validate our approach digital. We note that the experimental values of the instantaneous thermal efficiency of the solar collector is not very far from the values from the theoretical models. However, for low flows, the model overestimates the performance. We believe that this discrepancy can explained on one hand by measurement errors but mostly by theoretical approaches made at the equated the theoretical model are the underestimation of losses in the theoretical model amounted for low airflow rates, because of the higher temperature difference between the absorber and the outside. We could introduce empirical correction coefficient of overall losses, from experimental data, as was done in **[12]**. In this work, the evolution of the overall losses U_L coefficient was plotted against the temperature difference $(T_f - T_a)$ which is all the more important that the flow rate is low, T_f being the average air temperature within the air gap and the ambient temperature T_a. The proposed equation is of the form $U_L = a\,(T_f - T_a)^b$. The evolution of the instantaneous thermal efficiency of the solar collector shown in **Figure 8**. We note that the addition of GEM_{PEHD} absorber made a remarkable increase is about 25% relative to a solar collector with another type absorber. Analysis of the various results obtained show that while the duration of the capture of solar radiation near room temperature, the solar collector has a good performance, which is around 55%. This performance explained by lower optical and thermal losses through the glazing system. Taking into account the means of realization of our system, its simplicity and its cost price is quite low and compared to other more sophisticated solar collector performance found in the literature, we can conclude that our prototype still competitive with a quality / price very valuable. Based on experimental tests on our system and the different results obtained, we find that more solar illumination is perpendicular to the greater the amount of heat absorbed collecting area is considerable, because the glazing transitivity and absorptivity of the absorbing plate are inversely proportional to the angle of incidence of solar radiation. It is therefore very important to choose the inclination of the solar collector with the seasons to take better advantage of the system.

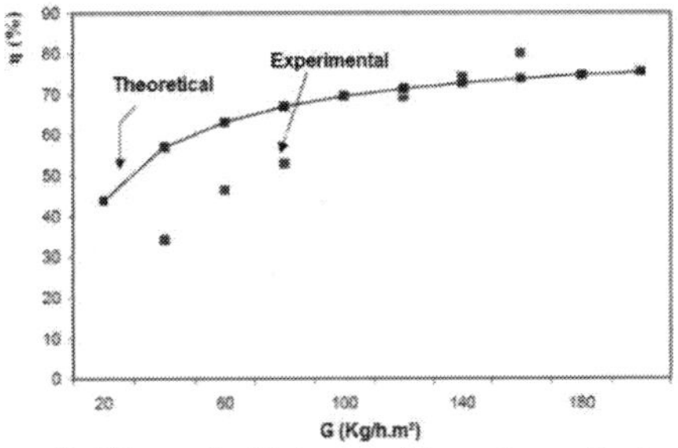

Figure 7: Thermal efficiency of the solar collector as a function of airflow

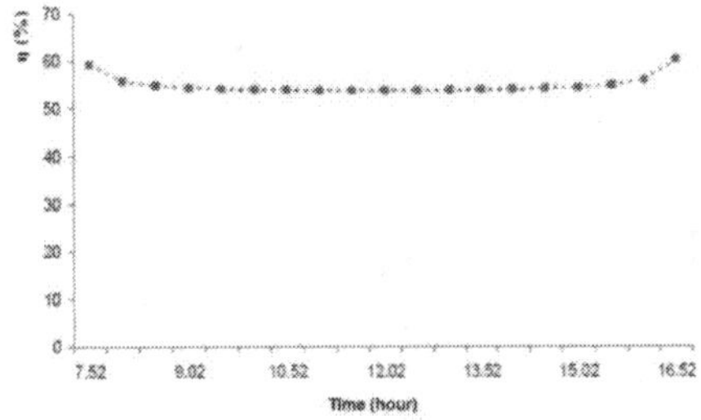

Figure 8: Change thermal efficiency of the solar collector with a clear sky and a throughput of 35 kg / h.

As regards the development of in and out air temperatures, the results confronted with experimentally provides those delivered digitally. **Figure 9** shows the temporal variation of the temperature of each sensor element. In this figure, the absorber and the inner surface of the insulator have higher temperatures, this is due to the power absorbed by the absorber is important and is greater than that absorbed by the glass, and the insulation is placed beneath the absorber plate, the latter has a high thermal conductivity. The lowest temperature is that of the insulation of the rear side, indicating that the insulation has low thermal conductivity. The increase in indoor and outdoor temperatures glazing is due to absorption by the window of the incident solar radiation and the heat released by the absorber. The temperature of the inner surface is slightly greater than that of the outer face; this is due to the trapping of solar infrared ray inside the sensor thereby increasing the temperature inside of the glass with respect to outside, which assigned to the action of wind. The coolant outlet temperature is slightly lower than that of the absorber, due to losses. The analysis results successively shows good agreement between the results provided by the model and those found in the literature. It is reported that due to the existing deviations from the experimental and theoretical work cited, compared with those of our modeling, are due to the aging phenomenon of the constituent elements the solar collector when it comes to enter the corresponding values thermophysical and optical parameters, on the other hand the illuminations seem unfitted models considered under specific conditions.

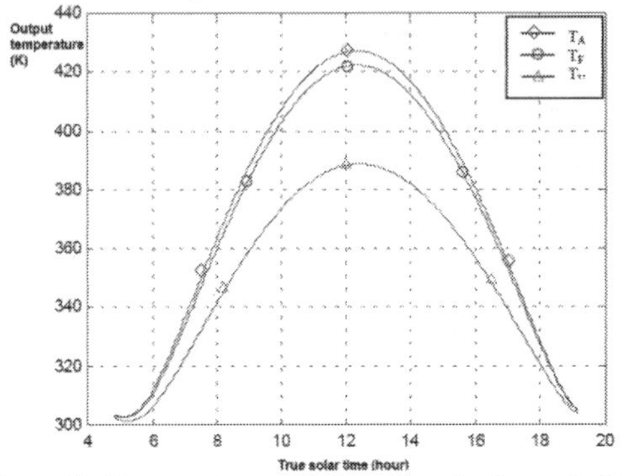

Figure 9: Temperature variation of each element of the sensor

3.2. *Study of aging GEM_{PEHD}*: The samples were prepared after summers spent summers in the solar collector. Each of these techniques giving us access to the results of different origins (chemical, physical), the confrontation of all these data allow us to envisage a better interpretation of the results. Infrared spectroscopy is the first additional technique that we used. A test campaign showed different behavior during the first hours of irradiation. This difference is a greater amount of carbonyls present in the samples prior to irradiation and for other campaigns. The amount of vinyl product remains the same as for other campaigns, and has a similar evolution during irradiation, with two separate steps before and after 6 h of irradiation (**Figure 10**).

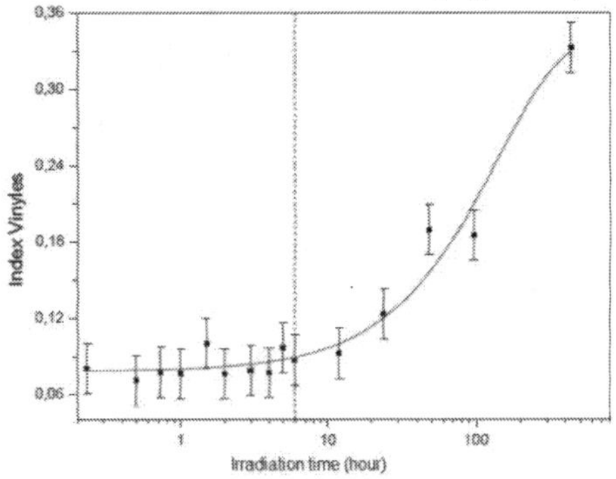

Figure 10: Evolution of vinyl degradation products as a function of UV irradiation time.

https://doi.org/10.1051/matecconf/201814902046

The reason for the presence of strong carbonyl degradation product in samples prior to irradiation from the manufacturing protocol. Shaping done on a hot plate at a temperature of 150 °C, to which the GEM_{PEHD} undergoes a major thermal oxidation, which causes formation of carbonyl-type degradation products if the exposure time is too long. We chose to keep this trial campaign since the first moments of exposure to UV show interesting behavior (**Figure 11**) **[13]**.

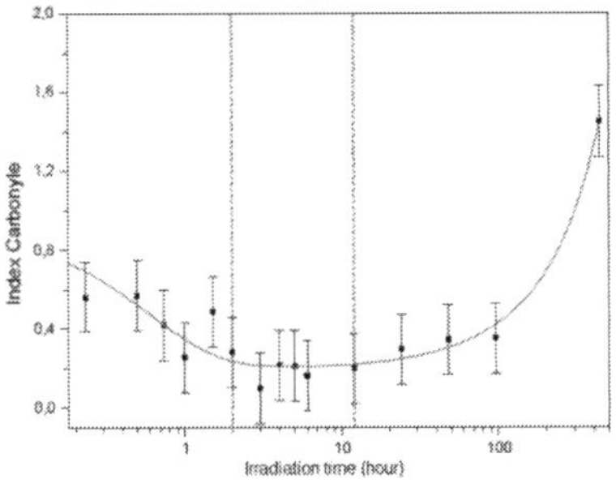

Figure 11: Evolution of carbonyl degradation products as a function of UV irradiation time.

The differential scanning calorimeter used in the following part will allow us to have access to this information. This technique allows us to obtain information on physical samples and especially the degree of crystallinity (% Cr) and the melting temperature (T_f). **Figure 12** shows the change in the rate of crystallinity of the samples according to the UV ray exposure time. As in the case of the infrared spectroscopy analysis, two distinct phases can observed. Crystallinity rate remains stable at around 72% during the first 6 hours, then increases exponentially in the semi logarithmic representation to reach the value of 79% after 18 days of irradiation.

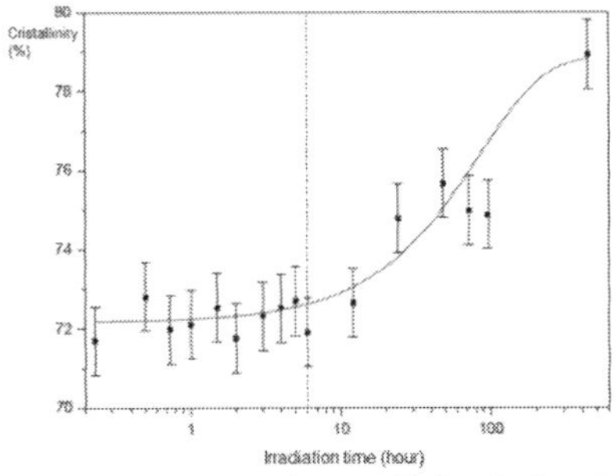

Figure 12: Evolution of crystallinity during the UV irradiation.

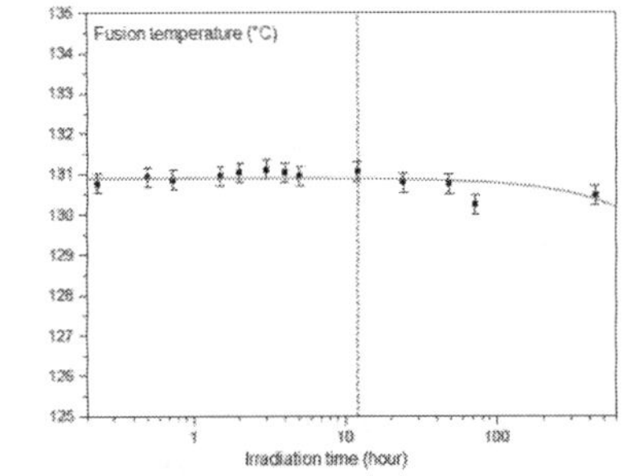

Figure 13: Evolution of the melting temperature of the samples as a function of UV irradiation.

This increase in the degree of crystallinity accompanied by a slight decrease in the melting temperature (**Figure 13**). In the literature, different authors interpret these results by changing the macromolecular chains of GEM_{PEHD} due to UV irradiation. UV radiation causes breaks chains with consequent increases in crystallinity **[14]**. Indeed, in the amorphous phase, chains see their length reduced, making them more mobile **[15]** and thus favoring the phenomenon called secondary crystallization **[16]**, illustrated here by increasing the degree of crystallinity after 6 h of exhibition.

From the melting temperature values and crystallinity, it is possible to calculate the size of the crystal lamellae. Preliminary calculations performed on irradiated samples led to size strips of the order of 30.1 nm. The evolution of this value during the deterioration (**Figure 14**) may also presented in two stages, in the first moments with a stability followed by a decrease. This development is in line with the first results of degree of crystallinity and melting temperature. The rupture of the macromolecular chains caused by UV radiation leading to lamellae smaller dimensions which plays on the melting temperature and the crystallinity. However, even if all the results obtained in AED are well in line with the already established models.

Monitoring of weight average molecular weight of the GEM_{PEHD} during UV irradiation performed by GPC and shows a rapid increase them from an hour of exposure (**Figure 15**). These results are consistent with an increase in the average length of the macromolecular chains, which can attributed to cross-linking phenomena.

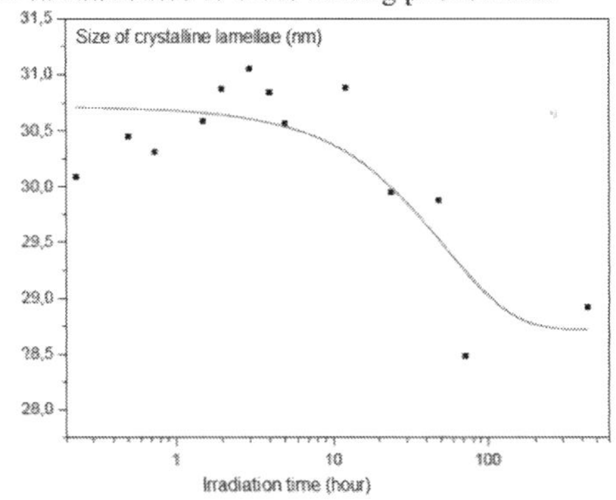

Figure 14: Evolution of the size of crystalline lamellae as a function of UV irradiation time.

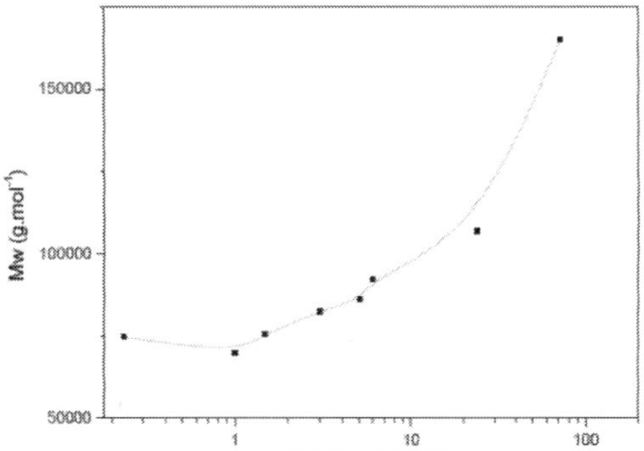

Figure 15: Changes in the average molar mass by weight of the GEM_{PEHD} as a function of UV irradiation time.

4. Conclusion:

The results obtained show that the outlet temperature varies depending on the solar flux as well as geometric and climatic parameters. Among the main results, we quote:

- Large temperature value achieved in the summer solstice, which can reach 420 K in true solar noon. It mainly depends on the power consumption that is a function of the optical and geometrical parameters of the sensor. What say we leave this type of collector can favorably recommended in the water-heating sector.
- The temperature of the absorber is closest to T_F, which can be justified by its high absorptivity for visible solar radiation and a high emissivity for the infrared radiation of long wavelength with its coating. It is possible to keep the greater part of the incident solar energy and not lose as little heat by long wavelength radiation when the absorbent surface becomes hot. The glazing temperature is lower than that of TF when dependent optical, climatic, and specific parameters of the wind speed.

Analyzes performed by FTIR and DSC provide information about aging GEM_{PEHD}. It is found that thermal aging has an important influence on the volumetric strain due to increased crystallinity. The X-ray diffraction analysis showed the post crystallization phenomenon by revelation of the increased crystallinity of the aged sample. While FTIR analysis has shown that, this phenomenon is due only to the change in the structural morphology of the GEM_{PEHD}. Investigations conducted by DSC shows that the GEM_{PEHD} has a melting temperature of 131 °C for an aged sample while valley 135 °C for a blank sample characteristic of increasing the crystallinity of confirming the findings with the same further analyzes.

Each analysis technique having its own sensitivity to this or that phenomenon, aggregation and correlation of all results obtained are essential for a comprehensive understanding of degradation mechanisms GEM_{PEHD}. However, a question remains open on the choice of the molecule inserted into the GEM_{PEHD}, this choice that can condition the type of information obtained. In our case, with Irganox B225, the results show a high sensitivity to rearrangements of the macromolecular chains of the GEM_{PEHD}.

After various tests, we conclude that the value of using the GEM_{PEHD} as absorber in the solar collector is doubly beneficial. First, given its very attractive cost compared to other as efficient absorbers (such as metals) gives good thermal performance of the solar collector even reaching 72%. While the life of the GEM_{PEHD} is rather long (several years).

Reference:

[1] OCDE. Énergie : les cinquante prochaines années. OECD Publishing, 2010.

[2] A. Bidart and L. Dubois, Les énergies fossiles et renouvelables, Dossier Pédagogique de la Fondation Polaire Internationale, 2003.

[3] L. Ibos, S. Datcu and S. Matteï, Mesure d'émissivité totale à température ambiante de matériaux inhomogènes à l'aide d'un émissomètre portable, 5ème Colloque Interuniversitaire Franco-Québécois Thermique des Systèmes, Lyon, 73-80, 2001.

[4] L. Ibos, M. Marchetti, A. Boudenne, S. Datcu, J. Livet and Y. Candau, Infrared emissivity measurement device : Principle and applications, Measurement Science and Technology, 17, 2950-2956, 2006.

[5] M. Marchetti, L. Ibos, A. Boudenne, S. Datcu, Y. Candau, J. Dumoulin and J. Livet, Analyse du comportement radiatif de matériaux de l'infrastructure routière : Principes et mesures de l'émissivité infrarouge, Bulletin des Laboratoires des Ponts et Chaussées, 272, 45-55, 2008.

[6] E. Tang-Kwor, Contribution au développement de méthodes périodiques de mesure de propriétés thermophysiques des matériaux opaques, Thèse Université Paris 12, 1998.

[7] Ministère de l'Energie et des Mines, Bilan énergétique nationale de l'année 2010, www.mem-algeria.org, 2011.

[8] N. M. Nahar and P. G. Jagdish, Studies on gap spacing between absorber and cover glazing in flat plate solar collectors, International journal of energy research, 13, 1989.

[9] S. W. Churchill and H. S. Chu, Correlating Equations for Laminar and Turbulent Free Convection from a Vertical Plate, *International Journal of Heat and Mass Transfer,* 18, 1323-1329, 1975.

[10] B. Cheron, *Transferts Thermiques - Résumé De Cours, Problèmes Corrigés*, Ellipses marketing, 1999.

[11] I. Tari, Natural convection simulations and numerical determination of critical tilt angles for a parallel plate channel, Energy conversion and management, 51, 685-695, 2010.

[12] R. Alvarado, J. Xamán, J. Hinojosa and G. Álvarez, Interaction between natural convection and surface thermal radiation in tilted slender cavities, International Journal of thermal sciences, 47, 355–368, 2008.

[13] Q. Wu and al., *Surface photo-oxidation and photostabilization of photocross-linked olyethylene,* Polymer Degradation and Stability, 68(1), 97-102, 2000.

[14] R. Yang and al., *Effects of inorganic fillers on the natural photo-oxidation of high-density Polyethylene,* Polymer Degradation and Stability, 88(2), 333-340, 2005.

[15] L. Guadagno and al., *Chemical and morphologial modifications of irradiated linear low density polyethylene,* Polymer Degradation and Stability, 72(1), 175-186, 2001.

[16] I. H. Craig and al., *Photo-induced scission and crosslinking in PELD, PELLD and PEHD,* Polymer Engineering and Science, 45(4), 579-587, 2005.

Evaluation of landfill site choice using AHP and GIS case study: Oum Azza, morocco

A. AIT ERROUHI[1], L. BAHI[1], L. OUADIF[1], A. AKHSSAS[1], Y. BOUROUMINE[1] and A. BAHI[1]
[1]Laboratory of Applied Geophysics, Geotechnics and Geology of Engineers and Environment, Mohammadia School of Engineers, Mohammed V University, Rabat, Morocco

Abstract The amount of household waste produced by the mainly urban population in Morocco continues to increase year after year. Morocco is thus confronted with the problems of storage and disposal of this waste, which are composed largely of organic matter, bio-waste. In the recent past, these wastes have been buried in wild dumps or in old quarries located in areas close to urban centers without any treatment or control. This can constitute significant risks both for public health and the environment and for the future of the socio-economic activities of the country. For which evaluating municipal landfills is vital due to the importance of health and environmental issues and the possibility of contravening environmental and engineering criteria and principles. The present research is conducted using new techniques of approach among others, the analytical hierarchy process AHP, which will allow us to weight and combine all criteria and to classify their respective weights, as well as the use of GIS which will contribute to the evaluation of these criteria in order to determine whether the choice of the current site according to the results of our study meets the requirements of a controlled landfill respecting the environment.

1 Introduction

Waste management solids remained debatable up to now in the near total of the countries of the world. The capital concern relates to the evacuation of the refuse produced by the households with an aim of improving cleanliness and protection of public health. This waste finishes for most of the time in waste grounds. Nearly 98% of cities in developing countries have only not controlled discharges which, in the averages and small towns summarize simple wild dumps into cubes in which waste is often burned for the dry periods to reduce the quantity [1]. The access to these sites is controlled little and there exists an important informal sector of recovery involving health risks and embarrassments of exploitation. It should be noted that land filling is the most widespread waste treatment solution in the world but also the least expensive. However, while this practice is relatively controlled and restraint in western countries, it is still a problem in developing countries due to lack of controls, inadequate or non-functional arrangements and funding difficulties.

In Morocco, in order to face the problem of wild dumps and their eminent character, land filling is certainly a good choice, but this option is far from being ecologically sustainable where the waste is very wet and very rich in organic matter and generate large volumes of leachate. Like the majority of the provinces of the Kingdom, the province of Temara Sekhirat includes a controlled landfill in Oum Azza, which covers an area of 110 hectares and receives 1,800 tons of solid household and similar waste. This landfill raises the dissatisfaction of the neighboring population by obvious nuisances such as odors, groundwater pollution and road degradation due to truck passages.

The objective of our study is to evaluate the site using the geographic information system GIS, which plays a crucial role in planning, to see if the landfill meets legislative requirements and respects the environment, Possibility of identifying other better sites that could well house the latter.

GIS can also play a key role in maintaining data to facilitate collection operations, customer service, analysis of optimal locations for transfer stations, planning of the route for the transport of waste to transfer stations and stations transferring to landfills, and long-term landfill monitoring [2], As it can be a decision-making tool, simplifying the search for sites suitable for a particular use because of its ability to extract spatial characteristics and classification. The process of site selection is complicated, hence the need to integrate several decision support tools such as multi-criteria evaluation (MCE) for optimal selection [3].

Among MCE techniques, the Analytic Hierarchy Process (AHP) is one of the most common analytical techniques for complex decision-making problems and is widely used because of its agility and easy to use. It helps decision-makers discern which is best suited to their goal and problem design. It provides a comprehensive and rational framework for structuring a decision problem, representing and quantifying its elements to link these elements to the overall objectives and to evaluate alternative solutions [2]. To that end, a hierarchical tree is created: the decision-making hierarchy tree represents the decision strategy graphically; it uses a multi-level hierarchical structure of objectives, criteria, sub-criteria, and alternatives. The pertinent data are derived by using a set of pairwise comparisons. These comparisons are used to obtain the weights of importance of the decision criteria, and the relative performance measures of the alternatives in terms of each individual decision criterion [4].

The evaluation of this waste disposal site is a complex process because it requires considerable expertise in various social and environmental fields, such as soil science, distance to homes, engineering, hydrogeology, and topography, Land use, leaching of contaminants, sociology and economics [5].

Therefore, the site selection of a solid waste landfill must also involve the processing of a significant amount of spatial data, regulations and acceptance criteria, as well as an effective correlation between them. Some promising approaches to landfill assessment have been discussed by Gebhardt [6], Swallow [7], Chen [8] and Sasao [9].

The present study's purpose is to develop a methodology to evaluate the landfill site choice using AHP and GIS for the study area Oum Azza, Morocco.

2 Study area

The region of Rabat, Sale, Zemmour and Zaers (RSZZ) is located in the north-west of the country, covering an area of 9,580 km2 and 2,134,533 inhabitants (RGPH 2014). It is bounded on the north by the region of Rharb-Cherrda-Beni hssen, on the east by the region of Fes-Meknes, on the south by the region of Chaouiya Ouedrigha and on the West by the Atlantic Ocean.

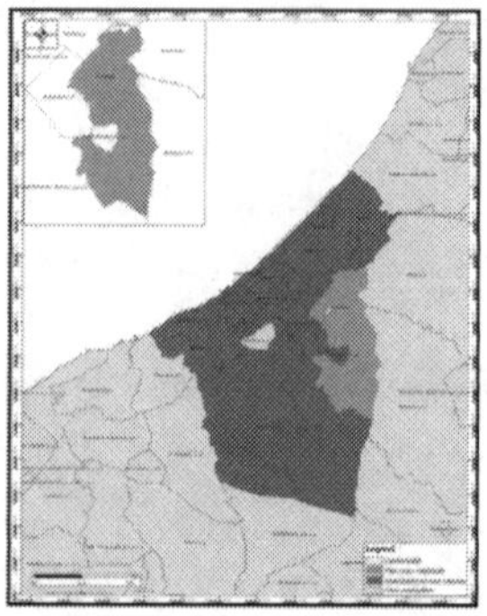

Fig. 1: Location of study area

This region is characterized by relatively favorable climatic conditions. The geographical distribution of temperatures and precipitation is influenced by the latitude, distance from the ocean and altitude.

In terms of temperature, three different thermal domains are categorized: an oceanic domain in the coastal zone, a semi-continental domain (Tifelt, Roummani and Khemisset) and a middle mountain area (Oulmes area). In addition, the average temperature varies between 17 and 18 ° C. In terms of rainfall, there are two regimes according to altitude: Plains and plateaus with a rainfall of 500 mm / year (Rabat-Tifelt, Tiddas), the mid-mountain zone with a rainfall of 660 mm / Year (Oulmes).

The RSZZ region houses a landfill in the rural commune of Oum Azza on the Aïn Aouda plateau. It is located between Wadi Akrach to the west and the reservoir of the Sidi Mohamed Ben Abdellah dam in the east, with elevation between 160 and 200 NGM. It receives household and similar waste from 13 communes: Rabat, Sale, Bouknadel, Temara, Skhirate, Harhoura, Ain Aouda, Mers Kheir, Al Manzeh, Ain Attig, Sebbah, S.Yahya Zaers and Oum Azza, and serves a population of 980,000 inhabitants. Its area is about 111 ha.

3 Methods and materials

3.1 Geographic Information System

Available information on geology, hydrogeology, hydrology, soil, climatology and eco-sociology has been assembled and compiled in digital form. Several source maps helped to prepare the geographic database. These maps were at different scales and included different types and quantities of information. Thus, several maps have been produced, namely: topographic maps of the region Rabat Kenitra, especially topographic map of the city of Temara with a scale of 1/50000, municipality of Rommani with a scale of 1/50000 and rural communities of Larba't As-Sehoul, Sidi Bettach and Sidi Bouqnadel with a scale of

1/50000, geologic map of the study area with a scale of 1/100000 and World Imagery from GIS. All these data allowed the preparation of a database in the information system with spatial reference of the region of Rabat Sale and Kenitra.

The allocated model to evaluate the current site of the Oum Azza landfill for the storage of waste, consists of a compilation of relative geographical data in GIS as a decision support system. In each version, various localization techniques have allowed the creation of basic spatial documents in raster or vector format [10 and 11].

3.2 Analytic Hierarchy Process AHP

Analytic hierarchy process (AHP) is one of the most popular analytical techniques for complex decision making problems and is widely used due to its flexibility and easy to use. An AHP hierarchy can have many levels to characterize a decision condition. The selected factors governing the suitability of the site are weighted using this analytical method which is aided by pairwise comparison matrix that uses a scale of relative importance [12]. Once the weights are determined, the AHP calculates the relation between the weights of criteria and then calculates the total value of each option based on the weight obtained [13]. To make a decision in an organized way to generate priorities we need to decompose the decision into the following steps:

1- To define the problem and determine the kind of knowledge sought.
2- To structure the decision hierarchy from the top with the goal of the decision, then the objectives from a broad perspective, through the intermediate levels (criteria on which subsequent elements depend) to the lowest level.
3- Construct a set of pairwise comparison matrices. Each element in an upper level is used to compare the elements in the level immediately below with respect to it. The scale for comparison consists of values ranging from 1 to 9 which describe the intensity of importance, by which a value of 1 expresses equal importance and a value of 9 is given to those factors having an extreme importance over another factor.
4- Use the priorities obtained from the comparisons to weigh the priorities in the level immediately below. Do this for every element. Then for each element in the level below add its weighed values and obtain its overall or global priority [14].

In AHP, at first the decision problem was decomposed into simpler decision problems to form a decision hierarchy. Then, pairwise comparisons were done. Pairwise comparison method is mainly used because it is hierarchical, statistical and easy to use. It is a precise method with high credibility.

In the second part of our study, after the maps obtained using GIS, AHP will allow us to evaluate the current landfill site based on four criteria: economic impact, environmental impact, social impact and technical impact. In order to identify the criterion to be taken into consideration to ensure that the landfill meets legislative and environmental requirements. To do this we have developed an algorithm that will help us to make these calculations for n criteria.
The matrix of the ratios of all weights

$$W=[W_i/W_j]=\begin{pmatrix} W_1/W_1 & \cdots & W_1/W_n \\ \vdots & \ddots & \vdots \\ W_n/W_1 & \cdots & W_n/W_n \end{pmatrix}$$

With n : the number of compared alternative
W_1, Wn : their current weights

The matrix of pairwise comparisons $A=[a_{ij}]$ represents the intensities of the stakeholders preference between individual pairs of alternatives. They are usually chosen from a given scale (1/9,1/7,...,7,9). Given n alternatives, a decision maker compares pairs of alternatives for all the possible pairs, and a comparison matrix A is obtained, where the element aij shows the preference weight of Ai obtained by comparison with Aj [15].

$$A=[a_{ij}]=\begin{bmatrix} 1 & a_{1j} & a_{1n} \\ 1/a_{1j} & 1 & a_{in} \\ 1/a_{1n} & 1/a_{in} & 1 \end{bmatrix} \quad (1)$$

The a_{ij} elements estimate the ratios w_i/w_j where w is the vector of current weights of the alternative.

$$Aw = \lambda_{\max} . w \; (2)$$

With λ_{max} : eigenvalue, [16].
The consistency index (CI) as follows:

$$CI = \frac{\lambda_{max} - n}{n-1} \quad (3)$$

Consistency Ratio (CR) to measure how consistent the judgements have been relative to large samples of purely random judgements.

If CR ≤ 0.1 is true for matrix A, the assessments of the relative importance of the criteria are considered as acceptable.

If the CR is much in excess of 0.1 the judgements are untrustworthy because they are too close for comfort to randomness and we need to consider revising our subjective judgments (the exercise is valueless or must be repeated). For Saaty [15], defined the consistency ratio (CR) as :

$$CR = \frac{CI}{RI} \quad (4)$$

With RI is the random index: consistency index for the n row matrixes of randomly generated comparisons in pairs.

$$RI = \frac{\lambda_{max} - n}{n - 1} \quad (5)$$

With $\bar{\lambda}_{max}$, [17] :

$$\bar{\lambda}_{max} = (2.7699 * n) - 4.3513 \quad (6)$$

Figure 2 shows the algorithm that we have done to calculate the weight of criteria.

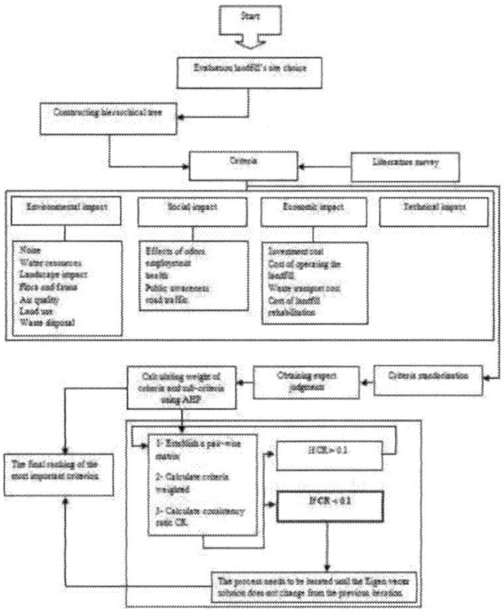

Fig.2: The flow chart illustrating the methodology used for evaluating the most important criteria

4 Results and discussion

After the digitization of the collected data, it was possible to extract the main data relating to the study area of Oum Azza. We observe at the vegetation cover, the majority of land is occupied by food crops, which are largely self-consumed by peasants and the local population. Also, Trees and shrubs are mostly identified along hedges along roads and certain plots (figure 4). As far as the infrastructure of the region is concerned, there is the Bouregreg complex, including the dam and the reservoir of Sidi Mohamed Ben Abdellah, the roadways P 22 and S 202, the Rabat bypass highway which has just been put into service and which crosses the town of Rommani, High and medium voltage lines, and the site of Sidi Mohammed and of Sidi M'bark belong to the National cultural heritage (holy places) (figure 3). (Figure 5), shows that the municipality of Oum Azza is crossed by several permanent and non-permanent watercourses. This municipality belongs to the watershed of the Bou-Regreg, the rainfall on the plateau is directed to the east, towards the wadi Korifla and the Sidi Mohamed Ben Abdellah reservoir, to the west, to the wadi Akrach. Wadi Akrach is a direct tributary of the Oued Bou-Regreg. It is not the same for the wadi Korifla which first conflates to the Oued Mechraa. The wadis draining the plateau, and particularly the Hachichia river, are relatively little used for

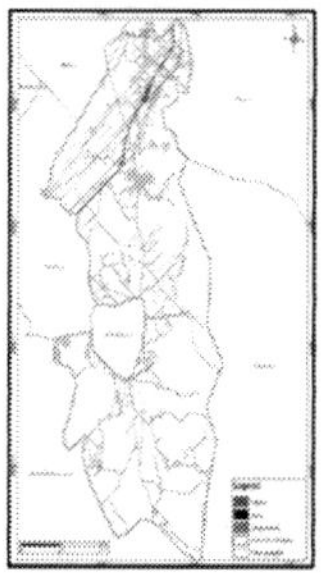

Fig. 3: Infrastructure map of Oum Azza

Fig. 4: Map of vegetation cover of Oum Azza

Fig. 5: Hydrological map of Oum Azza

irrigation of crops. The same is true for daya.

In order to evaluate the current site of the landfill, the table 1 shows a list of exclusion criteria on which we based ourselves for the rest of our study.

Table. 1: List of exclusion criteria and areas

Exclusion criteria	Exclusion area
Living areas	500 m
Roads	300 m
Highway	500 m
Electrical networks	200 m
Cemeteries and religious institutions	500 m
Green spaces	500 m
Protected areas	500 m
Agricultural areas	300 m
Forests	300 m
Natural parks	1000 m
Oueds, chaabats	500 m
Fountains	500 m
Water resources	400 m

From the table 1, we have inserted the corresponding exclusion areas for each exclusion criterion in our GIS database. The software allowed us to conclude that the current site of the Oum Azza landfill does not meet certain exclusion criteria (figures 6, 7 and 8).

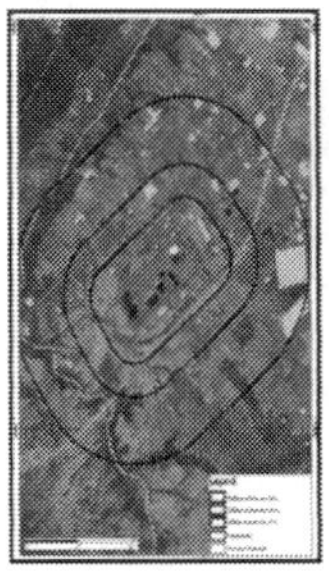

Fig.6 : Map showing the exclusion criteria of living areas

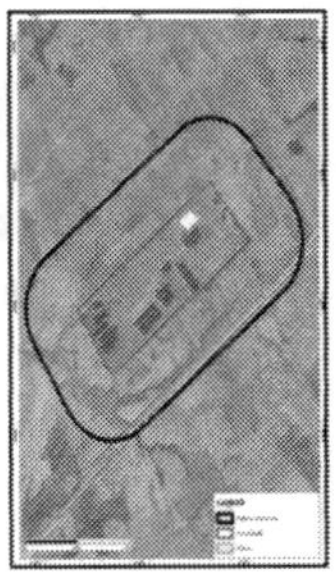

Fig.7 : Map showing the exclusion criteria of the highway

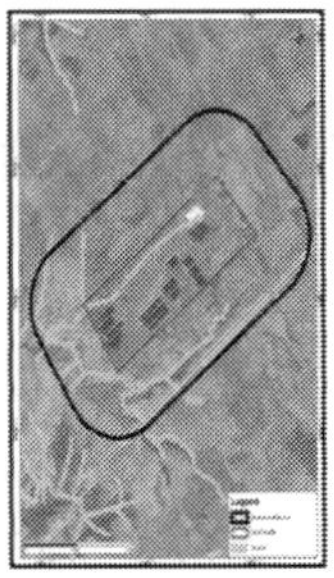

Fig.8 : Map showing the exclusion criteria of water resources

When assessing the current landfill site, we note that the current site does not meet certain exclusion criteria; as has been shown in the preceding figures, for the (figure 6) concerning living areas; it can be seen that the exclusion area of 500 m was not taken into account and there are indeed households, also for (figure 8) shows that the criterion of exclusion for water resources at a distance of 200 m was not respected, which could harm the health of the population and the same case for the criterion of exclusion of the highway to an exclusion area of 500 m (figure 7).

In the present descriptive-analytic study, to determine the weight of each of the criteria, economic, environment, social and technical impact, we used the method of decision-making (AHP). These criteria will be calculated by the application that we have developed. Then the criteria, evaluation options of the Oum Azza landfill will be ranked and prioritized.

Table. 2: Pairwise comparison matrix of the main criteria

Environment (En), social (S), economical (Eco), technical (T).

Criteria	En	S	Eco	T	Weight
En	1	2	9	5	0.518
S	0.5	1	9	3	0.316
Eco	0.11	0.11	1	0.2	0.038
T	0.2	0.33	5	1	0.128

Table. 3: Pairwise comparison matrix for the sub-criteria with respect to environment

Noise (N), water resources (WR), landscape impact (LI), flora and fauna (F), air quality (A), land use (LU), waste disposal (WD).

En	N	WR	LI	F	A	LU	WD	Weight
N	1	0.11	3	0,33	0,11	0,14	3	0,043
WR	9	1	7	3	0,33	2	7	0,227
LI	0,3	0,14	1	0,2	0,11	0,2	2	0.029
F	3	0,33	5	1	0,2	0,33	5	0,092
A	9	3	9	5	1	5	9	0,425
LU	7	0,5	5	3	0,2	1	5	0,158
WD	0,3	0,14	0,5	0,2	0,11	0,2	1	0,024

Table. 4: Pairwise comparison matrix for the sub-criteria with respect to social

Effects of odors (EO), employment (E), health (H), public awareness (P), road traffic (R).

S	E	EO	H	P	R	Weight
E	1	3	0.2	2	5	0.173
EO	0.33	1	0.11	0.2	0.5	0.04
H	5	9	1	7	9	0.6
P	0.5	5	0.14	1	5	0.139
R	0.2	2	0.11	0.2	1	0.048

Table. 5: Pairwise comparison matrix for the sub-criteria with respect to economy

Investment cost (IC), cost of operating the landfill (CO), waste transport cost (WT), cost of landfill rehabilitation (CR).

Eco	IC	CO	WT	CR	Weight
IC	1	0.14	0.125	0.33	0.048
CO	7	1	0.5	5	0.367
WT	8	2	1	3	0.467
CR	3	0.2	0.33	1	0.118

Table. 6: final weights

Criteria		Sub-criteria		Final weight
Description	Weight	Description	Weight	
ENVIRON MENT	51.8 %	Noise	4.3 %	2.22 %
		Water resources	22.7 %	11.7 %
		Landscape impact	2.9 %	1.5 %

Criteria		Sub-criteria		Final weight
Description	Weight	Description	Weight	
		Flora and fauna	9.2 %	4.76 %
		Air quality	42.5 %	22.02 %
		Land use	15.8 %	8.18 %
		Waste disposal	2.4 %	1.24 %
SOCIAL	31.6 %	Effects of odors	4 %	1.26 %
		Employment	17.3 %	5.4 %
		Health	60 %	18.9 %
		Public awareness	13.9 %	4.39 %
		Road traffic	4.8 %	1.51 %
ECONOMICAL	3.8 %	Investment cost	4.8 %	0.18 %
		Cost of operating the landfill	36.7 %	1.39 %
		Waste transport cost	46.7 %	1.77 %
		Cost of landfill rehabilitation	11.8 %	0.44 %
TECHNICAL	12.8 %	-	-	12.8 %

From tables 2, 3, 4 and 5, the CR obtained for the assessment of the current discharge site for each criterion and sub-criteria is 0.042, 0.074, 0.086 and 0.055 respectively. The consistency ratios are less than 0.1 indicating constant judgment and a reasonable level of consistency. According to the results of table 6, it was concluded that the most important criterion is the environment with a weight of 0.581, followed by the social and technical criteria which had as weights 0.316 and 0.128 respectively, on the other hand the economic criterion is ranked as the least important with 0.038. At the level of sub-criteria we distinguish that air quality is more vulnerable with a weight of 0.220 and that the least vulnerable is the investment cost with a weight of 0.018, given the region, where the landfill site is located, is coastal, it undergoes a sea-breeze phenomenon, the latter during the day, the earth absorbs the heat emitted by the sun quickly and heats the air above it. On the other hand, the solar radiation on the sea is absorbed more slowly because of the higher thermal capacity of the latter. During the night, the earth cools faster than the sea, because of their difference in thermal capacity (the earth accumulates less heat from the sun than the sea, it cools faster at night), and inverse phenomenon occurs: the air above the earth is colder than that above the sea, which generates the spread of harmful odors from the landfill that detract the air quality which causes the dissatisfaction of the neighboring population.

Given that most of the criteria considered in this study may not have been considered in the selection of the current landfill site, this site is vulnerable and inhibited to be located in this area. The criteria considered in this study should be taken into account by decision makers to assess existing landfills and select the most suitable sites for future landfills. In addition, buffer zone distances for the above criteria should be established in the selection of future sites and maintained thereafter to avoid undesirable environmental impacts. In addition, the tool used in this study provides critical support to the decision-making process.

5 Conclusion

The present study studies the evaluation of the current site of the Oum Azza on the basis of different criteria. The results highlight the importance of multicriteria decision analysis to avoid vulnerability to the landfill and GIS environment that requires the collection of data from different sources with different formats to create a database unified. Thus, GIS data should be updated frequently to reflect the current situation in the study area. The criteria used in this study should be taken into account by decision makers in order to carry out evaluation studies at different existing landfill sites and to select the most suitable sites for future landfills.

References

[1]http://www.plateforme-re-sources.org/wp-content/uploads/2015/10/08-FS-Decharge.pdf http://ma.chm-cbd.net/rabat-sale kenitra/region/monographie-de-la-region-rabat-sale-kenitra

[2] B. Kaoutar, B. Lahcen A Decision Support Approach for Optimized Sitting Of Municipal Solid Waste Landfill Case Study Tangier Morocco, International Journal of Engineering Research and Applications, **Vol 2**, Issue 6, pp 1676-1684, (2012).

[3] M. Oyinloye, J. Fasakin, Geographical information system (GIS) for sitting and management of solid waste disposal in Akure,

Nigeria, Academia Journal of Scientific Research, **Vol 1**, pp 090-103, (2013).

[4] E. Triantaphyllou, Stuart H. Mann, Using the analytic hierarchy process for decision making in engineering applications: some challenges, International Journal of Industrial Engineering: Applications and Practice, **Vol 2**, pp 35-44, (1995).

[5] V. Sumathi, U. Natesan and C. Sarkar, GIS-based approach for optimized siting of municipal solid waste landfill, ELSEVIER Waste Management, **Vol 28**, pp 2146–2160, (2008).

[6] K. Gebhardt, J. Jankowski, Preliminary landfill siting and related analysis using simple modeling techniques, Journal of Engineering Geology, **Vol 23**, pp 291–306, (1987).

[7] S.K. Swallow, J.J. Opaluch, T.F. Weaver, Siting noxious facilities: an approach that integrates technical, economics, and political consideration. Land Economics, **Vol 68** (3), pp 283–301, (1992).

[8] W.Y. Chen, J.J. Kao, Fuzzy drastic for landfill siting. In: *Proceedings of the 13th International Conference on Solid Waste Technology and Management*, **Vol 15**, pp 239-253, (1997).

[9] T. Sasao, An estimation of the social costs of landfill siting using a choice experiment. Waste Management, **Vol 24**, pp 753–762, (2004).

[10] B. Sener, M. Süzen, and V. Doyuran, "Landfill site selection by using geographic information systems." Environmental Geology, **Vol 49**, pp 376-388, (2006).

[11] S. Sener, E. Sener, and R. Karagüzel, "Solid waste disposal site selection with GIS and AHP methodology: A case study in Senirkent–Uluborlu (Isparta) Basin, Turkey." Environmental Monitoring and Assessment, **Vol 173**, (2011).

[12] S. A. Al Raisi2014, *msc,* landfill suitablity analysis using ahp method and state of heavy metals pollution in selected landfills in Oman, European Scientific Journal, **Vol 10**, (2014).

[13] S. Zakerian, M. Mokhtari, Evaluation of a Landfill Site Using AHP and TOPSIS Models: A Case Study of Ardakan Landfill, Iran, Avicenna Journal of Environmental Health Engineering, **Vol 2**, 11-12 (2015).

[14] T.L. Saaty, Decision making with the analytic hierarchy process, Int. J. Services Sciences, **Vol 1**, No. 1, (2008).

[15] T.L. Saaty, and L.G. Vargas, Models, Methods, Concepts and Applications of the Analytic Hierarchy Process, Boston: *Kluwer Academic Publishers*, (2000).

[16] Perron-Frobenius, *Theorie der Matrizen, Math. Ann.*, **Vol 64**, (1907).

[17] J. A. ALONSO, Consistency in the Analytic Hierarchy Process: A NEW APPROACH International Journal of Uncertainty, Fuzziness and Knowledge-Based Systems **Vol 14**, pp 445-459, (2006).

Environmental and energy assessment of the use of Andesite in cement

Evaluation environnementale et énergétique de l'utilisation de l'Andésite dans le ciment

Moussa HAMIDI[1], Larbi KACIMI[2], Pierre CLASTRES[3]
[1]Université de Djilali Bounaama,, Faculté des sciences et de la technologie, Khemis Miliana W. Ain Defla, Algérie
[2]Université USTO, Faculté de Chimie, Oran, Algérie.
[3]Université de Toulouse, Laboratoire LMDC, INSA-UPS, Toulouse, France.

Abstract. The present study is a contribution to the research of new sources of supply of additions and raw materials in order to valorize local materials in the manufacture of cement. It consists in checking the environmental validity of the industrial use of andesite by evaluating the energy consumption of manufacturing andesitic cement. This evaluation was made on CEMI 52.5R type cement mortars whose basic constituent is clinker, with substitution rates of raw and calcined andesitic rock at 800 °C varying from 0-40%. and this through mechanical performance tests at several deadlines (2 days, 7 days, 28 days and 90 days). The results show that the analysis of the influence on the environment of this addition, in its raw state or calcined at 800 ° C, made it possible to say that globally the impact of andesite on the matrices seems positive to long term. The replacement of the cement by andesite systematically induces a reduction in the energy required to produce one ton of binder, due to the lower energy demand for grinding than that required for the thermal treatment of the clinker. Overall, the energy cost of andesite cement (in kWh per MPa compressive strength) is lower than that of ordinary cement for low rates of clinker replacement and obtaining cements with moderate strengths.
Keywords: Andesite; Cement; Impact; Environment; Energy.

Résumé. La présente étude est une contribution à la recherche de nouvelles sources d'approvisionnement en ajouts et en matières premières en vue de valoriser des matériaux naturels locaux dans la fabrication du ciment. Elle consiste à vérifier la validité environnementale de l'utilisation industrielle de l'andésite en évaluant la consommation énergétique de fabrication du ciment andésitique. Cette évaluation a été faite sur des mortiers de ciment de type CEMI 52.5R dont le constituant de base est le clinker, avec des taux de substitution de roche andésitique brute ou calcinée à 800°C variant de 0-40%, et ce par le biais de tests de performances mécaniques à plusieurs échéances (2 jours, 7 jours, 28 jours et 90 jours). Les résultats montrent que l'analyse de l'influence sur l'environnement de cette addition, à l'état brut ou calcinée à 800°C, a permis de dire que globalement l'impact de l'andésite sur les matrices semble positif à long terme. Le remplacement du ciment par l'andésite induit systématiquement une diminution de l'énergie nécessaire pour produire une tonne de liant, en raison de la demande d'énergie de broyage plus faible que celle nécessaire au traitement thermique du clinker. Globalement, le coût énergétique du ciment d'andésite (en kWh par MPa de résistance à la compression) est inférieur à celui du ciment ordinaire pour les faibles taux de remplacement de clinker et l'obtention de ciments à résistances modérées.
Mots-Clés : Andésite ; Ciment ; Impact ; Environnement ; Energie.

1 Introduction

La recherche de nouvelles sources d'approvisionnement en ajouts et en matières premières en vue de valoriser des matériaux naturels locaux dans la fabrication du ciment, est devenue une préoccupation écologique majeure pour les cimentiers, en termes de préservation de l'environnement et développement durable. Pour limiter l'impact environnemental relatif au procédé de production du ciment, l'emploi de ciment partiellement substitué par des additions minérales est envisagé comme liant pour l'industrie cimentaire. Cette substitution fait actuellement partie des développements les plus récents dans la production du ciment, en apportant une amélioration sur les propriétés mécaniques des matériaux cimentaires et en réduisant la consommation du clinker, en contribuant de manière simple et économique à résoudre les problèmes liés à l'environnement. La présente étude consiste à vérifier la validité environnementale de l'utilisation industrielle de l'andésite sur l'environnement et ce en évaluant la consommation énergétique de fabrication du ciment à base de roche andésitique. Les andésites sont des dépôts volcaniques riches en silice (53-63%) et de composition chimique et minéralogique variable [1-2]. Les propriétés physiques et mécaniques de l'andésite sont généralement satisfaisantes aux exigences acceptables pour leur utilisation dans divers domaines [3-5]. Ces propriétés lui permettent notamment la possibilité d'être utilisée comme matière première pour la fabrication du verre [6-12]. Les travaux antérieurs sur l'andésite Algérienne, ont montré qu'il s'agit d'andésites appartenant à la série calco-alcaline, à texture microlitique fortement porphyrique. L'évaluation des impacts environnementaux liés à l'utilisation de cette roche dans la fabrication du ciment a été faite sur des mortiers à ciment de type CEMI 52.5R et d'un mortier dont le constituant de base est le clinker, avec des taux de substitution de roche andésitique brute et calcinée à 800°C variant de 0-40%, et ce par le biais des tests de performances mécaniques à plusieurs échéances (2, 7, 28 et 90 jours). L'andésite utilisée est de provenance de la région de la wilaya de Tipaza-Algériee.

2 Expérimentation

2.1 Matériaux utilisés

Les matériaux utilisés dans la formulation des mortiers et des pâtes pour tous les modèles expérimentaux sont: le ciment, le clinker, le gypse, l'andésite, le sable et l'eau. Selon la norme européenne [EN 197-1] [15], trois types de ciment, nommés A, B et C ont été utilisés:

- Ciment A : CEM I 52,5R, fabriqué à l'usine de Martres- France de la compagnie Lafarge.
- Ciment B : CEM I 52,5 R de l'usine Teil de Lafarge en France.
- Ciment C : CEM II/A 42,5 N de la cimenterie de Chlef (ECDE), Algérie.

Le sable est choisi selon la norme EN 196-1 [16], commercialisé en sacs plastiques de 1350 ± 5g. L'eau de gâchage est une eau potable de robinet. La roche volcanique andésitique est de provenance la région de Tipaza, appartenant à la partie centrale du Nord Algérien, de réserve géologique d'environ 54 milliards m^3 [9, 12-13]. Elle est composée principalement de Feldspath, Cristobalite, Quartz, Montmorillonite, Calcite et Dolomite (Figure 1), sa masse volumique apparente est de 2,30 g/cm3. Les compositions chimiques et minéralogiques des différents constituants utilisés sont données au

Compositions Chimiques des matériaux utilisés						
Matériaux	**SiO_2**	**Al_2O_3**	**Fe_2O_3**	**CaO**	**MgO**	**K2O**
CEMI52.5R	22,84	2,70	1,84	67,41	0,81	0,23
Clinker	21.50	5.18	3.76	66.36	0.95	0.66
Andésite	57,25	14,97	5,96	8,95	1,97	1,24
Matériaux	**Na_2O**	**TiO_2**	**P_2O_5**	**SO_3**	**PF**	
CEMI52.5R	0,14	-	-	2,23	1,72	
Clinker	0.11	-	-	0.78	0.48	
Andésite	1,76	0,62	0,40	2,67	3,27	
Composition minéralogique des ciments utilisés						
Phases	**C_3S**	**C_2S**	**C_3A**	**C_4AF**	**Gypse**	
CEM I 52,5R	73,6	10,4	4,0	5,6	~ 3	
Clinker	61,02	15,63	7,37	11,43	~ 5	

Tableau 1. Compositions Chimiques et minéralogiques des matériaux utilisés

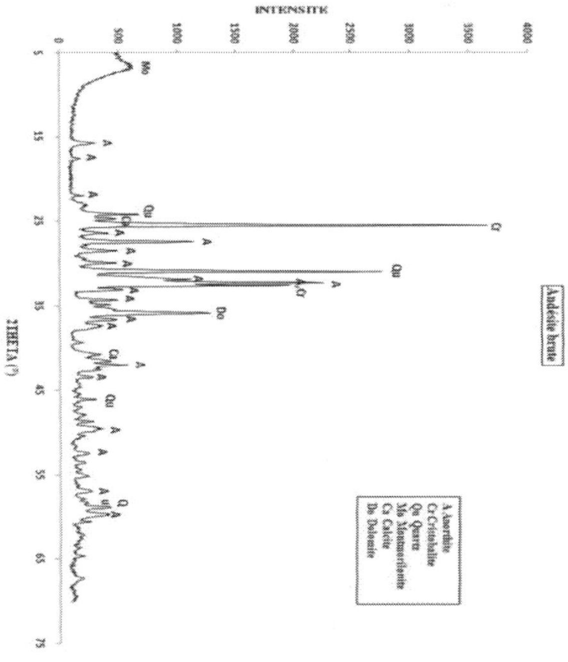

Fig. 1. Diffractogramme RX de l'andésite brute

2.2 Méthodes d'essais

https://doi.org/10.1051/matecconf/201814902048

Pour la caractérisation mécanique, l'étude expérimentale a été menée sur des éprouvettes de mortiers de 40x40x160 mm3, conformément à la norme EN 196-1 [16]. Les mortiers sont élaborés suivant les rapports massiques (S/C = 3 et E/C= 0,5 avec C : Ciment, S : Sable et E : Eau).

A chaque échéance d'âge, une série de trois (03) éprouvettes est soumise à l'essai de flexion et les six (06) demi-prismes obtenues sont soumis à l'essai de compression.

by the publisher.

3 Résultats et discussions

3.1 Evolution des résistances mécaniques

Les résultats des essais mécaniques obtenus pour les différentes éprouvettes des mortiers à base des trois ciments A, B et C à ajout andésitique brut et calciné à différentes températures (700°C, 800°C et 900°C) après 2, 7, 28 et 90 jours de durcissement sont présentés dans le tableau 2.

Tableau 2. Résistances de mortiers à base des trois ciments A, B et C à ajout andésitique

	Age 2	Age 7	Age 28	Age 90		Age 2	Age 7	Age 28	Age 90		Age 2	Age 7	Age 28	Age 90
P%	Effet du type de ciment A/80µm					Effet du type de ciment B/80µm					Effet du type de ciment C/80µm			
0	34,6	57,5	70,8	77,5		28,2	51,4	66,1	68,6		22,2	34,6	44,0	46,1
10	30,9	52,1	66,4	65,1		29,5	53,9	64,4	69,4		18,1	32,1	43,2	45,1
20	27,4	47,4	54,7	57,2		26,7	46,9	55,9	63,1		14,9	28,5	37,2	41,1
30	21,6	38,4	47,6	50,8		22,6	39,8	47,7	53,1		13,3	25,8	34,1	38,3
40	16,9	29,3	37,0	41,3		17,6	32,3	38,4	45,6		12,0	23,2	30,7	33,3
	Effet de calcination Ciment A/80µm, 700°C					Effet de calcination Ciment A/80µm, 800°C					Effet de calcination Ciment A/80µm, 900°C			
10	29.6	49.6	64.5	n.a		32.0	54.0	67.0	71.1		33.6	53.3	63.2	n.a
20	26.5	45.9	59.5	n.a		29.8	47.8	60.1	65.3		28.9	46.5	58.5	n.a.
30	23.7	38.4	50.1	n.a		23.1	43.9	54.3	57.1		22.4	39.6	50.0	n.a.
40	17.0	31.2	38.7	n.a		17.7	33.7	45.0	48.2		17.9	33.3	42.9	n.a.

3.2 Utilisation de l'andésite dans les ciments comme constituant principal

Les travaux effectués par [14], ont montré que l'andésite a une activité pouzzolanique modérée et pourrait probablement être considérée comme une pouzzolane naturelle au sens de la norme EN 197-1 [15]. Le Tableau 3 présente quelques exemples de ciments qui pourraient être produits, selon les exigences mécaniques et la norme EN 197-1 [15] pour les ciments A, B et C utilisés dans cette étude.

Tableau 3. Exemples de ciments pouvant être produits en utilisant l'andésite, sur la base des exigences mécaniques et de composition de la norme EN 197-1 [15], et sur les travaux présentés dans [14]

Exigences de composition EN 197-1			Formulations étudiées satisfaisant aux exigences mécaniques		Exigences mécaniques EN 197-1		Teneur de l'andésite répondant aux exigences mécaniques et de composition (%)
Classe de résistance	Résistance en compression (MPa) 2 Jours	Résistance en compression (MPa) 28 Jours	Ciment	Taux d'andésite (%)	Type de ciment	Taux de pouzzolane (%)	
32,5 N 32,5 R 42,5 N	≥ 16 (7 Jrs) ≥ 10 ≥ 10	≥ 32,5 ≥ 42,5	A, B C A, B C	40 20-30 30 10	CEM II/B-P CEM II/B-P CEM II/B-P CEM II/A-P	21–35 21–35 21–35 6–20	35 25 30 10
42,5 R	≥ 20	≥ 42,5	A, B C	30 <10	CEM II/B-P -	21–35 6–20	30 -
52,5 N	≥ 20	≥ 52,5	A, B C	20 <10	CEM II/A-P -	6–20	20 -
52,5 R	≥ 30	≥ 52,5	A, B C	10 <10	CEM II/A-P -	6–20	10 -

Nous pouvons voir que, selon le clinker choisi, toutes les classes de résistance pourraient être fabriquées et ce en allant jusqu'à 35% d'andésite. Le présent travail ne permet de proposer que le ciment de type CEM II, mais il est possible de fabriquer des ciments de type CEM IV (par exemple, à base d'andésite et de cendres volantes) et CEM V (à base d'andésite et de laitier, d'andésite ou de laitier et de cendres volantes), à condition d'effectuer des tests supplémentaires. De toute évidence, des essais supplémentaires doivent être effectués comme, par exemple, le broyage simultané du clinker et de l'andésite, qui doit être effectué dans les mêmes conditions (même appareillage et même temps). En outre, le liant doit être optimisé, par exemple en ajustant la teneur en sulfate.

3.3 L'andésite est-elle un matériau éco-efficace?

L'andésite peut-elle être considérée comme un matériau éco-efficace lorsqu'elle est utilisée dans les matériaux cimentaires? A première vue, nous pourrions avancer que le remplacement du clinker ou du ciment par un ajout minéral est une bonne alternative pour la production de liants économiques et écologiques acceptables. En vue de répondre à cette question, un bilan complet du cycle de vie serait nécessaire, mais c'est au-delà du cadre de la présente étude. Néanmoins, en vue de proposer des pistes de réflexion, nous présentons une analyse reposant sur un calcul simple de demande en énergie quand l'andésite est utilisée en tant qu'ajout de remplacement de ciment. Ce calcul est basé sur l'énergie consommée pour la production de ciment et d'andésite, qui comprend le procédé de fabrication (par exemple, le traitement thermique de la poudre crue du ciment, la calcination de l'andésite) et le broyage final des matériaux (clinker et andésite). L'énergie nécessaire (E, exprimée en kWh/t de liant) est donnée par l'équation (1), en fonction de la composition du liant et de l'énergie nécessaire au procédé de fabrication et de broyage des matériaux (ciment et andésite).

$$E\ (kWh/t) = C.(E_{pc} + E_{bc}) + A.(E_{pa} + E_{ba}) \qquad (1)$$

Le calcul de l'énergie de performance $E_{perf.}$ *(en kW h/t/MPa)* se fait en utilisant l'équation (2), en divisant l'énergie nécessaire E par la résistance à la compression du mortier contenant le liant de l'étude. $E_{perf.}$, donne le coût de l'énergie de chaque 1 MPa de résistance.

$$E_{perf.}\ [(kWh/t)/MPa] = E/R_c \qquad (2)$$

E : l'énergie de consommation (en *kWh*) pour la production d'une tonne de liant.
C et A : les proportions de ciment et d'andésite dans le liant respectivement.
E_{pc} et E_{bc} : les consommations d'énergie (en *kWh/t*) du procédé de fabrication et du broyage final du ciment, respectivement.
E_{pa} et E_{ba} : Les consommations d'énergie (en *kWh/t*) du procédé de calcination et du broyage final de l'andésite, respectivement.
$E_{perf.}$: La performance de consommation énergétique pour la production d'une tonne de liant, par rapport à la résistance en compression du mortier contenant le liant considéré.
R_c : La résistance en compression du mortier de composition de C+A.
Trois cas ont été considérés :

1. L'andésite naturelle, n'ayant pas besoin de calcination mais devant être broyée (l'andésite a été considérée comme déjà dans un état granuleux, c'est à dire qu'aucune énergie n'a été prise en compte pour l'extraction de la roche).
2. L'andésite calcinée, qui doit être calcinée et broyée.
3. Les fines d'andésite naturelle provenant du procédé de broyage des agrégats de carrière. Dans ce cas, les fines ont été considérées comme des sous-produits, sans besoin de calcination ou de broyage supplémentaire.

Les données utilisées pour le calcul ont été prises de la littérature et sont reportées dans le tableau 4 [17-25].

Tableau 4. Données utilisées pour le calcul de la consommation d'énergie

	Consommation d'énergie (kWh/t).			
	Ciment	Référence	Andésite	Référence
Procédé de fabrication E_{pc}=950=950[(a)] Broyage final E_{bc}=50		[17, 20-21, 24] [19, 23]	Brute [(b)] ou Calcinée E_{pa} = 0 ou 500[(c)] Fines de broyage ou Agrégats de sol E_{ba}= 0 ou 75	- [18, 22]

[(a)] Les valeurs d'énergie dans la littérature se situent entre 800 et 1200 kWh/t pour le procédé sec (les données initiales en GJ/t, avec 3,6 GJ = 1000 kWh).
[(b)] Sans tenir compte du procédé d'extraction de la carrière.
[(c)] Évaluation pour une calcination à 800°C.

L'énergie totale requise pour produire du ciment a été évaluée dans la plage de 800-1200 kWh par tonne de ciment, dont environ 50 kWh/t pour le broyage final du clinker. Dans le cas de d'andésite, l'énergie nécessaire pour le procédé de calcination a été évaluée à 500 kWh/t (température d'environ 800°C). L'énergie de broyage d'andésite est admise comme étant plus élevée que celle du clinker, en raison de la présence de minéraux durs. Selon de nombreux auteurs [18, 22], plus de 50% d'énergie supplémentaire est nécessaire pour que la finesse de l'andésite atteigne celle du clinker. Les résultats du calcul sont donnés dans la Figure 2 pour des teneurs d'andésite jusqu'à 40%.

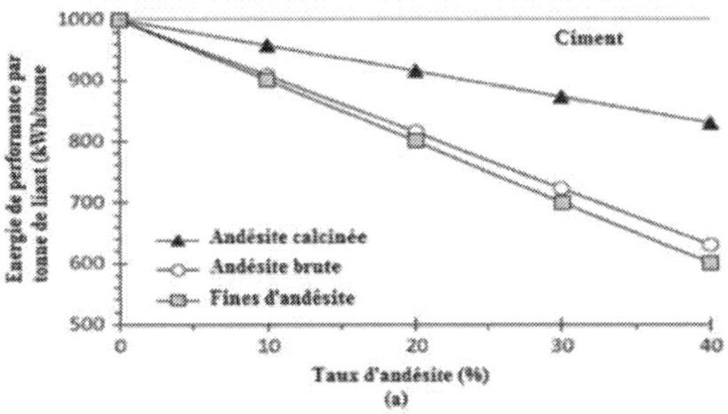

(a)

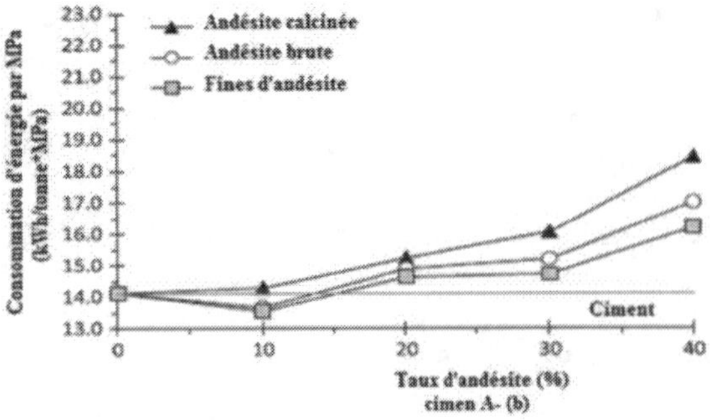

Fig. 2. Consommation énergétique de fabrication des ciments contenant de l''andésite.
(a) Besoin énergétique de liant (E, en kWh / t de liant).
(b) Energie de performance (E performance, en kWh/t/MPa) pour le ciment à 28 jours.

3.4 Besoin énergétique du liant

L'examen de l'énergie nécessaire pour une tonne de liant montre que la consommation d'énergie par le ciment d'andésite a toujours été inférieure à celle de la référence (ciment sans ajout) (Figure 2 (a)), bien que le broyage de l'andésite exige plus d'énergie que celui du clinker. L'andésite reste moins consommatrice d'énergie, en raison de la demande énergétique plus faible du procédé de fabrication (sans prendre en compte l'extraction de la roche). Il est évident que l'andésite calcinée augmente la demande en énergie d'une valeur qui se situe entre celle provoquée par l'andésite naturelle et le témoin (un liant de 100% ciment). Dans le cas des fines d'andésite naturelle, sous-produit provenant du procédé de broyage d'agrégats, un faible gain d'énergie a été enregistré comparativement à l'andésite naturelle.

3.5 Energie de performance

Lorsque la consommation d'énergie est considérée par rapport à la performance des mortiers (par exemple par rapport à la résistance en compression à 28 jours), les conclusions sont très différentes de celles tirées de la partie du « besoin énergétique » car, selon le ciment utilisé et le taux d'andésite, des points se trouvent sous ou au-dessus de la courbe de référence (ciment sans ajout). Dans le cas du ciment A (Figure 2 (b)), la plupart des liants d'andésite exigent plus d'énergie comparativement au ciment ordinaire, en particulier pour des taux de remplacement d'andésite élevés. Cela signifie qu'une résistance d'un MPa exige plus d'énergie, en raison de l'activité pouzzolanique modérée de l'andésite. L'andésite calcinée a fait le cas le plus défavorable, avec une demande d'énergie du ciment supérieure à celle du ciment témoin et les ciments à andésite brute. L'augmentation de la résistance du ciment due à la calcination de l'andésite n'a pas été suffisamment élevée pour compenser la hausse de consommation énergétique de fabrication. Parmi les formulations testées du ciment A, seul le liant avec 10% d'andésite a conduit à de meilleurs résultats par rapport à la référence de point de vue consommation énergétique-performance des mortiers. Les besoins énergétiques pour les liants à 20 et 30% d'andésite ont été légèrement supérieurs mais restent assez proches de la référence. Comme nous l'avons remarqué, les fines d'andésite naturelle provenant du procédé de broyage d'agrégats ont été le meilleur choix pour un taux de substitution de 30% d'andésite. Il est à noter que les résultats de performance pourraient être améliorés si l'efficacité de l'andésite était améliorée, par exemple en choisissant un couple de ciment/andésite compatible. La Figure 2 (c) présente une comparaison de l'énergie de la performance pour des liants fabriqués composé d'andésite naturelle et de ciments A, B ou C. Nous pouvons constater que le type de ciment a une grande influence sur les résultats. Les consommations d'énergie (par rapport à la résistance en compression) des ciments d'andésite ont été inférieures ou égales à celles du ciment seul pour les différentes teneurs d'andésite : 10, 30 et plus de 40% pour les ciments A, B et C respectivement. Ces résultats étaient prévisibles puisque les performances d'andésite ont été très différentes d'un ciment à autre. Cela signifie que, d'un point de vue rendement énergétique, il est préférable de n'utiliser l'andésite qu'avec des ciments ayant une efficacité modérée, comme le ciment C dans la présente étude. Pour ce ciment, chaque 1 MPa vaut moins de 1 kWh par tonne de liant, quelle que soit la teneur de l'andésite. Ce résultat peut être intéressant pour des pays en développement, où les ciments ont généralement de faibles performances. C'est le cas ici avec le ciment C, qui est de provenance algérienne. En résumé, nous pouvons dire que l'énergie nécessaire pour produire du liant (Figure 2 (a)) est systématiquement la plus faible lorsque l'andésite est utilisée, mais de point de vue rendement énergétique (pour atteindre un niveau donné de résistance, Figure 2 (b) et (c)), l'andésite peut être qualifiée comme écologiquement acceptable seulement dans certains cas. Suivant le point de vue adopté, la conclusion peut être très différente. Il est probable que l'application d'une approche similaire pour d'autres paramètres environnementaux, tels que la production de CO_2, conduirait à des conclusions similaires, car les émissions de CO_2 suivent la même tendance : une forte émission de CO_2 pour la production de clinker contre une plus faible émission de CO_2 pour l'andésite.

4 Conclusions

Dans la présente étude, nous avons étudié l'impact environnemental de la valorisation de l'andésite en substitution du ciment dans la matrice cimentaire. L'analyse de l'influence de cette addition, à l'état brut et calcinée à 800°C, sur l'environnement a permis de tirer les conclusions suivantes :

- Les principaux constituants de l'andésite utilisée sont : Feldspath, Cristobalite, Quartz, Montmorilonite, Calcite et Dolomite et la somme de ces oxydes majeurs est de 78,18%. Selon la norme ASTM C 618, cette andésite peut être considérée comme matériau pouzzolanique.
- L'incorporation progressive de la roche andésitique entraîne une légère diminution de résistances

mécaniques. Toutefois, à l'âge de 28 et 90 jours, les mortiers contenant jusqu'à 10% d'andésite peuvent se comparer avec le produit témoin.

- Globalement l'impact de l'andésite sur les matrices à long terme semble positif.
- Le remplacement du ciment par l'andésite induit systématiquement une diminution de l'énergie nécessaire pour produire une tonne de liant, en raison de la demande d'énergie de broyage moins faible par rapport au traitement thermique du clinker.
- Par rapport à la performance du mortier, le coût énergétique du ciment d'andésite (en kWh par MPa de résistance à la compression) est inférieur à celle du ciment ordinaire pour les faibles taux de remplacement et les ciments présentant une efficacité modérée.

Références

1. D. Bondar, C. J. Lynsdale, N. B. Milestone, N. Hassan, Geopolymer Cement from Alkali-Activated Natural Pozzolans : Effect of Addition of Minerals, 2nd International Conference On Sustainable Construction Materials and Technologies, Università Politecnica delle Marche, Ancona, Italia, 28-30 June, (2010).
2. Y. Tatsumi et T. Akahashi, Operation of subduction factory and production of andesite, (JAMSTEC), Vol.101 (1), P.145-153, (2006).
3. P. M. Black, Andesites as resources for roading and concrete industries, North Island of New Zealand, Department of Geology, University of Auckland, Auckland 1020, P. 328-332, (2009).
4. Takanori, Y. Kawabata, H. Hamada, Y. Sagawa, Alkali-Silica Reactivity of Andesite in NaCl Saturated Solution, International Conference on Durability of Concrete Structures, China, P. 1-8, (2008).
5. U. Chinje melo, N. Billong, Activité pouzzolanique des déchets de briques et tuiles cuites. Laboratoire des Matériaux et du Contrôle de Qualité, (AJST), vol. **5** N°1, P. 92-100, (2004).
6. Y. Kawabata, 2007. Petrological study on evaluating of alkali-silica reactivity and expansion analysis of andesite, Doboku Gakkai Ronbunshuu, **E (63)**, P. 689-703, (2007).
7. M. Saidi, 2010. Development and Structural Study of Cements Containing Additions of Industrial Waste. European Journal of Scientific Research Vol.**40** (2), P. 297-306 (2010).
8. S. Martínez-Ramírez, M.T. Blanco Varela, I. Ereña, M. Gener, Pozzolanic reactivity of zeolitic rocks from two different Cuban deposits: Characterization of reaction products, Applied Clay Science vol. **32**, P. 40–52, (2006).
9. F. Ait Hamou, Etude pétrographique et géochimique du volcanisme d'âge miocène de la région de Hadjout (Ouest Algérois), Thèse de magister en géologie, (1987).
10. G. G. Gvazava, O. I. Chokhonelidze, G. Sh. Kurtanidze, V. I. Kharazishvili, N. D. Natbeladze, Use of Bakury Andesite in the production of dark-green bottles, Glass and Ceramics, vol. **47** N°10, P. 371-374, (1990).
11. A. DAA, et H.M. El-Kaliouby Utilisation d'andésite dans la transformation des organismes électro-céramiques, Les progrès de la céramique appliquée, vol.**106** (3), P. 113-119, (2007).
12. D. Belhai Evolution tectonique de la zone ouest-Algéroise, Approche Stratigraphique et structurale, Thèse de Doctorat d'Etat, USTHB, Algérie, (1996).
13. H. Benali, Les minéralisations associées aux roches magmatiques tertiaires du nord de l'Algérie, Thèse de Doctorat d'Etat, U.S.T.H.B, Algérie, (2003).
14. M. Hamidi Elaboration d'un éco-ciment composé à base d'un ajout andésitique brut et activé, Thèse Docteur ès Sciences, USTO, Oran Algérie, (2014).
15. Norme AFNOR. NF EN 197-1, Cement. Composition, specifications and conformity criteria for common cements, (2011).
16. Norme EN 196-1 Méthodes d'essais des ciments - Partie 1 : Détermination des résistances mécaniques, (2005).
17. IZ. Bribián et al., Life cycle assessment of building materials: Comparative analysis of energy and environmental impacts and evaluation of the eco-efficiency improvement potential. Build Environ, **46**(5), 1133-40, (2011).
18. JN. De La Vergne The hard rock miner's handbook. 3rd Ed. McIntosh Engineering: North Bay, (2003).
19. C. Hosten, B. Fidan An industrial comparative study of cement clinker grinding systems regarding the specific energy consumption and cement properties. Powder Technol. 221:183-8, (2012).
20. NA. Madlool et al., A critical review on energy use and savings in the cement industries. Renew Sustainable Energy Rev **15**(4): 2042-60, (2011).
21. JA. Moya et al., Pardo N, Mercier A. The potential for improvements in energy efficiency and CO2 emissions in the EU27 cement industry and the relationship with the capital budgeting decision criteria. J Clean Prod **19**(11), 1207-15, (2011).
22. RH. Snow et al., Size reduction and Size Enlargement. In Perry RH, Green DW, editors. Perry's Chemical Engineers, Handbook. 7th Edition. McGraw-Hill, (1997).
23. D. Touil, S. Belaadi, C. Frances Energy efficiency of cement finishes grinding in a dry batch ball mill. Cement Concrete Research, 36(3) : 416-21, (2006).
24. World Business Council for Sustainable Development (WBCSD). Cement Sustainability Initiative (CSI). Global Cement Database on CO2 and Energy Information, <http://wbcsdcement.org/index.php?Itemid=74>.
25. E. Worrell et al., Potentials for energy efficiency improvement in the US cement industry. Energy **25**(12) :1189-214, (2000).

The Rehabilitation effects on the thermal behaviour of Algerian earthen built heritage.

Bencheikh Darda[1], Bederina Madani[2].
[1]University Saad Dahleb, ETAP laboratory, Architecture and urbanism institute, Blida , Algeria.
[2]University Ammar Telidji, Civil engineering laboratory, Department of civil engineering, Laghouat, Algeria.

Abstract. The earthen built heritage presents today a state of deterioration due to the climate changes that put it seriously at risk, through the degradation of its envelope building materials. These degradations lead its occupants to rehabilitate it by using modern building materials. This paper focuses on the rehabilitation of vernacular earthen architecture and its influence on the thermal behaviour of ksourian houses in sub-arid zones. It is based on a comparative study that combines the site measurements and the numerical simulation by Energyplus. Two courtyard houses, constructed of an adobe brick, were evaluated, one in its initial shape and the other rehabilitated. The results shows that the rehabilitation of the envelope made on this house did not influence its thermal qualities significantly; a very small difference justified on its internal thermal behaviour was recorded in cold period and almost minor during summer.

1 Introduction

The sustainable approach reorients the foundations of architecture towards a greater consideration of the environment and the impact of any construction on nature [1], while ensuring comfort for the occupants and reducing their energy consumption, especially in the residential and tertiary set; an archetypal energy-consuming sector [2]. According to the National Agency for the Promotion and Rational Use of Energy (*APARUE*), its consumption in Algeria presents more than 40% of final energy consumption (2012), with an annual average of equal accommodation to 1,303 PET [3].

Indeed, the problem of energy consumption has been one of the major concerns for decades, particularly, in the hot and dry climate, due to the absence of thermal comfort, which obliged the residents to use active heating and air-conditioning systems, that account for 51% of the total energy consumed [4]. This problem has, in fact, favored the vernacular architecture, since these inhabitants have endeavored to treat the comfort passively, by the good adaptation of the primitive forms to the context in which it fits [5], without resorting to the usual modern technical installations. The use of a building material characterized by a high thermal inertia is one of the main rules that lead to the creation of thermal comfort in this architecture [6] (a result which was demonstrated in a study of the effect of the thermal inertia on the hygrothermal comfort under the sub-humid climate), and consequently to the reduction of energy consumption, according to a study made by *Nayaran* [7]. It should be underlined that this energy can be reduced to 62% with an own selection of systems and building materials [8].

Among these materials, we can cite the rammed earth, a building material that all civilizations had adopted at one time or another. According to the United Nations, nearly 50% of the world's population lives in earthen houses [9]. It has been the most widely used building material in hot and dry climates [10], which has in fact endowed the vernacular architecture with a hygrothermal comfort without resorting to the active system, thanks to its inertia and its excessive thickness, which varies from 0.5 to 1 meter [11].

Traditionally, a regular maintenance of earthen houses was sufficient, but today most of these houses (34% of the world's sites [12]) show a significant deterioration caused by the climate change, which have placed them seriously at risk, and have lead their inhabitants to ignore them, and sometimes to rehabilitate them. The studies of *Kadri* (2011), *Sarbu* (2011) and that of *Sam* (2012) have reported that the rehabilitation of stone and concrete structures improves the internal quality of the houses in terms of thermal comfort and reduces its energy consumption factor [13, 14, 15]. For that purpose, the present paper seeks to examine the impact of earthen vernacular building rehabilitation, especially adobe houses rehabilitation , on the thermal comfort (as criteria of built heritage assessment) under sub-arid climate, through evaluating two houses, one degraded and the other rehabilitated, by site measurements, in order to validate the results by numerical simulations.

This study may be an opportunity to highlight the thermal comfort aspect in the vernacular architecture or to make it reappear.

2 Material and methods

The Temperature measurements were taken under winter and summer conditions (January and July, when temperature drops and peaks were predicted) by using a thermo-hygrometer with Self-contained mechanical recording. This instrument has enabled to measure the temperatures by a metal bimetallic strip (Accuracy: ± 1%).

In order to validate the results recorded in site measurements, a numerical simulation using the Energyplus software (version 1.2.2) have been developed. Previously, a laboratory tests were conducted on a sample of an adobe. In fact, these tests permit to determine the characteristics required for the study (density, thermal conductivity and specific heat) to execute these simulations.

3 Study case: vernacular ksar of Zgag El-Hadjadj

The studied ksar is an example of vernacular architecture of southern Algeria. It is located in the region of Laghouat (33.8 °N, 2.9 °E), which is characterized by a pre-Saharan climate (Figure 1) , with a very hot summer, where temperatures can reach 40°C, And a cold winter marked by a low temperature (which can decrease at 6 °C during the day and at -2°C during the night) .

The Ksar of Zgag El-Hadjadj is a set of compact habitat built in height, developed over an area of three hectares, in arborescent frame. It comprises 135 houses spread over eleven islets of irregular pentagonal shape oriented North-South (Figure 2). Its urban fabric is structured by squares, streets, alleys and impasses [16-17]. This organization enables to limit the effects of heat by minimizing the walls exposed to the sun, thus contributes to summer thermal comfort [18]. This ksar was classified as a national cultural heritage in 2007.

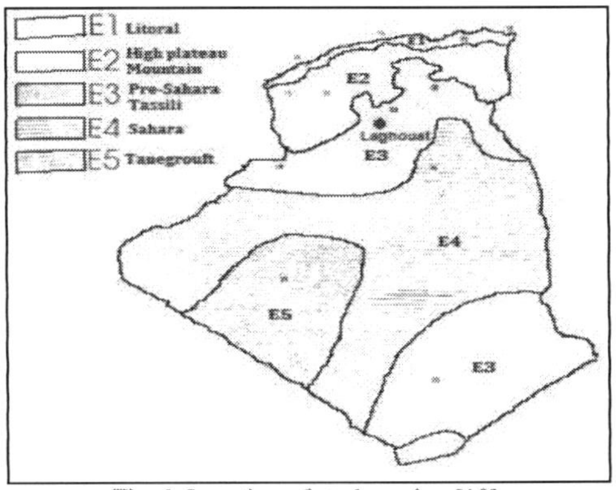

Fig. 1. Location of study region [19].

The Ksar's houses are developed on square plans; they are characterized by introversion and centrality. The construction system used is composed of adobe brick walls with stone basement (with an average thickness of 50 to 70 cm) and palm's trunk or poplar's joist beams.

3.1 Pathology and disorder

The houses of Ksar Zgag El-Hadjadj are currently in a state of a very advanced degradation threatening the hygiene, comfort and safety of their inhabitants. The visit of the houses allowed to examine the visible disorders at the level of the structure and to identify all the potential sources of the deterioration observed. The pathologies encountered are inherent to the adobe's high vulnerability to erosion.

For the delivery of a few houses in habitation conditions, these disorders have been treated by tackling the causes of symptoms. The rehabilitation was carried out by several works, shown in the table below.

Table 1. Rehabilitation work made at rehabilitated houses.

Problem	Solution
Damaged supports	The functional replacement or the reinforcement by intermediate posts arranged across the beam to be strengthened, in order to mask it by false ceilings.
cracks	The resumption by the intercalation with more resistant elements (pieces of brick), and sometimes by the grouting with earth to completely seal the vacuum.
Unstuck fillers	The scratching and the repair using the Sand-lime mixture, which promotes the water vapour permeability required for breathing the old masonry.
Altered floor coverings	The Rebuilding with the stone floors, in order to be covered first with an insulated concrete slab (while blocking the rising of moisture).
Rain run-offs	The addition of gutters and gargoyles to immobilize the water runoff.

4 Results and discussion

The investigation have been carried out on two houses (one in its initial shape and the other rehabilitated), which were carefully selected in order to evaluate the effect of rehabilitation on the thermal behaviour of earthen architecture. The study settled the parameter of centrality with regard to the ksar (Figure 2), which is common between both samples, and plays a key role in the thermal performance of the ksourian houses.

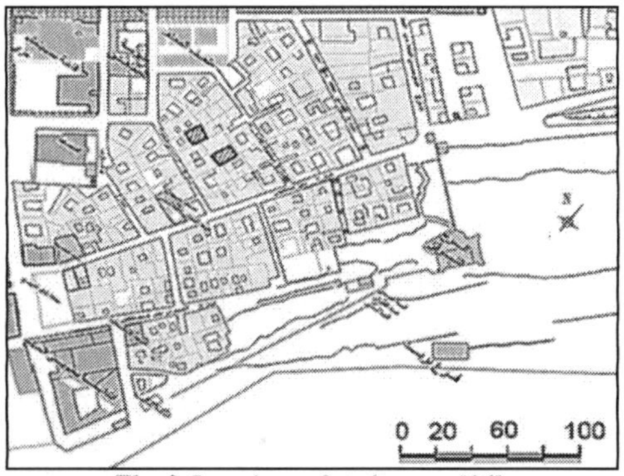

Fig. 2. Locations of study cases [16].

4.1 Site measurements

The temperature recordings were taken in the most exposed areas to hostile climatic conditions, which do not contain active heating and air-conditioning systems, as it has been seen above that the old inhabitants did not use it. For the rehabilitated case, the choice was mainly focused on the space where the rehabilitation affected the entire envelope. Rooms on the first floor with discontinuous occupancy have been selected (Figure 3-4), taking into account the parameters previously targeted. The courtyard is the chosen space control, to compare the results maintained inside the houses, since it presents the not covered space, which totally exposed to climatic conditions.

Note that the external hourly temperature recordings were obtained on the days of measurements (5-6 January and 9-10 July 2017), from the city meteorological station [20].

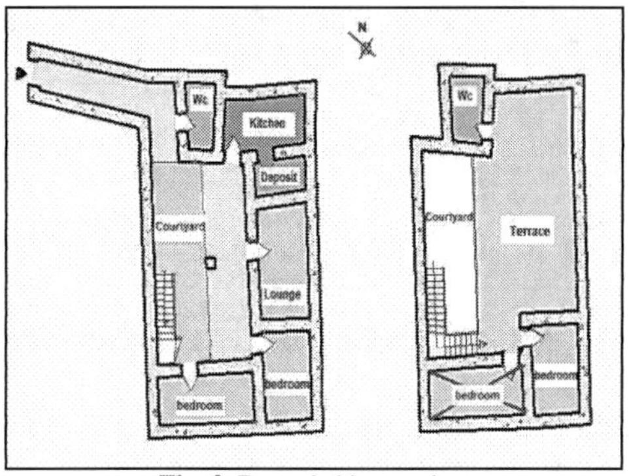

Fig. 3. Degraded house plan.

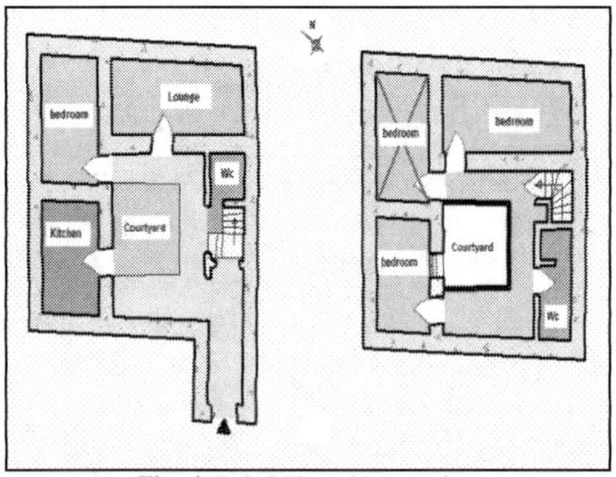

Fig. 4. Rehabilitated house plan.

4.1.1 Degraded house

The reading of the following graphs (Figure 5) shows that the internal temperature is stable throughout the day with an amplitude of 2.7°C between the maximum temperature (9,9°C) marked about 5pm and the minimum one (7,2°C) marked at 7am . This stability is also remarkable in the courtyard, where the amplitude is equal to 2.33°C, while outdoor temperatures have a high amplitude of 13°C. This is translated by the amortization of the external contributions insured by the adobe [21], with delaying the cold transfers through the walls. This amortization had also a negative impact, which is obviously determined in the room temperature recordings, during the first eight hours, in which a more pleasant external environment was recorded, with a temperature difference that reached 6°C. Indeed, the ventilation plays a major role in this phenomenon, by evacuating the cold air accumulated in the space during the night. This is clearly established from the results obtained in the courtyard, with an average temperature of 16.96°C. The winter thermal comfort was not reached the inhabitants to use the fire placed in the bedroom, and the need of heating was only for no more than two months.

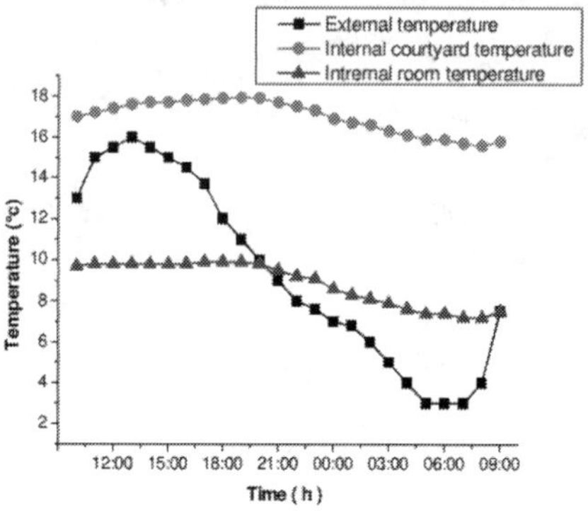

Fig. 5. Variation of winter temperature.

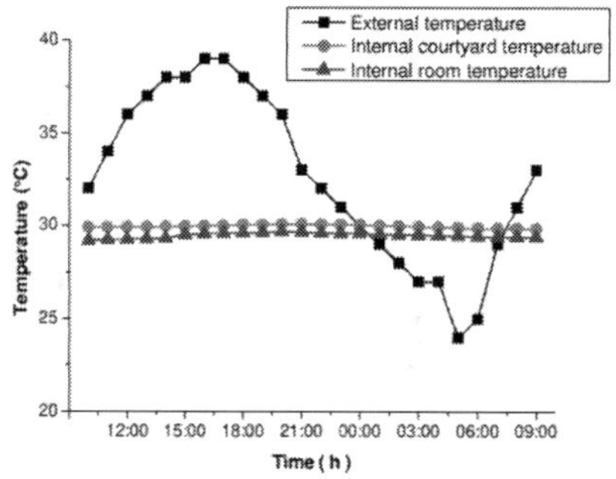

Fig. 6. Variation of summer temperature.

The range of thermal comfort is limited according to Givoni between 20 and 27°C. For this purpose, it can be seen that thermal comfort is almost guaranteed in the warm period, inside the house (room and courtyard), where temperatures fluctuate slightly from 29 to 30.76°C in the shade. (Figure 6).The thermal inertia and the night ventilation are the recommended solutions, under the climatic conditions of July. The night ventilation was neglected by the inhabitants during measurements, which justify the summer results obtained.

4.1.2 Rehabilitated house

The results obtained from the graph presented below (Figure 7) shows that the winter interior temperature is almost constant (it varies between 7.5 and 11°C), compared to the outside temperatures, where the amplitude is equal to 13°C. This is justified by the thermal mass of the materials constituting the house envelope and the low ratio of openings in relation to the total area of the envelope, which limits the thermal losses. Nevertheless, these temperatures remain below the range of comfort determined by Givoni. However, at the courtyard , the temperatures recorded were approximately pleasant (with an average temperature of 13.29°C) comparing it with those of the room ; This is due to the direct solar contributions and the greenhouse effect provided by the nylon cover placed by the inhabitants during the cold period. On the other hand, its suppression during the warm period allows to ventilate the space naturally and consequently lowers the ambient temperatures, in order to stabilize it, thanks to the thickness of the wall (which is equal to 57cm) that permits to delay the transfer of heat received from a radiant heat source by storing it in the wall mass (Figure 8). Note that, compactness of the urban fabric limits the walls exposed to winds and solar radiations, which has further helped to achieve these results.

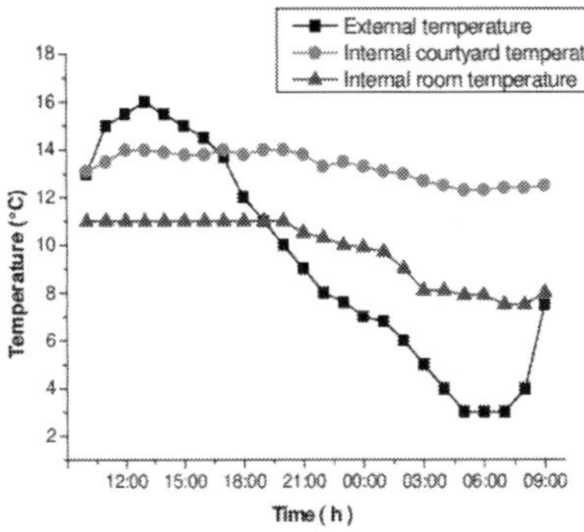

Fig. 7. Variation of winter temperature.

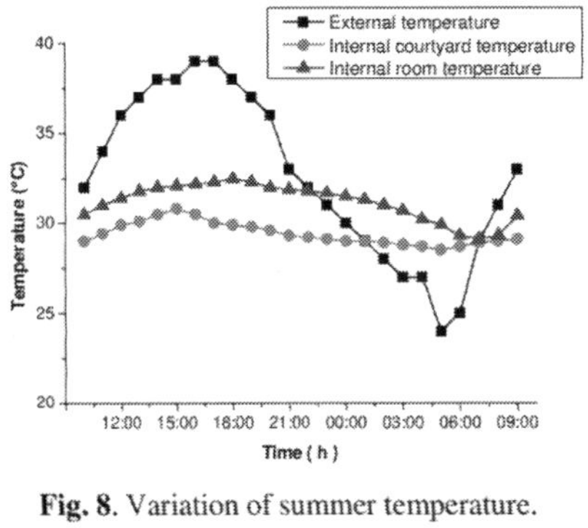

Fig. 8. Variation of summer temperature.

4.1.3 Comparative analysis of thermal behaviour results

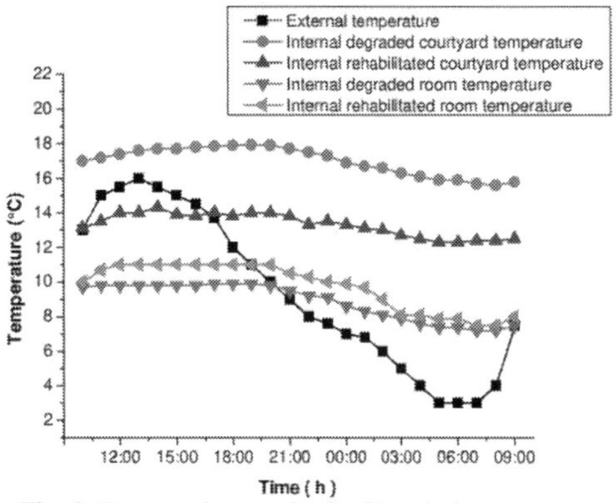

Fig. 9 .Comparative graph of cold period temperature recordings.

The site measurement results, shown graphically above (Figure 9), demonstrate that during the cold period, the courtyard thermal behaviour in the degraded house is more pleasant than that in the rehabilitated house, with an average difference of 3.67°C. This is explained by the natural ventilation, which allows to release the cold air stoked in the walls during the night. On the other hand, the presence of nylon cover in the rehabilitated house courtyard slows down the evacuation of this air, which leads to its accumulation in space after its restitution. This coverage also has a positive impact on the thermal environment, which is summarized by the passive heating provided by the greenhouse effect. This is clearly established from the results, where an increase in temperatures was observed during the afternoon (direct solar radiation).

The resulting temperatures in the rooms have an average difference of 0.85°C, which is reflected by the presence of a wall exposed to climatic conditions in the degraded case, the northeast orientation of this wall has negatively influenced the winter thermal comfort in the space.

The thermal inertia causes difficulties in terms of thermal regulation in intermittent occupancy spaces [22], which explains the low temperatures (out of comfort range) observed in the rooms of the two cases studied.

Figure 10 illustrates that the courtyards of the two houses show approximately the same results in the warm period. The small marked difference is justified by the surface, the shape and the location of the courtyard: the smaller and more centralized the courtyard, the more shadow is observed, the faster refreshment and a long-lasting air pool puddle is marked [23].

Oppositely, a large difference of room temperatures between the two cases was recorded; this is expressed by the wall thickness, which varies from a house to another (57cm in the rehabilitated house and 65cm in the degraded one). This thickness makes it possible to dampen the temperature variations resulting in the amplitude reductions [24]. The thicker the wall, the slower the heat transfers. These results are consistent with those obtained by other studies in the same field [25].

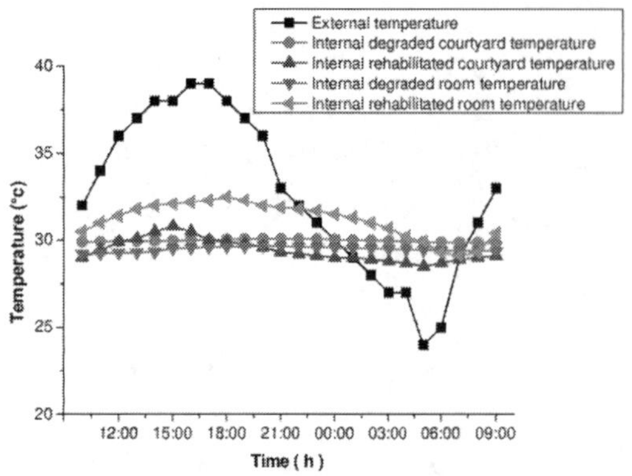

Fig. 10. Comparative graph of warm period temperature recordings.

From this comparison, it can be seen that the rehabilitation did not influence the house thermal qualities significantly; a very small difference justified on its internal thermal behaviour was recorded in cold period and almost minor during summer.

4.2 Numerical simulations

Energyplus is a software that has shown its performance and has been the subject of extensive validations [26]. The simulation by this software takes into account most of the parameters affecting the thermal balance, in particular the characteristics of the building materials; density, thermal conductivity, specific heat and thermal diffusivity, which have been determined, by taking an adobe sample results extracted from the ksar.

Table 2. Adobe characterization results.

Characteristics	Values
Density (kg/m^3)	2567,6
Thermal Conductivity (w/m.K)	0.817
Specific heat (J/kg. °C)	1075,1
Thermal diffusivity (m^2/s)	0,7596

The Ksar's organization was also taken into consideration during the simulation, in order to ensure the compactness of the urban fabric and the centrality of the studied houses in relation to their location. The architectural characteristics are often taken into account by this software in a simplified way, while guaranteeing a precision excellent level of houses behaviour.

The simulations are made without heating and air conditioning systems, using the meteorological data of the sub-arid climate of Laghouat.

Table 3. Site measurements /numerical simulations comparison (Degraded house).

		Site measurements		
		Ext	courtyard	room
Average Temperature (°C)	Cold period	9,57	16,96	8,86
	Warm Period	32,71	29,96	29,46
		Numerical Simulations		
		Ext	courtyard	room
Average Temperature (°C)	Cold period	11,81	15,36	9,83
	Warm Period	34,43	29,37	29,25

Table 4. Site measurements /numerical simulations comparison (Rehabilitated house).

		Site measurements		
		Ext	courtyard	room
Average Temperature (°C)	Cold period	9,57	13,29	9,71
	Warm Period	32,71	29,39	31,16
		Numerical Simulations		
		Ext	courtyard	room
Average Temperature (°C)	Cold period	11,81	13,45	10,09
	Warm Period	34,43	28,86	28,64

The measurement/simulation comparison presented in the tables above shows a small average relative error of 1.07% between the measured and simulated data. This is reflected in the homogeneity of the building materials, which was assumed by the software. However, reality shows a significant state of aging. The uncertainty obtained can also be related to the positioning of the measuring instrument and its accuracy.

Lastly, in order to verify numerically the outcomes earlier revealed concerning the rehabilitation effects on the thermal behaviour, another simulation was made on the house before its rehabilitation; the results show a small average difference of 0.4°C in the room. On the other hand, no improvement had been made on the courtyard results, which is contrasted with the statement announced previously.

5 Conclusion

This work concretizes a comparative study relative to the thermal comfort, between two vernacular earthen houses, one in the initial shape and the other rehabilitated. The site measurements have made it possible to determine the thermal quality of the earthen built heritage during the cold and the warm season, and to check the influence of the envelope rehabilitation on its interior thermal behaviour.

The experimental results show that the summer thermal comfort of sub-arid zones can be assured by the thermal inertia of building materials and the natural ventilation; two complementary and indispensable parameters for effective results.

It should be recalled, in this regard, that spatial organization of the old houses also favours their thermal behaviour; the courtyard centrality makes it possible to control sunshine by protecting from the summer sun, and also, to limit the overheating and lower the temperature by provoking a fresh air circulation, thus renewing the warm indoor air.

The comparison of the results obtained from two houses, showed that the rehabilitation of the envelope had a negligible effect on the thermal comfort. The reduction of the wall thickness and the presence of a wall exposed to climatic conditions were the parameters justifying the difference between the internal temperatures of the studied cases. This was also verified by a numerical simulation using the Energyplus software, which showed an average difference of 1.07%. This concordance has as well permitted to affirm that this software adequately predicted the thermal comfort.

To summarize, it is important to emphasize that the vernacular earthen architecture refers to sustainable architecture; it is the object of a renewed interest, which must be exerted by a reinforced involvement of construction's actors. Its rehabilitation presents a symbolic act that demonstrates the cultural value of the goods to future generations.

References

1. A .Diemer, Développement durable plutôt qu'écodéveloppement : le nouveau gadget idéologique de l'occident. *Colloque francophone ; Les représentations Nord Sud du développement durable*, Université Blaise Pascal – IUFM Auvergne (2012).
2. Transition to Sustainable Buildings. Strategies and Opportunities to 2050. IEA (2013). [ISBN : 978-92-64-20241-2].
3. Ministère de l'Energie et des Mines, 'Consommation Energétique Finale de l'Algérie, Chiffre Clé -Année 2012-', in APRUE 'Données et Indicateurs'. Ed (2014).
4. G. Kang, D. Song. Optimized blind control method to minimize heating, cooling and lighting energy. *Energy Procedia* **78**, pp.2857–2862 (2015).
5. C. Rubio-Bellido, P. Arcos, C. Lainez, M. José. Adaptation Strategies and Resilience to Climate Change of Historic Dwellings. *Sustainability* **7**, pp.3695-3713, (2015).
6. D. Medjelakh, S.Abdou. Impact de l'inertie thermique sur le confort hygrothermique et la consommation énergétique du bâtiment. *Revue des Energies Renouvelables* **3** , pp. 329-341 (2008).
7. T. Narayan. A Passive Courtyard Home in Jaipur, India: Design Analysis for Thermal Comfort in a Hot Desert Climate. Arizona State University, [online] Thulasi.Narayan@asu.edu . Last saved 03/02/05.
8. B-V-V. Reddy, K-S. Jagadish , Embodied energy of common and alternative building materials and technologies *.energy Build* **35**, pp129-37 (2003).
9. H. Guillaud. Architectures de terre Utilisée de tous temps et en tous lieux, la terre crue a de plus en plus d'adeptes dans le monde. Retour sur un matériau respectueux de l'environnement. *Les matériaux de construction*, TDC **977** (2009).
10. M. Gernot. *Building with earth: Design and Technology of a Sustainable Architecture*. Basel: Birkhäuser - Publishers for Architecture (2006).
11. K. Heathcote. The thermal performance of earth buildings. *informes de la construccion*; **62**, 523 pp.117-126, (2016).
12. Word Heritage, inventory of earthen architecture. WHEAP (2012). [ISBN: 978-2-906901-70-4]
13. N. Kadri, A. Mokhtari. Contribution à l'étude de réhabilitation thermique de l'enveloppe du bâtiment *.Revue des Energies Renouvelables* **14**, 2 pp.301-311, (2011).
14. F. Sam. *Réhabilitation thermique d'un local dans une zone aride – Cas de Ghardaïa-*. thèse de magistère, UMMTO (2012).
15. I. Sarbu, C. Sebarchievici , Thermal rehabilitation of building *.International journal of energy* **5**, 2 (2011).
16. Plan permanent de sauvegarde et de mise en valeur du secteur sauvegardé ; Le vieux ksar de Laghouat N°11-141 du 28 mars 2011/ JO N°20 du 30/03/2011.
17. A.N. Si Amer. La prise en compte des attributs touristiques dans le processus de patrimonialisation de l'espace oasien : cas de l'Oasis de Zgag El Hadjadj à Laghouat, Algérie. *Colloque internationale ; Ressources patrimoniales et alternatives touristiques : entre oasis et montagnes*, Ouarzazate, Maroc. Nov. 2011.
18. M. Khoukhi. N. Fezzioui. Thermal comfort design of traditional houses in hot dry region of Algeria». *International Journal of Energy and Environmental Engineering* (2012).
19. Ministère de l'habitat, *Recommandations architecturales.* Ed ENAG, Alger (1993).
20. R.O.M, Regional office of meteorology, Laghouat, (2017).
21. N. Matari , A .Mahi et M. Lachi , Effet de l'enveloppe du bâtiment sur le confort thermique. Application au climat aride. *Congrès Français de Thermique SFT 2015, La thermique de l'habitat et de la ville*, La Rochelle pp.1-8 (2015).
22. D. Pajani. *La thermographie du bâtiment ; principes et applications du diagnostic thermographique.* Paris, Eyrolles, 216p, (2010).
23. P. Supic. L'aspect bioclimatique de l'habitat vernaculaire. Revue *Architecture et Comportement.* **10**, pp.27-47 (1994).
24. P. Oliver. *Built to meet needs; cultural issues in vernacular architecture.* Oxford, Architectural press, 475p (2006)
25. S. Serrano, A. De Garcia, L-F. Cabeza. Adaptation of rammed earth to modern construction systems: comparative study of thermal behavior under summer conditions. *Applied energy* **175**, pp.180-188 (2016).
26. C. Spitz. *Analyse de la fiabilité des outils de simulation et des incertitudes de métrologie appliquée à l'efficacité énergétique des bâtiments.* Thèse du doctorat, Université de Grenoble (2012).

Proposal of industrialization process of dredged sediments

Proposition d'un processus d'industrialisation des sédiments de dragage

Hadj Sadok Rachid[1], Belas Nadia[1], Tahlaiti Mahfoud[2], Belaribi Omar[1], Taieb Fatima[1], Belguesmia Khalil[1], Safer Omar[1]
[1]Laboratoire Construction Transport et Protection de l'Environnement, Université de Mostaganem, Département génie civil et architecture, Mostaganem, Algérie
[2]Laboratoire de l'institut de recherche en génie civil et mécanique (GeM) et l'institut catholique d'art et des métiers (ICAM) à Nantes, Ecole centrale de Nantes, France

Abstract. Our study will focus on life cycle assessment (LCA) of dredged sediments in its environment from sediment extraction to waste treatment. This tool is part of an environmental management approach, to compare the environmental loads of the different stages of the life cycle of the same product and to deduce which stage of the scenario is the most polluting in environmental terms. Optimize the modeling of industrialization process of dredged sediments using the SimaPro 8.2.3 software to propose a model that is the most respectful of the environment. We will focus on the environmental impacts; we will try to propose the most environmentally friendly scenario to exploit these dredged sediments in the field of building construction.

Résumé. Notre étude portera essentiellement sur l'analyse de cycle de vie (ACV) des sédiments dragués dans son environnement depuis l'extraction des sédiments jusqu'au traitement des déchets. Cet outil s'inscrit dans une démarche de management environnemental, pour comparer les charges environnementales des différentes étapes du cycle de vie du même produit ; et en déduire quelle étape du scenario est la plus polluante en termes environnementaux. Optimiser la modélisation d'un processus d'industrialisation des sédiments de dragage à l'aide du logiciel SimaPro 8.2.3 pour proposer un modèle le plus respectueux de l'environnement. On se focalisera sur les impacts environnementaux ; On essaiera de proposer le scénario le plus respectueux de l'environnement, afin d'exploiter ces sédiments dragués dans le domaine de la construction des bâtiments.

1 Introduction

Cet article s'inscrit dans une longue étude de recherche au sein du laboratoire LCTPE portant sur le domaine de la valorisation des sédiments de dragage. Dans cette étape de recherche, on complètera le travail par une étude environnementale par analyse du cycle de vie des sédiments de dragage, pour faire face au problème d'envasement des barrages dû à l'érosion hydraulique mettant en danger ses potentialités. L'Algérie a investi des moyens colossaux pour alimenter en eau potable sa population et permettre l'irrigation des terres agricoles. De grands efforts ont été déployés pour assurer l'autosuffisance en eau potable et fournir l'eau aux industriels ; Des réalisations de barrages de toutes dimensions à travers tout le territoire national ; ont vu et vont pouvoir voir le jour.

Plusieurs études de recherches récentes [1][2][3][4][5][6][7] sur la valorisation des sédiments issus du dragage des barrages ont été menées dans le domaine de caractérisation physico-chimique, mécanique et de durabilité. Elles ont démontré la possibilité de leur utilisation en tant que matière première dans le domaine de génie civil, et spécialement en substitution partielle des sédiments de dragage dans le ciment.

L'Algérie dispose de 72 grands barrages, d'une capacité totale de 7692,24 millions m³ selon les dernières mesures en 2014. Leur envasement est d'environ 1085,94 millions m³. Cela représente un pourcentage d'envasement de 14,12 % de la capacité totale [8].

De grandes opérations de dragage essaient d'éliminer une partie des sédiments dès leur arrivée dans les réservoirs des barrages. La troisième campagne de dévasement est en cours depuis 2014. Trois grandes opérations de dragage des barrages de Bouhanifia et Foum El Gherza et K'sob sont en cours. Elles visent à récupérer 21,3 millions m³ de capacité de stockage dans un délai moyen de 30 mois [8].

2 La description du système de production des sédiments de dragage

Dans cette partie d'industrialisation des sédiments de dragage, nous allons détailler le procédé de production et décrire les différents processus de traitement des sédiments ; afin de les valoriser comme matière première dans la substitution du ciment. Pour cela deux scenarios sont envisageables :

2.1 Scénario (SN°1)

Il consiste à utiliser les sédiments après la décantation totale et le séchage à l'air libre dans les zones de rejet. Là on extrait ces sédiments par un chargeur et on les transporte avec un dumper jusqu'à la station de traitement des sédiments de dragage ; l'installation de l'usine étant au niveau des zones de rejet.

2.1 Scénario (SN°2)

On exploite les sédiments après dragage dans les zones de rejets ou au niveau des conduites terrestres, est incorporé un sécheur directement pour récupérer l'eau et les sédiments séchés, afin d'optimiser le temps de décantation à l'air libre. Le temps de séjour sera de 10 à 60min, avec une consommation énergétique réduite 0,6-0,7 kWh/kg d'eau évaporée. On suppose que le taux d'humidité est de 30% pour le scénario N°2 [9].

3 Méthodologie de l'analyse du cycle de vie

L'analyse de cycle de vie (ACV) est la méthodologie de référence la plus complète pour réaliser l'inventaire des flux quantifies et évaluer les impacts environnementaux d'un produit ou service ; Elle distingue 4 étapes [10] :

a) La définition des frontières du système étudié, de l'unité fonctionnelle et du périmètre prise des informations et des données.
b) La réalisation de l'inventaire du cycle de vie à collecter et compiler.
c) La transformation de cet inventaire sous forme d'indicateur des impacts.
d) L'interprétation des résultats.

4 L'analyse du cycle de vie de production des sédiments de dragage

4.1 Le champ de l'étude (Scope)

4.1.1 Fonction

L'analyse portera sur les quatre étapes de production, le séchage, le concassage, la calcination et le stockage des sédiments de dragage. Il s'agira de modéliser les différentes étapes de production afin de produire une tonne de sédiments de dragage calcinés prêts à être substituée dans le ciment.

4.1.2 Unité fonctionnelle

Production d'une (1) tonne de sédiments de dragage calcinés pour la substitution partielle dans le ciment (pour une finesse ≥ 3000 cm²/g).

4.1.3 Hypothèses et limitations

Pour toute cette analyse de processus de production des sédiments de dragage, on exclut l'étape de dragage ; car cette étape est faite par l'ANBT [8]. On exploite les sédiments après dragage dans les zones de rejet.

Pour les différents procédés de production, référence est faite aux données collectées au niveau de la recherche bibliographique effectuée dans ce travail sur la valorisation des sédiments de dragage. Nous allons détailler toutes ces données dans l'inventaire du cycle de vie.

La fin de vie n'a pas été prise en compte dans cette analyse car l'objectif final de cette recherche est l'application de l'analyse du cycle de vie des sédiments de dragage dans le domaine bâtiment.

4.1.4 Frontière du système

On va calculer les flux élémentaires entrants dans le processus de production des sédiments de dragage, comme l'utilisation des ressources naturelles et l'énergie ; ainsi que l'utilisation d'espace au niveau des zones de rejet. Les flux élémentaires sortants dans l'environnement sont des émissions dans l'air, l'eau et le sol.

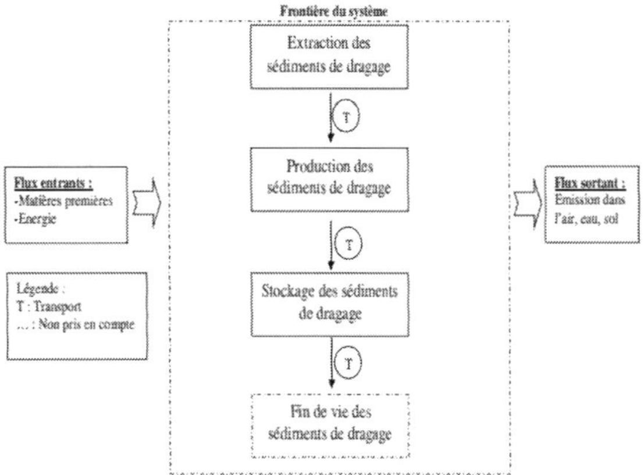

Fig. 1. Principe de définition de la frontière du système

4.2 L'inventaire du cycle de vie :

4.2.1 La méthode de collecte de données :

Les données ont été collectées on deux étapes :

1) Une recherche bibliographique sur la valorisation des sédiments de dragage.

2) L'exploitation de la base de données Ecoinvent par le logiciel SimaPro 8.2.3 multi-user de l'école d'ingénieurs ICAM de Nantes.

4.2.2 L'inventaire du procédé de production des sédiments de dragage

Il est donné par le tableau 1

Table 1. Inventaire du procédé de production des sédiments de dragage

Procédé	Consommation	Unité	Scénario N°1	Scénario N°2	Réf
Extraction	Gasoil	MJ/t	76,9	76,9	*[11]*
Séchage	Electricité	KWh/t	-	300	*[9]*
Concassage	Electricité	KWh	0,57	0,57	Estimation
Broyage	Electricité	KWh	10,90	10,90	Estimation
Tamisage	Electricité	KWh	5,45	5,45	Estimation
Calcination flash	Gaz naturel	MJ/t	1894,7	1894,7	*[11]*
Stockage en sac	Papier kraft	Kg	3	3	Estimation

4.2.3 Matières premières

L'estimation de la composition minéralogique des sédiments de dragage du barrage est donnée par le tableau 2.

Table 2. Estimation minéralogique des sédiments de dragage du barrage de Fergoug.

Minéraux	Composition	Classification	Sédiments
Quartz	SiO_2	Oxyde de silicium	42,62
Calcite	$CaCO_3$	Carbonate anhydre	31,90
Dolomite	$CaMg(CO_3)_2$	Carbonate anhydre	12,06
Kaolinite	$(K,H_3O)(Al,Mg,Fe)_2(Si,Al)_4O_{10}(OH)_2,H_2O$	Phyllosilicate	13,40

4.2.4 Extraction des sédiments (Engins) :

Les émissions dans l'air sont présentées dans le tableau 3.

Table 3. Hypothèse de calcul d'après : Hugrel et Joumard, 2006[11].

Emission dans l'air	Unité	Quantité	Données
CO2	Kg/t	5,599	bibliographiques [Hugrel et Joumrad, 2006]
CO		0,016	
COGNM		0007	
NOx		0071	
CH4		0	
Particules PM10		0,004	
N2O		0,001	
H2O		0	

4.3 Méthodes de calcul des indicateurs d'impacts :

Nous avons choisi la méthode CML baseline afin de réaliser l'évaluation des impacts. La réflexion sera basée sur 11 indicateurs qui sont : Epuisement des ressources, épuisement des ressources (combustibles fossiles), réchauffement climatique, destruction de la couche d'ozone, toxicité humaine, Ecotox aquatique d'eau douce, photochimique, acidification atmosphérique, eutrophisation [10].

4.4 Evaluation des impacts environnementaux :

Apres modélisation par le logiciel SimaPro 8.2.3 multi-user de l'école d'ingénieurs ICAM de Nantes, les figures suivantes représentent les impacts estimés pour la production d'une tonne de sédiments de dragage avec les deux scenarios envisagés. Cette modélisation nous permet d'avoir une première idée pour appréhender le développement d'un processus de production des sédiments de dragage.

a) Modélisation du scénario N°1 de production des sédiments de dragage :

La figure 2 représente les impacts sur l'ensemble des indicateurs de la modélisation du scénario N°1 de production des sédiments de dragage avec la méthode CML-IA.

La figure 2 nous permet de constater que la phase de calcination flash est la phase la plus impactante sur l'ensemble des indicateurs de la méthode CML-IA. Elle est suivie par le transport par tapis vibrant et le broyage qui sont aussi observés sur les huit catégories de dommages de la méthode CML-IA baseline. Ceci confirme les résultats faits lors de l'analyse de l'arborescence de la modélisation du scénario N°1 de production des sédiments de dragage.

b) Modélisation du scénario N°2 de production des sédiments de dragage :

La figure 3 représente les impacts sur l'ensemble des indicateurs de la modélisation du scénario N°2 de production des sédiments de dragage avec la méthode CML-IA.

La figure nous permet de constater que la phase de calcination flash et le séchage sont les phases les plus impactantes sur l'ensemble des indicateurs de la méthode CML-IA.

5 Comparaison et discussion des résultats :

L'analyse du cycle de vie (ACV) des deux processus de production des sédiments de dragage permet d'envisager une comparaison environnementale des indicateurs d'impacts des deux procèdes envisagés afin d'optimiser le procédé le plus respectant de l'environnement.

L'évaluation environnementale des deux procédés envisagés est représentée dans le tableau 4.

Le scénario N°1 présente un excellent bilan environnemental dans l'ensemble des indicateurs de la méthode CML-IA par rapport au scénario N°2.

La figure 4 représente l'histogramme de comparaison des impacts sur l'ensemble des indicateurs de la modélisation des deux scénarios de production des sédiments de dragage avec la méthode CML-IA.

a) scénario N°1 :

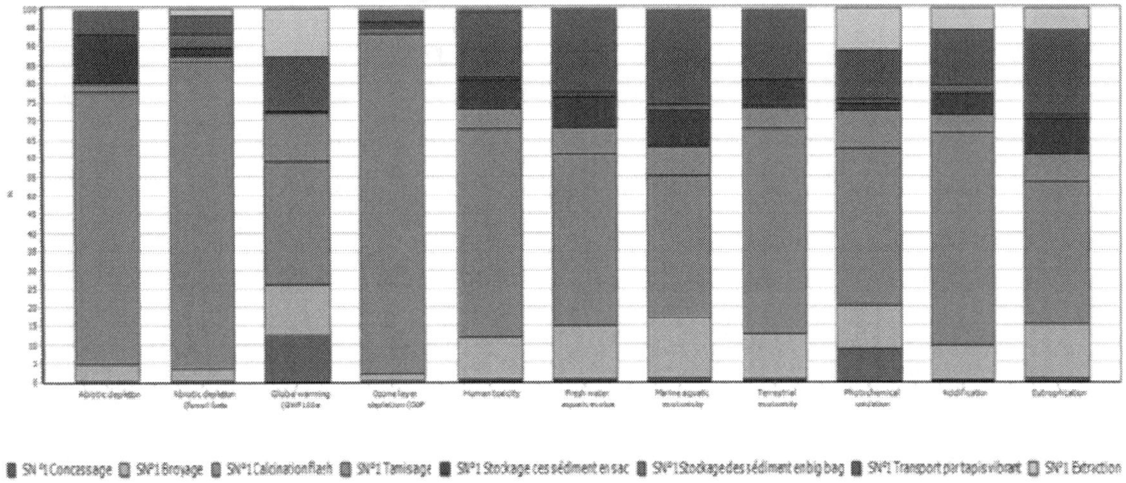

Fig. 2. Histogramme des impacts sur l'ensemble des indicateurs de la modélisation du scénario N°1 de production des sédiments de dragage avec la méthode CML-IA

b) scénario N°2 :

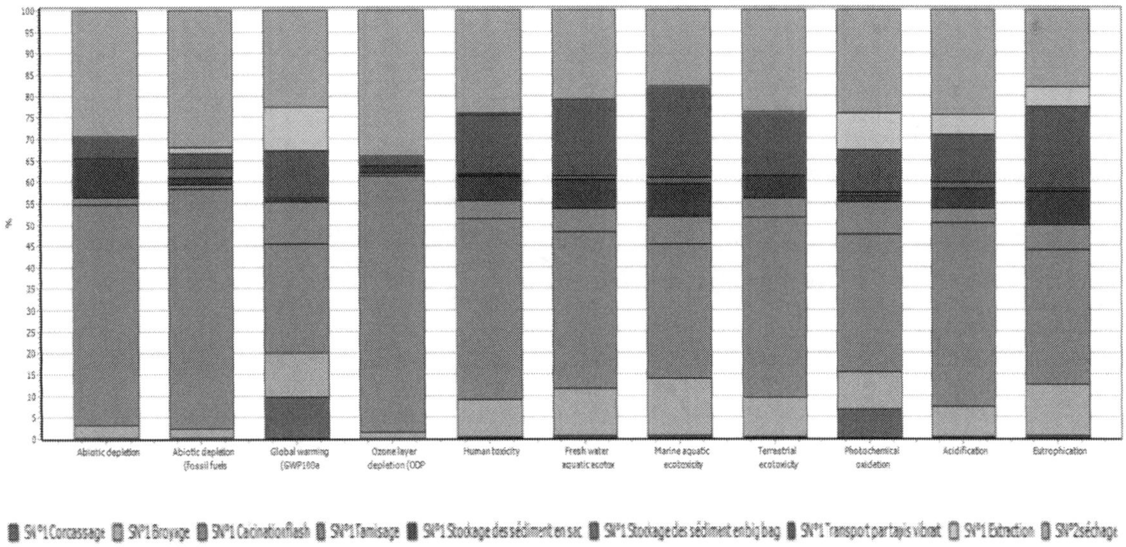

Fig. 3. Histogramme des impacts sur l'ensemble des indicateurs de la modélisation du scénario N°2 de production des sédiments de dragage avec la méthode CML-IA

Table 4. Indicateurs d'impacts environnementaux des deux scénarios de production des sédiments de dragage

Catégorie d'impact	Unité	Scénario N°1	Scénario N°2
Epuisement des ressources	Kg Sb eq	7,56E-05	0,000107
Epuisement des ressources (combustibles fossiles)	MJ	2,91E+03	4,28E+03
Changement climatique	Kg CO2 eq	726	939
Destruction de la couche d'ozone	Kg CFC-11 eq	1,95E-05	2,97E-05
toxicité humaine	Kg 1,4-DB eq	21,4	28,2
Ecotox aquatique d'eau douce	Kg 1,4-DB eq	19	24,1
Ecotoxicité aquatique marine	Kg 1,4-DB eq	6,75E-04	8,22E+04
Toxicité terrestre	Kg 1,4-DB eq	0,133	0,175
Oxydation photochimique	Kg C2H4 eq	0,0644	0,0848
Acidification atmosphérique	Kg SO2 eq	0,461	0,611
Eutrophisation	Kg PO4--- eq	0,114	0,138

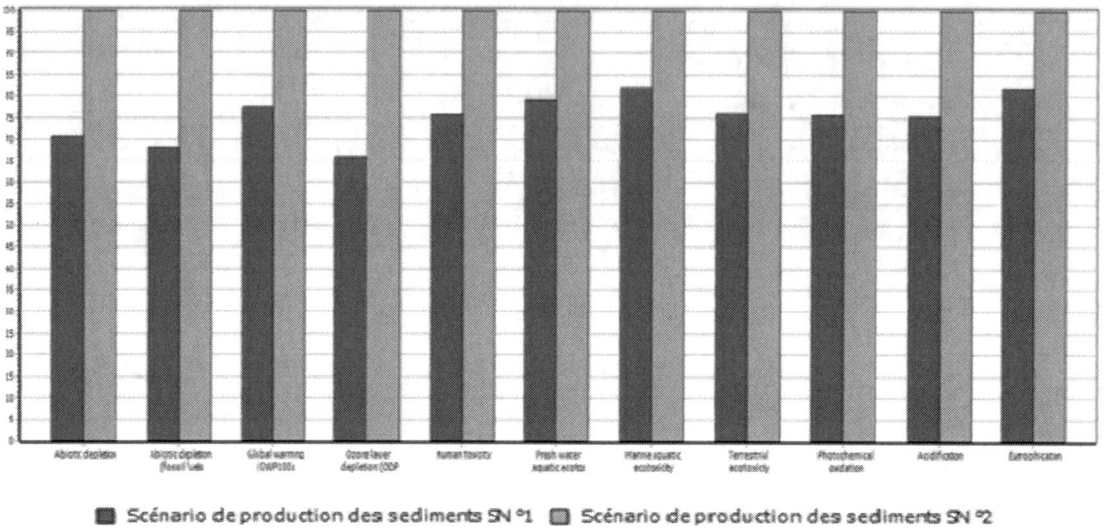

Fig. 4. Histogramme de comparaison des impacts sur l'ensemble des indicateurs de la modélisation des deux scénarios de production des sédiments de dragage avec la méthode CML-IA

Le scénario N°1 de production des sédiments de dragage est le plus respectueux de l'environnement sur l'ensemble des indicateurs de la méthode CML-IA baseline offre un bénéfice moyen de l'ordre de 30% d'amélioration sur l'ensemble des indicateurs par rapport au scénario N°2.

La figure 5 représente l'histogramme des processus à l'origine des impacts des deux scénarios N°1 et N°2 de production des sédiments de dragage.

Sur l'ensemble du processus de production des sédiments de dragage, on observe que la consommation du gaz naturel, avec un gain de 1,78E-5kg CFC-11eq de la phase calcination flash, est le principal contributeur. Le second est la consommation de l'électricité avec un gain de 1,4E-6 KgCFC-11eq dû à l'ensemble des opérations de traitement des sédiments de dragage. Le troisième est la consommation du papier Kraft avec un gain de 0,5E-6 KgCFC-11eq dû au stockage des matériaux.

La figure 6 représente l'histogramme des substances à l'origine des impacts des deux scénarios de production des sédiments de dragage.

Sur l'ensemble du processus de production des deux scénarios N°1 et N°2 des sédiments de dragage, on observe que les substances « Méthane, bromochlorodifluoro-, Halon 1211 », « Méthane, bromotrifluoro-,Halon1301 » et « Ethane, 1,2-dichloro-1,1,2,2-tetrafluoro », sont les principaux contributeurs. Les autres couleurs correspondent à d'autres substances.

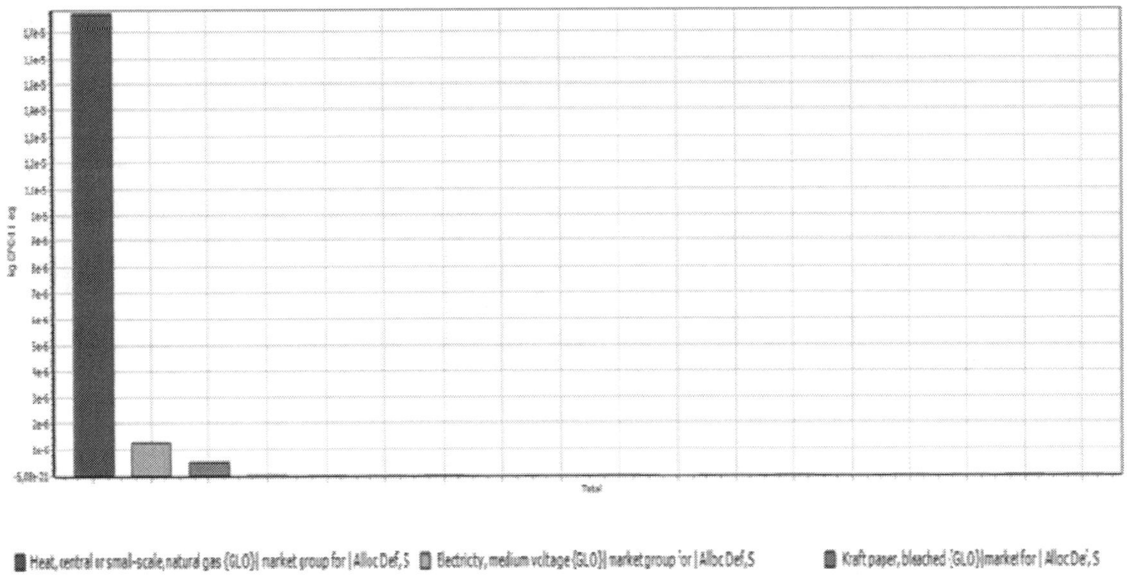

Fig. 5. Histogramme des processus à l'origine des impacts des deux scénarios N°1et N°2 de production des sédiments de dragage

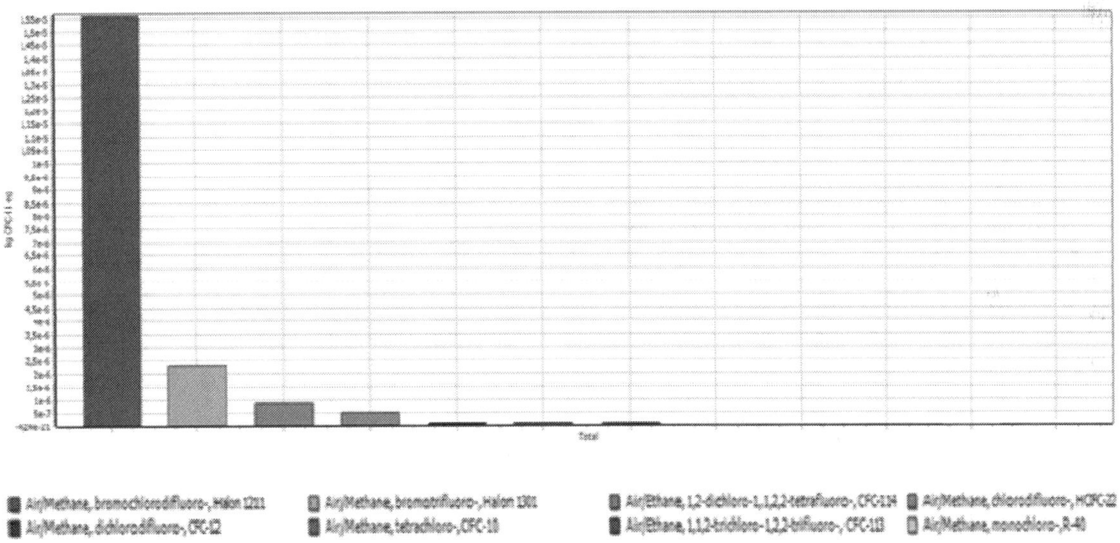

Fig. 6. Histogramme des substances à l'origine des impacts des deux scénarios N°1et N°2 de production des sédiments de dragage

6 Conclusion

Les scénarios étudies ont porté sur le développement d'un nouveau procédé de production des sédiments de dragage en Algérie. Le scénario N°1 consistait à utiliser les sédiments à l'état sec. Cette opération a demandé un temps de séchage à l'air libre assez important. Le scénario N°2 a été fait sur des sédiments à l'état humide. Le temps de séchage sera plus réduit. Le choix se portera sur le scénario N°1, les impacts environnementaux ont été plus faibles ; d'où une nette amélioration sur les indicateurs environnementaux.
L'étape de calcination est la plus impactante sur l'environnement. Il Ya lieu d'envisager une recherche plus détaillée sur un nouveau procédé, avec une optimisation de consommation d'énergie.

References

1. F. KAZI AOUAL-BENSLAFA, (2014). *Caractérisation des sédiments de dragage du barrage de Bouhanifia pour une réutilisation.* XIIIème Journées nationales génie côtier, Dunkerque, (2014)

2. Y. SENHADJI, *Evaluation de la vase de quatre barrages algériens*, Conférence internationale des

matériaux innovants & leurs applications. Les 21et 22 avril 2016, Oujda-Maroc, (2016)

3. M.A. CHIKOUCHE M.A, *Effet des ajouts à base d'argiles gréseuses et vaseuses sur les propriétés des matériaux cimentaires.* Mémoire magister Université de M'sila, (2008)
4. A. SEMCHA, *Valorisation des sédiments de dragage : Applications dans le BTP, cas du barrage de Fergoug.* Thèse Université de Reims Champagne-Ardenne, (2008)
5. M. SILINE, *Valorisation d'une argile gréseuse comme ajout pour la fabrication d'un ciment à faible impact environnemental.* 31emes Rencontres de l'AUGC ENS, Cachan, (2013)
6. N.E. BOUHAMOU, *Etude du comportement d'un béton autoplaçant à base de vase de dragage vis-à-vis du retrait.* Colloque éco matériau, Montpellier, France, (214)
7. O. BELARIBI, *Durabilité des bétons autoplaçants à base de vase et de pouzzolane.* Thèse en cotutelle université de Mostaganem et l'université Cergy-pontoise, (2015)
8. ANBT, *Agence nationale des barrages et transferts,* (2015)
9. ADEME, DABEE, Département Industrie et Agriculture. *Le séchage thermique ADEME*, (2006)
10. J. OLIVIER, S. MYRIAM, *Analyse du cycle de vie comprendre et réaliser un écobilan.* **2eme édition**, (2010)
11. LABORATOIRE CENTRAL DES PONTS ET CHAUSSEES LCPC, *Evaluation environnementale d'un procédé de fabrication de méta kaolin (calcination flash),* (2010)

Parametric Modelling and Traditional Architecture: Improving the thermal comfort of the traditional courtyard house in Morocco

Khalid EL HARROUNI, Mouhcine BEN AICHA and Rime EL HARROUNI
Ecole Nationale d'Architecture, Laboratoire Energie, Durabilité et Maitrise des Ambiances, Rabat, Morocco

Abstract. The traditional courtyard house of the Mediterranean Basin has been viewed as a complex regulating system that creates a microclimate which historically worked, and still works, in a *passive* way to provide acceptable thermal comfort in summer. The internal courtyard is generally described as a positive factor that can moderate extreme outdoor climatic conditions. However, some researches have shown that the courtyard could become a negative factor from the energy efficiency point of view. For this purpose, this paper is based on a research study exploring sustainable characteristics of Moroccan traditional housing and its climatic adaptation, delving into the Rabat-Salé case study. A traditional courtyard model is used as a case study to analyze the indoor thermal comfort without using mechanical heating and cooling systems. The thermal behavior of the rooms surrounding the courtyard is analyzed under a temperate and humid climate of Rabat-Salé medina. The simulation modelling is carried out to analyze the effectiveness of different parameters to improve the indoor climate during summer and winter, including the façade orientation, the air infiltration, the surroundings, the ceiling height, the walls and roof/ceiling insulation and the shading devices. Tools for climatic design, Mahoney's tables, Givoni and Szokolay bio climatic diagrams have been also used to improve design strategies in terms of thermal comfort.

1 Introduction

The courtyard architectural *type* transforms a one or two facades building into a four solar orientations architecture. This is one of the main advantages of including a courtyard in the plant of a building inserted in a compact urban fabric. Whether to gain natural light, to create an open private space or to control the building thermal behavior, in the past the different *types* of courtyard house design used to follow the same principle: a central courtyard surrounded by rooms aligned with four interior facades.

The climatic behavior of the courtyard is especially important in arid areas that are characterised by hot, dry days and cool clear nights to radiate heat away to the sky, -climates with high temperatures during the day and low temperatures during the night and high temperature amplitude between the seasons-. Also for climates with cool but sunny days at lower latitudes where sun penetration is both: easy and welcome.

Some researches [1-2] have shown that the courtyard could become a negative factor from the climatic point of view. It can be hotter than its surroundings while, during the winter, the temperature difference remains smaller. It seems that for courtyard comfort, the worst climates are hot and humid, where little wind is available to relieve stuffiness; or cold and cloudy ones at higher latitudes, where little winter sun can penetrate to provide warmth.

Morocco has recently adopted new thermal regulation regarding energy efficiency in buildings [3]. The objective is to introduce minimum requirements in new residential and tertiary buildings in order to optimize their needs for heating and air conditioning, while improving their thermal comfort. The aim of this study is to simulate and to examine the thermal behaviour of a traditional courtyard house model under a temperate and humid climate in Rabat-Sale and to explain the findings, to compare them with similar research and to evaluate the architectural parameters that create a positive climatic adaptation to the environmental climatic conditions.

2 General

2.1 The traditional house in the medina

As a first phase of the research, we have explored the values of the medina within an analytical investigation of domestic and public spaces [4]. The traditional house can be defined by the following elements: the entrance, the central yard or patio and its surrounding rooms. These elements make it possible to describe the components of the domestic spaces (Fig. 1):

- the entrance: the corridor from the front door and into the house itself is often angled to minimize the view from outside;

- the central yard and its surrounding rooms: the central courtyard has no roof or can have an opening in the roof. The ceiling height in the rooms varies from 3 to 5 meters and the rooms are often quite elongated. The use of the rooms is often non-specific and can be used for various purposes. The family life and the housekeeping take place in these domestic spaces, which can be used interchangeably for eating, sleeping, recreation and other domestic tasks. The rooms are organized around the patio which is the only space bringing in light and air to the rooms. There are usually very few or no windows facing the street, almost all the openings and windows are facing the inner courtyard.

The opening is important because it modifies the indoor climate and contributes to natural ventilation. It also works as a protection from sun and wind, walls are high enough to protect from the sun rays, that usually reach the courtyard until the afternoon. Then the warm air rises, and convention is created, ventilating the rooms. However, some studies have shown that the roof opening in the inner courtyard can be negative, especially in hot and humid climates. In hot humid climates, the temperature in the courtyard can be higher than outdoors because there is little wind and it can become stuffy inside the house.

Common traditional building materials are stone, earth, and wood, because of capillary effect, absorb water, which can then evaporate from their surfaces and thus hinder the interior air from being rewarmed by convection.

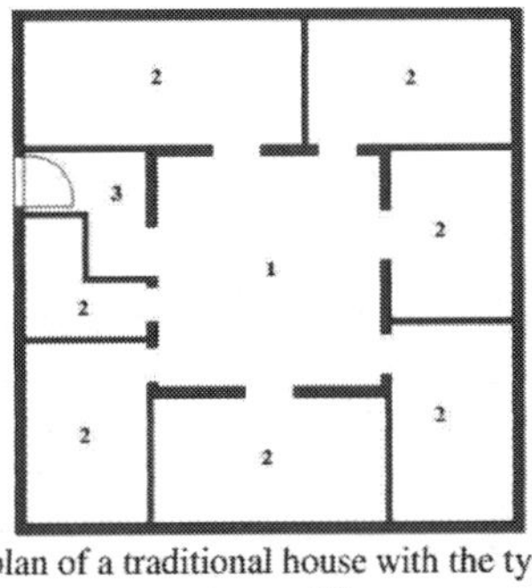

Fig. 1. A plan of a traditional house with the typical angle entrance [5].

The large residential area surrounding the city centre is formed by the composition of the residential quarters. A residential quarter is formed by the junction of neighborhood units along quarter streets. The residential area and the public area are enclosed all together by the city walls, and thus form a closed physical and social unit, the city or the medina.

A street in the medina is a long space between two walls which have no openings facing the street. The ratio between the height of the walls and the width of the street is big: about 10/1. Because of the streets depth and narrow, the sun rays can't reach the pavement or facades during the day.

2.2 The problem

Contemporary buildings and constructions in Morocco are often designed without taking the climate into consideration. Consequently, a great part of the built environment has a poor climate adaptation. However, as mentioned before, Morocco has now its thermal regulation and energy efficiency buildings code. The implementation of the thermal regulation is related to the building envelope with some technical specifications: 20°C in the winter for heating and 26°C in the summer for air conditioning. Those are the ideal temperatures for a good adequacy between thermal comfort of the room and energy consumption of the devices and equipments.

A building's indoor climate depends largely on the architecture and design of the building. Layout of the building, location and color of walls, location, size and orientation of windows and the buildings ventilation ability largely determines the climate in the building but also the building materials matters. Building materials can affect through thermal capacity and thermal conductivity.

One of the most frequent complains among the families living in the Moroccan traditional houses, regardless of the age of the house, involves moisture. The moisture that creates musty odors at the base of the walls, and the condensations that concentrate on the ceiling and other cold surfaces, affect approximately 50% [6] of the traditional houses.

Also extreme temperatures varying from winter to summer are a difficulty. These thermal discomfort problems are primarily due to the following factors: the climate, the thermal quality of the building materials, the ventilation and the renewal of the air, the orientation of the frontages and the lack of sun shining in the patio during the winter season.

The inquire housing in Morocco [6] confirms this thermal discomfort for the traditional houses that have the problem of low temperatures (77% in Rabat-Salé). The moisture in more than 30% of the traditional houses affects the interior basement walls, and the ceiling during winter season.

3 Case study

3.1.The site and the climate

Rabat-Salé is located in latitude 34°03'n, longitude 6°46'w and at an altitude of 75.3m above sea level. The climate is temperate, Mediterranean climate, the Atlantic ocean has a cooling effect that transforms the climate along the coast into relatively mild.

The temperature data given by the local meteorological station show that the coldest month is January: the mean temperature is 12.5°C with an extreme minimum of -0.5°C. July is the hottest month with mean air temperature of 22.3°C and extreme maximum of 47.3°C. The maximum and the minimum relative humidity are also given for every month, the maximum value is 96% and the minimum is 42%. The maximum rainfall in Rabat-Salé is 110mm in December. The maximum sunshine hours is evaluated to 13,85 hours per day in June; the highest global radiation is observed in June-July and the lowest in December.

3.2 Hypothesis

The aim of the study is to analyze the climatic behavior of the rooms surrounding the courtyard in a temperate and humid zone. It seems that one of the important factors is the ratio between the courtyard length to height of the rooms surrounding it. In hot and arid region for example [1], when the room walls are higher than the courtyard length, the patio climate is more comfortable than the outdoor climate, due to better shading that is provided by the surrounding walls. In contrast, it seems that when the length of the courtyard is longer than the height of its surrounding walls, the temperatures inside the patio are higher than the outdoor temperatures.

The internal courtyard can then affect the indoor climate. Unfortunately, this space is not studied because of the limitation of the computer program, DEROB-LTH [7] which cannot simulate an open space, a courtyard for example. Secondly, due to the lack of the courtyard climatic measurements during the summer and winter seasons -including the temperature, the humidity and the wind speed-, the main hypothesis of the present study is to consider the courtyard as an open space and to treat it as an outdoor space. Third, the DEROB-LTH software used in this study does not take into account the humidity of the air, which can be the principal source of the discomfort in the Rabat-Salé region.

3.3 Analysis methods

3.3.1 Architectural bioclimatic tools

Givoni/Szokolay bio climatic diagrams and Mahoney's tables [8-9] have been used in this study. There are still useful tools to give general considerations of the comfort zone with some recommendations for passive design of building.

According to the Rabat-Salé Givoni/Szokolay bioclimatic diagrams (Fig. 2-3), the comfort zone can be extended by using heating in January, in the rest of the winter months one can obtain a comfortable indoor climate with passive methods -internal gains-. The diagram shows that, because of the mild climate in Rabat-Sale, ventilation is enough during the summer although the buildings must be designed with good ventilation possibilities and protection against solar radiation during the summer months.

Based on a diagnosis of some climatic indicators, the Mahoney tables give the performance specifications and the design recommendation for buildings in Rabat-Salé in order to have a best indoor climate:

-layout: orientation north and south (long axis east-west);
-spacing: compact layout of estates;
-rooms should be double banked, with temporary provision for air movement;
-openings: medium openings, 25-40 %;
-position of openings: in north and south walls, at body height on windward side, openings also in internal walls;
-walls and floors: light, short time-lag, low thermal capacity;
-roofs: light, well insulated roofs.

The most important design parameters affecting indoor thermal comfort and energy conservation in building scale are building form and layout, building envelope design, building materials and optical and thermo-physical properties of the building envelope. Among these parameters, building envelope design, as it separates the outdoor and indoor environment, is the most important.

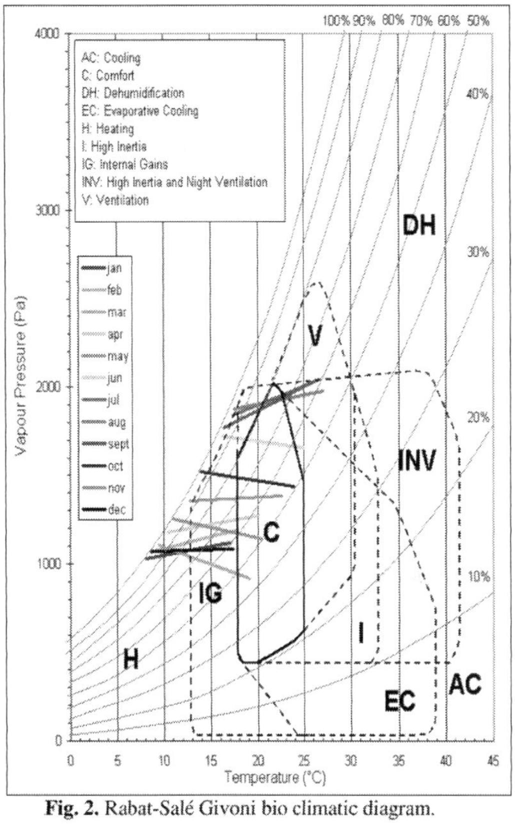

Fig. 2. Rabat-Salé Givoni bio climatic diagram.

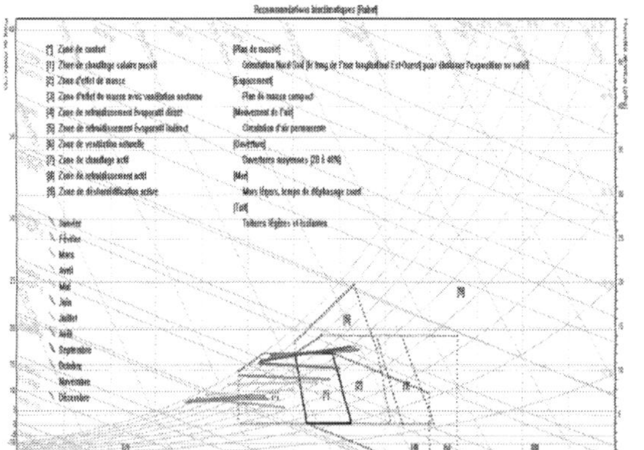
Fig. 3. Rabat-Salé Szokolay bio climatic diagram (Awrach).

3.3.2 Parametric and simulations modelling

DEROB, which is an acronym for dynamic energy response of buildings, is a MS Windows based flexible simulation tool using a RC-network for thermal model design. It was originally developed at the numerical simulation laboratory of the school of architecture of the university of Texas at Austin. The DEROB-LTH modules are further developed to suit the local needs at the department of building science at Lund Institute of Technology. The calculations are performed in a dynamic way for each hour during a specified period of simulation, and they are influenced by climatic factors such as outdoor temperature, solar radiation and the sky temperature. Properties for the indoor climate of the building can be calculated based on these simulated results.

The Plot size surface of the traditional house model used as a case study is 88.0 m² (11.0 m of width and 8.0 m of length) and the height is of 3.5 m but it is variable for simulation modelling.

Orientation of all façades (facing street, backyard or adjacent to other buildings) is also variable.

Regarding the openings, the total area of the window size is 2,16 m² and the percentage of the façade is about 6.85 %. The total area of the doors is of 8,40 m².

Figure 4 and Table 1 give the description of the traditional house model used as a case study.

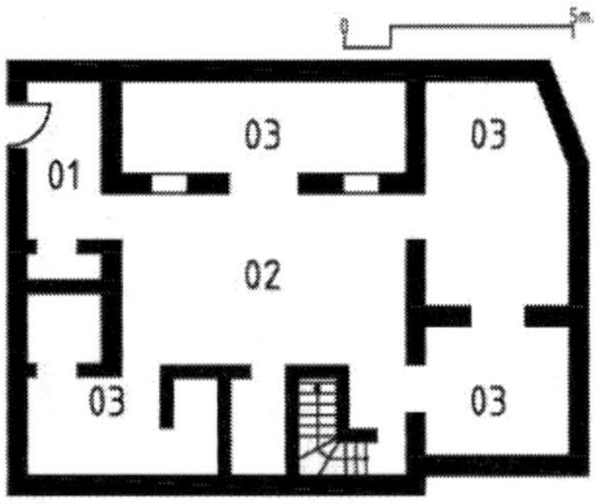

Fig. 4. Plan of the traditional house model.

Other data include air changes and internal heat loads during winter and summer (day and night ventilations) for both main rooms of the model house.

To perform DEROB modelling, the thermal behaviour of the volumes surrounding the courtyard is analyzed under a temperate and humid climate (Rabat-Salé) using passive design of building. Different simulations by DEROB-LTH were carried out using different parameters to improve the indoor climate. The operative temperatures in volumes 2 and 3 (DEROB model in Figure 5) are simulated during summer and winter using the climatic data and all information given in the project description.

Table 1. Description of materials.

	NAME	CONDUCTIVITY (W/MK)	SPECIFIC HEAT (WH/KGK)	DENSITY (KG/M³)
OPAQUE MATERIALS	RAMMED EARTH / PISE	0.80	0.20	2000
	EARTH	1.40	0.22	1300
	SAND	0.40	0.24	1700
	CLAY MIXED WITH 40% SAND	1.20	0.23	2000
	LIME PLASTER	0.80	0.24	1700
	PLASTER	0.35	0.30	900
	LIME CEMENT PLASTER	1.00	0.24	1800
	CEMENT MORTAR	0.93	0.29	1800
	ZELLIGE / CERAMIC	0.80	0.24	1900
	WOOD	0.16	0.70	700
	MINERAL WOOL	0.04	0.24	50

	EXISTING WALLS/FLOOR/ ROOF	THICKNESS (MM)	NAME
WALLS/ FLOORS / ROOM	FOUNDATION GROUND	500	EARTH
		15	LIME PLASTER
		200	CLAY MIXED WITH 40% SAND
		15	CEMENT MORTAR
		15	ZELLIGE / CERAMIC **ABSORPTIVITY** 40 %
	FAÇADE AND ROOM WALL	15	LIME PLASTER **ABSORPTIVITY** 20 %
		500	RAMMED EARTH / PISÉ
		20	PLASTER **ABSORPTIVITY** 20 %
	ADIABATIC WALL	500	MINERAL WOOL
		250	RAMMED EARTH / PISÉ
		20	PLASTER **ABSORPTIVITY** 20 %
	FLOOR AND ROOF STRUCTURE	15	ZELLIGE / CERAMIC
		15	CEMENT MORTAR
		150	CLAY MIXED WITH 40% SAND
		25	SAND
		15	LIME PLASTER
		130	WOOD **ABSORPTIVITY** 60 %
	DOOR	35	WOOD

WINDOWS	GLASS TYPES	**DATA TYPE**	**EMITT. FRONT**	**EMITT. BACK**	**TRANS-MITT.**	**REFL.**
		FRESNEL GLASS : NORMAL TRANSMITTANCE AND REFLECTIVITY.	83.7 %	83.7%	83.0%	7.0%
	WINDOW TYPES	**PANE**	**GLASS**		**THICKNESS**	
		1	CLEAR 4		4 MM	

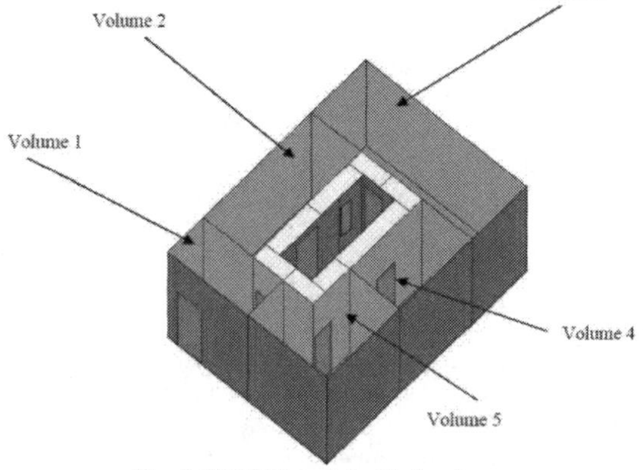

Fig. 5. DEROB model with 5 volumes

The operative temperatures in these rooms (Volumes 2 and 3) are first simulated during winter and the variants are:

- The façade orientation: the orientation for solar access is important, especially in the winter. The four orientation cases, the façade facing to south (case 0), east (case 1), north (case 2) and west (case 3) were analyzed separately. Next simulation cases, the parameters were studied in combination in order to improve the indoor climate.
- The shading devices: the gallery in the traditional house is a covered space of the circulation and the transition between the courtyard or patio and the rooms. In winter, the galleries make it possible to circulate when it is raining. In summer, the galleries create shade and become an obstacle for the intense sun rays access. The shading device in the courtyard has a width of 0.7 m in the baseline case. In cases 4, 5, 6 and 7, the shading device is out in the courtyard.
- The ceiling height: the traditional house rooms have generally a high ceiling height (more than 3.0 m). In the baseline case, the ceiling height h=3.5 m, which is reduced to h=3.0 m in cases 8, 9, 10 and 11.
- The surroundings: the site layout in terms of the width street and the surroundings height were analyzed in combination with the orientation. The surrounding variation facing building height and the small width street are examined in the case 12.
- The roof/ceiling and walls insulation: thermal insulation of the roof and the walls is also important for better comfort. Composite insulating panels (wood-wool/mineral-wool) for walls and wood-wool slabs for roof (Table 2) were used (cases 13 and 14).
- The air infiltration: the ventilation is an important factor for comfort, especially during the night and day ventilations. The air change hour (ach) is reduced from 4 to 2 ach during evening (between 19 and 24), and from 3 to 1 ach during night (between 01 and 07). This corresponds to the case 15.

Table 2. Walls and roof thermal insulation.

Insulation material	Conductivity (W/mK)	Thickness (cm)	Density (Kg/m³)
Composite Insulating panel (wood wool / Mineral wool)	0.04	5	200
Wood wool slab	0.07	3	300

The winter parametric modelling process starts from case 0 which is the baseline case and consists in varying the above parameters to reach case 15 which is the best possible solution obtained by reducing the air changes hour during evening and night. Combining case 10 corresponding to reduction of the ceiling height with case 14 using roof and walls insulation.

The operative temperature in the volumes 2 and 3 (the main rooms of the house) are displayed in figures 6-7 for the baseline case (case 0) and the best solution (case 15).

The winter indoor climate has also shown only small difference between the orientation cases (façade facing to the south, east, north and west) and the shading devices of the courtyard. Indeed, these orientations are equal for this house model while north orientation combining with the ceiling height reduction give higher indoor temperatures. The north orientation in the present case means that the rooms (volume 2 and volumes 3) have west and north windows.

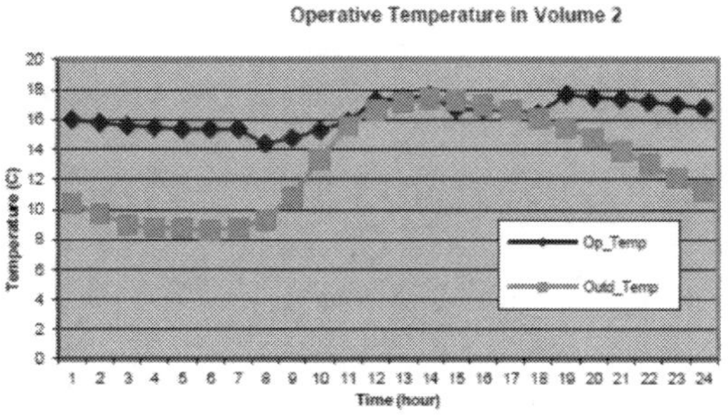

Fig. 6. Operative temperature in volume 2 during December (case 0)

The optimal orientation, the façade facing to the north, is then selected in order to have a moderate indoor climate in both main rooms.

Thermal insulation of the roof and the walls is important for a better comfort. We considered walls and roof ceiling with additional insulation: an inner insulated layer of 50 mm using composite insulating panels (wood wool / mineral wool) for walls and an outer insulated layer of 30 mm of wood wool slabs for the roof. In wintertime, this type of insulation gives a better performance than the previous cases.

Another important factor related to comfort is the air infiltration, especially the day ventilation which is useful to improve the indoor climate during the winter. Operative temperatures in figure 8 show the effect of the ventilation when reducing the ach during evening and night. An important increasing of 3 degrees is observed. Finally, the case 15 is the best possible solution obtained in the winter parametric modelling process.

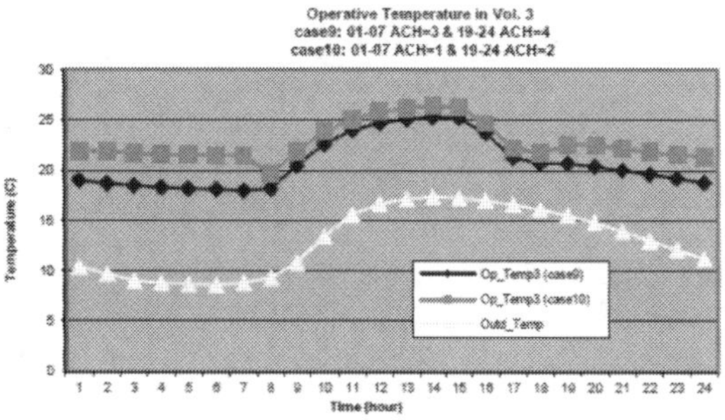

Fig. 7. Operative temperature in volume 3 and air changes hour reduction during December (case15)

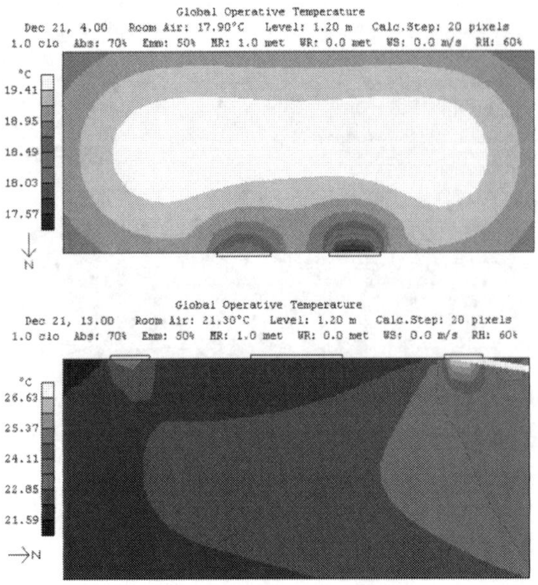

Fig. 8. Global operative temperature in volumes 2 and 3 during night and day times (winter)

The operative temperatures in volumes 2 and 3 are then simulated during summer and the influence of the following parameters is studied: the ceiling height (case 17), the shading devices of the courtyard (case 18) and the air infiltration in terms of day (case 19) and night (case 20) ventilations. During winter, the main rooms should be closed at night and open during the day to maximize the indoor temperatures. The reversed strategy (night ventilation) should be used in the summer

Case 20 seems to be the best possible solution obtained by increasing the ceiling height and the dimension of the shading device in the courtyard and by adopting the option of the night ventilation.

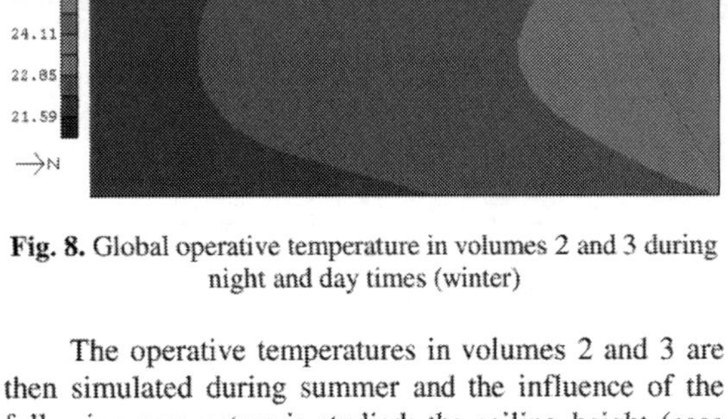

Fig. 9. Operative temperature in volume 3 and ventilation, August (cases 19 and 20)

4 Conclusions

We can conclude through all simulations that the courtyard can be considered a positive factor in the passive design of dwellings in the Rabat-Salé region.

In order to keep traditional courtyard house in this area, while improving its energy efficiency and meeting the new energy regulation requirements, some strategies should be taken into account by designers:

-Lengthening the courtyard in the axis parallel to the north and south façades.

-Increasing (more than 3.0 m.) the ceilings height in the south and west oriented internal rooms.

-Diminishing (less than 3.0) the ceilings height in the north and east oriented internal rooms.

-Increasing the insulation in the roof, north and east internal facades.

-Moderating the insulation in the south and west internal facades.

As general considerations to be implemented by owners:

-Reducing the air changes hour during evening and night during the winter, maintaining only the minimum time required for the air renovation and moisture ventilation.

-Increasing the dimensions of the shading devices of the courtyard during the summer.

-Increasing the night ventilation as much as possible and a low air changes rate is recommended during daytime in the hot season.

References

1. Y . Etzion, The Thermal Behaviour of Non-Shaded Closed Courtyard in Hot, Arid Zones, Architectural Science Review, **33**, 79-83 (1999)
2. I.M. Meir and D. Pearlmutter, Attached Courtyard – Asset or Debit, *Proceeding of International Conference on Energy and Buildings in the Mediterranean Area*, Saloniki, Greece, 245-252 (1992)
3. Royaume du Maroc, Bulletin Officiel, Décret N° 2.13.874 Approuvant le Règlement Général de Construction Fixant les Règles de Performance Energétique des Constructions et Instituant le Comité National de l'Efficacité Energétique dans le Bâtiment, **6306**, 4256-4269 (2014)
4. A. Abarkan and A.M. Salama, Courtyard Housing in Northern Africa: Changing Paradigms, *ENHR 2000 Conference*, Gavle, Sweden, *Housing in the 21st Century, Fragmentation and Reorientation*, 26-30 (2000)
5. T. Brunzell and S. Duric, Moroccan Architecture, Traditional and Modern - A Field Study in Casablanca, Morocco, *Bachelor Thesis*, LTH School of Engineering, Housing Development & Management, Lund University (2012)
6. Royaume du Maroc, MATEUH, SEH/DPI/OH, *Enquête Logement 2000*, L'Observatoire de l'Habitat (2000)
7. H. Kvist, DEROB-LTH for MS Windows, *User Manual*, Department of Building Science, Lund Institute of Technology, Lund University, Sweden (1999)
8. B. Givoni, *L'homme, l'Architecture et le Climat.* Éditions du Moniteur, Paris (1978)
9. S.V. Szokolay, *Introduction to Architectural Science. The Basis of Sustainable Design.* Architectural Press (2008)

Stability of tillians and impact on the road network: case of the outside basin of Ouergha (Morocco)

Stabilite des talus et impact sur le reseau routier: cas du bassin versant d'Ouergha (Maroc)

I. JAOUDA[1], A. AKHSSAS[1], L. OUADIF[1], L. BAHI[1], A. LAHMILI[1]
[1] Mohammadia Engineering School, 3GIE Laboratory, Mohammed V University in Rabat, Morocco

Abstract. The study concerns the cartography of the spatio-temporal evolution of the embankments of the road network of the Oued Ouergha watershed using data from the Landsat images on the one hand, and the modeling of soil stability processes. On the other hand. This region experiences an intense destabilization linked to the combination of several natural factors, such as the brutality and the abundance of the precipitations and the outcrop on the marls and limestones of Cretaceous. Human intervention in this vulnerable environment accentuates its fragility through the clearing and degradation of vegetation cover. The analysis of satellite data has shown that the most degraded soils are not necessarily those with the highest rate of erosion over the past fifteen years and that some well-developed soils are unstable areas after clearing and their non-rational exploitation. Human intervention and the choice of road routes sometimes, combined with climate change, remain the most important factors in weakening and increasing the vulnerability of the slopes. The results obtained made it possible to identify and monitor vulnerable zones at the Ouergha watershed scale, where interventions are needed to limit the processes of slope degradation and their impact on the various road access routes in the region.
Key words: slope stability, road network, climate change, land use, Ouergha, Morocco.

Résumé. L'étude concerne la cartographie de l'évolution spatio-temporelle des talus du réseau routier du bassin versant d'Ouergha à partir des données des images Landsat d'une part, et la modélisation des processus da la stabilité des versants d'autre part. Cette région connait une déstabilisation intense liée à la conjugaison de plusieurs facteurs naturels, tels que la brutalité et l'abondance des précipitations et l'affleurement sur les marnes et calcaires du crétacé. L'intervention humaine dans ce milieu vulnérable accentue sa fragilité par le défrichement et la dégradation du couvert végétal. L'analyse des données satellitaires a permis de montrer, que les versants les plus dégradés ne sont pas forcément ceux qui ont le plus grand taux d'érosion au cours des quinze dernières années et que certains sols bien développés constituent des zones instables après défrichement et leur exploitation non rationnelle. L'intervention humaine et le choix parfois des tracés routiers, conjuguée avec les changements climatiques, restent les facteurs les plus importants dans la fragilisation et l'accentuation de la vulnérabilité des versants. Les résultats obtenus ont permis l'identification, le suivi des zones vulnérables à l'échelle du bassin versant d'Ouergha où des interventions sont nécessaires pour limiter les processus de la dégradation des versants et leur impact sur les différents accès routiers de la région.
Mots clés: Stabilité des versants, réseau routier, changements climatiques, occupation du sol, Ouergha, Maroc.

1 Introduction

Les marnes constituent plus de 70% des substrats qui affleurent dans le Pré-rif marocain et la majorité des sols fertiles de cette région se développent sur ce type de substrat. Ces sols constituent un capital important pour les habitants de cette. Les marnes sont des roches relativement tendres, elles subissent une géodynamique très active à leur surface et leur fragilité les rend très vulnérables aux aléas de la nature et à l'activité anthropique. Une combinaison de facteurs naturels et anthropiques a provoqué une érosion hydrique intense qui s'est manifesté par une dégradation des sols et des versants.

L'instabilité des versants dans le bassin versant de l'oued Ouergha est un phénomène complexe, tous les processus de dégradation s'y opèrent. La dégradation des versants est un processus qui se produit de façon inaperçue et qui est accéléré par l'activité anthropique. Elle dégrade le sol et affecte les la stabilité des versants et il a un impact direct sur les routes et les pistes.

Les routes et les pistes de la zone d'étude montrent différents types d'instabilité dues à la conjugaison de phénomène naturelles comme la pluviométrie importante essentiellement au niveau des zones qui peuvent être qualifiées de zone éclaires, et des caractéristiques lithologiques des formations géologiques de la région. Le diagnostique intégré et approfondi mené dans le bassin versant d'Ouergha a permis d'élaborer des documents cartographiques qui délimitent les zones vulnérables à l'instabilité des versants; pour orienter les priorités d'intervention et d'aménagement et aussi pour atténuer ou même éviter la dégradation des versant et protéger le milieu naturel.

1.1 Présentation du périmètre d'étude

Le bassin versant du barrage El WAHDA est drainé par l'Oued de l'Ouergha. Il s'étend sur une superficie de 6190 km^2. Il se situe dans le Nord du Maroc entre les latitudes 34°25' et 35°25', et les longitudes -5°24' et -3°45', il s'étend sur la majeure partie de la montagne rifaine.

Il est limité au Nord par la crête rifaine passant par Ketama, Bab Berred et Bab Taza, distant de la méditerranée d'environ 30 Km. Au Sud, par les bassins versants des affluents pré-rifains du Sebou et à l'Est par les affluents de la Moulouya et du Sebou du Périf oriental (Fig. 1).

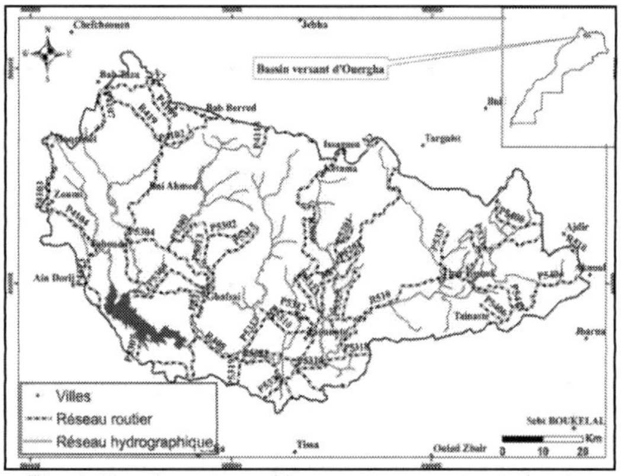

Fig. 1. Situation géographique et réseau routier du bassin versant du l'Ouergha.

Le bassin d'Ouergha est constitué en grande en grande partie par la nappe d'Aknoul et la nappe de Ketama (fig.2). Au nord affleurent, par endroits, des terrains géologiques appartenant aux zones méso-rifaine et pré-rifaine ([1]; [2]).

La nappe d'Aknoul constituent l'essentiel des affleurements surtout dans la moitié orientale du bassin, il s'agit surtout des marnes et calcaires du crétacé supérieur (Sénonien), auxquelles sont superposés localement des calcaires détritiques et marnes noires du Paléocène et de l'Eocène ([3]). Dans la moitié occidentale du bassin, s'étend le synclinal de Dhar Souk, occupé par des dépôts post-nappe marneux localement gréseux et conglomératiques du Tortonien et du Messinien.

Les formations du Jurassique et du Crétacé inférieur (grès et pélites, calcaires et dolomies) appartiennent à la nappe de Senhaja qui fait partie de la zone méso-rifaine. Par contre, les marno-calcaires du Miocène moyen font partie de la zone pré-rifaine. L'unité de Ketama comporte un flysch schisto-gréseux du Crétacé inférieur épais de quelques Kilomètres ([4]).

L'unité de Tanger est considérée comme la couverture plus au moins décollée, de l'unité de Ketama, elle est formée par des schistes et des calcaires du Crétacé moyen et supérieur ([5]). La zone Méso-Rifaine apparaît au sud du bassin versant sous forme de fenêtre sous les nappes intra-rifaines. Les terrains de cette unité sont constitués par une série jurassico-crétacée, constituée dans la région de Taounate par des marnes surmontées par une formation molassique composée de grés et de conglomérats ([6]).

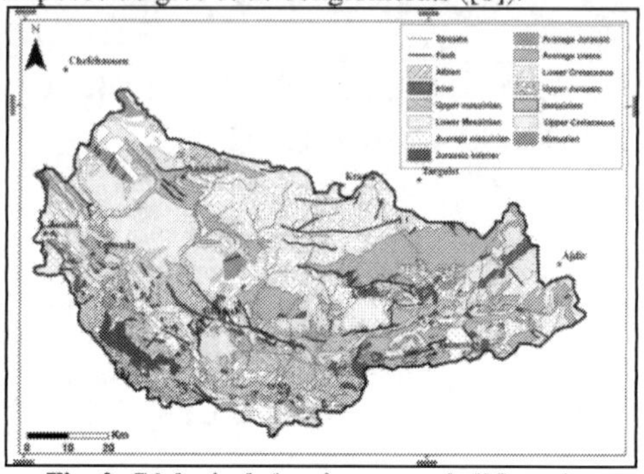

Fig. 2. Géologie du bassin versant du l'Ouergha.

Les lignes du relief sont fortement dépendantes des structures géologiques. Le relief est partout accidenté et accuse des pentes fortes à très fortes et n'est relativement atténué qu'en faveur des vallées. Le ruissellement a façonné des collines douces et arrondies sur les substrats marneux du Miocène supérieur et marneux ou marno-schisteux du crétacé supérieur ([7]).

Les lignes de crête les plus élevées ainsi que les hauts massifs sont façonnés sur les calcaires ou grès calcaires ou gréseux de l'Eocène et de l'Oligocène (fig. 3) du Miocène inférieur, et sur les calcaires du Jurassique. Ces hauts niveaux dominent des versants à pentes fortes. Par endroits, ils dominent aussi des versants à topographie ondulée témoignant de glissements successifs ayant sévis sur les versants marno-schisteux du Sénonien.

Dans les fonds des vallées des Oueds importants, notamment celle de l'Ouergha, se trouve des terrasses du Quaternaire récent et à topographie plane. Par endroits, sur les flancs de versants des vallées se rencontrent à des niveaux supérieurs aux précédents, des terrasses et replats datant du Quaternaire et d'époques plus anciennes ([7]).

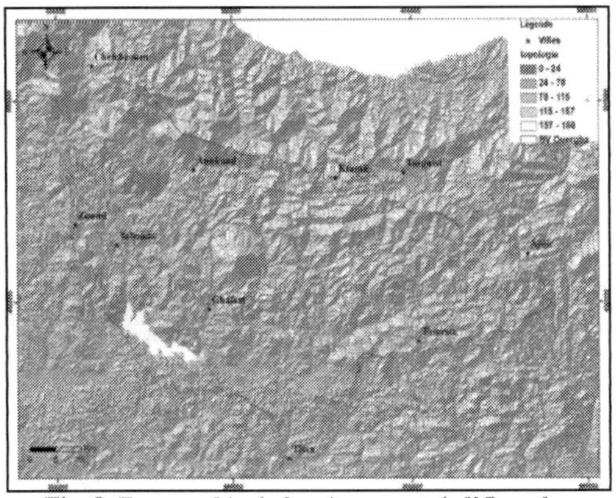

Fig. 3. Topographie du bassin versant de l'Ouergha.

En raison de son relief montagneux et sa situation géographique, la pluviométrie annuelle moyenne est de 650 mm. Elle peut excéder les 1200 mm en année humide surtout sur les reliefs environnants situés à des altitudes plus élevées que celle d'Ouergha (500m).

Les précipitations se répartissent principalement entre les mois d'octobre et mai. Décembre est le mois le plus pluvieux avec 87,5 mm ([8]). Les mois de juillet et août sont chauds et secs et les pluies ne dépassent pas 1,4 mm en moyenne.

L'agressivité des pluies, synthétisée à partir de la spatialisation des stations hydrologiques, montre que la valeur du facteur R varie de 66,7 à 128,7 (fig. 4).

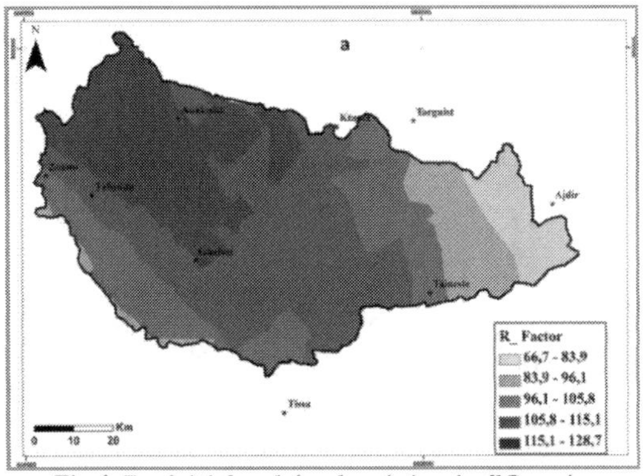

Fig.4. Erosivité des pluies dans le bassin d'Ouergha

La répartition géographique des différentes classes du facteur d'érosivité des pluies par rapport à la superficie totale du bassin est la suivante: 95% de la superficie du bassin est soumise à une érosivité de 70 à 119 et 5% de la surface du bassin versant est soumise à une faible érosivité.

Ces résultats sont proches de ceux obtenus par la DREF-NE ([9]) dans une étude réalisée sur le bassin versant d'Oued Beht limitrophe au bassin d'Ouergha ([10]).

Ainsi, il est rapporté, dans cette étude que l'agressivité pluviale est très concentrée dans l'intervalle 80 à 120 avec 86%. La classe de forte érosivité dépassant 100 représente 24% du bassin versant d'Ouergha. Aussi, les valeurs obtenues sont élevées comparativement à celles du bassin versant de l'oued Boussouab dans le Rif oriental rapportées par Sadiki ([11]), qui ont trouvé que R varie de 31,2 à Anguied au Sud-Est du bassin à environ 60 au Nord-Ouest et à celles du Rif central (bassin versant de l'oued Leben). Enfin, selon la classification de Marnique ([12]), le risque d'érosivité au niveau du bassin versant d'Ouergha est fort à très forte. Par conséquent, le facteur R pourrait entraîner l'apparition et le développement du phénomène de l'érosion linéaire.

Le réseau hydrographique du bassin versant Ouergha (Fig.5) s'écoule du NE vers le SW. L'oued Ouergha est le deuxième affluant principal de l'oued Sebou après l'oued Baht.

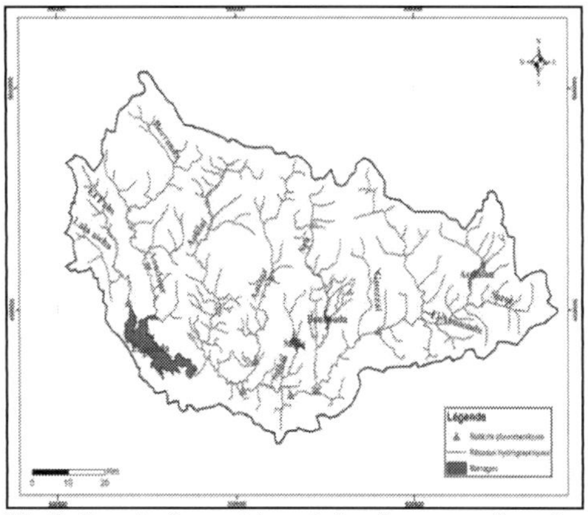

Fig.5: Carte hydrographique du bassin versant Ouergha.

L'oued Ouergha est forme sur la rive gauche par la réunion de l'oued Asfalou venant de l'Est, l'oued Teheris et l'oued Ras Ouergha venant du nord.

Il passe a la station hydrometrique de Bab-Ouender, en aval de laquelle l'oued Ouergha reçoit de gros affluents sur sa rive droite de l'aval versl'amont:

- L'oued Sra: contrôlé par la station hydrométrique de pont du Sker;
- L'oued Amzez: controlé par la station hydrométrique de Galez;
- L'oued Aoulai: contrôlé par la station hydrométrique de Rhafsai.
- L'oued Aoudour: contrôlé par la station hydrométrique de Tarfant ;
- L'oued Aoudai: peu avant de passe au site de barrage Al Wahda et a la station hydrométrique de M'Jara, immédiatement a l'aval du site.

Le bassin versant Ouergha renferme quatre barrages, de l'amont vers l'aval: Asfalou, Bouhouda, Sahla el Al-Wahda, ce dernier est le plus grand aménagement hydraulique du Maroc et le deuxième en Afrique.

De point de vue pente, sept classes ont été identifiées. La classe 6-16%, éparpillée sur tout le

 https://doi.org/10.1051/mateconf/201814902052

territoire, représente 43% de la surface du bassin versant (fig. 6).

La classe 15-25 est localisée surtout dans l'amont et au centre du bassin versant sur 23% de la superficie du bassin versant.

Les pentes dépassant 25% sont localisées dans l'amont sur une superficie de 18,5% du bassin versant. Ces résultats montrent le caractère accidenté du bassin versant de l'Ouergha.

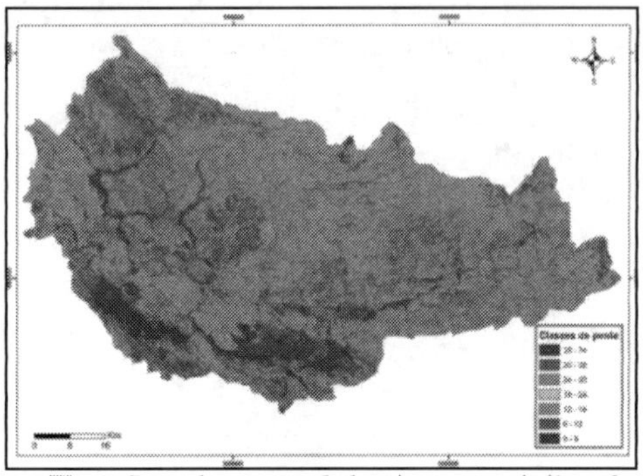

Fig.6. Carte des pentes du bassin versant de l'oued Ouergha.

L'occupation actuelle des terres: La carte présentant les classes d'occupation du sol du périmètre étudié (fig. 7) a été obtenue suite aux résultats de traitement des images satellitaires et de classification des bondes.

En effet, six principaux types d'occupation du sol (forêt, terrains fortement dégradées ...).répartis sur l'ensemble de la zone d'étude.

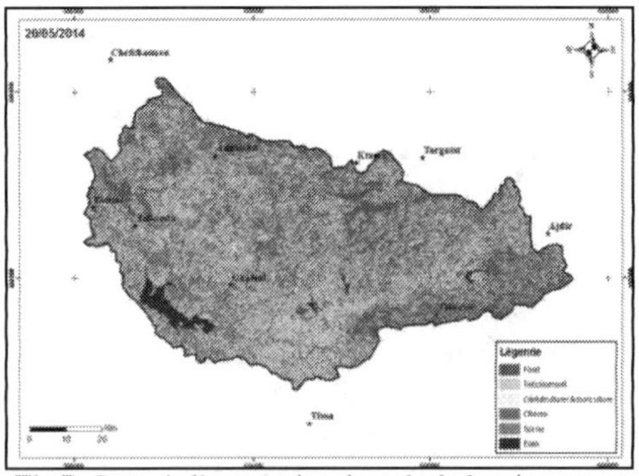

Fig.7. Carte de l'occupation des sols du bassin versant Ouergha.

L'érosion comme par ailleurs dans le Rif, la dégradation des versants au niveau du périmètre est importante et y constitue une contrainte majeure (fig. 8).

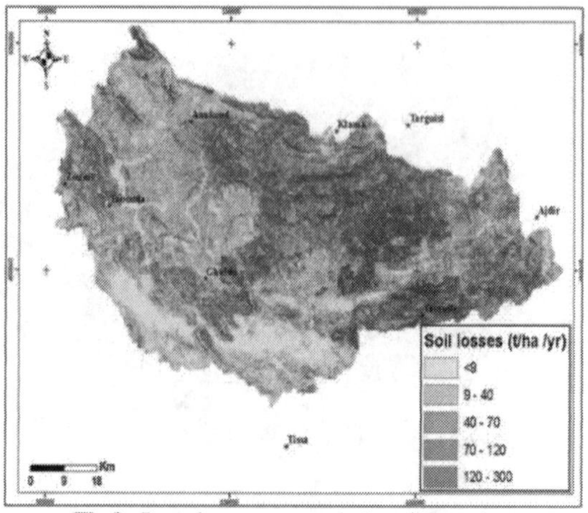

Fig.8. Carte des pertes en sols dans l'Ouergha.

Elle s'y manifeste par des formes et intensités variables. Et cela pour des raisons lithologiques et structurales, climatiques et anthropiques.

La carte géologique (Fig. 2) montre l'extension importante des substrats tendres de différents âges et qui sont sensibles à la dégradation telles que les marnes, les pélites, les marno-schistes et dans une certaine mesure les marno-calcaires.

Les pentes moyennes à fortes favorisent les processus d'érosion. En effet, les pentes dépassant 20% totalisent 50 % de l'ensemble des classes de pentes. La couverture végétale peu dense, protège mal les versants contre le ruissellement provoqué par des pluies agressives. Enfin, le surpâturage et l'usage inadéquat des terres sont aussi des facteurs qui favorisent la dégradation des terres.

2 Méthodologie

Toutes les cartes (fig. 9) ont été saisies, traitées et analysées à l'aide de l'outil SIG, notamment le logiciel Arc-gis.

La combinaison des ces couches thématiques à l'aide de l'outil précité a abouti à la réalisation de la carte de stabilité des versants.

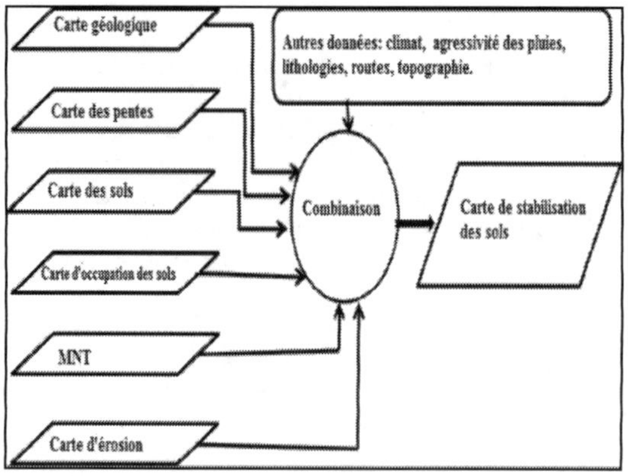

Fig. 9. Combinaison des différentes cartes thématiques à l'aide du SIG.

 https://doi.org/10.1051/matecconf/201814902052

3 Résultats et discutions

Du fait qu'elles sont toutes géo-référencées dans le même système de coordonnées, leur superposition et confrontation ont permis donc la délimitation des unités cartographiques de la carte résultante précitée. Cette dernière est obtenue suite à la combinaison thématique de toutes ces couches analysées en mode raster.

En effet, une fois la superposition réalisée, les valeurs des pixels homologues dans chacune des couches sont analysés et regroupés dans des classes aussi homogènes que possible répondant en fonction de leur caractéristiques à des besoins spécifiques en matière de gestion et de stabilisation.

Ces classes correspondent donc aux différentes unités cartographiques de la carte de figure 9. Une méthodologie similaire été adoptée dans l'étude et l'évaluation des terres d'un versant situé dans la plaine de Tadla ([14]). Les unités de la carte de la figure 10 correspondant donc à des catégories de terre dont les recommandations tiennent compte de la variabilité de leurs caractéristiques spatiales locales présentées dans les cartes ci-dessous cartes géologique, lithologie, pédologique, relief et occupation du sol et routes.

L'analyse des facteurs du milieu, surtout le climat et les sols fait ressortir que les principales contraintes des ces derniers sont la pente et la dégradation des versants.

En plus des facteurs précités, les recommandations prennent aussi en considération l'aptitude des terres ([15]; [16]).

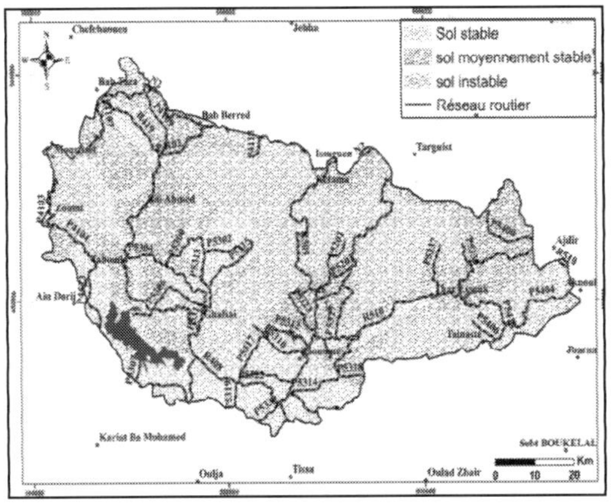

Fig. 10. Carte de stabilité des versants de l'Ouergha.

La carte de stabilité des versants reforme trois types des versants:

Versant stable: Les terrains délimités dans cette catégorie concernent une grande partie du périmètre. Ils s'étendent au nord et à l'ouest de la ville de Dhar souk sur les dépôts du Miocène post-nappe et sur les marno-schistes de la nappe d'Aknoul. Les pentes sont inférieures à 20% et le relief dans l'ensemble est collinaire. Les sols sont essentiellement des sols peu évolués d'apport colluviaux sur colluvions de marnes miocène et marno-schistes du crétacé supérieur et secondairement des marnes noires de l'Eocène et du Crétacé inférieur.

Cette catégorie de terre englobe aussi et d'une manière très localisée, les sols alluviaux irrigués des basses terrasses quaternaire, à relief plat, occupant le fond des vallées des principaux oueds, notamment de Oued Ouergha. Illustrant la situation inquiétante dans la zone 1 tout en impactant les routes (Photo1, Fig. 13).

Versant moyennement stable: Les pentes des versants varient de 20 à 35 %. Les terrains sont surtout des marno-calcaires du Miocène moyen, des marnes de l'Eocène moyen et parfois des calcaires et dolomies du Lias inférieur. Les sols dominants sont des sols d'érosion régosoliques. Les causes principales de l'instabilité des versants sont la topographie, l'érosion hydrique et la pente (Photo 2, Fig.12).

Versant instable: Les pentes de cette catégorie de terre sont fortes et dépassent 35. L'altitude est plus importante en moyenne, elle est de 1200 m. Les terrains sont caractérisés par les marnes du Miocène supérieur, et grès et pélites du Jurassique supérieur. Les sols sont surtout des sols peu évolués.

L'érosion est active par ravinement et donne parfois des badlands, ce phénomène provoque l'instabilité des versants, la fragilité des tracés routiers et la perturbation des routes dans les cas extrêmes (Photo 2, Fig.12).

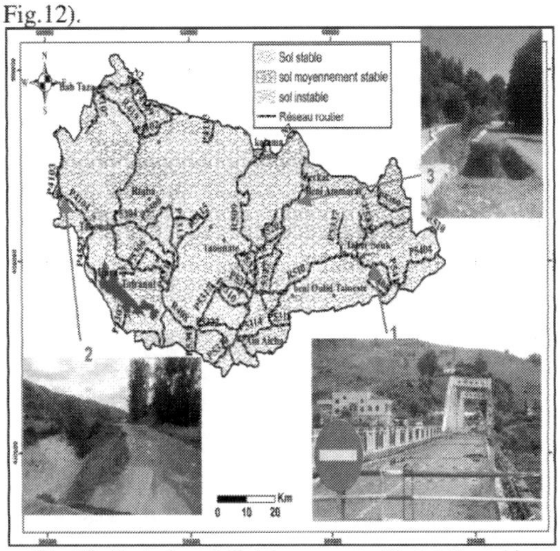

Fig. 12. Carte de stabilité des versants; Photo1 Pont Ain Aaicha sur Oued Ouergha (affaissement); photo2: Ain Aicha S302 (tassement); Photo 3: Zoumi P4103 (glissement).

Conclusion

L'utilisation de l'imagerie satellitaire et du SIG a permis de délimiter les zones où les versants constituent un danger sur les tracés routiers inventoriés et sur les pistes qui relient les différentes agglomérations de la zone d'étude. Les versants instables se situent dans la partie centrale du bassin (couleur rouge fig.10), les zones moyennement stables se trouve dans la partie Est du bassin versant et aussi sur la partie Ouest de bassin.

Cette instabilité dépend essentiellement de La nature des terrains qui sont constitués des marnes et calcaires du crétacé supérieur, auxquelles sont superposés localement des calcaires détritiques et marnes noires du Paléocène. La forte pente qui dépasse les 35%. La pluviométrie importante, elle peut excéder les 1200 mm en année humide surtout sur les reliefs environnants situés à des altitudes plus élevées que celle d'Ouergha (500m) et la lithologie des sols en plus de la dégradation des couvert végétal qui protège les versants conte l'instabilité et rendre les zones stable naturellement.

La réalisation de la carte de stabilité des versant découle l'analyse et l'intégration des différentes caractéristiques morpho-géo-pédologiques influençant la stabilité des terres et constituant autant de thèmes organisés dans la base de données spatiales. Cette dernière peut être actualisée et alimentée au besoin par d'autres informations complémentaires. Elle peut ainsi aider les responsables du développement de la zone dans l'orientation et la prise de décisions en matière d'utilisation rationnelle et de conservation des versants soumis à des processus d'instabilité très importants.

Bibliographie

1. **J.**C Vidal. Carte géologique du Rif: Taïneste - Echelle 1/50 000, échelle 1/50 000 Direction de la géologie. Ministère de l'Énergie et des Mines. Royaume du Maroc (1983).

2. Leblanc D., Feinberg H. coll., Lorenz H.GCarte géologique du Rif: Taïneste, échelle 1/50000. Direction de la géologie. Ministère de l'Énergie et des Mines. Royaume du Maroc (1983).

3. Asebriy L., Bourgois J., Cherkaoui T.E, Azsimousa A. Evolution tectonique récente de la zone de faille du Nékor: importance paléogéographique et structurale dans le Rif externe, Maroc. Journal of African Earth Sciences (and the Middle East), volume 17, Issue 1, July Pages 65-74 (1993).

4. Asebriy L. Bourgois J. Leiking M. Sur l'âge du métamorphisme anchi-épizonal de l'unité de Ketama, Rif central (Maroc). C. R. Acad. Sci. Paris, t. 313, Série II, p. 787-793 (1991).

5. Asebriy L., Azdimousa1 A., Jabaloy A., Booth-Rea G., González-Lodeiro F. and Bourgois J. Lithostratigraphy and structure of the temsamane unit (eastern external rif, morocco). Revista de la Sociedad Geológica d'España, 20 (3-4) (2007).

6. Leblanc D. Etude géologique du Rif externe oriental au Nord de Taza (Maroc). Notes et Mémoires, Service Géologique Marocain, Rabat, 281: 1-159 (1979).

7. Asebriy L., Azdimousa A., Bourgois J., Poupeau G. & Montigny R. Histoire thermique et surrection du Rif externe et des nappes de flyschs associées (Nord Maroc). Trav. Inst. Sci., Rabat, sér. Géologie & Géogr. phys., 21, 15-26 (2003).

8. Ihissou Y. Évaluation de l'impact de l'arboriculture fruitière en zones de montagnes sur la conservation des sols et des eaux: cas de la zone de Tahar-souk (Province de Taounate). PFE, ENA Meknès, 95p (2014).

9. DREF du Nord-Est: Etude d'aménagement du bassin versant d'Oued Beht en amont du barrage El Kansera, Direction Régionale des Eaux et Forêts, Fès (2007).

10. El Gaatib R., Larabi A. et Mohamed Faouzi1 M. Integrated elaboration of priority planning of vulnerable areas to soil erosion hazard using Remote Sensing and GIS techniques: A pilot case of the Oued Beht Watershed (Morocco), J. Mater. Environ. Sci. 6, 11, 3110-3127, ISSN: 2028-2508 CODEN: JMESCN (2015).

11. Sadiki A, S. Bouhlassa, J. Auajjar, A. Faleh and J. Macaire, Utilisation d'un SIG pour l'évaluation et la cartographie des risques d'érosion par l'Equation universelle des pertes en sol dans le Rif oriental (Maroc) : cas du bassin versant de l'oued Boussouab, Bulletin de l'Institut Scientifique, Rabat, section Sciences de la Terre, 26, 69-79, (2004).

12. Manrique L.A. Land Erodibility assessment methodology, LEAM, Using soil survey data based on soil taxonomy, University of Hawaii, Editorial and publication Shop, Honolulu, Hawaii USA, 28p (1988).

13. Ministère de l'Agriculture et de la Mise en valeur Agricole. Étude de reconnaissance au 1/100 000 en vue de la mise en valeur agricole dans les cercles de Taounate et Ghafsai, Province de Taounate (1994).

14. Loukili M., Bock L., Engels P., Mathieu L. Approche géomorpho-pédologique et Système d'Information Géographique (SIG) pour la gestion des terres au Maroc. EGS 7: 37- 52 (2000).

15. FAO (1976). Cadre pour l'évaluation des terres. Bulletin pédologique de la FAO, 32, Rome, 64 p.

16. FAO Guidelines for land evaluation for rainfed aqriculture. FAO soils bulletin, 52, Rome, 237p (1983).

17. Logiciel Google earth

18. (https://www.google.com/intl/fr/earth/).

Thermal characterization of a new effective building material based on clay and olive waste

Lamrani Mohamed[1], Khalfaoui Mohamed[1] , Laaroussi Najma[2] and Khabbazi Abdelhamid[2]
[1]Mohammed V University in Rabat, EST de Salé, LASTIMI, Salé, Maroc
[2]Mohammed V University in Rabat, EST de Salé, MEAT-EMDD, Salé, Maroc

Abstract. The influence of thermophysical properties of wall materials on energy performance and comfort in traditional building was investigated. The clay is the most commonly used sustainable building material. The study looked at the effects of the addition of pomace olive on the thermophysical properties of clay bricks to improve the energy efficiency of this ecological material. An experimental measurement of thermal properties of clay mixed with pomace olive was carried out by using the transient and steady state hot-plate and flash methods. The experimental methods are applied to measure the thermal properties of the composite material. The estimation of these thermal characteristics is based on a one dimensional model and the experimental errors are found less than 3%. The composite samples were prepared with different granular classes and mass fractions of the pomace olive in the mixture. The results show that the density of the new material was not substantially influenced by the size of the pomace olive. However, the thermal conductivity and diffusivity decrease from 0.65 $W.m^{-1}.K^{-1}$ and 4.21×10^{-7} $m^2.s^{-1}$ to 0.29 $W.m^{-1}.K^{-1}$ and $2.47\times10^{-7} m^2.s^{-1}$, respectively, according to the variation of the volume fraction of pomace olive from 0 (pure clay) to 71% showing that the olive pomace can be used as effective secondary raw materials in the making of clay bricks.
Key words: Ecological material, building efficiency, insulation, thermophysical characterization.

1 Introduction

Actuellement, la valorisation des matériaux locaux utilisés dans le secteur du bâtiment et l'économie d'énergie constituent un champ de recherche très dynamique au Maroc. Plusieurs études ont été menées sur des matériaux locaux tels que l'argile et le plâtre consolidés avec des produits de nature végétale ou animale. Cherki et al. [1] ont étudié les propriétés thermiques d'un matériau isolant à base du liège et du plâtre. L'effet des fibres d'alfa sur les caractéristiques thermophysiques de l'argile a été mis en évidence par Elhamdouni et al. [2]. Mounir et al. [3] ont réalisé une étude expérimentale des propriétés thermiques de l'argile consolidée avec du liège utilisant la méthode du plan chaud en régime permanent et transitoire et la méthode flash. L'efficacité énergétique d'un matériau composite à base d'argile et de paille a été montrée par El Azhary et al. [4]. Les propriétés thermiques des briques en terre cuite ont été mesurées en utilisant différentes techniques de métrologie thermique par Laaroussi et al. [5-6].

Notre étude expérimentale est basée sur la caractérisation d'un matériau composite réalisé par l'incorporation des résidus des grignons d'olive consolidés avec de l'argile qui constitue l'élément de base du pisé au sud marocain.

Le but de ce travail est consacré à mettre en évidence l'influence de la taille et des proportions occupées par les grignons dans le mélange sur les propriétés du matériau testé. L'identification de ces propriétés thermophysiques est faite à l'aide de la méthode du plan chaud asymétrique en régimes permanent et transitoire développée par Jannot et al. [7] pour la mesure de la conductivité et l'effusivité thermiques et la méthode Flash pour la mesure de la diffusivité thermique.

2 Matériaux utilisés

Argile : L'argile utilisée dans cette étude est extraite de la région d'Errachidia au sud marocain.

Grignon d'olive : Le grignon d'olive est le rejet d'une huilerie traditionnelle de la région. Il est séché pendant quelques jours à l'air libre. Les résultats d'analyse chimique sont présentés dans le tableau 1.

Tableau 1. Analyse chimique du grignon d'olive.

Cellulose	Hémi-cellulose	Lignine	Matière grasse	Cendres
34.25%	12.40%	23.36%	4.33%	3.01%

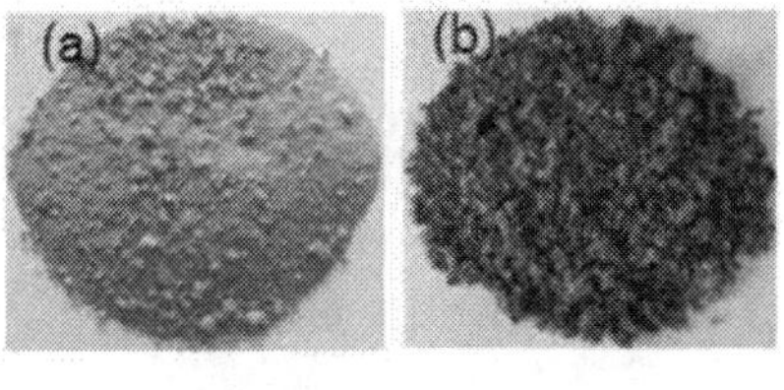

Fig. 1. (a) Argile, (b) Grignons d'olive.

3 Préparation des échantillons

La préparation des échantillons consiste à répartir les grignons d'olive en classes avec différentes tailles par un procédé de tamisage normalisé et à préparer la pâte d'argile avec un taux de gâchage massique w/g = 0.25. Les échantillons fabriqués ont des épaisseurs de 20 mm et sont élaborés selon deux approches différentes :

1- Des échantillons dont la variable est la taille des grignons d'olive. Trois échantillons correspondants à cinq classes avec différentes tailles (d_1-D_1=0.63-1.25mm, d_2-D_2=1.25-2.5 mm, d_3-D_3=2.5-5 mm, d_4-D_4=5-8 mm et d_5-D_5=8-12.5 mm) ont été élaborés dans des moules identiques dont les dimensions sont 100 × 100 × 20 mm^3. Le moule est entièrement rempli par un volume des grignons d'olive correspondant à une teneur X=100%, ensuite la pâte d'argile est ajoutée afin d'occuper l'espace inter-granulaire. La distribution des grains dans le mélange est assurée par une bétonnière pour homogénéiser le matériau composite.

2- La préparation d'échantillons dont la variable est la teneur, X, des grignons d'olive dans le mélange pour la classe d_2-D_2=1.25-2.5 mm. On rempli le volume apparent par des fractions volumiques des grignons décroissantes variantes de 100%, 80% , 60% , 40% , 20% à X=0% représentant un échantillon d'argile brute sans additifs afin de comparer la variation de ses propriétés thermiques avec ceux du matériau composite. La figure 2 représente l'aspect des échantillons fabriqués avec différentes tailles des grignons d'olive. Les échantillons ont été séchés à l'étuve pour éliminer l'humidité présente dans les pores. Ensuite, leurs masses ont été mesurées puis rangées dans des sacs en plastique.

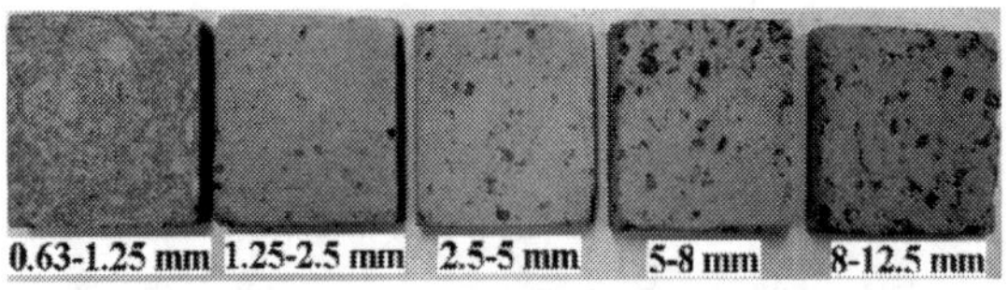

Fig. 2. Echantillons argile-grignons d'olive avec différentes tailles étudiées.

4 Approche théorique de mesure

4.1 Méthode du Plan chaud asymétrique

La méthode du plan chaud asymétrique en régime permanent permet d'identifier la conductivité thermique d'un matériau compris entre deux plans isothermes "Figure 3".

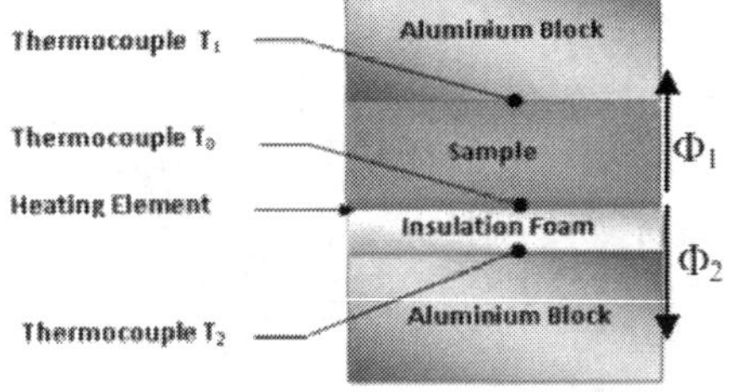

Fig. 3. Dispositif expérimental de la méthode Plan chaud asymétrique en régime permanent.

L'échantillon étudié est placé sur un élément chauffant, sous lequel est placée une mousse isolante en polyéthylène ayant les dimensions de 10×100×100 mm^3 et de conductivité 0,04 $W.m^{-1}.K^{-1}$, pour que la majorité du flux de chaleur, émis par l'élément chauffant, passe à travers l'échantillon. Cet ensemble est placé ensuite entre deux blocs d'aluminium de dimensions 50×100× 100 mm^3 qui ont une grande conductivité et jouent le rôle de plaques isothermes pour que le système atteint l'équilibre thermique le plus vite possible.

La mesure des températures est faite à l'aide des thermocouples mesurant T_0, T_1 et T_2 placés aux centres des faces de l'échantillon et de la mousse isolante. La conductivité thermique λ_1 de l'échantillon, une fois que le système ait atteint le régime permanent, est donnée par l'équation (1) :

$$\lambda_1 = \frac{e_1}{T_0 - T_1}\left[\frac{U^2}{R_\Omega . S} - \frac{\lambda_2}{e_2}(T_0 - T_2)\right] \quad (1)$$

Où e_1 est l'épaisseur de l'échantillon ; λ_2=0.04 $W.m^{-1}K^{-1}$ et e_2=10 *mm* sont successivement la conductivité thermique et l'épaisseur de la mousse isolante ; U est la tension imposée aux bornes de l'élément chauffant de résistance électrique R_Ω=40 Ω et de surface S=0.0105 m^2.

L'étalonnage du système expérimental a été réalisé à l'aide d'un échantillon de référence à conductivité connue.

4.2 Méthode Flash

La caractérisation de la diffusivité thermique est effectuée par la méthode Flash par résolution de l'équation de la chaleur dans l'espace de Laplace en utilisant le modèle quadripolaire [8].

Une impulsion thermique $\emptyset_0$ ($W.m^{-2}$) de courte durée a été envoyée sur la face supérieure de l'échantillon parfaitement isolé à la périphérie. Les pertes de chaleur des deux faces de l'échantillon (100×100×20 mm^3) sont représentées par le même coefficient d'échange convectif h. L'élévation de température est mesurée par un thermocouple placé au centre de la face non irradiée "Figure 4".

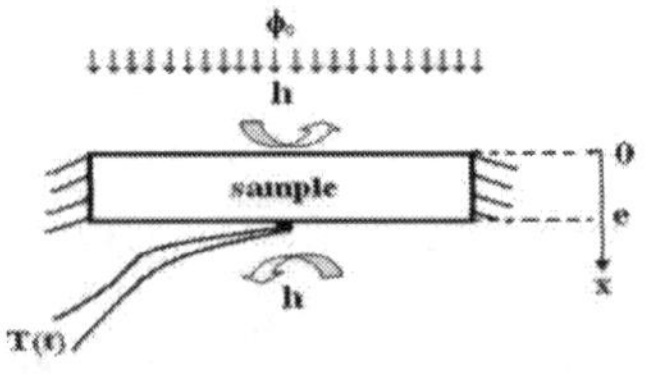

Fig. 4. Principe de la méthode Flash.

L'élévation de la température dans l'espace de Laplace est donnée par l'expression suivante :

$$\theta(p) = \frac{\frac{\emptyset_0 \, (1 - e^{-\tau . p})}{p}}{h^2 \frac{\sinh(ke)}{\lambda k} + 2h\cosh(ke) + \lambda k . \sinh(ke)} \quad (2)$$

τ est la durée de l'impulsion thermique, $= \sqrt{\frac{p}{a}}$, p est le paramètre de Laplace et a est la diffusivité thermique.

λ et e sont succéssivement la conductivité thermique et l'épaisseur de l'échantillon.

Après l'inversion numérique de cette expression par l'algorithme de De Hoog [9], on obtient l'expression théorique de la température. L'algorithme de Levenberg-Marquardt [10-11] est utilisé pour l'estimation des paramètres a, $\emptyset_0$ et h par minimisation de l'erreur quadratique entre le thermogramme expérimental représentant l'évolution de la température de la face arrière de l'échantillon et l'expression théorique de la température.

4.3 Méthode du Plan chaud asymétrique en régime transitoire

La méthode du plan chaud asymétrique en régime transitoire permet d'identifier l'effusivité thermique de l'échantillon "Figure 5". L'échantillon est posé sur un élément chauffant de même surface (100×100 mm^2), un thermocouple de type K est installé au centre de la face inférieure de l'élément chauffant. Le dispositif est placé entre deux blocs de polystyrène ayant chacun une épaisseur de 50 mm. L'ensemble est ensuite placé entre deux blocs d'aluminium d'épaisseur 40 mm chacun. Un échelon de flux de chaleur est émis par l'élément chauffant et l'élévation de température T(t) au centre de l'échantillon est enregistrée.

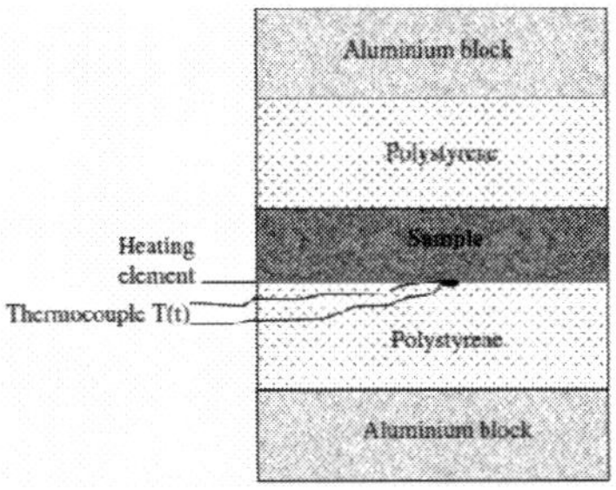

Fig. 5. Dispositif expérimental de la méthode Plan chaud asymétrique en régime transitoire

Le système est modélisé avec l'hypothèse d'un transfert de chaleur unidirectionnel (1D) au centre de l'échantillon. Compte tenu de la faible valeur du flux de chaleur atteignant les blocs d'aluminium à travers le polystyrène et leurs grandes capacités thermiques, leurs températures sont supposées égales et constantes. En appliquant le formalisme des quadripôles sur le modèle représenté par la figure 3, on peut écrire :

$$\begin{bmatrix} \theta \\ \Phi_{01} \end{bmatrix} = \begin{bmatrix} 1 & 0 \\ C_h & 1 \end{bmatrix} \begin{bmatrix} 1 & R_c \\ 0 & 1 \end{bmatrix} \begin{bmatrix} A & B \\ C & D \end{bmatrix} \begin{bmatrix} A_i & B_i \\ C_i & D_i \end{bmatrix} \begin{bmatrix} 0 \\ \Phi_1 \end{bmatrix} = \begin{bmatrix} A_1 & B_1 \\ C_1 & D_1 \end{bmatrix} \begin{bmatrix} 0 \\ \Phi_1 \end{bmatrix} \quad (3)$$

$$\begin{bmatrix} \theta \\ \Phi_{02} \end{bmatrix} = \begin{bmatrix} A_i & B_i \\ C_i & D_i \end{bmatrix} \begin{bmatrix} 0 \\ \Phi_2 \end{bmatrix} \quad (4)$$

$$\Phi_0 = \frac{\emptyset}{p} = \Phi_{01} + \Phi_{02} \quad (5)$$

θ est la transformée de Laplace de l'élévation de température T(t) ; Φ_{01} est la transformée de Laplace de la densité de flux de chaleur dissipée par l'élément chauffant vers l'échantillon (vers le haut) ; Φ_{02} est la transformée de Laplace de la densité de flux de chaleur dissipée par l'élément chauffant vers le polystyrène (vers le bas) ; Φ_0 est la transformée de Laplace de la densité de flux de chaleur totale produite par l'élément chauffant ; Ø est la densité de flux de chaleur totale produite par l'élément chauffant et C_h sa capacité thermique ($C_h = \rho_h c_h e_h$) ; Φ_1 est la transformée de Laplace de la densité de flux de chaleur qui atteint le bloc d'aluminium du haut ; Φ_2 est la transformée de Laplace de la densité de flux de chaleur qui atteint le bloc d'aluminium du bas.

avec :

$$A = D = cosh\left(\frac{\rho c}{E} e\sqrt{p}\right);\; B = \frac{sinh\left(\frac{\rho c}{E} e\sqrt{p}\right)}{E\sqrt{p}};$$

$$C = E\sqrt{p}\, sinh\left(\frac{\rho c}{E} e\sqrt{p}\right) \quad (6)$$

$$A_i = D_i = cosh\left(\sqrt{\frac{p}{a_i}} e_i\right);\; B_i = \frac{sinh\left(\sqrt{\frac{p}{a_i}} e_i\right)}{\lambda_i \sqrt{\frac{p}{a_i}}};$$

$$C_i = \lambda_i \sqrt{\frac{p}{a_i}}\, sinh\left(\sqrt{\frac{p}{a_i}} e_i\right) \quad (7)$$

E est l'effusivité thermique de l'échantillon $E = \sqrt{\lambda \rho c}$; ρc est la capacité thermique volumique de l'échantillon ; e est l'épaisseur de l'échantillon ; λ_i, a_i et e_i sont respectivement la conductivité thermique, la diffusivité thermique et l'épaisseur du polystyrène.

En combinant ces cinq dernières équations, on en déduit la valeur de θ par la relation :

$$\theta(p) = \frac{\Phi_0(p)}{\frac{D_1}{B_1} + \frac{D_i}{B_i}} \quad (8)$$

Le principe de cette méthode est d'estimer les valeurs des paramètres (E, ρc, R_c, C_h) en minimisant l'erreur quadratique $\Psi = \sum_{j=0}^{N} [T_{exp}(t_j) - T_{mod}(t_j)]^2$ entre les points expérimentaux et les valeurs théoriques calculées par la transformée de Laplace inverse de la relation (8) en utilisant l'algorithme de Levenbeg-Marquart [10-11]. La transformée de Laplace inverse de la relation (8) est réalisée en utilisant l'algorithme de De Hoog [9].

5 Résultats et discussions

Pour chaque échantillon, trois tests ont été réalisés pour tenir compte de l'erreur de mesure et pour que les résultats de caractérisation obtenus soit statistiquement validés. La moyenne arithmétique de ces trois essais est adoptée comme résultat final de caractérisation pour chaque échantillon testé.

5.1 Masse volumique des échantillons

Connaissant les dimensions et les masses sèches des échantillons, la masse volumique de chaque échantillon a été calculée "Tableau 2".

La loi des mélanges pour un milieu à deux composants a été utilisée pour déduire la fraction volumique, notée *y*, des grignons dans le mélange.

$$\rho_{Go-arg} = y.\rho_{Go} + (1-y)\rho_{arg} \quad (9)$$

avec: $\rho_{Go\text{-}arg}$, ρ_{arg} et ρ_{Go} sont respectivement les masses volumiques apparentes du matériau composite, de l'argile et des grignons d'olive.

Tableau 2. Masse volumique apparente moyenne et fraction volumique des grignons d'olive des échantillons fabriqués.

N°	Echantillon	X (%)	ρ_{moy} (kg.m^{-3})	$\frac{\Delta\rho}{\rho_{moy}}$ %)	y
1	Go-ag 0.63-1.25mm	100	1138.55	0,088	0.502
2	Go-ag 1.25-2.5mm		1142.42	0.121	0.499
3	Go-arg 2.5-5 mm		1139.14	0.079	0.502
4	Go-arg 5-8 mm		1141.55	0.102	0.500
5	Go-arg 8-12.5 mm		1140.19	0.192	0.501
6	Go-arg 1.25-2.5mm	80	1239.17	0.206	0.423
7	Go-arg 1.25-2.5mm	60	1398.3	0.219	0.298
8	Go-arg 1.25-2.5mm	40	1553.97	0.148	0.176
9	Go-arg 1.25-2.5mm	20	1642.59	0.147	0.107
10	Argile	0	1778.68	0.099	------
	Grignons d'olive		504	------	------

Nous constatons que la masse volumique apparente ne varie pratiquement pas en fonction de la taille. Cependant, plus la teneur en grignons augmente dans mélange plus le composite devient de plus en plus léger.

5.1 Conductivité thermique

La conductivité thermique des échantillons mesurée par la méthode du Plan chaud asymétrique en régime permanent est représentée dans le tableau 3, l'erreur moyenne issue de l'expérimentation, calculée par la formule $\frac{\Delta\lambda}{\lambda_{moy}}$, ne dépasse pas 3%.

Tableau 3. Résultats de caractérisation issus des trois méthodes expérimentales utilisées

N°	Echantillon Go-arg	X (%)	λ (W.m^{-1}.K^{-1})	a×10^{-7} (m^2.s^{-1})	E (W.s$^{1/2}$.m^{-2}.K^{-1})
1	0.63-1.25mm	100	0.298	2.47	595.56
2	1.25-2.5mm		0.296	2.49	594.59
3	2.5-5 mm		0.297	2.48	594.88
4	5-8 mm		0.296	2.49	594.42
5	8-12.5 mm		0.298	2.50	596.10
6	1.25-2.5mm	80	0.351	2.91	653.85
7	1.25-2.5mm	60	0.401	3.15	715.28
8	1.25-2.5mm	40	0.463	3.31	805.58
9	1.25-2.5mm	20	0.553	3.75	904.07
10	Argile	0	0.651	4.21	1001.12
	Erreur maximale		2.37 %	2.41%	0.86 %

λ : *Conductivité thermique par la méthode du Plan chaud asymétrique en régime permanent.*
***a**: Diffusivité thermique par la méthode Flash.*
***E**: Effusivité thermique par la méthode Plan chaud asymétrique en régime transitoire.*

Nous constatons l'influence de l'augmentation de la proportion volumique des grignons d'olive sur la conductivité thermique du matériau, ce qui montre que plus le matériau est léger plus sa qualité d'isolation thermique est améliorée.

5.2 Diffusivité thermique

5.2.1 Résultats de mesure

La figure 6 illustre le résultat de minimisation des deux thermogrammes expérimental et théorique avec leurs résidus pour l'échantillon de classe 1.25-2.5mm.

La courbe théorique reste en très bon accord avec les points expérimentaux jusqu'au t=1400s, ceci montre la validité du modèle 1D au centre de l'échantillon. Cette constatation est validée par l'observation des résidus qui restent faibles et plats le long de l'expérience.

Le tableau 3 regroupe aussi les résultats de la diffusivité thermique moyenne obtenus pour chaque échantillon. L'erreur de mesure moyenne ($\frac{\Delta a}{a_{moy}}$) correspondant à la diffusivité ne dépasse pas 3%. Cela montre que le dispositif expérimental utilisé donne une bonne reproductibilité des mesures.

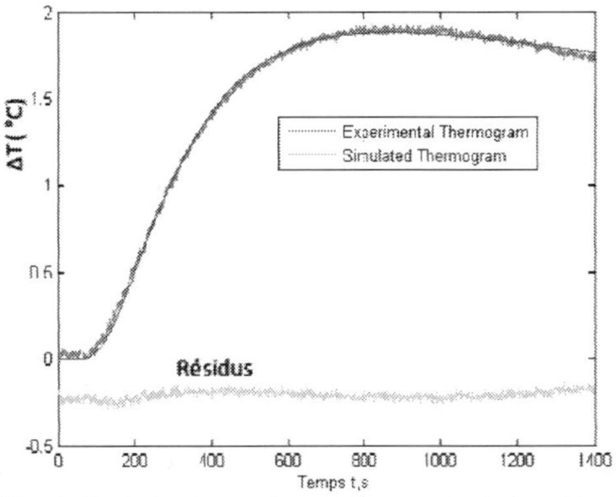

Fig. 6. Minimisation par le modèle complet de la méthode Flash pour l'échantillon de classe 1.25-2.5 mm

5.2.2 Analyse de la sensibilité

La figure 7 montre l'allure des courbes des sensibilités réduites $\left(a\frac{\partial T}{\partial a}\,;\,\emptyset_0\frac{\partial T}{\partial \emptyset_0};\,h\frac{\partial T}{\partial h}\right)$ relatives aux trois paramètres d'identification. La sensibilité d'un paramètre est le degré d'influence qui nous informe sur la fiabilité des résultats obtenus.

L'analyse des courbes des sensibilités pour les trois paramètres (a, $\emptyset_0, h$) à estimer par la méthode Flash montre que la diffusivité peut être identifiée à t=300s car sa sensibilité est élevée. Pour h et $\emptyset_0$, ils sont estimés à des temps plus longs. Le calcul des sensibilités réduites pour chaque paramètre a été programmé sous MATLAB.

L'analyse du thermogramme réalisé sur cet exemple est similaire aux thermogrammes trouvés pour les autres séries d'échantillons.

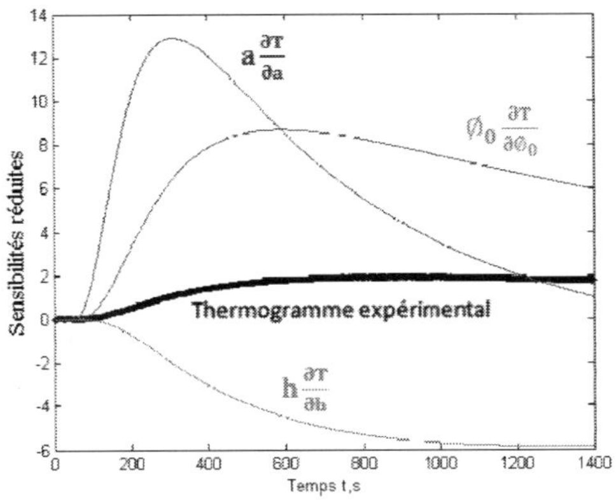

Fig. 7. Courbes des sensibilités réduites des trois paramètres à estimer de l'échantillon de classe 1.25-2.5 mm

5.3 Effusivité thermique

La méthode du plan chaud asymétrique transitoire a été appliquée aux échantillons fabriqués du composite argile-grignons d'olive. Pour chaque thermogramme obtenu, l'algorithme de Levenberg-Marquart est appliqué pour identifier les paramètres (E, ρc, Rc, C_h) en minimisant l'écart quadratique entre la courbe expérimentale et la courbe théorique simulée.

La figure 8 montre les courbes expérimentale et simulée après minimisation, avec les résidus correspondants, pour l'un des échantillons de classe granulaire 1.25-2.5mm.

La figure 9 illustre le tracé des courbes des sensibilités réduites ($E\frac{\partial T}{\partial E}$; $(\rho c)\frac{\partial T}{\partial(\rho c)}$; $Rc\frac{\partial T}{\partial Rc}$ et $C_h\frac{\partial T}{\partial C_h}$) relatives aux paramètres d'identification.

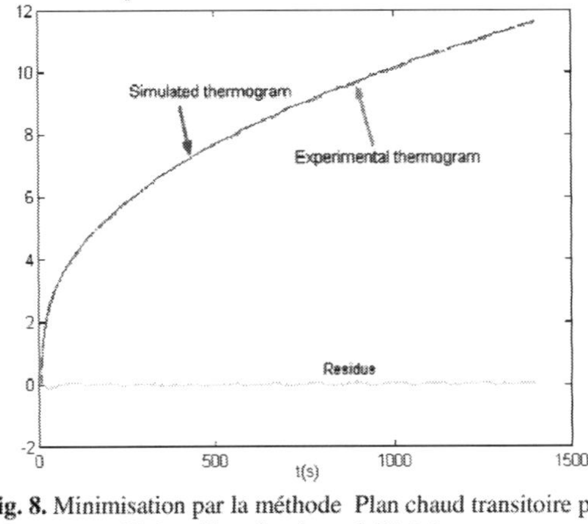

Fig. 8. Minimisation par la méthode Plan chaud transitoire pour l'échantillon de classe 1.25-2.5 mm

D'après la figure 8, on peut constater que le modèle théorique simulé est en très bon accord avec la courbe expérimental et que le modèle 1D reste valable jusqu'au t=1400s. Les paramètres E et ρc à estimer donnent des valeurs fiables puisque leurs sensibilités sont élevées.

Le Tableau 3 regroupe les résultats d'identification de l'effusivité thermique des échantillons par la méthode du plan chaud transitoire avec une erreur moyenne $\frac{\Delta E}{E_{moy}}$ qui ne dépasse pas 1%.

On peut noter que l'éffusivité ne varie pratiquement pas en fonction de la taille des grignons d'olive et décroit significativement en fonction de la teneur en grignons.

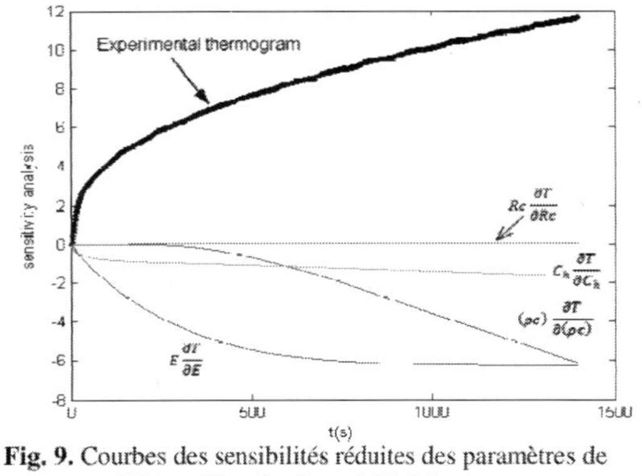

Fig. 9. Courbes des sensibilités réduites des paramètres de l'échantillon de classe 1.25-2.5 mm

Les erreurs de mesure relevées ne dépassent pas 1% pour toutes les mesures effectuées, ce qui confirme la fiabilité et la robustesse de la méthode adoptée pour la caractérisation du matériau étudié.

6 Intérêt pratique du composite étudié

On considère deux murs composés par le composite argile-grignons d'olive et par l'argile brute sans grignons avec la même épaisseur et soumis au même gradient de température en régime permanent ce qui nous a permis de déduire le ratio des deux flux de chaleur traversant ces deux murs en fonction de leur conductivité thermique :

$$\frac{\varnothing_{Go-arg}}{\varnothing_{arg}} = \frac{\lambda_{Go-arg}}{\lambda_{arg}} \quad (10)$$

En se basant sur les résultats expérimentaux obtenus, on peut calculer le facteur de légèreté et le gain énergétique direct par transmission de chaleur:

$$\text{Gain\_énergétique} = 100 \times \left(1 - \frac{\varnothing_{Go-arg}}{\varnothing_{arg}}\right) \quad (11)$$

$$\text{Le facteur de légèreté} = 100 \times \left(1 - \frac{\rho_{Go-arg}}{\rho_{arg}}\right) \quad (12)$$

Tableau 4. Gain énergétique et facteur de légèreté pour les différents échantillons par rapport à l'argile brute sans grignons.

N°	Echantillon	X(%)	Gain énergétique (%)	Facteur de légèreté (%)
1	Go-arg 0.63-1.25mm	100	54.22	35.99
2	Go-arg 1.25-2.5mm		54.53	35.77
3	Go-arg 2.5-5 mm		54.38	35.96
4	Go-arg 5-8 mm		54.53	35.82
5	Go-arg 8-12.5 mm		54.22	35.90
6	Go-arg 1.25-2.5mm	80	46.08	30.33
7	Go-arg 1.25-2.5mm	60	38.40	21.39
8	Go-arg 1.25-2.5mm	40	28.88	12.63
9	Go-arg 1.25-2.5mm	20	15.05	7.65

Le tableau 4 regroupe les résultats obtenus de ces deux paramètres pour les différents échantillons par rapport à l'argile brute sans grignons.

Ces résultats montrent que l'incorporation des grignons d'olive dans l'argile permet d'avoir un matériau qui nous permet d'avoir un gain énergétique de 54% environ avec une légèreté de 36% que l'argile brute, ce qui améliorera le confort intérieur des locaux et réduira la facture énergétique de conditionnement d'air et contribuera par conséquent à l'amélioration de l'efficacité énergétique dans le bâtiment traditionnel en pisé.

7 Conclusions

S'agissant d'un matériau local, naturel, abondant et à faible coût, le grignon d'olive améliore la qualité d'isolation thermique de l'argile et offre une meilleure efficacité énergétique. Le composite ainsi développé peut constituer une alternative intéressante aux matériaux traditionnellement utilisés en construction dans les régions rurales du sud marocain.

Les propriétés thermophysiques du matériau composite sont nettement améliorées par l'ajout progressif des additifs proposés. Ainsi, la teneur en grignons influe positivement sur les performances énergétiques de l'argile en offrant une légèreté de 36% et un gain énergétique de 54% environ.

Cependant, les propriétés thermiques du matériau ne constituent pas le seul critère de choix pour la construction, il faut également tenir compte de ses propriétés mécaniques qui constituent un paramètre important pour la stabilité du bâtiment.

Références

1. A. Cherki, B. Remy, A. Khabbazi, Y. Jannot, B. Baillis, *Experimental thermal properties characterization of insulating cork–gypsum composite, Constr.Build. Mater.* (2014); 54: 202–209.
2. Y. Elhamdouni, A. Khabbazi, C. Benayad, A. Dadi, O. Idriss Ahmid. *Effect of fiber alfa on thermophysical characteristics of a material based on clay. Energy Procedia* 74 (2015) 718 – 727
3. S. Mounir, Y. Maaloufa, A. Cherki, A. Khabbazi, *Thermal properties of the composite material clay/granular cork, Constr. Build. Mater.*(2014); 70 :183–190.
4. K. El Azhary, Y. Chihab, N. Laaroussi, M. Garoum, M. Mansour. *Energy efficiency and thermal properties of the composite material clay-straw. Energy Procedia* 00 (2017) 000–000
5. N. Laaroussi, G. Lauriat, M. Garoum, A. Cherki, Y. Jannot, *Measurement of thermal properties of brick materials based on clay mixtures, Constr. Build.Mater.*(2014); 70:351–361.
6. N. Laaroussi, A. Cherki, M. Garoum, A. Khabbazi, A. Feiz, *Thermal properties of a sample prepared using mixtures of clay bricks, Energy Proc.*(2013); 42:337–346.
7. Y. Jannot, V. Felix, A. Degiovanni, *A centred Hot Plate method for measurement of thermal properties of thin insulating materials, Meas.Sci.Technol.* (2010); 21 : 035106.
8. D. Maillet, S. André, J.C.Batsale, A.Degiovanni, C. Moyne. *Thermal quadrupoles: solving the heat equation through integral transforms. Chichester (PA): John Wiley & Sons Ltd*; (2000)
9. F.R. De Hoog, *An improved method for numerical inversion of Laplace transforms, Society for Industrial and Applied Mathematics*, (1982); 3(3): 357–66.
10. D.W. Marquardt, *An Algorithm for Least-Squares Estimation of Nonlinear Parameters*, J. Soc. Ind. Appl. Math. 11 (1963) 431–441. doi:10.1137/0111030.
11. K. Levenberg, *A method for the solution of certain nonlinear problems in least squares.* Q. Appl. Math. 2164-8, (1944).

Bioaccumulation of heavy metals by the flora and benthic macrofauna of the Bouregreg estuary wetland

M. Khamar, E. Cherkaoui et A. Nounah

Laboratory of Civil Engineering and Environment (LGCE),
Higher School of Technology, SALE, Mohammed V University in Rabat.

e-mail:m_khamar@yahoo.fr

Abstract: Estuarine systems account for a high proportion of wetlands in Morocco due to the development of the river system. These estuarine and coastal Moroccan environments are as rich in fauna and flora as their European equivalents and present much originality. However, these coastal areas are generally highly urbanized and industrialized, and therefore subject to domestic and industrial discharges. The Bouregreg Estuary represents a good example of this situation, in view of this pollution (organic, chemical and biological) and the absence of fresh water supply upstream of the estuary after the dam was built, is increasingly unable of restoring the equilibrium of this ecosystem. This tide can instead ensure the polluting discharge downstream or upstream. Metallic contamination affects the different compartments of the estuarine ecosystem (sediment, water) and benthic species (fauna and flora). The objective of this study is to assess the accumulation and bioaccumulation of five heavy metals (Cu, Fe, Zn, Cr and Pb) at the sediment level of two dominant plant species (Artiplex portulacoides and Sarcocornia fructicosa) and three species of benthic macrofauna (Scrobicularia plana, Venerupis decussata and Hediste diversicolor) from the wetland of this estuary. This work will enable us to clearly understand the nutritional relationships between plant, animal, water and sediment species. The results of the analysis revealed that bioaccumulation varies from one species to another and from one metal to another. Thus, the levels of Lead and Chromium at Sarcocornia fructicosa are higher than those found at Artiplex portulacoides. Nevertheless, the latter accumulates better the other metals: Cu, Zn and Fe. While, the macrofauna shows a fairly large variation depending on the life style and sensitivity of species. Thus, Hediste diversicolor showed high levels of the various metals analyzed compared to Scrobicularia plana and Venerupis decussata.

Key words: Estuary of Bouregreg, Heavy metals, Bioaccumulation, flora, macrofauna, water, sediment

1 INTRODUCTION

The estuaries are environments of great ecological and economic importance. Urban and industrial pressure, as well as the development of estuaries, leads to an increase in their pollution [9]. On the Atlantic coast of Morocco, the estuary of Bouregreg is an edifying case of this situation. Indeed this estuary drains in its passage the polluted rejections of many industrial units and the untreated domestic rejections of some districts of the two agglomerations Rabat and Salé which do not cease developing in its surroundings.

Anthropogenic activities are the source of many pollutants disseminated in the environment, in particular heavy metals. The latter result either from discharges directly into ecosystems or from indirect flows such as industrial discharges, landfill leachate, agricultural run-off . The estuary of Bouregreg does not escape this scourge. Several studies have revealed the contamination of this estuary in these different compartments ([15, 20]).

Among the main pollutants generated by industrial activities are heavy metals (Cu, Pb, Cr, etc.) which formally pose particular problems of concern. Thus the presence of a fraction of heavy metals in the aquatic environment is reversibly associated with superficial sediments. These elements, by nature non-biodegradable, accumulate in the sediment and present a high toxicity to the fauna and flora of this ecosystem [14]. This bioaccumulation of toxic metals will cause neffffects in biological chains is one of the most pernicious effects of marine pollution.
The objective of this study is to assess the accumulation and bioaccumulation of these five heavy metals (Cu, Fe, Zn, Cr and Pb) in the sediment and two dominant plant species (Artiplex portulacoides and Sarcocornia fructicosa) And three animal species belong to the zoobenthic groups (Scrobicularia plana, Venerupis decussata and Hediste diversicolor) of the wetland of the downstream part of this Bouregreg estuary. This work will allow us to clearly identify the nutritional relationships between the living part (plant and animal species) on the one hand and their biotopes (water and sediment) on the other hand. In addition to the role of these benthic species as bioindicators of the aquatic ecosystem, this makes it possible to perceive their impacts on consumer health (man).

2 MATERIALS AND METHODS

The estuary of Bouregreg is located on the Moroccan Atlantic between the two cities Rabat and Salé at 34 ° north and 6 ° 50 'west. It has a length of 23 km, limited upstream by the Sidi Mohammed Ben Abdellah dam and an average width of 150 m (Fig1). This estuarine ecosystem has several original features linked, on the one hand, to its disturbed state by anthropogenic action (increase in urban effluents and reduction of terrigenous and freshwater effluents in the watershed) and, on the other hand, to its Natural situation, isolated on a long straight coast, beaten by the great swells of the Atlantic. Indeed, this estuary has a natural vegetation cover composed of indigenous species specific to wetlands and terrestrial (eucalyptus, ..). After the development of the Bourgegreg estuary with the closure and rehabilitation of the Oulja landfill, the construction of the Bettana collector and the construction of the banks for the construction of the New Hassan II bridge and the My bridge. Youssef in addition to the closuture of the Akreuch dump. The wetland has suffered considerable degradation by limiting its area (construction of the marina on the right bank of the river and the development of the banks). Although these species have adapted very well to these changes. This leaves a greater value for diversity and species richness. In addition to this wetland estuary provides a refuge for many species of fish, birds, amphibians, mammals and reptiles and in addition a fairly high diversity of benthic macrofauna [5].

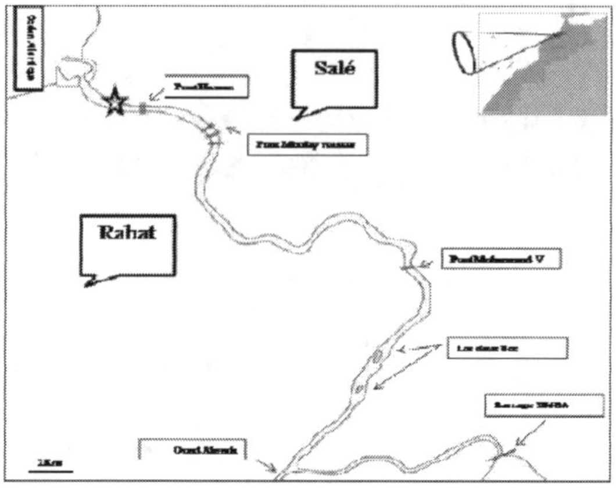

Fig.1: Location of the study area (★).

The study of the impact of discharges of wastewater and waste at the estuary level is approached by the accumulation of five heavy metals: Cu, Zn, Pb, Cr, Fe in two plant species (*Artiplex portulacoides and Sarcocornia fructicosa*) And three animal species belong to the zoobenthic groups (*Scrobicularia plana, Venerupis decussata* and *Hediste diversicolor*) of the Bouregreg estuary. This choice has been dictated by the fact that these metals are commonly used in industrial units and also because they are characterized by high toxicity to humans and aquatic organisms.

To carry out this work, sampling of plant species was done by hand. Once in the laboratory, we dry the harvested plants. This is an indispensable operation for the determination of species. We introduce the plants carefully, flat in a newspaper that will have to be changed every day. The determination is made at the Rabat Institute of Science in the Botanical Laboratory. For the analysis of heavy metals, both species are rinsed with tap water and then with distilled water. They are dried at 60 ° C. to constant weight and then crushed.
For the macrofauna, sampling was carried out in the slikkes at low tide by the flat drill at a depth of 15 to 20 cm and then sieved on a sieve with a diameter of 1 mm of opening of the mesh. The sampled species (*Scrobicularia plana, Venerupis decussata and Hediste diversicolor*) are washed under tap water and then rinsed with distilled water. The pulp of ten individuals of each species collected separately in boxes, dried in an oven at 60 ° C to constant weight. It is then finely ground. Water and sediments are taken from the same area of collection of animal and plant species.

The five metals were assayed at the UATRS at the National Center for Scientific and Technical Research (CNRST).

3 RESULTS AND INTERPRETATIONS

3.1 Water and Sediment

According to the classification of Chassé & Glémarec (1976), the sediments of the study area are dominated by fine sands (74.745%) and pelites (17.16%), the other coarse sand and gravel fractions do not exceed Not ??? the 10%, coarse sand and gravel represent 4.75% and 3.15% respectively (Table 1).

Table 1: Formation of the sediment of the study area.

The sedimentary fraction	The percentage
Pelites(<63µm),	17,165%
Fine sand (63-500µm)	74,745%
Coarse sands (>500µm),	4,575%
gravels	3,515%

The sediment of this area of the estuary is well classified with the Trask index (1.58). The median is 200 µm, which is justified by a fine characteristic [14]. On the other hand, the loss on fire revealed that the organic matter in the sediments oscillates between 3% and 3.5%.
Analysis of the five metallic elements (Cr, Cu, Zn, Pb and Fe) in the waters of the Bouregreg estuary showed that, with the exception of iron, concentrations of other elements (Cr, cu, Zn and Pb) remain below the unit. The chromium remains below the limit of detection, whereas the iron constituting the soil is at a concentration of 0.8 g / l. The comparison of its contents with the codex standards shows that Fe exceeds the norm by 0.48 times, and 0.05 times for Cu, and 0.1 times for Pb (Table 2).

Table 2 : Heavy metal content in the waters, sediments of the study area of the Bouregreg estuary.

Compartiment	Cr	Cu	Pb	Zn	Fe
Water of the estuary (mg / l)	<0,04	0,02	0,715	0,01	800
CODEX Standards		0,400	0,100		1,500
sediment (mg/kg)	84,3	16,86	146,12	59,01	16500

While concentrations of these metals are appreciable in sediments. The concentration varies from metal to metal. The average tensides of metallic elements ranging from 16.86 mg / kg for copper to 146.12 mg / kg for lead and 59.01 mg / kg. With the exception of iron with an average grade of 16,500 g / kg, this is related to the geological nature of the region (iron-rich land); Nevertheless, these elements (Cr, Cu and Pb), which are found in quantities less than one in water, show significant sediment concentrations. This highlights the importance of the sediment compartment of this estuarine ecosystem in the accumulation of these elements of the aquatic environment. Although the lead content in the Bouregreg estuary is comparable to that observed in the same estuary before the development project ([3], [21]) and lower than in the estuary of Loukkos and Oum Errbia [11] [18].

3.2 The Macrofauna

Aquatic environments are colonized by animal and plant populations, whose structure under normal conditions responds to a certain balance. The bioaccumulation of metallic elements is linked to the metabolism of the species to ensure its growth and maintains its ionic balance.
For the three species studied (*Hediste diversicolor Scrobicularia plana and Venerupis decussata*) the accumulation of different metallic elements from one species to another according to the element.

Venerupis decussata:
The association of Fe in Venerupis decussata is very high, followed by Zn, Pb, Cu and Cr. However, it remains the weakest in comparison with the other two species (Hediste diversicolor, Scrobicularia plana). From the comparison of the heavy metal values at this level with the codex standards, we notice an increase of almost 1067 times for The Fe, 33 times for Cu and 262 times for the Pb. These high concentrations can be judged by the existence of industrial discharges near the estuary, and it can be said that this species is unfit for human consumption. (Fig. 2)

Hediste diversicolor:
The concentration of Fe in H. diversitor is very high, followed by Zn, Pb, Cu and Cr.
Based on the comparison of the values of heavy metals in this species with Codex standards, we note an increase of 2600 times for Fe, 332 times for Pb, and 55 times for Cu. These high concentrations are not only due to the existence of industrial discharges near the estuary, but also to the nature of the species. It can be said that the species is not suitable for human consumption (Fig 2)

Scrobicularia plana:
The iron concentration in Scrobicularia plana is also very high, followed by Zn, Pb, Cu, and Cr. However, it remains weaker than that of Hediste diversicolor. When we compare these values with those of Codex, we notice

an increase of 1800 times for Fe, 281 times for Pb and 59 times for Cu. With a Zn content which is the highest compared to the two other species. This species is also unsuitable for human consumption. (Fig. 2)

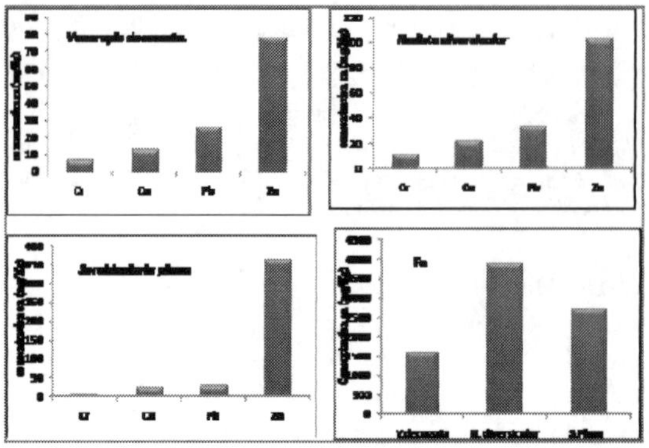

Figure 2: Variations of the Zn, Pb, Cu, Cr and Fe contents in the benthic macrofauna Bouregreg estuary

The rate of bioaccumulation varies greatly from one species to another. It is also observed that the concentration of Fe within Hediste is much higher than that found in the other two species, with a multiplication factor of 2.44 with respect to the clam, and 1.44 with respect to Scrobicularia plana. The concentration of iron, a major element of the parent rock, remains very high in H. diversicolor (3.9 g / kg). This accumulation in the three species results from the bioavailability of the metals, in parallel with the role that the physico-chemical parameters of the pH, salinity and temperature may play in addition to the physiological processes specific to the species. However, Scrobicularia plana showed strong accumulations in zn with a multiplication factor of 4.61 with respect to the clam, and 3.55 with respect to Hediste diversicolor. This result remains comparable to that found in the same species by Kaimoussi et al.1998 [15], and remains less than that noted in Chamelea gallina at the mouth of the Oued Moulouya [1] (Fig. 2). While, for lead, chromium and copper, the three species show no significant differences in accumulation.

However, the levels of chromium and lead remain lower than those recorded in the sediments of the three species. While zinc is found at high concentrations in all three species compared to concentrations analyzed in sediment. Similarly for copper in S.plana and H. diversicolor (Table 3). This could be related to the mode of nutrition of these species. In fact, these molluscs are filtering species. While polychaetes are psamnivores (eat sand and assimilate nutrient particles), these polychaetes can consume highly degraded halophytic plants (either the animal prospects around its burrow without leaving it completely, or it emits a net of mucus trapping particles Out of his gallery).

Table 3: Comparison of heavy metal content in animal species and WHO standards and Codex standards.

Compartiment	Cr	Cu	Pb	Zn	Fe
V.decussata (mg/kg)	6,55	13,1	26,21	78,62	1600
S.plana (mg/kg)	6,47	23,74	28,05	362,55	2700
H. diversicolor (mg/kg)	11,06	22,11	33,17	102,27	3900
Sediment (mg/kg)	84,3	16,86	146,12	59,01	16500
S.marginatus(mg/kg)	1.44	12.5	4.66	63.01	662.5
M.galloprovincialis (mg/kg)	1.93	12.4	8.31	232.7	416.7
Standards WOH 1982	-	10	5	100	-
Standards CODEX		0,40	0,100		1,50

The concentration of the metallic elements (Cu, Pb and Fe) for the three species exceeds the standards allowed by WHO 1982 with the exception of Zn in *V. decussata* [29], and those of Codex with reports in the order of 59 , 35 times (*S. plana*) and 32.75 (*V.decussata*) and 55 times for *H.diversicolor* for copper. This variation in concentrations may be related to the lifestyle of the three species. Indeed *V. decussata* life buried to a few centimeters (maximum 15cm) in the substrate to the infralittoral stage. It enjoys varied substrates of sand, small muddy gravel and mud, while *Scrobicularia plana* species of the mediolateral is present up to a few meters deep while *Hediste diversicolor* is a maple tree species that lives in a U-shaped gallery or Of Y in muddy or sandy muddy sediments. This habitat situation will influence their diets (water and sediment) and consequently the concentration of the different heavy metals. This is evident when comparing these concentrations with those found in *M. galloprovincialis*, which is fixed to the rocks of the mouth of the estuary of Bouregreg, and which replenishes only water, and shows concentrations much lower than those Recorded in our species (Table 3) [6].

3.3 Plant species

The Bouregreg Estuary revealed the presence of 14 species in 9 families, most of which are located in the wetland area of the marina. In this zone the family of CHENOBODIACEAE remains dominant by two species *Artiplex portulacoides* and *Sarcocornia fructicos.*) The presence of these species results from biotope that favors their development. Indeed, for *Artiplex portulacoides* is the typical plant of the highest areas of salt areas, out of reach of regular floods. However, it can withstand immersion for a short time; While Sarcocornia fructicosa prefers well drained, sandy and saline soils. It easily tolerates spray and drought. For good growth, it must be exposed to the sun, to the partial shade [12].

Analysis of the five metallic elements (Cr, Cu, zn, Pb and Fe) in the two plant species shows that: iron has two relatively high values in both species (200 mg / Kg for *Atiplex portulacoides* and 300 mg / Kg For *Sarcocornia fructicosa*). For zinc and copper the highest values are present in *Artiplex portulacoides*. As regards chromium and lead, they are characterized by a significant decrease compared to other metals in both species, especially chromium. (Figure 3). The comparison between these two species shows that *Artiplex portulacoides* accumulates more copper, zinc, with contents of the order of 129.38 and 98.44 mg / kg respectively, while *Sarcocornia fructicosa* accumulates more chromium of lead and iron, with Values of 12.75; 35.05 and 300 mg / kg respectively. This deference of accumulation between the two species could be related to the physiology and morphology of each species and mainly their root system and the shape and leaf area.

These values are lower than those recorded in sewage irrigated bean (Cu 1.775 g / kg, Zn 1.4 g / kg [12] and in the tomato stem 17.7 g / kg of copper.17 However, the concentration Of Zn is less than that of lettuce (194 mg / kg). [19] However, the high concentrations of these two metals (Cu and Zn) in addition to iron are due to the metabolic requirements of the plant. These elements are part of the composition of certain enzymes or serve as cofactors for the development of the plant [28].

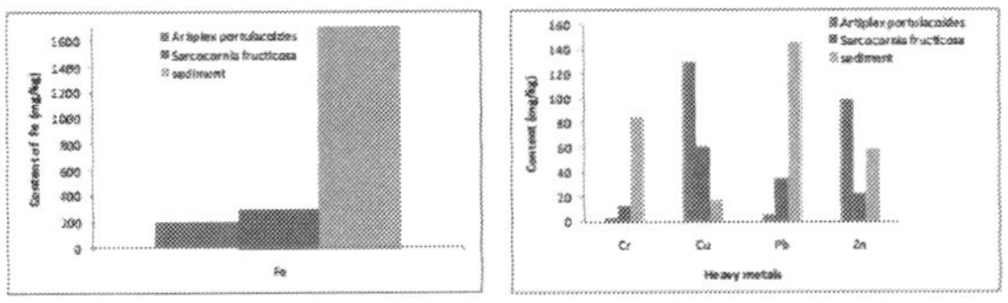

Fig.3: Cr, Cu, Pb Zn and Fe content in both plant species

Although most of the heavy metals detected in water and sediments are found to accumulate in appreciable quantities in the different aerial parts (stem, leaf) of the two plant species (Table 4). Certainly the accumulation of metals by plants is linked to different mechanisms. Indeed, plants have a very extensive root system that they use to extract water and minerals from the soil necessary for their growth. During this extraction, they also absorb other compounds dissolved in the aqueous phase. These mechanisms are also used for toxic compounds in the environment such as heavy metals. The transfer of metal in contaminated soil to plants depends on its bioavailability and solubility and the pH of the biotope [25]. Elevated metal concentrations were observed in the tissues of the Brassicajuncea plant (for Zn, Cu and Pb), particularly in plant leaves in low pH (2029 mg / kg, 71 mg / kg and 55 Mg / kg, respectively) [8]. Other authors have linked the accumulation of heavy metals in plants to the concentration of these metals in the soil [27], plant genetics [10], soil physico-chemical (pH, Eh, organic matter, Amount of clay) ([24] - [16]).

Table 4: Heavy metal content in the waters, sediments and plants of the Bouregreg estuary wetland.

Compartiment	Cr	Cu	Pb	Zn	Fe
V.decussata (mg/kg)	6,55	13,1	26,21	78,62	1600
S.plana (mg/kg)	6,47	23,7	28,05	362,5	2700
H. diversicolor (mg/kg)	11,06	22,1	33,17	102,2	3900
Artiplexportulacoides(mg/kg)	2,81	129,3	5,63	98,44	200
Sarcocornia fructicosa (mg/kg)	12,7	60,5	35,05	22,3	300
Sediment (mg/kg)	84,3	16,8	146,12	59,01	16500
Water of the estuary (mg/l)	<0,04	0,02	0,715	0,01	800

4 CONCLUSION

The spatial distribution of the demographic structure of *M. galloprovincialis* fluctuates widely from one area to another. It seems that it depends mainly on the physical and chemical factors, edaphic and hydrological coastline are: the nature of the substance, the tidal phenomenon, the availability of food and behavior of species. Different regression lines show that there is good correlation between the total length to the width of the shell, with significant correlation coefficients. This result demonstrates the use of this species as a bio indicator for monitoring the water quality of coastal Rabat -Sale.

Analysis of water, sediment, vegetation and macrofauna in the Bouregreg Estuary reveals that the entire compartment of this ecosystem is heavily contaminated by heavy metals (Cr, Cu, Zn, Pb and Fe) . Similarly, we observe that there is accumulation and bioaccumulation from the water to the sediment and the biological part (benthic plants and animals). Concentrations that remain low in water are found in remarkable quantities in sediments and flora and fauna.

All the results of the analyzes carried out showed us that the macrofauna studied (edible bivalve molluscs) is not in conformity with the standards and therefore unsuitable for human consumption. The levels of lead and chromium at *Sarcocornia fructicosa* are significantly higher than those found at *Artiplex portulacoides*, it seems that this one has a relatively strong intake of these two metals which could be associated with the shape of its leaves Which promote storage. Plants have a very extensive root system that they use to extract water and minerals from the soil for their growth. During this extraction, they also absorb other compounds dissolved in the aqueous phase (heavy metals) and consequently the depollution of their biotope.

The ability of these two species to tolerate and accumulate these metals will open up a new avenue of research on the treatment of these areas (phytoremediation). In addition the accumulation of these metals in the macrofauna will play an important role as biological indicators of metal pollution

REFERENCES

[1] Basraoui Y, Zegmout M, El addouli j, Demnati S, Chahlaoui A et Chafi A 2010. SCIENCE 06(3) (2010) 64 - 74

[2] Cheggour M., Texier H., Moguedet G. &Elkaîm B. 1990. Hydrobiologia, 207, 209-219.

[3]Cheggour M., 1999.Thèse de Doct. Etat, Université Cadi Ayyad, Marrakech,315 pp.

[4] Cherkaoui E., Bayed A. et Hily C. 2003. Bio. Mar. (2003) 44 :339-352.

[5]Cherkaoui E., 2006. Thèse Doctorat, Université Mohammed V, Rabat. 305p.

[6]Cherkaoui E., Nounah A. and Khamar M. 2013.3. Journal of Environmental Science and Engineering B 2 (2013) 432-435

[7]Ciéslinski G. Vanrees J., Huang P.M. Kozak L.M. Rostad H.P. w. et Knott d. R. 1996. Plant and soil (182) pp: 115-124.

[8] Clemente R., J. Walker David J., Bernal Pilar M.2005. The effect of soil amendments.Environmental Pollution 138 (2005) 46-58.

[9]-Djouai 2005. Journée d'études sur : « eau : vecteur du développement humain » 1 Octobre 2005 pp.9

[10]Ebbs S.D., Last M.m., Bradyd j., Cornich J., Gordon R. And Kochian L.V. 1997. J. Environm. Qual. (26) pp : 1424-1430.

[11]El Morhit M., Fekhaoui M., Serghini A., El blidi S., El abidi A., Bennaakam R., Yahyaoui A. & Jbilou M. 2008. Bulletin de l'Institut Scientifique, Rabat, section Sciences de la Terre, 2008, n°30,39-47.

[12] Fellah A. 2009. Thèse Doctorat, Université AbouBekrBelkaid, Tlemcen.169p.

[13] Förstner U et Wittman GTW, 1981.).Springer-Verlag :Berlin; 486.

[14]Fournier j. , Bonnot-Courtois, paris R. Voldoire O. & Levot M. 2012 : analyses granulométriques, principes et méthodes CNRST, Dinard, 99p.

[15] Gillert P. 1987. cahier de biologie Marine, 28 : 339-350

[16] Hinsinger P. 2000. Topical Days on phytoanagement of contaminated Environments, Mol, Belgium, may 15 - 16.

[17] Kassaoui H., Lebkiri M., Lebkiri A., Rifi E., Badoc A. & Douira A. 2009. Bull. Soc. Pharm. Bordeaux, 2009, 148, 77-92.

[18]Kaimoussi A., mouzdahi A., Bakkas S.& Chafik A. 1998. Bulletin de l'Institut Scientifique, Rabat, 1997-1998, n°21, pp. 95-102.

[19] Khamar M., Bouya D. &Ronneau C. 1999. Africancrops science conferenceproceedings, Vol. 4.pp 263-267.

[20] Khamar M., Cherkaoui E. &Nounah A. 201. Journées Internationales Oiseaux d'Eau et Zones Humides Faculté des Sciences Ben M'sik Casablanca 23-25 février 2012.

[21]Kourradi R., 2007. Thèse Doctorat ès Sciences, Université Mohamed V – Agdal Faculté des Sciences, Rabat, Maroc

[22] Lemine M. 1993. Thèse docorat 3 ème cycle, ENS, takaddoumrabat 337p.

[23] Leval K.T. and Haselwandter K. 1997. Mycorrhizol (7), pp :139-153.

[24]Losi M. E., Amrhein C. and Frankenberg W. 1994..J. Environ. Qual. 23, pp: 1141-1150.

[25] Mengel K. et Kirkby E.A. 1987. Principles of plant nutrition. 4th edition 686p.

[26] Sbaa M., Chergui H., Melhaoui M. MELHAOUI &Bouali A. 2001.Actes Inst. Agron. Vet. (Maroc) 2001, Vol. 21 (1), 45-52.

[27] Smolders 2000 Topical Days on phytoanagement of contaminated Environments, Mol, Belgium, may 15 -16.

[28] Wang W., Vinocur B., Shoseyov O., Altman A. 2004. TRENDS in Plant Science 9 : 244-252.

[29]WHO, 1982. World Health Organization, Geneva, ISBN: 978-92-4-120940-3, pp: 106.

Susceptibility assessment of landslides: A comparison of two GIS-based methods in the region of Al Hoceima (Eastern Rif, Morocco).

A. N. EL FAHCHOUCH [1], L. AIT BRAHIM[2], O. RAJI[3]
[1,2]Mohammed V University, faculty of science, department of earth sciences, GEORISK research unit: Geological risks, remote sensing and sustainable development. Ibn Battuta street, PO Box 1014, RABAT, MOROCCO.
[3]Mohammed VI Polytechnic University, Geology & Sustainable Mining, BENGUERIR, MOROCCO.

Abstract. The evaluation of the degree of susceptibility to landslides has become a major concern in mountainous areas, it is a key component of manager policies efforts in disaster prevention, mitigate risk and manage the consequences. The region of Al Hoceima is one of most mountain regions in Morocco, and is highly exposed landslides events. For this reason, the area was selected in order to determine its susceptibility to landslides using two methods. The purpose of this study is to evaluate and to compare the results of multivariate (logical regression) and bivariate (landslide susceptibility) methods in Geographical Information System (GIS) based landslide susceptibility assessment procedures. In order to achieve this goal, the selected methods were compared by the Seed Cell Area Indexes (SCAI) and by the spatial locations of the resultant susceptibility pixels. We found that both of the methods converge in 80% of the area; however, the weighting algorithm in the bivariate technique (landslide susceptibility method) had some severe deficiencies, as the resultant hazard classes in overweighed areas did not converge with the factual landslide inventory map. The result of the multivariate technique (logical regression) was more sensitive to the different local features of the test zone and it resulted in more accurate and homogeneous susceptibility maps. This information may have direct applications in landslides susceptibility research programs and can assist decision-makers in the implementation of preventive management strategies in the most sensitive areas.
Keywords: Landslide susceptibility, Multivariate, Bivariate, Al Hoceima, Morocco.

1 Introduction

Mass movements, In particular the landslides are very destructive factors, which cause every year lots of financial losses in mountainous area [1]. They can interact in very different ways and scattered in different places [2]. They are in general a manifestation of landmasses gravity displacement destabilized because of natural conditions.

Landslide susceptibility expresses the spatial probability that a landslides, which exist or may potentially occurs in an area for different local environmental conditions [3]. The most important conditions for a landslide susceptibility zonation include geomorphology, geology, geotechnical parameters, hydrology, topography, meteorology and botany [1].

The susceptibility of an area may be determined through very diverse, their comparison has largely fueled the scientific literature [4]. GIS and remote sensing has contributed largely to the development of these methods. Among the most recent methods used, there is the bivariate analysis based on Bayes' theorem named theory of evidence [5-7]. Actually, statistical methods seem to be more applicable and the preferred by the scientific community [8].

In Morocco, the identification, description and mapping of landslide hazard is a relatively new field of study. In the 1960's, [9] Avenard (1965) have mapped out the erosion in the Sebou basin. Since then, several studies have followed [10-15]. Most of these studies were conducted in Rif region, which is highly exposed to landslides hazards. However, the evaluation of the susceptibility to landslides in this area using probabilistic methods remains insufficient.

The objective of this research is to: (i) systematically inventory region, characterize and map the landslides in the Al Hoceima region, (ii) Build a geographic information systems (GIS) database of these hazard events and its causative factors, (iii) Assess the susceptibility to landslides through a bivariate and multivariate probability model. (iv) Compare the results of the two selected calculation methods

https://doi.org/10.1051/matecconf/201814902055

2 Study area

The study area (Al Hoceima region) is located in the central part of the (Rif northern Morocco), between latitudes 34°00" to 35°20" N and longitudes 4°20" to 3°50" W "figure 1". The total surface area of the selected area is 414 km2. This Rif in general and the Al Hoceima region is located in an area geologically unstable, subject to earthquakes [16] and is considered as the area most affected by different ground instabilities [15].

Geologically, the Al Hoceima area is part of the internal domain of the Rif Belt related to the Alpine Orogeny. The area is formed by a range of diverse facies and is characterised by the stacking of several structural units that extends from Paleozoic to Cenozoic. Among these units, there is the external dorsal limestone, which supports, as a tectonic klippe, the terrains of the internal dorsal limestone and those of the Palaeozoic Ghomarides nappes. These units overlay the terrains of the pre-dorsal zone (Tertiary sole) "figure 1". From a geomorphologic point of view, Al Hoceima region is dominated by rugged and mountainous terrain characterized by very steep slopes and by diverse topography. Elevations range from a few meters to 1200m [34-35]. The Al Hoceima region has a large forest area and under Mediterranean climate influences, it receives an annual rainfall of between 360 and 700 mm.

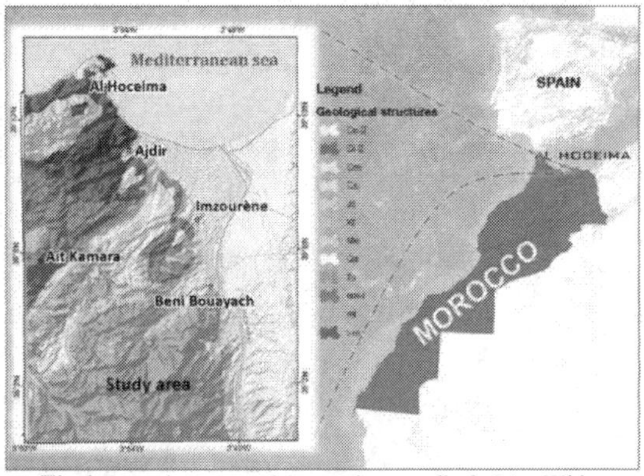

Fig.1. Location map of the study area, Ce-2: Lias White limestone; Ci-2: Marl-sandstone from Priabonian-Oligocene; Cm: Cretaceous Early Marl; Cs: Cretaceous late Marl; Js: Jurassic limestone; Kt: Schist from ketama unit; Me: Flysch siltstone; Qa: Quaternary; Ts: Marl of Tisirene unit; em-i: Ultrarif Jurassic Marl; mi: Triasic marl-clay of mezzo-rif unit; t-m: Marl-dolomitic.

3 Data and methodological approach

This study is based on a comparison of two methods of susceptibility assessment using GIS tools. The framework, which was adopted to achieve this goal, consists of three steps "figure 2": (i) Data sets and database creation (ii) Landslide susceptibility assessment by bivariate and multivariate methods and (iii) Comparison of methods.

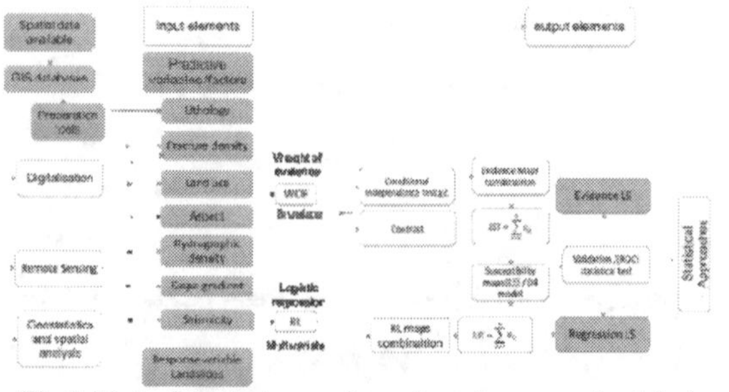

Fig.2. Methodological procedure adopted to assess landslind succptibility

3.1 Data sets and database creation

The selection of causal factors is crucial to determine the susceptibility to landslides events. In this context, we reviewed multiple physical parameters used throughout the world for the development of susceptibility mapping [17-19]. A preliminary analysis was necessary to reduce the number of parameters and refine these accordingly. The choice of variables is based on two vital factors: i) applicability at the scale of the study area; and ii) data availability. Consequently, the seven following quantifiable variables were chosen to evaluate the susceptibility of study area to landslides:

Land cover: The spatial distribution of landslides is also related to land uses/land cover. Indeed, it is recognized that bare soils favor terrain instability [20]. In our case, the land-cover map was produced by the analysis of a Landsat ETM+ image (year 2000).

Fracture density: lineation has an effective role in a landslide occurrence [21], to obtain the fracture density map, we used the results of the analysis of satellite images supplemented by data from geological maps and fields data; Which allowed us to establish a fracturing density map.

Lithology: Represents a major conditioning factor, the lithological map is based on the main lithological units obtained by the segmentation of the landscape into homogeneous macro-areas closely associated with sediment facies [22].

Slope: An important parameter in stability considerations and, more generally, slopes greater than 5°, are the most able to produce instabilities. However, it has to be taken into account that high slope angles do not always is synonym of largest landslide occurrence [23].

Aspect: is referred to as orientation from north of the terrain surface slopes. It is determined by measuring, in a clockwise direction, the angle between the north and the horizontal projection of the maximum slope. Both slope and aspect maps were prepared in ArcGIS from DEM.

Hydrographic network density: The disturbance of the slopes by the torrents is considered as a criterion of

https://doi.org/10.1051/matecconf/201814902055

instability [1]. The erosion of stream banks can cause the triggering of a landslides or its reactivation.

Seismicity: Using a global catalog of landslides, [24] Keefer had identified that the area where landslides are likely to occur, increases with magnitude of earthquakes and that a minimum magnitude of 4 is required to trigger a landslide.

3.2 Bivariate method

The weight of evidence method is based on a statistical bayesian bivariate approach. Actually is being used in several geology fields, and the interest for this method extended in the field of landslide hazard assessment is constantly increasing [25, 6, 7].

A detailed description of the mathematical formulation of the method is available in [26] and mathematical relationships for landslide hazard mapping is available in [27]. The method calculates the weight for each predictive factor based on the presence (positive) or absence (negative) of the training point theme units (D) within the area of each binary predictor theme (B), as indicated in [28]:

$$W_i^+ = \ln \frac{P\{B/Vd\}}{P\{B/\overline{Vd}\}} \quad (1)$$

$$W_i^- = \ln \frac{P\{\overline{B}/Vd\}}{P\{\overline{B}/\overline{Vd}\}}$$
(2)

A positive weight (Wi+) indicates that the causative factor is present at the landslide location and there is a positive correlation between presence of the causative factor and landslides. A negative weight (Wi-) indicates an absence of the causative factor and shows the level of negative correlation. The difference between the two weights is known as the weight contrast, Wf (W+ - W-), and the magnitude of contrast reflects the overall spatial association between the causative factor and landslides. In weights-of-evidence model, the combination of causative factors assumes that the factors are conditionally independent of one another with respect to the landslides [26].

The calculation of weights of evidence assumes conditional independence among the causative factor input to the model. One technique to assess the conditional independence between pairs of binary causative factor is to calculate a Chi-squared statistic (X2) to assess the variation between the expected and observed occurrences on and off the patterns in the two factor.

The pairwise test between two evidential factors involves a contingency table calculation, applicable only to locations at which landslides occur. The rows of the contingency table are the classes of one evidential factor, and the columns of the contingency table are the classes of the second factor. Each cell (i,j) of the table records the number of landslides occurring for a specific overlap of the i-th class of factor 1 and the j-th class of factor 2.

The null hypothesis of conditional independence is tested by determining if the measured chi-squared value exceeds a theoretical chi-squared value, given the level of significance. The (α) is the number of rows – 1 times the number of columns - 1. So for binary themes, the (α) =1. The level of significance for most tests is taken as 95%, equal to (1-probability) or (α) =0.05 [5, 29] Probability values (α) <0.05 indicates some conditional probability, or the failure of the conditional independence test at the 95% level. Low values of probability indicate conditional independence and the lower the value the greater the indication of conditional independence. The main problem of this test is a sensitivity in the number of landslides taken into account, the results varying by 30% [30, 31].

3.3 Multivariate (logical regression) method

As with the weight of evidence method, the logical regression has been largely used to study the susceptibility to landslides. Indeed, it allows forming a multivariate regression relation between causative factors that might affect the probability of the landslide occurrence [32]. The predicted values, which range from 0 to 1, can be defined by the following formulas [33], 32]:

$$p = ez/(1+ ez) \quad (3)$$

$$z= b0+ b1\,x1 + b2\,x2 + \ldots\ldots +bn\,xn \quad (4)$$

where P is the probability of landslide occurrence (landslide susceptibility index), z is the linear logistic model, b0 is the intercept of the model, n is the number of landslide–causative factors, b is the weight of the each factor, and x is the landslide conditional factors.

Logistic regressions have several advantages, namely [34]:

1. They work with quantitative and / or categorical variables

2. No analysis of factor conditional independence (X2 test) is necessary to obtain consistent results;

3. The method of calculation does not necessarily require a partition of the study area by homogeneous units

3.4 Comparison of methods

The results susceptibility maps produced from bivariate analysis and logical regression analyses. Both of them produced acceptable results will be compared depending on how their high susceptibility classes differ or converge. Therefore, the seed cell area index (SCAI) was used which is the area percent values divided with the landslide seed cell percent values. The high and very high susceptibility classes should have very small SCAI values and low and very low susceptibility classes to have higher SCAI values.

4 Results and discussion

4.1 Landslides inventory and predictive factors mapping

At a scale of 1:50,000 to 1:30,000 through interpretation of multiple data sources (satellite image, geological and geomorphological maps) combined with field data, a landslide inventory maps was constructed "figure 3". It shows the spatial distribution of existing landslides that have moved in the past. 187 landslides of various dimensions have been distinguished, they cover an area of 30 km2 accounting for 7 % of the study area. The rotational and translational landslides represents the majority with 80% of the total number of observed landslides, followed by flows (12%) and falls (8%).

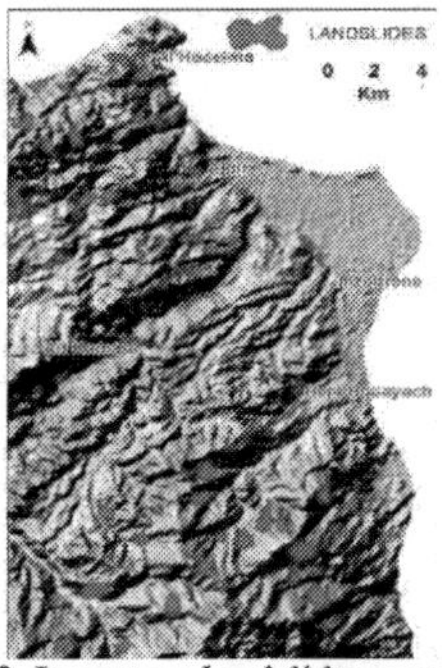

Fig.3. Inventery landslides map

Figure 4 shows the predictive variables maps for the study area: (i) the land use "figure 4a", shows a dominance of bare soils with about 57% followed by agricultural areas and urban area with about 33%. It is noted that the majority of inventoried landslides (60%) occur in areas with bare soil. (ii) the orientation of the lineaments is dominated by the NE-SW direction. The highest fracturing density is found in the NW of the study area "figure.4b". The main lithological units of the study area have been grouped into four categories "figure 4c", which according to their percentages are: 43.05% of quartzitic sandstone, 37.78% of shale, 14.05% of land Marl-schist, and finally 5.09% of the limestones. In this sense, it should be noted that most of inventoried landslides in our study area are related to the presence of clays, marls and shales. With regard to slope, the majority of the territory has a slope higher than 6% "figure 4d". The maximum values are naturally found in mountain areas where the slope usually exceeds 20%. The Aspect map shows that the most exposed areas are in the mountainous areas on the western side of the study area "figure 4e". The study area is traversed by a very dense hydrographic network but the rivers are non-permanent limited to periods of high floods "figure 4e". It has a strong seismicity at the NW but the magnitudes do not exceed 5 "figure 4g".

Generally, in case of landslides, evaluation of a prediction result is based on predictive ability of the prediction result. The best validation method is to see if the prediction result may be identical with inventory landslides.

Table 1. Shi-sequar result for each predictive factor; Probability values depend on chi-squared values (number of degrees of freedomX2) and Values α < .05 indicate some conditional dependence.

Parameters	Test	Fracture density	Seismicity	Lithology	Hydrographic network density	Aspect	Land use
Slope	X^2	109	**6**	30	43	117	19
	α	0.00	**0.99**	0.00	0.00	0.00	0.00
Fracture density	X^2		34	87	47	**12**	**12**
	α		0.00	0.00	0.00	**0.76**	**0.71**
Seismicity	X^2			25	24	**10**	**18**
	α			0.00	0.00	**0.84**	**0.35**
Lithology	X^2				**14**	**18**	27
	α				**0.59**	**0.33**	0.00
Hydrographic network density	X^2					**6**	**15**
	α					**0.99**	**0.56**
Aspect	X^2						26
	α						0.00

In trying to determine the best combination of predictive factors and landslide occurrence, the conditional independence was tested for each parameter. The table 1 shows chi-square for all possible pairs. As a result, for the degree of dependency, which exceeds the 76 per cent, it evidence a dependence between Hydrographic network density-Aspect (99%) and between Slope-Seismicity (99%) and Aspect-Seismicity (84%). Based on these results, we selected four models: a benchmark model combining all parameters and three other models by combining only independent factors "table 2".

Table 2. The variables combinations

Model	Combination
M1	Slope **(S)** + Fracture density **(Frd)** + Seismicity **(Ss)** + Lithology **(Lth)** + Hydrographic network density **(Hyd)** + Aspect **(As)** + Land use **(Lu)**
M2	Frd + Lth + Hyd + Lu + Ss
M3	Frd + Lth + Hyd + Lu + S
M4	Frd + Lth + As + Lu + S

4.2 Application of weights of evidence mode and logistic Regression

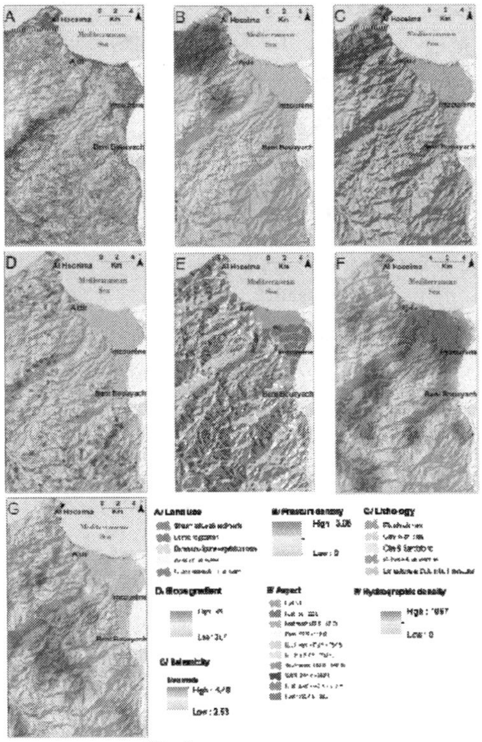

Fig.4. parameters maps

The maps in figure 5 show spatial variability of susceptibility to landslides from model. To test the success and prediction rates of models, ROC curve was used "figure 7". Each resulting map was reclassified into four susceptibility classes based on the quartile ranges.

According to the results, all selected models provide generally good performance but the map produced by the model 4 exhibited the better one with 0.7395 (73.95 %) of predictive accuracy. The north of the study area is highly susceptible to landslides, whilst the Al Hoceima plain have low susceptibility to landslides.

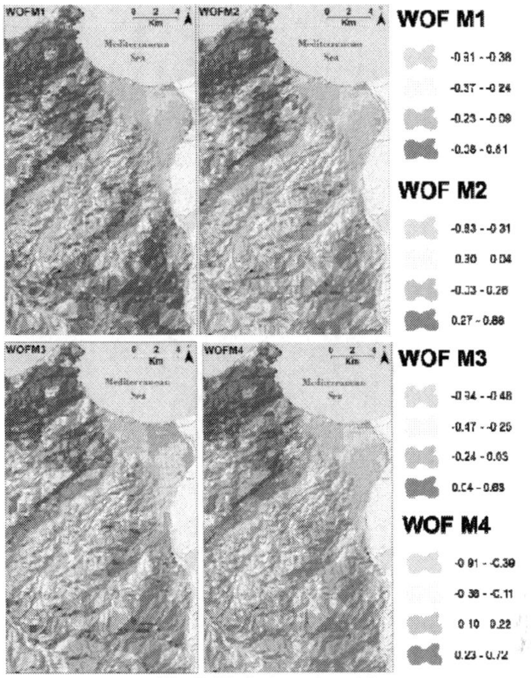

Fig.5. Wof adopted models

LR model was employed using the same models as for weight of evidence method. The generated Landslide susceptibility map was also reclassified into four susceptibility classes "figure 6". Compared to WOF susceptibility maps, more area seem to be susceptible to landslides and only Al Hoceima plain show the lowest susceptibility values. Unlike WOF results, models 1 and 3 posters the best prediction performance of the landslide susceptibility maps produced with respectively 0.78 and 0.75 "figure 7".

5 Conclusion

Landslides are very tragic disasters that can occurs regularly in a variety of places. During the last years, many quantitative methodologies have been developed to study the relationship between landslides and causative factors. In this paper, two methods were selected and tested in the Al Hoceima area: The weight of evidence and the logistic regression.

Seven predictive factors were identified in order to apply these statistic methods, some of them were derived from DEM and the rest from available data of the study area. This study shows that the approach taken using GIS is appropriate for study area and that all selected models present good performances. However, the landslide susceptibility map produced using the logistic regression method predicts higher percentages of landslides, especially in very high zones than the logistic regression method. Furthermore, it should be noted that the best model (best combination of predictive factor) is different from one method to the other.

Generally, all the models tested show that the northern zone of the study area and the high-altitude zones at the south are the areas which have the high susceptibility values in comparison with the Al Hoceima plain which have the lowest values.

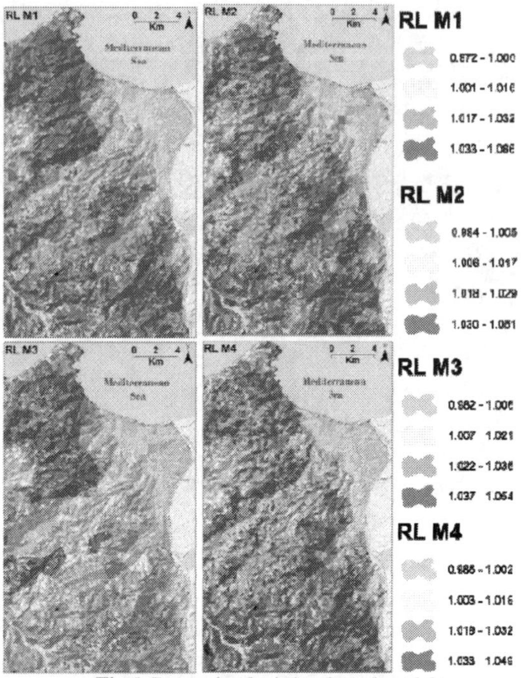

Fig.6. Regression logistic adopted models

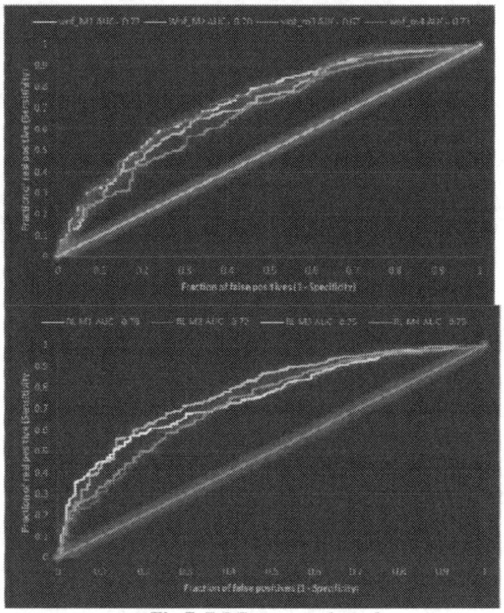

Fig.7. ROC compared graphs

References

1. Y. Thiery, J. P.Malet, S. Sterlacchini, A. Puissant, O. Maquaire, *Revue Internationale de Géomatique*, 15(2), **227-245** (2005)
2. S. Gruber, C. Huggel, R. Pike, *Developments in Soil Science*, 33, **527-550** (2009)
3. S. Lee, J. S. Won, S. W. Jeon, I. Park, M. J. Lee, *Mathematical Geosciences*, 47(5), **565-589**, (2015)
4. C.J. Van Westen, T.W.J. Van Asch, R. Soeters, *Bulletin of Engineering Geology and the Environment*, 65, **167–184**. (2006)
5. G. Bonham-Carter, *Pergamon Press, Oxford*, **398**, (1994)
6. C-J. Van Westen, G.F. *Lulie Geomorphology* 54 (1–2), **77–89**, (2003)
7. Y. Thiery, Strasbourg, **16**, (2006)
8. E. Yesilnacar, T. Topal, *Engineering Geology*, 79(3), **251-266**, (2005)
9. J.M. Avenard, INRA, **114** (1965)
10. G. Maurer, *Université de Paris*, **500**. (1968)
11. A. El Gharbaoui, *série Géologie et Géographie Physique*. n° 15. **439**, (1981)
12. Kh. Margaa, *Université de Franche-Comté*, **196**, (1994)
13. A. Fares, *Université de Franche Comté*. **177**, (1994)
14. A. Sossey, *Université Mohamed V, Faculté des Sciences*, **175**. (2005)
15. A.N. El-Fahchouch, L. Ait-Brahim, O. Raji, A. Khouakhi, *Afrique science* 11(2), **44-57**, (2015)
16. A. Talhaoui, M. Aberkan, I. Brahim, A. El Mouraouah, Pangea, 43, **3-18** (2005)
17. E. Yesilncar, T. Topal, *Engineering Geology, Vol. 79*, **251-266**(2005)
18. L. J. Wang, K. Sawada, S. Moriguchi, *Computers & Geosciences*, 57, **81-92**, (2013)
19. E. K. Sangchini, S. N. Emami, N. Tahmasebipour, H. R. Pourghasemi, S. A. Naghibi, S. A. Arami, B. Pradhan, *Arabian Journal of Geosciences*, 9(3), **1-15** (2016)
20. P. Reichenbach, C. Busca, A.C. Mondini, M. Rossi *Environmental Management*, 54(6):**1372-1384**. (2014)
21. H. Shahabi, B. B. Ahmad, S. Khezri, *Arabian Journal of Geosciences*, 6(10), **3885-3907** (2013)
22. VAN WESTEN, C. Johannes. (1993).
23. A. Donnarumma, P. Revellino, G. Grelle, F. M. Guadagno, *Landslide Science and Practice*. **425-433** (2013)
24. D. K. Keefer, *Surveys in geophysics*, 23(6), **473-510**, (2002)
25. S. Lee, K. Min, *Environmental Geology*, 43: **120-131**. (2002)
26. I. M. Kettles, , G. F. Bonham-Carter, *Environment, Analysis*, 2(2), **99-110** (2002)
27. R. K. Dahal, S. Hasegawa, A. Nonomura, M. Yamanaka, T. Masuda, K. Nishino, *Environmental Geology*, 54(2), **311-324** (2008)
28. G. F. Bonham-Carter, F. P. Agterberg, D. F. Wright, *Statistical applications in the earth sciences*, 89(9), **171-183** (1989)
29. M. Davis, *Neuron*, 36(3), **340-343**, (2002).
30. A. Pistocchi, L. Luzi, P. Napolitano, *Environmental geology*, **41**(7), (2002).
31. P. Dumolard, N. Dubus, L. Charleux, *Les statistiques en géographie, coll. Atouts, Belin, Paris*. (2003)
32. L. Ayalew, H. Yamagishi, *Geomorphology*, 65(1), **15-31** (2005).

https://doi.org/10.1051/matecconf/201814902055

33. S. Menard, *Sage University Paper Series on Quantitative Applications in Social Sciences, vol. 106. Thousand Oaks, California.* **98**. (1995)
34. Y.Thiery, J. P.Malet, S. Sterlacchini, A. Puissant, O. Maquaire, *Geomorphology*, 92(1), **38-59**, (2007).
35. L. Ait Brahim, C. Nakhcha, B. Deffontaines, H. Sami, P. Chotin (2004)- Revue Télédétection- photo interprétation. Editions ESKA, France N° 4, pp.**13-49.**
36. L. Ait Brahim, B. Tadili, C. Nakhcha, I. Mouayn, M. Ramdani, M. Limouri, (2002) - Pure and Applied Geophysics,: Volume 161, Numbers 5-6, pp. **1081-1091**

Non-linear finite element analysis of reinforced concrete members and punching shear strength of HSC slabs

Kernou Nassim[1], Belakhdar Khalil[2] and Benyamina Abdelrahmane Bekaddour[3]
[1] Université Docteur Tahar Moulay Saida , Département de génie civil et hydraulique, Saida, Algérie
[2] Université de Tamanrasset, Département de science et technologie, Algérie
[3] Université Ibnkhaldoun de Tiaret, Département de génie civil, Tiaret, Algérie

Abstract. A rational three-dimensional nonlinear finite element model (NLFEAS) is used for evaluating the behavior of high strength concrete slabs under monotonic transverse load. The non-linear equations of equilibrium have been solved using the incremental-iterative technique based on the modified Newton-Raphson method. The convergence of the solution was controlled by a load convergence criterion. The validity of the theoretical formulations and the program used was verified, through comparison with results obtained using ANSYS program and with available experimental test results. A parametric study was conducted to investigate the effect of different parameters on the behavior of slabs which was evaluated in terms of load-deflection characteristics, concrete and steel stresses and strains, and failure mechanisms. Also, punching shear resistance of slabs was numerically evaluated and compared with the prediction specified by some design codes.

1 Introduction

Reinforced concrete structures present significant nonlinear and accordingly complex behavior and consequently nonlinear analysis of this kind of structure has been subject of research for many years [2]. In the last decades very sophisticated and accurate structural models were proposed, that can take into account most aspects of the nonlinear behavior of concrete structures [7,9].

The complex behavior of concrete slabs under real conditions leads to strong uncertainties related to material properties of slabs and the loading conditions. These uncertainties must be taken into account to be included in the design guidelines of such structures especially when some structure does not fit exactly in any standard due to its size, new material properties, complexity or multidisciplinary nature. In these cases probabilistic analysis can be pursued since design standards are not able to cover the full range of application that engineers are able to conceive.

Most research on the punching shear strength of slabs has been concerned with the generation of experimental data on simply supported slabs and the development of empirical equations. A few analytical analyses have also been proposed by various investigators based on different models. The most known empirical formula for predicting the shear strength of slabs as result of experimental and analytical studies are those suggested by the ACI-318 code approach.

However, In order to ensure the serviceability requirement of slabs it is necessary to predict the cracking and the deflections of slab structures under service loads. In order to assess the margin of safety of slab structures against failure an accurate estimation of the ultimate load is also essential, in addition to the prediction of the load-deformation behavior of the slab throughout the range of elastic and inelastic response is desirable. The main objectives of this research are: 1-To calibrate and validate the program by comparing the predicted results with available experimental tests and other known software-programs. 2-To study Punching shear behavior and the major material and geometrical parameters that strongly affect it.

2 Finite Element Formulation

For the nonlinear analysis of concrete slabs a nonlinear finite element code NLFEAS (Non-Linear Finite Element Analysis of Slabs) has been used. NLFEAS has been developed to predicate and study the behavior of normal and high strength concrete [9]. Its concept and modeling characteristics are described in the following section.

2.1Modeling of concrete material in compression

The behavior of concrete in compression is simulated by an elasto-plastic work hardening model up to onset of crushing. The plasticity model and its constituent, and the modeling of concrete under triaxial state of stress, will be discussed in terms of the following elements: 1) Uniaxial stress-strain relationship, 2) Failure criterion, and 3) Crushing condition.

2.1.1. Uniaxial stress-strain relationship

Frequently, the widely used stress-strain relationship of concrete up to the peak stress is given as a parabolic relationship as follows:

$$\sigma = fc'\left[\frac{2\varepsilon}{\varepsilon_0} - \left(\frac{\varepsilon}{\varepsilon_0}\right)^2\right] \quad (1)$$

Where ε_0 is the strain at peak stress f_c given by

$$\varepsilon_0 = 2.\frac{fc'}{E_c} \quad (2)$$

In the presence of orthogonal cracks which are caused by shear or transverse tensile stresses, concrete exhibits lower compressive strength and stiffness than uniaxially compressed state. Such degradation or softening in compressive strength of concrete is taken into consideration in the present study by multiplying the uniaxial compressive concrete stress defined in Equation (1) by a softening factor λ, as shown in Fig. 1. Among various compression reduction models available in literature, the model suggested by Vecchio, and Aspiotis (1994), is implemented in the present finite element formulation. The model can be expressed as:

$$\lambda = \frac{1}{1 + K_c.K_f} \quad (3)$$

Where: K_c, represents the effect of the transverse cracking and straining, K_f represents the dependence on the strength of the concrete (f_c), K_c and K_f are given by:

$$K_c = 0.27\left(\frac{\varepsilon_r}{\varepsilon_0} - 0.37\right) \quad (4)$$

$$K_f = 2.55 - 0.2629\sqrt{fc'} \le 1.11 \quad (5)$$

Where: ε_r is the tensile strain normal to the cracked plane given by:

For cracked sampling point in the principal direction "1"

$$\varepsilon_r = \varepsilon_1 \quad (6)$$

For doubly cracked sampling point in both directions "1"and "2"

$$\varepsilon_r = \left(\varepsilon_1^2 + \varepsilon_2^2\right)^{\frac{1}{2}} \quad (7)$$

Where ε_1 is the transverse tensile strain in the principal direction "1" normal to the cracked plane, ε_2 is the tensile strain in the second direction "2" normal to the second crack plane.

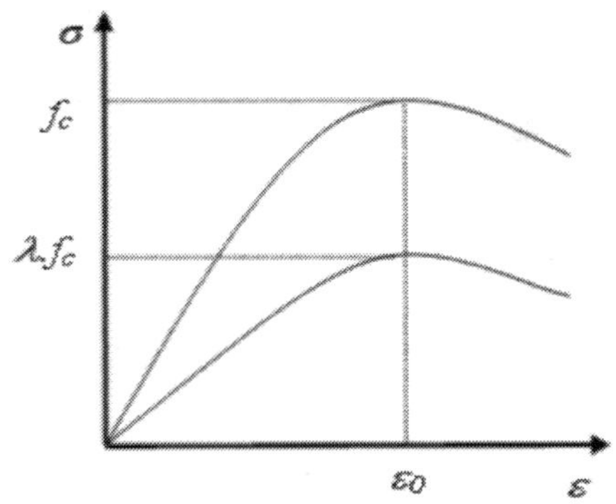

Fig.1. Compression reduction of transversely cracked concrete

2.1.2 Failure criterion

Under triaxial state of stress the failure criterion for concrete is generally assumed to be dependent on three stress invariants. However, the failure criterion used in this study is dependent on two stress invariants which has been proved to be adequate for most practical situations and has been successfully used by many investigators for analyzing reinforced concrete plate and shells which can be expressed as:

$$f(\sigma) = f(I_1, J_2) = \sqrt{\alpha I_1 + 3\beta J_2} - \sigma_0 = 0 \quad (8)$$

Where:

I_1 is the first stress invariant given by:

J_2 is the second deviatoric stress invariant

σ_0 is the equivalent effective stress at the onset of plastic deformation,

$$\alpha = \sigma_0(\beta - 1) \quad (9)$$

$$\beta = \frac{1 - 2.\gamma}{\gamma^2 - 2.\gamma} \quad (10)$$

Where, γ is a constant to be determined from equal biaxial compression state, which is given by:

$$\sigma_x = \sigma_y = -\gamma\sigma_0 \quad (11)$$

The adopted values of γ are taken according to experimental work conducted by Hussein and Marzouk [4]where they found that the constant γhas the values of 19, 14 and 9% for concrete strength of 42.7, 73.7 and 96.5 MPa, respectively.

2.1.3. The crushing condition

The experimental tests of concrete under multiaxial loading indicate that the crushing is a strain related phenomenon, so concrete is considered to crush when the strain reaches a specified ultimate value, after that the current stresses drop suddenly to zero and the concrete is assumed to lose completely its resistance against further deformation. Hence the crushing criterion is directly obtained by using the same form of yield criterion but in terms of strains, as follow:

$$C.I_1 + \sqrt{(C.I_1)^2 + 3.\beta.J_2} = \varepsilon_{cu} \qquad (12)$$

Where ε_{cu} the ultimate concrete strain that can be obtained from the uniaxial compression test. Frequently, the ultimate concrete crushing strain ε_{cu} is estimated to be in the range of 0.0030 as suggested by ACI-318-02 [10] as given by BS8110 [11] and CSA-94 codes [12].

2.2. Modeling of concrete materials in tension

When tensile stress exceeds a limiting value a crack is assumed to form in the plane perpendicular to the direction of that stress and concrete behaves no longer isotropic, and therefore, the normal stiffness is reduced through tension-stiffening concept. Once concrete has cracked, fixed smeared cracking model is used in the current study to model the crack. The gradual release of tensile stresses normal to the cracked plane is represented by bilinear average stress-strain curves to simulate the tension stiffening behavior (Fig.2).

The shear stiffness is also reduced when cracking occurs because it retains the two major mechanisms by which shear is transferred across the crack (the aggregate interlock of the rough crack surfaces and dowel action of the reinforcing bars crossing the crack planes). A bilinear shear retention model is used as shown in Fig. 3. The material parameters used in the analysis are as follow:

Tension stiffening: $\alpha_1 = 5$, $\alpha_2 = 0.6$

Shear retention: $\gamma_1 = 10$, $\gamma_2 = 0.5$, $\gamma_3 = 0.04$

Besides, at onset of cracking Poisson's ratio (υ_c) is set to zero.

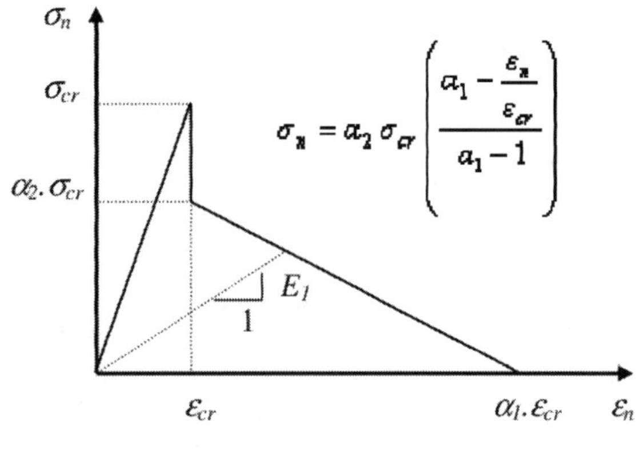

Fig.2. Bilinear average tensile stress-strain of concrete

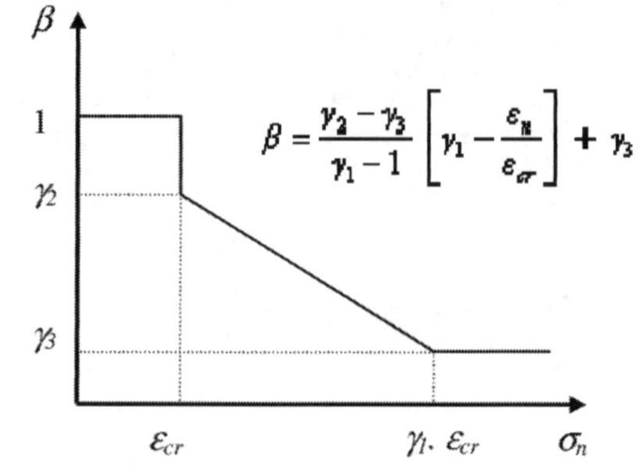

Fig.3. Bilinear shear retention model

2.3. Material modeling of reinforcement

In contrast to concrete, the material modeling of steel is rather simple. Frequently, the steel is modeled using linear elastic-full plastic model, as shown in Fig.4.

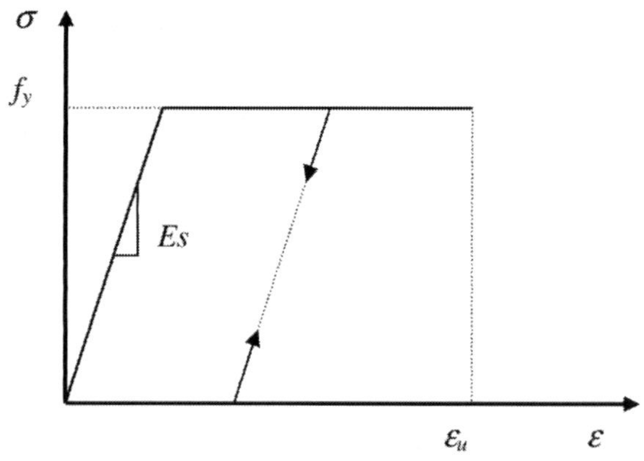

Fig.4. Modeling of steel reinforcing bars

2.4 Finite element idealization

The concrete is represented by using 8-noded brick element. The reinforcing bars are modeled as one dimensional element subjected to axial force only. The non-linear equations of equilibrium have been solved using the incremental-iterative technique based on the modified Newton-Raphson method. The convergence of the solution was controlled by a load convergence criterion. NLEFAS program interface as shown in Fig.5.

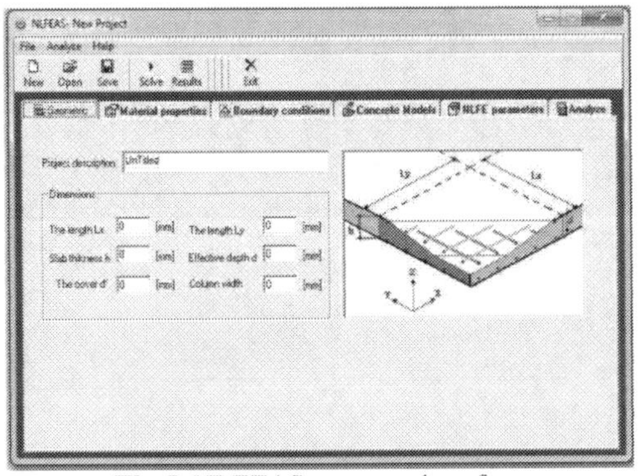

Fig.5. NLEFAS program interface

4. Description of selected experimental slab

A simply supported square slabs were selected from an experimental test of reinforced concrete slabs made of HSC, to be used to validate the predicted analytical results and to be used for the reliability analysis. The selected experimental slab was one of eleven slabs carried out by Tmaszewicz [2]. The selected slab was named "ND95-2-3D", it has span of 2200 mm, with thickness of 240mm. The concrete strength f_c' was 80.3 MPa while the steel yielding stress f_y was500 MPa. The steel ratio was 2.34 %

The following material properties are assumed in the analysis: Es = 200.000MPa, $Ec = 3320.\sqrt{f_c'} + 6900$, $f_t = 0.33\sqrt{f_c'}$, and ν_c=0.24.

3 Finite element mesh

The concrete is represented by using 8-noded brick element. The reinforcing bars are modeled as one dimensional element subjected to axial force only, and perfect bond is assumed to occur between the two materials,

By taking advantage of symmetry, a segment representing one quarter of the slabs has been considered in the finite element analysis, noting that the slab was modeled upside down and the loading is applied on the column as shown in Fig.6.

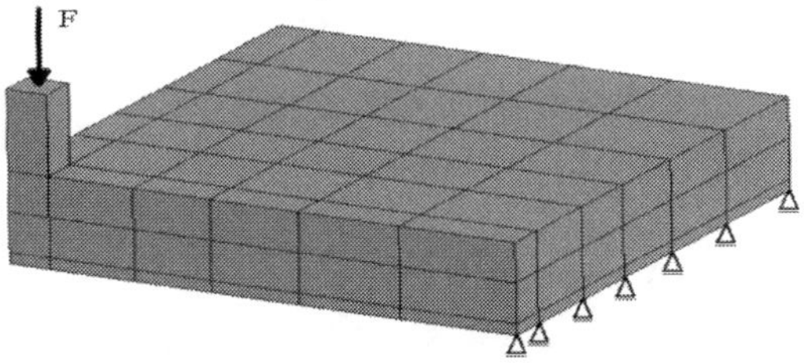

Fig.6. Finite element mesh

4 Validation of the NLFEM program

The results of the present nonlinear finite element analysis of the investigated slabs in terms of ultimate load are compared against the experimental measurements and listed in Table 1. Fig. 6 shows load-deflection curves of selected slabs of the present finite element analysis and experimental results. According to Fig 6 and Table 1, it can be observed that the present finite element model performs satisfactory and it predicts accurately the real behavior of slabs.

Table 1. Comparison of the predicted and experimental results

	Experimental		FE Analysis		FEA / Exp	
Slab	Ultimate load P_u(KN)	Ultimate deflection D_u(mm)	Ultimate load P_u(KN)	Ultimate deflection D_u(mm)	P_u/P_u	D_u/D_u
ND95 2-3D	1250	7.05	1180	5.84	0.944	0.828
ND115-1-1	2450	11.65	2540	10.29	1.037	0.883

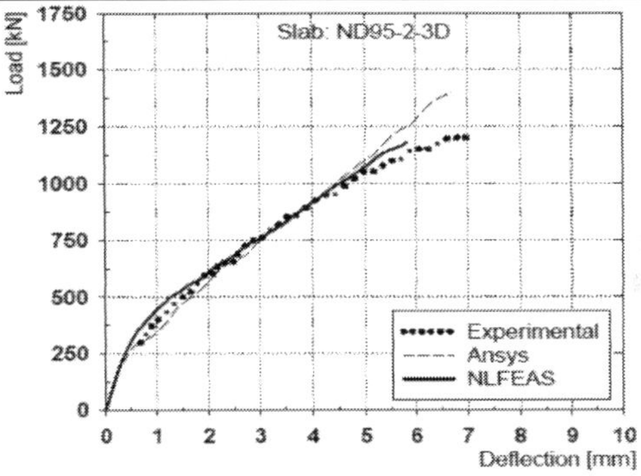

Fig.6. Comparison of predicted and experimental load-deflection curves.

4.1Effect of compressive strength grade

The concrete compressive strength considered as one of the important primary variables that was studied in the present investigation. The influence of *fc'* on the selected slabs' behavior was studied through altering the values of *fc'* by different values, and the obtained results are illustrated in Fig.8. It can be seen that generally, the higher the value of *fc'*, the higher the expected stiffness and the ultimate load of any slab. It was stated previously that increasing the value of *fc'* influences the load-deflection behavior particularly after cracking, but the shape of the load-deflection curve was not highly influenced.

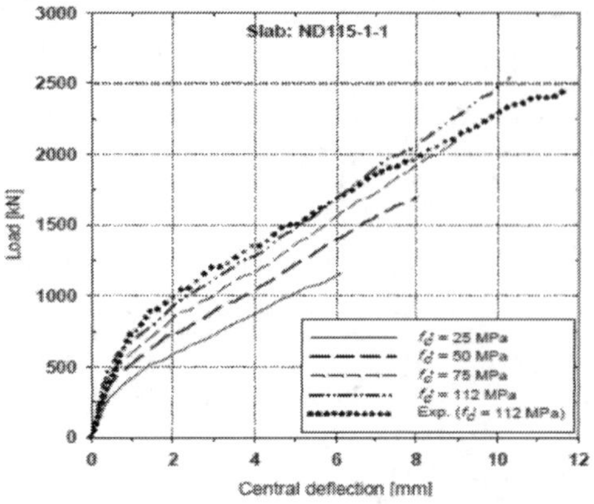

Fig.8. Effect of compressive strength *fc'* on the slab response

4.2 Comparison of the experimental punching strength versus present prediction and codes predictions

Experimental results were used to check the validity of the punching shear strength formula given in AS3600 [1], CEB-FIP [3]and NLFEAS [9]. Marzouk and Hussein [4] tested 17 square specimens to investigate the punching shear behaviour of high-strength concrete slabs. Tomaszewicz [2] tested 19 square flat slabs with orthogonal, equally spaced flexural reinforcement and without shear reinforcement. A total of 29 test results from four research studies conducted by Ramdane [8], Hallgren and Kinnunen [6], Marzouk and Hussein [4] and Tomaszewicz [2], were compared to values of punching strength calculated using AS3600. the punching shear strength values specified in different codes vary with concrete compressive strength f'_c and is usually expressed in terms of $f_c^{'n}$. In AS3600 (Cl. 9.2.3) the punching shear strength is expressed as proportional to $\sqrt{f_c^{'}}$. The ultimate shear strength for slabs without prestress is given by $V_{uo} = ud\left(f_{cv}\right)$ where: *u*= length of the critical perimeter.

f_{cv} = punching shear strength (MPa)

$$f_{cv} = 0.17\left(1 + \frac{2}{\beta_h}\right)\sqrt{f_c^{'}} \le 0.34\sqrt{f_c^{'}} \qquad (13)$$

β_η: ratio of longest column dimension to shorter column dimension

In CEB-FIP the punching shear resistance, F_{sd} is expressed as proportional to $\left(f_{ck}\right)^{\frac{1}{3}}$, Where f_{ck} is the characteristic compressive strength of concrete. The highest concrete grade considered in CEB-FIP, which corresponds to f_{ck} equal to 80 MPa. For the calculation of punching load capacity.

$$F_{sd} = 0.18\xi\left(100\rho f_{ck}\right)^{\frac{1}{3}} u_1 d \qquad (14)$$

In this study, measured concrete strength is taken as f_{ck}. Fig.9 shows the ratios between test results and the failure loads predicted by different formulae plotted with respect to the concrete strength. The experimental ultimate loads (P_{test}) of the slabs to the values predicted by AS3600 and CEB-FIP is compared. two points from AS3600 fall below the safety margin with one result for a slab with a concrete strength of 112 MPa. The AS3600 formula can be considered to be applicable up to 100 MPa. However the ratios between observed and calculated loads clearly show that AS3600 is less conservative for the HSC slabs and high scatter is found. Therefore, all above codes and present FE model may be considered applicable for predicting the punching strength of HSC up to 118 MPa.

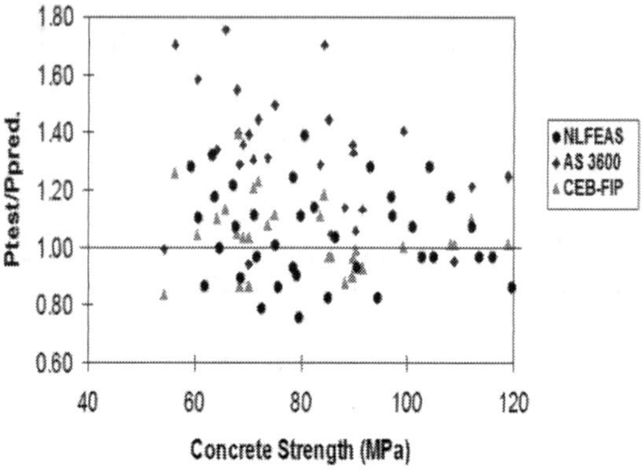

Fig.9. Ratios of experimental and predicted shear strengths

5 Conclusion

The main conclusion extracted from the current study is that the nonlinear finite element method based on advanced 3D models is a powerful and relatively economical tool which can be effectively used to simulate the true behavior of reinforced concrete even under complicated conditions.

The presented finite element program (NLFEAS) is used for the analysis of high strength reinforced concrete slabs subjected to monotonic concentrated loads after calibration using many experimental data. Predictions of the present FE program are compared with valuable experimental data, and the comparisons were judged good. Moreover, the model was able to predict the total response and capacity of the high strength concrete slabs and to indicate the type of failure (flexural or shear).

In AS3600 the punching shear strength is expressed as proportional to $\sqrt{f_c^{'}}$. However in CEB-FIP punching shear strength is assumed to be proportional to $\sqrt[3]{f_c^{'}}$.

The comparison of experimental results show that the AS3600 formula is applicable up to 100 MPa. However the ratios between observed and calculated loads clearly show that AS3600 is less conservative for the HSC slabs. The code of CEB-FIP formula is less conservative for HSC slabs and may be unsafe for some cases.

References

1. AS3600: Concrete Structures Standard. Standards Association of Australia, (1994).
2. A. Tomaszewicz. High-strength Concrete SP2 - Plates and Shells. Report 2.3, Punching Shear Capacity of Reinforced Concrete Slabs. Report No. STF70A93082, SINTEF, Trondheim, (1993).
3. CEB-FIP State-of-the-art report on high-strength concrete. 90/1/1, Bulletin d'Information No. 197, (1990).
4. H.Marzouk, A.Hussein. Experimental Investigation on the Behavior of High Strength Concrete Slabs. ACI structural journal November-December 1991; 88(6) 701-513;(1991).
5. H.Marzouk ,Z. Chen, Z. "Finite Element Analysis of High Strength Concrete Slabs," ACI Structural Journal, V. 90, No. 5, September-October 1993, pp. 505-513; (1993).
6. M.Hallgren, and S.Kinnunen,. Increase of Punching Shear Capacity by using High-Strength Concrete. 4th International Symposium on Utilization of High-strength/High-performance Concrete. Paris, 1996, pp. 1037-1046;(1996).
7. N.Kernou, B.Youcef, K. Belakhdar, K. . Assessment of reliability and punching shear resistance of slabs, GRAĐEVINAR, 67 (11), 1051-1062, (2015) doi: https://doi.org/10.14256/JCE.1295.2015.
8. K.E.Ramdane, Punching Shear of High Performance Concrete Slabs. 4th International Symposium on Utilization of High-strength/High-performance Concrete. Paris, 1996, pp. 1015-1026; (1996).
9. M. M. Smadi and K. A. Belakhdar, "Nonlinear finite element analysis of high strength concrete slabs", Computers and Concrete, Vol. 4, No. 3, pp 187-206, (2007).
10. ACI 318-2002. Building Code Requirements for Structural Concrete (ACI 318-02) and Commentary (2002).
11. BS-8110. Structural Use of Concrete, Part 1:Code Practice For Design and Construction,. British Standards Institution, Londan (1985).
12. CSA, Design of concrete structures for buildings, Standard A23.3- 94, Canadian Standards Association, Rexdale, Ontario,(2004).

Elemental Chemical Analysis by X-ray Fluorescence of Calcarenite Stones Used in Historical Monuments Building of Rabat (Morocco)

Y. El Rhaffari[1], M. Hraita[1], A. Rahmouni[1], A. Samaouali[1], M. Boukalouch[1], Y. Geraud[2]
[1]Thermodynamic-Energy Team, Center of Research in Energy,Faculty of Science, Mohammed V University, P.O. Box 1014,Rabat, Morocco, Email: ayounes1@hotmail.com
[2] Université de Lorraine, Ecole Nationale Supérieure de Géologie, UMR 7359-Géo Ressources rue du Doyen Marcel Roubault TSA 70605, 54518 Vandoeuvre Les Nancy Cedex, France

Abstract. A good knowledge of chemical properties, thermophysical and petrophysical parameters, makes it possible to understand the mechanisms of alteration of historical monuments, and to predict solutions limiting their evolution.
The main objective of this work is chemical analysis by X-ray fluorescence spectrometry on calcarenite stones, often used as a building material for historical monuments of Rabat (Morocco), in order to identify the nature of the rock.
The elemental chemical composition shows high concentrations of CaO and SiO_2, which confirms that the main phases of the calcarenite rock studied are calcite and quartz. These are the characteristic minerals of calcareous rocks.

1 Introduction

The main objective of this work is to determine the mineralogical properties of calcarenite rocks, often used as building materials for historical monuments of Rabat.
Calcarenite was the material used to build the historical monuments of the city of Rabat, it was used to build the Roman and Islamic buildings of Chellah [1,2].
The experimental approach consists in studying X-ray fluorescence spectrometry (XRF), in order to identify the nature of the rock and to better understand the processes of heat and fluid transfer in material.
Chemical analyzes of several samples of calcarenites Plio-Quaternary in the city of Rabat indicate that the main component of these deposits is calcium carbonate ($CaCO_3$), its content varies between 50 and 57%, reflecting theimportance of bioclasts marines. In contrast, the content of silica (SiO_2), which is in the form of quartz grains, is very low [3,4,5].

2 Description of method of work

2.1 Sampling

The material tested sample is a calcarenite stone,sampled from a quarry located near the historicalmonument of Chellah "figure1".

Fig. 1. Main gate of the Chellah monument

Geologically, it is a calcareous limestone, ofPlio-Quaternary age in the form of a system ofdune ridges running parallel along the Atlanticcoast. The deposit calcarenite took place under theaction of a substantially aeolian dynamics withplenty of round quartz grains mats whichremobilized sand beach, and by the largehorizontal stratifications [6].
Ten cylindrical samples were extracted from the calcarenite material by an Diamond Drill with Water Cooled Bits "figure 2". The samples were placed in a forced ventilation oven, where they were dried at a

temperature of 55 ° C, until their mass became constant.

Fig. 2. Ten cylindrical samples extracted by coring

2.2 X-ray fluorescence spectrometry

Chemical analysis by X-ray fluorescencespectrometry was carried out for the identification of the mineral species of the stone studied. The apparatus used is an Axios of Panalytical, which requires samples in powder form. It is a sequential spectrometer with a measuring channel on a single goniometer covering the complete range of measurement of Beryllium to Uranium elements. It is controlled by a microprocessor for maximum flexibility and the entire system is controlled by a computer with spectrum analysis and processing software.

During X-ray fluorescence spectroscopy (XRF), the atoms contained in the reference and analytical substances are excited by the spectra of lines and braking of an X-ray tube to emit their characteristic X-ray fluorescence. Radiation emitted by the sample is fractionated by monochromatization on crystal analyzers in the spectrometer, so that it is possible to measure the intensities of the different spectral lines or spectral regions (wavelengths dispersive).

After correction of matrix effects and line superposition, the measured intensity of an elementary line serves as a gauge for the concentration of the element.

3 Results and interpretations

We have analyzed the powders of ten samples by X-ray fluorescence spectrometry (XRF). The elemental chemical composition "Table 1" shows high contents of CaO with percentages between 45.12% and 50.96%, and a non-negligible amount of SiO2 with percentages between 4.10% and 11.39%. In presence of traces of Al_2O_3 (between 0.45% and 1.65%), MgO (between 0.54% and 0.78%), Fe_2O_3 (between 0.38% and 1.27% Na_2O (between 0.13% and 0.19%), SrO (between 0.05% and 0.15%), P_2O_5 (between 0.07% and 0.11%), SO_3 (0.05% and 0.09%), Cl (between 0.03% and 0.04%), TiO_2 (between 0% and 0.08% %) and MnO_2 (between 0% and 0.02%).

4 CONCLUSION

Chemical elemental composition shows that the principal phases of the calcarenite rock of Rabat are calcite and quartz; they are characteristic minerals of calcareous rocks. This result is in agreement with the results found by Samaouali [7].

A good knowledge of the mineralogical composition, which plays an important role in phenomena of thermal transfers, capillary and evaporation, therefore makes it possible to better understand the mechanisms of alteration of historic monuments and to envisage solutions limiting their progress.

Table 1. Results of chemical analysis by (XRF)

	CaO	P,a,F	SiO_2	$A_{l2}O_3$	MgO	Fe_2O_3	Na_2O	SrO	P_2O_5	SO_3	K_2O	Cl	TiO_2	MnO_2	Sum
Ech1	50,83	41,37	5,40	0,66	0,71	0,49	0,16	0,10	0,08	0,07	0,05	0,04	0	0	100
Ech2	50,96	41,1	5,42	0,74	0,68	0,51	0,19	0,09	0,07	0,09	0,06	0,04	0	0	100
Ech3	49,23	40,8	7,08	1,01	0,62	0,72	0,15	0,06	0,08	0,05	0,06	0,03	0,04	0	100
Ech4	45,12	39,11	11,3	1,65	0,75	1,27	0,18	0,05	0,11	0,05	0,17	0,03	0,08	0	100
Ech5	50,19	43,52	4,37	0,45	0,55	0,38	0,13	0,15	0,08	0,04	0,02	0,03	0,02	0	100
Ech6	45,68	46,52	5,77	0,61	0,54	0,41	0,13	0,08	0,07	0,06	0,05	0,03	0	0	100
Ech7	49,88	41,64	5,26	1,27	0,78	0,59	0,16	0,06	0,08	0,06	0,09	0,03	0,04	0	100
Ech8	47,44	42,19	7,50	0,93	0,61	0,70	0,16	0,09	0,10	0,05	0,07	0,03	0,05	0,02	100
Ech9	50,9	43,05	4,10	0,47	0,56	0,39	0,15	0,14	0,09	0,05	0,02	0,03	0	0	100
Ech10	50,88	41,3	5,45	0,68	0,60	0,49	0,16	0,10	0,10	0,06	0,04	0,03	0,03	0,02	100

5 References

1. Bellitir D., 1998. Etude géologique et techniques des calcarénites de la côte atlantique et des calcaires mésozöques des rides prérifaines, utilisés au Maroc dans les constructions anciennes et actuelles. Thèse de Doctorat, Université de GAND, Belgique, 268p.
2. Bellitir D., Nijs R., Asebriy L et Aberkan M., 1998. Evolution de la calcarénite dans les constructions; vitesse d'altération en fonction des facteurs naturels et industriels. Mines, Géologie et Energie, 57 : 83-88.
3. H. Azouaoui, N. El Hatimi, and N. El Yamine, Plio-Quaternary formations of the Casablanca area (Morocco): sedimentological and geotechnical aspects, Bull Eng Geol Environ, 2000, **59**, 59.
4. R. Benboughaba, Phd thesis, Etude Sédimentologique et Géophysique des formations littorales plio-quaternaires de l'axe Rabat-Kénitra(Maroc), Université Mohammed V of Rabat, Morocco, 2001.
5. A. Samaouali, L. Laânab, Y. Geraud, A. Nounah, and M. Boukalouch, Experimental study of chemical deterioration of chellah monument stones, Phys. Chem. News, 2008, **44**, 103.
6. Y. El Rhaffari, M. Hraita , A. Samaouali, M. Boukalouch and Y. Géraud, Thermal and petrophysical characteristics of calcarenite rocks used in the construction of monuments of Rabat, Romanian Journal of Materials 2014, **44**(2), 153.
7. A. Samaouali, Processus d'altération et de transfert de fluides dans les pierres calcarénites du monument Chellah– Rabat, Phd thesis, University Mohammed V of Rabat, Morocco, 2011.

Analysis of the Behaviour of Semi Rigid Steel End Plate Connections

A. Bahaz[1], S. Amara[1], J.P. Jaspart[2] and J.F. Demonceau[2]
[1]University Amar Telidji, SREML laboratory, Laghouat, Algeria
[2]Liège University, ArGEnCo Department, Liège, Belgium

Abstract. The analysis of steel-framed building structures with full strength beam to column joints is quite standard nowadays. Buildings utilizing such framing systems are widely used in design practice. However, there is a growing recognition of significant benefits in designing joints as partial strength/semi-rigid. The design of joints within this partial strength/semi-rigid approach is becoming more and more popular. This requires the knowledge of the full nonlinear moment-rotation behaviour of the joint, which is also a design parameter. The rotational behaviour of steel semi rigid connections can be studied using the finite element method for the following three reasons: i) such models are inexpensive; ii) they allow the understanding of local effects, which are difficult to measure accurately physically, and iii) they can be used to generate extensive parametric studies. This paper presents a three-dimensional finite element model using ABAQUS software in order to identify the effect of different parameters on the behaviour of semi rigid steel beam to column end plate connections. Contact and sliding between different elements, bolt pretension and geometric and material non-linearity are included in this model. A parametric study is conducted using a model of two end-plate configurations: flush and extended end plates. The studied parameters were as follows: bolts type, end plate thickness and column web stiffener. Then, the model was calibrated and validated with experimental results taken from the literature and with the model proposed by Eurocode3. The procedure for determining the moment–rotation curve using finite element analysis is also given together with a brief explanation of how the design moment resistance and the initial rotational stiffness of the joint are obtained.

1 Introduction

Extended end-plate connections are widely used in steel structures as moment resistance connections and as an alternative to fully welded ones that have been considered for use in steel frames. These connections consist of end-plates welded to the end of beams and field bolted to the connecting columns as illustrated in Fig. 1. The understanding of the behaviour of joints is enhanced by developing analytical models. Various forms of analysis and mathematical modelling methods have been suggested in order to study the semi-rigid characteristics of beam-to-column joints and their influence on the response of the rest of structural members. Eurocode 3, Part 1.8 [1] includes a simplified analytical model to represent the semi-rigid connection behaviour based on the moment- rotation (M-φ) characteristics of the joint. The component method [1] is not applicable to end-plate joints with pre-tensioned bolts and research into these types of end-plate connections is therefore much needed [2].

Many researches on end-plate steel connections have been carried out analytically and experimentally in order to determine an accurate method for predicting the connection rotational behaviour. Pertold et al. [3] performed two experimental tests on beam-to-beam and beam-to-column joints. For both tests, failure occurred at the column flange in the compression zone. Lima et al. [4] tested three specimens to enable the prediction of the moment resistance and the rotation capacity about the minor axis of the beam-to-column semi-rigid joints. These investigations motivated the development of a mechanical model to assess the structural response of joints.

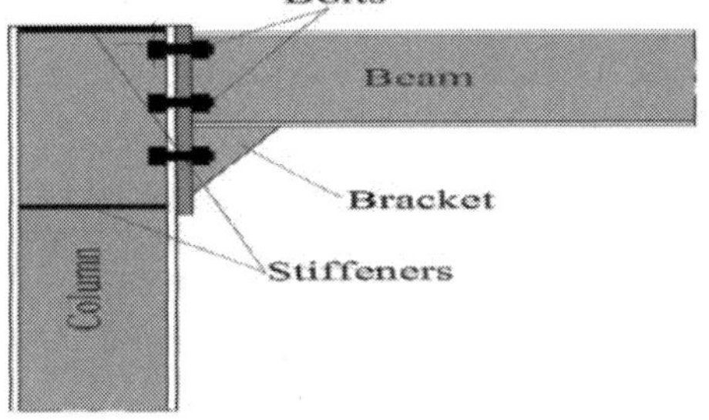

Fig. 1. Typical end-plate connection

Furthermore, connection types and details are numerous and innovative with many parameters, that must be accounted collectively to characterise the behaviour of the joint. Such parameters include whether the end plate is flush or extended, whether the end plate extends beyond one or both of the beam flanges, the diameter and the grade of the bolts, the number of bolt rows, the vertical and horizontal spacing of the bolts, the end-plate thickness, the stiffening of steel and the coefficient of friction at the contact surface. Because of this, it is almost impossible to study the behaviour of these joints comprehensively except by physical tests [5].

However, due to limited number of tests, the numerical simulations are widely used currently. The more accurate finite element model for parametric analysis is particularly important [6]. The first study into joint behaviour using FEM has been carried out by Bose et al [7], which is related to welded beam-to-column joints, including: plasticity, strain hardening and buckling. Maggi et al. [8] performed parametric analyses on the behaviour of bolted extended end-plate connections using 3D models. Dai et al. [9] made a simulation study of ten (10) fire tests on restrained steel beam-column assemblies using five different types of joints: fin plate, flexible end- plate, flush end-plate, web cleat and extended end-plate.

The aim of this paper is to present a full 3D ABAQUS [16] FE model to obtain the behaviour of steel beam-to-column bolted extended and flush end-plate joints. This model includes: contact and sliding between different elements; bolt pre-tension; geometric and material non- linearity. The results from the FE Analysis (FEA) are verified by comparing the moment–rotation curve with those obtained from experimental results found in the literature and with the results achieved using the model proposed by EC3 [1]. A brief explanation of how the design of the moment resistance and the initial rotational stiffness of the joint are obtained.

2 End-plate connections

End-plate connections have become more popular in steel building constructions due to their economy, simplicity of fabrication, and good structural performance. In the literature end-plate connections have been presented with three typologies: header, flush and extended end-plate connections. From which five observations [9] can be made:

- The most studied are the extended and double end-plate connections.
- Most models use four bolts and no stiffeners.
- 3D FE are mostly used for bolts, although truss, beams and plane stress elements have been used. The nut is less commonly modelled, however, when it is modelled, a 3D FE are used.

The extended and the flush end-plate 6-bolts unstiffened connections was analysed in this work as shown in Fig. 2. The analysis was conducted using ABAQUS/Standard software [16]. Both materials and geometry nonlinearities were considered. All parts of the models are presented in detail as follows.

3 Finite element modelling

ABAQUS software[16] offers a wide range of options regarding element types, material behaviour and numerical solution controls, as well as graphic user interfaces (known as Abaqus/CAE), auto-meshers, and sophisticated postprocessors and graphics to speed the analyses. In this paper, the structural system modelling is based on the use of this software.

3.1. Element types and meshing

A suitable type of element needs to be selected for the finite element model to obtain reliable results [10]. The element C3D8I (an 8-node linear brick, incompatible mode) is adopted for H-shaped steel beams, columns, high strength bolts and plate.

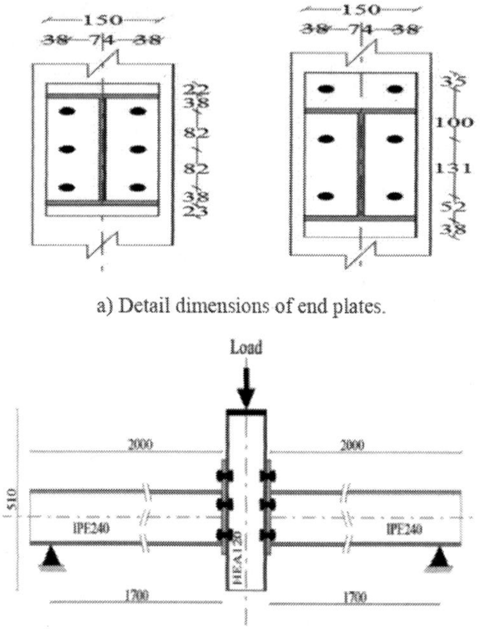

a) Detail dimensions of end plates.

b) The specified dimensions.

Fig. 2. Model description.

This element can effectively avoid shear locking phenomenon (comparing with element C3D8R), which will significantly affect the initial stiffness of the connection [6]. Since this solid element has no rotational degree of freedom, the number of elements through the thickness of each component plate plays a critical role [10]. For the regions where the hexahedra formulation was not possible to be used the wedge approach was used with an element designated by C3D6 which is a 6-node linear triangular prism and constant pressure element [11].

To reduce the number of contact planes and the complexity of the model [12], the bolt nut forms an integral component with the bolt shank rather than as an individual part, as shown in Fig. 3.

Different mesh sizes have been examined to determine a reasonable mesh that provides both reliable results with less computational time. The results show that, if the mesh is too coarse, a convergence problem will occur as the contact element was used between the column flange and the endplate surface. However, if the mesh is too fine, the computational time will be excessive. The finite element mesh adopted for all joint components is shown in Fig.3, with the smallest and largest element sizes being 5 mm and 50 mm, respectively. The fine mesh was created at the region around the bolts and studs to achieve reliable results. The finite element mesh of each specimen contains approximately 12300 elements.

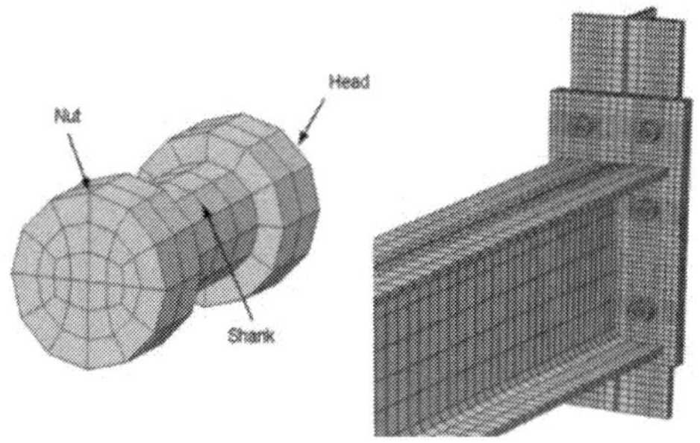

Fig. 3. Finite element model

3.2. Material proprieties

The material behaviour used for the joint was represented by the bilinear plus nonlinear stress-strain curve. The plasticity behaviour in the connection was represented by the isotropic work hardening assumption and the Von Mises criteria. The yield ultimate tensile strength and the elongation of the connection components are given in Table 1. These values were selected based on tensile coupon tests conducted on structural steel specimens [13].

Table 1. Mechanical properties.

	$f_y\,(MPa)$	$f_u\,(MPa)$	$A\%$
Beam web	343	456	27.0
Beam flange	356	480	33.5
Column web	345	456	27
Column flange	338	435	33.5
End-plate	310	464	27
Bolts M16/8.8	893	110	5

The true stress and true strain relationship is then obtained and tabulated for the use in ABAQUS [16] as shown in Fig. 4.

3.3. Boundary conditions

Due to the symmetry of geometry, one-half of a real specimen was modelled with double symmetry conditions. All nodes at the mid-span section of the column were prevented from translating in the X direction as shown in Fig. 5. For the application of the end support boundary conditions, the end of the steel beam was converted to a rigid body.

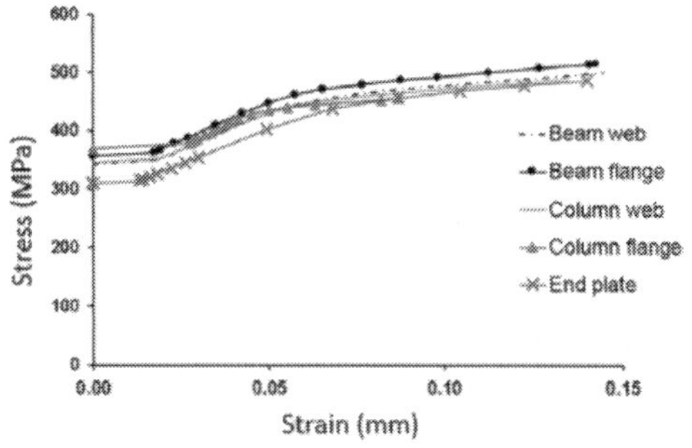

Fig. 4. True plastic stress-strain curves.

A reference point was specified on the rigid body and it was prevented from translation in the X direction to avoid the buckling effect. A support with only its translational displacements in the Y and Z directions restrained was defined herein to simulate a simply supported boundary condition.

3.4. Contact interaction and constrain conditions

The interaction between the different components of joint should be taken into consideration in the modelling. The contact between these components, viz. end plate-column flange, bolt-column flange and bolt-end plate were represented using the surface-to-surface contact interaction technique.

One surface was taken as a master surface and the other as slave one. The normal behaviour is assumed to be a ''hard'' contact pressure-over closure relationship. Because of a tangential behaviour, penalty frictional formulation is used and according to Shi G et al. numerical simulation results [5], the coefficient of friction was considered as 0.44. The welds between the beam end plate and the column stiffeners were defined as a Tie constraint.

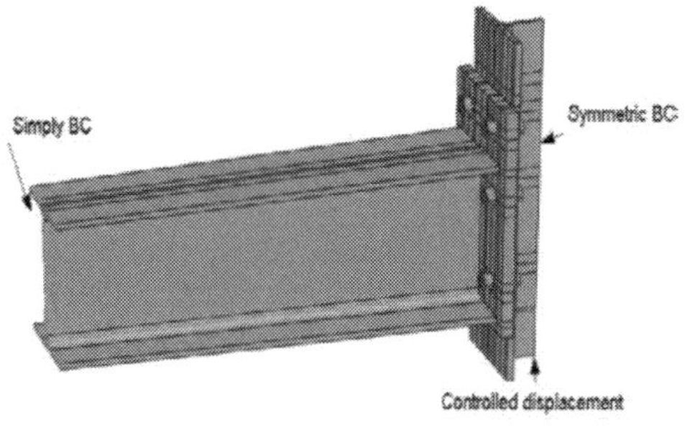

Fig. 5. Boundary conditions.

3.5. Load application and analysis method

A nonlinear analysis is carried out in two steps. In the first step the boundary conditions and contact are defined, loads are applied on the contact surfaces to make convergence easier, in the same step the pretension load was applied to bolts connecting the beam to the column.

The bolt load feature available in ABAQUS [16] was invoked in order to include the bolt pretension in this

https://doi.org/10.1051/matecconf/201814902058

step and in the second step a displacement-controlled load was applied to the column using the arc length method (Riks) as the convergence algorithm.

This method is generally used to predict the unstable and nonlinear collapse of a structure. It is an implicit load control method. In the RIKS method, the load is applied proportionally in several load steps. In each load step, the equilibrium iteration is performed and the equilibrium path is tracked in the load–displacement space. This method is often used in static analysis and shows to be a strong method for nonlinear analysis.

4 Validation of the proposed FE model

4.1. Comparison of moment-rotation curves

Figure 6 shows the schematic of how the joint rotation 'θ' was calculated based on the predicted vertical deflection 'u' at a distance of L = 850 mm using the expression [15] given below:

$$\theta = tan^{-1}\left(u/l\right) \tag{1}$$

The validation of the proposed finite element model is examined by comparing its numerical results with those obtained experimentally by Abidelah et al. [13] in terms of moment capacity.

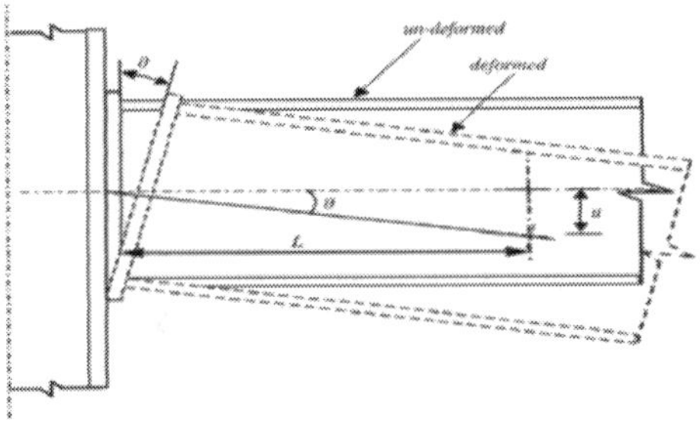

Fig. 6. A sketch of the deflected beam to estimate the connection rotation.

From Table 2 and Fig. 7, it can be seen that finite element model results show good correlation with the experimental results qualitatively and quantitatively and have proven to simulate the behaviour of the connection effectively.

Table 1. Comparison between experimental and FE moment capacity.

Test configuration	Experimental test (kN.m)	FE model (kN.m)	Ratio FE/EXP
Test EEP	54	62.06	1.19
Test FEP	48	52.03	1.08
		average	**1.13**

4.2. Comparison of failure modes

A comparison between the failure mode observed in tests and the one obtained by numerical simulation is shown in Fig. 8. It is clear that, as the displacement increases, an increase of the gap between column flange and end plate in the tensile zone occurs. In compressive zone, the beam flange is subjected to a local buckling. Moreover, the shear deformation of the panel zone was evident. As shown in Fig. 8, the plastic deformation, gap between endplate and column flange, panel zone deformation and local buckling were well simulated. It is clear from the above comparisons that the FE model simulates, with acceptable accuracy, the behaviour of steel endplate bolted connections.

5 Parametric study

Based on the verified model, 3D solid models of steel beam-to-column connection (model description shown in Fig. 2) were constructed to investigate the behaviour of two connection types; extended and flush end plate connections. The studied parameters are as follows: end-plate thickness, bolts type and presence of column stiffeners.

The naming convention used for sample identification (XX–BT-S) is a combination of the connection configuration (XX), bolt type (BT) and the presence of column stiffeners (S).
The connection configurations are designated as follows:
- (FEOS) flush stiffened end plate with ordinary bolts.
- (FEP) flush end plate with pretension bolts.
- (EEOS) extended stiffened end plate with ordinary bolts.
- (EEP) extended end plate with pretension bolts.

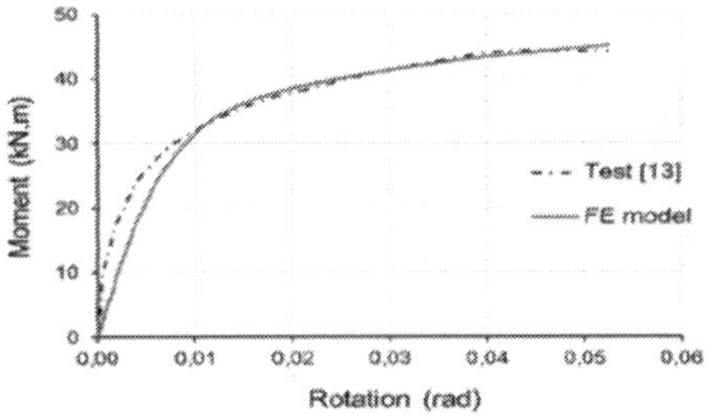

a) Flush end plate (FEP).

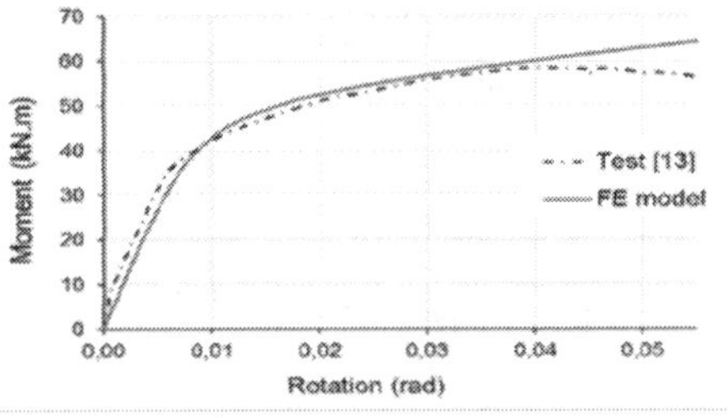

b) Extended end plate (EEP).

Fig. 7. Comparison of moment–rotation curves for two end-plate joints specimens.

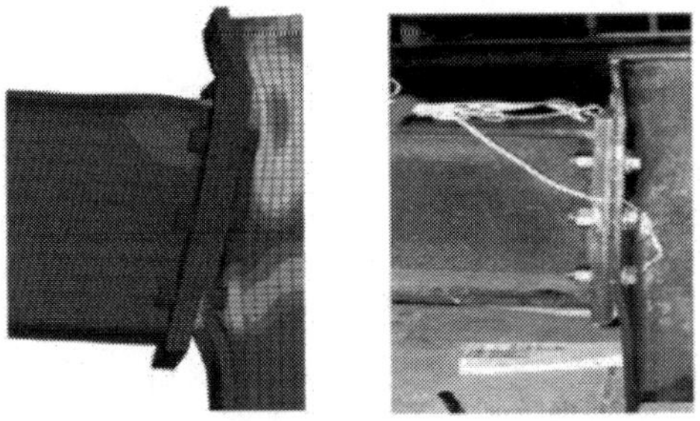

Fig. 8. Experimental and predicted failure modes of the FEP 15.

6 Results and discussions

6.1. Effect of end-plate thickness

Three end-plate thicknesses 8 mm, 15 and 20 mm for both connections have been used. The result comparison of the extended end plate is shown in Fig. 9. It can be seen that when the endplate thickness decreases to 8 mm, low moment and rotation capacities are achieved. It is also shown from Fig. 9 and Fig. 10 that the thickness of the endplate has a great influence on the initial rotation stiffness of the bolted extended end plate connections compared to the flush end plate connections.

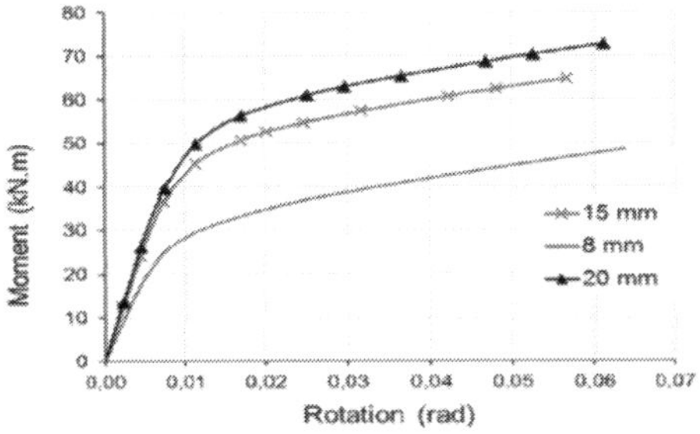

Fig. 9. Moment–rotation curve variation for EEP, with different end plate thickness (t_p).

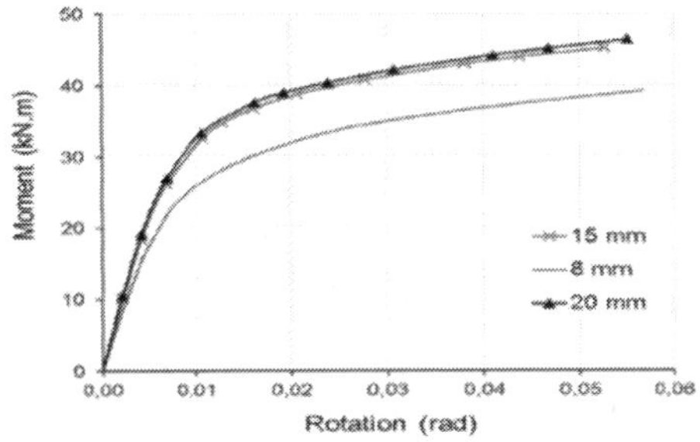

Fig. 10. Variation moment–rotation curve of FEP with different end plate thickness (t_p).

As the thickness increases from 8 to 15 mm, the initial stiffness increases, but when the thickness is greater than 15 mm, little difference is observed between the moment–rotation curve of EEP15 and EEP20. The reason for this is that as the thickness of the endplate increases to a certain value, the strength of the bolt becomes the influential factor, most areas of the endplate remain elastic and the bolt will yield first. This permits higher tensile forces in the bolts which will increase the bending moment. However, the brittle bolt failure was observed when the endplate increased to 15 and 20 mm, which caused the low rotation capacity as the brittle failure of the bolt decreased the rotation capacity of the connection. As the thickness of the plate increased, the mode of failure changed from complete yielding of the flange (end plate 8) to bolt fracture with flange yielding (end plate 15) and to bolt fracture (end plate 20). EC3—part1.8 [1] has named these 3 modes of failure of bolted endplate connection as mode 1, mode 2 and mode 3 respectively.

From the above discussion, it is suggested that for the endplate connections in bending, it is necessary to limit the thickness of the endplate in order to avoid brittle failure of the bolts and to obtain ductile joints by the yielding of the endplate in bending. The endplate should also have a certain thickness to avoid the brittle failure of the endplate itself.

6.2. Effect of the presence of column stiffeners with ordinary bolts

The influence of the presence of column stiffeners with ordinary bolts on the behaviour of the end plate connections is studied in this section. Four models are analysed, as shown in Fig. 12. It can be observed that using column stiffeners with ordinary bolts instead of pre-tensioned bolts increases considerably the ultimate moment and the rotation capacity of the connections. In addition, it can be noticed that stiffening the column web increases slightly the initial stiffness of the connection, since the flange column without stiffeners bends easily as shown in the Fig. 13.

The presence of column stiffeners minimizes the deformation in the panel zone of the connection and it has a significant effect on the ultimate moment and rotation capacities than that of unstiffened pre-tensioned bolted connections.

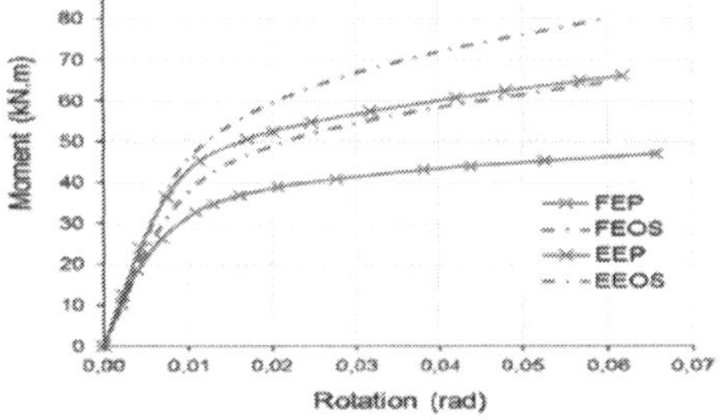

Fig. 12. Moment–rotation curves of the pretensioned bolted and column stiffness joints (t_p=15mm).

Furthermore, the design of end plate connections without column stiffeners will produce column web buckling, leading to a premature failure in the joint. Figure 13 shows that an increase in capacities is rather significant in stiffened connections with ordinary bolts than with pre-tensioned ones.

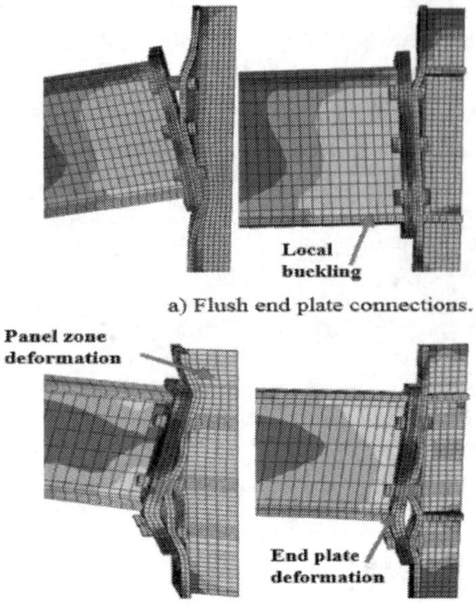

a) Flush end plate connections.

b) Extended end plate connection.

Fig. 13. Distribution of Von Mises stress in EEP and EEOS connections.

Conclusion

The results presented herein focused on the behavioural variations of bolted extended and flush end plate connections due to changes in plate thickness, bolts type and presence of column stiffeners. Based on the studied results, the following conclusions can be drawn:

- The presence of column stiffeners with ordinary bolts instead of the pre-tensioned bolts alone, gives an important moment capacity but a slight increase on the initial stiffness. For the flush end plate, the effect is more significant.
- The thickness of the end plate has a great influence on the behaviour of extended plate joints compared to the flush ones.
- The comparison between the use of ordinary bolts with column stiffeners and pre-tensioned bolts alone, cannot be achieved, because the economic parameter must be taken into account (design cost, cost of reparation and erection of each type).

Finally, the finite element results, which have been well validated against test results, can provide extra valuable results for the mechanical behaviour of joints, which are difficult to measure in physical tests, such as pressure distribution caused by bolt pretension, friction between the end-plate and the column flange and the principal stress in the connection.

The finite element method allows for further parametric analyses of bolt end-plate connections to be implemented so as to obtain comprehensive results that can be used to propose a complete analytical design procedure that is consistent with the component method of joint design used in Eurocode 3.

References

1. CEN. Eurocode 3, *Design of steel structures. part 1.8: Design of joints*, (2005).
2. C. H. Shi G, Shi YJ, Wang YQ, Li S, Experimental study on semirigid end-plate connections in multi-story steel frames, J. Tsinghua University (Science Technol., no. **44**, p. 391–4, (2004).
3. J. Pertold, R. Y. Xiao, and F. Wald, Embedded steel column bases: I. experiments and numerical simulation," J. Constr. Steel Res., vol. **56**, pp. 253–270, (2000).
4. L. R. O. De Lima, S. A. L. De Andrade, P. C. G. Da, and L. S. Da Silva, Experimental and mechanical model for predicting the behaviour of minor axis beam-to-column semi-rigid joints, Int. J. Mech. Sci., vol. **44**, no. 6, pp. 1047–1065, (2002).
5. G. Shi, Y. Shi, Y. Wang, and M. A. Bradford, Numerical simulation of steel pretensioned bolted end-plate connections of different types and details, vol. **30**, pp. 2677–2686, (2008).
6. M. Wang, Y. Shi, Y. Wang, and G. Shi, Numerical study on seismic behaviors of steel frame end-plate connections, JCSR, vol. **90**, pp. 140–152, (2013).
7. S. K. Bose, G. M. McNeice, and A. N. Sherbourne, Column webs in steel beam-to-column connexions part II-Design recommendations, Comput. Struct., vol. **2**, no. 1–2, pp. 281–301, (1972).
8. Y. I. Maggi, R. M. Gonçalves, R. T. Leon, and L. F. L. Ribeiro, Parametric analysis of steel bolted end plate connections using finite element modeling,vol. **61**, pp. 689–708, (2005).
9. C. Díaz, M. Victoria, P. Martí, and O. M. Querin, FE model of beam-to-column extended end-plate joints, J. Constr. steel Res., vol. **67**, no. 10, pp. 1578–1590, (2011).
10. H. Thai and B. Uy, Finite element modelling of blind bolted composite joints, JCSR, vol. **112**, pp. 339–353, (2015).
11. H. Augusto, C. Rebelo, and L. S. Silva, Modelling of the Dissipative Behaviour of Partial-Strength Beam-to-Column Steel Connections, 15th World Conf. Earthq. Eng., no. 1994, (2012).
12. E. R. Romero, Finite element simulation of a bolted steel joint in, Tampere university of technology Elena, (2010).
13. A. Abidelah, Analyse numérique du comportement d'assemblages métalliques. Approche Numérique Et Validation Expérimentale, Université Blaise Pascal – Clermont II, 2009.
14. B. Gil and E. Bayo, An alternative design for internal and external semi-rigid composite joints. Part II : Finite element modelling and analytical study, Eng. Struct., vol. **30**, pp. 232–246, (2008).
15. K. S. Al-Jabri, A. Seibi, and A. Karrech, Modelling of unstiffened flush end-plate bolted connections in fire, J. Constr. Steel Res., vol. **62**, no. 1–2, pp. 151–159, (2006).
16. *ABAQUS user'smanual version **12.1***. Pawtucket, R.I.: Hibbit. Karlsson and Sorenson, (2012).

Simplified Model for the Hybrid Method to Design Stabilising Piles Placed at the Toe of Slopes

Dib M.[1] and Kouloughli S.[1]
[1]Université des frères Mentouri, Civil Engineering department, Constantine, Algeria

Abstract. Stabilizing precarious slopes by installing piles has become a widespread technique for landslides prevention. The design of slope-stabilizing piles by the finite element method is more accurate comparing to the conventional methods. This accuracy is because of the ability of this method to simulate complex configurations, and to analyze the soil-pile interaction effect. However, engineers prefer to use the simplified analytical techniques to design slope stabilizing piles, this is due to the high computational resources required by the finite element method. Aiming to combine the accuracy of the finite element method with simplicity of the analytical approaches, a hybrid methodology to design slope stabilizing piles was proposed in 2012. It consists of two steps; (1): an analytical estimation of the resisting force needed to stabilize the precarious slope, and (2): a numerical analysis to define the adequate pile configuration that offers the required resisting force. The hybrid method is applicable only for the analysis and the design of stabilizing piles placed in the middle of the slope, however, in certain cases like road constructions, piles are needed to be placed at the toe of the slope. Therefore, in this paper a simplified model for the hybrid method is dimensioned to analyze and design stabilizing piles placed at the toe of a precarious slope. The validation of the simplified model is presented by a comparative analysis with the full coupled finite element model.

1 Introduction

Nailing unstable slopes using piles is one of the successful reinforcement techniques. Piles were proven to be efficient to stabilize precarious slopes through the additional resisting force that they offer. Many authors studied this subject, among them: De Beer and Walleys (1972), Ito and Matsui (1975), Sommer (1977), Fukuoka (1977), Wang *et al.* (1979), Nethero (1982), Oakland and Chameau (1984), Poulos (1995), and Poulos and Chen (1997).

In general, the Design methods for slope stabilizing piles are: pressure or displacement-based method, and numerical methods. In 2012, Kourkoulis *et al.* proposed a hybrid method for the design of slope-stabilizing piles. This method merges the accuracy of the finite element method with simplicity of the analytical approaches. However, the hybrid method was only developed to analyze piles placed in the middle of unstable slopes.

2 The hybrid method for analysis and design of slope stabilizing piles

The hybrid method for the design of slope-stabilizing piles was proposed by Kourkoulis *et al.* in 2012. It consists of two steps; (1): an analytical estimation of the resisting force needed to stabilize the precarious slope, and (2): a numerical analysis to define the adequate pile configuration that offers the required resisting force for a desired pile head displacement.

In the analytical part, the estimations of the needed resisting force to stabilize the precarious slope is done by the conventional slope stability techniques. However, in the numerical part, Kourkoulis *et al.* proposed a new approach. This approach aims to reduce the numerical cost by reducing the size of the model; so instead of modeling the entire slope-piles system, only two piles and the surrounded soil are modeled as presented in figure 1.

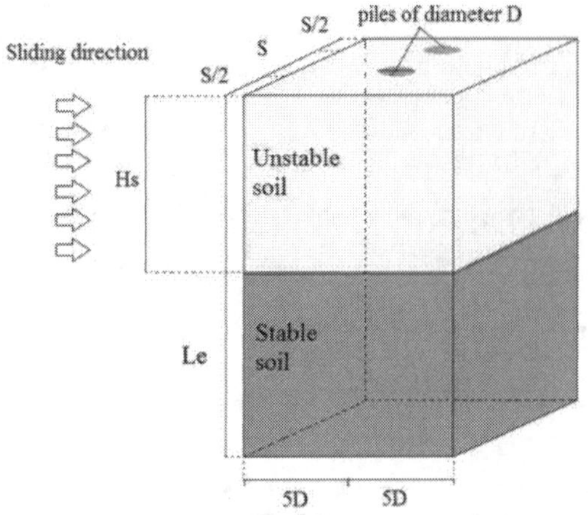

Fig. 1. The simplified model for the hybrid method [12].

The length of the simplified model is 10D, and its width is 2S (two times the spacing between piles). In full scale, it is assumed that the slope is reinforced by a row of piles with spacing S repeated indefinitely. Hs is the sliding depth, and Le represents the pile length.

For piles placed in the middle of precarious slopes, it was found that beyond a distance of 5D up-slope of the piles row, the soil displacement profile becomes almost uniform [12]. Therefore, 10D (2 times 5D) was chosen to be the length of the simplified model.

The Hybrid method was developed to design piles placed in the middle of the slope, but in certain cases like road constructions, piles are needed to be placed at the toe of the slope. Therefore, in this paper a simplified model for the hybrid method is dimensioned to analyse and design stabilizing piles placed at the toe of a precarious slope.

3 Investigation of the uniformity of soil displacement profile for piles placed at the toe of the slope

3.1 Properties of slopes and piles

In order to investigate the uniformity of soil displacement profile for piles placed at the toe of the slope, different slope geometries with pre-existing potential sliding interface (PEPSI) were tested, with inclination angle ranging from 21° to 26° and pile spacing of 3D and 4D.

For the unstable and the stable parts of the slopes, soil properties are resumed in table 1. Soil is simulated with an elastoplastic model (Mohr-Coulomb failure criterion).

Table 1. Soil properties.

	Unit Wight (KN/m^3)	Undrained shear strength Cu (KPa)	Young modulus E (KPa)
Unstable part	16	72	48000
Stable part	19.2	191	482000

Concrete piles with 1.2 m diameter were used in the analysis, where an elastic behaviour was chosen in the simulations. Materials properties of piles are resumed in table 2.

Table 2. Piles properties.

	Unit Wight (KN/m^3)	Young modulus E (KPa)	Poisson's ratio
Concrete	25	25 000 000	0.2

3.2 The numerical analysis

Fully coupled slopes with pre-existing potential sliding interface (PEPSI) were analyzed using finite element code Abaqus, an example is presented in figure 2.

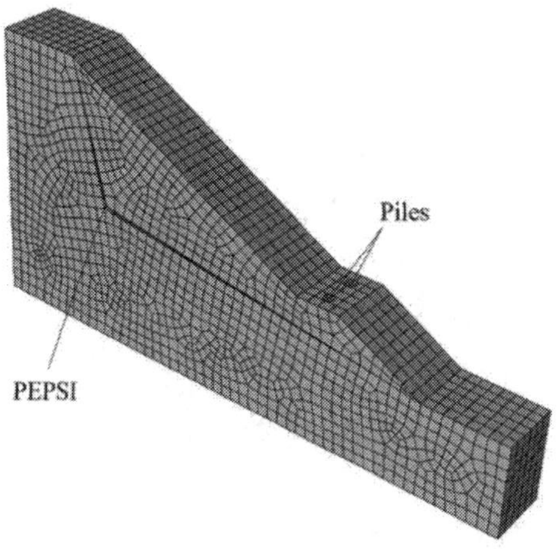

Fig. 2. Finite element mesh of the fully coupled slope reinforced by two piles with preexisting potential sliding interface (PEPSI).

3.3 Results and discussions

By analysing the displacement profiles, the following results were obtained: Figure 3 represents the displacement profiles at a distance of 5D up-slope of the piles.

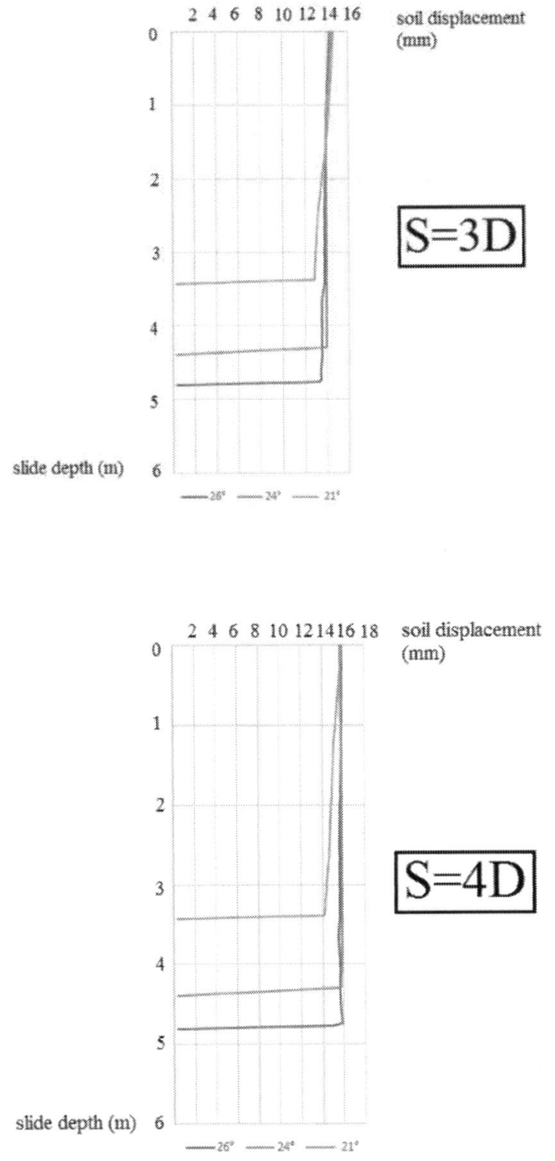

Fig.3. Soil displacement profiles at a distance of 5D up-slope of the piles row for different slopes inclinations, and pile spacings of 3D and 4D.

One can notice that either for pile spacing of 3D or 4D, the soil displacement profile at a distance of 5D up-slope of the piles is almost uniform. Figure 4 represents a cut at a distance of 5D upslope of piles with soil displacement contours.

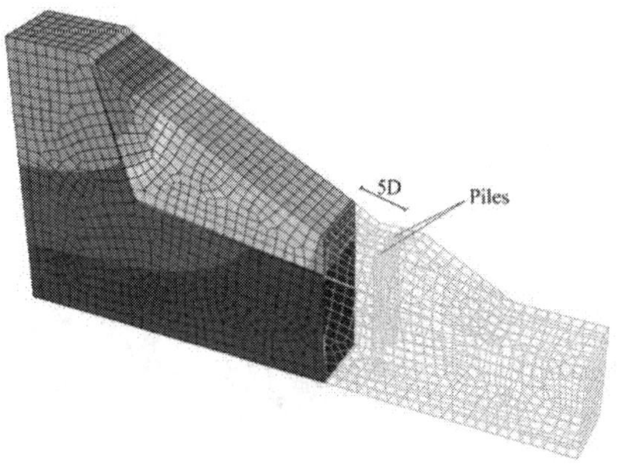

Fig. 4. Cut at a distance of 5D up-slope of piles with soil displacement contours.

4 The simplified model for the hybrid method to design stabilizing piles placed at the toe of slopes

As found above, the soil displacement profile at a distance of 5D up-slope the row of piles is almost uniform. Therefore, the simplified model for the hybrid method to design stabilizing piles placed at the toe of slopes is identical to that developed by Kourkoulis *et al.* in 2012, The length of simplified model is 10D, and its width is 2S, as shown in figure 1.

5 Validation of the simplified model against fully coupled reinforced slope analysis

In order to validate the simplified model, a fully coupled analysis was conducted first, it consists of a slope with 24° inclination reinforced by a row of concrete piles of 1.2 m diameter, and spacing S=3D. Then a simplified model was analysed as presented in figure 5.

https://doi.org/10.1051/matecconf/201814902059

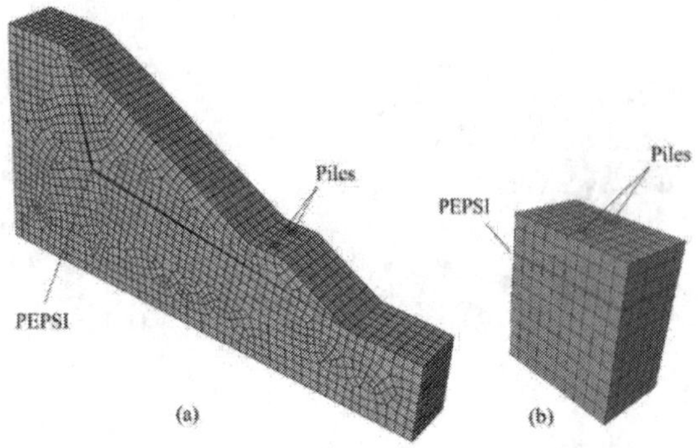

Fig. 5. The validation of the simplified model, finite element mesh for: (a) the fully coupled model, (b) the simplified model.

6 Results and discussions

For the fully coupled model, the pile head deflection necessary to stabilize the slope by coupled methodology was found to be 6.3 mm.

Using the method of slices, it was found that the required resisting force (RF) needed to stabilize the slope is 341 KN/m. After the analysis of the simplified model, the needed RF (341 KN/m) was given by the piles at a pile head deflection of 7.4 mm, with an error of 17%. (figure 6)

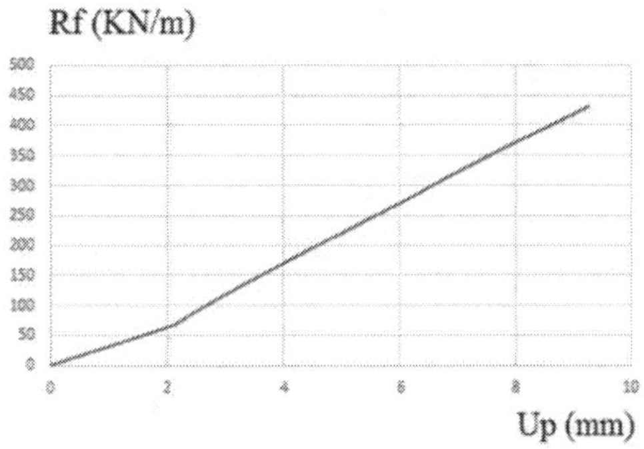

Fig. 6. Resisting force (Rf) with respect to pile head deflection (Up) given by the simplified model.

7 Conclusions

In this research about the simplified model for the hybrid method to design stabilizing piles placed at the toe of slopes, the following conclusions were reached:

The displacement profile at a distance of 5D up-slope the pile location may be considered as uniform.

The simplified model for the analysis and the design of slope stabilizing piles placed at the toe of the slope is identical to that of piles located in the middle of the slope.

References

1. De Beer, E. E. and Wallays, M., Forces Induced in Piles by Unsymmetrical Surcharges on the Soil Around the Pile, *International Proceedings of the 5th Euro Conference on Soil Mechanics and Foundation Engineering, Madrid,* **1**, 154-176 (1972)

2. Ito, T. and Matsui, T., Methods To Estimate Lateral Force Acting on Stabilizing Piles, Soils and Foundations, **15**, 4, 43-59 (1975)

3. Sommer, H., Creeping slope in a stiff clay, *Proc., 9th Int. Conf. on Soil Mechanics and Foundation Engineering*, A. A. Balkema, Rotterdam, Netherlands, 113–118 (1977)

4. Fukuoka, M., The effects of horizontal loads on piles due to landslides, *Proc., 9th Int. Conf. on Soil Mechanics and Foundation Engineering, Japanese Society of Soil Mechanics and Foundation Engineering, Tokyo*, 27–42 (1977)

5. Wang, M. C., Wu, A. H., and Scheessele, D. J., Stress and deformation in single piles due to lateral movement of surrounding soils, Behavior of deep foundations: ASTM special technical publication, **670**, Raymond Lunggren, ed., ASTM, West Conshohocken, PA, 578–591 (1979)

6. Nethero, M. F., Slide control by drilled pier walls, Application of Walls to Landslide Control Problems, *Proc., ASCE National Convention, R. B. Reeves, ed., ASCE, New York*, 19–29 (1982)

7. Oakland, M. W. and Chameau, J. L. A., Finite-Element Analysis of Drilled Piers Used for Slope Stabilization: In Laterally Loaded Deep Foundations: Analysis and Performance, ASTM STP 835, J.A. Langer, E.T. Mosley and C.D. Thompson, Eds., ASTM, 182-193 (1984)

8. Poulos, H. G., Design of Reinforcing Piles to Increase Slope Stability, Canadian Geotechnical Journal, **32**, 808-818 (1995)

9. Poulos H. G. and Chen L. T., Pile Response Due to Excavation-Induced Lateral Soil Movement, Journal of Geotechnical and Geoenvironmental Engineering, **123**, 2, 94-99 (1997)

 https://doi.org/10.1051/matecconf/201814902059

10. Kourkoulis, R. *Interplay of Mat Foundations and Piles With a Failing Slope. PhD thesis,* National Technical University of Athens, Greece (2009)

11. Kourkoulis, R. Gelagoti, F. Anastasopoulos, I. and Gazetas, G., Slope Stabilizing Piles and Pile-Groups: Parametric Study and Design Insights, Journal of Geotechnical and Geoenvironmental Engineering, **137**, 7, 663-678 (2011)

12. Kourkoulis, R. Gelagoti, F. Anastasopoulos, I. and Gazetas, G., Hybrid Method for Analysis and Design of Slope Stabilizing Piles, Journal of Geotechnical and Geoenvironmental Engineering, **138**, 1, 1-14 (2012)

13. Gazetas, G., and Uddin, N., Permanent deformation of preexisting sliding surfaces in dams, J. Geotech. Eng., **120**, 11, 2041–2061 (1994)

Analysis of the restoration of mosque Ali Betchine of Algiers

Analyse de la restauration de la mosquée Ali Betchine d'Alger

Attari Nassereddine[1], Mechouari Zineb[1], Laoues Souad[1]
[1] Ecole Polytechnique d'Architecture et d'Urbanisme (epau), Alger, Algérie.
(Laboratoire Ville Architecture et Patrimoine, LVAP)

Abstract: Ali Betchine is one of the oldest mosques in the city of Algiers of the Ottoman period. Classified as a national monument in 1947. In 1986, UNESCO gave Ali Betchine its universal cultural dimension. This classification of the mosque has resulted in the building currently being protected from any major architectural shuffle in order to protect and preserve it. The Ali Betchine mosque underwent several transformations during the French colonization, it was converted into a church under the name "Notre Dame des Victoires". After the independence of the country, this church was converted into a mosque. The evolution of heritage protection in Algeria can be monitored after a period of follow-up by the country's first regulations on heritage protection. It is noticed that an awareness of the architectural heritage of the country was beginning to emerge. The Ali Betchine mosque underwent restoration work which ended in September 2010 after more than twelve years of closure to the faithful. The paper traces the different stages of its restoration.

Résumé : Ali Betchine est l'une des plus anciennes mosquées de la ville d'Alger de la période ottomane. Classée monument historique en 1947 à l'échelle nationale. En 1986, l'UNESCO a donné à Ali Betchine sa dimension culturelle universelle. Ce classement de la mosquée a fait que l'édifice bénéficie actuellement d'une protection qui empêche tout remaniement architectural majeur dans le but de le protéger et le conserver. La mosquée Ali Betchine a subi plusieurs transformations pendant la colonisation française, elle était convertie en église sous l'appellation Notre Dame des Victoires. Après l'indépendance du pays, cette église a été reconvertie en mosquée. On peut suivre l'évolution de la protection du patrimoine en Algérie, et cela après une période de suivi par les premières règlementations propres au pays en matière de protection du patrimoine. On remarque qu'une prise de conscience du patrimoine architecturale du pays commençait à voir le jour. La mosquée Ali Betchine a subi des travaux de restauration qui ont prirent fin en septembre 2010 après plus de douze années de fermeture aux fidèles. Le papier retrace les différentes étapes de sa restauration.

Introduction:

En termes de patrimoine classé, l'Algérie est un pays riche, mais qui se trouve altéré par l'usure du temps, par les interventions insolites des habitants, les affectations et les modifications de l'occupant français pendant la période coloniale, auquel s'ajoute le manque d'entretien. Face à cet état, des opérations de réhabilitation et de restauration ont étaient entreprises. La restauration est une opération qui a pour but la réinsertion de l'édifice concerné dans son contexte urbain, leur réappropriation en tant que mémoire collective et la restitution de leur architecture originelle dans les limites des possibilités de l'évolution urbaine du contexte physique. En se basant sur des documents fiables, et dans certain cas sur des études de toutes les traces archéologiques selon le cas à restaurer, les restaurateurs doivent respecter le plus proche possible les règles émises par les différentes chartes et conventions internationales.

Nous avons abordé la restauration en tant que mode d'intervention sur le patrimoine bâti, qui par son activation, peut assurer leur conservation et jouer un rôle sur leur durabilité, et l'impact qui va générer sur l'environnement. L'article présente une analyse et une évaluation de la restauration de la mosquée Ali Betchine.

La mosquée Ali Betchine porte le nom du corsaire vénitien « *Piccinini* », un grand négociant italien converti à l'islam au mois de radjeb de l'an 1032 de l'hégire (1622) après J-C. Selon **Klein**, c'était un grand amiral de la flotte algérienne entre 1630 et 1646.

Il est venu volontairement à Alger poursuivre son métier d'armateur de navires destinés à la course qu'il pratiquait déjà dans l'Adriatique. D'autres sources signifient qu'il aurait été d'abord

un captif chrétien, affranchi et converti ensuite à l'islam par *Hadj Mustafa BenQara Ali*. Différents documents établissent qu'il était le

chef de *Ta'ifat al-Riyyas*. Ce poste le plaçait au premier rang sur la scène politique, puisqu'il était considéré comme « *gouverneur et capitaine général de la mer et terre* ».

Après avoir embrassé la religion musulmane, Piccinini, devenu « Ali Betchine », fait construire cette mosquée sur l'emplacement d'un ancien bagne d'esclaves chrétien sil se fait aider par l'un de ses amis armateur, « Fath Allah Ben Khodja ». 1 Durant la période coloniale, la mosquée Ali Betchine a subi plusieurs modifications et d'autres affectations.

La mosquée a été utilisée comme pharmacie par l'armée française. Puis, servira d'église sous le nom de «***Notre Dame des Victoires***»,de 1843 à 1962. On y aménagea alors une lourde cloche qui ne manqua pas d'endommager le minaret.

Fig1. la mosquée Ali Betchine, la porte Bab-El-Oued en 1834. Source http://jf.vinaccio.free.fr/site0201/alg02013

Fig2. L'intérieur de l'église Notre Dames des Victoires (1953)
Source :http://alger-roi.fr/Alger/rue_bab_el_oued

Description de la mosquée : La mosquée Ali Betchine possède une grande salle de prière de forme carrée, située à cinq mètres du niveau des rues voisines, repose sur un groupe important de magasins voûtés. On accède, au niveau de la mosquée par deux escaliers donnant ; l'un sur la rue Bab-el-Oued, l'autre sur la rue la Casbah.

La salle de prière est surmontée d'une coupole centrale octogonale construite sur quatre piliers épais et des colonnes en pairs (sous chaque arc, il existe deux colonnes, une en tuf et une en pierre), se trouvant dans la cour centrale. Cette cour est entourée de trois galeries couvertes de vingt petites coupoles.

Le minaret qui forme l'angle des rues Bab el Oued et de la Casbah s'élevait autrefois à une hauteur d'environ 26m ; il n'en subsiste au lendemain de la conquête française que la tour de base, qui ne dépasse plus guère 15m de haut. Sa face ouest est encore flanquée de nos jours de la fontaine dite d'*Ain- ech-Châra*.

Par sa forme et sa construction ; le minaret se rattache à un type local des plus simples : une tour carrée qu'aurait surmontée un lanternon carré, qui furent défigurés dans leur ensemble lors de leur reconversion coloniale.

Les points d'appui, entre la coupole et les galeries, sont constitués par les piliers cruciformes, situés aux quatre angles de la salle de prière. Des coupolettes, à base octogonale épaulent la coupole centrale et complètent la couverture du carré central et en occupant les quatre angles de la calotte octogonale. Entre les piliers d'angles et sur chaque face de la salle carrée, se trouvent quatre colonnes en ciment groupées par deux. Les piliers et les colonnes reçoivent des arcs brisés, situés au-dessus de la grande coupole. Les arcs séparent la salle de prière des galeries qui l'entourent sur ces quatre côtés. Les galeries sont couvertes de coupolettes, à la base octogonale, reliées au carré de base par des pendentifs, et qui sont séparées par des doubleaux.

On note simplement que les piliers d'angles sont remplacés par seize colonnes, groupés par quatre à chaque angle. D'autre part, les galeries qui bordent le carré central ne sont pas couvertes totalement par des coupolettes. Celles-ci n'apparaissent qu'aux angles des galeries ; et en avant du mihrab, le reste est couvert par des voûtes d'arêtes. Des berceaux surbaissés remplacent les coupolettes et complètent la couverture du carré central.

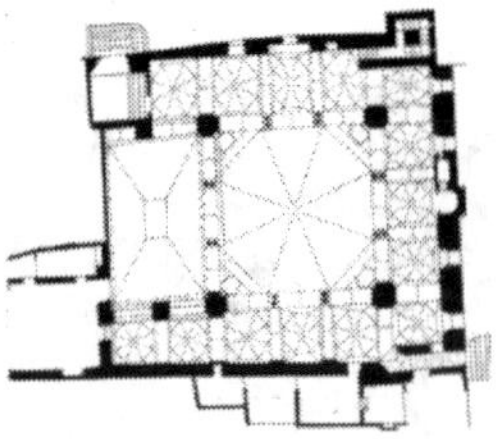

Fig 3. Plan de la mosquée Ali Betchine **[2)**

Fig 4. Photo de la mosquée pendant les travaux.

Fig 5. Photo de la mosquée 2016

Cet intérieur était blanchi à la chaux et n'offrait aucune décoration. A droite, en entrant par la rue de la Casbah, en face du mihrab et sur le dépourvu de galerie, se trouvait une cour à ciel ouvert renfermant quelques arbres et un jeu d'eau. C'est dans cette partie, qu'ont été installés le chœur et la sacristie de l'église N. D. des Victoires.

Si le minaret n'é été que partiellement dérasé en 1860, les latrines et les lieux d'ablutions ont été en revanche entièrement démolis. Leur emplacement a été englobé, selon Devoulx, dans l'habitation portant le n°2 de la rue de la Casbah.

La façade septentrionale, donnant sur la rue de la Casbah, et percée de trois fenêtres ogivales, contient au rez-de-chaussée nef boutiques dont sept seulement sont apparentes aujourd'hui, et la grande porte d'entrée, que quatre marches assez larges relient à la nef. Cette issue est actuellement clôturée par une porte sculptée, d'un assez joli travail, qui appartenait primitivement à la mosquée Ketchaoua, et qui a été placée là par les Français eu 1843. Au-dessus de la rosace médiale de chacun des deux battants de cette porte, on lit la phrase suivante, qui se détache en relief comme le reste de l'ornementation: « ماشاءالله », que la volonté de Dieu s'accomplisse. On attribue ces sculptures au Maître « Ahmed ben Lablabtchi » le responsable de la corporation des menuisiers. **(3)**

Sur le plan architectural, l'édifice religieux est considéré parmi les premières et authentiques typologies de mosquées ottomanes actuellement subsistantes. Sa superficie s'étend sur environ 500m².

1. Les travaux de restauration de la mosquée Ali Betchine :

a. Consolidation :

Les fissures présentes sur les murs, les arcs et les coupoles de la mosquée ont été réparées selon deux procédés différents en fonction de leur degré d'importance :

- Les fissures ayant une ouverture inférieure à 3mm de largeur ont été réparées avec des agrafes métalliques associées avec des injections sous pression,
- Les fissures ayant une ouverture supérieure à 3mm de largeur ont été réparées avec du mortier armé en treillis soudés avec injection sous pression.

Pour cela, il faudra noter qu'avant tout, ils ont décapé les enduits intérieurs et extérieurs pour permettre des mesures précises de l'ouverture des fissures.

La réparation avec les agrafes métalliques a été exécutée de la manière suivante :

1- Après le décapage des enduits, ils ont procédé à des percements dans la maçonnerie de 30cm de part et d'autre de la fissure, à une profondeur d'au moins 20cm, et un espacement de 15cm dans le sens de la longueur de la fissure. Les trous avaient un diamètre de 25mm.
2- Ensuite, ils ont mis en place des agrafes en acier maxydable de diamètre 8 et 1.00m de longueur, dont les parties encastrées ont été enrobées de mortier bâtard.
3- Les fissures ont été ensuite colmatées par injection sous pression d'un lait de chaux et ciment avec des proportions équivalentes (01 volume de chaux, 01 volume de ciment 250), et l'injection a été faite après le nettoyage des fissures à l'eau.

La chaux était de type hydraulique, la chaux aérienne est formellement proscrite pour les injections.

La réparation des fissures avec le mortier armé a été exécutée comme suit :

1- Apres décapage de l'enduit et piquage des joints à 1cm de profondeur, il été procédé à la mise en place de l'armature en treillis soudés de Ø4 ou Ø5 qui doit déborder de 1m de chaque côté de la fissure et maintenue à 0.5cm de la surface de la maçonnerie.
2- L'armature en treillis soudés a été accrochée à son appui à l'aide de tiges métalliques de 8Ø en acier inoxydable, disposées chaque 50cm et encastrées dans toute l'épaisseur de la maçonnerie. Elles été enrobées de mortier bâtard dans des trous de 25mm de diamètre.
3- Ils ont exécuté ensuite l'enduit en trois couches (ou deux couches pour l'extérieur), dont la composition et le dosage étaient précis.

4- Les fissures ont été colmatées par injection sous pression d'un mortier bâtard après avoir été nettoyées à l'eau. Le dosage du mortier pour l'injection est comme suit :
   - ✓ Ciment : 01 volume,
   - ✓ Chaux hydraulique : 01 volume,
   - ✓ Sable : 07 volumes.

5- Des chainages horizontaux ont été exécutés au-dessus des murs et arcs de la salle de prière. Ils ont des dimensions de 90 x 20cm (ferraillés en 8 T12, cadre Ø6 chaque 20cm), et 30 x 20cm (ferraillés en 4 T12, cadre Ø6 chaque 20cm).
Les coupolettes restituées reposeront sur des chainages horizontaux de dimensions 20 x 20 (ferraillés en 4 T12, cadre Ø6 chaque 20cm).
Des chainages verticaux et horizontaux ont été exécutés dans les locaux. Ils ont des dimensions de 20 x 20cm et étaient ferraillés en 4T12, cadres Ø6 chaque 20cm.

Des étaiements traditionnels sont effectués pour la reconstruction de planchers d'autres étaiements de type chevalement sont fait pour soutenir les naissances d'arcs dans le but est de reconstituer les chapiteaux endommagés. Des mesures à l'ultrason aux scléromètres tendres et sures pour déterminer la résistance des colonnes composées des tuf et de pierre.

b. Intervention dans les locaux- voûtes :

Un décapage a été nécessaire pour réaliser la nature et l'état réel des murs et voûtes, où plusieurs interventions ont été déjà remarquées. Une attention particulière a été portée aux voûtes, celles-ci étaient mal aérées avec la présence du phénomène de condensation qui humidifie les murs et par conséquent, rend le crépi friable. La présence d'un puits au sein même de l'épaisseur d'un mur porteur laisse à considérer les remontées capillaires susceptibles d'altérer la base des murs.

6.3.1. e. Planchers :

- Au-dessus du voligeage, ils ont procédé au coulage d'une chape en béton de 5cm d'épaisseur armée en treillis soudés avec interposition d'un film polyane.
- Ils ont procédé à l'exécution de planchers collaborant en poutre de bois (solives) et chape armée en treillis soudés dans la salle de prière et les locaux et au-dessus de l'accès à la terrasse à partir du minaret.
- Les poutres ont des dimensions de 7 x 14cm posées chaque 20cm entraxes. Elles sont encastrées dans les chainages horizontaux du plancher.
- Au-dessus des poutres, un voligeage est posé, composé de planches en bois de 2cm d'épaisseur.
- Les éléments en bois ont reçu au préalable un traitement avec un produit hydrofuge

La salle de prière après la restauration :

Fig 6. Vue sur l'ensemble de la salle de prière, 15/09/2016. Source : Auteur

Conclusion :

La restauration d'un patrimoine bâti ancien, à l'exemple de notre cas d'étude la mosquée Ali

Betchine, et cela en vue, de garantir sa conservation dans de bonnes conditions, n'est pas une mince affaire, il s'agit surtout, d'une opération compliquée, délicate et exigeante. La décision de restaurer un édifice suppose une volonté de prise en charge à long terme. La restauration n'est que le premier pas d'un processus de conservation et de mise en valeur qui ne devra jamais cesser.

Référence :

1. Chergui Samia, 2011, *Les mosquées d'Alger ; construire, gérer et conserver (XVI-XIX* siècles), 1ere éd., P U De Paris-Sorbonne, Paris, 371p .
2. Dokali Rachid, 1974, *Les Mosquées de la période turque à Alger*, 1ere éd, 1ere éd., SNED, Alger, 127p.
3. Kerdoune Aicha, 2011, *Les Mosquées historiques de la ville d'Alger,* 1ere éd., Alpha Editions, Telmcen, 256 p.
4. Duval George, 1990, Restauration et réutilisation des monuments anciens, techniques contemporaines, 1ere éd., Mardage, Belgiun, 1990, 290p.

Crack identification based on the nonlinear response of plates with variably oriented surface crack.

H. Chouiyakh[1], L. Azrar[2, 3] K. Alnefaie[3] and O.Akourri[1]
[1] Abdelmalek Esaadi University, Mathematical modelling and control, Tangier,Morocco
[2] Mohammed V University; School of Technical Education of Rabat (ENSET), Research Centre STIS, Rabat, Morocco
[3] King Abdulaziz University, Mechanical Engineering Department, Faculty of Engineering, Jeddah, Saudi Arabia

Abstract. In order to secure structural and operational safety of structures, it is important to implement a structural health monitoring (SHM) strategy to issue early warnings on damage or deterioration prior to costly repair or even catastrophic collapse. Developing a SHM strategy for structures enables evaluating structural integrity, durability and reliability of the monitored structure. Hence, the main objective of this work is to develop a damage detection procedure based on a plate's dynamic response and the Hilbert transform. Rectangular plates are considered and assumed to contain a surface crack which is centrally located, with a depth of h_0, a length of 2C and inclined with an angle β. Von Karman plate theory is adopted herein, and the crack is modeled through the line spring model given by fracture mechanics. The plate is assumed to behave nonlinearly due to large deformation. The differential quadrature method is used to investigate the linear and nonlinear dynamic behaviors of cracked plates. The influence of crack's parameters on modal properties is discussed. The eigenfrequencies of cracked plates with respect to crack half length C and orientation β are performed. For crack characterization, Hilbert transform is applied to the obtained linear and nonlinear time responses. It is shown throughout this paper that identified backbones describe changes in crack orientation.

1 Introduction

Vibration based structural health monitoring is the process of identifying the health of structures based on changes in dynamic behavior caused by damage [1]. Many techniques have been extensively discussed in literature such as: optimization techniques, regression models, signal processing, etc. Thanks to its use for vibration signal's identification [2], the Hilbert Transform is adopted in this work to detect crack presence. The dynamic behavior of cracked structures was described by many models, for 1D solids, i.e. beams containing one or multiple open cracks [3, 4], or breathing ones. Vibration of damaged 2D solids such as plates and shells was investigated in case of horizontal crack [5,6]. The oriented crack case was considered in [7,8] where the line spring model was adopted and the flexibilities were calculated based on fracture mechanics developments [9,10].
To the best of the authors' knowledge crack identification of plates is not well addressed as beams, hence Hilbert transform is investigated in this paper, where performed nonlinear forced responses are Hilbert transformed. The free vibration of the cracked plate is first investigated numerically by using the differential quadrature method. It is shown that the developed numerical procedure is accurate since it matches well with the semi-analytical results and the identified backbones describe changes in crack orientation.

2 Vibration of plates with variably oriented surface crack

A rectangular cracked plate of dimensions L_x, L_y and H and material properties of E, ρ, υ, is considered that is subjected to bending and tensile stresses. The plate is assumed to contain a crack at the centre of its top surface, inclined with an angle β with respect to x axis, having as depth h, and length 2C (see figure 1).

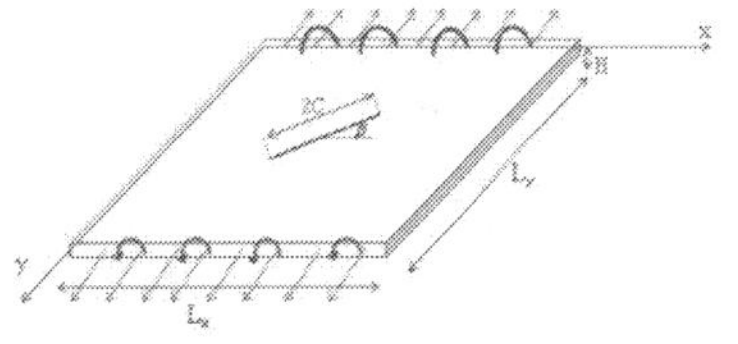

Fig. 3. Rectangular plate with inclined surface crack

For transverse vibration analysis, the considered problem is modeled through the following partial differential equation

$$D\nabla^4 w = -\rho H\frac{\partial^2 w}{\partial t^2} + N_x\frac{\partial^2 w}{\partial x^2} + F$$
$$+\frac{CD(1+\cos(2\beta))}{((\alpha_{bt}/6)+\alpha_{bb})(3+\upsilon)(1-\upsilon)h+2C}\left(\frac{\partial^4 w}{\partial y^4}+\upsilon\frac{\partial^4 w}{\partial x^2\partial y^2}\right) \quad (1)$$
$$+\frac{2CD\sin(2\beta)}{((C_{bt}/6)+C_{bb})(1+\upsilon)h+2C}\left(\frac{\partial^4 w}{\partial x\partial y^3}+\upsilon\frac{\partial^4 w}{\partial x^3\partial y}\right)$$
$$-\frac{C(1+\cos(2\beta))}{(6\alpha_{bt}+\alpha_{tt})(1-\upsilon^2)h+2C}N_0\frac{\partial^2 w}{\partial y^2}-\frac{2C\sin(2\beta)}{(6C_{bt}+C_{tt})(1+\upsilon)h+2C}N_0\frac{\partial^2 w}{\partial y\partial x}$$

where $w = w(x,y,t)$ is the transverse displacement response, ∇^4 is the bi-harmonic operator and $D = \frac{Eh^3}{12(1-\upsilon^2)}$ is the plate's flexural rigidity. N_x is the in plane force per unit length and N_0 is the in plane force caused by the presence of the crack. The compliance coefficients α_{bb}, α_{tt} and $\alpha_{bt} = \alpha_{tb}$ are used to match the stretching and bending resistance for symmetric loading and they can be found in [9]. The other compliance coefficients C_{bb}, C_{tt} and $C_{bt} = C_{tb}$ can be found in [10], they are chosen to take into account the anti-symmetric mode loading.

The boundary conditions that must be satisfied by an edge parallel to **x** edge for example are as follows:

$$\text{Clamped edge}: \ w = 0 \ ; \ \frac{\partial w}{\partial y} = 0 \quad (2\text{-a})$$

$$\text{Pinned edge}: \ w = 0 \ ; \ M_y = -D\left(\frac{\partial^2 w}{\partial y^2} + \nu\frac{\partial^2 w}{\partial x^2}\right) = 0 \quad (2\text{-b})$$

$$\text{Free edge}: M_y = 0 \ ; \ \left(\frac{\partial^3 w}{\partial y^3} + (2-\nu)\frac{\partial^3 w}{\partial y\partial x^2}\right) = 0 \quad (2\text{-c})$$

3 Free vibrations of cracked plates based on the differential quadrature method (DQM)

Thanks to the efficiency of the differential quadrature method for solving partial and differential equations, it is used here for plate's problem. A short overview is first given and the DQM formulation for the inclined cracked plate problem is presented.

3.1 An overview of the DQM

The DQM requires to discretize domain of the problem into N points. The derivatives at any point are approximated by a weighted linear summation of all the functional values along the discretized domain, as follows [3]

$$\left.\frac{d^m f(x)}{dx^m}\right|_{x=\bar{z}_k} = \frac{d^m}{dx^m}\begin{bmatrix} f(\bar{z}_1) \\ f(\bar{z}_2) \\ f(\bar{z}_3) \\ \cdot \\ \cdot \\ \cdot \\ f(\bar{z}_N) \end{bmatrix} = c_{kj}^{(m)}\begin{bmatrix} f(\bar{z}_1) \\ f(\bar{z}_2) \\ f(\bar{z}_3) \\ \cdot \\ \cdot \\ \cdot \\ f(\bar{z}_N) \end{bmatrix} \quad (3)$$

where 'm' is the order of the highest derivative appearing in the problem, $f(\bar{z}_k)$ are the values of the function at the sampling points $\bar{z}_k$ relating the mth derivative to the functional values at $\bar{z}_k$. These coefficients can be determined by making use of Lagrange interpolation formula as follows:

$$f(x) = \sum_{i=1}^{N}\frac{L(x)}{(x-\bar{z}_i)L_1(\bar{z}_i)} \quad (4)$$

where: $L(x) = \prod_{j=1}^{N}(x-\bar{z}_j), \quad L_1(\bar{z}_i) = \prod_{j=1}^{N}(\bar{z}_i-\bar{z}_j)$

The weighting coefficients for the first order derivative to the functional values at $\bar{z}_k$ can be obtained as:

$$c_{kj}^{(1)} = \begin{cases} \dfrac{L_1(\bar{z}_i)}{(\bar{z}_k-\bar{z}_j)L_1(\bar{z}_j)} & k \neq j \\ -\displaystyle\sum_{j=1, j\neq k}^{N} c_{kj}^{(1)} & k = j \end{cases} \quad (5)$$

The second, third and higher derivatives can be calculated as:

$$c_{kj}^{(m)} = \sum_{l=1, j\neq k}^{N} c_{kl}^{(1)} c_{lj}^{(m-1)} \quad (6)$$

For accurate results, we adopt the Chebychev-Gauss-Lobatto mesh distribution given for the interval [a, b] by:

$$\bar{z}_k = \frac{(b-a)}{2}\left[1-\cos\left(\frac{k-1}{N-1}\pi\right)\right] \quad k = 1,2,\ldots,N \quad (7)$$

It should be noted that this method has been used and deeply tested in case of multi-cracked beams [3, 4].

3.2 Free vibration analysis using DQM

The free vibration problem is first solved by using 2D differential quadrature method. By assuming a harmonic motion, the time-space partial differential equation (1) is reduced to the following partial differential equation :

$$D\nabla^4 w = -\rho H\omega^2 w + \frac{CD(1+\cos(2\beta))}{((\alpha_{bt}/6)+\alpha_{bb})(3+\upsilon)(1-\upsilon)h+2C}\left(\frac{\partial^4 w}{\partial y^4}+\upsilon\frac{\partial^4 w}{\partial x^2\partial y^2}\right) \quad (8)$$
$$+\frac{2CD\sin(2\beta)}{((C_{bt}/6)+C_{bb})(1+\upsilon)h+2C}\left(\frac{\partial^4 w}{\partial x\partial y^3}+\upsilon\frac{\partial^4 w}{\partial x^3\partial y}\right)$$

In order to use the DQM, the plane of the plate is discretized by an (n_e x m_e) Gauss-Lobatto-Chebychev grid. At a given point **(x_i ,y_j)**, of the plate, each partial derivative in Eq.(10) is written, in the DQM analog, as follows:

$$\left.\frac{\partial^p w(x,y)}{\partial x^p}\right|_{x=x_i} = \sum_{k_x=1}^{ne} c_{ik_x}^{(p)} w(x_{k_x}, y_j) \quad (9\text{-a})$$

$$\left.\frac{\partial^q w(x,y)}{\partial y^q}\right|_{y=y_j} = \sum_{k=1}^{me} b_{jk_y}^{(q)} w(x_i, y_{k_y}) \tag{9-b}$$

$$\left.\frac{\partial^{p+q} w(x,y)}{\partial x^p \partial y^q}\right|_{y=y_j} = \sum_{k_x=1}^{ne} c_{ik_x}^{(m)} \sum_{k_y=1}^{me} b_{jk}^{(m)} w(x_{k_x}, y_{k_y}) \tag{9-c}$$

After some mathematical developments and using a matrix form, the following eigenvalue problem is obtained

$$[K]\{w\} = \Omega^2 \{w\} \tag{10}$$

After some mathematical developments and boundary conditions incorporation, one gets:

$$\begin{bmatrix} A & A_b^t \\ A_b & 0 \end{bmatrix} \begin{bmatrix} \{w\} \\ 0 \end{bmatrix} = \Omega^2 \begin{bmatrix} \{w\} \\ 0 \end{bmatrix} \tag{11}$$

The matrix A_b is the matrix generating the considered boundary conditions. This resulting eigenvalue problem is to be solved; and $\{w\}$ is a (n_e x m_e) row vector given by :

$$\{w\}^T = \{w(x_1,y_1) \ \ldots \ w(x_1,y_{me}) \ \ldots.. \ w(x_{ne},y_1) \ \ldots w(x_{ne},y_{me})\} \tag{12}$$

The numerical solution of this eigenvalue problem allows getting the eigenfrequencies and eigenmodes of plates with various crack characteristics. The impact of both crack length and orientation on natural frequencies and eigenmodes are investigated.

4 Nonlinear free response of cracked plates with inclined crack

In order to solve the nonlinear vibration problem, Galerkin method is first used:

$$w(x, y, t) = \sum_{i=1}^{\infty} \sum_{j=1}^{\infty} u_i(x) v_j(y) q_{ij}(t) \tag{13}$$

where $u_i(x)$ and $v_j(y)$ are the eigenmodes parts in x and y directions of the cracked rectangular plate, respectively. By inserting Eq.(13) into Eq(1), the following equation is obtained:

$$\begin{aligned} D\nabla^4 w = {} & \rho H \omega^2 w(x,y) + \frac{CD(1+\cos(2\beta))}{((\alpha_{bt}/6)+\alpha_{bb})(3+\upsilon)(1-\upsilon)h + 2C} \frac{\partial^4 w}{\partial y^4} \\ & + \frac{CD\upsilon(1+\cos(2\beta))}{((\alpha_{bt}/6)+\alpha_{bb})(3+\upsilon)(1-\upsilon)h + 2C} \frac{\partial^4 w}{\partial x^2 \partial y^2} \\ & + \frac{2CD\sin(2\beta)}{((C_{bt}/6)+C_{bb})(1+\upsilon)h + 2C} \frac{\partial^4 w}{\partial x \partial y^3} \\ & + \frac{2CD\upsilon\sin(2\beta)}{((C_{bt}/6)+C_{bb})(1+\upsilon)h + 2C} \frac{\partial^4 w}{\partial x^3 \partial y} \end{aligned} \tag{14}$$

The resulting in-plane forces are formulated according to Berger formulation as follows [7]:

$$\begin{cases} N_x = \dfrac{6D}{h^2 L_x L_y} q_{ij}^2(t) \displaystyle\int_0^{L_x}\int_0^{L_y} \left[\left(\frac{\partial u_i(x)}{\partial x}\right)^2 v_j^2(y) + \upsilon \left(\frac{\partial v_j(y)}{\partial y}\right)^2 u_i^2(x) \right] dxdy \\ N_\theta = \dfrac{6D}{h^2 L_x L_y} q_{ij}^2(t) \displaystyle\int_0^{L_x}\int_0^{L_y} \left[\left(\frac{\partial v_j(y)}{\partial y}\right)^2 u_i^2(x) + \upsilon \left(\frac{\partial u_i(x)}{\partial x}\right)^2 v_j^2(y) \right] dxdy \end{cases} \tag{15}$$

Based on the 1-mode analysis and by multiplying both sides of Eq(14) by $u_i(x)$ and $v_i(y)$ and integrating over the plate's area, the following nonlinear differential equation is obtained:

$$M_{ii}\ddot{q}_{ii}(t) + \mu \dot{q}_{ii}(t) + K_{ii} q_{ii}(t) + G_{ii} q_{ii}^3(t) = 0 \tag{16}$$

Note that a weak classical linear viscous damping μ is introduced. M_{ii} is the modal mass and K_{ii}, G_{ii} are modal rigidity of the cracked plate respectively. They are given by:

$$M_{ii} = \frac{\rho H}{D} \int_0^{L_x}\int_0^{L_y} u_i^2(x) v_i^2(y) dxdy \tag{17-a}$$

$$K_{ii} = \int_0^{L_x}\int_0^{L_y} \begin{pmatrix} u_i^{IV}(x)v_i(y) + 2u_i''(x)v_i''(y) + u_i(x)v_i^{IV}(y) \\ -\dfrac{CD(1+\cos(2\beta))}{((\alpha_{bt}/6)+\alpha_{bb})(3+\upsilon)(1-\upsilon)h + 2C}\left(u_i(x)v_i^{IV}(y) + \upsilon u_i^{IV}(x)v_i(y)\right) \\ -\dfrac{2CD\sin(2\beta)}{((C_{bt}/6)+C_{bb})(1+\upsilon)h + 2C}\left(u_i'(x)v_i'''(y) + \upsilon u_i'''(x)v_i'(y)\right)dxdy \end{pmatrix} u_i(x)v_i(y)dxdy \tag{17-b}$$

$$G_{ii} = \int_0^{L_x}\int_0^{L_y} \begin{pmatrix} -P_{1ii}u_i''(x)v_i(y) + \dfrac{C(1+\cos(2\beta))}{(6\alpha_{bt}+\alpha_{tt})(1-\upsilon^2)h + 2C} P_{2ii}u_i(x)v_i''(y) \\ + \dfrac{2C\sin(2\beta)}{(6C_{bt}+C_{tt})(1+\upsilon)h + 2C} P_{2ii}u_i'(x)v_i'(y) \end{pmatrix} u_i(x)v_i(y)dxdy \tag{17-c}$$

Numerical solution of the obtained nonlinear differential equation is performed for different crack parameters and various initial conditions. The resulting nonlinear time response is analyzed using the Hilbert transform.

5 Crack detection using Hilbert transform

The Hilbert Transform (HT) of a signal s(t), is an integral transformation, from time domain to time domain, defined by [2] :

$$H(s(t)) = \frac{1}{\pi} \int_{-\infty}^{+\infty} \frac{s(\tau)}{t-\tau} d\tau \tag{18}$$

The real signal s(t) and its HT h(t), form an analytical complex signal $\tilde{S}(t)$ of the form :

$$\tilde{S}(t) = s(t) + i\,h(t) = A(t)e^{i\theta(t)} \tag{19}$$

The instantaneous amplitude $A(t)$ and phase $\theta(t)$ change with time. They are given by:

$$A(t) = \sqrt{(s(t))^2 + (h(t))^2} \quad ; \quad \theta(t) = \text{Arctan}\left(\frac{h(t)}{s(t)}\right) \tag{20}$$

Instantaneous frequency (IF) is defined as the derivative of the phase:

$$\omega(t) = \frac{d\theta(t)}{dt} \tag{21}$$

It measures the rate and direction of a phase in the complex plane. It can be estimated by different algorithms [2], the one adopted in this work is based on the following formulae:

$$\omega(t) = \frac{s(t)\dot{h}(t) - \dot{s}(t)h(t)}{2\pi(s^2 + h^2)} \tag{22}$$

The 'Freevib' method is developed by Feldman [2] to analyze systems' vibrations. Based on the Hilbert Transform, it offers a direct way to plot the system skeleton curve, which includes modal frequency and non-linearity in spring characteristic. The 'Freevib' method proposes equations for identifying instantaneous characteristics of the dynamic system (Eq.20 and Eq.22)

5 Numerical results and discussion

In this section a rectangular simply supported (SSSS) cracked plate is considered. The isotropic material properties are taken as E=7.03 10^9 N/m^2, ρ=2660 kg/m^3, υ=0.3. In order to prove accuracy of the developed numerical method, linear eigenfrequencies are first calculated by using DQM and compared with literature. Note that the plate is discretized into (12x12) points. According to Table 1, it is shown that the method is accurate, and the results match very well with the semi-analytical ones given in [7].

Table 1: Calculated ω_1 for different β and C

	Lx=0.3 Ly=0.3			
	C=0.003 (m)		C=0.0075 (m)	
	Present (DQM)	[7]	Present (DQM)	[7]
β=0	251.81	251.81	244.649	244.65
β=20	252.607	252.61	246.291	246.32
β=40	254.618	254.64	250.440	250.51
β=60	256.905	256.92	255.135	255.18
β=80	258.396	258.40	258.184	258.19
β=90	258.601	258.60	258.601	258.60

By calculating eigenfrequencies of the cracked plate, it is judicious to investigate influence of crack parameters on the modal frequency. The first normalized eigenfrequency is depicted in figure 2 with respect to crack length and orientation. It is shown that a plate with a crack, either of length C=0 or β=90°, behaves as an undamaged one. The horizontal crack (β=0°) is the most severe one. In order to illustrate the applicability of the HT to crack identification, the former plate is considered. The plate's dimensions are L_x=L_y=1; H=0.01. The crack is assumed to have a length of 2C=0.4, and a depth h_0 = 0.6H. The obtained forced responses are analyzed by using the Hilbert transform, and then the instantaneous frequency and amplitude are calculated. By relating the instantaneous frequencies to instantaneous amplitudes, the backbone curves corresponding to the nonlinear free vibration behavior are obtained.
It is shown that the identified backbones by the Hilbert Transform describe the cracked plate nonlinear free vibration behavior. The presence of the crack is indicated since the plate's identified linear eigenfrequency is decreasing. According to figure 3, the horizontal crack (β=0°) is the most severe and the plate nonlinear behavior is clearly affected by the crack orientation. According to obtained curves, it is shown that a plate with vertical crack (β=90°) behaves as an undamaged one.

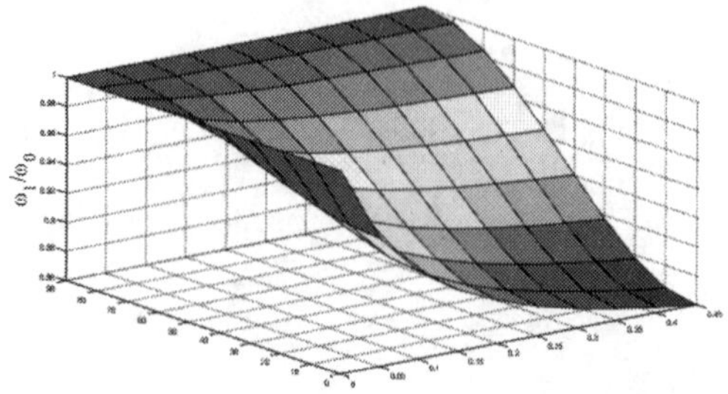

Fig. 2. First eigen frequency with respect to crack half length C and orientation β

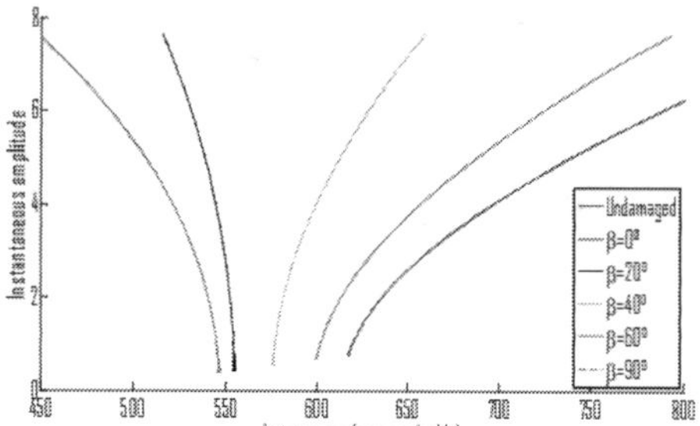

Fig. 3. Identified backbones for different crack angles

6 Conclusion

Through this work, the numerical study of cracked plate's vibrations is performed based on the differential quadrature method. The nonlinear free response of the cracked plate is investigated as well. Based on the obtained time responses and the Hilbert transform, identified backbones can describe the cracked plate nonlinear behavior due to the in plane forces. The obtained backbones indicate the presence of damage since the identified modal frequency is decreasing. Moreover, the plate's nonlinear behavior is affected by the presence of the crack. In a further work, dynamic behavior of plates with different excitation types can be investigated, as well as the quantification of crack parameters based on the nonlinear response of the plate.

References

1. C. Boller, Chang F-K, Fujino Y. Encyclopedia of Structural Health Monitoring. John Wiley & Sons, 2009.
2. M. Feldman. Hilbert transform in vibration analysis. Mechanical systems and signal processing **25**, 3 (2011).
3. H. Chouiyakh, L. Azrar, K. Alnefaie and O. Akourri. Multicracks identification of beams based on moving harmonic excitation. Structural Engineering and Mechanics, **58**, 6, (2016).
4. H. Chouiyakh, L. Azrar, K. Alnefaie and O. Akourri. Vibration and multi-crack identification of Timoshenko beams under moving mass using the

differential quadrature method. International Journal of Mechanical sciences, **120**, (2017).

5. J. R. Rice, N. Levy. Journal of Applied Mechanics, **39**, 1, (1972).
6. S. E. Khadem, M. Rezaee. Introduction of modified comparison functions for vibration analysis of a rectangular cracked plate. Journal of Sound and Vibration, **236**,2, (2000).
7. R. Ismail., M. P. Cartmell. An investigation into the vibration analysis of a plate with a surface crack of variable angular orientation. Journal of Sound and Vibration **331**,12 (2012).
8. T. Bose, A. R. Mohanty. Vibration analysis of a rectangular thin isotropic plate with a part-through surface crack of arbitrary orientation and position. Journal of Sound and Vibration **332**, 26 (2013).
9. J. Rice, N. Levy, The part-through surface crack in an elastic plate. Journal of Applied Mechanics **39** (1972).
10. P.F. Joseph, F. Erdogan. A surface crack in shells under mixed-mode loading conditions. ASME, Transactions, Journal of Applied Mechanics **55** (1988).

Mechanical behavior of the reinforced concrete frame with masonry filling Comportement mécanique des portiques en béton armé avec remplissage en maçonnerie

Jalal KETTAR[1], Khadija BABA[1], Abderrahman NOUNAH[1] and, Lahcen BAHI[2]
[1]GCE Laboratory, High School of Technology –Sale-, Mohammed V University in Rabat, Morocco
[2]3GIE Laboratory, Mohammedia School of Engineering, Mohammed V University in Rabat, Morocco

Abstract. Masonry is often used in the most of reinforced concrete structure constructions as a filling material that is important on structural characteristics. Structural contribution in the calculations was neglected or misunderstood, mainly due to the lack of a practical calculation methods and an appropriate regulatory tool.
The analysis of frames filled with masonry is very complex. This complexity is linked from one part to the difference in the nature of elements and its behavior that make up the masonry itself (brick and mortar) and their interaction, and on the other part, for the large dispersion that characterizes the bricks as well as the execution's quality parameters which make it difficult to define reliable criteria for the masonry. The objective of this work is to experimentally highlight the influence of the hollow brick masonry filler, commonly used in Morocco, on reinforced concrete frames subject to lateral stresses, to deepen understanding the seismic behavior of the masonry structures by evaluating the structural performance of a specimen wall. These experimental results will be compared to those found by modeling prototypes, using SAP 2000 software, based on various approaches and models as well as other results deduced from the other researchers. The experimental study was carried out according to standard NF EN 1052-3 on two reinforced concrete frames, of dimensions (2m X 1.6m), the one with the masonry filling, and the other without filling in order to determine the initial characteristic resistance to the shearing of the masonry walls. The obtained results showed that a filling has a beneficial effect on rigidity which can be doubled compared to an empty frame. in the same way the lateral resistance. But this effect is much contrasted; it depends a lot on the characteristics essentially of the materials (bricks and concrete). This is the main reason, which justifies the divergence of the results deduced from the nine models that we used.

Résumé. Dans la majorité des constructions type structure en béton armé, la maçonnerie est souvent utilisée comme matériau de remplissage dont l'influence sur les caractéristiques des structures est importante. Dans les calculs, sa contribution structurale a été négligée ou méconnue et ce, principalement, en raison de l'absence d'une méthode de calcul pratique ainsi que d'un outil réglementaire approprié. L'analyse des ossatures remplies de maçonnerie est très complexe. Cette complexité est liée d'une part à la différence de nature et de comportement des éléments qui composent la maçonnerie elle-même et à leur interaction et d'autre part, à la grande dispersion qui caractérise les briques ainsi que la qualité d'exécution qui rendent difficile la définition de critères fiables pour la maçonnerie. L'objectif de ce travail est de mettre en évidence expérimentalement l'influence du remplissage en maçonnerie de briques creuses, couramment utilisées au Maroc, sur des portiques en béton armé soumis à des sollicitations de type latéral, en vue d'approfondir la compréhension du comportement sismique des structures en maçonnerie, et ce par l'évaluation de la performance structurelle d'un spécimen de mur. Ces résultats expérimentaux vont être confrontés à ceux trouvés par modélisation de ces prototypes, en utilisant logiciel SAP 2000, s'appuyant sur diverses approches et modèles, et aussi d'autres déduites à partir des résultats d'autres chercheurs. L'étude expérimentale a été effectuée selon la norme NF EN 1052-3 sur deux cadres en béton armé de dimensions (2m x 1.6m), l'un, avec un remplissage en maçonnerie, et l'autre sans remplissage afin de déterminer la résistance caractéristique initiale au cisaillement des murs en maçonnerie. Les résultats obtenus, montrent qu'un remplissage entraine un effet bénéfique sur la rigidité, qui peut être doublé par rapport à un portique vide. Ainsi que la résistance latérale. Mais cet effet est très contrasté, il dépend beaucoup des caractéristiques essentiellement des matériaux (briques et béton). C'est la raison principale, qui justifie la divergence des résultats déduits à partir des neufs modèles qu'on a utilisé.

Keyswords: Reinforced concrete frame, Walls, Masonry, Hollow bricks, Tests, shearing, Rigidity, Modeling

1 Introduction

La maitrise du comportement des constructions en maçonnerie lors d'un séisme est devenue une nécessité; surtout après les pertes humains et les dégâts matériels très importants, provoqués par le séisme d'AL Hoceima au Maroc le 24 Février 2004, dégâts qui étaient d'autant plus amplifiés que la plupart de ces constructions étaient de type remplissage par maçonnerie. La recherche de méthodes pratiques de calcul, tenant compte du rôle des panneaux de remplissage dans la réponse sismique des structures, passe par la modélisation de la structure sollicitée à échelle 1, ce qui va permettre de diminuer les imprécisions de calcul. Cette modélisation ne peut se faire sans une bonne compréhension et une évaluation des effets, et principalement l'effort tranchant, que peut causer l'existence du remplissage en maçonnerie sur les structures sollicitées aux séismes et aussi la maitrise des caractéristiques des matériaux constituants les cadres remplis par la maçonnerie.

Cette étude propose une étude expérimentale présentant l'influence des murs de remplissage en maçonnerie sur le comportement des structures en béton armé soumises à un effort latéral. Les résultats expérimentaux seront confrontés à ceux obtenus par modélisations numériques à l'aide du logiciel SAP2000 et basées sur la méthode des bielles. Ce travail vise à améliorer la réglementation en vigueur, à savoir RPS 2011.

2 Matériels et méthodes

Nous avons analysé le comportement de murs en maçonneries sous une pression latérale et en quasi-statique. Ces essais doivent permettre d'évaluer la capacité portante et le mécanisme de rupture d'un mur en maçonnerie. Dans le montage expérimental, les matériaux de construction les plus courants et qui respectent les standards de la construction au Maroc, ont été utilisés. Cependant, nous avons testé deux spécimens de corps d'épreuve : Le premier, indexé ND1, représente un cadre vide, chargé unidirectionnellement, qui servira de référence. Le deuxième corps d'épreuve, réferencé ND2, représente un cadre identique au premier, mais rempli de maçonnerie de briques creuses (figure 1).

Les deux corps d'épreuve sont soumis à une charge verticale répartie de 6 t/m, constante durant tout l'essai, simulant le poids des étages supérieures. Ils sont en outre soumis à une charge latérale croissante

Fig.1a. Cadre vide (ND1) **Fig.1b.** Cadre rempli (ND2)

Fig.1. Corps d'épreuve

2.1 Description des corps d'épreuve

Les dimensions du remplissage sont de 160 cm pour la largeur (L) et de 120 cm pour la hauteur (h), soit un élancement de 0,75.

Le cadre est dimensionné et ferraillé de la même façon, que s'il appartenait au rez-de-chaussée d'un bâtiment courant (le béton est de classe 25/30 et l'acier est de type FeE50). Les poteaux ont une section carrée de 20 x 20 cm, comportant un ferraillage de 4T12 et les poutres ont la même section de 20 x 20 cm avec un ferraillage constitué de 3 barres T12 en bas et 3 barres T8 en haut; et un ferraillage transversal consitué de cadres T6.

Les panneaux de remplissage sont constitués par une maçonnerie de briques creuses en terre cuite (6 trous) , 28x15x7cm, assemblées par un mortier standard de ciment d'une épaisseur moyenne de 15 mm.

2.2 Instrumentation et Mesures

Les écrasements des panneaux sont réalisés par la machine de cisaillement des murs (Marque 3R°), suivant la Norme NF EN 1052-3. Cette machine est équipée de deux vérins de force identique (200KN), ces derniers, travaillent avec un système hydraulique, et sont pilotés par un pupitre Quantium, équipés d'un capteur de déplacement (figure 2).

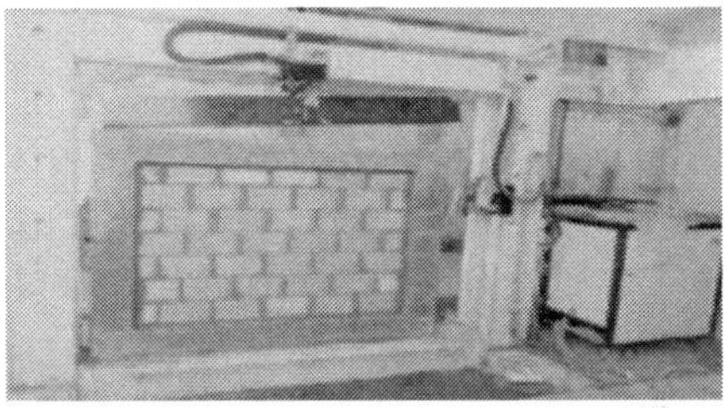

Fig.2. Appareil de cisaillement

2.3 Modelisation des murs en maçonnerie

La modélisation des murs en maçonnerie sous chargement latéral se fait selon deux approches soit locale, et bien globale. l'approche locale consiste en la modélisation par éléments finis très complexe censés, représenter la forme réelle des blocs de brique et de mortier, en se basant sur un modèle de l'endommagement, de plasticité, de fissuration ou bien mixte. La complexité de la modélisation rend cette approche très demandeuse en temps de calcul et nécessite une précision au niveau des caractéristiques des matériaux et leurs mise en œuvre. (figure 3). Quant à l'approche globale elle consiste à remplacer le mur en maçonnerie par des éléments simples type barres, soumis à un effort de compression sous forme de bielle, suivant le sens de chargement. (Figure 4).

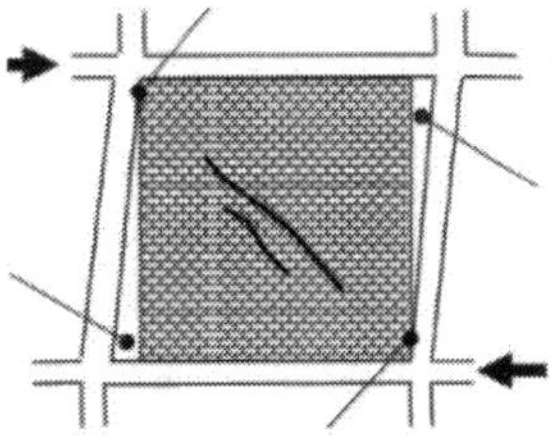

Fig.3. Modélisation par

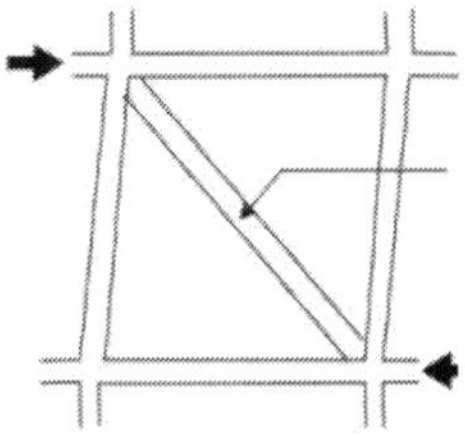

Fig.4. Bielle comprimée

Les caractéristiques mécaniques des éléments barres, à savoir le module de Young, le coefficient de Poisson, la résistance à la compression correspondent à ceux de la maçonnerie point sur lequel s'accordent tous les experts, par contre en ce qui concerne les caractéristiques géométriques, et principalement la largeur, de ces bielles, les avis divergent. En effet, plusieurs modèles, basés sur des études expérimentales, ont été établies (Stafford-Smith, 1963[1], Mainstone, 1971[2], klingner et Bertero, 1976[3]).

2.3.1 *Modèle de Mainstone* [2]

Cette approche est la plus utilisée, Mainstone a réalisé plusieurs essais en compression diagonale sur les modèles réduits de cadres et à grande échelle ; utilisant le concept de la bielle diagonale ainsi que le paramètre de la rigidité relative pour étudier la rigidité et la résistance de remplissage. La largeur de la diagonale équivalente *a* est en fonction de la raideur flexionnelle relative du remplissage et des poteaux constituant le cadre λ

$$\lambda = \sqrt[4]{\frac{E_m t \sin 2\theta}{4 E_c I_c \square}}$$

<u>Pour λ_h inférieur à 5 on a :</u>

* pour remplissage en brique

$$a=0{,}175.\lambda^{-0{,}4}.\sqrt{H^2+L^2}$$

* Pour remplissage en béton

$$a=0{,}115.\lambda^{-0{,}4}.\sqrt{H^2+L^2}$$

<u>Pour λ_h supérieur à 5 on a :</u>

* pour remplissage en brique

$$a=0{,}16.\lambda^{-0{,}3}.\sqrt{H^2+L^2}$$

* Pour remplissage en béton

$$a=0{,}11.\lambda^{-0{,}3}.\sqrt{H^2+L^2}$$

Avec :
Ec : module d'Young du portique,
Ic : inertie de la section du portique,
Em : module d'Young du remplissage,
t : épaisseur du remplissage,
L : longueur du remplissage,
h; D; H : sont définis sur la figure 5

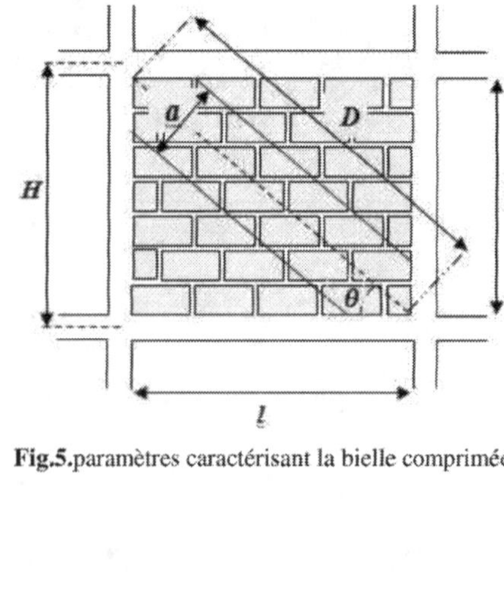

Fig.5.paramètres caractérisant la bielle comprimée équivalente.

2.3.2 *Modèle Elouali* [4]

Il a appliqué les formules de Mainstone [2] sur des structures dont le paramètre de rigidité est inférieur à 5, en corrigeant le coefficient pour l'adapter à ces types de remplissage.

*.Pour les briques d'argiles et de béton :

- $a/D=0.135\ (\lambda H)^{-0.4}$

*Pour les briques de chaux et sable :

- $a/D=0.175\ (\lambda H)^{-0.4}$

2.3.3 *Modèle de Dawe (Dawe et Al, 1989)* [5]

- $$a = \frac{\pi}{1.5x\lambda_h}\cos\theta + \frac{\pi}{1.5x\lambda_t}\sin\theta$$

Avec : $$\lambda_t = \left(\frac{E_m t_i \sin(2\theta)}{4 E_c I_b l_i}\right)^{1/4}$$

$$\lambda_{\square} = \left(\frac{E_m t_i \sin(2\theta)}{4 E_c I_c \square_i}\right)^{1/4}$$

I_b et Ic sont les inerties respectivement des poutres et des poteaux.

2.3.4 *Modèle de Durrani (Durrani & Al)* [6]

$$a = \gamma \sin(2\theta)\ D$$

$$\gamma = 0.32\sqrt{\sin(2\theta)}\left(\frac{h^4 E_m t_i}{m E_c I_c h_i}\right)^{-0.1}$$

$$m = \left(1 + \frac{6 E_h I_h h}{\pi \varepsilon_c I_c L}\right)$$

2.3.5 *Modèle de Hrennikoff (Hrennikoff,1941)* [7]

Le modèle de Hrennikoff est un modèle adéquat pour la modélisation d'un milieu homogène, isotrope, élastique linéaire (Hrennikoff, 1941). Ce modèle a été appliqué pour le béton armé (Kotronis, 2000, Mazars *et Al.*, 2002, Kotronis *et al*, 2003[8,9 et 10]), nous proposons d'étudier la possibilité d'étendre cette approche à la maçonnerie.

$$A_v = \frac{3}{8}\frac{3k^2-1}{k}\ at$$

$$A_{\square} = \frac{3}{8}(3-k^2)\ at$$

$$A_d = \frac{3}{16}\frac{(1-k^2)^{3/2}}{k}\ at$$

Avec t l'épaisseur du voile.

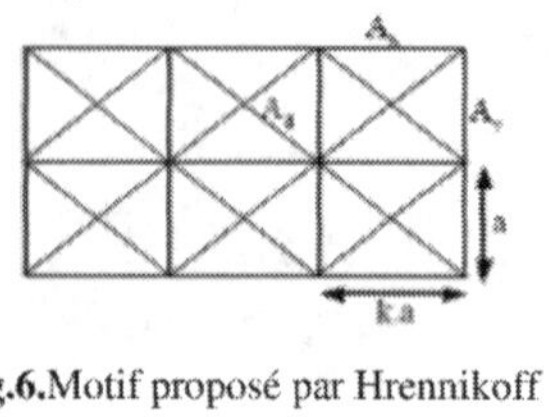

Fig.6. Motif proposé par Hrennikoff

2.3.6 *Modèle de Liauw et Kwan 1984 [11]*

Il a travaillé sur l'analyse des portiques remplis par une approche par éléments finis incrémentale et itérative, qui compte des non-linéarités des matériaux et du

comportement structural de l'interface cadre-remplissage. La méthode utilise une procédure itérative à déplacement incrémental basée sur la rigidité sécante.
La largeur de la diagonale est :

$$a = \frac{0{,}95.\,H.\cos\ \theta}{\sqrt{\lambda_{\square}}}$$

2.3.7 Modèle de Priestly 1992 [12]

Il a supposé que la largeur de la diagonale est constante sur toute sa largeur sans prendre en compte les propriétés de cadre de remplissage. La largeur de la diagonale est donnée par la formule suivante :

$$a = 0.25d$$

Avec d : longueur de la diagonale de remplissage.

2.3.8 Modèle de Cruz-Diaz 2002 [13]

La méthode utilisée par Cruz-Diaz met en évidence la corrélation entre la largeur équivalente de la diagonale et la longueur des briques de maçonnerie.

$$a = l\left(\frac{l}{l_c}\right)^r$$

Avec l : longueur des briques
r, lc : paramètres de calage, on peut prendre pour :
*. Joints verticaux remplis : r=0.545 ; lc=2.247
*. Joints verticaux secs : r=0.876 ; lc=0.384

2.3.9 Modèle adopté par le règlement PS92[14]

Le principe de calcul de résistance consiste à assimiler l'ensemble formé par les panneaux de maçonnerie et par les chaînages en béton armé qui les encadrent à un système triangulé dont les éléments diagonaux sont constitués par les bielles actives susceptibles de se former dans la maçonnerie. La largeur a de ces bielles est prise, dans les calculs, égale à la plus petite des deux valeurs d/6 et 4e, soit :

$$a = \min\,(d/6\ ;\ 4e)$$

Avec : d est la longueur de la bielle (diagonale du panneau), e est l'épaisseur brute de la maçonnerie

2.3.10 Modèle de FEMA [15]

$$a = 0.201\,\frac{\sqrt{H'^2 + L'^2}}{H^{0.4}(E_d.\,s.\sin(2\theta))^{0.1}}\,(E_f I_c H')^{0.1}$$

$$Avec\ \theta = \tan^{-1}\left(\frac{H'}{L'}\right)$$

Avec s et E_d : la largeur et le module de rigidité du remplissage, E_f : le module de Young du Portique
H et L sont la hauteur et la longueur du portique,
H'et L' la hauteur et la longueur du remplissage,
I : moment d'inertie de la section du poteau

3 Résultats et discussions

3.1 Largeur de la bielle du cadre ND2

Lors de suivi du chargement horizontal du corps d'épreuve ND2 nous avons noté que le cadre se séparait du remplissage au niveau de la diagonale tendue, avec une largeur mesurée de l'ordre de 37cm, (figure 8), valeur qui est en concordance avec la théorie de la diagonale comprimée équivalente de largeur a et qui varie, pour des panneaux carrés, entre 0.3 à 0.4 (Stafford Smith 1963[1])

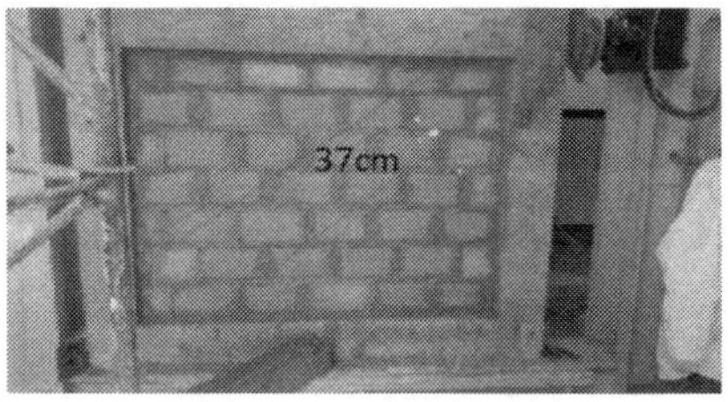

Fig.8: Diagonale comprimée équivalente

L'histogramme ci-dessous (figure 9), synthétise les valeurs obtenues à partir des modèles énumérés ci-dessus. Nous pouvons dire que le modèle de Durrani [6] coïncide presque parfaitement avec les résultats expérimentaux avec une erreur de 5%, pour le modèle adopté par le règlement parasismique Français PS 92 [12], donne des résultats relativement éloignés avec une incertitude de 24%, par contre les autres modèles sont un peu loin avec des erreurs variant de 35% à 75%, sauf pour le modèle de Dawe 1989[5], donne des erreurs excessives de l'ordre de 200%, (cette valeur de Dawe[5] est en concordance avec les résultats obtenus par Samouh [16]). Ceci est dû principalement à la diversité des corps d'épreuves utilisés. Et les matériaux utilisés.

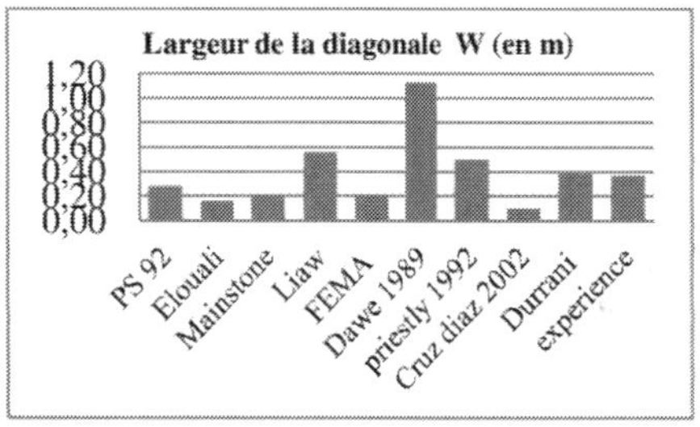

Fig.9: Largeur de la bielle diagonale a (en m)

3.2 Etude comparative du comportement des corps d'épreuve (ND1 & ND2)

Lors de l'interprétation de ces deux diagrammes (figure 10), qui représentent la loi de comportement de mur rempli (ND2) et celui du mur vide (ND1), on peut conclure que le remplissage améliore nettement la rigidité, ainsi que la résistance à la rupture

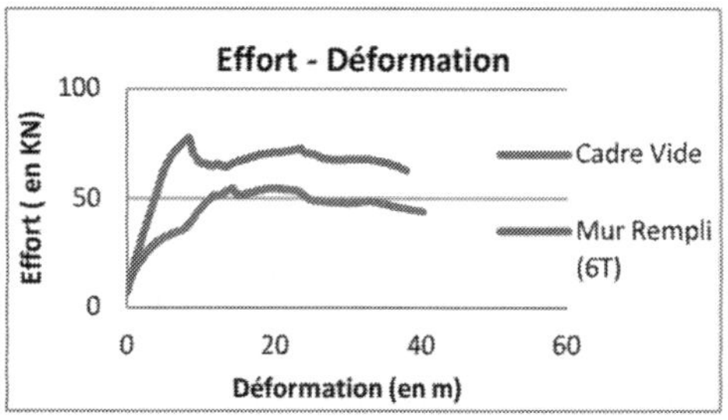

Fig.10 : Loi de comportement (Effort- Déformation)

En effet : La résistance à la rupture, s'est améliorée de presque 42%, grâce à l'ajout du remplissage (de 54 à 77 KN). Ainsi que de la même manière, la première fissuration s'est retardée avec la présence du remplissage,

ce qui agit sur le déplacement limite élastique qui s'est apprécié presque de 80%, par conséquent, la rigidité du portique s'est doublée, pour leND2. Ces résultats sont conformes à ceux obtenus par d'autres chercheurs, en particulier Irwin (1979) dont les valeurs étaient comprises entre 2,7 et 8,65.
Cette conclusion peut être justifiée analytiquement par les formules théoriques prononcées par le règlement Marocain Parasismique RPS 2011[17].

La rigidité totale est composée selon le RPS 2011 de deux termes, à savoir La rigidité apportée par le cadre (Kc) et celle apportée par le remplissage (Kr), ces derniers sont déduites à partir des deux formules suivantes : $K = Kp+Kr$

Avec Kp : Rigidité latérale du portique

$$Kp = 12 \sum (Ec.Ic) / h^3(L + 2.\lambda)$$

$$\lambda = Ls.\ Ic / hs.I$$

$\sum$ sur le nombre des travées.

Kr : La rigidité latérale du panneau :

$$\mathbf{Kr = 0.045\ .m.\ (Er.e\ cos^2\alpha)}$$

m: Nombre des travées
Er : Module d'élasticité
.e : épaisseur du panneau
α: est l'angle de la diagonale avec l'horizontale du panneau

Les résultats trouvés par les différents modèles précités ci-dessous en appliquant ces dernières formules de la rigidité, sont résumés dans l'histogramme ci-dessous, (figure 11) ;

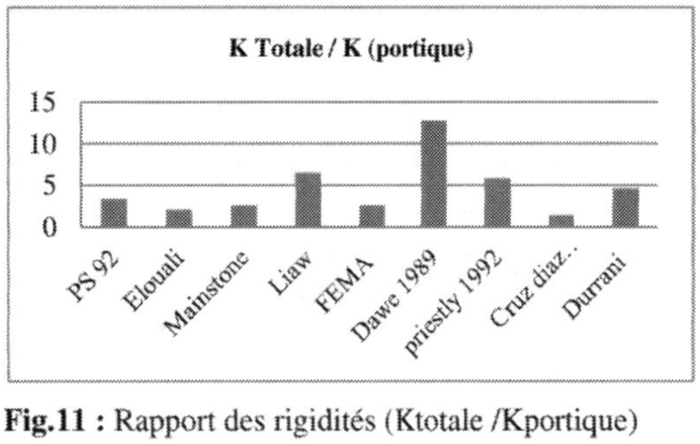

Fig.11 : Rapport des rigidités (Ktotale /Kportique)

Les modèles de Mainstone[2], d'Elouali[4] et FEMA[15], confirme ce résultat intéressant, que le rapport des rigidités est de l'ordre de 2. Cruz diaz [13], a obtenu 1.4 ; par contre les autres ont trouvé des valeurs comprises entre 3 et 6, sauf bien sûr, le modèle de Dawe[5], qui diverge complètement comme pour la largeur de a bielle. Ceci peut s'expliqué par le calibrage des modèles, en effet : Le modèle d'Elouali[4] s'est basé sur celui de Mainstone[2] , et de la même manière FEMA[15], et en plus La formule de calcul sur laquelle on s'est basée est déduite du RPS 2011, édité par T Elouali.[4] Mais ce qui est sure, que le remplissage augmente d'une manière significative

3.3 Étude comparative entre un mur vide modélisé et après expérience

Sur le tableau n°1, nous avons reporté les valeurs obtenus suite à la modélisation numérique par le logiciel éléments finis SAP2000, On peut valider notre modèle car l'erreur trouvée par l'expérience, pour le déplacement, est de presque 3%, par contre on a trouvé une erreur de l'ordre 9.5% concernant la rigidité

Tab.1:Résultats expérimentaux vs par modélisation

	la rigidité K (MN/m)	Le déplacement (mm)
Résultats expérimentaux	4.01	11.22
Valeurs SAP 2000	4.39	10.86

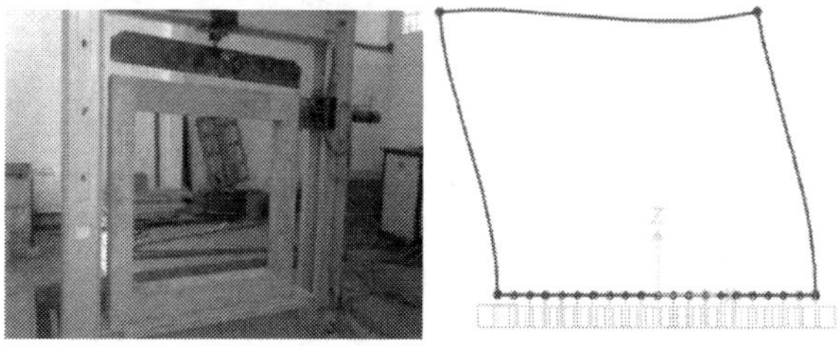

Fig.12 : Modèle du Cadre vide

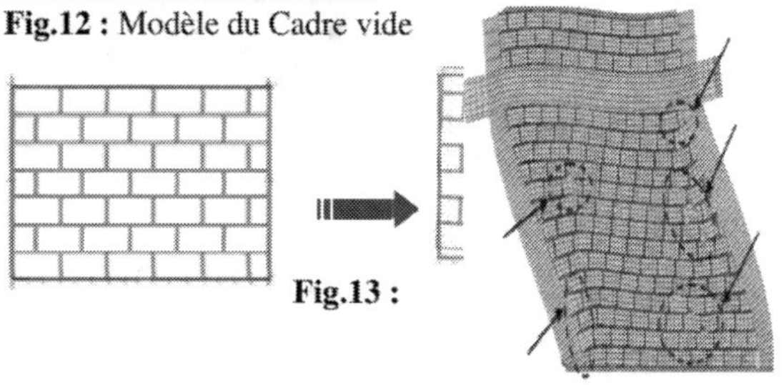

Fig.13 : Modèle du Cadre Rempli

On peut noter également que la contrainte limite élastique n'a augmenté que de 7% alors que la contrainte de rupture à augmenter de 150%, chose que l'on peut justifier par le comportement non linéaire de ce type de structure, et que de ce fait on peut dire que l'effort est repris en premier lieu par le cadre, jusqu'à fissuration et par la suite le chargement est transmis au remplissage en maçonnerie prend la relève pour équilibrer l'effort supplémentaire.

D'après l'essai de cisaillement du mur rempli, nous avons reporté une cartographie de fissuration très proche du modèle homogénéisé, ceci confirme la validité de ce modèle.

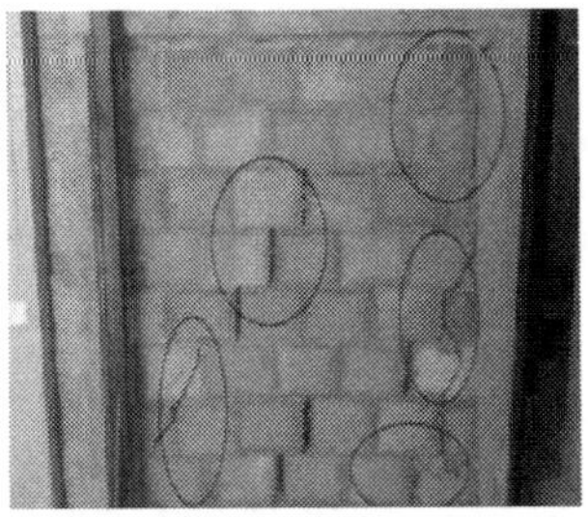

Fig.14 : Mode de rupture du mur rempli

4 Conclusion

Au terme de cette étude, et d'après l'exploitation des résultats obtenus, nous pouvons déduire que :

- La méthode des bielles est bien adaptée à ce genre de structures, elle est aussi justifiée par l'expérience.
- La présence du remplissage en maçonnerie peut jouer un rôle bénéfique en réduisant les déplacements horizontaux et les contraintes de flexion.
- l'effort tranchant et le moment fléchissant se trouvent réduits par la présence du remplissage.
- la présence du remplissage change le mode de déformation et de déplacement.
- Le remplissage entraine un effet bénéfique sur la rigidité, qui peut être doublée par rapport à un portique vide.
- De la même manière la résistance latérale. Mais cet effet est très contrasté, il dépend beaucoup des caractéristiques essentiellement des matériaux (briques et béton). C'est la raison principale, qui justifie la divergence des résultats déduits à partir des neufs modèles qu'on a utilisé.
- L'utilisation des modèles déjà connus pour dimensionner des structures, en particulier PS 92, à utiliser avec méfiance, car ces derniers sont calés sur des matériaux, et des techniques de mise en œuvre, différents par rapport qu'on utilise chez nous au Maroc. D'où la nécessité d'approfondir cette recherche, en multipliant les essais, afin d'élaborer un nouveau modèle tenant compte de ces contraintes
- A partir des diagrammes des lois de comportement (contraintes-déformation), on peut conclure que le mur rempli a un comportement non linéaire, et par conséquent, toute modélisation doit tenir compte de ce comportement ; nous recommandons l'utilisation de la méthode non linéaire.

5 Références

1. Stafford Smith, B., "Lateral Stiffness of Infilled Frames."Journal of Structural Division, ASCE, 1962, Vol. 88, No.ST6, pp. 183-199.
2. R. J. Mainstone, [1971], On the stiffnesses and strengths of infilled frames,Proc. of the Institution of Civil Engineers, Supplement (IV), Paper 73605:57–90.
3. R. E. Klingner, V. V. Bertero; Infilled Frames in Earthquake-resistant Construction, University of California, 1976 - 548 pages
4. T. Elouali (1998) Thesis Ph.D. "Effects of infill panels on the seismic response of frame buildings.
5. JL. Dawe, CK. Seah. Behavior of masonry infilled steel frames. Can J Civ Eng 1989;16:865–76.
6. A.J Durrani, Y.H. Luo, Seismic retrofit of flat slab buildings with masonry infill, Nceer Workshop on Seismic response of masonry infills, SanFrancisco, 1994
7. A. Hrennikoff A., «Solution of Problems of Elasticity by the Frame-Work Method ». ASME J. Appl. Mech., n° 8, 1941, p. A619– A715.
8. P. Kotronis, Cisaillement dynamique de murs en béton armé. Modèles simplifiés 2D et 3D. Matériaux. École normale supérieure de Cachan, ENS Cachan, 2000.
9. J. Mazars, L. Davenne, P. Kotronis, F. Ragueneau, Casaux G.A 3D beam element analysis for R/C structural walls'. 12th European Conference on Earthquake Engineering, cd paper no 751, London 9-13 September 2002
10. P, Kotronis, J. Mazars, L. Davenne, 'The equivalent reinforced concrete model for simulating the behavior

of shear walls under dynamic loading '. Engineering Fracture Mechanics , vol. 70, issues 7-8, pp. 1085-1097, 2003.

11.. T. C. Liauw and K. H. Kwan, "Nonlinear Behavior Of Non-Integral Infilled Frames," Computers & Structures, vol. 18, pp. 551-560, 1984.

12. T. Paulay and M. Priestley, Seismic Design of Reinforced Concrete and Masonry Buildings. New York: Jhon Wiley & Sons, 1992.

13. J.I.Cruz Diaz, Etude des murs de contreventement en maçonnerie d'éléments de terre cuite Thèse l'Université de Marne-la-Vallée (2002)

14. DTU NF P 06-013. Règles de construction parasismique - Règles PS applicables aux bâtiments, dites règles PS 92

15. «FEMA 273, 306 et 356» Federal Emergency Management Agency, USA, 1997,1998 et 2000

16. H. Samouh et All : Modélisation simplifiée des portiques avec remplissage en maçonnerie soumise à l'action sismique :conference paper sept 2011 ; DOI : 10.13140/2.1.4756.1761

17. «Regalement parasismique Marocain»(RPS2011) ;

18. M. Holmes, Steel frames with brickwork and concrete filling. Proceedings of the Institute of Civil Engineers 1961: 19(6501): 473–478.

19. H. Cherif, Modélisation simplifiée des portiques avec remplissage en maçonnerie soumis à l'action sismique test. Mast-Gc.

20. K, Lefebvre, Etude du comportement sous charges latérales des ossatures de béton armé avec murs de remplissage de maçonnerie, construites avant les années 1960 (2012) ; thèse présentée à l'école de technologie supérieure canada

21. T. Telford, Rc Frames under earthquake loading, state of the art report, Comité Euro-International Du Béton.

22. NF EN 1996 Eurocode 6 : Calcul des ouvrages en maçonnerie

23. NF EN 1052-3 Avril 2003 Méthodes d'essai de la maçonnerie - Partie 3 : détermination de la résistance initiale au cisaillement

24. F. B. HOUTI, Comportement des structures en maçonnerie sous sollicitation uni axiale, Universite Abou Bekr Belkaid – Tlemcen, 2002

25. N. Mezigheche, Comportement des remplissages dans les structures sous différentes sollicitations: cas de la maçonnerie, Mémoire de Magister, Faculté Des Sciences de l'Ingénieur,2010.

26. T. Bouchakour, Interaction entre ossature en béton armé et maçonnerie de briques creuses sous sollicitations de type sismique, Thèse de Doctorat, Mechanics. Ecole Nationale des Ponts et Chaussees, 1984.

27. A. Ghanem, Analyse non linéaire des structures planes en béton armé tenant compte du remplissage en maçonnerie, Mémoire de Magister, Université Mouloud Mammeri De Tizi-Ouzou, 2013.

28. I. Ait Ramdane, Comportement sismique d'ossatures en portiques béton armé en tennant compte de l'interaction murs de remplissage en maçonnerie-cadre des portiques, Mémoire de Magister, Université Mouloud Mammeri De Tizi-Ouzou, Faculté du Génie de la Construction, 2014

Numerical analysis of a reinforced concrete beam under blast loading

Yehya Temsah[1], Ali Jahami[1], Jamal Khatib[1,2], M Sonebi[3]
[1]Faculty of Engineering, Beirut Arab University, Beirut, Lebanon
[2]Faculty of Science and Engineering, University of Wolverhampton, Wolverhampton, UK
[3]School of Natural and Built Environment, Queens University of Belfast, Belfast, UK

Email: j.khatib@bau.edu.lb

Abstract. The types of Dynamic loads that might face an engineer during any design procedure vary. One of these loads is the explosion's pressure on buildings which is in other words the blast load. This research has examined the possibility of using a finite element method as a tool for predicting the dynamic response of blast loaded reinforced concrete beams. In this study, the advanced software, ABAQUS is used in order to model materials and consider the material nonlinearity, stiffness degradation and strain rate effects. Experimental results for several beams under explosion are chosen to be modeled and verified using ABAQUS. These experiments were carried out at the National University of Defense Technology in China. The results show that the material properties of concrete under impact loads (high strain rates) can be well defined in ABAQUS. Also the built in model CONWEP for blast load in ABAQUS can be used in the simulation process with an acceptable error.

1 Introduction

Many types of concrete structures may be subjected to blast and shock waves due to different types of blast loading. Some of these explosions are occurring naturally like the lightning phenomenon, and some of them are made by humans such as terrorist attacks. After World War one and two, people started to understand the importance of protecting buildings from blast effect in order to save human lives, and this presented a particular challenge to design. Nowadays this domain became one of most important when designing structures to resist impact loading or explosion in many parts of the world including the Middle East.

An explosion[1,2] can scientifically be defined as a sudden dissipation of energy which leads to a sudden rise in pressure. This pressure rise forms a wave called a "Blast Wave" that travels starting from a supersonic speed, which decreases as the wave front propagates. The pressure's amplitude has two regions: the positive phase region and the negative phase region as illustrated in Figure 1. Where "ΔP" is the over pressure,"t^+" is the duration of positive phase and "t^-" is the duration of negative phase. The impulse "I" as known is the area under curve. The positive phase pressure will cause positive impulse density (I^+/A) and the negative phase pressure will cause negative impulse density (I^-/A), where "A" is unit area. Therefore:

$$I^+ = A\int_{t_a}^{t_a+t^+}\left(P(t)-P_o\right)dt \quad (1)$$

$$I^- = A\int_{t^++t^-}^{t_a+t^++t^-}\left(P_o-P(t)\right)dt \quad (2)$$

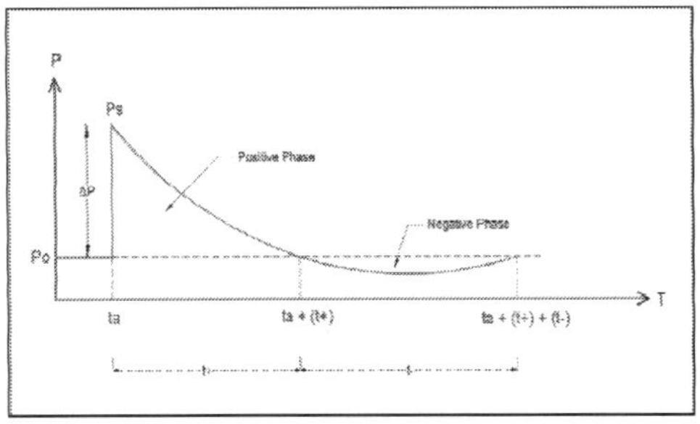

Figure 1. Ideal blast wave behavior (Ardila-Giraldo 2010)

Scaling of the properties of blast waves can be done by Hopkinson-Cranz Scaling[3]. This scaling principle states that when two explosive charges have the same geometry and explosive type but of different size are detonated in the same atmosphere and at equal scaled distances, then they will produce the same overpressure stated by:

$$Z = \frac{R}{Q^{1/3}} \quad (3)$$

Where "Z" is the scaled distance, R is the distance from the center of the explosive charge (in meter), and Q is the equivalent TNT mass (Kg).

As the wave reaches the object, part of it is reflected back and another part is transmitted through the object. The reflection and transmission of waves depend on a mechanical property of objects called the mechanical impedance or (shock impedance of a body)[1]. The mechanical impedance (z) is defined as the ratio of pressure change "P'" to the velocity change "u'" at the interface between the two mediums. For concrete structures and since the shock impedance of concrete is much higher than the shock impedance of air, the reflected blast wave from the interface of the two mediums will be magnified and reversible in direction (expansion wave "T"). This will lead into a resultant tension pressure which would cause the spalling of concrete as shown in Figure 2.

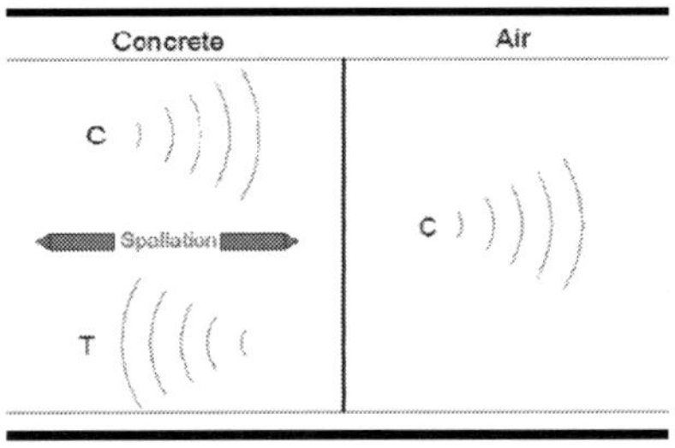

Figure 2. Spalling Phenomenon

The behavior of concrete under high strain loading is different from that under static loading. The strength and stiffness of concrete under dynamic conditions will be higher than under static conditions as illustrated in figure 3. Many researches were conducted to study the tensile and compressive strength of concrete at different strain rates. Some of the published results is presented in Malvar and Crawford (1998)[4]. The behavior of reinforcing steel is also affected by dynamic load conditions. The yield strength will be increased due to strain rate effect as well as the ultimate strength. However, the elastic modulus of steel won't be affected.

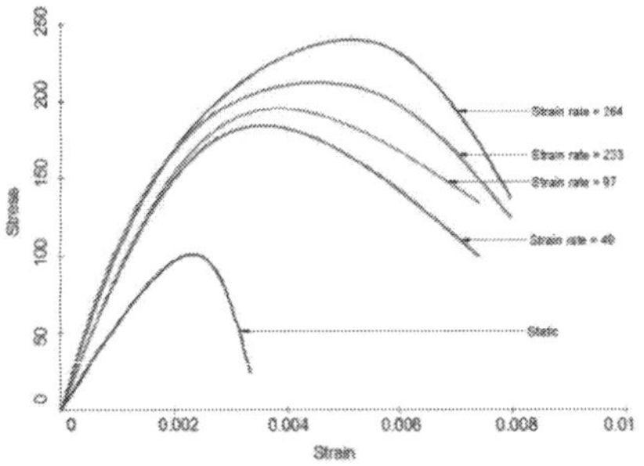

Figure 3. Uniaxial compressive behavior of concrete at different strain rates (Malvar and Crawford 1998)

2 Aim and Objectives

The aim of the research was to model the behavior of a Reinforced Concrete Beam under blast loading using ABAQUS. The specific objectives were to:

1- Model the material property under under high strain rate load.

2- Model the blast load using "CONWEP" Built-in model in ABAQUS and check the validity and accuracy of this model.

3- Show the effect of strain rate on the responses of reinforced concrete beams.

3 Data collection

The data were collected from the experiment that was carried out by Zhang et al[5] National University of Defense Technology. Three different sizes of beams were investigated with dimensional ratios of 3:4:5. The dimensions (mm) were (850*75*75), (1100*100*100) and (1350*125*125) mm. 6mm bars were used for the compressive. Tensile and hoop reinforcement. The hoops were placed at 60mm centres. Table 1 presents the number and dimensions of beams. The design compressive strength of concrete was 40 MPa. The yield stress and ultimate strength of reinforcement steel are 395 MPa and 501 MPa respectively. The explosive used in this experiment were compressed to a cakey cylinder suspended over the center of the reinforced concrete (RC) beam and` detonated electronically. The beam was supported using a steel frame as shown in Figure 4. The mid span displacement of the beam is measured using steel needles which were allowed to move in the vertical direction.

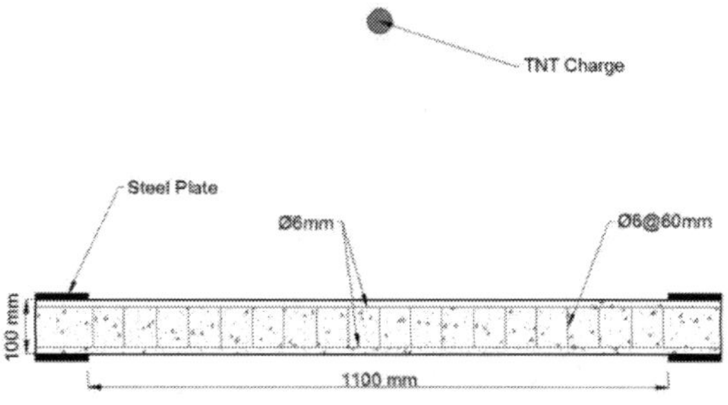

Figure 4. Experimental setup (Zhang et al 2013)

Beam	*Dimension (mm)*	*TNT (Kg)*	*Standoff distance (m)*
B2-1	100X100X1100	0.36	0.4
B2-2		0.75	0.4

Table 1. Experimental program (Zhang et al 2013)

4 Numerical Modeling

The volumetric finite element program "ABAQUS" was used to perform nonlinear analysis of the structural element (beam). The model is composed of two main parts: the concrete body that was modeled using a solid element, and steel reinforcement that was modeled as a rebar element. A Dynamic explicit analysis step was chosen for the case with a reasonable time step close to the experimental interaction time.

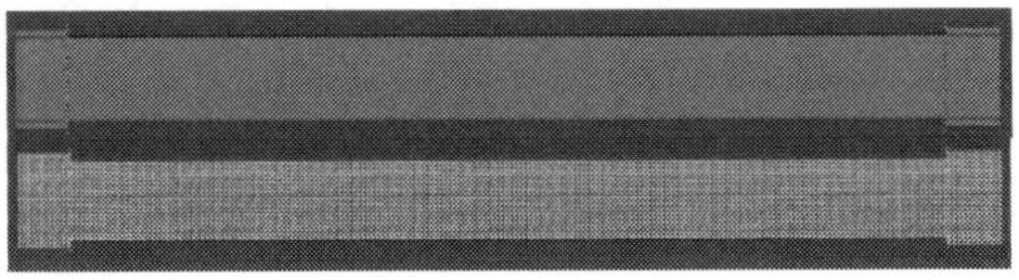

Figure 5. Finite Element Modeling of beam

The Concrete Damage Plasticity Model [6] will be used to define the behavior of concrete. This model works with static and dynamic load conditions. It was derived by Lubliner [7] in 1989 and developed by Lee and Fenves [8] in in 1998. It represents the nonlinear behavior of concrete using different input parameters such as: inelastic strain, cracking strain, stiffness degradation and recovery, and other parameters.

The behavior of concrete in compression and tension according to CDP method in ABAQUS is illustrated in Figures 6 and 7 respectively. If the load on the concrete specimens is released during the softening stage, the elastic stiffness is reduced and this can range from zero (undamaged) to one (total damage).

As for reinforcing steel, they can be modeled using several methods. For this simulation we will consider the elasto-plastic behavior of reinforcing steel, and we will assume a perfect bond between concrete and steel which is not accurate; since there is some slipping between the two materials (concrete and steel). But for now we are not interesting in the interaction between these materials.

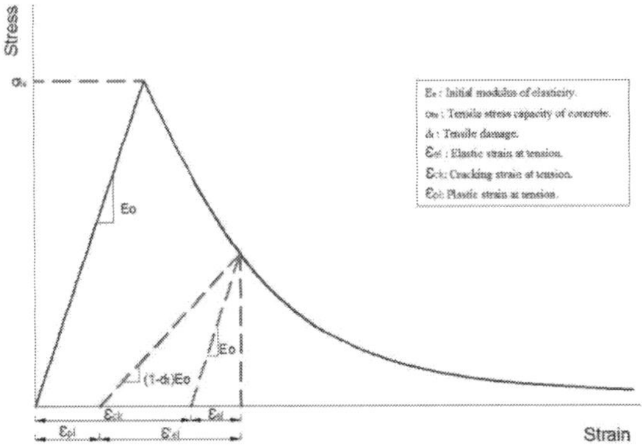

Figure 6. Uniaxial behavior of concrete under tension

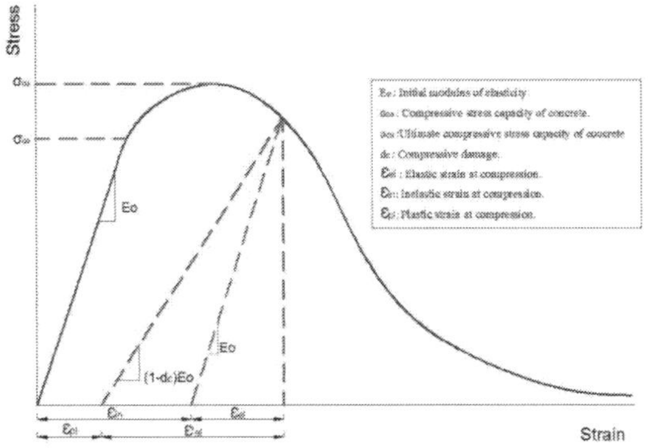

Figure 7. Uniaxial behavior of concrete under compression (Hibbitt et al 2011)

In this work, the built-in CONWEP model was used to define the blast load. The model provides the values of maximum overpressure, arrival time, positive phase duration and the exponential decay coefficient. Using the above parameters, the variation of incident pressure and reflected pressure can be constructed (Figure 1).

5. Model calibration

The model was calibrated until reaching an acceptable error. Two verifications were done for this process: comparing mid-span displacements and damage zone length. Table 2 and Figure 8 show the Mid-Span displacements for Beams B2-1 to B2-2. As shown the difference between experimental and numerical analysis in small. Therefore the first verification is acceptable.

Beam Sample	Mid-Span Displacement (mm)		Absolute error %
	Experimental	Numerical	
B2-1	25	23.5	6
B2-2	40	36.3	9.25

Table 2. Mid-Span Displacement for B2 samples

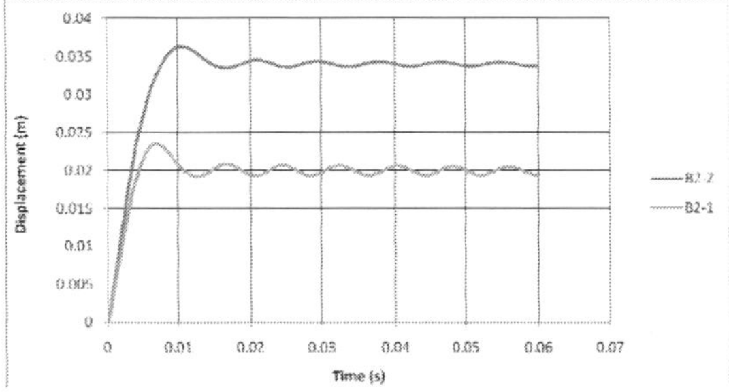

Figure 8. Time-Displacement Curves

As for the damage pattern, Tables 3 and 4 show the compression fracture zone width and the tensile fracture zone width (spalling area) in both Experimental work and Numerical analysis. Also Figures 9 and 10 presents graphically the damage pattern in both compression and tension zones for beams B2-1 and B2-2 respectively. According to these data the second verification is achieved, and as a final conclusion the experiment is verified and the model therefore is ready for further studies.

Beam Sample	Compression Fracture Zone Width (cm)	
	Experimental	Numerical
B2-1	8	6
B2-2	12	10.25

Table 3. Compression Fracture Zone width

Beam Sample	Tensile Fracture Zone width (cm)	
	Experimental	Numerical
B2-1	7	5.9
B2-2	15	13.75

Table 4. Tensile Fracture Zone width

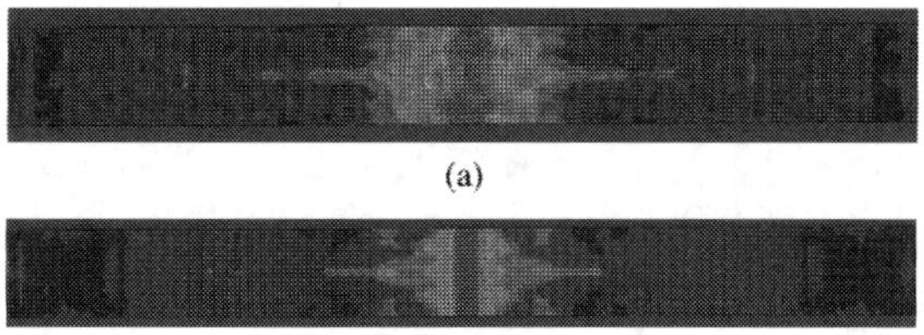

(a)

(b)

Figure 9. B2-1 zone damage.
a) Compression damage b) Tensile damage

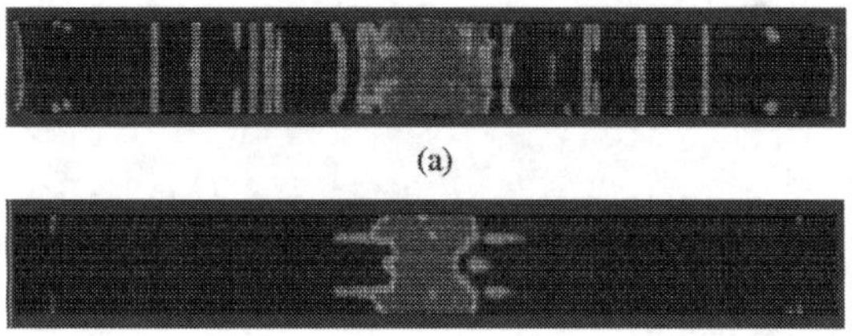

(a)

(b)

Figure 10. B2-2 zone damage.
a) Compression damage b) Tensile damage

6. Further Study

After Calibration process, the Effect of strain rate on structural response was studied. First the effect of strain rate on Mid-Span displacement is presented in Figures 11 to 12 as the following:

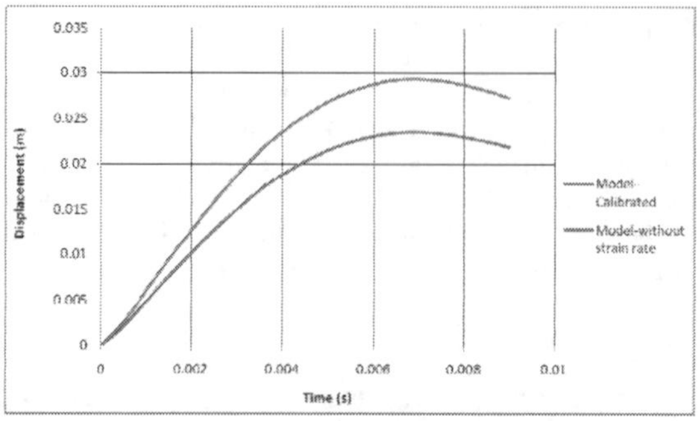

Figure 11. Strain rate effect on beam B2-1

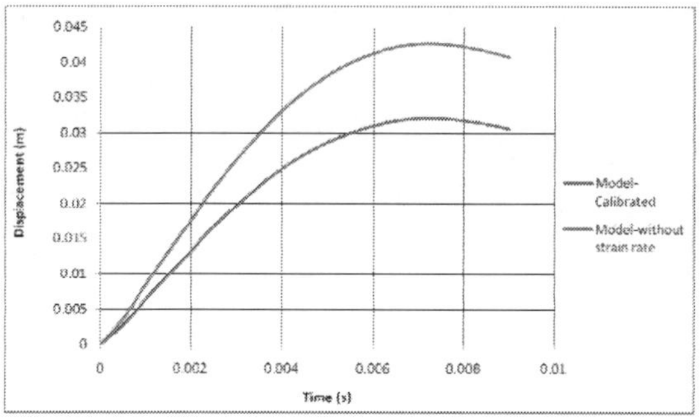

Figure 12. Strain rate effect on beam B2-2

In addition, the effect of strain rate on support reaction force is illustrated in Figures 13 and 14:

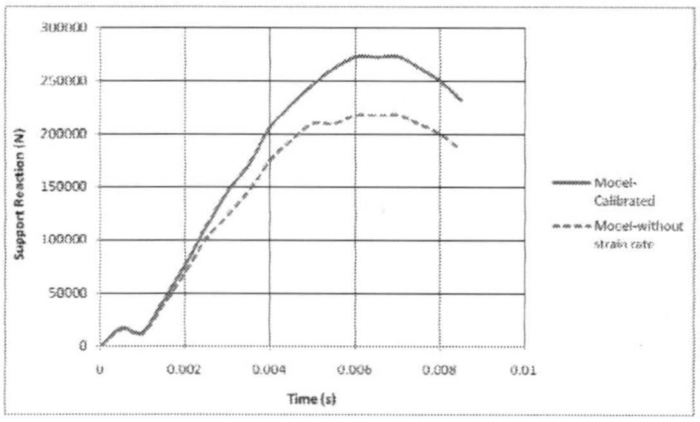

Figure 13. Strain rate effect on beam B2-1

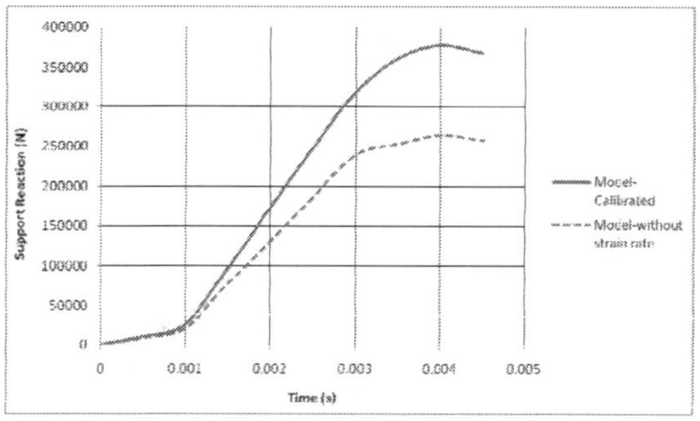

Figure 14. Strain rate effect on beam B2-2

All previous figures proved that the beam has a softer behavior in the absence of strain rate effects.

7. Conclusion

The following conclusions can be drawn from the current investigation:

1- The prediction of the the response of blast loaded reinforced concrete beams using a finite element program such as ABAQUS is possible.

2- The concrete properties must be defined using a nonlinear model (CDP). As for steel reinforcement it must be defined considering the elasto-plastic behavior.

3- The strain rate has a great effect on beam response which makes it important to be considered.

4- The Built-in CONWEP model successfully simulates the real explosion wave in the case of single structural elements without reflection phenomenon.

8. Further research

Further research may include the following:

1- Modeling the interaction between concrete and reinforcing steel and its effects on the crack pattern.

2- Modeling the effect of reflection and transmission of blast waves on the damage pattern of RC elements.

3- Modeling the thermal effect of an explosion on RC elements.

References

1. K. Ramamurthi, 2014, An Introduction to Explosions and Explosion Safety, Department of Mechanical Engineering,IIT Madras, from: http://nptel.ac.in/

2. Ardila-Giraldo, O. A., 2010. Investigation on the Initial Response of Beams to Blast and Fluid Impact, *PhD thesis, Purdue University.*

3. Magnusson, J., 2007. Structural Concrete Elements subjected to Air Blast Loading, *Licentiate thesis, Royal Institute of Technology, Division of Concrete Structures, Stockholm, Sweden.*

4. Malvar, L.J., Crawford, J.E., 1998. Dynamic increase factors for concrete, *Department of Defense Explosives Safety Seminar (DDESB), Orlando FL, USA.*

5. Duo Zhang , ShuJian Yao, Fangyun Lu, XuGuang Chen, Guhui Lin, Wei Wang & Yuliang Lin. (July 5, 2013). *Experimental study on scaling of RC beams under close-in blast loading*. Retrieved December 7, 2015 from: http: // www.sciencedirect.com /science/ article/pii/ S1350630713002203.

6. Hibbitt, Karlsson, Sorensen, 2011, ABAQUS User's Manual, *Pawtucket, 6th Edition.*

7. Lublinear, J., Oliver, J., Oller, S., Onate, E., 1989. A plastic-damage model for concrete, *Solids and Structures, Vol. 25, No. 3, pp. 299-326.*

8. Lee, J., Fenves, G., 1998. Plastic-damage model for cyclic loading of concrete structure, *Engineering Mechanics, Vol. 124, No. 8, pp. 892-900.*

Seismic behaviour of reinforced concrete frame with infill panels

Mrs. Sanaa Elmalyh [1], Mr. Azzeddine Bouyahyaoui [1] and Mr. Taoufik Cherradi[1]
[1] Mohammadia School of Engineers, University of Mohammed V, Dept. of Civil Engineering, Agdal, Rabat, Morocco

Abstract. Infill panels are usually used in structures; therefore, it is very important to study their behaviour and their influence on bare frames. Several models had been proposed to understand the effect of the infill panels on structure. The aim of this work is to identify the effect of the presence of infill panels on bare frames. The diagonal stiffness and shear compression failure of the structure, the nonlinear behaviour of the infilled frame under axial loading and the relation between the strength, ductility and stress-strain of different previous experiments will be presented. Experimental research on the interaction between the steel frames and infill panels has been done to determine the behaviour of steel frames with different types of infill walls. Many frames with composite panels subjected to cyclic-loading were realized to determine the damping ratio, the static and dynamic response, strength and stiffness degradation of infill elements. The objective of this paper is to represent and analyze the hysteretic behaviour, failure modes, and the variation of different parameters of the infilled bare frame like stiffness, strength, ductility and loading displacement curves.

1 Introduction

The non-linear behaviour of the infilled frame under seismic loading is very complex, thus it is very difficult to predict it just by analytical models. Therefore, experimental studies are necessary to predict this behaviour.

The presence of infill frame affect in a direct way, stiffness, strength and energy dissipation of the structure. In seismic zones, the interaction between the infill panels and bare frames influence the seismic response of the structure. The collapse of infilled reinforced concrete frames exposed to earthquake loading is caused in the most of the cases by the presence of infill panels and by the insufficiency of shear resisting capacity due to the lack of ductility and resistance of reinforced concrete bares. Finite element analysis can simulate the non-linear comportment of reinforced concrete elements of the model even if the non-linear response of each material has a complicated behaviour. The presence of the infill panels has an important effect on the behaviour of the bare frame, because the infill panels alter the mechanical characteristics of the portico under seismic loading, and their presence can influence the strength, stress, stiffness and ductility of the structure.

The paper presents a review of different experiments that had been done by many researchers to analyze the effect of the presence of infill panels on the behaviour of the Infilled RC frame under seismic cyclic loading.

2 Objective and scope

The goal of this research is to evaluate the risks of infill panels on the behaviour of bare fames. Also among the objectives of this work is to:

- Identify the mechanism failure of masonry walls, the stiffness, ductility and energy dissipation.
- Classify the behaviour mechanism of infill panels.
- Compare experimentally the response of model with fully and partially infill walls.
- Examine the influence of infill panels on the seismic response of the structure.

For that purpose, analytical and quasi-static cyclic experiments had been done by many researchers to evaluate the influence of infilled walls on bare frames, through different physical parameters of the model like the height, mortar formulation and the constituent of the wall, to study the influence of a cyclic loading on in-plane and out-of-plane drift. In this part, the description of the previous experiments program, materials, the connections between different elements, the method of function of the system used for test, the measure of the model, the test setup and loading are summarized. Moreover, the results of the experimental test such as, displacement, strength, stiffness, hysteresis response, failure mechanism, drift ratio, and the different types of damage in terms of different type of models will be defined.

3 Literature review

Infilled frame had been studied by many researchers to understand the behaviour of widespread strength of the Infilled RC-frame. [1] Had tested three-dimensional surfaces to determine the failure surface of masonry subjected to biaxial stress using a cubic tensor polynomial. Besides, [2] had reviewed the most criteria present in the literature and codes to interpret, the possible failure modes for masonry piers under in plane loading. Furthermore, [3] had evaluated the possibility of achieving a high level of ductility by flexural yielding in fully grouted reinforced concrete masonry shear walls; for that, six full-scale walls were tested to failure under reversed cyclic lateral loading to investigate the effect of the amount and distribution of vertical reinforcement and the level of axial compressive stress of the inelastic behaviour and ductility. Results showed that yielding of the outermost vertical bars extended to a height equivalent to half the wall length. In addition, the top wall displacement at the onset of yielding of the vertical reinforcement was highly dependent on the amount of reinforcement, but only minimally affected by the level of axial compressive load. However, at maximum loads, the displacements were less sensitive to either the amount of vertical reinforcement or the level of axial compression. Correspondingly, the displacement ductility was found to be very sensitive to the amount of vertical reinforcement, but was not dependent on the level of axial compression. In general, high levels of ductility and energy dissipation capabilities accompanied by relatively small strength degradation were observed for the test specimens.

The irregularity related with undefined location of infill walls could make undesirable effects on the general behaviour of structures [4].

[5] Had proposed a macro-model to simulate the comportment of infill panels, which were presented by two equivalent struts to study the global and local responses. Finite element analyses had been done to compare the results to experimental data, in order to evaluate the local effects on the frame and underline the influence of the coefficient of friction at the infill-frame interface. It was concluded that the presence of infill increases the stiffness and modifies the dynamic response of the structure.

The internal forces produced by the interaction of the infilled frame, were not supported by the bare-frame, caused by an additional shear force arising at the ends of beams and columns. This additional force can cause brittle collapse mechanisms. A parametric study was carried out addressed to the prediction of the shear forces. The results can be used as a support when the simplified model is adopted, consisting in substitution of infill with an equivalent in jointed concentric strut. Equivalent strut models do not present exactly the physics of the original systems; furthermore, information about the local shear transfer in contact regions can be ignored [6].

According to [7], the distribution of the original construction cost of the nonstructural elements of three types of buildings (offices, hospitals, and hotels) it is about 48% to 70% of the total cost. Therefore, the economic loss caused by nonstructural components is more of that due to structural components.

3.1 Seismic provisions related to mortar type

Mortars had been used to fill cracks and provide uniform bedding for masonry units. Such mortars were typically composed of clay, bitumen, or clay-straw mixtures. [8] Had studied the influence of grouting on the compressive strength of masonry prisms, results showed that grouted masonry prisms have failure loads, lower than those predicted using superposition to combine the capacity of the grouted area and the capacity of the unit area. Results indicate that the compression strength of grouted masonry can be expected to be less influenced by mortar properties than that of ungrouted masonry due to the dominance of the grout with respect to the failure mechanism. It was also noted that fully grouted specimens typically display more uniform behaviour with less scatter than ungrouted specimens.

Currently, the Masonry Standards Joint Committee Code [9] prohibits the use of masonry cement (MC) mortars in the construction of lateral force resisting systems for structures that fall into Seismic Design Categories. Two factors have likely contributed to establishing and maintaining this ban: (1) MC mortars were not common in high seismic regions where seismic codes were developed and (2) research into the behaviour of the ungrouted (solid or hollow) masonry assemblages has shown that MC mortars typically display lower bond strength than PCL mortars. Clearly, the first factor is social in nature rather than related to the actual seismic performance of shear walls constructed with masonry cements. The second factor, while well established, may not have as large an influence on seismic performance as expected.

3.2 Models subjected to in-plane actions

The in-plane behaviour of shear walls is so complex to lateral force resisting of the structure, as consequence their failure cause ruin of the structure. Many researchers have studied the reaction of infilled panels to quasi-static reversed-cyclic displacement. This part presents a summary of existing researchers and their results founded in the literature.

[10] Have reported the structural behaviour of a frame, with masonry wall subjected to in-plane monotonic loading by full-scale test and the method of discontinuous deformation analysis.

[11] Had studied reinforced concrete (RC) frame with the middle bay brick infilled, representing a five stories, three bay building in quarter-scale. As a result, the frame developed beam and column hinging near beam column interface before they reached their maximum story shear force and they eventually failed due to joint shear, exhibiting successive strength drops. The frame exhibited diagonal crack failure at brick infill in the bottom story at a relatively slow rate of increase before they started to break down. After the collapse of brick infill in the bottom story, it was acting as a soft story. The contribution of the infill brick was significant up to breaking. The strain measured in infilled beams and columns are 20% lesser than bare frame beams up to failure brick infill. For, [12] results show that the response of the total walls is not sensitive to

the reinforcement pattern, but it is sensitive to the aspect ratio of the wall.

It is clear that only if the infill is separated from the frame and a sufficient gap exists, then it can be considered that the frame and the panel do not interact. This solution is rarely used in common practice due to the difficulties related to connecting the walls to the frame. Out-of plane loads need to be withstood, while allowing free in-plane deformations and ensuring a good thermal and sound insulation. Moreover, if there is a gap only at the top of the wall, between the infill and the frame, a negative effect could occur: the compression strut acts more on the columns, which leads to a faster increase of shear forces [13].

3.3 Models subjected to out-of-plane actions

Several research had been done to evaluate the out-of-plane behaviour of partially and fully infill panels, such as [14], this research was based on the experimental testing approach where the walls were loaded out-of-plane using inflatable air bags. Testing was performed on ten unreinforced masonry walls, with cavity walls in two separated buildings. The test walls were isolated into vertically spanning panels. The walls are in reinforced concrete block and walls having a rigid top edge restraint consisting of a RC slab or beam were assumed to be subjected to the arching action during loading. The retrofit cavity ties used in this study included both mechanical and adhesive ties and was installed in each test specimen along two vertical lines separated by approximately 600 mm horizontally. Cavities between brick wythes were typically 50 mm. All the test walls were loaded semi-cyclically at a quasi-static loading rate. The maximum in-test g-force value for each wall was determined by dividing the maximum total test load by the weight of the test wall. Where the walls could be tested to complete collapse, the instability drift was measured using photogrammetry. It was concluded that, the top rigid restraint from the building frame causing "arching action" could greatly increase the OOP capacity of vertically spanning infill walls. For relatively low slenderness ratios (i.e., ratios of wall height to effective thickness), the ratio of strength-based capacity between a cavity wall with arching action and a similarly retrofitted cavity wall without arching action (e.g., Has-W1 and Has-W3, respectively) was observed to be as high as 4, "figure 1".

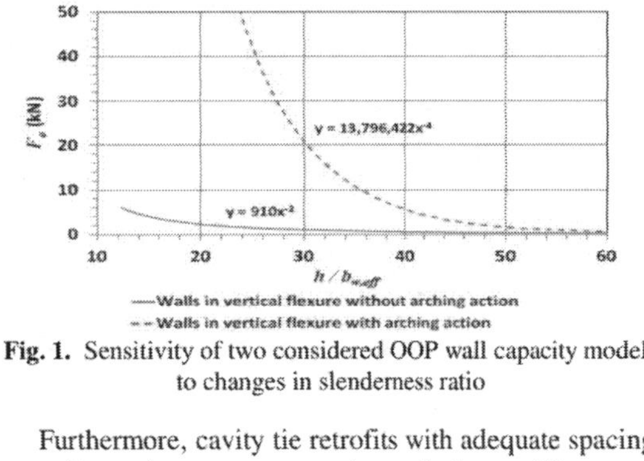

Fig. 1. Sensitivity of two considered OOP wall capacity models to changes in slenderness ratio

Furthermore, cavity tie retrofits with adequate spacing, as well as adequate compressive and shear stiffness, can greatly improve the OOP capacity of URM cavity walls, especially for walls with arching action. For a force-based analysis in one case-study, the cavity wall OOP strength was more than doubled when comparing the existing condition (i.e., Has-W5) to the condition in which new ties were vertically spaced at approximately 150 mm (i.e., Has-W6). The adhesive cavity ties considered in this study were not notably more useful to improving OOP cavity wall performance than were more easily installed mechanical cavity ties. In addition, [15] had evaluated the out-plane response of masonry infill panels when damaged due to in-plane forces. Two half-scaled specimens of large slenderness ratio (h/t=23) were subjected to simulated seismic load through a shake table. The out-of-plane objective was chosen as N21E component of the 1952 Taft earthquake with PGA of 0.156g and total duration of 54.16 s "figure 2a". The first 30 s of ground motion was considered for the simulation, which includes the strong motion portion, and the time axis of the accelerogram was scaled by a factor 1/2 to satisfy the similitude relations.

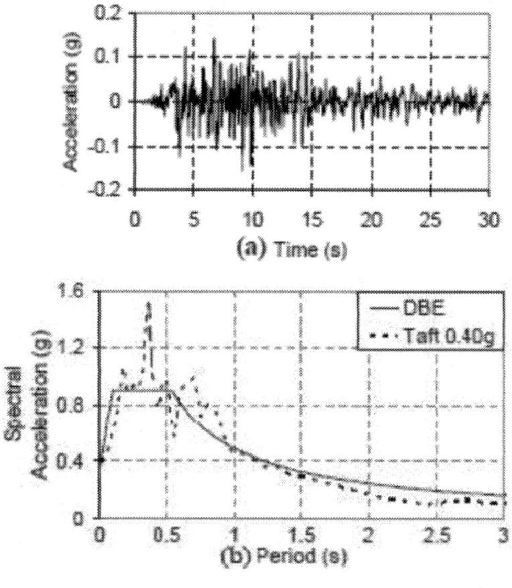

Fig. 2. (a) TAFT N21E ground motion (b) Comparison of response spectra of DBE and TAFT motion

As results, the big parts of the cracks were produced by the in-plane loading, the cracks form at initial stages of the in-plane loading developed and energy dissipation caused principally by sliding along the bed-joints. The first specimens indicate a diagonal bed joint cracking pattern, while the second specimen showed a sliding along the parallel bed joints and uniformly distributed cracks "figure 3" & "figure 4". The two models attain their maximum in-plane strengths of 32.2KN and 39.1KN at 7.50mm and 11.25mm displacement.

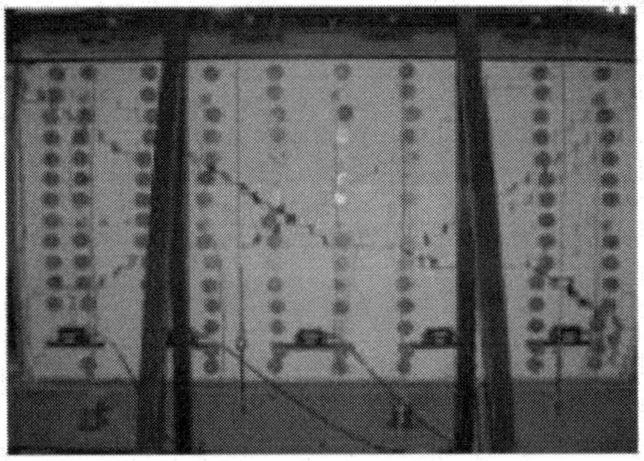

Fig. 3. Cracking pattern in the first specimen after 1.0 % in-plane drift cycles and arching phenomenon before failure under out-of-plane shake table motion

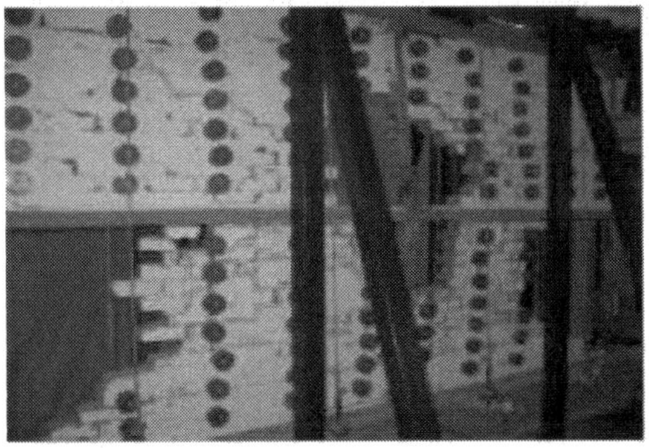

Fig. 4. Cracking pattern in the second specimen after 1.0 % in-plane drift cycles and partial collapse of masonry due to out-of-plane motion after 2.2 %in-plane drift cycles

For the out-of-plane loading, very few cracks were created. The first model collapsed during out-of-plane loading after 1.2% in-plane drift cycles, showed large out-of plane deflection, and arching before the failure. For the second model with interior grid elements, a partial collapse of masonry sub-panels were observed after 2.20% in-plane drift cycles, along with the failure of the interior vertical grid element at mid height.

As results, the two experimental models reached attained their maximum peak accelerations in the undamaged state, but after the damage was introduced, and a small deflection was observed in the second model, which could be caused by the confining effect that the interior grid generated which help in reducing out-of-plane deflections. The acceleration response decreased and was observed to saturate to a lower value with continued in-plane damage "figure 5".

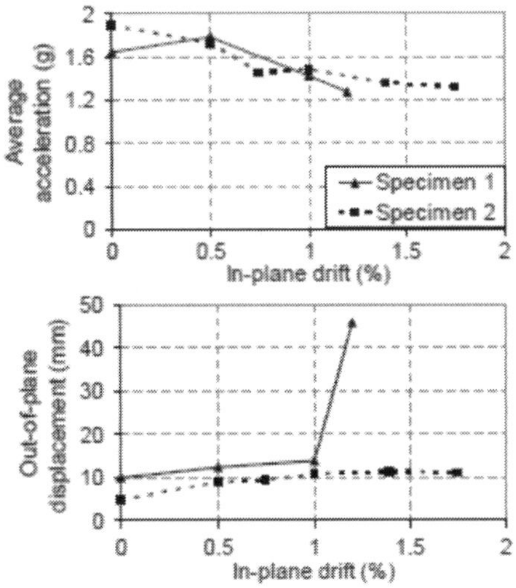

Fig. 5. Variation of average acceleration and out-of-plane displacement with in plane drift (Damage)

In addition, the variation of the acceleration en function of the height of the models shown in "figure 6", the interior grid elements provided in the second model seems not very important to modify the acceleration response. Therefore, the maximum acceleration response for two specimens was nearly similar.

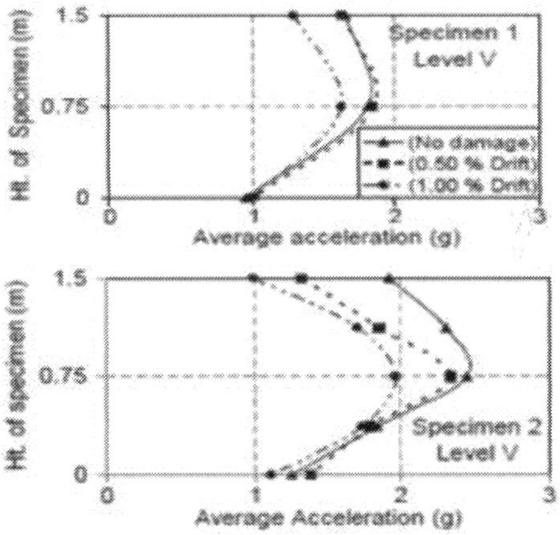

Fig. 6. Variation of average accelerations along the height of specimen

The sliding movement of sub-panels with the grid elements allowed better energy dissipation and grater deformability, without much reduction in lateral resistance. In addition, the in-plane load-displacement response presented enhanced the capacity of infill panel with interior grid elements "figure 7".

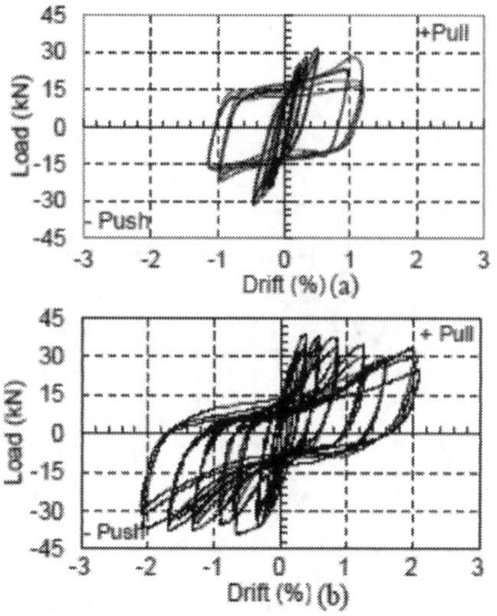

Fig. 7. Load-displacement curve (a) Specimen 1 (b) Specimen 2

Many National and International Standards have recommended different empirical formulations to tackle the above problem by dividing it into two parts; (i) Bare framed structure and (ii) Framed structures with masonry infill. They do not give due consideration to the location and the percentage of infill in the structures [16].

Extensive researches have been carried out worldwide in the last five decades. Many methods have been developed for the dynamic analysis of RC frames with masonry infill incorporating one or more simplifying assumptions regarding its stiffness. Unreinforced Masonry (URM) infill panels have a very high initial in-plane lateral stiffness and low deformability. Therefore, under seismic loads, the existence of URM walls change the whole lateral force transfer mechanism of the structure from a predominant frame action to predominant truss action [17].

The behaviour of 3D RC frames with various configurations & amount of masonry infill, under dynamic loading conditions has been studied by [18], components of the structure i.e., structural sub-assemblages of columns, beams and slabs were casted from the same materials. A series of various ground motion tests were performed in the building models using servo controlled hydraulic actuator. As a conclusion the masonry infill although do not interfere in the vertical load resisting system for the RC frame structures, they significantly affect the lateral load-resisting system. Also, the effect of de-bonding of masonry infill with RC frames as the base acceleration is increased, moreover, the natural frequencies and damping reduces. For the natural frequencies and damping values had been high in the presence of masonry infill. In addition, the decrease in the natural frequency is more, when the masonry infill is not constructed in the lower floor as compared to the upper floors without masonry infill. This is due to the contribution of walls towards the stiffness being more in the lower floors as compared to the upper floors. Moreover, there is an increase in the natural frequency when the masonry infill is not constructed in the upper floors. From this, it can be concluded that the contribution of masonry infill towards the mass is more in the upper floors as compared to the lower floors.

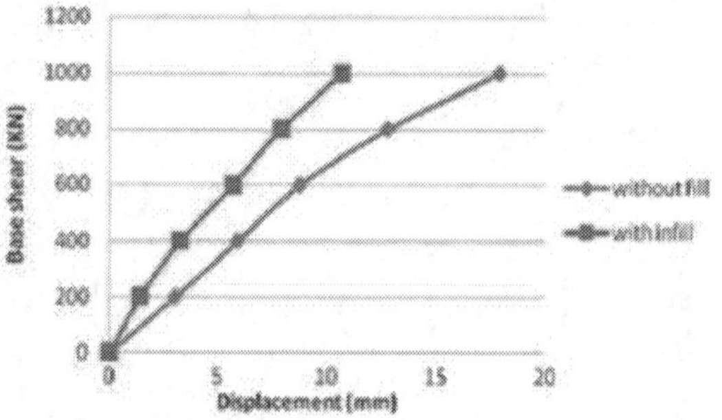

Fig. 8. Pushover Curve

[19] Had tested three full-scale wall specimens with different schemes, to improve the seismic comportment of shear-type damage of lightly RC walls. Two specimens were improved using supplementary RC panel and ultra-high strength fiber reinforced concrete panel to avoid shear failure. The other specimen had been improved reinforcement details to enhance shear capacity and flexural ductility. The test results had been compared with past lightly RC wall test as a reference wall, which failed prematurely in shear reproducing the failure observed in the field. The variations of the eventual failure mode and the damage process were detected. The experimental specimens indicated ductile and stable behaviour with less damage compared with the reference wall. Also, this test, prove that RC and UFC panels enhanced the behaviour of lightly RC wall and stopped crack creation in the central part of the wall panel. Also, increasing the amount of horizontal reinforcement and providing confinement in the boundary regions controlled the opening of crack width and made the structure more ductile.

Conclusion

Stiffness, strength, energy dissipation and mechanical behaviour of bare frames are more significant enhanced when the infill panels were taken into consideration in buildings. The contribution of infilled panels, steel not taken into account in the design, because of the insufficient information about the interaction between infill panels and RC frames. However, their interaction needs more investigation. From those experimental researches, we can conclude that:

- The presence of Infill panels in the structure modify in a direct way the stiffness, the strength, deformation, failure mechanisms and energy dissipation of RC frame under lateral loads .
- The damage caused like shear sliding in the joints, the crush of infill, diagonal tensile cracking, and plastic hinges, depends on the physical characteristics of the different element, the ratio between the infilled and the RC frame and the infilled RC frame interface.
- A number of models have been used to predict the component of infilled frames. However, there is a less of experimental studies, which work on RC frame with the presence of infilled. In addition, the existing codes

considered that infilled do not have an important influence on the RC frame. For that additional studies and experimental investigation must be done.

- The negligence of the presence of infilled walls can highly influence the mechanical characteristic of RC frame that can lead to unacceptable economical damage.
- The remarkable strength and stiffness of frames with and without infills prove the importance role of infill panels in the structure, consequently the infill panels must be taken into consideration in the design of the structures.
- The presence of infill panels in frame building makes significant modifications in the mechanical characteristics of the structure.
- The presence of infilled reduces the displacement capacity of the structure.
- The response of infilled frames had been simulated by the application of loads in compression in structure with equivalent diagonal struts, which might not consider the evolution forces in frame members.
- Brittle failure produced by shear forces was undervalued.
- When the interaction between infill panels and bare frame is taking into consideration the mechanical characteristic of the structure changes.
- The infill panels has an important role in resisting to lateral forces like earthquake and wind.
- The interaction between in-plane and out-of-plane panels was not taken into consideration.

Different models evaluated the contribution of infill-panels on the behaviour of bare frames. The present work presented a review of many experimental studies. The future work will be focused on a finite element study on Infilled RC frame with different storey subjected to cyclic loading.

References

1. C. A. Syrmakezis, P.G. Asteris, *"Masonry failure criterion under biaxial stress state"*, Journal of materials in Civil Engineeting, ASCE, **13**, No 1, Pages 58-64, (2001)
2. C. Calderini, S. Cattari, S.Lagomarsino, *"In-plane strength of unreinforced masonry piers"*, Earthquake Eng. Struct. Dyn., **38**, Issue 2, Pages 243-267, (2009)
3. M. T. Shedid; R. G. Drysdale; and W. W. El-Dakhakhni, *",Behavior of Fully Grouted Reinforced Concrete Masonry Shear Walls Failing in Flexure: Experimental Results"*, M.ASCE, Journal of Structural Engineering, **134**, Issue 11, November, (2008)
4. B. Kaushik, DC. Rai, SK. Jain, *"Code approaches to seismic design of masonry-infilled reinforced concrete frames: a state-ofthe-art review"*. Earthq Spectra, **22**, 4, Pages 961–983, (2006)
5. F. Alessandra, A. Netti, P. Monaco, *"The influence of masonry infill on the seismic behaviour of RC frame buildings"*, Engineering Structures, **44**, Pages 133-145, Novembre (2012)
6. L. Cavaleri, F. D. Trapani, "Prediction of the additional shear action on frame members due to infills*", Bulletin of Earthquake Engineering*, **13**, Issue 5, Pages 1425-1454, May, (2015)
7. S. Taghavi, E. Miranda, *"Response assessment of nonstructural building elements"*, PEER Report 2003/05, The Pacific Earthquake Engineering Research Center, Berkeley, CA, (2003)
8. R. G. Drysdale, A. A. Hamid, L. R. Baker, *"Masonry structures: Masonry Structures Behaviour and Design"*, 2nd Edition. The Masonry Society. Boulder, CO. (1999)
9. *"Masonry Standards Joint Committee. Building Code Requirements for Masonry Structures"*, ACI 530/ASCE 5/TMS 402. American Concrete Institute. American Society of Civil Engineers. The Masonry Society. Detroit, New York, Boulder, (2005)
10. Y.J. Chiou, J. Cherng and Y.W. Liou,, "Experimental and analytical study of masonry infilled frames", *J. Struct. Eng.*, 125, Pages 1109-1117, (2006)
11. S. Arulselvan, K. Subramanian, E.B. P. Pillai and A.R Santhakumar, *"RC infilled Frame-RC Plane Frame Interactions for Seismic Resistance"*, Journal of Applied Sciences 7(7), Pages 942-950, (2007)
12. M. Maleki1 , R. G. Drysdale , A. A. Hamid , and A. A. El-Damatty, *" Behaviour Of Partially Grouted Reinforced Masonry Shear Walls-Experimental Study"*, 11th Canadian Masonry Symposium, Toronto, Ontario, May 31- June 3, (2009)
13. TUCEB Code for seismic design - Part I – Design rules for buildings. P100-1, (2013)
14. K. Walsh, D. Dizhur, J. Shafaei, H. Derakhshan, and J. Ingham, *" Out-of-Plane In-Situ Testing of Masonry Cavity Walls in As-Built and Improved Conditions"*, Australian Earthquake Engineering Society Conference, Nov 21-23, Lorne, Victoria, (2014)
15. D. C. Rai, S. Komaraneni and V. Singhal, *"Out-Of-Plane Seismic Behaviour Of Brick Masonry Infilled Panels With Prior In-Plane Damage"*, 11th ,Canadian Masonry Symposium, Toronto, Ontario, May 31- June 3, (2009)
16. H. R. Tamboli, and U. N. Karadi , *"Seismic Analysis of RC frame Structure with and without Masonry Infill Walls"*, Indian Journal of Natural Sciences, Vol. **3**, Issue 14, Pages 1137-1194, (2012)
17. CVR. Murty, SK. Jain, *" Beneficial influence of masonry infill walls on seismic performance of RC frame buildings*", In: 12th World Conference on Earthquake Engineering, Auckland, (2000)
18. R. Srinivasan, S. Suresh Babu, and Sidramappa V. Itti, *"A Study on Performance of 3D RC Frames with Masonry in-fill under Dynamic Loading Conditions"*, KSCE Journal of Civil Engineering (0000) 00(0):1-7
19. E. Yuniarsyah, S. Kono, M. Tani, R. Taleb, H. Watanabe, T. Obara, T. Mukai, *" Experimental study of lightly reinforced concrete walls upgraded with various schemes under seismic loading*", Engineering Structures 138,Pages 1-15, (2017)

Effect of quick lime on physicochemical properties of clay soil

Mohammed Mustapha BESSAIM[1,3], Aicha BESSAIM[2], Hanifi MISSOUM[1,3], Karim BENDANI[1,3]
[1]University Abdelhamid Ibn Badis, Civil Engineering and Architecture Department, Mostaganem, Algeria
[2]University Mustapha Stambouli, Civil Engineering Department, Mascara, Algeria
[3]Construction, Transport and Protection of Environment Laboratory

Abstract. Clay soils are known for their water sensitivity, which causes irreparable damage to any structure built on this type of soil. In order to avoid such problem, it is necessary to use various improvement and stabilization methods such as treatment with lime. This process has been used successfully in the field for decades. The addition of lime generates various physicochemical reactions within the soil such as cation exchange and pozzolanic reactions which are largely responsible for the improvement of the soil in question. This paper presents a study concerning the variation of physicochemical properties of clayey soil with the addition of quicklime at different percentages. Experiments were performed on two clayey soils (CL type) in order to investigate the influence of quicklime on Atterberg limits and pH. These tests were carried out in an attempt to study and follow the development and progression of various reactions occurred within the soil with various lime percentages. The results show that the addition of quicklime causes a significant improvement in soil properties by reducing plasticity and thereby improves the soil workability. It can also be found that the addition of lime increase pH of soil, which allow activating pozzolanic reactions who tend to stabilize the soil in question by formation of cementitious compounds. Finally, the pH can be considered as a relevant parameter who allows a better understanding of the reactions that occur in the soil matrix.

Keywords: Soil; Quicklime; Chemical reactions; Consistency limits; pH.

1 Introduction

Clayey soils have undesirable properties such as low bearing capacity, high compressibility and dispersive behavior. Such properties cause several damage on construction built on this type of soil. Hence, these soils must be treated prior to construction operations, so that desired properties can be achieved [1]. In this regard, improvement methods must be used in order to remedy such problems. Various treatment techniques exist which aim to enhance and stabilize the soil properties. Among these improvement methods, there is treatment with the addition of cementing agents such as lime. This remediation mode is often used. On the one hand, because of its facility and economic appearance and on the other hand due to its effectiveness who allow to obtain engineering materials having superior properties [2-6]. Several studies on the basic lime-clay-reaction mechanisms have been widely reported in literature [7-10] allowing a better understanding of the physicochemical evolution of the system and micro structural features induced after the addition of lime [11-15].

The addition of quicklime (CaO) leads to a modification in the pore fluid chemistry. The first reaction taking place is hydration reaction, highly exothermic, leading to the formation of hydrated lime in the clay water system. The dissociation reaction of the hydrated lime formed due to the quicklime hydration leads to an increase of the pH and to a high concentration of calcium ions in the pore water [15]. These phenomena cause several reactions in the soil matrix, creating a cation exchange effect and pozzolanic reactions [16]. Firstly, cation exchange lead to a reduction of the thickness of the double layer surrounding clay particles due to calcium ions released from lime that tend to replace the exchangeable cations from clay [17], which also permits to stabilize the clay particles causing an improvement of soil characteristics such plasticity and compaction parameters [18]. Moreover, this exchange causes clay minerals flocculate and agglomerate leading to a quick change in plasticity, workability and particle size distribution. In addition to cationic exchange, the high pH environment induced by lime enables the dissolution of the alumino-silicate constituents of clay minerals, promoting the development of the pozzolanic reactions [15]. These reactions result in the formation of calcium silicate hydrate (CSH), calcium aluminate hydrate (CAH), and calcium alumino silicate hydrate (CASH) gels [19,20]. These amorphous phases crystallise with time, causing the formation of cementitious compounds bonding clay particles together [21-24]. Similarly, this reaction will lead to decrease of liquid limit, plasticity index, and swelling potential of clay [25-28]. However, all of these interactions are highly dependent on pH of soil. The presence of a high pH environment, namely pH value of 12.4–12.6 (the pH of a saturated lime solution) or higher, is essential to guarantee the effectiveness of lime (CaO) stabilisation of clays by means of pozzolanic reactions. Thus, as long as the lime is available in the system and the pH remains high enough to maintain the solubility, pozzolanic reactions continue over time to form bonding compounds that cause long-term strength increase [16-32].

The current study investigated the effect of quicklime on the physicochemical properties of two clayey soils. In the first hand we studied the effect of quicklime in the consistency limits of this soil; on the other hand, we studied the influence of this binder in the variation of the pH of the soil. Comprehension of this parameter is crucial because it governs the various reactions which are responsible for soil modification and improvement. So this may help to better understand the interactions occurring within the soil matrix.

2 Materials and experimental methods

2.1 Materials

The soils used in this study soil are residual soils. The studied soils were collected at a depth of about 0.5 to 1 m, where the first soil was a light green and the second was brown in color. The both (soil 1 and 2) are normally consolidated. After that, the soils were excavated, placed in plastic bags and transported to the laboratory for identification and preparation testing. Extreme precautions were taken during sampling to keep the clay in its natural conditions. Table 1 gives a summary of the geotechnical properties of the soils. The soils can be classified as the clay with low plasticity (CL) according to the Unified Classification System (USCS).

2.2 Lime

The used lime for this study is quicklime (CaO). Table 2 shows the physicochemical properties of the lime. These results show that the lime used in this study contains a high amount of quicklime (CaO), which confirms the purity of the tested lime.

2.3 Samples preparation

In this study, a number of specimens from the clay samples were investigated. To study the influence of lime on physicochemical properties of these soils, quicklime was added to each specimen at room temperature. Initially, all materials were dried thoroughly and then sieved through a #40 sieve. Then, the quicklime was thoroughly mixed until homogeneity was reached. Distilled water was used in all tests. The amount of soil-admixture was ranging from 0% to 8% of quicklime. This established mixture aim to investigate the effects of lime on Atterberg limits and the pH of the samples.

2.4 Laboratory tests

The experiments conducted in the present study include Atterberg limits and pH test.

2.4.1 Consistency limits tests

The Atterberg limit test was performed on the non treated and treated soil samples with different amount of quicklime. From these tests, the effects of lime on the consistency limits of samples were studied. Atterberg limits test was determined in accordance with [33].

2.4.2 pH measurements

The pH tests were determined using a combined glass electrode, the apparatus has been calibrated for each test using standard solution and preparation of specimens as well as the measurement has been done at room temperature according to [36].

Table 1. Basic properties of the tested soils

Properties	Soil 1	Soil 2
Soil color	Green	Brown
Water content (%)	16	7.8
Liquid limit (%)	42	34
Plastic limit (%)	21	18
Plastic index (%)	21	16
Specific gravity	2.70	2.65
ρd max (Mg/m3)	1.80	1.72
O.M.C (%)	14	22
Sand (%)	12	8
Silt (%)	48	67
Clay (%)	40	25
Soil classification (USCS)	CL	CL
pH	8.9	8.4

Table 2. Properties of quicklime

Properties	Values
Physical properties	
Bulk density (g/l)	600-900
Over 630 μ (%)	0
Over 90 μ (%)	< 10
Chemical properties (%)	
CaO	> 83.30
MgO	< 0.50
Fe_2O_3	< 2.00
Al_2O_3	< 1.50
SiO_2	< 2.50
SO_3	< 0.50
$Na_2O + K_2O$	0.4 – 0.5
CO_2	< 5.00
$CaCO_3$	< 10.00
Insoluble in HCl	< 1.00

3 Results and discussion

3.1 Effect of quicklime on Atterberg limits of soil samples

In order to investigate the effect of quicklime on the consistency limits, measures of liquid and plastic limits were carried out on soil samples at different percentages ranging from 0% to 8%. The results are shown in Fig. 1. The results showed that there is a slight increase in liquid and plastic limit at 2% of lime and thereafter both limits start to decrease for the both, soil 1 and 2. This increase in consistency limits can be explained due to a flocculation of soil particles after adding lime. In this phenomenon, the clay particles tend to agglomerate which tend to create bigger aggregates. These latter, have the ability to retain water within the flocculated structure which leads to an increase in the water content. The same result has also been reported by [34,35]. However, increasing lime percent more than 2% leads to decrease the liquid limit with an increase of plasticity for the two treated samples. The reason of such behavior can be explained by cation exchange, where calcium ions released from the addition of lime which tends to replace other cations around the clay particle surface leading to reducing the thickness of the double layer. Following these results, plasticity index undergoes a relatively large decrease of about 15% with 8% lime as shown in Fig.1.a. For the second soil, the plasticity index decrease from 16% to 10.5% with the same lime content as represented in Fig.1.b. These results are in agreement with those published by [16]. Phenomena such cation exchange and flocculation occur shortly after adding the lime which generate a decrease in the soil plasticity [37-40]. Furthermore, the reduction in plasticity can be attributed to the increasingly granular nature of the lime treated clay soil induced by the short term reactions (cation exchange and flocculation–agglomeration). Moreover, the lime addition causes a significant improvement in soil properties by reducing plasticity and eliminating swelling [16].

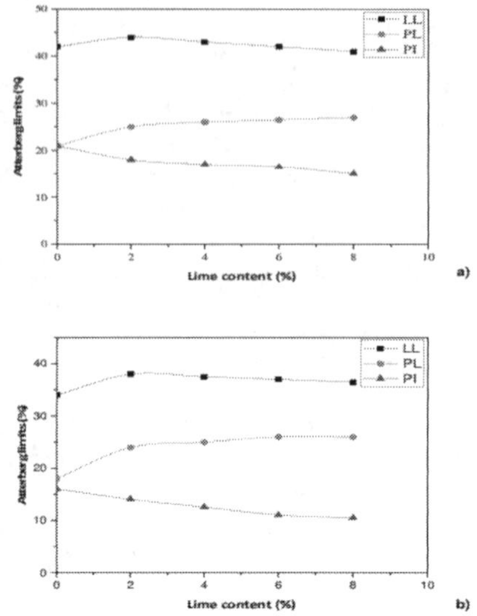

Fig. 1. Atterberg limits of lime-treated samples: a) Soil 1, b) Soil 2.

3.2 Effect of lime on pH of soils samples

Fig. 2 shows the variation of pH with the percentage of added lime in a range of (0 to 8%). In the soil 1, the natural pH is estimated to 8.9. As can be seen, the pH value rises rapidly to 11.9 with 2% lime. Then, there is an increase of pH value until reaches 12.5 with the lime percentage of 4%. After that, the pH value increases slowly to reach a constant value equal to 12.6 with 8% lime as show in Fig.2.a. However, a tiny effect of lime addition above 4% is noted in term of pH. The same results were observed in the soil 2, where the initial value of pH soil is 8.4. With 2% lime, the pH of quicklime-soil mixture reached 11.7 and increased to about 12.4 with lime percentage of 4%. After that, a lower increase of pH until remained constant value of 12.5 with 8% as presented in Fig.2.b. The same behavior has been seen by [16] and it in accord with our results. As can be seen the "lime fixation point" was reached by 4% lime for the two soils. These results are in agreement with the results of several researchers [16,41]. According to [41] a pH of 12.4 is the optimum value or the lime fixation point how allows activating the pozzolanic reaction, according to pH measurements. The increase of pH value can be explained to a variation in the pore fluid chemistry following the addition of quicklime. Firstly, there is the hydration reaction of quicklime, highly exothermic (allow to reduce the water content), leading to the formation of hydrated lime ($Ca(OH)_2$) in the clay water system as follow [42]:

$$CaO+H_2O \rightarrow Ca(OH_2)+Heat(\Delta) \quad (1)$$

$$Ca(OH)_2 \rightarrow Ca^{2+} + 2(OH)^- \quad (2)$$

The dissociation reaction of the hydrated lime formed due to the quicklime hydration leads to an increase of the pH and to a high concentration of calcium ions in the pore water [15]. This high pH leads the dissolution of the silica and the alumina released from clay minerals, which react with lime (combination with the calcium) creating new cementitious compounds such as calcium silicate hydrates (CSH), calcium aluminate hydrates (CAH) and calcium alumino-silicate hydrates (CASH) as shown in the following equations [42]:

$$Ca^{2+} + 2(OH)^{-} + SiO_2 \rightarrow CSH \quad (3)$$

$$Ca^{2+} + 2(OH)^{-} + Al_2O_3 \rightarrow CAH \quad (4)$$

This high alkalinity environment (pH=12.4) produced by the addition of lime allows activating pozzolanic reaction. Maintaining this rate is essential to guarantee the effectiveness of quicklime (CaO) stabilization of clays by means of pozzolanic reactions [15]. These latter allow to bind soil particles together and result in the improvement of the soil properties [23, 24].

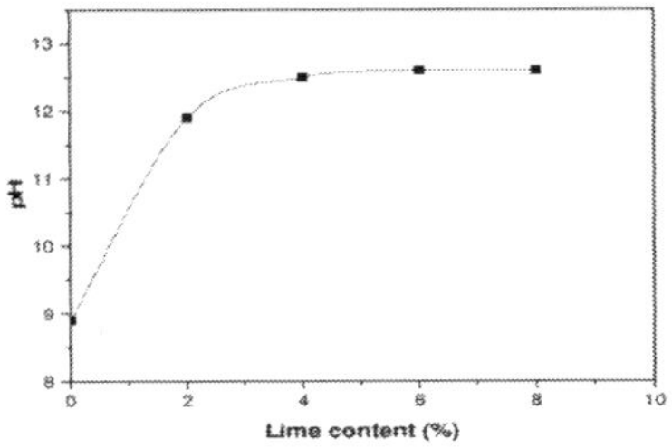

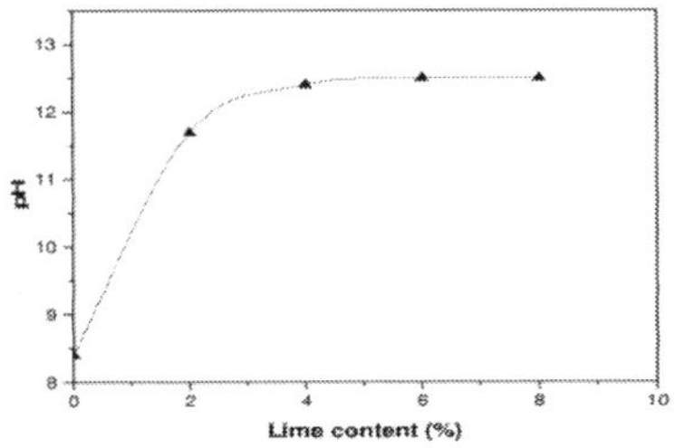

Fig. 2. pH values of lime-soil mixtures: a) Soil 1, b) Soil 2.

Conclusion

A laboratory investigation was undertaken to study the effect of quicklime on physicochemical properties of two clayey soils (CL type). The main findings allow the following conclusions to be drawn:

The addition of quicklime in clayey soils leads to a significant modification in soil properties with reducing plasticity and thereby improving the soil workability.

The lime addition increase pH of the soil due to the dissociation reaction of the hydrated lime formed by the quicklime hydration.

The addition of quicklime reduces plasticity index and increase the pH. This is in agreement with the results of most of the researchers.

The change in soil behavior (decrease in plasticity and an increase in pH) is directly linked to the reactions and to the change of pore fluid chemistry of the medium following the addition of quicklime.

The pH is a relevant parameter that allows a better understanding and can be used to study and follow the development and progression of various reactions such as cation exchange and pozzolanic reactions.

A percentage of 4% lime was found sufficient to reach the lime fixation point.

The high pH is essential to activate pozzolanic reactions for the formation of the cementitious compounds and thereby improving soil properties.

To guarantee the effectiveness of quicklime stabilization of the clayey soils, a pH above or equal to lime fixation point value is primordial.

References

1. K.A. Saeed, K.A. Kassim, H. Nur., Physicochemical characterization of cement treated kaolin clay. Gradevinar, 66, 513-521 (2014).
2. D. Castro-Fresno, D. Movilla-Quesada, A. Vega-Zamanillo, M.A. Calzada-Pérez., Lime stabilization of bentonite sludge for tunnel boring. Appl. Clay Sci. 51, 250–257 (2011).
3. M. Di Sante, E. Fratalocchi, F. Mazzieri, E. Pasqualini., Time of reaction in a lime treated clayey soil and influence of curing conditions on its microstructure and behaviour. Appl. Clay Sci. 99, 100–109 (2014).
4. S.Z. George, D.A. Ponniah, J.A. Little., Effect of temperature on lime-soil stabilization. Constr. Build. Mater. 6, 247–252 (1992).
5. S. Kolias, V. Kasselouri-Rigopoulou, A. Karahalios., Stabilization of clayey soils with high calcium fly ash and cement. Cem. Concr. Compos. 27, 301–313 (2005).
6. A. Modarres, Y.M. Nosoudy., Clay stabilization using coal waste and lime-Technical and environmental impact. Appl. Clay Sci. 116-117, 281–288 (2015).
7. J.L Eades, R.E. Grim., Reactions of hydrated lime with pure clay minerals in soil stabilization. Highw. Res. Board Bull. 262, 51–53 (1960).
8. E. Pomakhina, D. Deneele, A.C. Gaillot, M. Paris, G. Ouvrard., 29Si solid state NMR investigation of pozzolanic reaction occurring in lime treated Ca-bentonite. Cem. Concr. Res. 14, 626–632 (2012).
9. L. Dewindt, D. Deneele, N. Maubec., Kinetics of lime/bentonite pozzolanic reactions at 20 and 50°C: batch tests and modeling. Cem. Concr. Res. 59, 34–42 (2014).
10. Y.C. Chemeda, D. Deneele, G.E. Christidis, G. Ouvrard., Influence of hydrated lime on the surface properties and interaction of kaolinite particles. Appl. Clay Sci. 107, 1–13 (2015).
11. G. Russo, S. Dal Vecchio, G. Mascolo., Microstructure of a lime stabilised compacted Silt. In: Schanza, Dans In T. (Ed.), Experimental

Unsaturated Soil Mechanics. Springer, Berlin Heidelberg, pp. 49–56 (2007).
12. Z. Metelkova, J. Bohac, R. Prikryl, I. Sedlarova., Maturation of loess treated with variable lime admixture: pore space textural evolution and related phase changes. Appl. Clay Sci. 61, 37–43 (2012).
13. G. Russo, G. Modoni., Fabric changes induced by lime addition on a compacted alluvial soil. Geotechnique Lett. 3, 93–99 (2013).
14. E. Vitale, D. Deneele, G. Russo., Multiscale analysis on the behaviour of a lime treated bentonite, VI Italian Conference of Researchers in Geotechnical Engineering-Geotechnical Engineering in Multidisciplinary Research: from Microscale to Regional Scale, CNRIG2016. Procedia Eng. 158, 87–91(2016).
15. E. Vitale, D. Deneele, M. Paris, G. Russo., Multi-scale analysis and time evolution of pozzolanic activity of lime treated clays. Applied Clay Science, 141, 36–45 (2017).
16. M. Al Mukhtar, A. Lasledj, J.F, Alcover., Behavior and mineralogy changes in lime treated expansive soil at 20 °C. Applied Clay Science, 50, 191–198 (2010).
17. C.D. Rogers, S. Glendinning., Modification of clay soils using lime. In: CDF Rogers, Dans S.G. (Ed.), Lime Stabilisation. Thomas Telford, London, pp. 99–112 (1996).
18. S. Leroueil, J.P Le Bihan., Liquid limits and falling cones. Can. Geotech. J. 33, 793–798 (1996).
19. J.K. Mitchell., Practical problems from surprising soil behaviour. J. Geotech. Eng., ASCE 112, 259–289 (1986).
20. J.B. Croft., The pozzolanic reactivates of some New South Wales fly ashes and their application to soil stabilization. Proc. ARRB, Australia 2, 144–167 (1964).
21. O.G. Ingles, J.B. Metcalf., Soil Stabilisation: Principles and Practice. Butterworths, Sidney, Australia (1972).
22. F.G. Bell., Lime stabilisation of clay minerals and soils. Eng. Geol. 42, 223–237 (1996).
23. D.N. Little., Fundamentals of the Stabilization of Soil With Lime: National Lime Association. Bulletin vol. 332. Arlington, USA, pp. 1–20 (1996).
24. R. James, A.H. Kamruzzaman, A. Haque, A. Wilkinson., Behaviour of lime-slag-treated clay. Proceedings of the ICE-Ground Improv. vol. 161 (4), pp. 207–216 (2008).
25. M. Khemissa, A. Mahamedi., Cement and lime mixture stabilization of an expansive overconsolidated clay. Appl. Clay Sci. 95, 104–110 (2014).
26. A.R. Goodarzi, M. Salimi., Stabilization treatment of a dispersive clayey soil using granulated blast furnace slag and basic oxygen furnace slag. Appl. Clay Sci. 108, 61–69 (2015).
27. J.M. Kinuthia, S. Wild., Effects of some metal sulphates on the strength and swelling properties of lime-stabilised kaolinite. Int. J. Pavement Eng. 2 (2), 103–120 (2001).
28. R.W. Grimshaw., The Chemistry and Physics of Clays. Fourth Editing. Ernest Benn (1971).
29. J. Locat, H. Tremblay, S. Leroueil., Mechanical and hydraulic behaviour of a soft inorganic clay treated with lime. Can. Geotech. J. 33 (4), 654–669 (1996).
30. P.V. Sivapullaiah, A. Sridharan, A.N. Ramesh., Strength behaviour of lime treated soils in the presence of sulphate. Can. Geotech. J. 37, 1358–1367 (2000).
31. H. Tremblay, S. Leroueil, J. Locat., Mechanical improvement and vertical yield stress prediction of clayey soil from eastern Canada treated with lime or cement. Can. Geotech. J. 38, 567–579 (2001).
32. S.M. Rao, P. Shivananda., Role of curing temperature in progress of lime-soil reactions. Geotech. Geol. Eng. 23 (1), 79–85 (2005).
33. ASTM D-4318., Standard test method for liquid limit, plastic limit and plasticity index of soils. Annual Book of ASTM Standards, vol. 04.08. Easton, PA: American Society for Testing and Materials (2000).
34. J. Locat, H. Tremblay, S. Lerouil., "Mechanical and hydraulic behavior of a soft inorganic clay treated with lime". Journal of Geotechnical 33, 654-669 (1966).
35. S.H. Chew, A.H.M. Kamruzzaman, F.H. Lee, "Physicochemical and engineering behavior of cement treated clayey". Journal of Geotechnical and Geo-Environmental Engineering 130, 696-706 (2004).
36. R.C. Bates., Electrometric pH determination, John Wiley and sons Inc. New York (1954).
37. B. Le Runigo, O. Cuisinier, Y.J. Cui, V. Ferber, D. Deneele., Impact of initial state on the fabric and permeability of a lime-treated silt under long-term leaching. Canadian Geotechnical Journal, 46, 1243–1257 (2009).
38. T. Umesha, S. Dinesh, P. Sivapullaiah., Control of dispersivity of soil using lime and cement. International Journal of Geology, 3, 8–16 (2009).
39. J.C. Dong., Investigation of aggregates size effect on the stiffness of lime and/or cement treated soils: From laboratory to field conditions (Ph.D thesis). Université Paris Est, France (2013).
40. S.A. Bourokba, A. Hachichi, H. Souli, S. Taibi, J.M. Fleureau., Effect of lime on some physical parameters of a natural expansive clay from Algeria, European Journal of Environmental and Civil Engineering (2015).
41. J.L. Eades, R.E. Grim., A quick test to determine lime requirements for lime stabilization. Highw. Res 139: 61–72 (1966).
42. J.M. Kinuthia, S. Wild, G.I. Jones., Effects of monovalent and divalent metal sulphates on consistency and compaction of lime-stabilized kaolinite. Appl. Clay Sci. 14, 27–45 (1999).

Analysis the dynamic response of earth dam in free vibration and forced by introducing the effect of the interaction dam foundation

Boumaiza Malika[1], Mohamadi Sadika[2] and Ait Ahmed Fatiha[3]
[1]Université de Bouira, Université Houari Boumediene, laboratoire LBE, Algérie.
[2] Université de Bouira, Ecole nationale supérieure des travaux publics, Laboratoire TPiTE. Algérie.
[3]Université de Bouira, Uuniversité Mouloud Mammeri, Tizi Ouzou. Algérie.

Abstract: The present study concerns the analysis of the dynamic response of earth dam, in free and forced vibration (under the effect of earthquake) using the finite element method. The analysis is carried out at the end of dam construction without filling. The behavior of the dam materials and the foundation is linear elastic. In free vibration, to better understand the effect of the dam foundation interaction, we will take into account different site conditions and see their influence on the free vibration characteristics of the dam. In forced vibration, to study the seismic response of the dam, the system is subjected to the acceleration of the Boumerdes earthquake of May 21, 2003 recorded at the station n ° 2 of the dam of Kaddara in the base, with a parametric study taking into account the influence of the main parameters such as the mechanical properties of the soil: rigidity, density.

1 Introduction

À l'état naturel un massif en terre est en équilibre sous l'effet des contraintes en place, et la moindre perturbation de cet état naturel donne naissance à un nouvel état de déformations contraintes.
Ces perturbations peuvent être d'origine naturelle, les séismes par exemple, comme elles peuvent être dues à l'activité de l'homme ; exploitation des carriers ou par exemple élévation d'une structure de retenue pour palier aux problèmes de l'envasement de ces retenues et des inondations de régions [1].
Quelques 245 barrages en remblai – principalement de petits barrages étaient endommagés suites au séisme de Bhuj d'une magnitude de 7.7 le 26 janvier 2001 à Gujarât en Inde. D'autres barrages ont reçu de sérieuses détériorations pendant les séismes de Kolinski le 17 août 1999 en Turquie et de Chi Chi à Taiwan [2].
Ces évènements récents ont montré que les tremblements de terre présentent toujours des dangers pour les barrages que les ingénieurs doivent bien prendre en charge pendant l'analyse de la sécurité sismique de ces ouvrages.
La stabilité sismique était déjà évoquée en 1955 au 5ième congrès du comité international des grands barrages – ICOLD à Paris.
Désormais dans le monde, il n'y a pas de zones qu'on peut considérer comme non sismiques. Le calcul sismique de barrages étaient toujours jusqu'à 1989 par de méthodes pseudo statiques avec des accélérations horizontales du sol égale à 0.1g, alors qu'aujourd'hui tout le monde s'accorde à dire que les tremblements de terres peuvent induire des accélérations beaucoup plus élevées que cette valeur et dans les deux sens, et que les méthodes à utiliser doivent effectivement reproduire le comportement réelle des ouvrages, pour pouvoir bien se prononcer sur la sécurité sismique des ouvrages [3,4].

2 Données de l'analyse

L'exemple de référence concerne un barrage en terre à noyau construit sur une couche homogène de sol.la figure.1 montre la géométrie de barrage.

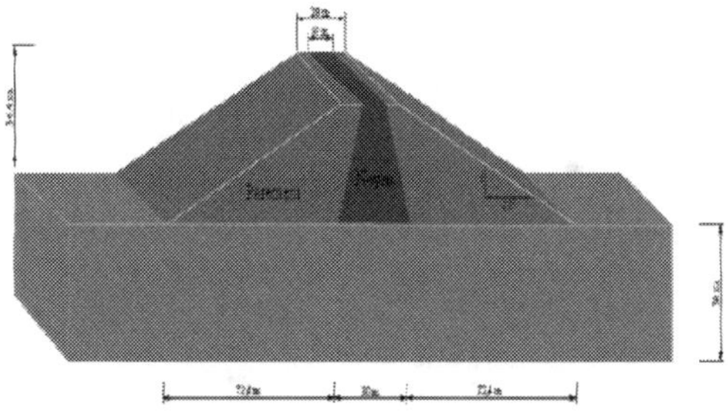

Fig. 1. La géométrie de barrage.

Les propriétés mécaniques du barrage sont récapitulées dans le tableau.1

Tableau.1 : Les propriétés mécaniques du barrage.

	Densité ρ (t/m3)	Coefficient de poisson υ	Module d'élasticité *E* (MPa)
Fondation	2.2	0.25	1000
Noyau (argile)	1.98	0.3	30.7
Parement	2,13	0.3	60

3 Analyse des vibrations libres

Commençant par l'étude du comportement dynamique du barrage en vibrations libres, en tenant compte de l'effet de l'interaction barrage-fondation

3.1 Barrage seul

Le tableau suivant récapitule les périodes naturelles de barrage seul pour les cinq premiers modes de vibration.

Tableau.2 : Les périodes naturelles pour les cinq premiers modes de vibration.

Modes	1	2	3	4	5
Périodes(Sec)	1.147	0.810	0.661	0.654	0.535

Les figures suivantes représentent les cinq modes de déformations :

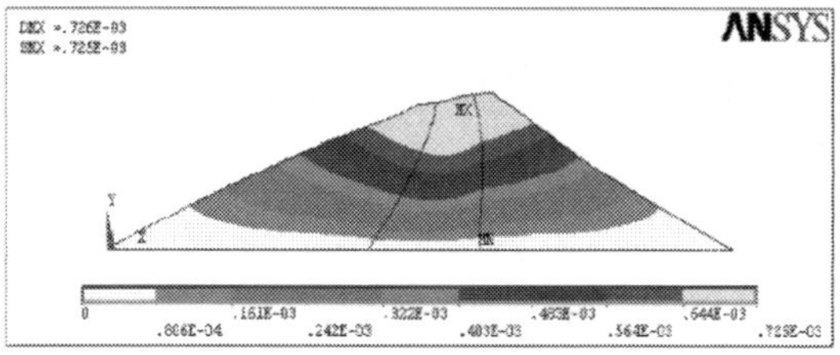

Premier mode de vibration et la répartition des contraintes σ xx (Pa).

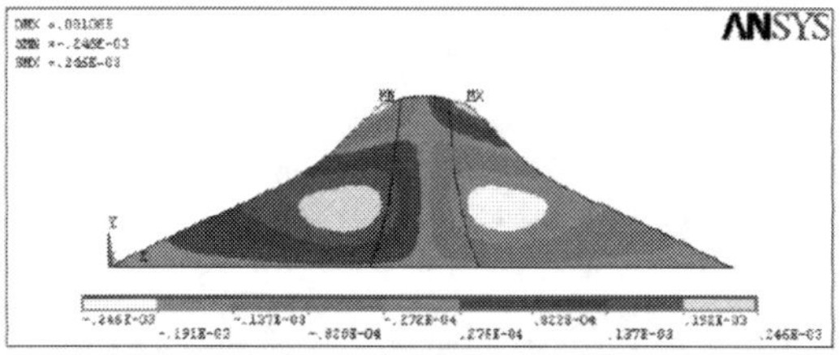

Deuxieme mode de vibration et la répartition des contraintes σ xx (Pa).

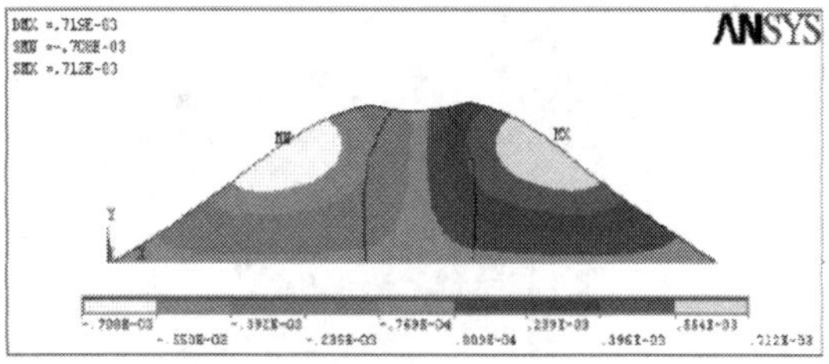

Troisième mode de vibration et la répartition des contraintes σ xx (Pa).

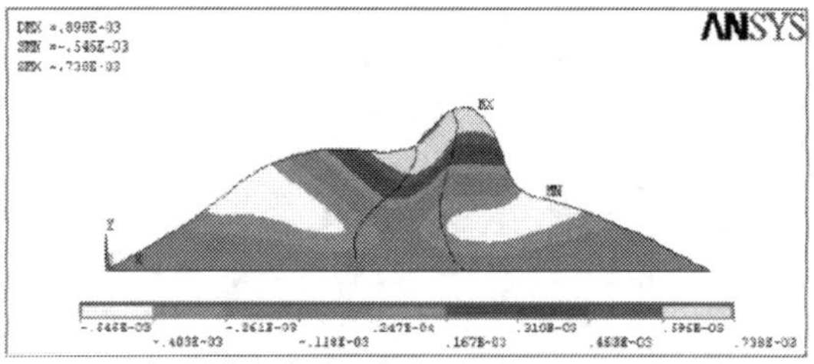

Quaterième mode de vibration et la répartition des contraintes σ xx (Pa).

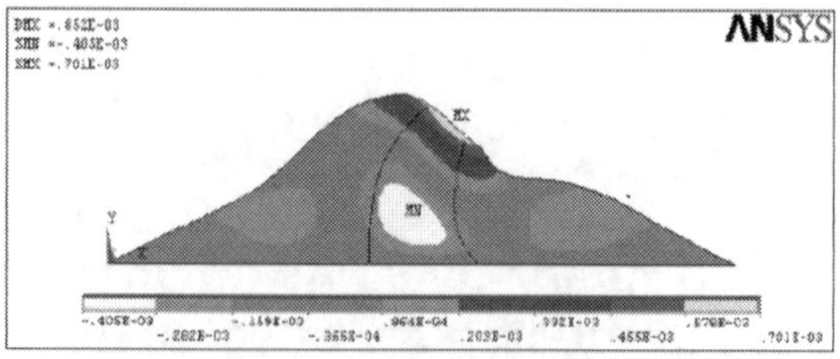

Cinquième mode de vibration et la répartition des contraintes σ xx (Pa).

3.2 Système barrage-fondation

Tableau.3 : Les périodes naturelles pour les cinq premiers modes de vibration.

Modes	1	2	3	4	5
Périodes(Sec)	1.147	0.810	0.661	0.654	0.535

Les figures suivantes représentent les cinq modes de déformations :

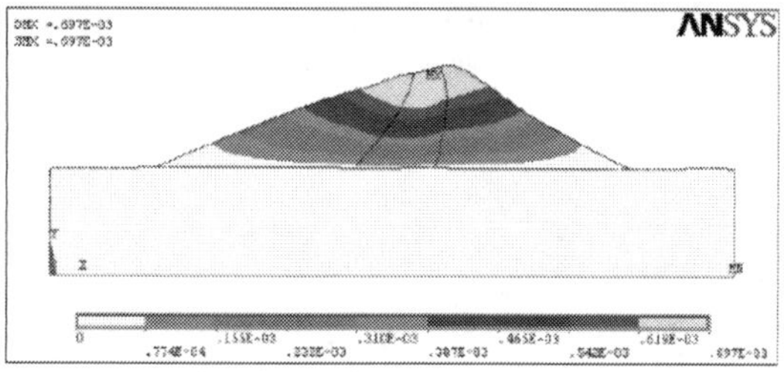

Premier mode de vibration et la répartition des contraintes σ xx (Pa).

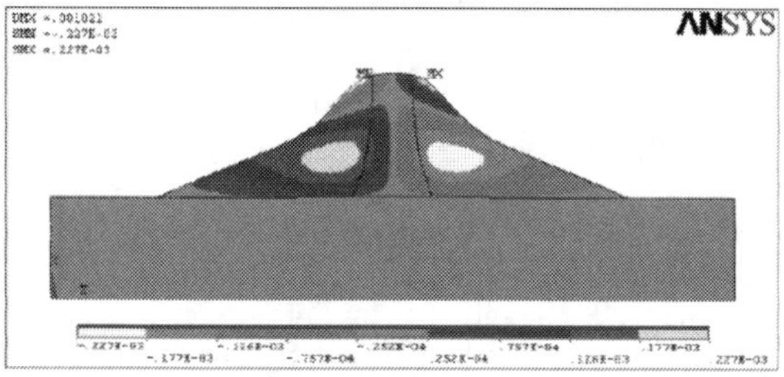

Deuxième mode de vibration et la répartition des contraintes σ xx (Pa).

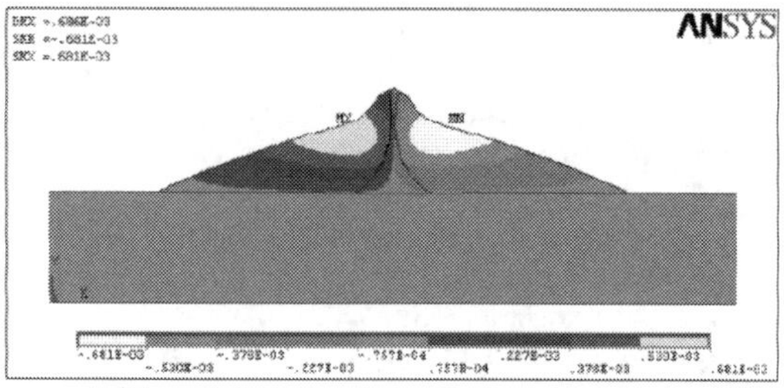

Troisième mode de vibration et la répartition des contraintes σ xx (Pa).)

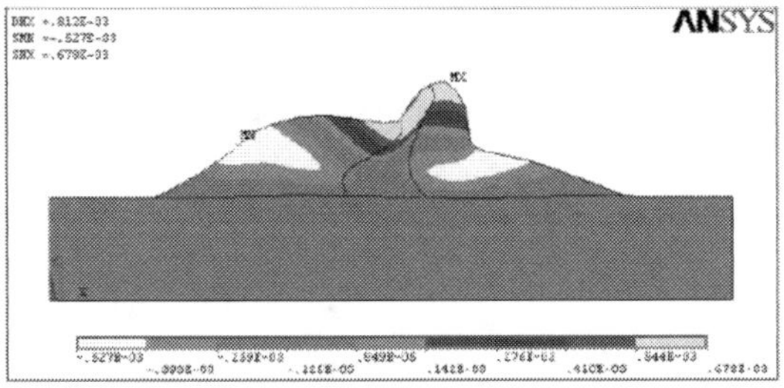

Quatrième mode de vibration et la répartition des contraintes σ xx (Pa).

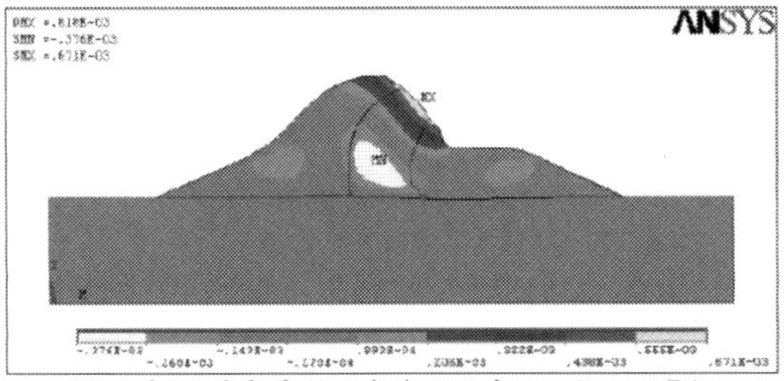

Cinquième mode de vibration et la répartition des contraintes σ xx (Pa).

En comparant les résultats obtenus pour le cas du barrage seul et ceux obtenus pour le système barrage-fondation, on remarque que la prise en compte de la fondation a tendance d'allonger les périodes de vibration du barrage. Pour mieux comprendre l'effet de l'interaction barrage fondation, on va tenir compte de différentes conditions de site et voir leur influence sur les caractéristiques de vibration libre du barrage. Les résultats obtenus sont représentés dans le tableau suivant :

Tableau.4 : Les périodes naturelles pour les cinq premiers modes de vibration en fonction de variation de rapport de rigidité E_f/E_b .

	Les rapports de rigidité E_f/E_b					
Modes	4	2	1	1/2	1/4	1/8
1	1.289	1.453	1.776	1.258	3.131	4.308
2	0.895	1.001	1.237	0.875	2.225	3.073
3	0.740	0.835	1.053	0.723	1.908	2.621
4	0.720	0.831	1.017	0.704	1.686	2.249
5	0.648	0.758	0.886	0.619	1.453	1.932

On constate que lorsque la fondation est très rigide $E_f/E_b>4$, la période du système barrage-fondation est presque identique à celle du barrage seul (sans interaction).

-Par contre, dans le cas où le rapport de rigidité $E_f/E_b<4$, l'influence de l'interaction apparaît d'une manière significative.

-Pour le cas $E_f/E_b=1/8$ on observe une augmentation qui dépasse le double pour le mode fondamental et presque le double pour les modes supérieurs.

Donc on peut conclure que :

-Pour le cas d'un sol flexible les périodes de vibrations sont supérieures à celles obtenues pour un cas de sol rigide.

-Avec l'augmentation de la rigidité de la fondation, les périodes de vibrations convergent vers les périodes propres pour le cas d'un barrage considéré encastré à sa base.

4 Analyse de comportement sismique du barrage

L'analyse du comportement dynamique du barrage sous l'effet du séisme (sans pression d'eau) dans le cas élastique est faite, avec une étude paramétrique tenant compte de l'influence des principaux paramètres tels que les propriétés mécaniques du sol : rigidité, densité.

L'analyse du comportement dynamique du barrage est faite avec la méthode des éléments finis en utilisant des éléments quadratiques à deux degré de liberté u_x *et* u_y.

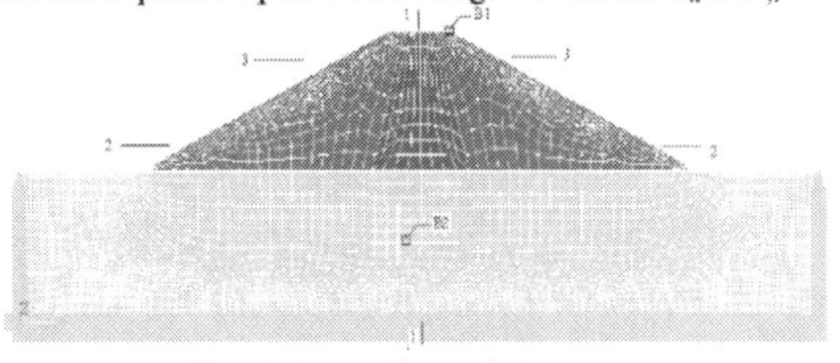

Fig. 2. Le maillage de barrage.

Pour étudier la réponse sismique du barrage, le système est soumis aux accélérations du séisme de Boumerdes du 21 mai 2003 enregistrées à la station n°2 du barrage de Kaddara.

Le rapport d'amortissement visqueux ξ =5%. Les valeurs de coefficients de Rayleigh α_R= 0.32 et β_R=0.0078. La méthode de Newmark est utilisée. Avec les valeurs des paramètres d'intégration sont γ = 0.5 et β=0.25 pour avoir la solution inconditionnellement stable. Le pas de temps est pris égal à Δt = 0.005.

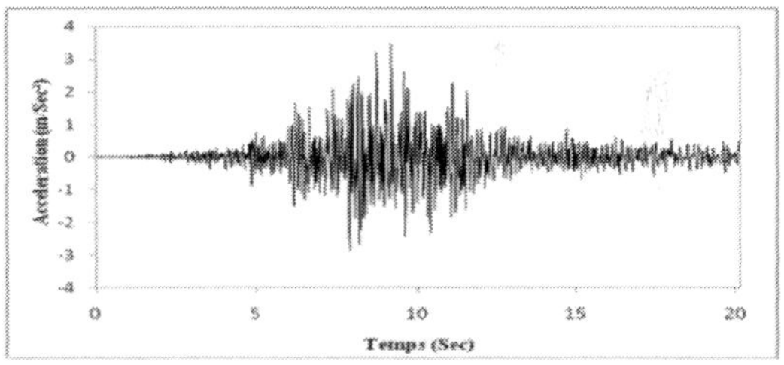

Fig. 3. *Enregistrement de la composante N.S du séisme de boumerdes (21Mai 2003). Accélération.*

4.1 La réponse du barrage au tremblement de terre

La réponse du barrage à l'excitation maximale est représentée dans la figure.4, on remarque une déformation latérale importante dans le barrage en s'éloignant de la fondation, tel que le déplacement maximal est observé à la crêté du barrage et atteint une valeur de u_{max} ≈0.12 *m*. On peut observer que l'amplification est négligeable à la fondation tandis quelle augmente en allant vers la crête du barrage.

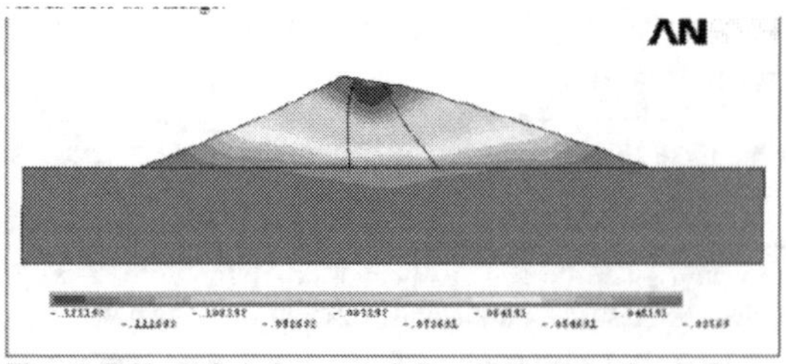

Fig. 4. Déformé du barrage à l'excitation maximale (u_{max} =0.12 m à la crête de barrage)

La figure.5, montrent la variation de l'accélération en fonction de temps pour le nœud B_1 situé à la crête du barrage, on remarque une amplification des ($a_{max} \approx 11$ *m/ Sec²*).

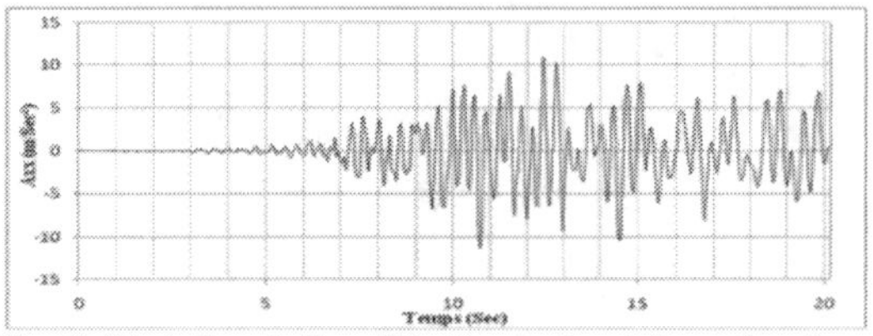

Fig. 5. *Variation de l'accélération horizontale en fonction de temps à la crête du Barrage (Nœud B1).*

4.2 Influence de la densité et de la rigidité

Pour étudier l'influence de la densité et de la rigidité, on procède de la manière suivante :
En premier, nous varions la densité de noyau et des recharges amont et aval (tableau.5),
Ensuite la rigidité de noyau, recharges amont et aval et fondation (tableau.6).

Tableau.5 : Valeurs de la densité pour le noyau et les recharges amont et aval

La densité de noyau	(Kg/m^3)	1980	3000	3200
La densité des recharges amont et aval		1700	1980	2130

Tableau.6 : Valeurs de la rigidité pour le noyau et les recharges amont et aval

La densité de noyau	(Kg/m^3)	1980	3000	3200
La densité des recharges amont et aval		1700	1980	2130

4.2.1 Influence de la densité de noyau et des recharges amont et aval

Les figures .6 et 7 montrent l'influence de la densité du noyau et des recharges amont et aval sur les contraintes, on remarque que l'augmentation de la densité du noyau conduit à une augmentation négligeable des contraintes (la masse du noyau représente une petite partie de la masse de barrage), tandis que la diminution de la densité des recharges amont et aval implique une augmentation des contraintes.

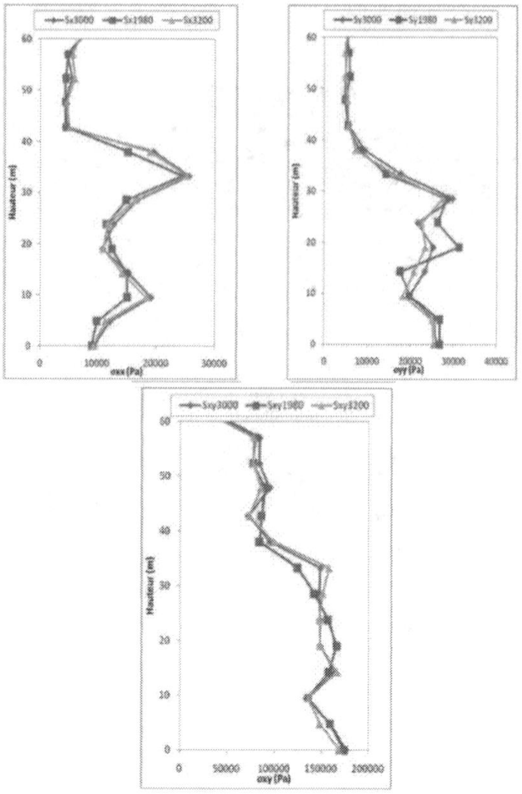

Fig. 6. Influence de la densité du noyau sur la distribution des contraintes (σ xx), (σ yy) et (σ xy).

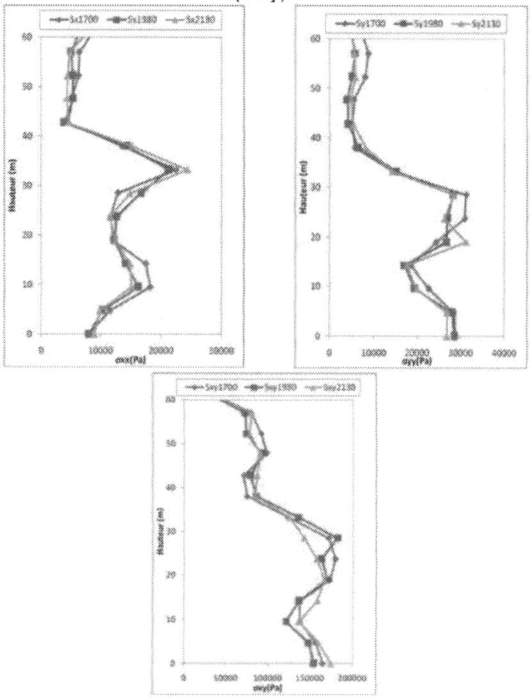

Fig. 7. Influence de la densité des recharges amont et aval sur la distribution des contraintes (σ xx), (σ yy) et (σ xy).

4.2.2 Influence de la rigidité de noyau et des recharges amont et aval

En analysant les figures.8, 9 et 10, on remarque :
La diminution de la rigidité de la fondation implique une diminution des contraintes à la base de la fondation du barrage et en s'éloignant de la base on aura une augmentation considérable des contraintes, donc quand la fondation est flexible on a l'accroissement des contraintes horizontales, verticales et de cisaillement dans le barrage.
Si la rigidité des recharges amont et aval diminue on aura une augmentation importante des contraintes horizontales, verticales et de cisaillement, par contre celle de noyau n'a pas une grande influence sur les contraintes. Tandis que la variation de la rigidité de la fondation, influe modérément les contraintes du barrage.

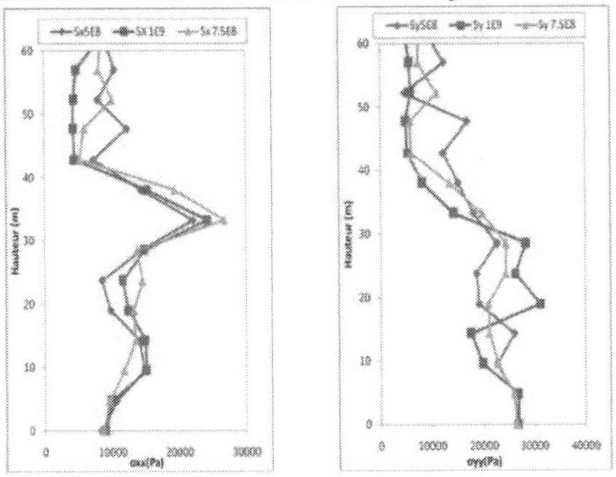

Fig. 8. Influence de la rigidité de la fondation sur la distribution des contraintes (σ xx), (σ yy) et (σ xy).

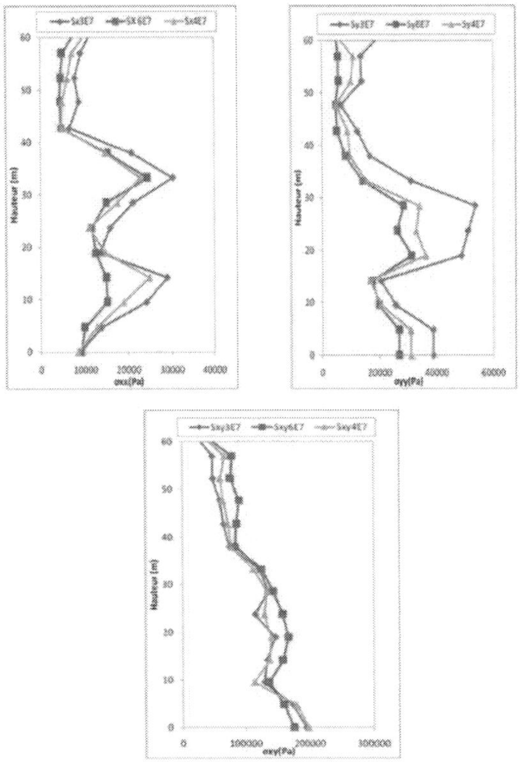

Fig. 9. Influence de la rigidité des recharges amont et aval sur la distribution des contraintes (σxx), (σ yy) et (σ xy).

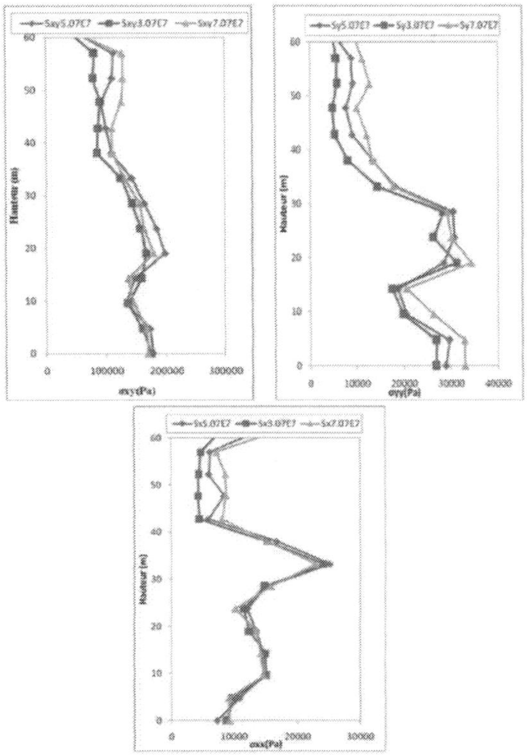

Fig. 10. *Influence de la rigidité de noyau sur la distribution des contraintes (σ xx), (σ yy) et (σ xy).*

L'analyse élastique prouve que le chargement sismique induit principalement le déplacement latéral, ce qui augmente avec la distance de la base du barrage. On observe le maximum près de la crête. Les propriétés mécaniques du noyau (rigidité, densité) affectent modérément la réponse élastique, alors que ceux des recharges amont et aval affectent de manière significative la réponse du barrage.

5 Conclusion

La présente étude concerne l'analyse du comportement de barrage en terre libre et forcé, basé sur la méthode des éléments finis en tenant compte du comportement linéaire des matériaux constituant le barrage.
Les résultats obtenus montrent que :
1. En vibrations libres:
· La prise en compte de l'interaction barrage-fondation à tendance de prolonger les périodes de vibration.
· L'augmentation de la rigidité de la fondation implique que les périodes de vibration se rapprochent des périodes d'un barrage encastré à sa base, et on a le cas contraire dans le cas où la fondation est flexible.
2. En vibration forcées :
· L'analyse montre que le séisme induit de grands déplacements latéraux, ces derniers augmentent en s'éloignant de la fondation et allant vers la crête de barrage.
En plus on a une amplification des accélérations.
La densité des recharges amont et aval influe d'une façon considérable la distribution des contraintes dans le barrage, et la diminution de leurs rigidité à tendance d'augmenté les contraintes dans le barrage.

References

1. Robin Fell, Patrick MacGregor, David Stapledon and Graeme Bell (2005). Geotechnical engineering of dams. A.A.BALKEMAPUBLISHERS Leiden/London/New York/Philadelphia/Singapore.
2. ANCOLD (1998). Guidelines for design of dams for earthquake. Australian National Comminittee on Large dams, Melbourne.
3. US Corps of Engineers (1984). Rationalising the seismic coefficient method. Miscellaneous Paper GL84-13.
4. Seed, H.B., (1979). "Considerations in the earthquake-resistant design of earth and rock fill dams". Geotechnique, Vol.29, No.3.

Effect of swelling mineral on geotechnical characteristics of clay soil

Mechri Mohammed Cherif[1], Medjnoun Amal[1], Bahar Ramdane[1]
[1]University of Sciences and Technology Houari Boumediene (USTHB), Faculty of Civil Engineering, LEEGO, Bab Ezzouar, Algiers, Algeria

Abstract. The shrinking and swelling is a very common phenomenon in the world, mainly in arid and semi-arid areas. The construction on these soils often requires a study of the risk of swelling, in advance. The emergence of this phenomenon in soils is caused primarily by the presence of swelling clays and water. The risk is most significant when the rate of the expansive minerals, such as bentonite, is important. Several researchers have studied the relationship between the percentages of swelling minerals and soil swelling potential. The researchers have shown the difficulty of characterization, because of the complexity of the phenomenon under natural conditions, where several other phenomena occur, and which are not considered in laboratory experimentation. This article addresses the characterization of several soils, with different swelling rates from the physical, chemical and mechanical tests. It shows the relationship between the quantity of swelling minerals and some site conditions, such as water content and consolidation stress, on the ability of soil to swell. This work aims to establish empirical relationships between the percentage of swelling minerals and soil swelling potential.

1 Introduction

The shrinking - swelling of soils is one of the geotechnical problems. It is observed on the plastical clay soils under the various environmental conditions. This phenomenon affects all the weight construction realized on these soils without the knowledge of their properties. The damages resulting from this phenomenon are expensive in repairing damaged structures [1].

The swelling characteristics of expansive soils are due to several parameters associated to the initial soil conditions (initial dry density and initial moisture content) [2, 3], the granularity and clay content. Nonetheless, the first factor which conditions the swelling of a clay soil and affects these physical-chemical and mechanical properties is the mineralogical composition; the quantity and type of minerals present in clay soils. It's known for example, that the phenomenon of swelling depends not only on the granularity, but also on the superficial activity of the clay itself which will be all the more intense as the specific area of particles will be high.

A montmorillonite will swell more than illite and this exhibits for such phenomenon more than the kaolinite; the properties of the soil will be all more marked as the percentage of expansive minerals (montmorillonite) will be higher [4, 5, 6, 7, 8].

But in nature, the soil is very heterogeneous; its composition depends on other elements in addition to swelling minerals. Several mineralogical researches on natural soils show the considerable presence of swelling clay and non-swelling clay in addition to the presence of bentonite and silt [4, 5, 6, 7, 8, 9].

Therefore, the objective of this study is to highlight and quantify the influence of swelling minerals on the different properties of studied soils.

2 Test procedures

The materials used in this study are: Bentonite delivered by the treatment company "ENOF", the Kaolinite delivered by the treatment company "SOALKA", the silt and the fine sand from the Algiers region. Table 1 and 2 summarize the physical-chemical properties of used materials.

The preparing of tested mixtures was performed following the same procedures in order to optimize the repeatability between the prepared samples. The four materials were dried in oven at 105 °C for 24 h, then samples were prepared by manual mixing in different weight proportions, by varying percentage of bentonite, kaolinite and keeping the rate of the Silt and fine sand fixed as shown in table 3.

The mixtures prepared were kept in closed plastic bags and conserved more than 24 h; this allowed moisture in the soil to distribute evenly throughout the mass of soil (for compaction and oedometric tests). The granulometric distribution by sedimentation of studied mixtures is presented in figure 1, soil geotechnical and physical-chemical properties are presented in table 4.

Table 1. Physical - chemical properties of used materials.

Properties	Materials			
	Bentonite	Kaolinite	Silt	fine sand
Liquid Limit (%) (ASTM D 4318 00)	149,39	63,40	30,84	/
Index Plastic (%) (ASTM D 4318 00)	119,59	26,94	9,98	/
Bleu Value (g/100g)	12.84	2,00	0,24	0.12
Activity of Skempton	2.84	0.84	/	/
pH	8,58	6,01	6,75	6,03
Sst (m^2/g)	269,50	42,00	4,98	2,63

Table 2. Chemical composition of the materials used.

	SiO_2	Al_2O_3	Fe_2O_3	CaO	MgO	K_2O	Na_2O	SO_3	Cl
Bentonite	60,80	17,33	1,72	3,62	3,00	1,89	4,45	0,29	0,06
Kaolinite	53,04	31,98	1,83	2,93	0,00	3,31	0,01	0,00	0,02
Silt	49,43	16,19	11,46	4,83	7,82	2,14	1,55	0,16	0,06
Fine sand	48,78	7,39	8,49	15,36	0,83	0,87	0,91	0,09	0,04

Table 3. Mineralogical composition of the studied soils.

Index of soil / Material	5B/60K	10B/55K	20B/45K	30B/35K	40B/25K	50B/15K
Bentonite (%)	5	10	20	30	40	50
Kaolinite (%)	60	55	45	35	25	15
Silt (%)	20	20	20	20	20	20
Fine sand (%)	15	15	15	15	15	15

Table 4. Physical - chemical and geotechnical properties of the studied soils.

Properties \ Index of soil	5B/60K	10B/55K	20B/45K	30B/35K	40B/25K	50B/15K
Liquid Limit (%)	49,60	55,80	75,44	88,85	92,54	101,81
Index Plastic (%)	20,34	26,34	45,91	56,46	57,21	69,07
Activity	0,48	0,61	1,00	1,04	1,08	1,17
Classification (UCSC)	CL	CH	CH	CH	CH	CH
Max. Dry unit weight (KN/m^3) (ASTM D 698 12)	15,06	14,95	14,90	14,90	14,72	14,47
Optimum Moisture Content (%) (ASTM D 698 12)	14,33	15,13	17,38	17,66	21,82	23,78
Specific Area (m^2/g)	45,50	59,50	77,00	115,50	136,50	154,00
Swelling Potential (%) (ASTM D 4546 03)	1,02	2,08	8,10	11,25	15,60	17,90
Pressure swelling (KPa) (ASTM D 4546 03)	34,54	77,47	193,72	214,94	321,49	470,80

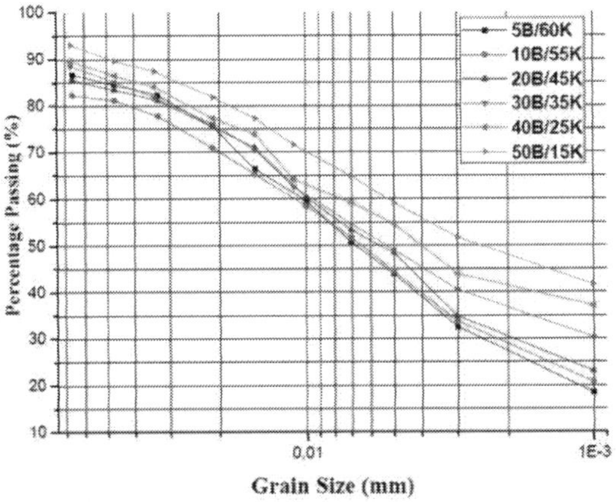

Fig. 1. Grain size distribution of the studied mixtures.

3 Results and discussion

3.1. Effect of minerals on the Atterberg limits

The results presented in figure 2 show that the liquidity limit increases from 50 to 100%, with an increase of nearly 50% for the rate of bentonite ranging from 5 to 50%, due to the increase in fine particles of the bentonite and their swelling characters, which further favors water absorption.

The plasticity index has been evaluated for all studied mixtures and the results represented in of figure 2. A proportional relationship between bentonite content (between 5 and 50 %) and the plasticity index of studied mixtures were observed, rising from 20 % to 70 %, an increase of nearly 70 %. This fact due to the total surface area of water adsorption of samples that depending on grains size of bentonite.

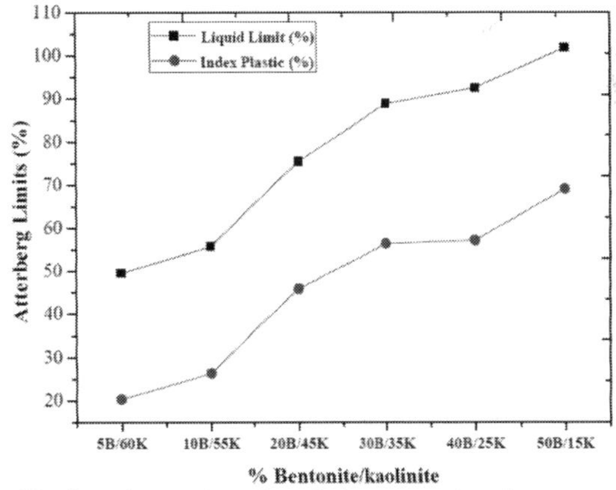

Fig. 2. Variation of Atterberg limits versus mineral contents (bentonite / kaolinite).

3.2 Effect of minerals on the activity and total surface area

From the figure 3, it can be seen that the activity and the total surface area are proportional to the bentonite rate and evolving at the same intensity. The lowest values are obtained for mixtures with high levels of kaolinite (5B/60K and 10B/55K) with an activity ranging from 0,48 to 0,61 and a specific area between 45,50 and 59,50 m^2/g, respectively. On the other hand, when bentonite contents increase from 5 to 50%, the activity rises from 0,48 to 1,17 and the specific surface area from 45,50 to 154 m^2/g respectively. The increase of more than 60 % is observed. These results show the activity and the development of specific area is much higher when the rate of active clay (bentonite) is important. There is a direct impact on the swelling potential of these mixtures, which is in accordance with other researchers work studies [10, 11, 12, 13, 14] between mixtures and swelling potential.

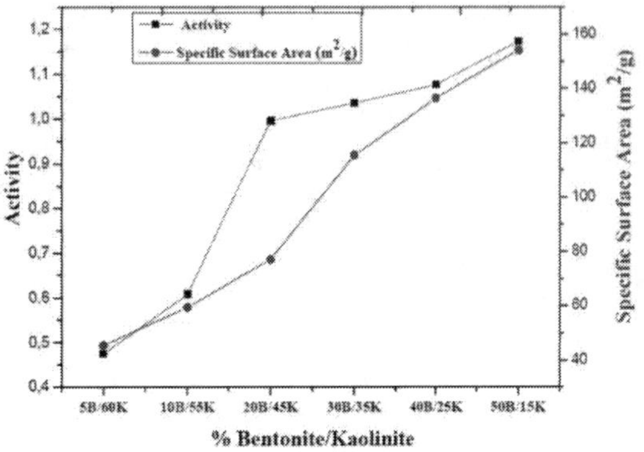

Fig. 3. Variation of the activity and total surface area versus mineral contents (bentonite / kaolinite).

3.3 Minerals effect on compaction characteristics

The effect of minerals percentage (bentonite / kaolinite) in studied mixtures on compaction parameters (optimal water content and dry density) are shown in figure 4. The dry density decreases slightly for a high rate of bentonite from 5 to 50%, a reduction of 4% is observed which is not significant.

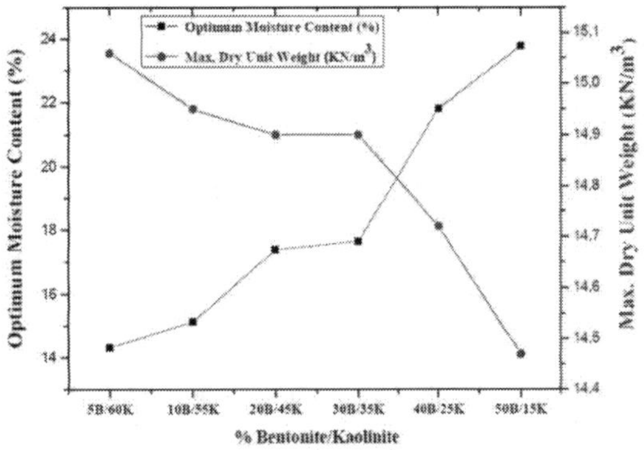

Fig. 4. Variation of compaction parameters as a function of mineral content (bentonite / kaolinite).

The optimal water content is very sensitive to the percentage of bentonite, a proportional relationship is found between the optimal water content and the bentonite rate, which increases from 14,33% for the mixture with 5% of bentonite compensate by 65 % Kaolinite (5B/65K) up to 23,78% for mixing with 50% bentonite compensate by 15% kaolinite (50B/15K), this variation is caused by the increase in the amount of water adsorbed by the bentonite particles present in the mixtures.

3.4 Effect of minerals on swelling parameters

Figure 5 shows the kinetics of free swelling as a function of logarithm of time. It is noted that the swelling kinetics have a conventional form with an initialization phase, a primary swelling phase and a secondary swelling phase, which evolve differently as a function of the minerals percentages (bentonite / kaolinite);

reconstituted soils (5B / 60K and 10B / 55K) with high rates of kaolinite (60 and 55% respectively) and low percentages of bentonite (5 and 10% respectively) show a rapid initialization phase; at the moment that studied mixtures are in contact with water they swell immediately, this phase corresponds to swelling of macrostructure that depends on the permeability and suction of tested samples, In the second swelling phase, minor variations in volume are noted, the equilibrium is established rapidly (after 48 h).

Reconstituted soils (20B / 45K, 30B / 35K, 40B / 25K, and 50B / 15K) with considerable percentages of bentonite 20, 30, 40, and 50%, respectively, indicated very slow kinetics of swelling initialization phase, this may be due to a low permeability [15, 16, 17]. Then primary swelling phase begins. It can be noticed that an important deformation occurs which correspond to swelling of microstructure, that depending on the degree of hydration of bentonite minerals and cation exchanges.

In the secondary phase of swelling low deformations were recorded, this phase is slow and gradual; the stability is established only after 360 hours.

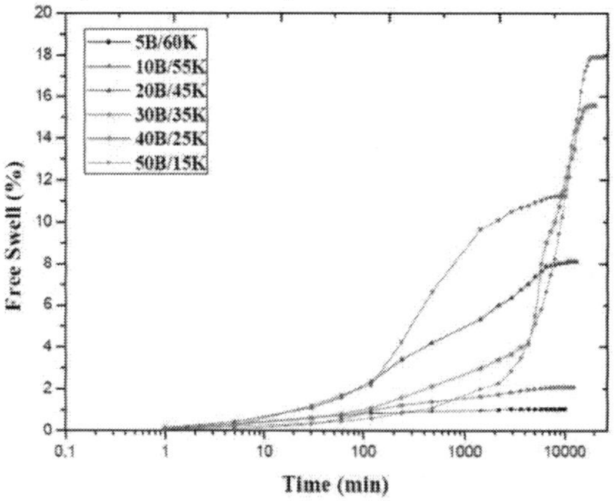

Fig. 5. Evolution of free swelling as a function of time.

The variation of swelling parameters as a function of minerals percentage contained in the studied mixtures is shown in figure 6. A proportional relationship between the bentonite content and swelling parameters is observed, and inversely proportional to kaolinite percentages. This is associated to the nature and rates of active minerals present in mixtures, which is in accordance with work of many researchers [2, 4, 5, 8, 14], The swelling potential increases from (1,02 to 17,90%) for bentonite contents ranging from (5 to 50%), at the same time swelling pressure increases from (34,54 to 470,80 KPa), reaching a value of more than 90%.

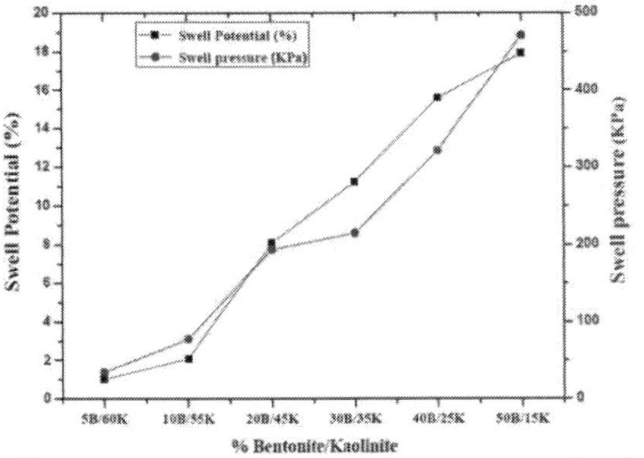

Fig. 6. Variation of swelling rate and pressure versus mineral contents (bentonite / kaolinite).

4 Conclusion

The present researchers show the impact of nature and rate of minerals present in various reconstituted soils. The following conclusions can be drawn from these results:

Limits of Atterberg are strongly influenced by the variation of mineral contents (bentonite / kaolinite), they have a proportional relationship with the percentage of bentonite, an increase of nearly 50% for the Bentonite contents ranging from 5 to 50% is noticed.

The development of the specific area is as higher as the level of active clay (bentonite) contained in studied mixtures is greater; There is an increase of 70% when bentonite contents range from 5 to 50%.

Results obtained for dry density of studied soils as a function of mineral content show a slight variation, there was a decrease of only 4% for bentonite rate ranging from 5 to 50%. However, the optimal water content is highly sensitive and proportional to bentonite levels, where an increase of 40 % is recorded for bentonite contents varying from 5 to 50%.

Kinetics of swelling progress differently according to the rate and type of minerals; For mixtures with high rate of kaolinite compensated by low bentonite content (5B / 60K, 10B / 55K and 20B / 45K) the primary swelling phase occurs at the instantant of contact with water and stabilization is established, swelling is due only to deformations caused by the macrostructure, in case of mixtures with high bentonite contents compensated by a low percentage of kaolinite (30B/35K, 40B/25K and 50B/15K), swelling tendency is very slow and progressive, the greater deformations are recorded during the second swelling phase which correspond to deformations caused by microstructure hydration.

Swelling parameters are highly sensitive to the bentonite content; swelling potential is higher at rates of 1,02 to 17,90% for bentonite percentages ranging from 5 to 50%, at the same time swelling pressure increases from 34,54 to 470,80 KPa, where an increase of more than 90% is noted.

We find that with a bentonite level beyond 30% the swelling potential becomes dangerous with demanding values (Liquid Limit = 88%, Ac = 1,04, Specific Area = 115,05 m2/g and Pressure swelling = 214,94 KPa).

References

1. Vincent M., Hédou F., Chirouze M., Plat E., Leroy S. (2008). Cartographie à l'échelle départementale de l'aléa retrait-gonflement des sols argileux à des fins préventives. Symposium International Sécheresse et Construction, Pages 55-62.
2. Komine H. & Ogata N. (1994). Exprimental study on swelling characteristics of compacted bentonite. Canadian Geotechnical Journal, vol. 31, p. 478-490.
3. Sànchez M., Gens A., Guimarães In & Olivella S. (2005). Adouble structure generalized plasticity model for expansive material. International journal for Numerical and Analytical Methods in Geomechanics, in press.
4. Tabani P. (1999). Transfert hydrique dans des sols déformables. Thèse de doctorat, Institu National Polytechnique de Lorraine, École Supérieure de Géologie, Nancy, 173p.
5. Xu Y.F., Matsuoka H. & Sun D. A. (2003). Swelling characteristics of fractal-textured bentonite and its mixtures. Epplied clay science. Vol 22. p 197-209.
6. Anil Kumar Mishra., Sarita Dhawan., Sudhakar M. Rao. (2008). Analysis of Swelling and Shrinkage Behavior of Compacted Clays, Geotech Geol Eng. 26, 289–298
7. Maison T., Laouafa F., Fleureau J.M., Delalain P. (2010). Analyse au niveau microscopique du comportement hydrique d'agrégats argileux, Journées nationales de géotechnique et de géologie de l'ingénieur (JNGG 2010).
8. Binod Tiwari, Beena Ajmera. (2011). Consolidation and swelling behavior of major clay minerals and their mixtures, Applied Clay Science. 54, 264–273
9. F. Homand, T. Belem, M. Souley. (2001). Friction and degradation of rock joint surface under shear loads International Journal for Numerical and Analytical Methods in Geomechanics, 25 (10), pp. 973–999
10. Tran N.L. (1977). Un nouvel essai d'identification des sols : l'essai au bleu de méthylène. Bulletin de liaison des Laboratoires des Ponts et Chaussées, no. 88, pp. 136-137.
11. Van olphen H., Fripiat J.J. (1979). Data Handbook for Clay Minerals and Other Non metallic Materials, Pergamon, Oxford, UK.
12. Lautrin D. (1989). Utilisation pratique des paramètres dérivés de l'essai au bleu de méthylène, Bull. de liaison des Ponts et Chaussées, n°88, pp. 136-147
13. Bedin A., Jacquard C. (2004). Pathologie des fondations superficielles sur sols argileux – retour d'expérience en Midi-Pyrénées, 10 pp.
14. Mohie Eldin Elmashad a & Alaa A Ata. (2014). Effect of seawater on consistency, infiltration rate and swelling characteristics of montmorillonite clay. HBRC Journal. 12, pp 175–180

 https://doi.org/10.1051/matecconf/201814902067

15. Tay YY, Stewart DI, Cousens TW (2001) Shrinkage and desiccation cracking in bentonite–sand landfill liners. Eng Geol 60:263–274
16. Kockar MK, Akgu¨n H, Aktu¨rk O¨ (2005) Preliminary evaluation of a compacted bentonite/sand mixture as a landfill liner material. J Solid Waste Technol Manag 31(4):187–192
17. Stewart DI, Cousens TW, Studds PG, Tay YY (1999) Design parameters for bentonite-enhanced sand as a landfill liner. Proceedings of the Institution of Civil Engineers—Geotechnical engineering 137(4):189–195.

Study of the interaction of the rupture zones of contiguous anchor plates in analogical medium

H. Abbad[1], N. Elmeiche[1 2], I. Mechab[2 3] and C. Zemri[2 1]
[1]Université Djillali Liabes, LGCE, Département de Génie Civil et Travaux Publics, Sidi Bel Abbes, Algérie
[2]Université Mustapha Stambouli, Département de Génie Civil, Mascara, Algérie
[3]Université Djillali Liabes, LMPM, Département de Génie Mécanique, Sidi Bel Abbes, Algérie

Abstract. Knowing what occurs above anchor plates is certainly instructive. In this general vision of the interaction soil-anchor plates that our work was directed. An experimental study which required the design and implantation of a model containing plastic granules powder to similate a natural environment, is presented. The latter is subjected to the removal of anchor paltes. For each test, digital photographs are taken to materialize different deformed configurations during the pullout process. These photos processed in couples by the 7D software (image correlation) giving the evolution of the displacement field and plane strain analogical environment. Particular attention is paid to the discussion of the interference of rupture zones of contiguous anchors by reducing the distance between plates.

1 Introduction

Les plaques d'ancrage sont des éléments structuraux légers employés pour résister à des forces d'arrachement « traction » sollicitant les pylônes, les mâts de transmission et les structures soumises à la poussée d'Archimède telle que les stations offshore. Des études théoriques 'Meyerhof [1]; Kumar et Kouser [2]', et expérimentales sur modèle réduits sur des plaques contigües ancrées dans un sol pulvérulent effectuées par 'Hanna et Sparks [3] ; Geddes et Murray [4] ' ont mis en évidence l'influence du sol d'ancrage, de la géométrie des plaques, de la profondeur de d'ancrage et de la distance entre deux plaques voisines sur leur comportement. Voir de plus près ce qui se passe dans le milieu environnant deux plaques d'ancrage voisines, reste cependant toujours instructif. Cela est d'autant plus vrai que le comportement du milieu est conditionné par l'existence des configurations particulières des plaques 'Abbad [5]'. L'analyse cinématique de ces milieux, moyennant une modélisation physique, permet de se rendre compte du comportement réel du milieu étudié et surtout répondre à des interrogations des problématiques précises. L'utilisation d'un milieu analogique dans ce genre de modélisation rapproche à la compréhension de la phénoménologie des sols réels tant que les deux matériaux ont des points communs tels que l'incompressibilité des grains et leur caractère frottant.

2 Modèle expérimental

Le modèle réduit sur lequel nous avons réalisé différents essais est un dispositif parallélépipédique, de dimensions égales à (100x80x5) cm, conçu pour contenir des granulés en plastique (matériau analogique) simulant un milieu granulaire. La paroi d'avant du modèle est constituée d'une plaque de verre de 6mm d'épaisseur, avec un indice réfraction constant sur toute sa surface. Les autres parois sont réalisées avec des plaques en bois rigides. Des parallélépipèdes en bois lissé de dimensions (5x5x1,5) cm, modélisant des plaques d'ancrage, sont en contact ponctuel avec des tiges fileteé. Des écrous sont soudés à un système d'attache au châssis du modèle, ce qui permet de contrôler l'arrachement des deux plaques en dévissant les tiges. Un déplacement strictement vertical est imposé aux plaques d'ancrage du fait que ces dernières sont guidées par des rails de 5 mm d'épaisseur, placés sur la paroi arrière, empêchant toute rotation (figure 1).

Les grains en plastique, ronds de diamètre moyen 1mm, de différentes couleurs (rouge, bleue, jaune, verte et noire) sont mélangés pour matérialiser un milieu visuellement contrasté. Afin d'obtenir un milieu compact et homogène, le matériau est déposé uniformément sur toute la largeur du bac par petites couches de 2cm d'épaisseur, chaque couche est compactée moyennant une dame de compactage métallique. Le milieu à un indice des vides de 0,321, correspondant à une densité relative Dr = 96% (e_{min} = 0.302, e_{max}= 0.855). Vu les très faibles contraintes mises en jeu, les grains en plastique

https://doi.org/10.1051/matecconf/201814902068

constituant le milieu étudié sont considérés indéformables.

Afin de décrire le champ de déplacement et de déformation au dessus d'une plaque d'ancrage, plusieurs séries d'essais ont été réalisées. Pour chaque essai réalisé, une série de photos numériques, de résolution 3200 x 2400 pixels, est prise de face, l'axe de la prise de vue étant perpendiculaire au plan du modèle réduit. Ces photos matérialisent différentes configurations déformées du milieu granulaire, engendrées par l'arrachement des plaques d'ancrage. Les images enregistrées sont traitées ultérieurement pour restituer le champ de déplacement pour déterminer le champ de déformation par le logiciel 7D (Vacher et al [6]). Il s'agit d'une méthode de corrélation entre deux images, ces dernières seront comparées en utilisant le coefficient de corrélation du niveau de gris. En d'autres termes, les points de l'image initiale sont retrouvés sur les images successives avec une incertitude de mesure de l'ordre de 0,1pixel, voire 0,01 pixel en petite déformation. L'image pour laquelle la déformation sera analysée est divisée en un certain nombre d'éléments de grille, carrés de côtés 20 pixels.

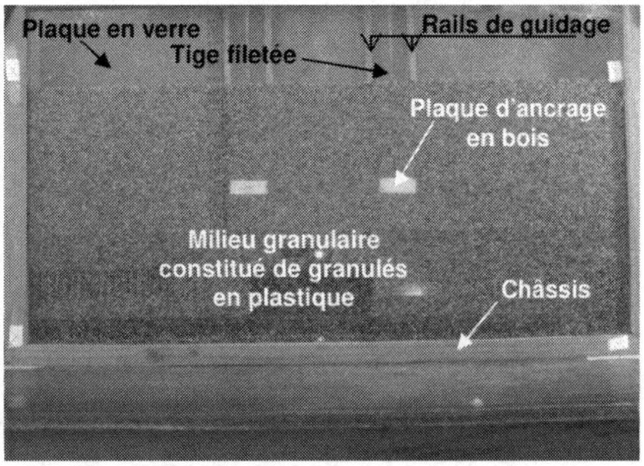

Fig. 1. Vue du modèle réduit.

3 Déplacements et déformations au-dessus des plaques d'ancrage

La mise en évidence de l'influence du ratio d'ancrage (H/B) sur le comportement du milieu granulaire au-dessus des plaques d'ancrage est volontairement cinématique, faisant abstraction de toute mesure de l'état de contrainte. Les essais en question ont été effectués à des profondeurs d'ancrage de la plaque carrée de H = 3B, H = 5B et H = 7B, H et B étant respectivement la profondeur d'ancrage de la plaque et la largeur de celles-ci.

Lors des traitements d'images, un domaine d'étude est défini par l'opérateur. On suivra à volonté l'évolution des déplacements et des déformations dans ce domaine. L'arrachement ou soulèvement imposé à la plaque d'ancrage se fait à des pas constants de l'ordre de 0,266 mm correspondant à 1pixel.

3.1 Champ de déplacement

Les figures (2a, 2b et 2c) reproduisent graphiquement les trajectoires des grains pour un arrachement de 20 pixels et pour des distances entre nus des plaques d'ancrage respectives de (3B), (4B) et (6B). On remarque que le déplacement des grains s'accentue verticalement suivant l'axe de la plaque, et horizontalement vers les bords de celle-ci. Les vecteurs de déplacement sont tracés avec un coefficient d'amplification de 7 (700%). En premier lieu, nous constatons que les vecteurs de déplacement sont globalement verticaux et orientés vers le haut, les trajectoires des grains s'amortissent, à peu près, à l'approche d'un plan horizontal parallèle à la surface libre du massif, suivant un arc incurvé vers le haut.

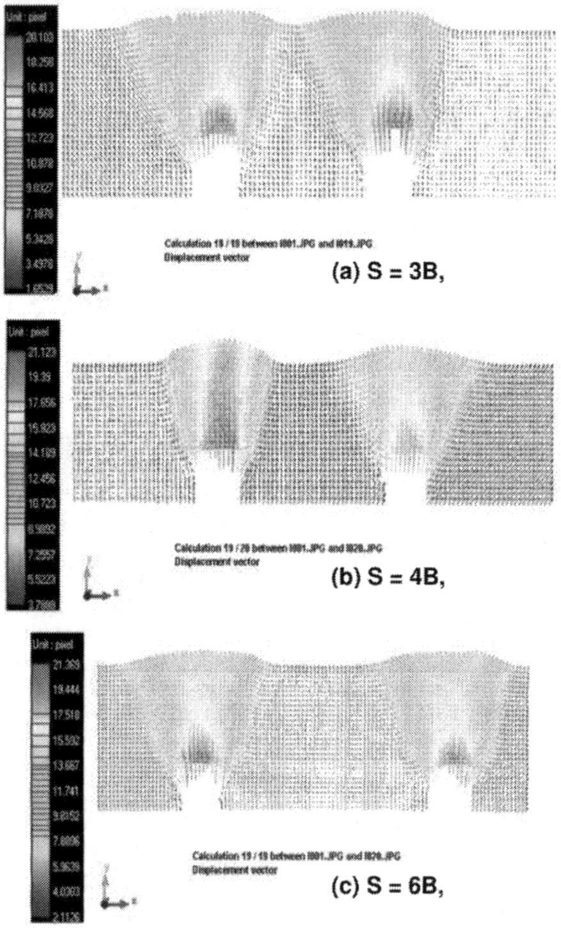

Fig. 2. Vecteurs de déplacements pour différentes distances entre nus de plaques.

Les figures (3a, 3b et 3c) représentent les champs de déplacement verticaux du matériau soumis à l'arrachement des deux plaques d'ancrage écartées entre nus respectivement de (3B), (4B) et 6B, l'enfoncement est de 20 pixels. Localement, au-dessus de chaque plaque, le milieu se comporte de manière similaire à une plaque isolée. Remarquons que la formation d'un « coin rigide » au-dessus chaque plaque confirme la compacité élevée du matériau.

En fonction de l'intensité de l'arrachement des plaques, un fort gradient de déplacement se met en place sur un contour triangulaire légèrement incurvé confirmant l'existence des coins rigides au-dessus des plaques.

On remarque qu'au-delà d'un écartement S = 4B les plaques agissent indépendamment sur le matériau. A une distance S = 3B, l'interaction des zones de rupture de chaque plaque crée un bulbe de déplacement vertical qui atteint la valeur de 5.17 pixels au niveau de la surface du massif à mi distance entre nus de plaques.

Le refoulement des grains est limité par des droites inclinées passantes par les nus des plaques, l'inclinaison de ces droites est d'environ 31°.

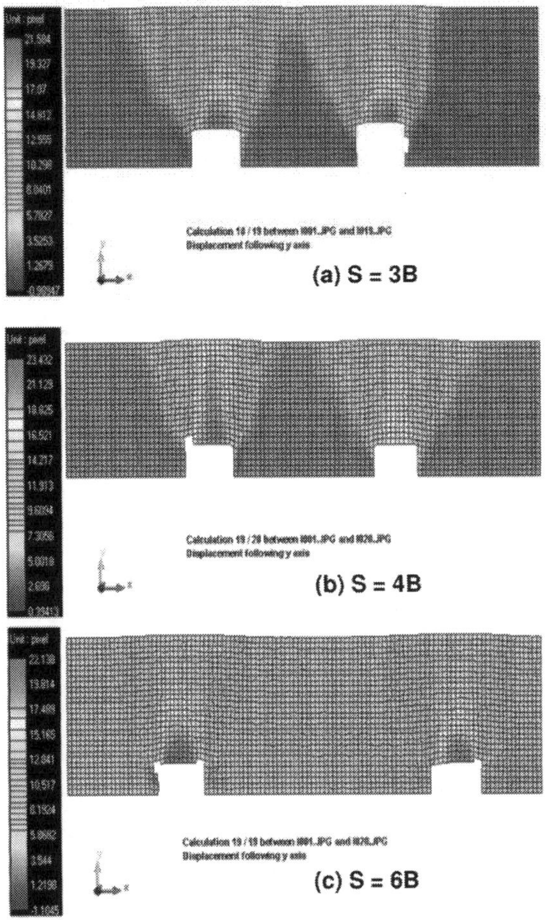

Fig. 3. Champ de déplacement vertical pour différentes distances entre nus de plaques.

La figure 4 schématise la superposition entre les isovaleurs du déplacement vertical pour un arrachement de la plaque de 20 pixels et le mode de rupture préconisée par Ilamparuthi et al en 2002 [7], nous constatons une similitude de forme entre les isovaleurs de déplacement et le mode de rupture généralisé pour un ancrage de H = 3B, en effet ce mode de rupture concerne les milieux granulaires denses soumis à l'arrachement d'une plaque d'ancrage superficielle.

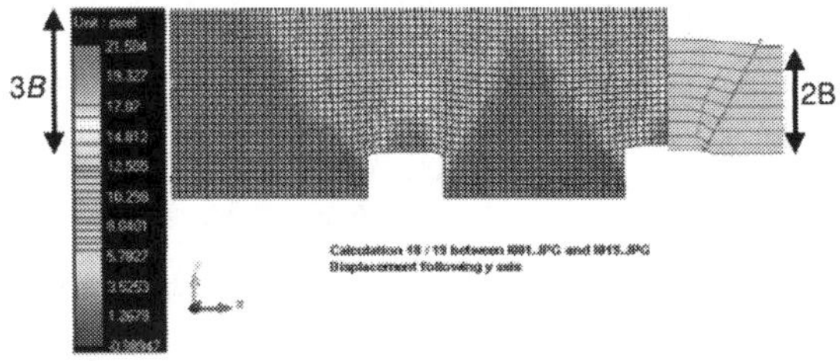

Fig. 4. Superposition entre les isovaleurs du déplacement vertical et les résultats de Ilamparuthi et al. 2002

3.2 Champ de déformation

Les figures (5a, 5b et 5c) représentent les champs de déformation (cisaillement maximal) dans le milieu granulaire pour un arrachement de 20 pixels et pour des écartements entre nus de plaques respectifs de (3B), (4B) et (6B). On remarque que les bulbes de cisaillement au niveau de la surface du massif à mi distance entre nus de plaques sont d'autant plus grands que les plaques se rapprochent l'une de l'autre avec une plus grande intensité de cisaillement qui atteint la valeur de 3% pour un écartement de 3B alors qu'elle ne dépasse pas les 0.76% pour un écartement de 4B et les 0.38% pour un écartement de 6B.

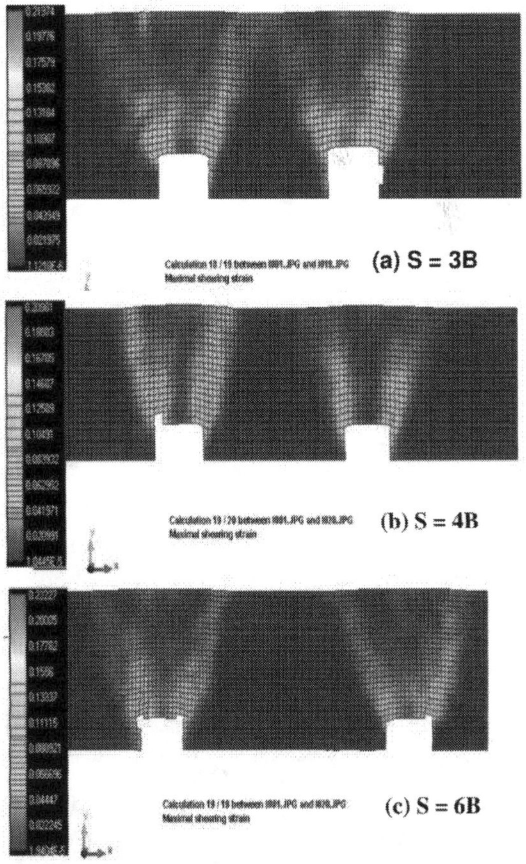

Fig. 5. Champ de déformation (cisaillement maximal) pour différentes distances entre nus de plaques.

3 Conclusion

L'étude expérimentale que nous avons menée, s'est basée principalement sur l'évolution du milieu granulaire soumis à l'arrachement des plaques d'ancrage.

Les essais d'arrachement effectués ont permis de quantifier le champ de déplacement vertical et celui des déformations du milieu granulaire au-dessus des « plaques d'ancrage » grâce à la technique de corrélation d'image. Le choix d'un arrachement de 20 pixels (déplacement imposé) se justifie par le souci de présenter des champs de déformations correspondant à un état de rupture. Une multitude de couple de photos ont été traité, allant jusqu'à un arrachement de 33 pixels et plus. Au-delà de 18 pixels, la configuration déformée est pratiquement inchangée, ce qui montre qu'à ce stade de sollicitation, le milieu a atteint sa rupture et qu'un état de plasticité s'installe.

L'interaction des zones de ruptures des plaques voisines est étudiée cinématiquement en faisant varier le paramètre distance entre nus de plaques. Il s'avère qu'un écartement minimal d'environ quatre fois la largeur de la semelle (*4B*) est nécessaire pour que deux plaques d'ancrage voisines agissent indépendamment sur le milieu.

La corrélation entre la phénoménologie de l'évolution des champs de déplacement et de déformation au-dessus de la plaque d'ancrage et le mode de rupture du massif d'ancrage, faisant objet d'études antérieures a été mise en évidence grâce à la superposition entre les isovaleurs du déplacement vertical (pour un enfoncement de 20 pixels) et le mode de rupture généralisé défini par Ilamparuthi *et al.* (2002) [7].

Références

1. G.C Meyerhof, S.I. Adams, *The ultimate uplift capacity of foundations*, Can Geotech J **4**, 225-244 (1968)
2. J Kumar, K.M Krouzer, *Interference effect on the vertical uplift capacity of two shallow horizontal anchors*, Géotechnique, 58 **10**, 821-824 (2008)
3. T.H Hanna, R Sparks, M Yilmaz, *Anchor behaviour in sand,* J Soil Mech Found Div, ASCE 98 **11** pp1187 (2007)
4. J.D Geddes, E.J Murray, *Plate anchor groups pulled vertically in sand*, J Geotech Eng, ASCE, 122 **7**, 509-516 (1996)
5. H Abbad, *Mise au point d'un dispositif d'analyse des déformations planes par corrélation d'images*, Thèse de doctorat, Université de Sidi Bel Abbès, Algérie (2011)
6. P Vacher, S Dumoulin , Morestin and SMguil-Touchal, *Bidimensional strain measurement using digital images*, Poc. Instn. Mech. Engrs, **213** 811-817 (1999)
7. Ilamparuthi K, Dickin EA, Muthukrishnaiah K, *Experimental investigation of the uplift capacity of circular plate anchors in sand*, Can Geotech J **39**, 644-648 (2002)

Study and monitoring of the stability of a rock mass at the Dar-El-Oued tunnel, wilaya of Jijel – Algeria

Étude et suivi de la stabilité d'un massif rocheux au niveau du tunnel de Dar-El-Oued, wilaya de Jijel – Algérie

Chahra YELLAS [1], Riad BENZAID [2], Mustapha TEKKOUK [3]
[1, 2, 3] Laboratoire de Génie Géologique, Université Mohammed Seddik Benyahia, Jijel, Algérie.

Abstract. The construction of a civil engineering work requires many preliminary studies as well as various calculation methods to give an assessment or a forecast of the state of the rock in place and the behavior of the structure in its real environment and under various conditions (construction phase, service and operating phase). The present work concerns the access zone of the Dar-El-Oued tunnel, located on the national road n°43 about 40 km West of the city of Jijel (northeastern Algeria). The geotechnical study of this tunnel was carried out following the classification methods relating to underground structures. Through the results of this study we have determined the influence of land pressures on the mechanical behavior of the ground, which allowed us to measure the state of stress in the rock and to make a comparison between the results of the measurements of convergence and that of numerical modeling. On the other hand, the influence of the velocity of vibration induced by explosive shots has been discussed in order to check the stability of the face as the digging progresses. From the results obtained, it can be said that the surface displacements of the soil are relatively small compared to those determined by the numerical model at the level of the vault and the slab. Note the good agreement between the results provided by our numerical model and those obtained in-situ.

Résumé. La construction d'un ouvrage de génie civil nécessite de nombreuses études préliminaires ainsi que diverses méthodes de calcul pour donner une évaluation ou une prévision de l'état de la roche en place et du fonctionnement de l'ouvrage dans son environnement réel et sous diverses sollicitations (phase de construction, phase de service et d'exploitation). Le présent travail concerne le front de taille du tunnel de Dar-El-Oued, situé sur la route nationale n°43 à environ 40 km à l'ouest de la ville de Jijel (Nord-Est Algérien). L'étude géotechnique de ce tunnel a été réalisée en suivant les méthodes de classifications relatives aux ouvrages souterrains. A travers les résultats de cette étude nous avons déterminé l'influence des pressions des terres sur le comportement mécanique du terrain, ce qui nous a permis de mesurer l'état de contrainte dans la roche et de faire une comparaison entre les résultats des mesures de convergence et celle de la modélisation numérique. D'autre part, l'influence de la vitesse de vibration induite par les tirs d'explosifs a été discutée dans le but de vérifier la stabilité du front de taille au fur et à mesure de l'avancement des travaux de creusement. A partir des résultats obtenus, on peut dire que les déplacements en surface du sol sont relativement faibles par rapport à ceux déterminés par le modèle numérique au niveau de la voûte et du radier. Notons ainsi la bonne concordance entre les résultats prévus par notre modèle numérique et ceux des mesures obtenues in-situ.

1 Introduction

Le projet du tunnel de Dar El-Oued s'inscrit dans les travaux d'adaptation fonctionnelle de la route côtière qui relie la ville de Jijel à la ville de Bejaia (Nord-Est Algérien). Il est né de l'exigence d'éviter le tronçon de l'actuelle route côtière dont la section est sensiblement réduite à proximité des fameuses grottes merveilleuses (Figure 1).

L'ajustement de la Grotte Merveilleuse, ainsi que tout le versant à pic sur la mer, à lui seul de grande valeur

 https://doi.org/10.1051/matecconf/201814902069

touristique. D'où la décision de créer un nouveau tunnel, en amont du tracé actuel, qui perce le Djebel Hamra du côté mer, en évitant ainsi les Grottes Merveilleuses et le tronçon de côte attenant.

Les travaux de creusement du tunnel ont provoqué une modification de l'état de contraintes dans la roche autour de la galerie. Ce changement d'état de contraintes est à l'origine des déplacements, qui affectent le terrain avoisinant et les structures préexistantes, particulièrement sensible, sachant que ce passage d'un état d'équilibre à un autre n'est pas instantané.

Les travaux de creusement ont provoqué une succession d'instabilités intermédiaires, qui influencent l'état d'équilibre final. Quelle que soit la technique de construction employée, il se produit toujours des déplacements autour de la galerie, qui peuvent se propager dans le massif et atteindre la surface du sol. Ces déplacements dépendent des caractéristiques du sol dans chaque section, des conditions géométriques et de la technique de creusement. C'est pourquoi il est important de localiser les déformations et les déplacements associés au creusement.

Dans le souci d'assurer une sécurité optimale pour le projet de construction du tunnel de Dar El-Oued, les pressions sur le soutènement ont étaient calculées par plusieurs méthodes théoriques, en plus d'une analyse basée sur une modélisation numérique par éléments finis en utilisant le logiciel SAP 2000.

Fig. 1. L'ancien tracé de la route à proximité du site des Grottes Merveilleuses

2 Description du tunnel de Dar El-Oued

Cet ouvrage se situe sur la route nationale n°43 (RN43) à environ 40 km à l'Ouest de la ville de Jijel (Nord-Est Algérien) au niveau du massif carbonaté de Dar-El-Oued, bordé au Nord par la mer Méditerranée et à l'Est par l'oued de Dar-El-Oued (Figure 2). Le tracé du tunnel se prolonge sur une longueur de 620 m creusé dans les calcaires dolomitiques d'âge Jurassique. Il est constitué de deux voies de 3,50 m de large chacune et des accotements de 1,40 m de large, avec un gabarit en hauteur de 5 m environ. Un système de drainage rigoureux été réalisé sur toute la longueur de la galerie pour assurer une bonne étanchéité du tunnel. A l'entrée Est, un faux tunnel sur 5 ml de long été réalisé pour faciliter le début des travaux. Par ailleurs l'embouchure Ouest est protégée par une berlinoise en micro pieux renforcée avec des tirants d'ancrages.

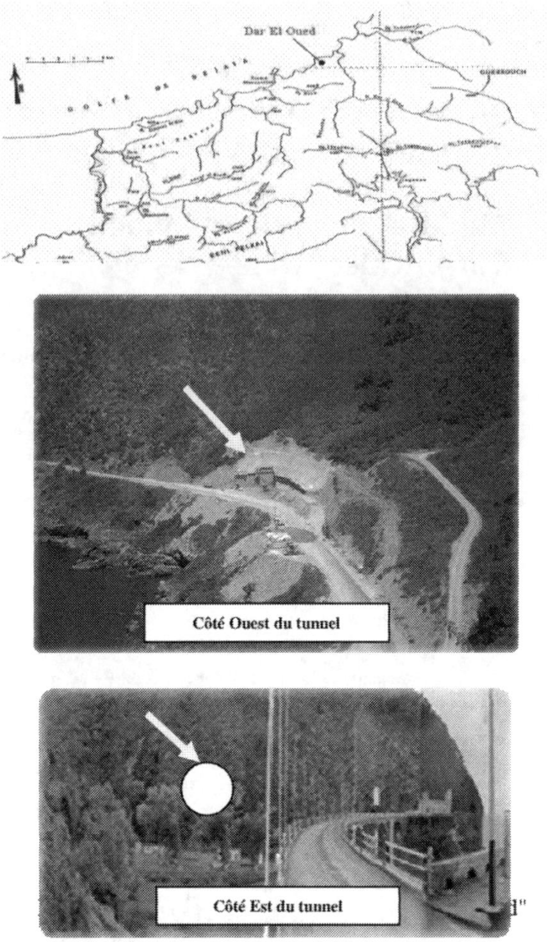

3 Influence de la vitesse de vibration sur la stabilité du front de taille

Pour prédire les effets d'un tir en un point donné de la section du tunnel, il convient de déterminer la relation entre la quantité d'explosif utilisée et la vitesse de vibration résultante au point considéré, ce qui permet d'adapter le plan de tir à l'environnement. La seule façon appropriée de définir cette relation et de prévoir sur le site des tirs expérimentaux où l'on fait croître la charge toute en mesurant la vitesse de vibration correspondante [1,2]. En avant-projet et à défaut de tirs expérimentaux, il est toujours possible d'utiliser des relations empiriques. Les formules publiées sont diverses et donnent des résultats souvent très différents. En effet, les relations établies ne font pas apparaître les facteurs pourtant importants que sont la nature du rocher et la géométrie du plan de tir. On peut néanmoins retenir en toute première approximation la formule proposée par Chapot (1981) [3] pour les tirs en tunnels :

$$V = K.\ Q^{0,9} / D^{1,8} \qquad (1)$$

Avec, V vitesse particulaire en (cm/s) ; D distance entre le point de mesure et le point de tir en (m) et Q la quantité d'explosif en (gr).

Le paramètre K varie suivant le site et la nature de la roche. Pour les tirs en tunnel, on peut retenir à titre indicatif les valeurs suivantes : K = 1,2 (valeur moyenne) et K = 2,4 (valeur maximale). Le Tableau 1 présente les résultats obtenus sur site, avec une quantité d'explosif égale à 22,5 kg. On a pris des valeurs de la constante K supérieures à celles indiquées ci-dessus afin d'évaluer son influence sur la vitesse de vibration.

Tableau 1. Vitesses de vibration en (m/s) à partir de la distance du front de taille [4]

	Distance (m)						
	40	**50**	**60**	**70**	**80**	**90**	**100**
K = 1,2	0,12	0,08	0,06	0,04	0,037	0,030	0,024
K = 1,4	0,15	0,10	0,07	0,05	0,043	0,035	0,029
K = 1,6	0,17	0,11	0,08	0,06	0,049	0,040	0,033
K = 1,8	0,19	0,13	0,09	0,07	0,05	0,048	0,037
K = 2	0,21	0,14	0,10	0,078	0,062	0,05	0,04
K = 2,2	0,23	0,15	0,11	0,08	0,068	0,055	0,045
K = 2,4	0,25	0,17	0,12	0,09	0,07	0,06	0,049
K = 2,6	0,28	0,18	0,13	0,10	0,08	0,065	0,05
K = 2,8	0,3	0,2	0,14	0,12	0,09	0,075	0,06
K = 3	0,32	0,21	0,15	0,11	0,086	0,07	0,058

En examinant les résultats obtenus sur la Figure 3, on déduit que pour les valeurs les plus probables de la constante K (1,6 à 1,8) les vitesses de vibration atteignent 0,17 à 0,19 m/s quand la distance du front de taille est égale à 40 m et elles descendent à des valeurs proches de 0,04 m/s pour la distance maximale de 100 m.

A partir d'un tir, une fraction de l'énergie cédée au terrain se propage sous forme vibratoire, tant dans les fluides (air, eau) que dans les solides. Cette vibration transitoire sollicite les structures sur son passage et y déclenche divers types de vibrations. Lorsque les efforts engendrés par ces vibrations excèdent localement la résistance de ces structures, il peut y avoir apparition de dommages c'est-à-dire la rupture de la roche [3]. On retiendra à ce sujet les propositions qui nous semblent intéressantes :

- toutes les limites ont un caractère probabiliste ;
- les seuils inferieurs à 1cm/s, tel qu'ils ont pu être proposés dans certains projets de norme, ne semblent pas être à retenir, la probabilité d'apparition de dégâts étant alors très réduite ;
- la zone de 1 à 3 cm/s constitue une fourchette applicable aux immeubles courants habités et aux chantiers de travaux publics classiques.
- à partir de 3cm/s, on ressent intensément les vibrations.

En examinant le tracé du tunnel de Dar-El-Oued on a constaté l'absence des édifices privés ou publique, qui peuvent être menacés par les vibrations engendrées par l'usage d'explosif, au voisinage de l'ouvrage réalisé. Une seule exception été faite pour les grottes (un site touristique important) qui se trouve à une distance effective de 200 m de la cavité du tunnel. Pour cette raison particulière, la vitesse de vibration a été diminuée.

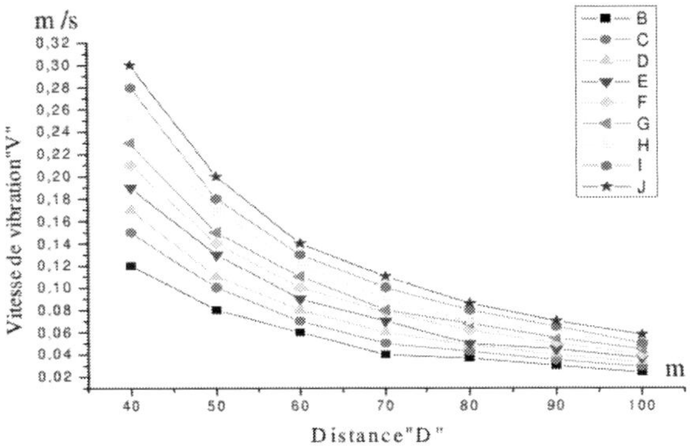

Fig. 3. Influence de la vitesse de vibration en fonction de la distance du front de taille [4]

4. Mesures de convergence

Les mesures les plus fréquentes effectuées dans le cas des tunnels sont les mesures de convergence entre plots scellées aux parois. On mesure, généralement à l'aide d'un fil invar maintenu à une tension constante par un dynamomètre, le rapprochement de deux points situés sur des parois opposées de l'ouvrage.

Comme le déplacement d'un point de la paroi intègre la déformation de l'ensemble du massif, les résultats ne sont ainsi pas trop dispersés. C'est pourquoi les mesures de convergence constituent le moyen de contrôle par excellence du bon comportement d'une galerie souterraine. Le suivi des résultats de l'auscultation en fonction de l'éloignement du front de taille permet de juger la bonne adéquation du soutènement mis en place aux exigences de déformation prévues dans le projet.

Dans le tunnel de Dar El-Oued, des mesures de convergence ont été réalisées, elles consistent à relever et à rendre sous forme graphique et numérique les déplacements dans le plan transversal du tunnel (5 points de mesures pour chaque station) comme le montre la Figure 4.

L'évolution de la déformation relative d'une zone de terrain dans le temps est l'indication fondamentale des courbes de convergence. Des valeurs stables dans le temps indiquent une stabilité mécanique de l'ouvrage, même si ces valeurs sont importantes.

Une convergence négative apparaît durant les premiers jours d'ouverture du terrassement de la section, mais elle reste faible en moyenne pendant les sept premiers jours. Sur certains profils, la stabilisation s'amorce naturellement, il faut noter que cette stabilisation ne peut être obtenue qu'après fermeture de la section (bétonnage des banquettes et du radier).

Les mesures d'auscultation réalisées consistent à relever et à restituer sous forme graphique et numérique l'état des contraintes entre les revêtements provisoire et définitif, sous le pied des cintres, à travers des capteurs de charge et des extensomètres.

En particulier, les efforts sur le cintre sont mesurés avec des tiges du type à corde vibrante placées sur les ailes, dans le béton de revêtement définitif avec des tiges placées sur l'extrados et sur l'intrados de la section surveillée.

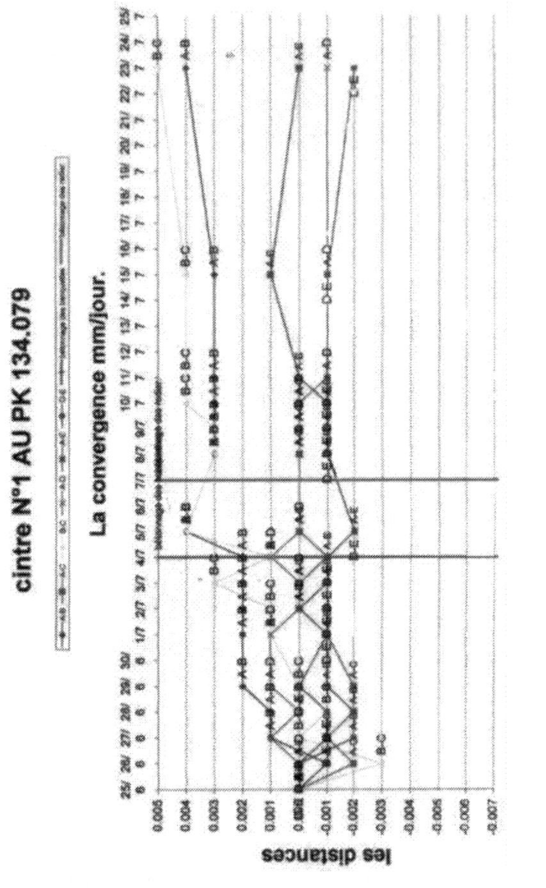

Fig. 4. Mesures de convergence sur le cintre n°1 au PK 134,079 [4]

L'amplitude des déplacements prévisibles est déterminée par les calculs effectués au stade du projet. On retient rarement une valeur maximale admissible pour les déplacements, car l'amplitude des déplacements mesurés ne permet pas à elle seule de juger la stabilité de l'excavation. La vitesse de déplacement est maximale au passage du front dans le plan de mesure. On peut admettre que le déplacement journalier admissible est de l'ordre de 1/4 du déplacement total prévu au passage du front et doit descendre du 1/20 après un délai de l'ordre d'une semaine.

Dans ce type de projet, l'accélération est le critère de base utilisé pour l'interprétation des mesures de déplacement. Généralement, il est admis qu'une accélération des déformations conduit à la rupture. D'autre part, l'influence du soutènement doit également être prise en compte pour une bonne interprétation de son installation et son renforcement. Une décélération non marquée ne peut survenir qu'après un ou deux jours du passage du front.

Les calculs sur le modèle numérique on été effectués au moyen du logiciel SAP 2000, en schématisant la section du tunnel par des tiges rectilignes soumises aux valeurs des pressions déduites à partir des expressions de Terzaghi [1,2]:

- Pression verticale :

$$\mathbf{P_v} = \frac{B\left(\gamma - \frac{2C}{B}\right)}{2\tan\theta} \cdot \left[1 - e^{\frac{-2.H\tan\theta}{B}}\right] \qquad (2)$$

- La poussée latérale :

$$\mathbf{P_h} = 0{,}3.\gamma\left(0{,}5m + h_p\right) \qquad (3)$$

Le résultat des pressions sur les trois sections étudiées sont résumés dans le Tableau 2 :

Tableau 2. Valeurs des pressions calculées

Section	Pression verticale (t/m²)	Pression latérale (t/m²)
Section 1	41,79	8,9
Section 2	20,80	9,77
Section 3	20,51	9,34

En ce qui concerne le revêtement définitif, les déformations de la coupe 2 étaient moins intenses que celle des coupes 1 et 3 représentée dans la Figure 5. On remarque que ces déformations se concentrent au niveau des piédroits, du radier et beaucoup plus au niveau de la calotte. Les valeurs de ces déformations diminuent lors de la mise en place du soutènement provisoire. D'autre part, on à constaté que le terrain est moins déformable dans la section 2 constitué par une dolomie massive. La plus part des déplacements sont situées en clé de la voûte et en radier. Au final, et après avoir réalisé le soutènement définitif (section entièrement soutenue), les déplacements et les déformations sont moins importantes au niveau de la section 2 par rapport aux sections 1 et 3.

5. Conclusion

A partir de cette étude, on peut conclure que dans la construction des tunnels, les déplacements en surface du sol sont plus faibles par rapport à ceux déterminés au niveau de la voûte et du radier.

Aussi, on note la bonne concordance entre les résultats prévus par notre modèle numérique et les résultats des mesures obtenues sur le site du tunnel. Pour cela et pour assurer la stabilité de l'ouvrage, l'exécution du radier qui s'oppose à la pression ascendante est indispensable.

Sur site, la convergence est mesurée durant toute la période de réalisation du tunnel, quel que soit la position du front de taille (même en position d'arrêt des travaux), d'autant plus que la convergence est fonction de deux temps : le temps de creusement (vitesse de creusement et délais de mise de soutènement) et le temps caractéristique du comportement différé du terrain (relaxation).

Enfin, une connaissance de la convergence finale est très intéressante pour prévoir le comportement différé du terrain.

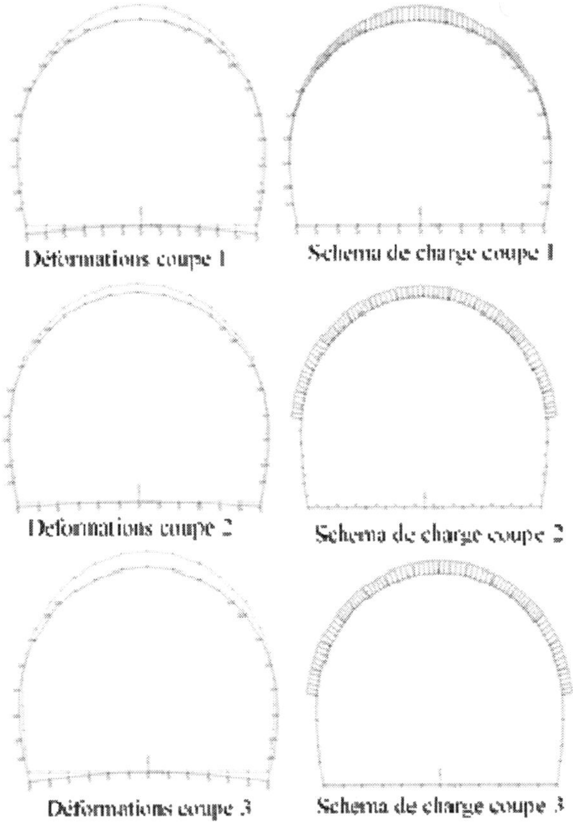

Fig. 5. Représentation des déformations sur les trois sections étudiées

Références

[1] A. Bouvard, G. Colombet, Esteulle, F. Ouvrage souterrain : Conception, Réalisation, Entretien, Presse de L'ENPC, 2^{éme} édition, Paris, pp.61-64 (1992).

[2] K. Széchy. The art of tunneling. Akadémiai Kiadó, 1097p (1973).

[3] P. Chapot, Etude des vibrations provoquées par les explosifs dans les massifs rocheux, Rapport de recherche LPC n°**105** (1981).

[4] C. Yellas, Etude géologique et géotechnique du Tunnel de Oued Dar-El-Oued (Grottes Merveilleuses) Wilaya de Jijel, Mémoire de Magister, Université de Jijel- Algérie (2008).

Use of reinforced earth to improve the stability of railway sleepers

Utilisation de la terre armée pour l'amélioration de la stabilité des traverses ferroviaires

Souad ATOUB[1], Riad BENZAID[2], Mustapha TEKKOUK[3]
[1, 2, 3] Laboratoire de Génie Géologique, Université Mohammed Seddik Benyahia, Jijel, Algérie.

Abstract. Soil reinforcement with inclusion has been used since the dawn of time, however, decisive progress in this area dates back only a few decades, When the techniques of soil reinforcement are passed to an industrial development with metallic inclusions then geosynthetic and for applications as well in embankments as in cuttings. Reinforced earth is one of the most used methods for soil strengthening. It is a relatively recent technique whose principle consists in associating with a cohesionless soil tensile reinforcements, which thus give the material an anisotropic cohesion in the direction of the reinforcements. Among the areas where the reinforced earth is currently experiencing a remarkable expansion, railway infrastructure, traditional means for the transport of passengers and goods, and which has never ceased to be the subject of technical progress. This progress based for a long time on the empiricism, allowed the progressive increase, as well speeds as tonnage by axle. However, the railway remains a privileged place of deformation under the effect of the repetitive application of loads. These deformations are largely related to the problem of ballast settlement, a problem whose causes are diverse and that we will try to study in the laboratory on a physical model in real size. The present work consists in experimentally analyzing the load-settlement relationships that exist during the introduction of stainless steel reinforcement into an unstable rail ballast under static vertical loading and to establish the conditions that lead to the reduction of recorded settlements in order to improve the stability of railway sleepers.

Résumé. Le renforcement des sols à l'aide d'inclusion a été utilisé depuis l'aube du temps cependant, les progrès décisifs dans ce domaine ne datent que de quelques dizaines d'années, lorsque les techniques de renforcement des sols sont passées à un développement industriel avec des inclusions métalliques puis géosynthètiques et à des applications aussi bien en remblais qu'en déblais. La terre armée est l'un des procédés les plus utilisés pour le renforcement des sols. C'est une technique relativement récente dont le principe consiste à associer à un sol pulvérulent des armatures résistantes à la traction, qui confèrent ainsi au matériau une cohésion anisotrope, dans la direction des armatures. Parmi les domaines où la terre armée connaît actuellement une remarquable expansion, les infrastructures des chemins de fer, moyen traditionnel pour le transport des voyageurs et des marchandises, et qui n'a jamais cessé d'être l'objet de progrès techniques. Ce progrès basé longtemps sur l'empirisme, a permis l'augmentation progressive, tant des vitesses que du tonnage par essieu. Cependant, la voie ferrée reste un siège privilégié de déformations sous l'effet de l'application répétitive des charges. Ces déformations sont liées en grande partie au problème de tassement du ballast, un problème dont les causes sont diverses et qu'on va essayer de l'étudier en laboratoire sur un modèle physique en grandeur réelle. Le présent travail consiste à analyser par voie expérimentale, les relations charge-tassement qui existent lors d'une introduction des armatures inoxydable dans un ballast ferroviaire instable sous chargement vertical statique et d'établir les conditions qui mènent à la réduction des tassements enregistrés afin d'améliorer la stabilité des traverses ferroviaires.

https://doi.org/10.1051/matecconf/201814902070

1 Introduction

La voie ferrée connaît actuellement une profonde évolution et tend à s'adapter aux besoins de notre époque. Des progrès considérables ont été réalisés ces dernières années tant dans le domaine des performances pures que dans celui de l'exploitation courante. Cela s'est entre autres traduit par l'utilisation de plus en plus fréquente des grandes vitesses, une augmentation du confort des voyageurs, du tonnage par essieu, et par une diminution des dépenses d'entretien [1,2].

Cependant, sous l'effet de l'application répétitive de la charge, la voie ferrée reste un siège continu de déformations. Ces déformations ou dégradations de la voie sont en grande partie liées aux problèmes de tassement du ballast [3].

Le maintien de la géométrie de la voie ferrée est directement lié au comportement du ballast. Ce dernier, a pour rôle essentiel de transmettre au sol la charge qu'exerce le rail sur le blochet en veillant à s'opposer aux mouvements longitudinaux et transversaux des traverses [4, 5].

2 Principe général du comportement du ballast armé

Il est admis que sous l'effet d'une charge verticale, l'armature introduite dans le ballast ferroviaire diminue les déformations latérales et verticales du milieu pulvérulent. Ainsi admise, cette hypothèse est vérifiable au cas où les forces de frottement entre les grains du milieu pulvérulent et la surface de l'armature seraient suffisantes. Un tel cas limitera le glissement latéral, par conséquent, les déformations verticales du ballast sont à leur tour atténuées et le tassement du sol minimisé.

En empêchant le glissement latéral des grains du milieu pulvérulent, l'armature (sous forme de treillis soudé) transmet les efforts de traction et assure une cohésion locale de la couche de ballast [6, 7]. Ce phénomène correspond à une augmentation locale de l'angle de frottement interne du sol.

L'interaction sol-armature développée est de deux types [6, 7] :
- Un frottement latéral entre les éléments orientés dans le sens de la traction ;
- Une résistance en butée le long des éléments transversaux.

Lorsque le sol remplit les ouvertures des treillis soudés, une portion de la résistance du mécanisme est désorientée aux fils d'armatures transversales des treillis. Les contraintes dans les fils d'armatures transversales sont transférées aux fils longitudinaux qui résiste aux contraintes imposées, ces contraintes vont être transférées à travers les nœuds (d'où on remarque leur importance). C'est l'effet de "raquette" où le système de renforcement bidimensionnel répartit les charges sur le sol, tout comme les raquettes soutiennent le poids d'un homme sur la neige (Figure 1).

3 Dimensionnement des renforcements

La procédure de dimensionnement applique depuis longtemps le concept d'équilibre plastique de Rankine qui utilise les conditions des pressions des terres actives pour les charges permanentes et le concept de Boussinesq pour les charges de service.

Notons que la force admissible du treillis soudé considérée à la moindre apparition des dommages ou glissements doit être inclue dans le dimensionnement.

La valeur de la force de traction maximale dans le treillis soudé (T_{max}) est calculée en fonction de sa déformation maximale (ε_{max}) par la formule suivante :

$$T_{max} = (T_a / \varepsilon_a)\, \varepsilon_{max} \qquad (1)$$

Avec (T_a / ε_a) le module de traction du treillis soudé.

L'équilibre interne du massif de sol est satisfait lorsque la composante horizontale de la poussée latérale totale est égale à la somme des tractions dans tous les fils de renforcements :

$$\sum T_{max} = 0{,}5\, K_a \cdot \gamma \cdot H^2 + K_q \cdot q \cdot H \qquad (2)$$

K_a : le coefficient des poussées des terres actives.
K_q : le coefficient de transmission des surcharges.
H : l'épaisseur du ballast au-dessus du renforcement.
q : les surcharges ferroviaires.

En connaissant la résistance maximale fournie par le treillis soudé (RT_{max} unitaire), on peut déduire le choix et le nombre de treillis adéquat par mètre linéaire ainsi que la détermination des déformations maximales (ε_{max}).

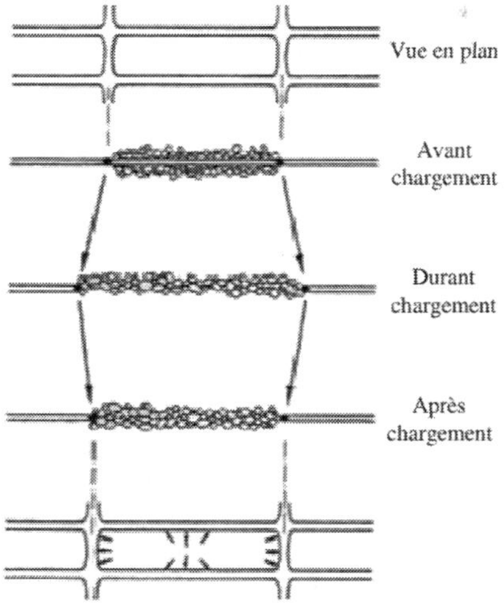

Fig. 1. Mécanisme de mise en tension des treillis soudés.

4 Objectif des essais

On se propose d'analyser par voie expérimentale les relations qui apparaissent lors de l'introduction d'armatures de renforcement dans un ballast ferroviaire instable et, d'établir les conditions qui permettraient de réduire le tassement de ce dernier. Notre étude se limite

 https://doi.org/10.1051/matecconf/201814902070

aux déformations du ballast dans les sections de voies ferroviaires d'accès au garage ou d'entretien et aux autres tronçons où des déformations sont observées au niveau de la couche du ballast ferroviaire.

Par l'expérience menée au laboratoire, nous essayerons de connaître :

- L'effet du renforcement sur le tassement de la couche de ballast et par conséquent, l'influence sur la stabilité des traverses ferroviaires.
- L'effet du changement de la position en profondeur des armatures de renforcement sur la valeur du tassement.

5 Mode d'exécution de l'expérience

Le dispositif de laboratoire ayant servi à la réalisation de nos essais de simulation est représenté en Figure 2.

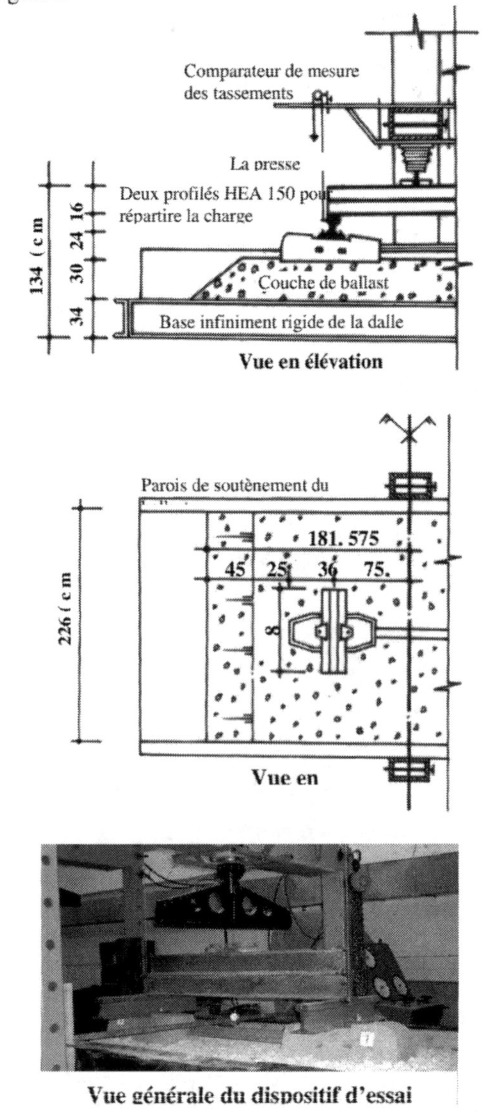

Fig. 2. Banc d'essai

Composé d'une base infiniment rigide, le banc d'essai supporte la couche de ballast étudiée. Cette couche de ballast d'épaisseur 30 cm et de dimensions en plan de (363,15 cm x 226 cm) est chargée d'une traverse ferroviaire en béton armé de type SL. Les deux blochets en béton armé ont chacun les dimensions suivantes : B = 29 cm, L = 68 cm et h = 24 cm [4].

L'armature de renforcement utilisée est un treillis soudé constitué de fil de fer cylindrique de diamètre 4 mm, il se présente en mailles carrées de (15 cm x 15 cm), disposé horizontalement. Ce type d'armature a été choisi sur la base des remarques suivantes : il doit être suffisamment résistant à la traction, peu extensible (à cause de l'assemblage des fils au nœud), durable (acier inoxydable).

Notons que la forme en mailles carrées du treillis soudé ne divise pas le sol en deux couches séparées, son seul but est d'augmenter localement l'effet de l'angle de frottement interne du sol et de résister aux forces de tractions latérales.

La couche de ballast de granulométrie 25/50 mm, est soumise à l'aide d'une presse hydraulique à une sollicitation maximale de deux forces verticales statiques de 10 tonnes chacune distantes de 151,15 cm (ce qui représente l'entre axe de la traverse SL). Le pas de chargement appliqué est de 1 tonne jusqu'à une valeur de 10 tonnes par blochet, ce qui correspond à une charge totale de 20 tonnes par essieu (catégorie C des voies classées par l'Union Internationale des Chemins de fer en ce qui concerne les efforts supportés par la voie ferrée).

Un comparateur de déplacement placé au-dessus de chaque blochet de la traverse ferroviaire chargée mesure les tassements au cours des essais (voir Figure 1).

Deux séries d'essais ont été réalisés pour chaque disposition du programme expérimental (ballast de référence et ballast renforcé). Après chaque série d'essai, le ballast est remis à son état initial.

6 Programme d'essais

Essai n° 1 :
C'est l'essai de référence (Figure 3), la traverse ferroviaire est placée sur une couche de ballast ordinaire.

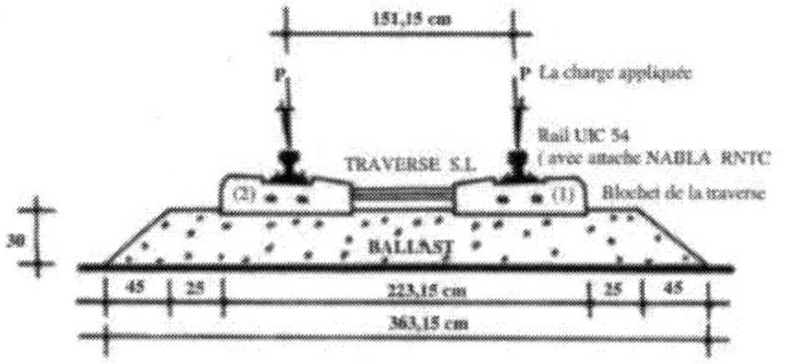

Fig. 3. Essai de référence

Essai n° 2 :
Essai renforcé 1 (Figure 4), la traverse ferroviaire est placée sur une couche de ballast renforcé avec un treillis soudé introduit à mi-hauteur.

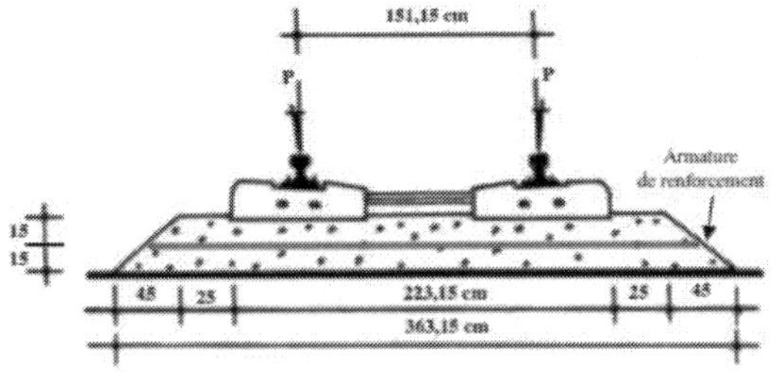

Fig. 4. Essai renforcé (1)

Essai n°3 :
Essai renforcé 2 (Figure 5), la traverse ferroviaire est placée sur une couche de ballast renforcé avec un treillis soudé introduit à 7,5 cm de la surface.

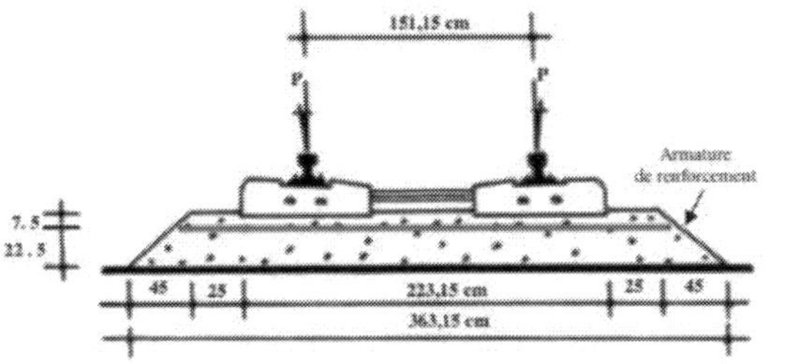

Fig. 5. Essai renforcé (2)

7 Etude et analyse des résultats expérimentaux

7.1 Tassements de la couche du ballast

La couche de ballast est chargée d'une traverse ferroviaire soumise à une sollicitation de deux forces verticales symétriques appliquées sur les rails au niveau des blochets. Le pas de chargement simultanée est de 1 tonne et la charge maximale exercée par blochet est de 10 tonnes.

7.1.1 Cas du ballast de référence

La courbe de la Figure 6 donne l'évolution du tassement le long de l'expérience. Dans ce cas de figure la valeur du tassement moyen est de l'ordre de 2,69 cm.

7.1.2 Cas du ballast renforcé

Renforcé avec un treillis soudé à mi-hauteur et, à 7,5 cm de sa surface, la couche de ballast tasse moins (Figure 6). Le tassement est nettement réduit par rapport à l'essai de référence, avec des valeurs moyennes de l'ordre de 1,255 cm et 1,074 cm respectivement.

Les différences entre les tassements de la traverse sur ballast renforcé et non renforcé sont considérables. Les tassements de la traverse surmontant le ballast renforcé à mi-hauteur, sont deux fois plus faibles que ceux du ballast de référence. En effet, le tassement moyen enregistré pour les deux essais de référence a été réduit d'environ 53% après le renforcement de la couche de ballast avec un treillis soudé disposé à sa mi-hauteur.

La réduction du tassement est nettement plus marquée quand l'armature de renforcement est introduite à 7,5 cm de la surface du ballast. En effet, le tassement moyen obtenu sur le ballast de référence est réduit de 60 %.

Les réductions des tassement ainsi obtenues sont dues aux frottements développés entre les grains du ballast et la surface des armatures de renforcement. Les mailles du treillis soudé opposent une butée qui empêche le glissement des grains "effet de raquette". L'interaction sol-armature contribue à améliorer l'adhérence des deux composants du ballast renforcé. Sous ces conditions (frottement mobilisé est adhérence améliorée), la couche de ballast aura un glissement latéral limité et par conséquent des déformations verticales atténuées. Le glissement latéral des grains ainsi empêché, l'armature aidera à transmettre les efforts de traction tout en assurant la cohésion locale de la couche du ballast.

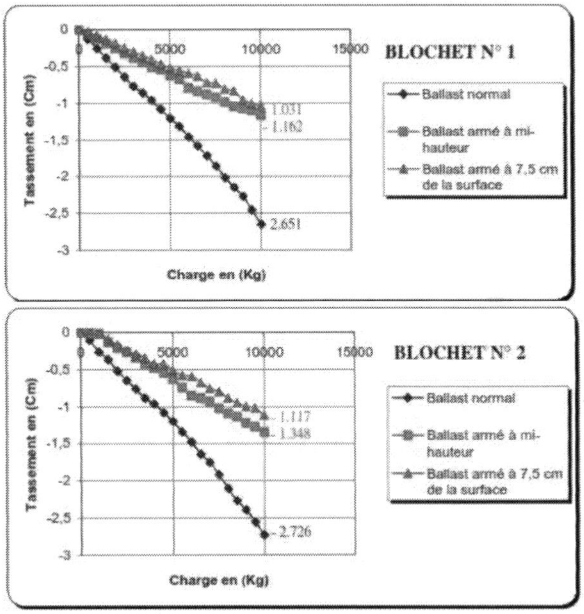

Fig. 6. Tassement sous le blochet n°1 et n°2 de la traverse chargée sur du ballast armé et non armé

Notons que l'intérêt de l'utilisation des treillis soudé comme armatures de renforcement réside dans le fait que la couche renforcée garde son unicité, elle n'est nullement stratifiée ou divisée en deux une fois l'armature introduite en son sein. Cette même armature participe à augmenter l'effet de l'angle de frottement interne du ballast par la transmission des efforts de traction ce qui induit la diminution des déformations latérales dues aux charges verticales appliquées.

7.2 Influence de la disposition des armatures de renforcement au niveau du ballast

Le changement de la disposition en profondeur des armatures de renforcement à plus ou moins influencé les valeurs des tassements de la couche du ballast (Figure 6). En effet, placée à 7,5 cm de la surface, l'armature de renforcement a contribué à réduire d'avantage le tassement par rapport à la disposition à mi-hauteur.

Ainsi donc, le tassement moyen enregistré à mi-hauteur est réduit de 14% une fois que les armatures été placées plus haut, soit à 7,5 cm de la surface du ballast. Cela s'explique par le fait que le coefficient de frottement apparent entre le sol et les armatures diminue en fonction de la profondeur [8].

D'autre part, le sol sera moins dilatant si la pression de confinement est plus élevée. Ceci explique pourquoi l'effet favorable de la dilatance sur le coefficient de frottement apparent diminue lorsque la contrainte verticale moyenne (γ.H) exercée par le ballast sur les armatures augmente. Le coefficient de frottement apparent sera diminué et les forces de frottement sol-armature réduites.

Tous ces éléments expliquent parfaitement les conditions de l'amélioration du renforcement du sol étudié.

Par le biais de ces résultats, nous confirmons que le ballast renforcé avec des treillis soudés a gardé le même principe de comportement que celui des massifs en terre armée.

8 Conclusions

Le travail présenté est une étude expérimentale sur un modèle physique en grandeur réelle, il consiste à introduire des treillis soudés dans un ballast ferroviaire sollicité par un chargement vertical dans le but de minimiser les tassements de ce dernier.

A travers cette recherche, nous pouvons tirer les conclusions suivantes :

- Les résultats obtenus montrent que le ballast renforcé est très utile dans les superstructures ferroviaires instables.
- L'utilisation du treillis soudé peut améliorer la stabilité de la superstructure des tronçons de voies où le ballast manifeste d'une façon permanente des déformations importantes.
- Les efforts de tractions au sein du ballast renforcé sont repris par la mobilisation de frottement entre le milieu pulvérulent et l'armature, ce qui a pour effet de limiter les déformations horizontales et verticales de la couche considérée.
- Le tassement du ballast renforcé est d'environ 50 % plus faible que celui du ballast de référence.
- L'utilisation des treillis soudés permet d'obtenir un renforcement bidimensionnel. Ce qui favorise d'une part, la répartition des contraintes dans la couche renforcée et d'autre part, permettra d'avoir un frottement sol-armature dans les deux directions du renforcement "effet de raquette".

En perspectives, le frottement sol-armature peut être amélioré par l'augmentation de la surface de contact. Pour cela l'utilisation de lanières métalliques plus larges est conseillée. D'autre part, ces essais menés en laboratoire peuvent servir de base à d'autres études sur des ouvrages réels à caractère expérimental.

Le système de renforcement proposé dans ce travail peut trouver application dans les cas pratiques suivants :

(a) Les passages à niveau entre les routes et les chemins de fer, où la différence de rigidité entre les deux voies est à l'origine des déformations souvent rencontrées.
(b) La décompression des dalles de transitions à l'entrée des ouvrages d'art ferroviaires.
(c) Les sections des voies ferroviaires construites sur sol compressible.
(d) Les sections des voies ferroviaires servant comme voies de garage ou d'entretien.
(e) Les voies anciennes.

Références bibliographiques

[1] Renforcement des Sols : Expérimentation En vraie grandeur des Années 80', Symposium international sur le renforcement des sols, Paris 18-19 novembre 1993, presse de l'ENPC (1993).
[2] V. Profillidis, P. Humbert. Etude en élastoplasticité par la méthode des éléments finis du comportement de la voie ferrée et de sa fondation – bull. LPC, **141**, réf. 3051 (1986)
[3] A. Al-Shaer. Permanent settlement analysis of ballasted railway tracks - Dynamical approach. Engi Sci. Ecole des Ponts, ParisTech, (2005).
[4] SNTF. Entretien de la voie : Le Ballast. Document technique destiné au personnel d'encadrement de la SNTF (Algérie), Réalisé par : SOFRERAIL-SOFRETU- France (2000).
[5] S. Malab Mankour, A. Mansouri, A. Benaissa. Comportement du ballast sous chargement vertical pour différentes structures de l'assise ferroviaire. Séminaire national de génie civil, Sidi Bel Abbés, Algérie. 15 -16 mai (2001).
[6] R. Benzaid. Stabilité des traverses ferroviaires sur un ballast armé sous chargement statique sur un support rigide, étude expérimentale - Mémoire de magister, Université badji-mokhtar, Annaba, Algérie (2002).
[7] M. Tekkouk, R. Benzaid. La terre armée, une solution d'amélioration des fondations pour remblais en terrain meuble. XXIXeme Rencontres Universitaires de Génie Civil. Tlemcen, 29 au 31 Mai (2011).
[8] F. Schlosser, M. Bastick. Reinforced earth, In Foundation Engineering Hand Book, second edition, edited by: Hsai-Yang Fang. Chapman and hall, New York, pp. 778-795 (1991).

Landslides susceptibility assessment using AHP method in Kanyosha watershed (Bujumbura-Burundi): Urbanisation and management impacts

KUBWIMANA Désiré[1], AIT BRAHIM Lahsen[2], BOUSTA Mahfoud[2], DEWITTE Olivier[3], ABDELOUAFI Abdellah[2] and BAHAJ Tarik[2]

[1]University of Burundi, Department of Earth Sciences, Bujumbura, Burundi

[2]Université Mohammed V, Faculté Sciences, Département Sciences de la Terre, Unité de recherche GEORISK. Avenue Ibn Battouta Rabat – Agdal, Boîte Postale 1014, Rabat, Maroc

[3] Musée Royal pour l'Afrique Centrale, Bruxelles, Belgique.

Abstract. The Kanyosha watershed is unstable due to the presence of several landslides, which occupy about 3% of the study area. They are causing major damage which costs expensive to the Government of Burundi as well as to the population residing there and their properties. Roads, schools, irrigation canals, houses, crop fields, etc., are in danger of collapse. These landslides are mostly naturally occurring but can sometimes be reactivated by heavy rains or human activities during the excavation of building materials from the river bed.In order to carry out this study, we used the multivariate statistical classification with weighting of the responsible parameters of landslides risk to reach the susceptibility map of mass movements in the Kanyosha watershed. Remote sensing, geology, morphometry and bibliography were the data sources for the different parameters. Google Earth images, ortho-photos and field prospecting helped us to identify the landslides needed to validate the susceptibility map.During the fieldwork, we observed 34 landslides of different types, which were superimposed on the mass movements susceptibility map obtained using the Analytic Hierarchy Process (AHP) and compared to previous studies in which the matrix indexing method was used. We found approximately similar results with the consideration of different scales of work. These reasons confirm the validity of the susceptibility map at the level of the Kanyosha watershed, a map which is an essential document for urban planning and land management.

Keywords: Landslides, ortho-photos, remote sensing, Kanyosha watershed, susceptibility map, Burundi.

1 Introduction

Bujumbura city, which is the political and economic capital of Burundi, is affected by many types of natural hazards, including landslides. The latter is responsible for loss of life among the population, the destruction of roads and houses, distrurbance in electrical power supply and loss of agricultural production [7]. . Unfortunately, areas prone to landsliding are currently being built up due to rapid extension of Bujumbura city.Therefore, to provide information needed to limit this hazard one needs to extend investigations to the region covered by watersheds of the five main rivers accross Bujumbura city, which will be referred to as Bujumbura region hereafter.In this respect, our contribution was limited to the watershed of the Kanyosha River.The aim of this study was to derive a map of susceptibility to landsliding in Kanyosha watershed from auxilliary information using the Analytic Hierarch Process (AHP).

2 Materials and methods

To prepare the datat needed to achieve this study, we used three sources of information, namely: (1) Literature on the lithology and pedology; (2) 10-m Digital Elevation Model (DEM) of Burundi from which we derived slope, exposition and drainage maps; (3) Landsat 7+ imagery to derive a map of rock fractures. After creating the maps of the predisposing factors, they were all standardized by classifying them into five classes based on expert knowledge. Then they were assigned weights obtained by pairwise comparison according to the AHP approach. Finally, the weighted factors were summarized using GIS operations to obtain the landslide susceptibility map which was validated using landslides identified from detailed satellite images and field work. Our results were also compared to results obtained by Van Western and Soeters.

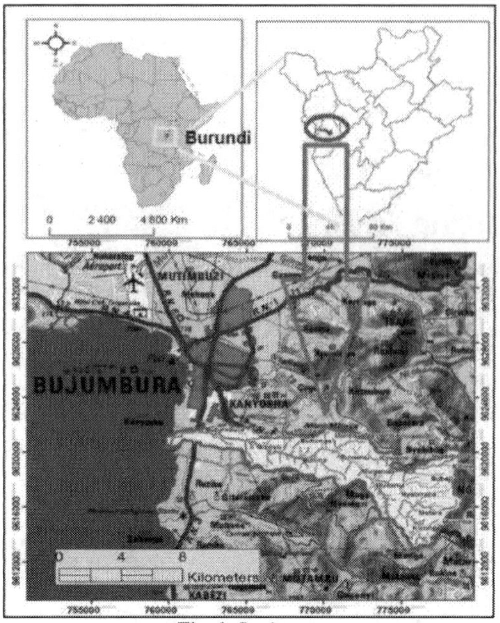

Fig. 1. Study area

Fig. 2. Examples of observed landslides

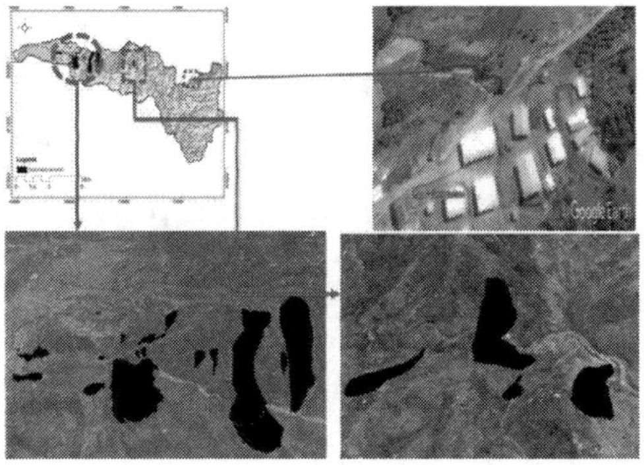

Fig. 3. Landslides identified at Kanyosha watershed

2.1 Predisposing factors

2.1.1 Lithology

Lithology is one of the main factors controlling the development of deep rotational landslides. [2, 3, 10]. Landslides result from the disruption of the equilibrium between forces acting on a slope, which in turn depend on lithology and the structural pattern of rock types. The stability of slopes is affected by weathering processes and by the fluctuation of the water table. In this study the different facies were classified into five categories according to their susceptibility to landslides (Figure4).

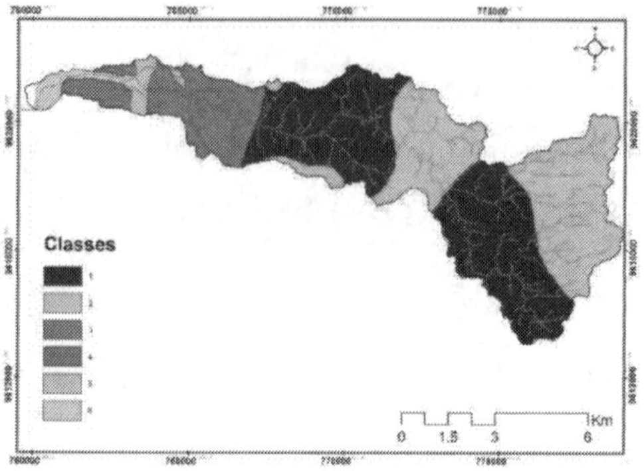

Fig. 4. Lithological map

2.1.2 Fracturation

Schistosity, fracturation, stratification or fault planes are weak zones where landslides are often initiated. These features were identified using band 5 of Landsat 7 ETM+. Multi-directional filters were applied to identify linear features that could correspond to fracturation zones. Linear features other than fracturation zones, e.g. roads and river channels were identified and discarded using detailed ortho-photos and Google Earth images.A five-class fracture density map was made using GIS operations (Figure5). The susceptibility to landsliding increases as the density of fracturation increases.

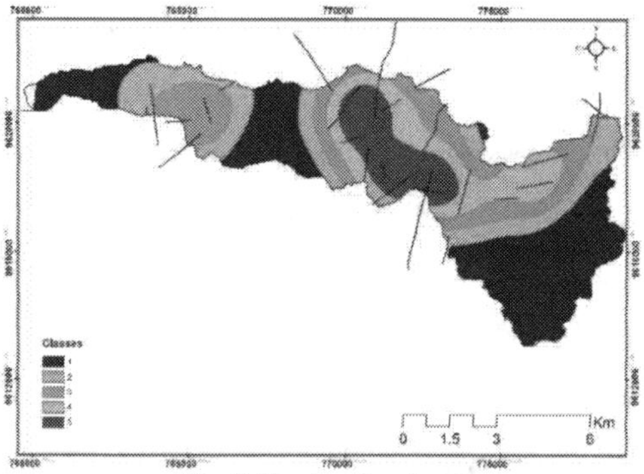

Fig. 5. Fracturation density map

2.1.3 Pedology

Soil type is one of the key parameters that influence landsliding. Landslides occur mainly in clayey soils whereas sandy soils resist against landsliding. Deep landslides occur in deep soils and vice-versa.The map on Figure6 was derived from the 1 :50 000 soil map of Burundi (Bujumbura sheet). It is composed of five classes of susceptibility to landsliding, from the least to the most susceptible to landsliding.

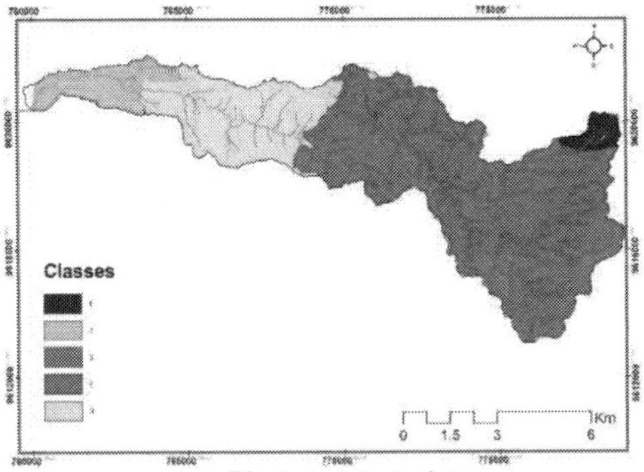

Fig.6. Pedological map

2.1.4 Drainage

Water plays a critical role in landsliding by producing hydrostatic pressure within soil and rock voids, fractures and faults and due to percolation forces. It also causes expansion of some clay minerals. In this study the effect of water on landsliding was assessed based on drainage density (Figure7).Areas with high drainage density are also areas under high risk of landsliding.

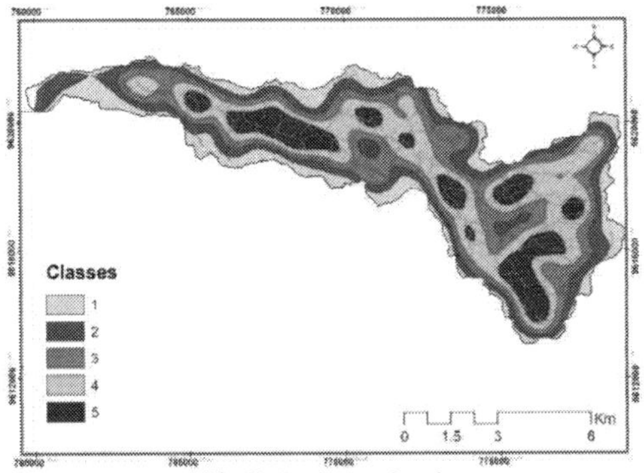

Fig.7. Drainage density map

2.1.5 Slope

Slope steepness is certainly the mostly used variable in landslide studies **[11]**.It is directly linked to the shear strength**[10]**.Slope steepness was derived from the 10-m DEM using the "3D analysist"of ArcGIS 10.2.2. The resulting slope steepness map (in degrees) was subsequently classified into a five-class slope map which is shown at Figure8.The susceptibility to landsliding increases as the slope steepness increases.

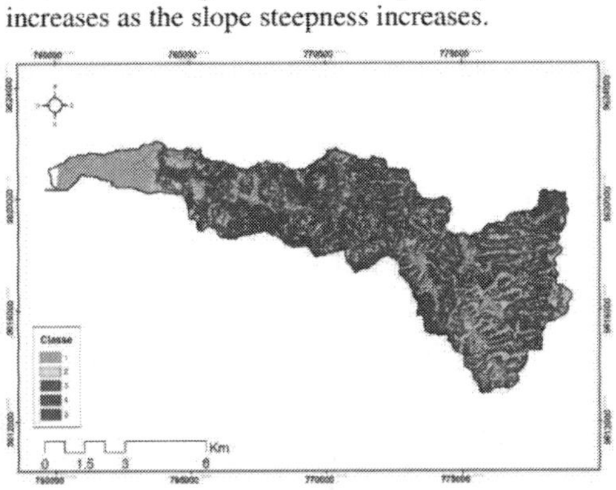

Fig.8. Slope map

2.1.6 Aspect

Slope orientation influences the amount of solar radiation and, thus soil humidity Slope orientation also controls the amount of precipitations when these are related to dominant winds with pronounced directional component (9). The slope orientation map was derived from the 10-m DEM using the function « Aspect » of ArcGIS 10.2.2. Slope orientations are defined taking the north as reference and using 0 – 360° as clockwise value range. The slope orientation map was classified into an 8-class map based on the four cardinal points and on the intermediate directions (Figure9).

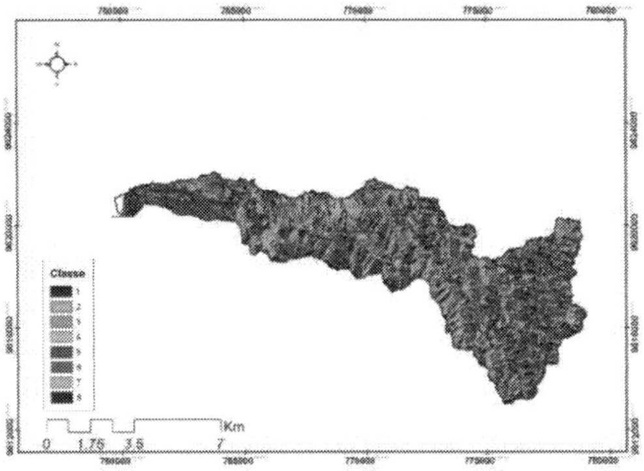

Fig 9. Aspect map

2.1.7 Elevation

Elevation is often accounted for in landslides studies as it is linked to precipitations. **[5]**. Elevation was readily obtained from the 10-m DEM. In our study area, elevation varies from 758 à 2537 m. This elevation map was subsequently classified into a five-class elevation map (Figure10). Low-elevation areas receive small amount of precipitations and, thus have low risk to landsliding.

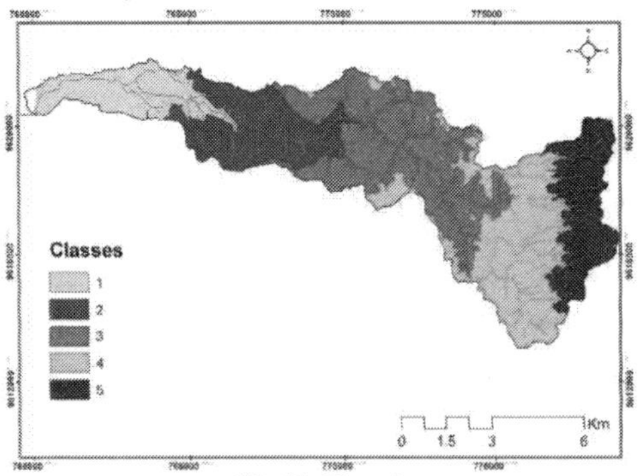

Fig.10. Elevation map

2.2 Approach principles

The AHP is a multi-criteria decision support tool that was developed by[12] and that helps to set priorities in order to make the best decision for complex decision making situations involving alternative decisions, decision criteria and trade-offs. After defining the problem you want to solve, four steps are followed to perform this method. :

1- Determine significant layers.
2- Reclassify the layers.
3- Weight the input layers.
4- Sum up the weighted layers

My concern was always about how I can assign a weight to each layer in a kind of a scientific way. However, Analytic Hierarchy Process (AHP) sorted it out. The main concept of AHP is to pair-wise compare all criteria. A value of 1 expresses "equal importance" and a value of 9 is given for those factors having an "extreme importance" over another factor in a comparison matrix.

3 Results and validation

The result of AHP method of the seven factors considered in this study is shown in Table 1.

Table 1. AHP matrix

	Lith	Frac	Pedo	Drai	Slop	Aspe	Elev	W
Lith	1.00	2.00	3.00	4.00	5.00	6.00	7.00	0.35
Frac	0,50	1.00	2.00	3.00	4.00	5.00	6.00	0.24
Pedo	0,33	0,50	1.00	2.00	3.00	4.00	5.00	0.16
Drai	0,25	0,33	0,50	1.00	2.00	3.00	4.00	0.11
Slop	0,20	0,25	0,33	0,50	1.00	2.00	3.00	0.07
Aspe	0,16	0,20	0,25	0,33	0,50	1.00	2.00	0.04
Elev	0,14	0,16	0,20	0,25	0,33	0,50	1.00	0.03

The consistency ratio (RC) is 0.0247, i.e. is smaller than the threshold value of 0.1, which means that there are small and tolerable inconsistencies [12]. Using map algebra [1] we obtained a landslide susceptibility map for Kanyosha watershed (Figure 11).

The landsliding susceptibility map was validated using field observations (Fig. 4) which are represented by triangles in Fig. 13 and also by comparing our map with the map obtained by [13]and cited by[8].
In total we observed 34 landslides. Their frequences are given in Table 2 and suggest that:

1. Areas with very high susceptibility to landsliding contain 61.8% of observed landslides ;
2. Areas with high susceptibility to landsliding contain 26.5% of observed landslides ;
3. Areas with moderate susceptibility to landsliding contain 11.8% of observed landslides ;
4. No landslide is observed in areas with small to very small susceptibility to landsliding.

Weights in Table.1 indicate that four out of the seven parameters are more important as they occupy more than 86%. They are ranked as follows : Lithology (35%), Fracturation density (24%), Pedology (11%), Drainage network density (11%).

Our results are approximately similar to those obtained by [13] using the "Matrix index" approach and a case study larger than Kanyosha watershed. Together with results from the 34 observed landslides, this similarity supports the validity of our landsliding susceptibility map. Results from field work indicate that about 3% of Kanyosha watershed correspond to landslides. Areas with very high susceptibility to landsliding correspond to weathered gneissic facies. This is because the weathering product of this facies is composed of clays (silts), which store water in raining season and shrink or collapse in dry season due to the evaporation of the stored water. However, no landslides were observed in fluvio-lacustrine clayey and/or sandy deposits located in the alluvial plain and piedmont. This is explained by small values of slope steepness that characterize these landforms (<17°).

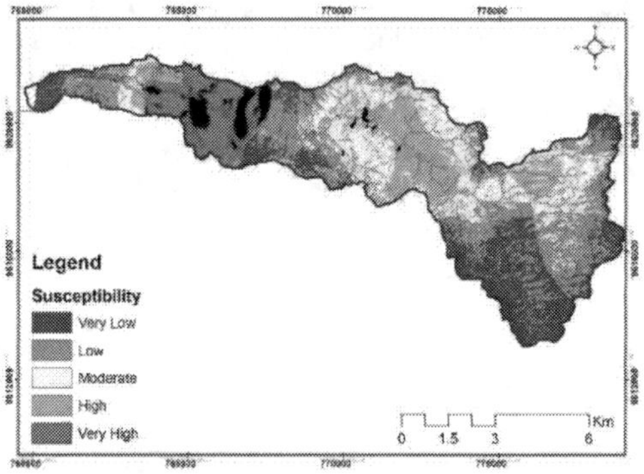

Fig.11. Landslide susceptibility map

4 Conclusions

Kanyosha watershed is an unstable zone due to landslides which occupy about 3% of the area. These landslides are causing important dammages, e.g. roads, schools, irrigation channels, houses, crops, etc.Most of these landslides are of natural origin but can be reactivated human activities, e.g. extraction of building materials in river beds.

In this study we applied the Analytic Hierarch Process (AHP) to derive a landsliding susceptibility map for Kanyosha watershed using remote sensing data, geology, pedology, morphometry and information from literature. The obtained map was validated based on landslides identified from field work, Google Earth images and detailed ortho-photos.

The landsliding susceptibility map is of paramount importance for landuse planning in general and urban planning in particular. Further studies need to be conducted to improve our landsliding susceptibility map, e.g. by applying so-called Weight of evidence method which is more appropriate for any townplanning project.

References

1. C.D. Tomlin. Cartographic Modelling In M. Goodchild, D. Maguire, D. Rhind, (Eds.). Geographical Information Systems: Principles and Application. Harlow, Essex, UK: Longman Group, Ltd., 361-370. (1991)
2. C.I. Fernandez, T.F. Del Castillo, R. El Hamdouni, J.C. Montero. Verification of landslide susceptibility mapping: A case study. Earth Surface Processes and Landforms, 24(6): 537-544, (1999)
3. Carrara. Multivariate models for landslide hazard evaluation. Mathematical Geology, 15(3): 403-427 (1983)
4. D. Kubwimana. *Contribution à l'évaluation de la susceptibilité aux mouvements de terrains dans le bassin versant de Kanyosha Bujumbura (Burundi) Impacts sur l'Aménagement et l'Urbanisme*, (Université Mohammed V, Département des Sciences de la Terre, 2017). Unpublished master dissertation.
5. Gallart, & N., Clotet. Some aspects of the geomorphic processes triggered by an extreme rainfall event: the November 1982 flood in Eastern Pyrenees. Catena Supplement, 13: 79-85 (1988)
6. I. D. Moore, R.B. Grayson, A.R. Ladson. Digital terrain modeling. A review of hydrological, geomorphological, and biological applications. Hydrological Processes, 5(1): 3-30. (1991)
7. J. Moeyersons, P. Trefois, La protection du site CURGO - Kabezi –Bujumbura contre les risques hydrologiques dans le bassin de la rivière Nyabage. Etude de formulation d'une opération de conservation, (2008)
8. L. Nibigira, S. Draidia, H. Havenith. GIS-based landslide susceptibility mapping in the Great Lakes region of Africa, Case study of Bujumbura Burundi. Engineering Geology for Society and Territory 2, 985 – 988.
9. M. Mastere, L. Aït Brahim, E.Guelzim, N.El Fahchouch, Landslide assessment by using digital photogrammetric techniques. European journal of remote sensing, (2009)
10. O. Dewitte. Cinématique de glissements de terrain et prédiction de leur réactivation : approche probabiliste dans la région d'Oudenaarde, Université de Liège,(2006)
11. P.M. Atkinson, & R. Massari. Generalised linear modelling of susceptibility to landsliding in the central Apennines, Italy. Computers & Geosciences, 24(4): 373-385. (1998)
12. Saaty, The Analytic Hierarchy Process. McGraw-Hill, New York. (1980).

Rock stability analysis – A case study

A. Lahmili[1], L. Ouadif[1], A. Akhssas[1] and L. Bahi[1]
[1]Université Mohammed V, Ecole Mohammadia d'Ingénieurs, Laboratoire de Géophysique Appliquée, de Géotechnique et de Géologie de l'Ingénieur et de l'environnement, Rabat, Maroc

Abstract: The problems of the stability of the mineral-bearing structure ST2 at 560 m of depth in the east zone of Bou-azzer mine disturbs the advance of the exploitation. The geological and structural study based on field observations and the analysis of core drilling shows the presence of altered and fractured diorite surmounted by cobalt mineralization. Based on the empirical methods of Barton (Q-system) and Bieniawski (RMR) the bed rock is classified as poor quality. The analytical study made it possible to dimension supporting by bow-pieces and bolting. The existence of several types of discontinuities (fault, diaclases and joints) has made the realization of numerical simulation by the finite elements method very difficult. These discontinuities create a network of natural fractures which cut out the blocks in various forms likely to be detached or slip into the excavation, thus encouraging the infiltration of water creating pressure on the massif. The classical studies show their limits in practice for installation of supporting because they must take into account the characteristics of discontinuities. Hence a structural analysis of the massif is essential. The cracking survey of ST2 at 560 m of depth in the east zone of Bou-azzer mine at 560m of depth, and their processing by the DIPS software, showed the existence of three main families of discontinuity NW-SE with a dip of 75SW, NS subverticale and NE-SW with a dip of 57NW, and two families of minor joints NW-SE and NE-SW with successive dips of 40SW and 75SE. The analysis of fracturing surveys allowed us to evaluate the risks of falling blocks and the families of discontinuity responsible for them, and to limit the zones presenting a risk of slip and the families responsible for them. The importance of this study is of knowing how and where to put supporting to be opposed to the risk of fall and tilting of the blocks, caused by the network of discontinuities of the massif.

Résumé : Les problèmes de la stabilité de la mine de Bou-azzer zone Est au niveau de la structure minéralisée ST2 à 560 m de profondeur perturbe l'avancement de l'exploitation. L'étude géologique et structurale basées sur les observations de terrain et l'analyse des sondages carottés, montre la présence de la diorite altérée et fracturée surmontée de la minéralisation de Cobalt constituée de Skuttérudite et la Safflorite, et parfois Loellingite qui entre en contact avec la serpentine. En se basant sur les méthodes empiriques Barton (Q-système), ou Bieniawski (RMR) le terrain est classé de qualité médiocre, l'étude analytique a permis de dimensionner le soutènement par cintres et boulons d'encrages. L'existence de plusieurs type de discontinuités (faille, diaclases et joints) ont rendu la réalisation de simulation numérique par la méthode des éléments finis très difficile, ces discontinuités créent un réseau de fractures naturelles qui découpent les blocs en différentes formes susceptibles de se détacher ou de glisser dans l'excavation, favorisant ainsi l'infiltration des eaux créant une pression sur le massif. Les études classiques entamées précédemment montrent leurs limites en pratique pour la pose du soutènement, parce qu'ils doivent prendre en considération les caractéristiques des discontinuités. D'où une analyse structurale du massif s'avère indispensable. Les levées de la fissuration au niveau -560 dans la ST2 de la mine de Bou-azzer Est, et leurs traitement par le logiciel DIPS, a montré l'existence de trois familles de discontinuité principales NW-SE avec un pendage de 75SW, N-S subverticale et NE-SW avec un pendage de 57NW et deux familles de joints mineurs NW-SE et NE-SW avec des pendages successive de 40SW et 75SE. Le traitement des relevées de la fracturation nous a permis d'évaluer les risques de chute de blocs et les familles de discontinuité qui en sont responsables, et de limiter les zones présentant un risque de glissement et les familles qui en sont responsable. L'importance de cette étude est de savoir comment et où mettre le soutènement pour s'opposer au risque de chute et de basculement des blocs, provoqué par le réseau de discontinuité du massif.

https://doi.org/10.1051/matecconf/201814902072

Introduction

La minéralisation cobaltifères dans la mine de Bou-azzer se présente essentiellement sous forme de filons à fort pendage et de puissance variable qui se développent dans les serpentines au contact avec les diorites quartziques du Précambrien II inférieur. L'exploitation est régulièrement confrontée à des problèmes de stabilité de terrain, ce qui présente évidemment un risque pour la sécurité du personnel et du matériel, et un impact négatif sur la productivité des chantiers.

Dans la présente étude nous allons caractérisé les massifs rocheux de la ST2 de la mine de Bou-azzer Est, au niveau -560m pour le choix du mode de soutènement approprié, en utilisant des méthodes empiriques comme les classifications de Deer, Hoek et Brown, de Barton, de Bienswiski et L'AFTES, complétée par une étude de la fracturation pour évaluer les risques de basculement de blocs et le risque de glissement et limiter les familles de fractures qui en sont responsable.

1 Contexte géologique

1.1 Situation géographique

Le district de Bou-azzer, se situe dans L'Anti-Atlas central dans la boutonnière précambrienne de Bou azzer-El Graara, à 140 kilomètres au sud-ouest d'Ouarzazate (figure 1).

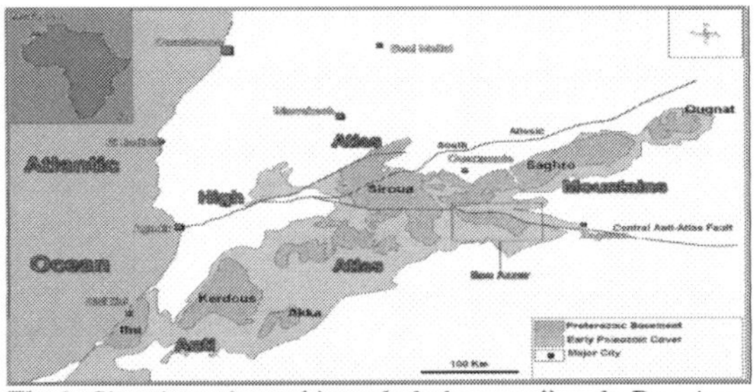

Fig 1: Situation géographique de la boutonnière de Bou-Azzer El Graara [1]

1.2 Géologie régionale

La boutonnière de Bou-azzer El Graara est formée essentiellement d'un socle d'âge Protérozoïque et d'une couverture Infracambrienne (figure 2). Une série ophiolitique encaisse principalement les minéralisations cobaltifères et chromifères.

Précambrien I : constitué de gneiss, d'amphibolites, de migmatites et de leucogranites déformés et polymétamorphiques ayant subi plusieurs phases tectoniques (éburnéenne, panafricaine)

Précambrien II représenté par une série volcano-sédimentaire de Tachdamt-Bleida, et un complexe ophiolitique de 4 à 5Km, auquel sont liées les minéralisations d'arséniures de cobalt et nickel de Bou azzer Est du précambrien II inférieur, et la série de Tiddiline à dominance de matériaux détritiques formés de molasses, arkose, de la carapace d'altération météorique discontinue d'Ambed qui repose sur le substratum serpentineux du Précambrien II supérieur.

Infracombrien : La couverture Infracambrienne repose en discordance sur le socle. Elle comprend à sa base une puissante série volcano-sédimentaire (formation d'Ouarzazate d'âge néo-protérozoïque supérieur ou PIII) bien stratifiée et subhorizontale, constituée par des andésites, des rhyolites avec beaucoup d'ignimbrites. Les laves intercalées dans cette série sont de nature calco-alcaline surmontée par une série sédimentaire gréso-dolomitique, dont la puissance ne dépasse pas 1000m dans la région de Bou-azzer.

La tectonique postérieure au PIII est de style cassant. C'est l'orogenèse hercynienne, qui a fait rejouer verticalement et horizontalement des blocs du socle et donnent naissance, dans la couverture, à des plis disharmoniques et parfois chevauchants.

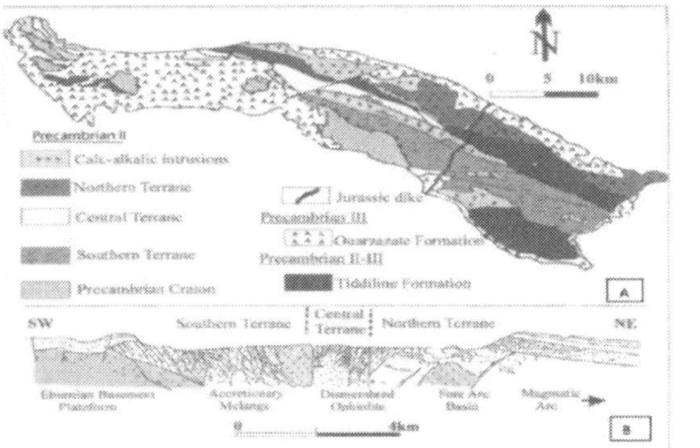

Fig 2: Carte géologique de la boutonnière de Bou Azzer, [2].

1.3 Géologie du chantier Bou-Azzer Est

Le gisement de Bou azzer Est, actuellement en exploitation, est situé à environ 2Km à l'Est du centre minier de Bouazzer. Le contexte géologique local du secteur est caractérisé par l'affleurement des terrains du Précambrien II inférieur, représentés par les serpentines en contact avec les diorites quartziques. La structure ST2 orientée NE-SW et subvertical, actuellement exploitée a une extension moyenne de 160m de longueur et une puissance qui varie entre 0.5cm et 2.7m, est reconnue jusqu'à une profondeur de -560m.

2 Etude Géotechnique du massif de la ST2

Au niveau -560m la minéralisation de la structure ST2 est de direction NW-SE avec un pendage de 60° vers le NE. Les serpentines constituent le toit et les diorites le mur (figure. 3). L'étude géologique montre que le sens du pendage des serpentines a changé entre les niveaux -510m et -560m, et la schistosité qui les affectent a pendage fort vers le SE, ce qui a engendré des problèmes d'éboulements dans les premières phases d'exploitation du niveau -560m.

 https://doi.org/10.1051/matecconf/201814902072

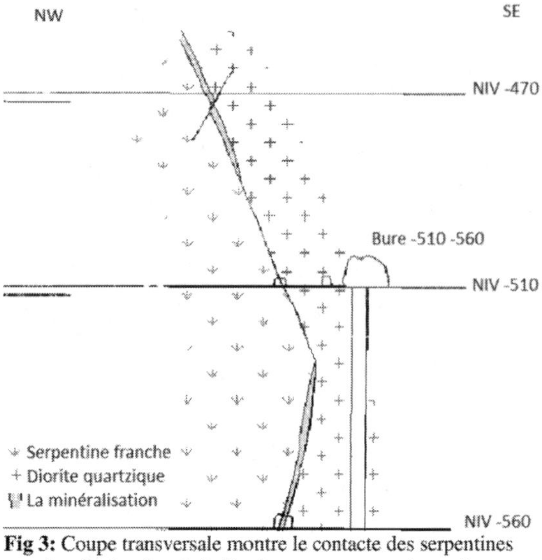

Fig 3: Coupe transversale montre le contacte des serpentines avec les diorites la minéralisation se trouve au niveau du contact.

2.1 Identification des roches

Quatre sondages carottés ont été réalisé à partir du niveau -440m jusqu'au -720m pour caractériser le massif rocheux de la ST2, et ont permis d'identifier les différents facies et de localiser la minéralisation.

Il ressort que les infiltrations sont remarquables, des petits blocs chutent facilement lors du purgage. Les serpentines sont friables schistosées, les diorites sont tendres et fracturés, les chlorites témoignent de l'existence de l'altération hydrothermal et les discontinuités existantes sont de différentes types (faille, diaclases, fentes).

2.2 Classifications empiriques du massif rocheux.

Les facies concernés par les classifications basées sur approche empirique, sont les Diorites quartziques altéré chloritisées, Diorites quartziques altérées chloritisés fracturées, Diorites quartziques très fracturées, Serpentine franche et Caisse minéralisée avec chlorite.

*2.2.1 Classification de DEERE : RQD (*Rock Quality Designation) [3]

Basée sur l'indice RQD, Deere classe les massifs rocheux de très médiocre à excellent

Le calcule de la moyenne des RQD pour chaque faciès, en utilisant les sondages, nous a permis de classer le massif de la ST2 selon Deer de faible (tableau 1)

Tableau 1. Classification de Deere (RQD)

Roche	RQD	Classification de Deer
Diorite quartzique altéré fracturé + Chlorite		
Caisse Minéralisée+ chlorite	**36**	**Faible**
Serpentine Franche	**26**	**Faible**

Note: the first data row is split across columns: "Diorite quartzique altéré" with **40** / **Faible** at the bottom of the left column, then "fracturé + Chlorite" with empty cells at the top of the right column. Printed values per row part:

Roche	RQD	Classification de Deer
Diorite quartzique altéré	**40**	**Faible**
fracturé + Chlorite		

2.2.2 Classification de Hoek et Brown GSI

Introduit par Hoek (1995) et Hoek, Kaiser et Bawden [4] (1995), le GSI (Geological Strength Index) est un nombre sans dimension, déterminé empiriquement, qui varie entre 5 (très mauvaise qualité) et 85 (excellente qualité), et qui peut être estimé à partir d'un examen de la qualité de la masse rocheuse in situ. L'observation du terrain et l'étude des sondages carottés, ont permis de déterminer une valeur GSI moyenne approximative pour les différentes roches.

Le GSI modifié proposé par Sonmez et Ulusay (1999) [5], est un perfectionnement de la GSI estimait trop approximative.

Le calcul du GSI et GSI modifié pour le massif rocheux de la ST2 de Bou-azzer EST a permis de classer les différents faciès tableau 2.

Tableau 2. Classification GSI et GSI modifié

Roche	GSI	GSI Modifié	Classification
Diorite quartzique altéré chloritisé fracturé	65	43,00	Moyenne
Caisse Minéralisée+ Chlorite	65	42,50	Moyenne
Serpentine	40	33,67	Mauvaise

2.2.3 Classification de l'AFTES [7]

L'Association Française des Travaux en Souterrain (AFTES), précise clairement les divers facteurs influençant la conception d'un projet souterrain dans un massif rocheux.

Pour chaque critère, l'AFTES fournit des recommandations propres afin de permettre le choix du soutènement, après étude des trois faciès de la ST2 (-560) les recommandations sont regroupées dans le tableau 3

Tableau 3. Le soutènement retenu après l'utilisation de l'AFTES pour la ST2 (-560 m).

Roche	Diorite quartzique altéré chloritisé fracturé
Les Soutènements données par l'AFTES	-Le béton projeté - Boulons à ancrage ponctuel - Boulons à ancrage réparti - Cintres lourds.
Diorite quartzique altéré	-Le béton projeté - Boulons à ancrage ponctuel - Boulons à ancrage réparti

chloritisé fracturé	- Cintres lourds.
Serpentine franche	-Le béton projeté -Boulons à ancrage ponctuel -Boulons à ancrage réparti -Cintres lourds. -Voussoirs constitués de plaques de métal assemblées.
Caisse minéralisée+ chlorite	- Le béton projeté - Boulons à ancrage ponctuel - Boulons à ancrage réparti - Cintres lourds.

2.2.4 Classification de Barton [8,9] :

On utilisant la méthode de Barton ou Q-System nous avons pu classer le massif de la ST2 (-560) de mauvais à très mauvais et de proposer le soutènement adapté (tableau 4).

Tableau 4 : soutènement pour chaque faciès donné par le Q-système.

Roche	Qsys	Classification de Barton	De	Soutènement
Diorite quartzique altéré chloritisé fracturé	2,67	Mauvaise	2	Boulonnage Lb=1,60m/Es=1,35m
Caisse minéralisée + chlorite	1,60	Mauvaise	2	Boulonnage Lb=1,60m/Es=1,35m
Serpentine	0,22	Très Mauvaise	2	Grillage + Béton projeté 50 - 90 mm avec boulonnage

2.2.5 Classification de Bieniawski [10, 11]

La classification de Bieniawski ou la RMR est la somme de cinq notes représentant la quantification de cinq paramètres caractérisant le rocher (la résistance en compression simple; le RQD; espacement de famille des discontinuités; la rugosité et l'altération des discontinuités, la présence d'eau; l'orientation de la galerie par rapport à l'orientation de la famille de discontinuités critiques). La somme des cinq notes est ajustée par une note relative au sens du creusement par rapport au sens du pendage des discontinuités. Bieniawski, par cette classification, propose la méthode d'excavation envisageable et le type de soutènement le mieux adapté.

D'après la RMR, le massif de la ST2 (-560m) est classé de médiocre pour les Serpentines franches à moyen pour les Diorites quartziques altérés chloritisés fracturés et pour la caisse minéralisée et Chlorite, le soutènement proposé est de type Béton projeté et Cintres métalliques pour les trois type de facies et Boulon à ancrage uniquement pour les Diorites quartziques altérés chloritisés fracturés et la caisse minéralisée et Chlorite.

3 Etude de la fracturation

L'existence de plusieurs type de discontinuités (failles, diaclases et joints) ont rendu la réalisation de simulation numérique très difficile, ces discontinuités créent un réseau de fractures naturelles qui découpent les blocs en différentes formes susceptibles de se détacher ou de glisser dans l'excavation, favorisant ainsi l'infiltration des eaux et créant une pression sur le massif. Les études classiques entamées précédemment montrent leurs limites en pratique pour la pose du soutènement, parce qu'ils doivent prendre en considération les caractéristiques des discontinuités, d'où une analyse structurale du massif s'avère indispensable

Les levées de la fracturation ont été traitées par Dips (Logiciel de projection stéréographique sur le canevas de Wolf-Smith).

La projection stéréographique montre une large concentration des pôles vers l'Est (figure 4).

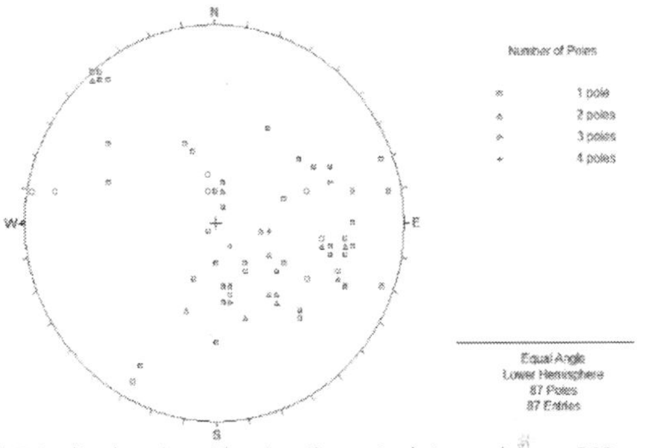

Fig 4: Distribution des pôles des discontinuités au niveau -560m

Les discontinuités se répartissent autour de trois familles principales NNE-SSW, NNW-SSE, NE-SW et une famille secondaire NW-SE très diffuse, elle est de moindre importance) (figure 5).

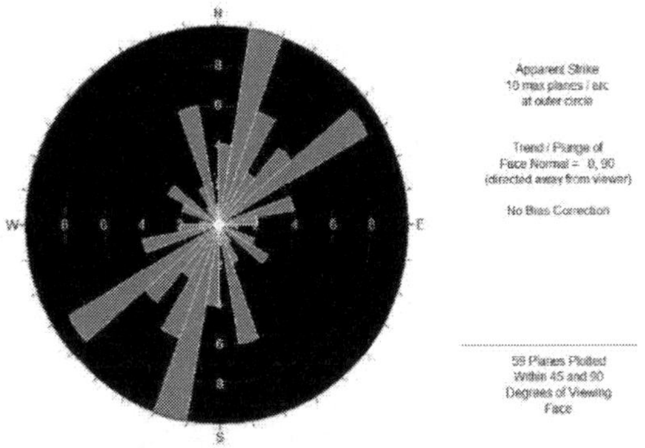

Fig 5 : Diagramme Rosette des lignes des discontinuités Niv-560m.

La concentration des pôles montre quatre à cinq zones de concentration avec une concentration maximale de 10.25%. (figure 6)

https://doi.org/10.1051/matecconf/201814902072

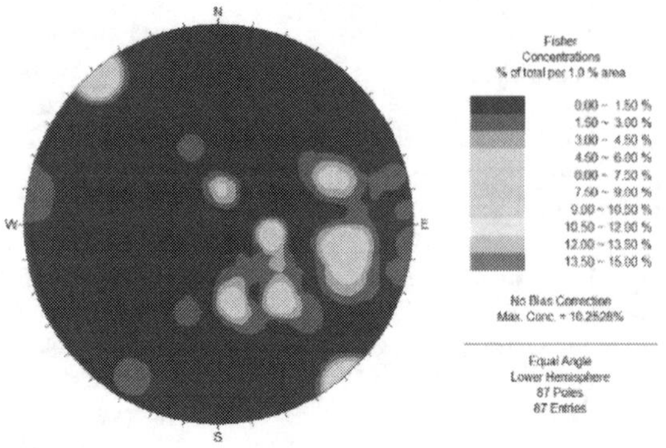

Fig 6: Contours des concentrations des pôles au niveau -560m

Le regroupement des pôles qui définissent la même famille de discontinuités (figure 7) , nous a permis d'identifier trois familles principales avec des joints mineurs, de directions variable, et de pendages faible à moyen et subvertical (tableau 5)

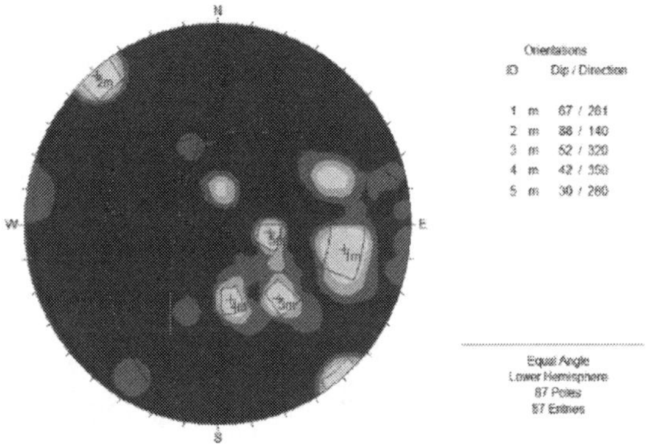

Fig 7: Regroupement des pole au niveau -560m.

Tableau 5: Les principales familles de discontinuités au niveau -560m

Famille	Pendage	Sens du pendage	Description
1	67	283	Famille principale
2	88	140	Famille principale
3	52	320	Famille principale
4	42	350	Joints Mineurs
5	30	280	Joints Mineurs

L'étude de la fracturation montre que le massif est caractérisé par un état de fracturation important, pour l'évaluation des risques engendrés par ces familles de discontinuités, nous avons utilisé les techniques Kinematics recommandés par Goodman, en ajoutant les variabilités standard 1 et 2 ainsi que les plans des épontes diorite et serpentine

3.1 Risque au basculement :

Nous avons considéré les deux épontes de la galerie, la Diorite Quartzique de pendage subvertical (85°) et de direction de pendage 160 et la Serpentine de pendage moyen 60° et de direction de pendage 340. Les plans limitant le risque de basculement sont caractérisés par les mêmes directions que les épontes, leurs pendages est calculé par la formule : $PL = |P - \varphi|$

Avec PL : pendage du plan limite, P : pendage du plan de l'éponte et φ l'angle de frottement interne.

Cette analyse nous a permis de visualiser les zones de risques de chute de blocs, ainsi que les familles responsables de ces risques (figure 8), ainsi il y a un risque de basculement dans la partie entre la minéralisation et l'éponte de la serpentine engendré par la deuxième famille (88/140).

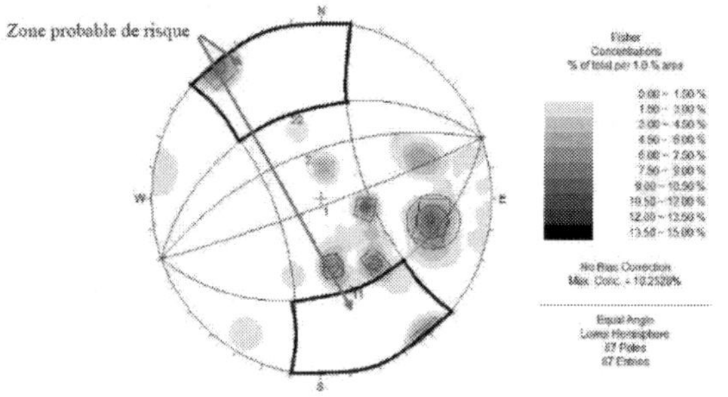

Fig 8 : Risque de Basculement.

3.2 Risque de glissement.

Nous avons considéré des enveloppes engendrées par les plans des deux épontes et des connes limites verticaux de direction N00 et d'angles égale à l'angle de frottement interne de la roche

Par cette analyse nous avons pu visualiser les zones de risques de glissement, ainsi que les familles responsables de ces risques (Figure 9)

Le risque de glissement est localisé dans la partie entre la minéralisation et la serpentine engendré par la famille principale 3 et à moindre degré par les familles 4 et 5

https://doi.org/10.1051/matecconf/201814902072

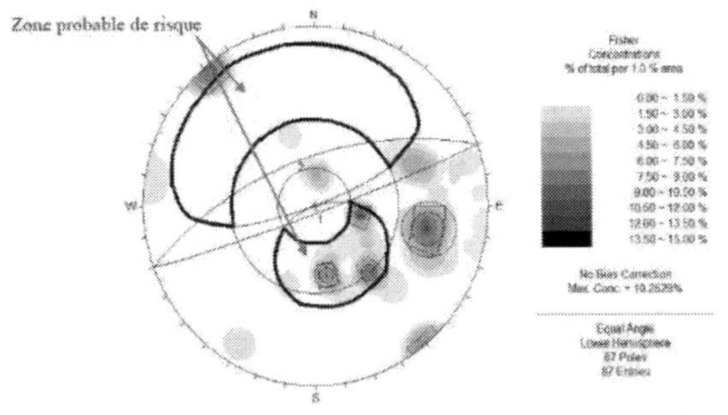

Fig 9 : Risque de glissement

Les risques engendrés par l'intersection des différentes familles de discontinuité est déterminé en tenant compte des plans des épontes et des plans des familles de discontinuités, et en ajoutant des connes verticaux de direction N00 d'angle égale à 90-φ (φ : angle de frottement interne de la roche encaissante). Nous avons donc pu localiser les zones de risques d'intersections des familles de discontinuité, engendré par l'intersection des familles 1-3, 1-4, 5-4 et 4-3 dans la partie entre la minéralisation et la serpentine et 1-2 dans la partie entre la minéralisation et les diorites. Les familles 4 et 5 ne sont pas principales et constituent de ce fait un risque négligeable; la famille 2 qui caractérise les plans de schistosités des serpentines n'affecte pas les diorites qui se trouve épargnée du risque d'instabilité (Figure 10).

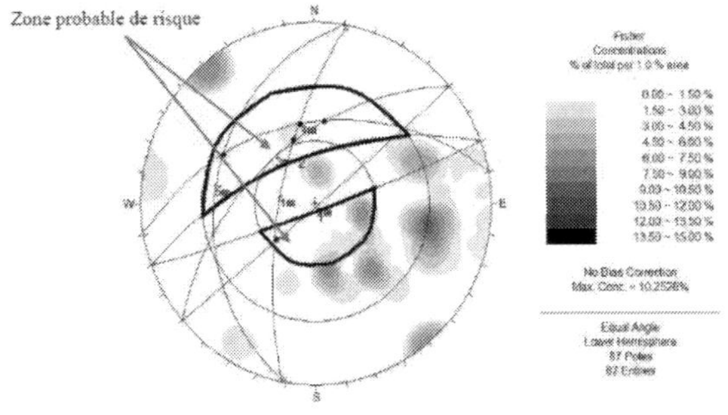

Fig 10: Risques de glissement engendrés par l'intersection des discontinuités.

Conclusion

Les filons minéralisé cobaltifères de la mine de Bou-azzer se développent dans les serpentines au contact avec les diorites quartziques du Précambrien II inférieur.

La classification empirique (Deer, Hoek et Brown, Barton, Bienswiski et L'AFTES), du massif rocheux de la ST2 à Bou-Azzer Est, le qualifié de mauvais terrain et propose un soutènement de type Béton projeté et Cintres métalliques pour les trois type de facies Diorites quartziques altérés chloritisés fracturés, Serpentines et la caisse minéralisée et Chlorite et Boulon à ancrage avec précaution pour les Serpentines

L'analyse avancée de la fracturation du massif rocheux a permis le relevé des principales directions des discontinuités NNE-SSW, NNW-SSE, NE-SW, et de déterminer et de localiser les zones à risque de basculement et de glissement, puis de proposer un soutènement surfacique (Grillage) ou bien porteur (Cintres métalliques, Cadres en bois) pour soutenir les épontes fracturées et instable. Cependant l'étude des risques engendrés par l'intersection des discontinuités nous a permis de localiser les intersections des familles de discontinuités pour la mise en place du soutènement.

Références et bibliographie

[1] A Ait Bendra & R Bkali, Exploitation des filons cobaltifères de Bou-Azzer au contact des serpentines franches (cas d'étude: Aghbar et Bou-Azzer Est),CTT, 2013.

[2] Leblanc, M., 1975. Ophiolites précambriennes et gites arséniés de Cobalt (Bou Azzer - Maroc), Université Paris VI, Paris.

[3] Deere, D.U. and Deere, D.W. 1988. The rock quality designation (RQD) index in practice. In Rock classification systems for engineering purposes, (ed. L. Kirkaldie), ASTM Special Publication 984, 91-101. Philadelphia: Am. Soc. Test. Mat.

[4] Hoek, E., Kaiser, P.K. and Bawden. W.F. 1995. Support of underground excavations in hard rock. Rotterdam: Balkema.

[5] Sonmez, H., and Ulusay, R. (1999). Modification to the geological strength index (GSI) and their applicability to stability of slopes. International Journal of Rock Mechanics and Mining Science, 36: 743-760.

[6] A. Palmström, Journal of Rock Mechanics and Tunnelling Technology, The Rock Mass Index (RMi) applied in rock mechanics and rock engineering, 11, No.2, 1-40, (1996)

[7] AFTES : Recommandations du Groupe de Travail n°1. Caractérisation des massifs rocheux utile à l'étude et à la réalisation des ouvrages souterrains. Tunnels et ouvrages souterrains, (177) :pp 138-170, mai-juin 2003.

[8] N. Barton, R. Lien, and J. Lunde, Engineering classification of rock masses for the design of tunnel support, (1974)

[9] N. Barton, International Journal of Rock Mechanics & Mining Sciences, Some new Q-value correlations to assist site characteristics and tunnel design, 185–216, (2002)

[10] Z.T. Bieniawski, Rock Mechanics Design in Mining and Tunnelling, 272,(1984)

[11] Z.T. Bieniawski, Classification of Rock Masses for Engineering: *The RMR System and Future Trends, Comprehensive rock engineering*, 3,522-542, (1993)

Concrete lozenges impact on the slope erodibility

Latifa ELBOUANANI[1], Khadija BABA[1], Latifa OUADIF[2], Abdelaziz LAHMILI[2] and Lahcen BAHI[2]

[1]GCE Laboratory, High School of Technology-Sale, Mohammed V University in Rabat-Morocco.
[2]3GIE Laboratory, Mohammadia Engineering School, Mohammed V University in Rabat-Morocco.

Abstract. Water erosion is a natural phenomenon, the severity of which is related to several parameters depending on the raindrop impact, the characteristics of the soil and the level of human intervention. The risk of water erosion occurs in both drainage basins and road and rail slopes. Indeed, this phenomenon can be aggravated causing major gullies and landslides. The PK 94 of the Taourirte-Nador railway line, located in the oriental region of Morocco, whose slopes are formed by silt-clay formations experiencing intense erosion. The Moroccan experience for the slopes protection of water erosion knows the use of several methods in particular; concrete arcades and methods based on biological engineering. The objective is to contribute to this subject by proposing the use of concrete lozenges technique. It have the advantage of reducing both the area at risk of erosion and especially the length of slope which is an important factor in the Revised Universal Soil Loss Equation RUSLE used to quantify the quantities of soil eroded.

1 Introduction

Nowadays, climate change (a too hot summer and a too rainy winter) and the construction of more and more road, highway and railway networks influenced the slope stability and their resistance to the rainfall and runoff aggressiveness; which gave rise to water erosion.

Water erosion is a natural phenomenon, which is defined as the total process of detachment, transport and deposition of solid particles from the soil surface due to rainfall, runoff or both.

The Rif in general and the Prerif in particular are zones characterized by the predominance of friable lithological formations such as marls, marl-limestone and schist [1].

The drainage basins of the eastern Rif, although less watered, show a much accelerated soil erosion linked to the regime of limited rains in time and space, but which are most often violent and stormy [2]. In the same direction, the slopes surrounding the Taourirte-Nador railway line, more precisely PK 94, due to their location in the oriental Rif, are subject to intense erosion, which necessitates periodic maintenance, generating significant costs.

In Morocco, several approaches were used to map the soil erodibility in different regions in order to estimate erosion and to predict the necessary preventive or curative methods.

In this context, concrete lozenges are a new preventive technique; used in order to protect the slopes subject to water erosion.

The AMERCAIN approach, in particular the Revised Universal Soil Loss Equation RUSLE, is used to quantify the quantities of soil eroded before and after the use of this technique.

2 Study zone

2.1 Situation

The PK 94 slope of the Taourirte-Nador railway line is located in the North-West of Morocco in the eastern Rif, especially in the city of Nador, which is subject to intense water erosion.

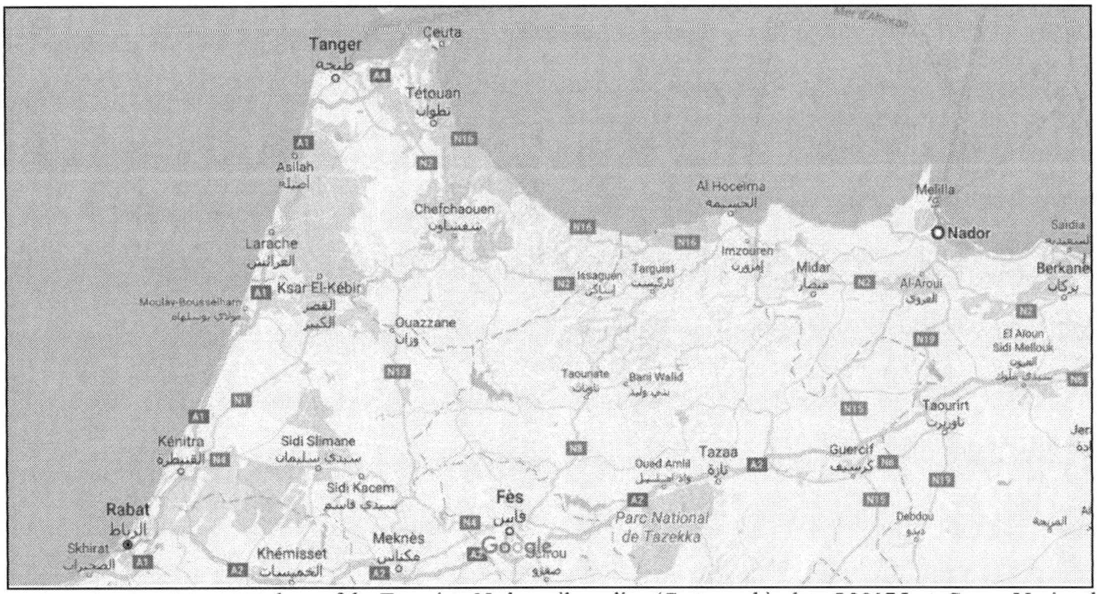

Fig.1. Situation of the PK94 slope of the Taourirte-Nador railway line (Cartographic data ©2017 Inst. Geogr. Nacional. Monde)

2.2 Geology

In the geological context, the oriental Rif presents formations characterized by a superior Miocene, and an inferior Pleistocene belonging to the Villafranchian. These lands consist mainly of pink silts with gravelly beds, and a small band of a powerful quaternary; belonging to the middle quaternary [3].

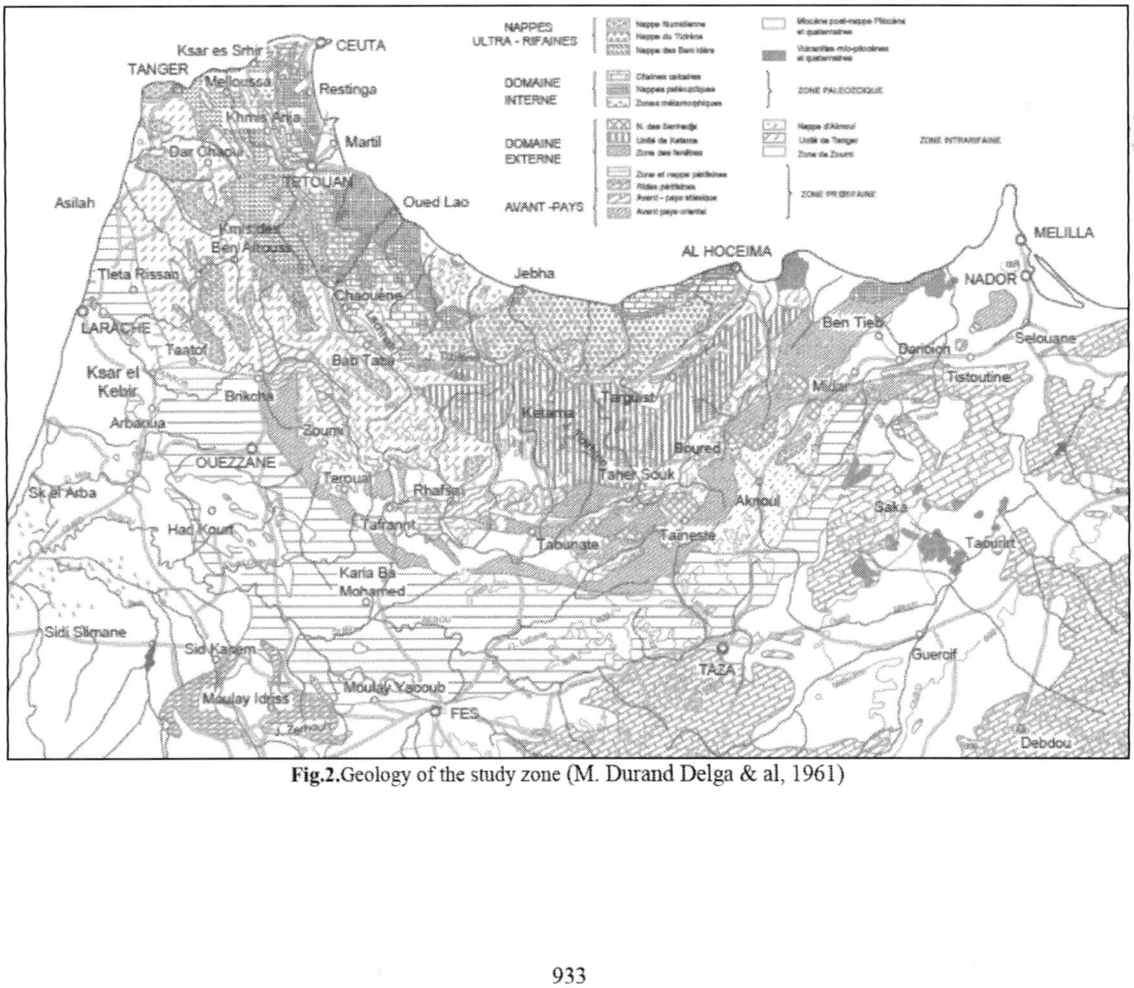

Fig.2. Geology of the study zone (M. Durand Delga & al, 1961)

2.3 Precipitations

The province of Nador is characterized by a Mediterranean climate, including alternating dry and wet seasons, from June to September, and from October to April. The average annual rainfall is approximately 373,5mm/ year. The rains are characterized by seasonal variability [3].

The following figure gives the average of total monthly rainfall in mm of 12 years (2001 to 2012)[3]:

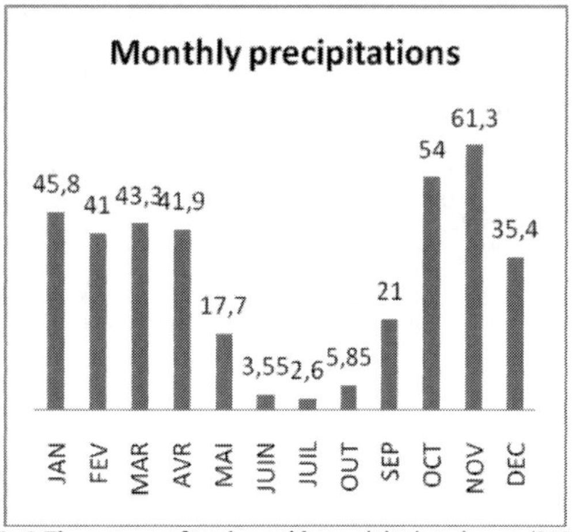

Fig.3. The average of total monthly precipitations in mm (2001 to 2012)

2.4 Granulometric characteristics

The soil is characterized by a strong dominance of silt-clay formations and fine sand [3]:

The silt-clay formations have an average of 90% of particles with a diameter of less than 80µm and 10% of fine sand.

The fine sand formation comprises an average of 25% of particles with a diameter of less than 80µm.

3. Methods

RUSLE is an erosion model designed to predict the long-time average annual soil loss (A) carried by runoff from specific field slopes in specified cropping and management systems as well as from rangeland. It is also applicable to non-agricultural conditions such as construction sites [4].

Soil erosion can be calculated using the Revised Universal Soil Loss Equation such as:

$$A= R * K * LS * C * P \qquad (1)$$

With:

A= Predicted soil loss (ton.ha-1.year-1)

R= Rainfall and runoff factor (MJ .mm/ha.h.year);

K= Soil erodibility factor (t.h/MJ.mm);

LS= Slope length and steepness factor (a dimensionless factor);

C= Crop management factor (a dimensionless factor);

P= Support practices factor (a dimensionless factor).

3.1 Rainfall and runoff factor (R)

The rainfall and runoff factor is the average annual of the sum of the storm EI value; which is the product of total storm energy by the maximum 30-min intensity (I30) (MJ. Mm / ha.h. y).

Given that, we have only the annual and monthly precipitation amounts data, we will use the Rango and Arnoldus equation [5], for the definition of the R factor.

Therefore, the R factor can be expressed as follows:

$$\text{Log } R= 1{,}74 * \text{Log } \Sigma\, (Pi^2/P) + 1{,}29 \qquad (2)$$

With:

Pi: the monthly precipitation in mm;

P: the annual precipitation in mm.

3.2 Soil erodibility factor (K)

The soil-erodibility factor (K) is the rate of soil loss caused by the effect of rainfall, runoff and infiltration. It's measured on a plot which has 22.1m long, a 1.83 m minimum width and a 9% slope.

Soil-erodibility factor can be defined by the soil-erodibility nomograph, or by the algebraic approximation used when the silt fraction does not exceed 70% [4].

Both the graphic and the algebraic method are based on five parameters: % modified silt, % modified sand, % organic matter (OM), classes for structure (s) and permeability (p).

K can be calculated using the algebraic approximation such us [4, 6]:

$$K= [2{,}1.10^{-4}(12–OM)\, M1.14 + 3{,}25(s–2) + 2{,}5(p-3)]/300 \qquad (3)$$

M is the product of the primary particles size.

M= (%modified silt or the 0.002-0.1 mm size fraction) x (%silt+%sand).

To defining the s factor, there is a nomographic classification of the soil structure based on the particles size [4]. It is given in the following table:

Table 1: Classification of the soil structure

Class	Type of soil structure	Size (mm)
1	Very fine granular or without structure	-
2	Fine granular	<2
3	Medium granular Coarse granular	2-5 5-10
4	Polyhedral, lamellar, massive, prismatic	>10

According the soil structure and the soil permeability, Cook & al(1985) [7] are defined six permeability class. They are presented in the following table:

Table 2: permeability class

Structural class	Permeability class	Permeability cm/s	Permeability code
Gravel, coarse sand	Fast	>4,4.10-3	1
Loam sand and loams sandy	Moderate to Fast	(1,4 – 4,4).10-3	2
Fine sand loams, loams	in moderation	(0,4 – 1,4).10-3	3
Loams, loam loams, clay loams	Slow to moderate	(0,14 – 0,4).10-3	4
Clay loams, clays	slow	(4 – 14).10-5	5
Tight, compacted	Very slow	<4.10-5	6

3.3 Slope length and steepness factor (LS)

LS are a dimensionless factor that represents the topography effect on erosion. LS combines both slope length factor (L in m) and the slope steepness factor (S in %).

3.3.1 Slope length factor (L)

Slope length factor can be calculated using the following equation:

$$L=(\lambda/72.6)^m \quad (4)$$

Where λ is the horizontal projection, 72.6 (ft) is the RUSLE unit plot length and m is a dimensionless variable slope length exponent [4, 6].

$$m=\beta/(1+\beta) \quad (5)$$

To calculate the slope length exponent (m) we should defining the β variable (Foster & al.1977) [4, 8] by the following equation:

$$\beta = \left(\frac{\sin\theta}{0.0896}\right)/(3 * (\sin\theta)^{0.8} + 0.56) \quad (6)$$

This value for the ration β of rill to interril erosion for conditions when the soil is moderately susceptible to both rill and interill erosion were computed fromthe equation (6).Where θ is the slope angle.

3.3.2 Slope steepness factor (S)

The slope steepness factor S is defined by the following equation (McCool & al.1987) [9]:

$$S = 10.8 \sin\theta + 0.03 \quad s < 9\% \quad (7)$$

$$S = 16.8 \sin\theta - 0.5 \quad s \geq 9\% \quad (8)$$

LS can be defined also using the following table using the slopes angles in % and the slopes length in ft:

Table 3: Slope length and steepness factor (LS) [4]

Slope %	Slope length in feet																
	<3	6	9	12	15	25	50	75	100	150	200	250	300	400	600	800	1000
0,20	0,05	0,05	0,05	0,05	0,05	0,05	0,05	0,05	0,05	0,05	0,06	0,06	0,06	0,06	0,06	0,06	0,06
0,50	0,07	0,07	0,07	0,07	0,07	0,07	0,08	0,08	0,09	0,09	0,10	0,10	0,10	0,11	0,12	0,12	0,13
1,00	0,09	0,09	0,09	0,09	0,09	0,10	0,13	0,14	0,15	0,17	0,18	0,19	0,20	0,22	0,24	0,26	0,03
2,00	0,13	0,13	0,13	0,13	0,13	0,16	0,21	0,25	0,28	0,33	0,37	0,40	0,43	0,48	0,56	0,63	0,69
3,00	0,17	0,17	0,17	0,17	0,17	0,21	0,30	0,36	0,41	0,50	0,57	0,64	0,69	0,80	0,96	1,10	1,23
4,00	0,20	0,20	0,20	0,20	0,20	0,26	0,38	0,47	0,55	0,68	0,79	0,89	0,98	1,14	1,42	1,65	1,86
5,00	0,23	0,23	0,23	0,23	0,23	0,31	0,46	0,58	0,68	0,86	1,02	1,16	1,28	1,51	1,91	2,25	2,55
6,00	0,26	0,26	0,26	0,26	0,26	0,36	0,54	0,69	0,82	1,05	1,25	1,43	1,60	1,90	2,43	2,89	3,30
8,00	0,32	0,32	0,32	0,32	0,32	0,45	0,70	0,91	1,10	1,43	1,72	1,99	2,24	2,70	3,52	4,24	4,91
10,00	0,35	0,37	0,38	0,39	0,40	0,57	0,91	1,20	1,46	1,92	2,34	2,72	3,09	3,75	4,95	6,03	7,02
12,00	0,36	0,41	0,45	0,47	0,49	0,71	1,15	1,54	1,88	2,51	3,07	3,60	4,09	5,01	6,67	8,17	9,57
14,00	0,38	0,45	0,51	0,55	0,58	0,85	1,40	1,87	2,31	3,09	3,81	4,48	5,11	6,30	8,45	10,40	12,23
16,00	0,39	0,49	0,56	0,62	0,67	0,98	1,64	2,21	2,73	3,68	4,56	5,37	6,15	7,60	10,26	12,69	14,96
20,00	0,41	0,56	0,67	0,76	0,84	1,24	2,10	2,86	3,57	4,85	6,04	7,16	8,23	10,24	13,94	17,35	20,57
25,00	0,45	0,64	0,80	0,93	1,04	1,56	2,67	3,67	4,59	6,30	7,88	9,38	10,81	13,53	18,57	23,24	27,66
30,00	0,48	0,72	0,91	1,08	1,24	1,86	3,22	4,44	5,58	7,70	9,67	11,55	13,35	16,77	23,14	29,07	64,71
40,00	0,53	0,85	1,13	1,37	1,59	2,41	4,24	5,89	7,44	10,35	13,07	15,67	18,17	22,95	31,89	40,29	48,29
50,00	0,58	0,91	1,31	1,62	1,91	2,91	5,16	7,20	9,13	12,75	16,16	19,42	22,57	28,60	39,95	50,63	60,84
60,00	0,63	1,07	1,47	1,84	2,19	3,36	5,97	8,37	10,63	14,89	18,92	22,78	26,51	33,67	47,18	59,93	72,15

3.4 Crop management factor (C)

C is a dimensionless factor that represents the effect of cropping and management practices.

The determination of the C factor is related to the density of the vegetation cover of the soil surface.

3.5 Support practices factor (P)

P factor, a dimensionless factor, is a ratio that expresses the effect of a specific support practice on the erosion.

The P factor varies of 0 to 1 as well as the agricultural or anti-erosive practices adopted.

The value 1 corresponds to the ground without anti-erosion practices.

Table 4: Value of support practices factor (Shin, 1999) [10].

Slope steepness (%)	P
0.0 – 7.0	0.55
7.0 – 11.3	0.60
11.3 – 17.6	0.80
17.6 – 26.8	0.90
> 26.8	1.00

4. Results and discussion

The concrete lozenges are inclined concrete drainage channels in the form of lozenges limiting the formation of deep runoff paths increasingly eroding the slope (Figure4).

They have two mains advantages: reducing the area exposed to erosion and collecting and transporting water.

In the sense of the RUSLE, this method directly and mainly influences two factors: LS and P in addition to area exposed to erosion.

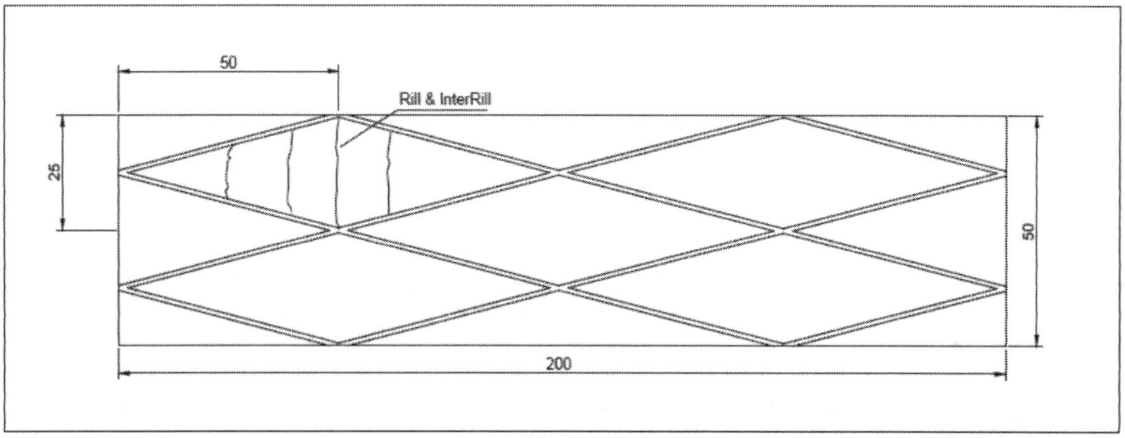

Fig.4. Arrangement of 4 concrete lozenges on an initially unprotected slop

In discussing the concrete lozenges impact on the slope erodibility, we will apply it to the PK 94 slope of the Taourirte-Nador railway line located in the North-West of Morocco in the oriental Rif.

Our study is carried out for a slope area of 50 x 200 = 10000 m2 = 1 ha.

4.1 Rainfall and runoff factor (R)

Using the equation (2) R factor has the following value: R= 150MJ .mm/ha.h.year

4.2 Soil erodibility factor (K)

K factor is computed according to the equation (3). It has the following value: K=0.107mm.t.ha/MJ.

Table 5: K factor

% clay	59	%
% Limon	30	%
% fin sand	10	%
M	1640	-
OM	1	%
s	2	-
p	3	-
K	**0,107**	**t.h/MJ.mm**

4.3 Slope length and steepness factor (LS)

The slope steepness is: 3H / 1V

The slope has: 3H/1V, H = 50m and L = 200m.

So on an area of 50 x 200 = 10000 m^2; we calculate the quantities of eroded soil with and without concrete lozenges.

For a single lozenge, the runoff length varies from 0 to Lh with Lh = 25m. We therefore take an average value of Lr equivalent to Lh / 2 = 17,5 m.

The following table presents the results of the LS factor calculation according to the table 3:

Table 6: LS Factor

	Area without lozenges	Area with 4 lozenges	
Slope steepness (3H / 1V)	33	33	%
The inclination of the slope	18,45	18,45	°
Runoff length Lh	50	17,5	m
Horizontal projection of runoff length	151,02	52,86	ft
LS	**7,7**	**3,22**	-

4.4 Predicted soil loss A

The table 7 contains the data required for the Predicted soil loss (A) calculation both before and after the use of concrete lozenges technique.

Table 7: Predicted soil loss (A)

	Area without lozenges	Area with 4 lozenges	
Exposed area	1	0,93	ha
LS	7,7	3,22	-
Erodibility factor (K)	0,107	0,107	mm.t.h/ MJ
Erosivity factor (R)	150	150	MJ.mm/h a.h.year
The slope is devoid of vegetation cover and support practice (C = 1 and P = 1)	1	1	-
Predicted soil loss (A)	**123,58**	**48,06**	**ton/year**

5. Conclusion

It is concluded that water erosion is a natural risk that can be accentuated or attenuated by human intervention.

Concrete lozenges are a technique to protect the slopes from erosion (by more than 150 %) and to reduce maintenance interventions, because it allows to:

Reduce the area exposed to the raindrops impact and runoff (7% for 4 lozenges).

Reduce a length factor LS by more than 150%.

Indeed, the number of lozenges required is conditioned by the nature of the soil, the climate conditions, slope length and steepness, crop management and support practices.

A large number of lozenges can have a significant financial impact on the project.

References

1. Ennadifi Y. *Etude géologique du Prérif oriental et son avant-pays (région comprise entre Mezguitem, Ain Zora et Tizroutine). Notes &Mém.Serv. géol.Maroc,* (1974) 253, 66 p.

2. Abdelhamid SADIKI, Saïdati BOUHLASSA, Jamal AUAJJAR, Ali FALEH & Jean-Jacques MACAIRE. *Utilisation d'un SIG pour l'évaluation et la cartographie des risques d'érosion par l'Equation universelle des pertes en sol dans le Rif oriental (Maroc) : cas du bassin versant de l'oued Boussouab.* Bulletin de l'Institut Scientifique, Rabat, section Sciences de la Terre, 2004, n°26, p. 69-79, (2004).

3. Novec, Tanger Med Ingeneering. *Etude d'impact sur l'environnement du nouveau port Nador West Med, mission 01: Etude d'Impact Envionnemental Novec, version définitive -N714-13c,* (Juillet 2014).

4. Renard, K.G. G.R. Foster,G.A. Weesies,D.K. McCool,andD.C.Yoder, coordinators. *Predicting Soil Erosion by Water: A Guide to Conservation Planning With the Revised Universal Soil Loss Equation (RUSLE).USDA*; United State Department of Agriculture, Agricultural Research Service agriculture handbook number **703,** 404pp, (1997).

5. Rango A. &Arnoldus H.M.J. *Aménagement des bassins versants. Cahiers techniques de la FAO*, (1987).

6. Wischmeier, W.H. & Smith D.D. *Prediction rainfall erosion losses, a guide to conservation planning Science .U.S. Dept. Agriculture. Agric.* Handbook **537**, 60p, (1978).

7. A.CHEHLAFI, A.KCHIKACH, A. DERRADJI; *Protection des talus autoroutiers par arcades bétonnées ou maçonnées.* Rock Slope Stability, Marakech - Maroc, 2 au 4 Avril (2014).

8. Foster,G.R, L.D.Meyer and C.A.Onstad. *A runoff erosivity factor and variable slope length exponents for soil loss estimates.*Trans. ASAE 20:683-687, (1977).

9. McCool, D.K, L.C.Brown, G.R.Foster, et al. *Revised slope steepness factor for the Universal Soil Loss Equation.*Trans.ASAE **30:1387.1396** (1987).

10. Shin, G. J. *The analysis of soil erosion analysis in watershed using GIS*, Ph.D. Dissertation, Department of Civil Engineering, Gang-won National University, (1999).

Spatial correlation between the predictor variables and the weighting values calculated during the mapping of the environmental factors of mass movements in the Beni Idder region (northern Rif)

Ait Brahim L., Jemmah A. I., Bousta M., El Hamdouni I., Abdelouafi A., El Mahsani A., Lallout I. Labraimi A.Et Merouni F.Z

:

Université Mohammed V, Faculté des sciences, Unité de recherche GEORISK, LGRN, 4 Avenue Ibn Battouta Rabat-Agdal, BP 1014, Rabat, Maroc.

laitbrahim@gmail.com

Abstract. The Tleta of Beni Ider region located in the SW of Tetouan (Rif Septentrional) knows many mass instabilities. The diagnostic via the inventory, the mapping and the characterization of mass movements was made by using satellite imagery, aerial photography and field data coupled with existing documents (geological, geomorphological,...). The understanding of both their spatial distribution and the mechanism generating them, is very complex because of the existence of an important number of natural factors (geological, geomorphological, hydrological) in a relative mountainous landscape with deep valleys, steep slopes and significant elevation changes. Thus, a multidisciplinary approach was adopted to elaborate the landslide susceptibility map of the region taking into account interactions and causal relationships between the various natural parameters that tend to accentuate and aggravate the setting of landslides. The multidisciplinary database allowed us to evaluate the susceptibility thanks to a bivariate probabiliste model (Weight of Evidence). The obtained landslide susceptibility map is a major contribution to the development of urban development plans in the region.

Introduction

L'objectif de cette partie est d'élaborer le zonage de la carte de susceptibilité aux mouvements de terrain à méso-échelle (1/50 000ème) au niveau de la région de Béni Idder. Le choix de cette échelle d'analyse pour cette région a été conditionné par un certain nombre de paramètres essentiels : La présence d'une densité relativement importante et diversifiée des mouvements de terrain ; c'est une zone centrale qui regroupe les cinq classes de susceptibilité identifiées. A cela, s'ajoute l'accroissement progressif de l'installation de la population le long de la route nationale N°2 (Tetouan-chechaouen) et des routes secondaires (P4704, P4702, P4405, P4403 etc.) qui ne cessent de s'accroitre dans le cadre du désenclavement de cette région.

L'approche probabiliste statistique utilisée pour le zonage de la susceptibilité aux mouvements de terrain à méso-échelle (1/50 000ème) est la théorie de l'évidence (TDE) qui est une méthode bayésienne bivariée testée en contextes montagneux [1] [2] [3] [4] [5]. La démarche adoptée est représentée dans la figure.1, et peut être résumée en cinq grandes étapes : **i.** La sélection des différents paramètres à intégrer dans le modèle ; **ii.** La mise en œuvre de tests pour s'assurer de la validité des paramètres sélectionnés ; **iii.** La réalisation d'un nombre de simulations par la théorie de l'évidence ; **iiii.** La comparaison des résultats qui permettra d'en sélectionner la meilleure. **(Fig.1)**

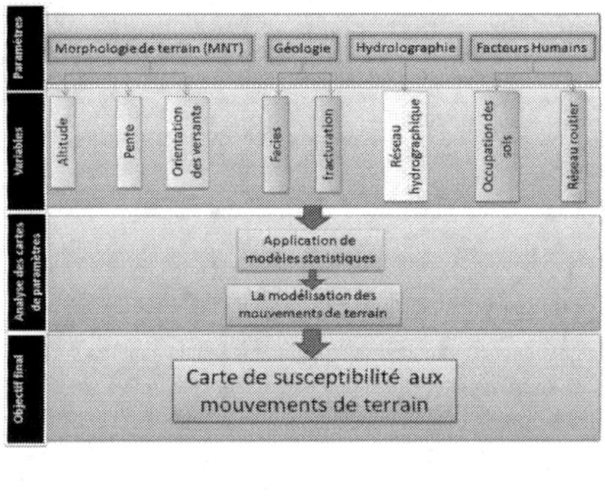

Fig.1 la démarche adoptée pour le zonage de la susceptibilité aux mouvements de terrain à méso-échelle (1/50 000ème) par la théorie de l'évidence (TDE) de la région de Béni Idder

1 Secteur d'étude

2.1 Cadre géographique

La région d'étude est celle de Beni Idder (Rif septentrional), qui se trouvent à 35°24'0" N et 5°30'36" E. Administrativement, elle est délimitée au Nord par Tétouan au Sud par Chefchaouene, à l'Est par Talembote et à l'Ouest par Khmis El Arous. Avec une superficie d'environ 635 km^2. (**Fig.2**)

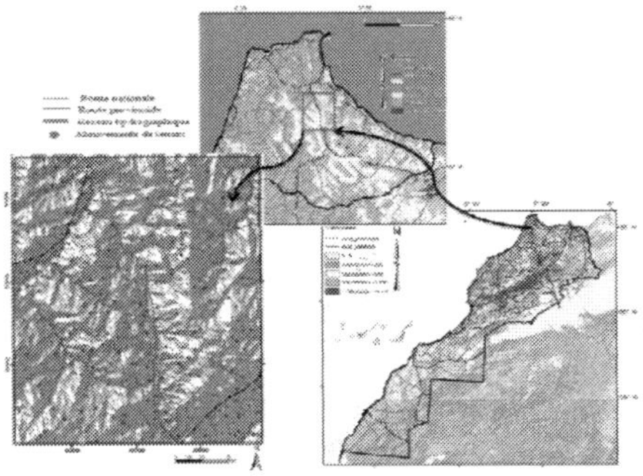

Fig.2 Situation géographique du secteur d'étude

2.2 Cadre géomorphologique et géologique

L'hypsométrie de la région de Tleta de Beni Ider lui imprime un caractère moyennement montagneux (1900m). Secteur caractérisé par des reliefs très accidentés et des pentes très abruptes notamment au niveau des grands contacts chevauchants entre les nappes de charriage, caractérisées par des vallées fortement encaissées, et des dénivellations relativement fortes. Une telle morphologie est le résultat de la conjugaison de plusieurs facteurs, à savoir : la lithologie complexe, le système climatique, la dynamique érosive, le style de déformations et l'activité tectoniques récente. Les unités géomorphologiques distinctes dans le secteur, ont été reconnues par [6].
De point de vue géologique ; la zone d'étude appartienne au Rif septentrional. Elle est situé entre la dorsale calcaire constitué par un empilement d'écaille calcairo-dolomitique à l'est, l'unité d'Amettrasse-Bettara qui englobe des marnes d'âge éocène-oligocène au sud, et vers l'ouest apparaissent des unités de Tisirène (nappes des flysch) constitués des argilites et des marnes gréseux d'âge crétacé inférieur. La zone et couverte d'un manteau de dépôts récents constitué de coulées boueuses ,d'éboulis et d'alluvions [7].

2 Présentation de la variable à modélisé

Nous avons recensé et cartographié les différents mouvements de terrain de Beni Idder, en consultant les documents disponibles et en utilisant les données satellitaires (Landsat7 ETM+ haute résolution, Spot) et une série de photo-aériennes à l'échelle de 1/ 50 000eme, les résultats ont été vérifié sur le terrain (Fig.3). Nous avons classé les mouvements de terrains par typologie dans le Rif (malgré la diversité lithologique, et la complexité morphologique) en quatre classes [8]: **i.** Les glissements qui occupent 43,07 % de la totalité des mouvements de terrain. Ils se localisent surtout dans des formations gréso-pélitiques ou des intercalations marno-calcaires. **ii.** Les éboulements (écroulements+ coulée de débris) qui présentent 31, 46 % du nombre total des mouvements recensés. Ils regroupent un ensemble de phénomènes rapides et brutaux qui affectent des roches rigides et fracturées tels que les calcaires, les grès, ou les roches cristallines. **iii.** Les coulées boueuses constituent 8,24 % ; elles se localisent au niveau des terrains argileux, ou argileux sableux. **iiii.** Les effondrements constituent 17,23 % du nombre total des mouvements développés surtout au niveau des calcaires de la dorsale Les mouvements de terrain dans la région de Beni Idder, présent 10.10 % de la surface total de la région **(Tab. 1)**.

Tableau 1. Pourcentages en nombre et en surface des quatre types de mouvements de terrain de la région d'étude

Types	*Nom-bre*	*% en Nom-bre*	*Superf icie en Km²*	*% en Super-ficie*
Glissements	115	43,07	28,65	44,66
Eboulements	84	31, 46	10,92	17,01
Coulées Boueuses	22	8,24	7,68	11.97
Effondrements	46	17,23	16,91	26,36

3 Outils et méthode ;

3.1 La théorie de l'évidence

La théorie de l'évidence (Weight of evidence) est l'approche que nous avons adoptée dans notre étude. C'est une version log-linéaire du théorème général de Bayes utilisant les principes de calcul de probabilité fondés sur les notions de probabilité a priori et de probabilité a posteriori [9]. Le principe de la méthode consiste à définir une relation mathématique entre les facteurs de prédisposition (variable prédictive, *Vp*) et l'occurrence spatiale des glissements de terrain à travers ses données historiques (variable dépendante, *Vd=VM)* [1]. Elle suppose trois hypothèses : **1.** Les glissements de terrain potentiels se déclencheront selon les mêmes conditions que dans le passé, **2.** Les facteurs de prédisposition *(Ve=VP)* sont connus *apriori* et introduits dans l'analyse, **3.** L'ensemble des glissements de terrain est inventorié sur l'aire d'étude.

La méthode est fondée sur le calcul d'un poids positif (1 :*W+*) et d'un poids négatif (2 :*W-*), dont les valeurs dépendent de l'association entre *Vp* et *Vd.* Ce ci pour calculer la probabilité a priori et la probabilité a posteriori. La probabilité a priori, qui est la probabilité qu'une unité de terrain (maille de calcul, un pixel) contienne une variable dépendante (*Vd=VM)*, dans notre cas les glissements de terrain), est calculée suivant la densité sur la zone d'étude. La probabilité a posteriori est estimée après le calcul de la probabilité a priori selon la densité de *Vd (VM)* pour chaque variable considérée comme prédictive (*Vp*, dans notre cas chaque facteur de prédisposition) [1] [2] [3]. .

(1) $$W^+ = \ln\frac{P\{B \mid E\}}{P\{B \mid \bar{E}\}}$$ (2) $$W^- = \ln\frac{P\{\bar{B} \mid E\}}{P\{\bar{B} \mid \bar{E}\}}$$

B étant la classe de Vp

Le ratio représente la probabilité de présence sur la probabilité d'absence d'un phénomène. Les pondérations sont additionnées en utilisant le logarithme naturel des ratios appelé logit. Pour chaque VP, deux poids sont calculés, un positif (W^+) et un autre négatif (W^-) dont les valeurs dépendent de la relation spatiale entre les VP et la VM survenu dans le passé [9]. Lorsque plusieurs *Vp* sont combinées, les zones (mailles de calculs) qui ont un poids respectivement élevé ou faible correspondent respectivement à une forte ou faible probabilité de présence de *VM*.

3.2 Outils de travail ;

Les données nécessaires à cette étude ont été obtenues à partir des cartes topographiques existantes, des cartes géologiques et des images satellitaires en utilisant des techniques de traitement d'image. Toutes les fonctions de localisation, géologiques et géomorphologiques fournis par les différentes cartes thématiques ont été numérisées en utilisant le logiciel ArcGIS, puis transférées dans ArcView pour les analyses ultérieures. Ces données ont été regroupées en 5 groupes de variables : **1.** un inventaire des mouvements de terrain, **2.** Des paramètres topographiques (MNT, gradient de pente, expositions), **3.** Des paramètres géologiques (lithologie, failles), **4.** Le réseau hydrographique (densité du réseau de drainage), **5.** L'occupation du sol.

4 Facteurs prédictifs

La base de données multi-sources élaborée (sous forme de cartes) des facteurs de prédisposition de l'occurrence spatiale des mouvements de terrain comporte deux variables géologiques : lithologie, fracturation ; trois variables géométriques : cartes hypsométrique, des pentes et d'exposition un variable topographique : le réseau hydrographique et l'occupation du sol. Toutes les cartes élaborées ont été rastérisées avec la même taille du Pixel qui est de l'ordre de 25m.

4.1 La lithologie

La carte lithologique qu'on a réalisée présente cinq ensembles lithologiques ou cinq classes à savoir : les dolomies et calcaires, les marnes, les grès, les schistes et les alternances marno-calcaires. **(Fig.4)**

La fréquence des glissements au niveau des Grès, des marno-calcaires et Marnes montre une fréquence relativement élevée, et au niveau des marno-calcaires elle est jugée très élevée.

4.2 Occupation des sols

L'occupation du sol est obtenue par analyse et interprétation d'images satellitaires et de photographies aériennes. Nous avons définit cinq classes : **1.** La forêt qui présente 33,2% de la surface totale, **2.** La végétation naturelle avec une surface de 26%, **3.** terres arables de 22,9%, **4.** terrains agricoles avec 17,3% de la surface, **5.** terrains nus de 0,6% **(Fig.5)**.

La corrélation entre l'occupation du sol et la fréquence des glissements de terrain montre que la fréquence des glissements de terrain est relativement faible sur les terrains nus et plus élevé sur des terres arables et forêts.

4.3 Le gradient de pente

On a réalisé la carte des pentes à partir du MNT de la Région de Beni Idder avec cinq classes bien déterminer : classe 1 (0°- 5°), classe 2 (5°- 15°), classe 3 (15°- 25°), classe 4 (25°- 45°) et classe 5 (45°- 89°) **(Fig.6).**

L'examen de la répartition des glissements de terrain avec les classes de gradient de pente, montre une augmentation de la fréquence des glissements de terrain dans trois classes de gradient de pente où la fréquence maximale est atteinte dans la classe 5°-15°.

4.4 Hypsométrie

On a réalisé la carte d'hypsométrie avec les classes suivantes : 33m- 400m, 400m- 650m, 650m-900m, 900m- 1250m et 1250m- 1913m **(Fig 7)**. Au nveau des altitudes de 400m à 900m la fréquence des mouvements de terrains est très élevée, on note aussi que la fréquence des mouvements de terrain est assez importante aux altitudes inférieures à 400 m.

4.5 Exposition des versants

La carte d'exposition a été reclassifiée en cinq classes importantes (**Fig.8**). La répartition des glissements de terrain cartographiées par rapport aux classes des expositions, montre que les glissements sont relativement plus fréquents au niveau des orientés Nord et Ouest.

4.6 La fracturation

L'activité tectonique récente et héritée peut intervenir, en conditionnant les mouvements de terrain dans les zones où les terrains sont fragilisés [10] [11]. Ainsi la cartographie des mouvements de terrain dans la région d'étude, montre qu'un nombre important de mouvements s'est développé le long des principaux accidentsde direction NS et les fronts de chevauchements de direction, , etc). Ces diverses failles guident un certain nombre d'écoulements de surface. Elles montrent aussi une relation avec la répartition des sources d'eau [10] [11] [12]. **(Fig.9)**

La relation entre la fréquence des glissements de terrain et la distance aux failles, montre que la fréquence des glissements de terrain diminue lorsque la distance aux failles augmente. Avec le maximum des mouvements de terrain se trouve à 100 m des failles.

4.7

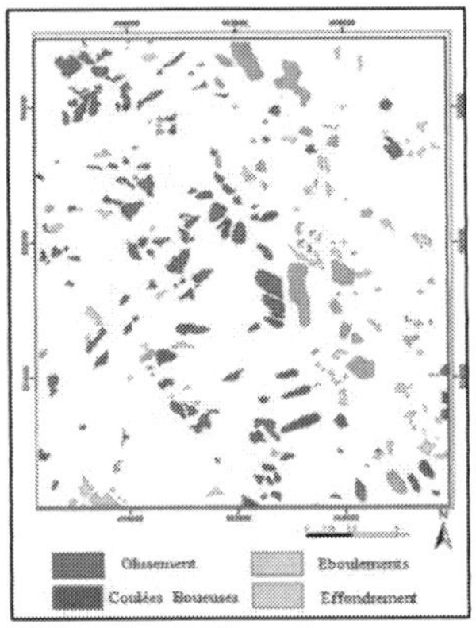

Fig.3 carte d'inventaire des mouvements de terrain.

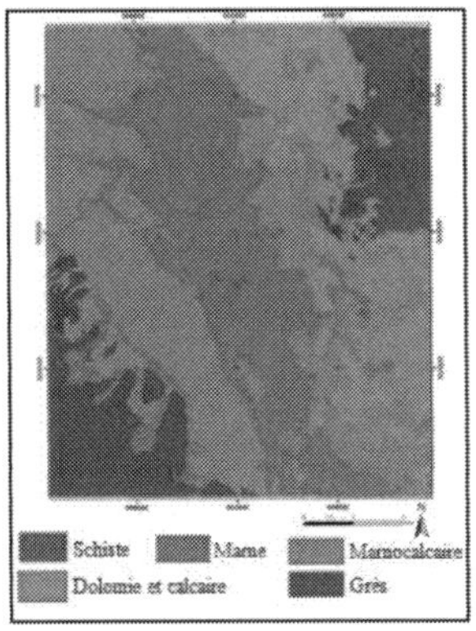

Fig.4 Carte des différentes formations lithologiques.

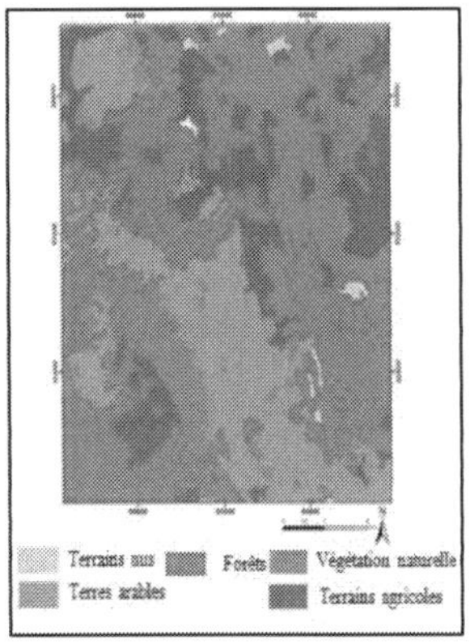

Fig.5 Carte d'occupation des sols.

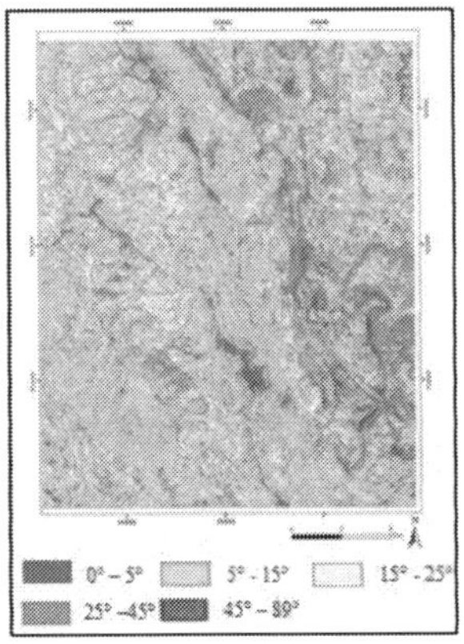

Fig.6 Carte des gradients de pente.

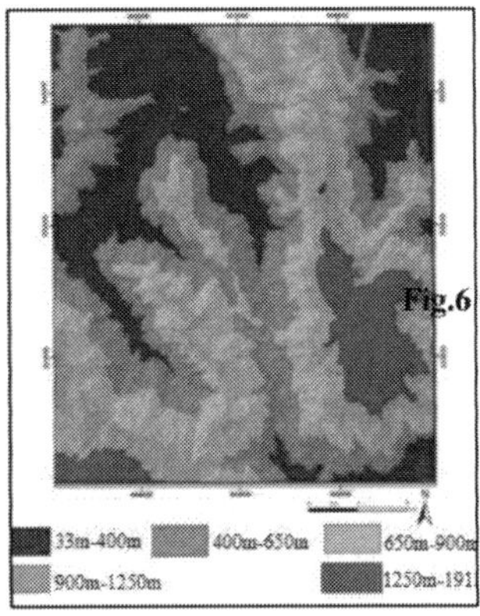

Fig.7 Carte d'hypsométrie (en mètre).

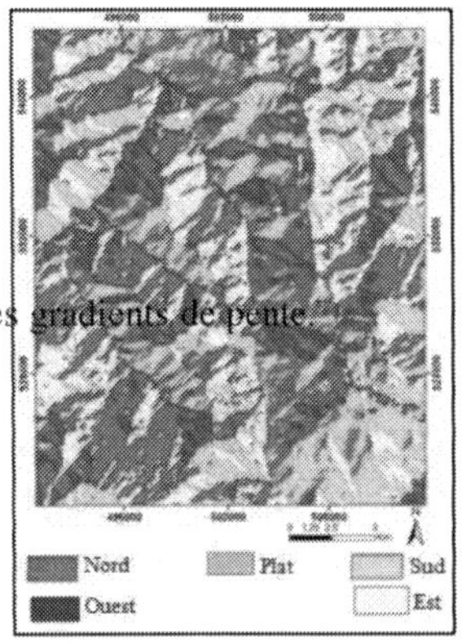

Fig.8 Carte d'expositions.

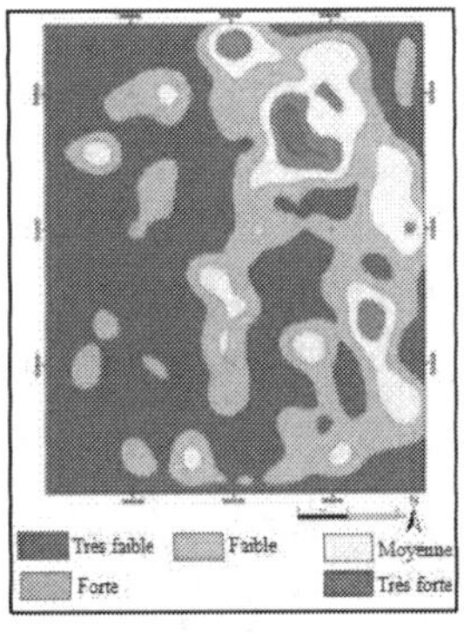

Fig.10 Carte de la densité du réseau hydrographique.

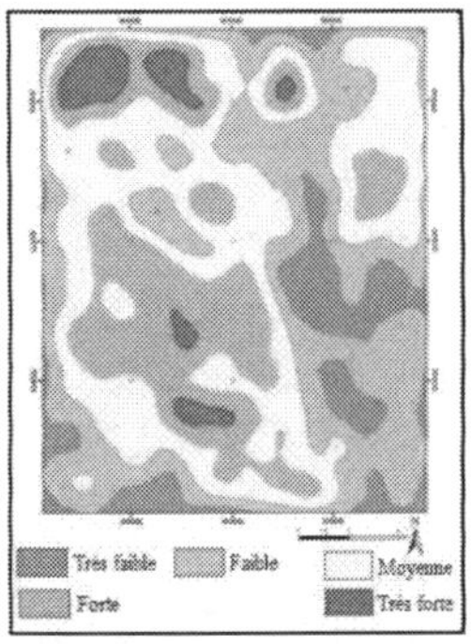

Fig.11 Carte de densité du réseau hydrographique

Réseau hydrographique

Le réseau hydrographique est pris en considération pour deux raisons : **i.** La nature imperméable de la majorité des terrains dans le Rif (à l'exception de la dorsale calcaire), se traduit par un écoulement de surface très important lors de la période pluviale et à chaque orage ce qui conduit soit à une érosion soit à une infiltration à travers des zones fragiles (stratification inclinée, schistosité, failles etc…) contribuant à une diminution de la résistance au cisaillement des matériaux ; **ii.** Son analyse en terme d'anomalies et de perturbation peut nous renseigner dans les zones dépourvues de terrain plio-quaternaire de notre secteur d'étude de la réactivation récente des failles [10] [12]. **(Fig.10)** La relation entre la fréquence des glissements de terrain et la densité du réseau hydrographique, montre que la fréquence des glissements de terrain augmente dans la classe ou la distance entre 200 et 400 m.

5 Résultats et analyses

5.1 Calcul du poids ;

Les cartes des paramètres de prédisposition combinées avec la carte de glissement de terrain permettent de calculer les poids positifs et négatifs et les contrastes, comme indiqué dans **le tableau.2**.

Les valeurs positives du contraste **(C) enregistrés** pour les paramètres lithologiques (Marno-calcaire, Marne et dolomie-calcaire), l'occupation du sol (terres arables et terrains nus), distance aux failles (inférieure à 300 m), attestent de leur rôle significatif dans l'apparition des mouvements de terrain.

Tableau 2 : Corrélation spatiale entre les variables prédictives et les valeurs de pondération calculées par la théorie de l'évidence, (W+) : la valeur la probabilité a priori, (W-) : la valeur de la probabilité à posteriori, (C) : le contraste qui est la différence entre W+ et W-,S(C) : la variance du contraste

Formations géologiques	*Surface en Km^2*	*W+*	*W-*	*C(Contraste)*	*S(C)*
Grès	73,443	-0,4067	0,0436	-0,4503	0,5246
Dolomie et calcaires	130,64	0,0368	-0,0099	0,0466	0,3441
Marno-calcaires	212,39	0,0430	-0,0227	0,0657	0,2961
Schistes	37,478	-1,1347	0,0431	-1,1778	1,0135
Marnes	173,90	0,1915	-0,0836	0,2751	0,3005
Distance par rapport aux Failles	*Surface en Km^2*	*W+*	*W-*	*C(Contraste)*	*S(C)*
100	332,40	0,0498	-0,0621	0,1120	0,2856
200	196,47	0,0650	-0,0310	0,0960	0,3002
300	70,891	-1,0694	0,0813	-1,1508	0,7242
400	16,292	0,4266	-0,0141	0,4407	0,7326
> 500	4,4508	1,0577	-0,0130	1,0707	1,0392

Gradient de pente	*Surface en Km^2*	*W+*	*W-*	*C(Contraste)*	*S(C)*
0 - 5°	67,686	0,6162	-0,1063	0,7225	0,3587
5 - 15°	219,90	0,0598	-0,0339	0,0938	0,2927
15 - 25°	213,09	-0,5493	0,1967	-0,7459	0,3553
25 - 45°	118,28	0,1250	-0,0318	0,1568	0,3443
45 - 89°	8,0862	0,3906	-0,0064	0,3971	1,0252
Exposition des versants	*Surface en Km^2*	*W+*	*W-*	*C(Contraste)*	*S(C)*
Plat	37,194	0,9925	-0,1107	1,1033	0,3946
Nord	241,45	-0,3411	0,1670	-0,5082	0,3165
Est	83,763	-0,3121	0,0406	-0,3527	0,4746
Sud	77,692	0,1046	-0,0157	0,1202	0,4115
Ouest	187,01	0,1115	-0,0514	0,1628	0,3003
Distance au réseau hydro-graphique	*Surface en Km^2*	*W+*	*W-*	*C(Contraste)*	*S(C)*
100	69,838	0,2128	-0,0300	0,2429	0,4119
200	190,34	-0,2555	0,0939	-0,3494	0,3330
300	205,51	-0,1038	0,0467	-0,1505	0,3103
400	137,97	0,0737	-0,0216	0,0953	0,3337
>500	26,157	0,8868	-0,0620	0,9487	0,4819
Occupation des sols	*Surface en Km^2*	*W+*	*W-*	*C(Contraste)*	*S(C)*
Forêts	218,00	-0,3595	0,1475	-0,5070	0,3211
Végétation naturelle	161,96	0,0152	-0,0053	0,0205	0,3139
Terres arables	143,20	0,2774	-0,0971	0,3745	0,3019
Terrains agricoles	107,76	-0,0204	0,0041	-0,0245	0,3690
Terrains nus	3,6054	1,9718	-0,0326	2,0044	0,7732
L'hypso-métrie	***Surface en Km^2***	***W+***	***W-***	***C(Contraste)***	***S(C)***
33 – 400 m	148,20	0,0073	-0,0022	0,0095	0,3335
400 – 650 m	190,28	0,2217	-0,1116	0,3334	0,2929
650 – 900 m	191,37	-0,0259	0,0110	-0,0369	0,3104
900 – 1250 m	70,632	-1,0567	0,0797	-1,1364	0,7242
1250 – 1913 m	33,887	0,0985	0,0124	-0,2608	0,3255

5.2 Calcul de l'indépendance conditionnel (IC)

5.2.1 Degré de liberté entre chaque paire de variables de predisposition

Parmi les sept variables de predisposition traités, six ont été retenues (occupation des sols, la pente, le facies, hypsométrie, distance aux failles et le réseau hydrographique) étant donné que les valeurs du degré de liberté **(α)** entre chaque paire de variables obtenue par la méthode de WofE (tableau.4) sont supérieurs à **0,05 (α>0,05)** [1] [2] [3]. La variable exposition a été retirée (α< 0,05) étant donné sa dépendance avec la variable pente. Ce résultat a été confirmé par le test de **chi-carrée** (X^2) etant donné que l'ecart n'est pas important entre l'écart de X^2 observé etX^2 théorique pour chaque couple de facteurs et pour l'ensemble des variables à modéliser.

Tableau 3 : Valeurs du degré de liberté α entre chaque paire de variables

	Occupa-tion des sols	Pente	Facies	Hypso-métrie	Réseau hydro-graphique
Distance au faille	0,991 9	0,8358	0,3607	0,9996	0,9330
Occupa-tion des sols		0,9841	0,8133	0,3840	0,9611
Pente			0,8392	0,9736	0,9952
Facies				0,8091	0,6982
Hypso-métrie					0,7724

5.2.2 La carte de susceptibilité aux mouvements de terrain de Beni Idder (WofE)

La carte obtenue permet de distinguer 5 classes de susceptibilité : très forte, forte, moyenne, faible, et très faible (**Fig.11**).

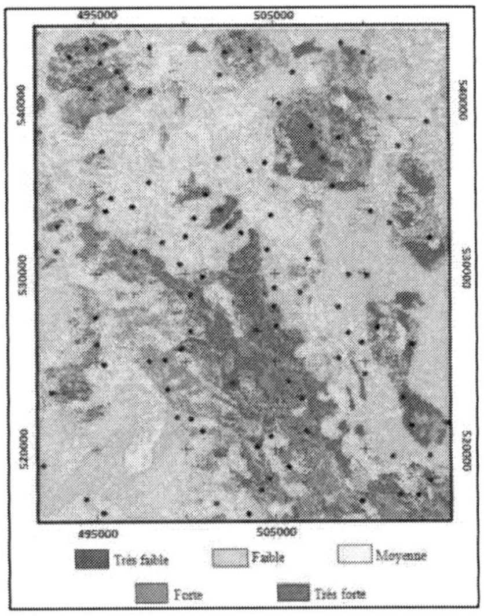

Fig.11 La carte de susceptibilité aux mouvements de terrain par la théorie de l'évidence de la région de Tleta de Beni Ider

Pour la validation de la carte de susceptibilité aux mouvements de terrain de la région de Tleta de Beni Ider, on a superposé les mouvements de terrain cartographiés sur la carte de susceptibilité obtenue afin de définir la proportion des mouvements de terrain par classes de susceptibilité. Le résultat statistique en proportion des mouvements de terrain par classes de susceptibilité a montré que 53% des mouvements de terrain cartographiés se situent dans la zone à très forte à forte susceptibilité, 24% des mouvements de terrain dans la zone moyenne.

Tableau. 4 : Les pourcentages des mouvements de terrain par rapport les classes de susceptibilité

Susceptibilité	Pourcentage des MT (%)
Très forte	29
Forte	24
Moyenne	20
faible	15
très faible	12

Conclusion

L'analyse de la susceptibilité aux mouvements de terrain dans la région montagneuse de Beni Idder a été facilitée grâce à l'utilisation d'outils de la géomatique (Télédétection, SIG, GPS, etc.).

La théorie de l'évidence a été testée dans la région de Tétouan-Chefchaouen (Rif septentrional) pour évaluer la susceptibilité liée aux quatre types de mouvements de terrain. Les facteurs conditionnellement indépendants ont été retenus et intégrés dans le processus de modélisation afin de produire les cartes de probabilité d'occurrence de mouvements de terrain.

Les facteurs de prédisposition de l'occurrence spatiale des mouvements de terrain, ont été rastérisés avec la même taille du Pixel qui est de l'ordre de 25m, puis classés en 5 classes. Les mouvements de terrain significatifs et représentatifs ont été identifiés et cartographiés.

Le test chi-carré X^2 a permis de définir les facteurs prédictifs conditionnellement indépendants vis à vis aux mouvements de terrain VM. Les poids positifs et négatifs ainsi que leurs contrastes ont été calculés pour chaque classe de variables. Ce calcul a permis d'identifier les variables qui sont positivement corrélées ($W^+ > 0$ et $W^-<0$) ou négativement corrélées ($W^+ <0$ et $W^->0$).

La carte de susceptibilité élaborée fait ressortir que : **i.** Les glissements de terrain sont les plus fréquents et se localisent sur des terrains marneux à marno-calcaires ; **ii**. Les éboulements viennent par la suite en terme de fréquence et se localisent au niveau de la dorsale calcaire (dolomie calcaire, marno-calcaire) ; **iii**. Les coulées boueuses se localisent dans les zones où dominent les marnes ; **iii**. Les effondrements sont situés essentiellement au centre de la dorsale calcaire ce qui est justifié par sa nature lithologique. Ainsi les zones les plus exposées aux instabilités de terrains se localisent au niveau de la dorsale calcaire et des nappes des flyschs.

Les résultats statistiques montrent que 53% des mouvements de terrain cartographiés se situent dans la zone à très forte à forte susceptibilité, 24% des mouvements de terrain dans la zone moyenne.

La carte de susceptibilité aux mouvements de terrain élaborée, servira dans l'aménagement et le développement durable de la région, ceci pour une éventuelle extension urbaine, ou lors des planifications de nouveaux tracés de routes et autoroutes. Elle constitue un document utile pour la prévention et l'information des populations aux risques de mouvements de terrain dans la région de Tleta de Beni Ider.

References

1. A. Carrara , M. Cardinali, F. Guzzetti, P. Reichenbach, *GIS-based techniques for mapping landslide hazard.* Geographical Information System in assessing natural hazard(1995)

2. C.J. Van Westen, E.A. Castellanos Abella, L.K. Sekhar, *Spatial data for landslide susceptibility, hazards and vulnerability assessment: an overview.* In: Engineering geology. (2008)

3. Y. Thiery , J-P. Malet, S. Sterlacchini, A. Puissant, O. Maquaire, *Landslide susceptibility assessment by bivariate methods at large scales: Application to a complex mountainous environment.* Geomorphology. (2007)

4. M. El Moulat, L. Ait Brahim, M. Mastere and A. I. Jemmah, *"Mapping of Mass Movements Susceptibility in the Zoumi Region Using Satellite Image and GIS Technology (Moroccan Rif),"* International Journal of Scientific & Engineering Research, vol. **6(2)**, pp. 210-217, (2015)

5. M. Mansour, L. Ait Brahim, P. Potherat, Z. Sekkat. *"Etude du glissement de Bouhalla par une approche multisource (commune de Chaouene, Rif, Maroc)."* Revue de Génie civil, **81**, Mai 1999, pp.34-41, (1999).

6. A.El Gharbaoui, *La terre et l'Homme dans la péninsule tingitane. Etude sur l'Homme et le milieu dans le Rif occidental.* Travaux de l'Institut Scientifique, série Géologie et Géographie Physique, Rabat, **15**, 439 p. (1981)

7. M. Moukhchane, S. Bouhlassa, A. Chalouan ; *Approche cartographique et magnétique pour l'identification des sources de sédiments : cas du bassin versant Nakhla (Maroc).* Sécheresse, 9 :227-32. (1998)

8. D.J. Varnes, *Landslide Hazard Zonation, a review of principles and practice.* IAEG Commission on Landslides. UNESCO, Paris. (1984)

9. G.F. Bonham-Carter, F.P. Agterberg, *Weights of evidence modeling: a new approach to mapping mineral potential.* Geological Survey of Canada Paper, **89-9**. (1989)

10. L. Aït Brahim - *Tectonique cassante et état de contaraintes récentes du Maroc Nord, contibution à l'étude du risque sismo-tectonique au Maroc.* Thèse de doctorat d'état, Université Mohamed V, Faculté des Sciences, Rabat. (1991)

11. L. Aït Brahim, P. Chotin , S. Hinaj, A. Abdelouafi, A. El Adraoui , C. Nakcha, D. Dhont, M. Charroud, F. Sossey Alaoui, M. Amrhar, A. Bouaza, H. Tabyaoui, A. Chaouni. *Paleostress evolution in the Moroccan African margin from Triassic to Present.* Tectonophysics. (2002)

12. Deffontaines et al., *Investigation of active faults in Moroco using morphometric methods and pattern analyses sonderdruck an Geologische Rundschau*, **8**, Helt 1, Stuttgart, Deutschland, P. 199-210, (1992).

Simple Evaluation of the Influence of an Inert Additive on the Swelling Characteristics of Clay Soil

Bahia LOUAFI[1], Ramdane BAHAR[2]
Laboratory of Environment, Water Geomechanics and Structures (LEEGO),
Faculty of Civil Engineering USTHB, Algiers, Algeria

Abstract: In this study, an experimental investigation carried out to assess the performance of addition of an inert material on stabilization of an expansive soil in order to solve the lifting structures caused by this phenomenon. The results showed that an addition of glass is capable of reducing the actions of swelling. From the global analysis of the porosity using an ultrasonic device, we measure the propagation velocities of ultrasonic pulse through the tested clay samples. This reveals a change occurring in the porosity of the clay when glass is added to it. As a result some interesting conclusions concerning the expansive soil-material amendment are drawn.
Keywords: clay, swelling, glass, soil-additive

1. Introduction

En génie civil, le phénomène de retrait –gonflement de certains sols argileux et des formations géologiques argileuses induisent des tassements différentiels qui se manifestent par des désordres affectant principalement les structures construites en surfaces (bâtiments, fondations superficielles, ouvrages de soutènement, remblai…). Les exemples de désordres liés à la présence d'argile gonflante sont nombreux et variés.

Malgré ces inconvénients, les propriétés des argiles gonflantes; faible perméabilité, potentiel important de rétention, et bonne capacité d' auto cicatrisation en font des matériaux très intéressants dans de nombreuses applications, en particulier, en géotechnique de l'environnement pour la construction de barrières ouvragées destinées à confiner des colis de déchets nucléaires stockés en profondeur, pour l'élaboration de noyau de barrages en terre. Cependant ces matériaux peuvent être soumis à des variations de température et contrainte mécanique et hydraulique importantes qui peuvent mettre en cause leur bon fonctionnement.

Les argiles gonflantes sont donc présentes dans de nombreux contextes leurs propriétés pouvant être à la fois profitables et ou dommageables. Il est donc important d'appréhender le comportement hydromécanique de ces matériaux afin de mieux maîtriser leur utilisation.

En Algérie qui est un pays à climat contrasté, plusieurs cas de désordres liés au gonflement ont été recensés, la raffinerie de In-Amenas, l'hopital N'gaous (batna) [1,2], les lignes de chemin de fer (Jijel) [3], … etc.

Plusieurs études ont été consacrées à ce phénomène afin de le maitriser. Néanmoins sa complexité n'a pas encore permis d'atteindre jusqu'à présent des solutions pouvant le résoudre.

De nombreuses relations empiriques ont été proposées pour caractériser l'expansibilité et la pression de gonflement lors de son hydratation. Ces relations expriment le gonflement sur la base de quelques caractéristiques du sol qui sont en principe simples à déterminer en laboratoire ([4]). Des études aussi nombreuses se sont intéressées aux facteurs intervenant dans le gonflement, ([4, 5, 6])

Un bon nombre de solutions a été proposé afin de résoudre ce problème. Parmi ces solutions, il y a les techniques de stabilisation des sols. Parmi ces techniques, on a la stabilisation chimique, thermique, et la stabilisation des sols par additifs tels que l'apport de chaux et de ciment.

Plusieurs auteurs ont étudié l'influence de l'ajout de sable sur le gonflement ([7, 8, 9, 10]). Ces études confirment le rôle positif qu'a le sable sur la réduction du gonflement des sols expansifs et ce, par sa capacité d'amortir le gonflement à travers les vides importants qu'il contient.

L'étude présentée ici a comme objectif d'analyser en laboratoire l'effet de la performance d'un ajout d'un matériau inerte similaire au sable -le verre-, sur le comportement d'un sol potentiellement gonflant utilisé dans des aspects différents.

2. Caractéristiques géotechniques des matériaux utilisés

Dans cette étude on a utilisé de la bentonite et des billes de verre.

2.1 La bentonite

La bentonite est un matériau argileux composé de plusieurs minéraux, principalement de la montmorillonite et d'un faible pourcentage de Kaolinite et d'Illite avec en général quelques traces de quartz.
La bentonite utilisée dans cette investigation provient de M'zila. C'est un matériau non traité, de couleur grise et se présente sous forme de poudre finement broyée. Le gisement de la bentonite de M'zila est situé au Nord-est de la ville de Mostaganem et à environ 10 Km à l'est du village Achastas.

2.1.1 Quelques propriétés de la bentonite

L'analyse granulométrique a nécessité un tamisage à sec pour la distribution dimensionnelle en poids des particules supérieures ou égales à 80 µm, et une sédimentométrie pour la distribution dimensionnelle en poids des particules inférieures à 80 µm. Les résultats montrent que cette bentonite a été broyée très finement, le pourcentage des éléments de diamètre inférieur à 80 µm est égale à 92% et les éléments de diamètre inférieur à 2µm sont évalués à 42%.

Ses limites d'Atterberg qui sont la limite de liquidité, la limite de plasticité sont: W_l = 187%, W_p = 47%. Son activité de skempton est de 3.33. On remarque la valeur très élevée de la limite de liquidité.
Cette bentonite a une grande affinité à l'eau. Elle est composée essentiellement de la silice (58%) et d'alumine (19%) qui sont des minéraux argileux favorisant le gonflement.

2.2 Le verre

Les billes utilisées dans cette étude sont en verre de forme sphérique de 1mm de diamètre.

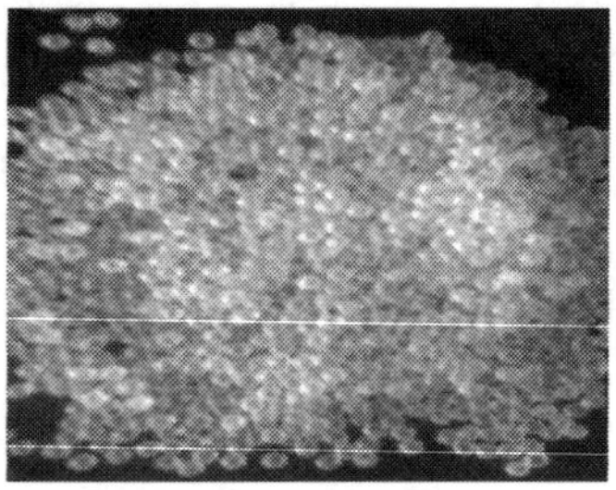

Fig. 1. Billes de verre utilisé

Des essais de compression à l'œdomètre effectués sur des billes de 5mm de diamètre, ont montré que les billes ne tassent pas jusqu'à ce que la charge appliquée dépasse la limite de résistance à la compression des billes provoquant leur cassure et leur dégradation.

3. Procédure expérimentale et tests réalisés

3.1La consistance

La consistance est un paramètre très important dans la reconnaissance d'un matériau, il nous informe sur son état et son comportement en fonction de son humidité.
Dans cette partie, on étudie l'influence de l'ajout de la poudre de verre sur la plasticité de la bentonite.
On a à déterminer les limites d'Atterberg (la limite de liquidité W_L, la limite de plasticité W_P et l'indice de plasticité I_P.) de plusieurs échantillons correspondant à des dosages différents de poudre de verre.
La détermination de ces paramètres s'est faite suivant la NF P 94-051.

3.1 Gonflement

Une série d'essais a été menée afin d'étudier la performance d'un ajout de billes de verre (d=1mm) sur les caractéristiques de gonflement de la bentonite. Les échantillons ont été préparés à la température ambiante à une même teneur en eau de 26.5 % et une même densité sèche 1.3 par malaxage et compactés statiquement et directement dans la trousse oedométrique à l'aide d'une presse manuelle.

3.2 Porosité

Le but de cette investigation est de tester la porosité des mélanges argile-poudre de verre à travers la vitesse de propagation des pulsations ultrasonores. On essaiera d'aboutir à quelques caractéristiques des pulsations liant le changement du milieu aux variations de la vitesse des pulsations ultrasonores.

Pour dérouler cette analyse, on a confectionné des échantillons cylindrique de bentonite seule et de bentonite + billes de verre de hauteur h= 140mm et de diamètre d=70 mm à une teneur en eau optimale (W_{opt}= 26,5%) et une densité égale à 1.5. Les échantillons ont été préparés par malaxage à l'aide d'un pétrin.

Les tests réalisés commencent par l'étalonnage de l'analyseur,

barreau de calibrage dont on connait les caractéristiques au préalable. Ensuite on mesure la vélocité d'un train d'onde ultrasonique, qui traverse les éprouvettes de sol testées

Pour garantir une bonne transmission des ondes dans le corps de l'éprouvette et avant d'ajuster le système de mesure, on applique des minces couches de graisses de contact sur les deux faces des transducteurs (émetteur et récepteur). Sur l'écran de l'analyseur sont représentés le temps ou la vitesse de transit de l'onde, cela selon la configuration de l'analyseur.

4. Présentation et discussion des résultats

2.1. Variation de la limite de liquidité

Les résultats de la variation de la limite de liquidité de la bentonite avec la teneur en poudre de verre sont représentés dans la courbe suivante.

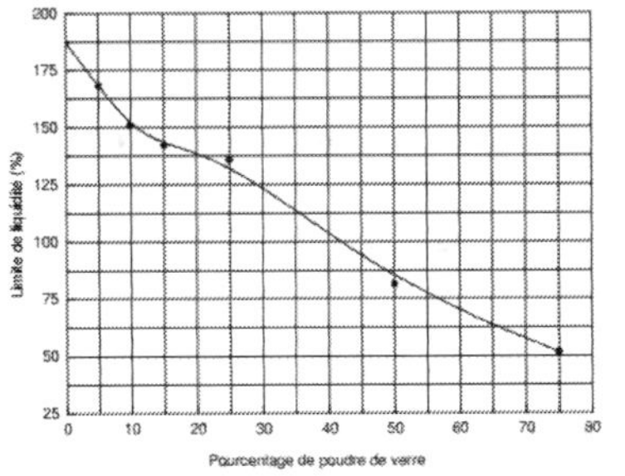

Fig.2. Variation de la limite de liquidité en fonction de la teneur en poudre de verre

En analysant ces résultats, on constate une diminution de la limite de la liquidité avec l'augmentation de la teneur en poudre de verre. Cette chute est d'abord très rapide pour de faibles teneurs en poudre, puis elle est régulière pour des teneurs moyennes mais moins importantes que la précédente et devient enfin très faible pour de très fortes teneurs en poudrette de verre des mélanges. Ces phases de variation peuvent être le résultat du changement du comportement des mélanges par ajout de matériaux inertes d'un sol argileux très plastique à un sol faiblement plastique pour de très fortes quantités de poudrette de verre ajoutée. Cette réduction est due à la diminution de la teneur en éléments fins actifs qui contribuent à la plasticité.

En outre, ces résultats montrent qu'il n'y a pas une liaison claire entre le dosage en poudre de verre et la limite de liquidité, ce qui dicte qu'il existe d'autres facteurs autres que celui de la diminution de la fraction argileuse, il s'agit aussi de la diminution de la surface spécifique d'adsorption d'eau et la réduction du nombre de liaisons entre les particules argileuses qui contribuent à la réduction de cette limite de liquidité.

2.2. Variation de la limite de plasticité

Les résultats de la variation de la limite de plasticité de la bentonite avec la teneur en poudre de verre sont représentés dans la courbe de la figure 3 suivante.

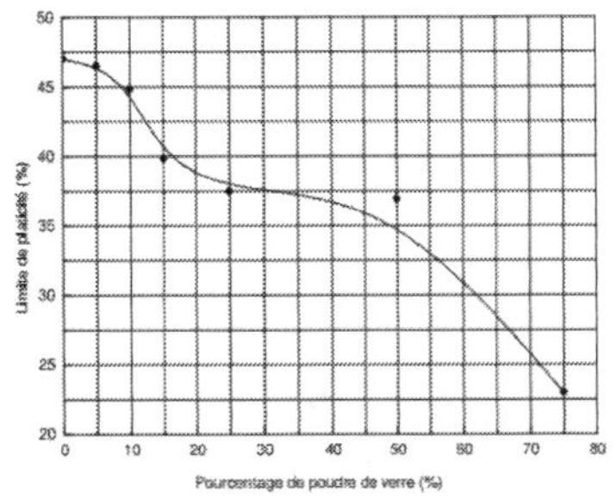

Fig.3. Variation de la limite de plasticité en fonction de la teneur en poudre de verre

On constate de cette courbe une diminution de la limite de la plasticité avec la teneur en poudre de verre. Cette réduction est serrée comparée à la limite de liquidité, mais elle est aussi due à la diminution de la teneur en éléments fins actifs et l'augmentation de la quantité du matériau inerte. Elle confirme aussi le changement de comportement du matériau à très fortes teneurs en poudrette de verre.

2.3. Variation de l'indice de plasticité

Les résultats de la variation de l'indice de plasticité de la bentonite avec la teneur en poudre de verre sont représentés dans la courbe suivante.

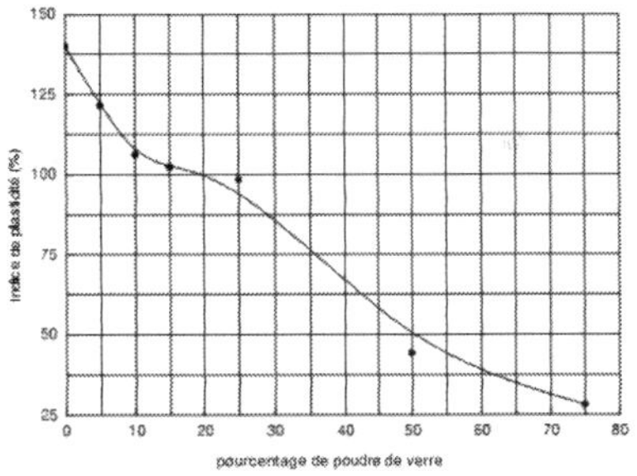

Fig.4. Variation de l'indice de plasticité en fonction du pourcentage en poudre de verre

La courbe de la figure 4 montre une diminution de l'indice de plasticité avec la teneur en poudre de verre. Cette réduction est due à la diminution des limites de liquidité et de plasticité. Son taux de réduction a été important du fait que la diminution de la limite de liquidité est plus importante que celle de la limite de plasticité, et puisque l'indice de plasticité n'est rien d'autre que la différence entre ces deux limites, alors sa décroissance a été importante.

$$W_L \searrow \text{ et } W_P \searrow \Longrightarrow I_P = W_L - W_P \searrow$$

Cette diminution de l'indice de plasticité engendre la réduction de l'étendue du domaine plastique, ce qui confirme le passage d'un matériau très plastique vers un matériau moins plastique et moins gonflants par conséquent.

4.1 Gonflement

Les résultats des tests effectués dans cette investigation, donnée dans les figures 5, 6, 7 et 8 appellent les observations suivantes : On constate que les deux paramètres de gonflement diminuent avec le pourcentage de billes additionné

Le taux de réduction de la pression de gonflement est de 24%, 66% et 94% pour dosages de 25% ,50% et 75% respectivement., On constate aussi que la variation du gonflement avec la teneur en verre n'est pas linéaire, montrant que la diminution du gonflement n'est pas que le résultat de la diminution de la fraction argileuse, d'autres facteurs contribuent à cette réduction qui est la réduction des liaisons bentonite-bentonite engendrée par le surplus de billes et la dispersion des particules de bentonite dans le mélange.

En revanche, la figure 8 montre l'évolution de la teneur en eau finale avec la teneur en billes de verre qui est due à la réduction de la surface spécifique globale d'adsorption d'eau. On remarque que le taux de réduction n'est pas aussi important comparée à celui de la pression et le potentiel de gonflement, cela est dû à la participation des billes de verre dans la préservation de l'eau par adsorption.

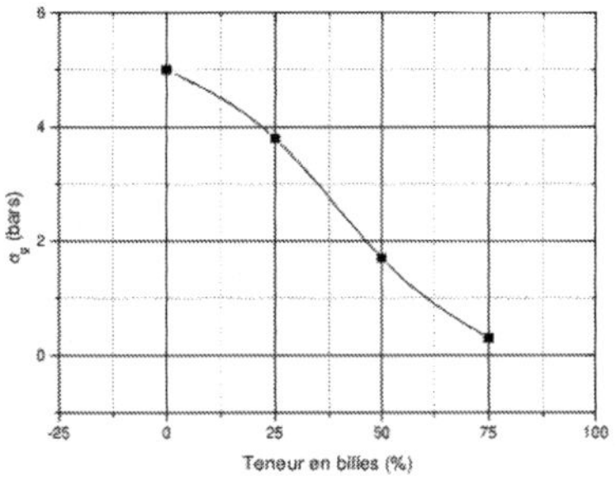

Fig.5. Variation de la pression avec la teneur en billes de verre

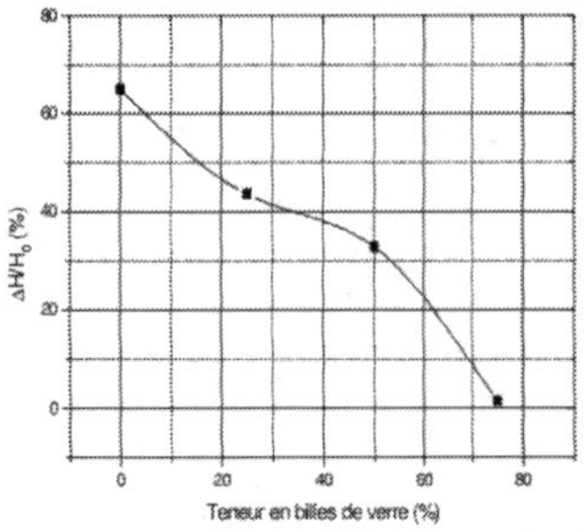

Fig.6. Variation du taux de gonflement en fonction de la teneur en billes de verre

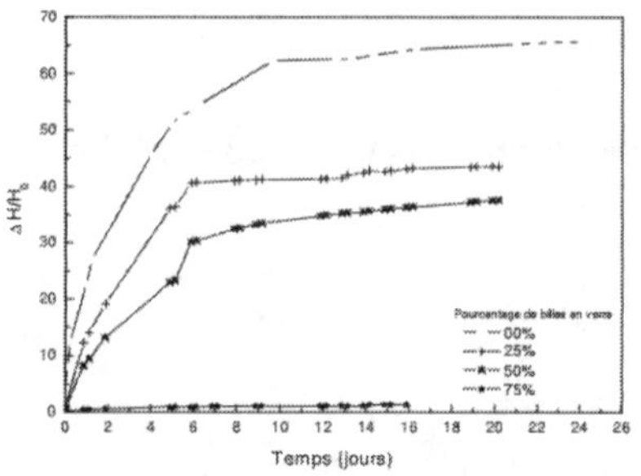

Fig.7. Evolution de la cinétique de gonflement de la bentonite seule et des mélanges bentonite - billes en verre

Les courbes de gonflement en fonction du temps de la bentonite mélangée aux billes de verre sont représentées par la figure 7. On remarque que la tendance de la variation du gonflement dans le temps est la même pour les différentes concentrations. Ces courbes montrent un potentiel de gonflement plus élevé pour la bentonite seule et qui diminue avec l'augmentation du dosage en billes de verre des mélanges.

Sur chaque courbe, on distingue que l'évolution du potentiel de gonflement se produit en deux temps. Dans un premier temps le gonflement est très rapide, il démarre dès que l'échantillon est inondé. Cette première étape de gonflement correspond à un gonflement macroscopique régi principalement par la perméabilité du milieu et le gradient de succion. Dans un deuxième temps le gonflement devient lent et progressif (gonflement microscopique) jusqu'à atteindre la durée nécessaire à la stabilisation. Cette dernière se raccourcit avec l'augmentation de la teneur en billes de verre des mélanges. La deuxième phase de gonflement constituerait le processus d'hydratation progressive des minéraux argileux et des cations adsorbés, cette hydratation s'effectuant à partir de l'eau qui est disponible dans les pores. On remarque aussi que l'ajout des billes de verre réduit aussi le temps de stabilisation du gonflement des mélanges. Plus exactement la première phase qui correspond au gonflement primaire a tendance à disparaitre à 75%. Ce changement dans la cinétique du gonflement peut être dû en grande partie, au changement dans la perméabilité des échantillons. Celle-ci augmente suite au traitement par ajout de billes. Cette nouvelle structure du sol aide l'eau à s'infiltrer rapidement et remplir les gros pores.

En revanche, la figure 8 montre l'évolution de la teneur en eau finale avec la teneur en billes de verre qui est due à la réduction de la surface spécifique globale d'adsorption d'eau. On remarque que le taux de réduction n'est pas aussi important comparée à celui de la pression et le potentiel de gonflement, cela est dû à la participation des billes de verre dans la préservation de l'eau par adsorption.

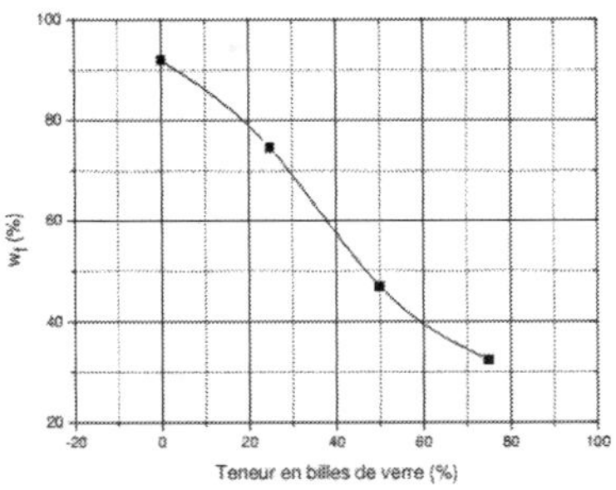

Fig.8. Variation de la teneur en eau finale en fonction de la teneur en billes de verre

4.3 Evolution de la vitesse de propagation des ondes ultrasonores

La figure 9 montre la diminution de la vitesse de propagation des ondes ultrasonores par ajout de billes de verre, pour devenir très faible et incontrôlable à des teneurs élevées. On signale qu'à 50% d'ajout, le sol étudié devient très poreux et la vitesse devient très variable, l'appareil affiche en effet plusieurs valeurs de vitesse. Ces dernières sont tous faibles tendant vers zéro.

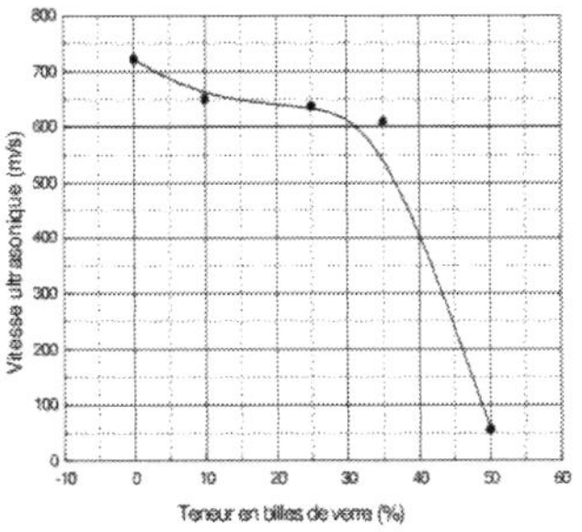

Fig. 9. Vitesse et teneur en billes de verre

Cette variation dans le sens décroissant exprime l'augmentation de la porosité des mélanges par ajout de billes en verre. Ce résultat permet de mettre en évidence la variation dans la structure du sol traité et de répondre au pourquoi de la diminution du gonflement par ajout de ce matériau.

5. Conclusion

La présente étude comporte des essais géotechniques standards, a étudié comment agit l'additif (le verre) pour réduire le gonflement d'une bentonite. Elle a permis d'asseoir des conclusions intéressantes qu'on résume dans ce qui suit.

1. La poudre de verre a un effet réducteur sur tous les paramètres de consistance. Cette réduction est élevée pour de faibles teneurs en poudre et diminue pour des teneurs élevées.
2. Le potentiel et la pression de gonflement de la bentonite traitée avec billes de verre diminuent avec le dosage en billes. Cette réduction semble être plus importante que le dosage ajouté. Le non linéarité de la diminution est dû à la diminution au départ la fraction argileuse par ajout de la poudre du verre et à la diminution de la densité de l'argile.
3. Enfin l'analyse globale de la porosité par propagation de pulsations ultrasoniques, a montré une diminution de la vitesse. Cette variation dans le sens décroissant exprime l'augmentation de la porosité des mélanges par ajout des billes de verre. Ce résultat met en évidence la diminution de la densité de l'argile par ajout de billes de verre.
4. Ces constatations confirment les relations déjà établies entre la consistance, la porosité et le comportement des sols vis-à-vis du gonflement. Il en ressort que la réduction du gonflement d'un sol est engendrée par la rupture des liaisons entre les particules argileuses et la création des pores suite à l'incorporation du matériau d'amendement qui est le verre.
5. Les résultats obtenus encouragent à approfondir l'étude et à réfléchir peut-être à d'autres techniques pour lutter contre le phénomène de gonflement des sols, par exemple, partant d'un esprit écologique, l'exploitation des déchets industriels dans la stabilisation du gonflement des sols.

References

[1] Z. Derriche et M. Kebaili, "Prévision du gonflement des argiles d'In Amenas," *Bulletin des Laboratoires des ponts et Chaussées*, pp. 15-23, (1998)
[2] A. Hachichi and J. M. Fleureau, "Caratérisation de Quelques Sols Gonflants d'Algérie," *Revue Française de Géotechnique*, No. 86, pp. 37-51, (1999).
[3] M. Lamara and A. Merah, "Analyse d'un Cas Pat logique de Gonflement de Sol (Université de Laghouat)," *Journée d'Études sur les Sols Gonflants*, Tlemcen, 27-28 Octobre 2002, pp. 101-110.
[4] M.A. El-Sobhy & A. El-Sayed '*'Some factors affecting swelling of clayey soils''*. Faculty of Engineering Department, University of Al-Azhar, Cairo, Egypt (1981).
[5] F.H. Chen, *Foundation on expansive soil. Elsevier edition. (1988)*
[6] A. Djedid, and N. Ouadah,, *Paramètres d'état et gonflement des sols argileux, exemple de la teneur en eau et de la densité sèche initiale*. Colloque international sols non saturés et environnement. pp.436-444 (2009)..
[7] B. Louafi,, Analyse de Quelques *Mécanismes physiques de la stabilisaton des sols expensifs par additifs à aspect inerte*. Thèse de doctorat , Université des Sciences et la Technologie Houari Boumediènne, Alger, Algerie (2013).
[8] H.A. Alawadiji,. Swell and compressibility characteristics of sand-bentonite mixtures inundated with liquids. Applied Clay Science, 15(3-4), 411-430 (1999).
[9] L. Bengraa , A. Hachichi A., S.A Bourokba., & A. Benaissa, *Etude de la stabilisation des argiles gonflantes par ajout de sable de carrière*. 2éme Journée d'études sur les sols gonflants. Université de Tlemcen (Algérie), pp. 101- 112 (2005).
[10] B. Louafi & R. Bahar, *SAND: An Additive for Stabilzation of Swelling Clay Soils,* International Journal of Geosciences, 2012, Vol.3.pp.719-725 (2012).

Contribution of geophysics for the characterization of ore deposits, case of the Bouarfa mine, Morocco.

Bahi Anas, Akhssas Ahmed, Ouadif Latifa, Ahmed Ait Errouhi and Kawtar Ouatiki
Laboratory of Applied Geophysics, Geotechnics, Engineering Geology and Environment, Mohammadia School of Engineers, Mohammed V University, Rabat,Morocco

Abstract. Geophysical methods are one of the best ways to determine the potential of a mining area. Indeed, for the prospection of metalliferous deposits, the magnetic method remains essential as it makes it possible to characterize the extent and the depth of the anomalous bodies present in the prospected zone,mainly due to their reaction with the values of the total magnetic field. Thus, for the manganese mining zone of Bouarfa and more exactly Hammraouet, the genesis of the deposit is such that the manganese is generally associated with ferruginous minerals and clearly affect the total magnetic field. That said, the raw results obtained must imperatively be subjected to corrections to only keep the effect of the anomalous bodies on the magnetic field. Hence, we must proceed with the correction of the daytime effect then we eliminate the regional effects to get the map of the residual anomaly and finally we perform a reduction to the pole that will allow us to have well individualized and unipolar anomalies and thus allowing us to locate precisely the potentially mineralized axes. The synthesis of the various maps obtained by spatial interpolation of the data recorded in situ and corrected allows us to locate areas of mining interest and to determine the different tectonic phenomena that took place in our study area. In time, these maps will serve us to guide the subsequent mechanical surveys to confirm the results obtained by geophysical methods.

Introduction

Geophysical methods have become unavoidable for anyone interested in medium and deep geology. Even more so today for the mining geologist, since the easiest deposits to discover have already been and because we are seeking deeper and deeper depths. Thus, in the case of metalliferous deposits, the so-called geophysical methods of potential constitute the most interesting means for their characterization. As a matter of fact, the magnetic method is the most appropriate tool especially when we are dealing with a Manganese mineralization, because it is usually associated with iron. This makes it possible to measure the total magnetic field and thus calculate the effect that the presence of useful ore can have on said field. Hence, to highlight this phenomenon, we focused our interest on the mining area of Bouarfa located in eastern Morocco, north of Bouarfa, at a faulty corridor with an East-West direction.

The Manganese mine in Bouarfa has long been considered one of the largest in Morocco since 1912, reaching more than 2,000,000 tons in 1967 when the production stopped. This sudden halt could be explained by the fact that ore prices had fallen and that exploitation was becoming increasingly difficult given the means implemented at that time.

This research project proposes to lay the foundations necessary for resuming exploitation of this deposit by conducting a geophysical campaign at the formerly operated Hammraouet area. This campaign will be preceded, as usual, by a geological study of the area in question, which will take into account both the various description works carried out during the exploitation of the deposit and the observations in situ.

Therefore, first, we will describe the geological and mining context of our study area and then look into more details about the geophysical prospecting methods used and the extent of work done on the ground. Finally, a synthesis of the different results obtained will be carried out to determine the structural scheme of the mineralization in the Hammraouet mining area.

I- Geological and mining context

The Bouarfa manganese deposit is located in the province of Figuig two kilometers north of the town of Bouarfa as shown in the processed radar satellite image in Figure 1.

https://doi.org/10.1051/matecconf/201814902076

Fig.1:Radar satellite image locating the study area

The study area thus belongs to the northern border chain of the high eastern atlas, it is bordered on the north by Jbel Bou Arfa and on the south by Jbel Bou Mokhta as represented on the extract of the geological map of the region in figure 2. It is a pleated structure dated from the Jurassic until the Cretaceous which ensures the junction between the Central High Atlas and the Saharan Atlas in Algeria. Its extensive tectono-sedimentary style in tilted blocks has generated a structure in wrinkles and depocentres.

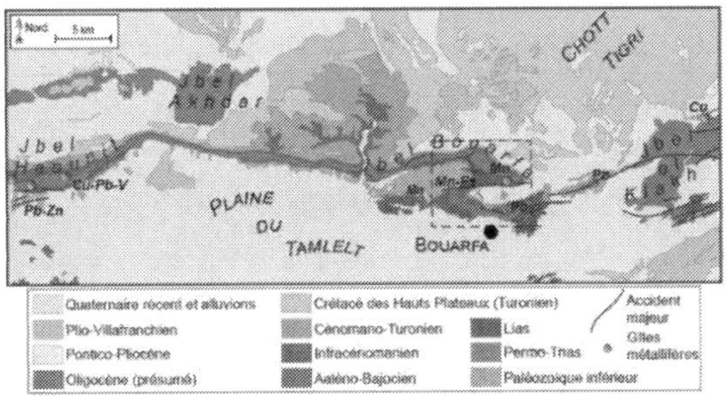

Fig.2:Geological map of Tamlelt [1]; 1/200 000, Study area is located in the frame.

The stratigraphic log shown in Figure 3 shows that the Cambro-Ordovician is confined to the Hamda formation which is outcropping to the west of our study area. It is mainly composed of quartzo-phyllades, grauwackes and some veins of siderosis.
Then there is the basic formation which is deposited in major discordance on the Cambro-Ordovician in the form of conglomerates of all kinds as well as some limestones in platelets.
The main area of interest that concerns us in terms of this unit is the buttonhole Hamda-Hammraouet. In this buttonhole, we distinguish the formation of Bou Mokhta dated lower Lias which contains three more or less distinct members as represented in figure 4.
This formation begins firstly, by the arrival of the first carbonated levels marking the liasic transgression, by a marine regime having experienced phases of emersion, then on the other hand, by a lagoon deltaic epicontinental environment, marked by evaporites deposits and by a rather pronounced phase of dolomitization. The age of the formation is attributed to the lower Lias (Hettangien - Sinemurian), its power is variable from West to East and towards the South, and it varies from 50 to 150m [2].
The first member of the Bou Mokhta Formation is characterized by the presence of carbonate levels in the form of small irregular brownish to black centimetric beds marking the first Liase marine transgression on the Basic Formation and characterizing the lower limit of this member. The upward sedimentation of the limb becomes more carbonated and the power of the layers becomes larger to reach decametric values. At the top of the limb is a succession of two limestone benches forming a bar of 4 to 5 m of power nicknamed thanks to its color "chocolate bar". This "chocolate bar" is characteristic and clearly visible in the landscape, it is the upper limit of the Member. The total power of the Lower Member is 8 to 20 m, it can reach 50 m in the South-West of the study area.
The lower limit of the Intermediate Member begins at the top of the "Chocolate Bar", a new sequence begins with withdrawals and returns from the sea which results in a detrital terrigenous sedimentation. Then calcareous banks are inserted, in which a lagoon environment predominates and results in a deposit of gypsum in metric lenses. Towards the summit of this limb, reddish, oolitic, bioclastic, and dolomitic limestones alternate. The power of this Member is reduced to a few meters at the northwest end, to the Southwest, it can reach 50 m.
The upper limb "Hammraouet" finishes the formation of Bou Mokhta, it marks the landscape by its brown to blackish cliff 8 to 15 m high, and its lower limit begins with argillites and limestones. This member consists of two bars, the lower bar of 10 to 15m of power, consisting of dolomitic limestone beds. Towards the top of this limb, the second, thinner bar settles through stratification joints in the form of reddish to brownish argillites and dolomitic limestones with biological laminae. These two bars together form the "Hammraouet Bar".
Ensuite, on retrouve la formation d'Ain Beida datée du lias inférieur moyen formée par une alternance de calcaires et d'argilites avec des minéralisations de Mn sous forme de canaux et de lentilles plus ou moins étendus.
Next, we find the Ain Beida formation dated from the lower middle lias formed by an alternation of limestones and argillites with mineralization of Mn in the form of channels and lenses more or less extended.
The Middle Lias happens subsequently through training Bouarfa with outcrops in the form of alternating conglomerates and shales rich in granite elements. It should also be noted that the layer of arkoses is quite important in this formation.
Finally, the slab of the high plateaus overhangs our study area in the north with a succession of dolomitic benches with bioclastic limestones and oolitic baselimestones.

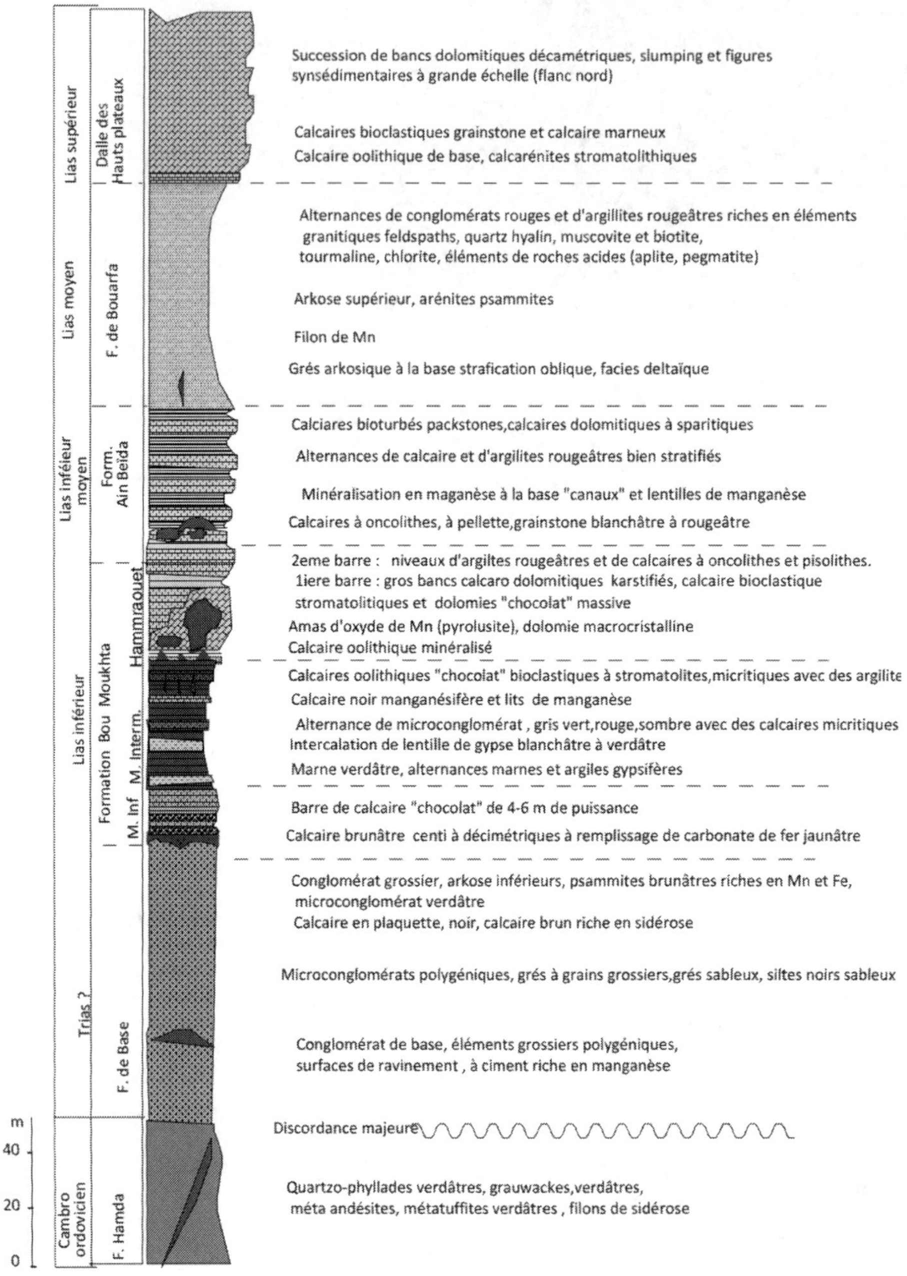

Fig 3: Stratigraphic log of the study area.

https://doi.org/10.1051/matecconf/201814902076

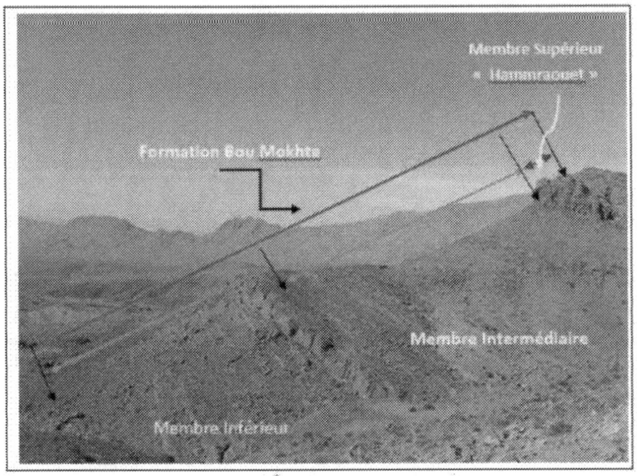

Fig.4:Schematic representation of the Bou Mokhta Formation

The mineralization at the Hammraouet zone is a substitution type in the chocolate dolomite layers and in direct relation with tectonics. Pouit and Jouravsky [2] report the presence of hausmannite, polianite (former name of pyrolusite or nsutite), pyrolusite and manganite. Several Fe mineralizations are also notable. The polianite is covered with a layer of goethite with inclusions of polianite. Thus at Hamaraouet, Fe is mineralogically separated from Mn [3]. This type of mineralization is the second largest in the sector after Ain Beida. It is arranged in metric heaps very rich in Mn [4-5].

Figure 5 shows the simplified structural scheme which synthetizes the geological context.

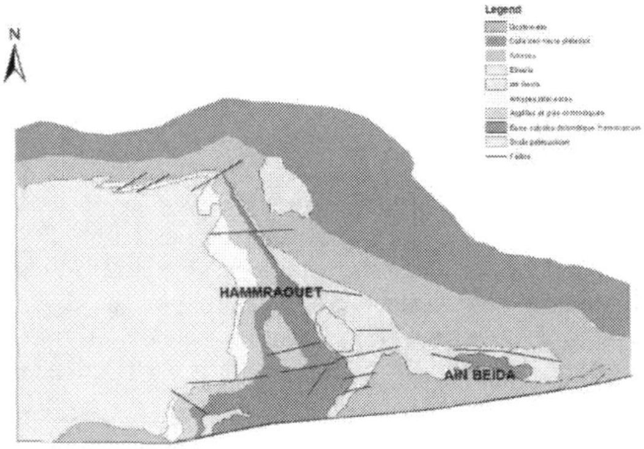

Fig.5:Simplified structural scheme of the Bouarfa Domain

II- Geophysical prospecting method and work performed

Method characteristics:

Given the nature of the mineralization and its genesis, the magnetic prospecting method has emerged as the most interesting method for recognizing the Manganese deposit at Hammraouet [6].The magnetic method is based on the measurement of the total field, which in the presence of manganese ore, in our case study, associated with iron oxides, will undergo a disturbance created by the contrast of magnetic susceptibility between the mineralization and the host [7-8].Magnetic prospecting thus consists of searching for and locating rocks, formations and deposits by anomalies or local variations that they produce in the terrestrial field [7-8].

Work carried out in situ:

The measurements carried out on our study site consisted of:

- 6 magnetic profiles spaced 25 to 50 m depending on the difficulties posed by the rugged terrain.
- A cumulative coverage distance of 3.48 linear km.
- Measuring stations spaced 25 m apart.

The details of this work are shown in Figure 6.

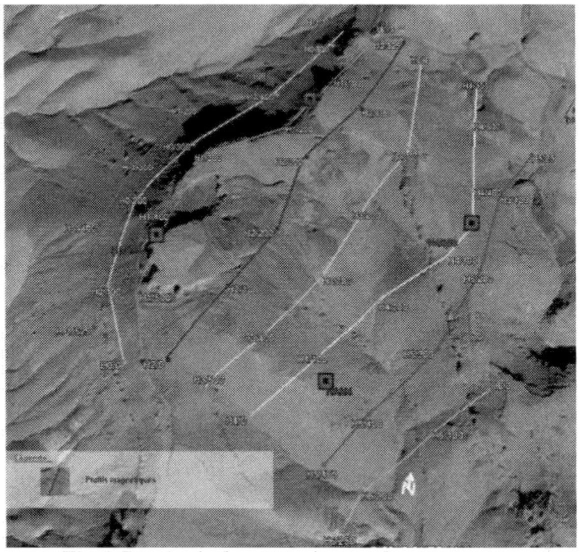

Fig.6:Geophysical prospection work map.

Data correction:

In situ data are subject to the effects of regional magnetism and other distortions that may affect the distribution of magnetic anomalies and their equilibrium. Thus, it is mandatory to correct the values of the measured magnetic field in order to obtain values representative of the magnetism of the structures in place.

The first correction to be applied is due to the diurnal variations of the magnetic field, which are themselves the result of variations in the current in the ionosphere in view of the solar activity.

Then, we apply the correction of the regional field. The measured total field values are affected by regional effects. Thus, these values will be filtered so as to obtain a map of the residual magnetic field.

Once the map of the residual field obtained, the anomalies which appear there are not necessarily localized within the plumbs of the different structures and are in most cases asymmetrical and bipolar. These distortions are due to the effects of the directions of the inductive field and the magnetization, which are not always vertical. The reduction at the pole is a transformation that makes it possible to put the anomalies back in line with the structures. Its calculation requires knowing not only the direction of the inductive field, but also that of the magnetization which is generally that of the magnetization induced and thus parallel to the current field. [9-11]

III- Results and discussions

The results from this magnetic study lead to the production of iso-gamma maps more commonly known as the total field map (see Figure 7). This will show us the gradients of magnetic anomalies that will inform us about the probability of existence of the mineralized zones and plumbs.

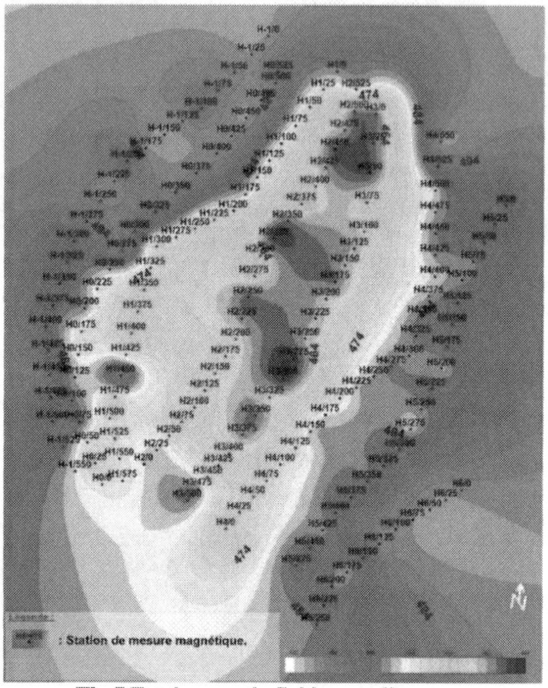

Fig.7:Total magnetic field anomalies map.

This map of the total field is the result of the graphical representation of the values of the measured total field to which the correction due to diurnal variation has been applied.

The appearance of the magnetic anomalies that appear in this map depends essentially on:

- The direction of the body relative to magnetic North.
- The magnetic susceptibility of the material forming the desired body.
- Geometry and extent.
- The depth.
- The lithology of the bedrock.

In our case, we notice the presence of a large minimum traversed by the profiles H1, H2, H3 and H4; surrounded by maxima that are located individually on the profile H1 and south of H3. This allows us to say that we are dealing with NW-SE oriented bipolar anomalies.

In order to eliminate the effects of the regional magnetic value on the measured gross values, which disturb the geometry of the anomalous zones and even sometimes cause false anomalies.

To do this, we subtract the values of the regional magnetic field from those of the Total field. This results in the creation of the residual magnetic map (see Figure 8). It represents the local magnetic effects and shows all the anomalies caused by the local existence of substances of different magnetic susceptibilities.

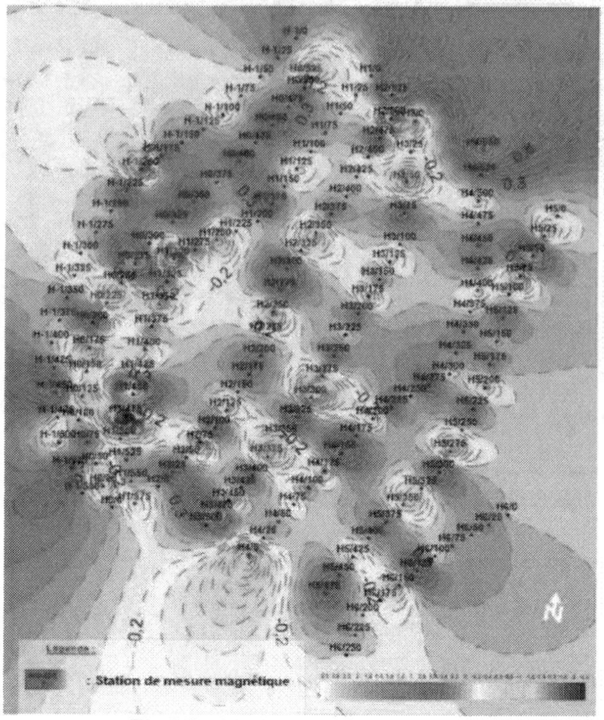

Fig.8:Residual anomalies map.

The observation of the residual map shows five alignments of bipolar anomalies intersecting perpendicularly all the 6 profiles. These alignments are parallel to the direction of the limestone bar of Hammraouaet. In the same way, each anomaly is discontinuous, which shows that we are dealing with discontinuous mineralization concentrations. To better reveal the plumb of suspected mineralized zones, we proceed to the reduction to the pole to obtain the reduced to pole magnetic map [9-11], (see Figure 9), which will allow us to have anomalies well Individualized and unipolar and thus allow us to locate with precision the potentially mineralized axes.

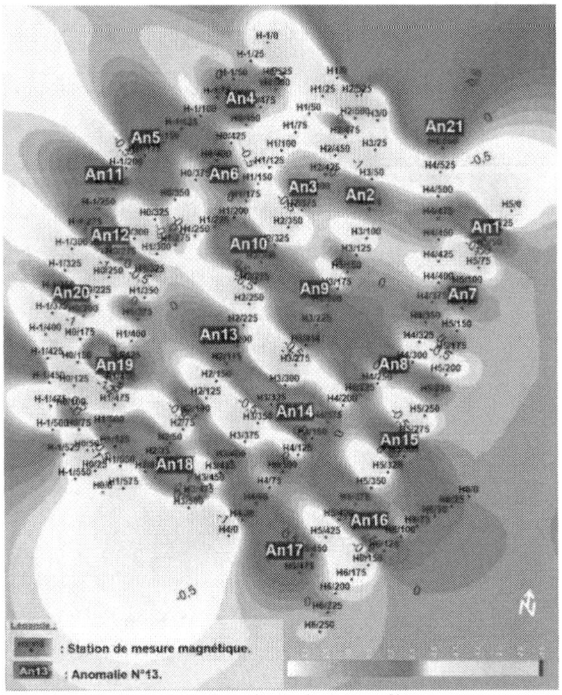

Fig.9:Reduced to the pole anomalies map

As in the residual magnetic field map, the magnetic map reduced to the pole shows a succession of discontinuous positive anomalies aligned NW-SE along five parallel axes.

The reduced to the pole magnetic map, reflects the existence of 21 generally discontinuous anomalies and distributed along NW-SE axes. These anomalies do not have the same volumes or the same gradient. The anomalies An14, An16 and An18 seem to be the most important with respect to their extensions and their magnetic gradients.

Most of these anomalies appear to consist of a single mineralized source, but perhaps due to earlier faulting effects, this source has been subdivided into several mineralized clusters. For example, the anomalies An14, An13 and An16 constitute the same cluster but it has undergone recoil forces.

The map in Figure 10 shows the probable faults that affected the potentially mineralized zones. The location of these faults will be of great importance in all direct research such as drilling and other prospecting galleries.

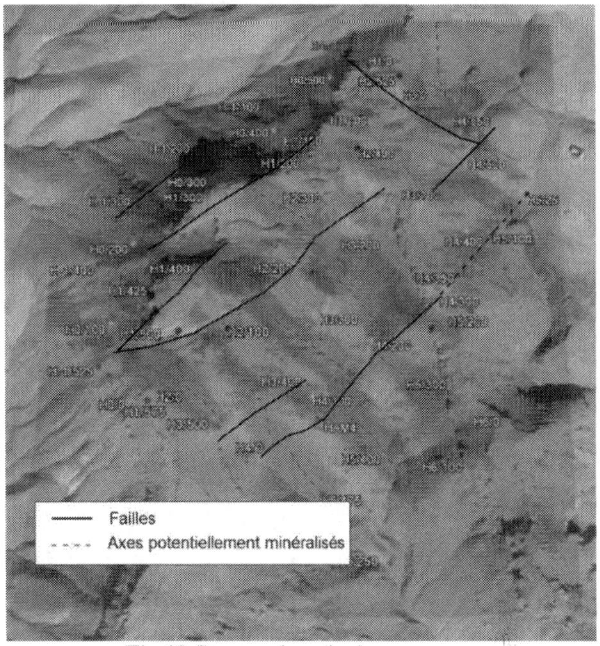

Fig.10:Structural synthesis map

Conclusion

The geophysical prospecting campaign using the magnetic method allowed us to highlight the different mineralized axes as well as the faults that affect them. This has helped us locating different areas of interest that will serve as a base to guide the mechanical survey campaigns which, in time, will give a clearer idea of the ore reserves in the study area.

That being said, magnetic study should be extended to the north and east of the initial study area to address other anomalies that may have eluded us.

In addition, the magnetic method alone cannot be sufficient to evaluate the economic potential of a deposit. Therefore, a complementary prospecting method must be used. We will be able to choose the grounding resistivity method to confirm the plumbs of the anomalies and the directions of the mineralized axes deduced from the magnetic survey.

Finally, it would be interesting to implement electrical soundings above the anomalies detected in order to have a clearer idea about the depths of the mineralized levels.

References:

1. **Du Dresnay R (1965)** Notice géologique sur la région de Bou-Arfa. Colloque sur les gites stratiformes de plomb, zinc, et manganèse du Maroc (2 mai - 14 mai 1962). Notes et mémoires du service géologique 181:107-112.

2. **Pouit G, Jouravsky G (1965)** Les minéralisations manganésifères de la région de Bou-Arfa. Colloque sur les gites stratiformes de plomb, zinc, et manganèse du Maroc (2 mai - 14 mai 1962). *Notes et mémoires du service géologique 181:113-131*

3. **Caillère S (1938)** Quelques observations sur la composition minéralogique et la structure des minerais de manganèse de Bouarfa. (Maroc oriental). Bull Soc Sc Nancy 2:221-250

4. **Ludovic Lafforgue.** Place de la minéralisation de manganèse de Bouarfa dans l'évolution mésocénozoïque de l'oriental marocain. Minéralogie. Université Paris-Saclay, **2016**. Français. < NNT: 2016SACLS277 >.

5. **Ludovic Lafforgue 2015**. Origin of the Bouarfa Manganese Ore Deposit (Eastern High Atlas, Morocco): Insights from Petrography and Geochemistry of the Mineralization.*MINERAL RESOURCES IN A SUSTAINABLE WORLD • 13th SGA Biennial Meeting 2015. Proceedings, Volume 5.*

6. **M. Jaffal1,** Interpretation of magnetic data of the Laachach gossan (central Jebilets, Morocco): mining implications. *Estudios Geológicos, 66(2) julio-diciembre* **2010**, *171-180*

7. **Debeglia N. (2005)** – Estimation de la direction d'aimantation pour une réduction au pôle optimale du champ magnétique. *BRGM/RP-54059-FR, 34 p., 15 fig.*

8. **Pierre Edouard Dousset, Jean Yves Tollardo.** Etude métallogénique du val d'Aoste (Italie): essai de valorisation des ressources minières par les méthodes de prospection géophysique, géochimique ou alluvionnaire. Minéralogie. Université Scientifique et Médicale de Grenoble, **1981**. Français. <tel-00749389>

9. **Baranov V. and Naudy H.; 1964:** Numerical calculation of the formula of reduction to the magnetic pole.*Geophysics, 29, 67-79.*

10. **Gunn P.J.; 1975:** Linear transformation of gravity and magnetic fields. *Geophysical prospecting, 23, 300-312.*

11. **G.R.J Cooper, D.R. Cowan (2005).**Differential reduction to the pole. *Computer and Geosciences, 31, 989-999*

Influence of seismic isolation systems and soil-structure interaction on the response of structures

Influence des systèmes d'isolation parasismique et de l'interaction sol-structure sur la réponse des structures

Hasrouri Samah[1] et Ounis Abdelhafid[2]
[1] Laboratoire de Recherche en Hydraulique Appliquée, LRHYA, Université Hadj Lakhder, Batna, Algérie

ABSTRACT. The reduction of cyclic loading triggering major damage in urban areas is a major challenge in earthquake engineering. The processes of structural control especially control structures for passive isolation systems and earthquake sinks of energy, which consists in superimposing on the structure a device which modifies the rigidity or the damping of the structural system without the demand for an external energy source and without introducing energy for its operation, these devices with taking account the effect of soil-structure interaction are currently regarded as effective solutions to these problem by reducing the level of acceleration imposed on the structure and consequently forces shear and the relative displacements in the superstructure. This reduction of shear forces and displacements will limit the structural damage.

1 Introduction

L'isolation sismique à la base est une méthode de conception parasismique qui consiste à introduire entre les fondations et la superstructure, des dispositifs qui ont une déformabilité horizontale très importante et une rigidité verticale très élevée. Ces dispositifs permettent à découpler le mouvement du sol de la structure dans le but de réduire les forces transmises à cette dernière. L'isolateur capte les déformations (inélastiques) et filtre les accélérations (hautes fréquences) de sorte que la superstructure isolée se déplace essentiellement selon un mode rigide subissant de faibles accélérations et presque pas de déformations.

Les isolateurs parasismiques en élastomère font partie des types d'isolateur les plus utilisés à cause de leur facilité de fabrication, d'installation et d'entretien. Toutefois, ces isolateurs présentent un comportement linéairement élastique et visqueux à grandes déformations de cisaillement dépendant de plusieurs facteurs qui varient durant la phase d'utilisation (température, déformation interne, fréquence de chargement...).

Il s'ensuit que, pour assurer la sécurité de l'ouvrage isolé, il est important de connaître l'impact de la variation de chacun des facteurs sur les propriétés mécaniques des isolateurs en caoutchouc.

L'objectif de ce travail est d'étudier l'influence des systèmes d'isolation parasismique à base d'élastomères comme moyen de contrôle passif sur la réponse sismique des structures avec tenir compte l'interaction sol-structure.

2 procédures de simulation numérique

2.1 Présentation du modèle

Une structure sera modélisée avec et sans prise en compte de l'effet d'interaction sol-structure en considérant l'isolation parasismique à la base de type LRB (lead rubber bearing), et l'analyse de l'influence de ces derniers sur le comportement des structures avec le logiciel SAP2000.

IL s'agit d'un portique en béton armé de 5,0 m d'ouverture " figure .1". Les poteaux de 3.14 m de hauteur, ont une section de 0,3×0, 4 m². Le plancher est constitué d'une dalle en corps creux de 16+4 cm d'épaisseur et dimensions en plan de 4,0×5,0 m. La structure repose sur des semelles superficielles en béton de section carrée de 2 m de côté.

 https://doi.org/10.1051/matecconf/201814902077

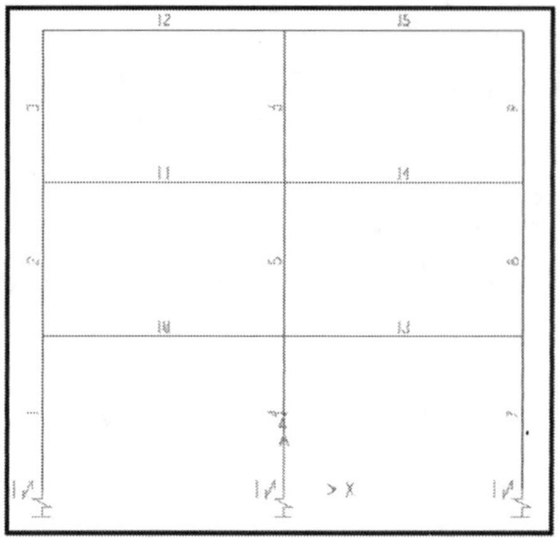

Fig. 1. Portique sur des appuis élastiques

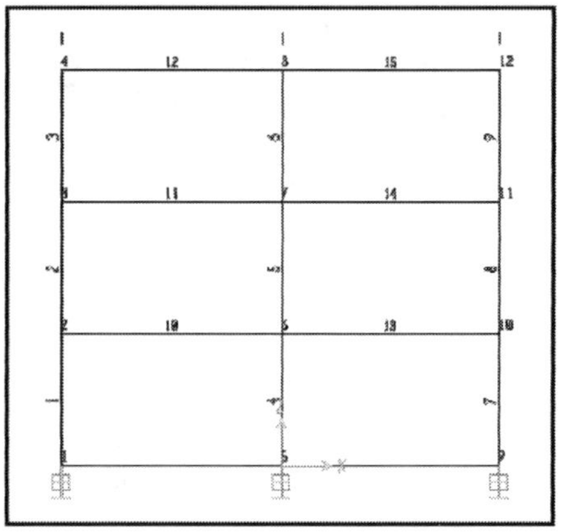

Fig. 2. Portique avec des isolateurs à la base(LRB)

2.2 Caractéristiques des matériaux

2.2.1 Caractéristiques géotechniques du sol

Pour l'étude de l'interaction sol-structure (ISS) [1]. Le sol est modélisé par un ressort horizontal, un ressort vertical et un ressort de basculement en chaque nœud (Capra et al. 84), (Oudjene et al. 01).

Le modèle ressort de sol " figure .1" définit les réactions élastiques de ce dernier à l'égard des composantes de déplacement statique de la fondation.

Les valeurs des raideurs des ressorts de sol sont calculées à l'aide des relations données dans le tableau 1.

Les valeurs des coefficients sans dimension β x,y, β z et βφ,ψ, du tableau 2 sont données par des abaques spécifiques en fonction du rapport des dimensions des semelles et la direction de l'action sismique considérée (Zacek, 96).

Le sol (sable) est supposé homogène avec un comportement élastique linéaire et ses propriétés sont résumées dans le tableau 2.

Table 1. Raideurs équivalentes pour un milieu semi-infini [1]

Mouvement	Semelle rigide rectangulaire
Horizontal x ou y	$K_{x,y} = \frac{E}{2(1-v^2)}\beta_{x,y}\sqrt{b.a}$
Vertical	$K_z = \frac{E}{2(1-v^2)}\beta_z\sqrt{b.a}$
Basculement	$K_{\psi,\varphi} = \frac{E}{2(1-v^2)}\beta_{\psi,\varphi}b^2\sqrt{b.a}$

Table 2. Caractéristiques des sols

Sites		Site meuble (sable lâche)	Site très meuble (sable très lâche)
Module de Young E(Mpa)		10	5
Coefficient de poisson v		0.5	0.5
Densité volumique (KN/cm3		17	19
Angle de frottement (deg)		35	35
Cohésion (KN/M2)		0	0
Paramètre sans dimension	βx, y	1.42	1.42
	Bz	2.12	2.12
	βψ,φ	1.65	1.65
raideurs des ressorts de sol (kn/m)	Kx, y	18930	11833.33
	Kz	28260	17666.66
	Kψ,φ	88000	85937.5

2.2.2 Caractéristiques de l'isolateur LRB

Les isolateurs à base d'élastomère " figure .2", modélisés par des éléments Link, ont été fournis à tous les supports des poteaux à la base. Les propriétés des Link ont été fournis par des calculs du modèle d'isolateur de base LRB.

Les calcule détaillé de l'isolateur de base " figure .3 et 4" ont été obtenus par les codes IBC2000 et IBC2006 pour finaliser la conception des isolateurs de base.

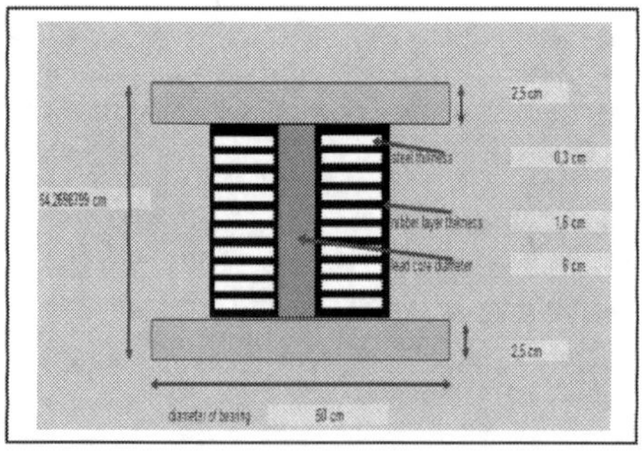

Fig. 3. Les isolateurs utilisés à base d'élastomère (Site meuble)

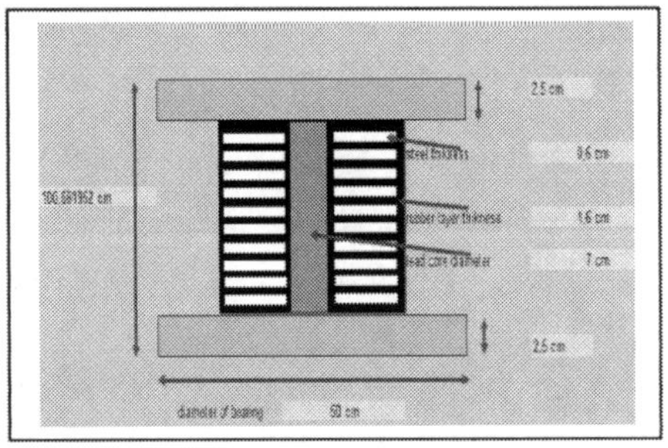

Fig. 4. Les isolateurs utilisés à base d'élastomère (Site très meuble)

Pour une analyse linéaire modale " figure .5"

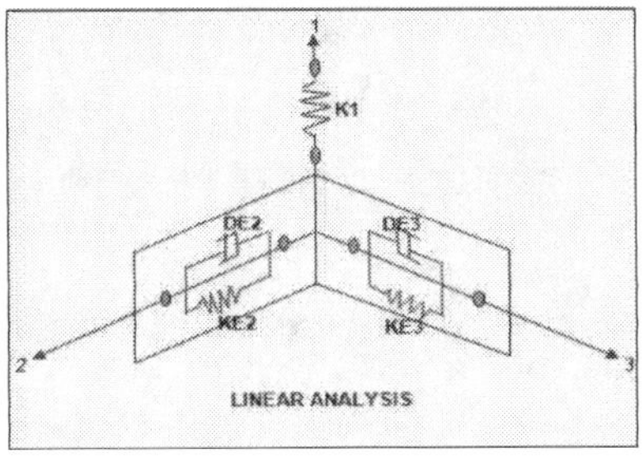

Fig. 5. Modèle du système d'isolation LRB pour une analyse linéaire

Le modèle utilisé pour caractériser le comportement hystérèse du LRB " figure .6" est exigé généralement la définition de trois paramètres à savoir :
La rigidité post-élastique Kp, la force élastique Fy et le déplacement élastique Dy [6].

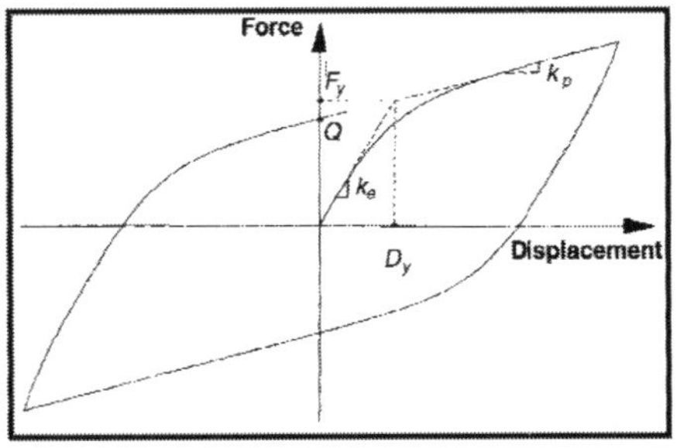

Fig. 6. Model Relation idéale hystérèse force-déplacement d'un appui en caoutchouc de plomb [2]

-La rigidité effective $Keff = Kp + \frac{Q}{D} \quad D \geq D_y$

-La rigidité élastique est approximativement égale à 6,5 à 10 Kp

-La période naturelle

$$T = 2\Pi\sqrt{\frac{W}{K_{eff}\, g}}$$

-Le taux effectif d'amortissement effectif

$$Beff = \frac{4Q(D - Q/9Kp)}{2\Pi(Kp\, D + Q)\, D}$$

-Le déplacement au centre de rigidité

$$D = \left(\frac{g}{4\Pi 2}\right)\frac{S_{d1}\, T_d}{B_d}$$

$$S_{d1} = \frac{2}{3} SM1$$

Où : g est l'accélération,
Sd1 est le coefficient spectral,
Td est la période isole,
Bd est le coefficient d'amortissement.

Et

$$S_{M1} = Fv\, S_1$$

Où

- S1 : est 5% d'amortissement d'accélération spectrale pour le site disponible sur les cartes accompagnant à IBC-2000.
- Fv : le coefficient de site défini pour les classes des divers sites et des niveaux d'accélération.

Les caractéristiques d'isolateur LRB [9, 10, 11] sont données dans le tableau 3.

Table 3. Caractéristiques d'isolateur LRB

Classe de site	Site meuble	Site très meuble
	D	E
Coefficient spectral Sd	0.56	0.89
K1=Kv (kN/m2)	687637,21	432670,61
KE2 (kN/m2)	350,86	350,45
DE2	0,0945	0,094

3 Résultats

3.1 Effet de l'interaction sol structure (ISS) et d'isolations parasismiques à la base (LRB) sur la valeur de l'effort tranchant à la base

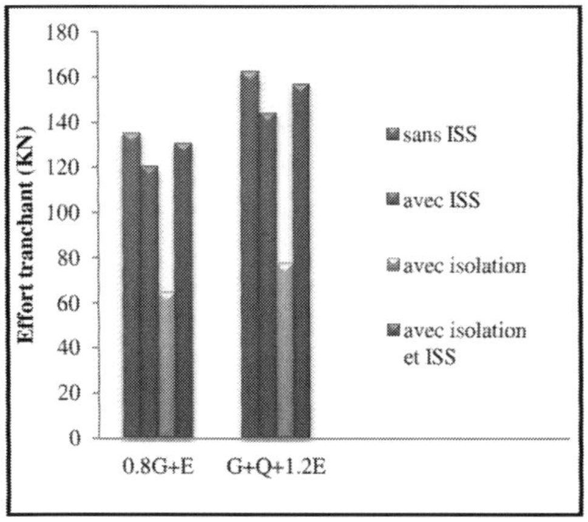

Fig. 7. Effort tranchant à la base sans et avec isolation et ISS des sites meubles

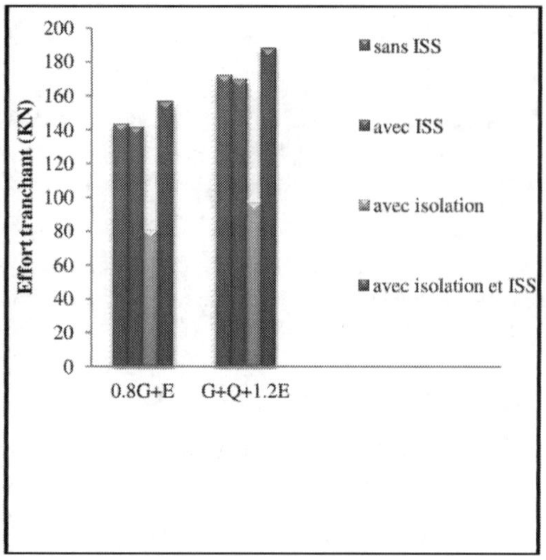

Fig. 8. Effort tranchant a la base sans et avec isolation et ISS des sites très meubles

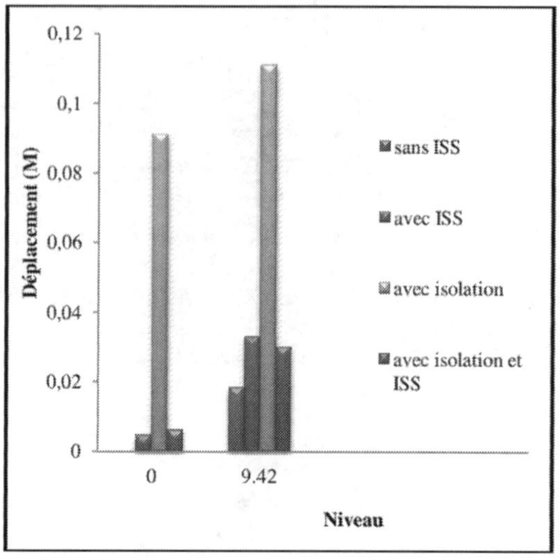

Fig. 10. Déplacements horizontaux avec isolation et ISS des sites très meubles

3.2 Effet d'isolation parasismique et ISS sur les déplacements horizontaux des nœuds

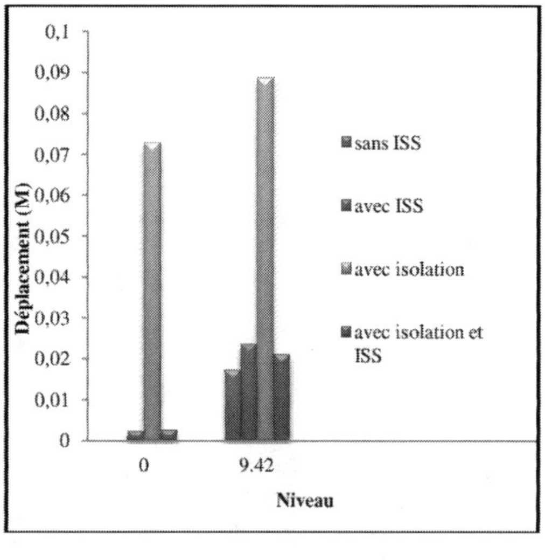

Fig. 9. Déplacements horizontaux avec isolation et ISS des sites meubles

3.3 Effet d'isolation parasismique et ISS sur les déplacements verticaux des nœuds

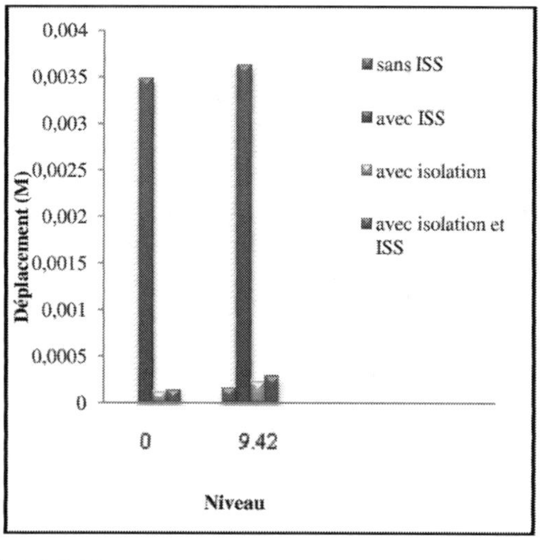

Fig. 11. Déplacements verticaux avec isolation et ISS des sites meubles

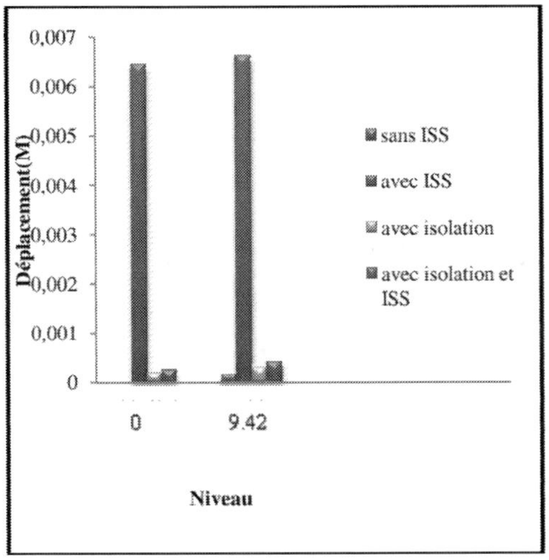

Fig. 12. Déplacements verticaux avec isolation et ISS des sites très meubles

4 Conclusion

Les conclusions principales tirées de la présente étude sont :

-La modélisation des structures avec prise en compte de l'effet ISS est le plus proche de la réalité car sa dépendance des caractéristiques des sols est plus importante.

-La modélisation des structures encastrées (sans tenir compte l'effet ISS) n'est pas à recommander car elle donne des résultats trop conservatifs.

-Les systèmes d'isolation parasismique (appui en caoutchouc à faible- amortisseurs avec un noyau de plomb (LRB)), peuvent être un moyen efficace pour limiter les effets dévastateurs d'un séisme sur une structure donnée

-les efforts tranchants de la structure munie de systèmes d'isolation et avec prise en compte l'effet ISS sont supérieurs à ceux de la structure soumise à l'effet d'interaction. Cependant les déplacements prennent des valeurs inferieurs.

-Les déplacements horizontaux des structures munies des systèmes d'isolation et soumis à l'effet d'interaction sont inférieurs à ceux des structures avec l'effet d'interaction.

-Les déplacements verticaux des structures et des fondations munies des systèmes d'isolation et soumis à l'effet d'interaction sont inférieurs à ceux des structures avec l'effet d'interaction sol structure.

-Les structures munies des systèmes d'isolation et soumise à l'effet d'interaction limiteront les dommages structurales des sites lâches et très lâches par la réduction des efforts de cisaillement et des déplacements.

Il est nécessaire de tenir compte de certains aspects actuellement non pris en compte dans le RPA 2003 :

1-Les règles RPA20003 doivent donner plus de détails concernant les différents aspects inhérents à une analyse de l'influence de l'interaction sol structure.

2-Les règles RPA20003 doivent donner des études spécifier concernant les différents systèmes d'isolation parasismique et pour l'utilisation de ces derniers doivent être adoptées.

3-Spécifier le type de structures pour lesquelles l'interaction sol-structure est importante et donner des directives pour les calculs.

4-le règlement algérien stipule qu'en cas de non disponibilité d'étude de sol, un site 3 sera pris en compte dans l'étude sismique, il est noté que cette considération conduit à un surdimensionnement coûteux plus que les frais qui seraient engagés pour une étude de sol. De ce fait il faudrait insister plus sur bonne étude de sol qui permet une économie et une sécurité.

5 Références

1. Alain Pecker .*Comportement des sols interaction sol-structure.*
2. Anil K. Chopra *University of Califomia at Berkeley, dynamics of structures theory and applications to Earthquake Engineering.*
3. Davidovici V. *La construction en zone sismique, Paris, Editions Le Moniteur* (1999).
4. Michael D. Symans, PhD, *seismic protective systems passive energy dissipation.*
5. Trevor E Kelly, S.E.Holmes Consulting Group Ltd Revision 0 (July 2001 DESIGN GUIDELINES).
6. F Naeim-*Design of structures with seismic isolation.*
7. A.Pecket, Dynamique des sols (Presse de l'école des nationales des ponts et Chausses 1984).
8. KHALIL Louay, SADEK Marwan & SHAHROUR Isam Influence de l'interaction sol-structure (ISS) sur la fréquence fondamentale.
9. (2006) *INTERNATIONAL BUILDING CODE®*
10. (2000)*IBC*
11. Paul W Meisel Member VISCMA, *Understanding the* (2000) *IBC Code* (Architectural Components and Equipment Restraint)
12. KINETICS™ *Guide to Understanding IBC Seismic for MEP*, required basic project information
13. *Règle parasismiques algériennes* (RPA99/version 2003), Ministère de l'habitat Centre national de recherché appliqué en génie parasismique.

Developed approach to determine the seismic vulnerability of reinforced concrete structures

SERRAYE Mahmoud [1], AMRI Salima [2]
[1] Geo-materials Laboratory, Civil Engineering Department, Faculty of Technology, Saad Dahleb university of Blida, ALGERIA.
[2] Faculty of Technology, university of Medea, ALGERIA

Abstract. Several evaluation methods of the seismic vulnerability have been developed around the world. Which are very use ful from humanitarian and socioeconomic point of view. Generally these methods use knowledge obtained from previous earthquakes and they are basing on seismic intensity scales and on buildings direct observation. But the macroseismic intensity expresses the consequences of the seism, and not its physical characteristics of the structures. Contrary to this type of methods, an approach based on a nonlinear analysis (Push-Over method) is proposed in this work. It consists in modeling the excitation of the earthquake by a response spectrum and building's behavior by capacity curves. These capacity curves are obtained from numerical modeling performed by Opensees software. The superposition of the two curves, response spectra and capacity curve, makes it possible to determine the performance point and consequently to deduce the state of expected damage. To estimate the probability of damage of a building at a given level of solicitation (defined by S_d), we excites a group of buildings characterized by different parameters related to the geometry of the building and those are related to the materials used (concrete, steel) by seismic solicitation (Response spectrum - RPA 99). The performance point for each building is determined by a procedure defined in FEMA 440. We classifies the buildings according to the position of performance point on their curve which defines a damage state of ds (Mild, Moderate, Important or Ruin) according to the damage levels of Risk-UE. A statistical analysis is then made for each class to build the fragility curves.

1. INTRODUCTION

Aujourd'hui, nombre important des modèles d'estimation de dommages sismiques ont pu être réalisés à travers le monde, permettant la réalisation des scénarii catastrophe, très utiles d'un point de vue humanitaire, social ou économique pour la gestion d'une éventuelle crise. Généralement les méthodes d'évaluation de la vulnérabilité sismiques actuelles correspondent à des approches basées sur l'utilisation des relations entre intensité macrosismique et les dommages post sismiques, telle que l'EMS 98, GNDT, RISK-UE "level I".... Mais la valeur de l'intensité macrosismique exprime les conséquences du séisme, et fonction de l'ampleur des dommages.

Contrairement à ce type des méthodes, l'estimation de l'état de dommages sur la base du comportement du bâtiment (courbes de capacité) prend directement en compte les paramètres liés au mouvement du sol, tels que l'accélération et le déplacement, en plus le comportement (réponse) des structures dépend fortement avec les dimensions géométries des constructions (hauteur d'étage, nombre d'étages, nombre portées aux deux directions et taille...). L'approche HAZUS représente la seule méthode présentant à ce jour une approche de ce type, son caractère exhaustif et notamment l'explicitation des relations utilisées ont imposé ce modèle comme une référence au niveau international, mais son accessibilité uniquement aux spécialistes, influer à son utilisation, reste limité (moins publique). Le projet européen RISK-UE "level II" a comme objectif le développement un modèle spécifique au contexte européen. C'est cette caractéristique de méthodes existantes qui a poussé la communauté scientifique à développer une approche s'appuyant sur la méthode de capacité spectrale (capacity spectrum method of ATC-40, 1996) [1].

Cette approche a été développée dans cette communication, basée sur des courbes de capacités spécifiques au contexte du bâti urbain algérien. Les courbes de capacités ont la propriété d'exprimer un niveau d'endommagement directement en fonction de la sollicitation sismique spécifique (spectre de réponse RPA99) à l'enjeu étudié. Dans ce qui suit les différentes étapes de l'approche développée.

2 PROCEDURE L'APPROCHE DEVELOPPEE

L'approche développée pour l'estimation de dommages repose sur la méthode push-over, elle se décompose en quatre étapes principales:

2.1 La première étape: construction les courbes de capacité

Une analyse non linéaire (analyse push over) des structures à l'aide une modélisation numérique est réalisée par logiciel Opensees pour construire les courbes de capacité. Les courbes tirant en terme effort tranchant (V) à la base en fonction du déplacement (D) au sommet. Ces courbes sont transformées en termes d'accélération spectrale *'Sa'* et de déplacement spectral *'Sd'* (voir figure 01et 02) par les relations [1] suivantes:

$$S_d = \frac{D}{\Gamma_1 \Phi_{r,1}} \quad (1)$$

$$S_a = \frac{V / W}{\alpha_1} \quad (2)$$

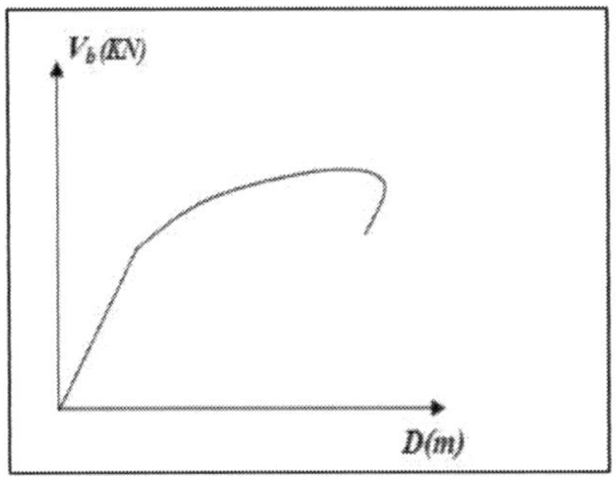

Fig. 1. La courbe de capacité au format V-D

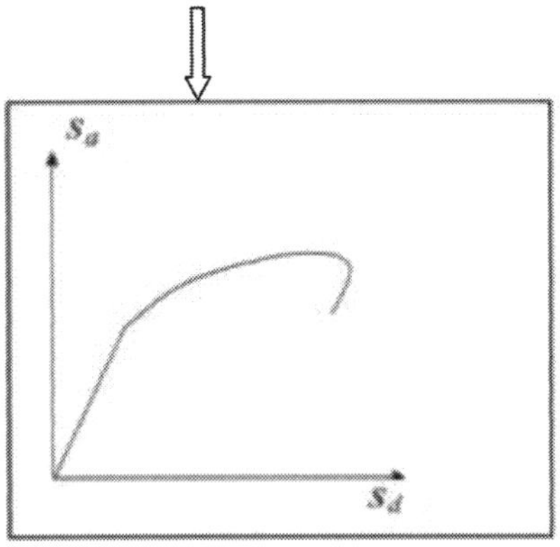

Fig. 2. La courbe de capacité au format Sa-Sd

Comme recommandé dans le document ATC40 [1], on adopte pour les actions sismiques appliquées à la structure de forme triangulaire et en supposant que déformée modale est normalisée à $\Gamma_1\phi_{r,1}$ =1.4, et les facteurs de charges appliquer lors d'une analyse Push-Over $\alpha = 0.8$ [1].

Avant la modélisation numérique elle consiste de caractériser les bâtiments études selon une classe typologique définie par des paramètres lié à la géométrie du structure:

- Nombres et hauteur d'étage,
- Nombres et longueur de travées en deux sens zz et xx,
- Section des poteaux,

Et par des paramètres lié aux matériaux (béton, acier) telle que :

- La résistance à la compression de béton (fc).
- Limite élastique de l'acier (fy).

Dans cette étape, nous avons générés les paramètres des structures par la simulation de Monte-Carlo.

2.2 La seconde étape : Choix de l'action sismique

Dans cette approche, la sollicitation du séisme est représentée par un spectre de réponse la norme RPA99 version 2003 [2] transformé en terme d'accélération spectrale Sa, et déplacement spectral Sd, (voir figure 3) selon les formules suivantes [1] :

$$S_{ai} = \left(\frac{4\pi^2}{T_i^2} S_{di}\right)/g \quad (3)$$

$$S_{di} = \left(\frac{T_i^2}{4\pi^2} S_{ai}\right) \quad (4)$$

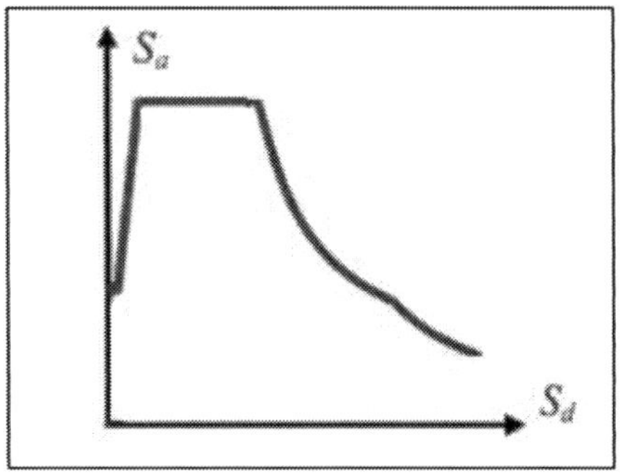

Fig. 3. Définition de l'action sismique.

Nous avons tenu compte des variabilités aux actions sismiques (spectre de réponse) est intégrée au effet de site S [2] et le facteur de qualité Q, (voir figure 04)

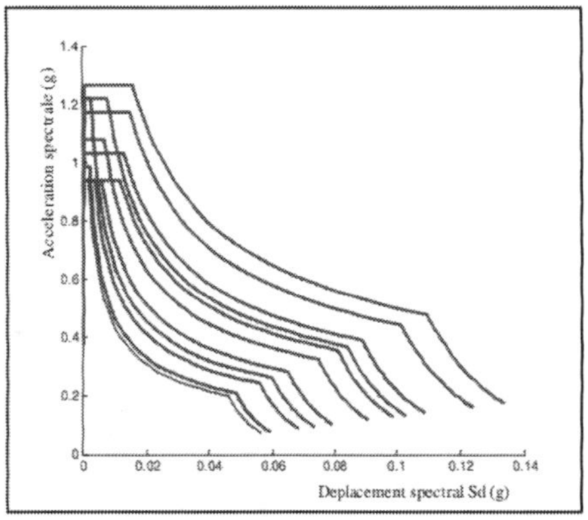

Fig. 4. Variabilités des spectres de réponse.

2.3 Troisième étape : Obtention des points de performance

Ils existent plusieurs techniques pour obtenir la valeur de point de performance de structure. Dans l'approche développée, les coordonnées des points de performance sont calculés à travers la superposition de la courbe de spectre de réponse au format ADRS et les courbes de capacité transformées au format (Sa, Sd) par une procédure itérative présentée au FEMA 440 (2009) [4].

En fonction la position le point de performance sur la courbe de capacité détermine un certain niveau de dommage. Ainsi, plusieurs niveaux d'endommagement peuvent être distingués à travers cette représentation graphique.

Exemple de position point de performance

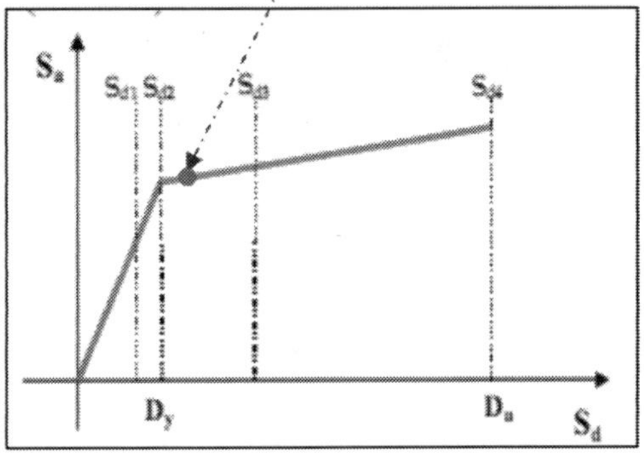

Fig. 5. Signification la position de performance sur la courbe de capacité.

Dans cet approche, on va utiliser les niveaux d'endommagement fournies par programme Risk-UE [3] qui définie quatre niveaux ou degrés de dégâts: faibles, modérés, importants et très importants. Les niveaux d'endommagements proposés par le projet Risk-UE sont présentés au Tableau 01 [3].

Tableau .1 Définitions des niveaux d'endommagement selon Risk-UE [3].

Damage Grade	Definition	Spectral Displacement Limits (UNIGE, CIMNE)
0	No damage	D < 0.7 Dy
1	Slight damage	0.7Dy ≤ D < 1.0 Dy
2	Moderate damage	1.0 Dy ≤ D < Dy+Duy
3	Extensive damage	Dy+Duy ≤ D < Du
4	Very heavy damage	Du ≤ D
		Duy = 0.25*(Du-Dy)

2.4 Quatrième étape : Construction des courbes de fragilité

Dans ce travail le modèle de distribution statistique retenu pour représenter les fonctions de fragilité de la structure est un modèle de type Log-normal qui représente convenablement la combinaison de variables dont les effets ont multiplicatifs (Chintanapakdee et Chopra, 2003). Cette fonction est contrôlée par deux paramètres: la valeur moyenne Sd et son écart type β_{Sd}. Sa densité de probabilité cumulée s'exprime par la relation suivante [5]:

$$P[ds|S_d] = \Phi\left[\frac{1}{\beta_{ds}}\ln\left(\frac{S_d}{\overline{S}_{d,ds}}\right)\right] \tag{5}$$

$S_{d,ds}$ est la valeur médiane du déplacement spectral pour laquelle la structure atteint le seuil de l'état de dommages ds ; β_{ds} est l'écart type du déplacement spectral pour l'état de dommages ds, et Φ est la fonction de répartition de la loi normale centrée réduite. La valeur médiane du déplacement spectral Sd est déterminée pour chaque niveau de dommage ds à partir des déformations inter-étages moyennes Δds est calculé par l'expression [1] :

$$S_{d,ds} = \frac{\Delta_{ds} \times H}{\Gamma_1} \tag{6}$$

Avec H hauteur de la structure et $\Gamma 1$ facteur de participation du premier mode, selon la forme des déformations du structures, le document ATC40 donne des valeurs simplifiés pour $\Gamma 1$, sur cette base en supposant que $\Gamma 1 = 1.4$ [1] qui correspondant notre cas.

3. EXEMPLE D'APPLICATION ET RESULTATS OBTENUES

Dans cette étude, nous nous intéressons à une seule typologie, c'est-à-dire nous avons fait le choix de traiter les constructions en béton armé type portique «poutres-poteaux». Ce choix de typologie est basé sur l'importance et la prépondérance de ces constructions dans le milieu urbain Algérien. Les structures étudiés ont été différenciées en fonction des paramètres liés aux géométries des structures, tel que le nombre d'étages "low-Rise" (entre 1 et 3 étages) ainsi que nombre de travée. Etc. des paramètres liés aux matériaux utilisés de la construction (béton, acier).

3.1 Construire les courbes de capacité

Les 1000 bâtiments étudies de classe typologique "low-Rise" (entre 1et 3 étages) et sont différenciées par caractères liés à la géométrie de structure (hauteur d'étage, nombre d'étages, nombre trames et section des poteaux …) et par des paramètres liés aux matériaux de construction.

Le tableau 2. Donne les différents paramètres lié à la géométrie de la structure et paramétrés lié au matériau de construction.

	Paramètre	*Nombre des bâtiments*	
Paramètres lies aux dimensions géométriques de la structure	*Nombres d'étages*	*R*	*130*
		R+1	*800*
		R+2	*70*
	Hauteur d'étage (m)	*Moyenne*	*Ecart type*
		3.20	*0.20*
		Nombre des bâtiments	
	Nombre de travée zz	*2*	*70*
		3	*180*
		4	*750*
	4) Longueur de trames zz (m)	*Moyenne*	*Ecart type*
		4.00	*0.20*
		Nombre des bâtiments	
	5) Nombre de travée xx	*2*	*55*
		3	*820*
		4	*125*
	6) Longueur de trames xx (m)	*Moyenne*	*Ecart type*
		4.00	*0.20*
	7) Section des poteaux (cm)	*Moyenne)*	*Ecart type*
		30x30	*5.00*
matériau De construction	*La résistance à la compression de béton (fc) (Mpa)*	*Moyenne*	*Ecart type*
		20.00	*3.25*
	Limite élastique de l'acier fy (Mpa)	*Moyenne*	*Ecart type*
		420.00	*20.00*

Nous présentons les différentes courbes de capacité de bâtiments étudiés dans notre travail à la figure 06.

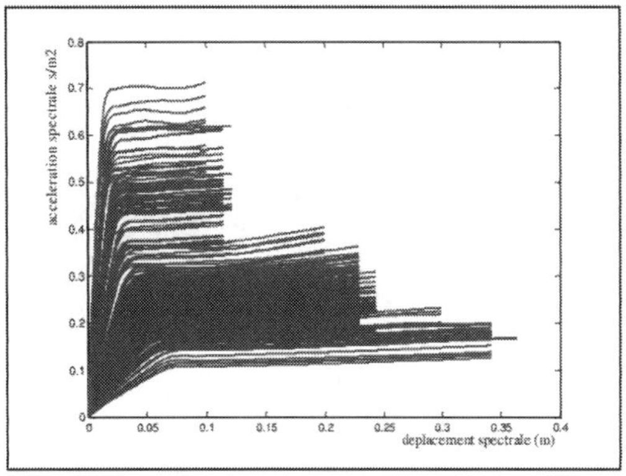

Fig. 6. Courbes de capacité en format ADRS (exemple de 1000 bâtiments).

3.2 Générer les spectres de réponse

Dans l'approches développée, nous allons utiliser une sollicitation sismique est considérée à travers un spectre de réponse de Règlement Parasismique Algérienne (RPA 99 version 2003) [37] qui définie par:

- la zone sismique III.
- le groupe d'usage groupe1B (A=0.30).

32 combinaisons sont formés par des différentes valeurs de facteur de qualité Q (1, 1.05, 1.10, 1.15, 1.20, 1.25, 1.35) avec le site T2 (0.30, 0.40, 0.50, 0.70) permettant de générés le spectre de réponse dans le plan (Sa-Sd) comme présenté à la figure 07.

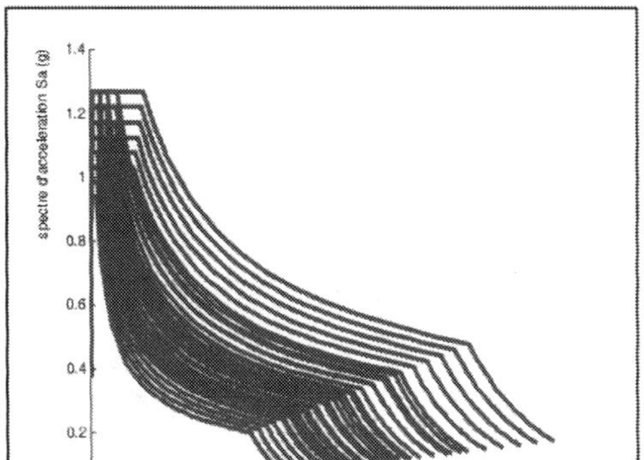

Fig. 7. Génération spectres de réponse (32 combinaisons entre Q et T2).

3.3 Déterminer les points de performances

Dans le présent de travail, on utilise une procédure itérative définie par le code FEMA440 (2009) pour la détermination les points de performance comme présentée aux figures 08. Nous avons considéré dans cet exemple 1000 bâtiments (courbes de capacité) soumise aux différentes agressions sismiques (32 spectre de réponse) (voir la figure 09).

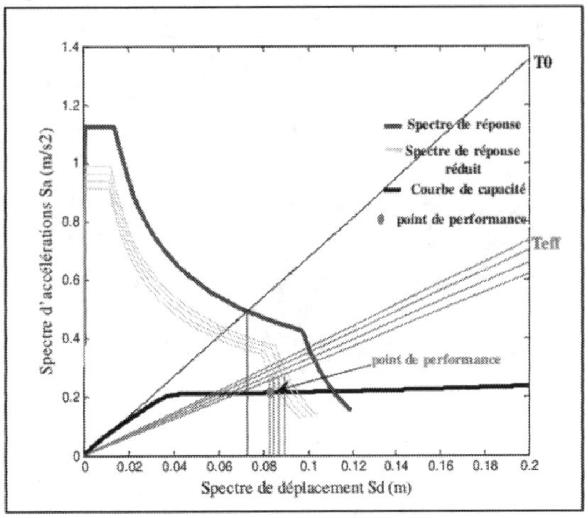

Fig. 8. Schéma pour d'obtention du point de performance selon procédure « FEMA440 ».

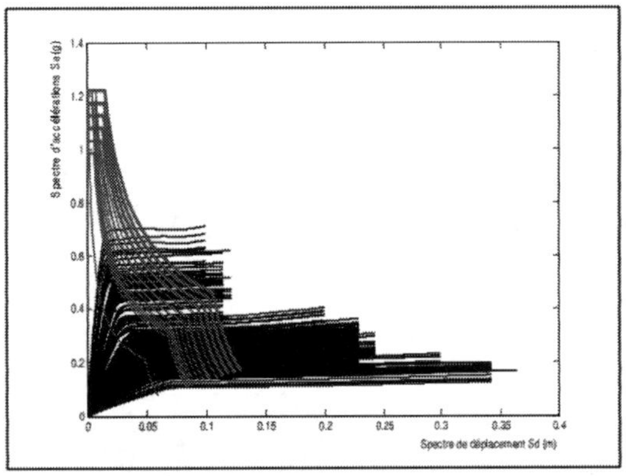

Fig. 9. Superposition entre courbes de capacité (1000 bâtiments) et spectres de réponse générée par 32 combinaisons.

3.4 Construire des courbes de fragilité

D'après la classification des bâtiments selon l'état d'endommagent (faibles, modérés, importants et très importants). On détermine la valeur moyenne de S_d et l'écart type de chaque classe ds. Une fois les paramètres de la fonction de fragilité Sd,ds obtenus, on peut construire les courbes de fragilité à partir de l'équation (04). Les courbes de fragilité sont représentées à la figure 10.

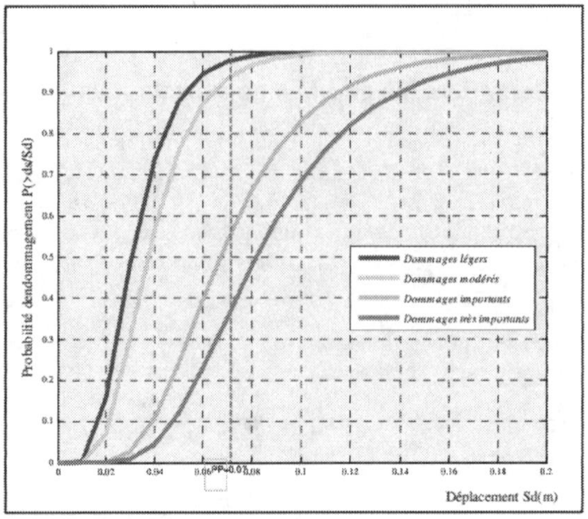

Fig. 10. Construction courbes de fragilité.

Connaissant le déplacement spectral maximum Sd il est possible de déterminer la probabilité d'endommagement, en fonction des quatre niveaux de dommages. Ainsi, pour un déplacement subi de X cm, nous pouvons quantifier les probabilités d'avoir respectivement P1% d'endommagement de niveau 1 (dommages légers), P2% d'endommagement du niveau 2 (dommages modérés), P3% d'endommagement du niveau 3 (dommages importants), P4% d'endommagement du niveau 4 (dommages très importants allant jusqu'à la ruine). A titre illustratif, pour un déplacement spectral de 7 cm, les probabilités que la structure se trouve aux niveaux d'endommagement 1, 2, 3 ou 4 sont respectivement de 97%, 93%, 55% et 37%.

4. CONCLUSION

Par cette communication nous avons essayé de contribuer au développement d'un outil (courbes de fragilité) permettant l'amélioration de la gestion du risque sismique en Algérie.

Une méthodologie développée pour la détermination des courbes de fragilité a été présentée. Cette méthodologie est basée sur de la méthode push-over et l'utilisation des simulations Monte Carlo. Des courbes de fragilité ont été établies pour des structures de type portique en béton peuvent être utilisées pour la mise en conformité sismiques de structures existantes. Les résultats (courbes de fragilité) présentés ici constituent un exemple d'utilisation pour l'évaluation des dommages et constituent une première estimation des niveaux de risques dans la définition du risque sismique des constructions. Les perspectives directes de ce travail, sont l'application de la méthodologie développée et sa validation sur d'autres typologies de structures

De nombreux études de recherches dans le risque naturel et notamment sismique se sont réalisés dans les dernières années (HAZUS, RISK-EU). Ils ont apporté ainsi une quantité importante d'information, concernant essentiellement le développement de nouvelles solutions techniques pour les constructions et la protection de différents enjeux d'une part, et l'estimation des conséquences potentielles suite à la manifestation des différents scénarii de catastrophes naturelles d'autre part. Les décideurs disposent par conséquent aujourd'hui d'une large gamme d'outils, dont certains présentés dans ce travail, qui doivent servir comme un appui pour la mise en place d'une meilleure politique de mitigation. Les résultats obtenus nous semblent cohérents avec ceux déjà obtenus par d'autres chercheurs, Tout fois nous devons procéder à une validation de ces résultats en les comparant à des endommagements enregistrés lors de séisme (exemple ceux de Boumerdes).

5. REFERENCES

[1] Applied Technology Council ATC.40, (1996).

[2] Règlement Parasismique Algérienne RPA (99 version 2003).

*[3] RISK-UE, An advanced approach to earthquake risk scenarios with applications to different european towns, Projet Européen, EVK4-CT-**2000-00014**, (2003).*

*[4] Federal Emergency Management Agency FEMA**440**,(2009).*

*[5] HAZUS. Direct physical damage general building stock, chapter **5**, HAZUS99 Technical Manual, (1999).*

Assessment of the tsunami hazard on Moroccan coasts using numerical modeling

Meriem Amine[1], Latifa Ouadif[1], Khadija Baba[2] & Lahcen Bahi[1]
[1]Laboratory of applied Geophysics, Geotechnics, Geology of the Engineer and Impact on the Environment (L3GIE), Mohammadia School of Engineers, Mohamed V University, Rabat, Morocco.
[2]Laboratory of civil engineering and environment (GCE), High School of Technology-Sale, Mohamed V University, Rabat, Morocco.

Abstract. Caused by earthquakes in the marine environment, tsunamis are often of particular violence. Pests, their damage far exceeds those caused by earthquakes earthquakes. In Morocco, seismic sensors in the national network often detect earthquakes in the marine environment. However, specialists minimize their threat given the peculiarity of the Moroccan ocean zone. The latter is characterized by the importance of the length of the faults and the depth of the seismic focus.

This work can be considered as a first approach to the study and understanding tsunamis. Although Morocco has known since historical times, earthquakes and tsunamis along its coasts.

This preliminary study is mainly intended to show the contribution of the numerical simulation of tsunamis, with a concrete application of the 1755 Tsunami of Lisbon based on the four source zones involving the five potentially tsunamigenic faults that are generally mentioned in the literature for this major event.

1 Introduction

Le tsunami est une catastrophe naturelle parmi les plus destructrices de l'histoire, il est perçu comme une série de vagues océaniques destructrices déferlantes de très grande hauteur et ayant été provoquée par de brutales perturbations de la couche d'eau de l'océan, causant de nombreux dégâts.

La majorité des tsunamis est générée par des séismes se produisant en domaine maritime, mais seuls les séismes de forte magnitude et de faible profondeur hypocentrale génèrent des tsunamis, en effet le séisme doit générer une déformation suffisante du fond de l'océan (de l'ordre de quelques mètres). C'est pour cette raison que la plupart des séismes générant des tsunamis ont lieu le long de grandes zones de subduction. À l'approche des côtes, l'amplitude augmente au détriment de la vitesse, qui baisse avec la profondeur d'eau. En conséquence le train d'onde sera comprimé spatialement, et son amplitude deviendra plus importante. Dans les cas les plus forts, l'onde peut atteindre quelques dizaines de mètres de hauteur et inonder sur des distances de plusieurs kilomètres. Lors de l'inondation, lorsque l'onde pénètre dans les terres, les dégâts peuvent être considérables: arbres coupés, infrastructures détruites, déplacement de masses importantes (bateaux, conteneurs,...). Les caractéristiques des ondes tsunamis lors de l'inondation dépendent de la morphologie des terrains.

Dans la région de l'Atlantique Nord-Est, les pays les plus exposés aux tsunamis sont le Maroc, le Portugal et l'Espagne [1; 2; 3] (Figure 1). Dans ces pays, les structures côtières et de défense de la mer ne sont pas conçues pour résister à l'impact du tsunami et ne servent que des brise-lames qui protègent les marinas et les installations portuaires des vagues de l'océan.

Au Maroc, l'aléa tsunami peut être considéré comme étant à faible probabilité, mais à forte gravité du fait de l'importance des enjeux présents, et surtout à venir, sur la bande littorale.

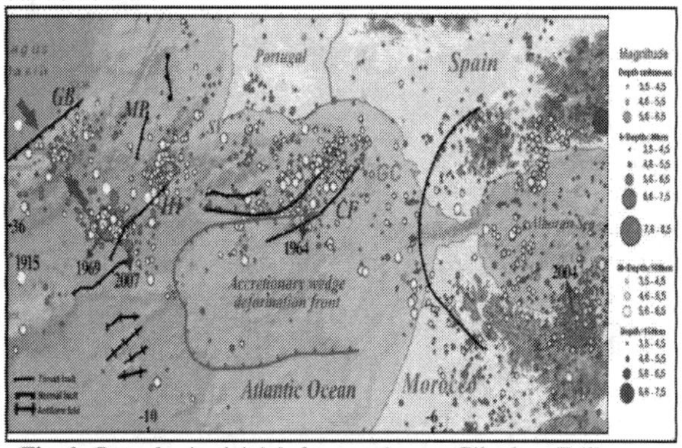

Fig. 1. Carte de sismicité de la zone Açores-Gibraltar. Failles majeures (de Gutsher et al., 2009 et Rosas et al., 2012) et distribution d'hypocentres sismiques de M ≥ 3,5 pendant la période 1901-2010.

Dans cette étude nous considérons cinq scénarios tsunamigéniques, tous d'origine sismique. Ces scénarios, de grandeur allant de 8,0 à 8,6, représentent les principaux défauts de tremblement de terre actifs dans l'océan atlantique qui peuvent déclencher de gros tsunamis dans la région. La valeur inférieure (Mw = 8.0) est compatible avec l'événement du 28 février 1969 [1; 4], tandis que les grandeurs supérieures (Mw = 8.2-8.6) sont considérées comme représentant l'événement du 1er novembre 1755 [1]. Nous effectuons une modélisation numérique de la propagation du tsunami pour tous les scénarios considérés et l'inondation pour le scénario qui a causé le pire impact dans la zone étudiée.

Les résultats prévus montrent différents niveaux d'impact que les différentes conditions de l'onde de tsunami pourraient générer dans la région. L'impact dramatique du tsunami dans la région montre la nécessité de nouvelles normes de tsunami non seulement pour les structures de défense de la mer, mais aussi pour les habitations côtières et les maisons afin de fournir une évacuation potentielle sur place.

La conception et la construction d'installations côtières capables d'arrêter l'inondation du tsunami dans les zones côtières à risque demeurent l'une des mesures recommandées pour atténuer les risques liés aux tsunamis. D'autre part, les normes de construction des défenses maritimes doivent répondre aux besoins de sécurité des zones vulnérables, qui pourraient découler des estimations et des prévisions d'impact et de dégâts.

L'inondation du tsunami et la propagation des ondes terrestres dépendent de manière critique de la qualité du terrain numérique et des modèles numériques de propagation du tsunami. Le modèle numérique du terrain doit inclure une description complète de la bathymétrie, de l'infrastructure topographique et côtière [5].

2 Evaluation du risque tsunami sur le littoral marocain

Le Maroc, par son contexte géologique particulier et sa proximité de la limite de plaque Nubie-Eurasie (NEPB) est le littoral ouest africain qui est le plus exposé aux tsunamis induits par les séismes [6] (figure1). Les tremblements de terre sous-marins les plus graves ressenties au Maroc ont été ceux générés en mer le long de la côte atlantique [7; 8; 9; 10]. Certains de ces événements étaient tsunamigène, comme ce fut le cas de l'événement du 1er Novembre 1755, qui était certainement le tsunami le plus dévastateur jamais enregistré au Maroc. Ses vagues ont ravagé les côtes ibériques et atlantique marocaine, avec des hauteurs qui ont atteint 5-15 m et ont été observées à travers l'Atlantique dans les Antilles [11], mais il persiste encore de nombreuses incertitudes sur la source exacte du séisme et ses effets sur les côtes, en particulier marocaines. Il s'agit du seul tsunami connu en Atlantique nord et demeure en cela un événement de référence pour établir des scénarios de risque sur cet espace.

3 Modélisation de l'aléa tsunami sur les côtes marocaines

La modélisation numérique, développée depuis les années 1960, consiste à construire un ensemble de fonctions mathématiques décrivant le phénomène. Elle repose sur la programmation de modèles théoriques ou mathématiques (intégration des équations du mouvement, d'équations différentielles, etc…) qui sont adaptés aux moyens numériques. En modifiant les variables de départ, on peut ainsi prédire les modifications du système physique.

Dans le cas des études de tsunamis, la modélisation numérique permet d'évaluer l'aléa et de proposer, sur la base d'analyses sismotectoniques, des scénarios crédibles dans des zones où les données historiques sont incomplètes [12].

La modélisation de tsunami consiste à reproduire des événements passés selon une approche déterministe ou en imaginer de nouveaux. Elle comprend trois phases distinctes: 1) la génération du tsunami, 2) la propagation au large et 3) l'interaction des ondes avec le plateau continental et l'inondation de la côte.

3.1 Code de simulation

Les connaissances sur les tsunamis historiques ayant affecté le Maroc sont insuffisantes pour estimer l'aléa. Par conséquent, la modélisation numérique est le seul moyen de déterminer, au moins approximativement, les zones qui seront inondées, les hauteurs d'eau et les temps d'arrivée d'un tsunami d'origine sismique. Afin de sensibiliser les décideurs sur la réalité de ce risque, l'événement de référence retenu pour ce modèle est le tsunami du 1er novembre 1755. Les simulations numériques du tsunami effectuées dans le cadre de ce travail ont été réalisées avec le code SWAN (Simulating WAves Nearshore) sous le logiciel Mirone. SWAN est un modèle de vague de troisième génération, qui calcule des ondes aléatoires, à crête courte, générées par le vent dans les régions côtières et les eaux intérieures.

La rupture sismique est censée être instantanée et le déplacement du fond marin est calculé en utilisant la théorie du demi-espace élastique [13], mise en œuvre dans le code MIRONE. Le déplacement vertical du fond de la mer est ensuite transféré à la surface libre de l'océan, selon l'hypothèse que les deux déformations, du fond de la mer et de la surface de l'océan, seraient égales. Ce modèle consiste dans un premier temps à calculer la déformation initiale associée à une rupture sismique en faisant appel aux algorithmes développés par Okada (1985) qui correspondent à la méthode couramment utilisée par les différents modèles (la théorie demi-espace élastique) [14]. Ces algorithmes permettent de calculer la déformation du plancher océanique à partir des caractéristiques de la faille à l'origine du séisme. Le tsunami initial est ensuite déduit en admettant que l'intégralité de cette déformation est instantanément transmise à la surface de l'eau.

3.2 Scénarios tsunamigènes

La source exacte du tremblement de terre du 1er Novembre 1755 demeure encore à ce jour incertaine [15]. Quatre zones sources impliquant cinq failles potentiellement tsunamigènes sont généralement évoquées dans la littérature pour cet événement majeur. R. OMIRA et al.(2009, 2010) ont modélisé les tsunamis générés par chacune de ces sources ; À savoir la faute de la Banque Gorringe (GB) [16], la faute de Horseshoe (HS), la faute de Marques de Pombal (MP) [17], la faute de la Banque Portimao (PB), et la faute de Cadiz Wedge (CW) [18] (figure 2). La relation de mise à l'échelle de Scholz (1982) qui repose sur les longueurs de défaut et la relation longueur / largeur a été utilisée pour obtenir l'ampleur du glissement de chaque scénario.

En ce qui concerne le module de cisaillement (μ) qui caractérise la rigidité de la zone de rupture du tremblement de terre, une valeur de 3×10^{10} Pa est adoptée pour les cinq failles.

Les paramètres de défaut pour les scénarios étudiés sont illustrés dans le tableau 1. Dans la Figure 2, les scénarios sont tracés.

Tableau 1. Paramètres de fautes typiques des scénarios tsunamigènes candidats dans la région du golfe de Cadix. L: la longueur de la faute en kilomètres; W: la largeur de la faille en kilomètres; D: la profondeur depuis le fond de la mer jusqu'au sommet de la faille en kilomètres; μ: le module de cisaillement et Mw: l'amplitude du moment.

Scenario name	L (km)	W (km)	Epicenter coordinates		D (km)	Slip (m)	Strike (°)	Dip (°)	Rake (°)	μ (Pa)	Mw	Source Zone
			Lon	**Lat**								
GBF	127	60	-11.332	36.665	5.0	8.3	233.0	35	90	3.0×10^{10}	8.1	GBZ
HSF	165	70	-9.913	35.796	4.0	10.7	42.1	35	90	3.0×10^{10}	8.3	HMZ
MPF	129	70	-9.890	36.574	4.0	8.0	20.0	35	90	3.0×10^{10}	8.1	HMZ
PBF	105	55	-8.585	36.314	2.5	7.2	266.3	24	90	3.0×10^{10}	8.0	AZ
CWF	168	200	-8.059	35.407	5.0	10	349.0	5	90	3.0×10^{10}	8.6	ABZ

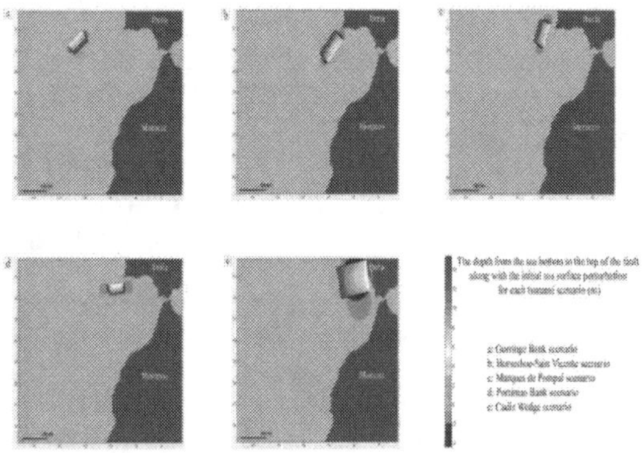

Fig. 2. Perturbation initiale de la surface de la mer pour chaque scénario de tsunami dans le golfe de Cadix; (a) Scénario de la Banque de Gorring, (b) Scénario de Horseshoe; (c) Scénario de Marques de Pombal, (d) Scénario de la Banque de Portimao, (e) Scénario de Cadix Wedge.

3.3 Modélisation numérique du Tsunami

Pour simuler la propagation du tsunami, nous utilisons un code non linéaire et un modèle de terrain numérique à haute résolution qui intègre des données d'infrastructures bathymétriques, topographiques et côtières.

Une grille de données bathymétriques / topographiques est générée pour couvrir l'océan atlantique et les côtes marocaines, cette grille a servit comme base pour la réalisation de la modélisation du tsunami.

Le résultat de ces processus a été la génération d'une couche de grille (figure 3), permettant une meilleure représentation des structures bathymétriques, topographiques et côtières afin de mener une modélisation plus réaliste de l'impact du tsunami dans les côtes du Maroc.

Les données bathymétriques/ topographiques utilisées proviennent de la base de données internationale NOAA, National Oceanic and Atmospheric Administration.

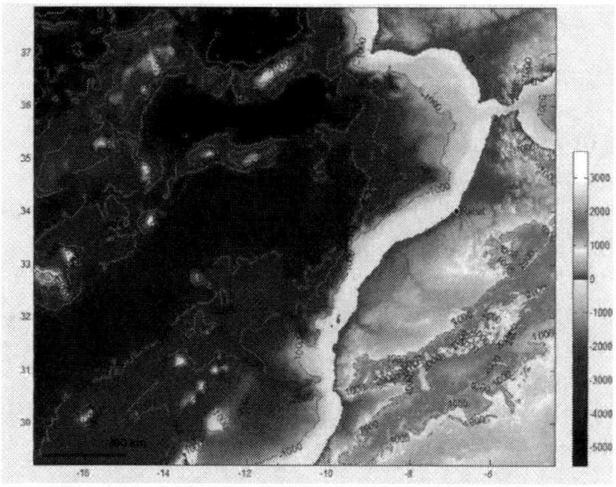

Fig. 3. Modèle d'élévation numérique intégrant les caractéristiques bathymétriques et topographiques de la côte marocaine

4 Résultats

La simulation de la propagation des vagues de tsunami a été réalisé dans le but d'analyser leur comportement en s'approchant du littoral afin de comparer les résultats pour tous les scénarios considérés et pouvoir tirer celui qui affecte le plus les côtes marocaines.

4.1 Scénario Gorringe Bank (GB) Mw = 8.1

Le scénario Mw = 8,1 GB, est considéré comme une source candidate pour l'événement Lisbonne de 1755 [16]. La simulation numérique montre une direction de l'énergie du tsunami vers le nord, car la caractéristique bathymétrique de la Gorringe Bank empêche une propagation efficace vers les côtes du golfe de Cadix (figure 4). L'analyse de la propagation des vagues de tsunami près du littoral marocain montre des similitudes dans le comportement des vagues avec celles déclenchées par les scénarios HS et MP.

Ces similitudes sont dues à la proximité de ces défauts sismiques ainsi qu'à l'azimut de défaut presque similaire.

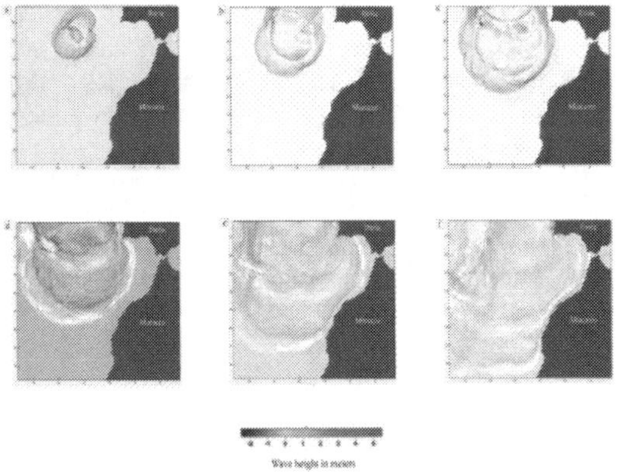

Fig. 4. Simulation de la propagation des vagues du tsunami, depuis la zone épicentrale, suite au scénario Mw = 8.1 GB

4.2 Scénario Horseshoe (HS) Mw = 8.3

L'examen des résultats des simulations indique que l'énergie du tsunami du scénario Mw = 8.3 HS est principalement orientée vers les côtes du nord-ouest du Maroc (figure 5).

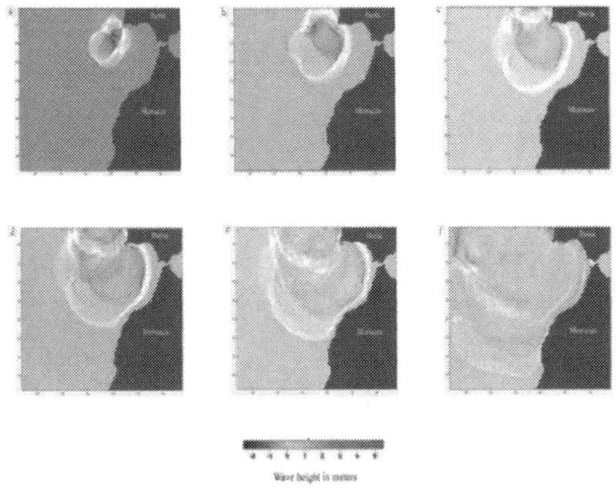

Fig. 5. Simulation de la propagation des vagues du tsunami, depuis la zone épicentrale, suite au scénario Mw = 8.3 HS

4.3 Scénario Marques de Pombal (MP) Mw = 8.1

Le scénario Mw = 8.1 MP, est basé sur une grande structure tectonique active qui est considérée comme l'une des sources candidates pour l'événement de 1755 [17]. Les simulations numériques montrent que le scénario Mw = 8.1 MP a déclenché un grand tsunami dans la région du Golfe de Cadix (figure 6). La profondeur de l'eau moins profonde dans la zone hors SW Iberia entraîne une diminution de l'impact du tsunami le long des côtes atlantiques du Maroc.

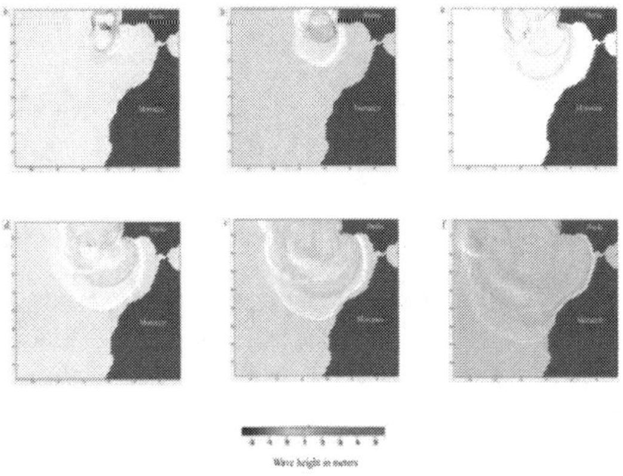

Fig. 6. Simulation de la propagation des vagues du tsunami, depuis la zone épicentrale, suite au scénario Mw = 8.1 MP

4.4 Scénario Portimao Bank (PB) Mw = 8.0

Le scénario Mw = 8.0 PB, par son emplacement très proche du Golfe de Cadix, présente un impact principalement sur cette région.

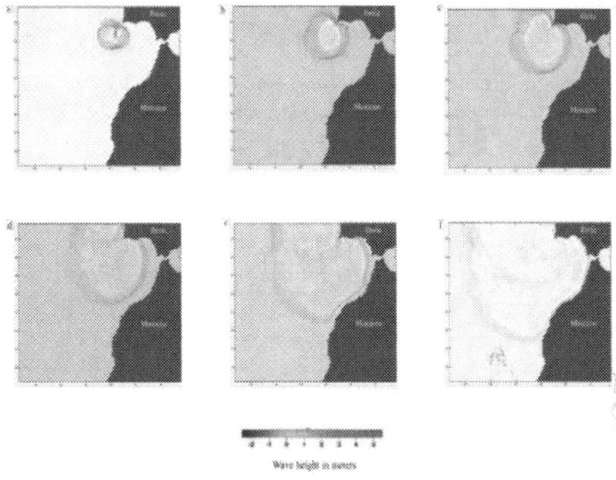

Fig. 7. Simulation de la propagation des vagues du tsunami, depuis la zone épicentrale, suite au scénario Mw = 8.0 PB

4.5 Scénario Cadiz Wedge (CW) Mw = 8.6

La dalle de subduction Cadiz Wedge a été proposée par Gutscher (2004) comme responsable de la génération de l'événement de 1755. La modélisation numérique indique que la localisation et l'azimut du scénario CW contraignent les modèles d'énergie du tsunami et la plupart des énergies sont dirigées vers le détroit de Gibraltar (figure 8).

 https://doi.org/10.1051/matecconf/201814902079

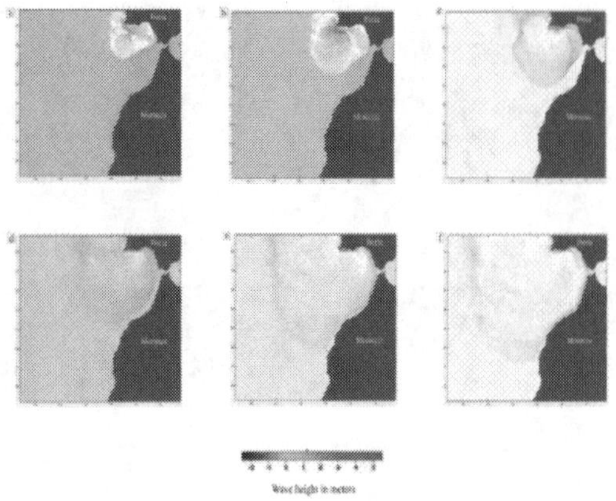

Fig. 8. Simulation de la propagation des vagues du tsunami, depuis la zone épicentrale, suite au scénario Mw = 8.6 CW

5 Discussion

Il est clair que la bathymétrie du domaine de propagation, l'emplacement de la faute et son azimut sont les principaux facteurs qui limitent la distribution de l'énergie du tsunami et la direction des hauteurs des ondes. Pour les scénarios HS, MP et GB ayant des sites proches, les propagations du tsunami près des rives marocaines présentent certaines similitudes dans les caractéristiques et la directivité. Concernant le scénario PB, son emplacement et sa directivité induisent une propagation vers la région du Golfe de Cadix. L'exception concerne le scénario CW, pour lequel la propagation du tsunami à proximité de la côte a une directivité différente des autres scénarios considérés. Cela s'explique principalement par l'emplacement de cette source provoquant une propagation du tsunami et des modes d'énergie totalement différents par rapport aux autres sources.

Ce travail montre le haut niveau de destruction que les zones côtières du Maroc, du Portugal et de l'Espagne pourraient ressentir suite à l'apparition de grands événements tsunami régionaux, et donc le besoin de systèmes robustes pour la préparation et l'alerte au tsunami dans la région pour atténuer ce risque.

6 Conclusions

La revue des données historiques, instrumentales et géologiques disponibles sur le risque de tsunami au Maroc permettent de conclure que, comme pour les pays voisins, ce risque est réel et que par conséquent des mesures de prévention et de protection des côtes sont nécessaires d'autant plus que les sources les plus dangereuses se trouvent à proximité ce qui ne permet pas un temps de réaction supérieur à une heure. Ceci implique:

- la connaissance du risque qu'encourt une région donnée en fonction de l'aléa et de la vulnérabilité, y compris la délimitation des zones potentiellement exposées au risque et également des endroits les plus vulnérables; pour ce faire, l'étude doit intégrer dans un système SIG de multiples facteurs relatifs à la structure de l'habitat et de l'environnement physique ainsi que des données socio-économiques;

- éviter de nouvelles constructions dans les zones identifiées comme exposées au déferlement des tsunamis pour minimiser les pertes lors de futurs tsunamis; ici, le rôle de l'aménagement du territoire et de son application stricte est de première importance;

- appliquer des normes de construction aux nouveaux bâtiments (résistance au déferlement des vagues, fondations consolidées, etc.) pour minimiser les dommages dus aux tsunamis;

- protéger des tsunamis les constructions et infrastructures existantes dans les zones exposées via des digues ou des murs anti-tsunamis qui, certes, sont des ouvrages très coûteux et peu esthétiques mais qui ont prouvé leur efficacité;

- prendre des précautions spécifiques pour l'installation d'infrastructures nécessaires durant les crises (hôpitaux, pompiers, protection civile, etc.).

Si la désignation des institutions et des personnes chargées de l'alerte est pratiquement achevée, reste le problème de la gestion de l'alerte au niveau des populations et de la transmission rapide de l'information. Ici, deux solutions existent: d'une part, l'évacuation horizontale qui consiste à se retirer à l'intérieur des terres sur un point haut et, d'autre part, l'évacuation verticale qui réside dans le fait de se réfugier aux étages supérieurs des bâtiments. Cette dernière solution est préconisée lorsque le temps d'évacuation horizontale est très court, lorsque la densité de population est élevée et les moyens d'évacuation horizontale (routes, ponts, etc.).

References

1. Baptista, M.A. & Miranda, J.M. (2009). Revision of the Portuguese catalog of tsunamis. Natural Hazards and Earth System Sciences, 9, 25-42.
2. Kaabouben, F., Baptista, M.F., Iben Brahim, A., El Mouraouah, A. & Toto, A. (2009). On the Moroccan tsunami catalogue. Natural Hazards and Earth System Sciences, 9, 1227-1236.
3. Carmona, P. & Ruiz, J.M. (2009). Geomorphological evolution of the river Loukkos estuary around the Phoenician city of Lixus on the Atlantic littoral of Morocco. Geoarcheology, 24(6), 821 –845.
4. Gjevik, B., Moe, H., Ommundsen, A., 1997. Sources of the Maelstrom. Nature 388 (28), 837–838
5. Mellas, S., Leone, F., Omira, R., Gherardi, M., Baptista, M.A., Zourarah, B., Péroche, M. & Lagahé, E. (2012). Le risque tsunamique au Maroc: modélisation et évaluation au moyen d'un premier jeu d'indicateurs d'exposition du littoral atlantique. Physio-Géo, 6, 119-139.
6. El Alami, S.O. & Tinti, S. (1991). A preliminary evaluation of the tsunami hazards in the Moroccan coasts. Science of Tsunami Hazards, 9(1), 31-38.
7. Elmrabet, T. (2005) The great earthquakes in the Maghreb region and their consequences on man and environment. Imprimerie Beni Snassene, Rabat (en Arabe).
8. Omira, R., Baptista, M.A., Miranda, J.M., Toto, E., Catita, C. & Catalao, J. (2010). Tsunami vulnerability assessment of Casablanca-Morocco using numerical modeling and GIS tools. Natural Hazards, 54(1), 75-95.

9. Omira, R., Baptista, M.A., Leone, F., Matias, L., Mellas, S., Zourarah, B., Miranda, J.M., Carrilho, F. & Cherel, J.P. (2013). Performance of coastal sea-defense infrastructure at El Jadida (Morocco) against tsunami threat: lessons learned from the Japanese 11 March 2011 tsunami. Natural Hazards and Earth System Sciences, 13, 1779-1794.
10. Mellas, S. (2012). Evaluation du risque tsunamique sur le littoral atlantique marocain. Thèse de Doctorat, Université Chouaib Doukkali, Faculté des Sciences d'El Jadida.
11. Ambraseys, N.N. (1962). Data for the investigation of the seismic sea-waves in the Eastern Mediterranean. Bulletin of the Seismological Society of America, 52(4), 895-913.
12. Abe, K. (1979). Quantification of tsunamigenic earthquakes by the Mt scale. Tectonophysics, 166(1), 27-34.
13. Mansinha, L., Smylie, D.E., (1971). The displacement fields of inclined faults. Bulletin of the Seismological Society of America. 61 (5): 1433-1440.
14. Okada, Y., (1985). Surface deformation due to shear and tensile faults. In a half-space. Bulletin of the Seismological Society of America. 75 (4): 1135-1154.
15. Gutscher, M. A., Dominguez, S., Graham, K., Westbrook, Leroy, P., (2009). Deep structure, recent deformation and analog modeling of the Gulf of Cadiz accretionary wedge: Implications for the 1755 Lisbon earthquake. Tectonophysics, Elsevier, 475 (1), pp.85-97.
16. Johnston, A., (1996). Seismic moment assessment of earthquakes in stable continental regions – III. New Madrid, 1811-1812, Charleston 1886 and Lisbon 1755. Geophys. J. Int. 126, 314-344.
17. Zitellini, N., et al., (2001). Source of 1755 Lisbon earthquake and tsunami investigated. Eos (Transactions, American Geophysical Union) 82, 285-291.
18. Gutscher, M. A., Dominguez, S., Westbrook, G., Gente, P., Babonneau, N., Mulder, T., Gonthier, E., Bartolome, R., Luis, J., Rosas, F., Terrinha, P., and the Delila and DelSis Scientific Teams. (2008). Tectonic shortening and gravitational spreading in the Gulf of Cadiz accretionary wedge: observations from multi-beam bathymetry and seismic profiling. Journal of Marine and Petroleum Geology, Sp. Vol. on Submarine instabilities. doi:10.1016/j.marpetgeo.2007.11.008.
19. Ammar, A., Mauffret, A., Gorini, C. & Jabour, H. (2007). The tectonic structure of the Alboran margin of Morocco. Revista de la Sociedad Geologica de España, 20, 247-271.
20. Atillah, A., El Hadani, D., Moudni, H., Lesne, O., Renou, C., Mangin, A. & Rouffi, F. (2011). Tsunami vulnerability and damage assessment in the coastal area of Rabat and Salé, Morocco. Natural Hazards and Earth System Sciences, 11, 3397-3414.
21. Blanc, P.-L. (2009). Earthquakes and tsunami in November 1755 in Morocco: a different reading of contemporaneous documentary sources. Natural Hazards and Earth System Sciences, 9, 725-738.
22. Debrach, J. (1946). Raz de marée d'origine séismique enregistré sur le littoral atlantique du Maroc. Annales SPGM, 1946, pp. 59-71.
23. Medina, F., Mhammdi, N., Chiguer, A., Akil, M. & Jaaidi, E.B. (2011). The Rabat and Larache boulder fields; new examples of high-energy deposits related to extreme waves in north-western Morocco. Natural Hazards, 59, 725-747.
24. Mhammdi, N., Medina, F., Kelletat, D., Ahmamou, M. & Aloussi, L. (2008). Large boulders along the Rabat coast (Morocco); possible emplacement by the November, 1st, 1755 a.d. tsunami. Science of Tsunami Hazards, 27(1), 17-30.
25. Mhammdi, N., Medina, F., Trentesaux, A., Font, E., Belkhayat, Z., & Geawhari, M. A. (2015). Sedimentary evidence of palaeo-tsunami deposits along the Loukkos estuary (Moroccan Atlantic coast). Science of Tsunami Hazards, 34(2), 83-100.
26. Mouzzoun, M., Moustachi, O., Taleb, A., (2012). Evaluation de la vulnérabilité sismique des bâtiments en béton armé par l'utilisation des courbes de fragilités (Seismic vulnerability assessment of reinforced concrete Buildings using fragility curves). J. Mater. Environ. Sci. 3 (6) (2012) 1037-1044. ISSN:2028-2508 CODEN:JMESCN.

Superficial structures cartography of the Essaouira basin under ground (Morocco), by small refraction seismic: contribution of the static corrections in the reinterpretation of the speeds variations.

M. DAHAOUI[1], L. AIT BRAHIM[3], A. El ADRAOUI[2], A. ABDELOUAFI[6], M. DAKKI[4] and A. El IMRANI[5]
[1,3] Université Mohammed V, Faculté des Sciences, laboratoire Géophysique et Risques naturels, structure Géorisk, Avenue Ibn Battouta, BP 1014, Rabat, Maroc
[2,6] Université Mohammed V, Faculté des Sciences, laboratoire GEOS, structure GEORISK, Avenue Ibn Battouta, BP 1014, Rabat, Maroc
[4,5] Office National d'Hydrocarbures et des Mines, (ONHYM), Rabat, Maroc

Abstract. The static corrections are a necessary step in the sequence of the seismic processing. This paper presents a study of these corrections in the Essaouira basin. The main objective of this study is to calculate the static corrections by exploiting the seismic data acquired in the field to improve the deep structures imaging. It is to determine the roof and the basis of the superficial layers which constitute the weathered zone while calculating the delays of seismic wave's arrivals in these layers. The purpose is to cancel the effect of the topography and the weathered zone, in order to avoid any confusion when the seismic and geological interpretation. The results obtained show the average values of the static corrections varying between - 127 and 282 ms (double time), with existence of high values by location, particularly in the Eastern and North-Eastern of the basin, which meant the presence of altered zone with irregular topography and whose thickness and speeds vary laterally. In effect the variations of velocities in the fifty meters from the surface may introduce significant anomalies in seismic refraction, with heavy consequences when the interpretation or the drilling establishment. These variations are mainly due to lateral changes in facies and variations in the formations thickness. The calculation of the static corrections, revealed high values at certain areas (East and North-East), which will enable us to better orient the future campaigns in these zones. It is therefore necessary to concentrate the seismic cores drillings and the small refraction seismic profiles by tightening the seismic lines meshes in order to have the maximum values of static corrections and thereafter a better imaging of the reflectors.
Keywords: Essaouira basin, Static corrections, seismic refraction, weathered zone, seismic waves, lateral variation of lithology.

1 Introduction

Les bassins marocains ont été depuis longtemps l'objet de plusieurs recherches pétrolières justifiées par l'intérêt économique que présentent certaines formations géologiques. Or ces recherches n'ont découvert que de modestes réserves dans le bassin d'Essaouira et le bassin de Rharb. Cependant plusieurs efforts ont été déployés dans l'exploration pétrolière au Maroc au cours des dernières années. Les bassins sédimentaires ont fait l'objet de plusieurs campagnes d'exploration, réalisées par différents établissements spécialisés dans le domaine [1].

La prospection pétrolière consiste à explorer le sous-sol en utilisant l'imagerie sismique, fondée sur le principe physique de transmission des ondes dans les milieux solides dans le but d'analyser les structures géologiques en profondeur, et la découverte des zones de piégeage des hydrocarbures, avant de recourir au forage, qui présente un outil extrêmement coûteux [2]. Dans ce cadre, d'importantes méthodes géophysiques sont mises en œuvre : la sismique réflexion à couverture multiple est la plus utilisée actuellement en acquisition des données de terrain. Cette méthode fournit une échographie du sous-sol en deux ou trois dimensions selon la mise en œuvre de chacune sur le terrain [8].

Les couches superficielles existantes aux dessus d'un réservoir (Zone Altérée) constituent un problème non négligeable, étant donné que la propagation des ondes sismique dans ces couches engendre un retard sur le temps d'arrivée [9].

Dans ce travail, on se concentre sur l'étude de la zone altérée (Weathered Zone : WZ) par le calcul des corrections statiques à appliquer aux données sismiques en utilisant deux approches différentes, à savoir : la méthode des carottages sismiques (Upholes) et celle de la sismique petite réfraction [7].

2 Cadre géologique

Le bassin d'Essaouira qui fait partie du bassin côtier d'El Jadida-Agadir [8], se localise entre 9°46'11" W et 31°30' 44 " N et s'étend sur une superficie d'environ 20 000 km². Il est limité au Nord par le bassin de Doukkala au Nord-Est par les Jebilet occidentales, à l'Est par la plaine du Haouz, au Sud-Est par le Haut Atlas occidental, au Sud par les plateaux de Haha et à l'Ouest par l'Océan Atlantique (figure 1).

Le bassin d'Essaouira est caractérisé par des dépôts paléozoïques qui ont été largement affectés, au Carbonifère, par l'orogénèse hercynienne. En effet, durant le Trias et le début du Lias, les dépôts ont été contrôlés par des failles normales extensives entrainant la formation de blocs basculés et d'épaisses séries détritiques rouges, de sel et de dolomies. Durant le Jurassique et le Crétacé, la sédimentation a été caractérisée par des carbonates marins de plate-forme et des anhydrites qui furent interrompus par plusieurs intervalles silicoclastiques [1].

Sur presque tout l'arrière-pays affleurent les formations du Crétacé; des formations du Tertiaire (couches phosphatées) et du Quaternaire (dépôts superficiels) s'y rencontrent localement dans des cuvettes synclinales. La zone côtière est presque partout recouverte par les formations dunaires du Pliocène et du Quaternaire, sur une bande parallèle à l'Océan et large d'une vingtaine de kilomètres [4]. (Figure1).

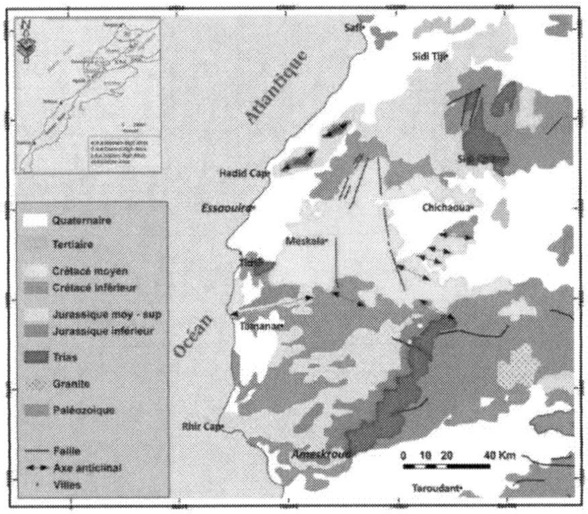

Fig. 1. Carte de localisation et contexte géologique du bassin d'Essaouira.

3 Outils et méthodes

Pour atteindre les objectifs fixés, nous avons utilisé deux méthodes complémentaires : 1) Nous avons commencé par l'analyse des résultats de calculs par la méthode de la petite réfraction afin d'évaluer les caractéristiques de la couche altérée, cerner l'architecture du bassin en deux dimensions, et suivre sont évolution spatio-temporelle (temps en fonction de profondeur). Cette méthode permet d'avoir l'épaisseur et la vitesse moyenne de la couche entre le tir directe et le tir inverse [3], mais pour avoir des valeurs de la correction statique entre ces couplets de points de mesures l'interpolation par les carottages sismiques (Upholes) s'avère nécessaire comme solution pour une étude complète de la zone altérée [5].

Les corrections statiques correspondent généralement à des valeurs utilisées pour compenser les retards sur le temps d'arrivée en sismique dues aux effets de l'altimétrie et de l'altération des couches superficiels. Le nom de statique provient de la constance des corrections en fonction de temps c'est-à-dire que pour une trace brute donnée, tout le train d'onde est déclaré lors de ces corrections de la même valeur de temps. Les valeurs de la correction statique peuvent être positives ou négative selon le positionnement des traces par rapport au plan de référence (Datum Plane : DP) [3, 5]. Ci-dessous la formule (1) pour calculer les corrections statiques [3, 7].

$$\mathbf{CS = \{[\textstyle\sum (hi/vi) + (Ed - Es + \sum hi)] / Vc\}} \quad (1)$$

Avec: Vi et hi: sont respectivement la vitesse et l'épaisseur de la couche i déduites de la dromochronique.

Zs: Elévation de point de mesure par rapport au niveau zéro de la mer obtenue par les levées topographiques.

Ed: Elévation du plan de référence (DP) qui égale à 400 m dans la zone d'Essaouira.

Vc: La vitesse de remplacement estimée à partir de la connaissance préalable (Vc = 3000m/s dans le bassin d'Essaouira).

2) Dans un deuxième temps, et dans le cadre de l'évolution de la propagation des vitesses au niveau du bassin nous avons étudié la répartition lithologique à

partir des données de forages réels les plus représentatives (figure 4) pour estimer d'une part le taux de variation des vitesses en fonction de la variation lithologique [6], et d'autre part pour comparer l'harmonisation des résultats obtenus par cette méthode avec celle de la petite réfraction. Il nous semble obligatoire avant toute modélisation de préparer les données nécessaires de la zone d'étude. Ces dernières sont collectées à partir des études faites sur la zone altérée par la méthode de la petite réfraction et les carottages sismiques.

En utilisant des logiciels de modélisation (Petrel & ArcGis 10.5), on a pu cartographier et modéliser les différents paramètres de la zone altérée, sous forme des cartes des iso-vitesses, d'épaisseurs (isopâques) et des carottages vitesses-lithologie pour chacune des couches mise en évidences dans cette étude, où chaque point est placé selon ses coordonnées XY.

4 Résultats

Bien que les points des carottages sismiques ne sont pas trop nombreux et ne couvrent pas une surface importante, il apparaît que les résultats obtenus pour cette méthode concordent dans l'ensemble avec ceux obtenus par la méthode de la petite réfraction (figure 5A et 5B). D'autre part la fréquence de la présence ou de l'absence de ces couches (H0, H1 et H2) est aussi identique pour les deux méthodes. Seulement on note l'absence total d'épaisseur de la couche H3 par la méthode de la petite réfraction (PT), par contre sa présence au niveau de la méthode des carottages sismiques (UPH).

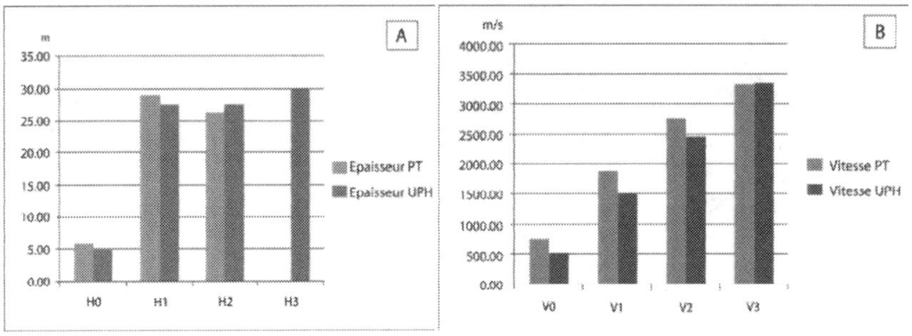

Fig. 2. A) Comparaison de la moyenne des épaisseurs des couches entre les deux méthodes. B) Comparaison de la moyenne des vitesses dans les couches entre les deux méthodes.

Les profils de corrélation montrent une variation latérale et verticale des vitesses, faisons apparaître l'existence d'une hétérogénéité lithologique du site étudié. En effet, les carottages sismiques ont permis d'identifier plusieurs formations lithologiques avec une dominance remarquable des calcaires et des argiles une fois et des calcaires et marnes une autre fois. Les deux premières couches sont généralement de même lithologie mais, parfois, la limite entre ces deux correspond à une variation lithologique.

Les résultats obtenus concernant le calcule des corrections statiques, montrent des valeurs moyennes varient entre – 127 et + 282 ms avec l'existence des valeurs élevées par endroit notamment dans la partie Nord-Est du bassin (Figure 3).

La présente carte (figure 3) a pour objectif d'indiquer les zones où les ondes sismiques sont affectées par le maximum de retard sur les temps d'arrivée. Ces zones sont caractérisées par des valeurs de corrections statiques (Cs) importantes, qui peuvent êtres positives ou négatives. Pour remédier à ce problème, on prévoit pour la prochaine campagne sismique, de concentrer les carottages sismiques et les petites réfractions en resserrant les mailles des lignes sismiques au niveau de ces zones afin d'y avoir le maximum de valeurs de la correction statique, ce qui sera très utile afin de fournir des images assez correctes de la géométrie des réflecteurs.

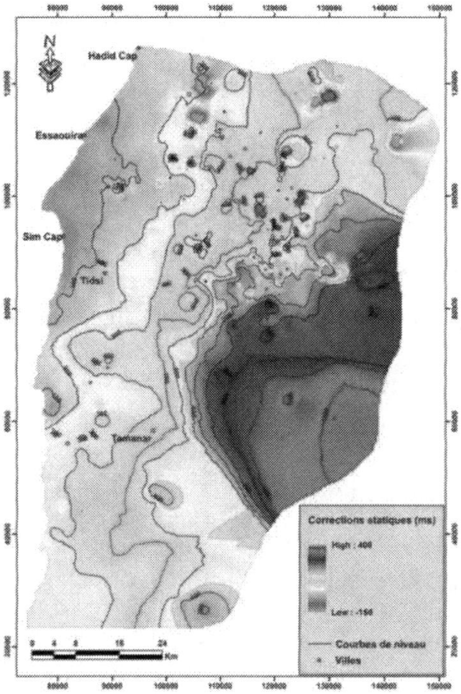

Fig. 3. Carte des corrections statiques "Cs" du bassin d'Essaouira.

5 Discussion & Interprétation

5.1 Couches altérées (H0, H1, H2)

Les cartes des iso-pâques de l'ensemble des couches (H0, H1 et H2), montrent que l'épaisseur varie latéralement dans la majeure partie des cartes. Toutefois, on remarque quelques zones très réduites et ponctuelles, indiquent une transition des couches altérées et hétérogènes vers autres qui ont des épaisseurs plus ou moins stables et homogènes (figure 4.C). La présence d'épaisseur dominant relativement la majeure partie de la carte, peut être expliqué par l'abondance de données dans ces zones par rapport aux autres, où il y a moins de données. Il est à noter que dans l'ensemble, les épaisseurs de la deuxième couche (Figure 4. B) sont plus élevées que celles de la première (Figure 4. A) (h2>h1).

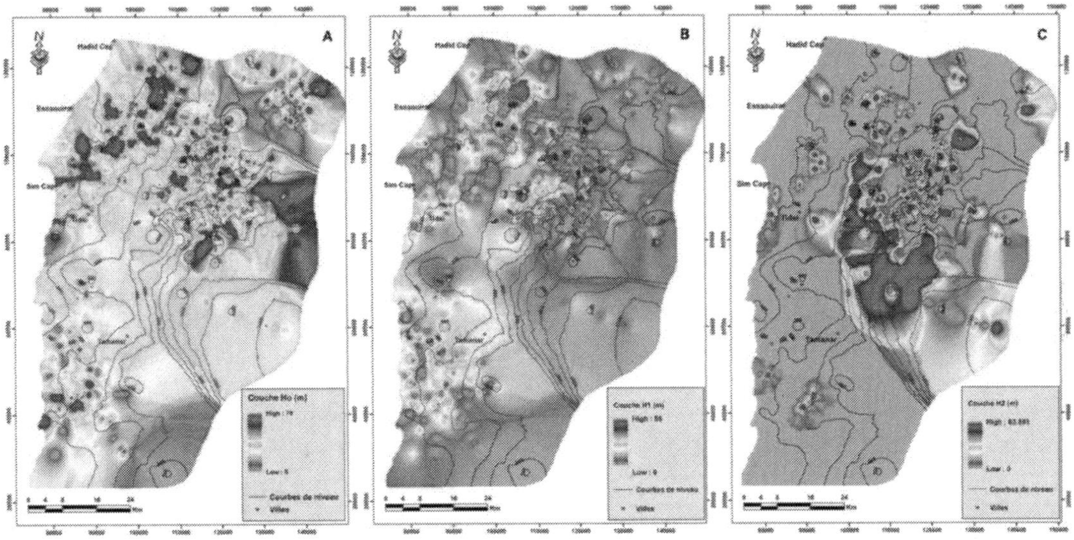

Fig. 4. A. Carte des iso-pâques de la première couche altérée « H0 ». B. Carte des iso-pâques de la deuxième couche altérée « H1 ». C. Carte des iso-pâques de la troisième couche altérée « H2 ».

En effet dans les zones de base altitude l'érosion implique une désagrégation superficielle de la roche ou du sol appelée météorisation. Cette érosion se produit sur place, et produit des débris qui restent sur place. A ces débris s'ajoutent les matériaux issus de l'érosion des zones de haute altitude. Ceci explique les épaisseurs élevées de la couche altérée dans les zones de basses altitudes. Contrairement, dans les zones de haute altitude, on distingue généralement destruction du matériel rocheux par érosion (ablation du matériel) puis son transport et son accumulation dans les zones de plus basse altitude. De ce fait, l'épaisseur de la zone altérée à hautes altitudes est relativement faible.

La présence des domaines d'épaisseurs bien concentrés au niveau de la carte, figure 4.C par rapport aux autres (figure 4.A et 4.B), indiquent une transition des couches altérées et hétérogènes vers autres qui ont des épaisseurs plus ou moins stables et homogènes. On peut dire que ces couches sont probablement liées à une zone en dehors de la zone altérée.

La variation de la somme des épaisseurs des trois couches ne montre pas une corrélation visible avec la topographie (épaisseur élevée dans les basses altitudes et épaisseur faible dans les hautes altitudes). L'explication la plus logique est que les reliefs montagneux escarpés sont des zones de dénudations, alors que les plaines et les piémonts sont des zones d'accumulation.

5.2 Variation des Vitesses (V0, V1, V2 et V3)

Les cartes des iso-vitesses, ressortent qu'il existe trois niveaux V0, V1 et V2, ceci indique qu'on est en présence de trois couches superficielles ayant respectivement les épaisseurs H0, H1 et H2. Dans l'analyse de quelques dromochroniques, on peut rencontrer une quatrième vitesse V3 d'une épaisseur non déterminée.

Au niveau de la première couche, les vitesses (Figure 5. A) sont largement inférieures à celles de la deuxième (figure 5.B) (V1>V0). Cette différence correspond probablement, soit à une différence de lithologie entre la couche H0 et H1, soit aux roches de même lithologie mais peu altérées en H1 par rapport à H0. Etant donné que la vitesse de remplacement de la zone d'Essaouira est fixée à 3000 m/s, on peut considérer que les vitesses relativement élevées de la troisième couche (figure 5.C), correspondent à celles des roches dures non altérées, de même pour la quatrième couche (figure 6) qu'on peut la considérée en de hors de la zone altérée. La distribution de la vitesse dans ce dernier cas montre une assez bonne corrélation avec la topographie, ou les basses altitudes correspondent aux zones de faibles vitesses et les hautes altitudes correspondent aux zones de fortes vitesses.

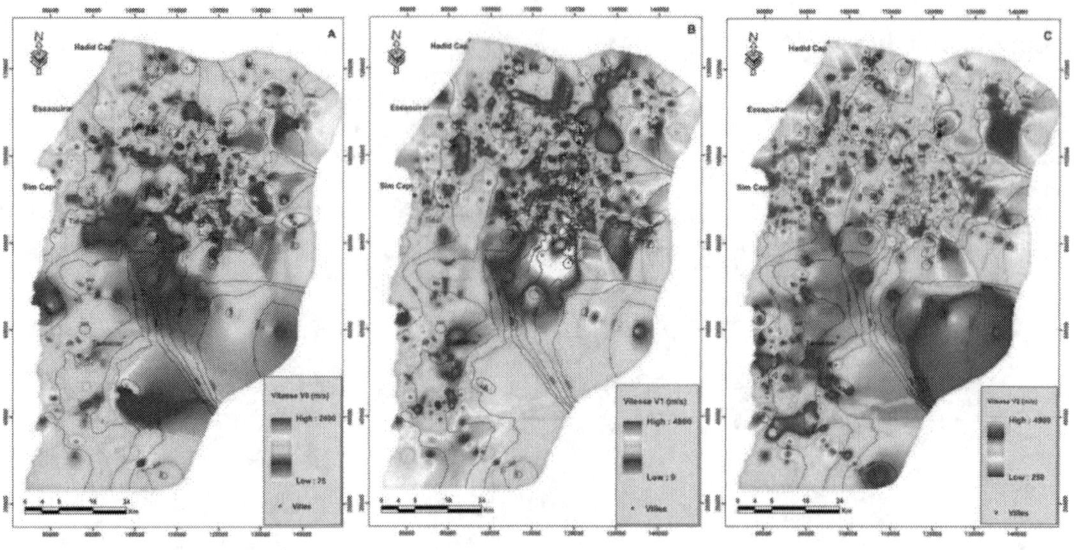

Fig. 5. A. Carte des iso-vitesses de la première couche "H0". B. Carte des iso-vitesses de la deuxième couche "H1". C. Carte des iso-vitesses de la troisième couche "H2".

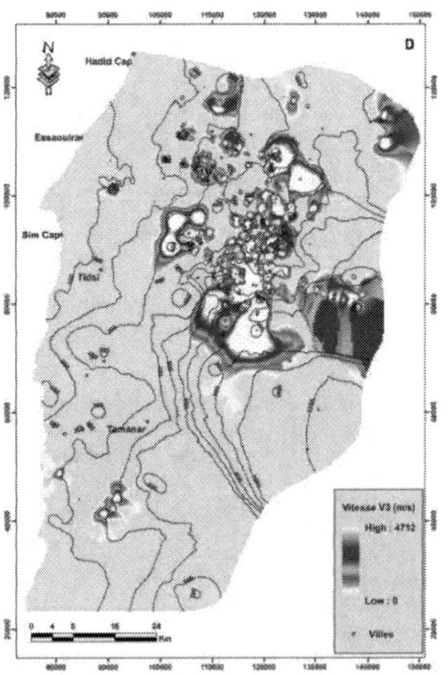

Fig. 6. Carte des iso-vitesses de la quatrième couche "H3".

D'après les trois vitesses enregistrées, il apparaît que V0<V1<V2. Donc, la vitesse augmente avec la profondeur car les roches deviennent plus consolidées et très peu altérées. On peut obtenir aussi sur ce site que le recouvrement du substratum représente un profil complexe et hétérogène (altéré).

5.3 Corrélation vitesses-lithologies

Les méthodes de calcules présentées précédemment, nous ont permis de suivre la variation des épaisseurs des différentes couches de la zone altérée en se basant sur la variation des vitesses des ondes sismiques qui y propagent. D'autre part, les valeurs des vitesses peuvent être corrélées avec la lithologie probable et la lithologie réelle obtenus par les carottages.

L'analyse des carottages obtenus ont permis d'identifier plusieurs formations lithologiques, ces formations, nous ont permis de réaliser deux coupes de différentes directions (NS, EW) afin de montrer la relation entre la vitesse sismique et la variation lithologique.

Le profile Nord-Sud (Figure 7.A) montre une élévation progressive du terrain vers le Sud, ainsi la lithologie de la zone altérée est variable verticalement et discontinues latéralement avec des passages latéraux fréquents, ceci peut être expliqué par la grande distance séparant les carottages (2 à 5 Km). Le profil de corrélation montre que les vitesses varient latéralement et verticalement, ainsi apparaît l'existence d'une hétérogénéité lithologique du site. Par endroits, les terrains où les vitesses sont inférieures à 700 m/s correspondent à des argiles, des marnes ou des calcaires qui sont très altérés (puits UH41, UH19). Par contre les vitesses voisinent les 3000 m/s, sont des marnes et des calcaires très tendres. Ces vitesses sont celles des roches non altérées.

Les deux puits situant à l'Est de la coupe orientée Est-Ouest (U20 et U21) (Figure 7.B), sont constitués essentiellement par des calcaires avec toutes ses variétés et c'est au niveau de ces puits où les vitesses atteignent son maximum : 2500 et 2700 m/s vers une profondeur de 470 m. Au niveau de cette coupe (figure 7.B), la valeur la plus élevée (2700 m/s), se présente au niveau de la base de la couche altérée du puits UH21, ceci est dû à une alternance des couches où la lithologie est dominée par les calcaires et parfois avec des marnes. Cependant la valeur minimale est de l'ordre de 440 m/s, apparaît au sein de la première couche altérée dans les formations du calcaire et du marne au niveau du puits UH29, ceci indique l'influence de la densité sur la vitesse de propagation des ondes sismiques.

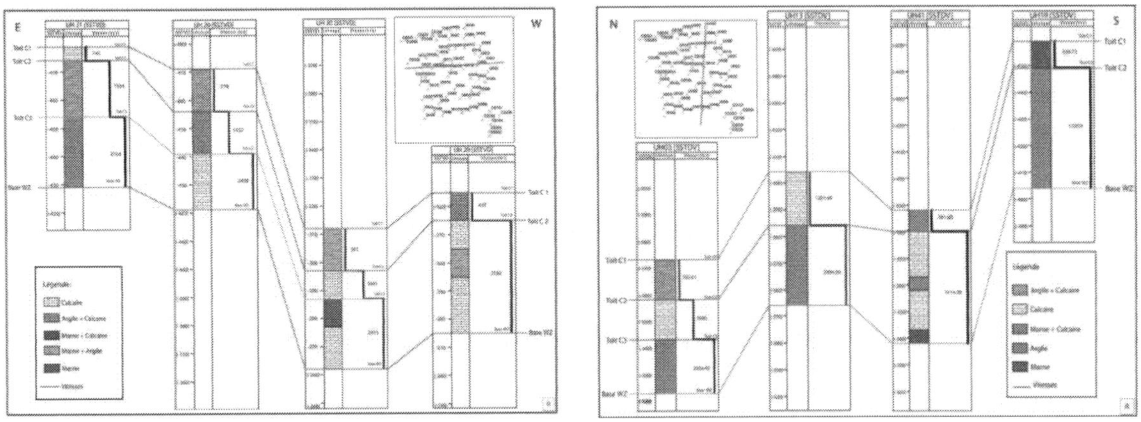

Fig. 7. A. Profil de corrélation 'vitesses-lithologie' de la zone altérée (transect Nord-Sud). B. Profil de corrélation 'vitesses-lithologie' de la zone altérée (transect Est-Ouest).

6 Conclusion

La sismique réflexion associée à la petite réfraction est un outil d'investigation extrêmement puissant joignant une grande profondeur de pénétration à une bonne résolution. Son coût est relativement élevé par rapport aux autres méthodes géophysiques, mais il demeure très compétitif par la précision de son diagnostic et la finesse de ses informations.

L'analyse détaillée des résultats de calcul des corrections statiques a dégagé la présence d'une zone altérée de topographie irrégulière dont l'épaisseur et les vitesses varient latéralement. Ces facteurs conduisent à souligner l'importance et la minutie à l'application des corrections statiques au cours du traitement des données sismiques issues du secteur d'étude. En effet les variations de vitesses dans les cinquante premiers mètres peuvent introduire des anomalies importantes sur la sismique (entre -127 et 282 ms : temps double de retard) pouvant avoir de lourds conséquences lors de l'interprétation ou de l'implantation de forage. Ces variations sont principalement dues à des changements latéraux de faciès et des variations d'épaisseur des formations.

Le calcul des corrections statiques obtenues par les deux méthodes, a mis en évidence des valeurs élevées au niveau de certaines zones (Nord-Est), ce qui nous ramène à faire un certain nombre de recommandations qui permettront d'orienter les futures campagnes au niveau de ces zones. A savoir d'y concentrer les carottages sismiques et les petites réfractions, en resserrant la mailles des lignes sismiques, afin d'y avoir le maximum de valeurs de la correction statique et par la suite une meilleur imagerie des réflecteurs.

Remerciement

Nous remercions l'ONHYM (DEP) pour avoir mis à notre disposition les enregistrements et les carottages sismiques. Cette étude a été réalisée dans le cadre du Master : « Géologie, Océanologie, Eau et SIG», Laboratoire Risques Géologiques, Télédétection et Développement Durable (GEORISK), Faculté des Sciences de Rabat. Nous tenons à remercier les professeurs de ce Master et les évaluateurs pour leurs commentaires et suggestions.

Références

1. R. Bouatmani, Le bassin d'Essaouira (avantpays atlasique, Maroc) : évolution géodynamique méso-cénozoïque et modélisation de la maturation thermique des hydrocarbures. Thèse Doctorat, Université Mohammed V Agdal, Faculté Sciences Rabat, **206 p**. (2002)
2. P. Broughton, A. Trepanier, Hydrocarbon generation in the Essaouira basin of western Morocco. A.A.P.G. Bull., 77, 6, 999-1015. (1993)
3. Bertin, CGG. *Sismique réfraction*. La sismique réfraction appliquée aux problèmes de la mécanique des roches. Bull. BRGM. Section 2. N°4 p 19-29. (1970)
4. A. COCHET & M. COMBE, Bassin d'Essaouira-Chichaoua et zone côtière d'Essaouira. Rapp. Notes Et Mémoires Du Service Géologique. N° 231. Inéd. MTPC/DH/DRE. (1975)
5. F. Coppens, Les corrections statiques de base. Fascicule réf 29870. IFP (France). (1982)
6. M. Djeddi, *Carottage sismique et Sismosondage*. Extrait de http://djeddimabrouk.fr.gd/. Date d'accès: Juillet 2016 (2013)
7. M. Djeddi, Cours de prospection sismique «*Les Corrections Statiques*» Département de Géophysique, Université M'Hamed Bougara de Boumerdes. Algérie. Nov. **22 p**. (2015)
8. P. Le Roy, Les bassins ouest marocains, leur formation et leur évolution dans le cadre de l'ouverture et du développement de l'Atlantique central (marge africaine). Thèse de Doctorat, Université de Bretagne occidentale, Brest, France, **327 p**. (1997)
9. ONAREP, Triassic depositional system: geology and play concept. Second Seminar on Petroleum exploration in Morocco, Rabat, 24-25 november 1998 (inédit), **105 p**. (1998)

 https://doi.org/10.1051/matecconf/201814902081

Probabilistic Landslide risk assessment - Application to Eastern Rif (Morocco)

Evaluation probabiliste du risque glissement de terrain - Application au Rif oriental du Maroc

RFIFI Mohamed [1,a], AIT BRAHIM Lahsen [1]
[1] Université Mohammed V, Faculté des sciences, Unité de recherche GEORISK, LGRN, 4 Avenue Ibn Battouta -Agdal, BP 1014, Rabat, Maroc.

Abstract. The present study is consecrated to the probabilistic mapping of the landslide risk at the local scale of an area that belongs to Al Hoceima city in the western Rif of Morocco. The study focuses mainly on the spatial analysis of multi sources data by using an environment GIS (Geographic Information Systems), and the application of the bivariate probabilistic model to qualify the risk susceptibility. The employed methodology is based on three stages. First, the evaluation of landslide susceptibility (S) by the analysis model cited before. Second, the identification and the estimation of the potential consequences (C) for the existing issues. Finally, the landslide risk (R) is evaluated by combining the susceptibility and the potential consequences map.
This study requires the use of spatial analysis techniques. It also refers to the risk maps scale, generally reduced and being inappropriate at the urban project area. The obtained risk map defines four risk intensities with a spatial resolution of two meters.
Keywords: Landslide - Probabilistic mapping - Susceptibility - Consequences - Risk – GIS.

1 Introduction

Les risques naturels constituent l'un des principaux enjeux d'aménagement, en particulier dans la région du Rif oriental marocain, zone caractérisée par des paramètres de prédisposition au risque. Cette étude a été initiée à partir d'un constat terrain et un intérêt scientifique initié par des travaux de recherches antérieurs.

Les mouvements de terrain dans le Rif ont fait l'objet de plusieurs travaux, dont ceux de [1] qui demeure l'un des pionniers dans l'étude de ces mouvements. Dans d'autres travaux [2, 3] consacrés à des régions précises du Rif, ont invoqué la combinaison de plusieurs facteurs provoquant cet instabilité.

L'objectif principal du présent travail est de mettre en évidence les relations entre des paramètres de susceptibilité (conditions géologiques, topographiques locales…) et les enjeux existants à l'échelle d'une zone urbanisée du Rif menacées par le risque de glissement de terrain.

2 Site d'étude

La zone d'étude est située à l'ouest de la ville d'Al Hoceima (figure 1), elle s'étend sur une superficie approximative de 1 800 000 m^2 et elle appartient au Rif oriental, une chaine montagneuse dont la structure est chevauchante avec une superposition de nappes (divisées en plusieurs domaines : les Ghomarides, les Sebtides et la dorsale calcaire).

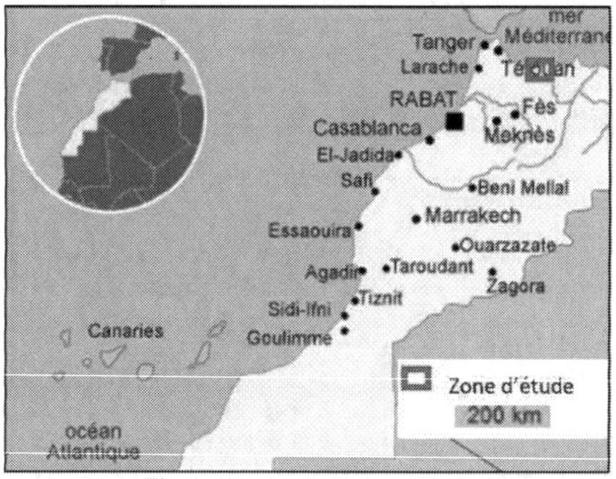

Fig. 1. Situation de la zone d'étude

Le secteur est marqué par l'affleurement des calcaires et des marnes avec un relief accidenté allant jusqu'à 70° de pente.

La zone est caractérisée également par une activité sismique remarquable. Plusieurs séismes ont secoués la région, notamment celui de 1994, 2004 et les derniers de 2016.

Le climat dominant la région est de nature semi-aride avec une moyenne de précipitations de 300 mm/an et une température moyenne annuelle égale à 18°C. Plusieurs facteurs ont favorisé le développement du réseau hydrographique dans la zone (formations géologiques, pente du terrain et surface de drainage).

3 Approche d'analyse du risque employée

3.1 Concepts de l'approche

- L'évaluation de l'ampleur du risque glissement de terrain consiste à exprimer le lien entre la susceptibilité et les conséquences du risque sur les enjeux exposés ;
- La susceptibilité est la probabilité spatiale qu'un glissement de terrain se produit sur un territoire pour différentes conditions environnementales locales. Les techniques d'analyse spatiale [4,5] permettent d'identifier des zones susceptibles selon deux approches : directe (qualitative) et indirecte (probabiliste). L'approche indirecte est la plus objective et permet d'obtenir des cartes de susceptibilité plus fiables et réalistes. Elle est retenue dans le cadre de ce travail.
- Le principe de l'approche indirecte « bi-variée » consiste à définir les relations mathématiques entre les facteurs de prédisposition (Variables indépendantes) et l'occurrence spatiale du glissement (variable dépendante).
- La définition des conséquences du risque consiste à évaluer une valeur d'endommagement pour chaque élément exposé.
- Le calcul du risque est effectué par croisant, à l'aide d'un tableau à double entrées [6], des classes de susceptibilité et les classes de conséquences potentielles simulées.

3.2 Données de l'étude

Les données pour l'analyse de la susceptibilité peuvent être regroupées en cinq groupes de variables (table 1) : inventaire géomorphologique des glissements de terrain observés, pente topographiques, géologie, formations superficielles, hydrographique et occupation du sol.

Table 1. Données analysées pour l'évaluation de la susceptibilité

Variable dépendante/explicative		Source
Vd	Glissement de terrain (GT)	Observations, cartes
Ve	Pente (PE)	Modèle numérique de terrain
	Lithologie (LI) Formations superficielles (FS)	Carte géologique et rapports
	Réseau hydrographique (HY)	Cartes et modèle numérique de terrain
	Occupation du sol (OS)	Cartes

L'inventaire géomorphologique des glissements de terrain a été réalisé à partir de l'observation terrain et les rapports techniques archivés (figure 2).

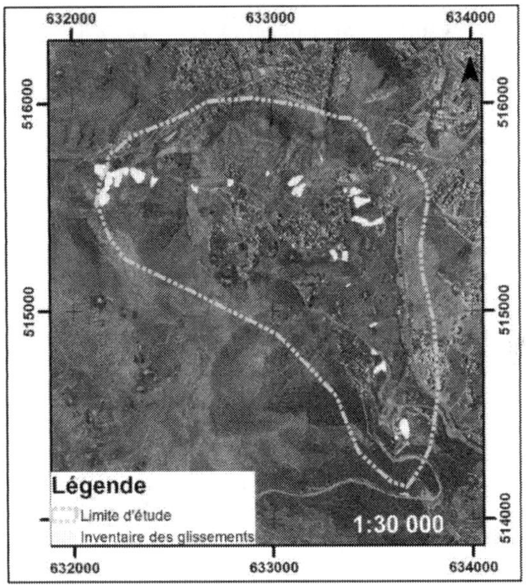

Fig. 2. Carte de l'inventaire géomorphologique des glissements de terrain actifs

Les valeurs attribuées pour qualifier le degré de l'endommagement (CEP) en cas de glissement reflète le cout des dommages monétaires et/ou humains engendrés par le phénomène. Les valeurs ont été quantifiées par retour d'expérience dans des travaux antérieurs [7].

3.3 Cartographie de la susceptibilité

La cartographie de la susceptibilité consiste à :

- Etablir un inventaire cartographique des zones d'occurrence des glissements ;
- Tester l'indépendance conditionnelle (test $\chi 2$) entre les cinq variables indépendantes pour in seuil de signification $\alpha = 0.05$ [8] ;
- Définir les combinaisons entre les variables indépendantes (au sens statistique). Suivant le protocole défini par [9] ;
- Valider statistiquement la combinaison employée en introduisant les 2/3 de la population totale des glissements de terrain, à partir d'un tirage aléatoire ;
- Caler le modèle obtenu en utilisant le 1/3 de la population des glissements restante ;
- Interpréter les résultats du modèle sous forme de classes de susceptibilité (S). Le choix des classes est établi par discrétisation manuelle, basée sur l'existence de seuils naturels visibles sur la courbe cumulative des valeurs de susceptibilité.

3.4 Cartographie de conséquences

L'évaluation est faite par identification des catégories des enjeux existants (structurels, corporels et fonctionnels) et par attribution des coefficients d'endommagement (CEP) à chaque détail exposé à l'enjeu. La valeur du coefficient est comprise entre 0 et 1 (table 2).

Table 2. Enjeux identifiés au niveau de la zone et valeurs du coefficient d'endommagement (CEP) attribuées

ENJEU CORPOREL ET STRUCTUREL (E1)	
Type de zone	**CEP**
Urbaine	1,0
Naturelle	0,3
Affleurement rocheux et sol nu	0,1
ENJEU FONCTIONNEL (E2)	
Type de l'infrastructure	**CEP**
Route principale	1,0
Route secondaire	0,8
Piste	0,6
Aménagement hydraulique	1,0
Confortement	1,0
Murs de soutènement	1,0

La combinaison des valeurs du CEP pour les enjeux étudiés (E1 et E2) permet de déduire 5 classes de dommages.

3.5 Cartographie du risque total

Cette cartographie est obtenue par le croisement des classes de la carte de susceptibilité et de conséquences potentielles suivant une matrice à double entrée (table 3).

Table 3. Croisement de la carte de susceptibilité et la carte de dommages

CLASSES DE SUSCEPTIBILITES					
S3	S2	S1	S0		CLASSES DE CONSEQUENCES POTENTIELLES
R3	R3	R2		C4	
R3	R2	R1		C3	
R2	R2	R1		C2	
R2	R1	R1		C1	
				C0	

4 Résultats et interprétation

4.1 Carte de susceptibilité finale modélisée

Les tests effectués pour le calage des variables indépendantes indiquent que la meilleure combinaison est celle qui regroupe les cinq variables, par ordre hiérarchique, la pente, la lithologie, l'occupation du sol, la géologie et la distance aux cours d'eau (FS + LI + OS + PE + HY).

Une combinaison de trois variables seulement (FS, LI et OC) produit une erreur relative sur les surfaces modélisées de 40% (figure 3). L'ajout des variables PE et HY permet une réduction de 29% d'erreur.

La cartographie de la susceptibilité finale en 4 classes a finalement été retenue. Elle identifie respectivement des aires de susceptibilité de 96 000 m^2, 664 000 m^2, 983 000 m^2 et 57 000 m^2 pour S0, S1, S2 et S3 (figure 4) :

- Susceptibilité forte (S3) : coïncide avec des terrains marneux de pentes moyennes et fortes exposés à l'action des eaux de surface (3% de l'aire d'étude) ;
- Susceptibilité modérée (S2) : corresponde généralement à des formations marneuses et des encroutements calcaire ayant des pentes moyennes (55% du secteur) ;
- Susceptibilité faible (S1) et nulle (S0) coïncide avec des terrains de faibles pentes occupant 42%.

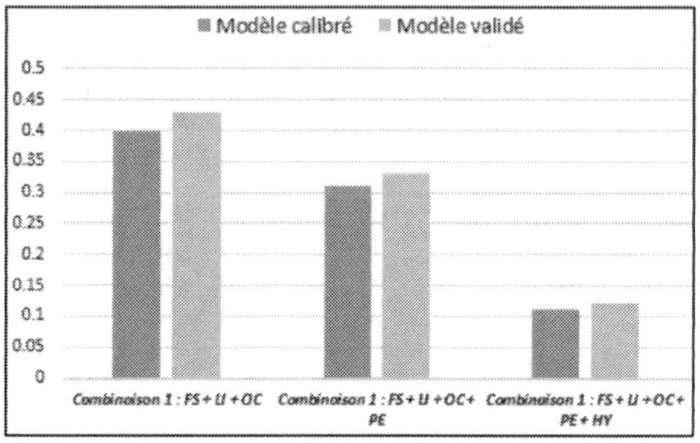

Fig. 3. Combinaison des paramètres de prédispositions au risque et résultats du modèle retrouvés

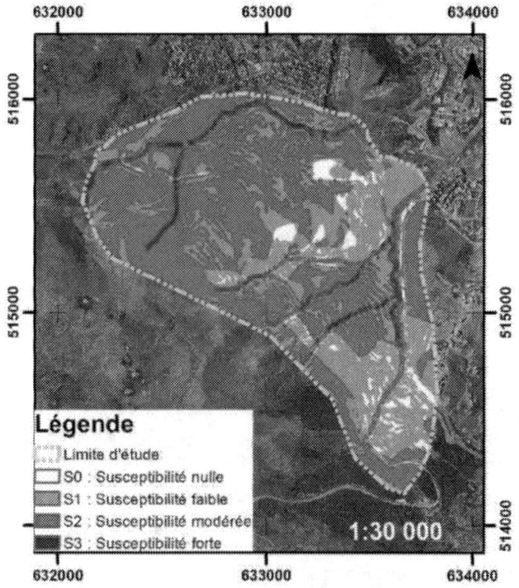

Fig. 4. Carte de susceptibilité aux glissements de terrain

4.2 Carte des conséquences potentielles

La carte de dommage met en évidence des enjeux liés aux infrastructures et réseau du transport existant, les zones urbanisées et en urbanisation progressive. La carte identifie respectivement des aires de conséquences potentielles de 927 000 m², 4 000 m², 733 000 m², 1 000 m² et 135 000 m² pour les classes C0, C1, C2, C3 et C4 (figure 5).

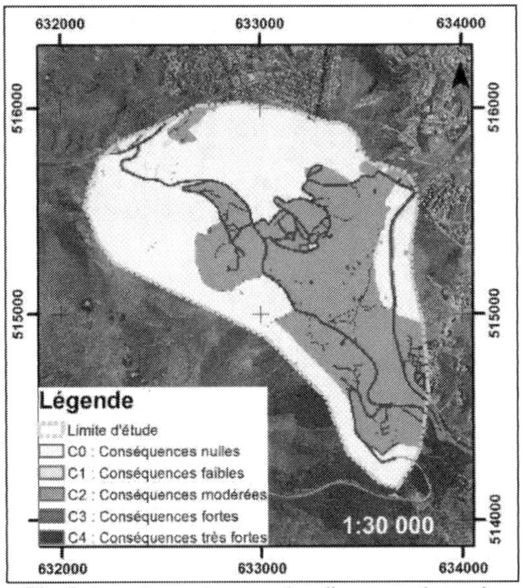

Fig. 5. Carte de conséquences des glissements de terrain

4.3 Carte du risque total

Le croisement des cartes de susceptibilité et de conséquences potentielles indique que les zones de risque fort (R3) sont très peu développées sur le territoire étudié (figure 6). Les zones R3 correspondent à des secteurs en urbanisation ayant une pente forte combinée à un écoulement de l'eau de ruissèlement. Le niveau de risque négligeable représente 60% de la zone (1 081 000 m²) et correspond aux zones non engagées dans des activités d'aménagement. Cependant, le risque faible et modéré se concentre à l'intérieur de la zone d'étude et ils représentent respectivement 406 000 m² et 294 000 m².

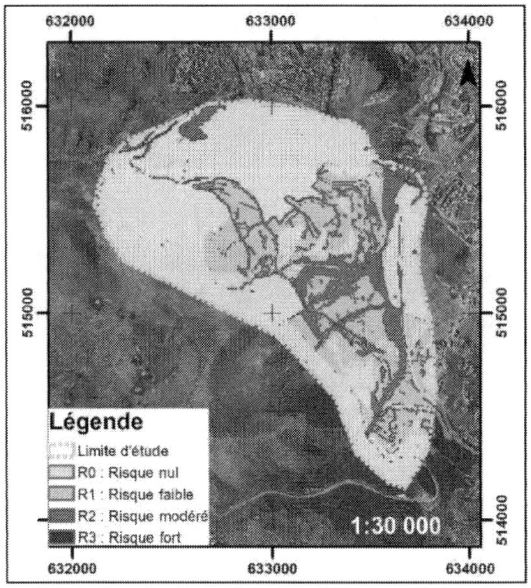

Fig. 6. Carte du risque global des glissements de terrain

5 Conclusion

Cette étude s'inscrit dans le cadre de l'analyse du risque « glissement de terrain » et l'évaluation probabiliste de son intensité à l'échelle d'une zone située à l'ouest de la ville d'Al Hoceima. L'étude démontre des niveaux de risques variables déterminés par les techniques d'analyse spatiale. A partir d'une démarche en 3 étapes, le risque est identifié et cartographié en associant plusieurs modèles d'analyse : (1) la susceptibilité de la zone au glissement de terrain est analysée par un modèle probabiliste bi-varié de la théorie d'évidence ; (2) les conséquences potentielles sont estimées par approche semi-empirique ; (3) le risque est évalué en combinant la carte de susceptibilité et celle des conséquences potentielles.

La cartographie du risque de glissement nécessite un inventaire géomorphologique des glissements au niveau de la zone. C'est un travail qui nécessite une bonne connaissance du terrain, une exploitation des rapports techniques disponibles et la mise en place d'une base de données utiles. Le bon modèle retrouvé par analyse probabiliste a démontré des résultats satisfaisants et une bonne adéquation aux particularités du secteur d'étude.

Références

1. Millies-Lacroix, *Les glissements de terrain, présentation d'une carte prévisionnelle des mouvements de terrain dans le Rif*. Mine et Géologie, **27**, pp. 45-54 (1968)
2. K. Margua, *Essai de cartographie des risques naturels : Application à l'aménagement de la région d'Al Hoceima*. Thèse universitaire, Univ. Franche-Comté, 196 p (1994)
3. A. Talhaoui, M. Aberkan, A. Iben Brahim, A. El Mouraouah, N. Jabour, *Contribution à l'étude des risques naturels. Cas de la région d'Al Hoceima.*

Oujda, Colloque des Bassins Sédimentaires Marocains (1999)

4. F. Guzzetti, A. Carrara, M. Cardinali, P. Reichenbach, *Landslide hazard evaluation: a review of current techniques and their application in a multi-scale study, Central Italy*. Geomorphology, **31**, pp. 181-216 (1999)
5. C.J. Van Westen, *The modelling of landslide hazard using GIS*. Surveys in Geophysics, **21**, pp. 241-255 (2000)
6. C. Ko Ko, P. Flentje, R. Chowdhury. *Landslides qualitative hazard and risk assessment method and its reliability*. Bulletin of Engineering Geology and the Environment, **63**, n°. 2, pp. 149-165 (2004)
7. F. Leone, *Concept de vulnérabilité appliqué à l'évaluation des risques générés par les phénomènes de mouvements de terrain*. Environnement et société. s.l: Université de Grenoble (1996)
8. G.F. Bonham-Carter, *Geographic Information System for Geoscientifists: modelling with GIS*. Computer methods in the Geosciences. Pergamon Press, **13** (1994)
9. Y. Thiery, S. Sterlacchini, J.P Malet, A. Puissant, A. Remaître, O. Maquaire, *Strategy to Reduce Subjectivity in Landslide Susceptibility Zonation by GIS in Complex Mountainous Environments*. Heraklion, Greece: 7th AGILE Conference on GIScience (2004)

Application Of Logistic Regression Method To Produce Landslide Susceptibility Map: A Case Study Of Tetouan Mazari, Morocco

L. Ait Brahim [1] And M. Elmoulat [1].
[1] University Mohamed V, Faculty of Science, Research Unit GeoRisk: Geological Risks, Remote Sensing and sustainable development, LGRN Ibn Battouta Avenue. Zip code 1014, Rabat – Agdal, Morocco.
laitbrahim@gmail.com

Abstract. The main purpose of this study is to use logistic regression (RL) model to map landslide susceptibility in and around the area of Tetouan Mazari in the Northern Morocco. Parameters, such as lithology, slope gradient, slope aspect, faults, drainage lines, and hillshade, were considered. Landslide susceptibility map was produced using RL method and then compared and validated. Before the modeling and validation, the observed landslides were separated into two groups. The first group was for training, and the other group was for validation steps. The accuracy of the model was measured by fitting them to a validation set of observed landslides. For validation process, the half landslides remaining was used. The final map was classified into five classes: Very High (32%), High (40%), Medium (7%), Low (7%) and Nil (15%). According to these values logistic regression was determined to be one of the most accurate method to generate landslide susceptibility map. Last but not least, logistic regression model can be used to manage and mitigate hazards related to landslides and to aid in land-use planning for the city of Tetouan..

1 Introduction

Landslides are significant natural hazards in many areas of the world. Every year they cause more than 100,000 deaths and injuries and more than one billion USD in damage [1] . Mapping the areas that are susceptible to be prone to landslides is mandatory for effective land-use management. Thus, there is an increasing need for methodologies to guide managers in choosing the best management strategies, including minimizing impacts from land-use activities in vulnerable slope areas [2]. A landslide susceptibility map indicates areas likely to undergo landslides in the future by correlating some of the factors that contribute to landslides with historic distributions of slope failures [3]. Many methods and techniques, often involving geographic information systems (GIS), have been proposed to produce landslide susceptibility maps [4]). Logistic regression has been applied for susceptibility mapping by various researchers including and [5].

The main objective of the present study is to create susceptibility map by using Logistic Regression (LR). For this purpose, initially, a detailed landslide inventory was prepared [6]. Then, the landslide locations mapped were separated into two groups for training and validation purposes. In addition to that six different controlling factors have been mapped to generate the susceptibility map.

2 Study area

Our site test is Mazari located in the province of Tetouan (Latitude: 35°45'N - 35°30'S; Longitude: 5°30'W - 5°9'E). It covers a total area of about 590 Km2. It is limited by Fnideq, the Mediterranean Sea, Chefchaouen, and Asilah on the North, East, South and West, respectively "Fig. 1". The study area is inserted within a hazardous and mountainous zone called the Western Rif (Northern Morocco). The Rif has been subject to a complex structural and Palaeogeographical evolution since the Triassic period until the Quaternary [7]. This tectonic zone is made up by three different paleogeographic domains.

2.1 The geological and structural sitting

The geological and structural sitting of the study area have been detailed in the geological map of Tetouan

https://doi.org/10.1051/matecconf/201814902082

sheet by (Kornprobst & Durand Delga 1985), which consist broadly of three main domains: (1) Internal domain: represented by the Ghomarides, which are metamorphic nappes of Paleozoic from Silurian to Carboniferous and Sebtides that are composed of polymetamorphic terrain and Paleozoic-Triassic cover called the unit of Federico. The Sebtides-Ghomarides are essentially thrust over calcareous units, namely, the Dorsale Calcaire that composed the third part of this domain. (2) Flyschs: formed by Cretaceous and Tertiary material. We distinguish three types of nappes: Numidian Flysch, Flysch of Tisirène and Flysch of Melloussa. (3) External zone: Represented by Tangier Unit.

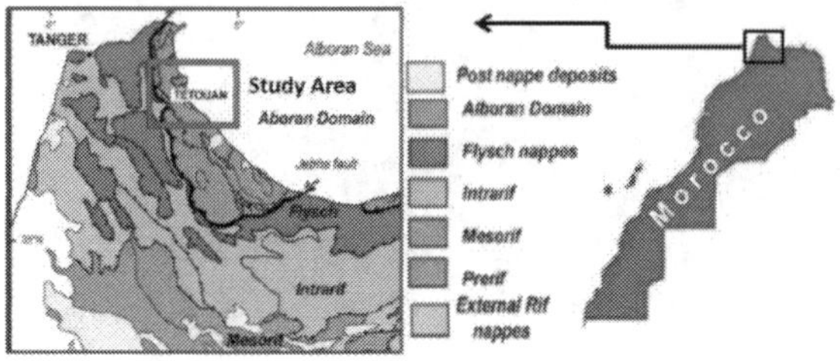

Fig 1. Location map showing the study area in the Province of Tetouan.

2.2 Seismic activity

The Tetouan region, although located in a moderate seismic zone, remains under the effect of the very active neighboring seismic zones (Alboran and Acores). However, no recent events of sufficient intensity (magnitude greater than or equal to 6, intensity VII or more MSK) occurred to slope instability.

2.3 Climate, Precipitation and Temperature

Climate: Our study area knows the succession of two different seasons in one year; to a warm and dry summer, which lasts for several months, succeeds a rainy and colder season corresponding to the end of autumn, winter and early spring. The rains are not continued but alternate during the whole year with periods of fine weather of long and short duration. The study area is part of an area which forms a transition between a Mediterranean and an Atlantic slope; in fact, the climate of the region benefits from two maritime influences, on the one hand a Mediterranean influence because the opening of the limestone chain at Tetouan allows to penetrate Mediterranean perturbations, on the other hand an Atlantic influence because it is exposed to the Atlantic winds of the West and South West. The rains are prolonged and very violent with a maximum in November, December and January, causing intense erosion on the slopes to flash flooding. For precipitation, 80% is recorded between November and April.

Precipitation: has a significant inter-annual variation, suggesting that the precipitation system is not stable from year to year, but also within a given year, which is an important hydrological system and erosion of the watershed.

Temperature: The temperatures are high, at the end of Spring until the end of the Autumn, especially in Summer when one attends a very strong Summer drought which causes desiccations into the soils. According to the doctoral work of [7] , the hottest months of the year are July and August with a high average (38°C for the two months), the coldest months are December and January (2.6°C and 2.2°C respectively). A thin blanket of snow falls in this area (4 to 5 days/year) and lasts few hours. The study area is part of sub-humid zone. There is a good relationship between mass movement and the height of precipitation level (Winter and Spring), itself dependent on altitude and the aspect; but the main feature of mass movements distribution remains the lithological factor, because the majority of these phenomena start in marly and schistose formations.

3 Methodology

3.1 Generation of database for LR model

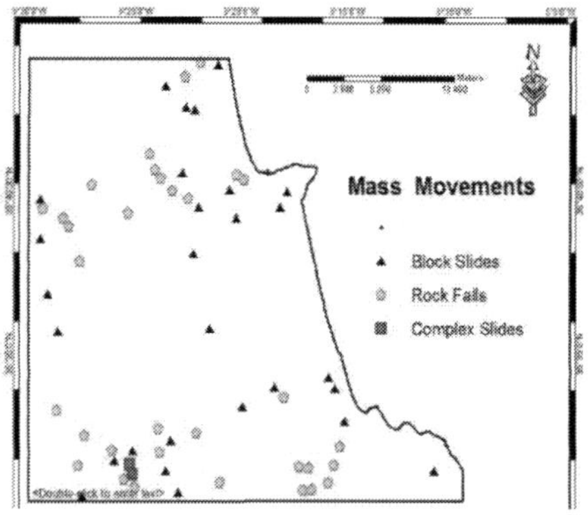

Fig 2. Mass Movements of Tetouan Mazari Area

A total of 126 landslides were mapped. According to the classification made by Varnes [8], the type of landslides movement encounter in our case study were rock falls, and translational and rotational slides. A landslide inventory map was prepared by mapping the landslide locations in Mazari [9] using interpretation of aerial photographs, high resolution satellite imagery, previous reports and extensive field surveys [10]. A random partitioning of the training data was carried out in the area of Mazari, 50% of the whole landslides were selected randomly to calibrate the logistic regression analysis and the half remaining was used for validation purposes "Fig. 2". The pixel size of the landslide map and all the independent variable maps was 100m. Six parameters were created for the landslide susceptibility map of the study area. These parameters are lithology, fault density, hydrographic network density, slope gradient, slope aspect, hypsometry "Fig. 3". The previous description of the geological sitting shows that the study area consists predominantly of marsh sandstone Flyschs, detrital limestones, mudstones and conglomerates. All these formations are almost alternative, which have an effect on slope stability.

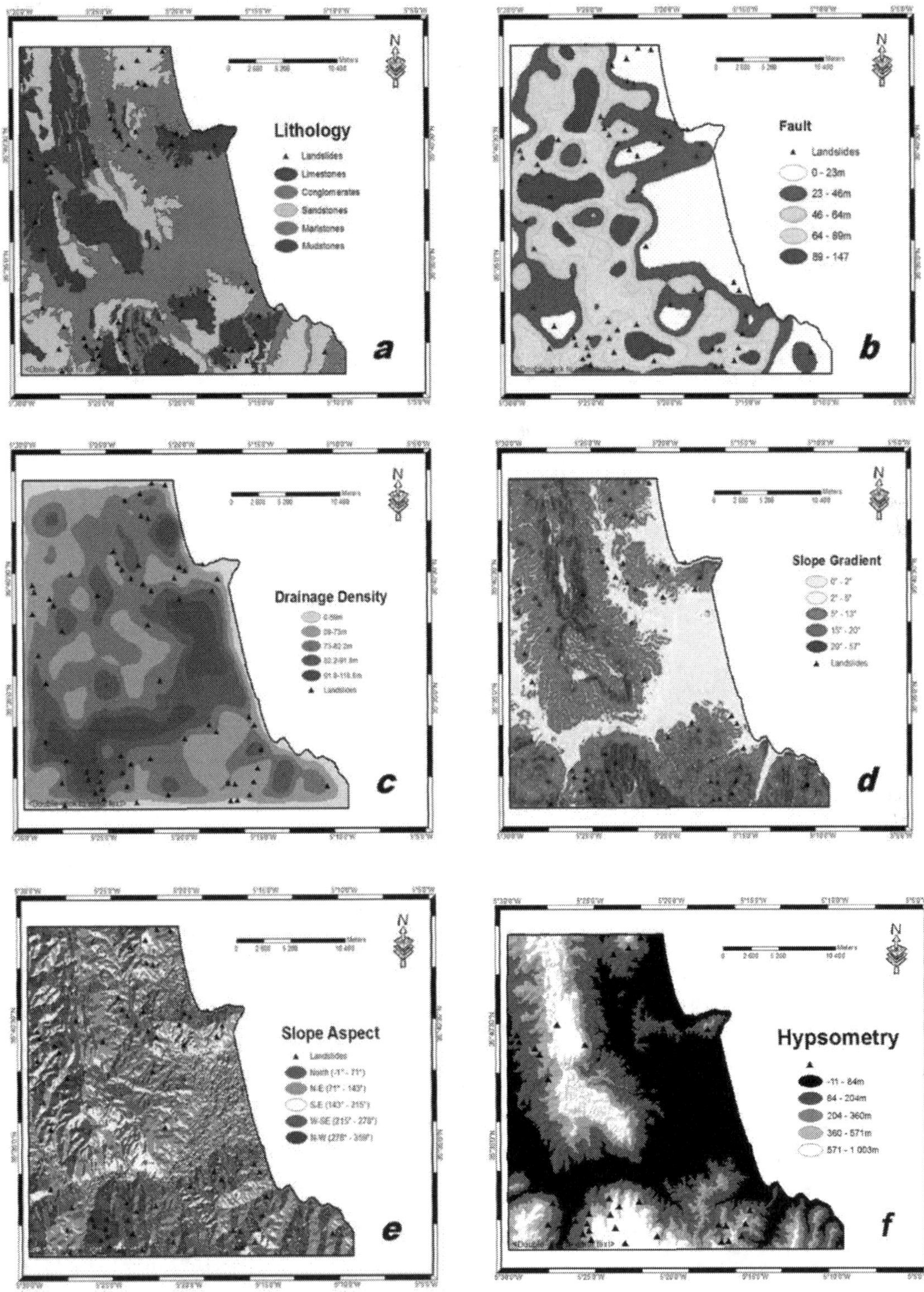

Fig 3. Various thematic data layers prepared in GIS for Tetouan-Ras-Mazari area

For example, mudstones become fluid in contact with water; hence, they are easily discharged when they are on surface. The alternation of limestones and sandstones with marlstones and clays is another unfavorable factor in terms of terrain stability [7].The lithological map used in this study has been derived form the geological map of the study area [11] and topographical map [12], combined to frequencies of landslide occurrence, and then were classified into five classes to create the lithology map "Fig. 3(a)".

According to previous work that has been done in the past by many researchers and experts in Western Rif [13], our case study is a faulted zone which presents favorable conditions for landslide occurrence. All faults are highlighted by the consistent movement caused within the features of different kinds of tectonic layering; they are posterior to thrust that accommodates horizontal shortening e geological and overthrusting. Hence, fault map of 341 types of tectonic movements were mapped and analyzed using geological map and Landsat 7 ETM+ 2015 satellite imagery observation and validated by ground truth studies [11] "Fig. 3(b)".

Water is considered as the primary factor to trigger landslide mechanisms. Rivers may induce failure of the banks due to slope undercutting. Many studies have shown that the proximity to the drainage lines is an important factor controlling the occurrence of landslides. This can be attributed to the fact that terrain modification caused by gully erosion may influence the initiation of a landslide. In order to map this parameter, we have used a DEM to extract the automatically The hydrographic network using ArcHydro tools. This extension was developed by ESRI along with a hands-on document that explained step-by-step how to perform drainage analysis on a terrain model. DEM Raster analysis is performed to generate data on flow direction, flow accumulation and watershed delineation. These data are then used to develop a vector representation of catchments and drainage lines. Using this information, a geometric network is constructed [14]. The final map of distance to drainage was validated using the topographic map of Tetouan scale 1:50,000 [12]"Fig. 3(c)".

Moreover, A digital elevation model (DEM) A DEM with 30×30m resolution representing our study area was a great source to generate various topographic factors, which influence the landslide activities in this zone. hence, the dem was modified to correct the artifacts associated with the aster sensor and to improve its visual quality. three thematic data layers were extracted which are slope (0° - 54°). aspect (0° - 360°), and hypsometry:

- Slope: The slope gradient is an important component of the slope stability analysis and is frequently used in landslide susceptibility studies. The more the slope gradient increases, the more it will correlate with increased likelihood of failure. However, soil thickness and strength are two factors that vary over a wide range between sites [15]. The slope gradient map of the study area was divided into 5 slope categories "Fig. 3(d)".
- Aspect: can be defined as the compass direction that a slope faces measured in degrees from north in a clockwise direction, ranking from 0° to 360°. The values of each cell in the output raster indicate the compass direction that the surface faces at that location. We especially used the aspect in relation to the effect of the solar radiation and the degree of humidity: The orientation of the slopes inthe region of Tetouan influences especially the degree of instability on the exposed slopes (Wet) North, NNW and NNE for the mudflows and landslides. The slopes (Dry) of the direction South, SSE, SSW show especially an erosion dominated by the streaming "Fig.3(e)".
- Hillshade: creates a shaded relief from a surface raster by considering the illumination source angle and shadows. It is obtained by determining illumination values for each cell in relation to neighboring cells "Fig.3(f)".

3.1.1 Overview of LR model

The binary logistic regression used in this study is one of the most popular statistical methods for determining landslide susceptibility and has often been applied for this purpose [16]. The application of logistic regression requires the inclusion of landslide triggering and/or conditioning parameters as independent variables. In general, the more independent variables that are included, the more complete the model will be, given that the consideration of variables plays a major role in determining the dependent variable. As it has been mentioned by Pham [17] in his paper that LR is another known popular method for landslide susceptibility assessment. Many authors [18], define LR is a mathematical model that used to predict the presence or absence of an event based on values of a set of controlling factors. It can be beneficial for either continuous or discrete, or any combination of both types, and they do not necessarily have normal distributions. The LR has been proven more efficient than other methods such as certainty factor, likelihood ration, artificial neural networks, and multi-criteria decision analysis for landslides susceptibility assessment [19].

$$p=1/(1+□^{(-z)}) \quad (1)$$

Where p is the probability of landslide occurrence, z is a parameter.

$$z = \alpha + \beta 1x2 + \beta 2x2 + \ldots + \beta nxn \quad (2)$$

Where α is the intercept of the equation; x1, x2,….,xn are the predictor factors; β1, β2,…., βn are the coefficients needed to be estimated for the predictor factors [20].

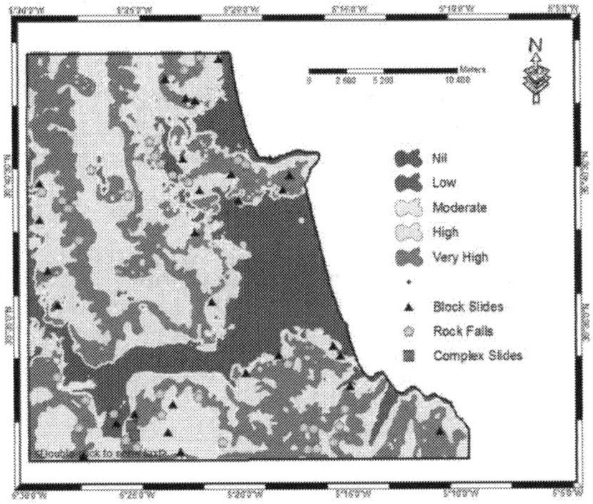

Fig 4. Final Susceptibility map for Tetouan Mazari area using RL model

4 Results and validation

Using the six independent variables as the input factors and the 63 landslide points as a dependent variable, forward stepwise logistic regression was used to analyze the data using the Logistic Regression tool implemented in The ArcGIS along with SDM utilities. An investigation of the susceptibility map highlights a high probability of landslides where mudstones and conglomerates outcrop. The susceptibility maps show also that the landslide is more prone in highly elevated areas greater than 57° with the Northeast, West and Southeast-facing slopes. This map reveals that landslides preferentially occur close to drainage lines and where illumination is present.

Validation of the results is considered to be one of the most important tasks, and without validation, the prediction model will have no scientific significance. The results of the landslide susceptibility map were validated using the test subset of other half remaining (63 landslides points) using the results from the Area Frequency Table and the receiver operating characteristic technique (ROC). The ROC curve is one of those statistical techniques that can be used to provide predictions of the performance and to compare the different models. The overall predicted accuracy of our model is 95.3% which is considered as an excellent according to many authors [21]. Finally, the susceptibility map has been classified into five classes: Very High (32%), High (40%), Medium (7%), Low (7%) and Nil (15%) by using natural breaks methods available in ArcGIS.

According to the results of the Table 1, we notice that the highest value of standard deviation (Sd) for lithology class is 4. This value is found in the formation of marlsones and mudstones or also called argillaceous limestone, which is an intimate mixture of clay and calcium carbonate. These formations regard as homogeneous from the geomorphological and engineering geological point of viewin terms of shallow slope instability. The Table 2 emphasizes that the highest Standard Deviation of the Logistic Regression is 0.315; which can be explained by shallow landslides that has been developed in argillaceous formations of the area. Thus, the instability causative factors are represented by the poor geotechnical properties of the marl-argillaceous material such as high plasticity, high clay fraction and high content of expansive lattice-type minerals that affect the strength properties of the rocks. In addition to that, the swelling and/or shrinkage of marl-argillaceous formations appear to be a very important factor in the localization of landslide phenomena.

Table 1. Summary statistics of The Logistic Regression Model

Factor	Class	Area (Km^2)	Landslides points	Std. C	Class Gen.
Lithology	Conglomerates	120.07	0	0	0
	Sandstones	234.35	21	0.569	99
	Mudstones	57.67	2	-1.308	99
	Marlstones	142.45	21	3.154	4
	Limestones	121.22	11	0.416	99
Fault	Very low	155.91	6	-2.168	1
	Low	95.77	10	0.876	99
	Midium	158.87	25	3.838	3
	High	148.45	12	-0.04	99
	Very high	114.84	2	-2.441	5
Slope	0° - 2°	202.7	15	-0.474	99
	2° - 5°	60.13	10	2.438	2
	5° - 13°	149.65	11	-0.411	99
	13° - 21°	217.62	17	-0.229	99
	21° - 50°	43.75	2	-0.882	99
Aspect	F-N	62.93	4	-0.548	99
	NE-E-SE	235.93	25	1.682	99
	S-E	84.11	6	-0.368	99
	W - SE	172.92	12	-0.68	99
	N - W	117.96	8	-0.602	99
Drainage	Very low	234.04	12	-2.064	1
	Low	223.7	17	-0.376	99
	Medium	162.74	18	1.539	99
	High	48.01	8	2.167	4
	Very high	5.35	0	0	99
Hypsometry	(-) 14 - 84 m	285.55	25	0.482	99
	84 - 204 m	133.62	15	1.434	99
	204 - 360 m	127.43	12	0.574	99
	360 - 572 m	111.8	3	-2.168	4
	572 - 1006 m	15.45	0	0	99

Table 2. The Logistic Regression coefficients obtained for the six independent factors

Factor	LR_Std_Dev
Lithology	0.559
Fault	0.559
Slope	0.39
Aspect	0
Drainage	0.362
Hypsometry	0.636

Regarding the fault parameter, the Table 2 shows that the highest value of the Standard Deviation of RL is 0.559, which is found in the low distance of 0-23m. Thus, we conclude that the areas to be prone to landslides are close to the faulted zone.

The drainage is considered as primary factor to trigger landslide mechanisms. According to the results of the Tables 1 & 2, we observe that the highest probability of the RL (0.362) are at low distances (0 – 59m). Consequently, the proximity to the drainage lines is an important factor controlling the occurrence of landslides.

For the slope, the landslide probability (0.390) is the highest in 0°-5° category. This result can be explained by the number of landslides occurred (10) in the slope's category.

Aspect is an important factor influencing the slope instability, since aspect related parameters, such as exposure to sunlight and drying winds, control the concentration of the soil moisture, which influence landslide occurrence. Based on the result of the final map, is it clear that the landslide percentage is relatively low and that it increases with the orientation angle reaching the maximum on the southwest aspect and then decreases.

Hillshade plays a significant role in landslide susceptibility studies. The results of the table 2 and the final susceptibility map display that the high value of the RL (0.646) is developed in moderate altitudes 204 – 360m and 360 – 572m.

5 Conclusion

The preparation of landslide susceptibility map is a major step forward in risk management. Nowadays, such maps can be prepared by GIS-based techniques and remote sensing data. There are many qualitative and quantitative techniques useful to analyze the relationship between landslides and their causative factors. This study used Logistic Regression (LR) to prepare a susceptibility map for the Tetouan Mazari area, Northern Morocco. Despite the diversity of suggested approaches to landslide susceptibility mapping, researchers still debate the performance and the accuracy of each method. The difficulty in obtaining data and the issue of selecting independent data to analyze, i.e., the parameters that are thought to be causally related, remain challenging in implementing the more data-based approaches. This problem has been discussed in detail by various researchers [21]. Otherwise, an alternative method should be identified because of the time, labor, and cost associated with the collection of inventory data. From the perspective of engineering, the pressure to solve a problem in the shortest period of time, with the most reliable data and in the most economical way requires that a given method should be selected carefully. Similar problems should be expected in the use of numerical methods based on expert opinions. The expert must propose the parameters that are causally related to landslides, and detailed land observations are needed to select the optimal parameters. Recently, the development of large-scale aerial photographs and high-resolution satellite images has considerably decreased the time and labor required for land studies. A total of 126 landslides were mapped using satellite imagery observation and field surveys. The causative factors considered are lithology, slope gradient, slope aspect, faults, drainage lines, and hillshade. After considering all of the potential parameter inputs, the problem solver should also consider the available opportunities and temporal constraints in selecting the optimal method. This study defined the problem, collected data possibly required for the solution, and identified the issues of how and from where these data could be obtained quickly and economically. This change might be attributed to the opportunity to perform inventory mapping more quickly and more easily due to the supply of high resolution images from satellite technologies, which overcome difficulties in time and labor in identifying and evaluating 126 of landslide locations in hundreds of square kilometers. LR-based susceptibility map was clearly more accurate. Consequently, the susceptibility map was successfully generated, validated and has a predicted accuracy of 95.3%. Then it has been classified to five classes: Very High (32%), High (40%), Medium (7%), Low (7%) and Nil (15%). Finally, the results of this study will be used in future for potential construction to mitigate landslide susceptibility.

Acknowledgement

The authors would like to express their sincere gratitude to the National Center of Scientific Research and Technology (CNRST) of Rabat-Morocco, for providing the financial support for this work through a scholarship referenced #H004/11.

References

1. R. Schuster, "Landslides: investigation and mitigation. Transportation Research Board National Research Council Socioeconomic significance of landslides," Turner A K, Schuster R L, Washington, 1996.
2. P. Gorsevski, P. Gessler, R. Folt and W. Elliot, "Spatial prediction of landslide hazard using logistic regression and ROC analysis," *Trans GIS,* vol. **10(3)**, p. 395–415, 2006

3. C. Van Westen, "The modelling of landslide hazards using GIS," *Surveys in Geophysics,* vol. **21** (2–3), p. 241–255, 2000.
4. T. Van Westen, T. Van Asch and R. Soeters, "andslide hazard and risk zonation - why is it still so difficult?," *Bull Eng Geol Environ,* vol. **65**, p. 167–184, 65.
5. R. Bernknopf, R. Campbell, D. Brookshire and C. Shapiro, "A probabilitic approach to landslide hazard mapping in Cincinnati, Ohio, with applications for economic evaluation," *Bulletin American Association of Engineering Geologists,* vol. **25** (I), pp. 39-56, 1988.
6. M. El Moulat, L. Ait Brahim, M. Mastere and A. I. Jemmah, "Mapping of Mass Movements Susceptibility in the Zoumi Region Using Satellite Image and GIS Technology (Moroccan Rif)," *International Journal of Scientific & Engineering Research,* vol. **6**(2), pp. 210-217, 2015.
7. M. Kirat, "Essai de cartographie géomorphologique et étude des mouvements de terrain dans la vallée de l'Oued El Kbir (province de Tetouan, Rif occidental:Maroc Septentrional)," *University of Sciences and Technology of Lille I,* no. PhD diss, 1993.
8. D. Varnes, "Slope movement types and processes," Schuster RL, Krizek RJ (eds) Landslides, analysis and control, 1978.
9. L. Ait Brahim , P. Chotin, S. Hinaj , A. Abdelouafi , A. El Adraoui , C. Nakhcha, D. Dhont , M. Charroud , F. Sossey Alaoui , M. Amrhar, A. Bouaza, H. Tabyaoui and A. Chaouni , "Paleostress evolution in the Moroccan African margin from Triassic to Present," Tectonophysics, vol. **357**, no. 1-4, pp. 187-205, 2002.
10. L. Ait Brahim, F. Sossey Alaoui, H. Siteri, M. Tahri and B. Baghdada, "Prise en compte des glissements de terrain dans la quantification des pertes en sols dans le bassin versant Nakhla," *Bulletin du Réseau d'Erosion,* vol. **21**, pp. 50-64, 2002.
11. J. Kornprobst and M. Durand Delga, "Carte Géologique du Rif Tetouan-Ras-Mazari," Ministère de l'énergie et des mines, Direction de la Géologie, Vols. Editions de Service Géologique du Maroc, Notes et Mémoires (No. 292)., 1985.
12. D. Map, Topographical Map of Tetouan, Rabat: Ministry of Agriculture and Agrarian Reform, Department of land conservation and topographic works, 1970.
13. M. Mastere, B. Van Vliet Lanoë and L. Aït Brahim L, "Land use mapping and its relation to mass wasting and gullying in NorthWestern Rif (Morocco). " *Geomorpholog, relief, process, environment*, vol. **3**, pp. 335–352, 2013.
14. ESRI, "Arc Hydro Tools-Tutorial," 2011. [Online]. Available: http://downloads.esri.com/archydro/archydro/tutorial/doc/arc%20hydro%20tools%202.0%20-%20tutorial.pdf. [Accessed 26 March 2017].
15. F. Dai, C. Lee, J. Li and Z. Xu, "Assessment of landslide susceptibility on the natural terrain of Lantau Island, Hong Kong," *Environmental Geology,* vol. **40**, pp. 381-391, 2001.
16. M. Ercanoglu and F. Temiz, "Application of logistic regression and fuzzy operators to landslide susceptibility assessment in Azdavay (Kastamonu, Turkey," *Environ Earth Sci,* vol. **64**, p. 949–964, 2011.
17. B. Pham, D. Bui, I. Prakash, L. Nguyen and M. Dholakia, "A comparative study of sequential minimal optimization-based support vector machines, vote feature intervals, and logistic regression in landslide susceptibility assessment using GIS," Environ Earth Sci , vol. **76**, p. 371, 2017.
18. S. Lee, "Application of logistic regression model and its validation for landslide susceptibility mapping using GIS and remote sensing data," *Remote Sens ,* vol. **26**, pp. 1477-1491, 2005.
19. K. Devkota, "Landslide susceptibility mapping using certainty factor, index of entropy and logistic regression models in GIS and their comparison at Mugling-Narayanghat road section in Nepal Himalaya," *Nat Hazards,* vol. **65**, p. 135–165, 2013
20. C. Chang and A. Fabbri, "Validation of Spatial Prediction Models for Landslide Hazard Mapping," Natural Hazards, vol. **30**, p. 451–472, 2003.
21. J. Chacon , C. Irigaray , T. Fernandez and R. El Hamdouni R, " Engineering geology maps: landslides and geographical information systems," Bull Eng Geol Environ , vol. **65**, p. 341–411, 2006.

Contribution of geophysical methods to the detection of underground cavities in salt series. Development in different projects in Morocco (roads, rail, etc...)

Tlemçani M.K.[1], Ait Brahim L.[2], El Mahsani A.[3]et Labrihmi A.[4]

[1,2,3]Université Mohammed V, Faculté des sciences, Unité de recherche GEORISK, LGRN, 4 Avenue Ibn Battouta Rabat-Agdal, BP 1014, Rabat, Maroc.

Abstract. As part of the geotechnical studies of a road project, a geophysical investigation by electrical tomography, microgravimetry, and boreholes, was carried out in order to locate if the areas can present or not a risks of ground instability along the road alignment. The road is located near a rock salt mine. This geophysical survey had demonstrated that the area may present major risks for the stability of the road platform. This risk is related to the existence of large expanses of saliferous rocks. The dissolution and exploitation of salt provoked underground voids under the future road. The results of this geophysical survey have allowed the designer office to propose more suitable solutions to solve this problem or even the possibility of shifting the road to another site.

1 Introduction

Le secteur d'étude est caractérisé en plus de son relief accidenté et difficile par la présence de zones d'effondrement qui sont certainement en relation étroite avec l'activité d'exploitation du sel gemme par la mine qui se trouve dans la région et qui occupe une partie du secteur étudié.

2 SITUATION ET CONTRAINTES GEOGRAPHIQUES

Situé à 2,5km au Nord-Ouest de la ville de Taza, dans le domaine rifain, près d'une mine de sel gemme.

Le tronçon routier empreinte le couloir de contact entre deux principaux domaines géologiques, à savoir le domaine atlasique au Sud et le domaine rifain au Nord. Le tronçon traverse des terrains à topographie sinueuse et accidentée, faisant partie du domaine du Prérif externe. Le paysage rencontré est géographiquement plus ou moins homogène et comprend un ensemble de collines généralement de nature marneuse.

3 CONTEXTE GEOLOGIQUE

Le site objet de la reconnaissance géophysique fait partie de la zone du contact entre le prérif charrié et l'avant-pays autochtone, Ce contact frontal contient souvent du gypse associé à du sel gemme sous forme de lits dans la masse charriée (complexe pré rifain). Il s'agit d'un matériel triasique remanié (il est nécessaire d'admettre qu'il a été totalement dissous avant d'être resédimenté (Leblanc, 1979).

Les formations salifères sous forme de lits sont parallèles à la surface de charriage et elles prennent une grande ampleur dans le secteur d'étude formant ainsi le gisement de sel gemme. Ces niveaux évaporitiques sont bien cristallisés et non tectonisés formant des masses irrégulières noyées au sein des sédiments marneux (surtout d'âge Miocène) expliqué aussi par desphénomènes diapiriques contemporains des charriages et qui ont amené le Trias en surface, en provoquant des ruptures dans la couverture tertiaire "figure.1".

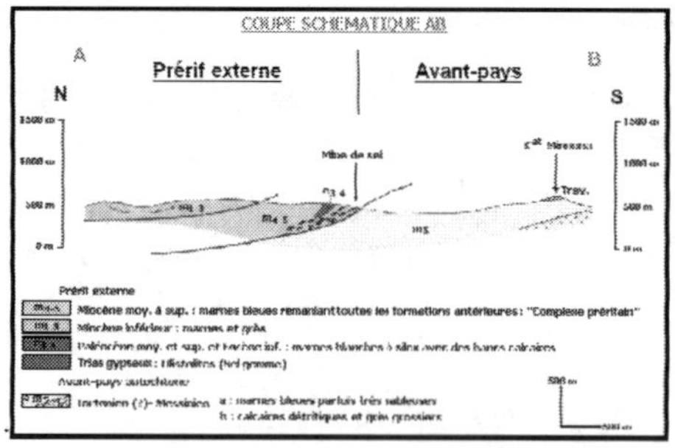

Fig. 1. Coupe géologique AB (Fig. II-7) d'après la carte géologique de Bab El Mrouj Nord Au 1/50 000, 1978

Les tranchées et les galeries creusées par les travaux de la mine montrent généralement trois formations d'épaisseurs variables, à la surface des marnes suivies par des intercalations marno-salifères et des masses dures de sel gemme"figure.2".

Fig. 2. Dispositif de mesure de tomographie électrique avec quatre câbles

4 DESCRIPTION DES METHODES GEOPHYSIQUE UTILISEES

4.1, Tomographie électrique

La tomographie électrique (ou imagerie) de résistivité électrique est une technique dérivant des méthodes classiques du sondage et du traîné électrique et permettant une imagerie continue 2D ou 3D des variations latérales et verticales des résistivités électriques du sous-sol grâce à des électrodes implantées en surface. Elle permet d'obtenir une coupe de la résistivité en fonction de la profondeur en mesurant le profil de résistivités pour différentes combinaisons d'électrodes de courant et de potentiel "figure.3".

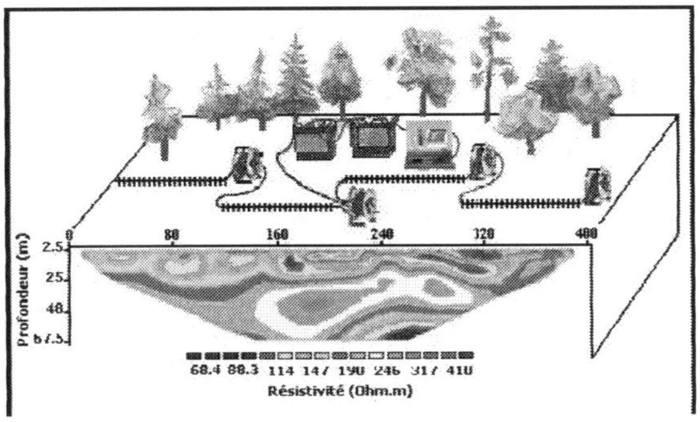

Fig. 3. Dispositif de mesure de tomographie électrique avec quatre câbles

4.2 La microgravimetrie

La méthode de prospection micro gravimétrique consiste à mesurer les variations de l'accélération de la pesanteur puis à appliquer aux valeurs brutes un certain nombre de corrections pour éliminer l'influence des autres corps célestes, de la rotation et de la forme de la Terre, afin de ne faire ressortir que les anomalies imputables seulement aux variations de densité du sous-sol. Autrement dit L'analyse des variations de la pesanteur en surface permet d'appréhender la répartition des masses à l'intérieur de la terre à toutes les échelles. L'anomalie de Bouguer est la réponse gravimétrique de toutes les hétérogénéités situées à toutes les profondeurs.

L'anomalie de Bouguer est donnée par la formule suivante:

$$\Delta g_B = \Delta g_{(mesurée)} \pm les\ 5\ corrections \qquad (1)$$

1-Correction de dérive de l'appareil y compris la correction de marée
2- Correction de latitude Δ_L= 0,081sin2Ø mgals/100m (Øest latitude en degré).
3- Correction d'altitude Δ_h= -0,3086 h mgals/m(h est la côte par apport à la référence)
4- Correction de plateau Δ_P= 0,4191ρ_Bh mgals/m($\rho_B$densité de plateau)
5- Correction de terrain Δ_T
Où

$$\Delta g_{\ (mesurée)} = g_{\ mesurée} - g_{th} \qquad (2)$$

Donc

$$\Delta g_B = g_{\ mesurée} - (g_{th} \pm \Delta g) \qquad (3)$$

Avec:$g_{\ mesurée}$ pesanteur mesurée
g_{th}la pesanteur théorique :
g_{th}= 978.0490 (1+0.0052884 $\sin^2$Ø -0.0000059 $\sin^2$2Ø)en gal

L'anomalie de Bouguer est représentée sous forme de carte dite de l'anomalie de Bouguer qui est le document de base en gravimétrie.

5 PROSPECTION GEOPHYSIQUE

5.1, Présentation

La campagne géophysique, avait comme objectifs :

- La délimitation des horizons salifères ;
- La détection des cavités souterraines crées par l'exploitation du minerai le long de l'axe routier ;
- La localisation des points singuliers et des zones pouvant présenter des risques d'instabilité de terrain.

Au total, six (6) profils tomographiques parallèles ont été réalisés avec des longueurs de 295m. Chaque profil est réalisé avec deux câbles colinéaires et formé par 60 électrodes avec un espacement constant de 5m. Les profils sont orientés sensiblement SW-NE avec une distance variable entre eux, elle est de l'ordre de 40m entre les profils successifs 1, 2, 3 et 4 et de l'ordre de 30m entre les profils 4, 5 et 6.

Les mêmes profils tomographiques sont exploités pour les mesures microgravimétriques; l'emplacement de chaque électrode est pris comme une station microgravimétrique en ajoutant entre deux électrodes successives une station (donc 2,5m entre deux stations voisines appartenant au même profil). Au total, 720 stations de mesure microgravimétrique réparties sur 6 profils avec 120 stations par profil.

L'ensemble de toutes les stations de mesures tomographiques et microgravimétriques ainsi que les sondages mécaniques exécutés sont présenté sur la "figure.4".

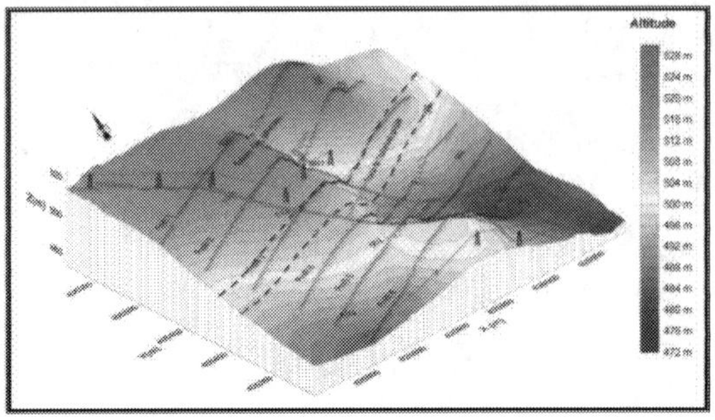

Fig.4. Carte de situation des profils de tomographieélectrique,microgravimétrique, et des sondages carottés par rapport au tracé de l'autoroute en mode 3D

5.2 Discussion des résultats

✓ Tomographie électrique

Les résultats après inversion des données de résistivité apparente donnent des images électriques de résistivités vraies du sous sol et de profondeurs, ce qui permet une interprétation quantitative des données. Mais ces images restent insuffisantes et l'espace des solutions demeure assez vaste. Cependant les divers relevés d'affleurements et les caractéristiques géologiques du terrain le long des profils ont permis de préciser et d'interpréter les images électriques obtenues du sous-sol.

On a systématiquement observé, sur chacune des images électriques, la présence de zones très conductrices dont la résistivité est très faible (représentées en bleu sur toutes les images électriques) et dont l'extension latéraleet verticale est variable. Ces zones, faiblement résistantes correspondent avec forte probabilité aux niveaux marneux. Le niveau représentant l'alternance des marnes et du sel (représenté en tranches vert-marron) apparaît dans les images électriques comme une plage de valeurs moyennes de résistivité. Ce niveau est aussi caractérisé par une extension variable latéralement et verticalement.

Les zones d'anomalies de résistivité (représentées par des « noyaux » rouges-violets sont généralement localisées entre 150 et 170m mais avec des profondeurs variables qui restent généralement dans les quinze premiers mètres, ces zones présentent aussi des variations latérales importantes ; elles sont largement allongées sous les profils 1, 2, 3 et 5 et très réduites sous les profils 4 et 6. A partir de ces résultats, nous pouvons admettre que les zones résistantes forment une continuité du profil 1 au profil 6 et que son extension verticale et latérale s'est avérée variable. Les indices géologiques et les relevés d'affleurement permettent de rattacher la zone d'anomalie au niveau du profil 6 aux masses de sel gemme. Les autres zones sont rattachées soit aux masses de sel gemme ou aux cavités. Cette interprétation reste dans le rang d'une hypothèse qui sollicite une confirmation par une autre méthode géophysique, à savoir la microgravimétrie"figure.5".

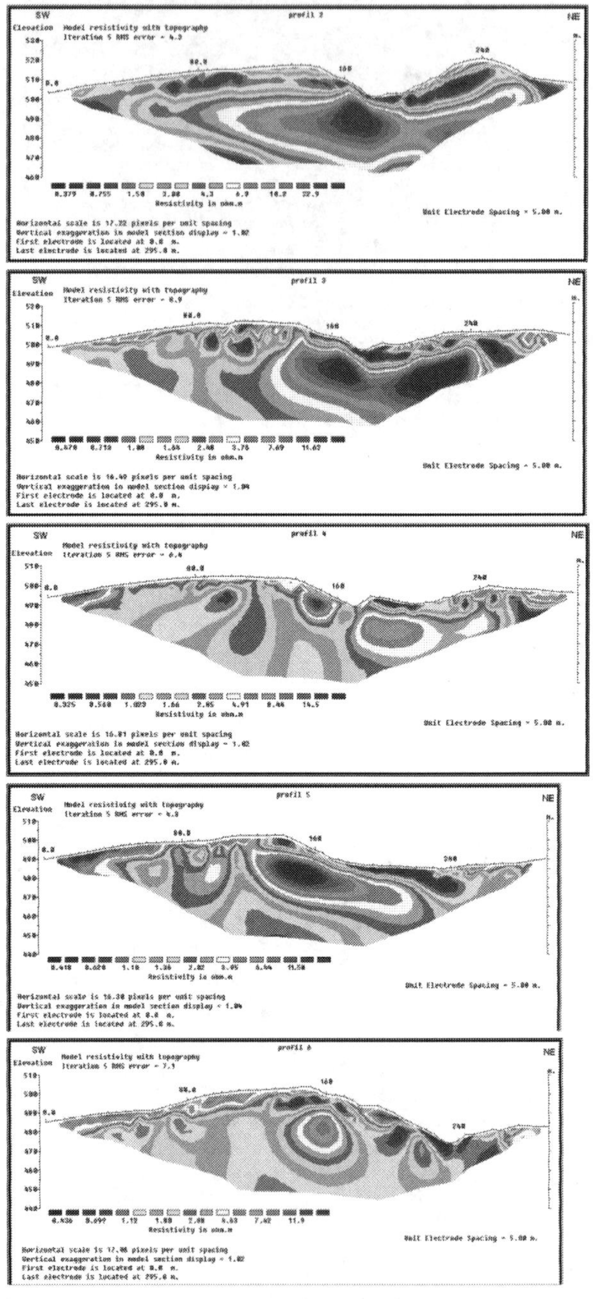

Fig.5. Modèle de résistivité électrique obtenu par inversion 2D des données des profils

5.3, Synthèse des résultats

5.3.1, La microgravimétrie

https://doi.org/10.1051/matecconf/201814902083

La campagne de microgravimétrie vise à confirmer les résultats de la tomographie électrique d'une part, et d'autre part, de répondre le mieux possible aux hypothèses déjà posées en ce qui concerne la présence ou non des cavités et des masses de sel gemme

Les mesures ont été effectuées à l'aide d'un gravimètre Scintrex CG5.Une large gamme de densités a été employée pour comparer les déférentes cartes d'anomalie de Bouguer qui en résultent, afin de choisir la carte qui présente le mieux le contraste en terme de densité (entre les marnes, le sel gemme et/ou les cavités) et donc qui permet au mieux de révéler les zones d'anomalies.

La carte d'anomalie de Bouguer pour les densités 2.0g/cm3 "figure.6" montre un bon contraste et par conséquent elles dévoilent certaines anomalies locales ; on cite particulièrement les anomalies négatives au niveau du profil 1 (de station 1068 à 1075), du profil 2 (de station 2072 à 2073 et de station 2094 à 2099), du profil 4 (de station 4092 à 4095 et de station 4088 à 4091) et du profil 6 (de station 6092 à 6098).

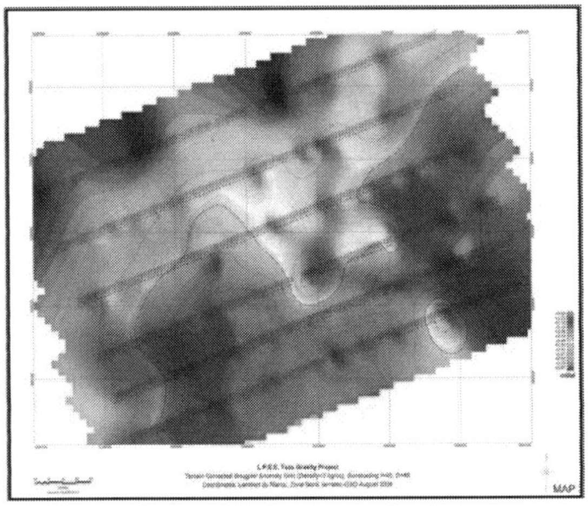

Fig.6. Carte d'anomalie de Bouguer (densité = 2,0g/cm3)

Pour une meilleure analyse de cette carte on procède àlatechnique de calcul de la dérivée première ($\delta\Delta g_B / \delta z$) dite aussi le gradient vertical. Cette technique permet de réduire l'effet de coalescence des sources d'anomalies et d'amplifier les anomalies dues aux structures superficielles par rapport à celles liées aux structures profondes. L'application de cette technique permet d'élaborer la carte de la première dérivée de l'anomalie de Bouguer "figure.8". L'analyse de cette carte montre clairement des anomalies (zones bleues) qui s'individualisent par endroits sur tous les profils.

Toutes ces cartes traitées indiquent la présence du sel gemme de faible densité au Nord et à l'Ouest et des sédiments marneux de densité plus élevée au Sud-Est du secteur d'étude. On observe en plus un certain nombre d'anomalies sur la carte du gradient verticale qui montrent de possibles cavités dans le sel. D'autres anomalies semblent être associées aux effets de terrain ceci est justifié par le fait que la correction de terrain est basée sur le modèle numérique de terrain qui n'a pas été assez précis pour se débarrasser de tous ces effets. Des annotations sur les figures 6 et 7 ont été faites pour montrer les interprétations adoptées.

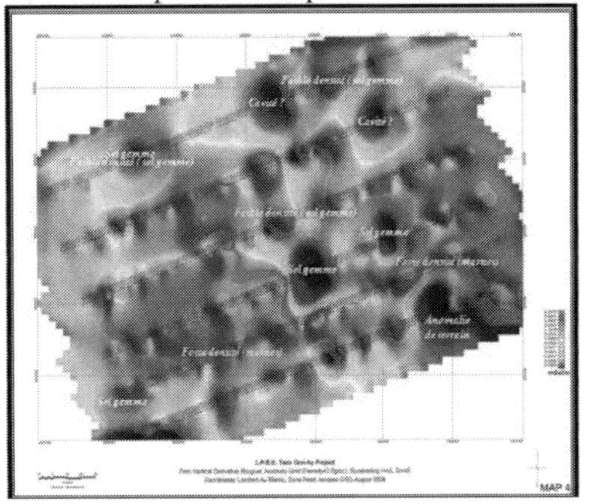

Fig.7. Carte de la première dérivée de l'anomalie de Bouguer (densité = 2,0g/cm3)

5.3.2 Les sondages carottés

Au total 11 sondages carottés de 30m de profondeur ont été exécutés et qui sont répartis sur tous les profils de tomographie électrique et de microgravimétrie. Ces sondages sont exécutés pour caler les résultats de la campagne géophysique et de vérifier les interprétations adoptées a fin de juger un certain nombre d'anomalies Les coupes lithologiques des sondages carottés exécutés confirment la vision lithologique concernant la répartition des formations géologiques dans le secteur d'étude. Ils montrent formellement que les terrains de recouvrement sont essentiellement marneux (et par endroits des masses de sel gemme en affleurement) qui reposent généralement sur une alternance marno-salifères et en profondeur on trouve généralement des marnes grisâtres. Ces coupes montrent également qu'il y a une importante variation latérale et verticale des formations géologiques dans le secteur étudié.Les résultats de ces sondages sont représentés dans les figures suivantes"figure.8,9,10,11,12,13" :

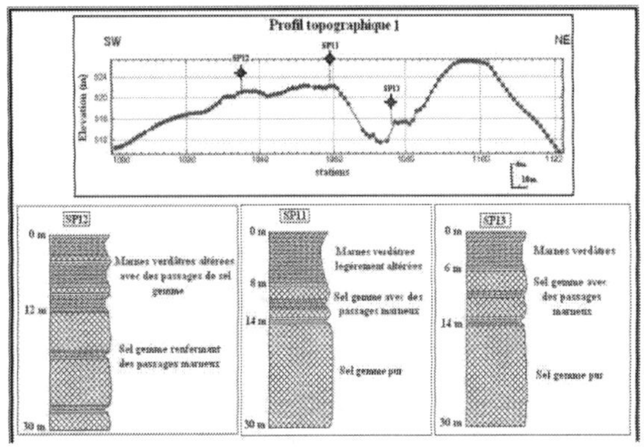

Fig.8. Coupes lithologiques des sondages carottés SP11, SP12 et SP13, leurs emplacements par rapport aux stations gravimétriques, et topographie du profil 1.

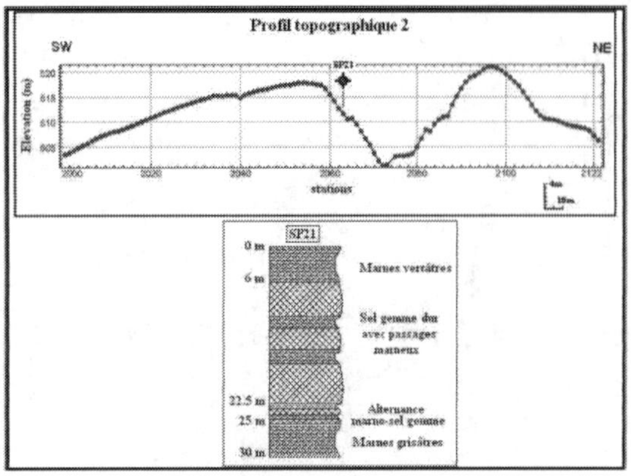

Fig.9. Coupe lithologique du sondage carotté SP21, son emplacement par rapport aux stations gravimétriques, et topographie du profil 2

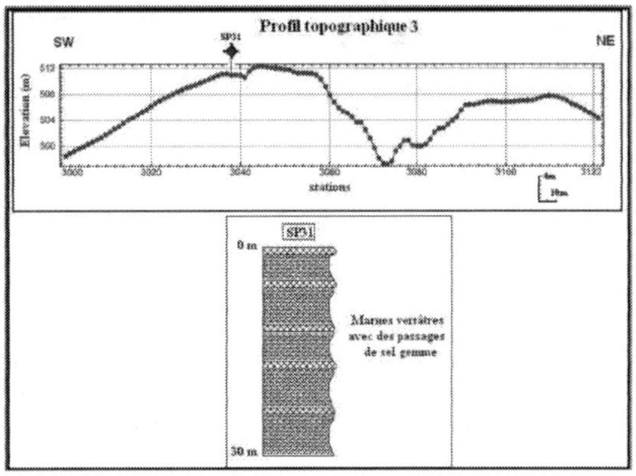

Fig.10. Coupe lithologique du sondage carotté SP31, son emplacement par rapport aux stations gravimétriques, et topographie du profil 3

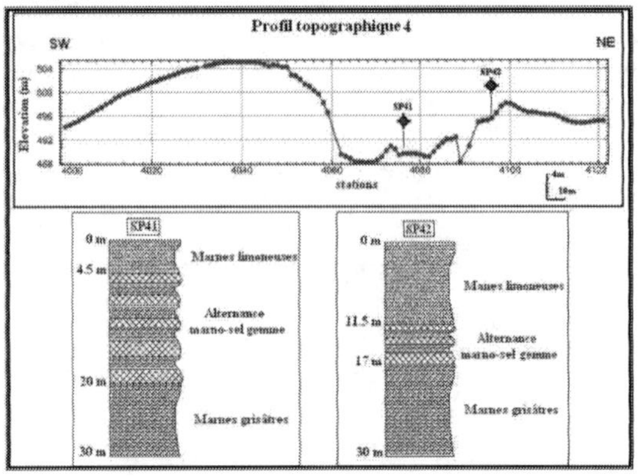

Fig.11. Coupes lithologiques des sondages carottés SP41 etSP42, leurs emplacements par rapport aux stationsgravimétriques, et topographie du profil 4

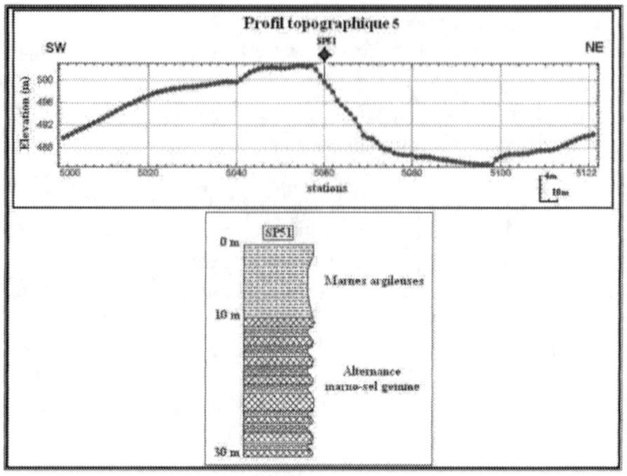

Fig.12. Coupe lithologique du sondage carotté SP51, son emplacement par rapport aux stations gravimétriques, et topographie du profil 5.

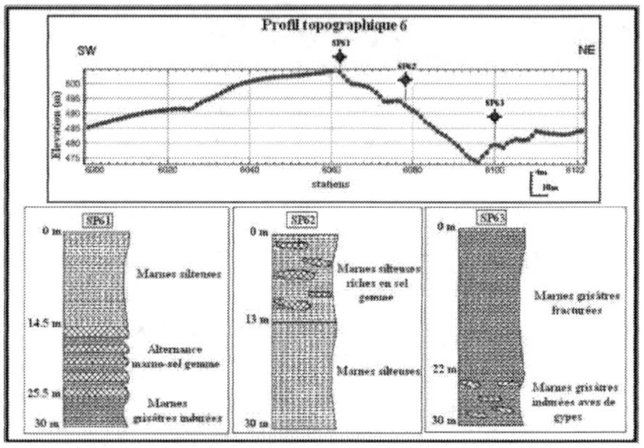

Fig.13. Coupes lithologiques des sondages carottés SP61, SP62 et SP63, leurs emplacements par rapport aux stations gravimétriques, et topographie du profil 6.

6 SYNTHESE GENERALE DES RESULTATS

A la lumière des données de la prospection de la tomographie électrique et de la microgravimétrie couplées par des sondages carottés, il a été possibled'élaborermodèle géologique de terrain faisant sortir les anomalies pouvant être en relation avec les cavités.

L'examen de la carte de la dérivée première de l'anomalie de Bouguer révèle l'existence de plusieurs anomalies positives interprétées comme formations marneuses et négatives interprétées par l'existence de masses de sel gemme et/ou de cavités. Ces anomalies sont bien corrélées aux résultats de tomographie électrique du secteur étudié. Leur confrontation aux données géologiques a permis de faire les observations suivantes :

- Seules les anomalies de résistivité et de densité au niveau des profils 1 et 2 semblent être des zones de vides possibles puisqu'elles présentent des valeurs de résistivités elevés par rapport aux autres anomalies des autres profils. En plus, les sondages mécaniques SP11,

SP13 et SP21 ne touchent pas le centre des anomalies, ils sont juste passés par la périphérie de ces anomalies. Ces sondages montrent aussi qu'il y'a des vastes masses de sel gemme.

- La carte de la première dérivée de l'anomalie de Bouguer ne montre pas des anomalies bien visibles au niveau de la ligne gravimétrique 3, mais le profil 3 de tomographie électrique montre qu'il y a une importante anomalie mais avec de valeurs de résistivités inférieures à celles rencontrées au niveau des profils 1 et 2. Pour ces raisons on a interprété cette zone comme une zone de concentration de sel gemme éliminant ainsi la possibilité d'existence de vides.
- La carte de la première dérivée de l'anomalie de Bouguer montre clairement des anomalies gravimétriques au niveau des profils 4, 5 et 6 ce qui est confirmé aussi au niveau des profils de tomographie électrique par des zones de résistivités peu élevées. Ces zones d'anomalies font l'objet des sondages mécaniques SP41, SP42, SP51, SP61 et SP62. Ces derniers montrent qu'il s'agit uniquement d'une concentration de sel gemme.
- Enfin, la zone d'anomalie gravimétrique localisée à l'extrémité du profil 6 ne trouve pas sa ressemblance en termes de résistivité par la tomographie électrique. Le sondage carotté SP63 montre qu'il s'agit de formations marneuses ce qui est concordant avec les valeurs de résistivités obtenues. L'anomalie gravimétrique est attribuée donc aux effets de terrain c'est-à-dire à des corrections de terrains qui probablement ne sont pas bien faites.

7 CONCLUSION GENERALE

La prospection géophysique utilisant la méthode de tomographie électrique et la microgravimétrie pour la reconnaissance du sous sol du tracé d'un projet routier près d'une mine de sel gemme a permis de mettre en évidence des zones d'anomalies de résistivité et de densité. Ces zones sont corrélées avec les sondages carottés, les conditions géologiques et avec les relevés d'affleurement du terrain afin de les interpréter. Cette corrélation a permis de délimiter d'importantes formations salifères et de désigner quelques zones qui peuvent contenir des vides.

Le couplage des données de la géophysique et des autres levés (sondages carottés, données géologiques) s'est donc avéré efficace, aboutissant ainsi à un modèle géologique satisfaisant. Cependant les résultats de cette prospection sont nettement défavorables pour le passage de tracée routier, car en plus du relief accentué, on note l'existence possible de cavités qui présentent des menaces d'effondrement et l'existence de vastes masses de sel gemme qui présentent des dangers non négligeables liés à leur dissolution en présence de l'eau et donc un risque d'instabilité de terrain.

En se basant sur ces résultats, le secteur étudié est jugé non valable pour le passage de la route et il est préconisé de trouver une autre variante vers le Sud du terrain prospecté où les masses de sel gemme sont moins importantes.

8 REFERENCES BIBLIOGRAPHIQUES

1. M.A. Annecchione, La contribution de la gravimétrie à l'étude hydrogéologique de la Moraine OakRidges, mémoire de maîtrise ès sciences appliquées (génie minéral), Ecole Polytechnique de Montréal, 124p (2004).
2. R.D. Barker, Depth of investigation of collinear symmetrical four- electrode arrays, Geophysics, 54(8):1031-1037 (1989).
3. G. Bernard , Techniques géophysiques de haute résolution, Microgravimétrie, cours en ligne, Ecole Polytechnique de Montréal, 48p (2005).
4. http://geo.polymtl.ca/~giroux/cours/gml6201a/documents/gravi.pdf
5. G. Bernard, Techniques géophysiques de haute résolution, Tomographie géophysique en forage, cours en ligne, Ecole Polytechnique de Montréal, 47p (2005).
6. M. Chouteau, Géophysique Appliquée I, gravimétrie, cours en ligne, Ecole Polytechnique de Montréal, 108p (2002).
7. http://geo.polymtl.ca/www_Cours/pdffiles/Gravimetrie.pdf
8. M. Chouteau. et E.Gloaguen, Tomographie électrique en génie et en environnement, cours de tomographie électrique, 64p (2003).
9. http://geo.polymtl.ca/~chouteau/documents/GML6201_tomoelec.ppt
10. M. Combe, Le couloir de Fès-Taza (les ressources en eau du Maroc Tomme 2), Notes et Mémoires du service géologique n° 231 Rabat (1975).
11. I D'amours, Modélisation et interprétation de tomographie de résistivité et de polarisation provoquée entre trous de forage, mémoire de maître ès sciences appliquées (Génie Minéral), Ecole polytechnique de Montréal, 156p (1998).
12. A Dridri. et B.Fedan, Rôle du contrôle structural dans la mise en place du réseau hydrographique de Sebou et d'Inaouène entre Fès et Oued Amlil, Bulletin de l'Institut scientifique, Rabat, Section sciences de la Terre, n°23, 10p, pp 67-77 (2001).
13. http://www.israbat.ac.ma/PublicationsIS/BIS_ST_23%5Cdridri_morpho.pdf.
14. L.S. Edwaerds, A modified pseudosection for resistively and induced-polarization. Geophysics, 42p (1977).
15. B. El hassani, Etude hydrogéologique du couloir Fès-Taza, Mémoire de fin d'études, ENIM, Rabat, 116p (1994).
16. A. Gartet, J.Gartet. et CC García C. C., Exploitation des salines au massif diapirique de tissa et impacts.

Landslide susceptibility mapping using AHP method and GIS in the peninsula of Tangier (Rif-northern morocco)

Ait Brahim L, Bousta M, Jemmah I A, El Hamdouni I, ElMahsani A, Abdelouafi A,. Sossey alaoui F. and Lallout I

Université Mohammed V, Faculté des Sciences, Département sciences de la terre, Unité de recherche GEORISK, LGRN, 4 Avenue Ibn Battouta Rabat-Agdal, BP 1014, Rabat, Maroc.

Abstract. The peninsula of Tangier (Northern Morocco) is submitted to a significant number of landslides each year due to its lithological, structural and morphological complexity; which cause a lot of damage to the road network and other related infrastructure. The main objective of this study is to create a landslide indexed susceptibility map of Tangier peninsula, by using AHP (Analytical Hierarchical Processes) model to calculate each factor's weight. The work is made via GIS by using an ArcGIS AHP extension. In the current research, First of all, the four main types of landslides were identified and mapped from existing documents, works and new data which came from either remote sensing or fieldwork. Lithology, land use, slope, hypsometry, exposure, fault density and drainage network density were used as main parameters controlling the occurrence of the selected landslides. Then, afterward, each parameter is classified into a number of significant classes based on their relative influence on gravitational movement genesis. The validity of the susceptibility zoning map which is obtained through linear summation of indexed maps was tested and cross-checked by inventoried and studied landslides. The obtained landslide susceptibility map constitutes a powerful decision-making tool in land-use planning, i.e. New highways, secondary highways, railways, etc. within the national development program in the Northern provinces. It is a necessary step for the landslides hazard assessment in the Tangier peninsula in northern Morocco.

Key words: Landslides, remote sensing, indexed parameters, susceptibility map, land-use planning, peninsula of Tangier, Rif, Morocco.

1 Introduction

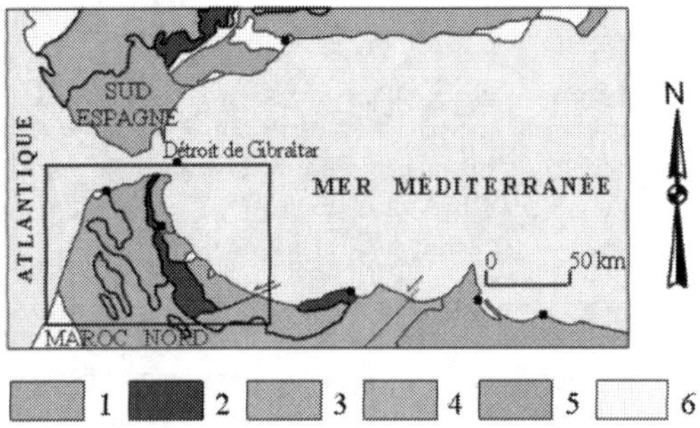

Fig. 1. Secteur d'étude : 1 : domaine d'Alboran, 2 : dorsale calcaire, 3: nappes de flysch, 4 : intrarif, 5 : avant pays, 6 : dépôt post nappes.

La chaîne du Rif et plus particulièrement la péninsule de Tanger (figure 1), présente la fréquence la plus importante, au Maroc, en termes d'instabilités de versants du fait de sa complexité géologique, morphologique et climatique [1-8]. Ces instabilités entraînent chaque année des dégâts en vies humaines, au niveau des habitations, des aménagements hydrauliques, électriques, des terres agricoles, du réseau ferroviaire et surtout le réseau routier. Actuellement, les villes de la péninsule de Tanger (Tétouan, Chefchaouen, Tanger, etc.) subissent une pression démographique (exode rural). En effet, le périmètre urbain bâti a déjà comblé toutes surfaces aménageables stables et facilement constructibles au niveau de ses trois villes. La croissance urbaine a entraîné et entraînera le débordement de l'urbanisme sur de nouveaux versants et plateaux adjacents. Le démarrage du plan d'aménagement des provinces du nord dont le but, est le désenclavement de cette région, a nécessité actuellement le lancement de constructions de nouveaux barrages (Martil), unités de productions d'énergie (centrales électriques, éoliennes) et d'infrastructures portuaires (MED1, MED2 etc.). Il a surtout permis la mise en place d'un important réseau de communication (autoroutes, rocade méditerranéenne, chemin de fer, lignes grande vitesse, aéroports, routes nationales, secondaires, etc.). Tous ses grands projets nécessitent l'établissement d'une nouvelle cartographie pour mettre en évidence les zones les plus exposées aux instabilités de terrain. L'objectif principal de cette étude est de cartographier la susceptibilité des mouvements de masse en utilisant l'AHP (approche heuristique) [9], [21].

2 Matériels et méthodologie

Dans le cadre de cette étude, nous avons eu recours aux cartes géologiques 1/100 000 et topographiques 1/50 000 du Rif septentrional ; aux données satellitaires Landsat7 ETM+ (résolution 30m, Panchro 15m), Spot 5 THR (Très haute résolution spatiale de 2,5 m, Google Earth professionnel), Radar ERS-1(résolution 12.5m). Les données des précipitations des stations météorologiques du Rif (1941-2016 ; DGH, 2017). Les données de l'occupation des sols en termes d'essences forestières (HCEFLCD, 2016). Les images satellitaires ont été fournies dans le cadre de l'action intégrée (AI 219 STU) entre la Faculté des Sciences de Rabat et l'Université Pierre et Marie Curie (Paris VI).

L'approche adoptée lors de cette étude comprend trois étapes principales : l'acquisition de l'information, l'élaboration d'une base de données numérique à l'aide d'un SIG et la zonation de la susceptibilité à échelle régionale et moyenne (méso-échelle)

1. La première étape, a été l'acquisition de l'information : Inventaire, caractérisation et cartographie des mouvements de masses à partir de cartes thématiques, de photographies aériennes, d'images satellitaires, d'identification et de contrôle sur le terrain.
2. Elaboration d'une base de données multi-sources et multi-échelles des facteurs de prédisposition de l'occurrence spatiale des mouvements de terrain, sélectionnés de part leur disponibilité et leur pertinence (variable à modéliser) ; Cette base de données a nécessité la consultation et la synthèse des données bibliographiques du secteur d'étude, l'utilisation de l'imagerie satellitaires disponibles (cf supra) et l'identification et le contrôle sur le terrain dans la mesure du possible.
3. Élaboration de la carte de susceptibilité des mouvements de masse à l'échelle régionale en combinant des cartes d'indices [21] après l'utilisation de l'approche AHP pour l'organisation des différents paramètres et l'attribution de poids relatifs à chaque classe [12].

2.1 Cartographie des mouvements de terrain : typologie, classification et inventaire

Les données extraites des images satellitaires associées à nos observations de terrain ont été d'un apport important pour la cartographie et l'inventaire des mouvements de masse de la péninsule de Tanger. La carte numérique géoréférencée de synthèse de la localisation et de la distribution spatiale des mouvements de terrains que nous avons élaborés présente quatre grands classes (Figure2) suivant les traits morphologiques les plus dominants : les écroulements (2.5%), les glissements (11.7%), les coulées boueuses (19.4%) et les glissements complexes (66.4%). Les mouvements sont représentés en tenant compte de l'échelle par des polygones qui délimitent les surfaces affectées. Cependant, pour plus de lisibilité des cartes présentées dans le présent travail, les 4 classes ont été représentées par des points de formes et de couleurs différentes. Cette synthèse a permis d'élaborer une carte de densité des mouvements de terrain qui correspond à une zonation préliminaire (hiérarchisée par des courbes d'isovaleurs) d'où se dégage les secteurs de plus grande occurrence. Cette carte va permettre aussi, de quantifier l'abondance spatiale des mouvements inventoriés en termes de fréquences. Le croisement de la densité des mouvements de terrain de la péninsule de Tanger avec les différents facteurs individuellement, permet à ce stade d'avoir une idée même globale sur le degré d'importance de chaque paramètre dans l'occurrence des mouvements de terrain. Cette carte servira de base pour tester le degré de validité de la carte de susceptibilité obtenue.

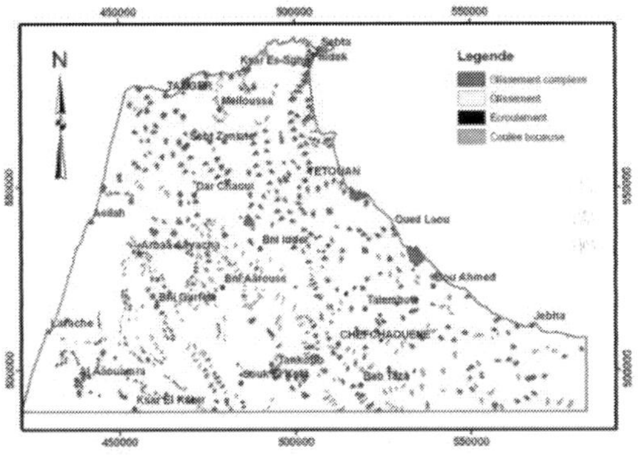

Fig. 2. Carte d'inventaire des instabilités de terrain

2.1.1 Facteurs de prédisposition

L'identification et la spatialisation des variables prédictives (paramètres de prédisposition) des mouvements de terrain de la péninsule de Tanger a nécessité la consultation et la synthèse des données bibliographiques disponibles, l'utilisation de l'imagerie satellitaires et des missions sur le terrain. La base de données multi-sources et multi-échelles des paramètres de prédisposition (lithologie, occupation de sol, fracturation, hypsométrie, pente, exposition) élaborée va permettre le croisement et la superposition entre eux et avec la variable à modéliser les mouvements de terrain, ainsi que leur mise à jour en fonction des nouvelles données acquises. Elle servira surtout à l'élaboration de la carte de susceptibilité.

2.1.1.1 Lithologie

L'occurrence et la distribution spatiale des glissements que nous avons identifiés et cartographiés montre qu'un nombre important de mouvements de masses se développent sur des affleurements de faciès assez diversifié. Pour l'extraction de l'information lithologique à l'échelle régionale, on a eu recourt à la partie couverte par les cartes géologiques 1/100 000 et 1/50 000 et le reste à partir de la carte géologique du Rif 1/500 000ème et les données satellitaires Landsat7 ETM+ et Spot 5 THR (Google Earth professionnel) grâce à plusieurs rehaussements spectraux (l'indice lithologique de Bardinet[17] ; la composition colorée minéralogique avec le triplet de Jenssen[18], l'ACP etc.). La carte de faciès élaborée (Figure3) montre une variation lithologique des affleurements que l'on peut regrouper en six grandes classes :1-marnes-argils, 2-grès, 3-dolomies-calcaires, 4-calcaires marneux, 5: les alluvions, 6-grès marneux. Les faciès marneux (l'unité de Tanger et de Beni Ider) sont caractérisés par l'aspect mécanique médiocre des argiles (riches en smectite) et leur sensibilité aux variations de la teneur en eau [13] ,[7] .

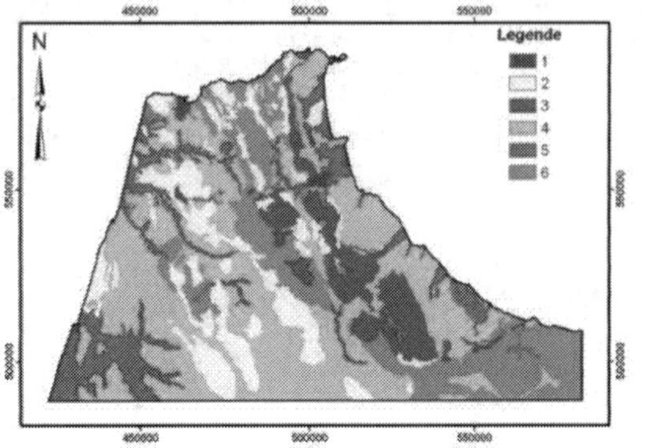

Fig. 3. Carte lithologique

2.1.1.2 Hypsométrie

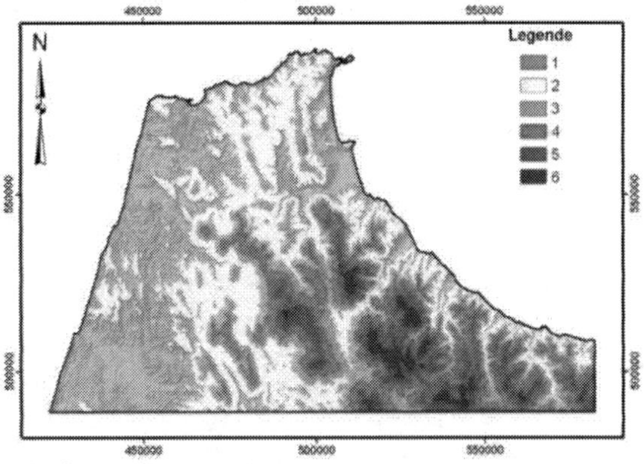

Fig. 4. Carte hypsométrique

La carte d'hypsométrie a été dérivée du MNT GDEM Aster 30m. Les altitudes varient entre le "zéro" (nivellement général marocain) et 2159 m point culminant du (Jbel Tissouka), ce qui a conditionné le choix de l'équidistance des courbes de niveaux pour mettre en relief l'ampleur des phénomènes morphologiques. La superposition de la carte des mouvements de terrain cartographiés à la carte hypsométrique montre que d'une façon générale : plus l'altitude augmente, plus la fréquence d'occurrence de mouvement de terrain est présente. Cependant, nous avons remarqué que certains affleurements de roches carbonatées massifs (à 1926 m et 2159 m), et de grès (1658 m) sont le siège de ruptures par écroulements et coulées de débris mais avec des taux inférieurs à la moyenne générale. Ainsi, le relief, par sa composante altimétrique constitue l'un des facteurs les plus déterminants dans la genèse des glissements de terrains de par sa relation directe avec la répartition de la pluviométrie (moyenne annuelle oscille entre 500 mm en basse altitude à plus 1200 mm sur les crêtes). Nous avons défini six classes d'altitudes (1 : 0-100 m; 2 :100-300m ; 3 :300-400m ; 4 :400-600m ;5 :600-1200m ; 6 : >1200m).

2.1.1.3 Pente

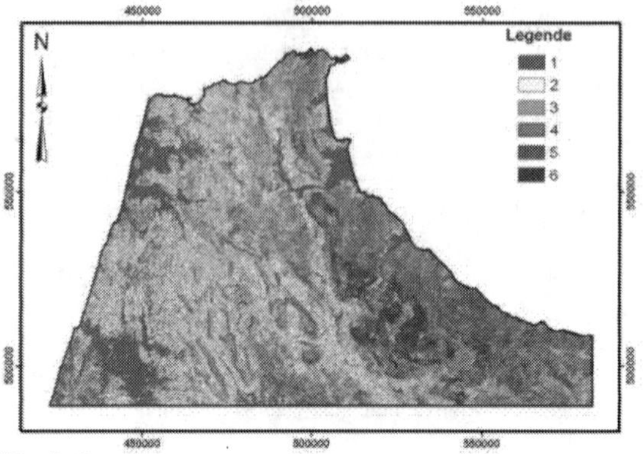

Fig. 5. Carte des pentes

Lors de la cartographie des glissements dans la péninsule de Tanger, les versants les plus propices à l'instabilité sont ceux dont l'inclinaison dépasse 5°, avec un maximum de susceptibilité entre 10° et 35°. Par typologie, les versants montrent la même tendance de comportement vis-à-vis de l'instabilité due aux mouvements complexes et aux coulées, mais avec des valeurs différentes. Le taux de rupture par glissement s'accroît avec l'inclinaison des versants. Sous un degré d'instabilité relativement faible, les pentes très fortes paraissent les plus sensibles aux éboulements et aux écroulements. Ainsi, pour le calcul de l'aléa des instabilités de versants, six classes (1: (0°-2.5°) ; 2: (2.5°-5°) ; 3: (5°-10°) ;4: (10°-20°) ; 5: (20°-30°) ; 6: (>30°)) de pentes (Figure 5) dérivées du MNT correspondant à six indices d'instabilité allant de 5° à 80° ont été établies sur la base d'une étude statistique comparative entre la densité des surfaces occupées par le phénomène de glissement et les pentes.

2.1.1.4 Exposition

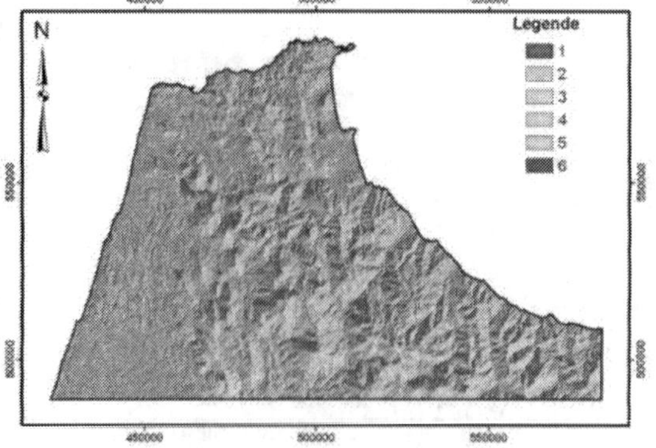

Fig. 6. Carte d'exposition

La carte d'exposition élaborée (Figure 6) montre 6 classes d'orientation des versants (1 :N ; 2 :NE ; 3 :E à

SE ; 4 :S ; 5 :SW ; 6 : W à NW). A l'échelle régionale, les expositions E, W et NW, SE, sont largement dominantes. Mais, à l'échelle du versant, une relative homogénéité de distribution des terrains, selon leur exposition par rapport au nord, est mise en évidence. A l'exception des terrains plats couverts principalement par des alluvions récents.

2.1.1.5 Fracturation

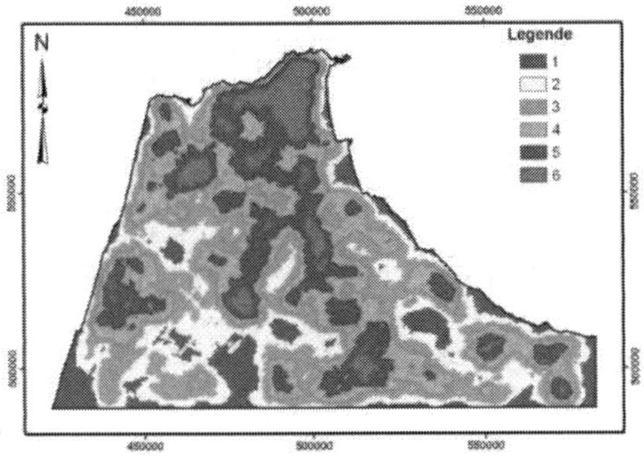

Fig. 7. Carte de densité de fracturation

La carte de synthèse de la fracturation de la péninsule de Tanger est le résultat de l'utilisation de 3 sources. La digitalisation des failles extraites des documents existants (cartes géologiques 1/50 000,1/100 000 et 1/500 000). La fracturation obtenue à partir de l'analyse de l'imagerie satellitaire Radar ERS1,Spot XS et Landsat7 ETM+[15] [24], et la carte de la fracturation récente obtenue à partir de l'analyse du réseau hydrographique en terme d'anomalie et de perturbation du réseau hydrographique. La carte synthétique de la fracturation s'organise selon 3 directions NE-SW, ENE-WSW et NW-SE. Ceci nous a permis d'établir une carte de densité de la fracturation (Figure 7) qui tient compte aussi bien de la longueur de la faille que de la présence de nœuds tectoniques. Les travaux effectués sur la tectonique mio-pliocène et quaternaire de la péninsule de Tanger montrent l'importance de ces directions dans la poursuite de la structuration de cette dernière [15],[19],[20]. Ceci est attesté aussi par le fait que ses failles ont d'une part guidé le tracé du réseau hydrographique [23] et d'autre part montrent une relation avec la répartition de l'activité sismique dans la péninsule de Tanger [22]. Sur le terrain, on a remarqué que des mouvements de terrain se situent le long de certains failles, à proximité de certains nœuds tectoniques (NE-SW et NW-SE) et de zone de broyage des failles (cisaillements et/ou décrochements) formée de matériaux fragilisés et de résurgences. Ce paramètre nécessite l'introduction de la notion de distance des mouvements de terrain par rapport aux failles [5]. Un certain nombre de failles permettent à l'eau de s'infiltrer entraînant ainsi l'accroissement de pressions interstitielles et la réduction de la résistance au cisaillement des sols [15]. Une fois que ces derniers arrivent à saturation, le mouvement de masse est déclenché avec formation d'un plan de cisaillement profond.

2.1.1.6 Occupations de sol

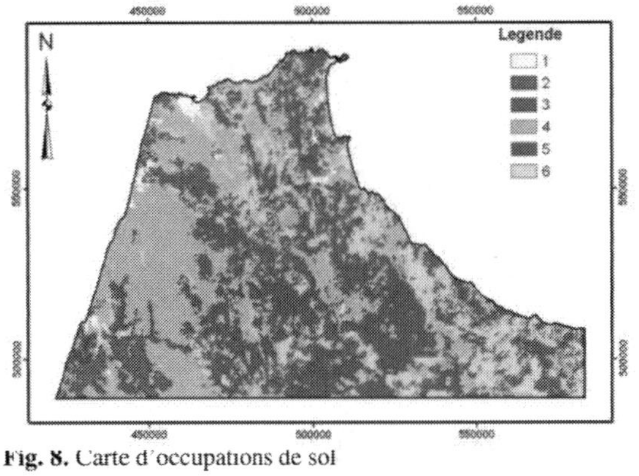

Fig. 8. Carte d'occupations de sol

La couverture végétale de la péninsule de Tanger exerçait jusqu'au années 1960 une action phyto-stabilisante sur les versants à substrat argileux ou à substrat rocheux [1-3]. Cette action se manifeste par l'atténuation de l'énergie cinétique des eaux de ruissellement et des précipitations, l'évapotranspiration, l'amélioration de la qualité physique du sol par la structure des racines et surtout la fixation du manteau d'altération susceptible de glisser par rapport au substratum. Depuis 1970, l'augmentation de la fréquence des années de sécheresse, les feux de forêts et l'action anthropique par le défrichement et surtout le développement accéléré du réseau routier (routes nationales, secondaires, rocade méditerranéenne, chemin de fer, autoroutes, lignes grande vitesse, etc.) ont accentué les différentes formes d'érosion superficielles et profondes rendant ainsi le sol plus conditionné à l'occurrence des mouvements de masses. Pour la cartographie de l'occupation du sol, nous avons utilisé les données de l'occupation des sols en termes d'essences forestières fourni par le HCEFLCD (2016). L'actualisation des données a été effectuée par interprétation visuelle des données satellitaires Spost5 très haute résolution. Nous avons aussi appliqué des classifications supervisées par maximum de vraisemblance sur une mosaïque de données Landsat ETM+. Ensuite, une vérification de la carte obtenue, par un certain nombre de contrôles sur le terrain. La carte de l'occupations du sols (Figure 8) a été classifiée en 6 classes: 1: plans d'eau, bâti et espaces urbains - 2 : périmètres irrigués et terrains nus - 3 : reboisements - 4 : terres arables - 5 : forêts et végétation naturelle - 6 : terrains agricoles.

2.1.1.7 Drainage

Le réseau hydrographique a été digitalisé à partir des cartes topographiques 1/100 000. La carte de sa densité élaborée comporte six classes (Figure 9). L'action érosive du réseau hydrographique peut causer la déstabilisation du terrain et donc participer fortement à l'occurrence des glissements de terrain. Plusieurs instabilités de terrains cartographiées sont réactivées le long de certains oueds surtout lors des périodes de crues (plus particulièrement au niveau des méandres).

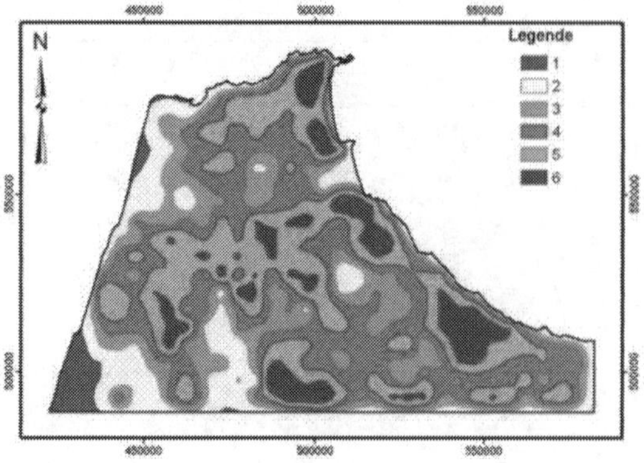

Fig. 9. Carte de densité de drainage

3 Cartographie de la susceptibilité aux instabilités de terrain par l'approche AHP

La cartographie de la susceptibilité aux instabilité de terrain par combinaison des cartes d'index adoptée dans le présent travail a été proposée initialement par Carrara et al., [9]. Il s'agit d'une méthode heuristique indirecte et qualitative pour l'évaluation de la susceptibilité aux mouvements de terrain. Elle a nécessité la hiérarchisation [12] des différents paramètres significatifs prédictifs sélectionnés (lithologie, fracturation, la pente, hypsometrie, exposition, occupation du sol). La valeur des poids relatifs attribuée est basée sur nos connaissances du contexte géodynamique responsable du développement des instabilités de terrain dans la péninsule de Tanger [14],[15] et[5]. Pour générer la matrice d'AHP, les six facteurs de prédisposition ont été utilisés dans une comparaison " pair-par-pair ". Ces facteurs ont d'abord été standardisés et classés en six classes. Nous avons également assigné des poids pour chaque classe de paramètres prédictifs en fonction du pourcentage de glissements de terrain inventoriés présents dans cette classe [16]. Finalement, une seule et unique carte d'index de l'AMT (IAMT) a été élaborée à l'aide de la formule ci-dessous proposée par Voogd [21].

$$IA_{MT} = \sum_{j=1}^{n} W_j w_{ij}$$

Avec :

IA_{MT} :Index de susceptibilité aux mouvements de terrain

W_j : Poids assigné au paramètre 'j'

W_{ij} : Poids assigné la classe 'i' paramètre 'j '

n : Nombre de paramètres

Table 1. Matrice AHP des poids

	Hyps	*Lith*	*Occu*	*Expo*	*Pent*	*Frac*	*Hydr*
Hyps	1	2	3	4	5	6	7
Lith	0,5	1	1	3	4	5	6
Occu	0,33	1	1	3	4	5	6
Expo	0,25	0,33	0,33	1	1	2	3
Pente	0,2	0,25	0,25	1	1	2	3
Frac	0,17	0,2	0,2	0,5	0,5	1	2
Hydr	0,14	0,17	0,17	0,33	0,33	0,5	1
Wi	0,36	0,21	0,2	0,08	0,07	0,05	0,03

4 Résultats et conclusions

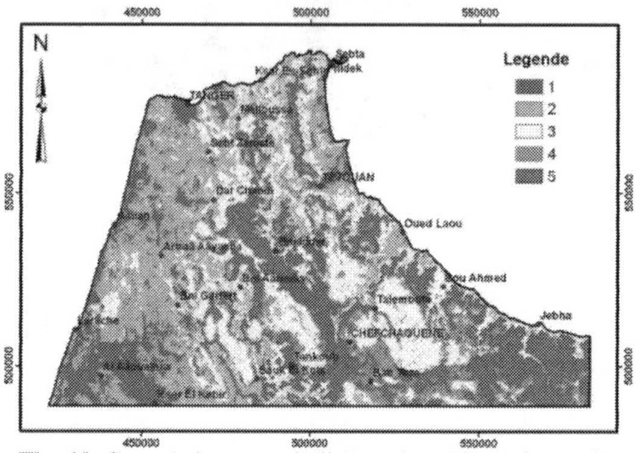

Fig. 10. Carte de la susceptibilité aux instabilités de terrain

La carte de susceptibilité de glissement de terrain obtenue par la méthode AHP montre 5 classes : très faible, faible, moyen, fort et très fort. Elle a été croisée avec les mouvements de masse inventoriés pour tester et évaluer sa validité. Les résultats montrent que : 61,09% des mouvements de masse identifiés sont situés dans des régions à susceptibilité forte à très forte. 20,32% des mouvements de masse sont situés dans une zone de susceptibilité moyenne et 18,58% des mouvements de masse appartiennent à des zones de susceptibilité très faible à faible. Ces derniers s'expliquent par l'apparition de glissements de terrain autour des principaux centres urbains de la péninsule de Tanger et le long des voies de communication déjà construites ou en cours de construction. En conclusion, les valeurs obtenues sont très satisfaisantes pour valider la carte de susceptibilité des mouvements de masse élaborée et le modèle AHP utilisé. La carte de susceptibilité au glissement de terrain obtenue fourni des informations utiles et efficaces sur les instabilités actuelles et leur éventuelle évolution dans le futur, en accord avec l'évolution morphologique de la région. Ce document constitue une contribution à la gestion et à la planification territoriale urbaine et régionale dans le cadre du programme national du développement de la péninsule de Tanger.

Références

1. A. M. Lacroix, *L'instabilité des versants dans le domaine rifain. Revue Géomorphologie Dynamique.* **Vol. XV**, n° 7, 8, 9, p. 97-109 (1965).
2. G. Maurer. *Les montagnes du Rif central, étude géomorphologique*. Thèse d'état, Université de Paris, 500 p (1968).
3. A. El Gharbaoui. *La terre et l'Homme dans la péninsule tingitane. Etude sur l'Homme et le milieu dans le Rif occidental.* Travaux de l'Institut Scientifique, série Géologie et Géographie Physique, Rabat, n° 15, 439 p (1981).
4. Y. El Kharim. *Etude des mouvements de versants dans la région de Tétouan (Rif occidental) : Inventaire, analyse et cartographie.* Thèse de Doctorat Es-Sciences, Université Abdelmalek Essaadi, Tétouan, Maroc, 250 p (2002).
5. L. Ait Brahim, M. Mansour, F. Sossey Alaoui et I. El Hammdouni. Rôle de la fracturation dans la désagrégation mécanique des calcaires et la mise en mouvement de la coulée de pierres d'Amtrasse (Rif, Maroc). Edition ISBN 9954-8407-0-2 "L'implication de la géotechnique dans le développement des infrastructures en Afrique ", pp 69-79 (2003).
6. F. Sossey Alaoui. *Traitement et intégration des données satellitaires optiques et Radar dans un SIG en vue de l'obtention de carte de l'aléa lié aux instabilités de terrain dans la péninsule de Tanger (Rif septentrional, Maroc).* Doctorat, Université Mohamed V, Faculté des Sciences, Rabat. 175p (2005).
7. M. Mansour et L. Ait Brahim. Utilisation de la télédétection pour l'analyse de la fracturation du domaine interne rifain (Maroc): relation avec la répartition des sources. Télédétection, **vol. 5**, p. 1-2 (2005).
8. M. Mastere, B. Van Vliet Lanoë & L. Aït Brahim. "Land use mapping and its relation to mass wasting and gullying in NorthWestern Rif (Morocco)." Geomorpholog, relief, process, environment, 3, 335–352, (2013).
9. A. Carrara, E. Catalano, M. Sorriso-Valvo, et al. Digital terrain analysis for land evaluation. Geologia Applicata e Idrogeologia, 1978, **vol. 13**, p. 69-127.Voogd, (1983)
10. M.J. Selby. *Hillslope Materials and Processes.* Oxford Univ. Press, Oxford (1993).
11. R. Dikau, A. Cavallin, et S. Jäger. Databases and GIS for landslide research in Europe. Geomorphology, **vol. 15**, no 3-4, p. 227-239 (1996).
12. A.K. Saha, R.P. Gupta, et M.K. Arora. GIS-based landslide hazard zonation in the Bhagirathi (Ganga) valley, Himalayas. International journal of remote sensing,, **vol. 23**, no 2, p. 357-369 (2002).
13. M. Kabbaj, M. Himmi, J. Benbouziane, S. Leroueil, A. Maatouk. Talus du Rif (Analyse générale du phénomène des glissements de terrains au nord du Maroc. L'ampleur, les causes et les remèdes pour le réseau routier). Rapport définitif (Coopération maroco- canadienne). Free Pub. Casablanca, 257p (1995).
14. L. Ait Brahim. *Tectoniques cassantes et états des contraintes récents au Nord du Maroc. Contribution al'étude du risque sismotectonique.* Thèse Doctorat d'Etat, Université Mohammed V, Rabat, (1991).
15. L. Ait Brahim, et F. Sossey Alaoui. Utilisation de la télédétection pour l'analyse de la fracturation du domaine interne rifain (Maroc) : relation avec la répartition des sources. Revue canadienne : Télédétection n22 pp 123-135 (2002).
16. LEE, Saro et TALIB, A. Jasmi. Analyse probabiliste de la susceptibilité aux glissements de terrain et de l'effet des facteurs. Géologie de l'environnement, **vol. 47**, no 7, p. 982-990 (2005).
17. CL. Bardinet. Méthodes et réalisations en cartographie automatique des données satellitaires. Bulletin de la Société Belge de Géologie, Tomme 97, fascicule 2, p. 101-113 (1988).
18. J.R. Jensen. *Introductory digital image processing: a remote sensing perspective.* 2nd edition in geographic information science (ed. k. c. clarke), upper saddle river, nj: prentice hall, 316 p (1996).
19. L. Ait Brahim, C. Nakhcha, B. Tadili, A. El Mrabet et N. Jabour. Structural analysis and interpretation of the surface deformations of the feb 24th 2004 Al Hoceima earthquake. Newsletter N°21, Avril, CSEM-EMSC, N° ISSN: 1607-1980, pp 10-12 (2004).
20. A. Tahayt, T. Mourabit, A. Rigo, et al. Mouvements actuels des blocs tectoniques dans l'arc Bético-Rifain à partir des mesures GPS entre 1999 et 2005. Comptes Rendus Geoscience , **vol. 340**, no 6, p. 400-413 (2008).
21. H. Voogd. *Evaluation multicritère pour la planification urbaine et régionale.* Londres: Pion, (1983).
22. L. AIT Brahim. Cadre sismotectonique et risque sismique de la chaîne rifaine, Travaux de l'Institut Scientifique, série géographique physique et géologie, n° 397 pp.121-135. (2002).
23. B. Deffontaines, P Chotin, L. Ait Brahim, M. Rozanov. Investigation of active faults in Morocco using morphonetic methods and pattern analysis. Sonderdruck an Geologische Rundschau, Band 81, Helt 1, Stuttgart, pp. 119-210 Allemagne (1992).
24. P. Chotin, L. Ait Brahim, B. Deffontaines, J.P. Rudant J.P, A. Chaouni. Nouvelles données ERS1-SAR sur la reconnaissance du réseau de failles dans la péninsule de Tanger Maroc : Télédétection- photo interprétation n°1995/2, p. p129-152, Editions ESKA,France (1995).

In situ measurement of CBR concrete modulus. Choice of the type of dam in morocco (static loads and earthquake cases).

TlemçaniM.K.[1], Ait Brahim L.[2], TardieuB.3

[1,3] Université Mohammed V, Faculté des sciences, Unité de recherche GEORISK, LGRN, 4 Avenue Ibn Battouta Rabat-Agdal, BP 1014, Rabat, Maroc.

[2]Expert international-France

Abstract. The Young's modulus of roller compacted concrete is rarely measured. However, it has a direct influence on the dam behavior and especially for static loads during earthquakes. In many cases, static and dynamic calculations are done with modules copied from other study. This dam was studied using in situ module measurements that contribute to a choice of the type of dam and the choice of materials from borrow pits near the dam.

Dans le cas d'un site de barrage au Maroc, il a été nécessaire de mesurer le module d'Young de façon à améliorer la qualité de la prévision des modèles statiques et dynamiques utilisés pour prévoir le comportement du barrage.

La médiocrité des caractéristiques de la fondation a conduit ce barrage ainsi que d'autres barrages récents au Maroc, au choix du type du barrage, il s'agit d'un barrage poids rectiligne de profil quasi symétrique. Un profil quasi symétrique permet d'améliorer l'angle d'incidence des forces appliquées par la gravité et par l'eau et de diminuer l'état de contrainte dans la fondation.

Les calculs montrent que le niveau de contrainte est faible dans la fondation, elle est également faible dans le corps du barrage. Ce qui justifie le choix d'un béton faiblement dosé en ciment et en matériaux fins et dont les agrégats proviennent de zones d'emprunt situées à proximité immédiate du barrage, malgré des caractéristiques mécaniques moyennes dues à un contenu schisteux notable.

Pour évaluer le module du BCR, différents types de mesures ont été effectués durant la construction du barrage : des mesures directes sur éprouvettes réalisées au laboratoire, des essais sur carottes extraites du corps du barrage, des essais cross-hole réalisés dans le corps du barrage. Ces différents essais et les résultats obtenus sont exposés après une brève description du projet.

1 DESCRIPTION DU PROJET

Il s'agit d'un barrage poids en BCR au profil à parement amont avec un fruit de 0.5H/1V et à parement aval avec un fruit de 0.7H/1V. La hauteur maximale sur fond de fouilles est d'environ 98.20m et la longueur en crête d'est de 383.30 m. Le volume du barrage est d'environ de 1 000 000 m3 de béton.

La coupe type originale du barrage est définie sur la base des caractéristiques géomécaniques au contact béton –fondation. La figure ci-après donne la coupe type du barrage au niveau du fond de vallée "figure1".

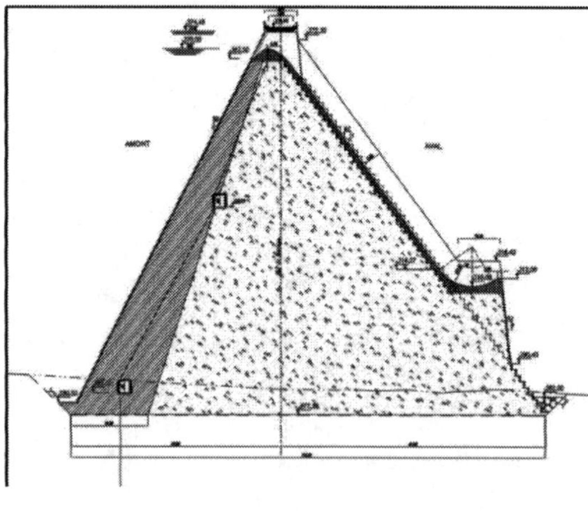

Fig.1.Coupe type au niveau de fond de vallée

La préparation des agrégats (concassage criblage) est faite sur le site. L'expérience montre que toutes les manipulations et les stockages ont tendance à générer des fines plastiques qui ne sont pas favorables à la résistance du béton.

Le BCR est mis en place selon la technique des couches inclinées rive-rive à 7%. Les opérations sont continuées, depuis la préparation de la banquette de 20 cm d'épaisseur au contact de l'ancien BCR jusqu'à la mise en place de la couche elle-même. Chaque levée présente une épaisseur de 3 mètres. Cette hauteur correspond à la hauteur des coffrages qui déterminent les marches amont et aval. La technique de construction par couches inclinées élimine en grande partie le risque d'occurrence d'une couche horizontale faible, mince et continue ce qui est toujours à craindre accidentellement dans une région où les agrégats sont schisteux.

Le nettoyage de la surface de BCR est réalisé à la soufflette eau-air comprimé. Les agrégats sont juste dégagés en prévision du mortier de liaison. L'entrepreneur met en place environ 1500 m3 /jour à raison d'environ 8 heures par couche inclinée en partie inférieure.

2 FORMULATION DU BETON

Le béton est réalisé en exploitant au maximum les ressources du site. Le bassin versant est géologiquement homogène avec une dominante de schistes argileux. Il était illusoire de rechercher des gîtes à granulats qui seraient pauvres en fines argileuses. La préférence a été donnée à l'exploitation des zones à proximité immédiate du lit mineur de l'oued en évitant les pieds des versants. Il y a donc des matériaux schisteux dans les alluvions de la rivière. De plus, l'examen des matériaux sortant de la du malaxeur montre que ceux-ci présentent fréquemment une pellicule de fines argileuses qui n'est pas favorable à une bonne adhérence du liant. Leur élimination nécessite un lavage efficace et couteux si l'objectif est exagérément exigeant.

La formule retenue est la suivante :
Graves calcaire 0/16 343 Kg dont 16 % de fines soit 55 Kg (carrière El Hajeb).
Graves de l'oued Bath 0/16 937 Kg dont 13% de fines soit 122 Kg.
16/31.5 de l'oued 478 Kg.
31,5/63 de l'oued 529 Kg.
Ciment CPJ 45 90 Kg (usine HOLCIM de fès).
Eau de l'oued Bath 134 Kg.
Adjuvant Plastiroute M 0,7%.

L'objectif de résistance est fixé à 7 MPa valeur minimum. La contrainte maximale de poids propre, qui est aussi sensiblement la contrainte de service est égale à 2,36 MPa, soit 33.6% de la résistance minimale exigée du BCR.

3 LES ESSAIS DE LABORATOIRE SUR ECHANTILLON ET SUR CAROTTES.

Les échantillons étaient prélevés régulièrement et soumis à des essais de compression simple et de traction (essai brésiliens).

Pour évaluer le module du BCR, une campagne de 9 échantillons a été programmée sur des BCR de plus de 90 jours. Tous les prélèvements ont été effectués le 13 Novembre 2014.Trois carottes de diamètre 175 mm extraite du massif de BCR entre la levée 19 et la levée 21, entre la cote 334,60 et la cote 341,50 NGM.Trois carottes de diamètre 137 mm extraite du massif de BCR entre la levée 19 et la levée 21, entre la cote 334,44 et la cote 339,50 NGM.

Les échantillons proviennent des forages FCB16 et FCB18.Trois carottes de diamètre 153 mm carottés dans des éprouvettes de laboratoire (diamètre 25 mm, hauteur 50 mm) conservées sur le chantier du barrage et correspondant au BCR entre la levée 19 et la levée 20.

Des carottes de BCR ont été prélevées en parallèle pour chaque diamètre et chaque levée de la mise en place pour déterminer la résistance à la compression.Trois essais ont été réalisés sur des carottes de diamètre 137 prélevées le 13/11/2014, trois essais ont été réalisés sur des carottes de diamètre 175 prélevées le 12/11/2014 et trois essais ont été réalisés par carottage en diamètre 153 d'échantillons de dimension plus importante réalisés et conservés au laboratoire du chantier.

La photo ci-dessous montre une des carottes qui ont servi aux essais.

Les essais à 90 jours ont été réalisés en mars et juin 2014. Les essais réalisés le 1 décembre 2014 correspondent à des bétons coulés entre décembre 2013 et mars 2014. Ils ont donc un âge compris entre 8 mois et 12 mois.

Les essais suivent la norme ASTM C469-02. Des cycles de chargement et de déchargement en compression sont effectués sur des éprouvettes cylindriques d'élancement 2, par augmentation de la charge selon une vitesse constante de 0,05 MPa/s pour atteindre 60% de la charge maximale de rupture. Les déformations correspondantes sont déterminées par enregistrement des déplacements par des extensomètres électroniques.

Un essai de compression simple est d'abord réalisé pour déterminer la charge maximale à la rupture.Une adaptation du processus de mesure a été effectué pour tenir compte du fait que les éprouvettes testées montrent lors du premier cycle de chargement une déformation résiduelle qui varie de 50 à 400 µm/m. Cette déformation résiduelle est apparue pour toutes les éprouvettes testées. Pourtant les cycles 2 et 3 montrent une réversibilité quasi parfaite avec des déformations résiduelles inférieures à 20µm/m. Les valeurs demodule donnée ci- dessous correspondent à 40% de la charge à la rupture pour le 3°cycle.

Un échantillon, le 24/1, donne un résultat de 10,5 GPa. Un échantillon, le 21/1 donne un résultat de 14,1 GPa. Tous les autres essais donnent des résultats entre 17 et 21 GPa, avec une moyenne de 19 GPa.

Les essais sont effectués pour des déformations comprises entre 10-4 et 5. 10-4.

Si on adopte une résistance maximale moyenne de Rc = 9 MPa et le module E = 19 GPa alors on peut en déduire que la rupture se produit pour une déformation limite de ϵ = 9/19000 = 4,7 10-4, ce qui est cohérent avec le champ de déformation utilisé pour les essais alternés permettant la détermination du module.

Le tableau suivant indique les résultats (la pente des courbes) pour les différents types d'éprouvettes, en distinguant les chargements successifs.

éprouvette	Premier chargement	Deuxième chargement	Troisième chargement
20/1	14,5	17,5	18,3
20/3	14,7	17,6	17,8
21/1	9,1	13,9	14,1
21/2	13	16,8	17,3
21/5	17,6	20,7	20,8
24/1	5,8	10	10,5
moy	13,8	17,3	17,66
436/13	16,1	20,8	21
440/14	12,8	17,1	17,2
454/14	17,6	20,7	20,8
moy	15,5	19,53	19,67

Dans tous les cas, les cycles alternés témoignent d'un comportement parfaitement élastique, les cycles 2 et 3 donnent des résultats très voisins. Le premier cycle témoigne d'un comportement irréversible qui peut correspondre soit à l'ajustement de l'échantillon dans la machine, soit à un comportement réel correspondant à un premier chargement. Entre le premier chargement et les chargements suivants, on constate une augmentation de 25%.

Les carottes prélevées dans le corps du barrage étaient déjà chargées par le poids propre du barrage. Le BCR a été mis en place entre décembre 2013 et Mar 2014 entre la cote 334,6 et la cote 341,5 et le prélèvement par carottage a été fait le 13 novembre 2014 alors que le béton était en cours de mise en place à la cote 353,50. La charge de poids propre est donc comprise entre 12 mètres et 18 mètres. On constate que le module mesuré dans les échantillons carottés en place est inférieur de 11% à celui mesuré sur carottage d'échantillon en laboratoire.

Les mesures sur échantillons carottés in situ donnent des résultats un peu plus faibles et correspondent probablement mieux à la réalité physique du barrage que les mesures sur échantillon de laboratoire.

La différence entre le premier chargement et les suivants mérite réflexion. Dans la mesure où les chargements alternés sont relativement rapides, on pourrait considérer que le module mesuré est assez proche du module dynamique qui s'établirait ainsi vers 17 à 18 GPa.

Le module statique est très vraisemblablement supérieur au module de premier chargement, soit 13 à 14 GPa.

4 ANALYSE DES ESSAIS CROSS-HOLE

Trois doublets de forage ont été réalisés dans le corps du barrage. Leur implantation est schématisée sur la figure ci-dessous.

Trois mesures de vitesse des ondes de compression et de cisaillement ont été réalisées du 20 au 23 mai 2015 dans ces doublets par la méthode du cross-hole. Cette méthode consiste à émettre un train d'onde dans un forage et à mesurer le temps qu'il met pour atteindre l'autre forage en plaçant un géophone à la même cote. Les doublets de forages sont éloignés de 4,20 mètres pour les essais CH1 dans le plot 16 à la cote 293,5 et CH2 dans le plot 14 à la cote 296,5, et 3,02 mètres pour le CH3, dans le plot 13 à la cote 429,5.

La masse volumique est supposée égale à ρ=2400 Kg/m3.

Les essais cross-hole exigent des forages bien réalisés équipés de tube PVC parfaitement étanches (bouchon de pied), scellés très soigneusement au coulis de ciment.

La sonde d'émission est constituée de deux marteaux placés de part et d'autre du dispositif de placage, dans un corps de sonde. Le dispositif de placage est constitué d'un obturateur pneumatique disposé à l'intérieur d'un tube fendu (type sonde Ménard) assurant le placage et le couplage à la paroi du tubage. Une fois descendue à la cote d'essai dans le forage, la sonde est plaquée contre la paroi par l'obturateur. L'opérateur augmente la pression dans le circuit. Lorsqu'elle atteint une valeur d'environ 25 à 35 bars, le premier marteau est libéré et frappe l'enclume qui transmet l'onde sismique au terrain par l'obturateur. Le déclenchement du premier marteau arme le second marteau qui est libéré à son tour et émet une onde inverse. Grace à la propriété de polarisation de l'onde de cisaillement, cette frappe inverse provoque, sur la composante verticale, une onde S inversée par rapport à celle émise par le premier marteau. Cela permet d'identifier l'arrivée de l'onde S avec plus de précision.

La sonde de réception est munie de capteurs tri-directionnels (Un capteur vertical et deux capteurs horizontaux). Le capteur vertical mesure principalement l'onde de cisaillement S. Les capteurs horizontaux mesurent principalement l'onde de compression P.

La déviation des forages est mesurée à l'aide d'une sonde inclinométrique Götzl ou équivalent afin de déterminer la distance entre le point d'émission et le point de réception pour chaque profondeur.

On appelle onde P l'onde de compression et onde S l'onde de cisaillement. Les vitesses de propagation sont respectivement Vp et Vs.

Vp et Vs sont reliées par le coefficient de poisson :

$$\upsilon = (Vp2 - 2Vs2)/(2(Vp2-Vs2)) \qquad (1)$$

Les modules sont respectivement Module de compression

$$E = 2.\rho.Vs2.\,(1+\upsilon) \qquad (2)$$

Module de cisaillement G= ρ. Vs2

Le fond de fouille est situé à la cote 277,30. Les levées sont espacées de 3 mètres.L'identification des ondes P ne présente pas de difficulté alors que les ondes S sont plus difficilement identifiables car elles sont parasitées par les terrains encaissants qui provoquent une atténuation importante du signal, notamment pour le récepteur le plus éloigné de la source. Cela conduit à une incertitude sur la détermination du coefficient de Poisson. Comme la valeur de ce dernier est assez bien connu, et vraisemblablement assez constante dans le massif, on peut le prendre comme donnée et en déduire une autre valeur possible pour la vitesse Vs.

Les résultats sont :

<u>Cross- hole CH1 :</u>

Dans le BCR, les vitesses Vp varient entre 2917 m/s et 3200 m/s, soit une valeur moyenne de 3056 m/s. les vitesses VS varient entre 1400 m/s et 1780 m/s soit une valeur moyenne de 1580 m/s.

Dans le rocher proche du contact béton rocher, donc injecté par les injections de consolidation, Vp varie entre 3500 m/s et 4038 m/s soit une valeur moyenne de 3829 m/s, Vs varie entre 1500 m/s et 1850 m/s soit une valeur moyenne de 1700 m/s.

<u>Cross-hole CH2 :</u>

Dans le BCR, les vitesses Vp varient entre 2950 m/s et 3141 m/s, soit une valeur moyenne de 3077 m/s. les vitesses VS varient entre 1450 m/s et 1820 m/s soit une valeur moyenne de 1635 m/s.

Dans le rocher proche du contact béton rocher, donc injecté par les injections de consolidation, Vp varie entre 3500 m/s et 4038 m/s soit une valeur moyenne de 3829 m/s, Vs varie entre 1850 m/s et 1900 m/s soit une valeur moyenne de 1875 m/s.

<u>Cross-hole CH3 :</u>

Dans le BCR, les vitesses Vp varient entre 2860 m/s et 3300 m/s, soit une valeur moyenne de 3041 m/s. les vitesses VS varient entre 1500 m/s et 1780 m/s soit une valeur moyenne de 1660 m/s. Quelques valeurs de Vs sont faibles (1540 m/s).

Le tableau suivant indique les différentes valeurs et calcul Vs en faisant l'hypothèse que ν est égal à 0,2, ce qui ne change pas significativement Vs et reste dans le champ de l'incertitude de la mesure.

	VP	VS	VS/VP	VS/VP carré	nu	ro	E	Vs pour nu 0,2
CH1 min	2917	1400	0,47994515	0,23034735	0,3503563	2400	1,2704E+10	1326,80
CH1 max	3200	1780	0,55625	0,30941406	0,27597715	2400	1,9405E+10	1782,23
CH1 moy	3056	1580	0,51701571	0,26730524	0,31758758	2400	1,5788E+10	1539,68
CH2 min	2950	1450	0,49152542	0,24159724	0,3407197	2400	1,3531E+10	1391,60
CH2 max	3141	1820	0,5794333	0,33574295	0,24727935	2400	1,9831E+10	1933,88
CH2 moy	3077	1635	0,53136172	0,28234527	0,30328613	2400	1,6723E+10	1626,31
CH3 min	2860	1500	0,52447552	0,27507458	0,31027388	2400	1,4151E+10	1584,43
CH3 max	3300	1780	0,53939394	0,29094582	0,29483527	2400	1,9692E+10	1675,85
CH3 moy	3041	1660	0,54587307	0,29797741	0,28777221	2400	1,7033E+10	1716,35
rocher CH1max	4038	1850	0,4581476	0,20989922	0,36716934	2400	2,246E+10	1209,02
rocher CH1min	3500	1500	0,42857143	0,18367347	0,3875	2400	1,4985E+10	1057,96
rocher CH2 max	4038	1900	0,47052997	0,22139845	0,35782301	2400	2,3528E+10	1275,26
rocher CH2 min	3500	1850	0,52857143	0,27938776	0,30614557	2400	2,1457E+10	1609,27

Il est à noter que la vitesse Vp dans le rocher schisteux est relativement élevée, entre 3500 m/s et 4000 m/s. Ce schiste est relativement peu compressible. C'est son angle de frottement assez faible qui a conduit à dessiner un projet quasi symétrique qui sollicite peu la fondation en cisaillement. Nous avons déjà constaté dans des formations géologiques analogues des déplacements très faibles à la mise en eau du barrage, donc des modules du rocher de fondation plutôt élevés. Par ailleurs, les injections de consolidation ont été réalisés avec beaucoup de soin et à relativement haute pression d'injection, et donc, semble-t-il avec efficacité. Les injections améliorent notablement le module de déformation des rochers fissurés.

Les figures suivantes montrent que le module croit avec la profondeur pour les Crosshole CH1 et CH2.

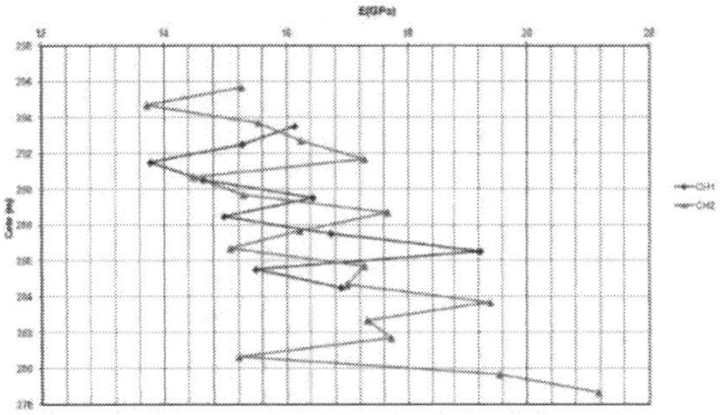

Cette tendance est sensiblement linéaire. Si on suppose que cette évolution est due à l'âge du béton, alors la progression du BCR est d'environ une levée de 3 mètres par mois et passe de 14,5 GPa à 18 GPa entre la cote 280 et la cote 295, soit en 5 mois. La valeur du module serait donc de la forme :

$$E = 14,5 + nj/150 \times 3,5 \quad (3)$$

nj étant le nombre de jours
14,5 est le module à la cote 296, c'est-à-dire le 19 juin 2012, soit pour un béton dont l'âge est de presque 3 ans, précisément 1068 jours, et 18 GPa serait le module pour un béton plus vieux d'au moins 5 mois.

Une telle variation du module pour un béton déjà âgé est peu vraisemblable. Le tableau et la courbe suivante donne l'évolution de la résistance à la compression des carottes extraites des éprouvettes avec l'âge qui donne une idée de l'évolution du module avec l'âge :
A 7 jours σ= 6,3 MPa
A 28 jours σ= 8,1 MPa
A 90 jours σ= 9,4 MPa

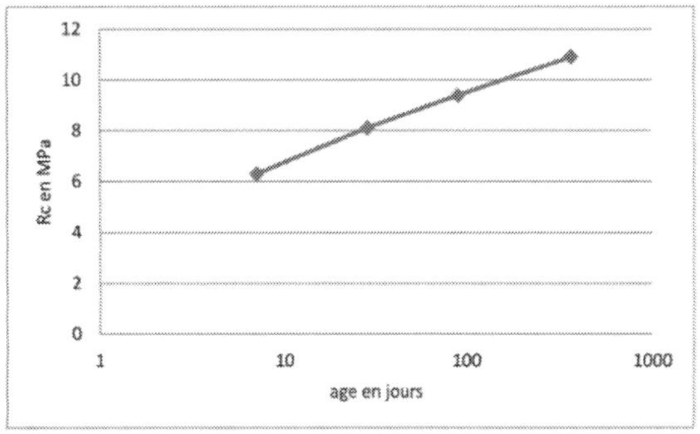

Au-delà de 1000 jours, la variation est très lente.

Cette progression du module est probablement due à la proximité croissante du rocher dont la vitesse est plus élevée que dans le béton.

On peut admettre en fonction des résultats des mesures, que le module du rocher est voisin de 20 GPa et que le module du BCR est supérieur à 14 GPa et compris entre 15 et 16 GPa. Ce module dynamique prend en compte le principe de mise en place du BCR, par couches de 0,30 m d'épaisseur inclinées à 7%, sans traitement du joint, ce qui peut conduire à une diminution du module pour l'ensemble de la masse du béton.

5 MESURES DE VITESSES SUR EPROUVETTES

En novembre 2013, des essais de mesures de vitesses d'onde avaient été réalisés sur le chantier, sur des éprouvettes de BCR.Différents essais ont été réalisés :

a) Vitesses de propagation d'onde mesurées sur carottes extraites du BCR in situ :

BCR in situ	Vitesse en m/s
Nombre	5
Max	3540
Moyenne	3000
Min	2300

b) Vitesses de propagation d'onde mesurées sur des carottes extraites sur des blocs réalisés au laboratoire.

BCR laboratoire	Vitesse en m/s
Nombre	5
Max	4090
Moyenne	3960
Min	3820

Les mesures faites sur les carottes du 21/03/2013, levée 14/couche 1, ont une résistance à la compression Rc = 10,2 MPa à 90 jours. Le test a été réalisé le 02/09/2013 soit après 160 jours. La moyenne est égale à 3320 m/s avec une dispersion assez faible.

Les mesures faites sur les carottes du10/04/2013, levée 14/couche 1, ont une résistance à la compression Rc = 9,8 MPa à 90 jours. Le test a été réalisé le 02/09/2013 soit après 140 jours. La moyenne est égale à 2515 m/s. Cette valeur basse est un peu surprenante d'autant que la résistance à la compression y est nettement supérieure à la moyenne. Dans le cas des mesures sur carottes extraites du corps du barrage, les valeurs mesurées, donnent des valeurs proches des valeurs mesurées par les essais cross-hole.

6 MODULE STATIQUE ET MODULE DYNAMIQUE

En principe, le module mesuré en laboratoire par un essai de compression simple est le module statique et le module mesuré au cross-hole est le module dynamique. Il est très généralement admis que le module dynamique est plus élevé que le module statique. Des anciens travaux montrent que sur des bétons beaucoup plus dosés que notre BCR. Il donne cependant des informations très utiles. Les éprouvettes saturées en eau libre (ce qui est le cas du BCR des barrages) présentent une augmentation de résistance à la traction importante en fonction de la vitesse de chargement. En revanche, le module d'Young varie très peu avec la vitesse de chargement, pas plus de 10% à 15% pour une multiplication de la vitesse de chargement de 5 ordres de grandeurs (page 75).

Dans le cas du BCR, la structure d'hydrate est lâche. Le squelette n'a été constitué qu'après la formation des cristaux d'hydrates entre les grains assez lointains

Le module mesuré par la méthode du cross-hole fournit une valeur du module dynamique pour la masse du béton prenant en compte le système de construction de cette masse qui n'est pas parfaitement homogène, car elle est formée de couches inclinées, de joints entre couches non traités au mortier et de joints entre levées de 3 mètres traités au mortier.

Les essais alternés en laboratoire montrent des valeurs un peu plus élevées pour le béton conservé en laboratoire que pour le béton carotté en place et qui intègre les spécificités de la mise en place (construction par couches inclinées et joints non traités).

Les essais alternés sont relativement rapides et dans une gamme de déformation plutôt supérieur aux déformations dynamiques lors du séisme de dimensionnement.

On pourrait donc retenir pour la masse du barrage, un module statique de 13 à 14 GPa et un module dynamique de 15 à 16 Gpa, mais il n'est pas impossible que les deux soient très voisins.

Il n'y a pas d'argument pour adopter un module statique plus bas que 13 GPa et il n'y a pas d'arguments pour adopter un module dynamique supérieur à 16 GPa.

Pour caractériser le module dynamique sous très faible déformation, on peut mesurer les vibrations naturelles du barrage, éventuellement sollicités, et en déduire les fréquences des premiers modes ce qui donne les modules du béton. Une telle mesure faite dans un pays voisin pour un barrage de profil et de hauteur voisins et une fondation assez similaire, indique un module d'environ 17 à 18 GPa.

7 REFERENCES BIBLIOGRAPHIQUES

1. J.M Medes et al. La technique du béton compacté au rouleau (BCR)-Possibilité d'application pour les grands barrages en Afrique –Sud ScienceTechnologie-56-62 (Janvier 1998).
2. Géologie et barrages en béton compacté au rouleau (BCR)- Bulletin of Engineering Geology and the Environment-The official journal of the IAEG-Volume 55, Issue 1, pp 77–81 (April 1997)
3. André GOUBET-Béton compacté au rouleau-REVUEFRANÇAISE DE GÉOTECHNIQUE-P47-53- (1998).
4. C. Bobard, J.L.Mattiuzzo- Essais cross hole, nouveau développement pour la réalisation et l'interprétation-journées Nationales de Géotechnique et de Géologie de l'Ingénieur -Nancy (2016).
5. Standard Test Methods for Crosshole Seismic Testing- ASTM D 4428/D 4428M (Janvier 2014).
6. G Benrnard, Techniques géophysiques de haute résolution, Tomographie géophysique en forage, cours en ligne, Ecole Polytechnique de Montréal, 47p (2005).

Underground Cemented Backfill, a Design Procedure for an Integrated Mining Waste Management.

KHALDOUN Abdelhadi[1], OUADIF Latifa[1], BAHI Lahcen[1], BABA Khadija[1].
[1] Mohamed V University, Mohammadia Engineering School, Department of Mineral Engineering, 3GIE Laboratory, Rabat, Morocco.

Abstract. From several case studies around the world, it is well known that the binder represents the major part of backfilling operation cost. Therefore, in the case of Imiter operation, research were mainly focused on the optimization of binder content. To this end, the definition of the physical and chemical properties of the future formula ingredients, specifically: tailings, waste material and hydraulic binder, was necessary. Analytical verifications were conducted to predict the UCB mechanical strength according to the defined underground functions and delivery network. Experimental testing, including: uniaxial compression, Immediate Bearing Index (IBI) and slump test, were then conducted to evaluate the possibility of reaching the required strength with the selected materials. The obtained results show that the tailings and mining wastes can be used as backfilling material with a specific binder content depending on each underground application. The followed approach can be applied for a prefeasibility evaluation for a backfilling facility.

Keywords: underground mining, waste management, cemented backfill, design optimization

1 Introduction

Based on the case of study of Imiter operation, this paper presents a simplified method for the design procedure of Underground Cemented Backfill (UCB). Unlike conventional cemented backfills, the proposed UCB includes both total mill tailing and crashed waste rock mixed with a specific binder content. The proposed method is an iterative process comprising tree main steps: 1) material characterization, 2) mechanical strength prediction 3) Recipe optimization. This last consist of an iterative process comprising laboratory investigation, full-size testing and in-situ monitoring.

2 Materials and methods

2.1 Waste material characterization

2.1.1 Waste dumps material

Table 1 presents results of conventional laboratory testing for crashed sand and aggregates from the mine waste dumps. These results confirm acceptable characteristics according to Moroccan standards.

Although **UCB** materials are not at the same requirement level of conventional concrete, this results show the quality of used materials which will eventually improve quality of final product.

Table 1. Laboratory characterization of crushed rock waste. (Testing standard: NM10.1.27, Morocco)

Test	**Flatness Coefficient (%)**	**Superficial Cleanliness (%)**	**Los Angeles Hardness (%)**	**Sand Equivalent (%)**
Results	23	0.3	15	56
Standard	≤ 25	≤ 3	≤ 30	≥ 60

2.1.2 Tailings

The methylene blue test gives an index value of **3.4 g/Kg**, it indicates a silty medium plasticity material according to **GTR*** Guide. At least **15%** of the particles in the fill material should be finer than **20 µm**; if the tailings contain insufficient fines, it may be necessary to add fly ash or additional cement **[1]**. In this case study, particle size investigation by sedimentation test gives a fines content for about **56%.**

*GTR : Guide des Terrassements Routiers, Réalisation des remblais et des couches de forme, LCPC, SETRA, Paris, 1992.

Potential health and safety hazards of tailing were evaluated by performing metal analysis. To this end, inductively coupled plasma with mass spectrometer detector (ICP-MS) was used. The concentration of heavy elements and total cyanide ranges with typical values of international guidelines. **Table** 2 and **Table** 3 present summarized results respectively for main metals and total cyanide analysis.

Table 2. Results of main heavy elements analysis by ICP method (Analyzing Standard: SW846-6020, USA)

Parameter	Antimony	Arsenic	Cadmium	Lead	Manganese	Nickel
MDL* (mg/l)	0.005	0.001	0.0001	0.001	0.001	0.001
Result (mg/l)	0.034	0.189	0.00064	0.014	0.223	0.167

*Minimum Detection Limit (method limit)

Table 3. Results of total cyanide analysis (Analyzing Standard: APHA-4500-CL, USA)

Parameter	Unit	WHO*	CCME**	USEPA***	Result
Total Cyanide	mg/L	0,070	0.200	0.200	0.022

*World Health Organisation, Directive for Cynaide Concentration in Drinking Wtaer 1996,
**Canadian Council of Ministers of the Environment, Maximum Acceptable Concentration of Cyanide in Drinking Water, 2007
***United States Environmental Protection Agency, Maximum Contaminant Level Goals in Drinking Water, 1997.

2.2 Mining Backfill Design

In addition to the weight of the backfill, the stress components in a filled stope are also affected by the walls-fill interactions. This allow horizontal stress transfer to the sidewalls due to frictional and/or cohesive interactions. Thus, the vertical pressure at the bottom of filled stope, is less than the weight of overlying fill. This phenomenon is known as the Arching Effect **[2, 3]**.

Uniaxial Compressive Strength (UCS), easily obtained from laboratory testing, is generally used to evaluate UCB stability. Many mine operations still using two-dimensional limit equilibrium analyses to determine fill exposure stability with no consideration of the arching effect. These analyses can result in an overestimation of the design UCS **[4]** which increase the backfilling costs. Analytical methods developed to account for arching phenomenon include mainly the initial Arching Theory and its modifications **[5-8]** and the 3D Sliding Wedge Failure **[9]**.

Mitchell et al. (1982) proposed an arching model based on wedge limit equilibrium analysis and physical modeling. The Equation 1 describing Mitchell solution was obtained by considering the followings assumptions:

- Stope geometry of a height largely superior to its width (H>>B),
- Null fill friction angle ($\phi = 0$),
- Fill cohesion equal to that of interface between the fill and the side wall.

$$\mathbf{UCS_{design} = \frac{\gamma L\ H}{H+L} FS} \quad (1)$$

where γ = fill bulk unit weight (kN/m3); L = strike length of stope (m); H = total height of fill (m); FS = factor of safety.

Despite limitations of the Mitchell et al. (1982) model related to the main above assumptions, this solution has been commonly used in mine backfill design, and has reduced significantly cement consumption compared to

that required by unconfined-model solutions. Some updates have been proposed by authors to overcome these limitations and develop a more general solution for estimating the required strength of the cemented fill material **[10-12].**
Unlike the 3D wedge model described above, models developed from the initial aching theory **[2]** take into consideration the frictional property of the backfill material ($\phi \neq 0$). Equation 2 and Equation 3 presents a modified two-dimensional solution for vertical and horizontal stress estimating **[6].**

$$\sigma_{vh} = \frac{\gamma B}{2K\tan\delta'}\left[1 - \exp\left(\frac{-2KH}{B\tan\delta'}\right)\right] \quad (2)$$

$$\sigma_{hh} = K\sigma_{vh} \quad (3)$$

where σ_{vh}(KPa) and σ_{hh}(KPa), are the vertical and horizontal stresses at depth H (m), respectively; γ (MN/m3) represents the unit weight of the backfill; δ' (°) is the effective friction angle between the wall and backfill, generally taken as 2/3 of the backfill frictional angle ϕ'_{bf}(°) **[13].**
K is the ratio of the horizontal stress σ_{hh} to vertical stress σ_{vh}. It's a reaction coefficient depending on the horizontal wall movement and material properties. When there is no relative displacement of the walls, the fill is said to be at rest, and the reaction coefficient is usually given by Equation **4 [14]**.

$$K = 1 - \sin(\varphi'_{bf}) \quad (4)$$

The design UCS depends on the underground function of the backfill and stope geometry. In Imiter Mine, typically stopes of 5 m width and 5 m height are mined with fill-and-cut instead 20 m height for sublevel stoping method with a stope length of 30 m. The principal UCB function for cut-and-fill stoping is to give a working platform for next tranches and walls support for the global stability of the mine. With sublevel stoping, UCB allows face freestanding exposure for pillar recovery. UCB ensure also the sill mat function for both method to minimize mining panels interfering and allow undercut mining.

2.2.1 Pillar recovery

Design considering the arching effect
For Imiter case, application of **Equation 1** and **Equation 2** gives respectively a required strength of 480 and 549 kPa with consideration of a fill unit weight of 20 kN/m3, a minimal fiction angle of 30° and a typical safety factor of 2.0.
The friction angle of the backfill is an important parameter when evaluating the stability with **Equation 2.** This parameter can be obtained with laboratory (triaxial, direct shear) or in-situ testing. The friction angle of the studied Imiter backfill is assumed to be greater than 30° because of the angular shape of the crashed sand and aggregates used at a percentage of approximately 60% from the total weight of the backfill solid material.
Design with no arching effect
In case of stope with no arching effect, which can correspond to a backfill with more than two exposed faces, the fill block is subjected to gravity loading similar to a uniaxial compression sample. The design UCS can be estimated by **Equation 5 [15],** which gives a design UCS of 800 kPa for Imiter case study.

$$UCS_{design} = (\gamma_f H_f)FS \quad (5)$$

where γ_f = fill bulk unit weight (kN/m3); H_f = total fill height (m); FS = Factor of Safety.

2.2.2 Working platform

In cut-and-fill stoping, backfilling is performed with cyclic operations. Each filled tranche serves as a working platform for workers and mining equipment for next level. UCB must so develop sufficient strength (bearing capacity) at the short term to allow operations sequencing. A strength limit of 100 KPa is generally admitted to avoid UCB liquefaction **[16]** and a bearing capacity of 30 KN for a working platform **[17].** The bearing capacity for fresh UCB can be estimated using equations Equation 6, Equation 7 and Equation 8 developed for foundation design in civil engineering:

$$q_f = cN_c + \gamma DN_q + \frac{1}{2}\gamma BN_\gamma \quad (6)$$

$$N_q = e^{(\pi\tan\varphi)}\tan^2\left(45 + \frac{\varphi}{2}\right) \quad (7)$$

$$N_c = (N_q - 1)\cot\varphi \quad (8)$$

Where c = cohesive strength of fill (kPa), ϕ = intern frictional angle of fill (degree). γ = bulk unit weight of the fill (kN/m 3), D = foundation footing depth from surface (m), B = foundation breadth (m); $N\gamma$ = unit weight bearing capacity factor; N c = cohesion bearing capacity factor; Nq = surcharge bearing capacity factor;
In case of backfilled stope, the B dimension is equivalent to the width of tire surface contact on the fill. This can be determined by Equation 9 **[16].**

$$B = \sqrt{\frac{F_t}{p}} \quad (9)$$

Where F_t is tire loading force (kN) and p is tire air pressure (kN/m^2).

In civil engineering, value of $N\gamma$ factor can be calculated by Equation 10 **[18].**

$$N_\gamma = 2(N_q - 1)\tan\varphi \quad (10)$$

Considering that fresh UCB has nil cohesion and an estimated friction angle of 30°, Equation 6 gives a bearing capacity of 100 kN/m^2. The limit bearing capacity of 30 kN/m^2 giving an adequate working platform **[17]** is reached with a friction angle of 23°.
To evaluate the trafficability of UCB in term of curing

time, Immediate Bearing Index (IBI) test, also known as Immediate CBR Index, can be used to predict the strength behavior of the fill. Based on IBI results, the UCB ability to be used as working plate-form is evaluated according the IBI standard classification.

2.2.3 Sill mat

To minimize mining panels interfering and improve the global strength of the rock mass, UCB is used as sill mat fill where it ensure a supporting function. However, the main function of the sill mat is to allow underhand cut-and-fill mining. Ideally, the design of a sill mat fill take into consideration the vertical and lateral loading, stope geometry, failure modes, seismic effects, and mechanical properties. It is critical to know the sill mat loading to determine the required strength of the fill. An under-estimation can cause premature failure when the sill mat is exposed by mining extraction. The overestimation implies additional cement which increases the sill mat cost. The vertical loading onto the sill mat can be estimated by the arching theory equations presented in section 2.5. Considering a vertical stope with an infinite backfill thickness, the vertical stress exercised by the overlying backfill on the UCB sill mat can be calculated by Equation11 **[19].**

$$\sigma_v = \frac{\gamma B}{2K\tan\varphi} \tag{11}$$

Where σ_v (kPa) is the vertical stress applied by the overlying backfill on the sill mat, B (m) is the span of the sill mat (or width of the stope), γ (kN/m 3) and ϕ (°) are, respectively, the unit weight and friction angle of the overlying backfill, and K (-) is the lateral earth pressure coefficient in the backfilled stope above the sill mat. As described by Mitchell (1991), four failure modes condition the sill mat stability; the failure can be a result of a flexural, sliding, rotational or caving mechanism. Assuming that there is a good and controlled quality of the contact fill-walls, the flexural failure is more critical than rotational or sliding failure. Flexural failure of the sill mat is mainly related to backfill loading. However, stresses generated by wall convergence can lead to compressive failure if the limit strength of the material is exceeded. Mott (2002) **[20]** proposed Equation 12 to estimate the limit flexural strength for a fixed sill mat. Flexural failure, referred to as tensile failure, is the most common and requiring the highest strength to prevent failure, as such it is used as the critical design criteria.

$$\sigma_t = \frac{3}{4}\left(\frac{\gamma L^2}{d}\right) \tag{12}$$

The corresponding factor of safety is given by Equation13; σ_t is assumed to be 1/10th of UCS as suggested by many authors **[21]**.

$$SF = \frac{4}{3}\left(\frac{\sigma_t h}{\gamma L^2}\right) \tag{13}$$

2.2.4 Underground delivery

Generally pumping system is required for underground delivery of UCB. Gravity cannot ensure the fill flow in case of important horizontal mining development.
Transfer pumps should be able to handle a flow rate approximately 20% greater than the design flow rate for the system (solids and water) **[1].** The decreasing in flow rate to the critical sedimentation value directly causes complete filling of the pipe. The UCB flowing is principally related to the consistency and so, the viscosity of the fresh backfill.
Due to the complexity of measuring the viscosity which depends on several factors, the slump cone test, originally developed to monitor the consistency of unhardened concrete, gives an indication of the backfill yield stress. The slump is the measure of the drop of the backfill material when it is released from a standard metallic cone. Determination of the slump provides a way to characterize a material's consistency that can be related to its transportability **[22, 23]**. For a successful pumping system, the ideal slump height range from 150 mm (5'') to 300 mm (12'') **[24, 25]**.

In term of particle size, the backfill material must also contain at least 15 % by weight passing 20 μm for a typical flow and retain sufficient colloidal water to create a non-segregating mixture **[26].**

2.3 Recipe optimization

In order to optimize the backfill mixture for Imiter mine, five recipes were defined with variation of the cement content from 1 to 10% as presented in **Table 4**. Mixtures presented in **Table 4** were reconstituted in laboratory for a volume of 1 m^3 and were subjected of the following testing: Uniaxial Compressive Strength (UCS); Immediate Bearing Index (IBI); Slump test.

Table 4. Laboratory UCB formulation

Component (Kg)	R1	R2	R3	R4	R5
Agregate	513	513	490	480	460
Sand	620	610	580	570	550
Dry tailings	513	500	490	490	460
Fresh water	341	341	341	325	335
Cement	22	44	100	140	200
Total weight	2009	2008	2001	2005	2005
% Cement	1	2	5	7	10

Uniaxial Compressive Strength
UCB samples were crafted as cylindrical specimens of 10 cm diameter and 20 cm length. Mechanical strength of the analyzed formulations was evaluated by UCS testing for curing time ranging from 1 to 28 days. The obtained results **(Fig. 1)** show that UCS values range from 0.14 to 2.44 MPa with an exponential trend from day 1 to day 28.

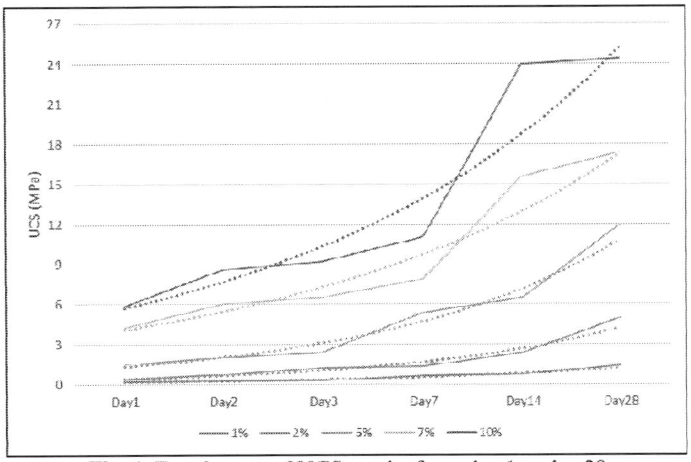

Fig. 1. Trend curve of UCS results from day 1 to day 28

Immediate Bearing Index (IBI)
The IBI test is generally used to evaluate trafficability of soil foundation during road construction. The test is originally realized on compacted soil simples with the optimum Proctor density. For our case, the test was conducted on crafted cylindrical specimen for the same backfill recipes used for UCS testing. **Figure 2** shows that evolution of IBI index for recipe R1, R2, R3 and R4 for a curing age until four days. The objective of this test is to define the curing age after which the backfill can be trafficable by mining equipment.

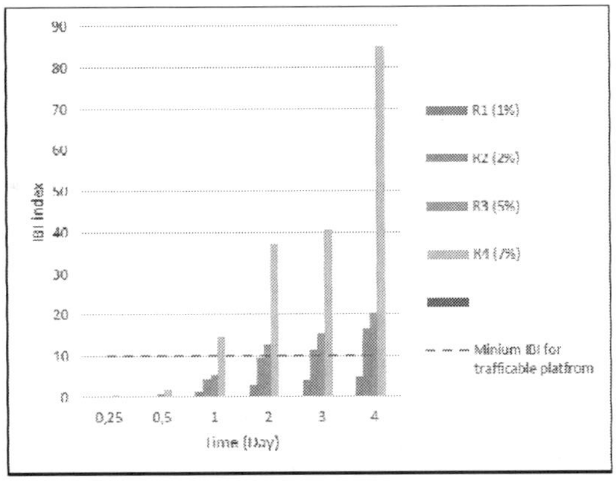

Fig. 2. Evolution of IBI index as function of curing time for recipe R1, R2, R3 and R4.

3 Discussion

For the backfill mixture optimization, five recipes were proposed with cement content ranging from 1 to 10%. Mechanical properties were evaluated through Uniaxial Compressive Strength testing. Trafficalbility and pumpbility were respectively analyzed by Immediate Bearing Index and slump test.
According to the UCB design requirements and experimental results, it appears that the adequate proportion of cement in the UCB mixture is situated between 2 and 5 % (recipes R2 and R3) in case of pillar recovery, and 4% in the case of self-supporting backfill. Except recipe R1, IBI index results show that all backfill recipes allow trafficable platform after four days of curing time. Recipe R3 and R4 can make possible equipment circulation only after a curing age of 2 days. However, if the recipe R3 is chosen as the optimal formulation, tree days of waiting time will be required to adequately upturn the extraction activity in the filled stope. For the concern of underground delivery, the realized cone tests give slumps from 210 to 220 mm, which indicate a fluid slurry easily manipulated by conventional pumping system.

4 Conclusion

This paper presented a simplified procedure for mining waste and tailing reutilization through Underground Cemented Backfill. Following this procedure will help operation to develop the adequate backfill mixture based on specificities of each extraction site. The case study of Imiter operation presented in this paper, presented an integrated solution to using both waste rock and tailings as an engineered filling material ensuring different underground functions. Despite the knowing environmental benefits of mining waste reusing, improved understanding of cementitious binding mechanisms will be important in improving and reducing potential heavy element release to the environment. Other important issues also remain to be resolved, including the long term evolving properties of the fill material, the seismic impact related to blast-induced vibrations. It must also be noted that the important quantity of available wastes implies evaluation of other reusing solutions, especially in building and large civil works.

Acknowledgements

The authors would like to thank the SMI mining operation, Morocco, for the support of this work.

References

1. Hambley D F, Backfill Mining. In: Darling P, editor. *Mining Engineering Handbook, 3rd ed.* USA, Society for Mining, Metallurgy, and Exploration, pp. 1375-1384 (2011).
2. Marston A. *The theory of external loads on closed conduits in the light of latest experiments*. Bulletin 96, Iowa Engineering Experiment Station, Ames Iowa (1930).
3. Terzaghi K, *Theoretical soil mechanics*. New York: John Wiley & Sons (1943).
4. Stone DMR., The optimization of mix designs for cemented rock fill. In: Glen H W, editor. MINEFILL'93 – The 5th International Symposium on Mining with Backfill. Johannesburg, South Africa, SAIMM, pp. 249-253 (1993).
5. Winch C, Geotechnical characteristics and stability of paste backfill at BHP Cannington mine. PhD, James Cook University, Townsville, Australia (1999).
6. Aubertin M, Li L, Arnoldi S, Belem T, Bussière B, Benzaazoua M, Simon R., Interaction between backfill and rock mass in narrow stopes. In: Culligan PJ, Einstein HH, Whittle AJ (eds) Soil and rock america, vol 1. Verlag Glückauf Essen (VGE), Essen, pp 1157–1164 (2003).
7. Li L, Aubertin M, Belem T, Formulation of a three dimensional analytical solution to evaluate stress in backfilled vertical narrow openings. Can Geotech J **42**: 1705–1717 (2005).
8. Pirapakaran K, Sivakugan N., Arching within Hydraulic Fill Stopes. Geotech Geol Eng 25: 25-35.
9. Mitchell RJ, Olsen RS, Smith JD 1982. Model studies on cemented tailings used in mine backfill. Can Geotech J **19**: 14–28 (2007).
10. Zou S, Nadarajah N, Optimizing backfill design for ground support and cost saving. In: Yale D P, editor. The 41st US symposium on rock mechanics USRMS) Golden: American Rock Mechanics Association, Colorado, USA, p. 7–21 (2006).
11. Li, L., and Aubertin, M. A modified solution to assess the required strength of exposed backfill in mine stopes. Canadian Geotechnical Journal, **49** (8):994–1002 (2012).
12. Li L, Aubertin M, An improved method to assess the required strength of cemented backfill in underground stopes with an open face. Int J Min Sci Technol **24**: 549–558 (2014).
13. Becker DE, Moore ID, Canadian Manual of Foundation Engineering. 4th ed. Canada: Canadian Geotechnical Society (2006).
14. Kumar, S, Design of Pile Foundations, in: Braja M. Das. (Ed.) *Geotechnical Engineering Handbook*, J. Ross Publishing, Floride, pp. 189-258 (2010).
15. Yu T R, Mechanisms of Fill Failure and Fill Strength Requirements. In: Canadian Association of Rock Mechanics, editor. The 6th Canadian Rock Mechanics Symposium, Sudbury, Canada, pp. 1-6 (1992).
16. Hassani F, Bois D, Economic and technical feasibility for backfill design in Quebec underground mines Final report 1/2, Canada-Quebec Mineral Development Agreement, Research & Development in Quebec Mines Contract no EADM 1989-1992, File no 71226002 (1992).
17. Craig RF, *Craig's Soil Mechanics*. In: Craig RF, editor. Bearing capacity. 7th ed. London, Spon Press, Taylor & Francis, pp. 294-295 (2004).
18. European Committee for Standardization, Eurocode 7, Geotechnical Design – Part 1: General Rules, Brussels: European Committee for Standardization (2004).
19. Mitchell R J, Sill mat evaluation using centrifuge models, Mining Science and Technology **13**: 301-313 (1991).
20. Mott, R, Applied strength of materials. Ontario: Prentice Hall (2002).
21. Regal, X., Thèse de Doctorat, Caractérisation du comportement en traction du béton sous fortes sollicitations: essais de flexion trois points aux barres de Hopkinson. Autre. Universitée d'Orléans (2016).
22. Clark CC, Vickery JD, Backer RR, Transport of total tailings paste backfill: results of full-scale pipe test loop pumping tests, report of investigation RI 9573. Washington, Spokane: National Institute for Occupational Safety and Health (1993).
23. Belem T, Benzaazoua M, An overview on the use of paste backfill technology as a ground support method in cut-and-fill mines. In: The 5th Int Symp on Ground support in Mining and Underground Construction: Villaescusa E,Potvin Y. editors., Perth, Australia, pp. 637 – 650 (2004).
24. Landriault D.A., Verburg R., Cincilla W. and Welch D, Paste technology for underground backfill and surface tailings disposal applications. Short course
25. Benzaazoua M, Ouellet J, Servant, S, Newman P, Verburg R, Cementitious mine backfill with high sulfur content: physical, chemical and mineralogical characterization. Cem Concr Res **29**: 719–725 (1999).
26. Hassani F., Archibald, J, *Mine backfill Hand Book.* Montreal: Canadian Institute of Mining, Metallurgy and Petroleum (1998).

Vibration analysis of carbon nanotubes-based zeptogram masses sensors and taking into account their rotatory inertia

A. Azrar[1], L. Azrar[2,3], and A. A. Aljinaidi[3]

[1] MMC, Faculty of Sciences and Techniques of Tangier, Abdelmalek Essaâdi University; Tangier; Morocco.
[2] Research Center STIS, Higher School of Technical Education of Rabat (ENSET), Mohammed V University in Rabat, Morocco.
[3] Department of Mechanical Engineering, Faculty of Engineering, Kind Abdulaziz University, Jeddah, Saudi Arabia

Abstract. In this research work, the transverse vibration behaviour of single-walled carbon nanotubes (SCNT) based mass sensors is studied using the Timoshenko beam and nonlocal elasticity theories. The nonlocal constitutive equations are used in the formulations and the CNT with different lengths, attached mass (viruses and bacteria) and the general boundary conditions are considered. The dimensionless frequencies and associated modes are obtained for one and two attached masses and different boundary conditions. The effects of transverse shear deformation and rotatory inertia, nonlocal parameter, length of the carbon nanotubes, and attached mass and its location are investigated in detail for each considered problem. The relationship between the frequencies and mode shapes of the sensor and the attached zeptogramme masses are obtained. The sensing devices for biological objects including viruses and bacteria can be elaborated based on the developed sensitivity and frequency shift methodological approach.

1 Introduction

Detection of the mass of bio-molecules has become an increasing growing field in the biological and biomedical sciences. It is recognized as one of the key technologies for predictive and preventive medicine [1]. The CNT is ultra light and is highly sensitive to its environment changes. Therefore, many researchers have explored the potential of using CNT as nano mechanical resonators in atomic-scale mass sensor [2,3]. Length scale effect analysis on vibration behavior of single walled carbon nanotubes with generalized boundary conditions and CNT conveying fluid have been elaborated by Azrar et al [4-7].

The aim of this paper is analyzing the dynamic behavior of a CNT-based biosensor where the rotary inertia of the attached bio-object and of the CNT is taken into account. The analytical solutions are obtained based on the on iterative-incremental procedures. The effect of the physical and geometrical parameters are investigated.

2 Mathematical formulation

Let us consider a slender single walled carbon nanotube of length L, diameter d, thickness h, moments of inertia I, cross section area A, the mass density ρ, Young modulus E and shear modulus G. The nonlocal elasticity theory combined with the Timoshenko beam model is adopted. The dynamic governing equations are formulated.

2.1 Dynamic equation of the SWCNT sensor based on nonlocal TBT

Based on the nonlocal elasticity and Timoshenko beam theories (TBT), the governing equations of the nonlocal CNT can be obtained as [4-7]

$$\begin{cases} ksAG\left(\dfrac{\partial \boldsymbol{\theta}}{\partial x}+\dfrac{\partial^2 \mathbf{w}}{\partial x^2}\right)-\rho A\left[1-(e_0a)^2\dfrac{\partial^2}{\partial x^2}\right]\dfrac{\partial^2 \mathbf{w}}{\partial t^2}=0 & (1)\\ EI\dfrac{\partial^2 \boldsymbol{\theta}}{\partial x^2}-ksAG\left(\theta+\dfrac{\partial \mathbf{w}}{\partial x}\right)-\rho I\left[1-(e_0a)^2\dfrac{\partial^2}{\partial x^2}\right]\dfrac{\partial^2 \boldsymbol{\theta}}{\partial t^2}=0 & (2)\end{cases}$$

where w denotes the displacement, $\boldsymbol{\theta}$ the rotation of the cross section and e_0a is the nonlocal parameter.

For the free vibrations, harmonic motion is assumed

$$\mathbf{w}(x,t)=\mathbf{W}(x)e^{i\omega t};\quad \psi(x,t)=\boldsymbol{\Psi}(x)e^{i\omega t} \tag{3}$$

where ω is the natural vibration frequency parameter. For dimensionless equations, the following variables are used.

$$y=\frac{x}{L};\ W=\frac{w}{L};\ \mu=\frac{e_0a}{L};\ \xi=\frac{EAL^2}{EI};\ \zeta=\frac{EI}{k_sAGL^2};\ \beta=\sqrt{\frac{\rho AL^4}{EI}}; \tag{4}$$

 https://doi.org/10.1051/matecconf/201814902087

$$\begin{cases} \dfrac{d\mathbf{\Psi}}{dy}+\dfrac{d^2\mathbf{W}}{dy^2}+\beta^2\zeta\left[\mathbf{W}-\mu^2\dfrac{d^2\mathbf{W}}{dy^2}\right]=0 & (5)\\ \zeta\left(1-\mu^2\xi\beta^2\right)\dfrac{d^2\mathbf{\Psi}}{dy^2}-\dfrac{d\mathbf{W}}{dy}+\left(\zeta\xi\beta^2-1\right)\mathbf{\Psi}=0 & (6)\end{cases}$$

Eliminating **W** or **Ψ** from eqs. (5) and (6) yields to the following uncoupled differential equations:

$$\left(\mu^2\beta^2(\zeta+\xi)-\mu^4\zeta\xi\beta^4-1\right)\frac{d^4\mathbf{W}}{dy^4}+\left(2\mu^2\zeta\xi\beta^4\right.$$
$$\left.-(\mu^2+\zeta+\xi)\beta^2\right)\frac{d^2\mathbf{W}}{dy^2}+\left(1-\zeta\xi\beta^2\right)\beta^2\mathbf{W}=0 \quad (7)$$

$$\left(\mu^2\beta^2(\zeta+\xi)-\mu^4\zeta\xi\beta^4-1\right)\frac{d^4\mathbf{\Psi}}{dy^4}+\left(2\mu^2\zeta\xi\beta^4\right.$$
$$\left.-(\mu^2+\zeta+\xi)\beta^2\right)\frac{d^2\mathbf{\Psi}}{dy^2}+\left(1-\zeta\xi\beta^2\right)\beta^2\mathbf{\Psi}=0 \quad (8)$$

The analytical solution of the ordinary differential equation Eq. (7) is sought in the following form

$$\mathbf{W}(y)=\mathbf{A}e^{i\lambda y} \quad (9)$$

where *A* is a constant.

Substituting Eq. (9) in Eq.(7), the following characteristic equation is obtained:

$$\left(\mu^2\xi\beta^2-1\right)\left(\mu^2\zeta\beta^2-1\right)\lambda^4+\left((2\zeta\xi\beta^2-1)\mu^2-\xi-\zeta\right)\beta^2\lambda^2$$
$$+\left(\zeta\xi\beta^2-1\right)\beta^2=0 \quad (10)$$

The solutions of this equation are given by:

$$\lambda_1=\sqrt{\frac{\left((1-2\zeta\xi\beta^2)\mu^2+\xi+\zeta\right)\beta^2+\sqrt{\Delta}}{2}};\quad \lambda_2=-\lambda_1;$$
$$\lambda_3=i\sqrt{\frac{\sqrt{\Delta}-\left((1-2\zeta\xi\beta^2)\mu^2+\xi+\zeta\right)\beta^2}{2}};\quad \lambda_4=-\lambda_3; \quad (11)$$

where $\Delta=\left(\mu^4+(\xi-\zeta)^2-2(\xi-\zeta)\mu^2\right)\beta^4+4\beta^2$

The transverse displacement and rotation solutions of (5, 6) are given by:

$$\mathbf{W}_1=A_1e^{i\lambda_1 y}+A_2e^{i\lambda_2 y}+A_3e^{i\lambda_3 y}+A_4e^{i\lambda_4 y};\quad 0\le y<y_1$$
$$\mathbf{W}_2=A_5e^{i\lambda_1 y}+A_6e^{i\lambda_2 y}+A_7e^{i\lambda_3 y}+A_8e^{i\lambda_4 y};\ y_1\le y<y_2 \quad (12)$$
$$\mathbf{W}_3=A_9e^{i\lambda_1 y}+A_{10}e^{i\lambda_2 y}+A_{11}e^{i\lambda_3 y}+A_{12}e^{i\lambda_4 y};\ y_2\le y<1$$

$$\mathbf{\Psi}_i=\frac{\left((\mu^2\xi\beta^2-1)\zeta^2\beta^2-1\right)}{1-\zeta\xi\beta^2}\left(\frac{d\mathbf{W}_i}{dy}\right)+\frac{\zeta\left(1-\mu^2\xi\beta^2\right)\left(\mu^2\zeta\beta^2-1\right)}{1-\zeta\xi\beta^2}\left(\frac{d^3\mathbf{W}_i}{dy^3}\right) \quad (13)$$

where A_i are constants to be determined with the boundary conditions at y= 0 and 1 and the continuity equations at y_1 and y_2 while $\mathbf{W}_1$, $\mathbf{W}_2$ and $\mathbf{W}_3$ and $\mathbf{\Psi}_1$, $\mathbf{\Psi}_2$ and $\mathbf{\Psi}_3$ are, respectively, the left, central and right transverse displacements and rotations divided at the points where the concentrated masses are attached.

Introducing the following non-dimensional coordinates:

$$y=x/L;\ y_1=x_1/L_1;\ y_2=x_2/L_2 \quad (14)$$

The boundary conditions are: (15)

$$\mathbf{M}_1(0)-K_r^L\frac{d\mathbf{W}_1(0)}{dy}=0 \qquad \mathbf{M}_3(1)+K_r^R\frac{d\mathbf{W}_3(1)}{dy}=0$$
$$\mathbf{Q}_1(0)+K_t^L\ \mathbf{W}_1(0)=0 \qquad \mathbf{Q}_3(1)-K_t^R\ \mathbf{W}_3(1)=0$$

where K_r^L, K_t^L *and* K_r^R, K_t^R are the translational and the rotational spring constants at the left and right ends of the CNT at y=0 and y=1, see ''figure 1''. The classical boundary conditions can be simply obtained as special cases when the stiffness's of the springs take some extreme values such as zero and infinity.

The continuity equations at the position y_1 are:

$$\mathbf{W}_1(y)\big|_{y=y_1}=\mathbf{W}_2(y)\big|_{y=y_1} \quad (16-a)$$
$$\left.\frac{d\mathbf{W}_1}{dy}\right|_{y=y_1}=\left.\frac{d\mathbf{W}_2}{dy}\right|_{y=y_1} \quad (16-b)$$
$$\mathbf{M}_1+c_1^2\bar{m}_1\beta^4\mathbf{W}_1\big|_{y=y_1}=\mathbf{M}_2\big|_{y=y_1} \quad (16-c)$$
$$\mathbf{Q}_1+\bar{m}_1\beta^4\mathbf{W}_1\big|_{y=y_1}=\mathbf{Q}_2\big|_{y=y_1} \quad (16-d)$$

and at y_2

$$\mathbf{W}_2(y)\big|_{y=y_2}=\mathbf{W}_3(y)\big|_{y=y_2} \quad (17-a)$$
$$\left.\frac{d\mathbf{W}_2}{dy}\right|_{y=y_2}=\left.\frac{d\mathbf{W}_3}{dy}\right|_{y=y_2} \quad (17-b)$$
$$\mathbf{M}_2+c_2^2\bar{m}_2\beta^4\mathbf{W}_2\big|_{y=y_2}=\mathbf{M}_3\big|_{y=y_2} \quad (17-c)$$
$$\mathbf{Q}_2+\bar{m}_2\beta^4\mathbf{W}_2\big|_{y=y_2}=\mathbf{Q}_3\big|_{y=y_2} \quad (17-d)$$

where

$$\mathbf{M}_i=\beta^2\left(\mu^2(\zeta\xi\beta^2-1)-\zeta\right)\mathbf{W}_i(y)-\left(\mu^2\xi\beta^2-1\right)\left(\mu^2\zeta\beta^2+1\right)\frac{d^2\mathbf{W}_i(y)}{dy^2}$$

$$\mathbf{Q}_i=\frac{\beta^2\left((1-2\zeta\xi\beta^2)\mu^2+\xi+\zeta\right)}{\zeta\xi\beta^2-1}\frac{d\mathbf{W}_i(y)}{dy}+\frac{\left(\mu^2\xi\beta^2-1\right)\left(\mu^2\zeta\beta^2-1\right)}{\zeta\xi\beta^2-1}\frac{d^3\mathbf{W}_i(y)}{dy^3}$$

$$K_r^L=\frac{k_r^LL}{EI};\ K_r^R=\frac{k_r^RL}{EI};\ K_t^L=\frac{k_t^LL^3}{EI};\ K_t^R=\frac{k_t^RL^3}{EI};$$
$$\bar{m}_i=\frac{m_i}{\rho AL};\ c_i=\frac{r_i}{L};$$

 https://doi.org/10.1051/matecconf/201814902087

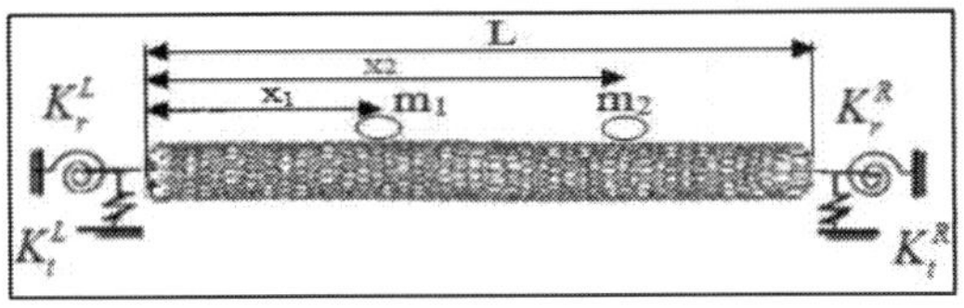

Fig. 1. Elastically restrained CNT at both ends.

3 Frequency equation

Substituting Eq. (12) into Eqs. (15)–(17), taking into account Eq. (14) one obtains the following system of equations expressed as

$$\begin{bmatrix} a_{1,1} & a_{1,2} & \text{L} & a_{1,12} \\ a_{2,1} & a_{2,2} & \text{L} & a_{2,12} \\ \text{M} & \text{M} & \text{L} & \text{M} \\ a_{12,1} & a_{12,2} & \text{L} & a_{12,12} \end{bmatrix} \begin{bmatrix} A_1 \\ A_2 \\ \text{M} \\ A_{12} \end{bmatrix} = \begin{bmatrix} 0 \\ 0 \\ \text{M} \\ 0 \end{bmatrix} \qquad (18)$$

The non-triviality condition is established by solving:

$$\det(\mathbf{H}) = 0 \qquad (19)$$

where H is the matrix of the coefficients $a_{i,j}$ of the system and the roots β_n are the eigenvalues of the problem. Eq.(18) is a hardly nonlinear function of β_n. A numerical technique based on the Newton–Raphson algorithm has been elaborated for numerical solutions.

4 Numerical results and discussions

Numerical results are presented using the following geometrical and material properties.

$$\rho = 2{,}3e^3\, kg/m^3,\ E = 1e^3\, Gpa,\ G = 420GPa,\ \nu = 0.19,$$
$$h = 0.34e^{-9}m,\ L = 20a, A = 7.85e^{-19}m^2, a = 1.5\times e^{-9}m,$$
$$d = 10^{-9}m,\ I = \pi d^4/64,\ ks = 0.877,\ \Omega_0 = EI/ks\,AG,$$

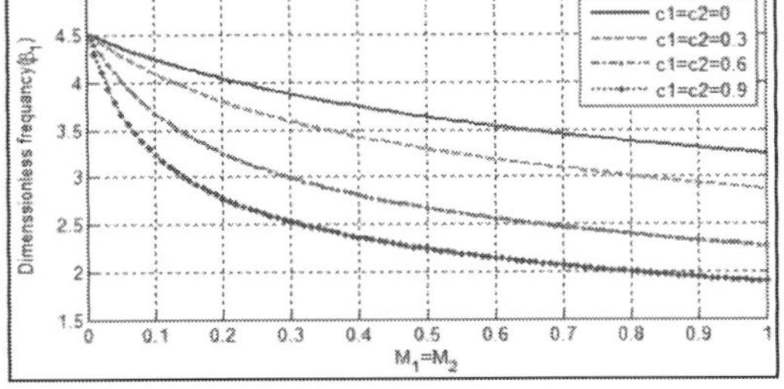

Fig. 2. Dimensionless frequencies as a function of dimensionless masses

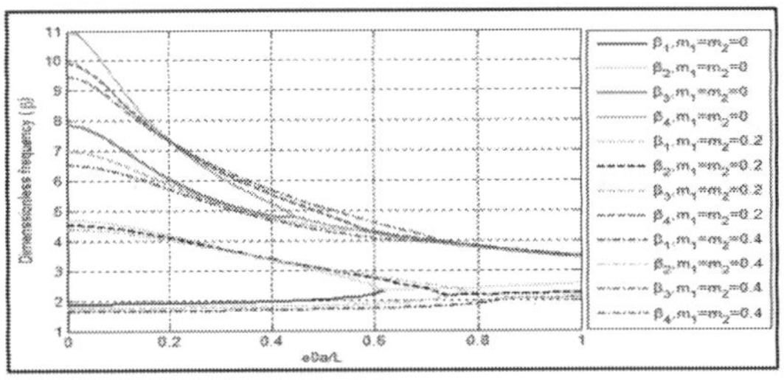

Fig. 3. Small length scale effect on the dimensionless frequency of a cantilever single walled CNT for different masses

Figure 2 shows the effect of mass inertia of the first dimensionless frequency of the clamped $K_t = +\infty$ sensor with attached mass. It can be seen that the frequency of the sensor decreases with increasing the attached mass and its inertia.

Figure 3 shows the variations of the eigenfrequency parameter associated to the first fourth modes shapes for a cantilever CNT with the nonlocal parameter e_0a/L for dimensionless masses. It is found that increasing the nonlocal parameter decreases the frequency and instability behaviors can be reached.

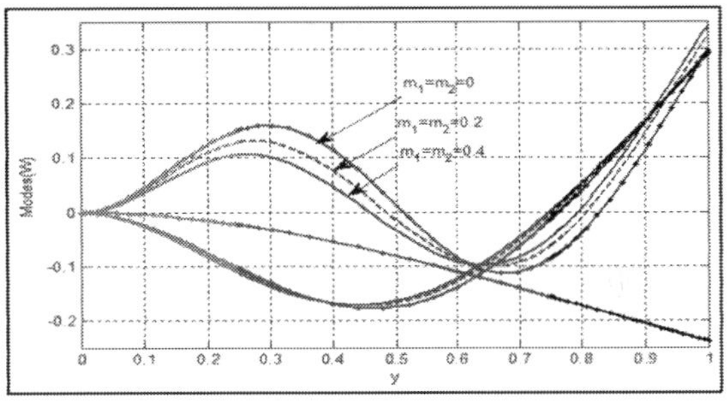

Fig. 4. The effect of zeptogram mass on the first three modes of cantilever SWCNT.

Figure 4 shows the effect of zeptogram mass on the first three modes of cantilever SWCNT. It can be seen that the attached mass has the most significant effect on mode shapes

Acknowledgements

The authors would like to acknowledge the technical and financial supports from the DSR and the King Abdulaziz University, Jeddah, Saudi Arabia.

5 Conclusions

The rotary inertia of the attached bio-object has a strong effect on the natural frequencies and cannot be simply neglected. The nonlocal Timoshenko beam model is more adequate than the nonlocal Euler- Bernoulli beam model for short CNT biosensors. The transverse frequency and mode shape distortion due to attached mass have been used as parameters for characterizing SWCNT as zeptogram mass sensors.

REFERENCES

[1] L. Hood, J.R. Heath, M.E. Phelps, B. Lin, Science **306**, 640-643, (2004).

[2] W. Shi, Z.-B. Shen, X.-L. Peng, X.-F. Li, International Journal of Mechanical Sciences **115-116**, 406–415, (2016)

[3] Z. B Shen, L. P. Sheng, X.F. Li, G. J. Tang, Physica E **44**, 1169–1175, (2012).

[4] A. Azrar, L. Azrar and A. A. Aljinaidi. Journal Mechanics of Advanced Materials and Structures, **23**, 1245-1262, (2016).

[5] A. Azrar, L. Azrar and A. A. Aljinaidi. Journal Composite Structures, **125**, 127-143, (2015).

[6] A. Azrar, L. Azrar and A. A. Aljinaidi. MATEC Web of Conferences 1, 09002, (2012).

[7] A. Azrar, L. Azrar and A. A. Aljinaidi, Revue de Mécanique Appliqué et Théorique, **2, 5**, 475-484, (2011).

Adsorption of cationic dye onto fly ash-based geopolymer: Batch and fixed bed column studies

Marouane El Alouani, Saliha Alehyen, Mohammed El Achouri and M'hamed Taibi

Mohammed V University in Rabat, Centre des Sciences des Matériaux (CSM), (LPCMIO), Ecole Normale Supérieure Rabat, Morocco

Abstract. Cationic dye adsorption from aqueous solution onto synthesized geopolymer was investigated by batch and fixed bed column experiments. The geopolymer material was elaborated by alkali solution and fly ash supplied by Jorf Lasfar power plant of Morocco. Physical and chemical characteristics of samples were determined by FX, DRX, SEM, ^{29}Si MAS NMR and Zeta potential methods. The Brunauer, Emmett and Teller (BET) technique is used to determine the surface area. The Barrett-Joyner-Halenda (BJH) method was performed to obtain pore size distribution curves and average pore diameter. Kinetics data were analyzed using pseudo-first-order, pseudo-second-order and intraparticle diffusion models. To predict the breakthrough curves and determine the main fixed bed column parameters, three kinetic models; Tomas, Bohart–Adams and Yoon-Nelson models are applied to fitting the experimental data. The kinetic study showed that the pseudo-second-order can be used to describe the methylene blue (MB) adsorption process on the geopolymer matrix. The kinetic models of the adsorption in dynamic column are suitable to describe the continuous adsorption process of dyestuff by the geopolymer. The results of this study indicated that geopolymer derived from fly ash can be used as a low cost effective adsorbent for cationic dye removal from industrial aqueous effluent.

1 Introduction

Synthetic organic dyes are used in many industrial sectors such as paper, leather plastic manufacture, automotive industry and textile industry. This various industrial activities are an important source of environmental pollution due to their high content of many chemical dyes.

Among the textile dyes most used in industry, methylene blue (MB) or basic blue 9. It is a water-soluble cationic dye and can reveal very harmful effects on living things such as difficulties in breathing, vomiting, diarrhea, nausea and several negative impacts on the aquatic environment [1]. Therefore it is very important to confirm the water quality, since even just 1.0 mg/L of dye concentration in drinking water can impart a significant color, making it unfit for human consumption [2]. Therefore, it is necessary to reduce dyes concentration in aqueous media and wastewater.

Recently, many different technologies have been utilized for removal of hazardous pollutants from water and wastewater, including biochemical methods [3], membrane separation [4], ion-echange [5], ultrafiltration [6], photocatalysis [7], coagulation/flocculation [8], and adsorption [9]. Among the mentioned methods, adsorption has some advantage when compared aforementioned conventional methods in terms the simplicity of utilization, effectiveness, low cost, no sludge formation, etc. However, natural and synthetic materials with low cost treatment such a bentonite, kaolinite, silicate, natural phosphate, activated carbon and fly ash based geopolymer have been applied for removal the organic pollutants from water contaminated by pollutants [10-13]. In recent years, many authors are interested in the synthesis of the adsorbents powdered and granulated goepolymer for removing the organic pollutants from aqueous media and wastewaters by the adsorption method [14-16]. The batch and column adsorption of organic pollutants from aqueous solutions by goepolymer has been studied by various scientists [17-19].

In this work, the geopolymer developed by fly ash and alkali solution has been tested for adsorption the dye (MB) in aqueous solutions by batch and fixed-bed column.

2 Materials and methods

2.1 Materials and chemicals

– Methylene Blue dye (MB) with the molecular formula $C_{16}H_{18}ClN_3S$ and molecular weight 319,852 g/mol, purchased from Sigma Aldrich.

– Alkaline solution: sodium silicate powder (Honeywell Riedel-de Haën®, Germany; 18 wt.% Na_2O, 63 wt.% SiO_2, 18wt.% loss on ignition) and sodium hydroxide (ACS AR Analytical Reagent Grade Pellets).

– Fly ash was obtained from Jorf Lasfar power plant in EL Jaddida, Morocco. The chemical composition (wt.%) of fly ash measured by X-ray Fluorescence (XRF) was: (SiO_2 (52.5%), Al_2O_3 (30.2), CaO (6.27%), MgO (1.23%), Fe_2O_3 (2.94%), SO_3 (0.787%), and LOI ("loss on ignition" 7.12%)) The total mass (SiO_2+Al_2O_3 = 82.7%) indicated the Jorf Lasfar fly ash was Class F.

X-ray diffraction (XRD) analysis was carried out using (Xpert Pro model) diffractometer equipped with a source operating Cu-Kα (1.54060 Å).

Nitrogen adsorption/desorption isotherms were recorded for powdered samples using a Micrometrics model 3Flex 3500 analyzer.

Surface areas were obtained by the Brunauer–Emmett–Teller (BET) equation, pore volume and pore size distributions were calculated from the desorption data using the Barrett–Joyner–Halenda (BJH) method. Malvern Zetasizer (nano series) was used to measure the zeta potential of geopolymer matrices well dispersed in deionized water (5% weight suspension). Before each set of measurements, the instrument was calibrated using potassium tungstosilicate solution, and all calculations were performed using the Zetasizer software. The method used is the Dynamic Light Scattering (DLS).

^{29}Si MAS –NMR of samples materials was carried out by MAS NMR Silicon (Magic Angle Spinning).^{29}Si MAS-NMR spectroscopic characterization was conducted with a Bruker apparatus, model Avance III 600 MHz.

2.2 Batch and fixed-bed column adsorption experiments

Batch experiments were used to determine the effect of contact time on the MB adsorption capacity. After each batch experiment, the adsorbent was separated by centrifugation. The initial concentration and the concentration of MB in the solution at the time (t) were determined by UV/VIS spectrophotometer (JASCO V-630 UV/VIS). The amount of adsorption at time (t), q_t (mg/g) was calculated using the equation:

$$q_t = \frac{(C_i - C_t)}{m} V \qquad (1)$$

Where C_i is the initial concentration ($mg.L^{-1}$) and C_t, the concentration ($mg.L^{-1}$) at the time t. V is the volume of solution (L) and m is the mass of the adsorbent (g).

Fixed bed column experiments were carried out using a glass column (diam, 1.2 cm; length, 30 cm) for adsorption of MB onto geopolymer at various bed height (Z=1cm (2.28 g); 3 cm (5.82 g); 5 cm (8.04 g)) by fixing influent concentration (40 mg/L) and flow rate of 6 ml/min.

3 Results and Discussion

3.1 Synthesis of the geopolymer and characterization of samples

3.1.1 Geopolymer Synthesis

The alkali activator solution is prepared by mixing sodium hydroxide (12M) and sodium silicate powder in a weight ratio of 2.5 for 24 h. The geopolymer was elaborated by mixing fly ash with alkaline activator solution in a solid to liquid ratio of 2.5 for 15 min. After mixing, the sample paste was cast in a cylindrical plastic mould for curing at the temperature of 60°C for 24 h. After the curing period, the geopolymer obtained was dried in at the ambient temperature for 3 days and crushed to particle size of (100-200 μm for powdered geopolymer and 0.98-1.85 mm for granulated geopolymer). The obtained samples are stored in a desiccator before their use.

3.1.2 Characterizations of samples

X-ray diffraction (XRD) analysis

The XRD pattern of fly ash and the geopolymer shown in Fig.1 illustrates the crystalline phases present in the fly ash and geopolymer namely, quartz and mullite. After geopolymerization process, the shift of peak towards lower angles is due to the formation of new product and the large band between 20° and 28° (2 Theta) indicates the presence of amorphous aluminosilicate.

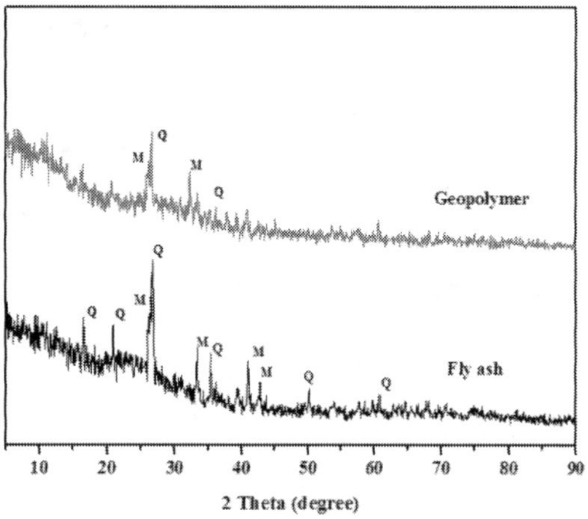

Fig.1. XRD patterns of fly ash and geopolymer (Q–quartz, M–mullite).

The surface area, pore structure properties and zeta potential of geopolymer

The surface area and porous structure of geopolymer are summarized in Table 1. The BET surface areas, pore volumes and pore sizes of the sample are 2.5080 m^2/g, 0.001567 m^3/g and 1.1738 nm, respectively. The geopolymer was classified as microporous materials with average pore size up to 2 nm.

Table 1. Textural properties by N_2 adsorption/ desorption studies

Typical parameters	FA- based geopolymer
S_{BET} (m^2/g)	2.5080
S_{ext} (m^2/g)	2.1338
S_{mic} (m^2/g)	0.3743
V_t (m^3/g)	0.001567
V_{mic} (m^3/g)	0.001037
D_p (nm)	1.1738

S_{BET}: The specific surface area; **S_{ext}**: The external area; **S_{mic}**: The micropore specific surface; **V_t**: The total pore volume; **V_{mic}**: The micropore volume; **D_p**: The average pore size

Fig.2 shows the zeta potential obtained for the geopolymer. As can be observed in Fig.2, the negative value (-41.6 mV) was observed in the surface zeta potential of geopolymer. This behaviour indicates that the surface of geopolymer is negatively charged and the compound is a condition for the adsorption of the cationic dye existing in aqueous solution.

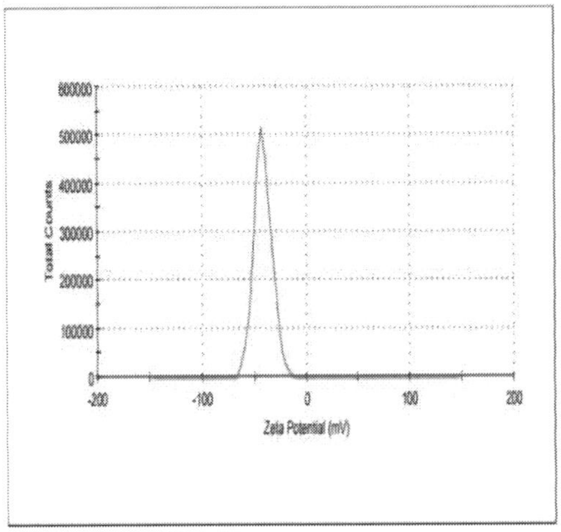

Fig.2. Zeta potential of geopolymer in water

^{29}Si RMN MAS analysis

^{29}Si RMN MAS spectrum of fly ash and geopolymer are shown in Fig.3. ^{29}Si RMN MAS spectrum of fly ash has strong peak at -107 ppm, which demonstrate tetrahedral $[SiO_4]^{4-}$ coordination SiQ_4 [20-22]. According to the ^{29}Si RMN MAS spectra of geopolymer, the peak at -94,66 ppm indicates the presence of SiQ_4 (2-3Al) in the geopolymer [23]. This result indicates that the chemical structure of fly ash is completely different from the geopolymer one.

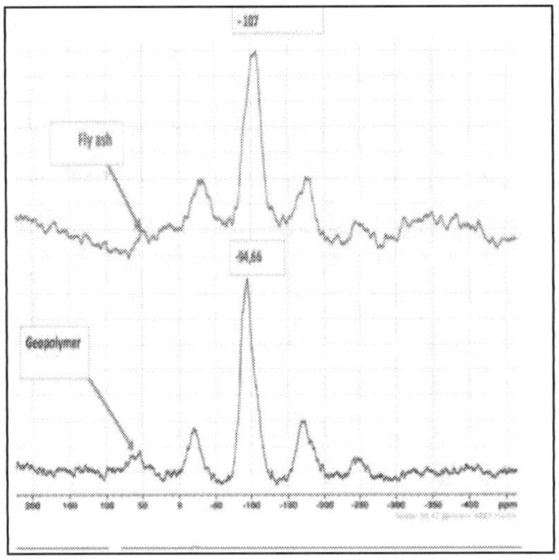

Fig.3. ^{29}Si MAS–RMN spectra of fly ash and geopolymer.

Scanning electron microscope (SEM) analysis

Scanning electron microscopy (SEM) was used to observe the surface texture of fly ash and geopolymer. The SEM image of fly ash (fig 4.a) shows that the sample particles have spherical forms and different sizes. The changes in the morphology of fly ash and geopolymer are illustrated in Fig.4. The results indicate the formation of the new material after activation the fly ash by alkali solution and SEM images shows the porous structure of geopolymer.

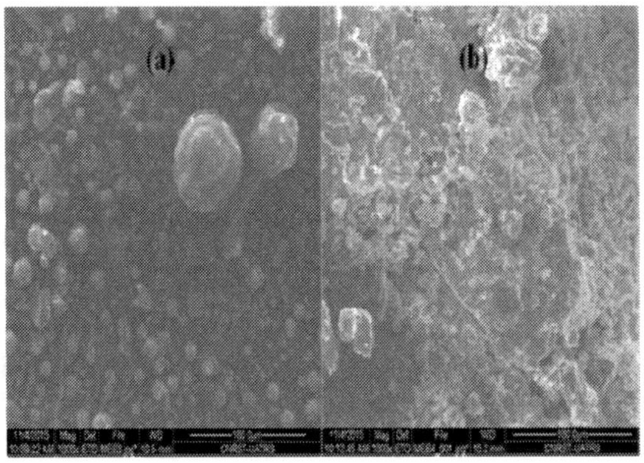

Fig.4. SEM images of the surface the fly ash (a) and geopolymer (b)

3.2 Adsorption studies

Effect of the contact time

The effect of the contact time on MB removal was considered by changing the contact time from 30 min to 210 min. The MB initial concentration was fixed at 40 mg/L with a fixed geopolymer dosage (1 g/L) and pH of solution was fixed at 6. The effect of contact time of MB onto geopolymer is displayed in Fig.5. The results indicated that MB capacity adsorption occurred in the

https://doi.org/10.1051/matecconf/201814902088

first 120 min and the adsorption equilibrium was reached at 150 min.

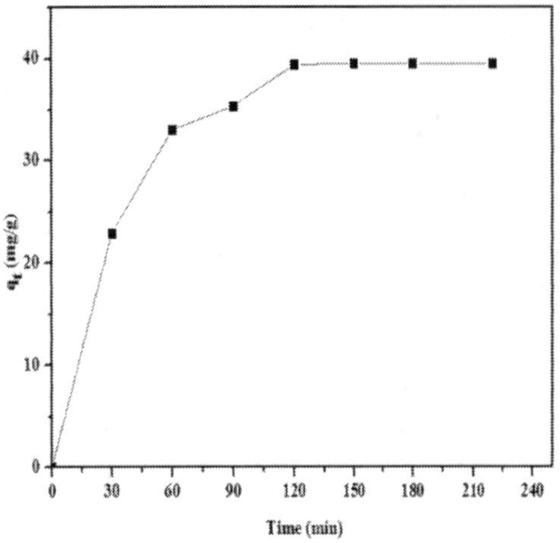

Fig.5. Effect of contact time on adsorption of MB onto geopolymer

Adsorption kinetics

Several models have been established to describe the adsorption kinetics and the rate-limiting step of the process. They include models of pseudo-first, second-order kinetic model, model intra-particle diffusion and sorption model Weber and Morris, the relationship of Adam-Bohart Thomas, etc [24]. The adsorption kinetics data of MB using the geopolymer were analyzed with pseudo-first-order [25], pseudo-second-order [26] and intraparticle diffusion kinetic models [27].

The experimental data of MB adsorption on geopolymer were simulated with three models and the results were presented in Table 2. The correlation coefficients (R_1^2) for pseudo-first-order kinetic model are between 0.718 and 0.926, the correlation coefficients (R_2^2), for the pseudo–second-order kinetic model are between 0.99 and 1, and the correlation coefficients (R_3^2) for intraparticle diffusion model are between 0.513 and 0.993. On the basis of R^2 value, the pseudo second order rate model fit best with experimental data. On other hand, the q_e (43.47 mg/g) calculated value from the pseudo-second-model is close with the experimental value (39.55 mg/g). This confirmed that the adsorption processes are better represented by pseudo-second-model and the adsorption process is controlled by the chemisorption process. These results indicated that the adsorption occurred probably by surface complexation reaction between cationic dye and negative charged sites on the geopolymer. Similar kinetic results were reported for the adsorption of MB onto spent tea leaves [28], oil palm (Elaeis guineesis) [29] and coal-bearing kaolinite [30].

Table 2. Simulated parameters by Pseudo-first-order, Pseudo-second-order and intraparticle diffusion kinetic for adsorption of MB onto geopolymer

Dye		**Pseudo -first-order**			**Pseudo-second -order**			**Intra-particule diffusion model**		
		$Ln(q_e - q_t) = Lnq_e - k_1 t$			$\frac{t}{q_t} = \frac{1}{k_2 q_e^2} + \frac{t}{q_e}$			$q_t = k_I t^{1/2} + I$		
C_0 mg/L	**q_{exp}**	**q_e (mg/g**	**k_1 (1/min)**	**R_1^2**	**q_e (mg/g)**	**k_2 (g/mg min)**	**R_2^2**	**I (mg/g)**	**k_{id} (mg/g min$^{0.5}$)**	**R_3^2**
40	39.55	87	0.051	0.926	43.47	0.002	0.99	17.84	1.678	0.796

Where:
q_e: The mass of adsorbate per unit mass of adsorbent at equilibrium; **q_t**: the mass of adsorbate per unit mass of adsorbent at time; **k_1**: Pseudo-first order rate constant (min^{-1}). **k_2**: pseudo-second order rate constant (g/(mg.min)), and **I**: The intraparticle diffusion constants (mg/g).

3.3 Colum adsorption studies

Influence of contact time was studied at different bed height and the concentration of MB was fixed at 40 mg/L. The breakthrough curves have been determined at various bed heights at 25°C. The results are illustrated in Fig.6. It can be seen that the breakthrough time and exhaustion times increased with increasing bed height. This can be explained with the interaction between of cationic dye (positively charge) and geopolymer (negatively charge of surface) is also greater. Similar kinetics behaviour for the adsorption of MB onto various sorbent materials was also reported in the literature [31-33]. Various mathematical equations were used to describe column adsorption process. The adsorption kinetics data of MB using geopolymer were analyzed with Thomas [34], Adams-Bohart [35] and Yoon-Nelson equations [36]. These models tested for MB adsorption in the column study were summarized in Table 3. For Thomas parameter model, rate constant k_{Th} decreased simultaneously with increasing bed height. However, the adsorption capacity increases with the increase in geopolymer mass in height. From these results, it is found that the adsorption capacity usually increases with the increase of adsorbent ratio. For the Yoon-Nelson model, the rate constant k_{YN} decreased and the time required for 50 % breakthrough increased with increasing bed height. This phenomenon can be explained by the number of active sites increases with the increases in adsorbent doses in height bed. Similar results were reported by other researchers [37-40]. For the values of Adams-Bohart model, the kinetic constant k_{AB} decreases simultaneously and the concentration at saturation (N_0) of column decreases with increasing bed height. This result can be explained by that the decrease the mass transfer resistance between the adsorbate (MB) and adsorbent surface (geopolymer). The kinetic models are found to describe well the breakthrough curves of methylene blue removal. The overall study results showed that the granulated geopolymer have high capacity and potential for removal the organic pollutants from aqueous solutions.

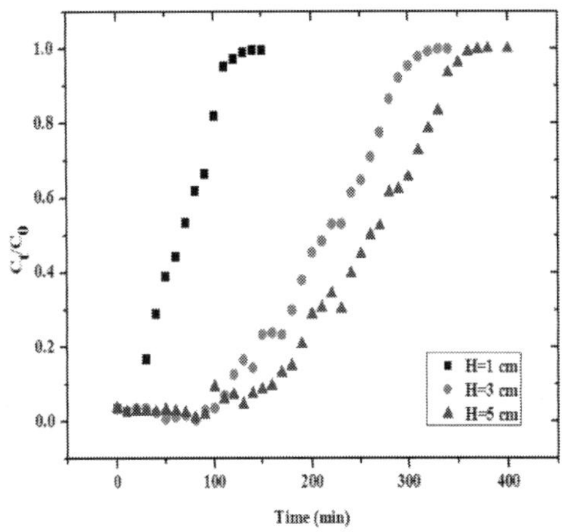

Fig.6. Break through curve of bed height (Initial concentration= 40 mg/L, flow rate 6 mL/min)

Table 3. Kinetic parameters for adsorption of MB onto geopolymer in the column study

	Thomas model			**Adams-Bohart**			**Yoon-Nelson**		
	$Ln(\frac{C_0}{C}-1)=k_{th}q_0\frac{m}{F}-k_{th}C_0t$			$Ln\frac{C}{C_0}=k_{AB}C_0t-k_{AB}N_0\frac{H}{v}$			$Ln\frac{C}{C_0-C}=k_{YN}t-k_{YN}t_{50}$		
Bed height	**q_0 (mg/g)**	**k_{TH}**	**R^2**	**k_{AB}**	**N_0**	**R^2**	**K_{YN}**	**t_{50}**	**R^2**
1 cm	6.475	$2.075.10^{-3}$	0.857	6.10^{-4}	5240	0.782	0.083	59.20	0.857
3 cm	7.820	$7.5.10^{-4}$	0.816	$3.5.10^{-5}$	4070	0.828	0.03	189.56	0.816
5 cm	6.496	$6.5.10^{-4}$	0.764	$2.75.10^{-5}$	2875	0.915	0.026	217.04	0.764

Where:
Thomas model: **C_0:** The initial concentration (mg/L), **C:** effluent concentration at the time t (mg/L), **m:** Mass of the adsorbent (g), **F:** flow rate (mL/min), **k_{th}:** Thomas model rate constant (mL/min g), **q_0:** The column adsorption capacity (mg/g).
Adams-Bohart model: **C_0:** The initial concentration (mg/L), **C:** effluent concentration at the time t (mg/L). **k_{YN}:** the Yoon-Nelson rate constant (min^{-1}), **H:** height of adsorbent in fixed bed in the column (cm) , **N_0:** the volumetric sorption capacity (mg/L) , **v:** The velocity (cm/min).
Yoon-Nelson model: **C_0:** The initial concentration (mg/L), **C:** effluent concentration at the time t (mg/L). **k_{YN}:** is the rate constant. **t_{50}:** The time required for 50 % Adsorbate breakthrough (min).

4 Conclusions

This study thoroughly reviewed the application of geopolymer for the treatment of dye-containing aqueous solution. A physico-chimical characterization of the samples indicates the formulation of geopolymer by fly ash and alkaline activator solution, and suggests that the adsorbent have functional groups suitable to remove MB. The kinetic studies for batch experiment reveal that MB adsorption on geopolymer was rapid and equilibrium was reached within 120 min. The kinetics of MB adsorption onto adsorbent was followed pseudo-second-order kinetic model. The breakthrough curves of column both showed that the geopolymer favors the adsorption of MB. The results of this research indicated that the geopolymer synthesized by fly ash and alkaline activator could be used adequately for the treatment of aqueous solution contaminated with cationic dye.

References

1. S. Nawar, H. Doma, Sci. Total Environ. **79**, 271-279 (1989)
2. M. A. M. Salleh, D. K. Mahmoud, W. A. Karim, A. Idris, Desalination. **280**, 1-13 (2011)
3. C. Tocchi, E. Federici, L. Fidati, E. Manzi, V. Vincigurerra, M. Petruccioli, Water Res. **46**, 3334-3344 (2012)
4. Y. Zhang, C. Causserand, P. Aimar, J.P. Cravedi, Water Res. **40**, 3793-3799 (2006)
5. S. Stopic, B. Friedrich, A. Widigdo, Metalurgia. **13**, 27-34 (2007)
6. D.J. Ennigrou, L. Gzara, M.R. Ben Romdhane, M. Dhahbi, Desalination. **246**, 363-369 (2009)
7. A. Sandoval, C. Hernández-Ventura, T. E. Klimova, Fuel. **198**, 22-30 (2017)
8. Y. Al-Ani, Y. Li, J. Taiwan Ins. Chem. Eng. **43**, 942-947 (2012)
9. Y. Y. Sia, I.A.W. Tan, M.O. Abdullah, MATEC Web Confer. **87**, 03010 (2017)
10. N. Barka, A. Assabane, A. Nounah, L. Laanab, Desalination. **235**, 264-275 (2009)
11. L. Li, S. Wang, Z. Zhu, J. Colloid Interface Sci. **300**, 52–59 (2006)
12. K. Bhattacharyya, S. Sengupta, G. K. Sarma, Appl. Clay Sci. **99**, 7-17 (2014)
13. D. Pathania, S. Sharma, P. Sing, Arab. J. Chem. **10**, S1445–S1451 (2017)
14. Y. Zhang, L. Liu, Particuology. **11**, 353-358 (2013)
15. T.W. Chenga, M.L. Leea, M.S. Koa, T.H. Uenga, S.F. Yang, Appl. Clay Sci. **56**, 90–96 (2012)
16. L. Li, S. Wang, Z. Zhu, J. Colloid Interface Sci. **300**, 52-59 (2006)
17. A. Jusoh, Y. K. Tam, A.G. Liew, M. J. Megat, K. Saed, Int. J. Eng. Tech. **1**, 58-63 (2004)
18. T.W. Cheng, M.L. Lee, M.S. Ko, T.H. Ueng, S.F. Yang, Appl. Clay Sci. **56**, 90-96 (2012)
19. C. Smaranda, M.C. Popescu, D. Bulgariu, T. Malutan, M. Gavrilescu, Process Saf. Environ. Prot. Process (2016), http://dx.doi.org/10.1016/j.psep.2016.06.027
20. F. Škvára, L. Kopecký, J. Němeček, Z. Bittnar, Ceramics-Silikáty. **50**, 208-215 (2006)
21. N.P. Bansal, J.P. Singh, W.M. Kriven, H. Schneider, Ceram. Trans. **153**, 175-186 (2003)
22. J.G.S. van Jaarsveld, J.S.J. van Deventer, L. Lorenzen, Miner. Eng. **10**, 659-669 (1997)
23. P.S. Singh, T. Bastow, T. Trigg, J. Mater. Sci. **40**, 3951-3961 (2005)
24. J. Febrianto, A.N. Kosasih, J. Sunarso, Y. Ju, N. Indraswati, S. Ismadji, J. Hazard. Mater. **162**, 616-645 (2009)
25. S.S. Gupta, K.G. Bhattacharyya, Adv. Colloid Interface Sci. **162**, 39-58 (2011)
26. Y.S. Ho, G. McKay, Process Biochem. **34**, 451–465 (1999)
27. T. Furusawa, J.M. Smith, Aiche J. **20**, 88-93 (1974)
28. B.H. Hameed, J. Hazard. Mater. **16,** 753-759 (2009)
29. H.D. Setiabudi, R. Jusoh, S. Suhaimi, S.F. J. Masrur, Taiwan. Inst. Chem. Eng. **63**, 363-370 (2016)
30. W. Goa, S. Zhao, H. Wu, W. Deligeer, S. Asuha. Appl. Clay Sci. **126**, 98-106 (2016)
31. R. Han, Y. Wang, X. Zhao, Y. Wang, F. Xie, J. Cheng, M. Tang, Desalination. **245**, 284-297 (2009)
32. O. Hamdaoui, J. Hazard. Materi. **B138**, 293-303 (2006)
33. Y. Zhuang, F. Yu, J. Chen, J. Ma, J. Environ. Chem. Eng. **4**, 147-156 (2016)
34. Z. Aksu, F. Gönen, Process Biochem. **39**, 599–613 (2004)
35. M. Calero, F. Hernáinz, G. Blázquez, G. Tenorio, M.A. Martín- Lara, J. Hazard. Mater. **171**, 886-893 (2009)
36. Z. Aksu, S.S. Cagatay, F. Gonen, J. Hazard. Mater. **143**, 362-371 (2007)
37. M. Ratnamala, M. Rahul, S. Sameer, M. Vaani, Omkar, T. Devdatt, Asian. J. Chem. **29**, 191-195 (2017)
38. N. Kannan, S. Murugavel, Elctron. J. Environ. Agric. Food Chem. **6**, 1860-1868 (2007)
39. P. Sivakumar, P.N. Palanisamy, International Journal of Chem. Tech. Research. **1**, 502-510 (2009)
40. Z. Zulfadhly, M.D. Mashitah, S. Bhatia, Envion. Pollut. **112**, 463-470 (2001)

Study of indoor performances of a building using Rammed earth

Wiçal CHEIKHI[1]; Khadija BABA[1]; Sidi Mohamed LAMRANI[2]; Abderrahmane NOUNAH[1] ; Mohamed KHALFAOUI[2] ; Lahcen BAHI[3]

[1] GCE Laboratory, High School of Technology –Sale-, Mohammed V University in Rabat, Morocco
[2] ASTIMI Laboratory, High School of Technology –Sale-, Mohammed V University in Rabat, Morocco
[3] 3GIE Laboratory, Mohammedia School of Engineering, Mohammed V University in Rabat, Morocco

Abstract. Nowadays, many assumptions hold that returning to rammed earth construction could be a solution to the energy challenges of the time. These assumptions are based on the ability of the material to maintain a comfortable indoor temperature regardless of the outdoor climatic variations. However, its influence on the energy performance of buildings remains little explored so far. This paper comes in this order of ideas. It is devoted to study and evaluate the energy efficiency of rammed earth buildings compared to those in concrete structure and masonry envelope. It takes as case study two buildings: one is traditional, entirely built with the technique of rammed earth; the other is constructed of concrete. In order to compare between their performances, we have performed a simulation of the thermal behavior of each by the dynamic thermal simulation software Design Builder.

1 Introduction

Since the beginning of its use in the construction industry, the rammed earth always proved it reliability as much as local, ecological material implementing excellent thermal properties. Indeed, whatever the climatic context of their establishment, rammed earth constructions are able to get freshness in summer and tepidity in winter. This property makes it possible to reduce the recourse to the systems of air-conditioning and heating, leading to an abusive consumption of energy.

In a world context centered on sustainable development and energy efficiency, working to reduce the energy consumption of the buildings becomes a major requirement. Indeed, on the national scale, construction industry, formed in majority of modern buildings, is today responsible for 36% of the total energy intake and 51% of the electricity's consumption [1]. Alarming figures, which invite all the speakers in the sector of the building to suggest effective and accessible solutions in order to improve the thermal performances of the envelope, regarded as the primary source of the energy losses. In this direction, several research tasks were undertaken, suggesting capable solutions to improve energy quality of the envelope. Soebarto & al. [2] carried out a study about the indoor performance of rammed earth houses. The study compared between the performances of two existing houses in Willunga, South Australia: the one uses rammed earth as the sole wall material and the second uses insulated rammed earth walls. The results showed that the insulated house could be 5 degrees warmer than the uninsulated one, and would potentially use less energy for heating.

The renovation of social accommodation which gives an occasion interesting for the reduction of the power consumption and the increase of the comfort of the inhabitants was studied by Gagliano &al. [3]. The improvement of the thermal insulation of the building envelope and the use of renewable energy's sources was offered, what drove to an economy of energy and to an increase of comfort. From his part,

Jaher &al. [4] led a dynamic thermal simulation aiming at the valuation of the thermal quality of the rammed earth used as an external cladding of a low-cost house in Bangladesh. The results of simulation showed that net improvements concerning thermal comfort and power consumption could result from the usage of the rammed earth applied as external cladding. Following the same logic, the present paper consists of establishing a comparative study between the thermal performances of rammed earth buildings and those in concrete; with an objective to show the interest of the use of the rammed earth that could be considered as energy-performance improvement in the contemporary building. Comparison is led on a building in rammed earth and a second variable of the same building constructed in concrete. The studied building is a residential type, constructed in two levels. The objective of this job is to show the influence of the material's envelope used to improve the thermal comfort of the building and as a result, the

 https://doi.org/10.1051/matecconf/201814902089

lightering of heating and air-conditioning costs. Simulation registers as part of Moroccan thermal regulation and is insured by both software BLOCK LOAD [5] and DESIGN BUILDER [6].

Fig. 1. 3D view of the studied building

2 Case study description

The studied building is a project under construction located in the peripheral zone of Marrakech city (Morocco, climatic zone 5), at about fifteen kilometers of the city center of this one. The building is a residential type family house (five persons) and composed of a ground floor and a first floor (see figure1). The ground floor spreads out on an area of 143Sqm and articulates around a patio which distributes the kitchen on the one side, the living area on the other one. The first floor has a surface of 90 Sqm, composed mainly of the master bedroom, which opens on a patio and is provided with a balcony (Figure2).

Fig. 2. Orientation and building layout

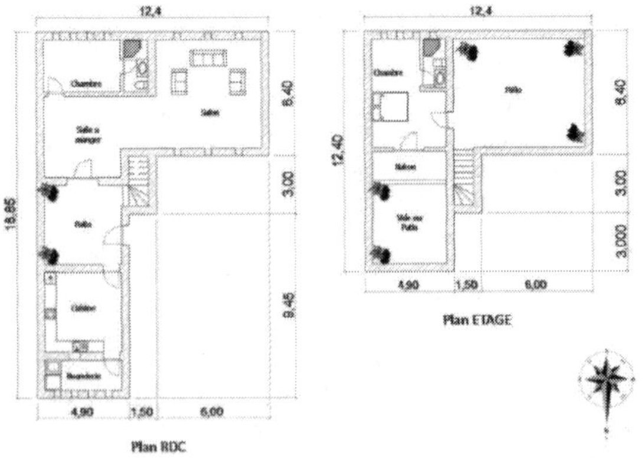

The building has an interesting orientation: the living rooms (bedrooms and living room) are facing the south, while the laundry room is located in the opposite direction (north) (Figure 2). This arrangement gives the possibility to enjoy the passive solar contributions in the living room and to create a buffer zone (laundry) protecting the building from thermal losses to the north.
The organization of the building around the patio gives the possibility to improve the natural lighting and thus reduce the use of artificial lighting devices. From an air circulation point of view, the building benefit from two natural ventilation modes: horizontal ventilation through the different windows; another vertical being made through the opening of the building on the Patio. The latter allows the air circulation by chimney effect. Effective ventilation device in arid regions, known for its ability to minimize overheating in summer.

3 Walls composition study

3.1 Initial Option (Rammed earth)

The original building is entirely built with the traditional technique of Rammed earth. All the vertical walls (façade walls and partitions) are composed of a mixture of 2/3 clay for 1/3 straw. The roof is made up of wooden beams, joists and a reed grid. The outer walls have a thickness of 0,4m, which is considered to be the minimum thickness for earth constructions. The thermophysical characteristics of the building walls have previously been the subject of an experimental study. It was based on the steady state asymmetric hot plate method [7] to determine the thermal conductivity (λ), the asymmetrical transient Hot Plate method to measure the effusivity (e), and the flash method [8] to determine the diffusivity (a) of the mixture used in the construction of our building. The experimental values found and used in our study are summarized in the table1.

Table 1. Thermophysical characteristics of the used Rammed earth

Thermal conductivity λ (W.m-1.K-1)	**0,463**
Thermal diffusivity a.10-7 (m2.s-1)	3,13
Thermal effusivity a (w. s1/2.m-2.K-1)	648,07
Specific heat C (J.Kg-1.K-1)	1128,59
Density ρ (kg/m3)	1310,69

3.1 Comparison Option (Masonry)

In order to compare the thermal performance of rammed earth building with masonry buildings, we have created a second variant of the studied building: it is a reinforced concrete structure and a masonry envelope. It corresponds to the model of construction in masonry usually used in the Moroccan context. The composition of this variant's walls is summarized in the table below.

Table 2. Envelope composition of the Masonry option

Envelope	External walls	Roof slab and Intermediate slab	Ground slab
Width (m)	0,30	0,25	0,40
Structure (From external to internal)	Lime Plaster/cement Hollow block 6 holes Air chamber Hollow block 12 holes Lime Plaster/cement	Interior plaster Slab Waterproof bitumen Screed to fall Finishing screed Floor tiles	floor tiles Finishing screed Screed to fall Concrete slab Polyane Layer Sand Gravel layer
U Value ($W/m^2.K$)	1,96	2,04	1,60

To calculate the U value, we took account of the external and internal surface resistances, Rse and Rsi

4 Energy simulation of the Buildings

4.1. Building Thermal summary

The thermal summary of the building is made using the software package Block Load. This calculates summer/winter loads for buildings. Heating and cooling requirements can be calculated also. This package uses the response factor method to calculate the loads. This method is approved by ASHRAE as a suitable method for the calculation of hourly loads. For our study, the objective behind the installation of the thermal balance is to quantify the energy consumption of the two variants to be compared (rammed earth variant and masonry variant). The results are shown in the attached tables.

Table 3. Characteristics of air conditioning equipment for both variants

Option	Masonry	Rammed earth
Heating capacity (kw)	32,2	29,3
Cooling Capacity (kw)	13,2	10,6
Equipement Characteristics	EER=3,05 COP=4,09	EER=3,05 COP=4,06

It is important to note that the coefficient of performance (COP) of the equipment required to heat the masonry variant of the building is slightly higher than the rammed earth version. While the cooling equipment EER is equal for both variables. In the light of this observation, it is possible to argue that rammed earth buildings allow a better conservation of heat compared to those in masonry.

4.2. Dynamic thermal simulation

Dynamic thermal simulation (DTS) simulates the behavior of the building hour by hour. To do this, the simulation is carried out using the Design Builder software which uses the dynamic simulation engine EnergyPlus to model the thermal performances of the building. This takes into account the following parameters:

- Location of the building, local weather

- The orientation of the building, the distribution of parts and glazing

- The materials structure of the building, its thermal inertia, its thermal bridges, its solar protection, etc.

- Heating, air conditioning, DHW production, ventilation, lighting

- The occupation of the parts and the behavior of the users

- Regulation of heating and possibly other equipment.

For each of the two studied variables, our building was subdivided into several thermal zones according to the scope of each, their location and whether they are covered or not (in the case of the patio and the balcony). Six (6) areas resulted from the ground floor and Four (4) from the first. The wall compositions used for each variable are summarized in Table 1 for the rammed earth construction variant and in Table 2 for the masonry variant. The heating setpoint temperature was set at 20 ° C and the cooling temperature was set at 26 ° C. We then defined the number of users of the building (five people) and an annual schedule for the occupancy and use of heating and air-conditioning equipment.

We then proceeded to define the outputs of the simulation. These include the internal temperature of each studied space, radiant temperatures, thermal losses, heating and cooling requirements and the total energy consumption of the building.

5 Results of the simulation

To compare the energy requirements in (kWh / m2 / year) of the heating and cooling of rammed earth buildings and concrete ones, we proceeded to simulate the energy requirements for the following two cases:

- Walls of 30cm in double wall of masonry.

- Walls of 40cm built in rammed earth.

The results of the dynamic energy simulation carried out on the Design Builder software show the following:

- Improved operating and radiant temperatures minimizing the "cold wall" sensation due to the temperature difference between the air and the wall surface (table 4).

Table 4. Temperatures in both options

Option	Masonry	Rammed earth
Air temperature (°C)	19,96	
Radiant temperature (°C)	15,12	16,65
Operative temperature (°C)	17,54	18,30

- A significant reduction in thermal losses through walls, the top and bottom floors are recorded in the rammed earth construction variable compared with the masonry (Figures 3 and 4 and table 5).

Fig. 3. Heat losses in the rammed earth option

Fig. 4. Heat losses in the masonry option

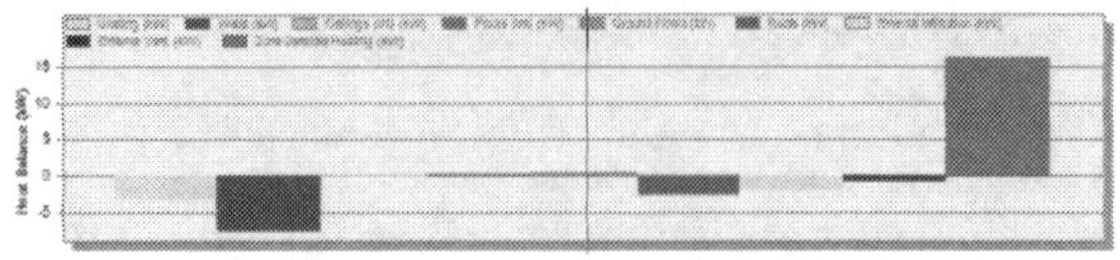

Table 5. Heat losses in both options

Heat loss (KW)	Masonry	Rammed earth
Walls	7,75	4,11
Roofs	2,73	1,94

- A decrease in heating requirements is noted for the building constructed in rammed earth. Table 6 shows a reduction in heating requirements which is as high as 32% compared to the masonry variant.

- A 26% increase in air conditioning needs is noted in the rammed building (Table 6) compared to the masonry variant. The air conditioning loads are increased because their walls with high thermal inertia characterize the rammed constructions. This inertia lead the heat coming from the diurnal solar radiation to be absorbed, stored and then trapped inside the walls. Its evacuation will require more time and will induce the use of air conditioning to compensate the sensation of thermal discomfort caused by this trend.

Table 6. Annual Heating and Cooling demand

Option	Heating demand [kWh/m2]	Cooling demand [kWh/m2]
Rammed earth	33.96	668.68
Masonry	49.94	507.32

6 Conclusion

Dynamic thermal simulation of the two variants (rammed earth and masonry) of the studied building allowed us to obtain temperature graphs, a quantification of the heating and cooling needs, a thermal loss reading ... All these results, to compare the thermal performance of rammed earth and masonry constructions and to identify the strengths and weaknesses of each. The conclusions drawn from the study are as follows:

- Rammed earth is an ecological, renewable and energy-saving building material. It fits perfectly into the energy efficiency approach.

- The use of rammed earth in construction makes it possible to reduce certain trends observed in masonry construction. We enumerate as an example the phenomenon of cold walls.

- The rammed earth is a material with high thermal inertia. Nevertheless, this feature should be considered with caution: If it reduces heating requirements considerably in the winter, it causes an increase in the air conditioning load in summer. Indeed, the walls with high thermal inertia absorb the heat of the solar radiation, store it and restore it in phase shift (in offset with the absorption time). This makes it possible in winter to extend the feeling of thermal comfort and reduce the use of heating systems, resulting in a gain of more than 30% in the energy load. While in the summer, the restitution of the phase-shifted heat flux on the contrary extend the sensation of overheating, inducing an additional use of the cooling systems. In order to reduce the consequences of this phenomenon it is recommended to:

- Decrease the overall rate of bay windows in rammed earth constructions. The more windows there are, the more the building will absorb heat. It is remarkable that the traditional rammed earth constructions of the South of Morocco are opaque buildings, benefiting only from very limited surfaces of openings. However, they are nonetheless sheltered from external climatic variations, providing freshness in summer and tepidity in winter.

- Promote the use of fixed shading devices (awning) for south-facing openings. This allows the building to benefit from passive solar contributions in winter and to reduce overheating in summer.

- Promote overnight ventilation by creating air chimneys, transverse ventilation, ventilation grilles, etc. This action allows the surplus heat accumulated during the day to be evacuated more quickly, thus reducing the need for air conditioning during the summer days.

References

1. Agence internationale de l'énergie, Morocco (2009).
2. Soebarto Veronica., Analysis of Indoor Performance of Houses Using Rammed Earth Walls, *Eleventh International IBPSA Conference Glasgow, Scotland,* 1530-1537, (2009).
3. A.Gagliano, F.Nocera, F.Patania, G.Capizzi, A case study of EnergyEfficiency Retrofit in social housing units; *EnergyProcedia* 42, 289 – 298, (2013).
4. Jaher Wasim, The effects of rammed earth on thermal comfort and energy consumption of a bungalow in Bangladesh, *Procedia environmental sciences* (2017).
5. Software package Block Load, for calculating the thermal summary of the building.
6. Design Builder Software Ltd, 2009. DesignBuilder 2.1 User's Manual. Available at: http://www.designbuilder.co.uk/component/option,com_docman/task,cat_view/gid,20/Itemid,30/.
7. Y Jannot, V Felix, A Degiovanni, A centered hot plate method for measurement of thermal properties of thin insulating materials, *Measurement Science and Technology, Volume 21, Number 3* (2010).
8. Parker, W. J.; Jenkins, W. J.; Butler, C. P. and Abbott, G. L., Flash method of determining thermal diffusivity, heat capacity and thermal conductivity. *J. App. Physics, 32: (9), 1679-1684.* (1961).
9. T.T. Bui, A. Mesbah, S. Maximillien, A. Limam, Behavior of rammed earth walls under compression or shear stress, *J. Mater. Environ. Sci. 7 (10),* 3584-3594, (2016)
10. Lucile Soudani, Antonin Fabbri, Monika Woloszyn, Pierre-Antoine Chabriac, Anne-Cécile Grillet, Jean-Claude Morel, Étude de la pertinence des hypothèses dans la modélisation hygrothermique du pise, *proceeding of Conférence IBPSA France-Arras,* (2014)
11. David Allinson, Matthew Hall, Hygrothermal analysis of a stabilised rammed earth test building in the UK, *Energy and Buildings 42, 845–852,* (2010).
12. G. Minke, Earth Construction Handbook: The Building Material Earth in Modern Architecture, *Southampton, WIT Press* (2000).
13. J.D. Revuelta-Acosta, A. Garcia-Diaz, G.M. Soto-Zarazua, E. Rico-Garcia, Adobe as a sustainable material: a thermal performance, *Journal of applied sciences 10 (19), 2211-2216,* (2010).
14. Julien Borderon, Richard Cantin, Joseph Virgone, Etude du comportement thermique dynamique de l'enveloppe non isolée du bâtiment : cas des murs inertes, *proceeding of Conférence IBPSA France -Lyon* (2008).
15. Lucile Soudani, Antonin Fabbri, Monika Woloszyn, Pierre-Antoine Chabriac, Anne-Cécile Grillet, Jean-Claude Morel, Étude de la pertinence des hypothèses dans la modélisation hygrothermique du pise, *proceeding of Conférence IBPSA France-Arras* (2014).
16. Morel J C, Mesbah A, Oggero M and Walker P. Building houses with local materials: means to drastically reduce the environmental impact of construction, *Building and Environment 36, 1119-1126,* (2001).
17. Paul W L, Taylor P A. A comparison of occupant comfort and satisfaction between a green building and a conventional building, *Building and Environment 43, 1858-1870,* (2008).
18. Ranime El Nabouch, Q.-B. Bui, P. Perrotin, O. Ple, J.-P. Plassiard, Modélisation numérique de structures en pisé : analyses et recommandations, *Rencontres Universitaires de Génie Civil, Bayonne, France. Hal 01167665* (2015).
19. Taylor P, Fuller R J, Luther M B. Energy use and thermal comfort in a rammed earth office building, *Energy and Building 40, 793-800,* (2008).

The Urban Heat Island phenomenon modelling and analysis as an adaptation of Maghreb cities to climate change

Kaoutar Ouali[1], Khalid El Harrouni[2], Moulay Larbi Abidi[1], Youssef Diab[3]
[1]Laboratory of Civil Engineering, Ecole Mohammadia d'Ingénieurs, Université Mohammed 5, Rabat, Morocco
[2]Laboratory Energy, Sustainability and Ambiances, Ecole Nationale d'Architecture, Rabat, Morocco
[3] Lab'Urba, Université Paris-Est-Marne-la-Vallée, Champs-sur-Marne, France

Abstract: The modeling of the urban microclimate, in particular the phenomenon of the Urban Heat Island (UHI), is becoming increasingly essential for city planning and urban design. The phenomenon analysis is henceforth possible thanks to the increase in computational power, the link between simulation tools and urban databases, which allow to represent explicitly the characteristics of the urban microclimate and to better understand its effects, through the analysis and evaluation of the different impacts of the urban climatic or anthropogenic contributors (urban morphology, land use, building sites, albedo, ...). However, the choice of the scaleof the study depends on a balance between the precision of the modeling, the capacities of calculation and the availability and reliability of the data.The UHI phenomenon has been the subject of several research studies in the European countries since the 2000s. Thispaper focuses mainly on the description of the phenomenon, the different methodsused to evaluate and modeled its impacts, using some approaches for mitigating these ones.The contribution aims to highlight the phenomenon of the UHI based on a bibliographic study of the latest research on this topic in Maghreb cities.The state of art focuses on the progress made during the last 15 years taking into account the UHI in the different strategies for adapting cities to climate change and for improving their resilience.

1 Introduction

The build environment is place of radiative, thermal, dynamic and hydric processes which modify the climate of the city. Urban environment stores some heat during day but releases little during night contrary to rural areas.In a current context where the consequences of the urban development must be mastered, the knowledge and the modeling of these various urban physical phenomena became an important stake. The controlof the energy consumptions of cities, characterization and quantification of the Urban Heat Island (UHI) impact, the study of the internal and external thermal comfort of the space urban users, are widely boundto the development of the knowledgeon the interactions between the city, the climate and the energy.

The UHI is one of the most documented phenomena in the scientific literature; it contributes to global warming and leads to an adaptation of the city's sustainable planning policies. The evaluation of the maximum amplitude of the empirical as the case of the large North American and European agglomerations [1] or with the models based on the energy balances surface or volume.It is very difficult to compare UHI studies because of the multiplicity of used methodologies, the simplistic historical vision of the single city center and the clear delimitation of urban space. Nowadays, the reality is more complex.We cannot quantify the difference in temperature due to urbanization alone without knowing the original conditions of the site; natural microclimatic phenomena must make the comparison even more difficult.

The present proposed overview cannot be exhaustive. Nevertheless, it provides an illuminating insight into the state of the issue and the progress made in recent years about the scientific knowledge of this phenomenon. This research carried out also the approaches to study the UHI, and some expected results of these studies, particularly in the Maghreb countries, notably in Morocco.

2 Literature Review

2.1.Characterization of the UHI

UHI refers to the excess air temperature in the urban canopy relative to the surrounding rural areas. The Urban Canopy Layer (UCL) is the layer of air closest to the surface in cities, extending upwards to approximately the mean building height.See schematic representation at figure 1.Each UHI has its own unique features and the quantification of this phenomenon depends on the specific city. Several studies have subsequently demonstrated that the level temperature measurement at the streets did not matter because the air temperature varies little in the street when it is more than one meter from the walls or the road [2, 3].In the literature, the UHI observation methods listed are based on: collecting statistics data of the temperature difference between urban and rural meteorological stations [4, 5, 6]; tendency study of temperature for a meteorological station whose environment is urbanized [7, 8, 9]; comparison of the temperature trend observed for several stations whose environment evolves differently [10]; observation by means of a network of fixed stations [11], observation along a path using vehicle [12].

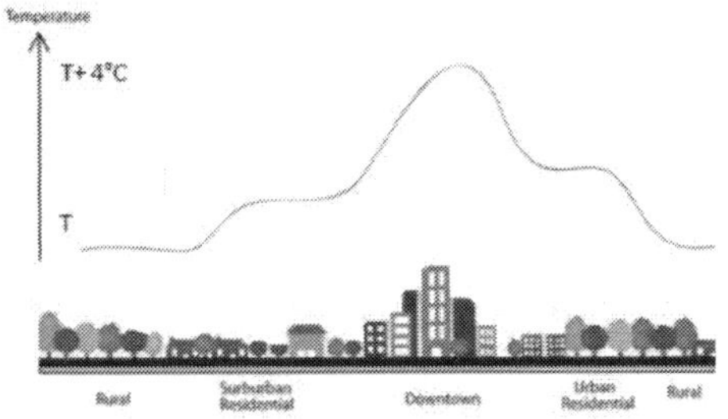

Fig.1. Schematic representation of the urban heat island profile

Numerous studies show that city centers that concentrate large quantities of buildings, economic activities and leave less space for vegetation and stretch of water are more affected by the UHI impacts [13].

2.2 Energy balance of an urban surface element

The development of the energy balance of the city requires the modeling of energy and mass transfers which are the basis of microclimatic processes. The atmosphere is a thermal machine whose energy state depends on the balance between these fluxes [14]. The overall surface of the soil is a patchwork of irrigated or not irrigated plant, mineral and synthetic surfaces with completely different radiative, thermal, aerodynamic and hydric properties [15]. Moreover, it is difficult to calculate the urban energy balance within the urban canopy, because of the accentuate "pleating" of surface, the geometric heterogeneity, the diversity of the thermo-physical properties of urban developments, and the various unknowns such as the uncertainty of the composition underground or the unpredictability of the quantity and quality of the energy equipment inside the buildings [16].

A modeling of urban surface elements or urban facets is required depending on the targeted spatial and temporal scale [16].While the first one is more used to the energy comparison of neighborhoods or to study energy transfers of the larger climate scale, the second is more suitable for the treatment of energy interactions between building and its immediate urban environment.

The energy balance of a system is the algebraic sum of the energy fluxes that exchange with its environment. Simply, it is the spatial scale of modeling that modifies the terms of the balance sheet. On a large scale, the usual practice is to aggregate the physical variables of the field, the fluxes, in order to be interested only in their average values. While on a small scale, we are interested in the singular phenomena, specific to the human scale, the comfort, the safety, and the energy consumption.The energy balance of a real surface mesh of an urban object in a steady state (figure2),is expressed in flux density [W.m^{-2}][17, 18, 19, 20]according to equation (1):

$$Q^* + Q_F = Q_H + Q_E + \Delta Q_S (+ \Delta Q_A) (1)$$

where:
Q* is the net all wave radiation;
$Q_F$is the anthropogenic heat flux corresponds to the sum of the heat outputs of motor vehicles, buildings, industries, urban facilities as well as human metabolism;
$Q_H$is the turbulent flow of sensible heat due to the aerodynamic heat transfer between the solid surfaces and the air mass, due to their temperature difference;
$Q_E$is the turbulent flow of latent heat due to the release of water vapor by natural soils, plant surfaces and porous surfaces;
$\Delta Q_S$is the change in the amount of heat stored by the urban area;$\Delta Q_A$is the net energy (sensible and latent) advection (horizontal transport).

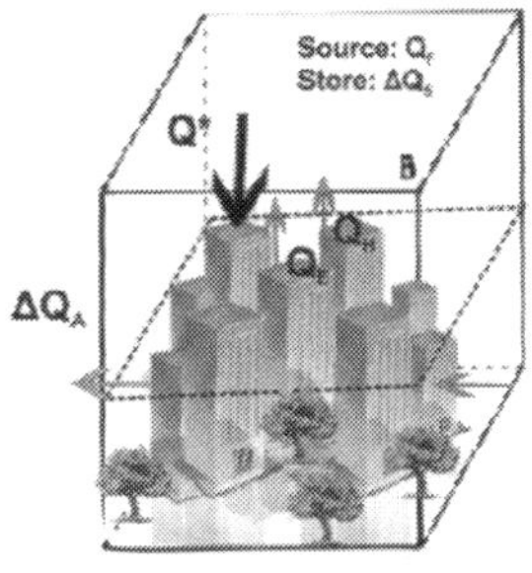

Fig.2. The energy balancean urban building air volume(OKE, 1978).

The term Q* represents the net flux of radiation over the entire electromagnetic spectrum. It is decomposed into two bands of wavelengths: solar radiation of 0.3 to 3μm (visible from 0.38 to 0.76μm) and thermal infrared (IRT) of 3 to 100μm.
The radiative balance (Q*) in [W.m^{-2}]:

$$Q^* = L\downarrow - L\uparrow + (1 - \alpha) K\downarrow \quad (2)$$

where:
$\alpha$is the albedo defined as the ratio of the amount of light or energy that is reflected back into the atmosphere by any particular surface,
L↑ is the net IRT flux exchanged with the surfaces of buildings, soil and all other solid surfaces belonging to the urban canopy,
L↓ is the net IRT flux exchanged with the atmosphere,
K↓ is the overall solar flux incident.
The incident global solar flux K ↓ is the sum of three terms, the flux coming directly from the sun, the diffuse solar flux coming from the reflection and multidirectional diffusion of the direct solar rays by the atmospheric particles and the diffuse solar flux coming from the multi-reflections on theothers urban surfaces and trapped in the canopy.

2.3 Factors affecting UHI

Many controllable and uncontrollable factors which contribute to UHI have been categorized [21]. This include the temporary effect variables, such as air speed and cloud cover and permanent effect variables like green areas, building material, and sky view factor and cyclic effect variables such as solar radiations and anthropogenic heat sources.Controllable factors include urban parameters such as city size, influenced by urban populations and densities, density of built-up areas including land use, distance between buildings and average height of buildings, urban geometry including street orientations, H/W ratio of street (H and W refer to the height and the width of the street canyon, and sky view factor (SVF), which is the visible area of the sky from a given surface. Apart of urban geometry, surfaces including buildings and pavements that are generally composed of dark materials that readily absorb and store the sun's heat. Most building materials are also impermeable thus further exacerbating the warming trend in cities. The high levels of urban air pollution as well as the urban heat generations are an others reason for the heat island formation.Natural parameters such as temperature, cloud cover and wind have a significant impact on the UHI. The natural location of a city also plays a role since its physical characteristics that include topography, mountain ranges and hills, rivers and/or other water bodies can determine the extent to which UHI can be affected.

2.3.1 Materials andcoatings

The ability of an urban material to capture heat and restore it at night is characterized by its inertiadepending on its thermal capacity, conductivity, diffusivity and effusiveness.Another significant anthropogenic factor that has a major impact on the UHI phenomenon is the albedo property of urban surface materials. A low albedo means a surface reflects a small amount of the incoming radiation and absorbs the rest. On the other hand, ahigh albedo means a surface absorbs a small amount of the incoming radiation and reflects the rest. The albedo of urban materials is generally lower than of materials found in the countryside. These materialstherefore increase the temperature and their inertia is higher, so they absorb the heat to restore it during the night period, preventing the refreshing of the city [22].Rough and dark surfaces absorb more radiation than smooth, flat and clear surfaces and reach a higher temperature. Smooth, flat surfaces are warmer than those that are rough and curved. [23]. Materials such as marble, mosaic and stone reach lower surface temperatures and thus contribute less to the effect of UHI.By studying the variation in the solar reflectivity of concrete, their composition, aging and climatic exposure have been shown to significantly alter their albedo [24].

2.3.2 Urban morphology

The urban morphology refers to the shape, dimension and spacing of buildings within a city, which together add to the amount of radiation received, reflected and emitted by the urban infrastructure. In urbanized areas, spaces and built environments are often obstructed, either fully or partially, by building blocks that quickly become large thermal masses triggering the build-up of UHI.

Shading is the second parameter affecting the radiative balance. Indeed, the shadows that the buildings make on the outside spaces limit the rise in temperature of the materials. These shadows are all the more important as the prospect of the urban construction is high [25].The orientation of the streets is also of particular importance. The orientation of urban streets and islets in relation to prevailing winds influences wind speed and turbulence [26]. The link between the wind speed inside and above a canyon street has also been shown: the wind inside the street is slowed down, its direction is changed and turbulent flows are generated. This link is largely influenced by the orientation of the street in relation to the wind and by the prospect of the street.Toudert[27] carried out an investigation on the contribution of street design toward the development of a comfortable microclimate for pedestrians during the hot and dry summertime of Ghardaia in Algeria. The main factors investigated in the study comprised: height-width ratio (H/W), street orientation, overhanging facades, and the use of galleries and rows of trees; they were able to determine that street orientation and H/W ratio of streets have the most significant impact on the microclimate in urban canyons.Radiative trapping is the third factor, also linked to urban geometry, for a highly prospective urban form, the solar rays undergo multiple reflections and heat the surfaces that compose the street before partially emerging towards the atmosphere. The temperature inside the canyon streetwill therefore be increased (Figure 3). Infrared rays emanating from heated urban materials are trapped in the same way [28].

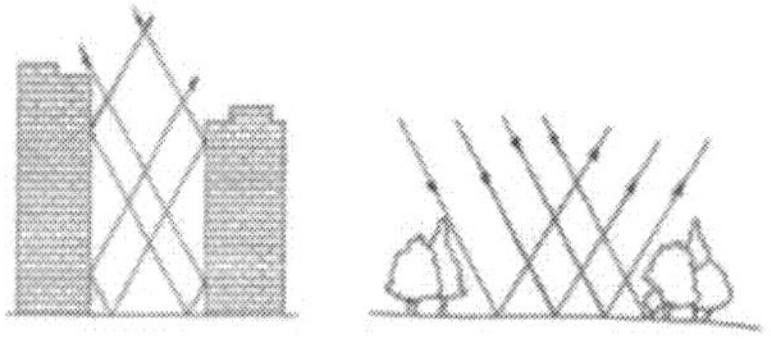

Fig.3. Illustration of radiativetrapping (Griffiths, 1976)

2.3.3 The natural canopy

The landscape in rural areas is typically dominated bylarge vegetated surfaces or open spaces. Rural temperatures are therefore lower andcooler than their surrounding urban areas due to shading provided by the trees andvegetation, and the process of evapo-transpiration in which plants release water fromthe earth's surface and accordingly dissipate ambient heat into the air. In contrast, urban areas are characterized by dry, impermeable and darkmineralized surfaces consisting primarily of roofs, pavements, roads and parking lots thattypically reflect less and absorb more of the sun's energy. As urbanization expandstherefore, more vegetation is lost, and more surfaces become paved and covered withbuildings thus resulting in lower levels of moisture and accordingly heat losssuch us in (figure 4). Indeed, the phase change of water by evaporation is accompanied by heat absorption. For example, during a summer rainfall that supplies 5mm/m² of water, 8 times less water evaporates in urban areas than in rural areas within the next twenty-four hours. The difference in the amount of heat taken from the environment corresponds to 2.5KWh/m^2 in twenty-four hours, slightly more thantwice the heat resulting from human activities in summer [29].

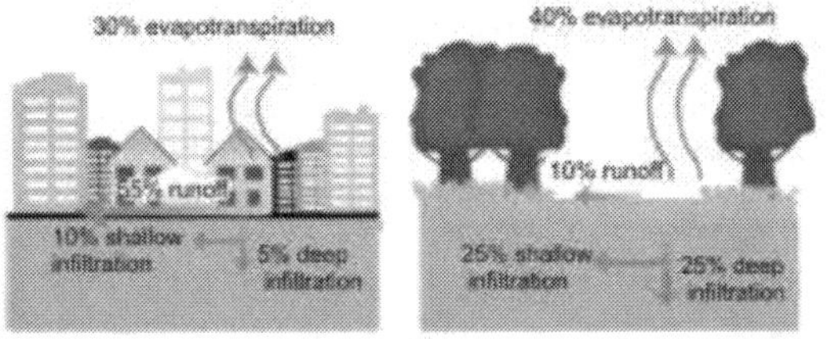

Fig.4. Evapotranspiration in urban areas (USEPA, 2008)

In the study of the mostoptimal urban street designs on thermal microclimate for example, Toudert[27] highlighted the importance of having rows of shade trees in urban canyonsdue to the important role they play in enhancing thermal comfort levels of urban streets.

2.3.4 Anthropogenic heat emissions

In urban areas, the high density of concentration on the human activities increases the anthropogenic heat releases of heating, air conditioning, transport and industrial activities.These anthropogenic heat fluxes are different from the city to another: Oke[18] has been able to highlight annual values, ranging from 3W/m^2 to 117W/m^2. While the amount of these releases depends on the climate (Montreal has one of the highest amounts of anthropogenic releases), it also depends heavily on building density and population of the city [18, 28, 30].

2.4 Effects of urban heat islands

The consequences of the UHI remain difficult to predict, and very variable according to climates.It is difficult to decide between reducing heating consumption and increasing consumption of air conditioning, especially in climate change context.Among the consequences of the urban heat island effect are the reduction in the efficiency of passive refreshing modes, air pollution, increased discomfort, and significant risks to human health and biodiversity [28] as well as the increase in energy consumption [31].

2.4.1 Impact on energy consumption

Numerous studies show that the UHI is both the cause and the consequence of the increase in energy consumption at the city [16].The case of Athens city is pertinent, the maximum intensity of the UHI can in fact reach 12°C, and an increaseof 3 to 5% in electrical consumption is observed when the temperature increases by 1°C [32].Santamouris [33] carried out a study in Athens in 1996, showing that the maximum intensity of the UHI can reach on average for the city center of Athens near 12°C, and even 16°C for the hyper center. It can be seen thatfor the city of Athens, where the mean heat island intensity exceeds 10°C, the cooling load of urbanbuildings can be doubled, the maximum power ofelectricity for cooling purposes may be tripled in particular at higher set point temperatures, whereas the minimum coefficient of performance COP value of the air conditioners may be reduced by up to 25% due to higher ambient temperatures. According to Akbari [34] the analysis of temperature trends in several large cities in the United States indicatesthat temperature increases are responsible for 5-10% of the energy consumption for air conditioning.

In 2005, Bozonnet[31] also revealed the vicious circle linked to the mutual influences of air conditioning systems and the UHI (Figure 5).

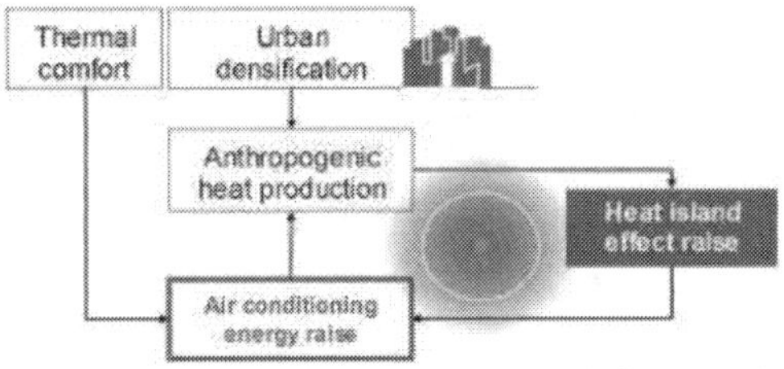

Fig.5. The vicious circle linked to the mutual influences of air conditioning systems and the UHI Bozonnet (2005)

2.4.2 Impact on outdoor comfort

The main external physical parameters that occur in thermal comfort are air temperature, humidity, solar and infrared radiation received, and air movements or convective influences (wind) [28]. All these elements are influenced by the form, the urban structure. However, a purely physiological and quantitative approach is inadequate to characterize external comfort. Furthermore, parameterthat strongly influence the feeling of comfort, there is a subjective assessment of the comfort that depends on the person and this psychological adaptation to the outside seems to be important [35]. The thermo-radiative energy balance of the external spaces is closely linked to the nature and morphology of the building arrangements. Athamena[36] studied through modeling and numerical simulation the relationship between specific urban morphology Eco-neighborhoods, microclimate and thermal comfort in outdoor spaces resulted in recommendations for the design of a comfortable outdoor urban space, for example, future neighborhoods should be constructed as self-contained porous buildings opening the internal public spaces to the incident winds, thus favoring the ventilation of spaces and limiting the corner effect which generates unpleasant heat sensations. Moreover, theorganization of the district around an open internal space connected to the outside by a crosswalk advances the dissipation of the heat accumulated during the day.

3 Approaches to Mitigation of UHI in Maghreb Cities

3.1 Case of Maghreb cities

There are few studies on the microclimate and human comfort of urban areas in hot dry climates.A study carried out in Constantine in Algeria with ENVI-met a pronostic three-dimentional climate model simulating the interactions between atmosphere, soil, vegetation and building on micro-scale level demonstrated that the urban canyon with a large H/W ratio is most advantageous for the outdoor thermal comfort in the summer provided that to choose the orientation offeringmore ventilation[37].The wind also represents a ratherperceptible element of the urban microclimate; it is one of the factors which determine the success or the failure of a public space, asit strongly influences the thermal comfort. It can be a source of ventilation in summer, or a source of nuisances in winter.Also ofJijel in Algeria; a simulations highlighted the effect of urban geometry on wind flow and natural outdoor ventilation in the city. The study has shown that the geometry of the ground plane and the arrangement of the building determine and model the wind flow. Because of humid climate and strong winds, an exploded geometry is not recommended for a site at high altitude. On the other hand, it will be appropriate for low lying areas [38].

The Sfax agglomeration in Tunisia was also studied on the trend of rising air temperatures, the spatial variability of the UHI and the impact of inadequate development choices on the local climate on the intensification of global warming. Using the satellite tools, the study showed that hot surfaces correspond to densely built spaces. The spreading, densification of the built-up areas between 1987 and 2007 and the vertical evolution that is known in the city are in line with the expansion of the hot surfaces. The temperature measurements in downtown Sfax, have shown that some recent developments such as the vertical trend and the use of curtain walls have worsened the warming. The urban design of the northern countries such as the curtain walls, neglect of the aerodynamics of the city and the scarcity of green spaces are inadequate to the local climate.

In future developments, urban planners should have relatively large streets oriented obliquely to the prevailing wind so that it benefits both facades and streets [39]. It seems sensible to impose, for new constructions, standards better suited to a warm climate.

3.2 Case of Moroccan cities

The topic of the UHIin Morocco was first discussed in 2006, as part of a cooperation program between LPEE (Laboratoire publicd'essais et d'études)and the University of Lund (Sweden), in Fez city comparative study between two districts with a different urban fabric: the urban traditionalarea in the medina (Seffarine), and the modern district on the outskirts (Adarissa), a standard of urban density and a low density urban neighborhood. The aim was to study the influence of the urban fabric and shape on the microclimate of the two neighborhoods.Johansson [40] proved that in hot dry climates a compact urban design with very deep canyons is preferable.

The study investigated theinfluence of urban geometry on outdoor thermal comfort by comparing an extremely deep and a shallow street canyon in Fez city. Continuous measurements during the hot summer and cool winter seasons show that, by day, the deep canyon wasconsiderably cooler than the shallow one. In summer, the maximum difference was on average 6°C and as great as 10°C during thehottest days. However, during winter, the shallow canyon is the more comfortable as solar access is possible [40].The main difference between the two districts is the level of the urban form: the configuration and density of the neighborhoods on the one hand, and the architecture and walls thermal characteristics of the building on the other. The urban form of modern districts has a heating effect on the microclimate. This influence can be seen mainly during the night and afternoon periods; the climate then being out of the comfort zone and it is sometimes necessary to air-condition. On the other hand, traditional neighborhoods with a high urban density have a regulatory impact on the microclimate in summer because they compensate for the large differences in daytime and nighttime temperatures which make it possible to maintain the climate for the most part in the comfort zone.In winter, the urban form of modern districts has a heating effect on the microclimate.

Casablanca, the largest industrial and commercial center in Morocco with rapid urbanization and explosive population growth, has also been the subject of two recent studies on the characterization and evaluation of the UHI by means of satellite tools [41, 42].

Landsat and MODIS data were combined in the simple biosphere model to assess the impact of urbanization on surface climate in Marrakech. A complete urbanization of the area would decrease its carbon uptake by 0.13 tons and increase its daytime surface temperature by 1.3°C, with 5.72% increase in energy consumption [43].

4 Conclusion

In Maghreb cities notably in Morocco, urban planning is being revised; new approaches are being implemented, and it is the appropriate time to sensitize to take into account of these mitigation measures.The last urban development plan incorporated measures (eco-neighborhoods, urban agriculture, etc.) but it is not a question here of proposing only strategies that can be envisaged in the short and medium terms (increase in vegetated areas, the albedo of the surfaces ...), but also to understand if longer-term strategies have their place because of their potentially important effect on the intensity of the urban climate and more particularly of the UHI.Significant progress has been made in recent years in taking into account the UHI in the different strategies for adapting cities to climate change and in improving their resilience. Several studies have focused on this subject in cities of different climatic zones and have allowed the implementation of an adaptation strategy specific to each one: Tokyo, Barcelona, Seville, Montreal, Paris and Bayroute. The most advanced city in the field of adaptation is Tokyo where a systematic plan for watering streets is proposed and a strategy to open public parks is advocated during a heat wave. Paris else, takes the issue seriously following the 2003 heat episode.Moreover, consider this problem within the framework of a systemic approach incorporating the interactions between the different elements of the urban ecosystem has not yet been addressed in Maghreb cities.

Bibliography

1. Oke,T.R.Initial guidance to obtain representative meteorological observations at urban sites. *Instruments and Observing Methods Report* No. **81**, WMO/TD No. 1250. (2006).
2. Nakamura, Y., et T. R. Oke, Wind, temperature and stability conditions in an east-west oriented urban canyon, Atmos. Environ**.**, **22**, 2691–2700.. (1988)

3. Eliasson,I.Urban-suburban-rural air temperature differences related to street geometry. Physical Geography, **15**(1), 1–22., (1994)
4. Ackerman, B., Temporal march of the chicago heat island. Climate Appl. Meteor**24**, 547–554, (1985)
5. Nasrallah, H. A., A. J. Brazel, et R. C. Balling Analysis of the Kuwait City urban heat island, Int. J. Climatol. **10**, 401–405. (1990)
6. Moreno-Garcia, M. C. Intensity and form of the urban heat island in Barcelona, Int. J. Climatol., **14**, 705–710. (1994),
7. Tarleton, L. F., et R. W. Katz Statistical explanation for trends in extreme summer temperatures at Phoenix, Arizona, Journal of Climate, **8**, 1704–1708 (1995),
8. Montávez, J. P., A. Rodríguez, et J. I. Jiménez A study of the urban heat island of Granada, Int. J. Climatol., **20**, 899–911. (2000)
9. Tereshchenko, I.E., et A.E.Filonov Air temperature fluctuations in Guadalajara, Mexico, from 1926 to 1994 in relation to urban growth., Int. J. Climatol.,**21**, 483–494. (2001),
10. Magee, N., J. Curtis, et G. Wendler The urban heat island effect at Fairbanks, Alaska, Theor. And Appl. Climatol., **64**, 39–47. (1999)
11. Kuttler, W., A.-B. Barlag, et F. Roßmann Study of the thermal structure of a town in a narrow valley, Atmos. Environ.,**30**, 365–378. (1996)
12. Yamashita, S., allOn relationships between heat island and sky view factor in the cities of Tama River basin, Japan, Atmos. Environ., **20**, 681–686. (1986)
13. Grégoire Pigeon. Les échanges surface-atmosphère en zone urbaine - projets CLU-ESCOMPTE et Capitoul *Thèse : Physique de l'Atmosphère* Université Toulouse III (2007)
14. Vinet, J. "Contribution à la modélisation thermo-aéraulique du microclimat urbain : Caractérisation de l'impact de l'eau et de la végétation sur les conditions de confort en espaces extérieurs". *Thèse Ecole Polytechnique de Nantes.* (2000).
15. Arnfield, A. Two decades of urban climate research: a review of turbulence, exchanges of energy and water, and the urban heat island. International Journal of Climatology,**23**:1–26. (2003).
16. Bouyer J., Modélisation et simulation des microclimats urbains : Étude de l'impact de l'aménagement urbain sur les consommations énergétiques des bâtiments. École NationaleSupérieured'Architecture de Nantes (2009).
17. Brutsaert, W. Evaporation into the atmosphere: theory, history and applications. Environmental Fluid Mechanics.Dordrecht, Holland. (1982).
18. Oke, T. R. *Boundary layer climates.Londres*, **2**édition. (1987).
19. Mestayer, P. G.,Anquetin, S. Climatology of cities. In Diffusion and transport of pollutants in atmospheric mesoscale flow fields, volume **1** de *ERCOFTAC series*, pages 165–189.Netherlands. (1994).
20. Guyot, G. *Climatologie de l'environnement : Cours et exercices corrigés.* Dunod, Paris, **2e** édition. (1999).
21. RIZWAN Ahmed Memon. A review on the generation, determination and mitigation of Urban Heat Island, Journal of Environmental Sciences**20**p/120–128 (2008).
22. Liebard, A. all. Concevoir, édifier et aménager avec le développement durable.*Traité d'architecture et d'urbanisme bioclimatiques. Le Moniteur*, 776 p. (2006)
23. Doulos, L.; Santamouris, M. &Livada, I.Passive cooling of outdoor urban spaces: The role of materials.Solar Energy**77**, 231-249 (2004)
24. *Levinson*H.F.,*et al.* The Mass Disruption of Oort Cloud Comets *Science*296, 2212 (2002)
25. ADOLPHE, L. all. SAGA Cités, Système d'Aide à la Gestion des Ambiances urbaines, *Rapport final. MENRT, Action Concertée Incitative Ville*, février (2002).
26. SACRE, Christian. Le confort dans les espaces extérieurs : analyse microclimatique. Nantes : Centre Scientifique et Technique du Bâtiment,138 p. (1983)
27. Ali- Toudert F., Mayer H. Numerical study on the effects of aspect ratio and orientation of an urban street canyon on outdoor thermal confort in hot and dry climate. *Building* and environement,**41** (2), pp. 94-108 (2006).
28. Colombert, M. "Contribution à l'analyse de la prise en compte du climat urbain dans les différents moyens d'intervention sur la ville". Thèse Université Paris-Est, (2008).
29. DETTWILLER, J. L'évolution séculaire de la température à Paris. La Météorologie, n°**13**, p. 95-130. (1978)
30. RINGENBACH, Nicolas. Bilan radiatif et flux de chaleur en climatologie urbaine : mesures, modélisation et validation sur Strasbourg. Thèse Strasbourg I, 167 p. (2004)
31. Bozonnet, E. Impact des microclimats urbains sur la demande énergétique des bâtiments – Cas de la rue canyon. Thèse :Génie Civil, La Rochelle, 175 p. (2005)
32. Santamouris, M. ADNOT, J. ALVAREZ, S. et al. Cooling the cities – rafraichir les villes. Paris : Ecole des Mines de Paris-*Les Presses*, 263 p. (2004)
33. Santamouris, M., Papanikolaou, N., Livada, I., Koronakis, I., Georgakis, C. and Assimakopoulos, D. N. 'On the impact of urban climate to the energy consumption of buildings', Solar Energy, vol**70**, no 3, p201–216 (2001)
34. Akbari, H. POMERANTZ, M. TAHA, H. Cool surfaces and shade trees to reduce energy use and improve air quality in urban areas. Solar Energy, vol.**70**, n°3, p. 295-310 (2001).
35. NIKOLOPOULOU, Marialena. STEEMERS, Koen. Thermal comfort and psychological adaptation as a guide for designing urban spaces. Energy and Buildings, vol.**35**, p. 95- 101 (2003)
36. Athamena K., Modelisation et simulation des microclimats urbains : étude de l'impact de la morphologie urbaine sur le confort dans les espaces extérieurs. thèse de doctorat de l'Ecole *Centrale de Nantes* (2012).
37. Bouchriba F., Impact de la géométrie des canyons urbains sur le confort thermique extérieur - Cas du Coudiat de Constantine – *Thése. Universitementouri* (2005)
38. Bouketta, Bouchahm and Y. Bouchahm, L'effet de la géométrie urbaine sur l'écoulement du vent et la ventilation naturelle extérieure Revue des Energies RenouvelablesV.**15** N°04 (2012)
39. Ali-Toudert F., Weidhaus J., Numerical assessment and optimization of a low-energy residential building for Mediterranean and Saharan climates using a pilot project in Algeria Renewable Energy**101** 327- 346 (2017)
40. Erik Johansson Influence of urban geometry on outdoor thermal comfort in a hot dry climate: A study in Fez, Morocco. Building and Environment**41** 1326–1338 (2006)
41. Bahi H. all. Effects of Urbanization and Seasonal Cycle on the Surface Urban Heat Island Patterns in the Coastal Growing Cities: A Case Study of Casablanca, Morocco, Remote Sens., *8*, 829. (2016)
42. Azmi R., Saadane A., and Kacimi I. Estimation of spatial distribution and temporal variability of land surface temperature over Casablanca and the surroundings of the city International Journal of Innovation and Applied StudiesVol. **11** No. 1 Apr., pp. 49-57(2015)
43. Lachir A., Bounoua L., Ping Zhang Kurtis Messouli M. Modeling the Urban Impact on Semiarid Surface Climate: A Case Study in Marrakech, Morocco Canadian Journal of Remote Sensing, **42**:379–395, (2016)

Effect of pH and time on the treatment by coagulation from slaughterhouse of the city of Rabat

Nisrine. Boughou*, Imane .Majdy, Essediya. Cherkaoui, Mohamed. Khamar, Abderrahman. Nounah

Civil Engineering and Environment Laboratory (CEEL), Team Materials, Water and Environment Higher School of Technology, Av. Crown Prince BP 227, SALE .Mohammed V University in Rabat

**For correspondence: Email: nisrine.boughou@gmail.com (N.Boughou); Phone:* +212668832535

Abstract. Water is an essential element in the agri-food industry. The resulting wastewater is often returned to the natural environment without prior treatment; which can be an important source of environmental pollution. Among these food processing industries, there are slaughterhouses that are classified as dangerous, unhealthy and inconvenient, so the treatment of their discharges is necessary before their spill. This article presents the results of the waste water treatment study of the Rabat slaughterhouse by coagulation-flocculation using ferric chloride and the effect of pH and retention time on the elimination of turbidity, Chemical oxygen demand (COD) and organic matter (OM). The results show that pH and retention time plays a major role in the elimination efficiency of these parameters. Thus, with a pH = 5.83, a significant reduction in COD and turbidity are observed after 90 min and 120 min of decantation.

1 Introduction

In recent years, the red meat slaughtering industry has expanded enormously around the world. All the conditions are there to allow this industry to continue its development. Nevertheless, this industry is among the most polluting because of the large quantities of wastewater discharged. The effluent from these slaughterhouses is characteristic and requires appropriate treatment (separation of solid wastes and fats, specific treatments).Among the many proposed physico-chemical methods[1,2-4,5], coagulation-flocculation-sedimentation remains a relatively simple technique [6-8,9], mainly applied to the removal of heavy metals and organic compounds [10,11,12], and widely used for the treatment of water that contains high organic matter [8]. Numerous factors can influence the effectiveness of such a treatment: type of coagulant, pH, temperature, mixing velocity, retention time [13]. A laboratory study was conducted to determine the performance of the treatment according to different parameters: Coagulant type, pH, retention time, COD, turbidity, conductivity and OM content were performed. Several tests have been carried out using various coagulants such as ferric sulfate Fe2 $(SO_4)_2$, ferric chloride ($FeCl_3$), aluminum sulfate (Al_2 $(SO4)_3$). The choice of ferric chloride coagulant has been dictated by its effectiveness in reducing turbidity and organic matter in wastewater, in addition to its validation by previous studies, iron salts appear to be more effective than aluminum [1, 2]. This article reproduces the results of a coagulation-flocculation treatment followed by decantation on a different pH series. The performance tests of this treatment as a function of the ferric chloride content showed that the optimization was at a concentration of 0.6 g of $FeCl_3$/L.

2 Materials and methods

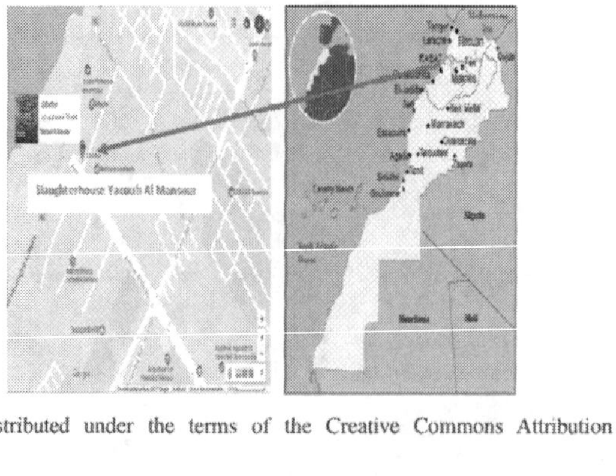

2.1 Study site

The town of Rabat has a municipal slaughterhouse located in Yacoub Al Mansour. Inaugurated in 1956, it covers an area of 1800 m^2, it is part of the traditional municipal slaughterhouses, and it is located in one of the most popular districts of the capital "Figure 1".

Fig.1. Geographical location of the Rabat slaughterhouse

2.2 Sample

The wastewater from the slaughterhouse studied was taken from the rejected slaughterhouse at Yacoub Al Mansour in Rabat, following the precautions of sampling [14].

2.3 Characterization of Releases

The Slaughterhouse wastewater from Rabat has been characterized by different physicochemical parameters (T °C, pH, Conductivity, Suspended matter, Biochemical Oxygen Demand, Chemical Oxygen Demand). The pH and temperature were determined by a Lutron pH meter equipped with a temperature probe. The electrical conductivity was measured by a WTW LF90 type conductivity meter, while the turbidity was measured by a HACH 21009 turbidimeter. The SM (Suspended Materials) is determined by filtering on a filter. COD is determined by acid oxidation by excess potassium dichromate at the temperature of 148 ° C of the oxidizable materials under the test conditions in the presence of silver sulphate as catalyst and sulphate of mercury [15] and BOD_5 is determined by the respiratory method using a WTW OXITOP DBO-meter, according to the technique described by DIN [16].

The organic matter content is determined by evaporation of the samples in porcelain capsules weighed first and then in an oven at 105 °C. and then weighed again and then after conditioning these capsules in the oven at 550 ° C. This is cooled and weighed again.

The flocculation-coagulation tests were carried out according to the Jar-Test protocol (JLT6 Leaching VELP Scientifica), which comprises 6 agitators, is equipped with 6 beakers which simultaneously and simultaneously agitate the waste water of the slaughterhouse and the coagulant. The coagulant used is ferric chloride ($FeCl_3$, $6H_2O$) as a conventional coagulant at a concentration of 0.6 g $FeCl_3$ / l. In the four beakers, the same volume of slaughterhouse water (752 ml), the same volume of $FeCl_3$ (48 ml) and different pH values (3.90, 5.83, 6.28 and 9, 28), the pH is adjusted by addition of sulfuric acid H_2SO_4 and sodium hydroxide NaOH. These beakers are subjected to rapid stirring of 120 rpm for 30 seconds. The speed is then reduced to 40 rpm for a period of 20 minutes (In order to promote the encounter and aggregation of the colloids but without destroying the already formed flocs). Subsequently, the mixture was subjected to a decantation of 15 min, 30 min, 45 min, 60 min and 90 min to monitor the pH effect as a function of the retention time. The performance of the treatment is expressed using the following equation: removal efficiency (%) = $((C_0\text{-}C_f) / C_0) \times 100$ where C_0 and Cf the initial and final concentrations of each of the parameters (COD, Turbidity, pH) [17].

3 Results and Discussion

3.1 Characteristics of Slaughterhouse Wastewater

The characteristics of the wastewater sample from Slaughterhouse are summarized in the following table.

Table 1. Physico-chemical characteristics of wastewater from the Rabat slaughterhouse [18].

Parameters	Value	Direct Releases Standards
T°C	19,06	30
pH	7,3	6,5-8,51
Turbidity (NTU)	853	-
E.C (µs/cm)	1711	2700
BOD_5 (mg d'O_2/l)	640	100
COD (mg d'O_2/l)	1386	500
Suspended matter (mg/l)	835	50

Photo 1: Treatment of slaughterhouse water by coagulation as a function of pH and retention time

3.2 Effect of pH

The "Figure 2" shows the variation of the turbidity as a function of the settling time, for the various analyzes as a function of the pH. It is important to note a very significant reduction in turbidity at pH 5.83. It goes from 853 NTU to 180 NTU after 120 min of decantation. This could be related to the chemical nature of the organic compounds. Indeed, the organic compounds tested seem to be better eliminated in a pH range between 5 and 6. And means that the Fe $(OH)_3$ formation zone. In view of the simultaneous presence of several hydrolysed Fe $(OH)_3$ iron species And Fe $(OH)_2$ (soluble and insoluble) at these pHs, the reactions involved are probably numerous. PH may be an important parameter in that it promotes the reaction of the functional groups of ferric chloride and of the organic matter in question. At acid pH (5 and 6), the cationic species of Iron are preponderant. The possible reactions would be the formation of complexes, soluble or insoluble according to the structure of the compound and the nature of the organic functions present. Admittedly at these pHs the organic matter is in its dissolved form which also increases the adsorption surface of the coagulant.

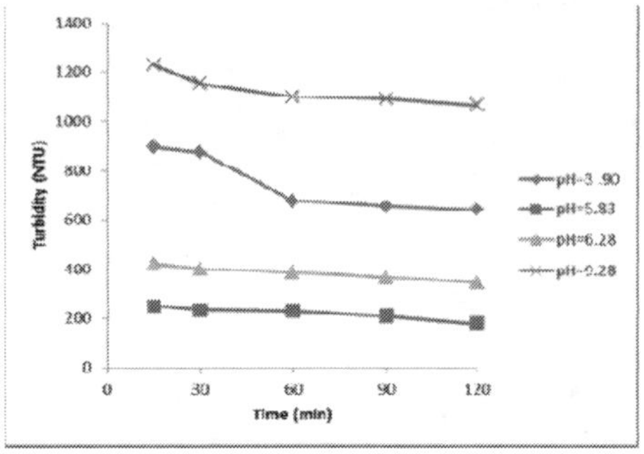

Fig. 2. Evolution of the turbidity according to the pH as a function of retention time

3.3 Effect of pH on conductivity

When $FeCl_3$ is added, it is noted that the conductivity shows high values between 5.83 and 6.28, which can be interpreted by the formation of a complex of Fe $(OH)_3$ from Fe (III) And also by the destabilization of the colloidal particles.

A decrease in conductivity from pH 9.28 to 1323 µs / cm at 120 min due to the formation of Fe $(OH)_3$ In view of the simultaneous presence of several hydrolyzed Fe $(OH)_3$ and Fe $(OH)_2$ (soluble and insoluble) at these pH values "Figure 3 ".

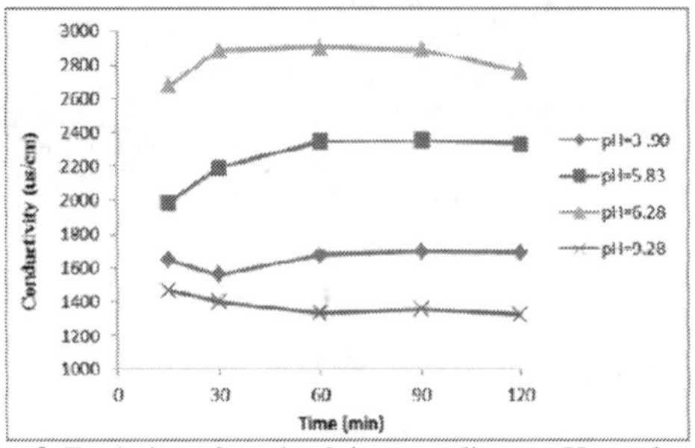

Fig. 3. Evolution of conductivity according to pH as a function of retention time

3.4 Effect of pH on COD removal

The best COD reduction yields were observed at pH 5.83 with a significant reduction from 1386 mg/l to 632 mg/l after 120 min of settling in a yield of 54.40%, this reduction can be interpreted by the removal of the organic matter under the effect of ferric chloride.

For the pH 9.28 can be interpreted by the presence of the fulvic acids (FA) which are derived from the decomposition of the organic plant matter "Figure 4".

PH can be an important parameter insofar as it conditions the dissociation of organic compounds. In general, the organic compounds appear to be better eliminated in a pH range between 5 and 6. Similarly, the coagulation of the organic matter is much more effective at acidic pH than near neutral pH.

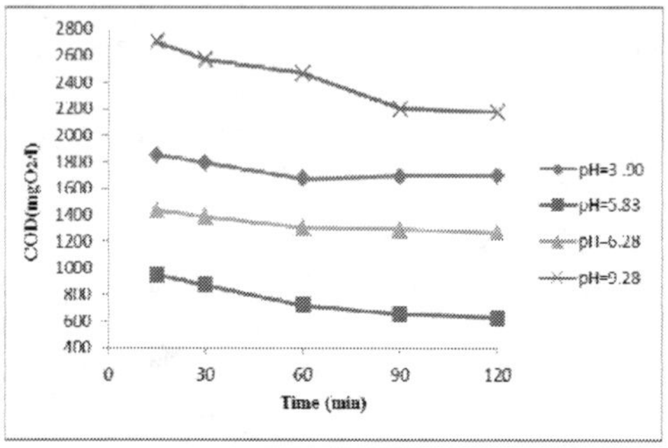

Fig.4. Evolution of the COD according to the pH as a function of retention time

Conclusion

Slaughterhouses are places where the consumption of water is very important because of the multiple uses that are made of it. Wastewater from these industries is not without impact on the environment. The results of this study confirm the choice of ferric chloride for the coagulation-flocculation of slaughterhouse wastewater. This method allows, after adjustment of the pH to a value of 5.83 to obtain the first 15 minutes of decantation, important elimination efficiency greater than 78.89% of the turbidity and 54.40% of the COD. These results show that the pH and the retention time have an effect on the elimination efficiency of these parameters.

References

1. Amokrane A., Comel C., Veron J., Water Research, 31 (1997) 2775-2782.
2. Yilmaz T., Apaydin S., B erktay A., The Open Environmental Engineering Journal, 3 (2010) 42-48.
3. MelikeYalılıKılıç, KadirKestioglu, TanerYonar, Journal of Biodiversity and Environmental Sciences, 1 (1) (2007) 37-43.
4. Marttinen S.K., Kettunen R.H., Sormunen K.M., Soimasuo R.M., Rintala J.A., Chemosphere, 46 (2002) 851-858.
5. Foo K.Y., Hameed B.H., Journal of Hazardous Materials, 171 (2009) 54-60.
6. Rivas F.J., Beltran F., Carvalho F., Acedo B., Gimeno O., Journal of Hazardous Materials, B116 (2004) 95-102.
7. He, P.J., Qu X., Shao L.M., Li G.J., Lee D.J., Journal of Hazardous Materials, 142 (2007) 288-296. 8. Lee Mao Rui, ZawawiDaud, Abd Aziz Abdul Latif, International Journal of Engineering Research and Applications, 2 (2012) 1929-1934.

9. Abdulhussain A. A., Guo J., Liu Z.P., Pan Y. Y., Al-Rekabi W. S., American Journal of Applied Sciences, 6 (4) (2009) 672-684.
10. Li S., Fang S., He S., Hao Y., Xiong G., IERI Procedia, 9 (2014) 64-70.
11. Gnandi K., Changbedji G., Kili K., Baba G., SalimO., Mine Water and the Environment, 24 (2005) 215-221.
12. Tchobanoglous G., Kreith F., McGraw-Hill Handbook of solid waste management, second edition, (2002).
13. Ozkan A., Yekeler M., Chemical Engineering and Processing, 43 (2004) 873-879.
14. ONEP, 1998.Approach to the typology of urban wastewater in Morocco. ONEP and GTZ. Rabat
15. (COD) according to DIN 38409-H52. Laboratory National Environment - Rabat. 1992b.
16. DIN (Deutsche Industry standards). Determination of the Biological Oxygen Demand (BOD) to DIN National Laboratory Environment - Rabat.1992a
17. Benradi F, El yahyaoui A, Bouhlassa S, Nounah N, Khamar M, Ghrissi F., J. Mater. About. Sci. 7 (3) (2016) 1001-1007.
18. Ministry of the Environment of Morocco: "Moroccan Standards, Official Bulletin of Morocco", No. 5062 of 30 Ramadan 1423. Rabat, 2002.

Solar Desalination by Humidification-Dehumidification of Air

J.Moumouh[1], M.Tahiri [1], L.Balli[1]
[1]Mohammadia School of Engineering, Mohammed V University-Agdal, Rabat, Morocco

Abstract. The importance of supplying potable water can hardly be overstressed. In many arid zones, coastal or inlands, seawater or brackish water desalination may be the only solution to the shortage of fresh water. The process based on humidification-dehumidification of air (HDH) principle mimic the natural water cycle. HDH technique has been subjected to many studies in recent years due to the low temperature, renewable energy use, simplicity, low cost installation and operation. An experimental test set-up has been fabricated and assembled. The prototype equipped with appropriate measuring and controlling devices. Detailed experiments have been carried out at various operating conditions. The heat and mass transfer coefficients have been obtained experimentally. The results of the investigation have shown that the system productivity increases with the increase in the mass flow rate of water through the unit. Water temperature at condenser exit increases linearly with water temperature at humidifier inlet and it decreases as water flow rate increases. HDH desalination systems realised on also work at atmospheric pressure; hence they do not need mechanical energy except for circulation pumps and fans. These kinds of systems are suitable for developing countries. The system is modular, it is possible to increase productivity with additional solar collectors and additional HDH cycles.

1 Introduction

The continued growth of demand for water and increasing shortage of supplies are two of the most certain and predictable scenarios of the 21st century. Population growth is threatening the availability of fresh water in many regions of the world.

Over a billion people do not have access to safe supply of water and the number is growing. Large areas of the world already suffer from drought while deserts and populations increase in size. Rainfall remains broadly constant, yet demand for water has doubled in the last 20 years. As demand outstrips renewable supply, the depletion of ground water is accelerated [1].

Over the last three decades, of available brackish sources and/or seawater has been used to supply potable water resources in arid and semi-arid regions, as well as islands with limited storage capacity. Typical examples in the Mediterranean are Morocco, Egypt and Israel. The standard methods of desalination such as multi-stage flash evaporation and multi-effect evaporation, vapour compression and reverse osmosis are reliable in the range of some100-500,000 m3/d fresh water production.

They are not used in regions with a poor infrastructure either for the supply in decentralized regions due to their permanent need of qualified maintenance and electricity supply. Smaller scale, decentralized solar desalination a system is desirable and makes economic sense. The present desalination system shall fill this gap [2].

2 Principle of the HDH Process

The HDH process is based on the fact that air can be mixed with quantities of vapor. Air ability to carry water vapor increases by increasing its temperature. For example, 1 kg of dry air can carry 0.5 kg of vapor and about (2814 kJ) when its temperature increases from 30° to 80° as reported in Ref. [3]. When air flow is in contact with salt water, air extracts a certain quantity of vapor at the expense of the sensible heat of salt water (providing cooling). On the other hand, the distilled water is recovered by maintaining humid air at contact with cooling surface, causing the condensation of a part of the vapor mixed with the air. The condensation occurs in another heat exchanger in which salt water is preheated by latent heat recovery.

The HDH process operates, as any distillation process, with only one cross of salt water (once through). It can also operate with water recirculation. In the first case, the evaporation of 1 kg of water causes a decrease of about 60°C in 10 kg of salt water. The amount of recovered distilled water varies from 5% to 20% of the quantity of salt water in circulation. This leads to a very low concentration of the salt water but a high sensitive heat loss. In the second case, more concentrated solution and lower energy consumption are obtained [4].

2.1 Description of the physical model of HDH open air-open water, Water heated

Salt water enters the system at the dehumidifier and is used to cool and dehumidify a hot, humidified air stream. Salt water gets preheated as the enthalpy of vaporization is transferred to it from the moist air (condensation process). The condensate is removed as product water and the salt water is then further heated in a solar collector or water heater before it enters the humidifier. The hot salt water is used to humidify a colder air stream, so mass and heat are transferred from the hot salt water to the moist air stream (evaporation process) and the remaining salt water is extracted as brine. The air stream circulates continuously in an open loop (fig.1) [5].

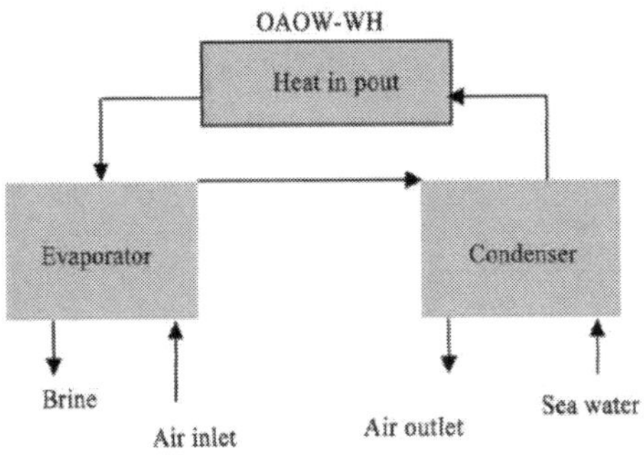

Fig. 1. Open air open water air-heated (OAOW-WH) system.

3 Experimental details

The purpose of this work is to verify experimentally the saltwater desalination process humidification-dehumidification of the air and to determine the effectiveness of this configuration. In fact, the experimental work was performed on a prototype installed in laboratry in Mohammadia School of engineering used to simulate experimentally the desalination process [6].

Fig.2 shows an image of the equipment before installing the thermal insulation. The condensation tower occupies is one compartment of the desalination unit.

The dimensions of the condensation tower are 72 cm height, 33 cm length, and 33 cm width. A copper tube coil is used in the condenser it as 50 m length and 1.27 cm outer diameter.

The other compartment of the desalination unit is the humidifying tower. Its dimensions are 170 cm height, 33 cm length, and 30 cm width. A packing material is fixed inside the humidifier; the specific area of packing material is $200m^2/m^3$.The water is sprayed on the packing material using a hydraulic grid.

Fig. 2. Desalination prototype by Humidification Dehumidification of Air.

A 200watts electric fan is used to blow the air through the unit. The fan is supported at the bottom of the tower. Water circulates by 500 watts centrifugal pump. A filter is used in the suction side to separate any impurities coming from the elevated tank water feeding.

A flow meter is connected on the delivery side of the pump to measure the water flow rate. A group of isolating valves are used to separate each part of the saline water loop from the system when necessary.

The external heat is provided via two electrical resistances 2400 watts that can be used to simulate the solar input during a day, and the actual heat input is monitored via voltage and amperage. Temperatures of air are measured with wet thermometers. The air flow is measured by Extech digital anemometer. The rate of distillate is measured with a 1000ml graduated test tube and a chronometer.

4 Results and Discussion

The variation of the distillate production with temperature in two different air flow rate is shown in fig.3. The amount of water production increases with increases in air flow rate.

The airflow rate increases, the productivity is increased at T_w = 60, 70 and 80 °C except for T_w =50°C is decreased. This is due to the increase in the diameter and the density distribution and kinetic energy of air bubbles in hot water layer.

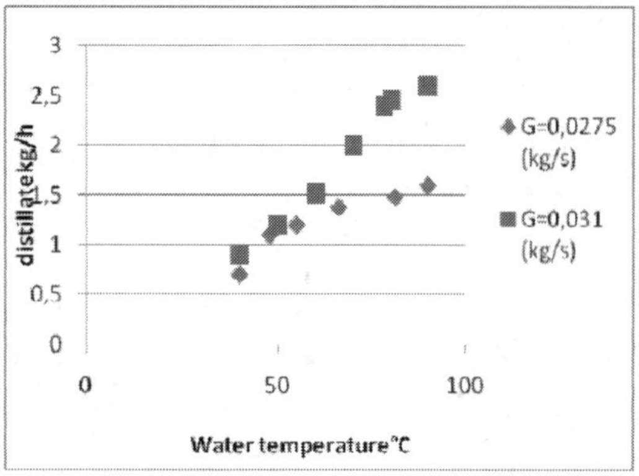

Fig. 3. Effect of flow air in distillate production.

The effect of water and air temperature on the productivity of the system is shown in Fig .4. The distillate flow rate increases with the increase of water temperature in evaporator chamber on the productivity of the system at Tw = 45, 50, 60 and 70◦C.

It found that the humidity of air increases appreciably, with the temperature of water at the inlet of the tower. The rate of distilled water varies linearly as a function of the air temperature at the humidifier outlet. This means that the increase of the hot water temperature at the entrance of the evaporation tower produces more heat inside the installation.

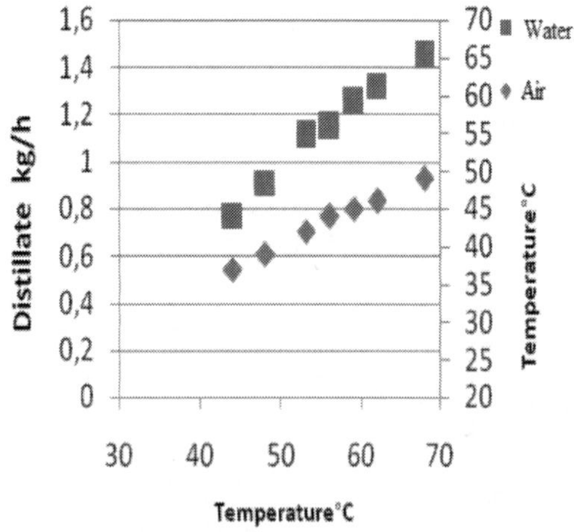

Fig.4. Distillate Evolution in function of T_{winth} ,T_{aouth}

Growing branch of the curve of variation of the water content of the air in function of the flow ratio is illustrated in Fig .5 .shows that the power to evaporation of the water increases by increasing the amount of water in the humidifier, it is similar to the literature in Ref. [7].

This implies that the heat conveyed is still sufficient to allow the exchange of heat and material between the water and air throughout this branch. Decreasing branch of the curve shown in Fig.5, shows that the vapor diffusion of water in air decreases with increasing amount of water in the humidifier, this implies that the heat conveyed by air decreases also.

Decreasing the air transfer of heat to the water is due to the decrease of the air-water mixture temperature by increasing the amount of water in the humidifier.

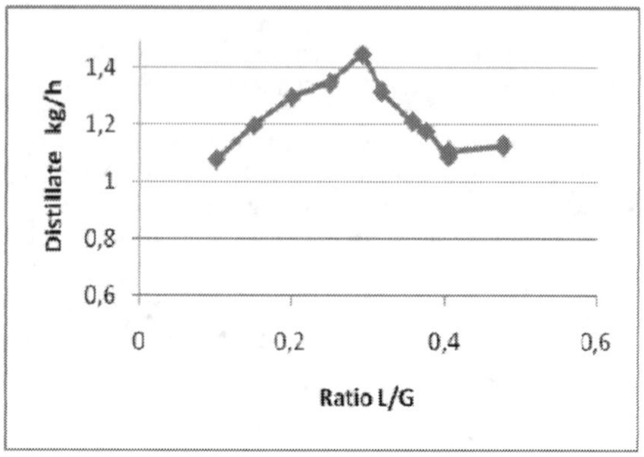

Fig. 5. Variation of production rate in function of water flow rate.

Fig. 5. indicates the variation of the quantity with the change of the normalized mass flow rate of the saltwater. It shows that there exists an optimum mass flow rate ratio corresponding to a maximum flesh water production.

5 Conclusion

The present work concerns an experimental study of a water desalination system. The system based on the humidification-dehumidification principle is a modular one and has the following independent components: water heater, an evaporator and a condenser. Some experimental results are analysed. They show the evolutions of water productivity in function of the temperature of air and water and the mass flow rate of water and air at the inlet and the outlet of each component of the system. The results show that there exists an optimum mass flow rate ratio corresponding to a maximum fresh water product.

References

1. Karim Bourouni et al.water humidification dehumidification of air, Solar Energy Conversion,(2103).
2. A. M. A. Dayem et M. Fatouh. Experimental and numerical investigation of humidification-dehumidification solar water desalination systems ,DES **247**, 594-609, (2009).
3. A. E. Kabeel, M. H. Hamed, Z. M. Omara, et S. W. Sharshir, Water Desalination ,DES, 286-305, (2013).
4. D. D. E. E. Geográfica, G. E. Energia, F. Diogo, A. Santos, et M. Azevedo, Renewable Energy Powered Desalination Systems : Technologies and Market Analysis Renewable Energy Powered Desalination Systems : Technologies and Market Analysis , (2014).

5. Prakash G, Narayan, Sharqawy MH, Summers EK, John H. The potential of solar-driven humidification-dehumidification desalination for small-scale decentralized water production.Desalination;(2010).

6. J.Moumouh et al. Theoretical and experimental study of a solar desalination unit based on humidificatione-dehumidification of air. IJHE **41** , 20818-822, (2016)

Nomenclature

G	[kg/s]	*Air flow rate*
L	[kg/s]	Water flow rate
T	[C]	Temperature

Subscripts

HDH	Humidification,dehumidification
w	water
a	air
outh	External of humidifier
inth	Internal of humidifier
WH	Water heater

Analysis of the recent Plio-Quaternary deformation and the correlation with the focal mechanisms of the significant earthquakes in the Bokoya, Ras Tarf (Eastern Rif)

A. El Mahsani [1] and L. Ait Brahim[1]
[1] Université Mohammed V, Faculté des sciences Département sciences de la terre, Unité de recherche GEORISK et LGRN, 4 Avenue Ibn Battouta Rabat-Agdal, BP 1014, Rabat, Maroc.

Abstract. Relative movements of microplates developed between the African and Eurasian plate characterize the continuity of the tectonic movements in Westerner of the Mediterranean Sea. The convergence movements between these plates, at the origin of the current structuring in oriental Rif. The recent tectonics in the south ripple of Alboran Sea, translated by an important seismic activity, In the North of Morocco, exactly in the Al Hoceima region, considered as zone of plates border with important seismicity. The structural analysis made in this paper and the interpretation of focals mechanisms earthquakes characteristics in the Al Hoceima region, has allows to deduct the main directions of faults, which we compared with the results of the satellite images. This study we has provides to demonstrate the directions of faults which are able to generate earthquakes in the future. The direction NE-SW represents the overall faults dominant in oriental Rif, this major direction coincides with the breaks of surface. A reduced number of faults shows a generall direction NNE-SSO to N-S through the Quaternary formations in the NO of Al Hoceima city, the direction NO does not show a good correlation with the directions of surface rupture, as well as with the deducted directions from focals mechanisms. This study drives us to propose that faults of direction NNE-SSO to NE-SO, represent the major directions of the faults in the Al Hoceima region, which represents the main cause of the seismic activity in the region.
Key words: Al Hoceima, Quaternary, Active faults, Focal mechanisms.

1 Introduction

La chaine alpine du Rif est issue du rapprochement entre les plaques Africaines et Eurasiatique qui subissent un rapprochement oblique dextre avec un taux de déplacement de 4.5 mm/année [1]. Dans le cadre de ce rapprochement la région d'Al Hoceima a connu de nombreux séismes violents dans le passé [2]. Récemment des séismes destructeurs ont eu lieu : 26 mai1994, 24 février 2004 et celui du 25 janvier 2016. Par conséquent, la région d'Al Hoceima est considérée comme la zone la plus active sismiquement au Maroc.

Les différents travaux effectués et méthodes utilisées (Mécanismes au foyer, Radar insar etc) pour déterminer les directions des failles responsables des séismes destructeurs dans la région d'Al Hoceima de 1994, 2004 et 2016 proposent des solutions différentes entre elles et surtout avec les déformations cosismiques de surface développées lors de ses séismes [3-7]. Dans le but de lever cette ambiguïté, nous avons d'une part repris l'analyse de nos relevés de terrain cartographiés lors des séismes de 1994 et 2004 et d'autre part effectué de récentes missions de terrain dans les formations géologiques récentes plio-quaternaire des régions limitrophes de la région d'Al Hoceima.

2 Présentation de la zone d'étude

La zone d'étude se situe au nord du Maroc, elle correspond à l'extrémité orientale de la chaine du Rif. Elle est limitée au nord par la partie méridionale de la mer d'Alboran. Les formations géologiques dans la région se présentent sous forme d'empilement de nappes de charriages à contacts anormaux E-W à vergence Sud [8]. Les principaux domaines structuraux présents dans la région d'étude sont : le massif des Bokoya marqué par un empilement de plusieurs unités structurales séparées les unes des autres par des contacts anormaux E-W [8]. Il est dominé par les formations carbonatées de la dorsale externe, sur lesquelles reposent sous forme de klippes tectoniques des terrains appartenant à la dorsale calcaire interne ainsi que les nappes Paléozoïques Ghomarides [8]. L'ensemble du massif chevauche vers le sud la série des flyschs gréseux de Tizirène par le biais d'une couche marneuse du Prédorsalien (semelle tertiaire), elle-même charriée sur les unités du Rif externe au sud [9]. Les formations post nappes correspondent aux bassins post nappes de Boudinar (tortono-messinien) avec le volcanisme de Ras Tarf qui est un massif andésitique d'âge miocène (première manifestation volcanique qui affecte la marge méditerranéenne marocaine postérieurement aux phases majeures de l'orogenèse alpine lié à un modèle de subduction) et celui du bassin d'Al Hoceima (baie de nekor) à remplissage quaternaire.

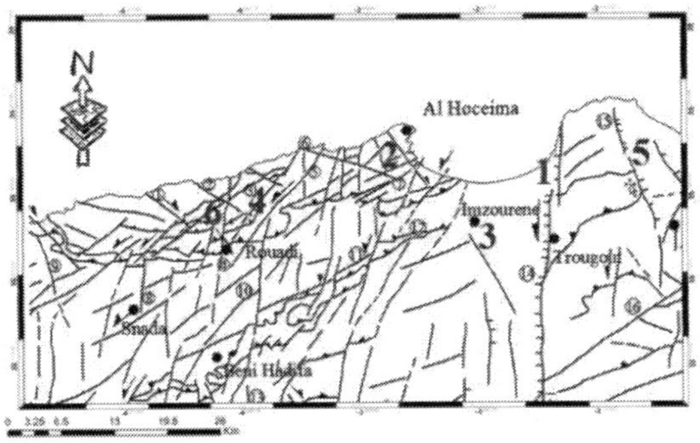

Fig. 1. Principales failles dans la région d'Al Hoceima (1: Faille Trougout, 2: Faille Boujibar, 3: Faille d'Imzouren, 4: Faille de Boussekour, 5: Faille Boudinar, 6 : Faille Rouadi).

Ces différentes unités structurales sont recoupées par des failles majeures transverses : du Nekor (NE-SW), d'Al-Hoceima (ENE-WSW), d'Imzouren (NNW-SSE), d'El Hammam (NNW-SSE), de Rouadi, (NNE-SSW), de Boussekour (NNE-SSW et celle d'ajdir WNW-ESE) [10, 4, 8, 11,12,13, 5, 14,6].

La région d'Al Hoceima est la zone la plus active sismiquement au Maroc avec une forte concentration d'épicentres (figure 2). Cette sismicité est la conséquence de l'activité tectonique complexe associée à la convergence Nord-Sud entre les plaques Afrique et Eurasie [3,15].

Fig. 2. Carte de la distribution des séismes, dans la région d'Al Hoceima entre 1900 et 2016.

Les principaux accidents cartographiques recensés responsables de la structuration de la région d'Al Hoceima qui recoupent les principaux fronts de chevauchement (E-W) sont (figure 1): Trougout (N-S), Imzouren (NNW-SSE), Hammam (NNW-SSE), Rouadi, (NNE-SSW), Boussekour (NNE-SSW), Nekor (NE-SW). [4, 5, 6, 8, 11, 12, 14].

3 Matériels et méthodes

La méthodologie adoptée dans cette étude est basée sur le traitement des images satellitaires de haute résolution et surtout l'analyse de la déformation cassante dans les formations plio-quaternaires (séries marines et continentales) de la région d'Al Hoceima suite aux nouveaux affleurements dégagées par la route côtière (la rocade). L'analyse des images satellitaires pour l'extraction des linéaments s'est faite par une série de traitements afin d'améliorer la qualité des résultats. La prise en compte de l'analyse du réseau hydrographique en termes d'anomalies et de perturbation était d'un apport considérable dans la détermination des segments des failles actifs ayant guidés le réseau hydrographique de la région d'Al hoceima [16, 17].

4 Carte des linéaments issus de l'image satellitaire

Les résultats issus de l'analyse des images satellitaires dans la région d'Al Hoceima montre une prédominance des directions NE-SW à ENE-WSW à secondairement la direction submeridienne.

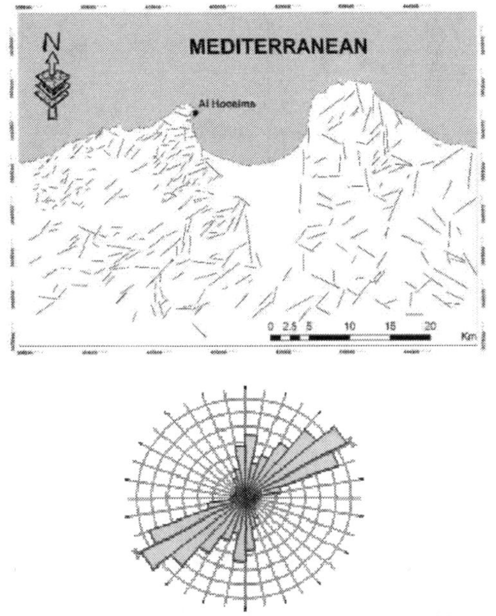

Fig. 3. Carte des linéaments issus de l'image satellitaire Landsat 8 Oli. La rosace des directions des linéaments montre les pourcentages suivants : N170-N010 25 %, N030-N050 27 %, N060-N070 33 %.

5 Analyse de la déformation récente dans la région d'Al Hoceima

La reprise de nos données de terrain (que nous n'avons pas publiées), associées aux récentes observations effectuées dans les formations géologiques récentes plio-quaternaire grâce aux nouveaux affleurements dégagées par la route côtière (la rocade), nous ont permis d'identifier et d'analyser de nouveau la déformation cassante dans les régions limitrophes de la région d'Al Hoceima.

5.1. Les formations quaternaires marines situées au NE de la ville d'Al-Hoceima

Le secteur d'Al Hoceima se caractérise par la présence des failles normales et des fractures de directions NNW-SSE à NE-SW (perpendiculaire à l'orientation de la côte). La majorité des failles montre un pendage sub-vertical (70° à 90°). Nous avons observé dans cette zone de nombreuses manifestations d'une activité tectonique récente, voire actuelle, de la faille d'Ajdir. Des fractures de direction N-S affectent les sables quaternaires et les schistes du socle paléozoïque (figure 4. B). Dans les sables, ces fractures N-S forment un réseau de failles normales néoformées à pendage opposé (figure 4. A). En surface, nous avons observé aussi, des fractures de directions subméridiennes au niveau de calcrètes récentes qui s'organisent sous forme de fentes de tension, parfois disposées en echelon compatibles avec un composante décrochante dextre. Le long de la côte nous avons identifiées des failles normales subméridiennes (N20 à N160) qui décalent les sables quaternaires de plusieurs dizaines de cm (figures 5 et 6). Parmi ces failles normales, celle de direction N160 qui décale avec un rejet vertical de l'ordre de 60 cm un niveau conglomératique à gros galets côtiers (figure 5).

Fig. 4.A. Formation sableuse quaternaire affectée par des fractures de direction N170 à N190 à pendage opposé F1, F2.

5.2. Les formations aux WNW d'Imzouren

Des niveaux récents de colluvions au niveau de **l'oued Rhis au NW d'Imzouren**, ainsi que des terrains villafranchiens nous ont permis d'observer des indices de déformations quaternaires à actuelles, liées à l'activité des failles actives dans le Rif oriental. Parmi ces indices, les plus représentatifs, **les nombreuses failles normales de direction N140 à N150** que nous avons observé à l'ouest du tracé cartographique de la faille d'Ajdir-Imzouren (figure 1).

Fig. 4.B. faille normale de direction N145, affectant une dune sableuse et le socle Paléozoïque.

Fig. 5. Le long de la cote, une faille normale N160, 80W décale un niveau conglomératique quaternaire avec un rejet vertical de 60cm.

Ces failles qui présentent aussi bien des pendages vers le SW que vers le NE (figure 7), affectent les terrains mésozoïques de la dorsale calcaire, les formations flyschoïde de Tizirène, et se prolongent dans les colluvions quaternaires sans atteindre systématiquement la surface. Dans plusieurs cas, elles affectent les calcrètes à la base des sols actuels. Il est important de noter ici que ce type de structure n'a pas été rencontré à proximité de la zone de faille d'Ajdir, mais dans une large zone située plus loin à l'Est (de la faille d'Ajdir). Ceci nous amène à penser que l'activité de cette structure n'est pas aujourd'hui cantonnée à un plan de faille cartographiquement bien défini, mais à des structures secondaires parallèles qui se distribuent sur un large couloir de faille de plusieurs centaines de mètres, voire quelques kilomètres.

5.3. Les régions de Trougout et Ras-tarf

Le soulèvement récent de l'extrémité Nord du massif volcanique de Ras Tarf a déjà été décrit grâce à l'étagement de terrasses marines étagées [5]. Ce soulèvement est guidé par la faille NS du massif de Trougout (qui est responsable aussi de l'effondrement du bassin d'Al Hoceima). **Les formations récentes au NE de la faille de Trougout**, montrent les traces d'une activité tectonique récente, grâce aux récents affleurements dégagés lors de la construction de la route (la rocade) qui longe la côte méditerranéenne. Ainsi, nous avons observé des plans de failles, de directions NNW-SSE à NNE-SSW (figure 8), qui mettent en contact des colluvions quaternaires avec les séries volcaniques tortono-messiniennes du Ras Tarf. La présence de facies mylonitique pulvérulent au niveau des plans de ses failles atteste de leur récente activité. Les miroirs de ces failles récemment dégagés montrent des stries subhorizontales et des tectoglyphes qui attestent du rejeu en décrochement senestre récent postérieurement au jeu en faille normale à fort escarpement vertical dans la topographie lors des phases tortono-messiniennes.

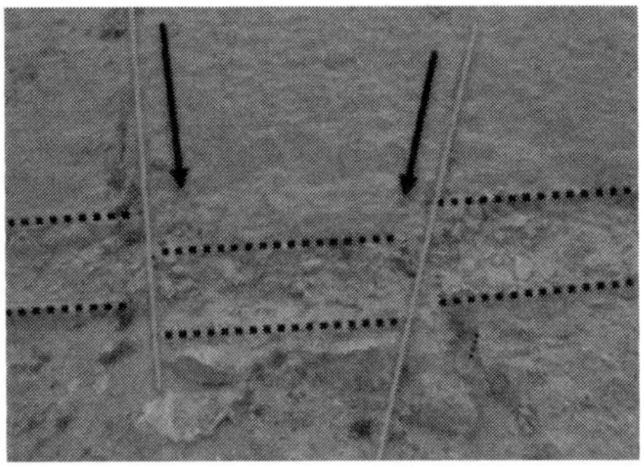

Fig. 6. Deux failles normales de direction N20 à regard opposé affectent une formation sableuse en décalant un niveau conglomératique

Fig. 7. Failles normales à pendage opposé de direction N100, affectant une formation du Villafranchien.

Fig. 8. Miroir de la faille de Trougout, dans la région d'Ahadad, B. l'escarpement de la terrasse de l'oued par le jeu de la faille de Trougout.

6 Ruptures de surface développées lors des séismes de 1994 et 2004

Les déformations de surface que nous avons cartographiées lors du séisme de 1994 se répartissent sur plus de 20 kms le long de la faille de Boussekour (de la côte méditerranéenne jusqu'à Rouadi). Elles correspondant à des fractures de direction NE-SW à légère composante inverse avec localement des fentes de tensions en échelon qui indiquent un jeu décrochant senestre [18].

Fig. 9. Le tracé de la faille de Boussekour-Aghbar qui montre une direction (NE-SW), responsable de l'évènement sismique du 26 Mai 1994.

Pour le séisme. L'analyse des mécanismes au foyer de ce séisme, propose [19] que la meilleure solution correspond au mouvement de décrochement à composante inverse du premier plan nodal sénestre orienté N22E. [6] proposent N20°E, [20] proposent une rupture complexe sur des failles de direction N175°E et N150°E. La direction déduite par les données InSAR (interférogrammes) réalisées par [21], proposent un plan de faille de direction N23E avec un pendage de 77° vers le SE. Les déformations de surface que nous avons cartographiées lors du séisme de 2004 se répartissent sur plus de 60 kms (entre Beni-Abdellah et Ajdir) (figure 11) dans la nappe de Tizirene, l'unité de kétama et les alluvions recents de l'oued Rhis (avec développement de liquéfactions locales). Elles correspondant à des fractures de direction NE-SW avec localement des fentes de tensions en échelon qui indiquent clairement un jeu decrochant senestre [4]. Les mécanismes au foyer montrent tous un glissement en décrochement sur des plans nodaux subverticaux orientés NNE-SSW et WNW-ESE (SED, HRV, IGN, IAG, USGS, EMSC) (figure 10).

Le plan nodal de faille de décrochement senestre, N11°E, 74° a été préféré par [22]. La répartition des répliques montrent deux essaims de direction NE-SW et NW-SE. Leurs mécanismes au foyer indiquent un jeu décrochant senetre pour les NE-SW et dextre pour les NW-SE [23]. Cette solution des deux failles a été reprise et retenue par [24] avec distance entre les deux failles estimée à 1,5 km. La direction déduite par les données InSAR (interférogrammes) réalisées par [21], proposent un plan de faille de direction N23E avec un pendage de 77° vers le SE. Devant cette complexité de la distribution des répliques et des solutions proposés par les mécanismes au foyer, l'analyse des données InSAR [25] n'a pas pu retenir le mouvement de décrochement N-S a N15E, comme unique plan responsable du séisme de 2004, car les interférogrammes cosismiques font ressortir dans le même pourcentage aussi la direction NW-SE.

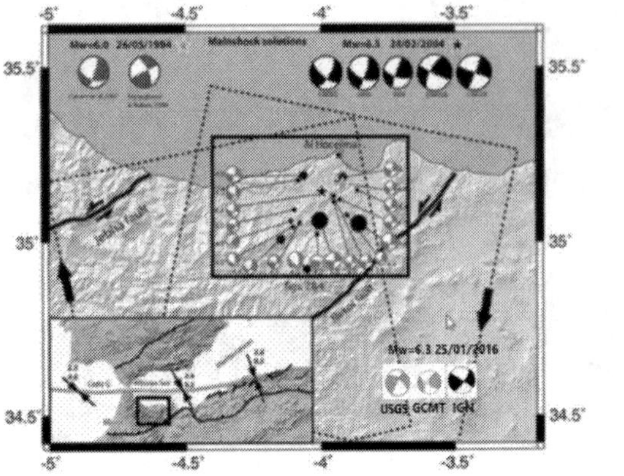

Fig. 10. Mécanismes au foyer calculés pour les évènements sismiques du 1994, 2004 et 2016 (EMSC, HRV, IGN, SWISS, USGS, GCMT). Modifié [23].

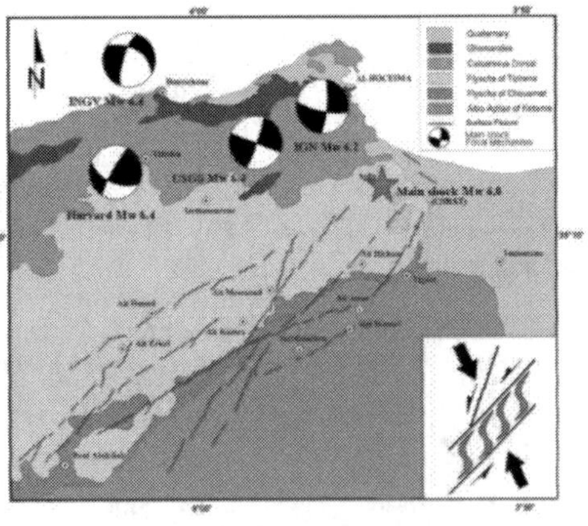

Fig. 11. Ruptures de surface de direction NE-SW cartographiées après le séisme du 2004, [17].

7 Résultats et discussions

Les résultats obtenus donnent ainsi trois grandes catégories de directions, NS à NNE-SSW, NE-SW et NNW-SSE à NW-SE. La direction NE-SW coïncide avec les résultats issus de l'analyse des images satellitaires et correspond à la direction la plus abondante dans la région d'Al Hoceima (figure 3). Cette direction est en accord avec la direction de la faille de Boussekour Aghbar (figure 9) responsable de l'événement sismique de 1994, ainsi que les fracturations de surface causées par le séisme du 2004 (figure 11). L'analyse des mécanismes au foyer des séismes majeurs (figure 10), a permis de mettre en évidence deux directions de failles en profondeur, NE-SW et NW-SE [24]. On note que les données issues de la méthode de sismique réflexion haute résolution ont permis de mettre en évidence l'existence des failles dans le large de la mer d'Al Hoceima de directions NE-SW [26]. Cela nous ramène à proposer que l'activité sismique dans la région d'Al Hoceima est guidée par des failles de direction allant de N-S a NE-SW (avec une prédominance des failles de direction NE-SW qui restent les plus abondantes dans la région d'Al Hoceima et de la mer d'Alboran).

8 Conclusion

La région d'Al Hoceima a enregistré des séismes majeurs ayant entrainé des dégâts importants aussi bien sur le plan humain que matériel. Les mécanismes au foyer de ses séismes proposent deux plans nodaux de direction NE-SW à NNE-SSW et NW-SE à WNW-ESE. Les déformations de surface cartographiées lors des séismes de 1994 et 2004 sous forme de fentes de tension organisées en échelon présentent une direction NNE-SSW à NE-SW [4]. Les données GPS indiquent un mouvement de convergence des plaques Afrique-Eurasie de direction NNW-SSE à NW-SE, avec un échappement du bloc du Rif central vers le SSW le long d'une faille de direction NE-SW à composante décrochante senestre [27]. Pour avoir plus d'information afin de trancher sur la direction de la direction des failles responsable en profondeur du déclanchement des séismes de 1994 et 2004, nous avons analysé les déformations cassantes dans les formations plio_quaternaires de la région d'Al Hoceima. La rosace des failles mesurées indique que la direction NNE-SSW à NE-SW est la plus prédominante. Nous proposons de retenir cette direction comme celle des failles responsables des séismes de 1994 et 2004. Ces failles se prolongent en mer avec une légère déviation pour rejoindre les accidents actifs de la ride d'Alboran responsables du séisme de 2016.

References

1. D. F. Argus, R. G. Gordon, C. DeMets & S. Stein. Closure of the Africa□Eurasia□North America plate motion circuit and tectonics of the Gloria fault. Journal of Geophysical Research: Solid Earth, **94**(B5), 5585-5602. (1989)
2. T. El Mrabet. *Les grands séismes dans la région maghrébine (Doctoral dissertation*, Thèse d'état, Faculté des lettres et des sciences humaines, Université Mohammed V. Rabat, **435 pp.**, {(in arabe)}). (2005)
3. M. Meghraoui. La faille sismique et ses implications pour l'évaluation des risques sismiques le long de la limite des plaques en Afrique du Nord. En conférence euro-méditerranéenne pour l'intégration de l'environnement(pp. 37-40). Springer, Cham. (2017, novembre).
4. L. Ait Brahim et al. Structural analysis and interpretation of the surface deformations of the February 24th, 2004 Al Hoceima earthquake. Centre Sismologique Euro-Méditerranéen European-

Mediterranean Seismological Centre www. emsc-csem. (2004).

5. A. Poujol, J.F. Ritz, A. Tahayt, P. Vernant, M. Condomines, P. H. Blard, P. H & A. K. Idrissi. Active tectonics of the Northern Rif (Morocco) from geomorphic and geochronological data. *Journal of Geodynamics*, *77*, 70-88. (2014).
6. A. Calvert et al. An integrated geophysical investigation of recent seismicity in the Al-Hoceima region of north Morocco.. Bulletin of the Seismological Society of America **87**, p. 637–651. (1997).
7. C. Ziyadin, M. Meghraoui, M. Akoglu, M., et al. Surface deformation associated with the Mw 6.4, 24 February 2004 Al Hoceima, Morocco, earthquake deduced from InSAR: implications for the active tectonics along North Africa. Bulletin of the Seismological Society of America, vol. **96**, no 1, p. 59-68. (2006).
8. O. Azzouz. *Lithostratigraphie et tectonique hercynienne des terrains paléozoïques ghomarides du Massif des Bokoya (Rif Interne, Maroc),* Thèse 3ème cycle, Univ Mohammed V, Fac. Sci. Rabat. (1992).
9. T. Mourier, Etude géologique et structurale du massif des Bokoya, Trav. Lab. Géol. de l'Afrique, 6. Univ. Paris Sud. (1982).
10. L. Ait Brahim L, B. Tadili, C. Nakhcha, I. Mouayn, M. Ramdani, M. Limouri. Using active faults and seismicity for the strong motion modeling in the Eastern Rif (Northern Morocco). *Pure and Applied Geophysics,: Volume 161, Numbers 5-6, pp.* 1081-1091. (2002)
11. A. Chalouan, Ouazani-Touhami, Les failles normales à faible pendage du Rif interne (Maroc) et leur effet sur l'amincissement crustal du domaine d'Alboran. Geogaceta 17, 107–109. (1995).
12. L. Ait Brahim. *Tectoniques cassantes et états des contraintes récents au Nord du Maroc. Contribution al'étude du risque sismotectonique*. These Doctorat d'Etat, Université Mohammed V, Rabat. (1991).
13. L .Ait Brahim,P. Chotin, M. Ramdani. Failles actives dans le Rif central et oriental (Maroc) - C. R. Acad. Sci. Paris, t. 310, série II, pp. 1123-129. (1990)
14. J. Galindo-Zaldívar, O. Azzouz, A. Chalouan, A. Pedrera, P. Ruano, A. Ruiz-Constán & M. Benmakhlouf. Extensional tectonics, graben development and fault terminations in the eastern Rif (Bokoya–Ras Afraou area). *Tectonophysics*, *663*, 140-149. (2015).
15. L. Ait Brahim. Cadre sismotectonique et risque sismique de la chaîne rifaine,. Travaux de l'Institut Scientifique, série géographique physique et géologie, n° 397 pp.121-135. (2002)
16. B. Deffontaines, P. Chotin, L. A. Brahim & M. Rozanov. Investigation of active faults in Morocco using morphometric methods and drainage pattern analysis. *Geologische Rundschau*, *81*(1), 199-210. (1992)
17. L. Ait Brahim, C. Nakhcha, B. Deffontaines, H. Sami, P. Chotin. Apports de l'analyse conjointe du réseau hydrographique et d'un modèle numérique de terrain pour l'étude de la néotectonique du Rif oriental, (région d'Al Hoceima – Nador, Maroc NE). *Revue Télédétection- photo interprétation. Editions ESKA, France* N° 4, pp.13-49. (2004)
18. L. Ait brahim, B. Tadili, M. Ramdani, O. Ibn ibrahim. Cadre structural et manifestation de surface du séisme d'Al Hoceima du 26/5/94. poster, 33 rd Meeting of the General committe of the international council of scientific union (ICSU) 11-17 Octobre-Rabat. (1994)
19. S. O. El Alami, B. Tadili, T. E. Cherkaoui, F. Medina, M. Ramdani, L.A. Brahim & M. Harnafi. The Al Hoceima earthquake of May 26, 1994 and its aftershocks: a seismotectonic study. (1998)
20. M. Bezzeghoud & E. Buforn. Source parameters of the 1992 Melilla (Spain, MW= 4.8), 1994 Alhoceima (Morocco, MW= 5.8), and 1994 Mascara (Algeria, MW= 5.7) earthquakes and seismotectonic implications. *Bulletin of the Seismological Society of America*, *89*(2), 359-372. (1999)
21. A. M. Akoglu, Z. Cakir, M. Meghraoui, S. Belabbes, S. O. El Alami, S. Ergintav,& H. S. Akyüz. The 1994–2004 Al Hoceima (Morocco) earthquake sequence: Conjugate fault ruptures deduced from InSAR. *Earth and Planetary Science Letters*, *252*(3), 467-480. (2006)
22. D. Stich, FDL. Mancilla, D. Baumont et J. Morales. Analyse des sources du tremblement de terre Mw 6.3 2004 Al Hoceima (Maroc) en utilisant des fonctions de temps source apparentes régionales. *Journal of Geophysical Research: Terre solide* , *110* (B6). (2005)
23. L. Dorbath, Y. Hahou, B. Delouis, C. Dorbath, J. Van Der Woerd, S. Bardrane, P. Tapponnier. Études sismologiques sur le séisme D'al Hoceima: localisation, et mécanisme du choc principal et des répliques, contraintes et structure de la zone epicentrale. In *Colloque Intenational Seisme d'Al Hoceima: bilan et perspectives* (pp. 24-26). (2005)
24. J. Van Der Woerd, C. Dorbath, F. Ousadou, L. Dorbath, B, E. Jacques & H. Haessler. The Al Hoceima Mw 6.4 earthquake of 24 February 2004 and its aftershocks sequence. Journal of Geodynamics, **77**, 89-109. (2014).
25. C. Ziyadin, M. Meghraoui, M. Akoglu et al. Surface deformation associated with the Mw 6.4, 24 February 2004 Al Hoceima, Morocco, earthquake deduced from InSAR: implications for the active tectonics along North Africa. Bulletin of the Seismological Society of America, vol. **96**, no 1, p. 59-68. (2006).
26. M. Lafosse, E. d'Acremont, A. Rabaute, B. Mercier de Lépinay, A. Tahayt, A. Ammar & C. Gorini. Evidence of Quaternary transtensional tectonics in the Nekor basin (NE Morocco). *Basin Research*, *29*(4), 470-489. (2017)
27. A. Koulali, D. Ouazar, A. Tahayt, R. W. King, P. Vernant, R. E. Reilinger & N. Amraoui. New GPS constraints on active deformation along the Africa–Iberia plate boundary. Earth and Planetary Science Letters, **308**(1), 211-217. (2011).

Mass movement susceptibility mapping - A comparison of logistic regression and Weight of evidence methods in Taounate-Ain Aicha region (Central Rif, Morocco)

JEMMAH A I [1] ET AIT BRAHIM L [1]
[1] Université Mohammed V, Faculté des sciences Département sciences de la terre, Unité de recherche GEORISK et LGRN, 4 Avenue Ibn Battouta Rabat-Agdal, BP 1014, Rabat, Maroc

laitbrahim@gmail.com

Abstract. Taounate region is known by a high density of mass movements which cause several human and economic losses. The goal of this paper is to assess the landslide susceptibility of Taounate using the Weight of Evidence method (WofE) and the Logistic Regression method (LR). Seven conditioning factors were used in this study: lithology, fault, drainage, slope, elevation, exposure and land use. Over the years, this site and its surroundings have experienced repeated landslides. For this reason, landslide susceptibility mapping is mandatory for risk prevention and land-use management. In this study, we have focused on recent large-scale mass movements. Finally, the ROC curves were established to evaluate the degree of fit of the model and to choose the best landslide susceptibility zonation. A total mass movements location were detected; 50% were randomly selected as input data for the entire process using the Spatial Data Model (SDM) and the remaining locations were used for validation purposes. The obtained WofE's landslide susceptibility map shows that high to very high susceptibility zones contain 62% of the total of inventoried landslides, while the same zones contain only 47% of landslides in the map obtained by the LR method. This landslide susceptibility map obtained is a major contribution to various urban and regional development plans under the Taounate Region National Development Program.

Introduction

La susceptibilité aux mouvements de mass est une composante importante dans l'équation de ce type de risque, celui-ci étant défini comme la probabilité d'occurrence du phénomène dans un certain secteur.
L'évaluation de celui-ci suppose l'appréciation quantitative et qualitative de la classification, du volume, de la superficie et de la distribution spatiale des mouvements de mass qui existent et/ou qui peuvent apparaître [1, 8].

La cartographie de la susceptibilité aux glissements de terrain met en évidence la distribution spatiale des versants potentiellement instables, fondée sur l'analyse des paramètres favorables aux mouvements de mass, par la considération de la superposition spatio-temporelle des facteurs qui ont généré de tels processus dans le passé [2]. A partir du principe « le passé et le présent représentent la clé de l'avenir », on peut formuler les hypothèses sur lesquelles est fondée l'estimation des risques de glissements de terrain :
- la distribution actuelle des glissements reflète les conditions géologiques, géomorphologiques, hydrogéologiques, climatiques, le degré de recouvrement de la végétation, l'impact des activités humaines ;
- les futurs glissements de terrain vont apparaître dans les mêmes conditions ;
- les facteurs de contrôle des glissements peuvent être identifiés, classés, hiérarchisés et évalués par différentes méthodes.

Le degré de fiabilité des cartes de susceptibilité aux glissements de terrain est influencé par plusieurs facteurs, comme la précision, la quantité et la qualité des données acquises, l'échelle de travail, la méthodologie et la modélisation.

Cet article propose ainsi une cartographie de la susceptibilité aux mouvements de mass dans la région de Taounate _ Ain Aicha, prenant en considération des différents aspects et limites méthodologiques. Après une brève présentation du secteur d'étude, la démarche méthodologique ainsi que les résultats seront présentés et discutés.

1 Contexte de l'étude

La présente étude s'inscrit dans le cadre d'une étude environnementale portant sur la région de Taounate _ Ain Aicha. Notre étude vise principalement à développer une approche méthodologique menant à l'élaboration d'une carte de susceptibilité aux mouvements de mass, en appliquant les méthodes de la théorie de l'évidence et la régression logistique et un système d'information géographique (SIG). L'un des objectifs de notre étude est donc de fournir une cartographie simple mais détaillée des zones sensibles aux mouvements de mass. Par ailleurs, cette cartographie peut servir de document de référence aux municipalités régionales pour l'identification des zones

 https://doi.org/10.1051/matecconf/201814902094

à risques naturels, lesquelles zones doivent s'inscrire dans les schémas d'aménagement du territoire. Enfin, la région de Taounate _ Ain Aicha a été choisie pour cette étude puisqu'elle constitue l'une des régions du Rif et qu'elles présentent des zones à risques élevés de mouvements de mass.

2 Zone d'étude et méthodologie

2.1 Zone d'étude

Le secteur étudié forme un rectangle de 640 km² de surface découpé selon le carroyage rectangulaire de la carte géologique (1/50.000) (Fig. 1).

La région de Taounate _ Ain Aicha au sein du Rif central est caractérisée par une alternance de reliefs montagneux et de vallées dont la plus importante est celle de l'Oued Ouerrhe dont le lit présente par endroits d'importants replats où se développent une agriculture d'irrigation. Cette topographie très contrastée est le résultat des déformations tectoniques de la région au cours du Tortono-messinien et du plio-quaternaire [1, 12] qui a porté en relief des rides de roches résistantes de calcaires et de grès qui sont bordées sur les versants par les roches tendres à moyennement résistantes de marnes, marno-calcaires et schisto-gréseuses. Les constructions, en général se situent sur les crêtes calcaires saturées. Ceci oblige l'expansion anarchique sur les versants argileux nord et sud inaptes à la construction (en plus de mouvements de terrain) sauf aux abords de l'Ouerrha et de ses affluents.

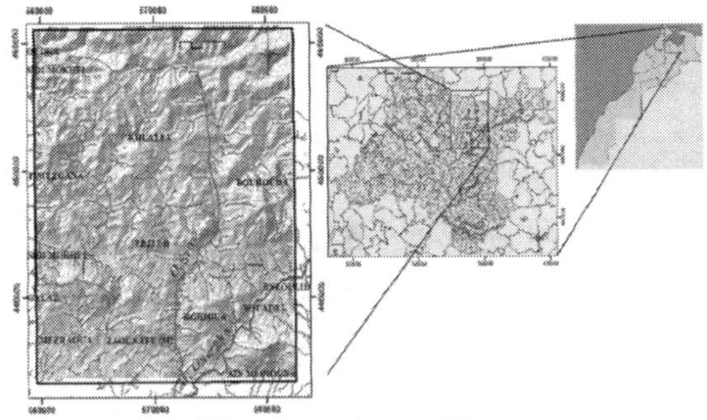

Fig.1 Situation géographique de la zone d'étude.

2.2 Méthodologie

2.2.1 Inventaire de mouvements de mass

L'inventaire des mouvements de mass été conduit via l'utilisation de la base de données existante, complétée la photo-interprétation systématique de l'ensemble du terrain d'étude selon une grille d'analyse. Ceci permet [1] d'éliminer les doublons et erreurs cartographiques de la base de données existante, [2] d'obtenir une représentation de glissements sous forme de polygones séparant la zone d'ablation de la zone d'accumulation pour chacun des phénomènes (fig.2). Quelques campagnes de validation sur le terrain ont permis de valider la qualité de l'inventaire.

La base de données contient finalement 463 phénomènes cartographiés. 339 d'entre eux sont des Eboulis (73%) et 124 sont des Solifluxion (27%). Pour le calcul juste 50% des mouvements de mass sont utilisés pour la modélisation de la susceptibilité. Alors que les 50 % restantes pour la vérification.

2.2.2 Constitution de la base de données des facteurs de prédisposition

Les bases de données (cartes géologiques, cartes topographiques, MNT et image satellitaire) ont été utilisées pour dériver les principaux facteurs de prédisposition (lithologie, réseau hydrographique, fracturation, pente, exposition, hypsométrie et occupation du sol,). L'ensemble des données thématiques a été converti en format raster.

– La lithologie représente un facteur conditionnant majeur dans la stabilité des versants aussi bien par, la nature des matériaux (qualité géotechnique des roches) que par la structure (direction et pendage des couches, plans de schistosité, etc.) (fig.3).

– La pente est l'un des facteurs déterminant dans la genèse des glissements de terrain. En effet, pour qu'une rupture puisse se produire sur un versant, il faut qu'il y ait l'intervention de la gravité accompagnée d'un un appel au vide. Ce phénomène ne pouvant prendre naissance sur des terrains plats, la prise en compte du degré de pente des versants et des couches est d'une importance capitale. (fig.4)

– L'hypsométrie, par sa composante altimétrique représente aussi l'un des facteurs les plus déterminants dans la genèse des glissements de terrains. Ainsi, plus l'altitude augmente, plus le risque d'occurrence de mouvement de mass est présent. (fig.5)

- La carte d'exposition (fig.6) à été générée à partir du MNT. Elle Exprime l'orientation des versants par rapport au nord. Elle a été déterminée par mesure, dans le sens des aiguilles d'une montre, de l'angle entre le nord et la projection horizontale de la ligne de pente maximale d'un versant. En se basant sur la connaissance de notre secteur d'étude, nous avons élaborée la carte en cinq classes selon les différentes directions (Plat, N, E, S, et W). Le changement de l'exposition des versants équivaut une grande variabilité de l'obliquité par rapport aux rayons solaire, la durée d'ensoleillement des versants, ainsi que des apports énergétiques (vents, température…) inégaux. Cela rend très différent la prédisposition des versants aux mouvements de mass. Les versants exposés vers le N et l'W montrent les fréquences les plus importantes d'instabilités de terrain.

– Le réseau hydrographique est généralement liée au phénomène d'érosion hydrique parmi lesquels on retrouve les sapements de berges. Ce phénomène se développe généralement dans les portions des cours d'eau où la vitesse d'écoulement est suffisamment puissante pour éroder la partie basale d'un versant. Dans la région de Taounate Ce sont essentiellement les berges des cours d'eau supérieur allant du 3ème au 5ème ordre (selon la hiérarchisation de Strahler), qui sont les plus propices aux instabilités. (fig.7)

– La fracturation peut conditionner ou déclencher un mouvement de masse au niveau de la zone de broyage des failles [3] (cisaillements et/ou décrochements) formée de matériaux fragilisés et permettre aussi (fig.8) l'infiltration des eaux en profondeur qui en contact avec les argiles une fois saturées vont déclencher des instabilités de terrain.

–L'occupation des sols (fig.9) : la cartographie a été réalisée par interprétation des images satellites couplée aux photographies aériennes (classification multi-spectrale supervisée) et la base de données de la couverture végétale et surtout des essences forestières mises à notre disposition par le HCEFLCD (Haut-Commissariat des Eaux et Forêts et de la Lutte contre la Désertification). Cette dernière couche exerce au niveau de la région de Taounate essentiellement une action phytostabilisante sur les versants à substrat argileux ou à substrat rocheux

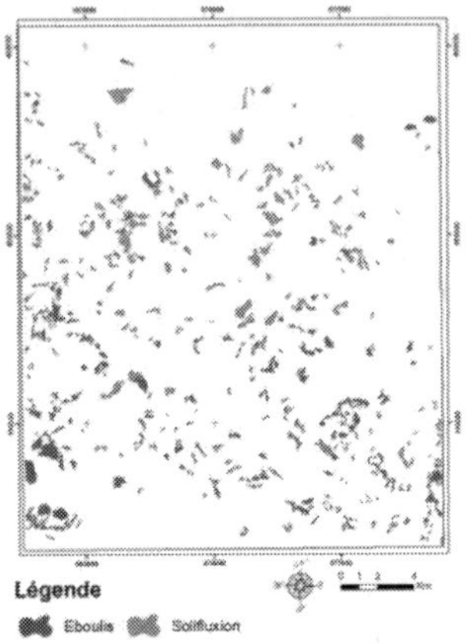

Fig.2 carte d'inventaire des mouvements de terrain

Fig.3 Carte des différentes formations lithologiques.

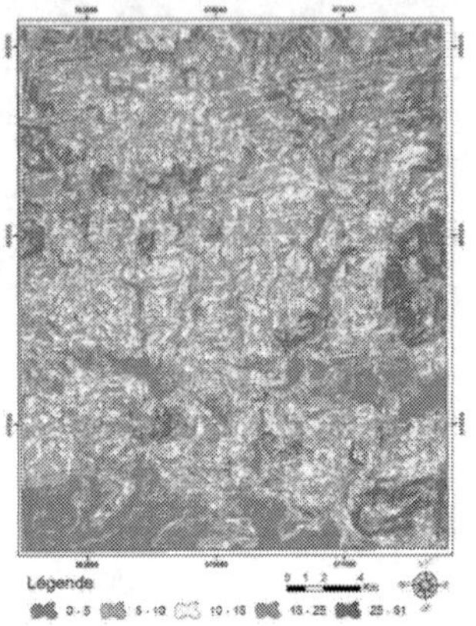

Fig.4 Carte des gradients de pente.

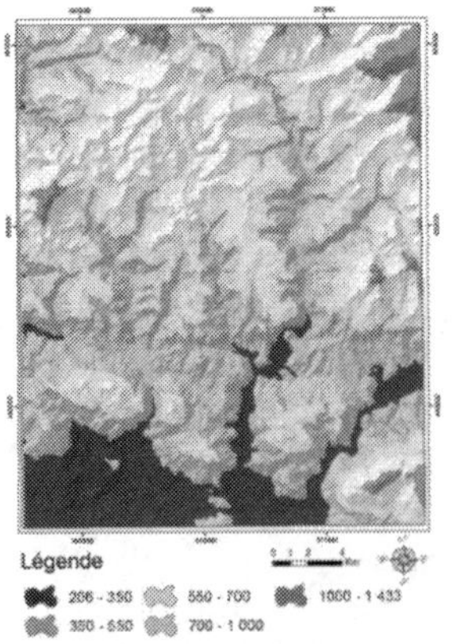

Fig.5 Carte d'hypsométrie (en mètre).

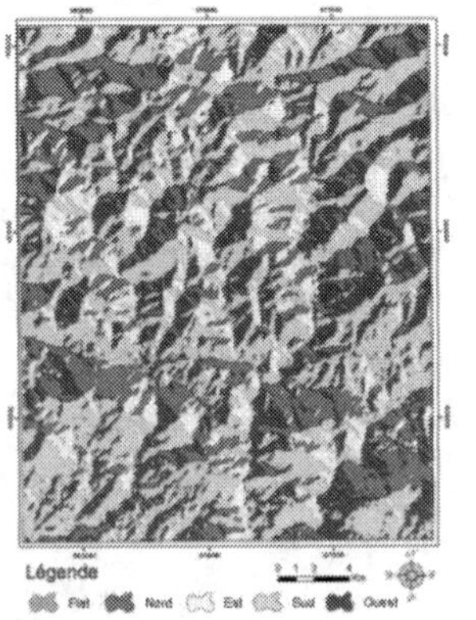

Fig.6 Carte d'expositions.

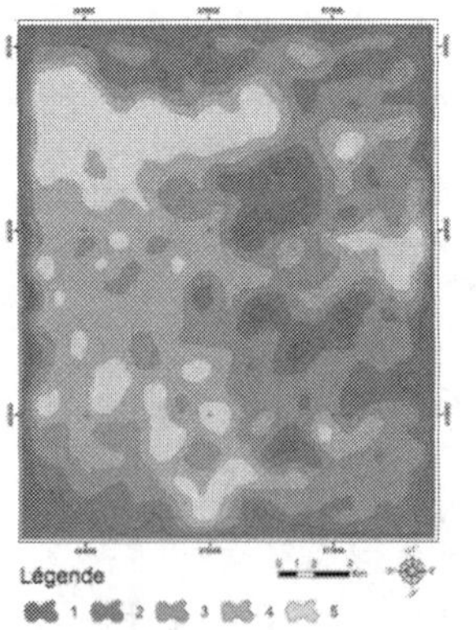

Fig.7 Carte de la densité du réseau hydrographique

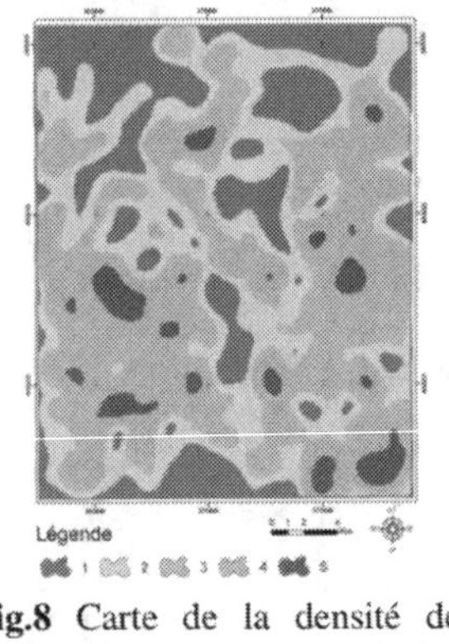

Fig.8 Carte de la densité de la fracturation.

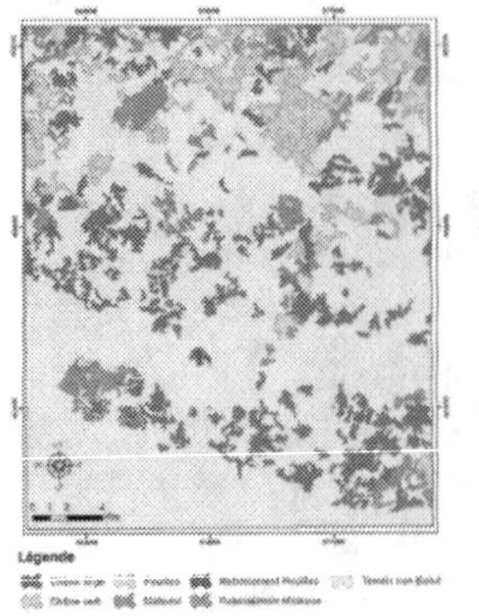

Fig.9 Carte d'occupation du sol

développant une couverture d'altération argileuse peu épaisse

2.2.3 *La théorie de l'évidence et la Régression Logistique*

Le concept fondamental de ces modèles est d'étudier le comportement d'une variable dépendante (localisation des mouvements de mass) à partir d'une combinaison de variables indépendantes prédictives pour des unités géomorphologiques homogènes.

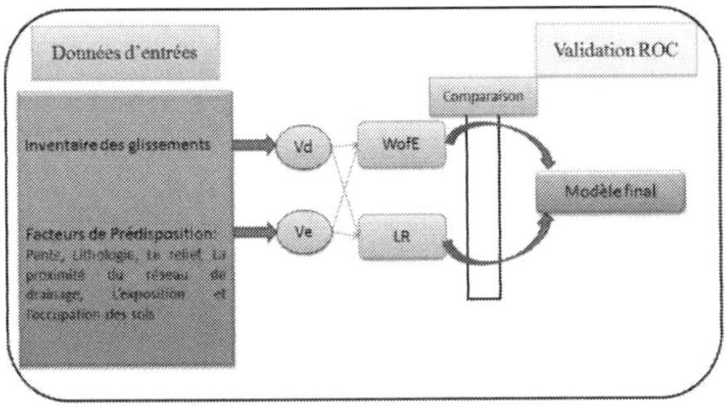

Fig.10 Organigramme de la démarche adoptée à méso-échelle (1/50 000ème) par la théorie de l'évidence (TDE) et la Régression Logistique

2.2.3.1 la théorie de l'évidence

L'analyse spatiale bivariée par la théorie de l'évidence, déjà appliquée en médecine [4, 8, 9, 10] et en géologie [4], est une version log-linéaire du théorème général de Bayes utilisant les principes de calcul de probabilité fondés sur les notions de probabilité a priori et de probabilité a posteriori. La probabilité a priori, qui est la probabilité qu'une unité de terrain (un pixel) contienne une variable dépendante (Vd, dans notre cas les glissements de terrain), est calculée suivant la densité sur la zone d'étude. La probabilité a posteriori est estimée après le calcul de la probabilité a priori selon la densité de Vd pour chaque variable considérée comme prédictive (Vp, dans notre cas chaque facteur de prédisposition). La méthode est fondée sur le calcul d'un poids positif (W+ ; (1)) et d'un poids négatif (W- ; (2)), dont les valeurs dépendent de l'association entre Vp et Vd. [5; 9, 10,11]

$$(1)\quad W_{+} = \ln \frac{P(B|Vd)}{P(B|\bar{V}d)} \qquad (2)\quad W_{-} = \ln \frac{P(\bar{B}|Vd)}{P(\bar{B}|\bar{V}d)}$$

Dans les équations (1) et (2), B est la classe de Vp et le symbole « _ » représente l'absence de Vp et/ou de Vd. Le ratio représente la probabilité de présence sur la probabilité d'absence d'un phénomène. Les pondérations sont additionnées en utilisant le logarithme naturel des ratios appelé logit. Les calculs des valeurs de W+ et de W- pour toutes les Vp permettent de calculer la probabilité a posteriori, qui actualise la probabilité a priori. Lorsque plusieurs Vp sont combinées, les zones qui ont un poids respectivement élevé ou faible correspondent respectivement à une plus grande ou une plus petite probabilité de présence de Vd.

2.2.3.2 La régression Logistique

L'analyse spatiale bivariée par régression logistique est un modèle non-linéaire, fondée sur une fonction (3) de répartition sigmoïdale entre la probabilité de présence d'un glissement de terrain (Vd) et différents facteurs de prédisposition (Vd), [4, 8]. Le modèle logistique pour la présence ou l'absence de glissements de terrain est défini par :

$$Pr\ (Vd) = \frac{1}{1+e^{-z}} \qquad (3)$$

où Pr (Vd) est la probabilité que la variable dépendante soit présente et Z est une combinaison linéaire de Vp. Comme Z varie de $-\infty$ à $+\infty$, le codage de la probabilité varie de 1 à 0, et la probabilité de présence de Vd correspond à son espérance conditionnelle. Comme pour l'analyse bivariée par théorie de l'évidence, lorsque plusieurs Vp sont combinées, les zones qui ont un poids respectivement élevé ou faible correspondent respectivement à une plus grande ou une plus petite probabilité de présence de Vd. [8, 9, 10]

2.2.3.3 Calage et validation des modèles

Le calage des modèles a été effectué en ne retenant initialement que 50% de la population totale de Vd (présence/absence des mouvements de masse) pour la meilleure combinaison de Vp. Les résultats sont analysés à partir d'une courbe prédictive représentant l'aire cumulée en fonction de la probabilité, sur laquelle des sauts permettent d'identifier trois classes de susceptibilité (faible, moyenne, forte).

La validation des modèles s'effectue d'un point de vue statistique et d'un point de vue expert. Pour la validation statistique, la même procédure que pour le calage est utilisée, en introduisant les 50% restants de la population totale de Vd. Pour la validation experte, les résultats des modèles en termes de surfaces de glissements de terrain modélisés sont comparés à l'inventaire géomorphologique. [9, 10,11]

3 RESULTATS ET DISCUSSION

Pour la théorie de l'évidence, les cartes des paramètres de prédisposition combinées avec la carte de mouvements de mass permettent de calculer les poids positifs et négatifs et les contrastes, comme indiqué dans le tableau 1.

Tableau 1 : Corrélation spatiale entre les variables prédictives et les valeurs de pondération calculées par la théorie de l'évidence, (W+) : la valeur la probabilité a priori, (W-) : la valeur de la probabilité à posteriori, (C) : le contraste qui est la différence entre W+ et W-, S(C) : la variance du contraste

Formations géologiques	*Surface en Km^2*	*W+*	*W-*	*C(Contraste)*	*S(C)*
calcaires	8,7663	0,9677	-0,0218	0,9895	0,3770
Schisto-gréseux	8,0880	0,9095	-0,0183	0,9278	0,4012
Marne	403,51	0,2858	-0,8223	1,1081	0,1793
Schistes	213,57	-1,2542	0,3207	-1,5749	0,2213
Distance par rapport aux Failles	***Surface en Km^2***	***W+***	***W-***	***C(Contraste)***	***S(C)***
> 500	135,26	-0,7003	0,1300	-0,8304	0,2139
400	183,34	-0,1318	0,0489	-0,1807	0,1533
300	154,93	0,1595	-0,0566	0,2162	0,1484
200	133,25	0,2349	-0,0716	0,3065	0,1523
100	30,750	0,7087	-0,0516	0,7603	0,2321
Gradient de pente	***Surface en Km^2***	***W+***	***W-***	***C(Contraste)***	***S(C)***
0 - 5°	94,264	-0,8657	0,0995	-0,9652	0,2693
5 - 10°	151,37	-0,0381	0,0118	-0,0498	0,1581
10 - 15°	160,30	0,1873	-0,0722	0,2595	0,1454
15 - 25°	181,99	0,1119	-0,0490	0,1610	0,1431
25 - 51°	41,676	0,1609	-0,0124	0,1733	0,2508
Exposition des versants	***Surface en Km^2***	***W+***	***W-***	***C(Contraste)***	***S(C)***
Nord	130,49	0,2088	-0,0620	0,2708	0,1540
Est	138,03	-0,2025	0,0506	-0,2531	0,1732
Sud	187,95	-0,1704	0,0651	-0,2355	0,1532
Ouest	171,89	0,1434	-0,0592	0,2025	0,1442
Distance au réseau hydrographique	***Surface en Km^2***	***W+***	***W-***	***C(Contraste)***	***S(C)***
>500	48,199	-0,6885	0,0406	-0,7291	0,3433
400	128,99	-0,3359	0,0706	-0,4066	0,1882
300	190,2162	0,0941	-0,0426	0,1367	0,1425
200	180,1351	0,2681	-0,1267	0,3947	0,1390
100	90,8756	-0,1390	0,0214	-0,1604	0,2018

Occupation des sols	*Surface en Km^2*	*W+*	*W-*	*C(Contraste)*	*S(C)*
Chêne liège	82,1165	0,4044	-0,0746	0,4790	0,1717
Chêne vert	51,6982	-0,5528	0,0373	-0,5900	0,3124
Feuillus	1,4756				
Reboisement Feuillus	24,3778	-0,2454	0,0087	-0,2540	0,3895
Matorral	11,6754	-0,7738	0,0102	-0,7840	0,7164
Terrain non Boisé	468,6057	-0,0099	0,0265	-0,0364	0,1496
'hypsométrie	***Surface en Km^2***	***W+***	***W-***	***C(Contraste)***	***S(C)***
206 – 400 m	130,5278	-0,1952	0,0457	-0,2409	0,1766
400 – 600 m	247,4623	0,3252	-0,2787	0,6039	0,1340
600 – 800 m	167,3002	-0,2848	0,0869	-0,3717	0,1658
800 – 1000 m	64,8983	-0,1894	0,0198	-0,2092	0,2376
1000 – 1433 m	19,4222	-2,0049	0,0281	-2,0330	1,0048

3.1 Calcul de l'indépendance conditionnel (IC)

3.1.1 Degré de liberté entre chaque paire de variables de predisposition

Parmi les sept variables de prédisposition traités, six ont été retenues (occupation des sols, la pente, le facies, l'exposition, distance aux failles et le réseau hydrographique) étant donné que les valeurs du degré de liberté (α) entre chaque paire de variables obtenue par la méthode de WofE (tableau.2) sont supérieurs à 0,05 (α>0,05). La variable hypsométrie a été retirée (α< 0,05) étant donné sa dépendance avec tous les variables. Ce résultat a été confirmé par le test de chi-carrée (X^2) étant donné que l'écart n'est pas important entre l'écart de X2 observé et X2 théorique pour chaque couple de facteurs et pour l'ensemble des variables à modéliser. [4, 9, 10]

Tableau 2 : Valeurs du degré de liberté α entre chaque paire de variables : (OS,occupation du sol ;Lt,lithologie ; Rh,réseau hydrographique ;Exp, exposition ; DF distance des failles)

	OS	Lt	R H	Exp	DF
Pente	0,8969	0,6467	0,4972	0,9999	0,7373
OS		0,8172	0,9999	1,0000	1,0000
Lt			0,0000	0,9948	0,0000
RH				0,9721	0,9285
Exp					0,7736

 https://doi.org/10.1051/matecconf/201814902094

L'analyse bivariée par régression logistique est plus souple à mettre en oeuvre car elle ne suppose pas une hypothèse d'indépendance conditionnelle [7, 10], ce qui peut réduire considérablement le temps d'analyse en amont.

D'après les valeurs des coefficients de régression logistique qu'on a réalisée à partir des paramètres de prédisposition (occupation des sols, la pente, le facies, exposition, distance aux failles et le réseau hydrographique), on peut dire que la pente et le réseau hydrographique sont les paramètres les plus importants dans l'apparition des mouvements de mass. [8, 9, 10]

Tableau3: Les valeurs des coefficients de régression obtenus pour les paramètres indépendants

Paramètres	Coefficients	Standard Déviation
Constant	-2,242407	0,519553
Pente	0,175012	0,060177
Occupation des sols	-0,014301	0,035736
Lithologie	-0,772629	0,105654
Densité du réseau hydrographique	0,175952	0,064487
Exposition	-0,026988	0,060661
Densité aux failles	0,142481	0,061142

Viens par la suite la densité aux failles, alors que l'occupation des sols, la lithologie et l'exposition avec ce coefficients négatif ne présente pas une grande importance dans l'apparition des mouvements de mass. La carte de susceptibilité aux mouvements de mass la région de Taounate_Ain Aicha permet de distinguer 5 classes de susceptibilité : très forte, forte, moyenne, faible, et très faible (Fig.11et 12).

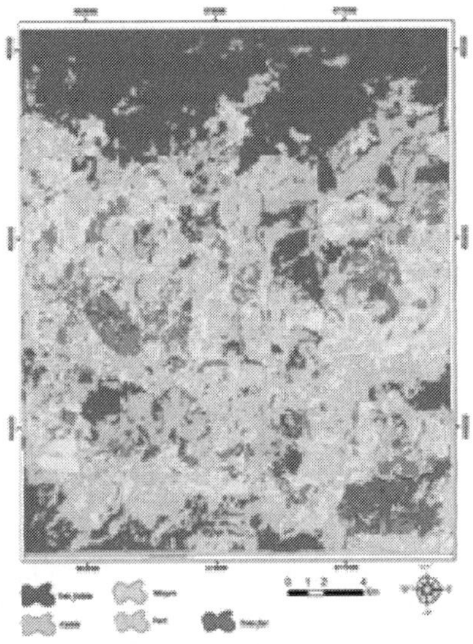

Fig.11 carte de susceptibilité aux mouvements de mass par la théorie de l'évidence de la région de Taounate_Ain Aicha

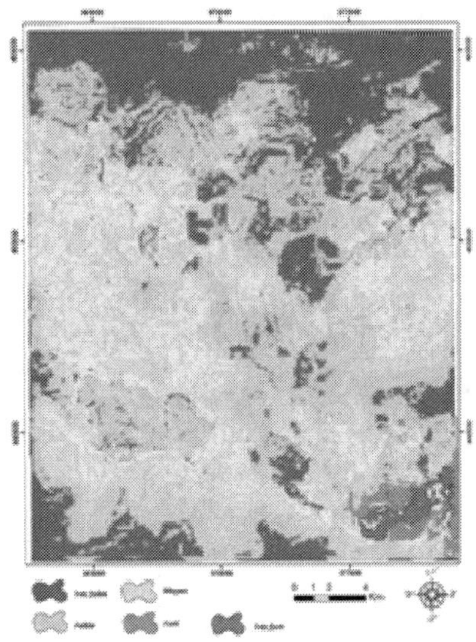

Fig.12 carte de susceptibilité aux mouvements de mass par la Régression logistique de la région de Taounate_Ain Aicha

Conclusion

Dans cet article, deux méthodes ont été utilisées pour l'analyse de la susceptibilité aux mouvements de terrain dans la région de Taounate - Ain Aicha : la théorie de l'évidence et la régression logistique. La théorie de l'évidence appliquée s'est avéré plus performante et adaptée que celle de la régression logistique à la prédiction des instabilités de terrain dans la région de Taounate - Ain Aicha. La précision de prédiction du modèle par la théorie de l'évidence est de 75% par contre celle obtenue pour la régression logistique n'est que 56%. De plus, la carte de susceptibilité aux glissements de terrain qui en résulte montre que les zones de susceptibilité élevée à très élevée élaborée par la théorie de l'évidence présentent le pourcentage le plus élevé (62%) de tous les glissements de terrain inventoriés par rapport aux autres zones par contre ce pourcentage n'est que 47% pour les mêmes zones sur la carte de susceptibilité obtenue par la régression logistique.
La carte de susceptibilité au glissement de terrain obtenue par la théorie de l'évidence (Wof E) constitue une contribution à la gestion et à la planification territoriales urbaines et régionales dans le cadre du programme national de développement de la province de Taounate.

References

1 R. Fell, J. Corominas, C. Bonnard, L. Cascini, E. Leroi, W. Z Savage; *Guidelines for Landslide Susceptibility, Hazard and Risk Zoning for Land Use Planning;* On behalf of the JTC-1 Joint Technical Committee on Landslides and Engineered Slopes:, Eng. Geol., 102, 85–98, (2008).

2 F. Guzzetti, A. Carrara, M. Cardinali, P. Reichenbach, *Landslide hazardevaluation:areview of current techniques and their application in a multi-scalestudy, central Italy. Geomorphology 31,* 181–216. (1999)

3 L. Ait Brahim, M. Mansour, F. Sossey Alaoui, I. EL Hamdouni, *Rôle de la fracturation dans la désagrégation mécanique des calcaires et la mise en mouvement de la coulée de pierres d'Amtrasse (Rif, Maroc),* Edition ISBN 9954-8407-0-2 "L'implication de la géotechnique dans le développement des infrastructures en Afrique ", 69-79 (2003).

4 D Spiegelhater., R.P. Kill-Jones, *Statistical and Knowledge Approaches to Clinical Decision-support Systems, with an Application in Gastroenterology*, Journal of the Royal Statistical Society, vol. **147**, 1, 1984 p. 35-77.

5 G.F Bonham-Carter, F.P Agterberg, D.F Wright. *Weights of evidence modeling: a new approach to mapping mineral potential.* Statistical Applications in the Earth Science: Geological Survey of Canada Paper, vol. **89-9**, pp. 171–183. (1989)

6 D.G. Kleinbaum, *Logistic Regression A Self-Learning Text. Springer-Verlag, New York.* (1994) https://doi.org/10.1007/978-1-4757-4108-7

7 F.P.Agterberg, Q.Cheng; *Conditional independence test for weights-of- evidence modeling*, Natural Resources Research, v. **11**, 4, p. 249-255. (2002)

8 S. Lee, B. Pradhan, *Landslide hazard mapping at Selangor, Malaysia using frequency ratio and logistic regression models,* Landslides *4*, 33–41(2007).

9 Y. Thiery, J-P. Malet, S. Sterlacchini, A. Puissant, O. Maquaire, *Landslide susceptibility assessment by bivariate methods at large scales: Application to a complex mountainous environment,* Geomorphology (2007).

10 A. Fares ; *Essai méthodologique de la cartographie des risques naturels liés aux mouvements de terrain. Application à l'aménagement de la ville de Taounate (Rif, Maroc).* Thèse Doctorat. Université de Franche Comté, (1994).

11 C.J. Van Westen, *GIS in landslide hazard zonation: a review, with examples from the Andes of Colombia,* Mountain Environments and Geographic Information Systems (1994).

12 L. Aït Brahim, *Tectonique cassante et état de contaraintes récentes du Maroc Nord, contibution à l'étude du risque sismo-tectonique au Maroc.* Thèse de doctorat d'état, Université Mohamed V, Faculté des Sciences, Rabat. (1991)

Structural behaviour of square RC columns confined with CFRP wraps

A. Benredjem[1], N. Chikh[2], H.A. Mesbah[3] and R. Benzaid[4]
[1,2] University of Mentouri Constantine, LMDC - Department of Civil Engineering, Constantine, Algeria
[3]University of Rennes 1, L.G.C.G.M - Department of Civil Engineering , Rennes, France
[4]University of Jijel, LGG - Department of Geology, Jijel, Algeria

Abstract. Loading capacity and strains of square reinforced concrete (RC) columns, strengthened with external carbon fiber reinforced polymer (CFRP) sheets, were tested and evaluated. The experimental parameters include: number of wrap layers, concrete strength and the slenderness of the columns (L/a). All test specimens were loaded to failure in axial compression. Compressive stress, axial and hoop strains have been recorded to evaluate the stress-strain relationship, ultimate strength, stiffness, and ductility of the specimens. Results clearly demonstrate that composite wrapping can enhance the structural performance of RC columns in terms of both maximum strength and ductility. The effects of test parameters are evidenced and compared.

1 Introduction

An increasing number of reinforced concrete structures have reached the end of their service life, either due to deterioration of the concrete and reinforcements caused by environmental factors, or due to an increase in applied loads. These deteriorated structures may be structurally deficient or functionally obsolete, and most are now in serious need of extensive rehabilitation. Carbon fiber reinforced plastics sheets or plates are well suited to this application because of their high strength-to-weight ratio, good fatigue properties, and excellent resistance to corrosion. Their application in civil engineering structures has been growing rapidly in recent years, and is becoming an effective and promising solution for strengthening deteriorated concrete members. Because CFRPs are quickly and easily applied, their use minimizes labour costs and can lead to significant savings in the overall costs of a project.

During the last decade, the use of FRP composites has been successfully promoted for external confinement of reinforced concrete (RC) columns all over the world. Several studies on the performance of FRP wrapped columns have been conducted, using both experimental and analytical approaches [1-9]. Such strengthening technique has proved to be very effective in enhancing their ductility and axial load capacity. However, most of the available studies on the behavior of FRP confined concrete columns have concentrated on circular shaped columns with normal strength. The data available for columns of square or rectangular cross sections have increased over recent years but are still limited [10-14]. Also the validation of these results and their applicability to large-scale RC columns is of great practical interest. This field remains in its infancy stages and more research investigation is needed on this subject to study the effect of slenderness and that of concrete strength.

This study deals with a series of tests on square plain concrete (PC) and reinforced concrete (RC) columns strengthened with CFRP sheets. A total of 48 concrete specimens were tested under axial compression. The data recorded included the compressive loads, axial strains, and radial strains. The parameters considered are the number of composite layers (1 and 3), the compressive strength of the unconfined concrete (25MPa and 60MPa) and the columns' slenderness ratio L/a (2; 4 and 7.4). To comply with existing RC members in practice, where reduced cover is often present, the corners for all prismatic specimens were almost kept sharp for CFRP application.

2 Experimental program

2.1 Materials

The Two kind of concrete mix have been realised to investigate the influence of concrete strength, normal strength concrete (NSC- 25 MPa) and high strength concrete (HSC- 60 MPa). The carbon-fiber sheets used were the SikaWrap-230C product, a unidirectional wrap. The manufacturer's guaranteed tensile strength for this CFRP is 4300 MPa, with a tensile modulus of 238 GPa, an ultimate elongation of 1.8 % and a fiber thickness of 0.13mm. The Sikadur-330 epoxy resin was used to bond the carbon fabrics over the square columns.

Table 1 summarizes the specimens involved in the experimental program. Eight series of experiments were performed to investigate the behavior of PC and RC square columns confined by CFRP composite. For all RC specimens the diameter of longitudinal and transverse reinforcing steel bars were respectively 12 mm and 8 mm. The longitudinal steel ratio was constant for all specimens and equal to 2.25%.The yield strength of the longitudinal and transversal reinforcement was 500 MPa and 235 MPa; respectively. The specimen notations are as follows. The specimen notations are as follows. The first two letters refer to the type of concrete: PC for plain concrete and RC for reinforced concrete, followed by the concrete mixture: I for normal strength (24.77MPa) and II for high strength (59.53MPa). The next letter indicates the slenderness ratio: x for L/a=2, y for L/a=4 and z for L/a=7.14. The last number specifies the number of layers.

Table 1. Details of test specimens

Specimen designation	Concrete mixture	Nominal dimensions (side x side x height) [mm]	Number of layers	Unconfined concrete strength [MPa]	Number of specimens
PCI. x0	I	140x140x280	-	24.77	2
PCI. x1		140x140x280	1		2
PCI. x3		140x140x280	3		2
RCI. x0		140x140x280	-		2
RCI. x1		140x140x280	1		2
RCI. x3		140x140x280	3		2
RCI. y0		140x140x560	-		2
RCI. y1		140x140x560	1		2
RCI. y3		140x140x560	3		2
RCI. z0		140x140x1000	-		2
RCI. z1		140x140x1000	1		2
RCI. z3		140x140x1000	3		2
PCII. x0	II	140x140x280	-	59.53	2
PCII. x1		140x140x280	1		2
PCII. x3		140x140x280	3		2
RCII. x0		140x140x280	-		2
RCII. x1		140x140x280	1		2
RCII. x3		140x140x280	3		2
RCII. y0		140x140x560	-		2
RCII. y1		140x140x560	1		2
RCII. y3		140x140x560	3		2
RCII. z0		140x140x1000	-		2
RCII. z1		140x140x1000	1		2
RCII. z3		140x140x1000	3		2

2.2 Specimen preparation

After concrete columns were fully cured, FRP wrapping procedure was performed according to the procedure specified by the manufacturer. The CFRP jackets were applied to the specimens by manual wet lay-up process. The concrete specimens were cleaned and completely dried before the resin was applied. The epoxy resin was directly applied onto the substrate. The fabric was carefully placed into the resin with gloved hands and smooth out any irregularities or air pockets using a plastic laminating roller. The roller was continuously used until the resin was reflected on the surface of the fabric, an indication of fully wetting. A second layer of resin was applied to allow the impregnation of the CFRP. The following layer is applied in the same way. Finally, a layer of resin was applied to complete the operation. Each layer was wrapped around the column with an overlap of ¼ of the perimeter to avoid sliding or deboning of fibers during tests. The wrapped specimens were left at room temperature for 1 week before testing.

2.3 Test procedures

Specimens were loaded under a monotonic uni-axial compression load up to failure. The load was applied at a rate corresponding to 0.24 MPa/s and was recorded with an automatic data acquisition system. Axial and lateral strains were measured using appreciable extensometer. The instrumentation included one lateral linear variable differential transducer (LVDT) placed in the form of a square frame at the mid-height of the specimens. Measurement devices also included three vertical LVDTs to measure the average axial strains. Prior to testing, all CFRP-wrapped columns were capped with sulfur mortar at both ends.

3 Test results and discussion

Figures Compression behavior of the CFRP wrapped specimens was mostly similar in each series in terms of stress-strain curves and failure modes of the specimens. All confined concrete columns failed by fracture of the composite wrap at one of the corners, because of the high stress concentration at these locations, Figure. 1. The collapse occurred in a sudden and explosive way, though some popping noises were heard during various stages of loading and were attributed to microcracking of the concrete. The strain values observed for the jacket tensile failure were substantially lower than the FRP failure strain, as many authors have already published. For short specimens (L/a =2), the fiber rupture starts mainly in their central zone, then propagates towards both ends. Regarding slender specimens, the collapse was mostly concentrated in their end regions, indicating that the greater the slender ratio, the smaller the area of CFRP ruptured.

For these columns at ultimate load, when confinement action was no longer provided due to FRP fracture, the internal steel started buckling and the crushed concrete fell down between the fractured FRP. Hence, this indicates that the concrete core is significantly damaged (but yet confined) even before reaching ultimate load. For all confined specimens, delamination was not observed at the overlap location of the jacket, which confirmed the adequate stress transfer over the splice.

Fig. 1. Failure of CFRP confined specimens

The average experimental results are reported in Table 2, with the increase in terms of compressive strength (f'_{cc}/f_{co}) and ductility ($\varepsilon_{cc}/\varepsilon_{co}$), intended as ultimate axial displacement.

Table 2. Test results

Specimen designation	f'_{co} [MPa]	f'_{cc} [MPa]	f'_{cc}/f_{co}	ε_{cc} [‰]	$\varepsilon_{cc}/\varepsilon_{co}$	$\varepsilon_{h,rup}$ [‰]	$\varepsilon_{h,rup}/\varepsilon_{ho}$
PCI. x0	24.77	24.77	1.00	2.20	1.00	3.88	1.00
PCI. x1		27.66	1.11	6.58	2.99	24.33	6.27
PCI. x3		32.03	1.29	5.89	2.67	17.01	4.38
RCI. x0	33,59	33.59	1.00	7.61	1.00	17.00	1.00
RCI. x1		39.52	1.17	15.06	1.97	20.58	1.21
RCI. x3		49.12	1.46	15.66	2.05	25.50	1.50
RCI. y0	30.49	30.49	1.00	1.67	1.00	9.78	1.00
RCI. y1		36.73	1.20	3.02	1.80	8.69	0.88
RCI. y3		41.85	1.37	5.61	3.35	7.77	0.79
RCI. z0	24.69	24.69	1.00	0.96	1.00	-	-
RCI. z1		33.92	1.37	2.05	2.13	-	-
RCI. z3		39.17	1.58	3.64	3.79	-	-
PCII. x0	59.53	59.53	1.00	3.66	1.00	3.31	1.00
PCII. x1		61.30	1.02	2.46	0.67	5.43	1.64
PCII. x3		70.35	1.18	3.09	0.84	13.39	4.04
RCII. x0	63.79	63.79	1.00	3.05	1.00	10.30	1.00
RCII. x1		74.84	1.17	3.87	1.26	16.36	1.58
RCII. x3		79.59	1.24	5.29	1.73	7.96	0.77
RCII. y0	63.62	63.62	1.00	2.05	1.00	0.35	1.00
RCII. y1		80.78	1.26	2.82	1.37	0.76	2.17
RCII. y3		82.44	1.29	2.79	1.36	0.76	2.17
RCII. z0	69.98	69.98	1.00	2.08	1.00	0.49	1.00
RCII. z1		75.77	1.08	2.53	1.21	0.82	1.67
RCII. z3		81.51	1.16	2.70	1.29	1.36	2.77

Representative stress-strain curves for each series of tested CFRP-wrapped specimens are reported in figure 2 (a-c) for NSC and in figure 3 (a-c) for HSC. They give the axial stress versus the axial and lateral strains for specimens with zero, 1 and 3 layers of CFRP wrap considering various slenderness ratio L/a (2, 4 and 7.14).

3.1 Stress-Strain Response

For NSC, all CFRP strengthened specimens showed a typical bilinear trend with a transition zone. Three zones can be observed for the stress-strain curves of the CFRP-confined specimens. The first zone is essentially a linear response governed by the stiffness of the unconfined concrete, which indicates that no confinement is activated in the CFRP wraps since the lateral strains in the concrete are very small. The unconfined concrete specimens show a sudden drop in stiffness and strength after reaching the maximum load point. In the second zone, a nonlinear transition occurs as the concrete expands, thus producing larger lateral strains. The CFRP wrap reacts accordingly and a confining action is created on the concrete core. During this stage a loss of stiffness occurs due to the rapidly growing network of cracks in the concrete. Finally, in the third zone, the concrete is fully cracked and the CFRP confinement is activated to provide additional load carrying capacity by keeping the concrete core intact. The stress-strain curve here increases linearly up to failure. However, no distinct post behaviour is observed for specimens with higher slenderness ratio. On overall, both ultimate compressive strength and ultimate strain are variably enhanced depending on the number of layers and the slenderness ratio.

As for the previous case of NSC, the first slope of the curve, regarding specimens with HSC, is also not substantially altered by the presence of CFRP. In this initial elastic zone, the confined and the unconfined specimens behave in the same manner, irrespective of the number of layers. The strengthening effect of the CFRP layers begins only after the concrete has reached the peak strength of the unconfined concrete: transversal strains in the concrete activate the CFRP jacket. The increase of load would produce large lateral expansions, and consequently a higher confining pressure, provided that the number of composite layers is quite sufficient. With low levels of confinement (one CFRP layer), the second part of the bilinear curve shifts from strain hardening to a flat plateau with a drastically reduced ductility.

No distinct post behaviour is observed as the slenderness ratio increases and little improvement is achieved in both strength and ductility.

3.2 Effect of unconfined concrete strength

To investigate the effects of concrete quality, different concrete strength (24.77 MPa and 59.35 MPa) have been used. Considering specimens confined with 3 CFRP layers, it can be seen from stress strain curves in figures 2 and 3 a varying strength increase of 37 to 58% for NSC specimens and of 16 to 29% for HSC specimens. This effect is even more pronounced on the axial deformations, where the relative variation increases are of 105 to 280% for NSC specimens and 25 to73% for HSC specimens. This clearly indicates that the effect on the strength and ductility capacities decreases with increasing concrete strength, as found by other authors. Mechanical effects of different concrete quality are also evident in the first branch of the curves, where stronger concrete shows higher stiffness with respect to concrete with lower strength.

3.3 Effect of CFRP Strengthening Ratio

In all cases the increase of the numbers of sheets generated an increase of compressive strength as well as axial deformation capacity. The level of increase is important for NSC specimens. Considering the cases of 1 and 3 CFRP layers, from results displayed in Table 2 and figure 2, it can be evaluated that the increase in the bearing capacity varies roughly from 17% to 58% as compared to the relative unconfined specimens, while the ultimate vertical deformations increase on average from 80% to 280%. From these findings, it is possible to assert that the increase in the number of CFRP sheets has a significant influence even though the increase in terms of strength is not as important as that of axial deformations which increase almost proportionally to the FRP strengthening ratio.

The effect of the number of CFRP layers on HSC specimens is relatively moderate compared to the case discussed previously. In this situation, the confinement pressure is activated at higher load (around 80% of the ultimate value). Consequently, the enhancement in the load carrying capacity is reduced and varies roughly from 8% to 29%, whereas the ultimate axial deformations undergo a significant reduction displaying an increase on average from 21 to 125%, as illustrated in Table 2 and in figure 3. It should be emphasized that the presence of quite sharp corners in all tested CFRP jacketed columns produced a cutting effect on confining sheets and hence affected the rate of enhancement in their load carrying and deformation capacities.

3.4 Effect of slenderness ratio

The comparison of results recorded for a slenderness ratio varying from 2 to 7.14 shows for NSC wrapped RC specimens a moderate decrease in the load carrying capacity and an important reduction in the axial deformation. However, in the case of HSC jacketed specimens, the strength was almost not affected whereas the ductility underwent a moderate decrease. This may be explained by the late activation of the confinement pressure which occurred at higher load (around 80% of the ultimate value).

On overall, the efficiency of the confinement provided by composite wraps was greatly affected by the premature damage of the CFRP fabric at the sharp column corner.

4 Conclusions

An experimental study on the axial compression behaviour of square RC columns confined externally with CFRP was presented. The following conclusions can be drawn:

• The failure of all CFRP wrapped specimens occurred in a sudden and explosive way preceded by typical creeping sounds. For short specimens (L/a =2), the fiber rupture

 https://doi.org/10.1051/matecconf/201814902095

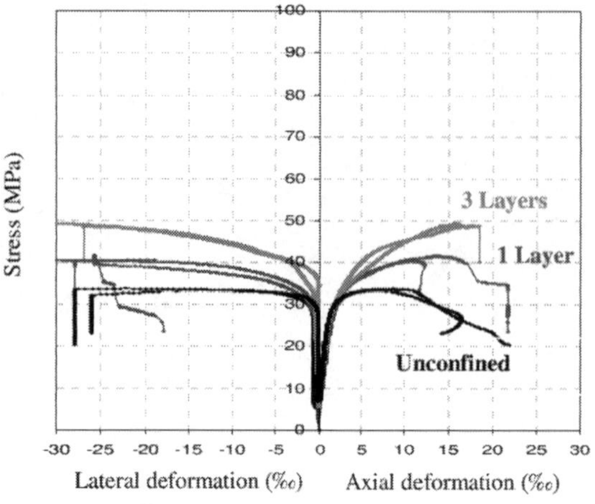

(a) RCI.x series

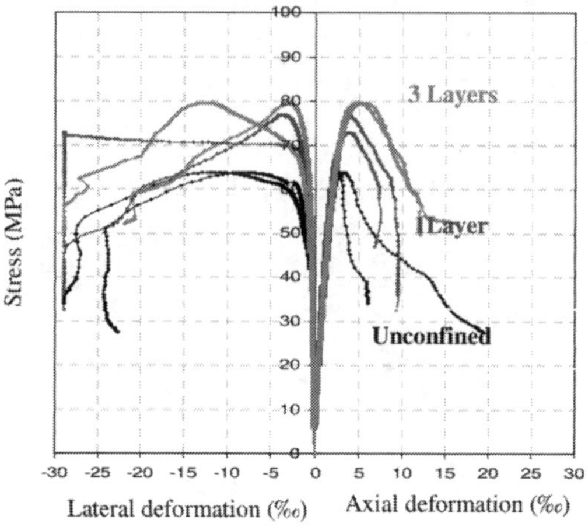

(a) RCII.x series

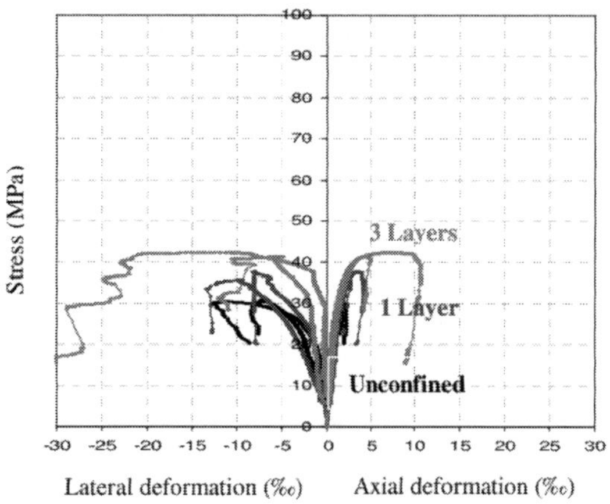

(b) RCI.y series

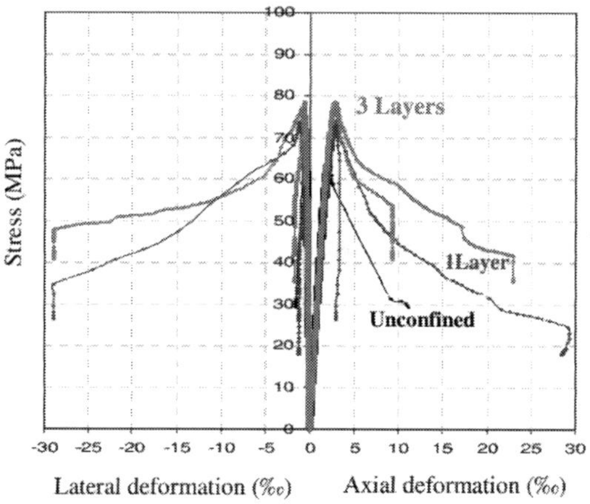

(b) RCII.y series

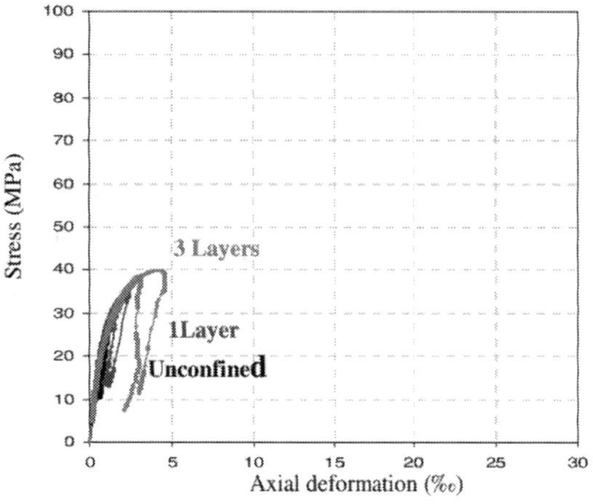

(c) RCI.z series

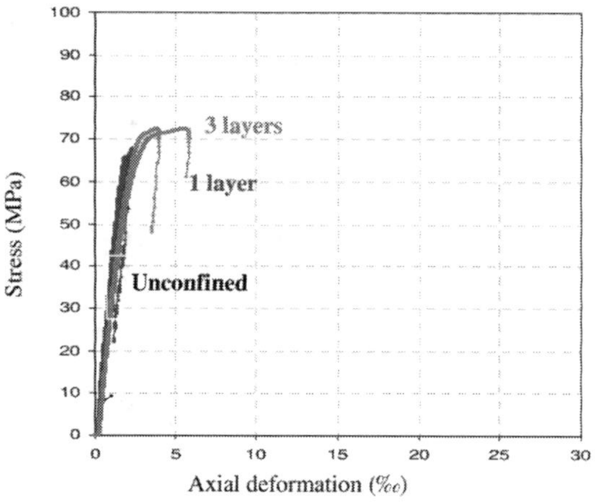

(c) RCII.z series

Fig. 2. Stress strain curves of NSC CFRP specimens

Fig.3. Stress strain curves of HSC FRP specimens

starts mainly in their central zone, then propagates towards both ends. Regarding slender specimens, the collapse was mostly concentrated in their upper or lower regions.

• Increasing the amount of CFRP sheets produce an increase in the compressive strength of the confined column but with a rate lower compared to that of the deformation capacity which almost proportional to the CFRP strengthening ratio;

• The CFRP confinement on low-strength concrete specimens produced higher results in terms of strength and strains than for high-strength concrete similar specimens. Therefore, the effect of CFRP confinement on the bearing and deformation capacities decreases with increasing concrete strength;

• The effect of increasing the slenderness ratio results in a decrease of the strengthening effect on strength and ductility. The rate of decrease is more important for NSC specimens.

References

1. H. Saadatmanesh, M.R. Ehsani, W. Li, ACI Stru. J., **91**, 4, (1994)
2. A. Nanni, N.M. Bradford, Constr. Build. Materials, **9**, 2, (1995)
3. R V.M. Karbhari, Y. Gao, ASCE J. Mat. Civ. Eng., **9**, 4, (1997)
4. P. Rochette, P. Labossière, ASCE J. Comp. Constr., **4**, 3, (2000)
5. O. Chaallal, M. Hassen, M. Shahawy, ACI Stru. J., **100**, 2, (2003)
6. J.L. Pan, T. Xu, Z.J. Hu, Constr. Build. Materials, **21**, 1, (2007).
7. A. Chakrabarti, A.M. Chandra, P. Bharagava, ASCE J. Reinf. Plas. Comp., **27**, 12, (2008)
8. C. Pellegrino, C. Modena, ASCE J. Comp. Constr., **14**, 6, (2010)
9. J. Piekarczyk, W. Piekarczyk, S. Blazewicz, Constr. Build. Materials, **25**, 5, (2011)
10. Y.A. Al-Salloum, J. Comp. Part B, **38**, 5–6, (2007)
11. H. Toutanji, M. Han, J. Gilbert, S. Matthys, ASCE J. Comp. Constr., **14,** 1, (2010)
12. T. Turgay, Z. Polat, H. O. Koksal, B. Doran, C. Karakoc, Mat. Design, **31**, 1, (2010)
13. A. De Luca, A. Nanni, ASCE J. Comp. Constr., **15,** 3, (2011)
14. Z. Wang, D. Wang, S.T. Smith, D. Lu, ASCE J. Comp. Constr., **16,** 2, (2012)

Soil reinforcement with geosynthetics

Mohammed Mustapha BESSAIM[1,3], Aicha BESSAIM[2], Hanifi MISSOUM[1,3], Karim BENDANI[1,3]
[1]University Abdelhamid Ibn Badis, Department of Civil Engineering, Mostaganem, Algeria
[2]University Mustapha Stambouli, Department of Civil Engineering, Mascara, Algeria
[3]Laboratory Construction, Transport and Environmental Protection

Abstract. The proportionality of existence of land with good bearing to erect any building or building is very small, to remedy this deficiency it is necessary to resort to techniques of reinforcement of the soils which can constitute a very important development. Among these methods of remediation, there is reinforcement by the geosynthetics which constitute an effective solution to these constraints. This process tends to stabilize the soil in question with increased load bearing capacity in civil engineering and geotechnical works such as embankments, slopes, embankments and hydraulic structures, with an inestimable gain in time, economy and durability while preserving the natural and environmental aspect.

Keywords: Soil, Geosynthetiques, Reinforcement, Improvement,

1 Introduction

Avec le développement et l'accroissement urbains et industriels, des terrains avec de bonnes conditions de construction sont moindres. Ces sols présentent des propriétés inadéquates pour une construction sûre et durable. Parmi ces sols, on peut distinguer les sols fins expansifs et de très grande compressibilité à l'instar des limons et les sols argileux. De même, les sols grenus et les sables fins lâches sont également des matériaux qui nécessitent une étude minutieuse avant de procéder à tous travaux de constructions. L'utilisation et la mauvaise étude de ces derniers engendrent des dégâts et des défauts irrémédiables entrainant usuellement la ruine de l'ouvrage en question. A cet effet, les chercheurs ont pensés à mettre en place des modalités de traitement, parmi lesquelles on peut citer le renforcement par les géosynthétiques dont le but d'exploiter un maximum de tels sols. Ces techniques constituent des avantages inestimables et permettent de résoudre les problèmes liés à l'instabilité ainsi qu'aux différents désordres rencontrés dans le domaine de génie civil car elles permettent d'une part d'améliorer les caractéristiques de l'ouvrage et d'autre part elles de préserver l'aspect esthétique et cela avec un coût réduit [1,2]. Ce processus a suscité l'intérêt de différents chercheurs qui ont étudiés la performance de ce procédé vis-à-vis la portance des fondations renforcée par ces matériaux [3-6] ou même pour le renforcement des pentes [7,8]. Après leurs applications avec succès sur le terrain, ces produits ont vite immergés le domaine de génie civil notamment le domaine de la géotechnique car ils ont pu satisfaire aux différents recommandations exigés dans les divers projets. Par ailleurs, les géosynthétiques fournissent une variété d'avantages tels qu'un gain dans la durée de vie des constructions, une amélioration des caractéristiques géotechniques du sol tout en augmentant les caractéristiques mécaniques, la possibilité de construction dans des milieux défavorables, sans oublier l'aspect économique car ils permettent de limiter les coûts de transport et d'installation en termes de structure et d'emprise au sol. En outre, le traitement des sols avec les géosynthétiques permet de préserver l'aspect écologique et environnemental de l'ouvrage.

Dans le présent article, nous allons entamés le renforcement des sols par les géosynthétiques tout en abordant également les différentes applications de ces premiers. De même, nous allons cités les différents types de géosynthétiques, leurs domaines d'utilisation notamment dans le domaine de la géotechnique.

https://doi.org/10.1051/matecconf/201814902096

2 Les différents types de géosynthétique

La forte demande de renforcement par les géosynthétiques ne cesse de s'amplifier et cela en raison des différents types de ces matériaux qui permettent une réalisation rapide, facile et moins coûteuse et cela dépend des différents contextes de l'édifice et cela c'est à cause dans la variété de choix et types qu'offre ces produits. Ils existent différent type de géosynthétiques tels que les géotextiles, les géogrilles, les géomembranes, les géofilets et les géocomposites [9].
Le choix du type de géosynthétique approprié, nécessitent une compréhension de chaque type qui correspond à une fonction principale pour une application spécifique considérée comme le montre le tableau suivant :

Tableau 1. Les types de géosynthétiques en fonction de leurs applicabilités [10].

Type	Différentes fonctions		
	Séparaion	**Renforcement**	**Filtration**
Géotextile	Fonction principale ou Fonction secondaire	Fonction principale ou Fonction secondaire	Fonction principale ou Fonction secondaire
Géogrilles	Fonction secondaire	Fonction principale	Non applicable
Géofilets	Fonction secondaire	Non applicable	Non applicable
Géomembranes	Fonction secondaire	Non applicable	Non applicable
Géocomposites	Fonction principale ou Fonction secondaire	Fonction principale ou Fonction secondaire	Fonction principale ou Fonction secondaire

Type	Différentes fonctions	
	Drainage	**Barrière**
Géotextile	Fonction principale ou Fonction secondaire	Non applicable
Géogrilles	Non applicable	Non applicable
Géofilets	Fonction principale	Non applicable
Géomembranes	Non applicable	Fonction principale
Géocomposites	Fonction principale ou Fonction secondaire	Fonction principale ou Fonction secondaire

3 Le renforcement des ouvrages par les géosynthétiques

Les géosynthétiques permettent d'améliorer les caractéristiques du sol en question, car ces premiers possèdent une grande capacité à la traction ce qui est quasi nul dans les sols. Ce type de traitement pourra être opté dans différents contextes tels que les chaussées, les pentes et les murs de soutènement [11,12].
Les différentes applications de renforcement avec les géosynthétiques sont réalisées dans le cas des remblais, les talus ainsi que pour les murs de soutènement comme le montre la figure suivante:

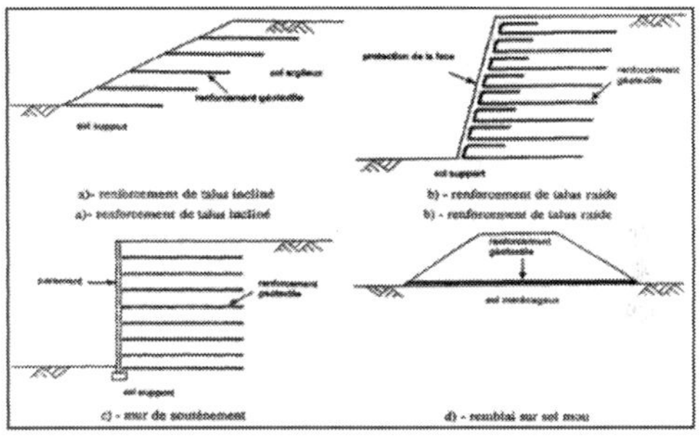

Fig. 1. Divers applications de renforcement des ouvrages avec géosynthétiques [13].

L'exemple de renforcement des pentes avec les géosynthétiques a été abordé par [14]. La stabilité des pentes peut grandement être améliorée avec les géosynthétiques suivant la formule suivante :

$$F.S = \sum \frac{moments\ résis \tan t}{moments\ moteur} \tag{1}$$

$$F.S = \frac{(\tau_e L_{ab} + \tau_f L_{bc})R + T_a Y}{WX} \tag{2}$$

Avec :

τ_e : La résistance de cisaillement de remblai
La-b : Longueur de l'arc a-b
τ_f : La résistance au cisaillement de la fondation du sol
L_{b-c} : Longueur de l'arc b-c
R : rayon depuis le centre critique de rupture.
T : la force de tension admissible du géotextile
Y : moment de géotextile
W : Poids du sol dans la zone de rupture.
X : moment depuis le centre de gravité vers le centre de rupture.

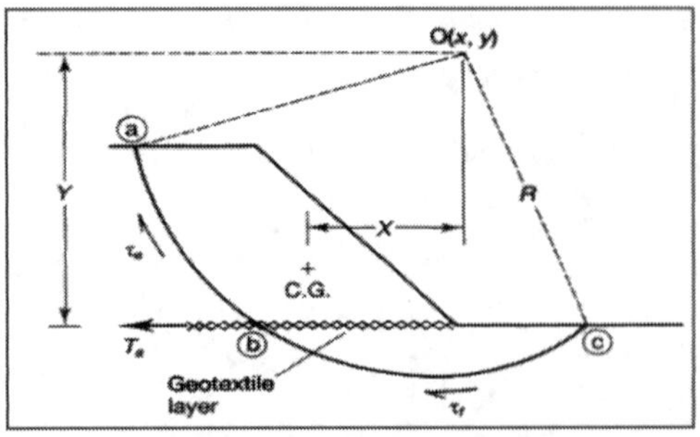

Fig. 2. Configuration générale utilisée pour modifier la stabilité des pentes avec les géosynthétiques [14].

Où il a été remarqué d'après l'équation précédente et depuis cette figure que la résistance du géotextile peut être augmentée si nécessaire pour conduire le facteur de sécurité jusqu'à une valeur acceptable [14]. Il est à connaitre que l'utilisation des géosynthétiques change selon le contexte d'étude prédéfinie. En terme de fonction mécanique, il faut assurer la reprise de forces de traction et de les reporter dans le sol par le biais de forces de cisaillement. Du point de vue hydraulique, il s'agit d'assurer une perméabilité suffisante afin d'éviter une retenue d'eau au-dessus du géotextile [13]. Les propriétés déterminantes des géosynthétiques afin assurer leurs rôles de renforcement. Ces derniers sont illustrés dans le tableau suivant:

Tableau 2. Propriétés déterminantes du géotextile pour un rôle de renforcement [13].

Functions partielles	Proprieties	Parameters caractéristiques
Function mécanique	Déformation Fluage Frottement sol-géotextile	Courbe « force-allongement » Augmentation de la déformation sous force de traction constante φ_{sg} Angle de frottement sol-géotextile
Fonction hydraulique	Perméabilité normale au plan du géotextile	ψ Permittivité k_n Perméabilité

Conclusion

Le renforcement des sols par les géosynthétiques représente un moyen d'apport inestimable. L'utilisation de ce procédé est connue par son efficacité depuis des décennies. Cette méthode s'est largement développée pour une vaste gamme d'applications dans divers domaines de géotechniques à l'instar du domaine mécanique et hydraulique. Cette technique est envisageable dans le cas des sols dont les propriétés sont inadéquates et qui répondent mal aux exigences et aux conditions des ouvrages et structures à supporter, d'où la nécessité d'un traitement au préalable. Ce processus permet un gain de la durée de vie des constructions et cela grâce à une méthode relativement facile à mettre en œuvre à des prix moins élevés vis-à-vis d'autres techniques de traitement. Le renforcement par les géosynthétiques est considéré comme étant une alternative avec des avantages prodigieux qui permettent d'une part d'améliorer les propriétés géotechniques du sol et d'autre part de préserver l'aspect naturel et environnemental.

Références

1. Leshchinsky, D., Han, J., Geosynthetic reinforced multitiered walls. J. Geotech. Geoenviron. Eng. 130 (12), 1225e1235 (2004).
2. Ruan, X., Leshchinsky, D., Leshchinsky, B.A., Global stability of bilinear reinforced slopes. Transp. Infrastruct. Geotechnol. 2, 34e46 (2015).
3. Yetimoglu, T.,Wu, J.T.H., Saglamer, A., Bearing capacity of rectangular footings on geogrid-reinforced sand. ASCE J. Geotechnical Eng. 120 (12), 2083e2099 (1994).
4. Dash, S., Sireesh, S., Sitharam, T., Model studies on circular footing supported on geocell reinforced sand underlain by soft clay. Geotext. Geomembranes 21 (4), 197e219 (2003).
5. Boushehrian, J.H., Hataf, N., Experimental and numerical investigation of the bearing capacity of model circular and ring footings on reinforced sand. Geotext. Geomembranes 21 (4), 241e256 (2003).
6. Ghosh, A., Ghosh, A., Bera, A.K., Bearing capacity of square footing on pond ash reinforced with jute-geotextile. Geotext. Geomembranes 23 (2), 144e173 (2005).
7. El Sawwaf, M.A., Strip footing behavior on pile and sheet pile stabilized sand slope. J. Geotech. Geoenvironmental Eng. 131 (6), 705e715 (2005).
8. El Sawwaf, M.A.,Behavior of strip footing on geogrid-reinforced sand over a soft clay slope. Geotext. Geomembranes 25 (1), 50e60 (2007)..
9. Koerner, R. M. Designing with Geosynthetics, 2nd Edition, Prentice-Hall, Inc., Englewood Cliffs, N.J (1990).
10. Koerner, R. M., A note on geotextile design methods, Geotechnical Fabrics Report, 2, No.2, pp. 28-29 (1984).
11. Hausmann, M. R., Fabric reinforced unpaved road design methods-Parametric studies, Proceedings of the Third International Conference on Geotextiles, Vienna, Austria, IFAI, pp. 19-24 (1986).
12. Yako, M. A. and Christopher, B. R., Polymerically reinforced retaining walls and slopes in North America, Workshop on Applications of Polymeric Reinforcement in Soil Retaining Structures, ed. P. Jarrett, Royal Military College, Kingston, Ontario, Canada, pp.213-226(1987).
13. Sere, A. "Ouvrages renforcés par des géotextiles chargés en tête : Comportement et Dimensionnement". Thèse de Doctorat de l'Ecole Nationale des Ponts et Chaussées, 327 pages (1995).
14. Koerner, R. M., Geosynthetics in geotechnical engineering, Chapter 22, Foundation Engineering Handbook, 2nd ed. Kluwer Academic Publishers, Boston, MA, pp. 796-813(1991).

Parametric study of the thermal performance of a typical administrative building in the six thermal zones according to the RTCM, using TRNSYS

LAMRANI Abdelghafour[1], SAFAR Said[1], ROUGUI Mohamed[1]*
1 LGCE, Ecole Supérieure de Technologie Salé, Mohammed V University in Rabat 10 000, Morocco.
*Corresponding author. Email: rougui93@yahoo.fr

Abstract. In this work, we present a parametric study of a new administrative building, located in El-Ksar El Kebir region (Morocco). In order to have a building that complies with the RTCM in a technically and economically sound manner, we have carried out a number of interventions to insulate the components of the building, namely external walls, exposed roofs and openings. In this perspective, we have modelled the building envelope as a multi-zone building in TRNSYS and we have adopted an occupation scenario for this type of building. After determining the optimal insulation solutions, we simulated the administrative building in the five other thermal zones, to determine its feasibility in the latter.

1. Introduction

In recent years, the world experienced a terrible climate change, which everyone is talking about, so some countries have begun to apply different solutions to remedy this problem. The COP 21, in Paris, was closed by an international agreement on climate., setting as a target by 2100 the limitation of global warming between 1.5 ° C and 2 ° C. witch was applied in the COP 22, in Marrakech, so Morocco has been working for a few years in a program that converges with global climate objectives, through the implementation of laws on energy efficiency and sustainable development, such as the Moroccan Building Regulation in Morocco RTCM, targeting the building.Morocco's energy consumption continues to rise rapidly each year.

Currently, the building sector accounts for 28% of final energy consumption, with a contribution of one third of total CO_2 emissions, given that the potential for energy savings is estimated at 40% in this sector worldwide whole. A well thought-out construction design will always be more efficient and less costly since it is characterized by a long duration.

Energy consumption in Morocco continues to grow rapidly every year. Growth in demand for electricity exceeds the most accurate forecasts, with a growth rate of 7% to 8% in recent years. The country must import more than 97% of its energy needs (more than 16% of electricity was imported directly in 2014) with the construction sector as it is the largest consumer of energy, about 25% of The total energy used. Energy consumed in the residential sector represents 72% of the total energy absorbed by the building sector and the rest is devoted to higher education. Thus, the building sector represents the largest depository of savings; about 40% could be saved by combining several factors (effective lighting system, effective envelope and efficient electrical equipment).

In the perspective of the involvement of scientific research in development of technical solutions for energy efficiency, Benhamou et al [4] carried out a transient thermal simulation of a villa type house in the suburb of Marrakech in order to evaluate the effects of insulating external walls and roofs and integration of overhangs. The study concerns a new project and the results showed that the effect of insulating external walls has small beneficial effects on heating load while it increases the cooling load during summer, in this specific climate conditions. This works was conducted before the implementation of the RTCM; thus, they did not show if the integration of those solutions would reduce the energy demand for heating and cooling to meet the requirements values of the RTCM.

Although the issue of energy efficiency in existing buildings is very important given the scale of existing buildings in Morocco, the first version of the RTCM, which was published in the Official Journal of Morocco in November 2014 and becomes a mandatory regulation by November 2015 [5], it concerns only new projects. In this work, we tried to address those issues by modeling a new typical administrative building's envelope and providing solutions of insulation and integration of passive techniques (Overhangs, Double glazing, Insulating external walls and roofs) through nine cases in order to meet the requirements values of the Moroccan Thermal Regulation standards. The first case is the real building; the others correspond to the lack of one or several solutions. We modeled the building's envelope as a multi-zone building in TRNSYS and we adopted a scenario of occupation for this type of building.

Nomenclature

- RTCM : Règlement Thermique de Construction au Maroc;
- ADEREE : Agence National pour le Développement des Energies Renouvelables et de l'Efficacité Energétique ;
- DMN : Direction de la Météorologie National ;
- TGBV : Taux Général de Bais Vitrées ;
- ACH: Air Change per Hour (dimensionless);
- K1: constant coefficient of infiltration (1/hr);
- K2: Temperature coefficient of infiltration (ACH/°C);
- K3: Wind Speed coefficient of infiltration (ACH.s/m);
- WS: Wind Speed (m/s);
- T: Temperature (°C).

2. Region climate and building description

2.1. Region Climate

El-Ksar El Kebir is a city in northwest of Morocco with 126,523 inhabitants[7], about 160 km from Rabat, 32 km from Larache and 110 km from Tangier.

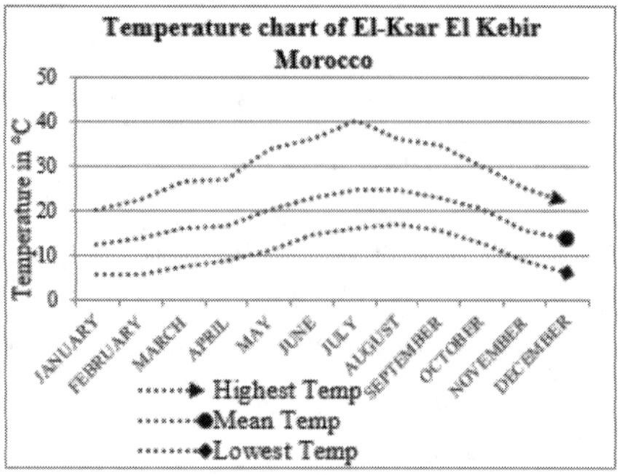

Fig.1. Temperature chart of El-Ksar El Kebir Morocco

The climate of the region is Mediterranean with a peak of temperature reaching 40°C during July as shown in the figure 1. The lowest temperature is recorded during January with a value of 5°C. The region is also known by its variation in wind speed with a mean value of 16 Km/h and 148 Km/h as maximum.

According to the RTCM, El-Ksar EL Kebir is located in the second thermal zone. The climate zoning work of Morocco was carried out in close coordination between the DMN (The National Meteorology Directorate) and ADEREE (National Agency for the Development of Renewable Energy and Energy Efficiency), with the support of international expertise. Moroccan territory was divided into six homogeneous climatic zones based on the analysis of climate data recorded by 37 meteorological stations over the period 1999-2008 (10 years) [1].

2.2. Thermal zones of RTCM

Two zoning types were established by the DMN:
- Heating degree-day, base 18 °C.
- Cooling degree-day, base 21 °C.

In this paper we present the cooling and heating degree-day chart of El-Ksar El-Kebir (Figure 2) based on the analysis of climate data, taken from Meteonorm [7], of this region. The heating degree-day in El-Ksar El Kebir reaches 157.59 in January as a peak value, whereas the cooling degree-days reach 113.07 in August.

The figure 2 shows that the total amount of heating's degree-day is 591, while the total of cooling's degree-day is 402; thus buildings in El-Ksar EL-Kebir require more energy for heating than cooling.

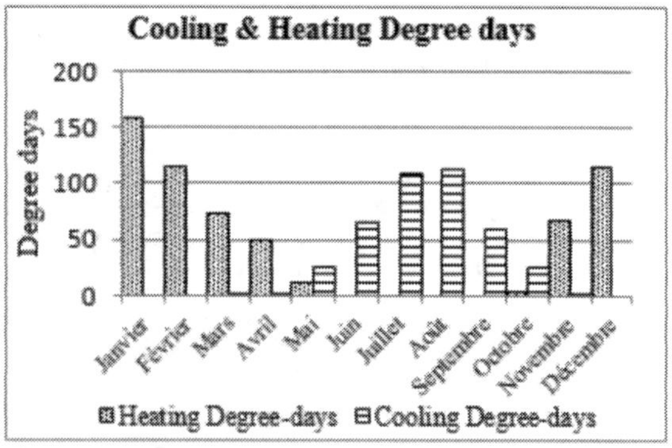

Fig.2. Cooling & heating Degree Days of El-Ksar El Kebir

2.3. Building description:

This analysis concerns a municipal building in the city of El-Ksar El Kebir with a built area of 186.86 m², and consists of two levels; the effective area of both levels is 266.16 m², divided into several offices. The main façade is oriented south. Glass panels surface represent 15% of the external façades, which leads to the perspective approach as described in the RTCM.

The following figures (3, 4) show the space allocation of the two levels.

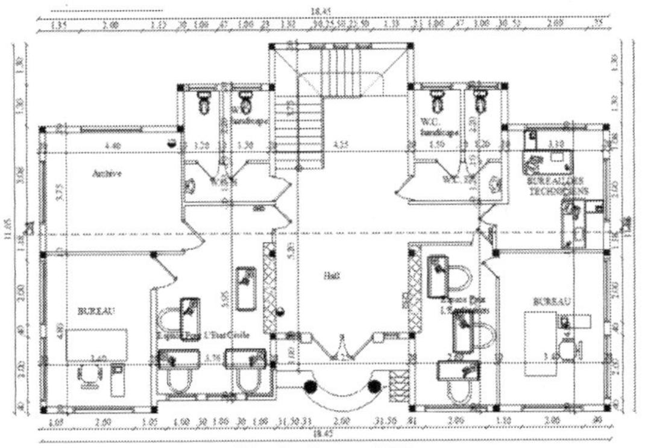

Fig.3. Ground Floor

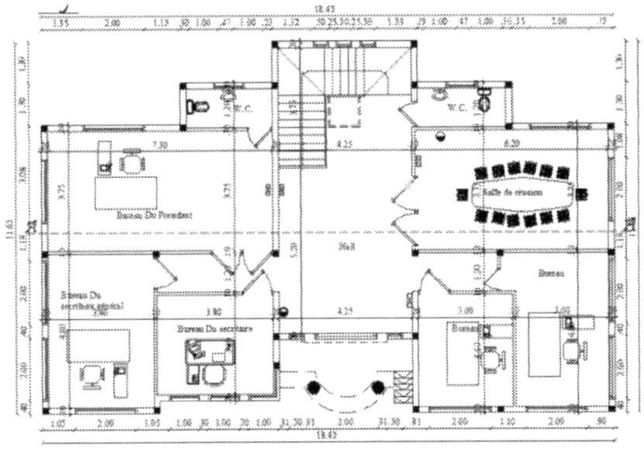

Fig.4. Second floor

Spatial delimitation of the building is presented in table 1 which presents the scenarios adopted for the communal building's occupation. Spaces are named respectively to the occupying functions. Knowing that the working hours of the administrators is limited between 8am and 4pm for 5 days, eliminating weekends and holidays. The set point temperature is 18 °C in winter and 21° C in summer.

Table1: Building offices and occupation scenario (part 1)

	workrooms	Workday/week	Duration	Set Temp (Cooling/heating) °C
Ground floor	Bureau 1	5 days	08h→16h	21/18
	Etat Civil	5 days	08h→16h	21/18
	Archive	5 days	08h→16h	21/18
	Hall	5 days	08h→16h	21/18
	Bureau Techniciens	5 days	08h→16h	21/18
	Bureau 2	5 days	08h→16h	21/18
	Egalisation	5 days	08h→16h	21/18
	WC H	5 days	08h→16h	#
	WC F	5 days	08h→16h	#
	Escalier	5 days	08h→16h	21/18

Table1: Building offices and occupation scenario (Part 2)

	workrooms	Workday/ week	Duration	Set Temp (Cooling/heating) °C
Floor	Sec-Générale	5 days	08h→16h	21/18
	Secretaire	5 days	08h→16h	21/18
	President	5 days	08h→16h	21/18
	Hall	5 days	08h→16h	21/18
	Salle Réunion	2 days	08h→20h	21/18
	Bureau 3	5 days	08h→16h	21/18
	Bureau 4	5 days	08h→16h	21/18
	WC President	5 days	08h→16h	#
	WC Réunion	5 days	08h→16h	#
	Escalier	5 days	08h→16h	21/18

The compositions of the structural element from outside to inside and their properties are given as [8]:

- Ewternal Wall: 1.5 cm Cement, 15 cm brick (12trous), 0.4 cm blade of air, 6.5 cm brick (8trous), 1.5 cm cement;
- Exposed roof: 2 cm Terracotta tiles, 2 cm tightness, 8/12 Insulation, 5 Shape, 16 cm hollow slab, 1.5 cm cement;
- Intermediate roof: 2 cm Terracotta tiles, 5 Shape, 16 cm hollow slab, 1.5 cm cement;
- Low floor: 2 cm Terracotta tiles, 5 cm Shape, 10 cm Rocks, 2 cm cement, 0.2 cm Vapor barrier;
- Internal Wall: 1.5 cm Cement, 15 cm brick (12trous), 1.5 cm cement;

3. Thermal simulation procedures:

The work was carried out in the simulation studio of TRNSYS, which is a complete simulation package containing several tools, from simulation engine programs and graphical connection programs to plotting and spreadsheet software. It is an integrated tool which can be used from the design of a project to its simulation [9].

In this paper, two important parameters were handled carefully as following:

- The effect of shading devices on the total radiation received on a glazed surface. The overhang and wing-walls play an important role in shading windows and reducing the solar radiation. In order to model these phenomena in TRNSYS we used the TYPE 34 (Overhang and Wingwall shading). Buildings directly heated radiation often include a shading device to shield receiver surfaces from direct radiation in summer months.

This component computes the solar radiation on a vertical receiver shaded by an overhang and/or Wingwall [10].

- Infiltration has an important effect on the balance heat transfer. A semi empirical approximation can be used in TRNSYS to compute the influence of infiltration on the cooling and heating balance. Thermal Energy System Specialists TESS implemented a TRNSYS type called TYPE 571 (Humidity ratio input), which uses a semi empirical model recommended by the ASHRAE for the calculation of infiltration to a conditioned zone [11]. The correlation is shown in equation 1.

$$ACH = K_1 + K_2 * (T_{zone} - T_{ambient}) + K_3 * WS \quad (1)$$

Suggested values for K1, K2, K3 taken from the ASHRAE Handbook of Fundamentals 1989 are provided below in Table 2 for varying levels of construction.

Table 2: Values for K1, K2, K3

Construction	K_1 [ach]	K_2 [ach/°C]	K_3 [ach.s/m]	Description
Tight	0.1	0.011	0.034	New building in which special precautions have been taken to prevent infiltration.
Medium	0.1	0.017	0.049	Building constructed using conventional construction procedures.
Loose	0.1	0.023	0.070	Evidence of poor construction on older buildings where joints have separated.

4. Results and discussion

4.1. Result of simulation without insulation:

Table 3 presents the simulation results in the case 0, where the building is in a normal state without any intervention, these results show that the annual heating needs is 10.68 kWh/m²/year, and 130.91 kWh/m²/year for cooling.

Table 3: Results of the simulation in the case 0, with a monthly demand of heating and cooling.

MONTH	HEATING DEMAND (kWh)	HEATING RATIO (kWh/M²/Y)	COOLING DEMAND (kWh)	COOLING RATIO (kWh/M²/Y)
CASE 0	2842.34	10.68	34842.42	130.91

According to the RTCM, it is not required to make a performance-based approach, since the building TGBV (General rate of glass panels in percentage, which is the total surface of glass panels by the total surface of external facades) doesn't exceed 45% it is not mandatory to use the performance approach, so we use the perspective approach that requires this conditions cited hereafter:

- The U of the exposed roof: ≤0.65 W / m².K instead of the initial value 3.72 W / m².K (Not Conform);
- The external walls U: ≤ 0.8 W / m².K instead of the initial value 0.59 W / m².K (Conform);
- The U of glazing: ≤ 5.8 W / m².K instead of the initial value 5.74 W / m².K (Conform);
- The SHGC (solar heat gain coefficient): NA.

4.2. Optimization of energy balance:

In view to reduce the energy bill of the building and to meet the requirement values of RTCM, 9 cases have been proposed:

- Case 1: exposed roof insulation with 4 cm of extruded polystyrene;
- Case 2: exposed roof insulation with extruded polystyrene 8 cm;
- Case 3: exposed roof insulation with extruded polystyrene 8 cm, 10 cm overhang;
- Case 4: exposed roof insulation with extruded polystyrene 8 cm, 20 cm overhang;
- Case 5: insulation of the exposed roof with Extruded polystyrene 8 cm, 30 cm overhang;
- Case 6: insulation of the exposed roof with Extruded polystyrene 8 cm, 60 cm overhang;
- Case 7: insulation of the exposed roof with Extruded polystyrene 8 cm, 80 cm overhang, and 10 cm Wingwall;
- Case 8: insulation of the exposed roof with Extruded polystyrene 8 cm, 10 cm of overhang, and 10 cm of Wingwall with TGBV reduced to 10%.
- Case 9: insulation of the exposed roof with Extruded polystyrene 12 cm, 10 cm of overhang, and 10 cm of Wingwall with TGBV reduced to 10%.

Figure 5 shows the ratio of the energy required for heating and cooling, in different cases.

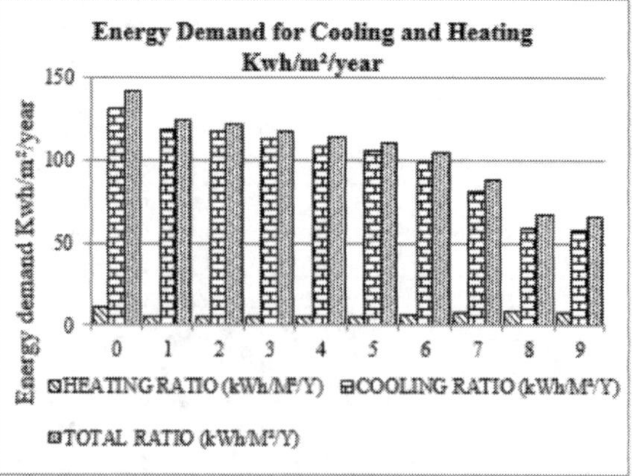

Fig.5. Energetic demand for heating and cooling.

Two variants have given the best technical-economic results for the El-Ksar El Kebir region, which are:

- Variant 8: Insulation of the roof Exposed with 8 cm extruded polyethylene, 10 cm of advanced cap and 10 cm of the lateral forward.
- Variant 9: Insulation of the roof Exposed with 12 cm extruded polyethylene, 10 cm forward of the cap and 10 cm of the lateral forward.

Table 4: Results of the simulation of the adopted cases, with the total ratio of demand in (Kwh/m²/Y).

EL KSAR EL KEBIR 1	Hating demande (kWh)	Hating ratio (kWh/ M²/Y)	Cooling Demande (kWh)	Coolin g ratio (kWh/ M²/Y)	Total ratio (kWh/ M²/Y)
VARIANTE 0	2842.34	10.68	34842.42	130.91	**141.59**
VARIANTE 8	1915.48	7.20	14806.84	55.63	**62.83**
VARIANTE 9	1890.49	7.10	14710.29	55.27	**62.37**

These two variants tightening duplicate for the other 5 zones, namely (Zone 2: Agadir, Zone 3: Fes, Zone 4: Ifrane, Zone 5: Marrakech, Zone 6: Er-rachidia), to determine the feasibility of this economic building In these thermal zones.

The results obtained for the five other zones using the same variants are shown in the following figure:

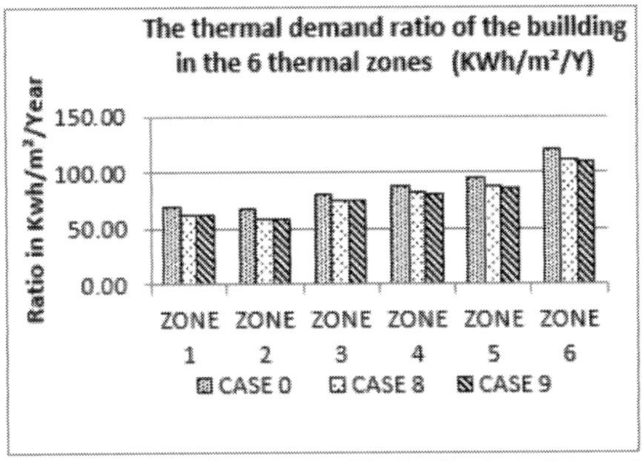

Fig.6. The thermal demand ratio of the building in the 6 thermal zones.

4.3. Discussion

The results obtained by the simulation of the building without intervention, and with a glazing rate of 15% (the initial state according to the architect), shows that the ratio of the heating demand is 10.68 Kwh / m² / Year, and 130.91 Kwh / m² / year for cooling ones, thus a total demand of 141.59 Kwh / m² / year. Table 9 shows the impact of the TGBV reduction on the thermal budget since we have a 50% reduction in energy demand, whereas the application of roof insulation combined with the overhangs allows a reduction of 55% since the total ratio reaches 62.37 Kwh / m² / year, equivalent to an energy bill of 7018 Dhs / year. The previous figure is the subject of the integration of this typical building in the other five zones. The results obtained for Agadir are the best with the lowest demand 59 Kwh / m² / year, cost the expensive monk of 6475 Dhs / year, Fes also represents conditions that can admit this building since the energy demand of that 74.5 Kwh / m² / year equivalent to an annual invoice of 9100 Dhs / year, but for the last three zones (Ifran, Marrakech and Er-rachidia) can not admit this building since it consumes more energy this year. Which pushes to seek other suitable solutions for this type of climate.

Although the building envelope complies with the Moroccan Building Regulation, the building consumes more energy in extremely cold, extremely hot areas.

5. Conclusion:

In this work, we have established a parametric study of thermal simulation of building office during summer and winter. In order to minimize the consumption, we proposed 9 variants. The final solution has achieved the objectives. A reduction in the annual bill of energy was obtained amounting to 6 887 kWh per year, with 8 264 MAD/year of benefices. The main conclusion of this work is that the insulation of the exposed roof allows a significant reduction in the total energy demand, and that this communal building can be the subject of a standardized structure for the 3 zones, Tangier, Agadir and Fes, in view of the energy performance that it showed, specifically for Agadir, in order to apply to the requirements of the RTCM on the one hand and to follow the global approach to fight against global warming.

References:

1. RTCM, Règlement Thermique de Construction au Maroc (2014)
2. The GEF CEO Endorsement Request – PIMS 3230 Morocco EE Building Code4, (May 1, 2009)
3. Rapport des activités annuelles 2014 ONE
4. Hassan Bouhess, Hassan Hamdi, Brahim Benhamou, Amín Bennouna, Pierre Hollmuller And Karim Limam, dynamic simulation of an earth-to-air heat exchanger connected to a villa type house in marrakech; Proceedings Of BS2013:13th Conference Of International Building Performance Simulation Association, Chambéry, France, August 26-28 (2013)
5. Bulletin officiel du Maroc 6360
6. Haut Commissariat au Plan RGPH (2014)
7. Meteonorm Global Meteorological databas Version 7
8. Bibliothèque BINAYAT perspective 2014
9. TRNSYS Manuel: Volume 6, Multizone modeling with TYPE 56 and TRNBuild~
10. Mathematical reference of TRNSYS 16, 5.7.4. Type 34: Overhang and Wingwall Shading
11. ASHRAE Handbook of Fundamentals (1989), chapter 22: Ventilation and Infiltration

12. N. Fezzioui, B. Droui, M. Benyamine, et S. Larbi, Influence des caractéristiques dynamiques de l'enveloppe d'un bâtiment sur le confort thermique au sud Algérien, Revue des Energies Renouvelables Vol. **11** N°1 (2008) 25 – 34.

 https://doi.org/10.1051/matecconf/201814902098

Introduction of inclined open channels for the control of surface runoff of slopes in road structures

O. Hniad[1], L. Bahi[1] and S. Amgaad[1]
[1]Laboratoire de Géophysique appliquée, de Géotechnique, de Géologie de l'Ingénieur et de l'Environnement, Ecole Mohammadia d'Ingénieurs, Rabat, Morocco

Abstract. The phenomenon of water erosion induced by runoff speeds at the surface of the embankments causes their instability. Particularly in road environments, gullying on the slope's surface due to runoffs causes landslides, which in turn cause considerable damage and consequent disorders to the road network. The aim of this research is to put in place a new technology for superficial water drainage on slope surfaces. Our study has developed a methodology involving the change of the geometric configuration of the water flow, aiming at velocity control of the flows by choosing slanting waterways with small slopes coupled to vertical drains. A modelling of the proposed solution will evaluate its effectiveness as to prevent the erosive factor and to identify other factors that are responsible for slope disorders.

1 Introduction

Erosive phenomena have long been a studies' priority in civil engineering in order to protect constructions (road embankments and roadside slopes in particular) from the effects of water erosion.

Several parameters are known to contribute to the accentuation or mitigation of the erosion phenomenon caused by runoff. A study started at the University of Missouri in 1914 was conducted to measure the amount of soil lost by erosion, leading to the Wischmeier and Smith model established in 1957.

Rainfall in some parts of Morocco, such as the Rabat-Tangier and Fez-Taza highways, has greatly favoured soil displacement on roadside slopes, leading to grain deposit accumulation in rainwater collection ditches at the lower parts of slopes. Potential solutions include geometric land management to divert water by reducing its erosive impact, or planting the embankment to reduce thrusting and runoff. A draining network of concrete arcades is one of the solutions already undertaken on the Rabat-Tangier motorway in order to introduce an artificial factor reducing the erosive phenomenon [1].

The purpose of this study is to present an alternative protection measure, to evaluate its efficiency and to obtain its evolution functions according to the various parameters involved in the RUSLE equation. The solution will then be evaluated to determine reduction rates of eroded quantities versus a reference slope using the RUSLE model parameters [2].

2 Presentation of study zone

The area concerned by the first part of this study is located in the Tangier-Assilah province, located in the Tangérois watershed located north-west of Morocco. It is bound on the north by the Strait of Gibraltar, on the west by the Atlantic Ocean, on the south by the Lower Loukoss watershed and the province of Larache, and on the east by the Côtiers Méditerranéens watershed and the Province of Tétouan. The province's area is 1195 km^2.

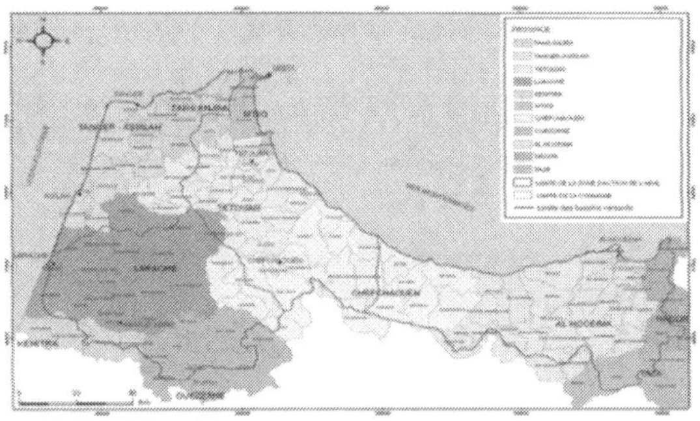

Fig. 1. Location of the Tangiers Province and the Tangérois watershed.

Outside the coastal plains, areas with steep or strongly undulating geomorphology cover more than 80% of the region's territory. The Tangérois presents an alternation of valleys mainly covered with quaternary alluvium and marly sandstone hills (figure 2).

 https://doi.org/10.1051/matecconf/201814902098

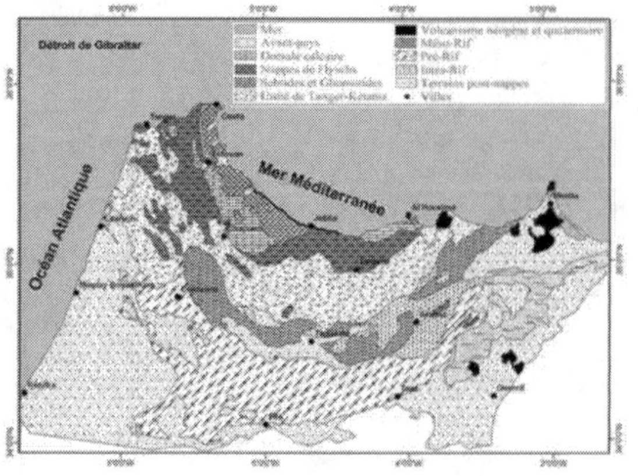

Fig. 2. Geological mapping of the Rif's chain (Rampnoux & al., 1979).

The Tangérois climate is subhumid. The region has an average rainfall of 800 millimeters per year. Eastern winds are dominant in the region. Temperatures generally range from 30°C to 40°C, with few exceptions. East or west winds are common almost every day, and average 7 to 20 km/h.

The main Oueds crossing the region are:
- Descending into the Atlantic Ocean: Hachef, Boukhalf, Bougadou and Mharhar;
- Descending towards the Straits of Gibraltar: Lihoud, Souani, Moghogha and Mellaleh.

The Tanger-Tétouan region benefits from a highway infrastructure composed of these main roads: (1) The A1 linking Rabat to Tangier and the latter to Tanger Med Port, and (2) the A6 linking Tétouan to Fnideq. The highly rugged terrain of the region favors large amounts of soil cuttings, but also the development of the runoff erosion phenomenon, which is a great inconvenience for roadside slopes. Said phenomenon has negative impacts on other aspects such as water quality in retention structures [3].

Other catalytic factors add to the relief: (1) abundant rainfall in the region, (2) dominant marl substratum, (3) degraded natural vegetation, and (4) land misuse.

In addition to water erosion, there is also strong wind erosion, but such phenomena remain localized: we note the formation of dunes on coastal sands; there is significant wind erosion east of Assilah in cultivated sandy soils as well.

3 The Revised Universal Soil Loss Equation

3.1 The RUSLE equation

In order to mathematically determine soil loss amounts by runoff erosion, it was first necessary to determine the parameters involved in the phenomenon. (Cook, 1936) has regrouped them in three categories: (1) rainfall conditions, (2) soil nature and parameters and (3) natural or artificial protection measures for the slopes [4].

The U.S.L.E (Universal Soil Loss Equation) model developed by Wischmeier and Smith involves a set of variables influencing water erosion, which belong to aforementioned categories and are regrouped in the following equation [2]:

$$A = RKLSCP \quad (1)$$

Where:
- A is the amount of soil lost per unit area and time. The R and K units are conventionally selected in order to obtain A in tons per hectare per year;
- R is the rainfall and runoff factor. It is to be completed by a factor taking into account snowmelt runoff or potentially significant artificial water that may induce erosive phenomena;
- K is the soil erodibility factor taking into account composition and nature of the soil subjected to water erosion;
- L is the slope length factor, defined as the ratio of the soil loss of the length of the slope of land to that of a 22.13-meter-long slope under identical conditions;
- S is the slope effect factor, defined as the ratio of soil loss of the slope to that of a 9% slope under identical conditions;
- C is the cover and crop management factor, defined as the ratio of soil loss of an area with specified cover and management to that of an identical area kept fallow;
- P is the factor of conservation measures and anti-erosion support practices, which is the ratio of soil loss with a support practice against erosion to that of the same parcel without an adequate conservation measure.

3.2 Rainfall/runoff factor R

The rainfall and runoff factor should not take into account a single meteorological factor, but include several of them. Indeed, it is necessary to quantitatively evaluate the impact of a raindrop depending on its fall on the slope surface. It can be highlighted as follows: a high intensity rainfall over a short period can lead to a quantitative erosion equivalent to that of a low rainfall but maintained over a longer period or with higher impact velocity. This lead (Wischmeier & al., 1958) to introduce the rainfall energy parameter [5].

The initial soil moisture can also be introduced in a similar illustration: a previously wet soil promotes runoff, and thus increases the erosive factor compared to said soil in its dry state.

The best estimate of the R-factor is based on the EI_{30} method developed by Wischmeier, taking into account rainfall energy E and maximum rainfall intensity over a period of 30 minutes. This computation approach implies that thunderstorms have already been partitioned into intervals (not necessarily of equal duration) in which intensity is almost constant.

Application of the EI_{30} method, according to (Renard & al., 1991) [6], is impossible in most cases due to unavailability of instantaneous rainfall measurements (in the form of pluviographs). This gave rise to several R-factor approximations based on more readily acquired parameters such as daily [7,8], monthly [9,10] or annual rainfall.

The PNUD/FAO MOR 87/002 [11] project has established an R-factor estimation based on available

https://doi.org/10.1051/matecconf/201814902098

monthly and annual rainfall averages, and is expressed as follows:

$$Log(R) = 1.299 + 1.744\ Log(\Sigma p_i^2/p_{tot}) \quad (2)$$

Where p_i and p_{tot} are monthly and annual rainfall averages respectively, all taken in inches.

3.3 Soil erodibility factor K

The K-factor reflects the soil's susceptibility to rainfall erosion according to its nature. Indeed, two different soils will not lose equal quantities of matter in similar conditions. *K*-factor is the amount of soil lost on a 9% slope and a 22.13-meter-long unitary parcel [2].

Evaluation of this factor, according to the nomogram of Wischmeier [12], is based on three main factors of the studied soil, including its physical and chemical properties: (1) its proportions in silt, clay and very fine sands, (2) its organic matter content and (3) its permeability (figure 3).

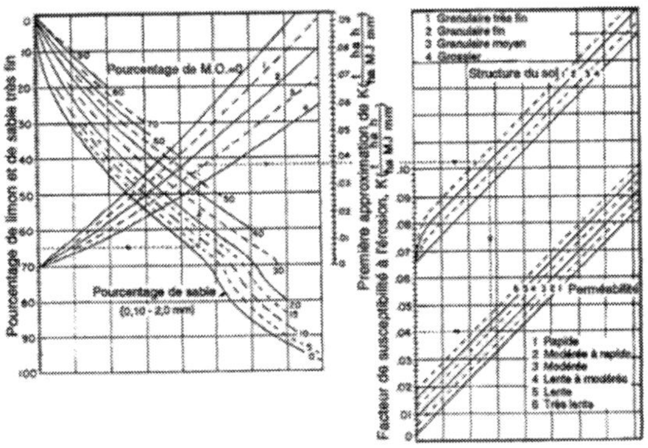

Fig. 3. Wischmeier's nomogram used in determining the *K*-factor [12].

From the nomogram, the following equation is extracted [2]:

$$100K = 2.1M^{1.14}10^{-4}(12-a) + 3.25(b-2) + 2.5(c-3) \quad (3)$$

Where:
- M is the grain size factor, being: $M = (\%_{silt} + \%_{very\ fine\ sands})\ (100 - \%_{clay})$;
- *a* is the organic matter content expressed in %;
- *b* is the soil's structural class index, which depends on its particle size;
- *c* is the soil's permeability class index, which depends on its textural class.

Table 1. *b*-index used in Wischmeier's nomogram and equation.

Structure type	Size (mm)	Class (b)
Very fine or structureless	---	1
Fine	< 2	2
Medium	2 to 5	3
Coarse	5 to 10	
Polyhedral, lamellar, prismatic	> 10	4

Table 2. *c*-index used in Wischmeier's nomogram and equation.

Textural class	Permeability level	Class (c)
Gravel, coarse sands	Fast	1
Silty sands, sandy silts	Moderate to fast	2
Fine sandy silts, silts	Moderate	3
Silts, clayey silts	Slow to moderate	4
Clayey silts, clays	Slow	5
Tight texture	Very slow	6

It should be noted that (Salehi,, 1990) [13] was able to illustrate that the validity of Wischmeier's nomogram was limited to the United States, and that it was affected by soil differences: After studying similar soil types in Canada, it was found that the textural difference between studied soils and those of reference induced differences between the values of *K* obtained using the nomogram and those measured directly on Canadian test plots.

3.4 Topographic factors LS

The two topographic factors *L* and *S* respectively characterize the length and slope of the terrain in their influence on the runoff-eroded quantities. In erosion studies, the disadvantage of considering them separately is eliminated, hence the determination of a single LS factor.

The *L*-factor is based on the ratio between the magnitude λ and the distance of a unit parcel, which equals 22.13m [2]. λ is the distance between the runoff's point of origin and, as the case may be, the point where the slope becomes low enough for solid material deposit, or the point where the water runoff is introduced into a channel purposefully built for drainage (in the case of a drainage network or a built channel).

The *S*-factor represents the influence of slope inclination on the amount of eroded soil. This factor can be determined in case of constant gradient slopes.

The *L*-factor is as follows:

$$L = (\lambda/22.13)^m \quad (4)$$

Where m is a variable characterizing the favored mode of erosion in the studied slope (if furrows favor the erosive factor or not). Said magnitude involves a factor β. The exponent m is given by the following formula, θ being the slope's angle:

$$M = \beta/(1+\beta)\ where\ \beta = \sin\theta/[0.0896(3\sin\theta+0.56)] \quad (5)$$

(Wischmeier & Smith, 1978) gave arbitrary values of m to be chosen according to slope inclination, without taking into account the effect of erosion mode: (1) 0.5 for slopes greater than 5%, (2) 0.4 for slopes between 3.5% and 4.5%, (3) 0.3 for slopes between 1% and 3%, and (4) 0.2 for slopes less than 1%.

The slope-effect factor S is given by the following undermentioned formula [2]:

$$S = 65.41(\sin\theta)^2 + 4.56\sin\theta + 0.065 \quad (6)$$

Thus, the LS-factor can be given in the formula:

 https://doi.org/10.1051/matecconf/201814902098

$$LS = (\lambda/22.13)^m [65.41(\sin\theta)^2 + 4.56\sin\theta + 0.065] \quad (7)$$

3.5 Cover and management factor C

C-factor, as defined by (Wischmeier & Smith, 1978), is the ratio of soil loss from a crop area (or any crop management program) to loss of the same fallow area under identical conditions.

In case where the soil has not been cultivated, factor *C* is not taken into account in the RUSLE equation, thus giving eroded amounts equal to *RKLS* in the absence of protective measures. In other words, value of *C* is taken equal to one for fallow land. In case where crops or a management program is put in place, the value of *C* is much lower.

Apart from the values in the USDA Manual N°537 (Wischmeier & Smith, 1978), research on *C* values has remained insufficient. Some comparisons were made between the manual's values and measured values for the Lennoxville area in Canada (Salehi & al., 1991), indicating good comparability for soils outside the United States.

3.6 Support practice and conservation factor P

By definition, the *P*-factor in the revised USLE equation is, according to (Wischmeier & Smith, 1978), the soil loss ratio of a parcel with conservation measures to an identical parcel without said measures, cultivated in the slope's direction.

Existing anti-erosion support practices include contour cultivation, contour ridging, stripcropping, terracing [2], and natural or fabricated mulching.

Anti-erosion techniques break down into two families: biological and mechanical. In arid zones, mechanical techniques take over and are more developed and recommended because of the difficulty of adopting biological methods, although a vegetation cover multiplies the support practice's effectiveness.

4 Describing the designed method

The introduction of a new superficial drainage method mainly affects the *P*-factor value of the RUSLE equation. (Shin, 1999) gave *P*-factor values for different agricultural practices and for different slope values in table 3:

Table 3. *P*-factor values for different agricultural practices.

	Contouring	Stripcropping	Terracing
< 7.0	0.55	0.27	0.10
7.0 – 11.3	0.60	0.30	0.12
11.3 – 17.6	0.80	0.40	0.16
17.6 – 26.8	0.90	0.45	0.18
> 26.8	1.0	0.50	0.20

The purpose of surface runoff drainage is to collect runoff water and convey it to a collection point through a defined flow network, which will also reduce the LS value according to the definition of (Wischmeier \& Smith, 1978). Reducing runoff's travel distance on the soil surface will also contribute to reducing furrow-induced erosion, a phenomenon likely to occur in freshly built slopes.

The suggested solution is about realizing a runoff drainage network composed of: (1) concrete channels quasi-perpendicular to the runoff's direction with an inclination varying between a 10 and a 20-degree angle, along with (2) equidistant vertical water descents linking aforementioned channels. Points of intersection of channels and descents will be out of phase in order to induce a flow discontinuity in vertical elements, thus limiting flow velocities in channels and risk of degradation of concrete matter constituting the solution's elements (figure 4).

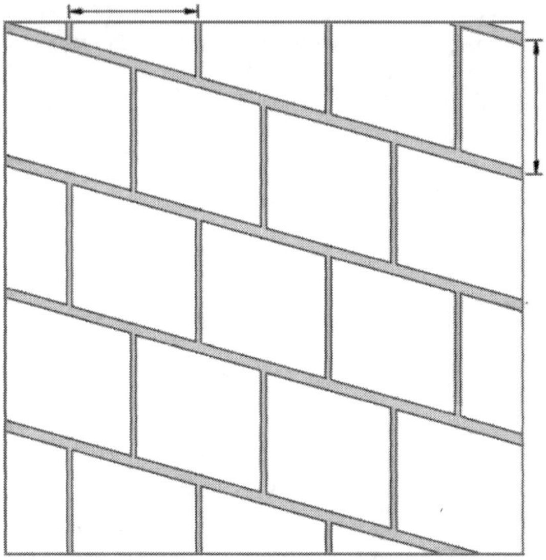

Fig. 4. Scheme of the suggested drainage solution on a one-hectare square experimental parcel (Channel inclination is 15 degrees and spacing between channels and descents equal to 25 meters).

Disposition of a runoff diversion ditch perpendicular to the slope renders our approach similar to contour ridging. The main objective of inclined channels, as for soil ridges, is to reduce the runoff's energy by nullifying runoff velocity and facilitating rainwater runoff's catchment. Choice of concrete for inclined channels and descents has several advantages: (1) canceling water infiltration, avoiding saturation of the upper part of the ground, (2) reducing landslide risk of superficial layers, (3) avoiding ditches' clogging and (4) offering the possibility of exploiting collected water in other potential uses.

The geometric similarity of the inclined channels and ridges allows us to estimate an anti-erosion factor close to that considered for the ridging method. The actual effectiveness of the method can only be evaluated by an experimental study quantifying eroded soils in the presence of the designed method compared to a similar quantitative on reference plot without anti-erosion measures. A value of 0.5 would be appropriate, since contour ridging is known to reduce by half the eroded quantities in crops perpendicular to slope direction; and for slopes exceeding 25% (which is the major case of road embankments), the presence or absence of vegetation does not have a significant contribution (value of *P* between 0.9 and 1 without ridging (Wischmeier & Smith, 1978)).

5 Results and discussions

5.1 Rainfall runoff factor

Rainfall data for calculation of *R* are obtained from the rainfall pluviogram shown in figure 5, showing rainfall distribution over a typical year as well as temperature variations during the same year. For such a year, the average annual rainfall is around 700 millimeters. The station is located at coordinates 35°43'43' 'north, 5°55'01" west.

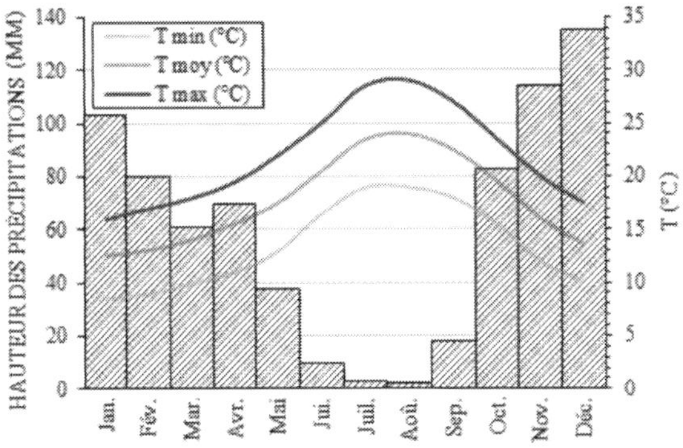

Fig. 5. Rainfall in millimeters and temperatures per month in the Tangier region (INRA, 2007).

In this application, we will use the Rango & Arnoldus model [11] involving monthly and annual rainfall averages instead of rainfall energy. The value obtained for the rainfall factor in our case study is 186.34 $MJ.mm.ha^{-1}.h^{-1}.year^{-1}$.

Multiple rainfall samples from several years (including dry and wet ones) will allow a better estimate of the R-factor. (Ouallali & al., 2016) evaluated the erosivity factor in the Oued Arbaa Ayacha watershed at the frontier of the communes of Larache and Tangier-Assilah between 100 and 125 $MJ.mm.ha^{-1}.h^{-1}.year^{-1}$ [14]. (Khali Issa & al., 2016) estimated the factor in the Kalaya watershed between 94.7 and 96.6 $MJ.mm.ha^{-1}.h^{-1}.year^{-1}$ [15]. (Modeste & al., 2016) studied the R-factor on the Ourika watershed and found values between 55 and 100.5 $MJ.mm.ha^{-1}.h^{-1}.year^{-1}$ [16]. (Yjjou & al., 2014) also estimated the R-factor for the upper Oum Er-Rbia watershed between 70 and 119 $MJ.mm.ha^{-1}.h^{-1}.year^{-1}$over most of the watershed's area [17].

5.2 Soil erodibility factor

Soils concerned by the study were taken from sites near the commune of Gzenaya in the Tanger-Assilah province. The green clays taken were kept in their natural state as best as possible, and a paraffin shell served as hygrometric insulation allowing minimal alteration and accidental water content variation [18].

Geotechnical study of the collected samples underlines the highly plastic nature of soils north of the country and the preponderance of clayey particles. A percentage of fine sands and silts of 49% and a clay particle content estimated at 35% were found. According to the USCS classification, this soil is classified as CH (inorganic clays with high plasticity).

(Edahbi & al., 2014), in their study of soils in the Loukkos region south of the study area, calculated the organic matter content of Dehs (with particle size distribution close to the studied soil) using organic carbon content by multiplying it by the universal coefficient of 1.724, resulting in a percentage of organic matter equal to 1.01% [19].

We will use a value of 1% for this first part of our study. Calculating the value of *K* gives a value of 0.245 $t.h.MJ^{-1}.mm^{-1}$. (Khali Issa & al., 2016) found values of up to 0.724 $t.h.MJ^{-1}.mm^{-1}$ for the most erodible soils in the Kalaya watershed [15]. (Modeste & al., 2016) find *K* values between 0.15 and 0.69 $t.h.MJ^{-1}.mm^{-1}$ for the Ourika region [16]. (Markhi & al., 2015) find that this value varies from 0.23 to 0.85 $t.h.MJ^{-1}.mm^{-1}$ for the N'Fis watershed [20].

5.3 Cover factor

In our study, we must consider the slope's initial state at the very commissioning of the road project including it. In general, freshly constructed embankments are not vegetated, which will lead us to a cover factor *C* equal to 1 [2].

Planting of the protected slope is one of the strong recommendations for improving this solution. As local planting and use of local plants is common practice in the country, it is recommended in our case study to reduce runoff quantities and kinetic energy of the rains favoring the erosive water factor.

5.4 Support practice factor

The conservation method factor *P* is taken as 0.5 in our case. A comparison of two test plots will allow us to have a value specific to the solution, but will nonetheless be inferior to than 0.5 since this value considers a very unfavorable state of the solution, making it almost similar to the agricultural ridging solution.

5.5 Topographic factors

As part of this study, we took a simplified square study plot of one hectare to show the effect of changing the geometrical parameters of our solution (in particular, the spacing between the inclined channels and the vertical descents) on the gain in eroded quantities. Calculations were made on channels inclined by a 15-degree angle and spacing values of 50 m, 33 m, 25 m, 20 m, and a tight mesh of 10 m of spacing to illustrate the usefulness of channels' number. The aforementioned spacings correspond to a number of inclined channels equal to 2, 3, 4, 5 and 10 channels respectively.

In the case of clay or clayey soils, a suitable terrain slope would be between 1V:3H and 1V:1.5H, equivalent to slope percentages between 33% and 67%. Both layouts and lower slopes (10% and 25% slope) were studied in order to evaluate eroded amounts' reduction from a steep slope to a weaker one. A comparison between cleared quantities for two slope percentages and gain in the form of non-eroded soils was performed.

5.6 Eroded soil amounts

Our simulations' results include area reductions provided by the designed drainage method, eroded soil quantities per year on the one-hectare square test plot considered and a relative comparison between different geometric parameter changes of the method. They are summarized in the following table:

Table 4. Erodible soil surface reduction and *A*-values for a square one hectare parcel.

	50 m	33 m	25 m	20 m	10 m
S_b : covered area by concrete channels and descents					
m^2/ha	177.03	267.29	375.06	467.82	945.14
%	1.77	2.67	3.75	4.68	9.45
A : eroded amounts in tons per year					
10%	39.89	31.87	27.30	24.08	15.98
25%	189.62	144.04	119.27	102.41	62.45
33%	305.97	229.77	188.81	161.13	96.40
67%	921.52	677.47	548.86	463.07	267.39

It is found that with the introduction of two inclined channels and appropriate descents connecting them, and maintaining the approximation of *P* to the 0.5 value of the homologous approach of contour ridging in the protections of agricultural parcels, cutting the runoff path allowed the reduction of eroded soil amounts to percentages between 30% and 34.5% of that with no runoff drainage solution on the four simulated slopes (figure 6).

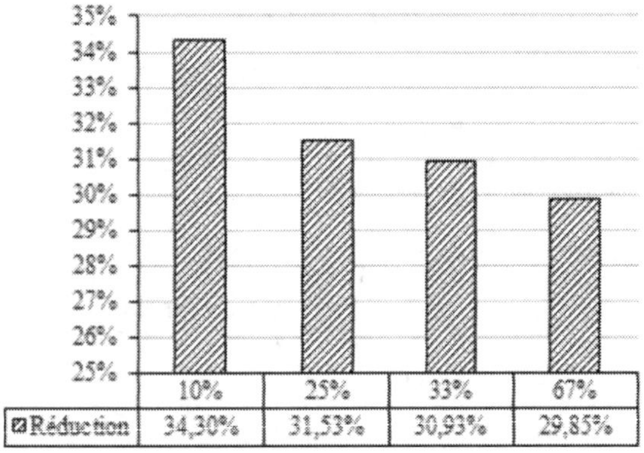

Fig. 6. Reduction percentages of *A* with the contribution of the inclined two-channel drainage solution versus embankment slopes.

Eroded quantities are greatly reduced by reducing slope or multiplying water interceptors. (Chehlafi & al., 2014) draw an identical conclusion for the solution in concrete arcades arranged by the Autoroutes Du Maroc on the A1 highway linking Rabat to Tanger [1].

A considerable reduction (between 35.2% and 38%) in the amount of land lost by water erosion is possible by moving from an inclination of 33% to 25%. this amounts to a larger quantity of cuttings during the realization of the slope, but remains limited to 21% of additional cut. Higher reductions occur when lowering slope value from 25% to 10% slope (between 74% and 79%), but a slope of 10% remains extremely low, prompting excessive excavation.

Going from a number of inclined channels to a greater number also reduces eroded quantities: with a passage from two channels to three, the value of A is reduced to 75% of the quantity (evaluation carried out on 33% and 25% slopes). Higher reductions are possible and it is found that these are almost identical throughout the slopes taken, and in particular between 25% and 33%, favoring again the choice of the lower slope.

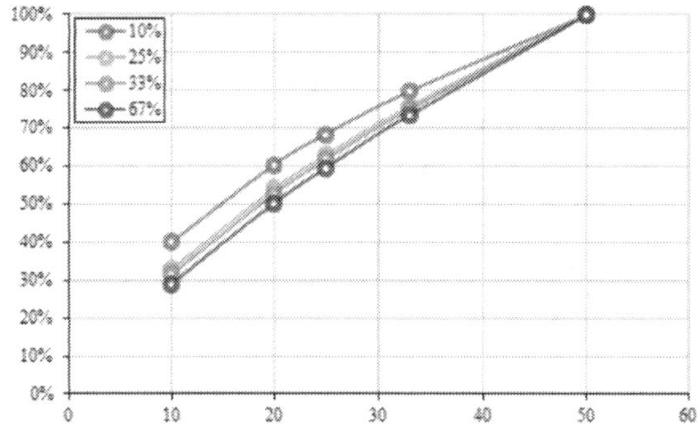

Fig. 7. Reduction percentages of *A* versus channel spacing by slope inclination (relative to two-channel layout).

5.7 Dimensioning of collection ditches on embankment foot

Dimensioning of the collector ditch should allow the flow and transport of solid particles without any deposited volume affecting said transport. Our test plot is characterized by a thrust with an average density on the samples of 1.74 t/m^3 in the Gzenaya region [18].

It can be considered that the flow does not allow the transport of solid grains beyond a ditch filling greater than half of its section. This will allow us to apply the formula:

$$A_{max} = 0.5\rho_h L_{ditch} S_{ditch} \quad (8)$$

For ditch sections, we referred to the STRADAL catalog for opencast hydraulic ditches. The following ditch areas will be studied: 0.5 m^2, 0.8 m^2, 1.125 m^2 and 1.301 m^2. The maximum section of the supplier will also be taken into account, but it should be noted that its own weight per meter (more than one ton per meter of concrete ditch) and its installation would require heavy machinery if not cast on site.

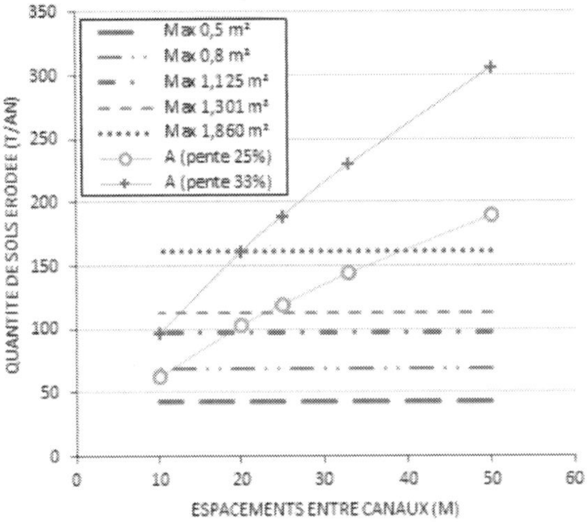

Fig. 8. Tolerance of ditches in relation to eroded soil amounts at 25% and 33% inclination (Eroded quantities versus channel spacing).

According to the graph in figure 8, it is necessary to consider higher section areas for the collecting ditches to be realized. The design of a particular trapezoidal section would be the best approach: a section that can be reconstituted from concrete slabs easily moved on site and assembled to form the ditch while having adjustable geometric configuration depending on the selected area.

5.8 Minimum spacing between inclined channels

In order to allow a direct choice of the collection ditch area and the appropriate spacing for the designed solution, a numerical solution based on the factors of the RUSLE equation has been established.

In this sense, the maximum capacity supported by a main collection ditch at the foot of the embankment was taken equal to half of its section and the study was conducted on the same square 1 ha test section as previously defined.

Minimum spacing necessary for the channels to allow them to ensure a correct drainage according to the fixed collection ditch area was evaluated on 7 different soils: two samples of the clays of the Tangier region including Gzenaya [18], Gharb tirs [21], Loukoss tirs [19], Doukkala tirs [22], and two soils collected from the Lemhaya area near Fès [23], as well as the Ifrane region.

The established formula for this study is as follows:

$$e_{min} = \alpha(0.01R)^{\beta} K^{\gamma} S_{ditch}^{\lambda} \quad (9)$$

Where:

- $\alpha = 0.136(\sin\theta)^{-2.271}$
- $\beta = -1.286(\sin\theta)^{-0.133}$
- $\gamma = -2.0734(\sin\theta)^2 + 2.272\sin\theta - 2.1418$
- $\lambda = 1.2078(\sin\theta)^2 - 1.5027\sin\theta + 1.8554$

In the majority of the simulations carried out, an error margin of around 5% between values of *A* obtained with spacings calculated by this formula and the maximum capacity supported by the embankment's collection ditch is ascertained, confirming the validity of this approach for direct calculation of the spacings adapted to each chosen ditch.

6 Conclusions

The contribution of the surface inclined channels drainage solution reduced the amount of soil eroded on the studied test plot to nearly 35%.

A change from 33% to 25% was found to have a considerable effect on the reduction of *L* and *S* factors, hence the lesser eroded soil quantity. *A* reduction in channel and descent spacing, considered equidistant in our study, also contributes to the reduction of the runoff path, lowering the *A*-values with a trend independent of slope value.

The evaluation of the spacings required for a set of ditch areas at the collection ditches made it possible to predict a layout of the drainage network by directly exploiting the same parameters of the RUSLE equation, thus facilitating studies to the optimization of the channel sections and the arrangement of the vertical descents according to the desired in-channel velocity adjustments.

A more accurate estimate of the value of the anti-erosion measurement factor will give us greater precision in our simulations, and will confirm this solution as suitable for the surface drainage of highly erodible road embankments. The participation of a revegetation of erodible surfaces will have a dual role: (1) improving the efficiency of the solution by reducing the value of *C* (and of *A* as a consequence), and (2) a better integration of the method in the landscape and the environment.

The study of this solution by varying the dimensions and the inclinations of the channels will be able to orient the solution towards an optimization of the spacings and dimensions of the channels and descents according to the type of soil. The flexibility of this solution will also allow its implementation on slopes with more than one type of soil.

References

1. A. Chehlafi, A Kchikach, A. Derradji. Protection des talus autoroutiers par arcades betonnees ou maçonnees. *Journées nationales de géotechnique et de géologie de l'ingénieur* (2014)
2. W.H. Wischmeier, D.D. SMITH, et al. Predicting rainfall erosion losses - a guide to conservation planning (1978)
3. Direction Générale des Collectivités Locales - Ministère de l'Intérieur - Royaume du Maroc. La région de Tanger-Tétouan-Al Hoceima : monograghie générale (2015)
4. H.L. Cook. The nature and controlling variables of the water erosion process. *Soil Sci. Soc. AM*, **1**, p.60-64 (1936)
5. W.H. Wischmeier, D.D. Smith, & R.E. Uhland. Evaluation of factors in the soil loss equation. *Agric. Eng.* **39**, p.458-462 (1958)
6. K.G. Renard, G.R. Foster, G.A. Weesies, J.P. Porter. RUSLE: Revised universal soil loss equation. *Journal of soil and water conservation* **46**, 30-33 (1991)

7. V. Bagarello, F. D'Asaro. Estimating single storm erosion index. Trans. Am. Soc. Agric. Engrs 37, p.785–791 (1994)
8. G. Petkovsek, M. Mikos. Estimating the R factor from daily rainfall data in the sub-Mediterranean climate of southwest Slovenia. Hydrological sciences journal, 49, p.869-877 (2004)
9. M. Grimm, R.J.A. Jones, E. Rusco, L. Montanarella. Soil erosion risk in Italy: a revised USLE approach. EUR 20677 EN, Office for Official Publications of the European Communities, Luxemburg, Luxemburg, 26p (2003)
10. K.G. Renard, J.R. Freimund. Using Monthly Precipitation Data to Estimate the R factor in the Revised USLE. *J. Hydrol*, **157**, p.287–306 (1994)
11. A. Rango, H.M.J. Arnoldus. Aménagement des bassins versants. *Cahiers techniques de la FAO*, p.1-11 (1987).
12. W.H. Wischmeier, C.B. Johnson, B.V. Cross. A soil erodibility nomograph for farmland and construction sites. *J. Soil and Water Conservation.* **26**, p.189-193 (1971)
13. F. Salehi, A.R. Pesant, R. Lagace. Validation of the universal soil loss equation for three cropping systems under natural rainfall in Southeastern Québec. *Canadian Agricultural Engineering,* **33**, p.11-16 (1991)
14. A. Ouallali, M. Moukhchane, H. Aassoumi, et al. Evaluation et cartographie des taux d'érosion hydrique dans le bassin versant de l'Oued Arbaa Ayacha (Rif occidental, Nord Maroc). *Journal of Geoscience and Environment Protection*, **4** (2016)
15. L. Khali Issa, K.B.H Lech-hab, A. Raissouni, & al. Quantitative Mapping of Soil Erosion Risk Using GIS/USLE Approach at the Kalaya Watershed (North Western Morocco).
16. M. Modeste, A. Khattabi, N. Mhammdi, et al. Cartographie Des Risques De L'érosion Hydrique Par L'équation Universelle révisée des pertes En Sols, La Télédétection Et Les SIG Dans Le Bassin Versant De L'Ourika (Haut Atlas, Maroc). *European Scientific Journal, ESJ*, **12** (2016)
17. M. Yjjou, R. Bouabid, A. El Hmaidi, et al. Modélisation de l'érosion hydrique via les SIG et l'équation universelle des pertes en sol au niveau du bassin versant de l'Oum Er-Rbia. *The International Journal of Engineering and* Science (IJES), **3**, p.83-91 (2014)
18. T. El Bahlouli & L. Bahi. Study on the swelling behavior of the green clay of Gzenaya (Morocco) compared with test conditions. *International Journal of Engineering Research and Applications*, **2**, p.1413-1423 (2012)
19. M. Edahbi, M. Khaddor, F. Salmoun. Caractérisation des sols du Nord du Maroc (Bassin Loukkos). *J. Mater. Environ. Sci*, **5**, p.2133-2138 (2014)
20. A. Markhi, N.E. Laftouhi, A. Soulaimani, et al. Quantification et évaluation de l'érosion hydrique en utilisant le modèle RUSLE et déposition intégrés dans un SIG. Application dans le bassin versant N'fis dans le Haut Atlas de Marrakech (Maroc). *European Scientific Journal*, **11** (2015)
21. T. El Bahlouli & L. Bahi. Analysis of the swelling behavior of the Tirs of Gharb (Morocco) compared with factors related to the change of soil-moisture. *International Journal of Engineering and Technology*, **5**, p.3649-3659 (2013)
22. J. Wilbert. Tirs et sols tirsifiés au Maroc. *Les Cahiers de la recherche agronomique*, **20**, p.23 (1965)
23. D. Touhtouh, Y. Moujahid, E.M. El Faleh, et al. Physical and chemical characterization of three types of soils of Saïs, Morocco.

AUTHOR INDEX

AUTHOR INDEX

AUTHOR INDEX

AUTHOR INDEX

AUTHOR INDEX